REFERENCE
DO NOT REMOVE
FROM LIBRARY

WITHDRAWN FROM
TSC LIBRARY

Mechanical Engineers' Handbook

Mechanical Engineers' Handbook
Third Edition

Materials and Mechanical Design

Edited by
Myer Kutz

JOHN WILEY & SONS, INC.

This book is printed on acid-free paper. ♾

Copyright © 2006 by John Wiley & Sons, Inc. All rights reserved.

Published by John Wiley & Sons, Inc., Hoboken, New Jersey.
Published simultaneously in Canada.

No part of this publication may be reproduced, stored in a retrieval system, or transmitted in any form or by any means, electronic, mechanical, photocopying, recording, scanning, or otherwise, except as permitted under Section 107 or 108 of the 1976 United States Copyright Act, without either the prior written permission of the Publisher, or authorization through payment of the appropriate per-copy fee to the Copyright Clearance Center, Inc., 222 Rosewood Drive, Danvers, MA 01923, (978) 750-8400, fax (978) 750-4470, or on the web at www.copyright.com. Requests to the Publisher for permission should be addressed to the Permissions Department, John Wiley & Sons, Inc., 111 River Street, Hoboken, NJ 07030, (201) 748-6011, fax (201) 748-6008, or online at http://www.wiley.com/go/permission.

Limit of Liability/Disclaimer of Warranty: While the publisher and author have used their best efforts in preparing this book, they make no representations or warranties with respect to the accuracy or completeness of the contents of this book and specifically disclaim any implied warranties of merchantability or fitness for a particular purpose. No warranty may be created or extended by sales representatives or written sales materials. The advice and strategies contained herein may not be suitable for your situation. The publisher is not engaged in rendering professional services, and you should consult a professional where appropriate. Neither the publisher nor author shall be liable for any loss of profit or any other commercial damages, including but not limited to special, incidental, consequential, or other damages.

For general information on our other products and services, please contact our Customer Care Department within the United States at (800) 762-2974, outside the United States at (317) 572-3993 or fax (317) 572-4002.

Wiley also publishes its books in a variety of electronic formats. Some content that appears in print may not be available in electronic books. For more information about Wiley products, visit our web site at www.wiley.com.

Library of Congress Cataloging-in-Publication Data:
Mechanical engineers' handbook/edited by Myer Kutz.—3rd ed.
p. cm.
Includes bibliographical references and index.
ISBN-13 978-0-471-44990-4
ISBN-10 0-471-44990-3 (cloth)
1. Mechanical engineering—Handbooks, manuals, etc. I. Kutz, Myer.
TJ151.M395 2005
621—dc22

2005008603

Printed in the United States of America.

10 9 8 7 6 5 4 3 2

To Sol, Dorothy, and Jeanne, in Blessed Memory

Contents

Preface

The first volume of the third edition of the *Mechanical Engineers' Handbook* is comprised of two major parts. The first part, Materials, which has 14 chapters, covers metals, plastics, composites, ceramics, and smart materials. The metals covered are carbon, alloy, and stainless steels; aluminum and aluminum alloys; copper and copper alloys; titanium alloys; nickel and its alloys; magnesium and its alloys; and superalloys. Chapters on some of these materials, such as ceramics, smart materials, and superalloys, are updated versions of chapters that have appeared in the *Handbook of Materials Selection* (Wiley, 2002), and they are entirely new to the *Mechanical Engineers' Handbook.* The intent in all of the materials chapters is to provide readers with expert advice on how particular materials are typically used and what criteria make them suitable for specific purposes. This part of Volume I concludes with a chapter on sources of materials data, the intent being to provide readers with guidance on finding reliable information on materials properties, in addition to those that can be found in this volume, and a chapter on analytical methods of materials selection, which is intended to give readers techniques for specifying which materials might be suitable for a particular application.

The second part of Volume I, Mechanical Design, which has 22 chapters, covers a broad range of topics, including the fundamentals of stress analysis, the finite-element method, vibration and shock, and noise measurement and control and then moving into modern methodologies that engineers use to predict failures, eliminate defects, enhance quality and reliability of designs, and optimize designs. There are chapters on failure analysis and design with all classes of materials, including metals, plastics and ceramics, and composites. I should point out that, to a large extent, the two parts of Volume I go hand in hand. After all, it is useful to know about the properties, behavior, and failure mechanisms of all classes of materials when faced with a product design problem. Coverage in the second part of Volume I extends to lubrication of machine elements and seals technology. Chapters in this part of Volume I provide practitioners with techniques to solve real, practical everyday problems, ranging from nondestructive testing to CAD (computer-aided design) to TRIZ (the acronym in Russian for Theory of Inventive Problem Solving), STEP [the Standard for the Exchange of Product Model Data is a comprehensive International Organization for Standardization standard (ISO 10303) that describes how to represent and exchange digital product information], and virtual reality. Topics of special interest include physical ergonomics and electronic packaging.

While many of the chapters in Volume I are updates or entirely new versions of chapters from the second edition of the *Mechanical Engineers' Handbook* and the *Handbook of Materials Selection,* a number of chapters—on Six Sigma, TRIZ, and STEP—are new additions to this family of handbooks. Contributors of the chapters in Volume I include professors, engineers working in industry and government installations, and consultants, mainly from North America, but also from Egypt, the Netherlands, and Germany.

Vision for the Third Edition

Basic engineering disciplines are not static, no matter how old and well established they are. The field of mechanical engineering is no exception. Movement within this broadly based discipline is multidimensional. Even the classic subjects on which the discipline was founded, such as mechanics of materials and heat transfer, continue to evolve. Mechanical engineers continue to be heavily involved with disciplines allied to mechanical engineering, such as industrial and manufacturing engineering, which are also constantly evolving. Advances in other major disciplines, such as electrical and electronics engineering, have significant impact on the work of mechanical engineers. New subject areas, such as neural networks, suddenly become all the rage.

In response to this exciting, dynamic atmosphere, the *Mechanical Engineers' Handbook* is expanding dramatically, from one volume to four volumes. The third edition not only is incorporating updates and revisions to chapters in the second edition, which was published in 1998, but also is adding 24 chapters on entirely new subjects as well, incorporating updates and revisions to chapters in the *Handbook of Materials Selection,* which was published in 2002, as well as to chapters in *Instrumentation and Control,* edited by Chester Nachtigal and published in 1990.

The four volumes of the third edition are arranged as follows:

Volume I: *Materials and Mechanical Design*—36 chapters
- Part 1. Materials—14 chapters
- Part 2. Mechanical Design—22 chapters

Volume II: *Instrumentation, Systems, Controls, and MEMS*—21 chapters
- Part 1. Instrumentation—8 chapters
- Part 2. Systems, Controls, and MEMS—13 chapters

Volume III: *Manufacturing and Management*—24 chapters
- Part 1. Manufacturing—12 chapters
- Part 2. Management, Finance, Quality, Law, and Research—12 chapters

Volume IV: *Energy and Power*—31 chapters
- Part 1: Energy—15 chapters
- Part 2: Power—16 chapters

The mechanical engineering literature is extensive and has been so for a considerable period of time. Many textbooks, reference works, and manuals as well as a substantial number of journals exist. Numerous commercial publishers and professional societies, particularly in the United States and Europe, distribute these materials. The literature grows continuously, as applied mechanical engineering research finds new ways of designing, controlling, measuring, making and maintaining things, and monitoring and evaluating technologies, infrastructures, and systems.

Most professional-level mechanical engineering publications tend to be specialized, directed to the specific needs of particular groups of practitioners. Overall, however, the mechanical engineering audience is broad and multidisciplinary. Practitioners work in a variety of organizations, including institutions of higher learning, design, manufacturing, and consulting firms as well as federal, state, and local government agencies. A rationale for an expanded general mechanical engineering handbook is that every practitioner, researcher, and bureaucrat cannot be an expert on every topic, especially in so broad and multidisciplinary a field, and may need an authoritative professional summary of a subject with which he or she is not intimately familiar.

Starting with the first edition, which was published in 1986, our intention has always been that the *Mechanical Engineers' Handbook* stand at the intersection of textbooks, research papers, and design manuals. For example, we want the handbook to help young engineers move from the college classroom to the professional office and laboratory where they may have to deal with issues and problems in areas they have not studied extensively in school.

With this expanded third edition, we have produced a practical reference for the mechanical engineer who is seeking to answer a question, solve a problem, reduce a cost, or improve a system or facility. The handbook is not a research monograph. The chapters offer design techniques, illustrate successful applications, or provide guidelines to improving the performance, the life expectancy, the effectiveness, or the usefulness of parts, assemblies, and systems. The purpose is to show readers what options are available in a particular situation and which option they might choose to solve problems at hand.

The aim of this expanded handbook is to serve as a source of practical advice to readers. We hope that the handbook will be the first information resource a practicing engineer consults when faced with a new problem or opportunity—even before turning to other print sources, even officially sanctioned ones, or to sites on the Internet. (The second edition has been available online on knovel.com.) In each chapter, the reader should feel that he or she is in the hands of an experienced consultant who is providing sensible advice that can lead to beneficial action and results.

Can a single handbook, even spread out over four volumes, cover this broad, interdisciplinary field? We have designed the third edition of the *Mechanical Engineers' Handbook* as if it were serving as a core for an Internet-based information source. Many chapters in the handbook point readers to information sources on the Web dealing with the subjects addressed. Furthermore, where appropriate, enough analytical techniques and data are provided to allow the reader to employ a preliminary approach to solving problems.

The contributors have written, to the extent their backgrounds and capabilities make possible, in a style that reflects practical discussion informed by real-world experience. We would like readers to feel that they are in the presence of experienced teachers and consultants who know about the multiplicity of technical issues that impinge on any topic within mechanical engineering. At the same time, the level is such that students and recent graduates can find the handbook as accessible as experienced engineers.

Contributors

Brian A. Baker
Special Metals Corporation
Huntington, West Virginia

H. Barry Bebb
ASI
San Diego, California

Bruce L. Bramfitt
Research Laboratories
International Steel Group, Inc.
Bethlehem, Pennsylvania

Robert E. Brown
Magnesium Monthly Review
Prattville, Alabama

J. A. Collins
The Ohio State University
Columbus, Ohio

John G. Cowie
Copper Development Association
New York, New York

Robert L. Crane
Wright Patterson Air Force Base
Dayton, Ohio

S. R. Daniewicz
Mississippi State University
Starkville, Mississippi

B. S. Dhillon
University of Ottawa
Ottawa, Ontario, Canada

Jaap H. van Dieën
Vrije Universiteit
Amsterdam, The Netherlands

George M. Diehl
Machinery Acoustics
Phillipsburg, New Jersey

Matthew J. Donachie
Rennselaer at Hartford
Hartford, Connecticut

Stephen J. Donachie
Special Metals Corporation
Huntington, West Virginia

Warren C. Fackler
Telesis Systems
Cedar Rapids, Iowa

Mahmoud M. Farag
The American University in Cairo
Cairo, Egypt

Franklin E. Fisher
Loyola Marymount University
Los Angeles, Calaifornia
and
Raytheon Company
El Segundo, California

S. W. Freiman
National Institute of Standards and Technology
Gaithersburg, Maryland

E. R. Fuller, Jr.
National Institute of Standards and Technology
Gaithersburg, Maryland

Bernard J. Hamrock
The Ohio State University
Columbus, Ohio

Martin Hardwick
STEP Tools, Inc.
Troy, New York

James A. Harvey
Under the Bridge Consulting, Inc.
Corvallis, Oregon

R. Nathan Katz
Worcester Polytechnic Institute
Worcester, Massachusetts

J. G. Kaufman
Kaufman Associates
Columbus, Ohio

James Kelly
Rochester, Michigan

Jeremy S. Knopp
Wright Patterson Air Force Base
Dayton, Ohio

Konrad J. A. Kundig
Metallurgical Consultant
Tucson, Arizona

Ming C. Leu
University of Missouri-Rolla
Rolla, Missouri

James E. McMunigal
MCM Associates
Long Beach, California

Ruth E. McMunigal
MCM Associates
Long Beach, California

Dietrich Munz
University of Karlsruhe
Karlsruhe, Germany

Maury A. Nussbaum
Virginia Polytechnic Institute and State University
Blacksburg, Virginia

Xiaobo Peng
University of Missouri-Rolla
Rolla, Missouri

Edward N. Peters
General Electric Company
Selkirk, New York

Singiresu S. Rao
University of Miami
Coral Gables, Florida

A. Ravi Ravindran
Pennsylvania State University
University Park, Pennsylvania

G. V. Reklaitis
Purdue University
West Lafayette, Indiana

Michael Slocum
Breakthrough Management Group
Longmont, Colorado

Vishu H. Shah
Consultek
Brea, California

Gaylord D. Smith
Special Metals Corporation
Huntington, West Virginia

Bruce M. Steinetz
NASA Glenn Research Center at Lewis Field
Cleveland, Ohio

Steven Ungvari
Strategic Product Innovations, Inc.
Columbus, Ohio

G. S. White
National Institute of Standards and Technology
Gaithersburg, Maryland

Emory W. Zimmers, Jr..
Lehigh University
Bethlehem, Pennsylvania

Tarek I. Zohdi
University of California
Berkeley, California

Carl Zweben
Devon, Pennsylvania

Mechanical Engineers' Handbook

PART 1

MATERIALS

CHAPTER 1

CARBON AND ALLOY STEELS

Bruce L. Bramfitt
International Steel Group, Inc.
Research Laboratories
Bethlehem, Pennsylvania

1 INTRODUCTION

Steel is the most common and widely used metallic material in today's society. It can be cast or wrought into numerous forms and can be produced with tensile strengths exceeding 5 GPa. A prime example of the versatility of steel is in the automobile where it is the material of choice and accounts for over 60% of the weight of the vehicle. Steel is highly formable as seen in the contours of the automobile outerbody. Steel is strong and is used in the body frame, motor brackets, driveshaft, and door impact beams of the vehicle. Steel is corrosion resistant when coated with the various zinc-based coatings available today. Steel is dent resistant when compared with other materials and provides exceptional energy absorption in a vehicle collision. Steel is recycled and easily separated from other materials by a magnet. Steel is inexpensive compared with other competing materials such as aluminum and various polymeric materials.

In the past, steel has been described as an alloy of iron and carbon. Today, this description is no longer applicable since in some very important steels, e.g., interstitial-free (IF) steels and type 409 ferritic stainless steels, carbon is considered an impurity and is present in quantities of only a few parts per million. By definition, steel must be at least 50% iron and must contain one or more alloying elements. These elements generally include carbon, manganese, silicon, nickel, chromium, molybdenum, vanadium, titanium, niobium, and aluminum. Each chemical element has a specific role to play in the steelmaking process or in achieving particular properties or characteristics, e.g., strength, hardness, corrosion resistance, magnetic permeability, and machinability.

Reprinted from *Handbook of Materials Selection,* Wiley, New York, 2002, by permission of the publisher.

2 STEEL MANUFACTURE

In most of the world, steel is manufactured by integrated steel facilities that produce steel from basic raw materials, i.e., iron ore, coke, and limestone. However, the fastest growing segment of the steel industry is the "minimill" that melts steel scrap as the raw material. Both types of facilities produce a wide variety of steel forms, including sheet, plate, structural, railroad rail, and bar products.

Ironmaking. When making steel from iron ore, a blast furnace chemically reduces the ore (iron oxide) with carbon in the form of coke. Coke is a spongelike carbon mass that is produced from coal by heating the coal to expel the organic matter and gasses. Limestone (calcium carbonate) is added as a flux for easier melting and slag formation. The slag, which floats atop the molten iron, absorbs many of the unwanted impurities. The blast furnace is essentially a tall hollow cylindrical structure with a steel outer shell lined on the inside with special refractory and graphite brick. The crushed or pelletized ore, coke, and limestone are added as layers through an opening at the top of the furnace, and chemical reduction takes place with the aid of a blast of preheated air entering near the bottom of the furnace (an area called the bosh). The air is blown into the furnace through a number of water-cooled copper nozzles called tuyeres. The reduced liquid iron fills the bottom of the furnace and is tapped from the furnace at specified intervals of time. The product of the furnace is called pig iron because in the early days the molten iron was drawn from the furnace and cast directly into branched mold configurations on the cast house floor. The central branch of iron leading from the furnace was called the "sow" and the side branches were called "pigs." Today the vast majority of pig iron is poured directly from the furnace into a refractory-lined vessel (submarine car) and transported in liquid form to a basic oxygen furnace (BOF) for refinement into steel.

Steelmaking. In the BOF, liquid pig iron comprises the main charge. Steel scrap is added to dilute the carbon and other impurities in the pig iron. Oxygen gas is blown into the vessel by means of a top lance submerged below the liquid surface. The oxygen interacts with the molten pig iron to oxidize undesirable elements. These elements include excess carbon (because of the coke used in the blast furnace, pig iron contains over 2% carbon), manganese, and silicon from the ore and limestone and other impurities like sulfur and phosphorus. While in the BOF, the liquid metal is chemically analyzed to determine the level of carbon and impurity removal. When ready, the BOF is tilted and the liquid steel is poured into a refractory-lined ladle. While in the ladle, certain alloying elements can be added to the steel to produce the desired chemical composition. This process takes place in a ladle treatment station or ladle furnace where the steel is maintained at a particular temperature by external heat from electrodes in the lid placed on the ladle. After the desired chemical composition is achieved, the ladle can be placed in a vacuum chamber to remove undesirable gases such as hydrogen and oxygen. This process is called degassing and is used for higher quality steel products such as railroad rail, sheet, plate, bar, and forged products. Stainless steel grades are usually produced in an induction or electric arc furnace, sometimes under vacuum. To refine stainless steel, the argon–oxygen decarburization (AOD) process is used. In the AOD, an argon–oxygen gas mixture is injected through the molten steel to remove carbon without a substantial loss of chromium (the main element in stainless steel).

Continuous Casting. Today, most steel is cast into solid form in acontinuous-casting (also called strand casting) machine. Here, the liquid begins solidification in a water-cooled copper mold while the steel billet, slab, or bloom is withdrawn from the bottom of the mold. The partially solidified shape is continuously withdrawn from the machine and cut to length for further processing. The continuous-casting process can proceed for days or weeks as ladle after ladle of molten steel feeds the casting machine. Some steels are not continuously cast but are poured into individual cast-iron molds to form an ingot that is later reduced in size by forging or a rolling process to some other shape. Since the continuous-casting process offers substantial economic and quality advantages over ingot casting, most steel in the world is produced by continuous casting.

Rolling/Forging. Once cast into billet, slab, or bloom form, the steel is hot rolled through a series of rolling mills or squeezed/hammered by forging to produce the final shape. To form hot-rolled sheet, a 50–300-mm-thick slab is reduced to final thickness, e.g., 2 mm, in one or more roughing stands followed by a series of six or seven finishing stands. To obtain thinner steel sheet, e.g., 0.5 mm, the hot-rolled sheet must be pickled in acid to remove the iron oxide scale and further cold rolled in a series of rolling stands called a tandem mill. Because the cold-rolling process produces a hard sheet with little ductility, it is annealed either by batch annealing or continuous annealing. New casting technology is emerging where thin sheet (under 1 mm) can be directly cast from the liquid through water-cooled, rotating rolls that act as a mold as in continuous casting. This new process eliminates many of the steps in conventional hot-rolled sheet processing. Plate steels are produced by hot rolling a slab in a reversing roughing mill and a reversing finishing mill. Steel for railway rails is hot rolled from a bloom in a blooming mill, a roughing mill, and one or more finishing mills. Steel bars are produced from a heated billet that is hot rolled in a series of roughing and finishing mills. Forged steels are produced from an ingot that is heated to forging temperature and squeezed or hammered in a hydraulic press or drop forge. The processing sequence in all these deformation processes can vary depending on the design, layout, and age of the steel plant.

3 DEVELOPMENT OF STEEL PROPERTIES

In order to produce a steel product with the desired properties, basic metallurgical principles are used to control three things:

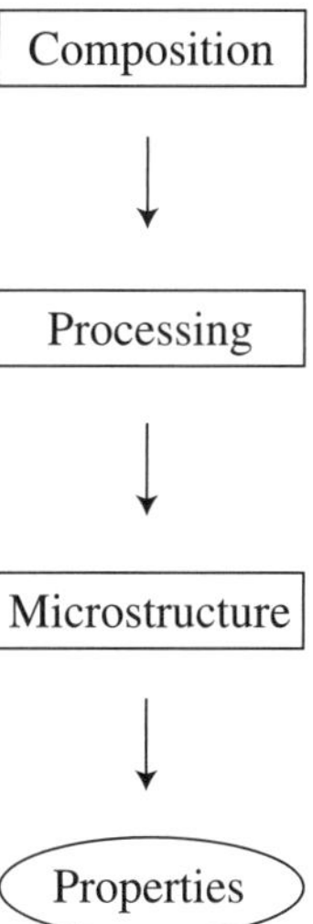

This means that the steel composition and processing route must be closely controlled in order to produce the proper microstructure. The final microstructure is of utmost importance in determining the properties of the steel product. This section will explore how various microstructures are developed and the unique characteristics of each microstructural component in steel. The next section will discuss how alloy composition also plays a major role.

Iron–Carbon Equilibrium Diagram. Since most steels contain carbon, the basic principles of microstructural development can be explained by the iron–carbon equilibrium diagram. This diagram, shown in Fig. 1, is essentially a map of the phases that exist in iron at various carbon contents and temperatures under equilibrium conditions. Iron is an interesting chemical element in that it undergoes three phase changes when heated from room temperature to liquid. For example, from room temperature to 912°C, pure iron exists as ferrite (also called alpha iron), from 912 to 1394°C, it exists as austenite (gamma iron), from 1394 to 1538°C it exists as ferrite again (delta iron), and above 1538°C it is liquid. In other words, upon heating, iron undergoes allotropic phase transformations from ferrite to austenite at 912°C, austenite to ferrite at 1394°C, and ferrite to liquid at 1538°C. Each transformation undergoes a change in crystal structure or arrangement of the iron atoms in the crystal lattice. It must be remembered that all chemical elements in their solid form have specific arrangements of atoms that are essentially the basic building blocks in producing the element in the form that we physically observe. These atomic arrangements form a latticework containing billions of atoms all aligned in a systematic way. Some of these lattices have a cubic arrangement, with an atom at each corner of the cube and another atom at the cube center. This arrangement is called body-centered-cubic (bcc). Others have an atom at each corner of the cube and atoms at the center of each face of the cube. This is called face-centered-cubic (fcc). Other arrangements are hexagonal, some are tetragonal, etc. As an example, pure iron as ferrite has a bcc arrangement. Austenite has a fcc arrangement. Upon heating, bcc ferrite will transform to fcc austenite at 912°C. These arrangements or crystal structures impart different properties to steel. For example, a bcc ferritic stainless steel will have properties much different from a fcc austenitic stainless steel, as described later in this chapter.

Since pure iron is very soft and of low strength, it is of little interest commercially. Therefore, carbon and other alloying elements are added to enhance properties. Adding carbon to pure iron has a profound effect on ferrite and austenite, discussed above. One way to understand the effect of carbon is to examine the iron–carbon diagram (Fig. 1). This is a binary (two-element) diagram of temperature and composition (carbon content) constructed under near-equilibrium conditions. In this diagram, as carbon is added to iron, the ferrite and austenite phase fields expand and contract depending upon the carbon level and temperature. Also, there are fields consisting of two phases, e.g., ferrite plus austenite.

Since carbon has a small atomic diameter when compared with iron, it is called an interstitial element because it can fill the interstices between the iron atoms in the cubic lattice. Nitrogen is another interstitial element. On the other hand, elements such as manganese, silicon, nickel, chromium, and molybdenum have atomic diameters similar to iron and are called substitutional alloying elements. These substitutional elements can thus replace iron atoms at the cube corners, faces, or center positions. There are many binary phase diagrams (Fe–Mn, Fe–Cr, Fe–Mo, etc.) and tertiary-phase diagrams (Fe–C–Mn, Fe–C–Cr, etc.) showing the effect of interstitial and substitutional elements on the phase fields of ferrite and austenite. These diagrams are found in the handbooks listed at the end of the chapter.

Being an interstitial or a substitutional element is important in the development of steel properties. Interstitial elements such as carbon can move easily about the crystal lattice whereas a substitutional element such as manganese is much more difficult to move. The movement of elements in a crystal lattice is called diffusion. Diffusion is a controlling factor

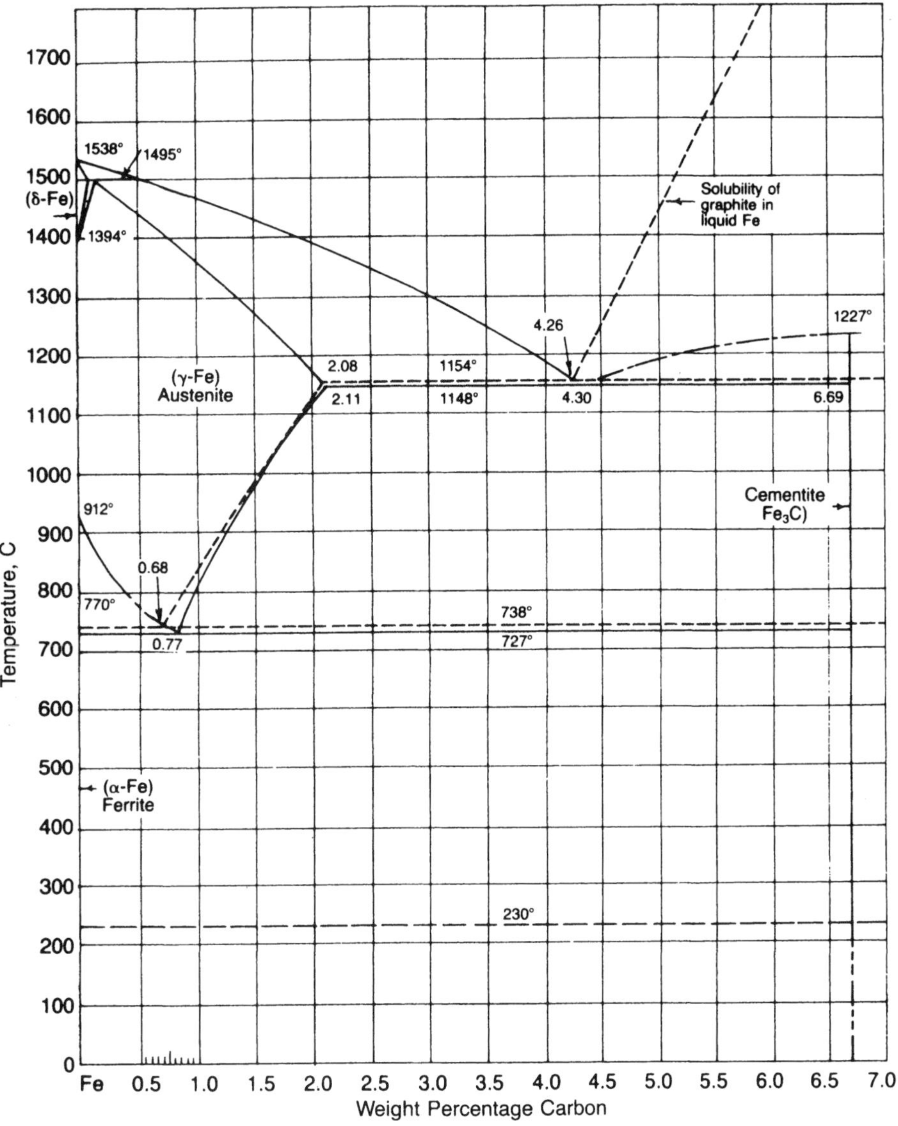

Figure 1 Iron–carbon binary-phase diagram. (*Source: Steels: Heat Treatment and Processing Principles,* ASM International, Materials Park, OH 44073-0002, 1990, p. 2.)

in the development of microstructure. Another factor is solubility, which is a measure of how much of a particular element can be accommodated by the crystal lattice before it is rejected. In metals when two or more elements are soluble in the crystal lattice, a solid solution is created (somewhat analogous to a liquid solution of sugar in hot coffee). For example, when added to iron, carbon has very limited solubility in ferrite but is about 100 times more soluble in austenite, as seen in the iron–carbon diagram in Fig. 2 (a limited version of the diagram in Fig. 1). The maximum solubility of carbon in ferrite is about

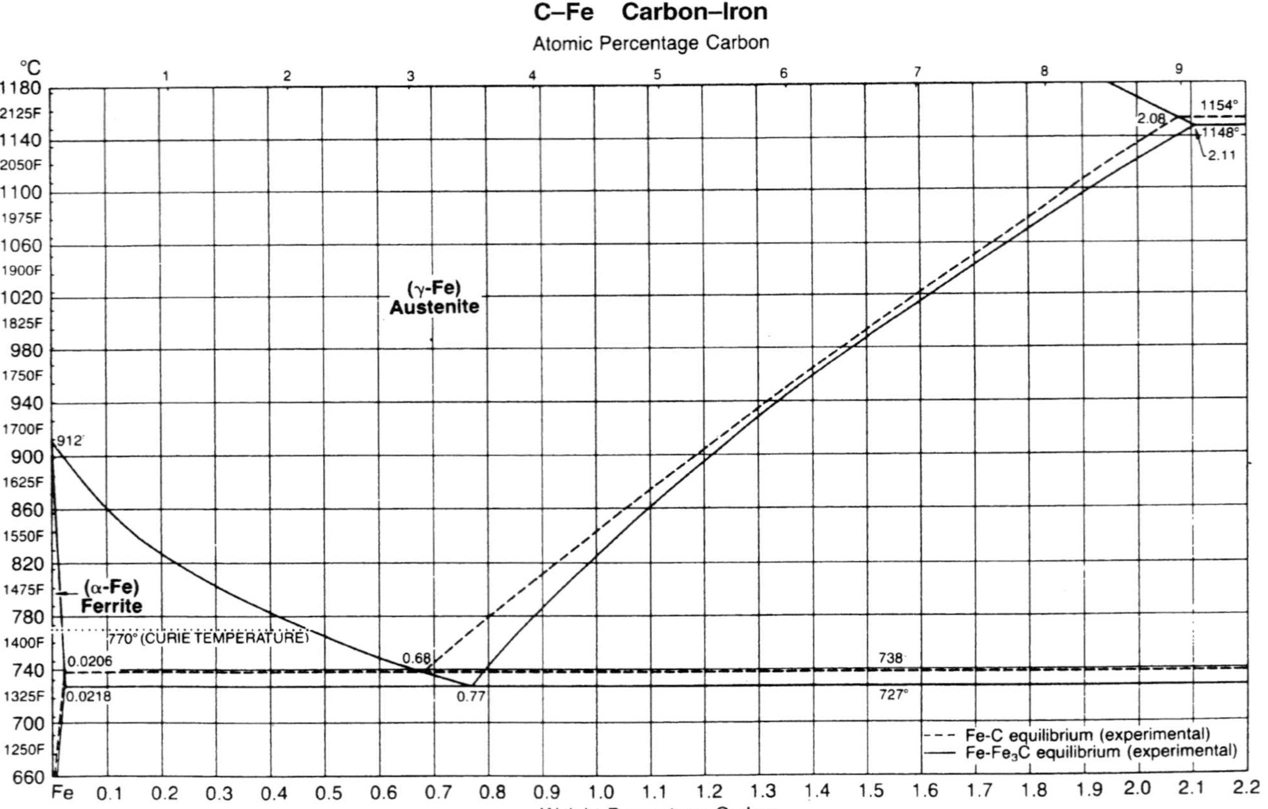

Figure 2 Expanded portion of the iron–carbon binary-phase diagram in Fig. 1. (*Source: Steels: Heat Treatment and Processing Principles,* ASM International, Materials Park, OH 44073-0002, 1990, p. 18.)

0.022% C at 727°C while the maximum solubility of carbon in austenite is 100 times more, 2.11% C at 1148°C. At room temperature the solubility of carbon in iron is only about 0.005%. Any amount of carbon in excess of the solubility limit is rejected from solid solution and is usually combined with iron to form an iron carbide compound called cementite. This hard and brittle compound has the chemical formula Fe_3C and a carbon content of 6.7%. This is illustrated in the following two examples. The first example is a microstructure of a very low carbon steel (0.002% C), shown in Fig. 3*a*. The microstructure consists of only ferrite grains (crystals) and grain boundaries. The second example is a microstructure of a low-carbon steel containing 0.02% C, in Fig. 3*b*. In this microstructure, cementite can be seen as particles at the ferrite grain boundaries. The excess carbon rejected from the solid solution of ferrite formed this cementite. As the carbon content in steel is increased, another form of cementite appears as a constituent called pearlite, which can be found in most carbon steels. Examples of pearlite in low-carbon (0.08% C) and medium-carbon (0.20% C) steels are seen in Figs. 4*a* and 4*b*. Pearlite has a lamellar (parallel-plate) microstructure, as shown at higher magnification in Fig. 5, and consists of layers of ferrite and cementite. Thus, in these examples, in increasing the carbon level from 0.002 to 0.02 to 0.08 to 0.20%, the excess carbon is manifested as a carbide phase in two different forms, cementite particles and cementite in pearlite. Both forms increase the hardness and strength of iron. However, there is a trade-off; cementite also decreases ductility and toughness.

Pearlite forms on cooling austenite through a eutectoid reaction as seen below:

$$\text{Austenite} \leftrightarrow Fe_3C + \text{ferrite}$$

A *eutectoid* reaction occurs when a solid phase or constituent reacts to form two different solid constituents on cooling (a *eutectic* reaction occurs when a liquid phase reacts to form two solid phases). The eutectoid reaction is reversible on heating. In steel, the eutectoid reaction (under equilibrium conditions) takes place at 727°C and can be seen on the iron–carbon diagram (Fig. 1) as the "V" at the bottom left side of the diagram. A fully pearlitic microstructure forms at 0.77% C at the eutectoid temperature of 727°C (the horizontal line on the left side of the iron–carbon diagram). Steels with less than 0.77% C are called *hypoeutectoid* steels and consist of mixtures of ferrite and pearlite with the amount of pearlite increasing as the carbon content increases. The ferrite phase is called a *proeutectoid* phase because it forms prior to the eutectoid transformation that occurs at 727°C. A typical example of proeutectoid ferrite is shown in Fig. 6. In this photomicrograph, the ferrite (the white-appearing constituent) formed on the prior austenite grain boundaries of hypoeutectoid steel with 0.60% C. The remaining constituent (dark appearing) is pearlite. Steels between 0.77% C and about 2% C are called *hypereutectoid* steels and consist of pearlite with proeutectoid cementite. Cementite forms a continuous carbide network at the boundaries of the prior austenite grains. Because there is a carbide network, hypereutectoid steels are characterized as steels with little or no ductility and very poor toughness. This means that in the commercial world the vast majority of carbon steels are hypoeutectoid steels.

Thus, according to the iron–carbon diagram, steels that are processed under equilibrium or near-equilibrium conditions can form (a) pure ferrite at very low carbon levels generally under 0.005% C, (b) ferrite plus cementite particles at slightly higher carbon levels between 0.005% C and 0.022% C, (c) ferrite plus pearlite mixtures between 0.022% C and 0.77% C, (d) 100% pearlite at 0.77% C, and (e) mixtures of pearlite plus cementite networks between 0.77% C and 2% C. The higher the percentage of cementite, the higher the hardness and strength and lower the ductility and toughness of the steel.

Departure from Equilibrium (*Real World*). Industrial processes do not occur at equilibrium, and only those processes that take place at extremely slow heating and cooling rates can be considered near equilibrium, and these processes are quite rare. Therefore, under real con-

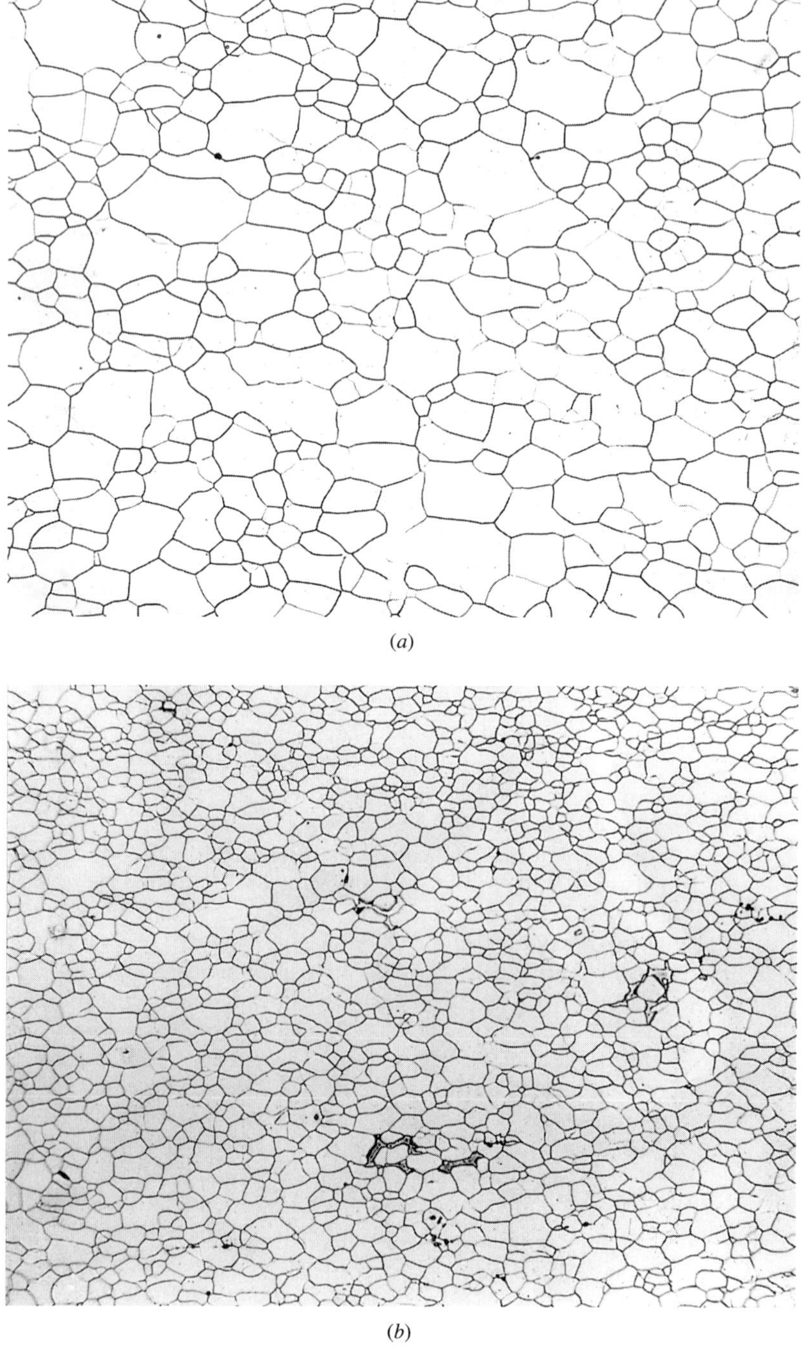

Figure 3 (*a*) Photomicrograph of a very low carbon steel showing ferrite grains and (*b*) photomicrograph of a low-carbon steel showing ferrite grains with some cementite on the ferrite grain boundaries. (*a*) 500X and (*b*) 200X. Marshalls etch.

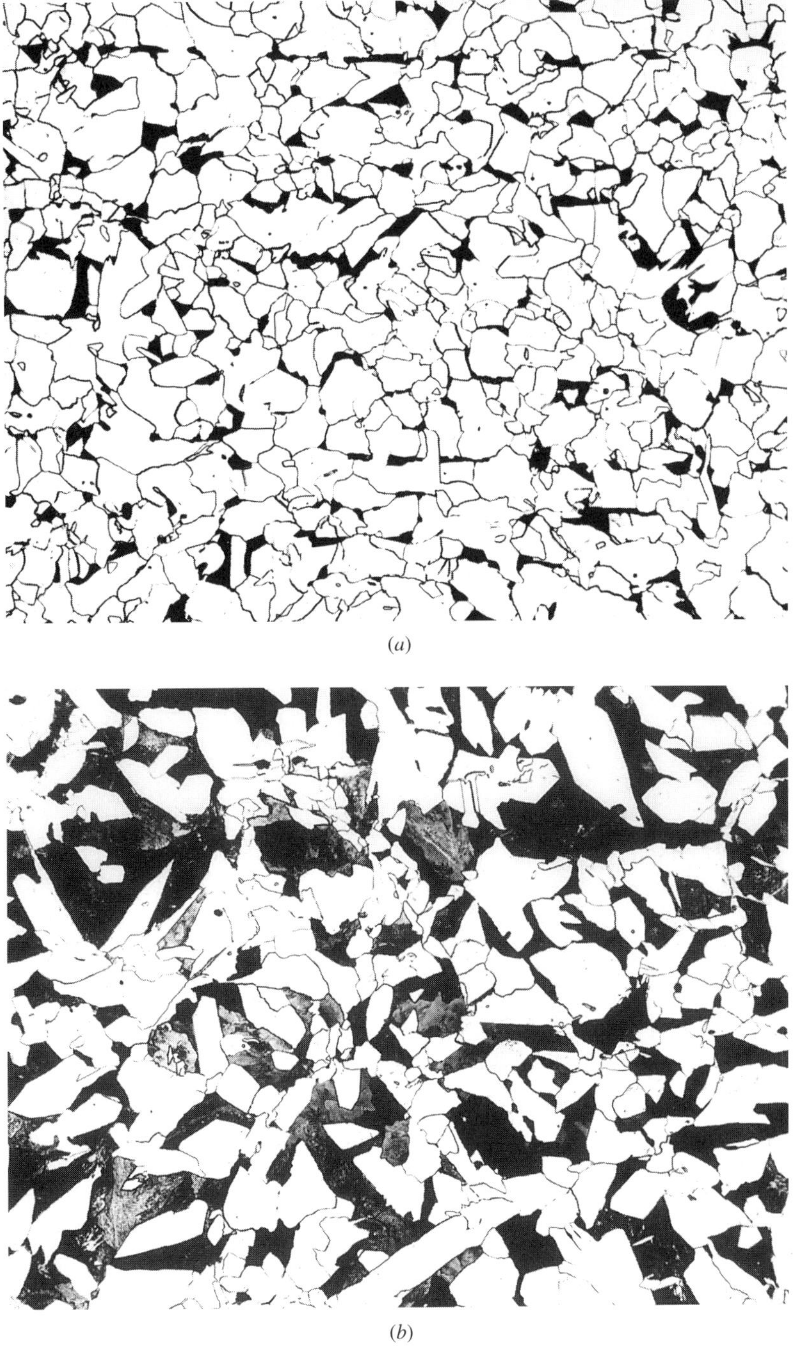

Figure 4 (*a*) Photomicrograph of an SAE/AISI 1008 steel showing ferrite grains and pearlite (dark) and (*b*) photomicrograph of an SAE/AISI 1020 steel showing ferrite grains with an increased amount of pearlite. (*a*) and (*b*) both 200X. 4% picral + 2% nital etch.

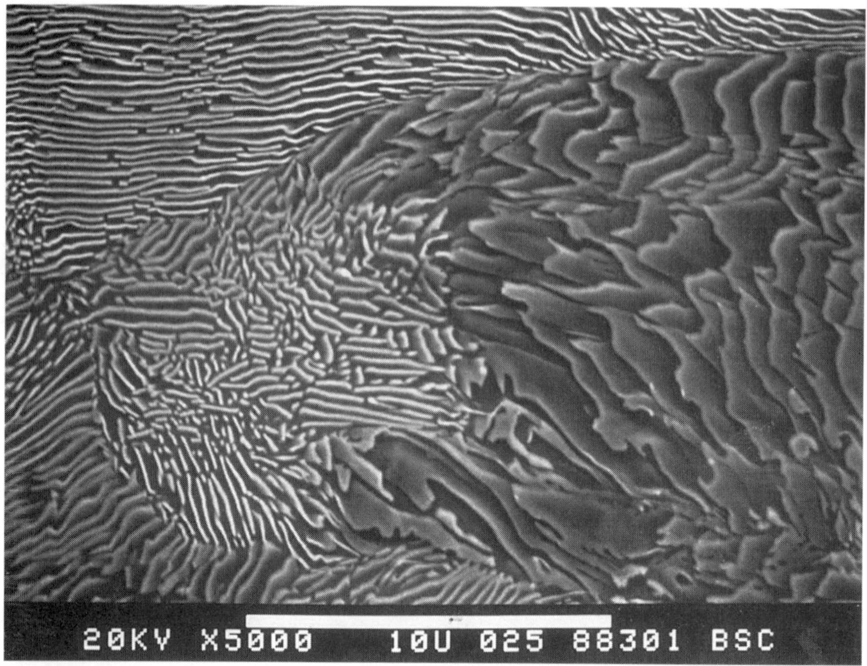

Figure 5 Scanning electron micrograph of pearlite showing the platelike morphology of the cementite. 5000X. 4% picral etch.

ditions, the iron–carbon diagram can only be used as a rough guideline since the equilibrium transformation temperatures shift to lower temperatures on cooling and to higher temperatures on heating. If steels are cooled at very fast rates, e.g., quenching in water, the iron–carbon diagram can no longer be used since there is a major departure from equilibrium. In fact, during the quenching of steel, new constituents form that are not associated with the iron–carbon diagram. Therefore, at fast cooling rates the concept of time–temperature transformation (TTT) diagrams must be considered. These diagrams are constructed under isothermal (constant) temperature (called IT diagrams) or continuous-cooling conditions (called CT diagrams). It is important to know how these diagrams are constructed so that we can understand the development of nonequilibrium microstructures, which are so important in carbon and alloy steels.

Isothermal Transformation Diagram. This diagram is formed by quenching very thin specimens of steel in salt baths set at various temperatures. For example, thin specimens of 0.79% C steel can be quenched into seven different liquid salt baths set at 650, 600, 550, 500, 450, 400, and 200°C. The specimens are held for various times at each temperature then pulled from the bath and quickly quenched in cold water. The result will be a diagram called an isothermal transformation (IT) diagram, as shown in Fig. 7. The diagram is essentially a map showing where various constituents form. For example, at 650°C, austenite (A) begins to transform to pearlite if held in the bath for 10 s. The curve drawn through this point is the pearlite transformation start temperature and is labeled beginning of transformation in Fig. 7. At about 100 s the pearlite transformation is finished. The second curve represents the pearlite transformation finish temperature and is labeled the end of transfor-

Figure 6 Photomicrograph of a medium-carbon hypoeutectoid steel showing a pearlite matrix and proeutectoid ferrite nucleating on the original (prior) austenite grain boundaries. 200X. 4% picral + 2% nital etch.

mation in Fig. 7. In this steel, pearlite forms at all temperatures along the start of the transformation curve from 727°C (the equilibrium temperature of the iron–carbon diagram) to 540°C, the "nose" of the curve. At the higher transformation temperatures, the pearlite interlamellar spacing (the spacing between cementite plates) is very coarse and decreases in spacing as the temperature is decreased, i.e., nose of the IT diagram is approached. This is an important concept since a steel with a coarse pearlite interlamellar spacing is softer and of lower strength than a steel with a fine pearlite interlamellar spacing. Commercially, rail steels are produced with a pearlitic microstructure, and it has been found that the finer the interlamellar spacing, the harder the rail and the better the wear resistance. This means that rails will last longer in track if produced with the finest spacing allowable. Most rail producers employ an accelerated cooling process called head hardening to obtain the necessary conditions to achieve the finest pearlite spacing in the rail head (the point of wheel contact).

If the specimens are quenched to 450°C and held for various times, pearlite does not form. In fact, pearlite does not isothermally transform at transformation temperatures (in this case, salt pot temperatures) below the nose of the diagram in Fig. 7. The new constituent is called bainite, which consists of ferrite laths with small cementite particles (also called precipitates). An example of the microstructure of bainite is shown in Fig. 8. This form of bainite is called upper bainite because it is formed in the upper portion below the nose of the IT diagram (between about 540 and 400°C). Lower bainite, a finer ferrite–carbide microstructure, forms at lower temperatures (between 400 and about 250°C). Bainite is an important constituent in tough, high-strength, low-alloy steel.

If specimens are quenched into a salt bath at 200°C, a new constituent called martensite will form. The start of the martensitic transformation is shown in Fig. 7 as M_s (at 220°C).

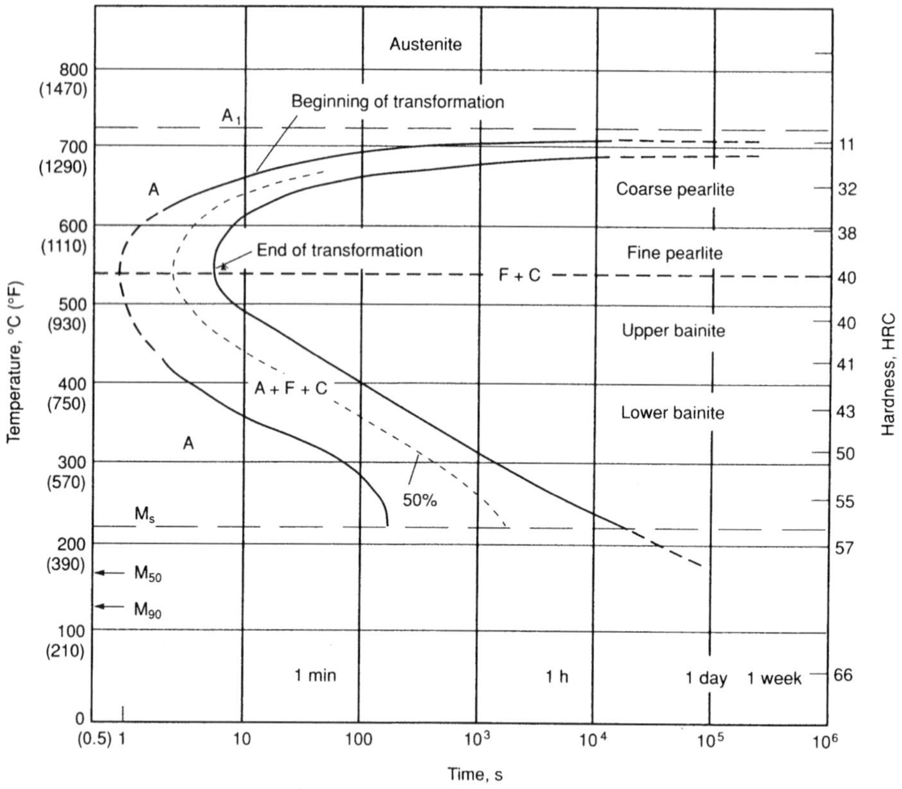

Figure 7 Isothermal transformation diagram of SAE/AISI 1080 steel showing the beginning and end of transformation curves with temperature and time. (*Source: ASM Handbook, Vol. 1, Properties and Selection: Irons, Steels, and High-Performance Alloys,* ASM International, Materials Park, OH 44073-0002, 1990, p. 128.)

Martensite is a form of ferrite that is supersaturated with carbon. In other words, because of the very fast cooling rate, the carbon atoms do not have time to diffuse from their interstitial positions in the bcc lattice to form cementite particles. An example of martensite is shown in Fig. 9. Steel products produced with an as-quenched martensitic microstructure are very hard and brittle, e.g., a razor blade. Most martensitic products are tempered by heating to temperatures between about 350 and 650°C. The tempering process allows some of the carbon to diffuse and form as a carbide phase from the supersaturated iron lattice. This softens the steel and provides some ductility. The degree of softening is determined by the tempering temperature and the time at the tempering temperature. The higher the temperature and the longer the time, the softer the steel. Most steels with martensite are used in the quenched and tempered condition.

Continuous-Cooling Transformation Diagram. The other more useful form of a time–temperature transformation diagram is the continuous-cooling transformation (CT) diagram. This differs from the IT diagram in that it is constructed by cooling small specimens at various cooling rates and measuring the temperatures at which transformations start and finish using a device called a dilatometer (a machine that measures dilation). Each phase transfor-

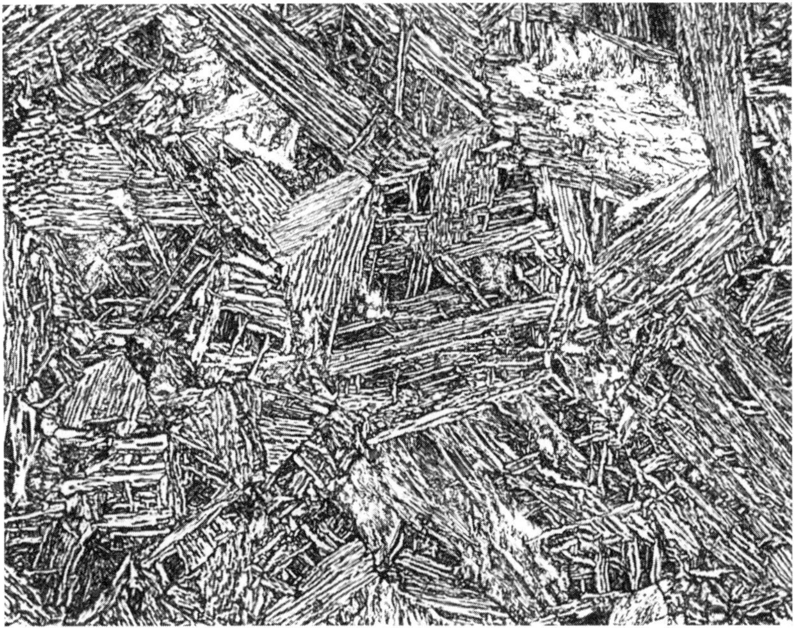

Figure 8 Photomicrograph of a low-alloy steel showing a bainitic microstructure. 500X. 4% picral + 2% nital etch.

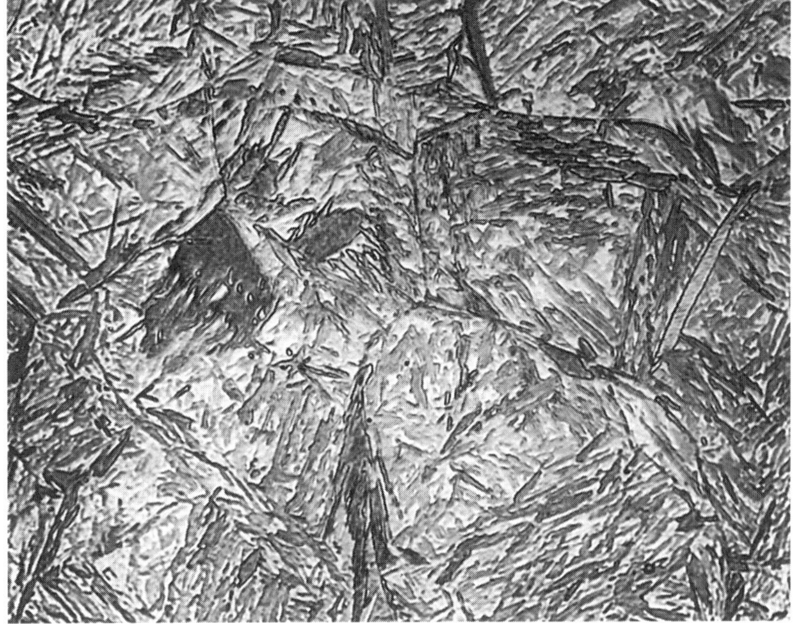

Figure 9 Photomicrograph of a low-alloy steel showing a martensitic microstructure. 1000X. 4% picral + HCl and 10% sodium metabisulfate etch.

mation undergoes a distinct volume change (positive on cooling and negative on heating) that can be measured by a sensitive length-measuring device in the dilatometer. A CT diagram has similar features to the IT diagram shown in Fig. 7 but is produced by continuous cooling rather than isothermal conditions. A continuous-cooling diagram is applicable for most industrial processes and should be used in lieu of an IT diagram. A CT diagram can also be constructed by quenching one end of a Jominy bar described below.

Hardenability Concept. In thick products, e.g., large-diameter bars, thick plates, and heavy forgings, the through-thickness properties are achieved through hardenability. Hardenability is the ability to induce depth of hardness in a steel product. The hardness level is obtained by controlling the amount of martensite in the microstructure. To increase the depth of hardness, certain alloying elements are added to the steel for increased hardenability. Elements, such as nickel, chromium, and molybdenum, shift the pearlite nose of the IT and CT diagrams to the right (longer times). With the nose out of the way on cooling, martensite can be formed over a wider range of cooling rates when compared with a steel without alloying elements.

There is a fairly simple test to measure the hardenability of steel called the Jominy test. A 25.4-mm-diameter and 102-mm-long bar is austenitized to 845°C for 1 h and then water quenched at one end of the bar. The quenching takes place in a specially designed fixture where the bar is suspended in the vertical position and water is directed against the machined bottom end face of the bar. After quenching, parallel flats 0.38 mm deep are machined on opposite sides of the bar. Hardness is measured at 1.6-mm ($\frac{1}{16}$-in.) intervals from the quenched end. The hardness is plotted against depth from the quenched end to produce a hardenability curve or band. A hardenability band for medium-carbon Society of Automotive Engineers/American Iron and Steel Institute (SAE/AISI) 1045 steel is shown in Fig. 10*a*. The two curves that form the band represent the maximum and minimum hardness values from many Jominy tests. To illustrate the concept of hardenability, compare the hardenability band for SAE/AISI 1045 steel to low-alloy SAE/AISI 4145 steel in Fig. 10*b*. These steels are similar except that the low-alloy steel has chromium and molybdenum additions as shown below:

C	Mn	Si	Cr	Mo
0.42/0.51	0.50/1.00	0.15/0.35	—	—
0.42/0.49	0.65/1.10	0.15/0.35	0.75/1.20	0.15/0.25

As can be seen from the hardenability bands, the higher manganese, chromium, and molybdenum additions in the SAE/AISI 4145 steel produced a much greater depth of hardness than the plain-carbon steel. For example, a hardness of HRC 45 (Rockwell *C* scale) was achieved at a depth of only 3–6.5 mm in the SAE/AISI 1045 steel compared with a hardness of HRC 45 at a depth of 21–50 mm in the SAE/AISI 4145 steel. This low-alloy steel has many times the depth of hardness or hardenability of the plain-carbon steel. This means that a hardness of HRC 45 can be achieved in the center of a 100-mm-diameter bar of SAE/AISI 4145 steel compared to a 10-mm-diameter bar of SAE/AISI 1045 steel (both water quenched). The depth of hardness is produced by forming martensite near the quenched end of the bar with mixtures of martensite and bainite further in from the end and eventually bainite at the maximum depth of hardness. Hardenability is important since hardness is roughly proportional to tensile strength. To convert hardness to an approximate tensile strength, the conversion table in the American Society for Testing and Materials (ASTM) E140 can be used. A portion of this table is

Hardness			
Rockwell C Scale	Vickers	Brinell 3000-kg Load	Approximate Tensile Strength (MPa)
60	697	(654)	—
55	595	560	2075
50	513	481	1760
45	446	421	1480
40	392	371	1250
35	345	327	1080
30	302	286	950
25	266	253	840

Heat-treating temperatures recommended by SAE
Normalize (for forged or rolled specimens only): 870 °C (1600 °F)
Austenitize: 845 °C (1550 °F)

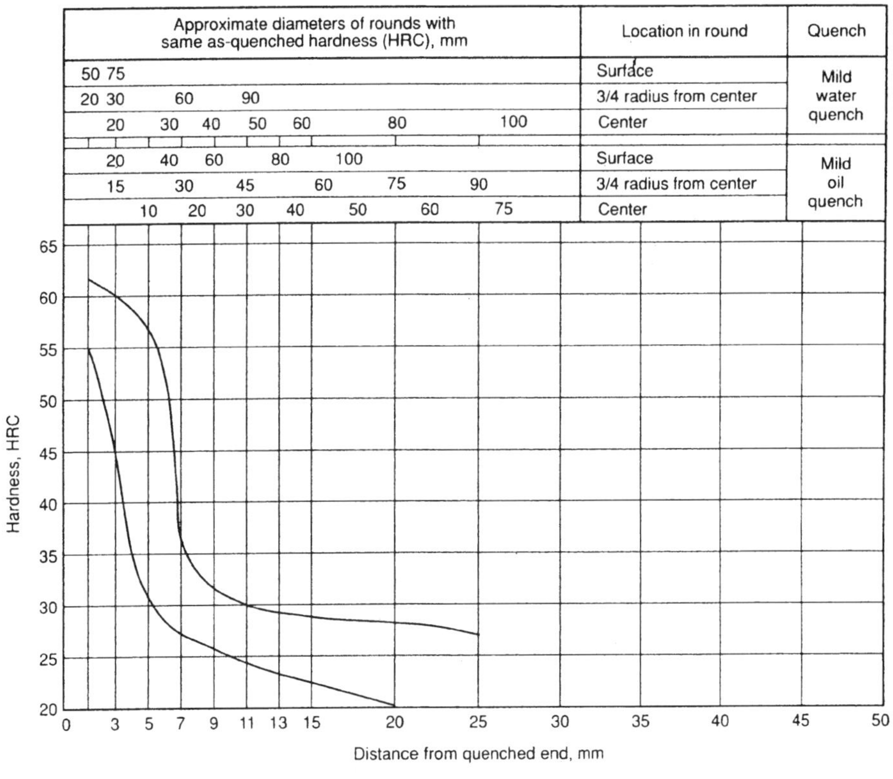

(a)

Figure 10 Hardenability curves for (*a*) SAE/AISI 1045 and (*b*) SAE/AISI 4145 showing depth of hardness with distance from the quenched end of a Jominy bar (*Source: ASM Handbook, Vol. 1, Properties and Selection: Irons, Steels, and High-Performance Alloys,* ASM International, Materials Park, OH 44073-0002, 1997, p. 487.)

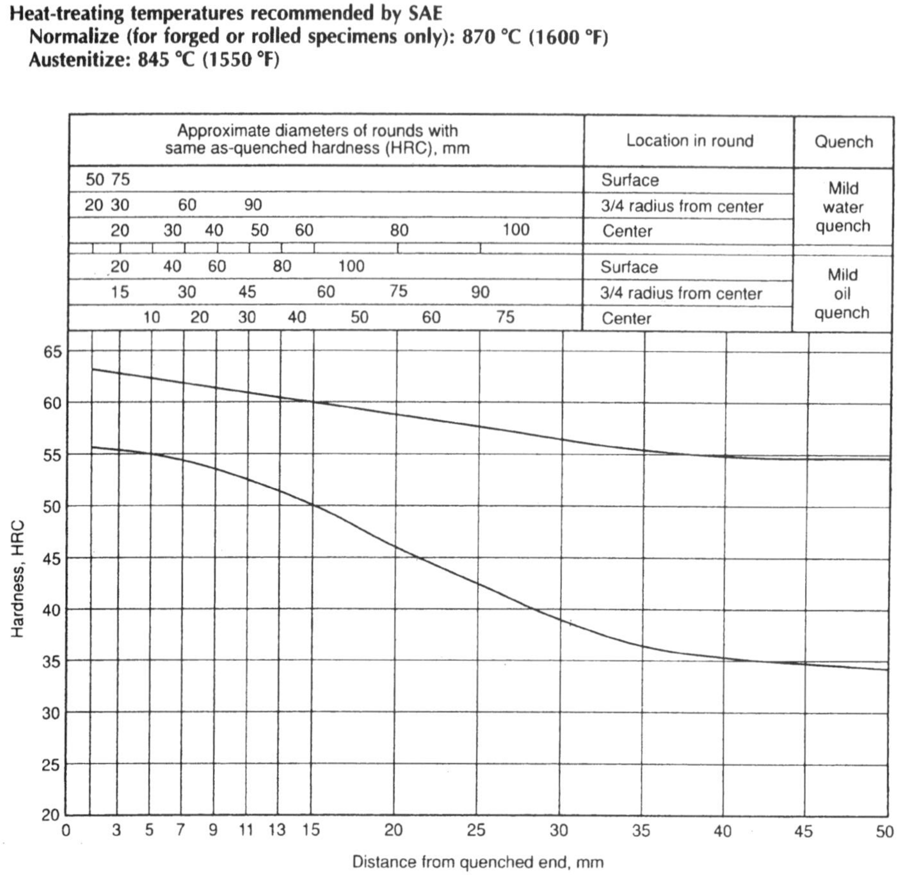

Heat-treating temperatures recommended by SAE
Normalize (for forged or rolled specimens only): 870 °C (1600 °F)
Austenitize: 845 °C (1550 °F)

(b)

Figure 10 *(Continued)*

This table also lists Vickers and Brinell hardness values, which are different types of hardness tests. It can be seen that a hardness of HRC 45 converts to an approximate tensile strength of 1480 MPa.

4 ROLE OF ALLOYING ELEMENTS IN STEEL

In the hardenability concept described in the previous section, alloying elements have a profound effect on depth of hardness. Alloying elements also change the characteristics of the iron–carbon diagram. For example, in the iron–carbon diagram (see Fig. 1) austenite cannot exist below the eutectoid temperature of 727°C. However, there are steels where austenite is the stable phase at room temperature, e.g., austenitic stainless steels and austenitic manganese steels. This can only be achieved through alloying. There are, however, special conditions where small amounts of austenite can be retained at room temperature during rapid quenching of low-alloy steel. When this occurs, the austenite is too rich in alloying elements to transform at room temperature and is thus retained as small regions in a martensitic microstructure. Because of this, it is called retained austenite. The retained austenite can be transformed through tempering the steel.

In austenitic stainless steels, when nickel is added with chromium, the austenite phase field is expanded, allowing austenite to be stable at room temperature. The popular SAE/ AISI 304 austenitic stainless steel contains 18% Cr and 8% Ni. Austenitic manganese steel (Hadfield steel) contains 12% Mn with 1% C. The Mn and C allow austenite to be stable at room temperature. Because of this ability, nickel and manganese are, therefore, called austenite stabilizers. Other elements are ferrite stabilizers, e.g., chromium, silicon, and molybdenum. A ferrite-stabilizing element expands the ferrite phase field, and the austenite phase field is restricted within what is called a gamma loop (gamma, γ, is the symbol for austenite). A gamma loop can be seen in the iron–chromium equilibrium diagram in Fig. 11. The gamma loop is shown at the left side of the diagram. According to this diagram, iron–chromium alloys with 12.7% Cr or higher, the transformation from austenite (γ) to ferrite (α) does not occur and ferrite exists from room temperature to melting. Iron–chromium alloys make up an important class of stainless steels called ferritic and martensitic stainless steels.

Each particular alloying element has an influence on the structure and properties of steel. The following elements are important alloying elements in steel:

Carbon. Carbon is the most common alloying element in steel. It is inexpensive and has a strong influence on hardness and strength. It is the basic and essential alloying element in all plain-carbon, low-alloy, and tool steels. Carbon is an interstitial element that occupies sites between the larger iron atoms in the bcc and fcc lattices. The influence of carbon on the strength of iron can be seen in Fig. 12. Carbon can increase yield strength of pure iron (0% C) with a strength of about 28–190 MPa. At 0.005% C, the maximum solubility of carbon at room temperature the sevenfold increase in strength is due to interstitial solid solution strengthening. Any excess carbon, above 0.005% C, will form an iron carbide com-

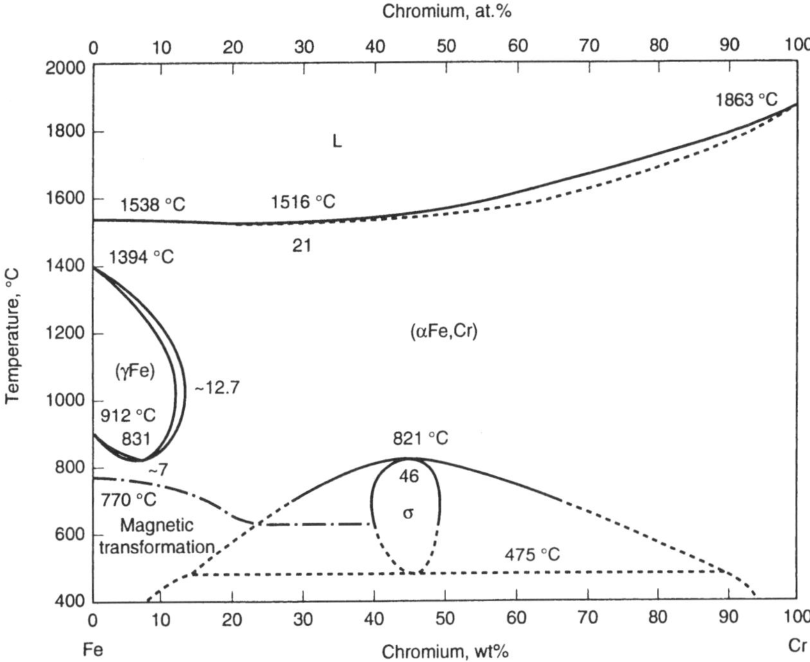

Figure 11 Iron–chromium equilibrium phase diagram. (*Source: ASM Handbook, Vol. 20, Materials Selection and Design,* ASM International, Materials Park, OH 44073-0002, 1997, p. 365.)

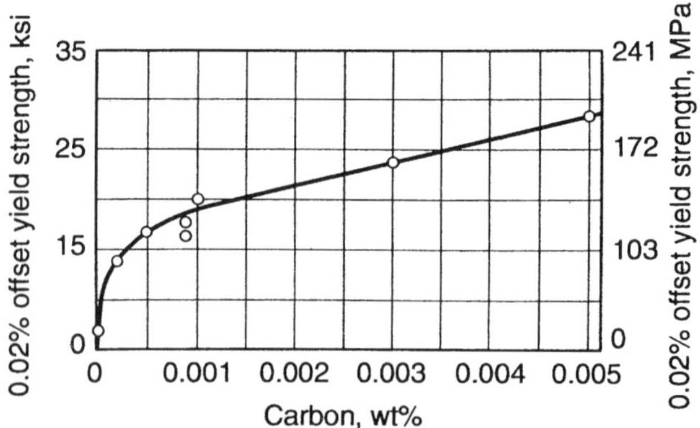

Figure 12 Effect of carbon in solid solution on the yield strength of iron. (*Source: ASM Handbook, Vol. 20, Materials Selection and Design,* ASM International, Materials Park, OH 44073-0002, 1997, p. 367.)

pound called cementite (Fe_3C). Cementite can exist as a particle, as a component of lamellar pearlite, or as a proeutectoid network on prior austenite grain boundaries in hypereutectoid steel. Thus, carbon in the form of cementite has a further influence on the strength of steel, as seen in Fig. 13. In this plot, the steels between 0.1% C and 0.8% C contain about 10–100% pearlite. Yield strength peaks at about 425 MPa at 0.6% C whereas tensile strength (ultimate strength) increases to 790 MPa at 0.8% C. These properties are for carbon steels in the air-cooled condition. In a 0.8% C steel, a further increase in strength can be achieved if faster cooling rates are used to produce a finer pearlite interlamellar spacing. In a fully pearlitic, head-hardened rail steel (accelerated cooled), the yield strength can increase to 860 MPa and tensile strength to 1070 MPa. Carbon also has a negative effect on properties, as seen in Fig. 13. For example, the percent reduction in area (as well as total elongation not shown) decreases with increasing carbon. The percent reduction in area is a measure of the cross-sectional area change in a tensile specimen before and after fracture. Notch toughness also decreases with carbon content, as seen in the decrease in upper shelf energy and the increase in transition temperature. Shelf energy is the upper portion or upper shelf of a curve of absorbed energy plotted from a Charpy test.

Manganese. Manganese is also an essential element in all carbon, low-alloy, and alloy steels. Manganese has several roles as an alloying element. One role is to assure that all residual sulfur is combined to form manganese sulfide (MnS). Manganese is generally added to steel with a minimum manganese–sulfur ratio of 20:1. Without manganese the sulfur would combine with iron and form iron sulfide (FeS), which is a brittle compound that lowers toughness and ductility and causes a phenomenon called hot shortness. Hot shortness is a condition where a compound (such as FeS) or insoluble element (such as copper) in steel has a low melting point and thus forms an unacceptable cracklike surface condition during hot rolling. Another role of manganese is in strengthening steel. Manganese is a substitutional element and can replace iron atoms in the bcc or fcc lattice. Each 0.1% Mn added to iron will increase the yield strength by about 3 MPa. Manganese also lowers the eutectoid transformation temperature and lowers the eutectoid carbon content. In large amounts (12% or higher), manganese is an austenite stabilizer in alloy steels and forms a special class of steels called austenitic manganese steels (also called Hadfield manganese steels). These steels are

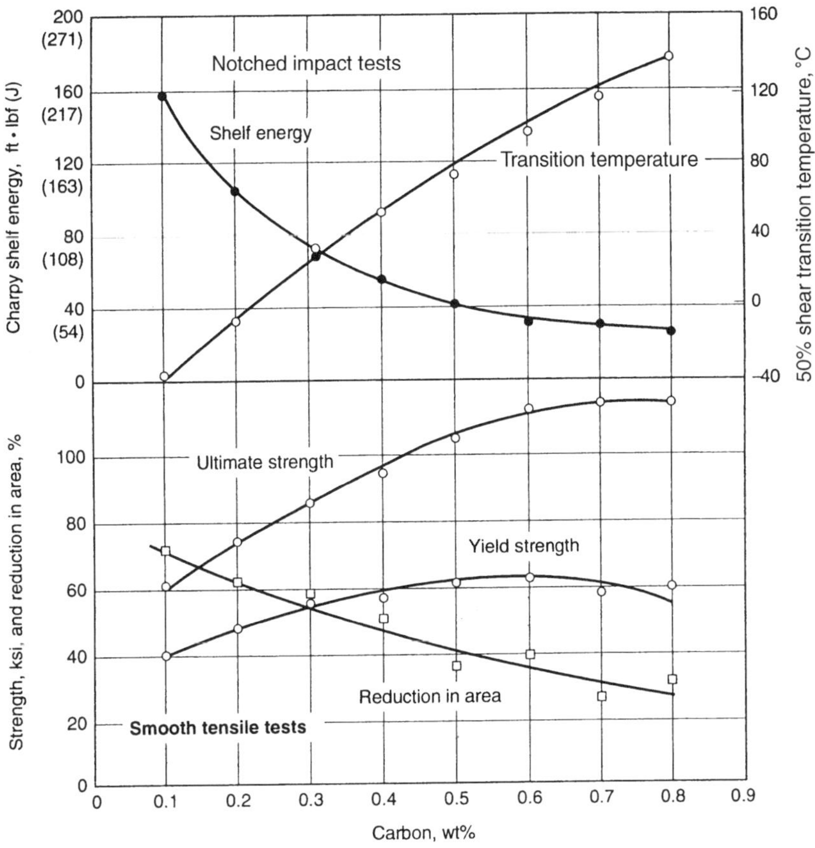

Figure 13 Effect of carbon on the tensile and notched impact properties of ferrite–pearlite steels. (*Source: ASM Handbook, Vol. 20, Materials Selection and Design,* ASM International, Materials Park, OH 44073-0002, 1997, p. 367.)

used in applications requiring excellent wear resistance, e.g., in rock crushers and in railway track connections where two rails meet or cross.

Silicon. Silicon is added to many carbon and low-alloy steels as a deoxidizer, i.e., it removes dissolved oxygen from molten steel during the steel-refining process. Oxygen is an undesirable element in steel because it forms oxide inclusions, which can degrade ductility, toughness, and fatigue resistance. Silicon has a moderate effect on strengthening steel but is usually not added for strengthening. Each 0.1% Si increases the yield strength of iron by about 8 MPa. It is a ferrite stabilizer and is found in some stainless steels. Silicon is also added to steel for enhanced electrical properties, e.g., iron–silicon transformer steels at 3.25% Si. These carbon-free steels have high magnetic permeability and low core loss.

Phosphorus. Phosphorus is considered a tramp or residual element in steel and is carefully restricted to levels generally below 0.02%. However, like carbon, phosphorus is an interstitial element that can substantially strengthen iron. For this reason, phosphorus is added to a special class of steels called rephosphorized steels for strength. Rephosphorized steels also have enhanced machinability.

Sulfur. Sulfur is also considered a tramp element in steel and is usually restricted to below about 0.02%. Although an element with a small atomic diameter, sulfur is not considered an interstitial alloying element because it is insoluble in iron. However, as in the case of phosphorus, sulfur is added to a special class of steels called resulfurized steels that have improved machinability. These steels are called free-machining steels.

Copper. In most steels copper is considered a tramp (residual) element and is restricted to levels below 0.04%. Copper, having a much lower melting point than iron, can create a detrimental steel surface condition known as hot shortness. Although not generally added to steel, there is a very special class of steels that contain high levels of copper to take advantage of the precipitation of copper particles during aging (a tempering process). These copper particles increase strength and hardness. Copper is also added to low-alloy steels for atmospheric corrosion protection (these steels are called weathering steels). One problem with copper in steel is that it cannot be oxidized and removed during steel refining. Thus, over time, the copper level of steel produced from steel scrap is slowly increasing.

Nickel. Nickel is an important element because of its positive effect on hardenability. Many important low-alloy steels contain nickel for this reason. Nickel, being a substitutional element in the iron lattice, has a small effect on increasing yield strength. Nickel, being an austenite stabilizer, is also a vital element in austenitic stainless steels. Nickel is also important in steels for cryogenic applications and storage tanks for liquefied hydrocarbon gases. Nickel does not form a carbide and remains in solid solution.

Chromium. Like nickel, chromium has a positive effect on hardenability and is an important alloying element in many low-alloy steels. For corrosion resistance, chromium is present in all stainless steels as a solid solution element. In addition to hardenability and solid solution effects, chromium forms several important chromium carbides that are necessary for wear resistance in many tool steels and steels used for rolls in hot- and cold-rolling mills.

Molybdenum. Molybdenum is a potent hardenability element and is found in many low-alloy steels. Molybdenum, like chromium, forms several types of carbides that are important for wear-resistant applications, e.g., tool steels. Molybdenum is added to minimize temper embrittlement in low-alloy steels. Temper embrittlement occurs when low-alloy steels are tempered in the temperature range of 260–370°C. The embrittlement is caused by tramp elements such as phosphorus that accumulate at the prior austenite grain boundaries and thus weaken the boundaries. Adding molybdenum prevents the accumulation of these undesirable elements at the boundaries. Molybdenum also enhances the creep strength of low-alloy steels at elevated temperatures and is used in rotors and other parts of generators in electric power plants. Creep is an undesirable process that allows steel to slowly elongate or creep under load. Eventually the component will fail.

Vanadium. Although vanadium is a potent hardenability element, its most useful role is in the formation of a vanadium nitride and vanadium carbide (it can also be in a combined form of vanadium carbonitride). A very important role of vanadium is in microalloyed steels, also called high-strength, low-alloy (HSLA) steels. These steels are strengthened by precipitation of vanadium nitrides and vanadium carbides (vanadium carbonitrides). The formation of vanadium carbide is important for wear resistance. Vanadium carbide is much harder than iron carbide, chromium carbide, and molybdenum carbide. Vanadium is thus important in high-speed tool steels, which are used as drill bits that retain their hardness as the tool heats by friction.

Tungsten. Tungsten is not an addition to low-alloy steels but is a vital alloying element in high-speed tool steels where it forms hard tungsten carbide particles.

Aluminum. Aluminum is employed as a deoxidizer in steel and is generally used in conjunction with silicon (also a deoxidizer). A deoxidizer removes undesirable oxygen from molten steel. Once removed, the steel is called "killed." Aluminum–silicon deoxidized (killed) steels are known as fine-grain steels. Another important role of aluminum is the formation a aluminum nitride (AlN) precipitate. Many steels depend upon the formation of AlN, especially steels used for sheet-forming applications requiring a high degree of formability such as parts that require deep drawing. These steels are called drawing-quality special-killed (DQSK) steels. The AlN precipitates help in the formation of an optimum crystallographic texture (preferred orientation) in low-carbon sheet steels for these deep-drawing applications. When aluminum combines with nitrogen to form AlN, the dissolved interstitial nitrogen is lowered. Lower interstitial nitrogen (interstitial nitrogen is also called free nitrogen) provides improved ductility. Aluminum can also substitute for silicon in electrical steels for laminations in electric motors and transformer cores.

Titanium. Titanium is a strong deoxidizer but is usually not used solely for that purpose. Titanium is important in microalloyed steels (HSLA steels) because of the formation of titanium nitride (TiN) precipitates. Titanium nitrides pin grain boundary movement in austenite and thus provide grain refinement. Another role of titanium is in steels containing boron where a titanium addition extracts nitrogen from liquid steel so that boron, a strong nitride former, remains in elemental form to enhance hardenability. Because of its affinity for both carbon and nitrogen, titanium is important in IF steels. Interstitial-free steels are an important class of steels with exceptional formability. Titanium, being a strong carbide former, is used as a carbide stabilizer in austenitic stainless steels (AISI type 321), ferritic stainless steels (AISI type 409, 439, and 444), and precipitation hardening stainless steels (AISI type 600 and 635).

Niobium (Columbium). Niobium is also important in microalloyed (HSLA) steels for its precipitation strengthening through the formation of niobium carbonitrides. Some microalloyed steels employ both vanadium and niobium. Because of its affinity for both carbon and nitrogen, niobium is an element found in some IF steels. Niobium is also added as a carbide stabilizer (prevents carbides from dissolving and re-forming in undesirable locations) in some austenitic stainless steels (AISI type 347, 348, and 384), ferritic stainless steels (AISI type 436 and 444), and precipitation hardening stainless steels (AISI type 630).

Tantalum. Because of its affinity for carbon, tantalum, like niobium, is added as a carbide stabilizer to some austenitic stainless steels (AISI type 347 and 348).

Boron. On a weight percent basis, boron is the most powerful hardenability element in steel. A minute quantity of boron, e.g., 0.003%, is sufficient to provide ample hardenability in a low-alloy steel. However, boron is a strong nitride former and can only achieve its hardenability capability if in elemental form. Thus, boron must be protected from forming nitrides by adding a sufficient amount of titanium to first combine with the nitrogen in the steel.

Calcium. Calcium is a strong deoxidizer in steel but is not used for that purpose. In an aluminum-deoxidized (killed) steel, calcium combines with sulfur to form calcium sulfide particles. This form of sulfide remains as spherical particles as compared with manganese sulfide, which is soft and elongates into stringers upon hot rolling. Thus, steels properly treated with calcium do not have the characteristics associated with MnS stringers, i.e., property directionality or anisotropy.

Zirconium. Although expensive and rarely added to steel, zirconium acts like titanium in forming zirconium nitride participates.

Nitrogen. Nitrogen is added to some steels, e.g., steels containing vanadium, to provide sufficient nitrogen for nitride formation. This is important in microalloyed steels containing vanadium. Nitrogen, being an interstitial element like carbon, strengthens ferrite. A number of austenitic stainless steels contain nitrogen for strengthening (AISI type 201, 202, 205, 304N, 304LN, 316N, and 316LN).

Lead. Lead is added to steel for enhanced machinability. Being insoluble in iron, lead particles are distributed through the steel and provide both lubrication and chip breaking ability during machining. However, leaded steels are being discontinued around the world because of the environmental health problems associated with lead.

Selenium. Selenium is added to some austenitic stainless steels (AISI type 303Se), ferritic stainless steels (AISI type 430Se), and martensitic stainless steels (AISI type 416Se) for improved machined surfaces.

Rare Earth Elements. The rare earth elements cerium and lanthanum are added to steel for sulfide shape control, i.e., the sulfides become rounded instead of stringers. It is usually added in the form of mish metal (a metallic mixture of the rare earth elements).

Residual Elements. Tin, antimony, arsenic, and copper are considered residual elements in steel. They are also called tramp elements. These elements are not intentionally added to steel but remain in steel because they are difficult to remove during steelmaking and refining. Steels made by electric furnace melting employing scrap as a raw material contain higher levels of residual elements than steels made in an integrated steelmaking facility (blast furnace–BOF route). Some electric furnace melting shops use direct-reduced iron pellets to dilute the effect of these residuals.

Hydrogen. Hydrogen gas is also a residual element in steel and can be very deleterious. Hydrogen is soluble in liquid steel and somewhat soluble in austenite. However, it is very insoluble in ferrite and is rejected as atomic hydrogen (H^+). If trapped inside the steel, usually in products such as thick plate, heavy forgings, or railroad rail, hydrogen will accumulate on the surfaces of manganese sulfide inclusions. When this accumulation takes place, molecular hydrogen (H_2) can form and develop sufficient pressure to create internal cracks. As the cracks grow, they form what is termed hydrogen flakes and the product must be scrapped. However, if the product is slow cooled from the rolling temperature, the atomic hydrogen has sufficient time to diffuse from the product, thus avoiding hydrogen damage. Also, most modern steelmakers use degassing to remove hydrogen from liquid steel.

5 HEAT TREATMENT OF STEEL

One of the very important characteristics of steel is the ability to alter the microstructure through heat treatment. As seen in the previous sections, many different microstructural constituents can be produced. Each constituent imparts a particular set of properties to the final product. For example, by quenching a steel in water, the steel becomes very hard but brittle through the formation of martensite. By tempering the quenched steel, some ductility can be restored with some sacrifice in hardness and strength. Also, superior wear properties can be obtained in fully pearlitic microstructures, particularly if an accelerated cooling process is employed to develop a fine interlamellar spacing. Complex parts can be designed by taking advantage of the formability and ductility of ferritic sheet steel through cold rolling and annealing. The amount of pearlite in ferritic steel can be adjusted by carbon content and cooling rate to produce a wide range of hardness and strength. In quenched and tempered steels, a bainitic microstructure has a unique combination of high strength and toughness. Thus steel, more than any other metallic material, can be manipulated through heat treatment to provide a multiplicity of microstructures and final properties. The common types of heat treatment are listed below:

Annealing (Full Annealing). One of the most common heat treatments for steel is annealing. It is used to soften steel and to improve ductility. In this process, the steel is heated into the lower regions of the austenite phase field and slow cooled to room temperature. The resulting microstructure consists of coarse ferrite or coarse ferrite plus pearlite, depending upon carbon and alloy content of the steel.

Normalizing. Steel is normalized by heating into the austenite phase field at temperatures somewhat higher than those used by annealing followed by air cooling. Many steels are normalized to establish a uniform ferrite plus pearlite microstructure and a uniform grain size.

Spheroidizing. To produce a steel in its softest possible condition, it is usually spheroidized by heating just above or just below the eutectoid temperature of 727°C and holding at that temperature for an extended time. This process breaks down lamellar pearlite into small spheroids of cementite in a continuous matrix of ferrite, as seen in Fig. 14. To obtain a very uniform dispersion of cementite spheroids, the starting microstructure is usually martensite. This is because carbon is more uniformly distributed in martensite than in lamellar pearlite. The cementite lamella must first dissolve and then redistribute the carbon as spheroids whereas the cementite spheroids can form directly from martensite.

Process Annealing (Recrystallization Annealing). Process annealing takes place at temperatures just below the eutectoid temperature of 727°C. This treatment is applied to low-carbon, cold-rolled sheet steels to restore ductility. In aluminum-killed steels, the recrystallized ferrite will have an ideal crystallographic texture (preferred orientation) for deep drawing into complex shapes such as oil filter cans and compressor housings. Crystallographic texture is produced by developing a preferred orientation of the ferrite grains, i.e., the crystal axes of the ferrite grains are oriented in a preferred rather than random orientation.

Stress Relieving. Steel products with residual stresses can be heated to temperatures approaching the eutectoid transformation temperature of 727°C to relieve the stress.

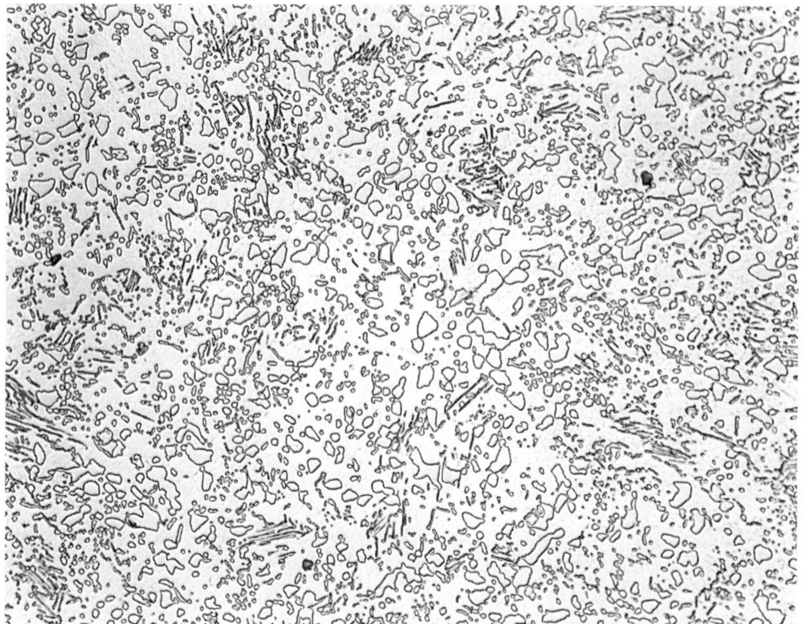

Figure 14 Photomicrograph of a medium-carbon steel in the spheroidized condition. 500X. 4% picral + 2% nital etch.

Quenching. To produce the higher strength constituents of bainite and martensite, the steel must be heated into the austenite phase field and rapidly cooled by quenching in oil or water. High-strength, low-alloy steels are produced by this process followed by tempering. It must be noted that employing microalloying additions such as Nb, V, and Ti can also produce HSLA steels. These microalloyed steels obtain their strength by thermomechanical treatment rather than heat treatment.

Tempering. When quenched steels (martensitic steel) are tempered by heating to temperatures approaching the eutectoid temperature of 727°C, the dissolved carbon in the martensite forms cementite particles, and the steels become more ductile. Quenching and tempering are used in a variety of steel products to obtain desired combinations of strength and toughness.

6 CLASSIFICATION AND SPECIFICATIONS OF STEELS

Since there are literally thousands of different steels, it is difficult to classify them in a simple straightforward manner. However, some guidelines can be used. For example, steels are generally classified as carbon steel or alloy steel. A classification system was developed by the SAE as early as 1911 to describe these carbon and alloy steels. The AISI collaborated with SAE to refine the compositional ranges of the classification that are used today. Recently, a Unified Numbering System (UNS) was established that incorporates the SAE/AISI number.

Many steel products are purchased by specifications describing specific compositional, dimensional, and property requirements. Specification organizations such as ASTM have developed numerous specifications for steel products and the testing of steel products. Some specific product user groups in the United States have developed their own specifications, e.g., the American Bureau of Ships (ABS) for ship plate and other marine products, Aerospace Materials Specifications (AMS) for aerospace applications, the American Railway Engineering and Maintenance of Way Association (AREMA) for rail and rail products, the SAE for automotive applications, and the American Society of Mechanical Engineers (ASME) for steels produced to boiler code specifications. In Japan, there are standards developed by the Japanese Industrial Standards (JIS) Committee of the Ministry of International Trade and Industry. In the United Kingdom, there are the British Standards (BS) developed by the British Standards Institute. In Europe, Germany has the Deutsches Institut für Normung (DIN) standards, France the Association Francaise de Normalisation (AFNOR) standards, and Italy the Ente Nazionale Italiano di Unificazione (UNI) standards.

Specifications can be as simple as a hardness requirement, i.e., ASTM A1 for rail steel to elaborate compositional and property requirements as in ASTM A808, "High-strength low-alloy carbon–manganese–niobium–vanadium steel of structural quality with improved notch toughness." Describing specific specifications is beyond the scope of this handbook, but many of the key sources can be found in the Bibliography at the end of this chapter.

6.1 Carbon Steels

Carbon steels (also called plain-carbon steels) constitute a family of iron–carbon–manganese alloys. In the SAE/AISI system, the carbon steels are classified as follows:

Nonresulfurized carbon steels	10xx series
Resulfurized steels	11xx series
Rephosphorized and resulfurized steels	12xx series
High-manganese carbon steels	15xx series

A four-digit SAE/AISI number is used to classify the carbon steels with the first two digits being the series code and the last two digits being the nominal carbon content in points of carbon (1 point = 0.01% C). For example, SAE/AISI 1020 steel is a carbon steel containing 0.20% C (actually 0.18–0.22% C). The chemical composition limits for the above SAE/AISI 10xx series of carbon steels for semifinished products, forgings, hot- and cold-finished bars, wire, rods, and tubing are listed in *SAE Materials Standards Manual* (SAE HS-30, 1996). There are slight compositional variations for structural shapes, plates, strip, sheet, and welded tubing (see SAE specification J403). The *SAE Manual* gives the SAE/AISI number along with the UNS number. The carbon level spans the range from under 0.06% C to 1.03% C.

Because of the wide range in carbon content, the SAE/AISI 10xx carbon steels are the most commonly used steels in today's marketplace. All SAE/AISI 10xx series carbon steels contain manganese at levels between 0.25 and 1.00%. For a century, manganese has been an important alloying element in steel because it combines with the impurity sulfur to form manganese sulfide (MnS). MnS is much less detrimental than iron sulfide (FeS), which would form without manganese present. Manganese sulfides are usually present in plain and low-alloy steels as somewhat innocuous inclusions. The manganese that does not combine with sulfur strengthens the steel. However, with the development of steelmaking practices to produce very low sulfur steel, manganese is becoming less important in this role.

The SAE/AISI 11xx series of resulfurized steels contain between 0.08 and 0.33% sulfur. Although in most steel, sulfur is considered an undesirable impurity and is restricted to less than 0.05%, in the SAE/AISI 11xx and 12xx series of steels, sulfur is added to form excess manganese sulfide inclusions. These are the free-machining steels that have improved machinability over lower sulfur steels due to enhanced chip breaking and lubrication created by the MnS inclusions.

The SAE/AISI 12xx series are also free-machining steels and contain both sulfur (0.16–0.35%) and phosphorus (0.04–0.12%). The SAE/AISI 15xx series contain higher manganese levels (up to 1.65%) than the SAE/AISI 10xx series of carbon steels.

Typical mechanical properties of selected SAE/AISI 10xx and 11xx series of carbon steels are listed in the first part of the table on pp. 20–23, Section 4, of the *ASM Metals Handbook,* Desktop Edition, 1985, for four different processing conditions (as rolled, normalized, annealed, and quenched and tempered). These properties are average properties obtained from many sources, and thus this table should only be used as a guideline. The as-rolled condition represents steel before any heat treatment was applied. Many applications utilize steel in the as-rolled condition. As can be seen from the aforementioned ASM table, yield and tensile strength is greater for steel in the normalized condition. This is because normalizing develops a finer ferrite grain size. Yield and tensile strength is lowest for steels in the annealed condition. This is due to a coarser grain size developed by the slow cooling rate from the annealing temperature. In general, as yield and tensile strength increase, the percent elongation decreases. For example, in the ASM table, annealed SAE/AISI 1080 steel has a tensile strength of 615 MPa and a total elongation of 24.7% compared with the same steel in the normalized condition, with a tensile strength of 1010 MPa and a total elongation of 10%. This relationship holds for most steel.

Special Low-Carbon Steels

These are the steels that are not classified in the aforementioned SAE table or listed in the aforementioned ASM table. As mentioned earlier, carbon is not always beneficial in steels. These are special steels with carbon contents below the lower level of the SAE/AISI 10xx steels. There are a number of steels that are produced with very low carbon levels (less than 0.002% C), and all the remaining free carbon in the steel is tied up as carbides. These steels are known as IF steels, which means that the interstitial elements of carbon and nitrogen are no longer present in elemental form in the iron lattice but are combined with elements such as titanium or niobium as carbides and nitrides (carbonitrides). Interstitial-free steels are required for exceptional formability, especially in applications requiring deep drawability. Drawability is a property that allows the steel to be uniformly stretched (or drawn) in thickness in a closed die without localized thinning and necking (cracking or breaking). An example of a deep-drawn part would be a compressor housing for a refrigerator. With proper heat treatment, IF steels develop a preferred crystallographic orientation that favors a high plastic anisotropy ratio or *r* value. High *r*-value steels have excellent deep-drawing ability and these steels can form difficult parts. Another type of low-carbon steel is a special class called DQSK steel. This type of aluminum-treated steel also has a preferred orientation and high *r* value. The preferred orientation is produced by hot rolling the steel on a hot strip mill followed by rapid cooling. The rapid cooling keeps the aluminum and interstitial elements from forming aluminum nitride particles (i.e., the Al and N atoms are in solid solution in the iron lattice). After rolling, the steel is annealed to allow aluminum nitride to precipitate. The aluminum nitride plays an important role in the development of the optimum crystallographic texture. The DQSK steel is used in deep-drawing applications that are not as demanding as those requiring IF steel.

A new family of steels called bake-hardening steels also have a low, but controlled carbon content. These steels gain strength during the paint–bake cycle of automotive production. Controlled amounts of both carbon and nitrogen combine with carbonitride-forming elements such as titanium and niobium during the baking cycle (generally 175°C for 30 min). The precipitation of these carbonitrides during the paint–bake cycle strengthens the steel by a process called aging.

Enameling steel is produced with as little carbon as possible because during the enameling process, carbon in the form of carbides can react with the frit (the particles of glasslike material that melts to produce the enamel coating) to cause defects in the coating. Thus, steels to be used for enameling are generally decarburized in a special reducing atmosphere during batch annealing. In this process, the carbon dissipates from the steel. After decarburization, the sheet steel is essentially pure iron. Enamel coatings are used for many household appliances such as washers and dryers, stovetops, ovens, and refrigerators. Also, steel tanks in most hot-water heaters have a glass (or enameled) inside coating.

Electrical steels and motor lamination steels are also produced with as low a carbon content as possible. Dissolved carbon and carbides in these steels are avoided because the magnetic properties are degraded. The carbides, if present in the steel, inhibit the movement of the magnetic domains and lower the electrical efficiency. These steels are used in applications employing alternating current (AC) in transformers and electric motors. Most electric motors for appliances and other applications have sheet steel stacked in layers (called laminations) that are wound in copper wire. Electrical steels used for transformers contain silicon, which is added to enhance the development of a specific crystallographic orientation that favors electrical efficiency.

6.2 Alloy Steels

Alloy steels are alloys of iron with the addition of one or more of the following elements; carbon, manganese, silicon, nickel, chromium, molybdenum, and vanadium. The alloy steels cover a wide range of steels, including low-alloy steels, stainless steels, heat-resistant steels, and tool steels. Some alloy steels, such as austenitic stainless steels, do not contain intentional additions of carbon. Silicon, when required, is added as a deoxidizer to the molten steel. Nickel provides strength and assists in hardening the steel by quenching and tempering heat treatment. This latter effect is called hardenability, which has been described earlier. Chromium is found in stainless steels for corrosion resistance. Chromium and molybdenum also assist in hardenability of the low-alloy steels. Vanadium strengthens the steel by forming precipitates of vanadium carbonitride. Vanadium is also a potent hardenability element.

Low-Alloy Steels

There is an SAE/AISI four-digit classification system for the low-alloy steels. As in the carbon steels, the first two digits are for the alloy class and the last two (or three) digits are for the carbon content. Because of the various combinations of elements, the system is more extensive than that used for the carbon steels. The general SAE/AISI classification system for low-alloy steels is as follows:

Manganese steels	13xx series
Nickel steels	23xx, 25xx series
Nickel–chromium steels	31xx, 32xx, 33xx, and 34xx series
Molybdenum steels	40xx, 44xx series
Chromium–molybdenum steels	41xx series
Nickel–chromium–molybdenum steels	43xx and 47xx series 81xx, 86xx, 87xx, and 88xx series 93xx, 94xx, 97xx, and 98xx series
Nickel–molybdenum steels	46xx and 48xx series

Chromium steels	50xx and 51xx series 50xxx, 51xxx, and 52xxx series
Chromium–vanadium steels	61xx series
Tungsten–chromium steels	71xxx, 72xx series
Silicon–manganese steels	92xx series
Boron steels	xxBxx series
Leaded steels	xxLxx series

The boron-containing steels are low-alloy steels with boron added in the amount of 0.0005–0.003%. Boron is a strong hardenability element. The leaded steels contain 0.15–0.35% lead for improved machinability (however, lead is no longer favored as an alloying addition because of health concerns).

A table in the aforementioned SAE HS-30 lists the composition limits of most of the families of the SAE/AISI low-alloy steels listed above. These steels are supplied in the form of bar, plate, and forged products and are usually heat treated to obtain specific mechanical properties, especially high strength and toughness. Mechanical properties of selected SAE/AISI low-alloy steels in the as-rolled, annealed, normalized and quenched and tempered conditions are listed in the aforementioned ASM table. These properties are average properties and should only be used as a guideline. For example, SAE/AISI 4340 steel is usually heat treated by heating the component to 950–1000°C followed by quenching in oil. The component is then tempered at a temperature between 205 and 650°C. According to the aforementioned ASM table, this nickel–chromium–molybdenum steel in the quenched and tempered condition (tempered at 205°C) can achieve a yield strength of 1675 MPa and a tensile strength of 1875 MPa. Quenched and tempered low-alloy steels are used in a large number of applications requiring high strength and good toughness. Note that in the annealed condition, SAE/AISI 4340 steel has a yield strength of only 745 MPa.

Other Low-Alloy Steels

There are a number of important steels that do not fit into the SAE/AISI classification system described above. Such classes are the microalloyed steels also called HSLA steels, dual-phase steels, trip steels, and high-performance steels.

Microalloyed (High-Strength, Low-Alloy) Steels. Microalloying is a term applied to steels that contain small additions of alloying elements that retard austenite recrystallization and pin austenite grain boundary movement by the formation of small carbide and/or nitride precipitates. These elements include vanadium, niobium, and titanium. These HSLA steels are produced for a variety of plates, structural shapes, bars, and sheet applications with yield strength varying from 290 to 690 MPa. These steels are covered under numerous SAE and ASTM specifications. The SAE high-strength, low-alloy steels are covered under specifications J410, J1392, and J1442 and the ASTM high-strength, low-alloy steels are covered under various specifications, including A242, A440, A441, A572, A588, A606, A607, A618, A633, A656, A690, A709, A714, A715, A808, A812, A841, A860, and A871. These HSLA steels have found wide application in areas such as bridge construction (structural beams), off-shore oil and natural gas platforms, ship hull and deck plate, and electrical transmission towers and poles. In the automobile, HSLA steels are used for safety (ultrahigh-strength impact door beams and energy-absorbing bumper assemblies) and for increasing fuel economy through thinner (lighter weight) chassis structural sections. Microalloyed HSLA steels are also employed in large-diameter gas transmission pipelines.

Dual-Phase Steels. A relatively recent development, dual-phase steels are produced by rapidly cooling a medium-carbon steel, containing vanadium or molybdenum, from the two-phase ferrite plus austenite region. The austenite transforms into islands of martensite (stabilized at room temperature by the V and Mo additions) in a ferrite matrix. Depending upon the alloy content, the martensite islands can also contain austenite that is retained below the transformation temperature (called retained austenite). Thus, dual-phase steel may contain both martensite and austenite as a second phase (called MA constituent). The unique characteristic of dual-phase steels is the continuous yielding behavior during deformation; i.e., there is a lack of a yield point during deformation. This provides increased uniform elongation and work hardening so that those components or parts produced from dual-phase steel actually gain strength during the forming operation. Dual-phase steels are being applied in applications such as automobile wheel rims and wheel disks. Because of their energy-absorbing characteristics, dual-phase steels are being used in critical locations of the automobile for safety to protect the occupants in the event of a crash.

Trip Steels. Similar to dual-phase steels, trip steels have emerged as an energy-absorbing high-strength steel for the automobile. The term "trip" is derived from the mechanism of *tr*ansformation *i*nduced *p*lasticity. These steels contain a high percentage of retained austenite (10–15%). The austenite transforms to martensite during the forming of the part, thus providing enhanced formability or transforms upon impact in a crash.

High-Performance Steels. There are a number of high-performance steels that are used in critical applications. These low-alloy steels, such as HY80 and HY100, are used in applications requiring high strength and excellent toughness. The "80" and "100" in the codes represent the minimum yield strength in ksi units. Another family of low-alloy steels is used in heat exchangers, high-temperature piping, and boiler applications. These steels, like $2\frac{1}{4}$% Cr–1% Mo, find wide use in these applications. Other high-performance steels are the Ni–Cr–Mo steels used as rotors for large steam generators and motors in electric power plants. These steels must withstand temperatures of superheated steam and must maintain high strength, superior toughness, as well as high fatigue strength and creep resistance. The Ni–Cr–Mo–V steels are also used in pressure vessels in nuclear reactors.

Higher Alloy Steels

There is a distinction between the low-alloy steels described above and the higher alloy steels (usually containing over 8% alloying elements). The higher alloy steels include stainless steels, tool steels, heat-resistant steels, wear-resistant steels, and ultrahigh-strength steels.

Stainless Steels. Stainless steels are corrosion-resistant steels that contain at least 10.5% chromium. Chromium is unique in that it forms a passive layer on the steel surface that provides protection from corrosion. There are basically five types of stainless steels: austenitic, ferritic, duplex, martensitic, and precipitation hardening steels. These five types of stainless steel have a somewhat simplified classification system as follows:

Austenitic stainless steels with low nickel	2xx series
Austenitic stainless steels	3xx series
Ferritic stainless steels	4xx series
Duplex stainless steel	329
Martensitic stainless steels	4xx series
Precipitation strengthening stainless steels	6xx (xx-x PH)

The classification system is different for the stainless steels than the system for SAE/AISI low-alloy steels in that the last two digits (xx) do not represent the carbon content and have no particular compositional meaning. Unfortunately, the classification system has some confusion with the ferritic and martensitic stainless steels both of the 4xx series. The 2xx series of austenitic stainless steels were developed during the 1950s when nickel became scarce. In these steels, manganese and nitrogen were substituted for the lower nickel level in order to maintain strength. Each type of stainless steel is expanded upon below:

AUSTENITIC STAINLESS STEELS. Austenitic stainless steels have sufficient alloying to stabilize austenite at room temperature. These steels being austenitic are nonmagnetic. Austenitic stainless steels have excellent low-temperature toughness, weldability, and corrosion resistance. On the other hand, they have relatively low yield strength and can only be strengthened by cold working the steel, by precipitation hardening, or by interstitial or substitutional solid solution strengthening.

The table on pp. 15.1–15.4 of the *ASM Metals Handbook,* Desktop Edition, 1985, lists the composition limits of the austenitic stainless steels. In general, the 3xx series are iron–chromium–nickel alloys that contain 16–26% chromium and 6–22% nickel. The popular type 304 austenitic stainless steel contains 18–20% Cr and 8–12% Ni and is often referred to as "18-8" stainless steel for the chromium and nickel content. There are many compositional variations of austenitic stainless steels. The following list summarizes these variations:

201	Low nickel replaced with manganese and nitrogen
202	Higher Mn than 201
205	Higher Mn and N than 202
301	Lower Ni and Cr to increase work-hardening ability
302	General-purpose 18–8 stainless steel
302B	Scaling resistance improved with Si
303	Enhanced machinability with a S addition
303Se	Improved machined surfaces with a selenium addition
304	Popular 18–8 stainless steel, lower C than 302
304L	Low-carbon 304 for improved corrosion resistance
304LN	Low-carbon 304 with nitrogen added for strength
304H	Higher carbon 304
304Cu	Copper added for improved cold working
304N	Nitrogen added for strength
305	Higher Ni for reduced work hardening
308	Higher Cr and Ni for weldability
309	High Cr and Ni for heat resistance
309S	Lower carbon 309
309Cb	Niobium (columbium) added
310	Higher Cr and Ni than 309 for improved heat resistance
310S	Lower carbon 310
310Cb	Niobium (columbium) added
314	Higher Si for improved heat resistance

316	Mo added for improved corrosion resistance
316F	Higher S and P for machinability
316L	Lower C for improved corrosion resistance and weldability
316LN	Lower C and higher nitrogen (for strength)
316H	Higher carbon 316
316N	Nitrogen added for strength
316Ti	Titanium added
316Cb	Niobium (columbium) added
317	Higher Cr and Mo for improved corrosion resistance
317L	Low-carbon 317 for improved weldability
321	Titanium added to minimize Cr carbide precipitation
330	High Ni to minimize carburization and improve thermal shock
347	Nb and Ta added to minimize Cr carbide precipitation
347H	Higher carbon 347
348	Ta and Co added for restricted nuclear applications
348H	Higher carbon 348
384	Higher Ni for decreased work hardening

The limiting of carbon is important in austenitic stainless steels. When heated, carbon forms chromium carbide that precipitates on the austenite grain boundaries and produces a condition known as sensitization. Because the chromium is tied up as carbide, the chromium adjacent to the boundaries will be depleted in chromium and corrosion can take place. Sensitization is reversible by heating the steel to temperatures between 1040 and 1150°C followed by rapid cooling to room temperature. The high temperature dissolves the carbides and the rapid cooling prevents reprecipitation of the carbides. More on austenitic stainless steel can be found in Chapter 2.

FERRITIC STAINLESS STEELS. The ferritic stainless steels are basically iron–chromium alloys with chromium ranging from 10.5 to 27%. The compositional limits for the ferritic stainless steels are listed in the aforementioned ASM table. Nickel is absent in ferritic stainless steels except for minor amounts, i.e., less than 1%, in some alloys. These steels have a microstructure of ferrite at room temperature and are magnetic. Type 409 stainless steel with the lowest chromium level (10.5–11.75%) is the least expensive of the ferritic stainless steel series and is used for automotive exhaust systems because it far outlasts carbon steel in that application. There are fewer variations of ferritic stainless steels than austenitic stainless steels. The ferritic stainless steels are listed below:

405	Low Cr with Al added
409	Low Cr, for automotive exhaust applications
429	Slightly less Cr, better weldability
430	General-purpose ferritic stainless steel
430F	Free machining with higher S and P
430Se	Selenium added for improved machined surfaces
434	Mo added for improved corrosion resistance

436	Mo, Nb, and Ta added for corrosion and heat resistance
439	Low C, Ti added to minimize sensitization
442	Higher Cr for improved oxide scaling resistance
444	Low C, Mo for corrosion resistance, Ti and Nb for sensitization
446	Highest Cr for improved scaling resistance

Ferritic stainless steels are described in more detail in Chapter 2.

DUPLEX STAINLESS STEELS. Type 329 is an iron–chromium alloy with 2.5–5% nickel and 1–2% molybdenum that has a mixed (duplex) microstructure of approximately equal percentages of ferrite and austenite. There are many more duplex stainless steels that have priority compositions and trade names (see Bibliography at the end of chapter). The corrosion characteristics of these duplex stainless steels are similar to austenitic stainless steels. However, they have higher strength and better resistance to stress–corrosion cracking than austenitic stainless steels. Duplex stainless steels are discussed in Chapter 2.

MARTENSITIC STAINLESS STEELS. To produce martensite in a stainless steel, the alloy must be transformed from the austenite phase field. According to the equilibrium phase diagram, this means that they have restricted chromium levels within the range required to form the gamma loop where austenite exists (see Fig. 11). The gamma loop is the region between 800 and 1400°C and 0 and 12.7% Cr in Fig. 11. Since austenite only exists in this restricted region, the steel must be heated within this temperature range and quenched to room temperature to form martensite. Martensitic stainless steels contain added carbon, which expands the gamma loop to allow higher chromium contents to be used. Because they can be heat treated, the martensitic stainless steels generally have higher strength than the austenitic and ferritic stainless steels. The martensitic stainless steels are listed below:

403	Select quality for highly stressed parts
410	General-purpose martensitic stainless steel
414	Ni added for improved corrosion
416	Higher P and S for improved machinability
416Se	Se added for improved machined surfaces
420	Higher C for increased strength
420F	Free machining with higher P and S
422	Mo, V, and W added for increased strength and toughness
431	Higher Cr, Ni added for improved corrosion resistance
440A	Highest Cr, C added for increased hardness
440B	Highest Cr, more C added for increased hardness/toughness
440C	Highest Cr, highest C for increased hardness/toughness
501	Low Cr, Mo added
502	Low C, Mo added

The compositional ranges for the martensitic stainless steels are shown in the aforementioned ASM table. Martensitic stainless steels are discussed in the next chapter.

PRECIPITATION HARDENING STAINLESS STEELS. The precipitation hardening stainless steels are iron–chromium–nickel alloys that develop high strength and toughness through additions of Al, Ti, Nb, V, and/or N, which form precipitates during an aging heat treatment. The base microstructures of precipitation hardening stainless steels can be either martensitic or austenitic depending upon composition and processing. Some selected grades are listed below:

600	Austenitic grade with Mo, Al, Ti, V, and B added
630	Martensitic grade with Cu and Nb added
631	Austenitic grade with Al added
633	Austenitic grade with Mo and N added
635	Martensitic grade with Al and Ti added

The compositional ranges of the precipitation hardening stainless steels are listed in the aforementioned ASM table.

OTHER STAINLESS STEELS. There are many stainless steels that do not fall within the AISI classification system. These steels have proprietary compositions and trade names. Details of many of these steels can be found in Chapter 2 and in the Bibliography at the end of this chapter.

Tool Steels. Tool steels are alloy steels that are used to cut or machine other materials. Tool steels contain various levels of Cr, Ni, Mo, W, V, and Co. The categories of tool steels are:

M series	Molybdenum high-speed steels
T series	Tungsten high-speed steels
Cr series	Chromium hot-work steels
H series	Molybdenum hot-work steels
A series	Air-hardening medium-alloy cold-work steels
D series	High-carbon high-chromium cold-work steels
O series	Oil-hardening cold-work steels
S series	Shock-resistant steels
L series	Low-alloy special-purpose tool steels
P series	Low-carbon mold steels
W series	Water-hardening tool steels

The compositional ranges of the various tool steels are listed in the table on pp. 758–759 of the *ASM Metals Handbook,* Vol. 20, 10th Edition, 1997. The high-speed steels are used in high-speed cutting tools such as drill bits. The hot-work tool steels are used in operations that utilize dies for punching, shearing, and forming materials at elevated temperatures, and the cold-work steels are used in similar operations at room temperature.

Heat-Resistant Steels. The upper temperature limit for use of carbon steels is about 370°C because of excessive oxidation and loss of strength. However, there are a number of alloy steels, called heat-resistant steels, that can be used at temperatures of 540–650°C. These steels include some of the ferritic stainless steels (405, 406, 409, 430, 434, and 439), quenched and tempered martensitic stainless steels (403, 410, 416, 422, and 431), precipitation hardening martensitic stainless steels (15-5 PH, 17-4 PH, and PH 13-8 Mo), precipitation hardening semiaustenitic stainless steels (AM-350, AM-355, 17-7 PH, and PH 15-7 Mo), and austenitic stainless steels (404, 309, 310, 316, 317, 321, 347, 202, and 216). In addition to the stainless steels, there are a number of proprietary alloys containing various levels of Cr, Ni, Mo, Nb, Ti, Cu, Al, Mn, V, N, or Si. The properties that are important to heat-resistant steels are creep and stress rupture. Creep is time-dependent strain occurring under stress. In more common terms creep is elongation or sagging of a part over time at elevated temperature. Stress rupture is a measure of the elevated temperature durability of material. These steels are generally specified under the ASME Boiler and Pressure Vessel Code.

Wear-Resistant Steels (*Austenitic Manganese Steels*). An important series of alloy steels are the austenitic manganese steels that contain 1.2% carbon and a minimum of 11% manganese. These steels, invented by Sir Robert Hadfield in 1882, are wear resistant and tough. Because they are difficult to hot work, these steels are usually cast into the final product form. The chemical compositional ranges for some selected austenitic manganese steels (ASTM A128) are listed below:

Grade	*C*	*Mn*	*Cr*	*Mo*	*Ni*	*Si*
A	1.55–1.35	11 (min)	—	—	—	1 (max)
B1	0.9–1.05	11.5–14	—	—	—	1 (max)
C	1.05–1.35	11.5–14	1.5–2.5	—	—	1 (max)
D	0.7–1.3	11.5–14	—	—	3–4	1 (max)
E1	0.7–1.3	11.5–14	—	0.9–1.2	—	1 (max)

The carbon addition is necessary to maintain an austenitic microstructure at room temperature. All grades must be heat treated by solution annealing at 1010–1090°C for 1–2 h per 25 mm of thickness followed by rapid water quenching. Because these alloys work harden during use, they are used in applications involving earthmoving (bucket blades), mining and quarrying (rock crushers), and railway trackwork (frogs, switches, and crossings).

Ultrahigh-Strength Steel

MARAGING STEEL. Another important series of alloy steels are the maraging steels. They are considered ultrahigh-strength steels because they can be heat treated to yield strength levels as high as 2.5 GPa. They also have excellent ductility and toughness. There are basically four grades that are produced to strength levels between 1.4 and 2.5 GPa.

Grade	*Ni*	*Co*	*Mo*	*Ti*	*Al*
18Ni (200)	18	8.5	3.3	0.2	0.1
18Ni (250)	18	8.5	5.0	0.4	0.1
18Ni (300)	18	9.0	5.0	0.7	0.1
18Ni (350)	18	12.5	4.2	1.6	0.1

The numbers in parentheses represent the nominal yield strength level in ksi. All maraging steels must be heat treated to achieve the desired properties. The heat treatment cycle for the 18Ni (200), 18Ni (250), and 18Ni (300) grades requires a solution treatment at 820°C for 1 h per 25 mm of thickness, cooling to room temperature, and an aging treatment at 480°C for 4 h. The 18Ni (350) grade has an extended aging time of 12 h. The heat treatment develops a martensitic microstructure on cooling from austenite at 820°C. The aging step precipitates intermetallic compounds, e.g., Ni_3Mo, that strengthen the martensitic matrix. Maraging steels can be machined before the aging treatment and have good weldability and excellent fracture toughness. They have found applications in missile and aircraft parts that require exceptional performance.

Music Wire. One of the strongest steel products commercially available is music wire. These wires can achieve levels of tensile strength approaching 5 GPa. The steel is basically SAE/AISI 1080. To obtain the ultrahigh-strength levels, rods of SAE/AISI 1080 are isothermally transformed to fine pearlite in a process known as patenting. The rods are then cold drawn to wire using large reductions in wire diameter through each die. The cold reduction forces the ferrite and cementite in the microstructure to align in a preferred orientation or fiber texture. The wires are used in musical instruments where they can be stretched under high tension to achieve certain musical notes. Ultrahigh-strength wires are also used to strengthen the rubber in automobile tires.

7 SUMMARY

Steel is one of the most versatile materials in today's society. It can be produced with a wide range of properties and is used in millions of applications. For example, stainless steels are used for their corrosion resistance, interstitial-free steels are used for their excellent formability characteristics, iron–silicon alloys are used for their electrical properties, austenitic manganese steels are used for their wear and abrasion resistance, microalloyed steels are used for their high strength, patented and cold-drawn eutectoid steel wires are used for their ultrahigh strength, dual-phase and trip steels are used for their energy absorption capability in a vehicle collision, and tool steels are used for their outstanding ability to cut and machine other materials. No other material can span such a range of properties and characteristics.

BIBLIOGRAPHY

Handbooks

ASM Metals Handbook, Properties and Selection: Irons, Steels and High Performance Alloys, Vol. 1, 10th ed., ASM International, Materials Park, OH, 1990.

ASM Metals Handbook, Materials Selection and Design, Vol. 20, 10th ed., ASM International, Materials Park, OH, 1997.

ASM Metals Handbook, Desk Edition, 2nd ed., ASM International, Materials Park, OH, 1998.

ASM Specialty Handbook®—Stainless Steels, ASM International, Materials Park, OH, 1994.

ASM Specialty Handbook®—Carbon and Alloy Steels, ASM International, Materials Park, OH, 1996.

Engineering Properties of Steel, ASM International, Materials Park, OH, 1982.

Stahlschlüssel (*Key to Steel*), 18th ed., Verlag Stahlschlüssel Wegst GMBH, Marburg, 1998.

Worldwide Guide to Equivalent Irons and Steels, 4th ed., ASM International, Materials Park, OH, 2000.

General References

Beddoes, J., and J. G. Parr, *Introduction to Stainless Steels,* 3rd ed., ASM International, Materials Park, OH, 1999.

Brooks, C. R., *Principles of the Heat Treatment of Plain Carbon and Low Alloy Steels,* ASM International, Materials Park, OH, 1996.

Krauss, G., *Steels—Heat Treatment and Processing Principles,* ASM International, Materials Park, OH, 1990.

Honeycombe, R., and H. K. D. H. Bhadeshia, *Steels: Microstructures and Properties,* 2nd ed., Wiley, New York, 1996.

Roberts, G., G. Krauss, and R. Kennedy, *Tool Steels,* 10th ed., ASM International, Materials Park, OH, 1998.

Specifications on Steel Products

Annual Book of ASTM Standards, Vol. 01.01, Steel—Piping, Tubing, Fittings, ASTM, West Conshohocken, PA, 2001.

Annual Book of ASTM Standards, Vol. 01.02, Ferrous Castings, Ferroalloys, Shipbuilding, ASTM, West Conshohocken, PA, 2001.

Annual Book of ASTM Standards, Vol. 01.03, Steel—Plate, Sheet, Strip, Wire, ASTM, West Conshohocken, PA, 2001.

Annual Book of ASTM Standards, Vol. 01.04, Steel—Structural, Reinforcing, Pressure Vessel, Railway, ASTM, West Conshohocken, PA, 2001.

Annual Book of ASTM Standards, Vol. 01.05, Steel—Bars, Forgings, Bearing, Chain, Springs, ASTM, West Conshohocken, PA, 2001.

Annual Book of ASTM Standards, Vol. 01.06, Coated Steel Products, ASTM, West Conshohocken, PA, 2001.

SAE Materials Standards Manual, SAE HS-30, Society of Automotive Engineers, Warrendale, PA, 2000.

CHAPTER 2

STAINLESS STEELS

James Kelly
Rochester, Michigan

Stainless steels are those alloys of iron and chromium, with or without other elements, containing at least 11% chromium. This is the minimum amount of chromium necessary to form a stable, passive chromium oxide film. It is this film that is the basis for the corrosion resistance of all stainless, and most nickel-base, corrosion-resistant alloys.

There are six basic classifications of stainless steels: ferritic, martensitic, martensitic age hardening, duplex austenitic–ferritic, and austenitic. The most commonly produced of these are the ferritics 409 for automotive applications and 430 for corrosion-resistant/decorative uses, the martensitic grade 410, and the age-hardening martensitic 17-4PH®. Of the austenitic–ferritic duplex grades, alloy 2205 is the most broadly available. The two most used austenitic stainless grades are 304L and 316L. A number of "superaustenitics" use nitrogen to maintain an austenitic structure with relatively high molybdenum, some 6%, and moderate nickel, 18–25%. Of the higher nickel grades alloys 825 and 20 are used for sulfuric acid and general chemical processing. The most commonly used of the very high nickel alloys is C-276. The austenitic stainless steels form a continuum with the nickel-base heat- and corrosion-resistant alloys. They are distinguished on the basis of nickel content by arbitrary or commercial definitions. There is no recognized metallurgical definition of where stainless ends and nickel base begins.

1 EFFECT OF ALLOYING ELEMENTS

The corrosion behavior of the alloying elements in pure form influences the corrosion properties of the alloys made from them.

Chromium is the first example, with outstanding corrosion resistance in the passive state. In solutions of neutral pH, dissolved oxygen from the air is sufficient to maintain passivity. But in low-pH solutions, stronger oxidizing agents must be present and halogen or sulfuric acids absent in order to stabilize the passive condition. Chromium metal is not resistant to corrosion by reducing acids.[1]

Some examples, from Uhlig,[1] of corrosion resistance of electrodeposited chromium are as follows:

Acid or Salt	Concentration (%)	Temperature		Corrosion Rate	
		°C	°F	mm/yr	mils/yr
Acetic	10	58	136	0.39	15
Ferric chloride	10	58	136	0.41[a]	16[a]
Fomic	10	58	136	30	1,200
Hydrobromic	10	58	136	4.7	186
Hydrofluoric	10	12	54	25	1,000
Perchloric	10	58	136	0.86	42
Phosphoric	10	58	136	0.28	34
Sulfuric	10	12	54	0.86	11
Sulfuric	10	58	136	250	10,000
Sulfuric	100	12	34	0.76	30
Sulfuric	100	58	136	1.8	69

[a] Pitting occurred; this number does not reflect uniform corrosion.

Three points can be made from these data. First, chromium as an alloying element is not particularly effective in promoting resistance to reducing or halogen acids. Second, in solutions of some halogen salts the passive layer was maintained by oxygen dissolved in the solution. Third, sulfuric acid behaves as a reducing acid in lower concentrations but as an oxidizing acid in concentrated form. When selecting alloys to resist sulfuric acid, one must bear this in mind.

Stainless steels containing only chromium and iron, specifically the ferritic and martensitic stainless steels, likewise have poor resistance to sulfuric acid solutions but may resist nitric acid. These chromium–iron alloys are not resistant to corrosion by halogen acids or chloride salts. Those ferritic alloys that do have good to excellent chloride pitting resistance, such as E-Brite® and AL29-4C®, gain that resistance by the addition of 1 and 4% Mo, respectively.

Austenitic nickel alloys with resistance to concentrated (oxidizing) sulfuric acid require high chromium, such as the ThyssenKrupp VDM alloy 33, or silicon, as in Haynes® nickel alloy D-205, and the stainless grades A610, A611, and Sandvik® SX.

There are a few highly corrosion resistant nickel alloys with little or no chromium, including the various Ni–Mo "B" grades and the 67Ni–31Cu alloy 400. Excellent in reducing environments, they have almost no tolerance to oxidizing compounds in the environment. A newly developed nickel–molybdenum alloy, B-10, includes 8% Cr in its composition for limited resistance to low levels of oxidizers.

Molybdenum, in contrast to chromium, has very low resistance to oxidizing solutions but does resist reducing and halogen acids. Oxidizing acids such as nitric, aqua regia, and concentrated sulfuric acids readily dissolve molybdenum metal. Hydrofluoric acid does not affect Mo, and hot (110°C) hydrochloric acid attacks molybdenum metal only slowly.[1] In both stainless steels and nickel-base alloys, molybdenum as an alloying element is required for resistance to halogen acids and to pitting by acid or oxidizing chlorides. The amount used ranges from 2% in 316L stainless steel up to 24–30% in alloys B through B-10. As an alloying addition, molybdenum improves the stability of the passive layer in the presence of halogens.

Tungsten at the 2–4% level is used, along with molybdenum, to improve chloride pitting corrosion resistance.

Nickel metal in general is attacked by oxidizing solutions, while reducing solutions are less aggressive. Some examples[1] follow:

Acid	Notes	Concentration (%)	Temperature °C	Temperature °F	Corrosion Rate mm/yr	Corrosion Rate mils/yr
Hydrochloric	Air saturated	10	30	86	2	80
	N_2 saturated	10	30	86	0.25	10
Sulfuric	Aeration by convection	10	77	170	0.31	12.1
	Air saturated	10	82	180	4	160

Nickel metal is strongly attacked by phosphoric acid solutions containing ferric (oxidizing) salts, whereas it resists phosphoric acid solutions that are free of oxidizing compounds. Nickel metal resists neutral chloride solutions, such as sodium chloride, but is attacked by acid or oxidizing chloride salts. Alloys for use in reducing acids invariably have a considerable nickel content, ranging from as little as 8% in 304L to as much as 71% in alloy B-2.

Copper is generally resistant to reducing acid solutions containing only low levels of oxygen but is readily attacked by oxidizing acids. These include nitric, sulfurous, and concentrated sulfuric acids as well as solutions containing oxidizing salts, such as ferric chloride. Copper in solution tends to reduce the corrosion rate of stainless alloys in (reducing) sulfuric acid. Those alloys intended for use in environments containing much sulfuric acid invariably have some copper as an alloy addition. These include 904L, 20Cb-3®, 825, Nicrofer® 3127 hMo (alloy 31), and the Hastelloy® alloys G, G-3, and G-30.

Additions of some 4 or 5% silicon increase resistance to corrosion in oxidizing environments. The silicon is primarily used in alloys meant to withstand concentrated, hence oxidizing, sulfuric acid. Such alloys in current production include A610, A611, Sandvik SX, and Haynes D-205™. The use of silicon as an alloying element for corrosion resistance dates back to before World War I. Although not a stainless steel, one of the oldest and most generally corrosion resistant alloys ever developed is the 14.5% silicon cast iron, Duriron®. A silicon oxide film is believed responsible for this grade's useful resistance to environments ranging from oxidizing to reducing. This includes seawater, and organic and many inorganic acids, though not halogen acids. A further addition of 3% molybdenum provides resistance to hot hydrochloric acid. Lack of strength and ductility limit the cast iron's range of use.

When one speaks about the effect of this or that pure chemical on an alloy, one must emphasize that real industrial environments are complex mixtures of chemicals. These mixtures may behave in surprising ways, quite unlike what one might expect from the behavior of alloys in pure, laboratory-controlled environments. Corrosion rates depend not only upon the concentrations of various chemicals but also on the temperature. The temperature of liquid inside a vessel is one point that can be measured, but the temperature at the surface of submerged heating coils in that vessel is another and higher value. Likewise the concentration of, say, an acid in the vessel is not the same as the concentration at the point where that acid is introduced to the mixture.

The most commonly used corrosion-resistant alloys are the stainless steels 304 (18% Cr–8% Ni, commonly known as 18-8 stainless steel) and 316 (17% Cr–11% Ni–2% Mo). The more corrosion resistant nickel alloys, such as C-276, have much higher levels of nickel, 57%, and molybdenum, 15.5%. Commercially pure nickel and nickel–copper alloys are used for special environments.

Oxidizing and reducing environments are defined chemically with respect to whether hydrogen is oxidized or reduced under the environment in question. In an oxidizing environment, hydrogen will only be present chemically combined with some other element, for example with oxygen to form H_2O. In a reducing environment that H^+ will be reduced to hydrogen gas, H_2.

Common oxidizing chemicals are nitric acid, HNO_3, and certain salts such as ferric chloride, $FeCl_3$, and cupric chloride, $CuCl_2$. The ferric and cupric ions are at relatively high valences, +3 and +2, respectively, and readily accept electrons from, or oxidize, other materials to get their own valences reduced to a more stable level. Sulfuric acid, H_2SO_4, is normally a reducing acid. At high concentrations, above about 95%, sulfuric acid changes its character and becomes an oxidizing acid. Of course, dissolved oxygen contributes to the oxidizing character of an environment. To some extent so does dissolved elemental sulfur.

To resist oxidizing conditions, an alloy must contain some amount of chromium. In an oxidizing acid, simple materials such as 304 (18% Cr–8% Ni) or 310 (25% Cr–20% Ni) are often used. An unusually high level of chromium, 33%, is present in a newly developed alloy, UNS R20033, meant to resist very oxidizing acids. In any of these alloys the nickel content is necessary to make a stable austenitic alloy, but it does not contribute specifically to oxidizing acid resistance. Small additions of molybdenum or copper may be tolerated in these alloys to enhance resistance to chlorides or sulfuric acid. But neither Mo nor Cu is helpful in resisting strongly oxidizing chemicals.

A common and severe test for resistance to oxidizing acids is boiling 65% nitric acid. The test is run for five periods of 48 h each, specimens being weighed after each test period, and the results averaged. This test is a good measure of resistance to intergranular corrosion in a sensitized alloy as well as to general corrosion in nitric acid. Test results[2] show 2205 at 0.13–0.20 mm/yr, which is good; 304 at 0.23 mm/yr; and RA333®, which has been stabilize annealed at 1700°F, at 0.29 mm/yr. In the case of RA333 it is the high chromium that helps, in spite of the 3% Mo. Other molybdenum-bearing grades do not fare so well, at 316L (2% Mo) at 0.87 mm/yr after only 24 h, AL-6XN® (6.3% Mo) at 0.74 mm/yr, 625 (9% Mo) at 0.76 mm/yr, and C-276 (15.5% Mo) at 0.74 mm/yr. These results do not mean that one cannot successfully use a higher Mo alloy in the presence of any nitric acid at all. They do indicate that high molybdenum alloys may not behave at all well in hot, concentrated oxidizing industrial environments.

One cannot readily find boiling 65% nitric acid (ASTM A 262 C) data for the 66% Ni–31% Cu alloy 400 (Monel® 400) or for the assorted B alloys—B, B-2, B-3, or B-4. Their corrosion rates in nitric acid are simply too high for the test to have any practical value. Alloys 400, B, and B-2 have no deliberate chromium addition, B-3 and B-4 only about 1.3% Cr. These grades may have excellent resistance to various reducing environments, but because there is essentially no chromium present, they will literally dissolve in nitric acid. Likewise, they are attacked by ferric, cupric and chlorate ions and even dissolved oxygen in HCl.

The common "reducing" acids are sulfuric under about 95%, phosphoric (H_3PO_4), and hydrochloric. Of these by far the most corrosive is HCl, phosphoric being the less troublesome. Because reducing industrial environments often do contain some oxidizing salts or oxygen from the air, most alloys used to withstand reducing chemical environments will contain chromium, at least 15%. The alloy additions used to resist the reducing components of the environment are nickel (Ni), molybdenum (Mo), and copper (Cu).

In sulfuric acid some amount of copper is usually used, such as in 20Cb-3® stainless steel, 904L, or 825. Even copper salts in the acid will reduce corrosion attack on stainless steel. 20Cb-3 uses carefully balanced proportions of Cu and Mo to resist sulfuric acid corrosion.

2 SOME FORMS OF CORROSION

2.1 General Corrosion

This is the most common form of corrosion, accounting for the greatest tonnage loss of metal. It is characterized by relatively uniform attack of the entire area exposed to the corrosive environment. The passive film slowly dissolves but continually re-forms. Since the attack is linear with time, the life of equipment subject to general corrosion is reasonably predictable. If the passive film is locally disrupted, as by chlorides, corrosion modes such as pitting, crevice, and stress corrosion may occur. These are more difficult to predict and tend to cause premature equipment failures. Erosion may also remove the passive film and contribute to much higher than expected general corrosion rates.

Stainless steel passivates simply by being exposed to air. A metallographic specimen of AL-6XN, for example, must be etched immediately after polishing. Otherwise it will passivate in air so that a uniform etch cannot be achieved. Passivation in acid is not required. But, during normal fabrication practices, enough iron is picked up to cause surface rusting in damp weather. A treatment in nitric–hydrofluoric acid may be used to remove this surface iron contamination.

Uniform corrosion rates may be stated as an average metal thickness loss with time, in mils per year or millimeters per year. A convenient rating for metals subject to uniform attack based on corrosion rates is as follows:

Excellent—rate less than 5 mils/year (0.13 mm/yr). Metals suitable for making critical parts.

Satisfactory—rate 5–50 mils/yr (0.13–1.3 mm/yr). Metals generally suitable for noncritical parts where a higher rate of attack can be tolerated

Unsatisfactory—rates over 50 mils/yr (1.3 mm/yr). Metals usually not acceptable in the environment.

An approximate ranking of a few common alloys by increasing resistance to general corrosion would be 304L, 316L, 2205, 20Cb-3/825, RA333, AL-6XN, 625, and C-276. Alloy selection does depend upon the exact corrosive environment in question. Some specific examples include hot concentrated caustic, where commercially pure nickel or the 76% nickel alloy 600 is used. Alloy 33 may be required for mixtures of caustic with oxidizing chemicals. For sulfuric acid alloys 20Cb-3 or 825 are usually chosen; however, if chlorides are present in the acid, one of the 6% molybdenum grades such as AL-6XN would be preferred. AL-6XN is used for organic acids, such as napthenic acid in refinery service. For nitric acid service chromium is beneficial, molybdenum not. Alloys commonly selected include 304L or a low-carbon version of 310. RA333 is used when the same piece of equipment must see very high temperatures, in the red heat range, in one zone and aqueous corrosion in another.

2.2 Stress–Corrosion Cracking

For just about every alloy, there is some chemical environment that, combined with stress, will cause cracking. For brass that environment is ammonia or other nitrogen compounds. The source of stress is usually residual forming and welding stresses, which may reach the yield point of the material. Operating stress is rarely the issue.

For austenitic stainless steels chlorides are the major cause of stress–corrosion cracking (SCC). An example is hot potable water under heat transfer conditions which permit chlorides to concentrate locally. Susceptible alloys include 304L, 316L, 321, and 347. Some 95% of 316L chemical plant equipment failures may be attributed to chloride SCC. The chlorides concentrate from trace amounts present in steam for heating or the cooling water in heat exchangers as well as from the product

Chloride SCC occurs most quickly in austenitic stainless steels with about 8–10% nickel, alloys with much lower or much higher nickel content being less susceptible. As thermal stress relief is rarely practical with stainless fabrications, the metallurgical solution is a change in alloy. Nickel-free ferritic steels, such as E-BRITE, are highly resistant to chloride SCC but impractical to fabricate into a vessel.

The traditional solution in North America has been to go to a higher nickel alloy. Alloys with about 30% or more nickel are generally considered to be good engineering solutions to most chloride SCC problems, although they will crack under very severe conditions. 20Cb-3 at about 34% and 825 at 40% nickel have long been chosen for this service. Likewise the fine-grained Incoloy® 800, UNS N08800, 31% Ni, had been used in years past. However, this low-carbon, fine-grained version of 800 is now rarely available. Regardless of what it is called, "800" today usually is 800HT®, UNS N08811, a higher carbon, coarse-grained version. This grade is designed to maximize creep rupture strength for high-temperature applications.

Since the mid-1980s the 6% molybdenum superaustenitics have become available. Grades such as 254 SMO® with only 18% nickel, or AL-6XN at 24% nickel, have been used effectively to resist chloride SCC. The material cost of the "6-moly" grades is between that of 316L stainless and nickel alloy C-276. Although lower in nickel, molybdenum contents above 2% tend to decrease susceptibility of austenitic stainless steel to chloride SCC.[3]

For greater resistance to both corrosion and chloride SCC the most used grade has been C-276, at 57% Ni 15.5% Mo. There are now a number of alloys in this class, including C-22, Inconel® 686, C-2000, 59, and a new Japanese grade, MAT 21. These very high nickel alloys can easily reach five times or more the cost of 316L stainless steel. They are metallurgically excellent solutions to chloride SCC, particularly in severe environments. However, they are expensive choices for service conditions under which 316L lasted a few years before cracking. There is a less expensive choice, one that has long been used in Europe—a duplex stainless steel which is about half austenite and half ferrite. Duplex stainless steel is a practical solution to most 304L or 316L SCC failures. The most commonly available duplex in North America is 2205, at a cost roughly 20% above that of 316L.

Other forms of SCC in stainless steels include caustic cracking and polythionic acid SCC. Caustic may crack carbon as well as stainless steel. High-nickel alloys, such as alloy 600 or, better, commercially pure nickel (UNS N02201), are used for hot concentrated caustic.

Polythionic acid stress–corrosion cracking (PASCC) is caused by sulfur compounds in the environment and most often encountered in refineries. Any stainless or nickel alloy that has been sensitized can be subject to PASCC. High-nickel alloys do not help; even 600 alloy will crack when sensitized.[4] To resist this form of SCC the alloy must contain a strong carbide-forming element, or "stabilizing" element, such as columbium or titanium. Examples include 321, 347, 20Cb-3, 825, and 625. In addition, the alloy must be given a stabilizing anneal so that the carbon is effectively combined with the Cb or Ti. RA333, because of its tungsten and molybdenum content, resists PASCC when stabilize annealed about an hour at 1700°F.

Normally, 304H or 316H would be quite sensitive to polythionic acid stress cracking, as these higher carbon, solution-annealed alloys readily sensitize. The matter has been addressed at one refinery by fabricating the equipment from one of these H-grade stainless steels, then stabilize annealing the completed fabrication. A temperature of about 1650°F (900°C) for a minimum of 1 h is used. This does precipitate carbides at the grain boundaries, but the temperature is high enough to permit chromium to diffuse back into the Cr-depleted grain boundary zone. In addition, this treatment relieves over half of the residual fabricating stress, thus reducing susceptibility to chloride SCC as well.

2.3 Pitting Corrosion

Pitting is an extremely localized form of corrosion that results in holes in the metal. Although total metal loss may be small, the equipment may be rendered useless because of perforation. Pitting usually requires a long initiation period before attack is visible. Once a pit has begun, the attack continues at an accelerating rate. Pits tend to grow in a manner that undermines or undercuts the surface. Typically a very small hole is seen on the surface. Poking at this hole with a sharp instrument may reveal a rather cavernous hole under what had looked like solid metal. In effect, a pit may be considered a self-formed crevice. Pitting attack increases with temperature.

Chloride solutions are the most common cause of pitting attack on stainless steels and nickel alloys. The alloying additions molybdenum, nitrogen, and, to some extent, chromium all contribute to pitting resistance. A laboratory measure of resistance to pitting corrosion is the critical pitting temperature, or CPT, which is the highest temperature at which an alloy resists pitting in a given environment. Alloy ranking with respect to chloride pitting resistance would be 304L (0% Mo, poor), followed by 316L (2% Mo), then four austenitics each with about 3% molybdenum, 20Cb-3/825/317L/RA333, the duplex 2205 (3% Mo–0.16% N), AL-6XN (6.3% Mo–0.22% N), 625 (9% Mo) and C-276 (15.5% Mo). Alloys AL-6XN and higher have sufficient chloride resistance to be used in hot seawater service. The lower molybdenum grades, including 2205 with 3% Mo, are unsuitable for use in seawater.

2.4 Crevice Corrosion

Crevice corrosion, more so than pitting, is the limiting condition which often prevents the use of conventional austenitic stainless steel in chloride environments. The attack usually occurs in small volumes of stagnant solution under gasket surfaces, lap joints, marine fouling, and solid deposits and in the crevices under bolt heads and the mating surfaces of male and female threads. The mechanism involves oxygen depletion in the crevice followed by chloride ion concentration and increase in acidity (decrease in pH) within the crevice. In a neutral, pH 7, chloride solution service the liquid within a crevice may contain 3–10 times as much chloride as the bulk solution and have a pH of only 2–3.

Susceptibility to crevice corrosion increases rapidly with temperature. Molybdenum and nitrogen additions to nickel–chromium–iron alloys improve their resistance to crevice corrosion. Together with the use of appropriate materials, design practice to minimize crevices and maintenance procedures to keep surfaces clean are required to combat the problem.

The usual laboratory measure of resistance to crevice corrosion is the critical crevice corrosion temperature, or CCCT, which is the highest temperature at which an alloy resists crevice corrosion in a given environment. For a given environment the CCCT is usually significantly lower than the CPT. Crevice corrosion resistance as measured by the ferric chloride test relates, to a degree, to performance in seawater. The results for a number of alloys[2] given the temperature for initiation of crevice corrosion in 10% ferric chloride ($FeCl_3 \cdot 6H_2O$) per ASTM G 48 Practice B and pitting reseistance equivalent (PRE) N = Cr = 3.3% Mo + 30(%N) are as follows:

Alloy	Mo (%)	Temperature °C	Temperature °F	Pitting Resistance Equivalent, N	Reference
316L	2.1	−3	27	23	5
825	2.7	−3	27	30	5
317L	3.2	2	35	29	5
2205	3.1	20	68	38	5
317 LXN™	4.4	20	68	34	5
28	3.5	24	75	38	6
904L	4.4	24	75	35	5
904L	4.4	25	77	35	7
G	6.5	30	86	43	5
28	3.5	35	95	39	7
2507	4.0	35	95	47	7
1925hMo	6.2	40	104	47	7
33	1.4	40	104	50	7
AL-6XN	6.2	43	110	48	5
27-7MO	7.3	45	113	57	8
625	9.0	45	113	51	5
625	9.0	55	131	51	6
31	6.5	55	131	54	6
G-30	5.5	50	122	48	7
C-276	15.4	55	130	66	5

2.5 Intergranular Corrosion

Intergranular corrosion consists of localized attack along the grain boundaries of the metal. Sensitization to this attack in stainless or nickel alloys is caused by precipitation of chromium-rich carbides in the grain boundaries at a temperature low enough that a chromium-depleted zone forms. This precipitation most commonly occurs from the heat of welding. It may also result from a slow cool after annealing or from prolonged exposure to intermediate temperatures, roughly 850–1470°F (450–800°C), in service.

A most effective means of combating intergranular corrosion is to restrict the carbon content of the alloy. In the stainless "L" grades 0.03% maximum is considered sufficient. High chromium and molybdenum additions, as in AL-6XN, also reduce the chance of intergranular attack.

Another approach is to add columbium or titanium to tie up the carbon, the same as is done to resist PASCC. 20Cb-3 stainless steel takes both approaches, being melted to low carbon as well as having a columbium addition.

2.6 Galvanic Corrosion

An electrical potential, or voltage, difference usually exists between two different metals that are in electrical contact and immersed in a corrosive solution. This potential difference causes current to flow and the less noble, or more anodic, metal suffers increased corrosion rate. The severity of attack depends upon the relative voltage difference between the metals, the relative exposed areas of each, and the particular corrosive environment.

The most common example is the old-fashioned flashlight battery, or dry cell. It has a shell of zinc metal (less noble, or anodic) filled with a moist corrosive paste that conducts electricity. The center post is made of graphite, which is quite noble (cathodic, does not tend to corrode). The potential (voltage) difference between zinc and graphite is about 1.5 V. When an electric connection is made in a flashlight, the zinc corrodes, giving up electrons, which flow through the lightbulb toward the graphite cathode, the positive pole. In this case, because generating electricity is the point, no one minds that the zinc corrodes and gets used up.

The ratio of cathodic (noble) to anodic areas is an important factor in galvanic corrosion. An undesirable situation is a large cathode connected to a small anode, or less noble metal. This can develop high current density, hence severe corrosion, at the anode. In that common zinc dry cell the zinc anode has a much larger area than the graphite cathode, so it has some useful working life before corroding through the zinc case.

For example, a large area of stainless steel in contact with a small surface area of carbon steel is undesirable. The potential difference will tend to corrode the carbon steel, and the very large area of stainless steel will make that corrosion occur quickly. The reverse condition is preferred. That is, a small area of stainless (or more noble) metal may be coupled with a much larger area of carbon steel (anodic) without significant problems. "Significant" depends upon the application. In the past, when ferritic stainless trim was used on carbon steel automobile bodies, the steel would tend to corrode most severely underneath the trim. In part that was because the crevice trapped salt, but it was accentuated by the galvanic difference between ferritic stainless and carbon steel.

There is some potential difference among the various stainless and nickel alloys. In practice, galvanic corrosion is rarely a problem among these various alloys. There is, however, a significant potential difference between copper and stainless alloys. So long as the stainless steel is passive (not actively corroding), it is more noble enough than copper to corrode the copper alloy. An example is when a heat exchanger with a Muntz metal (60% Cu–40% Zn) tubesheet is retubed with AL-6XN® alloy instead of the original copper–nickel tubes. The potential difference is enough to corrode the copper alloy tubesheet. One ought to either replace the tubesheet with stainless steel, or retube using a copper-base alloy.

Graphite is at the noble end of the galvanic series. If graphite is in contact with stainless or nickel alloys in a corrosive environment, those alloys may corrode preferentially.

Galvanic effects have a positive side and may be used to protect equipment from corrosion, a common example being a zinc coating on steel. The zinc corrodes preferentially, and in doing so protects the steel from corrosion (rusting). Zinc or magnesium anodes are often connected to equipment ranging from chemical process to steel ship hulls to protect them from corrosion.

3 AOD, DUAL CERTIFICATION, AND CHEMISTRY CONTROL

Most stainless steels, and a few higher nickel alloys, are available with different levels of carbon. For resistance to intergranular corrosion, a low carbon is preferred, usually 0.03%

carbon maximum in stainless steel. This is referred to as an "L" grade stainless steel, e.g., 304L and 316L. With respect to aqueous corrosion resistance, the lower the carbon, the better. For high-temperature service the opposite is true, and some minimum amount of carbon is required for both tensile and creep rupture strength.

The argon-oxygen decarburization (AOD) process for refining stainless steel was introduced in the 1970s. This made profound changes in how existing grades were produced as well as permitting totally new grades to be developed. Three of these changes are worth discussing—carbon, sulfur, and precise control of chemistry.

Prior to the AOD, carbon could not be removed in the refining process without also removing chromium. Low-carbon grades could only be produced by starting with low-carbon raw materials, specifically low-carbon ferrochrome. The expense of low-carbon ferrochrome meant that the L grades were inherently more expensive. The AOD now permits refining carbon to very low levels, even with starting stock of higher carbon.

Industrywide specifications such as the American Society for Testing and Materials (ASTM) were written prior to the introduction of this new melting process. For example, ASTM A 240 for 304 stainless steel, UNS S30400, calls out 0.08% carbon maximum, no minimum, 30,000 psi minimum yield strength. Low carbon 18-8, 304L, S30403 is limited to 0.03% carbon, with a consequent lower limit for yield strength, 25,000 psi minimum. In addition, there is a 304H, meant for high-temperature use, with carbon specified as a range, 0.04–0.10%, and annealing and grain size requirements. This constitutes three separate grades. It is more economical if the mills can melt steel to only two, not three, different levels of carbon and dual certify. Consider 304, UNS S30400. As the carbon is specified only as a maximum, it might be possible to melt 304 to 0.03 maximum carbon. Lower carbon would also result in lower than the 30,000 psi yield strength required. However, using the AOD it is now possible to add a very small, precisely controlled amount of nitrogen. This does not harm intergranular corrosion resistance, but it does tend to increase room temperature tensile properties. With care in annealing practice, it is possible to produce 304 with low enough carbon to meet the 304L specification yet with high enough yield strength to meet 304 requirements. As this metal meets all specified requirements of both 304L and 304, the mill test report will show both S30403 and S30400, i.e., dual certified.

S30403/S30400 is appropriate for corrosion service but not for high-temperature mechanical properties. For useful creep rupture strength some minimum amount of carbon is required, typically 0.04%. The situation was addressed a few years ago by adding a number of H grades to ASTM A 240, with controlled carbon for high-temperature strength. The stainless steel 304H, S30409 has carbon specified at 0.04–0.10% for high-temperature strength. In addition, there are grain size and minimum anneal temperature requirements. The 304, S30400 has no requirement for minimum carbon, control of grain size, or annealing temperature. Therefore any 304H containing no more than 0.08% carbon will meet 304 requirements and may be dual-certified with 304. One should note that dual-certified 304L/304 is suited only for aqueous corrosion service but would have rather low strength at high temperature. Likewise dual-certified 304/304H is meant for high-temperature service but may be unsatisfactory for welded construction in a wet corrosive environment. In practice, there is rather little actual S30400 produced as sheet or plate at this time. Most is dual certified, one way or another.

Like carbon, sulfur can now readily be refined to very low levels, typically less than 0.005%. Compare this with typical ASTM A 240 levels of 0.030% S maximum. Usually, stainless steel intended for plates is refined to a low-sulfur level to improve hot workability. The plate is generally formed and welded, with little machining by the customer. Low sulfur is quite detrimental to machinability. As bar products are commonly meant to be machined, most stainless bar actually must be resulfurized to some level, about 0.02%, for improved machinability. When a plate is intended to be drilled for a tubesheet, machinability becomes important and a resulfurized grade, still within the old 0.030 sulfur maximum, may be chosen.

It is the precise control of chemistry, in particular nitrogen, that has permitted development of the superaustenitic 6% molybdenum grades. The ability to closely control nitrogen as an alloying addition has also tremendously improved the weldability of duplex stainless steels. Whereas formerly the only duplex stainless steel used in North America was 3RE60, and that only in tubing, today several grades of duplex stainless plate, pipe, and bar products are used in significant and growing amounts.

Chemistry control also means that the producer will minimize the use of expensive alloying elements. To remain competitive, mills now melt to the bottom end of the allowable range. One consequence for the user is that 316L, specified as 2.00–3.00% Moly, now has a melt range of about 2.00–2.10% Mo. Through the 1970s the typical Mo level of 316L was about 2.3%. In addition, the average nickel content has dropped from just below 12% in the 1970s down to around 10.2% Ni as currently produced. The result has occasionally been that 25-yr-old 316L equipment, replaced in kind, gave unexpectedly short life. In part, of course, this is due to increased corrosive conditions from recycling, rather than dumping, water.

4 AVAILABILITY

The metallurgical aspects of alloy selection are well covered in this and other works. Along with corrosion resistance, strength, and fabricability, one of the most important material properties is availability. On a large project most alloys can be made in all product forms, in full mill heat lots. A contemporary example is 317LMX®, used as flat-rolled product in mill heat lot quantities for flue gas desulfurization scrubbers. This grade is rarely, if ever, carried in stock by mills or service centers.

Few mills produce all the forms, such as sheet, plate, bar, pipe, pipe fittings, and appropriate weld fillers, needed to fabricate a chemical process vessel. The minimum quantity a mill requires per product form is large, and lead times can be significant. This may limit flexibility in the event of error, last-minute design changes, or later maintenance needs.

Consider what alternate, if more expensive, grade may be used to fill in for unavailable product forms or sizes. For example, AL-6XN has occasionally been used to fill out a bill of materials specified 317LMX. The stainless steel grades 304L and 316L are used in such quantity that availability is unlikely to be an issue. The same may not be true for more highly alloyed or specialized materials.

5 FERRITIC STAINLESS STEEL

The ferritic grades listed below exhibit corrosion resistance over a very wide range. All are subject to severe embrittlement after prolonged exposure in the 700–1100°F (370–600°C) temperature range. Alloy 409 is weldable and is the alloy used for automotive exhaust systems and catalytic converter shells. It has some useful high-temperature oxidation resistance to about 1200°F (650°C). In plate gauges it may be welded with ERNiCr-3 for better toughness. Alloy 409 is not suited for decorative applications as it may rust from exposure to weather. Alloy 416 is free machining, available both as a ferritic and a martensitic version; 430 is commonly used for decorative purposes as well as food-handling equipment. E-BRITE has chloride-pitting resistance equal or superior to 316L and practical immunity to chloride SCC. E-BRITE is currently available only as tubing for heat exchangers. AL29-4-C® is extremely resistant to chloride pitting and is used for heat exchanger applications, available only as tubing or thin strip.

		Percent								
Alloy	UNS	Cr	Ni	Mo	Si	Mn	Cu	C	Fe	Other
409	S40900	11	—	—	0.4	0.3	—	0.015	87	—
405	S40500	12	—	—	0.3	0.5	—	0.06	86	0.3 Al
416	S41600	13	—	—	0.5	1	—	0.05	85	0.2 S, 0.04 P
430	S43000	16.5	—	—	0.5	0.5	—	0.08	83	—
439	S43035	17.2	—	—	0.5	0.4	—	0.015	81	0.5 Ti
E-BRITE®	S44627	26	—	1	0.2	0.1	—	0.002	72	0.1 Cb
AL29-4C®	S44735	29	—	4	0.35	0.3	—	0.02	66	0.5 Ti

6 MARTENSITIC STAINLESS STEELS

The only one of these that is readily welded is 410S. The higher carbon 410 may require significant preheat and postweld anneal. Consider the nickel alloy filler ERNiCr-3 for maximum resistance to weldment cracking at a sacrifice in strength and hardness. The corrosion resistance of 410 depends upon the heat-treated condition. For maximum corrosion resistance temper either below 750°F (400°C) or above 1100°F (600°C). Both the 410 and the martensitic 416 are used in sporting arms construction. The 440 series are used for cutting tools, 440A and B for pocket knives; 440C and 154CM have the best edge retention and are used for fine custom knives and surgical tools.

		Percent							
Alloy	UNS	Cr	Ni	Mo	Si	Mn	C	Fe	Other
410	S41000	12	—	—	0.3	0.4	0.14	87	—
410S	S41008	12	—	—	0.3	0.4	0.05	97	—
416	S41610	13	—	—	0.4	0.2	0.1	84	0.2 S, 0.05 P
420	S42000	13	—	—	0.4	0.4	0.2	86	—
431	S43100	15.6	1.9	—	0.2	0.4	0.1	81	—
440A	S44002	17	—	0.5	0.2	0.4	0.67	81	—
440B	S44003	17	—	0.5	0.2	0.4	0.8	81	—
440C	S44004	17	—	0.5	0.2	0.4	1.07	80	—
154CM®	—	14	—	4	0.3	0.5	1.05	80	—

7 AGE-HARDENING MARTENSITIC STAINLESS STEELS

		Percent									
Alloy	UNS	Cr	Ni	Mo	Cu	Mn	Al	Ti	C	Fe	Other
455	S45500	11.5	8.5	—	2	0.2	—	1.1	0.2	76	0.3 Cb
PH15-7Mo®	S15700	15	7	2.5	—	0.4	1	—	0.04	74	—
17-4PH®	S17400	15.5	4.7	—	3.3	0.5	—	—	0.05	75	0.3 Cb
17-7PH®	S17780	17	7.3	—	—	0.7	1.2	—	0.06	74	—
PH13-8Mo®	S13800	23	8	2.5	—	—	1.2	—	0.03	76	—

8 DUPLEX STAINLESS STEELS

Duplex stainless steels are characterized by high strength and good resistance to chloride SCC. They represent a more economical solution to the problem of stainless SCC than does the use of a high-nickel alloy. Some of the newer lean duplex grades, such as LDX 2101, are replacing carbon steels as well as 304L and 316L stainless steel. AL 2003 is intended to offer a cost reduction compared with 2205. Carbon steel may be replaced in part because the higher strength permits such weight reduction as well as eliminating maintenance costs. The grades most commonly available from North American service centers are alloys 2205 and 255. LDX 2101 is beginning to be used in Europe. Other grades are available from steel mills in heat lot quantities, appropriate for large projects.

		Percent									
Alloy	UNS	Cr	Ni	Mo	Cu	Mn	Si	N	C	Fe	Other
LDX 2101®	S32101	21.5	1.5	0.3	0.3	5	0.7	0.22	0.03	70	—
AL 2003™	S32003	21	3.5	1.6	—	1	0.5	0.17	0.02	73	—
SAF 2304	S32304	23	4	0.3	0.3	2	—	0.1	0.03	70	—
3RE60	S31500	18.4	4.8	2.7	—	1.6	1.7	0.07	0.03	70	—
Nitronic® 19D	S32001	21.2	1.3	—	—	5	0.4	0.15	0.02	71	—
2205	S32205	22.1	5.6	3.1	—	1.5	0.5	0.16	0.02	67	—
Zeron®100	S32760	25	7	3.5	0.5	0.5	0.3	0.22	0.02	62	0.7 W
2507	S32750	25	7	4	—	0.1	0.2	0.3	0.02	63	—
329	S32900	26.5	4.2	1.5	—	1	0.8	—	0.08	66	—
7-MoPLUS®	S32950	26.5	4.8	1.5	—	0.4	0.3	0.2	0.02	66	—
255	S32550	25.5	5.7	3.1	1.8	0.8	0.5	0.17	0.02	62	—

9 AUSTENITIC STAINLESS AND NICKEL ALLOYS

These nickel-bearing alloys are used for the majority of corrosive environments. Alloy 316L is by far the most common choice for chemical process equipment. Formerly, when 316L was inadequate, the designer switched to the high end of the spectrum, alloy C-276, which is still considered the most broadly useful of the high-nickel alloys. Today there are numerous other choices, ranging from those of intermediate cost to alloys superior to C-276 in specific environments.

Types S30600, S32615, and alloy D-205™ are specialized high-silicon materials intended for concentrated sulfuric acid heat exchanger service. Weldability is limited. A very high chromium grade, R20033, is also intended for hot concentrated sulfuric acid. It is weldable and has a broader range of potential uses than do the high-silicon alloys.

Materials in the 5–7% molybdenum range are used for seawater service and chemical process vessels in general. Alloys N08367, S31254, and N08926 constitute the "6 moly" alloys. These have both corrosion resistance and cost intermediate between 316L and alloy C-276. All are readily fabricated and broadly available. Nitrogen in these alloys retards sigma formation, in spite of rather high Mo and moderate Ni levels. Prolonged exposure in the 1000–1800°F (540–980°C) temperature range may cause sigma-phase precipitation.

Alloys in the 13–16% Mo range include C-276, C-22, 59, C-2000, and 686. The latter four are regarded as improvements over C-276, with higher Cr for oxidizing environments.

The newest, and possibly best of the very high end nickel alloys, MAT 21, makes use of a tantalum addition as well as 19% Mo.

Alloy selection for corrosive process environments is a complex process. It should include experience with similar equipment, extensive testing in the exact corrosive environment of interest, and detailed knowledge of the various alloys to be considered. Oftentimes, apparently minor contaminants can cause major changes in corrosion rates. One example is contamination of organic chlorides with small amounts of water. This can permit the organic compound to hydrolyze, forming hydrochloric acid. The HCl in turn may aggressively pit or stress corrode the standard 18-8 stainless steels. Other examples include the alloys B-2, 200, and 400, which contain no chromium. While they have excellent corrosion resistance in reducing environments, they have little or no resistance to oxidizing environments. Unexpected failures may therefore arise from contamination by small amounts of oxidizing salts (e.g., $FeCl_3$, $CuCl_2$, or $NaClO_3$) or even dissolved oxygen. Titanium behaves in the opposite manner and requires the presence of oxidizing species to develop its protective oxide film.

		Percent								
Alloy[a]	UNS	Cr	Ni	Mo	Si	Mn	Cu	C	Fe	Other
302	S30200	18.5	8.2	—	0.5	0.75	—	0.1	72	—
304L	**S30403**	**18.3**	**9**	**—**	**9.5**	**1.7**	**—**	**0.02**	**70**	**—**
321	S32100	17.3	9.3	—	0.7	1.8	—	0.01	70	0.2 Ti
347	S34700	17	9.5	—	0.7	1.5	—	0.04	70	0.5 Cb
316L	**S31603**	**16.4**	**10.2**	**2.1**	**0.5**	**1.6**	**—**	**0.02**	**69**	**—**
317L	S31703	18	11.6	3.1	0.4	1.5	—	0.02	65	—
317LMN	S31726	17	13	4.2	9.5	1.5	—	0.03	62	0.15 N
A610	S30600	18	15	—	4	0.7	—	0.01	62	—
254 SMO®	S31254	20	18	6.1	0.4	0.7	0.7	0.015	54	0.2 N
SX	S32615	18	20.5	0.9	5.5	1.5	2	0.04	51	—
654 SMO®	S32654	24	22	7.3	—	3	0.5	0.01	42	0.5 N
B66	S31266	24.5	22	5.6	—	3	1.5	0.02	44	2 W
904L	N08904	21	25	4.5	0.5	1.7	1.6	0.015	45	—
1925 hMO, 25-6MO	**N08926**	**20**	**25**	**6.2**	**0.4**	**0.7**	**0.9**	**0.01**	**46**	**0.2 N**
AL-6XN®	**N08367**	**20.5**	**24**	**6.3**	**0.4**	**0.3**	**—**	**0.02**	**48**	**0.22 N**
27-7Mo™	N08927	22	27	7.3	0.2	1	—	0.01	42	0.35 N
28	N08028	27	31	3.5	0.2	1.8	1	0.01	35	—
31	**N08031**	**27**	**31**	**6.5**	**0.2**	**1**	**1.2**	**0.01**	**33**	**0.2 N**
33	R20033	33	31	1.4	0.3	0.7	0.7	0.01	32	—
20Cb-3	N08020	20	33	2.2	0.4	0.4	3.3	0.02	40	0.5 Cb
3620 Nb	N08020	20	37	2.1	0.5	1.6	3.4	0.02	35	0.6 Cb
825	**N08825**	**21.5**	**40**	**2.8**	**0.3**	**0.6**	**2**	**0.01**	**29**	**0.8 Ti**
RA333®	**N06333**	**25**	**45**	**3**	**1**	**1.5**	**—**	**0.05**	**18**	**3 Co, 3 W**
G-30	**N06030**	**29.5**	**45**	**5.5**	**—**	**—**	**1.9**	**0.01**	**15**	**0.7 Cb, 2.5 W**
G	N06007	22	45	6.5	0.3	1.5	2	0.01	20	2.1 Cb, 0.8 W
G-3	N06985	22	48	7	0.4	0.8	2	0.01	18	0.3 Cb, 0.8 W
625	N06625	21.5	61	9	0.1	0.1	—	0.05	4	3.6 Cb
C-276	N10276	15.5	57	15.5	0.05	0.5	—	0.005	5.5	0.2 V,4 W
686	N06686	20.5	57	16.3	0.1	0.2	—	0.005	1	3.9 W, 0.2 Al
C-2000®	N06200	23	58.5	16	0.02	0.2	1.6	0.003	0.3	0.25 Al
C-22	N06022	21	57	13	0.05	0.3	—	0.003	4	0.2 V, 3 W
59	N06059	23	59	16	0.05	0.4	—	0.005	0.3	0.2 Al
MAT 21	—	19	60	19	—	—	—	—	—	1.8 Ta

Alloy[a]	UNS	Percent								
		Cr	Ni	Mo	Si	Mn	Cu	C	Fe	Other
D-205	—	20	64	2.5	5	—	2	0.02	6	—
B-2	N10665	—	71	28	0.01	0.15	—	0.002	0.8	—
B	N10001	—	66	30	—	0.5	—	0.06	3	0.3 V
B-3	N10675	1.3	67.7	28	0.03	0.5	—	0.002	1.5	—
B-4	N10629	1.3	67	28	—	—	—	0.005	3	—
400	N04400	—	66	—	0.02	1	31	0.1	1.5	—
K500	N05500	—	66	—	—	1	28	0.1	1.5	2.7 Al, 0.6 Ti
600	N06600	15.5	76	—	0.2	0.3	—	0.08	8	0.2 Ti
201	N02201	—	99.5	—	0.2	—	—	0.01	0.2	—

[a] Alloys listed in boldface are those most commonly available from stocking distributors.

10 WELDING[9]

There are some important differences between welding carbon steel and welding stainless steel. Alloys containing more than about 20% nickel have somewhat different requirements than the lower nickel austenitic stainless steels. There are specific issues regarding duplex stainless steels and weld fillers for the high-molybdenum grades.

10.1 Carbon versus Stainless Steel

Six important differences between welding the carbon or low-alloy structural steels and the austenitic stainless and nickel alloys are surface preparation, shielding gases, cold cracking versus hot cracking, distortion, penetration, and fabrication time.

1. *Surface Preparation.* When fabricating carbon steel, it is common practice to weld right over the scale (a so-called mill finish is a layer of blue-black oxide, or scale, on the metal surface), red rust, and even paint. The steel weld fillers normally contain sufficient deoxidizing agents, such as manganese and silicon, to reduce these surface iron oxides back to metallic iron. The resultant Mn–Si slag floats to the weld surface. Iron oxide, or scale, melts at a lower temperature, 2500°F (1371°C), than does the steel itself. One can see this in a steel mill when a large ingot is removed from the soaking pit for forging—the molten scale literally drips off the white-hot steel

Stainless steel, by contrast, must be clean and free of any black scale from hot-rolling, forging, or annealing operations. Such scale is predominately chromium oxide. Normally, stainless steel is produced with a descaled white or bright finish. Stainless steel melts at a lower temperature than does its chromium oxide scale, and the stainless steel weld filler chemistry is not capable of reducing this scale back to metallic chromium. As a result, with gas shielded processes it is difficult to get the weld bead to even "wet," or stick to, a scaled piece of stainless steel.

The need to clean or grind down to bright metal is more likely to cause trouble when stainless steel is being joined to carbon steel. In this dissimilar metal joint it is necessary to grind that carbon steel to bright metal, on both sides of the joint, free of all rust, mill scale, grease, and paint. The preferred weld fillers for this particular joint, to minimize the potential for any hard martensitic layer on the steel side, are ENiCrFe-3 covered electrodes or ERNiCr-3 bare filler wire; E309 electrodes are commonly used but are not the optimum choice.

Both stainless and high-nickel alloys designed for corrosion resistance are produced to very low carbon contents, less than 0.03% and sometimes less than 0.01% carbon. Any higher carbon content will reduce the metal's corrosion resistance. For this reason it is necessary to clean the alloy thoroughly of all grease and oil before welding. Also the very high nickel alloys, such as Monel® 400 or commercially pure nickel 200/201, are very sensitive to weld cracking from the sulfur in oil or marking crayons.

Metallic zinc paint is a common way to protect structural steel from corrosion. Even a small amount of that zinc paint overspray on stainless steel will cause it to crack badly when welded. Consider finishing all stainless steel welding before painting the structural steel in the area.

2. *Shielding Gases.* For gas–metal arc welding (GMAW, also called MIG) of the carbon steel the shielding gases are usually 95% Ar–5% O, 75% Ar–25% CO_2 (carbon dioxide), or 100% CO_2. While suitable for carbon or low-alloy steel welding wire, such gases are far too oxidizing for use with stainless or nickel alloys. It is not unknown to hear the complaint "Clouds of red smoke are coming off when I weld this stainless. . .heavy spatter" and then learn that the shielding gas used was 75% Ar–25% CO_2—a fine gas for carbon steel but not for any stainless steel.

One exception to this high CO_2 prohibition is when using flux-cored wire, either stainless or nickel alloy. Some of these wires are specifically formulated to run best with 75% Ar–25% CO_2.

Stainless and nickel alloys may be GMAW spray-arc welded using 100% argon shielding gas, though it is not necessarily the ideal choice. Stainless steels possess a passive chromium oxide film. The basis of stainless steel corrosion resistance, this film is not desirable when welding. Some 80% argon is necessary to get into a spray transfer mode. Beyond this, a helium addition, up to about 20%, gives a hotter arc and helps break up the oxide film. A very slight amount of CO_2, perhaps 1%, will prevent arc wander. Too much CO_2 may begin to add carbon to the weld, undesirable for corrosion resistance. Nitrogen may be added in gases specifically designed for welding the duplex stainless steels.

Pulse-arc welding is usually done with a 75% Ar–25% He mix. Short-circuiting arc transfer may be done with this 75% Ar–25% He mix but a 90% He–$7\frac{1}{2}$% Ar–$2\frac{1}{2}$% CO_2 "trimix" is more commonly used.

3. *Cold Cracking versus Hot Cracking.* Carbon steel weldments may harden and crack as they cool from welding. High hardness and the resulting cracking may occur when the steel contains more than 0.25% carbon. Alloying elements which increase hardenability, such as manganese, chromium, molybdenum, etc., can make steels of lower carbon content also harden. Hydrogen pickup from moisture in the air causes underbead cracking in steels that harden as they cool from welding. To prevent such cracking, the steel is usually preheated before welding. This retards the cooling rate of the weld and avoids martensite formation. Postweld heat treatment, or stress relief, is also applied to some steels or for certain applications. The martensitic stainless steels behave in this fashion and, because of their high hardenability, are quite difficult to weld.

Austenitic stainless and nickel alloys do not harden, no matter how fast they cool from welding. So, it is not necessary to preheat austenitic stainless, or to postweld heat treat. Indeed preheating austenitic alloys beyond what may be necessary to dry the metal can be harmful. Stress relief at 1100–1200°F (600–650°C) as applied to carbon steel is ineffective with stainless or nickel alloys and may damage the corrosion resistance of some grades.

Stainless steel weldments are usually quite resistant to cracking, unless contaminated, possibly by zinc or copper, more rarely by aluminum. A small amount of ferrite in the austenitic weld bead provides this hot-cracking resistance.

High-nickel alloys are less forgiving and may be susceptible to cracking in restrained joints, including heavy sections. This is a hot tearing, not a cold crack. That is, the weld bead tears rather than stretches as the bead contracts upon solidifying.

This hot tearing/hot cracking has nothing to do with hardness. The faster a nickel alloy weld freezes solid, the less time it spends in the temperature range where it can tear. For this reason preheating, which slows down the cooling rate, is actually harmful, as it permits more opportunity for hot tearing to occur.

4. *Distortion.* Stainless steel has poor thermal conductivity, only about one-fourth that of ASTM A 36 structural steel. This means the welding heat tends to remain concentrated, rather than spread out. Stainless steel also expands with heat about half again as much as does carbon steel. The combination of these two factors means that stainless or nickel alloy fabrications distort significantly more than similar designs in carbon steel.

Tack welds should be closer than with carbon steel and sequenced in a pattern, left side, right side, middle, etc. If the tacks are simply done in order from one end, the plate edges close up. In order to balance stresses, weld runs should be done symmetrically about the joint's center of gravity. Back step welding is helpful. This subject is well covered in Refs. 10 and 11.

Reducing heat input reduces the stresses and distortions from the welding operation. Heat into the workpiece is controlled by welding current, arc voltage, travel speed, and the specific welding process used. For the same amperes, volts, and speed, submerged arc welding (SAW) transfers the most heat; shielded-metal arc welding (SMAW) and GMAW with argon shielding next and roughly equivalent. Gas–tungsten arc welding (GTAW) can put the least heat into the work.

5. *Penetration.* There is some tendency for welders to want to increase heat input with the stainless and high-nickel alloys. First, the weld fillers tend to be sluggish and not flow well, as compared with carbon steel. Second, the arc simply will not penetrate stainless steel as it does carbon steel. The higher the nickel content, the less the penetration. Increasing welding current will not solve the problem! Stainless, and especially nickel, alloy joint designs must be more open. The base metal should be single or double beveled, with a root gap, so that the weld metal may be placed in the joint.

6. *Fabrication Time.* Cleanliness, distortion control measures, maintaining low interpass temperatures, and even machining add up to more time spent fabricating stainless than carbon steel. A shop experienced with stainless steel may require 1.6 times as long to complete the same fabrication in stainless as in carbon steel. A good carbon steel shop encountering stainless or nickel alloys for the first time may easily spend twice as long, maybe even three times as long, to do the stainless steel fabrication as it would for the same job in carbon steel.

10.2 Austenitic Alloys

The fundamental problem to be overcome in welding austenitic nickel-bearing alloys is the tendency of the weld to hot tear upon solidification from the melt. This matter is readily handled in stainless steels of up to about 15% or so nickel. In these stainless steel grades the weld metal composition is adjusted, usually by slightly higher chromium and reduced nickel, to form a small amount of ferrite upon solidification. The amount of ferrite in the weld may be measured magnetically and is reported as a ferrite number (FN). This ferrite acts to nullify the effects of the elements responsible for hot cracking in the Ni–Cr–Fe austenitics. These elements are chiefly phosphorus, sulfur, silicon, and boron. In higher nickel grades, about 20% nickel and over, it is metallurgically not possible to form any measurable amount of ferrite. Therefore other means of minimizing hot cracking must be used. Foremost among these is to use high-purity raw materials in the manufacture of weld fillers. Simply reducing the amounts of harmful P, S, Si, and B in the weld metal improves its ability to make a sound weld. Phosphorus, in particular, must be kept below 0.015% in the weld wire itself. Certain alloy additions such as manganese, columbium (niobium), and molybdenum serve in one way or another to reduce the austenitic propensity for weld hot cracking. Manganese ranges from about 2% in AWS E310-15 covered electrodes to 8% in ENiCrFe-3 nickel alloy covered electrodes. Columbium at the 0.5% level, as in 347 or 20Cb-3 stainless steel, is harmful whereas 2–4% Cb is quite beneficial in many nickel-base weld fillers. Molybdenum is not added with the intention of increasing weldability; nevertheless it does so. The 2% Mo contributes to 316L being the most weldable of the stainless steels, and 15% in C-276 accounts for the popularity of the various "C-type" electrodes in repair welding.

The distinction between the lower nickel stainless steel grades, which depend upon ferrite to ensure weldability, and the high-chnickel alloys, which require high-purity weld fillers, is an important one to remember. Most ferrite-containing (stainless) weld fillers are useless with nickel alloy base metal, as dilution of the weld bead with nickel from the base metal eliminates this ferrite. Likewise a high-purity nickel alloy weld filler, such as ER320LR, may not be so crack resistant when contaminated by phosphorus from use on 316L or carbon steel base metal.

Alloys under 20% Nickel

Most austenitic grades containing less than 20% nickel are joined with weld fillers that utilize some ferrite, perhaps 4-12 FN, to ensure freedom from hot cracks. In practice, this means stainless steels up through 317LMN or 309S, both about 13% nickel. The 310S type stands in an odd position between the stainless and the nickel alloys, having neither ferrite nor any particular alloy addition to the weld metal to ensure sound welds. Not surprisingly, 310S welds have a reputation for fissuring, unless the ER310 weld filler contains less than 0.015% P.

Alloys over 20% Nickel

Corrosion-resistant alloys in this category begin with the 18% nickel 254 SMO®, for this and other "6 moly" grades are welded with an overmatching nickel alloy filler such as ERNiCrMo-3.

Other nickel alloys are joined with matching composition weld fillers modified only by restrictions on P, S, Si, and B. A minor amount of titanium may be added for deoxidation. Other fillers contain significant manganese or columbium to improve resistance to fissuring and hot cracking.

Such chemistry modifications are rarely as effective as is the use of ferrite in the lower nickel stainless weld fillers. Welding technique and attention to cleanliness, then, become increasingly important to ensure the soundness of fully austenitic welds. These techniques include bead contour and low interpass temperature. Reinforced, convex stringer beads are much more resistant to centerbead cracking than are shallow, concave beads. Interpass temperature for most nickel alloy weldments is kept below 300°F (150°C). Cleanliness includes *not* using oxygen additions to the GMAW shielding gases for nickel alloys.

It is worth repeating here that high-nickel alloys cannot be reliably welded using stainless steel weld fillers. Stainless steel fillers (308, 309, etc.) depend upon a chemistry which will solidify from the melt as a duplex structure, containing a small amount of ferrite in with the austenite.

10.3 Duplex Stainless Steels[12]

When welding duplex stainless steel, the issues are maintaining the austenite–ferrite balance in the weldment and avoiding precipitation of nitrides and sigma. High heat input is used to weld duplex, similar to welding 316L stainless steel. This is in contrast to the lower heat preferred with the high-nickel alloys. Low heat input is *not* desirable with a duplex, as it may not permit sufficient transformation of ferrite to austenite upon cooling. Modern duplex weld fillers have a nitrogen addition as well as slightly higher nickel content than the base metal. This is to ensure that enough weld metal ferrite transforms to austenite to maintain a balance of the two phases.

Toughness and critical pitting temperature of duplex weldments vary with the choice of welding process in opposite directions—in order of increasing notch impact toughness, SMAW AC/DC, FCAW (flux-cored arc welding), SMAW DC-basic, SAW, GMAW argon shielding, GMAW 95% Ar–3% He–2% N_2 shielding, and GTAW; in order of increasing pitting resistance, as measured by critical pitting temperature, GTAW, GMAW, SAW, FCAW, SMAW DC-basic, and SMAW AC/DC.

10.4 High-Molybdenum Alloys[13]

High-molybdenum stainless and nickel alloys are welded with an overmatching filler metal. This is necessary to maintain corrosion resistance in the weld metal at least equal to the base metal. The reason is that molybdenum and chromium segregate as the weld metal solidifies from the melt. This leaves local areas with high and low molybdenum content. Pitting corrosion can start in the low-Mo areas, with the pits eventually growing even into metal with high molybdenum content. This occurs in alloys ranging from 316L to C-276, for the most part being more severe at higher alloy contents

This matter began to receive attention when the 6% Mo stainless steels came on the market. If any of these 6% Mo grades are welded without filler metal, the result is a weld bead that may be as low as 3% Mo in areas. The end result can be that this weld has only the pitting corrosion resistance of 317L stainless. In the case of tubular products autogenously welded in production, a high-temperature anneal is used to homogenize the metal. In addition, a small amount of nitrogen, 3–5%, is added to the torch gas. Fabrications of thin sheet, which cannot be annealed after welding, should have this nitrogen addition to minimize the loss of corrosion. Even so, because thin-sheet welds solidify more quickly, the segregation is less severe.

In normal fabrication of a 6% Mo grade, alloy 625 (ERNiCrMo-3) filler metal is used. The weld metal contains 9% Mo. After welding, segregation causes some areas to have as little as 6% Mo. The result is that the alloy 625 weld bead has approximately the same corrosion resistance as the 6% Mo base metal. Higher alloy weld fillers, such as ERNiCrMo-10 or ERNiCrMo-14, may also be used, though the benefit may be more theoretical than real. ERNiCrMo-4 is not suggested, as it has 5% less chromium than does AL-6XN, for example. Since the mid-1980s nearly all of the 6% Mo alloy fabrications have been made, and put into service, using a 9% Mo weld filler.

ERNiCrMo-3 weld filler is widely available and is appropriate for welding lower alloys such as 317L, 317LMN, and 904L for chloride service. The problem of reduced weld bead corrosion resistance from molybdenum and chromium segregation exists with most of the 13–16% Mo nickel alloys as well. Filler 686 CPT (ERNiCrMo-14) does appear to be markedly less susceptible to this effect than other high-molybdenum alloys.

WEB SITES

Detailed technical information on heat and corrosion resistant alloys is available at www.rolledalloys.com. Complete stainless steel data is at www.alleghenyludlum.com. Nickel alloys in general are covered at www.nidi.org.

REFERENCES

1. H. H. Uhlig, *Corrosion Handbook,* Wiley, New York, 1948
2. J. Kelly, *Corrosion Resistant Alloy Specifications & Operating Data,* Bulletin 151, Rolled Alloys, Temperance, MI, 2003.
3. M. Ueda, H. Abo, T. Sunami, T. Muta, and H. Yamamoto, "A New Austenitic Stainless Steel Having Resistance to Stress Corrosion Cracking," Technical Report, Overseas No. 2, Nippon Steel, Tokyo, January, 1973, p. 66.
4. C. H. Samans, "Stress Corrosion Cracking Susceptibility of Stainless Steels and Nickel Base Alloys in Polythionic Acids and Acid Copper Sulfate Solution," *Corrosion* 20, NACE International, Houston, TX, 1964.
5. J. Kelly, *AL-6XN® Alloy Physical, Mechanical and Corrosion Properties,* Bulletin No. 210, Rolled Alloys, Temperance, MI, 2002.
6. R. Kirchheiner, H. Portisch, R. Solomon, M. Jahudka, and J. Ettere, *Designing Components for Radioactive Waste Liquids in a Modern NiCrMo-Alloy,* Paper No. 166, Corrosion 98, NACE International, Houston, TX, 1998.
7. Paper No. 338, Corrosion 95, NACE International, Houston, TX, 1995.
8. Special Metals Publication No. SMC-092, Huntington, WV.
9. J. Kelly, *Heat Resistant Alloy Welding,* Bulletin No. 200, Rolled Alloys, Temperance, MI, 2002.
10. *Avesta, Handbook for the Welding of Stainless Steel,* Inf. 8901, Avesta Welding AB, Avesta, Sweden.
11. B. Lundqvist, *The Sandvik Welding Handbook,* AB Sandvik Steel, Sandviken, Sweden, 1977.
12. J. Kelly, *RA2205 Duplex Stainless,* Bulletin No. 1071, Rolled Alloys, Temperance, MI, 2003.
13. J. Kelly, *AL-6XN Alloy Fabrication,* Bulletin No. 203, Rolled Alloys, Temperance, MI, 2001.

TRADEMARKS

RA333 and RA330 are registered trademarks of Rolled Alloys, Inc.

AL-6XN, AL 2003, E-BRITE, and AL 29-4C are registered trademarks, and 317LXN a trademark, of ATI Properties, Inc.

7MoPLUS and 20Cb-3 are registered trademarks of Carpenter Technology Corporation.

254 SMO and 654 SMO are registered trademarks of Outokumpu AB.

Nicrofer is a registered trademark of ThyssenKrupp VDM.

Duriron is a registered trademark of Duraloy Technologies, Inc.

Hastelloy, Haynes, G-30, and C-2000 are registered trademarks, and D-205 a trademark, of Haynes International.

Incoloy, Inconel, Monel, 800HT, 25-6MO, and 27-7MO are registered trademarks of Special Metals, Inc.

Nintronic, 17-4PH, PH15-7MO, 17-7PH, and PH13-8MO are registered trademarks of AK Steel Corporation.

Oakite is a registered trademark of Oakite Products, Inc.

Sandvik SX and SAF are registered trademarks of Sandvik AB.

CHAPTER 3
ALUMINUM ALLOYS

J. G. Kaufman
Kaufman Associates
Columbus, Ohio

1 NATURE OF ALUMINUM ALLOYS

Aluminum is the most abundant metal and the third most abundant chemical element in the earth's crust, comprising over 8% of its weight. Only oxygen and silicon are more prevalent. Yet, until about 150 years ago aluminum in its metallic form was unknown to humans. The reason for this is that aluminum, unlike iron or copper, does not exist as a metal in nature. Because of its chemical activity and its affinity for oxygen, aluminum is always found combined with other elements, mainly as aluminum oxide. As such it is found in nearly all clays and many minerals. Rubies and sapphires are aluminum oxide colored by trace impurities, and corundum, also aluminum oxide, is the second hardest naturally occurring substance on earth—only a diamond is harder. It was not until 1886 that scientists learned how to economically extract aluminum from aluminum oxide via electrolytic reduction. Yet in the more than 100 years since that time, aluminum has become the second most widely used of the approximately 60 naturally occurring metals, second only to iron.

Aluminum alloys are broadly used in products and applications that touch us regularly in our daily lives, from aluminum foil for food packaging and easy-open aluminum cans for your beverages to the structural members of the aircraft in which we travel. The broad use of aluminum alloys is dictated by a very desirable combination of properties, combined with the ease with which they may be produced in a great variety of forms and shapes. In this chapter, we will review the characteristics of aluminum alloys that make them so attractive and note the variety of applications in which they are used.

For readers who are interested in a broader look at the aluminum industry as a whole, the publications of the Aluminum Association are highly recommended, especially:

- *Aluminum: Technology, Applications, and Environment*[1]
- *Aluminum Standards and Data (Standard and Metric Editions)*[2]
- *Standards for Aluminum Sand and Permanent Mold Castings*[3]
- *The Aluminum Design Manual*[4]

See the References at the end of the chapter for a more complete listing.

2 ADVANTAGES OF ALUMINUM ALLOYS

The first step in becoming familiar with the opportunities to utilize aluminum alloys advantageously is to briefly note some of the basic characteristics of wrought and cast aluminum alloys that make them desirable candidates for such a wide range of applications as well as their limitations. Wrought alloys (those mechanically formed by rolling, forging, and extrusion into useful products) are addressed first, then cast alloys (those cast directly to the near-final finished shape).

2.1 Wrought Aluminum Alloys

Low Density/Specific Gravity. One property of aluminum that everyone is familiar with is its light weight or, technically, its low density or specific gravity. The specific gravity of aluminum is only 2.7 times that of water and roughly one-third that of steel and copper. An easy number to remember is that 1 cubic inch (in.^{3}) of aluminum weighs 0.1 lb; 1 cubic foot (ft^3) weighs 170 lb compared to 62 lb for water and 490 lb for steel.

High Strength–Weight Ratio. The combination of relatively high strength with low density means a high strength efficiency for aluminum alloys and many opportunities for replacement of heavier metals with no loss (and perhaps a gain) in load-carrying capacity. This characteristic, combined with the excellent corrosion resistance and recyclability, has led to aluminum's broad use in containers, aircraft, and automotive applications.

Excellent Corrosion Resistance. As a result of a naturally occurring tenacious surface oxide film, many aluminum alloys provide exceptionally resistance to corrosion in many atmosphere and chemical environments. Alloys of the 1xxx, 3xxx, 5xxx, and 6xxx systems are especially favorable in this respect and are even used in applications where they are in direct contact with seawater and antiskid salts. With some electrocoating enhancements (e.g., anodizing), the oxide coating can be thickened for even greater protection. Aluminum and a number of its alloys possess excellent resistance to corrosive attack by many foods and chemicals as well as natural environments.

High Thermal Conductivity. Aluminum and its alloys are good conductors of heat, and, while they melt at lower temperatures than steels, about 1000°F (about 535°C), they are slower to reach very high temperatures than steel in fire exposure.

High Electrical Conductivity. Pure aluminum and some of its alloys have exceptionally high electrical conductivity (i.e., very low electrical resistivity), second only to copper among common metals as conductors.

Excellent Reflectivity. Aluminum with appropriate surface treatment becomes an excellent reflector and does not dull from normal atmospheric oxidation.

High Fracture Toughness and Energy Absorption Capacity. Many aluminum alloys are exceptionally tough and excellent choices for critical applications where resistance to unstable crack growth and brittle fracture are imperatives. Alloys of the 5xxx series, for example, are prime choices for liquefied natural gas (LNG) tankage. And special high-toughness versions of aerospace alloys, such as 2124, 7050, and 7475, replace the standard versions of these alloys for critical bulkhead applications.

Superior Cryogenic Toughness. Aluminum alloys, especially of the 3xxx, 5xxx, and 6xxx series, are ideal for very low temperature applications because the ductility and toughness as well as strength of many alloys at subzero temperatures are as high as or higher than at room temperature, even down to near absolute zero. As noted above, the 5xxx series alloys are regularly used for liquefied gas tankage operating at temperatures from −150 to −452°F (−65 to −269°C)

Fatigue Strength. On an efficiency basis (strength to density) the fatigue strengths of many aluminum alloys are comparable to those of steels.

Low Modulus of Elasticity. Aluminum alloys have elastic moduli about one-third those of steels (about 10×10^6 psi vs. about 30×10^6 psi), so they absorb about three times as much elastic energy upon deformation to the same stress. They also deflect three times more under load (see Section 2.3).

Superior Workability. Aluminum alloys are readily workable by virtually all metalworking technologies and especially amenable to extrusion (the process of forcing heated metal through shaped dies to produce specific shaped sections). This enables aluminum to be produced in a remarkable variety of shapes and forms in which the metal can be placed in locations where it can most efficiently carry the applied loads.

Ease of Joining. Aluminum alloys may be joined by a very broad variety of commercial methods, including welding, brazing, soldering, riveting, bolting, and even nailing in addition to an unlimited variety of mechanical procedures. Welding, while considered difficult by those familiar only with joining steel and who try to apply the same techniques to aluminum, is particularly easy when performed by proven techniques such as gas–metal arc welding (GMAW or MIG) or gas–tungsten arc welding (GTAW or TIG).

Versatile Array of Finishes. Aluminum can be finished in more ways than any other metal used today, including a variety of techniques that build upon its strong oxide coating and employ coloring, plus more conventional means such as painting and enameling.

Ease of Recyclability. Aluminum and its alloys are among the easiest to recycle of any structural materials. And they are recyclable in the truest sense, unlike materials which are reused but in lower quality products: Aluminum alloys may be recycled *directly* back into the same high-quality products like rigid container sheet (cans) and automotive components. Through such recycling, the life-cycle assessment advantages for aluminum are optimized.

2.2 Cast Aluminum Alloys

The desirable characteristics of wrought alloys are also generally applicable to cast alloys, but in fact the choice of one casting alloy or another tends to be more often made on the basis of their relative abilities to meet one or more of the four following characteristics:

- Ease of casting
- Quality of finish
- High strength, especially at high temperatures
- High toughness

1. *Ease of Casting.* Many aluminum casting alloys have relatively high silicon contents that provide excellent flow characteristics during casting, enabling them to be utilized for large and complex castings (e.g., even complete engines). Relatively minute details in the shape of the casting can be accurately and reliably replicated.

2. *Quality of Finish.* By proper selection of aluminum casting alloy, extremely fine surface quality can be achieved. While such alloys typically require more attention to casting practice, they are widely used in applications where the finished casting surface mirrors the finish needed in surface.

3. *High Strength.* Many aluminum casting alloys respond to heat treatment following casting and achieve relatively high levels of strength and excellent strength–weight ratios. With careful design of molds, high chill rates can be assured and both high strength and high toughness can be achieved.

4. *High Toughness.* With careful selection of alloy and heart treatment combined with process technology often referred to as premium engineered casting, optimizing metal flow and chill rated, the toughness of castings may be comparable to that of wrought alloys at comparable strength levels. Techniques such as hot isostatic pressing (HIP) are available to further reduce porosity and improve performance.

Unfortunately few casting alloys possess all four characteristics, but some generalizations useful in alloy selection for specific applications may be made:

1. *Ease of Casting.* The high-silicon 3xx.x series are outstanding in this respect as the relatively high silicon contents lend a characteristic of good flow and mold-filling capability. As a result, the 3xxx.x series are the most widely used and are especially chosen for large and very complex casting.

2. *Finish.* The 5xx.x and 7xx.x series are noteworthy for the fine finish they provide, but they are more difficult to cast than the 3xx.x series and so are usually limited to those applications where that finish is paramount. A good example is the use of 7xx.x alloys for bearings.

3. *Strength and Toughness.* The 2xx.x alloys typically provide the very highest strengths, especially at high temperatures, but are among the more difficult to cast and lack good surface characteristics. Therefore their use is usually limited to situations where sophisticated casting techniques can be applied and where strength and toughness are at a premium, as in the aerospace industry. Relatively high strengths and superior toughness can also be achieved with some of the higher purity 3xx.x alloys (e.g., A356.0, A357.0) in heat-treated conditions.

2.3 Limitations of Wrought and Cast Aluminum Alloys

There are several characteristics of aluminum alloys that require special attention in alloy selection or design:

Moduli of Elasticity. As noted earlier, the elastic moduli of aluminum alloys are about one-third those of steel. In applications such as bridges, where some designs may be deflection critical, this is a disadvantage and consideration should be given to the fact that aluminum alloys will deflect about three times more than comparably sized steel members. To compensate for this, aluminum members subject to bending loads are usually made deeper or thicker in their upper and lower extremities to reduce stresses and/or deflections.

Melting Temperature. Aluminum Alloys melt at about 1000°F (535°C), well below where steels melt, and so they should not be selected for applications such as flue pipes and fire doors where the low melting point may result in unsatisfactory performance. The useful limit of high-temperature structural application of aluminum alloys is about 500°F (about 260°C) for conventional alloys or about 600°F (315°C) for alumina-enhanced alloys. It is useful to note, however, that even in the most intense fires, aluminum alloys do not burn; they are rated noncombustible in all types of fire tests and achieve the highest ratings in flame-spread tests. Further, as noted earlier, they are slower to reach high temperatures in fires than other metals such as steels because of their higher thermal conductivity and emmissivity. Nevertheless, they should not be used where service requirements include structural strength above 500°F (260°C) or exposure above 700°F (370°C).

Stress–Corrosion Susceptibility of Some Alloys. Some aluminum alloys, notably the 2xxx and 7xxx alloys, when stressed perpendicular to the major plane of grain flow (i.e., in the short-transverse direction) may be subject to intergranular stress–corrosion cracking (SCC) unless they have been given a special thermal treatment to reduce or eliminate this type of behavior. If short transverse stresses are anticipated in relatively thick components, 2xxx alloys should only be utilized in the T6- or T8-type tempers (not T3- or T4-type tempers), and 7xxx alloys should only be used in the T7-type tempers (not T6- or T8-type tempers). Similarly, 5xxx alloys with more than 3% Mg should not be used in applications where a combination of high stress and high temperature will be experienced over a long period of time (more than several hundred hours at or above 150°F, or 65°C) because some susceptibility to SCC may be encountered (i.e., the alloys may become "sensitized"); for applications where temperatures above about 150°F (65°C) are likely to be encountered for long periods, the use of 5xxx alloys with 3% or less Mg, e.g., 5454, is recommended.

Mercury Embrittlement. Aluminum alloys should never be used when they may be in direct contact with liquid or vaporized mercury; severe grain-boundary embrittlement may result. This is an unlikely exposure for the vast majority of applications, but in any instance where there is the possibility of mercury being present, the use of aluminum alloys should be avoided.

3 DESIGNATION SYSTEMS

One advantage in using aluminum alloys and tempers is the universally accepted and easily understood alloy and temper systems by which they are known. It is extremely useful for both secondary fabricators and users of aluminum products and components to have a working knowledge of those designation systems. The alloy system provides a standard form of alloy identification that enables the user to understand a great deal about the chemical composition and characteristics of the alloy, and similarly, the temper designation system permits one to understand a great deal about the way in which the product has been fabricated.

The alloy and temper designation systems in use today for wrought aluminum were adopted by the aluminum industry in about 1955, and the current system for cast system was developed somewhat later. The aluminum industry managed the creation and continues the maintenance of the systems through its industry organization, the Aluminum Association. The alloy registration process is carefully controlled and its integrity maintained by the Technical Committee on Product Standards of the Aluminum Association, made up of industry standards experts.[5] Further, as noted earlier, the Aluminum Association designation system is the basis of the American National Standards Institute (ANSI) standards, incorporated in ANSI H35.1[6] and, for the wrought alloy system at least, is the basis of the near-worldwide International Accord on Alloy Designations.[7] It is useful to note that the international accord does not encompass casting alloy designations, so more variation will be encountered in international designations for casting alloys than for wrought aluminum alloys.

The Aluminum Association Alloy and Temper Designation Systems covered in ANSI H35.1[6] and *Aluminum Standards and Data*[2] are outlined in this chapter.

3.1 Wrought Aluminum Alloy Designation System

The Aluminum Association Wrought Alloy Designation System consists of four numerical digits, sometimes with alphabetic prefixes or suffices, but normally just the *four* numbers:

- The *first* digit defines the major alloying class of the series starting with that number.
- The *second* defines variations in the original basic alloy; that digit is always a 0 for the original composition, a 1 for the first variation, a 2 for the second variation, and so forth; variations are typically defined by differences in one or more alloying elements of 0.15–0.50% or more, depending upon the level of the added element.
- The *third* and *fourth* digits designate the specific alloy within the series; there is no special significance to the values of those digits, and they are not necessarily used in sequence.

Table 1 shows the meaning of the first of the four digits in the wrought alloy designation system. The alloy family is identified by that number and the associated main alloying ingredient(s), with three exceptions:

- Members of the 1000 series family are commercially pure aluminum or special-purity versions and as such do not typically have any alloying elements intentionally added; however they do contain minor impurities that are not removed unless the intended application requires it.

Table 1 Main Alloying Elements in Wrought Alloy Designation System

Alloy	Main Alloying Element
1000	Mostly pure aluminum; no major alloying additions
2000	Copper
3000	Manganese
4000	Silicon
5000	Magnesium
6000	Magnesium and silicon
7000	Zinc
8000	Other elements (e.g., iron or tin)
9000	Unassigned

- The 8000 series family is an "other elements" series, comprised of alloys with rather unusual major alloying elements such as iron and nickel.
- The 9000 series is unassigned.

The major benefit of understanding this designation system is that one can tell a great deal about the alloy just from knowledge of which it is a member. For example:

- As indicated earlier, the 1xxx series are pure aluminum and its variations; compositions of 99.0% or more aluminum are by definition in this series. Within the 1xxx series, the last two of the four digits in the designation indicate the minimum aluminum percentage. These digits are the same as the two digits to the right of the decimal point in the minimum aluminum percentage specified for the designation when expressed to the nearest 0.01%. As with the rest of the alloy series, the second digit indicates modifications in impurity limits or intentionally added elements. Compositions of the 1xxx series do not respond to any solution heat treatment but may be strengthened modestly by strain hardening.
- The 2xxx series alloys have copper as their main alloying element and, because it will go in significant amounts into solid solution in aluminum, they will respond to solution heat treatment; they are referred to as heat treatable.
- The 3xxx series alloys are based on manganese and are strain hardenable; they do not respond to solution heat treatment.
- The 4xxx series alloys are based on silicon; some alloys are heat treatable, others are not, depending upon the amount of silicon and the other alloying constituents.
- The 5xxx series alloys are based on magnesium and are strain hardenable, not heat treatable.
- The 6xxx series alloys have both magnesium and silicon as their main alloying elements; these combine as magnesium silicide (Mg_2Si) following solid solution, and so the alloys are heat treatable.
- The 7xxx series alloys have zinc as their main alloying element, often with significant amounts of copper and magnesium, and they are heat treatable.
- The 8xxx series contain one or more of several less frequently used major alloying elements like iron or tin; their characteristics depend on the major alloying element(s).

The compositions of a representative group of widely used commercial wrought aluminum alloys are given in Table 2, from *Aluminum Standards and Data*[2] and other Aluminum Association publications.

3.2 Cast Aluminum Alloy Designation System

The designation system for cast aluminum alloys is similar in some respects to that for wrought alloys but has a few very important differences as noted by the following description.

Like the wrought alloy system, the cast alloy designation system also has four digits but differs from the wrought alloy system in that a decimal point is used between the third and fourth digits to make clear that these are designations to identify alloys in the form of castings or foundry ingot.

As for the wrought alloy designation system, the various digits of the cast alloy system convey information about the alloy:

Table 2 Nominal Compositions of Wrought Aluminum Alloy[a]

Alloy	Percent of Alloying Elements[b]								Notes[c]
	Silicon	Copper	Manganese	Magnesium	Chromium	Nickel	Zinc	Titanium	
1060	—	—	—	—	—	—	—	—	1
1100	—	—	—	—	—	—	—	—	1
1145	—	—	—	—	—	—	—	—	1
1350	—	—	—	—	—	—	—	—	2
2008	0.65	0.9	—	0.38	—	—	—	—	3
2010	—	1.0	0.25	0.7	—	—	—	—	
2011	—	5.5	—	—	—	—	—	—	4
2014	0.8	4.4	0.8	0.50	—	—	—	—	
2017	0.50	4.0	0.7	0.6	—	—	—	—	
2024	—	4.4	0.6	1.5	—		—	—	
2025	0.8	4.4	0.8		—		—	—	
2036	—	2.6	0.25	0.45	—	—	—	—	
2117	—	2.6	—	0.35	—	—	—	—	
2124	—	4.4	0.6	1.5	—	—	—	—	5
2195	—	4.0	—	0.50	—	—	—	—	5
2219	—	6.3	0.30	—	—	—	—	0.06	7
2319	—	6.3	0.30	—	—	—	—	0.15	7
2618	0.18	2.3	—	1.6	—	1	—	0.07	8
3003	—	.12	1.2	—	—	—	—	—	
3004	—	—	1.2	1	—	—	—	—	
3005	—	—	1.2	0.40	—	—	—	—	
3105	—	—	0.6	0.50	—	—	—	—	
4032	12.2	0.9	—	—	—	—	0.9	—	
4043	5.2	—	—	—	—	—	—	—	
4643	4.1	—	—	0.20	—	—	—	—	
5005	—	—	—	0.8	—	—	—	—	
5050	—	—	—	1.4	—	—	—	—	
5052	—	—	—	2.5	0.25	—	—	—	
5056	—	—	0.12	5.0	0.12	—	—	—	
5083	—	—	0.7	4.4	0.15	—	—	—	
5086	—	—	0.45	4.0	0.15	—	—	—	
5154	—	—	—	3.5	0.25	—	—	—	
5183	—	—	0.8	4.8	0.15	—	—	—	
5356	—	—	0.12	5.0	0.12	—	—	0.13	
5454	—	—	0.8	2.7	0.12	—	—	—	
5456	—	—	0.8	5.1	0.12	—	—	—	
5457	—	—	0.30	1.0	—	—	—	—	
5554	—	—	0.8	2.7	0.12	—	—	0.12	
5556	—	—	0.8	5.1	0.12	—	—	0.12	
5657	—	—	—	0.8	—	—	—	—	
5754	—	—	0.5	3.1	0.3	—	—	—	
6005	0.8	—	—	0.50	—	—	—	—	
6009	—	—	—	—	—	—	—	—	
6013	0.8	0.8	0.50	1.0	—	—	—	—	
6053	0.7	—	—	1.2	0.25	—	—	—	
6061	0.6	0.28	—	1.0	0.20	—	—	—	
6063	0.40	—	—	0.7	—	—	—	—	

Table 2 *(Continued)*

Alloy	Percent of Alloying Elements[b]								Notes[c]
	Silicon	Copper	Manganese	Magnesium	Chromium	Nickel	Zinc	Titanium	
6066	1.4	1.0	0.8	1.1	—	—	—	—	
6070	1.4	0.28	0.7	0.8	—	—	—	—	
6101	0.50	—	—	0.6	—	—	—	—	
6111	0.85	0.7	0.28	0.75	—	—	—	—	
6151	0.9	—	—	0.6	0.25	—	—	—	
6201	0.7	—	—	0.8	—	—	—	—	
6262	0.6	0.28		1.0	0.09				9
6351	1.0	—	0.6	0.6	—	—	—	—	
6951	0.35	0.28	—	0.6	—	—	—	—	
7005	—	—	0.45	1.4	0.13	—	4.5	0.04	10
7049	—	1.6	—	2.4	0.16	—	7.7	—	
7050	—	2.3	—	2.2	—	—	6.2	—	11
7072	—	—	—	—	—	—	1.0	—	
7075	—	1.6	—	2.5	0.23	—	5.6	—	
7116	—	0.8	—	1.1	—	—	4.7	—	12
7129	—	0.7	—	1.6	—	—	4.7	—	12
7175	—	1.6	—	2.5	0.23	—	5.6	—	5
7178	—	2.0	—	2.8	0.23	—	6.8	—	
7475	—	1.6	—	2.2	0.22	—	5.7	—	5
8017	—	0.15	—	0.03	—	—	—	—	13
8090	—	1.3	—	0.95	—	—	—	—	14
8176	0.09	—	—	—	—	—	—	—	13

[a] Based on industry handbooks.[2,3] Consult those references for specified limits. Values are nominal, i.e., middle range of limits for elements for which a composition range is specified.
[b] Aluminum and normal impurities constitute remainder.
[c]
1. Percent minimum aluminum—for 1060: 99.60%; for 1100: 99.00%; for 1145: 99.45%; for 1350: 99.50%. Also, for 1100, 0.12% iron.
2. Formerly designated EC.
3. Also contains 0.05% vanadium (max.).
4. Also contains 0.40% lead and 0.4% bismuth.
5. This alloy has tighter limits on impurities than does its companion alloy (2024 or 7075).
6. Also contains 1.0% lithium, 0.42% silver, and 0.12% zirconium.
7. Also contains 0.10% vanadium plus 0.18% zirconium.
8. Also contains 1.1% iron.
9. Also contains 0.55% lead and 0.55% bismuth.
10. Also contains 0.14% zirconium.
11. Also contains 0.12% zirconium.
12. Also contains 0.05% max. vanadium plus 0.03% max. gallium.
13. Also contains 0.7% iron.
14. Also contains 2.4% lithium plus 0.10% zirconium.

- The first digit indicates the alloy group, as can be seen in Table 3. For 2xx.x through 8xx.x alloys, the alloy group is determined by the alloying element present in the greatest mean percentage, except in cases in which the composition being registered qualifies as a modification of a previously registered alloy. Note that in Table 3 the 6xx.x series is shown last and for cast alloys is designated as the unused series.
- The second and third digits identify the specific aluminum alloy or, for the aluminum 1xx.x series, indicate purity. If the greatest mean percentage is common to more than

Table 3 Cast Alloy Designation System

Alloy	Main Alloying Element
1xx.x	Pure aluminum, 99.00% maximum
2xx.x	Copper
3xx.x	Silicon, with added copper and/or magnesium
4xx.x	Silicon
5xx.x	Magnesium
7xx.x	Zinc
8xx.x	Tin
9xx.x	Other elements
6xx.x	Unused series

one alloying element, the alloy group is determined by the element that comes first in sequence. For the 1xx.x group, the second two of the four digits in the designation indicate the minimum aluminum percentage. These digits are the same as the two digits to the right of the decimal point in the minimum aluminum percentage when expressed to the nearest 0.01%.

- The fourth digit indicates the product form: xxx.0 indicates castings and xxx.1 for the most part indicates ingot having limits for alloying elements the same as those for the alloy in the form of castings. A fourth digit of xxx.2 may be used to indicate that the ingot has composition limits that differ from but fall within the xxx.1 limits; this typically represents the use of tighter limits on certain impurities to achieve specific properties in the cast product produced from that ingot.
- A letter before the numerical designation indicates a modification of the original alloy or an impurity limit. These serial letters are assigned in alphabetical sequence starting with A, but omitting I, O, Q, and X, with X being reserved for experimental alloys. Note that explicit rules have been established for determining whether a proposed composition is a modification of an existing alloy or if it is a new alloy.

Experimental alloys of either the wrought or cast series are indicated by the addition of the prefix X. The prefix is dropped when the alloy is no longer experimental. However, during development and before an alloy is designated as experimental, a new composition may be identified by a serial number assigned by the originator. Use of the serial number is discontinued when the composition is registered with the Aluminum Association and the ANSI H35.1 designation is assigned.

The compositions of a representative group of widely used commercial cast aluminum alloys are given in Table 4, from *Standards for Aluminum Sand and Permanent Mold Castings*[3] and other aluminum casting industry publications.[5,8–10]

3.3 Aluminum Alloy Temper Designation System

The temper designation is always presented immediately following the alloy designation (Section 3.2), with a dash between the two, e.g., 2014-T6 or A356.0-T6.

The first character in the temper designation is a capital letter indicating the general class of treatment as follows:

F = as fabricated

O = annealed

Table 4 Nominal Compositions of Aluminum Alloy Castings[a]

Alloy	Percent of Alloying Elements[b]								Notes[c]
	Silicon	Copper	Manganese	Magnesium	Chromium	Nickel	Zinc	Titanium	
201.0	—	4.6	0.35	0.35	—	—	—	0.25	1
204.0	—	4.6	—	0.25	—	—	—	—	
A206.0	—	4.6	0.35	0.25	—	—	—	0.22	
208.0	3.0	4.0	—	—	—	—	—	—	
213.0	2.0	7.0	—	—	—	—	2.5	—	2
222.0	—	10.0	—	0.25	—	—	—	—	
224.0	—	5.0	0.35	—	—	—	—	—	3
240.0	—	8.0	0.5	6.0	—	0.5	—	—	
242.0	—	4.0	—	1.5	—	2.0	—	—	
A242.0	—	4.1	—	1.4	0.20	2.0	—	0.14	
295.0	1.1	4.5	—	—	—	—	—	—	
308.0	5.5	4.5	—	—	—	—	—	—	
319.0	6.0	3.5	—	—	—	—	—	—	
328.0	8.0	1.5	0.40	0.40	—	—	—	—	
332.0	9.5	3.0	—	1.0	—	—	—	—	
333.0	9.0	3.5	—	0.28	—	—	—	—	
336.0	12.0	1.0	—	1.0	—	2.5	—	—	
354.0	9.0	1.8	—	0.5	—	—	—	—	
355.0	5.0	1.25	—	0.5	—	—	—	—	
C355.0	5.0	1.25	—	0.5	—	—	—	—	4
356.0	7.0	—	—	0.32	—	—	—	—	
A356.0	7.0	—	—	0.35	—	—	—	—	4
357.0	7.0	—	—	0.52	—	—	—	—	
A357.0	7.0	—	—	0.55	—	—	—	0.12	4, 5
359.0	9.0	—	—	0.6	—	—	—	—	
360.0	9.5	—	—	0.5	—	—	—	—	
A360.0	9.5	—	—	0.5	—	—	—	—	4
380.0	8.5	3.5	—	—	—	—	—	—	
A380.0	8.5	3.5	—	—	—	—	—	—	4
383.0	10.5	2.5	—	—	—	—	—	—	
384.0	11.2	3.8	—	—	—	—	—	—	
B390.0	17.0	4.5	—	0.55	—	—	—	—	
413.0	12.0	—	—	—	—	—	—	—	
A413.0	12.0	—	—	—	—	—	—	—	
443.0	5.2	—	—	—	—	—	—	—	
B443.0	5.2	—	—	—	—	—	—	—	4
C443.0	5.2								6
A444.0	7.0	—	—	—	—	—	—	—	
512.0	1.8	—	—	4.0	—	—	—	—	
513.0	—	—	—	4.0	—	—	1.8	—	
514.0	—	—	—	4.0	—	—	—	—	
518.0	—	—	—	8.0	—	—	—	—	
520.0	—	—	—	10.0	—	—	—	—	
535.0	—	—	0.18	6.8	—	—	—	0.18	7
705.0	—	—	0.5	1.6	0.30	—	3.0	—	
707.0	—	—	0.50	2.1	0.30	—	4.2	—	
710.0	—	0.50	—	0.7	—	—	6.5	—	
711.0	—	0.50	—	0.35	—	—	6.5	—	8
712.0	—	—	—	0.58	0.50	—	6.0	0.20	

Table 4 (*Continued*)

713.0	—	0.7	—	0.35	—	—	7.5	—	
771.0	—	—	—	0.9	0.40	—	7.0	0.15	
850.0	—	1.0	—	—	—	1.0	—	—	9
851.0	2.5	1.0	—	—	—	0.50	—	—	9
852.0	—	2.0	—	0.75	—	1.2	—	—	9

[a]Based on industry handbooks,[3,5,8–10] consult those references for specified limits. Values are nominal, i.e., average of range of limits for elements for which a range is shown; values are representative of separately cast test bars, not of specimens taken from commercial castings.
[b]Aluminum and normal impurities constitute remainder.
[c]1. Also contains 0.7% silver.
2. Also contains 1.2% iron.
3. Also contains 0.10% vanadium and 0.18% zirconium.
4. Impurity limits are much lower for this alloy than for alloy listed above it.
5. Also contains 0.055% beryllium.
6. Also contains up to 2.0% total iron.
7. Also contains 0.005% beryllium and 0.005% boron.
8. Also contains 1.0 iron.
9. Also contains 6.2% tin.

H = strain hardened

W = solution heat treated

T = thermally treated

Further information on each of these designations is provided by the descriptions that follow:

F = as fabricated. Applies to wrought or cast products made by shaping processes in which there is no special control over thermal conditions or strain-hardening processes employed to achieve specific properties. For wrought alloys there are no mechanical property limits associated with this temper, though for cast alloys there may be.

O = annealed. Applies to wrought products that are annealed to obtain the lower strength temper, usually to increase subsequent workability, and to cast products that are annealed to improve ductility and dimensional stability. The O may be followed by a digit other than zero.

H = strain hardened. Applies to products that have their strength increased by strain hardening. They may or may not have supplementary thermal treatments to produce some reduction in strength. The H is always followed by two or more digits.

W = solution heat treated. Applies only to alloys that age spontaneously after solution heat treating. This designation is specific only when digits are used in combination with W to indicate the period of natural aging, i.e., $W\frac{1}{2}h$.

T = thermally treated to produce stable tempers other than F. Applies to products that are thermally treated with or without supplementary strain hardening to produce stable tempers. The T is always followed by one or more digits.

The most widely used temper designations above are the H and T categories, and these are always followed by from one to four numeric digits that provide more detail about how the alloy has been fabricated. It is beyond the scope of this volume to describe the system in further detail and the reader if referred to Ref. 2 and 6 for additional needed information.

4 MECHANICAL PROPERTIES OF ALUMINUM ALLOYS

The typical properties of a representative group of wrought aluminum alloys are shown in Table 5,[2] presented in engineering units in Table 5A and in metric units in Table 5B. The typical properties of a representative group of cast aluminum alloys are shown in Table 6[3,8–10] presented in engineering units in Table 6A and in metric units in Table 6B. In aluminum industry usage, the typical values are indicative of average or mean values for all products of the respective alloys. In the case of castings, the values represent average or mean values from tests of separately cast test bars for quality control purposes. *These are not intended and should not be used for design purposes.*

For design mechanical properties, readers are referred to the materials specifications,[2,3] including ASTM standards, and to design handbooks such as *The Aluminum Design Manual*[4] and MMPDS/MIL-HDBK-5.[11] For design rules and guidelines, readers are referred to *The Aluminum Design Manual*[4] and Ref. 12–16.

For the physical properties of wrought aluminum alloys, readers are referred to *Aluminum Standards and Data* (*Standard and Metric Editions*)[2] and for the physical properties of cast aluminum alloys to casting industry standards.[3,8–10]

Table 5A Typical Mechanical Properties of Wrought Aluminum Alloys (Engineering Units)[a]

	Tension				Hardness	Shear	Fatigue	Modulus[e]
	Ultimate	Yield	Elongation (%)		Brinell	Ultimate	Endurance	of
Alloy & Temper	Strength (ksi)	Strength (ksi)	in 2 in[b]	in 4D[c]	Number (500 kg/10 mm)	Strength (ksi)	Limit[d] (ksi)	Elasticity (10^3 ksi)
1060-O	10	4	43	—	19	7	3	10.0
1060-H12	12	11	16	—	23	8	4	10.0
1060-H14	14	13	12	—	26	9	5	10.0
1060-H16	16	15	8	—	30	10	6.5	10.0
1060-H18	19	18	6	—	35	11	6.5	10.0
1100-O	13	5	35	45	23	9	5	10.0
1100-H12	16	15	12	25	28	10	6	10.0
1100-H14	18	17	9	20	32	11	7	10.0
1100-H16	21	20	6	17	38	12	9	10.0
1100-H18	24	22	5	15	44	13	9	10.0
1350-O	12	4	23[f]	—	—	8	—	10.0
1350-H12	14	12	—	—	—	9	—	10.0
1350-H14	16	14	—	—	—	10	—	10.0
1350-H16	18	16	—	—	—	11	—	10.0
1350-H19	27	24	1.5[f]	—	—	15	—	10.0
2008-T4	36	18	28		—	21	—	10.2
2010-T4	35	19	25			21	8	10.2
2011-T3	55	43	—	15	96	32	18	10.2
2011-T8	59	45	—	12	100	35	18	10.2
2014-O	27	14	—	18	45	18	13	10.6
2014-T4, T451	62	42	—	20	105	38	20	10.6
2014-T6, T651	70	60	—	13	135	42	18	10.6
2017-O	26	10	—	22	45	18	13	10.5
2017-T4, T451	62	40	—	22	105	38	18	10.5

Table 5A (*Continued*)

Alloy & Temper	Tension				Hardness Brinell Number (500 kg/10 mm)	Shear Ultimate Strength (ksi)	Fatigue Endurance Limit[d] (ksi)	Modulus[e] of Elasticity (10^3 ksi)
	Ultimate Strength (ksi)	Yield Strength (ksi)	Elongation (%)					
			in 2 in[b]	in 4D[c]				
2024-O	27	11	20	22	47	18	13	10.6
2024-T3	70	50	18	—	120	41	20	10.6
2024-T4, T351	68	47	20	19	120	41	20	10.6
2024-T361	72	57	13	—	130	42	18	10.6
2025-T6	58	37	—	19	110	35	18	10.4
2036-T4	49	28	24	—	—	—	18[g]	10.3
2117-T4	43	24	—	27	70	28	14	10.3
2124-T851	70	64	—	8	—	—	—	10.6
2195-T351	52	36	17	—	—	—	—	10.6
2195-T851	66	51	10	—	—	—	—	10.6
2219-O	25	11	18	—	—	—	—	10.6
2219-T62	60	42	10	—	—	—	15	10.6
2219-T81, T851	66	51	10	—	—	—	15	10.6
2219-T87	69	57	10	—	—	—	15	10.6
2618-T61	64	54		10	115	38	18	10.8
3003-O	16	6	30	40	28	11	7	10.0
3003-H12	19	18	10	20	35	12	8	10.0
3003-H14	22	21	8	16	40	14	9	10.0
3003-H16	26	25	5	14	47	15	10	10.0
3003-H18	29	27	4	10	55	16	10	10.0
3004-O	26	10	20	25	45	16	14	10.0
3004-H32	31	25	10	17	52	17	15	10.0
3004-H34	35	29	9	12	63	18	15	10.0
3004-H36	38	33	5	9	70	20	16	10.0
3004-H38	41	36	5	6	77	21	16	10.0
3105-O	17	8	24	—	—	12	—	10.0
3105-H12	22	19	7	—	—	14	—	10.0
3105-H14	25	22	5	—	—	15	—	10.0
3105-H16	28	25	4	—	—	16	—	10.0
3105-H18	31	28	3	—	—	17	—	10.0
3105-H25	26	23	8	—	—	15	—	10.0
4032-T6	55	46		9	120	38	16	11.4
5005-O	18	6	25	—	28	11	—	10.0
5005-H32	20	17	11	—	36	14	—	10.0
5005-H34	23	20	8	—	41	14	—	10.0
5005-H36	26	24	6	—	46	15	—	10.0
5005-H38	29	27	6	—	51	16	—	10.0
5050-O	21	8	24	—	36	15	12	10.0
5050-H32	25	21	9	—	46	17	13	10.0
5050-H34	28	24	8	—	53	18	13	10.0
5050-H36	30	26	7	—	58	19	14	10.0
5050-H38	32	29	6	—	63	20	14	10.0

Table 5A *(Continued)*

Alloy & Temper	Tension				Hardness Brinell Number (500 kg/10 mm)	Shear Ultimate Strength (ksi)	Fatigue Endurance Limit[d] (ksi)	Modulus[e] of Elasticity (10^3 ksi)
	Ultimate Strength (ksi)	Yield Strength (ksi)	Elongation (%)					
			in 2 in[b]	in 4D[c]				
5052-O	28	13	25	30	47	18	16	10.2
5052-H32	33	28	12	18	60	20	17	10.2
5052-H34	38	31	10	14	68	21	18	10.2
5052-H36	40	35	8	10	73	23	19	10.2
5052-H38	42	37	7	8	77	24	20	10.2
5056-O	42	22	—	35	65	26	20	10.3
5056-H18	63	59	—	10	105	34	22	10.3
5056-H38	60	50	—	15	100	32	22	10.3
5083-O	42	21	—	22	—	25	—	10.3
5083-H116	46	33	—	16	—	—	23	10.3
5083-H321	46	33	—	16	—	—	23	10.3
5086-O	38	17	22	—	—	23	—	10.3
5086-H32	42	30	12	—	—	—	—	10.3
5086-H34	47	37	10	—	—	27	—	10.3
5086-H116	42	30	12	—	—	—	—	10.3
5154-O	35	17	27	—	58	22	17	10.2
5154-H32	39	30	15	—	67	22	18	10.2
5154-H34	42	33	13	—	73	24	19	10.2
5154-H36	45	36	12	—	78	26	20	10.2
5154-H38	48	39	10	—	80	28	21	10.2
5454-O	36	17	22	—	62	23	—	10.2
454-H32	40	30	10	—	73	24	—	10.2
5454-H34	44	35	10	—	81	26	—	10.2
5454-H111	38	26	14	—	70	23	—	10.2
5456-O	45	23	—	24		—	—	10.3
5456-H116	51	37	—	16	90	30	—	10.3
5456-H321	51	37	—	16	90	30	—	10.3
5657-H25	23	20	12	—	40	12	—	10.0
5657-H28, H38	28	24	7	—	50	15	—	10.0
5754-O	32	14	26	—	—	19	—	10.3
6009-T4	32	18	25	—	—	19	—	10.0
6022-T4	37	22	26	—	—	22	—	10.0
6061-O	18	8	25	30	30	12	9	10.0
6061-T4, T451	35	21	22	25	65	24	14	10.0
6061-T6, T651	45	40	12	17	95	30	14	10.0
6063-O	13	7	—	—	25	10	8	10.0
6063-T4	25	13	22	—	—	—	—	10.0
6063-T5	27	21	12	—	60	17	10	10.0
6063-T6	35	31	12	—	73	22	10	10.0
6063-T83	37	35	9	—	82	22	—	10.0
6066-O	22	12	—	18	43	14	—	10.0
6066-T4, T451	52	30	—	18	90	29	—	10.0
6066-T6, T651	57	52	—	12	120	34	16	10.0
6070-T6	55	51	10	—	—	34	14	10.0

Table 5A (*Continued*)

Alloy & Temper	Tension: Ultimate Strength (ksi)	Tension: Yield Strength (ksi)	Elongation (%) in 2 in[b]	Elongation (%) in 4D[c]	Hardness Brinell Number (500 kg/10 mm)	Shear Ultimate Strength (ksi)	Fatigue Endurance Limit[d] (ksi)	Modulus[e] of Elasticity (10^3 ksi)
6101-H111	14	11	—	—	—	—	—	10.0
6101-T6	32	28	15	—	71	20	—	10.0
6111-T4	42	22	26	—	—	25	—	10.0
6111-T41	39	22	26	—	—	23	—	10.0
6262-T9	58	55		10	120	35	13	10.0
6351-T4	36	22	20	—	—	44	—	10.0
6351-T6	45	41	14	—	95	29	13	10.0
7049-T73	75	65	—	12	135	44	—	10.4
7049-T7352	75	63	—	11	135	43		10.4
7050-T7351X	72	63	—	12	—	—	—	10.4
7050-T7451	76	68	—	11	—	44	—	10.4
7050-T7651	80	71	—	11	—	47	—	10.4
7075-O	33	15	17	16	60	22		10.4
7075-T6, T651	83	73	11	11	150	48	23	10.4
7175-T74	76	66		11	135	42	23	10.4
7178-O	33	15	16	—	—	—	—	10.4
7178-T6, T651	88	78	11	—	—	—	—	10.4
7178-T76, T7651	83	73	11	—	—	—	—	10.4
7475-T61	82	71	11	—	—	—	—	10.4
7475-T651	85	74	—	13	—	—	—	10.4
7475-T7351	72	61	—	13	—	—	—	10.4
7475-T761	75	65	12	—	—	—	—	10.4
7475-T7651	77	67	—	12	—	—	—	10.4
8176-H24	17	14	15	—	80	—	—	10.0

[a] Based on *Aluminum Standards and Data*.[2] Consult that reference for limits. For tensile yield strengths, offset = 0.2%.
[b] Elongation measured over 2 in. gauge length on $\frac{1}{16}$-in-thick sheet-type specimens.
[c] Elongation measured over 2 in. gauge length (4D) in $\frac{1}{2}$-in.-diameter specimens.
[d] Based on 500,000,000 cycles of completely reversed stress using R. R. Moore type of machines and specimens.
[e] Average of tension and compression moduli; compressive modulus is nominally about 2% greater than the tension modulus.
[f] Measured over 10 in. gauge length in wire.
[g] At 10^7 cycles with flexural fatigue specimens.

Table 5B Typical Mechanical Properties of Wrought Aluminum Alloys (Metric Units)[a]

Alloy & Temper	Tension: Ultimate Strength (MPa)	Tension: Yield Strength (MPa)	Elongation (%) in 50 mm[b]	Elongation (%) in 5D[c]	Hardness Brinell Number (500 kg/10 mm)	Shear Ultimate Strength (MPa)	Fatigue Endurance Limit[d] (MPa)	Modulus[e] of Elasticity (GPa)
1060-O	70	30	43	—	19	50	20	69
1060-H12	85	75	16	—	23	55	30	69
1060-H14	100	90	12	—	26	60	35	69
1060-H16	115	105	8	—	30	70	45	69
1060-H18	130	125	6	—	35	75	45	69

Table 5B *(Continued)*

Alloy & Temper	Tension				Hardness Brinell Number (500 kg/10 mm)	Shear Ultimate Strength (MPa)	Fatigue Endurance Limit[d] (MPa)	Modulus[e] of Elasticity (GPa)
	Ultimate Strength (MPa)	Yield Strength (MPa)	Elongation (%) in 50 mm[b]	Elongation (%) in 5D[c]				
1100-O	90	35	35	42	23	60	35	69
1100-H12	110	105	12	22	28	70	40	69
1100-H14	125	115	9	18	32	75	50	69
1100-H16	145	140	6	15	38	85	60	69
1100-H18	165	150	5	13	44	90	60	69
1350-O	85	30	23[f]	—	—	55	—	69
1350-H12	95	85	—	—	—	60	—	69
1350-H14	110	95	—	—	—	70	—	69
1350-H16	125	110	—	—	—	75	—	69
1350-H19	185	165	1.5[f]	—	—	105	—	69
2008-T4	250	125	28	—	145	70	—	70
2010-T4, T41	240	130	25	—	145	70	—	70
2011-T3	380	295	—	13	95	220	125	70
2011-T8	405	310	—	10	100	240	125	70
2014-O	185	95	—	16	45	125	90	73
2014-T4, T451	425	290	—	18	105	260	140	73
2014-T6, T651	485	415	—	11	135	290	125	73
2017-O	180	70	—	20	45	125	90	73
2017-T4, T451	425	275	—	20	105	260	125	73
2024-O	185	75	20	20	47	125	90	73
2024-T3	485	345	18	—	120	285	140	73
2024-T4, T351	472	325	20	17	120	285	140	73
2024-T361	495	395	13	—	130	290	125	73
2025-T6	400	255	—	17	110	240	125	72
2036-T4	340	195	24	—	—	205	125[g]	71
2117-T4	295	165	—	24	70	195	95	71
2124-T851	485	440	—	8	—	—	—	73
2195-T351	360	250	17	—	—	—	—	73
2195-T851	455	350	10	—	—	—	—	73
2219-O	170	75	18	—	—	—	—	73
2219-T62	415	290	10	—	—	—	105	73
2219-T81, T851	455	350	10	—	—	—	105	73
2219-T87	475	395	10	—	—	—	105	73
2618-T61	440	370		10	115	260	90	73
3003-O	110	40	30	37	28	75	50	69
3003-H12	130	125	10	18	35	85	55	69
3003-H14	150	145	8	14	40	95	60	69
3003-H16	175	170	5	12	47	105	70	69
3003-H18	200	185	4	9	55	110	70	69
3004-O	180	70	20	22	45	110	95	69
3004-H32	215	170	10	15	52	115	105	69
3004-H34	240	200	9	10	63	125	105	69
3004-H36	260	230	5	8	70	140	110	69
3004-H38	285	250	5	5	77	145	110	69

Table 5B *(Continued)*

Alloy & Temper	Tension				Hardness Brinell Number (500 kg/10 mm)	Shear Ultimate Strength (MPa)	Fatigue Endurance Limit[d] (MPa)	Modulus[e] of Elasticity (GPa)
	Ultimate Strength (MPa)	Yield Strength (MPa)	Elongation (%) in 50 mm[b]	Elongation (%) in 5D[c]				
3105-O	115	55	24	—	—	85	—	69
3105-H12	150	130	7	—	—	95	—	69
3105-H14	170	150	5	—	—	105	—	69
3105-H16	195	170	4	—	—	110	—	69
3105-H18	215	195	3	—	—	115	—	69
3105-H25	180	160	8	—	—	95	—	69
4032-T6	380	315		9	120	260	110	79
5005-O	125	40	25	—	28	75	—	69
5005-H32	140	115	11	—	36	95	—	69
5005-H34	160	140	8	—	41	95	—	69
5005-H36	180	165	6	—	46	105	—	69
5005-H38	200	185	5	—	51	110	—	69
5050-O	145	55	24	—	36	105	85	69
5050-H32	170	145	9	—	46	115	90	69
5050-H34	190	165	8	—	53	125	90	69
5050-H36	205	180	7	—	58	130	95	69
5050-H38	220	200	6	—	63	140	95	69
5052-O	195	90	25	27	47	125	110	70
5052-H32	230	195	12	16	60	140	115	70
5052-H34	260	215	10	12	68	145	125	70
5052-H36	275	240	8	9	73	160	130	70
5052-H38	290	255	7	7	77	165	140	70
5056-O	290	150	—	32	65	180	140	71
5056-H18	435	405	—	9	105	235	150	71
5056-H38	415	345	—	13	100	220	150	71
5083-O	290	145	—	20	—	170	—	71
5083-H116	315	230	—	14	—	—	160	71
5083-H321	315	230	—	14	—	—	160	71
5086-O	260	115	22	—	—	165	—	71
5086-H32	290	205	12	—	—	—	—	71
5086-H34	325	255	10	—	—	185	—	71
5086-H116	290	205	12	—	—	—	—	71
5154-O	240	115	27	—	58	150	115	70
5154-H32	270	205	15	—	67	150	125	70
5154-H34	290	230	13	—	73	165	130	70
5154-H36	310	250	12	—	78	180	140	70
5154-H38	330	270	10	—	80	195	145	70
5454-O	250	115	22	—	62	160	—	70
5454-H32	275	205	10	—	73	165	—	70
5454-H34	305	240	10	—	81	180	—	70
5454-H111	260	180	14	—	70	160	—	70
5456-O	310	160	—	22		—	—	71
5456-H116	350	255	—	14	90	205	—	71
5456-H321	350	255	—	14	90	205	—	71

Table 5B (*Continued*)

Alloy & Temper	Tension: Ultimate Strength (MPa)	Tension: Yield Strength (MPa)	Tension: Elongation (%) in 50 mm[b]	Tension: Elongation (%) in 5D[c]	Hardness Brinell Number (500 kg/10 mm)	Shear Ultimate Strength (MPa)	Fatigue Endurance Limit[d] (MPa)	Modulus[e] of Elasticity (GPa)
5657-H25	160	140	12	—	40	95	—	69
5657-H28, H38	195	165	7	—	50	105	—	69
5754-O	220	100	26	—	—	130	—	71
6009-T4	220	125	25	—	—	130	—	69
6061-O	125	55	25	27	30	85	60	69
6061-T4, T451	240	145	22	22	65	165	95	69
6061-T6, T651	310	275	12	15	95	205	95	69
6063-O	90	50	—	—	25	70	55	69
6063-T4	170	90	22	—	—	—	—	69
6063-T5	185	145	12	—	60	115	70	69
6063-T6	240	215	12	—	73	150	70	69
6063-T83	255	240	9	—	82	150	—	69
6066-O	150	85	—	16	43	95	—	69
6066-T4, T451	360	205	—	16	90	200	—	69
6066-T6, T651	395	360	—	10	120	235	110	69
6070-T6	380	350	10	—	—	235	95	69
6101-H111	95	75	—	—	—	—	—	69
6101-T6	220	195	15[f]	—	71	140	—	69
6111-T4	280	150	26	—	—	175	—	69
6111-T41	270	150	26	—	—	160	—	69
6262-T9	400	380	—	9	120	240	90	69
6351-T4	250	150	20	—	—	44	—	69
6351-T6	310	285	14	—	95	200	90	69
7049-T73	515	450	—	10	135	305	—	72
7049-T7352	515	435	—	9	135	295	—	72
7050-T7351X	495	435	—	11	—	—	—	72
7050-T7451	525	470	—	10	—	305	—	72
7050-T7651	550	490	—	10	—	325	—	72
7075-O	230	105	17	14	60	150		72
7075-T6, T651	570	505	11	9	150	330	160	72
7175-T74	525	455	—	10	135	290	160	72
7178-O	230	105	15	14	—	—	—	72
7178-T6, T651	605	540	10	9	—	—	—	72
7178-T76, T7651	570	505	11	9	—	—	—	71
7475-T61	565	490	11	—	—	—	—	70
7475-T651	585	510	—	13	—	—	—	72
7475-T7351	495	420	—	13	—	—	—	72
7475-T761	515	450	12	—	—	—	—	70
7475-T7651	530	460	—	12	—	—	—	72
8176-H24	160	95	15	—	—	70	—	10

[a] Based on *Aluminum Standards and Data*.[2] Consult that reference for limits. For tensile yield strengths, offset = 0.2%.
[b] Elongation measured over 500 mm gauge length on 1.60-mm-thick sheet-type specimens.
[c] Elongation measured over 500 mm gauge length (5D) in 12.5-mm-diameter specimens.
[d] Based on 500,000,000 cycles of completely reversed stress using R. R. Moore type of machines and specimens.
[e] Average of tension and compression moduli; compressive modulus is nominally about 2% greater.
[f] Measured over 250 mm gauge length in wire.
[g] At 10^7 cycles with flexural fatigue specimens.

Table 6A Typical Mechanical Properties of Wrought Aluminum Alloy Castings (Engineering Units)[a]

Alloy & Temper	Tension			Hardness Brinell Number (500 kg/10 mm)	Shear Ultimate Strength (ksi)	Fatigue Endurance Limit[b] (ksi)	Modulus[c] of Elasticity (10^3 ksi)
	Ultimate Strength (ksi)	Yield Strength (ksi)	Elongation in 2 in or 4D (%)				
				Sand Cast			
201.0-T6	65	55	8	130	—	—	—
201-0-T7	68	60	6	—	—	14	—
201.0-T43	60	37	17	—	—	—	—
204.0-T4	45	28	6	—	—	—	—
A206.0-T4	51	36	7	—	40	—	—
208.0-F	21	14	3	—	17	11	—
213.0-F	24	15	2	70	20	9	—
222.0-O	27	20	1	80	21	9.5	
222.0-T61	41	40	<0.5	115	32	8.5	10.7
224.0-T72	55	40	10	123	35	9	10.5
240.0-F	34	28	1	90	—	—	—
242.0-F	31	20	1	—	—	—	10.3
242.0-O	27	18	1	70	21	8	10.3
242.0-T571	32	30	1	85	26	11	10.3
242.0-T61	32	20	—	90–120	—	—	10.3
242.0-T77	30	23	2	75	24	10.5	10.3
A242.0-T75	31	—	2	—	—	—	—
295.0-T4	32	16	9	80	26	7	10.0
295.0-T6	36	24	5	75	30	7.5	10.0
295.0-T62	41	32	2	90	33	8	10.0
295.0-T7	29	16	3	55–85	—	—	10.0
319-F	27	18	2	70	22	10	10.7
319.0-T5	30	26	2	80	24	11	10.7
319.0-T6	36	24	2	80	29	11	10.7
328.0-F	25	14	1	45–75	—	—	—
328.0-T6	34	21	1	65–95	—	—	—
355.0-F	23	12	3	—	—	—	10.2
355.0-T51	28	23	2	65	22	8	10.2
355.0-T6	35	25	3	80	28	9	10.2
355.0-T61	35	35	1	90	31	9.5	10.2
355.0-T7	38	26	1	85	28	10	10.2
355.0-T71	35	29	2	75	26	10	10.2
C355.0-T6	39	29	5	85	—	—	—
356.0-F	24	18	6	—	—	—	10.5
356.0-T51	25	20	2	60	20	8	10.5
356.0-T6	33	24	4	70	26	8.5	10.5
356.0-T7	34	30	2	75	24	9	10.5
356.0-T71	28	21	4	60	20	8.5	10.5
A356.0-F	23	12	6	—	—	—	10.5
A356.0-T51	26	18	3	—	—	—	10.5
A356.0-T6	40	30	6	75	—	—	10.5

Table 6A (*Continued*)

Alloy & Temper	Tension: Ultimate Strength (ksi)	Tension: Yield Strength (ksi)	Tension: Elongation in 2 in or 4D (%)	Hardness Brinell Number (500 kg/10 mm)	Shear Ultimate Strength (ksi)	Fatigue Endurance Limit[b] (ksi)	Modulus[c] of Elasticity (10^3 ksi)
A356.0-T71	30	20	3	—	—	—	10.5
357.0-F	25	13	5	—	—	—	—
357.0-T51	26	17	3	—	—	—	—
357.0-T6	50	43	2	—	—	—	—
357.0-T7	40	34	3	60	—	—	—
A357.0-T6	46	36	3	85	40	12	—
359.0-T62	50	42	6	—	—	—	—
A390.0-F	26	26	<1.0	100	—	—	—
A390.0-T5	26	26	<1.0	100	—	—	—
A390.0-T6	40	40	<1.0	140	—	13	—
A390.0-T7	36	36	<1.0	115	—	—	—
443.0-F	19	8	8	40	14	8	10.3
B443.0-F	17	6	3	25–55	—	—	—
A444.0-F	21	9	9	44	—	—	—
A444.0-T4	23	9	12	45			
511.0-F	21	12	3	50	17	8	—
512.0-F	20	13	2	50	17	9	—
514.0-F	25	12	9	50	20	7	—
520.0-T4	48	26	16	75	34	8	—
535.0-F	35	18	9	60–90	—	—	—
535.0-T5	35	18	9	60–90	—	—	—
A535.0-F	36	18	9	65	—	—	—
707.0-T5	33	22	2	70–100	—	—	—
707.0-T7	37	30	1	65–95	—	—	—
710.0-F	32	20	2	60–90	—	—	—
710.0-T5	32	20	2	60–90	—	—	—
712.0-F	34	25	4	60–90	—	—	—
712.0-T5	34	25	4	60–90	—	—	—
713.0-F	32	22	3	60–90	—	—	—
713.0-T5	32	22	3	60–90			
771.0-T5	32	27	3	70–100	—	—	—
771.0-T52	36	30	2	70–100	—	—	—
771.0-T53	36	27	2	—	—	—	—
771.0-T6	42	35	5	75–105	—	—	—
771.0-T71	48	45	2	105–135	—	—	—
850.0-T5	20	11	8	45	14	—	10.3
851.0-T5	20	11	5	45	14	—	10.3
852.0-T5	27	22	2	65	18	10	10.3
Permanent Mold							
201.0-T6	65	55	8	130	—	—	—
201-0-T7	68	60	6	—	—	14	—
201.0-T43	60	37	17	—	—	—	—
204.0-T4	48	29	8	—	—	—	—

Table 6A (*Continued*)

Alloy & Temper	Tension: Ultimate Strength (ksi)	Tension: Yield Strength (ksi)	Tension: Elongation in 2 in or 4D (%)	Hardness Brinell Number (500 kg/10 mm)	Shear Ultimate Strength (ksi)	Fatigue Endurance Limit[b] (ksi)	Modulus[c] of Elasticity (10^3 ksi)
A206.0-T4	62	38	17	—	42	—	—
A206.0-T7	63	50	12	—	37	—	—
208.0-T6	35	22	2	75–105	—	—	—
208.0-T7	33	16	3	65–95	—	—	—
213.0-F	30	24	2	85	24	9.5	—
222.0-T551	37	35	<0.5	115	30	8.5	10.7
222.0-T52	35	31	1	100	25	—	10.7
238.0-F	30	24	2	100	24	—	—
242.0-T571	40	34	1	105	30	10.5	10.3
242.0-T61	47	42	1	110	35	10	10.3
A249.0-T63	69	60	6	—	—	—	—
296.0-T7	39	20	5	80	30	9	10.1
308.0-F	28	16	2	70	22	13	—
319.0-F	34	19	3	85	24	—	10.7
319.0-T6	40	27	3	95	—	—	10.7
324.0-F	30	16	4	70	—	—	—
324.0-T5	36	26	3	90	—	—	—
324.0-T62	45	39	3	105	—	—	—
332.0-T5	36	28	1	105	—	—	—
328.0-T6	34	21	1	65–95	—	—	—
333.0-F	34	19	2	90	27	15	—
333.0-T5	34	25	1	100	27	12	—
333.0-T6	42	30	2	105	33	15	—
333.0-T7	37	28	2	90	28	12	—
336.0-T551	36	28	1	105	28	14	—
336.0-T65	47	43	1	125	36	—	—
354.0-T61	48	37	3	—	—	—	—
354.0-T62	52	42	2	—	—	—	—
355.0-F	27	15	4	—	—	—	10.2
355.0-T51	30	24	2	75	24	—	10.2
355.0-T6	42	27	4	90	34	10	10.2
355.0-T61	45	40	2	105	36	10	10.2
355.0-T7	40	30	2	85	30	10	10.2
355.0-T71	36	31	3	85	27	10	10.2
C355.0-T6	48	28	8	90	—	—	10.2
C355.0-T61	46	34	6	100	—	—	10.2
C355.0-T62	48	37	5	100	—	—	10.2
356.0-F	26	18	5	—	—	—	10.5
356.0-T51	27	20	2	—	—	—	10.5
356.0-T6	38	27	5	80	30	13	10.5
356.0-T7	32	24	6	70	25	11	10.5
356.0-T71	25	—	3	60–90	—	—	10.5
A356.0-F	27	13	8	—	—	—	10.5
A356.0-T51	29	20	5	—	—	—	10.5

Table 6A *(Continued)*

	Tension						
Alloy & Temper	Ultimate Strength (ksi)	Yield Strength (ksi)	Elongation in 2 in or 4D (%)	Hardness Brinell Number (500 kg/10 mm)	Shear Ultimate Strength (ksi)	Fatigue Endurance Limit[b] (ksi)	Modulus[c] of Elasticity (10^3 ksi)
A356.0-T6	41	30	12	80	—	—	10.5
357.0-F	28	15	6	—	—	—	—
357.0-T51	29	21	4	—	—	—	—
357.0-T6	52	43	5	100	35	13	—
357.0-T7	38	30	5	70	—	—	—
A357.0-T6	52	42	5	100	35	15	—
359.0-T61	48	37	6	—	—	—	—
359.0-T62	50	42	6	—	—	16	—
A390.0-F	29	29	<1.0	110	—	—	—
A390.0-T5	29	29	<1.0	110	—	—	—
A390.0-T6	45	45	<1.0	145	—	17	—
A390.0-T7	38	38	<1.0	120	—	15	—
443.0-F	23	9	10	45	16	8	10.3
B443.0-F	21	6	6	30–60	—	—	—
A444.0-F	24	11	13	44	—	—	—
A444.0-T4	23	10	21	45	16	8	—
513.0-F	27	16	7	60	22	10	—
535.0-F	35	18	8	60–90	—	—	—
705.0-T5	37	17	10	55–75	—	—	—
707.0-T7	45	35	3	80–110	—	—	—
711.0-T1	28	18	7	55–85	—	—	—
713.0-T5	32	22	4	60–90			
850.0-T5	23	11	12	45	15	9	10.3
851.0-T5	20	11	5	45	14	9	10.3
851.0-T6	18	—	8	—	—	—	10.3
852.0-T5	32	23	5	70	21	11	10.3
Die Cast							
360.0-F	44	25	3	75	28	20	10.3
A360.0-F	46	24	4	75	26	18	10.3
380.0-F	46	23	3	80	28	20	10.3
A380.0-F	47	23	4	80	27	20	10.3
383.0-F	45	22	4	75	—	21	10.3
384.0-F	48	24	3	85	29	20	—
390.0-F	40.5	35	<1	—	—	—	—
B390.0-F	46	36	<1	120	—	20	11.8
392.0-F	42	39	<1	—	—	—	—
413.0-F	43	21	3	80	25	19	10.3
A413.0-F	42	19	4	80	25	19	—
C443.0-F	33	14	9	65	29	17	10.3
518.0-F	45	28	5	80	29	20	—

[a] Based upon industry handbooks.[3,8–10] Consult those references for limits. Values are representative of separately cast test bars, not of specimens taken from commercial castings. For tensile yield strengths, offset = 0.2%.

[b] Based on 500,000,000 cycles of completely reversed stress using R. R. Moore type of machines and specimens.

[c] Average of tension and compression moduli; compressive modulus is nominally about 2% greater than the tension modulus.

Table 6B Typical Mechanical Properties of Aluminum Alloy Castings (Metric Units)[a]

	Tension						
Alloy & Temper	Ultimate Strength (MPa)	Yield Strength (MPa)	Elongation in 5D (%)	Hardness Brinell Number (500 kg/10 mm)	Shear Ultimate Strength (MPa)	Fatigue Endurance Limit[b] (MPa)	Modulus[c] of Elasticity (GPa)
Sand Cast							
201.0-T6	450	380	8	130	—	—	—
201-0-T7	470	415	6	—	—	95	—
201.0-T43	415	255	17	—	—	—	—
204.0-T4	310	195	6	—	—	—	—
A206.0-T4	350	250	7	—	275	—	—
2008-0-F	145	655	3	—	115	75	—
213.0-F	165	105	2	70	140	60	—
222-0-O	185	140	1	80	145	65	
222.0-T61	285	275	<0.5	115	220	60	74
224.0-T72	380	275	10	123	240	60	73
240.0-F	235	195	1	90	—	—	—
242.0-F	145	140	1	—	—	—	71
242.0-O	185	125	1	70	145	55	71
242.0-T571	220	205	1	85	180	75	71
242.0-T61	220	140	—	90–120	—	—	71
242.0-T77	205	160	2	75	165	70	71
A242.0-T75	215	—	2	—	—	—	—
295.0-T4	220	110	9	80	180	50	69
295.0-T6	250	165	5	75	205	50	69
295.0-T62	285	220	2	90	230	55	69
295.0-T7	200	110	3	55–85	—	—	69
319-F	185	125	2	70	150	70	74
319.0-T5	205	180	2	80	165	75	74
319.0-T6	250	165	2	80	200	75	74
328.0-F	170	95	1	45–75	—	—	—
328.0-T6	235	145	1	65–95	—	—	—
355.0-F	160	85	3	—	—	—	70
355.0-T51	195	160	2	65	150	55	70
355.0-T6	240	170	3	80	195	60	70
355.0-T61	240	240	1	90	215	65	70
355.0-T7	260	180	1	85	195	70	70
355.0-T71	240	200	2	75	180	70	70
C355.0-T6	270	200	5	85	—	—	—
356.0-F	165	125	6	—	—	—	73
356.0-T51	170	140	2	60	140	55	73
356.0-T6	230	165	4	70	180	60	73
356.0-T7	235	205	2	75	165	60	73
356.0-T71	195	145	4	60	140	60	73
A356-0-F	160	85	6	—	—	—	73
A356.0-T51	180	125	3	—	—	—	73
A356.0-T6	275	205	6	75	—	—	73
A356.0-T71	205	140	3	—	—	—	73
357.0-F	170	90	5	—	—	—	—

Table 6B *(Continued)*

Alloy & Temper	Tension			Hardness Brinell Number (500 kg/10 mm)	Shear Ultimate Strength (MPa)	Fatigue Endurance Limit[b] (MPa)	Modulus[c] of Elasticity (GPa)
	Ultimate Strength (MPa)	Yield Strength (MPa)	Elongation in 5D (%)				
357.0-T51	180	115	3	—	—	—	—
357.0-T6	345	295	2	—	—	—	—
357.0-T7	275	235	3	60	—	—	—
A357.0-T6	315	250	3	85	275	95	—
359.0-T62	345	290	6	16	—	—	—
A390.0-F	180	180	<1.0	100	—	—	—
A390.0-T5	180	180	<1.0	100	—	—	—
A390.0-T6	275	275	<1.0	140	—	90	—
A390.0-T7	250	250	<1.0	115	—	—	—
443.0-F	130	55	8	40	95	55	71
B443.0-F	115	40	3	25–55	—	—	—
A444.0-F	145	60	9	43,400	—	—	—
A444.0-T4	23	60	12				
511.0-F	145	85	3	50	115	55	
512.0-F	140	90	2	50	115	60	
514.0-F	170	85	9	50	140	50	—
520.0-T4	330	180	16	75	235	55	—
535.0-F	240	125	9	60–90	—	—	—
535-0-T5	240	125	9	60–90	—	—	—
A535.0-F	250	125	9	65	—	—	—
707.0-T5	230	150	2	70–100	—	—	—
707.0-T7	255	205	1	65–95	—	—	—
710.0-F	220	140	2	60–90	—	—	—
710.0-T5	220	140	2	60–90	—	—	—
712.0-F	235	170	4	60–90	—	—	—
712.0-T5	235	170	4	60–90	—	—	—
713.0-F	220	150	3	60–90	—	—	—
713.0-T5	220	150	3	60–90			
771.0-T5	220	185	3	70–100	—	—	—
771.0-T52	250	205	2	70–100	—	—	—
771.0-T53	250	185	2	—	—	—	—
771.0-T6	290	240	5	75–105	—	—	—
771.0-T71	330	310	2	105–135	—	—	—
850.0-T5	140	75	8	45	95	—	71
851.0-T5	140	75	5	45	95	—	71
852.0-T5	185	150	2	n 65	125	60	71
Permanent Mold							
201.0-T6	450	380	8	130	—	—	—
201-0-T7	470	415	6	—	—	95	—
201.0-T43	415	255	17	—	—	—	—
204.0-T4	330	200	8	—	—	—	—
A206.0-T4	430	260	17	—	290	—	—

Table 6B (*Continued*)

Alloy & Temper	Tension Ultimate Strength (MPa)	Tension Yield Strength (MPa)	Tension Elongation in 5D (%)	Hardness Brinell Number (500 kg/10 mm)	Shear Ultimate Strength (MPa)	Fatigue Endurance Limit[b] (MPa)	Modulus[c] of Elasticity (GPa)
A206.0-T7	435	345	12	—	255	—	—
208.0-T6	240	150	2	75–105	—	—	—
208.0-T7	230	110	3	65–95	—	—	—
213.0-F	205	165	2	85	165	65	—
222.0-T551	255	240	<0.5	115	205	60	74
222.0-T52	240	215	1	100	170	—	74
238.0-F	205	165	2	100	165	—	—
242.0-T571	275	235	1	105	205	70	74
242.0-T61	325	290	1	110	450	70	74
A249.0-T63	475	415	6	—	—	—	—
296.0-T7	270	140	5	80	205	60	70
308.0-F	195	110	2	70	150	90	—
319.0-F	235	130	3	85	165	—	74
319.0-T6	275	185	3	95	—	—	74
324.0-F	205	110	4	70	—	—	—
324.0-T5	250	180	3	90	—	—	—
324.0-T62	310	270	3	105	—	—	—
332.0-T5	250	195	1	105	—	—	—
328.0-T6	235	145	1	65–95	—	—	—
333.0-F	235	130	2	90	185	105	—
333.0-T5	235	170	1	100	185	85	—
333.0-T6	290	205	2	105	230	105	—
333.0-T7	255	195	2	90	195	85	—
336.0-T551	250	193	1	105	193	95	—
336.0-T65	325	295	1	125	250	—	—
354.0-T61	330	255	3	—	—	—	—
354.0-T62	360	290	2	—	—	—	—
355.0-F	185	105	4	—	—	—	70
355.0-T51	205	165	2	75	165	—	70
355.0-T6	290	185	4	90	235	70	70
355.0-T61	310	275	2	105	250	70	70
355.0-T7	275	205	2	85	205	70	70
355.0-T71	250	215	3	85	185	70	70
C355.0-T6	330	195	8	90	—	—	70
C355.0-T61	315	235	6	100	—	—	70
C355.0-T62	330	255	5	100	—	—	70
356.0-F	180	125	5	—	—	—	73
356.0-T51	185	140	2	—	—	—	73
356.0-T6	260	185	5	80	205	90	73
356.0-T7	220	165	6	70	170	75	73
356.0-T71	170	—	3	60–90	—	—	73
A356.0-F	185	90	8	—	—	—	73
A356.0-T51	200	140	5	—	—	—	73

Table 6B *(Continued)*

Alloy & Temper	Tension: Ultimate Strength (MPa)	Tension: Yield Strength (MPa)	Tension: Elongation in 5D (%)	Hardness Brinell Number (500 kg/10 mm)	Shear Ultimate Strength (MPa)	Fatigue Endurance Limit[b] (MPa)	Modulus[c] of Elasticity (GPa)
A356.0-T6	285	205	12	80	—	—	73
357.0-F	195	105	6	—	—	—	—
357.0-T51	200	145	4	—	—	—	—
357.0-T6	360	295	5	100	240	90	—
357.0-T7	260	205	5	70	—	—	—
A357.0-T6	360	290	5	100	240	105	—
359.0-T61	330	255	6	—	—	—	—
359.0-T62	345	290	6	—	—	110	—
A390.0-F	200	200	<1.0	110	—	—	—
A390.0-75	200	200	<1.0	110	—	—	—
A390.0-T6	310	310	<1.0	145	—	115	—
A390.0-T7	260	260	<1.0	120	—	105	—
443.0-F	160	60	10	45	110	55	71
B443.0-F	145	40	6	30–60	—	—	—
A444.0-F	165	75	13	44	—	—	—
A444.0-T4	160	70	21	45	110	55	—
513.0-F	185	110	7	60	150	70	—
535.0-F	240	125	8	60–90	—	—	—
705.0-T5	255	115	10	55–75	—	—	—
707.0-T7	310	240	3	80–110	—	—	—
711.0-T1	195	125	7	55–85	—	—	—
713.0-T5	220	150	4	60–90			
850.0-T5	160	75	12	45	105	60	71
851.0-T5	140	75	5	45	95	60	71
851.0-T6	125	—	8	—	—	—	71
852.0-T5	220	160	5	70	145	75	71
Die Cast							
A360.0-F	315	165	4	75	180	124	71
380.0-F	315	160	3	80	195	140	71
A380.0-F	325	160	4	80	185	140	71
383.0-F	310	150	4	75	—	145	71
384.0-F	330	165	3	85	200	140	—
390.0-F	280	240	<1	—	—	—	—
B390.0-F	315	250	<1	120	—	140	81
392.0-F	290	270	<1	—	—	—	—
413.0-F	295	145	3	80	170	130	71
A413.0-F	290	130	4	80	170	130	—
C443.0-F	230	95	9	65	200	115	71
518.0-F	310	193	5	80	200	140	—

[a] Based upon industry handbooks.[3,8–10] Consult those references for limits. Values are representative of separately cast test bars, not of specimens taken from commercial castings. For tensile yield strengths, offset = 0.2%.

[b] Based on 500,000,000 cycles of completely reversed stress using R. R. Moore type of machines and specimens.

[c] Average of tension and compression moduli; compressive modulus is nominally about 2% greater.

5 CORROSION BEHAVIOR OF ALUMINUM ALLOYS

Although aluminum is a chemically active metal, its resistance to corrosion is attributable to an invisible oxide film that forms naturally and is always present unless it is deliberately prevented from forming. Scratch the oxide from the surface and, in air, the oxide immediately re-forms. Once formed, the oxide effectively protects the metal from chemical attack and also from further oxidation. Some properties of this natural oxide are as follows:

- It is very thin—200–400 billionths of an inch thick.
- It is tenacious. Unlike iron oxide or rust that spalls from the surface leaving a fresh surface to oxidize, aluminum oxide adheres tightly to aluminum.
- It is hard. Aluminum oxide is one of the hardest substances known.
- It is relatively stable and chemically inert.
- It is transparent and does not detract from the metal's appearance.

5.1 General Corrosion

The general corrosion behavior of aluminum alloys depends basically on three factors: (1) the stability of the oxide film, (2) the environment, and (3) the alloying elements; these factors are not independent of one another. The oxide film is considered stable between pH 4.5 and 9.0; however, aluminum can be attacked by certain anions and cations in neutral solutions, and it is resistant to some acids and alkalis.

In general, aluminum alloys have good corrosion resistance in the following environments: atmosphere, most fresh waters, seawater, most soils, most foods, and many chemicals. Since "good corrosion resistance" is intended to mean that the material will give long service life without surface protection, in support of this rating is the following list of established applications of aluminum in various environments:

In Atmosphere. Roofing and siding, truck and aircraft skin, architectural.

With Most Fresh Waters. Storage tanks, pipelines, heat exchangers, pleasure boats.

In Seawater. Ship hulls and superstructures, buoys, pipelines.

In Soils. Pipelines and drainage pipes.

With Foods. Cooking utensils, tanks and equipment, cans and packaging.

With Chemicals. Storage tanks, processing and transporting equipment.

It is generally true that the higher the aluminum purity, the greater is its corrosion resistance. However, certain elements can be alloyed with aluminum without reducing its corrosion resistance and in some cases an improvement actually results. Those elements having little or no effect include Mn, Mg, Zn, Si, Sb, Bi, Pb, and Ti; those having a detrimental effect include Cu, Fe, and Ni. Some guidelines for the different alloy groupings include the following:

Al–Mn Alloys. Al–Mn alloys (3xxx series) have good corrosion resistance and may possibly be better than 1100 alloy in marine environments and for cooking utensils because of a reduced effect by Fe in these alloys.

Al–Mg Alloys. Al–Mg alloys (5xxx series) are as corrosion resistant as 1xxx alloys and even more resistant to salt water and some alkaline solutions. In general, they offer the best combination of strength and corrosion resistance of all aluminum alloys.

Al–Mg–Si Alloys. Al–Mg–Si alloys (6xxx series) have good resistance to atmosphere corrosion but generally slightly lower resistance than Al–Mg alloys. They can be used unprotected in most atmospheres and waters.

Alclad Alloys. Alclad alloys are composite wrought products comprised of an aluminum alloy core with a thin layer of corrosion—protective pure aluminum or aluminum alloy metallurgically bonded to one or both surfaces of the core. As a class, alclad alloys have a very high resistance to corrosion. The cladding is anodic to the core and thus protects the core.

5.2 Pitting Corrosion

Pitting is the most common corrosive attack on aluminum alloy products. Pits form at localized discontinuities in the oxide film on aluminum exposed to atmosphere, fresh water, or saltwater, or other neutral electrolytes. Since in highly acidic or alkaline solutions the oxide film is usually unstable, pitting generally occurs in a pH range of about 4.5–9.0. The pits can be minute and concentrated and can vary in size and be widely scattered, depending on alloy composition, oxide film quality, and the nature of the corrodent. The resistance of aluminum to pitting depends significantly on its purity; the purest metal is the most resistant. The presence of other elements in aluminum, except Mn, Mg, and Zn, increases in susceptibility to pitting. Copper and iron have the greatest effect on susceptibility. Alclad alloys have greatest resistance to penetration since any pitting is confined to the more anodic cladding until the cladding is consumed.

5.3 Galvanic Corrosion

Aluminum in contact with a dissimilar metal in the presence of an electrolyte tends to corrode more rapidly than if exposed by itself to the same environment; this is referred to as galvanic corrosion. The tendency of one metal to cause galvanic corrosion of another can be predicted from a "galvanic series," which depends on environments. Such a series is listed below; the anodic metal is usually corroded by contact with a more cathodic one:

Anodic	Magnesium and zinc	Protect aluminum
	Aluminum, cadmium, and chromium	Neutral and safe in most environments
	Steel and iron	Cause slow action on aluminum except in marine environments
	Lead	Safe except for severe marine or industrial atmospheres
	Copper and nickel	Tend to corrode aluminum
Cathodic	Stainless steel	Safe in most atmospheres and fresh water; tends to corrode aluminum in severe marine atmospheres

Since galvanic corrosion is akin to a battery and depends on current flow, several factors determine the severity of attack:

Electrolyte Conductivity. The higher the electrical conductivity, the greater the corrosive effect.

Polarization. Some couples polarize strongly to reduce the current flow appreciably. For example, stainless steel is highly cathodic to aluminum, but because of polarization, the two can safely be used together in many environments.

Anode–Cathode Area Ratios. A high ratio minimizes galvanic attack; a low ratio tends to cause severe galvanic corrosion.

6 MACHINING ALUMINUM ALLOYS

Aluminum alloys are readily machined and offer such advantages as almost unlimited cutting speed, good dimensional control, low cutting force, and excellent life.

The cutting tool geometry used in machining aluminum alloys is described by seven elements: top or back rake angle, side rake angle, end-relief angle, side-relief angle, end cutting edge angle, and nose radius. The depth of cut may be in the range of $\frac{1}{16}$–$\frac{1}{4}$ in. (1.6–6.3 mm) for small work up to $\frac{1}{2}$–$1\frac{1}{2}$ in. (12.5–38 mm) for large work. The feed depends on finish. Rough cuts vary from 0.006 to 0.080 in. (0.15–2 mm) and finishing cuts from 0.002 to 0.006 in. (0.05–0.15 mm). Speed should be as high as possible, up to about 15,000 feet per minute (fpm) [about 5000 meters per minute (mpm)]. Cutting forces for an alloy such as 6061-T651 are 0.30–0.50 hp/in.^{3}/min for a 0° rake angle and 0.25–0.35 hp/in.^{3}/min for a 20° rake angle. Lubrication such as light mineral or soluble oil is desirable for high production. Alloys with a machinability rating of A or B may not need lubrication.

Cutting tool materials for machining aluminum alloys include water-hardening steels, high-speed steels, hard-cast alloys, sintered carbides, and diamonds:

1. Water-hardening steels (plain carbon or with additions of chromium, vanadium, or tungsten) are lowest in first cost. They soften if cutting edge temperatures exceed 300°F (150°C), have low resistance to edge wear, and are suitable for low cutting speeds and limited production runs.
2. High-speed steels are available in a number of forms, are heat treatable, permit machining at rapid rates, allow cutting edge temperatures of over 1000°F (540°C), and resist shock better than hard-cast or sintered carbides.
3. Hard-cast alloys are cast closely to finish size, are not heat treated, and lie between high-speed steels and carbides in terms of heat resistance, wear, and initial cost. They will not take severe shock loads.
4. Sintered carbide tools are available in solid form or as inserts. They permit speeds 10–30 times faster than for high-speed steels. They can be used for most machining operations. They should be used only when they can be supported rigidly and when there is sufficient power and speed. Many types are available.
5. Mounted diamonds are used for finishing cuts where an extremely high quality surface is required.

6.1 Single-Point Tool Operations

1. *Turning.* Aluminum alloys should be turned at high speeds with the work held rigidly and supported adequately to minimize distortion.

2. *Boring.* All types of tooling are suitable. Much higher speeds can be employed than for boring ferrous materials. Carbide tips are normally used in high-speed boring in vertical or horizontal boring machines.
3. *Planing and Shaping.* Aluminum permits maximum table speeds and high metal removal rates. Tools should not strike the work on the return stroke.

6.2 Multipoint Tool Operations

Milling. Removal rate is high with correct cutter design, speed and feed, machine rigidity, and power. When cutting speeds are high, the heat developed is retained mostly in the chips, with the balance absorbed by the coolant. Speeds are high with cutters of high-speed and cast alloys and very high with sintered carbide cutters. All common types of solid-tooth, high-carbon, or high-speed steel cutters can be employed. High-carbon cutters operating at a maximum edge temperature of 400°F are preferred for short-run production. For long runs, high-speed steel or inserted-tooth cutters are used. Speeds of 15,000 fpm (5000 mpm) are not uncommon for carbide cutters. Maximum speeds for high-speed and high-carbon steel cutters are around 5000 and 600 fpm (1650 and 200 mpm), respectively.

Drilling. General-purpose drills with bright finishes are satisfactory for use on aluminum. Better results may be obtained with drills having a high helix angle. Flute areas should be large; the point angle should be 118° (130°–140° for deeper holes). Cutting lips should be equal in size. Lip relief angles are between 12° and 20°, increasing toward the center to hold the chisel angle between 130° and 145°. No set rule can be given for achieving the correct web thickness. Generally, for aluminum, it may be thinner at the point without tool breakage. A 1Xs-Hi drill at 6000 rpm has a peripheral speed of 2000 fpm (680 mpm). For drilling aluminum, machines are available with speeds up to 80,000 rpm. If excessive heat is generated, hold diameter may be reduced even below drill size. With proper drills, feeds, speeds, and lubrication, no heat problem should occur.

For a feed of 0.008 inches per rotation (ipr) and a depth-to-diameter ratio of 4:1, the thrust value is 170 lb and the torque value is 10 lb-in. for a 1A-Ui drill with alloy 6061-T651. Aluminum alloys can be counterbored, tapped, threaded by cutting or rolling, and broached. Machining fluid should be used copiously.

Grinding. Resin-bounded silicon carbide wheels of medium hardness are used for rough grinding of aluminum. Finish grinding requires softer, vitrified-bonded wheels. Wheel speeds can vary from 5500 to 6000 fpm (1800–2000 mpm). Abrasive belt grinding employs belt speeds from 4600 to 5000 square feet per minute (sfpm). Grain size of silicon carbide abrasive varies from 36 to 80 for rough cuts and from 120 to 180 for finishing cuts. For contact wheel abrasive belt grinding, speeds are 4500–6500 sfpm. Silicon carbide or aluminum oxide belts (24–80 grit) are used for rough cuts.

Sawing, Shearing, Routing, and Arc Cutting Aluminum. Correct tooth contour is most important in *circular sawing.* The preferred saw blade has an alternate hollow ground side—rake teeth at about 15°. Operating speeds are 4000–15,000 fpm (1300–5000 mpm). Lower speeds are recommended for semi-high-speed steel, intermediate speeds for high-speed inserted-tooth steel blades, and high speeds for carbide-tipped blades. Band-sawing speeds should be between 2000 and 5000 fpm (655–1640 mpm). Spring-tempered blades are recommended for sheet and soft blades with hardened teeth for plate. Tooth pitch should not exceed material thickness: four to five teeth to the inch for spring tempered, six to eight teeth to the inch for flexible backed. Contour sawing is readily carried out. Lubricant should be applied to the back of the blade. Shearing of sheet may be done on guillotine shears. The clearance between blades is generally 10–12% of sheet thickness down to 5–6% for light gauge soft alloy sheet. Hold-down pads, shear beds, and tables should be covered to prevent marring. Routing can also be used with 0.188–0.50 in. (4.8–12.7 mm) material routed at feeds of 10–30 inches per minute (ipm) [25–75 centimeters per minute (cmpm)]. Heat-treated plates up to 3 in. (7.6 cm) in thickness material can be routed at feeds up to 10 ipm (25 cmpm). Chipless machining of aluminum can be carried out using shear spinning rotary swaging, internal swaging, thread rolling, and flame cutting.

7 FINISHING ALUMINUM

The aluminum surface can be finished in more different ways than any other metal. The normal finishing operations fall into four categories: mechanical, chemical, electrochemical, and applied. They are usually performed in the order listed, although one or more of the processes can be eliminated depending upon the final effect desired.

7.1 Mechanical Finishes

This is an important starting point since even with subsequent finishing operations a rough, smooth, or textured surface may be retained and observed. For many applications, the as-fabricated finish may be good enough. Or this surface can be changed by grinding, polishing, buffing, abrasive or shot blasting, tumbling or burnishing, or even hammering for special effects. Rolled surfaces of sheet or foil can be made highly specular by use of polishing rolls; one side or two sides bright and textures can be obtained by using textured rolls.

7.2 Chemical Finishes

A chemical finish is often applied after the mechanical finish. The most widely used chemical finishes include caustic etching for a matte finish, design etching, chemical brightening, conversion coatings, and immersion coatings. Conversion coatings chemically convert the natural oxide coating on aluminum to a chromate, a phosphate, or a combination chromate–phosphate coating, and they are principally used and are the recommended ways to prepare the aluminum surface for painting.

7.3 Electrochemical Finishes

These include electrobrightening for maximum specularity, electroplating of another metal such as nickel or chromium for hardness and wear resistance, and, most importantly, anodizing. Anodizing is an electrochemical process whereby the natural oxide layer is increased in thickness over a thousand times and made more dense for increased resistance to corrosion and abrasion resistance. The anodic oxide forms by the growth of cells, each cell containing a central pore. The pores are sealed by immersing the metal into very hot or boiling water. Sealing is an important step and will affect the appearance and properties of the anodized coating. There are several varieties of anodized coatings.

7.4 Clear Anodizing

On many aluminum alloys a thick, transparent oxide layer can be obtained by anodizing in a sulfuric acid solution—this is called clear anodizing. The thickness of the layer depends on the current density and the time in solution and is usually between 0.1 and 1 mil in thickness.

7.5 Color Anodizing

Color can be added to the film simply by immersing the metal immediately after anodizing and before sealing into a vat containing a dye or metallic coloring agents and then sealing the film. A wide range of colors have been imparted to aluminum in this fashion for many years. However, the colors imparted in this manner tend to fade from prolonged exposure to sunlight.

7.6 Integral Color Anodizing

More lightfast colors for outdoor use are achieved through integral color anodizing. These are proprietary processes utilizing electrolytes containing organic acids and, in some cases, small amounts of impurities are added to the metal itself to bring about the desired colors. This is usually a onstage process and the color forms as an integral part of the anodization. Colors are for the most part limited to golds, bronzes, grays, and blacks.

7.7 Electrolytically Deposited Coloring

Electrolytically deposited coloring is another means of imparting lightfast colors. Following sulfuric acid anodizing, the parts are transferred to a second solution containing metallic pigments that are driven into the coating by an electric current.

7.8 Hard Anodizing

Hard anodizing, or hardcoating as it is sometimes called, usually involves anodizing in a combination of acids and produces a very dense coating, often 1–5 mils thick. It is very resistant to wear and is normally intended for engineering applications rather than appearance.

7.9 Electroplating

In electroplating, a metal such as chromium or nickel is deposited on the aluminum surface from a solution containing that metal. This usually is done for appearance or to improve the hardness or abrasion resistance of the surface. Electroplating has a "smoothing-out" effect, whereas anodized coatings follow the contours of the base-metal surface, thus preserving a matte or a polished surface as well as any other patterns applied prior to the anodizing.

7.10 Applied Coatings

Applied coatings include porcelain enamel, paints and organic coatings, and laminates such as plastic, paper, or wood veneers. Probably as much aluminum produced today is painted as is anodized. Adhesion can be excellent when the surface has been prepared properly. For best results paint should be applied over a clean conversion-coated or anodized surface.

8 APPLICATIONS OF ALUMINUM ALLOYS

There are at least two approaches to overviewing important applications of aluminum alloys: *by alloy class,* as introduced in Section 3, and *by type of application*. We will consider both approaches in the information below, reviewing by alloy class in Section 8.1 and by application in Section 8.2.

All photographs are courtesy of the Aluminum Association unless otherwise indicated. For a more complete discussion of these and other applications of aluminum alloys, the reader is referred to Altenpohl's very complete treatise on the aluminum industry, *Aluminum: Technology, Applications, and Environment,*[1] and to *Aluminum Alloy Castings—Properties, Processes, and Applications.*[17]

8.1 Applications by Alloy Class

Wrought Alloys

1xxx—Pure Al. The major characteristics of the 1xxx series are:

- Strain hardenable
- Exceptionally high formability, corrosion resistance, and electrical conductivity
- Typical ultimate tensile strength range: 10–27 ksi (70–185 MPa)
- Readily joined by welding, brazing, and soldering

The 1xxx series represents the commercially pure aluminum, ranging from the baseline 1100 (99.00% minimum Al) to relatively purer 1050/1350 (99.50% minimum Al) and 1175 (99.75% minimum Al). The 1xxx series are strain hardenable but would not be used where strength is a prime consideration.

The primary uses of the 1xxx series include applications where the combination of extremely high corrosion resistance, formability, and/or electrical conductivity are required, e.g., foil and strip for packaging (Fig. 1), chemical equipment, tank car or truck bodies, spun hollowware, and elaborate sheet metal work.

Electrical applications are one major use of the 1xxx series, primarily 1350, which has relatively tight controls on impurities that would lower electrical conductivity. As a result, an electrical conductivity of 62% of the International Annealed Copper Standard (IACS) is guaranteed for this material which, combined with the natural light weight of aluminum, means a significant weight and therefore cost advantage over copper in electrical applications (Fig. 2).

2xxx—Al–Cu Alloys. The major characteristics of the 2xxx series are:

- Heat treatable
- High strength, at room and elevated temperatures
- Typical ultimate tensile strength range: 27–62 ksi (185–430 MPa)
- Usually joined mechanically but some alloys are weldable

The 2xxx series are heat treatable and possess in certain individual alloys (e.g., 2219) good combinations of high strength (especially at elevated temperatures), toughness, and

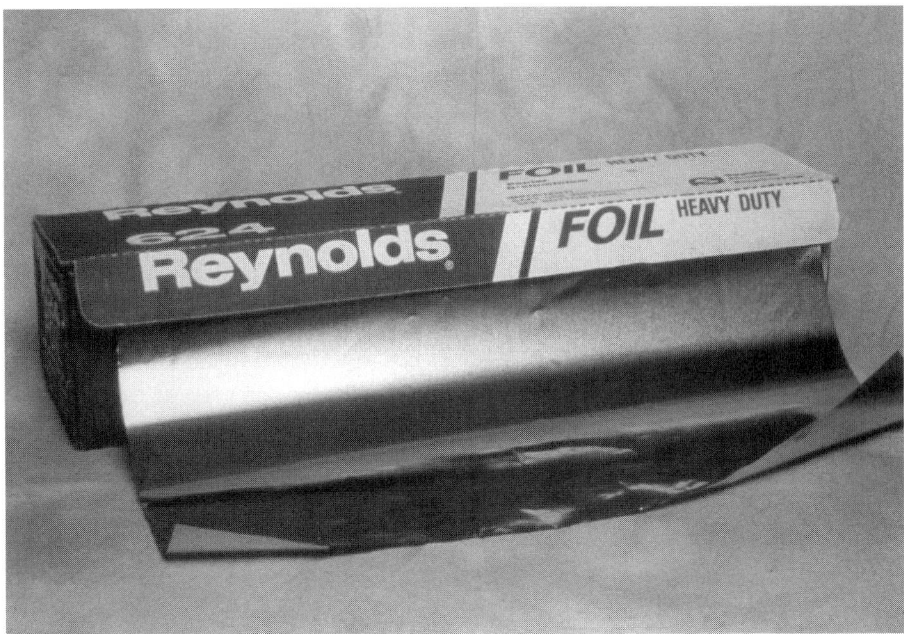

Figure 1 Reynolds Heavy Duty Foil product, an example of aluminum food wrapping products made of various grades of 1xxx commercially pure aluminum.

weldability. Aluminum–copper alloys are not as resistant to atmospheric corrosion as several other series and so are usually painted or clad for added protection.

The higher strength 2xxx alloys are widely used in aircraft wings and fuselage components (2X24; Fig. 3) and in truck body construction (2014; Fig. 4). Some specific alloys in the series (e.g., 2219 and 2048) are readily joined by gas-metal-arc (GMA) and gas-tungsten-arc (GTA) welding, and so are used for aerospace applications where that is the preferred method of joining.

Alloy 2195 is a new Li-bearing aluminum alloy providing very high modulus of elasticity along with higher strength and comparable weldability to 2219; it is being utilized for space applications (Fig. 5).

For applications requiring high strength plus high fracture toughness, there are high-toughness versions of several of the alloys (e.g., 2124, 2324, 2419, 2524) that have tighter control on the impurities that increases their resistance to unstable fracture, all developed specifically for the aircraft industry.

Alloys 2011, 2017, and 2117 are widely used for fasteners and screw-machine stock.

3xxx—Al–Mn Alloys. The major characteristics of the 3xxx series are:

- High formability and corrosion resistance with medium strength
- Typical ultimate tensile strength range: 16–41 ksi (110–285 MPa)
- Readily joined by all commercial procedures

The 3xxx series are strain hardenable, have excellent corrosion resistance, and are readily welded, brazed, and soldered.

Figure 2 Aluminum electrical bus bar installation with 1350 bus bar.

Alloy 3003 is widely used in cooking utensils and chemical equipment because of its superiority in handling many foods and chemicals and in builders' hardware because of its superior corrosion resistance. Alloy 3105 is a principal for roofing and siding.

Because of the ease and flexibility of joining, 3003 and other members of the 3xxx series are widely used in sheet and tubular form for heat exchangers in vehicles and power plants (Fig. 6).

Alloy 3004 and its modification 3104 are the principals for the bodies of drawn and ironed can bodies for beverage cans for beer and soft drinks (Fig. 7). As a result, they are among the most used individual alloys in the aluminum system, in excess of 3.5 billion pounds per year.

Figure 3 Aircraft wing and fuselage structure includes extrusions and plate of 2xxx alloys like 2024, 2124 and 2618 and 7xxx alloys like 7050 and 7475. External sheet skin may be alclad 2024, 2524, 2618 or 7475; the higher purity cladding provides corrosion protection to the AI–Cu and Al–Zn–Mg alloys that may darken with age otherwise.

Figure 4 Heavy dump and tank trucks and trailer trucks employ high-strength 2xxx or 6xxx extrusions for their structural members and tough 5xxx alloy sheet and plate for their load-carrying components.

4xxx—Al–Si Alloys. The major characteristics of the 4xxx series are:

- Heat treatable
- Good flow characteristics, medium strength
- Typical ultimate tensile strength range: 25–55 ksi (170–380 MPa)
- Easily joined, especially by brazing and soldering

There are two major uses of the 4xxx series, both generated by their excellent flow characteristics provided by their relatively high silicon contents. The first is for forgings: The workhorse alloy is 4032, a medium high-strength, heat-treatable alloys used principally in applications such as forged aircraft pistons. The second major application is as filler alloy; here the workhorse is 4043, used for GMA welding of 4xxx and 6xxx alloys for structural and automotive applications.

Figure 5 Fuel tanks and booster rockets such as those of the Space Shuttle are 2xxx alloys, originally 2219 and 2419, now Al–Li "Weldalite" alloy 2195.

As noted, the same characteristic leads to both types of application: good flow characteristic provided by the high silicon content. In the case of forgings, this ensures the complete and precise filling of complex dies; in the case of welding, it ensures complete filling of grooves in the members to be joined. For the same reason, other variations of the 4xxx alloys are used for the cladding on brazing sheet, the component that flows to complete the bond.

5xxx—Al–Mg Alloys. The major characteristics of the 5xxx series are:

- Strain hardenable
- Excellent corrosion resistance, toughness, weldability; moderate strength
- Building and construction, automotive, cryogenic, marine applications
- Representative alloys: 5052,5083,5754
- Typical ultimate tensile strength range: 18–51 ksi (125–350 MPa)

Aluminum–magnesium alloys of the 5xxx series are strain hardenable and have moderately high strength, excellent corrosion resistance even in salt water, and very high toughness even at cryogenic temperatures to near absolute zero. They are readily welded by a variety of techniques, even at thicknesses up to 8 in. (20 cm).

As a result, 5xxx alloys find wide application in building and construction (Fig. 8), highway structures including bridges (Fig. 9), storage tanks and pressure vessels, cryogenic tankage and systems for temperatures as low as −459°F (−270°C) near absolute zero, transportation (Fig. 10), and marine applications (Fig. 11), including offshore drilling rigs (Fig. 12).

Figure 6 Alloy 3003 tubing in large commercial power plant heat exchanger.

Alloys 5052, 5086, and 5083 are the workhorses from the structural standpoint, with increasingly higher strength associated with the higher Mg content. Specialty alloys in the group include 5182, the beverage can end alloy (Fig. 7), and thus among the largest in tonnage used; 5754 for automotive body frames and panels (Fig. 13); and 5252, 5457, and 5657 for bright trim applications, including automotive trim.

Care must be taken to avoid use of 5xxx alloys with more than 3% Mg content in applications where they receive continuous exposure to temperatures above 212°F (100°C). Such alloys may become sensitized and susceptible to stress–corrosion cracking. For this reason, alloys such as 5454 and 5754 are recommended for applications where high–temperature exposure is likely.

Figure 7 Bodies of most beverage cans are alloy 3004, the ends are 5182, making it the largest volume alloy combination in the industry.

6xxx—Al–Mg–Si Alloys. The major characteristics of the 6xxx series are:

- Heat treatable
- High corrosion resistance, excellent extrudibility; moderate strength
- Typical ultimate tensile strength range: 18–58 ksi (125–400 MPa)
- Readily welded by GMA and GTA welding methods

The 6xxx alloys are heat treatable and have moderately high strength coupled with excellent corrosion resistance. A unique feature is their great extrudability, making it possible to produce in single shapes relatively complex architectural forms and also to design shapes that put the majority of the metal where it will most efficiently carry the highest tensile and compressive stresses (Fig. 14). This is a particularly important advantage for architectural and structural members where stiffness criticality is important.

Alloy 6063 is perhaps the most widely used because of its extrudability; it is not only the first choice for many architectural and structural members but also has been the choice for the Audi automotive space frame members. A good example of its structural use was the all-aluminum bridge structure in Foresmo, Norway (Fig. 9); it was prefabricated in a shop and erected on the site in only a few days.

Higher strength alloy 6061 extrusions and plates find broad use in welded structural members such as automotive, truck (Fig. 4), and marine frames, railroad cars, and pipelines. Alloys like 6111 provide a fine combination of strength and formability, useful for external automotive panels (Fig. 13*b*).

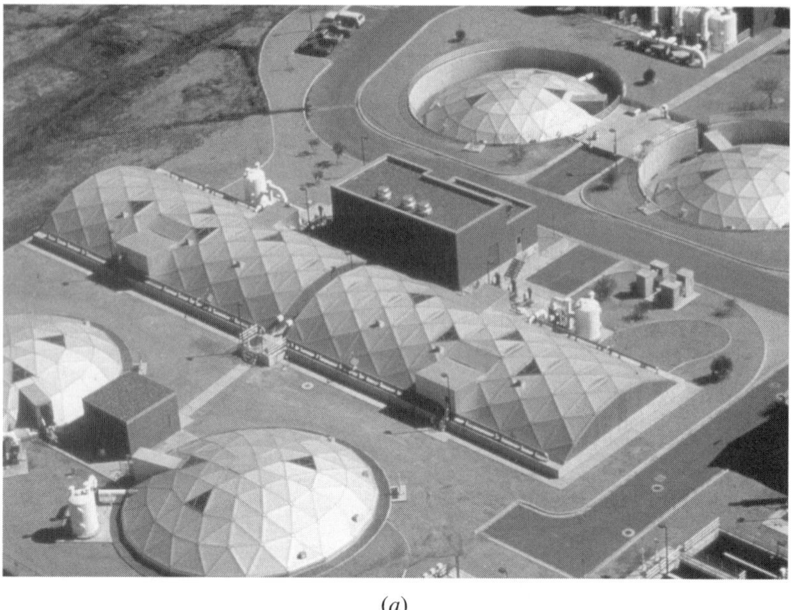

(*a*)

(*b*)

Figure 8 Sheet of 5xxx alloys often forms the surface of geodesic dome structures, as in these examples of (*a*) a water treatment plant and (*b*) a wide-span arena roof. The structural supports are typically 6061 or 6063 extruded shapes or tubular members.

(*a*)

(*b*)

Figure 9 The Foresmo bridge in northern Norway is an excellent example of the use of Al–Mg alloys for built-up girders systems; these photos illustrate a major advantage of replacement aluminum bridges: the ability to prefabricate the spans, (*a*) transport them, and (*b*) erect them in place quickly, minimizing the disruption to traffic.

Figure 10 Alloy 5454 has been widely used for railcar body construction where heavy loads, such as coal, and potentially high temperatures, may be involved.

Among specialty alloys in the series are 6066-T6, with high strength for forgings; 6071 for the highest strength available in 6xxx extrusions; and 6101and 6201 for high-strength electrical bus and electrical conductor wire, respectively.

7xxx—Al–Zn Alloys. The major characteristics of the 7xxx series are:

- Heat treatable
- Very high strength; special high toughness versions
- Typical ultimate tensile strength range: 32–88 ksi (220–605 MPa)
- Mechanically joined

The 7xxx alloys are heat treatable and, among the Al–Zn–Mg–Cu versions in particular, provide the highest strengths of all aluminum alloys. These alloys are not considered weldable by commercial processes and are regularly used with riveted construction.

The widest application of the 7xxx alloys has historically been in the aircraft industry (Fig. 3), where fracture-critical design concepts have provided the impetus for the high-toughness alloy development. There are several alloys in the series that are produced especially for their high toughness, notably 7050, 7055, 7150, 7175, and 7475; for these alloys, controlled impurity levels, particularly of Fe and Si, maximize the combination of strength and fracture toughness. Forgings of these alloys (Fig. 15) are often used for large structural members in aircraft.

The high strength–density combination for 7075-T73 (as well as 2014-T6) has made them choices for drill pipe, where the long lengths needed for deep wells require lightweight alloys.

The atmospheric corrosion resistance of the 7xxx alloys is not as high as that of the 5xxx and 6xxx alloys, so in such service they are usually coated or, for sheet and plate, used

(*a*)

(*b*)

Figure 11 High-speed single-hull ships (*a*) like the Proserio employ 5083 or 5383-H113/H321 machined plate for hulls, (*b*) internal hull stiffeners, decking and superstructure.

Figure 12 Demands of the superstructures of offshore oil rigs in high humidity and water exposure are met with 5454, 5086, and 5083 Al–Mg alloy welded construction.

in an alclad version. Also, special tempers have been developed to improve their resistance to exfoliation and stress–corrosion cracking, the T76 and T73 types, respectively. These tempers are especially recommended in situations where there may be high short transverse (through-the-thickness) stresses present during exposure to atmospheric or more severe environments.

The Cu-free 7xxx alloys have lower strength but are tougher and are both readily extrudable and weldable, so alloys like 7005 and 7029 find their way into applications like guard rail and automotive and truck bumpers.

8xxx—Alloys with Al+Other Elements (*Not Covered by Other Series*). The major characteristics of the 8xxx series are:

- Heat treatable
- High conductivity, strength, and hardness
- Typical ultimate tensile strength range: 17–35 ksi (115–240 MPa)

The 8xxx series is used for those alloys with lesser used alloying elements such as Fe, Ni, and Li. Each is used for the particular characteristics it provides the alloys.

Iron and Ni provide strength with little loss in electrical conductivity and so are used in a series of alloys represented by 8017 for conductors.

Lithium in alloy 8090 provides exceptionally high strength and modulus, and so this alloy is used for aerospace applications where increases in stiffness combined with high strength reduces component weight.

(*a*)

(*b*)

Figure 13 Automotive structures are likely to employ increasing amounts of 5754-0 formed sheet for parts such as internal door stiffeners or (*a*) the entire body in white; (*b*) external body panels are more likely to be higher strength 6111-T4.

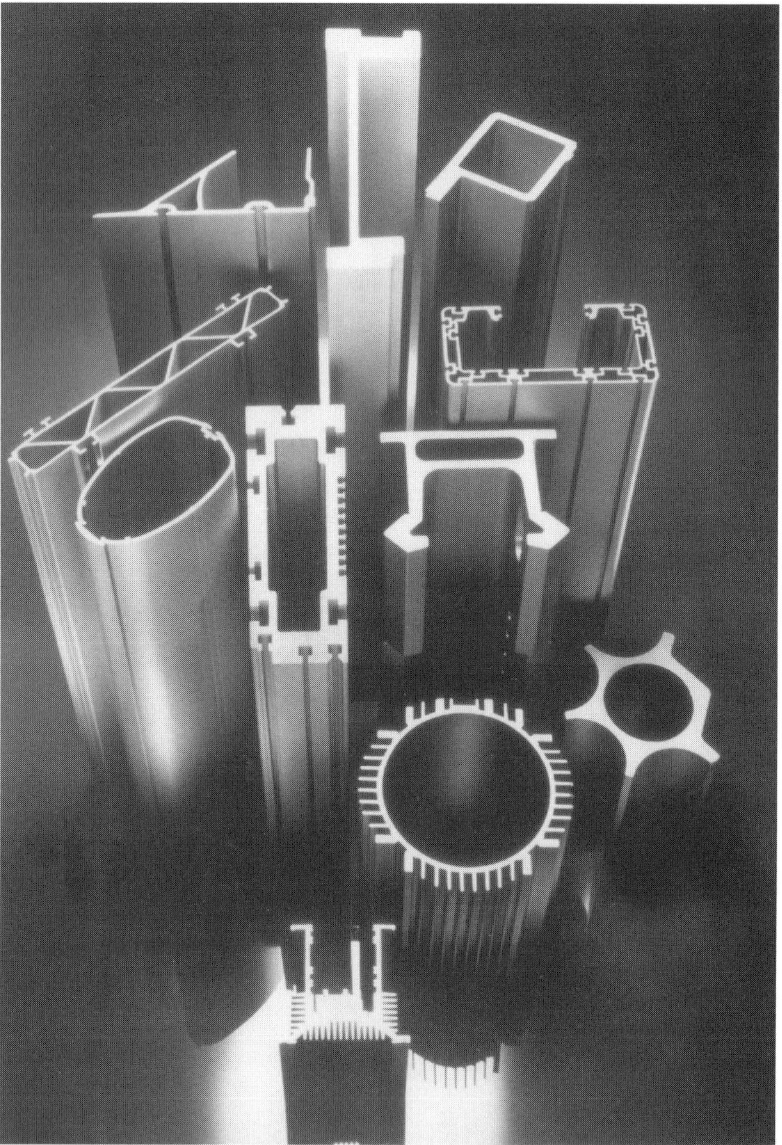

Figure 14 Power of extruded AI–Mg–Si alloys is the "put the metal where you need it" flexibility these alloys and the extrusion process provide.

Cast Alloys

In comparison with wrought alloys, casting alloys contain larger proportions of certain alloying elements such as silicon and impurity elements like iron, This results in a largely heterogeneous cast structure, i.e., one having a substantial volume of second phases. This second-phase material warrants careful study, since any coarse, sharp, and/or brittle con-

Figure 15 Example of a premium forged aircraft part, usually of alloys such as 7050 or 7175-T74.

stituent can create harmful internal notches and nucleate cracks when the component is later put under load. Fatigue performance is very sensitive to large heterogeneities, especially at or near the surface. As will be shown later, careful alloy selection combined with qood metallurgical and foundry practices can largely prevent such defects and provide exceptional quality castings for critical applications.

The elongation and strength, especially in fatigue, of most cast products are relatively lower than those of wrought products. This is because current casting practice is as yet unable to reliably prevent casting defects. In recent years, however, innovations in casting processes such as squeeze casting have brought about some significant improvement in the consistency and level of properties of castings, and these should be taken into account in selecting casting processes for critical applications. Another significant enhancement addressing porosity problems is HIP; its use has been shown to substantially improve the fatigue performance of aluminum castings.[17]

For applications where high ductility and toughness along with high strength are required, relatively high purity versions of casting alloys like A356.0-T6 (rather than 356.0-T6) and A357.0-T6 (rather than 357.0-T6) are recommended.

2xx.x—Al–Cu Alloys. The major characteristics of the 2xx.x series are:

- Heat treatable/sand and permanent mold castings
- High strength at room and elevated temperatures; some high toughness alloys
- Approximate ultimate tensile strength range: 19–65 ksi (130–450 MPa)

The strongest of the common casting alloys are heat-treated 201.0 and 204.0, which have found important application in the aerospace industry. Their castability is somewhat limited by a tendency to microporosity and hot tearing, so that they are best suited to investment casting. Their high toughness makes them particularly suitable for highly stressed components in machine tool construction, in electrical engineering (pressurized switchgear casings), and in aircraft construction.

Besides the standard aluminum casting alloys, there are special alloys for particular components, for instance, for engine piston heads, integral engine blocks, or bearings. For these applications the chosen alloy needs good wear resistance and a low friction coefficient as well as adequate strength at elevated service temperatures. A good example is the alloy 203.0, which to date is the aluminum casting alloy with the highest strength at around 400°F (200°C).

3xx.x—Al–Si+Cu or Mg Alloys. The major characteristics of the 3xx.x series are:

- Heat treatable/sand, permanent mold, and die castings
- Excellent fluidity/high strength/some high-toughness alloys
- Approximate ultimate tensile strength range: 19–40 ksi (130–275 MPa)
- Readily welded

The 3xx.x series of castings are one of the most widely used because of the flexibility provided by the high silicon content and its contribution to fluidity plus their response to heat treatment, which provides a variety of high-strength options. As a result, they are the best choice for large and complex castings (Fig. 16). Further the 3xx.x series may be cast by a variety of techniques ranging from relatively simple sand or die casting to very intricate permanent mold, investment castings, and the newer thixocasting and squeeze-casting technologies (Fig. 17).

Figure 16 Complex 3xx.x castings made by the investment casting processes provide the ability to obtain exceptionally intricate detail and fine quality.

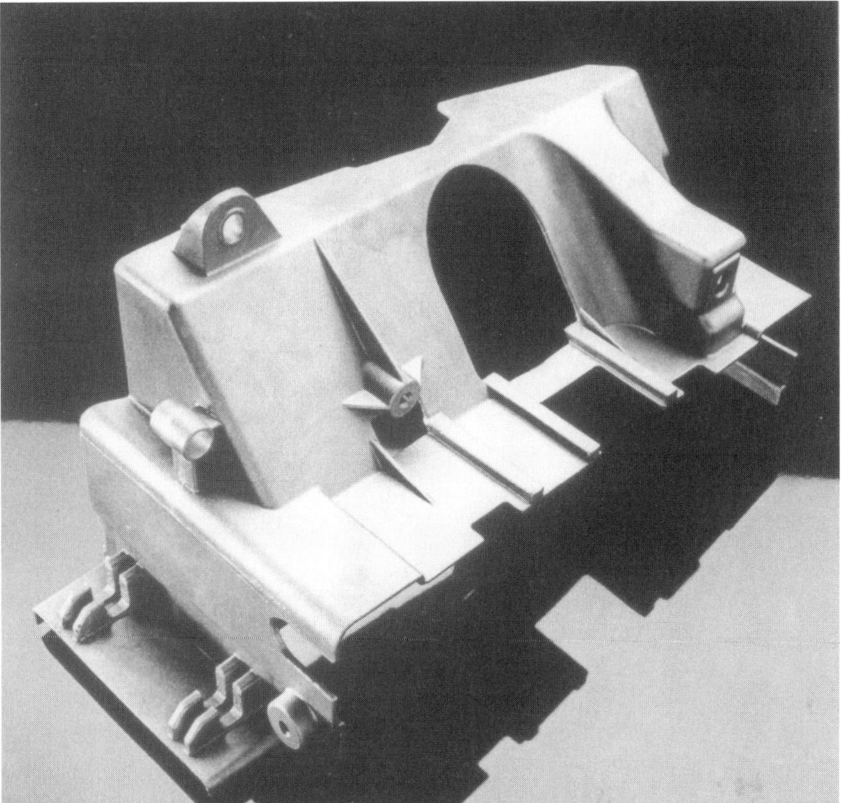

Figure 17 Thixoformed A356.0-T6 and A357.0-T6 may be used for critical aircraft components.

Among the workhorse alloys are 319.0 and 356.0/A356.0 for sand and permanent mold casting, with A356.0-T6 for such critical applications as automotive wheels (Fig. 18). For die castings, 360.0, 380.0/A380.0, and 390.0 are the most widely used. The newer squeeze-cast technologies have generally employed 357.0-T6 and A357.0-T6.

Alloy 332.0 is also one of the most frequently used aluminum casting alloys because it can be made almost exclusively from recycled scrap.

4xx.x—Al–Si Alloys. The major characteristics of the 4xx.x series are:

- Non–heat treatable/sand, permanent mold, and die castings
- Excellent fluidity/good for intricate castings
- Approximate ultimate tensile strength range: 17–25 ksi (115–170 MPa)

Alloy B413.0 is notable for its very good castability and excellent weldability, which are due to its eutectic composition and low melting point of 700°C. It combines moderate strength with high elongation before rupture and good corrosion resistance. The alloy is particularly suitable for intricate, thin-walled, leak-proof, fatigue-resistant castings.

Figure 18 Automotive wheels are often cast A356.0-T6.

These alloys have found applications in relatively complex cast parts for typewriter and computer housings and dental equipment and also for fairly critical components in marine and architectural applications.

5xx.x—Al–Mg Alloys. The major characteristics of the 5xx.x series are:

- Non–heat treatable/sand, permanent mold, and die
- Tougher to cast/provides good finishing characteristics
- Excellent corrosion resistance/machinability/surface appearance
- Approximate ultimate tensile strength range: 17–25 ksi (115–170 MPa)

The common feature of this group of alloys is good resistance to corrosion. Alloys 512.0 and 514.0 have medium strength and good elongation and are suitable for components exposed to seawater or to other similar corrosive environments. These alloys are often used for door and window fittings, which can be decoratively anodized to give a metallic finish or in a wide range of colors. Their castability is inferior to that of the Al–Si alloys because of their magnesium content and consequently long freezing range. For this reason they tend to be replaced by 355.0/AlSi5Mg, which has long been used for similar applications.

For die castings where decorative anodizing is particularly important, alloy 520.0 is quite suitable.

7xx.x—Al–Zn Alloys. The major characteristics of the 7xx.x series are:

- Heat treatable/sand and permanent mold cast (harder to cast)
- Excellent machinability/appearance
- Approximate ultimate tensile strength range: 30–55 ksi (205–380 MPa)

Because of the increased difficulty in casting 7xx.x alloys, they tend to be used only where the excellent finishing characteristics and machinability are important. Representative applications include furniture, garden tools, office machines, and farming and mining equipment.

8xx.x—Al–Sn Alloys. The major characteristics of the 8xx.x series are:

- Heat treatable/sand and permanent mold castings (harder to cast)
- Excellent machinability
- Bearings and bushings of all types
- Approximate ultimate tensile strength range: 15–30 ksi (105–210 MPa)

Like the 7xx.x alloys, 8xx.x alloys are relatively hard to cast and tend to be used only where their combination of superior surface finish and relative hardness are important. The prime examples are for parts requiring extensive machining and for bushings and bearings.

8.2 Applications by Market Area

In the paragraphs that follow, a review will be made of the alloys often selected for products in a number of the major markets in which aluminum is used

Electrical Markets

The major products for which aluminum is used in electrical applications are electric cable and bus conductor, where the high electrical conductivity (60% IACS) makes aluminum a cost-effective replacement for copper products:

- Electrical conductor wire—1350 where no special strength requirements exist and 6201 where a combination of high strength and high conductivity are needed
- Bus conductor—6101 (Fig. 2)
- Electrical cable towers—6063 or 6061 extruded shapes

Building and Construction Markets

Building and construction encompass those markets where architectural and/or structural requirements come together. Such applications include residential housing, commercial store fronts and structures, conference centers and areas (i.e., long roof bay requirements), highway bridges and roadside structures, and a variety of holding tanks and chemical structures (also considered below under Petroleum and Chemical Industry Components). Among the choices are the following:

- Bridges and other highway structures—6063 and 6061 extrusions (Figs. 8 and 9); 5083, 5083, and 5454 plate (Fig. 9)
- Bridge rail mounts—357.0-T6
- Store fronts, curtain wall—6063 extrusions
- Building sheet, siding—3005, 3105, and 5005 sheet
- Arena and convention center roofs—6061 extrusions with 5xxx alloy sheet panels (Fig. 8)
- Residential housing structures—6063 extrusions
- Architectural trim—5257, 5657, 6463
- Composite wall panels—5xxx alloy sheet plus expanded polymers

Automobile, Van, SUV, Bus, and Truck Applications

Automotive structures require a combination of aluminum castings, sheet, and extrusions to cover all good opportunities to increase gasoline mileage and reduce pollutants. Some examples are the following:

- Frame—5182 or 5754 sheet (Fig. 13*a*) or, for space frame designs, 6063 or 6061 extrusions
- External body sheet panels where dent resistance is important—2008, 6111 (Fig. 13*b*)
- Inner body panels—5083, 5754
- Bumpers—7029, 7129
- Air conditioner tubes, heat exchangers—3003 (Fig. 6)
- Auto trim—5257, 5657, 5757
- Door beams, seat tracks, racks, rails—6061 or 6063
- Hood, decklids—2036, 6016, 6111
- Truck beams—2014, 6070 (Fig. 4)
- Truck trailer bodies—5454, 5083, 5456 (Fig. 4)
- Wheels—A356.0 (Fig. 18), formed 5xxx sheet, or forged 2014-T6
- Housings, gear boxes—357.0, A357.0 (Fig. 16)

Aircraft and Aerospace Applications

Aircraft and aerospace applications require high strength combined with, depending upon the specific component, high fracture toughness, high corrosion resistance, and/or high modulus (sometimes all three). The result has been a great number of alloys and tempers developed specifically for this market, as illustrated by the examples below:

- Space mirror—high-purity aluminum
- Wing and fuselage skin—2024, alclad 2024, 7050 plate or extrusions (Fig. 3)
- Wing structures—2024, 2124, 2314, 7050 stiffened extrusions (Fig. 3)
- Bulkhead—2197, 7049, 7050, 7175
- Rocket tankage—2195, 2219, 2419 (Fig. 5)
- Engine components—2618
- Propellers—2025
- Rivets—2117, 6053
- If high modulus is critical—Li-bearing alloys 2090, 2091, 2195 or 8090
- If high fracture toughness is critical—2124, 2224, 2324, 7175, 7475
- For maximum fracture toughness—7475
- If stress–corrosion resistance is important—7X50 or 7X75 in the T73-type temper
- If resistance to exfoliation attack is vital—7xxx alloys in the T76-type temper
- For welded construction, as for shuttle tanks—5456, 2219, 2195

Marine Transportation

Many aluminum alloys readily withstand the corrosive attack of marine salt water and so find applications in boats, ships, offshore stations, and other components that are immersed in saltwater:

- Hull material—5083, 5383, 6063, 6061 (Fig. 11)
- Superstructure—5083, 5456
- Structural beams—5083, 5383, 6063, 6061 (Fig. 11)
- Off-shore stations, tanks—5083, 5456 (Fig. 12)

Rail Transportation

Much as for auto and truck bodies, aluminum lends itself to railcar structural and exterior panel applications:

- Beams—2014, 6061, 6070
- Exterior panels—5456, 6111
- Tank cars—5454, 5083
- Coal cars—5454, 5083, 5456 (Fig. 10)
- Cars for hot cargo—5454

Packaging Applications

Packaging applications require either great ductility and corrosion resistance for foil and wrapping applications or great strength and workability for rigid container sheet applications, i.e., cans. Alloy choices include the following:

- Aluminum foil for foods—1175 (Fig. 1)
- Rigid container (can) bodies—3004 (Fig. 7)
- Rigid container (can) ends—5182

Petroleum and Chemical Industry Components

The excellent combination of high strength combined with superior corrosion resistance plus weldability makes a number of aluminum alloys ideal for chemical industry applications, even some involving very corrosive fluids:

- Chemical piping—1060, 5254, 6063
- Pressure vessels (ASME Code)—5083, 5086, 6061, 6063
- Pipelines—6061, 6063, 6070
- Cryogenic tankage—5052, 5083, 5454, 6063, 6061
- Containers for hydrogen peroxide—5254, 5652

Other Markets

While not major markets in themselves, a variety of specialty products find great advantage in aluminum alloys:

- Screw machine products—2011, 6262
- Appliances—5005, 5052
- Tread plate—6061
- Weld wire—4043 (for welding 6xxx alloys), 5356, 5183, 5556 (for welding 5xxx alloys)

9 SUMMARY

Listed below are some of the characteristics of aluminum and its alloys that lead to their widespread application in nearly every segment of the economy. It is safe to say that no other material offers the same combination of properties.

- *Lightweight.* Very few metals have a lower density than aluminum, and they are not in common usage. Iron and copper are roughly three times as dense and titanium over 60% more dense than aluminum.
- *Good Formability.* Aluminum can be fabricated or shaped by just about every known method and is consequently available in a wide variety of forms.
- *Wide Range of Mechanical Properties.* Aluminum can be utilized as a weak, highly ductile material or, through alloying, as a material with a tensile strength approaching 100,000 psi.
- *High Strength-to-Weight Ratio.* Because of the combination of low density and high tensile strength, some aluminum alloys possess superior strength-to-weight ratios, equaled or surpassed only by highly alloyed and strengthened steels and titanium.
- *Good Cryogenic Properties.* Aluminum does not become brittle at very low temperatures; indeed, mechanical properties of most aluminum alloys actually improve with decreasing temperature.
- *Good Weatherability and General Corrosion Resistance.* Aluminum does not rust away in the atmosphere and usually requires no surface protection. It is highly resistant to attack from a number of chemicals.
- *High Electrical and Thermal Conductivity.* Pound for pound, aluminum conducts electricity and heat better than any other material except sodium, which can only be used under very special conditions. On a volume basis, only copper, silver, and gold are better conductors.
- *High Reflectivity.* Normally, aluminum reflects 80% of white light, and this value can be increased with special processing.
- *Finishability.* Aluminum is unique among the architectural metals with respect to the variety of finishes employed.

REFERENCES

1. D. G. Altenpohl, *Aluminum: Technology, Applications and Environment,* Aluminum Association, Washington, DC and TMS, Warrendale, PA, 1999.
2. *Aluminum Standards and Data* (*Standard and Metric Editions*), Aluminum Association, Washington, DC, published periodically.
3. *Standards for Aluminum Sand and Permanent Mold Casting,* Aluminum Association, Washington, DC, December 1992.
4. *The Aluminum Design Manual,* Aluminum Association, Inc., Washington, DC, 2005.
5. *The Aluminum Association Alloy and Temper Registrations Records:*
   (*i*) International Alloy Designations and Chemical Composition Limits for Wrought Aluminum and Aluminum Alloys, Aluminum Association, Washington, DC, January 2001.
   (*ii*) Designations and Chemical Composition Limits for Aluminum Alloys in the Form of Castings and Ingot, Aluminum Association, Washington, DC, January 1996.
   (*iii*) Tempers for Aluminum and Aluminum Alloy Products, Aluminum Association, Washington, DC, August 2001.

6. *American National Standard Alloy and Temper Designation Systems for Aluminum,* ANSI H35.1-1997, American National Standards Institute, Aluminum Association, Secretariat, Washington, DC, 1997.
7. *International Accord on Wrought Aluminum Alloy Designations,* Aluminum Association, Washington, DC, published periodically.
8. D. Zalenas (ed.), *Aluminum Casting Technology,* 2nd ed., American Foundrymens' Society, Des Plaines, IL, 1993.
9. *Product Design for Die Casting in Recyclable Aluminum, Magnesium, Zinc, and ZA Alloys,* Die Casting Development Council, La Grange, IL, 1996.
10. *The NFFS Guide to Aluminum Casting Design: Sand and Permanent Mold,* Non-Ferrous Founders Society, Des Plaines, IL, 1994.
11. *Strength of Aircraft Elements,* MMPDS-01 / MIL-HDBK-5J, Federal Aviation Administration, Washington, DC, 2004.
12. M. L. Sharp, *Behavior and Design of Aluminum Structures,* McGraw-Hill, New York, 1993.
13. M. L. Sharp, G. E. Nordmark, and C. C. Menzemer, *Fatigue Design of Aluminum Components and Structures,* McGraw-Hill, New York, 1996.
14. *Fatigue Design Handbook,* 2nd ed., SAE AE-10, Society of Automotive Engineers, Warrendale, PA, 1988.
15. J. R. Kissell and R. L. Ferry, *Aluminum Structures, a Guide to Their Specifications and Design,* Wiley, New York, 1995.
16. *Boiler & Pressure Vessel Code,* American Society of Mechanical Engineers, New York, updated Periodically.
17. J. G. Kaufman and E. Rooy, *Aluminum Alloy Castings—Properties, Processes, and Applications,* ASM International, Materials Park, OH, 2004.

Additional Selected Reading

18. *Aluminum Alloys—Selection and Application,* Aluminum Association, Washington, DC, December 1998.
19. J. R. Davis (ed.), *Aluminum and Aluminum Alloys,* ASM Specialty Handbook, ASM International, Materials Park, OH, February 1994.
20. H. Chandler (ed.), *Heat Treater's Guide—Practices and Procedures for Nonferrous Alloys,* ASM International, Materials Park, OH, February 1994.
21. J. G. Kaufman (ed.), *Properties of Aluminum Alloys—Tensile, Creep, and Fatigue Data at High and Low Temperatures,* Aluminum Association and ASM International, Materials Park, OH, December 1999.
22. *Aluminum Alloys for Cryogenic Applications,* Aluminum Association, Washington, DC, 1999.
23. *Life-Cycle Assessments for Aluminum Alloy Products,* Aluminum Association, Washington, DC, February 1996.
24. *Aluminum for Automotive Body Sheet Panels,* Publication AT3, Aluminum Association, Washington, DC, 1998.
25. *Aluminum Automotive Extrusion Manual,* Publication AT6, Aluminum Association, Washington, DC, 1998.
26. J. G. Kaufman, *Introduction to Aluminum Alloys and Tempers,* Aluminum Association and ASM International, Materials Park, OH, 2000.
27. J. G. Kaufman, *Fracture Resistance of Aluminum Alloys—Notch Toughness, Tear Resistance, and Fracture Toughness,* Aluminum Association and ASM International, Materials Park, OH, 2001.
28. *Aluminum and Aluminum Alloys,* American Society for Metals, Materials Park, OH.
29. *Aluminum Brazing Handbook,* Aluminum Association, Washington, DC.
30. *Aluminum Electrical Conductor Handbook,* Aluminum Association, Washington DC.
31. *Aluminum: Properties and Physical Metallurgy,* ASM International, Materials Park, OH.
32. *Aluminum Soldering Handbook,* Aluminum Association, Washington, DC.
33. *Forming and Machining Aluminum,* Aluminum Association, Washington, DC.

34. H. P. Goddard, et al., *The Corrosion of Light Metals,* Wiley, New York.

35. *Guidelines for the Use of Aluminum with Food and Chemicals,* Aluminum Association, Washington, DC.

36. *Handbook of Corrosion Data,* ASM International, Materials Park, OH.

37. J. D. Minford, *Handbook of Aluminum Bonding Technology and Data,* Marcel Dekker, New York.

38. *Metals Handbook Series,* ASM International, Materials Park, OH.

39. *The Surface Treatment and Finishing of Aluminum and Its Alloys,* American Society for Metals, Materials Park, OH.

40. *Welding Aluminum: Theory and Practice,* Aluminum Association, Washington, DC.

CHAPTER 4
COPPER AND COPPER ALLOYS

Konrad J. A. Kundig
Metallurgical Consultant
Tucson, Arizona

John G. Cowie
Copper Development Association
New York, New York

1 INTRODUCTION

Copper was the first metal humankind put into utilitarian service. Fabricated copper objects dating to approximately 8500 B.C. place the probable origin of the metal's use in Asia Minor, although prehistoric artifacts can be associated with civilizations on almost all inhabited continents. Copper is one of a few metals found in native (metallic) form in nature, and early "products" were undoubtedly made from nuggets. A solid copper mace head, found near Cata Huyuk, Anatolia (now Turkey), has been dated to 8000 B.C. Smelting is believed to have begun in what is now Israel as early as 3600 B.C., and a vigorous trade in the metal was well established within several hundred years. That prehistoric copper objects have survived through millennia attests to the metal's inherent chemical stability.[1,2]

Copper alloys appeared around 3500 B.C. in Mesopotamia, first, through the inclusion of tin (ushering in the Bronze Age) and later zinc (for brass). The resulting alloys were stronger than pure copper and were used for new applications, but the discovery of iron soon supplanted copper alloys in such items as tools and weaponry. It was the introduction of electricity in the nineteenth century that again brought copper into widespread use. Consumption grew quickly, reflecting the rate of electrification and industrialization and, more recently, the growth of global communications systems.

In 2003, worldwide consumption of refined copper was approximately 17.1 million short tons and in addition, more than 3.5 million tons of scrap were taken up by manufacturers, mainly for other-than-electrical products. The United States consumed more than 2.8 million tons of refined copper that year, somewhat less than one-half of which was mined domestically. Chile is the world's leading copper miner, but the United States has ample resources, and a newly discovered and extremely large orebody is currently under development in Arizona.

The properties that drive copper and copper alloy use include high electrical and thermal conductivities, favorable combinations of strength and ductility, ease of fabrication (machinability, castability, and welding and joining properties), resistance to corrosion, and aesthetic appeal. These properties are available in useful combinations through alloying and heat treatment. In the United States in 2003, 58.3% of all copper consumed found its way into electrical applications. Applications related to corrosion resistance (22.5%), heat transfer (11.2%), structural properties (6.3%), and aesthetics (1.7%) accounted for the remainder. In the future, a significant fraction of copper use might ensue from the metal's intrinsic biostatic properties. In terms of product form, wire and cable products make up 48.7% of total metallic copper consumption; plumbing and commercial tube, 17.5%; rod, bar, and mechanical wire, 15.3%; strip, sheet, and plate, 14.2%; castings, 3.7%; and powder, 0.6%.[3]

2 COPPER ALLOY FAMILIES

The Unified Numbering System (UNS) is the accepted alloy designation system in North America for wrought and cast copper and copper alloy products. Alloys are identified by five-digit numbers preceded by the letter C. The system is based on an earlier and still occasionally cited three-digit system developed by the U.S. copper and brass industry. For example, Copper Alloy No. 377 (forging brass) became C37700 under the UNS. The UNS is administered by the Copper Development Association (CDA). It is managed jointly by the American Society for Testing and Materials (ASTM) and the Society of Automotive Engineers (SAE), and alloy designations are common to standards issued by those organizations. Numbers from C10000 through C79999 denote wrought alloys, i.e., rolled, drawn, forged, or extruded. Cast alloys are numbered from C80000 through C99999. Descriptions of the copper alloy families are presented in Table 1.

2.1 Compositions

Compositions of currently accepted wrought and cast coppers and copper alloys, arranged by alloy families and similar alloys within families, are given in Tables 2–19 and 20–35, respectively.[4] (For example, leaded and nonleaded versions of alloys are tabulated separately.) Compositions are given as weight percent maximum unless shown as a range or minimum. These data are kept current on the CDA web site, www.copper.org.

3 PHYSICAL PROPERTIES OF COPPER AND COPPER ALLOYS

3.1 Pure Copper

With atomic number 29, copper is a member of subgroup IB on the periodic table of the elements. Other "noble metals" in this subgroup include silver and gold, with which copper

Table 1 Copper Alloy Families

Coppers	Metals which have a designated minimum copper content of 99.3% or higher.
High-copper alloys	For the wrought products, these are alloys with designated copper contents less than 99.3% but more than 96% which do not fall into any other copper alloy group. The cast high-copper alloys have designated copper contents in excess of 94%, to which silver may be added for special properties.
Brasses	These alloys contain zinc as the principal alloying element with or without other designated alloying elements such as iron, aluminum, nickel, and silicon. The wrought alloys comprise three main families of brasses: copper–zinc alloys, copper–zinc–lead alloys (leaded brasses), and copper–zinc–tin alloys (tin brasses). The cast alloys comprise five main families of brasses: copper–tin–zinc alloys (red, semired, and yellow brasses), "manganese bronze" alloys (high-strength yellow brasses), leaded "manganese bronze" alloys (leaded high-strength yellow brasses), copper–zinc–silicon alloys (silicon brasses and bronzes), and cast copper–bismuth and copper–bismuth–selenium alloys. Ingot for remelting for the manufacture of castings may vary slightly from the ranges shown.
Bronzes	Broadly speaking, bronzes are copper alloys in which the major alloying element is neither zinc nor nickel. Originally "bronze" described alloys with tin as the only or principal alloying element. Today, the term is generally used not by itself but with a modifying adjective. For wrought alloys, there are four main families of bronzes: copper–tin–phosphorus alloys (phosphor bronzes), copper–tin–lead–phosphorus alloys (leaded phosphor bronzes), copper–aluminum alloys (aluminum bronzes), and copper–silicon alloys (silicon bronzes). The cast alloys have four main families of bronzes: copper–tin alloys (tin bronzes), copper–tin–lead alloys (leaded and high-leaded tin bronzes), copper–tin–nickel alloys (nickel–tin bronzes), and copper–aluminum alloys (aluminum bronzes). The family of alloys known as "manganese bronzes," in which zinc is the major alloying element, is included in the brasses, above.
Copper–nickels	These are alloys with nickel as the principal alloying element, with or without other designated alloying elements.
Copper–nickel–zinc alloys	Known commonly as "nickel silvers" and occasionally "German silvers", these are alloys that contain zinc and nickel as the principal and secondary alloying elements, with or without other designated elements.
Leaded coppers	These comprise a series of cast copper alloys with 20% or more lead, sometimes with a small amount of silver, but without tin or zinc.
Special alloys	Alloys whose chemical compositions do not fall into any of the above categories are combined in "special alloys."

shares properties such as high ductility and chemical stability. Copper's atomic structure is $1s^2 2s^2 2p^6 3s^2 3p^6 3d^{10} 4s^1$. Its filled 3d state and loosely bound 4s electrons and the optical transitions between filled 3d and empty 4s states are responsible for copper's high electrical and thermal conductivity and its distinctive red color.

Copper's atomic weight is 63.546. Almost all copper found in nature exists as one of two stable isotopes, ^{63}Cu (occurring 69.09%) and ^{65}Cu (30.91%). Unstable isotopes, ranging from ^{58}Cu to ^{68}Cu, are β-emitters, although naturally occurring copper is not thought of as radioactive.

Table 36[5,6] contains a compilation of physical properties for the pure metal. Most important among these from an engineering standpoint are its high electrical and thermal conductivities. Copper is, in fact, the standard for electrical conductivity against which all other metals are compared. The International Annealed Copper Standard (IACS) is defined as the volume conductivity of an annealed pure copper wire 1 m long, weighing 1 g, having a density of 8.89 g/cm^3 at 298 K (25°C) and a conductivity of exactly 58 m/Ω mm^2 (i.e.,

Table 2 Compositions of Wrought Coppers, UNS C10100–C12000

Copper Number	Designation	Description	Cu (including Ag), min.	Ag % (min.)	Ag Troy oz (min.)	As	Sb	P	Te	Other Named Elements[a]
C10100[b]	OFE	Oxgen-free electronic	99.99[c]	—	—	0.0005	0.0004	0.0003	0.0002	1 0.0005 O
C10200[b]	OF	Oxygen-free	99.95[d]	—	—	—	—	—	—	0.0010 O
C10300	OFXLP	Oxygen-free copper	99.95[e]	—	—	—	—	0.001–0.005	—	—
C10400[b]	OFS	Oxygen-free with Ag	99.95[d]	0.027	8	—	—	—	—	0.0010 O
C10500[b]	OFS	Oxygen-free with Ag	99.95[d]	0.034	10	—	—	—	—	0.0010 O
C10700[b]	OFS	Oxygen-free with Ag	99.95[d]	0.085	25	—	—	—	—	0.0010 O
C10800	OFLP	—	99.95[e]	—	—	—	—	0.005–0.012	—	—
C10910[b]	—	—	99.95[d]	—	—	—	—	—	—	0.005 O
C10920	—	—	99.90	—	—	—	—	—	—	0.02 O
C10930	—	—	99.90	0.044	13	—	—	—	—	0.02 O
C10940	—	—	99.90	0.085	25	—	—	—	—	0.02 O
C11000[b]	ETP	Electrolytic tough pitch	99.90	—	—	—	—	—	—	2
C11010[b]	RHC	Remelted high conducivity	99.90	—	—	—	—	—	—	2
C11020[b]	FRHC	Fire-Refined high conductivity	99.90	—	—	—	—	—	—	2
C11030[b]	CRTP	Chemically refined tough pitch	99.90	—	—	—	—	—	—	2
C11040[b]	—	—	99.90	—	—	0.0005	0.0004	—	0.0002	3
C11100[b]	ETP	Electronic tough pitch, anneal resistant	99.90	—	—	—	—	—	—	4
C11300[b]	STP	Tough pitch with Ag	99.90	0.027	8	—	—	—	—	2
C11400[b]	STP	Tough pitch with Ag	99.90	0.034	10	—	—	—	—	2
C11500[b]	STP	Tough pitch with Ag	99.90	0.054	16	—	—	—	—	2
C11600[b]	STP	Tough pitch with Ag	99.90	0.085	25	—	—	—	—	2
C11700	—	—	99.9[f]	—	—	—	—	0.04	—	0.004–0.02 B
C12000	DLP	Phosphorus-deoxidized, low residual P	99.90	—	—	—	—	0.004–0.012	—	—

[a] 1. The following additional maximum limits shall apply: Bi, 1 ppm (0.0001%); Cd, 1 ppm (0.0001%); Fe, 10 ppm (0.0010%); Pb, 5 ppm (0.0005%); Mn, 0.5 ppm (0.00005%); Ni, 10 ppm (0.0010%); Se, 3 ppm (0.0003%); Ag, 25 ppm (0.0025%); S, 15 ppm (0.0015%); Sn, 2 ppm (0.0002%); Zn, 1 ppm (0.0001%). 2. Oxygen and trace elements may vary depending on the process. 3. The following additional maximum limits shall apply: Se, 2 ppm (0.0002%); Bi, 1.0 ppm (0.00010%); group total, Te + Se + Bi, 3 ppm (0.0003%); Sn, 5 ppm (0.0005%); Pb, 5 ppm (0.0005%); Fe, 10 ppm (0.0010%); Ni, 10 ppm (0.0010%); S, 15 ppm (0.0015%); Ag, 25 ppm (0.0025%); O, 100–650 ppm (0.010–0.065%). The total maximum allowable of 65 ppm (0.0065%) does not include oxygen. 4. Small amounts of Cd or other elements may be added by agreement to improve the resistance to softening at elevated temperatures.

[b] This is a high-conductivity copper which has, in the annealed condition, a minimum concuctivity of 100% IACS except for alloy C10100, which has a minimum conductivity of 101% IACS.

[c] Cu is determined by the difference between the impurity total and 100%. The Cu value is exclusive of Ag.

[d] Cu is determined by the difference between the impurity total and 100%.

[e] Includes P.

[f] Includes B + P.

Table 3 Compositions of Wrought Coppers, UNS C12100–C14181

Copper Number	Designation	Description	Cu (including Ag), min.	Ag % (min.)	Ag Troy oz (min.)	As	Sb	P	Te	Other Named Elements
C12100	DLPS	Phosphorus deoxidized, low residual P	99.90	0.014	4	—	—	0.005–0.012	—	—
C12200[a]	DHP	Phosphorus deoxidized, high residual P	99.9	—	—	—	—	0.015–0.040	—	—
C12210	—	—	99.90	—	—	—	—	0.015–0.025	—	—
C12220	—	—	99.9	—	—	—	—	0.040–0.065	—	—
C12300	DHPS	Phosphorus deoxidized, high residual P	99.90	—	—	—	—	0.015–0.040	—	—
C12500	FRTP	Fire-refined tough pitch	99.88	—	—	0.012	0.003	—	0.025[b]	0.003 Bi, 0.004 Pb, 0.050 Ni
C12510	—	—	99.9	—	—	—	0.003	0.03	0.025[b]	0.005 Bi, 0.05 Fe, 0.020 Pb, 0.050 Ni, 0.05 Sn, 0.080 Zn
C12900	FRSTP	Fire-refined tough pitch with Ag	99.88	0.054	16	0.012	0.003	—	0.025[c]	0.003 Bi, 0.004 Pb, 0.050 Ni
C13100	—	—	99.8	—	—	—	—	—	—	—
C14180	—	—	99.90	—	—	—	—	0.075	—	0.01 Al, 0.02 Pb
C14181	—	—	99.90	—	—	—	—	0.002	—	0.002 Cd, 0.005 C, 0.002 Pb, 0.002 Zn

[a] This includes oxygen-free Cu which contains P in an amount agreed upon.
[b] 0.025 Te + Se.
[c] Includes Te + Se.

having a resistance of exactly 0.15328 Ω). This value is designated "100% IACS" and corresponds to a volume resistivity of 16.70 nΩ·m. The conductivity of copper produced today often exceeds that of metal available at the time the international standard was established, and it is not unusual for today's electrical and electronic grades of copper to exhibit IACS values of at least 101%.

Copper's electrical conductivity is conventionally given on a volume basis. In these terms, aluminum's conductivity is about 62% IACS, but because of the large density difference in density between the two metals, it is roughly twice that of copper on a weight basis.

Table 4 Compositions of Wrought Coppers, UNS C14200–C15500

Copper Number	Designation	Description	Cu (including Ag), min.	Ag % (min.)	Ag Troy oz (min.)	As	Sb	P	Te	Other Named Elements[a]
C14200	DPA	Phosphorus deoxidized, arsenical	99.4	—	—	0.15–0.50	—	0.015–0.040	—	—
C14300	—	Cadmium copper, deoxidized	99.90[b]	—	—	—	—	—	—	0.05–0.15 Cd
C14410	—	—	99.90[c]	—	—	—	—	0.005–0.020	—	0.05 Fe, 0.05 Pb, 0.10–0.20 Sn
C14415	—	—	99.96[c]	—	—	—	—	—	—	0.10–0.15 Sn
C14420	—	—	99.90[d]	—	—	—	—	—	0.005–0.05	0.04–0.15 Sn
C14500[e]	PTE	Tellurium bearing	99.90[f]	—	—	—	—	0.004–0.012	0.40–0.7	—
C14510	—	Tellurium bearing	99.85[f]	—	—	—	—	0.010–0.030	0.30–0.7	0.05 Pb
C14520	DPTE	Phosphorus deoxidized, tellurium bearing	99.90[f]	—	—	—	—	0.004–0.020	0.40–0.7	—
C14530	—	—	99.90[c]	—	—	—	—	0.001–0.010	0.003–0.023[g]	1 0.003–0.023 Sn
C14700[e]	—	Sulfur bearing	99.90[h]	—	—	—	—	0.002–0.005[h]	—	2 0.20–0.50 S
C15000[i]	—	Zirconium copper	99.80	—	—	—	—	—	—	0.10–0.20 Zr
C15100[i]	—	—	99.80	—	—	—	—	—	—	0.05–0.15 Zr
C15150	—	—	99.90	—	—	—	—	—	—	0.015–0.030 Zr
C15500	—	—	99.75	0.027–0.10	8–30	—	—	0.040–0.080	—	0.08–0.13 Mg

[a] 1. Includes Te + Se. 2. Includes Cu + S + P.
[b] Includes Cd. Deoxidized with Li or other suitable elements as agreed upon.
[c] Includes Cu + Ag + Sn.
[d] Includes Te + Sn.
[e] Includes oxygen-free or deoxidized grades with deoxidizers (such as phosphorus, boron, lithium, or others) in an amount agreed upon.
[f] Includes Te.
[g] See note 1 in footnote *a*.
[h] See note 2 in footnote *a*.
[i] Cu + sum of named elements, 99.9% min.

Table 5 Compositions of Wrought Coppers, UNS C15715–C15815

Copper Number	Designation	Description	Cu (including Ag), min.	Al[a]	Fe	Pb	Oxygen[a]	Boron	Other Named Elements
C15715	—	—	99.62	0.13–0.17	0.01	0.01	0.12–0.19	—	—
C15720	—	—	99.52	0.18–0.22	0.01	0.01	0.16–0.24	—	—
C15725	—	—	99.43	0.23–0.27	0.01	0.01	0.20–0.28	—	—
C15760	—	—	98.77	0.58–0.62	0.01	0.01	0.52–0.59	—	—
C15815	—	—	97.82	0.13–0.17	0.01	0.01	0.19	1.2–1.8	—

[a] All aluminum present as Al_2O_3; 0.04% oxygen present as Cu_2O with a negligible amount in solid solution with copper.

The lighter metal is thus used for overhead transmission lines while copper is preferred for high-voltage underground lines to conserve space inside conduits and where reliable connectivity and corrosion resistance are important.

Copper's high thermal conductivity is useful in applications such as heat exchangers and thermal buffers in superconducting cables (copper itself is not a superconductor). Thermal conductivity decreases only about 7% between room temperature and the melting point. Alloying reduces thermal conductivity but improves strength and corrosion resistance.

3.2 Copper Alloys

Physical properties of wrought and cast copper alloys are listed in Tables 37A and 37B, respectively. From an engineering design standpoint, the most often sought physical properties are electrical and thermal conductivities, as is also the case with pure copper. Because these properties are based on electronic and lattice structures, both of which are affected by the presence of solute atoms, conductivity is strongly influenced by the presence of alloying (or impurity) elements. The effect varies widely with both element(s) in question and with their concentration. Metals such as silver, lead, cadmium, and zinc, for example, have a small influence at dilute concentrations (<0.10 at. %), although their effect at high concentrations is larger. Other elements, notably phosphorus, silicon, arsenic, and tin, are more potent at all concentrations. The magnitude of the effect also depends strongly on the state of the alloying element, i.e., whether it exists in solid solution (strongest effect) or is present in a combined form, such as an oxide or intermetallic compound.

However, it is useful to point out that copper alloys are noteworthy for their many *combinations* of beneficial physical and mechanical (as well as chemical, e.g., corrosion) properties. Alloying generally serves to improve mechanical properties and often has a positive influence on corrosion behavior; thus the selection of a copper alloy for a given application often involves a reasonable compromise among the various properties required. For example, zinc, tin, phosphorus, and beryllium degrade electrical conductivity while markedly increasing strength, yet brass (a copper–zinc alloy), phosphor bronzes (copper–tin–phosphorus), and beryllium coppers are widely used as electrical springs and connectors. Electrical conductivity in these alloys, while lower than that of pure copper, is nevertheless sufficiently high for the intended application, while the alloys' high mechanical strength ensures the required degree of contact force.

Table 6 Compositions of Wrought High-Copper Alloys, UNS C16200–C18090

Copper Alloy Number	Previous Trade Name	Cu (including Ag)	Fe	Sn	Ni	Co	Cr	Si	Be	Pb	Other Named Elements
C16200[a]	Cadmium copper	Rem.	0.02	—	—	—	—	—	—	—	0.7–1.2 Cd
C16500[a]	—	Rem.	0.02	0.50–0.7	—	—	—	—	—	—	0.6–1.0 Cd
C17000[a]	Beryllium copper	Rem.	—	—	—	0.20 min.[b]	—	0.20	1.60–1.79	—	0.20 Al
C17200[a]	Beryllium copper	Rem.	—	—	—	0.20 min.[b]	—	0.20	1.80–2.00	—	0.20 Al
C17300[a]	Beryllium copper	Rem.	—	—	—	0.20 min.[b]	—	0.20	1.80–2.00	0.20–0.6	0.20 Al
C17410[a]	Beryllium copper	Rem.	0.20	—	—	0.35–0.6	—	0.20	0.15–0.50	—	0.20 Al
C17450[a]	Beryllium copper	Rem.	0.20	0.25	0.50–1.0	—	—	0.20	0.15–0.50	—	0.20 Al, 0.50 Zr
C17455[a]	Beryllium copper	Rem.	0.20	0.25	0.50–1.0[c]	—	—	0.20	0.15–0.50	0.20–0.6	0.20 Al, 0.50 Zr
C17460[a]	Beryllium copper	Rem.	0.20	0.25	1.0–1.4	—	—	0.20	0.15–0.50	—	0.20 Al, 0.50 Zr
C17465[a]	Beryllium copper	Rem.	0.20	0.25	1.0–1.4[c]	—	—	0.20	0.15–0.50	0.20–0.6	0.20 Al, 0.50 Zr
C17500[a]	Beryllium copper	Rem.	0.10	—	—	2.4–2.7	—	0.20	0.4–0.7	—	0.20 Al

C17510[a]	Beryllium copper	Rem.	0.10	—	1.4–2.2	0.3	—	0.20	0.2–0.6	—	0.20 Al
C17530[a]	Beryllium copper	Rem.	0.20	—	1.8–2.5[c]	—	—	0.20	0.20–0.40	—	0.6 Al
C18000[a]	—	Rem.	0.15	—	1.8–3.0[d]	—	0.10–0.8	0.40–0.8	—	—	—
C18020[e]	—	Rem.	—	0.05–0.25	—	—	0.10–0.30	0.05	—	—	0.10–0.30 Zn
C18030[e]	—	Rem.	—	0.08–0.12	—	—	0.10–0.20	—	—	—	0.005–0.015 P
C18040[f]	—	Rem.	—	0.20–0.30	—	—	0.25–0.35	—	—	—	0.005–0.015 P, 0.05–0.15 Zn
C18045[e]	—	99.1 min.	—	0.20–0.30	—	—	0.20–0.35	0.05	—	—	0.15–0.30 Zn
C18050[g]	—	Rem.	—	—	—	—	0.05–0.15	—	—	—	0.005–0.015 Te
C18070[g]	—	99.0 min.	—	—	—	—	0.15–0.40	0.02–0.07	—	—	0.01–0.40 Ti
C18080[g]	—	Rem.	0.02–0.20	—	—	—	0.20–0.7	0.01–0.10	—	—	0.01–0.30 Ag, 0.01–0.15 Ti
C18090[h]	—	96.0 min.	—	0.50–1.2	0.30–1.2	—	0.20–1.0	—	—	—	0.15–0.8 Ti

Note: "Rem." = Remainder.

[a] Cu + sum of named elements, 99.5% min.

[b] Ni + Co, 0.20% min.: Ni + Fe + Co, .6% max.

[c] Ni value includes Co.

[d] Includes Co.

[e] Cu + sum of named elements, 99.9% min.

[f] Includes oxygen-free or deoxidized grades with deoxidizers (such as phosphorus, boron, lithium, or others) in an amount agreed upon.

[g] Cu + sum of named elements, 99.8% min.

[h] Cu + sum of named elements, 99.85% min.

Table 7 Compositions of Wrought High-Copper Alloys, UNS C18100–C188-35

Copper Alloy Number	Previous Trade Name	Cu (including Ag)	Fe	Sn	Ni	Co	Cr	Si	Be	Pb	Other Named Elements
C18100[a]	—	98.7 min.	—	—	—	—	0.40–1.2	—	—	—	0.03–0.06 Mg, 0.08–0.20 Zr
C18135[a]	—	Rem.	—	—	—	—	0.20–0.6	—	—	—	0.20–0.6 Cd
C18140[a]	—	Rem.	—	—	—	—	0.15–0.45	0.005–0.05	—	—	0.05–0.25 Zr
C18145[a]	—	Rem.	—	—	—	—	0.10–0.30	—	—	—	0.10–0.30 Zn, 0.05–0.15 Zr
C18150[b]	—	Rem.	—	—	—	—	0.50–1.5	—	—	—	0.05–0.25 Zr
C18200[a]	Chromium copper	Rem.	0.10	—	—	—	0.6–1.2	0.10	—	0.05	—
C18400[a]	Chromium copper	Rem.	0.15	—	—	—	0.40–1.2	0.10	—	—	0.005 As, 0.005 Ca, 0.05 Li, 0.05 P, 0.7 Zn
C18600[a]	—	Rem.	0.25–0.8	—	0.25	0.10	0.10–1.0	—	—	—	0.05–0.50 Ti, 0.05–0.40 Zr
C18610[a]	—	Rem.	0.10	—	0.25	0.25–0.8	0.10–1.0	—	—	—	0.05–0.50 Ti, 0.05–0.40 Zr
C18661[a]	—	Rem.	0.10	0.20	—	—	—	—	—	—	0.10–0.7 Mg, 0.001–0.02 P
C18665	—	99.0 min.	—	—	—	—	—	—	—	—	0.40–0.9 Mg, 0.002–0.04 P
C18700	Free-machining Cu	99.5 min.[c]	—	—	—	—	—	—	—	0.8–1.5	—
C18835	—	99.0 min.[a]	0.10	0.15–0.55	—	—	—	—	—	0.05	0.01 P, 0.30 Zn

Note: "Rem." = Remainder.
[a] Cu + sum of named elements, 99.5% min.
[b] Cu + sum of named elements, 99.7% min.
[c] Includes Pb.

Table 8 Compositions of Wrought High-Copper Alloys, UNS

Copper Alloy Number	Previous Trade Name	Cu (including Ag)	Fe	Sn	Ni	Co	Cr	Si	Be	Pb	Other Named Elements
C18100[a]	—	98.7 min.	—	—	—	—	0.40–1.2	—	—	—	0.03–0.06 Mg, 0.08–0.20 Zr
C18135[a]	—	Rem.	—	—	—	—	0.20–0.6	—	—	—	0.20–0.6 Cd
C18140[a]	—	Rem.	—	—	—	—	0.15–0.45	0.005–0.05	—	—	0.05–0.25 Zr
C18145[a]	—	Rem.	—	—	—	—	0.10–0.30	—	—	—	0.10–0.30 Zn, 0.05–0.15 Zr
C18150[b]	—	Rem.	—	—	—	—	0.50–1.5	—	—	—	0.05–0.25 Zr
C18200[a]	Chromium copper	Rem.	0.10	—	—	—	0.6–1.2	0.10	—	0.05	—
C18400[a]	Chromium copper	Rem.	0.15	—	—	—	0.40–1.2	0.10	—	—	0.005 As, 0.005 Ca, 0.05 Li, 0.05 P, 0.7 Zn
C18600[a]	—	Rem.	0.25–0.8	—	0.25	0.10	0.10–1.0	—	—	—	0.05–0.50 Ti, 0.05–0.40 Zr
C18610[a]	—	Rem.	0.10	—	0.25	0.25–0.8	0.10–1.0	—	—	—	0.05–0.50 Ti, 0.05–0.40 Zr
C18661[a]	—	Rem.	0.10	0.20	—	—	—	—	—	—	0.10–0.7 Mg, 0.001–0.02 P
C18665	—	99.0 min.	—	—	—	—	—	—	—	—	0.40–0.9 Mg, 0.002–0.04 P
C18700	Free-machining Cu	99.5 min.[c]	—	—	—	—	—	—	—	0.8–1.5	—
C18835	—	99.0 min.[a]	0.10	0.15–0.55	—	—	—	—	—	0.05	0.01 P, 0.30 Zn

Note: "Rem." = Remainder.
[a] Cu + sum of named elements, 99.5% min.
[b] Cu + sum of named elements, 99.7% min.
[c] Includes Pb.

4 MECHANICAL PROPERTIES OF COPPER AND COPPER ALLOYS

4.1 Strengthening Mechanisms

In the as-cast or annealed states, copper is relatively soft and quite ductile, one of the most readily formed of all engineering metals and easily capable of being rolled into sheet and cold drawn into fine wire and thin-walled tubes. Because copper exists in a single fcc crystal form at all temperatures below its melting point (Fig. 1), hardening (strengthening) can be achieved only through cold work or by grain refinement. Copper obeys the Hall–Petch $d^{1/2}$ proportionality with respect to grain size, although grain refinement per se is not commonly used for strengthening.

Alloying gives rise to one or more strengthening mechanisms depending on the constituent(s) added. For example, simple solid-solution hardening is produced by additions of nickel, which is one of several metals with unlimited solid solubility in copper. Silver, zinc, tin, and aluminum are also solid-solution strengtheners when added below their solid solubility limits. In such cases, the resulting alloy retains the face-centered α-phase of the pure metal. Higher concentrations of these elements lead to the formation of additional phases that also contribute to strengthening. Alloying elements such as chromium, zirconium, and beryllium exhibit retrograde solubility and are thus capable of producing precipitation hardening. In the case of beryllium, the effect is potent enough to raise tensile strengths above

Table 9 Compositions of Wrought High-Copper Alloys, UNS C

Copper Alloy Number	Previous Trade Name	Cu (including Ag)	Fe	Sn	Ni	Co	Cr	Si	Be	Pb	Other Named Elements
C18900[a]	—	Rem.	—	0.6–0.9	—	—	—	0.15–0.40	—	0.02	0.01 Al, 0.10–0.30 Mn, 0.05 P, 0.10 Zn
C18980[a]	—	98.0 min.	—	1.0	—	—	—	0.50	—	0.02	0.50 Mn, 0.15 P
C18990[b]	—	Rem.	—	1.8–2.2	—	—	0.10–0.20	—	—	—	0.005–0.015 P
C19000[a]	—	Rem.	0.10	—	0.9–1.3	—	—	—	—	0.05	0.15–0.35 P, 0.8 Zn
C19002[a]	—	Rem.	0.10	0.02–0.30	1.4–1.7[c]	—	—	0.20–0.35	—	0.05	0.01 Mg, 0.05 P, 0.02–0.50 Ag, 0.04–0.35 Zn, 0.005–0.05 Zr
C19010[a]	—	Rem.	—	—	0.8–1.8	—	—	0.15–0.35	—	—	0.01–0.05 P
C19015[d]	—	Rem.	—	—	0.50–2.4	—	—	0.10–0.40	—	—	0.02–0.15 Mg, 0.02–0.20 P
C19020[d]	—	Rem.	—	0.30–0.9	0.50–3.0	—	—	—	—	—	0.01–0.20 P
C19025[e]	—	Rem.	0.10	0.7–1.1	0.8–1.2	—	—	—	—	—	0.03–0.07 P, 0.20 Zn
C19030[e]	—	Rem.	0.10	1.0–1.5	1.5–2.0	—	—	—	—	0.02	0.01–0.03 P
C19100[a]	—	Rem.	0.20	—	0.9–1.3	—	—	—	—	0.10	0.15–0.35 P, 0.35–0.6 Te, 0.50 Zn
C19140[a]	—	Rem.	0.05	0.05	0.8–1.2	—	—	—	—	0.40–0.8	0.15–0.35 P, 0.50 Zn
C19150[a]	—	Rem.	0.05	0.05	0.8–1.2	—	—	—	—	0.50–1.0	0.15–0.35 P
C19160[a]	—	Rem.	0.05	0.05	0.8–1.2	—	—	—	—	0.8–1.2	0.15–0.35 P, 0.50 Zn

Note: "Rem." = Remainder.
[a] Cu + sum of named elements, 99.5% min.
[b] Cu + sum of named elements, 99.9% min.
[c] Ni value includes Co
[d] Cu + sum of named elements, 99.8% min.
[e] Cu + sum of named elements, 99.7% min.

200,000 psi (1379 MPa) and hardness above Rockwell C 40. Some high-strength aluminum bronzes undergo a diffusionless martensitic transformation similar to that in steel, although the phenomenon is not truly exploited for strengthening in these alloys.

Alloying with combinations of nickel and tin produces an order–disorder transformation capable of raising strength nearly to the levels achieved in copper–beryllium alloys. Like the beryllium coppers, these so-called spinodal alloys are used in applications such as electrical connectors, where favorable combinations of strength, formability, and electrical conductivity are needed.

4.2 Temper

In view of the complex strengthening mechanisms involved, it is easier to describe the copper alloys' mechanical/metallurgical state by the single term *temper.* That is, copper alloys are said to have a harder temper if they have been cold worked, precipitation heat treated, or

Table 10 Compositions of Wrought High-Copper Alloys, UNS C19200–C19900

Copper Alloy Number	Previous Trade Name	Cu	Fe	Sn	Zn	Al	Pb	P	Other Named Elements
C19200[a]	—	98.5 min.	0.8–1.2	—	0.20	—	—	0.01–0.04	—
C19210[a]	—	Rem.	0.05–0.15	—	—	—	—	0.025–0.04	—
C19215[a]	—	Rem.	0.05–0.20	—	1.1–3.5	—	—	0.025–0.050	—
C19220[a]	—	Rem.	0.10–0.30	0.05–0.10	—	—	—	0.03–0.07	0.005–0.015 B, 0.10–0.25 Ni
C19260[b]	—	98.5 min.	0.40–0.8	—	—	—	—	—	0.02–0.15 Mg, 0.20–0.40 Ti
C19280[a]	—	Rem.	0.50–1.5	0.30–0.7	0.30–0.7	—	—	0.005–0.015	—
C19400	—	97.0 min.	2.1–2.6	—	0.05–0.20	—	0.03	0.015–0.15	—
C19410[a]	—	Rem.	1.8–2.3	0.6–0.9	0.10–0.20	—	—	0.015–0.050	—
C19450[a]	—	Rem.	1.5–3.0	0.8–2.5	—	—	—	0.005–0.05	—
C19500[a]	—	96.0 min.	1.0–2.0	0.10–1.0	0.20	0.02	0.02	0.01–0.35	0.30–1.3 Co
C19520[a]	—	96.6 min.	0.50–1.5	—	—	—	0.01–3.5	—	—
C19700[a]	—	Rem.	0.30–1.2	0.20	0.20	—	0.05	0.10–0.40	0.05 Co, 0.01–0.20 Mg, 0.05 Mn, 0.05 Ni
C19710[c]	—	Rem.	0.05–0.40	0.20	0.20	—	0.05	0.07–0.15	0.03–0.06 Mg, 0.05 Mn, 0.10 Ni[d]
C19720[c]	—	Rem.	0.05–0.50	0.20	0.20	—	0.05	0.05–0.15	0.06–0.20 Mg, 0.05 Mn, 0.10 Ni[d]
C19750[a]	—	Rem.	0.35–1.2	0.05–0.40	0.20	—	0.05	0.10–0.40	0.05 Co, 0.01–0.20 Mg, 0.05 Mn, 0.05 Ni
C19800[a]	—	Rem.	0.02–0.50	0.10–1.0	0.30–1.5	—	—	0.01–0.10	0.10–1.0 Mg
C19810[a]	—	Rem.	1.5–3.0	—	1.0–5.0	—	—	0.10	0.10 Cr, 0.10 Mg, 0.10 Ti, 0.10 Zr
C19900[c]	—	Rem.	—	—	—	—	—	—	2.9–3.4 Ti

Note: "Rem." = Remainder.
[a] Cu + sum of named elements, 99.8% min.
[b] Cu + sum of named elements, 99.9% min.
[c] Cu + sum of named elements, 99.5% min.
[d] Ni value includes Co.

both and a softer temper when they are in an as-cast, as-wrought, or annealed state. Thus, when specifying copper alloys, it is necessary to call out the UNS designation as well as the product form (wire, rod, sheet, etc.) and temper. As usual, higher strength and hardness are gained at the cost of reduced ductility and, especially in the case of alloying, lower electrical conductivity. Tempers are defined in ASTM standard practice B 601, shown in Table 38.[7]

4.3 Mechanical Properties Data

Mechanical properties of selected coppers and copper alloys are listed in Tables 39 and 40. Note that the tables identify typical and/or minimum properties. Typical (nominal) properties

Table 11 Compositions of Wrought Copper-Zinc Alloys (Brasses), UNS C21000–C28000

Alloy Number	Previous Trade Name	Cu	Pb	Fe	Zn	Other Named Elements
C21000[a]	Gilding, 95%	94.0–96.0	0.05	0.05	Rem.	—
C22000[a]	Commercial bronze, 90%	89.0–91.0	0.05	0.05	Rem.	—
C22600[a]	Jewelry bronze, 87½%	86.0–89.0	0.05	0.05	Rem.	—
C23000[a]	Red brass, 85%	84.0–86.0	0.05	0.05	Rem.	—
C23030[a]	—	83.5–85.5	0.05	0.05	Rem.	0.20–0.40 Si
C23400[a]	—	81.0–84.0	0.05	0.05	Rem.	—
C24000[a]	Low brass, 80%	78.5–81.5	0.05	0.05	Rem.	—
C24080[a]	—	78.0–82.0	0.20	—	Rem.	0.10 Al
C25600[b]	—	71.0–73.0	0.05	0.05	Rem.	—
C26000[b]	Cartridge brass, 70%	68.5–71.5	0.07	0.05	Rem.	—
C26130[b]	—	68.5–71.5	0.05	0.05	Rem.	0.02–0.08 As
C26200[b]	—	67.0–70.0	0.07	0.05	Rem.	—
C26800[b]	Yellow brass, 66%	64.0–68.5	0.15	0.05	Rem.	—
C27000[b]	Yellow brass, 65%	63.0–68.5	0.10	0.07	Rem.	—
C27200[b]	—	62.0–65.0	0.07	0.07	Rem.	—
C27400[b]	Yellow brass, 63%	61.0–64.0	0.10	0.05	Rem.	—
C28000[b]	Muntz metal, 60%	59.0–63.0	0.30	0.07	Rem.	—

Note: "Rem." = Remainder.
[a] Cu + sum of named elements, 99.8% min.
[b] Cu + sum of named elements, 99.7% min.

may vary by as much as 5–10% in practice because of variations in processing and composition. They cannot be guaranteed unless by special arrangement between customer and supplier and should not be used for design or specification purposes, for which ASTM or other standards should instead be called out. Most data in Table 39 include annealed tempers plus representative hard or heat-treated tempers, as appropriate. Additional tempers are often available, and readers are encouraged to contact the Copper Development Association or its web site, www.copper.org, for all current data. Note also that yield strength is presented as either 0.2% offset or 0.5% extension under load, depending on yield behavior. Extension data are given for alloys whose stress–strain curves do not exhibit a well-defined yield point.

Fatigue strength in copper metals decreases to approximately one-half the static tensile strength in 10^8 cycles. The rate of decrease decreases with the number of cycles N and may become independent of N in some alloys, for which an absolute endurance ratio (fatigue strength at N cycles divided by static tensile strength) can be identified.

4.4 Effect of Temperature

Copper alloys used at elevated temperatures are selected on the basis of time-dependent deformation behavior. Slow deformation under constant load, or creep, is followed by failure, or stress rupture, which occurs at a stress considerably lower than short-term tensile strength. Conversely, applied stresses decrease when metals are held at constant strain at elevated temperatures. The phenomenon, known as stress relaxation, is an important design parameter in high-temperature bolts and electrical connectors.

High-conductivity coppers and leaded alloys are not chosen on the basis of elevated temperature properties. Nonleaded brasses and bronzes, chromium copper, and beryllium coppers have intermediate to high strengths at elevated temperatures and in some cases exhibit quite exceptional strength. Aluminum bronzes and copper–nickels have superior resistance to creep and stress rupture, while 90-10 copper–nickel, nickel–aluminum bronze,

Table 12 Compositions of Wrought Copper–Zinc–Lead Alloys (Leaded Brasses), UNS C30000–C39999

Copper Alloy Number	Previous Trade Name	Cu	Pb	Fe	Zn	Other Named Elements
C31200[a]	—	87.5–90.5	0.7–1.2	0.10	Rem.	0.25 Ni
C31400[a]	Leaded commercial bronze	87.5–90.5	1.3–2.5	0.10	Rem.	0.7 Ni
C31600[a]	Leaded commercial bronze (nickel bearing)	87.5–90.5	1.3–2.5	0.10	Rem.	0.7–1.2 Ni, 0.04–0.10 P
C32000[a]	Leaded red brass	83.5–86.5	1.5–2.2	0.10	Rem.	0.25 Ni
C33000[a]	Low-leaded brass (tube)	65.0–68.0	0.25–0.7	0.07	Rem.	—
C33200[a]	High-leaded brass (tube)	65.0–68.0	1.5–2.5	0.07	Rem.	—
C33500[a]	Low-leaded brass	62.0–65.0	0.25–0.7	0.15[b]	Rem.	—
C34000[a]	Medium-leaded brass, $64\frac{1}{2}$%	62.0–65.0	0.8–1.5	0.15[b]	Rem.	—
C34200[a]	High-leaded brass, $64\frac{1}{2}$%	62.0–65.0	1.5–2.5	0.15[b]	Rem.	—
C34500[a]	—	62.0–65.0	1.5–2.5	0.15	Rem.	—
C35000[a]	Medium-leaded brass, 62%	60.0–63.0[c]	0.8–2.0	0.15[b]	Rem.	—
C35300[d]	High-leaded brass, 62%	60.0–63.0[c]	1.5–2.5	0.15[b]	Rem.	—
C35330[d]	—	59.5–64.0	1.5–3.5[e]	—	Rem.	0.02–0.25 As
C35600[d]	Extra high leaded brass	60.0–63.0	2.0–3.0	0.15[b]	Rem.	—
C36000[d]	Free-cutting brass	60.0–63.0	2.5–3.7	0.35	Rem.	—
C36500[a]	Leaded Muntz metal, uninhibited	58.0–61.0	0.25–0.7	0.15	Rem.	0.25 Sn
C37000[a]	Free-cutting Muntz metal	59.0–62.0	0.8–1.5	0.15	Rem.	—
C37100[a]	—	58.0–62.0	0.6–1.2	0.15	Rem.	—
C37700[d]	Forging brass	58.0–61.0	1.5–2.5	0.30	Rem.	—
C37710[d]	—	56.5–60.0	1.0–3.0	0.30	Rem.	—
C38000[d]	Architectural bronze, low leaded	55.0–60.0	1.5–2.5	0.35	Rem.	0.50 Al, 0.30 Sn
C38500[d]	Architectural bronze	55.0–59.0	2.5–3.5	0.35	Rem.	—

Note: "Rem." = Remainder.
[a]Cu + sum of named elements, 99.6% min.
[b]For flat products, the iron shall be 0.10% max.
[c]Cu, 61.0% min. for rod.
[d]Cu + sum of named elements, 99.5% min.
[e]Pb may be reduced to 1.0% by agreement.

and 70-30 copper–nickel, based on their rated strengths in 100,000-h stress-rupture tests, are the most suitable alloys for use at temperatures above 662°F (350°C).

The ductility of copper and its alloys generally does not decrease with decreasing temperature, exceptions being alloys containing high-volume fractions of β-phase, which may show some effect. Charpy V-notch impact strengths of six copper alloys are listed as functions of temperature in Table 41. On the other hand, the ductility of β-containing alloys at elevated temperatures is useful in hot forging, extrusion, and rolling operations.

5 CORROSION BEHAVIOR

Copper and its alloys derive their inherent corrosion resistance from their tendency to form stable, adherent corrosion product films—tarnishes and patinas—when exposed to corrosive environments. Once established and left undisturbed, such films sharply limit the progress of corrosion reactions.

Copper and brass tubes, valves, and fittings are widely used for freshwater handling. Stable corrosion films typically form and coat the insides of the tubes. Catastrophic corrosion is not encountered in properly treated waters, although copper and low-zinc brasses are

Table 13 Compositions of Wrought Copper–Zinc–Tin Alloys (Tin Brasses), UNS C40000–C49999

Copper Alloy Number	Previous Trade Name	Cu	Pb	Fe	Sn	Zn	P	Other Named Elements
C40400[a]	—	Rem.	—	—	0.35–0.7	2.0–3.0	—	—
C40500[a]	Penny bronze	94.0–96.0	0.05	0.05	0.7–1.3	Rem.	—	—
C40810[a]	—	94.5–96.5	0.05	0.08–0.12	1.8–2.2	Rem.	0.028–0.04	0.11–0.20 Ni
C40820[b]	—	94.0 min.	0.02	—	1.0–2.5	0.20–2.5	0.05	0.10–0.50 Ni
C40850[a]	—	94.5–96.5	0.05	0.05–0.20	2.6–4.0	Rem.	0.01–0.20	0.05–0.20 Ni
C40860[a]	—	94.0–96.0	0.05	0.01–0.05	1.7–2.3	Rem.	0.02–0.04	0.05–0.20 Ni
C41000[a]	—	91.0–93.0	0.05	0.05	2.0–2.8	Rem.	—	—
C41100[a]	—	89.0–92.0	0.10	0.05	0.30–0.7	Rem.	—	—
C41120[a]	—	89.0–92.0	0.05	0.05–0.20	0.30–0.7	Rem.	0.01–0.35	0.05–0.20 Ni
C41300[a]	—	89.0–93.0	0.10	0.05	0.7–1.3	Rem.	—	—
C41500[a]	—	89.0–93.0	0.10	0.05	1.5–2.2	Rem.	—	—
C42000[a]	—	88.0–91.0	—	—	1.5–2.0	Rem.	0.25	—
C42200[a]	—	86.0–89.0	0.05	0.05	0.8–1.4	Rem.	0.35	—
C42220[a]	—	88.0–91.0	0.05	0.05–0.20	0.7–1.4	Rem.	0.02–0.05	0.05–0.20 Ni
C42500[a]	—	87.0–90.0	0.05	0.05	1.5–3.0	Rem.	0.35	—
C42520[a]	—	88.0–91.0	0.05	0.05–0.20	1.5–3.0	Rem.	0.01–0.20	0.05–0.20 Ni
C42600[a]	—	87.0–90.0	0.05	0.05–0.20	2.5–4.0	Rem.	0.01–0.20	0.05–0.20 Ni[c]
C43000[a]	—	84.0–87.0	0.10	0.05	1.7–2.7	Rem.	—	—
C43400[a]	—	84.0–87.0	0.05	0.05	0.40–1.0	Rem.	—	—
C43500[a]	—	79.0–83.0	0.10	0.05	0.6–1.2	Rem.	—	—
C43600[a]	—	80.0–83.0	0.05	0.05	0.20–0.50	Rem.	—	—
C44250[d]	—	73.0–76.0	0.07	0.20	0.50–1.5	Rem.	0.10	0.20 Ni
C44300[d]	Admiralty, arsenical	70.0–73.0	0.07	0.06	0.8–1.2[e]	Rem.	—	0.02–0.06 As
C44400[d]	Admiralty, antimonial	70.0–73.0	0.07	0.06	0.8–1.2[e]	Rem.	—	0.02–0.10 Sb
C44500[d]	Admiralty, phosphorized	70.0–73.0	0.07	0.06	0.8–1.2[e]	Rem.	0.02–0.10	—
C46200[d]	Naval brass, 63½%	62.0–65.0	0.20	0.10	0.50–1.0	Rem.	—	—
C46400[d]	Naval brass, uninhibited	59.0–62.0	0.20	0.10	0.50–1.0	Rem.	—	—
C46500[d]	Naval brass, arsenical	59.0–62.0	0.20	0.10	0.50–1.0	Rem.	—	0.02–0.06 As
C47000[d]	Naval brass welding and brazing rod	57.0–61.0	0.05	—	0.25–1.0	Rem.	—	0.01 Al
C47940[d]	—	63.0–66.0	1.0–2.0	0.10–1.0	1.2–2.0	Rem.	—	0.10–0.50 Ni[c]
C48200[d]	Naval brass, medium leaded	59.0–62.0	0.40–1.0	0.10	0.50–1.0	Rem.	—	—
C48500[d]	Naval brass, high leaded	59.0–62.0	1.3–2.2	0.10	0.50–1.0	Rem.	—	—
C48600[d]	—	59.0–62.0	1.0–2.5	—	0.30–1.5	Rem.	—	0.02–0.25 As

Note: "Rem." = Remainder.
[a] Cu + sum of named elements, 99.7% min.
[b] Cu + sum of named elements, 99.5% min.
[c] Ni value includes Co
[d] Cu + sum of named elements, 99.6% min.
[e] For tubular products, the minimum Sn content may be 0.9%.

Table 14 Compositions of Wrought Copper–Tin–Phosphorus Alloys (Phosphor Bronzes), UNS C50000–C52999

Copper Alloy Number	Previous Trade Name	Cu	Pb	Fe	Sn	Zn	P	Other Named Elements
C50100[a]	—	Rem.	0.05	0.05	0.50–0.8	—	0.01–0.05	—
C50200[a]	—	Rem.	0.05	0.10	1.0–1.5	—	0.04	—
C50500[a]	Phosphor bronze, 1.25% E	Rem.	0.05	0.10	1.0–1.7	0.30	0.03–0.35	—
C50510[b]	—	Rem.	—	—	1.0–1.5	0.10–0.25	0.02–0.07	0.15–0.40 Ni
C50580[a]	—	Rem.	0.05	0.05–0.20	1.0–1.7	0.30	0.01–0.35	0.05–0.20 Ni
C50590[a]	—	97.0 min.	0.02	0.05–0.40	0.50–1.5	0.50	0.02–0.15	—
C50700[a]	—	Rem.	0.05	0.10	1.5–2.0	—	0.30	—
C50705[a]	—	96.5 min.	0.02	0.10–0.40	1.5–2.0	0.50	0.04–0.15	—
C50710[a]	—	Rem.	—	—	1.7–2.3	—	0.15	0.10–0.40 Ni
C50715[c]	—	Rem.	0.02	0.05–0.15	1.7–2.3	—	0.025–0.04	—
C50725[a]	—	94.0 min.	0.02	0.05–0.20	1.5–2.5	1.5–3.0	0.02–0.06	—
C50780[a]	—	Rem.	0.05	0.05–0.20	1.7–2.3	0.30	0.01–0.35	0.05–0.20 Ni
C50900[a]	—	Rem.	0.05	0.10	2.5–3.8	0.30	0.03–0.30	—
C51000[a]	Phosphor bronze, 5% A	Rem.	0.05	0.10	4.2–5.8	0.30	0.03–0.35	—
C51080[a]	—	Rem.	0.05	0.05–0.20	4.8–5.8	0.30	0.01–0.35	0.05–0.20 Ni
C51100[a]	—	Rem.	0.05	0.10	3.5–4.9	0.30	0.03–0.35	—
C51180[a]	—	Rem.	0.05	0.05–0.20	3.5–4.9	0.30	0.01–0.35	0.05–0.20 Ni
C51190[a]	—	Rem.	0.02	0.05–0.15	3.0–6.5	—	0.025–0.045	0.15 Co
C51800[a]	Phosphor bronze	Rem.	0.02	—	4.0–6.0	—	0.10–0.35	0.01 Al
C51900[a]	—	Rem.	0.05	0.10	5.0–7.0	0.30	0.03–0.35	—
C51980[a]	—	Rem.	0.05	0.05–0.20	5.5–7.0	0.30	0.01–0.35	0.05–0.20 Ni
C52100[a]	Phospher bronze, 8% C	Rem.	0.05	0.10	7.0–9.0	0.20	0.03–0.35	—
C52180[a]	—	Rem.	0.05	0.05–0.20	7.0–9.0	0.30	0.01–0.35	0.05–0.20 Ni
C52400[a]	Phospher bronze, 10% D	Rem.	0.05	0.10	9.0–11.0	0.20	0.03–0.35	—
C52480[a]	—	Rem.	0.05	0.05–0.20	9.0–11.0	0.30	0.01–0.35	0.05–0.20 Ni

Note: "Rem." = Remainder.
[a] Cu + sum of named elements, 99.5% min.
[b] Cu + sum of named elements, 99.7% min.
[c] Cu + Sn + Fe + P, 99.5% min.

Table 15 Compositions of Wrought Copper–Tin–Lead–Phosphorus Alloys (Leaded Phosphor Bronzes), UNS C53000–C54999

Copper Alloy Number	Previous Trade Name	Cu	Pb	Fe	Sn	Zn	P	Other Named Elements
C53400[a]	Phosphor bronze B-1	Rem.	0.8–1.2	0.10	3.5–5.8	0.30	0.03–0.35	—
C53800[b]	—	Rem.	0.40–0.6	0.030	13.1–13.9	0.12	—	0.06 Mn, 0.03 Ni[c]
C54400[a]	Phosphor bronze B-2	Rem.	3.5–4.5	0.10	3.5–4.5	1.5–4.5	0.01–0.50	—

Note: "Rem." = Remainder.
[a] Cu + sum of named elements, 99.5% min.
[b] Cu + sum of named elements, 99.8% min.
[c] Ni value includes Co.

Table 16 Compositions of Wrought Copper–Phosphorus and Copper–Silver–Phosphorus Alloys (Brazing Alloys), UNS C55000–C55299

Copper Alloy Number[a]	Previous Trade Name	Cu	Ag	P	Other Named Elements
C55180	—	Rem.	—	4.8–5.2	—
C55181	—	Rem.	—	7.0–7.5	—
C55280	—	Rem.	1.8–2.2	6.8–7.2	—
C55281	—	Rem.	4.8–5.2	5.8–6.2	—
C55282	—	Rem.	4.8–5.2	6.5–7.0	—
C55283	—	Rem.	5.8–6.2	7.0–7.5	—
C55284	—	Rem.	14.5–15.5	4.8–5.2	—

Note: "Rem." = Remainder.

[a] Cu + sum of named elements, 99.85% min.

Table 17 Compositions of Copper–Silicon Alloys (Silicon Bronzes and Silicon Brasses) C64700–C66199

Copper Alloy Number[a]	Previous Trade Name	Cu (including Ag)	Pb	Fe	Sn	Zn	Mn	Si	Ni	Other Named Elements
C64700	—	Rem.	0.10	0.10	—	0.50	—	0.40–0.8	1.6–2.2[b]	—
C64710	—	95.0 min.	—	—	—	0.20–0.50	0.10	0.50–0.9	2.9–3.5[b]	—
C64725	—	95.0 min.	0.01	0.25	0.20–0.8	0.50–1.5	—	0.20–0.8	1.3–2.7[b]	0.01 Ca, 0.20 Cr, 0.20 Mg
C64730	—	93.5 min.	—	—	1.0–1.5	0.20–0.50	0.10	0.50–0.9	2.9–3.5[b]	—
C64740	—	95.0 min.	0.01½	0.25	1.5–2.5	0.20–1.0	—	0.05–0.50	1.0–2.0[b]	0.01 Ca, 0.05 Mg
C64750[c]	—	Rem.	—	1.0	0.05–0.8	1.0	—	0.10–0.7	1.0–3.0	0.10 Mg, 0.10 P, 0.10 Zr
C64760	—	93.5 min.	0.02½	—	0.30	0.20–2.5	—	0.05–0.6	0.40–2.5	0.05 Mg
C64770	—	Rem.	0.05½	0.10	0.05–0.50	0.30–0.8	0.10	0.40–0.8	1.5–3.0[b]	0.30 Mg
C64780	—	90.0 min.	0.02½	—	0.10–2.0	0.20–2.5	0.01–1.0	0.20–0.9	1.0–3.5	0.01 Cr, 0.01 Mg, 0.01 Ti, 0.01 Zr
C64785	—	Rem.	0.015	0.02	0.50–2.0	3.0–6.0	0.20–1.0	0.15	0.40–1.6[d]	3.0–6.0 Al, 0.015 P
C64790	—	Rem.	0.05	0.10	0.05–0.50	0.30–0.8	0.10	0.6–1.2	2.5–4.5[b]	0.05–0.50 Cr, 0.05–0.30 Mg
C64900	—	Rem.	0.05	0.10	1.2–1.6	0.20	—	0.8–1.2	0.10[b]	0.10 Al
C65100	Low-silicon bronze B	Rem.	0.05	0.8	—	1.5	0.7	0.8–2.0	—	—
C65400	—	Rem.	0.05	—	1.2–1.9	0.50	—	2.7–3.4	—	0.01–0.12 Cr
C65500	High-silicon bronze A	Rem.	0.05	0.8	—	1.5	0.50–1.3	2.8–3.8	0.6[b]	—
C65600	—	Rem.	0.02	0.50	1.5	1.5	1.5	2.8–4.0	—	0.01 Al
C66100	—	Rem.	0.20–0.8	0.25	—	1.5	1.5	2.8–3.5	—	—

Note: "Rem." = Remainder.

[a] Cu + sum of named elements, 99.5% min., except as noted.

[b] Ni value includes Co.

[c] Cu + sum of named elements, 99.92% min.

[d] Not including Co.

Table 18 Compositions of Other Wrought Copper–Zinc Alloys, UNS C66200–C69999

Copper Alloy Number[a]	Previous Trade Name	Cu (including Ag)	Pb	Fe	Sn	Zn	Ni	Al	Mn	Si	Other Named Elements
C66200	—	86.6–91.0	0.05	0.05	0.20–0.7	Rem.	0.30–1.0[b]	—	—	—	0.05–0.20 P
C66300	—	84.5–87.5	0.05	1.4–2.4[c]	1.5–3.0	Rem.	—	—	—	—	0.20 Co,[c] 0.35 P
C66400	—	Rem.	0.015	1.3–1.7[d]	0.05	11.0–12.0	—	—	—	—	0.30–0.7 Co[d]
C66410	—	Rem.	0.015	1.8–2.3	0.05	11.0–12.0	—	—	—	—	—
C66420	—	Rem.	—	0.50–1.5	—	12.7–17.0	—	—	—	—	—
C66430	—	Rem.	0.05	0.6–0.9	0.6–0.9	13.0–15.0	—	—	—	—	0.10 P
C66700	Manganese brass	68.5–71.5	0.07	0.10	—	Rem.	—	—	0.8–1.5	—	—
C66800	—	60.0–63.0	0.50	0.35	0.30	Rem.	0.25[b]	0.25	2.0–3.5	0.50–1.5	—
C66900[e]	—	62.5–64.5	0.05	0.25	—	Rem.	—	—	11.5–12.5	—	—
C66950	—	Rem.	0.01	0.50	—	14.0–15.0	—	1.0–1.5	14.0–15.0	—	—
C67000	Manganese bronze B	63.0–68.0	0.20	2.0–4.0	0.50	Rem.	—	3.0–6.0	2.5–5.0	—	—
C67300	—	58.0–63.0	0.40–3.0	0.50	0.30	Rem.	0.25[b]	0.25	2.0–3.5	0.50–1.5	—
C67400	—	57.0–60.0	0.50	0.35	0.30	Rem.	0.25[b]	0.50–2.0	2.0–3.5	0.50–1.5	—
C67420	—	57.0–58.5	0.25–0.8	0.15–0.55	0.35	Rem.	0.25[b]	1.0–2.0	1.5–2.5	0.25–0.7	—
C67500	Manganese bronze A	57.0–60.0	0.20	0.8–2.0	0.50–1.5	Rem.	—	0.25	0.05–0.50	—	—
C67600	—	57.0–60.0	0.50–1.0	0.40–1.3	0.50–1.5	Rem.	—	—	0.05–0.50	—	—
C68000	Bronze, low fuming (nickel)	56.0–60.0	0.05	0.25–1.25	0.75–1.10	Rem.	0.20–0.8[b]	0.01	0.01–0.50	0.04–0.15	—
C68100	Bronze, low fuming	56.0–60.0	0.05	0.25–1.2	0.75–1.10	Rem.	—	0.01	0.01–0.50	0.04–0.15	—
C68700	Aluminum brass, arsenical	76.0–79.0	0.07	0.06	—	Rem.	—	1.8–2.5	—	—	0.02–0.06 As
C68800	—	Rem.	0.05	0.20	—	21.3–24.1[f]	—	3.0–3.8[f]	—	—	0.25–0.55 Co
C69050	—	70.0–75.0	—	—	—	Rem.	0.50–1.5[b]	3.0–4.0	—	0.10–0.6	0.01–0.20 Zr
C69100	—	81.0–84.0	0.05	0.25	0.10	Rem.	0.8–1.4[b]	0.7–1.2	0.10	0.8–1.3	—
C69300	—	73.0–77.0	0.10	0.10	0.20	Rem.	0.10[b]	—	0.10	2.7–3.4	0.04–0.15 P
C69400	Silicon red brass	80.0–83.0	0.30	0.20	—	Rem.	—	—	—	3.5–4.5	—
C69430	—	80.0–83.0	0.30	0.20	—	Rem.	—	—	—	3.5–4.5	0.03–0.06 As
C69700	—	75.0–80.0	0.50–1.5	0.20	—	Rem.	—	—	0.40	2.5–3.5	—
C69710	—	75.0–80.0	0.50–1.5	0.20	—	Rem.	—	—	0.40	2.5–3.5	0.03–0.06 As
C69750	—	78.0–83.0	0.8–1.3	0.05	0.05	Rem.	0.01				

Note: "Rem." = Remainder.

[a] Cu + sum of named elements, 99.5% min., except as noted.

[b] Ni value includes Co.

[c] Fe + Co, 1.4–2.4%

[d] Fe + Co, 1.8–2.3%.

[e] Cu + sum of named elements, 99.8% min.

[f] Al + Zn, 25.1–27.1%.

Table 19 Compositions of Wrought Copper–Nickel Alloys (Copper–Nickels)

Copper Alloy Number	Previous Trade Name	Cu (including Ag)	Pb	Fe	Zn	Ni	Sn	Mn	Other Named Elements[a]
C70100[b]	—	Rem.	—	0.05	0.25	3.0–4.0[c]	—	0.50	—
C70200[b]	—	Rem.	0.05	0.10	—	2.0–3.0[c]	—	0.40	—
C70230[b]	—	Rem.	—	—	0.50–2.0	2.2–3.2	0.10–0.50	—	1, 0.40–0.8 Si, 0.10 Ag
C70240[b]	—	Rem.	0.05	0.10	0.30–0.8	1.0–4.0[c]	—	0.01–0.20	0.40–0.8 Si, 0.01–0.10 Ag
C70250[b]	—	Rem.	0.05	0.20	1.0	2.2–4.2[c]	—	0.10	0.05–0.30 Mg, 0.25–1.2 Si
C70260[b]	—	Rem.	—	—	—	1.0–3.0[c]	—	—	0.010 P, 0.20–0.7 Si
C70265[b]	—	Rem.	0.05	—	0.30	1.0–3.0[c]	0.05–0.8	—	0.01 P, 0.20–0.7 Si
C70270[b]	—	Rem.	0.05	0.28–1.0	1.0	1.0–3.0[a]	0.10–1.0	0.15	0.20–1.0 Si
C70280[b]	—	Rem.	0.02	0.015	0.30	1.3–1.7[c]	1.0–1.5	—	0.02–0.04 P, 0.22–0.30 Si
C70290[b]	—	Rem.	0.02	0.015	0.30	1.3–1.7[c]	2.1–2.7	—	0.02–0.04 P, 0.22–0.30 Si
C70310[b]	—	Rem.	0.05	0.10	2.0	1.0–4.0[c]	1.0	—	0.01 Mg, 0.05 P, 0.08–1.0 Si, 0.02–0.50 Ag, 0.005–0.05 Zr
C70350[b]	—	Rem.	0.05	0.20	1.0	1.0–2.0	—	0.20	1.0–2.0 Co, 0.04 Mg, 0.50–1.0 Si
C70370[b]	—	Rem.	0	0	1	1–2	—	0	1–2 Co, 0–1 Si, 0–1 Ag
C70400[b]	Copper–nickel, 5%	Rem.	0.05	1.3–1.7	1.0	4.8–6.2[c]	—	0.30–0.8	—
C70500[b]	Copper–nickel, 7%	Rem.	0.05	0.10	0.20	5.8–7.8[c]	—	0.15	—
C70600[b]	Copper–nickel, 10%	Rem.	0.05	1.0–1.8	1.0	9.0–11.0[c]	—	1.0	—
C70610[b]	—	Rem.	0.01	1.0–2.0	—	10.0–11.0[c]	—	0.50–1.0	0.05 C, 0.05 S
C70620[b]	—	86.5 min.	0.02	1.0–1.8	0.50	9.0–11.0[c]	—	1.0	0.05 C, 0.02 P, 0.02 S
C70690[b]	—	Rem.	0.001	0.005	0.001	9.0–11.0[c]	—	0.001	2
C70700[b]	—	Rem.	—	0.05	—	9.5–10.5[c]	—	0.50	—
C70800[b]	Copper–nickel, 11%	Rem.	0.05	0.10	0.20	10.5–12.5[c]	—	0.15	—
C71000[b]	Copper–nickel, 20%	Rem.	0.05	1.0	1.0	19.0–23.0[c]	—	1.0	—
C71100[b]	—	Rem.	0.05	0.10	0.20	22.0–24.0[c]	—	0.15	—
C71300[b]	—	Rem.	0.05	0.20	1.0	23.5–26.5[c]	—	1.0	—
C71500[b]	Copper–nickel, 30%	Rem.	0.05	0.40–1.0	1.0	29.0–33.0[c]	—	1.0	—
C71520[b]	—	65.0 min.	0.02	0.40–1.0	0.50	29.0–33.0[c]	—	1.0	0.05 C, 0.02 P, 0.02 S
C71580[b]	—	Rem.	0.05	0.50	0.05	29.0–33.0[c]	—	0.30	3
C71581[b]	—	Rem.	0.02	0.40–0.7	—	29.0–32.0[c]	—	1.0	4
C71590	—	Rem.	0.001	0.15	0.001	29.0–31.0[c]	0.001	0.50	2
C71640[b]	—	Rem.	0.05[d]	1.7–2.3	1.0[d]	29.0–32.0[c]	—	1.5–2.5	5, 0.06 C, 0.03 S
C71700[b]	—	Rem.	—	0.40–1.0	1.0	29.0–33.0[c]	—	1.0	0.30–0.7 Be
C71900[b]	—	Rem.	0.015	0.50	0.05	28.0–33.0[c]	—	0.20–1.0	0.04 C, 2.2–3.0 Cr, 0.02 P, 0.25 Si, 0.015 S, 0.01–0.20 Ti, 0.02–0.35 Zr
C72150[b]	—	Rem.	0.05	0.10	0.20	43.0–46.0[c]	—	0.05	0.10 C, 0.50 Si
C72200[e]	—	Rem.	0.05[d]	0.50–1.0	1.0[d]	15.0–18.0[c]	—	1.0	5, 0.30–0.7 Cr, 0.03 Si, 0.03 Ti
C72420[f]	—	Rem.	0.02	0.7–1.2	0.20	13.5–16.5[c]	0.10	3.5–5.5	1.0–2.0 Al, 0.05 C, 0.50 Cr, 0.05 Mg, 0.01 P, 0.15 Si, 0.15 S
C72500[e]	—	Rem.	0.05	0.6	0.50	8.5–10.5[c]	1.8–2.8	0.20	—
C72650[f]	—	Rem.	0.01	0.10	0.10	7.0–8.0[c]	4.5–5.5	0.10	—
C72700[f]	—	Rem.	0.02[g]	0.50	0.50	8.5–9.5[c]	5.5–6.5	0.05–0.30	0.15 Mg, 0.10 Nb
C72800[f]	—	Rem.	0.005	0.50	1.0	9.5–10.5[c]	7.5–8.5	0.05–0.30	0.10 Al, 0.02 Sb, 0.001 Bi, 0.001 B, 0.005–0.15 Mg, 0.10–0.30 Nb, 0.005 P, 0.05 Si, 0.0025 S, 0.01 Ti
C72900[f]	—	Rem.	0.02[g]	0.50	0.50	14.5–15.5[c]	7.5–8.5	0.30	0.15 Mg, 0.10 Nb
C72950[f]	—	Rem.	0.05	0.6	—	20.0–22.0[c]	4.5–5.7	0.6	—

Table 19 (*Continued*)

Note: "Rem." = Remainder.

[a] 1. Ag Includes B 2. The following additional maximum limits shall apply: 0.02% C, 0.015% Si, 0.003% S, 0.002% Al, 0.001% P, 0.0005% Hg, 0.001% Ti, 0.001% Sb, 0.001% As, 0.001% Bi, 0.05% Co, 0.10% Mg, and 0.005% oxygen. For C70690, Co shall be 0.02% max. 3. The following additional maximum limits shall apply: 0.07% C, 0.15% Si, 0.024% S, 0.05% Al, and 0.03% P. 4. 0.02% P max.; 0.25% Si max.; 0.01% S max.; 0.02–0.50% Ti. 5. The following additional maximum limits shall apply: When the product is for subsequent welding applications and is so specified by the purchaser, 0.50% Zn, 0.02% P, 0.02% Pb, 0.02% S and 0.05% C.

[b] Cu + sum of named elements, 99.5% min.

[c] Ni value includes Co.

[d] See note 5 in footnote a.

[e] Cu + sum of named elements, 99.8% min.

[f] Cu + sum of named elements, 99.7% min.

[g] 0.005% Pb max. for hot rolling.

affected by waters containing high concentrations of carbon dioxide. High iron concentrations can cause pitting, whereas low pH can lead to excessive copper dissolution and can also cause dezincification of uninhibited yellow brass. Hard, alkaline waters are generally benign toward copper.

NSF Standard 61[8] limits the amount of lead and copper permitted in potable waters contained in certain plumbing products. Metal enters the water due to leaching from exposed surfaces. Lead and copper levels can often be limited by redesign of the fixture, but where this option is not available, lead-free EnviroBrasses C89510, C89520, and C89550 can be specified in place of ordinary plumbing brasses. The alloys' bismuth and selenium contents provide the high machinability needed for cost-effective production.

Many copper alloys are well suited for use in seawater environments: Preferred marine alloys include copper–nickels and aluminum bronzes as well as high-strength beryllium copper, inhibited aluminum brass, phosphor bronzes, arsenical admiralty brass, and nickel–silvers. Copper alloys that perform well in seawater are also often acceptable for use with industrial and process chemicals. Copper–nickels, aluminum bronzes, silicon bronzes, manganese bronzes, and tin and phosphor bronzes should be considered for such environments. Table 42 provides general guidelines.

5.1 Forms of Corrosion

Copper alloys are widely used in aggressive environments, but they should be used with caution in certain media, where corrosion can take one or more of the following forms:

1. *Dealloying,* or *"parting," corrosion* is observed in brasses exposed to seawater and stagnant, neutral, or slightly acidic fresh waters, especially under sediments or biomass deposits. Additions of arsenic or antimony strongly reduce the phenomenon, and *inhibited* (dezincification-resistant) brasses, such as C31400, C31600, C35330, C69430, C69710,

Table 20 Compositions of Cast Coppers, UNS C80000–C81399

Alloy Number	Cu (including Ag)	P	Other Named Elements
C80100	99.95 min.	—	—
C80410	99.9 min.	—	—
C81100	99.70 min.	—	—
C81200	99.9 min.	0.045–0.065	—

Table 21 Compositions of Cast High-Copper Alloys, UNS C81400–C83299

Copper Alloy Number[a]	Cu	Ag	Be	Co	Si	Ni	Fe	Al	Sn	Pb	Zn	Cr	Other Named Elements
C81400	Rem.	—	0.02–0.10	—	—	—	—	—	—	—	—	0.6–1.0	—
C81500	Rem.	—	—	—	0.15	—	0.10	0.10	0.10	0.02	0.10	0.40–1.5	—
C81540	95.1 min.[b]	—	—	—	0.40–0.8	2.0–3.0[c]	0.15	0.10	0.10	0.02	0.10	0.10–0.6	—
C82000	Rem.	—	0.45–0.8	2.40–2.70[d]	0.15	0.20[d]	0.10	0.10	0.10	0.02	0.10	0.10	—
C82200	Rem.	—	0.35–0.8	0.30	—	1.0–2.0	—	—	—	—	—	—	—
C82400	Rem.	—	1.60–1.85	0.20–0.65	—	0.20	0.20	0.15	0.10	0.02	0.10	0.10	—
C82500	Rem.	—	1.90–2.25	0.35–0.70[d]	0.20–0.35	0.20[d]	0.25	0.15	0.10	0.02	0.10	0.10	—
C82510	Rem.	—	1.90–2.15	1.0–1.2	0.20–0.35	0.20	0.25	0.15	0.10	0.02	0.10	0.10	—
C82600	Rem.	—	2.25–2.55	0.35–0.65	0.20–0.35	0.20	0.25	0.15	0.10	0.02	0.10	0.10	—
C82700	Rem.	—	2.35–2.55	—	0.15	1.0–1.5	0.25	0.15	0.10	0.02	0.10	0.10	—
C82800	Rem.	—	2.50–2.85	0.35–0.70[d]	0.20–0.35	0.20[d]	0.25	0.15	0.10	0.02	0.10	0.10	—

Note: "Rem." = Remainder.
[a] Cu + sum of named elements, 99.5% min.
[b] Includes Ag.
[c] Ni value includes Co.
[d] Ni + Co.

C87800, or C87900, should be used where dealloying conditions are present. Certain aluminum-rich aluminum and nickel–aluminum bronzes are also vulnerable to dealloying, although susceptibility can be reduced through heat treatment and eliminated entirely by cathodic protection.

2. *Erosion–corrosion* and *cavitation* occur when protective corrosion films are removed by high-velocity fluids faster than the films can regenerate. Entrained particulates in the fluid accelerate the process. Critical flow velocities for selected copper alloys, below which erosion–corrosion does not occur, are listed in Table 43.

3. *Galvanic,* or *dissimilar-metal, corrosion,* can occur when two metals with significantly different electrochemical potentials are joined in a corrosive environment. The more

Table 22 Compositions of Cast Copper–Tin–Zinc and Copper–Tin–Zinc–Lead Alloys (Red and Leaded Red Brasses), UNS C83300–C83999

Copper Alloy Number[a]	Cu	Ag	Be	Co	Si	Ni	Fe	Al	Sn	Pb	Zn	Cr	Other Named Elements
C81400	Rem.	—	0.02–0.10	—	—	—	—	—	—	—	—	0.6–1.0	—
C81500	Rem.	—	—	—	0.15	—	0.10	0.10	0.10	0.02	0.10	0.40–1.5	—
C81540	95.1 min.[b]	—	—	—	0.40–0.8	2.0–3.0[c]	0.15	0.10	0.10	0.02	0.10	0.10–0.6	—
C82000	Rem.	—	0.45–0.8	2.40–2.70[d]	0.15	0.20[d]	0.10	0.10	0.10	0.02	0.10	0.10	—
C82200	Rem.	—	0.35–0.8	0.30	—	1.0–2.0	—	—	—	—	—	—	—
C82400	Rem.	—	1.60–1.85	0.20–0.65	—	0.20	0.20	0.15	0.10	0.02	0.10	0.10	—
C82500	Rem.	—	1.90–2.25	0.35–0.70[d]	0.20–0.35	0.20[d]	0.25	0.15	0.10	0.02	0.10	0.10	—
C82510	Rem.	—	1.90–2.15	1.0–1.2	0.20–0.35	0.20	0.25	0.15	0.10	0.02	0.10	0.10	—
C82600	Rem.	—	2.25–2.55	0.35–0.65	0.20–0.35	0.20	0.25	0.15	0.10	0.02	0.10	0.10	—
C82700	Rem.	—	2.35–2.55	—	0.15	1.0–1.5	0.25	0.15	0.10	0.02	0.10	0.10	—
C82800	Rem.	—	2.50–2.85	0.35–0.70[d]	0.20–0.35	0.20[d]	0.25	0.15	0.10	0.02	0.10	0.10	—

Note: "Rem." = Remainder.
[a] Cu + sum of named elements, 99.5% min.
[b] Includes Ag.
[c] Ni value includes Co.
[d] Ni + Co.

Table 23 Compositions of Cast Copper–Tin–Zinc and Copper–Tin–Zinc–Lead Alloys (Semired and Leaded Semired Brasses), UNS C84000–C84999

Copper Alloy Number[a]	Cu[b]	Sn	Pb	Zn	Fe	Sb	Ni[c]	S	P	Al	Si	Bi	Other Named Elements
C84200	78.0–82.0	4.0–6.0	2.0–3.0	10.0–16.0	0.40	0.25	0.8	0.08	0.05[d]	0.005	0.005	—	—
C84400	78.0–82.0	2.3–3.5	6.0–8.0	7.0–10.0	0.40	0.25	1.0	0.08	0.02[d]	0.005	0.005	—	—
C84410	Rem.	3.0–4.5	7.0–9.0	7.0–11.0	—	—	1.0	—	—	0.01	0.2	0.05	—[e]
C84500	77.0–79.0	2.0–4.0	6.0–7.5	10.0–14.0	0.40	0.25	1.0	0.08	0.02[d]	0.005	0.005	—	—
C84800	75.0–77.0	2.0–3.0	5.5–7.0	13.0–17.0	0.40	0.25	1.0	0.08	0.02[d]	0.005	0.005	—	—

[a] Cu + sum of named elements, 99.3% min.
[b] In determining Cu min., Cu may be calculated as Cu + Ni.
[c] Ni value includes Co.
[d] For continuous castings, P shall be 1.5% max.
[e] Fe + Sb + As shall be 0.8% max.

noble (cathodic) metal in the couple is protected at the expense of the more reactive (anodic) one. For example, copper alloys can accelerate corrosion in aluminum and mild steel, especially if the area ratio of copper to the anodic metal is large. Conversely, copper alloys can corrode when coupled to gold, silver, platinum-group metals, or carbon (or graphite). In either case, electrically insulating the coupled metals from one another prevents attack.

4. *Stress–corrosion,* or *environmentally assisted, cracking* occurs in many copper alloys exposed to aqueous ammonium ions, nitrites, mercury compounds, and moist atmospheres containing sulfur dioxide if the metals are under imposed or residual tensile stress. Yellow brasses are most vulnerable. Tin bronzes, red brasses, and aluminum bronzes are less sensitive, while pure copper and copper–nickels are essentially immune.

5.2 Corrosion, Copper, Health, and the Environment

Corrosion implies the dissolution of metal and the potential release to the environment or to a media from which it can be taken up by organisms, possibly with harmful consequences. However, It should be noted that copper is also an essential nutrient in most species, including humans, and that lack of adequate copper intake can result in deficiency and detri-

Table 24 Compositions of Cast Copper–Zinc Alloys (Yellow Brasses), UNS C85000–C85999

Copper Alloy Number	Cu[a]	Sn	Pb	Zn	Fe	Sb	Ni[b]	Mn	As	S	P	Al	Si	Other Named Elements
C85200[c]	70.0–74.0	0.7–2.0	1.5–3.8	20.0–27.0	0.6	0.20	1.0	—	—	0.05	0.02	0.005	0.05	—
C85400[d]	65.0–70.0	0.50–1.5	1.5–3.8	24.0–32.0	0.7	—	1.0	—	—	—	—	0.35	0.05	—
C85500[c]	59.0–63.0	0.20	0.20	Rem.	0.20	—	0.20	0.20	—	—	—	—	—	—
C85700[e]	58.0–64.0	0.50–1.5	0.8–1.5	32.0–40.0	0.7	—	1.0	—	—	—	—	0.8	0.05	—
C85800[e]	57.0 min.	1.5	1.5	31.0–41.0	0.50	0.05	0.50	0.25	0.05	0.05	0.01	0.55	0.25	—

[a] In determining Cu min., Cu may be calculated as Cu + Ni.
[b] Ni value includes Co.
[c] Cu + sum of named elements, 99.1% min.
[d] Cu + sum of named elements, 98.9% min.
[e] Cu + sum of named elements, 98.7% min.

Table 25 Compositions of Cast Manganese Bronze and Leaded Manganese Bronze Alloys (High-Strength Yellow Brasses), C86000–C86999

Alloy Number[a]	Cu[b]	Sn	Pb	Zn	Fe	Ni	Al	Mn	Si	Other Named Elements
C86100	66.0–68.0	0.20	0.20	Rem.	2.0–4.0	—	4.5–5.5	2.5–5.0	—	—
C86200	60.0–66.0	0.20	0.20	22.0–28.0	2.0–4.0	1.0[c]	3.0–4.9	2.5–5.0	—	—
C86300	60.0–66.0	0.20	0.20	22.0–28.0	2.0–4.0	1.0[c]	5.0–7.5	2.5–5.0	—	—
C86400	56.0–62.0	0.50–1.5	0.50–1.5	34.0–42.0	0.40–2.0	1.0[c]	0.50–1.5	0.10–1.5	—	—
C86500	55.0–60.0	1.0	0.40	36.0–42.0	0.40–2.0	1.0[c]	0.50–1.5	0.10–1.5	—	—
C86550	57.0 min.	1.0	0.50	Rem.	0.7–2.0	1.0[c]	0.50–2.5	0.10–3.0	0.10	—
C86700	55.0–60.0	1.5	0.50–1.5	30.0–38.0	1.0–3.0	1.0[c]	1.0–3.0	0.10–3.5	—	—
C86800	53.5–57.0	1.0	0.20	Rem.	1.0–2.5	2.5–4.0[c]	2.0	2.5–4.0	—	—

[a] Cu + sum of named elements, 99.0% min.
[b] In determining Cu min., Cu may be calculated as Cu + Ni.
[c] Ni value includes Co.

mental physiological consequences. Corrosion-based release potentially ensues simply from normal use of a copper product, as with plumbing goods and roofing sheet, or from disposal in a landfill. In an effort to avoid harmful impacts to the environment, regulations and, in some cases, design standards have been introduced to limit such releases.

Copper, once thought to be "toxic, persistent and bioaccumulative" by the U.S. Environmental Protection Agency (EPA), has been the subject of a large body of research over the past several decades that seeks to understand the metal's behavior with respect to health and the environment. Results of those studies show that, while *ionic* copper exhibits some degree of toxicity in organisms at elevated concentrations, the offending ionic copper species is easily transformed in nature via rapid and nearly complete binding of the copper by organic complexing agents or they change to far less bioavailable inorganic copper species. These forms of copper are not readily absorbed by organisms, and for this and other reasons, toxicity is sharply curtailed. Changes have been made in environmental regulations as a

Table 26 Compositions of Cast Copper–Silicon Alloys (Silicon Bronzes and Silicon Brasses), UNS C87000–C87999

Copper Alloy Number	Cu	Sn	Pb	Zn	Fe	Al	Si	Mn	Mg	Ni	S	Other Named Elements
C87300[a]	94.0 min.	—	0.20	0.25	0.20	—	3.5–4.5	0.8–1.5	—	—	—	—
C87400[b]	79.0 min.	—	1.0	12.0–16.0	—	0.8	2.5–4.0	—	—	—	—	—
C87500[a]	79.0 min.	—	0.50	12.0–16.0	—	0.50	3.0–5.0	—	—	—	—	—
C87600[a]	88.0 min.	—	0.50	4.0–7.0	0.20	—	3.5–5.5	0.25	—	—	—	—
C87610[a]	90.0 min.	—	0.20	3.0–5.0	0.20	—	3.0–5.0	0.25	—	—	—	—
C87800[a]	80.0 min.	0.25	0.15	12.0–16.0	0.15	0.15	3.8–4.2	0.15	0.01	0.20[c]	0.05	0.05 Sb, 0.05 As, 0.01 P
C87850[a]	74.0–78.0	0.30	0.10	Rem.	0.10	—	2.7–3.4	0.10	—	0.20[c]	—	0.10 Sb 0.05–0.20 P

[a] Cu + sum of named elements, 99.5% min.
[b] Cu + sum of named elements, 99.2% min.
[c] Ni value includes Co.

Table 27 Compositions of Cast Copper–Bismuth and Copper–Bismuth–Selenium Alloys (EnviroBrasses) UNS C88000–C89999

Copper Alloy Number	Cu	Sn	Pb	Zn	Fe	Sb	Ni[a]	S	P	Al	Si	Bi	Se	Other Named Elements
C89320[b]	87.0–91.0	5.0–7.0	0.09	1.0	0.20	0.35	1.0	0.08	0.30	0.005	0.005	4.0–6.0	—	—
C89325[c]	84.0–88.0	9.0–11.0	0.10	1.0	0.15	0.50	1.0	0.08	0.10	0.005	0.005	2.7–3.7	—	—[d]
C89510[b]	86.0–88.0	4.0–6.0	0.25	4.0–6.0	0.20	0.25	1.0	0.08	0.05	0.005	0.005	0.50–1.5[e]	0.35–0.75[e]	—
C89520[b]	85.0–87.0	5.0–6.0	0.25	4.0–6.0	0.20	0.25	1.0	0.08	0.05	0.005	0.005	1.6–2.2[f]	0.8–1.1[f]	—
C89550[b]	58.0–64.0	1.2	0.10	32.0–38.0	0.50	0.05	1.0	0.05	0.01	0.10–0.6	0.25	0.6–1.2	0.01–0.10	—
C89831[c]	87.0–91.0	2.7–3.7	0.10	2.0–4.0	0.30	0.25	1.0	0.08	0.050	0.005	0.005	2.7–3.7	—	—[d]
C89833[g]	87.0–91.0	4.0–6.0	0.10	2.0–4.0	0.30	0.25	1.0	0.08	0.050	0.005	0.005	1.7–2.7	—	—
C89835[c]	85.0–89.0	6.0–7.5	0.10	2.0–4.0	0.20	0.35	1.0	0.08	0.10	0.005	0.005	1.7–2.7	—	—[d]
C89837[c]	84.0–88.0	3.0–4.0	0.10	6.0–10.0	0.30	0.25	1.0	0.08	0.050	0.005	0.005	0.7–1.2	—	—[d]
C89844[g]	83.0–86.0	3.0–5.0	0.20	7.0–10.0	0.30	0.25	1.0	0.08	0.05	0.005	0.005	2.0–4.0	—	—
C89940[b]	64.0–68.0	3.0–5.0	0.01	3.0–5.0	0.7–2.0	0.10	20.0–23.0	0.05	0.10–0.15	0.005	0.15	4.0–5.5	—	0.20 Mn

[a] Ni value includes Co.
[b] Cu + sum of named elements, 99.5% min.
[c] Cu + sum of named elements, 99.0% min.
[d] 0.01–2.0% as any single or combination of Ce, La, or other rare earth elements, as agreed upon. According to the ASM International definition, *rare earth* refers to one of the group of chemically similar metals with atomic numbers 57–71, commonly refered to as lanthanides.
[e] Experience favors Bi:Se ≥ 2:1.
[f] Bi:Se ≥ 2:1.
[g] Cu + sum of named elements, 99.3% min.

Table 28 Compositions of Cast Copper–Tin Alloys (Tin Bronzes), UNS C90000–C91999

Copper Alloy Number[a]	Cu[b]	Sn	Pb	Zn	Fe	Sb	Ni[c]	S	P[d]	Al	Si	Mn	Other Named Elements
C90200	91.0–94.0	6.0–8.0	0.30	0.50	0.20	0.20	0.50	0.05	0.05	0.005	0.005	—	—
C90300	86.0–89.0	7.5–9.0	0.30	3.0–5.0	0.20	0.20	1.0	0.05	0.05	0.005	0.005	—	—
C90500[e]	86.0–89.0	9.0–11.0	0.30	1.0–3.0	0.20	0.20	1.0	0.05	0.05	0.005	0.005	—	—
C90700	88.0–90.0	10.0–12.0	0.50	0.50	0.15	0.20	0.50	0.05	0.30	0.005	0.005	—	—
C90710	Rem.	10.0–12.0	0.25	0.05	0.10	0.20	0.10	0.05	0.05–1.2	0.005	0.005	—	—
C90800	85.0–89.0	11.0–13.0	0.25	0.25	0.15	0.20	0.50	0.05	0.30	0.005	0.005	—	—
C90810	Rem.	11.0–13.0	0.25	0.30	0.15	0.20	0.50	0.05	0.15–0.8	0.005	0.005	—	—
C90900	86.0–89.0	12.0–14.0	0.25	0.25	0.15	0.20	0.50	0.05	0.05	0.005	0.005	—	—
C91000	84.0–86.0	14.0–16.0	0.20	1.5	0.10	0.20	0.8	0.05	0.05	0.005	0.005	—	—
C91100	82.0–85.0	15.0–17.0	0.25	0.25	0.25	0.20	0.50	0.05	1.0	0.005	0.005	—	—
C91300	79.0–82.0	18.0–20.0	0.25	0.25	0.25	0.20	0.50	0.05	1.0	0.005	0.005	—	—
C91600	86.0–89.0	9.7–10.8	0.25	0.25	0.20	0.20	1.2–2.0	0.05	0.30	0.005	0.005	—	—
C91700	84.0–87.0	11.3–12.5	0.25	0.25	0.20	0.20	1.2–2.0	0.05	0.30	0.005	0.005	—	—

Note: "Rem." = Remainder.
[a]Cu + sum of named elements, 99.4% min.
[b]In determining Cu min., Cu may be calculated as Cu + Ni.
[c]Ni value includes Co.
[d]For continuous castings, P shall be 1.5% max.
[e]Cu + sum of named elements, 99.7% min.

Table 29 Compositions of Cast Copper–Tin–Lead Alloys (Leaded Tin Bronzes), UNS C92000–C92900

Copper Alloy Number[a]	Cu[b]	Sn	Pb	Zn	Fe	Sb	Ni[c]	S	P[d]	Al	Si	Mn	Other Named Elements
C92200	86.0–90.0	5.5–6.5	1.0–2.0	3.0–5.0	0.25	0.25	1.0	0.05	0.05	0.005	0.005	—	—
C92210	86.0–89.0	4.5–5.5	1.7–2.5	3.0–4.5	0.25	0.20	0.7–1.0	0.05	0.03	0.005	0.005	—	—
C92220	86.0–88.0	5.0–6.0	1.5–2.5	3.0–5.5	0.25	—	0.50–1.0	—	0.05	—	—	—	—
C92300	85.0–89.0	7.5–9.0	0.30–1.0	2.5–5.0	0.25	0.25	1.0	0.05	0.05	0.005	0.005	—	—
C92310	Rem.	7.5–8.5	0.30–1.5	3.5–4.5	—	—	1.0	—	—	0.005	0.005	0.03	—
C92400	86.0–89.0	9.0–11.0	1.0–2.5	1.0–3.0	0.25	0.25	1.0	0.05	0.05	0.005	0.005	—	—
C92410	Rem.	6.0–8.0	2.5–3.5	1.5–3.0	0.20	0.25	0.20	—	—	0.005	0.005	0.05	—
C92500	85.0–88.0	10.0–12.0	1.0–1.5	0.50	0.30	0.25	0.8–1.5	0.05	0.30	0.005	0.005	—	—
C92600	86.0–88.5	9.3–10.5	0.8–1.5	1.3–2.5	0.20	0.25	0.7	0.05	0.03	0.005	0.005	—	—
C92610	Rem.	9.5–10.5	0.30–1.5	1.7–2.8	0.15	—	1.0	—	—	0.005	0.005	0.03	—
C92700	86.0–89.0	9.0–11.0	1.0–2.5	0.7	0.20	0.25	1.0	0.05	0.25	0.005	0.005	—	—
C92710	Rem.	9.0–11.0	4.0–6.0	1.0	0.20	0.25	2.0	0.05	0.10	0.005	0.005	—	—
C92800	78.0–82.0	15.0–17.0	4.0–6.0	0.8	0.20	0.25	0.8	0.05	0.05	0.005	0.005	—	—
C92810	78.0–82.0	12.0–14.0	4.0–6.0	0.50	0.50	0.25	0.8–1.2	0.05	0.05	0.005	0.005	—	—
C92900	82.0–86.0	9.0–11.0	2.0–3.2	0.25	0.20	0.25	2.8–4.0	0.05	0.50	0.005	0.005	—	—

Note: "Rem." = Remainder.
[a]Cu + sum of named elements, 99.3% min.
[b]In determining Cu min., Cu may be calculated as Cu + Ni.
[c]Ni value includes Co.
[d]For continuous castings, P shall be 1.5% max.

Table 30 Compositions of Cast Copper–Tin–Lead Alloys (High-Leaded Tin Bronzes), UNS C93000–C94500

Copper Alloy Number	Cu	Sn	Pb	Zn	Fe	Sb	Ni[a]	S	P[b]	Al	Si	Other Named Elements
C93100[c]	Rem.[d]	6.5–8.5	2.0–5.0	2.0	0.25	0.25	1.0	0.05	0.30	0.005	0.005	—
C93200[c]	81.0–85.0[d]	6.3–7.5	6.0–8.0	1.0–4.0	0.20	0.35	1.0	0.08	0.15	0.005	0.005	—
C93400[c]	82.0–85.0[d]	7.0–9.0	7.0–9.0	0.8	0.20	0.50	1.0	0.08	0.50	0.005	0.005	—
C93500[c]	83.0–86.0[d]	4.3–6.0	8.0–10.0	2.0	0.20	0.30	1.0	0.08	0.05	0.005	0.005	—
C93600[e]	79.0–83.0	6.0–8.0	11.0–13.0	1.0	0.20	0.55	1.0	0.08	0.15	0.005	0.005	—
C93700[c]	78.0–82.0	9.0–11.0	8.0–11.0	0.8	0.7[f]	0.50	0.50	0.08	0.10	0.005	0.005	—
C93720[c]	83.0 min.	3.5–4.5	7.0–9.0	4.0	0.7	0.50	0.50	—	0.10	—	—	—
C93800[c]	75.0–79.0	6.3–7.5	13.0–16.0	0.8	0.15	0.8	1.0	0.08	0.05	0.005	0.005	—
C93900[g]	76.5–79.5	5.0–7.0	14.0–18.0	1.5	0.40	0.50	0.8	0.08	1.5	0.005	0.005	—
C94000[h]	69.0–72.0	12.0–14.0	14.0–16.0	0.50	0.25	0.50	0.50–1.0	0.08[i]	0.05	0.005	0.005	—
C94100[h]	72.0–79.0	4.5–6.5	18.0–22.0	1.0	0.25	0.8	1.0	0.08[i]	0.50	0.005	0.005	—
C94300[c]	67.0–72.0	4.5–6.0	23.0–27.0	0.8	0.15	0.8	1.0	0.08[i]	0.08	0.005	0.005	—
C94310[c]	Rem.	1.5–3.0	27.0–34.0	0.50	0.50	0.50	0.25–1.0	—	0.05	—	—	—
C94320[c]	Rem.	4.0–7.0	24.0–32.0	—	0.35	—	—	—	—	—	—	—
C94330[c]	68.5–75.5	3.0–4.0	21.0–25.0	3.0	0.7	0.50	0.50	—	0.10	—	—	—
C94400[c]	Rem.	7.0–9.0	9.0–12.0	0.8	0.15	0.8	1.0	0.08	0.50	0.005	0.005	—
C94500[c]	Rem.	6.0–8.0	16.0–22.0	1.2	0.15	0.8	1.0	0.08	0.05	0.005	0.005	—

Note: "Rem." = Remainder.
[a] Ni value includes Co.
[b] For continuous castings, P shall be 1.5% max.
[c] Cu + sum of named elements, 99.0% min.
[d] In determining Cu min., Cu may be calculated as Cu + Ni.
[e] Cu + sum of named elements, 99.3% min.
[f] Fe shall be 0.35% max., when used for steel-backed.
[g] Cu + sum of named elements, 98.9% min.
[h] Cu + sum of named elements, 98.7% min.
[i] For continuous castings, S shall be 0.25% max.

Table 31 Compositions of Cast Copper–Tin–Nickel Alloys (Nickel–Tin Bronzes), UNS C94600–C94999

Copper Alloy Number	Cu	Sn	Pb	Zn	Fe	Sb	Ni	Mn	S	P	Al	Si	Other Named Elements
C94700[a]	85.0–90.0	4.5–6.0	0.10[b]	1.0–2.5	0.25	0.15	4.5–6.0[c]	0.20	0.05	0.05	0.005	0.005	—
C94800[a]	84.0–89.0	4.5–6.0	0.30–1.0	1.0–2.5	0.25	0.15	4.5–6.0[c]	0.20	0.05	0.05	0.005	0.005	—
C94900[d]	79.0–81.0	4.0–6.0	4.0–6.0	4.0–6.0	0.30	0.25	4.0–6.0[c]	0.10	0.08	0.05	0.005	0.005	—

[a] Cu + sum of named elements, 98.7% min.
[b] The mechanical properties of C94700 (heat treated) may not be attainable if the Pb content exceeds 0.01%.
[c] Ni value includes Co.
[d] Cu + sum of named elements, 99.4% min.

Table 32 Compositions of Cast Copper–Aluminum–Iron and Copper–Aluminum–Iron–Nickel Alloys (Aluminum Bronzes), UNS C95000—C95999

Copper Alloy Number	Cu	Pb	Fe	Ni	Al	Mn	Mg	Si	Zn	Sn	Other Named Elements
C95200[a]	86.0 min.	—	2.5–4.0	—	8.5–9.5	—	—	—	—	—	—
C95210[a]	86.0 min.	0.05	2.5–4.0	1.0[b]	8.5–9.5	1.0	0.05	0.25	0.50	0.10	—
C95220[c]	Rem.	—	2.5–4.0	2.5[b]	9.5–10.5	0.50	—	—	—	—	—
C95300[a]	86.0 min.	—	0.8–1.5	—	9.0–11.0	—	—	—	—	—	—
C95400[c]	83.0 min.	—	3.0–5.0	1.5[b]	10.0–11.5	0.50	—	—	—	—	—
C95410[c]	83.0 min.	—	3.0–5.0	1.5–2.5[b]	10.0–11.5	0.50	—	—	—	—	—
C95420[c]	83.5 min.	—	3.0–4.3	0.50[b]	10.5–12.0	0.50	—	—	—	—	—
C95500[c]	78.0 min.	—	3.0–5.0	3.0–5.5[b]	10.0–11.5	3.5	—	—	—	—	—
C95510[d]	78.0 min.	—	2.0–3.5	4.5–5.5[b]	9.7–10.9	1.5	—	—	0.30	0.20	—
C95520[c]	74.5 min.	0.03	4.0–5.5	4.2–6.0[b]	10.5–11.5	1.5	—	0.15	0.30	0.25	0.05 Cr, 0.20 Co
C95600[a]	88.0 min.	—	—	0.25[b]	6.0–8.0	—	—	1.8–3.2	—	—	—
C95700[c]	71.0 min.	—	2.0–4.0	1.5–3.0[b]	7.0–8.5	11.0–14.0	—	0.10	—	—	—
C95710[c]	71.0 min.	0.05	2.0–4.0	1.5–3.0[b]	7.0–8.5	11.0–14.0	—	0.15	0.50	1.0	0.05 P
C95720[c]	73.0 min.	0.03	1.5–3.5	3.0–6.0[b]	6.0–8.0	12.0–15.0	—	0.10	0.10	0.10	0.20 Cr
C95800[c]	79.0 min.	0.03	3.5–4.5[e]	4.0–5.0[e]	8.5–9.5	0.8–1.5	—	0.10	—	—	—
C95810[c]	79.0 min.	0.10	3.5–4.5[e]	4.0–5.0[e]	8.5–9.5	0.8–1.5	0.05	0.10	0.50	—	—
C95820[f]	77.5 min.	0.02	4.0–5.0	4.5–5.8[b]	9.0–10.0	1.5	—	0.10	0.20	0.20	—
C95900[c]	Rem.	—	3.0–5.0	0.50[b]	12.0–13.5	1.5	—	—	—	—	—

Note: "Rem." = Remainder.
[a] Cu + sum of named elements, 99.0% min.
[b] Ni value includes Co.
[c] Cu + sum of named elements, 99.5% min.
[d] Cu + sum of named elements, 99.8% min.
[e] Fe content shall not exceed Ni content.
[f] Cu + sum of named elements, 99.2% min.

result of our increased understanding of the behavior of copper in the environment. Significantly, copper was removed from the U.S. EPA list of persistent, bioaccumulative, and toxic (PBT) substances in 2004.

Copper and its alloys are among the most thoroughly recycled industrial metals. Scrap copper can be (and is) re-refined to primary purity standards, but it is more often used as feedstock for nonelectrical products and as a base for alloys. In 2003, the U.S. economy consumed approximately 26.4% as much scrap copper as it did primary refined copper. Recycling also conserves energy: Studies conducted in 1997 suggest that copper produced from ore consumes 62 MJ/kg, whereas clean scrap that requires only remelting requires approximately 8 MJ/kg for reuse; scrap that requires electrolytic refining consumes approximately 10 MJ/kg, and resmelted scrap requires approximately 28 MJ/kg.

6 BIOSTATIC AND ANTIMICROBIAL PROPERTIES

Copper and alloys high in copper are among the few metals that possess the ability to inhibit the growth of organisms on surfaces, a property that has been exploited since copper plates were first used to protect wooden ship hulls centuries ago. It is now generally held that the metal's mild toxicity makes surfaces inhospitable for the attachment and growth of marine

Table 33 Composition of Cast Copper–Nickel Alloys (Copper–Nickels), UNS C96000–C96999)

Copper Alloy Number	Cu	Pb	Fe	Ni[a]	Mn	Si	Nb	C	Be	Other Named Elements
C96200[b]	Rem.	0.01	1.0–1.8	9.0–11.0	1.5	0.50	1.0[c]	0.10	—	0.02 P, 0.02 S
C96300[b]	Rem.	0.01	0.50–1.5	18.0–22.0	0.25–1.5	0.50	0.50–1.5	0.15	—	0.02 P, 0.02 S
C96400[b]	Rem.	0.01	0.25–1.5	28.0–32.0	1.5	0.50	0.50–1.5	0.15	—	0.02 P, 0.02 S
C96600[b]	Rem.	0.01	0.8–1.1	29.0–33.0	1.0	0.15	—	—	0.40–0.7	—
C96700[b]	Rem.	0.01	0.40–1.0	29.0–33.0	0.40–1.0	0.15	—	—	1.1–1.2	0.15–0.35 Ti, 0.15–0.35 Zr
C96800[b,d]	Rem.	0.005	0.50	9.5–10.5	0.05–0.30	0.05	0.10–0.30	—	—	0.005–0.15 Mg, 7.5–8.5 Sn, 1.0 Zn
C96900[b]	Rem.	0.02	0.50	14.5–15.5	0.05–0.30	—	0.10	—	—	0.15 Mg, 7.5–8.5 Sn, 0.50 Zn
C96950[b]	Rem.	0.02	0.05	11.0–15.5	0.05–0.40	0.30	0.10	—	—	0.15 Mg, 5.8–8.5 Sn
C96970[b]	Rem.	0.02	0.50	8.5–9.5	0.30	—	0.10	—	—	0.15 Mg, 5.5–6.5 Sn, 0.50 Zn

Note: "Rem." = Remainder.
[a] Ni value includes Co.
[b] Cu + sum of named elements, 99.5% min.
[c] When product or casting is intended for subsequent welding applications, and so specified by the purchaser, the Nb content shall be 0.40% max.
[d] The following additional maximum impurity limits shall apply: 0.10% Al, 0.001% B, 0.001% Bi, 0.005% P, 0.0025% S, 0.02% Sb, 0.01% Ti.

life forms such as algae and barnacles while at the same time causing no harm to the surrounding environment. Marine vessels are now protected by copper-containing antifouling paints and coatings, but the biostatic properties of copper alloys themselves are used to maintain unobstructed flow in products such as condenser tubing, seawater piping, and inlet screens. Copper water tube and copper and brass fittings and plumbing goods are known to inhibit the growth of pathogens such as *Pneumophila legionella,* which causes Legionnaires'

Table 34 Compositions of Cast Copper–Lead Alloys (Leaded Coppers), UNS C98000—C98999

Copper Alloy Number	Cu	Sn	Pb	Ag	Zn	P	Fe	Ni	Sb	Other Named Elements
C98200[a]	Rem.	0.6–2.0	21.0–27.0	—	0.50	0.10	0.7	0.50	0.50	—
C98400[a]	Rem.	0.50	26.0–33.0	1.5	0.50	0.10	0.7	0.50	0.50	—
C98600	60.0–70.0	0.50	30.0–40.0	1.5	—	—	0.35	—	—	—
C98800	56.5–62.5[b]	0.25	37.5–42.5[c]	5.5[c]	0.10	0.02	0.35	—	—	—
C98820	Rem.	1.0–5.0	40.0–44.0	—	—	—	0.35	—	—	—
C98840	Rem.	1.0–5.0	44.0–58.0	—	—	—	0.35	—	—	—

Note: "Rem." = Remainder.
[a] Cu + sum of named elements, 99.5% min.
[b] Includes Ag
[c] Pb and Ag may be adjusted to modify alloy hardness.

Table 35 Compositions of Cast Special Alloys, UNS C99000—C99999

Copper Alloy Number[a]	Other Designation	Cu	Sn	Pb	Ni	Fe	Al	Co	Si	Mn	Other Named Elements
C99300	Incramet 800	Rem.	0.05	0.02	13.5–16.5	0.40–1.0	10.7–11.5	1.0–2.0	0.02	—	—
C99350	—	Rem.	—	0.15	14.5–16.0[b]	1.0	9.5–10.5	—	—	0.25	7.5–9.5 Zn
C99400	—	Rem.	—	0.25	1.0–3.5	1.0–3.0	0.50–2.0	—	0.50–2.0	0.50	0.50–5.0 Zn
C99500	—	Rem.	—	0.25	3.5–5.50	3.0–5.0	0.50–2.0	—	0.50–2.0	0.50	0.50–2.0 Zn
C99600	Incramute 1	Rem.	0.10	0.02	0.20	0.20	1.0–2.8	0.20	0.10	39.0–45.0	0.05 C, 0.20 Zn
C99700	—	54.0 min.	1.0	2.0	4.0–6.0	1.0	0.50–3.0	—	—	11.0–15.0	19.0–25.0 Zn
C99750	—	55.0–61.0	—	0.50–2.5	5.0	1.0	0.25–3.0	—	—	17.0–23.0	17.0–23.0 Zn

Note: "Rem." = Remainder.
[a]Cu + sum of named elements, 99.7% min.
[b]Ni value includes Co.

disease, and it was observed as recently as the 1980s that brass doorknobs and push plates used in hospitals are bacteriostatic.

Studies have shown that copper surfaces reduce the viability of methicillin-resistant *Staphylococcus aureus* (MRSA) by seven orders of magnitude in 90 min at room temperature. The effectiveness of copper alloys appears to be related to copper content, but all alloys tested show some bacteriostatic effect. Copper has also been shown to inhibit growth of the potentially deadly bacterium *Escherichia coli* O157:H7 and of *Listeria monocytogenes,* infection by which can be manifested as septicemia, meningitis, and encephalitis, among other diseases. Copper's ability to inhibit pathogens now promises to gain new applications for copper in products for health care and food-handling facilities.

7 FABRICATION

7.1 Machining

All copper alloys can be cut with standard machine tooling, and high-speed steel suffices for all but the hardest alloys. Carbide tooling can be used but is rarely necessary.

Chip Appearance and Machinability

The wide range of copper alloy physical and mechanical properties are illustrated in three types of machining chips (Fig. 2). Free-cutting alloys containing lead, sulfur, tellurium, or combinations of selenium and bismuth produce type I chips. The chips are short and fragmented and are easily cleared from the cutting zone. These alloys are intended for high-productivity machining operations. Alloys with multiphase microstructures tend to produce short, curly type II chips. The alloys are readily machinable but at slower rates than those of type I alloys. Copper and ductile single-phase alloys produce long and stringy type III chips. The alloys' low hardness facilitates cutting, but attention to tool design is important for optimum workpiece quality.

Machinability is commonly reported on a scale of 1–100, on which free-cutting brass, C36000, ranks 100. Other type I copper alloys are rated between 50 and 100, type II alloys between 30 and 50, and type III alloys between 20 and 40. When tested according to ASTM E 618,[9] leaded free-machining steel 12L14 has a machinability rating of approximately 21.

Type I alloys include leaded coppers, brasses, bronzes, nickel–silvers, architectural and forging brasses, tellurium, and sulfur-bearing coppers and alloys. Type II alloys include aluminum and nickel–aluminum bronzes, beryllium coppers, manganese bronze, Naval brass,

Table 36 Physical Properties of Pure Copper

Property	Value
Atomic weight	63.546
Atomic volume, cm^3/mol	7.11
Mass numbers, stable isotopes	63 (69.1%), 65 (30.9%)
Oxidation states	1, 2, 3
Standard electrode potential, V	$Cu/Cu^+ = 0.520$
	$Cu^+/Cu^{2+} = 0.337$
Electrochemical equivalent, mg/C[15]	0.3294 for Cu^{2+}
	0.6588 for Cu^+
Electrolytic solution potential, V (SCE)[15]	0.158 ($Cu^{2+} + e^- = Cu^+$)
	0.3402 ($Cu^{2+} + 2e^- = Cu$)
	0.522 ($Cu^+ + e^- = Cu$)
Density, g/m^3	8.95285 (pure, single crystal)
	8.94 (nominal)
Metallic (Goldschmidt) radius, nm	0.1276 (12-fold coordination)
Ionic radius, M^+, nm[3]	0.096
Covalent radius, nm[3]	0.138
Crystal structure	fcc, A1, Fm3m: cF4
Lattice parameter	0.361509 ± 0.000004 nm (25°C, 77 °F)
Electronegativity[3]	2.43
Ionization energy, kJ/mol[3]	
First	745
Second	1950
Ionization potential, eV[15]	7.724 Cu(I)
	20.29 Cu(II)
	36.83 Cu(III)
Hall effect[15]	
Hall voltage, V at 0.30–0.8116 T	-5.24×10^{-4}
Hall coefficient, mV·mA·T	−5.5
Heat of atomization, kJ/mol[3]	339
Thermal conductivity, W/m·K	394,[3] 398[15]
Electrical conductivity, MS/cm at 20°C	0.591
Electrical resistivity at 20°C, nΩ·m	16.70
Temperature coefficient of electrical resistivitiy, 0–100°C[15]	0.0068
Melting point	1358.03 K (1084.88 °C, 1984.79 °F)[15]
	1356 K (1083°C, 1981.4 °F)[3]
Heat of fusion, kJ/kg	205, 204.9, 206.8[15]
	212[3]
Boiling point	2868 K (2595 °C, 4703 °F), 2840 K (2567°C, 4652 °F)[15]
	2595 K (2868°C, 5194 °F)[15]
Heat of vaporization, kJ/kg	7369[3]
	4729, 4726, 4793[15]
Specific heat, kJ/kg·K	0.255 (100 K)[15]
	0.384 (293 K)[3]
	0.386 (293 K)[15]
	0.494 (2000 K)[15]
Coefficient of expansion, linear, $\mu m/m$	16.5
Coefficient of expansion, volumetric, $10^{-6}/K$	49.5
Tensile strength, MPa	230 (annealed)[3]
	209 (annealed)[15]
	344 (cold drawn)[15]

Table 36 (*Continued*)

Property	Value
Elastic modulus, GPa	125 (tension, annealed)[15]
	102–120 (tension, hard drawn)[3]
	128 (tension, cold drawn)[15]
	46.4 (shear, annealed)[15]
	140 (bulk)[15]
Magnetic susceptibility, 291 K, mks	-0.086×10^{-6} [3]
	-1.08×10^{-6} [15]
Emissivity	0.03 (unoxidized metal, 100°C)[3]
	0.8 (heavily oxidized surface)[15]
Spectral reflection coefficient, incandescent light	0.63[15]
Nominal spectral emittance, λ = 655 nm, 800°C	0.15[15]
Absorptivity, solar radiation	0.25[15]
Viscosity, mPa·s (cP)	3.36 (1085°C, 1985 °F)[15]
	3.33 (1100°C, 2012 °F)[15]
	3.41 (1145°C, 2093 °F)[3]
	3.22 (1150°C, 2102 °F)[15]
	3.12 (1200°C, 2192 °F)[15]
Surface tension, m·N/m (dyn/cm)	1300 (99.99999% Cu, 1084°C, 1983 °F, vacuum)[15]
	1341 (99.999% Cu, N_2, 1150°C, 2102 °F)[15]
	1104 (1145°C, 2093 °F)[3]
	(see Ref. 15 for additional data)
Coefficient of friction	4.0 (Cu on Cu) in H_2 or N_2 [15]
	1.6 (Cu on Cu) in air or O_2 [15]
	1.4 (clean)[15]
	0.8 (in paraffin oil)[15]
Velocity of sound, m/s	4759 (longitudinal bulk waves)[15]
	3813 (irrotational rod waves)[15]
	2325 (shear waves)[15]
	2171 (Rayliegh waves)[15]

Note: Superscript numbers in parentheses are reference numbers.

silicon red brass, and aluminum oxide dispersion-strengthened coppers. Type III alloys include nonleaded brasses, bronzes, and nickel–silvers as well as silicon bronzes, coppers, high-copper alloys, low-aluminum bronzes, and copper–nickels.

Recommended Machining Practices

Recommended tool geometries are illustrated in publications available from the CDA. The following guidelines can be used as starting points from which optimum conditions can be planned for various machining operations.

Single-Point Turning Tools. Type I alloys require little or no rake, and modest clearance angles can be used but are not necessary unless the tool tends to drag (Figs. 3*a* and 3*b*). Free-cutting brass can be cut at maximum attainable speed; speeds for other leaded alloys range from 300–1000 sfm (91–305 mpm). Straight light mineral oils are generally preferred, but soluble oils also give good results.

Type II alloys exhibit a wide range of properties, and cutting parameters vary accordingly. Ductile alloys should be cut with rake angles of 10° with reasonable clearance. Less ductile alloys require little or no rake to avoid chatter. Cutting speeds between 150 and 300

Table 37A Physical Properties of Selected Wrought Coppers and Copper Alloys

UNS Number	Melting Point,[a] Liquidus (°F, *°C*)	Melting Point, Solidus (°F, *°C*)	Density (lb/in.^{3}, 68°F; *g/cm^3, 20°C*)	Electrical Resistivity (Ω·cmil/ft, 68°F; *μΩ·cm, 20°C*)	Electrical Conductivity[a] (%IACS, 68°F; *MS/cm, 20°C*)	Thermal Conductivity (Btu·ft/(h·ft^2·°F), 68°F; *W/m·K, 20°C*)	Coefficient of Thermal Expansion (10^{-6}/°F, 68–212°F; *10^{-6}/°C, 20–100°C*)	Coefficient of Thermal Expansion (10^{-6}/°F, 68–392°F; *10^{-6}/°C, 20–200°C*)	Coefficient of Thermal Expansion (10^{-6}/°F, 68–572°F; *10^{-6}/°C, 20–300°C*)	Specific Heat Capacity (Btu/lb/°F, 68 F; *J/kg·K, 293 K*)	Modulus of Elasticity in Tension (ksi, *MPa*)	Modulus of Rigidity (ksi, *MPa*)
C10100–C10300	1,981	1,981	0.323	10.3	101	226.0	9.4	9.6	9.8	0.092	17,000	6,400
	1,083	*1,083*	*8.94*	*1.71*	*0.591*	*391.1*	*16.9*	*17.3*	*17.6*	*393.5*	*117,000*	*44,130*
C11000	1,981	1,949	0.322	10.3	101	226.0	9.4	9.6	9.8	0.092	17,000	6,400
	1,083	*1,065*	*8.91*	*1.71*	*0.591*	*391.1*	*16.9*	*17.3*	*17.6*	*393.5*	*117,000*	*44,130*
C12200	1,981		0.322	12.2	85	196.0	9.4	9.5	9.8	0.092	17,000	6,400
	1,083		*8.91*	*2.03*	*0.497*	*339.2*	*16.9*	*17.1*	*17.6*	*393.5*	*117,000*	*44,130*
C15000	1,976	1,796	0.321	11.2	93	212.3	9.4	9.8		0.092	18,700	
	1,080	*980*	*8.89*	*1.86*	*0.544*	*366.9*	*16.9*	*17.6*		*393.5*	*129,000*	
C17200	1,800	1,590	0.298	46.2	22	62.0			9.9	0.1	18,500	7,300
	982	*866*	*8.26*	*7.68*	*0.129*	*107.3*			*17.8*	*419.0*	*128,000*	*50,330*
C17500	1,955	1,885	0.311	22.8	45	120.0		9.8		0.1	19,000	7,500
	1,068	*1,029*	*8.61*	*3.79*	*0.263*	*207.7*		*17.6*		*419.0*	*131,000*	*51,710*
C18100	1,967		0.319	13.0	80	187.0	9.3	10.2	10.7	0.094	18,200	125,000
	1,075		*8.83*	*2.16*	*0.468*	*323.6*	*16.7*	*18.4*	*19.3*	*393.9*	*6,800*	*46,880*
C19100	1,983	1,973	0.32	15.0	65	150.0	*9.0*			0.09	17,000	6,400
	1,084	*1,078*	*8.86*	*2.49*	*0.38*	*259.6*	*16.2*			*377.1*	*117,000*	*44,130*
C19200	1,980	1,900	0.323	18.9	55	145			9.8	0.092	16,000	6,000
	1,082	*1,038*	*8.94*	*3.14*	*0.322*	*251.0*			*17.6*	*385.5*	*110,000*	*41,370*
C23000	1,880	1,810	0.316	28.0	37	92.0			10.4	0.09	17,000	6,400
	1,027	*988*	*8.75*	*4.65*	*0.216*	*159.2*			*18.7*	*377.1*	*117,000*	*44,130*
C26000	1,750	1,680	0.308	37.0	28	70.0			11.1	0.09	16,000	6,000
	954	*916*	*8.53*	*6.15*	*0.164*	*121.2*			*20.0*	*377.1*	*110,000*	*41,370*
C27000	1,710	1,660	0.306	38.4	27	67.0			11.3	0.09	15,000	5,600
	932	*904*	*8.47*	*6.38*	*0.158*	*116.0*			*20.3*	*377.1*	*103,400*	*38,610*
C28000	1,660	1,650	0.303	37.0	28	71.0			11.6	0.09	15,000	5,600
	904	*899*	*8.39*	*6.15*	*0.164*	*122.9*			*20.9*	*377.1*	*103,400*	*38,610*
C36000	1,650	1,630	0.307	39.9	26	67.0			11.4	0.09	14,000	5,300
	899	*888*	*8.5*	*6.63*	*0.152*	*116.0*			*20.5*	*377.1*	*96,500*	*36,500*
C37700	1,640	1,620	0.305	39.4	27	69.0			11.5	0.09	15,000	5,600
	893	*882*	*8.44*	*6.38*	*0.158*	*119.4*			*20.7*	*377.1*	*103,400*	*38,610*
C44400	1,720	1,650	0.308	41.5	25	64.0			11.2	0.09	16,000	6,000
	938	*899*	*8.53*	*6.90*	*0.146*	*110.8*			*20.2*	*377.1*	*110,000*	*41,370*
C46500	1,650	1,630	0.304	39.9	26	67.0			11.8	0.09	15,000	5,600
	899	*888*	*8.41*	*6.63*	*0.152*	*116.0*			*21.2*	*377.1*	*103,400*	*38,610*
C50500	1,970	1,900	0.321	21.6	48	120.0			9.9	0.09	17,000	6,400
	1,077	*1,038*	*8.89*	*3.59*	*10.281*	*207.7*			*17.8*	*377.1*	*117,000*	*44,130*

Table 37A (*Continued*)

UNS Number	Melting Point,[a] Liquidus (°F, *°C*)	Melting Point, Solidus (°F, *°C*)	Density (lb/in.^{3}, 68°F; *g/cm^3, 20°C*)	Electrical Resistivity Ω·cmil/ft, 68°F; *μΩ·cm, 20°C*)	Electrical Conductivity[a] (%IACS, 68°F; *MS/cm, 20°C*)	Thermal Conductivity (Btu·ft/(h·ft^2·°F), 68°F; *W/m·K, 20°C*)	Coefficient of Thermal Expansion (10^{-6}/°F, 68–212°F; *10^{-6}/°C, 20–100°C*)	Coefficient of Thermal Expansion (10^{-6}/°F, 68–392°F; *10^{-6}/°C, 20–200°C*)	Coefficient of Thermal Expansion (10^{-6}/°F, 68–572°F; *10^{-6}/°C, 20–300°C*)	Specific Heat Capacity (Btu/lb/°F, 68 F; *J/kg·K, 293 K*)	Modulus of Elasticity in Tension (ksi, *MPa*)	Modulus of Rigidity (ksi, *MPa*)
C51100	1,945	1,785	0.320	52.0	20	48.4			9.9	0.09	16,000	6,000
	1,063	*974*	*8.86*	*8.64*	*0.117*	*83.8*			*17.87*	*377.1*	*110,000*	*41,370*
C52100	1,880	1,620	0.318	79.8	13	36.0			10.1	0.09	16,000	6,000
	1,027	*882*	*8.80*	*13.27*	*0.076*	*62.3*			*18.2*	*377.1*	*110,000*	*41,370*
C61300	1,915	1,905	0.287	86.8	12	32.0			9.0	0.09	17,000	6,400
	1,046	*1,041*	*7.95*	*14.43*	*0.070*	*55.4*			*16.2*	*377.1*	*117,000*	*44,130*
C61800	1,913	1,904	0.272	79.8	13	37.0			9.0	0.09	17,000	6,400
	1,045	*1,040*	*7.53*	*13.27*	*0.076*	*64.0*			*16.2*	*377.1*	*117,000*	*44,130*
C62400	1,900	1,880	0.269	86.4	12	34.0			9.0	0.09	17,000	6,400
	1,038	*1,027*	*7.45*	*14.36*	*0.070*	*58.8*			*16.2*	*377.1*	*117,000*	*44,130*
C63200	1,940	1,905	0.276	148.0	7	20.0			9.0	0.09	17,000	6,400
	1,060	*1,041*	*7.64*	*24.6*	*0.041*	*34.6*			*16.2*	*377.1*	*117,000*	*44,130*
C65500	1,880	1,780	0.308	148.0	7	21.0			10.0	0.09	15,000	5,600
	1,027	*971*	*8.53*	24.6	*0.041*	*36.3*			*18.0*	*377.1*	*103,400*	*38,610*
C66700	2,000	1,920	0.308	61.0	17	56.0			11.1	0.09	16,000	
	1,093	*1,049*	*8.53*	*10.14*	*0.099*	*96.9*			*20.0*	*377.1*	*110,000*	
C67500	1,630	1,590	0.302	43.2	24	61.0			11.8	0.09	15,000	5,600
	888	*866*	*8.36*	*7.18*	*0.14*	*105.6*			*21.2*	*377.1*	*103,400*	*38,610*
C69400	1,685	1,510	0.296	167.0	6	15.0			11.2	0.09	16,000	
	918	*821*	*8.19*	*27.76*	*0.036*	*26.0*			*20.2*	*377.1*	*110,000*	
C70600	2,100	2,010	0.323	115.0	9	26.0			9.5	0.09	18,000	6,800
	1,149	*1,099*	*8.94*	*19.12*	*0.053*	*45.0*			*17.1*	*377.1*	*124,000*	*46,880*
C71500	2,260	2,140	0.323	225.0	4	17.0			9.0	0.09	22,000	8,300
	1,238	*1,171*	*8.94*	*37.4*	*0.027*	*29.4*			*16.2*	*377.1*	*152,000*	*57,230*
C72200	2,148	2,052	0.323	159.0	6	19.9			8.8	0.094	20,000	8,200
	1,176	*1,122*	*8.94*	*26.43*	*0.038*	*34.4*			*15.8*	*393.9*	*138,000*	*56,540*
C74500	1,870		0.314	115.0	9	26.0			9.1	0.09	17,500	6,600
	1,021		*8.69*	*19.12*	*0.053*	*45.0*			*16.4*	*377.1*	*121,000*	*45,510*
C77000	1,930		0.314	189.0	5	17.0			9.3	0.09	18,000	6,800
	1,054		*8.69*	*31.42*	*0.032*	*29.4*			*16.7*	*377.1*	*124,000*	*46,880*

[a] TFOO condition, TFOO = solution heat treated and aged (per ASTM B 601).

Table 37B Physical Properties of Selected Cast Copper Alloys

UNS Number	Melting Point, Liquidus (°F, *°C*)	Melting Point, Solidus (°F, *°C*)	Density (lb/in.^{3}, 68°F; *g/cm^3, 20°C*)	Electrical Resistivity Ω·cmil/ft, 68°F; *μΩ·cm, 20°C*)	Electrical Conductivity[a] (%IACS, 68°F; *MS/cm, 20°C*)	Thermal Conductivity (Btu·ft/(h·ft^2·°F), 68°F; *W/m·K, 20°C*)	Coefficient of Thermal Expansion (10^{-6}/°F, 68–392°F; *10^{-6}/°C, 20–200°C*)	Coefficient of Thermal Expansion (10^{-6}/°F, 68–572°F; *10^{-6}/°C, 20–300°C*)	Specific Heat Capacity (Btu/lb/°F, 68°F; *J/kg·K, 293 K*)	Modulus of Elasticity in Tension (ksi, *MPa*)	Relative Magnetic Permeability	Poisson's Ratio
C81100	1,981	1,948	0.323	11.3	92	200.0		9.4	0.09	17,000	1.0	
	1,083	*1,064*	*8.94*	*1.87*	*0.538*	*346.1*		*16.9*	*377.1*	*117,000*		
C81400	2,000	1,950	0.318	17.3	60	150.0		10.0	0.093	16,000		
	1,093	*1,066*	*8.81*	*2.87*	*0.348*	*259.0*		*18.0*	*389.7*	*110,000*		
82000	1,990	1,780	0.311[a]	23.1	45	150.0		9.9	0.1	*17,000*	1.001	0.33
	1,088	*971*	*8.61*	*3.85*	*0.263*	*259.6*		*17.8*	*419.0*	117,000		
C82500	1,800	1,575	0.297[b]	51.6	20	74.9	9.4		0.1	18,500	1.002	0.3
	982	*857*	*8.09*	*8.62*	*0.116*	*129.8*	*16.9*		*419.0*	*128,000*		
C83300	1,940	1,886	0.318	32.3	32				0.09	15,000		
	1,060	*1,030*	*8.8*	*5.38*	*0.187*				*377.1*	*10,340*		
C83450	1,860	1,580	0.319		20							
	1,015	*860*	*8.83*		*0.115*							
C83600	1,850	1,570	0.318	69.1	15	41.6	10.0		0.09	13,500	1.0	
	1,010	*854*	*8.83*	*11.49*	*0.087*	*72.0*	*18.0*		*377.1*	*93,100*		
C83800	1,840	1,550	0.312	69.1	15	41.8	10.0		0.09	13,300	1.0	
	1,004	*843*	*8.64*	*11.49*	*0.087*	*72.4*	*18.0*		*377.1*	*91,700*		
C84400	1,840	1,549	0.314	63.3	16	41.8	10.0[c]		0.09	13,000	1.0	
	1,004	*843*	*8.69*	*10.53*	*0.095*	*72.4*			*377.1*	*89,600*		
C85200	1,725	1,700	0.307	57.8	18	48.5	11.5		0.09	11,000		
	941	*927*	*8.50*	*9.62*	*0.104*	*83.9*	*20.7*[d]		*377.1*	*75,800*		
C85400	1,725	1,700	0.305	53.2	20	50.8	11.1		0.09	12,000	1.0	
	941	*927*	*8.44*	*8.85*	*0.113*	*87.9*	*20.2*[d]		*377.1*	*82,700*		
C85700	1,725	1,675	0.304	47.0	22	48.5		12.0	0.09	14,000	1.0	
	941	*913*	*8.41*	*7.81*	*0.128*	*83.9*		*21.6*	*377.1*	*87,000*		
C86200	1,725	1,650	0.288	136.7	8	20.5		12.0	0.09	15,000	1.24[e]	
	941	*899*	*7.97*	*22.73*	*0.044*	*35.5*		*21.6*	*377.1*	*103,400*		
C86300	1,693	1,625	0.283	130.8	8	20.5		12.0	0.09	14,200	1.09[e]	
	923	*885*	*7.83*	*21.74*	*0.044*	*35.5*		*21.6*	*377.1*	*97,900*		
C86400	1,616	1,583	0.301	54.2	19	51.0	11.0[c]		0.09	14,000		
	880	*862*	*8.33*	*9.01*	*0.111*	*88.3*	*19.8*		*377.1*	*87,000*		
C86500	1,616	1,583	0.301	47.0	22	49.6	11.3[d]		0.09	15,000	1.09[e]	
	880	*862*	*8.33*	*7.81*	*0.128*	*85.8*	*20.3*		*377.1*	*103,400*		
C87300	1,780	1,510	0.302		6	16.4		11.0		16,000		
	971	*821*	*8.36*		*0.035*	*28.4*		*19.8*		*110,000*		
C89510	1,871	1,500	0.313									
	1,021	*815*	*8.66*									
C89520	1,842	353										
	1,005	*196*										

Table 37B (*Continued*)

UNS Number	Melting Point, Liquidus (°F, °*C*)	Melting Point, Solidus (°F, °*C*)	Density (lb/in.^{3}, 68°F; *g/cm^3, 20°C*)	Electrical Resistivity Ω·cmil/ft, 68°F; *μΩ·cm, 20°C*)	Electrical Conductivity[a] (%IACS, 68°F; *MS/cm, 20°C*)	Thermal Conductivity (Btu·ft/(h·ft^2·°F), 68°F; *W/m·K, 20°C*)	Coefficient of Thermal Expansion (10^{-6}/°F, 68–392°F; *10^{-6}/°C, 20–200°C*)	Coefficient of Thermal Expansion (10^{-6}/°F, 68–572°F; *10^{-6}/°C, 20–300°C*)	Specific Heat Capacity (Btu/lb/°F, 68°F; *J/kg·K, 293 K*)	Modulus of Elasticity in Tension (ksi, *MPa*)	Relative Magnetic Permeability	Poisson's Ratio
C89550	1,638	1,588										
	892	*864*										
C87500	1,680	1,510	0.299	154.2	6	16.0		10.9	0.09	15,400		
	916	*821*	*8.28*	*25.64*	*0.039*	*27.7*		*19.6*	*377.1*	*106,000*		
C90300	1,832	1,570	0.318	87.2	12	43.2	10.0		0.09	14,000	1.0	
	1,000	*854*	*8.8*	*14.49*	*0.069*	*74.8*	*18.0*		*377.1*	*87,000*		
C90500	1,830	1,570	0.315	94.0	11	43.2	11.0		0.09	15,000	1.0	
	999	*854*	*8.72*	*15.63*	*0.064*	*74.8*	*19.8*		*377.1*	*103,400*		
C90700	1,830	1,528	0.317	107.4	10	40.8	10.2		0.09	15,000	1.0	
	999	*831*	*8.77*	*17.86*	*0.056*	*70.6*	*18.4*		*377.1*	*103,400*		
C91100	1,742	1,505		122.8	8				0.09	15,000		
	950	*818*		*20.41*	*0.049*				*377.1*	*103,400*		
C91300	1,632	1,505		150.4	7				0.09	16,000		
	889	*818*		*25.0*	*0.04*				*377.1*	*110,000*		
C91600	1,887	1,575	0.32	103.7	10	40.8	9.0		0.09	16,000		
	1,031	*857*	*8.86*	*17.24*	*0.058*	*70.6*	*16.2*		*377.1*	*110,000*		
C92200	1,810	1,518	0.312	72.5	14	40.2	10.0		0.09	14,000	1.0	
	988	*826*	*8.64*	*12.0*	*0.083*	*69.6*	*18.0*		*377.1*	*87,000*		
C92300	1,830	1,570	0.317	85.9	12	43.2	10.0		0.09	14,000		
	999	*854*	*8.77*	*14.29*	*0.069*	*74.8*	*18.0*		*377.1*	*87,000*		
C92600	1,800	1,550	0.315	115.7	9		10.0		0.09	15,000		
	982	*843*	*8.72*	*19.23*	*0.052*		*18.0*		*377.1*	*103,400*		
C92900	1,887	1,575	0.32	113.5	9	33.6	9.5		0.09	14,000		
	1,031	*857*	*8.86*	*18.87*	*0.052*	*58.2*	*17.1*		*377.1*	*87,000*		
C93200	1,790	1,570	0.332	85.9	12	33.6	10.0[d]		0.09	14,500		
	977	*854*	*8.91*	*14.29*	*0.07*	*58.2*	*18.0*		*377.1*	*100,000*		
C93700	1,705	1,403	0.32	102.0	10	27.1	10.3		0.09	11,000	1.0	
	929	*762*	*8.86*	*16.95*	*0.059*	*46.9*	*18.5*		*377.1*	*75,800*		
C93800	1,730	1,570	0.334	91.1	11	30.2	10.3		0.09	10,500	1.0	
	943	*854*	*9.25*	*15.15*	*0.066*	*52.3*	*18.5*		*377.1*	*72,400*		
C95200	1,913	1,907	0.276	94.0	11	29.1		9.0	0.09	15,000	1.1[e]	0.31
	1,045	*1,042*	*7.64*	*15.63*	*0.066*	*50.4*		*16.2*	*377.1*	*103,400*	*1.2*	

C95300	1,913	1,904	0.272	80.2	13	36.3	9.0	0.09	16,000	1.07[e]	
	1,045	*1,040*	*7.53*	*13.33*	*0.075*	*62.8*	*16.2*	*377.1*	*110,000*		
C95400	1,900	1,880	0.269	80.2	13	33.9	9.0	0.09	15,500	1.27[f]	
	1,038	*1,027*	*7.45*	*13.33*	*0.075*	*58.7*	*16.2*	*377.1*	*107,000*		
C95410	1,900	1,880	0.269	80.2	13	33.9	9.0	0.10	15,500		
	1,038	*1,027*	*7.45*	*13.33*	*0.075*	*58.7*	*16.2*	*419.0*	*107,000*		
C95500	1,930	1,900	0.373	122.8	8	24.2	9.0	0.10	16,000	1.32[f]	0.32
	1,054	*1,038*	*7.53*	*20.41*	*0.049*	*41.9*	*16.2*	*419.0*	*110,000*	1.2[g]	
C95800	1,940	1,910	0.276	146.7	7	20.8	9.0	0.105	16,500	1.05[f]	0.32
	1,060	*1,043*	*7.64*	*24.39*	*0.041*	*36.0*	*16.2*	*440.0*	*114,000*		
C96400	2,260	2,140	0.323	214.8	5	16.4	9.0	0.09	21,000		
	1,238	*1,171*	*8.94*	*35.71*	*0.028*	*28.5*	*16.2*	*377.1*	*145,000*		
C97300	1,904	1,850	0.321	182.3	6	16.5	9.0	0.09	16,000		
	1,040	*1,010*	*8.89*	*30.3*	*00.33*	*28.6*	*16.2*	*377.1*	*110,000*		
C97400	2,012	1,958	0.32	188.0	6	15.8	9.2	0.09	16,000		
	1,100	*1,070*	*8.86*	*31.25*	*0.032*	*27.3*	*16.6*	*377.1*	*110,000*		
C97600	2,089	2,027	0.321	207.4	5	13.0	9.3	0.09	19,000		
	1,143	*1,108*	*8.89*	*34.48*	*0.029*	*22.6*	*16.7*	*377.1*	*131,000*		
C97800	2,089	2,027	0.321	207.4	5	13.0	9.3	0.09	19,000		
	1,143	*1,108*	*8.89*	*34.48*	*0.029*	*22.6*	*16.7*	*377.1*	*131,000*		
C99400			0.3	85.9	16				19,300		
			8.3	*14.29*	*0.098*				*133,000*		
C99500			0.3	71.0	10		8.3		19,000		
			8.3	*11.64*	*0.057*		*14.9*		*131,000*		
C99700	1,655	1,615	0.296	353.8	3				16,500		
	902	*879*	*8.19*	*58.82*	*58.82*				*114,000*		
C99750	1,550	1,505	0.290	501.3	2	13.5		0.09	17,000		
	843	*818*	*8.03*	*83.3*	*0.012*	*24.3*		*377.1*	*117,000*		

[a] TFOO condition.
[b] Change in density upon aging = 0.6%.
[c] At 20°C, 68°F.
[d] 20–100°C, 68–212°F.
[e] Field strength 16 kA/m (200 Oe).
[f] As cast, field strength 16 kA/m (200 Oe).
[g] TQ50 temper, field strength 16 kA/m (200 Oe).

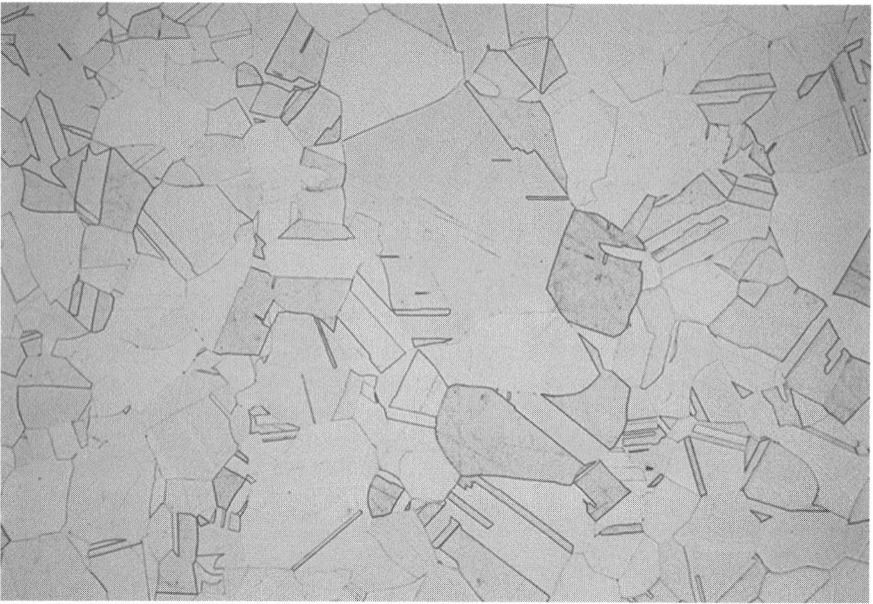

Figure 1 Microstructure of pure copper.

sfm (46 and 92 mpm) are recommended for high-speed steel single-point and form tools. Speeds can be increased to between 400 and 600 sfm (122 and 183 mpm) with carbide tooling. Begin with light cuts, 0.002 in./rev (0.05 mm/rev) and increase feed until surface finish and/or wear rates deteriorate. Alloys that tend to type III behavior require cutting fluids that provide effective lubrication and cooling. Mineral oil fortified with between 5 and 15% lard oil or a sulfurized fatty oil base thinned with a light mineral oil gives good lubrication. Alloys with higher machinability ratings can be cut with soluble oils.

Type III alloys require generous rake angles. Single-point tools should be ground with a back rake of 10°–20° and a side rake of as much as 20°–30°. Rake angles can be reduced somewhat with single-point carbide tools and dovetail or circular form tools. Coppers and copper–nickels tend to build up on the tool face, but burnishing the tool to a smooth finish reduces this tendency, as does the application of low-friction coatings such as titanium nitride. High-speed form tools should be ground with a front clearance angle between 7° and 12°; the clearance angle can be reduced slightly with carbide tooling.

Recommended straight-blade and circular cutoff tool geometries are illustrated in Fig. 3*c*; geometries for carbide and high-speed circular and dovetail form tools are illustrated in Fig. 3*d*.

Milling. Free-cutting type I alloys can be milled at speeds up to 500 sfm (152 mpm) (Fig. 4). Carbide tooling permits the highest speeds. Rake angles of 0°–10° are recommended, and land clearance should be between 5° and 15°. Side-milling cutters require a 12° clearance on each side. Cutting fluids are needed for heat removal at high speeds but are sometimes not used under moderate milling conditions.

Short-chip type II alloys can tolerate rake angles between 0° and 10°, but optimum conditions should be selected for individual alloys. Small-diameter cutters should be ground with 0° rake for hard aluminum bronzes. Milling speeds can be as high as 200 sfm (61

Table 38 Standard Temper Designations for Copper-Base Alloys per ASTM Standard B 601

Annealed Tempers—O	Temper Names
O10	Cast and annealed
O11	As cast and precipitation heat treated
O25	Hot rolled and annealed
O30	Hot extruded and annealed
O31	Hot extruded and annealed
O32	Extruded and precipitation heat treated
O30	Light anneal
O50	Soft anneal
O60	Annealed (also mill annealed)
O61	Soft anneal
O65	Annealed
O68	Drawing anneal
O70	Dead soft annealed
O80	Annealed to Temper—$\frac{1}{8}$ hard
O81	Annealed to Temper—$\frac{1}{4}$ hard
O82	Annealed to Temper—$\frac{1}{2}$ hard

Annealed Tempers with Grain Size Prescribed—OS	Nominal Average Grain Size, mm
OS005	0.005
OS010	0.010
OS015	0.015
OS020	0.020
OS025	0.025
OS035	0.035
OS050	0.050
OS060	0.060
OS070	0.070
OS100	0.100
OS120	0.120
OS150	0.150
OS200	0.200

Cold-Worked Tempers Based on Cold Rolling or Cold Drawing—H	Temper Names
H00	$\frac{1}{8}$ Hard
H01	$\frac{1}{4}$ Hard
H02	$\frac{1}{2}$ Hard
H03	$\frac{3}{4}$ Hard
H04	Hard
H06	Extra hard
H08	Spring
H10	Extra spring
H12	Special spring
H13	Ultra spring
H14	Super spring

Cold-Worked Tempers Based on Particular Products (Wire)—H	Temper Names
H50	Extruded and drawn
H52	Pierced and drawn
H58	Drawn general purpose
H60	Cold heading, forming
H63	Rivet
H64	Screw
H66	Bolt
H70	Bending
H80	Hard drawn
H85	Medium-hard drawn electrical wire
H86	Hard-hard drawn electrical wire
H90	As finned

Cold Worked and Stress Relieved—HR	Temper Names
HR01	$\frac{1}{4}$ Hard and stress relieved
HR02	$\frac{1}{2}$ Hard and stress relieved
HR04	Hard and stress relieved
HR08	Spring and stress relieved
HR10	Extra spring and stress relieved
HR12	Special spring and stress relieved
HR20	As finned

Cold Rolled and Order Strengthened	Temper Names
HT04	Hard temper and treated
HT08	Spring temper and treated

Hard Drawn End Annealed	Temper Names
HE80	Hard drawn and end annealed

Cold-Worked Tempers with Added Treatments—HR	Temper Names
HR50	Drawn and stress relieved

As-Manufactured Tempers—M	Temper Names
M01	As sand cast
M02	As centrifugal cast
M03	As plaster cast
M05	As pressure die cast
M06	As permanent mold cast
M07	As investment cast
M10	As hot forged—air cooled
M11	As hot forged—quenched
M20	As hot rolled
M25	As hot rolled and rerolled
M30	As extruded
M40	As hot pierced
M45	As hot pierced and rerolled

Table 38 (*Continued*)

Heat-Treated Tempers, T; Quench Hardened, TQ	Temper Names
TQ00	Quench hardened
TQ30	Quench hardened tempered
TQ50	Quench hardened temper annealed
TQ55	Quench hardened temper annealed, cold drawn and stress relieved
TQ75	Interrupted quench

Solution Heat Treated—TB	Temper Names
TB00	Solution heat treated (A)

Solution Treated and Cold Worked—TD	Temper Names
TD00	Solution treated and cold worked: $\frac{1}{8}$ hard
TD01	Solution treated and cold worked: $\frac{1}{4}$ hard
TD02	Solution treated and cold worked: $\frac{1}{2}$ hard
TD03	Solution treated and cold worked: $\frac{3}{4}$ hard
TD04	Solution treated and cold worked: hard (H)

Solution Heat Treated and Precipitation Heat Treated—TF	Temper Names
TF00	Precipitation hardened (AT)
TF01	Precipitation heat-treated plate—low hardness (ATLH)
TF02	Precipitation heat-treated plate—high hardness (ATHH)

Solution Heat Treated and Spinodal Heat Treated—TX	Temper Names
TX00	Spinodal hardened

Solution Heat Treated, Cold Worked and Precipitation Heat Treated—TH	Temper Names
TH01	$\frac{1}{4}$ Hard and precipitation heat treated ($\frac{1}{4}$ HT)
TH02	$\frac{1}{2}$ Hard and precipitation heat treated ($\frac{1}{2}$ HT)
TH03	$\frac{3}{4}$ Hard and precipitation heat treated ($\frac{3}{4}$ HT)
TH04	Hard and precipitation heat treated (HT)

Cold-Worked Tempers and Spinodal Heat Treated to Meet Standard Requirements Based on Cold Rolling or Cold Drawing—TS	Temper Names
TS00	$\frac{1}{8}$ Hard and spinodal hardened ($\frac{1}{8}$TS)
TS01	$\frac{1}{4}$ Hard and spinodal hardened ($\frac{1}{4}$TS)
TS02	$\frac{1}{4}$ Hard and spinodal hardened ($\frac{1}{4}$TS)
TS03	$\frac{3}{4}$ Hard and spinodal hardened ($\frac{3}{4}$TS)
TS04	Hard and spinodal hardened
TS06	Extra hard and spinodal hardened
TS08	Spring and spinodal hardened
TS10	Extra spring and spinodal hardened
TS12	Special spring and spinodal hardened
TS13	Special spring and spinodal hardened
TS14	Ultra spring and spinodal hardened

Mill Hardened Tempers—TM	Manufacturing Designation
TM00	AM
TM01	$\frac{1}{4}$ HM
TM02	$\frac{1}{2}$ HM
TM04	HM
TM06	XHM
TM08	XHMS

Precipitation Heat Treated or Spinodal Heat Treated and Cold Worked—TL	Temper Names
TL00	Precipitation heat treated or spinodal heat treated and $\frac{1}{8}$ hard
TL01	Precipitation heat treated or spinodal heat treated and $\frac{1}{4}$ hard
TL02	Precipitation heat treated or spinodal heat treated and $\frac{1}{2}$ hard
TL04	Precipitation heat treated or spinodal heat treated and hard
T008	Precipitation heat treated or spinodal heat treated and spring
TL10	Precipitation heat treated or spinodal heat treated and extra spring

Table 38 *(Continued)*

Precipitation Heat Treated or Spinodal Heat Treated, Cold Worked and Stress Relief Annealed—TR	Temper Names
TR01	Precipitation heat treated or spinodal heat treated, $\frac{1}{4}$ hard and stress relieved
TR02	Precipitation heat treated or spinodal heat treated, $\frac{1}{2}$ hard and stress relieved
TR04	Precipitation heat treated or spinodal heat treated, hard and stress relieved

Tempers of Welded Tube—W; As Welded, WM	Temper Names
WM50	As welded from annealed strip
WM00	As welded from annealed $\frac{1}{8}$ hard strip
WM01	As welded from annealed $\frac{1}{4}$ hard strip
WM02	As welded from annealed $\frac{1}{2}$ hard strip
WM03	As welded from annealed $\frac{3}{4}$ hard strip
WM04	As welded from annealed hard strip
WM06	As welded from annealed extra hard strip
WM08	As welded from annealed spring strip
WM10	As welded from annealed extra spring strip
WM15	As welded from annealed strip, thermal stress relieved
WM20	As welded from $\frac{1}{8}$ hard strip, thermal stress relieved
WM21	As welded from $\frac{1}{4}$ hard strip, thermal stress relieved
WM22	As welded from $\frac{1}{2}$ hard strip, thermal stress relieved

Welded Tube and Annealed—WO	Temper Names
WO50	Welded and light annealed
WO60	Welded and soft annealed
WO61	Welded and annealed

Welded Tube and Light Cold Worked—WH	Temper Names
WH55	Welded and light cold worked

Welded Tube and Cold Drawn—WH	Temper Names
WH00	Welded and drawn: $\frac{1}{8}$ hard
WH01	Welded and drawn: $\frac{1}{4}$ hard
WH02	Welded and drawn: $\frac{1}{2}$ hard
WH03	Welded and drawn: $\frac{3}{4}$ hard
WH04	Welded and drawn: hard
WH06	Welded and drawn: extra hard
WH55	Welded and cold reduced or light drawn
WH58	Welded and drawn, general purpose
WH80	Welded and reduced or hard drawn

Welded Tube, Cold Drawn and Stress Relieved—WR	Temper Names
WR00	Welded, drawn and stress relieved from: $\frac{1}{8}$ hard
WR01	Welded, drawn and stress relieved from: $\frac{1}{4}$ hard
WR02	Welded, drawn and stress relieved from: $\frac{1}{2}$ hard
WR03	Welded, drawn and stress relieved from: $\frac{3}{4}$ hard
WR04	Welded, drawn and stress relieved from: $\frac{1}{8}$ hard
WR06	Welded, drawn and stress relieved from: extra hard

mpm) with feed rates between 0.016 and 0.022 in./rev (0.4 and 0.56 mm/rev) for spiral cutters and from 0.010 to 0.022 in./rev (0.25–0.056 mm/rev) per tooth for end mills. Soluble oils are satisfactory, as are mineral oils containing about 5% lard oil.

Long-chip type III alloys require milling cutters with tooth spacing no finer than four to eight teeth per inch to facilitate chip removal. Spiral cutters are useful for wide cuts, for which helix angles as large as 53° have been found to be satisfactory. Generous rake angles (up to 15°) and adequate clearance (up to 15°) should be provided. Radial undercuts prevent tooth edges from dragging, and cutting faces should be finely polished to reduce buildup. Recommended milling speeds range from 50 to 100 sfm (15–45 mpm); feed rates range from 0.007 to 0.030 in./rev (0.18–0.76 mm/rev).

Table 39 Mechanical Properties of Selected Wrought Copper Alloys

Temper	Section Size in. mm.	Cold Work %	Typ/ Min.	Temp F C	Tensile Strength ksi MPa	Yield Strength (0.5% ext. under load) ksi MPa	Yield Strength (0.2% offset) ksi MPa	Yield Strength (0.05% offset) ksi MPa	El %	Rockwell Hardness B	Rockwell Hardness C	Rockwell Hardness F	Rockwell Hardness 30T	Vickers Hardness 500	Brinell Hardness 500	Brinell Hardness 3000	Shear Strength ksi MPa	Fatigue Strength[a] ksi MPa	Izod Impact Strength ft-lb J
C10100–C10300—Automotive Rectifiers, Conductors, Glass-to-Metal Seals, High-Resistance-Ratio Cryogenic Shunts, Bus Bars, Lead-in Wire, Vacuum Seals, Transistor Component Bases, Bus Conductors, Waveguides, Hollow Conductors, Anodes for Vacuum Tubes, Coaxial Cable, Coaxial Tube, Klystrons, Microwave Tubes																			
Rod																			
H04	2	16	TYP	68	45	40	—	—	20	45	—	85	—	—	—	—	26	—	—
	51			20	310	276	—	—	20	45	—	85	—	—	—	—	179	—	—
M20	1	0	TYP	68	32	10	—	—	55	—	—	40	—	—	—	—	22	—	—
	25.4			20	221	69	—	—	55	—	—	40	—	—	—	—	152	—	—
Tube																			
H55	0.065	0	TYP	68	40	32	—	—	25	35	—	77	45	—	—	—	26	—	—
	1.65			20	276	221	—	—	25	35	—	77	45	—	—	—	179	—	—
H80	0.065	0	TYP	68	55	50	—	—	8	60	—	95	63	—	—	—	29	—	—
	1.65			20	379	345	—	—	8	60	—	95	63	—	—	—	200	—	—
Wire																			
H04	−8	0	TYP	68	55	—	—	—	1	—	—	—	—	—	—	—	29	—	—
	2			20	379	—	—	—	1	—	—	—	—	—	—	—	200	—	—
H08	−8	0	TYP	68	66	—	—	—	1	—	—	—	—	—	—	—	33	—	—
	2			20	455	—	—	—	1	—	—	—	—	—	—	—	228	—	—
Shapes																			
H04	0.5	15	TYP	68	40	32	—	—	30	35	—	—	—	—	—	—	26	—	—
	12.7			20	276	221	—	—	30	35	—	—	—	—	—	—	179	—	—
Flat Products																			
H00	0.25	0	TYP	68	36	28	—	—	40	10	—	60	—	—	—	—	25	—	—
	6.35			20	248	193	—	—	40	10	—	60	—	—	—	—	172	—	—
H04	1	0	TYP	68	45	40	—	—	20	45	—	85	—	—	—	—	26	—	—
	25.4			20	310	276	—	—	20	45	—	85	—	—	—	—	179	—	—

C11000—Conductors, Electrical, Terminal Connectors, Busbars, Magnet Wire, Stranded Conductors, Wire, Electrical, Terminals, Switches, Radio Parts, Contacts, Trolley Wire, Gaskets, Radiators, Downspouts, Flashing, Roofing, Gutters, Building Fronts, Skylight Frames, Spouting, Ball Floats, Butts, Rivets, Nails, Cotter Pins, Soldering Copper, Tacks, Wire Screening, Fasteners, Heat Exchangers, Pans, Vats, Road Bed Expansion Plates, Rotating Bands, Kettles, Chimney Cap Screens, Chlorine Cells, Pressure Vessels, Anodes, Chemical Process Equipment, Printing Rolls

Rod																			
H04	0.25	40	TYP	68	55	50	—	—	10	60	—	94	—	—	—	—	29	—	—
	6.35			20	379	345	—	—	10	60	—	94	—	—	—	—	200	—	—
M20	1	0	TYP	68	32	10	—	—	55	—	—	40	—	—	—	—	22	—	—
	25.4			20	221	69	—	—	55	—	—	40	—	—	—	—	152	—	—
Tube																			
H80	0.065	40	TYP	68	55	50	—	—	8	60	—	95	63	—	—	—	29	—	—
	1.65			20	379	345	—	—	8	60	—	95	63	—	—	—	200	—	—
Wire																			
H04	0.08	0	TYP	68	55	—	—	—	1	—	—	—	—	—	—	—	29	—	—
	2			20	379	—	—	—	1	—	—	—	—	—	—	—	200	—	—
OS050	0.08	0	TYP	68	35	—	—	—	35	—	—	—	—	—	—	—	24	—	—
	2			20	241	—	—	—	35	—	—	—	—	—	—	—	165	—	—
Shapes																			
H04	0.5	15	TYP	68	40	32	—	—	30	35	—	—	—	—	—	—	26	—	—
	12.7			20	276	221	—	—	30	35	—	—	—	—	—	—	179	—	—

C12200—Gutters, Downspouts, Flashing, Roofing, Air Lines, Hydraulic Lines, Oil Lines, Air Conditioner Tubes and Condenser Sheets, Gas Lines, Heater Units, Heater Lines, Oil Burner Tubes, Air Conditioners, Refrigerators, Heater Elements, Wire Connectors, Tubing, LP Gas Service, Tubing, Medical Gas—Oxygen, Steam Lines, Paper Lines, Pulp Lines, Distiller Tubes, Dairy Tubes, Heat Exchanger Tubes, Evaporator Tubes, Condenser Tubes, Brewery Tubes, Oil Lines in Airplanes, Hydraulic Lines in Airplanes, Gasoline Lines in Airplanes, Air Lines in Airplanes, Oil Coolers in Airplanes, Tanks, Water Lines, Plating Hangers, Plating Anodes, Sugar House Refinery Lines, Print Rolls, Paper Rolls, Expansion Joint Tubes, Heat Exchanger Shells, Anodes for Electroplating, Kettles, Rotating Bands, Gage Lines, Casting Molds, Plating Anodes, Plating Racks, Plumbing Tube, Oil Coolers, Gasoline Lines, Plumbing Fittings, Plumbing Pipe

Pipe																			
H04	—	30	TYP	68	50	45	—	—	10	50	—	90	—	—	—	—	28	—	—
	—			20	345	310	—	—	10	50	—	90	—	—	—	—	193	—	—
Tube																			
H80	0.065	40	TYP	68	55	50	—	—	8	60	—	95	63	—	—	—	29	19	—
	1.65			20	379	345	—	—	8	60	—	95	63	—	—	—	200	131	—
OS050	0.065	0	TYP	68	32	10	—	—	45	—	—	40	—	—	—	—	22	11	—
	1.65			20	221	69	—	—	45	—	—	40	—	—	—	—	152	76	—
Flat Products																			
H00	0.25	0	TYP	68	36	28	—	—	40	10	—	60	—	—	—	—	25	—	—
	6.35			20	248	193	—	—	40	10	—	60	—	—	—	—	172	—	—
H04	1	0	TYP	68	45	40	—	—	20	45	—	85	—	—	—	—	26	—	—
	25.4			20	310	276	—	—	20	45	—	85	—	—	—	—	179	—	—
OS050	0.04	0	TYP	68	32	10	—	—	45	—	—	40	—	—	—	—	22	—	—
	1			20	221	69	—	—	45	—	—	40	—	—	—	—	152	—	—

Table 39 *(Continued)*

Temper	Section Size in. mm.	Cold Work %	Typ/ Min.	Temp F C	Tensile Strength ksi MPa	Yield Strength (0.5% ext. under load) ksi MPa	Yield Strength (0.2% offset) ksi MPa	Yield Strength (0.05% offset) ksi MPa	El %	Rockwell Hardness B	Rockwell Hardness C	Rockwell Hardness F	Rockwell Hardness 30T	Vickers Hardness 500	Brinell Hardness 500	Brinell Hardness 3000	Shear Strength ksi MPa	Fatigue Strength[a] ksi MPa	Izod Impact Strength ft-lb J
C15000—Pencil-Type and Light Soldering Guns: Tips, Rod Extensions, Resistance Welding Electrodes, Circuit Breaker Switches, Stud Bases for Power Transmitters and Rectifiers, Switches, Commutators, Resistance Welding Tips, Connectors, Solderless Wrapped, Welding Wheels																			
Rod																			
TL01	1.25	17	TYP	68	60	58	—	—	18	—	—	—	—	—	—	—	—	—	—
	31.8			20	414	400	—	—	18	—	—	—	—	—	—	—	—	—	—
TL06	0.204	76	TYP	68	62	56	—	—	8	—	—	—	—	—	—	—	—	—	—
	5.18			20	427	386	—	—	8	—	—	—	—	—	—	—	—	—	—
Wire																			
H02	0.5	30	TYP	68	53	49	—	—	23	—	—	90	—	—	—	—	—	—	—
	12.7			20	365	338	—	—	23	—	—	90	—	—	—	—	—	—	—
TF00	0.09	0	TYP	68	30	13	—	—	49	—	—	—	—	—	—	—	—	—	—
	2.29			20	207	90	—	—	49	—	—	—	—	—	—	—	—	—	—
TH04	0.04	98	TYP	68	76	72	—	—	1	—	—	—	—	—	—	—	—	—	—
	1			20	524	496	—	—	1	—	—	—	—	—	—	—	—	—	—
C17200—Electrical Switch and Relay Blades, Belleville Washers, Switch Parts, Fuse Clips, Spring Connectors, Current-Carrying Connectors, Relay Parts, Contact Bridges, Navigational Instruments, Clips, Fasteners, Lock Washers, Retaining Rings, Screws, Washers, Bolts, Roll Pins, Diaphragms, Valve Stems, Valve Seats, Shafts, Nonsparking Safety Tools, Flexible Metal Hose, Springs, Welding Equipment, Rolling Mill Parts, Spline Shafts, Pump Parts, Valves, Springs, Electrochemical, Pumps, Housings for Instruments, Wear Plates on Heavy Equipment, Bourdon Tubes, Bellows, Bushings, Bearings, Firing Pins																			
Rod																			
TD04	1	0	TYP	68	108	—	75	—	8	95	—	—	—	—	—	—	—	—	—
	25.4			20	745	—	517	—	8	95	—	—	—	—	—	—	—	—	—
TH04	3	0	TYP	68	195	—	145	—	4	—	41	—	—	—	—	—	—	—	—
	76.2			20	1344	—	1000	—	4	—	41	—	—	—	—	—	—	—	—
TH04	0.375	0	TYP	68	215	—	195	—	4	—	83	—	—	—	—	—	—	55	—
	9.53			20	1485	—	1345	—	4	—	83	—	—	—	—	—	—	380	—
TH04	3	0	TYP	68	195	—	145	—	4	—	41	—	—	—	—	—	—	50	—
	76.2			20	1344	—	1000	—	4	—	41	—	—	—	—	—	—	345	—

C17500—Fuse Clips, Switch Parts, Relay Parts, Conductors, Connectors, Fasteners, Resistance Welding Equipment, Seam Welding Dies, Resistance and Spot Welding Tips, Springs, Die-Casting Plunger Tips, Tooling for Plastic Molds, Stressed Parts

Rod																			
TB00	0.0	0	TYP	68	45	25	—	—	28	35	—	—	—	—	—	—	—	—	—
	0.0			20	310	172	—	—	28	35	—	—	—	—	—	—	—	—	—
TF00	0.0	0	TYP	68	110	90	—	—	18	96	—	—	—	—	—	—	—	40	—
	0.0			20	758	621	—	—	18	96	—	—	—	—	—	—	—	276	—
TH04	0.0	0	TYP	68	115	110	—	—	14	98	—	—	—	—	—	—	—	40	—
	0.0			20	793	758	—	—	14	98	—	—	—	—	—	—	—	276	—
Forgings																			
TB00	0.0	0	TYP	68	42	—	20	—	—	50	—	—	—	—	—	—	—	—	—
	0.0			20	293	—	138	—	—	50	—	—	—	—	—	—	—	—	—
TF00	4	0	TYP	68	115	—	90	—	—	92	—	—	—	—	—	—	—	—	—
	102			20	794	—	620	—	—	92	—	—	—	—	—	—	—	—	—
TB00	0.0	0	TYP	68	42	—	20	—	—	50	—	—	—	—	—	—	—	—	—
	0.0			20	293	—	138	—	—	50	—	—	—	—	—	—	—	—	—
TF00	4	0	TYP	68	115	—	90	—	—	92	—	—	—	—	—	—	—	—	—
	102			20	794	—	620	—	—	92	—	—	—	—	—	—	—	—	—
Flat Products																			
TB00	0.0	0	TYP	68	45	25	—	—	28	32	—	—	36	—	—	—	—	—	—
	0.0			20	310	172	—	—	28	32	—	—	36	—	—	—	—	—	—
TD02	0.0	0	TYP	68	68	60	—	—	8	70	—	—	64	—	—	—	—	—	—
	0.0			20	469	414	—	—	8	70	—	—	64	—	—	—	—	—	—
TD04	0.0	0	TYP	68	78	70	—	—	5	83	—	—	72	—	—	—	—	—	—
	0.0			20	538	483	—	—	5	83	—	—	72	—	—	—	—	—	—
TF00	0.0	0	TYP	68	110	90	—	—	12	96	—	—	80	—	—	—	—	—	—
	0.0			20	758	621	—	—	12	96	—	—	80	—	—	—	—	—	—
TH02	0.0	0	TYP	68	115	108	—	—	8	98	—	—	81	—	—	—	—	35	—
	0.0			20	793	745	—	—	8	98	—	—	81	—	—	—	—	241	—
TH04	0.0	0	TYP	68	115	110	—	—	8	98	—	—	81	—	—	—	—	—	—
	0.0			20	793	758	—	—	8	98	—	—	81	—	—	—	—	—	—

Table 39 (*Continued*)

Temper	Section Size in. mm.	Cold Work %	Typ/ Min.	Temp F C	Tensile Strength ksi MPa	Yield Strength (0.5% ext. under load) ksi MPa	Yield Strength (0.2% offset) ksi MPa	Yield Strength (0.05% offset) ksi MPa	El %	Rockwell Hardness B	Rockwell Hardness C	Rockwell Hardness F	Rockwell Hardness 30T	Vickers Hardness 500	Brinell Hardness 500	Brinell Hardness 3000	Shear Strength ksi MPa	Fatigue Strength[a] ksi MPa	Izod Impact Strength ft-lb J
C18100—Switches, Circuit Breakers, High-Temperature Wire, Contacts, Heat Sinks, Resistance Welding Wheels, Resistance Welding Tips, Semiconductor Bases, Resistance Welding Equipment, Fasteners, Fusion Energy Targets, Solar Collectors, Continuous Casting Molds, Stressed Parts																			
Wire																			
H04	0.8	90	TYP	68	158	—	141	—	17	—	—	—	—	—	—	—	—	—	—
	20.3			20	1089	—	972	—	17	—	—	—	—	—	—	—	—	—	—
H04	9.16	60	TYP	68	75	—	68	—	11	80	—	—	—	—	—	—	—	—	—
	233			20	517	—	469	—	11	80	—	—	—	—	—	—	—	—	—
Flat Products																			
H04	0.04	40	TYP	68	139	—	128	—	16	—	—	—	—	—	—	—	—	—	—
	1			20	958	—	882	—	16	—	—	—	—	—	—	—	—	—	—
C19100—Connectors, Components, Bolts, Turn Buckle Barrels, Nuts, Forgings, Screw Machine Parts, Bushings, Tie Rods, Gears, Pinions, Welding Torch Tips, Hardware																			
Rod																			
HR01	1	0	TYP	68	78	68	70	—	27	84	—	—	—	—	—	—	41	33	—
	25.4			20	538	469	483	—	27	84	—	—	—	—	—	—	283	228	—
HR01	0.125	75	TYP	68	104	77	92	—	6	95	—	—	—	—	—	—	56	—	—
	3.18			20	717	531	634	—	6	95	—	—	—	—	—	—	386	—	—
TH04	0.500	—			80	70	73		15	85								33	
	12.7	—			552	483	503											228	
C19200—Hydraulic Brake Lines, Gift Hollow Ware, Terminals, Circuit Breaker Parts, Fuse Clips, Cable Wrap, Contact Springs, Lead Frames, Connectors, Eyelets, Flexible Hose, Heat Exchanger Tubing, Gaskets, Air Conditioning Tubing																			
Tube																			
H80	—	40	TYP	68	56	52	52	—	7	—	—	—	—	—	—	—	—	—	—
	—			20	386	359	359	—	7	—	—	—	—	—	—	—	—	—	—
Flat Products																			
O60	0.0	0	TYP	68	45	—	20	—	25	38	—	—	—	—	—	—	—	—	—
	0.0			20	310	—	138	—	25	38	—	—	—	—	—	—	—	—	—
H08	0.04	0	TYP	68	74	—	71	—	2	76	—	—	—	—	—	—	—	—	—
	1			20	510	—	490	—	2	76	—	—	—	—	—	—	—	—	—

C23000—Weather Strip, Etching Parts, Trim, Kick Plates, Fire Extinguisher Cases, Costume Jewelry, Zippers, Medallions, Plaques, Tokens, Coinage, Rouge Boxes, Lipstick Containers, Dials, Compacts, Badges, Nameplates, Rotor Bars, Sockets, Conduit, Screw Shells, Rotor Bars, AC Motors, Sockets, Conduit, Screw Shells, Pump Cylinder Liners, Heat Exchanger Shells, Flexible Metal Hose, Fire Extinguishers, Pickling Crates, Tags, Radiator Cores, Condenser Tubes, Tubing for Heat Exchangers, Tubing for Instrumentation, Heat Exchangers, Fire Hose Couplings, Pipe Nipples, Pipe, Pump Lines, J-Bends, Pipe Service Lines, Traps, Fittings, Service Lines

Pipe																			
OS015	0.0	0	TYP	68	44	18	—	—	45	—	—	71	—	—	—	—	—	—	—
	0.0			20	303	124	—	—	45	—	—	71	—	—	—	—	—	—	—
Tube																			
OS050	0.065	0	TYP	68	40	12	—	—	55	—	—	60	15	—	—	—	—	—	—
	1.65			20	276	83	—	—	55	—	—	60	15	—	—	—	—	—	—
H80	0.065	35	TYP	68	70	58	—	—	8	77	—	—	68	—	—	—	—	—	—
	1.65			20	483	400	—	—	8	77	—	—	68	—	—	—	—	—	—
Wire																			
H00	0.08	0	TYP	68	50	—	—	—	25	—	—	—	—	—	—	—	35	20	—
	2			20	345	—	—	—	25	—	—	—	—	—	—	—	241	138	—
H08	0.08	0	TYP	68	105	—	—	—	—	—	—	—	—	—	—	—	54	—	—
	2			20	724	—	—	—	—	—	—	—	—	—	—	—	372	—	—
Flat Products																			
OS015	0.04	0	TYP	68	45	18	—	—	42	—	—	71	38	—	—	—	33	—	—
	1			20	310	124	—	—	42	—	—	71	38	—	—	—	228	—	—
H08	0.04	0	TYP	68	84	63	—	—	3	86	—	—	74	—	—	—	46	—	—
	1			20	579	434	—	—	3	86	—	—	74	—	—	—	317	—	—

Table 39 (*Continued*)

Temper	Section Size in. mm.	Cold Work %	Typ/ Min.	Temp F C	Tensile Strength ksi MPa	Yield Strength (0.5% ext. under load) ksi MPa	Yield Strength (0.2% offset) ksi MPa	Yield Strength (0.05% offset) ksi MPa	El %	Rockwell Hardness B	C	F	30T	Vickers Hardness 500	Brinell Hardness 500	3000	Shear Strength ksi MPa	Fatigue Strength[a] ksi MPa	Izod Impact Strength ft-lb J
C26000—Grillwork, Radiator Tube, Radiator Tanks, Odometer Contacts, Electrical Connectors, Tanks, Radiator Cores, Thermostats, Heater Cores, Door Knobs, Locks, Push Plates, Kick Plates, Decorative Hardware, Hinges, Finish Hardware, Shells—Electrical Sockets, Syringe Parts, Chain Links, Planters, Etched Articles, Bird Cages, Buttons, Snaps, Fireplace Screens, Coinage, Pen/Pencil Inserts and Clips, Costume Jewelry, Watch Parts, Lamps, Terminal Connectors, Reflectors, Screw Shells, Flashlight Shells, Lamp Fixtures, Fasteners, Grommets, Eyelets, Screws, Pins, Rivets, Sound-Proofing Equipment, Springs, Tubing for Instruments and Machines, Wire Screens, Pump Cylinders, Air Pressure Conveyer Systems, Liners, Chain, Bead Chain, Power Cylinders, Pumps, Heat Exchangers, Mechanical Housings for Lighters, Ammunition, Shells—Mechanical Housings for Ammunition, Ammunition Cartridge Cases, Washers, Stencils, Plumbing Accessories, Plumbing Brass Goods, Fittings, Traps, Bathroom Fixtures, Faucet Escutcheons																			
Rod																			
H00	1	6	TYP	68	55	40	—	—	48	60	—	—	—	—	—	—	36	—	—
	25.4			20	379	276	—	—	48	60	—	—	—	—	—	—	248	—	—
H02	1	20	TYP	68	70	52	—	—	30	80	—	—	—	—	—	—	42	22	—
	25.4			20	483	359	—	—	30	80	—	—	—	—	—	—	290	152	—
Tube																			
H80	0.0	35	TYP	68	78	64	—	—	8	82	—	—	73	—	—	—	—	—	—
	0.0			20	538	441	—	—	8	82	—	—	73	—	—	—	—	—	—
Wire																			
H00	0.08	0	TYP	68	58	—	—	—	35	—	—	—	—	—	—	—	38	—	—
	2			20	400	—	—	—	35	—	—	—	—	—	—	—	262	—	—
H08	0.08	0	TYP	68	130	—	—	—	3	—	—	—	—	—	—	—	60	22	—
	2			20	896	—	—	—	3	—	—	—	—	—	—	—	414	152	—
Flat Products																			
OS015	0.04	0	TYP	68	53	22	—	—	54	—	—	78	43	—	—	—	35	14	—
	1			20	365	152	—	—	54	—	—	78	43	—	—	—	241	97	—
H10	0.04	0	TYP	68	99	65	—	—	3	93	—	—	78	—	—	—	—	—	—
	1			20	683	448	—	—	3	93	—	—	78	—	—	—	—	—	—

C27000—Handrails, Grillwork, Radiator Cores, Tanks, Finish Hardware, Hinges, Locks, Stencils, Push Plates, Kick Plates, Flashlight Shells, Screw Shells, Socket Shells, Reflectors, Lamp Fixtures, Fasteners, Rivets, Pins, Screws, Grommets, Eyelets, Springs, Bead Chain, Chain, Fasteners, Plumbing Accessories, Sink Strainers

Rod																			
H00	1	6	TYP	68	55	40	—	—	48	55	—	—	—	—	—	—	36	—	—
	25.4			20	379	276	—	—	48	55	—	—	—	—	—	—	248	—	—
OS050	1	0	TYP	68	48	16	—	—	65	—	—	65	—	—	—	—	34	—	—
	25.4			20	331	110	—	—	65	—	—	65	—	—	—	—	234	—	—
Wire																			
H00	0.08	0	TYP	68	58	—	—	—	35	—	—	—	—	—	—	—	38	—	—
	2.0			20	400	—	—	—	35	—	—	—	—	—	—	—	262	—	—
H08	0.08	0	TYP	68	128	—	—	—	3	—	—	—	—	—	—	—	60	—	—
	2.0			20	883	—	—	—	3	—	—	—	—	—	—	—	414	—	—
Flat Products																			
OS015	0.04	0	TYP	68	53	22	—	—	54	—	—	78	43	—	—	—	—	—	—
	1			20	365	152	—	—	54	—	—	78	43	—	—	—	—	—	—
H08	0.04	0	TYP	68	91	62	—	—	3	90	—	—	76	—	—	—	47	20	—
	1			20	627	427	—	—	3	90	—	—	76	—	—	—	324	138	—

C28000—Door Frames, Large Architectural Trim, Architectural Panels, Sheet, Hardware, Decoration, Structural, Heavy Plate, Large Sheets, Decorative Hardware, Bolts, Brazing Rod, Valve Stems, Condenser Plates, Hot Forgings, Evaporator Tubes, Large Nuts and Bolts, Condenser Tube, Heat Exchanger Tube

Rod																			
M30	1	0	TYP	68	52	20	—	—	52	—	—	78	—	—	—	—	39	—	—
	25.4			20	359	138	—	—	52	—	—	78	—	—	—	—	269	—	—
H01	1	0	TYP	68	72	50	—	—	25	78	—	—	—	—	—	—	45	—	—
	25.4			20	496	345	—	—	25	78	—	—	—	—	—	—	310	—	—
Tube																			
O50	0.0	0	TYP	68	56	23	—	—	50	—	—	82	47	—	—	—	—	—	—
	0.0			20	386	159	—	—	50	—	—	82	47	—	—	—	—	—	—
H04	0.0	30	TYP	68	74	55	—	—	10	80	—	—	—	—	—	—	—	—	—
	0.0			20	510	379	—	—	10	80	—	—	—	—	—	—	—	—	—
Flat Products																			
O60	0.04	0	TYP	68	54	21	—	—	45	—	—	80	46	—	—	—	40	—	—
	1			20	372	145	—	—	45	—	—	80	46	—	—	—	276	—	—
H02	0.04	0	TYP	68	70	50	—	—	10	75	—	—	67	—	—	—	44	—	—
	1			20	483	345	—	—	10	75	—	—	67	—	—	—	303	—	—

Table 39 (*Continued*)

Temper	Section Size in. mm.	Cold Work %	Typ/ Min.	Temp F C	Tensile Strength ksi MPa	Yield Strength (0.5% ext. under load) ksi MPa	Yield Strength (0.2% offset) ksi MPa	Yield Strength (0.05% offset) ksi MPa	El %	Rockwell Hardness B	Rockwell Hardness C	Rockwell Hardness F	Rockwell Hardness 30T	Vickers Hardness 500	Brinell Hardness 500	Brinell Hardness 3000	Shear Strength ksi MPa	Fatigue Strength[a] ksi MPa	Izod Impact Strength ft-lb J
C36000—Terrazzo Strip, Fluid Connectors, Threaded Inserts for Plastic, Sensor Bodies, Thermostat Parts, Lock Bodies, Hardware, Fittings, Combs (to Straighten Hair), Bolts, Nuts, Screws, Valve Stems, Valve Seats, Valve Trim, Fluid Connectors, Nozzles, Unions, Adapters, Screw Machine Products, Automatic Screw Machine Parts, Pinions, Gears, Faucet Components, Pneumatic Fittings, Gauges, Plumbers' Brass Goods, Faucet Stems, Faucet Seats, Plumbing Fittings																			
Bar																			
H02	<0.50	0	TYP	68	—	—	—	—	—	65	—	—	—	—	—	—	—	—	—
	<12.7			20	—	—	—	—	—	65	—	—	—	—	—	—	—	—	—
Rod																			
H04	0.375	0	TYP	68	—	—	—	—	—	78	—	—	—	—	—	—	34	—	—
	12.7			20	—	—	—	—	—	78	—	—	—	—	—	—	234	—	—
Shapes																			
M30	0.5	0	TYP	68	49	18	—	—	50	—	—	68	—	—	—	—	30	—	—
	12.7			20	338	124	—	—	50	—	—	68	—	—	—	—	207	—	—
Flat Products																			
H02	0.25	11	TYP	68	56	45	—	—	20	62	—	—	—	—	—	—	33	—	—
	4.76			20	386	310	—	—	20	62	—	—	—	—	—	—	228	—	—
C37700—Furniture Hardware, Decorative Knobs, Door Handles, Valve Bodies for Refrigeration, Chemicals, Golf Putters, Valve Bodies for Scuba and Propane Spray Tanks, Valve Bodies for Agricultural Spray Tanks, Fuse Bodies, Covers, Valve Components, Forgings and Pressings of All Kinds																			
Rod																			
M30	1	0	TYP	68	52	20	—	—	45	—	—	78	—	—	—	—	—	—	—
	25.4			20	359	138	—	—	45	—	—	78	—	—	—	—	—	—	—
Shapes																			
M30	1	0	TYP	68	52	20	—	—	45	—	—	78	—	—	—	—	—	—	—
	25.4			20	359	138	—	—	45	—	—	78	—	—	—	—	—	—	—

C44000—Condenser Tube, Evaporator Tubing, Heat Exchanger Tubing, Condenser Tube Plates, Distiller Tubes, Ferrules, Strainers

Tube																			
OS025	0.0	0	TYP	68	53	22	—	—	65	—	—	75	37	—	—	—	—	—	—
	0.0			20	365	152	—	—	65	—	—	75	37	—	—	—	—	—	—
Wire																			
OS015	0.08	0	TYP	68	55	—	—	—	60	—	—	—	—	—	—	—	—	—	—
	2			20	379	—	—	—	60	—	—	—	—	—	—	—	—	—	—
Plate																			
M20	1	0	TYP	68	48	18	—	—	65	—	—	70	—	—	—	—	—	—	—
	25.4			20	331	124	—	—	65	—	—	70	—	—	—	—	—	—	—

C46500—Elevators, Architectural Metal, Bolts, Nuts, Rivets, Plater Bar for Jewelry, Plates, Baffles, Heat Sinks, Balls, Structural Uses, Welding Rod, Condenser Plates, Valve Stems, Aircraft Turn Buckle Barrels, Petrochemical Tanks, Heat Exchanger Tube, Marine Hardware, Propeller Shafts, Fittings

Rod																			
O60	1	0	TYP	68	57	25	—	—	47	55	—	—	—	—	—	—	40	—	—
	25.4			20	393	172	—	—	47	55	—	—	—	—	—	—	276	—	—
H02	1	20	TYP	68	75	53	—	—	20	82	—	—	—	—	—	—	44	—	—
	25.4			20	517	365	—	—	20	82	—	—	—	—	—	—	303	—	—
Tube																			
H04	0.0	35	TYP	68	88	66	—	—	18	95	—	—	—	—	—	—	—	—	—
	0.0			20	607	455	—	—	18	95	—	—	—	—	—	—	—	—	—
Flat Products																			
O60	0.25	0	TYP	68	58	25	—	—	49	56	—	—	55	—	—	—	40	—	—
	6.35			20	400	172	—	—	49	56	—	—	55	—	—	—	276	—	—
H01	0.04	0	TYP	68	70	58	—	—	17	75	—	—	68	—	—	—	43	—	—
	1			20	483	400	—	—	17	75	—	—	68	—	—	—	296	—	—
Wire																			
H04	0.08	80	TYP	68	79	—	—	—	1	—	—	—	—	—	—	—	—	32	—
Flat Products																			
OS025	0.04	0	TYP	68	40	14	—	—	48	—	—	68	—	—	—	—	—	—	—
	1			20	276	97	—	—	48	—	—	68	—	—	—	—	—	—	—
H08	0.04	0	TYP	68	75	—	—	—	4	79	—	—	70	—	—	—	—	—	—
	1			20	517	—	—	—	4	79	—	—	70	—	—	—	—	—	—

Table 39 (*Continued*)

Temper	Section Size in. mm.	Cold Work %	Typ/ Min.	Temp F C	Tensile Strength ksi MPa	Yield Strength (0.5% ext. under load) ksi MPa	Yield Strength (0.2% offset) ksi MPa	Yield Strength (0.05% offset) ksi MPa	El %	Rockwell Hardness B	C	F	30T	Vickers Hardness 500	Brinell Hardness 500	3000	Shear Strength ksi MPa	Fatigue Strength[a] ksi MPa	Izod Impact Strength ft-lb J
C51100—Bridge Bearing Plates, Resistance Wire, Electronic Connectors, Electrical Connectors, Electrical Flexing Contact Blades, Switch Parts, Wire Brushes, Fuse Clips, Electronic and Precision Instrument Parts, Electromechanical Spring Components, Fasteners, Cotter Pins, Lock Washers, Bellows, Springs, Chemical Hardware, Textile Machinery, Pressure Responsive Elements, Welding Rods, Perforated Sheets, Truss Wire, Sleeve Bushings, Beater Bar, Clutch Disks, Diaphragms, Bourdon Tubes																			
Flat Products																			
OS015	0.04	0	TYP	68	49	—	23	—	50	—	—	76	—	—	—	—	—	—	—
	1			20	338	—	159	—	50	—	—	76	—	—	—	—	—	—	—
HR08	0.0	0	TYP	68	98	—	88	—	8	—	—	—	—	—	—	—	—	—	—
	0.0			20	676	—	607	—	8	—	—	—	—	—	—	—	—	—	—
C52100—Bridge Bearing Plates, Thermostat Bellows, Power Conductor for Electrosurgical Pencil, Cymbals, Coinage, Electronic Connectors, Electrical Flexing Contact Blades, Electrical Connectors, Cold Headed Parts, Wire Brushes, Switch Parts, Fuse Clips, Fasteners, Heavy Duty, Lock Washers, Cotter Pins, Doctor Blades, Paper Industry, Welding Wire, Beater Bar, Clutch Disks, Bourdon Tubing, Diaphragms, Sleeve Bushings, Springs, Heavy Duty, Pneumatic Hammers, Well Drill Equipment, Thrust Bearings, Cold Headed Parts, Pinions, Gears, Clips, Heavy Duty, Springs, Helical Torsion, Springs, Helical Extension, Bellows, Truss Wire, Chemical Hardware, Perforated Sheets, Textile Machinery, Marine Parts																			
Rod																			
H02	0.5	20	TYP	68	80	65	—	—	33	85	—	—	—	—	—	—	—	—	—
	12.7			20	552	448	—	—	33	85	—	—	—	—	—	—	—	—	—
Wire																			
OS035	0.08	0	TYP	68	60	24	—	—	65	—	—	—	—	—	—	—	—	—	—
	2			20	414	165	—	—	65	—	—	—	—	—	—	—	—	—	—
H06	0.08	0	TYP	68	140	—	—	—	—	—	—	—	—	—	—	—	—	—	—
	2			20	965	—	—	—	—	—	—	—	—	—	—	—	—	—	—
Flat Products																			
OS015	0.04	0	TYP	68	65	—	—	—	60	—	—	85	—	—	—	—	—	—	—
	1			20	448	—	—	—	60	—	—	85	—	—	—	—	—	—	—
HR08	0.0	0	TYP	68	112	—	102	—	11	—	—	—	—	—	—	—	—	—	—
	0.0			20	772	—	703	—	11	—	—	—	—	—	—	—	—	—	—

C61300—Wire Screening, Fasteners, Bolts, Nuts, Gaskets, Tanks, Desalination Equipment, Seawater Piping, Machine Parts, Tube Sheets, Seamless Tubing and Pipe, Columns, Heat Exchanger Tube, Piping Systems, Acid-Resistant, Corrosion-Resistant Vessels, Condenser Tube, Water Boxes, Tie Rods, Mining Shovels, Marine Hardware, Protective Sheathing, Explosives Blending Chambers and Mixing Troughs, Ball Valve Seats

Rod																		
H04	1	25	TYP	68	82	55	—	—	35	90	—	—	—	—	—	—	45	—	—
	25.4			20	565	379	—	—	35	90	—	—	—	—	—	—	310	—	—
Flat Products																		
O60	1	0	TYP	68	76	33	—	—	42	79	—	—	—	—	—	—	45	25	—
	25.4			20	524	228	—	—	42	79	—	—	—	—	—	—	310	172	—

C61800—Welding Rod, Bushings, Bearing Liners, Pickling Hooks, Shafts, Welding Wire, Tie Rods, Bearings, Pump Impellers, Marine Hardware

Rod																		
H02	1	15	TYP	68	85	43	—	—	23	89	—	—	—	—	—	—	47	82	—
	25.4			20	586	293	—	—	23	89	—	—	—	—	—	—	324	565	—
Wire																		
H02	0.063	0	TYP	68	140	—	—	—	—	—	—	—	—	—	—	—	—	—	—
	1.60			20	965	—	—	—	—	—	—	—	—	—	—	—	—	—	—

C62400—Nuts, Hydraulic Bushings, Support Bushings, Valve Balls, Welding Wire, Tie Rods, Wear Plates, Drift Pins, Bushings, Gears, Cams

Bar																		
O50	1	0	TYP	68	100	50	—	—	14	92	—	—	—	—	—	—	—	—	—
	25.4			20	689	345	—	—	14	92	—	—	—	—	—	—	—	—	—
M30	3	0	TYP	68	90	40	—	—	18	87	—	—	—	—	—	—	—	—	—
	76			20	621	276	—	—	18	87	—	—	—	—	—	—	—	—	—

C63200—Marine Fasteners, Bolts, Nuts, Sleeve Bearings, Valve Stems, Seawater Pump Bodies, Bearing Cages, Shafts, Bushings, Cams, Seawater Valves, Marine Drive Shafts

Rod																		
O50	1	0	TYP	68	105	53	—	—	22	96	—	—	—	—	—	—	—	—	—
	25.4			20	724	365	—	—	22	96	—	—	—	—	—	—	—	—	—
Plate																		
O50	1	0	TYP	68	90	45	—	—	25	85	—	—	—	—	—	—	—	—	—
	25.4			20	621	310	—	—	25	85	—	—	—	—	—	—	—	—	—
Forgings																		
O50	5	0	TYP	68	100	45	—	—	22	92	—	—	—	—	—	—	—	—	—
	127			20	689	310	—	—	22	92	—	—	—	—	—	—	—	—	—

Table 39 (*Continued*)

Temper	Section Size in. mm.	Cold Work %	Typ/ Min.	Temp F C	Tensile Strength ksi MPa	Yield Strength (0.5% ext. under load) ksi MPa	Yield Strength (0.2% offset) ksi MPa	Yield Strength (0.05% offset) ksi MPa	El %	Rockwell Hardness B	C	F	30T	Vickers Hardness 500	Brinell Hardness 500	3000	Shear Strength ksi MPa	Fatigue Strength[a] ksi MPa	Izod Impact Strength ft-lb J
C65500—Sculpture, Motors, Rotor Bar, Pole Line Hardware, Bolts, Cotter Pins, Screws, Hinges, Burrs, Rivets, Nuts, Nails, Clamps, Oil Refinery Plumbing Tube, Wire, Screen Cloth, Piston Rings, Kettles, Heat Exchanger Tubes, Chemical Equipment, Channels, Cable, Bushings, Tanks, Butts, Shafting, Screen Plates, Welded Tanks, Doctor Blades, Paper Industry, Hydraulic Pressure Lines, Bearing Plates, Pressure Vessels, Welded Pressure Vessels, Wear Plates, Propeller Shafts, Hardware																			
Rod																			
OS050	1	0	TYP	68	58	22	—	—	60	60	—	—	—	—	—	—	43	—	—
	25.4			20	400	152	—	—	60	60	—	—	—	—	—	—	296	—	—
H06	1	50	TYP	68	108	60	—	—	13	95	—	—	—	—	—	—	62	—	—
	25.4			20	745	414	—	—	13	95	—	—	—	—	—	—	427	—	—
Tube																			
OS050	0.065	0	TYP	68	57	—	—	—	70	45	—	—	—	—	—	—	—	—	—
	1.7			20	393	—	—	—	70	45	—	—	—	—	—	—	—	—	—
OS050	0.065	0	TYP	68	57	—	—	—	70	45	—	—	—	—	—	—	—	—	—
	1.7			20	393	—	—	—	70	45	—	—	—	—	—	—	—	—	—
Wire																			
OS035	0.08	0	TYP	68	60	25	—	—	60	—	—	—	—	—	—	—	43	—	—
	2			20	414	172	—	—	60	—	—	—	—	—	—	—	296	—	—
H08	0.08	0	TYP	68	145	70	—	—	3	—	—	—	—	—	—	—	70	30	—
	2			20	1000	483	—	—	3	—	—	—	—	—	—	—	483	207	—
Flat Products																			
OS015	0.04	0	TYP	68	63	30	—	—	55	66	—	90	—	—	—	—	45	—	—
	1			20	434	207	—	—	55	66	—	90	—	—	—	—	310	—	—
H08	0.04	0	TYP	68	110	62	—	—	4	97	—	—	81	—	—	—	63	—	—
	1			20	758	427	—	—	4	97	—	—	81	—	—	—	434	—	—

C66700—Resistance Weldable Brass Products

Wire

H02	0.08	0	TYP	68	90	—	—	—	5	—	—	—	—	—	—	—	—	—	—
	2			20	621	—	—	—	5	—	—	—	—	—	—	—	—	—	—

Flat Products

OS015	0.04	0	TYP	68	55	24	24	—	44	—	—	83	47	—	—	—	—	—	—
	1			20	379	165	165	—	44	—	—	83	47	—	—	—	—	—	—
H10	0.04	0	TYP	68	100	—	93	—	2	93	—	—	78	—	—	—	—	—	—
	1			20	689	—	638	—	2	93	—	—	78	—	—	—	—	—	—

C67500—Clutch Disks, Pump Rods, Shafting, Bolts, Balls, Valve Stems, Valve Bodies, Aircraft Parts, Bushings, Hardware

Rod

H01	1	10	TYP	68	77	45	—	—	23	83	—	—	—	—	—	—	47	—	—
	25.4			20	531	310	—	—	23	83	—	—	—	—	—	—	324	—	—
H02	1	20	TYP	68	84	60	—	—	19	90	—	—	—	—	—	—	48	—	—
	25.4			20	579	414	—	—	19	90	—	—	—	—	—	—	331	—	—

C69400—Valve Stems

Rod

O60	1	0	TYP	68	85	43	—	—	25	85	—	—	—	—	—	—	—	—	—
	25.4			20	586	296	—	—	25	85	—	—	—	—	—	—	—	—	—
H04	0.75	0	TYP	68	100	57	—	—	21	95	—	—	—	—	—	—	—	—	—
	19			20	689	393	—	—	21	95	—	—	—	—	—	—	—	—	—

C70600—Power Steering Tube, Brake Lines, Screw Lamp Bases, Valve Bodies, Condensers, Evaporators, Pump Impellers for Oil Refining, Pressure Vessels, Condenser Plates, Weld Torch Tips, Distiller Tubes, Ferrules, Heat Exchanger Tubes, Evaporator Tubes, Salt Water Baffles, Salt Water Pipe Fittings, Water Hoses, Boat Hulls, Hot-Water Tanks, Ship Hulls, Tube Sheet for Salt Water Service, Salt Water Piping Systems, Salt Water Piling Wrap, Propeller Sleeves, Flanges

Tube

OS025	0.0	0	TYP	68	44	16	—	—	42	15	—	65	26	—	—	—	—	—	—
	0.0			20	303	110	—	—	42	15	—	65	26	—	—	—	—	—	—
H55	0.0	0	TYP	68	60	57	—	—	10	72	—	100	70	—	—	—	—	—	—
	0.0			20	414	393	—	—	10	72	—	100	70	—	—	—	—	—	—

Table 39 (*Continued*)

Temper	Section Size in. mm.	Cold Work %	Typ/ Min.	Temp F C	Tensile Strength ksi MPa	Yield Strength (0.5% ext. under load) ksi MPa	Yield Strength (0.2% offset) ksi MPa	Yield Strength (0.05% offset) ksi MPa	El %	Rockwell Hardness B	C	F	30T	Vickers Hardness 500	Brinell Hardness 500	3000	Shear Strength ksi MPa	Fatigue Strength[a] ksi MPa	Izod Impact Strength ft-lb J
C71500—Pump Impellers, Welding Backing Rings, Flexible Metal Hose, Boiler Parts, Evaporator Tubes, Distiller Tubes, Condenser Plates, Weld Wire, Heat Exchanger Tubes, Refrigerators, Process Equipment, Condenser Components, Heat Exchanger Components, Propeller Sleeves, Condensers, Ferrules, Seawater Pump Bodies and Internal Parts, Seawater Condensers, Valve Bodies, Salt Water Piping, Fittings, Salt Water Flanges, Water Boxes																			
Rod																			
H04	1	20	TYP	68	75	70	—	—	15	80	—	—	—	—	—	—	—	—	—
	25.4			20	517	483	—	—	15	80	—	—	—	—	—	—	—	—	—
Tube																			
OS025	0.0	0	TYP	68	60	25	—	—	45	45	—	80	—	—	—	—	—	—	—
	0.0			20	414	172	—	—	45	45	—	80	—	—	—	—	—	—	—
Flat Products																			
M20	1	0	TYP	68	55	20	—	—	45	35	—	—	—	—	—	—	—	—	—
	25.4			20	379	138	—	—	45	35	—	—	—	—	—	—	—	—	—
C72200—Condenser Tubing, Heat Exchanger Tubing, Salt Water Piping																			
Tube																			
H55	0.0	0	TYP	68	70	—	68	—	—	—	—	—	—	—	—	—	—	—	—
	0.0			20	483	—	469	—	—	—	—	—	—	—	—	—	—	—	—
O50	0.0	0	TYP	68	46	—	18	—	46	—	—	—	—	—	—	—	—	—	—
	0.0			20	317	—	124	—	46	—	—	—	—	—	—	—	—	—	—
Strip																			
H04	0.0	0	TYP	68	70	—	66	—	6	—	—	—	—	—	—	—	—	—	—
	0.0			20	483	—	455	—	6	—	—	—	—	—	—	—	—	—	—
O61	0.0	0	TYP	68	46	—	18	—	46	—	—	—	—	—	—	—	—	—	—
	0.0			20	317	—	124	—	46	—	—	—	—	—	—	—	—	—	—

C72500—Electronic Contacts, Telecommunications Connectors, Relays, Switch Springs, Connectors, Lead Frames, Control and Sensing Bellows, Brazing Alloy

Wire																			
OS015	0.08	0	TYP	68	60	25	—	—	—	—	—	—	—	—	—	—	—	—	—
	2			20	414	172	—	—	—	—	—	—	—	—	—	—	—	—	—
H14	0.08	99	TYP	68	120	—	—	—	—	—	—	—	—	—	—	—	—	—	—
	2			20	827	—	—	—	—	—	—	—	—	—	—	—	—	—	—
Flat Products																			
OS015	0.04	0	TYP	68	55	22	22	—	35	42	—	—	44	—	—	—	—	—	—
	1			20	379	152	152	—	35	42	—	—	44	—	—	—	—	—	—
H08	0.4	0	TYP	68	91	83	90	—	1	90	—	—	—	—	—	—	—	—	—
	1			20	627	572	621	—	1	90	—	—	—	—	—	—	—	—	—

C72650—Connectors for High Temperatures, Switches, Relays, Springs, Cladding and Inlay

Flat Products																			
TM00	0.0	0	TYP	68	110	—	80	75	18	—	—	—	—	—	—	—	—	—	—
	0.0			20	758	—	552	517	18	—	—	—	—	—	—	—	—	—	—
TM06	0.0	0	TYP	68	130	—	115	110	6	—	—	—	—	—	—	—	—	—	—
	0.0			20	896	—	793	758	6	—	—	—	—	—	—	—	—	—	—
TS04	0.0	0	TYP	68	147	—	—	122	4	—	—	—	—	—	—	—	—	—	—
	0.0			20	1013	—	—	841	4	—	—	—	—	—	—	—	—	—	—
TX00	0.0	0	TYP	68	130	—	—	78	6	—	—	—	—	—	—	—	—	—	—
	0.0			20	896	—	—	538	6	—	—	—	—	—	—	—	—	—	—

C72700—Connectors, Springs, Coil Springs

Flat Products																			
TB00	0.012	0	TYP	68	66	29	—	—	36	—	—	—	—	—	—	—	—	—	—
	0.3			20	455	200	—	—	36	—	—	—	—	—	—	—	—	—	—
TD04	0.027	75	TYP	68	126	—	100	—	—	—	—	—	—	—	—	—	—	—	—
	0.69			20	869	—	689	—	—	—	—	—	—	—	—	—	—	—	—
TS12	0.0	0	TYP	68	165	—	—	145	2	—	—	—	—	—	—	—	—	—	—
	0.0			20	1138	—	—	1000	2	—	—	—	—	—	—	—	—	—	—
TX00	0.0	33	TYP	68	115	—	—	77	15	—	—	—	—	—	—	—	—	—	—
	0.0			20	793	—	—	531	15	—	—	—	—	—	—	—	—	—	—

Table 39 (*Continued*)

Temper	Section Size in. mm.	Cold Work %	Typ/ Min.	Temp F C	Tensile Strength ksi MPa	Yield Strength (0.5% ext. under load) ksi MPa	Yield Strength (0.2% offset) ksi MPa	Yield Strength (0.05% offset) ksi MPa	El %	Rockwell Hardness B	C	F	30T	Vickers Hardness 500	Brinell Hardness 500	3000	Shear Strength ksi MPa	Fatigue Strength[a] ksi MPa	Izod Impact Strength ft-lb J
C72800—Springs, Extruded Shapes, Tubes, Bars, Rod, Wire, Forgings																			
Rod																			
TB00	0.25	0	TYP	68	76	—	36	—	23	—	—	—	—	—	—	—	—	—	—
	6.35			20	525	—	250	—	23	—	—	—	—	—	—	—	—	—	—
TF00	0.76	0	TYP	68	147	—	126	—	10	—	33	—	—	—	—	—	—	—	—
	19.3			20	1014	—	867	—	10	—	33	—	—	—	—	—	—	—	—
TH01	0.113	0	TYP	68	199	—	196	—	4	—	—	—	—	—	—	—	—	—	—
	2.87			20	1372	—	1351	—	4	—	—	—	—	—	—	—	—	—	—
Tube																			
TF00	0.0	0	TYP	68	294	—	250	—	5	—	67	—	—	—	—	—	—	—	—
	0.0			20	2024	—	1723	—	5	—	67	—	—	—	—	—	—	—	—
Strip																			
H01	0.089	0	TYP	68	89	—	76	—	13	89	—	—	—	—	—	—	—	—	—
	2.26			20	612	—	525	—	13	89	—	—	—	—	—	—	—	—	—
H08	0.063	0	TYP	68	139	—	132	—	3	—	30	—	—	—	—	—	—	—	—
	1.6			20	956	—	911	—	3	—	30	—	—	—	—	—	—	—	—
TH08	0.063	0	TYP	68	191	—	180	—	3	—	39	—	—	—	—	—	—	—	—
	1.6			20	1318	—	1242	—	3	—	39	—	—	—	—	—	—	—	—
Forgings																			
TF00	0.0	0	TYP	68	145	—	120	—	—	—	34	—	—	—	—	—	—	—	—
	0.0			20	1000	—	827	—	—	—	34	—	—	—	—	—	—	—	—

C72900—Connectors, Miniaturized Sockets, Relay Elements, Contacts, Controls, Wire, Marine Components, Springs, Wire, Marine Components

Tube																			
TF00	0.0	0	TYP	68	154	—	—	—	7	—	32	—	—	—	—	—	—	—	—
	0.0			20	1062	—	—	—	7	—	32	—	—	—	—	—	—	—	—
Wire																			
TF00	.030	0	TYP	68	160	—	174	—	—	—	—	—	—	—	—	—	—	—	—
	0.8			20	1103	—	1201	—	—	—	—	—	—	—	—	—	—	—	—
Flat Products																			
TD04	0.027	78	TYP	68	145	—	—	—	—	—	31	—	—	—	—	—	—	—	—
	0.69			20	1000	—	—	—	—	—	31	—	—	—	—	—	—	—	—
TM08	0.0	0	TYP	68	137	—	125	110	6	—	—	—	—	—	—	—	—	—	—
	0.0			20	944	—	862	758	6	—	—	—	—	—	—	—	—	—	—
TS12	0.0	0	TYP	68	181	—	—	163	—	—	—	—	—	—	—	—	—	—	—
	0.0			20	1248	—	—	1127	—	—	—	—	—	—	—	—	—	—	—
TX00	0.0	78	TYP	68	120	—	—	81	10	—	—	—	—	—	—	—	—	—	—
	0.0			20	827	—	—	558	10	—	—	—	—	—	—	—	—	—	—

C74500—Hollow Ware, Name Plates, Connectors, Rivets, Screws, Slide Fasteners, Optical Parts, Platers Bar, Etching Stock

Wire																			
OS015	0.08	0	TYP	68	63	—	—	—	35	—	—	—	—	—	—	—	—	—	—
	2			20	434	—	—	—	35	—	—	—	—	—	—	—	—	—	—
H04	0.08	84	TYP	68	130	—	—	—	1	—	—	—	—	—	—	—	—	—	—
	2			20	896	—	—	—	1	—	—	—	—	—	—	—	—	—	—
Flat Products																			
OS015	0.04	0	TYP	68	60	28	—	—	36	52	—	85	51	—	—	—	—	—	—
	1			20	414	193	—	—	36	52	—	85	51	—	—	—	—	—	—
H06	0.04	0	TYP	68	95	76	—	—	3	92	—	—	78	—	—	—	59	—	—
	1			20	655	524	—	—	3	92	—	—	78	—	—	—	407	—	—

C77000—Resistance Wire, Springs, Optical Goods, Eyeglass Frames

Wire																			
OS035	0.08	0	TYP	68	60	—	—	—	40	—	—	—	—	—	—	—	—	—	—
	2			20	414	—	—	—	40	—	—	—	—	—	—	—	—	—	—
H04	0.08	68	TYP	68	145	—	—	—	2	—	—	—	—	—	—	—	—	—	—
	2			20	1000	—	—	—	2	—	—	—	—	—	—	—	—	—	—
Flat Products																			
OS035	0.04	0	TYP	68	60	27	—	—	40	55	—	90	—	—	—	—	—	—	—
	1			20	414	186	—	—	40	55	—	90	—	—	—	—	—	—	—
H08	0.04	0	TYP	68	115	—	—	—	2	99	—	—	81	—	—	—	—	—	—
	1			20	793	—	—	—	2	99	—	—	81	—	—	—	—	—	—

[a] Fatigue Strength: 100×10^6 cycles.

Source: Copper Development Association.

Table 40 Mechanical Properties of Selected Cast Copper Alloys

Alloy Temper	Section Size in. mm	Cold Work %	Typ/ Min.	Temp °F °C	Tensile Strength ksi MPa	Yield Strength (0.5% ext. under load) ksi MPa	Yield Strength (0.2% offset) ksi MPa	Yield Strength (0.05% offset) ksi MPa	El %	Rockwell Hardness B	Rockwell Hardness C	Rockwell Hardness F	Rockwell Hardness 30T	Vickers Hardness 500	Brinell Hardness 500	Brinell Hardness 3000	Shear Strength ksi MPa	Fatigue Strength ksi MPa	Izod Impact Strength ft-lb J
C81100—Electrical and Thermal Conductor Applications, Bus Conductor Components, Furnaces for Steel Industry, Oxygen Nozzles, Blast Furnace Tuyeres, Bosh and Stack Cooling Plates, Smelters for the Steel Industry																			
As Sand Cast																			
M01	0.0	0	TYP	68	25	9	—	—	40	—	—	—	—	—	44	—	—	9	—
	0.0			20	172	62	—	—	40	—	—	—	—	—	44	—	—	62	—
C81400—Electrical Parts, Higher Hardness Electrical Conductors, Higher Hardness Thermal Conductors																			
As Sand Cast																			
M01	0.0	0	SMIN	68	45	—	15	—	15	—	—	—	—	—	—	—	—	—	—
	0.0			20	311	—	104	—	15	—	—	—	—	—	—	—	—	—	—
TF00	0.0	0	TYP	68	—	—	—	—	—	69	—	—	—	—	—	—	—	—	—
	0.0			20	—	—	—	—	—	69	—	—	—	—	—	—	—	—	—
TF00	0.0	0	SMIN	68	53	—	36	—	11	—	—	—	—	—	—	—	—	—	—
	0.0			20	366	—	248	—	11	—	—	—	—	—	—	—	—	—	—
C82000—Aerospace Components Requiring Nonsparking Action, Military Components Requiring Nonsparking Action, Contacts, Electrical Switches, Circuit Breaker Parts, Bearings, Resistance Welding Components, Switch Gear Parts, Contact and Switch Blades, Soldering Iron Components, Bushings																			
As Sand Cast																			
M01		0	TYP	68	50	—	20	—	20	52	—	—	—	—	—	—	—	—	—
				20	345	—	138	—	20	52	—	—	—	—	—	—	—	—	—
O11		0	TYP	68	65	—	37	—	12	70	—	—	—	—	—	—	—	—	—
				20	448	—	255	—	12	70	—	—	—	—	—	—	—	—	—
TB00		0	TYP	68	47	—	15	—	25	40	—	—	—	—	—	—	—	—	—
				20	324	—	104	—	25	40	—	—	—	—	—	—	—	—	—
TF00		0	TYP	68	96	—	75	—	6	96	—	—	—	—	—	—	—	18	—
				20	662	—	517	—	6	96	—	—	—	—	—	—	—	124	—
TF00	0.0	0	SMIN	68	90	—	70	—	3	—	—	—	—	—	—	—	—	—	—
	0.0			20	621	—	483	—	3	—	—	—	—	—	—	—	—	—	—

C82500—Sculpture, Safety Tools, Jewelry, Molds for Plastic Parts, Glass Molds, Aluminum Die Cast Plunger Tips, Valves, Business Machine Parts, Pressure Housings in Salt Water Environments, Sleeves, Permanent Molds for Brass and Aluminum Casting, Golf Clubs, Corrosion-Resistant Valves, Undersea Cable Housings, Ordnance Components

As Sand Cast																			
M01	0.0	0	TYP	68	80	—	45	—	20	82	—	—	—	—	—	—	—	—	—
	0.0			20	552	—	310	—	20	82	—	—	—	—	—	—	—	—	—
M01	0.0	0	SMIN	68	75	—	40	—	15	—	—	—	—	—	—	—	—	—	—
	0.0			20	518	—	276	—	15	—	—	—	—	—	—	—	—	—	—
TB00	0.0	0	TYP	68	60	—	25	—	35	63	—	—	—	—	—	—	—	—	—
	0.0			20	414	—	172	—	35	63	—	—	—	—	—	—	—	—	—
TF00	0.0	0	TYP	68	160	—	150	—	1	—	43	—	—	—	—	—	—	24	—
	0.0			20	1103	—	1035	—	1	—	43	—	—	—	—	—	—	165	—
TF00	0.0	0	SMIN	68	150	—	120	—	1	—	—	—	—	—	—	—	—	—	—
	0.0			20	1035	—	828	—	1	—	—	—	—	—	—	—	—	—	—

C83300—Adaptors, Electrical Hardware Parts, Connectors, Electrical Components, Terminal Ends for Electric Cables

As Sand Cast																			
M01	0.0	0	TYP	68	32	10	—	—	35	35	—	—	—	—	—	—	—	—	—
	0.0			20	221	69	—	—	35	35	—	—	—	—	—	—	—	—	—

C83450—Electrical Hardware, Low-Pressure Valve Bodies, Water Pump Parts, Iron Waterworks Valve Trim, Plumbing and Pipe Fittings

As Sand Cast																			
M01	0.0	0	TYP	68	37	15	—	—	31	—	—	—	—	—	62	—	—	—	—
	0.0			20	255	103	—	—	31	—	—	—	—	—	62	—	—	—	—
M01	0.0	0	SMIN	68	30	14	—	—	25	—	—	—	—	—	—	—	—	—	—
	0.0			20	207	97	—	—	25	—	—	—	—	—	—	—	—	—	—

C83600—Ornamental Fixtures, Hardware, Lightning Protection, Cooling Equipment, Heating Equipment, Trowels for Cement Working, Electrical Equipment, Switches, Electrical Hardware, Large Hold-Down Screws, Bearings, Small Gears, Low-Pressure Valves, Pumps, Valves for the Water Meter Industry, Valve Bodies for the Water Meter Industry, Transducer Housings, Valves, Rings, Printing Presses, Furnaces, Pump Fixtures, Impellers, Valve Bodies, Pumps, Couplings, Handles for Dental Equipment, Bearing Segments for Steel Industry, Pressure Blocks for Steel Industry, Air Actuators, Valve Bodies, Valves, Bushings, Pump Parts, Marine Products, Parts for Boats, Faucets, Pipe Fittings, Fixtures

As Sand Cast																			
M01	0.0	0	TYP	68	37	17	—	—	30	—	—	—	—	—	60	—	—	11	10.0
	0.0			20	255	117	—	—	30	—	—	—	—	—	60	—	—	76	14.0
M01	0.0	0	SMIN	68	30	14	—	—	20	—	—	—	—	—	—	—	—	—	—
	0.0			20	205	97	—	—	20	—	—	—	—	—	—	—	—	—	—
As Continuous Cast																			
M07	0.0	0	SMIN	68	36	19	—	—	15	—	—	—	—	—	—	—	—	—	—
	0.0			20	248	131	—	—	15	—	—	—	—	—	—	—	—	—	—
As Centrifugal Cast																			
M02	0.0	0	SMIN	68	30	14	—	—	20	—	—	—	—	—	—	—	—	—	—
	0.0			20	207	97	—	—	20	—	—	—	—	—	—	—	—	—	—

Table 40 (*Continued*)

Alloy Temper	Section Size in. mm	Cold Work %	Typ/ Min.	Temp °F °C	Tensile Strength ksi MPa	Yield Strength (0.5% ext. under load) ksi MPa	Yield Strength (0.2% offset) ksi MPa	Yield Strength (0.05% offset) ksi MPa	El %	Rockwell Hardness B	Rockwell Hardness C	Rockwell Hardness F	Rockwell Hardness 30T	Vickers Hardness 500	Brinell Hardness 500	Brinell Hardness 3000	Shear Strength ksi MPa	Fatigue Strength ksi MPa	Izod Impact Strength ft-lb J
C83800—Hardware, Electrical Components, Switches, Valves, Bushings, Railroad Catenary Fittings, Pump Fixtures, Air–Gas–Water Fittings, Plumbing Fixtures																			
As Sand Cast																			
M01	0.0	0	TYP	68	35	16	—	—	25	—	—	—	—	—	60	—	—	—	8.0
	0.0			20	241	110	—	—	25	—	—	—	—	—	60	—	—	—	11.0
M01	0.0	0	SMIN	68	30	13	—	—	20	—	—	—	—	—	—	—	—	—	—
	0.0			20	207	90	—	—	20	—	—	—	—	—	—	—	—	—	—
As Continuous Cast																			
M07	0.0	0	SMIN	68	30	15	—	—	16	—	—	—	—	—	—	—	—	—	—
	0.0			20	207	103	—	—	16	—	—	—	—	—	—	—	—	—	—
As Centrifugal Cast																			
M02	0.0	0	SMIN	68	30	13	—	—	20	—	—	—	—	—	—	—	—	—	—
	0.0			20	207	90	—	—	20	—	—	—	—	—	—	—	—	—	—
C84400—Ornamental Fixtures, Hardware, Cases for Dead Bolt Locks, Dead Bolt Locks, Door Hardware for Prisons, Heating Equipment, Cooling Equipment, Musical Instruments, Electrical Equipment, Valve Seat, Valves for Water Meters, Pump Fixtures, Valves, Valve Bodies for the Water Industry, Low-Pressure Fittings, Marine Fixtures, Pipe Fittings																			
As Sand Cast																			
M01	0.0	0	TYP	68	34	15	—	—	26	—	—	—	—	—	55	—	—	—	8.0
	0.0			20	234	103	—	—	26	—	—	—	—	—	55	—	—	—	11.0
M01	0.0	0	SMIN	68	29	13	—	—	18	—	—	—	—	—	—	—	—	—	—
	0.0			20	200	90	—	—	18	—	—	—	—	—	—	—	—	—	—
As Continuous Cast																			
M07	0.0	0	SMIN	68	30	15	—	—	16	—	—	—	—	—	—	—	—	—	—
	0.0			20	207	103	—	—	16	—	—	—	—	—	—	—	—	—	—
As Centrifugal Cast																			
M02	0.0	0	SMIN	68	29	13	—	—	18	—	—	—	—	—	—	—	—	—	—
	0.0			20	200	90	—	—	18	—	—	—	—	—	—	—	—	—	—

C85200—Ornamental Castings, Builders Hardware, Andirons, Chandeliers, Valves, Plumbing Fixtures, Plumbing Fittings

As Sand Cast																			
M01	0.0	0	TYP	68	38	13	—	—	35	—	—	—	—	—	45	—	—	—	—
	0.0			20	262	90	—	—	35	—	—	—	—	—	45	—	—	—	—
M01	0.0	0	SMIN	68	35	12	—	—	25	—	—	—	—	—	—	—	—	—	—
	0.0			20	241	83	—	—	25	—	—	—	—	—	—	—	—	—	—
As Centrifugal Cast																			
M02	0.0	0	SMIN	68	35	12	—	—	25	—	—	—	—	—	—	—	—	—	—
	0.0			20	241	83	—	—	25	—	—	—	—	—	—	—	—	—	—

C85400—Emblem, Ornaments, Builder Hardware, Window Hardware, Ornamental Fittings, Door Hardware for Prisons, Furniture, Decorative Furniture Legs, Musical Instruments, Chandeliers, Andirons, Mechanical Components Where Looks Are Important, Gas Cocks, Radiator Fittings, Battery Clamps, Ferrules, Valves, Marine Hardware, Ship Trimmings, Plumbing Parts

As Sand Cast																			
M01	0.0	0	TYP	68	38	13	—	—	35	—	—	—	—	—	50	—	—	—	—
	0.0			20	262	90	—	—	35	—	—	—	—	—	50	—	—	—	—
M01	0.0	0	SMIN	68	30	11	—	—	20	—	—	—	—	—	—	—	—	—	—
	0.0			20	207	76	—	—	20	—	—	—	—	—	—	—	—	—	—
As Centrifugal Cast																			
M02	0.0	0	SMIN	68	30	11	—	—	20	—	—	—	—	—	—	—	—	—	—
	0.0			20	207	76	—	—	20	—	—	—	—	—	—	—	—	—	—

C85700—Ornamental Hardware, Window Hardware, Door Hardware for Prisons, Musical Instruments, Mechanical Components Where Appearance Is Important, Ship Trim, Marine Hardware, Flanges, Fittings

As Sand Cast																			
M01	0.0	0	TYP	68	50	18	—	—	40	—	—	—	—	—	75	—	—	—	—
	0.0			20	345	124	—	—	40	—	—	—	—	—	75	—	—	—	—
M01	0.0	0	SMIN	68	—	14	—	—	—	—	—	—	—	—	—	—	—	—	—
	0.0			20	—	97	—	—	—	—	—	—	—	—	—	—	—	—	—
As Centrifugal Cast																			
M02	0.0	0	SMIN	68	40	14	—	—	15	—	—	—	—	—	—	—	—	—	—
	0.0			20	276	97	—	—	15	—	—	—	—	—	—	—	—	—	—

Table 40 (*Continued*)

Alloy Temper	Section Size in. mm	Cold Work %	Typ/ Min.	Temp °F °C	Tensile Strength ksi MPa	Yield Strength (0.5% ext. under load) ksi MPa	Yield Strength (0.2% offset) ksi MPa	Yield Strength (0.05% offset) ksi MPa	El %	Rockwell Hardness B	Rockwell Hardness C	Rockwell Hardness F	Rockwell Hardness 30T	Vickers Hardness 500	Brinell Hardness 500	Brinell Hardness 3000	Shear Strength ksi MPa	Fatigue Strength ksi MPa	Izod Impact Strength ft-lb J
C86200—Brackets, Structural Parts, Screw-Down Nuts, Shafts, Pressing Dies for Wood Pulp, Wear Rings for Pressing Dies for Wood Pulp Industry, Struts, Worm Gears, Hooks, Cams, Marine Racing Propellers, Valve Stems, Bushings, Frames, High-Strength Machine Parts, Gears, Boat Parts, Clamps, Rudders, Marine Castings. Gun Mounts																			
As Sand Cast																			
M01	0.0	0	TYP	68	95	48	—	—	20	—	—	—	—	—	—	180	—	—	12.0
	0.0			20	655	331	—	—	20	—	—	—	—	—	—	180	—	—	16.0
M01	0.0	0	SMIN	68	90	45	—	—	18	—	—	—	—	—	—	—	—	—	—
	0.0			20	621	310	—	—	18	—	—	—	—	—	—	—	—	—	—
As Continuous Cast																			
M07	0.0	0	SMIN	68	90	45	—	—	18	—	—	—	—	—	—	—	—	—	—
	0.0			20	621	310	—	—	18	—	—	—	—	—	—	—	—	—	—
As Centrifugal Cast																			
M02	0.0	0	SMIN	68	90	—	45	—	18	—	—	—	—	—	—	—	—	—	—
	0.0			20	621	—	310	—	18	—	—	—	—	—	—	—	—	—	—
C86300—																			
As Sand Cast																			
M01	0.0	0	TYP	68	119	62	—	—	18	—	—	—	—	—	—	225	—	25	15.0
	0.0			20	821	427	—	—	18	—	—	—	—	—	—	225	—	172	20.0
M01	0.0	0	SMIN	68	110	60	—	—	12	—	—	—	—	—	—	223	—	—	—
	0.0			20	760	415	—	—	12	—	—	—	—	—	—	223	—	—	—
As Continuous Cast																			
M07	0.0	0	SMIN	68	110	62	—	—	14	—	—	—	—	—	—	—	—	—	—
	0.0			20	758	427	—	—	14	—	—	—	—	—	—	—	—	—	—
As Centrifugal Cast																			
M02	0.0	0	SMIN	68	110	60	—	—	12	—	—	—	—	—	—	—	—	—	—
	0.0			20	758	414	—	—	12	—	—	—	—	—	—	—	—	—	—

C86400—Window Hardware, Door Hardware for Prison, Piano Keys, Musical Instruments, Electrical Equipment, Switches, Electrical Components, Screw-Down Nuts, Roller Bearings, Valve Stems, Machinery Parts, Propellers, Cams, Bearings, Light-Duty Gears, Brackets, Lever Arms, Fittings, Pump Fixtures, Bushings, Bearing Cage Blanks, Covers for Marine Hardware, Marine Hardware, Marine Fittings, Boat Parts, Plumbing Fixtures

As Sand Cast																			
M01	0.0	0	TYP	68	65	25	—	—	20	—	—	—	—	—	90	105	—	25	30.0
	0.0			20	448	172	—	—	20	—	—	—	—	—	90	105	—	172	41.0
M01	0.0	0	SMIN	68	60	20	—	—	15	—	—	—	—	—	—	—	—	—	—
	0.0			20	414	138	—	—	15	—	—	—	—	—	—	—	—	—	—
As Centrifugal Cast																			
M02	0.0	0	SMIN	68	60	20	—	—	15	—	—	—	—	—	—	—	—	—	—
	0.0			20	414	138	—	—	15	—	—	—	—	—	—	—	—	—	—

C86500—Welding Guns, Brackets, Electrical Hardware, Gears, Machinery Parts (Substituted for Steel and Malleable Iron), Wear Rings for Pressing Dies for Wood Pulp Industry, Forming Dies for Wood Pulp Industry, Machinery, Struts, Hooks, Lever Arms, Machinery Parts requiring High Strength, Frames, Compressors, Pressing Dies for Wood Pulp, Propellers for Salt and Fresh Water, Covers for Marine Hardware, Rudders, Boat Parts, Clamps

As Sand Cast																			
M01	0.0	0	TYP	68	71	29	—	—	30	—	—	—	—	—	100	130	—	20	—
	0.0			20	490	200	—	—	30	—	—	—	—	—	100	130	—	138	—
M01	0.0	0	SMIN	68	65	25	—	—	20	—	—	—	—	—	—	—	—	—	—
	0.0			20	448	172	—	—	20	—	—	—	—	—	—	—	—	—	—
As Continuous Cast																			
M07	0.0	0	SMIN	68	70	25	—	—	25	—	—	—	—	—	—	—	—	—	—
	0.0			20	483	172	—	—	25	—	—	—	—	—	—	—	—	—	—
As Centrifugal Cast																			
M02	0.0	0	SMIN	68	65	25	—	—	20	—	—	—	—	—	—	—	—	—	—
	0.0			20	448	172	—	—	20	—	—	—	—	—	—	—	—	—	—

C87300—(Silicon Bronze) Statuary and Plaques, Marine Hardware, Pumps, Boilers, Chemical Vessels

As Sand Cast																			
M02	0.0	0	SMIN	68	65	25	—	—	20	—	—	—	—	—	—	—	—	—	—
	0.0			20	448	172	—	—	20	—	—	—	—	—	—	—	—	—	—
As Centrifugal Cast																			
M02	0.0	0	SMIN	68	45	18	—	—	20	—	—	—	—	—	—	—	—	—	—
	0.0			20	310	124	—	—	20	—	—	—	—	—	—	—	—	—	—

Table 40 (*Continued*)

Alloy Temper	Section Size in. mm	Cold Work %	Typ/ Min.	Temp °F °C	Tensile Strength ksi MPa	Yield Strength (0.5% ext. under load) ksi MPa	Yield Strength (0.2% offset) ksi MPa	Yield Strength (0.05% offset) ksi MPa	El %	Rockwell Hardness B	Rockwell Hardness C	Rockwell Hardness F	Rockwell Hardness 30T	Vickers Hardness 500	Brinell Hardness 500	Brinell Hardness 3000	Shear Strength ksi MPa	Fatigue Strength ksi MPa	Izod Impact Strength ft-lb J
C87500—Window Hardware, Levers, Pump Fixtures, Fittings, Bearings, Impellers, Gears, Valve Bodies, Boat Parts, Small Boat Propellers, Plumbing Fixtures																			
As Sand Cast																			
M01	0.0	0	TYP	68	67	30	—	—	21	—	—	—	—	—	115	134	—	22	—
	0.0			20	462	207	—	—	21	—	—	—	—	—	115	134	—	152	—
M01	0.0	0	SMIN	68	60	24	—	—	16	—	—	—	—	—	—	—	—	—	—
	0.0			20	414	165	—	—	16	—	—	—	—	—	—	—	—	—	—
As Centrifugal Cast																			
M02	0.0	0	SMIN	68	60	24	—	—	16	—	—	—	—	—	—	—	—	—	—
	0.0			20	414	165	—	—	16	—	—	—	—	—	—	—	—	—	—
As Permanent Mold Cast																			
M05	0.0	0	SMIN	68	80	30	—	—	15	—	—	—	—	—	—	—	—	—	—
	0.0			20	552	207	—	—	15	—	—	—	—	—	—	—	—	—	—
C89510—Lead-Free Plumbing Castings																			
As Sand Cast																			
M01	0.0	0	TYP	68	30	20	—	—	12	—	—	—	—	—	37	—	—	—	—
	0.0			20	210	135	—	—	12	—	—	—	—	—	37	—	—	—	—
C89520—Lead-Free Plumbing Castings																			
As Sand Cast																			
M01	0.0	0	TYP	68	31	21	—	—	10	—	—	—	—	—	54	—	—	—	—
	0.0			20	210	140	—	—	10	—	—	—	—	—	54	—	—	—	—
C89550—Lead-Free Plumbing Castings																			
As Sand Cast																			
M01	0.0	0	MIN	68	35	21	20	—	5	—	—	—	—	—	—	—	—	—	—
	0.0			20	240	140	135	—	5	—	—	—	—	—	—	—	—	—	—
M01	0.0	0	TYP	68	48	29	28	—	8	—	—	—	—	—	—	—	—	—	—
	0.0			20	330	200	190	—	8	—	—	—	—	—	—	—	—	—	—

C90300—Heavy-Construction Equipment Bearings, Swivels, Bearings, Bushings, Pump Impellers, Piston Rings, Valve Bodies, Gear Blanks, Gears, Pump Bodies, Valves, Steam Fittings

As Sand Cast																			
M01	0.0	0	TYP	68	45	21	—	—	30	—	—	—	—	—	70	—	—	—	—
	0.0			20	310	145	—	—	30	—	—	—	—	—	70	—	—	—	—
M01	0.0	0	SMIN	68	40	18	—	—	20	—	—	—	—	—	—	—	—	—	—
	0.0			20	276	124	—	—	20	—	—	—	—	—	—	—	—	—	—
As Continuous Cast																			
M07	0.0	0	SMIN	68	44	22	—	—	18	—	—	—	—	—	—	—	—	—	—
	0.0			20	303	152	—	—	18	—	—	—	—	—	—	—	—	—	—

C90500—Clamps, Heavy-Construction Equipment, Electrical Connectors, Nuts, Bearings, Pump Impellers, Pump Bodies, Valves, Finishing Dies for Wood Pulp Industry, Expansion Bearings, Worm Gears, Seal Rings, Valve Bodies, Gear Blanks, Gears, Piston Rings, Bushings, Water Conditioners, Steam Fittings

As Sand Cast																			
M01	0.0	0	TYP	68	45	22	—	—	25	—	—	—	—	—	75	—	—	13	10.0
	0.0			20	310	152	—	—	25	—	—	—	—	—	75	—	—	90	13.0
M01	0.0	0	SMIN	68	40	18	—	—	20	—	—	—	—	—	—	—	—	—	—
	0.0			20	275	124	—	—	20	—	—	—	—	—	—	—	—	—	—
As Continuous Cast																			
M07	0.0	0	SMIN	68	44	25	—	—	10	—	—	—	—	—	—	—	—	—	—
	0.0			20	303	172	—	—	10	—	—	—	—	—	—	—	—	—	—
As Centrifugal Cast																			
M02	0.0	0	TYP	68	45	22	—	—	25	—	—	—	—	—	75	—	—	13	10.0
	0.0			20	310	152	—	—	25	—	—	—	—	—	75	—	—	90	13.0
M02	0.0	0	SMIN	68	40	18	—	—	20	—	—	—	—	—	—	—	—	—	—
	0.0			20	276	124	—	—	20	—	—	—	—	—	—	—	—	—	—

Table 40 *(Continued)*

Alloy Temper	Section Size in. mm	Cold Work %	Typ/ Min.	Temp °F °C	Tensile Strength ksi MPa	Yield Strength (0.5% ext. under load) ksi MPa	Yield Strength (0.2% offset) ksi MPa	Yield Strength (0.05% offset) ksi MPa	El %	Rockwell Hardness B	Rockwell Hardness C	Rockwell Hardness F	Rockwell Hardness 30T	Vickers Hardness 500	Brinell Hardness 500	Brinell Hardness 3000	Shear Strength ksi MPa	Fatigue Strength ksi MPa	Izod Impact Strength ft-lb J
C90700—Worm Wheels, Gears, Bearings for Heavy Loads and Relatively Low Speeds, Worm Gears, Gear Boxes, Speed Reducers, Valve Bodies, Restaurant Equipment, Bearings																			
As Sand Cast																			
M01	0.0	0	TYP	68	44	22	—	—	20	—	—	—	—	—	80	—	—	25	—
	0.0			20	303	152	—	—	20	—	—	—	—	—	80	—	—	172	—
M01	0.0	0	SMIN	68	35	17	—	—	10	—	—	—	—	—	65	—	—	—	—
	0.0			20	241	117	—	—	10	—	—	—	—	—	65	—	—	—	—
As Continuous Cast																			
M07	0.0	0	SMIN	68	40	25	—	—	10	—	—	—	—	—	—	—	—	—	—
	0.0			20	276	172	—	—	10	—	—	—	—	—	—	—	—	—	—
As Centrifugal Cast																			
M02	0.0	0	TYP	68	55	30	—	—	16	—	—	—	—	—	102	—	—	—	—
	0.0			20	379	207	—	—	16	—	—	—	—	—	102	—	—	—	—
M02	0.0	0	SMIN	68	50	28	—	—	12	—	—	—	—	—	95	—	—	—	—
	0.0			20	345	193	—	—	12	—	—	—	—	—	95	—	—	—	—
As Permanent Mold Cast																			
M05	0.0	0	TYP	68	55	30	—	—	16	—	—	—	—	—	102	—	—	—	—
	0.0			20	379	207	—	—	16	—	—	—	—	—	102	—	—	—	—
C91100—Hollow Electrical Conductors, Bearings, Piston Rings, Bushings																			
As Sand Cast																			
M01	0.0	0	TYP	68	35	25	—	—	2	—	—	—	—	—	—	135	—	—	—
	0.0			20	241	172	—	—	2	—	—	—	—	—	—	135	—	—	—
M01	0.0	0	SMIN	68	—	—	—	—	—	—	—	—	—	—	—	—	—	—	—
	0.0			20	—	—	—	—	—	—	—	—	—	—	—	—	—	—	—
C91300—Bells, Valve Bodies, Bushings, Piston Rings, Bearings																			
As Sand Cast																			
M01	0.0	0	TYP	68	35	30	—	—	0	—	—	—	—	—	—	170	—	—	—
	0.0			20	241	207	—	—	0	—	—	—	—	—	—	170	—	—	—
M01	0.0	0	SMIN	68	—	—	—	—	—	—	—	—	—	—	—	—	—	—	—
	0.0			20	—	—	—	—	—	—	—	—	—	—	—	—	—	—	—
As Continuous Cast																			
M07	0.0	0	SMIN	68	—	—	—	—	—	—	—	—	—	—	—	160	—	—	—
	0.0			20	—	—	—	—	—	—	—	—	—	—	—	160	—	—	—

C91600—Nuts, Gears, Fittings, Bearings, Pump Impellers, Steam Castings, Bushings, Piston Rings

As Sand Cast																			
M01	0.0	0	TYP	68	44	22	—	—	16	—	—	—	—	—	85	—	—	—	—
	0.0			20	303	152	—	—	16	—	—	—	—	—	85	—	—	—	—
M01	0.0	0	SMIN	68	35	17	—	—	10	—	—	—	—	—	65	—	—	—	—
	0.0			20	241	117	—	—	10	—	—	—	—	—	65	—	—	—	—
As Centrifugal Cast																			
M02	0.0	0	TYP	68	60	32	—	—	16	—	—	—	—	—	106	—	—	—	—
	0.0			20	414	221	—	—	16	—	—	—	—	—	106	—	—	—	—
M02	0.0	0	SMIN	68	45	25	—	—	10	—	—	—	—	—	85	—	—	—	—
	0.0			20	310	172	—	—	10	—	—	—	—	—	85	—	—	—	—
As Permanent Mold Cast																			
M05	0.0	0	TYP	68	60	32	—	—	16	—	—	—	—	—	106	—	—	—	—
	0.0			20	414	221	—	—	16	—	—	—	—	—	106	—	—	—	—

C92200—Ornamental Castings, Heating Equipment, Cooling Equipment, Nuts, Medium-Pressure Hydraulic Equipment, Valves for Water Meters, Pumps Used to 550°F, Gears, Bushings, Pump Impellers, Bearings, Piston Rings, Fittings Used to 550°F, Cryogenic Valves, Valve Components, Marine Castings, Medium-Pressure Steam Equipment to 550°F

As Sand Cast																			
M01	0.0	0	TYP	68	40	20	—	—	30	—	—	—	—	—	65	—	—	11	—
	0.0			20	276	138	—	—	30	—	—	—	—	—	65	—	—	76	—
M01	0.0	0	SMIN	68	34	16	—	—	22	—	—	—	—	—	—	—	—	—	—
	0.0			20	234	110	—	—	22	—	—	—	—	—	—	—	—	—	—
As Continuous Cast																			
M07	0.0	0	SMIN	68	38	19	—	—	18	—	—	—	—	—	—	—	—	—	—
	0.0			20	262	131	—	—	18	—	—	—	—	—	—	—	—	—	—
As Centrifugal Cast																			
M02	0.0	0	SMIN	68	34	16	—	—	22	—	—	—	—	—	—	—	—	—	—
	0.0			20	234	110	—	—	22	—	—	—	—	—	—	—	—	—	—
As Permanent Mold Cast																			
M05	0.0	0	SMIN	68	34	16	—	—	24	—	—	—	—	—	—	—	—	—	—
	0.0			20	234	110	—	—	24	—	—	—	—	—	—	—	—	—	—

Table 40 (*Continued*)

Alloy Temper	Section Size in. mm	Cold Work %	Typ/ Min.	Temp °F °C	Tensile Strength ksi MPa	Yield Strength (0.5% ext. under load) ksi MPa	Yield Strength (0.2% offset) ksi MPa	Yield Strength (0.05% offset) ksi MPa	El %	Rockwell Hardness B	C	F	30T	Vickers Hardness 500	Brinell Hardness 500	3000	Shear Strength ksi MPa	Fatigue Strength ksi MPa	Izod Impact Strength ft-lb J
C92300—Structural Castings, Nuts, High-Pressure Hydraulic Equipment, Valve Bodies, Bearings, Piston Rings, Pump Parts, Gears, Bushings, Pump Impellers, High-Pressure Steam Equipment																			
As Sand Cast																			
M01	0.0	0	TYP	68	40	20	—	—	25	—	—	—	—	—	70	—	—	—	12.0
	0.0			20	276	138	—	—	25	—	—	—	—	—	70	—	—	—	16.0
M01	0.0	0	SMIN	68	36	16	—	—	18	—	—	—	—	—	—	—	—	—	—
	0.0			20	248	110	—	—	18	—	—	—	—	—	—	—	—	—	—
As Continuous Cast																			
M07	0.0	0	SMIN	68	40	19	—	—	16	—	—	—	—	—	—	—	—	—	—
	0.0			20	276	131	—	—	16	—	—	—	—	—	—	—	—	—	—
As Centrifugal Cast																			
M02	0.0	0	SMIN	68	36	16	—	—	18	—	—	—	—	—	—	—	—	—	—
	0.0			20	248	110	—	—	18	—	—	—	—	—	—	—	—	—	—
C92600—Lead Screw Nuts, Nuts, Heavy-Duty Bearings, Pump Pistons, Valve Bodies, Bushings, Fittings, Valves, Gears, Piston Rings, Steam Castings, Pump Impellers																			
As Sand Cast																			
M01	0.0	0	TYP	68	44	20	—	—	30	—	—	78	—	—	70	—	—	—	7.0
	0.0			20	303	138	—	—	30	—	—	78	—	—	70	—	—	—	9.0
M01	0.0	0	SMIN	68	40	18	—	—	—	—	—	—	—	—	—	—	—	—	—
	0.0			20	276	124	—	—	—	—	—	—	—	—	—	—	—	—	—
C92900—Worm Gears, General Service Bearings, Pump Bodies, Impellers for Mine Water, Cams, Gears, Wear Plates																			
As Sand Cast																			
M01	0.0	0	TYP	68	47	26	—	—	20	—	—	—	—	—	80	—	—	—	12.0
	0.0			20	324	179	—	—	20	—	—	—	—	—	80	—	—	—	16.0
M01	0.0	0	SMIN	68	45	25	—	—	8	—	—	—	—	—	75	—	—	—	—
	0.0			20	310	172	—	—	8	—	—	—	—	—	75	—	—	—	—
As Continuous Cast																			
M07	0.0	0	SMIN	68	45	25	—	—	8	—	—	—	—	—	—	—	—	—	—
	0.0			20	310	172	—	—	8	—	—	—	—	—	—	—	—	—	—
As Permanent Mold Cast																			
M05	0.0	0	TYP	68	47	26	—	—	20	—	—	—	—	—	80	—	—	—	12.0
	0.0			20	324	179	—	—	20	—	—	—	—	—	80	—	—	—	16.0

C93200—Automotive Fittings, Washers, Insert Bearings, Bearings, Pumps, Machine Parts, Pump Fixtures, Machine Tool Bearings, Trunion Bearings, Rolling Mill Bearings, Pump Impellers, Hydraulic Press Main Lining, Hydraulic Press Stuffing Box, Forging Press Toggle Lever Bearings, Diesel Engine Wrist Pin Bushings, Water Pump Bushings, Fuel Pump Bushings, Linkage Bushings for Presses, Roll Neck Bearings, Bearings for Cranes, Main Spindle Bearings, Fittings, Bushings, Thrust Washers, General-Purpose Bushings

As Sand Cast																			
M01	0.0	0	TYP	68	35	18	—	—	20	—	—	—	—	—	65	—	—	16	6.0
	0.0			20	241	124	—	—	20	—	—	—	—	—	65	—	—	110	8.0
M01	0.0	0	SMIN	68	30	14	—	—	15	—	—	—	—	—	—	—	—	—	—
	0.0			20	207	97	—	—	15	—	—	—	—	—	—	—	—	—	—
As Continuous Cast																			
M07	0.0	0	SMIN	68	35	20	—	—	10	—	—	—	—	—	—	—	—	—	—
	0.0			20	241	138	—	—	10	—	—	—	—	—	—	—	—	—	—
As Centrifugal Cast																			
M02	0.0	0	SMIN	68	30	14	—	—	15	—	—	—	—	—	—	—	—	—	—
	0.0			20	207	97	—	—	15	—	—	—	—	—	—	—	—	—	

C93700—Brackets, Nuts, Washers for Engines, Bushings for high speed and heavy pressure, Bushings, Machine Parts, Slide Guides for Steel Mills, Pumps, Impellers, Bearing Plates, Corrosion-Resistant Castings, Pressure-Tight Castings, High-Speed, Heavy-Load Bearings, Parts for Steel Mill Maintenance, Applications Requiring Acid Resistance to Sulfite Fluids, Bearings, Crank Shafts, Large Marine Bearings

As Sand Cast																			
M01	0.0	0	TYP	68	35	18	16	—	20	—	—	—	—	—	60	—	18	13	5.0
	0.0			20	241	124	110	—	20	—	—	—	—	—	60	—	124	90	7.0
M01	0.0	0	SMIN	68	30	12	—	—	15	—	—	—	—	—	—	—	—	—	—
	0.0			20	207	83	—	—	15	—	—	—	—	—	—	—	—	—	—
As Continuous Cast																			
M07	0.0	0	SMIN	68	35	20	—	—	6	—	—	—	—	—	—	—	—	—	—
	0.0			20	241	138	—	—	6	—	—	—	—	—	—	—	—	—	—
As Centrifugal Cast																			
M02	0.0	0	SMIN	68	30	12	—	—	15	—	—	—	—	—	—	—	—	—	—
	0.0			20	207	83	—	—	15	—	—	—	—	—	—	—	—	—	—

Table 40 *(Continued)*

Alloy Temper	Section Size in. mm	Cold Work %	Typ/ Min.	Temp °F °C	Tensile Strength ksi MPa	Yield Strength (0.5% ext. under load) ksi MPa	Yield Strength (0.2% offset) ksi MPa	Yield Strength (0.05% offset) ksi MPa	El %	Rockwell Hardness B	Rockwell Hardness C	Rockwell Hardness F	Rockwell Hardness 30T	Vickers Hardness 500	Brinell Hardness 500	Brinell Hardness 3000	Shear Strength ksi MPa	Fatigue Strength ksi MPa	Izod Impact Strength ft-lb J
C93800—Bearings, Low-Friction, Moderate-Pressure Bushings, Railroad Applications, Acid-Resistant Applications, Pumps, General Service Bearings for Moderate Pressure, Wearing Material for Wedges, Pump Impellers for Acid Mine Water, Freight Car Bearings, Backs for Lined Journal Bearings for Passenger Cars, Backs for Lined Journal Bearings for Locomotives, Pump Bodies for Acid Mine Water, Wearing Material for Shoes, Wearing Material for Rod Bushings, Railroad Engine Casings, Industrial Centrifuges, Machine Parts, Large Marine Bearings																			
As Sand Cast																			
M01	0.0	0	TYP	68	30	16	—	—	18	—	—	—	—	—	55	—	15	10	5.0
	0.0			20	207	110	—	—	18	—	—	—	—	—	55	—	103	69	7.0
M01	0.0	0	SMIN	68	26	14	—	—	12	—	—	—	—	—	—	—	—	—	—
	0.0			20	179	97	—	—	12	—	—	—	—	—	—	—	—	—	—
As Continuous Cast																			
M07	0.0	0	SMIN	68	25	16	—	—	5	—	—	—	—	—	—	—	—	—	—
	0.0			20	172	110	—	—	5	—	—	—	—	—	—	—	—	—	—
As Centrifugal Cast																			
M02	0.0	0	TYP	68	33	20	—	—	12	—	—	—	—	—	—	—	—	—	—
	0.0			20	228	138	—	—	12	—	—	—	—	—	—	—	—	—	—
M02	0.0	0	SMIN	68	26	14	—	—	12	—	—	—	—	—	—	—	—	—	—
	0.0			20	179	97	—	—	12	—	—	—	—	—	—	—	—	—	—

C95200—Electrical Hardware, Nuts, Large Gear Parts, Pickling Tanks, Hydrant Parts, High-Strength Clamps, Valve Bodies, Mildalkali Applications, Wear Plates, Bearing Liners, Valves, Worm Wheels, Worms, Pump Rods, Plungers, Gears, Bushings, Bearings, Acid-Resistant Pumps, Valve Seats, Hot Mill Guides, Welding Jaws, Pickling Equipment, Pump Parts, Thrust Pads, Marine Engines, Marine Hardware, Propellers, Covers for Marine Hardware, Gun Mountings, Gun Slides

As Sand Cast																			
M01	0.0	0	TYP	68	80	27	—	—	35	—	—	—	—	—	—	125	40	22	30.0
	0.0			20	552	186	—	—	35	—	—	—	—	—	—	125	276	152	41.0
M01	0.0	0	SMIN	68	65	25	—	—	20	—	—	—	—	—	—	110	—	—	—
	0.0			20	448	172	—	—	20	—	—	—	—	—	—	110	—	—	—
As Continuous Cast																			
M07	0.0	0	SMIN	68	68	26	—	—	20	—	—	—	—	—	—	—	—	—	—
	0.0			20	469	179	—	—	20	—	—	—	—	—	—	—	—	—	—
As Centrifugal Cast																			
M02	0.0	0	SMIN	68	65	25	—	—	20	—	—	—	—	—	—	110	—	—	—
	0.0			20	450	170	—	—	20	—	—	—	—	—	—	110	—	—	—
As Permanent Mold Cast																			
M05	0.0	0	TYP	68	93	35	—	—	38	78	—	—	—	—	—	—	—	—	—
	0.0			20	640	240	—	—	38	78	—	—	—	—	—	—	—	—	—
As Permanent Mold Cast																			
M05	0.0	0	TYP	68	93	35	—	—	38	78	—	—	—	—	—	—	—	—	
	0.0			20	640	240	—	—	38	78	—	—	—	—	—	—	—	—	

Table 40 (*Continued*)

Alloy Temper	Section Size in. mm	Cold Work %	Typ/ Min.	Temp °F °C	Tensile Strength ksi MPa	Yield Strength (0.5% ext. under load) ksi MPa	Yield Strength (0.2% offset) ksi MPa	Yield Strength (0.05% offset) ksi MPa	El %	Rockwell Hardness B	Rockwell Hardness C	Rockwell Hardness F	Rockwell Hardness 30T	Vickers Hardness 500	Brinell Hardness 500	Brinell Hardness 3000	Shear Strength ksi MPa	Fatigue Strength ksi MPa	Izod Impact Strength ft-lb J
C95400—Welding Guns, Nuts, Large Hold-Down Screws, Bearings, Pawl, Pump Parts, Worm Gears, Valve Bodies, Machine Parts, Valves, Bushings, Pickling Hooks, Valve Guides, Valve Seats, Bearing Segments for the Steel Industry, Pressure Blocks for the Steel Industry, High-Strength Clamps, Gears, Landing Gear Parts, Spur Gears, Heavily Loaded Worm Gears, Covers for Marine Hardware, Ship Building, Ordnance Fittings																			
As Sand Cast																			
M01	0.0	0	TYP	68	85	35	—	—	18	—	—	—	—	—	—	170	47	28	16.0
	0.0			20	586	241	—	—	18	—	—	—	—	—	—	170	324	193	22.0
M01	0.0	0	SMIN	68	75	30	—	—	12	—	—	—	—	—	—	150	—	—	—
	0.0			20	517	207	—	—	12	—	—	—	—	—	—	150	—	—	—
As Continuous Cast																			
M07	0.0	0	SMIN	68	85	32	—	—	12	—	—	—	—	—	—	—	—	—	—
	0.0			20	586	221	—	—	12	—	—	—	—	—	—	—	—	—	—
TQ50	0.0	0	SMIN	68	95	45	—	—	10	—	—	—	—	—	—	—	—	—	—
	0.0			20	655	310	—	—	10	—	—	—	—	—	—	—	—	—	—
As Centrifugal Cast																			
M02	0.0	0	SMIN	68	75	30	—	—	12	—	—	—	—	—	—	150	—	—	—
	0.0			20	515	205	—	—	12	—	—	—	—	—	—	150	—	—	—
TQ50	0.0	0	SMIN	68	90	45	—	—	6	—	—	—	—	—	—	190	—	—	—
	0.0			20	620	310	—	—	6	—	—	—	—	—	—	190	—	—	—
As Permanent Mold Cast																			
M05	0.0	0	TYP	68	105	46	—	—	11	—	—	—	—	—	—	—	—	—	—
	0.0			20	725	320	—	—	11	—	—	—	—	—	—	—	—	—	—
M05	0.0	0	SMIN	68	100	40	—	—	10	—	—	—	—	—	—	—	—	—	—
M05	0.0	0	TYP	68	105	46	—	—	11	—	—	—	—	—	—	—	—	—	—
Cast and Annealed (Homogenized)																			
	0.0	0	SMIN	68	75	30	—	—	—	—	—	—	—	—	—	—	—	—	—
	0.0			20	517	207	—	—	—	—	—	—	—	—	—	—	—	—	—

C95410—Bearings, Spur Gears, Gears, Pickling Hooks, Worms, Valve Components, Bushings, Pickling Baskets

As Sand Cast																			
M01	0.0	0	TYP	68	85	35	—	—	18	—	—	—	—	—	—	170	47	28	14.0
	0.0			20	586	241	—	—	18	—	—	—	—	—	—	170	324	193	19.0
M01	0.0	0	SMIN	68	75	30	—	—	12	—	—	—	—	—	—	150	—	—	—
	0.0			20	515	205	—	—	12	—	—	—	—	—	—	150	—	—	—
TQ50	0.0	0	TYP	68	105	54	—	—	8	—	—	—	—	—	—	195	50	35	14.0
	0.0			20	724	372	—	—	8	—	—	—	—	—	—	195	345	241	19.0
TQ50	0.0	0	SMIN	68	90	45	—	—	6	—	—	—	—	—	—	190	—	—	—
	0.0			20	620	310	—	—	6	—	—	—	—	—	—	190	—	—	—
As Continuous Cast																			
M07	0.0	0	SMIN	68	85	32	—	—	12	—	—	—	—	—	—	—	—	—	
	0.0			20	586	221	—	—	12	—	—	—	—	—	—	—	—	—	
TQ50	0.0	0	SMIN	68	95	45	—	—	10	—	—	—	—	—	—	—	—	—	
	0.0			20	655	310	—	—	10	—	—	—	—	—	—	—	—	—	
As Centrifugal Cast																			
M02	0.0	0	SMIN	68	75	30	—	—	12	—	—	—	—	—	—	150	—	—	—
	0.0			20	515	205	—	—	12	—	—	—	—	—	—	150	—	—	—
TQ50	0.0	0	SMIN	68	90	45	—	—	6	—	—	—	—	—	—	190	—	—	—
	0.0			20	620	310	—	—	6	—	—	—	—	—	—	190	—	—	—
As Permanent Mold Cast																			
M05	0.0	0	SMIN	68	100	40	—	—	10	—	—	—	—	—	—	—	—	—	—
	0.0			20	690	275	—	—	10	—	—	—	—	—	—	—	—	—	—

Table 40 *(Continued)*

Alloy Temper	Section Size in. mm	Cold Work %	Typ/ Min.	Temp °F °C	Tensile Strength ksi MPa	Yield Strength (0.5% ext. under load) ksi MPa	Yield Strength (0.2% offset) ksi MPa	Yield Strength (0.05% offset) ksi MPa	El %	Rockwell Hardness B	Rockwell Hardness C	Rockwell Hardness F	Rockwell Hardness 30T	Vickers Hardness 500	Brinell Hardness 500	Brinell Hardness 3000	Shear Strength ksi MPa	Fatigue Strength ksi MPa	Izod Impact Strength ft-lb J
C95500—Window Hardware, Piano Keys, Musical Instruments, Electrical Hardware, Stuffing Box Nuts, Hand Gun Recoil Mechanisms, Valve Bodies, Bushings, Valve Components, Bearings, Sewage Treatment Applications, Pump Fluid Ends, Valve Seats, Piston Guides, Valve Guides, Pickling Equipment, Air Craft Components, Wear Plates, Welding Jaws, Glass Molds, Landing Gear Parts, Hot Mill Guides, Worms Wheels, Worms, Glands, Gears, Landing Gear Parts, Machine Parts, Covers for Marine Hardware, Ship Building, Marine Hardware, Marine Applications, Ordnance Fittings																			
As Sand Cast																			
M01	0.0	0	TYP	68	100	44	—	—	12	87	—	—	—	—	—	195	48	31	13.0
	0.0			20	689	303	—	—	12	87	—	—	—	—	—	195	331	214	18.0
M01	0.0	0	SMIN	68	90	40	—	—	6	—	—	—	—	—	—	190	—	—	—
	0.0			20	620	275	—	—	6	—	—	—	—	—	—	190	—	—	—
TQ50	0.0	0	TYP	68	120	68	—	—	10	—	—	—	—	—	—	230	70	38	15.0
	0.0			20	827	469	—	—	10	—	—	—	—	—	—	230	483	262	20.0
TQ50	0.0	0	SMIN	68	110	60	—	—	5	—	—	—	—	—	—	200	—	—	—
	0.0			20	760	415	—	—	5	—	—	—	—	—	—	200	—	—	—
As Continuous Cast																			
M07	0.0	0	SMIN	68	95	42	—	—	10	—	—	—	—	—	—	—	—	—	—
	0.0			20	655	290	—	—	10	—	—	—	—	—	—	—	—	—	—
TQ50	0.0	0	SMIN	68	110	62	—	—	8	—	—	—	—	—	—	—	—	—	—
	0.0			20	758	427	—	—	8	—	—	—	—	—	—	—	—	—	—
As Centrifugal Cast																			
M02	0.0	0	SMIN	68	90	40	—	—	6	—	—	—	—	—	—	190	—	—	—
	0.0			20	620	275	—	—	6	—	—	—	—	—	—	190	—	—	—
TQ50	0.0	0	SMIN	68	110	60	—	—	5	—	—	—	—	—	—	200	—	—	—
	0.0			20	760	415	—	—	5	—	—	—	—	—	—	200	—	—	—
As Permanent Mold Cast																			
	0.0	0	SMIN	68	110	60	—	—	5	—	—	—	—	—	—	—	—	—	—
	0.0			20	760	415	—	—	5	—	—	—	—	—	—	—	—	—	—

C95800—Nuts, Shafts, Propeller Hub, Propeller Blades, Pickling Equipment, Valve Bodies, Machinery, Worms, Bushings, Worm Wheels, Gears, Wear Plates, Valves in Contact with Seawater, Covers for Marine Hardware, Ship Building, Marine Hardware, Elbows

As Sand Cast																			
M01	0.0	0	TYP	68	95	38	—	—	25	—	—	—	—	—	—	159	58	31	20.0
	0.0			20	655	262	—	—	25	—	—	—	—	—	—	159	400	214	27.0
M01	0.0	0	SMIN	68	85	35	—	—	15	—	—	—	—	—	—	—	—	—	—
	0.0			20	585	240	—	—	15	—	—	—	—	—	—	—	—	—	—
As Continuous Cast																			
M07	0.0	0	SMIN	68	85	35	—	—	18	—	—	—	—	—	—	—	—	—	—
	0.0			20	586	241	—	—	18	—	—	—	—	—	—	—	—	—	—
As Centrifugal Cast																			
M02	0.0	0	SMIN	68	85	35	—	—	15	—	—	—	—	—	—	—	—	—	—
	0.0			20	585	240	—	—	15	—	—	—	—	—	—	—	—	—	—
As Permanent Mold Cast																			
M05	0.0	0	TYP	68	96	52	—	—	17	88	—	—	—	—	—	—	—	—	
	0.0			20	660	360	—	—	17	88	—	—	—	—	—	—	—	—	
M05	0.0	0	SMIN	68	90	40	—	—	15	—	—	—	—	—	—	—	—	—	
	0.0			20	620	275	—	—	15	—	—	—	—	—	—	—	—	—	

C96400—Pump Fixtures, Fittings, Pump Bodies, Steam Fittings, Valves, Pump Bodies, Flanges and Elbows Used for Seawater Corrosion Resistance, Boat Parts

As Sand Cast																			
M01	0.0	0	TYP	68	68	37	—	—	28	—	—	—	—	—	—	140	—	18	—
	0.0			20	469	255	—	—	28	—	—	—	—	—	—	140	—	124	—
M01	0.0	0	SMIN	68	60	32	—	—	20	—	—	—	—	—	—	—	—	—	—
	0.0			20	415	220	—	—	20	—	—	—	—	—	—	—	—	—	—
As Continuous Cast																			
M07	0.0	0	SMIN	68	65	35	—	—	25	—	—	—	—	—	—	—	—	—	—
	0.0			20	448	241	—	—	25	—	—	—	—	—	—	—	—	—	—

Source: Copper Development Association.

Table 41 Impact Properties of Cast Copper Alloys

UNS Alloy Number	Temperature		Charpy V-Notch Impact Strength		UNS Alloy Number	Temperature		Charpy V-Notch Impact Strength		UNS Alloy Number	Temperature		Charpy V-Notch Impact Strength	
	°F	°C	ft·lb	J		°F	°C	ft·lb	J		°F	°C	ft·lb	J
C83600	−305	−188	11	15	C92200	−320	−196	13	18	C95500	−305	−188	12	16
	−100	−74	13	18		−108	−78	14	19		−200	−130	16	22
	68	20	19	26		68	20	19	26		−78	−60	18	24
	392	200	15	20		212	100	14	19		68	20	18	24
	572	300	13	18		392	200	13	18		392	200	28	38
C86500	−305	−188	12	16		572	300	12	16		572	300	26	35
	−100	−74	19	26	C95200	−320	−196	25	34	C95700	−290	10	10	14
	68	20	19	26		68	20	30	41		−148	−100	16	22
	212	100	18	24		212	100	33	45		−58	−50	23	31
											68	20	32	41

Source: Copper Development Association.

Drilling. Stand twist drills are satisfactory for most copper alloys; however, high production rates might require special drill configurations (Fig. 5). Small-diameter holes in copper are best cut with fast-twist drills, while free-cutting grades can be cut very rapidly with straight-fluted "brass" drills. Full rake angles are normally retained in drills used on long-chip type III alloys. Drill-tip angles should be between 100° and 110° and lip clearance angles may have to be as steep as 20°. Notching the cutting edge helps break up stringy chips. Drills for short-chip type II alloys and free-cutting type I alloys should be flat ground to a 0° rake, with standard 118° tip angles and lip clearances between 12° and 15°. Speeds are similar to those for turning, and small-diameter holes can be drilled at maximum spindle speed. Recommended feed rates range from 0.002 to 0.030 in./rev (0.05–0.76 mm/rev) depending on alloy, i.e., higher rates for free-cutting alloys and lower rates for types II and III alloys. Cutting fluids are beneficial but not necessary with free-cutting alloys but may be required for others.

Boring. For best surface condition, free-cutting alloys should be bored with a 0° back rake angle, a 5° side rake angle, and speeds of 500–1000 sfm (152–305 mpm) (Fig. 6). Rake angles can be increased to between 0° and 5° in type II alloys and between 5° and 10° in type III alloys. Side rake angles will range between 5° and 10° and between 15° and 20° for types II and III alloys, respectively. Recommended cutting speeds will range between 400 and 600 sfm (122 and 183 mpm) for type II alloys and between 200 and 500 sfm (61 and 152 mpm) for type III alloys.

Reaming. Fluted reamers used with copper alloys are similar to those used with steel, except that clearances should be increased to 8°–10° (Fig. 7). A rake angle (hook) of 5° is used on all copper alloys except free-cutting alloys, which are generally reamed with zero or negative rake. Cutting tools must be lapped to a fine finish. Speeds for copper and other long-chip alloys will range from 40 to 90 sfm (12–27 mpm), 75 to 150 sfm (23–76 mpm) for short-chip alloys, and 100 to 200 sfm (30–61 mpm) for free-cutting type I alloys.

Table 42 Guidelines for Copper Alloy Selection in Corrosive Environments

Environment	Comments
Concentrated or dilute acids or alkalis not containing air or other oxidants (nitric acid, dichromates, chlorine and ferric salts), complexing agents [cyanides, ammonia, chlorides (when hot), amines], and compounds that react directly with copper (sulfur, hydrogen sulfide, silver salts, mercury and its compounds, acetylene)	Tin bronzes, silicon bronzes, nickel–silver, copper–aluminum bronzes and low-zinc brasses are generally acceptable.
Dilute or concentrated acids or acid salts, both organic and inorganic	Yellow brasses and other high-zinc alloys should not be specified.
Dilute or concentrated alkalis, neutral chloride or sulfate solutions or mild oxidizing agents such as calcium hypochlorite, hydrogen peroxide, and sodium nitrate	High-zinc alloys should not be specified.
Nonoxidizing acetic, hydrochloric, and phosphoric acids	Tin bronzes, aluminum bronzes, nickel–silver, copper and silicon bronzes recommended. High-zinc alloys should not be specified.
Hot and concentrated hydrochloric acid	May be aggressive toward alloys that resist attack when the acid is cold and dilute.
Nitric, chromic, and other oxidizing acids	Copper alloys not recommended.
Alkalis	70-30 copper nickel alloys recommended for concentrated bases; high-tin bronzes, nickel–silver, silicon bronzes, and most other alloys except high-zinc brasses can be used with dilute bases. Aluminum bronzes not containing tin are susceptible to dealuminification in hot dilute alkalis.
Ammonium hydroxide, substituted ammonium compounds, amines, and cyanides; mercury, mercury salts	Copper alloys should be avoided.
Neutral salt solutions	Generally safe with most copper alloys, although corrosion rates may vary. Chlorides are more aggressive than sulfates and carbonates, especially in aerated solutions. Copper–nickels and aluminum bronzes are preferred for use with concentrated brines.
Basic salts	Behavior similar to that of hydroxides but less aggressive.
Gases	Do not attack copper alloys when dry but can become aggressive when moist. All copper alloys are attacked by moist chlorine, but chlorinated water can be handled by high-tin bronze, nickel–silver, and copper. High-zinc brasses are attacked by moist carbon dioxide.
Organic compounds	Generally benign toward copper alloys. Exceptions include hot, moist, chlorinated hydrocarbons and aerated organic acids. Moist acetylene forms explosive compounds in contact with copper and alloys containing more than 65% copper.
Foodstuffs and beverages	Contact with acidic foodstuffs and beverages can impart a metallic taste and should be avoided. Copper and copper alloys are suitable for brewing and distilling, and nickel–silvers are commonly used with dairy products. Leaded alloys should be used with caution.

Table 43 Velocity Guidelines for Copper Alloys in Pumps and Propellers Operating in Seawater

	Peripheral Velocity	
UNS Alloy Number	ft/s	m/s
C83600	<30	<9.1
C87600	<30	<9.1
C90300, C92200	<45	<13.7
C95200, C86500	<75	<22.8
C95500, C95700, C95800	>75	>22.8

Source: Copper Development Association.

Threading and Tapping. Free-cutting brass accepts fine- to medium-pitch rolled threads quite well, but coarse or deep threads may call for alloys such as C34000, C34500, or C35300, which have lower lead contents and are more ductile than C36000. Ductile nonleaded alloys can be roll threaded; operations proceed best when alloys are in the annealed condition. Thread die cutting generally follows the recommendations for turning operations: soft, ductile long-chip alloys require large rake angles (17°–25° and up to 30° for pure copper); hard multiphase type II alloys require intermediate rake angles, generally 12°–17° but up to 25° for high-strength aluminum bronze. Free-cutting brass and other highly leaded alloys require zero rake. See Figs. 8*a* and 8*b* for recommended tool geometries.

Sawing. Coarse teeth are better for cutting soft or thick materials; finer teeth should be used with hard materials and thin sections (Fig. 9). Sawing conditions range from between 130 and 150 strokes/min for type I alloys to between 90 and 120 strokes/min for type II alloys and between 60 and 90 strokes/min for type III alloys. Thin rods of type I alloys less than 1½ in. (38 mm) in diameter can be circular sawed at rim speeds between 4000 and 8000

Figure 2 Three types of machining chips produced by copper and copper alloys: type I (free-cutting) chips, center; type II (short curly chips), right; type III (long stringy chips), left.

sfm (1220 and 2440 mpm) at feeds of 60 in./min (1524 mm/min). Speeds and feeds are reduced by between one-third and one-half for types II and III alloys, respectively. Thicker sections require proportionately slower speeds and lower feed rates.

Free-Cutting Brass Machining Economics

Free-cutting brass (C36000) and leaded free-machining steel (12L14) often contend equally (on the basis of mechanical properties) for large-volume screw-machined parts such as automotive fittings. Historic evaluations for these products based on raw material costs alone have been superceded in recent years by considerations of such environmental or quality-driven factors as recycling and the need for improved corrosion performance by the application of protective surface treatments. All of these factors favor free-cutting copper alloys, which are readily recycled and normally do not require corrosion protection. Components made from these alloys can, therefore, be less costly, despite their relatively higher raw materials costs, than leaded steels. The significantly higher production rates attainable with brass also reduce unit-part costs, with cost reductions increasing with the amount of metal removed during machining.

The gross raw material cost of C36000 is approximately 2½ times higher than leaded steel, but unlike steel, turnings from brass are routinely recycled for credits ranging between 75 and 80% of the cost of new rod. Since typical screw machine products generate 60–70% turnings, *net* raw material costs can be significantly reduced.

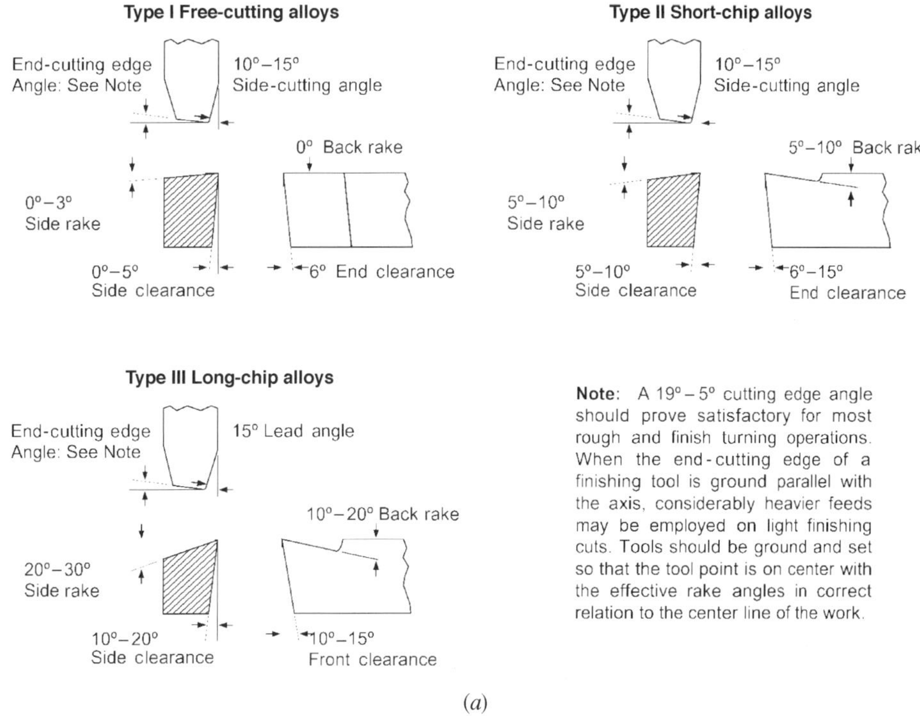

Figure 3 (*a*) Carbon and high-speed steel single-point turning tools. (*b*) Carbide-tipped single-point turning tools. (*c*) Straight-blade and circular cutoff tools. (*d*) Carbide and high-speed circular and dovetail form tools.

Type I Free-cutting alloys

End-cutting edge Angle 8° – 15°
10° – 15° Lead angle (or to suit)
0° Back rake
2° – 6° Side rake
4° – 6° Side clearance
4° – 6° Front clearance

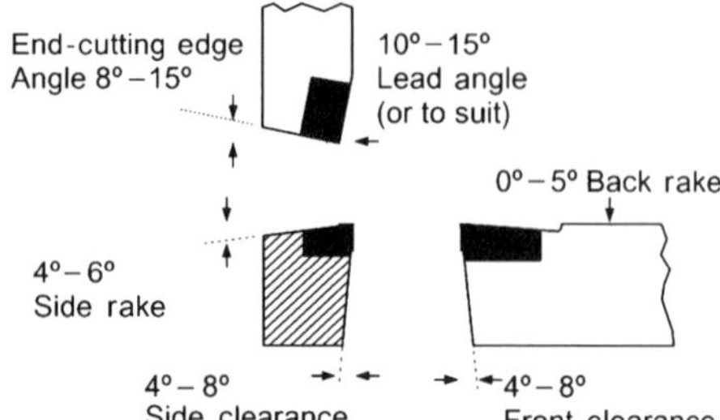

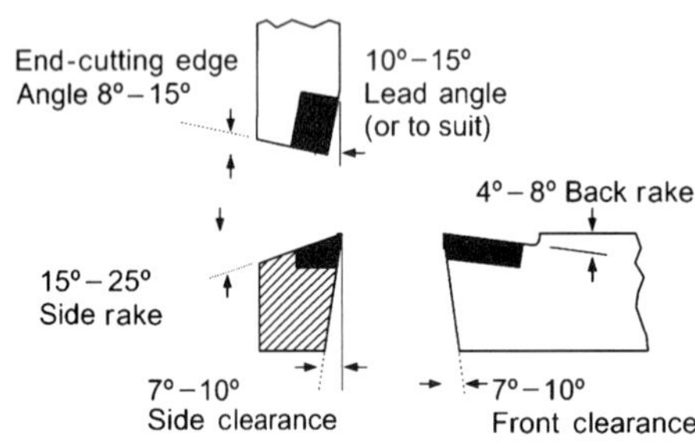

Note: Rake angles are based on the tool shank being set parallel with the center line of the work and with the tool point on center. Placing the tool point above or below center will change the effective rake angles appreciably, particularly on work of small diameter. On a setup where the tool holder is not parallel with the center line, the rake angles should be ground so that when the tool is mounted, they are in correct relation.

(*b*)

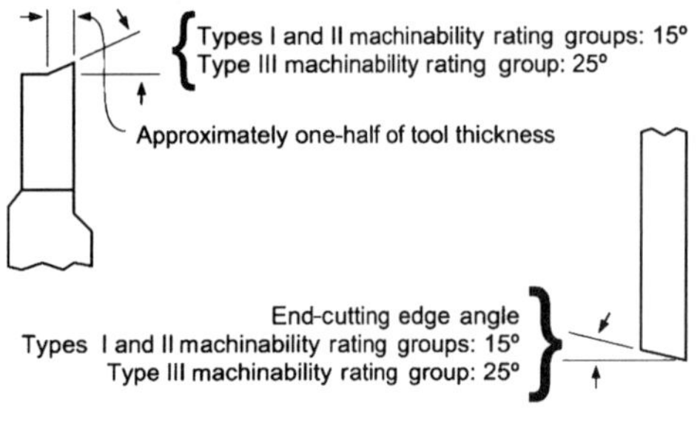

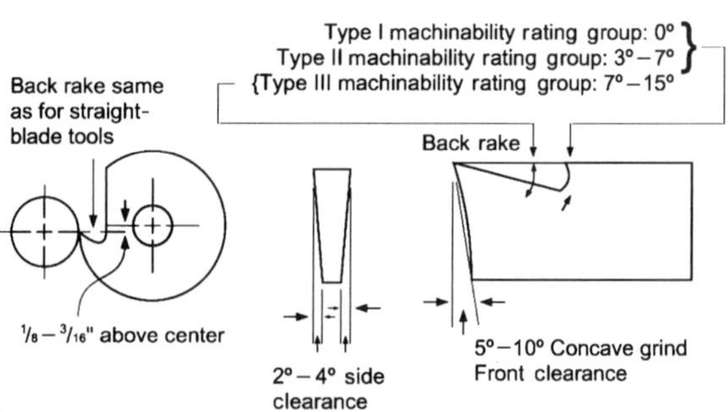

(*c*)

Figure 3 (*Continued*)

A typical hose fitting (Fig. 10) machined from a rod generates 71% scrap in the form of turnings. With allowance for recycling, the net raw material cost difference between brass and steel might be as small as 3%.

Higher cutting speeds attainable with brass reduce machining costs in brass to slightly more than 47% of the costs for steel, which, accounting for raw material, produces a finished product cost that is 17% lower in brass than in steel.

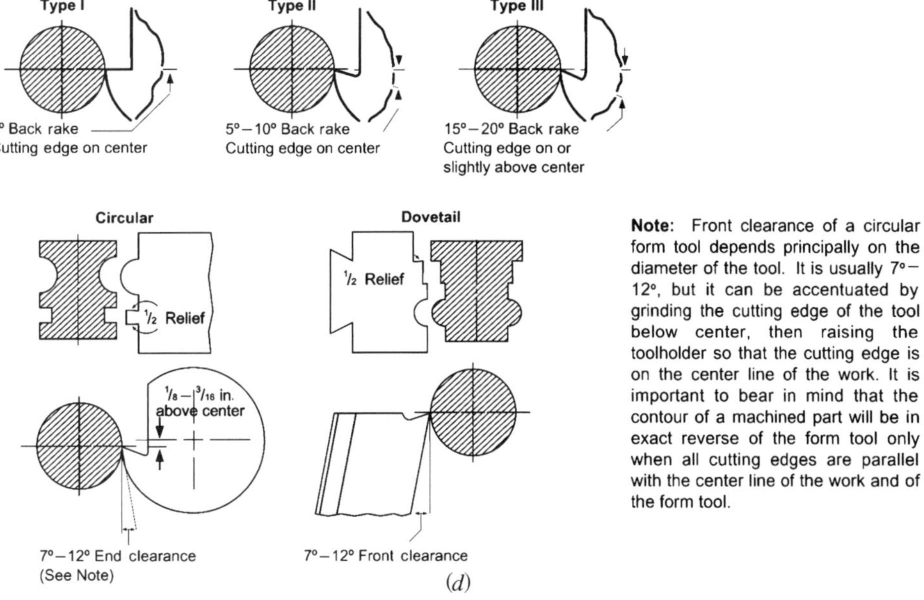

(*d*)

Figure 3 (*Continued*)

7.2 Casting

Copper casting alloys are widely used in the manufacture of plumbing, electrical, and mechanical products and for bearings and industrial valves and fittings. The wide selection of available products enables the designer to choose a combination of physical, mechanical, chemical (corrosion-resistant), and biostatic properties that best suit product requirements.

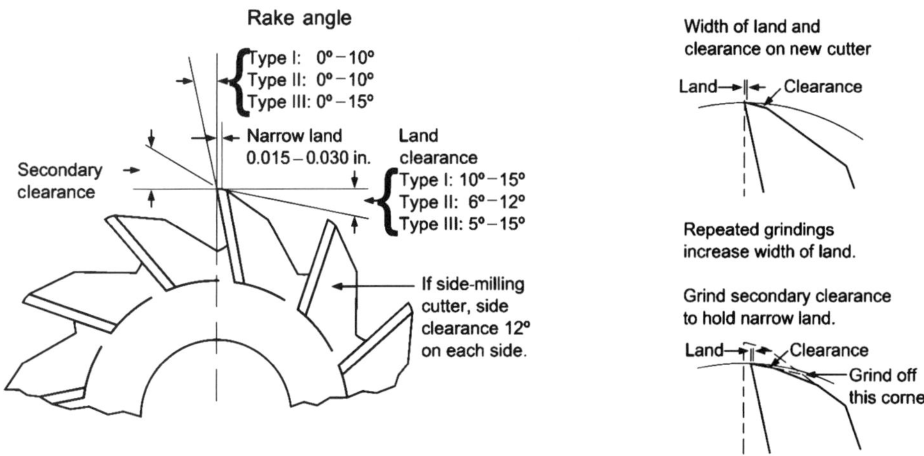

Figure 4 Milling cutter rakes and clearances.

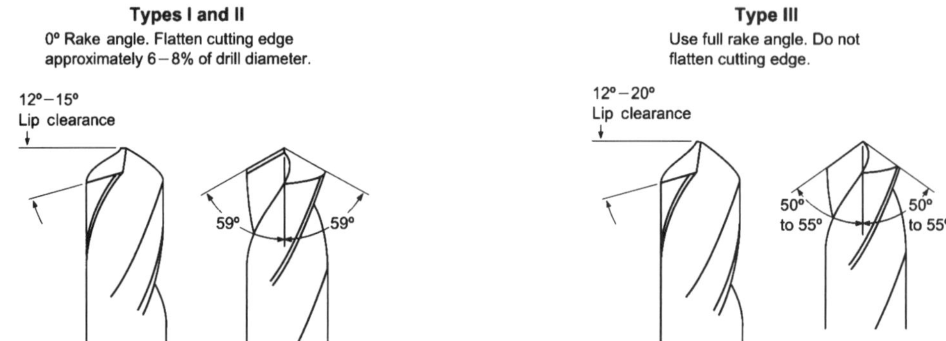

Figure 5 Drill point and clearance angles.

Sand casting accounts for approximately 75% of U.S. copper alloy foundry production. It is relatively inexpensive, acceptably precise, and highly versatile, being applicable to castings ranging in size from a few ounces to many tons. It can be used with all copper alloys.

Copper alloys can be cast by all other methods, including shell, plaster, and investment casting and the continuous and centrifugal processes. The latter process is also used for a large variety of industrial goods such as valves, bearings (which are also continuously cast), marine propeller housing hydraulic cylinders, flanges, fittings, and other tubular items. The choice among processes depends on product requirements, the number of castings to be produced, and cost.

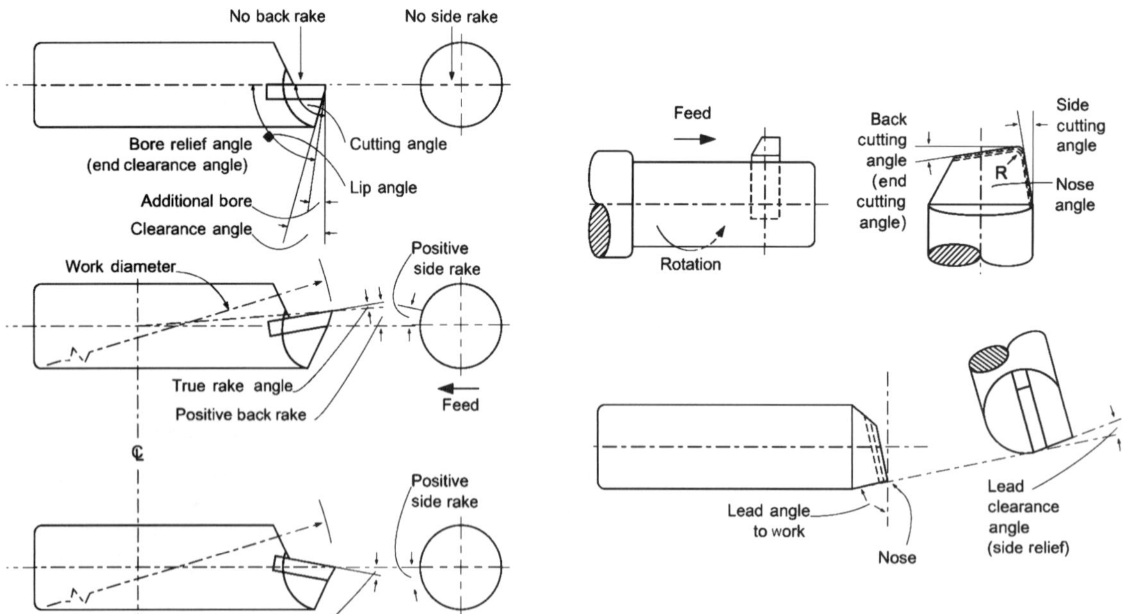

Figure 6 Boring tools.

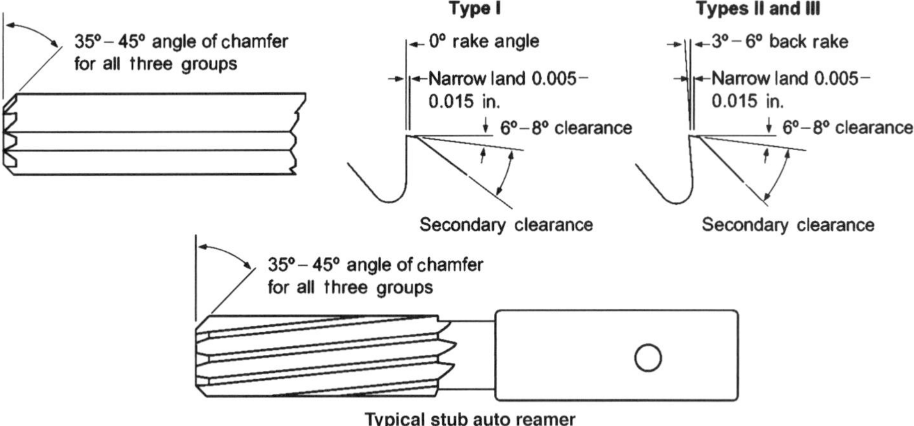

Figure 7 Reamer angles and clearances.

Of special interest are metal-mold processes, which include static, low-pressure, and vacuum permanent mold casting, pressure die casting, and semisolid casting similar to that used extensively with aluminum alloys. All of these processes are best applied to alloys with wide freezing ranges and are less well suited for (although not excluded from) pure copper, high-copper alloys, and other alloys that solidify over a narrow temperature span. Examples of acceptable alloys include red, semired, and yellow brasses, silicon brasses, manganese bronzes, nickel–silvers, and aluminum and nickel–aluminum bronzes.

Permanent-mold castings are characterized by good part-to-part dimensional consistency and very good surface finishes (about 70 μin., 1.8 μm). Rapid solidification tends to create finer grain sizes, which lead to somewhat higher mechanical properties than those for sand castings, as shown in Table 40. Die costs can be relatively high, but the overall costs of the processes are quite favorable for medium-to-large production quantities.

Squirrel cages for high-efficiency AC induction motors, products formerly limited exclusively to aluminum, are now produced as well in high-conductivity copper by the high-pressure die casting process. This long-sought technology depends on the maintenance of elevated die temperatures and the use of dies made from ductile nickel-base superalloys.

Another recent development is the semisolid casting of copper alloys. The process is similar to pressure die casting except that it is performed between the liquidus and solidus temperatures, i.e., while the alloy is in a semisolid, or "slushy," state. Castings exhibit freedom from porosity (pressure tightness), improved mechanical properties compared with sand-cast alloys, and excellent surface finish. As with other metal-mold processes, relatively high die costs can be offset by large production quantities and the process's ability to cast to near-net shapes.

7.3 Forging

Hot forging is used effectively for the manufacture of high-quality, near-net shape products requiring combinations of soundness, uniform mechanical properties and dimensional tolerances, and good surface finish. Medical gas-mixing valves, pressure regulator housings, and architectural and electrical hardware are typical examples. Forging brass, C37700, is the

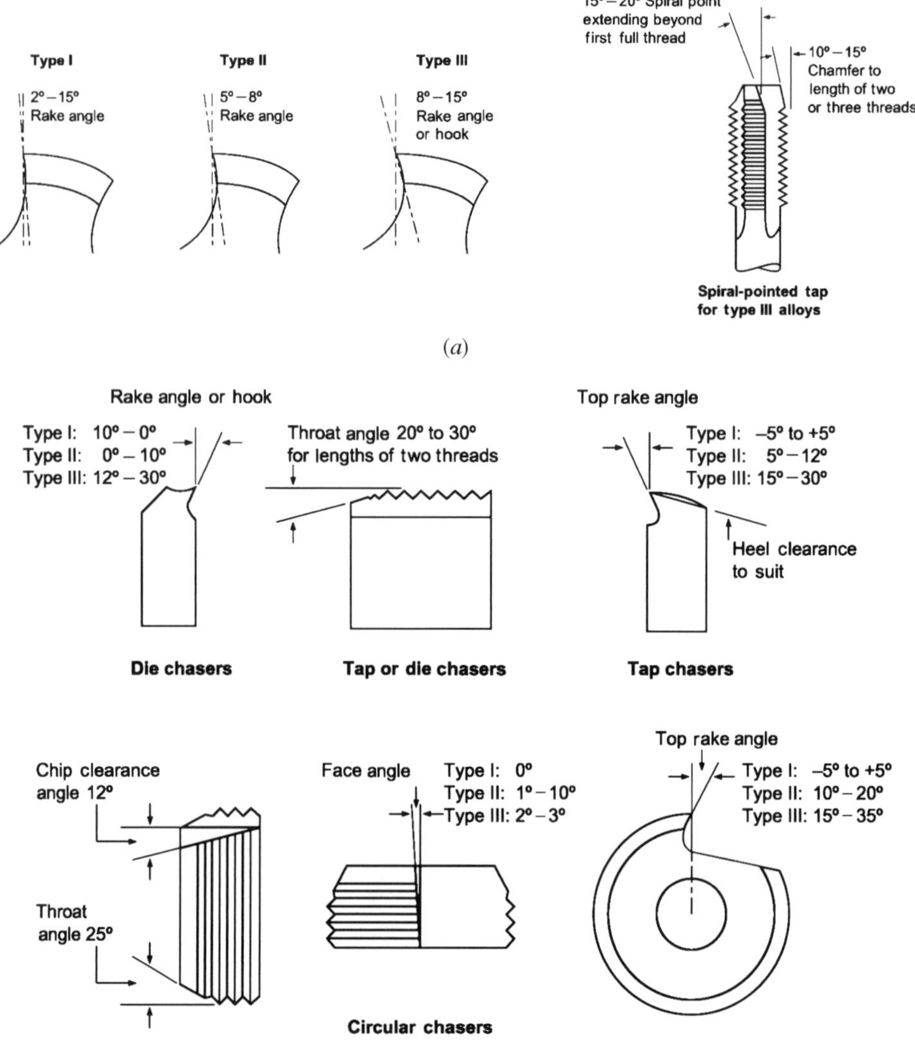

Figure 8 (*a*) Tap rake angle. (*b*) Chasers for die heads and collapsible taps.

workhorse hot-forging alloy (in addition to its good forging characteristics, the alloy contains lead for high machinability), but the process is applicable to other alloys, especially those containing sufficient β-phase to ensure hot ductility.

7.4 Welding, Brazing, and Soldering

Coppers and High-Copper Alloys

Most coppers and many copper alloys are readily weldable, the exceptions being undeoxidized oxygen-containing coppers and alloys containing lead, bismuth, or selenium. All copper alloys can be brazed and soldered.

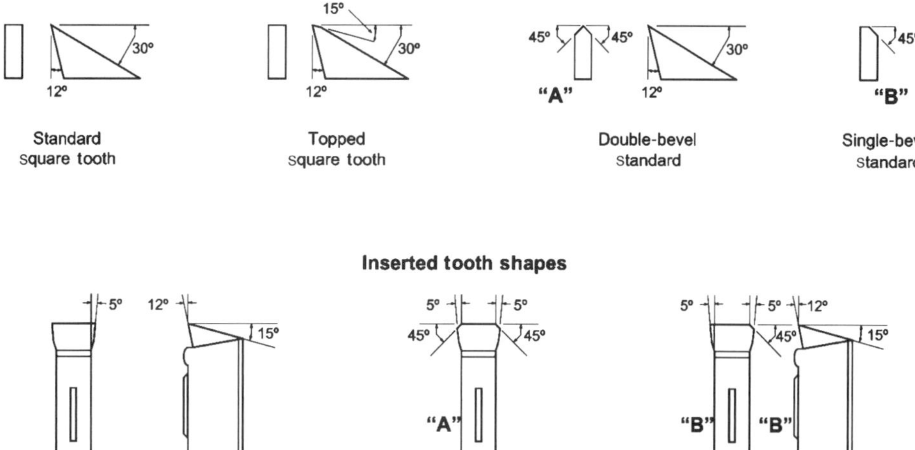

Figure 9 Solid and insert tooth shapes for circular saws.

Figure 10 Hose fitting, a typical automatic screw machine part made in alloy C36000 (left) and free-cutting brass and 12L14 leaded steel (right). The brass version typically provides lower finished cost.

Welding processes used include gas–tungsten arc (GTAW), gas–metal arc (GMAW), and, to a much lesser degree, oxyfuel (OFW) welding. Shielded metal arc welding (SMAW) and resistance welding are generally not recommended for copper. Less conventional processes such as friction welding, electron beam (EB) welding, and laser welding produce very high quality joints in copper and high-copper alloys. The suitability of joining processes for coppers and high-copper alloys is given in Table 44.

Oxygen-free grades such as C10100 are considered highly weldable. Their low dissolved oxygen content (<10 ppm) avoids the generation of porosity defects that can arise in, for example, electrolytic tough pitch copper, C11000, when hydrogen present as atmospheric moisture or surface contamination reacts with dissolved oxygen to generate water vapor in the liquid metal pool. Susceptible coppers, such as C11000, should be joined by brazing or soldering. Deoxidized coppers, such as C12200, are not vulnerable to hydrogen-related welding defects.

Pure copper's high thermal conductivity requires higher heat input during welding than do other materials, including copper alloys. Preheat and maintenance of high interpass temperatures are important for adequate penetration. Note that the high heating requirements will anneal out the effects of cold work, if any, in the base metal.

High-copper alloys are weldable, but the effect of prior heat treatment may be lost in the heat-affected zone (HAZ). For best results, these alloys should be welded in the solution-annealed condition and heat treated (aged) after welding. High-copper alloys have lower thermal conductivity than pure copper and, thus, require somewhat less preheat and lower interpass temperatures.

Dissimilar-Metal Combinations

Successful and safe dissimilar-metal weld joints can be made in coppers and copper alloys, and between copper metals and other metals, if a reasonable amount of precaution is taken before and during welding.

Aluminum bronze (AWS A5.6/ER CuAl-A2) and silicon bronze (AWS A5.6/ER CuSi) are used to join dissimilar metals. Deoxidized coppers, including AWS A5.6/ERCu, can also be used for many dissimilar-metal combinations, but they are not as versatile as either the silicon or aluminum bronze alloys. (Note that these metals are *welding* alloys, which are different from copper-base brazing alloys.) Aluminum bronze is the strongest of the four alloys. It has a slightly higher thermal conductivity and slightly lower coefficient of thermal expansion than does silicon bronze. All copper alloys have good corrosion resistance.

Both silicon bronze and aluminum bronze can be used to weld ferrous and nonferrous metals and alloys to each other and in various combinations (Table 45). Other combinations are also weldable.

Table 44 Welding Processes for Coppers and High-Copper Alloys[a]

Alloy	UNS Number	OFW	SMAW	GMAW	GTAW	RW	SSW	Brazing	Soldering	EBW
OF Cu	C10100-C10800	F	NR	G	G	NR	G	E	E	G
ETP Cu	C11000-C11900	NR	NR	F	F	NR	G	E	G	NR
Deox Cu	C12000-C12300-C17500	G	NR	E	E	NR	E	E	E	G
Be Cu	C17000-	NR	F	G	G	F	F	G	G	F
Cd/Cr Cu	C16200/C18200	NR	NR	G	G	NR	F	G	G	F

[a]E = excellent, G = good, F = fair, NR = not recommended.

Source: Copper Development Association.

Table 45 Dissimilar Metal Combinations Welded with Silicon Bronze and Aluminum Bronze

	Mild Steel	Galvanized Steel	Stainless Steel	Cast Iron	Copper	Copper–Nickel	Silicon Bronze	Aluminum Bronze	Brass
Mild steel		X	X	X	X	X	X	X	X
Galvanized steel	X		X	X	X		X	X	X
Stainless steel	X	X			X	X	X	X	X
Cast iron	X	X	X		X	X	X	X	X
Copper	X	X	X	X		X	X	X	X
Copper–nickel	X		X	X	X		X	X	X
Silicon bronze	X	X	X	X	X	X		X	X
Aluminum bronze	X	X	X	X	X	X	X		X

Source: Copper Development Association.

Shielding Gas Requirements

Argon, helium, or mixtures of the two are commonly used as shielding gases for GTAW and GMAW of copper and high-copper alloys. The choice of gas depends on the degree of heat input needed. Helium produces a higher heat input than argon and is normally used with thick sections. Helium–argon mixtures give intermediate heat input. As rule of thumb, use argon when manually welding sections less than $\frac{1}{8}$ in. (3.2 mm) thick, although a mixture of 75% He–25% Ar can be used for the automatic welding of thin sections.

With GMAW, straight argon requires approximately 100°F (55°C) higher preheat temperatures than when welding with 100% helium. With GTAW, preheat temperatures should be raised by 200°F (110°C) for pure argon.

Welding Processes

High-quality joints are best made using GTAW or GMAW processes. Plasma arc welding (PAW) is also used, and conditions used with GTAW also generally apply to this process. SMAW can be used for less critical joints, and it is not discussed here.

GTAW. Manual GTAW is generally preferred for thin sections. Some fabricators use automatic GTAW for relatively thick sections, claiming that the process affords better control than GMAW and is capable of producing X-ray-quality welds for especially critical applications. That preference is not meant to imply that GMAW is an inferior process. GMAW is widely and routinely used to weld copper at economically favorable speeds.

Preheat should be considered for all thicknesses, but it is absolutely necessary for work thicker than about 0.1 in. (0.25 mm). Preheat temperatures are about 100°F (55°C) higher than those used for GMAW.

GTAW can be used in all positions and is best for out-of-position welding. Pulsed current is recommended for vertical and overhead work. Use direct current, with electrode negative.

GMAW. GMAW is normally used when welding heavy sections in copper. Preheat is important, and straight helium or a 75% He–25% Ar mixture is normally used as a shielding gas. When welding with 100% argon, recommended preheat temperatures are raised by about 100°F (55°C) over what they would be with helium. GMAW can be used for vertical and overhead work, but pulsed current and small-diameter wire is recommended in such cases.

Joint Preparation

Joint designs for welding copper and copper alloys are based on the metals' high thermal conductivity and the need to gain good penetration. Joint angles are generally 10°–20° wider than those used with steels and low-conductivity alloys. A separation of $\frac{3}{32}$ in. (2.4 mm) should be used for square-groove butt joints. For double V-groove joints, use a total included angle between 80° and 90° and a $\frac{3}{32}$–$\frac{1}{8}$ in. (2.4–3.2 mm) joint separation. An 80°–90° included angle is also recommended for single V-groove joints, although no joint separation is necessary in this case.

All welded surfaces should be thoroughly cleaned, dried and degreased. Wire brushing normally suffices for pure coppers, but beryllium coppers and other copper alloys that form tightly adhering oxide films may require grinding or chemical cleaning to provide completely bare surfaces.

Filler Metals

The filler metal normally used to weld copper is deoxidized copper supplied as wire or bare electrodes meeting American Welding Society (AWS) 5.6/ERCu. It contains 0.05% phosphorus maximum. In addition to phosphorus, deoxidized copper contains manganese, tin, and silicon as deoxidizers and to increase fluidity and may also contain zinc and aluminum. Its conductivity ranges from about 15 to 40% IACS, which is considerably higher than other copper welding alloys.

Silicon bronze (AWS 5.6/ERCuSi) and aluminum bronzes (AWS 5.6/ERCuAl-1, AWS 5.6/ERCuAl-2, AWS 5.6/ERCuAl-3) also weld well with copper, and they are often used when joining copper to other metals (unless high conductivity is important). The alloys produce higher strength joints than are possible with deoxidized copper (strength of CuAl-A1 < CuAl-A2 < CuAl-A3), although strength is often less important than conductivity in welded copper assemblies.

On the other hand, strength may be an important consideration when welding high-copper alloys. In such cases, aluminum bronzes or silicon bronzes can be used as filler metals.

Distortion Control

The thermal expansion of copper metals is about 50% higher than that of carbon steels. Thermal conductivity can be as much as eight times higher. The potential for distortion during welding is, therefore, considerably greater when welding copper than with steel. Small or light-gage components should be firmly clamped or fixtured to minimize warping. Multiple tack welds are also helpful, especially with large items or thick sections. Preheating, which is necessary in any case, is likewise beneficial since it tends to reduce temperature differences across large areas.

The thermal stresses that cause distortion can also lead to cracking, another reason for correct preheat, which minimizes these stresses. Root passes should be large. High heat input during the initial pass, along with copper's high conductivity, creates more uniform temperatures around the weld zone and, therefore, avoids the sharp thermal gradients that can lead to cracking. Copper or ceramic backing rings, when used in conjunction with GTAW, help control root-pass penetration.

Avoiding Cracking

Weld cracking most often occurs when the base and filler metals (or both of the base metals) have widely different melting points, thermal conductivities, or thermal expansion coefficients. The chance for cracking grows worse when more than one of these conditions are present at the same time. To prevent cracking caused by widely different melting points, filler metal with a melting point that lies between those of the base metals should be used. In some cases, it may be helpful to butter the filler metal onto the lower melting base metal before laying down the rest of the passes. Buttering also reduces dilution.

Differences in thermal conductivity between two base metals produce different heating and cooling rates on the two sides of the weld joint. The metal with the higher conductivity will tend to draw heat away from the weld zone, and in severe cases uneven heat flow may prevent complete fusion of the low-conductivity metal. Uneven heat flow can also lead to distortion in the finished assembly. The situation can be mitigated by preheating the base metal with the higher conductivity, leading to a more even heating of the weld assembly. Heating the higher conductivity base metal also reduces the cooling rate after welding, reducing postweld thermal stresses.

Large differences in thermal expansion between two base metals (and between base metals and the filler metal) can arise during and after welding and during service. The stresses can be high enough to cause cracking in both cases. Cracking during welding is called hot cracking or hot tearing, and it results from the normal weakness of metals at high temperatures. Postweld cracking occurs when stresses are not adequately relieved. Cracking of this type is accelerated when the metals are cycled between low and high temperatures during service. Cracking can be avoided by selecting a filler metal with a thermal expansion coefficient that is intermediate between those of the two base metals and by applying preheat and, if necessary, postheat in order to reduce the level of thermal stresses during and after welding.

Weld Properties

As-cast copper is fairly soft, and ductility is high. Properties of welded joints are similar. That is, weld metal should have about the same strength and ductility as a fine-grained casting, and the properties of metal in the HAZ should resemble those of wrought and annealed metal. Any additional strength in the base metal due to prior cold work will be reduced, especially in regions in and near the HAZ. Welding reduces properties in heat-treated high-copper alloys.

Safety and Health

Copper and certain elements contained in high-copper alloys (chromium, beryllium, cadmium, arsenic, lead, manganese, and nickel) can cause serious health effects. Government regulations, therefore, impose strict limits on exposure to welding fumes, dust, and grinding particles when elements that are known to be especially harmful are likely to be present. Respirators and fume exhaust systems must be used if called for, and eating or the storage of food and beverages near welding operations should be avoided.

8 TUBE AND PIPE PRODUCTS

8.1 Plumbing Tube

In the United States, copper plumbing tube is manufactured to meet the requirements of specifications established by ASTM. Table 46 lists the six standard types, applicable standards, uses, tempers, and commercially available lengths of plumbing tube.

Types K, L, M, DWV, and medical gas tube are designated by ASTM standard sizes, with the actual outside diameter always $\frac{1}{8}$ in. larger than the standard size designation. Each type represents a series of sizes with different wall thicknesses. Type K tube has thicker walls than type L tube, and type L walls are thicker than type M, for any given diameter. All inside diameters depend on tube size and wall thickness. Copper tube for air conditioning and refrigeration (ACR) field service is designated by actual outside diameter.

"Temper" describes the strength and hardness of the tube. In the piping trades, drawn temper tube is often referred to as "hard" tube and annealed as "soft" tube. Tube in the hard-temper condition is usually joined by soldering or brazing using capillary fittings or by welding. Tube in the soft temper is commonly joined by the same techniques and also by

Table 46 Types and Applications of Copper Tube Products[a]

				Commercially Available Lengths[c]		
Tube Type	Color Code	Standard	Application[b]	Nominal or Standard Sizes (in.)	Drawn (ft)	Annealed (ft)
Type K	Green	ASTM B 88[d]		Straight lengths		
			Domestic water service and distribution	$\frac{1}{4}$–8	20	20
			Fire protection	10	18	18
			Solar	12	2	12
				Coils		
			Fuel/fuel oil	$\frac{1}{4}$–1	—	60
			HVAC	$\frac{1}{4}$ and $1\frac{1}{2}$	—	100
			Snow melting	2	—	60
			Compressed air		—	40
			Natural gas		—	45
			Liquified petroleum (LP) gas			
			Vacuum			
Type L	Blue	ASTM B 88		Straight lengths		
			Domestic water service and distribution	$\frac{1}{4}$–10	20	20
			Fire protection	12	18	18
			Solar			
				Coils		
			Fuel/fuel oil			
			Natural gas			
			Liquified petroleum (LP) gas	$\frac{1}{4}$–1	—	60
			HVAC		—	100
			Snow melting	$1\frac{1}{4}$–$1\frac{1}{2}$	—	60
			Compressed air	2	—	40
			Vacuum		—	45
Type M	Red	ASTM B 88		Straight lengths		
			Domestic water service and distribution	$\frac{1}{4}$–12	20	N/A
			Fire protection			
			Solar			
			Fuel/fuel oil			
			HVAC			
			Snow melting			
			Vacuum			
DWV	Yellow	ASTM B 306		Straight lengths		
			Drain, waste, vent	$1\frac{1}{4}$–8	20	N/A
			HVAC			
			Solar			
ACR	Blue	ASTM B 280		Straight lengths		
			Air conditioning	$\frac{3}{8}$–$4\frac{1}{8}$	20	[e]
			Refrigeration			
			Natural gas	$\frac{1}{8}$–$1\frac{5}{8}$	—	50
				Coils		
			Liquefied petroleum (LP) gas			
			Compressed air			

Table 46 (*Continued*)

Tube Type	Color Code	Standard	Application[b]	Commercially Available Lengths[c]: Nominal or Standard Sizes (in.)	Drawn (ft)	Annealed (ft)
OXY, MED, OXY/MED, OXY/ACR, ACR/MED	(K) Green, (L) Blue	ASTM B 819		Straight lengths		
			Medical gas Compressed medical air Vacuum	$\frac{1}{4}$–8	20	N/A

[a] Copper Development Association. See http://www.copper.org/applications/plumbing/homepage.html.
[b] There are many other copper and copper alloy tubes and pipes available for specialized applications. For information on these products, contact the Copper Development Association.
[c] Individual manufacturers may have commercially available lengths in addition to those shown in this table.
[d] Tube made to other ASTM standards is also intended for plumbing applications, although ASTM B 88 is by far the most widely used. ASTM standard classification B 698 lists six plumbing tube standards, including B 88.
[e] Available as special order only.

Source: The Copper Tube Handbook, Copper Development Association, New York, 2005.

the use of flare-type and compression fittings. It is also possible to expand the end of one tube so that it can be joined to another by soldering or brazing without a capillary fitting—a procedure that can be efficient and economical in many installations. Tube in both the hard and soft tempers can also be joined by a variety of mechanical joints that can be assembled without the use of the heat source required for soldering and brazing.

As for all materials, the allowable internal pressure for any copper tube in service is based on the formula used in the American Society of Mechanical Engineers Code for Pressure Piping (ASME B31):

$$P = \frac{2S(t_{\text{min}} - C)}{D_{\text{max}} - 0.8(t_{\text{min}} - C)}$$

where

P = allowable pressure, psi
S = maximum allowable stress in tension, psi
t_{min} = wall thickness (minimum), in.
D_{max} = outside diameter (maximum), in.
C = constant

For copper tube, because of copper's superior corrosion resistance, the ASME Code permits the factor C to be zero. Thus the formula becomes

$$P = \frac{2St_{\text{min}}}{D_{\text{max}} - 0.8t_{\text{min}}}$$

The value of S in the formula is the maximum allowable stress (ASME B31) for continuous long-term service of the tube material. It is only a small fraction of copper's ultimate tensile strength or of the burst strength of copper tube and has been confirmed to be safe by years of service experience and testing. The allowable stress value depends on the service temperature and on the temper of the tube, drawn or annealed.

In Tables 47–50, the rated internal working pressures are shown for both annealed (soft) and drawn (hard) types K, L, M, and DWV copper tube for service temperatures from 100

Table 47 Rated Internal Working Pressures for Copper Tube: Type K[a]

Part 1: 1/4–2 in.

	Annealed						
Nominal or Standard Size, in.	S = 6000 psi, 100°F	S = 5100 psi, 150°F	S = 4900 psi, 200°F	S = 4800 psi, 250°F	S = 4700 psi, 300°F	S = 4000 psi, 350°F	S = 3000 psi, 400°F
1/4	1074	913	877	860	842	716	537
3/8	1130	960	923	904	885	753	565
1/2	891	758	728	713	698	594	446
5/8	736	626	601	589	577	491	368
3/4	852	724	696	682	668	568	426
1	655	557	535	524	513	437	327
1 1/4	532	452	434	425	416	354	266
1 1/2	494	420	404	396	387	330	247
2	435	370	355	348	341	290	217

	Drawn[b]						
Nominal or Standard Size, in.	S = 10,300 psi, 100°F	S = 10,300 psi, 150°F	S = 10,300 psi, 200°F	S = 10,300 psi, 250°F	S = 10,000 psi, 300°F	S = 9700 psi, 350°F	S = 9400 psi, 400°F
1/4	1850	1850	1850	1850	1796	1742	1688
3	1946	1946	1946	1946	1889	1833	1776
1/2	1534	1534	1534	1534	1490	1445	1400
5/8	1266	1266	1266	1266	1229	1193	1156
3/4	1466	1466	1466	1466	1424	1381	1338
1	1126	1126	1126	1126	1093	1061	1028
1 1/4	914	914	914	914	888	861	834
1 1/2	850	850	850	850	825	801	776
2	747	747	747	747	726	704	682

Part 2: 2 1/2–12 in.

	Annealed						
Nominal or Standard Size, in.	S = 6000 psi, 100°F	S = 5100 psi, 150°F	S = 4900 psi, 200°F	S = 4800 psi, 250°F	S = 4700 psi, 300°F	S = 4000 psi, 350°F	S = 3000 psi, 400°F
2 1/2	398	338	325	319	312	265	199
3	385	328	315	308	302	257	193
3 1/2	366	311	299	293	286	244	183
4	360	306	294	288	282	240	180
5	345	293	281	276	270	230	172
6	346	295	283	277	271	231	173
8	369	314	301	295	289	246	184
10	369	314	301	295	289	246	184
12	370	314	302	296	290	247	185

Table 47 *(Continued)*

Nominal or Standard Size, in.	Drawn[b]						
	S = 10,300 psi, 100°F	S = 10,300 psi, 150°F	S = 10,300 psi, 200°F	S = 10,300 psi, 250°F	S = 10,000 psi, 300°F	S = 9700 psi, 350°F	S = 9400 psi, 400°F
$2\frac{1}{2}$	684	684	684	684	664	644	624
3	662	662	662	662	643	624	604
$3\frac{1}{2}$	628	628	628	628	610	592	573
4	618	618	618	618	600	582	564
5	592	592	592	592	575	557	540
6	595	595	595	595	578	560	543
8	634	634	634	634	615	597	578
10	634	634	634	634	615	597	578
12	635	635	635	635	617	598	580

[a]Based on maximum allowable stress in tension (psi) for the indicated temperatures (°F).
[b]When brazing or welding is used to join drawn tube, the corresponding annealed rating must be used.
Source: Copper Development Association.

Table 48 Rated Internal Working Pressures for Copper Tube: Type L[a]

Part 1: $\frac{1}{4}$–2 in.

Nominal or Standard Size, in.	Annealed						
	S = 6000 psi, 100°F	S = 5100 psi, 150°F	S = 4900 psi, 200°F	S = 4800 psi, 250°F	S = 4700 psi, 300°F	S = 4000 psi, 350°F	S = 3000 psi, 400°F
$\frac{1}{4}$	912	775	745	729	714	608	456
$\frac{3}{8}$	779	662	636	623	610	519	389
$\frac{1}{2}$	722	613	589	577	565	481	361
$\frac{5}{8}$	631	537	516	505	495	421	316
$\frac{3}{4}$	582	495	475	466	456	388	291
1	494	420	404	395	387	330	247
$1\frac{1}{4}$	439	373	358	351	344	293	219
$1\frac{1}{2}$	408	347	334	327	320	272	204
2	364	309	297	291	285	242	182

Nominal or Standard Size, in.	Drawn[b]						
	S = 10,300 psi, 100°F	S = 10,300 psi, 150°F	S = 10,300 psi, 200°F	S = 10,300 psi, 250°F	S = 10,000 psi, 300°F	S = 9700 psi, 350°F	S = 9400 psi, 400°F
$\frac{1}{4}$	1569	1569	1569	1569	1524	1478	1432
$\frac{3}{8}$	1341	1341	1341	1341	1302	1263	1224
$\frac{1}{2}$	1242	1242	1242	1242	1206	1169	1133
$\frac{5}{8}$	1086	1086	1086	1086	1055	1023	991
$\frac{3}{4}$	1002	1002	1002	1002	972	943	914
1	850	850	850	850	825	801	776
$1\frac{1}{4}$	755	755	755	755	733	711	689
$1\frac{1}{2}$	702	702	702	702	682	661	641
2	625	625	625	625	607	589	570

Table 48 (*Continued*)

Part 2: 2½–12 in.

Nominal or Standard Size, in.	Annealed: S = 6000 psi, 100°F	S = 5100 psi, 150°F	S = 4900 psi, 200°F	S = 4800 psi, 250°F	S = 4700 psi, 300°F	S = 4000 psi, 350°F	S = 3000 psi, 400°F
2½	336	285	274	269	263	224	168
3	317	270	259	254	248	211	159
3½	304	258	248	243	238	202	152
4	293	249	240	235	230	196	147
5	269	229	220	215	211	179	135
6	251	213	205	201	196	167	125
8	270	230	221	216	212	180	135
10	271	231	222	217	212	181	136
12	253	215	207	203	199	169	127

Nominal or Standard Size, in.	Drawn[b]: S = 10,300 psi, 100°F	S = 10,300 psi, 150°F	S = 10,300 psi, 200°F	S = 10,300 psi, 250°F	S = 10,000 psi, 300°F	S = 9700 psi, 350°F	S = 9400 psi, 400°F
2½	577	577	577	577	560	544	527
3	545	545	545	545	529	513	497
3½	522	522	522	522	506	491	476
4	504	504	504	504	489	474	460
5	462	462	462	462	449	435	422
6	431	431	431	431	418	406	393
8	464	464	464	464	451	437	424
10	466	466	466	466	452	439	425
12	435	435	435	435	423	410	397

[a] Based on maximum allowable stress in tension (psi) for the indicated temperatures (°F).
[b] When brazing or welding is used to join drawn tube, the corresponding annealed rating must be used.
Source: Copper Development Association.

to 400°F. The ratings for drawn tube can be used for soldered systems and systems using properly designed mechanical joints. Fittings manufacturers can provide information about the strength of their various types and sizes of fittings.

When welding or brazing is used to join tubes, the annealed ratings must be used, since the heating involved in these joining processes will anneal (soften) the hard tube. This is the reason that annealed ratings are shown in Table 49 for type M and Table 50 for DWV tube, although they are not furnished in the annealed temper. Rated internal working pressures for ACR copper are listed in Table 51, while pressure–temperature ratings of soldered and brazed joints are given in Table 52.

8.2 Nonflammable Medical Gas Piping Systems

Safety standards for oxygen and other positive-pressure medical gases require the use of type K or L copper tube (see ASTM B 819). Special cleanliness requirements are called for

Table 49 Rated Internal Working Pressures for Copper Tube: Type M[a]

Part 1: ¼–2 in.

Nominal or Standard Size, in.	Annealed[b]						
	$S =$ 6000 psi, 100°F	$S =$ 5100 psi, 150°F	$S =$ 4900 psi, 200°F	$S =$ 4800 psi, 250°F	$S =$ 4700 psi, 300°F	$S =$ 4000 psi, 350°F	$S =$ 3000 psi, 400°F
¼	—	—	—	—	—	—	—
⅜	**570**	**485**	**466**	**456**	**447**	**380**	**285**
½	**494**	**420**	**403**	**395**	**387**	**329**	**247**
⅝	—	—	—	—	—	—	—
¾	**407**	**346**	**332**	**326**	**319**	**271**	**204**
1	**337**	**286**	**275**	**270**	**264**	**225**	**169**
1¼	**338**	**287**	**276**	**271**	**265**	**225**	**169**
1½	**331**	**282**	**270**	**265**	**259**	**221**	**166**
2	**299**	**254**	**244**	**239**	**234**	**199**	**149**

Nominal or Standard Size, in.	Drawn[c]						
	$S =$ 10,300 psi, 100°F	$S =$ 10,300 psi, 150°F	$S =$ 10,300 psi, 200°F	$S =$ 10,300 psi, 250°F	$S =$ 10,000 psi, 300°F	$S =$ 9700 psi, 350°F	$S =$ 9400 psi, 400°F
¼	—	—	—	—	—	—	—
⅜	982	982	982	982	953	925	896
½	850	850	850	850	825	800	776
⅝	—	—	—	—	—	—	—
¾	701	701	701	701	680	660	639
1	580	580	580	580	563	546	529
1¼	582	582	582	582	565	548	531
1½	569	569	569	569	553	536	520
2	514	514	514	514	499	484	469

Part 2: 2½–12 in.

Nominal or Standard Size, in.	Annealed[b]						
	$S =$ 6000 psi, 100°F	$S =$ 5100 psi, 150°F	$S =$ 4900 psi, 200°F	$S =$ 4800 psi, 250°F	$S =$ 4700 psi, 300°F	$S =$ 4000 psi, 350°F	$S =$ 3000 psi, 400°F
2½	**274**	**233**	**224**	**219**	**215**	**183**	**137**
3	**253**	**215**	**207**	**203**	**199**	**169**	**127**
3½	**252**	**214**	**206**	**202**	**197**	**168**	**126**
4	**251**	**213**	**205**	**201**	**197**	**167**	**126**
5	**233**	**198**	**190**	**186**	**182**	**155**	**116**
6	**218**	**186**	**178**	**175**	**171**	**146**	**109**
8	**229**	**195**	**187**	**183**	**180**	**153**	**115**
10	**230**	**195**	**188**	**184**	**180**	**153**	**115**
12	**230**	**195**	**188**	**184**	**180**	**153**	**115**

Table 49 (*Continued*)

Nominal or Standard Size, in.	Drawn[c]						
	S = 10,300 psi, 100°F	S = 10,300 psi, 150°F	S = 10,300 psi, 200°F	S = 10,300 psi, 250°F	S = 10,000 psi, 300°F	S = 9700 psi, 350°F	S = 9400 psi, 400°F
$2\frac{1}{2}$	471	471	471	471	457	444	430
3	435	435	435	435	423	410	397
$3\frac{1}{2}$	433	433	433	433	421	408	395
4	431	431	431	431	419	406	394
5	400	400	400	400	388	377	365
6	375	375	375	375	364	353	342
8	394	394	394	394	382	371	359
10	394	394	394	394	383	371	360
12	395	395	395	395	383	372	360

[a] Based on maximum allowable stress in tension (psi) for the indicated temperatures (°F).
[b] Types M and DWV are not normally available in the annealed temper. Boldface values are provided for guidance when drawn temper tube is brazed or welded.
[c] When brazing or welding is used to join drawn tube, the corresponding annealed rating must be used.
Source: Copper Development Association.

because oxygen under pressure may cause the spontaneous combustion of some organic oils (the residual of lubricating oil used during manufacture) and for the safety of patients receiving medical gases.

8.3 Fuel Gas Distribution Systems

Until recently, nearly all interior fuel gas distribution systems in residential applications have used threaded steel pipe. Some areas of the United States, however, have had extensive experience using copper tube for distribution of natural and liquefied petroleum (LP) gas, ranging from single-family attached and detached to multistory, multifamily dwellings. In addition, copper gas distribution lines have been installed for many years in commercial buildings such as strip malls, hotels, and motels. Major code bodies in the United States. and Canada have approved copper tube for fuel gas systems. In 1989 in the United States, provisions for the use of copper tube and copper alloy fittings for interior distribution systems were incorporated in the National Fuel Gas Code (ANSI Z223.1/NFPA 54). Similar provision is made in the Canadian document CAN/CGA-B149.1, *Natural Gas Installation Code.* Copper and copper alloy tube (except tin-lined copper tube) should not be used if the gas contains more than an average of 0.3 grains of hydrogen sulfide per 100 standard cubic feet (scf) of gas (0.7 mg/100 L).

9 COPPER ALLOY SLEEVE BEARINGS

Copper alloy sleeve bearings are available in wrought, sintered, and cast forms, the latter being the most commonly used type since they offer the broadest range of applicability. Table 53 summarizes performance of the cast bearing bronze families with respect to five key bearing properties.

Tin bronzes exhibit relatively high thermal conductivity and low friction coefficients against steel, both of which result in low bearing operating temperatures. The alloys are hard

Table 50 Rated Internal Working Pressure for Copper Tube: Type DWV[a]

Part 1: $1\frac{1}{4}$–2 in.

Nominal or Standard Size, in.	Annealed[b]						
	S = 6000 psi, 100°F	S = 5100 psi, 150°F	S = 4900 psi, 200°F	S = 4800 psi, 250°F	S = 4700 psi, 300°F	S = 4000 psi, 350°F	S = 3000 psi, 400°F
$1\frac{1}{4}$	**330**	**280**	**269**	**264**	**258**	**220**	**165**
$1\frac{1}{2}$	**293**	**249**	**240**	**235**	**230**	**196**	**147**
2	**217**	**185**	**178**	**174**	**170**	**145**	**109**

Nominal or Standard Size, in.	Drawn[c]						
	S = 10,300 psi, 100°F	S = 10,300 psi, 150°F	S = 10,300 psi, 200°F	S = 10,300 psi, 250°F	S = 10,000 psi, 300°F	S = 9700 psi, 350°F	S = 9400 psi, 400°F
$1\frac{1}{4}$	566	566	566	566	549	533	516
$1\frac{1}{2}$	503	503	503	503	489	474	459
2	373	373	373	373	362	352	341

Part 2: 3–8 in.

Nominal or Standard Size, in.	Annealed[b]						
	S = 6000 psi, 100°F	S = 5100 psi, 150°F	S = 4900 psi, 200°F	S = 4800 psi, 250°F	S = 4700 psi, 300°F	S = 4000 psi, 350°F	S = 3000 psi, 400°F
3	**159**	**135**	**130**	**127**	**125**	**106**	**80**
4	**150**	**127**	**122**	**120**	**117**	**100**	**75**
5	**151**	**129**	**124**	**121**	**119**	**101**	**76**
6	**148**	**126**	**121**	**119**	**116**	**99**	**74**
8	**146**	**124**	**119**	**117**	**114**	**97**	**73**

Nominal or Standard Size, in.	Drawn[c]						
	S = 10,300 psi, 100°F	S = 10,300 psi, 150°F	S = 10,300 psi, 200°F	S = 10,300 psi, 250°F	S = 10,000 psi, 300°F	S = 9700 psi, 350°F	S = 9400 psi, 400°F
3	273	273	273	273	265	257	249
4	257	257	257	257	250	242	235
5	260	260	260	260	252	245	237
6	255	255	255	255	247	240	232
8	251	251	251	251	244	236	229

[a] Based on maximum allowable stress in tension (psi) for the indicated temperatures (°F).

[b] Types M and DWV are not normally available in the annealed temper. Boldface values are provided for guidance when drawn temper tube is brazed or welded.

[c] When brazing or welding is used to join drawn tube, the corresponding annealed rating must be used.

Source: Copper Development Association.

Table 51 Rated Internal Working Pressure for Copper Tube: ACR[a] (Air Conditioning and Refrigeration Field Service)

Part 1: $\frac{1}{8}$–$1\frac{5}{8}$ in.

Nominal or Standard Size, in.	Coils, Annealed (Coils, Drawn—Not Manufactured)						
	S = 6000 psi, 100°F	S = 5100 psi, 150°F	S = 4900 psi, 200°F	S = 4800 psi, 250°F	S = 4700 psi, 300°F	S = 4000 psi, 350°F	S = 3000 psi, 400°F
$\frac{1}{8}$	3074	2613	2510	2459	2408	2049	1537
$\frac{3}{16}$	1935	1645	1581	1548	1516	1290	968
$\frac{1}{4}$	1406	1195	1148	1125	1102	938	703
$\frac{5}{16}$	1197	1017	977	957	937	798	598
$\frac{3}{8}$	984	836	803	787	770	656	492
$\frac{1}{2}$	727	618	594	581	569	485	363
$\frac{5}{8}$	618	525	504	494	484	412	309
$\frac{3}{4}$	511	435	417	409	400	341	256
$\frac{3}{4}$	631	537	516	505	495	421	316
$\frac{7}{8}$	582	495	475	466	456	388	291
$1\frac{1}{8}$	494	420	404	395	387	330	247
$1\frac{3}{8}$	439	373	358	351	344	293	219
$1\frac{5}{8}$	408	347	334	327	320	272	204

Part 2: $\frac{3}{8}$–$4\frac{1}{8}$ in.

Nominal or Standard Size, in.	Straight Lengths						
	Annealed						
	S = 6000 psi, 100°F	S = 5100 psi, 150°F	S = 4900 psi, 200°F	S = 4800 psi, 250°F	S = 4700 psi, 300°F	S = 4000 psi, 350°F	S = 3000 psi, 400°F
$\frac{3}{8}$	914	777	747	731	716	609	457
$\frac{1}{2}$	781	664	638	625	612	521	391
$\frac{5}{8}$	723	615	591	579	567	482	362
$\frac{3}{4}$	633	538	517	506	496	422	316
$\frac{7}{8}$	583	496	477	467	457	389	292
$1\frac{1}{8}$	495	421	404	396	388	330	248
$1\frac{3}{8}$	440	374	359	352	344	293	220
$1\frac{5}{8}$	409	348	334	327	320	273	205
$2\frac{1}{8}$	364	309	297	291	285	243	182
$2\frac{5}{8}$	336	286	275	269	263	224	168
$3\frac{1}{8}$	317	270	259	254	249	212	159
$3\frac{5}{8}$	304	258	248	243	238	203	152
$4\frac{1}{8}$	293	249	240	235	230	196	147

and strong and have good corrosion resistance, especially against seawater. They are wear resistant and withstand pounding well. The alloys are also somewhat abrasive and should be used against hardened steel shafts (300–400 Brinnell hardness minimum). The alloys work well with grease lubricants and can function as boundary-type bearings.

Leaded tin bronzes are essentially free-cutting versions of the tin bronzes. Properties and applications are similar.

Table 51 *(Continued)*

Nominal or Standard Size, in.	Drawn[b] S = 10,300 psi, 100°F	S = 10,300 psi, 150°F	S = 10,300 psi, 200°F	S = 10,300 psi, 250°F	S = 10,000 psi, 300°F	S = 9700 psi, 350°F	S = 9400 psi, 400°F
$\frac{3}{8}$	1569	1569	1569	1569	1524	1478	1432
$\frac{1}{2}$	1341	1341	1341	1341	1302	1263	1224
$\frac{5}{8}$	1242	1242	1242	1242	1206	1169	1133
$\frac{3}{4}$	1086	1086	1086	1086	1055	1023	991
$\frac{7}{8}$	1002	1002	1002	1002	972	943	914
$1\frac{1}{8}$	850	850	850	850	825	801	776
$1\frac{3}{8}$	755	755	755	755	733	711	689
$1\frac{5}{8}$	702	702	702	702	682	661	641
$2\frac{1}{8}$	625	625	625	625	607	589	570
$2\frac{5}{8}$	577	577	577	577	560	544	527
$3\frac{1}{8}$	545	545	545	545	529	513	497
$3\frac{5}{8}$	522	522	522	522	506	491	476
$4\frac{1}{8}$	504	504	504	504	489	474	460

[a] Based on maximum allowable stress in tension (psi) for the indicated temperatures (°F).
[b] When brazing or welding is used to join drawn tube, the corresponding annealed rating must be used.
Source: Copper Development Association.

High-leaded tin bronzes comprise the most commonly used family of bronze bearing alloys. Although not quite as strong as tin bronzes and leaded tin bronzes, they operate well under moderate loads and medium-to-high speeds—applications that make up the bulk of sleeve bearing use. They are especially recommended for boundary-lubricated and mixed-film operation and provide a level of protection in the event of lubricant failure. They tolerate dirt in lubricants, can accommodate misaligned shafts, and can be run against unhardened shafts. Alloy C93200 is the workhorse of the series.

High-strength brasses (sometimes called manganese bronzes) are modifications of yellow brasses made by addition of aluminum, manganese, and iron to achieve higher strength, which can reach more than 115 ksi (800 MPa). Fatigue resistance is moderate, but the alloys can operate under very high loads and at moderately high speeds. They require hardened, well-aligned shafts and reliable, clean lubrication.

Aluminum bronzes are among the highest strength copper-base bearing alloys. Alloy C95500 can be heat treated (suffix HT) to tensile strengths of more than 115 ksi (800 MPa). The compressive strength of alloy C95400 at 500°F (260°C) is the same as that of tin bronzes at room temperature. The aluminum bronzes also have the highest fatigue strength among bearing alloys and resist heavy and repeated impact well. They must be used against shafts hardened to 500 HB or higher and require clean, reliable full-film lubrication. Both bearing and shaft surfaces should be machined to finishes finer than 15–20 μin. (0.4–0.5 μm). Correct shaft alignment is important.

Silicon brasses have good bearing characteristics at moderately high speeds. They are readily machinable, contain no lead, and are stronger than standard tin bearing alloys. Lubrication must be clean and reliable, and operation should be against hardened shafts.

Copper beryllium alloys can be heat treated to attain very high hardness. Alloy C82800, for example, exhibits a tensile strength in excess of 165 ksi (1100 MPa). The beryllium coppers have good bearing properties and can withstand extremely high stresses. They re-

Table 52 Pressure–Temperature Ratings of Soldered and Brazed Joints

Joining Material[b]	Service Temperature, °F		Maximum Working Gage Pressure (psi) for Standard Water Tube Sizes[a] by Nominal or Standard Size in Inches				
			$\frac{1}{8}$–1 in.	$1\frac{1}{4}$–2 in.	$2\frac{1}{4}$–4 in.	5–8 in.	10–12 in.
Alloy Sn50 50-50 tin–lead solder[c]	100	Pressure[d]	200	175	150	135	100
		DWV[e]	—	95	80	70	—
	150	Pressure[d]	150	125	100	90	70
		DWV[e]	—	70	55	45	—
	200	Pressure[d]	100	90	75	70	50
		DWV[e]	—	50	40	35	—
	250	Pressure[d]	85	75	50	45	40
		DWV[e]	—	—	—	—	—
	Saturated steam	Pressure	15	15	15	15	15
Alloy Sb5 95-5 tin–antimony solder	100	Pressure[d]	1090	850	705	660	500
		DWV[e]	—	390	325	330	—
	150	Pressure[d]	625	485	405	375	285
		DWV[e]	—	225	185	190	—
	200	Pressure[d]	505	395	325	305	230
		DWV[e]	—	180	150	155	—
	250	Pressure[d]	270	210	175	160	120
		DWV[e]	—	95	80	80	—
	Saturated steam	Pressure	15	15	15	15	15
Alloy E	100	Pressure[d]	710	555	460	430	325
		DWV[e]	—	255	210	215	—
	150	Pressure[d]	475	370	305	285	215
		DWV[e]	—	170	140	140	—
	200	Pressure[d]	375	290	240	225	170
		DWV[e]	—	135	110	115	—
	250	Pressure[d]	320	250	205	195	145
		DWV[e]	—	115	95	95	—
	Saturated steam	Pressure	15	15	15	15	15
Alloy HB	100	Pressure[d]	1035	805	670	625	475
		DWV[e]	—	370	310	315	—
	150	Pressure[d]	710	555	460	430	325
		DWV[e]	—	255	210	215	—
	200	Pressure[d]	440	340	285	265	200
		DWV[e]	—	155	130	135	—
	250	Pressure[d]	430	335	275	260	195
		DWV[e]	—	155	125	130	—
	Saturated steam	Pressure	15	15	15	15	15
Joining materials melting at or above 1100°F[f]	Pressure–temperature ratings consistent with the materials and procedures employed (see Tables 47–51, annealed)						
	Saturated steam	Pressure	15	15	15	15	15

Table 52 (*Continued*)

Note: For extremely low working temperatures in the 0 F to minus 200 F range, it is recommended that a joint material melting at or above 1100 F be employed (see footnote *f*).

[a] Standard water tube sizes per ASTM B 88.

[b] Alloy designations are per ASTM B 32.

[c] The Safe Drinking Water Act Amendment of 1986 prohibits the use in potable water systems of any solder having a lead content in excess of 0.2%.

[d] Ratings up to 8 in. in size are those given in ASME B16.22 Wrought Copper and Copper Alloy Solder Joint Pressure Fittings and ASME B16.18 Cast Copper and Copper Alloy Solder Joint Fittings. Rating for 10–12 in. sizes are those given in ASME B16.18 Cast Copper and Copper Alloy Solder Joint Pressure Fittings.

[e] Using ASME B16.29 Wrought Copper and Wrought Copper Alloy Solder Joint Drainage Fittings—DWV and ASME B16.23 Cast Copper Alloy Solder Joint Drainage Fittings—DWV.

[f] These joining materials are defined as brazing alloys by the American Welding Society.

Source: Copper Development Association.

quire hardened and precisely aligned shafts and clean, reliable lubrication. Their moderate impact and fatigue strengths require that bearings be properly supported.

Lead-free bearing bronze contains bismuth in place of lead. Properties are comparable. For example, lead-free alloy C89320 is very similar to the high-leaded tin bronze C93200.

10 STANDARDS AND SPECIFICATIONS

Copper and copper alloys designated according to the UNS are cited in AMS, ASME, ASTM, RWMA (Resistance Welding Manufacturers' Association), FED, and MIL specifications. In addition, ingot standards are listed for cast-metal feedstock by North American ingot manufacturers. Cross references to all relevant standards by UNS number and product form are available from the CDA and can be viewed on the *Standards and Properties* section of its web site, www.copper.org.

11 ADDITIONAL INFORMATION

Complete technical information regarding copper and its alloys is maintained by the CDA (www.copper.org) and the Canadian Copper and Brass Development Association (www.coppercanada.ca). Copper development centers are located in 24 other countries. Information about these centers can be found through the International Copper Association (www.copperinfo.com).

Table 53 Comparison of Bearing Selection Criteria for Cast Copper Bearing Alloys

Bearing Alloy Families	Load Capacity and Fatigue Resistance	Maximum Operating Temperature	Conformability and Embeddability	Resistance to Seizure	Hardness and Wear Resistance
Lead red brasses	Moderate	Moderate	Low	Low	Good
High-leaded tin bronzes	Moderate/high	High	Good	Good	High
Aluminum bronzes	Very high	Very high	Poor	Moderate	Very high
High-strength brasses	Moderate	Moderate	Poor	Moderate	High
Copper beryllium alloys	High	High	Poor	Poor	Very high
Leaded coppers	Moderate	High	Very good	Very good	Moderate

Source: Copper Development Association.

REFERENCES

1. G. Joseph, K. J. A. Kundig (eds.), *Copper: Its Trade, Manufacture, Use, and Environmental Status,* ASM International and International Copper Association, Materials Park, OH, 1999.
2. R. Raymond, *Out of the Fiery Furnace—the Impact of Metals on Mankind,* Macmillan Company of Australia, South Melbourne, Australia, 1984.
3. *Annual Data 2004, Copper Supply and Consumption,* Copper Development Association, New York, 2004. See also: http://www.copper.org/resources/market_data/pdfs/2004_Annual_Data.pdf.
4. *ASTM Standard Designations for Wrought and Cast Copper and Copper Alloys,* Copper Development Association Inc. and ASTM International, New York, 2004. See also http://www.copper.org/resources/properties/db/SDAPropertiesSelectionServlet.jsp?service=COPPERINTRA&Action=search.
5. A. Sutulov, *Copper at the Crossroads,* Internmet Publications, Santiago, Chile, 1985, pp. 55–57.
6. R. E. Bolz and G. I. Tuve (ed.), *Handbook of Tables of Applied Engineering Science,* 2nd ed., CRC Press, Boca Raton, FL, 1974.
7. *Temper Designations for Copper and Copper Alloys—Wrought and Cast,* ASTM B 601, ASTM International, West Conshohocken, PA, www.astm.org.
8. National Sanitation Foundation International, Ann Arbor, MI, www.nsf.org.
9. *Tentative Method for Evaluating Machining Performance Using an Automatic Screw/Bar Machine,* ASTM E 618-81, American Society for Testing and Materials.

CHAPTER 5
SELECTION OF TITANIUM ALLOYS FOR DESIGN

Matthew J. Donachie
Rensselaer at Hartford
Hartford, Connecticut

1 INTRODUCTION

1.1 Purpose

The purpose of this chapter is to create a sufficient understanding of titanium and its alloys so that selection of them for specific designs will be appropriate. Knowledge of titanium alloy types and their processing will give a potential user the ability to understand the ways in which titanium alloys (and titanium) can contribute to a design. The knowledge provided here should enable the user to ask the important questions of titanium alloy providers so as to evaluate the capability of primary producers (largely melt shops) and component producers while addressing the necessary mechanical property and corrosion/environmental behavior that will influence alloy selection. There is no cook book for titanium selection although there is a fair degree of standardization relative to generic alloy compositions. Proprietary/restricted processing leads to titanium alloy conditions and properties not listed in a handbook or catalog of materials. Some proprietary alloy chemistries exist.

Larger volume customers, particularly those in the aerospace industry, frequently dictate the resultant material conditions that generally will be available from a supplier. Proprietary alloy chemistries and/or proprietary/restricted processing required by such customers can lead to alloys or alloy variants that may not be widely available as noted above. In general, proprietary processing is more likely to be encountered than proprietary chemistry nowadays. With few exceptions, critical applications for titanium and its alloys will require the customer to work with one or more titanium producers to develop an understanding of what is available and what a selector/designer can expect from a chosen titanium alloy.

Properties of the titanium alloy families sometimes are listed in handbooks or vendor/supplier brochures. However, not all data will be available.

1.2 What Are Titanium Alloys?

For purposes of this chapter titanium alloys are those alloys of about 50% or higher titanium that offer exceptional strength-to-density benefits plus corrosion properties comparable to the excellent corrosion resistance of pure titanium. The range of operation is from cryogenic temperatures to around 538–595°C (1000–1100°F). Titanium alloys based on intermetallics such as gamma titanium aluminide (TiAl intermetallic compound which has been designated γ) are included in this discussion. These alloys are meant to compete with superalloys at the lower end of superalloy temperature capability, perhaps up to 700°C (~1300°F). They may offer some mechanical advantages for now but often represent an economic debit. Limited experience is available with the titanium aluminides.

1.3 Temperature Capability of Titanium Alloys

Although the melting point of titanium is in excess of 1660°C (3000°F), commercial alloys operate at substantially lower temperatures. It is not possible to create titanium alloys that operate close to their melting temperatures. Attainable strengths, crystallographic phase transformations, and environmental interaction considerations cause restrictions. Thus, while titanium and its alloys have melting points higher than those of steels, their maximum upper useful temperatures for structural applications generally range from as low as 427°C (800°F) to the region of about 538–595°C (1000–1100°F) dependent on composition. As noted, titanium aluminide alloys show promise for applications at higher temperatures, perhaps up to 700°C (~1300°F), although at one time they were expected to offer benefits to higher temperatures.

Actual application temperatures will vary with individual alloy composition. Since application temperatures are much below the melting points, incipient melting is not a factor in titanium alloy application.

1.4 Strength and Corrosion Capability of Titanium and Its Alloys

Titanium owes its industrial use to two significant factors:

- Titanium has exceptional room temperature resistance to a variety of corrosive media
- Titanium has a relatively low density and can be strengthened to achieve outstanding properties when compared with competitive materials on a strength-to-density basis.

Table 1 compares typical strength-to-density values for commercial purity (CP) titanium, several titanium alloys, and a high-strength steel. Figure 1 visually depicts the strength improvements possible in titanium alloys compared to magnesium, aluminum, or steel. In addition to the excellent strength characteristics, titanium's corrosion resistance makes it a desirable material for body replacement parts and other tough corrosion-prone applications. Modulus of elasticity (Young's modulus) is an important factor in addition to strength (specific strength) in some applications. Recently, after many years use of standard titanium α–β (such as Ti–6Al–4V) alloy composition as biomaterials, titanium β-alloys with lower moduli than the customary α–β alloys have been incorporated into biomedical orthopedic applications.

1.5 How Are Alloys Strengthened?

Metals are crystalline and, in the solid state, the atoms of a metal or alloy have various crystallographic arrangements, often occurring as cubic or hexagonal structures. Some crystal structures tend to be associated with better property characteristics than others. In addition, crystalline aggregates of atoms have orientation relationships. Unique crystalline aggregates are called grains and, in an alloy, there are usually many grains with random orientation directions from grain to grain. Metal alloys with multiple random grain directions are known as polycrystals, often having roughly equal dimensions in all directions (referred to as equiaxed). However, columnar-shaped grains (one long axis) are common as a result of casting operations and other elongated grains can result from plastic deformation during forming deformation of an alloy to a component. Grain size and shape are refined by deformation, application of heat and the transformation of existing grains into new ones of the same or different crystallographic structure. The occurrence of a particular crystal structure and chemistry (or range of chemistries) defines a phase, i.e., a chemically homogeneous, physically distinct, mechanically separable part of a system. The basic phases in titanium are α and β (see later sections for more information).

The peripheral surface of a grain is called a grain boundary. Aggregates of atoms without grain boundaries are rarely created in nature. The introduction of different atom types and

Table 1 Comparison of Typical Strength-to-Density Ratios at 20°C

Metal	Specific Gravity	Tensile Strength (lb/in.^{2})	Tensile Strength ÷ Specific Gravity
CP	4.5	58,000	13,000
Ti–6Al–4V	4.4	130,000	29,000
Ti–4Al–3Mo–1V	4.5	200,000	45,000
Ultrahigh-strength steel (4340)	7.9	287,000	36,000

Source: From *Titanium: A Technical Guide,* 1st ed., ASM International, Materials Park, OH, 1988, p. 158.

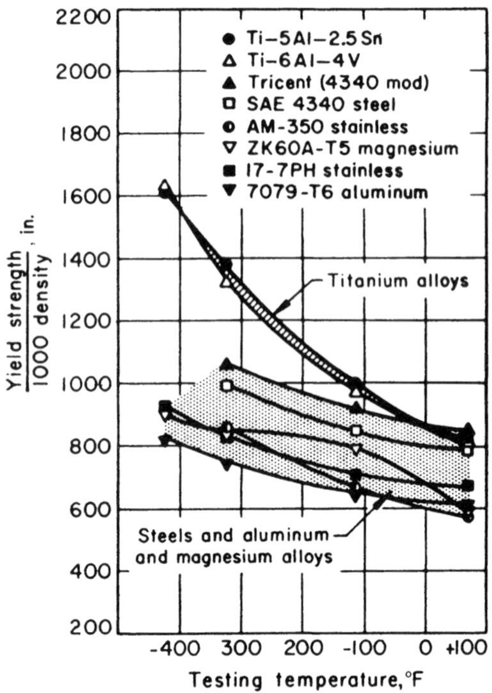

Figure 1 Yield strength-to-density ratio as a function of temperature for several titanium alloys compared to some steel, aluminum and magnesium alloys. (From *Titanium: A Technical Guide,* 1st ed., ASM International, Materials Park, OH, 1988, p. 158.)

additional crystal phases and/or the manipulation of grain boundaries enable inhibition of the movement through the crystal lattice or grains of imperfections that cause deformation to occur. Titanium alloys are basis titanium modified by changes in chemistry, in a polycrystalline form, with various phases present in the grains or at grain boundaries.

1.6 Manufacture of Titanium Articles

Appropriate compositions of all titanium alloy types can be wrought processed (forged, rolled to sheet, or otherwise formed) into a variety of shapes. Powder metallurgy also can be used to accomplish shape formation. Casting, particularly investment casting, is used to cast appropriate compositions in complex shapes usually with properties approaching those of the wrought forms. Fabricated titanium alloy structures can be built up by welding or brazing. Machining of titanium alloys requires forces about equal to those for machining austenitic stainless steel. Titanium alloys do have metallurgical characteristics that make them more difficult to machine than steels of comparable hardness.

In welding or machining of titanium, the effects of the energy input (heat energy, deformation energy) on the microstructure and properties of the final titanium alloy product must be considered. Many titanium alloys can combust if appropriate conditions of temperature are exceeded. Care is needed in machining and in storing scrap.

Many titanium alloys are available as wrought components in extruded, forged, or rolled form. Hot deformation is the preferred forming process. Cold rolling may be used to increase short-time strength properties for some lower temperature applications. Properties of titanium alloys generally are controlled by adjustments in chemistry (composition) and by modification of the processing (including heat treatment).

1.7 Titanium Alloy Information

Some chemistries and properties are listed in this chapter, but there is no substitute for consultation with titanium manufacturers about the forms (some cast, some powder, mostly wrought) which can be provided and the exact chemistries available. It should be understood that not all titanium alloys, particularly those with specific processing, are readily available as off-the-shelf items. Design data for titanium alloys are not intended to be conveyed in this chapter, but typical properties are indicated for some materials. Design properties should be obtained from internal testing if possible, from titanium metal or component producers, or from other validated sources if sufficient test data are not available in-house. Typical properties are merely a guide for comparison. Exact chemistry, section size, heat treatment, and other processing steps must be known to generate adequate property values for design.

The properties of titanium alloy compositions, although developed over many years, are not normally well documented in the literature. However, since many consumers actually only use a few alloys within the customary user groups, data may be more plentiful for certain compositions. In the case of titanium, the most used and studied alloy, whether wrought or cast, is Ti–6Al–4V.

The extent to which data generated for specific applications are available to the general public is unknown. However, even if such data were disseminated widely, the alloy selector needs to be aware that apparently minor chemistry changes or variations in processing operations such as forging conditions, heat treatment, etc., dramatically affect properties of titanium alloys. All property data should be reconciled with the actual manufacturing specifications and processing conditions expected. Alloy selectors should work with competent metallurgical engineers to establish the validity of data intended for design as well as to specify the processing conditions that will be used for component production.

Application of design data must take into consideration the probability of components containing locally inhomogeneous regions. For titanium alloys, such segregation can be disastrous in gas turbine applications. The probability of occurrence of these regions is dependent upon the melting procedures, being essentially eliminated by so-called triple melt. All facets of chemistry and processing need to be considered when selecting a titanium alloy for an application.

For sources of property data other than that of the producers (melters, forgers, etc.) or an alloy selector's own institution, one may refer to handbooks or to organizations such as ASM International which publish compilations of data that may form a basis for the development of design allowables for titanium alloys. Standards organizations such as ASTM publish information about titanium alloys, but that information does not ordinarily contain any design data. Additional information may be available from industry organizations (see Table 11 given later).

2 METALLURGY OF TITANIUM ALLOYS

2.1 Structures

As noted above, metals are crystalline and the atoms take various crystallographic forms. Some of these forms tend to be associated with better property characteristics than other crystal structures. Titanium, as does iron, exists in more than one crystallographic form. Titanium has two elemental crystal structures: In one, the atoms are arranged in a body-centered-cubic (bcc) array, in the other they are arranged in a hexagonal close-packed (hcp) array. The cubic structure is found only at high temperatures, unless the titanium is alloyed with other elements to maintain the cubic structure at lower temperatures. The bcc crystal structure is designated as β and the hcp structure as α. Thus titanium's two crystal structures are commonly known as α- and β-phases.

Not only crystal structure but also overall "structure," i.e., appearance, at levels above that of atomic crystal structure is important. Structure for our purposes will be defined as the macrostructure and microstructure (i.e., macro and microappearance) of a polished and etched cross section of metal visible at magnifications up to and including 10,000×. Two other microstructural features which are not determined visually but are determined by other means such as X-ray diffraction or chemistry are phase type (α, β, etc.) and texture (orientation) of grains.

Alpha actually means any hexagonal titanium, pure or alloyed, while β means any cubic titanium, pure or alloyed. The α and β "structures"—sometimes called systems or types—are the basis for the generally accepted classes of titanium alloys. These are α, near-α, α–β, and β. Sometimes a category of near-β is also considered. The preceding categories denote the general types of microstructure after processing.

Crystal structure and grain structure (a component of microstructure) are not synonymous terms. Both (as well as the chemical composition and arrangement of phases in the microstructure) must be specified to completely identify the alloy and its expected mechanical, physical, and corrosion behavior. The important fact to keep in mind is that, while grain shape and size do affect behavior, the crystal structure changes (from α to β and back again) which occur during processing play a major role in defining titanium properties.

2.2 Crystal Structure Behavior in Alloys

An α-alloy (so described because its chemistry favors the α-phase) does not normally form the β-phase on heating. A near-α- (sometimes called "superalpha") alloy forms only limited β-phase on heating, and so it may appear microstructurally similar to an α-alloy when viewed at lower temperatures. An α–β alloy is one for which the composition permits complete transformation to β on heating but transformation back to α plus retained and/or transformed β at lower temperatures. A near-β- or β-alloy composition is one which tends to retain, indefinitely at lower temperatures, the β-phase formed at high temperatures. However, the β that is retained on initial cooling to room temperature is metastable for many β alloys. Dependent on chemistry, it may precipitate secondary phases during heat treatment.

3 METALS AT HIGH TEMPERATURES

3.1 General

While material strengths at low temperatures usually are not a function of time, at high temperatures the time of load application becomes very significant for mechanical properties. Concurrently, the availability of oxygen at high temperatures accelerates the conversion of some of the metal atoms to oxides. Oxidation proceeds much more rapidly at high temperatures than at room or lower temperatures. For alloys of titanium there is the additional complication of titanium's high affinity for oxygen and its ability to "getter" oxygen (or nitrogen) from the air. Dissolved oxygen greatly changes the strength and ductility of titanium alloys. Hydrogen is another gaseous element which can significantly affect properties of titanium alloys. Hydrogen tends to cause hydrogen embrittlement while oxygen will strengthen but reduce the ductility yet not necessarily embrittle titanium as hydrogen does.

3.2 Mechanical Behavior

In the case of short-time tensile properties of tensile yield strength (TYS) and ultimate tensile strength (UTS), the mechanical behavior of metals at higher temperatures is similar to that at room temperature but with metals becoming weaker as the temperature increases. However, when steady loads below the normal yield or ultimate strength determined in short-time tests are applied for prolonged times at higher temperatures, the situation is different. Figure 2 illustrates the way in which most materials respond to steady extended-time loads at high temperatures. A time-dependent extension (creep) is noticed under load. If the alloy is exposed for a long time, the alloy eventually fractures (ruptures). The degradation process is called creep or, in the event of failure, creep rupture (sometimes stress rupture) and alloys may be selected on their ability to resist creep and creep-rupture failure. Cyclically applied loads that cause failure (fatigue) at lower temperatures also cause failures in shorter times (lesser cycles) at high temperatures. When titanium alloys operate for prolonged times at high temperature, they can fail by creep rupture. However, tensile strengths, fatigue strengths, and crack propagation criteria are more likely to dominate titanium alloy performance requirements.

In highly mechanically loaded parts such as gas turbine compressor disks, a common titanium alloy application, fatigue at high loads in short times, low-cycle fatigue (LCF), is the major concern. High-cycle fatigue (HCF) normally is not a problem with titanium alloys unless a design error occurs and subjects a component to a high-frequency vibration that forces rapid accumulation of fatigue cycles. While life under cyclic load (stress–cycle, S-N, behavior) is a common criterion for design, resistance to crack propagation is an increasingly

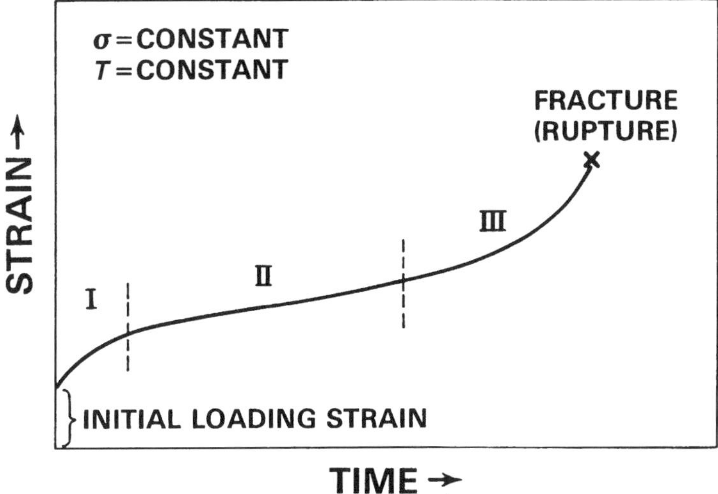

Figure 2 Creep rupture schematic showing time-dependent deformation under constant load at constant high temperatures followed by final rupture. (All loads below the short time yield strength. Roman numerals denote ranges of the creep rupture curve.)

desired property. Thus, the crack growth rate vs. a fracture toughness parameter is required. The parameter in this instance may be the stress intensity factor (K) range over an incremental distance which a crack has grown—the difference between the maximum and minimum K in the region of crack length (a) measured. A plot of the resultant type (da/dn vs. ΔK) is shown in Fig. 3 for several wrought titanium alloys.

Creep–fatigue interactions can play a role in titanium alloy response to loading. This type of fatigue in titanium alloys is sometimes called "dwell time fatigue" or may be called interrupted low-cycle fatigue (ILCF). In dwell or interrupted fatigue, the cyclic loading is interrupted so a steady load is imposed for a short time before cyclic loading is resumed. This process causes an interaction between creep (steady load) and fatigue (cyclic load) to take place, occurring at surprisingly low temperature levels—less than 200°C (390°F). The

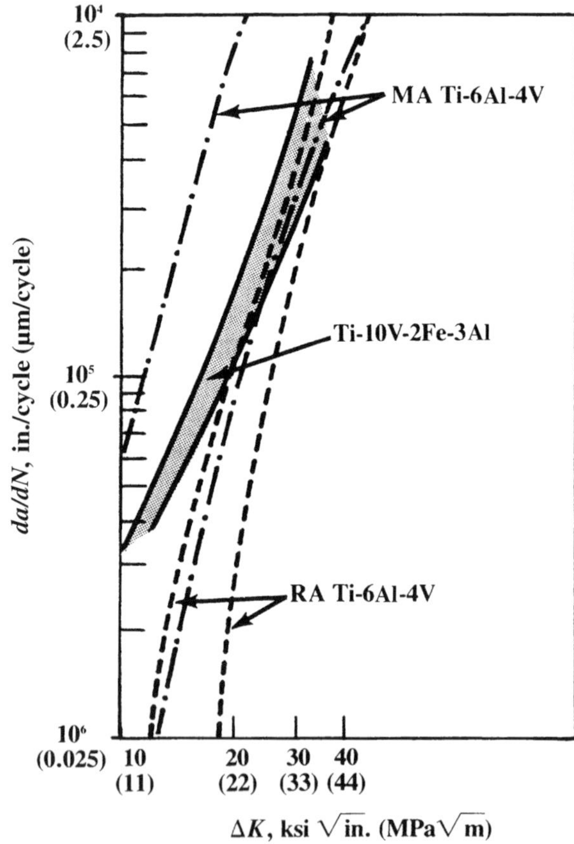

For Ti–10V–2Fe–3Al: $R = 0.05$; $F = 1$–30 Hz
For MA Ti–6Al–4V: $R = 0.08$; $F = 1$–25 Hz
For RA Ti–6Al–4V: $R = 0.08$; $F = 6$ Hz

Figure 3 Comparison of fatigue crack growth rate (da/dn) vs. toughness change (ΔK). Curves for several titanium alloys. Note that MA = mill annealed while RA = recrystallization annealed. (From *Titanium: A Technical Guide,* 1st ed., ASM International, Materials Park, OH, 1988, p. 184.)

fatigue life of α/β titanium alloys in dwell loads is substantially reduced compared to continuously fatigue loaded alloys.

4 MICROSTRUCTURE AND PROPERTIES OF TITANIUM AND ITS ALLOYS

The grain size, grain shape, and grain boundary arrangements in titanium have a very significant influence on mechanical properties, and it is the ability to manipulate the phases/grains present as a result of alloy composition that is responsible for the variety of properties that can be produced in titanium and its alloys. Transformed β-phase products in alloys can affect tensile strengths, ductility, toughness, and cyclic properties. To these effects must be added the basic strengthening effects of alloy elements.

4.1 Alloy Composition and General Behavior

The α titanium alloys usually have high amounts of aluminum which contribute to oxidation resistance at high temperatures. (The α–β-alloys also contain, as the principal element, high amounts of aluminum, but the primary reason is to control the α-phase.) The α-alloys cannot be heat treated to develop higher mechanical properties because they are single-phase alloys.

The addition of certain alloy elements to pure titanium provides for a wide two-phase region where α and β coexist. This behavior enables the resultant alloys to be heat treated or processed, if desired, in the temperature range where the alloy is two-phase. The two-phase condition permits the structure to be refined by the α–β-α transformation process on heating and cooling. The process of heating to a high temperature to promote subsequent transformation is known as solution heat treatment. By permitting some β to be retained temporarily at lower temperature, β-favoring alloy elements enable optimum control of the microstructure. The microstructure is controlled by subsequent transformation, after cooling alloys from the forging or solution heat treatment temperature, when the alloys are "aged" (reheated, after rapid cooling, to temperatures well below the β-transus). The α–β-alloys, when properly treated, have an excellent combination of strength and ductility. They generally are stronger than the α- or the β-alloys.

The β-alloys are metastable; that is, they tend to transform to an equilibrium, or balance of structures. The β-alloys generate their strength from the intrinsic strength of the β-structure and the precipitation of α and other phases from the alloy through heat treatment after processing. The most significant commercial benefit provided by a β-structure is the increased formability of such alloys relative to the hexagonal crystal structure types (α and α–β).

Although three principal categories or classes (α, α–β, β) of titanium have been mentioned, as noted earlier, the α category is sometimes subdivided into α and near-α and the β category is considered as near-β and β. Thus, when CP titanium is added to the list, we may find titanium materials to be listed under one of the following:

- Unalloyed (CP)
- The α and near-α
- The α–β
- The β and near-β

Figure 4 shows the main characteristics of the different titanium alloy family groupings. Commercial purity titanium is excluded from this figure. For purposes of alloy selection, separation as α-, α–β-, and β-alloys plus CP titanium is usually sufficient.

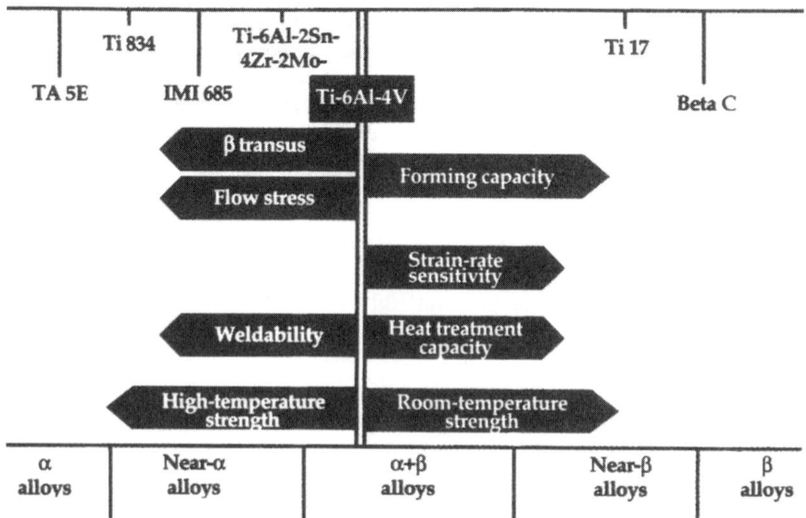

Figure 4 Main characteristics of different titanium alloy family groupings.

Titanium aluminides differ from conventional titanium alloys in that they are principally chemical compounds alloyed to enhance strength, formability, etc. The aluminides have higher operational temperatures than conventional titanium but at higher cost and, generally, have lower ductility and formability.

In addition to alloys, titanium is sold and used in CP forms usually identified as grades. Pure titanium usually has some amount of oxygen alloyed with it. The strength of CP titanium is affected by the interstitial (oxygen and nitrogen) element content. A principal difference among grades is the oxygen (and nitrogen) content, which influences mechanical properties. Small additions of some alloy elements such as palladium are added for increased corrosion resistance in certain grades. A summary of the compositions of many commercial and semicommercial titanium grades and alloys is given in Table 2.

4.2 Strengthening of Titanium Alloys

Desired mechanical properties such as yield or ultimate strength-to-density (strength efficiency), perhaps creep and creep-rupture strength as well as fatigue-crack growth rate, and fracture toughness are extremely important. Manufacturing considerations such as welding and forming requirements play a major role in generating titanium alloys' properties. They normally provide the criteria that determine the alloy composition, structure (α, α–β, or β), heat treatment (some variant of either annealing or solution treating and aging), and level of process control selected or prescribed for structural titanium alloy applications.

By introducing atoms, phases, grain boundaries, or other interfaces into titanium, the movement of imperfections that cause deformation to occur is inhibited. The process of modifying composition and microstructure enables titanium alloys to be strengthened significantly. Final strength is a function of composition and the various deformation processes used to form and strengthen alloys. It is quite important for the alloy selector to have a realistic understanding of the strengthening process in titanium alloys as the properties of titanium and its alloys can be modified considerably not only by chemistry modification but also by processing.

Table 2 Some Commercial and Semicommercial Grades and Alloys of Titanium

Designation	Tensile Strength (min.)		0.2% Yield Strength (min.)		Impurity Limits wt. % (max.)					Nominal Composition, wt. %				
	MPa	ksi	MPa	ksi	N	C	H	Fe	O	Al	Sn	Zr	Mo	Others
Unalloyed grades														
ASTM grade 1	240	35	170	25	0.03	0.08	0.015	0.20	0.18	—	—	—	—	—
ASTM grade 2	340	50	280	40	0.03	0.08	0.015	0.30	0.25	—	—	—	—	—
ASTM grade 3	450	65	380	55	0.05	0.08	0.015	0.30	0.35	—	—	—	—	—
ASTM grade 4	550	80	480	70	0.05	0.08	0.015	0.50	0.40	—	—	—	—	—
ASTM grade 7	340	50	280	40	0.03	0.08	0.015	0.30	0.25	—	—	—	—	0.2Pd
ASTM grade 11	240	35	170	25	0.03	0.08	0.015	0.20	0.18	—	—	—	—	0.2Pd
α and near-α alloys														
Ti–0.3Mo–0.8Ni	480	70	380	55	0.03	0.10	0.015	0.30	0.25	—	—	—	0.3	0.8Ni
Ti–5Al–2.5Sn	790	115	760	110	0.05	0.08	0.02	0.50	0.20	5	2.5	—	—	—
Ti–5Al–2.5Sn–ELI	690	100	620	90	0.07	0.08	0.0125	0.25	0.12	5	2.5	—	—	—
Ti–8Al–1Mo–1V	900	130	830	120	0.05	0.08	0.015	0.30	0.12	8	—	—	1	1V
Ti–6Al–2Sn–4Zr–2Mo	900	130	830	120	0.05	0.05	0.0125	0.25	0.15	6	2	4	2	0.08Si
Ti–6Al–2Nb–1Ta–0.8Mo	790	115	690	100	0.02	0.03	0.0125	0.12	0.10	6	—	—	1	2Nb, 1Ta
Ti–2.25Al–11Sn–5Zr–1Mo	1000	145	900	130	0.04	0.04	0.008	0.12	0.17	2.25	11	5	1	0.2Si
Ti–5.8Al–4Sn–3.5Zr–0.7Nb–0.5Mo–0.35Si	1030	149	910	132	0.03	0.08	0.006	0.05	0.15	5.8	4	3.5	0.5	0.7Nb, 0.35Si
α–β alloys														
Ti–6Al–4V[a]	900	130	830	120	0.05	0.10	0.0125	0.30	0.20	6	—	—	—	4V
Ti–6Al–4V–ELI[a]	830	120	760	110	0.05	0.08	0.0125	0.25	0.13	6	—	—	—	4V
Ti–6Al–6V–2Sn[a]	1030	150	970	140	0.04	0.05	0.015	1.0	0.20	6	2	—	—	0.7Cu, 6V
Ti–8Mn[a]	860	125	760	110	0.05	0.08	0.015	0.50	0.20	—	—	—	—	8.0Mn
Ti–7Al–4Mo[a]	1030	150	970	140	0.05	0.10	0.013	0.30	0.20	7.0	—	—	4.0	—
Ti–6Al–2Sn–4Zr–6Mo[b]	1170	170	1100	160	0.04	0.04	0.0125	0.15	0.15	6	2	4	6	—
Ti–5Al–2Sn–2Zr–4Mo–4Cr[b,c]	1125	163	1055	153	0.04	0.05	0.0125	0.30	0.13	5	2	2	4	4Cr
Ti–6Al–2Sn–2Zr–2Mo–2Cr[c]	1030	150	970	140	0.03	0.05	0.0125	0.25	0.14	5.7	2	2	2	2Cr, 0.25Si
Ti–3Al–2.5V[d]	620	90	520	75	0.015	0.05	0.015	0.30	0.12	3	—	—	—	2.5V
Ti–4Al–4Mo–2Sn–0.5Si	1100	160	960	139	[e]	0.02	0.0125	0.20	[e]	4	2	—	4	0.5Si

Table 2 (*Continued*)

Designation	Tensile Strength (min.)		0.2% Yield Strength (min.)		Impurity Limits wt. % (max.)					Nominal Composition, wt. %				
	MPa	ksi	MPa	ksi	N	C	H	Fe	O	Al	Sn	Zr	Mo	Others
β alloys														
Ti–10V–2Fe–3Al[a,c]	1170	170	1100	160	0.05	0.05	0.015	2.5	0.16	3	—	—	—	10V
Ti–13V–11Cr–3Al[b]	1170	170	1100	160	0.05	0.05	0.025	0.35	0.17	3	—	—	—	11.0Cr, 13.0V
Ti–8Mo–8V–2Fe–3Al[b,c]	1170	170	1100	160	0.03	0.05	0.015	2.5	0.17	3	—	—	8.0	8.0V
Ti–3Al–8V–6Cr–4Mo–4Zr[a,c]	900	130	830	120	0.03	0.05	0.20	0.25	0.12	3	—	4	4	6Cr, 8V
Ti–11.5Mo–6Zr–4.5Sn[a]	690	100	620	90	0.05	0.10	0.020	0.35	0.18	—	4.5	6.0	11.5	—
Ti–15V–3Cr–3Al–3Sn	1000[b]	145[b]	965[b]	140[b]	0.05	0.05	0.015	0.25	0.13	3	3	—	—	15V, 3Cr
	1241[f]	180[f]	1172[f]	170[f]										
Ti–15Mo–3Al–2.7Nb–0.2Si	862	125	793	115	0.05	0.05	0.015	0.25	0.13	3	—	—	15	2.7Nb, 0.2Si

[a] Mechanical properties given for the annealed condition; may be solution treated and aged to increase strength.

[b] Mechanical properties given for the solution-treated-and-aged condition; alloy not normally applied in annealed conditions.

[c] Semicommercial alloy; mechanical properties and composition limits subject to negotiation with suppliers.

[d] Primarily a tubing alloy; may be cold drawn to increase strength.

[e] Combined $O_2 + 2N_2 = 0.27\%$.

[f] Also solution treated and aged using an alternative aging temperature (480°C, or 900°F)

Source: From *Titanium: A Technical Guide,* 2nd ed., ASM International, Materials Park, OH, 2001, p. 8.

Titanium alloys derive their strength from a fineness of microstructure produced by transformation of crystal structures from β to α in grains plus dispersion of one phase in another, as in the case of precipitation of α phases from retained β in metastable β-alloys. The fine structure of titanium alloys often can be martensitic in nature, created by transformations as temperatures on the component being produced are reduced during cooling from deformation processing or solution treatment. The reader may recall that martensitic structures are produced in steels (and in other systems) and can create very strong and hard alloys. Martensitic reactions are found in titanium alloys; they are not as effective as those in steels in causing hardening but do bring about microstructure refinements and thus strength improvements in titanium alloys. Fine dispersions in the alloys usually are produced by "aging" through reheating and holding at an intermediate temperature after prior forging and heat treatment processing.

4.3 Effects of Alloy Elements

Alloy elements generally can be classified as α-stabilizers or β-stabilizers. The α-stabilizers, such as aluminum, oxygen, and nitrogen, increase the temperature at which the α-phase is stable. On the other hand, β-stabilizers, such as vanadium and molybdenum, result in stability of the β-phase at lower temperatures. The transformation temperature from $\alpha + \beta$ or from α to all β is known as the β-transus temperature. The β-transus is defined as the lowest equilibrium temperature above which the material is 100% β. The β-transus is critical in deformation processing and in heat treatment, as described below. By reference to the β-transus, heat treatment temperatures can be selected to produce specific microstructures during heat treatment. See, for example, the amount of α-phase that can be produced by temperature location relative to the β-transus for Ti–6Al–4V as shown in Fig. 5.

4.4 Intermetallic Compounds and Other Secondary Phases

Intermetallic compounds and transient secondary phases are formed in titanium alloy systems along with microstructural variants of the traditional β- and α-phases. The more important secondary phases, historically, have been ω and α_2 (chemically written as Ti_3Al). The ω-phase has not proven to be a factor in commercial systems using present-day processing practice. The α_2-phase has been a concern in some cases of stress–corrosion cracking (SCC). Alloys with extra-high aluminum were found to be prone to SCC. Some interest in α_2 centered on its use as a matrix for a high-temperature titanium alloy. However, the high-temperature alloy matrix of choice is γ-TiAl, mentioned previously. The γ-phase is not a factor in the property behavior of conventional titanium alloys.

4.5 Elastic Constants and Physical Properties

Titanium is a low-density element (approximately 60% of the density of steel and superalloys) which can be strengthened greatly by alloying and deformation processing. The physical and mechanical properties of elemental titanium are given in Table 3. Titanium is nonmagnetic and has good heat transfer properties. Its coefficient of thermal expansion is somewhat lower than that of steel's and less than half that of aluminum.

Titanium's modulus can vary with alloy type (β vs. α) and processing, from as low as 93 GPa (13.5×10^6 psi) up to about 120.5 GPa (17.5×10^6 psi). For reference, titanium alloy moduli on average are about 50% greater than the moduli for aluminum alloys but only about 60% of the moduli for steels and nickel-base superalloys. Wrought titanium alloys

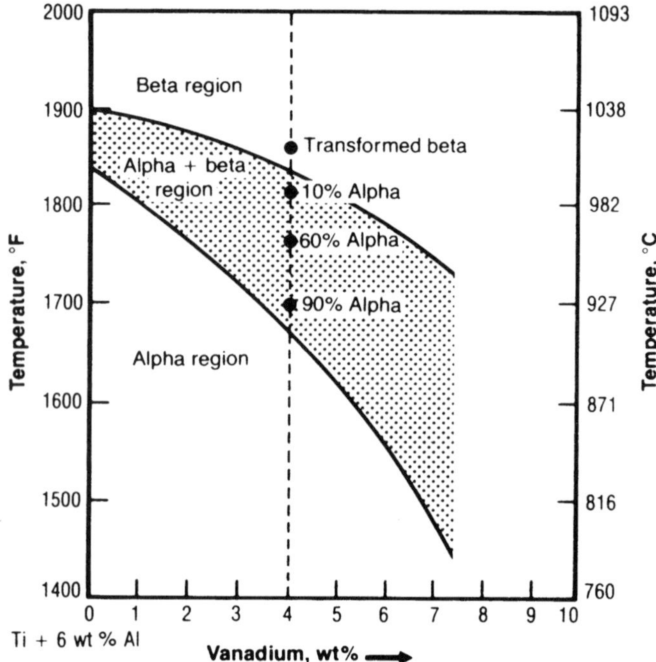

Figure 5 Phase diagram that predicts the results of heat treatment or forging practice. (From *Titanium: A Technical Guide,* 1st ed., ASM International, Materials Park, OH, 1988, p. 51.)

can have their crystals oriented by processing such that a texture develops. When that happens, instead of the usual random orientation of grains leading to uniformity of mechanical properties, a nonuniform orientation occurs and leads to a greater than normal range of property values. By appropriate processing, it is possible to orient wrought titanium for optimum elastic modulus at the high end of the modulus values quoted above. Although textures can be produced, processing that leads to directional grain or crystal orientation similar to directional solidification in castings or directional recrystallization in oxide dispersion-strengthened alloys is not practical in titanium alloy systems.

4.6 Effects of Processing

Properties of titanium alloys of a given composition generally are controlled by variations of the processing (including heat treatment) and are modified for optimum fatigue resistance by surface treatments such as shot peening. Process treatments can produce either acicular or equiaxed microstructures in most titanium alloys if phase transformations from α to β to α (or to related phases) are permitted to occur. Microstructures have been identified which show α as acicular or equiaxed and with varying amounts of α-phase. The platelike or acicular α produced by transformation from the β-phase has special aspects as far as properties are concerned. Table 4 shows the relative behavior of equiaxed vs. platelike α. No one microstructure is good for all applications.

Table 3 Physical and Mechanical Properties of Elemental Titanium

Atomic number	22
Atomic weight	47.90
Atomic volume	10.6 W/D
Covalent radius	1.32 Å
First ionization energy	158 kcal/g·mol
Thermal neutron absorption cross section	5.6 barns/atom
Crystal structure	• Alpha: close-packed, hexagonal ≤882.5°C (1620°F) • Beta: body-centered, cubic ≥882.5°C (1620°F)
Color	Dark gray
Density	4.51 g/cm^3 (0.163 lb/in.^{3})
Melting point	1668 ± 10°C (3035°F)
Solidus/liquidus	1725°C
Boiling point	3260°C (5900°F)
Specific heat (at 25°C)	0.518 J/kg·K (0.124 Btu/lb·°F)
Thermal conductivity	9.0 Btu/h·ft^2·°F
Heat of fusion	440 kJ/kg (estimated)
Heat of vaporization	9.83 MJ/kg
Specific gravity	4.5
Hardness	70–74 R_B
Tensile strength	35 ksi·min
Modulus of elasticity	14.9 × 10^6 psi
Young's modulus of elasticity	116 × 10^9 N/m^2 16.8 × 10^6 lbf/in.^{2} 102.7 GPa
Poisson's ratio	0.41
Coefficient of friction	0.8 at 40 m/min (125 ft/min) 0.68 at 300 m/min (1000 ft/min)
Specific resistance	554 $\mu\Omega$·mm
Coefficient of thermal expansion	8.64 × 10^{-6}/°C
Electrical conductivity	3% IACS (copper 100%)
Electrical resistivity	47.8 $\mu\Omega$·cm
Electronegativity	1.5 Pauling's
Temperature coefficient of electrical resistance	0.0026/°C
Magnetic susceptibility	1.25 × 10^{-6} 3.17 emu/g
Machinability rating	40

Source: From *Titanium: A Technical Guide,* 1st ed., ASM International, Materials Park, OH, 1988, p. 11.

Table 4 Relative Advantages of Equiaxed and Acicular Microstructures

Equiaxed
Higher ductility and formability
Higher threshold stress for hot-salt stress corrosion
Higher strength (for equivalent heat treatment)
Better hydrogen tolerance
Better low-cycle fatigue (initiation) properties
Acicular
Superior creep properties
Higher fracture-toughness values

Source: From *Titanium: A Technical Guide,* 1st ed., ASM International, Materials Park, OH, 1988, p. 168.

4.7 Hydrogen (in CP Titanium)

The solubility of hydrogen in α titanium at 300°C (572°F) is about 8 at % (about 0.15 wt %, or about 1000 ppm by weight). Hydrogen in solution has little effect on the mechanical properties. Damage is caused by hydrides which form. Upon precipitation of the hydride, titanium alloy ductility suffers. Hydrogen damage of titanium and titanium alloys, therefore, is manifested as a loss of ductility (embrittlement) and/or a reduction in the stress-intensity threshold for crack propagation. Figure 6 shows the effect of hydrogen on reduction of area.

No embrittlement is found at 20 ppm hydrogen. Twenty parts per million corresponds to about 0.1 at % of hydrogen. Other data show that, independent of the heat treatment, this low a concentration has little effect on the impact strength, a different measure of embrittlement. However, as little as 0.5 at % hydrogen (about 100 atom ppm) can cause measurable embrittlement. Slow cooling from the α-region—e.g., 400°C (752°F)—allows sufficient hydride to precipitate to reduce the impact energy. The only practical approach to control the hydrogen problem is to maintain a low concentration of the element. Hydrogen also can have a potent effect on other titanium alloy properties.

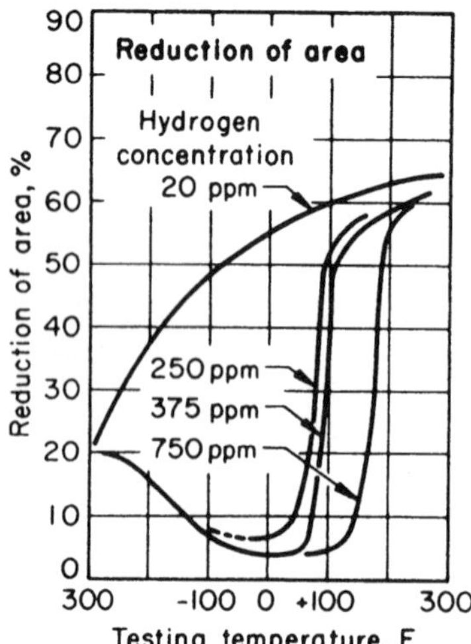

Figure 6 Ductility of alpha titanium vs. test temperature, showing embrittling effects of hydrogen. (From *Titanium: A Technical Guide,* 1st ed., ASM International, Materials Park, OH, 1988, p. 161.)

4.8 Oxygen and Nitrogen (in CP Titanium)

Oxygen and nitrogen have a significant effect on strength properties. As the amount of oxygen and nitrogen increases, the toughness decreases until the material eventually becomes quite brittle. Embrittlement occurs at a concentration considerably below the solubility limit. The allowed oxygen content is higher than the allowed nitrogen content. Yield and ultimate strengths increase as oxygen (and nitrogen) levels go up. the higher strengths in CP titanium grades come from higher oxygen levels. Oxygen (and nitrogen) can have a potent effect on other titanium alloy properties as well.

4.9 Mechanical Properties of Titanium Alloys

The grain size, grain shape, and grain boundary arrangements in titanium have a very significant influence on mechanical properties, and it is the ability to manipulate the phases/grains present as a result of alloy composition that is responsible for the variety of properties that can be produced in titanium and its alloys. Transformed β-phase products in alloys can affect tensile strengths, ductility, toughness, and cyclic properties. To these effects must be added the basic strengthening effects of alloy elements.

Interstitial elements are those elements such as oxygen that are significantly smaller than the titanium atom and so may dissolve in the titanium-phase crystal lattice as solid solutions without substituting for titanium atoms. Of course, some interstitial elements also may form second phases with titanium. For example, as indicated earlier, hydrogen can combine with titanium atoms to form a titanium hydride. As is the case for comparable-size elements, interstitial elements may have a preference for one phase over another in titanium. As indicated above, a significant influence on mechanical behavior of CP titanium is brought about by hydrogen, nitrogen, carbon, and oxygen, which dissolve interstitially in titanium and have a potent effect on mechanical properties. These effects carry over to titanium alloys in varying degrees.

The ELI (extra-low interstitial) levels specified for some titanium alloys implicitly recognize the effect of reduced interstitials on ductility. The ELI-type material is used for critical applications where enhanced ductility and toughness are produced by keeping interstitials at a very low level. Hydrogen is always kept at a low level to avoid embrittlement, yet there still remains concern about the most reasonable level to specify in both CP and alloyed titanium to protect against embrittlement but keep manufacturing cost low.

Although data are not provided here for grain size effects on titanium grades, it is generally accepted that fineness of structure (smaller particle size, grain size, etc.) is more desirable from the point of view of TYS in metallic materials. The UTS is not particularly affected by grain size, but ductility as represented by elongation or reduction in area generally is improved with smaller grain sizes. Ductility is a measure of toughness, but toughness is not normally at issue in CP titanium grades. Another measure of toughness is Charpy impact strength. The chemistry and minimum tensile properties for various specifications for CP and modified titanium grades at room temperature are given in Table 5.

Elevated temperature behavior of titanium grades has been studied, but titanium grades are not customarily used at high temperatures. The near-α- or α–β-alloys are the preferred materials where high-temperature mechanical properties are desired. With allowance for grain size effects and possible minor chemistry variations, cast CP titanium materials should behave in much the same way as wrought.

Alpha Alloys

The α-alloys such as Ti–5Al–2.5Sn, Ti–6Al–2Sn–4Zr–2Mo+Si and Ti–8Al–1Mo–1V (see Table 2) are used primarily in gas turbine applications. The Ti–8Al–1Mo–1V alloy and Ti–

Table 5 CP and Modified Ti: Minimum Room Temperature Tensile Properties for Various Specifications

Designation	Chemical Composition (% max.)				Tensile Properties[a]				
					Tensile Strength		Yield Strength		Minimum Elongation (%)
	C	O	N	Fe	MPa	ksi	MPa	ksi	
JIS Class 1	—	0.15	0.05	0.20	275–410	40–60	165[b]	24[b]	27
ASTM grade 1 (UNS R50250)	0.10	0.18	0.03	0.20	240	35	170–310	25–45	24
DIN 3.7025	0.08	0.10	0.05	0.20	295–410	43–60	175	25.5	30
GOST BT1-00	0.05	0.10	0.04	0.20	295	43	—	—	20
BS 19-27t/in.^{2}	—	—	—	0.20	285–410	41–60	195	28	25
JIS Class 2	—	0.20	0.05	0.25	343–510	50–74	215[b]	31[b]	23
ASTM grade 2 (UNS R50400)	0.10	0.25	0.03	0.30	343	50	275–410	40–60	20
DIN 3.7035	0.08	0.20	0.06	0.25	372	54	245	35.5	22
GOST BTI-0	0.07	0.20	0.04	0.30	390–540	57–78	—	—	20
BS 25-35t/in.^{2}	—	—	—	0.20	382–530	55–77	285	41	22
JIS Class 3	—	0.30	0.07	0.30	480–617	70–90	343[b]	50[b]	18
ASTM grade 3 (UNS R50500)	0.10	0.35	0.05	0.30	440	64	377–520	55–75	18
ASTM grade 4 (UNS R50700)	0.10	0.40	0.05	0.50	550	80	480	70	15
DIN 3.7055	0.10	0.25	0.06	0.30	460–590	67–85	323	47	18
ASTM grade 7 (UNS R52400)	0.10	0.25	0.03	0.30	343	50	275–410	40–60	20
ASTM grade 11 (UNS R52250)	0.10	0.18	0.03	0.20	240	35	170–310	25–45	24
ASTM grade 12 (UNS R53400)	0.10	0.25	0.03	0.30	480	70	380	55	12

[a] Unless a range is specified, all listed values are minimums.
[b] Only for sheet, plate, and coil.
Source: From *Materials Property Handbook—Titanium,* ASM International, Materials Park, OH, 1994, p. 224.

6Al–2Sn–4Zr–2Mo + Si are useful at temperatures above the normal range for the work horse α–β-alloy, Ti–6Al–4V. The Ti–8Al–1Mo-1V and Ti–6Al–2Sn–4Zr–2Mo + Si alloys have better creep resistance than Ti–6Al–4V and creep resistance is enhanced with a fine acicular (Widmanstatten) structure. In its normal heat-treated condition, Ti–6Al–2Sn–4Zr–2Mo + Si alloy actually has a structure better described as α–β.

The α- and near-α-alloys therefore are usually employed in the solution-annealed and stabilized condition. Solution annealing may be done at a temperature some 35°C (63°F) below the β-transus temperature while stabilization is commonly produced by heating for 8 h at about 590°C (1100°F). These alloys are more susceptible to the formation of ordered Ti_3Al, which promotes SCC.

Alpha–Beta Alloys

The most important titanium alloy is the α–β-alloy Ti–6Al–4V. This alloy has found application for a wide variety of aerospace components and fracture-critical parts. With a strength-to-density ratio of 25 × 106 mm (1 × 10^6 in.), Ti–6Al–4V is an effective lightweight structural material and has strength–toughness combinations between those of steel and aluminum alloys. High-strength α–β-alloys include Ti–6Al–6V–2Sn and Ti–6Al–2Sn–4Zr–6Mo. The dominant phase in all of these alloys is α, but it is dominant to a lesser extent in the high-strength alloys than in Ti–6Al–4V. These high-strength alloys are stronger and more readily heat treated than Ti–6Al–4V.

When α–β titanium alloys are heat treated high in the α–β range and then cooled, the resulting structure, because of the presence of globular (equiaxed) primary α in the transformed β (platelike) matrix, is called equiaxed. When a 100% transformed β-structure is achieved by cooling from above the β-transus, the structure may be called acicular, or needlelike. Generally speaking, α–β alloys would be annealed just below the β-transus to produce a maximum of transformed acicular β with approximately 10% of equiaxed α present. Some titanium alloys—e.g., Ti–6A1–2Sn–4Zr–2Mo—are given β heat treatments to enhance high-temperature creep resistance. (Castings and powder products may be given a β anneal, too, in order to break up the structure, although not necessarily for optimizing creep strength.)

In actual components, the structure of titanium α–β-type alloys is controlled not only by how much work is done and by how close to or above the β-transus the alloy is processed, but also by the section size of the component. Ideally, alloys should have good hardenability, i.e., ability to reach desired cooling rates and attendant microstructures in fairly thick sections. Many α–β-alloys do not have great hardenability. The Ti–6Al–4V alloy only has sufficient hardenability to be effectively heat treated to full property levels in sections less than 25 mm (1 in.) thick.

One of the least understood concepts in the behavior of α–β titanium alloys is that of aging. With few exceptions titanium alloys do not age in the classical sense: that is, where a secondary, strong intermetallic compound appears and strengthens the matrix by its dispersion. A dispersion is produced on aging of titanium α–β-alloys, but it is thought to be β dispersed in the α or in martensitic α'. Beta is not materially different from the α-phase with respect to strength; however the effectiveness of strengthening in titanium alloys appears to center on the number and fineness of α–β-phase boundaries. Annealing and rapid cooling, which maximize α–β boundaries for a fixed primary α-content, along with aging, which may promote additional boundary structure, can significantly increase alloy strength.

Beta Alloys

An alloy is considered to be a β-alloy if it contains sufficient β-stabilizer alloying element to retain the β-phase without transformation to martensite on quenching to room temperature. A number of titanium alloys (see Table 2) contain more than this minimum amount of β-stabilizer alloy addition. The more highly β-stabilized alloys are alloys such as Ti–3Al–8V–6Cr–4Mo–4Zr (Beta C) and Ti–15V–3Cr–3Al–3Sn. Solute-lean β-alloys are sometimes classified as β-rich α–β-alloys, and this class includes Ti–10V–2Fe-3Al and proprietary alloys such as Ti-17 (Ti–5Al–2Sn–2Zr–4Mo–4Cr) and Beta CEZ (Ti–5Al–2Sn–4Zr–4Mo–2Cr).

In a strict sense there is no truly stable β-alloy because even the most highly alloyed β will, on holding at elevated temperatures, begin to precipitate ω, α, Ti_3Al, or silicides, depending on temperature, time, and alloy composition. All β-alloys contain a small amount of aluminum, an α-stabilizer, in order to strengthen α that may be present after heat treatment. The composition of the precipitating α is not constant and will depend on the temperature of heat treatment. The higher the temperature in the α–β phase field, the higher will be the aluminum content of α.

The processing window for β-alloys is tighter than that normally used for the other alloy types (α- and α–β-alloys). For the less highly β-stabilized alloys, such as Ti–10V–2Fe–3Al, for example, the thermomechanical process is critical to the property combinations achieved as this has a strong influence on the final microstructure and the resultant tensile strength and fracture toughness that may be achieved. Exacting control of thermomechanical processing is somewhat less important in the more highly β-stabilized alloys, such as Ti–3Al–8V–6Cr–4Mo–4Zr and Ti–15V–3Cr–3Al–3Sn. In these alloys, the final microstructure, precipitated α in the β-phase, is so fine that microstructural manipulation through thermomechanical processing is not as effective.

Properties

Wrought Alloys. Typical minimum property values for titanium alloy mill products are listed in Table 6. Fractions of room temperature strength retained at elevated temperatures by the same titanium alloys are shown in Table 7. Data for unalloyed titanium is included in Table 7 to illustrate that alloys not only have higher room temperature strengths than unalloyed titanium but also retain much larger fractions of that strength at elevated temperatures. Typical tensile strengths and 0.1% creep strengths as functions of temperature of some selected alloys are shown in Figures 7 and 8, respectively.

Fatigue life in unalloyed titanium depends on grain size, interstitial (e.g., oxygen) level, and degree of cold work, as illustrated in Figure 9. A decrease in grain size in unalloyed titanium from 110 μm down to 6 μm improves the 10^7 cycle fatigue endurance limit by 30%. The HCF endurance limits of unalloyed titanium depend on interstitial contents just as do the TYS and UTS. The ratio of HCF endurance limit and TYS at ambient temperature appears to remain relatively constant as TYS changes with changing interstitial content but does show a temperature dependence.

There are significant differences among titanium alloys in fracture toughness, but there also is appreciable overlap in their properties. Table 8 gives examples of typical plane-strain fracture toughness ranges for α–β titanium alloys. From these data it is apparent that the basic alloy chemistry affects the relationship between strength and toughness. From Table 8 it also is evident, as noted earlier, that transformed microstructures may greatly enhance toughness while only slightly reducing strength. It is well known that toughness depends on thermomechanical processing (TMP) to provide the desired structure. However, the enhancement of fracture toughness at one stage of an operation—for example, a forging billet—does not necessarily carry over to a forged part.

Because welds in alloy Ti–6A1–4V contain transformed products, one would expect such welds to be relatively high in toughness. This is in fact the case. In addition to welding, many other factors such as environment, cooling rates occurring in large sections (i.e., hardenability), and hydrogen content may affect K_{1c}.

Titanium alloys may show less resistance to notches than other alloys. Notch strength in fatigue is significantly lower than smooth strength. Scratches on the surfaces of titanium alloy components can lead to reduced fatigue capability. High levels of favorable compressive residual stresses usually exist in titanium alloys as a result of machining. These levels are sometimes enhanced by surface processing such as glass bead or shot peening.

Cast Alloys. Cast titanium alloys generally are α–β-alloys. They are equal, or nearly equal, in strength to wrought alloys of the same compositions. Typical room temperature tensile properties of several cast titanium alloys are shown in Table 9 while creep strength of cast Ti–6Al–4V is shown in Table 10. Virtually all existing data have been generated from alloy Ti–6A1–4V; consequently, the basis for most cast alloy property data is Ti6–Al–4V. Because the microstructure of cast titanium alloy parts is comparable to that of wrought material, many properties of cast plus HIP parts are at similar levels to those for wrought alloys. These properties include tensile strength, creep strength, fracture toughness, and fatigue crack propagation.

Generally, castings of titanium alloys are hot isostatically pressed (HIPed) to close casting porosity. The HIP conditions may affect the resultant properties since HIP is just another heat treatment as far as microstructure is concerned. It also should be noted that test results are often on small separately cast test coupons and will not necessarily reflect the property level achievable with similar processing on a full-scale cast part. Property levels of actual

Table 6 Tensile Strengths of Several Commercial Titanium-Base Alloys: Typical Room Temperature Values

Alloy Name	Nominal Composition	Condition	Tensile Strength		Yield Strength		Elongation (%)
			ksi	10^8 N/m^2	ksi	10^8 N/m^2	
5-2.5	Ti–5Al–2.5Sn	Annealed (0.25–4 h/1300–1600°F)	120–130	8.3–9.0	115–120	7.9–8.3	13–18
3-2.5	Ti–3Al–2.5V	Annealed (1–3 h/1200–1400°F)	95	6.5	90	6.2	22
6-2-1-1	Ti–6Al–2Nb–1Ta–1Mo	Annealed (0.25–2 h/1300–1700°F)	125	8.6	110	7.6	14
8-1-1	Ti–8Al–1Mo–1V	Annealed (8 h/1450°F)	145	10.0	135	9.3	12
Corona 5	Ti–4.5Al–5Mo–1.5Cr	α–β annealed after β processing	140–160	9.7–11.0	135–150	9.3–10.3	12–15
Ti-17	Ti–5Al–2Sn–2Zr–4Mo–4Cr	α–β or β processed plus aged	165	11.4	155	10.7	8
6-4	Ti–6Al–4V	Annealed (2 h/1300–1600°F)	140	9.6	130	9.0	17
		Aged	170	11.7	160	11.0	12
6-6-2	Ti–6Al–6V–2Sn	Annealed (3 h/1300–1500°F)	155	10.7	145	10.0	14
		Aged	185	12.8	175	12.1	10
6-2-4-2	Ti–6Al–2Sn–4Zr–2Mo	Annealed (4 h/1300–1550°F)	145	10.0	135	9.3	15
6-2-4-6	Ti–6Al–2Sn–4Zr–6Mo	Annealed (2 h/1500–1600°F)	150	10.3	140	9.7	11
		Aged	175	12.1	165	11.4	8
6-22-22	Ti–6Al–2Sn–2Zr–2Mo–2Cr–0.25Si	α–β processed plus aged	162	11.2	147	10.1	14
10-2-3	Ti–10V–2Fe–3Al	Annealed (1 h/1400°F)	140	9.7	130	9.0	9
		Aged	180–195	12.4–13.4	165–180	11.4–12.4	7
15-3-3-3	Ti–15V–3Cr–3Sn–3Al	Annealed (0.25 h/1450°F)	115	7.9	112	7.7	20–25
		Aged	165	11.4	155	10.7	8
13-11-3	Ti–13V–11Cr–3Al	Annealed (0.5 h/1400–1500°F)	135–140	9.3–9.7	125	8.6	18
		Aged	175	12.1	165	11.4	7
38-6-44	Ti–3Al–8V–6Cr–4Mo–4Zr	Annealed (0.5 h/1500–1700°F)	120–130	8.3–9.0	113–120	7.8–8.3	10–15
		Aged	180	12.4	170	11.7	7
β-III	Ti–4.5Sn–6Zr–11.5Mo	Annealed (0.5 h/1300–1600°F)	100–110	6.9–7.6	95	6.5	23
		Aged	180	12.4	170	11.7	7

Source: From *Materials Property Handbook—Titanium,* ASM International, Materials Park, OH, 1994, p. 106.

Table 7 Fraction of Room Temperature Strength Retained at Elevated Temperature for Several Titanium Alloys[a]

Temperature		Unalloyed Ti		Ti–6Al–4V		Ti–6Al–6V–2Sn		Ti–6Al–2Sn–4Zr–6Mo		Ti–6Al–2Sn–4Zr–2Mo		Ti-1100[a]		IMI-834	
°C	°F	TS	YS	TS	YS	TS	YS	TS	YS	TS	YS	TS	YS	TS	YS
93	200	0.80	0.75	0.90	0.87	0.91	0.89	0.90	0.89	0.93	0.90	0.93	0.92	—	—
204	400	0.57	0.45	0.78	0.70	0.81	0.74	0.80	0.80	0.83	0.76	0.81	0.85	0.85	0.78
316	600	0.45	0.31	0.71	0.62	0.76	0.69	0.74	0.75	0.77	0.70	0.76	0.79	—	—
427	800	0.36	0.25	0.66	0.58	0.70	0.63	0.69	0.71	0.72	0.65	0.75	0.76	—	—
482	900	0.33	0.22	0.60	0.53	—	—	0.66	0.69	0.69	0.62	0.72	0.74	—	—
538	1000	0.30	0.20	0.51	0.44	—	—	0.61	0.66	0.66	0.60	0.69	0.69	—	—
593	1100	—	—	—	—	—	—	—	—	—	—	0.66	0.63	0.63	0.61

[a] Short-time tensile test with less than 1 h at temperature prior to test. TS = tensile strength; YS = yield strength.

Source: From *Fatigue Data Handbook: Light Structural Alloys,* ASM International, Materials Park, OH, 1995, p. 189.

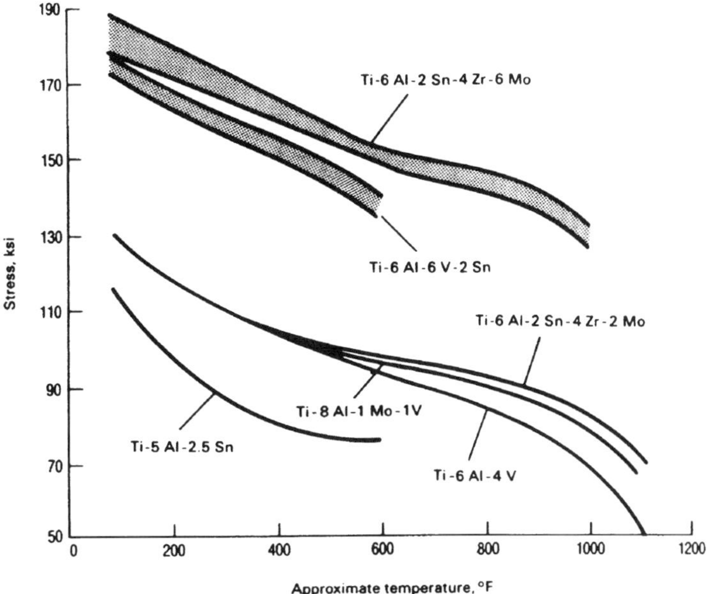

Figure 7 Comparison of typical ultimate tensile strengths of selected titanium alloys as a function of temperature. (From *Titanium: A Technical Guide,* 1st ed., ASM International, Materials Park, OH, 1988, p. 173.)

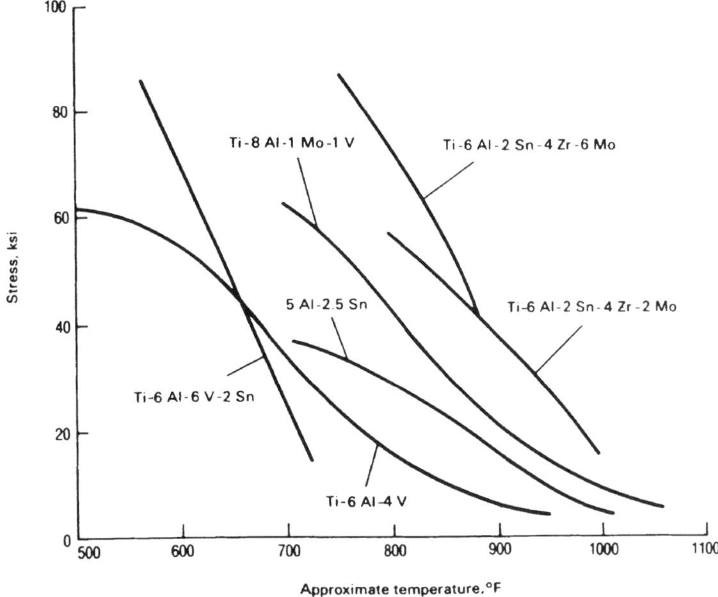

Figure 8 Comparison of typical 150-h, 0.1% creep strengths for selected titanium alloys. (From *Titanium: A Technical Guide,* 1st ed., ASM International, Materials Park, OH, 1988, p. 174.)

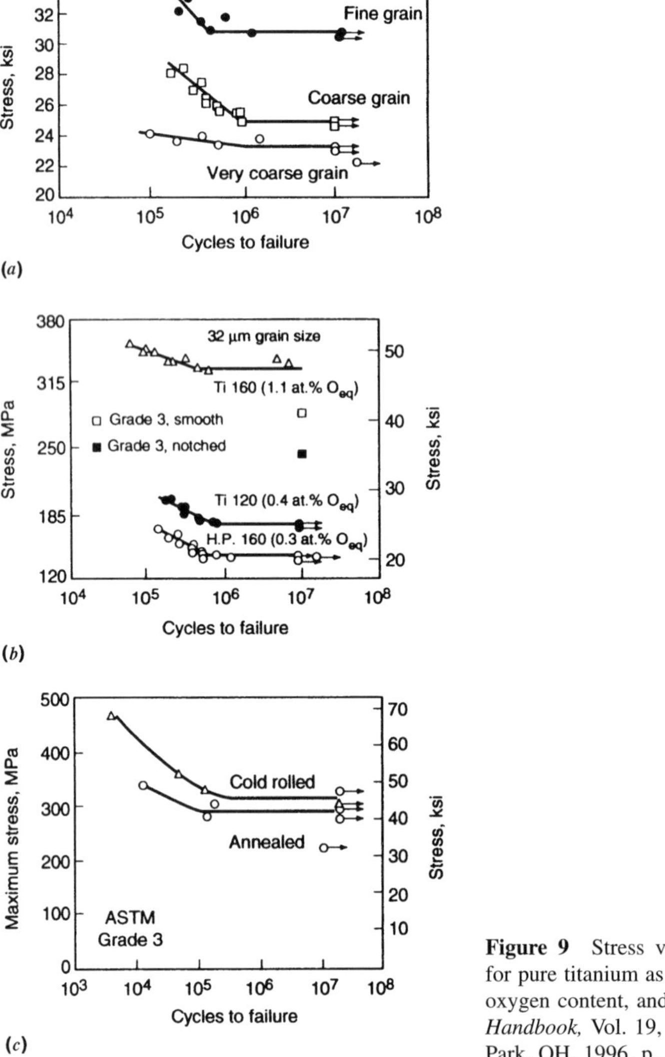

Figure 9 Stress versus cycles-to-failure curves for pure titanium as affected by (*a*) grain size, (*b*) oxygen content, and (*c*) cold work. (From *Metals Handbook,* Vol. 19, ASM International, Materials Park, OH, 1996, p. 837.)

cast parts, especially larger components, probably will be somewhat lower, the result of coarser grain structure or slower quench rates achieved.

Powder-Formed Alloys. It has been a goal of titanium alloy development to reduce costs by introducing powder metal processing. Very high purity powder is needed. Some applications for less demanding industries than the aerospace or biomedical markets may be able to use lower cost powder with lesser properties than conventional wrought alloys are capable of producing. High-purity powder is produced by special rotating electrode or similar processes under inert conditions. Subsequent handling and consolidation of the powder to form a net

Table 8 Typical Fracture Toughness Values of High-Strength Titanium Alloys

		Yield Strength		Fracture Toughness K_{Ic}	
Alloy	Alpha Morphology	MPa	ksi	MPa·m$^{1/2}$	ksi·in.$^{1/2}$
Ti–6Al–4V	Equiaxed	910	130	44–66	40–60
	Transformed	875	125	88–110	80–100
Ti–6Al–6V–2Sn	Equiaxed	1085	155	33–55	30–50
	Transformed	980	140	55–77	50–70
Ti–6Al–2Sn–4Zr–6Mo	Equiaxed	1155	165	22–23	20–30
	Transformed	1120	160	33–55	30–50

Source: From *Titanium: A Technical Guide,* 1st ed., ASM International, Materials Park, OH, 1988, p. 168.

or near-net-shape (NNS) component also require a highly inert environment. While powder-based components have shown comparable mechanical properties to wrought products, the costs of powder and the powder consolidation processes have not produced cost savings.

Summary: Wrought, Cast, and Powder Metallurgy Products. Powder metallurgy technology has been applied to titanium alloy processing with limited success, partially owing to economic issues. Wrought processing remains the preferred method of achieving shape and property control. Cast alloy processing is used but for a limited alloy base. Figure 10 shows fatigue scatterbands for wrought, cast, and powder metallurgy products of Ti–6Al–4V alloy for comparison of attainable properties.

Table 9 Typical Room Temperature Tensile Properties of Several Cast Titanium Alloys (bars machined from castings)[a]

	Yield Strength		Tensile Strength			
Alloy[b,c]	MPa	ksi	MPa	ksi	Elongation (%)	Reduction of area (%)
Commercially pure (grade 2)	448	65	552	80	18	32
Ti–6Al–4V, annealed	855	124	930	135	12	20
Ti–6Al–4V–ELI	758	110	827	120	13	22
Ti–1100, beta-STA[d]	848	123	938	136	11	20
Ti–6Al–2Sn–4Zr–2Mo, annealed	910	132	1006	146	10	21
IMI-834, beta-STA[d]	952	138	1069	155	5	8
Ti–6Al–2Sn–4Zr–6Mo, beta-STA[d]	1269	184	1345	195	1	1
Ti–3Al–8V–6Cr–4Zr–4Mo, beta-STA[d]	1241	180	1330	193	7	12
Ti–15V–3Al–3Cr–3Sn, beta-STA[d]	1200	174	1275	185	6	12

[a] Specification minimums are less than these typical properties.
[b] Solution-treated and aged (STA) heat treatments can be varied to produce alternate properties.
[c] ELI, extra low interstitial.
[d] Beta-STA, soluton treatment within β-phase field followed by aging.
Source: Metals Handbook, Vol. 2, ASM International, Materials Park, OH, 1990, p. 637.

Table 10 Ti–6Al–4V: Creep Strength of Cast Material

Test Temperature		Stress		Plastic Strain on Loading	Test Duration	Time, h, to Reach Creep of		
°C	°F	MPa	ksi	(%)	(h)	0.1%	0.2%	1.0%
455	850	276	40.0	0	611.2	2.0	9.6	610.0
425	800	276	40.0	0	500.0	15.0	60.0	—
425	800	345	60.0	0	297.5	3.5	11.0	291.5
400	750	448	65.0	0.7	251.4	7.5	22.0	—
370	700	414	60.0	0.3	500	240.0	—	—
315	600	517	75.0	2.04	330.9	0.02	0.04	0.1
260	500	534	77.5	2.1	307.9	0.01	0.02	0.1
205	400	552	80.0	0.56	138.0	0.1	0.13	1.5
205	400	531	77.0	0.8	18.2	0.02	0.04	0.16
175	350	517	75.0	0.01	1006.0	0.4	2.2	—
150	300	517	75.0	—	500	0.25	1.2	—
150	300	517	75.0	—	500	1.7	12.2	—
120	250	517	75.0	0.0	1006.1	9.8	160.0	—

Note: Specimens from hubs of centrifugal compressor impellers that were cast, HIPed (2 h at 900°C, or 1650°F), and 103.5 MPa, 15.0 ksi, and aged 1.5 h at 675°C (1250°F). Specimen blanks approximately 5.72 by 0.95 by 0.96 cm (2.25 by 0.37 by 0.37 in.) in section size, with the long axis oriented tangential to the hub section, were machined to standard-type creep specimens 3.81 mm (0.150 in.) in diameter. The specimens were lathe turned and then polished with 320-grit emery paper. The creep-rupture tests were performed at 120–455°C (250–850°F) using dead-load-type creep frames in air over a stress range of 276–552 MPa (40–80 ksi). The microstructure consisted of transformed β-grains with discontinuous grain boundary α and colonies of transformed β that contained packets of parallel-oriented α-platelets separated by a thin layer of aged β.

Source: A. Chakrabarti and E. Nichols, Creep Behavior of Cast Ti–6Al–4V Alloy, *Titanium '80: Science and Technology,* Proceedings of the 4th International Conference on Titanium, Kyoto, Japan, May 19–22, 1980, Vol. 2, H. Kimura and O. Izumi, Ed., TMS/AIME, New York, 1980, pp. 1081–1096.

5 MANUFACTURING PROCESSES

5.1 General Aspects of the Manufacture of Titanium Articles

Appropriate compositions of all titanium can be forged, rolled to sheet, or otherwise formed into a variety of shapes. Some compositions can be processed as large investment castings. Commercial large castings are made mostly in the titanium alloy Ti–6A–4V, which has been in production for about 50 years. Fabricated titanium structures can be built up by welding or brazing. Fabricated structures are primarily made with Ti–6A–4V, although stronger or more workable alloys are sometimes used. Fabricated structures may contain cast as well as wrought parts, although wrought parts are assembled in most applications.

Single-piece forged gas turbine fan and compressor disks are prime applications for titanium alloys. Titanium wrought, cast, and powder metallurgy products find use in the biomedical arena. Fan blades and compressor blades of titanium represent areas that continue to receive support despite the reported threat from composites. By and large, most titanium alloys are wrought, in particular forged.

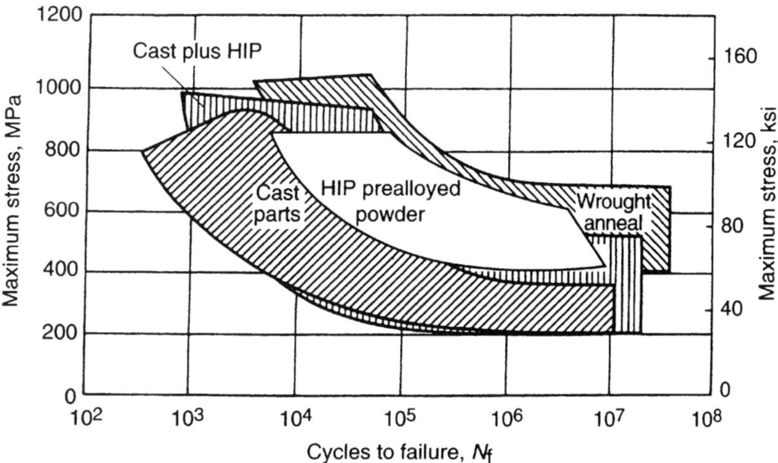

Figure 10 Fatigue scatter bands for ingot metallurgy, castings and powder metallurgy products of Ti–6Al–4V alloy. (From *Titanium: A Technical Guide,* 2nd ed., ASM International, Materials Park, OH, 2001, p. 116.)

The manufacture of titanium alloys consists of a number of separate steps of which the following represent the transfer of titanium from an ore to an ingot ready for either wrought or cast processing or to mill products:

- Production of titanium sponge (reduction of titanium ore to an impure porous form of titanium metal)
- Purification of the sponge
- Melting of sponge or sponge plus alloy elements or a master alloy to form an electrode
- Remelting and, possibly, remelting again to homogenize the first electrode and create an ingot for further processing
- Primary fabrication, in which ingots are converted into billets or general mill products such as bar, plate, sheet, strip, or wire
- Secondary fabrication where a billet or bar may be forged into an approximate final shape

5.2 Production of Titanium via Vacuum Arc Melting

Whether the final product is to be a forged or investment cast one, the essence of a titanium alloy's ability to create the properties desired hinges on the correct application of melting principles. Melting practices may be classified as either primary (the initial melt of elemental materials and/or scrap which sets the composition) or secondary (remelt, often more than once, of a primary melt for the purpose of controlling the solidification structure). The melt type or combination of melt types selected depends upon both the alloy composition (mill form and size) properties desired and sensitivity of the final component to localized inhomogeneity in the alloy.

The principal method for the production of titanium electrodes and ingots since commercial introduction of titanium alloys occurred in the 1950s has been the use of vacuum arc remelting (VAR). The purity of the titanium alloys produced is a function of the purity of the starting materials. Control of raw materials is extremely important in producing titanium and its alloys because there are many elements of which even small amounts can produce major, and at times undesirable, effects on the properties of the titanium alloys in finished form. To produce ingots of titanium or its alloys for commercial application, titanium from sponge is commonly alloyed with other pure elements, master melt of titanium plus alloy elements, and/or reclaimed titanium scrap (usually called "revert").

Because sponge is an uneven product consisting of a loose, granular mass, it does not compact as well as might be desired in some instances. Compacting is needed to make an electrode from which to melt the alloy. During melting, a piece of the sponge might fall unmelted into the solidifying electrode. Perhaps a chunk of revert or master melt might fall in. Whatever the situation, a gross inhomogeneity would result. Depending on the type and size of the inhomogeneity, a major structural defect could exist. Consequently, after some significant incidents in aircraft gas turbine engines 30–40 years ago, second and then third melts were instituted to provide almost certain homogenization of the alloy.

5.3 Cutting the Cost of Titanium Alloy Melting

A concept known as simultaneous nonconsumable arc-melting/plasma-refining process (SNAPP) has been developed which, in conjunction with a new kind of titanium melting furnace (VersaCast), is claimed to reduce the cost of melting titanium alloys. The process would lend itself to casting not just of round ingots but also to casting slabs or NNS small structures such as preforms and castings. The purpose of the SNAPP/VersaCast linkage is to cast closer to final dimensions, cutting steps from the melting and casting process for ingots and reducing yield losses.

Electron beam and plasma arc melting technologies are now available for the melting of titanium alloys or the remelting of scrap. The use of these technologies permits the controlled-hearth melting (CHM) of titanium alloys. Processes such as electron beam controlled-hearth melting (EBCHM) and plasma arc melting (PAM) have already demonstrated chemistry and quality improvements but not necessarily cost improvements for melting titanium alloys.

5.4 Defects and Their Control in Titanium Melting

Defects have been a concern for titanium ingot metallurgy production since the early days of the industry. Different types of defects were recognized, most stemming from sponge handling, electrode preparation, and melt practice. The principal characterization of these defects is as random hard and brittle particles such as titanium nitride (type I defect) or tungsten carbide particles (high-density inclusions, HDIs). Type II defects are a result of solidification segregation. Low-density inclusions (LDIs) are a variant of type II defects.

Over two dozen different defects have been cataloged. Defects prompted strict process controls which were agreed upon jointly by metal suppliers and customers alike. These controls have done much to attain either reduced-defect or defect-free materials. Despite the controls, occasional defects have been involved in significant titanium-alloy-related failures. The introduction of cold-hearth technologies will further reduce the incidence of defects in titanium ingots. Studies on EBCHM and PAM demonstrated the ability of hearth melting to remove HDIs with great confidence (the HDIs fall to the bottom of the molten hearth). Type II defects can be avoided by better process control and use of improved melting methods; Segregation defects are minimized by CHM.

5.5 Forging Titanium Alloys

Manufacturing processes such as die forging, hot and cold forming, machining, chemical milling, joining, and sometimes extrusion are all secondary fabrication processes used for producing finished parts from billet or mill products. Each of these processes may strongly influence properties of titanium and its alloys, either alone or by interacting with effects of processes to which the metal has previously been subjected. Titanium alloy forgings are produced by all the forging methods currently available. These include open die, closed die, rotary forging, and others. One of the main purposes of forging, in addition to shape control, is to obtain a combination of mechanical properties that generally does not exist in bar or billet. Tensile strength, creep resistance, fatigue strength, and toughness all may be better in forgings than in bar or other forms. Selection of the optimal forging method is based on the shape desired for the forged component as well as its desired mechanical properties and microstructure (which largely determines properties after alloy composition is set).

Open-die forging is used to produce some shapes in titanium when volume and/or size do not warrant the development of closed dies for a particular application. However, closed-die forging is used to produce the greatest amount of forged titanium alloys. Closed-die forging can be classified as blocker type (single die set), conventional (two or more die sets), or high definition (two or more die sets). Precision die forging is also conducted, usually employing hot-die/isothermal forging techniques. Conventional closed-die titanium forgings cost more than the blocker-type, but the increase in cost is justified because of reduced machining costs and better property control. The dies used in titanium forging are heated to facilitate the forging process and to reduce surface chilling and metal temperature losses which can lead to inadequate die filling and/or excessive surface cracking of the forged component. Hot-die/isothermal forging takes the die temperature to higher levels.

Forging is more than just a shape-making process. The key to successful forging and heat treatment is the β-transus temperature. Fundamentally there are two principal approaches to the forging of titanium alloys:

- Forging the alloy predominantly below the β-transus
- Forging the alloy predominantly above the β-transus

Conventional α–β forging is best described as a forging process in which all or most of the deformation is conducted below the β-transus. The α-, β-, and transformed β-phases will be present in the microstructure at some time. Structures resulting from α–β forging are characterized by deformed or equiaxed primary α-phase (α present during the forging process) and transformed β-phase (acicular in appearance). Beta forging is a forging technique in which the deformation of the alloy is done above the β-transus. In commercial practice, β forging actually involves initial deformation above the β-transus but final finishing with controlled amounts of deformation below the β-transus of the alloy. In β forging, the influences of mechanical working (deformation) are not necessarily cumulative because of the high temperature and because of the formation of new grains by recrystallization each time the β-transus is surpassed on reheating for forging. Beta forging, particularly of α- and α–β-alloys, results in significant reductions in forging pressures and reduced cracking tendency of the billet during forging.

An alternative titanium die-forging procedure involves the use of precision isothermal (sometimes superplastic) forging techniques. Precision isothermal forging produces a product form that requires much less machining than conventionally forged alloy to achieve final dimensions of the component. Precision-forged titanium is a significant factor in titanium usage in the aircraft and gas turbine engine field. Most precision forged-titanium is produced as NNS product, meaning that the forging is close to final dimensions but that some machining is required.

Superplastic forming, a variant of superplastic isothermal forging, currently is widely used in the aircraft industry and to a lesser extent in the gas turbine industry. Advantages of superplastic forming are, among others:

- Very complex parts can be formed.
- Lighter, more efficient structures can be designed and formed.
- It is performed in a single operation.
- More than one piece may be produced in a machine cycle.
- Pressure (force) is uniformly applied to all areas of the workpiece.

Superplastic forming coupled with diffusion bonding (SPFDB) has been used on titanium alloys to produce complex fabricated structures.

5.6 Casting

Cost factors associated with wrought alloy processing led to continual efforts to develop and improve casting methods for titanium and its alloys. The two principal casting methods have been investment casting or the use of a rammed graphite mold. Investment casting is preferred when close tolerances, thin sections, and better surface finishes are required.

The result of casting development has been a somewhat checkered application of titanium castings with a more widespread acceptance of the practice in the last 10 or 15 years. Titanium castings now are used extensively in the aerospace industry and to lesser measure in the chemical process, marine, biomedical, automotive, and other industries. While many investment cast parts are relatively small, the maximum pour weights of titanium alloy castings can reach 727 kg (1600 lb). The shipped weight of a titanium case for a gas turbine engine can be nearly 273 kg (600 lb). For example, the Pratt & Whitney PW4084 engine intermediate case is about 263 kg (578 lb). Cast components by the rammed graphite mold process have reached 2700 kg (5950 lb) in CP and alloyed grades. Investment casting is the rule in the aerospace and similar critical application areas.

The investment casting process uses a disposable mold to create a negative image of the desired component. Metal fills the mold and solidifies with the desired shape and dimensions very close to final desired values. Some machining is necessary. An α-case can be created during the casting process and must be prevented or removed.

Several alloy compositions were evaluated in early studies of investment cast titanium, but investigators soon concentrated on Ti–6Al–4V. Results supported the concept that cast titanium parts could be made with strength levels and characteristics approaching those of conventional wrought alloys. Subsequently, titanium components have been cast successfully from pure titanium, α–β, and β-alloys. Recently Ti–6Al–2Sn–4Zr–2Mo alloy thin-wall castings have been successfully produced with strength improvements over Ti–6Al–4V and comparable ductility. Nonetheless, the primary alloy used for casting of titanium components has been Ti–6Al–4V.

Some important casting concepts to remember are:

- Hot isostatic pressing may be required to close casting porosity.
- Heat treatment to develop properties may require close monitoring.
- Cast component properties will tend to fall in the lower end of the scatterband for wrought versions of the alloy chosen (usually Ti–6Al–4V, although other conventional alloys may be cast).
- Section thickness may affect properties generated in castings.

5.7 Machining and Residual Stresses

As noted earlier, machining of titanium alloys is similar to but more difficult than that of machining stainless steels. In welding or machining of titanium alloys, the effects of the energy input (heat energy, deformation energy) on the microstructure and properties of the final product must be considered, just as it must be done in forging. Favorable residual stresses have been generated on titanium surfaces for years. Properties measured will degrade dramatically if the favorable surface residual stresses are reduced, for example, by chemical polishing. Shot peening is a common method of increasing a titanium alloy's fatigue strength, at least in airfoil roots and other non-gas-path regions.

5.8 Joining

Components of titanium alloys are routinely welded. Titanium and most titanium alloys can be joined by the following fusion welding techniques:

- Gas–tungsten arc welding (GTAW)
- Gas–metal arc welding (GMAW)
- Plasma arc welding (PAW)
- Electron beam welding (EBW)
- Laser beam welding (LBW)

They can also be joined by brazing or such solid-state techniques as diffusion bonding inertia bonding, and friction welding.

Just as occurs in the heat treatment of titanium and its alloys, fusion welding processes can lead to pickup of detrimental gases. Alloys must be welded in such a way as to preclude interstitial gases such as oxygen from being incorporated in the weld or weld-heat-affected area. For successful arc welding of titanium and titanium alloys, complete shielding of the weld is necessary because of the high reactivity of titanium to oxygen and nitrogen at welding temperatures. Excellent welds can be obtained in titanium and its alloys in a welding chamber, where welding is done in a protective gas atmosphere, thus giving adequate shielding. When welding titanium and titanium alloys, only argon or helium and occasionally a mixture of these two gases are used for shielding. Since it is more readily available and less costly, argon is more widely used. Welding in a chamber, however, is not always practical. Open-air techniques can be used with fusion welding when the area to be joined is well shielded by an inert gas using a Mylar bag for gas containment. Such atmospheric control by means of a temporary bag, or chamber, is preferred. Because titanium alloy welds are commonly used in fatigue-critical applications, a stress-relief operation is generally required following welding to minimize potentially detrimental residual stresses. The essence of joining titanium and its alloys is adhering to the following conditions which need to be met:

- Detrimental interstitial elements must be excluded from the joint region.
- Contaminants (scale, oil, etc.) must be excluded from the joint region.
- Detrimental phase changes must be avoided to maintain joint ductility.

When proper techniques are developed and followed, the welding of thin- to moderate-section-thickness material in titanium alloys can be accomplished successfully using all of the processes mentioned. For welding titanium thicker than about 2.54 mm (0.10 in.) by the GTAW process, a filler metal must be used. For PAW, a filler metal may or may not be used for welding metal less than 12.7 mm (0.5 in.) thick.

Titanium and its alloys can be brazed. Argon, helium, and vacuum atmospheres are satisfactory for brazing titanium. For torch brazing, special fluxes must be used on the titanium. Fluxes for titanium are primarily mixtures of fluorides and chlorides of the alkali metals, sodium, potassium, and lithium. Vacuum and inert-gas atmospheres protect titanium during furnace and induction-brazing operations. Titanium assemblies frequently are brazed in high-vacuum, cold-wall furnaces. A vacuum of 10^{-3} Torr or more is required to braze titanium. Ideally, brazing should be done in a vacuum at a pressure of about 10^{-5}–10^{-4} Torr or in a dry inert-gas atmosphere if vacuum brazing is not possible.

6 OTHER ASPECTS OF TITANIUM ALLOY SELECTION

6.1 Corrosion

Although titanium and its alloys are used chiefly for their desirable mechanical properties, among which the most notable is their high strength-to-weight ratio, another important characteristic of the metal and its alloys is titanium's outstanding resistance to corrosion. CP titanium offers excellent corrosion resistance in most environments, except those media that contain fluoride ions. Unalloyed titanium is highly resistant to the corrosion normally associated with many natural environments, including seawater, body fluids, and fruit and vegetable juices. Titanium exposed continuously to seawater for about 18 years has undergone only superficial discoloration. Titanium is more corrosion resistant than stainless steel in many industrial environments, and its use in the chemical process industry has been continually increasing. Titanium exhibits excellent resistance to atmospheric corrosion in both marine and industrial environments.

Titanium and its alloys have found use in flue gas desulfurization as well as in the food, pharmaceutical, and brewing industries. Titanium and its alloys exhibit outstanding resistance to corrosion in human tissue and fluid environments (see below). Other applications which rely on the corrosion resistance of titanium and its alloys are chemical refineries, seawater piping, power industry condensers, desalination plants, pulp and paper mills, and marine piping systems.

The major corrosion problems for titanium alloys appear to be with crevice corrosion, which occurs in locations where the corroding media are virtually stagnant. Pits, if formed, may progress in a similar manner. Other problem areas are a potential for stress corrosion, particularly at high temperatures, resulting in hot-salt stress–corrosion cracking (HSSCC), which has been observed in experimental testing and an occasional service failure. Stress–corrosion cracking is a fracture, or cracking, phenomenon caused by the combined action of tensile stress, a susceptible alloy, and a specific corrosive environment. The requirement that tensile stress be present is an especially important characteristic of SCC. Aluminum additions increase susceptibility to SCC; alloys containing more than 6% Al generally are susceptible to SCC. Stress–corrosion cracking has been observed in salt water or other lower temperature fluid environments but is not a problem in most applications.

Hot-salt SCC of titanium alloys is a function of temperature, stress, and time of exposure. In general, HSSCC has not been encountered at temperatures below about 260°C (500°F). The greatest susceptibility occurs at about 290–425°C (about 550–800°F) based on laboratory tests. Time to failure decreases as either temperature or stress level is increased. All commercial alloys, but not unalloyed titanium, have some degree of susceptibility to HSSCC. The α-alloys are more susceptible than other alloys.

6.2 Biomedical Applications

Titanium alloys have become standards in dental applications and the orthopedic prosthesis industry where hip implants, for example, benefit from several characteristics of titanium:

- Excellent resistance to corrosion by body fluids
- High specific strength owing to good mechanical strengths and low density
- Modulus about 50–60% of that of competing cobalt-base superalloys

Corrosion resistance benefits are evident. High specific strength, however, enables a lighter implant to be made with attendant improvement in patient response to the device. Furthermore, the modulus of bone is very low, about 10% of that of stainless steel or cobalt-base alloys, and the degree of load transfer from an implant to the bony structure in which it is implanted (and which it replaces) is a direct function of the modulus. By reducing the elastic modulus, the bone can be made to receive a greater share of the load, which is important since bone grows in response to stress.

The result of using implants of titanium alloys of lower modulus than stainless steel and cobalt alloys is that the implant operates for a longer time before breakdown of the implant–bone assembly. The Ti–6Al–4V alloy continues to be the standard titanium alloy for biomedical applications. However, the recent introduction of β titanium alloys for orthopedic implant applications promises further modulus reductions. (The Ti–15Mo beta titanium alloy now is covered by ASTM F-2066 for surgical implant applications and alloys such as Ti–12Mo–6Zr–2Fe and Ti–13Nb–13Zr have been developed specifically for orthopedic implants.)

6.3 Cryogenic Applications

Many of the available α and α–β titanium alloys have been evaluated at subzero temperatures, but service experience at such temperatures has been gained only for a few alloys. The alloys Ti–5Al–2.5Sn and Ti–6A1–4V have very high strength-to-weight ratios at cryogenic temperatures and have been the preferred alloys for special applications at temperatures from −196 to −269°C (−320 to −452°F). Impurities such as iron and the interstitials oxygen, carbon, nitrogen, and hydrogen tend to reduce the toughness of these alloys at both room and subzero temperatures. For maximum toughness, ELI grades are specified for critical applications.

7 FINAL COMMENTS

Many titanium alloys have been developed, although the total is small compared to other metals such as steels and superalloys. A principal reason for this situation is the high cost of alloy development and of proving the worth and safety of a new material. In the sport world, titanium made a brief run at commercial non-gas-turbine applications when the golf club market virtually tied up titanium metal for a short time in the 1990s. Titanium bicycle frames are marketed but are quite expensive.

There continue to be areas of application that are attractive for titanium alloys. A low-cost β-alloy has been evaluated for automotive springs. For a drag racer, a titanium valve spring offered a 34% weight reduction, improved performance in the high-rpm range, and reduced fuel consumption. The question remains as to the cost effectiveness of titanium technology in normal automotive consumer markets.

New applications of old technology are being evaluated to produce elemental titanium at lower cost with the expectation of lowering titanium alloy costs as well. A new powder process is reportedly in the commercialization stages to produce lower pure titanium powder at costs comparable to titanium sponge. It is thought that such a product would make the powder metallurgy fabrication of components more economical.

Titanium alloy castings continue to make inroads with claims that cast alloy properties now rival those of wrought alloys. A 53-in. titanium investment cast fan frame hub casting supports the front fan section of a General Electric engine on several aircraft. The casting replaced 88 stainless steel parts that required welding and machining. In another application, a 330-lb titanium thrust beam was cast for a satellite launch vehicle. The casting replaced an aluminum part. Another application for Ti–6Al–4V is a casting for a lightweight and mobile 155-mm towed howitzer using thin-walled casting technology. While these are impressive applications, it is important to recognize that there are limits on the availability of cast titanium alloys.

The titanium market has been a roller coaster over the years, and gas turbine applications remain the most significant part of the application market. Within most aircraft gas turbine engine companies, only a few alloys have ever made it to production. Admittedly this list differs from company to company in the United States and somewhat greater with alloys used outside the United States. Nevertheless, it is apparent that, although the ability to push titanium's operating environment higher in temperature has resulted in significant gains, advances have tapered off. Since the mid 1950s, when Pratt & Whitney put the first titanium in U.S. gas turbines, much industrial and government funding has been used to increase alloy capabilities. It is obvious that the titanium alloy market for design is closely related to military or commercial high-end systems and will continue to be so for many years. Other applications are possible for titanium alloys. When cost is not a major stumbling block, particularly for high-end uses such as auto-racing vehicles or snowmobiles or for military applications such as body armor or for other people-safety-related applications, titanium alloys are available to do the job. A new alloy, ATI 425 titanium (Ti–4Al–2.5V–1.5Fe–0.25O) is available with good workability to produce body armor for the military. Alloy properties are reported to be comparable to those with Ti–6Al–4V alloy.

Titanium aluminides have been the subject of multidecades of study with interesting but scarcely commercially viable results. Barring a discovery of some unforeseen nature, the message is that, if an existing alloy works and a new alloy does not offer some benefit that overrides the development cost of proving up the alloy for its new use, do not change alloys. For the ever-shrinking cadre of developers in industry, the current status suggests that efforts to tailor existing alloys and "sell" them for new or existing applications may have the best return on investment.

If an alloy selector is starting from scratch to pick an alloy for an application, then any commercially available alloy may be fair game. On the other hand, the best alloy may not be available owing to corporate patent protection or insufficient market to warrant its continued production by the limited number of manufacturers. Then, selection of another alloy

Table 11 Associations Providing Titanium Information

Titanium Information Group	International Titanium Association
Trevor J. Glover, Secretary 5, The Lea Kidderminster, DY11 6JY United Kingdom	Jennifer Simpson, Executive Director 350 Interlocken Blvd. Suite 390 Broomfield, CO 80021-3485
TEL: +44 (0) 1562 60276 FAX: +44 (0) 1562 824851	TEL: 303 404 2221 FAX: 303 404 9111
WEB: www.titaniuminfogroup.co.uk E-MAIL: rayportman@talk21.com	WEB: www.titanium.org E-MAIL: jsimpson@titanium.org
Japan Titanium Society	
22-9 Kanda Nishiki-Cho Chiyoda-Ku, Tokyo ZIP 101 Japan	
TEL: 081 (3) 3293 5958 FAX: 081 (3) 3293 6187	
WEB: www.titan-japan.com	

from a producer may be necessary but may possibly require development costs to get the product in workable form and to determine design properties.

If possible, select a known alloy that has more than one supplier and more than one casting or forging source. In all likelihood, unless a special need (such as formability of sheet) or maximum high-temperature strength is required, Ti–6Al–4V might be the first choice. For special needs such as in marine applications or biomedical orthopedic situations, choice of other alloys may be warranted. In any event, one should work with the suppliers and others in the manufacturing chain to acquire typical or design properties for the alloy in the form it will be used. Generic alloys owned are best for the alloy selector not associated with one of the big corporate users of titanium alloys. User companies with proprietary interests usually have nothing to benefit from giving up a technological advantage by sharing design data or even granting a production release to use a proprietary alloy. Table 11 lists a few organizations chartered to provide assistance to users of titanium products. A list of suppliers should be available from them.

BIBLIOGRAPHY

R. Boyer, G. Welsch, and E. Collings, (Eds.), *Materials Property Handbook: Titanium Alloys,* ASM International Materials Park, OH, 1994.

Collings, E., Physical Metallurgy of Titanium Alloys, in *Materials Property Handbook: Titanium Alloys,* R. Boyer, G. Welsch, and E. Collings, (Eds.), ASM International, Materials Park, OH, 1994, pp. 1–122.

Donachie, M., *Titanium: A Technical Guide,* 2nd ed., ASM International, Material Park, OH, 2001.

Hanson, B., *The Selection and Use of Titanium,* Institute of Materials, London, England, 1995.

International Conferences on Titanium, Proceedings of a continuing series of conferences held periodically and published by various organizations since 1968.

Metals Handbook, 10th ed., ASM International, Material Park, OH, appropriate volumes on topics of interest.

The Effective Use of Titanium: A Designer and User's Guide, The Titanium Information Group, Kudderminster, England, 1992.

CHAPTER 6

NICKEL AND ITS ALLOYS

Gaylord D. Smith and Brian A. Baker
Special Metals Corporation
Huntington, West Virginia

1 INTRODUCTION

Nickel, the 24th element in abundance, has an average content of 0.016% in the outer 10 miles of the earth's crust. This is greater than the total for copper, zinc, and lead. However, few of these deposits scattered throughout the world are of commercial importance. Oxide ores commonly called laterites are largely distributed in the tropics. The igneous rocks contain high magnesium contents and have been concentrated by weathering. Of the total known ore deposits, more than 80% is contained in laterite ores. The sulfide ores found in the northern hemispheres do not easily concentrate by weathering. The sulfide ores in the Sudbury district of Ontario, which contain important by-products such as copper, cobalt, iron and precious metals, are the world's greatest single source of nickel.[1]

Nickel has an atomic number of 28 and is one of the transition elements in the fourth series in the periodic table. The atomic weight is 58.71 and density is 8.902 g/cm^3. Nickel has a high melting temperature (1453°C) and a ductile crystal structure (fcc). Nickel exhibits mild ferromagnetism at room temperature (saturation magnetization of 0.617 T and residual magnetism of 0.300 T) and has an electrical conductivity at 100°C of 82.8 W/m·K. The thermal expansion coefficient between 20 and 100°C is $13.3 \times 10^{-6}/C^{-1}$. The electrical resistivity of nickel at 20°C is 6.97 $\mu\Omega$·cm and the specific heat at 20°C is 0.44 kJ/kg·K. The modulus of elasticity in tension is 206 GPa and 73.6 GPa in shear. The Poisson ratio is 0.30.[2]

Nickel can be readily alloyed with other metallic elements to form a wide range of commercial alloys. As an alloying element, nickel is used in hardenable steels, stainless steels, special corrosion-resistant and high-temperature alloys, copper–nickel, "nickel–silvers," and aluminum–nickel. Nickel imparts ductility and toughness to cast iron. Nickel alloys are used in a multiplicity of consumer applications, such as household appliances, electronics, and automotive components. Selected nickel alloys are used in critical industrial technologies, including chemical processing, pollution control, and aircraft, missile, and ship production as well as electric power generation.

Unalloyed nickel is used for porous electrodes in batteries and fuel cells and as fine catalyst powders. It has long been employed for electroplating on more corrosive substrates. Unalloyed wrought nickel is very resistant to corrosion in marine and industrial atmospheres as well as natural and flowing seawater. It is resistant to corrosion by alkalis and nonoxidizing acids, by neutral and alkaline salt solutions, and by chlorine, hydrogen chloride, fluorine, and certain molten salts.

2 NICKEL ALLOYS

Most of the alloys listed and discussed are in commercial production. However, producers from time to time introduce improved modifications that make previous alloys obsolete. For this reason or for economic reasons, they may remove certain alloys from their commercial product line. Some of these alloys have been included to show how a particular composition compares with the strength or corrosion resistance of currently produced commercial alloys.

2.1 Classification of Alloys

Nickel and its alloys can be classified into the following groups on the basis of chemical composition.[3]

Nickel
(1) Pure nickel, electrolytic (99.56% Ni), carbonyl nickel powder and pellet (99.95% Ni); (2) commercially pure wrought nickel (99.6–99.97% nickel); and (3) anodes (99.3% Ni).

Nickel and Copper
(1) Low-nickel alloys (2–13% Ni); (2) cupronickels (10–30% Ni); (3) coinage alloy (25% Ni); (4) electrical resistance alloy (45% Ni); (5) nonmagnetic alloys (up to 60% Ni); and (6) high-nickel alloys, Monel (over 50% Ni).

Nickel and Iron
Wrought alloy steels (0.5–9% Ni); (2) cast alloy steels (0.5–9% Ni); (3) alloy cast irons (1–6 and 14–36% Ni); (4) magnetic alloys (20–90% Ni): (a) controlled coefficient of expansion (COE) alloys (29.5–32.5% Ni) and (b) high-permeability alloys (49–80% Ni); (5) nonmagnetic alloys (10–20% Ni); (6) clad steels (5–40% Ni); (7) thermal expansion alloys: (a) low expansion (36–50% Ni) and (b) selected expansion (22–50% Ni).

Iron, Nickel, and Chromium
(1) Heat-resisting alloys (40–85% Ni); (2) electrical resistance alloys (35–60% Ni); (3) iron-base superalloys (9–26% Ni); (4) stainless steels (2–25% Ni); (5) valve steels (2–13% Ni); (6) iron-base superalloys (0.2–9% Ni); (7) maraging steels (18% Ni).

Nickel, Chromium, Molybdenum, and Iron
(1) Nickel-base solution-strengthened alloys (40–70% Ni); (2) nickel-base precipitation-strengthened alloys (40–80% Ni).

The nominal chemical composition of nickel-base alloys is given in Table 1. This table does not include alloys with less than 30% Ni, cast alloys, or welding products. For these and those alloys not listed, the chemical composition and applicable specifications can be found in the *Unified Numbering System for Metals and Alloys,* published by the Society of Automotive Engineers, Inc.[4]

Table 1 Nominal Chemical Composition (wt %)

Material	Ni	Cu	Fe	Cr	Mo	Al	Ti	Nb	Mn	Si	C	Other Elements
Nickel												
Nickel 200	99.6	—	—	—	—	—	—	—	0.23	0.03	0.07	—
Nickel 201	99.7	—	—	—	—	—	—	—	0.23	0.03	0.01	—
Permanickel alloy 300	98.7	—	0.02	—	—	—	0.49	—	0.11	0.04	0.29	0.38 Mg
Duranickel alloy 301	94.3	—	0.08	—	—	4.44	0.44	—	0.25	0.50	0.16	—
Nickel–Copper												
Monel alloy 400	65.4	32	1.00	—	—	—	—	—	1.0	0.10	0.12	—
Monel alloy 404	54.6	45.3	0.03	—	—	—	—	—	0.01	0.04	0.07	—
Monel alloy R-405	65.3	31.6	1.25	—	—	0.1	—	—	1.0	0.17	0.15	0.04 S
Monel alloy K-500	65.0	30	0.64	—	—	2.94	0.48	—	0.70	0.12	0.17	—
Nickel–Chromium–Iron												
Inconel alloy 600	76	0.25	8.0	15.5	—	—	—	—	0.5	0.25	0.08	—
Inconel alloy 601	60.5	0.50	14.1	23.0	—	1.35	—	—	0.5	0.25	0.05	—
Inconel alloy 690	60	—	9.0	30	—	—	—	—	—	—	0.01	—
Inconel alloy 706	41.5	0.15	40	16	—	0.20	1.8	3	0.18	0.18	0.03	—
Inconel alloy 718	53.5	0.15	18.5	19	3.0	0.5	0.9	5.1	0.18	—	0.04	—
Inconel alloy X-750	73	0.25	7	15.5	—	0.70	2.5	1	0.50	0.25	0.04	—
Nickel–Iron–Chromium												
Incoloy alloy 800	31	0.38	46	20	—	0.38	0.38	—	0.75	0.50	0.05	—
Incoloy alloy 800H	31	0.38	46	20	—	0.38	0.38	—	0.75	0.50	0.07	—
Incoloy alloy 825	42	1.75	30	22.5	3	0.10	0.90	—	0.50	0.25	0.01	—
Incoloy alloy 925	43.2	1.8	28	21	3	0.35	2.10	—	0.60	0.22	0.03	—
Pyromet 860	44	—	Bal	13	6	1.0	3.0	—	0.25	0.10	0.05	4.0 Co
Refractaloy 26	38	—	Bal	18	3.2	0.2	2.6	—	0.8	1.0	0.03	20 Co
Nickel–Iron												
Nilo alloy 36	36	—	61.5	—	—	—	—	—	0.5	0.09	0.03	
Nilo alloy 42	41.6	—	57.4	—	—	—	—	—	0.5	0.06	0.03	—
Ni-Span-C alloy 902	42.3	0.05	48.5	5.33	—	0.55	2.6	—	0.40	0.50	0.03	—
Incoloy alloy 903	38	—	41.5	—	—	0.90	1.40	2.9	0.09	0.17	0.02	14 Co
Incoloy alloy 907	37.6	0.10	41.9	—	—	1.5	—	4.70	0.05	0.08	0.02	14 Co

Nickel–Chromium–Molybdenum												
Hastelloy alloy X	Bal[a]	—	19	22	9	—	—	—	—	—	0.10	—
Hastelloy alloy G	Bal	2	19.5	22	6.5	—	—	2.1	<1	1.5	<0.05	<1 W, <2.5 Co
Hastelloy alloy C-276	Bal	—	5.5	15.5	16	—	—	—	<0.08	<1	<0.01	2.5 Co, 4 W, 0.35 V
Hastelloy alloy C	Bal	—	<3	16	15.5	—	<0.7	—	<0.08	<1	<0.01	<2 Co
Inconel alloy 617	54	—	—	22	9	1	—	—	—	—	0.07	12.5 Co
Inconel alloy 625	Bal	—	2.5	21.5	9	<0.4	<0.4	3.6	—	—	0.03	—
MAR-M-252	Bal	—	—	19	10	1	2.6	—	<0.5	<0.5	0.15	10 Co, 0.005 B
Rene′ 41	Bal	—	—	19	10	1.5	3.1	—	—	—	0.09	11 Co, <0.010 B
Rene′ 95	Bal	—	—	14	3.5	3.5	2.5	3.5	—	—	0.15	8 Co, 3.5 W, 0.01 B, 0.05 Zr
Astroloy	Bal	—	—	15	5.3	4.4	3.5	—	—	—	0.06	15 Co
Udimet 500	Bal	—	<0.5	19	4	3.0	3.0	—	—	—	0.08	18 Co, 0.007 B
Udimet 520	Bal	—	—	19	6	2.0	3.0	—	—	—	0.05	12 Co, 1 W, 0.005 B
Udimet 600	Bal	—	<4	17	4	4.2	2.9	—	—	—	0.04	16 Co. 0.02 B
Udimet 700	Bal	—	—	15	5.0	4.4	3.5	—	—	—	0.07	18.5 Co, 0.025 B
Udimet 1753	Bal	—	9.5	16.3	1.6	1.9	3.2	—	0.1	0.05	0.24	7.2 Co, 8.4 W, 0.008 B, 0.06 Zr
Waspaloy	Bal	<0.1	<2	19	4.3	1.5	3	—	—	—	0.08	14 Co, 0.006 B, 0.05 Zr

[a] Balance.

2.2 Discussion and Applications

The same grouping of alloys used in Tables 1, 2, and 3, which give chemical composition and mechanical properties, will be used for discussion of the various attributes and uses of the alloys as a group. Many of the alloy designations are registered trademarks of producer companies.

Nickel Alloys

The corrosion resistance of nickel makes it particularly useful for maintaining product purity in the handling of foods, synthetic fibers, and caustic alkalies and also in structural applications where resistance to corrosion is a prime consideration. It is a general-purpose material used when the special properties of the other nickel alloys are not required. Other useful features of the alloy are its magnetic and magnetostrictive properties, high thermal and electrical conductivity, low gas content, and low vapor pressure.[5]

Typical *nickel 200* applications are food-processing equipment, chemical shipping drums, electrical and electronic parts, aerospace and missile components, caustic handling equipment and piping, and transducers.

Nickel 201 is preferred to nickel 200 for applications involving exposure to temperatures above 316°C (600°F). Nickel 201 is used as coinage, plater bars, and combustion boats in addition to some of the applications for nickel 200.

Permanickelalloy 300 by virtue of the magnesium content is age hardenable. But, because of its low alloy content, alloy 300 retains many of the characteristics of nickel. Typical applications are grid lateral winding wires, magnetostriction devices, thermostat contact arms, solid-state capacitors, grid side rods, diaphragms, springs, clips, and fuel cells.

Duranickel alloy 301 is another age-hardenable high-nickel alloy but is made heat treatable by aluminum and titanium additions. The important features of alloy 301 are high strength and hardness, good corrosion resistance, and good spring properties up to 316°C (600°P); it is on these mechanical considerations that selection of the alloy is usually based. Typical applications are extrusion press parts, molds used in the glass industry, clips, diaphragms, and springs.

Nickel–Copper Alloys

Nickel–copper alloys are characterized by high strength, weldability, excellent corrosion resistance, and toughness over a wide temperature range. They have excellent service in seawater or brackish water under high-velocity conditions, as in propellers, propeller shafts, pump shafts, and impellers and condenser tubes, where resistance to the effects of cavitation and erosion are important. Corrosion rates in strongly agitated and aerated seawater usually do not exceed 1 mil/year.

Monel alloy 400 has low corrosion rates in chlorinated solvents, glass-etching agents, sulfuric and many other acids, and practically all alkalies, and it is resistant to stress–corrosion cracking. Alloy 400 is useful up to 538°C (1000°F) in oxidizing atmospheres, and even higher temperatures may be used if the environment is reducing. Springs of this material are used in corrosive environments up to 232°C (450°F). Typical applications are valves and pumps; pump and propeller shafts; marine fixtures and fasteners; electrical and electronic components; chemical processing equipment; gasoline and freshwater tanks; crude petroleum stills, process vessels, and piping; boiler feedwater heaters and other heat exchangers; and deaerating heaters.

Monel alloy 404 is characterized by low magnetic permeability and excellent brazing characteristics. Residual elements are controlled at low levels to provide a clean, wettable surface even after prolonged firing in wet hydrogen. Alloy 404 has a low Curie temperature

Table 2 Mechanical Properties of Nickel Alloys

Material	0.2% Yield Strength (ksi)[a]	Tensile Strength (ksi)[a]	Elongation (%)	Rockwell Hardness
Nickel				
Nickel 200	21.5	67	47	55 Rb
Nickel 201	15	58.5	50	45 Rb
Permanickel alloy 300	38	95	30	79 Rb
Duranickel alloy 301	132	185	28	36 Rc
Nickel–Copper				
Monel alloy 400	31	79	52	73 Rb
Monel alloy 404	31	69	40	68 Rb
Monel alloy R-405	56	91	35	86 Rb
Monel alloy K-500	111	160	24	25 Rc
Nickel–Chromium–Iron				
Inconel alloy 600	50	112	41	90 Rb
Inconel alloy 601	35	102	49	81 Rb
Inconel alloy 690	53	106	41	97 Rb
Inconel alloy 706	158	193	21	40 Rc
Inconel alloy 718	168	205	20	46 Rc
Inconel alloy X-750	102	174	25	33 Rc
Nickel–Iron–Chromium				
Incoloy alloy 800	48	88	43	84 Rb
Incoloy alloy 800H	29	81	52	72 Rb
Incoloy alloy 825	44	97	53	84 Rb
Incoloy alloy 925	119	176	24	34 Rc
Pyromet 860	115	180	21	37 Rc
Refractaloy 26	100	170	18	—
Nickel–Iron				
Nilo alloy 42	37	72	43	80 Rb
Ni–Span–C alloy 902	137	150	12	33 Rc
Incoloy alloy 903	174	198	14	39 Rc
Incoloy alloy 907	163	195	15	42 Rc
Nickel–Chromium–Molybdenum				
Hastelloy alloy X	52	114	43	—
Hastelloy alloy G	56	103	48.3	86 Rb
Hastelloy alloy C-276	51	109	65	—
Inconel alloy 617	43	107	70	81 Rb
Inconel alloy 625	63	140	51	96 Rb
MAR-M-252	122	180	16	—
Rene′ 41	120	160	18	—
Rene′ 95	190	235	15	—
Astroloy	152	205	16	—
Udimet 500	122	190	32	—
Udimet 520	125	190	21	—
Udimet 600	132	190	13	—
Udimet 700	140	204	17	—
Udimet 1753	130	194	20	39 Rc
Waspaloy	115	185	25	—

[a] MPa = ksi × 6.895.

Table 3 1000-h Rupture Stress (ksi)[a]

Material	1200°F	1500°F	1800°F	2000°F
Nickel–Chromium–Iron				
Inconel alloy 600	14.5	3.7	1.5	—
Inconel alloy 601	28	6.2	2.2	1.0
Inconel alloy 690	16	—	—	—
Inconel alloy 706	85	—	—	—
Inconel alloy 718	85	—	—	—
Inconel alloy X-750	68	17	—	—
Nickel–Iron–Chromium				
Incoloy alloy 800	20	—	—	—
Incoloy alloy 800H	23	6.8	1.9	0.9
Incoloy alloy 825	26	6.0	1.3	—
Pyromet 860	81	17	—	—
Refractaloy 26	65	15.5	—	—
Nickel–Chromium–Moloybdenum				
Hastelloy alloy X	31	9.5	—	—
Inconel alloy 617	52	14	3.8	1.5
Inconel alloy 625	60	7.5	—	—
MAR-M-252	79	22.5	—	—
Rene′ 41	102	29	—	—
Rene′ 95	125	—	—	—
Astroloy	112	42	8	—
Udimet 500	110	30	—	—
Udimet 520	85	33	—	—
Udimet 600	—	37	—	—
Udimet 700	102	43	7.5	—
Udimet 1753	98	34	6.5	—
Waspaloy	89	26	—	—

[a]MPa = ksi × 6.895.

and its magnetic properties are not appreciably affected by processing or fabrication. This magnetic stability makes alloy 404 particularly suitable for electronic applications. Much of the strength of alloy 404 is retained at outgassing temperatures. Thermal expansion of alloy 404 is sufficiently close to that of many other alloys as to permit the firing of composite metal tubes with negligible distortion. Typical applications are waveguides, metal-to-ceramic seals, transistor capsules, and power tubes.

Monel alloy R-405 is a free-machining material intended almost exclusively for use as stock for automatic screw machines. It is similar to alloy 400 except that a controlled amount of sulfur is added for improved machining characteristics. The corrosion resistance of alloy R-405 is essentially the same as that of alloy 400, but the range of mechanical properties differs slightly. Typical applications are water meter parts, screw machine products, fasteners for nuclear applications, and valve seat inserts.

Monel alloy K-500 is an age-hardenable alloy that combines the excellent corrosion resistance characteristics of the Monel nickel–copper alloys with the added advantage of increased strength and hardness. Age hardening increases its strength and hardness. Still better properties are achieved when the alloy is cold worked prior to the aging treatment. Alloy K-500 has good mechanical properties over a wide temperature range. Strength is maintained up to about 649°C (1200°F), and the alloy is strong, tough, and ductile at temperatures as low as −253°C (−423°F). It also has low permeability and is nonmagnetic to −134°C (−210°F). Alloy K-500 has low corrosion rates in a wide variety of environments. Typical applications are pump shafts and impellers, doctor blades and scrapers, oil-well drill collars and instruments, electronic components, and springs.

Nickel–Chromium–Iron Alloys

This family of alloys was developed for high-temperature oxidizing environments. These alloys typically contain 50–80% nickel, which permits the addition of other alloying elements to improve strength and corrosion resistance while maintaining toughness.

Inconel alloy 600 is a standard engineering material for use in severely corrosive environments at elevated temperatures. It is resistant to oxidation at temperatures up to 1177°C (2150°F). In addition to corrosion and oxidation resistance, alloy 600 presents a desirable combination of high strength and workability and is hardened and strengthened by cold working. This alloy maintains strength, ductility, and toughness at cryogenic as well as elevated temperatures. Because of its resistance to chloride ion stress–corrosion cracking and corrosion by high-purity water, it is used in nuclear reactors. For this service, the alloy is produced to exacting specifications and is designated Inconel alloy 600T. Typical applications are furnace muffles, electronic components, heat exchanger tubing, chemical- and food-processing equipment, carburizing baskets, fixtures and rotors, reactor control rods, nuclear reactor components, primary heat exchanger tubing, springs, and primary water piping. Alloy 600, being one of the early high-temperature, corrosion-resistant alloys, can be thought of as being the basis of many of our present-day special-purpose high-nickel alloys, as illustrated in Fig. 1.

Inconel alloy 601 has shown very low rates of oxidation and scaling at temperatures as high as 1093°C (2000°F). The high chromium content (nominally 23%) gives alloy 601 resistance to oxidizing, carburizing, and sulfur-containing environments. Oxidation resistance is further enhanced by the aluminum content. Typical applications are heat-treating baskets and fixtures, radiant furnace tubes, strand-annealing tubes, thermocouple protection tubes, and furnace muffles and retorts.

Inconel alloy 690 is a high-chromium nickel alloy having very low corrosion rates in many corrosive aqueous media and high-temperature atmospheres. In various types of high-temperature water, alloy 690 also displays low corrosion rates and excellent resistance to stress–corrosion cracking—desirable attributes for nuclear steam generator tubing. In addition, the alloy's resistance to sulfur-containing gases makes it a useful material for such applications as coal gasification units, burners and ducts for processing sulfuric acid, furnaces for petrochemical processing, and recuperators and incinerators.

Inconel alloy 706 is a precipitation-hardenable alloy with characteristics similar to alloy 718, except that alloy 706 has considerably improved machinability. It also has good resistance to oxidation and corrosion over a broad range of temperatures and environments. Like alloy 718, alloy 706 has excellent resistance to postweld strain age cracking. Typical applications are gas turbine components and other parts that must have high strength combined with good machinability and weldability.

Inconel alloy 718 is an age-hardenable high-strength alloy suitable for service at temperatures from −253°C (−423°F) to 704°C (1300°F). The fatigue strength of alloy 718 is high, and the alloy exhibits high stress-rupture strength up to 704°C (1300°F) as well as oxidation resistance up to 982°C (1800°F). It also offers good corrosion resistance to a wide variety of environments. The outstanding characteristic of alloy 718 is its slow response to age hardening. The slow response enables the material to be welded and annealed with no

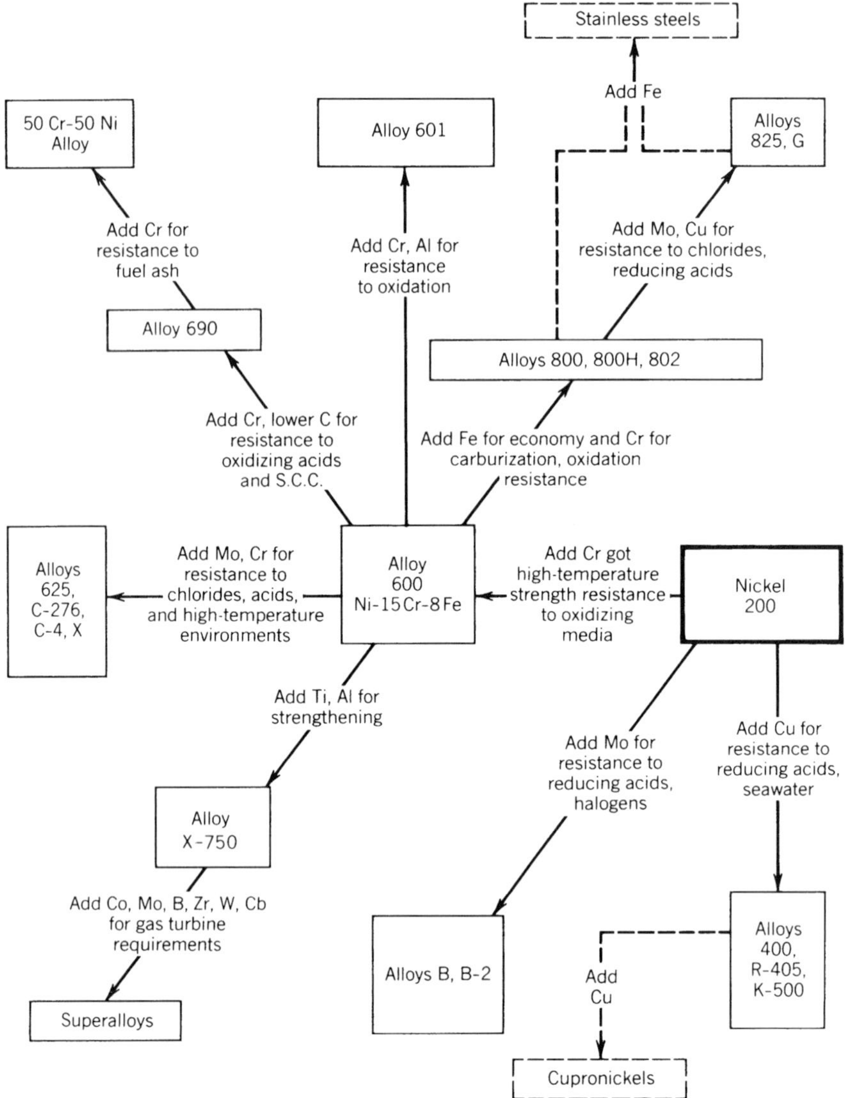

Figure 1 Some compositional modifications of nickel and its alloys to produce special properties.

spontaneous hardening unless it is cooled slowly. Alloy 718 can also be repair welded in the fully aged condition. Typical applications are jet engine components, pump bodies and parts, rocket motors and thrust reversers, and spacecraft.

Inconel alloy X-750 is an age-hardenable nickel–chromium–iron alloy used for its corrosion and oxidation resistance and high creep-rupture strength up to 816°C (1500°F). The alloy is made age hardenable by the addition of aluminum, columbium, and titanium, which combine with nickel, during proper heat treatment, to form the intermetallic compound $Ni_3(Al, Ti)$. Alloy X-750, originally developed for gas turbines and jet engines, has been adopted for a wide variety of other uses because of its favorable combination of properties. Excellent relaxation resistance makes alloy X-750 suitable for springs operating at temperatures up to about 649°C (1200°F). The material also exhibits good strength and ductility at temperatures as low as −253°C (−423°F). Alloy X-750 also exhibits high resistance to chloride ion stress–corrosion cracking even in the fully age-hardened condition. Typical applications are gas turbine parts (aviation and industrial), springs (steam service), nuclear reactors, bolts, vacuum envelopes, heat-treating fixtures, extrusion dies, aircraft sheet, bellows, and forming tools.

Nickel–Iron–Chromium Alloys

This series of alloys typically contains 30–45% Ni and is used in elevated- or high-temperature environments where resistance to oxidation or corrosion is required.

Incoloy alloy 800 is a widely used material of construction for equipment that must resist corrosion, have high strength, or resist oxidation and carburization. The chromium in the alloy imparts resistance to high-temperature oxidation and general corrosion. Nickel maintains an austenitic structure so that the alloy remains ductile after elevated-temperature exposure. The nickel content also contributes resistance to scaling, general corrosion, and stress–corrosion cracking. Typical applications are heat-treating equipment and heat exchangers in the chemical, petrochemical, and nuclear industries, especially where resistance to stress–corrosion cracking is required. Considerable quantities are used for sheathing on electric heating elements.

Incoloy alloy 800H is a version of Incoloy alloy 800 having significantly higher creep and rupture strength. The two alloys have the same chemical composition with the exception that the carbon content of alloy 800H is restricted to the upper portion of the standard range for alloy 800. In addition to a controlled carbon content, alloy 800H receives an annealing treatment that produces a coarse grain size—an ASTM number of 5 or coarser. The annealing treatment and carbon content are responsible for the alloy's greater creep and rupture strength.

Alloy 800H is useful for many applications involving long-term exposure to elevated temperatures or corrosive atmospheres. In chemical and petrochemical processing, the alloy is used in steam/hydrocarbon re-forming for catalyst tubing, convection tubing, pigtails, outlet manifolds, quenching-system piping, and transfer piping; in ethylene production for both convection and cracking tubes; in oxo-alcohol production for tubing in hydrogenation heaters; in hydrodealkylation units for heater tubing; and in production of vinyl chloride monomer for cracking tubes, return bends, and inlet and outlet flanges.

Industrial heating is another area of wide usage for alloy 800H. In various types of heat-treating furnaces, the alloy is used for radiant tubes, muffles, retorts, and assorted furnace fixtures. Alloy 800H is also used in power generation for steam superheater tubing and high-temperature heat exchangers in gas-cooled nuclear reactors.

Incoloy alloy 825 was developed for use in aggressively corrosive environments. The nickel content of the alloy is sufficient to make it resistant to chloride ion stress–corrosion cracking, and, with molybdenum and copper, alloy 825 has resistance to reducing acids. Chromium confers resistance to oxidizing chemicals. The alloy also resists pitting and intergranular attack when heated in the critical sensitization temperature range. Alloy 825 offers exceptional resistance to corrosion by sulfuric acid solutions, phosphoric acid solutions, and seawater. Typical applications are phosphoric acid evaporators, pickling-tank heaters, pickling hooks and equipment, chemical process equipment, spent nuclear fuel element recovery, propeller shafts, tank trucks, and oil-country cold-worked tubulars.

Incoloy alloy 925 was developed for severe conditions found in corrosive wells containing H_2S, CO_2, and brine at high pressures. Alloy 925 is a weldable, age-hardenable alloy having corrosion and stress–corrosion resistance similar to Incoloy alloy 825. It is recommended for applications where alloy 825 does not have adequate yield or tensile strength for service in the production of oil and gas, such as valve bodies, hanger bars, flow lines, casing, and other tools and equipment.

Pyromet 860 and *Refractaloy 26* are high-temperature precipitation-hardenable alloys with lower nickel content than Inconel alloy X-750 but with additions of cobalt and molybdenum. The precipitation-hardening elements are the same except the Al/Ti ratio is reversed with titanium content being greater than aluminum. Typical applications of both alloys are critical components of gas turbines, bolts, and structural members.[6]

Nickel–Iron

The nickel–iron alloys listed in Table 1 as a group have a low coefficient of expansion that remains virtually constant to a temperature below the Curie temperature for each alloy. A major application for *Nilo alloy 36* is tooling for curing composite airframe components. The thermal expansion characteristics of *Nilo alloy 42* are particularly useful for semiconductor lead frames and glass-sealing applications.

Ni-Span-C alloy 902 and *Incoloy alloys 903 and 907* are precipitation-hardenable alloys with similar thermal expansion characteristics to Nilo alloy 42 but having different constant coefficient of expansion temperature range. Alloy 902 is frequently used in precision apparatus where elastic members must maintain a constant frequency when subjected to temperature fluctuations. Alloys 903 and 907 are being used in aircraft jet engines for members requiring high-temperature strengths to 649°C (1200°F) With thermal expansion controlled to maintain low clearance.

Nickel–Chromium–Molybdenum Alloys

This group of alloys contains 45–60% Ni and was developed for severe corrosion environments. Many of these alloys also have good oxidation resistance and some have useful strength to 1093°C (2000°F).

Hastelloy alloy X is a non-age-hardenable nickel–chromium–iron–molybdenum alloy developed for high-temperature service up to 1204°C (2200°F). Typical applications are furnace hardware subjected to oxidizing, reducing, and neutral atmospheres; aircraft jet engine tail pipes; and combustion cans and afterburner components.[7,8]

Hastelloy alloy C is a mildly age-hardenable alloy similar in composition to alloy X except nearly all the iron is replaced with molybdenum and nickel. It is highly resistant to strongly oxidizing acids, salts, and chlorine. It has good high-temperature strength. Typical applications are chemical, petrochemical, and oil refinery equipment; aircraft jet engines; and heat-treating equipment.[9,10]

Hastelloy alloy C-276 is a modification of Hastelloy alloy C where the carbon and silicon contents are reduced to very low levels to diminish carbide precipitation in the heat-affected zone of weldments. Alloy C-276 is non–age hardenable and is used in the solution-treated condition. No postwelding heat treatment is necessary for chemical process equipment. Typical applications are chemical and petrochemical process equipment, aircraft jet engines, and deep sour gas wells.[9,10]

Hastelloy alloy G is a non-age-hardenable alloy similar to the composition of alloy X but with 2% copper and 2% columbium and lower carbon content. Alloy G is resistant to pitting and stress–corrosion cracking. Typical applications are paper and pulp equipment, phosphate fertilizer, and synthetic fiber processing.[9,10]

Inconel alloy 617 is a solid-solution-strengthened alloy containing cobalt that has an exceptional combination of high-temperature strength and oxidation resistance which makes alloy 617 a useful material for gas turbine aircraft engines and other applications involving exposure to extreme temperatures, such as steam generator tubing and pressure vessels for advanced high-temperature gas-cooled nuclear reactors.

Inconel alloy 625, like alloy 617, is a solid-solution-strengthened alloy but containing columbium instead of cobalt. This combination of elements is responsible for superior resistance to a wide range of corrosive environments of unusual severity as well as to high-temperature effects such as oxidation and carburization. The properties of alloy 625 that make it attractive for seawater applications are freedom from pitting and crevice corrosion, high corrosion fatigue strength, high tensile strength, and resistance to chloride ion stress–corrosion cracking. Typical applications are wire rope for mooring cables; propeller blades; submarine propeller sleeves and seals; submarine snorkel tubes; aircraft ducting, exhausts, thrust-reverser, and spray bars; and power plant scrubbers, stack liners, and bellows.

MAR-M-252, Rene' 41, *Rene'* 95, and *Astroloy* are a group of age-hardenable nickel-base alloys containing 10–15% cobalt designed for highly stressed parts operating at temperatures from 871 to 982°C (1600–1800°F) in jet engines. MAR-M-252 and Rene' 41 have nearly the same composition but Rene' 41 contains more of the age-hardening elements allowing higher strengths to be obtained. Rene' 95, of similar base composition but in addition containing 3.5% columbium and 3.5% tungsten, is used at temperatures between 371 and 649°C (700 and 1200°F). Its primary use is as disks, shaft retaining rings, and other rotating parts in aircraft engines of various types.[6,8,9]

Udimet 500, 520, 600, and 700 and *Unitemp* 1753 are age-hardenable, nickel-base alloys having high strength at temperatures up to 982°C (1800°F). All contain a significant amount of cobalt. Applications include jet engine gas turbine blades, combustion chambers, rotor disks, and other high-temperature components.[6,8,9]

Waspaloy is an age-hardenable nickel-base alloy developed to have high strength up to 760°C (14,000°F) combined with oxidation resistance to 871°C (1600°F). Applications are jet engine turbine buckets and disks, air frame assemblies, missile systems, and high-temperature bolts and fasteners.[6,8,9]

3 CORROSION

It is well recognized that the potential saving is very great by utilizing available and economic practices to improve corrosion prevention and control. Not only should the designer consider initial cost of materials, but he or she should also include the cost of maintenance, length of service, downtime cost, and replacement costs. This type of cost analysis can frequently show that more highly alloyed, corrosion-resistant materials are more cost effective. The National Commission on Materials Policy concluded that one of the "most obvious opportunities for material economy is control of corrosion."

Studies have shown that the total cost of corrosion is astonishing. The overall cost of corrosion is the United States was estimated by a study funded by the U.S. Federal Highway Administration. The report was published by NACE in July 2002. According to the report, metallic corrosion costs the United States about $276 billion a year. The report claims that about 25–30% of the costs of corrosion ($80 billion) is avoidable and could be saved by broader use of corrosion-resistant materials and the application of best anticorrosion technology from design through maintenance.

Since becoming commercially available shortly after the turn of the century, nickel has become very important in combating corrosion. It is a major constituent in the plated coatings and claddings applied to steel, corrosion-resistant stainless steels, copper–nickel and nickel–

copper alloys, high-nickel alloys, and commercially pure nickel alloys. Not only is nickel a corrosion-resistant element in its own right, but, owing to its high tolerance for alloying, it has been possible to develop many metallurgically stable, special-purpose alloys.[11]

Figure 1 shows the relationship of these alloys and the major effect of alloying elements. Alloy 500 with 15% chromium, one of the earliest of the nickel–chromium alloys, can be thought of as the base for other alloys. Chromium imparts resistance to oxidizing environments and high-temperature strength. Increasing chromium to 30%, as in alloy 690, increases resistance to stress–corrosion cracking, nitric acid, steam, and oxidizing gases. Increasing chromium to 50% increases resistance to melting sulfates and vanadates found in fuel ash. High-temperature oxidation resistance is also improved by alloying with aluminum in conjunction with high chromium (e.g., alloy 601). Without chromium, nickel by itself is used as a corrosion-resistant material in food processing and in high-temperature caustic and gaseous chlorine or chloride environments.

Of importance for aqueous reducing acids, oxidizing chloride environments, and seawater are alloy 625 and alloy C-276, which contain 9 and 16% molybdenum, respectively, and are among the most resistant alloys currently available. Low-level titanium and aluminum additions provide V strengthening while retaining good corrosion resistance, as in alloy X-750. Cobalt and other alloying element additions provide jet engine materials (superalloys) that combine high-temperature strength with resistance to gaseous oxidation and sulfidation.

Another technologically important group of materials are the higher iron alloys, which were originally developed to conserve nickel and are often regarded as intermediate in performance and cost between nickel alloys and stainless steels. The prototype, alloy 800 (Fe–33% Ni–21% Cr), is a general-purpose alloy with good high-temperature strength and resistance to steam and oxidizing or carburizing gases. Alloying with molybdenum and chromium, as in alloy 825 and alloy G, improves resistance to reducing acids and localized corrosion in chlorides.

Another important category is the nickel–copper alloys. At the higher nickel end are the Monel alloys (30–45% Cu, balance Ni) used for corrosive chemicals such as hydrofluoric acid and severe marine environments. At the higher copper end are the cupronickels (10–30% Ni, balance Cu), which are widely used for marine applications because of their fouling resistance.

Nickel alloys exhibit high resistance to attack under nitriding conditions (e.g., in dissociated ammonia) and in chlorine or chloride gases. Corrosion in the latter at elevated temperatures proceeds by the formation and volatilization of chloride scales, and high-nickel contents are beneficial since nickel forms one of the least volatile chlorides. Conversely, in sulfidizing environments, high-nickel alloys without chromium can exhibit attack due to the formation of a low-melting-point $Ni–Ni_3Si_2$ eutectic. However high chromium contents appear to limit this form of attacks.

Friend explains corrosion reactions as wet or dry (Ref. 10, pp. 3–5):

> The term wet corrosion usually refers to all forms of corrosive attack by aqueous solutions of electrolytes, which can range from pure water (a weak electrolyte) to aqueous solutions of acids or bases or of their salts, including neutral salts. It also includes natural environments such as the atmosphere, natural waters, soils, and others, irrespective or whether the metal is in contact with a condensed film or droplets of moisture or is completely immersed. Corrosion by aqueous environments is electrochemical in nature, assuming the presence of anodic and cathodic areas on the surface of the metal even though these areas may be so small as to be indistinguishable by experimental methods and the distance between them may be only of atomic dimensions.
>
> The term dry corrosion implies the absence of water or an aqueous solution. It generally is applied to metal/gas or metal/vapor reactions involving gases such as oxygen, halogens, hydrogen sulfide, and sulfur vapor and even to "dry" steam at elevated temperatures. ... High-temperature oxidation of metals has been considered to be an electrochemical phenomenon since it involves the diffusion

of metal ions outward, or of reactant ions inward, through the corrosion product film, accompanied by a flow of electrons.

The decision to use a particular alloy in a commercial application is usually based on past corrosion experience and laboratory or field testing using test spools of candidate alloys. Most often weight loss is measured to rank various alloys; however, many service failures are due to localized ttack such as pitting, crevice corrosion, intergranular corrosion, and stress–corrosion cracking, which oust be measured by other means.

A number of investigations have shown the effect of nickel on the different forms of corrosion. Figure 2 shows the galvanic series of many alloys in flowing seawater. This series gives an indication of the rate of corrosion between different metals or alloys when they are electrically coupled in an electrolyte. The metal close to the active end of the chart will behave as an anode and corrode, and the metal closer to the noble end will act as a cathode and be protected. Increasing the nickel content will move an alloy more to the noble end of the series. There are galvanic series for other corrosive environments, and the film-forming characteristics of each material may change this series somewhat. Seawater is normally used as a rough guide to the relative positions of alloys in solution of good electrical conductivity such as mineral acids or salts.

Residual stresses from cold rolling or forming do not have any significant effect on the general corrosion rate. However, many low-nickel-containing steels are subject to stress–corrosion cracking in chloride-containing environments. Figure 3 from work by LaQue and Copson[11] shows that nickel–chromium and nickel–chromium–iron alloys containing about 45% Ni or more are immune from stress–corrosion cracking in boiling 42% magnesium chloride.[10]

When localized corrosion occurs in well-defined areas, such corrosion is commonly called *pitting attack.* This type of corrosion typically occurs when the protective film is broken or is penetrated by a chloride–ion and the film is unable to repair itself quickly. The addition of chromium and particularly molybdenum makes nickel-base alloys less susceptible to pitting attack, as shown in Fig. 4, which shows a very good relationship between critical[11] pitting temperature in a salt solution. Along with significant increases in chromium and/or molybdenum, the iron content must be replaced with more nickel in wrought alloys to resist the fonnation of embrittling phases.[11,12]

Air *oxidation* at moderately high temperatures will form an intermediate subsurface layer between the alloy and gas quickly. Alloying of the base alloy can affect this subscale oxide and, therefore, control the rate of oxidation. At constant temperature, the resistance to oxidation is largely a function of chromium content. Early work by Eiselstein and Skinner has shown that nickel content is very beneficial under cyclic temperature conditions, as shown in Fig. 5.[13]

4 FABRICATION

The excellent ductility and malleability of nickel and nickel-base alloys in the annealed condition make them adaptable to virtually all methods of cold fabrication. As other engineering properties vary within this group of alloys, formability ranges from moderately easy to difficult in relation to other materials.

4.1 Resistance to Deformation

Resistance to deformation, usually expressed in terms of hardness or yield strength, is a primary consideration in cold forming. Deformation resistance is moderately low for the

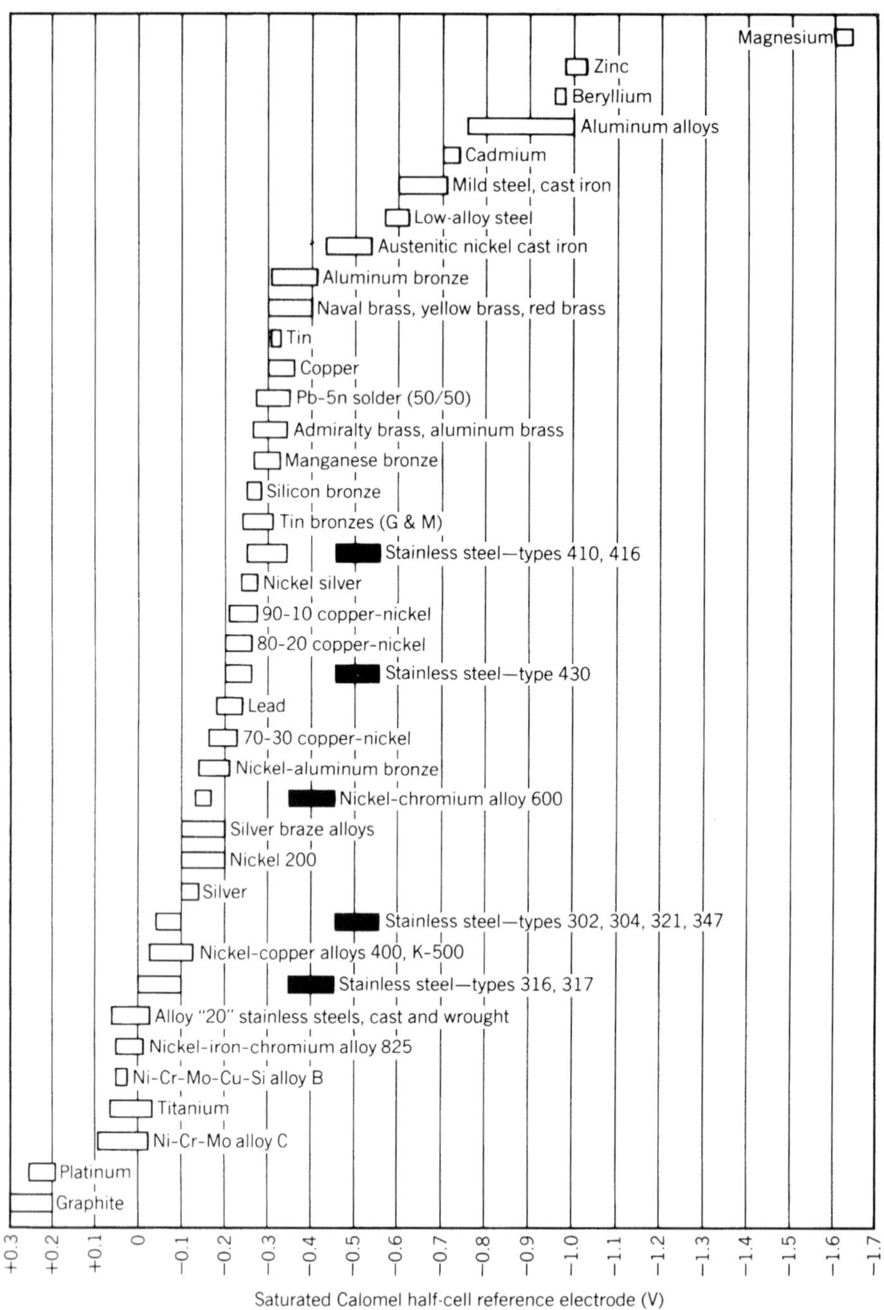

Figure 2 Corrosion potentials in flowing seawater (8–13 ft/s), temperature range 50–80°F. Alloys are listed in the order of the potential they exhibit in flowing seawater. Certain alloys, indicated by solid boxes, in low-velocity or poorly aerated water and at shielded areas may become active and exhibit a potential near −0.5 V.

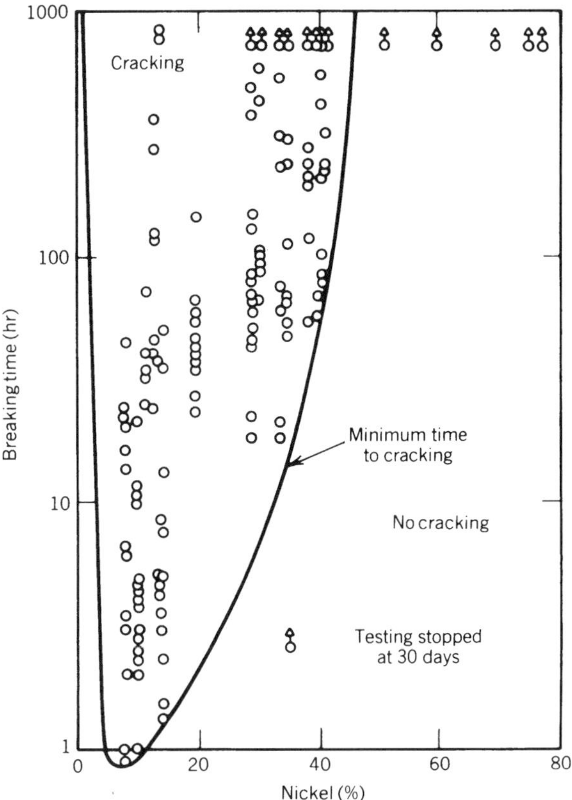

Figure 3 Breaking time of iron–nickel–chromium wires under tensile stress in boiling 42% magnesium chloride.

nickel and nickel–copper systems and moderately high for the nickel–chromium and nickel–iron–chromium systems. However, when properly annealed, even the high-strength alloys have a substantial range between yield and ultimate tensile strength. This range is the plastic region of the material and all cold forming is accomplished within the limits of this region. Hence, the high-strength alloys require only stronger tooling and more powerful equipment for successful cold forming. Nominal tensile properties and hardnesses are given in Table 2.

4.2 Strain Hardening

A universal characteristic of the high-nickel alloys is that they have face-centered-cubic crystallographic structures and, consequently are subject to rapid strain hardening. This characteristic is used to advantage in increasing the room temperature tensile properties and hardness of alloys that otherwise would have low mechanical strength or in adding strength to those alloys that are hardened by a precipitation heat treatment. Because of this increased strength, large reductions can be made without rupture of the material. However, the number of reductions in a forming sequence will be limited before annealing is required, and the percentage reduction in each successive operation must be reduced.

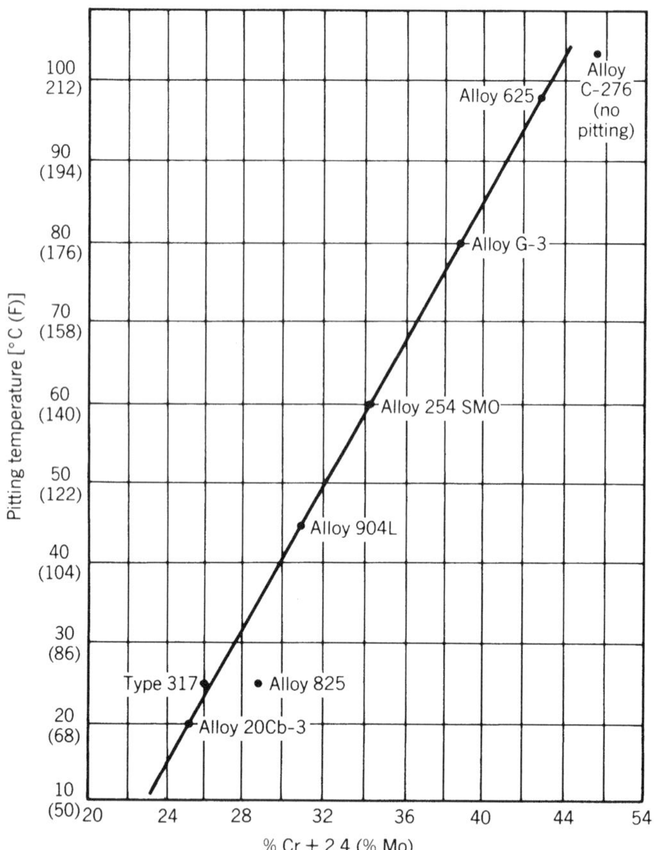

Figure 4 Critical temperature for pitting in 4% NaCl + 1% $Fe_2\ (SO_4)_3$ + 0.01 *M* HCl versus composition for Fe–Ni–Cr–Mo alloys.

Since strain hardening is related to the solid-solution strengthening of alloying elements, the strain-hardening rate generally increases with the complexity of the alloy. Accordingly, strain-hardening rates range from moderately low for nickel and nickel–copper alloys to moderately high for nickel–chromium and nickel–iron–chromium alloys. Similarly, the age-hardenable alloys have higher strain-hardening rates than their solid-solution equivalents. Figure 6 compares the strain-hardening rates of some nickel alloys with those of other materials as shown by the increase in hardness with increasing cold reduction.

Laboratory tests have indicated that the shear strength of the high-nickel alloys in double shear averages about 65% of the ultimate tensile strength (see Table 4). These values, however, were obtained under essentially static conditions using laboratory testing equipment having sharp edges and controlled clearances. Shear loads for well-maintained production equipment can be found in Table 5. These data were developed on a power shear having a 31-mm/m (3/8-in./ft) rake.

5 HEAT TREATMENT

High-nickel alloys are subject to surface oxidation unless heating is performed in a protective atmosphere or under vacuum. A protective atmosphere can be provided either by controlling

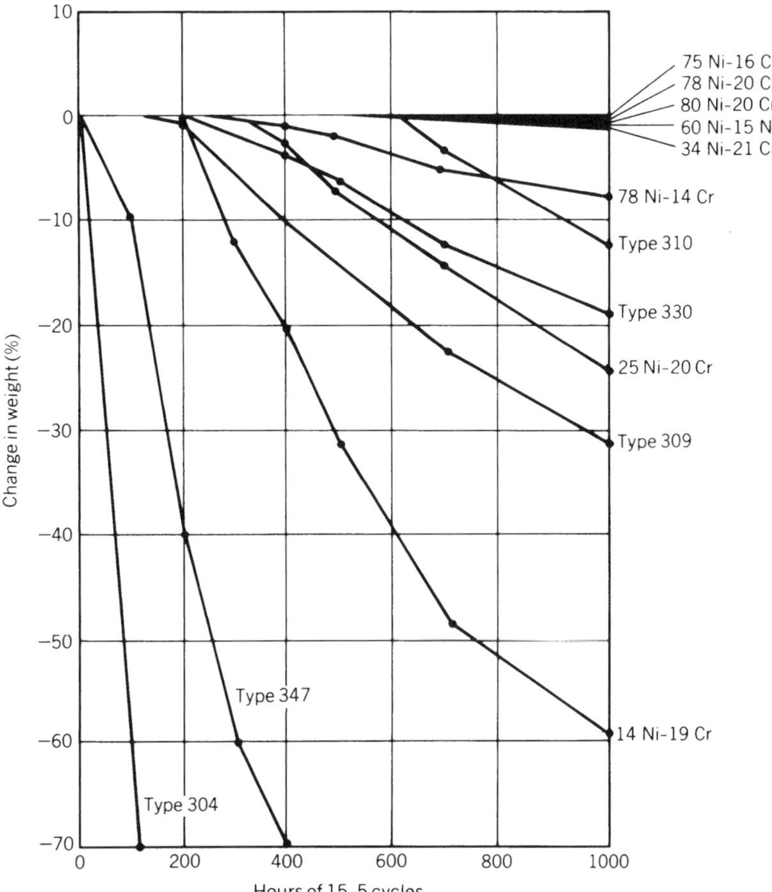

Figure 5 Effect of nickel content on air oxidation of alloys. Each cycle consisted of 15 min at 1800°F followed by a 5-min air cooling.

the ratio of fuel and air to minimize oxidation or by surrounding the metal being heated with a prepared atmosphere.

Monel alloy 400, Nickel 200, and similar alloys will remain bright and free from discoloration when heated and cooled in a reducing atmosphere formed by the products of combustion. The alloys that contain chromium, aluminum, or titanium form thin oxide films in the same atmosphere and, therefore, require prepared atmospheres to maintain bright surfaces.

Regardless of the type of atmosphere used, it must be free of sulfur. Exposure of nickel alloys to sulfur-containing atmospheres at high temperatures can cause severe sulfidation damage.

The atmosphere of concern is that in the immediate vicinity of the work, that is, the combustion gases that actually contact the surface of the metal. The true condition of the atmosphere is determined by analyzing gas samples taken at various points about the metal surface.

Furnace atmospheres can be checked for excessive sulfur by heating a small test piece of the material, for example, 13-mm- (½-in.-) diameter rod or 13 mm × 25 mm (½ in. ×

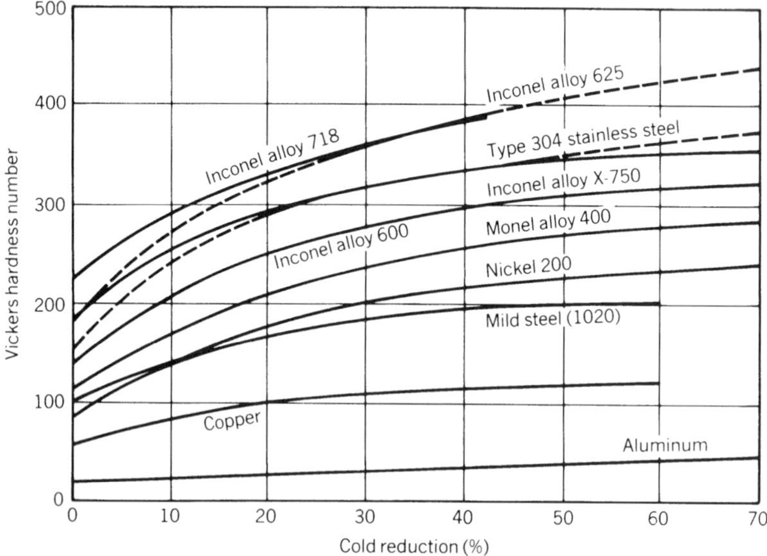

Figure 6 Effect of cold work on hardness.

1 in.) flat bar, to the required temperature and holding it at temperature for 10–15 min. The piece is then air cooled or water quenched and bent through 180° flat on itself. If heating conditions are correct, there will be no evidence of cracking.

5.1 Reducing Atmosphere

The most common protective atmosphere used in heating the nickel alloys is that provided by controlling the ratio between the fuel and air supplied to the burners. A suitable reducing

Table 4 Strength in Double Shear of Nickel and Nickel Alloys

Alloy	Condition	Shear Strength (ksi)[a]	Tensile Strength (ksi)	Hardness
Nickel 200	Annealed	52	68	46 Rb
	Half hard	58	79	84 Rb
	Full hard	75	121	100 Rb
Monel alloy 400	Hot rolled, annealed	48	73	65 Rb
	Cold rolled, annealed	49	76	60 Rb
Inconel alloy 600	Annealed	60	85	71 Rb
	Half hard	66	98	98 Rb
	Full hard	82	152	31 Rc
Inconel alloy X-750	Age hardened[b]	112	171	36 Re

[a] MPa = ksi × 6.895.
[b] Mill annealed and aged 1300°F (750°C)/20 h.

Table 5 Shear Load for Power Shearing of 6.35-mm (0.250-in.) Gauge Annealed Nickel Alloys at 31 mm/m (3/8 in./ft.) Rake as Compared with Mild Steel

Alloy	Tensile Strength (ksi)[a]	Hardness (Rb)	Shear Load (lb)[b]	Shear Load in Percent of Same Gauge of Mild Steel
Nickel 200	60	60	61,000	200
Monel alloy 400	77	75	66,000	210
Inconel alloy 600	92	79	51,000	160
Inconel alloy 625	124	95	55,000	180
Inconel alloy 718	121	98	50,000	160
Inconel alloy X-750	111	88	57,000	180
Mild steel	50	60	31,000	100

[a] MPa = ksi × 6.895.
[b] kg = lb × 0.4536.

condition can be obtained by using a slight excess of fuel so that the products of combustion contain at least 4%, preferably 6%, of carbon monoxide plus hydrogen. The atmosphere should not be permitted to alternate from reducing to oxidizing; only a slight excess of fuel over air is needed.

It is important that combustion take place before the mixture of fuel and air comes into contact with the work, otherwise the metal may be embrittled. To ensure proper combustion, ample space should be provided to burn the fuel completely before the hot gases contact the work. Direct impingement of the flame can cause cracking.

5.2 Prepared Atmosphere

Various prepared atmospheres can be introduced into the heating and cooling chambers of furnaces to prevent oxidation of nickel alloys. Although these atmospheres can be added to the products of combustion in a directly fired furnace, they are more commonly used with indirectly heated equipment. Prepared protective atmospheres suitable for use with the nickel alloys include dried hydrogen, dried nitrogen, dried argon or any other inert gas, dissociated ammonia, and cracked or partially reacted natural gas. For the protection of pure nickel and nickel–copper alloys, cracked natural gas should be limited to a dew point of −1–4°C (30–40°F).

Figure 7 indicates that at a temperature of 1093°C (2000°F), a hydrogen dew point of less than −30°C (−20°F) is required to reduce chromium oxide to chromium; at 815°C (1500°F) the dew point must be below −50°C (−60°F). The values were derived from the thermodynamic relationships of pure metals with their oxides at equilibrium and should be used only as a guide to the behavior of complex. alloys under nonequilibrium conditions. However, these curves have shown a close correlation with practical experience. For example, Inconel alloy 600 and Incoloy alloy 800 are successfully bright annealed in hydrogen having a dew point of −35 to −40°C (−30 to −40°F).

As indicated in Fig. 7, lower dew points are required as the temperature is lowered. To minimize oxidation during cooling, the chromium-containing alloys must be cooled rapidly in a protective atmosphere.

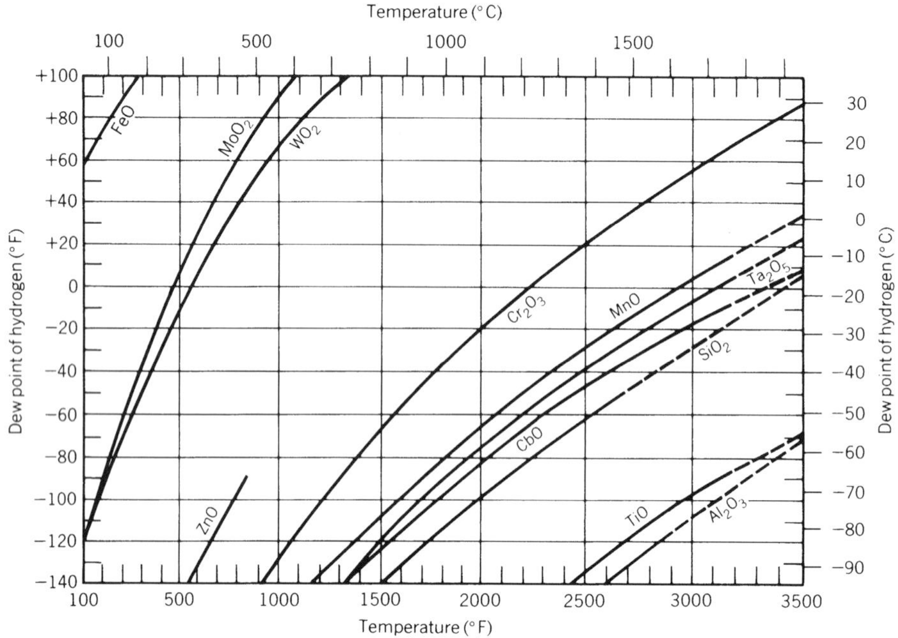

Figure 7 Metal/metal oxide equilibria in hydrogen atmospheres.

6 WELDING

Nickel alloys are characterized by good weldability by a variety of common joining processes. Thin sheet is generally joined without filler metal (autogenous) using gas–tungsten arc welding (GTAW) or plasma arc welding (PAW) processes. High-speed welding, using the GTAW process, requires magnetic arc deflection to maintain the arc in proper relation to the torch. While thin gauges can be welded at high speeds, the weld grain structure will be coarse and not exhibit the highest ductility. For optimum ductility, travel speed must be slow enough to produce an elliptical weld puddle rather the tear-drop shape which results from high-speed welding.

When filler metals are used, they usually are overmatching in composition to minimize galvanic corrosion effects in the weld metal. To minimize the amount of segregation which occurs in the fusion zone of the weld deposit, low heat input should be incorporated into the welding process when possible to optimize the corrosion resistance of the weld. Interpass weld temperature should not exceed 300°F (150°C) when joining thicker sections. Quality weld joints require careful consideration of the compatibility of weld metal with material being joined, the process used, dilution effects, joint design, service conditions, and the joint properties required. It is a good idea to involve a welding engineer where critical component requirements exist.

7 MACHINING

Nickel and nickel-base alloys can be machined by the same techniques used for iron-base alloys. However, higher loads will be imparted to the tooling requiring heavy-duty equipment to withstand the load and coolants to dissipate the heat generated. The cutting tool edge must be maintained sharp and have the proper geometry.

Table 6 Registered Trademarks of Producer Company

Trademark	Owner
Duranickel	Special Metals Corporation
Hastelloy	Haynes International
Incoloy	Special Metals Corporation
Inconel	Special Metals Corporation
MAR-M	Martin Marietta Corporation
Monel	Special Metals Corporation
Nilo	Special Metals Corporation
Ni-Span-C	Special Metals Corporation
Permanickel	Special Metals Corporation
Pyromet	Carpenter Technology Corporation
Rene	General Electric Company
Rene' 41	Allvac Metals Corporation
Udimet	Special Metals Corporation
Waspaloy	United Aircraft Corporation

8 CLOSURE

It has not been possible to give the composition of and discuss each commercial alloy and, therefore, one should refer to publications like Refs. 5, 7, and 8 for alloy listings, which are revised periodically to include the latest alloys available. (See Table 6 for the producer companies of some of the alloys mentioned in this chapter.)

REFERENCES

1. Joseph R. Boldt, Jr., *The Winning of Nickel,* Van Nostrand, New York, 1967.
2. *Nickel and Its Alloys,* NBS Monograph 106, May 1968.
3. Bassford, T. H., and Hosier, J., Nickel and Its Alloys, in M. Kutz (Ed.) *Mechanical Engineers' Handbook,* 2nd ed., Wiley, New York, 1989, pp. 71–89.
4. *Unified Numbering System for Metals and Alloys,* 9th ed., Society of Automotive Engineers, Warrendale, PA, and Conshohocken, PA, 2001.
5. *Alloy Handbooks and Bulletins,* Special Metals Corporation Publication, Huntington, WV, 2005.
6. *ASM Handbook,* 10th ed., Vol. 2, *Properties and Selection: Nonferrous Alloys and Special-Purpose Materials,* ASM International, Materials Park, OH, 1990.
7. *Hastelloy Alloy X Technical Bulletin,* Haynes International, Kokomo, IN.
8. *Alloy Digest,* ASM International, Materials Park, OH, 2003.
9. *Aerospace Structural Metals Handbook,* CINDA/USAF CRDA Handbook Operation, Purdue University, West Lafayette, IN, 2000.
10. W. Z. Friend, *Corrosion of Nickel and Nickel-Base Alloy,* Wiley, New York, 1980, pp. 3–5.
11. F. L. LaQue and H. R. Copson, *Corrosion Resistance of Metals and Alloys,* 2nd ed., Reinhold, New York, 1963.
12. J. Kolts et al., "Highly Alloyed Austenitic Materials for Corrosion Service," *Metal Prog.,* September 1983, pp. 25–36.
13. *High Temperature Corrosion in Refinery and Petrochemical Service,* Inco Publication, 1960.

CHAPTER 7
MAGNESIUM AND ITS ALLOYS

Robert E. Brown
Magnesium Monthly Review
Prattville, Alabama

1 INTRODUCTION

Magnesium, with a density of 1.74 g/cm^3, is the lightest of all structural metals. It is used mainly as an alloyed material in many forms, including castings, forgings, extrusions, rolled sheet, and plate. It is a plentiful element, representing 2.7% of the earth's crust. Magnesium is not found in metallic form but occurs mainly in nature as carbonates, dolomite, and magnesite. It is also found as carnallite salts in brines or in salt lakes in many areas of the world. A major source of magnesium is in seawater. There are 6 million tons of magnesium in 1 cubic mile of seawater.

Incorrect ideas prevail as to the fire risk and inflammability of magnesium. Many people first encounter magnesium metal as a powder or ribbon and assume that it is readily inflammable and hazardous to handle. This is not true as solid magnesium, such as in finished parts and coarse scrap, is completely harmless. A match cannot ignite it, and only if heated above the melting point will the metal burn, the solid remaining unburnt (Ref. 1, p. 1).

There are two major methods of producing magnesium, electrolytic and thermal reduction. The main electrolytic method of reduction is by electrolysis of magnesium chloride. Thermal reduction is principally achieved by reducing magnesium oxide by ferrosilicon. The world's largest magnesium-producing country is China, and it produces magnesium by the Pidgeon process, which uses ferrosilicon to reduce calcined dolomite.

The total production of magnesium metal in all areas of the world is relatively small. In 2003, estimated world production was about 500,000 metric tons.

2 USES

Magnesium is used in many forms and in many applications. However, the largest percent of primary magnesium produced each year is as an alloying element in aluminum, where it is added to improve strength and corrosion resistance. The next largest use is in die castings, most of which go to automotive uses. Desulfurization of iron and steel is a large user of magnesium also. A much smaller percent of the use goes to many other uses, including forging, sheet and plate, and sand casting.

2.1 Nonstructural Applications

Magnesium occupies a very high place in the electromotive series. This property allows magnesium to be used as a sacrificial anode to protect steel from corrosion. As an example, magnesium is used to protect buried pipelines and to protect the inside of household hot-water heaters. It is also used for protection of ship hulls, ballast tanks, seawater condensers, and steel pilings in marine environments. Alloys used for this service are either cast or extruded.

Magnesium is added to gray cast iron to produce nodular or ductile iron. Magnesium causes the carbon flakes in the cast iron to draw up into balls, thus giving the castings fewer breakage planes and strength that is equivalent of steel.

Magnesium, in powder or granule form, is added to molten iron to reduce the sulfur content prior to its conversion to steel. This improves the strength and toughness of the final steel products. Magnesium also is used as a deoxidizer or scavenger in the manufacture of copper-base alloys, such as brass and bronze, and the manufacture of nickel alloys. Magnesium is used with calcium to remove bismuth from lead.

Magnesium is also used as reducing agent in the production of titanium, zirconium, and beryllium. Very pure magnesium is used to reduce UF_4 into uranium metal.

Magnesium powder is also used to produce the Grignard reagents, an organic or organometallic intermediate used to produce fine chemicals and pharmaceuticals.

Magnesium sheet and plate are used for photoengraving and for large printing plates for fine-quality printing. Magnesium is also used for the construction of batteries, both dry cell and reserve-cell types such as seawater-activated cells.

2.2 Structural Applications

Magnesium has many attractive properties that have enabled the metal to be used in many structural applications. Its extreme lightness alone makes it attractive in all parts that are moved or lifted during their manufacture or use. Low inertia, which results from the low density of magnesium, is especially advantageous in rapidly moving parts. In addition, the low density of magnesium allows thicker sections to be used in parts, eliminating the need for a large amount of stiffening and simplifying the part and its manufacture (Ref. 2, p. 4).

Magnesium alloys have other attractive properties, such as elevated-temperature properties, for aircraft and missile applications, fatigue strength in wheels, and damping in electronic housings for aircraft and missiles as well as dimensional stability in electronic housings and in jigs and fixtures, machinability in tooling plate, dent resistance in magnesium sheet used for luggage, alkaline resistance to concrete, low resistance to the passage of X-rays, and thermal neutrons in X-ray cassettes and nuclear fuel cans (Ref. 2, p. 4).

Magnesium is easily fabricated by most traditional metal-forming processes, including casting of all types, forging, extrusion, rolling, and injection molding (Thixocasting). The base forms are transformed into finished products by machining, forming, and joining. Finishing for protective or decorative purposes is by chemical conversion coatings, painting, or electroplating.

Those properties mainly significant for structural applications are density (automotive and aerospace vehicle parts; portable tools such as chain saws; containers such as for computers, cameras, and briefcases; sports equipment such as catcher's masks and archery bows), high damping capacity (antivibration platforms for electronic equipment; walls for sound attenuation), excellent machinability (jigs and fixtures for manufacturing processes), and high corrosion resistance in an alkaline environment (cement tools).

Weight reduction of vehicles is an issue for many carmakers. Magnesium helps give a lighter weight car with improved fuel economy and reduced emissions regardless of the propulsion system used. The major parts made of magnesium alloys include instrument panel support beams (dashboards), steering wheels, steering columns, four-wheel drive gearboxes, and various support brackets and pieces. The most successful automobile ever built was the Volkswagen Beetle, with 21,000,000 cars sold. This car used magnesium for the crankcase and transmission housing and other parts averaging 22 kg per car or over 450,000 tons of magnesium during its life.

3 ALLOYS AND PROPERTIES

Many alloys have been developed to provide a range of properties and characteristics to meet the needs of a wide variety of applications. The most frequently used are listed in Table 1. There are two major classes—those containing aluminum as the principal alloying ingredient, others containing more exotic metals for alloying. Those containing aluminum

Table 1 Magnesium Alloys in Common Use

ASTM Designation	Ag	Al	Fe max	Mn	Ni max	Rare Earth	Si	Zn	Zr	Forms
AM50A		4.9	0.004	0.32	0.002			0.22		DC
AM60B		6.0	0.005	0.42	0.002			0.22 max.		DC
AS41B		4.2	0.0035	0.52	0.002		1.0	0.12		DC
AZ91D		9	0.005	0.33	0.002			0.7		DC
ACM522		5.3		0.17		2.6	E			DC
AJ52A		5		0.38		C		0.20		DC
AJ62A		6		0.38		D		0.20		DC
AZ31B		3	0.005	0.6	0.005			1		S, P, F, E
AZ61A		6.5	0.005	0.33	0.005			0.9		F, E
AZ80A		8.5	0.005	0.31	0.005			0.5		F, E
AZ81A		7.6		0.24				0.7		SC, PM, IC
AZ91E		9	0.005	0.26	0.0010			0.7		SC, PM
EZ33A						3.2		2.5	0.7	SC, PM
K1A									0.7	SC, PM
M1A				1.6						E
QE22A	2.5					2.2			0.7	SC, PM, IC
WE43A			0.01	0.15	0.005	A		0.20	0.7	SC, PM, IC
WE54A				0.15	0.005	B			0.7	SC, PM, IC
ZE41A				0.15		1.2		4.2	0.7	SC, PM, IC
ZE63A						2.6		5.8	0.7	SC, PM, IC
ZK40A								4	0.7	E
ZK60A								5.5	0.7	F, E

A = 4 Yttrium; 3 RE. C = 2.0 Strontium. E = 2.0 Calcium.
B = 5.1 Yttrium; 4 RE. D = 2.5 Strontium.
DC = die casting; E = extrusion; F = forging; IC = investment casting; P = plate; PM = permanent mold; S = sheet; SC = sand casting

are strong and ductile and have excellent resistance to atmospheric corrosion. Zirconium is an excellent grain refiner for magnesium but cannot be used with aluminum. Within this class, those alloys containing rare earth elements or yttrium are especially suited for applications at higher temperatures up to 300°C. Those that do not have zinc as a principal alloying ingredient and are strong and ductile.[3]

3.1 Mechanical Properties of Castings

The largest and most rapidly growing structural use of magnesium alloys is in die casting, and most die castings are for the automotive industry in all areas of the world. Both hot- and cold-chamber die-casting processes are being used to produce automotive parts of all sizes and shapes. Weight limits are usually restricted to parts with net weights under 25 lb. This method is frequently the most economical to produce a given part and is especially effective in producing parts with very thin sections. The acceptance of magnesium die casting has been greatly assisted by the development of high-purity Mg–Al alloys with very low impurity levels, i.e., Fe + Ni + Cu less than 0.005%. This low level of impurities enables magnesium alloy components to withstand standard salt spray corrosion testing with results equal to or better than those of competitive aluminum alloys.

AM 60B and AM 50A are being used for their higher elongation. These alloys are used for the instrument panel support beam (dashboard). This part extends from the right-door pillar to the left-door pillar; hence it is a vital structural part of the automobile frame.

New creep-resistant magnesium alloys have been developed to expand the use of magnesium in vehicles. In a car the front end is the most important with respect to weight saving. If magnesium parts such as a crankcase or transmission can be converted to magnesium, much of the weight can be reduced. Crankcases, oil pans, and transmissions all operate at temperatures over 150°C.

Development of sand-cast magnesium alloys that can operate effectively at higher temperatures has been ongoing for many years. These high-temperature sand-cast alloys are used for helicopter transmission housings and jet engine gearboxes. Most of these alloys are based on adding rare earths, zirconium, and yttrium. These alloys are not compatible with the high-pressure die-casting process.

Newer heat-resistant and high-temperature creep-resistant alloys for die casting are being developed using several new types of compositions. Noranda has a new alloy based on Mg–Al–Sr–Ca alloys. The Noranda alloy composition contains 5% Al–2% Sr (AJ52X, where J denotes strontium). General Motors is developing a similar alloy with large amounts of calcium. Dead Sea Magnesium and Volkswagen are developing a high-temperature resistant alloy based on Mg–Al–Ca–RE (rare earth) with optional additions of Sr and Zn. These alloys, while still under development, are able to give improved elevated-temperature properties, die castability, and excellent salt spray corrosion resistance. These types of alloys will expand the use of magnesium into automatic transmission casings, crankcases, oil pump body, oil pan, belt pulley, cylinder cover, and engine fan applications.

Other creep-resistant alloys being investigated include both high-pressure die-cast alloys and sand-cast alloys. The large size and complexity of engine blocks and transmission housings may create a need for additional higher temperature alloys that could be sand cast.

Mechanical properties of cast alloys are given in Table 2.

3.2 Mechanical Properties of Wrought Materials

Wrought products are produced as forgings, extrusions, sheet, and plate. Mechanical properties are given in Table 3.

Table 2 Typical Mechanical Properties for Castings

Alloy	Temper	Tensile Strength (MPa)	Yield Strength (MPa)	Elongation in 2 in. (%)
Sand and Permanent Mold Castings				
AZ81A	T4	276	85	15
AZ91E	F	165	95	3
	T4	275	85	14
	T6	275	195	6
EZ33A	T5	160	105	3
K1A	F	185	51	20
QE22A	T6	275	205	4
WE43A	T6	235	190	4
WE54A	T6	270	195	4
ZE63A	T6	295	190	7
Investment Castings				
AZ81A	T4	275	100	12
AZ91E	F	165	100	2
	T4	275	100	12
	T5	180	100	3
	T7	275	140	5
EZ33A	T5	255	110	4
K1A	F	175	60	20
QE22A	T6	260	185	4
Die Castings				
AM50A	F	200	110	10
AM60B	F	220	130	8
AS41B	F	210	140	6
AZ91D	F	230	160	3
Die Castings for Elevated Temperatures				
ACM522	F (20°C)	200	158	4
	F (175°C)	152	132	9
AJ52X	F (20°C)	228	161	13
	F (175°C)	141	100	18
AJ62X	F (20°C)	240	143	7
	F (175°C)	143	103	19

3.3 Physical Properties of Magnesium

A selection of physical properties of pure magnesium is given in Table 4. Most of these are insensitive to alloy addition, but melting point, density, and electrical resistivity may vary slightly.

4 FABRICATION

4.1 Machining

Magnesium is the easiest metal to machine. It requires less power for removing a given volume of metal by machining than any other commonly machined metal (Ref. 2, p. 3).

Table 3 Typical Mechanical Properties of Wrought Products

Alloy	Temper	Tensile Strength (MPa)	Yield Strength (MPa)		Elongation in 2 in. (%)
			Tensile	Compressive	
Sheet and Plate					
AZ31B	O	255	150	110	21
	H24	290	220	180	15
Extrusions					
AZ31B	F	260	200	95	15
AZ61A	F	310	230	130	16
AZ80A	F	340	250	140	11
	T5	380	275	240	7
M1A	F	255	180	125	12
ZK40A	T5	275	255	140	4
ZK60A	F	340	250	185	14
	T5	365	305	250	11
Forgings					
AZ31B	F	260	195	85	9
AZ61A	F	195	180	115	12
AZ80A	F	315	215	170	8
	T5	345	235	195	6
	T6	345	250	185	5
ZK60A	T5	305	205	195	16
	T6	325	270	170	11

4.2 Joining

Magnesium can be joined by the many traditional joining methods both mechanical and thermal. Joining methods that have been used successfully include mechanical fasteners such as rivets and treaded fasteners (bolts and screws). Welding is done by protected arc processes such as metal inert gas (MIG) and tungsten inert gas (TIG) welding. Other welding methods

Table 4 Physical Properties of Pure Magnesium

Latent heat of vaporization	5150–5400 kJ/kg
Latent heat of sublimation	6113–6238 kJ/kg
Density	1.738 g/cm^3
Melting point	650°C
Thermal expansion	25.2×10^{-6}/K
Specific heat	1.025 kJ/kg·K at 20°C
Latent heat of fusion	360–377 kJ/kg
Heat of combustion	25,020 kJ/kg
Boiling point	1090°C
Electrical resistivity	4.45 $\Omega\cdot m \times 10^{-8}$
Crystal structure	Close-packed hexagonal
	$a = 0.32092$
	$c = 0.52105$
	$c/a = 1.633$

Source: From Ref. 2, p. 9.

such as plasma, electron beam, friction stir welding, explosion joining, and ultrasonic joining can all be used.

Magnesium has been joined by adhesive bonding for many years. Large structural components have been joined by the use of adhesives. The large six-engine U.S. bomber, the B-36, used over 12,000 lb of magnesium sheet attached to the fuselage and wing frames by adhesives. Adhesive bonding permits the joining of magnesium with other metals with no electrical contact between them, thus eliminating galvanic corrosion. Adhesive systems that are acrylic based are being used in prototype vehicles.

Magnesium welding can also be used to repair casting defects in magnesium-cast parts.

4.3 Forming

Most common wrought processes such as forging, extrusion, deep drawing, bending, spinning, stretch forming, and dimpling can form magnesium. Magnesium has a hexagonal close-packed crystalline structure with little cold workability. Magnesium is best formed when hot (180–225°C) (Ref. 1, p. 584).

Forging is accomplished with hydraulic presses or slow-acting mechanical presses. Magnesium deforms laterally rather than longitudinally. Forging temperatures are relatively low (275–400°C), so conventional low-alloy, hot-work steels are satisfactory materials for the forging dies (Ref. 1, p. 559). The ease with which magnesium can be worked greatly reduces the number of forging operations needed to produce finished parts (Ref. 2, p. 47).

Magnesium alloys have been extruded for many years. Both indirect and direct extrusion have been used. Hydrostatic extrusion processes are being developed in Europe for magnesium alloys. Extrusion billet temperatures usually range from 320 to 400°C (Ref. 1, p. 555).

Magnesium is rolled to sheet and plate by a conventional method, reducing large, preheated slabs in a series of passes through a conventional rolling mill. Because the metal is hexagonal, it requires careful handling and reheating after a series of passes when the temperature of the product gets too cold. Starting slab temperatures for breakdown should be 425–450°C. Sheet rolling range should be 350–440°C with about 25–50% reduction before reheating (Ref. 1, p. 555).

Twin roll casting is also being used to produce thin magnesium sheet directly from molten magnesium alloy. Work is being done to produce sheet to a final thickness as it is cast between the rolls.

5 CORROSION AND FINISHING

Magnesium, as castings or as wrought products, has always faced corrosion problems. It is anodic to any other structural metal and will be preferentially attacked in the presence of an electrolyte. Galvanic contact must be avoided in structural assemblies by separating magnesium from other metals by the use of films and tapes.

5.1 Chemical Conversion Coatings

The development of high-purity alloys helped reduce the surface corrosion of finished magnesium products, but more protection is needed, especially for automotive parts. Many types of finishing and protective coatings have been tried over the years. Chromate treatments to fully clean magnesium surfaces have worked quite well in the past. The environmental hazards of elemental hexavalent chromium have dictated the elimination of many coatings that contain hazardous materials. New methods have been developed that do not contain chromium and are even more effective in providing protection to the magnesium.

5.2 Anodic Coatings

Conversion coatings without chromium are being used followed by other surface-modifying treatments. The most widely used is anodic coating applied by placing the part in a bath as an anode while the tank is the cathode. Anodic coatings (Mag-Oxide, Tagnite, Keronite) have been very successful. Also microplasmic coatings and others are being used.

5.3 Painting

If a good chemical conversion or anodic coating is present, any paint or powder coating will provide protection. The best protection results from the application of baked, alkaline-resistant paints.

5.4 Electroplating

Magnesium can be electroplated fairly easily. This can be the traditional electroplating or the new "electroless" type of plating. Once a base coating is applied chemically, standard electroplating procedures can be applied to magnesium to give decorative and protective finishes.

6 RECYCLING

Magnesium is a highly recyclable material consuming only 5% of the energy required to manufacture the primary metal.[4] The increasing demand for magnesium applications has created increased quantities of scrap. The treatment of the secondary material that becomes available not only is an ecological necessity but also offers economic potential.

Present die-casting methods have a yield of 50–75% based on total cast weight. That means that 25–50% must be recycled. There are two basic types of recycling processes: with the use of flux and fluxless refining. Both are being used effectively.

The clean die-casting scrap, consisting of gates and runners, is called type 1. This is the most highly sought after scrap. For many years it was sold to secondary magnesium plants for remelting and refining. Now, many die-casting producers are working to recycle the type 1 magnesium scrap in their own plants. Type 1 can be easily remelted without flux. Most other types of die casting and industry scrap require flux to melt it and refine out the impurities.[5]

REFERENCES

1. E. F. Emley, *Principles of Magnesium Technology,* Pergamon, New York, 1966.
2. M. Avedesian and H. Baker, *Magnesium and Magnesium Alloys,* ASM Specialty Handbook, ASM International, Materials Park, OH, 1999.
3. R. Busk, *Magnesium Products Design,* Marcel Dekker, New York, 1987.
4. L. Riopelle, *J. Metals,* **10,** 44–46 (1996).
5. G. Hanko, C. Lochbichler, et al., "Magnesium" in *Proceedings of the 6th International Conference on Magnesium Alloys and Their Applications,* K. Kainer (ed.), Wiley-VCH, Weinheim, Germany, 2004.

BIBLIOGRAPHY

Aghion, B. E., and D. Eliezer, *Magnesium 2000,* Magnesium Research Institute, Potash House, Beer-Sheva, Israel, 2000.

Aghion, D. E., and D. Eliezer, *Magnesium Alloys-Science, Technology, and Applications,* Israeli Consortium for Magnesium Technology Development, S. Neaman Institute, Technion City, Haifa, 2004.

Jenkins, E. D., *Magnesium Overcast-Story of the Convair B-36,* Specialty Press, North Branch, MN, 2001.

Beck, A., *The Technology of Magnesium and Its Alloys,* F. A. Hughes, Abbey House, London, 1940.

Kaplan, C. H., J. Hryn, and B. Clow, *Magnesium Technology 2000,* The Metallurgical Society, Warrendale, PA, 2000.

CHAPTER 8
SELECTION OF SUPERALLOYS FOR DESIGN

Matthew J. Donachie
Rensselaer at Hartford
Hartford, Connecticut

Stephen J. Donachie
Special Metals Corporation
Huntington, West Virginia

1 INTRODUCTION

1.1 Purpose

The purpose of this chapter is to create a sufficient understanding of superalloys so that selection of them for specific designs will be appropriate. Knowledge of superalloy types and their processing will give a potential user the ability to understand the ways in which superalloys can contribute to a design. The knowledge provided here should enable the user to ask the important questions of superalloy providers so as to evaluate the capability of primary melt shops and component producers while addressing the necessary mechanical property and corrosion/environmental behavior that will influence alloy selection. There is no cook book for superalloy selection. Proprietary alloys and/or proprietary/restricted processing lead to superalloy conditions and properties not listed in a handbook or catalog of materials. With few exceptions, critical applications will require work with one or more superalloy manufacturers to develop an understanding of what is available and what a superalloy selector/designer can expect from a chosen superalloy.

1.2 What Are Superalloys?

For purposes of this chapter superalloys are those nickel-, iron–nickel-, and cobalt-base corrosion-resistant alloys generally used above a nominal temperature of 540°C (1000°F). The iron–nickel-base superalloys are an extension of stainless steel technology and generally are melted and cast to electrode/ingot shapes for subsequent fabrication to components. The iron–nickel-base superalloys usually are wrought, i.e., formed to shape or mostly to shape by hot rolling, forging, etc. On the other hand, after primary production by melting and ingot casting, the cobalt-base and nickel-base superalloys may be used either in wrought or cast form depending on the application or the alloy composition involved. The stainless steels, nickel–chromium alloys, and cobalt dental alloys which evolved into the superalloys used chromium to provide elevated-temperature corrosion resistance. A Cr_2O_3 layer on the surface proved very effective in protection against oxidation. Eventually, cast superalloys for the highest temperatures were protected against oxidation by chromium and aluminum. In our opinion, superalloys must contain chromium, probably at the level of 5% (some would argue 8%) or higher for reasonable corrosion resistance.

1.3 How Are Superalloys Strengthened?

Metals are crystalline and, in the solid state, the atoms of a metal or alloy have various crystallographic arrangements, often occurring as cubic structures. Some crystal structures tend to be associated with better property characteristics than others. In addition, crystalline aggregates of atoms have orientation relationships. Uniform crystalline aggregates are called grains and, in an alloy, there are usually many grains with random a orientation direction for each grain. Metal alloys with multiple random grain directions are known as polycrystals (often referred to as equiaxed) having roughly equal dimensions in all directions, although columnar-shaped grains (one long axis) are common in cast products.

The peripheral surface of a grain is called a grain boundary. Aggregates of atoms without grain boundaries are rarely created in nature. However, metals without grain boundaries (single crystals) or metals with aligned boundaries (columnar-grained structures) can be produced in superalloys by appropriate manufacturing techniques. The introduction of different atom types, new crystal phases, and/or the manipulation of grain boundaries enable inhibition of the movement through the crystal lattice or grains of the imperfections that cause deformation to occur. Superalloys thus are the basis metals (iron, nickel, cobalt) modified by changes in chemistry, in (generally) a polycrystalline form with various phases present in the grains or grain boundaries.

Superalloys consist of an austenitic face-centered-cubic (fcc) crystal structure matrix phase, γ, plus a variety of secondary phases. Important secondary phases are γ' fcc ordered $Ni_3(Al, Ti)$ and various MC, $M_{23}C_6$, M_6C, and M_7C_3 (rare) carbides in nickel- and iron–nickel-base superalloys, where M represents various metal elements such as titanium, zirconium, hafnium, niobium, tungsten, molybdenum, and chromium. Carbides are the principal secondary phases in cobalt-base alloys. Also, γ', a body-centered tetragonal (bct) phase of ordered Ni_3Nb, a hexagonal ordered Ni_3Ti (η) phase, and the δ orthorhombic Ni_3Nb intermetallic phase can be found in nickel- and iron–nickel-base superalloys. Ordered structures are those in which specific metal atoms occur on defined locations in the crystal lattice, in contrast to disordered structures wherein the atoms of a phase occur randomly. An example of an ordered strengthening phase is $Ni_3(Al, Ti)$, which occurs in an fcc crystal form. The nickel atoms are found on the faces of the cube and the aluminum or titanium atoms occupy the corners of the cube.

It is quite important for the engineer selecting alloys to have a realistic understanding of the strengthening process in superalloys as the properties of superalloys can be modified considerably by processing to manipulate the strengthening level achieved. Superalloys derive their strength from solid-solution hardeners and from secondary precipitate phases that form in the γ-matrix and produce precipitation (age) hardening. Principal strengthening precipitate phases in nickel-base and iron–nickel-base superalloys are γ' and γ''. Carbides may provide limited strengthening directly (e.g., through dispersion hardening) or, more commonly, indirectly (e.g., by stabilizing grain boundaries against movement). The δ- and η-phases are useful (along with γ') in control of grain structure of wrought superalloys during wrought processing after melting. By controlling grain structure, strength can be significantly influenced. The extent to which the secondary phases contribute directly to strengthening depends on the alloy and its processing. It should be noted that improper distributions of carbides and precipitate phases can be detrimental to properties.

In addition to those elements that produce solid-solution hardening and/or promote carbide and γ' formation, other elements (e.g., boron, zirconium, hafnium) are added to enhance mechanical or chemical properties. Superalloy microstructure and property control can be complex. As many as 14 elements may be controlled in some superalloys. Some carbide- and γ'-forming elements may contribute significantly to corrosion properties as well. Table 1 gives a generalized list of the ranges of alloying elements in superalloys. Table 2 provides a tabulation of the specific effects of the various alloying elements.

1.4 Manufacture of Superalloy Articles

Appropriate compositions of all superalloy types can be forged, rolled to sheet, or otherwise formed into a variety of shapes. The more highly alloyed superalloy compositions normally are processed as small castings. Large castings are made principally in the weldable superalloy IN 718, although single-crystal turbine blades of complex nickel-base superalloys have been cast in lengths up to several feet for applications in land-based gas turbine frames. Fabricated superalloy structures can be built up by welding or brazing, but the more highly alloyed the superalloy composition (i.e., higher amount of hardening phase), the more difficult it is to weld a superalloy. Machining of superalloys is similar to but more difficult than that of machining stainless steels. In welding or machining of superalloys, the effects of the energy input (heat energy, deformation energy) on the microstructure and properties of the final superalloy product must be considered.

Table 1 Common Ranges of Major Alloying Additions in Superalloys

Element	Range (%)	
	Fe–Ni and Ni Base	Co Base
Cr	5–25	19–30
Mo, W	0–12	0–11
Al	0–6	0–4.5
Ti	0–6	0–4
Co	0–20	—
Ni	—	0–22
Nb	0–5	0–4
Ta	0–12	0–9
Re	0–6	0–2

Source: Metals Handbook Desk Edition, ASM International, Materials Park, OH, 1999, p. 395.

Iron–nickel- and nickel-base superalloys are readily available as wrought components in extruded, forged, or rolled form; the higher strength alloys generally are found only in the cast condition. Powder metallurgy processing is an accepted method to produce wrought components of certain higher strength alloys. Hot deformation is the preferred forming process, cold forming usually being restricted to thin sections (sheet). Cold rolling may be used to increase short-time strength properties for applications at temperatures below the nominal temperature of 540°C (1000°F) established in this chapter for superalloy use. Properties of superalloys generally are controlled by adjustments in chemistry (composition), control of grain size, and modification of the processing (including heat treatment).

1.5 Superalloy Information

While some chemistries and properties are listed in this chapter, there is no substitute for consultation with superalloy manufacturers about the forms (cast, wrought) which can be provided and the exact chemistries available. A few producers are noted at the end of this chapter. Also, it should be understood that not all superalloys are readily available as off-the-shelf items. While literally hundreds, perhaps thousands, of superalloy compositions have been evaluated since the mid-1930s, only a handful are routinely produced. Moreover, some superalloys are not available for use in all forms and sizes. Many of the highest strength alloys will be useful only as powder metal products or as castings.

Design data for superalloys are not intended to be conveyed here, but typical properties are indicated for some materials. Design properties should be obtained from internal testing, if possible, or from producers or other validated sources if sufficient test data are not available in-house. Typical properties are merely a guide for comparison. Exact chemistry, section size, heat treatment, and other processing steps must be known to generate adequate property values for design.

The properties of the extraordinary range of superalloy compositions which have been developed over the years are not normally well documented in the literature. However, since many consumers of superalloys actually only use a few alloys, within the customary user groups data may be more plentiful for certain compositions. The extent to which such data

Table 2 Role of Alloying Elements in Superalloys

Effect[a]	Iron Base	Cobalt Base	Nickel Base
Solid-solution strengtheners	Cr, Mo	Nb, Cr, Mo, Ni, W, Ta	Co, Cr, Fe, Mo, W, Ta, Re
fcc matrix stabilizer	C, W, Ni	Ni	—
Carbide form			
MC	Ti	Ti	W, Ta, Ti, Mo, Nb, Hf
M_7C_3	—	Cr	Cr
$M_{23}C_6$	Cr	Cr	Cr, Mo, W
M_6C	Mo	Mo, W	Mo, W, Nb
Carbonitrides: M(CN)	C, N	C, N	C, N
Promotes general precipitation of carbides	P	—	—
Forms γ' $Ni_3(Al,Ti)$	Al, Ni, Ti	—	Al, Ti
Retards formation of hexagonal η (Ni_3Ti)	Al, Zr	—	—
Raises solvus temperature of γ'	—	—	Co
Hardening precipitates and/or intermetallics	Al, Ti, Nb	Al, Mo, Ti,[b] W, Ta	Al, Ti, Nb
Oxidation resistance	Cr	Al, Cr	Al, Cr, Y, La, Ce
Improve hot corrosion resistance	La, Y	La, Y, Th	La, Th
Sulfidation resistance	Cr	Cr	Cr, Co, Si
Improves creep properties	B	—	B, Ta
Increases rupture strength	B	B, Zr	B[c]
Grain boundary refiners	—	—	B, C, Zr, Hf
Facilitates working	—	Ni_3Ti	—
Retard γ' coarsening	—	—	Re

[a] Not all these effects necessarily occur in a given alloy.
[b] Hardening by precipitation of Ni_3Ti also occurs if sufficient Ni is present.
[c] If present in large amounts, borides are formed.
Source: Metals Handbook Desk Edition, ASM International, Materials Park, OH, 1999, p. 395.

are available to the general public is unknown. However, even if such data were disseminated widely, the alloy selector needs to be aware that processing treatments such as forging conditions, heat treatment, coatings for corrosion protection, etc., dramatically affect properties of superalloys. All data should be reconciled with the actual manufacturing specifications and processing conditions expected. Alloy selectors should work with competent metallurgical engineers to establish the validity of data intended for design as well as to specify the processing conditions that will be used for component production.

Application of design data must take into consideration the probability of components containing locally inhomogenous regions. For wrought superalloys, the probability of occurrence of these regions (which are highly detrimental to fatigue life) is dependent upon the melting method selected (see the discussion on melting in Section 5 for details). For cast superalloys, the degree of inhomogeneity and the likelihood of defects such as porosity are related to the alloy composition, the investment casting technique used, and the complexity of the final component.

For sources of property data other than that of the producers (melters, forgers, etc.) or an alloy selector's own institution, one may refer to organizations such as ASM International which publish compilations of data that may form a basis for the development of design allowables for superalloys.

Standards organizations such as the American Society for Testing and Materials (ASTM) publish information about superalloys, but that information does not ordinarily contain design data. The great versatility of superalloys in property modification is also a detriment to the universal compilation of property values. The same nominal alloy may have some composition modifications made from one manufacturer or customer to another. Sometimes this extends from one country to another. Tweaking of the casting or forging process or the heat treatment, in addition to what seem like minor composition changes, can cause significant variations in properties. All facets of chemistry and processing need to be considered when selecting a superalloy for an application.

Some information is given on chemistry and properties of a few superalloys as follows: nominal compositions of selected superalloys in Tables 3 and 4; physical properties in Tables 5 and 6; short-time (tensile) properties in Tables 7 and 8: time-dependent (creep-rupture) properties in Tables 9 and 10; dynamic modulus values in Tables 11 and 12. It should be noted that time-dependent mechanical properties are particularly affected by grain size. Selected reference publications for superalloy technical information or properties are given in the Bibliography.

2 METALS AT HIGH TEMPERATURES

2.1 General

While material strengths at low temperatures are usually not a function of time, at high temperatures the time of load application becomes very significant for mechanical properties. Concurrently, the availability of oxygen at high temperatures accelerates the conversion of some of the metal atoms to oxides. Oxidation proceeds much more rapidly at high temperatures than at room or lower temperatures.

2.2 Mechanical Behavior

In the case of short-time tensile properties (yield strength, ultimate strength), the mechanical behavior of metals at higher temperatures is similar to that at room temperature, but with metals becoming weaker as the temperature increases. However, when steady loads below the normal yield or ultimate strength determined in short-time tests are applied for prolonged times at higher temperatures, the situation is different. Figure 1 illustrates the way in which most materials respond to steady extended-time loads at high temperatures.

Owing to the higher temperature, a time-dependent extension (creep) is noticed under load. If the alloy is exposed for a long time, the alloy eventually fractures (ruptures). The degradation process is called creep or, in the event of failure, creep rupture (sometimes stress rupture), and alloys are selected on their ability to resist creep and creep rupture failure. Data for superalloys frequently are provided as the stress which can be sustained for a fixed time (e.g., 100 h rupture) versus the temperature. Figure 2 shows such a plot with ranges of expected performance for various superalloy families. One of the contributory aspects of elevated temperature failure is that metals tend to come apart at the grain boundaries when tested for long times above about 0.5 of their absolute melting temperature. Thus, fine-grained alloys which are usually favored for lower temperature applications may not be the

Table 3 Nominal Compositions of Wrought Superalloys

Alloy	Composition, %										
	Cr	Ni	Co	Mo	W	Nb	Ti	Al	Fe	C	Other
Solid-Solution Alloys											
Iron–nickel base											
Alloy N-155 (Multimet)	21.0	20.0	20.0	3.00	2.5	1.0	—	—	32.2	0.15	0.15 N, 0.2 La, 0.02 Zr
Haynes 556	22.0	21.0	20.0	3.0	2.5	0.1	—	0.3	29.0	0.10	0.50 Ta, 0.02 La, 0.02 Zr
19-9 DL	19.0	9.0	—	1.25	1.25	0.4	0.3	—	66.8	0.30	1.10 Mn, 0.60 Si
Incoloy 800	21.0	32.5	—	—	—	—	0.38	0.38	45.7	0.05	—
Incoloy 800H	21.0	33.0	—	—	—	—	—	—	45.8	0.08	—
Incoloy 800HT	21.0	32.5	—	—	—	—	0.4	0.4	46.0	0.08	0.8 Mn, 0.5 Si, 0.4 Cu
Incoloy 801	20.5	32.0	—	—	—	—	1.13	—	46.3	0.05	—
Incoloy 802	21.0	32.5	—	—	—	—	0.75	0.58	44.8	0.35	—
Nickel base											
Haynes 214	16.0	76.5	—	—	—	—	—	4.5	3.0	0.03	—
Haynes 230	22.0	55.0	5.0 max.	2.0	14.0	—	—	0.35	3.0 max.	0.10	0.15 max. B, 0.02 La
Inconel 600	15.5	76.0	—	—	—	—	—	—	8.0	0.08	0.25 Cu
Inconel 601	23.0	60.5	—	—	—	—	—	1.35	14.1	0.05	0.5 Cu
Inconel 617	22.0	55.0	12.5	9.0	—	—	—	1.0	—	0.07	—
Inconel 625	21.5	61.0	—	9.0	—	3.6	0.2	0.2	2.5	0.05	—
RA333	25.0	45.0	3.0	3.0	3.0	—	—	—	18.0	0.05	—
Hastelloy B	1.0 max.	63.0	2.5 max.	28.0	—	—	—	—	5.0	0.05 max.	0.03 V
Hastelloy N	7.0	72.0	—	16.0	—	—	0.5 max	—	5.0 max	0.06	—
Hastelloy S	15.5	67.0	—	15.5	—	—	—	0.2	1.0	0.02 max.	0.02 La
Hastelloy W	5.0	61.0	2.5 max.	24.5	—	—	—	—	5.5	0.12 max.	0.6 V
Hastelloy X	22.0	49.0	1.5 max.	9.0	0.6	—	—	2.0	15.8	0.15	—
Hastelloy C-276	15.5	59.0	—	16.0	3.7	—	—	—	5.0	0.02 max.	—
Haynes HR-120	25.0	37.0	3.0	2.5	2.5	0.7	—	0.1	33.0	0.05	0.7 Mn, 0.6 Si, 0.2 N, 0.004 B
Haynes HR-160	28.0	37.0	29.0	—	—	—	—	—	2.0	0.05	2.75 Si, 0.5 Mn
Nimonic 75	19.5	75.0	—	—	—	—	0.4	0.15	2.5	0.12	0.25 max. Cu
Nimonic 86	25.0	65.0	—	10.0	—	—	—	—	—	0.05	0.03 Ce, 0.015 Mg
Cobalt base											
Haynes 25 (L605)	20.0	10.0	50.0	—	15.0	—	—	—	3.0	0.10	1.5 Mn
Haynes 188	22.0	22.0	37.0	—	14.5	—	—	—	3.0 max.	0.10	0.90 La
Alloy S-816	20.0	20.0	42.0	4.0	4.0	4.0	—	—	4.0	0.38	—
MP35-N	20.0	35.0	35.0	10.0	—	—	—	—	—	—	—
MP159	19.0	25.0	36.0	7.0	—	0.6	3.0	0.2	9.0	—	—
Stellite B	30.0	1.0	61.5	—	4.5	—	—	—	1.0	1.0	—
UMCo-50	28.0	—	49.0	—	—	—	—	—	21.0	0.12	—

Table 3 (*Continued*)

Alloy	Composition, %										
	Cr	Ni	Co	Mo	W	Nb	Ti	Al	Fe	C	Other
Precipitation-Hardening Alloys											
Iron–nickel base											
A-286	15.0	26.0	—	1.25	—	—	2.0	0.2	55.2	0.04	0.005 B, 0.3 V
Discaloy	14.0	26.0	—	3.0	—	—	1.7	0.25	55.0	0.06	—
Incoloy 903	0.1 max.	38.0	15.0	0.1	—	3.0	1.4	0.7	41.0	0.04	—
Pyromet CTX-1	0.1 max.	37.7	16.0	0.1	—	3.0	1.7	1.0	39.0	0.03	—
Incoloy 907	—	38.4	13.0	—	—	4.7	1.5	0.03	42.0	0.01	0.15 Si
Incoloy 909	—	38.0	13.0	—	—	4.7	1.5	0.03	42.0	0.01	0.4 Si
Incoloy 925	20.5	44.0	—	2.8	—	—	2.1	0.2	29	0.01	1.8 Cu
V-57	14.8	27.0	—	1.25	—	—	3.0	0.25	48.6	0.08 max.	0.01 B, 0.5 max V
W-545	13.5	26.0	—	1.5	—	—	2.85	0.2	55.8	0.08 max.	0.05 B
Nickel base											
Astroloy	15.0	56.5	15.0	5.25	—	—	3.5	4.4	<0.3	0.06	0.03 B, 0.06 Zr
Custom Age 625 PLUS	21.0	61.0	—	8.0	—	3.4	1.3	0.2	5.0	0.01	—
Haynes 242	8.0	62.5	2.5 max.	25.0	—	—	—	0.5 max.	2.0 max.	0.10 max.	0.006 max. B
Haynes 263	20.0	52.0	—	6.0	—	—	2.4	0.6	0.7	0.06	0.6 Mn, 0.4 Si, 0.2 Cu
Haynes R-41	19.0	52.0	11.0	10.0	—	—	3.1	1.5	5.0	0.09	0.5 Si, 0.1 Mn, 0.006 B
Inconel 100	10.0	60.0	15.0	3.0	—	—	4.7	5.5	<0.6	0.15	1.0 V, 0.06 Zr, 0.015 B
IN-100	10	60	15	3	—	—	4.7	5.5	<0.6	0.15	0.06 Zr, 1.0 V
Inconel 102	15.0	67.0	—	2.9	3.0	2.9	0.5	0.5	7.0	0.06	0.005 B, 0.02 Mg, 0.03 Zr
Incoloy 901	12.5	42.5	—	6.0	—	—	2.7	—	36.2	0.10 max.	—
Inconel 702	15.5	79.5	—	—	—	—	0.6	3.2	1.0	0.05	0.5 Mn, 0.2 Cu, 0.4 Si
Inconel 706	16.0	41.5	—	—	—	—	1.75	0.2	37.5	0.03	2.9 (Nb + Ta), 0.15 max. Cu
Inconel 718	19.0	52.5	—	3.0	—	5.1	0.9	0.5	18.5	0.08 max.	0.15 max. Cu
Inconel 721	16.0	71.0	—	—	—	—	3.0	—	6.5	0.4	2.2 Mn, 0.1 Cu
Inconel 722	15.5	75.0	—	—	—	—	2.4	0.7	7.0	0.04	0.5 Mn, 0.2 Cu, 0.4 Si
Inconel 725	21.0	57.0	—	8.0	—	3.5	1.5	0.35 max.	9.0	0.03 max.	—
Inconel 751	15.5	72.5	—	—	—	1.0	2.3	1.2	7.0	0.05	0.25 max. Cu
Inconel X-750	15.5	73.0	—	—	—	1.0	2.5	0.7	7.0	0.04	0.25 max. Cu
M-252	19.0	56.5	10.0	10.0	—	—	2.6	1.0	<0.75	0.15	0.005 B
MERL-76	12.4	54.4	18.6	3.3	—	1.4	4.3	5.1	—	0.02	0.35 Hf, 0.06 Zr
Nimonic 80A	19.5	73.0	1.0	—	—	—	2.25	1.4	1.5	0.05	0.10 max. Cu
Nimonic 90	19.5	55.5	18.0	—	—	—	2.4	1.4	1.5	0.06	—
Nimonic 95	19.5	53.5	18.0	—	—	—	2.9	2.0	5.0 max.	0.15 max.	+B, +Zr
Nimonic 100	11.0	56.0	20.0	5.0	—	—	1.5	5.0	2.0 max.	0.30 max.	+B, +Zr
Nimonic 105	15.0	54.0	20.0	5.0	—	—	1.2	4.7	—	0.08	0.005 B

Nimonic 115	15.0	55.0	15.0	4.0	—	—	4.0	5.0	1.0	0.20	0.04 Zr
C-263	20.0	51.0	20.0	5.9	—	—	2.1	0.45	0.7 max.	0.06	—
C-1023	15.5	bal	10	8.5			3.6	4.2		0.16	B 0.006
GTD 222	22.5	bal	19		2	0.8	2.3	1.2	—	0.1	Ta 1, Zr 0.012, B 0.005
Pyromet 860	13.0	44.0	4.0	6.0	—	—	3.0	1.0	28.9	0.05	0.01 B
Pyromet 31	22.7	55.5	—	2.0	—	1.1	2.5	1.5	14.5	0.04	0.005 B
Refractaloy 26	18.0	38.0	20.0	3.2	—	—	2.6	0.2	16.0	0.03	0.015 B
Rene 41	19.0	55.0	11.0	10.0	—	—	3.1	1.5	<0.3	0.09	0.01 B
Rene 88	16	56.4	13.0	4	4	0.7	3.7	2.1	—	0.03	0.03 Zr
Rene 95	14.0	61.0	8.0	3.5	3.5	3.5	2.5	3.5	<0.3	0.16	0.01 B, 0.05 Zr
Rene 100	9.5	61.0	15.0	3.0	—	—	4.2	5.5	1.0 max.	0.16	0.015 B, 0.06 Zr, 1.0 V
Rene 220	18	bal	12	3		5	1	0.5		0.02	B 0.01
Udimet 500	19.0	48.0	19.0	4.0	—	—	3.0	3.0	4.0 max.	0.08	0.005 B
Udimet 520	19.0	57.0	12.0	6.0	1.0	—	3.0	2.0	—	0.08	0.005 B
Udimet 630	17.0	50.0	—	3.0	3.0	6.5	1.0	0.7	18.0	0.04	0.004 B
Udimet 700	15.0	53.0	18.5	5.0	—	—	3.4	4.3	<1.0	0.07	0.03 B
Udimet 710	18.0	55.0	14.8	3.0	1.5	—	5.0	2.5	—	0.07	0.01 B
Udimet 720	18	55	14.8	3	1.25	—	5	2.5	—	0.035	0.03 Zr
Udimet 720LI	16	57	15.0	3	1.25	—	5	2.5	—	0.025	0.03 Zr
Unitemp AF2-1DA	12.0	59.0	10.0	3.0	6.0	—	3.0	4.6	<0.5	0.35	1.5 Ta, 0.015 B, 0.1 Zr
Waspaloy	19.5	57.0	13.5	4.3	—	—	3.0	1.4	2.0 max.	0.07	0.006 B, 0.09 Zr

Table 4 Nominal Compositions of Cast Superalloys

Alloy designation	Nominal composition, %												
	C	Ni	Cr	Co	Mo	Fe	Al	B	Ti	Ta	W	Zr	Other
Nickel base													
B-1900	0.1	64	8	10	6	—	6	0.015	1	4[a]	—	0.10	—
CM186	0.07	bal	6	9	0.5		5.7	0.015	0.7	3	8	0.005	Re 3, Hf 1.4
CMSX-2	—	66.2	8	4.6	0.6	—	56	—	1	6	8	6	—
CMSX-4	—	bal	6.5	9	0.6	—	5.6	—	1.0	6.5	6	—	
CMSX-6	—	bal	10	5	3	—	4.8	—	4.7	2	—	—	
CMSX-10	—	bal	1.8–4.0	1.5–9.0	0.25–2.0	—	5.0–7.0	—	0.1–1.2	7.0–10.0	3.5–7.5	—	
CMSX486	0.07	bal	5	9	0.7		5.7	0.015	0.7	4.5	9	0.005	Re 3, Hf 1
Hastelloy X	0.1	50	21	1	9	18	—	—	—	—	1	—	—
IN 939	0.15	bal	22.4	19			1.9	0.01	3.7	1.4	1.6	0.1	Nb 1
Inconel 100	0.18	60.5	10	15	3	—	5.5	0.01	5	—	—	0.06	1 V
Inconel 713C	0.12	74	12.5	—	4.2	—	6	0.012	0.8	1.75	—	0.1	0.9 Nb
Inconel 713LC	0.05	75	12	—	4.5	—	6	0.01	0.6	4	—	0.1	—
Inconel 738	0.17	61.5	16	8.5	1.75	—	3.4	0.01	3.4	—	2.6	0.1	2 Nb
Inconel 792	0.2	60	13	9	2.0	—	3.2	0.02	4.2	—	4	0.1	2 Nb
Inconel 718	0.04	53	19	—	3	18	0.5	—	0.9	—	—	—	0.1 Cu, 5 Nb
X-750	0.04	73	15	—	—	7	0.7	—	2.5	—	—	—	0.25 Cu, 0.9 Nb
M-252	0.15	56	20	10	10	—	1	0.005	2.6	—	—	—	—
MAR-M 200	0.15	59	9	10	—	1	5	0.015	2	—	12.5	0.05	1 Nb[b]
MAR-M 246	0.15	60	9	10	2.5	—	5.5	0.015	1.5	1.5	10	0.05	—
MAR-M 247	0.15	59	8.25	10	0.7	0.5	5.5	0.015	1	3	10	0.05	1.5 Hf
PWA 1426	0.10	bal	6.5	12	2		6	0.015		4	6	0.03	Re 3, Hf 1.5
PWA 1480	—	bal	10	5.0	—	—	5.0	—	1.5	12	4.0	—	—
PWA 1484	—	bal	5	10	2	—	5.6	—	—	9	6	—	
PWA 1484		bal	5	10	1.9		5.6			8.7	5.9		Re 3, Hf 0.1
PWA 1487		bal	5	10	1.9		5.6			8.4	5.9		Re 3, Hf 0.25, Y 0.013
Rene 41	0.09	55	19	11.0	10.0	—	1.5	0.01	3.1	—	—	—	—
Rene 77	0.07	58	15	15	4.2	—	4.3	0.015	3.3	—	—	0.04	—
Rene 80	0.17	60	14	9.5	4	—	3	0.015	5	—	4	0.03	—
Rene 80 Hf	0.08	60	14	9.5	4	—	3	0.015	4.8	—	4	0.02	0.75 Hf
Rene 100	0.18	61	9.5	15	3	—	5.5	0.015	4.2	—	—	0.06	1 V
René 142	0.12	bal	6.8	12	2		6.2	0.015		3	5	.02	Re 3, Hf 1.5
Rene N4	0.06	62	9.8	7.5	1.5	—	4.2	0.004	3.5	4.8	6	—	0.5 Nb, 0.15 Hf
RR 2000	—	bal	10	15	3	—	5.5	—	4.0	—	—	—	
SRR 99	—	bal	8	5	—	—	5.5	—	2.2	3	10	—	

Nickel base (Continued)													
Rene N5	—	bal	7	8	2	—	6.2	—	—	7	5	—	
Rene N6[c]	—	bal	4.25–6	10–15	0.5–2	—	5–6.25	—	—	7–9.25	5–6.5	—	
Udimet 500	0.1	53	18	17	4	2	3	—	3	—	—	—	—
Udimet 700	0.1	53.5	15	18.5	5.25	—	4.25	0.03	3.5	—	—	—	—
Udimet 710	0.13	55	18	15	3	—	2.5	—	5	—	1.5	0.08	—
Waspaloy	0.07	57.5	19.5	13.5	4.2	1	1.2	0.005	3	—	—	0.09	—
WAX-20(DS)	0.20	72	—	—	—	—	6.5	—	—	—	20	1.5	—
Cobalt base													
AiResist 13	0.45	—	21	62	—	—	3.4	—	—	2	11	—	0.1 Y
AiResist 213	0.20	0.5	20	64	—	0.5	3.5	—	—	6.5	4.5	0.1	0.1 Y
AiResist 215	0.35	0.5	19	63	—	0.5	4.3	—	—	7.5	4.5	0.1	0.1 Y
FSX-414	0.25	10	29	52.5	—	1	—	0.010	—	—	7.5	—	—
Haynes 21	0.25	3	27	64	—	1	—	—	—	—	—	—	5 Mo
Haynes 25; L-605	0.1	10	20	54	—	1	—	—	—	—	15	—	—
J-1650	0.20	27	19	36	—	—	—	0.02	3.8	2	12	—	—
MAR-M 302	0.85	—	21.5	58	—	0.5	—	0.005	—	9	10	0.2	—
MAR-M 322	1.0	—	21.5	60.5	—	0.5	—	—	0.75	4.5	9	2	—
MAR-M 509	0.6	10	23.5	54.5	—	—	—	—	0.2	3.5	7	0.5	—
MAR-M 918	0.05	20	20	52	—	—	—	—	—	7.5	—	0.1	—
NASA Co-W-Re	0.40	—	3	67.5	—	—	—	—	1	—	25	1	2 Re
S-816	0.4	20	20	42	—	4	—	—	—	—	4	—	4 Mo, 4 Nb, 1.2 Mn, 0.4 Si
V-36	0.27	20	25	42	—	3	—	—	—	—	2	—	4 Mo, 2 Nb, 1 Mn, 0.4 Si
Wi-52	0.45	—	21	63.5	—	2	—	—	—	—	11	—	2 Nb + Ta
X-40 (Stellite alloy 31)	0.50	10	22	57.5	—	1.5	—	—	—	—	7.5	—	0.5 Mn, 0.5 Si

[a] B-1900 + Hf also contains 1.5% Hf.

[b] MAR-M 200 + Hf also contains 1.5% Hf.

[c] Designated R′ 162 in U.S. patent 5,270,123. Also contains 0.02–0.07% C, 0.003–0.01% B, 0–0.3% Y, and 0–6% Ru.

Table 5 Effect of Temperature on Short-Time Mechanical Properties of Selected Wrought Superalloys

Alloy	Form	Ultimate Tensile Strength at 21°C (70°F)		540°C (1000°F)		760°C (1400°F)		Yield Strength at 0.2% offset at 21°C (70°F)		540°C (1000°F)		760°C (1400°F)		Tensile Elongation (%) at 21°C (70°F)	540°C (1000°F)	760°C (1400°F)
		MPa	ksi	MPa	ksi	MPa	ksi	MPa	ksi	MPa	ksi	MPa	ksi			
Nickel base																
Astroloy	Bar	1415	205	1240	180	1160	168	1050	152	965	140	910	132	16	16	21
Cabot 214	—	915	133	715	104	560	84	560	81	510	74	495	72	38	19	9
D-979	Bar	1410	204	1295	188	720	104	1005	146	925	134	655	95	15	15	17
Hastelloy C-22	Sheet	800	116	625	91	525	76	405	59	275	40	240	35	57	61	63
Hastelloy G-30	Sheet	690	100	490	71	—	—	315	46	170	25	—	—	64	75	—
Hastelloy S	Bar	845	130	775	112	575	84	455	65	340	49	310	45	49	50	70
Hastelloy X	Sheet	785	114	650	94	435	63	360	52	290	42	260	38	43	45	37
Haynes 230	[a]	870	126	720	105	575	84	390	57	275	40	285	41	48	56	46
Inconel 587	Bar	1180	171	1035	150	830	120	705	102	620	90	605	88	28	22	20
Inconel 597	Bar	1220	177	1140	165	930	135	760	110	720	104	665	96	15	15	16
Inconel 600	Bar	660	96	560	81	260	38	285	41	220	32	180	26	45	41	70
Inconel 601	Sheet	740	107	725	105	290	42	455	66	350	51	220	32	40	34	78
Inconel 617	Bar	740	107	580	84	440	64	295	43	200	29	180	26	70	68	84
Inconel 617	Sheet	770	112	590	86	470	68	345	50	230	33	230	33	55	62	59
Inconel 625	Bar	965	140	910	132	550	80	490	71	415	60	415	60	50	50	45
Inconel 706	Bar	1310	190	1145	166	725	105	1005	146	910	132	660	96	20	19	32
Inconel 718	Bar	1435	208	1275	185	950	138	1185	172	1065	154	740	107	21	18	25
Inconel 718 Direct Age	Bar	1530	222	1350	196	—	—	1365	198	1180	171	—	—	16	15	—
Inconel 718 Super	Bar	1350	196	1200	174	—	—	1105	160	1020	148	—	—	16	18	—
Inconel X750	Bar	1200	174	1050	152	—	—	815	118	725	105	—	—	27	26	—
M-252	Bar	1240	180	1230	178	945	137	840	122	765	111	720	104	16	15	10
Nimonic 75	Bar	745	108	675	98	310	45	285	41	200	29	160	23	40	40	67
Nimonic 80A	Bar	1000	145	875	127	600	87	620	90	530	77	505	73	39	37	17
Nimonic 90	Bar	1235	179	1075	156	655	95	810	117	725	105	540	78	33	28	12
Nimonic 105	Bar	1180	171	1130	164	930	135	830	120	775	112	740	107	16	22	25
Nimonic 115	Bar	1240	180	1090	158	1085	157	865	125	795	115	800	116	27	18	24
Nimonic 263	Sheet	970	141	800	116	650	94	580	84	485	70	460	67	39	42	21
Nimonic 942	Bar	1405	204	1300	189	900	131	1060	154	970	141	860	125	37	26	42
Nimonic PE 11	Bar	1080	157	1000	145	760	110	720	105	690	100	560	81	30	30	18
Nimonic PE 16	Bar	885	128	740	107	510	74	530	77	485	70	370	54	37	26	42
Nimonic PK 33	Sheet	1180	171	1000	145	885	128	780	113	725	105	670	97	30	30	18
Pyromet 860	Bar	1295	188	1255	182	910	132	835	121	840	122	835	121	22	15	18
Rene 41	Bar	1420	206	1400	203	1105	160	1060	154	1020	147	940	136	14	14	11
Rene 95	Bar	1620	235	1550	224	1170	170	1310	190	1255	182	1100	160	15	12	15

Udimet 400	Bar	1310	190	1185	172	—	—	930	135	830	120	—	—	30	26	—
Udimet 500	Bar	1310	190	1240	180	1040	151	840	122	795	115	730	106	32	28	39
Udimet 520	Bar	1310	190	1240	180	725	105	860	125	825	130	725	105	21	20	15
Udimet 630	Bar	1520	220	1380	200	965	140	1310	190	1170	170	860	125	15	15	5
Udimet 700	Bar	1410	204	1275	185	1035	150	965	140	895	130	825	120	17	16	20
Udimet 710	Bar	1185	172	1150	167	1020	148	910	132	850	123	815	118	7	10	25
Udimet 720	Bar	1570	228	—	—	1455	211	1195	173	—	—	1050	152	13	—	9
Unitemp AF2-IDA6	Bar	1560	226	1480	215	1290	187	1015	147	1040	151	995	144	20	19	16
Waspaloy	Bar	1275	185	1170	170	650	94	795	115	725	105	675	98	25	23	28
Iron base																
A-286	Bar	1005	146	905	131	440	64	725	105	605	88	430	62	25	19	19
Alloy 901	Bar	1205	175	1030	149	725	105	895	130	780	113	635	92	14	14	19
Discaloy	Bar	1000	145	865	125	485	70	730	106	650	94	430	62	19	16	—
Haynes 556	Sheet	815	118	645	93	470	69	410	60	240	35	220	32	48	54	49
Incoloy 800	Bar	595	86	510	74	235	34	250	36	180	26	150	22	44	38	83
Incoloy 801	Bar	785	114	660	96	325	47	385	56	310	45	290	42	30	28	55
Incoloy 802	Bar	690	100	600	87	400	58	290	42	195	28	200	29	44	39	15
Incoloy 807	Bar	655	95	470	68	350	51	380	55	255	37	225	32.5	48	40	34
Incoloy 825[b]	—	690	100	~590	~86	~275	~40	310	45	~234	~34	180	~26	45	~44	~86
Incoloy 903	Bar	1310	190	—	—	—	—	1105	160	—	—	—	—	14	—	—
Incoloy 907[c]	—	~1365	~198	~1205	~175	~655	~95	1110	161	~960	~139	~565	~82	~12	~11	~20
Incoloy 909	Bar	1310	190	1160	168	615	89	1020	148	945	137	540	78	16	14	34
N-155	Bar	815	118	650	94	428	62	400	58	340	49	250	36	40	33	32
V-57	Bar	1170	170	1000	145	620	90	83	120	760	110	485	70	26	19	34
19-9 DL	—	815	118	615	89	—	—	570	83	395	57	—	—	43	30	—
16-25-6	—	980	142	—	—	415	60	770	112	—	—	345	50	23	—	11
Cobalt base																
AirResist 213	—	1120	162	—	—	485	70	625	91	—	—	385	56	14	—	47
Elgiloy	—	690[d]–2480[e]	100[d]–360[e]	—	—	—	—	480[d]–2000[e]	70–290	—	—	—	—	34	—	—
Haynes 188	Sheet	960	139	740	107	635	92	485	70	305	44	290	42	56	70	43
L-605	Sheet	1005	146	800	116	455	66	460	67	250	36	260	38	64	59	12
MAR-M918	Sheet	895	130	—	—	—	—	895	130	—	—	—	—	48	—	—
MP35N	Bar	2025	294	—	—	—	—	1620	235	—	—	—	—	10	—	—
MP159	Bar	1895	275	1565	227	—	—	1825	265	1495	217	—	—	8	8	—
Stellite 6B	Sheet	1010	146	—	—	—	—	635	92	—	—	—	—	11	—	—
Haynes 150	—	925	134	—	—	—	—	317	46	—	—	—	—	8	—	—

[a] Cold-rolled and solution-annealed sheet, 1.2–1.6 mm (0.048–0.063 in.) thick.
[b] Annealed.
[c] Precipitation hardened.
[d] Varies with processing.
[e] Work strengthened and aged.

Table 6 Effect of Temperature on Short-Time Mechanical Properties of Selected Cast Superalloys (Numbers in Parentheses Are Estimated)

	Ultimate Tensile Strength at						0.2% Yield Strength at						Tensile Elongation, % at		
	21°C (70°F)		538°C (1000°F)		1093°C (2000°F)		21°C (70°F)		538°C (1000°F)		1093°C (2000°F)		21°C (70°F)	538°C (1000°F)	1093°C (2000°F)
Alloy	MPa	ksi	MPa	ksi	MPa	ksi	MPa	ksi	MPa	ksi	MPa	ksi			
Nickel base															
IN-713 C	850	123	860	125	—	—	740	107	705	102	—	—	8	10	—
IN-713 LC	895	130	895	130	—	—	750	109	760	110	—	—	15	11	—
B-1900	970	141	1005	146	270	38	825	120	870	126	195	28	8	7	11
IN-625	710	103	510	74	—	—	350	51	235	34	—	—	48	50	—
IN-718	1090	158	—	—	—	—	915	133	—	—	—	—	11	—	—
IN-100	1018	147	1090	150	(380)	(55)	850	123	885	128	(240)	(35)	9	9	—
IN-162	1005	146	1020	148	—	—	815	118	795	115	—	—	7	6.5	—
IN-731	835	121	—	—	275	40	725	105	—	—	170	25	6.5	—	—
IN-738	1095	159	—	—	—	—	950	138	—	—	—	—	—	—	—
IN-792	1170	170	—	—	—	—	1060	154	—	—	—	—	4	—	—
M-22	730	106	780	113	—	—	685	99	730	106	—	—	5.5	4.5	—
MAR-M200	930	135	945	137	325	47	840	122	880	123	—	—	7	5	—
MAR-M246	965	140	1000	145	345	50	860	125	860	125	—	—	5	5	—
MAR-M247	965	140	1035	150	—	—	815	118	825	120	—	—	7	—	—
MAR-M421	1085	157	995	147	—	—	930	135	815	118	—	—	4.5	3	—
MAR-M432	1240	180	1105	160	—	—	1070	155	910	132	—	—	6	—	—
MC-102	675	98	655	95	—	—	605	88	540	78	—	—	5	9	—
Nimocast 75	500	72	—	—	—	—	179	26	—	—	—	—	39	—	—
Nimocast 80	730	106	—	—	—	—	520	75	—	—	—	—	15	—	—
Nimocast 90	700	102	595	86	—	—	520	75	420	61	—	—	14	15	—
Nimocast 242	460	67	—	—	—	—	300	44	—	—	—	—	8	—	—
Nimocast 263	730	106	—	—	—	—	510	74	—	—	—	—	18	—	—
Rene 77	—	—	—	—	—	—	—	—	—	—	—	—	—	—	—
Rene 80	—	—	—	—	—	—	—	—	—	—	—	—	—	—	—

Udimet 500	930	135	895	130	—	—	815	118	725	105	—	—	13	13	—
Udimet 710	1075	156	—	—	240	35	895	130	—	—	170	25	8	—	—
CMSX-2[a]	1185	172	1295[b]	188[b]	—	—	1135	165	1245[b]	181[b]	—	—	10	17[b]	—
GMR-235	710	103	—	—	—	—	640	93	—	—	—	—	3	—	18[b]
IN-939	1050	152	915[b]	133[b]	325[c]	47[c]	800	116	635[b]	92[b]	205[c]	30[c]	5	7[b]	25[b]
MM 002[d]	1035	150	1035[b]	150[b]	550[c]	80[c]	825	120	860[b]	125[b]	345[c]	50[c]	7	5[b]	12[b]
IN-713 Hf[e]	1000	145	895[b]	130[b]	380[c]	55[c]	760	110	620[b]	90[b]	240[c]	35[c]	11	6[b]	20[b]
Rene 125 Hf[f]	1070	155	1070[b]	155[b]	550[c]	80[c]	825	120	860[b]	125[b]	345[c]	50[c]	5	5[b]	12[b]
MAR-M 246 Hf[g]	1105	160	1070[b]	155[b]	565[c]	82[c]	860	125	860[b]	125[b]	345[c]	50[c]	6	7[b]	14[b]
MAR-M 200 Hf[h]	1035	150	1035[b]	150[b]	540[c]	78[c]	825	120	860[b]	125[b]	345[c]	50[c]	5	5[b]	10[b]
PWA-1480[a]	—	—	1130[b]	164[b]	685[c]	99[c]	895	130	905[b]	131[b]	495[c]	72[c]	4	8[b]	—
SEL	1020	148	875[b]	127[b]	—	—	905	131	795[b]	115[b]	—	—	6	7[b]	—
UDM 56	945	137	945[b]	137[b]	—	—	850	123	725[b]	105[b]	—	—	3	5[b]	—
SEL-15	1060	154	1090[b]	158[b]	—	—	895	130	815[b]	118[b]	—	—	9	5[b]	—
Cobalt base															
AiResist 13[i]	600	87	420[b]	61[b]	—	—	530	77	330[b]	48[b]	—	—	1.5	4.5[b]	—
AiResist 215[i]	690	100	570[j]	83[j]	—	—	485	70	315[j]	46[j]	—	—	4	12[j]	—
FSX-414	—	—	—	—	—	—	—	—	—	—	—	—	—	—	—
Haynes 1002	770	112	560	81	115	17	470	68	345	50	95	14	6	8	28
MAR-M 302	930	135	795	115	150	22	690	100	505	73	150	22	2	—	21
MAR-M 322[i]	830	120	595[b]	86[b]	—	—	630	91	345[b]	50[b]	—	—	4	6.5[b]	—
MAR-M 509	785	114	570	83	—	—	570	83	400	58	—	—	4	6	—
WI-52	750	109	745	108	160	23	585	85	440	64	105	15	5	7	35
X-40	745	108	550	80	—	—	525	76	275	40	—	—	9	17	—

[a] Single crystal [001].
[b] At 760°C (1400°F).
[c] At 980°C (1800°F).
[d] RR-7080.
[e] MM 004.
[f] M 005.
[g] MM 006.
[h] MM 009.
[i] Data from *Metals Handbook,* Vol. 3, 9th ed., ASM International, Metals Park, OH, 1980.
[j] At 650°C (1200°F).

Source: Nickel Development Institute, except as noted

Table 7 Effect of Temperature on 1000 h Stress-Rupture Strengths of Selected Wrought Superalloys

		Rupture Strength at							
		650°C (1200°F)		760°C (1400°F)		870°C (1600°F)		980°C (1800°F)	
Alloy	Form	MPa	ksi	MPa	ksi	MPa	ksi	MPa	ksi
Nickel base									
Astroloy	Bar	770	112	425	62	170	25	55	8
Cabot 214	—	—	—	—	—	30	4	15	2
D-979	Bar	515	75	250	36	70	10	—	—
Hastelloy S	Bar	—	—	90	13	25	4	—	—
Hastelloy X	Sheet	215	31	105	15	40	6	15	2
Haynes 230	—	—	—	125	18	55	8	15	2
Inconel 587	Bar	—	—	285	41	—	—	—	—
Inconel 597	Bar	—	—	340	49	—	—	—	—
Inconel 600	Bar	—	—	—	—	30	4	15	2
Inconel 601	Sheet	195	28	60	9	30	4	15	2
Inconel 617	Bar	360	52	165	24	60	9	30	4
Inconel 617	Sheet	—	—	160	23	60	9	30	4
Inconel 625	Bar	370	54	160	23	50	7	20	3
Inconel 706	Bar	580	84	—	—	—	—	—	—
Inconel 718	Bar	595	86	195	28	—	—	—	—
Inconel 718 Direct Age	Bar	405	59	—	—	—	—	—	—
Inconel 718 Super	Bar	600	87	—	—	—	—	—	—
Inconel X750	Bar	470	68	—	—	50	7	—	—
M-252	Bar	565	82	270	39	95	14	—	—
Nimonic 75	Bar	170	25	50	7	5	1	—	—
Nimonic 80A	Bar	420	61	160	23	—	—	—	—
Nimonic 90	Bar	455	66	205	30	60	9	—	—
Nimonic 105	Bar	—	—	330	48	130	19	30	4
Nimonic 115	Bar	—	—	420	61	185	27	70	10
Nimonic 942	Bar	520	75	270	39	—	—	—	—
Nimonic PE.11	Bar	335	49	145	21	—	—	—	—
Nimonic PE.16	Bar	345	50	150	22	—	—	—	—
Nimonic PK.33	Sheet	655	95	310	45	90	13	—	—
Pyromet 860	Bar	545	79	250	36	—	—	—	—
Rene 41	Bar	705	102	345	50	115	17	—	—
Rene 95	Bar	860	125	—	—	—	—	—	—
Udimet 400	Bar	600	87	305	44	110	16	—	—
Udimet 500	Bar	760	110	325	47	125	18	—	—
Udimet 520	Bar	585	85	345	50	150	22	—	—
Udimet 700	Bar	705	102	425	62	200	29	55	8
Udimet 710	Bar	870	126	460	67	200	29	70	10
Udimet 720	Bar	670	97	—	—	—	—	—	—
Unitemp AF2-IDA6	Bar	885	128	360	52	—	—	—	—
Waspaloy	Bar	615	89	290	42	110	16	—	—
Iron base									
A286	Bar	315	46	105	15	—	—	—	—
Alloy 901	Sheet	525	76	205	30	—	—	—	—
Discaloy	Bar	275	40	60	9	—	—	—	—
Haynes 556	Sheet	275	40	125	18	55	8	20	3
Incoloy 800	Bar	165	24	66	9.5	30	4.4	13	1.9
Incoloy 801	Bar	—	—	—	—	—	—	—	—
Incoloy 802	Bar	170	25	110	15	69	10	24	3.5

Table 7 (*Continued*)

		Rupture Strength at							
		650°C (1200°F)		760°C (1400°F)		870°C (1600°F)		980°C (1800°F)	
Alloy	Form	MPa	ksi	MPa	ksi	MPa	ksi	MPa	ksi
Iron base (Continued)									
Incoloy 807	Bar	—	—	105	15	43	6.2	19	2.7
Incoloy 903	Bar	510	74	—	—	—	—	—	—
Incoloy 909	Bar	345	50	—	—	—	—	—	—
N-155	Bar	295	43	140	20	70	10	20	3
V-57	Bar	485	70	—	—	—	—	—	—
Cobalt base									
Haynes 188	Sheet	—	—	165	24	70	10	30	4
L-605	Sheet	270	39	165	24	75	11	30	4
MAR-M918	Sheet	—	—	60	9	20	3	5	1
Haynes 150	—	—	—	40[a]	5.8	—	—	—	—

[a] At 815°C (1500°F).

best materials for creep-rupture-limited applications at high temperatures. Elimination or reorientation/alignment of grain boundaries is sometimes a key factor in maximizing the higher temperature life of an alloy.

A factor in mechanical behavior not often recognized is that the static modulus (e.g., Young's modulus E) is affected by increases in temperature. This should be obvious from the above discussion about creep. Dependent on rate of load application, moduli determined by measurements from a tensile test tend to gradually fall below moduli determined by dynamic means. Since moduli are used in design and may affect predictions of life and durability, every effort should be made to determine the dynamic, not just static, moduli of superalloys for high-temperature applications. Moduli of cast alloys also may be affected dramatically by the orientation of grains or by use of single crystals. A more accurate method of depicting moduli would be to use the appropriate single-crystal elastic constants (three needed), but these are rarely available.

Cyclically applied loads that cause failure (fatigue) at lower temperatures also cause failures in shorter times (lesser cycles) at high temperatures. For example, Fig. 3 shows schematically how the cyclic resistance is degraded at high temperatures when the locus of failure is plotted as stress versus applied cycles (*S–N*) of load. From the *S–N* curves shown, it should be clear that there is not necessarily an endurance limit for metals and alloys at high temperatures. Cyclic loads can be induced not only by mechanical loads in a structure but also by thermal changes. The combination of thermally induced and mechanically induced loads leads to failure in thermal-mechanical fatigue (TMF), failure that occurs in a relatively low number of cycles. Thus TMF is a low-cycle fatigue (LCF) process that can be induced by high repeated mechanical loads, while lower stress repeated mechanical loads lead to fatigue failure in a high number of cycles (HCF). Dependent on application, LCF failures in structures can be either mechanically induced or the TMF type. In airfoils in the hot section of gas turbines, TMF is a major concern. In highly mechanically loaded parts such as gas turbine disks, mechanically induced LCF is the major concern. Normally HCF is not a problem with superalloys unless a design error occurs and a component is subjected to a high-frequency vibration that forces rapid accumulation of fatigue cycles. Although

Table 8 Effect of Temperature on Stress-Rupture Strengths of Selected Cast Superalloys

	Rupture Strength at											
	815°C (1500°F)				870°C (1600°F)				980°C (1800°F)			
	100 h		1000 h		100 h		1000 h		100 h		1000 h	
Alloy	MPa	ksi	MPa	ksi	MPa	ksi	MPa	ksi	MPa	ksi	MPa	ksi
Nickel base												
IN-713 LC	425	62	325	47	295	43	240	35	140	20	105	15
IN-713 C	370	54	305	44	305	44	215	31	130	19	70	10
IN-738 C	470	68	345	50	330	38	235	34	130	19	90	13
IN-738 LC	430	62	315	46	295	43	215	31	140	20	90	13
IN-100	455	66	365	53	360	52	260	38	160	23	90	13
MAR-M 247 (MM 0011)	585	85	415	60	455	66	290	42	185	27	125	18
MAR-M 246	525	76	435	62	440	63	290	42	195	28	125	18
MAR-M 246 Hf(MM 006)	530	77	425	62	425	62	285	41	205	30	130	19
MAR-M 200	495	72	415	60	385	56	295	43	170	25	125	18
MAR-M 200 Hf(MM 009)	—	—	—	—	—	—	305	44	—	—	125	18
B-1900	510	74	380	55	385	56	250	36	180	26	110	16
Ren 77	—	—	—	—	310	45	215	31.5	130	19	62	9.0
Rene 80	—	—	—	—	350	51	240	35	160	23	105	15
IN-625	130	19	110	16	97	14	76	11	34	5	28	4
IN-162	505	73	370	54	340	49	255	37	165	24	110	16
IN-731	505	73	365	53	—	—	—	—	165	24	105	15
IN-792	515	75	380	55	365	53	260	38	165	24	105	15
M-22	515	75	385	56	395	57	285	41	200	29	130	19
MAR-M 421	450	65	305	44	310	46	215	31	125	18	83	12
MAR-M 432	435	63	330	48	295	40	215	31	140	20	97	14
MC-102	195	28	145	21	145	21	105	15	—	—	—	—
Nimocast 90	160	23	110	17	125	18	83	12	—	—	—	—
Nimocast 242	110	16	83	12	90	13	59	8.6	45	6.5	—	—
Udimet 500	330	48	240	35	230	33	165	24	90	13	—	—
Udimet 710	420	61	325	47	305	44	215	31	150	22	76	11
CMSX-2	—	—	—	—	—	—	345	50	—	—	170	25
GMR-235	—	—	—	—	—	—	180	26	—	—	75	11
IN-939	—	—	—	—	—	—	195	28	—	—	60	9
MM 002	—	—	—	—	—	—	305	44	—	—	125	18
IN-713 Hf(MM 004)	—	—	—	—	—	—	205	30	—	—	90	13
Ren 125 Hf(MM 005)	—	—	—	—	—	—	305	44	—	—	115	17
SEL-15	—	—	—	—	—	—	295	43	—	—	75	11
UDM 56	—	—	—	—	—	—	270	39	—	—	125	18
Cobalt base												
HS-21	150	22	95	14	115	17	90	13	60	9	50	7
X-40 (HS-31)	180	26	140	20	130	19	105	15	75	11	55	8
MAR-M 509	270	39	225	33	200	29	140	20	115	17	90	13
PSX-414	150	22	115	17	110	16	85	12	55	8	35	5
WI-52	—	—	195	28	175	25	150	22	90	13	70	10

Table 9 Physical Properties of Selected Wrought Superalloys

			Melting Range		Specific Heat Capacity					
					At 21°C (70°F)		At 538°C (1000°F)		At 871°C (1600°F)	
Designation	Form	Density, g/cm[a]	°C	°F	J/kg·K	Btu/lb·°F	J/kg·K	Btu/lb·°F	J/kg·K	Btu/lb·°F
Nickel base										
Astroloy	Bar	7.91	—	—	—	—	—	—	—	—
D-979	Bar	8.19	1200–1390	2225–2530	—	—	—	—	—	—
Hastelloy X	Sheet	8.21	1260–1355	2300–2470	485	0.116	—	—	700	0.167
Hastelloy S	Bar	8.76	1335–1380	2435–2515	405	0.097	495	0.118	595	0.142
Inconel 597	Bar	8.04	—	—	—	—	—	—	—	—
Inconel 600	Bar	8.41	1355–1415	2470–2575	445	0.106	555	0.132	625	0.149
Inconel 601	Bar	8.05	1300–1370	2375–2495	450	0.107	590	0.140	680	0.162
Inconel 617	Bar	8.36	1330–1375	2430–2510	420	0.100	550	0.131	630	0.150
Inconel 625	Bar	8.44	1290–1350	2350–2460	410	0.098	535	0.128	620	0.148
Inconel 690	Bar	8.14	1345–1375	2450–2510	—	—	—	—	—	—
Inconel 706	Bar	8.08	1335–1370	2435–2500	445	0.106	580	0.138	670	0.159
Inconel 718	Bar	8.22	1260–1335	2300–2435	430	0.102	560	0.133	645	0.153
Inconel X750	Bar	8.25	1395–1425	2540–2600	430	0.103	545	0.130	715	0.171
Haynes 230	—	8.8	1300–1370	2375–2500	397	0.095	473	0.112	595	0.145
M-252	Bar	8.25	1315–1370	2400–2500	—	—	—	—	—	—
Nimonic 75	Bar	8.37	—	—	460	0.11	—	—	—	—
Nimonic 80A	Bar	8.16	1360–1390	2480–2535	460	0.11	—	—	—	—
Nimonic 81	Bar	8.06	—	—	460	0.11	585	0.14	670	0.16
Nimonic 90	Bar	8.19	1335–1360	2435–2480	460	0.11	585	0.14	670	0.16
Nimonic 105	Bar	8.00	—	—	420	0.10	545	0.13	670	0.16
Nimonic 115	Bar	7.85	—	—	460	0.11	—		—	—
Nimonic 263	Sheet	8.36	—	—	460	0.11	—	—	—	—
Nimonic 942	Bar	8.19	1240–1300	2265–2370	420	0.10	—	—	—	—
Nimonic PE 11	Bar	8.02	1280–1350	2335–2460	420	0.10	585	0.14	—	—
Nimonic PE 16	Bar	8.02	—	—	545	0.13	—	—	—	—
Nimonic PK 33	Sheet	8.21	—	—	420	0.10	545	0.13	670	0.16
Pyromet 860	Bar	8.21	—	—	—	—	—	—	—	—
Rene 41	Bar	8.25	1315–1370	2400–2500	—	—	545	0.13	725	0.173
Rene 95	Bar	—	—	—	—	—	—	—	—	—
Udimet 500	Bar	8.02	1300–1395	2375–2540	—	—	—	—	—	—
Udimet 520	Bar	8.21	1260–1405	2300–2560	—	—	—	—	—	—
Udimet 700	Bar	7.91	1205–1400	2200–2550	—	—	575	0.137	590	0.141
Unitemp AFI-IDA	Bar	8.26	—	—	420	0.100	—	—	—	—
Waspaloy	Bar	8.19	1330–1355	2425–2475	—	—	—	—	—	—
Iron base										
Alloy 901	Bar	8.21	1230–1400	2250–2550	—	—	—	—	—	—
A-286	Bar	7.91	1370–1400	2500–2550	460	0.11	—	—	—	—
Discaloy	Bar	7.97	1380–1465	2515–2665	475	0.113	—	—	—	—
Haynes 556	Sheet	8.23	—	—	450	0.107	—	—	—	—
Incoloy 800	Bar	7.95	1355–1385	2475–2525	455	0.108	—	—	—	—
Incoloy 801	Bar	7.95	1355–1385	2475–2525	455	0.108	—	—	—	—
Incoloy 802	Bar	7.83	1345–1370	2450–2500	445	0.106	—	—	—	—
Incoloy 807	Bar	8.32	1275–1355	2325–2475	—	—	—	—	—	—
Incoloy 825[a]	—	8.14	1370–1400	2500–2550	440	0.105	—	—	—	—
Incoloy 903	Bar	8.14	1320–1395	2405–2540	435	0.104	—	—	—	—
Incoloy 904	Bar	8.12	—	—	460	0.11	—	—	—	—
Incoloy 907	—	8.33	1335–1400	2440–2550	431	0.103	—	—	—	—
Incoloy 909	—	8.30	1395–1430	2540–2610	427	0.102	—	—	—	—
N-155	Bar	8.19	1275–1355	2325–2475	430	0.103	—	—	—	—
19-9 DL	—	7.9	1425–1430	2600–2610	—	—	—	—	—	—
16-25-6	—	8.0	—	—	—	—	—	—	—	—
Cobalt base										
Haynes	Sheet	8.98	1300–1330	2375–2425	405	0.097	510	0.122	565	0.135
L-605	Sheet	9.13	1330–1410	2425–2570	385	0.092	—	—	—	—
Haynes 150	—	8.05	1395	2540	—	—	—	—	—	—
MP35N	—	8.41	1315–1425	2400–2600	—	—	—	—	—	—
Elgiloy	—	8.3	1495	2720	—	—	—	—	—	—

Table 9 *(Continued)*

		Thermal Conductivity						Mean Coefficient of Thermal Expansion, 10^{-6} K		
		At 21°C (70°F)		At 538°C (1000°F)		At 871°C (1600°F)				
Designation	Form	W/m·K	Btu/ft²·in. h·°F	W/m·K	Btu/ft²·in. h·°F	W/m·K	Btu/ft²·in. h·°F	At 538°C (1000°F)	At 871°C (1600°F)	Electrical Resistivity, nΩ·m
Nickel base										
Astroloy	Bar	—	—	—	—	—	—	13.9	16.2	—
D-979	Bar	12.6	87	18.5	128	—	—	14.9	17.7	—
Hastelloy X	Sheet	9.1	63	19.6	136	26.0	180	15.1	16.2	1180
Hastelloy S	Bar	—	—	20.0	139	26.1	181	13.3	14.9	—
Inconel 597	Bar	—	—	18.2	126	—	—	—	—	—
Inconel 600	Bar	14.8	103	22.8	158	28.9	200	15.1	16.4	1030
Inconel 601	Bar	11.3	78	20.0	139	25.7	178	15.3	17.1	1190[a]
Inconel 617	Bar	13.6	94	21.5	149	26.7	185	13.9	15.7	1220[a]
Inconel 625	Bar	9.8	68	17.5	121	22.8	158	14.0	15.8	1290
Inconel 690	Bar	13.3	95	22.8	158	27.8	193	—	—	148
Inconel 706	Bar	12.6	87	21.2	147	—	—	15.7	—	—
Inconel 718	Bar	11.4	79	19.6	136	24.9	173	14.4	—	1250[a]
Inconel X750	Bar	12.0	83	18.9	131	23.6	164	14.6	16.8	1220[a]
Haynes 230	—	8.9	62	18.4	133	24.4	179	14.0	15.2	1250
M-252	Bar	11.8	82	—	—	—	—	13.0	15.3	—
Nimonic 75	Bar	—	—	—	—	—	—	14.7	17.0	1090[a]
Nimonic 80A	Bar	8.7	60	15.9	110	22.5	156	13.9	15.5	1240[a]
Nimonic 81	Bar	10.8	75	19.2	133	25.1	174	14.2	17.5	1270[a]
Nimonic 90	Bar	9.8	68	17.0	118	—	—	13.9	16.2	1180[a]
Nimonic 105	Bar	10.8	75	18.6	129	24.0	166	13.9	16.0	1310[a]
Nimonic 115	Bar	10.7	74	17.6	124	22.6	154	13.3	16.4	1390[a]
Nimonic 263	Sheet	11.7	81	20.4	141	26.2	182	13.7	16.2	1150[a]
Nimonic 942	Bar	—	—	—	—	—	—	14.7	16.5	—
Nimonic PE 11	Bar	—	—	—	—	—	—	15.2	—	—
Nimonic PE 16	Bar	11.7	81	20.2	140	26.4	183	15.3	18.5	1100[a]
Nimonic PK 33	Sheet	10.7	74	19.2	133	24.7	171	13.1	16.2	1260[a]
Pyromet 860	Bar	—	—	—	—	—	—	15.4	16.4	—
Rene 41	Bar	9.0	62	18.0	125	23.1	160	13.5	15.6	1308
Rene 95	Bar	8.7	60	17.4	120	—	—	—	—	—
Udimet 500	Bar	11.1	77	18.3	127	24.5	170	14.0	16.1	1203
Udimet 700	Bar	19.6	136	20.6	143	27.7	192	13.9	16.1	—
Unitemp AFI-IDA	Bar	10.8	75	16.5	114	19.5	135	12.4	14.1	—
Waspaloy	Bar	10.7	74	18.1	125	24.1	167	14.0	16.0	1240
Iron base										
Alloy 901	Bar	13.3	92	—	—	—	—	15.3	—	—
A-286	Bar	12.7	88	22.5	156	—	—	17.6	—	—
Discaloy	Bar	13.3	92	21.1	146	—	—	17.1	—	—
Haynes 556	Sheet	11.6	80	17.5	121	—	—	16.2	17.5	—
Incoloy 800	Bar	11.6	80	20.1	139	—	—	16.4	18.4	989
Icolony 801	Bar	12.4	86	20.7	143	25.6	177	17.3	18.7	—
Incoloy 802	Bar	11.9	82	19.8	137	24.2	168	16.7	18.2	—
Incoloy 807	Bar	—	—	—	—	—	—	15.2	17.6	—
Incoloy 825	—	11.1	76.8	—	—	—	—	14.0[a]	—	1130
Incoloy 903	Bar	16.8	116	20.9	145	—	—	8.6	—	610
Incoloy 904	Bar	16.8	116	22.4	155	—	—	—	—	—
Incoloy 907	—	14.8	103	—	—	—	—	7.7[b]	—	697
Incoloy 909	—	14.8	103	—	—	—	—	7.7[b]	—	728
N-155	Bar	12.3	85	19.2	133	—	—	16.4	17.8	—
19-9 DL	—	—	—	—	—	—	—	17.8	—	—
16-25-6	—	—	—	15	104	—	—	16.9	—	—

Table 9 *(Continued)*

		Thermal Conductivity						Mean Coefficient of Thermal Expansion, 10^{-6} K		
		At 21°C (70°F)		At 538°C (1000°F)		At 871°C (1600°F)				Electrical
Designation	Form	W/m·K	Btu/ft²·in. h·°F	W/m·K	Btu/ft²·in. h·°F	W/m·K	Btu/ft²·in. h·°F	At 538°C (1000°F)	At 871°C (1600°F)	Resistivity, nΩ·m
Cobalt base										
Haynes 188	Sheet	—	—	19.9	138	25.1	174	14.8	17.0	922
L-605	Sheet	9.4	65	19.5	135	26.1	181	14.4	16.3	890
Stellite 6B	—	14.7	101	—	—	—	—	15.0	16.9	910
Haynes 150	—	—	—	0.75[c]	5.2	—	—	—	16.8[d]	810
MP35N	—	—	—	—	—	—	—	—	15.7[d]	1010
Eligiloy	—	1.0	7.2	1.4	10	—	—	—	15.8[d]	995

[a] At 21–93°C (70–200°F).
[b] At 25–427°C (77–800°F).
[c] At 705°C (1300°F).
[d] At 980°C (1800°F).

S–N plots are common, the designer should be aware that data for LCF and TMF behavior frequently are gathered as plastic strain (epsilon plastic, ε_p or $\Delta\varepsilon_p$) versus applied cycles. The final component application will determine the preferred method of depicting fatigue behavior.

While life under cyclic load (*S–N* behavior) is a common criterion for design, resistance to crack propagation is an increasingly desired property. Thus, the crack growth rate (da/dn) versus a fracture toughness parameter is required. The parameter in this instance is the stress intensity factor (K) range over an incremental distance which a crack has grown—the difference between the maximum and minimum K in the region of crack length measured. Plots of the resultant type (da/dn versus ΔK) is shown in Fig. 4 for several wrought superalloys.

The nature of superalloys is that they resist the creep-rupture process better than other materials, have very good higher temperature short-time strength (yield, ultimate) and very good fatigue properties (including fatigue crack propagation resistance), and combine these mechanical properties with good to exceptional oxidation resistance. Consequently, superalloys are the obvious choice when structures are to operate at higher temperatures. Generally, the nominal temperature range of superalloy operation is broken up into the intermediate range of 540°C (1000°F) to 760°C (1400°F) and the high-temperature range which occurs above about 816°C (1500°F).

3 PROPERTIES OF SUPERALLOYS

3.1 Physical/Environmental

Iron, cobalt, and nickel, the bases of superalloys, are transition metals located in a similar area in the periodic table of the elements. The density, melting point, and physical properties of the superalloy base elements are given in Table 13. As can be seen, pure iron has a density of 7.87 g/cm³ (0.284 lb/in.³), while pure nickel and cobalt have densities of about 8.9 g/cm³ (0.322 lb/in.³). The superalloys are usually created by adding significant levels of the alloy elements chromium, aluminum, and titanium plus appropriate refractory metal elements such as tungsten and molybdenum to the base metal. Densities of superalloys are a function of the amounts of these elements in the final compositions. For example, iron–nickel-base

Table 10 Physical Properties of Selected Cast Superalloys

Alloy	Density, g/cm³	Melting Range °C	Melting Range °F	Specific Heat At 21°C (70°F) J/kg·K	Specific Heat At 21°C (70°F) Btu/lb·°F	Specific Heat At 538°C (1000°F) J/kg·K	Specific Heat At 538°C (1000°F) Btu/lb·°F	Specific Heat At 1093°C (2000°F) J/kg·K	Specific Heat At 1093°C (2000°F) Btu/lb·°F	Thermal Conductivity At 93°C (200°F) W/m·K	Thermal Conductivity At 93°C (200°F) Btu·in./h·ft²·°F	Thermal Conductivity At 538°C (1000°F) W/m·K	Thermal Conductivity At 538°C (1000°F) Btu·in./h·ft²·°F	Thermal Conductivity At 1093°C (2000°F) W/m·K	Thermal Conductivity At 1093°C (2000°F) Btu·in./h·ft²·°F	Mean Coefficient of Thermal Expansion(s),[a] 10^{-6} K At 93°C (200°F)	At 538°F (1000°C)	At 1093°C (2000°F)
Nickel base																		
IN-713 C	7.91	1260–1290	2300–2350	420	0.10	565	0.135	710	0.17	10.9	76	17.0	118	26.4	183	10.6	13.5	17.1
IN-713 LC	8.00	1290–1320	2350–2410	440	0.105	565	0.135	710	0.17	10.7	74	16.7	116	25.3	176	10.1	15.8	18.9
B-1900	8.22	1275–1300	2325–2375	—	—	—	—	—	—	(10.2)[b]	(71)[b]	16.3	113	—	—	11.7	13.3	16.2
Cast alloy 625	8.44	—	—	—	—	—	—	—	—	—	—	—	—	—	—	—	—	—
Cast alloy 718	8.22	1205–1345	2200–2450	—	—	—	—	—	—	—	—	—	—	—	—	—	—	—
IN-100	7.75	1265–1335	2305–2435	—	—	480	0.115	605	0.145	—	—	17.3	120	—	—	13.0	13.9	18.1
IN-162	8.08	1275–1305	2330–2380	—	—	—	—	—	—	—	—	—	—	—	—	12.2	14.1	—
IN-731	7.75	—	—	—	—	—	—	—	—	—	—	—	—	—	—	—	—	—
IN-738	8.11	1230–1315	2250–2400	420	0.10	565	0.135	710	0.17	—	—	17.7	123	27.2	189	11.6	14.0	—
IN-792	8.25	—	—	—	—	—	—	—	—	—	—	—	—	—	—	—	—	—
M-22	8.63	—	—	—	—	—	—	—	—	—	—	—	—	—	—	12.4	13.3	—
MAR-M-200	8.53	1315–1370	2400–2500	400	0.095	420	0.10	565	0.135	13.0	90	15.2	110	29.7	206	—	13.1	17.0
MAR-M-246	8.44	1315–1345	2400–2450	—	—	—	—	—	—	—	—	18.9	131	30.0	208	11.3	14.8	18.6
MAR-M-247	8.53	—	—	—	—	—	—	—	—	—	—	—	—	—	—	—	—	—
MAR-M-421	8.08	—	—	—	—	—	—	—	—	—	—	19.1	137	32.0	229	—	14.9	19.8
MAR-M-432	8.16	—	—	—	—	—	—	—	—	—	—	—	—	—	—	—	14.9	19.3
MC-102	8.84	—	—	—	—	—	—	—	—	—	—	—	—	—	—	12.8	14.9	—
Nimocast 75	8.44	1410[c]	2570[c]	—	—	—	—	—	—	—	—	—	—	—	—	12.8	14.9	—
Nimocast 80	8.17	1310–1380	2390–2515	—	—	—	—	—	—	—	—	—	—	—	—	12.8	14.9	—
Nimocast 90	8.18	1310–1380	2390–2515	—	—	—	—	—	—	—	—	—	—	—	—	12.3	14.8	—
Nimocast 242	8.40	1225–1340	2235–2445	—	—	—	—	—	—	—	—	—	—	—	—	12.5	14.4	—
Nimocast 263	8.36	1300–1355	2370–2470	—	—	—	—	—	—	—	—	—	—	—	—	11.0	13.6	—
Rene 77	7.91	—	—	—	—	—	—	—	—	—	—	—	—	—	—	—	—	—
Rene 80	8.16	—	—	—	—	—	—	—	—	—	—	—	—	—	—	—	—	—
Udimet 500	8.02	1300–1395	2375–2540	—	—	—	—	—	—	—	—	—	—	—	—	13.3	—	—
Udimet 710	8.08	—	—	—	—	—	—	—	—	12.1	84	18.1	126	—	—	—	—	—
Cobalt base																		
FSX-414	8.3	—	—	—	—	—	—	—	—	—	—	—	—	—	—	—	—	—
Haynes 1002	8.75	1305–1420	2380–2590	420	0.10	530	0.126	645	0.154	11.0	76	21.8	151	32.1	222	12.2	14.4	—
MAR-M-302	9.21	1315–1370	2400–2500	—	—	—	—	—	—	18.7	130	22.2	154	—	—	—	13.7	16.6
MAR-M-322	8.91	1315–1360	2400–2475	—	—	—	—	—	—	—	—	—	—	—	—	—	—	—
MAR-M-509	8.85	—	—	—	—	—	—	—	—	—	—	27.9	194	44.6	310	9.8	15.9	18.3
WI-52	8.88	1300–1355	2425–2475	420	0.10	—	—	—	—	24.8	172	27.4	190	40.3	280	—	14.4	17.5
X-40	8.60	—	—	—	—	—	—	—	—	11.8	82	21.6	150	—	—	—	15.1	—

[a] From room temperature to indicated temperature.
[b] Estimated.
[c] Liquidus temperature.
Source: Nickel Development Institute.

Table 11 Dynamic Moduli of Selected Wrought Superalloys

		Dynamic Modulus of Elasticity									
		At 21°C (70°F)		At 540°C (1000°F)		At 650°C (1200°F)		At 760°C (1400°F)		At 870°C (1600°F)	
Alloy	Form	GPa	10^6 psi	GPa	10^6 psi	GPa	10^6 psi	GPa	10^6 psi	GPa	10^6 psi
Nickel base											
D-979	Bar	207	30.0	178	25.8	167	24.2	156	22.6	146	21.2
Hastelloy S	Bar	212	30.8	182	26.4	174	25.2	166	24.1	—	—
Hastelloy X	Sheet	197	28.6	161	23.4	154	22.3	146	21.1	137	19.9
Haynes 230	[a]	211	30.6	184	26.4	177	25.3	171	24.1	164	23.1
Inconel 587	Bar	222	32.1	—	—	—	—	—	—	—	—
Inconel 596	Bar	186	27.0	—	—	—	—	—	—	—	—
Inconel 600	Bar	214	31.1	184	26.7	176	25.5	168	24.3	157	22.8
Inconel 601[b]	Sheet	207	30.0	175	25.4	166	24.1	155	22.5	141	20.5
Inconel 617	Bar	210	30.4	176	25.6	168	24.4	160	23.2	150	21.8
Inconel 625	Bar	208	30.1	179	25.9	170	24.7	161	23.3	148	21.4
Inconel 706	Bar	210	30.4	179	25.9	170	24.7	—	—	—	—
Inconel 718	Bar	200	29.0	171	24.8	163	23.7	154	22.3	139	20.2
Inconel X750	Bar	214	31.0	184	26.7	176	25.5	166	24.0	153	22.1
M-252	Bar	206	29.8	177	25.7	168	24.4	156	22.6	145	21.0
Nimonic 75	Bar	221	32.0	186	27.0	176	25.5	170	24.6	156	22.6
Nimonic 80A	Bar	219	31.8	188	27.2	179	26.0	170	24.6	157	22.7
Nimonic 90	Bar	226	32.7	190	27.6	181	26.3	170	24.7	158	22.9
Nimonic 105	Bar	223	32.3	186	27.0	178	25.8	168	24.4	155	22.5
Nimonic 115	Bar	224	32.4	188	27.2	181	26.3	173	25.1	164	23.8
Nimonic 263	Sheet	222	32.1	190	27.5	181	26.2	171	24.8	158	22.9
Nimonic 942	Bar	196	28.4	166	24.1	158	22.9	150	21.8	138	20.0
Nimonic PE 11	Bar	198	28.7	166	24.0	157	22.8	—	—	—	—
Nimonic PE 16	Bar	199	28.8	165	23.9	157	22.7	147	21.3	137	19.9
Nimonic PK 33	Sheet	222	32.1	191	27.6	183	26.5	173	25.1	162	23.5
Pyromet 860	Bar	200	29.0	—	—	—	—	—	—	—	—
Rene 95	Bar	209	30.3	183	26.5	176	25.5	168	24.3	—	—
Udimet 500	Bar	222	32.1	191	27.7	183	26.5	173	25.1	161	23.4
Udimet 700	Bar	224	32.4	194	28.1	186	27.0	177	25.7	167	24.2
Udimet 710	Bar	222	32.1	—	—	—	—	—	—	—	—
Waspaloy	Bar	213	30.9	184	26.7	177	25.6	168	24.3	158	22.9
Iron base											
A-286	Bar	201	29.1	162	23.5	153	22.2	142	20.6	130	18.9
Alloy 901	Bar	206	29.9	167	24.2	153	22.1	—	—	—	—
Discaloy	Bar	196	28.4	154	22.3	145	21.0	—	—	—	—
Haynes 556	Sheet	203	29.5	165	23.9	156	22.6	146	21.1	137	19.9
Incoloy 800	Bar	196	28.4	161	23.4	154	22.3	146	21.1	138	20.0
Incoloy 801	Bar	208	30.1	170	24.7	162	23.5	154	22.3	144	20.9
Incoloy 802	Bar	205	29.7	169	24.5	161	23.4	156	22.6	152	22.0
Incoloy 807	Bar	184	26.6	155	22.4	146	21.2	137	19.9	128	18.5
Incoloy 903	Bar	147[c]	21.3	152[d]	22.1	—	—	—	—	—	—
Incoloy 907[d]	—	165[c]	23.9	165[d]	23.9	159[c]	23	—	—	—	—
N-155	Bar	202	29.3	167	24.2	159	23.0	149	21.6	138	20.0
V-57	Bar	199	28.8	163	23.6	153	22.2	144	20.8	130	18.9
19-9 DL	—	203	29.5	—	—	152	22.1	—	—	—	—
16-25-6	—	195	28.2	—	—	123	17.9	—	—	—	—

Table 11 (Continued)

		Dynamic Modulus of Elasticity									
		At 21°C (70°F)		At 540°C (1000°F)		At 650°C (1200°F)		At 760°C (1400°F)		At 870°C (1600°F)	
Alloy	Form	GPa	10^6 psi	GPa	10^6 psi	GPa	10^6 psi	GPa	10^6 psi	GPa	10^6 psi
Cobalt base											
Haynes 188	Sheet	207	30	—	—	—	—	—	—	—	—
L-605	Sheet	216	31.4	—	—	185	26.8	—	—	166	24.0
MAR-M-918	Sheet	225	32.6	186	27.0	176	25.5	168	24.3	159	23.0
MP35N	Bar	231[c]	33.6	—	—	—	—	—	—	—	—
Haynes 150	—	217[c]	31.5	—	—	—	—	—	—	—	—

[a]Cold-rolled and solution-annealed sheet, 1.2–1.6 mm (0.048–0.063 in.) thick.
[b]Data for bar, rather than sheet.
[c]Annealed.
[d]Precipitation hardened.

superalloys have densities of around 7.9–8.3 g/cm^3 (0.285–0.300 lb/in.^{3}), not too dissimilar to densities of nickel-base superalloys, which may range from about 7.8 to 8.9 g/cm^3 (0.282–0.322 lb/in.^{3}). Cobalt-base superalloy densities, on the other hand, range from about 8.3 to 9.4 g/cm^3 (0.300–0.340 lb/in.^{3}). Aluminum, titanium, and chromium reduce superalloy density whereas the refractory elements such as tungsten, rhenium, and tantalum increase it.

The melting temperatures of the basis superalloy elements are nickel, 1452°C (2647°F); cobalt, 1493°C (2723°F); and iron, 1535°C (2798°F). When metals are alloyed, there is no

Table 12 Dynamic Modulus of Elasticity for a Few Cast Superalloys

	Dynamic Modulus of Elasticity					
	At 21°C (70°F)		At 538°C (1000°F)		At 1093°C (2000°F)	
Alloy	GPa	10^6 psi	GPa	10^6 psi	GPa	10^6 psi
Nickel base						
IN-713 C	206	29.9	179	26.2	—	—
IN-713 LC	197	28.6	172	25.0	—	—
B-1900	214	31.0	183	27.0	—	—
IN-100	215	31.2	187	27.1	—	—
IN-162	197	28.5	172	24.9	—	—
IN-738	201	29.2	175	25.4	—	—
MAR-M-200	218	31.6	184	26.7	—	—
MAR-M-246	205	29.8	178	25.8	145	21.1
MAR-M-247	—	—	—	—	—	—
MAR-M-421	203	29.4	—	—	141	20.4
Rene 80	208	30.2	—	—	—	—
Cobalt base						
Haynes 1002	210	30.4	173	25.1	—	—
MAR-M-509	225	32.7	—	—	—	—

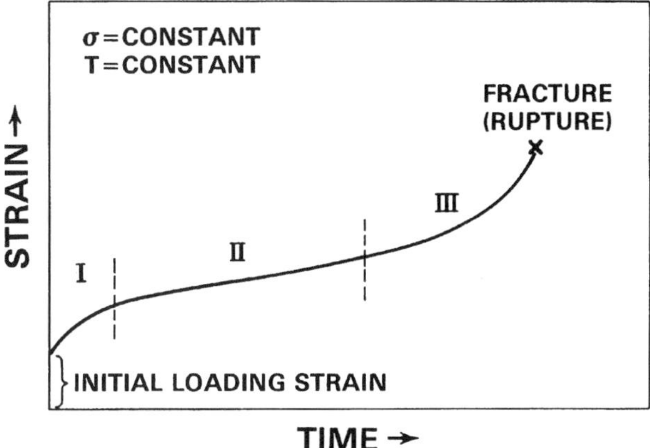

Figure 1 Creep-rupture schematic showing time-dependent deformation under constant load at constant high temperatures followed by final rupture. (All loads below the short time yield strength. Roman numerals denote ranges of the creep-rupture curve.)

longer a single melting point for a composition. Instead, alloys melt over a range. The lowest melting temperature (incipient melting temperature) and melting ranges of superalloys are functions of composition and prior processing. Just as the basis metal is higher melting, so generally are incipient melting temperatures greater for cobalt-base superalloys than for nickel- or iron–nickel-base superalloys. Nickel-base superalloys may show incipient melting

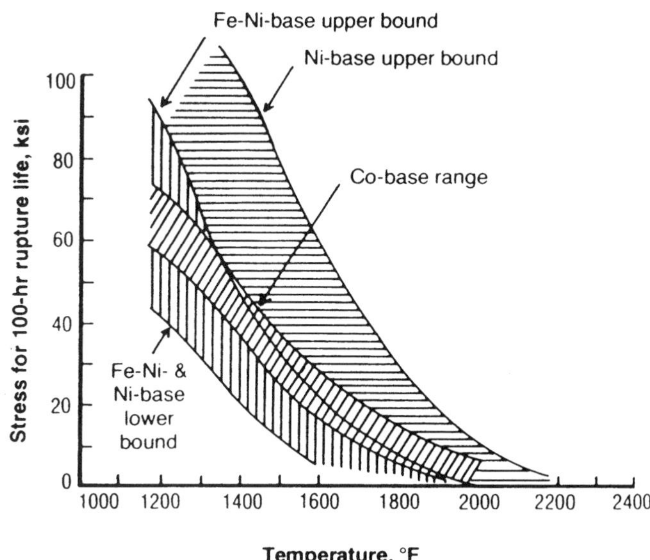

Figure 2 Creep-rupture curves showing ranges for superalloy families. (From *Superalloys Source Book,* ASM International, Materials Park, OH, 1984, p. 3.)

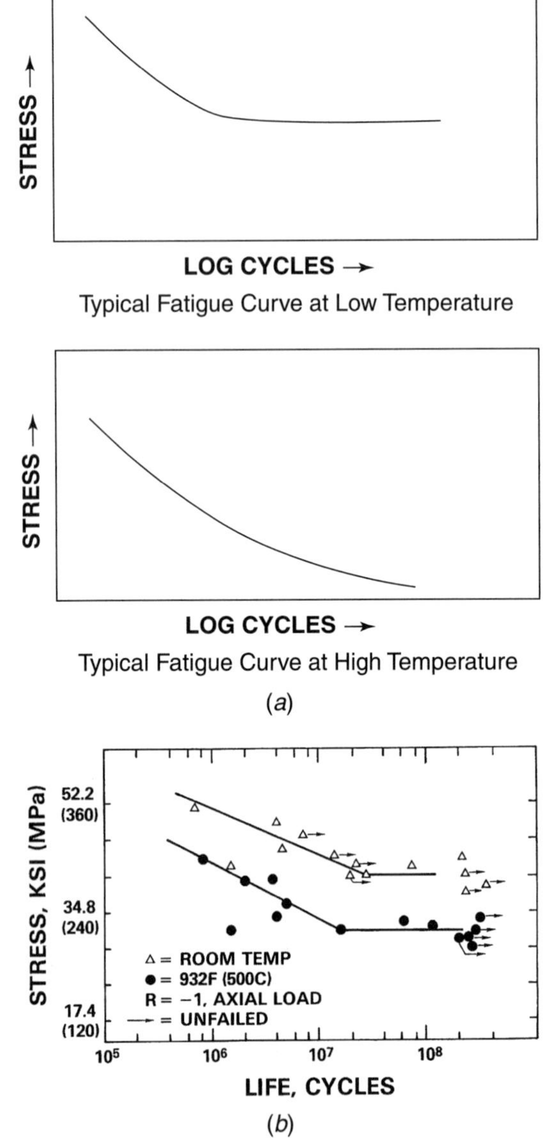

Figure 3 Stress versus cycles to failure (*S–N*) curves showing: (*a*) schematic typical fatigue response at high temperature versus that at lower temperature and (*b*) actual fatigue curves at room and elevated temperature for a specific nickel-base superalloy.

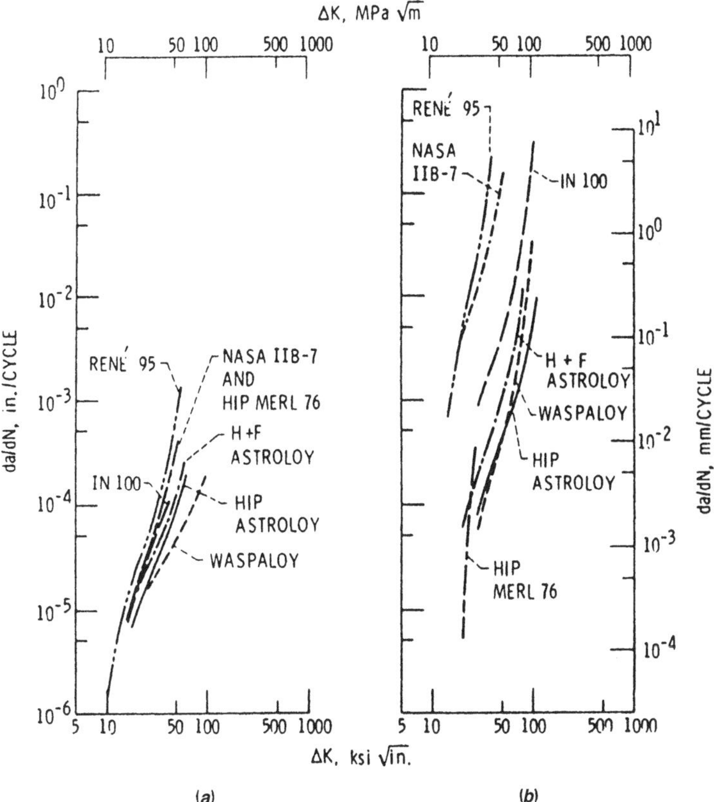

Figure 4 Crack growth rate (da/dn) versus toughness change (ΔK) curves for several superalloys at 649°C (1200°F). Note that HIP = hot isostatically pressed and H + F = HIP + forge. Unless otherwise noted, all alloys are forged. (From *Superalloys II,* Wiley, 1987, p. 284. Used by permission.)

Table 13 Some Physical Properties of Superalloy Base Elements

	Crystal Structure	Melting Point		Density		Expansion Coefficient[a]		Thermal Conductivity[a]	
		°F	°C	lb/in.^{3}	g/cm^3	°F × 10^{-6}	°C × 10^{-6}	Btu/ft^2/hr/°F/in.	cal/cm^2/s/°C/cm
Co	hcp	2723	1493	0.32	8.9	7.0	12.4	464	0.215
Ni	fcc	2647	1452	0.32	8.9	7.4	13.3	610	0.165
Fe	bcc	2798	1535	0.28	7.87	6.7	11.7	493	0.175

[a]At room temperature.

Source: From *Superalloys II,* Wiley, New York, 1987, p. 14.

at temperatures as low as 1204°C (2200°F). However, advanced nickel-base single-crystal superalloys having limited amounts of melting-point depressants tend to have incipient melting temperatures equal to or in excess of those of cobalt-base superalloys.

The physical properties, electrical conductivity, thermal conductivity, and thermal expansion of superalloys tend to be low (relative to other metal systems). These properties are influenced by the nature of the base metals (transition elements) and the presence of refractory-metal additions.

The corrosion resistance of superalloys depends primarily on the alloying elements added and the environment experienced. Contaminants in the atmosphere can cause unexpectedly high corrosion rates. The superalloys employed at the highest temperatures are coated to increase oxidation/corrosion resistance. More information follows later in this chapter.

3.2 Mechanical

The superalloys are relatively ductile; the ductilities of cobalt-base superalloys generally are less than those of iron–nickel- and nickel-base superalloys. Short-time tensile ductilities generally range from as low as 10% to as high as 70% but γ' hardened alloys are in the lower end, usually between 10 and 40%. Creep-rupture ductilities can range much lower. At the 760°C (1400°F) tensile ductility minimum area (see mention below for tensile testing), creep-rupture ductilities of castings have gone below 1.5%, although most current high-strength polycrystalline (PC) equiaxed cast alloys have rupture ductilities in excess of 2.0%. Single-crystal directionally solidified (SCDS) superalloy ductilities can vary with orientation of the single crystal relative to the testing direction.

Superalloys typically have dynamic moduli of elasticity in the vicinity of 207 GPa (30×10^6 psi), although moduli of specific PC alloys can vary from 172 to 241 GPa (25×10^6–35×10^6 psi) at room temperature depending on the alloy system. Processing that leads to directional grain or crystal orientation can result in moduli of about 124–310 GPa (about 18×10^6–45×10^6 psi) depending on the relation of grain or crystal orientation to testing direction. As noted above, dynamic measurement of modulus of elasticity at high temperatures is necessary because static modulus is greatly influenced by high temperatures and shows significant reductions over the dynamic value at a common high temperature. Static modulus may drop by around 25–30% as temperatures increase from room temperature to 871°C (1600°F).

Short-time tensile yield properties of γ' hardened alloys range from 550 MPa (80 ksi) to 1380 MPa (200 ksi) at room temperature. Actual values depend on composition and processing (cast versus wrought). Wrought alloys tend to have the highest values, with the highest hardener content alloys (e.g., Rene 95, IN 100) having the highest strengths. However, strength also is a function of grain size and stored energy. Alloys such as U630 or IN 718 can be produced with very high yield strengths. Solid-solution-hardened alloys such as the sheet alloy Hastelloy X have lower strengths. The ultimate strengths of superalloys range from 690 MPa (100 ksi) to 1520 MPa (230 ksi) at room temperature, with γ'-hardened alloys in the high end of the range.

Superalloys tend to show an increase of yield strength from room temperature up to a nominal temperature of 760°C (1400°F) and drop off thereafter. This is in contrast to non-superalloys, which tend to continuously decrease in short yield time strength as temperatures increase. Ultimate strengths generally do not show this trend. Concurrently, tensile ductility tends to decrease, with minimums in the range of 649°C (1200°F) to 760°C (1400°F). Many published data for alloys do not show significant increase in tensile yield strength over the range to 760°C (1400°F).

The highest tensile properties are found in the finer grain size wrought or powder metallurgy superalloys used in applications at the upper end of the intermediate-temperature regime, nominally 760°C (1400°F). The highest creep-rupture properties invariably are found in the coarser grain cast superalloys which are used in the high-temperature regime. Rupture strengths are a function of the time at which they are to be recorded. The 1000-h stress-capability rupture is obviously lower than the 100-h capability. Creep capability also is a function of the amount of creep permitted in a test. For example, the time to 0.5, 1.0, 2.0, or 5.0% might each be valuable for design, dependent on a component's intended use. It is much more difficult to find this information than it is to find creep-rupture capability information. Handbooks generally do not carry much creep information. Creep-rupture strengths for 100 h failure at 982°C (1800°F) may range from 45 MPa (6.5 ksi) for an older γ'-hardened wrought alloy such as U500 to 205 MPa (30 ksi) for the PC equiaxed cast superalloy Mar-M 246. Columnar grain and single-crystal alloys can be much stronger.

Cyclic properties are not commonly tabulated for superalloys. Properties of interest would be the 10^3–10^5 and 10^6–10^8 cycle fatigue strength capabilities. This could mean stress for a fixed-cyclic-life-to-a-particular-sized-crack or stress for a fixed-cyclic-life-to-fracture for LCF regimes or only stress for a fixed-cyclic-life-to-fracture for HCF regimes. Also, crack propagation rates versus toughness parameter (da/dn versus ΔK) are desired. The life values, when available, lend themselves to tabulation, but the da/dn values are best represented by graphs for wrought alloys used at intermediate temperatures. The LCF strengths are usually related to an alloy's yield strength while HCF strengths are usually related to an alloy's ultimate strength. For cast alloys used in the hottest sections of a gas turbine, there appears to be a relation of TMF strength to the creep strength of an alloy for a given alloy form, e.g., as for columnar grain directionally solidified (CGDS) nickel-base superalloys.

Superalloys usually are processed to optimize one property in preference to others. The same composition, if used in cast and wrought states, may have different heat treatments applied to the different product forms. Even when a superalloy is used in the same product form, process treatments may be used to optimize one property over others. For example, by adjusting the processing conditions (principally heat treatment) of wrought Waspaloy, substantial yield strength improvement can be made at the expense of creep-rupture strengths.

4 EVOLUTION OF SUPERALLOYS

4.1 Introduction

During the first quarter of the twentieth century, chromium was added at various times to bases of cobalt, nickel, and iron. The resulting products were remarkably resistant to atmospheric (moisture-based) environments and to oxidation at high temperatures. By the time of the World War II, some of these alloys, now containing additional alloy elements, had come into use for such applications as resistance wires, dental prostheses, cutlery, furnace and steam turbine components, etc. With the development of the gas turbine engine during the war, the need became apparent for corrosion-resistant materials to operate in demanding mechanical load conditions at high temperatures. At this point, the fledgling superalloy industry began to expand.

By modifying the stainless steels, higher strengths were achieved without the need for special high-temperature strengthening phases. Phases such as η (a nickel–titanium compound) or γ' (a nickel–aluminum compound) had been introduced into the nickel–chromium families of alloys just prior to the war to produce high strength at high temperatures. The increasing temperatures forced alloy developers to include these phases (η, γ') in the iron-base alloys to take the high-temperature strength characteristics beyond those of the modified stainless steels such as 19-9DL. Precipitation-hardened iron–nickel-base alloys were invented in Germany and, after modification, made their way to the United States and, as A-286 or V-57, are still in use today.

Nevertheless, the need for creep-rupture strength continually increased. Some of this need was met in the early years by adapting a cobalt-base corrosion-resistant alloy (Vitallium) for use in aircraft engine superchargers and, later, to airfoils in the hot sections of gas turbines. Similar cobalt-base superalloys are still in use today. However, creep-rupture requirements for aircraft gas turbine applications soon outstripped those of the cobalt-base and also the iron–nickel-base superalloys and the cobalt-base superalloys. Thus the use of nickel-base superalloys, modified to provide more of the hardening phase γ', increased.

Superalloys frequently are slightly modified for specific applications and a new name or specification is assigned. As an example, alloy PWA 1487 is PWA 1484 with a small addition of yttrium and slight changes in hafnium and tantalum (see Table 2). Sometimes chemistry modifications are made without changing the specification or name assigned to an alloy. Actual chemistry levels can vary since individual superalloy suppliers and/or patent holders make modifications to the original alloy compositions. The same nominal alloy name may result in modest to significantly differing chemistries from one superalloy user to another. Although not common now, proprietary additives once were added by some producers to assist in creating optimum properties of the resultant alloy product. For current chemistries, check with the appropriate superalloy producer or patent holder.

4.2 Improvement of Superalloys by Chemistry Control

The production of superalloy components initially requires some sort of melting process. The melting produces ingots which are remelted, converted to powder for subsequent consolidation to a component, or investment cast. Remelting is used to produce an ingot which can be processed to wrought mill forms (e.g., bar stock) or forged. Until the start of the second half of the twentieth century, melting of superalloys was conducted in air or under slag environments. The properties of modern superalloys derive principally from the presence of many elements which are reactive with oxygen and so were being lost to some or a great degree in the melting and casting process. When vacuum melting techniques were introduced to commercial production of articles, they were pioneered by superalloys. The vacuum enabled the melting of superalloys containing higher amounts of the hardeners aluminum and titanium. Furthermore, the concurrent reduction in gases, oxides, and other impurities created a significant improvement in the ductility of superalloys. Moreover, with more hardener content, strengths of superalloys began to increase dramatically. Figure 5 shows the improvement in creep-rupture life achieved with the reduction in oxygen content.

5 MELTING AND CASTING PRACTICES

5.1 General Aspects

The development of superalloys as they are employed today is largely a story of the development of modern melting technology. Whether the final product is to be a forged or investment cast one, the essence of a superalloy's ability to create the properties desired hinges on the correct choice and adaptation of melting principles.

Superalloy melting practices may be classified as either primary (the initial melt of elemental materials and/or scrap which sets the composition) or secondary (remelt of a

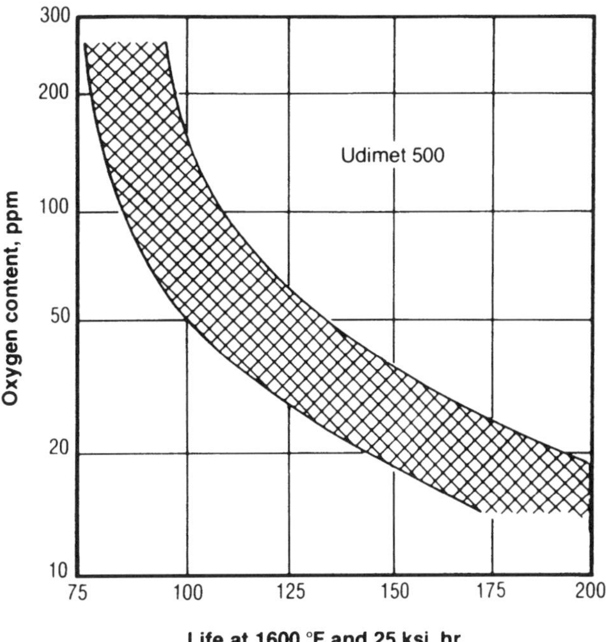

Figure 5 Improvement of rupture life at 871°C (1600°F) and 172 MPa (25 ksi) by reduced oxygen content produced by vacuum melting. (From *Superalloys Source Book,* ASM International, Materials Park, OH, 1984, p. 11.)

primary melt for the purpose of controlling the solidification structure). The melt type or combination of melt types selected depends upon the alloy composition, mill form and size desired, properties desired, and sensitivity of the final component to localized inhomogeneity in the alloy.

5.2 Melting and Refining Superalloys

The two most common primary melt practices are argon–oxygen decarburizing (AOD) treatment of electric arc processed metal (arc AOD) and vacuum induction melting (VIM). The two common secondary melt practices are vacuum arc remelting (VAR) and electroslag remelting (ESR). A few alloy mill products may be manufactured from cast ingots after only primary melting. For the majority of superalloys the common practice combinations are arc AOD + ESR, VIM + ESR, VIM + VAR, and VIM + ESR + VAR. Common material specifications, such as Aerospace Materials Specifications (AMS), will specify acceptable melting practices.

The arc AOD practice used for superalloys is little different from that used for stainless steels. An electric arc furnace is the recipient of the initial charge and uses power from an arc struck between graphite electrodes and the charge to heat and melt the charge. Once the charge melts, no further electric power is needed. Heat is input into the charge by injecting ("blowing") oxygen under the surface of the melt to react with elements such as aluminum, titanium, silicon, and carbon. A desulfurizing addition (usually CaO) is made. The slag formed by the oxidation products and the desulfurizing addition is physically removed from the furnace.

The deslagged charge is transferred to the AOD vessel, which is a refractory-lined steel shell with "tuyeres" in the bottom. The tuyeres are used to inject a mixture of argon and oxygen into the molten bath. By controlling the ratio of argon to oxygen, the selective oxidation or reduction of elements may be accomplished. The principal element reduced is carbon. Elements which need to be retained in the melt (particularly chromium) may also be oxidized into the slag. However, additions of aluminum made to the heat will react with the slag and reduce the chromium back into the molten charge. There is no external source of heat to an AOD vessel. The molten charge is heated by making additions of aluminum and oxidizing them (Al_2O_3 has a very high heat of formation). Cooling is accomplished by adding solid scrap to the bath. When the desired composition and pouring temperature are reached, the heat is poured off into a teeming ladle and transferred to the electrode/ingot pouring system. Precautions are taken (shrouding of the pour stream with argon) to minimize the reaction of the pour stream with atmospheric gases.

The other prevalent primary melt practice is VIM. The VIM furnace consists of a ceramic lined crucible built up around water-cooled induction coils. The crucible is mounted in a vacuum chamber. The vacuum chamber may have several vacuum ports (of various sizes) built into it so that, without breaking vacuum,

- charge material may be introduced into the crucible,
- molds may be introduced into the chamber,
- systems for removing slag from the pour stream (tundish) may be introduced, and
- sampling of the molten metal may be made for chemistry control.

The charge is separated into three types of components, virgin, reactive, and revert. Note that virgin material is usually elemental material but may also consist of any other type of material which has not been previously vacuum melted. Reactive elements refer to those elements which are strong oxide formers. For superalloys these are primarily aluminum, titanium, and zirconium. Operation of the VIM consists of charging the virgin portion of the charge (minus the reactive portion) into the crucible. The furnace is pumped down (if not already under vacuum) and a measurement made of the increase in pressure (as a function of time) when the vacuum ports are blanked off. This "leak-up rate" is a measurement of the baseline furnace vacuum integrity. Power is applied to the induction coils. The magnetic fields produced induce a current in the charge material, heating the charge. When the charge is molten, it will evolve gas (refining). The rate of gas evolution is measured by taking leak-up rates. When consecutive stable leak-up rates are obtained, then refining (degassing) is complete. This produces the lowest possible oxygen in the alloy. Reactive additions are then made and revert is added. (Revert is previously vacuum melted alloy which is of the same or nearly the same composition as the intended melt).

Additions of some form of calcium are usually made to reduce sulfur in the charge by forming a CaS slag. A chemistry sample is taken and analyzed. "Trim" additions are made to bring the heat to a precise composition. The metal is (top) poured into the molds through a system for metal transfer plus slag control and regulation of the pour rate. Cast product for direct processing (e.g., for investment casting) is referred to as ingot. Most often, the VIM product is intended for secondary melting and is referred to as an electrode, as both secondary processes consume this product in an electrically heated manner.

Compared to VIM, arc AOD uses the lowest cost charge material and has higher production rates. That is, arc AOD is the lowest cost process. However, arc AOD is limited in its ability to precisely control composition in a narrow range (particularly for superalloys with high titanium levels) and also produces much higher oxygen content material. However, the choice of ESR as the secondary melting process may compensate to some degree for the higher oxide inclusion content of the arc AOD process.

Electroslag remelting is the most frequently used secondary melt process for arc AOD electrodes. In ESR, AC is applied to an electrode situated inside a water-cooled crucible containing a molten slag charge. The intended circuit of current is from the electrode, through the molten slag, through the solidifying ingot, through the water cooled-stool, and by symmetrically located bus bars back to the electrode. The slag is generally a CaF_2 base modified by major additions of CaO, MgO, and Al_2O_3. Minor additions of TiO_2 and ZrO_2 may be made to counteract loss of titanium and zirconium during the melting process. (Note that minor compositional changes may occur during ESR.) The current passing through the slag keeps the slag molten. The molten slag melts the immersed face of the electrode. As the molten metal gathers into drops on the melt surface of the electrode, it is reduced in sulfur (through reaction with the CaF_2) and entrained oxides are incorporated into the slag. The fully formed drop falls through the slag and is collected in the water-cooled crucible. The slag, which has formed a solid layer against the cooled crucible wall, is partially remelted by the molten metal but remains as a layer between the ingot and the crucible. As the electrode is consumed, an ingot is built up in the crucible. At any given time the ingot contains a pool of molten metal at the ingot top and a zone of liquid + solid between the solid portion of the ingot and the molten pool.

However, solidifying consumable melted ingots are susceptible to the formation of unacceptable segregation. For a given alloy there is a critical thickness and angle (to the growth direction) of the liquid + solid zone at which the liquid in that zone may form self-perpetuating channels in the solidification structure. Such channels are known as "freckles" and often are of such high solute content that subsequent thermal treatment cannot eliminate the elemental concentration. Thus, the final product may contain channels of hard particles and be very detrimental to fatigue properties. The more highly alloyed the material being melted, the smaller and shallower must be the liquid + solid zone in order to avoid freckle formation. The shape and thickness of the liquid + solid zone may be modified to be more resistant to freckle formation by reducing the melt rate (the rate of deposition of molten metal into the pool), reducing the ingot diameter, or improving the heat extraction from the sides of the crucible.

Vacuum arc remelting is used in preference to ESR where larger ingot sizes of highly alloyed materials are needed. It is a DC process which is conducted in a vacuum. As occurs in ESR, the electrode is melted into a water-cooled crucible. The melting is accomplished by initially striking an arc between the electrode face and the crucible stool. The arc melts the electrode face and the molten metal collects into drops which fall into the crucible to form an ingot against which the arc is maintained. Because there is no additional source of heat (as, e.g., the cap of molten slag in ESR) and no insulating slag skin on the ingot, heat extraction in VAR is greater than in ESR. Thus the profile of the molten zone in VAR is generally shallower (than in ESR) for any given alloy, melt rate, and ingot diameter combination.

There is no compositional change in VAR with the exception of minor elements with high vapor pressures. Thus, residual amounts of detrimental elements such as bismuth and lead may be removed by VAR. Unfortunately, volatile elements such as magnesium (used for control of sulfide shape) are also removed. The removal of magnesium is generally compensated for by providing an elevated level in the primary melt process.

5.3 Pros and Cons of Remelted Ingot Processing

In VAR the interface between the ingot and the crucible does not contain the slag skin found in ESR. Instead, the surface of the ingot, being the first material to solidify, is low in solute elements. More importantly, any oxides or nitrides from the electrode are swept into this surface layer as they melt out of the electrode and onto the surface of the molten pool. This layer is called a "shelf." A disruption in the arc stability may cause the arc to undercut a small section of this shelf. If this section drops into the pool, it is unlikely to remelt and thus forms an alloy-lean "spot" which contains layers of entrapped oxide and nitride ("dirt"). These inhomogeneous regions are light or "white" etching and are known as "dirty white spots." Their generation is triggered by conditions which do not generally leave any electronic signature (at current levels of detection capability) in the VAR records. While robust VAR process parameters and good electrode quality may minimize the frequency of dirty-white-spot occurrence, their formation cannot be completely avoided. Thus, the possible presence of dirty white spots in a component (and the degree of their degradation of properties) must be considered in material/melt process selection for a component.

In selecting ESR as the secondary melt practice, positive considerations are that ESR does not generally form dirty white spots, that ESR products may also have better hot workability than VAR products, and that the process yield may be modestly higher (lower cost). Negative considerations are that ESR products cannot be made in large ingot sizes (compared to VAR) without the formation of freckles. The state of the art in ESR controls is such that, at sizes where ESR will normally not form freckles, some undetected abnormality in the melt may cause sufficient disruption that freckles will be produced anyway. Electroslag remelting is not a robust process (with regard to avoiding defect formation) when operated toward the extreme end of its size capability. However, advances in ESR controls have allowed commercially useful ingots of ESR Waspaloy to be produced while maintaining an adequate distance from the melt regimes where freckle formation would become a concern. More highly alloyed materials (such as IN 718) cannot yet be robustly produced by ESR in similar sizes.

In selecting VAR as the secondary melt practice, positive considerations are that (for a given alloy) VAR will produce a larger ingot (than ESR) without the presence of freckles and that at the selected size the process should be much more robust (in freckle resistance) than is ESR. The principal negative consideration is that dirty white spots will exist in the ingot and that they cannot be detected and their removal cannot be assured.

Several producers of critical rotating components in the gas turbine industry have adopted the use of a hybrid secondary melt process: VIM to ensure a low-oxygen, precise chemistry initial electrode followed by ESR. The ESRed electrode will be clean and sound but may contain freckles. The clean, sound electrode is then remelted by VAR. The improved cleanliness and soundness of the electrode facilitates VAR control. This product (referred to as "triple melt") has a much reduced frequency of dirty-white-spot occurrence compared to double-melt (VIM + VAR) product. However, even in triple melt, dirty white spots will occur and must be considered in the component design. The trade-off is the additional cost of an extra processing step versus the enhanced resistance to premature component failure from dirty white spots.

5.4 Melting Considerations in Alloy Selection

As noted earlier, superalloys are available not only in standard mill forms (plate, sheet, and bar—from which components may be machined) but may be produced as specific component shapes by the use of forging or casting. It is critically necessary that the property requirements of a component be fully understood. For instance, a component selected to resist corrosive attack may not need to be melted by a process that assures fatigue resistance.

The most critical feature of melt process selection is the size of the component to be manufactured from an ingot. The larger components require larger ingot sizes. As previously discussed, the more highly alloyed a material, the greater the restriction of maximum ingot size. For forged components consideration must be made not only of the final component size but also of the capability of the ingot to be sufficiently deformed to develop the properties characteristic of a forged structure. Internal cleanliness is also a consideration in selecting the melting route.

Generally, for fatigue-sensitive components a melt practice is selected which will guarantee the absence of freckles from the structure. For larger size components this favors a final melt by VAR rather than by ESR. However, while the absence of freckles may be guaranteed, there are no active VAR practices that can guarantee the elimination of dirty white spots. Thus, any component designed with a final melt practice of VAR must be designed with stress assumptions made assuming a specific flaw size is present in the new product. The flaw size is, of course, determined by knowledge of white-spot frequency, size, and location within the component (for a specific alloy composition and melt route). White-spot frequency and radial location in an ingot vary not only with ingot size but also with the precise melt conditions chosen to produce that size.

In forged applications the use of the triple-melt process (VIM/ESR/VAR) is increasing as it has been demonstrated that this process reduces the frequency of dirty-white-spot formation while it also may be inferred that the average size of the dirty white spot is decreased. Further improvement in reduction of white spots may be made by a fourth melt (VIM/ESR/VAR/VAR). This process is not currently used commercially but may eventually be found useful for very high strength, highly stressed, freckle-prone alloys not produced by a powder metallurgy route.

For alloys such as Waspaloy, which may be produced in useful (for disk forgings) freckle-free ingot sizes the process VIM/ESR is often chosen because of its complete freedom from formation of dirty white spots. In addition, VIM/ESR may be selected for production of mill products whose manufacture does not require large initial ingot sizes.

For alloys to be used in the chemical industry, principally as sheet and plate, the need to assure absence of freckles may not be critical, although possible pinhole corrosion at freckles must be considered. These alloys are commonly produced by VIM/ESR or arc AOD/ESR. A driving force for the use of ESR is that ESR of rectangular shapes (slabs) is common. This allows the direct rolling of such alloys for flat mill products rather than the introduction of an intermediate forging process to reduce round ingot to slab.

In summary it should be noted that no superalloy melting process produces a truly homogenous and uniform product. Specifying that the melted ingot *will* be uniform does not change the result. It is incumbent upon the component designer to understand the real and differing nature of each melt practice and select that practice which produces the properties needed while having the size capability to allow the component to be manufactured.

6 COMPONENT PRODUCTION

6.1 Casting Practices to Produce Parts

The principal casting practice is investment casting (also known as the lost-wax process). A reverse cast model of the desired component is made and wax is solidified in the resultant die. Then a series of these wax models are joined to a central wax-pouring stem. The assembly is coated (invested) with appropriate ceramic, processed to remove wax, and fired to strengthen the invested ceramic mold. A small percentage of a (VIM) master alloy is remelted and cast into the mold. Upon solidification, a series of components in the desired form are created, attached to the central pouring stem. These objects, frequently turbine hot-section airfoil components, are removed and then machined and processed to desired dimensions.

Superalloy investment castings now are available in sizes from a few inches in length up to about 48 in. diameter. Turbine airfoils can now be made not only for aircraft gas turbines but also with airfoil lengths of several feet for land-based power turbines.

Conventional investment castings are PC equiaxed with more or less randomly oriented grains. Mechanical properties are nominally random but may show some directionality. Increased property strength levels have been achieved by CGDS, which removes grain boundaries that are perpendicular to the applied principal load in turbine airfoils. Removal of these grain boundaries dramatically improves longitudinal (parallel to airfoil axis) creep-rupture properties of nickel-base cast superalloys. Transverse boundaries continue to be a problem but are minimized by the addition of hafnium, which enhances grain boundary ductility. The ultimate solution is to directionally solidify a superalloy as a single crystal, i.e., a superalloy with no grain boundaries. Maximum creep-rupture strength in nickel-base superalloys now is achieved with SCDS alloys. The SCDS process has been applied not only to aircraft gas turbine engines but also to large-frame land-based gas turbines.

An interesting benefit of directional solidification in nickel-base superalloys is the reduction of the elastic modulus in the primary solidification direction (parallel to the longitudinal axis). Reduced moduli mean less thermal stress on hot-section airfoils and thus longer life in TMF.

6.2 Casting Considerations in Alloy Selection

Selection of cast superalloys is best conducted in conjunction with an investment casting foundry which can indicate the likelihood of success in casting a given design in the alloy selected. Figure 6 shows shapes and surface appearance of turbine blades produced to the three standard casting conditions (equiaxed, columnar grain, and single crystal). These parts also contain complex internal cooling passages that effect the necessary cooling required for airfoil operation in ultrahot gas streams.

Not all alloys are equally good for casting. Casting defects occur and vary with the alloy composition and casting practice. Oxide (dross) carried over from the master melt may cause surface and internal inclusion problems. Porosity is another major concern, especially with large castings for the cases of gas turbines. Porosity may be controlled by mold design and pouring modifications. Non-surface-connected porosity may be closed by hot isostatic pressing (HIP) of cast articles. Surface porosity in large castings may be repaired by welding. Other casting concerns have included intergranular attack (IGA) caused during chemical removal of molding materials, selectively located coarse grains in PC materials, misorientation and spurious grains in DS alloys, etc. Alloys may be modified in chemistry to optimize product yield but with a possible compromise in properties. Product yield is an important determinator of final component cost.

6.3 Forging and Powder Metal Superalloys

Forging is the most common method of producing wrought components for superalloy applications. Mill products such as bar stock, wire, etc., are produced, but the most demanding

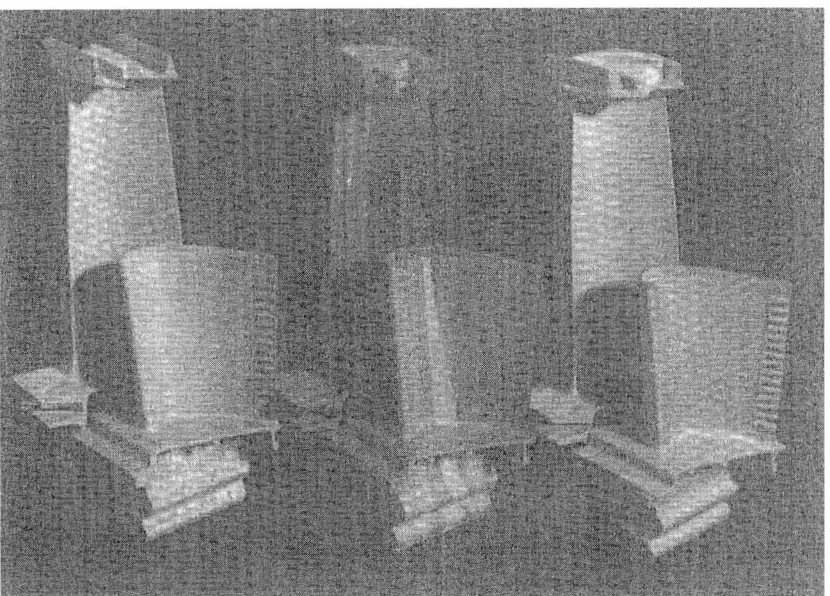

Figure 6 Turbine blades produced by equiaxed grain (left), directionally solidified polycrystal (center), and directionally solidified single-crystal (right) casting processes.

applications use forged wrought ingot metallurgy components. Disks are forged to near-net size or to approximate shape using large presses with input stock from billets previously forged from consumably remelted ingots. The forging process requires the application of heat and pressure to move the alloy from its billet or powder metallurgy preform stage to a stage ready for machining to a finished disk. One or more intermediate-shape stages usually are involved when conventional forging is practiced. Isothermal (usually superplastic) forging may go directly from billet to final stage in one step.

As alloy strengths increase, it becomes increasingly difficult to move the alloy around during forging. When higher pressures are required, defects became more probable. Superplastic forging (isothermal forging) became available in the time frame when the strengths of superalloys were bumping up against metallurgical limits. Simultaneously, the discipline of powder metallurgy opened the way to create billets of the highest strength alloys without the positive segregation defects (of alloy elements) that made casting and forging of billets to final parts so difficult.

In powder metallurgy, used principally for alloys such as Rene 95 or IN 100 or other advanced alloys, a preliminary ingot is remelted and then atomized to small droplets that solidify as "miniature ingots" with limited or no segregation of alloy elements. The powders are consolidated by HIP and/or by extrusion and processed to achieve 100% density. The aggregates are homogenous and display uniform properties. Sometimes the powder is pressed directly to form the final size/shape, but usually the powder is compacted to an intermediate stage (e.g., extruded billet) and isothermally forged to final form. Powder metallurgy is expensive, but the savings in subsequent machining costs, the ability to control defects, and the ability to use very high strength compositions make a reasonable cost trade. Possible defects in PM include carbide stringers, ceramic inclusions, and "reactive" inclusions (e.g., hair) from the powder production process.

Conventional alloys such as Waspaloy, U 720, and IN 718 are produced by ingot metallurgy and standard forging practices. High-strength alloys such as Rene 95, IN 100, and Rene 88 generally are produced using powder metallurgy techniques.

6.4 Forging/Working Considerations in Alloy Selection

The stronger the alloy to be utilized in design, the more difficult it will be to manipulate by mechanical working forces. Limits exist on the capability of an alloy to be worked without cracking, encountering other defects, or stalling a forge or extrusion press. Some shaping equipment such as rotary forging devices may be better able to change the shape of the stronger superalloys than other equipment. Powder metallurgy offers an alternative processing route for high-strength alloys used for rotating disks, shafts, hubs, etc., in aircraft and power generation turbines. Powder production reflects an art that is not always directly transferable from one producer or one alloy to another. Powder components are best made by producing an ingot from powder and then forging the ingot to component shape. Extensive work with an alloy melt shop and an associated powder producer may be necessary to create a satisfactory metal powder of an alloy selected for design.

6.5 Joining

Welding and brazing are used to manufacture some superalloy components, but superalloy castings have been employed to reduce component complexity and the need for welding in manufacture.

Components of superalloys often could be repaired by joining processes in the fledgling days of the superalloy industry. That is not necessarily true today. Cobalt-base superalloys, which do not depend on precipitated phases for strength, are generally amenable to fusion welding. Sheet metal structures are routinely fabricated by welding for cobalt-base superalloys. Repair welding of some cobalt-base superalloys is practiced. Wrought iron–nickel-base superalloys can be welded, but precipitation-hardened versions are welded with difficulty because properties of precipitation-hardened alloys are degraded by the heat of the joining process. This same principle applies to the precipitation-hardened nickel-base superalloys which become increasingly more difficult to weld as the amount of hardener in an alloy increases. Figure 7 shows a relationship between hardener content and weldability for nickel-base superalloys.

Inertia bonding is used to join nickel-base superalloys, but conventional fusion welding is not customarily used, although electron beam welding is possible. Other solid-state joining processes such as transient liquid-phase (TLP) bonding have been used with some measure of success.

The essence of the joining situation for superalloys is that nickel-base airfoils generally are not repair welded if cracks appear. Cobalt-base airfoils, on the contrary, may be welded to extend life. Sheet metal of cobalt-base, solid-solution-strengthened nickel-base, and lower hardener content nickel-base alloys can be joined as can iron–nickel-base alloys. The higher strength nickel-base alloy of choice for welding is IN 718, which, owing to the unique aspects of its hardening by γ'', can be fusion welded with relative ease.

6.6 Summary of Component Production

Figure 8 presents a view of the superalloy manufacturing process. Many producers have been involved in the business of producing superalloys over the years. During the past 50 years

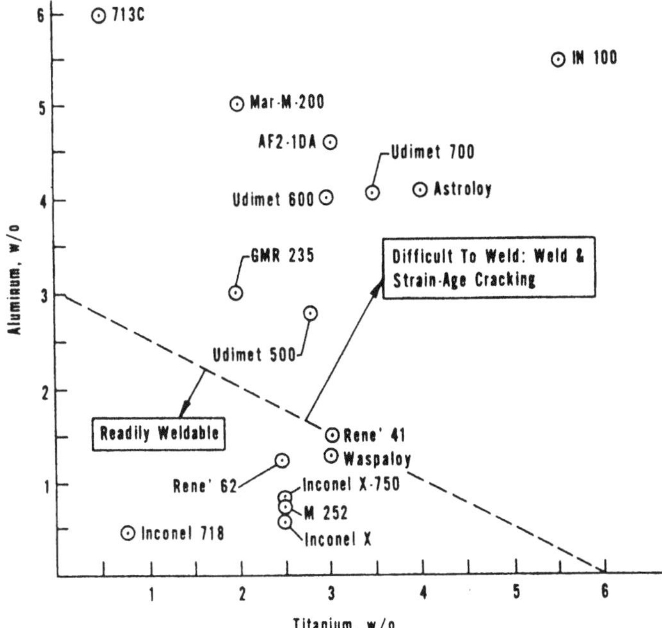

Figure 7 Weldability diagram for some γ'-strengthened nickel-base superalloys showing the influence of total age-hardening elements (Al plus Ti) on cracking tendency. (From *Superalloys II,* Wiley, 1987, p. 152. Used by permission.)

of progress, many advances have been made. New companies have formed and others have merged or gone out of business. Some sources of superalloy expertise in ingot melting, component forging, and article casting are listed below.

6.7 Some Superalloy Information Sources

The following is only a partial list. More information as to web sites, locations, etc., may be available. Not all companies or institutions active in superalloy technology are represented. No recommendation is made or implied by this list. The changing industry mix may make some parts of this list obsolete.

Product	Possible Source
GENERAL INFORMATION	
ASM International, Materials Park, OH	www.asminternational.org
Nickel Development Institute, Toronto, ON, Canada	www.nidi.org
Cobalt Development Institute, Guildford, Surrey, UK	www.cobaltdevinstitute.com
International Chromium Development Institute, Paris, France	www.chromium-assoc.com
Specialty Steel Industry of North America, Washington, DC	www.ssina.com

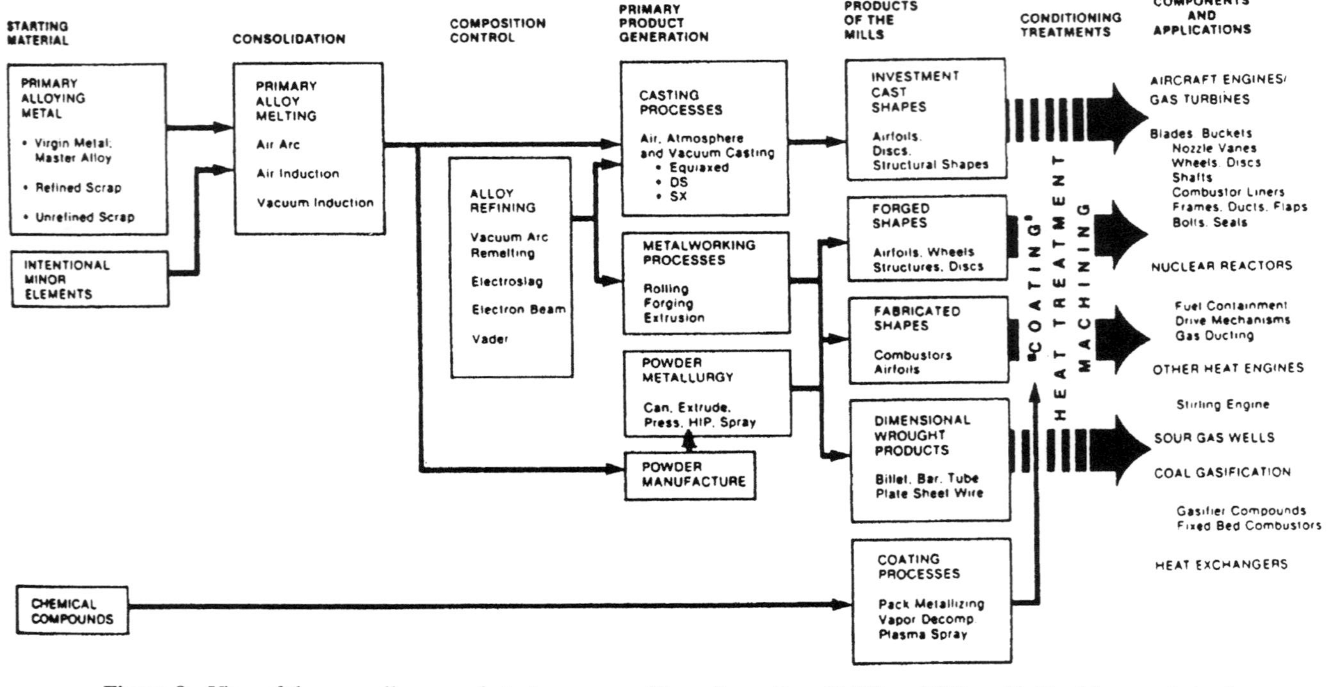

Figure 8 View of the superalloy manufacturing process. (From *Superalloys II,* Wiley, 1987, p. 23. Used by permission.)

Product	Possible Source
Minerals, Metals and Materials Society of the American Institute of Mining, Metallurgical and Petroleum Engineers, New York	www.tms.org
MELTING/INGOT PRODUCTION, FORMING AND/OR MILL PRODUCTS	
Special Metals Corporation, New Hartford, NY; Huntington, WV; Hereford, UK	www.specialmetals.com
Carpenter Technology Corporation, Wyomissing, PA	www.cartech.com
Haynes International, Kokomo, IN	www.haynesintl.com
Teledyne Allvac, An Allegheny Technologies Co, Monroe, NC	www.allvac.com
Cannon-Muskegon Corp., Subsidiary of SPS Technologies, Muskegon, MI	www.greenvillemetals.com/cmgroup.htm
Howmet, Dover, NJ; Devon, UK	www.howmet.com
Doncasters PLC, Sheffield UK	www.doncasters-bramah.com
Precision Rolled Products, Florham Park, NJ	Website unnkown
INVESTMENT CASTINGS	
Howmet Corporation, Hampton, VA; Whitehall, MI; LaPorte, IN; Wichita Falls, TX; Devon, England; Gennevilliers, France; Dives, France; Terai, Japan	www.howmet.com
Precision Cast Parts, Minerva, OH; Cleveland, OH; Mentor, OH; Douglas, CA	www.precast.com
Doncasters Precision Castings, Droitwich, Worcs, UK; Bochum, Germany; Groton, CT	www.doncasters-deritend.com, www.doncasters-bochum.com, www.doncasters.com
Hitchiner Manufacturing Co., Gas Turbine Division, Milford NH	www.hitchiner.com
FORGINGS	
Wyman Gordon Co., N. Grafton, MA; Livingston, Scotland; Houston, TX	www.wyman-gordon.com
Schlosser Forge, Cucamonga CA	www.aerospace-engine-parts.com
Ladish Co., Cudahy, WI	www.ladish.com
Carlton Forge Works, Paramount, CA	Website unknown
Carmel Forge, Tirat Carmel, Israel	Website unknown
Doncasters PLC, Monk Bridge, UK; Blaenavon, UK; Leeds, UK	www.doncasters-monkbridge.com, www.doncasters-blaenavon.com, www.doncasters.com
Thyssen Umformtechnik, Remscheid, Germany	www.tut-gmbh.com
Fortech Clermont, Ferrand, France	Website unknown
Firth Rixson, Monroe, NY; Verdi, NV	Website unknown
Forged Metals, Fontana, CA	Website unknown

Product	Possible Source
Coating and/or Refurbishment/Repair	
Chromalloy Gas Turbine Corporation, Carson City, NV; Gardena, CA; Orangeburg, NY; Harrisburg, PA; Middletown, NY; Columbus, IN; Manchester, CT; Phoenix, AZ	www.chromalloy-cnv.com, www.chromalloy-cla.com, www.chromalloyhit.com
Sermatech International, Limerick, PA; Muncie, IN; Houston, TX; Manchester, CT	www.sermatech.com

This list almost certainly will change and should only be used as a guide to locate potential producers. Most consumers deal with the forging and investment casting vendors who produce the product for subsequent final processing. For best control of properties, many consumers maintain liaisons with melters and frequently audit all aspects of the manufacturing process. It is vital to remember in superalloy selection that most superalloys are not off-the-shelf consumer items. They are made to user specifications which may vary from user to user and time to time. Diligence in working with the producers will pay dividends in obtaining optimum properties and reducing difficulties for users of superalloys. This concept is indigenous to the industry and superalloy users for mission-critical or human-life-critical applications. It is more costly than buying off the shelf. For some applications such as in oil development or coal gasification, there may be less stringent controls that permit off-the-shelf purchase of generic mill products. However, no forging or investment casting is an off-the-shelf item.

7 OTHER ASPECTS OF SUPERALLOY SELECTION

7.1 Corrosion and Coatings for Protection

Initial superalloys were intended to have both strength and adequate oxidation resistance, and this was accomplished with superalloys which contained upward of 20% chromium. Oxidation resistance up to temperatures around 982°C (1800°F) was excellent. However, to increase the design flexibility of nickel-base superalloys, chromium content was reduced so that more hardener could be added. Concurrently, some superalloys were put into service in environments (e.g., marine salts) that intensified oxidation or ion-induced attack such as hot corrosion. Also, the operating temperatures (surface environment temperatures) to which the alloys are exposed in the most demanding high-temperature conditions has increased with the strength increases of the available alloys.

At higher temperatures, the chromium oxide formed during prior heat treatment is less protective and does not regenerate with exposure to high temperatures. General oxidation and intergranular oxidation along the grain boundaries of superalloys is a problem with chromium-protected superalloys, However, the problem is not as great as initially anticipated owing to the protective nature of aluminum oxide formed in greater amounts by the higher aluminum values of second- and third-generation γ'-hardened superalloys. Nevertheless, general oxidation still occurs and causes reduced cross sections, thus effectively increasing stresses in the remaining material. Grain boundary oxidation creates notches. The combination of these events is of concern, and to protect against them, coatings are applied to superalloys.

The early coatings were diffusion coatings produced by pack aluminizing or slurry application. The chemistry of the coating was determined by the chemistry of the alloy. Later coatings were produced by overlaying a specific chemistry of a protective nature on the surface of the component using physical vapor deposition. Diffusion-type coatings can coat internal (non–line-of-sight) surfaces while the overlay coatings can only coat external surfaces that can be seen by the coating apparatus. Diffusion coatings are cheaper and, for a given thickness, probably nearly as protective as overlay coatings. However, overlay coatings can be made nearly twice as thick as the diffusion coatings so overlay coatings often are the coatings of choice for turbine airfoil applications. Diffusion coatings are used to coat internal cooling passages in hot-section airfoils. Some commercial diffusion coating processes are available, but most overlay coating processes are proprietary, having been developed by superalloy users such as the aircraft gas turbine manufacturers.

Overlay coatings are generally more expensive than diffusion coatings. Some diffusion coatings are deposited in conjunction with precious metals such as gold or platinum. There are significant benefits to this incorporation of the noble metals in the coating if the application can justify the increased cost.

Coating and corrosion technology are complex and do not lend themselves to a simple overall description and formula for protection. Hot-corrosion phenomena are found below a nominal temperature of 927°C (1700°F). Coatings and higher chromium content in an alloy inhibit surface attack. Coatings, in general, preserve the surface so that a component may be reused by removing and then restoring the coating.

Coating selection is based on knowledge of oxidation/corrosion behavior in laboratory, pilot plant, and field tests. Attributes that are required for successful coating selection include the following:

- High resistance to oxidation and/or hot corrosion
- Ductility sufficient to provide adequate resistance to TMF
- Compatibility with the base alloys
- Low rate of interdiffusion with the base alloy
- Ease of application and low cost relative to improvement in component life
- Ability to be stripped and reapplied without significant reduction of base-metal dimensions or degradation of base-metal properties

7.2 Special Alloys for Hot-Corrosion Resistance

As temperature decreases below 927°C (1700°F), the amount of hot-corrosion attack decreases until, at a nominal temperature below 760°C (1400°F), attack may begin to increase dramatically with decreasing temperature. In this regime, the province of land-based power gas turbines, attack is resisted best by forming chromium oxide on the surface. Consequently, alloys for this region and application are those such as IN 738 specifically designed to have higher chromium levels, sometimes above 20% (IN 939 is an alloy in this latter category). Other alloys have been devised for optimum hot-corrosion resistance at temperatures above 760°C (1400°F). Rene 80 and IN 792 are such alloys.

7.3 Thermal Barrier Coatings

Allied with superalloy protective coating technology is the development of ceramic, so-called thermal barrier, coatings (TBCs), which protect components by reducing surface temperature by several hundred degrees, enabling a superalloy to operate at lower temperatures in a region where the superalloy may have higher strength. Thermal barrier coatings have found use on a wide range of alloys. They use overlay coating technology. A thin overlay coat serves as the bond interface between the ceramic of the TBC and the base superalloy.

7.4 Postservice Refurbishment and Repair

One important aspect of modern superalloy use is the concern for extending service life of components. Cracking, erosion/corrosion of surfaces, and mechanical property loss are major economic factors in component applications. These factors have become of greater interest as the base cost of materials and subsequent components has risen dramatically over the past 40 years. Primary attention to alloy selection has been given with regard to initial producibility and material properties. However, in service, when public safety consideration limits or product integrity design limits are reached, costly components may be withdrawn from use. Some components may appear to be unaffected by service time. For economic reasons there may be incentives to return these components to service.

Other components after service may have visible changes in appearance. For example, a high-pressure turbine blade may be missing a coating in the hottest regions of the airfoil or a crack may develop in a vane airfoil. Seals may be worn. It is highly desirable that such damage can be repaired so that the costly parts can be returned to service.

If possible, superalloys would be selected on the basis of restoration of capability after initial capability has (apparently) been reduced by service exposure of the component. However, most applications do not permit alloy selection on that basis. The best alloy from a property and initial economic viewpoint is usually the choice. However, it is common practice to refurbish or repair many components (usually stationary parts) which have visible external changes. It is not practical to describe most of the varied refurbishment and repair schemes in effect for superalloys, but a brief summary will indicate the range of opportunities possible.

Stripping and recoating of turbine airfoils are one refurbishment and repair practice. Oxidation- and corrosion-resistant coatings and TBCs may be reapplied (after appropriate surface cleaning treatments) to restore resistance of the surface to heat and gaseous environments. In the case of missing, eroded, cracked, and/or routed material, welding traditionally has been used to fill the gaps. Welding of high-strength nickel-base superalloys is very difficult, if not impossible. Welding of solid-solution-hardened nickel-base alloys such as Hastelloy X and of lower strength precipitation-hardened nickel-base alloys is more readily accomplished. However, of the high-strength alloys, only IN 718 precipitation-hardened alloy (cast or wrought) may be considered truly weldable. Welding of (lower strength) cobalt-base turbine airfoil and sheet materials also is accomplished readily. Welding is followed by machining, possible heat treatment, and coating.

Recently, instead of fusion welding, some repair shops have been using HVOF (high-velocity oxy fuel) processes to deposit repair material and methods such as HIP are then used to densify the deposited material. Other processes which rely on laser buildup of material are claimed to be viable as repair processes for eroded surfaces. The main characteristic of refurbishment and repair processes is both the restoration of geometry and the provision of adequate mechanical properties in the repaired area. Care must be taken to assure that no additional alloy degradation occurs owing to the refurbishment and repair practice.

The restoration of mechanical properties degraded by creep and/or fatigue is not as clearcut as stripping and recoating or insertion of material by welding using HVOF or similar deposition processes. In the laboratory, reheat treatment has been shown to restore the mechanical properties of some superalloys after service exposure. The degree of restoration is a function of mechanical history of the component. Property degradation cannot be measured nondestructively. Extensive records of service operation and a statistical record of how a given alloy performs in a specific service application are required to make an educated guess as to the amount of prior damage and the potential for recovery of mechanical properties by reheat treatment. Results of reheat treatment of service exposed parts are variable. Most postservice procedures do not provide for mechanical property restoration.

Selection of refurbishment and repair processes for superalloy components must be as rigorous as the processes followed for original material selection. Indeed, refurbishment and repair processes may have to be held to even higher standards since the material being returned to service may have hidden damage and/or property loss. Development and selection of a particular refurbishment/repair procedure should be undertaken with a suitable supplier of such services. A robust process validated by a comprehensive test program should be followed before approval is granted to use any refurbishment and repair procedure. Appropriate metallurgical assistance should be sought during the development of such procedures.

8 ALLOY SELECTION SUMMARY

8.1 Alloys Available for Intermediate-Temperature Applications

Wrought alloys generally are used. Waspaloy and Astroloy were standard nickel-base superalloys selected in the past. Waspaloy is readily available but Astroloy may not be as readily procured. Wrought U-720 has found many applications.

Castings may be used for some applications of significant physical size. Large case structures for gas turbines are routinely cast in IN 718. However, in most instances, wrought alloys generally are used. Ductilities and fracture toughness of wrought alloys are better than those of cast alloys. Strength in tensile tests usually is better too. Creep may be of concern, but rupture life normally is not an issue. Iron–nickel alloys such as A-286 might be considered as they may have sufficient strength and will be considerably less costly than their nickel-base counterparts. The better alloy is the nickel-base (sometimes called nickel–iron-base) superalloy IN 718.

IN 718 is the most widely used superalloy today. As a wrought alloy, it can be made in various strength levels and in very large ingot sizes. It is the most weldable high-strength superalloy available. As a casting, IN 718's weldability is a significant factor in its application as large cases.

Costs of IN 718 are lower than some other superalloys because of the widespread use of the alloy. IN 718 also is unique in that it contains none of the strategic element cobalt. IN 718 received a significant application boost in the late 1970s when a cobalt "shortage" caused the price of cobalt to soar to astronomical levels. IN 718 at that time faced continued competition from alloys such as Waspaloy and Astroloy. The lack of cobalt in the alloy swung the pendulum to IN 718, and in the succeeding decades, IN 718 has solidified its position as the most used superalloy.

In lieu of IN 718, alloys such as Waspaloy, U-720, and others can be adapted to designs. Powder metal techniques allow IN 100, Rene 95, and other high-strength and damage-tolerant alloys to be fabricated and used. An important trend in the intermediate-temperature area is the development of dual material/property gas turbine disks. One of the major concerns for gas turbine disks to operate is the difficulty of getting all desired properties in one material. Extensive work has been done to validate the position that a disk may be made of more than one material. Demonstrations have shown that creep-resistant rims can be attached to the brute-strength and fatigue resistance cores. Selection of materials for such an application, however, requires extensive alloy/process development.

8.2 What to Look for in Wrought Alloys for Intermediate-Temperature Applications

For massive applications such as turbine disks, high tensile yield and ultimate strength are desired. Good tensile ductility is important and good mechanical LCF behavior with acceptable crack propagation rates at expected load is a must. If an alloy is to be used as sheet, then good formability is a must, coupled with good weldability. Massive parts such as disks can benefit from good forgeability, but such a quality does not exist when high tensile strengths are required. Powder metallurgy processing enables production of forged components not otherwise processable. Sonic inspectability of finished shapes is crucial. Cost is a very important factor, but one that may have to be subverted to properties and processing if a desired component is to be made. Of course, use of special processing techniques such as powder metallurgy may enable a part to be formed that could not be made in any other way, and so a high cost may be worth paying.

The essence of superalloy selection for intermediate-temperature applications is that there are standard alloys of capability similar to or less capable than Waspaloy that can be procured and forged by conventional means. Similarly, sheet alloys are available that can be manipulated in conventional ways. For the higher strength applications, there are no easy off-the-shelf technologies or alloys that can just be picked from a catalog and put to work. Selection of alloys is a preliminary step that must be expanded upon to obtain data and components in a reasonable time frame at acceptable costs.

8.3 Alloys Available for High-Temperature Applications

The highest temperature applications invariably require cast superalloys where the requirement of maximum high-temperature strength is concerned. However, some applications, such as combustion chambers, may require less strength and sheet alloys may be applied. Nickel- and cobalt-base sheet is available. Hastelloy X, IN 617, HA 188, and others have long experience records. HA 230 is also used extensively. Property data generally are available for these alloys. The sheet form of these alloys is readily available.

Turbine airfoil alloys or some combustor nozzles require stronger alloys than provided by the old standby cobalt-base and early nickel-base PC cast alloys. Many of these highest strength alloys are proprietary to various companies, usually aircraft engine providers. The strongest alloys are the single-crystal nickel-base superalloys.

Cast superalloys for the bottom end of the high-temperature application spectrum may be those standbys such as IN 713 or even cobalt-base alloys such as X-40. These alloys have decades of experience. IN 713 is a good general-purpose PC equiaxed casting alloy which also happens to have no cobalt. Alloys for higher temperatures include U-700, Rene 80, IN 792, IN 100 (Rene 100), Mar-M 246, Mar-M 247, CM 247 LC, Mar-M 200CG (PWA 1422), Rene 125, CM 186, and single-crystal alloys such as PWA 1480, PWA 1484, Rene N4, Rene N6, CMSX-4, and CMSX-486. Some alloys (PWA 1422, Rene 125) are used only as CGDS components and others only as SCDS components (PWA 1480, PWA 1484, Rene N4, Rene N6, CMSX alloys). Alloys such as Rene 80 have been available in CGDS as well as PC equiaxed form. Others such as IN 792 have been available not only as PC equiaxed components but also in single-crystal form. A defining feature of the short list above is that nearly half of the alloys mentioned are associated with specific aircraft gas turbine companies. Although non-U.S. alloys were omitted from the list, some non-U.S. alloys also tend to be associated with specific manufacturing company ownership. Thus, many of the alloys that satisfy the most demanding environments may not be available for applications outside of their corporate patent realm.

8.4 What to Look for in Cast Alloys for High-Temperature Applications

Alloys to be used for turbine airfoils require high creep and creep-rupture strengths. To maximize strength, the alloys for the most demanding applications in high-pressure turbine (HPT) sections should be SCDS materials. For the ultimate in creep-rupture strength, SCDS alloys with special crystallographic orientation (e.g., ⟨111⟩) may be used. In addition to maximizing creep-rupture strength through directionally solidified processing, TMF strength is optimized by the reduction in modulus achieved by orienting a specific direction (⟨001⟩) of the superalloy crystal parallel to the airfoil axis. An alloy for the most stringent turbine airfoil applications should have a high melting point and good to excellent oxidation resistance. Additionally, the alloy should be able to accept a coating and have good LCF strength at the temperature where the airfoil attachment is made to a disk. These attachment temperatures are nominally at or below 760°C (1400°F). The SCDS processing also will ensure that thin-section properties are optimized. As section thickness is reduced, for a fixed load, a superalloy ruptures in less time than a standard thick test bar would fail. The order of property reduction is PC equiaxed most reduced, CGDS possibly less, and SCDS least.

For turbine vanes where no centrifugal load exists, airfoils may be made from PC equiaxed high-strength cast cobalt-base alloys instead of directinality solidified processed nickel-base alloys. High incipient melting temperature is desired for first-stage turbine vanes. A special type of superalloy, an oxide-dispersion-strengthened (ODS) alloy, also has been used for turbine vanes in some applications. One such alloy, MA 754, relies on yttria dispersed in a corrosion-resistant nickel–chromium matrix to provide adequate creep-rupture capability. The ODS alloys are not common. MA 6000 is another such alloy which may have enough strength for a high-pressure turbine blade in aircraft gas turbines. A problem with PC equiaxed airfoils is that the thermal mechanical stresses are much higher than on CGDS or SCDS parts owing to the higher modulus of PC equiaxed parts. The modulus of the CGDS and SCDS parts may be only 60% of the value for the PC equiaxed nickel-base cast alloys. In the most demanding conditions, TMF problems can be minimized by using oriented grain or crystal structures to reduce stresses.

For low-pressure turbine (LPT) blade airfoils, alloys such as the IN 100 (Rene 100) or IN 792 and Rene 80 PC equiaxed alloys previously used for HPT airfoils may be chosen. If temperatures or stress conditions are sufficiently relaxed, IN 713, U700, or similar first-generation PC equiaxed cast nickel-base superalloys may suffice.

9 FINAL COMMENTS

Many superalloys have been developed, but not so many have been adopted for use. A principal reason for this situation is the high cost of proving the worth and safety of a new material. Within most aircraft engine companies, only a few airfoil alloys and a comparably short list of disk alloys have ever made it to production. Admittedly this list differs from company to company, but the message is the same. If an existing alloy works and a new alloy does not offer some benefit that overrides the development cost of proving up the alloy for its new use, one should not change alloys.

If an alloy selector is starting from scratch to pick an alloy for an application, then any commercially available alloy may be fair game. On the other hand, the best alloy may not be available owing to corporate patent protection. Then, selection of another alloy from a superalloy producer may be necessary but may possibly require extensive development costs to get the product in workable form and to determine design properties. If possible, a known alloy should be selected that has more than one supplier and more than one casting or forging source. Engineers should work with the suppliers and others in the manufacturing chain to acquire typical or design properties for the alloy in the form it will be used. Generic alloys owned by superalloy melters or developers are best for the alloy selector not associated with one of the big corporate users of superalloys. Companies with proprietary interests usually have nothing to benefit from giving up a technological advantage by sharing design data or even granting a production release to use a proprietary alloy.

BIBLIOGRAPHY

Betteridge, W., and J. Heslop, *The Nimonic Alloys,* 2nd ed., Crane, Russak and Co., New York, 1974.

Davis, J. R. (ed.), *Heat-Resistant Materials,* (an ASM Specialty Handbook) ASM International, Materials Park, OH, 1997.

Donachie, M. J. (ed.), *Superalloys Source Book,* American Society for Metals, Materials Park, OH, 1984.

Donachie, M. J., and S. J. Donachie, *Superalloys: A Technical Guide,* 2nd ed., ASM International, Materials Park, OH, 2002.

Lai, G. Y., *High Temperature Corrosion of Engineering Alloys,* ASM International, Materials Park, OH, 1990.

Sims, C. T., N. J. Stoloff, and W. C. Hager (eds.), *Superalloys II,* Wiley, New York, 1987.

Sullivan, C. P., M. J. Donachie, and F. R. Morral, *Cobalt-Base Superalloys 1970,* Cobalt Information Center, Brussels, 1970.

Tamarin, Y., *Protective Coatings for Turbine Blades,* ASM International, Materials Park, OH, 2002.

The proceedings of a continuing series of conferences in Europe, first held in 1978 and at four-year intervals thereafter, with emphasis on gas turbines, power engineering, and other applications (initial proceedings published as follows: *High Temperature Alloys for Gas Turbines,* Applied Science Publishers, 1978).

The proceedings of a continuing series of conferences in the United States, first held at Seven Springs Mountain Resort, Champion, PA, in 1968 and at four-year intervals thereafter with emphasis on high-temperature materials (initial proceedings published as follows: *International Symposium on Structural Stability in Superalloys,* Vols. 1 and 2, AIME, New York, 1968).

The proceedings of a continuing series of conferences in the United States, first held in 1989 and at irregular intervals thereafter, with emphasis on the metallurgy of IN 718 and related alloys (initial proceedings published as follows: *Superalloy 718 Metallurgy and Applications,* AIME, New York, 1989).

The proceedings of a series of conferences in the United States published as *Heat-Resistant Materials and Heat-Resistant Materials,* Vol. 2, ASM International, Materials Park, OH, 1991 and 1995.

Directory of Materials Property Databases, Advanced Materials and Processes, Special Supplement, August 2000. ASM International, Materials Park, OH.

Property data on CD-ROM, disk or in a handbook form from ASM International (e.g., *Atlas of Creep and Stress Rupture Curves,* ASM International, Materials Park, OH, 1988).

Property data handbooks published under the auspices of various government agencies.

CHAPTER 9

PLASTICS: INFORMATION AND PROPERTIES OF POLYMERIC MATERIALS

Edward N. Peters
General Electric Company
Selkirk, New York

1 INTRODUCTION

Plastics or polymers are ubiquitous. Through human ingenuity and necessity natural polymers have been modified to improve their utility and synthetic polymers developed. Synthetic polymers in the form of plastics, fibers, elastomers, adhesives, and coatings have come on the scene as the result of a continual search for man-made substances that can perform better or can be produced at a lower cost than natural materials such as wood, glass, and metal, which require mining, refining, processing, milling, and machining. The use of plastics can also increase productivity by producing finished parts and consolidating parts. For example, an item made of several metal parts that require separate fabrication and assembly can often be consolidated to one or two plastic parts. Such increases in productivity have led to fantastic growth in macromolecules. Indeed, the use of plastics has increased almost 20-fold in the last 30 years. Today there is a tremendous range of polymers and formulated plastics available which offer a wide range of performance. This chapter presents concise information on their synthesis, structure, properties, and applications.

1.1 Classification of Plastics

The availability of plastic materials for use in various applications is huge.[1–5] There are various ways to categorize or classify plastics which facilitate understanding similarities and difference between materials. The two major classifications are thermosetting material and thermoplastics materials.[5–7] As the name implies, thermosetting plastics, or thermosets, are set, cured, or hardened into a permanent shape.[5,6] The curing, which usually occurs rapidly under heat or UV light, leads to an irreversible crosslinking of the polymer. Thermoplastics differ from thermosetting materials in that they do not set or cure under heat.[5] When heated, thermoplastics merely soften to a mobile, flowable state where they can be shaped into useful objects. Upon cooling, thermoplastics harden and hold their shape. Thermoplastics can be repeatedly softened by heat and shaped.

Another major classification of thermoplastics is as amorphous or semicrystalline plastics.[5,6] Most polymers are either completely amorphous or have an amorphous component as part of a semicrystalline polymer. Amorphous polymers are hard, rigid glasses below a fairly sharply defined temperature, which is known as the glass transition temperature, or T_g. Above the T_g the amorphous polymer becomes soft and flexible and can be shaped. Mechanical properties show profound changes near the glass transition temperature.

Semicrystalline polymers have crystalline melting points, T_m, which are above their glass transition temperature. The degree of crystallinity and the morphology of the crystalline phase have an important effect on mechanical properties. Crystalline plastics will become less rigid above their glass transition temperature but will not flow until the temperature is above its T_m.

Many mechanical and physical properties of plastics are related to their structure. In general, at ambient temperatures, amorphous polymers have greater transparency and greater dimensional stability over a wide temperature range. In semicrystalline plastics the ordering and packing of the polymer chains make them denser, stiffer, stronger, and harder. Semicrystalline plastics also have better inherent lubricity and chemical resistance and during molding will tend to shrink more and have a greater tendency to warp.

Another important class of polymeric resins is elastomers. Elastomeric materials are rubberlike polymers with glass transition temperatures below room temperature. Below that glass transition temperature an elastomer will become rigid and lose its rubbery characteristics.

1.2 Chemical/Solvent Resistance

The solvent or chemical resistance of plastics is a measure of the resin's ability to withstand chemical interaction with minimal change in appearance, weight, dimensions, and mechanical properties over a period of time.[8] When polymers are exposed to chemical environments, they can dissolve in the solvent, absorb the solvent, become plasticized by the solvent, react with the solvent or chemical, or be stress cracked.[6] Hence mechanical properties can be reduced and dimensions of molded parts changed.

Qualitative comments are made in this chapter about the chemical resistance of various plastics. Generally, these comments are based on laboratory screening that the manufacturers have reported. The stress level of the test part has a very significant influence on the performance of the material. When the stress level increases, resistance to chemicals and solvents decreases. Sources of stress can occur from injection molding, applied loads, forming, and assembly operations.

When an application requires exposure to or immersion in chemicals or other harsh environments, it is of prime importance to use prototypes of suitable stressed material for testing under actual operating conditions.[8]

1.3 Plastics Additives

The wide variety of synthetic polymers offers a huge range of performance attributes.[9–13] Additives play a key role in optimizing the performance of polymers and transforms them into commercially viable products. Additives are materials that are added to a polymer to produce a desired change in material properties or characteristics. A wide variety of additives are currently used in thermoplastics to enhance material properties, expand processability, modify aesthetics, and/or increase environmental resistance. Enhancement of properties include thermal stability, flame retardancy, impact resistance, and UV light stability. Reinforcing fibers—i.e., high-strength, inert fibrous material such as glass and carbon fibers—can be incorporated in the polymer matrix to improve modulus and strength while lowering the coefficient of thermal expansion and unusually lowering impact and increasing density. Particulates of flaked fillers usually increase stiffness as well as lower costs.[13] Plasticizers lower modulus and enhance flexibility. Thus additives are found in most commercial plastics and are an essential part of the formulation in which performance is tailored for specific applications.[9–13]

This chapter focuses on properties of the base resins with very few additives other than stabilizers. Properties for plastics that contain glass fiber (GF) are listed for those plastics that routinely use such reinforcing fibers.

1.4 Properties

Plastic materials can encounter mechanical stress, impact, flexure, elevated temperatures, different environments, etc. Hence various properties are measured and reported to give an indication of a material's ability to perform under various conditions. Properties are determined on standard test parts and are useful for comparative purposes. A description of properties used in the chapter would be instructive and are listed below.

Density: The mass per unit volume of a material at 73°F (23°C).

Tensile Modulus: An indicator of the stiffness of a part. It is basically the applied tensile stress, based on the force and cross-sectional area, divided by the observed strain at that stress level.

Tensile Strength: The amount of force required to elongate the plastic by a defined amount. Higher values mean the material is stronger.

Elongation at Break: The increase in length of a specimen under tension before it breaks under controlled test conditions. Usually expressed as a percentage of the original length. Also called "strain."

Flexural Modulus: The ratio of applied stress to the deflection caused in a bending test. It is a measure of the stiffness of a material during the first part of the bending process.

Flexural Strength: The maximum stress which can be applied to a beam in pure bending before permanent deformation occurs.

Izod Impact Strength: The amount of impact energy that is required to cause a test specimen to break completely. The specimen may be notched or unnotched.

Heat Deflection Temperature (HDT, also called of deflection temperature under load, or DTUL): Gives an indication of a material's ability to perform at higher temperatures while supporting a load. It shows the temperature at which a test specimen will deflect a given distance under a given load in flexure under specified test conditions, usually at 1.82 and/or 0.42-MPa loads.

Vicat Softening Temperature: A measure of the temperature at which a plastic starts to soften at specified test conditions according to ISO 306. It gives an indication of a material's ability to withstand limited short-term contact with a heated object.

Relative Thermal Index (RTI, formerly called continuous-use temperature): The continuous operating temperature of plastics materials used in electrical applications, as specified by Underwriter's Laboratories (UL). It is the maximum temperature at which the material retains at least 50% of the original value of all properties tested after the specified amount of time. The RTI tests are important if the final product is to receive UL recognition.

UL94 VO: Classification of flammability of plastic materials that follows the UL 94 standard, vertical burning test. The VO rating is where the total burn time for the five specimens after two 10-s applications of flame is $\leq$ 50 s. Sample thicknesses on VO rated test specimens are reported. A rating on thin samples would suggest possible utility in applications requiring thin walls.

Oxygen Index: A flammability test that determines the minimum volumetric concentration of oxygen that is necessary to maintain combustion of a specimen after it has been ignited.

Hardness: The resistance of a material to indentation under standardized conditions. A hard indenter or standard shape is pressed into the surface of the material under a specified load. The resulting area of indentation or the depth of indentation is measured and assigned a numerical value. For plastics, the most widely used methods are Rockwell and Shore methods and Ball hardness.

Coefficient of Thermal Expansion (*CTE*): A measure of how much a material will lengthen (or shorten upon cooling) based on its original length and the temperature difference to which it is exposed. It becomes important when part dimensions are critical or when two different materials with different CTEs are attached to each other.

Shrinkage: The percentage of reduction in overall part dimensions during injection molding. The shrinkage occurs during the cooling phase of the process.

Water Absorption/Moisture Absorption: The percentage weight gain of a material after immersion in water for a specified time at a specified temperature. Low values are preferred and are important in applications requiring good dimensional stability. As water is absorbed, dimensions of the part tend to increase and physical properties can deteriorate. In addition, low water absorption is very important for most electrical properties.

Relative Permittivity (formerly called dielectric constant): A measure of the amount of electrical energy stored in a material. It is equal to the capacitance of a material divided by the capacitance of a dimensionally equivalent vacuum. Important for high-frequency or power applications in order to minimize power losses. Low values indicate a good insulator. Moisture, frequency, and temperature increases may have adverse effects.

Dissipation Factor (also called "loss tangent"): A measure of the dielectric characteristics of a plastic resin. Moisture, frequency, and temperature increases may have adverse effects.

2 POLYOLEFINIC THERMOPLASTICS

2.1 Polyethylene (PE, HDPE, LDPE, LLDPE)

Polyethylene plastics are lightweight, semicrystalline thermoplastics that are prepared by the catalytic polymerization of ethylene.[14–16] Depending on the temperature, pressure, catalyst, and use of a comonomer, three basic types of polyethylene can be produced: high-density polyethylene (HDPE), low-density polyethylene (LDPE), and linear low-density polyethylene (LLDPE). LDPE and LLDPE contain branching. This branching results in decreased crystallinity and lower density. Most properties of PEs are a function of their density and molecular weight. As density decreases, the strength, modulus, and hardness decrease and flexibility, impact, and clarity increase. Hence HDPE exhibits greater stiffness, rigidity, improved heat resistance, and increased resistance to permeability than LDPE and LLDPE.

LDPE is prepared under more vigorous conditions, which results in short-chain branching. The amount of branching and the density can be controlled by the polymerization conditions.

LLDPE is prepared by using an α-olefin comonomer during polymerization. Hence branching is introduced in a controlled manner and where the chain length of the branching is uniform. In general, the comonomers are 1-butene, 1-hexene, 1-octene, and 4-methyl-1-pentene (4M1P).

Polyethylene polymers are versatile and inexpensive resins that have become the largest volume use of any plastics. They exhibit toughness, near-zero moisture absorption, excellent chemical resistance, excellent electrical insulating properties, low coefficient of friction, and ease of processing. They are not high-performance resins. Their heat deflection temperatures are reasonable but not high. Specialty grades of PE include very low-density (VLDPE), medium-density (MDPE), and ultrahigh-molecular-weight PE (UHMWPE)

Some typical properties of HDPE and LDPE are listed in Table 1. Properties of the various copolymers of LLDPE appear in Table 2.

Uses. HDPE's major use is in blow-molded containers (milk, water, antifreeze, household chemical bottles), shipping drums, carboys, automotive gasoline tanks; injection-molded material-handling pallets, crates, totes, trash and garbage containers, and household and automotive parts; and extruded pipe products (corrugated, irrigation, sewer/industrial, and gas distribution pipes).

LDPE/LLDPEs find major applications in film form for food packaging as a vapor barrier film, including stretch and shrink wrap; for plastic bags such as grocery, laundry, and dry cleaning bags; for extruded wire and cable insulation; and for bottles, closures, and toys.

2.2 Polypropylene (PP)

Polypropylene is a versatile semicrystalline thermoplastic offering a useful balance of properties which are typically higher than for HDPE. It is prepared by the catalyzed polymerization of propylene.[17,18] Crystallinity is key to the properties of PP. The degree of crystallinity is a function of the degree of geometric orientation (stereoregularity) of the methyl groups on the polymer chain (backbone). There are three possible geometric forms of PP. Isotactic PP has predominantly all the methyl groups aligned on the same side of the chain (above or below the chain). In syndiotactic PP the methyl groups alternate above and below the polymer chain. Finally, atactic PP has the methyl groups randomly positioned along the chain. Both isotactic and syndiotactic PP will crystallize when cooled from the molten states; however, commercial PP resins are generally isotactic.

Isotactic PP is highly crystalline thermoplastic that exhibits low density; rigidity; good chemical resistance to hydrocarbons, alcohols, and oxidizing agents; negligible water absorption; excellent electrical properties; and excellent impact/stiffness balance. PP has the highest flexural modulus of the commercially available polyolefins. In general, PP has poor impact resistance. However, PP–elastomer blends have improved impact strength. Unfilled PP has poor flame resistance and is degraded by sunlight. Flame-retardant and UV-stabilized grades are available. Typical properties appear in Table 3.

Table 1 Typical Properties of HPDE, LDPE, LLDPE, and UHMWPE

Property	HDPE	LDPE	LLDPE	UHMWPE
Density (g/cm^3)	0.94–0.97	0.915–0.935	0.91–0.92	0.93
Tensile modulus (GPa)	0.76–1.0	0.14–0.31	0.13–0.19	110
Tensile strength (MPa)	22–32	7–17	14–21	30
Elongation at break (%)	200–1000	100–700	200–1100	300
Flexural modulus (GPa)	0.7–1.6	0.24–0.33	0.25–0.37	—
HDT, 0.45 MPa (°C)	65–90	43	—	79
Vicat softening temperature (°C)	120–130	90–102	80–94	136
Brittle temperature (°C)	<−75	<−75	<−75	—
Hardness (Shore)	D60–D69	D45–60	D45–50	—
CTE (10^{-5}/°C)	15	29	18	0.1.6
Shrinkage (in./in.)	0.007–0.009	0.015–0.035	—	—
Water absorption, 24 h (%)	<0.01	<0.01	<0.01	<0.01
Relative permittivity at 1 MHz	2.3	2.2	2.3	2.3
Dissipation factor at 1 MHz	0.0003	0.0003	0.0004	0.0005

Table 2 Typical Properties of LLDPE Copolymers

Property	1-Butene	1-Hexene	1-Octene	4M1P
Density (g/cm^3)	0.919	0.919	0.920	—
Tensile modulus (GPa)	0.185	0.206	0.200	0.277
Tensile strength (MPa)	41	39	58	42
Elongation at break (%)	430	430	440	510

Uses. End uses for PP are in blow-molding bottles and automotive parts, injection-molded closures, appliances (washer agitators, dishwasher components), house wares, automotive parts, luggage, syringes, storage battery cases, and toys. PP can be extruded into fibers and filaments for use in carpets, rugs, and cordage. In addition, PP can be extruded into film for packaging applications.

2.3 Polymethylpentane (PMP)

Polymethylpentane is prepared by the catalytic polymerization of 4-methyl-1-pentene.[14] PMP is a semicrystalline thermoplastic that exhibits transparency (light transmission up to 90%), very low density, chemical resistance, negligible water absorption, and good electrical properties. PMP resins are available as homopolymers and copolymers with higher α-olefins. The homopolymer is more rigid and the copolymers have increased ductility. PMP is degraded by sunlight and high-energy radiation. Moreover, strong oxidizing acids attack it. PMP grades are available with increased radiation resistance as well as with reinforcing fillers. Properties appear in Table 4.

Uses. Applications for PMP resins cover a wide spectrum of end uses. These include medical (hypodermic syringes, disposable cuvettes, blood-handling equipment, respiration equipment, laboratory ware), packaging (microwave and hot-air-oven cookware, service trays, coated

Table 3 Typical Properties of PP

Density (g/cm^3)	0.90–0.91
Tensile modulus (GPa)	1.4–1.8
Tensile strength (MPa)	22–35
Elongation at break (%)	10–60
Flexural modulus (GPa)	1.0–1.39
Notched Izod (kJ/m)	0.03–0.13
HDT at 1.81 MPa (°C)	60–65
HDT at 0.45 MPa (°C)	75–107
Vicat softening temperature (°C)	130–148
CTE (10^{-5}/°C)	9.0
Hardness (Shore)	D76
Hardness (Rockwell)	R60–R90
Shrinkage (in./in.)	0.01–0.02
Water absorption, 24 h (%)	<0.01
Relative permittivity at 1 MHz	2.25
Dissipation factor at 1 MHz	0.0003

Table 4 Typical Properties of PMP

Density (g/cm^3)	0.84
Tensile modulus (GPa)	1.75
Tensile strength (MPa)	20.5
Elongation at break (%)	18
Flexural modulus (GPa)	1.55
Flexural strength (MPa)	30.0
Notched Izod (kJ/m)	0.150
HDT at 1.81 MPa (°C)	49
HDT at 0.45 MPa (°C)	85
Vicat softening temperature (°C)	173
CTE (10^{-5}/°C)	3.8
Hardness (Rockwell)	L85
Relative permittivity at 100 Hz	2.12

paper plates), and miscellaneous (transparent ink cartridges for printers, light covers, sightglasses, lenses, liquid level and flow indicators, fluid reservoirs).

3 SIDE-CHAIN-SUBSTITUTED VINYL THERMOPLASTICS

3.1 Polystyrene (PS, IPS, HIPS)

The catalytic or thermal polymerization of styrene yields general-purpose or crystal polystyrene (PS).[19,20] The designation of "crystal" refers to its high clarity. PS is an amorphous polymer (atactic geometric configuration) and exhibits high stiffness, good dimensional stability, moderately high heat deflection temperature, and excellent electrical insulating properties. However, it is brittle under impact and exhibits very poor resistance to surfactants and a wide range of solvents.

The three main categories of commercial PS resins are crystal PS, impact PS, and expanded PS foam.

Copolymerization of styrene with a rubber, polybutadiene, results in the rubber being grafted onto the PS. This impact-modified PS has increased impact strength, which is accompanied by a decrease in rigidity and heat deflection temperature. Depending on the levels of rubber, impact polystyrene (IPS) or high-impact polystyrene (HIPS) can be prepared. These materials are translucent to opaque and generally exhibit poor weathering characteristics.

PS finds wide use in styrofoam. Typically styrofoam is produced from expandable styrene (EPS) beads, which contain a blowing agent. When heated, the blowing agent vaporizes and expands the PS and forms a low-density foam. The density of the foam is controlled by the amount of blowing agent.

Typical properties for PS, IPS, and HIPS appear in Table 5.

Uses. Ease of processing, rigidity, clarity, and low cost combine to support applications in toys, displays, consumer goods (television housings), medical (labware, tissue culture flasks), and house wares such as food packaging, audio/video consumer electronics, office equipment, and medical devices.

EPS can readily be prepared and is characterized by excellent low thermal conductivity, high strength-to-weight ratio, low water absorption, and excellent energy absorption. These

Table 5 Typical Properties of PS, IPS, and HIPS

Property	PS	IPS	HIPS
Density (g/cm^3)	1.04	1.04	1.04
Tensile modulus (GPa)	3.14	2.37	1.56
Tensile strength (MPa)	51.1	26.9	15.0
Elongation at break (%)	21	40	65
Flexural modulus (GPa)	3.54	2.61	1.68
Flexural strength (MPa)	102	56.5	30
Notched Izod (kJ/m)	0.021	0.112	0.221
HDT at 1.81 MPa (°C)	77	79	73
HDT at 0.42 MPa (°C)	89	88	79
Vicat softening temperature (°C)	105	104	96
Oxygen index (%)	17.8	—	—
Hardness (Rockwell)	R130	R110	R75
CTE (10^{-5}/°C)	9.0	9.0	9.0
Shrinkage (in./in.)	0.005	0.005	0.005
Relative permittivity at 1 kHz	2.53	—	—

attributes have made EPS of special interest as insulation boards for construction (structural foam sections for insulating walls), protective packaging materials (foamed containers for meat and produce), insulated drinking cups and plates, and flotation devices.

3.2 Syndiotactic Polystyrene (SPS)

Syndiotactic polystyrene is a semicrystalline polymer and is produced via metallocene-catalyzed polymerization of styrene.[21] By comparison to general-purpose, amorphous polystyrene (PS, HIPS), SPS has stereoregularity in its structure, which facilitates crystallization. In SPS the phenyl groups alternate above and below the polymer chain. Hence SPS has a high crystalline melt point and good chemical resistance.

The slow rate of crystallization and high T_g of SPS typically require an oil-heated tool when injection molding and longer cycle times to maximize physical propereties. Typical properties appear in Table 6.

Uses. SPS is targeted at automotive under-the-hood and specialty electronic applications. It will compete with polyamides, thermoplastic polyesters, and polyphenylene sulfide (PPS).

3.3 Styrene/Acrylonitrile (SAN) Copolymer

Copolymerization of styrene with a moderate amount of acrylonitrile (AN) provides a clear, amorphous polymer (SAN).[20] The addition of the polar AN group gives increased heat deflection temperature and chemical resistance compared to PS. Other benefits of SAN are stiffness, strength, and clarity. Like PS the impact resistance is still poor. SAN is chemically resistant to hydrocarbons, bases, most detergents, and battery acid. However, SAN has poor resistance to chlorinated and aromatic solvents, esters, ketones, and aldehydes. The composition and molecular weight can be varied to change properties. An increase in AN will increase SAN's physical properties but will make melt processing more difficult and will decrease the transparency. In general, the AN level in SAN does not exceed 30% for molding applications. Typical properties appear in Table 7.

Table 6 Typical Properties of SPS

Property	SPS	SPS + 30% GF
Density (g/cm^3)	1.05	1.25
Tensile modulus (GPa)	3.44	10.0
Tensile strength (MPa)	41	121
Elongation at break (%)	1	1.5
Flexural modulus (GPa)	3.9	9.7
Flexural strength (MPa)	71	166
Notched Izod (kJ/m)	0.011	0.096
HDT at 1.81 MPa (°C)	99	249
HDT at 0.42 MPa (°C)	—	263
CTE (10^{-5}/°C)	9.2	2.5
Water absorption, 24 h (%)	0.04	0.05
Relative permittivity, 1 MHz	2.6	2.9
Dissipation factor at 1 MHz	<0.001	<0.001

Uses. SAN is utilized in typical PS-type applications where a slight increase in heat deflection temperature and/or chemical resistance is needed. Such applications include appliances (refrigerator compartments, knobs, blender and mixer bowls), electronics (cassette cases, tape windows, meter lenses), packaging (bottle jars, cosmetic containers, closures), medical (syringe components, dialyzer housings), and other (glazing, battery cases, pen barrels).

3.4 Acrylonitrile/Butadiene/Styrene (ABS) Polymers

ABS is a generic name for a family of amorphous ter-polymer prepared from the combination of acrylonitrile, butadiene, and styrene monomers.[22] Acrylonitrile (A) provides chemical

Table 7 Typical Properties of SAN

Property	SAN	SAN + 30% GF
Density (g/cm^3)	1.07	1.22
Tensile modulus (GPa)	3.45	11.0
Tensile strength (MPa)	76	139
Elongation at break (%)	2.5	1.6
Flexural modulus (GPa)	3.65	9.60
Flexural strength (MPa)	128	—
Notched Izod (kJ/m)	0.016	0.060
HDT at 1.81 MPa (°C)	87	100
HDT at 0.42 MPa (°C)	103	108
Vicat softening temperature (°C)	111	—
RTI (°C)	50	74
Oxygen index (%)	19.0	—
Hardness (Rockwell)	R125	R123
CTE (10^{-5}/°C)	6.6	1.9
Shrinkage (in./in.)	0.005	—
Water absorption, 24 h (%)	—	0.15
Relative permittivity at 1 MHz	2.9	3.6
Dissipation factor at 1 MHz	0.009	0.008

resistance, hardness, rigidity, and fatigue resistance and increases the heat deflection temperature. Butadiene (B) provides toughness and low-temperature ductility but lowers the heat resistance and rigidity. Styrene (S) provides rigidity, hardness, gloss, aesthetics, and processing ease. This three-monomer system offers a lot of flexibility in tailoring ABS resins. Optimization of these monomers can enhance a desired performance profile. Because of the possible variations in composition, properties can vary over a wide range.

Typical resin grades in this product family consist of a blend of an elastomeric component and an amorphous thermoplastic component. Typically the elastomeric component is the polybutadiene-based copolymer (ABS with high B). The amorphous thermoplastic component is SAN.

ABS plastics offer a good balance of properties centering on toughness, hardness, high gloss, dimensional stability, and rigidity. Compared to PS, ABS offers good impact strength, improved chemical resistance, and similar heat deflection temperature. ABS is also opaque. ABS has good chemical resistance to acids and bases but poor resistance to aromatic compounds, chlorinated solvents, esters, ketones, and aldehydes. ABS has poor resistance to UV light. Typical properties are shown in Table 8.

Uses. ABS materials are suitable for tough consumer products (refrigerator door liners, luggage, telephones, business machine housings, power tools, small appliances, toys, sporting goods, personal care devices), automotive (consoles, door panels, various interior trim, exterior grills and lift gates), medical (clamps, stopcocks, blood dialyzers, check valves), and building and construction (pipes, fittings, faucets, conduit, shower heads, bathtubs).

3.5 Acrylonitrile/Styrene/Acrylate (ASA) Polymers

Acrylonitrile/styrene/acrylate (ASA) ter-polymers are amorphous thermoplastics and are similar to ABS resins. However, the butadiene rubber has been replaced by an acrylate-based elastomer, which has excellent resistance to sunlight. Hence ASA offers exceptional durability in weather-related environments without painting. In outdoor applications, ASA resins retain color stability under long-term exposure to UV, moisture, heat, cold, and impact. In addition, ASA polymers offer high gloss and mechanical properties similar to those of ABS

Table 8 Typical Properties of ABS Resins

Property	General Purpose	High Heat	High Impact
Density (g/cm^3)	1.05	1.05	1.04
Tensile modulus (GPa)	2.28	2.28	2.00
Tensile strength (MPa)	43	47	39
Elongation at break (%)	—	—	26
Flexural modulus (GPa)	2.48	2.41	2.07
Flexural strength (MPa)	77	83	65
Notched Izod (kJ/m)	0.203	0.214	0.454
HDT at 1.81 MPa (°C)	81	99	81
HDT at 0.42 MPa (°C)	92	110	97
Vicat softening temperature (°C)	—	—	99
RTI (°C)	60	—	60
Hardness (Rockwell)	—	R110	—
CTE (10^{-5}/°C)	8.82	7.92	9.45
Shrinkage (in./in.)	0.006	0.006	0.006

resins. ASA has good chemical resistance to oils, greases, and salt solutions. However, resistance to aromatic and chlorinated hydrocarbons, ketones, and esters is poor.

ASA resins exhibit good compatibility with other polymers. This facilitates its use in polymer blends and alloys and as a cost-effective, cap-stock (overlayer) to protect PS, polyvinyl chloride (PVC), or ABS in outdoor applications.

Various grades are available for injection molding, profile and sheet extrusion, thermoforming, and blow molding. Typical properties appear in Table 9.

Uses. ASA resins have applications in automotive/transportation (body moldings, bumper parts, side-view mirror housings, truck trailer doors, roof luggage containers), building/construction (window lineals, door profiles, downspouts, gutters, house siding, windows, mail boxes, shutters, fencing, wall fixtures), sporting goods (snowmobile and all-terrain vehicle housings, small water craft, camper tops, windsurfer boards), and consumer items (garden hose fittings and reels, lawnmower components, outdoor furniture, telephone handsets, covers for outdoor lighting, spa and swimming pool steps and pumps, housings for garden tractors).

3.6 Poly(methyl methacrylate) (PMMA)

The catalytic or thermal polymerization of methyl methacrylate yields PMMA. It is often referred to as acrylic. PMMA is a strong, rigid, clear, amorphous polymer. The optical clarity, rigidity, colorability, and ability to resist sunlight and other environmental stresses make PMMA ideal for glass replacement.

In addition, PMMA has low water absorption and good electrical properties.

Acrylics have fair chemical resistance to many chemicals. However, resistance to aromatic and chlorinated hydrocarbons, ketones, and esters is poor. PMMA properties appear in Table 10.

Table 9 Typical Properties of ASA Resins

	ASA	ASA/PVC	ASA/PC
Density (g/cm^3)	1.06	1.21	1.15
Tensile modulus (GPa)	1.79	—	–
Tensile strength (MPa)	41	39	62
Elongation at break (%)	40	30	25
Flexural modulus (GPa)	1.79	1.93	2.52
Flexural strength (MPa)	59	48	88
Notched Izod (kJ/m)	0.320	0.961	0.320
Notched Izod, −30°C (kJ/m)	0.059	0.107	0.080
HDT at 1.81 MPa (°C)	77	74	104
HDT at 0.42 MPa (°C)	88	82	116
Vicat softening temperature (°C)	99	—	—
RTI (°C)	50	50	—
Hardness (Rockwell)	R86	R102	R110
CTE (10^{-5}/°C)	9.0	8.46	7.2
Shrinkage (in./in.)	0.006	0.004	0.006
Water absorption, 24 h (%)	—	0.11	0.25
Relative permittivity at 1 MHz	3.2	—	—
Dissipation factor at 1 MHz	0.026	—	—

Table 10 Typical Properties of PMMA, SMMA, and SMA Resins

Property	PMMA	SMMA	SMA
Density (g/cm^3)	1.19	1.09	1.08
Tensile modulus (GPa)	3.10	3.50	—
Tensile strength (MPa)	70	57.2	48.3
Elongation at break (%)	6	2	—
Flexural modulus (GPa)	3.10	3.50	3.61
Flexural strength (MPa)	103	103	115.8
Notched Izod (kJ/m)	0.016	0.020	0.011
HDT at 1.81 MPa (°C)	93	98	96
HDT at 0.42 MPa (°C)	94	—	—
Vicat softening temperature (°C)	103	—	118
RTI (°C)	90	—	—
Hardness (Rockwell)	M91	M64	L108
CTE (10^{-5} 1/°C)	7.6	7.92	6.3
Shrinkage (in./in.)	0.004	0.006	0.005
Water absorption, 24 h (%)	0.3	0.15	0.10
Relative permittivity at 1 kHz	3.3	—	—

Uses. PMMA is used in construction (glazing, lighting diffusers, domed skylights, enclosures for swimming pools and buildings), automotive (exterior lighting lenses in cars and trucks, nameplate, medallions, lenses on instrument panels), household (laboratory, vanity and counter tops, tubs), medical (filters, blood pumps), and others (appliances, aviation canopies and window, outdoor signs, display cabinets).

3.7 Styrene/Maleic Anhydride (SMA) Copolymer

Styrene/maleic anhydride resins are copolymers of styrene and maleic anhydride (MA) and offer increased heat deflection temperatures, strength, solvent reistance, and density compared to PS.[20] SMA resins are usually produced via catalyzed bulk polymerization. There is a strong tendency to form the 1:1 styrene–MA copolymer. Random SMA resins containing 5–12% MA are produced via starve feeding, i.e., keeping the MA concentration low during the polymerization. SMA resins are brittle and have poor UV resistance. Impact-modified grades and ter-polymers (grafting rubber into polymer during polymerization) are available. Typical properties for SMA prepared with 9% MA appear in Table 10.

Uses. SMA copolymers have been used in automotive (instrument panels, headliners) food service items, plumbing, and electrical applications.

3.8 Styrene/Methyl Methacrylate (SMMA) Copolymer

Styrene/methyl methacrylate copolymers are prepared by the catalyzed polymerization of styrene and methyl methacrylate (MMA). The advantages of SMMA over PS include improved outdoor weathering and light stability, better clarity, increased chemical resistance, and improved toughness. Properties for SMMA prepared with 30% MMA appear in Table 10.

Uses. Applications for SMMA plastics are in small appliances and kitchen and bathroom accessories.

3.9 Polyvinyl Chloride (PVC)

The catalytic polymerization of vinyl chloride yields polyvinyl chloride.[23] It is commonly referred to as PVC or vinyl and is second only to PE in volume use. Normally, PVC has a low degree of crystallinity and good transparency. The high chlorine content of the polymer produces advantages in flame resistance, fair heat deflection temperature, and good electrical properties. The chlorine also makes PVC difficult to process. The chlorine atoms have a tendency to split out under the influence of heat during processing and heat and light during end use in finished products, producing discoloration and embrittlement. Therefore, special stabilizer systems are often used with PVC to retard degradation.

PVC has good chemical resistance to alcohols, mild acids and bases, and salts. However, PVC is attacked by halogenated solvents, ketones, esters, aldehydes, ethers, and phenols.

There are two major classifications of PVC: rigid and flexible (plasticized). In addition, there are also foamed PVC and PVC copolymers. Typical properties of PVC resins appear in Table 11.

Rigid PVC

PVC alone is a fairly rigid polymer, but it is difficult to process and has low impact strength. Both of these properties are improved by the addition of elastomers or impact-modified graft copolymers, such as ABS and ASA resins. These improve the melt flow during processing and improve the impact strength without seriously lowering the rigidity or the heat deflection temperature.

Uses. With this improved balance of properties, rigid PVCs are used in such applications as construction (door and window frames, water supply, pipe, fittings, conduit, building panels and siding, rainwater gutters and downspouts, interior molding and flooring), packaging (bottles, food containers, films for food wrap), consumer goods (credit cards, furniture parts), and other (agricultural irrigation and chemical processing piping)

Table 11 Typical Properties of PVC Materials

Property	General Purpose	Rigid	Rigid Foam	Plasticized	Copolymer
Density (g/cm^3)	1.40	1.34–1.39	0.75	1.29–1.34	1.37
Tensile modulus (GPa)	3.45	2.41–2.45	—	—	3.15
Tensile strength (MPa)	56.6	37.2–42.4	>13.8	14–26	52–55
Elongation at break (%)	85	—	>40	250–400	—
Notched Izod (kJ/m)	0.53	0.74–1.12	>0.06	—	0.02
HDT at 1.81 MPa (°C)	77	73–77	65	—	65
Brittle temperature (°C)	—	—	—	−60 to −30	
Hardness	D85 (Shore)	R107–R122 (Rockwell)	D55 (Shore)	A71–A96 (Shore)	
CTE (10^{-5}/°C)	7.00	5.94	5.58		
Shrinkage (in./in.)	0.003				
Relative permittivity at 1 kHz	3.39				
Dissipation factor at 1 kHz	0.081				

Plasticized PVC

The rigid PVC is softened by the addition of compatible, nonvolatile, liquid plasticizers. The plasticizers are usually used in >20 parts per hundred resins. It lowers the crystallinity in PVC and acts as internal lubricant to give a clear, flexible plastics. Plasticized PVC is also available in liquid formulations known as plastisols or organosols.

Uses. Plasticized PVC is used for construction (wire and cable insulation, interior wall covering), consumer goods (outdoor apparel, rainwear, upholstery, garden hose, toys, shoes, tablecloths, sporting goods, shower curtains), medical (clear tubing, blood and solution bags, connectors), and automotive (seat covers). Plastisols are used in coating fabric, paper, and metal and rotationally cast into balls, dolls, etc.

Foamed PVC

Rigid PVC can be foamed to a low-density cellular material that is used for decorative moldings and trim. Foamed PVC is also available via foamed plastisols. Foamed PVC adds greatly to the softness and energy absorption characteristics already inherent in plasticized PVC. In addition, it gives rich, warm, leatherlike material.

Uses. Upholstery, clothing, shoe fabrics, handbags, luggage, and auto door panels and energy absorption for quiet and comfort in flooring, carpet backing, auto headliners, etc.

PVC Copolymers

Copolymerization of vinyl chloride with 10–15% vinyl acetate gives a vinyl polymer with improved flexibility and less crystallinity than PVC, making such copolymers easier to process without detracting seriously from the rigidity and heat deflection temperature. These copolymers find primary applications in flooring and solution coatings.

3.10 Poly(vinylidene chloride) (PVDC)

Poly(vinylidene chloride) is prepared by the catalytic polymerization of 1,1-dichloroethylene. This crystalline polymer exhibits high strength, abrasion resistance, high melting point, better than ordinary heat resistance (100°C maximum service temperature), and outstanding impermeability to oil, grease, water vapor, oxygen, and carbon dioxide. It is used in films, coatings, and monofilaments.

When the polymer is extruded into film, quenched, and oriented, the crystallinity is fine enough to produce high clarity and flexibility. These properties contribute to widespread use in packaging film, especially for food products that require impermeable barrier protection.

PVDC and/or copolymers with vinyl chloride, alkyl acrylate, or acrylonitrile are used in coating paper, paperboard, or other films to provide more economical, impermeable materials. Properties appear in Table 12.

Table 12 Typical Properties of PVDC

Density (g/cm^3)	1.65–1.72
Tensile strength (MPa)	25
Elongation at break (%)	120
Notched Izod (kJ/m)	0.04
Hardness (Rockwell)	M50–M65

Uses. PVDC is used in food packaging were barrier properties are needed. Applications for injection-molding grades are fittings and parts in the chemical industry. PVDC pipe is used in the disposal of waste acids. PVDC is extruded into monofilament and tape that is used in outdoor furniture upholstery.

4 POLYURETHANE AND CELLULOSIC RESINS

4.1 Polyurethanes (TPU, PUR)

Polyurethanes (PUs) are prepared from polyols and isocyanates.[24,25] The isocyanate groups react with the hydroxyl groups on the polyol to form a urethane bond. The polyol can be a low-molecular-weight polyether or polyester. The isocyanate can be aliphatic or aromatic and in the preparation of linear PU is typically difunctional. However, isocyanates with greater functionality are used in preparing rigid foam PUs. The family of PU resins is very complex because of the enormous variation in the compositional features of the polyols and isocyanates. This variety results in a large number of polymer structures and performance profiles. Indeed, PUs can be rigid solids or soft and elastomeric or a have a foam (cellular) structure.

The majority of PU resins are thermoset (PUR). However, there are also important thermoplastic polyurethane resins (TPU). Polyurethanes offer high impact strength, even at low temperatures, good abrasion resistance, excellent heat resistance, excellent resistance to nonpolar solvents, fuels, and oils, and resistance to ozone, oxidation, and humidity.

TPUs are generally processed by injection-molding techniques. PURs are processed by reaction injection molding (RIM) and various foaming techniques. A major use of PUR is in the production of flexible, semirigid, and rigid foams. In addition, PUs can be used as fibers, sealants, adhesives, and coatings.

Typical properties of PUs appear in Table 13.

Uses. Typical applications for PUs are in consumer goods (furniture padding, bedding, skateboard and roller blade wheels, shoe soles, athletic shoes, ski booths, backing on carpets and tiles), automotive (padding, seals, fascias, bumpers, structural door panels), and miscellaneous (tubing, membranes, bearings, nuts, seals, gaskets).

Table 13 Typical Properties of PUs

Property	TPU Polyester	TPU Polyether	PUR–RIM Foam	PUR–RIM Solid	PUR–RIM Elastomer
Density (g/cm^3)	1.21	1.18	0.56	1.13	1.04
Tensile strength (MPa)	41	38	13.8	39	24
Elongation at break (%)	500	250	10	10	250
Flexural modulus (GPa)	0.14	0.54	0.86	0.14	0.36
Flexural strength (MPa)	—	—	27.6	60	—
Notched Izod (kJ/m)	—	—	—	0.05	0.6
HDT at 0.45 MPa (°C)	59	45	70	—	—
Vicat softening temperature (°C)	167	140	—	—	—
Brittle temperature (°C)	<−68	<−70	—	—	—
Hardness (Shore)	D55	D70	D65	D76	D58
CTE (10^{-5}/°C)	13	11.5	7.9	10	11
Shrinkage (in./in.)	0.008	0.008	0.8	1.1	1.3

4.2 Cellulosic Polymers (CA, CAB, CP)

Cellulose-based plastics are manufactured by the chemical modification of cellulose.[26–29] Cellulose does not melt and hence is not a thermoplastic material. However, esterification of cellulose gives organic esters of cellulose, which are thermoplastic. These include cellulose acetate (CA), cellulose acetate butyrate (CAB), and cellulose proprionate (CP). Cellulosics are noted for their wide range of properties, which include clarity (up to 80% light transmission), abrasion resistance, stress crack resistance, high gloss, and good electrical properties. CA offers good rigidity and hardness. CAB and CP offer good weatherability, low-temperature impact strength, and dimensional stability.

In general, cellulosic esters are resistant to aliphatic hydrocarbons, ethylene glycol, bleach, and various oils. However, alkaline materials attack them. Cellulosic esters have high water absorption and low continuous-use temperatures. Typical properties appear in Table 14.

Uses. Typical applications for cellulosic plastics are automotive (steering wheels, trim), films (photographic, audio tape, pressure-sensitive tape), home furnishings (table edging, Venetian blind wands), packaging (tubular containers, thermoformed containers for nuts, bolts, etc.), and miscellaneous (tool and brush handles, toys, filaments for toothbrushes, eye glass frames, lighting fixtures).

5 ENGINEERING THERMOPLASTICS: CONDENSATION POLYMERS

Engineering thermoplastics comprise a special-performance segment of synthetic plastics that offer enhanced properties.[2,3] When properly formulated, they may be shaped into mechanically functional, semiprecision parts or structural components. Mechanically functional implies that the parts may be subjected to mechanical stress, impact, flexure, vibration, sliding friction, temperature extremes, hostile environments, etc., and continue to function.

As substitutes for metal in the construction of mechanical apparatus, engineering plastics offer advantages such as transparency, light weight, self-lubrication, and economy in fabrication and decorating. Replacement of metals by plastic is favored as the physical properties

Table 14 Typical Properties of Cellulosic Materials

Property	CA	CAB	CP
Density (g/cm^3)	1.28	1.19	1.19
Tensile modulus (GPa)	2.17	1.73	1.73
Tensile strength (MPa)	40	34.5	35
Elongation at break (%)	25	50	60
Flexural modulus (GPa)	2.4	1.8	—
Flexural strength (MPa)	66	60	—
Notched Izod (kJ/m)	0.16	0.187	0.41
HDP at 1.81 MPa (°C)	61	65	72
HDP at 0.42 MPa (°C)	72	72	80
Vicat softening temperature (°C)	—	—	100
Hardness (Rockwell)	R82	R75	R70
CTE (10^{-5}/°C)	13.5	13.5	14
Relative permittivity at 1 kHz	3.6	3.6	3.6

and operating temperature ranges of plastics improve and the cost of metals and their fabrication increases.

5.1 Thermoplastic Polyesters

Thermoplastic polyesters are prepared from the condensation polymerization of a diol and typically a dicarboxylic acid. Usually the dicarboxylic acid is aromatic, i.e., terephthalic acid. As a family of polymers thermoplastic polyesters are typically semicrystalline and hence have good chemical resistance. An important attribute of a semicrystalline polymer is a fast rate of crystallization, which facilitates short injection-molding cycles.

Poly(butylene terephthalate) (PBT)

Poly(butylene terephthalate) is prepared from butanediol with dimethyl terephthalate.[29–31] PBT is a semicrystalline polymer which has a fast rate of crystallization and rapid molding cycles. PBT has a unique and favorable balance of properties between polyamides and polyacetals. PBT combines high mechanical, thermal, and electrical properties with low moisture absorption, extremely good self-lubricity, fatigue resistance, very good chemical resistance, very good dimensional stability, and good maintenance of properties at elevated temperatures. Dimensional stability and electrical properties are unaffected by high-humidity conditions.

PBT has good chemical resistance to water, ketones, alcohols, glycols, ethers, and aliphatic and chlorinated hydrocarbons at ambient temperatures. In addition, PBT has good resistance to gasoline, transmission fluid, brake fluid, greases, and motor oil. At ambient temperatures PBT has good resistance to dilute acids and bases, detergents, and most aqueous salt solutions. It is not recommended for use in strong bases or aqueous media at temperatures above 50°C.

PBT grades range from unmodified to glass-fiber-reinforced to combinations of glass fiber and mineral fillers which enhance strength, modulus, and heat deflection temperature. Both filled and unfilled grades of PBT offer a range of UL and other agency compliance ratings.

A high-density PBT combines the inherent characteristics of PBT with the advantages of high levels of mineral reinforcement. This combination provides a balance of mechanical, thermal, and electrical properties, broad chemical and stain resistance, low water absorption, and dimensional stability. In addition, the smooth, satin finish and heavy weight offer the appearance and quality feel of ceramics while providing design flexibility, injection-molding processing advantages, and recycling opportunities which are common in engineering thermoplastics. Properties appear in Table 15.

Uses. Applications of PBT include automotive components (brake system parts, fuel injection modules, grill-opening panels), electrical/electronic components (connectors, smart network interface devices, power plugs, electrical components, switches, relays, fuse cases, light sockets, television tuners, fiber-optic tubes), medical (check valves, catheter housings, syringes), consumer goods (hair dryer and power tool housings, iron and toaster housings, food processor blades, cooker-fryer handles), and miscellaneous (gears, rollers, bearing, housings for pumps, impellers, pulleys, industrial zippers). High-density PBT is being used in kitchen and bath sinks, countertops, wall tiles, shower heads, speaker housings, medical equipment, and consumer goods.

PBT/PC Alloy

PBT/PC resins are thermoplastic alloys of PBT and polycarbonate (PC). The amorphous PC provides impact resistance and toughness while the PBT provides enhanced chemical resis-

Table 15 Typical Properties of PBT

Property	PBT	PBT + 30% GF	PBT + 40% GF
Density (g/cm^3)	1.31	1.53	1.63
Tensile strength (MPa)	52	119	128
Elongation at break (%)	300	3	–
Flexural modulus (GPa)	2.34	7.58	8.27
Flexural strength (MPa)	83	190	200
Notched Izod (kJ/m)	0.053	0.085	0.096
Unnotched Izod (kJ/m)	1.602	0.801	0.961
HDT at 1.81 MPa (°C)	54	207	204
HDT at 0.42 MPa (°C)	154	216	216
RTI (°C)	—	140	—
Hardness (Rockwell)	R117	R118	R118
CTE (10^{-5}/°C)	8.1	2.5	2.7
Shrinkage (in./in.)	0.006	0.004	0.004
Water absorption, 24 h (%)	0.08	0.06	0.05
Relative permittivity at 1 MHz	3.1	3.7	4.0
Dissipation factor at 1 MHz	0.419	0.02	0.02

tance and thermal stability. Impact modification completes the balanced performance profile by providing both low- and high-temperature durability. Hence high levels of impact strength are achieved at low temperatures, below −40°C.

PBT/PC resins offer a balance of performance characteristics unique among engineering thermoplastics with its optimal combination of toughness, chemical resistance, dimensional stability, lubricity, and high heat distortion temperature. In addition, this family of resins offers very good aesthetics, lubricity, UV resistance, and color retention. Originally developed for the automotive industry, PBT/PC resins are designed to provide resistance to various automotive fluids—gasoline, greases, oils, etc. In general, the higher the amount of PBT in the resin blend, the higher the resin's chemical resistance. Hence resistance to gasoline may vary from grade to grade. PBT/PC resins generally are not very hydrolytically stable. Properties appear in Table 16.

Uses. Applications for PBT/PC resins include automotive bumpers/fascia, tractor hoods and panels, components on outdoor recreational vehicles, lawn mower decks, power tool housings, material-handling pallets, and large structural parts.

Poly(ethylene terephthalate) (PET)

Poly(ethylene terephthalate) is prepared from the condensation polymerization of dimethyl terephthalate or terephthalic acid with ethylene glycol.[29–31] PET is a semicrystalline polymer that exhibits high modulus, high strength, high melting point, good electrical properties, and moisture and solvent resistance. The crystallization rate of PET is relatively slow. This slow crystallization is a benefit in blow-molding bottles where clarity is important. Indeed, a small amount of a comonomer is typically added during polymerization of PET. The function of the comonomer is to disrupt the crystallinity and lower the rate of crystallization. However, in injection-molding applications the slow rate of crystallization will increase the molding cycle time.

For most injection-molding applications PET generally contains glass fiber or mineral filler to enhance properties. Typical properties appear in Table 17.

Table 16 Typical Properties of PBT/PC Alloy

Property	PBT/PC	PBT/PC + 30% GF
Density (g/cm^3)	1.21	1.44
Tensile strength (MPa)	59	92
Elongation at break (%)	120	4
Flexural modulus (GPa)	2.04	5.38
Flexural strength (MPa)	85	138
Notched Izod (kJ/m)	0.710	0.171
Notched Izod, −30°C (kJ/m)	3.204	0.112
Unnotched Izod (kJ/m)	0.299	0.641
HDT at 1.81 MPa (°C)	99	149
HDT at 0.42 MPa (°C)	106	204
Hardness (Rockwell)	R112	R109
CTE (10^{-5}/°C)	8.4	2.3
Shrinkage (in./in.)	0.009	0.005
Water absorption, 24 h (%)	0.12	0.09
Relative permittivity at 1 MHz	3.04	3.9
Dissipation factor at 1 MHz	0.019	0.02

Uses. Primary applications of PET include blow-molded beverage bottles; fibers for wash-and-wear, wrinkle-resistant fabrics; and films that are used in food packaging, electrical applications (e.g., capacitors), magnetic recording tape, and graphic arts. Injection-molding application of PET include automotive (cowl vent grills, wiper blade supports) and electrical (computer fan blades, fuse holders and connectors).

Poly(trimethylene terephthalate) (PTT)

Poly(trimethylene terephthalate) is prepared from 1,3-propanediol and terephthalic acid. PTT is a semicrystalline polymer that exhibits properties between PET and PBT. In particular, it

Table 17 Typical Properties of PET

Property	PET	PET + 30% GF
Density (g/cm^3)	1.41	1.60
Tensile modulus (GPa)	1.71	11.5
Tensile strength (MPa)	50	175
Elongation at break (%)	180	2
Flexural modulus (GPa)	2.0	—
Flexural strength (MPa)	—	225
Notched Izod (kJ/m)	0.090	—
HDT at 1.81 MPa (°C)	63	225
HDT at 0.42 MPa (°C)	71	—
Vicat softening temperature (°C)	—	260
Oxygen index (%)	—	21
Hardness (Rockwell)	R105	—
CTE (10^{-5}/°C)	9.1	2.0
Water absorption, 24 h (%)	—	0.15
Relative permittivity at 1 MHz	3.3	4.2
Dissipation factor at 1 MHz	—	0.018

offers high modulus, high strength, good electrical properties, and moisture and solvent resistance. Its crystallization rate is slower than PBT but faster than PET.[32] Properties appear in Table 18.

Uses. Initial applications for PTT were in carpet fiber, where it offers a softer feel in combination with stain resistance and resiliency. Other fiber markets include textiles and monofilaments. Injection-molding applications would be various automotive parts.

5.2 Polyamides (Nylon)

Polyamides, commonly called nylons, are produced by the condensation polymerization of dicarboxylic acids and diamines or the catalytic polymerization of a lactam monomer (a cyclic amide).[33,34] In general, polyamides are semicrystalline thermoplastics. Polyamides are a class of resins characterized by broad chemical resistance, high strength, and toughness. In addition, polyamides absorb high levels of water. Moisture from the atmosphere diffuses into the polymer and hydrogen bonds to the amide groups. This absorbed water causes dimensional changes where the molded part will increase in size and weight. The higher the amount of amide groups in the polymer, the greater the moisture uptake. In addition, the water acts as a plasticizer and lowers the rigidity and strength. Polar solvents such as alcohols are also absorbed into the nylon.

There are numerous dicarboxylic acids and diamines that can be used to make polyamides. The shorthand method for describing the various types of polyamides uses a number to designate the number of carbon atoms in each starting monomer(s). In the case of terephthalic and isophthalic acids T and I are used, respectively.

Polyamide 6/6 and 6 (PA6, PA6/6)

The two major types of polyamides are polyamide 6/6, or nylon 6/6 (PA6/6), and polyamide 6, or nylon 6 (PA6).[33,34] PA6/6 is made from a six-carbon dicarboxylic acid and a six-carbon diamine—i.e., adipic acid and hexamethylene diamine. PA6 is prepared from caprolactam. Both PA6/6 and PA6 are semicrystalline resins.

Key features of nylons include toughness, fatigue resistance, and chemical resistance. Nylons do exhibit a tendency to creep under applied load. Nylons are resistant to many chemicals, including ketones, esters, fully halogenated hydrocarbons, fuels, and brake fluids. Nylons have a relatively low heat deflection temperature. However, glass fibers or mineral fillers are used to enhance the properties of polyamides. In addition to increasing the heat deflection temperature, the fibers and fillers lessen the effect of moisture and improve the dimensional stability. Properties of PA66 and PA6 appear in Table 19.

Table 18 Typical Properties of PTT

Property	PTT	PTT + 30% GF
Density (g/cm^3)	1.35	1.55
Flexural modulus (GPa)	2.76	10.3
Tensile strength (Mpa)	67	159
Notched Izod (kJ/m)	0.05	0.11
HDT at 1.81 MPa (°C)	59	216
Shrinkage (in./in.)	0.020	0.002
Dielectric constant at 1 MHz	3.0	—
Relative permittivity at 1MHz	0.02	—

Table 19 Typical Properties of PA6 and PA6/6

Property	PA6	PA6 + 33% GF	PA6/6	PA6/6 + 40% GF
Density (g/cm^3)	1.13	1.46	1.14	1.44
Tensile modulus (GPa)	—	—	3.30	—
Tensile strength (MPa)	79	200	86	—
Elongation at break (%)	70	3	45	—
Flexural modulus (GPa)	2.83	9.38	2.90	9.3
Flexural strength (MPa)	108	276	—	219
Notched Izod (kJ/m)	0.053	0.117	0.059	0.14
HDT at 1.81 MPa (°C)	64	210	90	250
HDT at 0.45 MPa (°C)	165	220	235	260
RTI (°C)	105	120	—	—
Hardness (Rockwell)	R119	R121	—	M119
CTE (10^{-5}/°C)	8.28	2.16	8.10	3.42
Shrinkage (in./in.)	0.013	0.003	0.0150	0.0025
Water absorption, 24 h (%)	—	—	1.2	—
Relative permittivity at 1 MHz	—	3.8	3.6	—
Dissipation factor at 1 MHz	—	0.022	0.02	—

Uses. The largest application of nylons is in fibers. Molded applications include automotive components (electrical connectors, wire jackets, fan blades, valve covers, emission control valves, light-duty gears), electronic (connectors, cable ties, plugs, terminals, coil forms), related machine parts (gears, cams, pulleys, rollers, boat propellers), and appliance parts.

Polyamide/PPE Alloys

Polyamide/polyphenylene ether (PA/PPE) alloys are compatible blends of amorphous PPE and a semicrystalline PA which have a microstructure in which the PA is the continuous phase and the PPE is the discrete phase.[3,35–37] The PPE acts as an organic reinforcing material. This technology combines the inherent advantages of PPE (dimensional stability, very low water absorption, and high heat resistance) with the chemical resistance and ease of processing of the PA. This combination results in a chemically resistant material with the stiffness, impact resistance, dimensional stability, and heat performance required for automotive body panels that can undergo on-line painting.

PA/PPE alloys offer broad environmental resistance to commonly used automotive fuels, greases, and oils. In addition, this family of alloys is resistant to detergents, alcohols, aliphatic and aromatic hydrocarbons, and alkaline chemicals.

Since PPE does not absorb any significant amount of moisture, the effect of moisture on properties is reduced. Indeed, the moisture uptake in PA/PPE alloys is lower. Hence PA/PPE alloys minimize the effect of moisture on rigidity, strength, and dimensional stability vis-à-vis PA.[35,36] In addition, heat deflection temperatures have been enhanced by the PPE. Properties are shown in Table 20.

Uses. PA/PPE alloys are used in automotive body panels (fenders and quarter panels), automotive wheel covers, exterior truck parts, under-the-hood automotive parts (air intake resonators, electrical junction boxes and connectors), and fluid-handling applications (e.g., pumps).

Table 20 Typical Properties of PPE/PA6/6 Alloys

	Unfilled		10% GF		30% GF	
Property	PA	PPE/PA	PA	PPE/PA	PA	PPE/PA
Density (g/cm^3)	1.14	1.10	1.204	1.163	1.37	1.33
Flexural modulus (GPa)						
Dry as molded	2.8	2.2	4.5	3.8	8.3	8.1
100% Relative humidity	0.48	0.63	2.3	2.6	4.1	5.8
At 150°C	0.21	0.70	0.9	1.6	3.2	4.3
Flexural strength (MPa)						
Dry as molded	96	92	151	146	275	251
100% Relative humidity	26	60	93	109	200	210
At 150°C	14	28	55	60	122	128

Polyamide 4/6 (PA4/6)

Nylon 4/6 is prepared from the condensation polymerization of 1,4-diaminobutane and adipic acid.[33] PA4/6 has a higher crystalline melting point, greater crystallinity, and a faster rate of crystallization than other PAs. In comparison to PA6/6 and PA6, PA4/6 offers high tensile strength and heat deflection temperature; however, it has higher moisture uptake, which can affect properties and dimensional stability. Properties appear in Table 21.

Uses. Application for PA4/6 include under-the-hood automotive parts, gears, bearings, and electrical parts.

Semiaromatic Polyamide (PA6/6T, PA6I/6T)

Semiaromatic PAs have been developed in order to increase the performance over that of PA6/6 and PA6.[33] In general, these resins are modified copolymers based on

Table 21 Typical Properties of Nylon 4/6

Property	PA4/6	PA4/6 + 30% GF
Density (g/cm^3)	1.18	1.41
Melting point (°C)	295	—
Glass transition temperature (°C)	75	—
Tensile modulus (GPa)	3.3	10.0
Tensile strength (MPa)	100	210
Elongation at break (%)	15	4
Flexural modulus (GPa)	3.1	—
Flexural strength (MPa)	149.6	—
Notched Izod (kJ/m)	0.096	0.069
HDT at 1.81 MPa (°C)	190	290
HDT at 0.42 MPa (°C)	280	290
Vicat softening temperature (°C)	290	290
Hardness (Shore)	D85	—
CTE (10^{-5}/°C)	9.0	5.0
Water absorption, 24 h (%)	3.7	2.6
Relative permittivity at 1 kHz	3.83	—

poly(hexamethylene terephthalate), or PA6/T. Pure nylon 6/T exhibits a very high T_m of 370°C and a T_g of 180°C. This high T_m is above its decomposition temperature. PA6/T copolymers have been prepared using additional monomers such as isophthalic acid, adipic acid, caprolactam, or 1,5-hexyl diamine, which lower the crystalline melting point to useful ranges for melt processing. These ter-polymers exhibit T_m values of 290–320°C and T_g values of 100–125°C and offer enhanced performance (i.e., stiffer, stronger, greater thermal and dimensional stability) over PA6/6 and 6. Melt processing requires higher temperatures and often an oil-heated mold (>100°C). These semicrystalline PAs have good chemical resistance, good dielectric properties, and lower moisture absorption and are more dimensionally stable in the presence of moisture than PA6 and PA6/6. Properties appear in Table 22.

Uses. High heat application, automotive (radiator ventilation and fuel supply systems), electrical/electronic (housings, plugs, sockets, connectors), recreational (tennis rackets, gold clubs), mechanical (industrial and chemical processing equipment, bearings, gears), appliance and plumbing parts, and aerospace components.

Aromatic Polyamides (PPTA, MPIA)

Polyamides prepared from aromatic diamines and aromatic diacids give very high heat aromatic nylons or aramides. Examples are poly(*p*-phenyleneterephthalamide), or PPTA, and poly(*m*-phenyleneisophthalamide), or MPIA. These wholly aromatic PAs have high strength, high modulus, high chemical resistance, high toughness, excellent dimensional stability, and inherent flame resistance. MPIA has a T_g of 280°C and is difficult to melt process. Typically it is spun into fibers. PPTA has very high T_g and T_m of 425 and 554°C, respectively. In addition, PPTA exhibits liquid crystalline behavior and is spun into highly oriented, very high modulus, crystalline fibers. Properties appear in Table 23.

Uses. PPTA fibers have uses in bullet-resistant apparel, composites, brake and transmission parts, gaskets, ropes and cables, sporting goods, tires, belts, and hoses.

MPIA fibers have uses in heat-resistant and flame-retardant apparel, electrical insulation, and composite structures.

Table 22 Typical Properties of Semiaromatic Polyamides

Property	PA6/6T	PA6/6T +35% GF	PA6T/6I	PA6T/6I +40% GF
Density (g/cm^3)	1.16	1.43	1.21	1.46
Tensile modulus (GPa)	3.20	12.0	2.44	11.1
Tensile strength (MPa)	100	210	108	187
Elongation at break (%)	11.5	3	5	2
Flexural modulus (GPa)	—	—	3.43	10.8
Flexural strength (MPa)	—	—	157	284
Notched Izod (kJ/m)	0.070	—	0.049	0.079
Unnotched Izod (kJ/m)	—	—	0.395	0.592
HDT at 1.81 MPa (°C)	100	270	140	295
HDT at 0.45 MPa (°C)	120	—	—	—
RTI (°C)	125	—	120	120
CTE (10^{-5}/°C)	7.0	1.5	—	—
Shrinkage (in./in.)	0.0065	0.0035	0.006	0.002
Water absorption, 24 h (%)	1.8		0.3	0.2
Dielectric constant at 1 MHz	4.0	4.2	—	—
Dissipation factor at 1 MHz	0.030	0.020	—	—

Table 23 Typical Properties of Aromatic PAs

Property	PPTA	MPIA
Density (g/cm^3)	—	1.38
Melting point (°C)	554	—
Glass transition temperature (°C)	425	280
Tensile modulus (GPa)	80–125	—
Tensile strength (MPa)	1500–2500	61
Elongation at break (%)	2	25
Flexural modulus (GPa)	—	3.1
Oxygen index (%)	29	28
CTE (10^{-5}/°C)	−0.32	0.62

5.3 Polyacetals (POM)

Polyacetal homopolymer, or polyoxymethylene (POM), is prepared via the polymerization of formaldehyde followed by capping each end of the polymer chain with an ester group for thermal stability. Polyacetal copolymer (POM-Co) is prepared by copolymerizing trioxane with relatively small amounts of a comonomer such as ethylene oxide.[29] The comonomer functions to stabilize the polymer to reversion reactions. POM and POM-Co are commonly referred to as acetals and are also semicrystalline resins.

Polyacetals exhibit rigidity, high strength, excellent creep resistance, fatigue resistance, toughness, self-lubricity/wear resistance, and solvent resistance. Acetals are resistant to gasoline, oils, greases, ethers, alcohols, and aliphatic hydrocarbons. They are not recommended for use with strong acids.

Properties are enhanced by the addition of glass fiber or mineral fillers. Properties of POM and POM-Co appear in Table 24.

Uses. Applications of polyacetals include moving parts in appliances and machines (gears, bearings, bushings, rollers, springs, valves, conveying equipment), automobiles (door handles, fasteners, knobs, fuel pumps, housings), plumbing and irrigation (valves, pumps, faucet

Table 24 Typical Properties of Polyacetals

Property	POM	POM-Co	POM +25% GF	POM-Co +30% GF
Density (g/cm^3)	1.42	1.41	1.58	1.60
Tensile modulus (GPa)	3.12	2.83	9.50	9.20
Tensile strength (MPa)	68.9	60.6	140	135
Elongation at break (%)	50	60	3	2.5
Flexural modulus (GPa)	2.83	2.58	8.00	—
Flexural strength (MPa)	97	90	—	—
Notched Izod (kJ/m)	0.074	0.054	0.096	—
HDT at 1.81 MPa (°C)	136	110	172	160
HDT at 0.45 MPa (°C)	172	158	176	—
Vicat softening point (°C)	—	151	178	158
Hardness (Rockwell)	M94	M80	—	—
CTE (10^{-5}/°C)	11.1	11.0	5.6	6.0
Shrinkage (in./in.)	0.02	—	0.008	—
Water absorption, 24 h (%)	0.2	0.2	0.17	0.17
Relative permittivity at 1 MHz	3.7	4.0	—	4.3
Dissipation factor at 1 MHz	—	0.005	—	0.006

underbodies, shower heads, impellers, ball cocks), industrial or mechanical products (rollers, bearings, gears, conveyer chains, housings), consumer products (A/V cassette components, toiletry articles, zippers, pen barrels, disposable lighters, toy parts), and electronic parts (key tops, buttons, switches).

5.4 Polycarbonate (PC)

Most commercial PCs are derived from the reaction of bisphenol A and phosgene.[29,38–40] PCs are amorphous resins which have a unique combination of outstanding clarity and high impact strength. In addition, PCs offer high dimensional stability, resistance to creep, and excellent electrical insulating characteristics. Indeed, PCs are among the stronger, tougher, and more rigid thermoplastics available. PC is a versatile material and a popular blend material used to enhance the performance of ABS, ASA, and polyesters (PBT).

PCs offer limited resistance to chemicals. PC properties are shown in Table 25.

Uses. Applications of PC include glazing (safety glazing, safety shields, nonbreakable windows), automotive parts (lamp housings and lenses, exterior parts, instrument panels), packaging (large water bottles, reusable bottles), food service (mugs, food processor bowls, beverage pitchers), ophthalmic (optical lenses, corrective eyewear, sun wear lenses), medical/laboratory ware (filter housings, tubing connectors, eyewear, health care components), consumer (various appliance parts and housings, power tool housings, cellular phone housings, food processor bowls, lighting), media storage [compact discs (CDs), digital video discs (DVDs)], and miscellaneous (electrical relay covers, aircraft interiors, building and construction). Extruded PC film is used in membrane switches.

5.5 Polycarbonate/ABS Alloys (PC/ABS)

Polycarbonate/ABS alloys are amorphous blends of PC and ABS resins. They offer a unique balance of properties that combines the most desirable properties of both resins.[41] The ad-

Table 25 Typical Properties of PCs

Property	PC	PC + 30% GF
Density (g/cm^3)	1.20	1.43
Tensile modulus (GPa)	2.38	8.63
Tensile strength (MPa)	69	131
Elongation at break (%)	130	3
Flexural modulus (GPa)	2.35	7.59
Flexural strength (MPa)	98	158
Notched Izod (kJ/m)	0.905	0.105
Unnotched Izod (kJ/m)	3.20	1.06
HDT at 1.81 MPa (°C)	132	146
HDT at 0.45 MPa (°C)	138	152
Vicat softening point (°C)	154	165
RTI (°C)	121	120
Hardness (Rockwell)	R118	R120
CTE (10^{-5}/°C)	6.74	1.67
Shrinkage (in./in.)	0.006	0.002
Water absorption, 24 h (%)	0.15	0.14
Relative permittivity at 1 MHz	2.96	3.31

dition of ABS improves the melt processing of PC/ABS blends, which facilitates filling large, thin-walled parts. Moreover, the ABS enhances the toughness of PC, especially at low temperatures, while maintaining the high strength and rigidity of the PC. In addition, PC/ABS offers excellent UV stability, high dimensional stability at ambient and elevated temperatures, and the ability for chlorine/bromine-free flame retardance.

The properties are a function of the ABS-to-PC ratio. Properties appear in Table 26.

Uses. Automotive (interior and exterior automotive applications as instrument panels, pillar, bezel, grills, interior and exterior trim), business machines (enclosures and internal parts of products such as lap- and desk-top computers, copiers, printers, plotters, and monitors), telecommunications [mobile telephone housings, accessories, and smart cards (GSM SIM cards)], electrical (electronic enclosures, electricity meter covers and cases, domestic switches, plugs and sockets and extruded conduits), and appliances (internal and external parts of appliances such as washing machines, dryers, and microwave ovens).

5.6 Polyestercarbonates (PECs)

Polyestercarbonate resins have iso- and terephthalate units incorporated into standard bisphenol A polycarbonate.[38] This modification of the polymer enhanced the performance between that of PC and polyarylates. Thus PECs have properties similar to PC but with higher heat deflection temperature and better hydrolytic performance. Higher levels of iso- and terephthalate units result in higher heat deflection temperatures and continuous-use temperatures. Properties of PECs with different heat deflection temperatures appear in Table 27.

Uses. PEC is marketed into typical polycarbonate applications that require slightly higher heat deflection temperature and continuous-use temperatures, such as consumer goods, electrical/electronic, and automotive parts.

5.7 Polyarylates (PARs)

The homopolymers of bisphenol A and isophthalic acid or terephthalic acids are semicrystalline.[42] The semicrystalline PARs have a high crystalline melting point and very slow crystallization rates. Hence oil-heated molds and log cycle times would have made commercialization unattractive. Amorphous PARs are prepared from a mixture of isophthalic and terephthalic acids and bisphenol A and can be melt processed without difficulty. They are clear, slightly yellow in color, dimensionally stable, and resistant to creep, have excellent electrical properties, are rigid, and have good impact strength.

Table 26 Typical Properties of PC/ABS Blends

	At selected PC/ABS ratio (wt/wt)			
Properties	0/100	50/50	80/20	100/0
Density (g/cm^3)	1.06	1.13	1.17	1.20
Tensile modulus (GPa)	1.8	1.9	2.5	2.4
Tensile strength (MPa)	40	57	60	65
Elongation at break (%)	20	70	150	110
Notched Izod, 25°C (kJ/m)	0.30	0.69	0.75	0.86
Notched Izod, −20°C (kJ/m)	0.11	0.32	0.64	0.15
HDT at 1.81 MPa (°C)	80	100	113	132

Table 27 Typical Properties of PECs

Property	PEC-1	PEC-2
Density (g/cm^3)	1.20	1.20
Tensile strength (MPa)	71.8	78.0
Elongation at break (%)	120	75
Flexural modulus (GPa)	2.03	2.33
Flexural strength (MPa)	95	96.6
Notched Izod (kJ/m)	0.534	0.534
Unnotched Izod (kJ/m)	3.20	3.22
HDT at 1.81 MPa (°C)	151	162
HDT at 0.42 MPa (°C)	160	174
RTI (°C)	125	130
Hardness (Rockwell)	R122	R127
CTE (10^{-5}/°C)	5.1	4.5
Linear mold shrinkage (in./in.)	0.007	0.009
Water absorption, 24 h (%)	0.16	0.19
Relative permittivity at 1 MHz	3.19	3.45

PARs have poor chemical resistance to ketones, esters, and aromatic and chlorinated hydrocarbons. Typical properties appear in Table 28.

Uses. PARs are marketed into applications requiring a higher heat deflection temperature than PC. These include electrical/electronic and automotive applications.

5.8 Modified Polyphenylene Ether (PPE)

Poly(2,6-dimethylphenylene ether) or PPE, is produced by the oxidative coupling of 2,6-dimethyl phenol.[43–46] PPE is an amorphous thermoplastic. The PPE polymer has a very high heat deflection temperature, good inherent flame resistance, outstanding dimensional stability,

Table 28 Typical Properties of PARs

Property	PAR	PAR + 40% GF
Density (g/cm^3)	1.21	1.40
Tensile modulus (GPa)	2.0	8.28
Tensile strength (MPa)	66	149
Elongation at break (%)	50	0.8
Flexural modulus (GPa)	2.14	7.59
Flexural strength (MPa)	86	220
Notched Izod (kJ/m)	0.22	0.085
HDT at 1.81 MPa (°C)	174	—
Oxygen index (%)	34	—
UL94 VO (mm)	1.60	—
RTI (°C)	120	—
Hardness (Rockwell)	R100	R125
CTE (10^{-5}/°C)	5.6	3.0
Shrinkage (in./in.)	0.009	0.002
Water absorption (%)	0.27	0.18
Relative permittivity at 1 MHz	2.62	3.8

and outstanding electrical properties. In addition, PPE has one of the lowest moisture absorption rates found in any engineering plastic.

PPE by itself is difficult to process; hence, PPE is commonly blended with styrenics (e.g., HIPS, ABS) to form a family of modified PPE-based resins.[44–46] What is truly unique about the PPE/PS blends is that the PPE and PS form a miscible, single-phased blend. Most polymers have limited solubility in other polymers. Modified PPE resins cover a wide range of compositions and properties.

Modified PPE resins are characterized by ease of processing, high impact strength, outstanding dimensional stability at elevated temperatures, long-term stability under load (creep resistance), and excellent electrical properties over a wide range of frequencies and temperatures. Another unique feature of modified PPE resins is their ability to make chlorine/bromine-free flame-retardant grades.

Modified PPE resins are especially noted for their outstanding hydrolytic stability—they do not contain any hydrolyzable bonds. Their low water absorption rates—both at ambient and elevated temperatures—promote the retention of properties and dimensional stability in the presence of water and high humidity and even in steam environments. In addition, modified PPE resins are also virtually unaffected by a wide variety of aqueous solutions—slats, detergents, acids, and bases. PPE is a versatile material and is used in alloys with PA to enhance the performance and decrease the moisture absorbance of the PA.

The chemical compatibility with oils and greases is limited. It is not recommended for contact with ketones, esters, toluene, and halogenated solvents. Typical properties appear in Table 29.

Uses. Applications include automotive (instrument panels, trim, spoilers, under-the-hood components, grills), telecommunication equipment (TV cabinets, cable splice boxes, wire board frames, electrical connectors, structural and interior components in electrical/electronic

Table 29 Typical Properties of Modified PPE Resins

Property	190 Grade	225 Grade	300 Grade
Density (g/cm^3)	1.12	1.11	1.12
Tensile modulus (GPa)	2.70	—	—
Tensile strength (MPa)	61	67	76
Elongation at break (%)	18	17	20
Flexural modulus (GPa)	2.35	2.49	2.50
Flexural strength (MPa)	93	99	110
Notched Izod (kJ/m)	0.029	0.019	0.023
Unnotched Izod (kJ/m)	0.721	—	—
HDT at 1.81 MPa (°C)	88	99	145
HDT at 0.42 MPa (°C)	96	109	156
Vicat softening temperature (°C)	113	129	—
RTI (°C)	95	95	105
UL94 VO (mm)	1.5	1.5	1.5
Oxygen index (%)	39	—	—
Hardness (Rockwell)	R120	—	R119
CTE (10^{-5}/°C)	8.8	—	5.4
Shrinkage (in./in.)	0.006	0.006	0.006
Water absorption, 24 h (%)	0.08	—	0.06
Relative permittivity at 1 MHz	2.60	2.55	2.63
Dissipation factor at 1 MHz	0.0055	0.007	0.009

equipment), plumbing/water handling (pumps, plumbing fixtures), consumer goods (microwaveable food packaging, appliance parts), medical, and building and construction.

6 HIGH-PERFORMANCE MATERIALS

High-performance resins arbitrarily comprise the high end of engineering plastics and offer premium performance. Typically high-performance resins will be used in tough metal replacement applications or replacement of ceramic materials. Such materials offer greater resistance to heat and chemicals, high RTIs, high strength and stiffness, and inherent flame resistance. These resins have higher cost and the processing can be more challenging.

6.1 Polyphenylene Sulfide (PPS)

The condensation polymerization of 1,4-dichlorobenzene and sodium sulfide yields a semicrystalline polymer, PPS.[47] It is characterized by high heat resistance, rigidity, excellent chemical resistance, dimensional stability, low friction coefficient, good abrasion resistance, and electrical properties. PPS has good mechanical properties, which remain stable during exposure to elevated temperatures. Water absorption for PPS is very low and hydrolytic stability is very high. PPS resins are inherently flame resistant.

PPS has excellent chemical resistance to a wide range of chemicals. Indeed, even at elevated temperatures PPS can withstand exposure to a wide range of chemicals, such as mineral and organic acids and alkali. However, it is attacked by chlorinated hydrocarbons.

Depending on the polymerization process, PPS can be a linear or branched polymer. The branched polymer is somewhat more difficult to process due to the very high melting temperature and relatively poor flow characteristics.

PPS resins normally contain glass fibers or mineral fillers. Properties appear in Table 30.

Table 30 Typical Properties of PPS

Property	Branched	Branched +40% GF	Linear	Linear +40% GF
Density (g/cm^3)	1.35	1.60	1.35	1.65
Tensile modulus (GPa)	—	14.5	4.0	15.7
Tensile strength (MPa)	65	150	66	150
Elongation at break (%)	2	1.2	12	1.7
Flexural modulus (GPa)	3.85	15.0	3.90	15.0
Flexural strength (MPa)	104	153	130	230
Notched Izod (kJ/m)	0.080	0.578	0.139	0.241
Unnotched Izod (kJ/m)	0.107	0.482	0.167	0.589
HDT at 1.81 MPa (°C)	115	>260	115	265
RTI (°C)	220	—	220	—
UL94 VO (mm)	—	0.8	—	0.6
Oxygen index (%)	44	46.5	44	47
Hardness (Rockwell)	R120	R123	M95	M100
CTE (10^{-5}/°C)	4.9	4.0	5.3	4.1
Shrinkage (in./in.)	—	0.004	0	0.004
Water absorption, 24 h (%)	—	0.03	—	0.03
Relative permittivity at 1 MHz	—	3.9	—	4.7
Dissipation factor at 1 MHz	—	0.0014	—	0.020

Uses. Applications for PPS resins include industrial (parts requiring heat and chemical resistance, submersible, vane, and gear-type pump components), electrical/electronic (high-voltage electrical components), automotive (electrical connectors, under-the-hood components), appliance parts (hair dryers, small cooking appliances, range components), medical (hydraulic components, bearing cams, valves), and aircraft/aerospace.

6.2 Polyarylsulfones (PSU, PES, PPSU)

Polyarylsulfones are a class of amorphous, high-end-use-temperature thermoplastics that characteristically exhibit excellent thermo-oxidative stability, good solvent resistance, creep resistance, transparency, and high heat deflection temperatures.[48–50] Polyarylsulfones have excellent hydrolytic stability even in saturated steam; they can be repeatedly steam sterilized. Typically the polyarylsulfone families of resins have low resistance to weathering and are degraded by UV light. There are three major categories of aromatic polyarylsulfone resins: polysulfone (PSU), polyphenylsulfone (PPSU), and polyethersulfone (PES).

PSU is prepared by nucleophilic aromatic displacement of the chloride on bis(*p*-chlorophenyl)sulfone by the anhydrous disodium salt of bisphenol A. This amorphous polymer has a T_g of 190°C.

PPSU is higher performing than PSU. This wholly aromatic resin is prepared from biphenol and bis(*p*-chlorophenyl)sulfone.[50] The lack of aliphatic groups in this all-aromatic polymer and the presence of the biphenyl moiety impart enhanced chemical/solvent resistance, outstanding toughness, greater resistance to combustion, greater thermo-oxidative stability, and a T_g of 220°C. PPSU has excellent resistance to mineral acids, caustic, salt solution, and various automotive fluids. Exposure to esters, ketones, and polar aromatic solvents should be avoided.

PES consists of a diphenyl sulfone unit linked through an ether (oxygen) unit. Again the lack of aliphatic groups results in higher thermo-oxidative stability. PES offers high heat (T_g of 230°C), chemical/solvent resistance, and improved toughness over PSU. PES is chemically resistant to most inorganic chemicals, greases, aliphatic hydrocarbons, and gasoline. However, esters, ketones, methylene chloride, and polar aromatic solvents attack PES.

Typical properties of PSU, PPSU, and PES appear in Table 31.

Uses. Typical applications of polyarylsulfones include medical/laboratory (surgical equipment, laboratory equipment, life support parts, autoclavable tray systems, suction bottles, tissue culture bottles, surgical hollow shapes), food handling (microwave cookware, coffee makers, hot-water and food-handling equipment, range components), electrical/electronic (components, multipin connectors, coil formers, printed circuit boards), chemical processing equipment (pump housings, bearing cages), and miscellaneous (radomes, alkaline battery cases).

6.3 Liquid Crystalline Polyesters (LCPs)

Liquid crystalline polymers have a rigid rodlike aromatic structure. The rodlike molecules arrange themselves in parallel domains in both the melt and solid states. In the molten state the molecules readily slide over one another, giving the resin very high flow under shear. Most commercially important polyester LCPs are based on *p*-hydroxybenzoic acid (HBA). An example would be the copolyester of HBA and hydroxynaphthanoic acid (HNA) with a molar ratio of 73/27.

LCPs are highly crystalline, thermotropic (melt-orienting) thermoplastics. Because of the possibility of melt orientation during molding, LCPs can have anisotropic properties; i.e.,

Table 31 Typical Properties of Polyarylsulfones

Property	PSU	PES	PPSU
Density (g/cm^3)	1.24	1.37	1.29
Tensile modulus (GPa)	2.48	2.41	2.35
Tensile strength (MPa)	69	82	70
Elongation at break (%)	75	50	60
Flexural modulus (GPa)	2.55	2.55	2.42
Flexural strength (MPa)	102	110	91
Notched Izod (kJ/m)	0.080	0.075	0.694
Unnotched Izod, −40°C (kJ/m)	0.064	—	0.425
HDT at 1.81 MPa (°C)	174	203	204
HDT at 0.42 MPa (°C)	180	210	—
Vicat softening temperature (°C)	188	226	—
RTI (°C)	160	180	180
UL94 VO (mm)	—	0.46	0.58
Oxygen index (%)	—	38	—
Hardness (Rockwell)	M69	M88	—
CTE (10^{-5}/°C)	5.1	5.5	5.5
Shrinkage (in./in.)	0.005	0.006	0.006
Water absorption, 24 h (%)	0.3	0.43	0.37
Relative permittivity at 1MHz	3.19	3.45	3.5
Dissipation factor at 1 MHz	—	0.0076	—

properties can differ in the direction of flow and perpendicular to the flow direction. LCPs offer high strength, rigidity, dimensional stability, inherent flame resistance, and high heat resistance. LCPs have high flow and can deliver exceptionally precise and stable dimensions in very thin walled applications.

LCPs are resistant to weathering, burning, γ-radiation, steam autoclaving, and most chemical sterilization methods. They have outstanding strength at elevated temperatures.

LCPs are resistant to virtually all chemicals, including acids, organic hydrocarbons, and boiling water. It is attacked by concentrated, boiling caustic but is unaffected by milder solutions. Typical properties of the HBA-HNA LCPs appear in Table 32.

Uses. Typical applications of LCPs include electrical/electronic (stator insulation, rotors, boards for motors, burn-in sockets, interface connectors, bobbins, switches, chip carriers and sensors), medical (surgical instruments, needleless syringes, dental tools, sterilizable trays and equipment, drug delivery systems and diagnostics), industrial (chemical process and oil field equipment), and packaging (food packaging requiring barrier properties)

6.4 Polyimides (PMDA-ODA)

Polyimides are a class of polymers prepared from the condensation reaction of a carboxylic acid anhydride with a diamine.[51] Polyimides are among the most heat resistant polymers. Poly(pyromellitimide-1,4-diphenyl ether) is prepared from pyromellitic anhydride (PMDA) and 4,4′-oxydianiline (ODA). PMDA-ODA has a T_g of 360°C or higher. This very high T_g does not lend itself to standard melt-processing techniques. PMDA-ODA resins are available as films and direct-formed parts from formulated resin. PMDA-ODA has excellent thermal stability, useful mechanical properties over a very wide temperature range, creep resistance, high toughness, and excellent wear resistance. PMDA-ODA had excellent electrical prop-

Table 32 Typical Properties of LCP Resins (HBA/HNA)

Property	LCP	LCP + 15% GF	LCP + 30% GF
Density (g/cm^3)	1.40	1.50	1.62
Tensile modulus (GPa)	10.6	12.0	15.0
Tensile strength (MPa)	182	190	200
Elongation at break (%)	3.4	3.1	2.1
Flexural modulus (GPa)	9.10	12.0	15.0
Flexural strength (MPa)	158	240	280
Notched Izod (kJ/m)	0.75	0.35	0.17
Unnotched Izod (kJ/m)	2.0	0.48	0.23
HDT at 0.45 MPa (°C)	—	250	250
HDT at 1.81 MPa (°C)	187	230	235
Vicat softening point (°C)	145	162	160
Oxygen index (%)	—	—	45
Hardness (Rockwell)	—	M80	M85
CTE (10^{-5}/°C)			
Parallel to flow	0.4	1.0	0.6
Perpendicular to flow	3.62	1.8	2.3
Shrinkage (in./in.)			
Parallel to flow	0.0	0.001	0.002
Perpendicular to flow	0.007	0.004	0.004
Water absorption, 24 h (%)	0.03	—	—
Relative permittivity at 1 MHz	3.0	3.0	3.7
Dissipation factor at 1 MHz	0.02	0.018	0.018

erties to a broad range of chemicals. However, it will be attacked by 10% and stronger solutions of sodium hydroxide. Typical properties appear in Table 33.

Uses. PMDA-ODA films are used as wire and cable wrap, motor-slot liners, flexible printed circuit boards, transformers, and capacitors. Molded parts are used in application requiring resistance to thermally harsh environments such as automotive transmission parts, thermal

Table 33 Typical Properties of Polyimide

Property	PMDA-ODA	15% Graphite	40% Graphite
Density (g/cm^3)	1.43	1.51	1.65
Tensile strength (MPa)	86.2	65.5	51.7
Elongation at break (%)	7.5	4.5	3.0
Flexural modulus (GPa)	3.10	3.79	4.83
Flexural strength (MPa)	110	110	90
Notched Izod (kJ/m)	0.043	0.043	—
Unnotched Izod (kJ/m)	0.747	0.320	—
HDT at 2 MPa (°C)	360	360	—
Oxygen index (%)	53	49	—
Hardness (Rockwell)	—	M80	M85
CTE (10^{-5}/°C)	5.4	4.9	3.8
Water absorption, 24 h (%)	0.24	0.19	0.14
Relative permittivity at 1 MHz	3.55	13.4	—
Dissipation factor at 1 MHz	0.0034	0.0106	—

and electrical insulators, valve seats, rotary seal rings, thrust washers and discs, bushings, etc.

6.5 Polyetherimides (PEIs)

Polyetherimides are prepared from the polymerization of aromatic diamines and etherdianhydrides.[52,53] PEI is an amorphous thermoplastic which contains repeating aromatic imide and ether units. The rigid aromatic imide units provide PEI with its high-performance properties at elevated temperatures, while the ether linkages provide it with the chain flexibility necessary to have good melt flow and processability by injection molding and other melt-processing techniques.

PEI resins offer high strength, rigidity, high heat resistance, inherent flame resistance, low smoke generation, excellent processability with very tight molding tolerances, heat resistance up to 200°C, RTI of 170°C, excellent dimensional stability (low creep sensitivity and low coefficient of thermal expansion), superior torque strength and torque retention and stable dielectric constant and dissipation factor over a wide range of temperatures and frequencies.

PEIs have excellent hydrolytic stability, UV stability, and radiation resistance. They are well suited for repeated steam, hot air, ethylene oxide gas, and cold chemical sterilizations. In addition, PEIs have proven property retention through over 1000 cycles in industrial washing machines with detergents and compliancy with Food and Drug Administration (FDA), European Union (EU), and national food contact regulations.

In addition PEI resins are resistant to a wide range of chemicals, including alcohols, hydrocarbons, aqueous detergents and bleaches, strong acids, mild bases, and most automotive fuels, fluids, and oils. Chlorinated hydrocarbons and aromatic solvents can attack PEI resins. Typical properties appear in Table 34.

Table 34 Typical Properties of PEIs

Property	PEI	10% GF + PEI	20% GF + PEI	30% GF + PEI
Density (g/cm^3)	1.27	1.34	1.42	1.51
Tensile modulus (GPa)	3.59	4.69	6.89	9.31
Tensile strength (MPa)	110	116	131	169
Elongation at break (%)	60	6	4	3
Flexural modulus (GPa)	3.52	5.17	6.89	8.96
Flexural strength (MPa)	165	200	228	228
Notched Izod (kJ/m)	0.053	0.053	0.064	0.085
Unnotched Izod (kJ/m)	1.335	0.481	0.481	0.427
HDT at 1.81 MPa (°C)	201	209	210	210
HDT at 0.42 MPa (°C)	210	210	210	212
Vicat softening temperature (°C)	219	223	220	228
RTI (°C)	170	170	170	180
UL94 VO (mm)	—	0.4	0.4	0.3
Oxygen index (%)	47	47	50	50
Hardness (Rockwell)	M109	M114	M114	M114
CTE (10^{-5}/°C)	5.5	3.24	2.52	1.98
Shrinkage (in./in.)	0.006	0.005	0.004	0.003
Water absorption, 24 h (%)	0.25	0.21	0.19	0.16
Dissipation factor at 1 kHz	0.0012	0.0014	0.0015	0.0015
Relative permittivity at 1 kHz	3.15	3.5	3.5	3.7

Uses. Automotive (transmission components, throttle bodies, ignition components, sensors, thermostat housings), automotive lighting (headlight reflectors, fog light reflectors, bezels, light bulb sockets), telecommunications (molded interconnect devices, electrical control units, computer components, mobile phone internal antennae, RF duplexers or microfilters, fiber-optic connectors), electrical (lighting, connectors to reflectors), HVAC/fluid handling (water pump impellers, expansion valves, hot-water reservoirs, heat exchange systems), tableware/catering (food trays, soup mugs, steam insert pans or gastronome containers, cloches, microwavable bowls, ovenware, cooking utensils, reusable airline casseroles), medical (reusable medical devices like sterilization trays, stopcocks, dentist devices, and pipettes), and aircraft (air and fuel valves, food tray containers, steering wheels, interior cladding parts, semistructural components).

6.6 Polyamide Imides (PAIs)

PAIs are prepared from trimellitic anhydride and various aromatic diamines.[54] PAIs are tough, high-modulus thermoplastics capable of high continuous-use temperatures. In addition, they have inherent flame resistance and very high strength. PAIs have very broad chemical resistance but are attacked by aqueous caustic and amines. PAIs have high melt temperature and require very high processing temperatures. PAIs contain an amide group and hence will absorb moisture, which can affect dimensional stability. Properties appear in Table 35.

Uses. Automotive (housings, connectors, switches, relays, thrust washers, valve seats, bushings, wear rings, rollers, thermal insulators), composites (printed wiring boards, radomes), and aerospace (replacement of metal parts).

6.7 Aromatic Polyketones (PEK, PEEK)

Aromatic polyketones are a familty of semicrystalline high-performance thermoplastics that consist of a combination of ketone, ether, and aromatic units.[55] The absence of aliphatic

Table 35 Typical Properties of PAIs

Property	PAI	PAI + 30% Graphite Fiber
Density (g/cm^3)	1.38	1.42
Tensile modulus (GPa)	5.2	20.0
Tensile strength (MPa)	117	207
Elongation at break (%)	15	4
Flexural modulus (GPa)	3.59	17.9
Flexural strength (MPa)	189	316.7
Notched Izod (kJ/m)	0.136	0.07
Unnotched Izod (kJ/m)	1.088	—
HDT at 1.81 MPa (°C)	254	275
RTI (°C)	—	220
Hardness (Rockwell)	E78	E94
Oxygen index (%)	41	49
CTE (10^{-5}/°C)	3.6	4.9
Relative permittivity at 1 MHz	4.0	3.8
Dissipation factor at 1 MHz	0.009	—
Water absorption, 24 h (%)	1.0	0.8

groups results in outstanding high-temperature performance, exceptional thermal stability, excellent environmental resistance, high mechanical properties, resistance to chemical environments at elevated temperatures, inherent flame resistance, excellent friction and wear resistance, and impact resistance. In addition, aromatic polyketones have high chemical purity suitable for applications in silicon chip manufacturing.

There are two major aromatic polyketone resins: polyetherketone (PEK) and polyetheretherketone (PEEK). Typical properties appear in Table 36.

Uses. Applications are in the chemical process industry (compressor plates, valve seats, pump impellers, thrust washers, bearing cages), aerospace (aircraft fairings, radomes, fuel valves, ducting), and electrical/electronic (wire coating, semiconductor wafer carriers).

7 FLUORINATED THERMOPLASTICS

Fluoropolymers or fluoroplastics are a family of fluorine-containing thermoplastics that exhibit some unusual properties.[56,57] These properties include inertness to most chemicals, resistance to high temperatures, extremely low coefficient of friction, weather resistance, and excellent dielectric properties. Mechanical properties are normally low but can be enhanced with glass or carbon fiber or molybdenum disulfide fillers. Properties are shown in Table 37.

7.1 Poly(tetrafluoroethylene) (PTFE)

Poly(tetrafluoroethylene) is a highly crystalline polymer which is produced by the polymerization of tetrafluoroethylene. PTFE offers very high heat resistance (up to 250°C), exceptional chemical resistance, and outstanding flame resistance.[56] The broad chemical resistance includes strong acids and strong bases. In addition, PTFE has the lowest coefficient of friction of any polymer.

Table 36 Typical Properties of Aromatic Polyketones

Property	PEEK	PEEK +30% GF	PEEK +30% Carbon Fiber	PEK	PEK +30% GF
Density (g/cm^3)	1.32	1.49	1.44	1.30	1.53
Tensile modulus (GPa)	3.6	9.7	13.0	4.0	10.5
Tensile strength (MPa)	192	157	208	104	160
Elongation at break (%)	50	2.2	1.3	5	4
Flexural modulus (GPa)	3.7	10.3	13.0	3.7	9.0
Flexural strength (MPa)	170	233	318	—	—
Notched Izod (kJ/m)	0.08	0.1	0.09	0.08	0.1
Unnotched Izod (kJ/m)	—	0.73	0.75	—	—
HDT at 1.81 MPa (]C)	160	315	315	165	340
HDT at 0.42 MPa (°C)	—	—	315	—	—
RTI (°C)	—	—	—	260	260
UL94 VO (mm)	—	—	—	1.6	1.6
Oxygen index (%)	—	—	—	40	46
Hardness (Rockwell)	M126	R124	R124	R126	R126
CTE (10^{-5}/°C)	16	11.5	7.9	5.7	1.7
Shrinkage (in./in.)	0.004	0.005	0.001	0.006	0.004
Water absorption, 24 h (%)	0.5	0.11	0.06	0.11	0.08
Relative permittivity at 1 MHz	3.2	3.7	—	3.4	3.9

Table 37 Typical Properties of Fluoropolymers

Property	PTFE	PCTFE	FEP	PVDF	ECTFE
Density (g/cm^3)	2.29	2.187	2.150	1.77	1.680
Tensile modulus (GPa)	4.10	14.0	—	1.194	—
Tensile strength (MPa)	27.6	40	20.7	44	48.3
Elongation at break (%)	~275	150	~300	43	200
Flexural modulus (GPa)	5.2	1.25	—	—	–
Flexural strength (MPa)	—	74	—	—	—
Notched Izod (kJ/m)	0.19	0.27	0.15	—	—
HDT at 0.45 MPa (]C)	132	126	—	104	116
HDT at 1.81 MPa (°C)	60	75	—	71	77
RTI (°C)	260	199	204	—	150–170
Hardness	D42 (Shore)	D77 (Shore)	D55 (Shore)	D75 (Shore)	R93 (Rockwell)
Dielectric constant at 10^2 Hz	2.1	3.0	2.1	9	2.5
Dissipation factor at 10^3 Hz	0.0003	—	—	0.03	—
CTE (10^{-5}/°C)	12.0	4.8	9.3	11	—

PTFE has a high melting point and extremely high melt viscosity and hence cannot be melt processed by normal techniques. Therefore PTFE has to be processed by unconventional techniques (PTFE powder is compacted to the desired shape and sintered).

Uses. Typically PTFE is used in applications requiring long-term performance in extreme-service environments. PTFE applications include electrical (high-temperature, high-performance wire and cable insulation, sockets, pins, connectors), mechanical (bushings, rider rings, seals, bearing pads, valve seats, chemical resistance processing equipment and pipe, nonlubricated bearings, pump parts, gaskets, and packings), nonstick coatings (home cookware, tools, food-processing equipment), and miscellaneous (conveyor parts, packaging, flame-retardant laminates).

7.2 Poly(chlorotrifluoroethylene) (PCTFE)

Poly(chlorotrifluoroethylene) is less crystalline and exhibits higher rigidity and strength than PTFE.[57] Poly(chlorotrifluoroethylene) has excellent chemical resistance and heat resistance up to 200°C. Unlike PTFE, PCTFE can be molded and extruded by conventional processing techniques.

Uses. PCTFE applications include electrical/electronic (insulation, cable jacketing, coil forms), industrial (pipe and pump parts, gaskets, seals, diaphragms, coatings for corrosive process equipment and in cryogenic systems), and pharmaceutical packaging.

7.3 Fluorinated Ethylene–Propylene (FEP)

Copolymerization of tetrafluoroethylene with some hexafluoropropylene produces fluorinated ethylene–propylene polymer, FEP, which has less crystallinity, lower melting point, and improved impact strength than PTFE. FEP can be molded by normal thermoplastic techniques.[57]

Uses. FEP applications include wire insulation and jacketing, high-frequency connectors, coils, gaskets, and tube sockets.

7.4 Polyvinylidene Fluoride (PVDF)

Polyvinylidene fluoride is made by emulsion and suspension polymerization and has better ability to be processed but less thermal and chemical resistance than FEP, CTFE, and PTFE.[57] PVDF has excellent mechanical properties and resistance to severe environmental stresses and good chemical resistance.

Uses. Polyvinylidene fluoride applications include electronic/electrical (wire and cable insulation, films), fluid handling (solid and lined pipes, fittings, valves, seals and gaskets, diaphragms, pumps, microporous membranes), and coatings (metal finishes, exterior wall panels, roof shingles)

7.5 Poly(ethylene chlorotrifluoroethylene) (ECTFE)

The copolymer of ethylene and chlorotrifluoroethylene is ECTFE and has high strength and chemical and impact resistance. ECTFE can be processed by conventional techniques.

Uses. Poly(ethylene chlorotrifluoroethylene) applications include wire and cable coatings, chemical-resistant coatings and linings, molded laboratory ware, and medical packing.

7.6 Poly(vinyl fluoride) (PVF)

Poly(vinyl fluoride) is the least chemical resistant fluoropolymer.[57] PVF offers weathering and UV resistance, antisoiling, durability, and chemical resistance.

Uses. Poly(vinyl fluoride) uses include protective coatings (aircraft interior, architectural fabrics, presurfaced exterior building panels, wall coverings, glazing, lighting).

8 THERMOSETS

Thermosetting resins are used in molded and laminated plastics.[58] They are first polymerized into a low-molecular-weight linear or slightly branched polymer or oligomers, which are still soluble, fusible, and highly reactive during final processing. Thermoset resins are generally highly filled with mineral fillers and glass fibers. Thermosets are generally catalyzed and/or heated to finish the polymerization reaction, crosslinking them to almost infinite molecular weight. This step is often referred to as curing. Such cured polymers cannot be reprocessed or reshaped.

The high filler loading and the high crosslink density of thermoset resins result in high densities and low ductility but high rigidity and good chemical resistance.

8.1 Phenolic Resins

Phenolic resins combine the high reactivity of phenol and formaldehyde to form prepolymers and oligomers called resoles and novolacs. These materials can be used as adhesives and coatings or combined with fibrous fillers to give phenolic resins, which when heated undergo a rapid, complete crosslinking to give a highly cured structure. The curing typically requires heat and pressure. Fillers include cellulosic materials and glass and mineral fibers/fillers. The high-crosslinked aromatic structure has high hardness, dimensional stability, rigidity, strength, heat resistance, chemical resistance, and good electrical properties.

Phenolic resins have good chemical resistance to hydrocarbons, phenols, and ethers. However, they can undergo chemical reaction with acids and bases.

Uses. Phenolic applications include appliance parts (handles, knobs, bases, end panels), automotive uses (parts in electric motors, brake linings, fuse blocks, coil towers, solenoid covers and housings, ignition parts), electrical/electronic (high-performance connectors and coil bobbins, circuit breakers, terminal switches and blocks, light sockets, receptacles), and miscellaneous (adhesives in laminated materials such as plywood, coatings).

8.2 Epoxy Resins

Epoxy resins are materials whose molecular structure contains an epoxide or oxirane ring. The most common epoxy resins are prepared from the reaction of bisphenol A and epichlorohydrin to yield low-molecular-weight resins that are liquid either at room temperature or on warming. Other epoxy resins are available—these include novalac and cycloaliphatic epoxy resins. Each polymer chain usually contains two or more epoxide groups. The high reactivity of the epoxide groups with aromatic and aliphatic amines, anhydrides, carboxylic acids, and other curing agents provides facile conversion into highly crosslinked materials. The large number of variations possible in epoxy structure and curing agent coupled with their ability to be formulated with a wide variety of fillers and additives results in broad application potential. Cured epoxy resins exhibit hardness, strength, heat resistance, good electrical properties, and broad chemical resistance.

Most epoxies are resistant to numerous chemicals, including hydrocarbons, esters, and bases. However, epoxies can exhibit poor resistance to phenols, ketones, ethers, and concentrated acids.

Uses. Epoxy resins are used in composites (glass-reinforced, high-strength composites in aerospace, pipes, tasks, pressure vessels), coatings (marine coatings, protective coatings on appliances, chemical scrubbers, pipes), electrical/electronic (encapsulation or casting of various electrical and electronic components, printed wiring boards, switches, coils, insulators, bushings), and miscellaneous (adhesives, solder mask, industrial equipment, sealants).

8.3 Unsaturated Polyesters

Thermoset polyesters are prepared by the condensation polymerization of various diols and maleic anhydride to give a very viscous unsaturated polyester oligomer which is dissolved in styrene monomer. The addition of styrene lowers the viscosity to a level suitable for impregnation and lamination of glass fibers. The low-molecular-weight polyester has numerous maleate/fumarate ester units that provide facile reactivity with styrene monomer. Unsaturated polyesters can be prepared from a variety of different monomers.

Properly formulated glass-reinforced unsaturated polyesters are commonly referred to as sheet molding compound (SMC) or reinforced plastics. The combination of glass fibers in the cured resins offers outstanding strength, high rigidity, impact resistance, high strength-to-weight ratio, and chemical resistance. SMC typically is formulated with 50% calcium carbonate filler, 25% long glass fiber, and 25% unsaturated polyester. The highly filled nature of SMC results in a high density and a brittle, easily pitted surface.

Bulk molding compounds (BMCs) are formulated similar to SMCs except $\frac{1}{4}$-in. chopped glass is typically used. The shorter glass length gives easier processing but lower strength and impact.

Thermoset polyesters have good chemical resistance, which includes exposure to alcohols, ethers, and organic acids. They exhibit poor resistance to hydrocarbons, ketones, esters, phenols, and oxidizing acids.

Uses. The prime use of unsaturated polyesters is in combination with glass fibers in high-strength composites and in SMC and BMC materials. The applications include transportation markets (large body parts for automobiles, trucks, trailers, buses, and aircraft, automotive ignition components), marine markets (small- to medium-sized boat hulls and associated marine equipment), building and construction (building panels, housing and bathroom components—bathtub and shower stalls), and electrical/electronic (components, appliance housing, switch boxes, breaker components, encapsulation).

8.4 Vinyl Esters

Vinyl esters are part of the unsaturated polyester family. They are prepared by the reaction of an epoxy resin with methacrylic acid. Thus the epoxide group is converted into a methacrylate ester. Vinyl esters offer an enhancement in properties over unsaturated polyesters with greater toughness and better resistance to corrosion to a wide range of chemicals. This chemical resistance includes halogenated solvents, acids, and bases.

Uses. Applications for vinyl esters are similar to those for unsaturated polyesters but where added toughness and chemical resistance are required, i.e., electrical equipment, flooring, fans, adsorption towers, process vessels, and piping.

8.5 Alkyd Resins

Alkyd resins are unsaturated polyesters based on branched prepolymers from polyhydridic alcohols (e.g., glycerol, pentaerythritol, ethylene glycol), polybasic acids (e.g., phthalic anhydride, maleic anhydride, fumaric acid), and fatty acids and oils (e.g., oleic, stearic, palmitic, linoleic, linolenic, lauric, and licanic acids). They are well suited for coating with their rapid drying, good adhesion, flexibility, mar-resistance, and durability. Formulated alkyd molding resins have excellent heat resistance, are dimensionally stable at high temperatures and have excellent dielectric strength, high resistance to electrical leakage, and excellent arc resistance.

Alkyl resins can be hydrolyzed under alkaline conditions.

Uses. Alkyd resin applications include coatings (drying oils in paints, enamels, and varnish) and molding compounds when formulated with reinforcing fillers for electrical applications (circuit breaker insulation, encapsulation of capacitors and resistors, coil forms).

8.6 Diallyl Phthalate (DAP)

Diallyl phthalate is the most widely used compound in the allylic ester family. The neat resin is a medium-viscosity liquid. These low-molecular-weight prepolymers can be reinforced and compression molded into highly crosslinked, completely cured products.

The most outstanding properties of DAP are excellent dimensional stability and high insulation resistance. In addition, DAP has high dielectric strength and excellent arc resistance. DAP has excellent resistance to aliphatic hydrocarbons, oils, and alcohols but is not recommended for use with phenols and oxidizing acids.

Uses. DAP applications include electrical/electronic (parts, connectors, bases and housings, switches, transformer cases, insulators, potentiometers) and miscellaneous (tubing, ducting, radomes, junction bases, aircraft parts). DAP is also used as a coating and an impregnating resin.

8.7 Amino Resins

The two main members of the amino resin family of thermosets are the melamine–formaldehyde and urea–formaldehyde resins. They are prepared from the condensation reaction of melamine and urea with formaldehyde. In general, these materials exhibit extreme hardness, inherent flame resistance, heat resistance, scratch resistance, arc resistance, chemical resistance, and light fastness.

Uses. Melamine–formaldehyde resins find use in colorful, rugged dinnerware, decorative laminates (countertops, tabletops, furniture surfacing), electrical applications (switchboard panels, circuit breaker parts, arc barriers, armature and slot wedges), and adhesives and coatings.

Urea–formaldehyde resins are used in particleboard binders, decorative housings, closures, electrical parts, coatings, and paper and textile treatment.

9 GENERAL-PURPOSE ELASTOMERS

Elastomers are polymers that can be stretched substantially beyond their original length and can retract rapidly and forcibly to essentially their original dimensions (on release of the force).[59,60]

The optimum properties and/or economics of many rubbers are obtained through formulating with reinforcing agents, fillers, extending oils, vulcanizing agents, antioxidants, pigments, etc. End-use markets for formulated rubbers include automotive tire products (including tubes, retread applications, valve stems and inner liners), adhesives, cements, caulks, sealants, latex foam products, hose (automotive, industrial, and consumer applications), belting (V-conveyor and trimming), footwear (heels, soles, slab stock, boots, and canvas), and molded, extruded, and calendered products (athletic goods, flooring, gaskets, household products, O-rings, blown sponge, thread, and rubber sundries). A summary of general-purpose elastomers and properties is provided in Table 38.

10 SPECIALTY ELASTOMERS

Specialty rubbers offer higher performance over general-purpose rubbers and find use in more demanding applications.[61] They are more costly and hence are produced in smaller volume. Properties and uses are summarized in Table 39.

ACKNOWLEDGEMENT

The author wishes to thank John Wiley & Sons for permission to adapt the content on pages 115–129 from *Mechanical Engineer's Handbook,* 2nd ed. 1988, edited by M. Kutz.

Table 38 Properties of General-Purpose Elastomers

Rubber	ASTM Nomenclature	Outstanding Characteristic	Property Deficiency	Temperature Use Range (°C)
Butadiene rubber	BR	Very flexible; resistance to wear	Sensitive to oxidation; poor resistance to fuels and oil	−100–90
Natural rubber	NR	Similar to BR but less resilient	Similar to BR	−50–80
Isoprene rubber	IR	Similar to BR but less resilient	Similar to BR	−50–80
Isobutylene–isoprene rubber (butyl rubber)	IIR	High flexibility; low–permeability air		−45–150
Chloroprene	CR	Flame resistant; fair fuel and oil resistance; increased resistance toward oxygen, ozone, heat, light	Poor low-temperature flexibility	−40–115
Nitrile–butadiene	NBR	Good resistance to fuels, oils, and solvents; improved abrasion resistance	Lower resilience; higher hysteresis; poor electrical properties; poorer low-temperature flexibility	−45–80
Styrene–butadiene rubber	SBR	Relatively low cost	Less resilience; higher hysteresis; limited low-temperature flexibility	−45–80
Ethylene–propylene copolymer	EPDM	Resistance to ozone and weathering	Poor hydrocarbon and oil resistance	−50 to <175
Polysulfide	T	Chemical resistance; resistance to ozone and weathering	Creep; low resilience	−45–120

Table 39 Properties of Specialty Elastomers

Elastomer	ASTM Nomenclature	Temperature Use Range (°C)	Outstanding Characteristic	Typical Applications
Silicones (polydimethylsiloxane)	MQ	−100–300	Wide temperature range; resistance to aging, ozone, sunlight; very high gas permeability	Seals, molded and extruded goods; adhesives, sealants; biomedical; personal care products
Fluoroelastomers	CFM	−40–200	Resistance to heat, oils, chemical	Seals such as O-rings, corrosion-resistant coatings
Acrylic	AR	−40–200	Oil, oxygen, ozone, and sunlight resistance	Seals, hose
Epichlorohydrin	ECO	−18–150	Resistance to oil, fuels; some flame resistance; low gas permeability	Hose, tubing, coated fabrics, vibration isolators
Chlorosulfonated	CSM	−40–150	Resistance to oil, ozone weathering, oxidizing chemicals	Automotive hose, wire and cable, linings for reservoirs
Chlorinated polyethylene	CM	−40–150	Resistance to oils, ozone, chemicals	Impact modifier, automotive applications
Ethylene–acrylic	—	−40–175	Resistance to ozone, weathering	Seals, insulation, vibration damping
Propylene oxide	—	−6–150	Low temperature properties	Motor mounts

REFERENCES

1. E. N. Peters, "Plastics: Thermoplastics, Thermosets, and Elastomers," in *Handbook of Materials Selection,* M. Kutz (ed.), Wiley-Interscience, New York, Chapter 11, pp. 335–355, 2002.
2. D. W. Fox, E. N. Peters, "Engineering Thermoplastics: Chemistry and Technology" in *Applied Polymer Science,* 2nd ed.., R. W. Tess, and G. W. Poehlein (eds.), American Chemical Society, Washington, DC, Chapter 21, pp. 495–514, 1985.
3. E. N. Peters, R. K. Arisman, "Engineering Thermoplastics" in *Applied Polymer Science—21st Century,* C. D. Craver, C. E. Carraher (eds.), Elsevier, New York, pp. 177–196, 2000.
4. E. N. Peters, "Inorganic High Polymers" in *Kirk-Othmer Encyclopedia of Chemical Technology,* 3rd ed., Vol. 13, M. Grayson (ed.), Wiley, New York, pp. 398–413, 1981.
5. E. N. Peters, "Introduction to Polymer Characterization" in *Comprehensive Desk Reference of Polymer Characterization and Analysis,* R. F. Brady, Jr. (ed.), Oxford University Press: New York, Chapter 1, pp. 3–29, 2003.
6. S. H. Goodman (ed.), *Handbook of Thermoset Plastics,* 2nd ed., Plastics Design Library, Brookfield, CT, 1999.
7. O. Olabisi (ed.), *Handbook of Thermoplastics,* Marcel Dekker, New York, 1998.
8. E. N. Peters, "Behavior in Solvents" in *Comprehensive Desk Reference of Polymer Characterization and Analysis,* R. F. Brady, Jr. (ed.), Oxford University Press, New York, Chapter 20, pp. 535–554, 2003.
9. H. Zweifel, *Plastics Additives Handbook,* 5th ed. Hanser/Gardner, Cincinnati, OH, 2001.
10. J. Stepek, and H. Daoust, *Additives for Plastics,* Springer-Verlag, New York, 1983.
11. G. Pritchard, *Plastics Additives: An A-Z Reference,* Kluwer Academic Publishers, London, 1998.
12. J. T. Lutz, (ed.), *Thermoplastic Polymer Additives: Theory and Practice,* Marcel Dekker, New York, 1989.
13. G. Wypych, *Handbook of Fillers: The Definitive User's Guide and Databook on Properties, Effects, and Users,* 2nd ed., Plastics Design Library, Brookfield, CT, 2000.
14. G. M. Benedikt and B. L. Goodall (ed.), *Metallocene-Catalyzed Polymers—Materials, Properties, Processing and Markets,* Plastics Design Library, Brookfield, CT, 1998.
15. V. Shah (ed.), *Handbook of Polyolefins,* 2nd ed., Wiley, New York, 1998.
16. A. J. Peacock, *Handbook of Polyethylene: Structures, Properties, and Applications,* Marcel Dekker, New York, 2000.
17. H. G. Karian (ed.), *Handbook of Polypropylene and Polypropylene Composites,* Marcel Dekker, New York, 1999.
18. C. Maier and T. Calafut, *Polypropylene—The Definitive User's Guide,* Plastics Design Library, Brookfield, CT, 1998.
19. D. A. Smith (ed.), *Addition Polymers: Formation and Characterization,* Plenum Press, New York, 1968.
20. J. Scheirs and D. Priddy (eds.), *Modern Styrenic Polymers,* Wiley, New York, 2003.
21. D. Bank, R. Brentin, and M. Hus, *"SPS Crystalline Polymer: A New Material for Automotive Interconnect Systems,"* SAE Conference, SAE, Troy, MI, **71,** 305–309, 1997.
22. H. Vernaleken, "Polycarbonates" in *Interfacial Synthesis,* Vol. 2, F. Millich and C. Carraher (eds.), Marcel Dekker, New York, Chapter 13, pp. 65–124, 1997.
23. E. J. Wickson (ed.), *Handbook of PVC Formulating,* Wiley, New York, 1993.
24. D. Randall and S. Lee, *The Polyurethanes Book,* Wiley, New York, 2003.
25. K. Uhlig, *Discovering Polyurethanes,* Hanser Gardner, New York, 1999.
26. W. Pigman and D. Horton (eds.), *The Carbohydrates,* 2nd ed., Academic, New York, 1970.
27. G. O. Aspinall (ed.), *The Polysaccharides,* Vol. 1, Academic, New York, 1982.
28. R. Gilbert (ed.), *Cellulosic Polymers,* Hanser Gardner, Cincinnati, OH, 1993.
29. L. Bottenbruch (ed.), *Engineering Thermoplastics: Polycarbonates–Polyacetals–Polyesters–Cellulose Esters,* Hanser Gardner, New York, 1996.
30. S. Fakirov (ed.), *Handbook of Thermoplastic Polyesters,* Wiley, New York, 2002.
31. J. Scheirs and T. E. Long (eds.), *Modern Polyesters,* Wiley, New York, 2003.

32. B. J. Chisholm and J. G. Zimmer, "Isothermal Crystallization Kinetics of Commercially Important Polyalkylene Terephthalates" in *J. Appl. Polym. Sci.,* **76,** 1296–1307 (2000).

33. M. I. Kohan (ed.), *Nylon Plastics Handbook SPE Monograph,* Hanser Gardner, Cincinnati, OH, 1995.

34. S. M. Ahorani, *n-Nylons: Their Synthesis, Structure and Properties,* Wiley, New York, 1997.

35. R. R. Gallucci, "Polyphenylene Ether-Polyamide Blends" in *Conference Proceedings for the Society of Plastics Engineers, Inc., 44th Annual Technical Conference,* Society of Plastics Engineers, Washington, DC, pp. 48–50, 1986.

36. E. N. Peters, "High Performance Polyamides. I. Long Glass PPE/PA66 Alloys" in *Conference Proceedings for the Society of Plastics Engineers, Inc., 55th Annual Technical Conference,* Society of Plastics Engineers, Washington, DC, pp. 2322–2326, 1997.

37. B. Majumdar and D. R. Paul, "Reactive Compatibilization" in *Polymer Blends: Formulation and Performance,* Vol. 2, D. R. Paul and C. P. Bucknall (eds.), Wiley, New York, pp. 539–580, 1999.

38. D. G. LeGrand and J. T. Bendler (eds.), *Polycarbonates: Science and Technology,* Marcel Dekker, New York, 1999.

39. W. F. Christopher and D. W. Fox, *Polycarbonates,* Reinhold, New York, 1962.

40. H. Schnell, *Chemistry and Physics of Polycarbonates,* Wiley-Interscience, New York, 1964.

41. E. N. Peters, "Plastics and Elastomers" in *Mechanical Engineer's Handbook,* 2nd ed., M. Kutz (ed.), Wiley-Interscience, New York, Chapter 8, pp. 115–129, 1998.

42. L. M. Robeson, "Polyarylate" in *Handbook of Plastic Materials and Technology,* I. I. Rubin (ed.), Wiley-Interscience, New York, Chapter 21, pp. 237–246, 1990.

43. A. S. Hay, "Polymerization by Oxidative Coupling. II. Oxidation of 2,6-Disubstituted Phenols" in *J. Polym. Sci.,* **58,** 581–591 (1962).

44. D. M. White, "Polyethers, Aromatic" in *Kirk-Othmer Encyclopedia of Chemical Technology,* 5th ed., Wiley, New York, in press.

45. E. P. Cizek, U. S. Patent 3,338,435, 1968.

46. E. N. Peters, "Polyphenylene Ether Blends and Alloys" in *Engineering Plastics Handbook,* J. M. Margolis (ed.), McGraw-Hill, New York, in press.

47. J. W. Gardner and P. J. Boeke, "Poly(phenylene sulfide) (PPS)" in *Handbook of Plastic Materials and Technology,* I. I. Rubin (ed.), Wiley-Interscience, New York, Chapter 37, pp. 417–432, 1990.

48. R. N. Johnson, A. G. Farnham, R. A. Clendinning, W. F. Hale and C. N. Merriam, "Poly(aryl ethers) by nucleophilic aromatic substitution. I. Synthesis and properties" in *J. Polym. Sci., Part A-1,* **5,** 2375–2398 (1967).

49. J. E. Harris, "Polysulfone (PSO)" in *Handbook of Plastic Materials and Technology,* I. I. Rubin (ed.), Wiley-Interscience, New York, Chapter 40, pp. 487–500, 1990.

50. L. M. Robeson, "Poly(phenyl) Sulfone" in *Handbook of Plastic Materials and Technology* I. I. Rubin (ed.), Wiley-Interscience, New York, Chapter 34, pp. 385–394, 1990.

51. C. E. Sroog, "Polyimides" in *J. Polym. Sci., Macromol. Rev.,* **11,** 161–208 (1976).

52. T. Takekoshi, "Polyimides" in *Adv. Polym. Sci.,* **94,** 1–25 (1990).

53. I. W. Serfaty, "Polyetherimide (PEI)" in *Handbook of Plastic Materials and Technology,* I. I. Rubin (ed.), Wiley-Interscience, New York, Chapter 24, pp. 263–276, 1990.

54. J. L. Thorne, "Polyamid-Imid (PAI)" in *Handbook of Plastic Materials and Technology,* I. I. Rubin (ed.), Wiley-Interscience, New York, Chapter 20, pp. 225–236, 1990.

55. T. W. Haas, "Polyetheretherketone (PEEK)" in *Handbook of Plastic Materials and Technology,* I. I. Rubin (ed.), Wiley-Interscience, New York, Chapter 25, pp. 277–294, 1990.

56. S. Ebnesajjad, *Fluoroplastics,* Vol. I: *Non-Melt Processible Fluoroplastics,* Plastics Design Library, Brookfield, CT, 2000.

57. S. Ebnesajjad, *Fluoroplastics,* Vol. 2: *Mel-Processible Fluoroplastics,* Plastics Design Library, Brookfield, CT, 2002.

58. S. H. Goodman (ed.), *Handbook of Thermoset Plastics,* 2nd ed., Plastics Design Library, Brookfield, CT, 1999.

59. P. A. Ciullo and N. Hewitt, *The Rubber Formulary,* Plastics Design Library, Brookfield, CT, 1999.

60. M. Morton (ed.), *Rubber Technology,* Van Norstrand Reinhold, New York, 1973.

61. J. M. Zeigler and F. W. G. (eds.), in *Silicone-Based Polymer Science: A Comprehensive Resource,* American Chemical Society, Washington, DC, 1990.

CHAPTER 10
COMPOSITE MATERIALS

Carl Zweben
Devon, Pennsylvania

1 INTRODUCTION

The development of composite materials and related design and manufacturing technologies is one of the most important advances in the history of materials. Composites are multifunctional materials having unprecedented mechanical and physical properties that can be tailored to meet the requirements of a particular application. Many composites also exhibit great resistance to high-temperature corrosion and oxidation and wear. These unique characteristics provide the mechanical engineer with design opportunities not possible with conventional monolithic (unreinforced) materials. Composites technology also makes possible the use of an entire class of solid materials, ceramics, in applications for which monolithic versions are unsuited because of their great strength scatter and poor resistance to mechanical and thermal shock. Further, many manufacturing processes for composites are well adapted to the fabrication of large, complex structures, which allows consolidation of parts, reducing manufacturing costs.

Composites are important materials that are now used widely, not only in the aerospace industry, but also in a large and increasing number of commercial mechanical engineering applications, such as internal combustion engines; machine components; thermal control and electronic packaging; automobile, train, and aircraft structures and mechanical components, such as brakes, drive shafts, flywheels, tanks, and pressure vessels; dimensionally stable components; process industries equipment requiring resistance to high-temperature corrosion, oxidation, and wear; offshore and onshore oil exploration and production; marine structures; sports and leisure equipment; and biomedical devices.

It should be noted that biological structural materials occurring in nature are typically some type of composite. Common examples are wood, bamboo, bone, teeth, and shell. Further, use of artificial composite materials is not new. Straw-reinforced mud bricks were employed in biblical times. Using modern terminology, discussed later, this material would be classified as an organic fiber-reinforced ceramic matrix composite.

In this chapter, we consider the properties of reinforcements and matrix materials (Section 2) and the properties of composites (Section 3).

1.1 Classes and Characteristics of Composite Materials

There is no universally accepted definition of a composite material. For the purpose of this work, we consider a composite to be a material consisting of two or more distinct phases, bonded together.[1]

Solid materials can be divided into four categories: polymers, metals, ceramics, and carbon, which we consider as a separate class because of its unique characteristics. We find both reinforcements and matrix materials in all four categories. This gives us the ability to create a limitless number of new material systems with unique properties that cannot be obtained with any single monolithic material. Table 1 shows the types of material combinations now in use.

Composites are usually classified by the type of material used for the matrix. The four primary categories of composites are polymer matrix composites (PMCs), metal matrix composites (MMCs), ceramic matrix composites (CMCs), and carbon/carbon composites (CCCs). At this time, PMCs are the most widely used class of composites. However, there are important applications of the other types, which are indicative of their great potential in mechanical engineering applications.

Figure 1 shows the main types of reinforcements used in composite materials: aligned continuous fibers, discontinuous fibers, whiskers (elongated single crystals), particles, and numerous forms of fibrous architectures produced by textile technology, such as fabrics and braids. Increasingly, designers are using hybrid composites that combine different types of reinforcements to achieve more efficiency and to reduce cost.

A common way to represent fiber-reinforced composites is to show the fiber and matrix separated by a slash. For example, carbon fiber-reinforced epoxy is typically written "carbon/epoxy," or, "C/Ep." We represent particle reinforcements by enclosing them in parentheses followed by "p"; thus, silicon carbide (SiC) particle-reinforced aluminum appears as "(SiC)p/Al."

Composites are strongly heterogeneous materials; that is, the properties of a composite vary considerably from point to point in the material, depending on which material phase the point is located in. Monolithic ceramics and metallic alloys are usually considered to be homogeneous materials, to a first approximation.

Many artificial composites, especially those reinforced with fibers, are anisotropic, which means their properties vary with direction (the properties of isotropic materials are the same in every direction). This is a characteristic they share with a widely used natural fibrous composite, wood. As for wood, when structures made from artificial fibrous composites are required to carry load in more than one direction, they are used in laminated form.

Many fiber-reinforced composites, especially PMCs, MMCs, and CCCs, do not display plastic behavior as metals do, which makes them more sensitive to stress concentrations. However, the absence of plastic deformation does not mean that composites are brittle ma-

Table 1 Types of Composite Materials

	Matrix			
Reinforcement	Polymer	Metal	Ceramic	Carbon
Polymer	X	X	X	X
Metal	X	X	X	X
Ceramic	X	X	X	X
Carbon	X	X	X	X

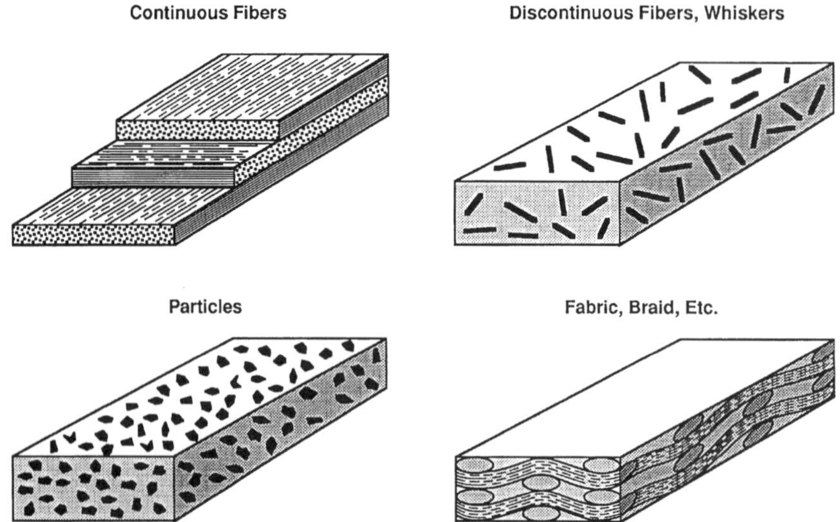

Figure 1 Reinforcement forms.

terials like monolithic ceramics. The heterogeneous nature of composites results in complex failure mechanisms that impart toughness. Fiber-reinforced materials have been found to produce durable, reliable structural components in countless applications. The unique characteristics of composite materials, especially anisotropy, require the use of special design methods.

1.2 Comparative Properties of Composite Materials

There are a large and increasing number of materials that fall in each of the four types of composites, making generalization difficult. However, as a class of materials, composites tend to have the following characteristics: high strength; high modulus; low density; excellent resistance to fatigue, creep, creep rupture, corrosion, and wear; and low coefficient of thermal expansion (CTE). As for monolithic materials, each of the four classes of composites has its own particular attributes. For example, CMCs tend to have particularly good resistance to corrosion, oxidation, and wear, along with high-temperature capability.

For applications in which both mechanical properties and low weight are important, useful figures of merit are specific strength (strength divided by specific gravity or density) and specific stiffness (stiffness divided by specific gravity or density). Figure 2 presents specific stiffness and specific tensile strength of conventional structural metals (steel, titanium, aluminum, magnesium, and beryllium), two engineering ceramics (silicon nitride and alumina), and selected composite materials. The composites are PMCs reinforced with selected continuous fibers—carbon, aramid, E-glass, and boron—and an MMC, aluminum containing silicon carbide particles. Also shown is beryllium–aluminum, which can be considered a type of metal matrix composite, rather than an alloy, because the mutual solubility of the constituents at room temperature is low.

The carbon fibers represented in Fig. 2 are made from several types of precursor materials: polyacrylonitrile (PAN), petroleum pitch, and coal tar pitch. Characteristics of the

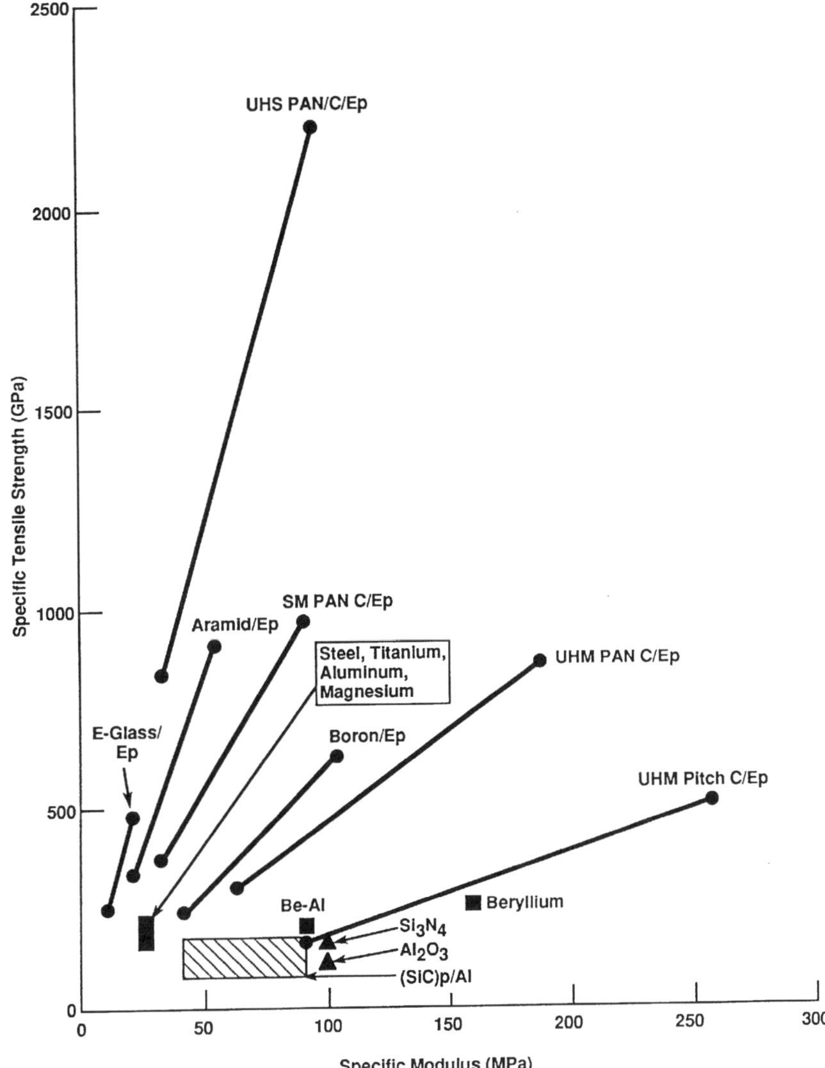

Figure 2 Specific tensile strength (tensile strength divided by density) as a function of specific modulus (modulus divided by density) of composite materials and monolithic metals and ceramics.

two types of pitch-based fibers tend to be similar but very different from those made from PAN. Several types of carbon fibers are represented: standard-modulus (SM) PAN, ultrahigh-strength (UHS) PAN, ultrahigh-modulus (UHM) PAN, and UHM pitch. These fibers are discussed in Section 2. It should be noted that there are dozens of different kinds of commercial carbon fibers, and new ones are continually being developed.

Because the properties of fiber-reinforced composites depend strongly on fiber orientation, fiber-reinforced polymers are represented by lines. The upper end corresponds to the axial properties of a unidirectional laminate, in which all the fibers are aligned in one direction. The lower end represents a quasi-isotropic laminate having equal stiffness and approximately equal strength characteristics in all directions in the plane of the fibers.

As Figure 2 shows, composites offer order-of-magnitude improvements over metals in both specific strength and stiffness. It has been observed that order-of-magnitude improvements in key properties typically produce revolutionary effects in a technology. Consequently, it is not surprising that composites are having such a dramatic influence in engineering applications.

In addition to their exceptional static strength properties, fiber-reinforced polymers also have excellent resistance to fatigue loading. Figure 3 shows how the number of cycles to failure (N) varies with maximum stress (S) for aluminum and selected unidirectional PMCs subjected to tension–tension fatigue. The ratio of minimum stress to maximum stress (R) is 0.1. The composites consist of epoxy matrices reinforced with key fibers: aramid, boron, SM carbon, high-strength (HS) glass, and E-glass. Because of their excellent fatigue resistance, composites have largely replaced metals in fatigue-critical aerospace applications, such as helicopter rotor blades. Composites also are being used in commercial fatigue-critical applications, such as automobile springs.

The outstanding mechanical properties of composite materials have been a key reason for their extensive use in structures. However, composites also have important physical properties, especially low, tailorable CTE and high thermal conductivity, that are key reasons for their selection in an increasing number of applications.

Many composites, such as PMCs reinforced with carbon and aramid fibers, and silicon carbide particle-reinforced aluminum, have low CTEs, which is advantageous in applications requiring dimensional stability. By appropriate selection of reinforcements and matrix materials, it is possible to produce composites with near-zero CTEs.

Coefficient of thermal expansion tailorability provides a way to minimize thermal stresses and distortions that often arise when dissimilar materials are joined. For example, Fig. 4 shows how the CTE of silicon carbide particle-reinforced aluminum varies with particle content. By varying the amount of reinforcement, it is possible to match the CTEs of a variety of key engineering materials, such as steel, titanium, and alumina (aluminum oxide).

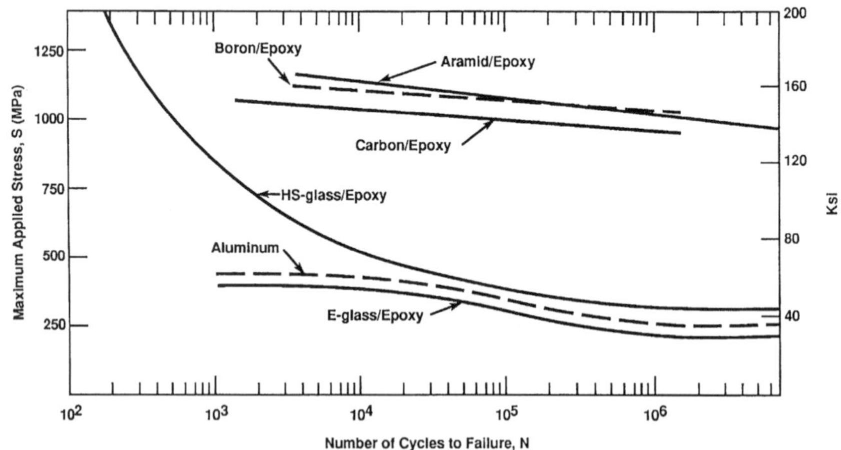

Figure 3 Number of cycles to failure as a function of maximum stress for aluminum and unidirectional polymer matrix composites subjected to tension–tension fatigue with a stress ratio $R = 0.1$. (From Ref. 2. Copyright ASTM. Reprinted with permission.)

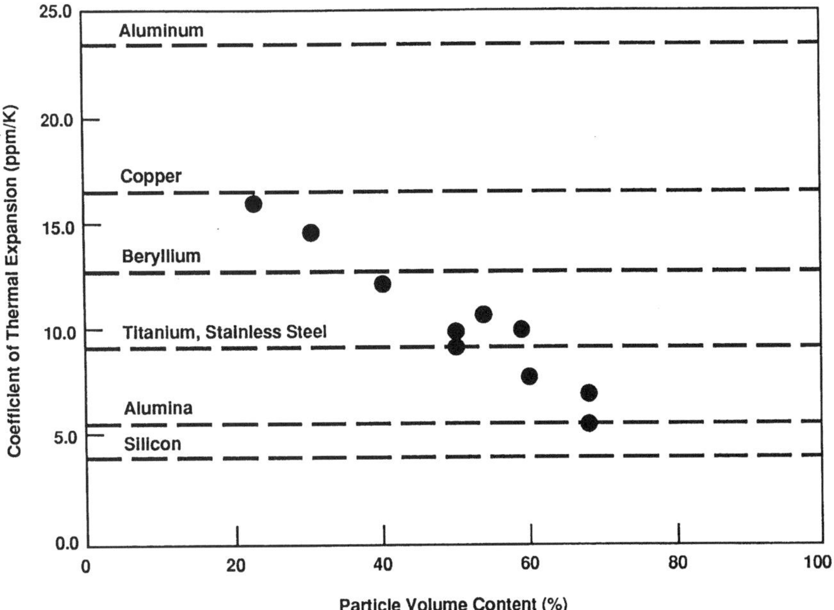

Figure 4 Variation of coefficient of thermal expansion with particle volume fraction for silicon carbide particle-reinforced aluminum. (From Ref. 3, p. 17. Copyright 1992 by TMS.)

The ability to tailor CTE is particularly important in applications such as electronic packaging, where thermal stresses can cause failure of ceramic substrates, semiconductors, and solder joints.

Another unique and increasingly important property of some composites is their exceptionally high thermal conductivity. This is leading to increasing use of composites in applications for which heat dissipation is a key design consideration. In addition, the low densities of composites make them particularly advantageous in thermal control applications for which weight is important, such as laptop computers, avionics, and spacecraft components, such as radiators.

There are a large and increasing number of thermally conductive composites, which are discussed in Section 3. One of the most important types of reinforcements for these materials is pitch fibers. Figure 5 shows how thermal conductivity varies with electrical resistivity for conventional metals and carbon fibers. It can be seen that PAN-based fibers have relatively low thermal conductivities. However, pitch-based fibers with thermal conductivities more than twice that of copper are commercially available. These reinforcements also have very high stiffnesses and low densities. At the upper end of the carbon fiber curve are fibers made by chemical vapor deposition (CVD). Fibers made from another form of carbon, diamond, also have the potential for thermal conductivities in the range of 2000 W/m·K (1160 Btu/h·ft·°F).

1.3 Manufacturing Considerations

Composites also offer a number of significant manufacturing advantages over monolithic metals and ceramics. For example, fiber-reinforced polymers and ceramics can be fabricated in large, complex shapes that would be difficult or impossible to make with other materials.

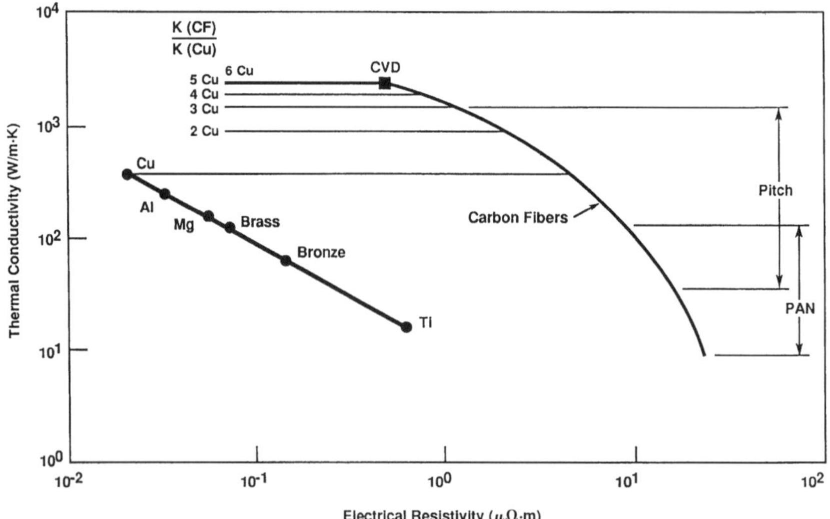

Figure 5 Thermal conductivity as a function of electrical resistivity of metals and carbon fibers (adapted from one of Amoco Performance Products).

The ability to fabricate complex shapes allows consolidation of parts, which reduces machining and assembly costs. Some processes allow fabrication of parts to their final shape (net shape) or close to their final shape (near-net shape), which also produces manufacturing cost savings. The relative ease with which smooth shapes can be made is a significant factor in the use of composites in aircraft and other applications for which aerodynamic considerations are important.

2 REINFORCEMENTS AND MATRIX MATERIALS

As discussed in Section 1, we divide solid materials into four classes: polymers, metals, ceramics, and carbon. There are reinforcements and matrix materials in each category. In this section, we consider the characteristics of key reinforcements and matrices.

There are important issues that must be discussed before we present constituent properties. The conventional materials used in mechanical engineering applications are primarily structural metals for most of which there are industry and government specifications. The situation is very different for composites. Most reinforcements and matrices are proprietary materials for which there are no industry standards. This is similar to the current status of ceramics. The situation is further complicated by the fact that there are many test methods in use to measure mechanical and physical properties of reinforcements and matrix materials. As a result, there are often conflicting material property data in the usual sources, published papers, and manufacturers' literature. The data presented in this chapter represent a carefully evaluated distillation of information from many sources. The principal sources are listed in the Bibliography and References. In view of the uncertainties discussed, the properties presented in this section should be considered approximate values.

Because of the large number of matrix materials and reinforcements, we are forced to be selective. Further, space limitations prevent presentation of a complete set of properties. Consequently, properties cited are room temperature values, unless otherwise stated.

2.1 Reinforcements

The four key types of reinforcements used in composites are continuous fibers, discontinuous fibers, whiskers (elongated single crystals), and particles (Fig. 1). Continuous, aligned fibers are the most efficient reinforcement form and are widely used, especially in high-performance applications. However, for ease of fabrication and to achieve specific properties, such as improved through-thickness strength, continuous fibers are converted into a wide variety of reinforcement forms using textile technology. Key among them at this time are two-dimensional and three-dimensional fabrics and braids.

Fibers

The development of fibers with unprecedented properties has been largely responsible for the great importance of composites and the revolutionary improvements in properties compared to conventional materials that they offer. The key fibers for mechanical engineering applications are glasses, carbons (also called graphites), several types of ceramics, and high-modulus organics. Most fibers are produced in the form of multifilament bundles called strands or ends in their untwisted forms and yarns when twisted. Some fibers are produced as monofilaments, which generally have much larger diameters than strand filaments. Table 2 presents properties of key fibers, which are discussed in the following subsections.

Fiber strength requires some discussion. Most of the key fibrous reinforcements are made of brittle ceramics or carbon. It is well known that the strengths of monolithic ceramics decrease with increasing material volume because of the increasing probability of finding strength-limiting flaws. This is called size effect. As a result of size effect, fiber strength typically decreases monotonically with increasing gauge length (and diameter). Flaw sensitivity also results in considerable strength scatter at a fixed test length. Consequently, there

Table 2 Properties of Key Reinforcing Fibers

Fiber	Density [g/cm^3 (Pci)]	Axial Modulus [GPa (Msi)]	Tensile Strength [MPa (ksi)]	Axial Coefficient of Thermal Expansion [ppm/K (ppm/°F)]	Axial Thermal Conductivity (W/m·K)
E-glass	2.6 (0.094)	70 (10)	2000 (300)	5 (2.8)	0.9
HS glass	2.5 (0.090)	83 (12)	4200 (650)	4.1 (2.3)	0.9
Aramid	1.4 (0.052)	124 (18)	3200 (500)	−5.2 (−2.9)	0.04
Boron	2.6 (0.094)	400 (58)	3600 (520)	4.5 (2.5)	—
SM carbon (PAN)	1.7 (0.061)	235 (34)	3200 (500)	−0.5 (−0.3)	9
UHM carbon (PAN)	1.9 (0.069)	590 (86)	3800 (550)	−1 (−0.6)	18
UHS carbon (PAN)	1.8 (0.065)	290 (42)	7000 (1000)	−1.5 (−0.8)	160
UHM carbon (pitch)	2.2 (0.079)	895 (130)	2200 (320)	−1.6 (−0.9)	640
UHK carbon (pitch)	2.2 (0.079)	830 (120)	2200 (320)	−1.6 (−0.9)	1100
SiC monofilament	3.0 (0.11)	400 (58)	3600 (520)	4.9 (2.7)	—
SiC multifilament	3.0 (0.11)	400 (58)	3100 (450)	—	—
Si–C–O	2.6 (0.094)	190 (28)	2900 (430)	3.9 (2.2)	1.4
Si–Ti–C–O	2.4 (0.087)	190 (27)	3300 (470)	3.1 (1.7)	—
Aluminum oxide	3.9 (0.14)	370 (54)	1900 (280)	7.9 (4.4)	—
High-density polyethylene	0.97 (0.035)	172 (25)	3000 (440)	—	—

is no single value that characterizes fiber strength. This is also true of key organic reinforcements, such as aramid fibers. Consequently, the values presented in Table 2 should be considered approximate values and are useful primarily for comparative purposes. Note that, because unsupported fibers buckle under very low stresses, it is very difficult to measure their inherent compression strength, and these properties are almost never reported. Instead, composite compression strength is measured directly.

Glass Fibers. Glass fibers are used primarily to reinforce polymers. The leading types of glass fibers for mechanical engineering applications are E-glass and HS glass. E-glass fibers, the first major composite reinforcement, were originally developed for electrical insulation applications (that is the origin of the "E"). E-glass is, by many orders of magnitude, the most widely used of all fibrous reinforcements. The primary reasons for this are its low cost and early development compared to other fibers. Glass fibers are produced as multifilament bundles. Filament diameters range from 3 to 20 μm (118–787 μin.). Table 2 presents representative properties of E-glass and HS glass fibers.

E-glass fibers have relatively low elastic moduli compared to other reinforcements. In addition, E-glass fibers are susceptible to creep and creep (stress) rupture. HS glass is stiffer and stronger than E-glass and has better resistance to fatigue and creep.

The thermal and electrical conductivities of glass fibers are low, and glass fiber-reinforced PMCs are often used as thermal and electrical insulators. The CTE of glass fibers is also low compared to most metals.

Carbon (Graphite) Fibers. Carbon fibers, commonly called graphite fibers in the United States, are used as reinforcements for polymers, metals, ceramics, and carbon. There are dozens of commercial carbon fibers, with a wide range of strengths and moduli. As a class of reinforcements, carbon fibers are characterized by high stiffness and strength and low density and CTE. Fibers with tensile moduli as high as 895 GPa (130 Msi) and with tensile strengths of 7000 MPa (1000 ksi) are commercially available. Carbon fibers have excellent resistance to creep, stress rupture, fatigue, and corrosive environments, although they oxidize at high temperatures. Some carbon fibers also have extremely high thermal conductivities—many times that of copper. This characteristic is of considerable interest in electronic packaging and other applications where thermal control is important. Carbon fibers are the workhorse reinforcements in high-performance aerospace and commercial PMCs and some CMCs. Of course, as the name suggests, carbon fibers are also the reinforcements in carbon/carbon composites.

Most carbon fibers are highly anisotropic. Axial stiffness, tension and compression strength, and thermal conductivity are typically much greater than the corresponding properties in the radial direction. Carbon fibers generally have small, negative axial CTEs (which means that they get shorter when heated) and positive radial CTEs. Diameters of common reinforcing fibers, which are produced in the form of multifilament bundles, range from 4 to 10 μm (160–390 μin.). Carbon fiber stress–strain curves tend to be nonlinear. Modulus increases under increasing tensile stress and decreases under increasing compressive stress.

Carbon fibers are made primarily from three key precursor materials: PAN, petroleum pitch, and coal tar pitch. Rayon-based fibers, once the primary CCC reinforcement, are far less common in new applications. Experimental fibers also have been made by chemical vapor deposition. Some of these have reported axial thermal conductivities as high as 2000 W/m·K, five times that of copper.

PAN-based materials are the most widely used carbon fibers. There are dozens on the market. Fiber axial moduli range from 235 GPa (34 Msi) to 590 GPa (85 Msi). They generally provide composites with excellent tensile and compressive strength properties, although compressive strength tends to drop off as modulus increases. Fibers having tensile strengths as high as 7 GPa (1 Msi) are available. Table 2 presents properties of three types of PAN-based carbon fibers and two types of pitch-based carbon fibers. The PAN-based fibers are SM, UHS, and UHM. SM PAN fibers are the most widely used type of carbon fiber reinforcement. They are one of the first types commercialized and tend to be the least expensive. UHS PAN carbon fibers are the strongest type of another widely used class of carbon fiber, usually called intermediate modulus (IM) because the axial modulus of these fibers falls between those of SM and modulus carbon fibers.

A key advantage of pitch-based fibers is that they can be produced with much higher axial moduli than those made from PAN precursors. For example, UHM pitch fibers with moduli as high as 895 GPa (130 Msi) are available. In addition, some pitch fibers, which we designate UHK, have extremely high axial thermal conductivities. There are commercial UHK fibers with a nominal axial thermal conductivity of 1100 W/m·K, almost three times that of copper. However, composites made from pitch-based carbon fibers generally are somewhat weaker in tension and shear, and much weaker in compression, than those using PAN-based reinforcements.

Boron Fibers. Boron fibers are primarily used to reinforce polymers and metals. Boron fibers are produced as monofilaments (single filaments) by chemical vapor deposition of boron on a tungsten wire or carbon filament, the former being the most widely used. They have relatively large diameters, 100–140 μm (4000–5600 μin.), compared to most other reinforcements. Table 2 presents representative properties of boron fibers having a tungsten core and diameter of 140 μm. The properties of boron fibers are influenced by the ratio of overall fiber diameter to that of the tungsten core. For example, fiber specific gravity is 2.57 for 100-μm fibers and 2.49 for 140-μm fibers.

Fibers Based on Silicon Carbide. Silicon-carbide-based fibers are primarily used to reinforce metals and ceramics. There are a number of commercial fibers based on silicon carbide. One type, a monofilament, is produced by chemical vapor deposition of high-purity silicon carbide on a carbon monofilament core. Some versions use a carbon-rich surface layer that serves as a reaction barrier. There are a number of multifilament silicon-carbide-based fibers made by pyrolysis of polymers. Some of these contain varying amounts of silicon, carbon and oxygen, titanium, nitrogen, zirconium, and hydrogen. Table 2 presents properties of selected silicon-carbide-based fibers.

Fibers Based on Alumina. Alumina-based fibers are primarily used to reinforce metals and ceramics. Like silicon-carbide-based fibers, they have a number of different chemical formulations. The primary constituents, in addition to alumina, are boria, silica, and zirconia. Table 2 presents properties of high-purity alumina fibers.

Aramid Fibers. Aramid, or aromatic, polyamide fibers are high-modulus organic reinforcements primarily used to reinforce polymers and for ballistic protection. There are a number of commercial aramid fibers produced by several manufacturers. Like other reinforcements, they are proprietary materials with different properties. Table 2 presents properties of one of the most widely used aramid fibers.

High-Density Polyethylene Fibers. High-density polyethylene fibers are primarily used to reinforce polymers and for ballistic protection. Table 2 presents properties of a common reinforcing fiber. The properties of high-density polyethylene tend to decrease significantly with increasing temperature, and they tend to creep significantly under load, even at low temperatures.

2.2 Matrix Materials

The four classes of matrix materials are polymers, metals, ceramics, and carbon. Table 3 presents representative properties of selected matrix materials in each category. As the table shows, the properties of the four types differ substantially. These differences have profound effects on the properties of the composites using them. In this section, we examine characteristics of key materials in each class.

Polymer Matrix Materials

There are two major classes of polymers used as matrix materials: thermosets and thermoplastics. Thermosets are materials that undergo a curing process during part fabrication, after which they are rigid and cannot be re-formed. Thermoplastics, on the other hand, can be repeatedly softened and re-formed by application of heat. Thermoplastics are often subdivided into several types: amorphous, crystalline, and liquid crystal. There are numerous types of polymers in both classes. Thermosets tend to be more resistant to solvents and corrosive environments than thermoplastics, but there are exceptions to this rule. Resin selection is based on design requirements, as well as manufacturing and cost considerations. Table 4 presents representative properties of common matrix polymers.

Polymer matrices generally are relatively weak, low-stiffness, viscoelastic materials. The strength and stiffness of PMCs come primarily from the fiber phase. One of the key issues in matrix selection is maximum service temperature. The properties of polymers decrease with increasing temperature. A widely used measure of comparative temperature resistance of polymers is glass transition temperature (T_g), which is the approximate temperature at which a polymer transitions from a relatively rigid material to a rubbery one. Polymers typically suffer significant losses in both strength and stiffness above their glass transition temperatures. New polymers with increasing temperature capability are continually being developed, allowing them to compete with a wider range of metals. For example, carbon fiber-reinforced polyimides have replaced titanium in some aircraft gas turbine engine parts.

An important consideration in selection of polymer matrices is their moisture sensitivity. Resins tend to absorb water, which causes dimensional changes and reduction of elevated-temperature strength and stiffness. The amount of moisture absorption, typically measured as percent weight gain, depends on the polymer and relative humidity. Resins also desorb moisture when placed in a drier atmosphere. The rate of absorption and desorption depends strongly on temperature. The moisture sensitivity of resins varies widely; some are very resistant.

In a vacuum, resins outgas water and organic and inorganic chemicals, which can condense on surfaces with which they come in contact. This can be a problem in optical systems and can affect surface properties critical for thermal control, such as absorptivity and emissivity. Outgassing can be controlled by resin selection and baking out the component.

Thermosetting Resins. The key types of thermosetting resins used in composites are epoxies, bismaleimides, thermosetting polyimides, cyanate esters, thermosetting polyesters, vinyl esters, and phenolics.

Epoxies are the workhorse materials for airframe structures and other aerospace applications, with decades of successful flight experience to their credit. They produce composites with excellent structural properties. Epoxies tend to be rather brittle materials, but toughened

Table 3 Properties of Selected Matrix Materials

Material	Class	Density [g/cm^3 (Pci)]	Modulus [GPa (Msi)]	Tensile Strength [MPa (ksi)]	Tensile Failure Strain (%)	Thermal Conductivity [W/m·K (BTU/h·ft·°F)]	Coefficient of Thermal Expansion [ppm/K (ppm/°F)]
Epoxy	Polymer	1.8 (0.065)	3.5 (0.5)	70 (10)	3	0.1 (0.06)	60 (33)
Aluminum (6061)	Metal	2.7 (0.098)	69 (10)	300 (43)	10	180 (104)	23 (13)
Titanium (6Al–4V)	Metal	4.4 (0.16)	105 (15.2)	1100 (160)	10	16 (9.5)	9.5 (5.3)
Silicon carbide	Ceramic	2.9 (0.106)	520 (75)	—	<0.1	81 (47)	4.9 (2.7)
Alumina	Ceramic	3.9 (0.141)	380 (55)	—	<0.1	20 (120)	6.7 (3.7)
Glass (borosilicate)	Ceramic	2.2 (0.079)	63 (9)	—	<0.1	2 (1)	5 (3)
Carbon	Carbon	1.8 (0.065)	20 (3)	—	<0.1	5–90 (3–50)	2 (1)

Table 4 Properties of Selected Thermosetting and Thermoplastic Matrices

	Density [g/cm^3 (Pci)]	Modulus [GPa (Msi)]	Tensile Strength [MPa (ksi)]	Elongation to Break (%)	Thermal Conductivity (W/m·K)	Coefficient of Thermal Expansion [ppm/K (ppm/°F)]
Epoxy[a]	1.1–1.4 (0.040–0.050)	3–6 (0.43–0.88)	35–100 (5–15)	1–6	0.1	60 (33)
Thermosetting polyester[a]	1.2–1.5 (0.043–0.054)	2–4.5 (0.29–0.65)	40–90 (6–13)	2	0.2	100–200 (56–110)
Polypropylene[b]	0.90 (0.032)	1–4 (0.15–0.58)	25–38 (4–6)	>300	0.2	110 (61)
Nylon 6-6[b]	1.14 (0.041)	1.4–2.8 (0.20–0.41)	60–75 (9–11)	40–80	0.2	90 (50)
Polycarbonate[b]	1.06–1.20 (0.038–0.043)	2.2–2.4 (0.32–0.35)	45–70 (7–10)	50–100	0.2	70 (39)
Polysulfone[b]	1.25 (0.045)	2.2 (0.32)	76 (11)	50–100	—	56 (31)
Polyetherimide[b]	1.27 (0.046)	3.3 (0.48)	110 (16)	60	—	62 (34)
Polyamideimide[b]	1.4 (0.050)	4.8 (0.7)	190 (28)	17	—	63 (35)
Polyphenylene sulfide[b]	1.36 (0.049)	3.8 (0.55)	65 (10)	4	—	54 (30)
Polyether etherketone[b]	1.26–1.32 (0.046–0.048)	3.6 (0.52)	93 (13)	50	—	47 (26)

[a] Thermoset.
[b] Thermoplastic.

formulations with greatly improved impact resistance are available. The maximum service temperature is affected by reduced elevated-temperature structural properties resulting from water absorption. A typical airframe limit is about 120°C (250°F).

Bismaleimide resins are used for aerospace applications requiring higher temperature capabilities than can be achieved by epoxies. They are employed for temperatures of up to about 200°C (390°F).

Thermosetting polyimides are used for applications with temperatures as high as 250–290°C (500–550°F).

Cyanate ester resins are not as moisture sensitive as epoxies and tend to outgas much less. Formulations with operating temperatures as high as 205°C (400°F) are available.

Thermosetting polyesters are the workhorse resins in commercial applications. They are relatively inexpensive, easy to process, and corrosion resistant.

Vinyl esters are also widely used in commercial applications. They have better corrosion resistance than polyesters but are somewhat more expensive.

Phenolic resins have good high-temperature resistance and produce less smoke and toxic products than most resins when burned. They are used in applications such as aircraft interiors and offshore oil platform structures, for which fire resistance is a key design requirement.

Thermoplastic Resins. Thermoplastics are divided into three main classes: amorphous, crystalline, and liquid crystal. Polycarbonate, acrylonitrile–butadiene–styrene (ABS), polystyrene, polysulfone, and polyetherimide are amorphous materials. Crystalline thermoplastics include nylon, polyethylene, polyphenylene sulfide, polypropylene, acetal, polyethersulfone, and polyether etherketone (PEEK). Amorphous thermoplastics tend to have poor solvent resistance. Crystalline materials tend to be better in this respect. Relatively inexpensive thermoplastics such as nylon are extensively used with chopped E-glass fiber reinforcements in countless injection-molded parts. There are an increasing number of applications using continuous fiber-reinforced thermoplastics.

Metals

The metals initially used for MMC matrix materials generally were conventional alloys. Over time, however, many special matrix materials tailored for use in composites have been developed. The key metallic matrix materials used for structural MMCs are alloys of aluminum, titanium, iron, and intermetallic compounds, such as titanium aluminides. However, many other metals have been used as matrix materials, such as copper, lead, magnesium, cobalt, silver, and superalloys. The *in situ* properties of metals in a composite depend on the manufacturing process and, because metals are elastic–plastic materials, the history of mechanical stresses and temperature changes to which they are subjected.

Ceramic Matrix Materials

The key ceramics used as CMC matrices are silicon carbide, alumina, silicon nitride, mullite, and various cements. The properties of ceramics, especially strength, are even more process sensitive than those of metals. In practice, it is very difficult to determine the *in situ* properties of ceramic matrix materials in a composite.

As discussed earlier, in the section on fiber properties, ceramics are very flaw sensitive, resulting in a decrease in strength with increasing material volume, a phenomenon called "size effect." As a result, there is no single value that describes the tensile strength of ceramics. In fact, because of the very brittle nature of ceramics, it is difficult to measure tensile strength, and flexural strength (often called modulus of rupture) is typically reported. It should be noted that flexural strength is also dependent on specimen size and is generally much higher than that of a tensile coupon of the same dimensions. In view of the great difficulty in measuring a simple property like tensile strength, which arises from their flaw sensitivity, it is not surprising that monolithic ceramics have had limited success in applications where they are subjected to significant tensile stresses.

The fracture toughness of ceramics is typically in the range of 3–6 $MPa \cdot m^{1/2}$. Those of transformation-toughened materials are somewhat higher. For comparison, the fracture toughnesses of structural metals are generally greater than 20 $MPa \cdot m^{1/2}$.

Carbon Matrix Materials

Carbon is a remarkable material. It includes materials ranging from lubricants to diamonds and structural fibers. The forms of carbon matrices resulting from the various carbon/carbon manufacturing processes tend to be rather weak, brittle materials. Some forms have very high thermal conductivities. As for ceramics, *in situ* matrix properties are difficult to measure.

3 PROPERTIES OF COMPOSITE MATERIALS

There are a large and increasing number of materials in all four classes of composites: PMCs, MMCs, CMCs, and CCCs. In this section, we present mechanical and physical properties of some of the key materials in each class.

Initially, the excellent mechanical properties of composites were the main reason for their use. However, there are an increasing number of applications for which the unique and tailorable physical properties of composites are key considerations. For example, the extremely high thermal conductivity and tailorable CTE of some composite material systems are leading to their increasing use in electronic packaging. Similarly, the extremely high stiffness, near-zero CTE, and low density of carbon fiber-reinforced polymers have made these composites the materials of choice in spacecraft structures.

Composites are complex, heterogeneous, and often anisotropic material systems. Their properties are affected by many variables, including *in situ* constituent properties; reinforcement form, volume fraction, and geometry; properties of the interphase, the region where the reinforcement and matrix are joined (also called the interface); and void content. The process by which the composite is made affects many of these variables. The same matrix material and reinforcements, when combined by different processes, may result in composites with very different properties,

Several other important things must be kept in mind when considering composite properties. For one, most composites are proprietary material systems made by proprietary processes. There are few industry or government specifications for composites, as there are for many monolithic structural metals. However, this is also the case for many monolithic ceramics and polymers, which are widely used engineering materials. Despite their inherently proprietary nature, some widely used composite materials made by a number of manufacturers have similar properties. A notable example is SM carbon fiber-reinforced epoxy.

Another critical issue is that properties are sensitive to the test methods by which they are measured, and there are many different test methods used throughout the industry. Further, test results are very sensitive to the skill of the technician performing the test. Because of these factors, it is very common to find significant differences in reported properties of what is nominally the same composite material.

In Section 2, we discussed the issue of size effect, which is the decrease in strength with increasing material volume that is observed in monolithic ceramic key reinforcing fibers. There is some evidence, suggestive but not conclusive, of size effects in composite strength properties, as well. However, if composite strength size effects exist at all, they are much less severe than for fibers by themselves. The reason is that the presence of a matrix results in very different failure mechanisms. However, until the issues are resolved definitively, caution should be used in extrapolating strength data from small coupons to large structures, which may have volumes many orders of magnitude greater.

As mentioned earlier, the properties of composites are very sensitive to reinforcement form, volume fraction, and geometry. This is illustrated in Table 5, which presents the properties of several common types of E-glass fiber-reinforced polyester composites. The reinforcement forms are discontinuous fibers, woven roving (a heavy fabric), and straight, parallel continuous fibers. As we shall see, discontinuous reinforcement is not as efficient as contin-

Table 5 Effect of Fiber Form and Volume Fraction on Mechanical Properties of E-Glass-Reinforced Polyester[4]

	Bulk Molding Compound	Sheet Molding Compound	Chopped Strand Mat	Woven Roving	Unidirectional Axial	Unidirectional Transverse
Glass content, wt. %	20	30	30	50	70	70
Tensile modulus, GPa (Msi)	9 (1.3)	13 (1.9)	7.7 (1.1)	16 (2.3)	42 (6.1)	12 (1.7)
Tensile strength, MPa (ksi)	45 (6.5)	85 (12)	95 (14)	250 (36)	750 (110)	50 (7)

uous. However, discontinuous fibers allow the composite material to flow during processing, facilitating fabrication of complex molded parts.

The composites using discontinuous fibers are divided into three categories. One is bulk molding compound (BMC), also called dough molding compound, in which fibers are relatively short, about 3–12 mm, and are nominally randomly oriented in three dimensions. BMC also has a very high loading of mineral particles, such as calcium carbonate, which are added for a variety of reasons: to reduce dimensional changes from resin shrinkage, to obtain a smooth surface, and to reduce cost, among others. Because it contains both particulate and fibrous reinforcement, BMC can be considered a type of hybrid composite.

The second type of composite is chopped strand mat (CSM), which contains discontinuous fibers, typically about 25 mm long, nominally randomly oriented in two directions. The third material is sheet molding compound (SMC), which contains chopped fibers 25–50 mm in length, also nominally randomly oriented in two dimensions. Like BMC, SMC also contains particulate mineral fillers, such as calcium carbonate and clay.

The first thing to note in comparing the materials in Table 5 is that fiber content, here presented in the form of weight percent, differs considerably for the four materials. This is significant, because, as discussed in Section 2, the strength and stiffness of polyester and most polymer matrices is considerably lower than those of E-glass, carbon, and other reinforcing fibers. Composites reinforced with randomly oriented fibers tend to have lower volume fractions than those made with aligned fibers or fabrics. There is a notable exception to this. Some composites with discontinuous-fiber reinforcement are made by chopping up composites reinforced with aligned continuous fibers or fabrics that have high-fiber contents.

Examination of Table 5 shows that the modulus of SMC is considerably greater than that of CSM, even though both have the same fiber content. This is because SMC also has particulate reinforcement. Note, however, that although the particles improve modulus, they do not increase strength. This is generally the case for particle-reinforced polymers, but, as we will see later, particles often do enhance the strengths of MMCs and CMCs, as well as their moduli.

We observe that the modulus of the BMC composite is greater than that of CSM and SMC, even though the former has a much lower fiber content. Most likely, this results from the high mineral content and also the possibility that the fibers are oriented in the direction of test and are not truly random. Many processes, especially those involving material flow, tend to orient fibers in one or more preferred directions. If so, then one would find the modulus of the BMC to be much lower than the one presented in the table if measured in other directions. This illustrates one of the limitations of using discontinuous-fiber reinforcement: it is often difficult to control fiber orientation.

The moduli and strengths of the composites reinforced with fabrics and aligned fibers are much higher than those with discontinuous fibers, when the former two types of materials are tested parallel to fiber directions. For example, the tensile strength of woven roving is more than twice that of CSM. The properties presented are measured parallel to the warp direction of the fabric (the warp direction is the lengthwise direction of the fabric). The elastic and strength properties in the fill direction, perpendicular to the warp, typically are similar to, but somewhat lower than, those in the warp direction. Here, we assume that the fabric is "balanced," which means that the number of fibers in the warp and fill directions per unit length are approximately equal. Note, however, that the elastic modulus, tensile strength, and compressive strength at 45° to the warp and fill directions of a fabric are much lower than the corresponding values in the warp and fill directions. This is discussed further in the sections that cover design.

As Table 5 shows, the axial modulus and tensile strength of the unidirectional composite are much greater than those of the fabric. However, the modulus and strength of the unidirectional composite in the transverse direction are considerably lower than the corresponding axial properties. Further, the transverse strength is considerably lower than that of SMC and CSM. In general, the strength of PMCs is weak in directions for which there are no fibers. The low transverse moduli and strengths of unidirectional PMCs are commonly overcome by use of laminates with fibers in several directions. Low through-thickness strength can be improved by use of three-dimensional reinforcement forms. Often, the designer simply assures that through-thickness stresses are within the capability of the material.

In this section, we present representative mechanical and physical properties of key composite materials of interest for a broad range of mechanical engineering applications. The properties represent a distillation of values from many sources. Because of space limitations, it is necessary to be selective in our choice of materials and properties presented. It is simply not possible to present a complete set of data that will cover every possible application. As discussed earlier, there are many textile forms, such as woven fabrics, used as reinforcements. However, we concentrate on aligned, continuous fibers because they produce the highest strength and stiffness. To do a thorough evaluation of composites, the design engineer should consider alternative reinforcement forms. Unless otherwise stated, room temperature property values are presented. We consider mechanical properties in Section 3.1 and physical properties in Section 3.2.

3.1 Mechanical Properties of Composite Materials

In this section, we consider mechanical properties of key PMCs, MMCs, CMCs, and CCCs that are of greatest interest for mechanical engineering applications.

Mechanical Properties of Polymer Matrix Composites

As discussed earlier, polymers are relatively weak, low-stiffness materials. To obtain materials with mechanical properties that are acceptable for structural applications, it is necessary to reinforce them with continuous or discontinuous fibers. The addition of ceramic or metallic particles to polymers results in materials that have increased modulus, but, as a rule, strength typically does not increase significantly and may actually decrease. However, there are many particle-reinforced polymers used in electronic packaging, primarily because of their physical properties. For these applications, ceramic particles, such as alumina, aluminum nitride, boron nitride, and even diamond, are added to obtain an electrically insulating material with higher thermal conductivity and lower CTE than the monolithic base polymer. Metallic particles such as silver and aluminum are added to create materials that are both electrically and thermally conductive. These materials have replaced lead-based solders in many applications. There are also magnetic composites made by incorporating ferrous or permanent magnet particles in various polymers. A common example is magnetic tape used to record audio and video.

We focus on composites reinforced with continuous fibers because they are the most efficient structural materials. Table 6 presents room temperature mechanical properties of unidirectional polymer matrix composites reinforced with key fibers: E-glass, aramid, boron, SM PAN carbon, UHS PAN carbon, UHM PAN carbon, UHM pitch carbon, and ultrahigh-thermal-conductivity (UHK) pitch carbon. We assume that the fiber volume fraction is 60%, a typical value. As discussed in Section 2, UHS PAN carbon is the strongest type of IM carbon fiber.

The properties presented in Table 6 are representative of what can be obtained at room temperature with a well-made PMC employing an epoxy matrix. Epoxies are widely used, provide good mechanical properties, and can be considered a reference matrix material.

Table 6 Mechanical Properties of Selected Unidirectional Polymer Matrix Composites

Fiber	Axial Modulus [GPa (Msi)]	Transverse Modulus [GPa (Msi)]	Inplane Shear Modulus [GPa (Msi)]	Poisson's Ratio	Axial Tensile Strength [MPa (ksi)]	Transverse Tensile Strength [MPa (ksi)]	Axial Compressive Strength [MPa (ksi)]	Transverse Compressive Strength [MPa (ksi)]	Inplane Shear Strength [MPa (ksi)]
E-glass	45 (6.5)	12 (1.8)	5.5 (0.8)	0.28	1020 (150)	40 (7)	620 (90)	140 (20)	70 (10)
Aramid	76 (11)	5.5 (0.8)	2.1 (0.3)	0.34	1240 (180)	30 (4.3)	280 (40)	140 (20)	60 (9)
Boron	210 (30)	19 (2.7)	4.8 (0.7)	0.25	1240 (180)	70 (10)	3310 (480)	280 (40)	90 (13)
SM carbon (PAN)	145 (21)	10 (1.5)	4.1 (0.6)	0.25	1520 (220)	41 (6)	1380 (200)	170 (25)	80 (12)
UHS carbon (PAN)	170 (25)	10 (1.5)	4.1 (0.6)	0.25	3530 (510)	41 (6)	1380 (200)	170 (25)	80 (12)
UHM carbon (PAN)	310 (45)	9 (1.3)	4.1 (0.6)	0.20	1380 (200)	41 (6)	760 (110)	170 (25)	80 (12)
UHM carbon (pitch)	480 (70)	9 (1.3)	4.1 (0.6)	0.25	900 (130)	20 (3)	280 (40)	100 (15)	41 (6)
UHK carbon (pitch)	480 (70)	9 (1.3)	4.1 (0.6)	0.25	900 (130)	20 (3)	280 (40)	100 (15)	41 (6)

Properties of composites using other resins may differ from these and have to be examined on a case-by-case basis.

The properties of PMCs, especially strengths, depend strongly on temperature. The temperature dependence of polymer properties differs considerably. This is also true for different epoxy formulations, which have different cure and glass transition temperatures. Some polymers, such as polyimides, have good elevated-temperature properties that allow them to compete with titanium. There are aircraft gas turbine engine components employing polyimide matrices that see service temperatures as high as 290°C (550°F). Here, again, the effect of temperature on composite properties has to be considered on a case-by-case basis.

The properties shown in Table 6 are axial, transverse and shear moduli, Poisson's ratio, tensile and compressive strengths in the axial and transverse directions, and inplane shear strength. The Poisson's ratio presented is called the major Poisson's ratio. It is defined as the ratio of the magnitude of transverse strain divided by axial strain when the composite is loaded in the axial direction. Note that transverse moduli and strengths are much lower than corresponding axial values.

As discussed in Section 2, carbon fibers display nonlinear stress–strain behavior. Their moduli increase under increasing tensile stress and decrease under increasing compressive stress. This makes the method of calculating modulus critical. Various tangent and secant definitions are used throughout the industry, contributing to the confusion in reported properties. The values presented in Table 6, which are approximate, are based on tangents to the stress–strain curves at the origin. Using this definition, tensile and compressive moduli are usually very similar. However, this is not the case for moduli using various secant definitions. Using these definitions typically produces compression moduli that are significantly lower than tension moduli.

Because of the low transverse strengths of unidirectional laminates, they are rarely used in structural applications. The design engineer uses laminates with layers in several directions to meet requirements for strength, stiffness, buckling, and so on. There are an infinite number of laminate geometries that can be selected. For comparative purposes, it is useful to consider quasi-isotropic laminates, which have the same elastic properties in all directions in the plane. Laminates are quasi-isotropic when they have the same percentage of layers every $180/n°$, where $n \geq 3$. The most common quasi-isotropic laminates have layers that repeat every 60°, 45°, or 30°. We note, however, that strength properties in the plane are not isotropic for these laminates, although they tend to become more uniform as the angle of repetition becomes smaller.

Table 7 presents the mechanical properties of quasi-isotropic laminates. Note that the moduli and strengths are much lower than the axial properties of unidirectional laminates made of the same material. In most applications, laminate geometry is such that the maximum axial modulus and tensile and compressive strengths fall somewhere between axial unidirectional and quasi-isotropic values.

The tension–tension fatigue behavior of unidirectional composites, discussed in Section 1, is one of their great advantages over metals (Fig. 6). In general the tension–tension *S–N* curves (curves of maximum stress plotted as a function of cycles to failure) of PMCs reinforced with carbon, boron, and aramid fibers are relatively flat. Glass fiber-reinforced composites show a greater reduction in strength with increasing number of cycles. Still, PMCs reinforced with HS glass are widely used in applications for which fatigue resistance is a critical design consideration, such as helicopter rotors.

Metals are more likely to fail in fatigue when subjected to fluctuating tensile rather than compressive load. This is because they tend to fail by crack propagation under fatigue loading. However, the failure modes in composites are very different and more complex. One consequence is that composites tend to be more susceptible to fatigue failure when loaded

Table 7 Mechanical Properties of Selected Quasi-Isotropic Polymer Matrix Composites

Fiber	Axial Modulus [GPa (Msi)]	Transverse Modulus [GPa (Msi)]	Inplane Shear Modulus [GPa (Msi)]	Poisson's Ratio	Axial Tensile Strength [MPa (ksi)]	Transverse Tensile Strength [MPa (ksi)]	Axial Compressive Strength [MPa (ksi)]	Transverse Compressive Strength [MPa (ksi)]	Inplane Shear Strength [MPa (ksi)]
E-glass	23 (3.4)	23 (3.4)	9.0 (1.3)	0.28	550 (80)	550 (80)	330 (48)	330 (48)	250 (37)
Aramid	29 (4.2)	29 (4.2)	11 (1.6)	0.32	460 (67)	460 (67)	190 (28)	190 (28)	65 (9.4)
Boron	80 (11.6)	80 (11.6)	30 (4.3)	0.33	480 (69)	480 (69)	1100 (160)	1100 (160)	360 (52)
SM carbon (PAN)	54 (7.8)	54 (7.8)	21 (3.0)	0.31	580 (84)	580 (84)	580 (84)	580 (84)	410 (59)
UHS carbon (PAN)	63 (9.1)	63 (9.1)	21 (3.0)	0.31	1350 (200)	1350 (200)	580 (84)	580 (84)	410 (59)
UHM carbon (PAN)	110 (16)	110 (16)	41 (6.0)	0.32	490 (71)	490 (71)	270 (39)	70 (39)	205 (30)
UHM carbon (pitch)	165 (24)	165 (24)	63 (9.2)	0.32	310 (45)	310 (45)	96 (14)	96 (14)	73 (11)
UHK carbon (pitch)	165 (24)	165 (24)	63 (9.2)	0.32	310 (45)	310 (45)	96 (14)	96 (14)	73 (11)

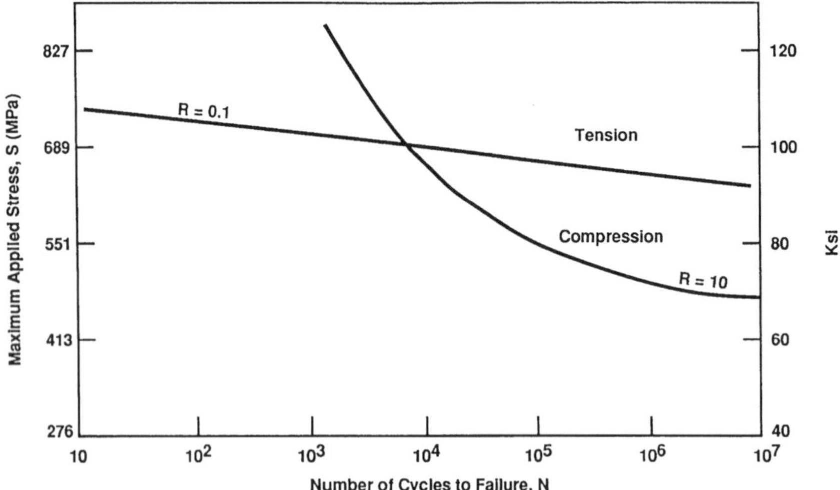

Figure 6 Cycles to failure as a function of maximum stress for carbon-fiber-reinforced epoxy laminates loaded in tension–tension ($R = 0.1$) and compression–compression ($R = -10$) fatigue (after Ref. 5).

in compression. Figure 6 shows the cycles to failure as a function of maximum stress for carbon fiber-reinforced epoxy laminates subjected to tension–tension and compression–compression fatigue. The laminates have 60% of their layers oriented at 0°, 20% at +45°, and 20% at −45°. They are subjected to a fluctuating load in the 0° direction. The ratios of minimum stress to maximum stress (R) for tensile and compressive fatigue are 0.1 and 10, respectively. We observe that the reduction in strength is much greater for compression–compression fatigue. However, the composite compressive fatigue strength at 10^7 cycles is still considerably greater than the corresponding tensile value for aluminum.

Polymer matrix composites reinforced with carbon and boron are very resistant to deformation and failure under sustained static load when they are loaded in a fiber-dominated direction. (These phenomena are called creep and creep rupture, respectively.) The creep and creep rupture behavior of aramid is not quite as good. Glass fibers display significant creep, and creep rupture is an important design consideration. Polymers are viscoelastic materials that typically display significant creep when they are not constrained with fibers. Therefore, creep should be considered when composites are subjected to significant stresses in matrix-dominated directions, such as the laminate through-thickness direction.

Mechanical Properties of Metal Matrix Composites

Monolithic metallic alloys are the most widely used materials in mechanical engineering applications. By reinforcing them with continuous fibers, discontinuous fibers, whiskers, and particles, we create new materials with enhanced or modified properties, such as higher strength and stiffness, better wear resistance, lower CTE, and so on. In some cases, the improvements are dramatic.

The greatest increases in strength and modulus are achieved with continuous fibers. However, the relatively high cost of many continuous reinforcing fibers used in MMCs has limited the application of these materials. The most widely used MMCs are reinforced with discontinuous fibers or particles. This may change as new, lower cost continuous fibers and processes are developed and as cost drops with increasing production volume.

Continuous Fiber-Reinforced MMCs. One of the major advantages of MMCs reinforced with continuous fibers over PMCs is that many, if not most, unidirectional MMCs have much greater transverse strengths, which allow them to be used in a unidirectional configuration. Table 8 presents representative mechanical properties of selected unidirectional MMCs reinforced with continuous fibers corresponding to a nominal fiber volume fraction of 50%. The values represent a distillation obtained from numerous sources. In general, the axial moduli of the composites are much greater than those of the monolithic base metals used for the matrices. However, MMC transverse strengths are typically lower than those of the parent matrix materials.

Mechanical Properties of Discontinuous Fiber-Reinforced MMCs. One of the primary mechanical engineering applications of discontinuous fiber-reinforced MMCs is in internal combustion engine components. Fibers are added primarily to improve the wear resistance and elevated-temperature strength and fatigue properties of aluminum. The improvement in wear resistance eliminates the need for cast iron sleeves in engine blocks and cast iron insert rings in pistons. Fiber-reinforced aluminum composites also have higher thermal conductivities than cast iron and, when fiber volume fractions are relatively low, their CTEs are closer to that of unreinforced aluminum, reducing thermal stresses.

The key reinforcements used in internal combustion engine components to increase wear resistance are discontinuous alumina and alumina–silica fibers. In one application, Honda Prelude engine blocks, carbon fibers are combined with alumina to tailor both wear resistance and coefficient of friction of cylinder walls. Wear resistance is not an inherent property, so that there is no single value that characterizes a material. However, in engine tests, it was found that ring groove wear for an alumina fiber-reinforced aluminum piston was significantly less than that for one with a cast iron insert.

Mechanical Properties of Particle-Reinforced MMCs. Particle-reinforced metals are a particularly important class of MMCs for engineering applications. A wide range of materials fall into this category, and a number of them have been used for many years. An important example is a material consisting of tungsten carbide particles embedded in a cobalt matrix that is used extensively in cutting tools and dies. This composite, often referred to as a cermet, cemented carbide, or simply, but incorrectly, "tungsten carbide," has much better fracture toughness than monolithic tungsten carbide, which is a brittle ceramic material. Another interesting MMC, tungsten carbide particle-reinforced silver, is a key circuit breaker contact pad material. Here, the composite provides good electrical conductivity and much greater hardness and wear resistance than monolithic silver, which is too soft to be used in this application. Ferrous alloys reinforced with titanium carbide particles, discussed in the

Table 8 Mechanical Properties of Selected Unidirectional Continuous Fiber-Reinforced Metal Matrix Composites

Fiber	Matrix	Density [g/cm^3 (Pci)]	Axial Modulus [GPa (Msi)]	Transverse Modulus [GPa (Msi)]	Axial Tensile Strength [MPa (ksi)]	Transverse Tensile Strength [MPa (ksi)]	Axial Compressive Strength [MPa (ksi)]
UHM carbon (pitch)	Aluminum	2.4 (0.090)	450 (65)	15 (5)	690 (100)	15 (5)	340 (50)
Boron	Aluminum	2.6 (0.095)	210 (30)	140 (20)	1240 (180)	140 (20)	1720 (250)
Alumina	Aluminum	3.2 (0.12)	240 (35)	130 (19)	1700 (250)	120 (17)	1800 (260)
Silicon carbide	Titanium	3.6 (0.13)	260 (38)	170 (25)	1700 (250)	340 (50)	2760 (400)

next subsection, have been used for many years in commercial applications. Compared to the monolithic base metals, they offer greater wear resistance and stiffness and lower density.

MECHANICAL PROPERTIES OF TITANIUM CARBIDE PARTICLE-REINFORCED STEEL. A number of ferrous alloys reinforced with titanium carbide particles have been used in mechanical system applications for many years. To illustrate the effect of the particulate reinforcements, we consider a particular composite consisting of austenitic stainless steel reinforced with 45% by volume of titanium carbide particles. The modulus of the composite is 304 GPa (44 Msi) compared to 193 GPa (28 Msi) for the monolithic base metal. The specific gravity of the composite is 6.45, about 20% lower than that of monolithic matrix, 8.03. The specific stiffness of the composite is almost double that of the unreinforced metal.

MECHANICAL PROPERTIES OF SILICON CARBIDE PARTICLE-REINFORCED ALUMINUM. Aluminum reinforced with silicon carbide particles is one of the most important of the newer types of MMCs. A wide range of materials fall into this category. They are made by a variety of processes. Properties depend on the type of particle, particle volume fraction, matrix alloy, and the process used to make them. Table 9 shows how representative composite properties vary with particle volume fraction. In general, as particle volume fraction increases, modulus and yield strength increase and fracture toughness and tensile ultimate strain decrease. Particle reinforcement also improves short-term elevated-temperature strength properties and fatigue resistance.

MECHANICAL PROPERTIES OF ALUMINA PARTICLE-REINFORCED ALUMINUM. Alumina particles are used to reinforce aluminum as an alternative to silicon carbide particles because they do not react as readily with the matrix at high temperatures and are less expensive. Consequently, alumina-reinforced composites can be used in a wider range of processes and applications. However, the stiffness and thermal conductivity of alumina are lower than the corresponding properties of silicon carbide, and these characteristics are reflected in somewhat lower values for composite properties.

Mechanical Properties of Ceramic Matrix Composites

Ceramics, in general, are characterized by high stiffness and hardness, resistance to wear, corrosion and oxidation, and high-temperature operational capability. However, they also have serious deficiencies that have severely limited their use in applications that are subjected to significant tensile stresses. Ceramics have very low fracture toughness, which makes them

Table 9 Mechanical Properties of Silicon Carbide Particle-Reinforced Aluminum

Property	Aluminum (6061–T6)	Titanium (6Al–4V)	Steel (4340)	Composite Particle Volume Fraction 25	55	70
Modulus, GPa (Msi)	69 (10)	113 (16.5)	200 (29)	114 (17)	186 (27)	265 (38)
Tensile yield strength, MPa (ksi)	275 (40)	1000 (145)	1480 (215)	400 (58)	495 (72)	225 (33)
Tensile ultimate strength, MPa (ksi)	310 (45)	1100 (160)	1790 (260)	485 (70)	530 (77)	225 (33)
Elongation (%)	15	5	10	3.8	0.6	0.1
Density, g/cm^3 (lb/in.^{3})	2.77 (0.10)	4.43 (0.16)	7.76 (0.28)	2.88 (0.104)	2.96 (0.107)	3.00 (0.108)
Specific modulus, GPa	5	26	26	40	63	88

very sensitive to the presence of small flaws. This results in great strength scatter and poor resistance to thermal and mechanical shock. Civil engineers recognized this deficiency long ago and, in construction, ceramic materials like stone and concrete are rarely used to carry tensile loads. In concrete, this function has been relegated to reinforcing bars made of steel or, more recently, PMCs. An important exception has been in lightly loaded structures where dispersed reinforcing fibers of asbestos, steel, glass, and carbon allow modest tensile stresses to be supported.

In CMCs, fibers, whiskers, and particles are combined with ceramic matrices to improve fracture toughness, which reduces strength scatter and improves thermal and mechanical shock resistance. By a wide margin, the greatest increases in fracture resistance result from the use of continuous fibers. Table 10 compares fracture toughnesses of structural metallic alloys with those of monolithic ceramics and CMCs reinforced with whiskers and continuous fibers. The low fracture toughness of monolithic ceramics gives rise to very small critical flaw sizes. For example, the critical flaw sizes for monolithic ceramics corresponding to a failure stress of 700 MPa (about 100 ksi) are in the range of 20–80 μm. Flaws of this size are difficult to detect with conventional nondestructive techniques.

The addition of continuous fibers to ceramics can, if done properly, significantly increase the effective fracture toughness of ceramics. For example, as Table 10 shows, addition of silicon carbide fibers to a silicon carbide matrix results in a CMC having a fracture toughness in the range of aluminum alloys.

The addition of continuous fibers to a ceramic matrix also changes the failure mode. Figure 7 compares the tensile stress–strain curves for a typical monolithic ceramic and a conceptual continuous fiber-reinforced CMC. The monolithic material has a linear stress–strain curve and fails catastrophically at a low strain level. However, the CMC displays a nonlinear stress–strain curve with much more area under the curve, indicating that more energy is absorbed during failure and that the material has a less catastrophic failure mode. The fiber–matrix interphase properties must be carefully tailored and maintained over the life of the composite to obtain this desirable behavior.

Although the CMC stress–strain curve looks, at first, like that of an elastic–plastic metal, this is deceiving. The departure from linearity in the CMC results from internal damage mechanisms, such as the formation of microcracks in the matrix. The fibers bridge the cracks, preventing them from propagating. However, the internal damage is irreversible. As the figure

Table 10 Fracture Toughness of Structural Alloys, Monolithic Ceramics, and Ceramic Matrix Composites

Matrix	Reinforcement	Fracture Toughness ($MPa \cdot m^{1/2}$)
Aluminum	None	30–45
Steel	None	40–65[a]
Alumina	None	3–5
Silicon carbide	None	3–4
Alumina	Zirconia particles[b]	6–15
Alumina	Silicon carbide whiskers	5–10
Silicon carbide	Continuous silicon carbide fibers	25–30

[a] The toughness of some alloys can be much higher.
[b] Transformation-toughened.

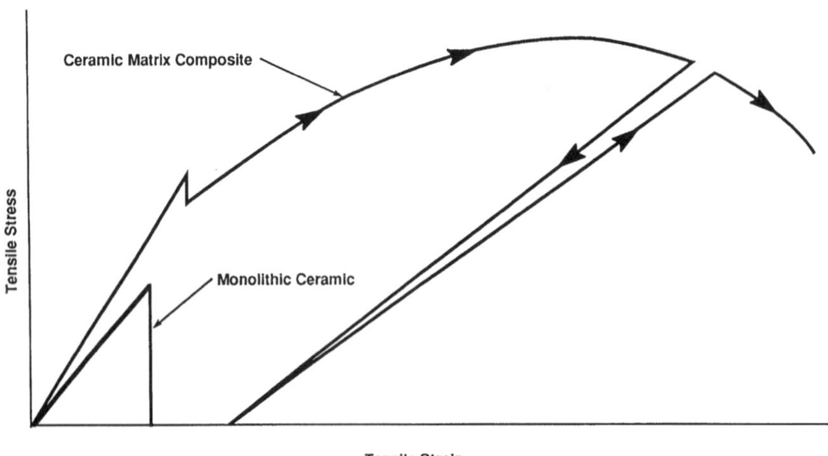

Figure 7 Stress–strain curves for a monolithic ceramic and ceramic matrix composite reinforced with continuous fibers.

shows, the slope of the stress–strain curve during unloading and subsequent reloading is much lower than that representing initial loading. For an elastic–plastic material, the slopes of the unloading and reloading curves are parallel to the initial elastic slope.

There are numerous CMCs at various stages of development. One of the most mature types consists of a silicon carbide matrix reinforced with fabric woven of silicon-carbide-based fibers. These composites are commonly referred to as SiC/SiC. We consider one version. Because the modulus of the particular silicon-carbide-based fibers used in this material is lower than that of pure silicon carbide, the modulus of the composite, about 210 GPa (30 Msi), is lower than that of monolithic silicon carbide, 440 GPa (64 Msi). The flexural strength of the composite parallel to the fabric warp direction, about 300 MPa (44 ksi), is maintained to a temperature of at least 1100°C for short times. Long-term strength behavior depends on degradation of the fibers, matrix, and interphase. Because of the continuous fiber reinforcement, SiC/SiC displays excellent resistance to severe thermal shock.

Mechanical Properties of Carbon/Carbon Composites

Carbon/carbon composites consist of continuous and discontinuous carbon fibers embedded in carbon matrices. As for other composites, there are a wide range of materials that fall in this category. The variables affecting properties include type of fiber, reinforcement form, and volume fraction and matrix characteristics.

Historically, CCCs were first used because of their excellent resistance to high-temperature ablation. Initially, strengths and stiffnesses were low, but these properties have steadily increased over the years. CCCs are an important class of materials in high-temperature applications such as aircraft brakes, rocket nozzles, racing car brakes and clutches, glass-making equipment, and electronic packaging, among others.

One of the most significant limitations of CCCs is oxidation, which begins at a temperature threshold of approximately 370°C (700°F) for unprotected materials. Addition of oxidation inhibitors raises the threshold substantially. In inert atmospheres, CCCs retain their properties to temperatures as high as 2800°C (5000°F).

Carbon matrices are typically weak, brittle, low-stiffness materials. As a result, transverse and through-thickness elastic moduli and strength properties of unidirectional CCCs are low. Because of this, two-dimensional and three-dimensional reinforcement forms are commonly used. In the direction of fibrous reinforcement, it is possible to obtain moduli as high as 340 GPa (50 Msi), tensile strengths as high as 700 MPa (100 ksi), and compressive strengths as high as 800 MPa (110 ksi). In directions orthogonal to fiber directions, elastic moduli are in the range of 10 MPa (1.5 ksi), tensile strengths 14 MPa (2 ksi), and compressive strengths 34 MPa (5 ksi).

3.2 Physical Properties of Composite Materials

Material physical properties are critical for many applications. In this category, we include, among others, density, CTE, thermal conductivity, and electromagnetic characteristics. In this section, we concentrate on the properties of most general interest to mechanical engineers: density, CTE, and thermal conductivity.

Thermal control is a particularly important consideration in electronic packaging because failure rates of semiconductors increase exponentially with temperature. Since conduction is an important method of heat removal, thermal conductivity is a key material property. For many applications, such as spacecraft, aircraft, and portable systems, weight is also an important factor, and consequently, material density is also significant. A useful figure of merit is specific thermal conductivity, defined as thermal conductivity divided by density. Specific thermal conductivity is analogous to specific modulus and specific strength.

In addition to thermal conductivity and density, CTE is also of great significance in many applications—for example, semiconductors and ceramic substrates used in electronics are brittle materials with coefficients of expansion in the range of about 3–7 ppm/K. Semiconductors and ceramic substrates are typically attached to supporting components, such as packages, printed circuit boards (PCBs), and heat sinks with solder or an adhesive. If the CTE of the supporting material is significantly different from that of the ceramic or semiconductor, thermal stresses arise when the assembly is subjected to a change in temperature. These stresses can result in failure of the components or the joint between them.

A great advantage of composites is that there are an increasing number of material systems that combine high thermal conductivity with tailorable CTE, low density, and excellent mechanical properties. Composites can truly be called multifunctional materials.

The key composite materials of interest for thermal control are PMCs, MMCs, and CCCs reinforced with UHK carbon fibers, which are made from pitch; silicon carbide particle-reinforced aluminum; beryllium oxide particle-reinforced beryllium; and diamond particle-reinforced aluminum and copper. There also are a number of other special CCCs developed specifically for thermal control applications.

Table 11 presents physical properties of a variety of unidirectional composites reinforced with UHK carbon fibers, along with those of monolithic copper and 6063 aluminum for comparison. Unidirectional composites are useful for directing heat in a particular direction. The particular fibers represented have a nominal axial thermal conductivity of 1100 W/m·K. Predicted properties are shown for four matrices: epoxy, aluminum, copper, and carbon. Typical reinforcement volume fractions (V/O) are assumed. As Table 11 shows, the specific axial thermal conductivities of the composites are significantly greater than those of aluminum and copper.

Figure 8 presents thermal conductivity as a function of CTE for various materials used in electronic packaging. Materials shown include silicon (Si) and gallium arsenide (GaAs) semiconductors; alumina (Al_2O_3), beryllium oxide (BeO), and aluminum nitride (AlN) ceramic substrates; and monolithic aluminum, beryllium, copper, silver, and Kovar, a nickel–iron alloy. Other monolithic materials included are diamond and pyrolitic graphite, which

Table 11 Physical Properties of Selected Unidirectional Composites and Monolithic Metals

Matrix	Reinforcement	V/O (%)	Density [g/cm^3 (Pci)]	Axial Coefficient of Thermal Expansion [ppm/K (ppm/°F)]	Axial Thermal Conductivity [W/m·K (Btu/h·ft·°F)]	Transverse Thermal Conductivity [W/m·K (Btu/h·ft·°F)]	Specific Axial Thermal Conductivity [W/m·K (Btu/h·ft·°F)]
Aluminum (6063)	—	—	2.7 (0.098)	23 (13)	218 (126)	218 (126)	81
Copper	—	—	8.9 (0.32)	17 (9.8)	400 (230)	400 (230)	45
Epoxy	UHK carbon fibers	60	1.8 (0.065)	−1.2 (−0.7)	660 (380)	2 (1.1)	370
Aluminum	UHK carbon fibers	50	2.45 (0.088)	−0.5 (−0.3)	660 (380)	50 (29)	110
Copper	UHK carbon fibers	50	5.55 (0.20)	−0.5 (−0.3)	745 (430)	140 (81)	130
Carbon	UHK carbon fibers	40	1.85 (0.067)	−1.5 (−0.8)	740 (430)	45 (26)	400

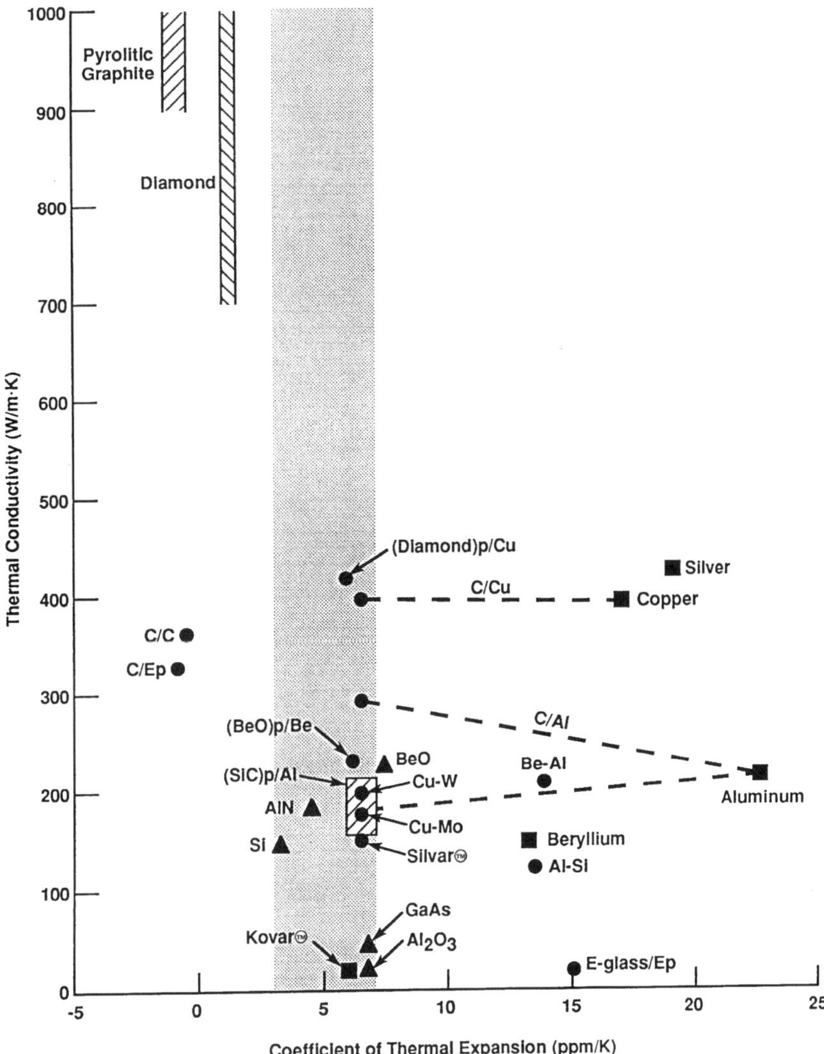

Figure 8 Thermal conductivity as a function of coefficient of thermal expansion for selected monolithic materials and composites used in electronic packaging.

have very high thermal conductivities in some forms. The figure also presents metal–metal composites, such as copper–tungsten (Cu–W), copper–molybdenum (Cu–Mo), beryllium–aluminum (Be–Al), aluminum–silicon (Al–Si), and Silvar, which contains silver and a nickel–iron alloy. The latter materials can be considered composites rather than true alloys because the two components have low solubility and appear as distinct phases at room temperature.

As Figure 8 shows, aluminum, copper, and silver have relatively high thermal conductivities but have CTEs much greater than desirable for most electronic packaging applications. By combining these metals with various reinforcements, it is possible to create new materials having CTEs isotropic in two dimensions (quasi-isotropic) or three dimensions in the desired range. The figure shows a number of composites: copper reinforced with UHK carbon fibers (C/Cu), aluminum reinforced with UHK carbon fibers (C/Al), carbon reinforced with UHK carbon fibers (C/C), epoxy reinforced with UHK carbon fibers (C/Ep), aluminum reinforced with silicon carbide particles [(SiC)p/Al], beryllium oxide particle-reinforced beryllium [(BeO)p/Be], diamond particle-reinforced copper [(Diamond)p/Cu], and E-glass fiber-reinforced epoxy (E-glass/Ep). With the exception of E-glass/Ep, C/Ep, and C/C, all of the composites have some configurations with CTEs in the desired range. The thermal conductivities of the composites presented are generally similar to, or better than, that of aluminum, while their CTEs are much closer to the goal range of 3–7 ppm/K. E-glass/Ep is an exception.

Note that although the CTEs of C/Ep and C/C are lower than desired for electronic packaging applications, the differences between their CTEs and those of ceramics and semiconductors are much less than the differences for aluminum and copper. Consequently, use of the composites can result in lower thermal stresses for a given temperature change.

The physical properties of the materials shown in Figure 8 and others are presented in Table 12.

The advantages of composites are even greater than those of conventional packaging materials when weight is considered. Figure 9 presents the specific thermal conductivities and CTEs of the materials appearing in Figure 8. Here, we find order-of-magnitude improvements. As discussed earlier, when a critical property is increased by an order of magnitude, it tends to have a revolutionary effect on technology. Several composites demonstrate this level of improvement; as a result, composites are being used in an increasing number of electronic packaging and thermal control applications.

Physical Properties of Polymer Matrix Composites

Table 13 presents physical properties of the polymer matrix composites discussed in Section 3.1. A fiber volume fraction of 60% is assumed. The densities of all of the materials are considerably lower than that of aluminum, and some are lower than that of magnesium. This reflects the low densities of both fibers and matrix materials. The low densities of most polymers give PMCs a significant advantage over most MMCs and CMCs, all other things being equal.

We observe that all of the composites have relatively low-axial CTEs. This results from the combination of low-fiber-axial CTE, high fiber stiffness, and low matrix stiffness. Note that the axial CTEs of PMCs reinforced with aramid fibers and some carbon fibers are negative. This means that, contrary to the general behavior of most monolithic materials, they contract when heated. The transverse CTEs of the composites are all positive and much larger than the magnitudes of the corresponding axial CTEs. This results from the high CTE of the matrix and a Poisson effect caused by constraint of the matrix in the axial direction and lack of constraint in the transverse direction. The transverse CTE of aramid composites is particularly high because the fibers have a relatively high positive radial CTE.

The axial thermal conductivities of composites reinforced with glass, aramid, boron, and a number of the carbon fibers are relatively low. In fact, E-glass and aramid PMCs are often used as thermal insulators. As Table 13 shows, most PMCs have relatively high thermal resistivities in the transverse direction as a result of the low thermal conductivities of the matrix and the fibers in the radial direction. Through-thickness conductivities of laminates tend to be similar to the transverse thermal conductivities of unidirectional composites.

Table 14 shows the inplane thermal conductivities and CTEs of quasi-isotropic laminates made from the same material as in Table 13. Here, again, a fiber volume fraction of 60% is assumed.

Table 12 Physical Properties of Isotropic and Quasi-Isotropic Composites and Monolithic Materials Used in Electronic Packaging

Matrix	Reinforcement	V/O (%)	Density [g/cm^3 (Pci)]	Coefficient of Thermal Expansion [ppm/K (ppm/°F)]	Thermal Conductivity [W/m·K (Btu/h·ft·°F)]	Specific Thermal Conductivity (W/m·K)
Aluminum (6063)	—	—	2.7 (0.098)	23 (13)	218 (126)	81
Copper	—	—	8.9 (0.32)	17 (9.8)	400 (230)	45
Beryllium	—	—	1.86 (0.067)	13 (7.2)	150 (87)	81
Magnesium	—	—	1.80 (0.065)	25 (14)	54 (31)	12
Titanium	—	—	4.4 (0.16)	9.5 (5.3)	16 (9.5)	4
Stainless steel (304)	—	—	8.0 (0.29)	17 (9.6)	16 (9.4)	2
Molybdenum	—	—	10.2 (0.37)	5.0 (2.8)	140 (80)	14
Tungsten	—	—	19.3 (0.695)	4.5 (2.5)	180 (104)	9
Invar	—	—	8.0 (0.29)	1.6 (0.9)	10 (6)	1
Kovar	—	—	8.3 (0.30)	5.9 (3.2)	17 (10)	2
Alumina (99% pure)	—	—	3.9 (0.141)	6.7 (3.7)	20 (12)	5
Beryllia	—	—	2.9 (0.105)	6.7 (3.7)	250 (145)	86
Aluminum nitride	—	—	3.2 (0.116)	4.5 (2.5)	250 (145)	78
Silicon	—	—	2.3 (0.084)	4.1 (2.3)	150 (87)	65
Gallium arsenide	—	—	5.3 (0.19)	5.8 (3.2)	44 (25)	8
Diamond	—	—	3.5 (0.13)	1.0 (0.6)	2000 (1160)	570
Pyrolitic graphite	—	—	2.3 (0.083)	−1 (−0.6)	1700 (980)	750
Aluminum–silicon	—	—	2.5 (0.091)	13.5 (7.5)	126 (73)	50
Beryllium–aluminum	—	—	2.1 (0.076)	13.9 (7.7)	210 (121)	100
Copper–tungsten (10/90)	—	—	17 (0.61)	6.5 (3.6)	209 (121)	12
Copper–molybdenum (15/85)	—	—	10 (0.36)	6.6 (3.7)	184 (106)	18
Aluminum	SiC particles	70	3.0 (0.108)	6.5 (3.6)	190 (110)	63
Beryllium	BeO particles	60	2.6 (0.094)	6.1 (3.4)	240 (139)	92
Copper	Diamond particles	55	5.9 (0.21)	5.8 (3.2)	420 (243)	71
Epoxy	UHK carbon fibers	60	1.8 (0.065)	−0.7 (−0.4)	330 (191)	183
Aluminum	UHK carbon fibers	26	2.6 (0.094)	6.5 (3.6)	290 (168)	112
Copper	UHK carbon fibers	26	7.2 (0.26)	6.5 (3.6)	400 (230)	56
Carbon	UHK carbon fibers	40	1.8 (0.065)	−1 (−0.6)	360 (208)	195

We observe that the CTEs of the quasi-isotropic composites are higher than the axial values of corresponding unidirectional composites. Note, however, that the CTEs of quasi-isotropic composites reinforced with aramid and carbon fibers are very small. By appropriate selection of fiber, matrix, and fiber volume fraction, it is possible to obtain quasi-isotropic materials with CTEs very close to zero. Note that through-thickness CTEs for these laminates typically will be positive and relatively large. However, this is not a significant issue for many applications.

Turning to thermal conductivity, we find that quasi-isotropic laminates reinforced with UHM pitch carbon fibers have an inplane thermal conductivity similar to that of aluminum alloys, while UHK pitch carbon fibers provide laminates with a conductivity more than 50% higher. Both materials have densities 35% lower than aluminum.

As mentioned above, through-thickness thermal conductivities of laminates tend to be similar to the transverse thermal conductivities of unidirectional composites, which are relatively low. However, if laminate thickness is small, this may not be a significant limitation.

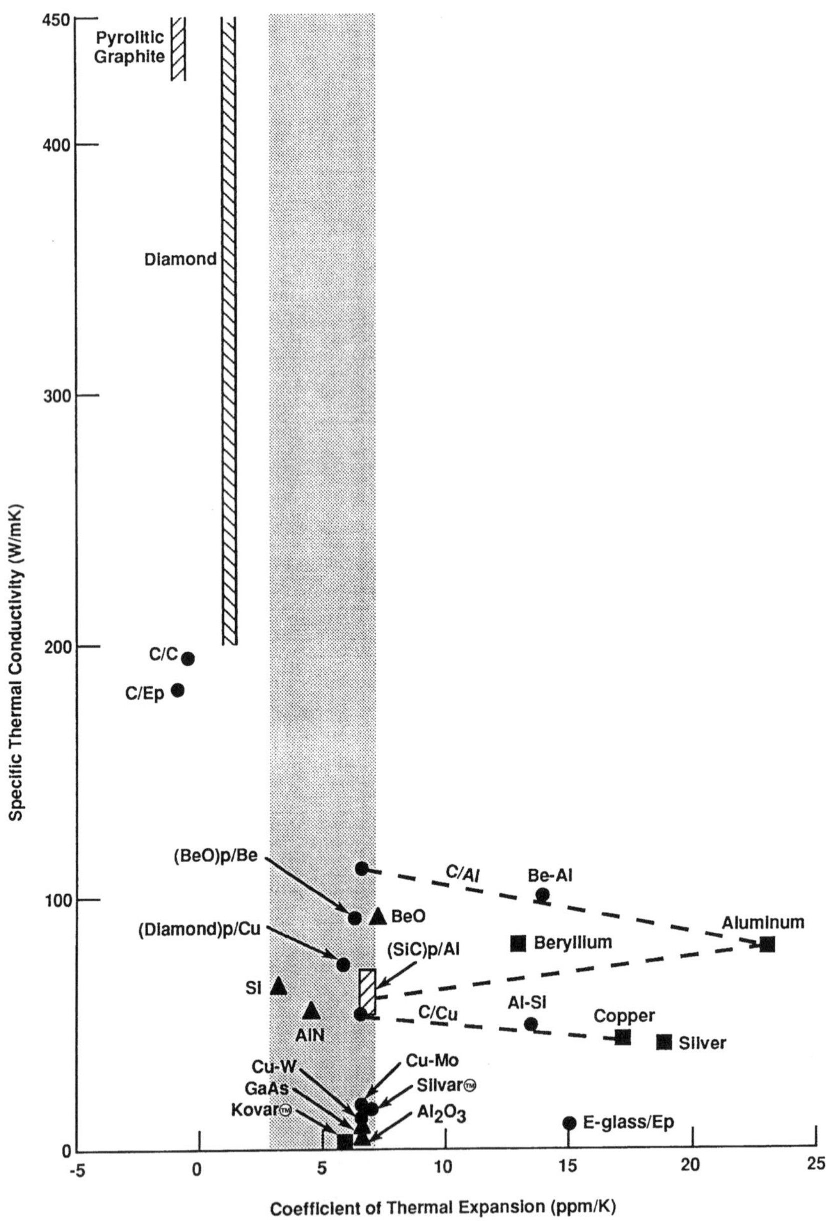

Figure 9 Specific thermal conductivity (thermal conductivity divided by specific gravity) for selected monolithic materials and composites used in electronic packaging.

Table 13 Physical Properties of Selected Unidirectional Polymer Matrix Composites

Fiber	Density [g/cm^3 (Pci)]	Axial CTE [10^{-6}/K (10^{-6}/°F)]	Transverse CTE [10^{-6}/K (10^{-6}/°F)]	Axial Thermal Conductivity [W/m·K (Btu/h·ft·°F)]	Transverse Thermal Conductivity [W/m·K (Btu/h·ft·°F)]
E-glass	2.1 (0.075)	6.3 (3.5)	22 (12)	1.2 (0.7)	0.6 (0.3)
Aramid	1.38 (0.050)	−4.0 (−2.2)	58 (32)	1.7 (1.0)	0.1 (0.08)
Boron	2.0 (0.073)	4.5 (2.5)	23 (13)	2.2 (1.3)	0.7 (0.4)
SM carbon (PAN)	1.58 (0.057)	0.9 (0.5)	27 (15)	5 (3)	0.5 (0.3)
UHS carbon (PAN)	1.61 (0.058)	0.5 (0.3)	27 (15)	10 (6)	0.5 (0.3)
UHM carbon (PAN)	1.66 (0.060)	−0.9 (−0.5)	40 (22)	45 (26)	0.5 (0.3)
UHM carbon (pitch)	1.80 (0.065)	−1.1 (−0.6)	27 (15)	380 (220)	10 (6)
UHK carbon (pitch)	1.80 (0.065)	−1.1 (−0.6)	27 (15)	660 (380)	10 (6)

Table 14 Physical Properties of Selected Quasi-Isotropic Polymer Matrix Composites

Fiber	Density [g/cm^3 (Pci)]	Axial CTE [10^{-6}/K (10^{-6}/°F)]	Transverse CTE [10^{-6}/K (10^{-6}/°F)]	Axial Thermal Conductivity [W/m·K (Btu/h·ft·°F)]	Transverse Thermal Conductivity [W/m·K (Btu/h·ft·°F)]
E-glass	2.1 (0.075)	10 (5.6)	10 (5.6)	0.9 (0.5)	0.9 (0.5)
Aramid	1.38 (0.050)	1.4 (0.8)	1.4 (0.8)	0.9 (0.5)	0.9 (0.5)
Boron	2.0 (0.073)	6.5 (3.6)	6.5 (3.6)	1.4 (0.8)	1.4 (0.8)
SM carbon (PAN)	1.58 (0.057)	3.1 (1.7)	3.1 (1.7)	2.8 (1.6)	2.8 (1.6)
UHS carbon (PAN)	1.61 (0.058)	2.3 (1.3)	2.3 (1.3)	6 (3)	6 (3)
UHM carbon (PAN)	1.66 (0.060)	0.4 (0.2)	0.4 (0.2)	23 (13)	23 (13)
UHM carbon (pitch)	1.80 (0.065)	−0.4 (−0.2)	−0.4 (−0.2)	195 (113)	195 (113)
UHK carbon (pitch)	1.80 (0.065)	−0.4 (−0.2)	−0.4 (−0.2)	335 (195)	335 (195)

Physical Properties of Metal Matrix Composites

In this section, we consider physical properties of selected unidirectional fiber-reinforced MMCs and of silicon carbide particle-reinforced aluminum MMCs.

Physical Properties of Continuous Fiber-Reinforced Metal Matrix Composites. Table 11 presents physical properties of unidirectional composites consisting of UHK pitch carbon fibers in aluminum and copper matrices. These materials both have very low, slightly negative axial CTEs for the assumed fiber volume fraction of 50%. As the table shows, the axial thermal conductivities for MMCs with aluminum and copper matrices are substantially greater than that of monolithic copper. A major advantage of having thermally conductive matrix materials is that the resulting composite transverse and through-thickness thermal conductivities are more than an order of magnitude higher than those of an epoxy–matrix composite.

Table 12 presents the properties of quasi-isotropic composites composed of aluminum and copper matrices reinforced with UHK pitch carbon fibers. Here, the fiber volume fraction of about 26% has been chosen to achieve an inplane CTE similar to that of aluminum oxide, 6.5 ppm/K (3.6 ppm/°F). The inplane thermal conductivities of the aluminum– and copper–matrix composites are 290 W/m·K (168 Btu/h·ft·°F) and 400 W/m·K (230 Btu/h·ft·°F), respectively. These values are considerably greater than those of any other material with a similar CTE, with the exception of diamond particle-reinforced copper, which is discussed later. Because of the lower fiber volume fractions, the through-thickness thermal conductivities of these composites will be somewhat higher than those of the unidirectional composites presented in Table 11.

Physical Properties of Particle-Reinforced Metal Matrix Composites. In this section, we consider the physical properties of silicon carbide particle-reinforced aluminum and diamond particle-reinforced copper.

PHYSICAL PROPERTIES OF SILICON CARBIDE PARTICLE-REINFORCED METAL MATRIX COMPOSITES. The physical properties of particle-reinforced composites tend to be isotropic (in three dimensions). As for all composites, the physical properties of silicon carbide particle-reinforced aluminum depend on constituent properties and reinforcement volume fraction. Figure 4 shows how the CTE of (SiC)p/Al varies with particle volume fraction for typical commercial materials. Table 15 presents density and CTE for several specific volume fractions, along with data for monolithic aluminum, titanium, and steel.

Thermal conductivity depends strongly on the corresponding properties of the matrix, reinforcement, and particle volume fraction. The thermal conductivity of very pure silicon carbide is slightly higher than that of copper. However, those of commercial particles are much lower. The thermal conductivities of silicon carbide particle-reinforced aluminum used

Table 15 Physical Properties of Silicon Carbide Particle-Reinforced Aluminum

Property	Aluminum (6061-T6)	Titanium (6Al–4V)	Steel (4340)	Composite Particle Volume Fraction 25	55	70
CTE, 10^{-6}/K (10^{-6}/°F)	23 (13)	9.5 (5.3)	12 (6.6)	16.4 (9.1)	10.4 (5.8)	6.2 (3.4)
Thermal Conductivity, W/m·K (Btu/h·ft·°F)	218 (126)	16 (9.5)	17 (9.4)	160–220 (92–126)	160–220 (92–126)	160–220 (92–126)
Density, g/cm^3 (Pci)	2.77 (0.10)	4.43 (0.16)	7.76 (0.28)	2.88 (0.104)	2.96 (0.107)	3.00 (0.108)

in electronic packaging applications tend to be in the range of monolithic aluminum alloys, about 160–218 W/m·K (92–126 Btu/h·ft·°F).

PHYSICAL PROPERTIES OF DIAMOND PARTICLE-REINFORCED COPPER METAL MATRIX COMPOSITES. Table 12 presents the physical properties of diamond particle-reinforced copper composites, which are developmental materials. As for other particle-reinforced composites, the properties can be expected to be relatively isotropic. This material has a thermal conductivity somewhat higher than that of monolithic copper, a much lower density, and a CTE in the range of semiconductors and ceramic substrates. This unique combination of properties makes this composite an attractive candidate for electronic packaging applications.

Physical Properties of Ceramic Matrix Composites

As discussed in Section 1, there are many CMCs and they are at various stages of development. One of the more mature systems is silicon carbide fiber-reinforced silicon carbide (SiC/SiC). For a fiber-reinforced composite with a fiber volume fraction of 40%, the density is 2.5 g/cm^3 (0.090 Pci), the CTE is 3 ppm/K (1.7 ppm/°F), the inplane thermal conductivity is 19 W/m·K (11 Btu/h·ft·°F), and the through-thickness value is 9.5 W/m·K (5.5 Btu/h·ft·°F).

Physical Properties of Carbon/Carbon Composites

The CTE of CCCs depends on fiber type, volume fraction, and geometry and matrix characteristics. In the fiber direction, CTE tends to be negative with a small absolute value. Perpendicular to the fiber direction, composite CTE is dominated by matrix properties. As a rule, the magnitude of transverse CTE is small. Both positive and negative values have been reported.

It is well known that in some forms carbon has exceptionally high thermal conductivities. For example, pyrolitic graphite can have a thermal conductivity as high as 2000 W/m·K (1160 Btu/h·ft·°F), five times that of copper. The conductivity of some types of diamond is much higher. Some CCCs also have very high thermal conductivities. Values have been reported as high as 400 W/m·K (230 Btu/h·ft·°F) for quasi-isotropic composites and 700 W/m·K (400 Btu/h·ft·°F) for unidirectional materials.

REFERENCES

1. A. Kelly (ed.), *Concise Encyclopedia of Composite Materials,* rev. ed., Pergamon, Oxford, 1994.
2. Z. L. H. Miner, R. A. Wolffe, and C. Zweben, "Fatigue, Creep and Impact Resistance of Kevlar® 49 Reinforced Composites," in *Composite Reliability,* ASTM STP 580, American Society for Testing and Materials, West Conshohocken, PA, 1975.
3. C. Zweben, "Overview of Metal Matrix Composites for Electronic Packaging and Thermal Management," *JOM,* **44**(7), 15–23 (July 1992).
4. A. F. Johnson, "Glass-Reinforced Plastics: Thermosetting Resins," in *Concise Encyclopedia of Composite Materials,* A. Kelly (ed.), rev. ed., Pergamon, Oxford, 1994.
5. J. Halpin, Lecture Notes, UCLA short course "Fiber Composites: Design, Evaluation, and Quality Assurance."

BIBLIOGRAPHY

Advanced Materials by Design, OTA-E-351, U.S. Congress Office of Technology Assessment, U.S. Government Printing Office, Washington, DC, June 1988.

Agarwal, B. D., and L. J. Broutman, *Analysis and Performance of Fiber Composites,* Wiley, New York, 1981.

Allen, H. G., *Analysis and Design of Structural Sandwich Panels,* Pergamon, Oxford, 1969.

Ambartsumyan, S. A., *Theory of Anisotropic Plates,* Technomic, Lancaster, PA, 1970.

Ashby, M. F., *Material Selection in Mechanical Design,* Pergamon, Oxford, 1992.

Ashton, J. E., and J. M. Whitney, *Theory of Laminated Plates,* Technomic, Lancaster, PA, 1970.

Bulletin 54, Chromalloy Metal Tectonics Company.

Calcote, L. R., *The Analysis of Laminated Composite Structures,* Van Nostrand Reinhold, New York, 1969.

Chou, T.-W., *Microstructural Design of Composite Materials,* Cambridge University Press, Cambridge, 1992.

Chou, T.-W. (ed.), *Materials Science Handbook,* Vol. 13, *Structural Properties of Composites,* VCH, Weinheim, Federal Republic of Germany, 1993.

Design Guide for Advanced Composite Applications, Advanstar Communications, Duluth, MN, 1993.

Deve, H. E., and C. McCullough, "Continuous-Fiber Reinforced Al Composites: A New Generation," *JOM,* 33–37 (July 1995).

Donomoto, T., et al., "Ceramic Fiber Reinforced Piston for High Performance Diesel Engines," SAE Technical Paper No. 830252, 1983.

Dvorak, G. J. (ed.), *Inelastic Deformation of Composite Materials,* Proceedings of the 19901 IUTAM Symposium, Springer, New York, 1991.

Engineered Materials Handbook, Vol. 1, *Composites,* American Society for Metals, Materials Park, OH, 1987.

Fisher, K., "Industrial Applications," *High-Performance Composites,* 46–49 (May/June 1995).

Grimes, G. G., et al., "Tape Composite Material Allowables Application in Airframe Design/Analysis," *Composites Eng.* **3**(7/8), 777–804 (1993).

Halpin, J. C., *Primer on Composite Materials: Analysis,* Technomic, Lancaster, PA, 1984.

Hayashi, T., H. Ushio, and M. Ebisawa, "The Properties of Hybrid Fiber Reinforced Metal and Its Application for Engine Block," SAE Technical Paper No. 890557, 1989.

Hoskin, B., and A. A. Baker, *Composite Materials for Aircraft Structures,* American Institute of Aeronautics and Astronautics, New York, 1986.

Hull, D., *An Introduction to Composite Materials,* Cambridge University Press, Cambridge, 1981.

Jones, R. M., *Mechanics of Composite Materials,* McGraw-Hill, New York, 1975.

Kedward, K., "Designing with Composites," Lecture Notes for Short Course "Composite Materials: Selection, Design and Manufacture for Engineering Applications," UCLA Extension, University of California, Los Angeles, 1996.

Kennedy, C. R., "Reinforced Ceramics Via Oxidation of Molten Metals," *Ceramic Industry,* 26–30 (Dec. 1994).

Kerns, J. A., et al., "Dymalloy, A Composite Substrate for High Power Density Electronic Components," in *Proceedings of the 1995 International Symposium on Microelectronics,* Los Angeles, CA, 1995.

Kliger, H. S., and E. R. Barker, "A Comparative Study of the Corrosion Resistance of Carbon and Glass Fibers," in 39th Annual Conference, Reinforced Plastics/Composites Institute, The Society of the Plastics Industry, Inc., January 1984.

Ko, F. K., "Three-Dimensional Fabrics for Composites," in *Textile Structural Composites,* T.-W. Chou and F. K. Ko (eds.), Elsevier Science, Amsterdam, 1989, pp. 129–171.

Ko, F. K., "Advanced Textile Structural Composites," in *Advanced Topics in Materials Science and Engineering,* J. I. Moran-Lopez and J. M. Sanchez (eds.), Plenum, New York, 1993.

Kulkarni, S. V., and C. Zweben (eds.), *Composites in Pressure Vessels and Piping,* American Society of Mechanical Engineering, New York, 1977.

Kwarteng K., and C. Stark, "Carbon Fiber Reinforced PEEK (APC-2/AS-4) Composites for Orthopaedic Implants," *SAMPE Quarterly,* 10–14 (Oct. 1990).

Lekhnitskii, S. G., *Anisotropic Plates,* Gordon & Breach, Philadelphia, PA, 1968.

Mallick, P. K., *Fiber-Reinforced Composites: Materials, Manufacturing and Design,* 2nd ed., Marcel Dekker, New York, 1993.

Marshall, A. C., *Composite Basics,* 4th ed., Marshall Consulting, Walnut Creek, CA, 1994.

Marshall, D. B., and A. G. Evans, "Failure Mechanisms in Ceramic-Fiber/Ceramic-Matrix Composites," *J. Am. Ceramic Soc.* **68**(5), 225–231 (May 1985).

McConnell, V. P., "Industrial Applications," *Adv. Composites,* 31–38 (Mar./Apr. 1992).

McConnell, V. P., "Fail-Safe Ceramics," *High-Performance Composites,* 27–31 (Mar./Apr. 1996).

Meyers, M. A., and O. T. Inal (eds.), *Frontiers in Materials Technologies,* Materials Science Monographs, 26, Elsevier, Oxford, 1985.

Military Handbook—5F, *Metallic Materials and Elements for Aerospace Vehicle Structures,* U.S. Department of Defense, Washington, DC, Dec. 1992.

Morrell, R., *Handbook of Properties of Technical and Engineering Ceramics,* Part 1: An Introduction for the Engineer and Designer, Her Majesty's Stationery Office, London, 1985.

Norman, J. C., and C. Zweben, "Kevlar® 49/Thornel® 300 Hybrid Fabric Composites for Aerospace Applications," *SAMPE Quarterly* **7**(4), 1–10 (July 1976).

Pfeifer, W. H., et al., "High Conductivity Carbon-Carbon Composites for SEM-E Heat Sinks," in Sixth International SAMPE Electronic Materials and Processes Conference, Baltimore, MD, June 22–25, 1992.

Premkumar, M. K., W. H. Hunt, Jr., and R. R. Sawtell, "Aluminum Composite Materials for Multichip Modules," *JOM,* 22–28 (July 1992).

"Properties of SiC/SiC Laminates (0/90 Fabric Layup)," Preliminary Engineering Data, Publication H-28488, Du Pont Composites.

Rawal, S. P., M. S. Misra, and R. G. Wendt, "Composite Materials for Space Applications," NASA CR-187472, National Aeronautics and Space Administration, Hampton, VA, 1990.

Savage, G., *Carbon–Carbon Composites,* Chapman and Hall, London, 1993.

Schmidt, K. A., and C. Zweben, "Advanced Composite Packaging Materials," in *Electronic Materials Handbook,* Vol. 1, *Packaging,* American Society for Metals, Materials Park, OH, 1989.

Schwartz, M. M., *Fabrication of Composite Materials,* American Society for Metals, Metals Park, OH, 1985.

Shih, W. T., F. H. Ho, and B. B. Burkett, "Carbon–Carbon (C–C) Composites for Thermal Plane Applications," in Seventh International SAMPE Electronics Conference, June 20–23, 1994, Parsippany, NJ.

Shtessel, V. E., and M. J. Koczak, "The Production of Metal Matrix Composites by Reactive Processes," *Mater. Technol.* **9**(7/8), 154–158 (1994).

Smith, D. L., K. E. Davidson, and L. S. Thiebert, "Carbon–Carbon Composites (CCC): A Historical Perspective," in *Proceedings, 41st International SAMPE Symposium,* March 24–28, 1996.

Smith, W. S., M. W. Wardle, and C. Zweben, "Test Methods for Fiber Tensile Strength, Composite Flexural Modulus and Properties of Fabric-Reinforced Laminates," in *Composite Materials: Testing and Design* (*Fifth Conference*), ASTM STP 674, S. W. Tsai (ed.), American Society for Testing and Materials, West Conshohocken, PA, 1979, pp. 228–262.

Strong, A. B., *Fundamentals of Composites Manufacturing: Materials, Methods, and Applications,* Society of Manufacturing Engineers, Dearborn, MI, 1989.

"Top Twenty Awards," *Mater. Eng.,* 28–33 (Nov. 1985).

Tsai, S. W., *Composites Design,* 4th ed., Think Composites, Dayton, OH, 1988.

Tsai, S. W., and H. T. Hahn, *Introduction to Composite Materials,* Technomic, Lancaster, PA, 1980.

Vasiliev, V. V., *Mechanics of Composite Structures,* R. M. Jones, English Edition Editor, Taylor and Francis, Washington, DC, 1988.

Vinson, J. R., and R. L. Sierakowski, *The Behavior of Structures Composed of Composite Materials,* Kluwer Academic, Dordrecht, 1993.

Warren, R. (ed.), *Ceramic–Matrix Composites,* Chapman & Hall, New York, 1992.

Zweben, C., "Advanced Composite Materials for Process Industries and Corrosion Resistant Applications," in *Proceedings of the Conference on Advances in Materials Technology for Process Industries' Needs,* National Association of Corrosion Engineers, Houston, TX, 1985, pp. 163–172.

Zweben, C., "Tensile Strength of Fiber-Reinforced Composites: Basic Concepts and Recent Developments," in *Composite Materials: Testing and Design,* ASTM STP 460, American Society for Testing and Materials, West Conshohocken, PA, 1970, pp. 528–539.

Zweben, C., "Advanced Composites—A Revolution for the Designer," in AIAA 50th Anniversary Annual Meeting and Technical Display, "Learn from the Masters" Lecture Series, Paper No. 81-0894, Long Beach, May 1981.

Zweben, C., "Thermomechanical Properties of Fibrous Composite Materials: Theory," in *Encyclopedia of Materials Science and Engineering,* M. B. Bever (ed.), Pergamon, Oxford, 1986.

Zweben, C., "Mechanical and Thermal Properties of Silicon Carbide Particle-Reinforced Aluminum," in *Thermal and Mechanical Behavior of Metal Matrix and Ceramic Matrix Composites,* ASTM STP 1080, L. M. Kennedy, H. H. Moeller, and W. S. Johnson (eds.), American Society for Testing and Materials, West Conshohocken, PA, 1989.

Zweben, C., "Is There a Size Effect in Composite Materials and Structures?" *Composites,* **25**, 451–454 (1994a).

Zweben, C., "Metal Matrix Composites: Aerospace Applications," in *Encyclopedia of Advanced Materials,* M. C. Flemings et al. (eds.), Pergamon, Oxford, 1994b.

Zweben, C., "Simple, Design-Oriented Composite Failure Criteria Incorporating Size Effects," in *Proceedings Tenth International Conference on Composite Materials,* ICCM-10, Whistler, BC, Canada, Aug. 1995a.

Zweben, C., "The Future of Advanced Composite Electronic Packaging," in *Materials for Electronic Packaging,* D. D. L. Chung (ed.), Butterworth-Heinemann, Oxford, 1995b.

CHAPTER 11

SMART MATERIALS

James A. Harvey
Under the Bridge Consulting, Inc.
Corvallis, Oregon

1 INTRODUCTION

The world has undergone two materials ages, the plastics age and the composite age, during the past centuries. In the midst of these two ages a new era has developed. This is the smart materials era. According to early definitions, smart materials are materials that respond to their environments in a timely manner.[1–4]

The definition of smart materials has been expanded to materials that receive, transmit, or process a stimulus and respond by producing a useful effect that may include a signal that the materials are acting upon it. Some of the stimuli that may act upon these materials are strain, stress, temperature, chemicals (including pH stimuli), electric field, magnetic field, hydrostatic pressure, different types of radiation, and other forms of stimuli.[5]

The effect can be caused by absorption of a proton, a chemical reaction, integration of a series of events, translation or rotation of segments within the molecular structure, creation and motion of crystallographic defects or other localized conformations, alteration of localized stress and strain fields, and others. The effects produced can be a color change, a change in index of refraction, a change in the distribution of stresses and strains, or a volume change.[5]

Another important criterion for a material to be considered smart is that the action of receiving and responding to stimuli to produce a useful effect must be reversible. Another important factor in determining if a material is smart pertains to its asymmetrical nature. This is primarily critical for piezoelectric materials. Other types of smart materials exhibit this trait. However, little research has been performed to verify this observation.

Also, it should be pointed out that the word "intelligent" is used to describe smart materials. The notation "smart" has been overused as a means to market materials and products.

From the purist point of view, materials are smart if at some point within their performance history their reaction to a stimulus is reversible. Materials that formally have the label of being smart include piezoelectric materials, electrostrictive materials, electrorheological materials, magnetorheological materials, thermoresponsive materials, pH-sensitive materials, UV-sensitive materials, smart polymers, smart gels (hydrogels), smart catalysts, and shape memory alloys. In this treatment of the subject we will be using some of these classifications; in some cases, however, the classification of a particular material may appear to be in error. This will be done to illustrate the rapid growth of the field of smart materials and the rediscovery of the smart behavior of materials known for centuries. As we continue to better understand smart materials, our definitions will change. In each material section there will be discussions pertaining to the material definition, types of materials that belong to that class, properties of the members, and applications of the materials. In some cases a more detailed discussion of application will be given to both illustrate the benefit of these materials and simulate the use of these materials in new applications.

Another important feature of smart materials is their inclusion in smart structures, which are simply structures with at least one smart material incorporated within its structure and from the effect of the smart material causes an action. A smart structure may have sensors (nerves), actuators (muscles), and a control (brain). Thus, the term *biomimetric* is associated with smart structures. Smart structures are being designed to make our life more productive and easy. With the number of sensors, actuators, and control systems available, coupled with the materials and the genius of scientists and engineers, these structures are becoming more commonplace. Reading nontechnical magazines, watching television, and going to stores can verify this.

In a comical manner the growth of smart structures can be illustrated by an article that appeared in the December 14, 2000, edition of the *New York Times.* This article describes the "Big Mouth Billy Bass," the singing fish, and how smart this novel toy is.

Examples of technical applications of smart structures are composite materials embedded with fiber optics, actuators, sensors, microelectromechanical systems (MEMSs), vibration control, sound control, shape control, product health or lifetime monitoring, cure monitoring, intelligent processing, active and passive controls, self-repair (healing), artificial organs, novel indicating devices, designed magnets, damping aeroelastic stability, and stress distributions. Smart structures are found in automobiles, space systems, fixed- and rotary-wing aircrafts, naval vessels, civil structures, machine tools, recreation, and medical devices.[5,6]

Another important feature related to smart materials and structures is that they encompass all fields of science and engineering. When searching for information on smart materials and structures, there are numerous sources, web sites, and professional societies that deal with this technology.

2 PIEZOELECTRIC MATERIALS

The simplest definition of piezoelectric materials can be obtained by first dividing the word into *piezo* and *electric.* Piezo is from the Greek word *piezein,* which means "to press tightly or squeeze." Combining *piezein* with electric, we have "squeeze electricity."

The history of piezoelectric materials is relatively simple and only the highlights will be presented. In 1880, Pierre and Paul-Jean Curie showed the piezoelectric effect in quartz and Rochelle salt crystals. Their first observations were made by placing weights on the faces of particular crystal cuts, like the X-cut quartz plate, detecting charges on the crystal surfaces, and demonstrating that the magnitude of the charge was proportional to the applied weight. This phenomenon has become known as the direct-pressure piezoelectric effect. In 1881 G. Lippmann predicted that a crystal such as quartz would develop a mechanical strain when an electric field was applied. In the same year the Curies reported the converse pressure piezoelectric effect with quartz and Rochelle salt. They showed that if certain crystals were subjected to mechanical strain, they became electrically polarized and the degree of polarization was proportional to the applied strain. The French inventor Langevin developed the first SONAR using quartz crystals in 1920. During the 1940s researchers discovered and developed the first polycrystal line, piezoelectric ceramic, barium titanate. A significant advantage of piezoelectric ceramics over piezoelectric crystals is their ability to be formed in a variety of shapes and sizes. In 1960, researchers discovered a weak piezoelectric effect in whalebone and tendon that led to intense search for other piezoelectric organic materials. And in 1969, Kawai found very high piezoelectric activity in polarized polyvinylidene fluoride (PVDF).[7,8]

The piezoelectric effect exists in a number of naturally occurring crystals, such as quartz, tourmaline, and sodium potassium tartrate. For a crystal to exhibit the piezoelectric effect, it must not have a center of symmetry. When a stress (tensile or compressive) is applied to a crystal with a center of symmetry, it will alter the spacing between the positive and negative sites in each elementary cell unit, thus causing a net polarization at the crystal surface. The effect is approximately linear. The polarization is directly related to the applied stress. It is direction dependent. Thus, compressive and tensile forces will generate electric fields and voltages of opposite polarity. The effect is also reciprocal; thus, when the crystal is exposed to an electric field, it undergoes an elastic strain changing its length based upon the field polarity.

As previously mentioned, the ceramic-type materials play an important role in the area of piezoelectrics. Piezoelectric ceramics are polycrystalline ferroelectric materials with a perovskite-crystal-type structure. The crystal structure is tetragonal/rhombohedral with a close proximity to cubic in nature. Piezoelectric ceramics have the general formula of $A^{2+}B^{4+}O_3^{2-}$, where A represents a large divalent metal ion such as barium or lead and B is one or more tetravalent metal ions such as titanium, zirconium, or manganese. These ceramics are considered to be masses of minute crystallites that change crystal forms at the Curie temperature. Above the Curie temperature, the ceramic crystallites have a simple cubic symmetry. This form is centrosymmetric with positive and negative charge sites coinciding; thus there are no dipoles present. The material is considered to be paraelectric. Below the Curie temperature, the ceramic crystallites have a tetragonal symmetry; this form lacks a center of symmetry with the positive and negative charge sites no longer coinciding, thus each unit cell has an electric dipole whose direction may be reversed and switched by the application of an electric field. The material is now considered to be ferroelectric.

The piezoelectric properties of ceramics can be enhanced by applying a large electric field at an elevated temperature, thus generating an internal remnant polarization that continues long after the removal of the electric and thermal fields. This technique is known as poling. The poling of piezoelectric ceramics has eliminated the use of piezoelectric crystals in many applications.

The vinylidene fluoride monomer $CH_2{=}CF$ yields the semicrystalline polymer (PVDF) upon polymerization. This polymer was found to be highly piezoelectric. Polyvinylidene fluoride is manufactured in sheet form from the nonpolar α-phase film extruded from the melt. The extruded film is uniaxially stretched. This process rotates the long polymer chains and forms the polar β-phase. The β-phase is needed for the high piezoelectricity. The final step consists of reorientation of the randomly directed dipoles associated with the stretched β-phase by applying a poling field in the direction normal to the plane of the film. The resulting piezoelectric PVDF has orthotropic symmetry.

Copolymers and rubber and ceramic blends of PVDF have been prepared for use as piezoelectric materials. The most common copolymer is one based on the polymerization reaction of vinylidene fluoride and trifluoroethylene. Polyvinylidene fluoride and its copolymers with trifluoroethylene have low mechanical quality factors and high hydrostatic-mode responses and their acoustic impedance is similar to water, making them ideal for underwater hydrophone applications.[9]

A series of polyimides has been developed at NASA Langley for use in piezoelectric applications. These polyimides have pendant trifluoromethyl (–CF3) and cyano (–CN) polar groups. Whenever these polyimides are exposed to applied voltages of the order of 100 MV/m at elevated temperatures, the polar groups develop a high degree of orientation, resulting in polymer films with high piezoelectric and pyroelectric properties. The piezoelectric response for these polyimides is in the same vicinity as those of PVDF at room temperature. However, the piezoelectric response of the polyimides is greater at elevated temperatures.[10]

Multiphase piezoelectric composites have been developed for their synergetic effect between the piezoelectric activity of monolithic ceramics and the low density of nonpiezoelectric polymeric materials. One class of piezoelectric composites that has been developed is the smart tagged composites. These piezoelectric composites are PZT-5A (soft piezoelectric ceramic) particles embedded into an unsaturated polyester polymer matrix and are used for structural health monitoring. Conductive metal-filled particles with polyimide films have been explored for microelectronic applications. The use of graphite-filled polymers has shown potential as microsensors and actuators.

Piezoelectric materials have been used in thousands of applications in a wide variety of products in the consumer, industrial, medical, aerospace, and military sectors. When a piezoelectric material is subjected to a mechanical stress, an electric charge is generated across the material. The ability of a material to generate a charge or electric field when subjected to a stress is measured by the piezoelectric voltage coefficient g.

The converse effect occurs when a piezoelectric material becomes strained when placed in an electric field. Under constant-stress conditions, the general equation for the piezoelectric charge coefficient d can be expressed as the change in the strain S of the piezoelectric material as a function of the applied electric field E.

In should be denoted that piezoelectric materials are also pyroelectric. They generate an electric charge as they undergo a temperature change. Whenever their temperature is increased, a voltage develops having the same orientation as the polarization voltage. Whenever the temperature is decreased, a voltage develops with orientation opposite to the polarization voltage.[11]

Some of the applications of materials that utilize direct and converse piezoelectric effects and the applications of PVDF piezoelectric film follow. Specific applications will be discussed.

One of the uses of piezoelectric ceramics and polymers is in ink-on demand printing. Several commercial available ink-jet printers are based upon this technology.[12,13] In this application the impulse ink jet is produced using a cylindrical transducer that is tightly bound to the outer surface of a cylindrical glass nozzle with an orifice ranging from mils to micrometers in diameter. Several industrious researchers have expanded upon this technique by using a printer as a chemical delivery system for the application of doped polymers for organic light-emitting displays. Researchers from Princeton University have used a color inkjet printer based upon piezoelectric technology with a resolution of 640 dots per line and replaced the inks with polymer solutions.[14] This technique has also been applied to the manufacture of color filters for liquid crystal displays.

Another application of piezoelectric polymer film involves the work of a group of researchers at the Thiokol Company and an earlier work[15,16] to monitor adhesive joints. The Thiokol study showed that during the bonding process, the PVDF piezoelectric film sensors monitored the adhesive cure ultrasonically and qualitatively determined the presence of significant void content. The Thiokol study also indicated that normal bond stresses were quantified during cyclic loading of single lap joints and electrometric butt joints. The researchers revealed that this was a significant step toward service life predictions. However, more and better understood monitoring techniques are needed.

An interesting application of piezoelectric ceramics can be illustrated by the hundreds of vertical pinball machines popular in the Pachinko parlors of Japan. The machines are assembled with stacks of piezoelectric disks that can act as both sensors and actuators. When a ball falls on the stack, the force of impact generates a piezoelectric voltage pulse that in turn generates a response from the actuator stack through a feedback control system. The stack expands, throwing the ball out of the hole and moving it up a spiral ramp through a sequence of events and then repeating the sequence.[17]

3 ELECTROSTRICTIVE MATERIALS

Piezoelectric materials are materials that exhibit a linear relationship between electric and mechanical variables. Piezoelectricity is a third-rank tensor. Electrostrictive materials also show a relationship between these two variables. However, in this case, it is a quadratic relationship between mechanical stress and the square of electrical polarization. Electrostriction can occur in any material and is a small effect. One difference between piezoelectric and electrostrictive materials is the ability of the electrostrictive materials to show a larger effect in the vicinity of its Curie temperature. Elecrostriction is a fourth rank tensor property observed both in centric and acentric insulators. This is especially true for ferroelectric materials such as the members of the perovskite family. Ferroelectrics are ferroic solids whose domain walls have the capability of moving by external forces or fields. In addition to ferroelectrics, the other principal examples of ferroic solids are ferromagnetics and ferroelastics, both of which have potential as smart materials. Other examples of electrostrictive materials include lead manganese niobate–lead titanate (PMN–PT) and lead lanthanium zirconate titanate (PLZT).

An interesting application of electrostrictive materials is in active optical applications. During the Cold War, the satellites that flew over the Soviet Union used active optical systems to eliminate atmospheric turbulence effects. Electrostrictive materials have the advantage over piezoelectric materials of being able to adjust the position of optical components due to the reduced hysteresis associated with the motion. Work on active optical systems has continued. Similar multilayer actuators were used to correct the position of the optical elements in the Hubble telescope. Supermarket scanners use actuators and flexible mirrors to read bar codes optically.[17]

Other examples of this family may be included in the electroactive polymers. This name has appeared in the technical literature for approximately a decade with an increasing interest in the last several years. Electroactive polymers include any polymer that is simulated by electricity and responds to its effect in a reversible manner. This classification is sort of a melting pot of smart materials.[18]

4 MAGNETOSTRICTIVE MATERIALS

Magnetorestrictive materials are materials that have the material response of mechanical deformation when stimulated by a magnetic field. Shape changes are the largest in ferromagnetic and ferromagnetic materials. The repositioning of domain walls that occurs when these solids are placed in a magnetic field leads to hysteresis between magnetization and an applied magnetic field. When a ferromagnetic material is heated above its Curie temperature, these effects disappear. The microscopic properties of a ferromagnetic solid are different than for a ferromagnetic solid. The magnetic dipoles of a ferromagnetic solid are aligned parallel. The alignment of dipoles in a ferromagnetic solid can be parallel or in other directions.[5,19,20]

Magnetorestrictive materials are usually inorganic in chemical composition and are alloys of iron and nickel and doped with rare earths. The most effective magnetorestrictive material is another alloy developed at the Naval Ordinance Laboratory, TERFENOL-D. It is an alloy of terbium, dysprosium, and iron. The full effect of magnetorestriction occurs in crystalline materials. One factor preventing magnetorestrictive materials from reaching commercial significance has been cost. Over the past three decades there has been a great deal of development of giant magnetorestrictive materials, colossal magnetorerestrictive materials, and organic and organometallic magnets.[5,21]

The giant magnetorestrictive effect was first observed in iron–chromium laminates in 1988. These laminates consisted of alternating layers of 50-Å-thick iron with chromium layers of various thicknesses. The iron layers oriented themselves with antiparallel magnetic moments. A magnetic field of 20 kOe applied in the plane of the iron layers will uncouple this antiferromagnetic orientation. Since the magnetic orientation of one layer is controlled, another layer is free to rotate with an applied field; this change in orientation of the two layers produces a macnetorestrictlve effect. This discovery of giant magnetorestrictive materials made the increased sensitivities in hard drive heads more cost effective.

In the 1990s, researchers further enhanced the field of magnetorestrictive materials with the development of colossal magnetorestrictive materials. Ratios of magnetorestrictive effect in excess of 100,000% were observed. These new materials were found to be epitaxial lanthanum–calcium–manganese–oxygen thin films, polycrystalline lanthanum–yttrium–calcium–manganese–oxygen, and polycrystalline lanthanum–barium–manganese–oxygen. These combinations enhanced the product performance at costs below that of original magnetorestrictive materials.[21]

Magnets that exhibit commercial potential should have magnetic saturation and coercive field properties that are operational at low temperatures. Magnets are useful below their critical temperature. Most inorganic magnets have critical temperatures well above room temperature. The first organometallic magnet was the ionic salt complex of ferric bis(pentamethylcyclopentadienide) and tetracyanoethylene. This complex was a ferromagnet below its critical temperature of 4.8 K. The highest effective critical temperature of an organometallic magnet was in the vicinity of 400 K.[22]

Organic magnets are different than organometallic magnets for two reasons. The first is obvious: Organic-based magnets contain metal atoms. This results in a rethinking of the principles of magnetism. The second difference refers to the fact that coupled spins of organic magnets reside only in the *p* orbitals while coupled spins of the organometallic magnets can reside in either the *p* orbitals or the *d* orbitals or a combination of the two. A series of organic magnets based upon the nitroxide chemistry was synthesized and their magnetic behavior studied between 0.6 and 1.48 K. The most studied compound of these nitroxide-based magnets is 4-nitrophenylnitronyl nitroxide. This compound showed a saturation magnetization equivalent to one spin per molecule. This is indicative of ferromagnetic behavior.[5,22]

5 ELASTORESTRICTIVE MATERIALS

This class of smart materials is the mechanical equivalent to electrorestrictive and magnetorestrictive smart materials. These smart materials exhibit high hysteresis between stress and strain. The motion of ferroelastic domain walls causes the hysteresis. This motion of the ferroelastic domain walls is very complex near a martensitic-phase transformation. At this phase change, two types of crystal structural changes occur. One is induced by mechanical stress and the other by domain wall motion. Martensitic shape memory alloys have wide, diffuse phase changes and the ability to exist in both high- and low-temperature phases. The domain wall movements disappear with total change to the high-temperature phase.[5,19,20] The elastorestrictive smart material family is in its infancy.

6 ELECTRORHEOLOGICAL MATERIALS

Rheological materials comprise an exciting group of smart materials. Electrorheological and magnetorheological materials can change their rheological properties instantly through the application of an electric or a magnetic field. Electrorheological materials (fluids) have been known for several centuries. The rheological or viscous properties of these fluids, which are usually uniform dispersions or suspensions of particles within a fluid, are changed with the application of an electric field. The mechanism of how these electrorheological fluids work is simple. In an applied electric field the particles orient themselves in fiberlike structures (fibrils). When the electric field is off, the fibrils disorient themselves. Another way to imagine this behavior is to consider logs in a river. If the logs are aligned, they flow down the river. If they are disordered, they will cause a log jam, clogging up the river. A typical example of an electrorheological fluid is a mixture of corn starch in silicone oil. Another fluid that has been experimented with as a replacement for silicone oil is chocolate syrup. Another feature of electrorheological systems is that their damping characteristics can be changed from flexible to rigid and vice versa. Electrorheological fluids were evaluated using a single-link flexible-beam test bed. The beam was a sandwich configuration with electrorheological fluids distributed along its length. When the beam was rapidly moved back and forth, the electrorheological fluid was used to provide flexibility during the transient response period of the maneuver for speed and made rigid at the end point of the maneuver for stability. A practical way of viewing this behavior is to compare it with fly fishing.[5] It has been suggested that rheological fluids be used in the construction of fishing rods and golf clubs.

7 MAGNETORHEOLOGICAL MATERIALS

Magnetorheological materials (fluids) are the magnetic equivalent of electrorheological fluids. These fluids consist of ferromagnetic or ferromagnetic particles that are either dispersed or suspended and the applied stimulus is a magnetic field. A simple magnetorheological fluid consists of iron powder in motor oil. The Lord Corporation provided a clever demonstration of magnetorheological fluids. It supplied an interlocking two-plastic-syringe system filled with a magnetorheological fluid and two small magnets. The fluid flows freely, without the magnets placed in the middle of the two syringes. With the two magnets in place, the fluid flows completely.

An interesting adaptation of magnetorestrictive fluids is a series of clastomeric matrix composites embedded with iron particles. During the thermal cure of the elastomer, a strong magnetic field was applied to align the iron particles into chains. These chains of iron particles were locked into place within the composite through a crosslinked network of the cured elastomer. If a compressive force stimulated the composite, it was 60% more resistant to deformation in a magnetic field. If the composite was subjected to a shear force, its magnetic-field-induced modulus was an order of magnitude higher than its modulus in a zero magnetic field.[5,23]

Recently attempts to enhance the properties of epoxies with magnetic fields showed that at low conversion rates of the epoxy with a hardener, economically generated magnetic fields had an effect on the properties of the final composite.[24] At the high conversion of the reactants there exists a need to drive the scarce unreacted glycidyl and amine functionalities together. Only magnetic fields generated by superconducting electromagnets are capable of this.[24]

There continues to be a great deal of research into magnets and magnetism. A new area of research involves magnetic nanocomposite films. Magnetic particles exhibit size effects. Below a critical size, magnetic clusters comprise single domains, whereas with bulk materials there are multiple domains. Nanomagnets show unusual properties of magnetism, such as superparamagnetism and quantum tunneling. These unique properties of magnetic nanoclusters can lead to applications in information storage, color imaging, magnetic refrigeration, ferrofluids, cell storage, medical diagnosis, and controlled drug delivery. A nanocomposite is considered to be the incorporation of these nanoclusters into polymeric matrices such as polyaniline.[25]

8 THERMORESPONSIVE MATERIALS

Amorphous and semicrystalline thermoplastic polymeric materials are unique due to the presence of a glass transition temperature. Changes in the specific volume of polymers and their rate of change occur at their glass transition temperatures. This transition affects a multitude of physical properties. Numerous types of indicating devices could be developed based upon the stimulus–response (temperature-specific volume) behavior. This chapter contains several examples of this type of behavior; however, its total impact is beyond the scope of this chapter. To take advantage of this behavior in product development or material selection, it is necessary to consult the many polymer references.

A few unique examples that illustrate thermoresponsive behavior are included in this chapter. One example refers to the polyethylene/poly(ethylene glycol) copolymers that were used to functionalize the surfaces of polyethylene films.[5,26] One may refer to this illustration as being a "smart surface" or functionally gradient surface. When the film is immersed in an aqueous dispersion of the copolymer, the ethylene glycol moieties attach to the polymer film surface, resulting in a film surface having solvation behavior similar to poly(ethylene glycol) itself. Due to the inverse temperature-dependent solubility behavior of poly(alkene oxide)s in water, surface-modified polymers are produced that reversibly change their hydrophilicity and solvation with changes in temperatures.[27] Similar behaviors have been observed as a function of changes in pH.[28–31]

Another interesting example of materials that respond smartly to changes in temperature includes cottons, polyesters, and polyamide/polyurethanes that are modified by poly(ethylene glycol)s. A combination of the thermoresponsiveness of these fabrics with a sensitivity to moisture resulted in a family of fabrics that can serve as smart pressure bandages. When exposed to an aqueous medium such as blood, these fabrics contract and apply pressure. Once the fabric dries, it releases the pressure.[23]

Polymers based upon the monomer vinyl methyl ether have the unique behavior of shrinking upon heating to approximately 40°C. In the right design with normal behaving polymers, one can construct a device that can grasp objects like a hand.

9 pH-SENSITIVE MATERIALS

By far, the widely known chemical classes of pH-sensitive materials are the acids, bases, and indicators. The indicators fit the definition of smart materials by changing color as a function of pH and the action is reversible.

Other examples of pH-sensitive materials include some of the smart gels and smart polymers mentioned in this chapter. There are a large number of pH-sensitive polymers and gels that are used in biotechnology and medicine. Usually these materials are prepared from various combinations of such monomers and polymers such as methacrylic acid, methyl methacrylate, carboxymethylethyl cellulose, cellulose acetate, cellulose phthalate, hydroxypropylmethylcellulose phthalate, hydroxypropylmethylcellulose acetate, hydroxypropylmethylcellulose succinate, diethylaminoethyl methacrylate, and butyl methacrylate.[32]

10 LIGHT-SENSITIVE MATERIALS

There are several different material families that exhibit different behavior to a light stimulus. Electrochromism is a change in color as a function of an electrical field. Other types of behavior for light-sensitive materials are thermochromism (color change with heat), photochromism (color change with light), and photostrictism (shape changes caused by changes in electronic configuration due to light).[5,19,20]

Electrochromic smart windows have been intensively researched over the past few decades. There have been over 1800 patents issued for optical switching devices, with the bulk of these being issued in Japan. A typical switchable glass is multilayered with an electrochromic device embedded inside. A window device may have glass with an interior conductive oxide layer both on the top and bottom. Inside the sandwich of glass and conductive oxide is the electrochromic device. This device consists of an electrochromic layer, an ion storage layer, and between these two layers an ion conductor.

An interesting light-sensitive material with both electro- and thermochromism behaviors, Li_xVO, was evaluated for a smart window application.[33]

Materials have been developed to exhibit both photochromic and photographic (irreversible behavior to light) behaviors. One such system is based upon a substituted indolinospirobenzopyran embedded in a polystyrene matrix. This system performs as a photochromic system at low exposure in the UV range and as a photographic system at high exposures. The image can be devisualized by heat and can be restored with UV irradiation many times.[34]

11 SMART POLYMERS

The term *smart polymers* was almost dropped as a classification for smart materials in this treatment of the subject. It is very confusing. Each field of science and engineering has its own definition of a smart polymer, each definition can fit in another classification, and its distinction in smartness can be confusing at times. The term is being included in this chapter because several excellent articles on the subject have "smart polymer" in the title.

In medicine and biotechnology, smart polymer systems usually pertain to aqueous polymer solutions, interfaces, and hydrogels. Smart gels, or hydrogels, will be treated separately. Smart polymers refer to polymeric systems that are capable of responding strongly to slight changes in the external medium: a first-order transition accompanied by a sharp decrease in the specific volume of the polymer. If the external medium is temperature, this transition is known as the glass transition temperature of the polymer and several properties of the polymer change. Among these properties are volume, coefficient of thermal expansion, specific heat, heat conductivity, modulus, and permeation. Manipulating the detection around the glass transition temperature of the polymer can develop in smart devices. There are numerous examples of product development that has resulted in failure because the glass transition temperature of the polymer was not considered. It should be noted that as the polymer cools down from high temperatures to below its glass transition temperature its below-glass-transition properties are returned and vice versa.

Smart polymers can respond to stimuli such as temperature, pH, chemical species, light, UV radiation, recognition, electric fields, magnetic fields, and other types of stimuli. The resulting response can be changes in phase, shape, optics, mechanical strength, electrical and thermal properties, reaction rate, and permeation rate.

12 SMART (INTELLIGENT) GELS (HYDROGELS)

In the literature you can find these smart materials under a variety of names, as reflected by the title of this section. The concept of smart gels is a combination of the simple concept of solvent-swollen polymer networks in conjunction with the material being able to respond to other types of stimuli. A partial list of these stimuli includes temperature, pH, chemicals, concentration of solvents, ionic strength, pressure, stress, light intensity, electric fields, magnetic fields, and different types of radiation.[35–39] The founding father of these smart gels, Toyochi Tanaka, first observed this phenomenon in swollen clear polyacrylamide gels. Upon cooling, these gels would cloud up and become opaque. Upon warming these gels regained their clarity. Upon further investigation to explain this behavior, it was found that some gel systems could expand to hundreds of times their original volume or could collapse to expel up to 90% of its fluid content with a stimulus of only a 1°C change in temperature. Similar behavior was observed with a change of 0.1 pH unit.

These types of behaviors led to the development of gel-based actuators, values, sensors, control-led release systems for drugs and other substances, artificial muscles for robotic devices, chemical memories, optical shutters, molecular separation systems, and toys. Other potential systems for the development of products with smart (intelligent) gels (hydrogels) include paints, adhesives, recyclable absorbents, bioreactors, bioassay systems, and display.

Numerous examples of the commercialization of these smart gels can be found in Ref. 35. This chapter will only include a few examples of smart gels. One such smart gel consists of an entangled network of two polymers, a poly(acrylic acid) (PAA) and a triblock copolymer of poly(propylene oxide) (PPO) and poly(ethylene oxide) (PEO) with a sequence of PEO–PPO–PEO. The PAA portion is a bioadhesive and is pH responsive, the PPO moieties are hydrophobic substances that assist in solubilizing lipophilic substances in medical applications, and the PEO functionalities tend to aggregate, resulting in gelation at body temperatures. Another smart gel system with a fairly complex composition consists of citosan, a hydrolyzed derivative of chitin (a polymer of *N*-acetylglucosamine that is found in shrimp and crab shells), a copolymer of poly(nisopropylacrylamide) and poly(acrylic acid), and a graft copolymer of poly(methacrylic aid) and poly(ethylene glycol). This gel system was developed for the controlled release of insulin in diabetics.

Polyampholytic smart hydrogels swell to their maximum extent at neutral pH values. When such gels, copolymers of methacrylic acid 2-(*N*,*N*-dimethylamino)ethyl methacrylate, are subjected to either acidic or basic media, they undergo rapid dehydration.[39]

One very unusual smart gel is based upon the polymerization of *N*-isopropylacrylamide, a derivative of tris(2,2′-bipyridyl)ruthenium(II) that has a polymerizable vinyl group, and

N,*N*′-methylenebisacrylamide. It is a self-oscillating gel that simulates the beating of the heart with color changes.[40]

13 SMART CATALYSTS

The development of smart catalysts is a new field of investigation and has shown a great deal of activity in universities of the oil-producing states. One such smart catalyst is rhodium based with a poly(ethylene oxide) backbone. Smart catalysts such as this one function opposite to a traditional catalyst; that is, as the temperature increases, they become less soluble, precipitating out of the reaction solution, thus becoming inactive. As the reaction solution cools down, the smart catalyst redissolves and thus becomes active again.[5,23] Other smart catalyst systems are being developed that dissociate at high temperatures (less active) and recombine at low temperatures (more active).[5,27]

14 SHAPE MEMORY ALLOYS

The shape memory effect in metals is a very interesting phenomenon. Imagine taking a piece of metal and deforming it completely and then restoring it to its original shape with the application of heat. Taking a shape memory alloy spring and hanging a weight on one end of the spring can easily illustrate this. After the spring has been stretched, heat the spring with a hot-air gun and watch it return to its original length with the weight still attached.

These materials undergo a thermomechanical change as they pass from one phase to another. The crystalline structure of such materials, such as nickel–titanium alloys, enters into the martensitic phase as the alloy is cooled below a critical temperature. In this stage the material is easily manipulated through large strains with a little change in stress. As the temperature of the material is increased above the critical temperature, it transforms into the austenitic phase. In this phase the material regains its high strength and high modulus and behaves normally. The material shrinks during the change from the martensitic to the austenitic phase,[5,19,20]

Nickel–titanium alloys have been the most used shape memory material. This family of nickel–titanium alloys is known as Nitinol, after the laboratory where this material was first observed (Nickel Titanium Naval Ordinance Laboratory). Nitinol has been used in military, medical, safety, and robotics applications. Specific applications include hydraulic lines on F-14 fighter planes, medical tweezers and sutures, anchors for attaching tendons to bones, stents for cardiac arteries, eyeglass frames, and antiscalding valves in water faucets and showers.[5,41,42]

In addition to the family of nickel–titanium alloys there are other alloys that exhibit the shape memory effect. These alloys are silver–cadmium, gold–cadmium, copper–aluminum–nickel, copper–tin, copper–zinc, combinations of copper–zinc with silicon or tin or aluminum, indium–thallium, nickel–aluminum, iron–platinum, manganese–copper, and iron–manganese–silicon.[43] Not all combinations of the two or three elements yield an alloy with the shape memory effect; thus it is recommended to review the original literature.

Several articles from Mitsubishi Heavy Industries describe the room temperature functional shape memory polyurethanes. To this writer, these papers only attest to the behavior of a polymer at its glass transition temperature. Thus if you wish to describe a polymer's behavior at its glass transition temperature and since the free-volume change is reversible, you may call it a smart polymer, a shape memory material, or a thermoresponsive material.

The unique characteristic of these polyurethanes is that their transition occurs in the vicinity of room temperature.[44,45]

15 UNUSUAL BEHAVIORS OF MATERIALS

As one researches the field of smart materials and structures, one realizes that there are many smart materials and there are many material behaviors that are reversible. The ability to develop useful products from smart materials is left up to one's imagination. For example, water is a very unique material. It expands upon freezing. As we know, the force generated by this expansion causes sidewalks and highways to crack. Now, what if you surround water pipes with a heating system that consists of a heater and water enclosed in a piezoelectric polymer or elastomer container that is in a fixed space. As the temperature drops to the freezing point of water, it expands and generates a force against the piezoelectric container which in turns generate electricity, thus powering the heater and keeping the pipes from freezing.

As previously mentioned, sometimes it is difficult to classify the material and its behavior. One such case involves a University of Illinois patent. The title of the patent is "Magnetic Gels Which Change Volume in Response to Voltage Changes for Magnetic Resonance Imaging."[46] The patent teaches the use of a matrix that has a magnetic and preferably superparamagnetic component and the capability of changing its volume in an electric field.

Fullerenes are spherically caged molecules with carbon atoms at the corner of a polyhedral structure consisting of pentagons and hexagons. The most stable of the fullerenes is a C60 structure known as a buckyball or buckminsterfullerene. The fullerenes have been under commercial development for the past decade. One application of the fullerenes as a smart material consists of embedding the fullerenes into sol–gel matrices for the purpose of enhancing optical limiting properties.[47]

A semiconducting material with a magnetic ordering at 16.1 K was produced from the reaction of buckyball with tetra(dimethylamino)ethylene. This organic-based magnet did not have the coercive or saturation magnetization to function totally as a ferromagnet. The replacement of buckyball with higher carbon number fullerenes in the reaction with tetra(diamethylamino)ethylene did not produce any complexes that showed magnetic ordering.[22]

16 COMMENTS, CONCERNS, AND CONCLUSIONS

In dealing with smart materials and structures there is still much confusion over the name of these materials and what makes a material or structure smart. Numerous products with "smart" in the name do not meet the definition of being smart, that is, responding to the environment in a reversible manner. The scientist/engineer should not fault advertising professionals for using the term smart in a product description. But I must admit I chuckle whenever I see the magazine *Smart Money*. Does this mean that I can spend money and it will return to me because it is reversible?

Confusion also exists within the smart materials community with the term smart polymers. Applications of a smart polymer center on the polymer's glass transition temperature. As design engineers become more familiar with that term and its significance to a design, in some applications the term will be eliminated and replaced with terms like "working smartly with a polymer."

We have not addressed the versatility of these smart materials. One such example may involve the smart shock absorbers. Two literature sources discuss current research on vibration suppression in automobiles using smart shock absorbers.[3,17] These smart shock absorbers were developed by Toyota and consist of multilayer piezoelectric ceramics. These multilayer stacks are positioned near each wheel. After analyzing the vibration signals, a voltage is fed back to the actuator stack, which responds by pushing on the hydraulic system to enlarge the motion. Signal processors analyze the acceleration signals from road bumps and respond with a motion that cancels the vibration.[3,17] Such active piezoelectric systems are used to minimize excess vibrations in helicopter blades and in the twin tail of F-18 fighter jets. An Internet source[48] has discussed work at the University of Rochester with smart shock absorbers using electrorheological fluids. The electrorheological fluids sense the force of a bump and immediately send an electric signal to precisely dampen the force of the bump, thus providing a smoother ride. In an engine mount, electrorheological fluids damp out the vibrations of the engine, thus reducing wear and tear on the vehicle. Another application of electrorheological fluids is in clutches to reduce the wear between the plates as a driver shifts gears. This can reduce the maintenance costs of high-duty trucks.[47] Or one can dampen the vibrations of the road, engine mounts, and other sources on an automobile or heavy-duty truck with magnetorheological fluids. The Lord Corporation has developed a series of trademarked fluids and systems known as Rheonetic Fluids that are commercially available for vibration control as well as for noise suppression.[49]

17 FUTURE CONSIDERATIONS

The future of smart materials and structures is wide open. The use of smart materials in a product and the type of smart structures that one can design are only limited by one's talents, capabilities, and ability to "think outside the box."

In an early work[5] and as part of short courses there were discussions pertaining to future considerations. A lot of the brainstorming that resulted from these efforts is now being explored. Some ideas that were in the conceptual stage are now moving forward. Look at the advances in information and comforts provided through smart materials and structures in automobiles.

Automobiles can be taken to a garage for service and be hooked up to a diagnostic computer that tells the mechanic what is wrong with the car. Or a light on the dashboard signals "maintenance required." Would it not be better for the light to inform us as to the exact nature of the problem and the severity of it? This approach mimics a cartoon that appeared several years ago of an air mechanic near a plane in a hanger. The plane says "Ouch" and the mechanic says "Where do you hurt?"

One application of smart materials is the work mentioned earlier of piezoelectric ink-jet printer that serves as a chemical delivery to print organic light-emitting polymers in a fine detail on various media. Why not take the same application to synthesize smaller molecules? With the right set one could synthesize smaller molecules in significant amounts for characterization and evaluation and in such a way that we could design experiments with relative ease.

A new class of smart materials has appeared in the literature. This is the group of smart adhesives. We previously mentioned that PVDF film strips have been placed within an adhesive joint to monitor performance. Khongtong and Ferguson developed a smart adhesive at Lehigh University.[50] They suggested that this new adhesive could form an antifouling coating for boat hulls or for controlling cell adhesion in surgery. The stickiness of the new adhesive can be switched on and off with changes in temperature. The smart adhesive also becomes water repellent when its tackiness wanes.[50] The term "smart adhesive" is appearing more frequently in the literature.

A topic of research that was in the literature a few years ago was "smart clothes" or "wearable computers" being studied at MIT. The potential of this concept is enormous. This sounds wonderful as long as we learn how to work smarter, not longer.

REFERENCES

1. Amato, *Sci. News,* **137**(10), 152–155 (Mar. 10, 1990).
2. Port, *Business Week,* No. 3224, 48–55 (Mar, 10, 1990).
3. Committee on New Sensor Technologies, Materials and Applications, National Materials Advisory Board, Commission on Engineering and Technical Systems, National Research Council Report, *Expanding the Vision of Sensor Materials,* National Academy Press, Washington, DC, 1995, pp. 33–45.
4. C. A. Rogers, *Sci. Am.,* **273**(3), 122–126 (Sept. 1995).
5. J. A. Harvey, *Kirk-Othiner Encyclopedia of Chemical Technology,* 4th ed., Supplement, Wiley, New York, 1998, pp. 502–504.
6. Chopra, in *Proceedings of the Smart Structures and Materials 1996—Smart Structures and Integrated Systems Conference for the Society of Photo-Optical Instrumentation Engineers,* Vol. 2717, San Diego, Feb. 26–29, 1996, pp. 20–26.
7. T. Thomas, Portland, OR, private communication, 1999.
8. Technical bulletins and notes, Sensor Technology Limited, Collingwood, Canada.
9. T. T. Wang, J. M. Herbert, and A. M. Glass (eds.), *Applications of Ferroelectric Polymers,* Blackle and Sons, Glasgow, 1988.
10. J. O. Simpson, http:hsti.1 arc. nasa. gov/randt/ I 995/S ection13 1. fm5 13. html.
11. Piezo Systems, *Technical Bulletin,* Cambridge, MA, 1993.
12. T. W. DeYoung and V. B. Maltsev, U.S. Patent 4,439,780, 1984.
13. S. D. Howkins, U.S. Patent 4,459,601, 1984.
14. T. R. Hebner, C. C. Wu, D. Marcy, M. H. Lu, and J. C. Sturm, *Appl. Phys. Lett.,* **72**(5), 519–521 (1998).
15. G. L. Anderson, J. Mommaerts, S. L. Tana, J. C. Duke, Jr., and D. A. Dillard, *J. Intell. Mater. Syst. Struct.,* **4,** 425–428 (1993).
16. G. L. Anderson, D. A. Dillard, and J. P. Wightman, *J. Adhesion,* **36,** 213 (1992).
17. R. E. Newnham and A. Amin, *CHEMTECH,* **29**(12), 38–46 (Dec. 1999).
18. *Proceedings for a Symposium entitled Electroactive Polymers* (*EAP*), Vol. 600, Boston, Nov. 29–Dec. 1, 1999, Materials Research Society, Warrendale, PA, 2000.
19. R. E. Newnham, *Mater. Res. Soc. Bull.,* **22**(5), 20–34 (May 1997).
20. B. Culshaw, *Smart Structures and Materials,* Artech House, Boston, 1996, pp. 43–45, 117–130.
21. K. Derbyshire and E. Korczynski, *Solid State Technol.,* 57–66 (Sept. 1995).
22. J. S. Miller and A. J. Epstein, *Chem. Eng. News,* 30–41 (Oct. 2, 1995).
23. R. Dagani, *Chem. Eng. News,* 30–33 (Sept. 18, 1995).
24. R. H. Gerzeski, *J. Adv. Mater.,* **33**(2), 63–69 (Apr. 2001).
25. B. Z. Tang, *CHEMTECH,* **29**(1), 7–12 (Nov. 1999).
26. D. E. Bergreiter, B. C. Ponder, G. Aguilar, and B. Srinivas, *Chem. Mater.,* **9**(2), 472–477 (1997).
27. R. Baum, *Chem. Eng. News,* 7 (Jan. 30, 1996).
28. Y. G. Takei, T. Aoki, K. Sanui, N. Ogata, Y. Sakurai, and T. Okano, *Macromolecules,* **27,** 6163–6166 (1994).
29. J. Jhan, R. Pelton, and Y. L. Den, *Langinuir,* **11,** 2301–2302 (1995).
30. J. L. Thomas, H. You, and D. A. Tirrell, *J. Am. Chem. Soc.,* **117,** 2949–2950 (1995).
31. D. H. Carey and G. S. Ferguson, *J. Am. Chem. Soc.,* **118,** 9780–9781 (1996).
32. I. Y. Galaev, *Russ. Chem. Rev.,* **65**(5), 471–489 (1995).

33. M. S. Khan, in *Proceedings of the Metal/Nonmetal Microsystems: Physics, Technology, and Applications Workshop,* Vol. 2780, Polanica Zdroj, Poland, Sept. 11–14, 1996, Society of Photo-Optical Instrumentation Engineers, 1996, pp. 56–59.

34. A. Vannikov and A. Kararasev, in *Proceedings of Smart Structures and Materials 1996—Smart Electronics and MEMS Conference,* Vol. 2722, San Diego, Feb. 28–29, 1996, Society of Photo-Optical Instrumentation Engineers, 1996, pp. 252–255.

35. R. Dagani, *Chem. Eng. News,* 26–37 (June 9, 1997).

36. R. S. Harland and R. K. Prud'homme, *Polyelectrolyte Gels, Properties and Applications,* American Chemical Society, Washington, DC, 1992.

37. D. DeRossi, K. Kajiwara, Y. Osada, and A. Yamauchi, *Polymer Gels Fundamental and Biomedical Applications,* Plenum, New York, 1991.

38. T. Okano (ed.), *Biorelated Polymers and Gels—Controlled Release and Applications in Biomedical Engineering,* Academic, Boston, 1998.

39. A. Dupommie, L. Merle-Aubry, Y. Merle, and E. Slany, *Makronmolek. Chem.,* **187,** 211 (1986).

40. R. Yoshida, T. Takahashi, T. Yamaguchi, and H. Ichijo, *Adv. Mater.,* **9,** 175 (1997).

41. G. Kauffman and I. Mayo, *Chem. Matters,* 4–7 (Oct. 1993).

42. D. Stoeckel and W. Yu, *Superelastic Nickel-Titanium Wire,* Technical Bulletin, Raychem, Menlo Park, CA.

43. K. Shimizu and T. Tadaki, in *Shape Memory Alloys,* H. Funakubo (ed.), Gordon and Breach, New York, 1987.

44. S. Hayashi, in *Proceedings of the US-Japan Workshop on Smart Materials and Structures,* Minerals, Metals and Materials Society, 1997, pp. 29–38.

45. S. Hayashi, S. Kondo, P. Kapadia, and E. Ushioda, *Plastics Eng.,* Feb. 29–31 (1995).

46. P. C. Lauterbur and S. Frank, U.S. Patent 5,532,006, July 2, 1996.

47. R. Signorini, M. Zerbetto, M. Meneghetti, R. Bozio, G. Brusatin, E. Menegazzo, M. Guglielmi, M. Maggini, G. Scorrano, and A Prato, in *Proceedings of the Fullerenes and Photoiniccs III Conference,* Vol. 2854, Society of Photo-Optical Instrumental Engineers, 1996, pp. 130–139.

48. T. Jones, http:www. rochester.edu/pr/releases/ce/j ones. htm.

49. Technical bulletins, Lord Corporation, Cary, NC.

50. S. Khongtong and G. Ferguson, *J. Am. Chem. Soc.,* **124,** 7254–7255 (2004).

CHAPTER 12

OVERVIEW OF CERAMIC MATERIALS, DESIGN, AND APPLICATION

R. Nathan Katz
Department of Mechanical Engineering
Worcester Polytechnic Institute
Worcester, Massachusetts

1 INTRODUCTION

Engineering ceramics possess unique combinations of physical, chemical, electrical, optical, and mechanical properties. Utilizing the gains in basic materials science understanding and advances in processing technology accrued over the past half century, it is now frequently possible to custom tailor the chemistry, phase content, and microstructure to optimize applications-specific combinations of properties in ceramics (which include glasses, single crystals, and coatings technologies, in addition to bulk polycrystalline materials). This capability in turn has led to many important, new applications of these materials over the past few decades. Indeed, in many of these applications the new ceramics and glasses are the key enabling technology.

Ceramics include materials that have the highest melting points, highest elastic moduli, highest hardness, highest particulate erosion resistance, highest thermal conductivity, highest optical transparency, lowest thermal expansion, and lowest chemical reactivity known. Counterbalancing these beneficial factors are brittle behavior and vulnerability to thermal shock and impact. Over the past three decades major progress has been made in learning how to design to mitigate the brittleness and other undesirable behaviors associated with ceramics and glasses. Consequently, many exciting new applications for these materials have emerged over the past several decades.

Among the major commercial applications for these materials are:

Reprinted from *Handbook of Materials Selection,* Wiley, New York, 2002, with permission of the publisher.

- Passive electronics (capacitors and substrates)
- Optronics/photonics (optical fibers)
- Piezoceramics (transducers)
- Mechanical (bearings, cutting tools)
- Biomaterials (hard-tissue replacement)
- Refractories (furnace linings, space vehicle thermal protection)
- Electrochemical (sensors, fuel cells)
- Transparencies (visible, radar)

This chapter will provide a brief overview of how ceramics are processed and the ramifications of processing on properties. Next a short discussion of the special issues that one encounters in mechanical design with brittle materials is provided. Short reviews of several of the above engineering applications of ceramics and glasses, which discuss some of the specific combinations of properties that have led design engineers to the selected material(s), follow. A section on how to obtain information on materials sources is provided. Tables listing typical properties of candidate materials for each set of applications are included throughout. Finally, some areas of future potential will be discussed.

2 PROCESSING OF ADVANCED CERAMICS

The production of utilitarian ceramic artifacts via the particulate processing route outlined in Fig. 1 actually commenced about 10,000 years ago.[1] Similarly, glass melting technology goes back about 3500 years, and as early as 2000 years ago optical glass was being produced.[1] While many of the basic unit processes for making glasses and ceramics are still recognizable across the millennia, the level of sophistication in equipment, process control, and raw material control have advanced by "light years." In addition, the past 50 years has

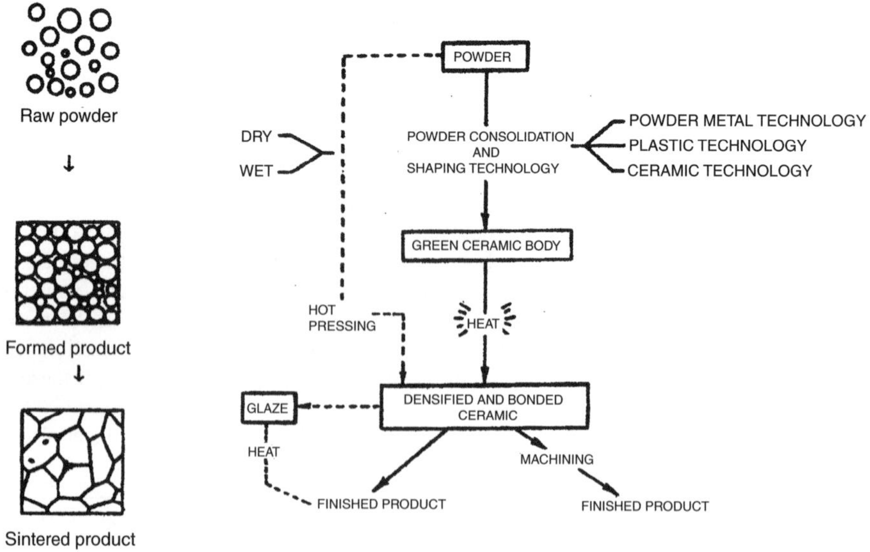

Figure 1 Processing of polycrystalline ceramics via the particular route.

created a fundamental understanding of the materials science principles that underlie the processing–microstrucure–property relationships. Additionally, new materials have been synthesized that possess extraordinary levels of performance for specific applications. These advances have led to the use of advanced ceramics and glasses in roles that were unimaginable 50 or 60 years ago. For example, early Egyptian glass ca. 2000 BC had an optical loss of $\sim 10^7$ dB/km, compared to an optical loss of $\sim 10^{-1}$ in mid-1980s glass optical fibers,[2] a level of performance that has facilitated the fiber-optic revolution in telecommunications. Similarly, the invention of barium titanate and lead zirconate titanate ceramics, which have much higher piezoelectric moduli and coupling coefficients than do naturally occurring materials, has enabled the existence of modern sonar and medical ultrasound imaging.[3]

The processing of modern ceramics via the particulate route, shown in Fig. 1, is the way that ~99% of all polycrystalline ceramics are manufactured. Other techniques for producing polycrystalline ceramics, such as chemical vapor deposition[4] or reaction forming,[5] are of growing importance but still represent a very small fraction of the ceramic industry. There are three basic sets of unit processes in the particulate route (and each of these three sets of processes may incorporate dozens of subprocesses). The first set of processes involves powder synthesis and treatment. The second set of processes involves the consolidation of the treated powders into a shaped preform, known as a "green" body. The green body typically contains about 50 vol % porosity and is extremely weak. The last set of unit processes utilizes heat, or heat and pressure combined, to bond the individual powder particles, remove the free space and porosity in the compact via diffusion, and create a fully dense, well-bonded ceramic with the desired microstructure.[6] If only heat is used, this process is called sintering. If pressure is also applied, the process is then referred to as hot pressing (unidirectional pressure) or hot isostatic pressing [(HIP), which applies uniform omnidirectional pressure].

Each of the above steps can introduce processing flaws that can diminish the intrinsic properties of the material. For example, chemical impurities introduced during the powder synthesis and treatment steps may adversely affect the optical, magnetic, dielectric, or thermal properties of the material. Alternatively, the impurities may segregate in the grain boundary of the sintered ceramic and negatively affect its melting point, high-temperature strength, dielectric properties, or optical properties. In green-body formation, platey or high-aspect-ratio powders may align with a preferred orientation, leading to anisotropic properties. Similarly, hot pressing may impose anisotropic properties on a material. Since ceramics are not ductile materials, they can (usually) not be thermomechanically modified after primary fabrication. Thus, the specific path by which a ceramic component is fabricated can profoundly affect its properties. The properties encountered in a complex shaped ceramic part are often quite different than those encountered in a simply shaped billet of material. This is an important point of which a design engineer specifying a ceramic component needs to be constantly mindful.

3 BRITTLENESS AND BRITTLE MATERIALS DESIGN

Even when ceramics are selected for other than mechanical applications, in most cases some levels of strength and structural integrity are required. It is therefore necessary to briefly discuss the issue of brittleness and how one designs with brittle materials before proceeding to discuss applications and the various ceramic and glass materials families and their properties.

The main issues in designing with a brittle material are that a very large scatter in strength (under tensile stress), a lack of capacity for mitigating stress concentrations via plastic flow, and relatively low energy absorption prior to failure dominate the mechanical behavior. Each of these issues is a result of the presence of one or more flaw distribution within or at the surface of the ceramic material and/or the general lack of plastic flow available in ceramics. As a consequence, ceramic and glass components that are subjected to tensile stresses are not designed using a single valued strength (*deterministic design*) as commonly done with metals. Rather, ceramic components are designed to a specified probability of failure (*probabilistic design*) that is set at acceptably low values. The statistics of failure of brittle materials whose strength is determined by a population of varying sized flaws are similar to modeling the statistics of a chain failing via its weakest link. The statistics utilized are known as Weibull statistics. A Weibull probability of failure distribution is characterized by two parameters, the characteristic stress and the Weibull modulus.[7] Computer programs for incorporating Weibull statistical distributions into finite-element design codes have been developed that facilitate the design of ceramic components optimized for low probabilities of failure.[8] The effectiveness of such probabilistic design methodology has been demonstrated by the reliable performance of ceramics in many highly stressed structural applications, such as bearings, cutting tools, turbocharger rotors, missile guidance domes, and hip prosthesis.

Flaws (strength-limiting features) can be intrinsic or extrinsic to the material and processing route by which a test specimen or a component is made. Intrinsic strength-limiting flaws are generally a consequence of the processing route and may include features such as pores, aggregations of pores, large grains, agglomerates, and shrinkage cracks. While best processing practices will eliminate or reduce the size and frequency of many of these flaws, it is inevitable that some will still persist. Extrinsic flaws can arise from unintended foreign material entering the process stream, i.e., small pieces of debris from the grinding media or damage (cracks) introduced in machining a part to final dimensions. Exposure to a service environment may bring new flaw populations into existence, i.e., oxidation pits on the surface of nonoxide ceramics exposed to high temperatures, or may cause existing flaws to grow larger as in the case of static fatigue of glass. In general, one can have several flaw populations present in a component at any time, and the characteristics of each population may change with time. As a consequence of these constantly changing flaw populations, at the present time the state of the art in life prediction of ceramic components for use in extreme environments significantly lags the state of the art in component design. As in most fields of engineering, there are some rules of thumb that one can apply to ceramic design.[9] While these are not substitutes for a carefully executed probabilistic finite-element design analysis, they are very useful in spotting pitfalls and problems when a full-fledged design cannot be executed due to financial or time constraints.

Rules of Thumb for Design with Brittle Materials

1. Point loads should be avoided to minimize stress where loads are transferred. It is best to use areal loading (spherical surfaces are particularly good); line loading is next best.
2. Structural compliance should be maintained by using compliant layers or springs or radiusing of mating parts (to avoid lockup).
3. Stress concentrators—sharp corners, rapid changes in section size, undercuts and holes—should be avoided or minimized. Generous radiuses and chamfers should be used.
4. The impact of thermal stresses should be minimized by using the smallest section size consistent with other design constraints. The higher the symmetry, the better (a cylinder will resist thermal shock better than a prism), and breaking up complex components into subcomponents with higher symmetry may help.

5. Components should be kept as small as possible—the strength and probability of failure at a given stress level are dependent on size; thus minimizing component size increases reliability.
6. The severity of impact should be minimized. Where impact (i.e., particulate erosion) cannot be avoided, low-angle impacts (20°–30°) should be designed for. Note this is very different than the case of metals, where minimum erosion is at 90°.
7. Avoid surface and subsurface damage. Grinding should be done so that any residual grinding marks are parallel, not perpendicular, to the direction of principal tensile stress during use. Machining-induced flaws are often identified to be the strength-limiting defect.

4 APPLICATIONS

The combinations of properties available in many advanced ceramics and glasses provide the designers of mechanical, electronic, optical, and magnetic systems a variety of options for significantly increasing systems performance. Indeed, in some cases the increase in systems performance is so great that the use of ceramic materials is considered an enabling technology. In the applications examples provided below the key properties and combinations of properties required will be discussed, as well as the resultant systems benefits.

4.1 Ceramics in Wear Applications

In the largest number of applications where modern ceramics are used in highly stressed mechanical applications, they perform a wear resistance function. This is true of silicon nitride used as balls in rolling element bearings, silicon carbide journal bearings or water pump seals, alumina washers in faucets and beverage dispensing equipment, silicon nitride and alumina-based metal-cutting tools, zirconia fuel injector components, or boron carbide sand blast nozzles, to cite some typical applications and materials.

Wear is a systems property rather than a simple materials property. As a systems property, wear depends upon what material is rubbing, sliding, or rolling over what material, upon whether the system is lubricated or not, upon what the lubricant is, and so forth. To the extent that the wear performance of a material can be predicted, the wear resistance is usually found to be a complex function of several parameters. Wear of ceramic materials is often modeled using an abrasive wear model where the material removed per length of contact with the abrasive is calculated. A wide variety of such models exist, most of which are of the form

$$V \propto P^{0.8} K_{\mathrm{Ic}}^{-0.75} H^{-0.5} N \tag{1}$$

where V is the volume of material worn away, P is the applied load, K_{Ic} is the fracture toughness, H is the indentation hardness, and N is the number of abrasive particles contacting the wear surface per unit length. Even if there are no external abrasives particles present, the wear debris of the ceramics themselves act as abrasive particles. Therefore, the functional relationships that predict that wear resistance should increase as fracture toughness and hardness increase are, in fact, frequently observed in practice.

Even though the point contacts that occur in abrasive wear produce primarily hertzian compressive stresses, in regions away from the hertzian stress field tensile stresses will be present and strength is, thus, a secondary design property. In cases where inertial loading or weight is a design consideration, density may also be a design consideration. Accordingly, Table 1 lists typical values of the fracture toughness, hardness, Young's modulus, four-point

Table 1 Key Properties for Wear-Resistant Ceramics

Material	K_{Ic} (MPa·m$^{1/2}$)	H (kg/mm^2)	E (GPa)	MOR (MPa)	ρ (g/cm^3)
Al_2O_3 99%	3.9–4.5	1900	360–395	350–560	3.9
B_4C	—	3000	445	300–480	2.5
Diamond	6–10	8000	800–925	800–1400	3.5
SiC	2.6–4.6	2800	380–445	390–550	3.2
Si_3N_4	4.2–7	1600	260–320	450–1200	3.3
TiB_2	5–6.5	2600	550	240–400	4.6
ZrO_2 (Y-TZP)	7–12	1000	200–210	800–1400	5.9

modulus of rupture (MOR) in tension, and the density for a variety of advanced ceramic wear materials. Several successful applications of ceramics to challenging wear applications are described below.

Bearings

Rolling element bearings, for use at very high speeds or in extreme environments, are limited in performance by the density, compressive strength, corrosion resistance, and wear resistance of traditional high-performance bearing steels. The key screening test to assess a material's potential as a bearing element is rolling contact fatigue (RCF). RCF tests on a variety of alumina, SiC, Si_3N_4, and zirconia materials, at loads representative of high-performance bearings demonstrated that only fully dense silicon nitride (Si_3N_4) could outperform bearing steels.[10] This behavior has been linked to the high fracture toughness of silicon nitride, which results from a unique "self-reinforced" microstructure combined with a high hardness. Additionally, the low density of silicon nitride creates a reduced centrifugal stress on the outer races at high speeds. Fully dense Si_3N_4 bearing materials have demonstrated RCF lives 10 times that of high-performance bearing steel. This improved RCF behavior translates into DN (DN = bearing bore diameter in millimeters × shaft rpm) ratings for hybrid ceramic bearings (Si_3N_4 balls running in steel races, the most common ceramic bearing configuration) about 50% higher than the DN rating of steel bearings. Other benefits of silicon nitride hybrid bearings include an order-of-magnitude less wear of the inner race, excellent performance under marginal lubrication, survival under lubrication starvation conditions, lower heat generation than comparable steel bearings, and reduced noise and vibration.

Another important plus for Si_3N_4 is its failure mechanism. When Si_3N_4 rolling elements fail, they do not fail catastrophically; instead they spall—just like bearing steel elements (though by a different microstructural mechanism). Thus, the design community only had to adapt their existing practices, instead of developing entirely new practices to accommodate new failure modes. The main commercial applications of silicon nitride bearing elements are listed in Table 2.

Cutting Tool Inserts

While ceramic cutting tools have been in use for over 60 years, it is only within the past two decades that they have found major application, principally in turning and milling cast iron and nickel-based superalloys and finishing of hardened steels. In these areas ceramics based on aluminum oxide and silicon nitride significantly outperform cemented carbides and coated carbides. High-speed cutting tool tips can encounter temperatures of 1000°C or higher. Thus, a key property for an efficient cutting tool is hot hardness. Both the alumina and Si_3N_4 families of materials retain a higher hardness at temperatures between 600 and 1000°C than either tool steels or cobalt-bonded WC cermets. The ceramics are also more chemically inert.

Table 2 Commercial Applications of Si_3N_4 Hybrid Bearings

Machine tool spindles	The first and largest application, its main benefits are higher speed and stiffness, hence greater throughput and tighter tolerances
Turbomolecular pump shaft	Presently the industry standard, the main benefits are improved pump reliability and marginal lubrication capability, which provide increased flexibility in pump mounting orientation
Dental drill shaft	The main benefit is sterilization by autoclaving
Aircraft wing flap actuators	Wear and corrosion resistance are the main benefits
In-line skates/mountain bikes	Wear and corrosion resistance are the main benefits
Space Shuttle main engine oxygen fuel pump	Here, the bearing is lubricated by liquid oxygen. Steel bearings are rated for one flight; Si_3N_4 hybrid bearings are rated for five.

The combination of hot hardness and chemical inertness means that the ceramics can run hotter and longer with less wear than the competing materials. Historic concerns with ceramic cutting tools have focused on low toughness, susceptibility to thermal shock, and unpredictable failure times. Improvements in processing together with microstructural modifications to increase fracture toughness have greatly increased the reliability of the ceramics in recent years

Alumina-based inserts are reinforced (toughened) with zirconia, TiC, or TiN particles or SiC whiskers. The thermal shock resistance of alumina–SiC_w is sufficiently high, so that cooling fluids can be used when cutting Ni-based alloys. Silicon-nitride-based inserts include fully dense Si_3N_4 and SiAlON's, which are solid solutions of alumina in Si_3N_4. Fully dense Si_3N_4 can have a fracture toughness of 6–7 $MPa \cdot m^{1/2}$, almost as high as cemented carbides ($\sim$9 $MPa \cdot m^{1/2}$), a high strength (greater than 1000 MPa), and a low thermal expansion that yields excellent thermal shock behavior. Silicon nitride is the most efficient insert for the turning of gray cast iron and is also used for milling and other interrupted cut operations on gray iron. Because of its thermal shock resistance, coolant may be used with silicon nitride for turning applications. SiAlON's are typically more chemically stable than the Si_3N_4's but not quite as tough or thermal shock resistant. They are mainly used in rough turning of Ni-based superalloys.

Ceramic inserts are generally more costly than carbides (1.5–2 times more), but their metal removal rates are $\sim$3–4 times greater. However, that is not the entire story. Ceramic inserts also demonstrate reduced wear rates. The combination of lower wear and faster metal removal means many more parts can be produced before tools have to be indexed or replaced. In some cases this enhanced productivity is truly astonishing. In the interrupted single-point turning of the outer diameter counterweights on a gray cast iron crankshaft a SiAlON tool was substituted for a coated carbide tool. This change resulted in the metal removal rate increasing 150% and the tool life increasing by a factor of 10. Each tool now produced 10 times as many parts and in much less time. A gas turbine manufacturer performing a machining operation on a Ni-based alloy using a SiAlON tool for roughing and a tungsten carbide tool for finishing required a total of 5 h. Changing to SiC-whisker-reinforced alumina inserts for both operations reduced the total machining time to only 20 min. This yielded a direct savings of $250,000 per year, freed up 3000 h of machine time per year, and avoided the need to purchase a second machine tool.

Ceramic Wear Components in Automotive and Light-Truck Engines

Several engineering ceramics have combinations of properties that make them attractive materials for a variety of specialized wear applications in automotive engines.

The use of structural ceramics as wear components in commercial engines began in Japan in the early 1980s. Table 3 lists many of the components that have been manufactured, the engine company that first introduced the component, the material, and the year of introduction. In some of these applications several companies have introduced a version of the component into one or more of their engines.

Many of these applications are driven by the need to control the emissions of heavy-duty diesels. Meeting current emissions requirements creates conditions within the engine fuel delivery system that increase wear of lubricated steel against steel. One of these conditions is increased injection pressure, another is an increase in the soot content of engine lubricating oils. Strategic utilization of ceramic components within the fuel delivery systems of many heavy-duty truck engines has enabled the engines to maintain required performance for warranties of 500,000 miles and more. The fuel injector link introduced by Cummins in 1989 is still in production. Well over a million of these components have been manufactured. And many of these have accumulated more than a million miles of service with so little wear that they can be reused in engine rebuilds. In a newer model electronic fuel injector, Cummins introduced a zirconia timing plunger. The part has proved so successful that a second zirconia component was added to the timing plunger assembly several years later. Increasingly stringent emissions requirements for heavy diesels has increased the market for ceramic components in fuel injectors and valve train components. Many of these heavy-duty engine parts are manufactured at rates of 20,000 up to 200,000 per month.

Perhaps the largest remaining problem for this set of applications is cost. Ceramic parts are still more expensive than generally acceptable for the automotive industry. Reluctance of designers to try ceramic solutions still exists, but it is greatly diminishing thanks to the growing list of reliable and successful applications of structural ceramic engine components.

4.2 Thermostructural Applications

Due to the nature of their chemical bond, many ceramics maintain their strength and hardness to higher temperatures than metals. For example, at temperatures above 1200°C, silicon carbide and silicon nitride ceramics are considerably stronger than any superalloy. As a consequence, structural ceramics have been considered and utilized in a number of demanding applications where both mechanically imposed tensile stresses and thermally imposed tensile stresses are present. One dramatic example is the ceramic (silicon nitride) turbocharger that has been in commercial production for automobiles in Japan since 1985. Over one

Table 3 Ceramic Wear Components in Automotive and Light-Truck Engines

Component	Engine Manufacturer	Engine Type	Ceramic	Year of Introduction
Rocker arm insert	Mitsubishi	SI	Si_3N_4	1984
Tappet	Nissan	Diesel	Si_3N_4	1993
Fuel injector link	Cummins	Diesel	Si_3N_4	1989
Injector shim	Yanmar	Diesel	Si_3N_4	1991
Cam roller	Detroit Diesel	Diesel	Si_3N_4	1992
Fuel injector timing plunger	Cummins	Diesel	ZrO_2	1995
Fuel pump roller	Cummins	Diesel	Si_3N_4	1996

SI = spark-ignited engine.

million of these have been manufactured and driven with no recorded failure. This is a very demanding application, as the service temperature can reach 900°C, stresses at ~700°C can reach 325 MPa, and the rotor must also endure oxidative and corrosive exhaust gases that may contain erosion inducing rust and soot particles. Silicon nitride gas turbine nozzle vanes have been flying for several years in aircraft auxiliary power units. Other applications include heat exchangers and hot-gas valving. Recently, ceramic matrix composites have been introduced as disks for disk breaks in production sports cars by two European manufacturers. A major future market for structural ceramics may be high-performance automotive valves. Such valves are currently undergoing extensive, multiyear fleet tests in Germany.

This class of applications requires a focus on the strength, Weibull modulus, m (the higher the m, the narrower the distribution of observed strength values), thermal shock resistance, and often the stress rupture (strength decrease over time at temperature) and/or creep (deformation with time at temperature) behavior of the materials. Indeed, as shown in Fig. 2, the stress rupture performance of current structural ceramics represents a significant jump in materials performance over superalloys.

The thermal shock resistance of a ceramic is a systems property rather than a fundamental materials property. Thermal shock resistance is given by the maximum temperature change a component can sustain, ΔT:

$$\Delta T = \frac{\sigma(1 - \mu)}{\alpha E} \frac{k}{r_m h} S \qquad (2)$$

where σ is strength, μ is Poisson's ratio, α is the coefficient of thermal expansion (CTE), E is Young's modulus, k is thermal conductivity, r_m is the half-thickness for heat flow, h is the heat transfer coefficient, and S is a shape factor totally dependent on component geometry.[11] Thus it can be seen that thermal shock resistance, ΔT, is made up of terms wholly dependent on materials properties and dependent on heat transfer conditions and geometry. It is the role of the ceramic engineer to maximize the former and of the design engineer to maximize the latter two terms. It has become usual practice to report the materials-related thermal shock resistance as the instantaneous thermal shock parameter, R, which is equal to

$$R = \frac{\sigma(1 - \mu)}{\alpha E} \qquad (3)$$

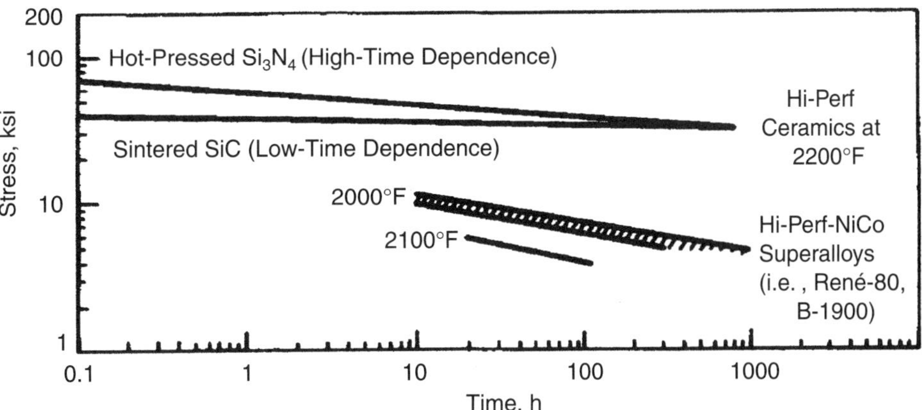

Figure 2 Stress rupture performance of nonoxide structural ceramics compared to superalloys (oxidizing atmosphere).

The value of R for selected ceramics is presented in Table 4. Another frequently used parameter is R', the thermal shock resistance where some heat flow occurs: R' is simply R multiplied by the thermal conductivity, k. For cases where heat transfer environments are complex, Ref. 12 lists 22 figures of merit for selecting ceramics to resist thermal stress.

4.3 Corrosion Resistance

Many advanced structural ceramics such as alumina, silicon nitride, or SiC have strong atomic bonding that yields materials that are highly resistant to corrosion by acidic or basic solutions at room temperature (the notable exception being glass or glass-bonded ceramics attacked by HF). This corrosion resistance has led to many applications. Carbonated soft drinks are acidic, and alumina valves are used to meter and dispense these beverages at refreshment stands. The chemical industry utilizes a wide variety of ceramic components in pumps and valves for handling corrosive materials. For example, the outstanding corrosion resistance of fully dense SiC immersed in a variety of hostile environments is given in Table 5. There are many cases where corrosion and particulate wear are superimposed, as in the handling of pulp in papermaking or transporting slurries in mineral processing operations, and ceramics find frequent application in such uses.

4.4 Passive Electronics

The role of passive electronics is to provide insulation (prevent the flow of electrons) either on a continuous basis (as in the case of substrates or packages for microelectronics) or on an intermittent basis, as is the case for ceramic capacitors (which store electric charge and hence need a high polarizability). These applications constitute two of the largest current markets for advanced ceramics. For electronic substrates and packages key issues include the minimization of thermal mismatch stresses between the Si (or GaAS) chip and the package material (so the CTE will be important) and dissipation of the heat generated as electrons flow through the millions of transistors and resistors that comprise modern microelectronic chips; hence the thermal conductivity is a key property. All other things being equal, the delay time for electrons to flow in the circuit is proportional to the square root of the dielectric constant of the substrate (or package) material. Additionally, the chip or package must maintain its insulating function, so resistivities of $>10^{14}$ are required. Most high-performance packages for computer chips are alumina. With the advent of microwave integrated circuits (e.g., cell phones) aluminum nitride substrates are beginning to be utilized

Table 4 Calculated Thermal Shock Resistance of Various Ceramics

Material	σ (MPa)	μ	CTE (cm/cm·K)	E (GPa)	R (K)
Al_2O_3 (99%)	345	0.22	7.4×10^{-6}	375	97
AlN	350	0.24	4.4×10^{-6}	350	173
SiC (sintered)	490	0.16	4.2×10^{-6}	390	251
PSZ	1000	0.3	10.5×10^{-6}	205	325
Si_3N_4 (sintered)	830	0.3	2.7×10^{-6}	290	742
LAS (glass CERAMIC)	96	0.27	0.5×10^{-6}	68	2061
Al-titanate	41	0.24	1.0×10^{-6}	11	2819

Table 5 Weight Loss of Fully Dense SiC in Acids and Bases[a]

Reagent (wt %)	Test Temperature (°C)	Weight Loss ($mg/cm^2 \cdot yr$)
98% H_2SO_4	100	1.5
50% NaOH	100	2.5
53% HF	100	<0.2
85% H_3PO_4	100	<0.2
45% KOH	100	<0.2
25% HCl	100	<0.2
10% HF + 57% HNO_3	25	<0.2

[a] Specimens submerged 125–300 h, continuously stirred.

Source: Data Courtesy of ESK-Wacker, Adrian, MI.

for high thermal conductivity. The environmental drawbacks to machining BeO have tended to favor the use of AlN to replace or avoid the use of BeO. Synthetic diamond is an emerging substrate material for special applications. Isotopically "pure" synthetic, single-crystal diamond has values of thermal conductivity approaching 10,000 W/mK. Typical values of the above properties for each of these materials are given in Table 6, along with selected properties of silicon for comparison. For design purposes exact values for specific formulations of the materials should be obtained from the manufacturers.

Over a billion ceramic capacitors or multilayer ceramic capacitors (MLCCs) are made every day.[13] Since electrons do not flow through capacitors, they are considered passive electronic components. However, the insulators from which ceramic capacitors are made polarize, thereby separating electric charge. This separated charge can be released and flow as electrons, but the electrons do not flow through the dielectric material of which the capacitor is composed. Thus, the materials parameter, which determines the amount of charge that can be stored, the dielectric constant k, is the key parameter for design and application. Table 7 lists the approximate dielectric constant at room temperature for several families of ceramics used in capacitor technology. The dielectric constant varies with both temperature and frequency. Thus, for actual design precise curves of materials performance over a relevant range of temperatures and frequencies are often utilized. Many ceramics utilized as capacitors are ferroelectrics, and the dielectric constant of these materials is usually a maximum at or near the Curie temperature.

Table 6 Key Properties for Electronic Substrates and Packages

Material	CTE (10^{-6}/K)	Thermal Conductivity (W/mK)	Resistivity	Dielectric Constant
Al_2O_3 (96%)	6.8	26	$>10^{14}$	9.5
Al_2O_3 (99%)	6.7	35	$>10^{14}$	10
AlN	4.5	140–240	$>10^{14}$	9
BeO	6.4	250	$>10^{14}$	6.5
Diamond	2	2000	$>10^{14}$	5.5
Silicon	2.8	150		

Note: CTE and thermal conductivity are at room temperature, and the dielectric constant is at 1 MHz.

Table 7 Dielectric Constants for Various Ceramic Capacitor Materials

Material	Dielectric Constant at RT
Tantalum oxide (Ta_2O_5)	~25
Barium titanate	~5,000
Barium–zirconium titanate	~20,000
Lead–zirconium titanate (PZT)	~2,000
PZT with W or Mg additives	~9,000
Lead magnesiun niobate (PMN)	~20,000
Lead zinc niobate (PZN)	~20,000

4.5 Piezoceramics

Piezoceramics are a multi-billion dollar market.[14] Piezoceramics are an enabling material for sonar systems, medical ultrasonic imaging, micromotors and micropositioning devices, the timing crystals in our electronic watches, and numerous other applications. A piezoelectric material will produce a charge (or a current) if subjected to pressure (the *direct* piezoelectric effect) or, if a voltage is applied, the material will produce a strain (the *converse* piezoelectric effect). Upon the application of a stress, a polarization charge, P, per unit area is created that equals $d\sigma$, where σ is the applied stress and d is the piezoelectric modulus. This modulus, which determines piezoelectric behavior, is a third-rank tensor[15] that is thus highly dependent on directions along which the crystal is stressed. For example, a quartz crystal stressed in the [100] direction will produce a voltage, but one stressed in the [001] direction will not. In a polycrystalline ceramic the random orientation of the grains in an as-fired piezoceramic will tend to minimize or zero out any net piezoelectric effects. Thus, polycrystalline piezoceramics have to undergo a postsintering process to align the electrically charged dipoles within the polycrystalline component. This process is known as poling and it requires the application of a very high electric field. If the piezoceramic is taken above a temperature, known as the Curie temperature, a phase transformation occurs and piezoelectricity will disappear. The piezoelectric modulus and the Curie temperature are thus two key materials selection parameters for piezoceramics.

The ability of piezoceramics to almost instantaneously convert electrical current to mechanical displacement, and vice versa, makes them highly useful as transducers. The efficiency of conversion between mechanical and electrical energy (or the converse) is measured by a parameter known as the coupling coefficient. This is a third key parameter that guides the selection of piezoelectric materials.

Although piezoelectricity was discovered by Pierre and Jacques Curie in 1880, piezoceramics were not widely utilized until the development of polycrystalline barium titanate in the 1940s and lead zirconate titanates (PZTs) in the 1950s. Both of these materials have high values of d and thus develop a high voltage for a given applied stress. PZT has become widely used because, in addition to a high d value, it also has a very high coupling coefficient. Sonar, in which ultrasonic pulses are emitted and reflected "echoes" are received, is used to locate ships and fish and map the ocean floor by navies, fishermen, and scientists all over the globe. Medical ultrasound utilizes phased arrays of piezoceramic transducers to image organs and fetuses noninvasively and without exposure to radiation. A relatively new application that has found significant use in the microelectronics industry is the use of piezoceramics to drive micropositioning devices and micromotors. Some of these devices can control positioning to a nanometer or less. Piezoceramic transducers are combined with sophisticated signal detection and generation electronics to create "active" noise and vibration damping devices. In such devices the electronics detect and quantify a noise spectrum and then drive the transducers to provide a spectrum 180° out of phase with the noise, thereby effectively canceling it.

Many of the current high-performance applications of piezoceramics are based on proprietary modifications of PZT, which contain additions of various dopants or are solid solutions with perovskite compounds of Pb with Mg, Mn, Nb, Sn, Mo, or Ni. Table 8 lists the range of several key piezoceramic selection parameters for proprietary PZT compositions from one manufacturer.

4.6 Transparencies

Transparent ceramics (which include glasses and single-crystal and polycrystalline ceramics) have been used as optical transparencies or lenses for millennia. Glass windows were in commercial production in first-century Rome, but it was not until the 1800s, with the need for precision optics for microscopes, telescopes, and ophthalmic lenses, that glasses and other optical materials became the object of serious scientific study. As noted in the introduction, progress in glass science and technology, coupled with lasers, has led to the current broadband digital data transmission revolution via optical fibers. Various ceramic crystals are used as laser hosts and specialty optical lenses and windows. A significant fraction of supermarket scanner windows combine the scratch resistance of sapphire (single-crystal alumina) with its ability to transmit the red laser light that we see at the checkout counter. While such windows are significantly more costly than glass, their replacement rate is so low that they have increased profitability for several supermarket chains. For the same reason the crystal in many high-end watches are scratch-resistant man-made sapphire. Polycrystalline translucent (as opposed to fully transparent) alumina is used as containers (envelopes) for the sodium vapor lamps that light our highways and industrial sites.

Not all windows have to pass visible light. Radar or mid- to far-infrared transparencies look opaque to the human eye but are perfectly functional windows at their design wavelengths. The most demanding applications for such transparencies is for the guidance domes of missiles. Materials that can be used for missile radomes include slip-cast fused silica, various grades of pyroceram (glass ceramics), and silicon-nitride-based materials. Infrared (IR) windows and missile domes include MgF_2 and ZnSe. Requirements exist for having missile guidance domes that can transmit in the visible, IR, and radar frequencies (multimode domes). Ceramic materials that can provide such functionality include sapphire and aluminum oxynitride spinel (AlON). In addition to optical properties missile domes must be able to take high aerothermal loading (have sufficient strength) and be thermal shock resistant (a high-speed missile encountering a rain cloud can have an instant ΔT of minus several hundred degrees kelvin).

Table 8 Key Properties for PZT-Based Piezoceramics

Material	Piezoelectric Modulus, d_{33}	Curie Temperature (°C)	Coupling Coefficent, k_{33}
A	226×10^{-12} m/V	320	0.67
B	635×10^{-12} m/V	145	0.68
C	417×10^{-12} m/V	330	0.73

Key properties for visible and IR optical materials include the index of refraction, n (which will be a function of wavelength), and absorption or loss. For radar transparencies key parameters are dielectric constant (which can be thought of as analogous to the index of refraction) and dielectric loss.

5 INFORMATION SOURCES

5.1 Manufacturers and Suppliers

There are hundreds of manufacturers of advanced ceramics and glasses. Locating ones that already have the material that is needed and can produce it in the configuration required can be a daunting task. There are two resources published annually that make this task much easier. The American Ceramic Society publishes a directory of suppliers of materials, supplies, and services that can help locate such information quickly. It is called *Ceramic Source.* This directory can also be accessed on the Web at www.ceramics.org. A similar *Buyers Guide* is published by *Ceramic Industry Magazine,* and this can also be viewed online at www.ceramicindustry.com. Once a likely source for your need has been identified, a visit to the supplier's web site can often provide a great deal of background information and specific data, which can make further contacts with the supplier much more meaningful and informative.

5.2 Data

Manufacturer's literature, both hard copy and posted on the Web, is an invaluable source of data. The handbooks, textbooks, and encyclopedias listed below are also excellent sources of data. However, before committing to a finalized design or to production, it is advisable to develop your own test data in conformance with your organization's design practice. Such data should be acquired from actual components made by the material, processing route, and manufacturer that have been selected for the production item.

ASM, *Ceramics and Glasses,* Vol. 4: *Engineered Materials Handbook,* ASM International, Materials Park, OH, 1991.

J. F. Shackelsford, W. Alexander, and J. Park (eds.), *Materials Science and Engineering Handbook,* 2nd ed., CRC Press, Boca Raton, FL, 1994.

C. X. Campbell and S. K. El-Rahaiby (eds.), *Databook on Mechanical and Thermophysical Properties of Whisker-Reinforced Ceramic Matrix Composites,* Ceramics Information Analysis Center, Purdue University, W. Lafayette, IN, and The American Ceramic Society, Westerville, OH, 1995.

R. J. Brook (ed.), *Concise Encyclopedia of Advanced Ceramic Materials,* Pergamon Press, Oxford, 1991.

5.3 Standards and Test Methods

To reliably design, procure materials, and assure reliability, it is necessary to have common, agreed-upon, and authoritative test standards, methods, and practices. Institutions such as the American Society for Testing of Materials (ASTM), the Japanese Institute for Standards (JIS), the German Standards Organization (DIN), and the International Standards Organization (ISO) all provide standards for their various constituencies. The following are a sampling of standards available from the ASTM and JIS for advanced ceramics and ceramic matrix composites. One can reach these organizations at the following addresses.

American Society for Testing Materials (ASTM), 100 Barr Harbor Drive, Conshohocken, PA 19428-2959:

- C-177-85(1993), Test Method for Steady State Heat Flux and Thermal Transmission by Means of the Gradient-Hot-Plate Apparatus
- C-1161-90, Test Method for Flexural Strength of Advanced Ceramics at Ambient Temperature
- C-1211-92, Test Method for Flexural Strength of Advanced Ceramics at Elevated Temperature
- C-1259-94, Test Method for Dynamic Young's Modulus, Shear Modulus and Poisson's Ratio for Advanced Ceramics by Impulse Excitation of Vibration
- C-1286-94, Classification for Advanced Ceramics
- C-1292-95A, Test Method for Shear Strength of Continuous Fiber-Reinforced Ceranic Composites (CFCCs) at Ambient Temperatures
- C-1337-96, Test Method for Creep and Creep-Rupture of CFFFs under Tensile Loading at Elevated Temperature
- C-1421-99, Standard Test Method for Determination of Fracture Toughness of Advanced Ceramics at Ambient Temperature
- C-1425-99, Test Method for Interlaminar Shear Strength of 1-D and 2D CFCCs at Elevated Temperature
- E-228-85(1989), Test Method for Linear Thermal Expansion of Solid Materials with a Vitreous Silica Dilatometer
- E-1269-94, Test Method for Determining Specific Heat Capacity by Differential Scanning Calorimetry
- E-1461-92, Test Method for Thermal Diffusion of Solids by the Flash Method

The Japanese Standards Association, 1-24, Akasaka 4, Minato-ku, Tokyo 107 Japan:

- Testing Methods for Elastic Modulus of High Performance Ceramics at Elevated Temperatures; JIS R 1605-(1989)
- Testing Methods for Tensile Strength of High Performance Ceramics at Room and Elevated Temperatures; JIS R 1606-(1990)
- Testing Methods for Fracture Toughness of High Performance Ceramics; JIS R 1607-(1990)
- Testing Methods for Compressive Strength of High Performance Ceramics; JIS R 1608-(1990)
- Testing Methods for Oxidation Resistance of Non-Oxide of High Performance Ceramics; JIS R 1609-(1990)
- Testing Methods for Vickers Hardness of High Performance Ceramics; JIS R 1610-(1991)
- Testing Methods of Thermal Diffusivity, Specific Heat Capacity, and Thermal Conductivity for High Performance Ceramics by Laser Flash Method; JIS R 1611-(1991)

In addition to testing standards, it is possible to obtain standard materials with certified properties to calibrate several of these new standards against your own tests. Such standard materials can be obtained from the National Institute of Science and Technology (NIST). For example, a standard material to calibrate ASTM C-1421-99 has just been made available. Materials standards are not available for all of the above tests.

5.4 Design Handbooks

It has been widely recognized that procedural handbooks that provide methodology on how to design with advanced ceramics and which can provide high-quality evaluated design data are sorely needed for ceramic materials. The ceramics matrix composites (CMCs) community has taken the initiative to begin the process of creating such a handbook for its constituency. The activity is sponsored by various U.S. governmental agencies, including the Department of Defense, the Department of Energy, the Federal Aviation Administration, and the National Aeronautics and Space Administration, and is entitled MIL-Handbook-17. This activity brings together materials suppliers, materials testers, designers, and end users who are engaged in developing a handbook that will provide design tools and guidance; provide guidelines on data generation, documentation, and use; and provide an authoritative source of design quality data. This is a work in progress and its completion is many years off, if ever. Nevertheless, much guidance in design and testing of advanced CMCs has already resulted from this activity. Progress can be followed by periodically accessing the handbook web sites at http://mil-17.udel.edu or http://www.materials-sciences.com/MIL17/. Unfortunately, no similar activity exists for monolithic ceramics.

6 FUTURE TRENDS

It has been estimated that in the United States advanced ceramics of the type discussed above are an over $8-billion-a-year industry with a growth rate of ~8% per year.[16]

The largest segment of this growth will come from the electronics area. Not only will there be significant growth in the "traditional" roles of ceramics as insulators, packages, substrates, and capacitors, but structural ceramics will play a major role in the equipment used in semiconductor manufacturing. This trend will be especially driven by the resistance of ceramics such as SiC, AlN, silicon nitride, and alumina to the erosive and corrosive environments within high-energy plasma chambers used in single-wafer processing operations.

The intertwined global issues of energy sufficiency and environmental protection will see commercial use of advanced ceramics in energy systems as diverse as solid oxide fuel cells and pebble-bed modular reactors (nuclear). As more and more industries move toward "green" (pollution-free) manufacturing, there will be growth in wear- and corrosion-resistant ceramics for industrial machinery. There will also be substantial growth potential for ceramic filters and membranes. One major environmentally driven opportunity will be particulate traps for diesel trucks and industrial power sources. This technology is just beginning to be commercialized, and it is certain to see rapid growth as emissions requirements for diesel engines grow more stringent. Not all progress in these areas will create increased markets for ceramics; some will reduce them. For example, the rapid growth of energy-efficient light-emitting diode technology for illumination will create a significant growth opportunity for producers of single-crystal SiC substrates and GaN materials. However, this will come at a cost to the ceramics industry of a significant decrease in glass envelopes for incandescent bulbs and fluorescent tubes. Another area of growth will be filters and membranes for filtration of hot or corrosive, or both, gases and liquids.

The explosive growth of fiber-optic- and microwave-based digital communications technology has produced significant opportunities and markets for advanced ceramics and glasses and will continue to do so into the foreseeable future.

Medical applications are sure to grow, in the areas both of diagnostics and prosthetics.

At the entrance to the Pohang Steel complex in Pohang, Republic of Korea, is a wonderful sign. It proclaims, "Resources are Limited—Creativity is Unlimited." This thought certainly applies to the global future of advanced ceramics. Creatively utilized advanced ceramics will effectively expand our resources, protect our environment, and create new technological opportunities. The potential opportunities go far beyond the few discussed in this chapter.

REFERENCES

1. P. B. Vandiver, "Reconstructing and Interpreting the Technologies of Ancient Ceramics," in *Materials Issues in Art and Archaeology,* Materials Res. Soc. Symposium Proceed., Vol. 123, Materials Research Socity, Pittsburgh, 1988, pp. 89–102.

2. *Materials Science and Engineering for the 1990's,* National Academy Press, Washington, DC, 1989, p. 24.

3. R. N. Katz, "Piezoceramics," *Ceramic Industry,* p. 20 (Aug. 20, 2000).

4. D. W. Richerson, *Modern Ceramic Engineering,* 2nd ed., Marcel Dekker, New York, 1992, pp. 582–588.

5. J. S. Haggerty and Y. M. Chiang, "Reaction-Based Processing Methods for Materials and Composites," *Ceramic Eng. Sci. Proc.,* **11**(7–8), 757–781 (1990).

6. See Ref. 4, Chapters 9–11.

7. A. F. McLean and D. Hartsock, "Design with Structural Ceramics," in *Treatise on Materials Science and Technology,* Vol. 29, J. B. Wachtman (ed.), Academic Press, Boston, 1989, pp. 27–95.

8. N. N. Nemeth and J. P. Gyekenyesi, "Probabilistic Design of Ceramic Components with the NASA/CARES Computer Program," in *Ceramics and Glasses,* Vol. 4, *Engineered Materials Handbook,* ASM International, Metals Park, OH, 1991, pp. 700–708.

9. R. N. Katz, "Application of High Performance Ceramics in Heat Engine Design," *Materials Sci. Eng.,* **71,** 227–249 (1985).

10. R. N. Katz, "Ceramic Materials for Roller Element Bearing Application," in *Friction and Wear of Ceramics,* S. Jahanmir (ed.), Marcel Dekker, New York, 1994, pp. 313–328.

11. W. D. Kingery, H. K. Bowen, and D. R. Uhlmann, *Introduction to Ceramics,* 2nd ed., Wiley, New York, 1976.

12. D. P. H. Hasselman, "Figures-of-Merit for the Thermal Stress Resistance of High Temperature Brittle Materials: A Review," *Ceramurgia International,* **4**(4), 147–150 (1998).

13. D. W. Richerson, *The Magic of Ceramics,* The American Ceramic Society, Westerville, OH, 2000, p. 141.

14. NSF Workshop Report, Fundamental Research Needs in Ceramics, Washington, DC, April, 1999, p. 9.

15. J. F. Nye, *Physical Properties of Crystals,* Oxford University Press, London, 1964.

16. T. Abraham, "US Advanced Ceramics Growth Continues," *Ceramic Industry,* 23–25 (Aug. 2000).

CHAPTER 13
SOURCES OF MATERIALS DATA

J. G. Kaufman
Kaufman Associates
Columbus, Ohio

1 INTRODUCTION AND SCOPE

It is the purpose of this chapter to aid engineers and materials scientists in locating reliable sources of high-quality materials property data. While sources in hard-copy form are referenced, the main focus is on electronic sources that provide well-documented searchable property data.

To identify useful sources of materials data, it is important to have clearly in mind at the outset (a) the intended use of data, (b) the type of data required, and (c) the quality of data required. These three factors are key in narrowing a search and improving its efficiency of a search for property data. Therefore, as an introduction to the identification of some specific potentially useful sources of materials data, we will discuss those three factors in some detail and then describe the options available in types of data sources.

Readers interested in a more comprehensive list of sources and of more discussion of the technology of material property data technology and terminology are referred to Westbrook's extensive treatment of these subjects in Refs. 1 and 2 and to the *ASM International Directory of Materials Property Databases.*[3]

2 INTENDED USES FOR DATA

Numeric material property data are typically needed for one of the following purposes by individuals performing the respective functions given in parentheses as part of their jobs:

- Modeling material or product performance
- Materials selection (finding candidate materials for specific applications)
- Analytical comparisons (narrowing the choices)
- Preliminary design (initially sizing the components)
- Final design (assuring performance; setting performance specifications)
- Material specification (defining specifications for purchase)
- Manufacturing (assuring processes to achieve desired product)
- Quality assurance (monitoring manufacturing quality)
- Maintenance (repairing deterioration/damage)
- Failure analysis (figuring out what went wrong)

It is useful to note some of the differing characteristics of data needed for these different functions.

2.1 Modeling Material and/or Product Performance

To an increasing extent, mathematical modeling is used to establish the first estimates of the required product performance and material behavior and even in some cases the optimum manufacturing processes that should be used to achieve the desired performance. The processes and/or performance analyzed and represented may include any of the issues addressed in the following paragraphs, and so the nature of the types of data described under the various needs is the same as that needed for the modeling process itself.

2.2 Materials Selection

The needs of materials specialists and engineers looking for materials data to aid in the selection of a material for some specific application are likely to be influenced by whether they (a) are in the early stages of their process or (b) have already narrowed the options down to two or three candidates and are trying to make the final choice. The second situation is covered in Section 2.3.

If the materials engineers are in the early stages of finding candidate materials for the application, they are likely to be looking for a wide variety of properties for a number of candidate materials. More often, however, they may decide to focus on two or three key properties that most closely define the requirements for that application and search for all possible materials providing relatively favorable combinations of those key properties. In either case, they may not be as much concerned about the quality and statistical reliability of the data at this stage as by the ability to find a wide variety of candidates and to make direct comparisons of the performance of those candidates.

Where there is interest in including relatively newly developed materials in the survey, it may be necessary to be satisfied with only a few representative test results or even educated guesstimates of how the new materials may be expected to perform. The engineers will need to be able to translate these few data into comparisons with the more established materials, but at this stage they are probably most concerned with not missing out on important new materials.

Thus, at this early stage of materials selection, the decision makers may be willing to accept data rather widely ranging in type and quality, with few restrictions on statistical reliability. They may even be satisfied with quite limited data to identify a candidate that may merit further evaluation.

2.3 Analytical Comparisons

If, on the other hand, the task is to make a final decision on which of two or three candidate materials should be selected for design implementation (the process defined here as "analytical comparison"), the quality and reliability of the data become substantially more important, particularly with regard to the key performance requirements for the application. It will be important that all of those key properties, e.g., density, tensile yield strength, and plane-strain fracture toughness, are available for all of the candidate materials make the next cut in the list.

The search will also be for data sources where the background of the data is well defined in terms of the number of tests made, the number of different lots tested, and whether the numbers included in the data source are averages or the result of some statistical calculation, such as that to define 3σ limits. It would not be appropriate at this stage to be uncertain whether the available data represent typical, average values or statistically minimum properties; it may not be important which they are, but the same quality and reliability must be available for all of the final candidates for a useful decision to be made.

In addition, the ability to make direct comparisons of properties generated by essentially the same, ideally standard, methods is very important. The decision maker will want to be able to determine if the properties reported were determined from the same or similar procedures and whether or not those procedures conformed to ASTM, ISO,[4] or other applicable standard test methods.

One final requirement is added at this stage: The materials for which the data are presented must all represent to the degree possible comparable stages of material production history. It would be unwise to base serious decisions on comparisons of data for a laboratory sample on one hand and a commercial-size production lot on the other. Laboratory samples have a regrettable history of promising performance that is seldom replicated in production-size lots.

Thus, for analytical comparisons for final candidate material selection, specialists need databases for which a relatively complete background of metadata (i.e., data about the data) are included and readily accessible.

Incidentally, it is not unusual at this point in the total process to decide that more data are needed for a particular candidate than are available in any existing database, and so a new series of tests is needed to increase confidence in the comparisons being made.

2.4 Preliminary Design

Once a decision is made on a candidate material (or sometimes two) for an application, the task of designing a real component out of that material begins. The requirement for statistical reliability steps up, and the availability of a data source that provides applicable metadata covering quality, reliability, and material history becomes even more important.

At this stage, the statistical reliability required includes not only a minimum value but one based upon a statistically significant sample size, ideally something comparable to the standards required in the establishment of MMPDS-01/MIL-HDBK-5J's A or B values.[5] In MMPDS/MIL-HDBK-5 terminology, an A value is one which would be expected to be equaled or exceeded by 99% of the lots tested with 95% confidence; the B value provides for 90% of lots tested equaling or exceeding the value with 95% confidence. Furthermore, the MIL-HDBK-5 guidelines require that A and B values be based upon predefined sample sizes, representing a minimum number of lots (normally 100 or more) and compositions (normally at least three) of a given alloy. The provision of such statistical levels needs to be a part of data sources used for design purposes, and the description of the statistical quality needs to be readily available in the data source.

For preliminary design, then, the data sources sought will include both statistically reliable data and well-defined metadata concerning the quality and reliability provided.

2.5 Final Design

Setting the final design parameters for any component or structure typically requires not only data of the highest level of statistical reliability but also, in many cases, data that have been sanctioned by some group of experts for use for the given purpose. It is also not unusual that at this stage the need is identified for additional test data generated under conditions as close as possible to the intended service conditions, conditions perhaps not available from any commercial database.

Databases providing the level of information required at this stage often contain what are characterized as "evaluated" or "certified" values. Evaluated data are those that, in addition to whatever analytical or statistical treatment they have been given, have been overviewed by an expert or group of experts who make a judgment as to whether or not the data adequately and completely represent the intended service conditions and if it is necessary to incorporate their own analysis into the final figures. This technique has been widely used in digesting and promulgating representative physical property data for many years; examples are the thermophysical property data provided by TPRC.[6]

Other databases may be said to provide "certified" data. In this case, the database or set of data going into a database have been evaluated by a group of experts and certified as the appropriate ones to be used for the design of a particular type of structure. Two examples are the aforementioned MMPDS/MIL-HDBK-5 values,[5] which are approved for aircraft design by the MMPDS/MIL-HDBK-5 Coordination Committee consisting of aerospace materials experts, and the ASME Boiler and Pressure Vessel Code,[7] with properties certified by materials experts in that field for the design of pressure vessels and companion equipment for high-temperature chemical processes.

As noted, it is often the case when designers reach this stage (if not the earlier preliminary design stage) that they find it necessary to conduct additional tests of some very specialized type to ensure adequate performance under the specific conditions the component or structure will see service but for which reliable databases have not previously been identified or developed. The net result is the creation of new materials databases to meet highly specialized needs and the need to do so in a manner that provides the appropriate level of statistical confidence.

2.6 Material Specification

Material specifications typically include specific property values that must be equaled or exceeded in tests of those materials that are being bought and sold. The properties that one requires in this case may differ from those needed for other purposes in two respects. First and foremost, they must be properties that will ensure that the material has been given the desired mechanical and thermal processing to consistently achieve the desired performance. Second, while in most cases there may be only one or a very few properties required (most often tensile properties), they are required at a very high level of precision and accuracy, similar to or better than that required for the MMPDS/MIL-HDBK-5 A properties defined earlier.

Examples are the material specifications required for the purchase of commercial aluminum alloy products.[8] These are usually only the chemical composition and the tensile properties. So while many of the other properties needed for design are not required as part of the purchase specification, those properties that are required are needed with very high reliability. In the case of aluminum alloys, the requirements for tensile strength, yield strength, and elongation are normally that 99% of lots produced must have properties that equal or exceed the published purchase specification values with 95% confidence, and they must have been defined from more that 100 different production lots from two or more producers.

In many cases, the databases needed to generate material specification properties are proprietary and are contained within individual companies or within the organizations that set industry specifications. However, the resultant statistically reliable specification properties are resident and more readily available in industry or ASTM material standards.[4,8]

2.7 Manufacturing

The properties required for manufacturing purposes may be the most difficult to find in commercially available databases because they typically involve the specialized treatments or processes utilized by specific producers or suppliers of the specific products in question. Sometimes these processes are proprietary and are closely held for competitive advantage. An added complication is that once some semifabricated component (e.g., aircraft sheet) has been purchased, it will require forming to very tight tolerances or finishing at some relatively high temperature. The fabricator may require data to enable the process to be carried through without otherwise damaging or changing the properties of the component but may have difficulty representing the fabricating conditions in meaningful tests. Fabricators may well have to carry out their own tests and build the needed database to provide the desired assurance of quality and to provide a source of information to which their employees can refer to answer specific questions. Typically such databases never become commercially available, and new situations will require compiling new data sources.

Some processing data sources are available, of course. The Aluminum Association, for example, provides to all interested parties a data source defining standard solution heat treating, artificial aging, and annealing treatments for aluminum alloys that will assure the proper levels of properties will be obtained.[8] In some cases, ASTM and ANSI material specifications will also contain such information.

2.8 Quality Assurance

Quality assurance may be considered to be the flip side of material specifications, and so the types of data and the data sources themselves required for the two functions are essentially the same.

Purchasers of materials, for example, may choose to do their own testing of the materials once delivered to their facilities by materials producers. If so, they will use exactly the same tests and refer to exactly the same data sources to determine compliance. The one difference may be that such purchasers may choose to gradually accumulate the results of such tests and build their own databases for internal use, not only by their quality control experts but also by their designers and materials experts, who must establish safe levels of performance of the structures. These types of databases also tend to be proprietary, of course, and are seldom made available to the outside world, especially competitors.

2.9 Maintenance

The principal value of material data in connection with maintenance concerns will likely be for reference purposes when problems show up with either deterioration of surface conditions or the suspicion of the development of fatigue cracks at local stress raisers.

In both cases, the important features of the types of data desired to address such issues are more likely to be those based upon individual exposures or prior service experience, and so the user may be more concerned about the degree of applicability than upon data quality and statistical reliability (though both features would be desired if available). Typically such data are hard to find in any event and once again are more likely to be buried in proprietary files than in published databases.

In the case of engineers and technicians needing databases comprised of service experience, they may well be faced with building their own data sources based upon their organization's production and service experience than expecting to be able to locate applicable external sources.

2.10 Failure Analysis

The occurrence of unexpected failure of components in a structure usually calls for some follow-up study to determine the cause and possible ways to avoid further loss. In such cases it is inevitable that such failure analysis will involve both (a) a review of the old databases used to design the part and (b) a search for or the development of new data sources that may shed more light on the material's response to conditions that developed during the life of the structure that had not been anticipated beforehand.

The types of databases sought in this case will likely be those containing statistically reliable data, but recognizing the unexpected nature of some problems, an interest in a wider range of data sources and a willingness to consider a lower level of data quality may result. Databases for failure analysis studies may need to be wider in scope and to cover subjects like corrosion that are not always easily treated by statistical means. In fact, sources covering failure experience may be the most valuable, though hardest to find, because historically engineers and scientists do not publish much detail about their mistakes.

The net result is that when dealing with failure analysis, the search may be quite broad in terms of data quality, and the focus most likely will be more on applicability to the problem than on the quality and structure of the compilation. As in the case of maintenance engineers and technicians needing databases comprised of service experience, failure analysts may well be faced with building new data sources based upon their organization's production and service experience.

3 TYPES OF DATA

It is useful at this stage to note that there are several basic types of materials databases, i.e., databases containing significantly different types of information and, hence, different data formats. Note that this is different from the type of platform or presentation format (e.g., hard copy, CD, online); these will be discussed in Section 6.

The two fundamental types of databases that will be discussed here are *textual data* and *numeric data*. In fact, many databases represent a combination of both types, but some basic differences worth noting are presented below.

The concept of metadata will also be described in more detail in this section.

3.1 Textual Data

The terminology "textual data" is generally applied to data entries that are purely alphabetic in nature with numbers used only as necessary to complete the thoughts. Textual databases are typically searched with alphabetical strings, e.g., by searching for all occurrences of a term like *aluminum* or *metallurgy* or whatever subject is of interest.

The subjects of textual databases are predominantly bibliographies and abstracts of publications, but they may reflect other specific subjects, such as textual descriptions of failures of components. Bibliographic databases seldom reflect the final answers to whatever inquiry is in mind but rather provide references to sources where the answers may be found.

The majority of all databases in existence in any form (see Section 6) are textual in nature, even many of those purporting to be property databases. Searchers of such textual property databases are searching based upon strings of alphanumeric characters reflecting their interest, not on numeric values of the properties, except as they are expressed as strings. This is quite different from the case for numeric property databases, as we shall see in the next section.

3.2 Numeric Databases

Databases classed as numeric (a) have data stored in them in numeric format and (b) are searched numerically. For example, numeric databases may be searched for all materials having a specific property equal to or greater than a certain values or within a certain specific range; this would not be possible in a textual database. To provide such searchability, almost all numeric databases are electronic in nature (see Section 4) though many hard-copy publications also contain extensive amounts of numeric data.

To accommodate numeric searching, the properties must be entered into a database digitally as numbers, not simply as alphanumeric strings. To be useful to material specialists and designers, they must have meaningful precision (i.e., numbers of significant figures) and units associated with each number. The numeric number of a property is of no value if it does not have the applicable unit(s) associated with it.

3.3 Metadata

The concept of metadata as "data about data" was introduced earlier, and it is appropriate at this stage to describe the concept and its importance in greater detail.[9] It is vital, especially for numeric databases and independent of their platform, to have ample background information on the numeric properties included in the database and closely associated with the individual properties.

Examples of metadata include the following:

- Original source of data (e.g., from tests at ABC laboratories)
- Test methods by which data were obtained (e.g., ASTM methods, size and type of specimen)
- Production history of the samples for which the data are applicable (e.g., annealed, heat treated, cold worked, special handling?)

- Exposure experienced by the samples tested, if any, prior to the test (e.g., held 1000 h at 500°F)
- Conditions under which the properties were determined (e.g., tested at 300°F, 50% humidity)
- Number of individual tests represented by and statistical precision of the values presented (e.g., individual test results, averages of *x* number of tests, statistical basis)
- Any subsequent evaluation or certification of the data by experts (e.g., ASME Boiler and Pressure Vessel Committee)

It should be clear that the value of information in any database, but especially in a numeric database, is greatly diminished by the absence of at least some and potentially all of these types of metadata. For example, a listing of the properties of an alloy is of no value at all if it is not clear at what temperature they were determined. It should be clear that metadata are an integral part of every material property database. Similarly the properties of a material are of virtually no value if it is not clear how the materials were mechanically and thermally processed before they were tested.

4 SUBJECTS OF DATA SOURCES

Next it is useful to note some of the major categories of available data and illustrate the manner in which they are likely to be classified or structured. From this point on, the discussion will focus primarily upon numeric materials property data of interest to the engineer and scientist.

The total breadth of properties and characteristics may reasonably be illustrated by the following four categories:

- Fundamental (atomic-level) properties
- Physical properties (atomic and macro/alloy levels)
- Mechanical properties (macro/alloy level)
- Application performance (macro/alloy or component levels)

To provide greater detail on the first three categories, it is helpful to utilize the taxonomy of materials information developed by Westbrook[1] and illustrated in Fig. 1. While the entire list of potential uses of data described above is not included, the taxonomy in Fig. 1 illustrates the variety of subject matter quite well.

The fourth major category identified above involves what is referred to as application performance but which in itself incorporates several areas:

1. Fabrication characteristics (sometimes called the "ilities")
   - Fabricability ("workability")
   - Forming characteristics ("formability")
   - Joining characteristics ("joinability")
   - Finishing characteristics ("finishability")
2. Service experience
   - Exposure conditions
   - Service history
   - Failures observed and their causes

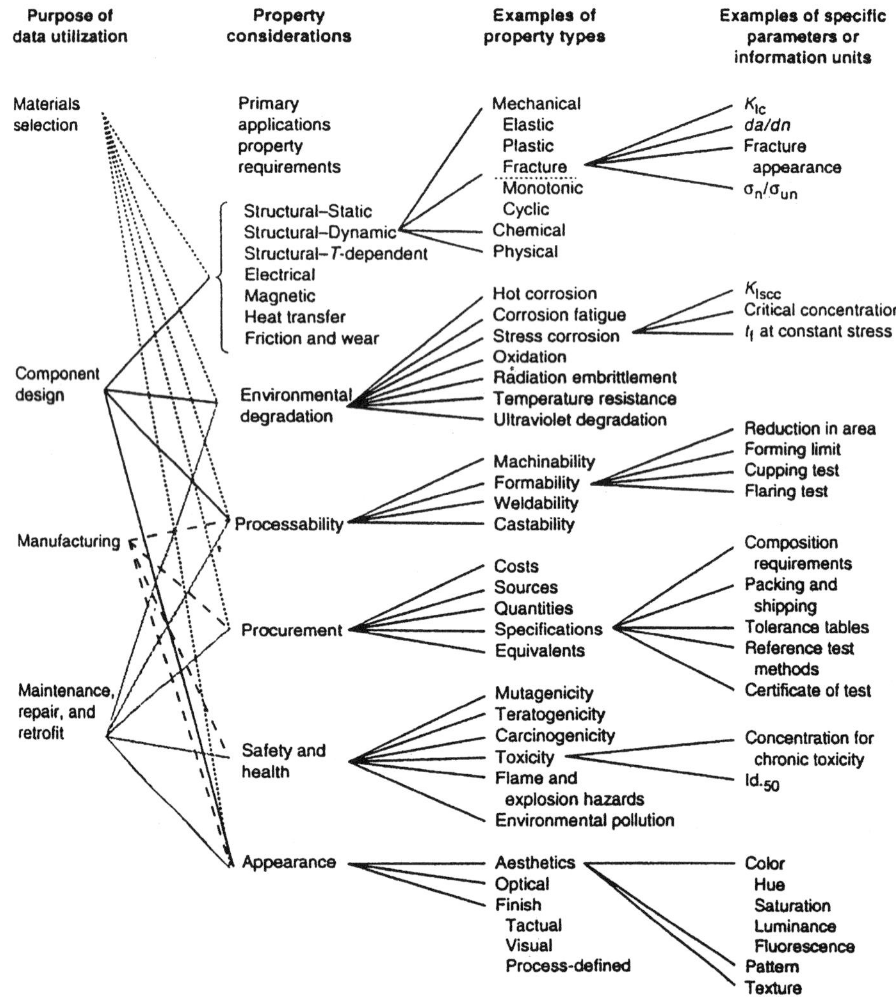

Figure 1 Taxanomy of materials information. From *ASM Handbook,* Vol. 20: *Materials Selection and Design (1997),* ASM International, Materials Park, OH, Fig. 1, p. 491.[1]

It is not necessary to discuss these categories individually; rather one simply needs to recognize that specific databases may focus on only one or a few of these subjects and rarely if ever would all subjects be included in any one database.

5 DATA QUALITY AND RELIABILITY

It was noted in Section 2 that individuals doing preliminary material selection or screening may have different needs with regard to quality and reliability than those doing design functions. It is appropriate at this stage to review the major factors that go into judgments of data quality and reliability.[10] There are two parts to such a discussion: (1) the two major factors affecting the data themselves and (2) the degree to which those factors are reflected in database content.

The two factors affecting quality and reliability include (a) the manner in which the data were obtained and (b) the statistical reliability of the data presented.

First and foremost, the users of a database will want to know that consistent standards were applied in assembling the data for that database and that the properties of one material may reliably be compared with those for another. They will also prefer that those properties have been generated by well-known standard methods such as those prescribed by organizations such as ASTM and ISO.[4] They may also be interested in knowing the specific source of the original data; realistically some laboratories (e.g., the National Institute of Standards and Technology, NIST) have a more widely recognized reputation than others.

The second major factor affecting how the user applies information from a specific database is the statistical reliability of the values included therein. The user will be interested in which of the following categories best describes the values presented:

- Raw data (the results of individual tests)
- Average of multiple tests (how many tests are represented?)
- Statistically analyzed (standard deviation, minimum value at what statistical definition, and with what confidence?)
- Evaluated/certified (by whom and for what purpose?)

Of equal importance to the user of database is the degree to which those factors affecting quality and reliability as noted above are presented in the database itself and, therefore, are fully understood by the user. In some instances there may be one or several screens of background information laying out the general guidelines upon which the database was generated. This is particularly effective if the entire database represents properties of a common lineage and character. On the other hand, if the user fails to consult this upfront information, some important delimiters may be missed and the data misinterpreted.

Another means of presenting the metadata concerning quality and reliability is within the database itself. This is especially true for such factors as time–temperature parameters that delimit the applicability of the data and units and other elements of information that may restrict its application (see Section 3.3).

It is also especially important in instances where the individual values may vary with respect to their statistical reliability. While the latter case may seem unlikely, it is actually quite common, as when "design" data are presented: In such cases, the strength values are likely to be statistical minimum values while moduli of elasticity and physical properties are likely to be average values, and the difference should be made clear in the database.

6 PLATFORMS: TYPES OF DATA SOURCES

The last feature we will consider before identifying specific data sources is the variety of platforms available for databases today. While it is not necessary to discuss them in detail, it will be obvious that the following choices exist:

- Hard copy (published books, monographs, etc.)
- Self-contained electronic (floppies, CDs, etc.)
- Internet sites (available online; perhaps downloadable)

The only amplification needed on these obvious options is that the last one, the Internet, has become an interesting resource from which to identify and locate specific sources of materials data, and that trend will likely continue to increase. Two large caveats go with the use of Internet sources, however: (1) A great deal of "junk" (i.e., unreliable, undocumented data) may be found on Internet web sites and (2) even those containing more reliable data seldom meet all users' needs with respect to covering the metadata. It is vital that the users themselves apply the guidelines listed earlier to judge the quality and reliability of sources located on the Internet.

7 SPECIFIC DATA SOURCES

It is beyond the scope of this chapter to provide an exhaustive list of data sources because there are thousands of them of varying presentation platforms, styles, and content. What we will do here is highlight a few of the potentially most useful sources, in the sense that their coverage is relatively broad and/or they represent good places to go to look for new and emerging sources of materials data.

The sources discussed below representing the variety of types of sources available will include the following:

- The Alloy Center on ASM International's website; ASM International is a technical society for materials engineers and specialists that produces and provides a wide range of materials databases in hard-copy and electronic form; this service is intended to be used by the engineers and material specialists themselves, the "end users" (see Section 7.1).
- STN International, a service of the American Chemical Society, also provides (for a fee) access to a very wide range of numeric databases as well as a great many textual/bibliographic databases on materials-related subjects; this service is focused on the professional researcher and technical librarian, not the "end user" (see Section 7.2).
- The Internet site knovel.com provides full-text searchable handbooks of many disciplines, including materials data. (see Section 7.3).
- The Internet as a whole is the most rapidly expanding source for materials data in a variety of forms (see Section 7.4).

For readers interested in more extensive listings, reference is made to the article "Sources of Materials Property Data and Information" by Jack Westbrook in Vol. 20 of the *ASM Handbook*[1] and to the *ASM Directory of Databases*.[3]

7.1 ASM International and the Alloy Center

ASM International has emerged as one of the strongest providers of numeric materials data sources, and those sources are generally in three formats or platforms: hard copy, disk (usually CD-ROM), and the Alloy Center on the Internet. As an example, one of the most extensive sources of high- and low-temperature data for aluminum alloys has recently been made available through a collaborative effort of ASM and the Aluminum Association in both the book *Properties of Aluminum Alloys—Tensile, Creep, and Fatigue Data at High and Low Temperatures*[11] and a searchable CD of the same title. Other representative data sources from ASM International include the following:

- ASM Handbook,[12] in Hard Copy and CD. Twenty volumes complete or available in a specific set covering material properties; the data sets are available for single workstations and also for local area network (LAN) arrangements. Other CDs are available covering heat treatment, testing and analysis, and manufacturing processes.

- Alloy Finder CD. Contains full alloy records from three ASM hard-copy reference standards: *Woldman's Engineering Alloys,*[13] the ASM *Worldwide Guide to Equivalent Irons and Steels.*[14] and the ASM *Worldwide Guide to Equivalent Nonferrous Metals and Alloys.*[15] The disk is searchable by composition as well as designation, so whether the user requires amplification of an alloy designation or to locate designations within specific composition ranges, the need is addressed.
- Alloy Digest on CD. Summaries of recently published data for new and emerging alloys are compiled periodically on disk as well as being made available in hard copy. The advantages include early warning of new materials; the limitation is the inability to provide consistent formats or data scope because such are not available for relatively new materials. More than 4200 data sheets are now available.
- Binary-Phase Diagrams on CD-ROM. The world's most extensive collection of binary-phase diagrams numbering in excess of 4700 is available on CD, in addition to ASM hard-copy publications such *Handbook of Ternary Alloy Phase Diagrams* and the monograph series on specific alloy groupings.
- Failure Analysis on CD-ROM. One of the most extensive collection of data expressly developed for failure analysis is available on searchable CD-ROM as well as hard copy.
- Materials Databases. Numeric materials databases covering a variety of classes of materials are available from ASM on disk in various formats, including within Mat.DB, MAPP, and Rover search software. The databases are organized and searchable by mechanical and physical property as well as alloy class: steel and stainless steel, aluminum, composites, copper, magnesium, plastics and polymers, nylon, and titanium, plus special disk covering corrosion data.
- Alloy Center on the ASM International Website. In collaboration with Granta Design, ASM International has established an extensive Alloy Center on their web site www.asminternational.org that provides searchable access to a wide and growing range of numeric materials property data. Where justified, some are presented in graphical format. The ASM Alloy Center is available for a reasonable subscription fee and, together with the availability of the ASM handbook series online, provides a formidable source of materials information.

7.2 STN International

STN International is the online worldwide scientific and technical information network and one of the most extensive sources of numeric materials property data.[16] Built and operated by the American Chemical Society at its Chemical Abstract Service site in Columbus, Ohio, STN International provides about 30 numeric databases, including many developed during the collaboration with the Materials Property Data Network.[17–19] Of great interest to materials data searchers, STN International has the most sophisticated numeric data search software available anywhere online. Thus the data sources may be searched numerically, i.e., using numeric values as ranges or with "greater or less than" types of operators, making it possible to search for alloys that meet required performance needs.

The disadvantages of using STN International to search for numeric data are twofold: (a) The search software is keyed to a command system best known to professional searchers and engineers and scientists will need patience and some training to master the technique and (b) use of the STN International system is billed via a time- and content-based cost accounting system, and so a private account is needed. Outright purchase of databases has not been a policy.

For those able to deal with those conditions, a number of valuable databases exist on STN International, including the following representative sources:

- AAASD—Aluminum Standards and Data (limited to 1990 information)
- ALFRAC—Aluminum Fracture Toughness Database (1968–1989 industry compilation)
- Aluminium—Aluminum Industry Abstracts (textual; 1868–present)
- ASMDATA—ASM Materials Databases online version
- BEILSTEIN—Beilstein Organic Compound Files (1779–1999)
- COPPERDATA—Copper and Copper Alloy Standards and Data
- DETHERM—Thermophysical Properties Database (1819–present)
- DIPPR—AIChE Design Institute Physical Property Data File
- GMELIN—GMELIN Handbook of Inorganic Chemistry (1817–present)
- ICSD—Inorganic Crystal Structure Database (1912–present)
- NUMERIGUIDE—Property Hierarchy and Directory for Numeric Files
- MDF—Metals Datafile (1982–1993)
- PIRA—PIRA and PAPERBASE Database (1975–present)
- PLASPEC—Plastics Material Selection Database
- RAPRA—Rubber, Plastics, Adhesives, and Polymer Composites (1972–present)

For more information on STN International, readers are referred to the web site, www.stn.org.

7.3 knovel.com

A rapidly growing and very useful scientific information web site is www.knovel.com. Produced by the Knovel subsidiary of William Andrew, Inc., it provides an impressive array of searchable versions of hard-copy engineering and scientific reference publications (e.g., the *Handbook of Chemistry and Physics*) as well as access to a wide range of materials data.

Unlike many online sources where such publications are reproduced only as readable print, Knovel subscription holders may work interactively with the various sources and search them individually, in selected groups, or in total. It is highly unusual to find such an array of highly respected sources accessible through a single site and with a specialized search software that caters to the needs of the searcher.

7.4 The Internet

As indicated earlier, there are many web sites on the Internet with materials information content in addition to those specifically identified above. In fact, there are so many, the challenge is to determine which have useful, reliable, and relatively easily accessible data.

Here we will focus on guidance on these latter points developed by Fran Cverna, Director of Electronic and Reference Data Sources at ASM International, who recently produced and presented a survey of scope and quality of materials data content provided at Internet web sites, and results of the most useful web sites screened by Cverna and her ASM resources are referenced here[20]:

www.about.com	Internet search engine—search Materials and Properties
www.nist.gov/public affairs/ database.htm	NIST Standard Reference Database
www.ecn.purdue.edu/MPHO	TPRL Thermophysical properties data
Id.inel.gov/shds	U.S. government solvent database
www.campusplastics.com	Campus Consortium plastics database
www.copper.org	Copper Development Association copper alloys database
www.brushwellman.com/homepage.htm	Brush Wellman supplier materials database
www.timet.com/overviewframe.html	Timet supplier materials database
www.specialmetals.com	Special Metals supplier materials database
www.cartech.com	Cartech supplier materials database (compositions)
www.macsteel.com/mdb	McSteel supplier materials database (limited)
www.aluminum.org	Aluminum Association applications, publications' limited data

Of these sources, two merit special mention: the Internet search engine site www. about.com and the NIST database site www.nist.gov/public affairs/database.htm, a part of thc Public Affairs menu at NIST's web site.

The About.com site provides an excellent means of locating materials data sites and provides a subcategory called Materials Properties and Data if you search for such information. Thirty-two sites are identified, some of which overlap the ASM survey, but to the author's experience others are unique to that site. Many included in the ASM survey are not included here, so the two are supplementary in scope. Among the specific types of information accessible via about.com are:

- Coefficient of thermal expansion
- Electrical conductivity
- Electrical resistivity
- Hardness conversion charts
- Metal temperature by color
- Metal weight calculator
- Periodic table of elements
- Specific gravity of metals
- Surface roughness comparison charts
- Thermal properties
- Wire gauge conversion tables
- Utilities to simplify tasks such as conversion charts for units, currency conversion, glossaries, and acronym definitions

In addition, About.com has links to many other materials sites, including to Aluminum.com, which also provides materials data for a variety of metals, but often without adequate citation and metadata.

The NIST database site provides direct online access to the highest quality, carefully evaluated numeric data from the following databases:

- Standard Reference Data—reliable scientific and technical data extracted from the world's literature, assessed for reliability and critically evaluated.
- Ceramics WebBook—evaluated data and access to data centers as well as tools and resources.
- Chemistry WebBook—chemical and physical property data for specific compounds.
- Fundamental Physical Constants—internationally recommended values of a wide range of often-used physical constants.
- Thermophysical Properties of Gases for the Semiconductor Industry.

In summary, there are many sources of numeric materials data available from Internet sites. It remains for the potential users of the data, however, to approach each site with caution, look for the pedigree of the data in terms of quality and reliability, and make certain that the source of interest meets the requirements of the intended use.

Acknowledgments

The contributions of Jack Westbrook, of Brookline Technologies, and Fran Cverna, of ASM international, are acknowledged and deeply appreciated.

REFERENCES

1. J. H. Westbrook, "Sources of Materials Property Data and Information," *ASM Handbook,* Vol. 20, ASM International, Materials Park, OH, 1997.
2. J. H. Westbrook and K. W. Reynard, "Data Sources for Materials Economics, Policy, and Management," in *Concise Encyclopedia of Materials Economics, Policy, and Management,* M. B. Bever (ed.), Pergamon, New York, 1993.
3. *ASM International Directory of Materials Property Databases,* ASM International, Materials Park, OH, published periodically.
4. ASTM and ANSI/ISO standards, *Annual Book of ASTM Standards,* American Society for Testing and Materials, Conshohocken, PA, American National Standards Institute and International Standards Organization standards, Brussels, published periodically.
5. *Metallic Materials and Elements for Aerospace Vehicle Structures,* MIL-HDBK-5H, U.S. Department of Defense, published periodically (now known as MMPDS-01).
6. Publications of the Thermophysical Properties Research Center (TPRC, previously known as CINDAS), Lafayette, IN.
7. *ASME Boiler and Pressure Vessel Code,* Section 2, Material—Properties, American Society of Mechanical Engineers, New York, published periodically.
8. *Aluminum Standards and Data, 2000* and *Aluminum Standards and Data 1998 Metric SI,* Aluminum Association, Washington, DC, published periodically.
9. J. H. Westbrook and W. Grattidge, "The Role of Metadata in the Design and Operation of a Materials Database," in *Computerization and Networking of Materials Databases,* ASTM STP 1106, J. G. Kaufman and J. S. Glazman, (eds.), American Socity for Testing and Materials, Philadelphia, PA, 1991.
10. J. G. Kaufman, "Quality and Reliability Issues in Materials Databases: ASTM Committee E49.05," in *Computerization and Networking of Materials Databases,* Vol. 3, ASTM STP 1140, T. I. Barry, and K. W. Reynard, (eds.), American Society for Testing and Materials, Philadelphia, PA, 1992, pp. 64–83.
11. J. G. Kaufman, *Properties of Aluminum Alloys—Tensile, Creep, and Fatigue Data at High and Low Temperatures,* ASM International, Materials Park, OH, 1999.
12. *ASM Handbook,* Vols. 1 and 2, *Properties and Selection,* ASM International, Materials Park, OH, published periodically.

13. *Woldman's Engineering Alloys,* 8th ed., Woldman's, London, published periodically.
14. *Worldwide Guide to Equivalent Irons and Steels,* ASM International, Materials Park, OH, published periodically.
15. *Worldwide Guide to Equivalent Nonferrous Metals and Alloys,* ASM International, Materials Park, OH, published periodically.
16. STN International, the Worldwide Scientific and Technical Information Network, Chemical Abstract Services (CAS), a division of the American Chemical Society, Columbus, OH.
17. J. G. Kaufman, "The National Materials Property Data Network Inc., The Technical Challenges and the Plan," in *Materials Property Data: Applications and Access,* PVP-Vol. 111, J. G. Kaufman, (ed.), MPD-Vol. 1, American Socity of Mechanical Engineers, New York, 1986, pp. 159–166.
18. J. G. Kaufman, "The National Materials Property Data Network, Inc.—A Cooperative National Approach to Reliable Performance Data," in *Computerization and Networking of Materials Data Bases,* ASTM STP 1017, J. S. Glazman and J. R. Rumble, Jr., (eds.), American Society for Testing and Materials, Philadelphia, PA, 1989, pp. 55–62.
19. J. H. Westbrook and J. G. Kaufman, "Impediments to an Elusive Dream," in *Modeling Complex Data for Creating Information,* J. E. DuBois and N. Bershon, (eds.), Springer-Verlag, Berlin, 1996.
20. F. Cverna, "Overview of Commercially Available Material Property Data Collections" (on the Internet), presented at the 2000 ASM Materials Solutions Conference on Materials Property Databases, ASM International, St Louis, MO, Oct. 10–12, 2000.

CHAPTER 14

QUANTITATIVE METHODS OF MATERIALS SELECTION

Mahmoud M. Farag
The American University in Cairo
Cairo, Egypt

1 INTRODUCTION

It is estimated that there are more than 40,000 currently useful metallic alloys and probably close to that number of nonmetallic engineering materials like plastics, ceramics and glasses, composite materials, and semiconductors. This large number of materials and the many manufacturing processes available to the engineer, coupled with the complex relationships between the different selection parameters, often make the selection of a material for a given component a difficult task. If the selection process is carried out haphazardly, there will be the risk of overlooking a possible attractive alternative material. This risk can be reduced by adopting a systematic material selection procedure. A variety of quantitative selection procedures have been developed to analyze the large amount of data involved in the selection process so that a systematic evaluation can be made.[1–11] Several of the quantitative procedures can be adapted to use computers in selection from a data bank of materials and processes.[12–17]

Experience has shown that it is desirable to adopt the holistic decision-making approach of concurrent engineering in product development in most industries. With concurrent engineering, materials and manufacturing processes are considered in the early stages of design and are more precisely defined as the design progresses from the concept to the embodiment and finally the detail stages. Figure 1 defines the different stages of design and shows the related activities of the material and manufacturing process selection. The figure illustrates the progressive nature of materials and process selection and defines three stages of selection: initial screening, comparing and ranking alternatives, and selecting the optimum solution. Sections 2, 3, and 4 discuss these three stages of material and process selection in more detail and Section 5 gives a case study to illustrate the procedure.

Although the materials and process selection is often thought of in terms of new product development, there are many other incidences where materials substitution is considered for an existing product. Issues related to material substitution are discussed in Section 6.

Unlike the exact sciences, where there is normally only one single correct solution to a problem, materials selection and substitution decisions require the consideration of conflicting advantages and limitations, necessitating compromises and trade-offs; as a consequence, different satisfactory solutions are possible. This is illustrated by the fact that similar components performing similar functions but produced by different manufacturers are often made from different materials and even by different manufacturing processes.

2 INITIAL SCREENING OF MATERIALS

In the first stages of development of a new product, such questions as the following are posed: What is it? What does it do? How does it do it? After answering these questions it is possible to specify the performance requirements of the different parts involved in the design and to broadly outline the main materials performance and processing requirements. This is then followed by the initial screening of materials whereby certain classes of materials and manufacturing processes may be eliminated and others chosen as likely candidates.

2.1 Analysis of Material Performance Requirements

The material performance requirements can be divided into five broad categories: functional requirements, processability requirements, cost, reliability, and resistance to service conditions.[1]

Functional Requirements

Functional requirements are directly related to the required characteristics of the part or the product. For example, if the part carries a uniaxial tensile load, the yield strength of a candidate material can be directly related to the load-carrying capacity of the product. However, some characteristics of the part or product may not have simple correspondence with measurable material properties, as in the case of thermal shock resistance, wear resistance, reliability, etc. Under these conditions, the evaluation process can be quite complex and may depend upon predictions based on simulated service tests or upon the most closely related mechanical, physical, or chemical properties. For example, thermal shock resistance can be related to the thermal expansion coefficient, thermal conductivity, modulus of elasticity, ductility, and tensile strength. On the other hand, resistance to stress–corrosion cracking can be related to tensile strength and electrochemical potential.

Processability Requirements

The processability of a material is a measure of its ability to be worked and shaped into a finished part. With reference to a specific manufacturing method, processability can be de-

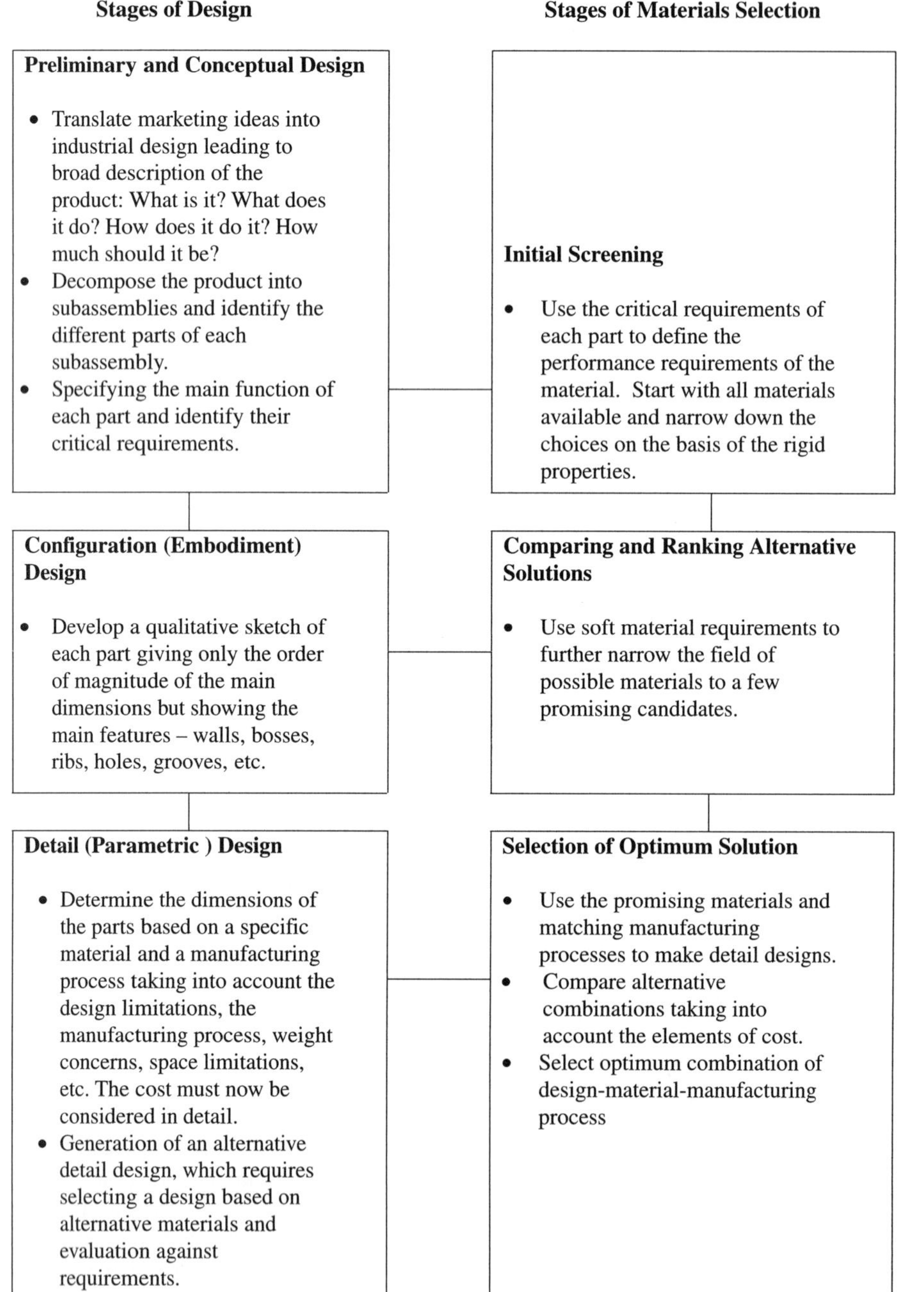

Figure 1 Stages of design and the related stages of materials selection.

fined as castability, weldability, machinability, etc. Ductility and hardenability can be relevant to processability if the material is to be deformed or hardened by heat treatment, respectively. The closeness of the stock form to the required product form can be taken as a measure of processability in some cases.

It is important to remember that processing operations will almost always affect the material properties so that processability considerations are closely related to functional requirements.

Cost

Cost is usually an important factor in evaluating materials, because in many applications there is a cost limit for a given component. When the cost limit is exceeded, the design may have to be changed to allow for the use of a less expensive material or process. In some cases, a relatively more expensive material may eventually yield a less expensive component than a low-priced material that is more expensive to process.

Reliability Requirements

Reliability of a material can be defined as the probability that it will perform the intended function for the expected life without failure. Material reliability is difficult to measure, because it is not only dependent upon the material's inherent properties, but it is also greatly affected by its production and processing history. Generally, new and nonstandard materials will tend to have lower reliability than established, standard materials.

Despite difficulties of evaluating reliability, it is often an important selection factor that must be taken into account. Failure analysis techniques are usually used to predict the different ways in which a product can fail and can be considered as a systematic approach to reliability evaluation. The causes of failure of a part in service can usually be traced back to defects in materials and processing, faulty design, unexpected service conditions, or misuse of the product.

Resistance to Service Conditions

The environment in which the product or part will operate plays an important role in determining the material performance requirements. Corrosive environments, as well as high or low temperatures, can adversely affect the performance of most materials in service. Whenever more than one material is involved in an application, compatibility becomes a selection consideration. In a thermal environment, for example, the coefficients of thermal expansion of all the materials involved may have to be similar in order to avoid thermal stresses. In wet environments, materials that will be in electrical contact should be chosen carefully to avoid galvanic corrosion. In applications where relative movement exists between different parts, wear resistance of the materials involved should be considered. The design should provide access for lubrication; otherwise self-lubricating materials have to be used.

2.2 Quantitative Methods for Initial Screening

Having specified the performance requirements of the different parts, the required material properties can be established for each of them. These properties may be quantitative or qualitative, essential or desirable. For example, the function of a connecting rod in an internal combustion engine is to connect the piston to the crank shaft. The performance requirements are that it should transmit the power efficiently without failing during the expected life of the engine. The essential material properties are tensile and fatigue strengths, while the desirable properties that should be maximized are processability, weight, reliability, and resistance to service conditions. All these properties should be achieved at a reasonable cost. The selection process involves the search for the material or materials that would best meet those requirements. In some cases none of the available materials can meet the requirements or the possible materials are too expensive or environmentally unsafe. In such cases, alternatives must be made possible through redesign, compromise of requirements, or development of new materials.

Generally, the starting point for materials selection is the entire range of engineering materials. At this stage, creativity is essential in order to open up channels in different directions, not let traditional thinking interfere with the exploration of ideas, and ensure that potential materials are not overlooked. A steel may be the best material for one design concept while a plastic is best for a different concept, even though the two designs provide the same function.

After all the alternatives have been suggested, the ideas that are obviously unsuitable are eliminated and attention is concentrated on those that look practical. Quantitative methods can be used for initial screening in order to narrow down the choices to a manageable number for subsequent detailed evaluation. Following are some of the quantitative methods for initial screening of materials.

Limits on Material Properties

Initial screening of materials can be achieved by first classifying their performance requirements into two main categories[1]:

- Rigid, or go–no go, requirements
- Soft, or relative, requirements

Rigid requirements are those which must be met by the material if it is to be considered at all. Such requirements can be used for the initial screening of materials to eliminate the unsuitable groups. For example, metallic materials are eliminated when selecting materials for an electrical insulator. If the insulator is to be flexible, the field is narrowed further as all ceramic materials are eliminated. Other examples of material rigid requirements include behavior under operating temperature, resistance to corrosive environment, ductility, electrical and thermal conductivity or insulation, and transparency to light or other waves. Examples of process rigid requirements include batch size, production rate, product size and shape, tolerances, and surface finish. Whether or not the equipment or experience for a given manufacturing process exists in a plant can also be considered as a hard requirement in many cases. Compatibility between the manufacturing process and the material is also an important screening parameter. For example, cast irons are not compatible with sheet-metal-forming processes and steels are not easy to process by die casting. In some cases, eliminating a group of materials results in automatic elimination of some manufacturing processes. For example, if plastics are eliminated because service temperature is too high, injection and transfer molding should be eliminated as they are unsuitable for other materials.

Soft, or relative, requirements are those which are subject to compromise and trade-offs. Examples of soft requirements include mechanical properties, specific gravity, and cost. Soft requirements can be compared in terms of their relative importance, which depends on the application under study.

Cost-per-Unit-Property Method

The cost-per-unit-property method is suitable for initial screening in applications where one property stands out as the most critical service requirement.[1] As an example, consider the case of a bar of a given length L to support a tensile force F. The cross-sectional area A of the bar is given by

$$A = \frac{F}{S} \tag{1}$$

where S is the working stress of the material, which is related to its yield strength divided by an appropriate factor of safety.

The cost of the bar (C') is given by

$$C' = C\rho AL = \frac{C\rho FL}{S} \tag{2}$$

where C = cost of material per unit mass
ρ = density of material

Since F and L are constant for all materials, comparison can be based on the cost of unit strength, which is the quantity

$$\frac{C\rho}{S} \tag{3}$$

Materials with lower cost per unit strength are preferable. If an upper limit is set for the quantity $C\rho/S$, then materials satisfying this condition can be identified and used as possible candidates for more detailed analysis in the next stage of selection.

The working stress of the material in Eqs. (1)–(3) is related to the static yield strength of the material since the applied load is static. If the applied load is alternating, it is more appropriate to use the fatigue strength of the material. Similarly, the creep strength should be used under loading conditions that cause creep.

Equations similar to (2) and (3) can be used to compare materials on the basis of cost per unit stiffness when the important design criterion is deflection in the bar. In such cases, S is replaced by the elastic modulus of the material. The above equations can also be modified to allow comparison of different materials under loading systems other than uniaxial tension. Table 1 gives some formulas for the cost per unit property under different loading conditions based on either yield strength or stiffness.

Ashby's Method

Ashby's material selection charts[4,5,9,10] are also useful for initial screening of materials. Figure 2 plots the strength versus density for a variety of materials. Depending upon the geometry and type of loading, different S–ρ relationships apply, as shown in Table 1. For simple axial loading, the relationship is S/ρ. For a solid rectangle under bending, $S^{1/2}/\rho$ applies, and for a solid cylinder under bending or torsion the relationship $S^{2/3}/\rho$ applies. Lines with these slopes are shown in Fig. 2. Thus, if a line is drawn parallel to the line $S/\rho = C$, all the

Table 1 Formulas for Estimating Cost per Unit Property

Cross Section and Loading Condition	Cost per Unit Strength	Cost per Unit Stiffness
Solid cylinder in tension or compression	$C\rho/S$	$C\rho/E$
Solid cylinder in bending	$C\rho/S^{2/3}$	$C\rho/E^{1/2}$
Solid cylinder in torsion	$C\rho/S^{2/3}$	$C\rho/G^{1/2}$
Solid cylindrical bar as slender column	—	$C\rho/E^{1/2}$
Solid rectangle in bending	$C\rho/S^{1/2}$	$C\rho/E^{1/3}$
Thin-walled cylindrical pressure vessel	$C\rho/S$	—

Note: From Ref. 1.

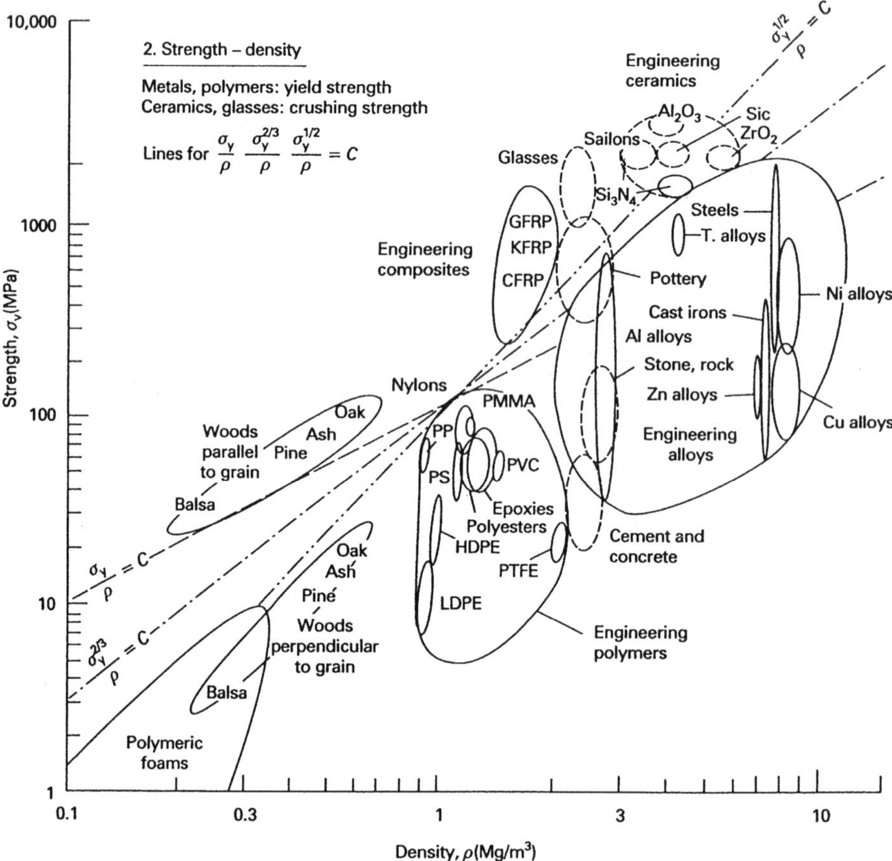

Figure 2 Example of Ashby's materials selection charts. (from Ref. 10, with permission from The Institute of Materials).

materials which lie on the line will perform equally well under simple axial loading conditions. Materials above the line are better and those below it are worse. A similar diagram can be drawn for elastic modulus versus density and formulas similar to those in Table 1 can be used to screen materials under conditions where stiffness is a major requirement

Dargie's Method

The initial screening of materials and processes can be a tedious task if performed manually from handbooks and supplier catalogs. This difficulty has prompted the introduction of several computer-based systems for materials and/or process selection.[12–15] As an illustrative example, the system MAPS 1 proposed by Dargie et al.[15] will be briefly described here. For this system, Dargie et al. proposed a part classification code similar to that used in group technology.

The first five digits of the MAPS 1 code are related to the elimination of unsuitable manufacturing processes. The first digit is related to the batch size. The second digit characterizes the bulk and depends on the major dimension and whether the part is long, flat, or compact. The third digit characterizes the shape, which is classified on the basis of being prismatic, axisymmetric, cup shaped, nonaxisymmetric, and nonprismatic. The fourth digit is related to tolerance and the fifth digit is related to surface roughness

The next three digits of the MAPS 1 code are related to the elimination of unsuitable materials. The sixth digit is related to service temperature. The seventh digit is related to the acceptable corrosion rate. The eighth digit characterizes the type of environment to which the part is exposed.

The system uses two types of databases for preliminary selection:

- Suitability matrices
- Compatibility matrix

The suitability matrices deal with the suitability of processes and materials for the part under consideration. Each of the code digits has a matrix. The columns of the matrix correspond to the value of the digit and the rows correspond to the processes and materials in the database. The elements of the matrix are either 0, indicating unsuitability, or 2, indicating suitability.

The compatibility matrix expresses the compatibility of the different combinations of processes and materials. The columns of the matrix correspond to the materials while the rows correspond to the processes. The elements of the matrix are 0 for incompatible combinations, 1 for difficult or unusual combinations, and 2 for combinations used in usual practice.

Based on the part code, the program generates a list of candidate combinations of materials and processes to produce it. This list helps the designer to identify possible alternatives early in the design process and to design for ease of manufacture.

Esawi and Ashby's Method

Another quantitative method of initial screening is proposed by Esawi and Ashby.[16] The method compares the approximate cost of resources of materials, energy, capital, time, and information needed to produce the component using different combinations of materials and manufacturing processes. The method can be used early in the design process and is capable of comparing combinations of materials and processes such as the cost of a polymer component made by injection molding with that of a competing design in aluminum made by die casting.

According to this method, the total cost of a component has three main elements: material cost, tooling cost, and overhead cost. The material cost is a function of the cost per unit weight of the material and the amount the material needed. Since the cost of tooling (dies, molds, jigs, fixtures, etc.) is normally assigned to a given production run, the tooling cost per component varies as the reciprocal of the number of components produced in that run. The overhead per component varies as the reciprocal of the production rate. Application of this method in initial screening requires a database, such as CES 4, which lists material prices, attributes of different processes, production rates, tool life, and approximate cost of equipment and tooling. CES 4 Software[17] contains records for 112 shaping processes such as vapor deposition, casting, molding, metal forming, machining, and composite forming. The software output can be in the form of graphs giving the variation of cost with the batch size for competing process/material combinations. Another type of output is relative cost per unit when a given component is made by different processing routes.

3 COMPARING AND RANKING ALTERNATIVE SOLUTIONS

After narrowing down the field of possible materials using one or more of the quantitative initial screening methods described in Section 2, quantitative methods can then also be used to further narrow the field of possible materials and matching manufacturing processes to a few promising candidates that have good combinations of soft requirements. Several such methods are described in Refs. 1 and 2 and following is a description of one of the methods.

3.1 Weighted-Properties Method

In the weighted-properties method each material requirement, or property, is assigned a certain weight, depending on its importance to the performance of the part in service.[1] A weighted-property value is obtained by multiplying the numerical value of the property by the weighting factor (α). The individual weighted-property values of each material are then summed to give a comparative materials performance index (γ). Materials with the higher performance index (γ) are considered more suitable for the application.

Digital Logic Method

In the cases where numerous material properties are specified and the relative importance of each property is not clear, determinations of the weighting factor α can be largely intuitive, which reduces the reliability of selection. The digital logic approach can be used as a systematic tool to determine α.[1] In this procedure evaluations are arranged such that only two properties are considered at a time. Every possible combination of properties or goals is compared and no shades of choice are required, only a yes or no decision for each evaluation. To determine the relative importance of each property or goal, a table is constructed, the properties or goals are listed in the left-hand column, and comparisons are made in the columns to the right, as shown in Table 2.

In comparing two properties or goals, the more important goal is given the number 1 and the less important is given as 0. The total number of possible decisions is $N = n(n - 1)/2$, where n is the number of properties or goals under consideration. A relative-emphasis coefficient or weighting factor α for each goal is obtained by dividing the number of positive decisions for each goal (m) into the total number of possible decisions (N). In this case $\Sigma\ \alpha = 1$.

To increase the accuracy of decisions based on the digital logic approach, the yes–no evaluations can be modified by allocating gradation marks ranging from 0 (no difference in importance) to 3 (large difference in importance). In this case, the total gradation marks for each selection criterion are reached by adding up the individual gradation marks. The weight-

Table 2 Determination of Relative Importance of Goals Using Digital Logic Method

Goals	Number of Positive Decisions $N = n(n - 1)/2$										Positive Decisions	Relative Emphasis Coefficient α
	1	2	3	4	5	6	7	8	9	10		
1	1	1	0	1							3	0.3
2	0				1	0	1				2	0.2
3		0			0			1	0		1	0.1
4			1			1		0		0	2	0.2
5				0			0		1	1	2	0.2
Total positive decisions											10	$\Sigma\alpha = 1.0$

Note: From Ref. 1.

ing factors are then found by dividing these total gradation marks by their grand total. A simple interactive computer program can be written to help in determining the weighting factors. A computer program will also make it easier to perform several runs of the process in order to test the sensitivity of the final ranking to changes in some of the decisions—sensitivity analysis.

Performance Index

In its simple form, the weighted-properties method has the drawback of having to combine unlike units, which could yield irrational results. This is particularly true when different mechanical, physical, and chemical properties with widely different numerical values are combined. The property with higher numerical value will have more influence than is warranted by its weighting factor. This drawback is overcome by introducing scaling factors. Each property is so scaled that its highest numerical value does not exceed 100. When evaluating a list of candidate materials, one property is considered at a time. The best value in the list is rated as 100 and the others are scaled proportionally. Introducing a scaling factor facilitates the conversion of normal material property values to scaled dimensionless values. For a given property, the scaled value B for a given candidate material is equal to

$$B = \text{scaled property} = \frac{\text{numerical value of property} \times 100}{\text{maximum value in list}} \tag{4}$$

For properties like cost, corrosion or wear loss, and weight gain in oxidation, a lower value is more desirable. In such cases, the lowest value is rated as 100 and B is calculated as

$$B = \text{scaled property} = \frac{\text{minimum value in list} \times 100}{\text{numerical value of property}} \tag{5}$$

For material properties that can be represented by numerical values, application of the above procedure is simple. However, with properties like corrosion, wear resistance, machinability, and weldability, numerical values are rarely given and materials are usually rated as very good, good, fair, poor, etc. In such cases, the rating can be converted to numerical values using an arbitrary scale. For example, corrosion resistance ratings excellent, very good, good, fair, and poor can be given numerical values of 5, 4, 3, 2, and 1, respectively. After scaling the different properties, the material performance index γ can be calculated as

$$\gamma = \sum_{i=1}^{n} B_i \alpha_i \tag{6}$$

where i is summed over all the n relevant properties.

Cost (stock material, processing, finishing, etc.) can be considered as one of the properties and given the appropriate weighting factor. However, if there is a large number of properties to consider, the importance of cost may be emphasized by considering it separately as a modifier to the material performance index γ. In the cases where the material is used for space filling, cost can be introduced on a per-unit-volume basis. A figure of merit M for the material can then be defined as

$$M = \frac{\gamma}{C\rho} \tag{7}$$

where C = total cost of material per unit weight (stock, processing, finishing, etc.)
ρ = density of material

When an important function of the material is to bear stresses, it may be more appropriate to use the cost of unit strength instead of the cost per unit volume. This is because higher strength will allow less material to be used to bear the load and the cost of unit strength may be a better indicator of the amount of material actually used in making the part. In this case, Eq. (7) is rewritten as

$$M = \frac{\gamma}{C'} \tag{8}$$

where C' is determined from Table 1 depending on the type of loading.

This argument may also hold in other cases where the material performs an important function like electrical conductivity or thermal insulation. In these cases the amount of the material and consequently the cost are directly affected by the value of the property.

When a large number of materials with a large number of specified properties are being evaluated for selection, the weighted properties method can involve a large number of tedious and time-consuming calculations. In such cases, the use of a computer would facilitate the selection process. The steps involved in the weighted-properties method can be written in the form of a simple computer program to select materials from a data bank. The type of material information needed for computer-assisted ranking of an alternative solution is normally structured in the form of databases of properties such as those published by ASM, as will be described in Section 8.

4 SELECTING THE OPTIMUM SOLUTION

Candidates that have the most promising performance indices can each now be used to develop a detail design. Each detail design will exploit the points of strength of the material, avoid the weak points, and reflect the requirements of the manufacturing processes needed for the material. The type of material information needed for detail design is different from that needed for initial screening and ranking. What is needed at this stage is detailed high-quality information about the highest ranking candidates. As will be shown in Section 8, such information is usually unstructured and can be obtained from handbooks, publications of trade organizations, and technical reports in the form of text, pdf files, tables, graphs, photographs, etc. There are instances where some of the desired data may not be available or may be for slightly different test conditions. In such cases educated judgment is required.

After completing the different designs, solutions are then compared, taking the cost elements into consideration in order to arrive at the optimum design–material–process combination, as will be illustrated in the following case study.

5 CASE STUDY IN MATERIALS SELECTION

The following case study illustrates the procedure for materials selection as described in Sections 2, 3, and 4 and is based on Ref. 18. The objective is to select the least expensive component that satisfies the requirements for a simple structural component for a sailing-boat mast in the form of a hollow cylinder of length 1000 mm which is subjected to compressive axial forces of 153 kN. Because of space and weight limitations, the outer diameter of the component should not exceed 100 mm, the inner diameter should not be less than 84 mm, and the mass should not exceed 3 kg. The component will be subjected to mechanical impact and spray of water. Assembly to other components requires the presence of relatively small holes.

5.1 Material Performance Requirements

Possible modes of failure and the corresponding material properties that are needed to resist failure for the present component include the following:

- Catastrophic fracture due to impact loading, especially near assembly holes, is resisted by the high fracture toughness of the material. This is a rigid-material requirement and will be used for initial screening of materials.
- Plastic yielding is resisted by high yield strength. This is a soft-material requirement but a lower limit will be determined by the limitation on the outer diameter.
- Local and global buckling are resisted by high elastic modulus. This is a soft-material requirement but a lower limit will be determined by the limitation on the outer diameter.
- Internal fiber buckling for fiber-reinforced materials is resisted by high modulus of elasticity of the matrix and high volume fraction of fibers in the loading direction. This is a soft-material requirement but a lower limit will be determined by the limitation on the outer diameter.
- Corrosion, which can be resisted either by selecting materials with inherently good corrosion resistance or by protective coating.
- Reliability of the component in service. A factor of safety of 1.5 is taken for the axial loading, i.e., the working axial force will be taken as 230 kN to improve reliability.

In addition to the above requirements the limitations set on dimensions and weight should be observed.

5.2 Initial Screening of Materials

The requirement for fracture toughness of the material is used to eliminate ceramic materials. Because of the limitations set on the outer and inner diameters, the maximum possible cross section of the component is about 2300 mm^2. To avoid yielding under the axial working load, the yield strength of the material should be more than 100 MPa, which excludes engineering polymers, woods, and some of the lower strength engineering alloys; see Fig. 2. Corrosion resistance is desirable but will not be considered a factor for screening, since the possibility of protection for less corrosive materials exists but will be considered as a soft requirement.

5.3 Comparing and Ranking Alternative Solutions

Table 3 shows a sample of materials that satisfy the conditions set in the initial screening stage. In a real-life situation the list in the table could be much longer, but the intent here is to illustrate the procedure. The yield strength, elastic modulus, specific gravity, corrosion resistance, and cost category are given for each material. At this stage, it is sufficient to classify materials into very inexpensive, inexpensive, etc. A better estimate of the material and manufacturing cost will be needed in making the final decision in selection. Because the weight of the component is important in this application, specific strength and specific modulus would be better indicators of the suitability of the material (Table 4). The relative importance of the material properties is given in Table 5 and the performance indices of the different materials, as determined by the weighted-properties method, are given in Table 6. The seven candidate materials with high performance indices ($\gamma > 45$) are selected for making actual component designs.

Table 3 Properties of Sample Candidate Materials

Material	Yield Strength (MPa)	Elastic Modulus (GPa)	Specific Gravity	Corrosion Resistance[a]	Cost Category[b]
AISI 1020 (UNS G10200)	280	210	7.8	1	5
AISI 1040 (UNS G10400)	400	210	7.8	1	5
ASTM A242 type1 (UNS K11510)	330	212	7.8	1	5
AISI 4130 (UNS G41300)	1520	212	7.8	4	3
AISI 316 (UNS S31600)	205	200	7.98	4	3
AISI 416 heat treated (UNS S41600)	440	216	7.7	4	3
AISI 431 heat treated (UNS S43100)	550	216	7.7	4	3
AA 6061 T6 (UNS A96061)	275	69.7	2.7	3	4
AA 2024 T6 (UNS A92024)	393	72.4	2.77	3	4
AA 2014 T6 (UNS A92014)	415	72.1	2.8	3	4
AA 7075 T6 (UNS A97075)	505	72.4	2.8	3	4
Ti–6Al–4V	939	124	4.5	5	1
Epoxy–70% glass fabric	1270	28	2.1	4	2
Epoxy–63% carbon fabric	670	107	1.61	4	1
Epoxy–62% aramid fabric	880	38	1.38	4	1

Note: Based on Ref. 18.
[a] 5, Excellent; 4, very good; 3, good; 2, fair; 1, poor.
[b] 5, Very inexpensive; 4, inexpensive; 3, moderate price; 2, expensive; 1, very expensive.

5.4 Selecting the Optimum Solution

As shown earlier, the possible modes of failure of a hollow cylinder include yielding, local buckling, global buckling, and internal fiber buckling. These four failure modes are used to develop the design formulas for the mast component. For more details on the design and optimization procedure [Eqs. (9)–(12)] refer to Ref. 18:

Condition for yielding:

$$\frac{F}{A} < \sigma_y \tag{9}$$

where σ_y = yield strength of material
F = external working axial force
A = cross-sectional area

Condition for local buckling:

$$\frac{F}{A} < \frac{0.121ES}{D} \tag{10}$$

Table 4 Properties of Sample Candidate Materials

Material	Specific Strength (MPa)	Specific Modulus (GPa)	Corrosion Resistance[a]	Cost Category[b]
AISI 1020 (UNS G10200)	35.9	26.9	1	5
AISI 1040 (UNS G10400)	51.3	26.9	1	5
ASTM A242 type1 (UNS K11510)	42.3	27.2	1	5
AISI 4130 (UNS G41300)	194.9	27.2	4	3
AISI 316 (UNS S31600)	25.6	25.1	4	3
AISI 416 heat treated (UNS S41600)	57.1	28.1	4	3
AISI 431 heat treated (UNS S43100)	71.4	28.1	4	3
AA 6061 T6 (UNS A96061)	101.9	25.8	3	4
AA 2024 T6 (UNS A92024)	141.9	26.1	3	4
AA 2014 T6 (UNS A92014)	148.2	25.8	3	4
AA 7075 T6 (UNS A97075)	180.4	25.9	3	4
Ti–6Al–4V	208.7	27.6	5	1
Epoxy–70% glass fabric	604.8	28	4	2
Epoxy–63% carbon fabric	416.2	66.5	4	1
Epoxy–62% aramid fabric	637.7	27.5	4	1

[a] 5, Excellent; 4, very good; 3, good; 2, fair; 1, poor.
[b] 5, Very inexpensive; 4, inexpensive; 3, moderate price; 2, expensive; 1, very expensive.

where D = outer diameter of cylinder
S = wall thickness of cylinder
E = elastic modulus of material

Condition for global buckling:

$$\sigma_y > \frac{F}{A}\left[1 + \left(\frac{LDA}{1000I}\right)\sec\left(\frac{(F/EI)^{1/2}\,L}{2}\right)\right] \tag{11}$$

where I = second moment of area
L = length of component

Table 5 Weighting Factors

Property	Specific Strength (MPa)	Specific Modulus (GPa)	Corrosion Resistance	Relative Cost
Weighting factor, α	0.3	0.3	0.15	0.25

Table 6 Calculation of Performance Index

Material	Scaled Specific Strength, ×0.3	Scaled Specific Modulus, ×0.3	Scaled Corrosion Resistance, ×0.15	Scaled Relative Cost, ×0.25	Performance Index (γ)
AISI 1020 (UNS G10200)	1.7	12.3	3	25	42
AISI 1040 (UNS G10400)	2.4	12.3	3	25	42.7
ASTM A242 type1 (UNS K11510)	2	12.3	3	25	42.3
AISI 4130 (UNS G41300)	9.2	12.3	6	15	42.5
AISI 316 (UNS S31600)	1.2	11.3	12	15	39.5
AISI 416 heat treated (UNS S41600)	2.7	12.7	12	15	42.4
AISI 431 heat treated (UNS S43100)	3.4	12.7	12	15	43.1
AA 6061 T6 (UNS A96061)	4.8	11.6	9	20	45.4
AA 2024 T6 (UNS A92024)	6.7	11.8	9	20	47.5
AA 2014 T6 (UNS A92014)	7	11.6	9	20	47.6
AA 7075 T6 (UNS A97075)	8.5	11.7	9	20	49.2
Ti–6Al–4V	9.8	12.5	15	5	42.3
Epoxy–70% glass fabric	28.4	12.6	12	10	63
Epoxy–63% carbon fabric	19.6	30	12	5	66.6
Epoxy–62% aramid fabric	30	12.4	12	5	59.4

Condition for internal fiber buckling:

$$\frac{F}{A} < \left[\frac{E_m}{4(1 + \nu_m)(1 - V_f^{1/2})}\right] \tag{12}$$

where E_m = elastic modulus of matrix material
ν_m = Poisson's ratio of matrix material
V_f = volume fraction of fibers parallel to loading direction

Figure 3 shows the optimum design range of component diameter and wall thickness as predicted by Eqs. (9)–(11) for AA 7075 aluminum alloy. Point O represents the optimum design. Similar figures were developed for the different candidate materials in order to determine the mast component's optimum design dimensions when made of the materials, and the results are shown in Table 7. Although all the materials in Table 7 can be used to make safe components that comply with the space and weight limitations, AA2024 T6 is selected since it gives the least expensive solution.

6 MATERIALS SUBSTITUTION

Substitution is an activity through which a product, a material, or a process is replaced by a more suitable alternative. An important consideration in making a substitution decision is

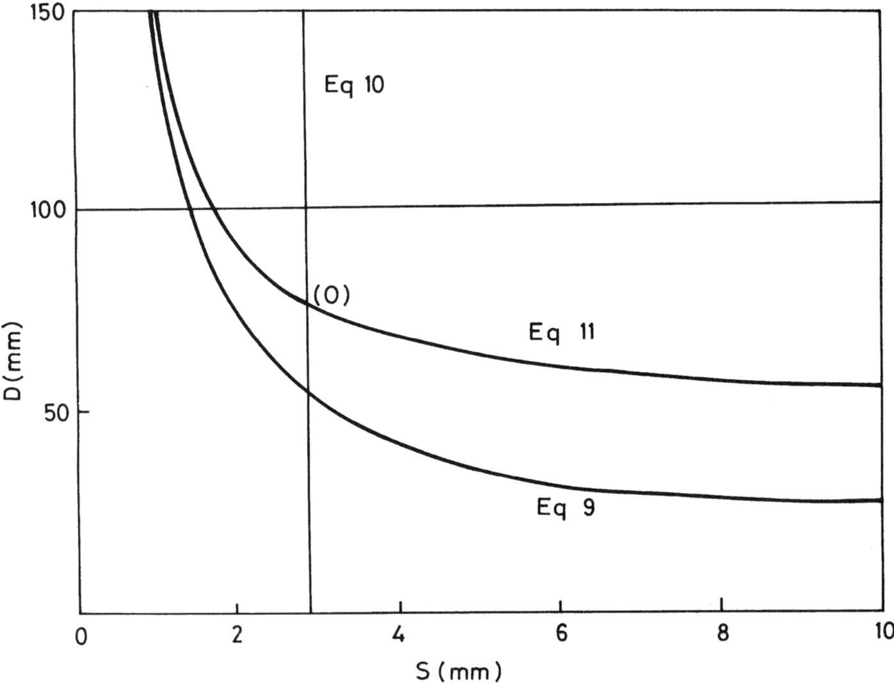

Figure 3 Design range as predicted by Eqs. 9–11 for AA 7075 aluminum alloy. (Reprinted from *Materials and Design,* **13,** M. M. Farag and E. El-Magd, An Integrated Approach to Product Design, Materials Selection, and Cost Estimation, 323–327, © 1992, with permission from Elsevier Science.)

Table 7 Designs Using Candidate Materials with Highest Performance Indices

Material	D_a (mm)	S (mm)	A (mm^2)	Mass (kg)	Cost/kg ($)	Cost of Component ($)
AA 6061 T6 (UNS A96061)	100	3.4	1065.7	2.88	8	23.2
AA 2024 T6 (UNS A92024)	88.3	2.89	801.1	2.22	8.3	18.4
AA 2014 T6 (UNS A92014)	85.6	2.89	776.6	2.17	9	19.6
AA 7075 T6 (UNS A97075)	78.1	2.89	709.1	1.99	10.1	20
Epoxy–70% glass fabric	78	4.64	1136.3	2.39	30.8	73.6
Epoxy–63% carbon fabric	73.4	2.37	546.1	0.88	99	87.1
Epoxy–62% aramid fabric	75.1	3.99	941.6	1.30	88	114.4

Note: Based on Ref. 18.

the relative value, which includes the price ratio, substitution costs, and customer's propensity to substitute. The common reasons for materials substitution include the following:

- Taking advantage of new materials or processes
- Improving service performance, including longer life and higher reliability
- Meeting new legal requirements
- Accounting for changed operating conditions
- Reducing cost and making the product more competitive

Generally, a simple substitution of one material for another does not produce an optimum solution. This is because it is not possible to realize the full potential of a new material unless the component is redesigned to exploit its strong points and manufacturing characteristics. Following is a brief description of some of the quantitative methods that are available for making decisions in materials substitution.

6.1 Pugh Method

The Pugh method (19) is useful as an initial screening method in the early stages of design. In this method, a decision matrix is constructed as shown in Table 8. Each of the properties of a possible alternative new materials is compared with the corresponding property of the currently used material and the result is recorded in the decision matrix as + if more favorable, − if less favorable, and 0 if the same. The decision on whether a new material is better than the currently used material is based on the analysis of the result of comparison, i.e., the total number of +'s, −'s and 0's. New materials with more favorable properties than drawbacks are selected as serious candidates for substitution and are used to redesign the component and for detailed analysis.

6.2 Cost–Benefit Analysis

The cost–benefit analysis is more suitable for the detailed analysis involved in making the final material substitution decision.[1] Because new materials are usually more complex and

Table 8 Example of Use of Pugh Decision Matrix for Materials Substitution

Property	Currently Used Material	New Material 1	New Material 2	New Material 3
1	C1	−	+	+
2	C2	+	+	+
3	C3	+	+	−
4	C4	0	+	−
5	C5	−	0	−
6	C6	0	0	0
7	C7	−	−	0
8	C8	−	+	0
9	C9	−	0	0
Total +		2	5	2
Total −		5	1	3
Total 0		2	3	4

often require closer control and even new technologies for their processing, components made from such materials are more expensive. This means that for materials substitution to be economically feasible, the economic gain as a result of improved performance ΔB should be more than the additional cost incurred as a result of substitution ΔC:

$$\Delta B - \Delta C > 1 \tag{13}$$

For this analysis it is convenient to divide the cost of materials substitution ΔC into the following:

- Cost Differences in Direct Material and Labor. New materials often have better performance but are more expensive. When smaller amounts of the new material are used to make the product, the increase in direct material cost may not be as great as it would appear at first. Cost of labor may not be an important factor in substitution if the new materials do not require new processing techniques and assembly procedures. If, however, new processes are needed, new cycle times may result and the difference in productivity has to be carefully assessed.
- Cost of Redesign and Testing. Using new materials usually involves design changes and testing of components to ensure that their performance meets the requirements. The cost of redesign and testing can be considerable in the case of critical components.
- Cost of New Tools and Equipment. Changing materials can have considerable effect on life and cost of tools, and it may influence the heat treatment and finishing processes. This can be a source of cost saving if the new material does not require the same complex treatment or finishing processes used for the original material. The cost of equipment needed to process new materials can be considerable if the new materials require new production facilities, as in the case of replacing metals with plastics.

Based on the above analysis, the total cost (ΔC) of substituting a new material, n, in place of an original material, o, in a given part is

$$\Delta C = (P_n M_n - P_o M_o) + f\left(\frac{C_t}{N}\right) + (T_n - T_o) + (L_n - L_o) \tag{14}$$

where P_n, P_o = price per unit mass of new and original materials used in part
M_n, M_o = mass of new and original materials used in part
f = capital recovery factor, can be taken as 15% in absence of information
C_t = cost of transition from original to new materials
N = total number of new parts produced
T_n, T_o = tooling cost per part for new and original materials
L_n, L_o = labor cost per part using new and old materials

The gains as a result of improved performance ΔB can be estimated based on the expected improved performance of the component, which can be related to the increase in performance index of the new material compared with the currently used material. Such increases include the saving gained as a result of weight reduction or increased service life of the component:

$$\Delta B = A\,(\gamma_n - \gamma_o) \tag{15}$$

where γ_n, γ_o = performance indices of new and original materials, respectively
A = benefit of improved performance of component expressed in \$ per unit increase in material performance index γ

7 CASE STUDY IN MATERIALS SUBSTITUTION

In the case study in materials selection that was discussed in Section 5, the aluminum alloy AA 2024 T6 was selected since it gives the least expensive solution. Of the seven materials in Table 7, AA 6061 T6, epoxy–70% glass fabric, and epoxy–62% aramid fabric result in components that are heavier and more expensive than those of the other four materials and will be rejected as they offer no advantage. Of the remaining four materials, AA2024 T6 results in the least expensive but the heaviest component. The other three materials—AA 2014 T6, AA 7075 T6, and epoxy–63% carbon fabric—result in progressively lighter components at progressively higher cost.

For the cases where it is advantageous to have a lighter component, the cost–benefit analysis can be used in finding a suitable substitute for AA 2024 T6 alloy. For this purpose Eq. (15) is used with the performance index γ being considered as the weight of the component and ΔC being the difference in cost of component and A is the benefit expressed in dollars of reducing the mass by 1 kg. Comparing the materials in pairs shows:

For $A <$ \$7/kg saved, AA2024 T6 is the optimum material.

For $A =$ \$7–\$60.5/kg saved, AA 7075 T6 is a better substitute.

For $A >$ \$60.5/kg saved, epoxy–63% carbon fabric is optimum.

8 SOURCES OF INFORMATION AND COMPUTER-ASSISTED SELECTION

8.1 Locating Materials Properties Data

One essential requisite to successful materials selection is a source of reliable and consistent data on materials properties. There are many sources of information, which include governmental agencies, trade associations, engineering societies, textbooks, research institutes, and materials producers. Locating the appropriate type of information is not easy and may require several cycles of iteration until the information needed is gathered. According to Kirkwood,[20] the steps involved in each cycle are: Define the question, set up a strategy to locate the needed information, use the resources you know best, go to less known sources when necessary, evaluate the quality of data/information resources, and start again if needed using the new information to help better define the question. To locate useful sources of materials data, it is important to identify the intended use of data, the type of data required, and the quality of data required.[21] In general, progressively higher quality of data is needed as the selection process progresses from initial screening to ranking and finally to selecting the optimum solution.

ASM International has recently published a directory of materials property databases,[22] which contains more than 500 data sources, including both specific databases and data centers. For each source, the directory gives a brief description of the available information, address, telephone number, email, web site, and approximate cost if applicable. The directory also has indices by material and by property to help the user in locating the most appropriate source of material information. Much of the information is available on CD-ROM or PC disk, which makes it possible to integrate the data source into computer-assisted selection systems. Other useful reviews of the sources of materials property data and information are also available.[23,24]

8.2 Types of Material Information

According to Cebon and Ashby,[25] materials information can be classified into structured information either from reference sources or developed in-house and unstructured information either from reference sources or developed in-house. Structured information is normally in the form of databases of properties and is most suited for initial screening (Section 2) and for comparing and ranking of materials (Section 3). Examples of structured reference sources of information material properties include ASM, Materials Universe, and Matweb databases. These databases, in addition to several others, are collected online under the Material Data Network.[26]

Unstructured information gives details about performance of specific materials and is normally found in handbooks, publications of trade organizations, and technical reports in the form of text, pdf files, tables, graphs, photographs, etc. Such information is most suited for detailed consideration of the top-ranking candidates that were selected in the initial screening and the ranking stages, as discussed in Section 4.

8.3 Computerized Materials Databases

Computerized materials databases are an important part of any computer-aided system for selection. With an interactive database, as in the case of ASM Metal Selector,[27] the user can define and redefine the selection criteria to gradually sift the materials and isolate the candidates that meet the requirements. In many cases, sifting can be carried out according to different criteria:

1. Specified numeric values of a set of material properties
2. Specified level of processability such as machinability, weldability, formability, availability, processing cost, etc.
3. Class of material, e.g., fatigue resistant, corrosion resistant, heat resistant, electrical materials
4. Forms like rod, wire, sheet, tube, cast, forged, welded
5. Designations: Unified Number System (UNS), American Iron and Steel Institute (AISI), common names, material group, or country of origin
6. Specifications which allow the operator to select the materials that are acceptable to organizations like the American Society for Testing and Materials (ASTM) and the Society of Automotive Engineers (SAE).
7. Composition, which allows the operator to select the materials that have certain minimum and/or maximum values of alloying elements

More than one of the above sifting criteria can be used to identify suitable materials. Sifting can be performed in the AND or OR modes. The AND mode narrows the search since the material has to conform to all the specified criteria. The OR mode broadens the search since materials that satisfy any of the requirements are selected.

The number of materials that survive the sifting process depends on the severity of the criteria used. At the start of sifting, the number of materials shown on the screen is the total in the database. As more restrictions are placed on the materials, the number of surviving materials gets smaller and could reach zero, i.e., no materials qualify. In such cases, some of the restrictions have to be relaxed and the sifting restarted.

The Material Data Network[26] is an online search engine for materials information and is sponsored by ASM International and Granta Design Limited. Souces of information that are linked to the network (member sites) include ASM International, the Welding Institute, National Physical Laboratory (U.K.), National Institute of Materials Science (Japan), U.K. Steel Association "Steel Specifications," Cambridge Engineering Selector, MatWeb, and IDES plastics data. Member sites can be searched simultaneously with one search string for all classes of materials. The information provided is both quantitative and qualitative, with tables, graphs, micrographs, etc. Some of the sites are freely available and do not require registration while others require registration in order to access the information.

8.4 Computer Assistance in Making Final Selection

Integrating material property database with design algorithms and computer-aided design/manufacturing (CAD/CAM) programs has many benefits, including homogenization and sharing of data in the different departments and decreased redundancy of effort and cost of information storage and retrieval. Several such systems have been cited in Ref. 22:

- The Computerized Application and Reference System (CARS) is developed from the AISI Automotive Steel Design Manual and performs first-order analysis of design using different steels.
- Aluminum Design System (ADS) is developed by the Aluminum Association (U.S.) and performs design calculations and conformance checks of aluminum structural members with the design specifications for aluminum and its alloys.
- Material Selection and Design for fatigue life predictions is developed by ASM International and aids in the design of machinery and engineering structures using different engineering materials.
- Machine Design's Materials Selection is developed by Penton Media (US) and combines the properties for a wide range of materials and the data set for design analysis.

8.5 Expert Systems

Expert systems, also called knowledge-based systems, are computer programs that simulate the reasoning of a human expert in a given field of knowledge. Expert systems rely on heuristics, or rules of thumb, to extract information from a large knowledge base. Expert systems typically consist of three main components:

- Knowledge base, which contains facts and expert-level heuristic rules for solving problems in a given domain. The rules are normally introduced to the system by domain experts through a knowledge engineer.
- Inference engine, which provides an organized procedure for sifting through the knowledge base and choosing applicable rules in order to reach the recommend solutions. The inference engine also provides a link between the knowledge base and the user interface.
- User interface, which allows the user to input the main parameters of the problem under consideration. It also provides recommendations and explanations of how such recommendations were reached.

A commonly used format for the rules in the knowledge base is in the form

IF (condition 1) and/or (condition 2)

THEN (conclusion 1)

For example, in the case of FRP selection

IF required elastic modulus, expressed in GPa, is more than 150 and specific gravity less than 1.7

THEN oriented carbon fibers at 60% by volume

Expert systems are finding many applications in industry, including the areas of design, trouble shooting, failure analysis, manufacturing, materials selection, and materials substitution.[12] When used to assist in materials selection, expert systems provide impartial recommendations and are able to search large databases for optimum solutions. Another important advantage of expert systems is their ability to capture valuable expertise and make it available to a wider circle of users. An example is the Chemical Corrosion Expert System, which is produced by the National association of Corrosion Engineers (NACE) in the United States.[22] The system prompts the user for information on the environmental conditions and configuration of the component and then recommends candidate materials.

REFERENCES

1. M. M. Farag, *Materials Selection for Engineering Design,* Prentice-Hall, London, 1997.
2. G. Dieter, "Overview of the Materials Selection Process," in *ASM Metals Handbook,* Vol. 20, *Materials Selection and Design,* Volume Chair G. Dieter, ASM International, Materials Park, OH, 1997, pp. 243–254.
3. J. Clark, R. Roth, and F. Field III, "Techno-Economic Issues in Materials Selection," in *ASM Metals Handbook,* Vol. 20, *Materials Selection and Design,* Volume Chair G. Dieter, ASM International, Materials Park, OH, 1997, pp. 255–265.
4. M. F. Ashby, "Materials Selection Charts," *ASM Metals Handbook,* Vol. 20, *Materials Selection and Design,* Volume Chair G. Dieter, ASM International, Materials Park, OH, 1997, pp. 266–280.
5. M. F. Ashby, "Performance Indices," in *ASM Metals Handbook,* Vol. 20, *Materials Selection and Design,* Volume Chair G. Dieter, ASM International, Materials Park, OH, 1997, pp. 281–290.
6. D. Bourell, "Decision Matrices in Materials Selection," in *ASM Metals Handbook,* Vol. 20, *Materials Selection and Design,* Volume Chair G. Dieter, ASM International, Materials Park, OH, 1997, pp. 291–296.
7. T. Fowler, "Value Analysis in Materials Selection and Design," in *ASM Metals Handbook,* Vol. 20, *Materials Selection and Design,* Volume Chair G. Dieter, ASM International, Materials Park, OH, 1997, pp. 315–321.
8. F. A. Crane and J. A. Charles, *Selection and Use of Engineering Materials,* Butterworths, London, 1984.
9. M. F. Ashby, *Materials Selection in Mechanical Design,* Pergamon, London, 1992.
10. M. F. Ashby, *Mater. Sci. Technol.,* **5,** 517–525 (1989).
11. R. Sandstrom, "An Approach to Systematic Materials Selection," *Mater. Design,* **6,** 328–338 (1985).
12. V. Weiss, "Computer-Aided Materials Selection," in *ASM Metals Handbook,* Vol. 20, *Materials Selection and Design,* Volume Chair G. Dieter, ASM International, Materials Park, OH, 1997, pp. 309–314.
13. P. A. Gutteridge, and J. Turner, "Computer Aided Materials Selection and Design," *Mater. Design,* **3,** 504–510 (Aug. 1982).
14. L. Olsson, U. Bengtson, and H. Fischmeister, "Computer Aided Materials Selection," in *Computers in Materials Technology,* T. Ericsson (ed.), Pergamon, Oxford, 1981, pp. 17–25.
15. P. P. Dargie et al., "MAPS 1: Computer Aided Design System for Preliminary Material and Manufacturing Process Selection," *Trans. ASME J. Mech. Design,* **104,** 126–136 (1982).
16. A. M. K. Esawi and M. F. Ashby, "Cost Estimates to Guide Pre-selection of Processes," *Mater. Design,* **24,** 605–616 (2003).
17. CES 4 Software, Granta Design Limited, Cambridge, U.K., www.Grantadesign.com, 2002.
18. M. M. Farag and E. El-Magd, "An Integrated Approach to Product Design, Materials Selection, and Cost Estimation," *Mater. Design,* **13,** 323–327 (1992).
19. S. Pugh, *Total Design: Integrated Methods for Successful Product Development,* Addison-Wesley, Reading, MA, 1991.
20. P. E. Kirkwood, "How to Find Materials Properties Data," in *Handbook of Materials Selection,* M. Kutz (ed.), Wiley, New York, 2002, pp. 441–456.
21. J. G. Kaufman, "Sources of Materials Data," in *Handbook of Materials Selection,* M. Kutz (ed.), Wiley, New York, 2002, pp. 457–473.

22. B. E. Boardman and J. G. Kaufman, *Directory of Materials Properties Databases,* Special Supplement to Advanced Materials and Processes, ASM International, Materials Park, OH, Aug. 2000.
23. J. H. Westbrook, "Sources of Materials Property Data and Information," in *ASM Metals Handbook,* Vol. 20, *Materials Selection and Design,* Volume Chair G. Dieter, ASM International, Materials Park, OH, 1997, pp. 491–506.
24. D. Price, "A Guide to Materials Databases," *Materials World,* 418–421 (July 1993).
25. D. Cebon and M. Ashby, "Data Systems for Optimal Materials Selection," *Adv. Mater. Process.,* **161,** 51–54 (2003).
26. www.matdata.net.
27. M. E. Heller, *Metal Selector,* ASM International, Materials Park, OH, 1985; also www.asminternational.org.

PART 2

MECHANICAL DESIGN

CHAPTER 15
STRESS ANALYSIS

Franklin E. Fisher
Professor Emeritus
Mechanical Engineering Department
Loyola Marymount University
Los Angeles, California
and
Raytheon Company
El Segundo, California

Revised from Chapter 8, *Kent's Mechanical Engineer's Handbook,* 12th ed., by John M. Lessells and G. S. Cherniak.

1 STRESSES, STRAINS, STRESS INTENSITY

1.1 Fundamental Definitions

Static Stresses

Total stress on a section *mn* through a loaded body is the resultant force S exerted by one part of the body on the other part in order to maintain in equilibrium the external loads acting on the part. Thus, in Figs. 1, 2, and 3 the total stress on section *mn* due to the external load P is S. The units in which it is expressed are those of load, that is, pounds, tons, etc.

Unit stress, more commonly called stress σ, is the total stress per unit of area at section *mn*. In general it varies from point to point over the section. Its value at any point of a section is the total stress on an elementary part of the area, including the point divided by the elementary total stress on an elementary part of the area, including the point divided by the elementary area. If in Figs. 1, 2, and 3 the loaded bodies are one unit thick and four units wide, then when the total stress S is uniformly distributed over the area, $\sigma = P/A = P/4$. Unit stresses are expressed in pounds per square inch, tons per square foot, etc.

Tensile stress or tension is the internal total stress S exerted by the material fibers to resist the action of an external force P (Fig. 1), tending to separate the material into two parts along the line *mn*. For equilibrium conditions to exist, the tensile stress at any cross section will be equal and opposite in direction to the external force P. If the internal total stress S is distributed uniformly over the area, the stress can be considered as unit tensile stress $\sigma = S/A$.

Compressive stress or compression is the internal total stress S exerted by the fibers to resist the action of an external force P (Fig. 2) tending to decrease the length of the material. For equilibrium conditions to exist, the compressive stress at any cross section will be equal and opposite in direction to the external force P. If the internal total stress S is distributed uniformly over the area, the unit compressive stress $\sigma = S/A$.

Shear stress is the internal total stress S exerted by the material fibers along the plane *mn* (Fig. 3) to resist the action of the external forces, tending to slide the adjacent parts in opposite directions. For equilibrium conditions to exist, the shear stress at any cross section will be equal and opposite in direction to the external force P. If the internal total stress S is uniformly distributed over the area, the unit shear stress $\tau = S/A$.

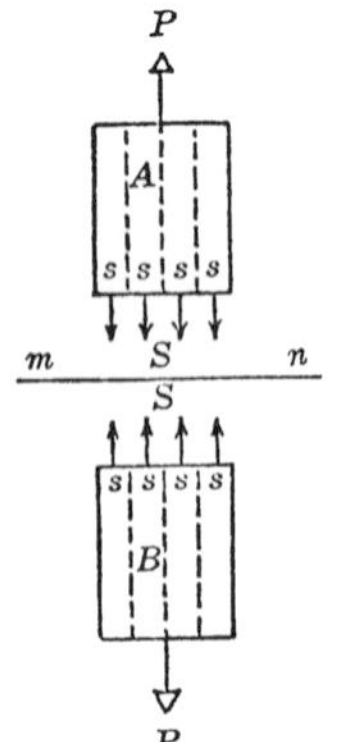

Figure 1 Tensile stress.

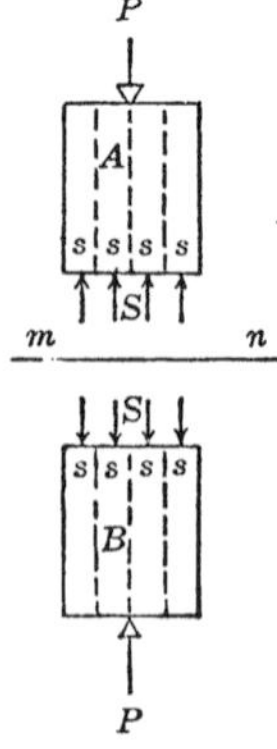

Figure 2 Compressive stress.

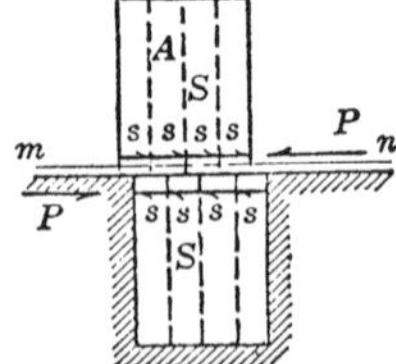

Figure 3 Shear stress.

Normal stress is the component of the resultant stress that acts normal to the area considered (Fig. 4).

Axial stress is a special case of normal stress and may be either tensile or compressive. It is the stress existing in a straight homogeneous bar when the resultant of the applied loads coincides with the axis of the bar.

Simple stress exists when tension, compression, or shear is considered to operate singly on a body.

Total strain on a loaded body is the total elongation produced by the influence of an external load. Thus, in Fig. 4, the total strain is equal to δ. It is expressed in units of length, that is, inches, feet, etc.

Unit strain, or deformation per unit length, is the total amount of deformation divided by the original length of the body before the load causing the strain was applied. Thus, if the total elongation is δ in an original gage length l, the unit strain $e = \delta/l$. Unit strains are expressed in inches per inch and feet per foot.

Tensile strain is the strain produced in a specimen by tensile stresses, which in turn are caused by external forces.

Compressive strain is the strain produced in a bar by compressive stresses, which in turn are caused by external forces.

Shear strain is a strain produced in a bar by the external shearing forces.

Poisson's ratio is the ratio of lateral unit strain to longitudinal unit strain under the conditions of uniform and uniaxial longitudinal stress within the proportional limit. It serves as a measure of lateral stiffness. Average values of Poisson's ratio for the usual materials of construction are:

Material	Steel	Wrought iron	Cast iron	Brass	Concrete
Poisson's ratio	0.300	0.280	0.270	0.340	0.100

Elasticity is that property of a material that enables it to deform or undergo strain and return to its original shape upon the removal of the load.

Hooke's law states that within certain limits (not to exceed the proportional limit) the elongation of a bar produced by an external force is proportional to the tensile stress developed. Hooke's law gives the simplest relation between stress and strain.

Plasticity is that state of matter where permanent deformations or strains may occur without fracture. A material is plastic if the smallest load increment produces a permanent

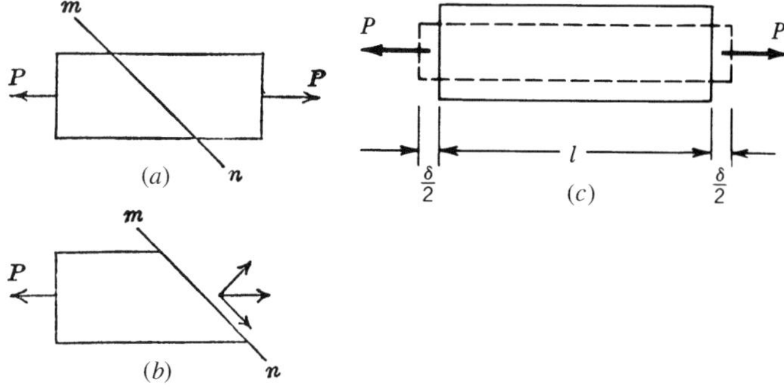

Figure 4 Normal and shear stress components of resultant stress on section *mn* and strain due to tension.

deformation. A perfectly plastic material is nonelastic and has no ultimate strength in the ordinary meaning of that term. Lead is a plastic material. A prism tested in compression will deform permanently under a small load and will continue to deform as the load is increased, until it flattens to a thin sheet. Wrought iron and steel are plastic when stressed beyond the elastic limit in compression. When stressed beyond the elastic limit in tension, they are partly elastic and partly plastic, the degree of plasticity increasing as the ultimate strength is approached.

Stress–strain relationship gives the relation between unit stress and unit strain when plotted on a stress–strain diagram in which the ordinate represents unit stress and the abscissa represents unit strain. Figure 5 shows a typical tension stress–strain curve for medium steel. The form of the curve obtained will vary according to the material, and the curve for compression will be different from the one for tension. For some materials like cast iron, concrete, and timber, no part of the curve is a straight line.

Proportional limit is that unit stress at which unit strain begins to increase at a faster rate than unit stress. It can also be thought of as the greatest stress that a material can stand without deviating from Hooke's law. It is determined by noting on a stress–strain diagram the unit stress at which the curve departs from a straight line.

Elastic limit is the least stress that will cause permanent strain, that is, the maximum unit stress to which a material may be subjected and still be able to return to its original form upon removal of the stress.

Johnson's apparent elastic limit. In view of the difficulty of determining precisely for some materials the proportional limit, J. B. Johnson proposed as the "apparent elastic limit" the point on the stress–strain diagram at which the rate of strain is 50% greater than at the origin. It is determined by drawing *OA* (Fig. 5) with a slope with respect to the vertical axis 50% greater than the straight-line part of the curve; the unit stress at which the line $O'A'$ which is parallel to *OA* is tangent to the curve (point *B*, Fig. 5) is the apparent elastic limit.

Yield point is the lowest stress at which strain increases without increase in stress. Only a few materials exhibit a true yield point. For other materials the term is sometimes used as synonymous with yield strength.

Yield strength is the unit stress at which a material exhibits a specified permanent deformation or state. It is a measure of the useful limit of materials, particularly of those whose stress–strain curve in the region of yield is smooth and gradually curved.

Ultimate strength is the highest unit stress a material can sustain in tension, compression, or shear before rupturing.

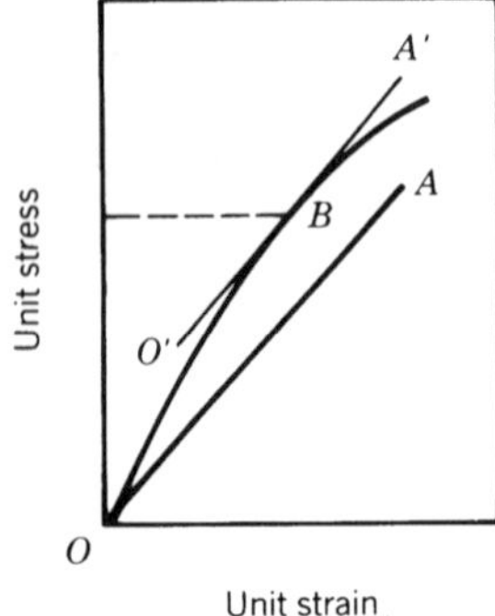

Figure 5 Stress–strain relationship showing determination of apparent elastic limit.

Rupture strength, or breaking strength, is the unit stress at which a material breaks or ruptures. It is observed in tests on steel to be slightly less than the ultimate strength because of a large reduction in area before rupture.

Modulus of elasticity (Young's modulus) in tension and compression is the rate of change of unit stress with respect to unit strain for the condition of uniaxial stress within the proportional limit. For most materials the modulus of elasticity is the same for tension and compression.

Modulus of rigidity (modulus of elasticity in shear) is the rate of change of unit shear stress with respect to unit shear strain for the condition of pure shear within the proportional limit. For metals it is equal to approximately 0.4 of the modulus of elasticity.

True stress is defined as a ratio of applied axial load to the corresponding cross-sectional area. The units of true stress may be expressed in pounds per square inch, pounds per square foot, etc.,

$$\sigma = \frac{P}{A}$$

where σ is the true stress, pounds per square inch, P is the axial load, pounds, and A is the smallest value of cross-sectional area existing under the applied load P, square inches.

True strain is defined as a function of the original diameter to the instantaneous diameter of the test specimen:

$$q = 2 \log_e \frac{d_0}{d} \text{ in./in.}$$

where q is true strain, inches per inch, d_0 is original diameter of test specimen, inches, and d is instantaneous diameter of test specimen, inches.

True stress–strain relationship is obtained when the values of true stress and the corresponding true strain are plotted against each other in the resulting curve (Fig. 6). The slope of the nearly straight line leading up to fracture is known as the coefficient of strain

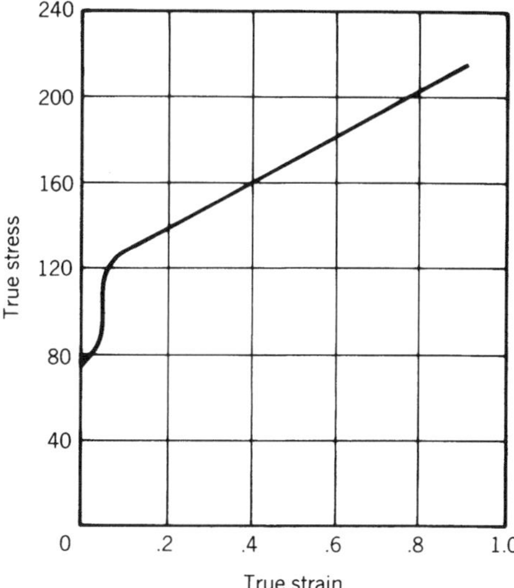

Figure 6 True stress–strain relationship.

hardening. It as well as the true tensile strength appears to be related to the other mechanical properties.

Ductility is the ability of a material to sustain large permanent deformations in tension, such as drawing into a wire.

Malleability is the ability of a material to sustain large permanent deformations in compression, such as beating or rolling into thin sheets.

Brittleness is that property of a material that permits it to be only slightly deformed without rupture. Brittleness is relative, no material being perfectly brittle, that is, capable of no deformation before rupture. Many materials are brittle to a greater or less degree, glass being one of the most brittle of materials. Brittle materials have relatively short stress–strain curves. Of the common structural materials, cast iron, brick, and stone are brittle in comparison with steel.

Toughness is the ability of the material to withstand high unit stress together with great unit strain without complete fracture. The area *OAGH,* or *OJK,* under the curve of the stress–strain diagram (Fig. 7), is a measure of the toughness of the material. The distinction between ductility and toughness is that ductility deals only with the ability to deform, whereas toughness considers both the ability to deform and the stress developed during deformation.

Stiffness is the ability to resist deformation under stress. The modulus of elasticity is the criterion of the stiffness of a material.

Hardness is the ability to resist very small indentations, abrasion, and plastic deformation. There is no single measure of hardness, as it is not a single property but a combination of several properties.

Creep, or flow of metals, is a phase of plastic or inelastic action. Some solids, as asphalt or paraffin, flow appreciably at room temperatures under extremely small stresses; zinc, plastics, fiber-reinforced plastics, lead, and tin show signs of creep at room temperature under moderate stresses. At sufficiently high temperatures, practically all metals creep under stresses that vary with temperature; the higher the temperature, the lower the stress at which creep takes place. The deformation due to creep continues to increase indefinitely and becomes of extreme importance in members subjected to high temperatures, as parts in turbines, boilers, superheaters, etc.

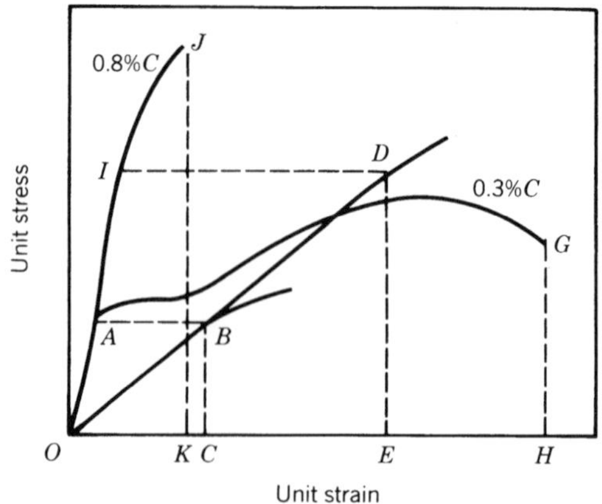

Figure 7 Toughness comparison.

Creep limit is the maximum unit stress under which unit distortion will not exceed a specified value during a given period of time at a specified temperature. A value much used in tests and suggested as a standard for comparing materials is the maximum unit stress at which creep does not exceed 1% in 100,000 h.

Types of fracture. A bar of brittle material, such as cast iron, will rupture in a tension test in a clean sharp fracture with very little reduction of cross-sectional area and very little elongation (Fig. 8*a*). In a ductile material, as structural steel, the reductions of area and elongation are greater (Fig. 8*b*). In compression, a prism of brittle material will break by shearing along oblique planes; the greater the brittleness of the material, the more nearly will these planes parallel the direction of the applied force. Figures 8*c*, 8*d*, and 8*e*, arranged in order of brittleness, illustrate the type of fracture in prisms of brick, concrete, and timber. Figure 8*f* represents the deformation of a prism of plastic material, as lead, which flattens out under load without failure.

Relations of elastic constants

Modulus of elasticity, E:

$$E = \frac{Pl}{Ae}$$

where P = load, pounds, l = length of bar, inches, A = cross-sectional area acted on by the axial load P, and e = total strain produced by axial load P.

Modulus of rigidity, G:

$$G = \frac{E}{2(1 + \nu)}$$

where E = modulus of elasticity and ν = Poisson's ratio.

Bulk modulus, K, is the ratio of normal stress to the change in volume.

Relationships. The following relationships exist between the modulus of elasticity E, the modulus of rigidity G, the bulk modulus of elasticity K, and Poisson's ratio ν:

$$E = 2G(1 + \nu) \qquad G = \frac{E}{2(1 + \nu)} \qquad \nu = \frac{E - 2G}{2G}$$

$$K = \frac{E}{3(1 - 2\nu)} \qquad \nu = \frac{3K - E}{6K}$$

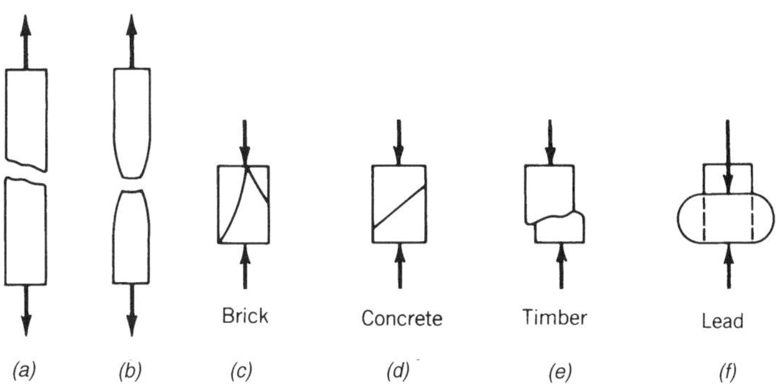

Figure 8 (*a*) Brittle and (*b*) ductile fractures in tension and compression fractures.

Allowable unit stress, also called allowable working unit stress, allowable stress, or working stress, is the maximum unit stress to which it is considered safe to subject a member in service. The term allowable stress is preferable to working stress, since the latter often is used to indicate the actual stress in a material when in service. Allowable unit stresses for different materials for various conditions of service are specified by different authorities on the basis of test or experience. In general, for ductile materials, allowable stress is considerably less than the yield point.

Factor of safety is the ratio of ultimate strength of the material to allowable stress. The term was originated for determining allowable stress. The ultimate strength of a given material divided by an arbitrary factor of safety, dependent on material and the use to which it is to be put, gives the allowable stress. In present design practice, it is customary to use allowable stress as specified by recognized authorities or building codes rather than an arbitrary factor of safety. One reason for this is that the factor of safety is misleading, in that it implies a greater degree of safety than actually exists. For example, a factor of safety of 4 does not mean that a member can carry a load four times as great as that for which it was designed. It also should be clearly understood that, even though each part of a machine is designed with the same factor of safety, the machine as a whole does not have that factor of safety. When one part is stressed beyond the proportional limit, or particularly the yield point, the load or stress distribution may be completely changed throughout the entire machine or structure, and its ability to function thus may be changed, even though no part has ruptured.

Although no definite rules can be given, if a factor of safety is to be used, the following circumstances should be taken into account in its selection:

1. When the ultimate strength of the material is known within narrow limits, as for structural steel for which tests of samples have been made, when the load is entirely a steady one of a known amount and there is no reason to fear the deterioration of the metal by corrosion, the lowest factor that should be adopted is 3.
2. When the circumstances of 1 are modified by a portion of the load being variable, as in floors of warehouses, the factor should not be less than 4.
3. When the whole load, or nearly the whole, is likely to be alternately put on and taken off, as in suspension rods of floors of bridges, the factor should be 5 or 6.
4. When the stresses are reversed in direction from tension to compression, as in some bridge diagonals and parts of machines, the factor should be not less than 6.
5. When the piece is subjected to repeated shocks, the factor should be not less than 10.
6. When the piece is subjected to deterioration from corrosion, the section should be sufficiently increased to allow for a definite amount of corrosion before the piece is so far weakened by it as to require removal.
7. When the strength of the material or the amount of the load or both are uncertain, the factor should be increased by an allowance sufficient to cover the amount of the uncertainty.
8. When the strains are complex and of uncertain amount, such as those in the crankshaft of a reversing engine, a very high factor is necessary, possibly even as high as 40.
9. If the property loss caused by failure of the part may be large or if loss of life may result, as in a derrick hoisting materials over a crowded street, the factor should be large.

Dynamic Stresses

Dynamic stresses occur where the dimension of time is necessary in defining the loads. They include creep, fatigue, and impact stresses.

Creep stresses occur when either the load or deformation progressively varies with time. They are usually associated with noncyclic phenomena.

Fatigue stresses occur when type of cyclic variation of either load or strain is coincident with respect to time.

Impact stresses occur from loads which are transient with time. The duration of the load application is of the same order of magnitude as the natural period of vibration of the specimen.

1.2 Work and Resilience

External work. Let P = axial load, pounds, on a bar, producing an internal stress not exceeding the elastic limit; σ = unit stress produced by P, pounds per square inch; A = cross-sectional area, square inches; l = length of bar, inches; e = deformation, inches; E = modulus of elasticity; $W = \frac{1}{2}Pe$ external work performed on bar, inch-pounds. Then

$$W = \frac{1}{2} A\sigma \left(\frac{\sigma l}{E}\right) = \frac{1}{2}\left(\frac{\sigma^2}{E}\right) Al \tag{1}$$

The factor $\frac{1}{2}(\sigma^2/E)$ is the work required per unit volume, the volume being Al. It is represented on the stress–strain diagram by the area ODE or area OBC (Fig. 9), which DE and BC are ordinates representing the unit stresses considered.

Resilience is the strain energy that may be recovered from a deformed body when the load causing the stress is removed. Within the proportional limit, the resilience is equal to the external work performed in deforming the bar and may be determined by Eq. (1). When σ is equal to the proportional limit, the factor $\frac{1}{2}(\sigma^2/E)$ is the *modulus of resilience,* that is, the measure of capacity of a unit volume of material to store strain energy up to the proportional limit. Average values of the modulus of resilience under tensile stress are given in Table 1.

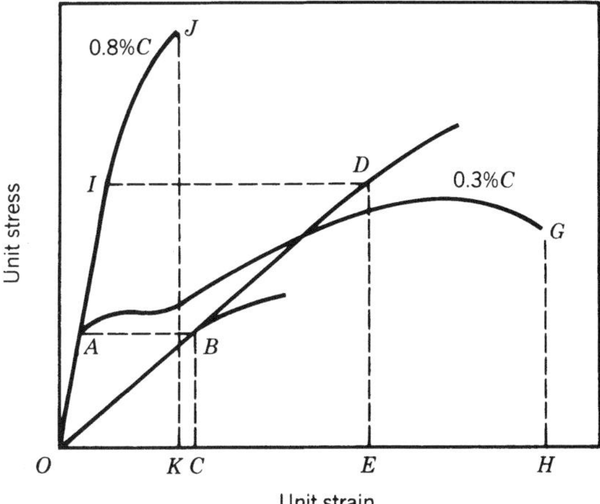

Figure 9 Work areas on stress–strain diagram.

Table 1 Modulus of Resilience and Relative Toughness under Tensile Stress (Average Values)

Material	Modulus of Resilience (in.·lb/in.^{3})	Relative Toughness (Area under Curve of Stress-Deformation Diagram)
Gray cast iron	1.2	70
Malleable cast iron	17.4	3,800
Wrought iron	11.6	11,000
Low-carbon steel	15.0	15,700
Medium-carbon steel	34.0	16,300
High-carbon steel	94.0	5,000
Ni–Cr steel, hot rolled	94.0	44,000
Vanadium steel, 0.98% C, 0.2% V, heat treated	260.0	22,000
Duralumin, 17 ST	45.0	10,000
Rolled bronze	57.0	15,500
Rolled brass	40.0	10,000
Oak	2.3[a]	13[a]

[a] Bending.

The total resilience of a bar is the product of its volume and the modulus of resilience. These formulas for work performed on a bar and its resilience do not apply if the unit stress is greater than the proportional limit.

Work required for rupture. Since beyond the proportional limit the strains are not proportional to the stresses, ½P does not express the mean value of the force. Equation (1), therefore, does not express the work required for strain after the proportional limit of the material has been passed and cannot express the work required for rupture. The work required per unit volume to produce strains beyond the proportional limit or to cause rupture may be determined from the stress–strain diagram as it is measured by the area under the stress–strain curve up to the strain in question, as *OAGH* or *OJK* (Fig. 9). This area, however, does not represent the resilience, since part of the work done on the bar is present in the form of hysteresis losses and cannot be recovered.

Damping capacity (hysteresis). Observations show that when a tensile load is applied to a bar, it does not produce the complete elongation immediately, but there is a definite time lapse which depends on the nature of the material and the magnitude of the stresses involved. In parallel with this it is also noted that, upon unloading, complete recovery of energy does not occur. This phenomenon is variously termed *elastic hysteresis* or, for vibratory stresses, damping. Figure 10 shows a typical hysteresis loop obtained for one cycle of loading. The area of this hysteresis loop, representing the energy dissipated per cycle, is a measure of the damping properties of the material. While the exact mechanism of damping has not been fully investigated, it has been found that under vibratory conditions the energy dissipated in this manner varies approximately as the cube of the stress.

2 DISCONTINUITIES, STRESS CONCENTRATION

The direct design procedure assumes no abrupt changes in cross section, discontinuities in the surface, or holes, through the member. In most structural parts this is not the case. The

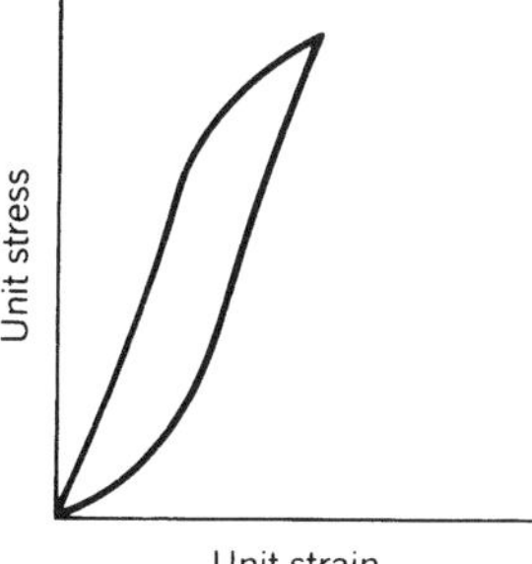

Figure 10 Hysteresis loop for loading and unloading.

stresses produced at these discontinuities are different in magnitude from those calculated by various design methods. The effect of the localized increase in stress, such as that caused by a notch, fillet, hole, or similar *stress raiser,* depends mainly on the type of loading, the geometry of the part, and the material. As a result, it is necessary to consider a stress concentration factor K_t, which is defined by the relationship

$$K_t = \frac{\sigma_{\text{max}}}{\sigma_{\text{nominal}}} \tag{2}$$

In general σ_{max} will have to be determined by the methods of experimental stress analysis or the theory of elasticity, and σ_{nominal} by a simple theory such as $\sigma = P/A$, $\sigma = Mc/I$, $\tau = Tc/J$ without taking into account the variations in stress conditions caused by geometrical discontinuities such as holes, grooves, and fillets. For ductile materials it is not customary to apply stress concentration factors to members under static loading. For brittle materials, however, stress concentration is serious and should be considered.

Stress-Concentration Factors for Fillets, Keyways, Holes, and Shafts
In Table 2 selected stress concentration factors have been given from a complete table in Refs. 1, 2, and 3.

Table 2 Stress Concentration Factors[a]

Type		K_t Factors					
Circular hole in plate or rectangular bar	$\frac{h}{a} = 0.67$		0.77	0.91	1.07	1.29	1.56
	$k = 4.37$		3.92	3.61	3.40	3.25	3.16
Square shoulder with fillet for rectangular and circular cross sections in bending	$\frac{h}{r} \Big/ \frac{r}{d}$	0.05	0.10	0.20	0.27	0.50	1.0
	0.5	1.61	1.49	1.39	1.34	1.22	1.07
	1.0	1.91	1.70	1.48	1.38	1.22	1.08
	1.5	2.00	1.73	1.50	1.39	1.23	1.08
	2.0		1.74	1.52	1.39	1.23	1.09
	3.5		1.76	1.54	1.40	1.23	1.10

[a]Adapted by permission from R. J. Roark and W. C. Young, *Formulas for Stress and Strain,* 6th ed., McGraw-Hill, New York, 1989.

3 COMBINED STRESSES

Under certain circumstances of loading a body is subjected to a combination of tensile, compressive, and/or shear stresses. For example, a shaft that is simultaneously bent and twisted is subjected to combined stresses, namely, longitudinal tension and compression and torsional shear. For the purposes of analysis it is convenient to reduce such systems of combined stresses to a basic system of stress coordinates known as principal stresses. These stresses act on axes that differ in general from the axes along which the applied stresses are acting and represent the maximum and minimum values of the normal stresses for the particular point considered.

Determination of Principal Stresses

The expressions for the principal stresses in terms of the stresses along the x and y axes are

$$\sigma_1 = \frac{\sigma_x + \sigma_y}{2} + \sqrt{\left(\frac{\sigma_x - \sigma_y}{2}\right)^2 + \tau_{xy}^2} \qquad (3)$$

$$\sigma_2 = \frac{\sigma_x + \sigma_y}{2} - \sqrt{\left(\frac{\sigma_x - \sigma_y}{2}\right)^2 + \tau_{xy}^2} \qquad (4)$$

$$\tau_1 = \pm\sqrt{\left(\frac{\sigma_x - \sigma_y}{2}\right)^2 + \tau_{xy}^2} \qquad (5)$$

where σ_1, σ_2, and τ_1 are the principal stress components and σ_x, σ_y, and τ_{xy} are the calculated stress components, all of which are determined at any particular point (Fig. 11).

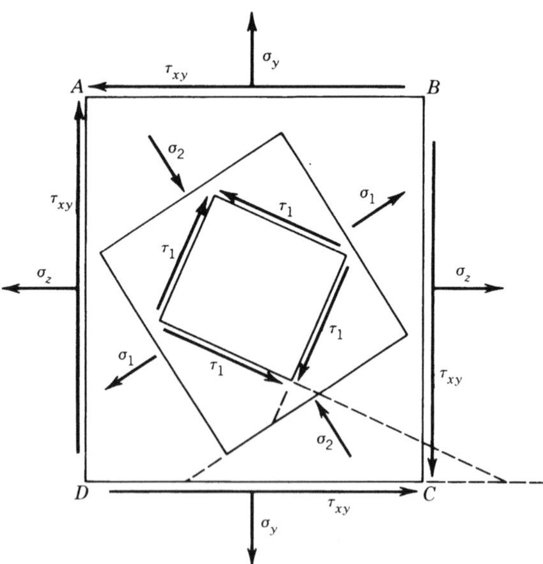

Figure 11 Diagram showing relative orientation of stresses. (Reproduced by permission from J. Marin, *Mechanical Properties of Materials and Design,* McGraw-Hill, New York, 1942.)

Graphical Method of Principal Stress Determination—Mohr's Circle

Let the axes x and y be chosen to represent the directions of the applied normal and shearing stresses, respectively (Fig. 12). Lay off to suitable scale distances $OA = \sigma_x$, $OB = \sigma_y$, and $BC = AD = \tau_{xy}$. With point E as a center construct the circle DFC. Then OF and OG are the principal stresses σ_1 and σ_2, respectively, and EC is the maximum shear stress τ_1. The inverse also holds—that is, given the principal stresses, σ_x and σ_y can be determined on any plane passing through the point.

Stress–Strain Relations

The linear relation between components of stress and strain is known as *Hooke's law.* This relation for the two-dimensional case can be expressed as

$$e_x = \frac{1}{E}(\sigma_x - \nu\sigma_y) \tag{6}$$

$$e_y = \frac{1}{E}(\sigma_y - \nu\sigma_x) \tag{7}$$

$$\gamma_{xy} = \frac{1}{G}\tau_{xy} \tag{8}$$

where σ_x, σ_y, and τ_{xy} are the stress components of a particular point, ν is Poisson's ratio, E is modulus of elasticity, G is modulus of rigidity, and e_x, e_y, and γ_{xy} are strain components.

The determination of the magnitudes and directions of the principal stresses and strains and of the maximum shearing stresses is carried out for the purpose of establishing criteria of failure within the material under the anticipated loading conditions. To this end several theories have been advanced to elucidate these criteria. The more noteworthy ones are listed below. The theories are based on the assumption that the principal stresses do not change

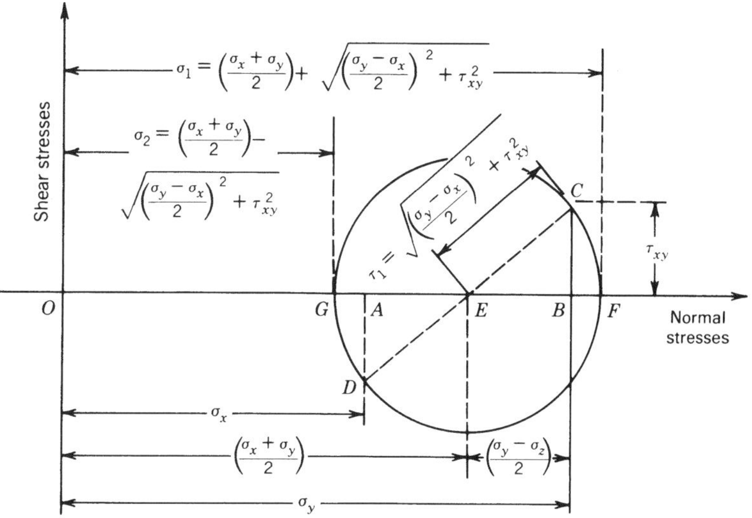

Figure 12 Mohr's circle used for the determination of the principal stresses. (Reproduced by permission from J. Marin, *Mechanical Properties of Materials and Design,* McGraw-Hill, New York, 1942.)

with time, an assumption that is justified since the applied loads in most cases are synchronous.

Maximum-Stress Theory (Rankine's Theory)

This theory is based on the assumption that failure will occur when the maximum value of the greatest principal stress reaches the value of the maximum stress σ_{max} at failure in the case of simple axial loading. Failure is then defined as

$$\sigma_1 \text{ or } \sigma_2 = \sigma_{max} \tag{9}$$

Maximum-Strain Theory (Saint Venant)

This theory is based on the assumption that failure will occur when the maximum value of the greatest principal strain reaches the value of the maximum strain e_{max} at failure in the case of simple axial loading. Failure is then defined as

$$e_1 \text{ or } e_2 = e_{max} \tag{10}$$

If e_{max} does not exceed the linear range of the material, Eq. (10) may be written as

$$\sigma_1 - \nu\sigma_2 = \sigma_{max} \tag{11}$$

Maximum-Shear Theory (Guest)

This theory is based on the assumption that failure will occur when the maximum shear stress reaches the value of the maximum shear stress at failure in simple tension. Failure is then defined as

$$\tau_1 = \tau_{max} \tag{12}$$

Distortion-Energy Theory (Hencky–Von Mises) (Shear Energy)

This theory is based on the assumption that failure will occur when the distortion energy corresponding to the maximum values of the stress components equals the distortion energy at failure for the maximum axial stress. Failure is then defined as

$$\sigma_1^2 - \sigma_1\sigma_2 + \sigma_2^2 = \sigma_{max}^2 \tag{13}$$

Strain-Energy Theory

This theory is based on the assumption that failure will occur when the total strain energy of deformation per unit volume in the case of combined stress is equal to the strain energy per unit volume at failure in simple tension. Failure is then defined as

$$\sigma_1^2 - 2\nu\sigma_1\sigma_2 + \sigma_2^2 = \sigma_{max}^2 \tag{14}$$

Comparison of Theories

Figure 13 compares the five foregoing theories. In general the distortion-energy theory is the most satisfactory for ductile materials and the maximum-stress theory is the most satisfactory for brittle materials. The maximum-shear theory gives conservative results for both ductile and brittle materials. The conditions for yielding, according to the various theories, are given in Table 3, taking $\nu = 0.300$ as for steel.

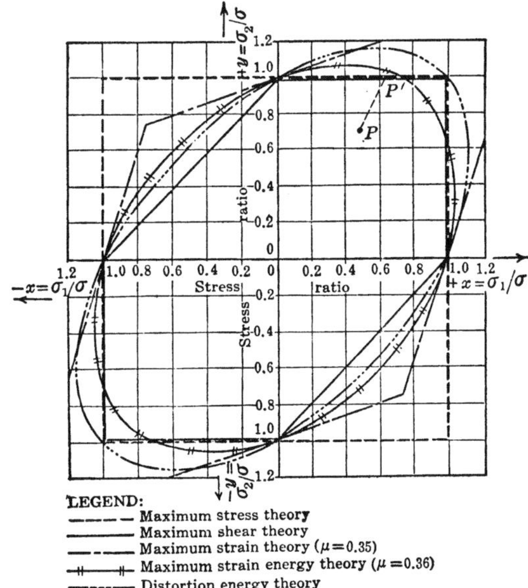

Figure 13 Comparison of five theories of failure. (Reproduced by permission from J. Marin, *Mechanical Properties of Materials and Design,* McGraw-Hill, New York, 1942.)

Static Working Stresses

Ductile Materials. For ductile materials the criteria for working stresses are

$$\sigma_w = \frac{\sigma_{yp}}{n} \qquad \text{(tension and compression)} \tag{15}$$

$$\tau_w = \frac{1}{2}\frac{\sigma_{yp}}{n} \tag{16}$$

Brittle Materials. For brittle materials the criteria for working stresses are

$$\sigma_w = \frac{\sigma_{\text{ultimate}}}{K_t \times n} \qquad \text{(tension)} \tag{17}$$

$$\sigma_w = \frac{\sigma_{\text{compressive}}}{K_t \times n} \qquad \text{(compression)} \tag{18}$$

where K_t is the stress concentration factor, n is the factor of safety, σ_w and τ_w are working stresses, and σ_{yp} is stress at the yield point.

Table 3 Comparison of Stress Theories

$\tau = \sigma_{yp}$	(from the maximum-stress theory
$\tau = 0.77\sigma_{yp}$	(from the maximum-strain theory)
$\tau = 0.50\sigma_{yp}$	(from the maximum-shear theory)
$\tau = 0.62\sigma_{yp}$	(from the maximum-strain-energy theory)

Working-Stress Equations for the Various Theories

Stress theory:

$$\sigma_w = \frac{\sigma_x + \sigma_y}{2} \pm \sqrt{\left(\frac{\sigma_x - \sigma_y}{2}\right)^2 + \tau_{xy}^2} \tag{19}$$

Shear theory:

$$\sigma_w = 2\sqrt{\left(\frac{\sigma_x - \sigma_y}{2}\right)^2 + \tau_{xy}^2} \tag{20}$$

Strain theory:

$$\sigma_w = (1 - \nu)\left(\frac{\sigma_x + \sigma_y}{2}\right) + (1 + \nu)\sqrt{\left(\frac{\sigma_x - \sigma_y}{2}\right)^2 + \tau_{xy}^2} \tag{21}$$

Distortion-energy theory:

$$\sigma_w = \sqrt{\sigma_x^2 - \sigma_x\sigma_y + \sigma_y^2 + 3\tau_{xy}^2} \tag{22}$$

Strain-energy theory:

$$\sigma_w = \sqrt{\sigma_x^2 - 2\nu\sigma_x\sigma_y + \sigma_y^2 + 2(1 + \nu)\tau_{xy}^2} \tag{23}$$

where σ_x, σ_y, τ_{xy} are the stress components of a particular point, ν is Poisson's ratio, and σ_w is working stress.

4 CREEP

Introduction

Materials subjected to a constant stress at elevated temperatures deform continuously with time, and the behavior under these conditions is different from the behavior at normal temperatures. This continuous deformation with time is called creep. In some applications the permissible creep deformations are critical, in others of no significance. But the existence of creep necessitates information on the creep deformations that may occur during the expected life of the machine. Plastic, zinc, tin, and fiber-reinforced plastics creep at room temperature. Aluminum and magnesium alloys start to creep at around 300°F. Steels above 650°F must be checked for creep.

Mechanism of Creep Failure

There are generally four distinct phases distinguishable during the course of creep failure. The elapsed time per stage depends on the material, temperature, and stress condition. They are: (1) Initial phase—where the total deformation is partially elastic and partially plastic. (2) Second phase—where the creep rate decreases with time, indicating the effect of strain hardening. (3) Third phase—where the effect of strain hardening is counteracted by the annealing influence of the high temperature which produces a constant or minimum creep rate. (4) Final phase—where the creep rate increases until fracture occurs owing to the decrease in cross-sectional area of the specimen.

Creep Equations

In conducting a conventional creep test, curves of strain as a function of time are obtained for groups of specimens; each specimen in one group is subjected to a different constant stress, while all of the specimens in the group are tested at one temperature.

In this manner families of curves like those shown in Fig. 14 are obtained. Several methods have been proposed for the interpretation of such data. (See Refs. 1 and 4.) Two frequently used expressions of the creep properties of a material can be derived from the data in the following form:

$$C = B\sigma^m \qquad \epsilon = \epsilon_0 + Ct \tag{24}$$

where C = creep rate, B and m = experimental constants (where $B = C/\sigma_1^m$ for C = 1%/hr), σ = stress, ϵ = creep strain at any time t, ϵ_0 = zero-time strain intercept, and t = time. See Fig. 15.

Stress Relaxation

Various types of bolted joints and shrink- or press-fit assemblies and springs are applications of creep taking place with diminishing stress. This deformation tends to loosen the joint and produces a stress reduction or stress relaxation. The performance of a material to be used under diminishing creep-stress condition is determined by a tensile stress relaxation test.

5 FATIGUE

Definitions

STRESS CYCLE. A stress cycle is the smallest section of the stress–time function that is repeated identically and periodically, as shown in Fig. 16.

MAXIMUM STRESS. σ_{max} is the largest algebraic value of the stress in the stress cycle, being positive for a tensile stress and negative for a compressive stress.

MINIMUM STRESS. σ_{min} is the smallest algebraic value of the stress in the stress cycle, being positive for a tensile stress and negative for a compressive stress.

RANGE OF STRESS. σ_r is the algebraic difference between the maximum and minimum stress in one cycle:

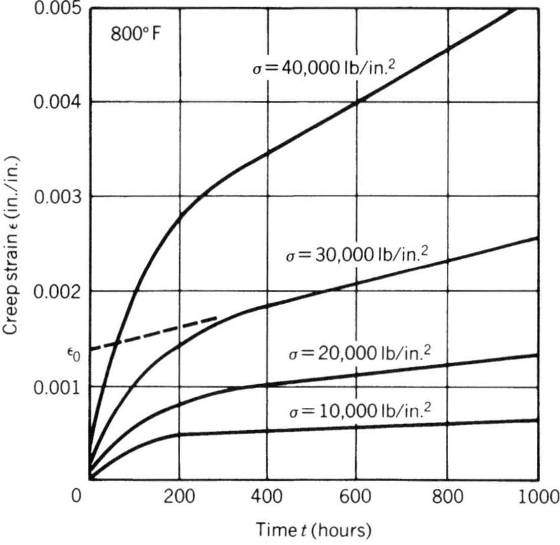

Figure 14 Curves of creep strain for various stress levels.

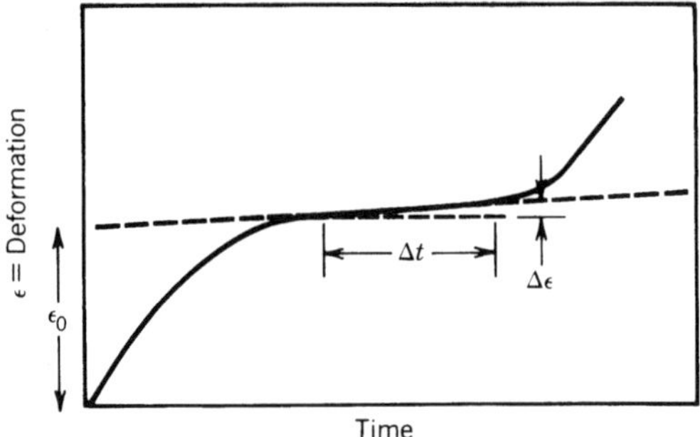

Figure 15 Method of determining creep rate.

$$\sigma_r = \sigma_{\max} - \sigma_{\min} \tag{25}$$

For most cases of fatigue testing the stress varies about zero stress, but other types of variation may be experienced.

Alternating-stress amplitude (variable-stress component). σ_a is one-half the range of stress, $\sigma_a = \sigma_r/2$.

Mean stress (steady-stress component). σ_m is the algebraic mean of the maximum and minimum stress in one cycle:

$$\sigma_m = \frac{\sigma_{\max} + \sigma_{\min}}{2} \tag{26}$$

Stress ratio. R is the algebraic ratio of the minimum stress and the maximum stress in one cycle.

5.1 Modes of Failure

The three most common modes of failure are*

$$\text{Soderberg's Law} \qquad \frac{\sigma_m}{\sigma_y} + \frac{\sigma_a}{\sigma_e} = \frac{1}{N} \tag{27}$$

$$\text{Goodman's Law} \qquad \frac{\sigma_m}{\sigma_u} + \frac{\sigma_a}{\sigma_e} = \frac{1}{N} \tag{28}$$

$$\text{Gerber's Law} \qquad \left(\frac{\sigma_m}{\sigma_u}\right)^2 + \frac{\sigma_a}{\sigma_e} = \frac{1}{N} \tag{29}$$

From distortion energy for plane stress

$$\sigma_m = \sqrt{\sigma_{xm}^2 - \sigma_{xm}\sigma_{ym} + \sigma_{ym}^2 + 3\tau_{xym}^2} \tag{30}$$

$$\sigma_a = \sqrt{\sigma_{xa}^2 - \sigma_{xa}\sigma_{ya} + \sigma_{ya}^2 + 3\tau_{xya}^2} \tag{31}$$

---

*This section is condensed from Ref. 1, Chap. 12.

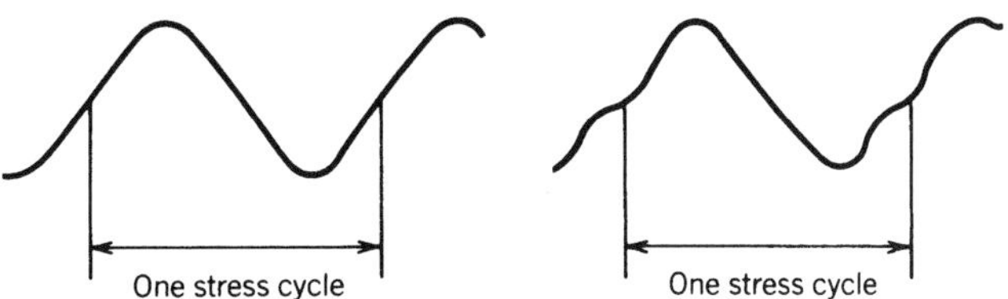

Figure 16 Definition of one stress cycle.

The stress concentration factor,[3] K_t or K_f, is applied to the individual stress for both σ_a and σ_m for brittle materials and only to σ_a for ductile materials. *N* is a reasonable factor of safety. σ_u is the ultimate tensile strength, and σ_y is the yield strength. σ_e is developed from the endurance limit σ_e' and reduced or increased depending on conditions and manufacturing procedures and to keep σ_e less than the yield strength:

$$\sigma_e = k_a k_b \cdots k_n \, \sigma_e'$$

where σ_e' (Ref. 1) for various materials is:

Steel	$0.5\sigma_u$ and never greater than 100 kpsi at 10^6 cycles
Magnesium	$0.35\sigma_u$ at 10^8 cycles
Nonferrous alloys	$0.35\sigma_u$ at 10^8 cycles
Aluminum alloys	$(0.16–0.3)\sigma_u$ at 5×10^8 cycles (Ref. 34)

and where the other *k* factors are affected as follows:

Surface Condition. For surfaces that are from machined to ground, the k_a varies from 0.7 to 1.0. When surface finish is known, k_a can be found[1] more accurately.

Size and Shape. If the size of the part is 0.30 in. or larger, the reduction is 0.85 or less, depending on the size.

Reliability. The endurance limit and material properties are averages and both should be corrected. A reliability of 90% reduces values 0.897, while one of 99% reduces 0.814.

Temperature. The endurance limit at −190°C increases 1.54–2.57 for steels, 1.14 for aluminums, and 1.4 for titaniums. The endurance limit is reduced approximately 0.68 for some steels at 1382°F, 0.24 for aluminum around 662°F, and 0.4 for magnesium alloys at 572°F.

Residual Stresses. For steel, shot peening increases the endurance limit 1.04–1.22 for polished surfaces, 1.25 for machined surfaces, 1.25–1.5 for rolled surfaces, and 2–3 for forged surfaces. The shot-peening effect disappears above 500°F for steels and above 250°F for aluminum. Surface rolling affects the steel endurance limit approximately the same as shot peening, while the endurance limit is increased 1.2–1.3 in aluminum, 1.5 in magnesium, and 1.2–2.93 in cast iron.

Corrosion. A corrosive environment decreases the endurance limit of anodized aluminum and magnesium 0.76–1.00, while nitrided steel and most materials are reduced 0.6–0.8.

Surface Treatments. Nickel plating reduces the endurance limit of 1008 steel 0.01 and of 1063 steel 0.77, but, if the surface is shot peened after it is plated, the endurance limit can be increased over that of the base metal. The endurance limit of anodized aluminum is in general not affected. Flame and induction hardening as well as carburizing increase the endurance limit 1.62–1.85, while nitriding increases it 1.30–2.00.

Fretting. In surface pairs that move relative to each other, the endurance limit is reduced 0.70–0.90 for each material.

Radiation. Radiation tends to increase tensile strength but to decrease ductility.

In discussions on fatigue it should be emphasized that most designs must pass vibration testing. When sizing parts so that they can be modeled on a computer, the designer needs a starting point until feedback is received from the modeling. A helpful starting point is to estimate the static load to be carried, to find the level of vibration testing in G levels, to assume that the part vibrates with a magnification of 10, and to multiply these together to get an equivalent static load. The stress level should be $\sigma_u/4$, which should be less than the yield strength. When the design is modeled, changes can be made to bring the design within the required limits.

6 BEAMS

6.1 Theory of Flexure

Types of Beams

A beam is a bar or structural member subjected to transverse loads that tend to bend it. Any structural members acts as a beam if bending is induced by external transverse forces.

A **simple beam** (Fig. 17*a*) is a horizontal member that rests on two supports at the ends of the beam. All parts between the supports have free movement in a vertical plane under the influence of vertical loads.

A **fixed beam, constrained beam,** or **restrained beam** (Fig. 17*b*) is rigidly fixed at both ends or rigidly fixed at one end and simply supported at the other.

A **continuous beam** (Fig. 17*c*) is a member resting on more than two supports.

A **cantilever beam** (Fig. 17*d*) is a member with one end projecting beyond the point of support, free to move in a vertical plane under the influence of vertical loads placed between the free end and the support.

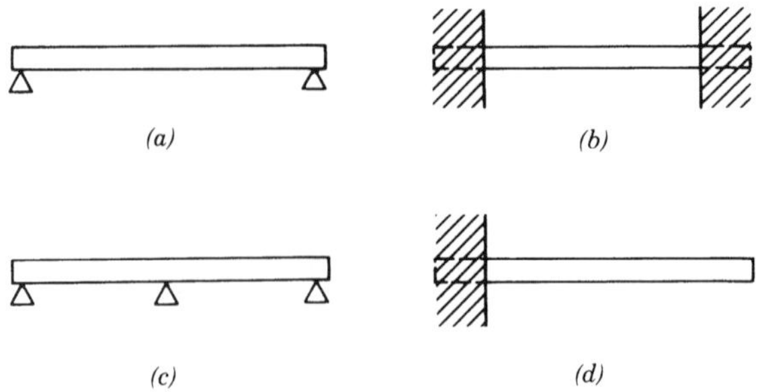

Figure 17 (*a*) Simple, (*b*) constrained, (*c*) continuous, and (*d*) cantilever beams.

Phenomena of Flexure

When a simple beam bends under its own weight, the fibers on the upper or concave side are shortened, and the stress acting on them is compression; the fibers on the under or convex side are lengthened, and the stress acting on them is tension. In addition, shear exists along each cross section, the intensity of which is greatest along the sections at the two supports and zero at the middle section.

When a cantilever beam bends under its own weight, the fibers on the upper or convex side are lengthened under tensile stresses; the fibers on the under or concave side are shortened under compressive stresses, the shear is greatest along the section at the support and zero at the free end.

The **neutral surface** is that horizontal section between the concave and convex surfaces of a loaded beam, where there is no change in the length of the fibers and no tensile or compressive stresses acting upon them.

The **neutral axis** is the trace of the neutral surface on any cross section of a beam. (See Fig. 18.)

The **elastic curve** of a beam is the curve formed by the intersection of the neutral surface with the side of the beam, it being assumed that the longitudinal stresses on the fibers are within the elastic limit.

Reactions at Supports

The reactions, or upward pressures at the points of support, are computed by applying the following conditions necessary for equilibrium of a system of vertical forces in the same plane: (1) The algebraic sum of all vertical forces must equal zero; that is, the sum of the reactions equals the sum of the downward loads. (2) The algebraic sum of the moments of all the vertical forces must equal zero. Condition 1 applies to cantilever beams and to simple beams uniformly loaded, or with equal concentrated loads placed at equal distances from the center of the beam. In the cantilever beam, the reaction is the sum of all the vertical forces acting downward, comprising the weight of the beam and the superposed loads. In the simple beam each reaction is equal to one-half the total load, consisting of the weight of the beam and the superposed loads. Condition 2 applies to a simple beam not uniformly loaded. The reactions are computed separately, by determining the moment of the several loads about each support. The sum of the moments of the load around one support is equal to the moment of the reaction of the other support around the first support.

Conditions of Equilibrium

The fundamental laws for the stresses at any cross section of a beam in equilibrium are: (1) Sum of horizontal tensile stresses = sum of horizontal compressive stresses. (2) Resisting shear = vertical shear. (3) Resisting moment = bending moment.

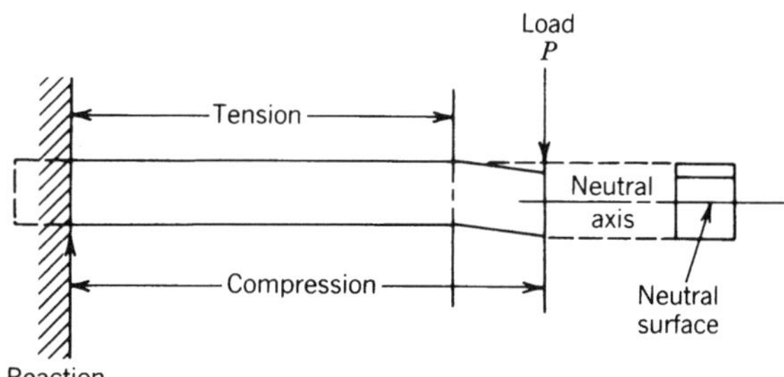

Figure 18 Loads and stress conditions in a cantilever beam.

Vertical Shear. At any cross section of a beam the resultant of the external vertical forces acting on one side of the section is equal and opposite to the resultant of the external vertical forces acting on the other side of the section. These forces tend to cause the beam to shear vertically along the section. The value of either resultant is known as the vertical shear at the section considered. It is computed by finding the algebraic sum of the vertical forces to the left of the section; that is, it is equal to the left reaction minus the sum of the vertical downward forces acting between the left support and the section.

A **shear diagram** is a graphic representation of the vertical shear at all cross sections of the beam. Thus in the uniformly loaded simple beam (Table 4) the ordinates to the line represent to scale the intensity of the vertical shear at the corresponding sections of the beam. The vertical shear is greatest at the supports, where it is equal to the reactions, and it is zero at the center of the span. In the cantilever beam (Table 4) the vertical shear is greatest at the point of support, where it is equal to the reaction, and it is zero at the free end. Table 4 shows graphically the vertical shear on all sections of a simple beam carrying two concentrated loads at equal distances from the supports, the weight of the beam being neglected.

Resisting Shear. The tendency of a beam to shear vertically along any cross section, due to the vertical shear, is opposed by an internal shearing stress at that cross section known as the resisting shear; it is equal to the algebraic sum of the vertical components of all the internal stresses acting on the cross section.

If V = vertical shear, pounds; V_r = resisting shear, pounds; τ = average unit shearing stress, pounds per square inch; and A = area of the section, square inches, then at any cross section

$$V_r = V = \tau A \qquad \tau = \frac{V}{A} \tag{32}$$

The resisting shear is not uniformly distributed over the cross section, but the intensity varies from zero at the extreme fiber to its maximum value at the neutral axis.

At any point in any cross section the vertical unit shearing stress is

$$\tau = \frac{VA'c'}{It} \tag{33}$$

where V = total vertical shear in pounds for section considered; A' = area in square inches of cross section between a horizontal plane through the point where shear is being found and the extreme fiber on the same side of the neutral axis; c' = distance in inches from neutral axis to center of gravity of area A'; I = moment of inertia of the section, inches^4; t = width of section at plane of shear, inches. Maximum value of the unit shearing stress, where A = total area, square inches, of cross section of the beam is

$$\text{For a solid rectangular beam:} \quad \tau = \frac{3V}{2A} \tag{34}$$

$$\text{For a solid circular beam:} \quad \tau = \frac{4V}{3A} \tag{35}$$

Horizontal Shear. In a beam, at any cross section where there is a vertical shearing force, there must be resultant unit shearing stresses acting on the vertical faces of particles that lie at that section. On a horizontal surface of such a particle, there is a unit shearing stress equal to the unit shearing stress on a vertical surface of the particle. Equation (33), therefore, also gives the horizontal unit shearing stress at any point on the cross section of a beam.

Table 4 Bending Moment, Vertical Shear, and Deflection of Beams of Uniform Cross Section under Various Conditions of Loading

P = concentrated loads, lb
R_1, R_2 = reactions, lb
w = uniform load per unit of length, lb/in.
W = total uniform load on beam, lb
l = length of beam, in.
x = distance from support to any section, in.
E = modulus of elasticity, psi
I = moment of inertia, in.^{4}
V_x = vertical shear at any section, lb
V = maximum vertical shear, lb
M_x = bending moment at any section, lb·in.
M = maximum bending moment, lb·in.
y = maximum deflection, in.

Simple Beam—Uniform Load

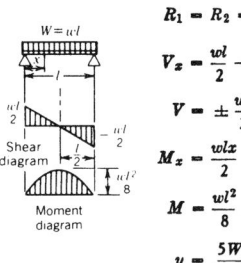

$$R_1 = R_2 = \frac{wl}{2}$$

$$V_x = \frac{wl}{2} - wx$$

$$V = \pm\frac{wl}{2}\left(\text{when}\begin{cases} x = 0 \\ x = l\end{cases}\right)$$

$$M_x = \frac{wlx}{2} - \frac{wx^2}{2}$$

$$M = \frac{wl^2}{8}\left(\text{when } x = \frac{l}{2}\right)$$

$$y = \frac{5Wl^3}{384EI} \text{ (at center of span)}$$

Simple Beam—Concentrated Load at Any Point

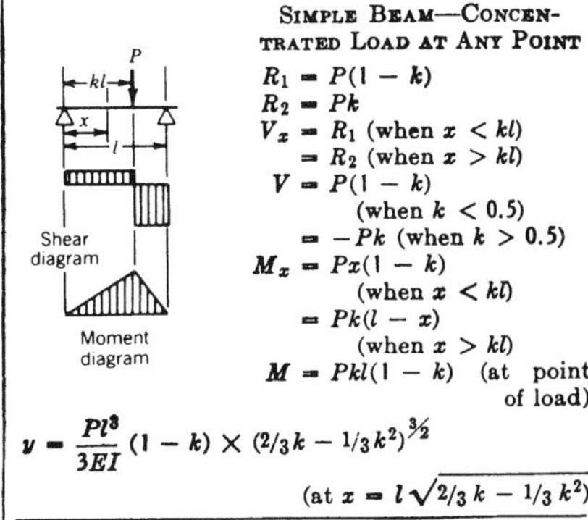

$R_1 = P(1 - k)$
$R_2 = Pk$
$V_x = R_1$ (when $x < kl$)
$= R_2$ (when $x > kl$)
$V = P(1 - k)$ (when $k < 0.5$)
$= -Pk$ (when $k > 0.5$)
$M_x = Px(1 - k)$ (when $x < kl$)
$= Pk(l - x)$ (when $x > kl$)
$M = Pkl(1 - k)$ (at point of load)

$$y = \frac{Pl^3}{3EI}(1 - k) \times (2/3k - 1/3k^2)^{3/2}$$

(at $x = l\sqrt{2/3\,k - 1/3\,k^2}$)

Simple Beam—Concentrated Load at Center

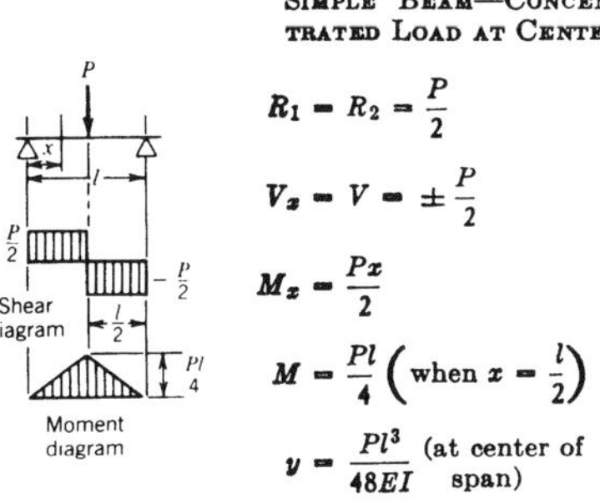

$$R_1 = R_2 = \frac{P}{2}$$

$$V_x = V = \pm\frac{P}{2}$$

$$M_x = \frac{Px}{2}$$

$$M = \frac{Pl}{4}\left(\text{when } x = \frac{l}{2}\right)$$

$$y = \frac{Pl^3}{48EI} \text{ (at center of span)}$$

Simple Beam—Two Equal Concentrated Loads at Equal Distances from Supports

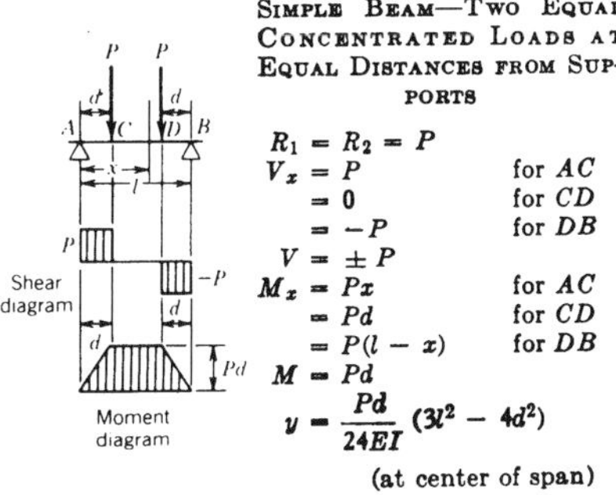

$R_1 = R_2 = P$		
$V_x = P$		for AC
$= 0$		for CD
$= -P$		for DB
$V = \pm P$		
$M_x = Px$		for AC
$= Pd$		for CD
$= P(l - x)$		for DB
$M = Pd$		

$$y = \frac{Pd}{24EI}(3l^2 - 4d^2)$$

(at center of span)

Simple Beam—Load Increasing Uniformly from Supports to Center of Span

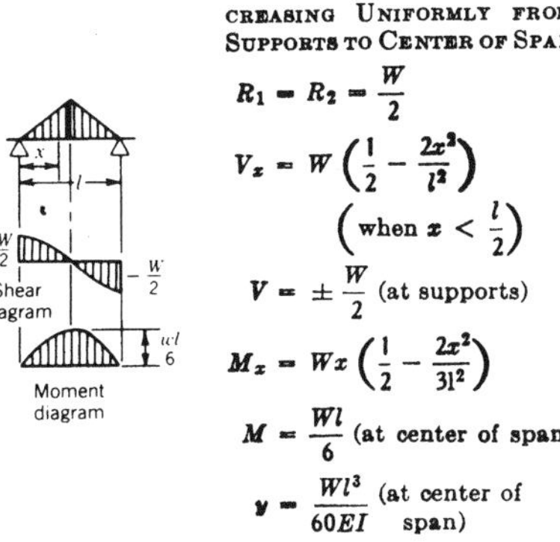

$$R_1 = R_2 = \frac{W}{2}$$

$$V_x = W\left(\frac{1}{2} - \frac{2x^2}{l^2}\right)\left(\text{when } x < \frac{l}{2}\right)$$

$$V = \pm\frac{W}{2} \text{ (at supports)}$$

$$M_x = Wx\left(\frac{1}{2} - \frac{2x^2}{3l^2}\right)$$

$$M = \frac{Wl}{6} \text{ (at center of span)}$$

$$y = \frac{Wl^3}{60EI} \text{ (at center of span)}$$

Cantilever Beam—Load Concentrated at Free End

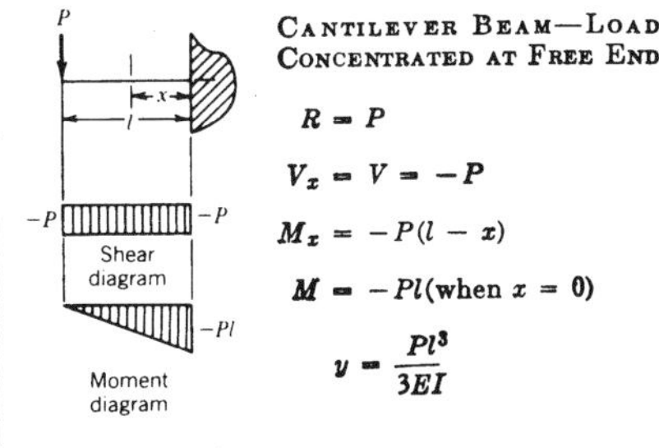

$$R = P$$

$$V_x = V = -P$$

$$M_x = -P(l - x)$$

$$M = -Pl(\text{when } x = 0)$$

$$y = \frac{Pl^3}{3EI}$$

Table 4 (*Continued*)

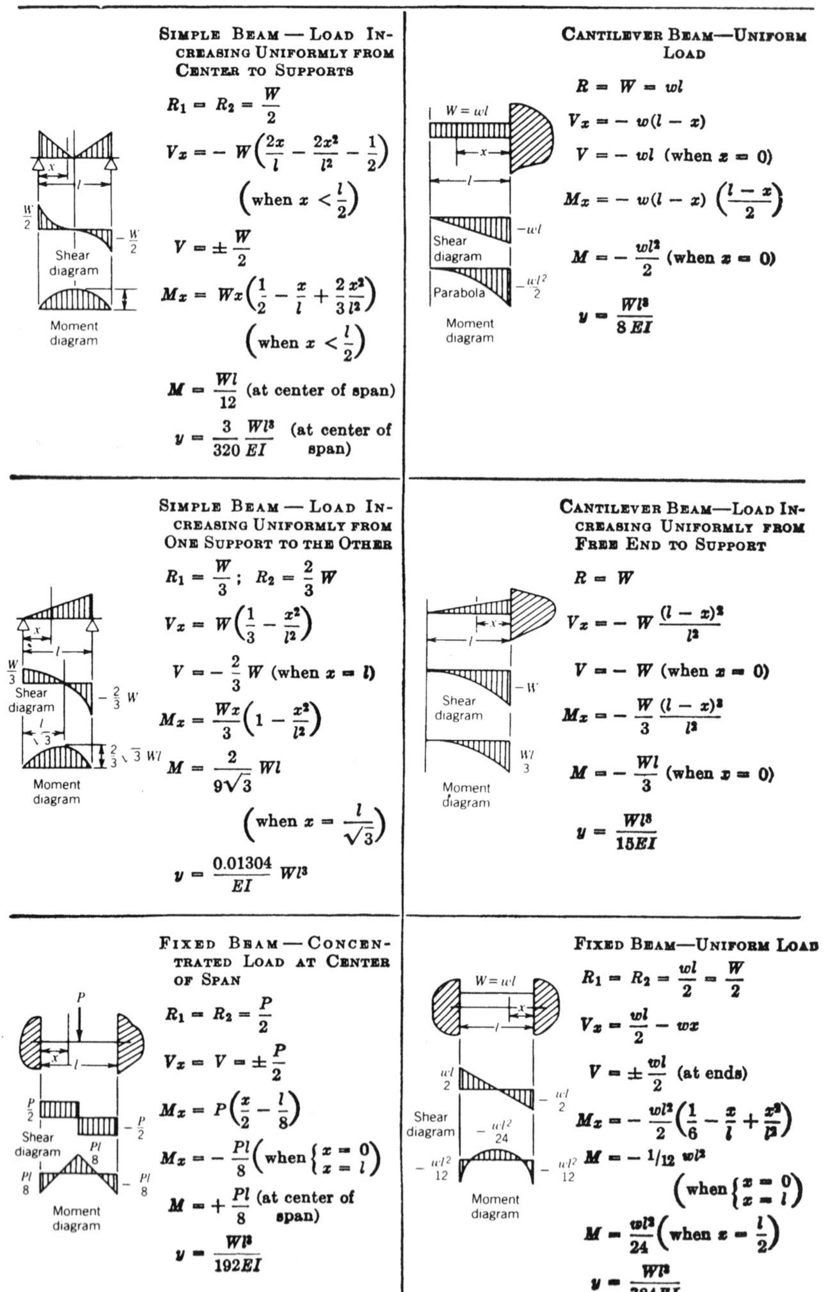

Simple Beam — Load Increasing Uniformly from Center to Supports

$$R_1 = R_2 = \frac{W}{2}$$

$$V_x = -W\left(\frac{2x}{l} - \frac{2x^2}{l^2} - \frac{1}{2}\right) \quad \left(\text{when } x < \frac{l}{2}\right)$$

$$V = \pm\frac{W}{2}$$

$$M_x = Wx\left(\frac{1}{2} - \frac{x}{l} + \frac{2}{3}\frac{x^2}{l^2}\right) \quad \left(\text{when } x < \frac{l}{2}\right)$$

$$M = \frac{Wl}{12} \text{ (at center of span)}$$

$$y = \frac{3}{320}\frac{Wl^3}{EI} \text{ (at center of span)}$$

Cantilever Beam—Uniform Load

$$R = W = wl$$

$$V_x = -w(l - x)$$

$$V = -wl \text{ (when } x = 0)$$

$$M_x = -w(l - x)\left(\frac{l - x}{2}\right)$$

$$M = -\frac{wl^2}{2} \text{ (when } x = 0)$$

$$y = \frac{Wl^3}{8EI}$$

Simple Beam — Load Increasing Uniformly from One Support to the Other

$$R_1 = \frac{W}{3}; \quad R_2 = \frac{2}{3}W$$

$$V_x = W\left(\frac{1}{3} - \frac{x^2}{l^2}\right)$$

$$V = -\frac{2}{3}W \text{ (when } x = l)$$

$$M_x = \frac{Wx}{3}\left(1 - \frac{x^2}{l^2}\right)$$

$$M = \frac{2}{9\sqrt{3}}Wl \quad \left(\text{when } x = \frac{l}{\sqrt{3}}\right)$$

$$y = \frac{0.01304}{EI}Wl^3$$

Cantilever Beam—Load Increasing Uniformly from Free End to Support

$$R = W$$

$$V_x = -W\frac{(l - x)^2}{l^2}$$

$$V = -W \text{ (when } x = 0)$$

$$M_x = -\frac{W}{3}\frac{(l - x)^3}{l^2}$$

$$M = -\frac{Wl}{3} \text{ (when } x = 0)$$

$$y = \frac{Wl^3}{15EI}$$

Fixed Beam — Concentrated Load at Center of Span

$$R_1 = R_2 = \frac{P}{2}$$

$$V_x = V = \pm\frac{P}{2}$$

$$M_x = P\left(\frac{x}{2} - \frac{l}{8}\right)$$

$$M_x = -\frac{Pl}{8}\left(\text{when} \begin{cases} x = 0 \\ x = l \end{cases}\right)$$

$$M = +\frac{Pl}{8} \text{ (at center of span)}$$

$$y = \frac{Wl^3}{192EI}$$

Fixed Beam—Uniform Load

$$R_1 = R_2 = \frac{wl}{2} = \frac{W}{2}$$

$$V_x = \frac{wl}{2} - wx$$

$$V = \pm\frac{wl}{2} \text{ (at ends)}$$

$$M_x = -\frac{wl^2}{2}\left(\frac{1}{6} - \frac{x}{l} + \frac{x^2}{l^2}\right)$$

$$M = -\tfrac{1}{12}wl^2 \quad \left(\text{when} \begin{cases} x = 0 \\ x = l \end{cases}\right)$$

$$M = \frac{wl^2}{24}\left(\text{when } x = \frac{l}{2}\right)$$

$$y = \frac{Wl^3}{384EI}$$

Table 4 (*Continued*)

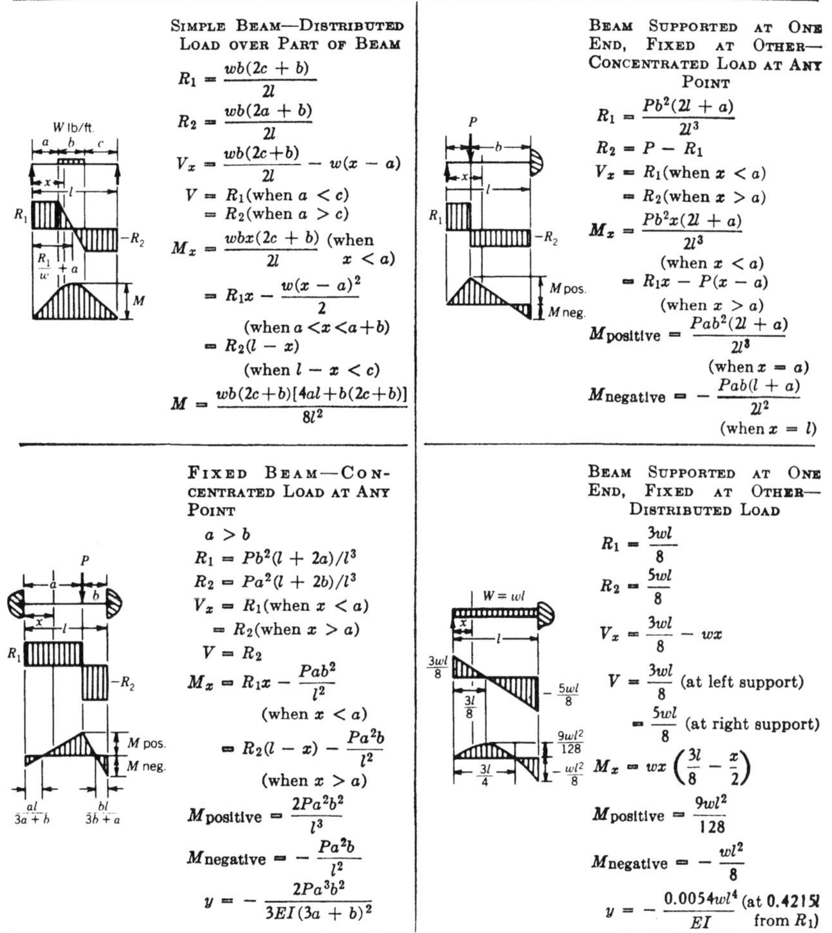

SIMPLE BEAM—DISTRIBUTED LOAD OVER PART OF BEAM $R_1 = \frac{wb(2c+b)}{2l}$ $R_2 = \frac{wb(2a+b)}{2l}$ $V_x = \frac{wb(2c+b)}{2l} - w(x-a)$ $V = R_1$ (when $a < c$) $= R_2$ (when $a > c$) $M_x = \frac{wbx(2c+b)}{2l}$ (when $x < a$) $= R_1x - \frac{w(x-a)^2}{2}$ (when $a < x < a+b$) $= R_2(l-x)$ (when $l - x < c$) $M = \frac{wb(2c+b)[4al+b(2c+b)]}{8l^2}$	BEAM SUPPORTED AT ONE END, FIXED AT OTHER—CONCENTRATED LOAD AT ANY POINT $R_1 = \frac{Pb^2(2l+a)}{2l^3}$ $R_2 = P - R_1$ $V_x = R_1$ (when $x < a$) $= R_2$ (when $x > a$) $M_x = \frac{Pb^2x(2l+a)}{2l^3}$ (when $x < a$) $= R_1x - P(x-a)$ (when $x > a$) $M_{\text{positive}} = \frac{Pab^2(2l+a)}{2l^3}$ (when $x = a$) $M_{\text{negative}} = -\frac{Pab(l+a)}{2l^2}$ (when $x = l$)
FIXED BEAM—CONCENTRATED LOAD AT ANY POINT $a > b$ $R_1 = Pb^2(l+2a)/l^3$ $R_2 = Pa^2(l+2b)/l^3$ $V_x = R_1$ (when $x < a$) $= R_2$ (when $x > a$) $V = R_2$ $M_x = R_1x - \frac{Pab^2}{l^2}$ (when $x < a$) $= R_2(l-x) - \frac{Pa^2b}{l^2}$ (when $x > a$) $M_{\text{positive}} = \frac{2Pa^2b^2}{l^3}$ $M_{\text{negative}} = -\frac{Pa^2b}{l^2}$ $y = -\frac{2Pa^3b^2}{3EI(3a+b)^2}$	BEAM SUPPORTED AT ONE END, FIXED AT OTHER—DISTRIBUTED LOAD $R_1 = \frac{3wl}{8}$ $R_2 = \frac{5wl}{8}$ $V_x = \frac{3wl}{8} - wx$ $V = \frac{3wl}{8}$ (at left support) $= \frac{5wl}{8}$ (at right support) $M_x = wx\left(\frac{3l}{8} - \frac{x}{2}\right)$ $M_{\text{positive}} = \frac{9wl^2}{128}$ $M_{\text{negative}} = -\frac{wl^2}{8}$ $y = -\frac{0.0054wl^4}{EI}$ (at $0.4215l$ from R_1)

Bending moment, at any cross section of a beam, is the algebraic sum of the moments of the external forces acting on either side of the section. It is positive when it causes the beam to bend convex downward, hence causing compression in upper fibers and tension in lower fibers of the beam. When the bending moment is determined from the forces that lie to the left of the section, it is positive if they act in a clockwise direction; if determined from forces on the right side, it is positive if they act in a counterclockwise direction. If the moments of upward forces are given positive signs and the moments of downward forces are given negative signs, the bending moment will always have the correct sign, whether determined from the right or left side. The bending moment should be determined for the side for which the calculation will be simplest.

In Table 4 let M be the bending moment, pound-inches, at a section of a simple beam at a distance x, inches, from the left support; w = weight of beam per 1 in. of length; l = length of the beam, inches. Then the reactions are $\frac{1}{2}wl$, and $M = \frac{1}{2}wlx - \frac{1}{2}xwx$. For the sections at the supports, $x = 0$ or $x = l$ and $M = 0$. For the section at the center of the span $x = \frac{1}{2}l$ and $M = \frac{1}{8}wl^2 = \frac{1}{8}Wl$, where W = total weight.

A moment diagram Table 4 shows the bending moment at all cross sections of a beam. Ordinates to the curve represent to scale the moments at the corresponding cross sections. The curve for a simple beam uniformly loaded is a parabola, showing $M = 0$ at the supports and $M = \frac{1}{8}wl^2 = \frac{1}{8}Wl$ at the center, M being in pound-inches.

The dangerous section is the cross section of a beam where the bending moment is greatest. In a cantilever beam it is at the point of support, regardless of the disposition of the loads. In a simple beam it is that section where the vertical shear changes from positive to negative, and it may be located graphically by constructing a shear diagram or numerically by taking the left reaction and subtracting the loads in order from the left until a point is reached where the sum of the loads subtracted equals the reaction. For a simple beam uniformly loaded, the dangerous section is at the center of the span.

The tendency to rotate about a point in any cross section of a beam is due to the bending moment at that section. This tendency is resisted by the **resisting moment,** which is the algebraic sum of the moments of all the horizontal stresses with reference to the same point.

Formula for Flexure

Let M = bending moment; M_r = resisting moment of the horizontal fiber stresses; σ = unit stress (tensile or compressive) on any fiber, usually that one most remote from the neutral surface; c = distance of that fiber from the neutral surface. Then

$$M = M_r = \frac{\sigma I}{c} \tag{36}$$

$$\sigma = \frac{Mc}{I} \tag{37}$$

where I = moment of inertia of the cross section with respect to its neutral axis. If σ is in pounds per square inch, M must be in pound-inches, I in inches^4, and c in inches.

Equation (37) is the basis of the design and investigation of beams. It is true only when the maximum horizontal fiber stress σ does not exceed the proportional limit of the material.

Moment of inertia is the sum of the products of each elementary area of the cross section multiplied by the square of the distance of that area from the assumed axis of rotation, or

$$I = \sum r^2 \,\Delta A = \int r^2 \, dA \tag{38}$$

where Σ is the sign of summation, ΔA is an elementary area of the section, and r is the distance of ΔA from the axis. The moment of inertia is greatest in those sections (such as I-beams) having much of the area concentrated at a distance from the axis. Unless otherwise stated, the neutral axis is the axis of rotation considered. I usually is expressed in inches^4. See Table 5 for values of moments of inertia of various sections.

Modulus of rupture is the term applied to the value of σ as found by Eq. (37), when a beam is loaded to the point of rupture. Since Eq. (37) is true only for stresses within the proportional limit, the value σ of the rupture strength so found is incorrect. However, the equation is used, as a measure of the ultimate load-carrying capacity of a beam. The modulus of rupture does not show the actual stress in the extreme fiber of a beam; it is useful only as a basis of comparison. If the strength of a beam in tension differs from its strength in compression, the modulus of rupture is intermediate between the two.

Section modulus, the factor I/c in flexure [Eq. (36)], is expressed in cubic inches. It is the measure of a capacity of a section to resist a bending moment. For values of I/c for

Table 5 Elements of Sections

A = area of section
I = moment of inertia about axis I–I
c = distance from axis I–I to most remote point of section
I/c = section modulus
r = radius of gyration

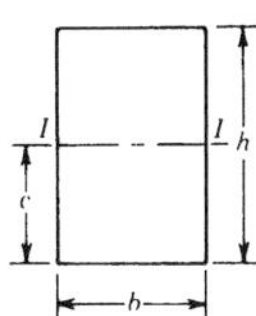

Rectangle

Axis through center

$A = bh$

$c = h/2$

$I = bh^3/12$

$I/c = bh^2/6$

$r = h/\sqrt{12} = 0.289h$

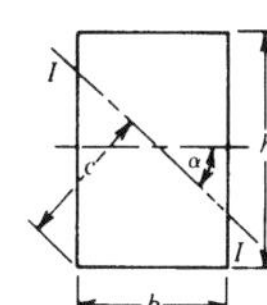

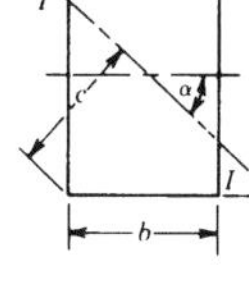

Rectangle

Axis any line through center of gravity

$A = bh$

$c = (b \sin \alpha + h \cos \alpha)/2$

$I = bh(b^2 \sin^2 \alpha + h^2 \cos^2 \alpha)/12$

$I/c = \dfrac{bh(b^2 \sin^2 \alpha + h^2 \cos^2 \alpha)}{6(b \sin \alpha + h \cos \alpha)}$

$r = \sqrt{(b^2 \sin^2 \alpha + h^2 \cos^2 \alpha)/12}$

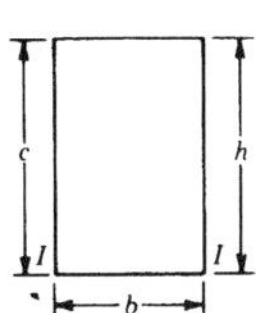

Rectangle

Axis on base

$A = bh$

$c = h$

$I = bh^3/3$

$I/c = bh^2/3$

$r = h/\sqrt{3} = 0.577h$

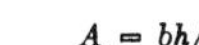

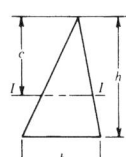

Triangle

Axis through center of gravity

$A = bh/2$

$c = 2/3\ h$

$I = bh^3/36$

$I/c = bh^2/24$

$r = h/\sqrt{18} = 0.236h$

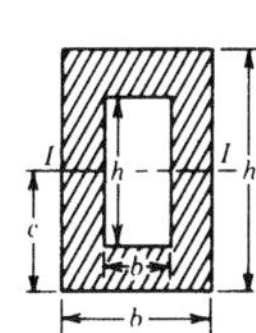

Hollow Rectangle

Axis through center

$A = bh - b_1h_1$

$c = h/2$

$I = (bh^3 - b_1h_1^3)/12$

$I/c = (bh^3 - b_1h_1^3)/6h$

$r = \sqrt{\dfrac{bh^3 - b_1h_1^3}{12(bh - b_1h_1)}}$

Triangle

Axis through base

$A = bh/2$

$c = h$

$I = bh^3/12$

$I/c = bh^2/12$

$r = h/\sqrt{6} = 0.408h$

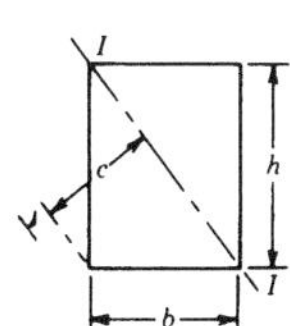

Rectangle

Axis on diagonal

$A = bh$

$c = bh/\sqrt{b^2 + h^2}$

$I = b^3h^3/6(b^2 + h^2)$

$I/c = b^2h^2/6\sqrt{(b^2 + h^2)}$

$r = bh/\sqrt{6(b^2 + h^2)}$

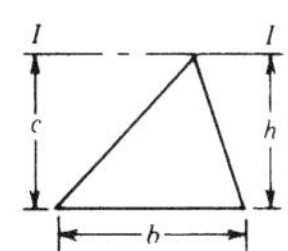

Triangle

Axis through apex

$A = bh/2$

$c = h$

$I = bh^3/4$

$I/c = bh^2/4$

$r = h/\sqrt{2} = 0.707h$

Table 5 *(Continued)*

EQUILATERAL POLYGON

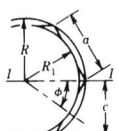

Axis through center, parallel to one side. n = number of sides

$A = nR_1^2 \tan\phi$

$c = a/2 \tan\phi = R_1$

$I = \{A(12R_1^2 + a^2)\}/48$

$I/c = \{A(12R_1^2 + a^2)\}/48R_1$

$r = \sqrt{(12R_1^2 + a^2)/48}$

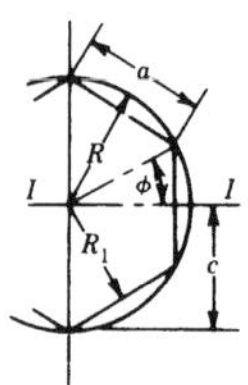

EQUILATERAL POLYGON

Axis through center, normal to side. n = number of sides

$A = nR_1^2 \tan\phi$

$c = a/(2\sin\phi) = R$

$I = \{A(6R^2 - a^2)\}/24$

$I/c = \{A(6R^2 - a^2)\}/24R$

$r = \sqrt{(6R^2 - a^2)/24}$

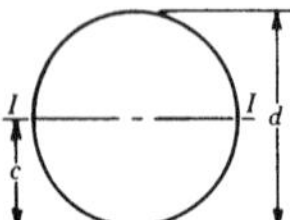

CIRCLE

Axis through center

$A = \pi d^2/4 = 0.7854d^2$

$c = d/2$

$I = \pi d^4/64 = 0.0491d^4$

$I/c = \pi d^3/32 = 0.0982d^3$

$r = d/4$

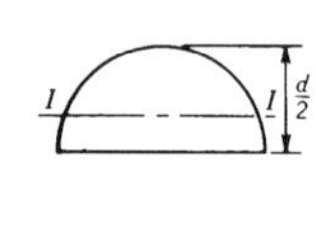

HALF CIRCLE

Axis through center of gravity

$A = \pi d^2/8 = 0.3927d^2$

$c = \{d(3\pi - 4)\}/6\pi = 0.2878d.$

$I = \{d^4(9\pi^2 - 64)\}/1152\pi = 0.0068d^4$

$I/c = \dfrac{\{d^3(9\pi^2 - 64)\}}{\{192(3\pi - 4)\}} = 0.0238d^3$

$r = \{d\sqrt{(9\pi^2 - 64)}\}/12\pi = 0.1322d$

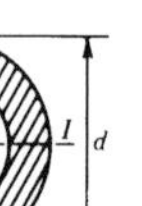

HOLLOW CIRCLE

Axis through center

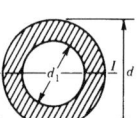

$A = \pi(d^2 - d_1^2)/4 = 0.7854(d^2 - d_1^2)$

$c = d/2$

$I = \pi(d^4 - d_1^4)/64 = 0.0491(d^4 - d_1^4)$

$I/c = \pi(d^4 - d_1^4)/32d = 0.0982(d^4 - d_1^4)/d$

$r = \sqrt{(d^2 + d_1^2)}/4$

ELLIPSE

Axis through center

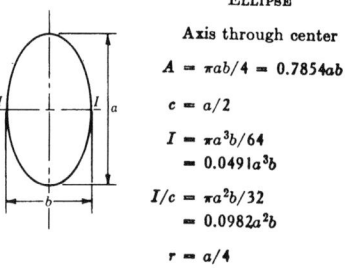

$A = \pi ab/4 = 0.7854ab$

$c = a/2$

$I = \pi a^3b/64 = 0.0491a^3b$

$I/c = \pi a^2b/32 = 0.0982a^2b$

$r = a/4$

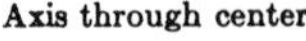

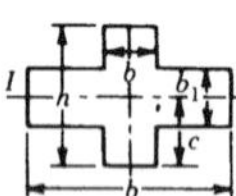

CROSSED RECTANGLES

Axis through center

$A = th + t_1(b - t)$

$c = h/2$

$I = \{th^3 + t_1^3(b - t)\}/12$

$I/c = \{th^3 + t_1^3(b - t)\}/6h$

$r = \sqrt{\dfrac{th^3 + t_1^3(b - t)}{12\{th + t_1(b - t)\}}}$

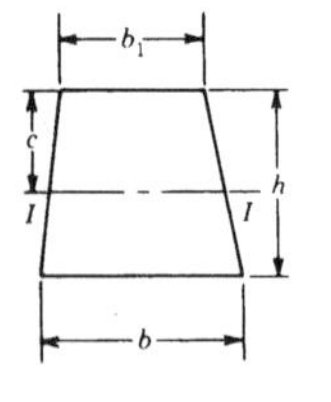

TRAPEZOID

Axis through center of gravity

$A = \{(b + b_1)h\}/2$

$c = \dfrac{\{(b_1 + 2b)h\}}{\{3(b + b_1)\}}$

$I = \dfrac{h^3(b^2 + 4bb_1 + b_1^2)}{36(b + b_1)}$

$I/c = \dfrac{h^2(b^2 + 4bb_1 + b_1^2)}{12(b_1 + 2b)}$

$r = \dfrac{h}{6(b + b_1)}\sqrt{2(b^2 + 4bb_1 + b_1^2)}$

simple shapes, see Table 5. See Refs. 5 and 6 for properties of standard steel and aluminum structural shapes.

Elastic Deflection of Beams

When a beam bends under load, all points of the elastic curve except those over the supports are deflected from their original positions. The radius of curvature ρ of the elastic curve at any section is expressed as

$$\rho = \frac{EI}{M} \tag{39}$$

where E = modulus of elasticity of the material, pounds per square inch; I = moment of inertia, inches^4, of the cross section with reference to its neutral axis; M = bending moment, pound-inches, at the section considered. Where there is no bending moment, ρ is infinity and the curve is a straight line; where M is greatest, ρ is smallest and the curvature, therefore, is greatest.

If the elastic curve is referred to a system of coordinate axes in which x represents horizontal distances, y vertical distances, and l distances along the curve, the value of ρ is found, by the aid of the calculus, to be $d^3l/dx \cdot d^2y$. Differential equation (40) of the elastic curve which applies to all beams when the elastic limit of the material is not exceeded is obtained by substituting this value in the expression $\rho = EI/M$ and assuming that dx and dl are practically equal:

$$EI \frac{d^2y}{dx^2} = M \tag{40}$$

Equation (40) is used to determine the deflection of any point of the elastic curve by regarding the point of support as the origin of the coordinate axis and taking y as the vertical deflection at any point on the curve and x as the horizontal distance from the support to the point considered. The values of E, I, and M are substituted and the expression is integrated twice, giving proper values to the constants of integration, and the deflection y is determined for any point. See Table 4.

For example, a cantilever beam in Table 4 has a length l inches and carries a load P pounds at the free end. It is required to find the deflection of the elastic curve at a point x inches from the support, the weight of the beam being neglected.

The moment $M = -P(l - x)$. By substitution in Eq. (40), the equation for the elastic curve becomes $EI(d^2y/dx^2) = -Pl + Px$. By integrating and determining the constant of integration by the condition that $dy/dx = 0$ when $x = 0$, $EI(dy/dx) = -Plx + \frac{1}{2}Px^2$ results. By integrating a second time and determining the constant by the condition that $x = 0$ when $y = 0$, $EIy = -\frac{1}{2}Plx^2 + \frac{1}{6}Px^3$, which is the equation of the elastic curve, results. When $x = l$, the value of y, or the deflection in inches at the free end, is found to be $-Pl^3/3EI$.

Deflection due to Shear

The deflection of a beam as computed by the ordinary formulas is that due to flexural stresses only. The deflection in honeycomb, plastic and short beams due to vertical shear can be considerable and should always be checked. Because of the nonuniform distribution of the shear over the cross section of the beam, computing the deflection due to shear by exact methods is difficult. It may be approximated by $y_s = M/AE_s$, where y_s = deflection, inches, due to shear; M = bending moment, pound-inches, at the section where the deflection is calculated; E_s = modulus of elasticity in shear, pounds per square inch; A = area of cross section of beam, square inches.[7] For a rectangular section, the ratio of deflection due to shear to the deflection due to bending will be less than 5% if the depth of the beam is less than one-eighth of the length.

6.2 Design of Beams

Design Procedure

In designing a beam the procedure is: (1) Compute reactions. (2) Determine position of the dangerous section and the bending moment at that section. (3) Divide the maximum bending moment (expressed in pound-inches) by the allowable unit stress (expressed in pounds per square inch) to obtain the minimum value of the section modulus. (4) Select a beam section with a section modulus equal to or slightly greater than the section modulus required.

Web Shear

A beam designed in the foregoing manner is safe against rupture of the extreme fibers due to bending in a vertical plane, and usually the cross section will have sufficient area to sustain the shearing stresses with safety. For short beams carrying heavy loads, however, the vertical shear at the supports is large, and it may be necessary to increase the area of the section to keep the unit shearing stress within the limit allowed. For steel beams, the average unit shearing stress is computed by $\tau = V/A$, where V = total vertical shear, pounds; A = area of web, square inches.

Shear Center

Closed or solid cross sections with two axes of symmetry will have a shear center at the origin. If the loads are applied here, then the bending moment can be used to calculate the deflections and bending stress, which means there are no torsional stresses. The open section or unsymmetrical section generally has a shear center that is offset on one axis of symmetry and must be calculated.[2,8,9] The load applied at this location will develop bending stresses and deflections. If any sizable torsion is developed, then torsional stresses and rotations must be accounted for.

Miscellaneous Considerations

Other considerations which will influence the choice of section under certain conditions of loading are: (1) Maximum vertical deflection that may be permitted in beams coming in contact with plaster. (2) Danger of failure by sidewise bending in long beams, unbraced against lateral deflection. (3) Danger of failure by the buckling of the web of steel beams of short span carrying heavy loads. (4) Danger of failure by horizontal shear, particularly in wooden beams.

Vertical Deflection

If a beam is to support or come in contact with materials like plaster, which may be broken by excessive deflection, it is usual to select such a beam that the maximum deflection will not exceed ($\frac{1}{360}$ × span). It may be shown that for a simple beam supported at the ends with a total uniformly distributed load W pounds, the deflection in inches is

$$y = \frac{30\sigma L^2}{Ed} \qquad (41)$$

where σ = allowable unit fiber stress, pounds per square inch; L = span of beam, feet; E = modulus of elasticity, pounds per square inch; d = depth of beam, inches.

If the deflection of a steel beam is to be less than $\frac{1}{360}$th of the span, it may be shown from Eq. (41) that, for a maximum allowable fiber stress of 18,000 psi, the limit of span in feet is approximately $1.8d$, where d = depth of the beam, inches.

For the deflection due to the impact of a moving load falling on a beam, see Section 6.6.

Horizontal Shear in Timber Beams

In beams of a homogeneous material which can withstand equally well shearing stresses in any direction, vertical and horizontal shearing stresses are equally important. In timber, however, shearing strength along the grain is much less than that perpendicular to the grain. Hence, the beams may fail owing to horizontal shear. Short wooden beams should be checked for horizontal shear in order that allowable unit shearing stress along the grain shall not be exceeded. (See the example below.)

Restrained Beams

A beam is considered to be restrained if one or both ends are not free to rotate. This condition exists if a beam is built into a masonry wall at one or both ends, if it is riveted or otherwise fastened to a column, or if the ends projecting beyond the supports carry loads that tend to prevent tilting of the ends which would naturally occur as the beam deflects. The shears and moments given in Table 4 for fixed-end conditions are seldom, if ever, attained, since the restraining elements themselves deform and reduce the magnitude of the restraint. This reduction of restraint decreases the negative moment at the support and increases the positive moment in the central portion of the span. The amount of restraint that exists is a matter which must be judged for each case in the light of the construction used, the rigidity of the connections, and the relative sizes of the connecting members.

Safe Loads on Simple Beams

Equation (42) gives the safe loads on simple beams. This formula is obtained by substituting in the flexure equation (36) the value of M for a simple beam uniformly loaded, as given in Table 4. Let W = total load, pounds; σ = extreme fiber unit stress, pounds per square inch; S = section modulus, inches^3; L = length of span, feet. Then

$$W = \frac{2}{3}\sigma\frac{S}{L} \tag{42}$$

If σ is taken as a maximum allowable unit fiber stress, this equation gives the maximum allowable load on the beam. Most building codes permit a value of σ = 18,000 psi for quiescent loads on steel. For this value of σ, Eq. (42) becomes

$$W = \frac{12{,}000\ S}{L} \tag{43}$$

If the load is concentrated at the center of the span, the safe load is one-half the value given by Eq. (43). If the load is neither uniformly distributed nor concentrated at the center of the span, the maximum bending moment must be used. The foregoing equations are for beams laterally supported and are for flexure only. The other factors which influence the strength of the beam, as shearing, buckling, etc., must also be considered.

Use of Tables in Design

The following is an example in the use of tables for the design of a wooden beam.

Example 1

Design a southern pine girder of common structural grade to carry a load of 9600 lb distributed uniformly over a 16-ft span in the interior of a building, the beam being a simple beam freely supported at each end.

Solution. From Table 4, the bending moment of a simple beam uniformly loaded is $M = wl^2/8$. Since $W = wl$ and $l = 12L$,

$$M = 9600 \times 16 \times {}^{12}\!/\!_{8} = 230{,}400 \text{ lb-in.}$$

If the allowable unit stress on yellow pine is 1200 psi,

$$\frac{I}{c} = \frac{230{,}400}{1200} = 192 \text{ in.}^3$$

From Table 5, the section modulus of a rectangular section is $bd^2/6$. Assume $b = 8$ in. Then $8d^2/6 = 192$, and $d = \sqrt{144} = 12.0$ in. A beam 8×12 in. is selected tentatively and checked for shear.

Maximum shearing stress (horizontal and vertical) is at the neutral surface over the supports. Equation (34) for horizontal shear in a solid rectangular beam is $\tau = 3V/2A$; $V = 9600/2 = 4800$, and $A = 8 \times 12 = 96$, whence $\tau = (3 \times 4800)/(2 \times 96) = 75$ psi.

If the safe horizontal unit shearing stress for common-grade southern yellow pine is 88 psi and since the actual horizontal unit shearing stress is less than 88 lb, the 8×12-in. beam will be satisfactory.

A **beam of uniform strength** is one in which the dimensions are such that the maximum fiber stress σ is the same throughout the length of the beam. The form of the beam is determined by finding the areas of various cross sections from the flexure formula $M = \sigma I/c$, keeping σ constant and making I/c vary with M. For a rectangular section of width b and depth d, the section modulus $I/c = \frac{1}{6}bd^2$, and, therefore, $M = \frac{1}{6}\sigma bd^2$. By making bd^2 vary with M, the dimensions of the various sections are obtained. Table 6 gives the dimensions b and d at any section, the maximum unit fiber stress σ, and the maximum deflection y of some rectangular beams of uniform strength. In this table, the bending moment has been assumed to be the controlling factor. On account of the vertical shear near the ends of the beams, the area of the sections must be increased over that given by an amount necessary to keep the unit shearing stress within the allowable unit shearing stress. The discussion of beams of uniform strength, although of considerable theoretical interest, is of little practical value since the cost of fabrication will offset any economy in the use of the material. A plate girder in a bridge or a building is an approximation in practice to a steel beam of uniform strength.

6.3 Continuous Beams

As in simple beams, the expressions $M = \sigma I/c$ and $\tau = V/A$ govern the design and investigation of beams resting on more than two supports. In the case of continuous beams, however, the reactions cannot be obtained in the manner described for simple beams. Instead, the bending moments at the various sections must be determined, and from these values the vertical shears at the sections and the reactions at the supports may be derived.

Consider the second span of length l_2 inches of the continuous beam (Fig. 19). Vertical shear V_x at any section x inches from the left support of the span is equal to the algebraic sum of all the vertical forces on one side of the section. Thus, if V_2 = vertical shear at a section to the right of but infinitely close to the left support, w_2x = uniform load, and ΣP_2

Table 6 Rectangular Beams of Uniform Strength[a]

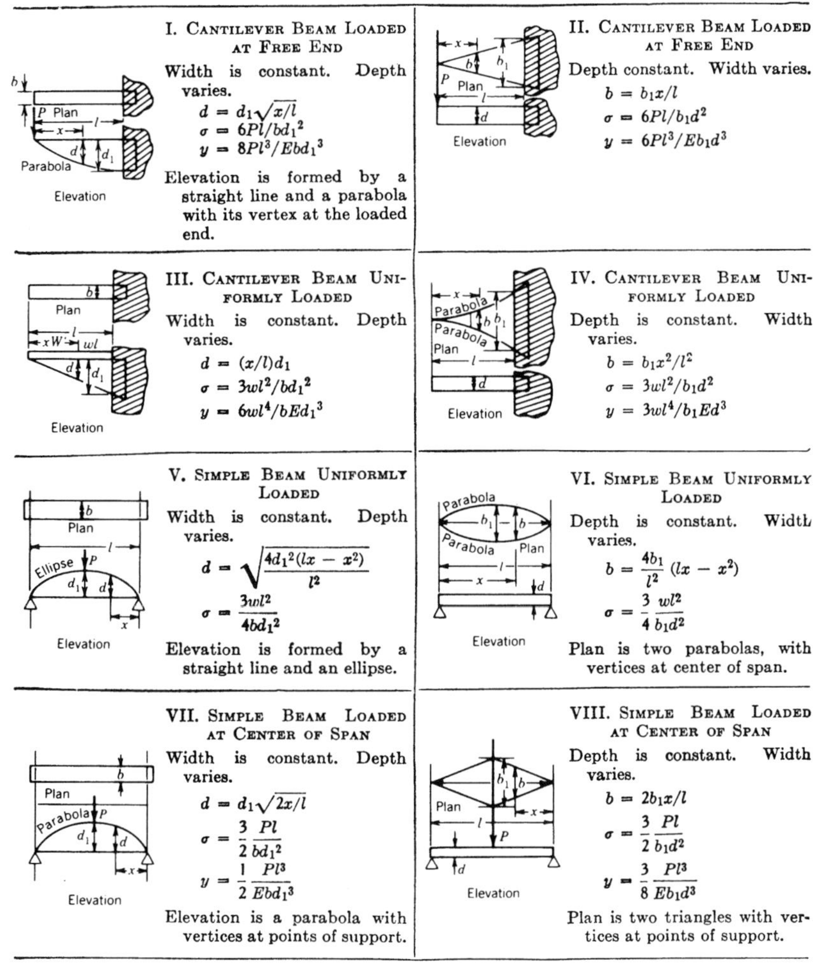

I. CANTILEVER BEAM LOADED AT FREE END Width is constant. Depth varies. $d = d_1\sqrt{x/l}$ $\sigma = 6Pl/bd_1^2$ $y = 8Pl^3/Ebd_1^3$ Elevation is formed by a straight line and a parabola with its vertex at the loaded end.	II. CANTILEVER BEAM LOADED AT FREE END Depth constant. Width varies. $b = b_1x/l$ $\sigma = 6Pl/b_1d^2$ $y = 6Pl^3/Eb_1d^3$
III. CANTILEVER BEAM UNIFORMLY LOADED Width is constant. Depth varies. $d = (x/l)d_1$ $\sigma = 3wl^2/bd_1^2$ $y = 6wl^4/bEd_1^3$	IV. CANTILEVER BEAM UNIFORMLY LOADED Depth is constant. Width varies. $b = b_1x^2/l^2$ $\sigma = 3wl^2/b_1d^2$ $y = 3wl^4/b_1Ed^3$
V. SIMPLE BEAM UNIFORMLY LOADED Width is constant. Depth varies. $d = \sqrt{\frac{4d_1^2(lx - x^2)}{l^2}}$ $\sigma = \frac{3wl^2}{4bd_1^2}$ Elevation is formed by a straight line and an ellipse.	VI. SIMPLE BEAM UNIFORMLY LOADED Depth is constant. Width varies. $b = \frac{4b_1}{l^2}(lx - x^2)$ $\sigma = \frac{3}{4}\frac{wl^2}{b_1d^2}$ Plan is two parabolas, with vertices at center of span.
VII. SIMPLE BEAM LOADED AT CENTER OF SPAN Width is constant. Depth varies. $d = d_1\sqrt{2x/l}$ $\sigma = \frac{3}{2}\frac{Pl}{bd_1^2}$ $y = \frac{1}{2}\frac{Pl^3}{Ebd_1^3}$ Elevation is a parabola with vertices at points of support.	VIII. SIMPLE BEAM LOADED AT CENTER OF SPAN Depth is constant. Width varies. $b = 2b_1x/l$ $\sigma = \frac{3}{2}\frac{Pl}{b_1d^2}$ $y = \frac{3}{8}\frac{Pl^3}{Eb_1d^3}$ Plan is two triangles with vertices at points of support.

[a] The sections of the beams near the ends must be increased over the amounts shown to resist the vertical shear expressed by the formula $\tau = 3/2\ V/A$.

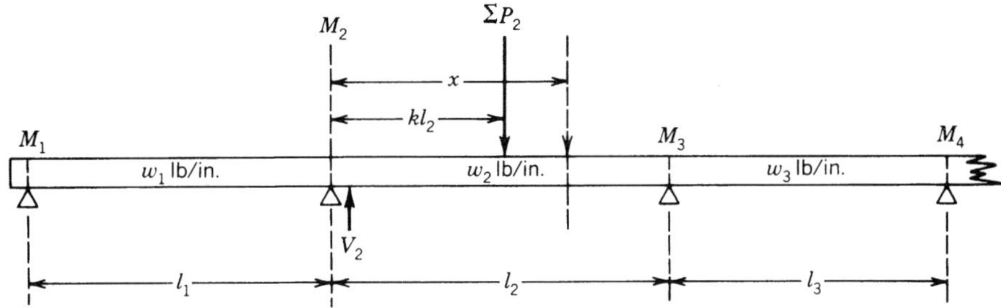

Figure 19 Continuous beam.

= sum of the concentrated loads along the distance x applied at a distance kl_2 from the left support, k being a fraction less than unity, then

$$V_x = V_2 - w_2 x - \Sigma P_2 \tag{44}$$

At any section x inches from the left support, the bending moment is equal to the algebraic sum of the moments of all forces on one side of the section. If M_2 is the moment in pound-inches at the support to the left,

$$M_x = M_2 + V_2 x - \frac{w_2 x^2}{2} - \Sigma P_2(x - kl_2) \tag{45}$$

Assume that $x = l_2$. Then M_x becomes the moment M_3 at the next support to the right, and the expression may be written

$$V_2 l_2 = M_3 - M_2 + \frac{w_2 l_2^2}{2} + \Sigma P_2(l_2 - kl_2) \tag{46}$$

From Eqs. (44), (45), and (46) it is evident that the bending moment M_x and the shear V_x at any section between two consecutive supports may be determined if the bending moments M_2 and M_3 at those supports are known.

To determine bending moments at the supports an expression known as the *theorem of three moments* is used. This gives the relation between the moments at any three consecutive supports of a beam. For beams with the supports on the same level and uniformly loaded over each span, the formula is

$$M_1 l_1 + 2M_2(l_1 + l_2) + M_3 l_2 = -\tfrac{1}{4} w_1 l_1^3 - \tfrac{1}{4} w_2 l_2^3 \tag{47}$$

where M_1, M_2, and M_3 are the moments of three consecutive supports; l_1 = length between first and second support; l_2 = length between second and third support; w_1 = uniform load per lineal unit over the first span; w_2 = uniform load per lineal unit over the second span. When both spans are of equal length and when the load on each span is the same, $l_1 = l_2$, $w_1 = w_2$, and Eq. (47) reduces to

$$M_1 + 4M_2 + M_3 = -\tfrac{1}{2} w l^2 \tag{48}$$

which applies to most cases in practice.

Equations (47) and (48) are used as follows: For any continuous beam of n spans there are $n + 1$ supports. Assuming the ends of the beam to be simply supported without any overhang, the moments at the end supports are zero, and there are, therefore, to be determined $n - 1$ moments at the other supports. This may be done by writing $n - 1$ equations of the form of Eqs. (47) and (48) for each support. These equations will contain $n - 1$ unknown moments, and their solution will give values of M_1, M_2, M_3, etc., expressed as coefficients of wl^2. The shear V_1 at any support may be determined by substituting values of M_1 and M_2 in Eq. (46), and the bending moment at any point in any span may be obtained by Eq. (45). The shear at any point in any span may be determined from Eq. (44).

Figure 20 gives values and diagrams for the reactions, shears, and moments at all sections of continuous beams uniformly loaded up to five spans. Note that the reaction at any support is equal to the sum of the shears to the right and to the left of that support.

6.4 Curved Beams

The derivation of the flexure formula, $\sigma = Mc/I$, assumes that the beam is initially straight; therefore, any deviation from this condition introduces an error in the value of the stress. If

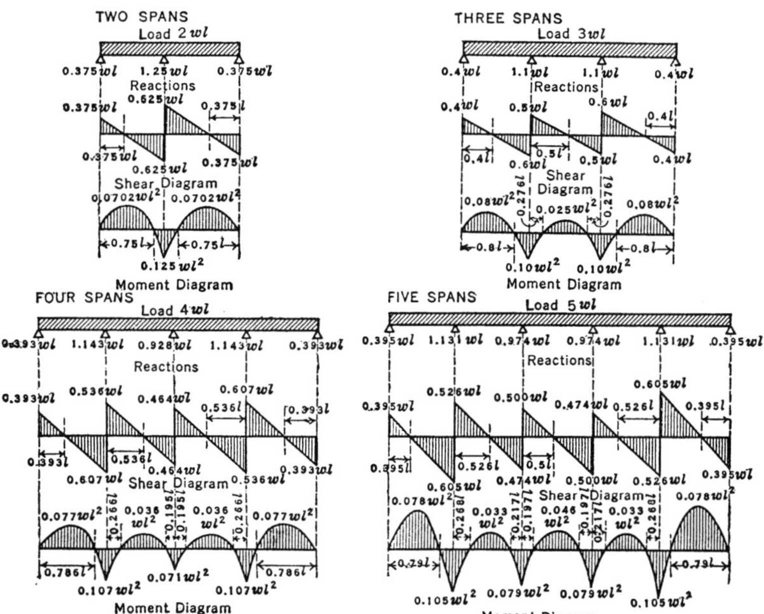

Figure 20 Shear and moment diagrams of continuous beams.

the curvature is slight, the error involved is not large, but in beams with a large amount of curvature, as hooks, chain links, and frames of punch presses, the error involved in the use of the ordinary flexure formula is considerable. The effect of the curvature is to increase the stress in the inside and to decrease it on the outside fibers of the beam and to shift the position of the neutral axis from the centroidal axis toward the concave or inner side.

The correct value for the unit fiber stress may be found by introducing a correction factor in the flexure formula, $\sigma = K(P/A \pm Mc/I)$; the factor K depends on the shape of the beam and on the ratio R/c, where R = distance, inches, from the centroidal axis of the section to the center of curvature of the central axis of the unstressed beam and c = instance, inches, of centroidal axis from the extreme fiber on the inner or concave side. Reference 8 has an analysis of curved beams, as does Table 7, which gives values of K for a number of shapes and ratios of R/c. For slightly different shapes or proportions K may be found by interpolation.

Deflection of Curved and Slender Curved Beams

The deflection of curved beams,[8,9] Fig. 21, in the curved portion can be found by

$$U = \int \frac{1}{2}\frac{P^2}{EA}\,ds + \int \frac{\phi V^2}{GA}\,ds + \int \frac{1}{2}\frac{M^2}{EAy_0R}\,ds + \int \frac{MP}{EAR}\,ds \tag{49}$$

$$\frac{\partial U}{\partial Q} = \delta_Q \tag{50}$$

where Q is a fictitious load of a couple where the deflection or rotation is desired or can be thought of as a 1-lb load or 1-in.-lb couple. y_0 is from Table 7, ϕ is a shape factor[2] often taken as 1, and ds is $R\,d\theta$. When $R/c > 4$, the last two terms condense to the integral of

Table 7 Values of Constant K for Curved Beams

Section	$\frac{R}{c}$	Values of K Inside Fiber	Values of K Outside Fiber	$\frac{Y_0}{R}$ [a]	Section	$\frac{R}{c}$	Values of K Inside Fiber	Values of K Outside Fiber	$\frac{Y_0}{R}$ [a]
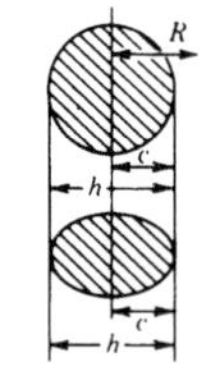	1.2	3.41	.54	.224	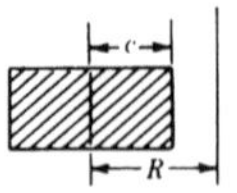	1.2	2.89	.57	.305
	1.4	2.40	.60	.151		1.4	2.13	.63	.204
	1.6	1.96	.65	.108		1.6	1.79	.67	.149
	1.8	1.75	.68	.084		1.8	1.63	.70	.112
	2.0	1.62	.71	.069		2.0	1.52	.73	.090
	3.0	1.33	.79	.030		3.0	1.30	.81	.041
	4.0	1.23	.84	.016		4.0	1.20	.85	.021
	6.0	1.14	.89	.0070		6.0	1.12	.90	.0093
	8.0	1.10	.91	.0039		8.0	1.09	.92	.0052
	10.0	1.08	.93	.0025		10.0	1.07	.94	.0033
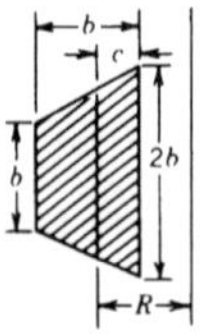	1.2	3.01	.54	.336	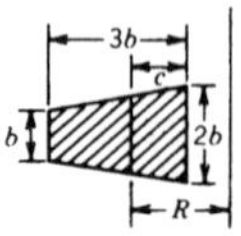	1.2	3.09	.56	.336
	1.4	2.18	.60	.229		1.4	2.25	.62	.229
	1.6	1.87	.65	.168		1.6	1.91	.66	.168
	1.8	1.69	.68	.128		1.8	1.73	.70	.128
	2.0	1.58	.71	.102		2.0	1.61	.73	.102
	3.0	1.33	.80	.046		3.0	1.37	.81	.046
	4.0	1.23	.84	.024		4.0	1.26	.86	.024
	6.0	1.13	.88	.011		6.0	1.17	.91	.011
	8.0	1.10	.91	.0060		8.0	1.13	.94	.0060
	10.0	1.08	.93	.0039		10.0	1.11	.95	.0039
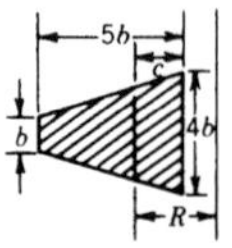	1.2	3.14	.52	.352	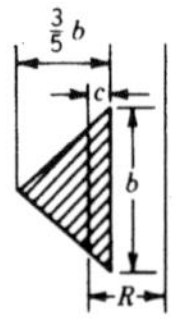	1.2	3.26	.44	.361
	1.4	2.29	.54	.243		1.4	2.39	.50	.251
	1.6	1.93	.62	.179		1.6	1.99	.54	.186
	1.8	1.74	.65	.138		1.8	1.78	.57	.144
	2.0	1.61	.68	.110		2.0	1.66	.60	.116
	3.0	1.34	.76	.050		3.0	1.37	.70	.052
	4.0	1.24	.82	.028		4.0	1.27	.75	.029
	6.0	1.15	.87	.012		6.0	1.16	.82	.013
	8.0	1.12	.91	.0060		8.0	1.12	.86	.0060
	10.0	1.11	.93	.0039		10.0	1.09	.88	.0039
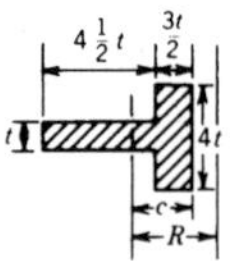	1.2	3.63	.58	.418	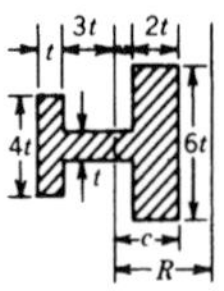	1.2	3.55	.67	.409
	1.4	2.54	.63	.299		1.4	2.48	.72	.292
	1.6	2.14	.67	.229		1.6	2.07	.76	.224
	1.8	1.89	.70	.183		1.8	1.83	.78	.178
	2.0	1.73	.72	.149		2.0	1.69	.80	.144
	3.0	1.41	.79	.069		3.0	1.38	.86	.067
	4.0	1.29	.83	.040		4.0	1.26	.89	.038
	6.0	1.18	.88	.018		6.0	1.15	.92	.018
	8.0	1.13	.91	.010		8.0	1.10	.94	.010
	10.0	1.10	.92	.0065		10.0	1.08	.95	.0065
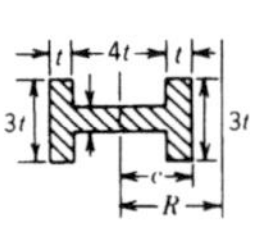	1.2	2.52	.67	.408	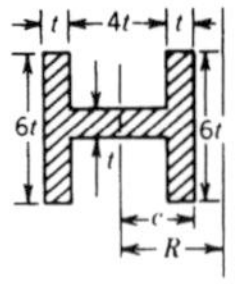	1.2	2.37	.73	.453
	1.4	1.90	.71	.285		1.4	1.79	.77	.319
	1.6	1.63	.75	.208		1.6	1.56	.79	.236
	1.8	1.50	.77	.160		1.8	1.44	.81	.183
	2.0	1.41	.79	.127		2.0	1.36	.83	.147
	3.0	1.23	.86	.058		3.0	1.19	.88	.067
	4.0	1.16	.89	.030		4.0	1.13	.91	.036
	6.0	1.10	.92	.013		6.0	1.08	.94	.016
	8.0	1.07	.94	.0076		8.0	1.06	.95	.0089
	10.0	1.05	.95	.0048		10.0	1.05	.96	.0057
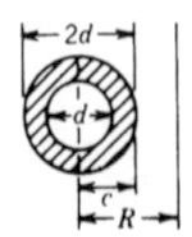	1.2	3.28	.58	.269	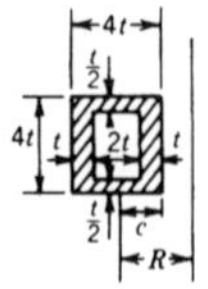	1.2	2.63	.68	.399
	1.4	2.31	.64	.182		1.4	1.97	.73	.280
	1.6	1.89	.68	.134		1.6	1.66	.76	.205
	1.8	1.70	.71	.104		1.8	1.51	.78	.159
	2.0	1.57	.73	.083		2.0	1.43	.80	.127
	3.0	1.31	.81	.038		3.0	1.23	.86	.058
	4.0	1.21	.85	.020		4.0	1.15	.89	.031
	6.0	1.13	.90	.0087		6.0	1.09	.92	.014
	8.0	1.10	.92	.0049		8.0	1.07	.94	.0076
	10.0	1.07	.93	.0031		10.0	1.06	.95	.0048

[a] Y_0 is distance from centroidal axis to neutral axis, where beam is subjected to pure bending.

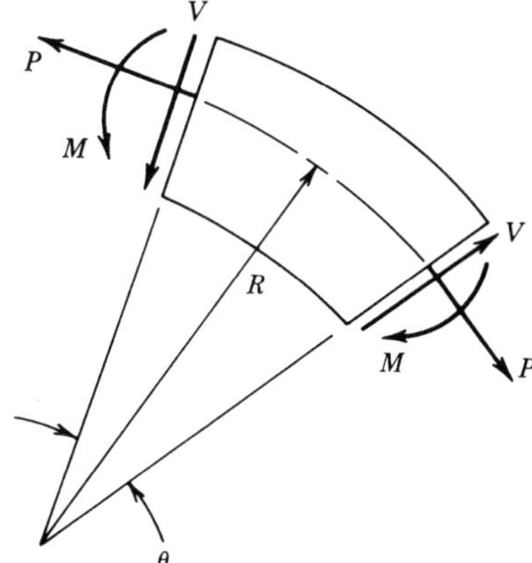

Figure 21 Positive sign convention for curved beams.

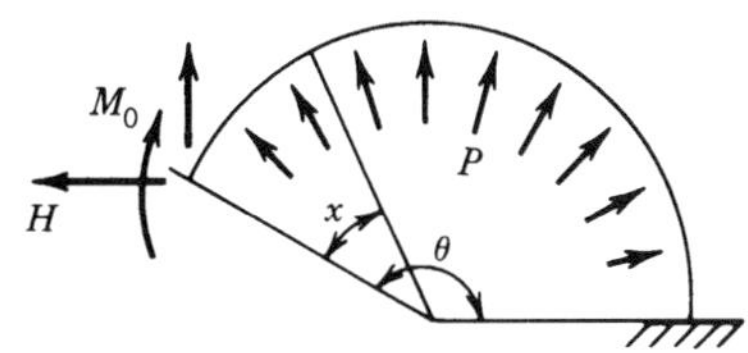

Figure 22 Circular cantilever with end loading and uniform radial pressure p pounds per linear inch.

$(M^2/2EI)\, ds$. When the length of the curved portion to the depth of the beam is greater than 10, the second term of Eq. (49) can be dropped. When in doubt, include all terms.

When beams are not curved (Fig. 22), such as some clamps, the following equations (used by permission of McGraw-Hill from the 4th ed. of Ref. 2) are useful:

$$M = M_0 + HR[\sin(\theta - x) - x] - VR\,[\cos(\theta - x) - c] + pR^2(1 - u) \tag{51}$$

Vertical deflection =

$$\frac{1}{EI}[M_0R^2(s - \theta c) + VR^3(\tfrac{1}{2}\theta + c^2\theta - \tfrac{3}{2}sc)$$
$$+ HR^3(\tfrac{1}{2} - c + sc\theta + \tfrac{1}{2}c^2 - s^2) + pR^4(s + sc - \tfrac{3}{2}\theta c - \tfrac{1}{2}s^3 - \tfrac{1}{2}c^2s)] \tag{52}$$

Horizontal deflection =

$$\frac{1}{EI}[M_0R^2(1 - \theta s - c) + VR^3(\tfrac{1}{2} - c + \theta sc + \tfrac{1}{2}c^2 - s^2)$$
$$+ HR^3(-2s + \theta s^2 + \tfrac{1}{2}\theta + \tfrac{3}{2}sc) + pR^4(1 - \tfrac{3}{2}\theta s + s^2 - c] \tag{53}$$

Rotation =

$$\frac{1}{EI}[M_0R\theta + VR^2(s - \theta c) + HR^2(1 - \theta s - c) + pR^3(\theta - s)] \tag{54}$$

where $u = \cos x$, $s = \sin\theta$, and $c = \cos\theta$.

6.5 Impact Stresses in Bars and Beams

Effect of Sudden Loads

If a sudden load P is applied to a bar, it will cause a deformation el, and the work done by the load will be Pel. Since the external work equals the internal work, $Pel = \sigma^2Al/2E$, and since $e = \sigma/E$, $P = \sigma A/2$, or $\sigma = 2P/A$. The unit stress and also the unit strain are double those obtained by an equal load applied gradually. However, the bar does not maintain equilibrium at the point of maximum stress and strain. After a series of oscillations, however, in which the surplus energy is dissipated in damping, the bar finally comes to rest with the same strain and stress as that due to the equal static load.

Stress due to Live Loads

In structural design two loads are considered, the dead load or weight of the structure and the live load or superimposed loads to be carried. The stresses due to the dead load and to the live load are computed separately, each being regarded as a static load. It is obvious that the stress due to the live load may be greatly increased, depending on the suddenness with which the load is applied. It has been shown above that the stress due to a suddenly applied load is double the stress caused by a static load. The term *coefficient of impact* is used extensively in structural engineering to denote the number by which the computed static stress is multiplied to obtain the value of the increased stress assumed to be caused by the suddenness of the application of the live load. If σ = static unit stress computed from the live load and i = coefficient of impact, then the increase of unit stress due to sudden loading is $i\sigma$, and the total unit stress due to live load is $\sigma + i\sigma$. The value of i has been determined by empirical methods and varies according to different conditions.

In the building codes of most cities, specified floor loadings for buildings include the impact allowance, and no increase is needed for live loads except for special cases of vibration or other unusual conditions. For railroad bridges, the value of i depends upon the proportion of the length of the bridge which is loaded. No increase in the static stress is needed when the mass of the structure, as in monolithic concrete, is great. For machinery and for unusual conditions, such as elevator machinery and its supports, each structure should be considered by itself and the coefficient assumed accordingly. It should be noted that the meaning of the word *impact* used above differs somewhat from its strict theoretical meaning and as it is used in the next paragraph. The use of the terms impact and *coefficient of impact* in connection with live load stresses is, however, very general.

Axial Impact on Bars

A load P dropped from a height h onto the end of a vertical bar of cross-sectional area A rigidly secured at the bottom end produces in the bar a unit stress which increases from 0 up to σ', with a corresponding total strain increasing from 0 up to e_1. The work done on the bar is $P(h + e_1)$, which, provided no energy is expended in hysteresis losses or in giving velocity to the bar, is equal to the energy $\frac{1}{2}\sigma' A e_1$ stored in the bar; that is,

$$P(h + e_1) = \tfrac{1}{2}\sigma' A e_1 \tag{55}$$

If e = strain produced by a static load P within the proportional limit

$$\frac{e}{e_1} = \frac{P/A}{\sigma'} \tag{56}$$

Combining this with Eq. (55) gives

$$\sigma' = \sigma + \sigma\sqrt{1 + 2\frac{h}{e}} \tag{57}$$

$$e_1 = e + e\sqrt{1 + 2\frac{h}{e}} \tag{58}$$

A wrought iron bar 1 in. square and 5 ft long under a static load of 5000 lb will be shortened about 0.012 in., assuming no lateral flexure occurs; but, if a weight of 5000 lb drops on its end from a height of 0.048 in., a stress of 20,000 psi will be produced.

Equations (57) and (58) give values of stress and strain that are somewhat high because part of the energy of the applied force is not effective in producing stress but is expended in overcoming the inertia of the bar and in producing local stresses. For light bars they give approximately correct results.

If the bar is horizontal and is struck at one end by a weight P moving with a velocity V, the strain produced is e_1. Then, as before, $\frac{1}{2}\sigma' A e_1 = Ph$. In this case $h = V^2/2g$ = height from which P would have to fall to acquire velocity V (g = acceleration due to gravity = 32.16 ft/s²). Combining with Eq. (56),

$$\sigma' = \sigma \sqrt{2\frac{h}{e}} \tag{59}$$

$$e_1 = e \sqrt{2\frac{h}{e}} \tag{60}$$

Impact on Beams

If a weight P falls on a horizontal beam from a height h producing a maximum deflection y and a maximum unit stress σ' in the extreme fiber, the values of σ' and y are given by

$$\sigma' = \sigma + \sigma \sqrt{1 + 2\frac{h}{y}} \tag{61}$$

$$y_1 = y + y \sqrt{1 + 2\frac{h}{y}} \tag{62}$$

where σ = extreme fiber unit stress and y = deflection due to P, considered as a static load. The value of σ may be obtained from the flexure formula [Eq. (37)]; that of y from the proper formula for deflection under static load.

If a weight P moving horizontally with a velocity V strikes a beam (the ends of which are secured against horizontal movement), the maximum fiber unit stress and the maximum lateral deflection are given by

$$\sigma' = \sigma \sqrt{2\frac{h}{y}} \tag{63}$$

$$y_1 = y \sqrt{2\frac{h}{y}} \tag{64}$$

where σ and y are as before and h is height through which P would have to fall to acquire the velocity V. These formulas, like those for axial impact on bars, give results higher than those observed in tests, particularly if the weight of the beam is great. For further discussion, see Ref. 7.

Rupture from Impact

Rupture may be caused by impact provided the load has the requisite velocity. The above formulas, however, do not apply since they are valid only for stresses within the proportional limit. It has been found that the dynamic properties of a material are dependent on volume, velocity of the applied load, and material condition. If the velocity of the applied load is kept within certain limiting values, the total energy values for static and dynamic conditions are identical. If the velocity is increased, the impact values are considerably reduced. For further information, see the articles in Ref. 10.

6.6 Steady and Impulsive Vibratory Stresses

For steady vibratory stresses of a weight W supported by a beam or rod, the deflection of the bar or beam will be increased by the dynamic magnification factor. The relation is given by

$$\delta_{\text{dynamic}} = \delta_{\text{static}} \times \text{dynamic magnification factor}$$

An example of the calculating procedure for the case of no damping losses is

$$\delta_{\text{dynamic}} = \delta_{\text{static}} \times \frac{1}{1 - (\omega/\omega_n)^2} \tag{65}$$

where ω is the frequency of oscillation of the load and ω_n is the natural frequency of oscillation of a weight on the bar.

For the same beam excited by a single sine pulse of magnitude A inches per second squared and a seconds duration, for $t < a$ a good approximation is

$$\sigma_{\text{dynamic}} = \frac{\delta_{\text{static}}(A/g)}{1 - [\omega/4\pi\omega_n]^2} \left[\sin \omega t - \frac{1}{4\pi^2}\left(\frac{\omega}{\omega_n}\right) \sin \omega_n t \right] \tag{66}$$

where A/g is the number of g's and ω is π/a.

7 SHAFTS, BENDING, AND TORSION

7.1 Definitions

Torsional stress. A bar is under torsional stress when it is held fast at one end, and a force acts at the other end to twist the bar. In a round bar (Fig. 23) with a constant force acting, the straight line *ab* becomes the helix *ad*, and a radial line in the cross section, *ob*, moves to the position *od*. The angle *bad* remains constant while the angle *bod* increases with the length of the bar. Each cross section of the bar tends to shear off the one adjacent

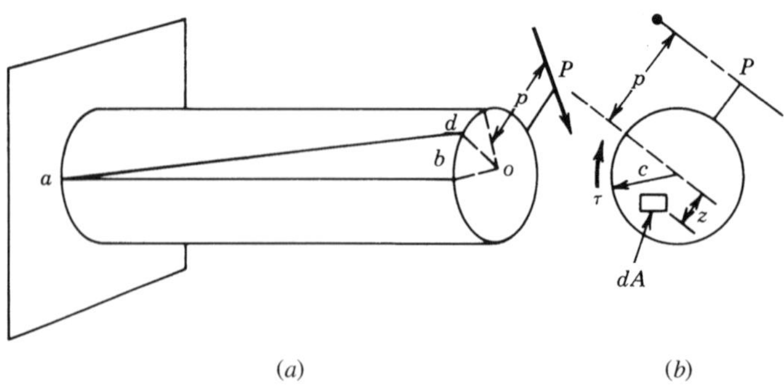

Figure 23 Round bar subject to torsional stress.

to it, and in any cross section the shearing stress at any point is normal to a radial line drawn through the point. Within the shearing proportional limit, a radial line of the cross section remains straight after the twisting force has been applied, and the unit shearing stress at any point is proportional to its distance from the axis.

Twisting moment, T, is equal to the product of the resultant, P, of the twisting forces multiplied by its distance from the axis, p.

Resisting moment, T_r, in torsion, is equal to the sum of the moments of the unit shearing stresses acting along a cross section with respect to the axis of the bar. If dA is an elementary area of the section at a distance of z units from the axis of a circular shaft (Fig. 23*b*) and c is the distance from the axis to the outside of the cross section where the unit shearing stress is τ, then the unit shearing stress acting on dA is $(\tau z/c)\ dA$, its moment with respect to the axis is $(\tau z^2/c)\ dA$, an the sum of all the moments of the unit shearing stresses on the cross section is $\int (\tau z^2/c)\ dA$. In this expression the factor $\int z^2\ dA$ is the polar moment of inertia of the section with respect to the axis. Denoting this by J, the resisting moment may be written $\tau J/c$.

The polar moment of inertia of a surface about an axis through its center of gravity and perpendicular to the surface is the sum of the products obtained by multiplying each elementary area by the square of its distance from the center of gravity of its surface; it is equal to the sum of the moments of inertia taken with respect to two axes in the plane of the surface at right angles to each other passing through the center of gravity. It is represented by J, inches^4. For the cross section of a round shaft,

$$J = \tfrac{1}{32}\pi d^4 \quad \text{or} \quad \tfrac{1}{2}\pi r^4 \tag{67}$$

For a hollow shaft,

$$J = \tfrac{1}{32}\pi(d^4 - d_1^4) \tag{68}$$

where d is the outside and d_1 is the inside diameter, inches, or

$$J = \tfrac{1}{2}\pi(r^4 - r_1^4) \tag{69}$$

where r is the outside and r_1 the inside radius, inches.

The polar radius of gyration, k_p, sometimes is used in formulas; it is defined as the radius of a circumference along which the entire area of a surface might be concentrated and have the same polar moment of inertia as the distributed area. For a solid circular section,

$$k_p^2 = \tfrac{1}{8}d^2 \tag{70}$$

For a hollow circular section,

$$k_p^2 = \tfrac{1}{8}(d^2 - d_1^2) \tag{71}$$

7.2 Determination of Torsional Stresses in Shafts

Torsion Formula for Round Shafts

The conditions of equilibrium require that the twisting moment, T, be opposed by an equal resisting moment, T_r, so that for the values of the maximum unit shearing stress, τ, within the proportional limit, the torsion formula for round shafts becomes

$$T_r = T = \tau \frac{J}{c} \tag{72}$$

if τ is in pounds per square inch, then T_r and T must be in pound-inches, J is in inches^4, and c is in inches. For solid round shafts having a diameter d, inches,

$$J = \tfrac{1}{32}\pi d^4 \qquad \text{and} \qquad c = \tfrac{1}{2}d \tag{73}$$

and

$$T = \tfrac{1}{16}\pi d^3 \tau \qquad \text{or} \qquad \tau = \frac{16T}{\pi d^3} \tag{74}$$

For hollow round shafts,

$$J = \frac{\pi(d^4 - d_1^4)}{32} \qquad \text{and} \qquad c = \tfrac{1}{2}d \tag{75}$$

and the formula becomes

$$T = \frac{\tau\pi(d^4 - d_1^4)}{16d} \qquad \text{or} \qquad \tau = \frac{16Td}{\pi(d^4 - d_1^4)} \tag{76}$$

The torsion formula applies only to solid circular shafts or hollow circular shafts, and then only when the load is applied in a plane perpendicular to the axis of the shaft and when the shearing proportional limit of the material is not exceeded.

Shearing Stress in Terms of Horsepower

If the shaft is to be used for the transmission of power, the value of T, pound-inches, in the above formulas becomes $63{,}030H/N$, where H = horsepower to be transmitted and N = revolutions per minute. The maximum unit shearing stress, pounds per square inch, then is

$$\text{For solid round shafts:} \qquad \tau = \frac{321{,}000H}{Nd^3} \tag{77}$$

$$\text{For hollow round shafts:} \qquad \tau = \frac{321{,}000Hd}{N(d^4 - d_1^4)} \tag{78}$$

If τ is taken as the allowable unit shearing stress, the diameter d, inches, necessary to transmit a given horsepower at a given shaft speed can then be determined. These formulas give the stress due to torsion only, and allowance must be made for any other loads, as the weight of shaft and pulley, and tension in belts.

Angle of Twist

When the unit shearing stress τ does not exceed the proportional limit, the angle *bod* (Fig. 23) for a solid round shaft may be computed from the formula

$$\theta = \frac{Tl}{GJ} \tag{79}$$

where θ = angle in radians; l = length of shaft in inches; G = shearing modulus of elasticity of the material; T = twisting moment in pound-inches. Values of G for different materials are steel, 12,000,000; wrought iron, 10,000,000; and cast iron, 6,000,000.

When the angle of twist on a section begins to increase in a greater ratio than the twisting moment, it may be assumed that the shearing stress on the outside of the section has reached the proportional limit. The shearing stress at this point may be determined by substituting the twisting moment at this instant in the torsion formula.

Torsion of Noncircular Cross Sections

The analysis of shearing stress distribution along noncircular cross sections of bars under torsion is complex. By drawing two lines at right angles through the center of gravity of a section before twisting and observing the angular distortion after twisting, it has been found from many experiments that in noncircular sections the shearing unit stresses are not proportional to their distances from the axis. Thus in a rectangular bar there is no shearing stress at the corners of the sections, and the stress at the middle of the wide side is greater than at the middle of the narrow side. In an elliptical bar the shearing stress is greater along the flat side than at the round side.

It has been found by tests[11,12] as well as by mathematical analysis that the torsional resistance of a section, made up of a number of rectangular parts, is approximately equal to the sum of the resistances of the separate parts. It is on this basis that nearly all the formulas for noncircular sections have been developed. For example, the torsional resistance of an I-beam is approximately equal to the sum of the torsional resistances of the web and the outstanding flanges. In an I-beam in torsion the maximum shearing stress will occur at the middle of the side of the web, except where the flanges are thicker than the web, and then the maximum stress will be at the midpoint of the width of the flange. Reentrant angles, as those in I-beams and channels, are always a source of weakness in members subjected to torsion. Table 8 gives values of the maximum unit shearing stress τ and the angle of twist θ induced by twisting bars of various cross sections, it being assumed that τ is not greater than the proportional limit.

Torsion of thin-wall closed sections (Fig. 24) is derived as

$$T = 2qA \tag{80}$$

$$q = \tau t \tag{81}$$

$$\theta_i = \frac{\theta}{L} = \frac{T}{2A}\frac{1}{2AG}\frac{S}{t} = \frac{T}{GJ} \tag{82}$$

where S is the arc length around area A over which τ acts for a thin-wall section; shear buckling should be checked. When more than one cell is used[1,13] or if a section is not constructed of a single material,[13] the calculations become more involved:

$$J = \frac{4A^2}{\oint ds/t} \tag{83}$$

Ultimate Strength in Torsion

In a torsion failure, the outer fibers of a section are the first to shear, and the rupture extends toward the axis as the twisting is continued. The torsion formula for round shafts has no theoretical basis after the shearing stresses on the outer fibers exceed the proportional limit, as the stresses along the section then are no longer proportional to their distances from the axis. It is convenient, however, to compare the torsional strength of various materials by using the formula to compute values of τ at which rupture takes place. These computed values of the maximum stress sustained before rupture are somewhat higher for iron and steel than the ultimate strength of the materials in direct shear. Computed values of the ultimate strength in torsion are found by experiment to be: cast iron, 30,000 psi; wrought iron, 55,000 psi; medium steel, 65,000 psi; timber, 2000 psi. These computed values of

Table 8 Formulas for Torsional Deformation and Stress

General formulas: $\theta = \frac{TL}{KG}$, $\tau = \frac{T}{Q}$, where θ = angle of twist, radians; T = twisting moment, in.·lb; L = length, in.; τ = unit shear stress, psi; G = modulus of rigidity, psi; K, in.^{4}, and Q, in.^{3}, are functions of the cross section.

Shape	Formula for K in $\theta = \frac{TL}{KG}$	Formula for Shear Stress
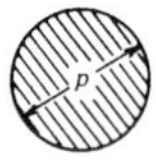 Solid circle, diameter d	$K = \frac{\pi d^4}{32}$	$\tau = \frac{16T}{\pi d^3}$
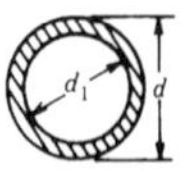 Hollow circle, d, d_1	$K = {}^1/_{32}\pi(d^4 - d_1{}^4)$	$\tau = \frac{16Td}{\pi(d^4 - d_1{}^4)}$
 Split thin tube, t, r	$K = {}^2/_3\,\pi r t^3$	$\tau = \frac{3T}{2\pi r t^2}$
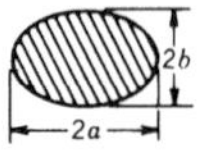 Solid ellipse, $2a$, $2b$	$K = \frac{\pi a^3 b^3}{a^2 + b^2}$	$\tau = \frac{2T}{\pi a b^2}$
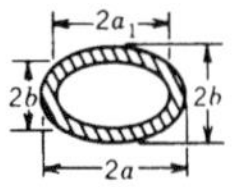 Hollow ellipse, $2a_1$, $2b_1$, $2a$, $2b$	$K = \frac{\pi a_1{}^3 b_1{}^3}{a_1{}^2 + b_1{}^2}[(1 + q)^4 - 1]$ $\quad q = \frac{a - a_1}{a_1}$ $\quad q = \frac{b - b_1}{b_1}$	$\tau = \frac{2T}{\pi a_1 b_1{}^2[(1 + q)^4 - 1]}$
 Equilateral triangle, b	$K = \frac{b^4\sqrt{3}}{80}$	$\tau = \frac{20T}{b^3}$
 Hexagon, b	$K = 2.69b^4$	$\tau = \frac{1.09T}{b^3}$
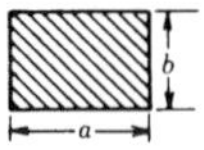 Rectangle, a, b	$K = \frac{ab^3}{16}\left[\frac{16}{3} - 3.36\frac{b}{a}\left(1 - \frac{b^4}{12a^4}\right)\right]$	$\tau = \frac{(3a + 1.8b)T}{a^2b^2}$
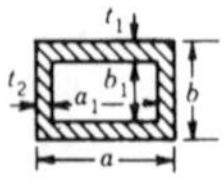 Hollow rectangle, a, b, a_1, b_1, t_1, t_2	$K = \frac{2t_1t_2(a - t_2)^2(b - t_1)^2}{at_2 + bt_1 - t_2^2 - t_1^2}$	$\tau = \frac{T}{2t_2(a - t_2)(b - t_1)}$
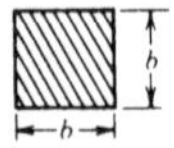 Square, b	$K = 0.1406b^4$	$\tau = \frac{4.8T}{b^3}$

Table 8 *(Continued)*

General formulas: $\theta = \frac{TL}{KG}$, $\tau = \frac{T}{Q}$, where θ = angle of twist, radians; T = twisting moment, in.·lb; L = length, in.; τ = unit shear stress, psi; G = modulus of rigidity, psi; K, in.^{4}, and Q, in.^{3}, are functions of the cross section.

Shape	Formula for K in $\theta = \frac{TL}{KG}$	Formula for Shear Stress
	r = fillet radius D = diameter largest inscribed circle $K = 2K_1 + K_2 + 2\alpha D^4$ $K_1 = ab^3\left[\frac{1}{3} - 0.21\frac{b}{a}\left(1 - \frac{b^4}{12a^4}\right)\right]$ $K_2 = cd^3\left[\frac{1}{3} - 0.105\frac{d}{c}\left(1 - \frac{d^4}{192c^4}\right)\right]$ $\alpha = \frac{b}{d}\left(0.07 + 0.076\frac{r}{b}\right)$	For all solid sections of irregular form the maximum shear stress occurs at or very near one of the points where the largest inscribed circle touches the boundary, and of these, at the one where the curvature of the boundary is algebraically least. (Convexity represents positive, concavity negative, curvature of the boundary.) At a point where the curvature is positive (boundary of section straight or convex) this maximum stress is given approximately by $\tau = G\frac{\theta}{L}c$ or $\tau = \frac{T}{K}c$ where $c = \frac{D}{1 + \frac{\pi^2 D^4}{16A^2}} \times \left[1 + 0.15\left(\frac{\pi^2 D^4}{16A^2} - \frac{D}{2r}\right)\right]$ where D = diameter of largest inscribed circle, r = radius of curvature of boundary at the point (positive for this case), A = area of the section.
	$K = 2K_1 + K_2 + 2\alpha D^4$ $K_1 = ab^3\left[\frac{1}{3} - 0.21\frac{b}{a}\left(1 - \frac{b^4}{12a^4}\right)\right]$ $K_2 = 1/3\, cd^3$ $\alpha = \frac{t}{t_1}\left(0.15 + 0.1\frac{r}{b}\right)$ $t = b$ if $b < d$ $t = d$ if $d < b$ $t_1 = b$ if $b > d$ $t_1 = d$ if $d > b$	
	$K = K_1 + K_2 + \alpha D^4$ $K_1 = ab^3\left[\frac{1}{3} - 0.21\frac{b}{a}\left(1 - \frac{b^4}{12a^4}\right)\right]$ $K_2 = cd^3\left[\frac{1}{3} - 0.105\frac{d}{c}\left(1 - \frac{d^4}{192c^4}\right)\right]$ $\alpha = \frac{b}{d}\left(0.07 + 0.076\frac{r}{b}\right)$	

twisting strength may be used in the torsion formula to determine the probable twisting moment that will cause rupture of a given round bar or to determine the size of a bar that will be ruptured by a given twisting moment. In design, large factors of safety should be taken, especially when the stress is reversed, as in reversing engines, and when the torsional stress is combined with other stresses, as in shafting.

7.3 Bending and Torsional Stresses

The stress for combined bending and torsion can be found from Eqs. (20) (shear theory) and (22) (distortion energy) with $\sigma_y = 0$:

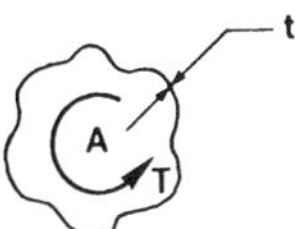

Figure 24 Thin-walled tube.

$$\frac{\sigma_w}{2} = \sqrt{\left(\frac{Mc}{2I}\right)^2 + \left(\frac{Tr}{J}\right)^2} \tag{84}$$

For solid round rods, this equation reduces to

$$\frac{\sigma_w}{2} = \frac{16}{\pi d^3} \sqrt{M^2 + T^2} \tag{85}$$

From distortion energy

$$\sigma = \sqrt{\left(\frac{Mc}{I}\right)^2 + 3\left(\frac{Tr}{J}\right)^2} \tag{86}$$

For solid round rods, the equation yields

$$\sigma = \frac{32}{\pi d^3} \sqrt{M^2 + \tfrac{3}{4}T^2} \tag{87}$$

8 COLUMNS

8.1 Definitions

A column or strut is a bar or structural member under axial compression which has an unbraced length greater than about eight or ten times the least dimension of its cross section. On account of its length, it is impossible to hold a column in a straight line under a load; a slight sidewise bending always occurs, causing flexural stresses in addition to the compressive stresses induced directly by the load. The lateral deflection will be in a direction perpendicular to that axis of the cross section about which the moment of inertia is the least. Thus in Fig. 25*a* the column will bend in a direction perpendicular to *aa*, in Fig. 25*b* it will bend perpendicular to *aa* or *bb*, and in Fig. 25*c* it is likely to bend in any direction.

Radius of gyration of a section with respect to a given axis is equal to the square root of the quotient of the moment of inertia with respect to that axis divided by the area of the section, that is

$$k = \sqrt{\frac{I}{A}} \qquad \frac{I}{A} = k^2 \tag{88}$$

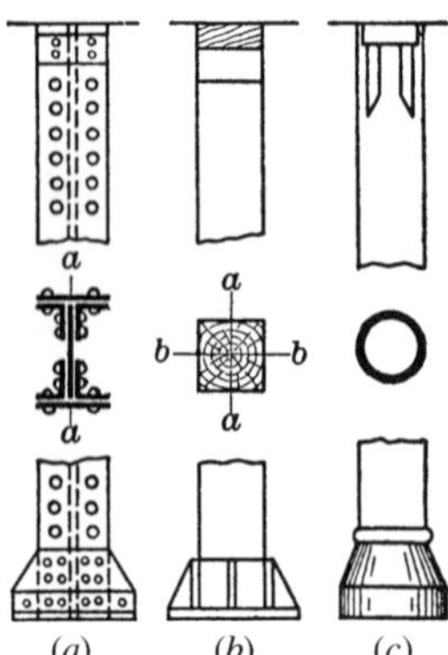

Figure 25 Column end designs.

where I is the moment of inertia and A is the cross-sectional area. Unless otherwise mentioned, an axis through the center of gravity of the section is the axis considered. As in beams, the moment of inertia is an important factor in the ability of the column to resist bending, but for purposes of computation it is more convenient to use the radius of gyration.

Length of a column is the distance between points unsupported against lateral deflection.

Slenderness ratio is the length l divided by the least radius of gyration k, both in inches. For steel, a *short column* is one in which $l/k < 20$ or 30, and its failure under load is due mainly to direct compression; in a *medium-length column,* $l/k =$ about 30–175, failure is by a combination of direct compression and bending; in a *long column,* $l/k >$ about 175–200, failure is mainly by bending. For timber columns these ratios are about 0–30, 30–90, and above 90, respectively. The load which will cause a column to fail decreases as l/k increases. The above ratios apply to round-end columns. If the ends are fixed (see below), the effective slenderness ratio is one-half that for round-end columns, as the distance between the points of inflection is one-half of the total length of the column. For flat ends it is intermediate between the two.

Conditions of ends. The various conditions which may exist at the ends of columns usually are divided into four classes: (1) Columns with round ends; the bearing at either end has perfect freedom of motion, as there would be with a ball-and-socket joint at each end. (2) Columns with hinged ends; they have perfect freedom of motion at the ends in one plane, as in compression members in bridge trusses where loads are transmitted through end pins. (3) Columns with flat ends; the bearing surface is normal to the axis of the column and of sufficient area to give at least partial fixity to the ends of the columns against lateral deflection. (4) Columns with fixed ends; the ends are rigidly secured, so that under any load the tangent to the elastic curve at the ends will be parallel to the axis in its original position.

Experiments prove that columns with fixed ends are stronger than columns with flat, hinged, or round ends and that columns with round ends are weaker than any of the other types. Columns with hinged ends are equivalent to those with round ends in the plane in which they have free movement; columns with flat ends have a value intermediate between those with fixed ends and those with round ends. It often happens that columns have one end fixed and one end hinged or some other combination. Their relative values may be taken as intermediate between those represented by the condition at either end. The extent to which strength is increased by fixing the ends depends on the length of bcolumn, fixed ends having a greater effect on long columns than on short ones.

8.2 Theory

There is no exact theoretical formula that gives the strength of a column of any length under an axial load. Formulas involving the use of empirical coefficients have been deduced, however, and they give results that are consistent with the results of tests.

Euler's Formula

Euler's formula assumes that the failure of a column is due solely to the stresses induced by sidewise bending. This assumption is not true for short columns, which fail mainly by direct compression, nor is it true for columns of medium length. The failure in such cases is by a combination of direct compression and bending. For columns in which $l/k > 200$, Euler's formula is approximately correct and agrees closely with the results of tests.

Let $P =$ axial load, pounds; $l =$ length of column, inches; $I =$ least moment of inertia, inches^4; $k =$ least radius of gyration, inches; $E =$ modulus of elasticity; $y =$ lateral deflection, inches, at any point along the column that is caused by load P. If a column has round ends, so that the bending is not restrained, the equation of its elastic curve is

$$EI \frac{d^2y}{dx^2} = -Py \tag{89}$$

when the origin of the coordinate axes is at the top of the column, the positive direction of x being taken downward and the positive direction of y in the direction of the deflection. Integrating the above expression twice and determining the constants of integration give

$$P = \Omega \pi^2 \frac{EI}{l^2} \tag{90}$$

which is Euler's formula for long columns. The factor Ω is a constant depending on the condition of the ends. For round ends $\Omega = 1$; for fixed ends $\Omega = 4$; for one end round and the other fixed $\Omega = 2.05$. P is the load at which, if a slight deflection is produced, the column will not return to its original position. If P is decreased, the column will approach its original position, but if P is increased, the deflection will increase until the column fails by bending.

For columns with value of l/k less than about 150, Euler's formula gives results distinctly higher than those observed in tests. Euler's formula is now little used except for long members and as a basis for the analysis of the stresses in some types of structural and machine parts. It always gives an *ultimate* and never an allowable load.

Secant Formula

The deflection of the column is used in the derivation of the Euler formula, but if the load were truly axial, it would be impossible to compute the deflection. If the column is assumed to have an initial eccentricity of load of e inches (see Ref. 7 for suggested values of e), the equation for the deflection y becomes

$$y_{\max} = e\left(\sec \frac{l}{2}\sqrt{\frac{P}{EI}} - 1\right) \tag{91}$$

The maximum unit compressive stress becomes

$$\sigma = \frac{P}{A}\left(1 + \frac{ec}{k^2}\sec \frac{l}{2}\sqrt{\frac{P}{EI}}\right) \tag{92}$$

where l = length of column, inches; P = total load, pounds; A = area, square inches; I = moment of inertia, inches^4; k = radius of gyration, inches; c = distance from neutral axis to the most compressed fiber, inches; E = modulus of elasticity; both I and k are taken with respect to the axis about which bending takes place. The ASCE indicates $ec/k^2 = 0.25$ for central loading. Because the formula contains the secant of the angle $(l/2)\sqrt{P/EI}$, it is sometimes called the *secant formula.* It has been suggested by the Committee on Steel-Column Research[14,15] that the best rational column formula can be constructed on the secant type, although of course it must contain experimental constants.

The secant formula can be used also for columns that are eccentrically loaded if e is taken as the actual eccentricity plus the assumed initial eccentricity.

Eccentric Loads on Short Compression Members

Where a direct push acting on a member does not pass through the centroid but at a distance e inches from it, both direct and bending stresses are produced. For short compression members in which column action may be neglected, the direct unit stress is P/A, where P = total load, pounds, and A = area of cross section, square inches. The bending unit stress is Mc/I, where $M = Pe$ = bending moment, pound-inches; c = distance, inches, from the centroid to the fiber in which the stress is desired; I = moment of inertia, inches^4. The total unit stress at any point in the section is $\sigma = P/A + Pec/I$, or $\sigma = (P/A)(1 + ec/k^2)$, since $I = Ak^2$, where k = radius of gyration, inches.

Eccentric Loads on Columns

Various column formulas must be modified when the loads are not balanced, that is, when the resultant of the loads is not in line with the axis of the column. If P = load, pounds, applied at a distance e inches from the axis, bending moment $M = Pe$. Maximum unit stress σ, pounds per square inch, due to this bending moment alone is $\sigma = Mc/I = Pec/Ak^2$, where c = distance, inches, from the axis to the most remote fiber on the concave side; A = sectional area, square inches; k = radius of gyration in the direction of the bending, inches. This unit stress must be added to the unit stress that would be induced if the resultant load were applied in line with the axis of the column.

The secant formula, Eq. (92), also can be used for columns that are eccentrically loaded if e is taken as the actual eccentricity plus the assumed initial eccentricity.

Column Subjected to Transverse or Cross-Bending Loads

A compression member that is subjected to cross-bending loads may be considered to be (1) a beam subjected to end thrust or (2) a column subjected to cross-bending loads, depending on the relative magnitude of the end thrust and cross-bending loads and on the dimensions of the member. The various column formulas may be modified so as to include the effect of cross-bending loads. In this form the modified secant formula for transverse loads is

$$\sigma = \frac{P}{A}\left[1 + (e + y)\frac{c}{k^2}\sec\frac{l}{2k}\sqrt{\frac{P}{AE}}\right] + \frac{Mc}{Ak^2} \tag{93}$$

In the formula, σ = maximum unit stress on concave side, pounds per square inch; P = axial end load, pounds; A = cross-sectional area, square inches; M = moment due to cross-bending load, pound-inches; y = deflection due to cross-bending load, inches; k = radius of gyration, inches; l = length of column, inches; e = assumed initial eccentricity, inches; c = distance, inches, from axis to the most remote fiber on the concave side.

8.3 Wooden Columns

Wooden Column Formulas

One of the principal formulas is that formerly used by the AREA, $P/A = \sigma_1(1 - l/60d)$, where P/A = allowable unit load, pounds per square inch; σ_1 = allowable unit stress in direct compression on short blocks, pounds per square inch; l = length, inches; d = least dimension, inches. This formula is being replaced rapidly by formulas recommended by the ASTM and AREA (American Railroad Engineering Association). Committees of these societies, working with the U.S. Forest Products Laboratory, classified timber columns in three groups (ASTM Standards, 1937, D245-37):

1. *Short Columns.* The ratio of unsupported length to least dimension does not exceed 11. For these columns, the allowable unit stress should not be greater than the values given in Table 9 under compression parallel to the grain.
2. *Intermediate-Length Columns.* Where the ratio of unsupported length to least dimension is greater than 10, Eq. (94), of the fourth-power parabolic type, shall be used

Table 9 Basic Stresses for Clear Material[a]

Species	Extreme Fiber in Bending or Tension Parallel to Grain	Maximum Horizontal Shear	Compression Perpendicular to Grain	Compression Parallel to Grain, $L/d = 11$ or Less	Modulus of Elasticity in Bending
Softwoods					
Bald cypress (Southern cypress)	1900	150	300	1450	1,200,000
Cedars					
Red cedar, Western	1300	120	200	950	1,000,000
White cedar, Atlantic (Southern white cedar) and northern	1100	100	180	750	800,000
White cedar, Port Orford	1600	130	250	1200	1,500,000
Yellow cedar, Alaska (Alaska cedar)	1600	130	250	1050	1,200,000
Douglas fir, coast region	2200	130	320	1450	1,600,000
Douglas fir, coast region, close grained	2350	130	340	1550	1,600,000
Douglas fir, Rocky Mountain region	1600	120	280	1050	1,200,000
Douglas fir, dense, all regions	2550	150	380	1700	1,600,000
Fir, California red, grand, noble, and white	1600	100	300	950	1,100,000
Fir, balsam	1300	100	150	950	1,000,000
Hemlock, Eastern	1600	100	300	950	1,100,000
Hemlock, Western (West Coast hemlock)	1900	110	300	1200	1,400,000
Larch, Western	2200	130	320	1450	1,500,000
Pine, Eastern white (Northern white), ponderosa, sugar, and Western white (Idaho white)	1300	120	250	1000	1,000,000
Pine, jack	1600	120	220	1050	1,100,000
Pine, lodgepole	1300	90	220	950	1,000,000
Pine, red (Norway pine)	1600	120	220	1050	1,200,000
Pine, southern yellow	2200	160	320	1450	1,600,000
Pine, southern yellow, dense	2550	190	380	1700	1,600,000
Redwood	1750	100	250	1350	1,200,000
Redwood, close grained	1900	100	270	1450	1,200,000
Spruce, Engelmann	1100	100	180	800	800,000
Spruce, red, white, and Sitka	1600	120	250	1050	1,200,000
Tamarack	1750	140	300	1350	1,300,000
Hardwoods					
Ash, black	1450	130	300	850	1,100,000
Ash, commercial white	2050	185	500	1450	1,500,000
Beech, American	2200	185	500	1600	1,600,000
Birch, sweet and yellow	2200	185	500	1600	1,600,000
Cottonwood, Eastern	1100	90	150	800	1,000,000
Elm, American and slippery (white or soft elm)	1600	150	250	1050	1,200,000
Elm, rock	2200	185	500	1600	1,300,000
Gums, blackgum, sweetgum (red or sap gum)	1600	150	300	1050	1,200,000
Hickory, true and pecan	2800	205	600	2000	1,800,000
Maple, black and sugar (hard maple)	2200	185	500	1600	1,600,000
Oak, commercial red and white	2050	185	500	1350	1,500,000
Tupelo	1600	150	300	1050	1,200,000
Yellow poplar	1300	120	220	950	1,100,000

[a]These stresses are applicable with certain adjustments to material of any degree of seasoning.

(For use in determining working stresses according to the grade of timber and other applicable factors. All values are in pounds per square inch. U.S. Forest Products Laboratory.)

to determine allowable unit stress until this allowable unit stress is equal to two-thirds of the allowable unit stress for short columns:

$$\frac{P}{A} = \sigma_1 \left[1 - \frac{1}{3} \left(\frac{l}{Kd} \right)^4 \right] \tag{94}$$

where P = total load, pounds; A = area, square inches; σ_1 = allowable unit compressive stress parallel to grain, pounds per square inch (see Table 9); l = unsupported length, inches; d = least dimension, inches; $K = l/d$ at the point of tangency of the parabolic and Euler curves, at which $P/A = \frac{2}{3}\sigma_1$. The value of K for any species and grade is $\pi/2\sqrt{E/6\sigma_1}$, where E = modulus of elasticity.

3. *Long Columns.* Where P/A as computed by Eq. (94) is less than $\frac{2}{3}\sigma_1$, Eq. (95) of the Euler type, which includes a factor of safety of 3, shall be used:

$$\frac{P}{A} = \frac{1}{36} \left[\frac{\pi^2 E}{(l/d)^2} \right] \tag{95}$$

Timber columns should be limited to a ratio of l/d equal to 50. No higher loads are allowed for square-ended columns. The strength of round columns may be considered the same as that of square columns of the same cross-sectional area.

Use of Timber Column Formulas

The values of E (modulus of elasticity) and σ_1 (compression parallel to grain) in the above formulas are given in Table 9. Table 10 gives the computed values of K for some common types of timbers. These may be substituted directly in Eq. (94) for intermediate-length columns or may be used in conjunction with Table 11, which gives the strength of columns of intermediate length, expressed as a percentage of strength (σ_1) of short columns. In the tables, the term "continuously dry" refers to interior construction where there is no excessive dampness or humidity; "occasionally wet but quickly dry" refers to bridges, trestles, bleachers,

Table 10 Values of K for Columns of Intermediate Length

ASTM Standards, 1937, D245–37						
	Continuously Dry		Occasionally Wet		Usually Wet	
Species	Select	Common	Select	Common	Select	Common
Cedar, western red	24.2	27.1	24.2	27.1	25.1	28.1
Cedar, Port Orford	23.4	26.2	24.6	27.4	25.6	28.7
Douglas fir, coast region	23.7	27.3	24.9	28.6	27.0	31.1
Douglas fir, dense	22.6	25.3	23.8	26.5	25.8	28.8
Douglas fir, Rocky Mountain region	24.8	27.8	24.8	27.8	26.5	29.7
Hemlock, west coast	25.3	28.3	25.3	28.3	26.8	30.0
Larch, western	22.0	24.6	23.1	25.8	25.8	28.8
Oak, red and white	24.8	27.8	26.1	29.3	27.7	31.1
Pine, southern	.....	27.3	.....	28.6	.....	31.1
Pine, dense	22.6	25.3	23.8	26.5	25.8	28.8
Redwood	22.2	24.8	23.4	26.1	25.6	28.6
Spruce, red, white, Sitka	24.8	27.8	25.6	28.7	27.5	30.8

Table 11 Strength of Columns of Intermediate Length, Expressed as a Percentage of Strength of Short Columns

ASTM Standards, 1937, D245–37
Values for expression $1 - \frac{1}{3}(l/Kd)^4$ in Eq. (94)

Ratio of Length to Least Dimension in Rectangular Timbers, l/d

K	12	13	14	15	16	17	18	19	20	21	22	23	24	25	26	27	28	29	30	31
22	97	96	95	93	91	88	85	81	77	72	67	...	...	...	...	...	...	...	...	...
23	98	97	95	94	92	90	87	84	81	77	72	67	...	...	...	...	...	...	...	...
24	98	97	96	95	93	92	89	87	84	80	76	72	67	...	...	...	...	...	...	...
25	98	98	97	96	94	93	91	89	86	83	80	76	72	67	...	...	...	...	...	...
26	99	98	97	96	95	93	92	91	89	86	83	80	76	72	67	...	...	...	...	...
27	99	98	98	97	96	95	93	92	90	88	85	82	79	74	71	67	...	...	...	...
28	99	98	98	97	96	95	94	93	91	89	87	85	82	79	75	71	67	...	...	...
29	99	99	98	98	97	96	95	94	92	91	89	87	84	82	79	75	71	67	...	...
30	99	99	98	98	97	97	96	95	94	92	90	88	86	84	81	78	75	71	67	..
31	99	99	99	98	98	97	96	95	94	93	92	90	88	86	84	81	78	75	71	67

Note. This table can also be used for columns not rectangular, the l/d being equivalent to $0.289l/k$, where k is the least radius of gyration of the section.

and grandstands; "usually wet" refers to timber in contact with the earth or exposed to waves or tidewater.

8.4 Steel Columns

Types

Two general types of steel columns are in use: (1) rolled shapes and (2) built-up sections. The rolled shapes are easily fabricated, accessible for painting, neat in appearance where they are not covered, and convenient in making connections. A disadvantage is the probability that thick sections are of lower strength material than thin sections because of the difficulty of adequately rolling the thick material. For the effect of thickness of material on yield point, see Ref. 15, p. 1377.

General Principles in Design

The design of steel columns is always a cut-and-try method, as no law governs the relation between area and radius of gyration of the section. A column of given area is selected, and the amount of load that it will carry is computed by the proper formula. If the allowable load so computed is less than that to be carried, a larger column is selected and the load for it is computed, the process being repeated until a proper section is found.

A few general principles should guide in proportioning columns. The radius of gyration should be approximately the same in the two directions at right angles to each other; the slenderness ratio of the separate parts of the column should not be greater than that of the column as a whole; the different parts should be adequately connected in order that the column may function as a single unit; the material should be distributed as far as possible from the centerline in order to increase the radius of gyration.

Steel Column Formulas

A variety of steel column formulas are in use, differing mostly in the value of unit stress allowed with various values of l/k. See Ref. 16 for a summary of the formulas.

Test on Steel Columns

After the collapse of the Quebec Bridge in 1907 as a result of a column failure, the ASCE, the AREA, and the U.S. Bureau of Standards cooperated in tests of full-sized steel columns. The results of these tests are reported in Ref. 17, pp. 1583–1688. The tests showed that, for columns of the proportions commonly used, the effect of variation in the steel, kinks, initial stresses, and similar defects in the column was more important than the effect of length. They also showed that the thin metal gave definitely higher strength, per unit area, than the thicker metal of the same type of section.

9 CYLINDERS, SPHERES, AND PLATES

9.1 Thin Cylinders and Spheres under Internal Pressure

A cylinder is regarded as thin when the thickness of the wall is small compared with the mean diameter, or $d/t > 20$. There are only tensile membrane stresses in the wall developed by the internal pressure p,

$$\frac{\sigma_1}{R_1} + \frac{\sigma_2}{R_2} = \frac{p}{t} \tag{96}$$

In the case of a cylinder where R_1, the curvature, is R and R_2 is infinite, the hoop stress is

$$\sigma_1 = \sigma_h = \frac{pR}{t} \tag{97}$$

If the two equations are compared, it is seen that the resistance to rupture by circumferential stress [Eq. (97)] is one-half the resistance to rupture by longitudinal stress [Eq. (98)]. For this reason cylindrical boilers are single riveted in the circumferential seams and double or triple riveted in the longitudinal seams.

From the equations of equilibrium, the longitudinal stress is

$$\sigma_2 = \sigma_L = \frac{pR}{2t} \tag{98}$$

For a sphere, using Eq. (96), $R_1 = R_2 = R$ and $\sigma_1 = \sigma_2$, making

$$\sigma_1 = \sigma_2 = \frac{pR}{2t} \tag{99}$$

In using the foregoing formulas to design cylindrical shells or piping, thickness t must be increased to compensate for rivet holes in the joints. Water pipes, particularly those of cast iron, require a high factor of safety, which results in increased thickness to provide security against shocks caused by water hammer or rough handling before they are laid. Equation (98) applies also to the stresses in the walls of a thin hollow sphere, hemisphere, or dome. When holes are cut, the tensile stresses must be found by the method used in riveted joints.

Thin Cylinders under External Pressure

Equations (97) and (98) apply equally well to cases of external pressure if P is given a negative sign, but the stresses so found are significant only if the pressure and dimensions are such that no buckling can occur.

9.2 Thick Cylinders and Spheres

Cylinders

When the thickness of the shell or wall is relatively large, as in guns, hydraulic machinery piping, and similar installations, the variation in stress from the inner surface to the outer surface is relatively large, and the ordinary formulas for thin-wall cylinders are no longer applicable. In Fig. 26 the stresses, strains, and deflections are related[1,18,19] by

$$\sigma_t = \frac{E}{1-\nu^2}(\epsilon_t + \nu\epsilon_r) = \frac{E}{1-\nu^2}\left[\frac{u}{r} + \nu\frac{\partial u}{\partial r}\right] \tag{100}$$

$$\sigma_r = \frac{E}{1-\nu^2}(\epsilon_r + \nu\epsilon_t) = \frac{E}{1-\nu^2}\left[\frac{\partial u}{\partial r} + \nu\frac{u}{r}\right] \tag{101}$$

where E is the modulus and ν is Poisson's ratio. In a cylinder (Fig. 27) that has internal and external pressures, p_i and p_o; internal and external radii, a and b; $K = b/a$; the stresses are

$$\sigma_t = \frac{p_i}{K^2-1}\left(1+\frac{b^2}{r^2}\right) - \frac{p_o K^2}{K^2-1}\left(1+\frac{a^2}{r^2}\right) \tag{102}$$

$$\sigma_r = \frac{p_i}{K^2-1}\left(1-\frac{b^2}{r^2}\right) - \frac{p_o K^2}{K^2-1}\left(1-\frac{a^2}{r^2}\right) \tag{103}$$

if $p_o = 0$, σ_t, σ_r are maximum at $r = a$; if $p_i = 0$, σ_t is maximum at $r = a$ and σ_r is maximum at $r = b$.

In shrinkage fits, Fig. 27, a hollow cylinder is pressed over a cylinder with a radial interference δ at $r = b$. p_f, the pressure between the cylinders, can be found from

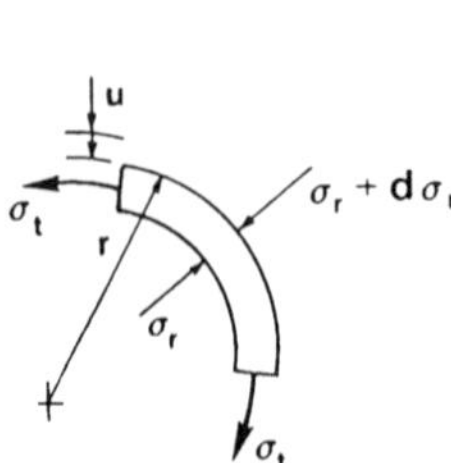

Figure 26 Cylindrical element.

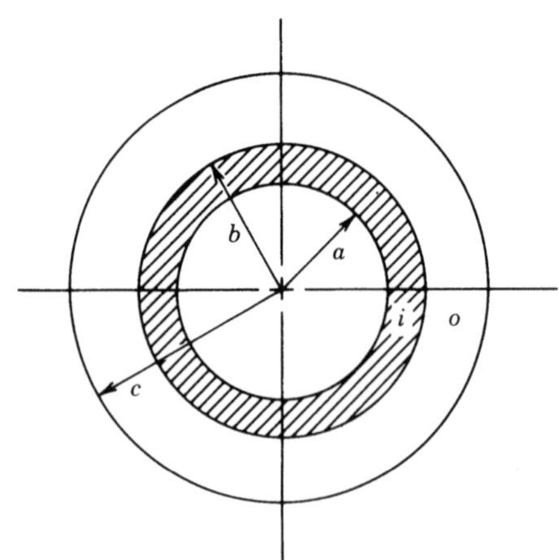

Figure 27 Cylinder press fit.

$$\delta = \frac{bp_f}{E_o}\left(\frac{c^2 + b^2}{c^2 - b^2} + \nu_o\right) + \frac{bp_f}{E_i}\left(\frac{a^2 + b^2}{b^2 - a^2} - \nu_i\right) \tag{104}$$

The radial deflection can be found at a, which shrinks, and c, which expands, by knowing σ_r is zero and using Eqs. (100) and (101):

$$u_a = \frac{\sigma_t}{E_i} a \qquad u_c = \frac{\sigma_t}{E_o} c \tag{105}$$

Spheres
The stress, strain, and deflections[19,20] are related by

$$\sigma_t = \frac{E}{1 - \nu - 2\nu^2}[\epsilon_t + \nu\epsilon_r] = \frac{E}{1 - \nu - 2\nu^2}\left[\frac{u}{r} + \nu\frac{\partial u}{\partial r}\right] \tag{106}$$

$$\sigma_r = \frac{E}{1 - \nu - 2\nu^2}[2\nu\epsilon_t + (1 - \nu)\epsilon_r] = \frac{E}{1 - \nu - 2\nu^2}\left[2\nu\frac{u}{r} + (1 - \nu)\frac{\partial u}{\partial r}\right] \tag{107}$$

The stresses for a thick-wall sphere with internal and external pressures p_i and p_o and $K = b/a$ are

$$\sigma_t = \frac{p_i(1 + b^3/2r^3)}{K^3 - 1} - \frac{p_o K^3(1 + a^3/2r^3)}{K^3 - 1} \tag{108}$$

$$\sigma_r = \frac{p_i(1 - b^3/r^3)}{K^3 - 1} - \frac{p_o K^3(1 - a^3/r^3)}{K^3 - 1} \tag{109}$$

If $p_i = 0$, $\sigma_r = 0$ at $r = a$, then

$$u_a = (1 - \nu)\frac{\sigma_t}{E} a \tag{110}$$

Conversely, if $p_o = 0$, $\sigma_r = 0$ at $r = b$, then

$$u_b = (1 - \nu)\frac{\sigma_t}{E} b \tag{111}$$

9.3 Plates

The formulas that apply for plates are based on the assumptions that the plate is flat, of uniform thickness, and of homogeneous isotropic material, thickness is not greater than one-fourth the least transverse dimension, maximum deflection is not more than one-half the thickness, all forces are normal to the plane of the plate, and the plate is nowhere stressed beyond the elastic limit. In Table 12 are formulas for deflection and stress for various shapes, forms of load, and edge conditions. For further information see Refs. 13 and 21.

9.4 Trunnion

A solid shaft (Fig. 28) on a round or rectangular plate loaded with a bending moment is called a trunnion. The loading generally is developed from a bearing mounted on the solid shaft. For a round, simply supported plate

Table 12 Formulas for Flat Plates[a]

Notation: W = total applied load, lb; w = unit applied load, psi; t = thickness of plate, in.; σ = stress at surface of plate, psi; y = vertical deflection of plate from original position, in.; E = modulus of elasticity; m = reciprocal of ν, Poisson's ratio. q denotes any given point on the surface of plate; r denotes the distance of q from the center of a circular plate. Other dimensions and corresponding symbols are indicated on figures. Positive sign for σ indicates tension at upper surface and equal compression at lower surface; negative sign indicates reverse condition. Positive sign for y indicates upward deflection, negative sign downward deflection. Subscripts r, t, a, and b used with σ denote, respectively, radial direction, tangential direction, direction of dimension a, and direction of dimension b. All dimensions are in inches. All logarithms are to the base e ($\log_e x = 2.3026 \log_{10} x$).

Type of Load and Support	Formulas for Stress and Deflection
	Circular Flat Plates
Outer edges supported. Uniform load over entire surface. $w\pi a^2 = W$	At center: $\max \sigma_r = \sigma_t = \frac{-3W}{8\pi m t^2}(3m+1)$ $\quad \max y = -\frac{3W(m-1)(5m+1)a^2}{16\pi E m^2 t^3}$ At q: $\sigma_r = -\frac{3W}{8\pi m t^2}\left[(3m+1)\left(1-\frac{r^2}{a^2}\right)\right]$ $\quad \sigma_t = -\frac{3W}{8\pi m t^2}\left[(3m+1)-(m+3)\frac{r^2}{a^2}\right]$ $y = -\frac{3W(m^2-1)}{8\pi E m^2 t^3}\left[\frac{(5m+1)a^2}{2(m+1)}+\frac{r^4}{2a^2}-\frac{(3m+1)r^2}{m+1}\right]$
Outer edges fixed. Uniform load over entire surface. $W = w\pi a^2$	At center: $\sigma_r = \sigma_t = -\frac{3W(m+1)}{8\pi m t^2}$ $\quad \max y = -\frac{3W(m^2-1)a^2}{16\pi E m^2 t^3}$ At q: $\sigma_r = \frac{3W}{8\pi m t^2}\left[(3m+1)\frac{r^2}{a^2}-(m+1)\right]$ $\quad \sigma_t = \frac{3W}{8\pi m t^2}\left[(m+3)\frac{r^2}{a^2}-(m+1)\right]$ $y = \frac{-3W(m^2-1)}{16\pi E m^2 t^3}\left[\frac{(a^2-r^2)^2}{a^2}\right]$
Outer edges supported. Uniform load over concentric circular area of radius r_0. $W = w\pi r_0^2$	At q, $r < r_0$: $\sigma_r = -\frac{3W}{2\pi m t^2}\left[m+(m+1)\log\frac{a}{r_0}-(m-1)\frac{r_0^2}{4a^2}-(3m+1)\frac{r^2}{4r_0^2}\right]$ $\sigma_t = -\frac{3W}{2\pi m t^2}\left[m+(m+1)\log\frac{a}{r_0}-(m-1)\frac{r_0^2}{4a^2}-(m+3)\frac{r^2}{4r_0^2}\right]$ $y = -\frac{3W(m^2-1)}{16\pi E m^2 t^3}\left[4a^2-5r_0^2+\frac{r^4}{r_0^2}-(8r^2+4r_0^2)\log\frac{a}{r_0}-\frac{2(m-1)r_0^2(a^2-r^2)}{(m+1)a^2}+\frac{8m(a^2-r^2)}{m+1}\right]$ At q, $r > r_0$: $\sigma_r = -\frac{3W}{2\pi m t^2}\left[(m+1)\log\frac{a}{r}-(m-1)\frac{r_0^2}{4a^2}+(m-1)\frac{r_0^2}{4r^2}\right]$ $\sigma_t = -\frac{3W}{2\pi m t^2}\left[(m-1)+(m+1)\log\frac{a}{r}-(m-1)\frac{r_0^2}{4a^2}-(m-1)\frac{r_0^2}{4r^2}\right]$ $y = -\frac{3W(m^2-1)}{16\pi E m^2 t^3}\left[\frac{(12m+4)(a^2-r^2)}{m+1}-\frac{2(m-1)r_0^2(a^2-r^2)}{(m+1)a^2}-(8r^2+4r_0^2)\log\frac{a}{r}\right]$ At center: $\max \sigma_r = \sigma_t = -\frac{3W}{2\pi m t^2}\left[m+(m+1)\log\frac{a}{r_0}-(m-1)\frac{r_0^2}{4a^2}\right]$ $\max y = -\frac{3W(m^2-1)}{16\pi E m^2 t^3}\left[\frac{(12m+4)a^2}{m+1}-4r_0^2\log\frac{a}{r_0}-\frac{(7m+3)r_0^2}{m+1}\right]$

[a] By permission from Ref. 22.

Table 12 *(Continued)*

Type of Load and Support	Formulas for Stress and Deflection
	Circular Flat Plates
Outer edges supported. Uniform load on concentric circular ring of radius r_0. W $r_0 \quad r_0$ 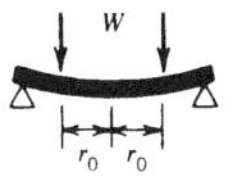	At q, $r < r_0$: $\max \sigma_r = \sigma_t = -\frac{3W}{2\pi m t^2}\left[\frac{1}{2}(m-1) + (m+1)\log\frac{a}{r_0} - (m-1)\frac{r_0^2}{2a^2}\right]$ $y = -\frac{3W(m^2-1)}{2\pi E m^2 t^3}\left[\frac{(3m+1)(a^2-r^2)}{2(m+1)} - (r^2+r_0^2)\log\frac{a}{r_0} + (r^2-r_0^2) - \frac{(m-1)r_0^2(a^2-r^2)}{2(m+1)a^2}\right]$ At q, $r > r_0$: $\sigma_r = -\frac{3W}{2\pi m t^2}\left[(m+1)\log\frac{a}{r} + (m-1)\frac{r_0^2}{2r^2} - (m-1)\frac{r_0^2}{2a^2}\right]$ $\sigma_t = -\frac{3W}{2\pi m t^2}\left[(m-1) + (m+1)\log\frac{a}{r} - (m-1)\frac{r_0^2}{2r^2} - (m-1)\frac{r_0^2}{2a^2}\right]$ $y = -\frac{3W(m^2-1)}{2\pi E m^2 t^3}\left[\frac{(3m+1)(a^2-r^2)}{2(m+1)} - (r^2+r_0^2)\log\frac{a}{r} - \frac{(m-1)r_0^2(a^2-r^2)}{2(m+1)a^2}\right]$
Outer edges fixed. Uniform load over concentric circular area of radius r_0. $W = w\pi r_0^2$ $r_0 \quad r_0$ 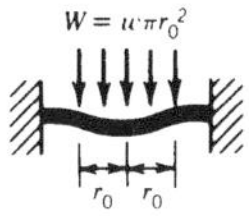	At q, $r < r_0$: $\sigma_r = -\frac{3W}{2\pi m t^2}\left[(m+1)\log\frac{a}{r_0} + (m+1)\frac{r_0^2}{4a^2} - (3m+1)\frac{r^2}{4r_0^2}\right]$ $\sigma_t = -\frac{3W}{2\pi m t^2}\left[(m+1)\log\frac{a}{r_0} + (m+1)\frac{r_0^2}{4a^2} - (m+3)\frac{r^2}{4r_0^2}\right]$ $y = -\frac{3W(m^2-1)}{16\pi E m^2 t^3}\left[4a^2 - (8r^2+4r_0^2)\log\frac{a}{r_0} - \frac{2r^2r_0^2}{a^2} + \frac{r^4}{r_0^2} - 3r_0^2\right]$ At q, $r > r_0$: $\sigma_r = -\frac{3W}{2\pi m t^2}\left[(m+1)\log\frac{a}{r} + (m+1)\frac{r_0^2}{4a^2} + (m-1)\frac{r_0^2}{4r^2} - m\right]$ $\sigma_t = -\frac{3W}{2\pi m t^2}\left[(m+1)\log\frac{a}{r} + (m+1)\frac{r_0^2}{4a^2} - (m-1)\frac{r_0^2}{4r^2} - 1\right]$ $y = -\frac{3W(m^2-1)}{16\pi E m^2 t^3}\left[4a^2 - (8r^2+4r_0^2)\log\frac{a}{r} - \frac{2r^2r_0^2}{a^2} - 4r^2 + 2r_0^2\right]$ At center: $\sigma_r = \sigma_t = -\frac{3W}{2\pi m t^2}\left[(m+1)\log\frac{a}{r_0} + (m+1)\frac{r_0^2}{4a^2}\right] = \max \sigma_r$ when $r_0 < 0.588a$ $\max y = -\frac{3W(m^2-1)}{16\pi E m^2 t^3}\left[4a^2 - 4r_0^2\log\frac{a}{r_0} - 3r_0^2\right]$
Outer edges fixed. Uniform load on concentric circular ring of radius r_0. 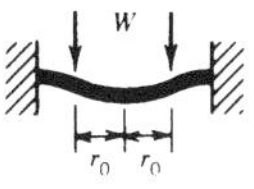 W $r_0 \quad r_0$	At q, $r < r_0$: $\sigma_r = \sigma_t = -\frac{3W}{4\pi m t^2}\left[(m+1)\left(2\log\frac{a}{r_0} + \frac{r_0^2}{a^2} - 1\right)\right] = \max \sigma$ when $r < 0.31a$ $y = -\frac{3W(m^2-1)}{2\pi E m^2 t^3}\left[\frac{1}{2}\left(1+\frac{r_0^2}{a^2}\right)(a^2-r^2) - (r^2+r_0^2)\log\frac{a}{r_0} + (r^2-r_0^2)\right]$ At q, $r > r_0$: $\sigma_r = -\frac{3W}{4\pi m t^2}\left[(m+1)\left(2\log\frac{a}{r} + \frac{r_0^2}{a^2}\right) + (m-1)\frac{r_0^2}{r^2} - 2m\right]$ $\sigma_t = -\frac{3W}{4\pi m t^2}\left[(m+1)\left(2\log\frac{a}{r} + \frac{r_0^2}{a^2}\right) - (m-1)\frac{r_0^2}{r^2} - 2\right]$ $y = -\frac{3W(m^2-1)}{2\pi E m^2 t^3}\left[\frac{1}{2}\left(1+\frac{r_0^2}{a^2}\right)(a^2-r^2) - (r^2+r_0^2)\log\frac{a}{r}\right]$ At center: $\max y = -\frac{3W(m^2-1)}{2\pi E m^2 t^3}\left[\frac{1}{2}(a^2-r_0^2) - r_0^2\log\frac{a}{r_0}\right]$

Table 12 *(Continued)*

Type of Load and Support	Formulas for Stress and Deflection
	Circular Flat Plates with Concentric Circular Hole
Outer edge supported. Uniform load over entire surface. $W = w\pi(a^2 - b^2)$	At inner edge: $\max \sigma = \sigma_t = -\frac{3w}{4mt^2(a^2-b^2)}\left[a^4(3m+1)+b^4(m-1)-4ma^2b^2-4(m+1)a^2b^2\log\frac{a}{b}\right]$ $\max y = -\frac{3w(m^2-1)}{2m^2Et^3}\left[\frac{a^4(5m+1)}{8(m+1)}+\frac{b^4(7m+3)}{8(m+1)}-\frac{a^2b^2(3m+1)}{2(m+1)} + \frac{a^2b^2(3m+1)}{2(m-1)}\log\frac{a}{b}-\frac{2a^2b^4(m+1)}{(a^2-b^2)(m-1)}\left(\log\frac{a}{b}\right)^2\right]$
Outer edge supported. Uniform load along inner edge. W	At inner edge: $\max \sigma = \sigma_t = -\frac{3W}{2\pi mt^2}\left[\frac{2a^2(m+1)}{a^2-b^2}\log\frac{a}{b}+(m-1)\right]$ $\max y = -\frac{3W(m^2-1)}{4\pi Em^2t^3}\left[\frac{(a^2-b^2)(3m+1)}{(m+1)}+\frac{4a^2b^2(m+1)}{(m-1)(a^2-b^2)}\left(\log\frac{a}{b}\right)^2\right]$
Supported along concentric circle near outer edge. Uniform load along concentric circle near inner edge. W, d, c	At inner edge: $\max \sigma = \sigma_t = -\frac{3W}{2\pi mt^2}\left[\frac{2a^2(m+1)}{a^2-b^2}\log\frac{c}{d}+(m-1)\frac{c^2-d^2}{a^2-b^2}\right]$
Inner edge supported. Uniform load over entire surface. $W = w\pi(a^2 - b^2)$	At inner edge: $\max \sigma = \sigma_t = \frac{3w}{4mt^2(a^2-b^2)}\left[4a^4(m+1)\log\frac{a}{b}+4a^2b^2+b^4(m-1)-a^4(m+3)\right]$ At outer edge: $\max y = \frac{3w(m-1)}{16Em^2t^3}\left[a^4(7m+3)+b^4(5m+1)-a^2b^2(12m+4) - \frac{4a^2b^2(3m+1)(m+1)}{(m-1)}\log\frac{a}{b}+\frac{16a^4b^2(m+1)^2}{(a^2-b^2)(m-1)}\left(\log\frac{a}{b}\right)^2\right]$
Outer edge fixed and supported. Uniform load over entire surface. $W = w\pi(a^2 - b^2)$	At outer edge: $\max \sigma_r = \frac{3w}{4t^2}\left[a^2-2b^2+\frac{b^4(m-1)-4b^4(m+1)\log\frac{a}{b}+a^2b^2(m+1)}{a^2(m-1)+b^2(m+1)}\right] = \nu \max \sigma$ At inner edge: $\max \sigma_t = -\frac{3w(m^2-1)}{4mt^2}\left[\frac{a^4-b^4-4a^2b^2\log\frac{a}{b}}{a^2(m-1)+b^2(m+1)}\right]$ $\max y = -\frac{3w(m^2-1)}{16m^2Et^3}\left[a^4+5b^4-6a^2b^2+8b^4\log\frac{a}{b} + \frac{\left\{[-8b^6(m+1)+4a^2b^4(3m+1)+4a^4b^2(m+1)]\log\frac{a}{b}-16a^2b^4(m+1)\left(\log\frac{a}{b}\right)^2 + 4a^2b^4-2a^4b^2(m+1)+2b^6(m-1)\right\}}{a^2(m-1)+b^2(m+1)}\right]$

Table 12 *(Continued)*

Type of Load and Support	Formulas for Stress and Deflection
	Circular Flat Plates with Concentric Circular Hole
Outer edge fixed and supported. Uniform load along inner edge.	At outer edge: $\max \sigma_r = \frac{3W}{2\pi t^2}\left[1 - \frac{2mb^2 - 2b^2(m+1)\log\frac{a}{b}}{a^2(m-1)+b^2(m+1)}\right] = \max \sigma \text{ when } \frac{a}{b} < 2.4$ At inner edge: $\max \sigma_t = \frac{3W}{2\pi m t^2}\left[1 + \frac{ma^2(m-1) - mb^2(m+1) - 2(m^2-1)a^2\log\frac{a}{b}}{a^2(m-1)+b^2(m+1)}\right]$ $= \max \sigma \text{ when } \frac{a}{b} > 2.4$ $\max y = -\frac{3W(m^2-1)}{4\pi m^2 E t^3}$ $\times\left[a^2 - b^2 + \frac{2mb^2(a^2-b^2) - 8ma^2b^2\log\frac{a}{b} + 4a^2b^2(m+1)\left(\log\frac{a}{b}\right)^2}{a^2(m-1)+b^2(m+1)}\right]$
Outer edge fixed. Uniform moment along inner edge.	At inner edge: $\max \sigma_r = \frac{6M}{t^2}$ $\max y = \frac{6M(m^2-1)}{mEt^3}\left[\frac{a^2b^2 - b^4 - 2a^2b^2\log\frac{a}{b}}{a^2(m-1)+b^2(m+1)}\right]$ At outer edge: $\sigma_r = -\frac{6M}{t^2}\left[\frac{2mb^2}{(m+1)b^2+(m-1)a^2}\right]$
Outer edge supported. Unequal uniform moments along edges.	At q: $\sigma_r = \frac{6}{t^2(a^2-b^2)}\left[a^2M_a - b^2M_b - \frac{a^2b^2}{r^2}(M_a - M_b)\right]$ $\sigma_t = \frac{6}{t^2(a^2-b^2)}\left[a^2M_a - b^2M_b + \frac{a^2b^2}{r^2}(M_a - M_b)\right]$ From outer edge level: $y = \frac{12(m^2-1)}{mEt^3(a^2-b^2)}\left[\frac{a^2-r^2}{2}\left(\frac{a^2M_a - b^2M_b}{m+1}\right) + \log\frac{a}{r}\left(\frac{a^2b^2(M_a-M_b)}{m-1}\right)\right]$

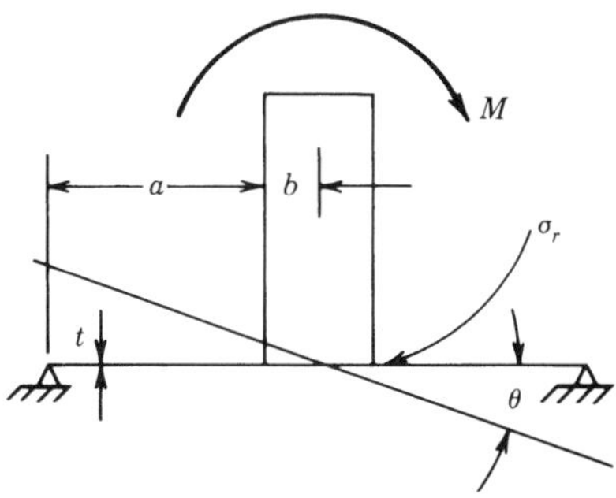

Figure 28 Simply supported trunnion.

$$\sigma_r = \frac{\beta M}{at^2} \tag{112}$$

$$\theta = \frac{\gamma M}{Et^3} \tag{113}$$

$$\left.\begin{aligned} \beta &= 10^{(0.7634-1.252x)} \\ \log \gamma &= 0.248 - \pi x^{1.5} \end{aligned}\right\} 0 < x = \frac{b}{a} < 1 \tag{114}$$

For the fixed-end plate

$$\left.\begin{aligned} \beta &= 10^{(1-1.959x)} \\ \log \gamma &= 0.179 - 3.75x^{1.5} \end{aligned}\right\} 0 < x = \frac{b}{a} < 1 \tag{115}$$

The equations for β, γ are derived from curve fitting of data (see, for example, Refs. 2, 4th ed., and 21).

9.5 Socket Action

In Fig. 29a, summation of moments in the middle of the wall yields

$$2\left[\left(\frac{\omega''}{2}\frac{l}{2}\right)\left(\frac{2}{3}\frac{l}{2}\right)\right] = F\left(a + \frac{l}{2}\right)$$

$$\omega'' = \frac{6}{l^2}\left[F\left(a + \frac{l}{2}\right)\right] \tag{116}$$

Summation of forces in the horizontal gives

$$\omega' = \frac{F}{l} \tag{117}$$

At B, the bearing pressure in Fig. 29c is

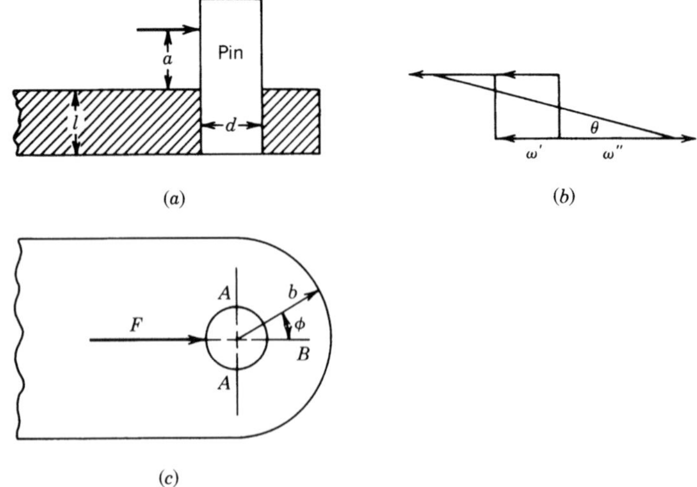

Figure 29 Socket action near an edge.

$$p_i = \frac{\omega' + \omega''}{d} \text{ psi} \tag{118}$$

In Eq. (102) $p_o = 0$ and

$$\sigma_t = \frac{p_i}{R^2 - 1}\left[1 + \left(\frac{b}{d/2}\right)^2\right]$$

At A in Fig. 29c

$$\sigma = \frac{\phi 8F}{\pi^2 bl}$$

where $2b/d = 2$, 4 and $\phi = 4.3$, 4.4;

$$F = (\omega' + \omega'')l$$

If a pin is pressed into the frame hole, σ_t created by p_f [Eq. (104)] must be added. Furthermore, if the pin and frame are different metals, additional σ_t will be created by temperature changes that vary p_f.

The stress in the pin can be found from the maximum moment developed by ω' and ω'' and then calculating the bending stress.

10 CONTACT STRESSES

The stresses caused by the pressure between elastic bodies (Table 13) are of importance in connection with the design or investigation of ball and roller bearings, trunnions, expansion rollers, track stresses, gear teeth, etc.

Contact Stress Theory

H. Hertz[23] developed the mathematical theory for the surface stresses and the deformations produced by pressure between curved bodies, and the results of his analysis are supported by research. Formulas based on this theory give the maximum compressive stresses which occur at the center of the surfaces of contact but do not consider the maximum subsurface shear stresses or the maximum tensile stresses which occur at the boundary of the contact area. In Table 13 formulas are given for the elastic stress and deformation produced by bodies in contact. Numerous tests have been made to determine the bearing strength of balls and rollers, but there is difficulty in interpreting the results for lack of a satisfactory criterion of failure. One arbitrary criterion of failure is the amount of allowable plastic yielding. For further information on contact stresses see Refs. 2, 24, and 25.

11 ROTATING ELEMENTS

11.1 Shafts

The stress[1] in the center of a rotating shaft or solid cylinder is

$$\sigma_r = \sigma_h = \frac{3 - 2\nu}{8(1 - \nu)}\left(\frac{\gamma\omega^2}{g}\right) r_o^2 \tag{119}$$

$$\sigma_z = \frac{\nu\gamma\omega^2}{4g(1 - \nu)} r_o^2 \tag{120}$$

Table 13 Areas of Contact and Pressures with Two Surfaces in Contact

Poisson's ratio = 0.3; P = load, lb; P_1 = load per in. of length, lb; E = modulus of elasticity.

Character of Surfaces	Maximum Pressure, s, at Center of Contact, psi	Radius, r, or Width, b, of Contact Area, in.
Two spheres 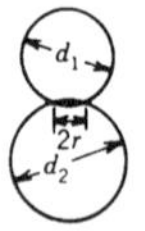	$s = 0.616 \sqrt[3]{PE^2 \left(\frac{d_1 + d_2}{d_1 d_2}\right)^2}$	$r = 0.881 \sqrt[3]{\frac{P}{E}\left(\frac{d_1 d_2}{d_1 + d_2}\right)}$
Sphere and plane 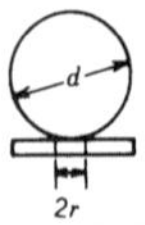	$s = 0.616 \sqrt[3]{\frac{PE^2}{d^2}}$	$r = 0.881 \sqrt[3]{\frac{Pd}{E}}$
Sphere and hollow sphere 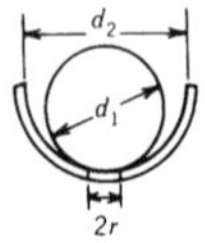	$s = 0.616 \sqrt[3]{PE^2 \left(\frac{d_2 - d_1}{d_1 d_2}\right)^2}$	$r = 0.881 \sqrt[3]{\frac{P}{E}\left(\frac{d_1 d_2}{d_2 - d_1}\right)}$
Cylinder and plane 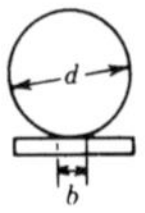	$s = 0.591 \sqrt{\frac{P_1 E}{d}}$	$b = 2.15 \sqrt{\frac{P_1 d}{E}}$
Two cylinders 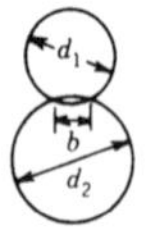	$s = 0.591 \sqrt{P_1 E \left(\frac{d_1 + d_2}{d_1 d_2}\right)}$	$b = 2.15 \sqrt{\frac{P_1}{E}\left(\frac{d_1 d_2}{d_1 + d_2}\right)}$
General case of two bodies in contact 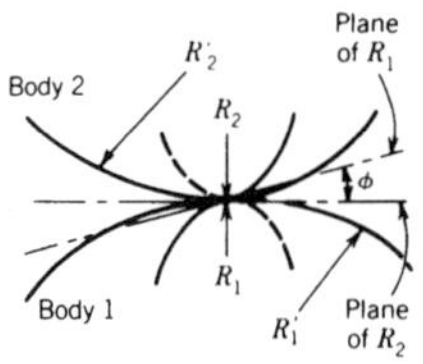	$s = \frac{1.5P}{\pi c d}$	$c = \alpha \sqrt[3]{\frac{P\delta}{K}}$ $d = \beta \sqrt[3]{\frac{P\delta}{K}}$ $\delta = \frac{4}{\frac{1}{R_1} + \frac{1}{R_2} + \frac{1}{R_1'} + \frac{1}{R_2'}}$ $K = \frac{8}{3} \frac{E_1 E_2}{E_2(1-\nu_1^2) + E_1(1-\nu_2^2)}$

$$\theta = \text{arc cos}\ \frac{1}{4}\delta \sqrt{\left\{\left(\frac{1}{R_1} - \frac{1}{R_1'}\right)^2 + \left(\frac{1}{R_2} - \frac{1}{R_2'}\right)^2 + 2\left(\frac{1}{R_1} - \frac{1}{R_1'}\right)\left(\frac{1}{R_2} - \frac{1}{R_2'}\right)\cos 2\phi\right\}}$$

θ	0°	10°	20°	30°	40°	50°	60°	70°	80°	90°
α	∞	6.612	3.778	2.731	2.136	1.754	1.486	1.284	1.128	1.00
β	0	0.319	0.408	0.493	0.567	0.641	0.717	0.802	0.893	1.00

where ν is Poisson's ratio, ω is in rad/sec, γ is the density in lb/in.^{3}, and g is 386 in./sec^2. The limiting ω can be found by using distortion energy; however, most shafts support loads and are limited by critical speeds from torsional or bending modes of vibration. Holzer's method and Dunkerley's equation are used.

11.2 Disks

A rotating disk[1,9,19] of inside radius a and outside radius b has $\sigma_r = 0$ at a and b, while σ_t is

$$\sigma_{ta} = \frac{3 + \nu}{4g} \gamma\omega^2 \left(b^2 + \frac{1 - \nu}{3 + \nu} a^2 \right) \tag{121}$$

$$\sigma_{tb} = \frac{3 + \nu}{4g} \gamma\omega^2 \left(a^2 + \frac{1 - \nu}{3 + \nu} b^2 \right) \tag{122}$$

Substitution in Eq. (105) gives the outside and inside radial expansions.

The solid disk of radius b has stresses at the center,

$$\sigma_t = \sigma_r = \frac{3 + \nu}{8g} \gamma\omega^2 b^2 \tag{123}$$

Substitution into the distortion energy [Eq. (22)] can give one the limiting speed.

11.3 Blades

Blades attached to a rotating shaft will experience a tensile force at the attachment to the shaft. These can be found from dynamics of machinery texts; however, the forces developed from a fluid driven by the blades develop more problems. The blades, if not in the plane, will develop additional forces and moments from the driving force plus vibration of the blades on the shaft.

12 DESIGN SOLUTION SOURCES AND GUIDELINES

Designs are composed of simple elements, as discussed here. These elements are subjected to temperature extremes, vibrations, and environmental effects that cause them to creep, buckle, yield, and corrode. Finding solutions to model these cases, as elements, can be difficult and when found the solutions are complex to follow, let alone to calculate. See Refs. 2, 21, and 26–32 for the known solutions. Always cross-check with another reference. The handbooks of Roark and Young[2] and Blevins[26] have been computerized using a TK solver and are distributed by UTS software. These closed-form solutions would ease some of the more complicated calculations and checks of finite-element solutions using a computer.

12.1 Computers

Most computer set-ups use linear elastic solutions where the analyst supplies mechanical properties of materials such as yield and ultimate strengths and cross-sectional properties like area and area moments of inertia. When solving more complex problems, some concerns to keep in mind:

Questions to Be Asked

1. Will I know if this model buckles?
2. Can one use a nonlinear stress–strain curve?
3. Is there any provision for creep and buckling?
4. How large and complex a structure can be solved? Look at a solved problem and relate it to future problems.

Things to Watch and Note

1. Press fit joints, flanges, pins, bolts, welds and bonds, and any connection interface present in modeling problems. The stress analysis of a single loaded weld is not a simple task. The stress solution for a trunnion with more complexity, such as seal grooves in the plate, requires many small finite elements to converge to a closed-form solution [Eq. (112)].
2. Vibration solutions with connection interfaces can give frequency solutions with 50% error with many connections and still have 10–15% error with no connections. The computer solution appears to be always on the high side.
3. Detailed fatigue stresses on elements can be derived out of the loads by printing out the force variation.
4. The materials have good operating range[33] and limitations for spring stress relaxation at higher temperatures or lower limits can be applicable to structural members.

Nickel alloys, Inconels, and similar materials	$-300°F \le T \le 1020°F$
300, 400, 17-4, 17-7 stainless or austenitic, martensite, and precipitation-hardening stainless steels	$-110°F \le T \le 570°F$
Spring steels	$-5°F \le T \le 430°F$
Patented cold-drawn carbon steels	$-110°F \le T \le 300°F$
Copper beryllium	$-330°F \le T \le 260°F$
Titanium alloys	
Bronzes	$-40°F \le T \le 175°F$
Aluminum	$-300°F \le T \le 400°F$
Magnesium	$-300°F \le T \le 350°F$

The high temperatures are for the onset of creep and stress relaxation and lower mechanical properties with higher temperature. The low temperatures show higher mechanical properties but are shock sensitive. Always examine for the mechanical properties for the temperature range and thermal expansion.[34–37] The mechanical properties at room temperature have predictable distributions with ample sample sizes, but if the temperature is varied, similar published results are not readily available.

Rubber, plastics, and elastomers have glassy transition temperatures below which the material is putty like and above which the material is rock like and brittle. All material mechanical properties vary a great deal due to temperature. This makes computer solutions much more complex. Testing is the final reliable check.

12.2 Testing

Most designs must pass some sets of vibration, environmental, and screen testing before delivery to a customer. It is at this time that design flaws show up and frequencies, stresses, and so on are verified. Some preliminary testing might help:

1. Compare impact hammer frequency test of part of or an entire system to the computer and hand calculations. The physical testing includes the boundary values sometimes difficult to simulate on a computer.
2. Spot bond optical parts to dissimilar-metal structural frame, which must be hot and cold soak tested to see if the bonding fractures the optical parts. Computers cannot predict a failure of this type well.
3. Check test joints and seal surfaces with pressure-sensitive gaskets to see if the developed pressures are sufficient to maintain the design to proper requirements. Then use operational testing to check for thermal warping of these critical surfaces.
4. Pressurize or load braze, weld, or solder part to check the process and its calculations for the pressures and loads.
5. Rapid Prototyping.[38] This method could be used to check a photoelastic model by vibrating it or freezing stresses in the model from static loads. It also could define areas of high stress for a smaller grid finite-element modeling. Stress coating on a regular plastic model could also point out areas of high stress.

REFERENCES

1. J. H. Faupel and F. E. Fisher, *Engineering Design,* 2nd ed., Wiley, New York, 1981.
2. R. J. Roark and W. C. Young, *Formulas for Stress and Strain,* 6th ed., McGraw-Hill, New York, 1989.
3. R. E. Peterson, *Stress Concentration Factors,* 2nd ed., Wiley, New York, 1974.
4. J. Marin, *Mechanical Properties of Materials and Design,* McGraw-Hill, New York, 1942.
5. *Aluminum Standards and Data,* 3rd ed., Aluminum Association, New York, 1972.
6. *AISC Handbook,* American Institute of Steel Construction, New York, 1989.
7. F. B. Seely and J. O. Smith, *Resistance of Materials,* 4th ed., Wiley, New York, 1957.
8. A. P. Boresi, O. Sidebottom, F. B. Seely, and J. O. Smith, *Advanced Mechanics of Materials,* 3rd ed., Wiley, New York, 1978.
9. S. P. Timoshenko, *Strength of Materials,* 3rd ed., Vols. I and II, Krieger, Melbourne, FL, 1958.
10. (a) H. C. Mann, "Relation of Impact and Tension Testing of Steels," *Metal Progress,* **27**(3), 36–41 (March 1935).
(b) H. C. Mann, "Relation between Tension Static and Dynamic Tests," *Proc. ASTM,* **35**(II), 323–340 (1935).
(c) H. C. Mann, "Fundamental Study of Design of Impact Test Specimens," *Proc. ASTM,* **37**(II), 102–118 (1937).
11. C. R. Young, *Bulletin 4,* School of Engineering Research, University of Toronto, 1923.
12. C. R. Bach and R. Baumann, *Elastizität und Festigkeit,* 9th ed., J. Springer, Berlin, 1924.
13. R. M. Rivello, *Theory Analysis of Flight Structures,* McGraw-Hill, New York, 1969.
14. B. G. Johnston (ed.), *Structural Research Council, Stability Design Criteria for Metal Structures,* 3rd ed., Wiley, New York, 1976.
15. *Trans. Am. Soc. Civil Eng.,* **xcviii** (1933).
16. S. P. Timoshenko and J. M. Gere, *Theory of Elastic Stability,* 2nd ed., McGraw-Hill, New York, 1961.
17. *Trans. Am. Soc. Civil Eng.,* **lxxxiii** (1919–20).
18. R. C. Juvinall, *Stress, Strain and Strength,* McGraw-Hill, New York, 1967.
19. S. P. Timoshenko and J. N. Goodier, *Theory of Elastic Stability,* 3rd ed., McGraw-Hill, New York, 1970.
20. M. Hetényi, *Handbook of Experimental Stress Analysis,* Wiley, New York, 1950.
21. W. Griffel, *Handbook of Formulas for Stress and Strain,* Frederick Ungar, New York, 1966.
22. R. J. Roark, *Formulas for Stress and Strain,* 2nd ed., McGraw-Hill, New York, 1943.

23. H. Hertz, *Gesammelte Werke,* Vol. 1, J. A. Berth, Leipzig, 1895.

24. R. K. Allen, *Rolling Bearings,* Pitman and Sons, London, 1945.

25. A. Palmgren, *Ball and Roller Bearing Engineering,* SKF Industries, Philadelphia, PA, 1945.

26. R. D. Blevins, *Formulas for Natural Frequency and Mode Shapes,* Krieger, Melbourne, FL, 1993.

27. R. D. Blevins, *Flow-Induced Vibration,* 2nd ed., Krieger, Melbourne, FL, 1994.

28. W. Flügge (ed.), *Handbook of Engineering Mechanics,* 1st ed., McGraw-Hill, New York, 1962.

29. A. W. Leissa, *Vibration of Plates NASA SP-160 (N70-18461) NTIS,* Springfield, VA.

30. A. W. Leissa, *Vibration of Shells NASA SP-288 (N73-26924) NTIS,* Springfield, VA.

31. A. Kleinlogel, *Rigid Frame Formulas,* 12th ed., Frederick Ungar, New York, 1958.

32. V. Leontovich, *Frames and Arches,* McGraw-Hill, New York, 1959.

33. M. O'Malley, "The Effect of Extreme Temperature on Spring Performance," *Springs* (May 1986), p. 19.

34. *Metallic Materials for Aerospace Structures,* 2 vols., Mil HDBK 5F, Department of Defense, 1990.

35. *Aerospace Structural Metals Handbook,* 5 vols., CINDAS/USAF CRDA, Purdue University, West Lafayette, IN, 1993.

36. *Structural Alloys Handbook,* 3 vols., CINDAS, Purdue University, West Lafayette, IN, 1993.

37. *Thermophysical Properties of Matter,* Vol. 12, *Metallic Expansion,* 1995; Vol. 13, *Non-Metallic Thermoexpansion,* 1977, IFI/Phenium, New York.

38. S. Ashley, "Rapid Prototyping Is Coming of Age," *Mechan. Eng.* (July 1995) pp. 62–69.

BIBLIOGRAPHY

Almen, J. O., and P. H. Black, *Residual Stresses and Fatigue in Metals,* McGraw-Hill, New York, 1963.

Di Giovanni, M., *Flat and Corrugated Diaphragm Design Handbook,* Marcel Dekker, New York, 1982.

Osgood, W. R. (ed.), *Residual Stresses in Metals and Metal Construction,* Reinhold, New York, 1954.

Proceedings of the Society for Experimental Stress Analysis.

Symposium on Internal Stresses in Metals and Alloys, Institute of Metals, London, 1948.

Vande Walle, L. J., *Residual Stress for Designers and Metallurgists,* 1980 American Society for Metals Conference, American Society for Metals, Metals Park, OH, 1981.

CHAPTER 16

AN INTRODUCTION TO THE FINITE-ELEMENT METHOD

Tarek I. Zohdi
Mechanical Engineering Department
University of California
Berkeley, California

The finite-element method (FEM), which has become a dominant numerical method in mathematical physics, applied mathematics, and engineering analysis, is a huge field of study. This chapter is intended to provide a concise review of basic field equations in solid mechanics and then illustrate how the FEM can be employed to solve them. The implementation, theory, and application of FEM is a subject of immense literature. We will not attempt to review this huge field.

To motivate the use of the FEM, we derive the classical equations of mechanical equilibrium. Throughout this analysis, boldface symbols imply vectors or tensors. The inner product of two vectors $\mathbf{u}$ and $\mathbf{v}$ is denoted $\mathbf{u}\cdot\mathbf{v}$. At the risk of oversimplification, we ignore the distinction between second-order tensors and matrices. Furthermore, we exclusively employ a Cartesian basis. Readers may consult the texts of Malvern[1] and Marsden and Hughes[2] for more background information. Hence, if we consider the second-order tensor $\mathbf{A}$ with its matrix representation

$$[\mathbf{A}] \overset{\text{def}}{=} \begin{bmatrix} A_{11} & A_{12} & A_{13} \\ A_{21} & A_{22} & A_{23} \\ A_{31} & A_{32} & A_{33} \end{bmatrix} \tag{1}$$

then the product of two second-order tensors $\mathbf{A}\cdot\mathbf{B}$ is defined by the matrix product $[\mathbf{A}][\mathbf{B}]$, with components of $A_{ij}B_{jk} = C_{ik}$. The second-order inner product of two tensors or matrices is $\mathbf{A} : \mathbf{B} = A_{ij}B_{ij} = \mathrm{tr}([\mathbf{A}]^T [\mathbf{B}])$. Finally, the divergence of a vector $\mathbf{u}$ is defined by $\nabla \cdot \mathbf{u} = u_{i,i}$; whereas for a second-order tensor $\mathbf{A}$, $\nabla \cdot \mathbf{A}$ describes a contraction to a vector with components $A_{ij,j}$.

1 DEFORMATION OF A SOLID

The term *deformation* refers to a change in the shape of the continuum between a reference configuration and current configuration. In the reference configuration, a representative particle of the continuum occupies a point **P** in space and has the position vector $\mathbf{X} = X_1\mathbf{e}_1 + X_2\mathbf{e}_2 + X_3\mathbf{e}_3$ (Fig. 1), where $(\mathbf{e}_1, \mathbf{e}_2, \mathbf{e}_3)$ is a Cartesian reference triad and X_1, X_2, X_3 (with center **O**) can be thought of as labels for the point. Sometimes the coordinates or labels (X_1, X_2, X_3, t) are called the referential coordinates. In the current configuration the particle originally located at point **P** is located at point **P′** and can be expressed also in terms of another position vector **x**, with the coordinates (x_1, x_2, x_3, t). These are called the current coordinates. It is obvious with this arrangement that the displacement is $\mathbf{u} = \mathbf{x} - \mathbf{X}$ for a point originally at **X** and with final coordinates **x**. When a continuum undergoes deformation (or flow), its points move along various paths in space. This motion may be expressed by $\mathbf{x}(X_1, X_2, X_3, t) = \mathbf{u}(X_1, X_2, X_3, t) + \mathbf{X}(X_1, X_2, X_3, t)$, which gives the present location of a point that occupied the position (X_1, X_2, X_3, t) at time $t = t_0$, written in terms of the labels X_1, X_2, X_3. The previous position vector may be interpreted as a mapping of the initial configuration onto the current configuration. In classical approaches, it is assumed that such a mapping is one to one and continuous, with continuous partial derivatives to whatever order that is required. The description of motion or deformation expressed previously is known as the Lagrangian formulation. Alternatively, if the independent variables are the coordinates **x** and t, then $\mathbf{x}(x_1, x_2, x_3, t) = \mathbf{u}(x_1, x_2, x_3, t) + \mathbf{X}(x_1, x_2, x_3, t)$, and the formulation is denoted as Eulerian (Fig. 1).

Partial differentiation of the displacement vector $\mathbf{u} = \mathbf{x} - \mathbf{X}$ produces the following displacement gradients: $\nabla_X \mathbf{u} = \mathbf{F} - \mathbf{1}$ and $\nabla_x \mathbf{u} = \mathbf{1} - \overline{\mathbf{F}}$, where

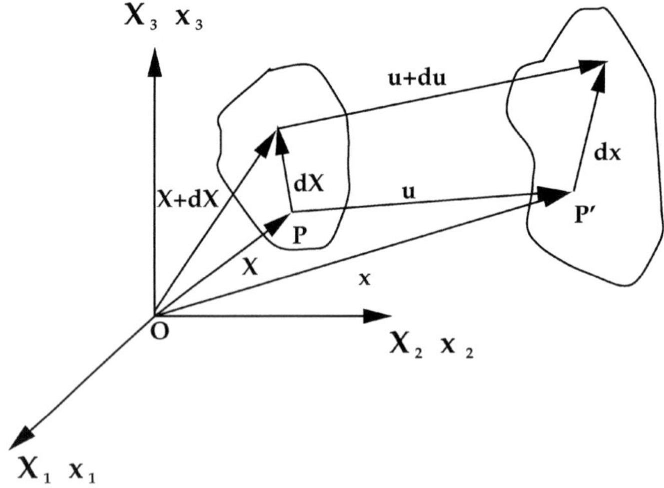

Figure 1 Different descriptions of a deforming body.

$$\nabla_{\mathbf{X}}\mathbf{x} \stackrel{\text{def}}{=} \frac{\partial \mathbf{x}}{\partial \mathbf{X}} = \mathbf{F} \stackrel{\text{def}}{=} \begin{bmatrix} \frac{\partial x_1}{\partial X_1} & \frac{\partial x_1}{\partial X_2} & \frac{\partial x_1}{\partial X_3} \\ \frac{\partial x_2}{\partial X_1} & \frac{\partial x_2}{\partial X_2} & \frac{\partial x_2}{\partial X_3} \\ \frac{\partial x_3}{\partial X_1} & \frac{\partial x_3}{\partial X_2} & \frac{\partial x_3}{\partial X_3} \end{bmatrix} \tag{2}$$

$$\nabla_{\mathbf{x}}\mathbf{X} \stackrel{\text{def}}{=} \frac{\partial \mathbf{X}}{\partial \mathbf{x}} = \overline{\mathbf{F}} \stackrel{\text{def}}{=} \begin{bmatrix} \frac{\partial X_1}{\partial x_1} & \frac{\partial X_1}{\partial x_2} & \frac{\partial X_1}{\partial x_3} \\ \frac{\partial X_2}{\partial x_1} & \frac{\partial X_2}{\partial x_2} & \frac{\partial X_2}{\partial x_3} \\ \frac{\partial X_3}{\partial x_1} & \frac{\partial X_3}{\partial x_2} & \frac{\partial X_3}{\partial x_3} \end{bmatrix} \tag{3}$$

where $\mathbf{F}$ is known as the material deformation gradient and $\overline{\mathbf{F}}$ is known as the spatial deformation gradient. Now, consider the length of a differential element in the reference configuration $d\mathbf{X}$ and $d\mathbf{x}$ in the current configuration, $d\mathbf{x} = \nabla_X\mathbf{x}\cdot d\mathbf{X} = \mathbf{F}\cdot d\mathbf{X}$. Taking the difference in the magnitudes of these elements yields

$$\begin{aligned} d\mathbf{x}\cdot d\mathbf{x} - d\mathbf{X}\cdot d\mathbf{X} &= (\nabla_{\mathbf{X}}\mathbf{x}\cdot d\mathbf{X})\cdot(\nabla_X\mathbf{x}\cdot d\mathbf{X}) - d\mathbf{X}\cdot d\mathbf{X} \\ &= d\mathbf{X}\cdot(\mathbf{F}^{\mathrm{T}}\cdot\mathbf{F} - \mathbf{1})\cdot d\mathbf{X} \stackrel{\text{def}}{=} 2\, d\mathbf{X}\cdot\mathbf{E}\cdot d\mathbf{X} \end{aligned} \tag{4}$$

Alternatively, we have with $d\mathbf{X} = \nabla_{\mathbf{x}}\mathbf{X}\cdot d\mathbf{x} = \overline{\mathbf{F}}\cdot d\mathbf{x}$ and

$$\begin{aligned} d\mathbf{x}\cdot d\mathbf{x} - d\mathbf{X}\cdot d\mathbf{X} &= d\mathbf{x}\cdot d\mathbf{x} - (\nabla_{\mathbf{x}}\mathbf{X}\cdot d\mathbf{x})\cdot(\nabla_{\mathbf{x}}\mathbf{X}\cdot d\mathbf{x}) \\ &= d\mathbf{x}\cdot(\mathbf{1} - \overline{\mathbf{F}}^{\mathrm{T}}\cdot\overline{\mathbf{F}})\cdot d\mathbf{x} \stackrel{\text{def}}{=} 2\, d\mathbf{x}\cdot\mathbf{L}\cdot d\mathbf{x} \end{aligned} \tag{5}$$

Therefore we have the so-called *Lagrangian* strain tensor

$$\mathbf{E} \stackrel{\text{def}}{=} \tfrac{1}{2}(\mathbf{F}^{\mathrm{T}}\cdot\mathbf{F} - \mathbf{1}) = \tfrac{1}{2}[\nabla_{\mathbf{X}}\mathbf{u} + (\nabla_{\mathbf{X}}\mathbf{u})^{\mathrm{T}} + (\nabla_{\mathbf{X}}\mathbf{u})^{\mathrm{T}}\cdot\nabla_{\mathbf{X}}\mathbf{u}] \tag{6}$$

Frequently, the Lagrangian strain tensor is defined in terms of the so-called right Cauchy–Green strain, $\mathbf{E} \stackrel{\text{def}}{=} \frac{1}{2}(\mathbf{C} - 1)$, $\mathbf{C} = \mathbf{F}^{\mathrm{T}}\cdot\mathbf{F}$. The Eulerian strain tensor is defined as

$$\mathbf{L} \stackrel{\text{def}}{=} \tfrac{1}{2}(\mathbf{1} - \overline{\mathbf{F}}^{\mathrm{T}}\cdot\overline{\mathbf{F}}) = \tfrac{1}{2}[\nabla_{\mathbf{x}}\mathbf{u} + (\nabla_{\mathbf{x}}\mathbf{u})^{\mathrm{T}} - (\nabla_{\mathbf{x}}\mathbf{u})^{\mathrm{T}}\cdot\nabla_{\mathbf{x}}\mathbf{u}] \tag{7}$$

In a similar manner as for the Lagrangian strain tensor, the Eulerian strain tensor can be defined in terms of the so-called left Cauchy–Green strain, $\mathbf{L} \stackrel{\text{def}}{=} \frac{1}{2}(\mathbf{1} - \mathbf{b}^{-1})$, $\mathbf{b} = \mathbf{F}\cdot\mathbf{F}^{\mathrm{T}}$.

REMARK. It should be clear that $d\mathbf{x}$ can be reinterpreted as the result of a mapping $\mathbf{F}\cdot d\mathbf{X} \rightarrow d\mathbf{x}$, or a change in configuration (reference to current), while $\overline{\mathbf{F}}\cdot d\mathbf{x} \rightarrow d\mathbf{X}$ maps the current to the reference system. For the deformations to be invertible and physically realizable, $\overline{\mathbf{F}}\cdot(\mathbf{F}\cdot d\mathbf{X}) = d\mathbf{X}$ and $\mathbf{F}\cdot(\overline{\mathbf{F}}\cdot d\mathbf{x}) = d\mathbf{x}$. We note that $(\det \overline{\mathbf{F}})(\det \mathbf{F}) = 1$ and the following obvious relation $(\partial\mathbf{X}/\partial\mathbf{x})\cdot(\partial\mathbf{x}/\partial\mathbf{X}) = \overline{\mathbf{F}}\cdot\mathbf{F} = \mathbf{1}$. It should be clear that $\overline{\mathbf{F}} = \mathbf{F}^{-1}$.

We now employ infinitesimal deformation theory. In infinitesimal deformation theory, the displacement gradient components being "small" implies that higher order terms like $(\nabla_{\mathbf{X}}\mathbf{u})^{\mathrm{T}}\cdot\nabla_{\mathbf{X}}\mathbf{u}$ and $(\nabla_{\mathbf{x}}\mathbf{u})^{\mathrm{T}}\cdot\nabla_{\mathbf{x}}\mathbf{u}$ can be neglected in the strain measure $\mathbf{L} = \frac{1}{2}[\nabla_{\mathbf{x}}\mathbf{u} + (\nabla_{\mathbf{x}}\mathbf{u})^{\mathrm{T}} - (\nabla_{\mathbf{x}}\mathbf{u})^{\mathrm{T}}\cdot\nabla_{\mathbf{x}}\mathbf{u}]$ and $\mathbf{E} = \frac{1}{2}(\nabla_{\mathbf{X}}\mathbf{u} + (\nabla_{\mathbf{X}}\mathbf{u})^{\mathrm{T}} + (\nabla_{\mathbf{X}}\mathbf{u})^{\mathrm{T}}\cdot\nabla_{\mathbf{X}}\mathbf{u}]$, leading to $\mathbf{L} \approx \boldsymbol{\epsilon}^{\mathrm{E}} \stackrel{\text{def}}{=} \frac{1}{2}[\nabla_{\mathbf{x}}\mathbf{u} + (\nabla_{\mathbf{x}}\mathbf{u})^{\mathrm{T}}]$ and $\mathbf{E} \approx \boldsymbol{\epsilon}^{\mathrm{L}} \stackrel{\text{def}}{=} \frac{1}{2}[\nabla_{\mathbf{X}}\mathbf{u} + (\nabla_{\mathbf{X}}\mathbf{u})^{\mathrm{T}}]$. If the displacement gradients are small compared with unity, $\boldsymbol{\epsilon}^{\mathrm{E}}$ and $\boldsymbol{\epsilon}^{\mathrm{L}}$ coincide closely to $\mathbf{L}$ and $\mathbf{E}$, respectively. If we assume that $\partial/\partial X \approx$

$\partial/\partial x$, we may use $\boldsymbol{\epsilon}^{\mathbf{E}}$ or $\boldsymbol{\epsilon}^{\mathbf{L}}$ interchangeably. Usually $\boldsymbol{\epsilon}$ is the symbol used for infinitesimal strains. Furthermore, to avoid confusion, when using models employing the geometrically linear infinitesimal strain assumption, we use the symbol of $\boldsymbol{\nabla}$ with no $\mathbf{X}$ or $\mathbf{x}$ subscript.

REMARK. The Jacobian of the deformation gradient, $\mathbf{F}$, is defined as

$$J \stackrel{\text{def}}{=} \det \mathbf{F} = \begin{vmatrix} \frac{\partial x_1}{\partial X_1} & \frac{\partial x_1}{\partial X_2} & \frac{\partial x_1}{\partial X_3} \\ \frac{\partial x_2}{\partial X_1} & \frac{\partial x_2}{\partial X_2} & \frac{\partial x_2}{\partial X_3} \\ \frac{\partial x_3}{\partial X_1} & \frac{\partial x_3}{\partial X_2} & \frac{\partial x_3}{\partial X_3} \end{vmatrix} \tag{8}$$

To interpret the Jacobian in a physical way, consider a reference differential volume given by $dS^3 = d\omega$, where $d\mathbf{X}^{(1)} = dS\mathbf{e}_1$, $d\mathbf{X}^{(2)} = dS\mathbf{e}_2$, and $d\mathbf{X}^{(3)} = dS\mathbf{e}_3$. The current differential element is described by $d\mathbf{x}^{(1)} = (\partial x_k/\partial X_1)\, dS\mathbf{e}_k$, $d\mathbf{x}^{(2)} = (\partial x_k/\partial X_2)\, dS\mathbf{e}_k$, and $d\mathbf{x}^{(3)} = (\partial x_k/\partial X_3)\, dS\mathbf{e}_k$, where $\mathbf{i}$ is a unit vector and

$$\underbrace{d\mathbf{x}^{(1)} \cdot (d\mathbf{x}^{(2)} \times d\mathbf{x}^{(3)})}_{\stackrel{\text{def}}{=} d\omega} = \begin{vmatrix} dx_1^{(1)} & dx_2^{(1)} & dx_3^{(1)} \\ dx_1^{(2)} & dx_2^{(2)} & dx_3^{(2)} \\ dx_1^{(3)} & dx_2^{(3)} & dx_3^{(3)} \end{vmatrix} = \begin{vmatrix} \frac{\partial x_1}{\partial X_1} & \frac{\partial x_2}{\partial X_1} & \frac{\partial x_3}{\partial X_1} \\ \frac{\partial x_1}{\partial X_2} & \frac{\partial x_2}{\partial X_2} & \frac{\partial x_3}{\partial X_2} \\ \frac{\partial x_1}{\partial X_3} & \frac{\partial x_2}{\partial X_3} & \frac{\partial x_3}{\partial X_3} \end{vmatrix} dS^3 \tag{9}$$

Therefore, $d\omega = J\, d\omega_0$. Thus, the Jacobian of the deformation gradient must remain positive definite; otherwise we obtain physically impossible "negative" volumes.

2 EQUILIBRIUM

We start with the following postulated balance law for an arbitrary part of a body Ω, ω, around a point P, with boundary $\partial\omega$:

$$\underbrace{\int_{\partial\omega} \mathbf{t}\, da}_{\text{surface forces}} + \underbrace{\int_{\omega} \mathbf{f}\, d\omega}_{\text{body forces}} = \underbrace{\frac{d}{dt}\int_{\omega} \rho\dot{\mathbf{u}}\, d\omega}_{\text{inertial forces}} \tag{10}$$

where ρ is the material density, $\mathbf{b}$ is the body force per unit mass ($\mathbf{f} = \rho\mathbf{b}$), and $\dot{\mathbf{u}}$ is the time derivative of the displacement. When the actual molecular structure is considered on a submicroscopic scale, the force densities, $\mathbf{t}$, which we commonly refer to as "surface forces," are taken to involve short-range intermolecular forces. Tacitly we assume that the effects of radiative forces, and others which do not require momentum transfer through a continuum, are negligible. This is a so-called local action postulate. As long as the volume element is large, our resultant body and surface forces may be interpreted as sums of these intermolecular forces. When we pass to larger scales, we can justifiably use the continuum concept.

Now consider a tetrahedron in equilibrium, as shown in Fig. 2. From Newton's laws, $\mathbf{t}^{(n)}\, \Delta A^{(n)} + \mathbf{t}^{(1)}\, \Delta A^{(1)} + \mathbf{t}^{(2)}\, \Delta A^{(2)} + \mathbf{t}^{(3)}\, \Delta A^{(3)} + \mathbf{f}\, \Delta\Omega = \rho\, \Delta\Omega\, \ddot{\mathbf{u}}$, where $\Delta A^{(n)}$ is the surface area of the face of the tetrahedron with normal $\mathbf{n}$ and $\Delta\Omega$ is the tetrahedron volume. Clearly, as the distance between the tetrahedron base [located at (0,0,0)] and the surface center, denoted h, goes to zero, we have $h \rightarrow 0 \Rightarrow \Delta A^{(n)} \rightarrow 0 \Rightarrow \Delta\Omega/\Delta A^{(n)} \rightarrow 0$. Geometrically, we have $\Delta A^{(i)}/\Delta A^{(n)} = \cos(x_i, x_n) \stackrel{\text{def}}{=} -n_i$, and therefore $\mathbf{t}^{(n)} + \mathbf{t}^{(1)} \cos(\mathbf{x}_1, \mathbf{x}_n) + \mathbf{t}^{(2)} \cos(\mathbf{x}_2, \mathbf{x}_n) +$

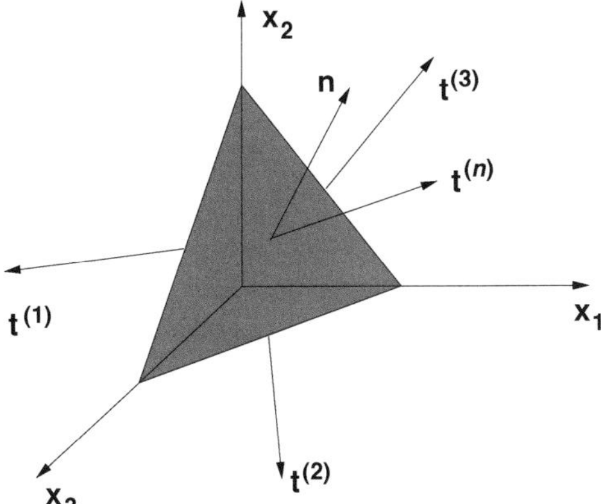

Figure 2 Cauchy tetrahedron: a "sectioned material point."

$\mathbf{t}^{(3)} \cos(\mathbf{x}_3, \mathbf{x}_n) = \mathbf{0}$. It is clear that forces on the surface areas could be decomposed into three linearly independent components. It is convenient to express the concept of stress at a point, representing the surface forces there, pictorially represented by a cube surrounding a point. The fundamental issue that must be resolved is the characterization of these surface forces. We can represent the force density vector, the so-called traction, on a surface by the component representation: $\mathbf{t}^{(i)} \stackrel{\text{def}}{=} (\boldsymbol{\sigma}_{i1}, \boldsymbol{\sigma}_{i2}, \boldsymbol{\sigma}_{i3})^{\mathrm{T}}$, where the second index represents the direction of the component and the first index represents the normal to the corresponding coordinate plane. From this point forth, we will drop the superscript notation of $\mathbf{t}^{(n)}$, where it is implicit that $\mathbf{t} \stackrel{\text{def}}{=} \mathbf{t}^{(n)} = \boldsymbol{\sigma}^{\mathrm{T}}\cdot\mathbf{n}$ or explicitly

$$\mathbf{t}^{(n)} = \mathbf{t}^{(1)}n_1 + \mathbf{t}^{(2)}n_2 + \mathbf{t}^{(3)}n_3 = \boldsymbol{\sigma}^{\mathrm{T}}\cdot\mathbf{n} = \begin{bmatrix} \boldsymbol{\sigma}_{11} & \boldsymbol{\sigma}_{12} & \boldsymbol{\sigma}_{13} \\ \boldsymbol{\sigma}_{21} & \boldsymbol{\sigma}_{22} & \boldsymbol{\sigma}_{23} \\ \boldsymbol{\sigma}_{31} & \boldsymbol{\sigma}_{32} & \boldsymbol{\sigma}_{33} \end{bmatrix}^{\mathrm{T}} \begin{bmatrix} n_1 \\ n_2 \\ n_3 \end{bmatrix} \tag{11}$$

where $\boldsymbol{\sigma}$ is the so-called Cauchy stress tensor.*

Substitution of Eq. (11) into Eq. (10) yields ($\omega \subset \Omega$)

$$\underbrace{\int_{\partial\omega} \boldsymbol{\sigma}\cdot\mathbf{n}\, da}_{\text{surface forces}} + \underbrace{\int_{\omega} \mathbf{f}\, d\omega}_{\text{body forces}} = \underbrace{\frac{d}{dt}\int_{\omega} \rho\dot{\mathbf{u}}\, d\omega}_{\text{inertial forces}} \tag{12}$$

A relationship can be determined between the densities in the current and reference configurations, $\int_{\omega} \rho\, d\omega = \int_{\omega_0} \rho J\, d\omega = \int_{\omega_0} \rho_0\, d\omega$. Therefore, the Jacobian can also be interpreted as the ratio of material densities at a point. Since the volume is arbitrary, we can assume that $\rho J = \rho_0$ holds at every point in the body. Therefore, we may write

$$\frac{d}{dt}(\rho_0) = \frac{d}{dt}(\rho J) = 0$$

when the system is mass conservative over time. This leads to writing the last term in Eq. (12) as

---

*Some authors follow the notation of the first index representing the direction of the component and the second index representing the normal to corresponding coordinate plane. This leads to $\mathbf{t} \stackrel{\text{def}}{=} \mathbf{t}^{(n)} = \boldsymbol{\sigma}\cdot\mathbf{n}$. In the absence of couple stresses, a balance of angular momentum implies a symmetry of stress, $\boldsymbol{\sigma} = \boldsymbol{\sigma}^{\mathrm{T}}$, and thus the difference in notations becomes immaterial.

$$\frac{d}{dt}\int_{\omega} \rho\dot{\mathbf{u}}\, d\omega = \int_{\omega_0} \frac{d(\rho J)}{dt}\dot{\mathbf{u}}\, d\omega_0 + \int_{\omega} \rho\ddot{\mathbf{u}}\, d\omega = \int_{\omega} \rho\ddot{\mathbf{u}}\, d\omega$$

From Gauss's divergence theorem and an implicit assumption that $\boldsymbol{\sigma}$ is differentiable, we have $\int_{\omega} (\nabla_{\mathbf{x}}\cdot\boldsymbol{\sigma} + \mathbf{f} - \rho\ddot{\mathbf{u}})\, d\omega = 0$. If the volume is argued as being arbitrary, then the relation in the integral must hold pointwise, yielding

$$\nabla_{\mathbf{x}}\cdot\boldsymbol{\sigma} + \mathbf{f} = \rho\ddot{\mathbf{u}} \tag{13}$$

Note: Invoking an angular momentum balance, under the assumptions that no infinitesimal "micromoments" or so-called couple stresses exist, it can be shown that the stress tensor must be symmetric,[1] i.e.,

$$\int_{\partial\omega} \mathbf{x}\times\mathbf{t}\, da + \int_{\omega} \mathbf{x}\times\mathbf{f}\, d\omega = \frac{\mathrm{d}}{\mathrm{dt}}\int_{\omega} \mathbf{x}\cdot\rho\dot{\mathbf{u}}\, d\omega$$

which implies $\boldsymbol{\sigma}^{\mathrm{T}} = \boldsymbol{\sigma}$. It is somewhat easier to consider a differential element, such as in Fig. 3, and to simply sum moments about the center. Doing this one immediately obtains $\boldsymbol{\sigma}_{12} = \boldsymbol{\sigma}_{21}$, $\boldsymbol{\sigma}_{23} = \boldsymbol{\sigma}_{32}$, and $\boldsymbol{\sigma}_{13} = \boldsymbol{\sigma}_{31}$. Therefore

$$\mathbf{t}^{(n)} = \mathbf{t}^{(1)}n_1 + \mathbf{t}^{(2)}n_2 + \mathbf{t}^{(3)}n_3 = \boldsymbol{\sigma}\cdot\mathbf{n} = \begin{bmatrix} \sigma_{11} & \sigma_{12} & \sigma_{13} \\ \sigma_{21} & \sigma_{22} & \sigma_{23} \\ \sigma_{31} & \sigma_{32} & \sigma_{33} \end{bmatrix}\begin{bmatrix} n_1 \\ n_2 \\ n_3 \end{bmatrix} = \boldsymbol{\sigma}^{\mathrm{T}}\cdot\mathbf{n} \tag{14}$$

3 INFINITESIMAL LINEARLY ELASTIC CONSTITUTIVE LAWS

The fundamental mechanism that produces forces in elastic deformation is the stretching of atomic bonds. If the deformations are small, then one can argue that we are dealing with only the linear portion of the response. Ultimately, this allows us to usually use linear relationships for models relating forces to deformations. The usual procedure to determine tensile properties of materials is to place samples of material in testing machines, apply the loads, and then measure the resulting deformations, such as lengths and changes in diameter in a portion of the specimen, of circular cross section, called the gage length. The location of the gage length is away from the attachments to the testing machine. The ends where the samples are attached to the machine are larger so that failure will not occur there first, which

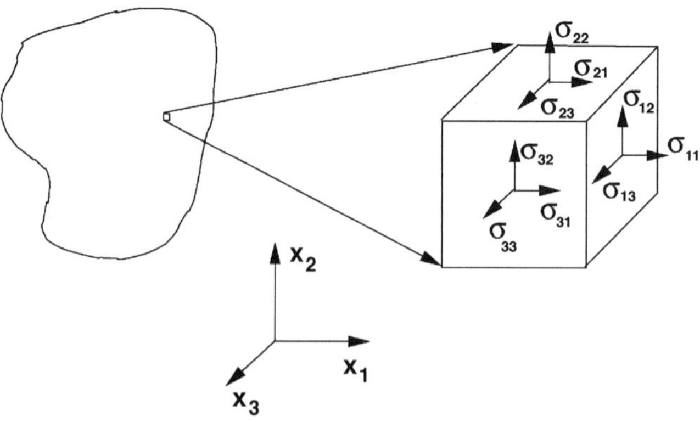

Figure 3 Stress at a point.

would ruin the experimental measurements. Slow rates of deformation are applied, and the response is usually measured with strain gauges or extensometers. For a metal, samples are usually 1.25 cm in diameter and 5 cm in length.

The immediate result of a tension test is the axial force (F) divided by the original area (A_0), denoted, loosely speaking, the "stress," and change in length ($\Delta L \stackrel{\text{def}}{=} L - L_0$) per unit length ($\epsilon_0$), or "engineering strain," measured by strain gauges. As a first approximation we define the tensile stiffness of the material, known as Young's modulus, denoted E, by $\sigma_0 \stackrel{\text{def}}{=} F/A_0 = E\,(\Delta L/L) \stackrel{\text{def}}{=} E\epsilon_0$. As we know, the terms σ_0 and ϵ_0 are, strictly speaking, not the true stress and strain, but simply serve our presentation purposes. Clearly, what we have presented is somewhat ad hoc; therefore, next we present the classical theory of isotropic elastic material responses for three-dimensional states of stress and strain.

We now discuss relationships between the stress and strain, so-called material laws or constitutive relations for geometrically linear problems (infinitesimal deformations).* The starting point to develop a constitutive theory is to assume a stored elastic energy function exists, a function denoted W, which depends only on the mechanical deformation. The simplest function that fulfills $\boldsymbol{\sigma} = \partial W/\partial \boldsymbol{\epsilon}$ is $W = \frac{1}{2}\boldsymbol{\epsilon} : \mathbb{E} : \boldsymbol{\epsilon}$. Such a function satisfies the intuitive physical requirement that, for any small strain from an undeformed state, energy must be stored in the material. Alternatively, a small-strain material law can be derived from $\boldsymbol{\sigma} = \partial W/\partial \boldsymbol{\epsilon}$ and $W \approx \mathbf{c}_0 + \mathbf{c}_1 : \boldsymbol{\epsilon} + \frac{1}{2}\boldsymbol{\epsilon} : \mathbb{E} : \boldsymbol{\epsilon} + \cdots$, which implies $\boldsymbol{\sigma} \approx \mathbf{c}_1 + \mathbb{E} : \boldsymbol{\epsilon} + \cdots$. We are free to set $\mathbf{c}_0 = \mathbf{0}$ (it is arbitrary) to have zero strain energy at zero strain, and furthermore we assume that no stresses exist in the reference state ($\mathbf{c}_1 = \mathbf{0}$); we obtain the familiar relation $\boldsymbol{\sigma} \approx \mathbb{E} : \boldsymbol{\epsilon}$. This is a linear (tensorial) relation between stresses and strains. The existence of a strictly positive stored energy function at the reference configuration implies that the linear elasticity tensor must have positive eigenvalues at every point in the body. Typically, different materials are classified according to the number of independent constants in $\mathbb{E}$. A general material has 81 independent constants, since it is a fourth-order tensor relating nine components of stress to strain. However, the number of constants can be reduced to 36 since the stress and strain tensors are symmetric. This is easily seen from the matrix representation of $\mathbb{E}$:

$$\underbrace{\begin{Bmatrix} \sigma_{11} \\ \sigma_{22} \\ \sigma_{33} \\ \sigma_{12} \\ \sigma_{23} \\ \sigma_{31} \end{Bmatrix}}_{\stackrel{\text{def}}{=}\{\boldsymbol{\sigma}\}} = \underbrace{\begin{bmatrix} E_{1111} & E_{1122} & E_{1133} & E_{1112} & E_{1123} & E_{1113} \\ E_{2211} & E_{2222} & E_{2233} & E_{2212} & E_{2223} & E_{2213} \\ E_{3311} & E_{3322} & E_{3333} & E_{3312} & E_{3323} & E_{3313} \\ E_{1211} & E_{1222} & E_{1233} & E_{1212} & E_{1223} & E_{1213} \\ E_{2311} & E_{2322} & E_{2333} & E_{2312} & E_{2323} & E_{2313} \\ E_{1311} & E_{1322} & E_{1333} & E_{1312} & E_{1323} & E_{1313} \end{bmatrix}}_{\stackrel{\text{def}}{=}[\mathbb{E}]} \underbrace{\begin{Bmatrix} \epsilon_{11} \\ \epsilon_{22} \\ \epsilon_{33} \\ 2\epsilon_{12} \\ 2\epsilon_{23} \\ 2\epsilon_{31} \end{Bmatrix}}_{\stackrel{\text{def}}{=}\{\boldsymbol{\epsilon}\}} \tag{15}$$

The symbol $[\cdot]$ is used to indicate standard matrix notation equivalent to a tensor form, while $\{\cdot\}$ to indicate vector representation. The existence of a scalar energy function forces $\mathbb{E}$ to be symmetric since the strains are symmetric, in other words $\mathrm{W} = \frac{1}{2}\boldsymbol{\epsilon} : \mathbb{E} : \boldsymbol{\epsilon} = \frac{1}{2}(\boldsymbol{\epsilon} : \mathbb{E} : \boldsymbol{\epsilon})^{\mathrm{T}} = \frac{1}{2}\boldsymbol{\epsilon}^{\mathrm{T}} : \mathbb{E}^{\mathrm{T}} : \boldsymbol{\epsilon}^{\mathrm{T}} = \frac{1}{2}\boldsymbol{\epsilon} : \mathbb{E}^{\mathrm{T}} : \boldsymbol{\epsilon}$, which implies $\mathbb{E}^{\mathrm{T}} = \mathbb{E}$. Consequently, $\mathbb{E}$ has only 21 free constants. The nonnegativity of W imposes the restriction that $\mathbb{E}$ remain positive definite. At this point, based on many factors that depend on the material microstructure, it can be shown that the components of $\mathbb{E}$ may be written in terms of anywhere between 21 and 2 independent parameters. Specifically, if there are an infinite number of planes where the material prop-

*Furthermore, we neglect all thermal effects.

erties are equal in all directions, there are two free constants, the Lamé parameters, and the material is of the familiar *isotropic* variety. An isotropic body has material properties that are the same in every direction at a point in the body, i.e., the properties are not a function of orientation at a point in a body. Accordingly, for isotropic materials, with two planes of symmetry and an infinite number of planes of directional independence (two free constants),

$$\mathbb{E} \stackrel{\text{def}}{=} \begin{bmatrix} \kappa + \frac{4}{3}\mu & \kappa - \frac{2}{3}\mu & \kappa - \frac{2}{3}\mu & 0 & 0 & 0 \\ \kappa - \frac{2}{3}\mu & \kappa + \frac{4}{3}\mu & \kappa - \frac{2}{3}\mu & 0 & 0 & 0 \\ \kappa - \frac{2}{3}\mu & \kappa - \frac{2}{3}\mu & \kappa + \frac{4}{3}\mu & 0 & 0 & 0 \\ 0 & 0 & 0 & \mu & 0 & 0 \\ 0 & 0 & 0 & 0 & \mu & 0 \\ 0 & 0 & 0 & 0 & 0 & \mu \end{bmatrix} \tag{16}$$

In this case we have

$$\mathbb{E} : \boldsymbol{\epsilon} = 3\kappa \frac{\text{tr}\,\boldsymbol{\epsilon}}{3} \mathbf{1} + 2\mu\boldsymbol{\epsilon}' \Rightarrow \boldsymbol{\epsilon} : \mathbb{E} : \boldsymbol{\epsilon} = 9\kappa\left(\frac{\text{tr}\,\epsilon}{3}\right)^2 + 2\mu\boldsymbol{\epsilon}' : \boldsymbol{\epsilon}' \tag{17}$$

where $\text{tr}\,\boldsymbol{\epsilon} = \boldsymbol{\epsilon}_{ii}$ and $\boldsymbol{\epsilon}' = \epsilon - \frac{1}{3}(\text{tr}\,\boldsymbol{\epsilon})\mathbf{1}$. The eigenvalues of an isotropic elasticity tensor are $(3\kappa, 2\mu, 2\mu, \mu, \mu, \mu)$. Therefore, we must have $\kappa > 0$ and $\mu > 0$ to retain positive definiteness of $\mathbb{E}$.

It is sometimes important to split infinitesimal strains into two physically meaningful parts $\boldsymbol{\epsilon} = (\text{tr}\,\epsilon/3)\,\mathbf{1} + [\boldsymbol{\epsilon} - (\text{tr}\,\epsilon/3)\,\mathbf{1}]$. The Jacobian, J, of the deformation gradient $\mathbf{F}$ is $\det(\mathbf{1} + \nabla_X \mathbf{u})$ and can be expanded as $J = \det \mathbf{F} = \det(\mathbf{1} + \boldsymbol{\nabla}_{\mathbf{X}}\mathbf{u}) \approx \mathbf{1} + \text{tr}\,\boldsymbol{\nabla}_{\mathbf{X}}\mathbf{u} + \mathcal{O}(\boldsymbol{\nabla}_{\mathbf{X}}\mathbf{u}) = \mathbf{1} + \text{tr}\,\boldsymbol{\epsilon} + \cdots$, and therefore with infinitesimal strains $(\mathbf{1} + \text{tr}\,\boldsymbol{\epsilon})\,d\omega_0 = d\omega \Rightarrow \text{tr}\,\boldsymbol{\epsilon} = (d\omega - d\omega_0)/d\omega_0$. Hence, $\text{tr}\,\epsilon$ is associated with the volumetric part of the deformation. Furthermore, since $\text{tr}[\boldsymbol{\epsilon} - (\text{tr}\,\boldsymbol{\epsilon}/3)\,\mathbf{1}] = 0$, the so-called strain deviator can only affect the shape of a differential element. *In other words, it describes distortion in the material.*

The stress $\boldsymbol{\sigma}$ can be split into two parts (dilatational and a deviatoric part): $\boldsymbol{\sigma} = (\text{tr}\,\boldsymbol{\sigma}/3)\,\mathbf{1} + [\boldsymbol{\sigma} - (\text{tr}\,\boldsymbol{\sigma}/3)\,\mathbf{1}] \stackrel{\text{def}}{=} -p\mathbf{1} + \boldsymbol{\sigma}'$, where we call the symbol p the hydrostatic pressure and $\boldsymbol{\sigma}'$ the stress deviator. This is one form of Hooke's law. The resistance to change in the volume is measured by κ. We note that $[(\text{tr}\,\boldsymbol{\sigma}/3)\,\mathbf{1}]' = 0$, which indicates that this part of the stress produces no distortion. Another fundamental form of Hooke's law is

$$\boldsymbol{\sigma} = \frac{E}{1 + \nu}\left(\epsilon + \frac{\nu}{1 - 2\nu}\,\text{tr}\,\boldsymbol{\epsilon}\mathbf{1}\right)$$

which implies $\boldsymbol{\epsilon} = [(1 + \nu)/E]\,\boldsymbol{\sigma} - (\nu/E)\,\text{tr}\,\boldsymbol{\sigma}\mathbf{1}$. To interpret the constants, consider a uniaxial test where $\sigma_{12} = \sigma_{13} = \sigma_{23} = 0 \Rightarrow \epsilon_{12} = \epsilon_{13} = \epsilon_{23} = 0$, $\sigma_{22} = \sigma_{33} = 0$. Under these conditions we have $\sigma_{11} = E\epsilon_{11}$ and $\epsilon_{22} = \epsilon_{33} = \nu\epsilon_{11}$. Therefore, Young's modulus E is the ratio of the uniaxial stress to the corresponding strain component. The Poisson ratio ν is the ratio of the transverse strains to the uniaxial strain.

Another commonly used set of stress–strain forms are the Lamé relations, $\sigma = \lambda\,\text{tr}\,\boldsymbol{\epsilon}\mathbf{1} + 2\mu\boldsymbol{\epsilon}$ or

$$\epsilon = -\frac{\lambda}{2\mu(3\lambda + 2\mu)}\,\text{tr}\,\boldsymbol{\sigma}\mathbf{1} + \frac{\boldsymbol{\sigma}}{2\mu}$$

To interpret the constants, consider a pressure test where $\sigma_{12} = \sigma_{13} = \sigma_{23} = 0$ and $\sigma_{11} = \sigma_{22} = \sigma_{33}$. Under these conditions we have

$$\kappa = \lambda + \frac{2}{3}\mu = \frac{E}{3(1 - 2\nu)} \qquad \mu = \frac{E}{2(1 + \nu)} \qquad \frac{\kappa}{\mu} = \frac{2(1 + \nu)}{3(1 - 2\nu)}$$

We observe that $\kappa/\mu \to \infty$, which implies that $\nu \to \frac{1}{2}$, and $\kappa/\mu \to 0$ implies $\nu \to -1$. Therefore, from the fact that both κ and μ must be positive and finite, this implies $-1 < \nu < 0.5$ and $0 < E < \infty$. For example, some polymeric foams exhibit $\nu < 0$, steels $\nu \approx 0.3$, and some forms of rubber have $\nu \to 0.5$. However, *no restrictions arise on* λ*, i.e., it could be positive or negative.*

4 FOUNDATIONS OF THE FINITE-ELEMENT METHOD

In most problems of mathematical physics the true solutions are nonsmooth, i.e., not continuously differentiable. For example, in the equation of static mechanical equilibrium,*

$$\nabla \cdot \boldsymbol{\sigma} + \mathbf{f} = \mathbf{0} \tag{18}$$

there is an implicit requirement that the stress $\boldsymbol{\sigma}$ is differentiable in the classical sense. Virtually the same mathematical structure holds for other partial differential equations of mathematical physics describing diffusion, heat conduction, etc. *In many applications, differentiability is too strong a requirement.* Therefore, when solving such problems we have two options: (1) enforcement of jump conditions at every interface or (2) weak formulations (weakening the regularity requirements). Weak forms, which are designed to accommodate irregular data and solutions, are usually preferred. *Numerical techniques employing weak forms, such as the FEM, have been developed with the essential property that whenever a smooth classical solution exists, it is also a solution to the weak-form problem.* Therefore, we lose nothing by reformulating a problem in a weaker way.

The FEM starts by rewriting the field equations in so-called "weak form." To derive a direct weak form for a body, we take the equilibrium equations (denoted the strong form) and form a scalar product with an arbitrary smooth vector-valued function $\mathbf{v}$ and integrate over the body, $\int_\Omega(\nabla\cdot\boldsymbol{\sigma} + \mathbf{f})\cdot\mathbf{v}\, d\Omega = \int_\Omega \mathbf{r}\cdot\mathbf{v}\, d\Omega = 0$, where $\mathbf{r}$ is called the residual. We call $\mathbf{v}$ a "test" function. If we were to add a condition that we do this for all ($\overset{\text{def}}{=}\forall$) possible test functions, then $\int_\Omega(\nabla\cdot\boldsymbol{\sigma} + \mathbf{f})\cdot\mathbf{v}\, d\Omega = \int_\Omega \mathbf{r}\cdot\mathbf{v}\, d\Omega = 0\ \forall\mathbf{v}$ implies $\mathbf{r} = \mathbf{0}$. Therefore, if every possible test function was considered, then $\mathbf{r} = \boldsymbol{\nabla}\cdot\boldsymbol{\sigma} + \mathbf{f} = \mathbf{0}$ on any finite region in Ω. Consequently, the weak and strong statements would be equivalent provided the true solution is smooth enough to have a strong solution. Clearly, $\mathbf{r}$ can never be zero over any finite region in the body, because the test function will "find" them. Using the product rule of differentiation, $\boldsymbol{\nabla}\cdot(\boldsymbol{\sigma}\cdot\mathbf{v}) = (\boldsymbol{\nabla}\cdot\boldsymbol{\sigma})\cdot\mathbf{v} + \boldsymbol{\nabla}\mathbf{v} : \boldsymbol{\sigma}$ leads to, $\forall\mathbf{v}$, $\int_\Omega(\nabla\cdot(\boldsymbol{\sigma}\cdot\mathbf{v}) - \boldsymbol{\nabla}\mathbf{v} : \boldsymbol{\sigma})\, d\Omega + \int_\Omega \mathbf{f}\cdot\mathbf{v}\, d\Omega = 0$, where we choose the $\mathbf{v}$ from an admissible set, to be discussed momentarily. Using the divergence theorem leads to, $\forall\mathbf{v}$, $\int_\Omega \boldsymbol{\nabla}\mathbf{v} : \boldsymbol{\sigma}\, d\Omega = \int_\Omega \mathbf{f}\cdot\mathbf{v}\, d\Omega + \int_{\partial\Omega} \boldsymbol{\sigma}\cdot\mathbf{n}\cdot\mathbf{v}\, dA$, which leads to $\int_\Omega \boldsymbol{\nabla}\mathbf{v} : \boldsymbol{\sigma}\, d\Omega = \int_\Omega \mathbf{f}\cdot\mathbf{v}\, d\Omega + \int_{\Gamma_t} \mathbf{t}\cdot\mathbf{v}\, dA$. If we decide to restrict our choices of $\mathbf{v}$'s to those such that $\mathbf{v}|_{\Gamma_u} = \mathbf{0}$, we have, where $\mathbf{d}$ is the applied boundary displacement on Γ_u, for infinitesimal strain linear elasticity,

*Here $\mathbf{f}$ are the body forces.

Find $\mathbf{u}$, $\mathbf{u}|_{\Gamma_u} = \mathbf{d}$ such that, $\forall \mathbf{v}$, $\mathbf{v}|_{\Gamma u} = \mathbf{0}$:

$$\underbrace{\int_\Omega \nabla\mathbf{v} : \mathbb{E} : \nabla\mathbf{u}\, d\Omega}_{\stackrel{\text{def}}{=}\mathcal{B}(\mathbf{u},\mathbf{v})} = \underbrace{\int_\Omega \mathbf{f}\cdot\mathbf{v}\, d\Omega + \int_{\Gamma_t} \mathbf{t}\cdot\mathbf{v}\, dA}_{\stackrel{\text{def}}{=}\mathcal{F}(\mathbf{v})} \tag{19}$$

This is called a weak form because it does not require the differentiability of the stress $\boldsymbol{\sigma}$. In other words, the differentiability requirements have been *weakened.* It is clear that we are able to consider problems with quite irregular solutions. We observe that if we test the solution with all possible test functions of sufficient smoothness, then the weak solution is equivalent to the strong solution. *We emphasize that provided the true solution is smooth enough, the weak and strong forms are equivalent, which can be seen by the above constructive derivation.*

5 HILBERTIAN SOBOLEV SPACES

A key question is the selection of the sets of functions in the weak form. Somewhat naively, the answer is simple, the integrals must remain finite. Therefore the following restrictions hold ($\forall \mathbf{v}$), $\int_\Omega \mathbf{f}\cdot\mathbf{v}\, d\Omega < \infty$, $\int_{\partial\Omega} \boldsymbol{\sigma}\cdot\mathbf{n}\cdot\mathbf{v}\, d\Omega < \infty$ and $\int_\Omega \nabla\mathbf{v} : \boldsymbol{\sigma}\, d\Omega < \infty$, and govern the selection of the approximation spaces. These relations simply mean that the functions must be square integrable. To make precise statements, one must have a method of book keeping. Such a system is to employ so-called Hilbertian Sobolev spaces. We recall that a norm has three main characteristics for any vectors $\mathbf{u}$ and $\mathbf{v}$ such that $\|\mathbf{u}\| < \infty$ and $\|\mathbf{v}\| < \infty$ and (1) $\|\mathbf{u}\| > 0$, $\|\mathbf{u}\| = 0$ if and only if $\mathbf{u} = 0$, (2) $\|\mathbf{u} + \mathbf{v}\| \leq \|\mathbf{u}\| + \|\mathbf{v}\|$, and (3) $\|\alpha\mathbf{u}\| \leq |\alpha|\|\mathbf{u}\|$, where α is a scalar. Certain types of norms, so-called Hilbert space norms, are frequently used in solid mechanics. Following standard notation, we denote $H^1(\Omega)$ as the usual space of scalar functions with generalized partial derivatives of order ≤ 1 in $L^2(\Omega)$, i.e., square integrable, in other words

$$u \in H^1(\Omega) \qquad \text{if} \qquad \|u\|^2_{H^1(\Omega)} \stackrel{\text{def}}{=} \int_\Omega \frac{\partial u}{\partial x_j}\frac{\partial u}{\partial x_j}\, d\Omega + \int_\Omega uu\, d\Omega < \infty$$

We define $\mathbf{H}^1(\Omega) \stackrel{\text{def}}{=} [H^1(\Omega)]^3$ as the space of vector-valued functions whose components are in $H^1(\Omega)$, i.e.,

$$\mathbf{u} \in \mathbf{H}^1(\boldsymbol{\Omega}) \qquad \text{if} \qquad \|\mathbf{u}\|^2_{\mathbf{H}^1(\Omega)} \stackrel{\text{def}}{=} \int_\Omega \frac{\partial u_i}{\partial x_j}\frac{\partial u_i}{\partial x_j}\, d\Omega + \int_\Omega u_i u_i\, d\Omega < \infty \tag{20}$$

and we denote $\mathbf{L}^2(\Omega) \stackrel{\text{def}}{=} [L^2(\Omega)]^3$. Using these definitions, a complete boundary value problem can be written as follows. The data (loads) are assumed to be such that $\mathbf{f} \in \mathbf{L}^2(\Omega)$ and $\mathbf{t} \in \mathbf{L}^2(\Gamma_t)$, but less smooth data can be considered without complications. Implicitly we require that $\mathbf{u} \in \mathbf{H}^1(\boldsymbol{\Omega})$ and $\boldsymbol{\sigma} \in \mathbf{L}^2(\boldsymbol{\Omega})$ without continually making such references. Therefore, in summary, we assume that our solutions obey these restrictions, leading to the following infinitesimal strain linear elasticity weak form:

Find $\mathbf{u} \in \mathbf{H}^1(\boldsymbol{\Omega})$, $\mathbf{u}|_{\Gamma_u} = \mathbf{d}$, such that, $\forall \mathbf{v} \in \mathbf{H}^1(\boldsymbol{\Omega})$, $\mathbf{v}|_{\Gamma_u} = \mathbf{0}$:

$$\int_\Omega \nabla\mathbf{v} : \mathbb{E} : \nabla\mathbf{u}\, d\Omega = \int_\Omega \mathbf{f}\cdot\mathbf{v}\, d\Omega + \int_{\Gamma_t} \mathbf{t}\cdot\mathbf{v}\, dA \tag{21}$$

We note that if the data in (21) are smooth and if (21) possesses a solution **u** that is sufficiently regular, then **u** is the solution of the classical linear elastostatic problem in strong form:

$$\boxed{\begin{aligned} \nabla\cdot(\mathbb{E} : \nabla\mathbf{u}) + \mathbf{f} = 0, &\qquad \mathbf{x} \in \Omega \\ \mathbf{u} = \mathbf{d} &\qquad \mathbf{x} \in \Gamma_u \\ (\mathbb{E} : \nabla\mathbf{u})\cdot\mathbf{n} = \mathbf{t} &\qquad \mathbf{x} \in \Gamma_t \end{aligned}} \tag{22}$$

6 FINITE-ELEMENT FORMULATION IN THREE DIMENSIONS

Consider the following general form:

$$\boxed{\begin{aligned} &\text{Find } \mathbf{u} \in \mathbf{H}^1(\Omega) \text{ such that, } \forall \mathbf{v} \in \mathbf{H}^1(\Omega), \\ &\int_\Omega \nabla\mathbf{v} : \mathbb{E} : \nabla\mathbf{u}\, d\Omega = \int_\Omega \mathbf{f}\cdot\mathbf{v}\, d\Omega + \int_{\Gamma_t} \mathbf{t}\cdot\mathbf{v}\, dA \end{aligned}} \tag{23}$$

It is convenient to write the bilinear form in the following (matrix) manner:

$$\int_\Omega ([\mathbf{D}]\{\mathbf{v}\})^T[\mathbb{E}]([\mathbf{D}]\{\mathbf{u}\})\, d\Omega = \int_\Omega \{\mathbf{v}\}^T\{\mathbf{f}\}\, d\Omega + \int_{\Gamma_t} \{\mathbf{v}\}^T\{\mathbf{t}\}\, dA \tag{24}$$

where $[\mathbf{D}]$, the deformation tensor, is

$$[\mathbf{D}] \stackrel{\text{def}}{=} \begin{bmatrix} \frac{\partial}{\partial x_1} & 0 & 0 \\ 0 & \frac{\partial}{\partial x_2} & 0 \\ 0 & 0 & \frac{\partial}{\partial x_3} \\ \frac{\partial}{\partial x_2} & \frac{\partial}{\partial x_1} & 0 \\ 0 & \frac{\partial}{\partial x_3} & \frac{\partial}{\partial x_2} \\ \frac{\partial}{\partial x_3} & 0 & \frac{\partial}{\partial x_1} \end{bmatrix} \qquad \{\mathbf{u}\} \stackrel{\text{def}}{=} \begin{Bmatrix} u_1 \\ u_2 \\ u_3 \end{Bmatrix} \qquad \{\mathbf{f}\} \stackrel{\text{def}}{=} \begin{Bmatrix} f_1 \\ f_2 \\ f_3 \end{Bmatrix} \qquad \{\mathbf{t}\} \stackrel{\text{def}}{=} \begin{Bmatrix} t_1 \\ t_2 \\ t_3 \end{Bmatrix} \tag{25}$$

It is clear that in an implementation of the FEM, the sparsity of **D** should be taken into account. It is also convenient to write

$$\boxed{\begin{aligned} u_1^h(x_1, x_2, x_3) &= \sum_{i=1}^{N} a_i\phi_i(x_1, x_2, x_3) \\ u_2^h(x_1, x_2, x_3) &= \sum_{i=1}^{N} a_{i+N}\phi_i(x_1, x_2, x_3) \\ u_3^h(x_1, x_2, x_3) &= \sum_{i=1}^{N} a_{i+2N}\phi_i(x_1, x_2, x_3) \end{aligned}} \tag{26}$$

or $\{\mathbf{u}^h\} = [\phi]\{\mathbf{a}\}$, where, for example, for trilinear shape functions

$$[\phi] \stackrel{\text{def}}{=} \begin{bmatrix} \phi_1\ \phi_2\ \phi_3\ \phi_4\ \phi_5\ \phi_6 \cdots \phi_N & 0\ 0\ 0\ 0\ 0\ 0\ 0 \cdots & 0\ 0\ 0\ 0\ 0\ 0\ 0 \cdots \\ 0\ 0\ 0\ 0\ 0\ 0\ 0 \cdots & \phi_1\ \phi_2\ \phi_3\ \phi_4\ \phi_5\ \phi_6 \cdots \phi_N & 0\ 0\ 0\ 0\ 0\ 0\ 0 \cdots \\ 0\ 0\ 0\ 0\ 0\ 0\ 0 \cdots & 0\ 0\ 0\ 0\ 0\ 0\ 0 \cdots & \phi_1\ \phi_2\ \phi_3\ \phi_4\ \phi_5\ \phi_6 \cdots \phi_N \end{bmatrix} \tag{27}$$

where $\mathbf{u}^h$ is the discrete (approximate) solution.

It is advantageous to write

$$\boxed{\{\mathrm{a}\} \stackrel{\text{def}}{=} \begin{Bmatrix} a_1 \\ a_2 \\ a_3 \\ \cdot \\ \cdot \\ \cdot \\ a_{3N} \end{Bmatrix} \quad \underbrace{\{\phi_i\} \stackrel{\text{def}}{=} \begin{Bmatrix} \phi_i \\ 0 \\ 0 \end{Bmatrix}}_{\text{for } 1 \le i \le N} \quad \underbrace{\{\phi_i\} \stackrel{\text{def}}{=} \begin{Bmatrix} 0 \\ \phi_i \\ 0 \end{Bmatrix}}_{\text{for } N+1 \le i \le 2N} \quad \underbrace{\{\phi_i\} \stackrel{\text{def}}{=} \begin{Bmatrix} 0 \\ 0 \\ \phi_i \end{Bmatrix}}_{\text{for } 2N+1 \le i \le 3N}} \tag{28}$$

and $\{\mathbf{u}^h\} = \sum_{i=1}^{N} \mathbf{a}_\mathbf{i}\{\phi_\mathbf{i}\}$. If we choose $\mathbf{v}$ with the same basis but a different linear combination $\{\mathbf{v}\} = [\phi]\{\mathbf{b}\}$, then we may write

$$\underbrace{\int_\Omega ([\mathbf{D}][\phi]\{\mathbf{b}\})^{\mathrm{T}}[\mathbb{E}]([\mathbf{D}][\phi]\{\mathbf{a}\})\, d\Omega}_{\{\mathbf{b}\}^{\mathrm{T}}[\mathbf{K}]\{\mathbf{a}\} \text{ stiffness}} = \underbrace{\int_\Omega ([\phi]\{\mathbf{b}\})^{\mathrm{T}}\{\mathbf{f}\}\, d\Omega}_{\text{body load}} + \underbrace{\int_{\Gamma_t} ([\phi]\{\mathbf{b}\})^{\mathrm{T}}\{\mathbf{t}\} dA}_{\text{traction load}} \tag{29}$$

Since $\{\mathbf{b}\}$ is arbitrary, i.e., the weak statement implies $\forall \mathbf{v} \Rightarrow \forall\{\mathbf{b}\}$, therefore

$$\boxed{\begin{aligned} \{\mathbf{b}\}^{\mathrm{T}}\{[\mathbf{K}]\{\mathbf{a}\} - \{\mathbf{R}\}\} = \mathbf{0} \Rightarrow [\mathbf{K}]\{\mathbf{a}\} &= \{\mathbf{R}\} \\ [\mathbf{K}] &\stackrel{\text{def}}{=} \textstyle\int_\Omega ([\mathbf{D}][\phi])^{\mathrm{T}}[\mathbb{E}]([\mathbf{D}][\phi])\, d\Omega \\ \{\mathbf{R}\} &\stackrel{\text{def}}{=} \textstyle\int_\Omega [\phi]^{\mathrm{T}}\{\mathbf{f}\}\, d\Omega + \int_{\Gamma_t} [\phi]^{\mathrm{T}}\{\mathbf{t}\}\, dA \end{aligned}} \tag{30}$$

This is the system of equations that is to be solved.

7 GLOBAL/LOCAL TRANSFORMATIONS

One strength of the FEM is that most of the computations can be done in an element-by-element manner. We define the entries of [**K**],

$$K_{ij} = \int_{\Omega} ([\mathbf{D}][\phi_i])^{\mathrm{T}}[\mathbb{E}]([\mathbf{D}][\phi_j])\, d\Omega \tag{31}$$

and

$$R_i = \int_{\Omega} [\phi_i]^{\mathrm{T}}\{\mathbf{f}\}\, d\Omega + \int_{\Gamma_t} [\phi_i]^{\mathrm{T}}\{\mathbf{t}\}\, dA \tag{32}$$

Breaking the calculations into elements, $K_{ij} = \Sigma_e K_{ij}^e$, where

$$K_{ij}^e = \int_{\Omega_e} ([\mathbf{D}][\phi_i])^{\mathrm{T}}[\mathbb{E}]([\mathbf{D}][\phi_j])\, d\Omega \tag{33}$$

To make the calculations systematic, we wish to use the generic or master element defined in a local coordinate system $(\zeta_1, \zeta_2, \zeta_3)$ (Fig. 4). Accordingly, we need the following mapping functions, from the master coordinates to the real space coordinates, $M : (x_1, x_2, x_3) \mapsto (\zeta_1, \zeta_2, \zeta_3)$ (for example, trilinear bricks):

$$\boxed{\begin{aligned} x_1 &= \sum_{i=1}^{8} X_{1i}\hat{\phi}_i \stackrel{\text{def}}{=} M_{x_1}(\zeta_1, \zeta_2, \zeta_3) \\ x_2 &= \sum_{i=1}^{8} X_{2i}\hat{\phi}_i \stackrel{\text{def}}{=} M_{x_2}(\zeta_1, \zeta_2, \zeta_3) \\ x_3 &= \sum_{i=1}^{8} X_{3i}\hat{\phi}_i \stackrel{\text{def}}{=} M_{x_3}(\zeta_1, \zeta_2, \zeta_3) \end{aligned}} \tag{34}$$

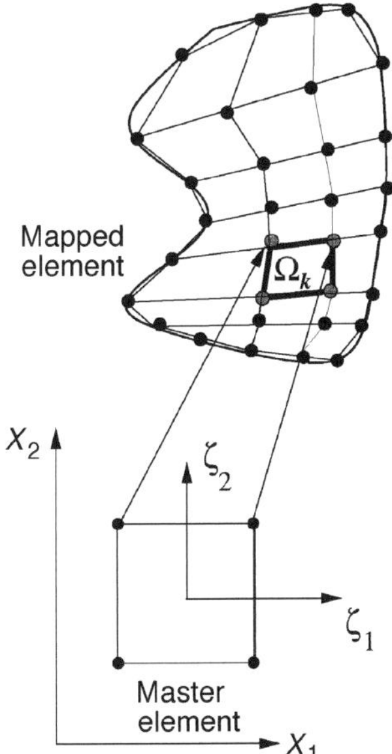

Figure 4 Two-dimensional finite-element mapping.

where (X_{1i}, X_{2i}, X_{3i}) are true spatial coordinates of the ith node and where $\hat{\phi}(\zeta_1, \zeta_2, \zeta_3) \stackrel{\text{def}}{=} \phi(x_1(\zeta_1, \zeta_2, \zeta_3), x_2(\zeta_1, \zeta_2, \zeta_3), x_3(\zeta_1, \zeta_2, \zeta_3))$. These types of mappings are usually termed parametric maps. If the polynomial order of the shape functions is as high as the element, it is an isoparametric map; lower, then a subparametric map; higher, then a superparametric map.

8 DIFFERENTIAL PROPERTIES OF SHAPE FUNCTIONS

The master element shape functions form a nodal base of trilinear approximation given by

$$\boxed{\begin{array}{ll} \hat{\phi}_1 = \frac{1}{8}(1 - \zeta_1)(1 - \zeta_2)(1 - \zeta_3) & \hat{\phi}_2 = \frac{1}{8}(1 + \zeta_1)(1 - \zeta_2)(1 - \zeta_3) \\ \hat{\phi}_3 = \frac{1}{8}(1 + \zeta_1)(1 + \zeta_2)(1 - \zeta_3) & \hat{\phi}_4 = \frac{1}{8}(1 - \zeta_1)(1 + \zeta_2)(1 - \zeta_3) \\ \hat{\phi}_5 = \frac{1}{8}(1 - \zeta_1)(1 - \zeta_2)(1 + \zeta_3) & \hat{\phi}_6 = \frac{1}{8}(1 + \zeta_1)(1 - \zeta_2)(1 + \zeta_3) \\ \hat{\phi}_7 = \frac{1}{8}(1 + \zeta_1)(1 + \zeta_2)(1 + \zeta_3) & \hat{\phi}_8 = \frac{1}{8}(1 - \zeta_1)(1 + \zeta_2)(1 + \zeta_3) \end{array}} \tag{35}$$

- For trilinear elements we have a nodal basis consisting of 8 nodes, and since it is vector valued, 24 total degrees of freedom, or three shape functions for each node. See Fig. 5.
- For triquadratic elements we have a nodal basis consisting of 27 nodes, and since it is vector valued, 81 total degrees of freedom, or three shape functions for each node. The nodal shape functions can be derived quite easily by realizing that it is a nodal basis, e.g., they are unity at the corresponding node and zero at all other nodes.

We note that the ϕ_i's are never really computed; we actually start with the $\hat{\phi}_i$'s. Therefore in the stiffness matrix and right-hand-side element calculations, all terms must be defined in terms of the local coordinates. With this in mind we lay down some fundamental relations, which are directly related to the concepts of deformation presented in our discussion in continuum mechanics. It is not surprising that a deformation gradient reappears in the following form:

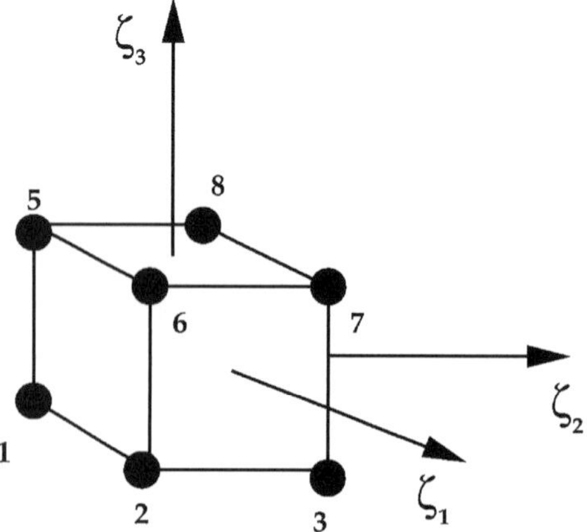

Figure 5 Trilinear hexahedron or "brick."

$$|\mathbf{F}| \stackrel{\text{def}}{=} \left|\frac{\partial \mathbf{x}(x_1, x_2, x_3)}{\partial \boldsymbol{\zeta}(\zeta_1, \zeta_2, \zeta_3)}\right| \qquad \text{where} \qquad \mathbf{F} \stackrel{\text{def}}{=} \begin{bmatrix} \frac{\partial x_1}{\partial \zeta_1} & \frac{\partial x_1}{\partial \zeta_2} & \frac{\partial x_1}{\partial \zeta_3} \\ \frac{\partial x_2}{\partial \zeta_1} & \frac{\partial x_2}{\partial \zeta_2} & \frac{\partial x_2}{\zeta_3} \\ \frac{\partial x_3}{\partial \zeta_1} & \frac{\partial x_3}{\partial \zeta_2} & \frac{\partial x_3}{\partial \zeta_3} \end{bmatrix} \tag{36}$$

The corresponding determinant is

$$|\mathbf{F}| = \frac{\partial x_1}{\partial \zeta_1}\left(\frac{\partial x_2}{\partial \zeta_2}\frac{\partial x_3}{\partial \zeta_3} - \frac{\partial x_3}{\partial \zeta_2}\frac{\partial x_2}{\partial \zeta_3}\right) - \frac{\partial x_1}{\partial \zeta_2}\left(\frac{\partial x_2}{\partial \zeta_1}\frac{\partial x_3}{\partial \zeta_3} - \frac{\partial x_3}{\partial \zeta_1}\frac{\partial x_2}{\partial \zeta_3}\right) + \frac{\partial x_1}{\partial \zeta_3}\left(\frac{\partial x_2}{\partial \zeta_1}\frac{\partial x_3}{\partial \zeta_2} - \frac{\partial x_3}{\partial \zeta_1}\frac{\partial x_2}{\partial \chi_2}\right) \tag{37}$$

The differential relations $\boldsymbol{\zeta} \rightarrow \mathbf{x}$ are

$$\boxed{\begin{aligned} \frac{\partial}{\partial \zeta_1} &= \frac{\partial}{\partial x_1}\frac{\partial x_1}{\partial \zeta_1} + \frac{\partial}{\partial x_2}\frac{\partial x_2}{\partial \zeta_1} + \frac{\partial}{\partial x_3}\frac{\partial x_3}{\partial \zeta_1} \\ \frac{\partial}{\partial \zeta_2} &= \frac{\partial}{\partial x_1}\frac{\partial x_1}{\partial \zeta_2} + \frac{\partial}{\partial x_2}\frac{\partial x_2}{\partial \zeta_2} + \frac{\partial}{\partial x_3}\frac{\partial x_3}{\partial \zeta_2} \\ \frac{\partial}{\partial \zeta_3} &= \frac{\partial}{\partial x_1}\frac{\partial x_1}{\partial \zeta_3} + \frac{\partial}{\partial x_2}\frac{\partial x_2}{\partial \zeta_3} + \frac{\partial}{\partial x_3}\frac{\partial x_3}{\partial \zeta_3} \end{aligned}} \tag{38}$$

The inverse differential relations $\mathbf{x} \rightarrow \boldsymbol{\zeta}$ are

$$\boxed{\begin{aligned} \frac{\partial}{\partial x_1} &= \frac{\partial}{\partial \zeta_1}\frac{\partial \zeta_1}{\partial x_1} + \frac{\partial}{\partial \zeta_2}\frac{\partial \zeta_2}{\partial x_1} + \frac{\partial}{\partial \zeta_3}\frac{\partial \zeta_3}{\partial x_1} \\ \frac{\partial}{\partial x_2} &= \frac{\partial}{\partial \zeta_1}\frac{\partial \zeta_1}{\partial x_2} + \frac{\partial}{\partial \zeta_2}\frac{\partial \zeta_2}{\partial x_2} + \frac{\partial}{\partial \zeta_3}\frac{\partial \zeta_3}{\partial x_2} \\ \frac{\partial}{\partial x_3} &= \frac{\partial}{\partial \zeta_1}\frac{\partial \zeta_1}{\partial x_3} + \frac{\partial}{\partial \zeta_2}\frac{\partial \zeta_2}{\partial x_3} + \frac{\partial}{\partial \zeta_3}\frac{\partial \zeta_3}{\partial x_3} \end{aligned}} \tag{39}$$

and

$$\begin{Bmatrix} dx_1 \\ dx_2 \\ dx_3 \end{Bmatrix} = \begin{bmatrix} \frac{\partial x_1}{\partial \zeta_1} & \frac{\partial x_1}{\partial \zeta_2} & \frac{\partial x_1}{\partial \zeta_3} \\ \frac{\partial x_2}{\partial \zeta_1} & \frac{\partial x_2}{\partial \zeta_2} & \frac{\partial x_2}{\partial \zeta_3} \\ \frac{\partial x_3}{\partial \zeta_1} & \frac{\partial x_3}{\partial \zeta_2} & \frac{\partial x_3}{\partial \partial \zeta_3} \end{bmatrix} \begin{Bmatrix} d\zeta_1 \\ d\zeta_2 \\ d\zeta_3 \end{Bmatrix} \tag{40}$$

and the inverse form

$$\begin{Bmatrix} d\zeta_1 \\ d\zeta_2 \\ d\zeta_3 \end{Bmatrix} = \begin{bmatrix} \frac{\partial \zeta_1}{\partial x_1} & \frac{\partial \zeta_1}{\partial x_2} & \frac{\partial \zeta_1}{\partial x_3} \\ \frac{\partial \zeta_2}{\partial x_1} & \frac{\partial \zeta_2}{\partial x_2} & \frac{\partial \zeta_2}{\partial x_3} \\ \frac{\partial \zeta_3}{\partial x_1} & \frac{\partial \zeta_3}{\partial x_2} & \frac{\partial \zeta_3}{\partial \partial x_3} \end{bmatrix} \begin{Bmatrix} dx_1 \\ dx_2 \\ dx_3 \end{Bmatrix} \tag{41}$$

Noting the following relationship

$$\mathbf{F}^{-1} = \frac{\text{adj }\mathbf{F}}{|\mathbf{F}|} \qquad \text{where} \qquad \text{adj }\mathbf{F} \stackrel{\text{def}}{=} \begin{bmatrix} A_{11} & A_{12} & A_{13} \\ A_{21} & A_{22} & A_{23} \\ A_{31} & A_{32} & A_{33} \end{bmatrix}^{\mathrm{T}} \tag{42}$$

where

$$A_{11} = \left[\frac{\partial x_2}{\partial \zeta_2}\frac{\partial x_3}{\partial \zeta_3} - \frac{\partial x_3}{\partial \zeta_2}\frac{\partial x_2}{\partial \zeta_3}\right] = |\mathbf{F}|\frac{\partial \zeta_1}{\partial x_1} \qquad \mathbf{A}_{12} = -\left[\frac{\partial x_2}{\partial \zeta_1}\frac{\partial x_3}{\partial \zeta_3} - \frac{\partial x_3}{\partial \zeta_1}\frac{\partial x_2}{\partial \zeta_3}\right] = |\mathbf{F}|\frac{\partial \zeta_2}{\partial x_1}$$

$$A_{13} = \left[\frac{\partial x_2}{\partial \zeta_1}\frac{\partial x_3}{\partial \zeta_2} - \frac{\partial x_3}{\partial \zeta_1}\frac{\partial x_2}{\partial \zeta_2}\right] = |\mathbf{F}|\frac{\partial \zeta_3}{\partial x_1} \qquad \mathbf{A}_{21} = -\left[\frac{\partial x_1}{\partial \zeta_2}\frac{\partial x_3}{\partial \zeta_3} - \frac{\partial x_3}{\partial \zeta_2}\frac{\partial x_1}{\partial \zeta_3}\right] = |\mathbf{F}|\frac{\partial \zeta_1}{\partial x_2}$$

$$A_{22} = \left[\frac{\partial x_1}{\partial \zeta_1}\frac{\partial x_3}{\partial \zeta_3} - \frac{\partial x_3}{\partial \zeta_1}\frac{\partial x_1}{\partial \zeta_3}\right] = |\mathbf{F}|\frac{\partial \zeta_2}{\partial x_2} \qquad \mathbf{A}_{23} = -\left[\frac{\partial x_1}{\partial \zeta_1}\frac{\partial x_3}{\partial \zeta_2} - \frac{\partial x_3}{\partial \zeta_1}\frac{\partial x_1}{\partial \zeta_2}\right] = |\mathbf{F}|\frac{\partial \zeta_3}{\partial x_2}$$

$$A_{31} = \left[\frac{\partial x_1}{\partial \zeta_2}\frac{\partial x_2}{\partial \zeta_3} - \frac{\partial x_2}{\partial \zeta_2}\frac{\partial x_1}{\partial \zeta_3}\right] = |\mathbf{F}|\frac{\partial \zeta_1}{\partial x_3} \qquad \mathbf{A}_{32} = -\left[\frac{\partial x_1}{\partial \zeta_1}\frac{\partial x_2}{\partial \zeta_3} - \frac{\partial x_2}{\partial \zeta_1}\frac{\partial x_1}{\partial \zeta_3}\right] = |\mathbf{F}|\frac{\partial \zeta_2}{\partial x_3}$$

$$A_{33} = \left[\frac{\partial x_1}{\partial \zeta_1}\frac{\partial x_2}{\partial \zeta_2} - \frac{\partial x_2}{\partial \zeta_1}\frac{\partial x_1}{\partial \zeta_2}\right] = |\mathbf{F}|\frac{\partial \zeta_3}{\partial x_3} \tag{43}$$

With these relations, one can then solve for the components of $\mathbf{F}$ and $\mathbf{F}^{-1}$.

9 DIFFERENTIATION IN REFERENTIAL COORDINATES

We now need to express $[\mathbf{D}]$ in terms of $(\zeta_1, \zeta_2, \zeta_3)$ via

$$[\mathbf{D}(\phi(x_1, x_2, x_3))] = [\hat{\mathbf{D}}(\hat{\phi}(M_{x_1}(\zeta_1, \zeta_2, \zeta_3), M_{x_2}(\zeta_1, \zeta_2, \zeta_3), M_{x_3}(\zeta_1, \zeta_2, \zeta_3))] \tag{44}$$

Therefore we write for the first column of $[\hat{\mathbf{D}}]$*

*This is for illustration purposes only. For computational efficiency, one should not program such operations in this way. Clearly, the needless multiplication of zeros is to be avoided.

$$
\begin{bmatrix}
\frac{\partial}{\partial \zeta_1}\frac{\partial \zeta_1}{\partial x_1} + \frac{\partial}{\partial \zeta_2}\frac{\partial \zeta_2}{\partial x_1} + \frac{\partial}{\partial \zeta_3}\frac{\partial \zeta_3}{\partial x_1} \\
0 \\
0 \\
\frac{\partial}{\partial \zeta_1}\frac{\partial \zeta_1}{\partial x_2} + \frac{\partial}{\partial \zeta_2}\frac{\partial \zeta_2}{\partial x_2} + \frac{\partial}{\partial \zeta_3}\frac{\partial \zeta_3}{\partial x_2} \\
0 \\
\frac{\partial}{\partial \zeta_1}\frac{\partial \zeta_1}{\partial x_3} + \frac{\partial}{\partial \zeta_2}\frac{\partial \zeta_2}{\partial x_3} + \frac{\partial}{\partial \zeta_3}\frac{\partial \zeta_3}{\partial x_3}
\end{bmatrix} \tag{45}
$$

for the second column

$$
\begin{bmatrix}
0 \\
\frac{\partial}{\partial \zeta_1}\frac{\partial \zeta_1}{\partial x_2} + \frac{\partial}{\partial \zeta_2}\frac{\partial \zeta_2}{\partial x_2} + \frac{\partial}{\partial \zeta_3}\frac{\partial \zeta_3}{\partial x_2} \\
0 \\
\frac{\partial}{\partial \zeta_1}\frac{\partial \zeta_1}{\partial x_1} + \frac{\partial}{\partial \zeta_2}\frac{\partial \zeta_2}{\partial x_1} + \frac{\partial}{\partial \zeta_3}\frac{\partial \zeta_3}{\partial x_1} \\
\frac{\partial}{\partial \zeta_1}\frac{\partial \zeta_1}{\partial x_3} + \frac{\partial}{\partial \zeta_2}\frac{\partial \zeta_2}{\partial x_3} + \frac{\partial}{\partial \zeta_3}\frac{\partial \zeta_3}{\partial x_3} \\
0
\end{bmatrix} \tag{46}
$$

and for the last column

$$
\begin{bmatrix}
0 \\
0 \\
\frac{\partial}{\partial \zeta_1}\frac{\partial \zeta_1}{\partial x_3} + \frac{\partial}{\partial \zeta_2}\frac{\partial \zeta_2}{\partial x_3} + \frac{\partial}{\partial \zeta_3}\frac{\partial \zeta_3}{\partial x_3} \\
0 \\
\frac{\partial}{\partial \zeta_1}\frac{\partial \zeta_1}{\partial x_2} + \frac{\partial}{\partial \zeta_2}\frac{\partial \zeta_2}{\partial x_2} + \frac{\partial}{\partial \zeta_3}\frac{\partial \zeta_3}{\partial x_2} \\
\frac{\partial}{\partial \zeta_1}\frac{\partial \zeta_1}{\partial x_1} + \frac{\partial}{\partial \zeta_2}\frac{\partial \zeta_2}{\partial x_1} + \frac{\partial}{\partial \zeta_3}\frac{\partial \zeta_3}{\partial x_1}
\end{bmatrix} \tag{47}
$$

For an element, our shape function matrix ($\stackrel{\text{def}}{=} [\hat{\phi}]$) has the following form for linear shape functions; for the first eight columns

$$\begin{bmatrix} \hat{\phi}_1 & \hat{\phi}_2 & \hat{\phi}_3 & \hat{\phi}_4 & \hat{\phi}_5 & \hat{\phi}_6 & \hat{\phi}_7 & \hat{\phi}_8 \\ 0 & 0 & 0 & 0 & 0 & 0 & 0 & 0 \\ 0 & 0 & 0 & 0 & 0 & 0 & 0 & 0 \end{bmatrix} \tag{48}$$

for the second eight columns

$$\begin{bmatrix} 0 & 0 & 0 & 0 & 0 & 0 & 0 & 0 \\ \hat{\phi}_1 & \hat{\phi}_2 & \hat{\phi}_3 & \hat{\phi}_4 & \hat{\phi}_5 & \hat{\phi}_6 & \hat{\phi}_7 & \hat{\phi}_8 \\ 0 & 0 & 0 & 0 & 0 & 0 & 0 & 0 \end{bmatrix} \tag{49}$$

and for the last eight columns

$$\begin{bmatrix} 0 & 0 & 0 & 0 & 0 & 0 & 0 & 0 \\ 0 & 0 & 0 & 0 & 0 & 0 & 0 & 0 \\ \hat{\phi}_1 & \hat{\phi}_2 & \hat{\phi}_3 & \hat{\phi}_4 & \hat{\phi}_5 & \hat{\phi}_6 & \hat{\phi}_7 & \hat{\phi}_8 \end{bmatrix} \tag{50}$$

which in total is a 3×24 matrix. Therefore the product $[\hat{\mathbf{D}}][\hat{\phi}]$ is a 6×24 matrix of the form, for the first eight columns

$$\begin{bmatrix} \frac{\partial \hat{\phi}_1}{\partial \zeta_1}\frac{\partial \zeta_1}{\partial x_1} + \frac{\partial \hat{\phi}_1}{\partial \zeta_2}\frac{\partial \zeta_2}{\partial x_1} + \frac{\partial \hat{\phi}_1}{\partial \zeta_3}\frac{\partial \zeta_3}{\partial x_1}, \ldots 8 \\ 0 \quad 0 \quad 0 \quad 0 \quad 0 \quad 0 \quad 0 \quad 0 \\ 0 \quad 0 \quad 0 \quad 0 \quad 0 \quad 0 \quad 0 \quad 0 \\ \frac{\partial \hat{\phi}_1}{\partial \zeta_1}\frac{\partial \zeta_1}{\partial x_2} + \frac{\partial \hat{\phi}_1}{\partial \zeta_2}\frac{\partial \zeta_2}{\partial x_2} + \frac{\partial \hat{\phi}_1}{\partial \zeta_3}\frac{\partial \zeta_3}{\partial x_2}, \ldots 8 \\ 0 \quad 0 \quad 0 \quad 0 \quad 0 \quad 0 \quad 0 \quad 0 \\ \frac{\partial \hat{\phi}_1}{\partial \zeta_1}\frac{\partial \zeta_1}{\partial x_3} + \frac{\partial \hat{\phi}_1}{\partial \zeta_2}\frac{\partial \zeta_2}{\partial x_3} + \frac{\partial \hat{\phi}_1}{\partial \zeta_3}\frac{\partial \zeta_3}{\partial x_3}, \ldots 8 \end{bmatrix} \tag{51}$$

and for the second eight columns

$$\begin{bmatrix} 0 \quad 0 \quad 0 \quad 0 \quad 0 \quad 0 \quad 0 \quad 0 \\ \frac{\partial \hat{\phi}_1}{\partial \zeta_1}\frac{\partial \zeta_1}{\partial x_2} + \frac{\partial \hat{\phi}_1}{\partial \zeta_2}\frac{\partial \zeta_2}{\partial x_2} + \frac{\partial \hat{\phi}_1}{\partial \zeta_3}\frac{\partial \zeta_3}{\partial x_2}, \ldots 8 \\ 0 \quad 0 \quad 0 \quad 0 \quad 0 \quad 0 \quad 0 \quad 0 \\ \frac{\partial \hat{\phi}_1}{\partial \zeta_1}\frac{\partial \zeta_1}{\partial x_1} + \frac{\partial \hat{\phi}_1}{\partial \zeta_2}\frac{\partial \zeta_2}{\partial x_1} + \frac{\partial \hat{\phi}_1}{\partial \zeta_3}\frac{\partial \zeta_3}{\partial x_1}, \ldots 8 \\ \frac{\partial \hat{\phi}_1}{\partial \zeta_1}\frac{\partial \zeta_1}{\partial x_3} + \frac{\partial \hat{\phi}_1}{\partial \zeta_2}\frac{\partial \zeta_2}{\partial x_3} + \frac{\partial \hat{\phi}_1}{\partial \zeta_3}\frac{\partial \zeta_3}{\partial x_3}, \ldots 8 \\ 0 \quad 0 \quad 0 \quad 0 \quad 0 \quad 0 \quad 0 \quad 0, \end{bmatrix} \tag{52}$$

and for the last eight columns

$$\begin{bmatrix} 0 & 0 & 0 & 0 & 0 & 0 & 0 & 0 \\ 0 & 0 & 0 & 0 & 0 & 0 & 0 & 0 \\ \multicolumn{8}{c}{\frac{\partial\hat{\phi}_1}{\partial\zeta_1}\frac{\partial\zeta_1}{\partial x_3}+\frac{\partial\hat{\phi}_1}{\partial\zeta_2}\frac{\partial\zeta_2}{\partial x_3}+\frac{\partial\hat{\phi}_1}{\partial\zeta_3}\frac{\partial\zeta_3}{\partial x_3}, \ldots 8} \\ 0 & 0 & 0 & 0 & 0 & 0 & 0 & 0 \\ \multicolumn{8}{c}{\frac{\partial\hat{\phi}_1}{\partial\zeta_1}\frac{\partial\zeta_1}{\partial x_2}+\frac{\partial\hat{\phi}_1}{\partial\zeta_2}\frac{\partial\zeta_2}{\partial x_2}+\frac{\partial\hat{\phi}_1}{\partial\zeta_3}\frac{\partial\zeta_3}{\partial x_2}, \ldots 8} \\ \multicolumn{8}{c}{\frac{\partial\hat{\phi}_1}{\partial\zeta_1}\frac{\partial\zeta_1}{\partial x_1}+\frac{\partial\hat{\phi}_1}{\partial\zeta_2}\frac{\partial\zeta_2}{\partial x_1}+\frac{\partial\hat{\phi}_1}{\partial\zeta_3}\frac{\partial\zeta_3}{\partial x_1}, \ldots 8} \end{bmatrix} \tag{53}$$

Finally with Gaussian quadrature for each element

$$K_{ij}^e = \underbrace{\sum_{q=1}^{g}\sum_{r=1}^{g}\sum_{s=1}^{g} w_q w_r w_s ([\hat{\mathbf{D}}]\{\hat{\phi}_i\})^{\mathrm{T}}[\hat{\mathbb{E}}]([\hat{\mathbf{D}}]\{\hat{\phi}_j\})|\mathbf{F}|}_{\text{standard}} \tag{54}$$

and

$$R_i^e = \underbrace{\sum_{q=1}^{g}\sum_{r=1}^{g}\sum_{s=1}^{g} w_q w_r w_s \{\hat{\phi}_i\}^{\mathrm{T}}\{\mathbf{f}\}|\mathbf{F}|}_{\text{standard}} + \underbrace{\sum_{q=1}^{g}\sum_{r=1}^{g} w_q w_r [\hat{\phi}_i|^{\mathrm{T}}\{\mathbf{t}\}|\mathbf{F}_s|}_{\text{for } \Gamma_t\cap\Omega_e\neq 0} \tag{55}$$

where w_q, w_r, w_s are Gauss weights and where $|\mathbf{F}_s|$ represents the (surface) Jacobians of element faces on the exterior surface of the body, where, depending on the surface on which it is to be evaluated, one of the ζ components will be $+1$ or -1. These surface Jacobians can be evaluated in a variety of ways, for example, using the Nanson formulas frequently encountered in the field of continuum mechanics.

10 POSTPROCESSING

Postprocessing for the stress, strain, and energy from the existing displacement solution, i.e., the values of the nodal displacements, the shape functions, is straightforward, namely, $[\mathbf{D}]\{\mathbf{u}^h\} = \{\boldsymbol{\epsilon}^h\}$. Therefore, for each element

$$
\begin{Bmatrix} \epsilon_{11}^h \\ \epsilon_{22}^h \\ \epsilon_{33}^h \\ 2\epsilon_{12}^h \\ 2\epsilon_{23}^h \\ 2\epsilon_{13}^h \end{Bmatrix} = \begin{bmatrix} \frac{\partial}{\partial x_1} & 0 & 0 \\ 0 & \frac{\partial}{\partial x_2} & 0 \\ 0 & 0 & \frac{\partial}{\partial x_3} \\ \frac{\partial}{\partial x_2} & \frac{\partial}{\partial x_1} & 0 \\ 0 & \frac{\partial}{\partial x_3} & \frac{\partial}{\partial x_2} \\ \frac{\partial}{\partial x_3} & 0 & \frac{\partial}{\partial x_1} \end{bmatrix} \underbrace{\begin{Bmatrix} \sum_{i=1}^{8} u_{1i}^h \phi_i \\ \sum_{i=1}^{8} u_{2i}^h \phi_i \\ \sum_{i=1}^{8} u_{3i}^h \phi_i \end{Bmatrix}}_{\text{known values}} \tag{56}
$$

where the global coordinates must be transformed to the master system in both the deformation tensor and the displacement representation. At each Gauss point, we add all eight contributions for each of the six components, then multiply by the corresponding nodal displacements that have previously been calculated. The following expressions must be evaluated at the Gauss points, multiplied by the appropriate weights and added together:

$$
\boxed{\begin{array}{lll}
\dfrac{\partial u_1^h}{\partial x_1} = \sum_{i=1}^{8} u_{1i}^h \dfrac{\partial \phi_i}{\partial x_1} & \dfrac{\partial u_2^h}{\partial x_1} = \sum_{i=1}^{8} u_{1i}^h \dfrac{\partial \phi_i}{\partial x_1} & \dfrac{\partial u_3^h}{\partial x_1} = \sum_{i=1}^{8} u_{1i}^h \dfrac{\partial \phi_i}{\partial x_1} \\
\dfrac{\partial u_1^h}{\partial x_2} = \sum_{i=1}^{8} u_{1i}^h \dfrac{\partial \phi_i}{\partial x_2} & \dfrac{\partial u_2^h}{\partial x_2} = \sum_{i=1}^{8} u_{1i}^h \dfrac{\partial \phi_i}{\partial x_2} & \dfrac{\partial u_3^h}{\partial x_2} = \sum_{i=1}^{8} u_{1i}^h \dfrac{\partial \phi_i}{\partial x_2} \\
\dfrac{\partial u_1^h}{\partial x_3} = \sum_{i=1}^{8} u_{1i}^h \dfrac{\partial \phi_i}{\partial x_3} & \dfrac{\partial u_2^h}{\partial x_3} = \sum_{i=1}^{8} u_{1i}^h \dfrac{\partial \phi_i}{\partial x_3} & \dfrac{\partial u_3^h}{\partial x_3} = \sum_{i=1}^{8} u_{1i}^h \dfrac{\partial \phi_i}{\partial x_3}
\end{array}} \tag{57}
$$

where u_{1i}^h denotes the x_1 component of the displacement of the ith node. Combining the numerical derivatives to form the strains we obtain $\epsilon_{11}^h = \partial u_1^h / \partial x_1$, $\epsilon_{22}^h = \partial u_2^h / \partial x_2$, and $\epsilon_{33}^h = \partial u_3^h / \partial x_3$ and $2\epsilon_{12}^h = \gamma_{12} = \partial u_1^h / \partial x_2 + \partial u_2^h / \partial x_1$, $2\epsilon_{23}^h = \gamma_{23} = \partial u_2^h / \partial x_3 + \partial u_3^h / \partial x_2$, and $2\epsilon_{13}^h = \gamma_{13} = \partial u_1^h / \partial x_3 + \partial u_3^h / \partial x_1$.

11 ONE-DIMENSIONAL EXAMPLE

Consider the general form

$$
\boxed{\begin{array}{l}
\text{Find } u \in H^1(\Omega) \text{ such that, } \forall v \in H^1(\Omega),\ v|_{\Gamma_u} = 0, \\
\displaystyle\int_\Omega \frac{dv}{dx} E \frac{du}{dx}\, dx = \int_\Omega fv\, dx + vt|_{\Gamma_t}
\end{array}} \tag{58}
$$

It is convenient to write the bilinear form in the following manner:

$$
\int_\Omega \frac{dv}{dx} E \frac{du}{dx}\, dx = \int_\Omega vf\, dx + vt|_{\Gamma_t} \tag{59}
$$

We write

$$u^h(x) = \sum_{j=1}^{N} a_j \phi_j(x) \tag{60}$$

If we choose v with the same basis but a different linear combination,

$$v(x) = \sum_{i=1}^{N} b_i \phi_i(x) \tag{61}$$

then we may write

$$\underbrace{\int_\Omega \left(\frac{d}{dx} \sum_{i=1}^{N} b_i \phi_i(x) \right) E \frac{d}{dx} \left(\sum_{j=1}^{N} a_j \phi_j(x) \right) dx}_{\{b\}^T\{\mathrm{K}\}\{a\}\ \text{stiffness}}$$

$$= \underbrace{\int_\Omega \left(\sum_{i=1}^{N} b_i \phi_i(x) \right) f\, dx}_{\text{body load}} + \underbrace{\left(\left(\sum_{i=1}^{N} b_i \phi_i(x) \right) t \right) \Bigg|_{\Gamma_t}}_{\text{traction load}} \tag{62}$$

Since the b_i are arbitrary, i.e., the weak form implies $\forall v \Rightarrow \forall b_i$, therefore

$$\boxed{\begin{aligned} &\sum_{i=1}^{N} b_i \left(\sum_{j=1}^{N} K_{ij} a_j - R_i \right) = 0 \Rightarrow [K]\{a\} = \{R\} \\ &K_{ij} \stackrel{\text{def}}{=} \int_\Omega \frac{d\phi_i}{dx} E \frac{d\phi_j}{dx}\, dx \\ &R_i \stackrel{\text{def}}{=} \int_\Omega \phi_i f\, dx + \phi_i t|_{\Gamma_t} \end{aligned}} \tag{63}$$

This is the system of equations that is to be solved.

One strength of the FEM is that most of the computations can be done in an element-by-element manner. Accordingly, we define the entries of $[K]$,

$$K_{ij} = \int_\Omega \frac{d\phi_i}{dx} E \frac{d\phi_j}{dx}\, dx \tag{64}$$

and

$$R_i = \int_\Omega \phi_i f\, dx + \phi_i t|_{\Gamma_t} \tag{65}$$

We can break the calculations into elements, $K_{ij} = \sum_e K_{ij}^e$, where

$$K_{ij}^e = \int_{\Omega_e} \frac{d\phi_i}{dx} E \frac{d\phi_j}{dx}\, dx \tag{66}$$

To make the calculations systematic, we wish to use the generic or master element defined in a local coordinate system (ζ). Accordingly, we need the following mapping functions, from the master coordinates to the real space coordinates, $M(x) \mapsto (\zeta)$ (Fig. 6):

$$x = \sum_{i=1}^{2} X_i \hat{\phi}_i \stackrel{\text{def}}{=} M_x(\zeta) \tag{67}$$

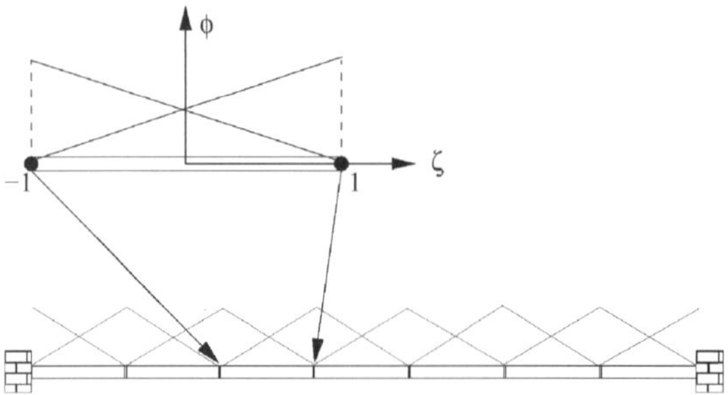

Figure 6 One-dimensional linear finite-element mapping.

where the X_i are true spatial coordinates of the ith node and where $\hat{\phi}(\zeta) \stackrel{\text{def}}{=} \phi(x(\zeta))$.

The master element shape functions form a nodal bases of linear approximation given by

$$\hat{\phi} = \tfrac{1}{2}(1 - \zeta) \qquad \hat{\phi}_2 = \tfrac{1}{2}(1 + \zeta) \tag{68}$$

- For linear elements we have a nodal basis consisting of two nodes and thus two degrees of freedom.
- The nodal shape functions can be derived quite easily by realizing that it is a nodal basis, e.g., they are unity at the corresponding node and zero at all other nodes.

We note that the ϕ_i's are never really computed; we actually start with the $\hat{\phi}_i$'s and then map them into the actual problem domain. Therefore in the stiffness matrix and right-hand-side element calculations, all terms must be defined in terms of the local coordinates. With this in mind, we introduce some fundamental quantities, such as the deformation gradient

$$F \stackrel{\text{def}}{=} \frac{dx}{d\zeta} \tag{69}$$

The corresponding "determinant" is $|F| = |dx/d\zeta|$. The differential relations $\zeta \rightarrow x$ are

$$\frac{d}{d\zeta} = \frac{d}{dx}\frac{dx}{d\zeta} \tag{70}$$

The inverse differential relations $x \rightarrow \zeta$ are

$$\frac{d}{dx} = \frac{d}{d\zeta}\frac{d\zeta}{dx} \tag{71}$$

We can now express d/dx in terms of ζ via

$$\frac{d\phi}{dx} = \frac{d}{dx}\,\phi(M(\zeta)) \tag{72}$$

Finally with quadrature for each element

$$K_{ij}^e = \sum_{q=1}^{g} w_q \underbrace{\left(\frac{d}{d\zeta}\,(\phi_i(M(\zeta))\right)\frac{d\zeta}{dx}\,E\left(\frac{d}{d\zeta}\,(\phi_j(M(\zeta))\right)\frac{d\zeta}{dx}\,|F|}_{\text{evaluated at } \zeta = \zeta_q}$$

and

$$R_i^e = \underbrace{\sum_{q=1}^{g} w_q \phi_i(M(\zeta)) f |F|_\zeta}_{\text{evaluated at } \zeta = \zeta_q} + \underbrace{\phi_i(M(\zeta))t}_{\text{evaluated on traction endpoints}}$$

where the w_q are Gauss weights.

Postprocessing for the stress, strain, and energy from the existing displacement solution, i.e., the values of the nodal displacements, the shape functions, is straightforward. Essentially the process is the same as the formation of the virtual energy in the system. Therefore, for each element

$$\frac{du}{dx} = \frac{d}{dx}\sum_{i=1}^{2} a_i\phi_i = \left(\frac{d}{d\zeta}\sum_{i=1}^{2} a_i\hat{\phi}_i\right)\frac{d\zeta}{dx} \tag{73}$$

REMARKS. On the implementation level, the system of equations to be solved are $[K]\{u^h\} = \{R\}$, where the stiffness matrix is represented by $K(I, J) = k(ELEM, i, j)$, where I, J are the global entries and i, j are the local entries. Here a global/local index relation must be made to connect the local entry to the global entry when solution time begins. This is a relatively simple and efficient storage system to encode. The element-by-element strategy has other advantages with regard to element-by-element system solvers. This is is trivial in one dimension; however, it can be extremely complicated in three dimensions. Major software exists which can automate these procedures. Some of the most widely used are:

- Abaqus, located at http://www.hks.com/
- Adina, located at http://www.adina.com/
- Ansys, located at http://www.ansys.com/
- LS-Dyna, located at http://www.lstc.com/
- Nastran, located at http://www.mscsoftware.com/

This is by no means a comprehensive list.

12 SUMMARY

Classical older techniques construct approximations from globally kinematically admissible functions. Two main obstacles arise: (1) It may be very difficult to find a kinematically admissible function over the entire domain and (2) if such functions are found, they lead to large, strongly coupled, and complicated systems of equations. These problems have been overcome by the fact that local approximations, i.e., over a very small portion of the domain, are possible that deliver adequate solutions and simultaneously lead to systems of equations which have an advantageous structure amenable to large-scale computation by high-speed computers. This piecewise or "elementwise" approximation technique was recognized at least 60 years ago by Courant[3]. There have been a variety of such approximation methods to solve equations of mathematical physics. The most popular is the finite-element method. The central feature of the method is to reduce the field equations in a physically sound and systematic manner to an assembly of discrete subdomains, or "elements." The process is designed to keep the resulting algebraic systems as computationally manageable and memory efficient as possible.

The implementation, theory, and application of the FEM is a subject of immense literature. We have not attempted to review this huge field. For general references on the subject see the standard books by Bathe,[4] Becker et al.,[5] Hughes,[6] Szabo and Babuska,[7] and Zienkiewicz and Taylor.[8]

REFERENCES

1. L. Malvern, *Introduction to the Mechanics of a Continuous Medium,* Prentice-Hall, Englewood Cliffs, NJ, 1968.
2. J. E. Marsden, and T. J. R. Hughes, *Mathematical Foundations of Elasticity,* Prentice-Hall, Englewood Cliffs, NJ, 1983.
3. R. Courant, "Variational Methods for the Solution of Problems of Equilibrium and Vibrations," *Bull. Am. Math. Soc.,* **49,** 1–23 (1943).
4. K. J. Bathe, *Finite Element Procedures,* Prentice-Hall, Englewood Cliffs, NJ, 1996.
5. E. B. Becker, G. F. Carey, and J. T. Oden, *Finite Elements: An Introduction,* Prentice-Hall, Englewood Cliffs, NJ, 1980.
6. T. J. R. Hughes, *The Finite Element Method,* Prentice-Hall, Englewood Cliffs, NJ, 1989.
7. B. Szabo, and I. Babúska, *Finite Element Analysis,* Wiley-Interscience, New York, 1991.
8. O. C. Zienkiewicz, and R. L. Taylor, *The Finite Element Method,* Vols. I and II, McGraw-Hill, New York, 1991.

CHAPTER 17

DESIGN FOR SIX SIGMA: A MANDATE FOR COMPETITIVENESS

James E. McMunigal
MCM Associates
Long Beach, California

H. Barry Bebb
ASI
San Diego, California

1 BACKGROUND

In contrast to the relatively standardized Six Sigma define, measure, analyze, improve, and control (DMAIC) process, an array of very different implementations of Design for Six Sigma (DFSS) has emerged. The different versions are identified by acronyms derived from the first character of the phase names, such as IDDOV, DIDOV, DMADV, DMEDI, ICOV, I^2DOV, and CDOV. While the various renditions of DFSS reveal significant differences, common themes can be found. In the broadest context, Six Sigma is characterized as an improvement process while DFSS is characterized as a creation process. At a somewhat lower level, most renditions of DFSS contain treatments of customer requirements, concept development, design and optimization, and verify and launch structured into the process phases like those mentioned above. For example, the five phases of IDDOV are identify project, define requirements, develop concept, optimize design, and verify and launch.

Conversely, the differences between implementations of DFSS can be large. The greatest difference resides in the choice of the foundation discipline, namely, statistical methods versus engineering methods. Renditions of DFSS that are tightly linked to Six Sigma process improvement methodologies are appropriately founded on statistical methods for consistency. Renditions of DFSS that are tightly linked to product development or improvement processes are best founded on engineering methods in order to capture the maximum possible benefits.

The IDDOV rendition of DFSS is described in the first book published on Design for Six Sigma[1] and is chosen as the basic rendition for this chapter. While Chowdhury's DFSS book[1] was written for managers, this chapter is written for engineers and engineering managers.

2 INTRODUCTION

Few executives would dispute the observation that global competitiveness of a manufacturing corporation is largely determined by its engineering capability. Curiously, few of the same executives would dispute the observation that they do not spend a major portion of their time focusing on how to improve the engineering capability within their corporation. The authors believe that *the best products come from the best engineering organizations* and that *implementation of a properly formulated Design for Six Sigma process is the fastest way to create "the best engineering organization."* These beliefs spawned the title of this chapter.

The Six Sigma DMAIC process has enjoyed unprecedented success, as illustrated by a recent article in *ASQ Six Sigma Forum Magazine.*[2] The Six Sigma process first designed by Mikel Harry and Richard Schroeder, the founders of the Six Sigma Academy, brings all of the key success factors together into a coherent process—top management commitment, a dedicated management structure of champions, and an effective DMAIC process that delivers rapid financial gains to the bottom line, as summarized in their book.[3] Chowdhury further emphasizes the combination of the power of people and the power of process in *The Power of Six Sigma*[4] and *Design for Six Sigma: The Revolutionary Process for Achieving Extraordinary Profits.*[1]

Design for Six Sigma would seem to be a natural sequel to Six Sigma. As Harry and Schroeder[3] point out, "Organizations that have adopted Six Sigma have learned that once they have achieved quality levels of roughly five sigma, the only way to get beyond the five sigma wall is to redesign the products and services from scratch using Design for Six Sigma (DFSS)." This and other broadly recognized and cited needs for DFSS have not yet led to the kind of pervasive application that Six Sigma continues to enjoy, perhaps because of a mixture of myths and truths.

Myths include:

- DFSS is only used when needed in the improve phase of Six Sigma.
- Six Sigma should be implemented prior to undertaking a DFSS initiative.
- DFSS is a collection of contemporary engineering methodologies.
- DFSS is not needed because all of the methods and tools are used in the product development process (PDP). (They may be available, but are they actually used?)

Truths include:

- There is no single process that one can say is DFSS. Multiple renditions of DFSS have been published, none of which has been broadly accepted as the basic model of DFSS.
- Payback from DFSS projects may be larger than from Six Sigma projects, but it takes longer.
- Credible renditions of DFSS encompass Taguchi robust optimization methods that appear to be in opposition to statistical methods such as classical design of experiments. The unfortunate controversy that results creates unnecessary confusion for executives.

An objective of this chapter is to dispel the myths and provide a fundamentally sound model of DFSS which the authors believe should serve as the basis for current applications and future renditions of DFSS in engineering environments. Tailoring DFSS to meet the particular needs of different industries and corporations is expected, encouraged, and accommodated.

The authors' experience has been that the IDDOV methodology described in Chowdhury's book on Design for Six Sigma[1] is the best methodology. IDDOV is designed to complement, not replace, the engineering processes indigenous to a corporation's technology and product development processes.

The design of this rendition of IDDOV is not ad hoc. The design is based on benchmarking an array of DFSS processes, research-and-development (R&D) processes, and engineering processes that support PDPs and the authors' experience in dozens of corporations in a broad range of industries, including medical instrumentation, heavy equipment, office equipment, fixed and rotary wing aircraft, automotive OEMs (original equipment manufacturers) and suppliers, chemical, and others in North America, Europe, and Asia.

Major advancements in engineering processes introduced to the Western world within the past 25 years include quality function deployment (QFD), Pugh concept generation and selection methods, TRIZ, Taguchi Methods®, axiomatic design, Six Sigma, and the various renditions of Design for Six Sigma. Many of these powerful methods and tools were not in the engineering curriculum when the more experienced engineers within corporations and other enterprises attended universities. Many, if not most, of the methodologies and tools are not yet taught in sufficient depth in most universities. It is left to hiring organizations to provide education and training in these contemporary methodologies. Six Sigma and Design for Six Sigma provide the "Trojan horse" for effectively bringing these methodologies into an organization.

The IDDOV process selected is consistent with and embodies Taguchi Methods®, known as quality engineering in Japan. The portion of quality engineering, known as robust engineering in the West, encompasses three phases—concept/system design, parameter design, and tolerance design—which correspond to the develop concept and optimize design phases of IDDOV. Indeed, robust engineering is the heart of IDDOV.

All of these factors strongly influenced Chowdhury's DFSS book[1] and the rendition of IDDOV presented in this chapter.

DMAIC and IDDOV Application Domains. DMAIC is an improvement process that digs down into a particular portion of an existing process or product. IDDOV is a creation process that parallels a PDP. From a logical perspective, DMAIC works vertically between a high-level problem area and a low-level root cause. IDDOV works horizontally along a time line through the various phases. This vertical versus horizontal logic leads to an unconventional representation of DMAIC, as illustrated in Fig. 1.

While IDDOV proceeds horizontally with a PDP, the Six Sigma DMAIC process works vertically down through the various system levels to find and correct root causes and then works back up through the system levels to complete the improve and control activities. DMAIC is a powerful process for improving existing processes. The intensity of DMAIC *find-and-fix* activities increases as products proceed toward production and launch.

The Big Picture. A system perspective of a PDP for a complex system is provided in Fig. 2 that shows how IDDOV fits with product development. The sequence of chevrons positioned over a winged-U system diagram represents a typical PDP. The left leg of the winged

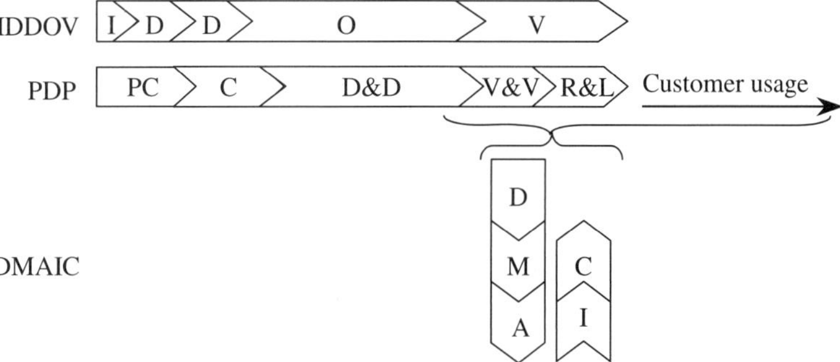

Figure 1 Unconventional representation of intersection of DMAIC and PDP shows DMAIC with a vertical orientation.

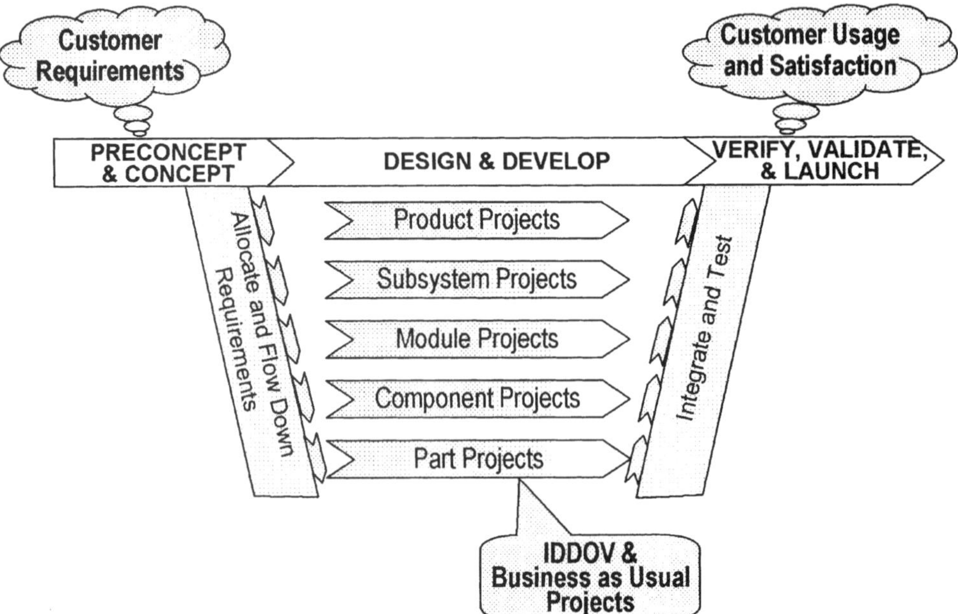

Figure 2 System perspective of a PDP for a complex system.

U depicts the process of allocating and flowing down requirements to guide the development of system architecture and the various lower level system elements within the architecture. Multiple IDDOV system element projects are depicted at the various system levels. As system element designs are completed, the process of integrating lower level system elements into higher and higher level system elements commences as depicted by the right leg of the winged U.

3 DFSS MANAGEMENT

The magic that has made Six Sigma and DFSS initiatives succeed where other quality initiatives have failed is largely attributable to the innovative management process created and published by Harry and Schroeder.[3] The management process features a hierarchy of "champions" spread through the organizations of an enterprise. The champions are dedicated to making the Six Sigma or Design for Six Sigma initiative succeed. To ensure long-term success, the CEO must serve as the chief champion with highly visible, unbridled exuberance. CEOs often appoint an executive champion to help lead the daily activities of the hierarchy of champions.

The management system contains many elements, including intense multiple-week, project-based training of "black belt" candidates who lead the projects. A hierarchy of belts is defined. It includes "master black belts" (MBBs) who lead the training and manage larger projects and lower level "green belts" who receive roughly half the training provided for black belts. The relationships of champions and belts are indicated in Fig. 3.

The management system helps a corporation achieve a lasting transformation rather than adding DFSS to a long string of failed quality initiatives—monthly fads. The management

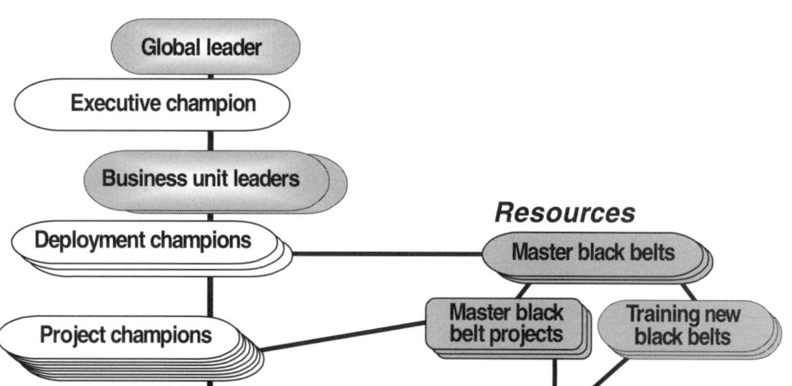

Figure 3 Relationships between line management, champions, and belts.

system contains two components: (1) a high-level enterprise system for planning, launching, deploying, growing, sustaining, and periodically revitalizing the initiative as outlined above and (2) a project management system to ensure excellence of execution of the DFSS process that leads to superior outcomes.

Project selection is a central management task. Selecting the right projects is a primary determinant of success. Project selection is an ongoing process that remains the same from the first wave of training black belt candidates and subsequent waves through the life of the current version of DFSS when many projects are selected for trained and experienced black belts. Project selection is typically led by MBBs working under the guidance of champions, executives, and managers at levels within a big, medium, or small enterprise. During the initial portion of deployment, the MBBs are outside, experienced experts. Over time, an enterprise develops internal MBBs who assume the leadership role.

Criteria are established by top management against which the projects are evaluated. The criteria should be derived from an enterprise's strategic and tactical needs combined with the technical knowledge of the project selection team. In manufacturing corporations, tactical criteria include warranty cost problems; product cost problems; safety problems; regulatory problems; published customer satisfaction ratings from organizations such as J. D. Powers, *Consumer Reports,* and trade publications; and complaints from customers, dealers, service technicians, retail outlets, and others. Strategic criteria include resolution of known problems but focus more on creating new, innovative products and services that cause current customers to repurchase from the same corporation and attract swarms of new customers away from competitive offerings.

Projects are sorted into IDDOV, OV, and other buckets. Projects selected against strategic criteria tend to be full IDDOV projects. Projects selected against tactical criteria normally require only the OV phases of IDDOV or some other quality process. The OV projects focus on robust optimization of an existing, troublesome subsystem to eliminate the problems and prevent their reoccurrence in the next development program or in the field. The verify-and-launch portion of OV warrants some emphasis. Typically, find-and-fix activities late in a program cycle tend to "patch" problems with changes that do not go through the normal verification and validation processes due to time constraints. It should not be surprising that such patches often turn into warranty, safety, or regulatory problems in the field. The V of OV strives to foster full verification and validation of all changes, including late "fixes" from firefighting activities.

IDDOV projects focus on creating new system elements ranging in scope from low-level system elements to the entire product. IDDOV projects should be started as early as possible in R&D and the early activities of a product development program.

Potential Pitfalls. Understandably, people within the enterprise generally do not know much about IDDOV. This lack of knowledge about the IDDOV process impedes deployment efforts. Even champions and managers who have received some overview training may not yet be comfortable with IDDOV. Considerable real-time guidance is necessary to help people within an enterprise understand and sincerely buy in to the IDDOV process. People need some level of understanding and buy-in in order to effectively support project selection activities.

Advocates and adversaries are created with the introduction of any major, new initiative in any organization. Experience suggests that in medium and large organizations the ratio of adversaries to advocates is about 10 to 1. Visible adversaries are not necessarily obstructionist. However, invisible adversaries are a real threat to successful deployment of a new initiative. Advocates and adversaries occupy the leading and lagging tails of the distribution. The majority of employees are the "wait and see-ers" who occupy the middle 70–80% of the population. This human element is real and must be dealt with by the MBBs upfront. Pronouncements of firm commitment to the new initiative from on high can help, but taken alone, pronouncements—and even directions—from the CEO do not solve the daily human problems that arise during project selection activities.

Pet projects are often motivated more by emotion than logic, which can blur objectivity. Owners of pet projects may be advocates pushing for their project to be selected or adversaries who want no part of IDDOV and hide their projects from the MBBs. Some are good IDDOV candidate projects and some are duds. It can be very difficult for outside MBBs to tell the difference. The only protection against selecting the wrong projects is for the MBBs to dig relatively deeply into the details with the project team members. It is often hard to get people to take time away from what they regard as their real job to work with MBBs to select projects for inclusion in a process that they know nothing about.

Secrecy is another barrier to ferreting out the right projects. Employees are naturally and properly dubious about revealing corporate secrets, even with knowledge that nondisclosure agreements have been signed. Adversaries will use secrecy to hide their projects. Even conscientious advocates may tend to reveal the less sensitive elements of a project when full disclosure is needed to understand if a project is a suitable candidate for IDDOV.

Inadequate time may be allocated for project selection. The project selection process can take a few weeks or a few months depending on the size of the enterprise and the complexity of its products—but it must be done well, however long it takes.

The above pitfalls are couched in the context of initial deployment. However, deployment is a continuing activity over a period of time measured in years. Initial deployment does not involve the entire organization. Deployment is rolled out sequentially to different portions of an enterprise over time. An aged initiative is still new to the people downstream from earlier roll-outs. Even as internal MBBs begin to displace external consultants, the cited challenges continue to exist until the initiative is pervasive throughout the enterprise. Then, the challenge shifts toward maintaining momentum and excitement.

Project selection must be completed prior to the initiation of the project-based training of black belt candidates. The output of project selection is a set of project charters delivered to the team leaders (black belt candidates) selected to participate in training. The charter contains the information needed to define the scope of the project and to help the team develop project plans.

4 DFSS PROCESS

4.1 Introduction

Storyboards have become a common means of conveying a process in Six Sigma and DFSS. The storyboard of the IDDOV process is displayed in Fig. 4.

Abridged descriptions of the five phases that follow are derived from Ref. 5.

Identify project	Select project, refine its scope, develop project plan, and form high-powered team.
Define requirements	Understand customer, corporate, and regulatory requirements and translate them into technical requirements using QFD.
Develop concept	Generate concept alternatives using Pugh concept generation, TRIZ, and other innovative methods. Select the best concept design and technology set using Pugh concept selection methods and conduct failure mode and effects analysis (FMEA) to strengthen selected concept/technology set. The concept may be developed at several levels starting with the system architecture for the entire product. Then, concepts are developed for the various system elements as needed.

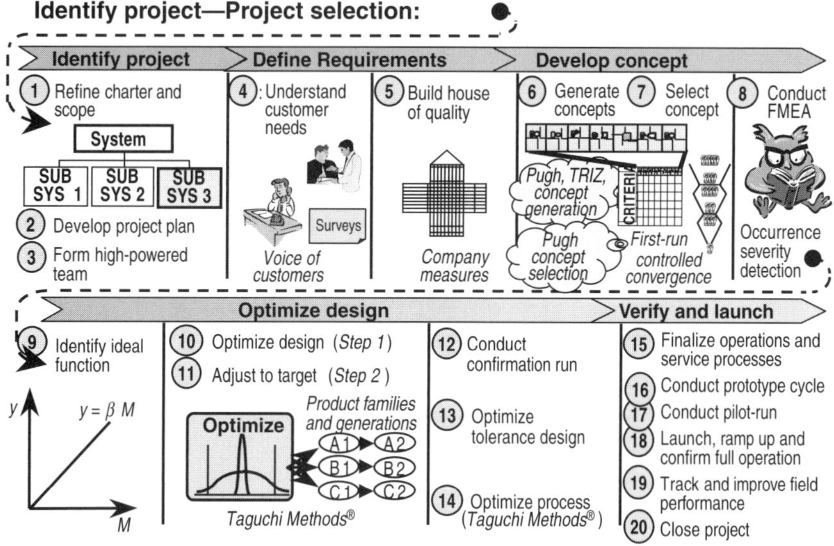

Figure 4 The visual storyboard of IDDOV showing the flow from project selection through the 20 process steps.

Optimize design	Optimize technology-set and concept design using robust optimization. Because Taguchi's robust engineering methodologies are central to the IDDOV formulation of DFSS, more detail is presented in this "O" phase.
Verify and launch	Finalize manufacturing process design, conduct prototype cycle, pilot run, ramp up to full production, and launch product.

The first two phases, identify project and define requirements, focus on getting the right product. The last two phases, optimize design and verify and launch, focus on getting the product right. The middle phase, develop concept, is the bridge between getting the right product and getting the product right.

As the bridge, conceptual designs should simultaneously respond to upstream requirements and downstream engineering, manufacturing, and service requirements. Optimization is one of many downstream requirements. If a concept design cannot be optimized to provide the required performance, it may be necessary to loop back and seek a new concept. Concept designs that cannot be optimized should be discarded. Marginal concepts are a major source of chronic problems that emerge in program after program. Concepts that do not optimize well are very sensitive to sources of variation. Hence, any deviation from ideal conditions causes a marginal concept to fail.

4.2 Process Details

The IDDOV process is described step by step through the five phases of IDDOV as outlined in the storyboard.

Identify Project. the "I" of IDDOV is the first phase of activities executed by the black belt candidate and project leader together with the team on the selected project. Intense project training commences. A MBB instructor may explain in the first day that "Projects are the means to get things done. The famous and overused list of action items seldom caused anybody to do anything that was important. Until action items are turned into projects with objectives, resources, and plans, very little can be accomplished." The instructor may emphasize that his or her first project is expected to deliver meaningful results to the enterprise with financial benefits measured in millions of dollars.

The identify project phase contains three storyboard steps: (1) Refine the charter and scope of the Project. (2) Develop the project plan. (3) Form a high-powered team.

Step 1: Refine charter and scope. Two scope tools are introduced—the *in–out of scope tool* and the *multigeneration plan (MGP) tool.*

The in–out of scope tool generally elicits a range of humor on first encounter since it is nothing but a circle drawn on a sheet of paper. However, it is magic when put to use. This simple visualization tool greatly facilitates team participation in understanding, refining, and reaching consensus on the scope of the project.

The MGP tool is a more involved tool to help the team determine what elements of scope should be included in the current project and what elements should be put off to future projects. The MGP typically contains three categories, such as product generation, platform generations, and technology generations, spread over three generations.

The primary purpose of both tools is to help define the scope of the current project, not to engage in long-range planning as the MGP tool might suggest.

Step 2: Develop project plan. Planning tools are more extensive. The foundation of all planning is the Shewhart–Deming plan–do–check–act (PDCA) circle. A special feature of the PDCA circle is the check–act step, which involves checking the progress against the plan and acting on the gaps on some predetermined cadence. The first use of the check–act pair should be to evaluate the completeness and doability of the completed project plan. Critical path/PERT (Program Evaluation Review Technique) is another important planning tool which combines the best elements of the two distinct methodologies into a single tool. A number of other planning tools like Gantt charts are introduced for use in the planning process.

Step 3: Form high-powered team. While project members should be identified prior to initiation of a project, some effort is usually required to bring them together as a team. Basic team-building methods for developing a high-powered team are introduced during training.

While this first critical phase, identify project, establishes the likelihood of the success of the project, the discussion is abbreviated since it is competently covered in broadly available Six Sigma and DFSS books and literature.

Define Requirements. The first "D" of IDDOV contains two critical-to-success actions, storyboard step 4, (understand customer requirements) and step 5 (build house of quality).

Step 4: Understand customer requirements. Customer, corporate, and regulatory requirements must be thoroughly understood if a team expects to deliver winning products to customers. T. Pare[6] suggests, "Don't overlook the most obvious way to get to know your customer—*talk to them.* Companies are exploring new ways of doing this." T. Peters[7] cites a study of 158 products in the electronics industry in which half were failures and half were successes:

- The unsuccessful products were technological marvels.
- The successes came from intense involvement with customers.

The successful firms had better and faster communications between customers and development teams. *Customer inputs were taken seriously!*

It is crucial to get the people responsible for delivering the products out and about with customers, with the support of marketing people. Throwing requirements over the wall from marketing to engineering is just as bad as engineering throwing drawings over the wall to manufacturing. Second-hand information cannot be as good as personal experience. Engineering and marketing people can be reluctant to have engineers talking with customers in prospect but are excitedly transformed in retrospect.

Don't Forget the Internal Customers. The primary internal customer is the recipient of the output of a project. Juran uses the phrase *Fit for use by next in line.* For the output of the project to be fit for use, the requirements of those next in line must be understood upfront. Stakeholders and sponsors are also customers. Treat internal customers as if they were external customers. In a formal sense, the boss is never the customer. Value-added activities flow horizontally. Vertical activities are overhead; they may be necessary, but they are not value adding.

The *Kano model*[8] was developed to classify requirements into basic needs, performance needs, and excitement needs. *Basic needs* are expected by customers and are often not spoken. Basic needs such as quality and reliability are established by industry standards. Customers also develop expectations about performance and features such as heating and cooling in automobiles. (Early automobiles did not have heaters or windshield defrosters. After all, buggies didn't!) *Performance needs* are items customers may desire that exceed the capability of their previous purchases. These needs are usually spoken in terms of more is better—larger flat-panel TV screens, easier to use alarm clocks in hotel rooms, more horsepower and less wind noise in vehicles. *Excitement needs* are seldom spoken because they are unknown. These needs are created by innovators within a corporation during concept development activities to gain competitive advantage.

Step 5: Build house of quality. The house of quality (HoQ) is the centerpiece of QFD. The HoQ provides an effective methodology for translating customer requirements into technical requirements. A completed HoQ contains an enormous amount of easy-to-access information that a team needs to guide the creation of concept designs that will create customer excitement. Technical requirements are often called company measures to indicate that they are the internal responses to external customer requirements.

The HoQ gets its name from the shape of the gabled matrix made up of different rooms for the different types of information contained in the HoQ as summarized below (Fig. 5). The HoQ is structured so that all customer information is entered horizontally and all company information is entered vertically. It is important to provide all customer information prior to undertaking the development of company information. The information is developed and entered into the HoQ in the order presented below.

Customer Information (Horizontal Entries)

1. Customer needs (*Gather the voice of the customer to identify customer requirements.*)
2. Critical customer requirements (CCRs) consolidated from "raw customer needs" (*To keep the HoQ from becoming too unwieldy, it is important to limit the number of customer requirements to those that are truly critical.*)
3. Company and regulatory requirements
4. Customer importance ratings for each CCR
5. Customer competitive comparisons of the company's product with competitive products on a CCR-by-CCR basis
6. Customer complaint history

Company Information (Vertical Entries)

7. Critical company measures (CCMs) consolidated from company measures

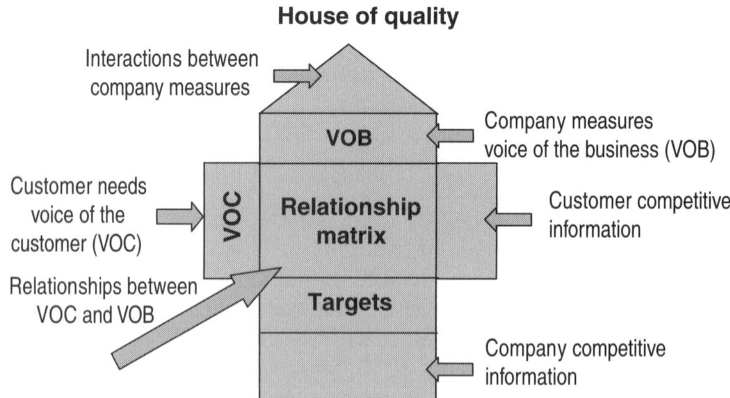

Figure 5 The HoQ with central rooms indicated.

8. Technical competitive comparisons for each CCM (*Tear down and compare competitive products.*)
9. Strengths of relationships between critical customer requirements and company measures (*A relationship matrix links company measures and customer measures. Strengths of the interactions between CCRs and CCMs are indicated by three weighng factors, typically 9, 3, 1 for strong, medium, and weak. The strength assigned each company measure indicates how much a change in the value of a company measure will change the degree of meeting customer needs.*)
10. Service history (*usually gathered at service centers*)
11. Target values for each company measure
12. Strengths of interactions between company measures (*The correlation matrix at the top of the HoQ indicates positive and negative interactions between the company measures. Negative interactions indicate a need for making trade-offs between company measures. A positive interaction indicates that the interacting company measures support each other in a way that can allow a reduction from ideal of one of the Company Measures without a significant impact on customer needs. Positive interactions sometimes provide opportunities for cost reductions.*)
13. Degree of organizational difficulty for each company measure (*Organizational difficulty relates to competency and capacity necessary to achieve the intent of the company measure.*)
14. Importance rating for each company measure (*This is a composite rating of the customer importance rating, the strength of relationships, and the degree of organizational difficulty to help guide the choice of company measures to be carried forward throughout the project. Selecting only the most critical company measures to carry forward is another opportunity to simplify the process.*)

Quality function deployment has fallen from favor in many corporations. The two main reasons for its fall from grace are as follows:

1. Teams try to develop so much information in the HoQ that it becomes unwieldy. It is difficult for people, especially engineers, to discard information even when it is pointed out that information not entered into the HoQ can be carried along with the HoQ.
2. The HoQ is best developed by a team of marketing and engineering people working together. In compartmentalized corporations, marketers and engineers seldom have either the opportunity or the desire to work together. Neither marketing nor engineering can gather customer information and build the HoQ without help from each other.

QFD remains the strongest method of gathering and interpreting customer information. The toughest competitors continue to pursue excellence in using QFD for competitive advantage. It is so important that most renditions of Six Sigma and Design for Six Sigma include QFD as a means of reintroducing it into Western world corporations.

Develop Concept. The second "D" of IDDOV—the develop concept phase—provides the maximum opportunity for innovation. Many pundits have asserted that the only way to achieve and sustain competitive advantage is to innovate faster than the toughest competitors. *In long races, whether leading or lagging, those who go fastest win!* Going fastest carries internal risk. Going slower than competitors carries external risk. During Jack Welch's long tenure as CEO of GE, he often emphasized, "Change or die."[9]

Innovations range from clever problem solving to breakthrough innovations. Innovation is the accelerator pedal within any enterprise (Fig. 6). Small and large innovations help to increase the rate of improvement within an enterprise. Innovations are created through the conceptual design process. Concept development utilizes a creativity toolkit that contains a broad array of tools and methods for generating, synthesizing, and selecting ideas for many purposes, including products, manufacturing processes, service processes, business processes, business strategies and plans, and innovative solutions for technical and nontechnical problems.

Winning products can only come from winning concepts. The creation of product concepts is simultaneously critically important, very difficult, highly exciting, and often neglected. Pugh,[10] the creator of Pugh concept generation and selection methods, states, "The wrong choice of concept in a given design situation can rarely, if ever, be recouped by brilliant detailed design." The concept design is really the first step of detailed product and process design. The quality of the concept design is a major determinant of the quality of the final design.

Some of the rationale behind Pugh's often-quoted observation is that concept development:

- Provides the greatest opportunity for innovation
- Places the greatest demands on the project team
- Is where *the most important decisions* are made
- Is where 80–90% of the cost and performance are locked in

A good concept design provides the foundation for all downstream endeavors, including optimization, detailed design, manufacturing, service, and customer usage. How well a concept design optimizes using Taguchi Methods® is a key measure of its quality. A concept design that does not optimize well is a poor concept. Poor concepts cannot be fixed by end-of-process firefighting. Indeed, poor concepts create chronic problems that seem to reappear in every new program and create disappointments in product performance, cost, quality, reliability, and useful life.

The huge amount of information that is generated during conceptual design places enormous demands on the team. Concept development is, indeed, a most difficult step in the PDP. In traditional development processes, the opportunity to achieve competitive superiority has been either won or lost by the end of concept development.

Conceptual design can be exciting since most engineers find the opportunity to create something new far more fun than the tedium of detailed design. Whether seeking better mousetraps, car door weather-strips, or flying machines, stretching one's imagination to cre-

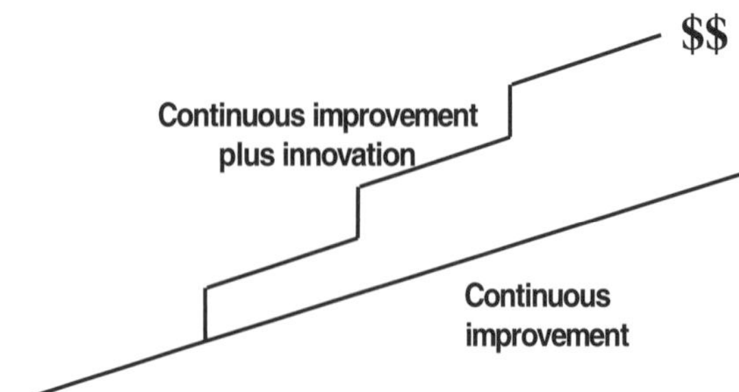

Figure 6 Innovation is accelerator pedal.

ate new widgets is fun, challenging, and rewarding. Where else in the PDP is a person likely to hear, "Eureka, we did it. I didn't think it was possible, but we did it!"

Conceptual design is often neglected because neither managers nor engineers appreciate its importance. Ever-increasing pressures to get product out the door on shorter and shorter schedules tend to disproportionately squeeze the time allocated for the unmanageable, "fluffy sandbox" of concept development. Managers seem to operate on the premise that good concepts can be developed in whatever abbreviated time is allocated to the task—the less time, the better.

Adequate time, resources, and support are needed to foster the pursuit of excellence of execution and the realization of meaningful results. The extra time needed to do things right up front is more than offset by time and resources saved later in the process. The hierarchy of champions can provide enormous support to ensure excellence of execution during the develop concept phase.

Step 6: Generate concepts. This step encompasses two very different innovation methodologies for generating concept alternatives:

(i) *Pugh concept generation* using a creativity toolkit. The toolkit contains familiar methods for creating ideas including brainstorming, brain-writing 6–3–5, painstorming, analogy, assumption busting, and other related methods.

(ii) *TRIZ or TIPS* (Theory of Inventive Problem Solving) created by the brilliant Russian engineer, Genrich Altshuller. TRIZ is a logical methodology to help anyone become creative, as suggested by the title of one of Altshuller's early books, Suddenly, an Inventor Was Born.

The output of the phrase define requirements—the HoQ—is the primary input into the develop concept phase. The HoQ information can be arranged and augmented as necessary to develop criteria for a good concept. Developing the criteria for a good concept should be completed prior to initiation of concept generation activities. The process flow for this phase is

HoQ >> criteria for a good concept >> concept generation (*Pugh and TRIZ*) >> concept selection >> controlled convergence >> system FMEA

Develop criteria for a good concept. A good concept satisfies upstream customer requirements (*get the right product*) and downstream engineering requirements (*get the product right*). Typically, criteria are not developed until concept selection. Developing criteria prior to concept generation has several benefits. First, the process of developing the criteria causes the team to collectively focus on the details in the HoQ and develop a common understanding of the upfront customer requirements. Second, development of the criteria requires consideration of the downstream technical requirements. Third, the criteria provide guidance during concept generation.

General, high-level criteria establish the framework for developing more detailed requirements specific to the project that a product concept must satisfy:

Upstream customer requirements: Satisfy internal and external customer requirements (HoQ).

Downstream engineering requirements: (1) Do not depend on unproven technologies, (2) can be optimized to provide high quality/reliability/durability at low cost, and (3) can be manufactured, serviced, and recycled at low cost.

The company measures with target values from the portion of HoQ for an automobile door system are displayed Fig. 7. The critical company measures—those selected to be carried forward against the criteria of new, important, or difficult—are boldface. Customer

Basic Door System

1	**Door closing effort outside**	**7.5 ft·lb**
2	Door opening effort outside	15 ft·lb
3	Door opening effort inside	8 ft·lb
4	Reach dist.-opening mech	26 in.
5	**Pull force inside**	**12 ft·lb**
6	Dynamic hold open force	15 ft·lb
7	Static hold open force	10 lb

Figure 7 Company measures for basic door system from HoQ.

importance for door-closing effort from outside and inside was rated 5 and 4, respectively. However, easy door-closing effort and good weather-strip sealing are in conflict, making the simultaneous achievement of easy closing and good sealing difficult. Hence, door closing effort was selected as the critical company measure to be carried forward because of both its importance and difficulty.

Consideration of the four, general, high-level requirements provides the basis for developing criteria to guide the generation and selection of alternative concepts. A team starts with the deployed upstream requirements and adds downstream technical criteria:

DEPLOYED REQUIREMENTS

Door-closing effort from outside	7.5 ft·lb	Closing effort
Pull force from inside	12 ft·lb	

TECHNICAL CRITERIA ADDED BY TEAM

No water leak	4-hr Soak/wind zero leak
Noise transmission	Test required
Wind noise	Test required
Design for assembly	Number of installation operations
Design for serviceability	Seal extraction/insertion forces
Robustness	Optimization experiment

Pugh's methodology involves three distinct activities:

(i) Concept generation is the process of creating a number (typically four to eight) of concept alternatives. TRIZ is an important complementary methodology for concept generation.

(ii) Concept selection is a process of synthesizing the best attributes of the alternatives into a smaller number of stronger alternatives and selecting a small number (one to three) of the stronger concepts.

(iii) Controlled convergence is the process of iterating between the first two activities with the objective of further synthesizing and winnowing down to the strongest one or two alternatives.

Concept generation, concept selection, and controlled convergence provide the framework for the develop concept phase of IDDOV.

For concept generation using Pugh Methods, concept generation should be planned as an exciting activity that draws out the highest possible levels of creativity from team members. People who choose product or manufacturing engineering as a profession tend to be more analytical than creative. Many, probably most, technical people within an organization do not regularly practice serious creativity and have little interest in starting now. Engineering environments seldom foster flamboyant creativity.

It is best to employ a trained facilitator who knows how to foster a free-flowing, creative environment and helps team members effectively select and use the tools within the creativity toolkit. Basic tools include brainstorming, brain-writing 6–3–5, pain-storming, analogy, and assumption-busting. With the exception of pain-storming, these tools are broadly known and treated in numerous creativity and quality books.

Pain-storming involves brainstorming the opposite of what you want to achieve. It is a way to help the team look at the problem from a different perspective. If the team gets stuck or reaches a dead point, the facilitator might suggest pain-storming to change the perspective by focusing on the "antisolution." Suppose your topic is how to speed up invoice preparation. Change the topic to the antisolution, how to slow down invoice preparation. Then, use brainstorming to generate ideas for the "anti" topic. This simple process of examining the topic from the opposite perspective usually stimulates a flood of new ideas. Pain-storming is sometimes called improvement's "evil twin."

Pugh's famous work hints at using three cycles for concept generation:

1. Group activity—gathering information and developing shared purpose
2. Individual activity—greating ideas
3. Group activity—combining, enhancing, improving, refining ideas

Organizing concept generation efforts around these three separate steps dramatically enhances the process.

The criteria for easy door closing and good sealing of an automobile weather-strip was developed above. The team used Pugh concept generation methods to create several alternatives, shown in Fig. 8.

Concept 7 was the most innovative. It consisted of an inflatable seal which was deflated to provide easy door closing and inflated after the door was closed to provide good sealing. This study was conducted in the mid 1980s. The inflatable seal was used by Mercedes about 10 years later.

For *concept generation using TRIZ methods,* the inflatable seal could have been derived using the TRIZ principles of separation. The separation principles include (1) separation in

Display drawings with descriptors

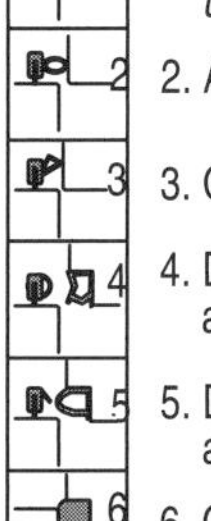

1. A basic compression strip of sponge rubber *(the company's current concept, which is used as the datum)*
2. A deflection strip of sponge rubber
3. Combination of compression/deflection using thinner walled section sponge rubber
4. Double-sealing strips with body side compression and sponge rubber. Body strip attached with plastic nails.
5. Double-sealing strips with body side deflection and sponge rubber. Body strip attached with plastic nails.
6. Compression/deflection using foam rubber pressed into metal carrier
7. Thin-wall low-compression strip which inflates with air after door is closed. This concept reduces door-closing effort casued by "trapped" internal air.

Figure 8 Concept alternatives for automobile door weather-strip.

space, (2) separation in time, and (3) separation in scale. The inflatable seal is an example of separation in time—deflated at one time and inflated at another time (*before and after door closing*). (For more information on the TRIZ methodology, refer to Chapter 18.)

Step 7: Concept selection. This is a process of synthesizing (*combining and separating*), enhancing the strengths, and attacking the weaknesses of the concept alternatives as an integral part of the process of selecting the best concept. Pugh's concept selection methodology is more than a process for simply selecting the best of the concept alternatives generated. It is as much a concept improvement methodology as it is a selection methodology. The steps in concept selection are as follows:

1. Prepare for First Run
   - **a.** Prepare characterizations of concepts—drawings, models, word descriptions, videos, working prototypes, etc.
   - **b.** Identify evaluation criteria for synthesis and selection of concepts.
   - **c.** Prepare evaluation matrix with drawings of concepts across the top row and criteria down the first column.
   - **d.** Select datum, usually the current concept. (*A better choice is the best competitive concept.*)
2. Conduct first run.
3. Conduct confirmation run.
4. Conduct controlled convergence runs (*additional runs to exhaustion of new concepts*).

(a) *Pugh Step 1: Prepare for first run.* The characterization of the concept alternatives is illustrated in Fig. 8. The second item, identify evaluation criteria, was done prior to concept generation. The next two items, prepare evaluation matrix and select datum, are combined with Pugh step 2 (conduct first run) in Fig. 8, which shows the information from the first run in the evaluation matrix.

(b) *Pugh Step 2: Conduct first run.* This step generates the matrix depicted in Fig. 9. The entries in the matrix, pluses, minuses, and sames (+'s, −'s, s's) are determined by comparing each concept with the datum concept for each criterion and determining whether the alternative concept is better, worse, or the same as the datum concept. The sums are used to evaluate the alternatives. There is no value in summing the s's. Pugh makes a number of suggestions about attacking the negatives and enhancing the positives in the synthesis process of striving to use the best attributes of strong concepts to turn negatives into positives in weaker alternatives and enhancing the positives of the stronger alternatives. Alternative 5, with the circled eight positives and four negatives, was selected as the strongest alternative.

(c) *Pugh Step 3: Conduct confirmation run.* This step is carried out by using the selected alternative as the datum and running the matrix a second time.
The output of steps 1–3 is typically one to three relatively strong concepts.

(d) *Pugh Step 4: Conduct controlled convergence runs.* This step involves multiple iterations of Steps 1–3 to further improve the concept. Often the most innovative ideas arise during controlled convergence runs when the team members have become very familiar with all of the concept alternatives and have generated new ideas during the synthesis process. It is a powerful methodology of intertwining divergent and convergent thinking to improve, often dramatically, product concepts. However, it is too often passed over as excessively time consuming under the pressures to get on with the real work of detailed design. Pugh countered arguments for shortcutting any of the concept selection process with "One thing is certain. It is extremely easy to select the wrong concept and very difficult to select the best one."[10]

Criteria \ Concept	1	2	3	4	5	6	7
Closing effort		+	+	+	+	S	+
Compression		+	+	+	+	−	+
Set		+	+	+	+	+	+
Meet freeze test	DATUM	S	S	−	S	+	S
Durability		S	S	S	S	−	−
Section change at radius		−	S	+	+	+	+
Squeak		S	S	S	S	S	S
Water leak		S	S	+	+	S	+
Wind noise		−	S	+	+	S	+
Pleasing to customer		S	S	S	+	S	S
Accommodate mfg. var.		S	S	+	S	S	−
Process capability		S	−	−	−	+	−
Cost		S	S	−	−	S	−
No. installation operations		S	S	−	−	S	S
R & R for repair		S	S	−	−	+	S
Robustness		−	S	+	+	S	+
Total +		3	3	8	8	5	7
Total −		3	1	5	4	2	4

Figure 9 Evaluation matrix with +'s, −'s, s's entered into matrix.

Some teams insist on using weighted matrices in the concept selection process. Weighting unnecessarily complicates the process. Numerical weighting of criteria is good practice when making a single-pass decision about known alternatives such as selecting automobiles, computers, cameras, or other items based on quantitative information about performance, features, price, etc. However, weighting criteria is not needed or advised when synthesizing attributes between alternatives to create improved or entirely new concepts, and weighting becomes unmanageable when conducting multiple runs in controlled convergence.

It is useful to rank order the criteria and look for concept alternatives with lots of +'s in the upper portion of the evaluation matrix.

Finally, note that robustness is the last criteria listed in the evaluation matrix. (*The team did not rank order the criteria.*) The final test of the strength of a concept selected is how well it optimizes. If the team selected more than one concept for further evaluation, conducting robust optimization experiments can be used to make a final selection.

The controlled convergence process is depicted in Fig. 10.

Step 8: Conduct FMEA. FMEA is a powerful methodology for strengthening the selected concept. At this stage the system FMEA is conducted. Refer to numerous books on FMEA for details.

Optimize Design. The "O" of IDDOV focuses on the Taguchi Methods® as the "heart" of IDDOV. Taguchi's robust optimization requires engineers and others to think differently in fundamental ways. Some of the premises in the Taguchi Methods® that require breaking traditional engineering thought patterns include the following:

A. *Work on the intended function, not the problems (unintended functions).* Problems such as heat, noise, vibration, degradation, soft failures (restart computer, clear paper jam in copier), high cost, etc., are only symptoms of poor performance of the intended function. Working on a problem such as audible noise does not necessarily improve the intended

Controlled convergence intertwines generation, synthesis, and selection in ways that strengthen both creativity and analysis.

It involves alternate convergent *(analysis)* and divergent *(synthesis)* thinking.

It helps the team to attack weaknesses and enhance strengths.

As team members learn more and gain new insights, they will consistently derive and create new, stronger concepts.

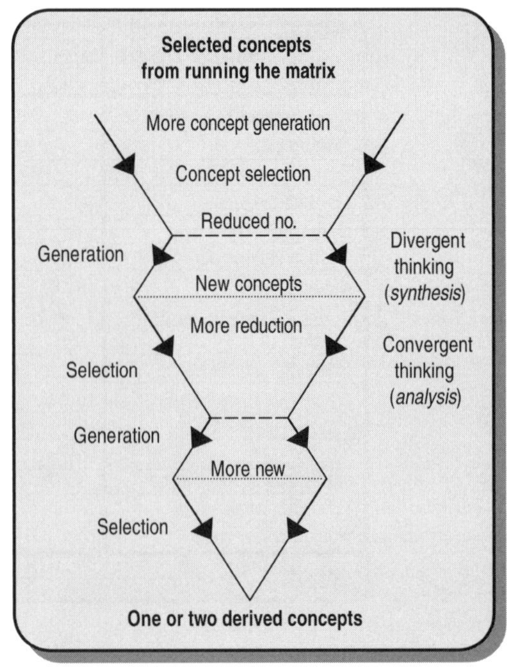

Figure 10 Visual depiction of controlled convergence process.

function. More often, another problem pops up. This is pejoratively called "whack-a-mole engineering," after the children's game. Whack one mole and another pops up somewhere else—whack one problem and another pops up somewhere else. The notion of optimizing the intended function to whack all moles with one whack rather than working directly on the problem (say, customer complaints about audible noise) is totally upside-down thinking for engineers who have spent much of their working life whacking problems.

B. *Strive for robustness rather than meeting requirements and specifications.* This is another notion that demands thinking differently. Traditional engineering measures of performance include meeting requirements and specifications, reliability data, warranty information, customer complaints, Cp/Cpk (Cp is the capability without respect to target, and Cpk is the capability with respect to target process capability), sigma level, scrap, rework, percent defective, etc.

C. *Use the signal-to-noise ratio as the measure of performance* (*robustness*). Increasing the signal-to-noise ratio simultaneously improves the performance against all traditional measures—whacks all moles with a single whack. The signal-to-noise ratio, borrowed from the communications industry, is simply a representation of the ratio of the power in the signal (the music) to the power in the noise (the static). The greater the signal-to-noise ratio, the greater the portion of total power that goes into making music compared to the power available to make static—by the law of conservation of energy. In an ideal AM/FM receiver, or more generally, the ideal function of any system, all of the power goes into making music, the intended function; no power remains to make static, the unintended function.

D. *Use multifactor-at-a-time* (*MFAT*) *experiments rather than the scientific method of changing one-factor-at-a-time* (*OFAT*). Engineering involves many factors that may interact with each other. OFAT experiments cannot reveal interactions. MFAT experiments are needed to characterize and optimize a system involving multiple interacting factors, $y = f(x_1, x_2, \ldots, x_n)$.

E. *Strive for the ideal function, not just meeting requirements and specifications.* The ideal function is a basic concept in robust engineering. Striving to get a system to perform as closely as possible to the ideal rather than striving to meet requirements and specifications requires thinking differently. Striving for the ideal function may seem like the impossible dream of striving for a perpetual-motion machine. Nevertheless, the measure of how far the value of the actual function differs from the value of the ideal function is a useful measure of robustness. The objective of robust optimization is to maximize the relative volumes of music and static by getting the actual function as close as possible to the ideal function.

The ideal system response, y, is a linear relation to the input M, $y = \beta M$. An actual response to some input M is some function of M together with various system parameters, $x_1, x_2, \ldots, x_n$, $y = f(M, x_1, x_2, \ldots, x_n)$. The actual function may be rearranged into two parts, the ideal function (useful part) and the deviation from the ideal function (harmful part), by adding and subtracting the ideal function, βM:

$$y = f(M, x_1, x_2, \ldots, x_n) = \underbrace{\beta M}_{\text{Ideal function}} + \underbrace{[f(M, x_1, x_2, \ldots, x_n) - \beta M]}_{\text{Deviation from ideal function}}$$

The ideal function represents all the radio signal energy going into the music with no energy available to cause static. The deviation from ideal is the portion of energy in the actual function that goes into causing static. Robust optimization is the process of minimizing the deviation from the ideal function by finding the values of controllable parameters that move the actual function, $f(M, x_1, x_2, \ldots, x_n)$, as close as possible to the ideal function, βM.

Some parameters are not controllable, such as environmental and usage conditions, variations in materials and part dimensions, and deterioration factors such wear and aging. Uncontrollable factors are called noise factors. When the value of the actual function remains close to the values of the ideal function for all anticipated values of the noise factors, the system is robust in the presence of noise. Such a system is insensitive to sources of variation, the noise factors.

Taguchi defines good robustness as "the state where the technology, product, or process performance is minimally sensitive to factors causing variability (either in manufacturing or user's environment) and aging at the lowest unit manufacturing cost."[11]

Figure 11 provides a graphical representation of the above discussion.

Working on the problems, the symptoms of poor function, does not necessarily improve the intended function. In energy terms, all functions are energy transformations. Reducing

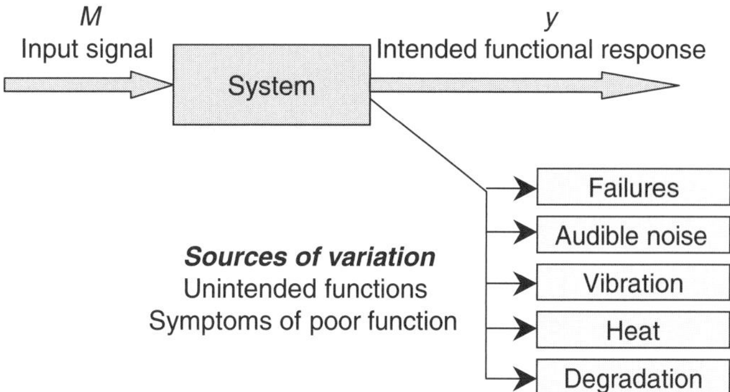

Figure 11 Energy transformations of a system into intended and unintended responses—*music and static.*

the energy in the unintended functions does not necessarily increase the energy transformed into the intended function. It may just pop up as another unintended function.

A simple example is provided by a case study[12] of an automotive timing belt. The problem was excessive audible noise. The team worked for more than a year to successfully reduce the audible noise. However, the solution also reduced the already short life of the belt by a factor of 2. Subsequent use of robust optimization reduced the audible noise by a factor of 20 and simultaneously doubled the life of the belt. Robust optimization focused on the intended function of the belt, namely, to transfer torque energy from one pulley to another pulley. Similar results have been achieved with more complex systems such as internal combustion engines by optimizing energy efficiency to reduce audible noise—once again, work on the intended functions, not the problems.

Step 9. Identify ideal function. This is the first, and often the most difficult, step in robust optimization. The representation of the ideal function as displayed in Fig. 12 is deceptively simple. The challenge is to determine what to measure. How should M and y be chosen for the optimization process? Said differently, what is the physics or chemistry of the system? What is the function to be optimized? Optimization is best performed on a single function. However, optimization can be conducted for systems involving multiple inputs and multiple outputs. The first step is to conduct functional analysis to break down the system to lower level, manageable elements which can range from the part level to complex subsystems. It is important to keep in mind that the focus is on variation of the function of the part or subsystem, not on variation in the part or parts that make up a subsystem.

Figure 12 illustrates the optimization process of striving to move as closely to the ideal function as possible as described above.

Step 10: Optimize design (Step 1 of Two-Step Optimization). This step focuses on maximizing the signal-to-noise ratio (S/N) to minimize variation of the function from the ideal function:

$$\mathrm{S/N} = 10 \log \left(\frac{\beta^2}{\sigma^2}\right)$$

where β^2 is a factor related to the energy in the signal and σ^2 is a factor related to the energy in the noise (sources of variation, not audible noise). Methods for calculating S/N from experimental data or math models are provided in numerous books.[13–15] Optimizing design involves a number of actions. A case study is used to illustrate the process.

Case Study. A published case study on wiper system chatter reduction[16] conducted by Ford Motor Company is used as an example. Windshield wiper chatter is the familiar, noisy skipping of the wiper blade that deteriorates the intended function of cleaning. The ideal

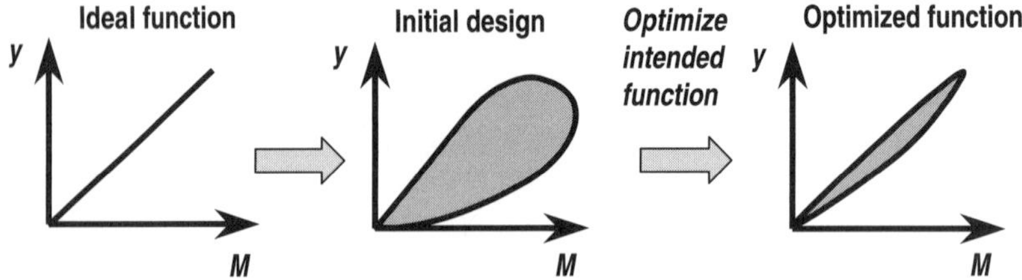

Figure 12 Comparisons between the ideal function and the variation from ideal in the actual function before and after optimization.

function is based on the assumption that for an ideal system the actual time to reach a point D_n during a wiping cycle should be the same as the designed target time:

$$Y_n = \beta M_n$$

where Y_n = actual time that blade reaches fixed point on windshield at nth cycle
M_n = designed target time that blade reaches fixed point on windshield at nth cycle
$\beta = n/\text{RPM}$ of motor

Three RPMs used in the experiments were M_n values of 40, 55, and 70 RPM.

Noise Strategy. Noise is generally categorized into (1) part-to-part variation, (2) deterioration and aging, (3) customer usage conditions (duty cycle), and (4) environmental conditions. The team used these general categories to define the specific noise factors entered into the lower portion of the P-diagram (*the customer usage space*) displayed in Fig. 13. The P-diagram, or parameter diagram, provides a convenient and orderly way to organize and display control factors, noise factors, input signal, and output response.

To simplify the experiment, the Ford team compounded the noises at two levels. The team deemed that wet or dry surface would be a dominant source of variation, so they kept it as a separate noise factor with the following definitions:

$$T_1 = \text{wet-surface condition}$$

$$T_2 = \text{dry-surface condition}$$

Several noise factors were compounded into two additional noise factors:

$$S_1 = -3^\circ \text{ at park/20°C/50\% humidity/before aging}$$

$$S_2 = 1^\circ \text{ at park/2°C/90\% humidity/after aging}$$

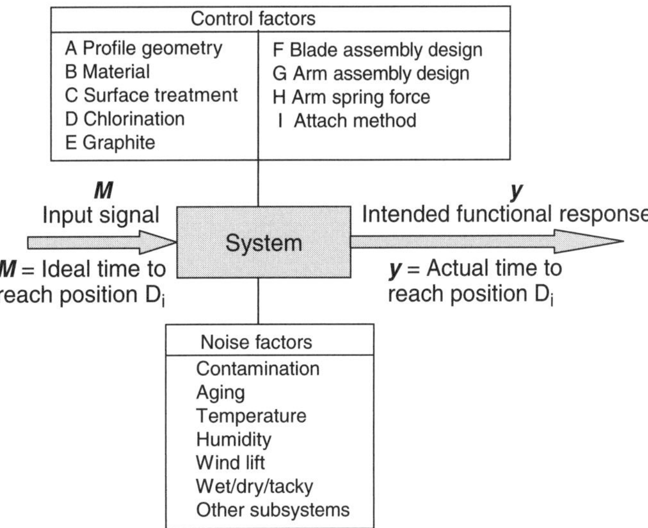

Figure 13 The P-diagram displays function inputs and outputs, and control and noise factors in an orderly manner.

At park refers to the parked position of the wiper blade. *Aging* is typically accomplished by using old parts, presumably old rubber blade inserts. *Before aging* means the use of new parts.

In robust optimization, it is not necessary to test at the extremes. All that is necessary is to ensure that the noise factors are significantly different to test the robustness of the system. This is a major advantage of robust optimization in shortening the length of time to conduct experiments and reducing the need to test to failure.

CONTROL FACTORS. The engineering design parameters are the control factors. The team brainstormed to select the design parameters that they believed had the largest impact on controlling wiping performance as the control parameters. The control parameters are entered into the upper portion of the P-diagram (the engineer's design space).

CONTROL FACTOR LEVELS. Robust optimization is a methodology for determining the values of the design parameters that optimize system performance. The team used their engineering knowledge and brainstorming to select values of the design parameters (control factors) that they believed covered the design space of optimum performance. A portion of the control factor levels is displayed in Table 1. The control factors not shown—*G*, graphite; *H*, chlorination; and *I*, attach method—have similar levels.

DESIGN OF THE EXPERIMENT. The team selected the popular L_{18} orthogonal array shown in Fig. 14. The popularity of this array stems from the fact that interactions between the control factors are reasonably balanced within the array. The standard L_{18} array contains up to eight control factors. The array in Fig. 14 was modified to accommodate nine control factors.

Each row of the 18 rows is one experimental run with the control factors set at the indicated levels—1, 2, or 3. When conducting experiments with a hardware/software system, 18 different sets with the indicated mixtures of levels are needed to conduct the 18 runs. For the wiper system, the only hardware changes were to park positions compounded with new and aged wiper blades. Two additional sets are needed to (1) establish the performance of the initial design and (2) experimentally confirm predictions made from the 18 runs.

Data are collected in the outer array for all combinations of wet/dry, S_1/S_2, and $M_1/M_2/M_3$ for a total of 12 data values for each of the 18 runs. If a particular combination in one of the runs is reasonably robust, the differences in the values of the data between different combinations of noise factors will be small. Less robust combinations of control factor levels will have larger differences in the values of data points. In any case, the trend in data values will track the RPM input values of 40, 55, and 70.

In this case study, a special test fixture was built with three sensors attached to the windshield to record a signal as the wiper blade passes by in order to determine the actual time to point D_n.

Table 1 Representative Portion of Control Factor Table

Control Factor	Level 1	Level 2	Level 3
A: Arm lateral rigidity	Design 1	Design 2	Design 3
B: Superstructure rigidity	Low	Medium	High
C: Vertebra shape	Straight	Concave	Convex
D: Spring force	Low	High	(None)
E: Profile geometry	Geo 1	Geo 2	Geo 3
F: Rubber material	Material 1	Material 2	Material 3

Run #	D	A	B	C	E	F	G	H	I	M_1 = 40 RPM				M_2 = 55 RPM				M_3 = 70 RPM			
										S_1		S_2		S_1		S_2		S_1		S_2	
										Wet	Dry	Wet	Dry	Wet	Dry	Wet	Dry	Wet	Dry	Wet	Dry
1	1	1	1	1	1	1	1	1	1												
2	1	1	2	2	2	2	2	2	1												
3	1	1	3	3	3	3	3	3	1												
4	1	2	1	1	2	2	1	3	2												
5	1	2	2	2	3	3	2	1	2												
6	1	2	3	3	1	1	2	2	2												
7	1	3	1	2	1	1	2	3	3												
8	1	3	2	3	2	2	3	1	3												
9	[illegible]	[illegible]	[illegible]	[illegible]	[illegible]	[illegible]	[illegible]	2	3												
[illegible]	[illegible]	[illegible]	[illegible]	[illegible]	[illegible]	[illegible]	[illegible]	[illegible]	2												
[illegible]	[illegible]	[illegible]	[illegible]	[illegible]	[illegible]	[illegible]	[illegible]	[illegible]	[illegible]												
[illegible]	[illegible]	[illegible]	[illegible]	[illegible]	[illegible]	[illegible]	[illegible]	[illegible]	2												
13	[illegible]	[illegible]	[illegible]	[illegible]	[illegible]	[illegible]	[illegible]	2	3												
14	2	2	2	3	1	2	1	3	3												
15	2	2	3	1	2	3	2	1	3												
16	2	3	1	3	2	3	1	2	1												
17	2	3	2	1	3	1	2	3	1												
18	2	3	3	2	1	2	3	1	1												

Figure 14 Modified L_{18} orthogonal array showing the inner array with levels 1, 2, or 3 indicated for each row representing one experimental run. The outer array is for collecting the data for each of the 18 runs under the inputs M_1, M_2, M_3, and combinations of the noise conditions S_1, S_2 and wet, dry, as shown at the top of the columns of the outer array.

Typical experimental results from one run are shown in Fig. 15. The graphical representation of the values of β and σ are indicated on the chart. Their actual values are of course calculated from least-square fits for the slope, β, and the square root of the variance for σ.

The values S/N and sensitivity (or efficiency) are calculated for each run using the following formulas:

$$\text{S/N} = 10 \log \left(\frac{\beta^2}{\sigma^2}\right) \qquad \text{Sensitivity} = 10 \log \beta^2$$

The S/N and β are calculated for each run using formulas that are not reproduced here. Columns for S/N and β are conveniently added to the right side of the outer array.

To determine the S/N and β for a particular control factor at a particular level, a small bit of work is needed. For example, consider D_1. To determine S/N and β for D_1, find all of the D's in the L_{18} array at level 1. All D_1's appear in the first nine rows of the D column. Then average the nine S/N values of D_1. Repeat the process for β. The S/N for D_1 shown in Fig. 16 is about 36 dB. Continue the process until S/N and β are calculated for all levels of all control factors. The process leads to a response table. The S/N response graph shown is created from the response table data (not shown).

With a modest amount of arithmetic, a team can make a prediction about the new S/N ratio for the system. The Ford wiper team predicted an S/N ratio of about 25 dB.

Step 11: Adjust to target (Step 2 of Two-Step Optimization). Two-step optimization is summarized as follows:

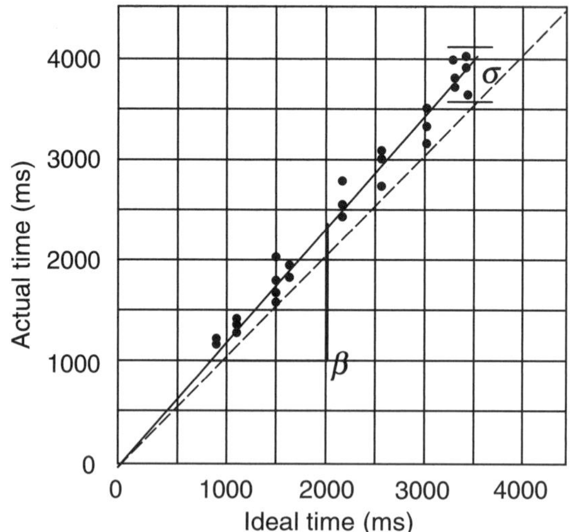

Figure 15 Typical data taken from one run. Solid line is the least square fit. The dotted line at 45° is the ideal function.

1. Maximize the S/N ratio.
2. Adjust β to target.

Maximizing the S/N ratio yields the configuration:

A2 B2 C D2 E1 F G3 H3 I

Adjusting to target yields the final configuration:

A2 B2 C1 D2 E1 F2 G3 H3 I3

Control factors that do not significantly impact the S/N are typically shown without their levels. These factors are often candidates for adjustment factors. Engineered systems usually

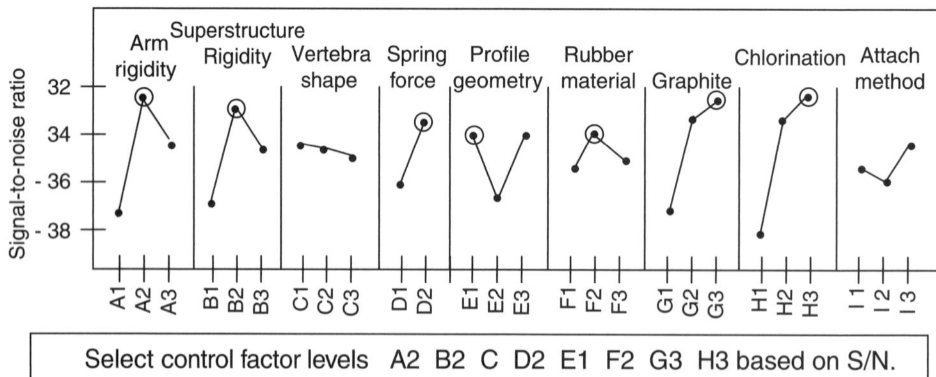

Figure 16 Response graph with selected control factor levels indicated.

provide a convenient control factor for adjusting β—one where the S/N remains relatively flat across the three levels while β shows significantly different values between the levels, such as *C, F,* and *I* in this experiment. Then β is adjusted to target by selecting the appropriate level for the adjustment control factor. If this convenient circumstance does not occur, it may be necessary to tinker with more than one control factor to set β to target. The windshield wiper team apparently chose to tinker with all three of the candidate adjustment factors.

Step 12: Conduct confirmation run. The prediction is followed by an experimental run, called the confirmation run, to validate the prediction. The team also ran an experiment with the original design to establish the baseline.

Baseline configuration A1 B1 C1 D1 E1 F1 G1 H1 I1

Optimized configuration A2 B2 C1 D2 E1 F2 G3 H3 I3

The confirmation test results are shown in Table 2.

The apparent small change in slope β from 1.082 to 1.01 is significant. It means that the actual time to point D_n is closer to the ideal time than the baseline (refer to Fig. 12).

Range of Variation. Figure 17 depicts the gain in S/N from a single robust optimization experiment assuming the average gain of 6 dB. The corresponding increase in σ level is indicated on the left side of the figure assuming that the baseline performance was at a σ level of 3.5. The formula, $\sigma_{OP}/\sigma_{BL} = (1/2)^{(gain/6)}$, can be written as $3.5\sigma_{OP} = 3.5\sigma_{BL}$ $(1/2)^{(gain/6)}$, or for a gain of 6 dB, as $7\sigma_{OP} = 3.5\sigma_{BL}$. The correlation of σ level and S/N is relative, not absolute. While the increase in σ level is a factor of 2 whatever the initial σ level, the baseline σ level is arbitrarily selected as 3.5 as an instructive example.

Returning to the case study of reducing windshield wiper system chatter, the range of variation of the baseline and optimized systems are related by $\sigma_{OP}/\sigma_{BL} = (1/2)^{(gain/6)}$. Hence, $\sigma_{OP} = \sigma_{BL}\ (1/2)^{(11.4/6)} = \sigma_{BL}/3.7$ or $\sigma_{BL} = 3.7\sigma_{OP}$.

Then, $3\sigma_{BL} = 11\sigma_{OP}$. If the baseline was at a 3σ level, the optimized system is at an 11σ level, an enormous improvement. Whatever the variation of the original wiper system, the range of variation was reduced by a factor of 3.7.

Cost Reduction. The S/N is relatively insensitive to the choice of the rubber material (factor *F*). Hence, the team is free to select the lowest cost material from the three tested. While this may not be very exciting for rubber wiper blades, it can be a significant cost opportunity in many situations.

Conclusions from Case Study. The study indicated that chlorination, graphite, and arm and superstructure rigidity have significant impacts on the wiper system while vertebra shape (load distribution) has minimal impact. Hence, a low-friction, high-rigidity wiper system provides a robust windshield wiper system that will remain chatter free for an extended

Table 2 Data from Prediction and Confirmation Run

	Predicted		Confirmed	
	S/N	β	S/N	β
Optimal	−24.71	1.033	−25.41	1.011
Baseline	−35.13	1.082	−35.81	1.01
Gain	10.42	—	11.4	—

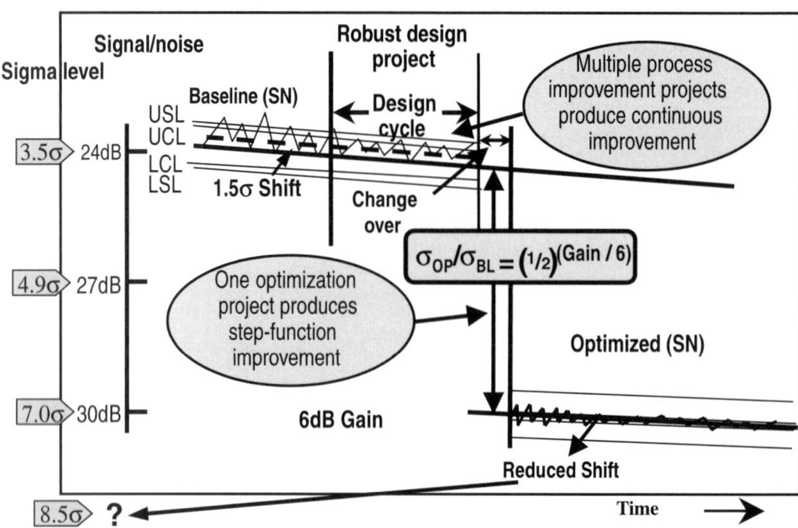

Figure 17 Depiction of improvement due to robust optimization experiment achieving 6 dB gain (30 dB − 24 dB = 6 dB), the average gain over thousands of experiments.

period of time. The increased durability of the system is achieved by including aged and worn wiper blades in the noise strategy.

Step 13: Optimize tolerance design. Tolerance design is not tolerancing. It is a methodology for balancing internal cost and external cost to either the customer or the company. It involves examination of the impact of the different control factors. A simple example that did not require tolerance design was the windshield wiper case study concerning the opportunity to use the lowest cost rubber material included in the study without significant impact on performance under customer usage conditions.

The methodology uses analysis of variance (ANOVA) to determine the percentage contribution of the various control factors to the performance of the system and Taguchi's quality loss function to determine internal and external costs. The combination facilitates the determination of whether to upgrade or degrade portions of the system. As another example of thinking differently, Taguchi goes even further to recommend starting with the lowest cost components and materials and upgrading only as necessary. This is opposite of the common practice of overspecifying things to be safe. Tolerance design can only be effectively conducted after optimization.

Step 14: Optimize process. The same methodologies discussed above are effectively used to optimize manufacturing and service processes in the spirit of concurrent engineering.

Benefits of Robust Optimization

Improved Quality, Reliability, and Durability. A major strength of Taguchi's robust optimization is the ability to make the function of the system insensitive to aged and worn parts as well as new parts. This capability distinguishes robust optimization from methods such as classical design of experiments that focus on reducing variability. Failure rate over time of usage is used to illustrate the impact of robust optimization on reliability. Taguchi shows that failure rate is reduced by the factor of $(\frac{1}{2})^{\text{gain}/3}$:

$$\text{Failure rate}_{\text{optimized}} = \text{failure rate}_{\text{initial}}\ (\tfrac{1}{2})^{\text{gain}/3}$$

The average gain over thousands of case studies is about 6 dB, which, on average, yields a factor of 4 reduction in failure rate. A more conservative prediction of failure rate after optimization is given by replacing gain/3 with gain/6, which yields a factor of 2 reduction in failure rate for a gain of 6 dB. The bathtub curves shown in Fig. 18 assume the more conservative improvement.

As emphasized in the develop concept phase, robust optimization is an important downstream measure of the quality of a conceptual design. Good concepts yield good gains such as the 11 dB gain in the wiper system case study or the average gain of about 6 dB. A concept that does not yield a good gain is a poor concept that will plague customers with problems and corporations with high warranty costs.

Reduce Development Time and Cost of Development and Product. The ability to optimize concept designs and technology sets combined with two-step optimization to increase the efficiency of product development is illustrated in Fig. 19.

After the function has been optimized, a subsystem can be set to different targets for different product requirements within a family of products. In addition, a good concept and final design can be carried over from generation to generation of product families. Multiple-use and carryover parts can dramatically reduce development cost and schedule.

Robust optimization enables some manufacturers to design only about 20% new parts and reuse about 80% of carryover parts in next-generation products for substantial competitive advantage. For other manufacturers, the percentages are reversed.

Verify and Launch. The "V" of IDDOV. The steps in verify and launch phase are traditional:

Step 15: Finalize operations and service processes.

Step 16: Conduct prototype cycle.

Step 17: Conduct pilot run.

Step 18: Launch, ramp up and confirm full operation.

Step 19: Track and improve field performance.

Step 20: Close project.

These steps are typical of any development process and are not discussed in detail. However, the steps 15 and 16 warrant brief elaboration, starting with step 16. This is a manufacturing

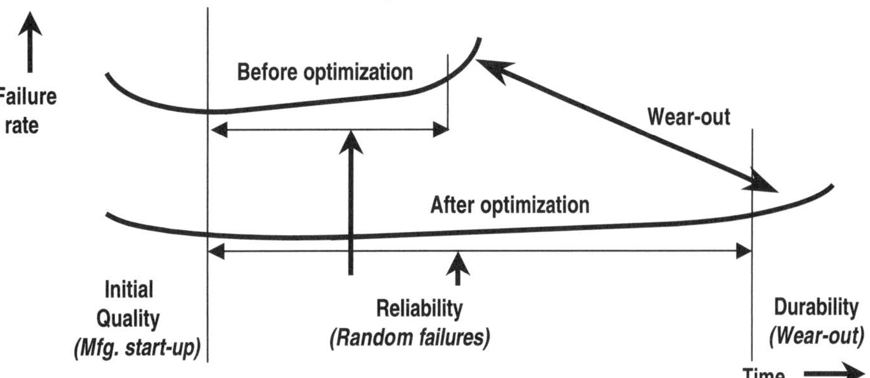

Figure 18 Good concepts can be optimized to provide superior quality, reliability, and durability. Poor concepts are not significantly improved through optimization.

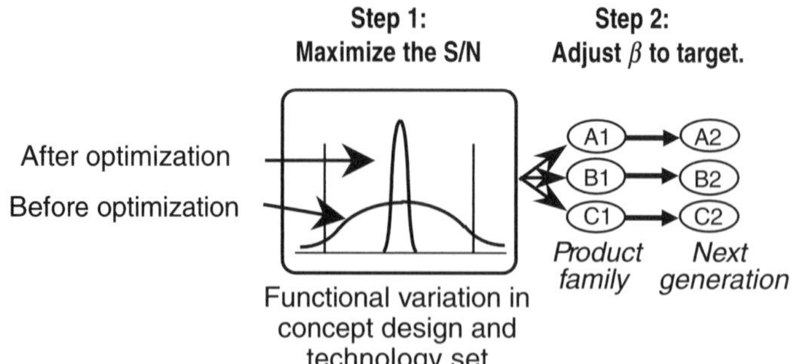

Figure 19 A good concept and technology set can be optimized and then set to different targets for different applications.

intent prototype cycle, in contrast with the many test rigs that have been built and tested throughout the IDDOV process.

The key recommendation in step 16 is to plan and execute a single manufacturing intent prototype cycle. Traditionally, teams use manufacturing intent prototype cycles to improve reliability to target by tinkering to correct problems as they emerge. The first cycle usually generates so many changes that configuration control is lost. Hence, the team redesigns and builds another set of manufacturing intent prototypes and repeats the test procedures. Often additional build–test–fix cycles are required to reach an acceptable level of performance and reliability at great expense in time and resources.

An upfront commitment that only a single prototype cycle will be executed fosters excellence of execution of all of the steps in the design process to complete the design prior to building the first set of prototypes. A strong concept design and robust optimization give credibility to the notion that performance and reliability requirements can be met with only a single manufacturing intent prototype cycle.

Step 15 is critical to the notion of a single prototype cycle. The manufacturing intent prototype should be assembled using full production processes and fully tested against all anticipated usage conditions, including completed operator and service documentation. This can only be done if all operations and service procedures are defined and documented.

An opportunity exists to improve manufacturing quality using Taguchi's on-line quality engineering. Unfortunately, on-line quality engineering has not received much attention in the Western world. A key attribute of on-line quality engineering is the focus on managing to target rather than to control limits. Managing the manufacturing process to target tends to yield a bell-shaped distribution where only a small number of parts and assemblies are near control limits. Managing to control limits tends to yield a flat distribution where a large portion of the parts and assemblies are near the control limits. Items near control limits are sometimes identified as latent defects just waiting to morph into active defects under customer usage. The portion of latent defects that morph into active defects depends on the ratio of specification limits (tolerance) to control limits, Cpk, or in contemporary terms, σ level. If the specification limits are significantly removed from control limits, say 6σ performance, the portion of latent defects that become active defects remains small. Obviously, at the other extreme of specifications set at the 3σ control limits, the portion of morphed defects becomes large. Managing to target leads to better products at lower overall cost.

5 SUMMARY AND CONCLUSIONS

The IDDOV process is all about increasing signal and decreasing noise, or in more traditional language, opening up design tolerances and closing down manufacturing variations in order to send superior products to market that droves of potential customers will purchase in preference to competitive offerings. The bottle model was created by the second author[17,18] in the early 1980s and published in 1988 to provide a visual representation of the role of robust engineering in the development process. The effort was motivated by numerous failed attempts to explain robust optimization to senior management. The bottle model is depicted in Fig. 20. It is presented here as a summary of the entire IDDOV process.

The interpretation of the bottle model curves differs between statistical language of capability indices, Cp and Cpk, and robust engineering language of S/N.

- The measure of statistical variation is the capability index, Cp, defined as the ratio of the design width (USL − LSL) and the process width (UCL − LCL):

$$\mathrm{Cp} = \frac{\mathrm{USL} - \mathrm{LSL}}{\mathrm{UCL} - \mathrm{LCL}}$$

- The measure of robustness of function is the S/N, defined as the ratio of the energy in the intended function to the energy in the unintended functions:

$$\mathrm{S/N} = 10 \log \left(\frac{\beta^2}{\sigma^2} \right)$$

Of course, the measurement scales and units are quite different between the two interpretations:

- Cp is a measure of variation of directly measured topics such as time, dimension, etc., in appropriate units like seconds and centimeters.
- S/N is a measure of functional performance in terms of the energy ratios in units of decibels.

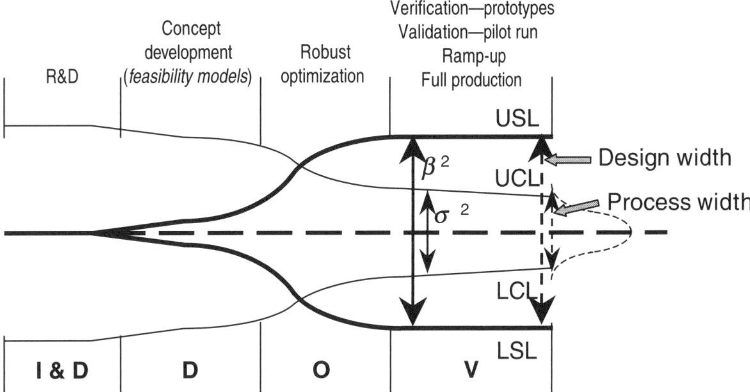

Figure 20 The bottle model provides a visual representation of the IDDOV process and engineering progress in opening up design tolerances (dark line) and closing down manufacturing variation (light line) or alternatively increasing the signal and increasing the impact of S/N.

The caption in Fig. 20 includes the phrase "impact of noise." Sources of variation of noise are not necessarily reduced. Only its impact on functional variation is reduced. Recall that the definition of Robustness is when technology, product, or process performance is minimally sensitive to factors which cause variability (either in manufacturing or a user's environment) and aging at the lowest unit manufacturing cost. Noise factors such as environment and customer usage conditions cannot be reduced. However, the system can be made minimally sensitive to such sources of variability. Part-to-part variation may even be allowed to increase to reduce manufacturing cost. Tolerance design is used to determine allowable variation that does not perceptibly impact the customer.

The distinction between statistical variation and robustness is critical. Statistics has to do with measuring and reducing statistical variation of parts and processes. Robustness has to do with measuring and minimizing functional variation in the presence of the statistical variation (i.e., sources of noise).

The robust optimization portion of the bottle model shows a rapid increase in signal (or design latitude) and rapid decrease in noise (or statistical variation) that results from conducting a series of robust optimization experiments on each of the system elements involved, as illustrated in Fig. 17 for a single system element.

The version of IDDOV described in Ref. 1 and in more technical detail herein has been deployed in a number of large, medium, and small corporations that have reaped *"extraordinary profits."*

IDDOV is an engineering process to support product development processes. It is not a PDP as some renditions of DFSS apparently strive for. The entire IDDOV process is used within technology development, within the advanced product concept development phase, within the design and development phase of typical PDPs, and in OEM manufacturing operations and suppliers to OEMs.

IDDOV is also used to parallel the entire PDP. Within any PDP, many IDDOV projects will be in progress through all phases.

IDDOV is much more than DFSS for breaking through the Five Sigma wall that the Six Sigma DMAIC process encounters. It is a customer-to-customer engineering methodology that carries the voice of the customer throughout the engineering process and delivers superior, low-cost manufactured products that droves of people around the world purchase in preference to competitive offerings.

Last, deployment of IDDOV is not a cost. It is an investment with high returns (ROIs). In tomorrow's competitive world, it will be difficult to compete without pervasive implementation of a competent form of DFSS such as IDDOV.

Acknowledgments

A special thanks for technical review of the chapter go to Ruth McMunigal, B.S., Green Belt, Black Belt candidate; Robert J. McMunigal, M.A.; and Kelly R. McMunigal, B.A., Yellow Belt, Green Belt candidate.

REFERENCES

1. S. Chowdhury, *Design for Six Sigma: The Revolutionary Process for Achieving Extraordinary Profits,* ASI, Dearborn, 2002.
2. A. Fornari and G. Maszle, "Lean Six Sigma Leads Xerox," *ASQ Six Sigma Forum Mag.,* **3**(4), 57 (Aug. 2004).

3. M. Harry and R. Schroeder, *Six Sigma: The Breakthrough Management Strategy Revolutionizing the World's Top Corporations,* Doubleday, New York, 2000.
4. S. Chowdhury, *The Power of Six Sigma: An Inspiring Tale of How Six Sigma is Transforming the Way We Work*, ASI, Dearborn, 2001.
5. B. Bebb, "Role of Taguchi Methods in Design for Six Sigma," *Taguchi's Quality Engineering Handbook,* in G. Taguchi, S. Chowdhury, and Y. Wu (eds.) Wiley, Hoboken, NJ, 2005, pp. 1492–1521.
6. T. P. Pare, "How to Find out What They Want," *Fortune,* **128**(13), 39 (Autumn/Winter 1993), www.fortune.com.
7. T. Peters, *Thriving on Chaos*, 1988, p. 185, www.fortune.com.
8. J. Terninko, *Step-by-Step QFD: Customer-Driven Product Design*, St. Lucie Press, Boca Raton, FL, 1997, pp. 67–74.
9. N. Tichy and S. Sherman, *Control Your Destiny or Someone Else Will,* Harper Business, New York, 1994, p. 8.
10. S. Pugh, *Total Design: Integrated Methods for Successful Product Engineering,* Addison-Wesley, Workingham, England, 1990.
11. G. Taguchi, S. Chowdhury, and S. Taguchi, *Robust Engineering: Learn How to Boost Quality While Reducing Cost and Time to Market,* McGraw-Hill, New York, 2000, p. 4.
12. B. Bebb, personal communication, ASI, 2005.
13. G. Taguchi, S. Chowdhury, and S. Taguchi, *Robust Engineering: Learn How to Boost Quality While Reducing Cost and Time to Market*, McGraw-Hill, New York, 2000.
14. Y. Wu and A. Wu, *Taguchi Methods for Robust Design*, ASME Press, New York, 2000.
15. G. Taguchi, *Taguchi on Robust Technology Development: Bringing Quality Engineering Upstream,* American Society of Mechanical Engineers, New York, 1993.
16. G. Taguchi, S. Chowdhury, and S. Taguchi, *Robust Engineering: Learn How to Boost Quality While Reducing Cost and Time to Market*, McGraw-Hill, New York, 2000, pp. 82–92.
17. H. B. Bebb, "Responding to the Competitive Challenge, the Technological Dimension," *The Bridge,* **18**(1), 4–6 (Spring 1988), published by the National Academy of Engineering.
18. H. B. Bebb, "Quality Design Engineering: The Missing Link in U.S. Competitiveness," in *Tolerance and Deviation Information*, V. A. Tipnis (ed.), New York, 1992, pp. 113–128.

BIBLIOGRAPHY

Creveling, C. M., J. L. Slutsky, and D. Aritis, *Design for Six Sigma*, Prentice-Hall, Englewood Cliffs, NJ, 2003.

ISO/IEC 15288 Systems Thinking, www.iso.org.

McMunigal, J. E., "The ABCs of Cost as Metric," in J. McMuniga (consulting ed.) *Quality Technology Asia,* A.P. Publications, Singapore, Mar., 1992.

McMunigal, J. E., Commissioned speaker and workshop director on "Malcolm Baldrige National Quality Award" for the Institute for International Research Conference on Benchmarking, Singapore, Feb. 1992.

McMunigal, J. E., "TQM and Japanese Optimization Techniques," in *Quality Technology Asia*, Singapore, Oct., 1991.

McMunigal, J. E., "TQM and Japanese Optimization Techniques," presented at the Third Annual European Conference on Japanese Optimization Technology, Wolfson College, Cambridge University, Cambridge, U.K., Oct. 1990.

McMunigal, J. E., "American Agents of Change: Total Quality Management and Taguchi Method," presented at the Second Annual European Conference on Taguchi Method, London, U.K., Sept. 1989.

Minitab® standard software for most Six Sigma models, www.minitab.com.

http://www.triz-journal.com.

Yang, K., El-Haik, *Design for Six Sigma,* McGraw-Hill, New York, 2003.

CHAPTER 18

TRIZ

James E. McMunigal
MCM Associates
Long Beach, California

Steven Ungvari
Strategic Product Innovations, Inc.
Columbus, Ohio

Michael Slocum
Breakthrough Management Group
Longmont, Colorado

Ruth E. McMunigal
MCM Associates
Long Beach, California

1 WHAT IS TRIZ?

TRIZ is the acronym for the Russian words *Teoriya Resheniya Izobretatelskikh Zadatch* (Theory of the Solution of Inventive Problems). TRIZ's development, evolution, and refinement covers over 50 years of rigorous, empirically based analysis.

The creativity and innovation mentioned within the context of science are rare. Typically, creativity and innovation are considered spontaneous phenomena occurring in a capricious and unpredictable way. Individuals such as Michelangelo, Leonardo da Vinci, and Thomas Edison appear to have possessed innate, natural ability for creative thought and inventiveness. What characteristics enabled them, or anyone, to perform as a highly creative thinker?

The term *theory to the solution of inventive problems* implies there is an innovation and/or creative thought process (supported by an underlying construct and architecture) that can be deployed on an as-needed basis. The implications of such a theory, if true, are enormous, suggesting that technicians can elevate their creative thinking abilities by orders of magnitude when the need arises.

2 ORIGINS OF TRIZ

The catalyst for TRIZ was a Russian named Genrich Altshuller (1926–1998). His interest in inventions began at an early age, patenting a device for generating oxygen from hydrogen peroxide by age 14. Altshuller's fascination with inventions and innovation continued through Stalin's regime and World War II. After the war, he was assigned as a patent examiner for the Department of the Navy. He found himself helping would-be inventors solve various problems with their inventions. Over time, Altshuller became fascinated with the study of inventions and understanding how their inventors' minds worked. His initial attempts were psychologically based; however, these probes provided little if any insight on how creativity could be "engineered."

Altshuller turned his attention to studying inventions and reverse engineering them to understand the essential engineering problem being solved and the elegance of the solution as described in the patent application. Patent applications, called Author Certificates (ACs) in the former Soviet Union, were concise documents of three to four pages. The AC consisted of a descriptive title of the invention, a schematic of the new invention, a rendering of the current design, the purpose of the invention, and a description of the solution.

2.1 Altshuller's First Discovery

The brevity of the ACs facilitated analysis, cataloguing, and mapping of solutions to the problems. As the number of inventors applying for an AC increased, Altshuller uncovered similar patterns of solutions for similar problems. He developed a scientific, standardized approach to a problem and incorporated a latent knowledge base as an integral element of the solution process when he recognized that similar technological problems gave rise to similar patents. This phenomenon was repeated in widely disparate engineering disciplines, in various geographical areas, during different time frames.

Altshuller postulated the possibility of creating a mechanism for describing "types" of problems and mapping them to types of solutions. This led to a mechanism naming the 39 typical engineering parameters, the contradiction matrix, and 40 inventive principles.

2.2 Altshuller's Second Discovery

As Altshuller assembled chronological technology maps, he uncovered regularity in the evolution of engineered systems. He described these time-based phenomena as "laws" and called them the *eight laws of engineered systems evolution*. The term laws does not imply that they conform to a strict scientific construction as one would describe in the field of physics or chemistry. The laws, though general in nature, are recognizable, and predictable and provide a road map to future derivatives. Today, these eight laws have been expanded into more than 400 "sublines" of evolution and are useful in technology development, product planning, and the establishment of defensible patent fences.

2.3 Altshuller's Third Discovery

The third truism that emerged was the realization that inventions are vastly different in their degree of "inventiveness." Indeed, many of the patents he studied were filed to describe a system and provide some degree of protection. These patents were useless to Altshuller's determination for discovering the secret of how an inventor reaches the highest order. To differentiate inventiveness, he devised a scale of 1–5 for categorizing the elegance of the solution. See Fig. 1.

Only levels 3 and 4 solutions are deemed "inventive." Within the body of TRIZ knowledge, "inventive" states that the solution was one that did not compromise conflicting requirements. For example, strength versus weight is an example of conflicting parameters. To increase strength, the engineer will typically make something thicker or heavier. An inventive solution would increase strength with no additional weight or even a reduction in weight.

2.4 Altshuller's Levels of Inventiveness

Level 1: Parametric Solution

A solution utilizing well-known methods and parameters within an engineering field of specialty is the lowest level solution and is not an inventive solution. For example, a problem with roads and bridges icing over can be solved with the application of salt or sand.

Level	Nature of solution	Number of trials to find the solution	Origin of the solution	% of patents at this level
1	Parametric	None to few	The designer's field of specialty	32%
2	Significant improvement in paradigm	10–50	Within a branch of technology	45%
3	Inventive solution in paradigm	Hundreds	Several branches of technology	18%
4	Inventive solution out of paradigm	Thousands to tens of thousands	From science—physical/chemical effects	4%
5	True discovery	Millions	Beyond contemporary science	1%

Figure 1 Altshuller's levels of inventiveness.

Level 2: Significant Improvement in the Technology Paradigm
A significant improvement in the system utilizing known methods possible from several engineering disciplines is a level 2 solution. It is a significant improvement over the previous system, but it is not inventive.

A level 2 solution of the icing problem would be required if conventional means were prohibited. This type of solution demands choosing between several variants that leave the original system essentially intact. The roadways or bridges, for example, could be formulated or coated with an exothermic substance that is triggered at a certain temperature.

Level 3: Invention within the Paradigm
The elimination of conflicting requirements within a system utilizing technologies and methods within the current paradigm is a level 3 solution. It is deemed to be inventive because it eliminates the conflicting parameters in such a way that both requirements are satisfied simultaneously.

A level 3 solution to the conflicting requirements of strength versus weight has been solved in aircraft by the use of honeycomb structures and composites.

Level 4: Invention Outside the Paradigm
Creation of a new generation of a system with a solution derived not in technology but in science is a level 4 solution. It integrates several branches of science. The invention of the radio, the integrated circuit, and the transistor are examples of level 4 solutions.

Level 5: True Discovery
A level 5 discovery in one that is beyond the bounds of contemporary science. It will often spawn entire new industries or allow for the accomplishment of tasks in radically new ways. Laser and the Internet are examples of level 5 inventions.

3 BASIC FOUNDATIONAL PRINCIPLES

Altshuller's *three discoveries* provide the construct for the formation of the foundational underpinnings upon which all TRIZ theory, practices, and tools are built. The three building blocks of TRIZ are *ideality, contradictions,* and the *maximal use of resources.*

3.1 Ideality

The notion of ideality is a simple concept. Ideality states that, in the course of time, systems move toward a state of increased ideality. Ideality is the ratio of useful functions F_U to harmful functions F_H:

$$\text{Ideality} = I = \frac{\sum F_u}{\sum F_H}$$

Useful functions embody all of the desired attributes, functions, and outputs from the system. From an engineering point of view, it is what is termed design intent. Harmful functions, on the other hand, include the expenses or fees that are associated with the system; the space it occupies, the resources it consumes, cost to manufacture, cost to transport, cost to maintain, etc.

Extrapolating the concept to its theoretical limit, one arrives at a situation where a system's output consists solely of useful functions with the complete absence of any harmful consequences. Altshuller called this state the ideal final result (IFR). The IFR is not calculated; it is a tool to define the ideal end state. Once the end state is defined, the question as to why it is difficult to attain it flushes out the real (contradictory) problems that must be overcome.

One might argue that it is absurd to think of solving problems from the theoretical notion of the IFR instead of explicitly defining the current dimensions of the problem. However, it is precisely this point of view that opens up innovative vistas by reducing prejudice, bias, and psychological inertia (PI).

Psychological inertia is analogous to thinking only within one's paradigm. An engineer competent in mechanics, for example, is unlikely to search for a solution in chemistry because it is outside his or her paradigm.

Problems with long duration yield an especially target-rich environment for TRIZ. Those intelligent folks who own the problem tend to work in their technical domain and the solution space often resides elsewhere. Some examples where discipline lines were successfully crossed are mechanical to microelectronic and composite lay-up to injection mold.

The notion of ideality postulates that a system, any system, is not a goal in itself. It is only a goal or design intent of any system—the useful function(s) that the system provides. Taken to its extreme, the most ideal system is one that does not exist but nevertheless one that produces its intended useful function(s). See Figs. 2 and 3.

In Fig. 3 the system has not reached a state of ideality because the useful interaction between *A* and *B* is accompanied by some type of unwanted (harmful) function. An ideal system *A*, on the other hand, is one that does not exist even when its design intent is fully accomplished.

In the abstract, this notion might seem fantastical or even absurd. There is, however, a very subtle yet very powerful heuristic embodied in ideality. First, ideality creates a mind set for finding a noncompromising solution. Second, it is effective in delineating all of the technological hurdles that need to be to overcome in order to invent the best solution possible. Third, it forces the problem solver to find alternative means or resources in order to provide the intended useful function. The latter outcome is similar to an organization reassigning key functions to the individuals that are retained after a reduction in force.

3.2 Contradictions

The second foundation principle is the full recognition that systems are inherently rife with various conflicts. Within TRIZ these conflicts are called contradictions. In TRIZ an "inventive" problem is one that contains one or more contradictions. Typically, when one is faced with a contradictory set of requirements, the easy resolution is to find a compromising solution. This type of solution, while it may be expedient, is not an inventive solution. If we return to the example of weight versus strength, an inventive solution satisfies both requirements. Another example would be speed versus precision. A TRIZ level 3 solution would satisfy both requirements utilizing available "in-paradigm" methods, while a level 4

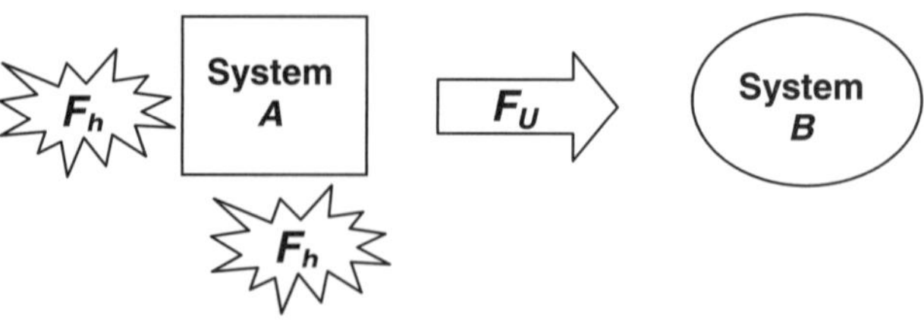

Figure 2 System *A* and system *B* interaction with useful output and harmful effects.

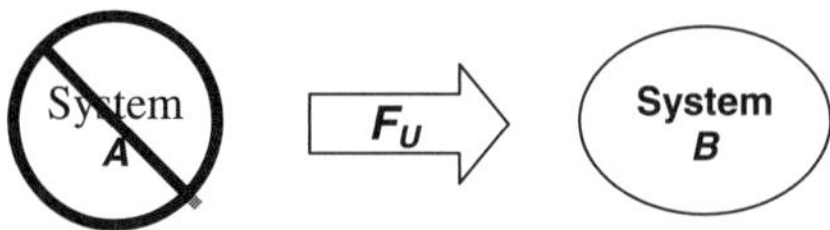

Figure 3 *A*'s effect of *A* when there is no system *A* yet system *A* function is carried out.

solution would incorporate technologies outside of the current paradigm. In both cases, however, speed and precision would be achieved at a quality level demanded by the contextual parameters of the situation. In TRIZ, two distinct types of contradictions are delineated: technical contradictions and physical contradictions. (Methods for solving technical contradictions will be discussed later in the chapter.)

3.3 Technical Contradictions

A technical contradiction is a situation where two identifiable parameters are in conflict: When one parameter is improved, the other is made worse. The two previously mentioned, weight versus strength and speed versus precision, are examples. See the Fig. 4.

3.4 Physical Contradictions

A physical contradiction is a situation where a single parameter needs to be in opposite physical states; e.g., it needs to be thin and thick, hot and cold, etc., at the same time. A physical contradiction is the controlling element or parameter linking the parameters of the technical contradiction. Figure 5 shows the pulley (*C*) upon which parameters *A* and *B* rotate as the physical contradiction.

The physical contradiction lies at the heart of an inventive problem; it is the ultimate contradiction. When the physical contradiction has been found, the process of generating an inventive solution has been greatly simplified. It stands to reason that when a physical contradiction is made to behave in two opposite states simultaneously, the technical contradiction is eliminated. For example, if by some means pulley *C* could rotate in opposite directions at the same time, both *A* and *B* would increase, hence eliminating the technical contradiction.

3.5 Maximal Use of Resources

The third foundation principle of TRIZ is the maximal utilization of any available resources before introducing a new component or complication into the system.

Resources are defined as any substance, space, or energy that is present in the system, its surroundings, or the environment. The identification and utilization of resources increase the operating efficiency of the system, thereby improving its ideality. In the former Soviet

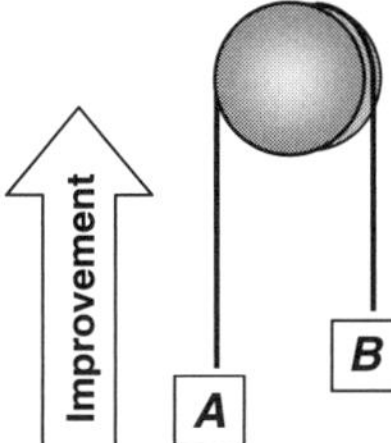

Figure 4 Relationship of parameters *A* and *B*; as one improves, the other worsens.

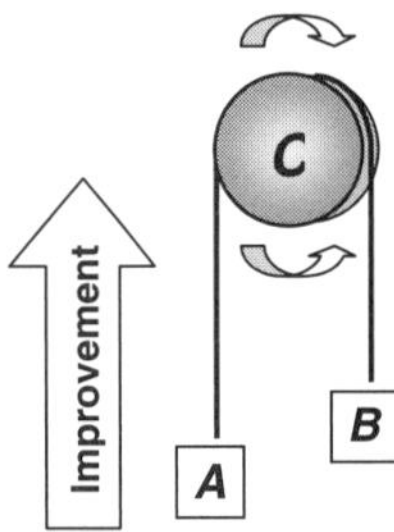

Figure 5 Physical contradictions example. For *A* and *B* to improve, *C* must rotate clockwise and counterclockwise simultaneoulsy.

Union, where money was scarce, necessity proved to be the mother of invention. In the West, on the other hand, system problems were often engineered out by "throwing money and complexity" at a system. The utilization of resources as an X agent to solve the problem was and still is not widely practiced.

A practiced TRIZ problem solver will marshal any in-system or environmental resources to assist in solving the problem. It is only when all resources have been exhausted or it is impractical to utilize one that the consideration for additional design elements comes into play. The mantra of a TRIZ problem solver is, *"Never solve a problem by making the system more complex."* More on this when the Algorithm for Problem Solving (ARIZ) is discussed. Table 1 lists the types of resources used in TRIZ.

4 A SCIENTIFIC APPROACH

TRIZ is comprised of a comprehensive set of analytical and knowledge-based tools that were heretofore buried at a subconscious level in the minds of creative inventors. If asked to explain specifically how they invent, most people are unable to provide a repeatable formula.

Table 1 Types of Resources

Substance—any material contained in the system or its environment, manufactured products or wastes
Energy—any kind of energy existing in the system
Space—any space available in the system and its environment
Time—time intervals before start, after finish, and between technological cycles, unused or partially used
Functional—possibilities of the system or its environment to carry out additional functions
Own—unused specific features and properties, characteristics of a particular system, such as special physical, chemical or geometrical properties; for example, resonance frequencies, magneto-susceptibility, radioactivity, and transparency at certain frequencies
System—new useful functions or properties of the system that can be achieved from modification of connections between the subsystems or a new way of combining systems
Organizational—existing but incompletely used structures or structures that can be easily built in the system, arrangement or orientation of elements or communication between them
Differential—differences in magnitude of parameters that can be used to create flux that carry out useful functions; for example, speed difference for steam next to a pipe wall versus in the middle, temperature variances, voltage drop across resistance, height variance
Changes—new properties or features of the system (often unexpected), appearing after changes have been introduced
Harmful—wastes of the system (or other systems) which become harmless after use

Through his work, Altshuller has codified the amorphous process of invention. His contribution to society is that he made the process of inventive thinking explicit. He has made it possible for anyone with a reasonable amount of intelligence to become an inventor. TRIZ makes it possible for people of average intelligence to access a large body of inventive knowledge and through analogical analysis formulate inventive "out-of-the-box" solutions.

4.1 How TRIZ Works

The general scheme in TRIZ is solution by abstraction. A specific problem is described into a more abstract form. The abstracted form of the problem has a counterpart solution at the level of abstraction. The connection between the problem and the solution is found through the use of various TRIZ tools. Once the solution analog is arrived at, the process is reversed, producing a specific solution. Figure 6 illustrates the process of solution by abstraction and Fig. 7 applies the process to an algebraic problem.

Assume that we were given the task of solving the problem found in the equation $3x^2 + 5x + 2 = 0$. Without a specific process, we would be reduced to the inefficient process of trial and error. An even more absurd method would be to try to arrive at the answer by brainstorming. Yet, brainstorming is often applied to problems that are much more complex than that shown above. This is what makes TRIZ so compelling—it provides a roadmap to highly creative and innovative solutions to seemingly impossible problems.

Figure 7 provides the general schema for how TRIZ works. The fundamental idea in TRIZ is to reformulate a problem into a more general (abstract) problem and then find an equivalent "solved" problem. These analogs, in theory, define the solution space that is occupied by one or several noncompromising alternative solutions.

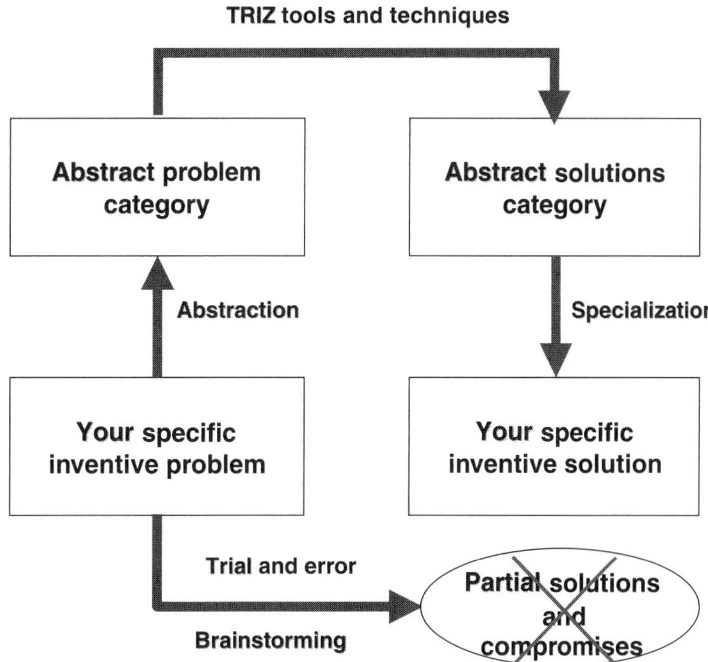

Figure 6 Solution by Abstraction using TRIZ.

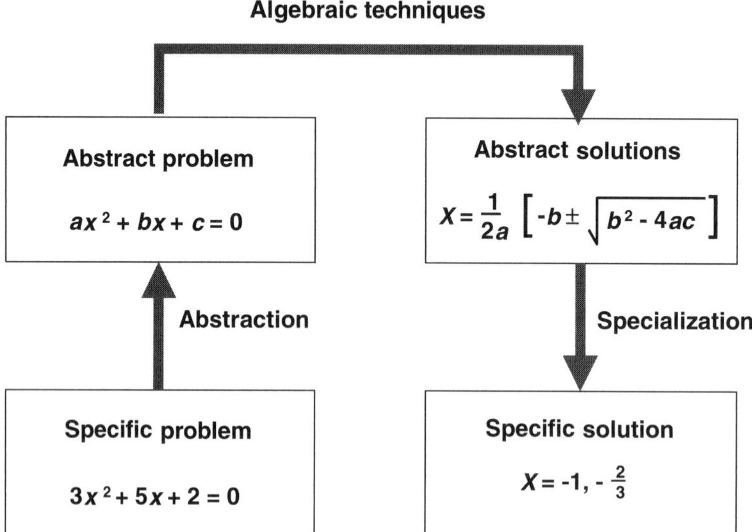

Figure 7 Solution by Abstraction using algebraic techniques.

The advantage of increasing the level of abstraction is that the solution space is expanded. Solving the equation illustrated in Fig. 7 is relatively simple, assuming knowledge of algebra. The correctness of the solution is also easier to verify because the solution space is very small, e.g., there is only one right answer! Inventive problems pose a greater challenge than the one shown above because the solution space is very large. Figure 8 illustrates this truism.

Figure 8 shows what happens when solving "inventive" versus noninventive problems. An inventive problem is often confused with problems of design, engineering, or a technological nature. For example, in constructing a bridge, the type of bridge to be built is largely an issue related to design. A cantilever bridge provides known design advantages over a suspension bridge in specific contexts and vice versa. This is an example of a noninventive design problem. The calculations of load and stress the bridge will have to withstand are an engineering problem. Coordinating the construction and assuring that materials meet specifications and the job is on time and within budget is a technical problem. While any of these problems are not insignificant by themselves, they are not inventive within the context of TRIZ because they are solvable by using known methods, formulas, schedules, etc. Furthermore, the path to the correct solution is defined and direct, and since the solution space is very small, verification of the answer is straightforward. This is not the case with inventive problems.

An inventive problem in the context of building a bridge would to be to make the bridge lighter and stronger, larger and less expensive, longer and more stable, etc. These problems are inventive because they have to overcome one or more contradictions. Therefore, to reiterate, a problem is an inventive one when one or several contradictions must be overcome in the solution and a compromise solution is not acceptable.

There are several distinguishing characteristics of an inventive versus typical problem, as shown in Fig. 8. The entire solution space can be quite large, containing both noninventive and inventive solutions. The two inner concentric circles represent level 3 and level 4 in-

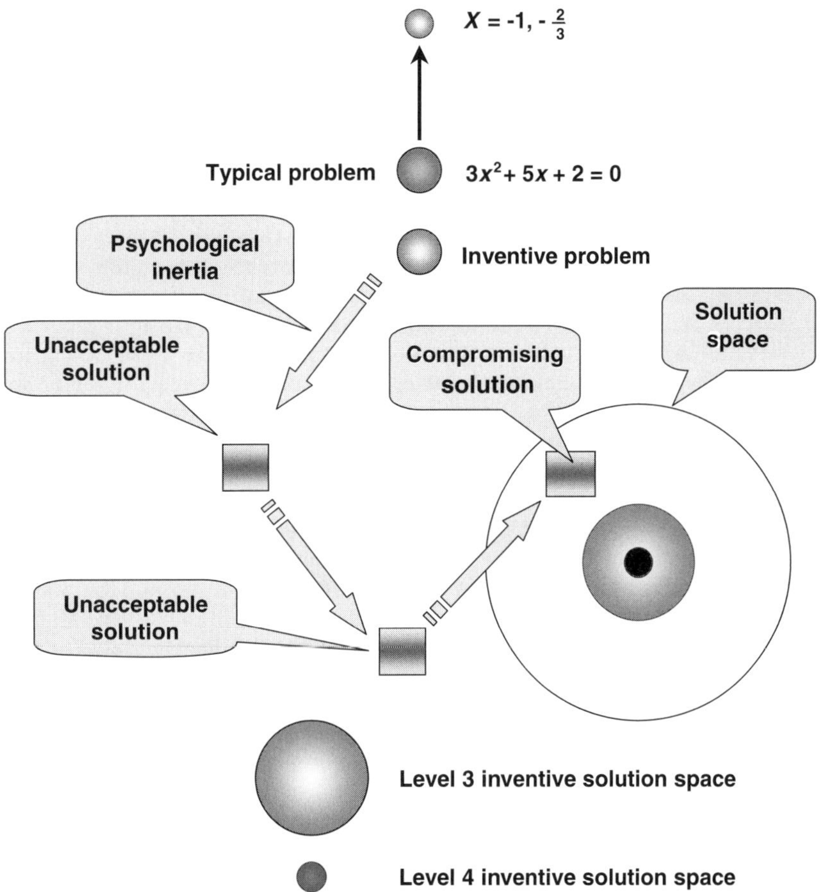

Figure 8 Solution space for inventive versus other problems.

ventive solutions, while the larger outer circle represents an area of noninventive solutions. Just as it is harder to hit the bull's-eye when shooting an arrow, so it is with hitting on an inventive solution. Why is this so?

The initial factor oftentimes driving one off the mark, PI, defined previously, presupposes a solution path as defined by one's individual paradigms. The route to a solution is often one of trial and error and strewn with several unacceptable solutions arrived at along the vector of one's PI. In a sense, the process of defining the current problem and then driving to a solution can be considered a "push" method for finding a solution. TRIZ is different because one of the initial steps of the TRIZ process is to define the ideal state, i.e., the solution space found in level 3 or 4 solutions. The articulation of the ideal solution acts to orient the problem solver and "pulls" him or her in that direction. Furthermore, TRIZ guides one to the ideal solution through the process of abstraction and finding analogs, as discussed previously. These two fundamental elements of TRIZ serve as a powerful magnet to draw one to an inventive solution and provide an example of how this has been accomplished by a previous inventor.

4.2 Five Requirements for a Solution to be Inventive

Within the context of TRIZ, before a proposed solution is labeled inventive, it must meet all of the stringent requirements outlined in Table 2.

5 CLASSICAL AND MODERN TRIZ TOOLS

In the course of his analytical work, Altshuller amassed a vast body of knowledge and invented analytical methods on how to access that body of knowledge. The subsequent evolution of TRIZ followed logical parallel paths. The creation of a body of "inventive" knowledge gave rise to various analytical tools making it easier to catalog and create more inventive knowledge that, in turn, spawned more sophisticated tools, and so on. The end result after more than 50 years of work is a complete set of sophisticated tools and an immense knowledge base of inventive ideas, methods, and solutions that can be mobilized to attack any inventive problem. To date, these tools have been used to solve problems related to product design and development, quality, manufacturing, cost reduction, production, warranty, and prevention of product failures, to name just a few applications.

The tools of TRIZ are subdivided into two major categories. The first division is by the nature of the tool, e.g., analytical versus knowledge base. The second differentiation is chronological: classical TRIZ versus I-TRIZ. The classical TRIZ tools span those derived from 1946 to 1985 with Altshuller as the primary inventive force. A protégée of Altshuller, Boris Zlotin of The Kishnev School (of TRIZ), continued developing the methodology, which for purposes of differentiation is called I-TRIZ.

5.1 Contradiction Matrix

The first of the classical TRIZ tools invented by Altshuller is the contradiction matrix. The objective of the matrix is to direct the problem-solving process to incorporate an idea that has been utilized before to solve an analogous "inventive" problem. The contradiction matrix accomplishes this by asking two simple questions: Which element of the system is in need of improvement? If improved, which element of the system is deteriorated? This is a technical contradiction. A portion of the 39×39 matrix is shown in Fig. 9.

The matrix is constructed by juxtaposing 39 engineering parameters along the vertical and horizontal axis. At the intersections Altshuller filled in from one to four numerical values hinting at ways to solve the problem. The numerical values identified one of the 40 inventive principles that were culled from the knowledge base as ways in which an analog to the specific problem had been solved previously. The 39 engineering parameters are general in nature and act as surrogates for the specific "real" parameters in conflict. The inventive principles are broad and nonspecific as the exact way in which they should be applied. In Fig. 9, the problem is trying to improve "convenience of use," but when this is attempted,

Table 2 Requirements of Inventive Solutions

Solution fully resolves the contradictory requirements.
Solution preserves all advantages of the current system.
Solution eliminates the disadvantages of the current system.
Solution does not introduce any new disadvantages.
Solution does not make system more complex.

	Deteriorated feature / Feature to improve	1 Weight of a moving object	2 Weight of nonmoving object	3 Length of a moving object	● ● ● ● ●	22 Waste of energy
1	Weight of a moving object			15,8 29,34	●	6, 2 34,19
2	Weight of a nonmoving object				●	18, 19 28, 15
3	Length of a moving object	8,15 29,34			●	7, 2 35,39
4	Length of a nonmoving object		35,28 40,29		●	6, 28
5	Area of a moving object	2,17 29,4		14,15 18,4	●	15, 17 30,26
6	Area of a nonmoving object		30,2 14,18		●	17, 7 30
7	Volume of moving object	2,26 29,40		1,7 35,4	●	7,15 13,16
●	● ● ● ● ● ●	●	●	●	●	●
33	Convenience of use	25,2 13,15	6,13,1 ,25	1,17 13,12	●	2,19 13
34	Repairability	2,27 35,11	2,27 35,11	1,28 10,25	●	15, 1 32,19
35	Adaptability	1,6 15,8	19,15 29,16	35,1 29,2	●	18, 15 1
36	Complexity of device	26,30 34,36	2,36 35,39	1,19 26,24	●	10,35 13,2
37	Complexity of control	27,26 28,13	6,13 28,1	16,17 26,24	●	35,3 15,19
38	Level of automation	28,26 18,35	28,26 35,10	14,13 17,28	●	23,28
39	Productivity	35,26 24,37	28,27 15,3	18,4 28,38	●	28,10 29,35

Figure 9 Contradiction matrix.

it results in "waste of energy." The matrix suggests that when this type of problem is encountered, principles 2, 9, and 13 have been utilized to resolve the contradiction. Table 3 provides details on these three principles.

The process for using the contradiction matrix follows the general schema shown in Fig. 7. The steps are as follows:

1. Describe the problem.
2. Select the parameter most closely aligned with one of the 39 engineering parameters from the feature-to-improve column.
3. State your proposed solution.

Table 3 Three of the Forty Inventive Principles

3. Local quality
   a. Change an object's structure from uniform (homogeneous) to nonuniform (heterogeneous) or change the external environment (or external iinfluence) from uniform to nonuniform
   b. Have different parts of the object carry out different functions
   c. Place each part of the object under conditions most favorable for its operation

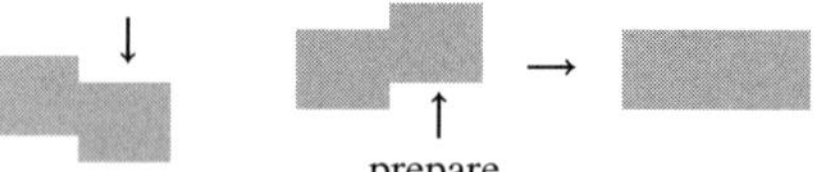

9. Preliminary antiaction
   a. If it is necessary to perform some action with both harmful and useful effects, consider a counter-action in advance that will negate the harmful effects
   b. Create stresses in an object that will counger known undesirable forces later on

13. The other way around
   a. Instead of an action dictated by the specifications of the problem, implement an opposite action
   b. Make a moving part of the object or the outside environment immovable and the nonmoving part movable
   c. Turn the object upside down and inside out; freeze it instead of boiling it

4. Select which feature will be deteriorated.
5. Note the inventive principle(s) at the intersection.
6. Apply the inventive principle(s).

5.2 Physical Contradictions

A physical contradiction (PC) is the controlling element in the system that links the two conflicting parameters in the technical contradiction; see Fig. 4. The PC expresses the most extreme form of contradictory requirements because the conflict must be resolved solely within a single entity. As Fig. 4 shows, the PC (pulley) is at the root of the inventive problem. If it were possible to make the pulley turn in opposite directions simultaneously, the technical contradiction would disappear. From a TRIZ standpoint, solving an inventive problem by satisfying the conflicting requirements of the PC results in elegant solutions with a greater degree of inventiveness.

5.3 Formulating and Solving Physical Contradictions

A PC is formulated according to the logic: "To perform function F_1, the object must exhibit property P, but to perform function F_2, it must exhibit property $-P$. The solution to PCs is accomplished by incorporating principles of separation. There are five separation principles that can be used to resolve a PC. See Table 4.

5.4 An Example

The principle of separation in time can be explained by a well-known illustration used by Altshuller. Assume that one is driving concrete piles for buildings into very hard ground. To

Table 4 Separation Principles

1. Separation in time
2. Separation in space
3. Separation between the system and its components
4. Separation upon condition
5. Coexistence of contradictory properties

facilitate ease of driving the piles, the tip profile should be sharp. Once in place, the pile should be stable, which means the profile should be blunt. In other words, the pile should be sharp and blunt—a PC. How can this be? The problem is solved by imbedding an explosive into the sharp end of the pile and, when it is in place, destroying the sharp profile by setting off the explosive. The tip profile is sharp (P) during time T_1 (driving into the ground) and it is blunt ($-P$) during time T_2 (in place).

5.5 Laws of Systems Evolution

The notion of predicting future technological patterns and derivatives has been recognized as a means of creating competitive leverage. Techniques such as technology forecasting, morphological analysis, trend extrapolation, and the Delphi process have been utilized since the World War II. All of these techniques are based on statistical probability modeling. In TRIZ, future derivatives are based on predetermined patterns of evolution that have been around since the invention of the wheel. Past evolutionary trends provide an "evolutionary crystal ball" for understanding how current technologies will morph over time. Altshuller termed these phenomena "laws of evolution."

These laws represent a stable and repeatable pattern of interactions between the system and its environment. These patterns occur because systems are subject to various cycles of improvement. When a new technological system emerges, it typically provides the minimum degree of functionality required to satisfy the inventor's intent. For example, the first powered flight by the Wright brothers occurred on December 17, 1903. The *Flyer,* with Orville Wright as the pilot, flew to a height of 10 ft and landed heavily after 12 s in the air. Today, jets are capable of flying at heights over 60,000 ft over thousands of miles at several times the speed of sound. What happened with airplanes has been repeated in other types of engineered systems.

The way in which systems evolve can be shown on life cycle, or "S," curves. Figure 10 shows the evolutionary picture.

From the time a system emerges to point *a,* its development is slow as it is unproven. At point *a,* the dominant design paradigm appears and the system is poised for commercialization. From point *a* to *b* the system experiences rapid improvement as commercialization and market pressures force cycles of continuous improvements. From point *b* to *c* the rate of improvement slows as the technology matures. As the system passes point *b,* the next system (*B*) is itself emerging. The abandonment of the original system in favor of the new one is governed by how much greater potential it possesses in comparison to the unrealized improvements remaining in system *A*.

Being a keen observer of inventive phenomena, Altshuller, through his analysis, uncovered eight describable, chronologically sequenced events. He called these events the laws of systems evolution. See Table 5.

Within these eight major laws, Altshuller and his students have found numerous "sub-lines" of evolution. Given the detail that is now captured in the evolutionary knowledge

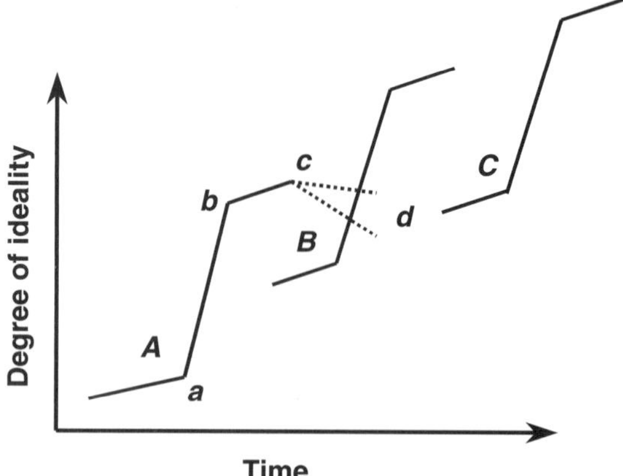

Figure 10 Life cycle, or "S" curves, evolutionary picture.

base, it is possible through the analysis of patents to fix where the technological system is positioned on its life-cycle curve.

Figure 11 shows a few of the sublines of Law 4, *increased dynamism.*

An analogy can be drawn between use of the laws of evolution and laws of motion. If the position of a moving object is known at a certain moment of time, any future position can be determined by solving equations containing velocity and direction. The laws of evolution serve as "equations" describing how the system will change as it travels through time. If the current position of the system is known, future derivatives can be "calculated" using the laws to indicate future positions. The implications to research-and-development initiatives, protection of intellectual assets, technology development strategy, patent strategies, and product development scenarios are profound.

5.6 Analytical Tools

In addition to the knowledge-based tools, Altshuller developed several analytical tools. The two most widely used are substance field modeling (Su-Field) and ARIZ.

5.7 Su-Field

The standard minimum system and its transformations (a generic formulation, according to the corollaries associated with Su-Field analysis) became the foundation of a set of standard

Table 5 Patterns of Technological Systems Evolution

1. Stages of evolution
2. Evolution toward increased indeality
3. Nonuniform development of systems elements
4. Evolution toward increased dynamism and controllability
5. Increased complexity, then simplification
6. Evolution with matching and mismatching components
7. Evolution toward microlevel and increased use of fields
8. Evolution toward decreased human involvement

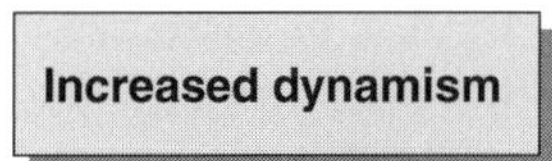

In the course of time, technological systems transition from rigid systems to flexible and adaptive ones

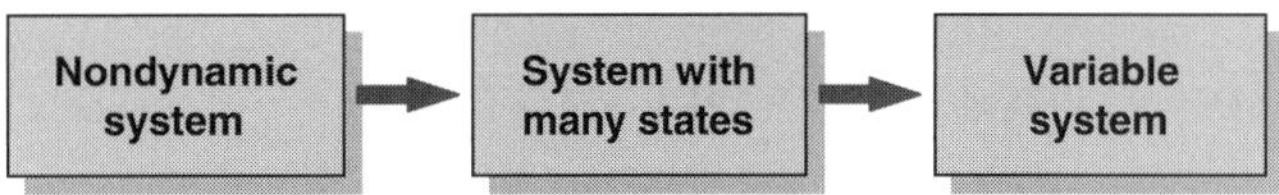

Evolution of Automotive Transmission

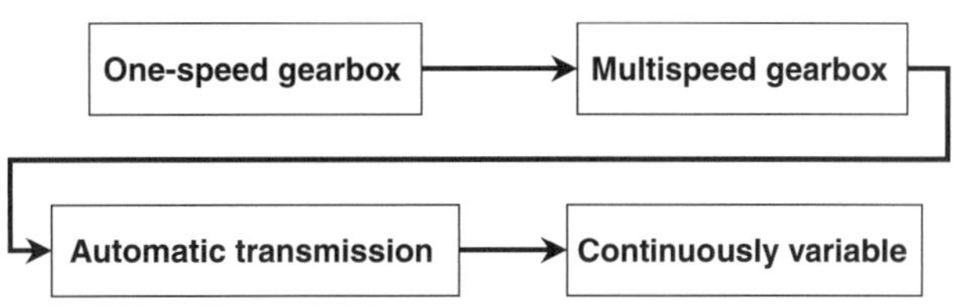

Figure 11 Sublines of Law 4, increased dynamism.

solutions (76 standard solutions) that is effectively utilized for manipulation, with the intent of the model transformations analogically resulting in solutions to a specific problem. These solutions, or standard transformations, are grouped into five classes:

Class 1. Composition and decomposition of Su-Field models (SFMs)

- Group 1-1: Synthesis of a SFM
- Group 1-2: Decomposition of SFMs

Class 2. Evolution of SFMS

- Group 2-1: Transition to complex SFMs
- Group 2-2: Evolution of SFM
- Group 2-3: Evolution by coorinating rhythms
- Group 2-4: Ferromagnetic SFMs (feSFMs)

Class 3. Transitions to supersystem and microlevel

- Group 3-1: Transitions to bisystem and polyststem
- Group 3-2: Transition to microlevel

Class 4. Measurement and detection standards

- Group 4-1: Instead of measurement and detection—system change
- Group 4-2: Synthesis of a measurement system
- Group 4-3: Enhancement of measurement systems
- Group 4-4: Transition to ferromagnetic measurement systems
- Group 4-5: Evolution of measurement systems

Class 5. Special rules of application

- Group 5-1: Substance introduction
- Group 5-2: Introduction of fields

Group 5-3: Use of phase transitions

Group 5-4: Physical effects use

Group 5-5: Substance particles obtaining

Su-Field analysis

6 PROBLEMS WITHOUT CONTRADICTIONS

Overcoming contradictions solves both simple and complex problems. Why do contradictions occur? Because, striving to improve the world around us, the inventor demands a lot from technical objects. This is logical, for in order to meet the increasing demand, technical systems (TSs) should constantly increase in efficiency (or decrease in harmful, or redundant, properties). This means that one group of inventive problems focuses on improving the existing technical systems. Once involved in the technological evolution process, they start facing contradictions. The increasing demand cannot always be met by improving the existing TS. This gives rise to a question: Are there problems where no contradiction can be defined?

Example. In the course of reconstruction, a match factory was equipped with high-performance machines that doubled the factory's production rate. Yet, there was an operation that slowed down the whole process: packing the ready matches into boxes. The old machines could not cope with twice as much production. Lack of space made it impossible to install two packing lines. Finally, a decision was made to remove the out-of-date packing equipment. The old equipment had some deficiencies, too. It was 'blind' and would often pack reject matches without heads or pack the wrong number of matches. Therefore, it became urgent to find an accurate method for packing millions of matches into boxes. There was a requirement for a system that would detect faulty matches.

There was no visible contradiction in this problem, but still there was the need to find a solution. The introduction of a small amount of ferromagnetic powder (application of a standard form Class 4, Group 4-4) to the ignition compound gave slight magnetic properties to each match. This was enough to orient the matches in a magnetic field and pack them faster and with higher accuracy (for a magnet of certain square surface attracts a fixed number of matches).

Let us analyze the problem and its solution in detail. First, as the conditions of the problem suggest, there was nothing to improve. The old TS was dismantled; therefore, a new system should be created. The matches were there, but what were we supposed to do with them? Should we count, orient, or package the matches? The problem was solved using the introduction of a ferromagnetic powder into the ignition compound of the match heads and using a magnetic field to create a system that could easily detect and control defect reduction in the packaged system.

In the beginning, there was one substance (the matches, S_1), and in the end there were two substances (the matches, S_1, and the ferromagnetic powder, S_2) and one field (magnetic, F_M). The system is depicted in Fig. 12.

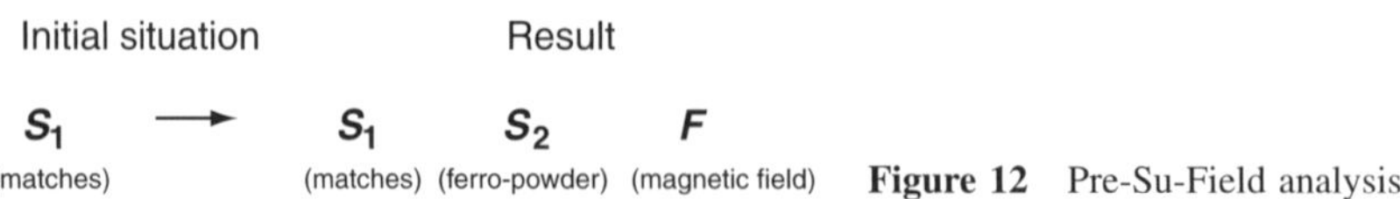

Figure 12 Pre-Su-Field analysis.

How does the system work? The magnetic field (F_M) acts on the ferro magnetic powder, S_2, which in turn acts on the matches (S_1). Graphically the operation can be represented as depicted in Fig. 13.

In other words, one should work from a single element (S_1) toward a system of interacting elements (S_1, S_2, and F_M). A double arrow (to avoid confusion with arrows that indicate the interaction between elements) indicates this transition. The entire process of transition is displayed in Fig. 14.

All this resembles the symbolic representations of a chemical reaction. Two elements (e.g., oxygen and nitrogen) are heated (i.e., an external thermal field is introduced). As a result of interaction, they form a molecule of water; but, if a single atom is withdrawn from the molecule, the water will disappear. Can we treat the right-hand triangle of this technical reaction, in Fig. 14, as a "molecule" of a TS? Let us validate this idea: Will the system work if we withdraw any of the substances? No, the system will fall apart and cease to be a system. The same holds true for the situation in which the field is withdrawn. Does this mean that the system's operation is secured by the presence of all three of the elements? Yes. This follows from the main principle of materialism: A substance can only be modified by material factors, i.e., by matter or energy (a field). With respect to a TS, this principle is as follows: A substance can only be modified as a result of a direct action performed by another substance (e.g., impact—mechanical field) or by field action of another substance (e.g., magnetic) or by an external field. As a consequence, the minimal number of elements any TS consists of is three: two substances and a field—thus the concept of a minimal TS was named a Su-Field.

7 RULES FOR THE INVENTOR: SU-FIELD SYNTHESIS

Discarding redundancies, SFMs shed light on the essence of transformations (synthesis and evolution) of technical systems and allow the use of universal technical language to represent the process of solving any inventive problem. That is why analysis of Su-Field structures in those parts of technical systems where contradictions occur under transformation is called Su-Field analysis**.** Su-Field analysis presents a general formula that shows the direction of solving the problem. This direction depends heavily on the initial conditions of the problem. Consider the example problem: Any slightest alteration of conditions will profoundly change the process of solving the problem. For example, no materials may be introduced into the match head, no cooling medium can be poured into the hollow boom of the robot, etc. How can you decide which step to take?

The SFM is defined as follows:

> A Su-Field model is a representation of the minimal, functioning and controllable technical system.

Quite often, conditions contain two substances and a field that have insufficient interaction and cannot be replaced with other substances or field. That is, the SFM is there (all

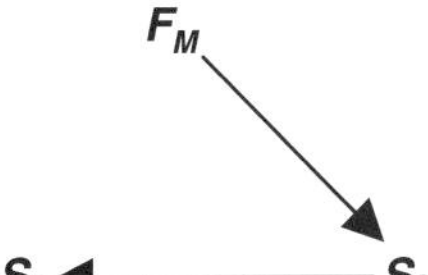

Figure 13 Su-Field model for the example system.

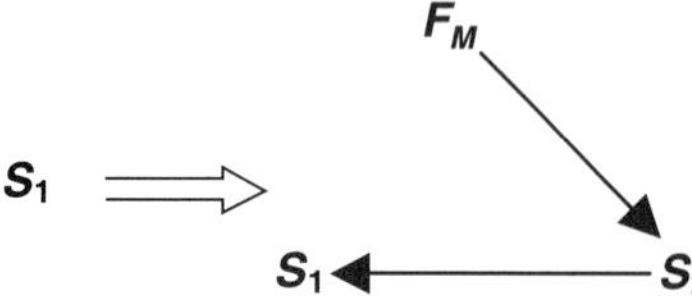

Figure 14 Incomplete system and the transformation to a solution model using Class 4, Group 4.4, from the 76 standard solutions.

three elements are present) and, at the same time, it is not there. It simply will not work. The same may happen after completing a SFM. That means that the SFM needs to be improved: The substances should become controllable, the field should have a desired effect, and the character of interaction of elements should proceed as required. There is a set of transformation rules for substances and fields in SFMs. The following is one such rule (see also Fig. 15):

> Formation of complex Su-Field by introducing an easily controllable admixture possessing desirable properties into the substance. The admixture can be introduced into the substance (internal complex Su-Field) or, where internal introduction is inadmissible, placed outside the substance (external complex Su-Field).

(a) *Internal Complex Su-Field.* Wetting of fabric; foaming of varnish (problem 3); emergence of multicolored inserts impressed at certain distance to the cutting edge indicates the wear of the cutting tool (Soviet patent 905,417).

(b) *External Complex Su-Field.* Admixing ferromagnetic powder to cereal, production of hollow metal porous balls: Polystyrene balls are given a metal coat and subsequently dissolved in organic solvent (U.S. patent 3,371,405). To avoid rumpling, the corrugations of the thin surface are filled with low-melting-point metal, which is withdrawn after treatment (Soviet patent 776,719) (see Fig. 16.)

8 CLASS 4: MEASUREMENT AND DETECTION STANDARDS

Group 4-1: Instead of Measurement and Detection—System Change

Standard 4-1-1. If we are given the problem of detection or measurement, it is proposed to change it such that there should be no need to perform detection or measurement at all.

EXAMPLE. To prevent a permanent electric motor from overheating, its temperature is measured by a temperature sensor. If the poles of the motor are made from an alloy with a Curie point equal to the critical value of the temperature, the motor will stop itself.

Standard 4-1-2. If we are given the problem of detection or measurement and it is impossible to change the problem to remove the need for detection or measurement, it is proposed to replace direct operations on the object with operations on its copy or picture.

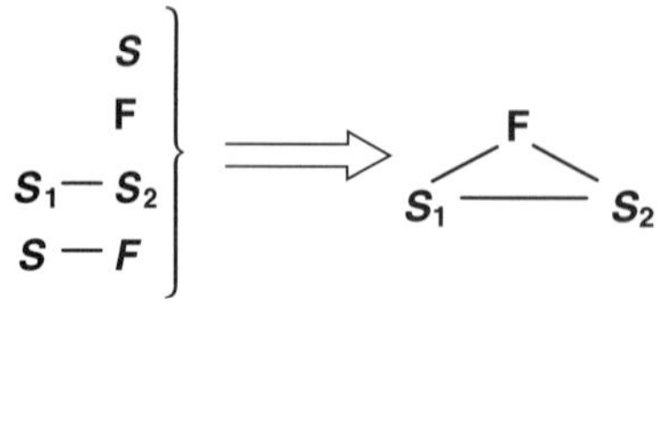

S
F
$S_1 — S_2$
$S — F$
F
S_1 S_2

Figure 15 Transformational rules for SFM.

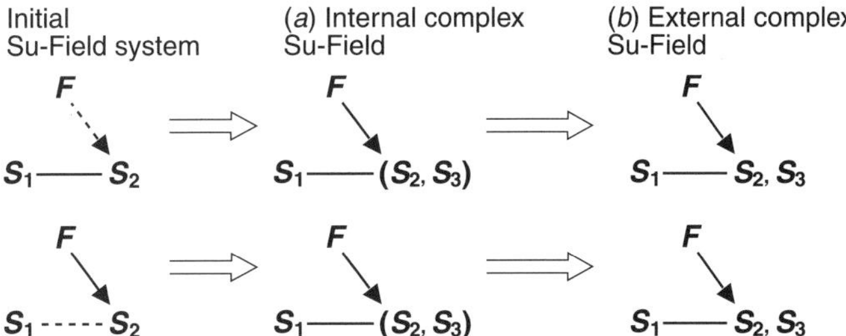

Figure 16 Complex SFM. Nonexistent interactions shown by dashed lines; parentheses indicate internal.

Example. It might be dangerous to measure the length of a snake. It is safe to measure its length on a photographic image of the snake and then recalculate the obtained result.

Standard 4-1-3. If we are given the problem of measurement and the problem cannot be changed to remove the need for measurement and it is impossible to use copies or pictures, it is proposed to transform this problem into a problem of successive detection of changes.

Note: Any measurement is carried out with a certain degree of accuracy. Therefore, even if the problem deals with continuous measurement, one can always single out a simple act of measurement involving two successive detections. This makes the problem considerably simpler.

Example. To measure a temperature, it is possible to use a material that changes its color depending on the current value of the temperature. Alternatively, several materials can be used to indicate different temperatures.

Group 4-2: Synthesis of Measurement Systems

Standard 4-2-1. If a non-SFM is not easy to detect or measure, the problem is solved by synthesizing a simple or dual SFM with a field at the output. Instead of direct measurement or detection of a parameter, another parameter identified with the field is measured or detected. Refer to Fig. 17.

If the conditions contain limitations on the introduction or attachment of substances, the problem has to be solved by synthesizing a Su-Field model using external environment as the substance:

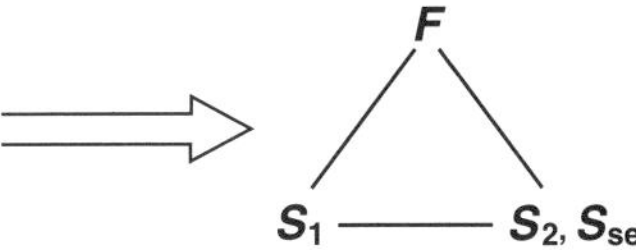

S_{se} is the substance from the surrounding environment. The left part of the formula coincides with that in the previous formulas.

Figure 17 Synthesizing SFM using external environment as the substance.

Example. To detect a moment when a liquid starts to boil, an electrical current is passed through the liquid. During boiling, air bubbles are formed; they dramatically reduce electrical resistance of the liquid.

Standard 4-2-2. If a system (or its part) does not provide detection or measurement, the problem is solved by transition to an internal or external complex measuring SFM, introducing easily detectable additives.

Example. To detect leakage in a refrigerator, a cooling agent is mixed with a luminophore powder.

Standard 4-2-3. If a system is difficult to detect or to measure at a given moment of time and it is impossible to introduce additives in the object, then additives that create an easily detectable and measured field should be introduced in the external environment and the changing state of the environment will provide an indication of the state of the object.

Example. To detect the wear of a rotating metal disc in contact with another disk, it is proposed to introduce luminophore into the oil lubricant, which already exists in the system. Metal particles collecting in the oil will reduce luminosity of the oil.

Standard 4-2-4. If it is impossible to introduce easily detectable additives in the external environment, they can be obtained in the environment itself, e.g., by decomposing it or by changing the aggregate state of the environment.

Note: Specifically, gas or vapor bubbles produced by electrolysis, cavitation, or by any other method are often used as additives obtained by decomposing the external environment.

Example. The speed of water flow in a pipe might be measured by the amount of air bubbles resulting from cavitation.

Group 4-3: Enhancement of measurement system

Standard 4-3-1. Efficiency of a measuring SFM is enhanced by the use of physical effects.

Example. The temperature of liquid media can be determined by measuring the change in the coefficient of retraction, which depends on the value of the temperature.

Standard 4-3-2. If it is impossible to detect or measure directly the changes that take place and if no field can be passed through the system, the problem is to be solved by exciting resonance oscillations (of the whole system or of its part), whose frequency change is an indication of the changes that take place. Refer to Figs. 18 and 19.

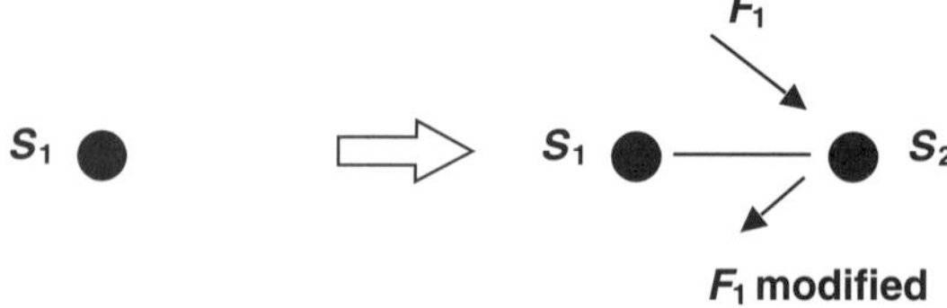

Figure 18

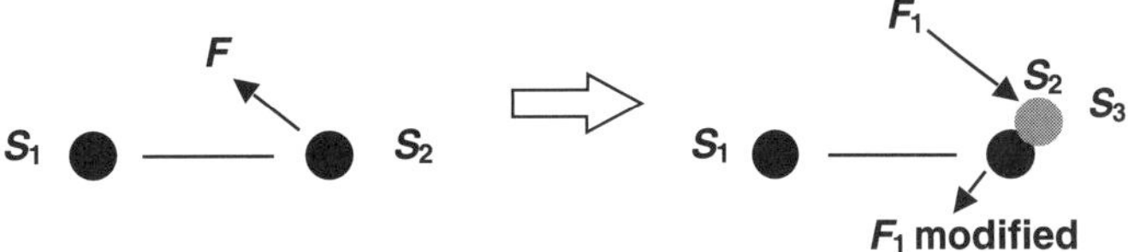

Figure 19

Example. To measure the mass of a substance in a container, the container is subjected to mechanically forced resonance oscillations. The frequency of the oscillations depends upon the mass of the system.

Standard 4-3-3. If no resonance oscillations can be excited in a system, its state can be determined by a change in the natural frequency of the object (external environment) connected with the system under control.

Example. The mass of boiling liquid can be determined by measuring the natural frequency of gas resulting from evaporation.

Group 4-4: Transition to Ferromagnetic Measurement Systems

Standard 4-4-1. The efficiency of a measuring SFM is enhanced by using a ferromagnetic substance and a magnetic field.

Note: The standard indicates the use of a ferromagnetic substance that is not crushed.

Example. A group of students developed a method of measuring speed, direction, time, and operating status of an operating system designed to unwind some type of material from one spool to another spool. To take mechanical rotations and put them in the form of analog pulses that could be analyzed by either a microprocessor or electronic component through a pulsed tachometer, the following detection method was developed. A pulsed tachometer can detect rotations of a rotating shaft that contain a ferromagnetic rotor comprised of "iron brushes" perpendicular to the axis. The magnet in the pickup sensor creates a magnetic field around the sensor. When the iron brushes on the rotor pass through the magnetic field, the flux change induces an electromotive force (EMF) in a coil sensor. These create analog pulses that can be used to determine operating speed, time, direction, and status.

Standard 4-4-2. Efficiency of detection or measurement is enhanced by transition to feSFMs, replacing one of the substances with ferromagnetic particles (or adding ferromagnetic particles), and by detecting or measuring the magnetic field.

Example. In an effort to orient or align numerous objects, ferromagnetic material can be added to the same portion of each object to be aligned. A magnet can then be used to attract the ferromagnetic portion of the object thus, orienting or aligning the objects.

Standard 4-4-3. If it is required to raise a system's efficiency of detection or measurement by going to a feSFM, while replacement of the substance with ferromagnetic particles is not allowed, the transition to the feSFM is performed by building a complex feSFM, introducing (or attaching) ferromagnetic additives to the substance.

Example. The addition of iron oxide (a ferromagnetic powder) is now included as a pigment in black ink to validate currency and other negotiable documents. This technology is in continual development as computers and high-quality color printers make counterfeiting an elementary process. The magnetic fields from these particles produce signatures that, when read by magnetic sensors, can also be used to determine denominations of currency by vending or change machines.

Standard 4-4-4. If it is required to enhance a system's efficiency of detection or measurement by going over to a feSFM, while introduction of ferromagnetic particles is not allowed, ferromagnetic particles are to be introduced in the external environment.

Example. The discovery of the electron resulted in extreme advances in the chemistry field. In 1927, Wolfgang Pauli developed a formal representation of the electron spin concept. Experimentation in 1967 produced data that indicated that electrons from ferromagnetic particles (Fe, Co, and Ni) were not spin polarized as had been previously theorized. To continue testing, an ultrahigh vacuum was constructed where photoemissions of electrons could be performed down to 4.2 K and in magnetic fields up to 50 kOe. This device obtained strikingly different results: The electrons photoemitted from various particles were highly spin polarized. Continued research allowed for the development of spin polarization spectroscopy, helping scientists to further understand magnetism. Recent testing utilizing thin ferromagnetic films indicates that the films may be useful in acting as a spin filter similar to plastic foils used with polarized light.

Standard 4-4-5. Efficiency of a feSFM measuring system is enhanced by the use of physical effects, such as going through the Curie point, Hopkins and Barkhausen effects, magnetoelastic effect, etc.

Example. Diagnosing and forecasting residual life of steel structures are important in determining the safety of large structures. Material magnetic memory (MMM) is effective in the assessment of the stressed–strained state of structures. This method envelops the theory that in zones of stress and strain concentration there are irreversible changes of the magnetic state of ferromagnetic items. Change of residual magnetization in tension, compression, torsion, and cyclic loading of ferromagnetic items is directly related to the maximal acting stress. The operator moves a sensor measuring the residual magnetic field intensity (H_p, A/m) along the weld over the entire perimeter and then transversely to the weld with the amplitude of deviation from the weld edge for 30–50 mm toward the base metal of the pipe element. The second operator records in the log book the data on residual magnetization of the metal, namely magnetic field intensity with the plus or minus sign. An abrupt change of the sign and value of H_p points to a concentration of residual stresses along the $H_p = 0$ line for a specific section of the welded joint. The main purpose of MMM is detection of the most critical sections and components in the controlled plant, which are characterized by strain concentration zones. After MMM, the traditional methods of nondestructive testing (ultrasonic test, X-ray, and eddy current inspection, etc.) are used to determine the presence of a particular defect.

Group 4-5: Evolution of Measurement Systems

Standard 4-5-1. Efficiency of a measuring system at any stage of its development is enhanced by transitioning to a measuring bi- or polysystem.

Note: For a simple formation of bi- and polysystems two or more elements are to be combined. The elements to be combined may be substances, fields, Su-Field pairs, and whole SFMs.

Example. It is difficult to accurately measure the temperature of a small beetle. However, if there are many beetles put together, the temperature can be measured easily.

Standard 4-5-2. Measuring systems are developed towards a transition to measuring the derivatives of the function under control. The transition is performed along the following line:

Measurement of a function → measurement of the first derivative of the function
→ measurement of the second derivative of the function

Example. Changes of stress in the rock are defined by the speed of changing the electrical resistance of the rock.

9 ALGORITHM FOR INVENTIVE PROBLEM SOLVING

ARIZ is the primary problem-solving tool in TRIZ. ARIZ was published in 1959 and revised many times: ARIZ-61, ARIZ-64, ARIZ-65, ARIZ-71 and ARIZ-85. Each revision improved the structure, language, and length of the algorithm. In its current state, we have a carefully crafted set of logical statements that transform a vaguely defined problem into an articulation of one with a clearly defined number of contradictions.

The assumptions designed into ARIZ are that the true nature of the problem is unknown and the process of finding a solution will follow the problem solver's vector of psychological inertia. It is why many of the steps in ARIZ are reformulations of the problem. With each reformulation, the problem is viewed from a different vantage point, yielding the possibility of new and novel ideas.

In mathematics, an algorithm is a precise set of steps designed to arrive at a single outcome. No consideration is given to the personality of the problem solver or to any changeable external conditions. The process is rote. In a broader context, an algorithm is a process following a set of sequential steps. ARIZ falls within that broader definition. ARIZ is a structured set of logic statements that guide the process of invention through a series of formulations and reformulations of the problem. If a chronic technological problem persists even after many attempts to solve it, the reason is often because the wrong problem is being solved. The selection of which problem to solve in an inventive situation is the starting point. It is critical that this selection is correct if there is any hope of arriving at an inventive solution in a timely manner.

As with any systematized process, ARIZ is dependent on the innate intelligence and knowledge of the subject matter expert and the skill with which he or she utilizes the tool. The strength of ARIZ, however, is that the process of thinking inventively is stripped of psychological inertia and regulated in a stepwise fashion toward the ideal solution, or in TRIZ terms, the IFR. The result is, the innate knowledge of the inventor is leveraged so that he or she is forced into thinking "inventively," e.g., into the solution space containing the most inventive ideas. Once the person is in the solution space, there are a number of inventive principles, analogs or Su-Field models that promote "thinking outside the box." See Fig. 20.

9.1 Steps in ARIZ

The architecture of ARIZ is composed of three major processes that are subdivided into nine high-level steps, each with their own substeps. ARIZ is designed to utilize all of the tools in TRIZ, including:

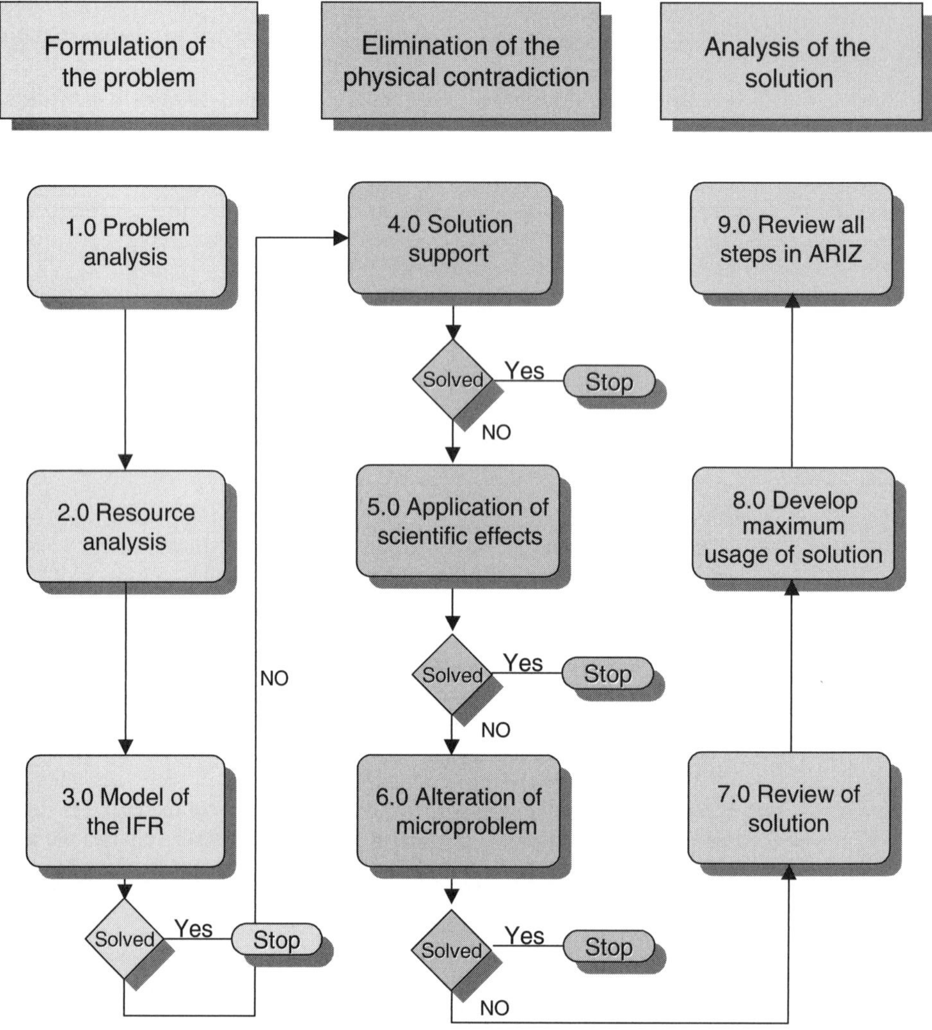

Figure 20 ARIZ flowchart.

- Ideality
- The ideal final result
- Elimination of physical and technical contradictions
- Maximal utilization of the resources of the system
- SFMs and standard solutions
- The 40 inventive principles

ARIZ is designed to manage the inventive process on two types of problems: micro- and macroproblems. A microproblem is focused on solving a contradiction contained within the system while a macroproblem is a redesign of the entire system. ARIZ is iterative in that the inventor is provided several alternative paths to solving a problem. If all the solutions generated at the microlevel are unsatisfactory, the problem must be solved at the macrolevel.

A portion of the algorithm (Stage 1—Formulation of the problem) is detailed below.

1. Problem analysis

1.1. Microproblem. Write down the conditions of the microproblem (*do not use technology-specific jargon*):

- A technological system for (*specify the purpose of the system*) that includes (*a list of the main elements of the system*). Technical contradiction 1: (*formulate*).
- Technical contradiction 2: (*formulate*).
- It is required to achieve (*specify desirable result*) without incurring (*specify the undesirable result*) with minimal changes or complications introduced into the system.

Note: Technical contradictions are defined using nouns for the elements in the system and actionable verbs describing the interaction between them.

1.2. Conflicting elements. Identify the conflicting elements: An article and a tool *Rules:*

1. If an element can be in two states, point out both of them.
2. An article is an element that is to be processed or improved. A tool is an element that has an immediate interaction with the article.
3. If there is more than one pair of the identical conflicting elements, it is sufficient to analyze just one pair.

1.3. Conflict intensification. Formulate the intensified technical contradiction (ITC) by showing an extreme state of the elements.

1.4. Conflict diagrams. Compile diagrams of the intensified technical contradictions.

1.5. Selection of the conflict. Select from two conflict diagrams, one for further analysis.

Rules:

1. Select a diagram which better emphasizes the main (primary) function.
2. If intensification of the conflicts resulted in impossibility of performing the main function, select a diagram that is associated with an absent tool.
3. If intensification of the conflicts resulted in elimination of the article, use a "95% principle."
4. Select a diagram which better emphasizes the main function but reformulate an associated technical contradiction by showing, not extreme, but very close to extreme, states of the elements.

1.6. Model of solution. Develop a model of solution by specifying actions of an X-resource capable of resolving the selected ITC:

- It is required to find such an X-resource that would preserve (*specify the useful action*) *while* eliminating (*specify harmful action*) with minimal changes or complications introduced into the system.

1.7. Model of solution diagram. Construct a diagram of the model of the solution.

1.8. Substance field analysis. Compile a Su-Field diagram that models the solution: *Compile a SFM representing a selected ITC.*

- *Compile a desirable SFM illustrating resolution of the conflict.*
- *Select the appropriate standard solution and compile the complete Su-Field transformation.*

2. Resource analysis

2.1. Conflict domain. Define the space domain within which the conflict develops.

2.2. Operation time. Define the period of time within which the conflict should be overcome:
- Operation time is associated with the time resources available:
  1. Pre-conflict time T_1
  2. Conflict time T_2
  3. Postconflict time T_3
- It is always preferable to overcome a conflict during T_1 and/or T_2

Substance and energy resources. List the substance and energy resources of the system and its environment:
- The substance and energy resources are physical substances and fields that can be obtained or produced easily within the system or its environment. These resources can be of three types:
  1. In-system resources:
    a. Resources of the tool
    b. Resources of the article
  2. Environmental resources:
    a. Resources of the environment that is specific to the system
    b. General resources that are natural to any environment such as magnetic or gravitation fields of the earth
  3. Overall system resources
    a. Side products: waste products of any system or any inexpensive or free foreign objects

3. Model of ideal solution

3.1. Selection of the X-resource. Select one of the resources from 2.3 for further modification. *Rules:*
1. Select in-system resources in the conflict domain first.
2. Modification of the tool is more preferable than modification of the article.

3.2. First ideal final result. The first IFR can be formulated as follows: The X-resource, without any complications or any harm to the system, terminates (*specify the undesirable action*) during the operation time within the conflict domain while providing the (*specify the useful action*).

3.3. Physical contradiction. Formulate a physical contradiction:
- To terminate (*specify the undesirable action*), the X-resource within the conflict domain and during the operation time must be (*specify the physical state P*).
- To provide (*specify the desirable action*), the X-resource within the conflict domain and during the operation time must be (*specify the opposite physical state −P*).

3.4. Elimination of physical contradiction macro. Use methods for elimination of physical contradictions:
- Separation of opposite physical properties in time
- Separation of opposite physical properties in space
- Separation of opposite physical properties between system and its components
- Separation of opposite properties upon conditions
- Combination of the above methods

Note: When applying the separation principles, use one or a combination of the following techniques:
- Separation in time:
  1. Think of ways to make the X-resource have property P before or after the conflict and property $-P$ during the conflict.
  2. Use "high-speed" processes.

3. Explore various phenomena possible for the X-resource developed during phase transitions.
4. Change the parameters or characteristics of the X-resource using a field.
5. Explore using phenomena associated with decomposition of the X-resource into its basic elementary structure and then its recovery, e.g., ionization, recombination, dissociation, association, etc.

- Separation in space:
  1. Divide the X-resource into two parts having properties P and $-P$ with one part in the conflict domain and the other outside the conflict domain.
  2. Combine the X-resource with a void, porosity, foam, bubbles, etc.
  3. Combine X-resource with other resources.
  4. Combine X-resource with a derivative of another resource (e.g., hydrogen and oxygen is a derivative of water).
- Separation between the system and its components. Divide the X-resource into several components in a way that one component has property P while the other has property $-P$.
  1. Decompose the X-resource into elementary particles, granules, flexible rods, shells, etc.
  2. Explore using the phenomena associated with the decomposition of the X-resource into its base elements.

10 CAVEAT

ARIZ is a highly developed complex tool and should not be used on typical straightforward engineering problems. Also, becoming proficient with ARIZ takes time and practice. As a general rule of thumb, it is recommended that an individual solve 10 problems with ARIZ before they claim a layman's level of competency with the tool.

11 CONCLUSION

TRIZ is a powerful comprehensive problem-solving tool. It is the product of a massive analytical study of the output of the world's best inventors and most creative inventions. The fundamental underlying principle of TRIZ is ideality. The ideality principle holds that over time systems evolve to higher levels of functionality through the elimination of internal contradictions and the efficient utilization of available resources.

In time, the study of inventions by Altshuller and others yielded a number of knowledge-based and analytical tools. Knowledge-based tools include the contradiction matrix, the 40 inventive principles, and the laws of systems evolution. Analytical tools include Su-Field analysis and ARIZ.

Contradiction as a goal is a tough sell to American engineers. We rely on "trades." For the TRIZ practitioner finding a contradiction is the answer. If something has to be on/off, hot/cold, liquid/solid, magnetic/nonmagnetic, or any other dichotomy, that contradiction *is* the answer. We are lucky to have the founding efforts of Jim Kowalick, the sustained efforts of Ellen Domb, and the addition of M. Michael Slocum at the *TRIZ Journal* (www.triz-journal.com). The online journal has archives also online which are readily available. Google TRIZ and you will find out a great deal of information. Some of the best case studies are locked in the vaults. The 300+ cases at Ontro by Michael Slocum are delineated in the *TRIZ journal*.

A number of these engineering tools and initiatives work together. The reader may note several *connects* in various chapters of this document. Another recent tool is Design for Six Sigma, (see Chapter 17). Where creativity or inductive reasoning is used, TRIZ may perform a positive service, especially those involving teams. A short list would include concurrent engineering, value engineering, ergonomic factors in design, processes, patents, total quality management, knowledge management, dimensional management, Six Sigma, and technical areas where teams are stalling or where a team needs to know if it is on the correct technical path or change is happening at the appropriate pace. Refer to the ABET certified course on Systems Engineering lecture notes of J. McMunigal.

Acknowledgments

A special "thank you" to John Opfell for collaboration utilizing engineering tools in the 1980s, including TRIZ starting in 1992 and for insights gleaned from direct Russian translation; Sam Brooks for long Sunday "chalk talks" on engineering tools and projects within Boeing featuring TRIZ and other engineering tools; Jeffrey A. Wolfe, Six Sigma black belt (BB); and Kelly R. McMunigal, Six Sigma yellow belt for technical support.

BIBLIOGRAPHY

Altshuller, G., *Creativity as an Exact Science,* Gordon & Breach, New York, 1984.

Altshuller, G., *The Innovation Algorithm,* Technical Innovation Center, Worcester, MA, 1999.

Batchelor, S., "Solving the Problems of Particle Filled Fibers Using the TRIZ Methodology," *TRIZ J.,* Oct. 1999, www.triz-journal.com.

Bosse, V., and J. E. McMunigal, book review of *Solving Problems with TRIZ* (An Exercise Book), *TRIZ J.,* May, 2004, www.triz-journal.com.

Carr, J., "Analysis of a Problem: Clogging of a Multi-Drum Filter Used in a Textile Application," *TRIZ J.* Aug. 1999, www.triz-journal.com.

Champa, R. and R. Handley, *Brainware for the Strategist,* Strategy Partners International, Mission Viejo, CA, 2003.

Clapp, T. G. and B. A. Dickinson, "Design and Analysis of a Method for Monitoring Felled Seat Seam Characteristics Utilizing TRIZ Methods," *TRIZ J.,* Dec. 1999, www.triz-journal.com.

Clapp, T. G., and M. S. Slocum, "Theory of Inventive Problem Solving Pedagogy in Engineering Education, Part I," *TRIZ J.,* Nov. 1998, www.triz-journal.com.

Gahide, S., "Smart Garment for Firefighters," *TRIZ J.,* June 1999, www.triz-journal.com.

Gasanov, A. M., B. M. Gochman, A. P. Yefimochkin, S. M. Kokin, and A. G. Sopelnyak, *Birth of an Invention,* Interpraks, Moscow, 1995.

Gibson, N., "The Determination of the Technological Maturity of Ultrasonic Welding," *TRIZ J.,* July 1999, www.triz-journal.com.

Heath, D., "Addressing Salt Issues in Textile Dyeing Using an ISQ and ARIZ," *TRIZ J.,* Jan. 2000, www.triz-journal.com.

Khona, V. J., "Increasing Speed of Yarn Spinning," *TRIZ J.,* Aug. 1999, www.triz-journal.com.

Kunst, B. and T. Class, "Automatic Boarding Machine Design Employing Quality Function Deployment, Theory of Inventive Problem Solving, and Solid Modeling," *TRIZ J.,* Jan. 2000, www.triz-journal.com.

McMunigal, J. E., "In Memory of Genrich Altschuller, " triz-viet nam, Jan., 1999, www.triz-vietnam.com.

McMunigal, James E., notes from Systems Engineering course, California State University—Long Beach, Spring 2000.

Rantanen, K. and E. Domb, *Simplified TRIZ,* CRC Press, Boca Raton, FL, 2002.

Raviv, D., "Introduction to Inventive Problem Solving in Engineering," *TRIZ J.,* Mar. 1997, www.triz-journal.com.

Rivin, E. I., "Use of the Theory of Inventive Problem Solving (TRIZ) In Design Curriculum," Innovations in Engineering Education, 1996 ABET Annual Meeting Proceedings, pp. 161–164; *TRIZ J.,* Mar. 1997, www.triz-journal.com.

Roberts, M., "B-cyclodextrin Molecules and Their Use in Breathable Barriers," *TRIZ J.,* Nov. 1999, www.triz-journal.com.

Salamatov, Y., *TRIZ: The Right Solution at the Right Time,* Insytec B.V., Hattem, Netherlands, 1999.

Savransky, S. D., *Engineering of Creativity,* CRC Press, Boca Raton, FL, 2000.

Slocum, M., and J. E. McMunigal, "TRIZ and the Deconstruction of the Major World Philosophies," No. 17, Altshuller Institute for TRIZ Studies, 2003, www.aitriz.org/2003/ABSTRACTS.htm.

Terninko, J., A. Zusman, and B. Zlotin, *Systematic Innovation,* CRC Press, Boca Raton, FL, 1998.

Ungvari, S., *TRIZ Two Day Workshop Manual,* Strategic Product Innovations, Columbus, OH, 1998.

Ungvari S., *TRIZ Refresher Course,* Strategic Product Innovations, Columbus, OH, 1999.

Ungvari, S., *TRIZ Problem Solving Guidebook,* Strategic Product Innovations, Columbus, OH, 1999.

Vijayakumar, S., "Maturity Mapping of DVD Technology," *TRIZ J.,* Sept. 1999, www.triz-journal.com.

CHAPTER 19
COMPUTER-AIDED DESIGN

Emory W. Zimmers, Jr. and Technical Staff
Enterprise Systems Center
Lehigh University
Bethlehem, Pennsylvania

1 INTRODUCTION TO CAD

Computers have a prominent, often controlling role throughout the life cycle of engineering products and manufacturing processes. Their role is becoming increasingly important as global competitive pressures call for improvements in product performance and quality coupled with significant reductions in product design, development, and manufacturing timetables.

Design engineers vastly improve their work productivity using computers. Performance of a product or process can be evaluated prior to fabricating a prototype using appropriate simulation software.

Computer-aided design (CAD) uses the mathematical and graphic processing power of the computer to assist the engineer in the creation, modification, analysis, and display of designs. Many factors have contributed to CAD technology becoming a necessary tool in the engineering world, such as the computer's speed at processing complex equations and managing technical databases. CAD combines the characteristics of designer and computer that are best applicable to the design process.

The combination of human creativity with computer technology provides the design efficiency that has made CAD such a popular design tool. CAD is often thought of simply as computer-aided drafting, and its use as an electronic drawing board is a powerful tool in itself. But the functions of a CAD system extend far beyond its ability to represent and manipulate graphics. However, geometric modeling, engineering analysis, simulation, and the communication of the design information can also be performed using CAD.

1.1 Historical Perspective on CAD

Graphical representation of data, in many ways, forms the basis of CAD. An early application of computer graphics was used in the SAGE (Semi-Automatic Ground Environment) Air Defense Command and Control System in the 1950s. SAGE converted radar information into computer-generated images on a cathode ray tube (CRT) display. It also used an input device, the light pen, to select information directly from the CRT screen.

Another significant advancement in computer graphics technology occurred in 1963, when Ivan Sutherland, in his doctoral thesis at MIT, described the SKETCHPAD system (Fig. 1). A Lincoln TX-2 computer drove the SKETCHPAD system. With SKETCHPAD, images could be created and manipulated using the light pen. Graphical manipulations such as translation, rotation, and scaling could all be accomplished on-screen using SKETCHPAD.

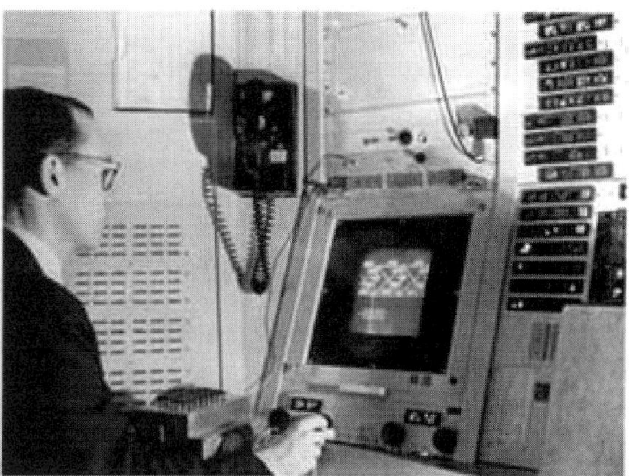

Figure 1 The SKETCHPAD was a project developed by Ivan Sutherland at MIT. Sutherland later went on to work on the first HMD, the precursor to virtual reality head displays.

Computer applications based on Sutherland's approach have become known as interactive computer graphics (ICG). The graphical capabilities of SKETCHPAD showed the potential for computerized drawing in design.

During his time as a professor of electrical engineering at the University of Utah, Sutherland continued his research on head-mounted displays (HMDs). The field of computer graphics, as we know it today, was born from among the many new ideas and innovations created by the researchers who made the University of Utah a hub for this kind of research. Together with the founder of the university's Computer Science Department, Sutherland cofounded Evans and Sutherland in 1968, which later went on to pioneer computer modeling systems and software. Today, almost 40 years later, the company remains an industry leader in developing both hardware and software to help professionals create realistic images used in a variety of applications such as simulation, training, and engineering (Fig. 2).

While at the California Institute of Technology, Sutherland served as the chairman of the Computer Science Department from 1976 to 1980. While he was there, he helped to introduce the integrated circuit design to academia. Together with Carver Mead, they developed the science of combining the mathematics of computing with the physics of real transistors and real wires and subsequently went on to make integrated circuit design a proper field of academic study.

In 1980, Sutherland left Caltech and launched the company Sutherland, Sproull, and Associates. Bought by Sun Labs in 1990, the acquisition formed the basis for Sun Microsystems Laboratories.

The high cost of computer hardware in the 1960s limited the use of ICG systems to large corporations such as those in the automotive and aerospace industries, which could justify the initial investment. With the rapid development of computer technology, computers became more powerful, with faster processors and greater data storage capabilities. Their physical size and cost decreased, and computers became affordable to smaller companies and personal users. Today it is rare to find an engineering, design, or architectural firm of any size without a working CAD system running on a personal computer (PC) or a workstation.

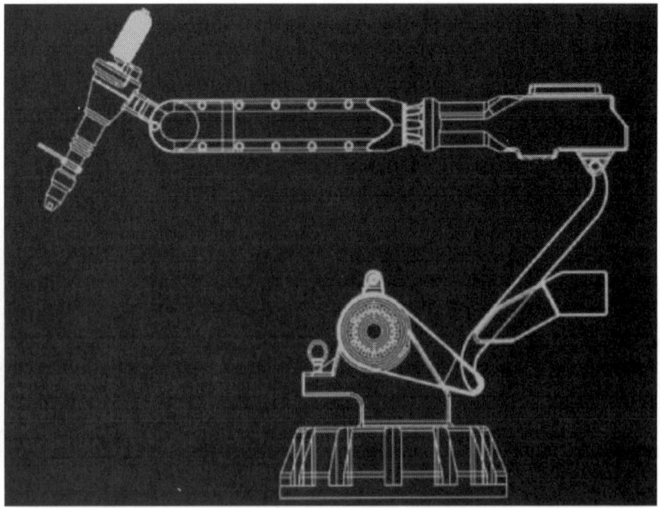

Figure 2 Image on a line-drawing graphics display.

1.2 Design Process

Before any discussion of CAD, it is necessary to understand the design process in general (Fig. 3). What are the series of events that lead to the beginning of a design project? How does the engineer go about the process of designing something? How does one arrive at the conclusion that the design has been completed? We address these questions by defining the process in terms of six distinct stages:

1. Customer input and perception of need
2. Problem definition
3. Synthesis
4. Analysis and optimization
5. Evaluation
6. Final design and specification

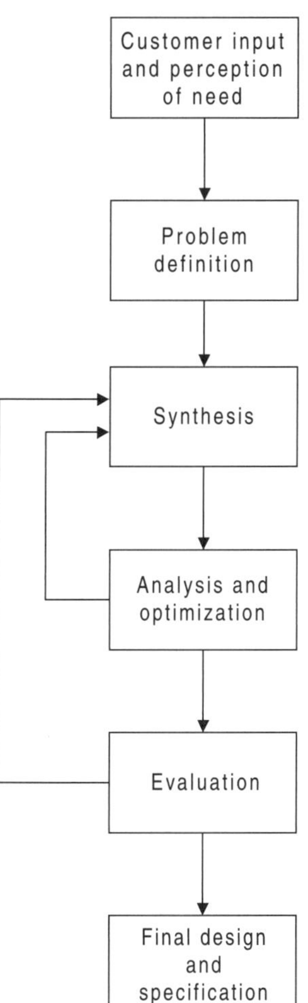

Figure 3 General design process.

A need is usually perceived in one of two ways. Someone must recognize either a problem in an existing design or a customer-driven opportunity in the marketplace for a new product. In either case, a need exists which can be addressed by modifying an existing design or developing an entirely new design. Because the need for change may only be indicated by subtle circumstances—such as noise, marginal performance characteristics, or deviations from quality standards—the design engineer who identifies the need has taken a first step in correcting the problem. That step sets in motion processes that may allow others to see the need more readily and possibly enroll them in the solution process.

Once the decision has been made to take corrective action to the need at hand, the problem must be defined as a particular problem to be solved such that all significant parameters in the problem are defined. These parameters often include cost limits, quality standards, size and weight characteristics, and functional characteristics. Often, specifications may be defined by the capabilities of the manufacturing process. Anything that will influence the engineer in choosing design features must be included in the definition of the problem. Careful planning in this stage can lead to fewer iterations in subsequent stages of design.

Once the problem has been fully defined in this way, the designer moves on to the synthesis stage, where knowledge and creativity can be applied to conceptualize an initial design. Teamwork can make the design more successful and effective at this stage. That design is then subjected to various forms of analysis, which may reveal specific problems in the initial design. The designer then takes the analytical results and applies them in an iteration of the synthesis stage. These iterations may continue through several cycles of synthesis and analysis until the design is optimized.

The design is then evaluated according to the parameters set forth in the problem definition. A scale prototype is often fabricated to perform further analysis and to assess operating performance, quality, reliability, and other criteria. If a design flaw is revealed during this stage, the design moves back to the synthesis/analysis stages for reoptimization, and the process moves in this circular manner until the design clears the evaluative stage and is ready for presentation.

Final design and specification represent the last stage of the design process. Communicating the design to others in such a way that its manufacture and marketing are seen as vital to the organization is essential. When the design has been fully approved, detailed engineering drawings are produced, complete with specifications for components, subassemblies, and the tools and fixtures required to manufacture the product and the associated costs of production. These can then be transferred manually or digitally using the CAD data to the various departments responsible for manufacture.

In every branch of engineering, prior to the implementation of CAD, design has traditionally been accomplished manually on the drawing board. The resulting drawing, complete with significant details, was then subjected to analysis using complex mathematical formulas and then sent back to the drawing board with suggestions for improving the design. The same iterative procedure was followed, and because of the manual nature of the drawing and the subsequent analysis, the whole procedure was time consuming and labor intensive. CAD has allowed the designer to bypass much of the manual drafting and analysis that was previously required, making the design process flow much more smoothly and efficiently.

It is helpful to understand the general product development process as a stepwise process. However, in today's engineering environment, the steps outlined above have become consolidated into a more streamlined approach called *concurrent engineering*. This approach enables teams to work concurrently by providing common ground for interrelated product development tasks. Product information can be easily communicated among all development processes: design, manufacturing, marketing, management, and supplier networks. Concurrent engineering recognizes that fewer iterations result in less time and money spent in moving from concept to manufacture and from manufacturing to market. The related processes of design for manufacturing (DFM) and design for assembly (DFA) have become integral parts of the concurrent engineering approach.

DFM and DFA methods use cross-disciplinary input from a variety of sources (e.g., design engineers, manufacturing engineers, suppliers, and shop-floor representatives) to facilitate the efficient design of a product that can be manufactured, assembled, and marketed in the shortest possible period of time. Often, products designed using DFM and DFA are simpler, cost less, and reach the marketplace in far less time than traditionally designed products. DFM focuses on determining what materials and manufacturing techniques will result in the most efficient use of available resources in order to integrate this information early in the design process. The DFA methodology strives to consolidate the number of parts, uses gravity-assisted assembly techniques, and calls for careful review and consensus approval of designs early in the process. By facilitating the free exchange of information, DFM and DFA methods allow engineering companies to avoid the costly rework often associated with repeated iterations of the design process.

In an attempt to define the stages or components of the design process, many definitions exist today which can vary from one to another. However, they all share a common thread that includes a needs statement by identifying the problem, a search for possible solutions, analysis and development of the solutions, testing, and finally usage of the final product. These kinds of descriptions of the design process are commonly called *models of the design process*. Figure 4 is an example of a model created by Pahl and Beitz in 1984 which was composed of four main phases:

- *Clarification of Task.* Involves collecting information about the design requirements and the constraints on the design and describing these in a specification.
- *Conceptual Design.* Involves establishment of the functions to be included in the design and identification and development of suitable solutions.
- *Embodiment Design.* The conceptual solution is developed in more detail, problems are resolved, and weak aspects are eliminated.
- *Detail Design.* The dimensions, tolerances, materials, and forms of individual components of the design are specified in detail for subsequent manufacture.

Figures 4 and 5 both depict two traditional methods of describing the design process, where there are sequential stages of design, with the manufacturing to follow. However, the current trend in manufacturing is to expedite the process by encouraging the design, development, analysis, and preparation of manufacturing information to be done simultaneously. This kind of engineering has previously been used in companies that produced established products or are constantly producing new products. The terms *concurrent engineering,* and *simultaneous engineering* have been coined to describe this type of engineering. This will be addressed in more depth in Chapter 17. Here, we will keep the sequential models of the design process during our discussion of modeling and communication in design.

1.3 Applying Computers to Design

Many of the individual tasks within the overall design process can be performed using a computer. As each of these tasks is made more efficient, the efficiency of the overall process increases as well. The computer is especially well suited to design in four areas, which correspond to the four stages of the general design process given above. Computers function in the design process through geometric modeling capabilities, engineering analysis calcu-

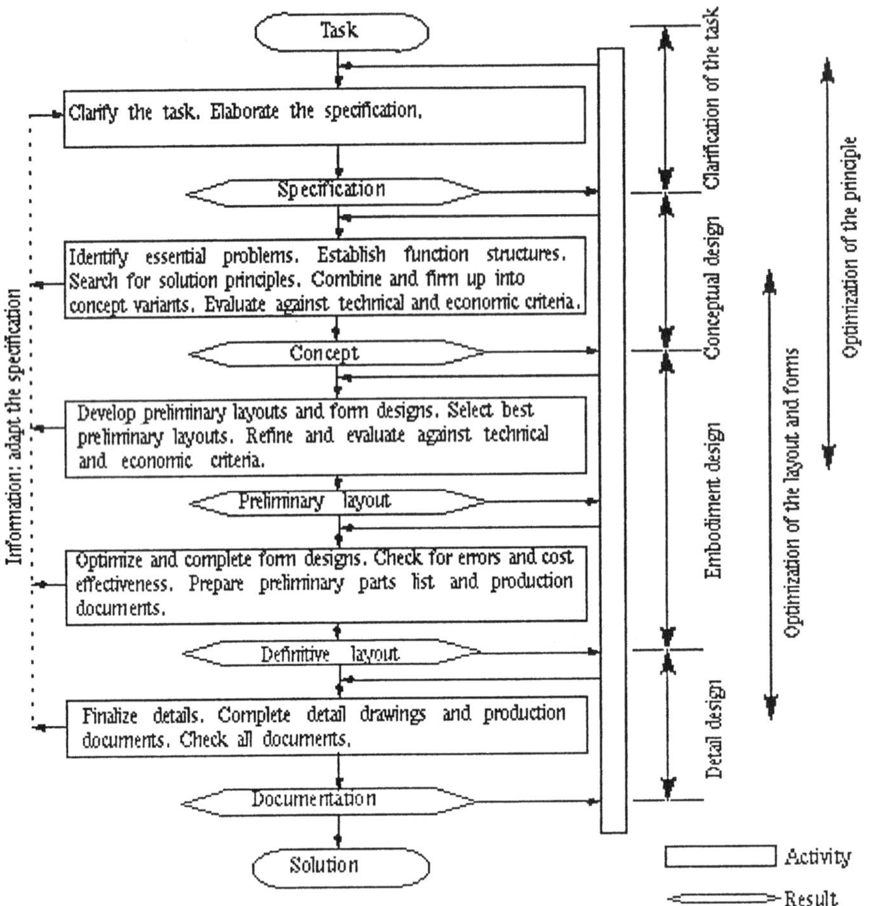

Figure 4 Pahl and Beitz design process model. Model presents a very clear, linear process. However, not all processes can be described in these terms, and there is no guarantee that the design team will not revisit earlier stages.

lations, automated evaluative procedures, and automated drafting. Figure 6 illustrates the relationship between CAD technology and the final four stages of the design process.

Geometric modeling is one of the keystones of CAD systems. It uses mathematical descriptions of geometric elements to facilitate the representation and manipulation of graphical images on a computer display screen. While the central processing unit (CPU) provides the ability to quickly make the calculations specific to the element, the software provides the instructions necessary for efficient transfer of information between the user and the CPU.

Three types of commands are used by the designer in computerized geometric modeling. The first type of command allows the user to input the variables needed by the computer to represent basic geometric elements such as points, lines, arcs, circles, splines, and ellipses. The second type of command is used to transform these elements. Commonly performed transformations in CAD include scaling, rotation, and translation. The third type of command allows the various elements previously created by the first two commands to be joined into a desired shape.

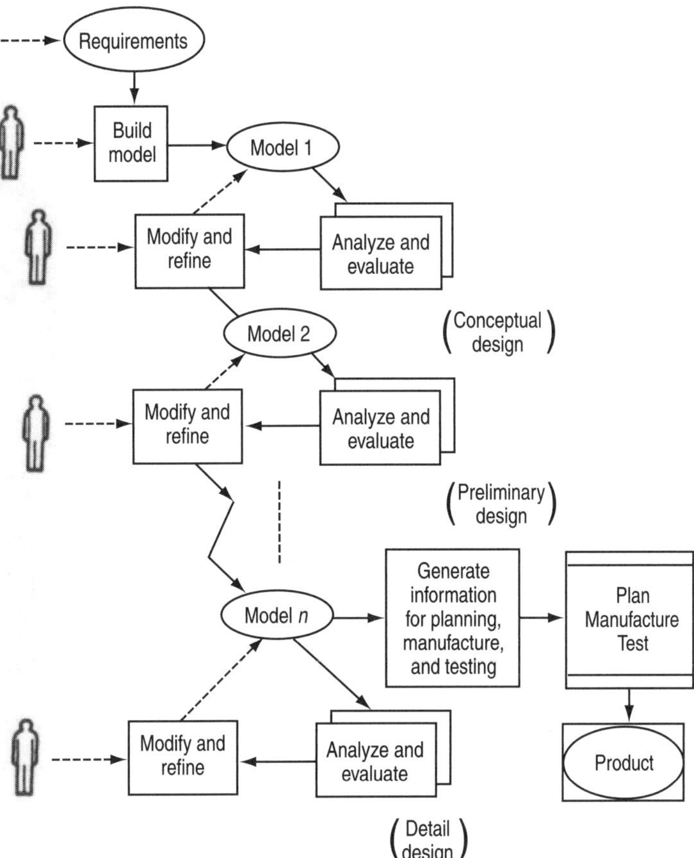

Figure 5 Oshuga design process.

During the whole geometric modeling process, mathematical operations are at work which can be easily stored as computerized data and retrieved as needed for review, analysis, and modification. There are different ways of displaying the same data on the CRT screen, depending on the needs or preferences of the designer. One method is to display the design as a two-dimensional (2D) representation of a flat object formed by interconnecting lines. Another method displays the design as a three-dimensional (3D) representation of objects. In 3D representations, there are four types of modeling approaches:

- Wireframe modeling
- Surface modeling
- Solid modeling
- Hybrid solid modeling

A *wireframe model* is a skeletal description of a 3D object. It consists only of points, lines, and curves that describe the boundaries of the object. There are no surfaces in a wireframe model. Three-dimensional wireframe representations can cause the viewer some confusion because all of the lines defining the object appear on the 2D display screen. This makes it hard for the viewer to tell whether the model is being viewed from above or below, inside or outside.

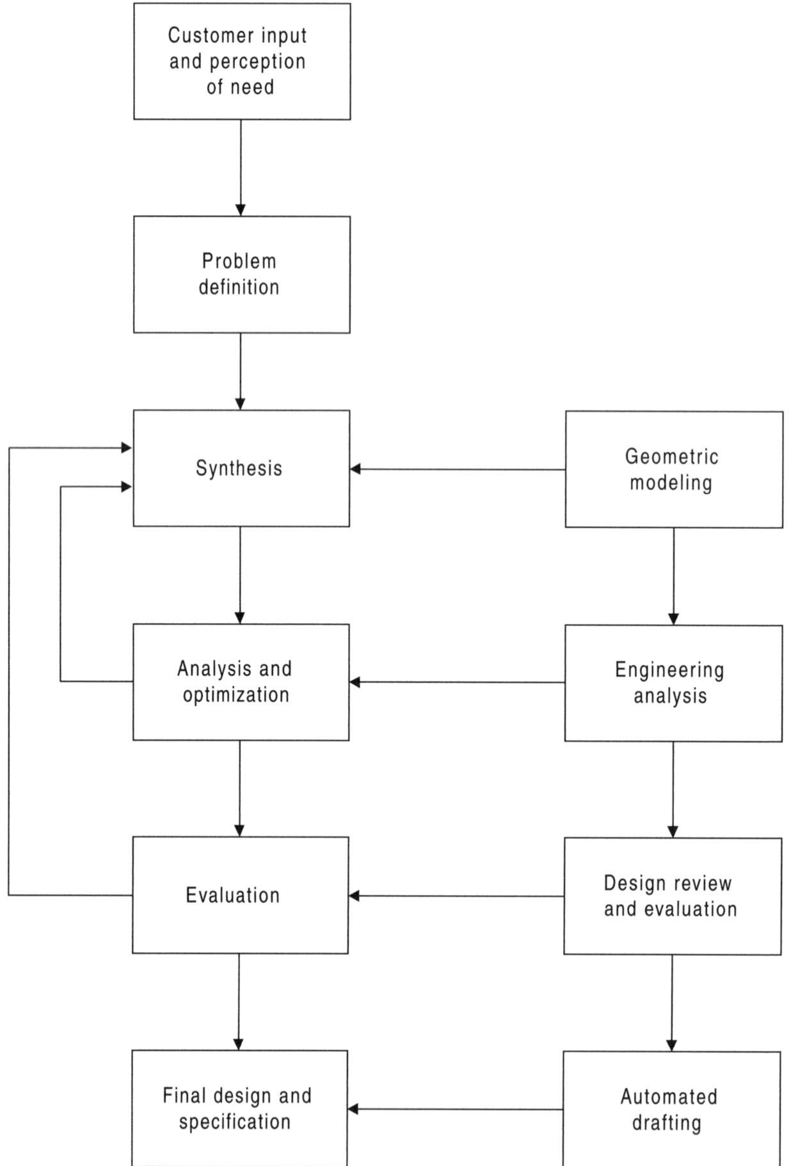

Figure 6 Application of computers to design process.

Surface modeling defines not only the edge of the 3D object but also its surface. In surface modeling, two different types of surfaces can be generated: faceted surfaces using a polygon mesh and true curve surfaces. NURBS (nonuniform rational B-spline) is a B-spline curve or surface defined by a series of weighted control points and one or more knot vectors. It can exactly represent a wide range of curves such as arcs and conics. The greater flexibility for controlling continuity is one advantage of NURBS. NURBS can precisely model nearly all kinds of surfaces more robustly than the polynomial-based curves that were used in earlier surface models. The surface modeling is more sophisticated than wireframe modeling. Here, the computer still defines the object in terms of a wireframe but can generate a surface "skin" to cover the frame, thus giving the illusion of a "real" object. However, because the computer has the image stored in its data as a wireframe representation having no mass, physical properties cannot be calculated directly from the image data. Surface models are very advantageous due to point-to-point data collections usually required for numerical control (NC) programs in computer-aided manufacturing (CAM) applications. Most surface modeling systems also produce the stereolithographic data required for rapid prototyping systems.

Solid modeling defines the surfaces of an object, with the added attributes of volume and mass. This allows image data to be used in calculating the physical properties of the final product. Solid modeling software uses one of two methods: constructive solid geometry (CSG) or boundary representation (B-rep). The CSG method uses Boolean operations (union, subtraction, intersection) on two sets of objects to define composite models. For example, a cylinder can be subtracted from a cube. B-rep is a representation of a solid model that defines an object in terms of its surface boundaries: faces, edges, and vertices.

Hybrid solid modeling allows the user to represent a part with a mixture of wireframe, surface modeling, and solid geometry. The I-DEAS Master Modeler offers this representation feature.

In CAD software, certain features have been developed to minimize the ambiguity of wireframe representations. These features include using dashed lines to represent the background of a view or removing those background lines altogether. The latter method is appropriately referred to as "hidden-line removal." The hidden-line removal feature makes it easier to visualize the model because the back faces are not displayed. Shading removes hidden lines and assigns flat colors to visible surfaces. Rendering adds and adjusts lights and textures to surfaces to produce realistic effects. Shading and rendering can greatly enhance the realism of the 3D image. Figures 7 and 8 show the same object represented as a pure wireframe and a wireframe with hidden-line removal.

Engineering analysis can be performed using one of two approaches: analytical or experimental. In the analytical method, the design is subjected to simulated conditions using any number of analytical formulas. By contrast, the experimental approach to analysis requires that a prototype be constructed and subsequently subjected to various experiments to yield data that might not be available through purely analytical methods.

There are various analytical methods available to the designer using a CAD system. Finite-element analysis and static and dynamic analysis are all commonly performed analytical methods available in CAD.

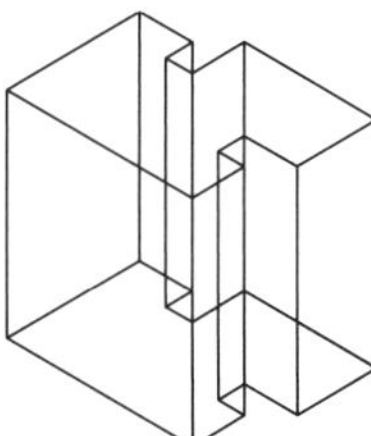

Figure 7 Pure wireframe model.

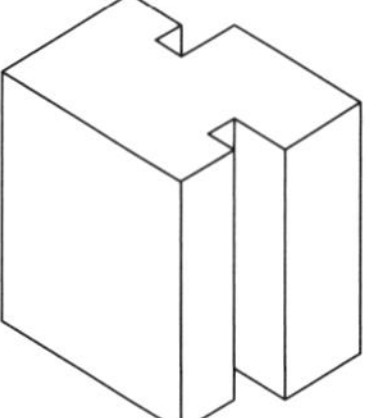

Figure 8 Wireframe model with hidden-line removal feature.

Finite-element analysis (FEA) is a computer numerical analysis program used to solve complex problems in many engineering and scientific fields, such as structural analysis (stress, deflection, vibration), thermal analysis (steady state and transient), and fluid dynamics analysis (laminar and turbulent flow) (Fig. 9).

The finite-element method divides a given physical or mathematical model into smaller and simpler elements, performs analysis on each individual element using required mathematics, and then assembles the individual solutions of the elements to reach a global solution for the model. FEA software programs usually consist of three parts: the preprocessor, the solver, and the postprocessor.

The program inputs are prepared in the preprocessor. Model geometry can be defined or imported from CAD software. Meshes are generated on a surface or solid model to form the elements. Element properties and material descriptions can be assigned to the model. Finally, the boundary conditions and loads are applied to the elements and their nodes. Certain checks must be completed before the analysis calculation. These include checking for duplication of nodes and elements and verifying the element connectivity of the surface elements so that the surface normals are all in the same direction. To optimize disk space and running time, the nodes and elements should usually be renumbered and sequenced. Many analysis options are available in the analysis solver to execute the model. The element stiffness matrices can be formulated and solved to form a global stiffness value for the model solution. The results of the analysis data are then interpreted by the postprocessor in an orderly manner. The postprocessor in most FEA applications offers graphical output and animation displays. Many vendors of CAD software are also developing pre- and postproc-

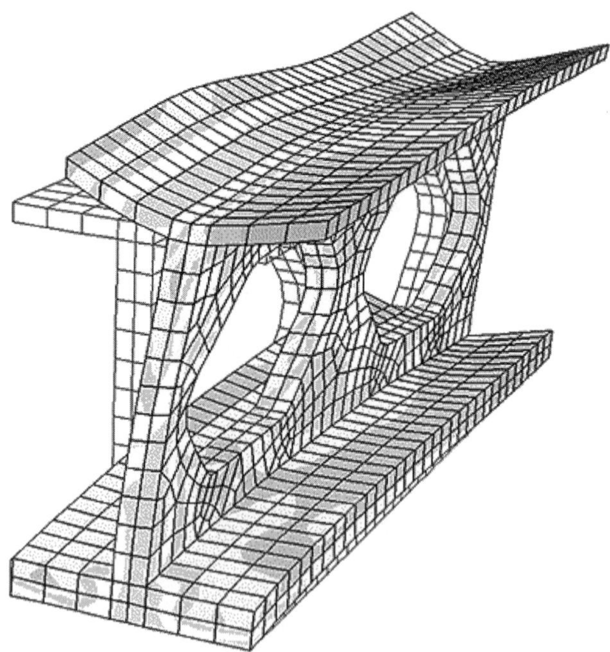

Figure 9 FEA of random vibration in a beam. Colors or gray scales are often used to show degrees of stress and deflection. The original shape is also outlined without shading for reference. (Courtesy of Algor, Inc.)

essors that allow users to graphically visualize their input and output. FEA is a powerful tool in effectively synthesizing a design into an optimized product.

Kinematic analysis and synthesis are used to study the motion or position of a set of rigid bodies in a system without reference to the forces causing that motion or the mass of the bodies (Fig. 10). It allows engineers to see how the mechanisms they design will function in motion. This luxury enables the designer to avoid faulty designs and also to apply the design to a variety of scenarios without constructing a physical prototype. Synthesis of the data extracted from kinematic analysis in numerous iterations of the process leads to optimization of the design. The increased number of trials that kinematic analysis allows the engineer to perform may have profound results in optimizing the behavior of the resulting mechanism before actual production.

Static analysis determines reaction forces at the joint positions of resting mechanisms when a constant load is applied. As long as zero velocity is assumed, static analysis can also be performed on mechanisms at different points of their range of motion. Static analysis allows the designer to determine the reaction forces on whole mechanical systems as well as interconnection forces transmitted to their individual joints. The data extracted from static analysis can be useful in determining compatibility with the various criteria set out in the problem definition. These criteria may include reliability, fatigue, and performance considerations to be analyzed through stress analysis methods.

Dynamic analysis combines motion with forces in a mechanical system to calculate positions, velocities, accelerations, and reaction forces on parts in the system. The analysis

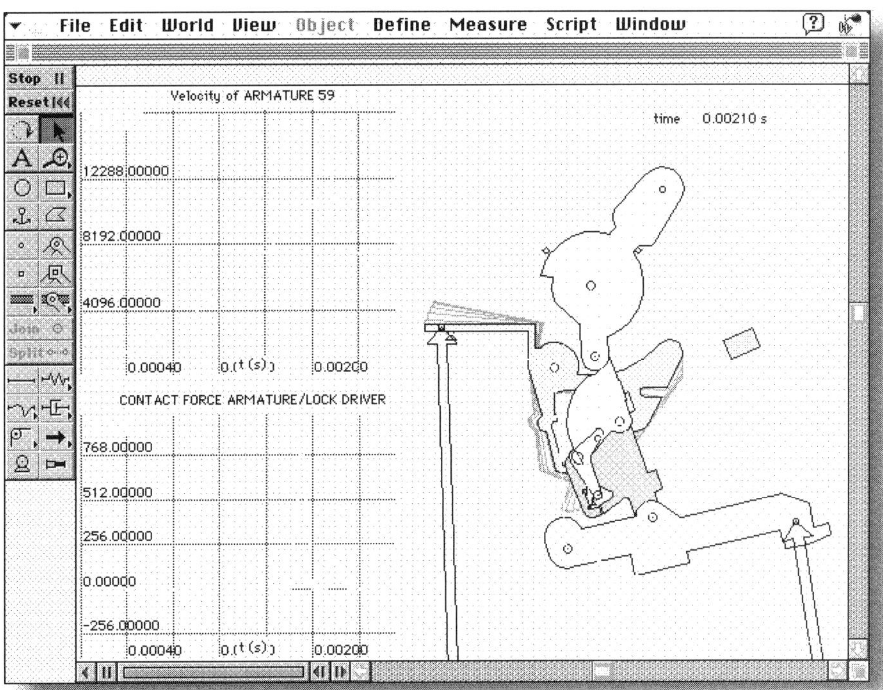

Figure 10 Kinematic analysis of a switch mechanism. (Courtesy of Knowledge Solutions, Inc.)

is performed stepwise within a given interval of time. Each degree of freedom is associated with a specific coordinate for which initial position and velocity must be supplied. Defining the system in various ways creates the computer model from which the design is analyzed. Generally, for data relating to individual parts, the user must supply geometry, forces, and overall system coordinates either directly or through a manipulation of data within the software.

The results of all of these types of analyses are typically available in many forms, depending on the needs of the designer. All of these analytical methods will be discussed in greater detail in Section 7 relating to CAD software.

Experimental analysis involves fabricating a prototype and subjecting it to various experimental methods. Although this usually takes place in the later stages of design, CAD systems enable the designer to make more effective use of experimental data, especially where analytical methods are thought to be unreliable for the given model. CAD also provides a useful platform for incorporating experimental results into the design process when experimental analysis is performed in earlier iterations of the process.

Design review can be easily accomplished using CAD. The accuracy of the design can be checked using automated tolerancing and dimensioning routines to reduce the possibility of error. Layering is a technique that allows the designer to superimpose images upon one another. This can be quite useful during the evaluative stage of the design process by allowing the designer to visually check the dimensions of a final design against the dimensions of stages of the design's proposed manufacture, ensuring that sufficient material is present in preliminary stages for correct manufacture.

Interference checking can also be performed using CAD. This procedure involves making sure that no two parts of a design occupy the same space at the same time.

Automated drafting capabilities in CAD systems facilitate presentation, which is the final stage of the design process. CAD data, stored in computer memory, can be sent to a pen plotter or other hard-copy device (see Section 6.2) to produce a detailed drawing quickly and easily. In the early days of CAD, this feature was the primary rationale for investing in a CAD system. Drafting conventions, including but not limited to dimensioning, crosshatching, scaling of the design, and enlarged views of parts of other design areas, can be included automatically in nearly all CAD systems. Detail and assembly drawings, bills of materials (BOM), and cross-sectioned views of design parts are also automated and simplified through CAD. In addition, most systems are capable of presenting as many as six views of the design automatically. Drafting standards defined by a company can be programmed into the system such that all final drafts will comply with the standard. Further, CAD systems can operate with other software applications to add to their presentation flexibility (Fig. 11).

Documentation of the design is also simplified using CAD. Product data management (PDM) has become an important application associated with CAD. PDM allows companies to make CAD data available interdepartmentally on a computer network. This approach holds significant advantages over conventional data management. PDM is not simply a database holding CAD data as a library for interested users. PDM systems offer increased data management efficiency through a client–server relationship among individual computers and a networked server. Benefits of implementing a PDM system include faster retrieval of CAD files through keyword searches and other search features; automated distribution of designs to management, manufacturing engineers, and shop-floor workers for design review; recordkeeping functions that provide a history of design changes; and data security functions limiting access levels to design files. PDM facilitates the exchange of information characteristic of the emerging agile workplace. As companies face increased pressure to provide clients with customized solutions to their individual needs, PDM systems allow an increased level of teamwork among personnel at all levels of product design and manufacturing, cutting the costs often associated with information lag and rework.

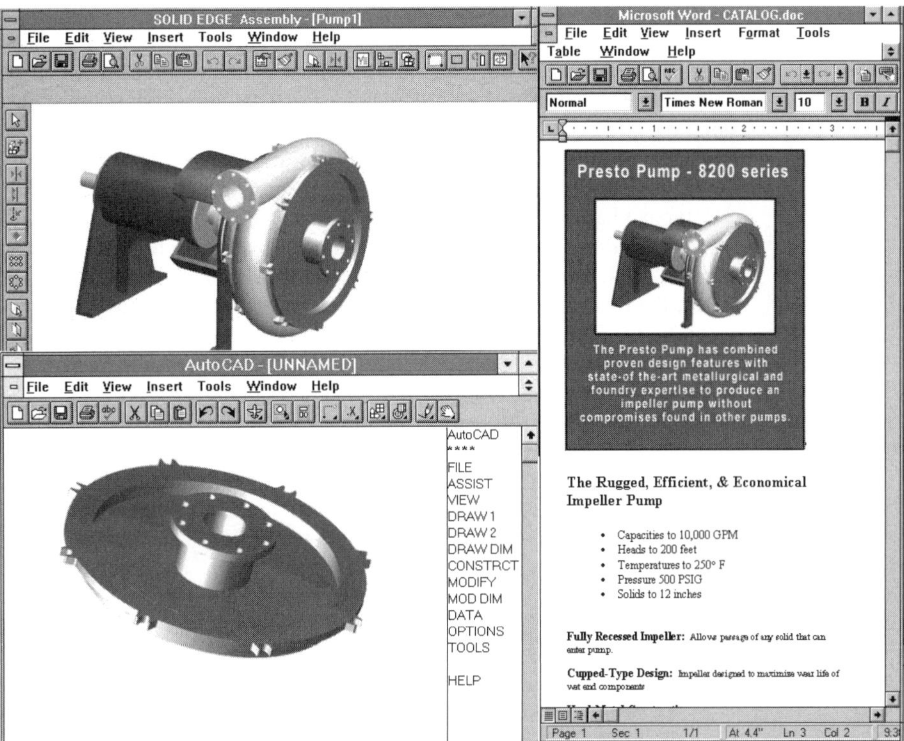

Figure 11 CAD files can be used in conjunction with other applications. Illustration shows Intergraph Corporation's Solid Edge software operating in conjunction with AutoCAD from Autodesk, and Microsoft Word. (Courtesy of Intergraph Corporation.)

Although CAD has made the design process less tedious and more efficient than traditional methods, the fundamental design process in general remains unchanged. It still requires human input and ingenuity to initiate and proceed through the many iterations of the process. Nevertheless, CAD design is such a powerful, time-saving design tool that it is now difficult to function in a competitive engineering world without such a system in place. The CAD system will now be examined in terms of its components: the hardware and software of a computer.

2 HARDWARE

Just as a draftsman traditionally requires pen and ink to bring creativity to bear on the page, there are certain essential components to any working CAD system. The use of computers for interactive graphics applications can be traced back to the early 1960s, when Ivan Sutherland developed the SKETCHPAD system. The prohibitively high cost of hardware made general use of interactive computer graphics uneconomical until the 1970s. With the development and subsequent popularity of personal computers, interactive graphics applications now are widespread in homes and workplaces.

CAD systems have become available for many hardware configurations. Most CAD systems have been developed for standard computer systems ranging from mainframes to microcomputers. Others, like turnkey CAD systems, come with all of the hardware and software required to run a particular CAD application and are supplied by specialized vendors.

Hardware is the tangible element of the computer system. Any physical component of a computer system, machinery, or equipment is considered to be hardware. In a standard PC, typical hardware components are the monitor, keyboard, mouse, and CPU, among others. The term *hardware* also extends beyond PCs to include any information technology (IT) devices, such as routers and hubs. It can also include parts of devices such as the telephone cable, the disk drives, or even the platters within the hard drive.

The term hardware is used to differentiate it from the intangible software. While you can physically pick up and move hardware as you please, software encompasses elements such as operating restrictions, programs, and documents.

Hardware is also used to explain the different stages of computer design or functionalities. An example of this is when an object has become hardware, it is considered to be moving into assembly, as opposed to when it was in the design stage. Terms such as *hardware engineer* and HAL (*hardware abstraction layer*) relate to the physical parts of the computer.

2.1 Central Processing Unit and Input/Output

The above systems all share a dependence on components that allow the actual interaction between computer and users. These electronic components are categorized under two general headings: CPU and input/output (I/O) devices. Input devices transfer information from the designer into the computer's CPU so that the data, encoded in binary sequencing, may be manipulated and analyzed efficiently. Output devices do exactly the opposite. They transfer binary data from the CPU back to the user in a usable (usually visual) format. Both types of devices are required in a CAD system. Without an input device, no information can be transferred to the CPU for processing, and without an output device, any information in the CPU is of little use to the designer because binary code is lengthy and tedious to interpret.

The main processing unit within the computer is known as the CPU, or microprocessor, or "processor" and handles all instructions and data. The difference between a microprocessor and a CPU is that a microprocessor can be the main processing chip in a computer as well as being any of the similar chips found in telephones or automobiles. The CPU, however, mostly refers only to the main processor and its related components within a computer. Nevertheless, many people today use these two terms interchangeably.

The make-up of the CPU chip is a tiny piece of silicon with an integrated circuit built on top of it. This whole unit then plugs into the computer's motherboard and works together with several other components, such as the hard drive and memory area, to operate the data [a motherboard is the printed circuit board platform that typically houses the CPU, random-access memory (RAM), and expansion slots].

The inside of the CPU is made up of a variety of different components, each with a particular function. Some examples of these functions include a cache, a bus, a control unit, a decode unit, an FPU (floating-point unit), and an ALU (arithmetic logic unit). Connecting all of these components is a network of circuits and wires, some of which are no larger than 0.13 μm in diameter, which is about one-thousandth the width of a human hair! To transfer the information, the CPU uses millions of electrical pulses per second to all of the different components.

Also contained within each CPU are millions of transistors that serve as the junction between the circuitry in the chip. To manipulate the data, these transistors either amplify or block the electrical pulses passing by. The calculations performed by the CPU can all take place on a chip no bigger than your thumbnail.

Many landmark events led to the creation of the CPU. First there was the creation of the transistor at Bell Labs in 1947, followed by the use of silicon to create transistors in 1954 at Texas Instruments. Then, at Texas Instruments and Fairchild Semiconductor, the first integrated circuit was created in 1958, all leading up to Intel's creation of the 4004 microprocessor in 1971.

By today's standards, the 4004 chip is meager compared to what is produced by Intel and AMD (Advanced Micro Devices), the leading manufacturers of CPUs. To clarify, modern CPUs run at speeds of around 2 GHz or more, while the 4004 chip ran at 108 kHz. That means the modern CPUs run about two thousand times faster than the 4004 did. Also, the 4004 chip only contained 2250 transistors, while the Pentium 4 CPU contains 42 million!

Devices such as floppy diskette drives, CD-ROM drives, printers, and LANs (local-area networks) are all a part of the information exchange with the CPU commonly known as "I/O." Information in the I/O process is carried via buses such as USB (universal serial bus), FireWire/IEEE (Institute of Electrical and Electronic Engineers) 1394, and SCSI (small-computer system interface).

The process of transferring data from the CPU to the memory is seemingly similar to an I/O activity. However, it is typically not considered as such. A majority of modern processors utilize a dedicated memory subsystem. Within the CPU, different types of events are differentiated by control lines; among these different types of events are memory cycles, I/O cycles, or some other type of event. To coordinate the various amounts of data trying to pass through the buses, the supporting hardware controls the external I/O and internal memory as needed. Without this management, bus contention issues can arise, where two devices attempt to use the same bus at the same time.

In its most elementary forms, I/O can be input only, output only, or both, though there are certainly many different types of I/Os. An example of an output-only channel would include turning on a status light to inform the user of the hard disk drive status. When it comes to input-only devices, voltage sensors on the power supply are good examples that can inform the processor if it is operating within the acceptable limits. I/O buses or devices that both send and receive signals from the computer are the most common and can include the mouse, keyboard, hard drives, network cards, audio cards, and video cards.

3 THE COMPUTER

The computer is a machine that accepts data as input, processes the data according to a set of programmable rules, and produces the result as output.

Broken down into its two most fundamental components, a computer consists of the physical machinery itself, known as the hardware, and the programs telling the computer what to do, known as the software.

Computer hardware is simply a large amount of switches with the software controlling how the switches should be set. The switches use values of 0 and 1 to represent whether they are open or closed, and through this, a single unit of information can be stored in one switch. This unit of information is known as a bit, which stands for "binary digit." Typically arranged into groups of eight, these eight bits make up one byte, which can hold 256 combinations, or a number from 0 to 255.

These combinations of switches, however, are not limited only to representing numbers. Based on the programmer's preference, a particular combination of bits can represent a specific number, letter, or even color for a graphics program.

Switches found inside a modern computer are not mechanical switches but electrical ones that comprise a portion of a larger electrical circuit. The flow of electricity through these switches is used to indicate the state of a switch and may, in turn, open or close other switches. When enough switches are present, very complex calculations can be computed. Within a modest modern computer, there are several million switches.

What is even more important than the sheer number of switches in a computer is the speed at which these switches can be used. In a single second, one switch could turn on and off more than a billion times, meaning that even complex tasks can be accomplished as fast as the user can input the commands. This is possible because at the heart of these complex commands are simple addition, subtraction, multiplication, and division commands utilizing the switches as counters.

Beyond these switches, there are many other components that help to make up the entire system. Broken down into its traditional components, the following are the hardware categories you would find in a computer:

- *Input Devices.* Receive the information and instructions from the user and transfer them into the computer.
- *Memory.* The array of switches that maintain an open or closed position until told to change. This enables the computer to remember programs and information.
- *Central Processing Unit.* Where a majority of the calculations and processing of information are completed utilizing a large array of switches.
- *Output Devices.* Display the results of programming after functions are performed.
- *Storage.* Maintains large amounts of information over a long duration of time.

Although this is the traditional breakdown of the components in a computer, these are very "soft" boundaries, as can be seen with a floppy disk. In one instance, it can be seen as an input device, delivering instructions to the computer. But in an entirely different instance, it could be seen as an output device when data or programs are being written to it. By the same token, there are also many devices that have fixed functions, such as the keyboard and mouse being input devices.

The most basic form of software is known as the OS (operating system) and lies underneath the most common software—those that are purchased or downloaded as special applications. These applications are programmed to perform a specific task, such as word processing or going onto the Internet. However, without the OS, these programs cannot run because it is the OS that actually runs the computer.

However, beneath the OS resides the BIOS (basic input/output system), which instructs the different components of the computer on how to communicate with each other. The BIOS is typically embedded within the computer and is loaded immediately after the computer is started.

Classes of Computers. There are four basic classes of computers that usually define a computer's size, power, and purpose:

- *Supercomputers.* These are the largest, fastest, and most expensive computers available. Although they technically fall under the mainframe class, their difference lies in the fact that supercomputers are designed to handle relatively few extremely complicated tasks in a short amount of time. Some applications of supercomputers lie in calculations involving intensive research and sophisticated applications like theoretical physics, turbulence calculations, weather forecasting, and advanced animated graphics.

- *Mainframes.* Designed to run as many simultaneous applications as possible, these computers are typically large and fast enough to handle many users when connected to multiple terminals. They are most commonly used by research facilities, large businesses, and the military.
- *Minicomputers.* Similar to mainframes in their function, minicomputers are smaller in size and power, though they are actually midrange computers. Still serving multiple users, minicomputers are often networked together with other minicomputers in business use.
- *Microcomputers.* Used almost synonymously with PCs, microcomputers are built around a one-chip microprocessor.

One must be careful not to take these classes as an absolute breakdown because there is some overlap and there are other classes that fall between the four mentioned. For example, a mini-supercomputer is simply one that falls between a supercomputer and a mainframe.

In addition, a computer's power and capabilities are not limited by only its class. Many advances in computer technology during the past few years have made new microcomputers more powerful than minicomputers. Even more surprising is that a new microcomputer can perform as well as some supercomputers from just a few decades past!

The influence of computer technology is a somewhat recent phenomenon due to the reduced cost of computers over the last two decades. However, the philosophical basis for the construction and employment of computing systems has a longer history than 20 years.

Charles Babbage, a nineteenth-century mathematician at Cambridge University in England, is often cited as a pioneer in the computing field. Babbage designed an "analytical engine," the capabilities of which would have surprisingly foreshadowed the same basic functions of today's computers had his design not been limited by the manufacturing capabilities of his time. The analytical engine was designed with considerations for input, storage, mathematical calculation, grouping results, and printing results in a typeface. Other less complex mechanical forms of computers include the slide rule and even the abacus.

The vast majority of contemporary computers are digital, although some analog computers do exist. These latter types have been relegated almost to a footnote in contemporary computing due to the overwhelming advances made in digital technology. The difference between digital and analog systems lies in the binary code.

Although digital computers vary in size, shape, price, and capabilities, all digital computers share four common features. First, the circuits used can exist in one of two states, either "on" or "off." This characteristic yields the basis for binary logic. Second, all share the ability to store data in binary form. Third, all digital computers can receive external input data, perform various functions relating to that data, and provide the user with the output or result of the performed function. Finally, digital computers can all be operated through the use of instructions organized into sets of separate steps. On a related note, many digital systems possess the ability to perform many different functions at the same time using a technique known as *parallel processing*.

3.1 Computer Evolution

Based on the advances leading to each stage of technological progress, computer systems have commonly been grouped into four generations:

- First generation: vacuum tube circuitry
- Second generation: transistors
- Third generation: small and medium integrated circuits
- Fourth generation: large-scale integration (LSI) and very large scale integration (VLSI)
- Fifth generation: near future

The *first generation* of computers, such as the Mark I and the ENIAC in the 1940s, were huge machines in terms of both size and mass. The ENIAC computer at the University of Pennsylvania in Philadelphia was constructed during World War II to calculate projectile trajectories. The circuitry of first-generation computers was composed of vacuum tubes and used very large amounts of electricity (it was said that whenever the ENIAC computer was turned on, the lights all over Philadelphia dimmed). ENIAC weighed 30 tons, occupied 15,000 ft^2 of floor space, and contained more than 18,000 vacuum tubes. It performed 5000 additions per second and consumed 40 kW of power per hour. Also, due to the vacuum tube circuitry, continuous maintenance was required to change the tubes as they burned out. Input and output functions were performed using punched cards and separate printers. Programming these computers was tedious and slow, usually performed directly in the binary language of the computer.

The *second generation* of computers was developed in the 1950s. These computers used transistors as a replacement to the vacuum tubes of their predecessors, decreasing maintenance requirements as well as electricity consumption. Information was stored using magnetic drums and tapes, and printers were connected online to the computer for faster hard-copy output. Unrelated to hardware considerations was the development of programming languages that could be written using more readily understandable commands and then separately converted into the binary data required by the computer.

Just as important as the development of new computer languages was allowing programmers to write with words, phrases, and mathematical formulas rather than in binary machine code. These languages featured the use of a compiler that used another program to translate the more human programming language into machine code the computer could understand. Between these improvements, the computer became a profitable tool for businesses. Based on these improvements, the computer industry, as we know it, was born with the second-generation computers. One problem that plagued second-generation computers was the dissipation of heat from its transistors.

The *third generation* of computers came about in 1964 with another major advancement in hardware and software. The first integrated circuit, which combined three separate electronic components into a single chip, was created at Texas Instruments and marked a major advance in hardware technology. Since then, both the number of components on a single chip and the chip's power have continued to rise, while the chip size has decreased.

The software advance of the third generation allowed users to run multiple programs at a time through the use of an OS. This OS would act as a master control program and coordinate the operation and use of system resources. This advancement came as a significant leap forward from the first- and second-generation computers that had to load each individual program before running them separately.

High-level software languages such as COBOL, FORTRAN, and BASIC were developed at this time and gained popularity. These languages were written in a way that the programmer could more readily understand and assemble automatically into a set of instructions for the computer to follow. The most significant development of this period was a downward cost spiral that precipitated the popularity of minicomputers—smaller computers designed for use by one user or a small number of users at a time, as opposed to the larger mainframes of previous generations.

The *fourth generation* of digital computers began around 1971. The steady decrease in processing time and cost for computer technology has continued with a corresponding increase in memory and computational capabilities. With LSI, more than 1000 components can be placed on a single integrated-circuit chip. VLSI chips contain more than 10,000 components, and current VLSI chips have 100,000 or more components on each chip. The semiconductor technology developed in the 1970s condensed whole computers into the size of a single chip known as a microprocessor. Semiconductors were responsible for the arrival of "personal computers" in the late 1970s and early 1980s.

The mark of this generation has been the ability to place more and more electronic components onto an increasingly smaller chip until today's circuits have millions of components fitting on a surface no larger than that of a shirt button. In the mid-1970s, computers had progressed enough to be reasonable for use in the common household as well as being capable of performing word processing, database manipulation, and financial work. By the 1980s, computer users no longer had to have prior programming knowledge. Today, they have become as common as the household TV or oven.

The *fifth-generation* computers have long been seen as "just around the corner." There are several technological factors indicating that the next generation is on our doorstep. The first is the concept of artificial intelligence in computers that would allow the computer to better mimic expert human thought. Next, there is the possibility of parallel processing, or the use of two independent central processors to work together to significantly increase the power of the computer. Also, the potential for superconducting materials presents a great opportunity for computers with their ability to offer extremely low resistance in electrical flow, thus increasing the speed of information flow throughout the computer.

The separation of the fourth- and fifth-generation computers is certain to be obscure, and it is still uncertain as to when the fifth generation will begin.

3.2 Classes of Computers

Computers, as briefly described above, can be divided into the following classes depending on their size, power, and capabilities: supercomputers, mainframes, minicomputers, and microcomputers

Supercomputers are the world's most powerful computers, often with processing speeds in excess of 20 million computations per second. The performance of the CRAY-2 supercomputer was rated at 100 million floating-point operations per second (MFLOPS). Supercomputers are often used to calculate extensive mathematical problems for scientific research purposes. These problems are characterized by the need for high precision and repetitive performance of floating-point arithmetic operations on large arrays of numbers.

Mainframes have large memory capabilities coupled with extremely fast processing speeds. These computers are less powerful than supercomputers, but they are used in large CAD systems where a significant amount of highly accurate analysis must occur. Mainframes are highly applicable to analytical methods and are often used in dynamic analysis, stress analysis, heat transfer analysis, and other analytical problems. This type of computer system is used most often in large engineering corporations, such as those in the automotive or aerospace industries, where centralized computing and data storage are essential. Mainframes support multiple users (some over 500) at terminals, giving them access almost instantaneously to the data required to design and share information among the project team. Because of their extensive memory capabilities, mainframes are also used for large database maintenance. Mainframe computers usually require a specialized support staff for maintenance and programming. The typical configuration of a mainframe system is a processor with 32- and 64-bit word addressing, 64 megabytes (MB) to 2 gigabytes (GB) of memory, and several gigabytes of storage space.

Minicomputers are somewhat smaller and less powerful than a mainframe, but they nevertheless offer a powerful, less expensive alternative to mainframe systems where a centralized computing environment is desired. They introduced the concept of distributed data processing. A typical minicomputer is available with 16–32-bit word addressing, several megabytes of memory, and multiple disk drives amounting to several megabytes to gigabytes of storage space. Turnkey CAD systems were offered as minicomputer systems in the late 1970s and early 1980s. A number of display terminals can be supported by minicomputer systems, and online printers for minicomputer systems are capable of delivering between two and three thousand lines of text per minute.

Microcomputers, which include the personal computer and engineering workstations, are desktop-size or smaller computers. These computers have seen the greatest growth in the number of systems being sold and used since the early 1980s. There are various reasons for this trend. Microcomputers are quickly becoming more powerful, with greater memory capabilities. A wide range of microcomputers are available with 8–32-bit word addressing, several megabytes of memory, and built-in hard-disk, floppy disk, CD-ROM, and tape backup systems.

Many companies operate best using a decentralized approach to computing; however, networks have become increasingly common in microcomputer environments in order to provide some of the advantages of centralized computing when desirable. Powerful servers that support massive client–server networks have largely replaced the huge mainframe computers. Even the power of a contemporary PC exceeds that of a mainframe from the 1960s and 1970s. Furthermore, the computational capability of engineering workstations today exceeds that of most minicomputers. The latest trend is to classify computers as supercomputers, servers, workstations, large PCs, and small PCs.

One common differential between types of computers is the word length. The term "word length" does not refer to words in human language; rather, it signifies the number of places in the base-2 units of the machine language in the various types of computers. Mainframes have traditionally run using 32-bit words, with minicomputers typically having 16-bit capabilities, and microcomputers an 8-bit word length. Word length influences processing speeds and memory-addressing capabilities in computer systems. Longer word length means that more information can be operated on or transferred to a different part of the system in fewer steps, thereby taking less time. The word length also influences memory capabilities by making virtual memory techniques available. While formerly applicable as a general rule for distinguishing the capabilities of the various types, various word lengths are now available in all types of computers. Even some home electronic game systems now employ 32-bit technology in a system about the size of a large textbook.

3.3 Central Processing Unit

A computer's CPU is the portion of the computer that retrieves and executes instructions. The CPU is essentially the brain of a CAD system. It consists of an ALU, a control unit, and various registers. The CPU is often simply referred to as "the processor." (See Section 2.1 for more information on the CPU.)

The ALU performs arithmetic operations, logic operations, and related operations according to the program instructions. The control unit controls all CPU operations, including

ALU operations, the movement of data within the CPU, and the exchange of data and control signals across external interfaces (e.g., the system bus). Registers are high-speed internal memory storage units within the CPU. Some registers are user visible, that is, available to the programmer via the machine instruction set. Other registers are dedicated strictly to the CPU for control purposes. An internal clock synchronizes all CPU components. The clock speed (the number of clock pulses per second) is measured in megahertz, or millions of clock pulses per second. The clock speed essentially measures how fast an instruction is processed by the CPU.

On August 27, 2001, Intel released its 2-GHz Pentium 4 chip, with a reduction in circuitry "wire" size to 0.13 μm, and an increase in the number of transistors to 55 million in 2002. This P4 chip uses NetBurst microarchitecture, which uses 20 stages in its pipeline. This increase in pipeline size allows the P4 to move more information more quickly and greatly increases the clock speed.

This increase in speed of the Pentium family of chips in just under a decade is quite amazing. These chips went from a 60 MHz clock speed to more than 2 GHz during this time. If this were translated into automobile speeds, travelers going from San Francisco to New York would be able to do so in just 13 seconds! In addition to these advancements in the Pentium chips, improved flexibility and microarchitecture have kept pace with the improving speeds.

3.4 CISC and RISC Computers

Computers can be divided into two classes depending on their method of using instructions:

- Complex instruction set computers (CISCs)
- Reduced instruction set computers (RISCs)

The reasons for designing CISCs are to simplify compilers and to improve performance. Underlying both of these reasons was the shift to high-level languages (HLLs) in computer programming. Computer architects attempted to design machines that provided better support for HLLs. The CISC was expected to yield smaller programs that would execute instructions more quickly.

The most basic command a computer CPU can understand and accept is an instruction set. Following any command is a two-step process for a computer. First, the computer must find the currently running program in order to figure out what step it is supposed to do next. Then it simply follows the instructions. Based on this, computers were originally designed to run as few instructions as possible in order to maintain higher operational speeds. The premise behind this idea was that the CPU would save time by having to retrieve instructions less frequently. As an example, imagine a parent telling a child to "clean your room" as opposed to saying, "pick up your clothes," waiting for the clothes to be picked up, then saying, "now put them in the hamper," and so forth.

Computer designers soon learned that giving a complex command like "clean your room" could often take longer to retrieve than several simple commands. This led to the development of RISCs.

Today, the actual architecture and styles of the chips determine the classification of chips because of the significant increase in chip speeds. Differentiation in instruction sets used within the chip are no longer the primary discerning factor because both CISC and RISC chips perform at a negligible speed difference. As an amusing point, many CISC chips actually have a smaller instructional set than many RISC chips.

In general, modern computer users will never notice a difference in usage from either the RISC or CISC chips. Intel's Pentium chip is a CISC chip, while Macintosh uses a RISC design. Some computers even use combinations of both RISC and CISC, further eliminating the differentiation between the two classes.

RISC technology is very new by comparison. However, RISC computers use fewer and simpler instructions than conventional CISC computers. Simpler instructions reduce the complexity of the circuits required to implement an instruction, thereby allowing individual instructions to execute quickly. RISC machines generally show a higher level of performance in comparison to a comparably complex CISC system, despite the fact that a RISC processor executes more instructions to accomplish a given task than does a CISC processor. The following characteristics are common on all RISC architecture computers: one instruction per machine cycle, unique register-to-register operations, and an instruction pipeline. RISC architecture often includes more general-purpose registers to maximize the number of operations that take place on the CPU. CISC computers, in contrast, employ more memory referencing. Studies have shown that the compilers on CISC machines tend to favor simpler instructions, such that the conciseness of the complex instruction sets seldom comes into play. The expectation that a CISC computer would produce smaller programs may not be realized because the more complex the instruction set, the more processor time is required to decode and execute each instruction. Longer opcodes required in CISC architecture produce longer instructions. In computationally intensive applications such as FEA, where calculation times are often measured in hours, RISC processors are generally more efficient at performing floating-point operations than CISC processors.

Most RISC architecture can be found in workstations which run on the UNIX operating system, such as Sparc MIPS, DEC Alpha, PA-RISC, and PowerPC. The Pentium architecture, however, is an excellent example of CISC design. It represents the result of decades of CISC architecture evolution. Pentium architecture incorporates the sophisticated design principles once found only on mainframes, supercomputers, and servers.

The RISC versus CISC debate continues to drive computer technology in new directions. There is a growing realization in the industry that RISC and CISC may benefit from each other. Notably, more recent designs, like the PowerPC line of the Apple Macintosh, are no longer "pure" RISC, and the more recent CISC designs, like the Pentium P4, have incorporated some RISC-like features.

3.5 Engineering PCs

Computer-aided design projects often range from simple 2D drawings to graphics-intensive engineering applications. Computationally intensive number crunching in 3D surface and solid modeling, photorealistic rendering, and FEA applications demand a great deal from a PC. Careful selection of a PC for these applications requires an examination of the capabilities of the CPU, RAM capacity, disk space, operating system, network features, and graphics capabilities. The industry advances quickly, especially in microprocessor capabilities. The following PC configurations list minimum requirements for various CAD applications. It should be noted that because of rapid advances in the industry, this listing may be dated by the time of publication.

For 2D drafting applications, a low-end PC is sufficient. This denotes a PC equipped with a Pentium II-266 MHz (or equivalent) processor running with 32 MB of RAM and Windows 95, 98, NT, or XP. A 4-GB hard drive is suggested by most software manufacturers, even though the software may only use 300 MB for the root program. A 16-inch high-resolution (1024 × 768) SVGA should also be considered a minimum requirement. CD-ROM drives are also essential.

For 3D modeling and FEA applications, a Pentium III 800-MHz processor running on a Windows XP or Windows 2000 operating system works well. The minimum memory requirement for Windows NT is 32 MB, and the operating system requires 160 MB of hard-drive space. However, most graphics programs will require at least 256 MB of RAM to run. The monitor for these applications should be 21 in. with 0.25 ultrafine dot pitch, high resolution (1600 × 1280), and 65,536 colors. Peripheral component interconnect (PCI) local bus (a high-bandwidth, processor-independent bus) and a minimum 4-GB hard drive are also required.

3.6 Engineering Workstations

The Intel Pentium CPU microprocessor reduced the performance gap between PCs and workstations. A current trend is the merging of PCs and workstations into "personal workstations." Pentium Pro and RISC processors, such as DEC Alpha, Sparc MIPS, Power PC, and PA-RISC, are all powerful personal workstations with high-performance graphics accelerators.

Until recently, operating systems were the main distinction between a low-end workstation and a high-end PC. The UNIX operating system, which supports multitasking and networking, usually ran on a workstation. DOS, Windows, and the Macintosh operating systems, which perform single tasks, usually ran on PCs. That distinction is beginning to disappear because of the birth of Microsoft Windows NT, Windows XP, and IBM's OS/2 operating systems. All of these operating systems support multitasking and networking as well. Other operating systems designed for workstations are being modified for use on a PC, such as IBM's AIX. UNIX now runs on laptop computers with the PowerPC microprocessor.

Many CAD and FEA software applications are traditionally UNIX-based applications. Since there are significant differences in price between UNIX and Windows NT and XP, more and more CAD and FEA vendors have released versions of their software for Windows. Memory capacity for a personal workstation can be up to 4–8 GB with up to 250 GB of disk space, such as those from Silicon Graphics and Hewlett-Packard's Indy Systems. Due to 64-bit technology and a scaleable modular platform, high-end workstation performance can now boast supercomputer-like performance at a fraction of the cost of a mainframe or supercomputer. The noted performance gain is a result of using more powerful 64-bit word addressing and up to 3.2-GHz clock speed. Dual processors available in some engineering workstations with the DEC Alpha system, Intel Pentium chip, or Motorola 68060 chip allow some degree of scaleable parallel processing within current engineering workstations.

The term "workstation" has colloquially become context dependent. However, technically speaking, a workstation is any networked computer that accesses or inputs data into the system.

Previously, workstations were known as "dumb terminals" because of their lack of processing and storage ability. This shortcoming required them to be linked to extremely large mainframes, which in turn only allowed them to see what the network would show them.

Today's workstations, or "clients," are no longer "dumb," since they are capable of storing their own information as well as a variety of programs. Nonetheless, these distributed client–server networks, as can be found in Novell's Netware or Microsoft's Windows NT or XP, still incorporate a single or many large and powerful computers.

For the Windows-based peer-to-peer network, there is no longer a need for a powerful central computer, as any of the computers in the network may serve as the server, a workstation, or both. The term workstation again changes meaning in the context of engineering, high-end graphics rendering, CAD, and other such lofty uses. In these contexts, the term typically refers to supermachines with the most memory, fastest screens, fastest processors, and, naturally, the highest prices.

3.7 Parallel Processing

To reach higher levels of productivity in the analysis of complex structures with thousands of components, such as in combustion engines or crash simulations, the application of parallel processing was introduced. Two parallel processing methods used in engineering applications are as follows (Fig. 12):

- Symmetric multiprocessing (SMP)
- Massively parallel processing (MPP)

Parallel processing is the use of multiple processors to complete complex tasks and calculations by dividing up the work. The ultimate goal of parallel processing is to constantly increase the power of the systems. This takes place in a parallel computer or supercomputer and is similar to having multiple workers work together to complete a single or many tasks in a shorter amount of time. SMP is the shared memory while MPP is the distributed memory used in these processes.

The difference between the SMP and the MPP systems is how they handle the memory relationship with the processor. In an SMP system, the more common of the two, the memory is shared between the two processors. This has the potential to lead to writing problems when both processors attempt to write to the memory at the same time. The MPP system,

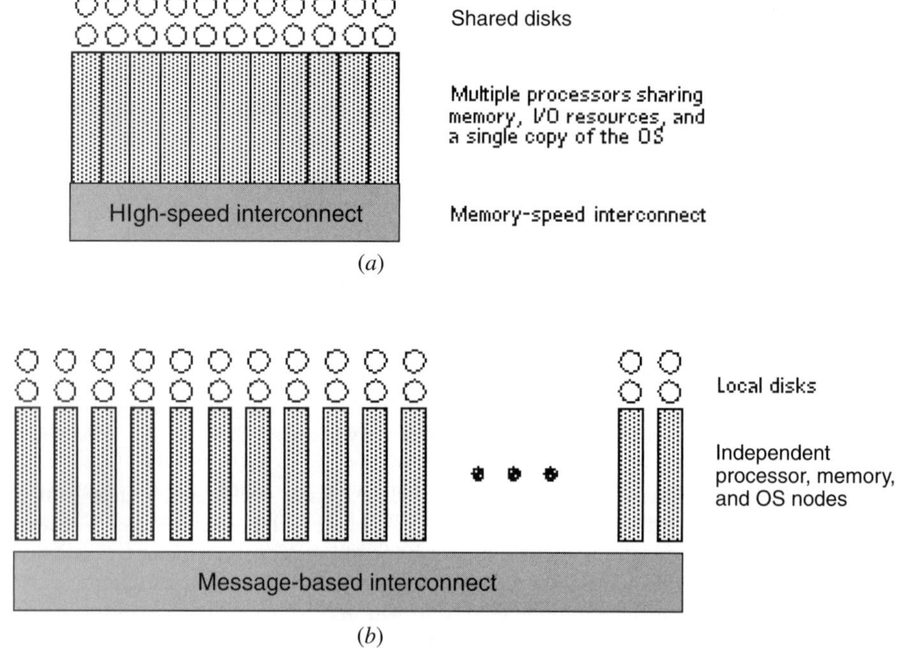

Figure 12 (*a*) SMP and (*b*) MPP layouts.

which is the more powerful of the two, has dedicated memory for each processor and utilizes an interconnection network to deal with the transfer of information from one processor to the next. MPP systems are traditionally more complex than the SMP systems and will usually be set up in large racks of interconnected processors working on a particular problem. Most supercomputer systems today are MPP systems and work on problems such as large simulations or weather forecasting.

MPP systems combine a number of processors into one machine and boast large amounts of processing power. While these machines were expected to revolutionize computerized engineering analysis, users could not simply upload existing applications to these half-billion dollar computers. On the contrary, programming MPP machines proved to be difficult even for experienced programmers, and few applications were ever developed for general use.

SMP, however, has shown extraordinary potential for use in general engineering analysis. SMP techniques essentially link PCs or workstations, each with existing memory and disk storage capacities and CPU capabilities, allowing the attributes of each individual machine to be applied to the computations. In addition, there is no lack of application software for SMP users. Applications for thermal, dynamic, stress (both linear and nonlinear), and fluid analyses are available for use with SMP computers.

To meet its needs for a more powerful supercomputer, the U.S. Department of Energy ran a project called ASCI (Accelerated Strategic Computing Initiative) which would require computational speeds of 100 TFLOPS (one trillion floating points per second, also known as teraflop). This project started in 1995 and had its first machine built for the project at the Sandia National Laboratory. This machine, called the ASCI Option Red Supercomputer, ran 9152 Pentium Pro processors and was the first computer to run more than 1 TFLOPS on a real application.

Another variation of parallel processing is through the use of PCs connected to the Internet. The SETI@Home project, for example, has its volunteer users download software that will run when the primary users are not using it. SETI (Search for Extraterrestrial Intelligence) analyzes radio telescope data in an effort to detect signals of intelligent beings from outer space.

4 MEMORY SYSTEMS

Memory systems store program information and other data in a specialized, high-speed component that facilitates efficient access. To accomplish this task, various technologies and organizational methods are employed. The typical computer system is equipped with a hierarchy of memory subsystems; some (internal to the system) can be directly accessed by the processor and some can be accessed by the processor via I/O devices (external). Internal memory consists of main memory, cache, and registers (the CPU's own local memory). External memory consists of secondary storage devices, such as magnetic disk and tape, and optical memory storage on CD-ROM.

One of the most fundamental components of a PC is its memory, and there are several components that go into the memory of a PC. Terms such as random access memory (RAM), dynamic RAM (DRAM), read-only memory (ROM), virtual memory, flash memory, erasable programmable ROM (EPROM), and many more all refer to memory.

The word *component* is used loosely due to the fact that while most memory is a physical object (created on a microchip), it can be created virtually through the use of the hard-disk drive or other unused media. However, physical memory is always faster than virtual memory, by definition. This is the reason why computers with a lot of physical memory are faster than computers with a lot of virtual memory.

Memory can be found in a wide variety of common everyday items where information from the user (or manufacturer) needs to be stored. These can range from computers, to cellular phones, to car radios, to VCRs. Of course, this information is only stored until the power source is removed from the memory.

There are two main categories of memory—volatile and nonvolatile—and within these two categories exists a large number of subcategories. Volatile memory is characterized by the fact that the information stored in the memory is only retained as long as there is a constant electricity flow to the memory. Once the energy flow is shut off, the stored information is lost. Computer RAM is a primary example of volatile memory as it is the primary memory used in a computer.

Nonvolatile memory, the counterpart to volatile memory, retains its information even when the flow of electricity is disrupted. This is the case for both physical and virtual memory. Within nonvolatile memory, there are two further classifications: permanent, where it can only be written once (read only), and erasable, where the memory can be reused. A good example of permanent nonvolatile memory is the ROM chips where the startup program and the BIOS are stored. Examples of erasable nonvolatile memory are flash memory, memory sticks, and smart media.

Technically, the first example of nonvolatile memory came about between 1948 and 1949 when Jay Forrester and other individuals in various locations developed core memory. The basic premise for core memory is that small iron doughnuts would have wires passed through them that would be able to carry a magnetic charge. These charges, either positive or negative, could represent either 0 or 1 states and would be transferred from the wires into the doughnuts. What was significant about this was that these doughnuts retained their charge even after the current to the wires was stopped. Thus the first nonvolatile memory was created.

The issue of combining temporary and permanent storage seemed to have found a solution in 1956 when IBM released its first hard drive. However, this was not the case. In 1958 the microchip, or integrated circuit, was demonstrated for the first time, which ultimately led to the foundation of many memory devices today. It was through the use of these circuits that two different states, 0 and 1, could be created and presented a great possibility for both permanent and temporary storage.

It was not until June 1968, however, that the first patent was granted to Robert Denard and IBM for a one-transistor DRAM cell. During the 10 years previous to this, the only applications of the microchip had been as a miniature engine inside of hearing aids, missiles, and eventually calculators. Denard's (and IBM's) DRAM was a taste of what would become much more complex memory chips.

From 1969 to 1971, the first generation of memory chips began to appear in both calculators and computers. First, Intel released its 1-kB RAM chip, a powerful one at the time, followed by its first commercially available DRAM chip, the Intel 1103, the following year. That same year (1970), Canon incorporated a Texas Instruments RAM chip into one of its calculators, soon to follow with a ROM chip in its Canon L100A and LE-10 calculators. From these humble beginnings, memory chips have grown in speed and power, though the fundamentals of the chips are still the same.

RAM is random-access memory because it appropriately utilizes byte storage in more of a random manner rather than a temporary one. In particular, any piece of storage can be used regardless of order and will not disturb any of the other bytes already stored there.

The process of writing and erasing in RAM is an automatic process that occurs many times during the device's operation. As new information is stored and used, other information will be erased and removed to free up memory for other applications. However, RAM cannot be manually erased at will, like a hard-disk drive or cassette tape. Nonetheless, the effects of erasing can be seen when the user instructs the device to erase a particular programming, such as the programmed recording time on a VCR.

There are two basic categories of RAM: DRAM and SRAM (static RAM). Since the 1970s, DRAM has helped run computers and continues to do so today. Compared to the newer SRAM, however, it can be up to 10 times slower. The reason for this is that DRAM requires that it be refreshed (recharged with electricity) several thousands of times per second whereas SRAM only needs to be refreshed whenever there is a write procedure. The time difference between these two different types of RAM is 5 ns (nanoseconds) versus 30–50 ns for the DRAM. A significant factor in the use of SRAM stems from the fact that in addition to being faster than DRAM, it is significantly more expensive. Because of this, DRAM is still the most popular kind of memory used in computers today. This means that SRAM is typically reserved for caches, which are naturally smaller (approximately 2 MB) and other electronic devices that do not require a lot of memory.

The first DRAM memory only worked at speeds of 300 ns, and as such, early computer processors were faster than their memory, wasting the potential utilization of the CPU. However, as processors have become faster, the increases in memory speed have not been able to keep up with the increases in CPU sppeds. As such, there has been an abundance of demand for faster DRAM.

To get around this issue, manufacturers have devised several solutions to help alleviate the problem. Some solutions are caching, as mentioned before, synchronizing with the computer's clock, bursting (receiving and processing multiple data addresses at once), and data rate doubling (data are sent twice during a clock cycle rather than once). Some variations of DRAM that use some or many of these techniques are EDO RAM (extended data output RAM), BEDO RAM (burst extended data output RAM), SDRAM (synchronous dynamic RAM), and DDR DRAM (double-data-rate SDRAM).

In addition to ROM, another type of permanent nonvolatile memory that needs to be mentioned is PROM (programmable ROM). PROM is similar to a CD in that the information is burned into the memory after it has been manufactured (as opposed to ROM where the information is incorporated while the memory is being manufactured).

The first PROM was invented by Texas Instruments in the late 1970s and was quickly hailed as a better version of memory than ROM because manufacturers would wait until the last moment to input their latest BIOS. In doing so, the manufacturers helped to maintain the shelf life of their product. However, even with this advantage, older BIOSs would be rendered obsolete when the newer ones were released, driving manufacturers to search for a reprogrammable BIOS chip.

The answer to the manufacturers' requests came with the advent of the EPROM (erasable PROM)—a special kind of PROM that would erase when subjected to ultraviolet light. Utilizing special equipment, the EPROM chip could then be reprogrammed. However, of even more significance than the EPROM was the creation of EEPROM (electrically erasable PROM)—another special case of PROM that could be erased with an electrical charge.

From the development of EEPROM came many of the flash memory, Smart Media, and Compact Flash based memory devices that all involve nonvolatile, erasable, and reprogrammable memory chips. A large majority of this technology can be found either installed or as an external component of digital cameras and audio players.

4.1 Memory Organizational Methods

Memory systems organize binary data into addressable words where data can be stored or retrieved. Some systems use a method called *interleaved memory,* which refers to an ability to access more than one word at a time. The advantage of interleaved memory is in situations where it is likely that the CPU will require the second word soon after the first.

The memory circuitry is a separate entity from the processing circuitry of the CPU. Therefore, for the two circuit systems to communicate, a memory controller is required. The controller deciphers the requests for memory information or storage access from the CPU and initiates the proper sequence of events. Because a controller can only control a certain amount of memory, multiple controllers are often used within a single system. The implementation of multiple memory controllers allows for interleaved memory transfer. Each controller, with its own memory domain within a system, increases efficient data transfer by allowing sequential data to be stored in different domains such that before one controller is finished transferring one part of the sequence, the next controller is already beginning its operation sequence. In some systems, memory can be shared among different CPUs of the same computer. In this case the controllers used in conjunction with a multiprocessor system would have access ports for each processor.

4.2 Internal Memory and Related Techniques

Registers

Memory within the CPU for data required to perform a specific task, such as the operands or result of a mathematical calculation, is stored in memory devices called registers. These registers are accessible by specific commands from the CPU. Other registers in the CPU are inaccessible for memory storage but are included as a part of the working system as a whole. Generally, registers hold the same number of bits, or binary digits, as the word length of the CPU. The number of registers in a CPU is variable, ranging from 1 to 16 or more.

Metal–Oxide–Semiconductor (MOS) RAM

This type of storage is semiconductor-based memory with wide-ranging applications in nearly all computer systems. RAM can be either dynamic or static. DRAM uses circuitry that must be periodically refreshed by rewriting the data in each block of memory. SRAM does not require the refreshing, but it is usually more expensive than DRAM. In general, SRAMs are faster than DRAMs. Both types of RAM are volatile; they lose their memory when power is turned off. Thus, RAM can be used only as temporary storage. To compensate, programs can be stored on magnetic disks or tapes (MRAM) or other forms of solid-state memory or a battery can be used to supply the necessary power to maintain semiconductor memory when the CPU is not in use.

Often, it is useful to separate memory into two types: RAM and ROM. RAM is used in memory blocks from which the CPU must be able to "read" (access stored information) and "write" (store new data). Because information in RAM can be accessed, subsequently modified, and possibly erased, it does not provide the needed security for important programs. For those applications, such as system programs, function tables, and library subroutines, data are best stored in ROM. Because access is usually limited to retrieval, data are not easily altered, and the integrity of key programs is ensured. Some ROMs can actually be erased and rewritten under certain conditions. These types are useful when a program needs periodic alterations but should be protected from general access and possible accidental erasure.

Cache Memory

Cache memory is designed to hold information relating to frequently used applications and subroutines in active memory. Cache memory, however, is usually a separate piece of hardware between the CPU and main memory. It provides faster data transfer to the CPU than does main memory but usually at a higher cost. This cost is usually well justified, especially if the computer is to be used with repetitive programs where the cache can dramatically increase the speed at which programs run and thus increase the efficiency of the user. Cache sizes of between 1K and 512K are usually offered for lower average cost per bit and faster average access time.

Virtual Memory

This technique addresses the problem of very large programs that use extensive address space and operate within a limited memory capacity. Programs use the registers of the CPU to keep the most active applications of the program available quickly. Other, less active applications are stored on magnetic disk space until needed. If needed, the application called for will be directed to occupy a less active register, and the data formerly in that register will be saved onto the disk in order to maintain any changes made to the data during execution of the program.

Memory Addressing

Some instructions from the CPU may require an operation involving one or more operands stored in memory. Operands of this type include those for logic, mathematical operations, and so forth. For example, an addressing mode might supply the ability to take operands from various locations in memory and store the result in a separate location. Most CPUs employ a variety of addressing modes depending on the operation. The variety of addressing modes for different tasks is a benefit in most CPUs. It provides a degree of flexibility in data management that increases efficiency and processing ability. Furthermore, memory addressing is made even more efficient in some systems by the ability to operate on single bits within an eight-digit byte, on the byte itself, and sometimes on words of 16 and 32 bits. A technique known as extended addressing can further increase the memory capabilities of the computer. Using this technique, the program running extended addressing functions considers memory as a number of pages. Each page is assigned a relocation constant, which is combined with the other addresses on that page to form a longer address than would normally be used. In an 8-bit computer, extended addresses might take the form of 10-bit words, which increases the number of addresses possible within the system.

4.3 External Memory

Most computer systems use some sort of magnetic storage system to store data after semiconductor memory has been erased due to a loss in power. Although computer systems of the 1990s have main memories much larger than those commonly used in previous decades, contemporary computer programs have many more capabilities; hence, they take up more memory space than programs of preceding decades. This leaves the main memory and cache memory still requiring external (with respect to the CPU) storage capabilities. Magnetic storage systems are one answer to this problem. These storage technologies provide the added benefit of an ability to copy data an almost unlimited number of times. Some copies can be stored away from the computer for security, while others may be kept nearby, perhaps online to facilitate access to the stored data. The two main magnetic storage configurations are disks and tapes.

4.4 Magnetic Media

The magnetic properties of some metals have allowed them to serve as the backbone of storage technology since before computers were invented. The first magnetic recording was made in Denmark. It was discovered in 1898 that when an electromagnet was applied to a steel wire, the wire would retain a magnetic charge based on the strength of the charge in the electromagnet, thus enabling the technology to create a "Telegrafoon" (Telegraphone). By varying the output of the electromagnet as the wire passed over it, sound patterns could be recorded onto the wire. The wire then passed over a special reader, which sensed the varying magnetic field in the wire and reproduced the original sounds.

The quality of the Telegraphone was not very good, but the idea was further developed in Germany during World War II to create an important invention called the Magnetophone. German engineers coated thin plastic tape with particles of iron ore, which are very sensitive to magnetic fields. The technology eventually evolved into tape recorders, reel-to-reel players, and other magnetic storage devices that are still in use today.

In the field of computers, magnetic storage technology is obviously of great importance. Magnetic tape was the most popular method of storing programs and data before diskette drives and hard drives came along. The advantage to disk-based magnetic media is that it can be accessed randomly, unlike tape, which is linear. The first computer hard drive was a magnetic drum that rotated to store information, and this concept eventually evolved into the disk platters that are used in modern hard drives. Advances in technology have led to the discovery of particles that are more sensitive to magnetic fields, which allow for higher data densities since more data can be packed into a smaller amount of space. More sensitive readers also have been developed that can detect even the smallest fluctuation in a magnetic field, allowing for more accuracy and, again, higher data densities.

As magnetic media have become more sophisticated, scientists and engineers are finally reaching the limits of the laws of physics and are facing problems such as the superparamagnetic effect. Under normal circumstances, magnetic particles can only become so small and densely packed before they become unstable due to temperature changes and other disturbances in the environment. Theoretically, this effect should kick in once data densities reach about 40 billion bits per square inch, but engineers at IBM have already come up with techniques that allow for 100 billion bits per square inch and beyond. Despite the threat of alternate storage technologies such as optical discs and holograms, magnetic storage is a tried-and-true, inexpensive, and reliable storage technology that should remain popular for some time to come.

4.5 Magnetic Disks

Magnetic disks are connected to the CPU through a controller attached to an I/O port. The controller can often control a number of disk drives, and it is usually programmed with a significant amount of data relating to error detection, data transfer, and other pertinent information.

Disk drives use a drive motor and one or more "heads" to read and write data on the disk. The drive motor turns a spindle on which the disk rests, rotating the disk at a controlled speed. A standard speed allows data written on one disk drive to be read on another drive. The head mechanism holds a read/write head for each recording surface on the disk. The heads are normally held away from the surface of the disk but placed extremely close to the surface or directly on the surface of the disk, depending on the type of drive, during data transfer. The magnetic disk is coated with a magnetic oxide, which forms the actual storage medium. The oxide must be organized, or "formatted," into closely spaced, spiral circular tracks. The tracks must be positioned accurately and consistently on all disks in order for the head mechanism to position itself accurately over a specific track. Data are recorded and subsequently read as analog variations in the magnetic field of the oxide medium. The CPU transfers these data to the disk drive controller, which converts the analog signal to a digital one for processing. Disks are further formatted into blocks, or "sectors," of equal area. On most disks, these blocks are established in the medium by the manufacturer. The seek time is the time needed to position the head at the desired track. Rotation latency time is the time elapsed for an appropriate sector to rotate and line up with the head. The sum of the seek time and the latency time gives the overall access time. This time consideration can be useful during the purchase of a disk drive, since the access time will affect the efficiency of the overall system if frequent storage access is required. Soft-sectoring, a technique in which the disk drive will establish blocks of unequal area on the disk, can also be used. Finding a particular track on a magnetic disk, where the distance between tracks can be as little as 0.01 mm, is no easy task. All hard-disk drives use a servocontrol mechanism to ensure accurate positioning over a desired track. Many floppy drives, where the distance between adjacent tracks is not as small, use a stepping motor system.

Floppy Disks

The floppy disk was designed as a cheap and simple device to provide quick, reliable access to information stored as a backup to computer memory. Physically, the floppy disk is a flexible diskette coated with magnetic oxide and contained within a square plastic housing. The housing has openings that allow the spindle of the drive motor to turn the disk and provide space for the head mechanism to make contact with the disk's surface. Floppy disks have continued to shrink in physical size and grow in terms of storage limits since their inception. The earliest floppy diskettes were 8-in. in diameter and capable of storing 256 kB of data. These disks typically used only one side of the diskette. The most commonly used diskettes today are 3.5 in. in diameter and are capable of storing 1.44 MB of data using both sides of the disk. Access times for floppy disks are 100–500 ms, and data transfer rates are generally lower than 300 kB/s. Floppy disk drive systems provide reliable data storage on a medium which can be removed from one computer and used on another. They are among the most common magnetic data storage systems currently in use.

Hard Disks

This term is a general heading for a category of disk drives where the disk remains within the drive. Because the disk and drive are housed within the same sealed unit, hard disks provide some significant advantages over floppy disks. The disk medium remains free of contaminants, resulting in greater reliability and data accuracy. The heads of a hard disk are lightweight and designed to aerodynamically hover extremely close to the disk surface without touching it. This virtually eliminates wear on either the disk or the drive heads. Shorter access times are also seen using hard-disk systems because two heads are normally associated with each surface.

Portable hard disks supplying 340 GB of storage space are now available. These disks provide extra security for important files and a convenient medium for storing large amounts of data for backup or other purposes.

Magneto-Optical Drive

A magneto-optical (MO) drive is the combination of magnetic media technology, such as hard drives, and optical laser technology, such as CD-ROM drives. It is the best of both worlds in that MO drives utilize the rewritability and reliability of the magnetic properties as well as the optical properties of higher storage density and a more rugged nature. This ruggedness allows MO media to withstand damage from heat, cold, or water.

MO drives store a higher storage density by using a laser installed within the drive to heat the portion of the magnetic disk that is being written. By heating the magnetic disk, the magnetic polarity of the disk is easier to change and results in a higher density of stable magnetic charge in the same size section of the disk. This, in turn, means that more information can be stored than on a nonheated disk.

In addition to the higher density of information, the disk, once it cools, also becomes much more resistant to a change in the magnetic layer, meaning that a laser must again be used if the information is needed to change. The natural advantage of this is that the information on the disk is safe from most external magnetic fields that had previously degraded or even erased information on other storage media, such as floppy disks. It would require a very high intensity magnetic field placed very close to the disk to erase the data written by an MO drive.

A low-intensity laser, one that is not powerful enough to affect the data it is reading, reads the information stored onto a disk from an MO drive. This laser is capable of reading the variations in polarity of the different areas within the disk and thus retrieves the information from the MO written disk.

A typical MO drive is separated into two parts: a fixed drive with the read/write components in it and a removable diskette that holds the MO storage media. The typical makeup of an MO disk is an optical storage media similar to a CD-ROM or CD-RW but sandwiched between two protective pieces of plastic. This makes for a unit that is slightly larger than a floppy diskette; however, the MO disks have significantly greater storage capacity and data transfer rates. Another significant advantage of the MO drive is that all of the diskettes are backward compatible, meaning that a drive that is capable of handling the newest MO disks can also handle the very first disks and everything in between as well.

As the technology of MO drives has grown, so has the storage capacity (as expected) as well as the technology to overwrite the data already stored on the MO disk. Originally, if information was to be overwritten on an MO disk, it required that all of the information be deleted first and then the new data could be placed on the disk. Today, however, new technology such as LIMDOW (laser intensify modulation direct overwrite) bypasses the deleting step by using a more intense laser beam to write over the preexisting data all in one step.

Magnetoresistive Heads

The capacity, data access speeds, reliability, and other attributes of a hard-disk drive are affected significantly by one specific component, the read–write head. The basic design of a hard-disk drive includes one or more magnetic disks which store the data and a read–write head that is attached to an arm. This arm has the ability to move in and out to access any part of the disk as well as up and down to get closer or farther away from the disk in order to read or write to it. It is the goal of the manufacturer of hard-disk drives to produce the most efficient read–write heads at the lowest cost.

When the hard drive was first introduced in 1956 by IBM, the read heads were simply wrapped in wire and electrically charged in order to create a magnetic field. To read the data on the disk, the magnetic fields of both the disk and the read head were analyzed, and the difference between the two was interpreted as data. However, the biggest setback in this design was that to produce a magnetic field large enough for the read head to detect, a large portion of the surface area of the hard disk needed to be used to store a minimal amount of data.

A breakthrough in technology occurred in 1991, led again by IBM, that allowed IBM to eventually market the first commercially available 1-GB hard drive. This breakthrough was the magnetoresistive (MR) head, which was used only for reading data. However, the significance of this new head was that it was made of a special material that was extremely sensitive to magnetic field changes yet emitted none of its own. This meant that there would be no interference created by the read head, and together with its increased sensitivity, information could be stored more densely on the surface of the hard disks.

There are also other significant advantages that the MR heads brought to hard-disk drives. The first was that the MR head allowed for the use of two different heads in the hard drive. Until this point, only one head had been used to both read and write the data, which limited the capabilities of this head. However, with the separation of the read and write heads, manufacturers could create more efficient heads refined to perform only one specific task. Another advantage that the MR heads offered was that the head did not actually need to touch the surface of the disk, meaning that there would be significantly less wear on the disk due to friction.

Another significant, recent advance in the read head technology was the discovery of the GMR (giant magnetoresistive) effect. This effect can occur when different metals are layered upon each other, resulting in a higher sensitivity to magnetic fields. This advance had the potential to yield more sensitive read heads than any others created with only a single metal. The two men credited with this simultaneous discovery are Albert Fert, from the University of Paris-Sud, and Peter Gruenberg, at the KFA Research Institute located in Jurich, Germany. Quickly, companies such as IBM and Seagate began trying to engineer read heads that would incorporate the GMR effect. When they succeeded, it allowed for more accurate readings from more densely packed magnetic data located on the hard drive platter. This made for even greater storage density than was possible with standard MR heads.

Today, to differentiate between the older and newer technology, hard-drive manufacturers sometimes refer to the MR heads as "anisotropic magnetoresistive heads," which was the original name. This would explicitly differentiate the old heads from the newer ones such as the GMR heads. An interesting note is that because of the extreme sensitivity of the MR heads, forensic scientists have begun using them to extract data from erased tapes and other magnetic media. In fact, a microscope based on MR technology was created for the purpose of reading the 18.5-min gap in tape 342 that had been erased during the Watergate scandal in the early 1970s. To win the contract to perform this task, the National Archives and Records Administration required that the company must use a data retrieval system that would leave the tape intact and the data as undisturbed as possible—certainly a task fit for MR technology.

4.6 Magnetic Tape

Magnetic tape was the first application of magnetic data storage employed with computer systems. This medium is effective, to a large extent, due to existing standards for data format. These standards, like those for magnetic disks, allow data to be used on different types of computers. Various forms of magnetic tape storage systems have been developed that satisfy specific user needs.

Magnetic tape has been in use for over half a century. One of the earlier forms in common use in computers was the punch-card paper tape; however, IBM transformed the industry in 1952 by developing the first widely successful magnetic information storage tape drive. This system, as well as its progenies, became the standard method for data storage for over 30 years and was visible in many buildings and movies worldwide.

Standard industrial tape drives use reels of 12.7-mm- (0.50-in.) wide tape at lengths of 731 or 365 m coated with an oxide medium similar to that used in magnetic disks. The tape moves past read/write heads that provide data transfer at 800–6250 bits per linear inch. The tape motion is servocontrolled for a high degree of accuracy, with the tape winding more than 180° around a capstan to provide sufficient physical control. The motors driving the tape reels must also be carefully controlled to ensure proper tape tension. Two mechanisms are typically used to ensure this control. In older tape drives, the tape moves over fixed- and variable-position pulleys. The variable pulleys are attached to a spring-loaded tension arm drawing the tape in a "W" form between the pulleys. The position of the tension arm gives the motor control mechanism the necessary information regarding tape winding and release to ensure the proper tension. Most current tape drive designs have abandoned the tension arm in favor of the following vacuum chamber technique. Between each reel and capstan, a vacuum chamber draws the tape into a loop 1–2 m long. The length of the tape in the chamber is detected using photoelectric sensors, and this information is used by the motor controller to govern the movement of the reels. Smaller magnetic tape drives are available for smaller systems such as microcomputers. In these drives, the tape is normally housed in a plastic cartridge that protects the tape medium from contamination. This provides excellent data integrity on a much smaller scale than industrial-type systems. An 8-mm tape is usually used with a workstation system, while PC-based tape drives are usually on a 4-mm format. Because magnetic tape moves in a linear fashion, it is inefficient for applications requiring rapid, random access to stored data. It does, however, provide an excellent means for back-up protection of important data.

Bit streams, sequences of 1s and 0s that merge to create discernable data, are stored on tape as magnetic patterns. Before the invention and implementation of the hard drives with which we are currently familiar, tape was the principal mode for computer data storage. Due to increasingly efficient technology improvements and the increased amount of time required for retrieving information stored on tape, tape has been consigned primarily to keeping annals. Even this role, however, is becoming less widely used, as there are increasingly more efficient ways to keep data records. Unlike other storage media such as Zip disks and CDs, you cannot run applications from magnetic tape. This is primarily due to the fact that the data are stored sequentially on the tape in one continuous fashion.

Terms commonly associated with tape are as follows:

Tape Archive (*TAR*). A compression format commonly used when storing and retrieving files on UNIX computers.

Tape Backup. A method of archiving computer data onto magnetic tape.

Tape Cartridge. A cartridge that contains magnetic tape and serves as the actual data storage medium.

Tape Drive. The mechanism that interfaces with a computer and reads and writes to a tape cartridge.

4.7 Optical Data Storage

While magnetic systems are a popular and reliable method for storing large amounts of data, they can become quite cumbersome in terms of physical size as the amount of data increases. There is also the limiting factor on speed imposed by the need to convert between the digital signal of the CPU and an analog signal. Digital data storage addresses both of these problems. First, the data storage capability of a compact disc (CD), essentially identical to those used in musical recordings, is 680 MB, and the technology for downloading the information stored on CDs keeps advancing. The initial transfer speed of CD drives was 150kB/s. Currently, 52X CD drives, capable of 7.8 MB/s are already on the market. Second, because the storage medium and the CPU use the same digital data format, there is no need for a controller to convert the signal from an analog signal.

Digital Versatile Disc

DVDs and CDs differ in their storage capacity because of many important reasons. In fact, DVDs will hold a minimum of 750% more data than a regular CD. First, both DVDs and CDs store data in a spiral fashion; however, the rings created by the DVD are more closely packed together, allowing for more data storage. Second, the indentations created that store the data on DVDs are much smaller than on CDs, further allowing for a larger storage capacity than CDs. DVDs also have the capability of being written to on both sides as well as having two recordable layers each. All of these advantages over regular CDs give the DVDs a maximum storage size of 17 GB per disc! There is no question then as to why DVDs are set to become the primary data storage device in the near future. Some commonly used terms are as follows:

- *DVD-Audio.* An audio format that can play sound at resolutions higher than a CD. CDs play 16-bit samples at 44.1 kHz while DVD-Audio plays 24-bit samples at 192 kHz. Despite the inability for humans to hear this change in intensity, the DVD-Audio can support other visual and audio systems (such as video, graphics, and Web links) and can provide surround sound for 2 h and conventional stereo acoustics for 4 h.
- *DVD-R* (*DVD Recordable*). Pioneer initiated DVD-R drives in the fall of 1997. At that point in time, DVD-Rs could hold only 3.9 GB. DVD-Rs are analogous to CD-R (CD recordable). A DVD-R is a write-once medium—meaning that valuable information cannot be accidentally erased—that can hold up to 4.7 GB of data using the same dye layer recording to burn information onto the disk. DVD-R can master DVD-Video and DVD-ROM disks.
- *DVD-RAM.* The first rewritable DVD. In 1998, Panasonic introduced the first DVD-RAM drive, with the capability to write 2.6 GB on a DVD-RAM disc and to burn 3.9 GB on a DVD-R disc. DVD-RAM drives can read DVD-Video and DVD-ROM and can maintain backward compatibility with CD media. Initially, DVD-RAM had compatibility problems with other drives and players as well as an inconvenient data-burning process, making the DVD-ROM difficult to market. Additionally, DVD-RW (DVD rewritable) and DVD+RW (DVD plus rewriteable) were introduced. These mediums were less confusing, stored more data, and eliminated the need for a caddy prior to recording data.
- *DVD-ROM.* Designed to replace the CD-ROM, the DVD-ROM was intended to be used for applications, specifically games and programs, that required a great deal of data storage. Most DVD-ROMs will play DVD-Video; however, DVD-ROM will not play in DVD-Video drives.
- *DVD-RW* (*DVD Rewritable*). This rewritable DVD is a substitute for the CD-RW. Both mediums use the same technology to read, write, and erase data. The DVD-RW drive was launched by Pioneer in 1999 and was initially known as DVD-R/W and DVD-ER. The DVD-RW is compatible with the original DVD-ROM drives but has significant compatibility problems with other players.
- *DVD+RW* (*DVD Plus Rewritable*). This is the third and most successful rewritable DVD medium, due to its compatibility successes with existing DVD drives and players. DVD+RW discs can be played in both DVD-ROM and DVD-Video mediums. Hewlett-Packard was the first company to make a DVD+RW drive available, followed by Philips and Sony.
- *DVD-Video.* Introduced in 1996, this format is used to replace VHS. It has the same movie-viewing function as does VHS; however, it is superior to VHS in several ways. DVD-Video not only provides laser-quality viewing but also remains in the same

condition after numerous viewings; it allows the viewer to operate five-channel digital surround-sound; and it has multiple viewing options (full screen, wide screen). Typically, the second side of the DVD is used to program the film in wide-screen (16 : 9 aspect ratio) format. DVD-Video also has a CSS (content scrambling system) to ban users from copying the discs.

5 INPUT DEVICES

Commonly used input devices in CAD systems include the alphanumeric keyboard, the mouse, the light pen, and the digitizer. All of these allow information transfer from the device to the CPU. The information being transferred can be alphanumeric, functional (in order to use command paths in the software), or graphic in nature. In each case, the devices allow an interface between the designer's thoughts and the machine which will assist in the design process.

5.1 Keyboard

The alphanumeric keyboard is one of the most recognizable computer input devices, as well as the principal method of text input in most systems. Rows of letters and numbers (typically laid out like a typewriter keyboard) with other functional keys (such as CTRL, ALT, and ESC), either dedicated to tasks such as control of cursor placement on a display screen or definable by the user, transfer bits of information to the CPU in one of several ways. The typical keyboard layout is almost identical to that of electronic and mechanical typewriters. Each key is comprised of a keycap (the plastic portion bearing the number, letter, or function of the key) and a small switch that resides under the keycap. Key depression can be detected through a simple mechanical switch, a change in magnetic coupling or a change in capacitance. The alphanumeric keyboard is dedicated to the input of alphanumeric information and special commands via function keys. Usually, keyboards are connected to computers through PS/2 or USB interfaces.

The keyboard controller is a small microchip that passes data entered into the computer from the keyboard. The keyboard buffer allows users to type more quickly than the computer can immediately process by storing the data in the computer's memory until the computer can manage it.

Most keyboards are organized so that the first six letters, from left to right on the top row of the keyboard are Q, W, E, R, T, and Y, lending the arrangement the name QWERTY—pronounced "KWEHR-tee." This format was first introduced for use in early mechanical typewriters in the 1870s. Two alternative forms of arranging the keyboard are the Dvorak and 101-key layouts. The Dvorak layout organizes the most common consonants and vowels on opposite sides of the home row, varying keystrokes between hands. The 101-key system has 101 keys and is also referred to as the AT (advanced technology) keyboard.

You can choose different layouts and languages for your keyboard. To do so in Windows, go to Start, select Settings, click Control Panel, and double-click Keyboard. Find the Keyboard Properties dialog box, go to the Language tab, and click Add to choose a different language. If you click on Properties in the Language tab, you will be able to pick different layouts for a specified language.

Keyboard shortcuts are time-saving alternatives to using the computer's mouse to carry out menu functions. For example, in most applications CTRL-I is the shortcut to italicize text in the Format menu under Font and CTRL-A is the shortcut to select an entire document under the Edit menu.

Special programmed-function keyboards with 16–32 buttons can also be used in conjunction with a CAD system. These keyboards are often separate from the alphanumeric keyboard, but the keys can similarly be dedicated or definable to specific CAD tasks. Some keyboards will employ cardboard or plastic overlays that show the function of each key. Where the keyboard is applicable to several tasks within the general CAD techniques, the overlays indicate which functions control the keys that command different techniques.

Maltron Keyboard

The Maltron ergonomic keyboard was designed by Lillian G. Malt and S. W. Hobday in 1976 in order to support and relieve stress on joints, tendons, and muscles for people who use computers on a daily basis. This keyboard rearranges the standard QWERTY key distribution to one that evenly disperses the typical typing burden across all of the fingers. Originally, the QWERTY system was developed to hinder typists' speed so they would not jam the typewriters.

Maltron keyboards are split into two halves with three main rows of keys in each half. Letters comprise these rows but are situated in a much different fashion than in the QWERTY system. The most commonly used letters are positioned directly under the fingers in the middle row:

QPYCB VMUZL
ANISF DTHOR
,?JG' _WK-X

There is a gap between the two hands that contains a numeric keypad with editing keys as well as two nine-key keypads that are controlled by the thumbs. The reason for these smaller keypads is that the thumb is the strongest part of the hand, so it makes sense that the keys used most frequently—such as spacebar, ENTER, CTRL, ALT, and BACKSPACE—are included in the area operated by the thumbs. This gap in between the two keyboard sections also situates the hands in separate areas so they can be further apart when typing, resulting in a decrease of stress on the wrists.

In addition to the unique layout of the keys on the Maltron keyboard, the keyboard's shape is distinct as well. Both halves of the keyboard are concave, with the keys arranged along the slopes of the curved surface. This shape allows the typist to move his or her fingers very slightly to reach the upper and lower rows, resulting in a lower stress impact for the typist. Furthermore, typists who switch from using the QWERTY keyboard to the Maltron keyboard experience an increased typing speed as well as make fewer mistakes after only a few weeks of practice.

5.2 Touch Pad

The touch pad is a device that allows command inputs and data manipulation to take place directly on the screen. The touch pad can be mounted over the screen of the display terminal, and the user can select areas or on-screen commands by touching a finger to the pad. Various techniques are employed in touch pads to detect the position of the user's finger. Low-resolution pads employ a series of light-emitting diodes (LEDs) and photodiodes in the x and y axes of the pad. When the user's finger touches the pad, a beam of light is broken between an LED and a photodiode, which determines a position. Pads of this type generally supply 10–50 resolvable positions in each axis. A high-resolution panel design generates high-frequency shock waves traveling orthogonally through the glass. When the user touches the panel, part of the waves in both directions are deflected back to the source. The coordinates of this input can then be calculated by determining how long after the wave was generated it was reflected back to the source. Panels of this type can supply resolution of up to 500 positions in each direction. A different high-resolution panel design uses two transparent panel layers. One layer is conductive while the other is resistive. The pressure of a finger on the panel causes the voltage to drop in the resistive layer, and the measurement of the drop can be used to calculate the coordinates of the input.

A touch pad is made of two different layers of material: the top, protective layer and the two layers of electrodes in a grid arrangement. The protective layer is designed to be smooth to the touch, allowing the user's finger to move effortlessly across the pad while still protecting the internals from dust and other harmful particles. Beneath the protective layer lies the two grids of electrodes that have alternating currents running through them. When the user touches the pad, the two grids interact and change the capacitance at that location. The computer registers this change and can tell where the touch pad was touched.

The touch pad was invented by George E. Gerpheide in 1988. In 1994, Apple Computer licensed and first used his design in its PowerBook notebook computers, and the touch pad has been the primary cursor control alternative to the mouse in notebook computers ever since. Other methods discussed later are the trackball and the TrackPoint pointing stick.

The input of graphical data is somewhat clumsy using a keyboard or touch pad. For this reason, various input devices that are specialized for graphics input have been created and are widely used in CAD.

5.3 Touch Screen

Touch screens are a special category of computer screens that allow the user to interface with the system other than with a traditional input device such as a keyboard or mouse, for example. However, a touch-sensitive interface must first be installed into the screen. This is done in three primary ways: resistive, surface wave, and capacitive.

The resistive method places an ultrathin metallic layer over the screen. This layer has both conductive and resistive properties, and touching the layer causes a variation in that conductivity and resistivity. The computer interprets this variation in the electrical charge of the metallic layer and the touched area becomes known. The advantage of resistive-based touch screens is that they are cheaper than the other two options and are not affected by dust or moisture. However, they also do not have the picture clarity that the others have.

The surface wave method has either ultrasonic waves or infrared light continually passed just over the surface of the screen. When an object interrupts the waves passing over the surface, the controller processes this information, and the location is determined. So far, this is the most advanced and expensive method used.

The final method, the capacitive method, requires that the screen be covered with an electrical-charge-storing material. When a user touches the screen, the charge is disrupted and can be interpreted as the location of the user's finger. The difficulty with the capacitive method is that it requires either a human finger or a special pointing device for the system to identify the touched portion of the screen. Nonetheless, this method is the least affected by external elements and can also boast the best clarity.

There are relatively few applications where the touch screen serves as a better interface than other devices, such as the mouse. The reasons for this are many. For example, users' productivity will decrease when using both the keyboard and the screen because they have to traverse an extra distance as well as may become tired more quickly due to raising their arms more frequently. In addition, the human finger is a very large pointing object and, as such, cannot be used for small and precise adjustments. Frequent use can also lead to a dirty screen, except in cases where a stylus is used.

There are, however, some key advantages to touch screens. Severe environments, such as factories, less affect a touch screen than a keyboard or mouse. Using software that utilizes large buttons on a touch screen both allows for efficiency of use in demanding environments, as and makes it easier to secure the data input method (as opposed to securing a mouse or a keyboard). For these reasons, touch screens are often used in public information workspaces. Additionally, the ability to reduce the size of touch screens makes them very useful for systems that do not allow the space for a mouse or keyboard, such as PDAs (personal digital assistants).

5.4 Mouse

The mouse is used for graphical cursor control and conveys cursor placement information in the *X–Y* coordinate plane to the CPU. A spherical roller is housed within the mouse such that the roller touches the plane upon which the mouse is resting. When the mouse is moved along a flat surface, the spherical roller simultaneously contacts two orthogonal potentiometers, each of which is connected to an analog-to-digital converter. The orthogonal potentiometers send *X*- and *Y*-axis vector information via a connecting wire to the CPU, which performs the necessary vector additions to allow cursor control in any direction on a 2D display screen. Often a mouse will be equipped with one to three pressure-sensitive buttons which assist in the selection of on-screen command paths. Since the mouse is inexpensive and simple to use, it has become a standard computer input device.

Douglas Engelbart presented the first mechanical mouse in 1968. A scientist at the Stanford Research Institute in California, he wanted to implement the same idea behind light pens and joysticks to create a device that would assist people who operated in computer workstations.

Wireless Mouse

The wireless mouse uses either radio frequencies or infrared signals to function. The radio-frequency version will work up to 6 ft away, anywhere in a room, but the infrared version requires a line-of-sight channel between it and the infrared port on the side of the computer. The technology employed for the infrared mouse is identical to the technology used in television remote controls. The radio signal mouse functions much the same as a wired mouse and has a transmitter that connects to the USB port on the computer.

Optical Mouse

The optical mouse was invented in response to user complaints that mechanical mice were too easily dirtied and were imprecise. The optical mouse uses instruments that produce light when currents pass through them—LEDs—and a mousepad that senses motion, instead of the conventional trackball system. Optical mice capture an image through navigation ICs (integration circuits) that show a pattern of light and shadows. Through the use of prediction, correlation, and interpolation, the mouse communicates to the computer the variables of the surface on which the mouse moves, and the computer can then determine the exact location of the mouse. These integrated circuits compute the movement of the mouse by evaluating subsequent images that the mouse tracks and sends to the circuits.

Optical navigation originated with military target tracking. Initially used in tracking cameras for following airplanes and other targets, optical navigation is now assisting with the manner in which we input data. Since an optical mouse has no moving parts, it does not require cleaning, unlike the mechanical version. The tracking is also much more accurate, and some manufacturers assert that the optical mouse may reduce computer-related stress injuries.

Advances in ASICs (application-specific integrated circuits), imaging arrays, and embedded mathematics have reduced the costs of optical navigation. We now have the capabilities to implant the above functions into one silicon circuit, making optical navigation technology reasonably priced for the average computer user.

5.5 Trackball

This device operates much like a mouse in reverse. The main components of the mouse are also present in the trackball. As in a mouse, the trackball uses a spherical roller which comes into contact with two orthogonally placed potentiometers, sending *X*- and *Y*-axis vector information to the CPU via a connecting wire. The difference between the mouse and the trackball lies in the placement of its spherical roller. In a trackball, the spherical roller rests on a base and is controlled directly by manual manipulation. As in a mouse, buttons may be present to facilitate the use of on-screen commands.

A trackball is immobile, so when the user maneuvers the ball on the top, the movement is almost identical to the movement a mouse ball makes as the user moves the mouse. Since it is stationary, a trackball can also be useful on surfaces where it would be difficult to use a mouse as well as being practical for users with minimal or limited desk space. Some desktop computers and notebook computers have trackballs built into their keyboards, both to save space and to allow the user to keep his or her hands on the keyboard at all times.

5.6 TrackPoint

A TrackPoint is a short joystick that is placed in the middle of the keyboard and controls the cursor. It is very space efficient and requires minimal movement to use. In addition to the TrackPoint, there are two buttons located on the bottom of the computer, allowing the TrackPoint system to perform the identical functions as does a mouse. Users use their finger to apply pressure to the TrackPoint in the direction they want the cursor to move. This device acts as a substitute for touchpads on some notebook computers.

Steve G. Steinberg at IBM invented the TrackPoint, and IBM first installed them on its IBM ThinkPad notebook computers. At this time, many notebook computer users were operating mice to control their cursors, as touchpads were not yet standard for notebook computers. The TrackPoint tripled IBM's notebook computer sales for the consecutive four months after launching the TrackPoint; however, production of its other computers without the TrackPoint was terminated because of low sales. TrackPoint is also available on both some Toshiba notebook computers, and limited desktop computer keyboards.

5.7 Light Pen

Another computer input device that can be used in CAD systems is the light pen. Somewhat of a misnomer, the light pen does not project light; rather it detects light from a raster-scan CRT screen (see Section 6). Light pens used in the SAGE radar system in the 1950s resembled guns which were pointed at the screen, with input delivered through a triggerlike device. Contemporary light pens are hand-held cylindrical instruments approximately the size of an ink pen. At one end of the cylinder is a lens and a photo-optical sensor. The other end of the cylinder is connected to the computer by a cable. The pen detects the timing of the screen's repeated illuminations (a process so fast that the constant flicker one would expect is absent and the screen appears to maintain a constant 2D image) by detecting the light pulse at the desired location when the screen is illuminated. The pulse is then transferred through the cable to the CPU, which uses the pulse to determine where the light pen is in contact with the screen and uses this information to continuously track its position. The location is determined by correlating the pulse with the graphical display data in the CPU to identify what graphical information was being displayed at the given time and location on the screen. The device can also send a second type of signal to the CPU indicating the selection of a point on the screen when a button on the pen is depressed. Light pens can thus be used to create lines and shapes which appear instantaneously to the human eye on the computer display. Light pens also select on-screen command paths in a manner similar to the mouse and trackball. Small graphical areas on the screen, sometimes called icons, can be associated with programmed software commands. If the user points the light pen at an icon and depresses the light pen button, the desired command will be executed on-screen, bypassing the need for keyboard inputs.

By using a light pen with a graphic tablet, artists are enabled to draw on a computer as easily as if they were using a pencil and paper. The light pen can also serve the same function on a touch screen as using one's finger. Pressing the light pen's tip against a touch-sensitive screen activates a switch in the screen, enabling the pen to control the cursor as would a mouse.

The light pen operates as follows*:

> The pen tip contains a combination of lenses, an aperture, and a photodiode detector. It makes time measurements, which are translated into *X* and *Y* grid coordinates on the monitor screen, and then acquires and magnifies the signal produced by the monitor's pixels. When the electron beam sweeps the back of the CRT (cathode-ray tube) face, it strikes the phosphor (at the pen's location), and the light emitted by the phosphor is focused through the lens and onto the photo detector. The accompanying software generates *X* and *Y* vectors, which correspond to a point on the grid (screen) and then communicates that information to the computer.

Light pens more accurately convey information between users and the computer system than do touch screens that are designed to work without light pens because they aim at particular pixels instead of finger-sized buttons. Light pens are ideal for environments in which weather, chemicals, foods, or liquids could threaten the delicate electronics used in a keyboard or a mouse. Light pens are a favorite with graphic artists, since the pens perform the same function as a mouse but are less difficult to handle.

5.8 Digitizer

A digitizer is an input device that consists of a large, flat surface coupled with an electronic tracking device, or cursor (Fig. 13). The cursor is tracked by the tablet underneath it and buttons on the cursor act as switches to allow the user to input position data and commands.

Digitizing tablets apply different technologies to sense and track cursor position. The three most common techniques use electromagnetic, electrostatic, and magnetostrictive methods to track the cursor. Electromagnetic tablets have a grid of wires underlying the tablet surface. Either the cursor or the tablet generates an alternating current that is detected by a magnetic receiver in the complementary device. The receiver generates and sends a digital signal to the CPU, giving the cursor's position. Despite their use of electromagnetism, these types of digitizers are not compromised by magnetic or conductive materials on their surface. Electrostatic digitizers generate a variable electric field which is detected by the tracking device. The frequency of the field variations and the time at which the field is sensed provide

---

* *Computing Encyclopedia,* Smart Computing Reference Series (Vol. 2, p. 210), Sandhills Publishing, Lincoln, NE, 2002.

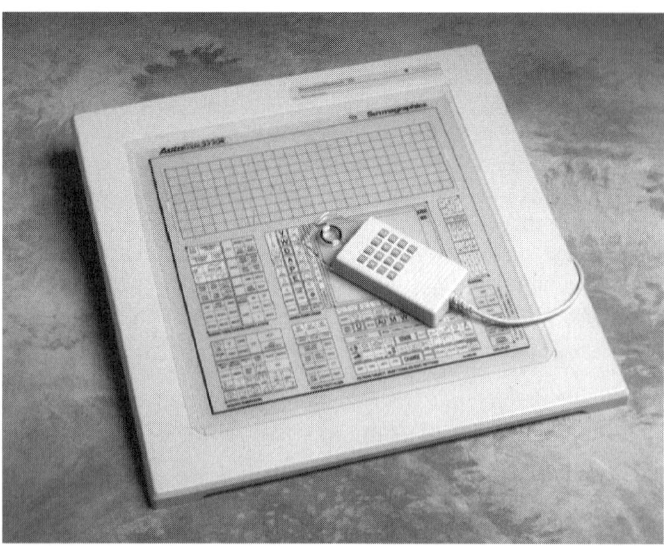

Figure 13 Digitizing tablet and cursor. (Courtesy of Summagraphics, Inc.)

the information necessary to give accurate coordinates. Electrostatic digitizers function accurately in contact with paper, plastic, or any other material with a small dielectric constant. They do lose accuracy, however, when even partially conductive materials are in close proximity to the tablet. Magnetostrictive tablets use an underlying wire grid, similar to that used in electromagnetic tablets. These tablets, however, use magnetostrictive wires (i.e., wires which change dimension depending on a magnetic field) in the grid. A magnetic pulse initiated at one end of a wire propagates through the wire as a wave. The cursor senses the wave using a loop of wire and relays a signal to the CPU, which then couples the time of the cursor signal with the time elapsed since the wave originated to give the position. These tablets require periodic remagnetization and recalibration to maintain their functional ability.

Digitizing tablets can usually employ various modes of operation. One mode allows the input of individual points. Other modes allow a continuous stream of points to be tracked into the CPU, either with or without one of the cursor buttons depressed, depending on the needs of the user. A digitizing-rate function, which enables the user to specify the number of points to be tracked in a given period of time, is also often present in CAD systems with digitizers. The rate can be adjusted as necessary to facilitate the accurate input of curves.

Whatever the type, digitizers are highly accurate graphical input devices and strongly suited to drafting original designs and to tracing existing designs from a hard-copy drawing. Resolution can be up to 1000 lines per linear inch. Tablet sizes typically range from 10 $\times$ 11 to 44 $\times$ 60 in. Often, plastic sheets with areas for command functions, such as switching between modes as discussed above, are laid over the tablet to allow the designer to access software commands directly through the tablet using one of the buttons on the tracking device. Many of these commands deal with the generation of graphical elements like lines, circles, and other geometries.

Digitizing tablets are commonly used in CAD, imaging, and illustration applications. Several different pointing devices are used, depending on the function for which the digitizing tablet is desired. Freehand artists often prefer a pen or stylus, due to the increased handling ease, as compared to a mouse. Instead of ink, generally a stylus either possesses sensors or is sensed when it makes contact with the digitizing tablet.

Another option more commonly used in CAD applications is a puck (or cursor). A puck appears to be very similar to a mouse, except for the clear plastic screen on its top. The screen has crosshairs on it that when used with the digitizing tablet create intersections that correspond to stationary locations on the computer monitor. The function of crosshairs, since the covering of a puck is clear, is to trace lines of a drawing when situated on top of the digitizing tablet.

One advantage of using a puck is that it leaves the user free to work without a keyboard. To facilitate ease of utilizing this device, pucks with up to 16 customizable buttons are available. For users who prefer a stylus, some digitizing tablets are available that allow users to designate areas of the board as buttons. Tapping these buttons enables the user to bypass a keyboard. For general computer use, several newer models of tablets facilitate a pointing system that can be used with a traditional mouse. Other commonly used terms for the digitizing tablet include touch tablets, digitizer tablets, digitizers, graphics tablets, and tablets.

5.9 Scanner

A scanner (or "optical scanner") is a tool that converts information from its physical form to digital signals through the use of light-sensing equipment. When one scans an object in computer technology, first the object is examined in small sections, then the data are converted to a computer-friendly digital format. Systems that are able to transfer the digital signal include graphics software or OCR (optical character recognition) software. Text scans use a scanner and OCR software, which converts text to a word processor file. Graphical scanners translate the visual patterns from photographs, magazines, and other pictorial sources to bit-mapped files, and a computer then edits or displays them. Most scanners are accompanied by imaging software, allowing the user to edit and resize the images, which is useful when one wishes to scan photographs or keep permanent records of graphics.

Imaging software that is used with scanners uses a program called TWAIN to read the images after they are scanned. TWAIN allows users to send images into applications such as Photoshop without having to first open the application and then transfer the image to image-editing software. TWAIN is typically installed in the imaging software program and mediates between the scanner and the imaging software, making it an industry standard.

Scanner terms are as follows:

Scan Head. The device that discriminates between levels of lightness and darkness, shading, and colors. The scan head moves over the scanned object and sends a bit-mapped representation of the object to the scanning system.

Scan line. The row of pixels that the scanner reads.

Different types of scanners are available, including flat-bed, sheet-fed, hand-held, and film scanners. All are able to scan in either black and white or color. High-resolution printed pictures are scanned using a high-resolution scanner (up to 2400 dots per inch), while graphics for a computer display are using a low-resolution scanner (600 dots per inch).

Flat-bed scanners are capable of scanning a variety of objects, which is probably why they are the most popular type of scanner. Flat-bed scanners can scan books, photographs, papers, and at least one side of a 3D object. The scanner looks like a copy machine and has a glass sheet on which the scanned objects are placed facedown. A scan head moves across the object being scanned to copy an image of the object. Some flat-bed scanners can scan multiple pages, and the average size of a flat-bed scanner is $12 \times 21 \times 4$ in.

Sheet-fed scanners scan individual pieces of paper at a high rate but, unlike flat-bed scanners, cannot scan books, bound papers, or 3D objects. The average sheet-fed scanner has measurements of $4.5 \times 12.3 \times 22.63$ in.

Film scanners were created to scan slides, transparencies, and negatives at a much higher resolution than flat-bed scanners. Film scanners cannot scan paper or 3D objects, and if a scanner is needed for both paper and occasional slides, then a transparency adapter is probably the most cost-effective way to scan both media. Film scanners are usually smaller than flat-bed and sheet-fed scanners, traditionally about $5.7 \times 3.6 \times 14.5$ in.

Hand-held scanners are the most inexpensive form of scanner, but since they are much smaller than the other types, they can only scan areas that are a few inches wide. Sometimes it is difficult for users to move the scanner head over the object at a constant rate, resulting in a compromised quality of the scanned image. If a user only needs to scan small amounts of text or images, then the hand-held scanner is ideal.

Scanners usually connect to a computer via a USB port, SCSI, or parallel port. SCSI allows computers to communicate with other electronic devices. Some popular companies that manufacture scanners include Canon, Hewlett-Packard, Panasonic, Epson, Microtek Labs, Visoneer, Acer, Umax, and Pacific Image Electronics.

6 OUTPUT DEVICES

Just as a CAD system requires input devices to transfer information from the user to the CPU, output devices are also necessary to transfer data in visual form back to the user. Electronic displays provide real-time feedback to the user, enabling visualization and modification of information without hard-copy production. Often, however, a hard copy is required for presentation or evaluation, and the devices which use the data in the CPU or stored in memory to create the desired copy are a second category of output devices.

6.1 Electronic Displays

Contemporary computer graphics displays use a CRT to generate an image on the display screen (Fig. 14). The CRT heats a cathode to project a beam of electrons onto a phosphor-coated glass screen. The electron beam energizes the phosphor coating at the point of contact, causing the phosphor to glow.

CRTs employ two different techniques, stroke writing and raster scan, to direct the beam onto the screen. A CRT using the stroke-writing technique directs the beam only along the vectors given by the graphics data in the CPU (Fig. 15). In a raster-scan CRT, the electron beam sweeps systematically from left to right and top to bottom at a continuous rate, employing what is known as "rasterization"(Fig. 16). The image is created by turning the electron beam on or off at various points along the sweep, depending on whether a light or a dark dot is required at those points to create a recognizable image.

Regardless of the technique used to create an image on a CRT screen, the phosphor glows for a very short period of time after being energized by the electron beam. Three types of graphics terminals represent the vast majority of those used in CAD systems. Each employs a different approach to create a continuous image on the screen and is discussed below.

Vector Refresh Terminals

These early graphics terminals (c. 1960s) use the stroke-writing approach in the CRT. The electron beam continuously refreshes the image at speeds around 40–50 cycles per second

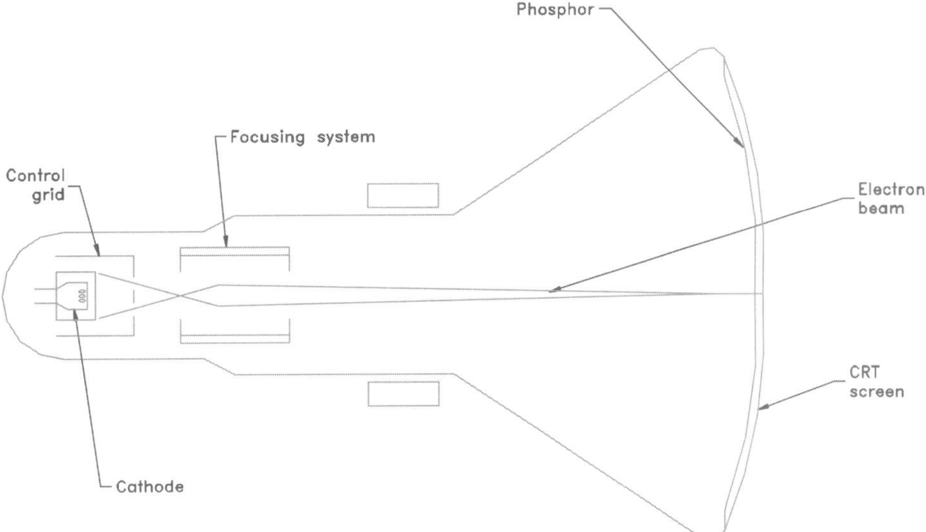

Figure 14 Diagrammatic representation of a CRT.

to avoid a noticeable flicker. Refresh terminals permit a high degree of movement in the displayed image as well as high resolution. Selective erasing or editing is possible at any time without erasing and repainting the entire image.

In a color CRT, there are three electron guns, each focusing only on the phosphors that have its own color (red, blue, or green). The gun hits each phosphor with varying intensities, until a pixel is formed by uniting one set of red, blue, and green phosphors. Inside the pixel, each color combines with the others to form the final color that is displayed on the screen. A shadow mask is a thin metal screen that assists in focusing the electron beams through the holes of the screen so that each electron gun color can only light its own phosphors.

Since the phosphors only light for less than a second, the electron gun is constantly shooting electrons at the screen to sustain a constant image. This process is referred to as "refreshing." The electron can refresh the screen approximately 100 times a second by constantly moving left to right and top to bottom across the screen. The computer's video

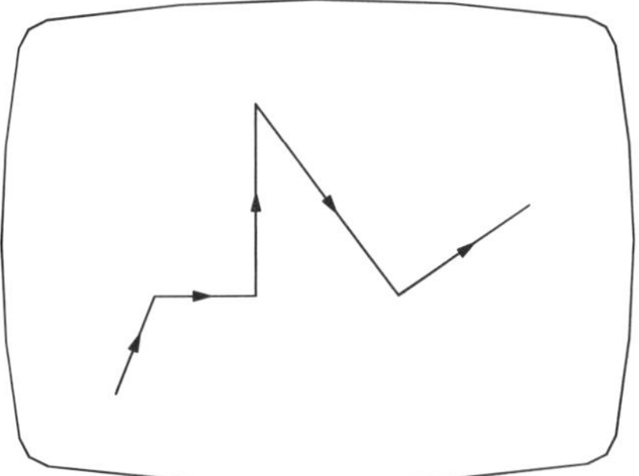

Figure 15 Electron beam path on a CRT display using stroke-writing technique

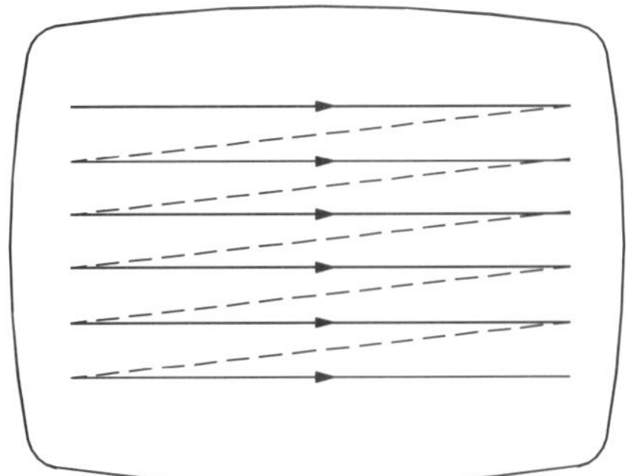

Figure 16 Electron beam path on a CRT display using raster-scan technique.

card communicates to the electron gun at what intensity it should strike specific pixels to create the proper image.

Ways to measure CRT monitor quality are as follows:

1. *Dot Pitch.* The measurement of the distance between pixels. A longer distance yields less sharp pictures.
2. *Resolution.* The measurement of the number of pixels a monitor can show at one time. Larger numbers typically produce sharper pictures.
3. *Refresh Rate.* Measured in hertz, it is the number of times per second the electron gun strikes each phosphor.

Two factors are primarily responsible for determining the price of a CRT monitor: sharpness and viewable screen size. The first commonly seen PC screens had 13-in. or smaller monitors (measured diagonally from corner to corner) and were primarily one color. In the mid-1990s, full-color monitor screens began to emerge in larger sizes (15–17-in. screens). Currently, CRT monitors are frequently available in sizes up to 21 in., although larger monitors are also offered. CRT monitors can be either vertical or horizontal, with the most popular form being horizontal. Graphical designers, however, often use vertical CRT monitors.

Direct-View Storage Tube

The direct-view storage tube (DVST) was one of the most widely used graphics display devices in the mid to late 1970s. In a DVST terminal, the CRT uses the stroke-writing technique to generate the desired image on the screen. One or more electron flood-guns continuously supply the necessary energy to maintain the image on the screen. DVST generates a long-lasting, flicker-free image with high resolution and no refreshing. It handles an almost unlimited amount of data. However, display dynamics are limited since DVST terminals do not permit selective erasure. The image also does not appear as bright as with a vector-refresh or raster-scan terminal.

Raster-Scan Terminals

As the name suggests, these displays use rasterization to create the image on the CRT screen. Much like a television set (the difference being that a TV input signal is analog while that of a computer terminal is digital), these terminals refresh the image by continuously sweeping over the screen in a constant pattern. This allows real-time changes to be made to the image. Raster scan is currently a dominant technology in CAD graphic display. Raster-scan terminal features include brightness, accuracy, selective erasure, dynamic motion capabilities, and the potential for unlimited color. The raster-scan terminal can display a large amount of information without flicker, although the resolution is not as high as with a DVST display. Raster-scan technology has moved rapidly since 1982, and with the introduction of graphic accelerators and additional add-on frame buffers to workstations and PCs, affordable, high-quality graphics has finally reached the end user.

Other forms of technology, such as the LCD (liquid crystal display), may ultimately replace the CRT. Due to its size, the CRT monitor is not the most efficient viewing technology for desktops. The CRT monitor is bulky, absorbs approximately three-fourths of the power a computer uses in a given day, and sometimes releases small amounts of X-ray radiation. The older CRT monitors also have curved screens to counteract shortcomings in the electron gun; however, many newer CRT monitors have flat screens. The LCD is the technology commonly seen in flat-screen televisions and computers and is becoming increasingly popular.

6.2 Hard-Copy Devices

Despite the ease with which design files can be managed using computer technology, a hard copy of the work is often required for recordkeeping and presentation. Output devices have been developed to interface with the computer and produce a hard copy of the requisite design or file. The processes used in hard-copy devices are analogous to those used in CRT displays. The various types of hard-copy devices are discussed next.

Vector Plotters

Just as storage tube and vector-refresh displays create images through an electron beam directed along vectors defined in the design file, vector plotters create an image using designated vectors. Vector plotters produce very high resolution hard copies. Two common kinds of vector plotters are the pen plotter and the COM plotter.

Pen Plotter. Pen plotters use mechanical ink pens directed along design vectors to create images on paper or similar media. Pen plotters are further divided into drum- and flat-bed types.

DRUM PLOTTER. The drum plotter consists of a cylindrical drum, most often mounted horizontally, and a pen mounted on a slide parallel to the surface of the drum (Fig. 17). The plotter matches the motion of the drum and that of the pen along the slide to draw the appropriate vectors and create the desired image. The drum plotter is able to produce hard copies that are limited in length only by thc amount of paper on a roll.

FLAT-BED PLOTTER. Flat-bed plotters act on the same vector-plotting concept, but the paper is attached to a flat surface. The writing tool is again movable along a metal slide. The slide provides one of the two coordinate axes upon which the design will be printed. The parallel tracks along which the slide itself moves provides the other axis. As the pen moves along the slide, the slide moves along the other axis, enabling the 2D vectors to be drawn.

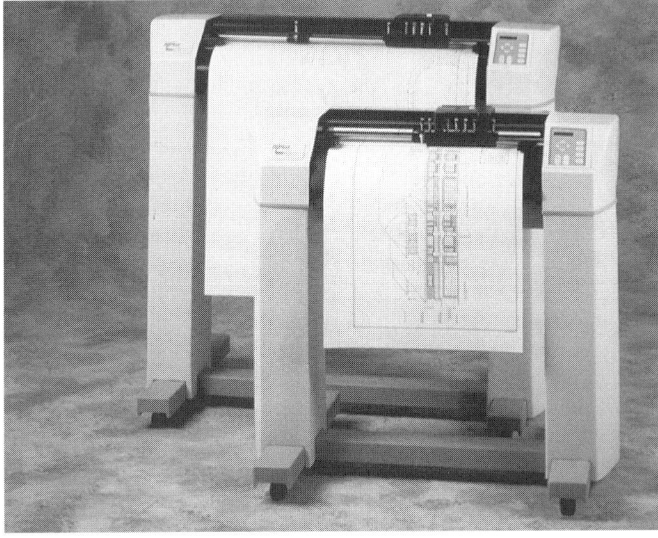

Figure 17 Two drum-type plotters. (Courtesy of Summagraphics, Inc.)

Computer-Output-to-Microfilm (COM) Plotter. COM plotters constitute a third type of vector plotter. These plotters produce images on film rather than on paper using light instead of ink. These expensive units facilitate efficient archiving of designs. Designs can be stored using a fraction of the space required for hard-copy filing and enlarged to original size when needed. The cost of producing a plot using a COM plotter can be significantly less compared to the cost associated with a flat-bed or drum plot, but the quality upon enlargement is generally poorer than that obtained using a pen plotter.

Raster Plotters

Many plotters require rasterization—breaking up the image into a series of dots—which will then be reconstituted to re-create the image in much the same way that a raster-scan CRT screen uses these dots to produce a recognizable image on the display terminal. Common raster plotters include the electrostatic plotter, inkjet plotter, and laser plotter. These types of plotters generally produce lower resolution images than those generated on a pen plotter. However, because the time to produce a rasterized plot is independent of complexity, these have the advantage of being significantly faster and are often used to create preliminary hard copies before a higher resolution plot is performed.

Electrostatic Plotter. Electrostatic or direct printing forms images for thermosensitive media by application of heat from nibs on thermal printing heads. The paper turns black at the precise points where heat is applied. Electrostatic plotters have few moving parts and thus are reliable and quiet. Resolution can be as high as 400 × 800 dpi, with gray scales ranging from 16 to 128 values. These are medium- to high-throughput devices, producing complex images in about a minute. On-board computing facilities such as RISC processors and fast hard-disk storage mechanisms contribute to rapid drawing and processing speeds. Expansion slots accommodate interface cards for LANs or parallel ports.

Inkjet Plotter. Inkjet plotters and printers fire tiny ink droplets at paper or a similar medium from thin nozzles in the printing head. Heat generated by a separate heating element almost instantaneously vaporizes the ink. The resulting bubble generates a pressure wave that ejects an ink droplet from the nozzle. Once the pressure pulse passes, ink vapor condenses and the negative pressure produced as the bubble contracts draws fresh ink into the nozzle. These plotters do not require special paper and can also be used for preliminary drafts. Inkjet plotters are available both as desktop units for 8.5 × 11-in. graphics and in wide format for engineering CAD drawings. Typical full-color resolution is 360 dpi, with black-and-white resolution rising to 700 × 720 dpi. These devices handle both roll-feed and cut sheet media in widths ranging from 8.5 to 36 in. Also, ink capacity in recently developed plotters has increased, allowing these devices to handle large rolls of paper without depleting any one ink color. Inkjet plotters are very user friendly, often including sensors for the ink supply and ink flow, which warn users of an empty cartridge or of ink stoppage, allowing replacement without losing a print. Other sensors eliminate printing voids and unwanted marks caused by bubbles in the ink lines. Special print modes typically handle high-resolution printing by repeatedly going over image areas to smooth image lines. In addition, inkjet plotters typically contain 6–64 MB of image memory and options such as hard drives, an Ethernet interface for networking, and built-in Postscript interpreters for faster processing. Inkjet plotters and printers are increasingly dominating other output technologies such as pen plotters in the design laboratory.

Laser Plotter. Laser plotters produce fairly high quality hard copies in a shorter period of time than pen plotters. A laser housed within the plotter projects rasterized image data in the form of light onto a photostatic drum. As the drum rotates further about its axis, it is dusted with an electrically charged powder known as toner. The toner adheres to the drum wherever the drum has been charged by the laser light. The paper is brought into contact with the drum and the toner is released onto the paper, where it is fixed by a heat source close to the exit point. Laser plotters can quickly produce images in black and white or in color and resolution is high.

7 SOFTWARE

Software is the collection of executable computer programs, including operating systems, languages, and application programs. All of the hardware described previously can do nothing without software to support it. In its broadest definition, software is a group of stored commands, sometimes known as a program, which provide an interface between the binary code of the CPU and the thought processes of the user. The commands provide the CPU with the information necessary to drive graphical displays and other output devices, to establish links between input devices and the CPU, and to define paths which enable other command sequences to operate. Software operates at all levels of computer function. Operating systems are a type of software that provides a platform upon which other programs may run. Likewise, individual programs often provide a platform for the operation of subroutines, which are smaller programs dedicated to the performance of specific tasks within the context of the larger program.

When software was first introduced, the general population was not familiar with the concept, and the term required further explanation. A common analogy was comparing hardware to a record and the software to the music the record played. The hardware, like the record, was tangible, and the software, like the music in the record, was intangible yet existed. A CD or floppy disk that you put into your computer is hardware, and the programs it loads is software.

System Software. System software is the OS that controls your machine.

Application Software. Application software is a conglomeration of the programs needed to carry out the tasks you need to complete. There are several types of application software. Different software is required for creating graphics, typing documents, adding spreadsheets, and creating slideshow presentations. There is also separate application software that focuses solely on industry, such as accounting, billing, retail business, and insurance companies.

Software is licensed and copyrighted and generally requires the user to agree to the specified terms of use. During the installation period, the user will be required to accept the user agreement, allowing the developers and manufacturers to continue to make profits on the software.

Shareware. The term "shareware" refers to programs that either are completely free or are free for a trial period, then obtainable by purchase after the specified time period. Shareware was introduced when programmers wanted to make their programs available to the public without profit or copyright restrictions. Sometimes there is an expiration date or a disabling function if the software is not purchased after a certain time period. Shareware can also refer to software that is free in an incomplete form but with available upgrades for a charge.

Freeware. Freeware is primarily a marketing term that is used to refer to shareware. Most commonly used in advertisements, "freeware" gives the impression that the software has no costs associated with it, which may or may not be true. Software that is entirely free has no copyright restrictions and is classified "in the public domain."

7.1 Operating Systems

Operating systems have developed over the past 50 years for two main purposes. First, operating systems attempt to schedule computational activities to ensure optimal performance of the computing system. Second, they provide a convenient environment for the development and execution of programs. An operating system may function as a single program or as a collection of programs that interact with each other in a variety of ways.

An operating system has four major components: process management, memory management, I/O operations, and file management. The operating system schedules and performs I/O, allocates resources and memory space, provides monitoring and security functions, and governs the execution and operation of various system programs and applications such as compilers, databases, and CAD software.

Operating systems that serve several users simultaneously (e.g., UNIX) are more complicated than those serving only a single user (e.g., MS-DOS, Macintosh OS). The two main themes in operating systems for multiple users are multiprogramming and multitasking.

Multiprogramming provides for the interleaved execution of two or more computer programs (jobs) by a single processor. In multiprogramming, while the current job is waiting for the I/O to complete, the CPU is simply switched to execute another job. Eventually the first job completes its I/O functions and is serviced by the CPU again. As long as there is some job to complete, the CPU remains active. Holding multiple jobs in memory at one time requires special hardware to protect each job, some form of memory management, and CPU scheduling. Multiprogramming increases CPU utilization and decreases the total time needed to execute the jobs, resulting in greater throughput.

Multitasking, or time sharing, is a logical extension of multiprogramming for situations where an interactive mode is essential. The techniques that use multiprogramming to handle multiple interactive jobs is referred to as multitasking or time sharing. The processor's time is shared among multiple users. Time sharing was developed in the 1960s, when most computers were large, costly mainframes. The requirement for an interactive computing facility could not be met by the use of a dedicated computer. An interactive system is used when a short response time is required. Time-sharing operating systems are very sophisticated, requiring extra disk management facilities, an online file system, and protective mechanisms.

The following sections discuss the two most widely used operating systems for CAD applications, UNIX and Windows NT. It should be noted that both of these operating systems can run on the same hardware architecture.

UNIX

The first version of UNIX was developed in 1969 by Ken Thompson and Dennis Ritchie of the Research Group of Bell Laboratories to run on a PDP-7 minicomputer. The first two versions of UNIX were created using assembly language, while the third version of UNIX was written using the C programming language. As UNIX evolved, it became widely used at universities and research and government institutions and eventually in the commercial world. UNIX quickly became the most portable of operating systems, operable on almost all general-purpose computers. It runs on PCs, workstations, minicomputers, mainframes, and supercomputers. UNIX has become the preferred program development platform for many applications such as graphics, networking, and databases. A proliferation of new versions of UNIX has led to a strong demand for UNIX standards. Most existing versions can be traced back to one of two sources: AT&T System V or 4.3 BSD (Berkeley UNIX) from the University of California, Berkeley (one of the most influential versions).

UNIX was designed to be a time-sharing, multiuser operating system. UNIX supports multiple processes (multiprogramming). A process can easily create new processes with the fork system call. Processes can communicate with pipes or sockets. CPU scheduling is a simple priority algorithm. Memory management is a variable-region algorithm with swapping supported by paging. The file system is a multilevel tree, which allows users to create their own subdirectories. In UNIX, I/O devices such as printers, tape drives, keyboards, and terminal screens are all treated as ordinary files (file metaphor) by both programmers and users. This simplifies many routine tasks and is a key component in extensibility of the systems. Certifiable security that protects users' data and network support are also two important features.

UNIX consists of two separable parts: the kernel and the system programs. The kernel is the collection of software that provides the basic capabilities of the operating system. In UNIX, the kernel provides the file system, CPU scheduling, memory management, and other operating system functions (I/O devices, signals) through system calls. System calls can be grouped into three categories: file manipulation, process control, and information manipulation. System programs use the kernel-supported system calls to provide useful functions, such as compilation and file manipulation. Programs, both system and user written, are normally executed by a command interpreter. The command interpreter is a user process called a shell. Users can write their own shell. There are, however, several shells in general use. The Bourne shell, written by Steve Bourne, is most widely available. The C shell, mostly by Bill Joy, is the most popular on BSD systems. The Korn Shell, by David Korn, has also become quite popular in recent years.

Solaris

Sun Microsystems developed a UNIX-based environment called Solaris. Currently, the Solaris OS, a product line marketed by Sun Microsystems, is the result of the company's goal to allow all computers to communicate with each other: "The Solaris OS is a product of that vision, and it has been a leader in what the company calls 'industrial strength' software for many years."*

Solaris manages the servers on which large-scale dot-coms run. For example, Solaris can link international offices for the same company on one server as well as connecting ISPs (Internet service providers) to the Internet. Solaris not only is capable of handling high-volume traffic across the globe but also is a highly reliable system that is the top UNIX-based OS currently on the market. Due to its high-volume and extremely stable network, Solaris is the system in highest demand for rapidly growing enterprises. Solaris also has a reputation for managing unpredictable levels of traffic.

In February 2000, Sun released Solaris 8. Solaris 8 was designed to react to the sharp increase of interest in the World Wide Web. The Sun Microsystems website (http://www.sun.com) has more information on the Solarsis product line.

Windows NT

The development effort for the new high-end operating system in the Microsoft Windows family, Windows NT (New Technology), has been led by David Culter since 1988. Market requirements and sound design characteristics shaped the Windows NT development. The architects of "NT," as it is popularly known, capitalized on the strengths of UNIX while avoiding its pitfalls. Windows NT and UNIX share striking similarities. There are also marked differences between the two systems. UNIX was designed for host-based terminal computing (multiuser) in 1969, while Windows NT was designed for client/server distributed computing in 1990. The users on single-user general-purpose workstations (clients) can connect to multiuser general-purpose servers with the processing load shared between them. There are two Windows NT-based operating systems: Windows NT Server and Windows NT

---

* *Computing Encyclopedia,* Smart Computing Reference Series (Vol. 4, p. 50), Sandhills Publishing, Lincoln, NE, 2002.

Workstation. The Windows NT Workstation is simply a scaled-down version of Windows NT Server in terms of hardware and software. Windows NT is a microkernel-based operating system. The operating system runs in privileged processor mode (kernel mode) and has access to system data and hardware. Applications run on a nonprivileged processor mode (user mode) and have limited access to system data and hardware through a set of digitally controlled application programming interfaces (APIs). Windows NT also supports both single-processor and SMP operations. Multiprocessing refers to computers with more than one processor. A multiprocessing computer is able to execute multiple threads simultaneously, one for each processor in the computer. In SMP, any processor can run any type of thread. The processors communicate with each other through shared memory. SMP provides better load balancing and fault tolerance. The Win32 subsystem is the most critical of the Windows NT environment subsystems. It provides the graphical user interface and controls all user input and application output.

Windows NT is a fully 32-bit operating system with all 32-bit device drivers, which paves the way for future development. It makes administration easy by providing more flexible built-in utilities and removes diagnostic tools. Windows NT Workstation provides full-crash protection to maximize uptime and reduce support costs. Windows NT is a complete operating system with fully integrated networking, including built-in support for multiple network protocols. Security is pervasive in Windows NT to protect system files from error and tampering. The NT file system (NTFS) provides security for multiple users on a machine.

Windows NT, like UNIX, is a portable operating system. It runs on many different hardware platforms and supports a multitude of peripheral devices. It integrates preemptive multitasking for both 16- and 32-bit applications into the operating system, so it transparently shares the CPU(s) among the running applications. More usable memory is available due to the advanced memory features of Windows NT. There are more than fourteen hundred 32-bit applications available for Windows NT today, including all major CAD and FEA software applications.

Hardware requirements for the Windows NT operating system fall into three main categories: processor, memory, and disk space. In general, Windows NT Server requires more in each of the three categories than does its sister operating system, Windows NT Workstation. The minimum requirements are an Intel-compatible microcomputer PC (IBM compatible) with a 486-DX2 33-MHz processor, such as an Intel Pentium, Apple Power-PC, or other supported RISC-based processor. The minimum memory requirement is 8 MB, though 32 MB is strongly recommended. The minimum disk space requirements for the operating system only are in the 100-MB range. NT Workstation requires 75 MB of x86 and 97 MB for RISC. For NT Server, 90 MB for x86 and 110 MB for RISC are required. There is no need to add additional disk space for any application that is run on the NT operating system.

7.2 Graphical User Interface and X Window System

DOS, UNIX, and other command-line operating systems have long been criticized for the complexity of their user interface. For this reason, the graphical user interface (GUI) is one of the most important and exciting developments of this decade. The emergence of the GUI revolutionized the methods of man–machine interaction used in the modern computer. GUIs are available for almost every type of computer and operating system on the market. A GUI is distinguished by its appearance and by the way an operator's actions and input options are handled. There are over a dozen GUIs. They may look slightly different, but they all share certain basic similarities. These include the following: a pointing device (mouse or digitizer), a bit-mapped display, windows, on-screen menus, icons, dialog boxes, buttons, sliders, check boxes, and an object–action paradigm. Simplicity, ease of use, and enhanced productivity are all benefits of a GUI. GUIs have rapidly become important features of CAD software.

GUI systems were first envisioned by Vannevar Bush in a 1945 journal article. Xerox was researching GUI tools at the Palo Alto Research Center throughout the 1970s. By 1983, every major workstation vendor had a proprietary window system. It was not until 1984, however, when Apple introduced the Macintosh computer that a truly robust window environment reached the average consumer. In 1984, a project called Athena at MIT gave rise to the X Window system. Athena investigated the use of networked graphics workstations as a teaching aid for students in various disciplines. The research showed that people could learn to use applications with a GUI much more quickly than by learning commands.

A GUI is the system of taskbars, icons, and other pictorial depictions that allow the user to communicate with the PC. Computers also use a GUI to exhibit and access information. Commonly used PCs have GUI systems, such as Windows or Macintosh. These OSs have icons that allow the user to start specific programs—such as locating the trash files by an icon of a garbage can—merely by clicking on the icons. GUI systems allow the user to more easily work with the computer. Instead of lists of commands, all the user has to do is click on a button icon to run specific programs. The goal is for users to reap greater efficiency by using a quick point-and-click method.

In earlier computers, users had to know specific sets of commands to start programs. Text-based languages such as COBOL (Common Business-Oriented Language) were typically used in operating systems. Today, users who have a GUI system do not have to be as computer literate to operate their PC. They need only to click on the appropriate icons to initiate certain programs, instead of having to remember long lists of commands. GUIs work with the concept that people's brains understand and remember pictures more easily than text.

The Windows and Macintosh OSs are the most popular GUIs. For the graphic icons, they use commonly recognized pictures from the office, such a file folder to indicate storage and a trash can to indicate where deleted files and programs will be located. Other aspects of GUIs can include toolbars and pull-down menus. A new form of GUI on which developers are currently working is a 3D GUI. This system would make it easier to view 3D designs, such as architectural plans. A different display form would be necessary, enabling users of the 3D GUI to view everything on the screen in a 3D environment.

Douglas Englebart's invention of the mouse in 1962 facilitated the switch to GUI systems. This early mouse was a small mechanism that was designed to point to and click on graphics in a GUI system. In 1963, a graduate student at MIT introduced Sketchpad as his thesis—a device that functioned similarly to a mouse but involved the use of a light pen on a computer screen. In 1973, Xerox incorporated the earlier mouse into another system similar to the GUI, called Alto.

Apple Computers was the first company to successfully incorporate the GUI system with a mouse. The company's cofounder, Steve Jobs, convinced people to work for his company and to develop a functioning GUI system in the new Apple system (LISA—Local Integrated Software Architecture). Despite Xerox's release of the same technology two years earlier in 1981, they did not meet with success at that time.

At MIT in 1984, the X Window system began to be developed. The project was called Athena, and the goal was to create a new teaching aid across all the academic disciplines using networked graphics workstations. Research showed that the students learned much more quickly with a GUI rather than by learning specific commands.

The X Window system is a non-vendor-specific window system. It was specifically developed to provide a common window system across networks connecting machines from different vendors. Typically, the communication is via TCP/IP protocol over an Ethernet network. The X Window system (commonly referred to as "X Windows" or "X") is not a

GUI; rather, it is a portable, network-transparent window system that acts as a foundation upon which to build GUIs (such as AT&T's OpenLook, OSF/Motif, and DEC Windows). The X Window system provides a standard means of communicating between dissimilar machines on a network and can be viewed in a window. The unique benefit provided by a window system is the ability to have multiple views showing different processes on different networks. Since the X Window system is in the public domain and not specific to any platform or operating system, it has become the de facto window system in heterogeneous environments from PCs to mainframes.

Unfortunately, a window environment does not come without a price. Extra layers of software separate the user and the operating system, including the window system, the GUI, and an Application Programming Interface (ToolKit) in a UNIX operating environment. GUIs also place extra demands on hardware. All visualization workstations require more powerful processing capabilities (>6 MIPS); large CPU memory and disk subsystems; built-in network I/O, typically Ethernet high-speed internal bus structures (at least 32 MB/s); high-resolution monitors (at least 1024 × 768); more colors (>256); etc. For PCs, both operating systems and GUIs are in a tremendous state of flux.

Some X Window system terms are *X Window, X terminal,* and *X server.*

X Window

X Window is a way for users to run applications on a computer other than their own and over a network while being able to see the data on their local computer. Developed in the mid-1980s, the primary function of the X Window was to allow users to utilize a GUI to make data look similar to how it would look on a Microsoft Windows format. X Window is not a Microsoft application. Where Microsoft Windows runs applications on a local machine, X Window "is designed to facilitate distributed processing and relies entirely on a client-server relationship, although the definition of the client and the server are the reverse of the typical relationship in a LAN (local area network) running Microsoft Windows."*

In X Window, the client is the application (X client), and the server is also software (X server). This application defines the server–client relationship in the software itself and enables the client and server to run on different computers, although it is possible to run both on the same computer. For this reason, X Window is exceptionally efficient for networks. Since X Window does not depend on one specific operating system or programming language, it is also easily movable. Other benefits to X Window include ease of installing new hardware or upgrading to a new server and cost reduction due to the lack of necessity for having high-powered PCs in terminals and workstations.

X Terminal

An X terminal is a workstation on an X network that does not actually run its own applications. While it appears that the computer is running its own applications, it is really displaying graphic images coming from the host computer that is really running the GUIs and graphical programs displayed. An X terminal relies solely on other computers on the network to run its software. To do this, an X server program is run on a local computer and is connected to an X client on what may be one of several host computers, or an application server.

The X terminal may be much less powerful than in high-powered LAN-connected computers. The X terminal, however, reduces hardware costs and enables older machines to remain functional at a constructive level for a much longer time.

---

* *Computing Encyclopedia,* Smart Computing Reference Series (Vol. 4, p. 218), Sandhills Publishing, Lincoln, NE, 2002.

X Server
The X server controls what the user on a local computer sees on the display. The X server is not the hardware that runs the primary applications; rather, it is the software that controls the mouse configuration, the screen resolution, and security on the local computer. The hardware that runs the applications is called the "host," and the client in the client–server relationship is also software that runs on the host.

X Window applications are not connected to the OS on the computer. These applications, as well as the client and the server, are independent of their physical setting and can run on almost any program, on the same or separate computers, and with or without a network. The strongest function of X Window and the X server–client relationship is its capacity for distributing areas of computing. This protocol formed a portion of the basis of thin-client architecture. This architecture also enables extremely efficient management of computing hardware resources.

7.3 Computer Languages

The computer must be able to understand the commands it is given to perform the desired tasks at hand. The binary code used by the computer circuitry is very easy for the computer to understand but can be tedious for and almost indecipherable to the human programmer. Languages for computer programming have developed to facilitate the programmer's job. Languages are often categorized as low- or high-level languages.

Low-Level Languages
The term *low-level* refers to languages that are easy for the computer to understand. These languages are often specific to a particular type of computer, so that programs created on one type of computer must be modified to run on another type. Machine language (ML) and assembly language (AL) are both considered low-level languages.

Machine language is the binary code that the computer understands. ML uses an operator command coupled with one or more operands. The operator command is the binary code for a specific function, for example, addition. The numbers to be added, in this example, are operands. Operators are also binary codes, arbitrary with respect to the machine used. For a hypothetical computer, all operator codes are established to be eight digits, with the operator command appearing after the two operands. If the operator code for addition then was 01100110, the binary (base-2) representation of the two numbers added would be followed by the code for addition. A command line to perform the addition of 21 and 14 would then be written as follows:

000101010000111001100110

The two operands are written in their binary forms (21_{10} as 00010101_2 and 14_{10} as 00001110_2) and are followed by the operator command (01100110 for addition). The binary nature of this language makes programming difficult and error correction even more so.

Assembly language operates in a similar manner to ML but substitutes words for machine codes. The program is written using these one-to-one relationships between words and binary codes and is separately assembled through software into binary sequences. Both machine and assembly languages are time intensive for the programmer, and because of the differences in logic circuitry between types of computers, the languages are specific to the computer being used. High-level languages address the problems presented by these low-level languages in various ways.

High-Level Languages

High-level languages give the programmer the ability to bypass much of the tediousness of programming involved in low-level languages. Often many ML commands will be combined within one HLL statement. The programming statements in HLL are converted to ML using a compiler. The compiler uses a low-level language to translate the HLL commands into ML and check for errors. The net gain in terms of programming time and accuracy far outweighs the extra time required to compile the code. Because of their programming advantages, HLLs are far more popular and widely used than low-level languages. The following commonly used programming languages are described below:

- FORTRAN
- Pascal
- BASIC
- C
- C++

FORTRAN (*FORmula TRANslation*). Developed at IBM between 1954 and 1957 to perform complex calculations, this language employs a hierarchical structure similar to that used by mathematicians to perform operations. The programmer uses formulas and operations in the order that would be used to perform the calculation manually. This makes the language very easy to use. FORTRAN can perform simple as well as complex calculations. FORTRAN is used primarily for scientific or engineering applications. CFP95 Suite, a software-benchmarking product by Standard Performance Evaluation Corp. (SPEC), is written in FORTRAN. It contains 10 CPU-intensive floating-point benchmarks.

The programming field in FORTRAN is composed of 80 columns, arranged in groups relating to a programming function. The label or statement number occupies columns 1–5. If a statement extends beyond the statement field, a continuation symbol is entered in column 6 of the next line, allowing the statement to continue on that line. The programming statements in FORTRAN are entered in columns 7–72. The maximum number of lines in a FORTRAN statement is 20. Columns 73–80 are used for identification purposes. Information in these columns is ignored by the compiler, as are any statements with a C entered in column 1.

Despite its abilities, there are several inherent disadvantages in FORTRAN. Text is difficult to read, write, and manipulate. Commands for program flow are complicated, and a subroutine cannot go back to itself to perform the same function.

Pascal. Pascal is a programming language with many different applications. It was developed by Niklaus Wirth in Switzerland during the early 1970s and named after the French mathematician Blaise Pascal. Pascal can be used in programs relating to mathematical calculations, file processing and manipulation, and other general-purpose applications.

A program written in Pascal has three main sections: the program name, the variable declaration, and the body of the program. The program name is typically the word "PROGRAM" followed by its title. The variable declaration includes defining the names and types of variables to be used. Pascal can use various types of data and the user can also define new data types, depending on the requirements for the program. Defined data types used in Pascal include strings, arrays, sets, records, files, and pointers. Strings consist of collections of characters to be treated as a single unit. Arrays are sequential tables of data. Sets define a data set collected with regard to sequence. Records are mixed data types organized into a hierarchical structure. Files refer to collections of records outside of the program itself, and pointers provide flexible referencing to data. The body of the program uses commands to execute the desired functions. The commands in Pascal are based on English and are arranged in terms of separate procedures and functions, both of which must have a defined beginning and end. A function can be used to execute an equation, and a procedure is used to perform sets of equations in a defined order. Variables can be either "global" or "local," depending on whether they are to be used throughout the program or within a particular procedure. Pascal is somewhat similar to FORTRAN in its logical operation, except that Pascal uses symbolic operators while FORTRAN operates using commands. The structure of Pascal allows it to be applicable to areas other than mathematical computation.

BASIC (Beginners All-purpose Symbolic Interactive Code). BASIC was developed at Dartmouth College by John Kemeny and Thomas Kurtz in the mid-1960s. BASIC uses mathematical programming techniques similar to FORTRAN and the simplified format and data manipulation capabilities similar to Pascal. As in FORTRAN, BASIC programs are written using line numbers to facilitate program organization and flow. Because of its simplicity, BASIC is an ideal language for the beginning programmer. BASIC runs in either the direct or programming mode. In the direct mode, the program allows the user to perform a simple command directly, yielding an instantaneous result. The programming mode is distinguished by the use of line numbers that establish the sequence of the programming steps. For example, a user who wishes to see the words "PLEASE ENTER DIAMETER" displayed on the screen immediately would execute the command PRINT "PLEASE ENTER DIAMETER." If, however, that phrase were to appear in a program, the above command would be preceded by the appropriate line number.

The compiler used in the BASIC language is unlike either FORTRAN or Pascal. Whereas other HLL compilers check for errors and execute the program as a whole unit, a BASIC program is checked and compiled line by line during program execution. BASIC is often referred to as an "interpreted" language as opposed to a compiled one since it interprets the program into ML line by line. This condition allows for simplified error debugging. In BASIC, if an error is detected, it can be corrected immediately; in FORTRAN and Pascal, the programmer must go back to the source program to correct the problem and then recompile the program as a separate step. The interpretive nature of BASIC does cause programs to run significantly slower than in either Pascal or FORTRAN.

C. C was developed from the B language by Dennis Ritchie in 1972. C was standardized by the late 1970s with Kernighan and Ritchie's book, *The C Programming Language*. C was developed specifically as a tool for the writing of operating systems and compilers. It originally became most widely known as the development language for the UNIX operating systems. C expanded at a tremendous rate over many hardware platforms. This led to many variations and a lot of confusion, and while these versions were similar, there were notable differences. This was a problem for developers who wanted to write programs that ran on several platforms. In 1989 the American National Standards Committee on Computers and Information Processing approved a standard version of C. This version is known as ANSI C, and it includes a definition of a set of library routines for file operations, memory allocation, and string manipulation.

A program written in C appears similar to Pascal. C, however, is not as rigidly structured as Pascal. There are sections for the declaration of the main body of the program and the declaration of variables. C, like Pascal, can use various types of data, and the programmer can also define new data types. C has a rich set of data types, including arrays, sets, records, files, and pointers. C allows for far more flexibility than Pascal in the creation of new data types and the implementation of existing data types. Pointers in C are more powerful than they are in Pascal. Pointers are variables that point, not to data, but to the memory location of data. Pointers also keep track of what type of data is stored there. A pointer can be defined as a pointer to an integer or a pointer to a character. CINT95 Suite, a software-benchmarking product, is written in C. It contains eight CPU-intensive integer benchmarks.

C++. C++ is a superset of the C language developed by Bjarne Stroustrup in 1986. C++'s most important addition to the C language is the ability to do object-oriented programming. Object-oriented programming places more emphasis on the data of a program. Programs are structured around objects. An object is a combination of the program's data and code. Like a traditional variable, an object stores data; unlike traditional languages, objects can also do things. For example, an object called "triangle" might store both the dimensions of the triangle and the instructions on how to draw the triangle. Object-oriented programming has led to a major increase in productivity in the development of applications over traditional programming techniques.

A program written in C++ no longer resembles C or Pascal. More emphasis is placed on a modular design around objects. The main section of a C++ code should be very small and may only call one or two functions; the declaration of variables in the main function should be avoided. Global variables and functions are avoided at all costs, and the use of variables in local objects is stressed. The avoidance of global variables and functions that do large amounts of work is intended to increase security and make programs easier to develop, debug, and modify.

Some computer languages have been developed or modified for use with software applications for the Windows NT operating system. These include languages such as Ada, COBOL, Forth, LISP, Prolog, Visual BASIC, and Visual C++.

Language

The two broad categories of language in the computer industry are natural language and machine language. Examples of natural languages are English, German, etc., and are defined as languages that are written or spoken by humans and evolve naturally. In the computer world, natural language is traditionally less difficult for people to understand and use and is therefore used in programming languages that incorporate vocabulary, grammar, and traditional sentence structure.

Natural language is sometimes extremely difficult to incorporate into AI (artificial intelligence) systems, for a number of reasons. The main difficulties occur when dealing with the complexities and irregularities as well as the issues with word meaning and context in natural language.

Machine language is written in a binary-encoded programming language consisting of 0s and 1s, which is the only language a computer understands, and is a representation of a computer program. Machine language is a logical and rational system that leaves little room for misreading or error. People usually write in programming languages such as C that an interpreter then translates into computer language so the computer can complete its task.

Machine language programs have a series of machine instructions that are composed of binary strings. Computer chip manufacturers call their products 32- or 64-bit processors because they are referring to the size string with which the processor works.

Language Processor. A language processor translates language from programming language into machine language. This software has two different components: interpreters and compilers.

Interpreters are programs that both translate and carry out instructions. These programs also launch other programs. *Interpretive overhead* is the term used when referring to the extra time required to run a program through an interpreter. The interpreter translates and executes only one line of source code at a time, and if the program loops back to a line that has already been translated, the process needs to be repeated again.

Sometimes, however, it is less time consuming to interpret a program than to first compile and then run the program. For programmers, a cycle run through an interpreting program will frequently save more time than an edit–compile–run–debug cycle.

Other advantages to interpreted environments include the adaptability and interactive nature of interpretation. Interpretable code is flexible and can be used with any machine type. It is also much easier to change, as there is no protracted time required for compilation.

Compilers, unlike interpreters, translate source code to machine code, but instead of executing the instructions, they save the entire translated code in a file. The compiler does not need to be loaded again to run the translated file, and the file can be executed without any further translation. Some compilers do not translate source code directly into machine code; rather, they translate into assembly language (low-level language), which is later translated to machine language by a different assembler.

Language Translation Program. A language translation program usually handles high-level languages and translates statements from one program to the other. To translate from one programming language to another, the source code is first converted into the target language; then the translated code goes through a target compiler until the code compiles with no errors. After the code is tested once again to ensure that it has no run time errors, the program is tested with sample tasks to make sure it functions properly.

Language translation programs are extremely valuable for several reasons. First, being able to convert older code to more modern languages both reduces the cost of run time on computers and functions as a time-saving device. Translation programs may take only a week to translate the same amount of code that it would typically take a year to translate manually. Additionally, software engineering advances are so rapid that language translation programs are vital to keep corporations and governments functioning.

8 CAD SOFTWARE

Contemporary CAD software is often sold in "packages" which feature all of the programs needed for CAD applications. These fall into two categories: graphics software and analysis software. Graphics software makes use of the CPU and its peripheral I/O devices to generate a design and represent it on-screen. Analysis software makes use of the stored data relating to the design and apply it to dimensional modeling and various analytical methods using the computational speed of the CPU.

8.1 Graphics Software

Traditional drafting has consisted of the creation of 2D technical drawings that operated in the synthesis stage of the general design process. However, contemporary computer graphics software, including that used in CAD systems, enables designs to be represented pictorially on the screen such that the human mind may create perspective, thus giving the illusion of three dimensions on a 2D screen. Regardless of the design representation, the drafting itself only involves taking the conceptual solution for the previously recognized and defined problem and representing it pictorially. It has been asserted above that this "electronic drawing-board" feature is one of the advantages of CAD. But how does that drawing board operate?

The drawing board available through CAD systems is largely a result of the supporting graphics software. That software facilitates graphical representation of a design on-screen by converting graphical input into Cartesian coordinates along x, y, and sometimes z axes. Design elements such as geometric shapes are often programmed directly into the software for simplified geometric representation. The coordinates of the lines and shapes created by the user can then be organized into a matrix and manipulated through matrix multiplication. The resulting points, lines, and shapes are relayed to both the graphics software and, finally, the display screen for simplified editing of designs. Because the whole process can take as little as a few nanoseconds, the user sees the results almost instantaneously. Some basic graphical techniques that can be used in CAD systems include scaling, rotation, and translation. All are accomplished though an application of matrix manipulation to the image coordinates. While matrix mathematics provides the basis for the movement and manipulation of a drawing, much of CAD software is dedicated to simplifying the process of drafting itself because creating the drawing line by line and shape by shape is a lengthy and tedious process in itself. CAD systems offer users various techniques that can shorten the initial drafting time.

Geometric Definition

All CAD systems offer defined geometric elements that can be called into the drawing by the execution of a software command. The user must usually indicate the variables specific to the desired element. For example, the CAD software might have, stored in the program, the mathematical definition of a circle. In the x–y coordinate plane, that definition is the following equation:

$$(x - m)^2 + (y - n)^2 = r^2$$

Here, the radius of the circle with its center at (m, n) is r. If the user specifies m, n, and r, a circle of the specified size will be represented on-screen at the given coordinates. A similar process can be applied to many other graphical elements. Once defined and stored as an equation, the variables of size and location can be applied to create the shape on-screen quickly and easily. This is not to imply that a user must input the necessary data in numerical form. Often, a graphical input device such as a mouse, trackball, digitizer, or light pen can be used to specify a point from which a line (sometimes referred to as a "rubber-band line" due to the variable length of the line as the cursor is moved toward or away from the given point) can be extended until the desired length is reached. A second input specifies that the desired endpoint has been reached, and variables can be calculated from the line itself. For a rectangle or square, the line might represent a diagonal from which the lengths of the sides could be extrapolated. In the example of the circle above, the user would specify that a circle was to be drawn using a screen command or other input method. The first point could be established on-screen as the center. Then, the line extending away from the center would define the radius. Often, the software will show the shape changing size as the line lengthens or shortens. When the radial line corresponds to the circle of desired size, the second point is defined. The coordinates of the two defined points give the variables needed for the program to draw the circle. The center is given by the coordinates of the first point, and the radius is easily calculated by determining the length of the line between points 1 and 2. Most engineering designs are much more complex than simple, whole shapes, and CAD systems are capable of combining shapes in various ways to create the desired design.

The combination of defined geometric elements enables the designer to create many unique geometries quickly and easily on a CAD system. The concepts involved in 2D combinations are illustrated before moving on to 3D combinations.

Once the desired geometric elements have been called into the program, they can be defined as "cells," individual design elements within the program. These cells can then be added as well as subtracted in any number of ways to create the desired image. For example, a rectangle might be defined as cell "A" and a circle might be defined as cell "B." When these designations have been made, the designer can add the two geometries or subtract one from the other using Boolean logic commands such as union, intersection, and difference. The concept for two dimensions is shown in Fig. 18. The new shape can also be defined as a cell and combined in a similar manner to other primitives or conglomerate shapes. Cell definition, therefore, is recognized as a very powerful tool in CAD.

8.2 Solid Modeling

Geometric or solid-modeling capabilities follow the same basic concept illustrated above, but with some other important considerations. First, there are various approaches to creating the design in three dimensions (Fig. 19). Second, different operators in solid-modeling software may be at work in constructing the 3D geometry.

In CAD solid-modeling software there are various approaches that define the way in which the user creates the model. Since the introduction of solid-modeling capabilities into the CAD mainstream, various functional approaches to solid modeling have been developed. Many CAD software packages today support dimension-driven solid-modeling capabilities, which include variational design, parametric design, and feature-based modeling.

Dimension-driven design denotes a system whereby the model is defined as sets of equations that are solved sequentially. These equations allow the designer to specify constraints, such as that one plane must always be parallel to another. If the orientation of the first plane is changed, the angle of the second plane will likewise be changed to maintain the parallel relationship. This approach gets its name from the fact that the equations used often define the distances between data points.

The *variational modeling* method describes the design in terms of a sketch that can later be readily converted to a 3D mathematical model with set dimensions (Fig. 20). If the designer changes the design, the model must then be completely recalculated. This approach is quite flexible because it takes the dimension-driven approach of handling equations sequentially and makes it nonsequential. Dimensions can then be modified in any order, making it well suited for use early in the design process when the design geometry might change dramatically. Variational modeling also saves computational time (thus increasing the run speed of the program) by eliminating the need to solve any irrelevant equations. Variational sketching involves creating 2D profiles of the design that can represent end views and cross sections. Using this approach, the designer typically focuses on creating the desired shape with little regard toward dimensional parameters. Once the design shape has been created, a separate dimensioning capability can scale the design to the desired dimensions.

Parametric modeling solves engineering equations between sets of parameters such as size parameters and geometric parameters. Size parameters are dimensions such as the di-

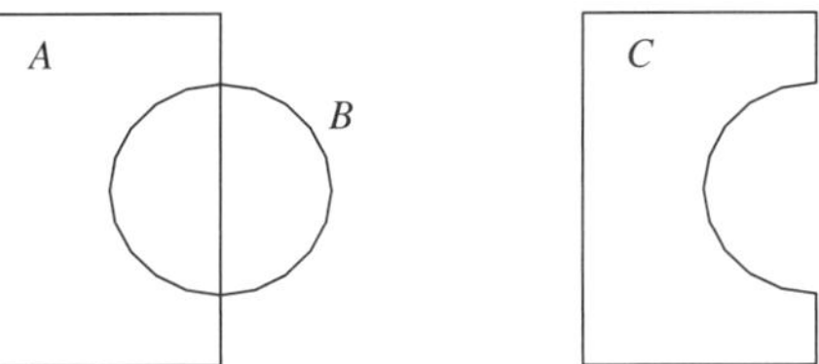

Figure 18 Two-dimensional example of Boolean difference.

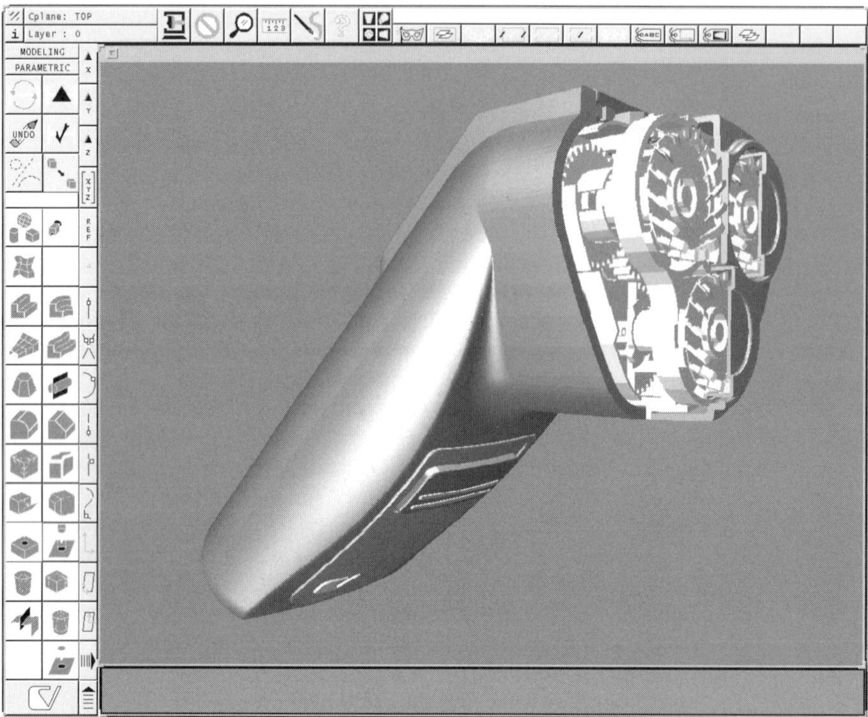

Figure 19 Solid model of an electric shaver design. (Courtesy of ComputerVision, Inc.)

ameter and depth of a hole. Geometric parameters are the constraints, such as tangential, perpendicular, or concentric relationships. Parametric modeling approaches keep a record of operations performed on the design such that relationships between design elements can be inferred and incorporated into later changes in the design, thus making the change with a certain degree of acquired knowledge about the relationships between parts and design elements. For example, using the parametric approach, if a recessed area in the surface of a design should always have a blind hole in the exact center of the area and the recessed portion of the surface is moved, the parametric modeling software will also move the blind hole to the new center. In the example in Fig. 21, if a bolt circle (BC) is concentric with a bored hole and the bored hole is moved, the bolt circle will also move and remain concentric with the bored hole. The dimensions of the parameters may also be modified using parametric modeling. The design is modified through a change in these parameters either internally, within the program, or from an external data source, such as a separate database.

Feature-based modeling allows the designer to construct solid models from geometric features that are industrial standard objects, such as holes, slots, shells, and grooves (Fig. 22). For example, a hole can be defined using a "thru-hole" feature. Whenever this feature is used, independent of the thickness of the material through which the hole passes, the hole will always be open at both sides. In variational modeling, in contrast, if a hole was created in a plane of specified thickness and the thickness was increased, the hole would be a blind hole until the designer adjusts the dimensions of the hole to provide an opening at both ends. The major advantage of feature-based modeling is the maintenance of design intent regardless of dimensional changes in design. Another significant advantage in using a feature-

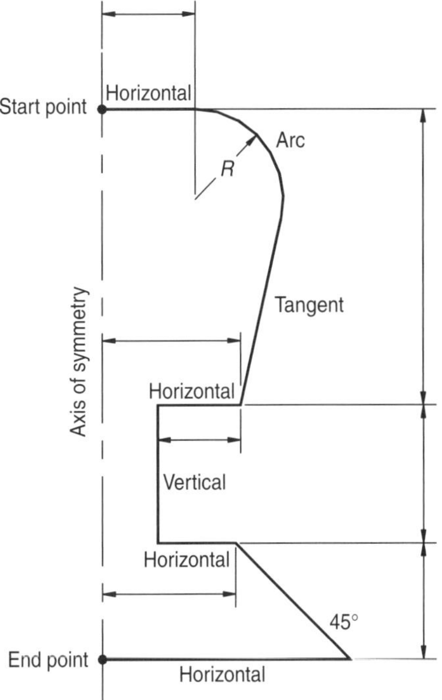

Figure 20 Variational sketch.

based approach is the capability to change many design elements relating to a change in a certain part. For example, if the threading of a bolt is changed, the threading of the associated nut would be changed automatically, and if that bolt design was used more than once in the design, all bolts and nuts could similarly be altered in one step. A knowledge base and inference engine make feature-based modeling more intelligent in some feature-based CAD systems.

Regardless of the modeling approach employed by a software package, there are usually two basic methods for creating 3D solid models. The first method is called constructive solid

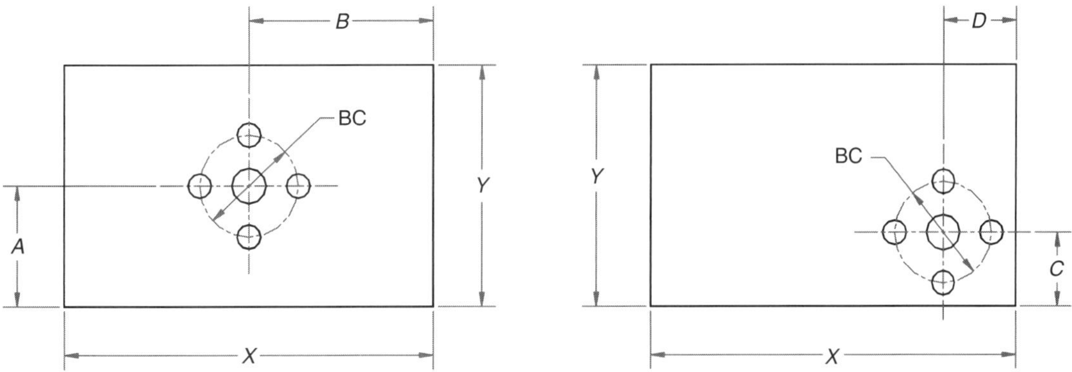

Figure 21 Parametric modeling.

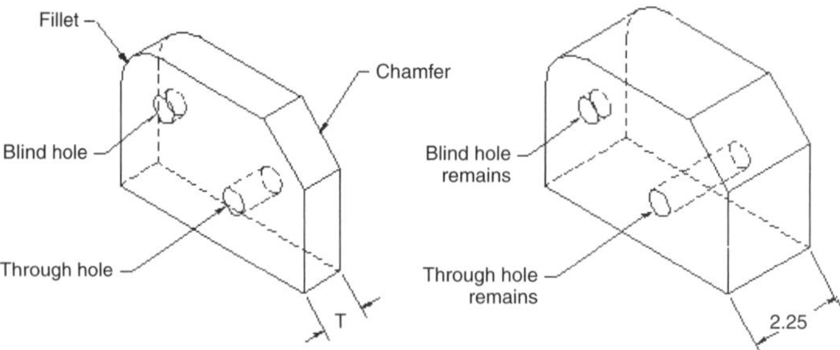

Figure 22 Feature-based modeling.

geometry (CSG). The second method is called boundary representation (B-rep). Most CAD applications offer both methods.

With the CSG method, using defined solid geometries such as those for a cube, sphere, cylinder, prism, etc., the user can combine them by subsequently employing a Boolean logic operator such as union, subtraction, difference, and intersection to generate a more complex part. In three dimensions, the Boolean difference between a cylinder and a torus might appear as follows in Fig. 23.

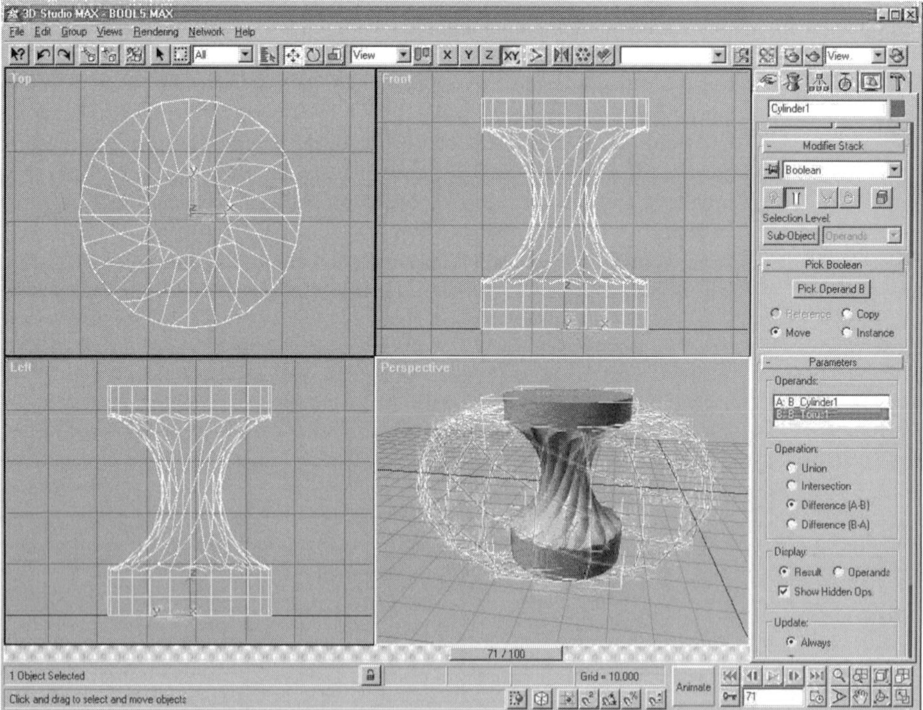

Figure 23 Boolean difference between a cylinder and a torus using Autodesk 3D StudioMax software. (Courtesy of Autodesk, Inc.)

The boundary representation method is a modeling feature in 3D representation. Using this technique, the designer first creates a 2D profile of the part. Then, using a linear, rotational, or compound sweep, the designer extends the profile along a line, about an axis, or along an arbitrary curved path, respectively, to define a 3D image with volume. Figure 24 illustrates the linear, rotational, and compound sweep methods.

Software manufacturers approach solid modeling differently. Nevertheless, every comprehensive solid modeler should have five basic functional capabilities. It should have interactive drawing, a solid modeler, a dimensional constraint engine, a feature manager, and assembly managers.

The drawing capabilities should indicate shapes and profiles quickly and easily, usually with one or two mouse clicks. The purpose of drawing interactively on the screen should be to capture the basic concept information in the computer as efficiently as possible. The solid modeler should be able to combine geometric elements using Boolean logic commands and transform 2D cross sections into 3D models with volume using linear, rotational, and compound sweep methods. A dimensional constraint engine controls relational variables associated with the model such that when the model is changed, the variables change correspondingly. It is the dimensional constraint engine that allows variables to be defined in terms of their relationship to other variables instead of as fixed geometric elements in the design file. The feature manager allows features such as holes, slots, and flanges to be introduced into the design. These features can save time in later iterations of design and represent a major advance in CAD system software in recent years. Assembly management involves the treatment of design units as conglomerate entities, often called cells, as a single functional unit. Assembly cells make management of the design a fairly easy task, since the user can essentially group any elements of interest into a cell and perform selective tasks on the cell as a whole.

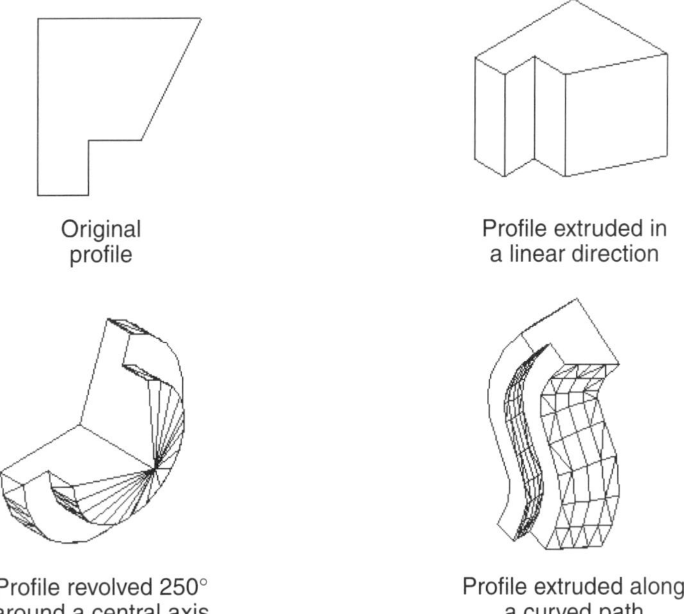

Figure 24 Various common sweep methods in CAD software.

Another significant advance in solid modeling over the past 20 years has been the creation of parts libraries using CAD data files. In early systems, geometries had to be created from within the program. Today, many systems will accept geometries from other systems and software. The Initial Graphics Exchange System (IGES) is an ANSI standard that defines a neutral form for the exchange of information between dissimilar CAD and CAM systems. Significant time can be saved when using models from differing sources. Often, corporations will supply magnetic disks or CD-ROMs with catalogued listings of various parts and products. In this way, the engineer can focus on the major design considerations without constantly redesigning small, common parts such as bearings, bolts, cogs, sprockets, etc. Libraries of parts are also becoming available on the Internet.

Editing Features

CAD systems also offer the engineer powerful editing features which reduce the design time by avoiding the traditionally required manual redrawing. Common editing features are performed on cells of single or conglomerate geometric shape elements. Most CAD systems offer all of the following editing functions as well as others that might be specific to a program being used:

- *Movement.* Allows a cell to be moved to another location on the display screen.
- *Duplication.* Allows a cell to appear at a second location without deleting the original location.
- *Rotation.* Rotates a cell at a given angle about an axis.
- *Mirroring.* Displays a mirror image of the cell about a plane.
- *Deletion.* Removes the cell from the display and the design data file.
- *Removal.* Erases the cell from the display but maintains it in the design data file.
- *Trimming.* Removes any part of the cell extending beyond a defined point, line, or plane.
- *Scaling.* Enlarges or reduces the cell by a specified factor along x, y, and z axes.
- *Offsetting.* Creates a new object that is similar to a selected object at a specified distance.
- *Chamfering.* Connects two nonparallel objects by extending or trimming them to intersect or join with a beveled line.
- *Filleting.* Connects two objects with a smoothly fitted arc of a specified radius.
- *Hatching.* Allows user to edit both hatch boundaries and hatch patterns.

Most of the editing features offered in CAD are transformations performed using algebraic matrix manipulations.

Transformations

Transformation in general refers to the movement or other manipulation of graphical data. Two-dimensional transformations are considered first in order to illustrate the basic concepts. Later, these concepts are applied to geometries with three dimensions.

Two-Dimensional Transformations. To locate a point in a two-axis Cartesian coordinate system, x and y values are specified. This 2D point can be modeled as a 1×2 matrix: (x,y). For example, the matrix $p = (3, 2)$ would be interpreted to be a point which is three units from the origin in the x direction and two units from the origin in the y direction.

This method of representation can be conveniently extended to define a line segment as a 2×2 matrix by giving the x and y coordinates of the two endpoints of the line. The notation would be

$$l = \begin{bmatrix} x_1 & y_1 \\ x_2 & y_2 \end{bmatrix}$$

Using the rules of matrix algebra, a point or line (or other geometric element represented in matrix notation) can be operated on by a transformation matrix to yield a new element.

There are several common transformations: translation, scaling, and rotation.

TRANSLATION. Translation involves moving the element from one location to another. In the case of a line segment, the operation would be

$$\begin{cases} x_1' = x_1 + \Delta x & y_1' = y_1 + \Delta y \\ x_2' = x_2 + \Delta x & y_2' = y_2 + \Delta y \end{cases}$$

where x', y' = coordinates of translated line segment
x, y = coordinates of original line segment
$\Delta x, \Delta y$ = movements in x and y directions, respectively

In matrix notation this can be represented as

$$l' = l + T$$

where

$$T = \begin{bmatrix} \Delta x & \Delta y \\ \Delta x & \Delta y \end{bmatrix}$$

is the translation matrix. Any other geometric element can be translated in space by adding Δx to the current x value and Δy to the current y value of each point that defines the element.

SCALING. The scaling transformation enlarges or reduces the size of elements. Scaling of an element is used to enlarge it or reduce its size. The scaling need not necessarily be done equally in the x and y directions. For example, a circle could be transformed into an ellipse by using unequal x and y scaling factors.

A line segment can be scaled by the scaling matrix as follows:

$$l' = l \times S$$

where

$$S = \begin{bmatrix} \alpha & 0 \\ 0 & \beta \end{bmatrix}$$

is the scaling matrix. Note that the x scaling factor α and y scaling factor β are not necessarily the same. This would produce an alteration in the size of the element by the factor α in the x direction and by the factor β in the y direction. It also has the effect of repositioning the element with respect to the Cartesian system origin. If the scaling factors are less than 1, the size of the element is reduced and it is moved closer to the origin. If the scaling factors are larger than 1, the element is enlarged and removed farther from the origin. Scaling can also occur without moving the relative position of the element with respect to the origin. In this case, the element could be translated to the origin, scaled, and translated back to the original location.

ROTATION. In this transformation, the geometric element is rotated about the origin by an angle θ. For a positive angle, the rotation is in the counterclockwise direction. This accomplishes rotation of the element by the same angle, but it also moves the element. In matrix notation, the procedure would be as follows:

$$l' = l \times R$$

where

$$R = \begin{bmatrix} \cos\theta & \sin\theta \\ -\sin\theta & \cos\theta \end{bmatrix}$$

is the rotation matrix.

Besides rotating about the origin point (0, 0), it might be important in some instances to rotate the given geometry about an arbitrary point in space. This is achieved by first moving the center of the geometry to the desired point and then rotating the object. Once the rotation is performed, the transformed geometry is translated back to its original position.

CONCATENATION. The previous single transformations can be combined as a sequence of transformations. This is called concatenation, and the combined transformations are called concatenated transformations.

During the editing process, when a graphic model is being developed, the use of concatenated transformations is quite common. It would be unusual that only a single transformation would be needed to accomplish a desired manipulation of the image. One example in which combinations of transformations would be required would be to uniformly scale a line l and then rotate the scaled geometry by an angle θ about the origin; the resulting new line is then

$$l' = l \times R \times S$$

where R is the rotation matrix and S is the scaling matrix. A concatenation matrix can then be defined as

$$C = RS$$

Concatenation is a unique feature used in many CAD functions in which a number of transformations are applied to a geometry. The advantage is in the amount of multiplication performed to get the desired picture. In the concatenation procedure, the transformation matrix C is first evaluated and then stored for future use. This eliminates the need of premultiplying the individual matrix to yield the desired transformed geometry.

The above concatenation matrix cannot be used in the example of rotating geometry about an arbitrary point. In this case the sequence would be translation to the origin, rotation about the origin, then translation back to the original location. Note that the translation has to be done separately.

Three-Dimensional Transformations Transformations by matrix methods can be extended to 3D space. The same three general categories defined in the preceding section are considered.

TRANSLATION. The translation matrix for a point defined in three dimensions would be

$$T = (\Delta x, \Delta y, \Delta z)$$

An element would be translated by adding the increments Δx, Δy, and Δz to the respective coordinates of each of the points defining the 3D geometry element.

SCALING. The scaling transformation is given by

$$S = \begin{bmatrix} \alpha & 0 & 0 \\ 0 & \beta & 0 \\ 0 & 0 & \gamma \end{bmatrix}$$

Equal values of α, β, and γ produce a uniform scaling in all three directions.

ROTATION. Rotation in three dimensions can be defined for each of the axes. Rotation about the z axis by an angle θ_z is accomplished by the matrix

$$R_z = \begin{bmatrix} \cos\theta_z & -\sin\theta_z & 0 \\ \sin\theta_z & \cos\theta_z & 0 \\ 0 & 0 & 1 \end{bmatrix}$$

Rotation about the y axis by the angle θ_y is accomplished similarly:

$$R_y = \begin{bmatrix} \cos\theta_y & 0 & \sin\theta_y \\ 0 & 1 & 0 \\ -\sin\theta_y & 0 & \cos\theta_y \end{bmatrix}$$

Rotation about the x axis by the angle θ_x is performed with an analogous transformation matrix:

$$R_x = \begin{bmatrix} 1 & 0 & 0 \\ 0 & \cos\theta_x & -\sin\theta_x \\ 0 & \sin\theta_x & \cos\theta_x \end{bmatrix}$$

All the three rotations about x, y, and z axes can be concatenated to form a rotation about an arbitrary axis.

Graphical Representation of Image Data

As discussed in the introduction to this section, one of the major advantages of CAD is in its ability to display the design interactively on the computer display screen. Wireframe representations, whether in two or three dimensions, can be ambiguous and difficult to understand. Because mechanical and other engineering designs often involve 3D parts and systems, CAD systems that offer 3D representation capabilities have quickly become the most popular in engineering design.

To generate a 2D view from a 3D model, the CAD software must be given information describing the viewpoint of the user. With this information, the computer can calculate angles of view and determine which surfaces of the design would be visible from the given point. The software typically uses surfaces that are closest to the viewer to block out surfaces that would be hidden from view. Then, applying this technique and working in a direction away from the viewer, the software determines which surfaces are visible. The next step determines the virtual distance between viewer and model, allowing those areas outside the boundaries of the screen to be excluded from consideration. Then, the colors displayed on each surface must be determined by combining considerations of the user's preferences for light source and surface color. The simulated light source can play a very important role in realistically displaying the image by influencing the values of colors chosen for the design and by determining reflection and shadow placements. Current high-end CAD software can simulate a variety of light sources, including spot lighting and sunlight, either direct or through some opening such as a door or window (Fig. 25).

Some systems will even allow surface textures to be chosen and displayed. Once these determinations have been made, the software calculates the color and value for each pixel

Figure 25 Vase design rendered with directed spot lighting and shadows in Autodesk 3D StudioMax software. (Courtesy of Autodesk, Inc.)

in the raster-scan display terminal. Since these calculations are computationally intensive, the choice of hardware is often just as important as the software used when employing solid-modeling programs with advanced surface representation features.

9 CAD STANDARDS AND TRANSLATORS

For CAD applications to run across systems from various vendors, four main formats facilitate this data exchange:

- Initial Graphics Exchange Specification
- Standard for the Exchange of Products
- Drawing Exchange Format
- American Committee for Interoperable Systems

9.1 IGES (Initial Graphics Exchange Specification)

IGES is an ANSI standard for the digital representation and exchange of information between CAD/CAM systems. Two-dimensional geometry and 3D CSG can be translated into IGES format. New versions of IGES also support B-rep solid-modeling capabilities. Common translator (IGES-in and IGES-out) functions available in the IGES library include IGES file parsing and formatting; general entity manipulation routines; common math utilities for matrix, vector, and other applications; and a robust set of geometry conversion routines and linear approximation facilities.

9.2 STEP (Standard for the Exchange of Products)

STEP is an international standard. It provides one natural format that can apply to CAD data throughout the life cycle of a product. STEP offers features and benefits that are absent from IGES. STEP is a collection of standards. The user can pull out an IGES specification and get all the data required in one document. STEP can also transfer B-rep solids between CAD systems. STEP differs from IGES in how it defines data. In IGES, the user pulls out the specification, reads it, and implements what it says. In STEP, the implementor takes the definition and runs it through a special compiler, which then delivers the code. This process assures that there is no ambiguous understanding of data among implementors. The conformance testing for STEP will eventually be built into the standard.

9.3 DXF (Drawing Exchange Format)

DXF, developed by Autodesk for AutoCAD software, is the de facto standard for exchanging CAD/CAM data on a PC-based system. Only 2D drawing information can be converted into DXF either in ASCII or in binary format.

9.4 ACIS (American Committee for Interoperable Systems)

The ACIS modeling kernel is a set of software algorithms used for creating solid-modeling packages. Software developers license ACIS routines from the developer, Spatial Technology, to simplify the task of writing new solid modelers. The key benefit of this approach is that models created using software based on ACIS should run unchanged with other brands of ACIS-based modelers. This eliminates the need to use IGES translators for transferring model data back and forth among applications. ACIS-based packages have become commercially available for CAD/CAM and FEA software packages. Output files from ACIS have the extension ".SAT."

9.5 Analysis Software

An important part of the design process is the simulation of the performance of a designed device. A fastener is designed to work under certain static or dynamic loads. The temperature distribution in a CPU chip may need to be calculated to determine the heat transfer behavior and possible thermal stress. Turbulent flow over a turbine blade controls cooling but may induce vibration. Whatever the device being designed, there are many possible influences on the device's performance.

The load types listed above can be calculated using FEA. The analysis divides a given domain into smaller, discrete fundamental parts called elements. An analysis of each element is then conducted using the required mathematics. Finally, the solution to the problem as a whole is determined through an aggregation of the individual solutions of the elements. In this manner, complex problems can be solved by dividing the problem into smaller and simpler problems upon which approximate solutions can be applied. General-purpose FEA software programs have been generalized such that users do not need to have detailed knowledge of FEA.

A finite-element model can be thought of as a system of solid blocks (elements) assembled together. Several types of basic geometric elements are available in the finite-element library. Well-known general-purpose FEA packages, such as NASTRAN and ANSYS, provide an element library.

To demonstrate the concept of FEA, a two-dimensional bracket is shown (see Fig. 26) divided into quadrilateral elements each having four nodes. Elements are joined to each other at nodal points. When a load is applied to the structure, all elements deform until all forces balance. For each element in the model, equations can be written which relate displacement and forces at the nodes. Each node has a potential of displacement in the x and y directions under F_x and F_y (x and y components of the nodal force) so that one element needs eight equations to express its displacement. The displacements and forces are identified by a coordinate numbering system for recognition by the computer program. For example, d^{I}_{xi} is the displacement in the x direction for element I at node i, while d^{I}_{yi} is the displacement in the y direction for the same node in element I. Forces are identified in a similar manner, so that F^{I}_{xi} is the force in the x direction for element I at the node i.

A set of equations relating displacements and forces for the element I should take the form of the basic spring equation, $F = kd$:

$$
\begin{aligned}
k_{11}d^{\mathrm{I}}_{xi} + k_{12}d^{\mathrm{I}}_{yi} + k_{13}d^{\mathrm{I}}_{xj} + k_{14}d^{\mathrm{I}}_{yj} + k_{15}d^{\mathrm{I}}_{xk} + k_{16}d^{\mathrm{I}}_{yk} + k_{17}d^{\mathrm{I}}_{xl} + k_{18}d^{\mathrm{I}}_{yl} &= F^{\mathrm{I}}_{xi} \\
k_{21}d^{\mathrm{I}}_{xi} + k_{22}d^{\mathrm{I}}_{yi} + k_{23}d^{\mathrm{I}}_{xj} + k_{24}d^{\mathrm{I}}_{yj} + k_{25}d^{\mathrm{I}}_{xk} + k_{26}d^{\mathrm{I}}_{yk} + k_{27}d^{\mathrm{I}}_{xl} + k_{28}d^{\mathrm{I}}_{yl} &= F^{\mathrm{I}}_{yi} \\
k_{31}d^{\mathrm{I}}_{xi} + k_{32}d^{\mathrm{I}}_{yi} + k_{33}d^{\mathrm{I}}_{xj} + k_{34}d^{\mathrm{I}}_{yj} + k_{35}d^{\mathrm{I}}_{xk} + k_{36}d^{\mathrm{I}}_{yk} + k_{37}d^{\mathrm{I}}_{xl} + k_{38}d^{\mathrm{I}}_{yl} &= F^{\mathrm{I}}_{xj} \\
k_{41}d^{\mathrm{I}}_{xi} + k_{42}d^{\mathrm{I}}_{yi} + k_{43}d^{\mathrm{I}}_{xj} + k_{44}d^{\mathrm{I}}_{yj} + k_{45}d^{\mathrm{I}}_{xk} + k_{46}d^{\mathrm{I}}_{yk} + k_{47}d^{\mathrm{I}}_{xl} + k_{48}d^{\mathrm{I}}_{yl} &= F^{\mathrm{I}}_{yj} \\
k_{51}d^{\mathrm{I}}_{xi} + k_{52}d^{\mathrm{I}}_{yi} + k_{53}d^{\mathrm{I}}_{xj} + k_{54}d^{\mathrm{I}}_{yj} + k_{55}d^{\mathrm{I}}_{xk} + k_{56}d^{\mathrm{I}}_{yk} + k_{57}d^{\mathrm{I}}_{xl} + k_{58}d^{\mathrm{I}}_{yl} &= F^{\mathrm{I}}_{xk} \\
k_{61}d^{\mathrm{I}}_{xi} + k_{62}d^{\mathrm{I}}_{yi} + k_{63}d^{\mathrm{I}}_{xj} + k_{64}d^{\mathrm{I}}_{yj} + k_{65}d^{\mathrm{I}}_{xk} + k_{66}d^{\mathrm{I}}_{yk} + k_{67}d^{\mathrm{I}}_{xl} + k_{68}d^{\mathrm{I}}_{yl} &= F^{\mathrm{I}}_{yk} \\
k_{71}d^{\mathrm{I}}_{xi} + k_{72}d^{\mathrm{I}}_{yi} + k_{73}d^{\mathrm{I}}_{xj} + k_{74}d^{\mathrm{I}}_{yj} + k_{75}d^{\mathrm{I}}_{xk} + k_{76}d^{\mathrm{I}}_{yk} + k_{77}d^{\mathrm{I}}_{xl} + k_{78}d^{\mathrm{I}}_{yl} &= F^{\mathrm{I}}_{xl} \\
k_{81}d^{\mathrm{I}}_{xi} + k_{82}d^{\mathrm{I}}_{yi} + k_{83}d^{\mathrm{I}}_{xj} + k_{84}d^{\mathrm{I}}_{yj} + k_{85}d^{\mathrm{I}}_{xk} + k_{86}d^{\mathrm{I}}_{yk} + k_{87}d^{\mathrm{I}}_{xl} + k_{88}d^{\mathrm{I}}_{yl} &= F^{\mathrm{I}}_{yl}
\end{aligned}
$$

The k parameters are stiffness coefficients that relate the nodal deflections and forces. They are determined by the governing equations of the problem using given material properties such as Young's modulus and Poisson's ratio and from element geometry.

The set of equations can be written in matrix form for ease of operation as follows:

$$
\begin{Bmatrix}
k_{11} & k_{12} & k_{13} & k_{14} & k_{15} & k_{16} & k_{17} & k_{18} \\
k_{21} & k_{22} & k_{23} & k_{24} & k_{25} & k_{26} & k_{27} & k_{28} \\
k_{31} & k_{32} & k_{33} & k_{34} & k_{35} & k_{36} & k_{37} & k_{38} \\
k_{41} & k_{42} & k_{43} & k_{44} & k_{45} & k_{46} & k_{47} & k_{48} \\
k_{51} & k_{52} & k_{53} & k_{54} & k_{55} & k_{56} & k_{57} & k_{58} \\
k_{61} & k_{62} & k_{63} & k_{64} & k_{65} & k_{66} & k_{67} & k_{68} \\
k_{71} & k_{72} & k_{73} & k_{74} & k_{75} & k_{76} & k_{77} & k_{78} \\
k_{81} & k_{82} & k_{83} & k_{84} & k_{85} & k_{86} & k_{87} & k_{88}
\end{Bmatrix}
\times
\begin{Bmatrix}
d^{\mathrm{I}}_{xi} \\ d^{\mathrm{I}}_{yi} \\ d^{\mathrm{I}}_{xj} \\ d^{\mathrm{I}}_{yj} \\ d^{\mathrm{I}}_{xk} \\ d^{\mathrm{I}}_{yk} \\ d^{\mathrm{I}}_{xl} \\ d^{\mathrm{I}}_{yl}
\end{Bmatrix}
=
\begin{Bmatrix}
F^{\mathrm{I}}_{xi} \\ F^{\mathrm{I}}_{yi} \\ F^{\mathrm{I}}_{xj} \\ F^{\mathrm{I}}_{yj} \\ F^{\mathrm{I}}_{xk} \\ F^{\mathrm{I}}_{yk} \\ F^{\mathrm{I}}_{xl} \\ F^{\mathrm{I}}_{yl}
\end{Bmatrix}
$$

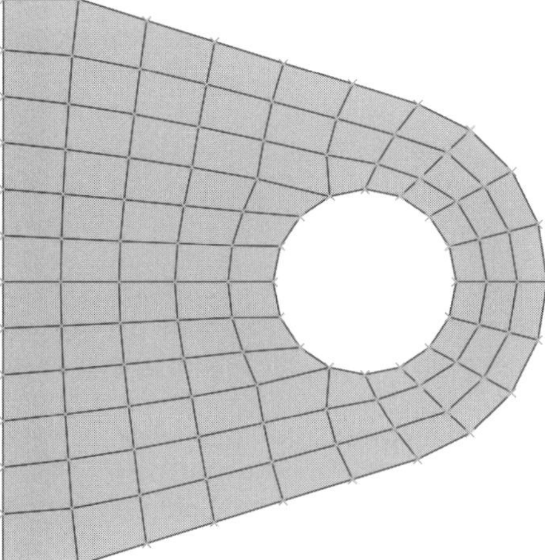

Figure 26 Two-dimensional FEA model of a bracket.

When a structure is modeled, individual sets of matrix equations are automatically generated for each element. The elements in the model share common nodes so that individual sets of matrix equations can be combined into a global set of matrix equations. This global set relates all of the nodal deflections to the nodal forces. Nodal deflections are solved simultaneously from the global matrix. When displacements for all nodes are known, the state of deformation of each element is known and stress can be determined through stress–strain relations.

For a 2D structure problem each node displacement has three degrees of freedom: one translational in each of x and y directions and one rotational in the (x,y) plane. In a 3D structure problem the displacement vector can have up to six degrees of freedom for each nodal point. Each degree of freedom at a nodal point may be unconstrained (unknown) or constrained. The nodal constraint can be given as a fixed value or a defined relation with its adjacent nodes. One or more constraints must be given prior to solving a structure problem. These constraints are referred to as boundary conditions.

FEA obtains stresses, temperatures, velocity potentials, and other desired unknown variables in the analyzed model by minimizing an energy function. The law of conservation of energy is a well-known principle of physics. It states that unless atomic energy is involved, the total energy of a system must be zero. Thus, the finite-element energy functional must equal zero.

The finite-element method obtains the correct solution for any analyzed model by minimizing the energy functional. Thus, the obtained solution satisfies the law of conservation of energy.

The minimum of the functional is found by setting to zero the derivative of the functional with respect to the unknown nodal point potential. It is known from calculus that the minimum of any function has a slope or derivative equal to zero. Thus, the basic equation for FEA is

$$\frac{dF}{dp} = 0$$

where F is the functional and p is the unknown nodal point potential to be calculated. The finite-element method can be applied to many different problem types. In each case, F and p vary with the type of problem.

9.6 Problem Types

Linear Statics

Linear static analysis represents the most basic type of analysis. The term *linear* means that the stress is proportional to strain (i.e., the materials follow Hooke's law). The term *static* implies that forces do not vary with time, or that time variation is insignificant and can therefore be safely ignored.

Assuming the stress is within the linear stress–strain range, a beam under constant load can be analyzed as a linear static problem. Another example of linear statics is a steady-state temperature distribution within a constant material property structure. The temperature differences cause thermal expansion, which in turn induces thermal stress.

Buckling

In linear static analysis, a structure is assumed to be in a state of stable equilibrium. As the applied load is removed, the structure is assumed to return to its original, undeformed position. Under certain combinations of loadings, however, the structure continues to deform without an increase in the magnitude of loading. In this case the structure has buckled or become unstable. For elastic, or linear, buckling analysis, it is assumed that there is no yielding of the structure and that the direction of applied forces does not change.

Elastic buckling incorporates the effect of differential stiffness, which includes higher order strain–displacement relationships that are functions of the geometry, element type, and applied loads. From a physical standpoint, the differential stiffness represents a linear approximation of softening (reducing) the stiffness matrix for a compressive axial load and stiffening (increasing) the stiffness matrix for a tensile axial load.

In buckling analysis, eigenvalues are solved. These are scaling factors used to multiply the applied load in order to produce the critical buckling load. In general, only the lowest buckling load is of interest, since the structure will fail before reaching any of the higher order buckling loads. Therefore, usually only the lowest eigenvalue needs to be computed.

Normal Modes

Normal-mode analysis computes the natural frequencies and mode shapes of a structure. Natural frequencies are the frequencies at which a structure will tend to vibrate if subjected to a disturbance. For example, the strings of a piano are each tuned to vibrate at a specific frequency. The deformed shape at a specific natural frequency is called the mode shape. Normal-mode analysis is also called real-eigenvalue analysis.

Normal-mode analysis forms the foundation for a thorough understanding of the dynamic characteristics of the structure. In static analysis the displacements are the true physical displacements due to the applied loads. In normal-mode analysis, because there is no applied load, the mode shape components can all be scaled by an arbitrary factor for each mode.

Nonlinear Statics

Nonlinear structural analysis must be considered if large displacements occur with linear materials (geometric nonlinearity), or structural materials that behave in a nonlinear stress–strain relationship (material nonlinearity), or a combination of large displacements and nonlinear stress–strain effects occurs. An example of geometric nonlinear statics is shown when a structure is loaded above its yield point. The structure will then tend to be less stiff and permanent deformation will occur, and Hooke's law will not be applicable anymore. In material nonlinear analysis, the material stiffness matrix will change during the computation. Another example of nonlinear analysis includes the contacting problem, where a gap may appear and/or sliding occurs between mating components during load application or removal.

Dynamic Response

Dynamic response in general consists of frequency response and transient response. Frequency response analysis computes structural response to steady-state oscillatory excitation. Examples of oscillatory excitation include rotating machinery, unbalanced tires, and helicopter blades. In frequency response, excitation is explicitly defined in the frequency domain. All of the applied forces are known at each forcing frequency. Forces can be in the form of applied forces and/or enforced motions. The most common engineering problem is to apply steady-state sinusoidally varying loads at several points on a structure and determine its response over a frequency range of interest. Transient response analysis is the most general method for computing forced dynamic response. The purpose of a transient response analysis is to compute the behavior of a structure subjected to time-varying excitation. All of the forces applied to the structure are known at each instant in time. The important results obtained from a transient analysis are typically displacements, velocities, and accelerations of the structure as well as the corresponding stresses.

10 APPLICATIONS OF CAD

Computer-aided design has been presented in terms of its applicability to design, the hardware and software used, and its capabilities as an entity unto itself. The use of CAD data in conjunction with specialized applications is now reviewed. These applications fall outside the realm of CAD software in a strict sense; however, they provide opportunities for the designer to use the data generated through CAD in new and innovative techniques that can similarly affect design efficiency. Many of the items discussed in this section apply to the evaluative stage of design. Some of the basic analytical methods that can be used in CAD to optimize designs have already been presented. Options open to the design engineer using information from a CAD database and special applications are now presented.

The following is a list of the minimum recommended system requirements to run two CAD packages:

ALGOR FEA

- CPU—P4 with multiple processors
- OS—Windows 2000, XP
- RAM—512 MB
- HDD—2 GB or more
- Graphics—32 MB
- Browser—IE 5.5 or 6
- Sound card and speakers
- CD-ROM

AutoCAD 2004

- CPU—P4 2 GHz
- OS—Windows 2000, XP
- RAM—256 MB

HDD—300 MB or more

Graphics—1024 X 768 True Color

Browser—IE 6

CD-ROM

10.1 Optimization Applications

As designs become more complex, engineers need fast, reliable tools. Over the last 20 years, FEA has become the major tool used to identify and solve design problems. Increased design efficiency provided by CAD has been augmented by the application of finite-element methods to analysis, but engineers still often use a trial-and-error method for correcting the problems identified through FEA. This method inevitably increases the time and effort associated with design because it increases the time needed for interaction with the computer, and solution possibilities are often limited by the designer's personal experiences.

Design optimization seeks to eliminate much of this extra time by applying a logical mathematical method to facilitate modification of complex designs. Optimization strives to minimize or maximize a characteristic, such as weight or physical size, which is subjected to constraints on one or more parameters. The size, shape, or both determine the approach used to optimize a design. Optimizing the size is usually easier than optimizing the shape of a design. The geometry of a plate does not significantly change by optimizing its thickness. On the other hand, optimizing a design parameter such as the radius of a hole does change the geometry during shape optimization.

Optimization approaches were difficult to implement in the engineering environment because the process was somewhat academic in nature and not viewed as easily applicable to design practices. However, if viewed as a part of the process itself, optimization techniques can be readily understood and implemented in the design process. Iterations of the design procedure occur as they normally do in design up to a point. At that point, the designer implements the optimization program. The objectives and constraints upon the optimization must first be defined. The optimization program then evaluates the design with respect to the objectives and constraints and makes automated adjustments in the design. Because the process is automatic, engineers should have the ability to monitor the progress of the design during optimization, stop the program if necessary, and begin again.

The power of optimization programs is largely a function of the capabilities of the design software used in earlier stages. Two- and three-dimensional applications require automatic and parametric meshing capabilities. Linear static, natural frequencies, mode shapes, linearized buckling, and steady-state analyses are required for other applications. Because the design geometry and mesh can change during optimization iterations, error estimate and adaptive control must be included in the optimization program. Also, when separate parts are to be assembled and analyzed as a whole, it is often helpful to the program to connect different meshes and element types without regard to nodal or elemental interface matches.

Preliminary design data are used to meet the desired design goals through evaluation, remeshing, and revision. Acceptable tolerances must then be entered along with imposed constraints on the optimization. The engineer should be able to choose from a large selection of design objectives and behavior constraints and use these with ease. Also, constraints from a variety of analytical procedures should be supported so that optimization routines can use the data from previously performed analyses.

Although designers usually find optimization of shape more difficult to perform than of size, the use of parametric modeling capabilities in some CAD software minimizes this difficulty. Shape optimization is an important tool in many industries, including shipbuilding, aerospace, and automotive manufacturing. The shape of a model can be designed using any number of parameters, but as few as possible should be used for the sake of simplicity. If the designer cannot define the parameters, neither design nor optimization can take place. Often, the designer will hold a mental note on the significance of each parameter. Therefore, designer input is crucial during an optimization run.

10.2 Virtual Prototyping

The creation of physical models for evaluation can often be time consuming and provide limited productivity. By employing kinematic and dynamic analyses on a design within the computer environment, time is saved, and often the result of the analysis is more useful than experimental results from physical prototypes. Physical prototyping often requires a great deal of manual work, not only to create the parts of the model, but also to assemble them and apply the instrumentation needed. Virtual prototyping uses kinematic and dynamic analytical methods to perform many of the same tests on a design model. The inherent advantage of virtual prototyping is that it allows the engineer to fine-tune the design before a physical prototype is created. When the prototype is eventually fabricated, the designer is likely to have better information with which to create and test the model.

Physical models can provide the engineer with valuable design data, but the time required to create a physical prototype is long and must be repeated often through iterations of the process. A second disadvantage is that through repeated iterations the design is usually changed, so that time is lost in the process when parts are reconstructed as a working model. Too often, the time invested in prototype construction and testing reveals less useful data than expected.

Virtual prototyping of a design is one possible solution to the problems of physical prototyping. Virtual prototyping employs computer-based testing so that progressive design changes can be incorporated quickly and efficiently into the prototype model. Also, with virtual prototyping, tests can be performed on the system or its parts in such a way that might not be possible in a laboratory setting. For example, the instrumentation required to test the performance of a small part in a system might disrupt the system itself, thus denying the engineer the accurate information needed to optimize the design. Virtual prototyping can also apply forces to the design that would be impossible to apply in the laboratory. For example, if a satellite is to be constructed, the design should be exposed to zero gravity to properly simulate its performance.

Prototyping and testing capabilities have been enhanced by rapid prototyping systems with the ability to convert CAD data quickly into solid, full-scale models that can be examined and tested. The major advantage of rapid prototyping is in the ability for the design to be seen and felt by the designer and less technically adept personnel, especially when aesthetic considerations must come into play. While rapid prototyping will be discussed further below, even this technology is somewhat limited in testing operations. For example, in systems with moving parts, joining rapid prototype model parts can be difficult and time consuming. With a virtual prototyping system, connections between parts can be made with one or two simple inputs. Since the goal is to provide as much data in as little time as possible, use of virtual prototyping before a prototype is fabricated can strongly benefit the design project.

Engineers increasingly perform kinematic and dynamic analyses on a virtual prototype because a well-designed simulation leads to information that can be used to modify design parameters and characteristics that might not have otherwise been considered. Kinematic and dynamic analysis methods apply the laws of physics to a computerized model in order to analyze the motion of parts within the system and evaluate the overall interaction and performance of the system as a whole. At one time a mainframe computer was required to perform the necessary calculations to provide a realistic motion simulation. Today, microcomputers have the computational speed and memory capabilities necessary to perform such simulations on the desktop.

One advantage of kinematic/dynamic analysis software is that it allows the engineer to deliberately overload forces on the model. Because the model can be reconstructed in an instant, the engineer can take advantage of the destructive testing data. Physical prototypes would have to be fabricated and reconstructed every time the test was repeated. There are many situations in which physical prototypes must be constructed, but those situations can often be made more efficient and informative by the application of virtual prototyping analyses.

10.3 Rapid Prototyping

One of the most recent applications of CAD technology has been in the area of rapid prototyping. Physical models traditionally have the characteristic of being one of the best evaluative tools for influencing the design process. Unfortunately, they have also represented the most time-consuming and costly stage of the design process. Rapid prototyping addresses this problem, combining CAD data with sintering, layering, or deposition techniques to create a solid physical model of the design or part. The rapid prototyping industry is currently developing technology to enable the small-scale production of real parts as well as molds and dies that can then be used in subsequent traditional manufacturing methods (Fig. 27). These two goals are causing the industry to become specialized into two major sectors. The first sector aims to create small rapid prototyping machines that one day might become as common in the design office as printers and plotters are today. The second branch of the

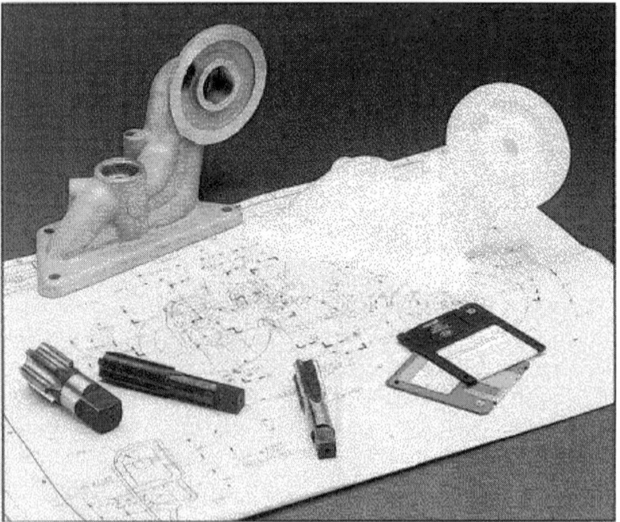

Figure 27 Rapid prototype (R) and part cast from prototype (L). (Courtesy of ProtoCam, Inc.)

rapid prototyping industry is specializing in the production of highly accurate, structurally sound parts to be used in the manufacturing process.

Stereolithography

A stereolithography machine divides a 3D CAD model into slices as thin as 0.0025 in. and sends the information to an ultraviolet laser beam. The laser traces the slice onto a container of photocurable liquid polymer, crosslinking (solidifying) the polymer into a layer of resin. The first layer is then lowered by the height of the next slice. The process is then repeated, and with each repetition, the solid resin is lowered by an increment equal to the height of the next slice until the prototype has been completed. The workspace on one large stereolithography machine has a workspace of 20 × 20 × 20 in.

Early stereolithographed parts were made of acrylate resins using the "Tri-Hatch" build style. This resulted in fabricated parts being very brittle, with rough surface finishes and significantly less accuracy than provided by today's stereolithography machines. Since 1990, advances in hardware, software, polymers, and processing methods have resulted in improved accuracy. The standard measurement for accuracy used in stereolithography applications is the $\varepsilon(90)$ value, which indicates the degree of 90th percentile error. Early machines were capable of an $\varepsilon(90)$ accuracy of about 400 μm. In 1990, a new technique called the "Weave" build style reduced the $\varepsilon(90)$ accuracy to 300 μm. The increased accuracy led the application of the models for verifying designs in checking for interference, tolerance buildups, and other potential design flaws. The next advance in stereolithography at 3-D Systems was the release of the "Star-Weave" structure, when $\varepsilon(90)$ reached 200 μm. At this accuracy, engineers could begin to use stereolithographed parts in iterations of the design process and for optimization routines. There are some applications for which computational simulations are simply not accurate enough. Airflow through an inlet manifold on an automobile engine is one example of such an application. Chrysler Corporation, for example, has used stereolithography to create manifold parts and subsequently tested them in the laboratory to identify the optimal design.

All of the benefits realized from the fabrication of plastic parts could potentially be greater with the ability to use the technology to create metal parts. The initial focus at 3-D Systems was to use the stereolithographed part in an investment casting system. The problem was that early stereolithographs were solid, causing thermal expansion to place stress on the ceramic shell, breaking the mold. The problem was addressed with the QuickCast build geometry. The structure of the stereolithograph is not solid, but about two-thirds hollow, with an open lattice structure. The internal lattice provides the structural strength, and the somewhat hollow properties of the model generate a smooth surface definition. The key to investment casting using this approach is to ensure that any liquid polymer left within the lattice structure of the pattern has an escape route to avoid its solidification to a thickness that would cause enough thermal expansion to break the ceramic mold. Therefore, a drain hole is usually provided to allow whatever resin is left in the mold to escape before the final UV curing process. A vent hole prevents a complete vacuum in the internal lattice.

A further advance in stereolithography technology was developed by Thomas Pang, an organic chemist at 3-D Systems, who developed an epoxy resin with significant advantages over the acrylate plastics originally used. The acrylates have high viscosity but not very high green strength. The lattice triangles in acrylates could not be too large or they would sag, and the high viscosity of the liquid meant that the resin flowed very slowly through the drain holes in the part. Epoxy resin offers one-tenth the viscosity of acrylates coupled with a fourfold increase in green strength. Also, the linear shrinkage effective upon hardening is decreased with the epoxy technology. Acrylates show 0.6–1.0% shrinkage, while epoxy resin offers only 0.06% linear shrinkage. With the epoxy resin, ε(90) values began to approach 100 μm. For these reasons, epoxy resin is viewed as a major improvement over acrylate systems.

The QuickCast system, first using acrylate plastics and then epoxy resin, has opened up the market to rapid manufacturing of accurate functional prototypes in aluminum, stainless steel, carbon steel, and others. The accuracy of the prototypes keeps growing greater, as shown by an ε(90) value of 91 μm achieved in December 1993 with the SLA-500/30 machine. Current research focus is on using rapid prototyping technology to create investment-cast steel tool molds for low-level manufacturing purposes. Ford Motor Co. implemented such technology in the production of a wiper-blade cover for its midyear 1994 Ford Explorer. Other rapid prototyping companies have also jumped into the rapid tooling market. DTM Corp. of Austin, Texas, has introduced a laser sintering process capable of sintering precursor powders into nylon, polycarbonate, and casting wax parts. The biggest advance at DTM is the use of laser sintering to create metal prototypes. In the process, called RapidTool at DTM, the powder used in its Sinterstation machines is a combination low-carbon steel and thermoplastic binder. The laser uses the data from a CAD file to trace the incremental slices of a part onto the powder, causing the plastic to bind with the metal, holding the shape of the part. A low-temperature furnace burns off the plastic binder, and then the temperature is raised to lightly fuse the steel particles, leaving an internal steel skeleton. The steel skeleton is subsequently infused with copper to provide a composite metal part slightly more than 50% iron by weight. The tools created using RapidTool technology use similar processes as those created using aluminum tooling. Accuracy using RapidTool is projected to reach 0.003 in. on features and 0.010 on any dimension.

Other materials for use with rapid prototyping are currently being tested and implemented by various other companies involved in the growing rapid prototyping industry. The technology presents not only the opportunity but also a realistic opportunity for further automatization in the design and implementation environments. One major factor impeding the implementation of such technology on a large scale is cost. Currently, even desktop systems, such as those from 3-D Systems range, from about $100,000 to $450,000, thus limiting their use in small and mid-size corporations. The cost associated with materials is also high, meaning that virtual prototyping, at least for the time being, can cut the cost of the prototyping and testing process before investing in the fabrication of even a rapid prototype. As with most technologies, however, prices are expected to drop with time, and when they do, it is expected that rapid prototyping technology will become as much of a fixture in the design engineering environment as CAD itself has become.

10.4 Computer-Aided Manufacturing

Although this chapter is primarily about CAD, we would be terribly remiss not to mention CAM in terms of our applications discussion. The two techniques are so integrally related in today's manufacturing environment that, often, one is not mentioned without the other. Acronyms such as CAD/CAM, CADAM (computer-aided design and manufacturing), and CIM (computer-integrated manufacturing) are often used to describe the marriage between the two. In essence, CAM uses data prepared through CAD to streamline the manufacturing process through the use of tools such as computer numerical control and robotics. CIM will be discussed in much further detail in a separate chapter of this handbook, and we refer the reader to Chapter 9 of the Manufacturing and Management volume to foster some basic understanding of this related subject.

BIBLIOGRAPHY

Ali, H., "Optimization for Finite Element Applications," *Mech. Eng.,* Dec. 1994, pp. 68–70.

Amirouche, F., *Principles of Computer Aided Design and Manufacturing,* Prentice-Hall, Englewood Cliffs, NJ, 2004.

Ashley, S., "Prototyping with Advanced Tools," *Mech. Eng.,* June 1994, pp. 48–55.

"Basics of Design Engineering," *Machine Design,* Apr. 6, 1995, pp. 83–126.

"Basics of Design Engineering," *Machine Design,* Apr. 4, 1996, pp. 47–83.

Bertoline, G. R., *Introduction to Graphics Communications for Engineers* (*B.E.S.T. Series*), 2nd ed., Purdue University—West Lafayette, McGraw-Hill Higher Education, New York, 2002.

Bertoline, G. R., E. N. Wiebe, and C. Miller, *Fundamentals of Graphics Communication,* 3rd ed., Purdue University—West Lafayette (Wiebe—North Carolina State University), McGraw-Hill Higher Education, New York, 2002.

Bertoline, G. R., and E. N. Wiebe, *Fundamentals of Graphics Communication,* 4th ed., Purdue University—West Lafayette (Wiebe—North Carolina State University), McGraw-Hill Higher Education, New York, 2005.

"CAD/CAM Industry Report," *Machine Design*, May 23, 1994, pp. 38–98.

Cary, H., and S. Helzer, *Modern Welding Technology,* Prentice-Hall, Englewood Cliffs, NJ, 2004.

Chang, T. C., H. P. Wang, and R. Wysk, *Computer Integrated Manufacturing,* Prentice-Hall, Englewood Cliffs, NJ, 2005.

Computing Encyclopedia, Smart Computing Reference Series, Vols. 1–5, Sandhills Publishing, Lincoln, NE, 2002.

Condoor, S., *Mechanical Design Modeling Using Pro/Engineer,* 1st ed., Parks College of St. Louis University, McGraw-Hill Higher Education, New York, 2002.

Cook, T., and R. Prater, *ABCs of Mechanical Drafting with an Introduction to AutoCAD 2000,* Prentice-Hall, Englewood Cliffs, NJ, 2002.

Deitz, D., "PowerPC: The New Chip on the Block," *Mech. Eng.,* Jan. 1996, pp. 58–62.

Dimarongonas, A. D., *Machine Design, A CAD Approach,* Wiley, New York, 2001.

Dix, M. and P. Riley, *Discovering AutoCAD 2004,* Prentice-Hall, Englewood Cliffs, NJ, 2004.

Duffy, B. A., *AutoCAD 2002 Assistant,* 1st ed., University of Louisville, McGraw-Hill Higher Education, New York, 2003.

Dvorak, P., "Engineering on the Other Personal Computer," *Machine Design,* Oct. 26, 1995, pp. 42–52.

Eggert, R., *Engineering Design,* Prentice-Hall, Englewood Cliffs, NJ, 2005.

"Engineering Drives Document Management," Special Editorial Supplement, *Machine Design,* June 15, 1995, pp. 77–84.

Ethier, S. and Ethier, C., *Instant AutoCAD: Essentials Using AutoCAD 2002.* Prentice-Hall, Englewood Cliffs, NJ, 2004.

Giesecke, F. E., Mitchell, A., Spencer, H., Hill, I. L., and Loving, R., *Engineering Graphics,* 7th ed., Prentice-Hall, Upper Saddle River, NJ, 2000.

Groover, M. P., and E. W. Zimmers, Jr., *CAD/CAM: Computer-Aided Design and Manufacturing,* Prentice-Hall, Englewood Cliffs, NJ, 1984.

Hanratty, P. J., "Making Solid Modeling Easier to Use," *Mech. Eng.,* Mar. 1994, pp. 112–114.

Harrington, D., *Inside AutoCAD 2005,* Prentice-Hall, Englewood Cliffs, NJ, 2004.

Heisel, J. D., D. Short, and C. H. Jensen, *Engineering Draw Fundamental Version w/CDrom 2002,* 5th ed., McGraw-Hill, New York, 2002.

Hodson, W. K. (ed.), *Maynard's Industrial Engineering Handbook,* 4th ed., McGraw-Hill, New York, 1992.

Jensen, C. H., J. D. Heisel, and D. Short, *Engineering Drawing And Design Student Edition 2002,* 6th ed., McGraw-Hill Higher Education, New York, 2002.

Kelley, D. S., *ProENGINEER 2001 Instructor with CD ROM,* 1st ed., Purdue University—West Lafayette, McGraw-Hill Higher Education, New York, 2002.

Kernighan, B. W. and Ritchie, D. M., *The C Programming Language,* 2nd ed., Prentice-Hall, Englewood Cliffs, NJ, 1988.

Krouse, J. K., *What Every Engineer Should Know about Computer-Aided Design and Computer-Aided Manufacturing,* Marcel Dekker, NewYork, 1982.

Leach, J. A., *AutoCAD 2002 Companion: Essentials of AutoCAD Plus Solid Modeling,* 1st ed., University of Louisville, McGraw-Hill Higher Education, New York, 2003a.

Leach, J. A., *MP AutoCAD 2004 Companion: Essentials of AutoCAD Plus Solid Modeling,* 1st ed., University of Louisville, McGraw-Hill Higher Education, New York, 2003b.

Leake, J., M., *Autodesk Inventor w/CD,* 1st ed., University of Illinois—Champaign, McGraw-Hill Higher Education, New York, 2004.

Lee, G., "Virtual Prototyping on Personal Computers," *Mech. Eng.,* July 1995, pp. 70–73.

Lee, K., *Principles of CAD/CAM/CAE Systems,* Addison Wesley Longman, Reading, MA, 1999.

Lueptow, R., and M. Minbiole, *Graphics Concepts with SolidWorks,* Prentice-Hall, Englewood Cliffs, NJ, 2004.

Madsen, D., and T. Shumaker, *Civil Drafting Technology,* Prentice-Hall, Englewood Cliffs, NJ, 2004.

Masson, R., "Parallel and Almost Personal," *Machine Design,* Apr. 20, 1995, pp. 70–76.

McMahon, C. and J. Browne, *CADCAM, Principles, Practice and Manufacturing Management,* 2nd ed, Addison Wesley Longman, Harlow, England, 1998.

Middlebrook, M., *AutoCAD LT2005 for Dummies,* Wiley, Hoboken, NJ, July 2004.

Mitton, M., *Interior Design Visual Presentation: A Guide to Graphics, Models, and Presentation Techniques,* 2nd ed., Wiley, Hoboken, NJ, July 2003.

Norton, R. L., Jr., "Push Information, Not Paper," *Machine Design,* Dec. 12, 1994, pp. 105–109.

Puttre, M., "Taking Control of the Desktop," *Mech. Eng.,* Sept. 1994, pp. 62–66.

Reis, R., *Electronic Project Design and Fabrication,* Prentice-Hall, Englewood Cliffs, NJ, 2005.

Rutenbar, R. A., Gielen, G. G. E., and Antao, B. A., eds., *Computer-Aided Design of Analog Integrated Circuits and Systems,* Wiley, Hoboken, NJ, Apr. 2002.

Saka, T., *AutoCAD for Architecture,* Prentice-Hall, Englewood Cliffs, NJ, 2004.

Sarfraz, Muhammad, ed., *Advances in Geometric Modeling*, Wiley, Hoboken, NJ, Mar., 2004.

Shigley, J. E. and C. R. Mischke, *Mechanical Engineering Design,* 5th ed., McGraw-Hill, New York, 1989.

Spencer, R., and M. Ghausi, *Introduction to Electronic Circuit Design,* Prentice-Hall, Englewood Cliffs, NJ, 2003.

Stallings, W., *Computer Organization and Architecture, Designing for Performance,* 4th ed., Prentice-Hall, Englewood Cliffs, NJ, 1996.

Teschler, L. (ed.), "Why PDM Projects Go Astray," *Machine Design,* Feb. 22, 1996, pp. 78–82.

Wallach, S., and J. Swanson, "Higher productivity with Scalable Parallel Processing," *Mechan. Eng.,* Dec. 1994, pp. 72–74.

CHAPTER 20
DATA EXCHANGE USING STEP

Martin Hardwick
Rensselaer Polytechnic Institute
Troy, New York
and
STEP Tools, Inc.
Troy, New York

1 WHAT IS STEP?

In design and manufacturing, many systems are used to manage technical product data. Each system has its own data formats so the same information has to be entered multiple times into multiple systems, leading to redundancy and errors. Although not unique to manufacturing, it is more acute there because design data are complex and three-dimensional (3D). The National Institute of Standards and Technology has estimated that data incompatibility is a $90 billion problem for the U.S. manufacturing industry.[1]

Over the years many solutions have been proposed. The most successful have been standards for data exchange. The first ones were national and focused on geometric data exchange. They include the Standard d'Exchange et de Transfer (SET) in France, the Verband des Automobilindustrie FlächenSchnittstelle (VDAFS) in Germany, and the Initial Graphics Exchange Specification (IGES) in the United States. Later a grand unifying effort was started by the International Organization for Standardization (ISO) to produce one international standard for all aspects of technical product data, named STEP for the Standard for Product Model Data.[2] The types of systems that use STEP are shown in Fig. 1.

Nearly every major CAD/CAM system now contains a module to read and write data defined by one of the STEP application protocols (APs). In the United States the most commonly implemented protocol is AP-203. This protocol is used to exchange data describing designs represented as solid models and assemblies of solid models. In Europe a very similar protocol called AP-214 performs the same function.

2 STEP APPLICATION PROTOCOLS

Following is a list of the STEP APs as of June 2004:

Part 201 Explicit Drafting
Part 202 Associative Drafting
Part 203 Configuration Controlled Design
Part 204 Mechanical Design Using Boundary Representation

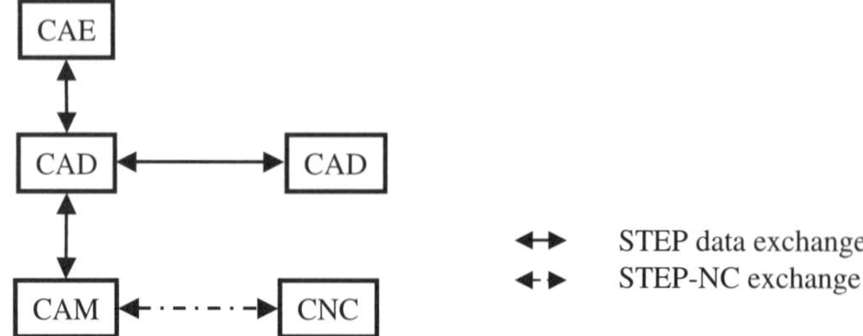

Figure 1 STEP application: CAD, computer-aided design; CAE, computer-aided engineering; CAM, computer-aided manufacturing; CNC, computerized numerical control.

Part 205 Mechanical Design Using Surface Representation
Part 206 Mechanical Design Using Wireframe Representation
Part 207 Sheet Metal Dies and Blocks
Part 208 Life Cycle Product Change Process
Part 209 Design through Analysis of Composite and Metallic Structures
Part 210 Electronic Printed Circuit Assembly, Design and Manufacturing
Part 211 Electronics Test Diagnostics and Remanufacture
Part 212 Electrotechnical Plants
Part 213 Numerical Control Process Plans for Machined Parts
Part 214 Core Data for Automotive Mechanical Design Processes
Part 215 Ship Arrangement
Part 216 Ship Molded Forms
Part 217 Ship Piping
Part 218 Ship Structures
Part 219 Dimensional Inspection Process Planning for CMMs
Part 220 Printed Circuit Assembly Manufacturing Planning
Part 221 Functional Data and Schematic Representation for Process Plans
Part 222 Design Engineering to Manufacturing for Composite Structures
Part 223 Exchange of Design and Manufacturing DPD for Composites
Part 224 Mechanical Product Definition for Process Planning
Part 225 Structural Building Elements Using Explicit Shape Rep
Part 226 Shipbuilding Mechanical Systems
Part 227 Plant Spatial Configuration
Part 228 Building Services
Part 229 Design and Manufacturing Information for Forged Parts
Part 230 Building Structure Frame Steelwork
Part 231 Process Engineering Data
Part 232 Technical Data Packaging
Part 233 Systems Engineering Data Representation
Part 234 Ship Operational Logs, Records and Messages
Part 235 Materials Information for Products
Part 236 Furniture Product and Project
Part 237 Computational Fluid Dynamics
Part 238 Integrated CNC Machining
Part 239 Product Life Cycle Support
Part 240 Process Planning

The ability to support many protocols within one framework is one of the key strengths of STEP. All the protocols are built on the same set of integrated resources (IRs) so they all use the same definitions for the same information. For example, AP-203 and AP-214 use the same definitions for 3D geometry, assembly data, and basic product information. Therefore CAD vendors can support both with one piece of code.

Each AP includes a scope describing its purpose, an activity diagram describing the functions that an engineer needs to perform within that scope, and an application requirement model describing the information requirements of those activities. These information requirements are then mapped into the common set of IRs and the result is a data exchange standard for the activities within the scope.

The ultimate goal is for STEP to cover the entire product life cycle, from conceptual design to final disposal, for all kinds of products. However, it will be a number of years before this goal is reached. The most tangible advantage of STEP to users today is the ability to exchange design data as solid models and assemblies of solid models. Other data exchange standards, such as the newer versions of IGES, also support the exchange of solid models, but less well.

STEP led the way with 3D data exchange by organizing an implementation forum for CAD vendors so that they could continually improve the quality of the solid-model data exchanges. The history of this success is relatively interesting because it shows that the initial reluctance of vendors to implement user-defined standards can be overcome with enough perseverance.

At first, in 1996, there was a significant body of opinion that solid-model geometric data could not be exchanged between systems using a neutral standard. However, in 1997 Ford, Allied Signal, and STEP Tools demonstrated the first successful data exchange of 3D geometry using STEP. Once this basic capability had been demonstrated, a pilot project, called AeroSTEP, was organized by Boeing and its aircraft engine vendors to test the first translators by exchanging data about where an engine fits onto the airframe. This project started out by exchanging simple faceted models but eventually demonstrated the exchange of models with great complexity.

The AeroSTEP project made it clear that the STEP data exchange of solid-model data was both feasible and valuable. As a result, vendor-neutral implementation forums were formed in Europe, the Far East, and the United States, and the quality of the translators was raised to the level that allowed anyone, including small organizations, to use STEP for data exchange of solid models after about 2001.

3 STEP FOR GEOMETRIC DIMENSIONING AND TOLERANCING

Manufacturing needs more than a geometric model to make a part. Many additional specifications are required, most importantly a description of the required tolerances, because this will drive the selection of the manufacturing process and the manufacturing tools used to manufacture the part.

A model of geometric tolerance and dimension (GD&T) is being added to STEP as part of an upgrade to AP-203 called Edition 2. Several iterations have already been made with significant feedback about functionality from the CAD vendors. The initial GD&T models were developed for several different APs with different scopes. A new harmonized model for all the known GD&T requirements was produced in September 2004 and used as the AP-203 Edition 2 Model.

GD&T information qualify geometric items already in the geometric model with plus/minus tolerances values, and reference data planes. There is a lot of sharing between data in the geometry model and data in GD&T model so the data definitions become quite intricate. This makes implementing the GD&T model almost as challenging as implementing the original geometric model. It was difficult to implement the geometry model because it is very object oriented with many advanced, inheritance relationships. The GD&T model is difficult to implement because there are many long data paths where simple numeric values become values with units, then values with units and tolerances, then values with units, tolerances and reference datums and so on.

Consequently, as of the summer of 2005 there are questions as to whether the AP-203 Edition 2 model can be implemented efficiently. Early implementation projects are being formed to show that the implementation is reasonable and feasible. The history of the original AP-203 implementation project is relevant here. It can be predicted that if the first experiments are successful, then an industry segment will work with the vendors to pilot the new capability to a level where it can be used for complex products, and then implementation forums will take it to a level where all users, including small job shops, can take advantage of the new capability.

4 STEP FOR CNC MACHINING

A tool path and material volume removal standard is being developed as an extension to STEP called STEP-NC (see Fig. 1).[3] Many authors have described how CNCs can become more intelligent if they are given more information about the product.[4] STEP is an attractive vehicle for communicating this information because it is so comprehensive and it already captures the necessary geometric data.

STEP-NC defines a CNC part program as a series of operations that remove material defined by features. The features supported include holes, slots, pockets, and volumes defined by 3D surfaces. Each operation contributes to the manufacture of a feature by defining the volume of material to be removed, the tolerances, the type of tool required, and some basic characteristics, such as whether this is a roughing or finishing operation. The operations are then sequenced into a work plan that converts the stock into the final part. The work plan may be sophisticated and include conditional operations that depend on the results of probing operations, and it may be divided into subplans to be executed concurrently on machines that have multiple cutting heads.

STEP-NC has the AP number AP-238 within the STEP framework. A key feature of STEP-NC AP-238 programs is that they are machine and organization independent. If a machine has the underlying capabilities (axes, table size, etc.), then a STEP-NC "compiler" should be able convert the part program into a sequence of tool movements for that machine. This has two significant consequences for industry:

- If parts can always be rapidly manufactured from an AP-238 description, then there is less requirement to keep copies of those parts in the inventory.
- If parts can be made independently of the axis codes, then the same CNC program can be run on many machines.

Figure 2 shows how design data are communicated to manufacturing in current practice. Design creates the specification for a product as a 3D model. Detailing decides the manufacturing requirements for the product by making a drawing. Path planning generates tools paths. Manufacturing controls production. For most industries, the figure is a simplification because other functions such as subcontracting, quality control and downstream maintenance must be considered. Nevertheless it shows the main pipeline for data that goes to the CNC control.

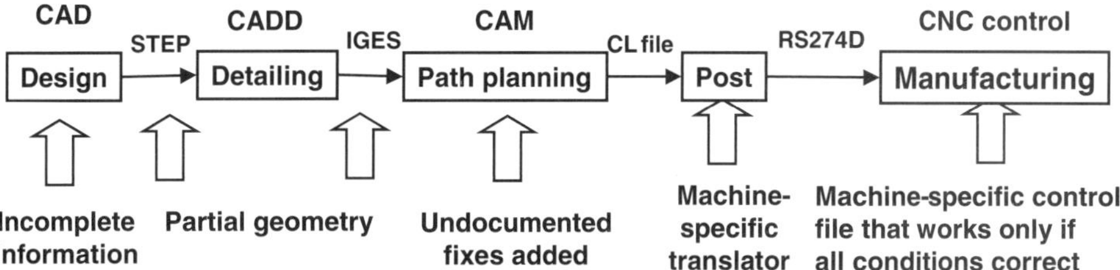

Figure 2 Current design to manufacturing pipeline.

The job of design is performed using a CAD system, the job of detailing is performed using a drawing CADD (CAD draughting) system, the job of path planning is performed using a CAM system, and the job of manufacturing is controlled using a CNC system. In many cases the CAD, CADD, and CAM functions are combined into a single integrated CAD/CAM system, but in all cases the CNC function is performed by a separate system.

Information can be lost in the pipeline because incomplete data are sent from the CAD to the CAM, because fixes to the geometry are made in the CAM and not communicated back to the CAD, because only the surface data are communicated to the post, and, most of all, because the RS274D standard only allows axis movement data to be communicated to the control. This means that no adjustments can be made on the control in response to changes in the available tooling, the control cannot optimize the machining process for the capabilities of the selected machine, and the operator cannot rely on software in the control to check the safety of the setup and the program.

In the new method enterprises can continue to use their existing systems for CAD, CADD, and CAM, but the end result is sent to the CNC as a STEP-NC AP-238 file instead of an RS274D file. Figure 3 shows the modified pipeline. The change is small, but the advantages are significant because all the information required to change the tooling or optimize the planning is included in the AP-238 data exchange file.

There has been considerable early testing of AP-238 as a new interface for defining CNC independent tool paths. This is the concept originally defined for Basic Control Language (BCL) (RS494). However, BCL did not have a means to display the design geometry and tolerances of the part or the features being manufactured in each operation. There is strong reason to believe that if these data are added, then AP-238 CNC machine-independent files can be used as the basis for contracting work between customers and suppliers, and if

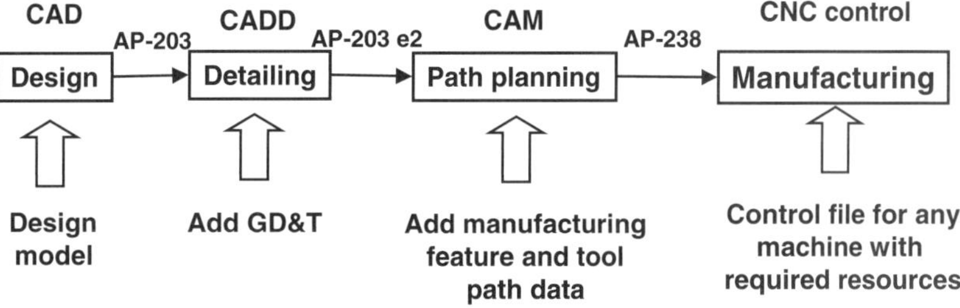

Figure 3 Modified design to manufacturing pipeline.

this is so, then the deployment of AP-238 may be driven by stronger economic forces than those that lead to the implementation of AP-203.

5 FUTURE OF STEP

Despite the many successes of STEP, there is still a question in users' minds about the speed of its development and deployment.[5] Many critics point out correctly that the Extensible Markup Language (XML) standards for e-commerce are being developed much more quickly.

Fundamentally, product model data are different to other kinds of e-commerce data, such as invoices, receipts, etc. The traditional method for communicating product model information is to make a drawing and the traditional method to communicate an invoice is to make a form. When you make a drawing or 3D model, you need to define information with many subtle and complex relationships, and this makes the STEP data exchange problem more difficult.

An XML data format is being developed for STEP, but the STEP architecture requires the information requirements of an AP to be mapped into the common set of IRs. This allows all of the protocols to share the same information and is essential if all of the interfaces shown in Fig. 3 will share and reuse the same set of data. However, the sharing necessarily divides the original data into multiple entities that are not so easy to understand in XML or any other format. This is disappointing because one of the attractions of XML is that it is self-documenting (at least for programmers and domain experts). Therefore, a new level of documentation is required in the STEP data to show how the information requirements have been mapped. The required structures are currently in development, and it is anticipated that STEP will have a self-documenting XML format in the near future.[5]

What has hindered STEP deployment is the commitment of those with the resources necessary to define the standards. Government does not want to take the lead, and industrial companies do not want to fund development of solutions that can also be used by their competitors. Consequently, much work only gets funded in situations of clear and desperate need, such as when the high cost of manufacturing is causing excessive job losses.

The Internet and the World Wide Web broke through this cycle when "killer" applications made the benefits of the new infrastructure clear and compelling for all users. AP-203 made STEP useful by allowing solid models to be exchanged between design systems. AP-238 will make STEP compelling for some users by allowing them to machine parts more efficiently. However, like the early Internet, there will be alternatives that are considered more reliable by other users. The killer application that makes STEP ubiquitous has yet to be identified.

REFERENCES

1. S. B. Brunnermeier and S. A. Martin, "Interoperability Cost Analysis of the U.S. Automotive Supply Chain," Research Triangle Institute, http://www.rti.org/publications/cer/7007-3-auto.pdf, Mar. 1999.
2. "ISO 10303-1:1994 Industrial Automation Systems and Integration Product Data Representation and Exchange—Overview and Fundamental Principles, International Organization for Standardization," Geneva, Switzerland, 1994.

3. "ISO 14649-1:2001 Industrial Automation Systems and Integration Physical Device Control—Part 1: Overview and Fundamental Principles, Draft International Standard," International Organization for Standardization, Geneva, Switzerland, 2001.

4. S. H. Suh, J. H. Cho, and H. D. Hong, "On the Architecture of Intelligent STEP-Compliant CNC," *Int. J. Computer Integrated Manufacturing,* **15**(2), 168–177 (Jan. 2002).

5. M. Hardwick, "On STEP-NC and the Complexities of Product Data Integration," *ACM/ASME Trans. Comput. Inf. Sci. Eng.,* **4**(1) (Mar. 2004), pp. 61–66.

CHAPTER 21
ENGINEERING APPLICATIONS OF VIRTUAL REALITY

Xiaobo Peng and Ming C. Leu
Department of Mechanical and Aerospace Engineering
University of Missouri-Rolla
Rolla, Missouri

1 INTRODUCTION

Since the first virtual reality simulator "Sensorama Simulator" was invented in 1956,[1] the concept of virtual reality has fascinated engineering researchers for over 40 years. Virtual reality systems have developed into a more and more practically useful entity in recent years due to their ability to fulfill the requirements of end users and the availability of more advanced computer hardware. Virtual reality has grown into an exciting field which has already seen the development of numerous, diverse engineering applications.[2]

1.1 What is Virtual Reality?

The term *virtual reality* (VR) has been used by many researchers and described in various ways. A broad definition of virtual reality comes from the Ref. 3: "Virtual Reality is a way for humans to visualize, manipulate and interact with computers and extremely complex data." Burdea and Coiffet[4] described virtual reality as "a high-end user interface that involves real-time simulation and interactions through multiple sensorial channels. These sensorial modalities are visual, auditory, tactile, smell, taste, etc." Jayaram et al.[5] stated that VR is often regarded as an extension of three-dimensional (3D) computer graphics with advanced input and output devices. Generally speaking, the features of VR are (1) it is a medium of communication, (2) it requires physical immersion, (3) it provides synthetic sensory stimulation, (4) it can mentally immerse the user, and (5) it is interactive.[1]

Virtual reality is closely associated with an environment commonly known as a virtual environment (VE). A VE is an interactive graphic display enhanced by special processing and by nonvisual display modalities, such as auditory and haptic feedback, to convince users that they are immersed in a real physical space.[6] According to Kalawsky,[7] VR is the same as VE, but it is more familiar to the public. VR can also be seen as a pinnacle of what is ultimately sought when implementing a VE system.

1.2 Types of VR Systems

Based on factors including display hardware, graphics-rendering algorithms, level of user involvement, and level of integration with the physical world, VR systems can be classified into different categories as described below.[8,9]

Immersive VR

The most common image of VR is that of an immersive VE—typically a user wearing a head-mounted display (HMD) or shutter glasses while viewing a computer monitor or standing inside a spatially immersive display. The concept of immersion is that the VE surrounds the user, wholly or partially. For example, a standard CAVE™ has screens on several (commonly four) sides of a cube to provide the sense of immersion, as shown in Fig. 1.[10]

Desktop VR

Nonimmersive VR systems normally run on standard desktop computers, hence the term *desktop VR*. Desktop VR systems use the same 3D computer graphics as immersive VR systems, but with two key differences. First, the VE does not surround the user; it is seen only on a single screen in front of the user. Second, the user typically navigates through and interacts with the environment using traditional desktop input devices such as a mouse and a keyboard (although specialized 3D input devices may also be used).

Image-Based VR

Displaying a realistic virtual world using only images (instead of computer-generated models) is called image-based rendering. Its basic approach is to manipulate the pixels in images

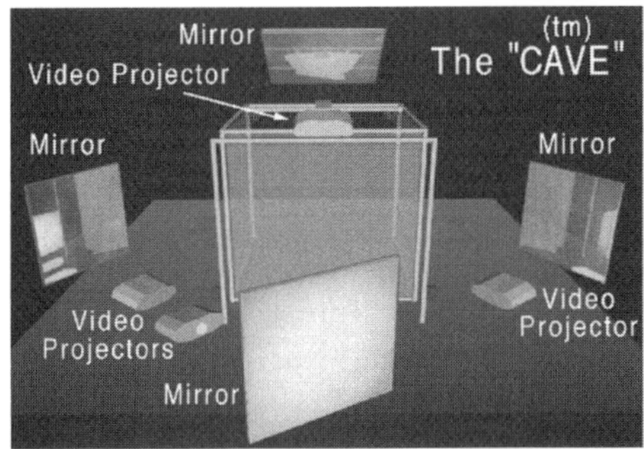

Figure 1 The CAVE system.[9]

to produce the illusion of a 3D scene, rather than to build the 3D scene on a computer. The simplest type of image-based VR is a panorama—a series of images taken with a camera at a single position pointed in multiple directions. Image-based rendering has not yet made a large impact on VR, but it is clearly a way to increase the realism of VEs.

Telepresence

Telepresence is a computer-generated environment consisting of interactive simulations and computer graphics in which a human experiences presence in a remote location. An example is a pilot in a sophisticated simulator which is used to control a real airplane 500 miles away and provides to the pilot visual and other sensory feedback as if the pilot were actually in the cockpit looking out through the windscreen and feeling the turbulence. Other applications of telepresence involve the use of remotely operated vehicles (e.g., robots) to handle dangerous conditions (e.g., nuclear accident sites) or for deep-sea and space exploration.

Augmented Reality

Augmented reality (AR) is the use of a transparent HMD to overlay computer-generated images onto the physical environment. Rapid head tracking is required to sustain the illusion. In the most popular AR systems, the user views the local real-world environment, but the system augments that view with virtual objects. For example, the Touring Machine system[11] acts as a campus information system, assisting a user in finding places and allowing him or her to pose queries about items of interest such as buildings and statues.

Aim of Chapter

The aim of this chapter is to introduce the techniques involved in the design and development of VR systems. The chapter will describe the main techniques and applications of the exiting VR technology, including hardware, software, and engineering applications.

2 VIRTUAL REALITY COMPONENTS

Virtual reality enables the user's immersion into a sensory-rich, interactive experience. A typical VR configuration is shown in Fig. 2. Using input and output sensors, a virtual world is generated and manipulated under the control of both the reality engine and the participant.[12] Software integrates various hardware elements into a coherent system that enables the user to interact with the VE. This chapter reviews the hardware and software tools commonly used for the creation of VEs and related issues. We will break down a VR system into its basic components and explain each one. This review is not intended to be exhaustive because VR hardware and software are constantly changing, with new developments taking place all the time.

2.1 VR Hardware

As shown in Fig. 2, VR hardware plays an extremely important role in a VR system. It provides physical devices that comprise an immersive environment such as CAVE, HMD, etc. Input and output devices transmit the information between the participant and the VE. They provide interactive ways for the information communication. This section will describe the hardware used in various VR systems, including tracking devices, input devices, visual stereo displays, haptic devices, and auditory displays.

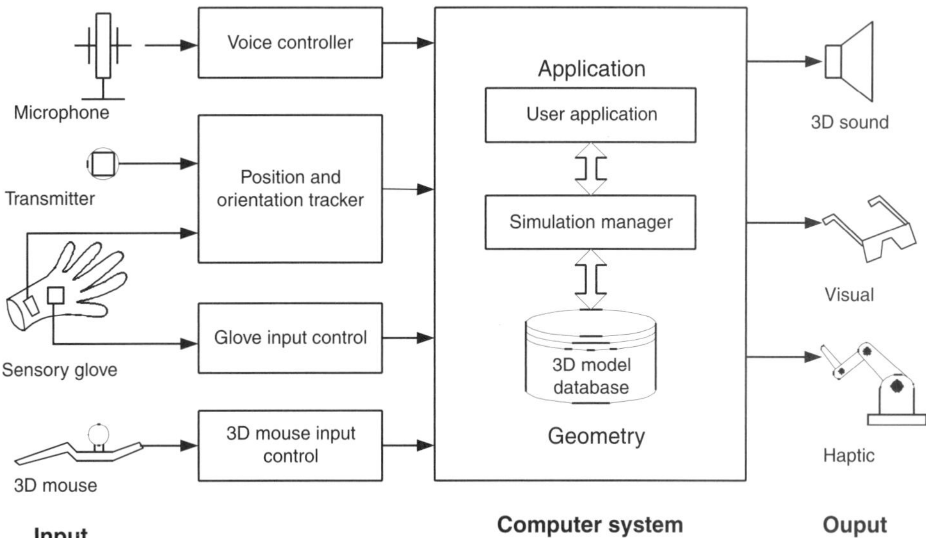

Figure 2 Virtual reality configuration.

Input Devices

The input device is the bridge to connect the real world and the virtual world. It gathers the input information from the user and transmits it to the virtual world. Using different input devices provides an intuitive and interactive environment for the user to immerse in the virtual world.

Tracking Devices. Two types of tracking devices are commonly used in VR systems: position trackers and body trackers. A position tracker is a sensor that reports an object's position (and possibly orientation) and maps it to the object's relative position in the VE. The current position-tracking devices used in VR include electromagnetic, mechanical, optical, videometric, ultrasonic, and neural position-sensing devices (see Table 1).[1]

A body-tracking device monitors the position and action of the participant. Commonly used body-tracking techniques include tracking the head, the hand and fingers, the eyes, the torso, the feet, etc.

Other Input Devices. In addition to the tracking devices described above, 3D mice, joysticks, sensory gloves, voice synthesizers, and force balls can be used as input devices for interaction with the user in the virtual world. Point input devices include the six-degree-of-freedom (6DOF) mouse and the force ball. In addition to functioning as a normal mouse, the 6DOF mouse can report height information as it is lifted in three dimensions. The force ball senses the forces and torques applied by the user in three directions. Figure 3 shows a SpaceMouse, a 6DOF mouse, and a SpaceBall, a force ball device, from 3Dconnexion.[13]

The CyberGlove® is a fully instrumented glove that uses resistive bend-sensing elements to accurately transform hand and finger motions into real-time digital joint-angle data. The CyberTouch® glove, which has tactors in the form of vibrators on each finger and the thumb, can provide the capability of tactile feedback. The CyberGrasp® is a lightweight, force-reflecting exoskeleton that fits over a CyberGlove® and adds resistive force feedback to each

Table 1 Types of Position-Tracking Devices

Type	Product Model	Company	Advantage	Disadvantage
Mechanical tracker	BOOM	FakeSpace	Using encoders and kinematic mechanisms provides very fast and accurate tracking.	Linkages restrict the user to a fixed location in the physical space.
	Phantom	SensAble Technologies		
Electromagnetic tracker	Flock of Birds	Ascension	No line of sight restriction. Wireless systems will reducing encumbrances on the paticipant.	Metal in the environment can cause distortion of data. The range of the generated magnetic field is short.
	IS-600	InterSencse		
	StarTrak8 (wireless)	Polhemus		
Optical tracker	HiBall Tracking System	UNC Tracker Research Group	Makes use of visual information like video camera to track the user.	Occlusion will happen if the tracked person or object and the camera are not clear.
Ultrasonic tracker	Logitech Ultrasonic Tracker	Logitech	Inexpensive because of use of commonly available speakers and microphones.	Limited range; easy to be interrupted in a noisy environment.

Source: Reference 1.

finger. With the CyberGrasp force feedback system, the user is able to feel the size and shape of a computer-generated 3D object in a simulated virtual world. The CyberForce® provides force feedback for the hand and arm. Figure 4 shows the four glove-type devices from Immersion.[14,15] Voice input devices and biocontrollers are also used to provide a convenient way for the user to interact with the VE.

Output Devices

How the user perceives the VR experience is based on what feedback the VE can provide. Output devices are applied in the VE to present the user with feedback about his or her actions. Three of the human perceptual senses, i.e., visual, aural, and haptic, are commonly presented to the VR user with synthetic stimuli through output devices.

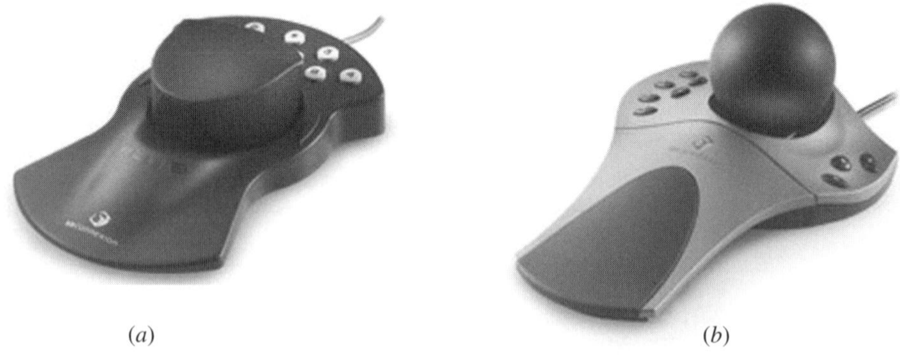

(*a*) (*b*)

Figure 3 (*a*) SpaceMouse and (*b*) SpaceBall from 3Dconnexion, Inc.

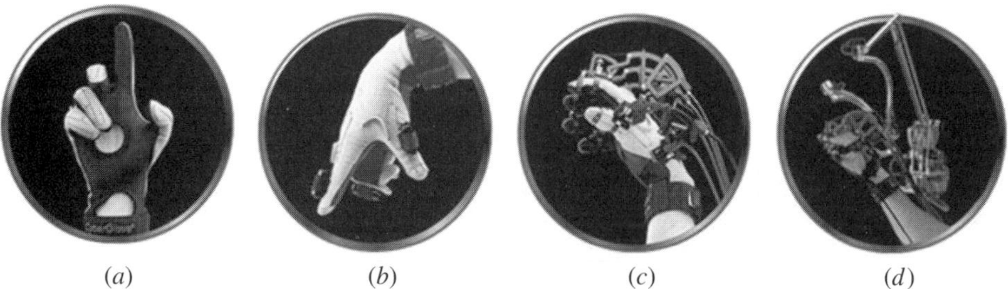

(*a*) (*b*) (*c*) (*d*)

Figure 4 (*a*) CyberGlove, (*b*) CyberTouch, (*c*) CyberGrasp, and (*d*) CyberForce. (Courtesy of Immersion.)

Visual Display. There are three general categories of VR visual displays. The cheapest method is using stereo graphics on a desktop monitor and a pair of liquid crystal display (LCD) shutter glasses. The HMD can provide a fairly immersive experience for the user. It is a helmet worn by the user with a small monitor in front of each eye to display images as a 3D view of the virtual world. Often the HMD is combined with a head tracker so as to allow the participant a full 360° view. Table 2 lists several commercially available HMDs. HMDs are less costly and more portable than projection systems. However, they have a lower resolution and can be cumbersome. Projection systems can be made up of a single large screen or several screens. By introducing stereoscopic images and with the application of stereo glasses, the image becomes 3D and the projection system gives the most realistic representation possible of graphical information. Immersadesk is the most popular single-screen projection system provided by Fakespace. The most commonly used multiscreen projection system is CAVE, which uses four to six large screens in the shape of a cube. It allows multiple viewers to physically immerse in the VE.[16] The screen can also be in the shape of cylinder, dome, or torus. An example is the Hemispherium™ shown in Fig. 5.[17]

Haptic Display. Haptics broadly refers to touch sensations that occur for the purpose of perception or physical contact with objects in the virtual world.[18] Haptics has played an important role in the application of VR. Haptic interface devices are becoming more and more mature. Figure 6 shows some example haptic devices. Force-feedback gloves such as the CyberTouch read finger-specific contact information and output finger-specific resistive forces. Exoskeleton mechanisms or body-based haptic interfaces, which a person wears on the arm or leg, present more complex multi-degree-of-freedom motorized devices. PHANToM™ is a ground-based device that provides the user a feel of force when he or she manipulates a virtual object with a physical stylus that can be positioned and oriented in six degrees of freedom.

Table 2 Head-Mounted Display Devices

HMD Model	Company	Display Type[a]	Resolution	Field of View
CyberEye CE-500	General Reality	LCD	800 × 600	30°
5DT HMD 800	5DT products	LCD	800 × 600	32°
Kiaser ProView XL35	TekGear	TFT	1024 × 768	30°
i-glasses 3D PRO	TekGear	LCoS	800 × 600	26°
Cy-Visor	TekGear	TFT	800 × 600	31°

[a] LCD = Liquid crystal display, TFT = Thin-film transistor, LCoS = Liquid Crystal on Silicon.

Figure 5 Birds-eye view from flight chair in Hemispherium.[15]

Auditory Display. Besides visual and touch information, sound can be incorporated in the VE to enhance the participant's sense of presence. Standard speakers or headphones are generally used as the auditory display hardware in the VE. An example is generating sound to confirm the selection when an object is selected. Another example is providing different pitches of sound to simulate the machining with different properties of material. The challenge of auditory display is to create sound that is realistic for a given application.

2.2 VR Software

Software is required to integrate various hardware devices into a coherent system that enables the user to interact with the VE. Three primary requirements are identified by Bierbaum and Just[19] for a VR system: performance, flexibility, and ease of use. When deciding which VR software is best suited to a particular application, various features of the VR software need to be considered. They include features to support cross-platform, VR hardware, and importing 3D models from other systems; 3D libraries; optimization of level of detail; interactivity of the virtual world; and multiuser networking. Valuable reviews of these features have been given by Vince[20] and Bierbaum et al.[19]

VR implementation software can be classified into two major categories: software development kits (SDKs) and authoring tools.[9] SDKs are programming libraries (generally in

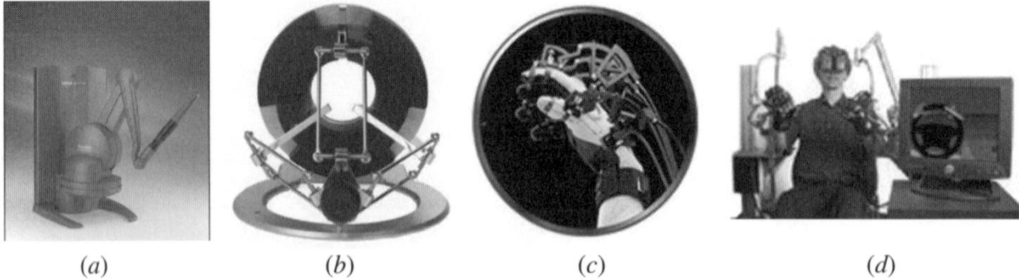

(*a*) (*b*) (*c*) (*d*)

Figure 6 Haptic devices: (*a*) PHANToM desktop device from SensAble; (http://www.sensable.com), (*b*) Omega haptic device from ForceDimension (http://www.forcedimension.com); (*c*) CyberGrasp from Immersion; (*d*) Haptic Workstation from Immersion.

C or C++) that have a set of functions for the developer to create VR applications. Authoring tools provide the graphical user interfaces (GUIs) for the user to develop the virtual world without requiring tedious programming. Table 3 contains a list of publicly available VR software packages for the creation of VEs.

3 VIRTUAL REALITY IN CONCEPT DESIGN

The fields of application of VR are extensive. This and the following sections present the VR applications in various engineering areas. The list of applications reviewed is not intended to be exhaustive; rather it provides some representations of current applications of VR in engineering.

In a conceptual design, the exact dimensions of the design part are not determined initially, and the designer is more interested in creating part shapes and features. Commercial computer-aided design (CAD) systems such as Unigraphics, Ideas, Catia, PRO/E, etc., are powerful geometric modeling tools, but they require precise data for designing objects and thus do not allow the users to implement their ideas on shape and feature design in an intuitive manner. Their user interface generally consists of windows, menus, icons, etc., which tend to prevent users from focusing on the design intent. Another limitation of the conventional CAD system is in the use of input devices. The designers use a 2D input device, usually the mouse, for the construction of 3D objects. Concept design using VR techniques enables the user to create, modify, and manipulate 3D CAD models intuitively and at the same time allows the user to visualize CAD models in a VE immersively. Following is a discussion of some systems that provide such capabilities.

3.1 3-Draw

3-Draw, a system for interactive 3D shape design, was introduced at MIT.[21] Its user interface is based on a pair of Polhemus six-degree-of-freedom tracking devices. One tracker is held as a palettelike sensor in the designer's left hand to specify a moving reference frame, and the other tracking device is used by the right hand as a styluslike sensor to draw and edit 3D curves in space. A graphic display is used to visualize the scene from a virtual camera position. After drawing and editing the 3D curves, which form wireframe models, the next steps are fitting surfaces to groups of linked curves and deforming the surfaces until the required shape is obtained.

Table 3 Virtual Reality Software

Type	Software	Website
SDK	CAVELib	http://www.vrco.com/products/cavelib/cavelib.html
	DI-Guy	http://www.bdi.com/content/sec.php?section=diguy
	DIVISION	http://www.ptc.com/products/index.htm
	WorldToolKit	http://www.sense8.com
	VR Juggler	http://www.vrjuggler.org/
Authoring tools	EON Studio	http://www.eonreality.com/
	WorldUp	http://www.sense8.com
	MultiGen Creator	http://www.multigen-paradigm.com/
	MotionBuilder	http://www.kaydara.com/products/

3.2 3DM

Built by researchers at the University of North Carolina,[22] the 3DM system uses an HMD to place the designer in a virtual modeling environment. An input device such as a Polhemus 3-space Isotrak held in one hand is used for all interactions, including selecting commands from a floating menu, selecting objects, scaling and rotating objects, or grabbing vertices to distort the surface of an object.

3.3 JDCAD

Researchers at the University of Alberta, Canada, developed a system called JDCAD.[23] It is a 3D modeling system which uses two 6DOF tracking devices, one to dynamically track the user's head and provide the kinetic 3D effect (e.g., correlation to the position and orientation of head) and the other used as a hand-held "bat"[24] to track hand movements. A bat is a tracker that reports 3D position and orientation data. It has three buttons mounted on it for signaling events. Figure 7 shows the bat that is used with JDCAD. By switching modes, the bat can be used to rotate and translate the model under construction to select objects for subsequent operations and to orient and align individual pieces of the model.

3.4 HoloSketch

Michael Deering[25] at Sun Microsystems created the HoloSketch system. It uses a head-tracked stereo display with a desktop cathode ray tube (CRT), unlike many VR systems which use HMDs. The user wears a pair of head-tracked stereo shutter glasses and manipulates the virtual world through a hand-held 3D mouse/wand. The HoloSketch system has a 3D multilevel fade-up circular menu. The fade-up menu is used to select the required drawing primitives or to perform one-shot actions such as cut or paste. The HoloSketch system supports several types of 3D drawing primitives, including rectangular solids, spheres, cylinders, cones, freeform tubes, and many more.

3.5 COVIRDS

Dani and Gadh[26] presented an approach for creating shape designs in a VE called COVIRDS (Conceptual VIRtual Design System), as shown in Fig. 8. This system uses VR technology

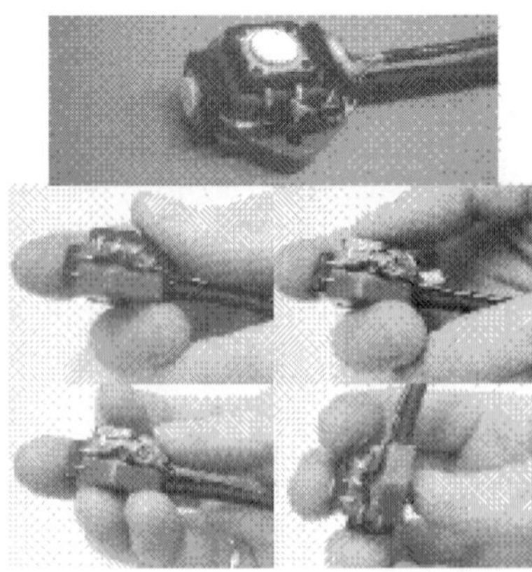

Figure 7 Bat input device.[24]

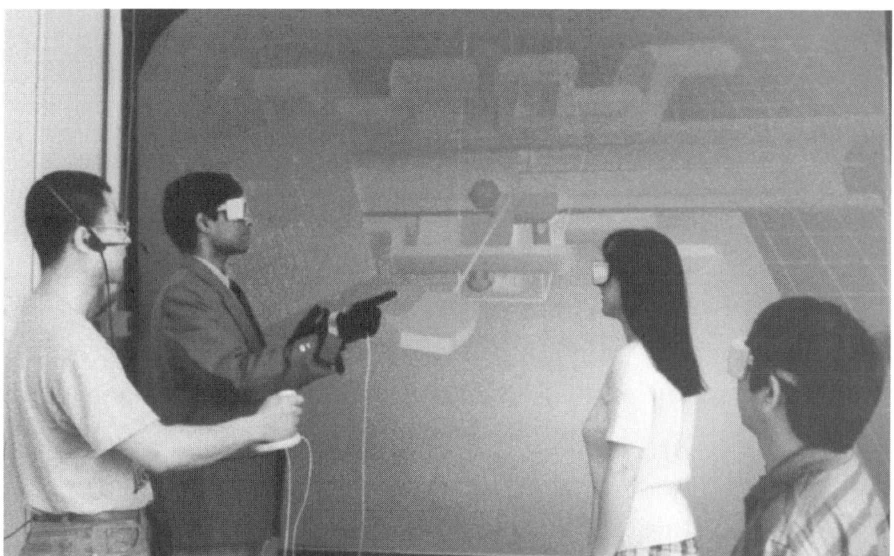

Figure 8 COVIRDS system.[27]

to provide a 3D virtual environment in which the designer can create and modify 3D shapes with an interface based on bimodal voice and hand tracking. A large-screen, projection-based system called the Virtual Design Studio (VDS) is used as an immersive VR-CAD environment. The designer creates 3D shapes by voice commands, hand motions, finger motions, and grasps and shape edits features with his or her hands. Tests have been conducted to compare the efficiency of the COVIRDS with the traditional CAD systems. It was claimed that the COVIRDS can achieve a productivity of 10–30 times over the conventional CAD systems.[28]

3.6 Virtual Sculpting with Haptic Interface

Peng and Leu[29,30] have developed a virtual sculpting system, as shown in Fig. 9. A VR approach has been taken to make use of modeling tools more intuitively and interactively. The virtual sculpting method is based on the metaphor of carving a solid block into a 3D

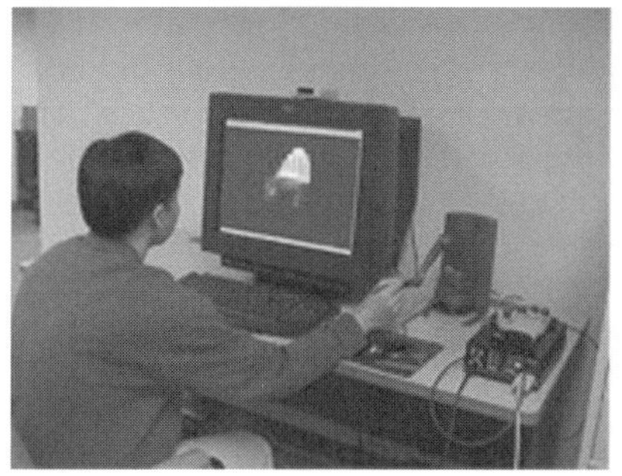

Figure 9 Chair generated using virtual sculpting system developed at University of Missouri-Rolla.[29]

freeform object workpiece like a real sculptor would do with a piece of clay, wax, or wood. The VR interface includes stereo viewing and force feedback. The geometric modeling is based on the sweep differential equation method[31] to compute the boundary of the tool swept volume and on the ray-casting method to perform Boolean operations between the tool swept volume and the virtual stock in dexel data to simulate the sculpting process.[30] Incorporating a haptic interface into the virtual sculpting system provides the user with a more realistic experience. Force feedback enables the user to feel the model creation process as if it were actual sculpting with physical materials. The PHANToM manipulator is used as a device to provide the position and orientation data of the sculpting tool and at the same time to provide haptic sensation to the user's hand during the virtual sculpting process. Multithreading is used in the geometry and force computations to address the different update rates required in the graphic and haptic displays.

4 VIRTUAL REALITY IN DATA VISUALIZATION

Virtual reality is beginning to have a major impact on engineering design by streamlining the process of converting analysis results into design solutions.[32] By immersing the user in the visualized solution, VR reveals the spatially complex structures in engineering in a way that makes them easy to understand and study.[33]

4.1 Finite-Element Analysis

There are two important procedures for finite-element analysis (FEA): preprocessing and postprocessing. The preprocessing sets up the model and the postprocessing displays the results. Applying VR in interactive FEA provides the preprocessor and postprocessor with a good way to describe the model and results in a 3D immersive environment.[34] VR offers a more intuitive environment and a real-time, full-scale interaction technique in the design.[35]

Yeh and Vance[36] incorporated finite-element data into a VE such that the designer can interactively modify the transverse force using an instrumented glove and view the resulting deformation and stress contour of a structural system. Methods are applied to allow the designer to modify design variables and immediately view the effects. A HMD and a BOOM (a fully immersive environment by Fakespare Systems, Inc.) are used with a 3D mouse as an interface device for interacting with the variables in the VE.

The university of Erlangen and BMW[37,38] developed a VE for car crash simulation, as shown in Fig. 10. The system was used at BMW to analyze a wide variety of time-dependent numerical simulations, ranging from crash worthiness to sheetmetal forming, vibrations, and acoustics. The fully immersive environment Fakespace BOOM has been used, in which the user can intuitively navigate and interact with different FEA methods relevant to the car-body development process. The system can visualize an average model consisting of 200,000 finite elements for each time step up to six time steps. The system can load simulation files into the VE directly, with on-the-fly polygon reduction but without any special preprocessing.

Connell and Tullberg[39,40] developed a framework for performing interactive finite-element simulation within a VE and a concrete method for its implementation. The developed software provides the ability to interchange large-scale components, including the FEA code, the visualization code, and the VR hardware support in the system. Standard VR-tools (dVISE) and FEA tools (ABAQUS) are used to build the VR module and FEA module, respectively. The real-time FEA results are visualized in a CAVE. Figure 11 shows a stress contour over the surface of the bridge deck under different loading conditions.

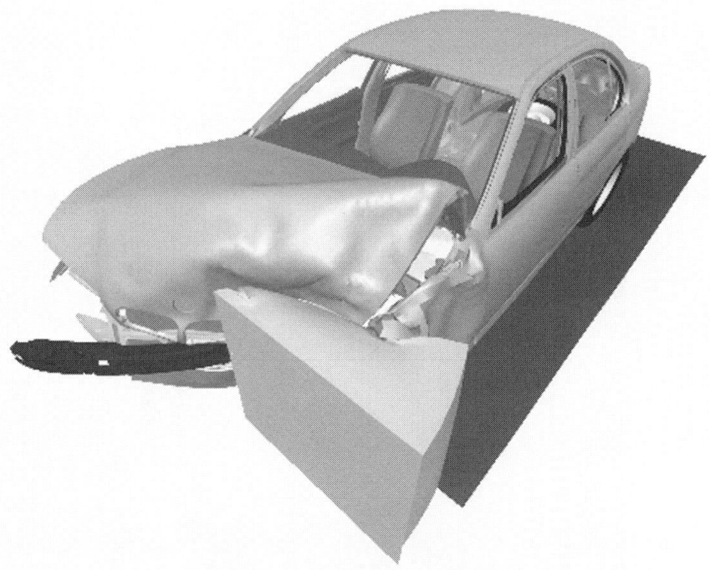

Figure 10 Full car crash simulation in a VE.[37]

4.2 Computational Fluid Dynamics

Computational fluid dynamics (CFD) software is a powerful tool that can be used to provide fluid velocity, temperature, and other relevant variables and is commonly used in industry to predict flow behavior. VR provides a fully interactive 3D interface in which users can interact with computer-generated geometric models in CFD analysis.

The NASA Virtual Wind Tunnel (VWT)[41] is a pioneering application of VR technology to visualize the results of modern CFD simulations. Figure 12 shows the user interacting

Figure 11 Concrete case screenshot.[39]

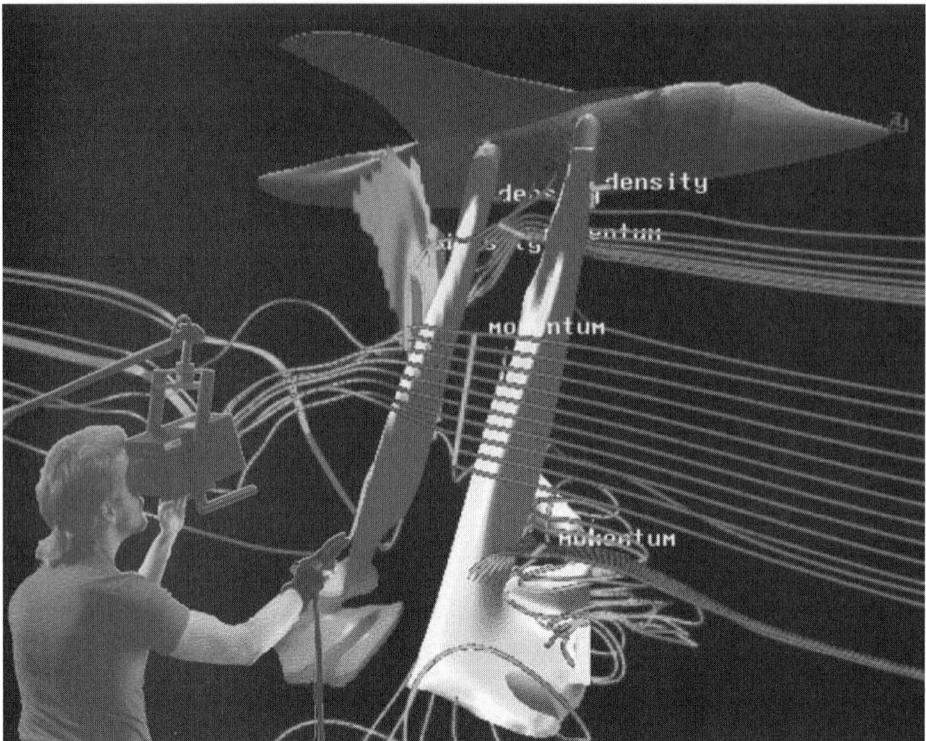

Figure 12 NASA Virtual Wind Tunnel.[41]

with the simulation via a BOOM display and a data glove. The data glove effectively acts as a source of "smoke," allowing the user to observe local flow lines. The computation of unsteady flow for visualization occurs in real time and the user can view the dynamic scene from many points of view and scales. The VR approach has an advantage that, unlike a real user in a flow field, the presence of the user does not disturb the flow.

CAVEvis[42] is a scientific visualization tool developed to study particle flow in vector and scalar fields using the CAVE. It is designed by the National Center for Supercomputing Applications. The scientific simulations include CFD simulations of tornados, hurricanes, and smog as well as the flow of gas and fluid around airplanes, cars, and other vehicles. To visualize very large and complex numerical datasets in real time, the system distributes visualization using asynchronous computation and rendering modules over multiple machines. The important issues related to the management of time-dependent data, module synchronization, and interactivity bottlenecks are addressed by the CAVEvis system. Figure 13 is a sample screen showing a number of different visualizations when a tornado touches down.

VR-CFD,[43] developed at Iowa State University, is a VR interface for visualization of (CFD) data; see Fig. 14. A CAVE-type VR facility is utilized to provide a collaborative environment where multiple users can investigate the entire flow field together. The system provides the features for creation and manipulation of flow visualization entities, including streamlines, rakes, cutting planes, isosurfaces, and vector fields in real time.

ViSTA FlowLib[45] is a framework which provides an interactive exploration of unsteady fluid flows in a VE. It combines efficient rendering techniques and parallel computations

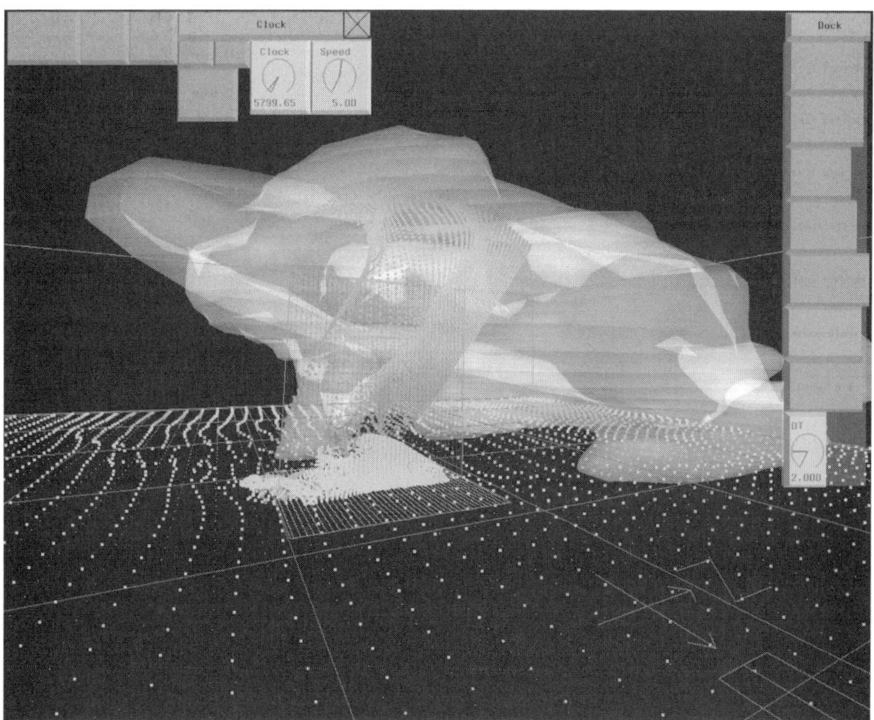

Figure 13 Sample screen of CAVEvis showing touch-down of tornado.[42]

Figure 14 Visualization and immersion in vector field.[44]

with intuitive multimodal user interfaces into a single powerful cross-platform CFD data visualization library. Special care was taken to achieve high scalability with respect to computing power, projection technology, and input–output device availability. Figure 15 shows the ViSTA environment. Van Reimersdahl et al.[47] introduced the haptic rendering subsystem of ViSTA FlowLib. The haptic rendering acts as a great assistance to visual rendering and facilitates the exploration process of CFD simulation data.

Other examples of data visualization and data representation in VEs include interactive FEA of shell structures called VRFEA,[48] stress analysis of a tractor lift arm,[49] a VE called COVISE for large-scale scientific data analysis,[50] and a virtual scientific visualization tool called MSVT.[51]

5 VIRTUAL REALITY IN DRIVING SIMULATION

A driving simulator is a VR tool that gives the user an impression that he or she is driving an actual vehicle. This is achieved by taking the driver inputs from the steering wheel and pedals and feeding back the corresponding visual, motion, and audio cues to the driver.[52] The advantages of a driving simulator vs. driving with a real vehicle may be a wide range of possible configurations, repeatable conditions, easy-to-change tasks and parameters, and good experimental efficiency. A driving simulator can be used to study design and evaluation of vehicles, highway systems, in-vehicle information and warning systems, traffic management systems, and many more.

The National Advanced Driving Simulator (NADS)[53] developed by the National Highway Traffic Safety Administration is the most advanced driving simulator existing today.

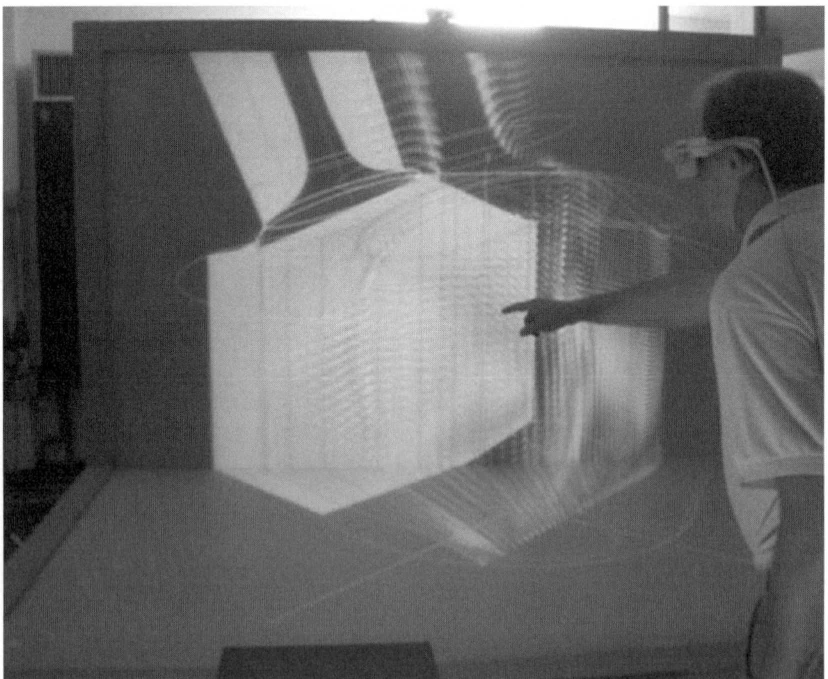

Figure 15 Interactive exploration of airflow into cylinder of spark ignition engine at HoloDesk.[46]

The applications of this driving simulator have included driver crash avoidance study, evaluation of advanced in-vehicle systems and control technologies, and highway design and engineering research related to traffic safety. NADS uses a simulation dome which is 24 ft in diameter with the interchangeable car cabs sitting inside of the dome. Different cabs that can be used currently include Ford Taurus, Chevy Malibu, Jeep Cherokee, and Freightliner. Within the dome are the projectors that provide 190° front and 65° rear field of view, as shown in Fig. 16. At the same time, the motion subsystem, on which the dome is mounted, provides 64 ft of horizontal and longitudinal travel and 330° of rotation providing a total of nine degrees of freedom. The VR effect is that the driver feels acceleration, braking, and steering cues as if he or she were driving a real car, SUV, truck, or bus. The system supports generation and control of traffic within the VE.

Ford Motor Company developed a motion-based driving simulator, VIRTTEX (VIRtual Test Track EXperience), to test the reactions of sleepy drivers.[55] This simulator can generate forces that would be experienced by a person while driving a car. The simulator dome houses five projectors, three for the forward view and two for the rear view, that rotate with the dome and provide a 300° computer-generated view of the road. Different car cabs in the simulator are attached to a hydraulic motion platform, called the hexapod, that can simulate the motion associated with more than 90% of the typical miles in the United States, including spinouts. A continuation of many years of driver drowsiness research is conducted by Volvo using VIRTTEX. The new safety technology developed through this research will be inte-

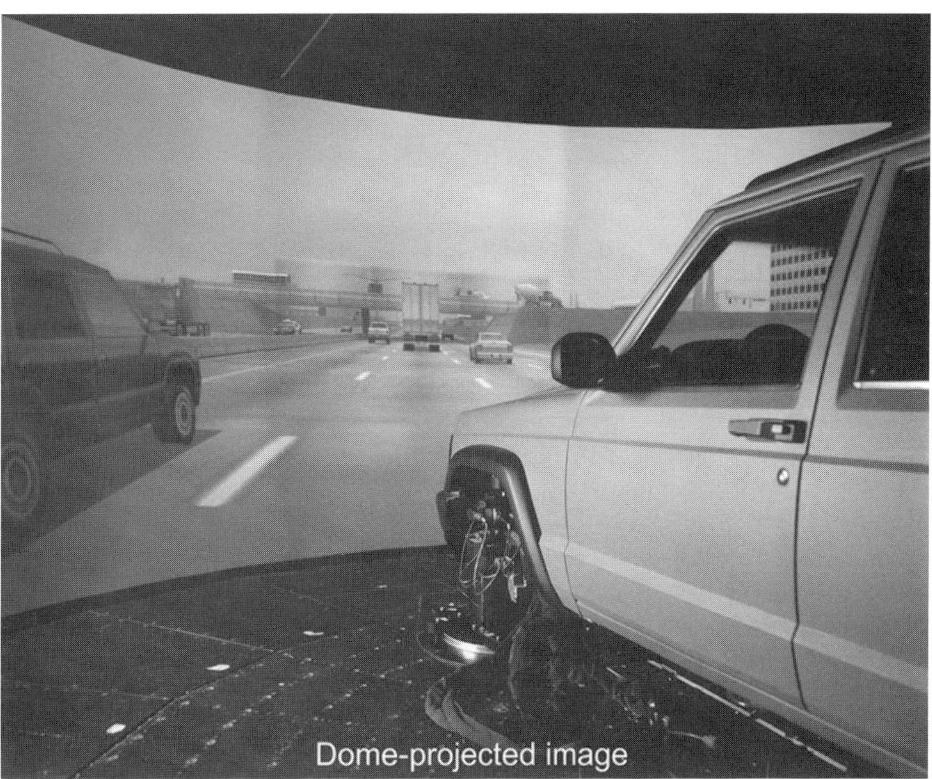

Figure 16 National Advanced Driving Simulator.[54]

grated into the new-car design. In Fig. 17, the left figure is the driving simulator dome inside which the actual car body is placed and the right figure is the hexapod system.

These driving simulators are the most advanced ones and make use of more than one linear or hexapod motion system. Other driving simulators are considered to be semiadvanced if they make use of one linear or hexapod motion system or use the vibrating platform for motion cues.

Kookmin University in Seoul, Korea, has developed a driving simulator,[57] which uses a 6DOF Stewart platform as the base. This platform helps the user to experience realistic force feedback. It has 32° of horizontal and vertical field of view for visualization. The applications of this simulator include vehicle system development, safety improvement, and human factor study. An advanced application is integration of an unmanned vehicle system with the driving simulator.

The State University of New York at Buffalo has conducted a project to study and enhance the ability of a driver to cope with inclement conditions using a VE.[58] The goal is to develop strategies for supplementing a driver's natural cues with synthetic cues derived from sensors and control strategies. To provide a system for the evaluation of driver assistive technologies for inclement conditions, a driver simulator has been developed.

The Drunk Driving Simulator developed at the University of Missouri-Rolla[59] is the initial development of a VR-based driving simulator that can be used as a tool to educate people, especially college and high school students, about the consequences of drunk driving. The VE is initially built on a desktop monitor with an HMD. The developed system consists of real-time vehicle dynamics, hardware control interface, force feedback, test environments, and driving evaluation. The drunk effects are simulated by implementing the control time lag effect and visual effect. Testing and evaluation techniques have been developed and carried out by students and the results are presented. The simulator has been extended to the CAVE system to provide sufficient immersion and realism.[60]

Other examples of a driving simulator include Iowa State University's Driving Simulator[52] and the UMTRI Driving Simulator.[61]

6 VIRTUAL REALITY IN MANUFACTURING

Application of VR technologies to modeling and simulation of manufacturing facilities and operations is an emerging area.[62] Generally speaking, there are four areas of VR in manu-

Figure 17 VIRTTEX Driving Simulator.[56]

facturing applications: process simulation, factory layout, assembly method prototyping, and part flow simulation.

6.1 Factory and Process Models

The biggest advantage of using VR for factory design is that it supports the user in planning space or logistical issues by allowing interactively moving and relocating the machines after the simulation has been carried out. By integrating into all levels of product development and production cycle, the virtual factory provides a unique environment for achieving these goals.[63]

VR-Fact!, developed at the University of Buffalo, provides an interactive virtual factory environment that explores the applications of VE in the area of manufacturing automation.[64] The VR-Fact! simulation as shown in Fig. 18 can be visualized using stereo HMDs and CrystalEye stereo glasses with a Silicon Graphics ONYX 2 computer. The system also supports viewing on a Fakespace Boom3C. VR-Fact! can be used to create digital mock-up of a real factory shopfloor for a given product mix and set of machines. By intuitively dragging and placing modular machines in the factory, designers can study issues such as plant layout, cluster formation, and part flow analysis. The VR walk-through environment

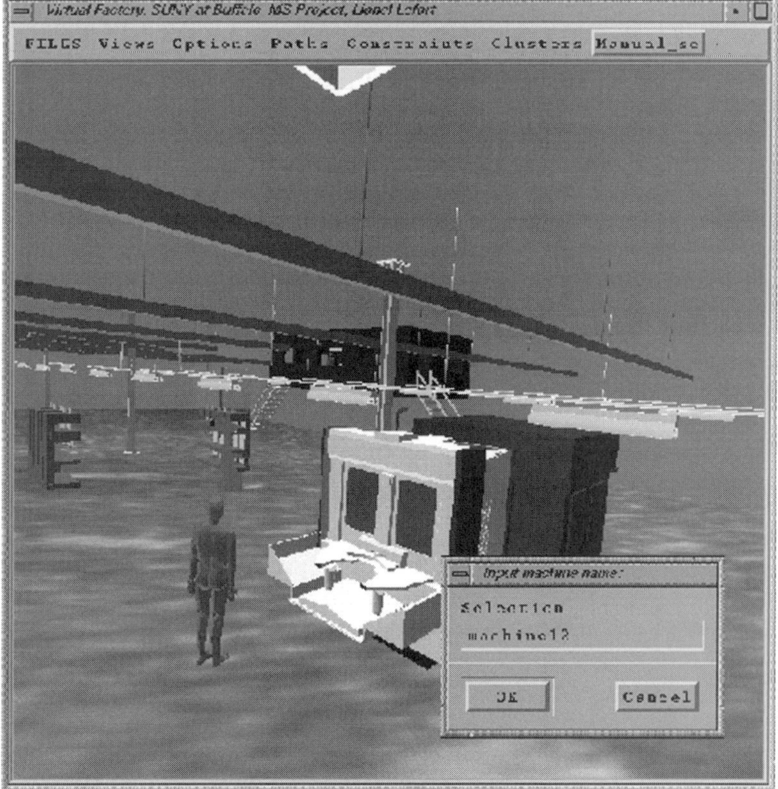

Figure 18 VR-Fact! simulation.[64]

of this software provides a unique tool for studying physical aspects of machine placements. It uses mathematical algorithms to generate independent manufacturing cells. This approach simultaneously identifies part families and machine groups and is particularly useful for large and sparse matrices.

Chawla and Banerjee[65,66] at Texas A&M University presented a 3D VE to simulate the basic manufacturing operations (unload, load, process, move, and store). It enables a 3D facility to be reconstructed using space-filling curves (SFCs) from a 2D layout. The facility layout is represented in 3D as a scenegraph structure, which is a directed acyclic graph. SGI's IRIS Performer is used for construction and manipulation of the scenegraph structure. Immersadesk™ of Fakespace Systems is used to visualize the VE developed with CAVELIB software. The scenegraph structure automatically encapsulates the static and dynamic behavior of the manufacturing system. These properties of the 3D objects include position and orientation, material properties, texture information, spatial sound, elementary sensory data, precedence relationships, event control lists, etc. There are two distinct types of manufacturing objects in terms of their ability to move within the facility: static and dynamic. The simulation of basic manufacturing process operations is represented by suitable object manipulation within the scenegraph.

Kibira and McLean[67] presented their work on the design of a production line for a mechanically assembled product in a VE. The simulation was developed as part of the Manufacturing Simulation and Visualization Program at the National Institute of Standards and Technology. Their research mainly focused on the partitioning and analysis of the assembly operation of the prototype product into different tasks and allocation of these tasks to different assembly workstations. The manufacturing process design is constructed using three software applications. The geometry of components is created by using AutoCAD™. IGRIP™ is used to provide the graphical ergonomic modeling of workstation operations. QUEST™ provides the discrete-event modeling of the overall production line. In addition, the difficulties of using simulation modeling for simultaneous graphical simulation of assembly operations and discrete-event analysis of a production process are addressed.

Kelsick et al.[68] presented an immersive virtual factory environment called VRFactory. This system was developed with an interface to the commercial discrete-event simulation program SLAM II to investigate how potential changes to a manufacturing cell affect part production. The results of the discrete-event simulation of a manufacturing cell are integrated with a virtual model of the cell. The VE is built in a CAVE-like projection screen–based facility called C2. The interaction device is a Fakespace PinchGlove equipped with a Flock of Birds tracker. This factory allows the user to explore the effect of various product mixes, inspection schedules, and worker experience on productivity while immersing in a visual, 3D space. The geometric models of all the machines are created using the Pro/Engineer CAD software and the WorldUP™ modeling software and then loaded in a virtual world. Figure 19 illustrates a snapshot of the models in VRFactory. The output of the simulation software AweSim is used to control part flow through the factory.

In cooperation with Ford Motor Company and the Lanner Group, Waller and Ladbrook[69] presented the work to automatically generated VR factory using the WITNESS VR simulation package. Mueller-Wittig et al.[70] presented the interactive visualization environment for an electronics manufacturing industry. The behavior of different manufacturing stages, including fabrication, assembly, testing, and packing, can be simulated in the VE.

6.2 VR Assembly

Virtual reality provides a means of immersing production engineers in a computer-synthesized working environment. Rather than using physical mock-ups or hand drawings,

Figure 19 VRFactory models.[68]

VR-based assembly can reduce the manufacturing lead time by simulating assembly sequences with virtual prototypes. It also provides an immersive interactive environment for planning assembly paths and part layout for assembly.[71] However, virtual assembly simulation is one of the most challenging applications of VR. This is mostly due to the very high degree of interactivity needed in the virtual assembly environment. Not only is the high amount of functionality needed, but also some of the interaction must be as natural as possible.

The Virtual Assembly Design Environment (VADE) was designed at Washington State University in collaboration with the National Institute of Standards and Technology (NIST).[72] It is a VR-based assembly design environment that allows engineers to plan, evaluate, and verify the assembly of mechanical systems. The VE was built with an SGI Onyx2 (with six processors) and two Infinite Reality pipes, a Flock of Birds motion tracker, a CyberGlove, and an HMD. The mechanical system can be designed using any parametric CAD software (e.g., Pro/Engineer) and automatically exported to the VADE system through a user-selected option. The user specifies the initial location and orientation of the parts and the assembly tools based on the actual setup being simulated. The system provides interactive dynamic simulation of parts using physical-based modeling techniques while the user performs the assembly. The assembly path can be interactively edited in the VE directly by the user. Then the assembly design information can be generated and made available to the designer in the CAD system. A research on heavy-machine virtual assembly[71] using VADE has been conducted by Komatsu in collaboration with Washington State University.

The Fraunhofer-Institute for Industrial Engineering in Germany has developed a planning system to carry out assembly planning operations.[73] The system applies a virtual model of a person (VirtualANTHROPOS) to execute assembly operations. It provides interfaces to connect with a CAD system and to present CAD objects in a virtual production system. After the assembly operations are decided by the user, a precedence graph can be generated with the assembly time and cost determined.

Zachmann and Rettig[74] at the University of Bonn, Germany, investigated robust and natural interaction techniques to perform assembly tasks in the VE. They identified that interaction metaphors for virtual assembly must be balanced among naturalness, robustness, precision, and efficiency. Multimodal input techniques were utilized to achieve robust and efficient interaction with the system, including speech input, gesture recognition, tracking, and menus. Precise positioning of parts was addressed by constraining interactive object motions and abstract positioning via command interfaces. A natural grasping algorithm was presented to provide intuitive interaction. In addition, a robust physically based simulation is proposed to create collision-free assembly paths. All of the algorithms were integrated into the VR assembly system. Figure 20 shows the interface of the virtual assembly simulation.

Yuan and Yang[76] presented a biologically inspired intelligence approach for virtual assembly. A biologically inspired neural network was incorporated into the development of a virtual assembly system. The neural network was shown useful in the planning of real-time optimal robot motion in dynamic situations without the need of any learning procedures. This work could lead to the improvement of flexible product manufacturing, i.e., automatically producing alternative assembly sequences with robot-level instructions for evaluation and optimization.

The virtual manufacturing group at Heriot-Watt University investigated the use of immersive VR in the design and planning of cable harnesses.[77] Fernando et al.[78] evaluated the use of constraint-based VEs to support assembly methods prototyping. McDermott and Bras[79] have developed a haptically enabled dis/reassembly simulation environment.

7 VIRTUAL REALITY IN CIVIL ENGINEERING AND CONSTRUCTION

Recently VR techniques have been widely and successfully explored in the area of civil engineering and construction. Virtual reality offers considerable benefits for many stages of

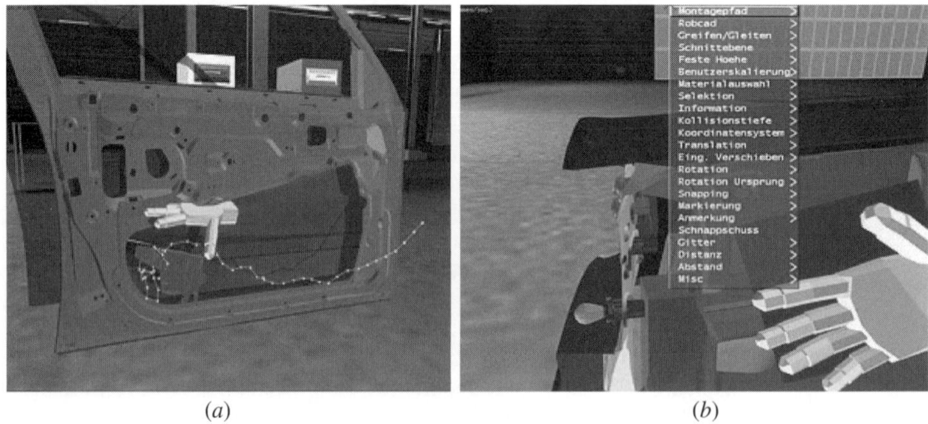

(*a*) (*b*)

Figure 20 Virtual assembly simulation: (*a*) multimodal user interface; (*b*) assembly path.[75]

the construction process in the following areas: (1) enabling a designer to evaluate the design and make modifications by immersing in a building, (2) virtually disassembling and reassembling the components to design the construction process, and (3) offering a "walk-through" view of the facility and experiencing the near-reality sense of the construction. Thabet et al.[8] provided a comprehensive survey of VR in construction. This section will present some recent success examples of VR applications in construction.

VIRCON[80,81] is a collaborative project between the University College London, Teeside University, the University of Wolverhampton, and 11 construction companies in the United Kingdom. The aim of this project was to provide a decision support system which allows planners to trade off the temporal sequencing of tasks with their spatial distribution using VR technologies. The research involved the development of a space-scheduling tool called "critical space analysis," its combination with critical path analysis in a space–time broker, and the development of advanced visualization tools for both of these analyses. This system applied visual 4D planning techniques that combine solid CAD models with the construction schedule (time); see Fig. 21. The 4D tools developed were not only to visualize the construction products but also to visualize movement of plants/temporary objects and highlight of spatial overload. This provides better evaluation and communication of activity dependency in spatial and temporal aspects.

DIVERCITY (Distributed Virtual Workspace for Enhancing Communication within the Construction Industry) is an undergoing project funded by the European Union Information Society Technologies.[82] The intention of this project is to investigate the use of VR in construction. Ten universities and research institutes are participating in this project. DIVERCITY aims to develop a "shared virtual construction workspace" that allows construction companies to conduct client briefing, design reviews, simulate what-if scenarios, test constructability of buildings, and communicate and coordinate design activities between teams. It allows the user to produce designs and simulate them in a VE. It also allows teams

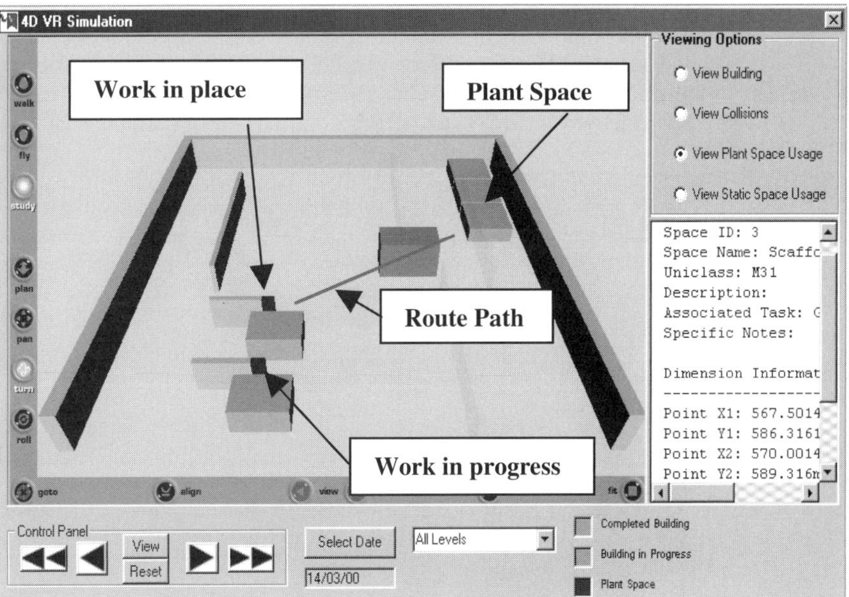

Figure 21 Visualization of construction in VIRCON.[80]

based in different geographic locations to collaboratively design, test, and validate shared virtual projects. Figure 22 illustrates a VE generated by DIVERCITY.

Waly and Thabet[84] presented a framework of the Virtual Construction Environment. The planning process using this environment contributes to the successful development and execution of a construction project. The virtual planning tool integrates a new planning approach that utilizes VR modeling techniques coupled with object-oriented technologies to develop an interactive collaborative environment. The VE enables the project team to undertake inexpensive rehearsals of major construction processes and test various execution strategies in a near-reality sense, prior to the actual start of construction. Cowden et al.[85] presented an immersive home design tool using CAVE, as shown in Fig. 23. The home design tool allows the user to navigate through the model of a residential home and redesign elements of the home in an immersive VE. The interactions provided by the tool are currently limited to hiding/showing individual elements.

Lipman[87] developed a VR system for steel structures using CAVE at NIST. This system can load the steel structures in the format of CIMsteel Integration Standards (CIS/2) and then convert the CIS/2 file to a VRML (Virtual Reality Modeling Language) file, which can then be visualized in CAVE. Yerrapathruni et al.[88] used a CAD system and CAVE to improve construction planning. A clean-room piping design using VR techniques was introduced by Dunston et al.[89] and Kamat and Martinez[90] described a VR-based discrete-event simulation of construction processes.

8 VIRTUAL REALITY IN OTHER DISCIPLINES (GEOLOGY AND CHEMICAL ENGINEERING)

The Center for Visualization at the University of Colorado has worked on exploring VR applications in the oil and gas industry. Dorn et al.[91] developed an Immersive Drilling Planner for platform and well planning in an immersive VE, as shown in Fig. 24. The planning of a well requires a wide variety of data, geometric accuracy of the data display, interaction with the data, and collaborative efforts of an interdisciplinary team. The system can import and visualize critical surface and subsurface data in the CAVE system. The data include seismic data, horizons, faults, rock properties, existing well paths and surveys, log data, bathymetry, pipeline maps, drilling hazards, etc. Using the data, the user can design the

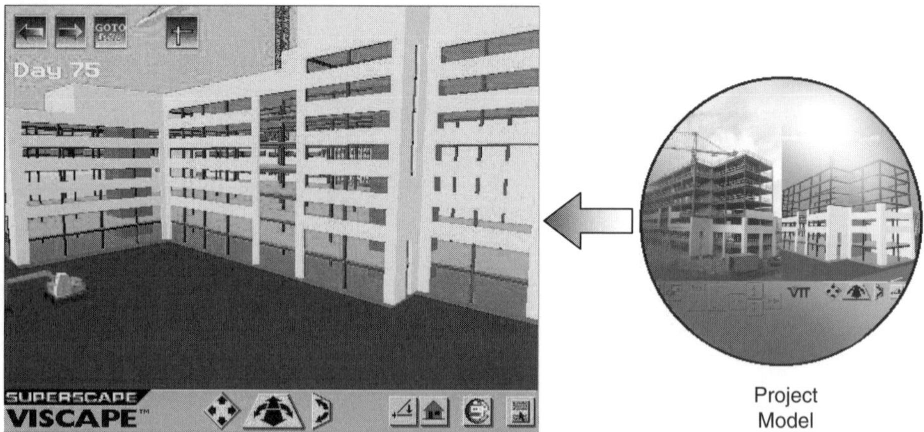

Figure 22 DIVERCITY construction planning.[83]

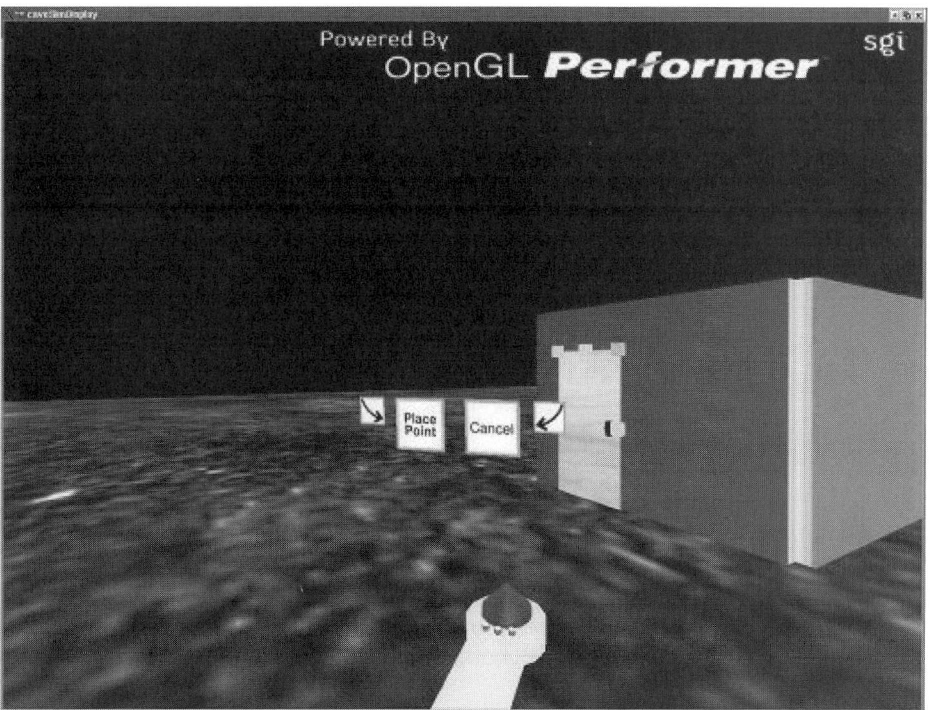

Figure 23 Interactions within virtual home design tool.[86]

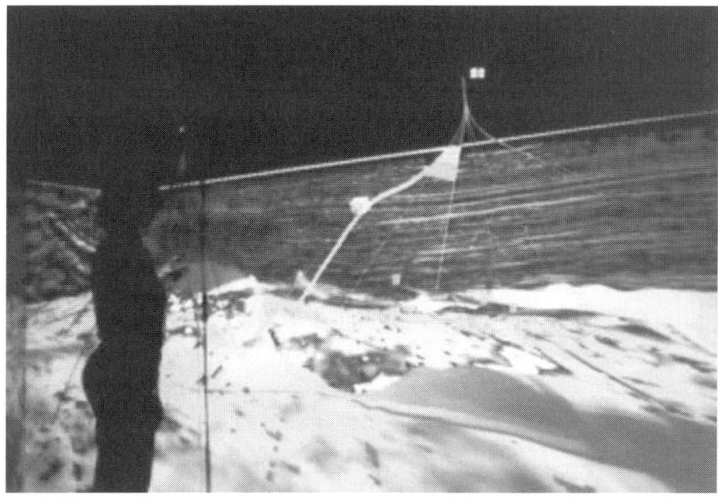

Figure 24 Immersive well path editing.[92]

platform and other properties of the well interactively. The Immersive Drilling Planner has been applied to several projects and has demonstrated development planning cycle time reductions from several months to a week or less. Furthermore, the immersive VE has enhanced the quality of results and has encouraged more effective communication among multidisciplinary team members.

The Virtual Reality Laboratory at Laurentian University in Canada[93] offers a collaborative immersive VE for mine planning and design. As shown in Fig. 25, the VE is built on a large spherical stereoscopic projection system with advanced earth modeling software. The system has a 9 × 22-ft curved screen and can contain up to 20 persons in the theater. Because of the constantly changing work environment and enormous overflow of data, mine design and planning are difficult, time-consuming, and expensive exercises. Large volumes of mine geometry, geology, geomechanics, and mining data can be interpreted and evaluated efficiently by utilizing a large-scale immersive visualization environment. The collaborative environment also makes the planning process quicker. The key benefits of using VR in mine design are data fusion, knowledge transfer, technical conflict resolution, and collaboration.

Bell and Fogler[95] presented a VR-based educational tool in chemical engineering. A series of interactive virtual chemical plant simulations have been developed to demonstrate the chemical kinetics and reactor design. Figure 26 shows one of the simulations focusing on nonisothermal effects in chemical kinetics. The simulations allow the user to explore domains which are otherwise inaccessible by the traditional education tool, such as the interiors of operating reactors and microscopic reaction mechanics. Different virtual laboratory accidents have been simulated to allow users to experience the consequences of not following proper laboratory safety procedures.

9 CONCLUSION

In the past decades, due to the availability of more advanced computers and other hardware and the increased interest from users and researchers, VR technology and application have made considerable progress. This chapter has reviewed VR software and hardware and their integration for a meaningful engineering application. Special devices such as HMDs, haptic devices, and data gloves used to interact with the VE that give the user the impression of being in the real world have been discussed.

VR has been applied to many aspects of the modern society: engineering, architecture, entertainment, art, surgical training, homeland security, etc. The various applications re-

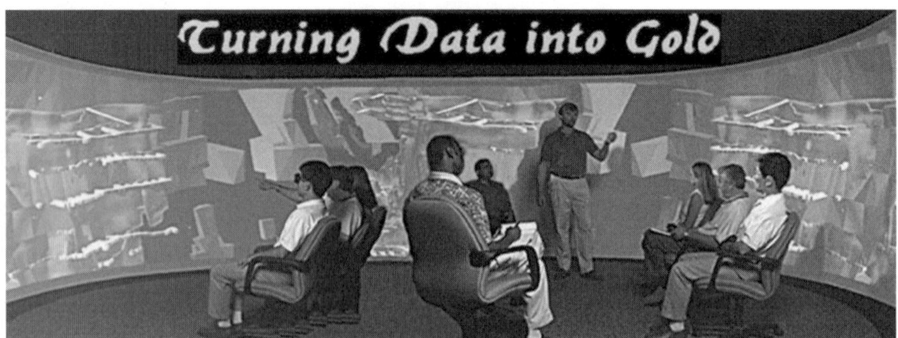

Figure 25 Project team discussing mine plan in the Virtual Reality Laboratory at Laurentian University.[94]

Figure 26 Vicher 2 snapshot.[96]

viewed using VR in engineering include concept design, data visualization, virtual manufacturing, driving simulation, construction engineering, and others. VR techniques provide intuitive interactions between the VE and the engineer. They enable the user to create, modify, and manipulate 3D models intuitively and at the same time allow the user to visualize and analyze data in an immersive VE. By immersing the user in the VE, VR reveals spatially complex structures in a way that makes them easy to understand and study. Visualizing large-scale models is imperative for some engineering applications, such as driving simulation and assembly layout design, which are difficult to achieve by traditional simulation methods. VR techniques conquer this hurdle by providing large-scale visualizing devices such as CAVE. This has been demonstrated in driving simulation, virtual manufacturing, construction engineering, and other applications.

REFERENCES

1. W. R. Sherman and A. B. Craig, *Understanding Virtual Reality: Interface, Application, and Design,* Morgan Kaufmann, San Francisco, CA, 2003, Chapter 1.
2. K. Warwick, J. Gray, and D. Roberts, *Virtual Reality in Engineering,* Institution of Electrical Engineers, London, 1993.
3. Aukstakalnis, S. and D. Blatner, *Silicon Mirage: The Art and Science of Virtual Reality*, Peachpit, Berkeley, CA, 1992.
4. G. Burdea and P. Coiffet, *Virtual Reality Technology,* Willey, New York, 1994, Chapter 1.
5. S. Jayaram, J. Vance, R. Gadh, U. Jayaram, and H. Srinivasan, "Assessment of VR Technology and Its Applications to Engineering Problems," *J. Comput. Info. Sci. Eng.,* **1**(1), 72–83, (2001).
6. P. Banerjee and D. Zetu, *Virtual Manufacturing,* Wiley, New York, 2001, Chapter 1.
7. R. Kalawsky, *The Science of Virtual Reality and Virtual Environments,* Addison-Wesley, Reading, MA, 1993.

8. W. Thabet, M. F. Shiratuddin, and D. Bowman, "Virtual Reality in Construction: A Review," in *Proceedings of International Conference on Engineering Computational Technology,* Prague, Czech Republic, Sep. 4–6, 2002, pp. 25–52.
9. J. Isdale, "What Is Virtual Reality?" http://vr.isdale.com/WhatIsVR.html, 1998.
10. VTCAVE Information Center, http://www.cave.vt.edu/.
11. S. Güven and S. Feiner, "Authoring 3D Hypermedia for Wearable Augmented and Virtual Reality," in *Proceedings of the ISWC '03 (Seventh International Symposium on Wearable Computers)*, White Plains, NY, Oct. 21–23, 2003, pp. 118–226.
12. K. Pimentel, and K. Teixeira, *Virtual Reality: Through the New Looking Glass,* Intel/Windcrest/McGraw-Hill, New York, 1993.
13. 3Dconnexion, http://www.3dconnexion.com/.
14. Immersion, http://www.immersion.com/.
15. B. Buxton, "A Directory of Sources for Input Technologies," http://www.billbuxton.com/InputSources.html, 2004.
16. C. Cruz-Neira, C. Sandin, and T. DeFanti, "Surround-Screen Projection-Based Virtual Reality: The Design and Implementation of the CAVE," *Proc. ACM SIGGRAPH,* **27,** 135–142 (1993).
17. J. Webster, "The Hemispherium Experience: Fasten Your Seat Belts," in *Proceedings of Trends, Technology, Theming and Design in Leisure Environments Conference,* London, May 11–13, 1999.
18. K. Salisbury, F. Conti, and F. Barbagli, "Haptic Rendering: Introductory Concepts," *IEEE Comp. Graph. Appl.,* **24**(2), 24–32 (2004).
19. A. Bierbaum and C. Just, "Software Tools for Virtual Reality Application Development," in *SIGGRAPH 98 Course 14: Applied Virtual Reality, SIGGRAPH,* Orlando, FL, July 19–24, 1998, pp. 3-2 to 3-22.
20. J. Vince, *Essential Virtual Reality Fast: How to Understand the Techniques and Potential of Virtual Reality?* Springer-Verlag, New York, 1998.
21. E. Sachs, A. Roberts, and D. Stoops, "3-Draw: A Tool for Designing 3D Shapes," *IEEE Comp. Graph. Appl.,* **11**(6), 18–26 (1991).
22. J. Butterworth, A. Davidson, S. Hench, and M. Olano, "3DM: A Three Dimensional Modeler Using a Head-Mounted Display," in *Proceedings 1992 Symposium on Interactive 3D Graphics,* Cambridge, MA, Mar. 30–Apr. 1, 1992, pp. 135–138.
23. J. Liang and M. Green, "JDCAD: A Highly Interactive 3D Modeling System," *Comp. Graph.,* **18**(4), 499–506 (1994).
24. C. Ware and D. R. Jessome, "Using the Bat: A Six Dimensional Mouse for Object Placement," *IEEE Comp. Graph. Appl.,* **8**(6), 65–70 (1988).
25. M. F. Deering, "The Holosketch VR Sketching System," *Commun. ACM,* **39**(5), 54–61 (1996).
26. T. H. Dani and R. Gadh, "Creation of Concept Shape Designs via a Virtual Reality Interface," *Computer-Aided Design,* **29**(8), 555–563 (1997).
27. R. Arangarasan, T. H. Dani, C. Chu, X. Liu, and R. Gadh, "Geometric Modeling in MUlti-Modal, Multi-Sensory Virtual Environment," in *Proceedings of 2000 NSF Design and Manufacturing Research Conference,* Vancouver, Canada, Jan. 3–6, 2000, CD-ROM.
28. C. Chu, T. Dani, and R. Gadh, "Evaluation of Virtual Reality Interface for Product Shape Design," *IIE Trans. Design Manufacturing,* **30,** 629–643 (1998).
29. X. Peng and M. C. Leu, "Interactive Virtual Sculpting with Force Feedback," in *Proceedings of 2004 Japan-USA Symposium on Flexible Automation,* S. K. Ong and A. Y. C. Nee (eds.), Denver, CO, July 19–21, 2004, UL-027.
30. X. Peng and M. C. Leu, "Interactive Solid Modeling in a Virtual Environment with Haptic Interface," in *Virtual and Augmented Reality Applications in Manufacturing,* S. K. Ong and A. Y. C. Nee (eds.), Springer-Verlag, London, 2004, pp. 43–61.
31. D. Blackmore and M. C. Leu, "Analysis of Swept Volume via Lie Groups and Differential Equations," *Int. J. Robot. Res.,* **11**(8), 516–537 (1992).
32. K. M. Bryden, "Virtual Reality Helps Convert Fluid Analysis Results to Solutions," http://www.memagazine.org/contents/current/webonly/webex722.html, *ME Magazine Online,* July 22, 2003.
33. S. Bryson, "Virtual Reality in Scientific Visualization," *Commun. ACM,* **39**(5), 62–71 (1996).
34. L. Lu, M. Connell, and O. Tullberg,, "The Use of Virtual Reality in Interactive Finite Element Analysis: State of the Art Report," paper presented at the AVR II and CONVR Conference, Gothenburg, Sweden, Oct. 4–5, 2002, pp. 60–68.

35. J. Berta, "Integrating VR and CAD," *IEEE Comp. Graph. Appl.,* **19**(5), 111–19 (1999).

36. T. P. Yeh and J. M. Vance, "Combining MSC/NASTRAN, Sensitivity Methods, and Virtual Reality to Facilitate Interactive Design," *Finite Elem. Anal. Design,* **26,** 161–169 (1997).

37. M. Schulz, T. Reuding, and T. Ertl, "Analyzing Engineering Simulation in a Virtual Environment," *IEEE Comp. Graph. Appl.,* **18**(6), 46–52 (1998).

38. M. Schulz, T. Ertl, and T. Reuding, "Crashing in Cyberspce—Evaluating Structural Behaviour of Car Bodies in a Virtual Environment," in *Virtual Reality Annual Interational Sysposium Proceedings,* Atlanta, Georgia, Mar. 14–18, 1998, pp. 160–166.

39. M. Connell and O. Tullberg, "A Framework for the Interactive Investigation of Finite Element Simulations within Virtual Environments," in *Proceedings of the 2nd International Conference on Engineering Computational Technology,* Leuben, Belgium, Sept. 6–8, 2000.

40. M. Connell and O. Tullberg, "A Framework for Immersive FEM Visualization Using Transparent Object Communication in a Distributed Network Environment," *Adv. Eng. Software,* **33,** 453–459 (2002).

41. S. Bryson and C. Levit, "The Virtual Wind Tunnel," *IEEE Comp. Graph. Appl.,* **12**(4), 25–34 (1992).

42. V. Jaswa, "CAVEvis: Distributed Real-Time Visualization of Time-Varying Scalar and Vector Fields Using the CAVE Virtual Reality Theater," in *Proceedings of IEEE Visualization,* Phoenix, AZ, Oct. 19–24, 1997, pp. 301–308.

43. V. Shahnawaz, J. Vance, and S. Kutti, "Visualization of Post-Process CFD Data in a Virtual Environment," in *Proceedings of the ASME Computers in Engineering Conference,* Las Vegas, NV, Sept. 12–15, 1999, DETC99/CIE-9042.

44. VRAC, Virtual Reality Applications Center, http://www.vrac.iastate.edu/~jmvance/CFD/CFD.html.

45. M. Schirski, A. Gerndt, T. van Reimersdahl, T. Kuhlen, P. Adomeit, O. Lang, S. Pischinger, and C. Bischof, "ViSTA FlowLib—A Framework for Interactive Visualizaiton and Exploration of Unsteady Flows in Virtual Environments," in *Proceedings of the 7th International Immersive Projection Technologies Workshop and the 9th Eurographics Workshop on Virtual Environments,* ACM Siggraph, Zurich, Switzerland, May 22–23, 2003, pp. 77–85.

46. Center for Computing and Communication, Aachen University, Germany, http://www.rz.rwth-aachen.de/vr/research/projects/VisIntCFD_e.php.

47. Van Reimersdahl, T., F. Bley, T. Kuhlen, and C. Bischof, "Haptic Rendering Techniques for the Interactive Exploration of CFD Datasets in Virtual Environments," in *Proceedings of the 7th International Immersive Projection Technologies Workshop and the 9th Eurographics Workshop on Virtual Environments,* ACM Siggraph, Zürich, Switzerland, May 22–23, 2003, pp. 241–246.

48. A. Liverani, F. Kuester, and B. Hamann, "Towards Interactive Finite Element Analysis of Shell Structures in VR," in *Proceedings of IEEE Information Visualization,* San Francisco, CA, Oct. 24–29, 1999, pp. 340–346.

49. Michael, J., and J. M. Vance, "Applying Virtual Reality Techniques to the Interactive Stress Analysis of a Tractor Lift Arm," *Finite Elem. Anal. Design,* **35**(2), 141–155 (2000).

50. D. Rantzau and U. Lang, "A Scalable Virtual Environment for Large Scale Scientific Data Analysis," *Future Generation Comp. Sys.,* **14**(3–4), 215–222 (1998).

51. J. J. Laviola, "MVST: A Virtual Reality-Based Multimodal Scientific Visualization Tool," in *Proceedings of the Second IASTED International Conference on Computer Graphics and Imaging,* Palm Springs, CA, Oct. 21–27, 1999, pp. 221–225.

52. X. Fang, A. P. Hung, and S. Tan, "Driving in the Virtual World," in *Proceedings of the 1st Human-Centered Transportation Simulation Conference,* Iowa City, IA, Nov. 4–7, 2001, CD-ROM.

53 L. Chen, Y. Papelis, G. Watson, and D. Solis, "NADS at the University of Iowa: A Tool for Driving Safety Research," in *Proc. of the 1st Human-Centered Transportation Simulation Conference,* Iowa City, IA, Nov. 4–7, 2001, CD-ROM.

54. NADS, http://www.nads-sc.uiowa.edu/.

55. P. Grant, B. Artz, J. Greenberg, and L. Cathey, "Motion Characteristics of the VIRTTEX Motion System," in *Proceedings of the 1st Human-Centered Transportation Simulation Conference,* Iowa City, IA, Nov. 4–7, 2001, CD-ROM.

56. Ford Motor Company, http://www.ford.com/en/innovation/safety/driverDistractionLab.htm.

57. D. Yun, J. Shim, M. Kim, Y. Park, and J. Kim, "The System Development of Unmanned Vehicle for the Tele-operated System Interfaced with Driving Simulator," in *Proceedings of IEEE International Conference on Robotics & Automation,* Seoul, Korea, May 21–26, 2001, pp. 686–691.

58. T. Singh, T. Kesavadas, R. M. Mayne, J. Kim, and A. Roy, "Design of Hardware/Algorithms for Enhancement of Driver-Vehicle Performance in Inclement Conditions Using a Virtual Environment," *SAE Trans. J. Passenger Car—Mechanical Systems,* **109,** 432–443 (2001).

59. M. Sirdeshmukh, "Development of Personal Computer Based Driving Simulator for Education to Prevent Drunk Driving," M.S. Thesis, University of Missouri-Rolla, Rolla, MO, 2004.

60. K. R. Nandanoor, "Development and Testing of Driving Simulator with a Multi-Wall Virtual Environment," M.S. Thesis, University of Missouri-Rolla, Rolla, MO, 2004.

61. P. Green, C. Nowakowski, K. Mayer, and O. Tsimhoni, "Audio-Visual System Design Recommendations from Experience with the UMTRI Driving Simulator," in *Proceedings of Driving Simulator Conference North America,* Dearborn, MI, Oct. 8–10, 2003, CD-ROM.

62. C. Cruz-Neira, "Virtual Reality Overview," in *ACM SIGGRAPH'93 Notes: Applied Virtual Reality, ACM SIGGRAPH'93 Conference,* Anaheim, CA, Aug. 1–3, 1993, pp. 1–18.

63. T. Kesavadas and M. Ernzer, "Design of an Interactive Virtual Factory Using Cell Formation Methodologies," in *Proceedings of the ASME Symposium on Virtual Environment for Manufacturing,* Nashville, TN, MH-Vol. 5, Nov. 14–19, 1999, pp. 201–208.

64. Virtual Reality Laboratory, University of Buffalo, http://www.vrlab.buffalo.edu/project_vfact/vfact.html.

65. R. Chawla and A. Banerjee, "A Virtual Environment for Simulating Manufacturing Operations in 3D," in *Proceedings of the 2001 Winter Simulation Conference,* Arlington, VA, Vol. 2, Dec. 9–12, 2001, pp. 991–997.

66. R. Chawla and A. Banerjee, "An Automated 3D Facilities Planning and Operations Model Generator for Synthesizing Generic Manufacturing Operations in Virtual Reality," *J. Adv. Manufacturing Sys.,* **1**(1), 5–17, 2002.

67. D. Kibira and C. McLean, "Virtual Reality Simulation of a Mechanical Assembly Production Line," in *Proceedings of the Winter Simulation Conference,* San Diego, CA, 2002, pp. 1130–1137.

68. J. Kelsick, J. M. Vance, L. Buhr, and C. Moller, "Discrete Event Simulation Implemented in a Virtual Environment," *J. Mech. Des.,* **125**(3), 428–433 (2003).

69. A. P. Waller, and J. Ladbrook, "Experiencing Virtual Factories of the Future," in *proceeding of the Winter Simulation Conference,* San Diego, CA, Dec. 8–11, 2002, pp. 513–517.

70. W. Mueller-Wittig, R. Jegathese, M. Song, J. Quick, H. Wang, and Y. Zhong, "Virtual Factory—Highly Interactive Visualization for Manufacturing," in *Proceedings of the Winter Simulation Conference,* San Diego, CA, Dec. 8–11, 2002, pp. 1061–1064.

71. F. Taylor, S. Jayaram, U. Jayaram, and T. Mitsui, "Functionality to Facilitate Assembly of Heavy Machines in a Virtual Environment," in *Proceedings of ASME Design Engineering Technical Conference,* Baltimore, MD, 2000 DETC2000/CIE-4590.

72. S. Jayaram, U. Jayaram, Y. Wang, K. Lyons, and P. Hart, "VADE: A Virtual Assembly Design Environment," *IEEE Computer Graphics and Applications,* **19**(6), 44–50 (1999).

73. H. J. Bullinger, M. Richter, K. A. Seidel,, "Virtual Assembly Planning," *Human Fact. Ergonomics Manufact.,* **10**(3) 331–341 (2000).

74. G. Zachmann and A. Rettig, "Natural and Robust Interaction in Virtual Assembly Simulation," in *Proceedings of the 8th ISPE International Conference on Concurrent Engineering: Research and Applications,* Anaheim, CA, July 28–Aug. 1, 2001, CD-ROM.

75. Assembly Simulation wtih Virtual Reality, University of Bonn, Germany, http://web.informatik.uni-bonn.de/II/ag-klein/people/zach/projects/fhg/ems.html.

76. X. Yuan and S. Yang, "Virtual Assembly with Biologically Inspired Intelligence," *IEEE Trans. Sys. Man Cybernet.,* Part C, **33**(2), 159–167 (2003).

77. F. M. Ng, J. M. Ritchie, and J. E. L. Simmons, "Cable Harness Design and Planning Using Immersive Virtual Reality—a Novel Concurrent Engineering Approach," in *Proceedings of the 15th National Conference on Manufacturing Research,* Bath, UK, Sept. 6–8, 1999, pp. 377–381.

78. T. Fernando, P. Wimalaratne, and K. Tan, "Constraint-Based Virtual Environment for Supporting Assembly and Maintainability Tasks," in *Proceedings of the 1999 ASME Design Engineering Technical Conferences,* Las Vegas, NV, Sept. 12–15, 1999, DETC99/CIE-9043.

79. S. D. McDermott and B. Bras, "Development of Haptically Enabled Dis/Reassembly Simulation Environment," in *Proceedings of the 1999 ASME Design Engineering Technical Conferences,* Las Vegas, NV, Sept. 12–15, 1999, DETC99/CIE-9035.

80. VIRCON demonstration CD, CCIR, University of Teeside, United Kingdom, http://sst.tees.ac.uk/ccir/projects/Specialisation/VIRCON/VirconCD/CCIR-VCS-Web_files/frame.htm.

81. N. Dawood, E. Sriprasert, Z. Mallasi, and B. Hobb, "Implementation of Space Planning and Visualisation in a Real-Life Construction Case Study: VIRCON Approach," paper presented at the Conference on Construction Applications of Virtual Reality, Virginia Tech, Blacksburg, VA, Sept. 24–26, 2003.

82. M. Sarshar and P. Christiansson, "Towards Virtual Prototyping in the Construction Industry: The Case Study of the DIVERCITY Project," in *Proceedings of Incite2004—International Conference on Construction Information Technology,* Langkawi, Malaysia, Feb. 18–21, 2004, pp. 581–588.

83. P. Christiansson, L. D. Dalto, J. O. Skjaerbaek, S. Soubra, and M. Marache, "Virtual Environments for the AEC Sector—The Divercity Experience," paper presented at the European Conference of Product and Process Modelling, Portoroz, Slovenia, Sept. 9–11, 2002.

84. A. F. Waly and W. Y. Thabet, "A Virtual Construction Environment for Preconstruction Planning," *Automation Construct.,* **12**(2), 139–154 (2003).

85. J. Cowden, D. Bowman, and W. Thabet, "Home Design in an Immersive Virtual Environment," in *Proceedings of CONVR Conference on Construction Applications of Virtual Reality,* Blackburg, VA, Sept. 24–26, 2003, pp. 1–7.

86. 3DI Group, Virginia Tech, http://research.cs.vt.edu/3di/.

87. R. R. Lipman, "Immersive Virtual Reality for Steel Structures," in *Proceedings of CONVR Conference on Construction Applications of Virtual Reality,* Blackburg, VA, Sept. 20–26, 2003, pp. 8–19.

88. S. Yerrapathruni, J. I. Messner, A. J. Baratta, and M. J. Horman, "Using 4D CAD and Immersive Virtual Environments to Improve Construction Planning," in *Proceedings of CONVR Conference on Construction Applications of Virtual Reality,* Blacksburg, VA, Sept. 24–26, 2003, pp. 179–192.

89. P. S. Dunston, X. Wang, T. Y. Lee, and L. M. Chang, "Benefits of CAD and Desktop Virtual Reality Integration for Cleanroom Piping Design," in *Proc. CONVR Conference on Construction Applications of Virtual Reality,* Blacksburg, VA, 2003.

90. V. R. Kamat and J. C. Martinez, "Interactive Discrete-Event Simulation of Construction Processes in Dynamic Immersive 3D Virtual Worlds," in *Proceedings of CONVR Conference on Construction Applications of Virtual Reality,* Blacksburg, VA, Sept. 24–26, 2003, pp. 197–201.

91. G. A. Dorn, K. Touysinhthiphonexay, J. Bradley, and A. Jamieson, "Immersive 3-D Visualization Applied to Drilling Planning," *Leading Edge,* **20**(12), 1389–1392 (2001).

92. BP Center for Visualization, University of Colorado at Boulder, http://www.bpvizcenter.com/Research_Consortia/Research_ImmersiveDrilling.php.

93. P. K. Kaiser, J. Henning, L. Cotesta, and A. Dasys, "Innovations in Mine Planning and Design Utilizing Collaborative Immersive Virtual Reality (CIVR)," in *Proceedings of the 104th CIM Annual General Meeting,* Vancouver, Canada, Apr. 28–May 1, 2002, CD-ROM.

94. MIRARCO, http://www.mirarco.org/aboutvr.php.

95. J. T Bell, and H. S. Fogler, "The Application of Virtual Reality to Chemical Engineering Education," in *Proceedings of 2004 IEEE Virtual Reality,* Chicago, IL, Mar. 27–31, 2004, pp. 217–218.

CHAPTER 22
PHYSICAL ERGONOMICS

Maury A. Nussbaum
Industrial and Systems Engineering
Virginia Polytechnic Institute and State University
Blacksburg, Virginia

Jaap H. van Dieën
Faculty of Human Movement Sciences
Vrije Universiteit
Amsterdam, The Netherlands

1 WHAT IS PHYSICAL ERGONOMICS?

1.1 Definitions and Brief History

Ergonomics is derived from the Greek *ergon* ("work") and *nomos* ("principle" or "law"). Original attribution is debated but is usually given to the Polish professor Wojciech Jastrzebowski in a treatise published in 1857 and intended to represent the "science of work." The same term seems to have been reinvented in 1949 by the British professor K. F. H. Murrell with the same general intended meaning. In the past decades ergonomics as a scientific and/or engineering discipline has seen a dramatic increase in research and application as well as attention from the general public. As such, the original term has, in cases, lost some of its original meaning. Returning to these origins and to provide a focus for this chapter, the following definition does well to encompass the current state of ergonomics as having a theoretical and multidisciplinary basis, being concerned with humans in systems and ultimately driven by real-work application (Ref. 1):

> Ergonomics produces and integrates knowledge from the human sciences to match jobs, products, and environments to the physical and mental abilities and limitations of people. In doing so it seeks to safeguard safety, health and well being whilst optimising efficiency and performance.

As a recognized professional domain, ergonomics is quite young, becoming formalized only after the World War II. Related efforts were certainly conducted much earlier, such as Borelli's mechanical analysis of physical efforts in the late-seventeenth century and the identification of work-related musculoskeletal illnesses by Ramazzini in the early-eighteenth century. It was only in the 1950s, however, that researchers with expertise in engineering, psychology, and physiology began to come together to realize their common goals and approaches. Ergonomics continues to be an inherently multidisciplinary and interdisciplinary field, which will be evident in the remaining material presented in this chapter.

1.2 Focus of This Chapter

As with any technical domain, the evolution and development of ergonomics have led to increasing specialization and branching. At a high level, three domains presently exist.[2] *Cognitive* ergonomics addresses human mental processes, including information presentation, sensation and perception, and memory. Related applications include mental workload, skill and training, reliability assessment, and human–computer interaction. *Organizational* ergonomics (also called organizational design or macroergonomics) focuses on humans in social–technical systems, addressing the design and evaluation of organizational structures and group processes.

The present chapter focuses solely on the third domain of *physical* ergonomics (PE), primarily because of its relevance to mechanical engineering and the design of mechanical processes and systems. PE addresses diverse physical characteristics of humans, including anatomical, dimensional, mechanical, and physiological, as they affect physical activity (or the consequences thereof). Among a variety of applications, PE is commonly used to address, for example, forceful and/or repetitive motions, workplace layout, tool and equipment design, and environmental stresses.

In the remainder of this section, the fundamental basis of PE is provided, along with its technical foundation and role in the design process, and a summary of key needs motivating PE research and application. In the subsequent section, several common PE analytical methods are described. The last section demonstrates two applications of the material addressing practical occupational design concerns.

This chapter is intended as both an introduction to and overview of PE, and three goals have motivated our presentation. Throughout, we wish to demonstrate the utility, if not the necessity, of PE in the mechanical design process, with a main focus on occupational applications and design for this context. We also intend to provide the reader with a level of familiarity with common contemporary methods and with sufficient information to allow for at least a rough understanding of application to actual problems. Finally, we have limited our formal citations to the literature in order to improve readability but have provided numerous sources, under Additional References, on the topics presented as well as related topics.

1.3 Basis of Physical Ergonomics

At its most basic level, PE is concerned with two aspects of physical effort: (1) the physical *demands* placed on the human and (2) the physical *capabilities* of the human in the situation where the demands are present. Simply, the goal is to ensure that demands do not exceed capacity, as is typical for the design of any mechanical system.

A major challenge in PE is the *measurement* of both demands and capacity in the wide variety of circumstances where human physical exertions are performed (or required, as in the occupational context). Physical requirements vary widely and include dimensions such as force, torque, repetition, duration, posture, etc. Similarly, human attributes vary widely across related dimensions (e.g., strength, endurance, mobility).

Along with the measurement problem is the related problem of *matching*. Given the variability in both demands and capacity, potential mismatches are likely (and prevalent in many cases). Occupationally, poor matching can lead to low productivity and quality, worker dissatisfaction and turnover, and in many cases musculoskeletal illness and injury.

Ongoing research is being widely conducted to address both the measurement and matching problems. In the material presented below, we have given examples of contemporary approaches, but the reader is advised that more advanced methods are currently available and the technology is rapidly advancing.

1.4 Disciplines Contributing to Physical Ergonomics

As noted earlier, ergonomics is inherently a multidisciplinary field. Indeed, it can be argued that ergonomics is essentially the intersection of several more primary fields, both receiving basic information from them and providing applications, tools, and procedures to them. Among the primary disciplines contributing to ergonomics and in which the ergonomist must have a fair degree of knowledge are anatomy, physiology, and mechanics. As will be seen in the subsequent presentation of methods, all are needed at some level to adequately address the measurement and matching problems earlier noted.

1.5 Ergonomics in the Design Process

Where are ergonomists and how do they contribute to the design process? As with any profession, the answer is necessarily diverse. In smaller enterprises, if any individual has responsibility for ergonomics, it is often along with safety and other related topics. In larger enterprises, there may be one or more persons with specific ergonomics training and expertise. In the largest, such as automotive manufacturers, whole departments may exist. Ergonomics consultants are also widely employed, as are ergonomists in academic centers.

Methods by which ergonomics is part of the design process are similarly quite varied. Broadly, ergonomics in design is either proactive or reactive. In the former, procedures such as those described below are applied early in a product or process life cycle, such as in the conceptual or prototype phases. In reactive situations, ergonomics methods are employed only after a design problem has been identified (e.g., a control cannot be reached, required torques cannot be generated, or work-related injuries are prevalent).

A natural question arises as to the need for ergonomics in design. Justification comes from a range of documented case studies, formal experiments, and economic evaluations. Though somewhat oversimplified, ergonomic design attempts to optimize design by minimizing adverse physical consequences and maximizing productivity or efficiency. It should be noted that these goals are at times in conflict, but ergonomic methods provide the tools whereby an optimal balance can be achieved.

An important justification for ergonomics is provided by the high costs of mismatches between demands and capabilities. Occupational musculoskeletal illnesses and injuries (e.g., sprains, strains, low-back pain, carpal tunnel syndrome) result in tens of billions of dollars (U.S.) in worker compensation claims, additional costs related to worker absenteeism and turnover, and unnecessary human pain and suffering. Increasingly, there are legal requirements and expectations for ergonomics in design (e.g., national standards or collective bargaining contracts). With respect to economics, it has been well documented (e.g., Refs. 3

and 4) that occupational ergonomic programs have led to reductions in overall injuries and illnesses and work days missed, with concurrent improvements in morale, productivity, and work quality. Further, ergonomic controls usually require small to moderate levels of investment and resources and do not drastically change jobs, tasks, or operations.

2 PHYSICAL ERGONOMIC ANALYSES

2.1 Overview

In this section we present several of the more common ergonomic methods, tools, and procedures. A focus is maintained on material that is directly relevant to design. In each section, a review of the underlying theory is given, followed by exemplary approaches (e.g., empirical equations or models). In some sections, sample applications to occupational tasks or scenarios are given. Sources of additional information are provided at the end of the chapter.

2.2 Anthropometry

Fundamentals of Anthropometry and Measurement

Anthropometry is the science that addresses the measurement and/or characterization of the human body, either individually or for populations. Engineering anthropometry is more application oriented, specifically incorporating human measures in design Examples include placement of a control so that most individuals can reach it, grip sizing for a hand-held tool, and height of a conveyor. Within ergonomics, anthropometric measures can be classified as either *static* or *functional*. The former are fundamental and generally fixed measures, such as the length of an arm or a body segment moment of inertia. Such static data are widely available from public and commercial sources. Functional measures are obtained during performance of some task or activity and may thus depend on several individual factors (e.g., training, experience, motivation). These latter measures are specific to the measurement situation and are hence relatively limited. Despite the availability of static measures, it is the functional measures that are directly relevant in design. The remainder of this section provides an overview of applied anthropometric methods. Results from anthropometry will also be critical in subsequent sections that address mechanical loading during task performance.

Static anthropometric measures are of four types: linear dimensions (e.g., body segment lengths), masses or weights, mass center locations, and moments of inertia. Linear dimensions can be obtained quite simply using tape measures or calipers, with more advanced recent approaches using three-dimensional (3D) laser scanning. A key issue with respect to linear dimensions is the differentiation between surface landmarks and underlying joint centers of rotation. The former are easily located (e.g., the lateral and medial boney "knobs" above the ankle joint), and methods have been developed to translate these to estimates of underlying joint centers that are required for biomechanical modeling (Section 2.4).

Mass (and/or volume) measures are often obtained using liquid immersion, though as noted above, recent scanning methods are also being employed. Locations of segment (or whole-body) center of mass can also be obtained using liquid immersion and a number of segmental balance methods. Segment moments of inertia are usually obtained using dynamical tests, where oscillatory frequencies are obtained during natural swinging or following a quick release. Representative geometric solids (e.g., a truncated cone) can also be used to model body parts and obtain analytical estimates.

Anthropometric Data and Use

Several large-scale anthropometric studies were conducted in the 1960s and 1970s, mostly in industrialized countries. Contemporary studies are typically of smaller scope, with the exception of the ongoing CAESAR (Civilian American and European Surface Anthropometry Resource) project (http://store.sae.org/caesar/). Anthropometric data are generally presented in tabular form, with some combination of means, standard deviations, and population percentiles. A normal statistical distribution is usually assumed, a simplification which is reasonable in most cases, though which also leads to larger magnitudes of errors at extremes of populations (e.g., the largest and smallest individuals).

Standard statistical methods can be employed directly for a number of applications. If, for example, we wish to design the height of a doorway to allow 99% of males to pass through unimpeded, we can estimate this height from the mean (μ) and standard deviation (σ) as follows (again assuming a normal distribution). Male stature has, roughly, $\mu = 175.58$ and $\sigma = 6.68$ cm. The standard normal variate, z, is then used along with a table of cumulative normal probabilities to obtain the desired value:

$$z_A = \frac{Y - \mu}{\sigma} \qquad \text{or} \qquad Y = \mu + z_A\sigma$$

where z_A is the z value corresponding to a cumulative area A and Y is the value to be estimated. Here, $z_{0.99} = 2.326$, and thus $Y = 191.1$ cm, or the height of a 99th percentile male. Clearly, however, further consideration is needed to address a number of practical issues. These include the relevance of the tabular values, whether this static value is applicable to a functional situation, and if/how allowances should be made for clothing, gait, etc.

Percentile calculations, as given above, are straightforward only for single measures. With multiple dimensions, such as several contiguous body segments, the associated procedures become more involved. To combine anthropometric measures, it is necessary to create a new distribution for the combination. In general, means add, but variances (or standard deviations) do not. Equations are given below for two measures, X and Y (a statistics source should be consulted for methods appropriate for $n > 2$ values):

$$\mu_{X \pm Y} = \mu_X \pm \mu_Y$$

$$\sigma_{X \pm Y} = [\sigma_X^2 + \sigma_Y^2 \pm 2 \operatorname{cov}(X, Y)]^{1/2} \qquad \text{or} \qquad \sigma_{X \pm Y} = [\sigma_X^2 + \sigma_Y^2 \pm 2(r_{XY})(\sigma_X)(\sigma_Y)]^{1/2}$$

where $\pm$ indicates addition if measures are to be added and subtraction otherwise, cov is the covariance, and r is the correlation coefficient. As can be seen from these equations, the variance (σ^2) of the combined measure reduces to the sum of the individual variances when the two measures are independent, or $\operatorname{cov}(X,Y) = r_{XY} = 0$. Human measures are generally moderately correlated, however, with r on the order of 0.2–0.8 depending on the specific measures.

Anthropometry in Design

Anthropometric data are often estimated using predictive formulas or standardized manikins. Most often, these approaches are intended as indicators of the "average" human. As such, its utility can be limited in that there is no individual who is truly average across multiple dimensions, and relationships between measures may not be linear or the same between people. For example, a person with a 50th percentile arm length likely does not have a 50th percentile leg length (it may be close or quite different). Further, many anthropometric tables only present average values (e.g., for center-of-mass location), making estimates of individual differences impossible.

A second limitation in the application of anthropometry arises from potential biases. As noted above, most of the larger datasets were derived several decades ago, thus not accounting for general and nontrivial secular trends toward larger body sizes across all populations. Many of these studies were also performed on military populations, and questions arise as to whether the values are representative in general. Additional biases can arise due to ethnic origins, age, and gender. Overall, application of anthropometric data requires careful attention to minimize such sources of bias.

Three traditional approaches have been employed when using anthropometry in design. Each may have value, depending on the circumstances, and differ in their emphasis on a portion of a population. The first, and most straightforward, is *design for extremes*. In this approach, one "tail" of the distribution in a measure is the focus. In the example above for door height, the tall males were of interest, since if those individuals are accommodated, then all shorter males and nearly all females will as well. Alternatively, the smaller individual may be of interest, as when specifying locations where reaching is required: If the smallest individual can reach it, so will the larger ones.

The second approach, *design for average*, focuses on the middle of the distribution. This has also been termed the "min–max" strategy, as it addresses the minimal dimension needed for small individuals and the maximal dimensions for large individuals. A typical nonadjustable seat or workstation is an example of designing for the average. In this case, both the smallest and largest users may not be accommodated (e.g., unable to find a comfortable posture).

Design for adjustability is the third approach, and this seeks to accommodate the largest possible proportion of individuals. For example, an office chair may be adjustable in height and/or several other dimensions. While this approach is generally considered the best among the three, with increasing levels or dimensions of adjustability comes increasing costs. In practice, designers must balance these costs with those resulting from failure to accommodate some users.

In all cases, the design strategy usually involves a goal or criterion for accommodation. Where the large individual is of concern (e.g., for clearance), it is common practice to design for the 95th percentile males. Similarly, the 5th percentile female is used when the smaller individual is of concern (e.g., for reaching). When the costs of failure to accommodate individuals is high, the tails are typically extended. From the earlier example, it might be desirable to ensure that 99.99% (or more) of the population can fit through a doorway.

Application of anthropometry in the design process usually involves a number of steps. Key anthropometric attributes need to first be identified, then appropriate sources of population data (or collect this if unavailable). Targets for accommodation are usually defined early (e.g., 99%) but may change as costs dictate. Mock-ups and/or prototypes are often built, which allow for estimating whether allowances are needed (e.g., for shoe height or gait in the doorway example). Testing may then be conducted, specifically with extremes of the population, to determine whether accommodations meet the targets.

2.3 Range of Motion and Strength

A number of measures are required to describe the capacity of an individual (or population) to achieve task performance (e.g., reach, lift, pull). Joint range-of-motion (ROM, also called mobility or flexibility) and joint (or muscle) strength begin to describe capacity and are especially relevant for tasks performed briefly or infrequently. Additional information will be required for highly demanding, prolonged, or frequent tasks, as well as additional types of measures (e.g., fatigue and environmental stress as described below).

Range of Motion

Joint ROM refers to the limits of joint motion and is represented as rotations about a given joint or of body segments (e.g., torso flexion). Two different forms of ROM are commonly measured. The first, passive (or assisted), involves external sources of force or moment to achieve joint motion. Examples include the use of gravity during a squat, to assess knee flexion, or forces/moments applied by an experimenter or device. The second, active ROM, requires muscle contraction to achieve joint motion and is associated with narrower motion limits than passive. In practice, the relevant type of ROM is determined by task requirements.

Measuring ROM from individuals is possible using a variety of equipment, from low-cost goniometers (for measuring included angles) to high-cost and sophisticated marker tracking systems. More often, population ROM data are obtained from a number of accessible sources (often in conjunction with anthropometric data). A number of factors can be expected to have an influence on ROM. Although ROM decreases with age, the changes are usually minimal in healthy individuals until the end of typical working life (i.e., 65). Women generally have higher ROM ranges, although gender differences are typically $<10\%$. Little association has been found between anthropometry and ROM, although ROM does decrease with obesity. In simple cases, such as those involving one joint, application of ROM data is straightforward and follows similar methods described in anthropometry (e.g., using percentiles). When multiple joints are involved, it is common to use human modeling software to assess the potential limitations due to ROM.

Strength

In common terminology, strength is an obvious concept. In practice, strength is not a single measure, but instead is specific to the kinematic requirements of a task. Depending on these requirements, strength might be measured as a force (e.g., capacity to pull) or a moment (e.g., shoulder elevation) for an isolated body segment, multiple segments, or the body as a whole. The capacity for an individual muscle to generate force depends on its length and velocity, with associated length–tension and velocity–tension relationships. Temporal dependencies (e.g., duration of effort) are discussed in Section 2.5.

Strength is typically measured using maximum voluntary efforts. An individual is placed in a specified posture or performs a specified motion while a load cell is used to record the relevant kinetics. Extensive data are available on population strengths, in many cases broken down by gender and/or age. Similar procedures as those described for anthropometry can again be used for design purposes (i.e., determining populations percentages with sufficient strength). These procedures usually involve comparing strength (as a measure of capacity) with task demands (determined using biomechanics, as described in the next section).

The vast majority of compiled population strength data are for static exertions, wherein postures are fixed (hence no length or velocity effects). Compared to dynamic exertions, data collection and interpretation are simpler, faster, and less expensive. Unfortunately, real-world efforts normally involve motion. Thus, there is a fundamental mismatch between available measures of capacity and task demands. Further, static tests usually have low associations with dynamic performance capabilities. When tasks do not involve substantial dynamics, use of static data should involve relatively minimal errors. When dynamics are substantial, assessing both capacity and demands becomes quite complex (and is currently a topic of extensive research).

Several factors can have important influences on strength. Women, on average, are weaker than males by approximately 35%, though the differences are highly joint and task specific. In addition, there is extensive overlap between the two gender distributions. Strength normally peaks in the late twenties to early thirties, and decreases with increasing rapidity afterward. By age 40, 50, and 70, typically strength reductions are on the order of 5, 20, and 40%, respectively. Strength has only a minor to moderate relationship with body size, and most attempts at predicting strength from anthropometry have been relatively inaccurate.

As noted above, there are length and velocity effects on muscle capacity, and these, in turn, are seen as posture and motion effects on strength. For most joints, strength is maximal near the middle of the ROM (e.g., an elbow angle of 90°), though several exceptions do exist. When muscles shorten during an exertion (a concentric effort), the capacity for force decreases with increasing velocity (there is a small increase in force capacity when muscles lengthen in an eccentric effort). When combined with the inertial effects of body segments, strength declines with increasing motion speed.

2.4 Biomechanics

Biomechanics in Ergonomics

Biomechanics is the application of classical mechanics to biological systems such as the human body. Although this comprises many branches of mechanics, in the context of ergonomics, we will deal only with dynamics and to a limited extent the mechanics of materials. In physical ergonomics, the human body can be seen as a mechanical system or rather a part of a mechanical system comprising also the tools, objects, and environment with which the human operator interacts. Inverse dynamics can be applied to the analysis of this mechanical system to estimate forces and moments being produced by or acting on the human body, whereas mechanics of materials contributes toward understanding the effects of these mechanical loads on the body. Direct measurement of forces and moments in the human is not feasible for practical and ethical reasons. Consequently the vast majority of available methods and data rely to some extent on inverse dynamical models.

Estimating Joint Moments

The principles of inverse dynamics will probably be known to any mechanical engineer and can be found in dynamics textbooks or specialty books on biomechanics (see Additional References). To summarize, application of the equations of motion based on Newton's second law ($\Sigma F = ma$) and Euler's extension of this law to angular motions ($\Sigma M = I\alpha$) to a system of linked segments is used to yield forces and moments acting at each of the segments in each of the links. In biomechanical analyses, moments about the joints of the human body are usually of interest, since these reflect the combined effect of all muscles spanning the joint. Note that physically these moments are thus the effect of muscle forces. In actual analysis, the moments of force of the muscles appear as a lumped moment in the moment equilibrium equation, and they do not appear in the force equilibrium equation. To perform this type of analysis, masses and moments of inertias of the segments and the accelerations of the segments need to be known. Finally, if two or more external forces act on the system, all but one of these need to be known. If all external forces are known, redundant information is available which can be used to validate the model with respect to anthropometric assumptions (see below) or to decrease the estimation errors that would result from assuming accelerations to be zero (see below).

A simplified example of a linked segment model that was used to estimate the moment on the knee while climbing ladders with different rung separations is given in Fig. 1. Video data were used to approximate the positions of centers of mass of the foot and lower leg as well as the joint rotation centers of the knee and ankle. Forces on a rung were measured with a force transducer. Segment masses were estimated on the basis of anthropometric data. First, a free-body diagram of the foot was created, and the reaction force at the ankle was calculated by equating the sum of the forces on this segment to its mass times acceleration. Next, the moment at the ankle was calculated from equating the sum of the moments to inertia times angular acceleration. Subsequently, the opposites of this force and moment were used as input for a free-body diagram of the lower leg and the reaction force and moment at the knee were calculated.

Segment masses and moments of inertias cannot be directly measured when dealing with the human body. Therefore estimations need to be made on the basis of anthropometric

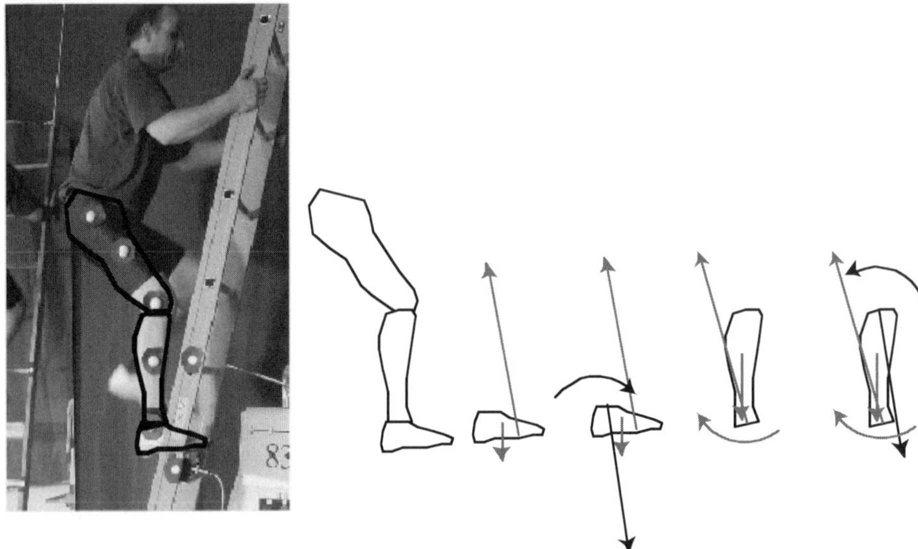

Figure 1 Example of linked segment model used to estimate moment on the knee while climbing ladders with different rung separations.

models (see the discussion of fundamentals in Section 2.2). Obviously these estimations may introduce errors, and the magnitude of such errors can be gauged by making use of redundant information when all external forces have been measured. For example, the moment about the low back in lifting can be calculated on the basis of a model of the lower body (legs and pelvis) using measured ground reaction forces on each foot as input. This same moment can also be estimated using a model of the upper body (arms and trunk) and the object lifted. It has been shown that with a careful choice of anthropometric assumptions, errors in moment estimates will generally be below 10 N·m.

Accelerations can be measured or calculated from position data by double differentiation. This involves labor-intensive measurements and is only feasible when at least a mock-up of the situation to be analyzed is available. Consequently, in many ergonomic applications, accelerations are assumed to be zero, in which case only the static configuration of the human body needs to be known or predicted. This simplification will lead to underestimation of mechanical loads on the human body in dynamic tasks, which in some cases can be substantial (e.g., in manual lifting, the moments around the low back may be underestimated by a factor of 2). Such errors may lead to questionable conclusions, even in a comparative analysis. For instance, earlier studies comparing stoop and squat lifting techniques appear to have been biased toward favoring the squat technique due to the application of static models. Consequently, early studies have often reported a lower low-back load in squat lifting as compared to stoop lifting, whereas more recent studies using dynamic models have reported the opposite.[5] If an analysis of a dynamic task is performed assuming acceleration to be zero but inputting the measured external forces on the body into the analysis—the so-called quasi-dynamic approach—reasonable estimates of joint moments result.

A second simplification often used is to assume that all movement and force exertion takes place in a single plane, which allows application of a 2D model. In analyses of asymmetric lifting tasks, this can cause significant and substantial errors in estimated moments around the low back (roughly 20% when the load is placed 30° outside of the primary plane of movement). At 10° of asymmetry, differences between 2D and 3D analyses have been found to be insignificant.

Data collection required for the estimation of net moments is usually not prohibitive in a comparative analysis of working methods and techniques, since this can be done in a laboratory mock-up setting. However, for monitoring and identification of the most stressful tasks or task elements, field measurements covering long periods are desirable. In this case, use of an inverse dynamics approach usually is prohibitive. Methods have been developed to estimate moments based on measurements of the electrical activity of muscles, the latter using electrodes applied on the skin overlying the muscle group of interest (electromyography, or EMG). However, it has been shown that additional kinematic data and extensive calibrations are needed to obtain valid estimates. Currently miniature kinematic sensors and efficient calibration procedures are being developed and tested to facilitate this type of measurements. Finally estimation of mechanical loads in the design stage can be done using inverse dynamics when external forces are known and postures (and movements) can be predicted. Several software programs which can in some cases be integrated with computer-aided design (CAD) applications allow for such analyses. Note that these models usually are static (assume accelerations to be zero), and the validity of the analysis will depend on the validity of the posture predictions made by the software or the user.

An indication of how load magnitude, as expressed by the moment about a joint, relates to the capacity of the musculoskeletal system can be obtained by comparison of the moments during a task to maximum voluntary moments. Usually such comparisons are made with the results of isometric strength tests (see Section 2.3). For example, lifting a 20-kg load manually has been predicted to exceed the shoulder strength of about 30% of the general population, whereas the same task performed with a hoist allowed over 95% to have sufficient strength.[6] Since many tasks are dynamic in nature and both joint angle and angular velocity strongly affect the moment capacity, dynamic reference data are needed. As noted earlier, however, such reference data are only partially available. Some commercial software packages provide a comparison of joint moments with population strength data, though the latter are usually static.

Several studies have shown that inverse dynamics of human movement can provide reliable and accurate estimates of joint moments. It should be noted, though, that joint moments do not always provide a definitive answer as to the actual extent of musculoskeletal loading, as will be discussed in the next paragraph. Further, since human motor behavior can be quite variable, the reliability of moment estimates derived from limited numbers of measurements, and more so when derived from model simulations, should be considered with care. When comparative analyses are done, substantial differences in net moments (e.g., >10%) will usually allow conclusions to be drawn with respect to musculoskeletal loading. For normative interpretation of joint moments with data on muscle strength, it is recommended that a margin of safety be included in view of the variability and sources of error both in moment estimates and strength data.

Estimating Muscle Forces

As discussed above, the joint moment reflects the combined effect of all muscle forces (and of elastic forces in passive structures like ligaments). A fundamental problem in biomechanics is that muscle forces cannot generally be calculated from the net moment, since the number of muscles spanning a joint creates an indeterminate problem. To relate mechanical loads imposed by a task to, for example, the mechanical tolerance of a joint, this problem needs to be solved.

Let us consider as an example the compression force acting on the contact surface of a joint. Since muscles act with small leverage about a joint relative to external forces and inertial forces on body segments, muscle forces greatly exceed the reaction forces calculated through inverse dynamics. Forces in ligaments (and other passive structures) are low, except at extreme joint positions. Consequently, the compression force on a joint is mainly determined by muscle forces. However, their magnitude will depend on which muscles (with what lever arms) produce the joint moment, and they can also be strongly affected by the level of cocontraction (i.e., activity of muscles producing moments opposite to the total moment around the joint).

Several types of models have been developed to obtain estimates of muscle force. Note that all of these models use the joint moment as a starting point of the estimation procedure. In addition, all rely on an anatomical model of the joint and surrounding muscles. Models can be differentiated on the basis of the way in which muscle activation is estimated. One class of models estimates muscle activation from measures of the electrical activity of muscles (EMG). Another class of models uses static optimization to simulate assumed optimal control of muscle activation. Models used in design software are mostly of the latter class. Usually these models rely on a cost function related to the efficiency of moment production. Consequently, antagonistic cocontraction is predicted to be absent. This has been shown not to produce large errors in predictions of forces on the spine for a wide range of different tasks, provided that a realistic cost function is used. However, effects of specific task conditions, which affect levels of cocontraction, will go unnoticed. At present, only EMG-based models account for modulations of the level of cocontraction, but other methods may be made to predict cocontraction, for instance by imposing constraints on stability of the joint.

Finally, it should be noted that close relationships between joint moments and joint compression force estimates have been found for the shoulder as well as the low back. This suggests that joint moments can be sufficient for comparative analyses of joint and muscle loading in many cases, though this conclusion may not hold for other joints.

Estimating Tissue Tolerances and Their Use in Design

To allow a normative interpretation of estimates of mechanical loads, such as joint compression forces, information on the strength of the loaded structures and tissues is needed. This will, in theory, also allow setting workload or design standards, but such an approach has had only limited application. Setting such standards is hampered by difficulties in obtaining estimates of tissue loading as well as difficulties in acquiring valid data on mechanical tissue strength. Tissue strength estimates obviously can be obtained only from cadaver material or nonhuman tissues. The validity thereof is fundamentally unsure, and availability is limited. In addition, biological materials display viscoelasticity, making strength dependent on loading rate and loading history. Furthermore, it has been a matter of debate whether normative data should be based on ultimate strength or values below that (e.g., yield strength). It is conceivable that subfailure damage can cause musculoskeletal disorders, suggesting that yield strength might be the correct criterion. Conversely, subfailure damage may be reparable and even a prerequisite for maintenance and adaptation of biological tissues. Finally, the strength of biological structures is highly variable between specimens, making it quite difficult to define thresholds.

For spinal loading, many ergonomic studies and some guidelines have proposed using spine compression to establish whether a certain load would be acceptable or not for occupational lifting. Compression of spinal motion segments (vertebrae and intervening disc) causes a fracture of the endplate, the boundary layer between the vertebrae and the intervertebral disc, at forces between 2 and 10 kN. Since extensive data are available on compression strength and the underlying properties of the spine, a population distribution of spinal strength can be estimated. This population distribution can be used to estimate the percentage of the population for which a given task, with a known peak compression force, is theoretically above the injury threshold. Based on this approach, a widely cited threshold (based on the 25th percentile of the general population) of 3.4 kN has been suggested by the National Institute for Occupational Safety and Health (NIOSH).[7] This threshold has been criticized because it does not account for the wide variability in compression strength. It thus on one hand is too limiting, especially for young males, and on the other hand may not provide enough protection for some groups, especially older females. Moreover, experiments using repetitive compression have shown that spinal motion segments fail at much lower forces versus one-cycle loading. Models based on creep failure and fatigue failure have been able to describe strength in repetitive loading, allowing a similar approach for cyclic tasks.

In view of the questionable validity of strength data, limiting the application to comparative analyses seems indicated. Specifically, tissue strength data can be used to assess the effects of a design toward reducing physical loading. For example, given the 2–10-kN range in compression strength of the human lumbar spine, a 50-N reduction in compression can be considered negligible. In contrast, spinal compression forces in manipulating a 20-kg load manually or with a hoist have been estimated at approximately 3 and 1.2 kN, respectively.[6] A reduction in compression of such magnitude as this is likely to substantially reduce the probability of injury (of course assuming that other task factors remain comparable).

2.5 Whole-Body and Localized Fatigue

In Section 2.3, strength was considered as the ability to generate a force or moment. This capacity, however, has a temporal dependency: The larger the effort, the smaller the duration over which it can be maintained. Fatigue, in turn, can be defined as the loss of capacity to generate a physical exertion as a result of prior exertions. Limits to physical exertion capacity can be more broadly discriminated as those involving the whole body (or large portions thereof), or more localized (e.g., a single joint).

Whole-Body Fatigue

For repetitive whole-body efforts, fatigue typically involves a loss of capacity in the body's metabolic system, or an inability to generate the necessary energy. Over a short duration, less than ~5 min, aerobic capacity (AC) provides a relevant measure and is defined as the maximum energy production rate using aerobic means (using oxygen, versus anaerobic metabolism). AC can be directly measured using a stress test (on a treadmill or bicycle ergometer) or indirectly estimated using submaximal tests. Average AC values for males and females are roughly 15 and 10.5 kcal/min, respectively, and both show consistent declines with age past ~30 years.

Of more relevance in practice is the ability to generate energy for prolonged periods, or physical work capacity (PWC). PWC is directly related to AC, so that the noted effects of gender and age also apply. Further, both AC and PWC are task dependent, since different muscle groups have different energetic capacities. As an example, AC is about 30% lower for tasks isolated to the upper versus lower extremities. PWC can be represented as a fraction of AC, with the fraction declining as the task duration is prolonged. Roughly 50% of AC can be maintained for up to 1 h and 25–33% of AC for up to 8 h. PWC can be estimated from individual or population AC as follows (units of minutes and kilocalories per minute; Ref. 7):

$$\mathrm{PWC} = \frac{[(\log 4400 - \log(\mathrm{duration})](\mathrm{AC})}{3.0}$$

With an estimate of individual or population PWC, design evaluation requires an assessment of task energetic demands. A range of approaches exist, different in both complexity and accuracy. Tabular values and subjective evaluations can provide quick-and-easy but rough estimations. Alternatively, direct measurement (of oxygen uptake), estimation from heart rate, or use of predictive equations can be used for more precise estimates.

Localized Muscle Fatigue

Localized muscle fatigue (LMF), in contrast to the above, involves an overload of the specific (or regional) capacity of a muscle or muscle group. For relatively simple efforts (e.g., static holds), the relationship between effort and endurance time has been well described. If the effort level is represented as a percentage of maximum voluntary capacity (MVC), $F =$ %MVC, then endurance time (T, in minutes) can be estimated as follows[8]:

$$T = \frac{1.2}{(0.01F - 0.15)^{0.618}} - 1.21$$

It should be noted, however, that the equation provided is only valid for efforts between roughly 20 and 95% MVC (being an empirical fit to a dataset). Endurance times at the lower end have not been well characterized, and endurance times at the higher end are quite brief. The relationship can also vary substantially,[9] including between individuals, muscle groups, with training, etc. Furthermore, it is important to note that endurance represents the culmination of an ongoing fatiguing process, specifically the time at which an effort can no longer be maintained because of declining muscle capacity. Nonetheless, this relationship has important design implications. If muscular efforts are required for more than brief times, the trade-off between effort level and duration on capacity can be estimated. If the duration is fixed, endurance times can be increased by reducing the relative effort (as a %MVC), by either reducing task kinetics or increasing capacity (e.g., using large muscle groups or keeping joints near the middle of the ROM).

It has proven to be difficult to expand the above results to more complex (and more realistic) exertions. Some research has led to guidelines for intermittent static tasks (periods of fixed efforts interrupted by fixed rest periods), but these have yet to be verified for general application. Little, if any, reliable guidelines exist for more complex time-varying exertions.

2.6 Environmental Stress

Human capacity can be both facilitated and compromised by environmental conditions. In this section, we focus primarily on the adverse consequences, with specific emphasis on exposure to heat and vibration and implications for design. A number of other important environmental exposures exist, such as light and sound, for which references are provided at the end of this chapter (see General References).

Heat Stress

The body has a complex control mechanism that maintains core temperature in a narrow range (near 37°C). This control is achieved except under conditions of extreme exposures, with adverse outcomes including heat exhaustion and fainting. Even with less extreme exposures, exposure to heat can be expected to compromise performance, on the order of 7 and 15% with prolonged exposure above 27 and 32°C, respectively.[10]

More quantitative estimates of the effects of heat are conducted by determining the rate of change in body heat content. Important sources to these "heat balance" calculations are metabolic, radiant, convective, and evaporative. The absolute and relative magnitudes of each heat source are dependent on the workload and specific environmental conditions (e.g., airflow, exposure surface areas, air vapor pressure). Individuals differ greatly in their tolerance to hot environments, with major effects due to physical fitness, age, gender, and level of obesity. In addition, individuals can become acclimated through repeated exposure, a process that takes roughly 1–2 weeks, with deacclimation occurring over a similar time period.

Assessment of environmental conditions requires measures that reflect the different sources of heat noted earlier. A common guideline[11] determines recommended exposure limits (in minutes per hour) based on metabolic heat and a compound measure of environmental heat, the wet-bulb-globe temperature (WBGT). The WBGT, in turn, is determined as the weighted average of dry-bulb (air temperature), wet-bulb (accounting for humidity and airflow), and globe (accounting for radiant heat) temperatures.

Additional quantitative assessments can be done by estimation of heat balance. Empirical equations have been developed (e.g., Ref. 11) to predict rates of body heat gain or loss from radiation, convection, and evaporation. Together with estimates (or direct measurement) of metabolic heat, the rate of net body heat storage can be obtained. Allowable levels of heat storage can be estimated (approximately 63 kcal for an individual with average body mass). From these estimates, one can determine an allowable exposure time as the ratio of maximum heat storage to rate of heat gain. If heat exposure is determined to be excessive, control measures can be applied based on the sources of heat storage (e.g., improve airflow to maximize evaporative heat loss, shielding to reduce radiant heat transfer).

Vibration

While historically used in a therapeutic mode (due to analgesic effects), exposure to vibration has been consistently associated with several adverse health outcomes. These include discomfort, reflexive muscle contraction, and decreased fine motor control in the short term and low-back pain and hand/finger circulation impairment in the long term.

Vibration exposure is quantified in both the temporal [peak, average, root-mean-square (rms)] and frequency domains (typically using octave or one-third octave bands). Spectral analysis is of particular utility, as the human response to vibration is strongly dependent on the frequencies of exposure. Different tissues, organs, and body segments also have distinct transmission and resonance responses as a function of frequency.

A number of vibration exposure guidelines have been developed by the International Organization for Standardization (ISO). The differing guidelines address the site of exposure (e.g., whole body or hand/arm), the direction of exposure (axis), posture (e.g., seated or standing), and vibration measures (e.g., acceleration or velocity). Exposure limits are obtained based on a spectral analysis of exposures. Controls for vibration exposure can be categorized as occurring at the source, path, or receiver. Examples of source control include avoiding resonances and using balancing to reduce vibration magnitudes. Control in the vibration path can be achieved by limiting exposure time and using vibration isolation. At the receiver (the human), vibration exposure can be reduced using damping apparel (e.g., gloves) and reducing contact forces and areas.

3 EXAMPLE APPLICATIONS

3.1 Overview

In this section, we present two practical applications of physical ergonomics in design. Each demonstrates the use of several of the methods and procedures presented above. Only the general approach and results are given, with references provided for additional detail.

3.2 Manual Material-Handling Systems

Associations between physically demanding manual handling of materials and work-related musculoskeletal injury have been well documented by ergonomists and epidemiologists. Mechanized assistive devices are widely used to control such problems. Material-handling systems can be generally categorized as positioners (e.g., lift tables, conveyors), used to place or orient objects, and manipulators (articulated arms, hoists), used to move and/or support objects. The latter were the focus of a detailed ergonomic/biomechanical analysis, with a goal of understanding the impact of using manipulators on task performance and the user's physical demands.

Only minimal investigation has been conducted on manipulators, mainly focusing on biomechanical modeling of the static gravitational component. While use of manipulators reduces the static component (i.e., object weight), substantial kinetic loads may result from body segment dynamics and the inertial dynamics of the manipulator and object handled. In addition, use of a manipulator frequently requires complex body postures and motions.

In the first study,[12] two mechanical manipulators (an articulated arm and an overhead hoist; Fig. 2) were used to move objects with moderate masses (10–40 kg) for several short-distance transfers. Differences between the two devices and manual (unassisted) movements were obtained for motion times, applied hand forces, and torso kinematics. Use of either manipulator increased motions times by 36–63% for symmetric motions (in the *y–z* plane)

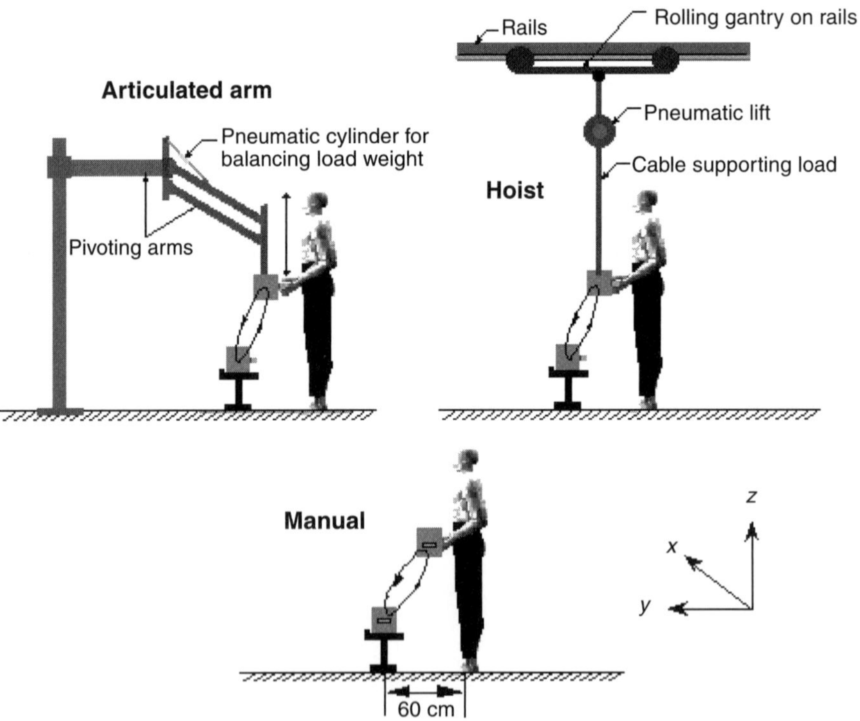

Figure 2 Illustration of object transfers performed manually or with a material-handling manipulator. (Reprinted with permission from *Human Factors,* Vol. 41, No. 2, 1999. Copyright 1999 by the Human Factors and Ergonomics Society.)

and by 62–115% for asymmetric motions (in the *x*–*z* plane). Hand forces were substantially lower when using either manipulator (by 40–50%), while torso motions were generally similar regardless of how the transfers were conducted. For self-paced job tasks (see below for effects of pacing), it was concluded that moderate-mass objects require significant increases in motion times, but with substantially reduced levels of upper extremity exertion.

A second study[6] examined physical loads in more detail using the same manipulators and motions. A 3D inverse dynamics biomechanical model was developed to allow for estimation of the strength demands (required moments versus joint strength) at the shoulders and low back as well as compressive and shear forces in the lower spine (the latter as estimates of the relative risk of low-back injury). Strength demands decreased substantially when using either manipulator in comparison to transfers done manually. Strength demands were also much less affected by increases in object mass. Spine compression and shear were reduced by roughly 40% when using either manipulator, primarily because of the decreased hand forces noted above and the resulting spinal moments. Estimates of trunk muscle activity (using EMG) suggested that the use of a hoist imposes higher demands on coordination and stability, particularly at extreme heights or with asymmetric motions. In contrast, the articulated arm imposed higher demands on strength and resulted in increased spinal compression, likely because of the higher system inertia. From both studies, it was concluded that use of either type of manipulator can reduce physical demands compared with manual methods, though performance (motion times) will be compromised. In addition, either a hoist or articulated arm may be preferable depending on the task requirements (e.g., height and asymmetry).

Two companion studies were conducted to address additional aspects of manipulators as used in practice. Effects of short-term practice (40 repetitions) were addressed in the first study,[13] using the two manipulators to lift and lower a 40-kg object. Nonlinear decreases in several measures were found (e.g., low-back moments and spine compression). These decreases, however, were fairly slow and comparable to rates of learning for manual lifting. Familiarization with material-handling manipulators is thus recommended, and the learning process may be somewhat lengthy. Effects of pacing (speed of object transfer) were determined in the second experiment[14] by having participants perform transfers 20% more rapidly than those presented above. This requirement led to roughly 10% higher hand forces, 5–10% higher torso moments, similar torso postures and motions, and 10% higher spine loads (compression and shear). When manipulators are used in paced operations (e.g., assembly lines) and the noted increases in required motion times are not accounted for, these results suggest that the risk of musculoskeletal injury may be increased.

3.3 Refuse Collection

Refuse collection can be considered a physically demanding job. Several studies have shown that many refuse collectors indeed suffer from musculoskeletal disorders. Especially low-back complaints and shoulder complaints appear to be prevalent. In The Netherlands, this has led to introduction of a job-specific guideline that sets limits on the amount of refuse a collector is allowed to collect per day. This guideline was based on energy demands of the task of refuse collecting, as discussed under Whole-Body Fatigue in Section 2.5. When this guideline was introduced, additional ergonomic measures to reduce workload were encouraged, and it was stated that insofar as these additional measures would lead to a reduction of energy consumption during the task, this could lead to adjustment of the guideline to allow a larger amount of refuse to be collected. A series of studies were performed to evaluate the efficacy of proposed ergonomic measures.

A first series of studies concentrated on mechanical workload in relation to the design of the two-wheeled containers used by Dutch refuse collectors. As a first step, a mechanical simulation model of pushing and pulling these containers was constructed. This 2D model served to evaluate the effect of design parameters on the moments acting around the shoulders and low back. It was predicted that mechanical loading could be strongly affected by the location of the center of mass of the container and the location of the handle. Furthermore the anthropometric characteristics of the user (body height), the direction of the force exerted on the container, and the angle over which it is tilted did strongly affect predictions of mechanical workload. Since the direction of the force exerted and the tilt angle cannot be predicted, it was considered necessary to do an experimental study in a mock-up situation. Four experienced refuse collectors with a wide range of body heights symmetrically tilted, pushed, and pulled a container of which the center of mass and handle positions were systematically varied over 9 and 11 positions, respectively. The position of the center of mass had substantial effect on workload, whereas changes in handle location had limited effect, because the study participants adjusted the tilt angle of the container to accommodate different handle positions. Furthermore, it was found that replacing the center of mass toward the axis could reduce workload relative to that with the standard container in use in The Netherlands.[15]

A second, more comprehensive experiment was performed with nine refuse collectors, again with a wide range of body heights, and including asymmetric tasks.[16] A standard container was compared to a redesigned container with an optimized handle position and larger wheels. In addition, for each of these containers three different positions of the center of mass were evaluated. Participants performed four tasks on a brick-paved road built in the laboratory. The tasks, selected on the basis of systematic observations of refuse collectors at work, were tilting the container and pulling it with one hand, tilting the container and pushing it with two hands, rotating the container, and pulling an empty container up the pavement. A 3D biomechanical model was used to estimate moments about the low back and shoulders (Fig. 3). In addition, models of the shoulder and the low back were used to estimate compression forces on these joints. Participants also gave a subjective rating of the ease of handling of each of the containers.

The use of the redesigned container resulted in a decrease of the force exerted on the container, a decrease of moments around the shoulder and low back, and a decrease of shoulder compression. No significant decrease in lower spine compression was found. Only when pulling the empty container up the pavement was an increase in shoulder joint moment found. In general, the participants rated the redesigned container as more easy to use. It was concluded that the redesign would reduce mechanical workload, especially when the task of pulling the container back up the pavement would be eliminated by leaving it placed against the curb, as is already often done in practice.

Energetic workload in refuse collecting could be reduced by gathering more containers at one collection site in a street than is currently the practice. To study the effect of this measure, a mock-up study was performed.[17] Gathering sites with 2, 16, and 32 containers were compared. Oxygen consumption, heart rate, and perceived exertion were measured in 18 men. In addition, the time it took to collect 32 containers was measured. The size of the gathering point had no effect on oxygen consumption, heart rate, or perceived exertion. However, the time it took to collect 32 containers was lower for the gathering points with 16 and 32 containers as compared to 2 containers. Since this would allow for more recovery during the shift, average oxygen consumption over the day would be decreased, and in the context of the guideline described above this would imply that more refuse can be collected.

Finally, job rotation between refuse collection and driving the refuse truck could be implemented as a measure to reduce workload in refuse collectors. A field study was per-

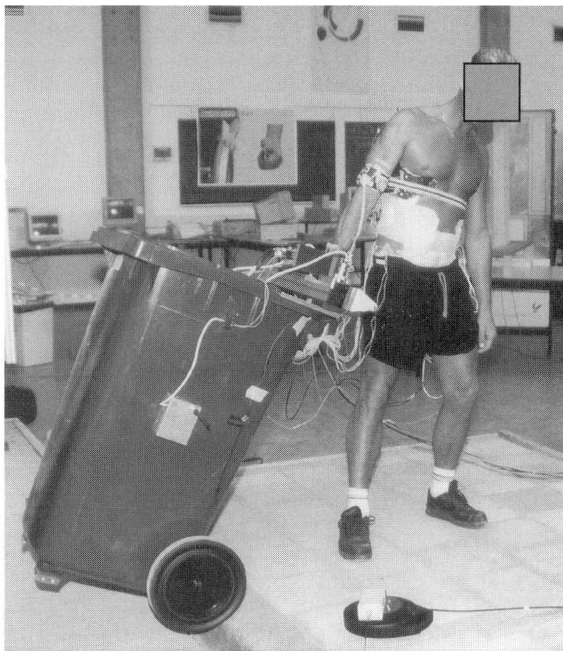

Figure 3 Refuse collector during data collection for estimation of mechanical workload in pulling refuse container up pavement.

formed to evaluate the efficacy of this measure.[18] Three teams of three workers participated in the study. Each team member worked at least one week as a truck driver only, one week as a refuse collector only, and two weeks rotating between these jobs. Of the latter two weeks, in one week the participants rotated between jobs on the same day; in the other week they rotated between days. Physical workload was quantified by measuring heart rate and subjective ratings. From the heart rate measurements the average oxygen consumption as a percentage of the individual maximum was estimated.

Job rotation yielded lower average oxygen consumption over the workday and subjective workload in comparison to refuse collecting only. In contrast, driving was associated with lower oxygen consumption, heart rate, and perceived exertion as compared to job rotation. No differences were found between rotating on or between days. It was concluded that job rotation can contribute to a reduction of physical workload in refuse collectors.

REFERENCES

1. *ISO Working Draft. Ergonomics Principles in the Design of Work Systems* (ISO/CD6385), TC/159/SC1, International Organization for Standardization, Geneva, 1999.
2. "The Discipline of Ergonomics," International Ergonomics Association, www.iea.cc/ergonomics, accessed July 27, 2004.
3. *Worker Protection: Private Sector Ergonomics Programs Yield Positive Results*. GAO/HEHS Publication 97-163, General Accounting Office, Washington, DC, 1997.
4. H. W. Hendrick, *Good Ergonomics Is Good Economics,* Human Factors and Ergonomics Society, Santa Monica, CA, 1996.

5. J. H. van Dieën, M. J. M. Hoozemans, and H. M. Toussaint, "Stoop or Squat: A Review of Biomechanical Studies on Lifting Technique," *Clin. Biomechan.,* **14,** 685–696 (1999).
6. M. A. Nussbaum, D. B. Chaffin, and G. Baker, "Biomechanical Analysis of Materials Handling Manipulators in Short Distance Transfers of Moderate Mass Objects: Joint Strength, Spine Forces and Muscular Antagonism," *Ergonomics,* **42,** 1597–1618 (1999).
7. *Work Practices Guide for Manual Lifting,* DHHS (NIOSH) Publication 81-122, National Institute for Occupational Safety and Health, Washington, DC, 1981.
8. B. W. Niebel, and A. Freivalds, *Methods, Standards, and Work Design,* 10th ed., McGraw-Hill, Boston, MA, 1999.
9. J. H. van Dieën and H. H. E. Oude Vrielink, "The Use of the Relation between Force and Endurance Time," *Ergonomics,* **37,** 231–243 (1994).
10. J. J. Pilcher, E. Nadler, and C. Busch, "Effects of Hot and Cold Temperature Exposure on Performance: A Meta-Analytic Review," *Ergonomics,* **45,** 682–698 (2002).
11. *Occupational Exposure to Hot Environments,* DHHS (NIOSH) Publication 86-113, National Institute for Occupational Safety and Health, Washington, DC, 1986.
12. M. A. Nussbaum, D. B. Chaffin, B. S. Stump, G. Baker, and J. Foulke, "Motion Times, Hand Forces, and Trunk Kinematics When Using Material Handling Manipulators in Short-Distance Transfers of Moderate Mass Objects," *Appl. Ergon.,* **31,** 227–237 (2000).
13. D. B. Chaffin, B. S. Stump, M. A. Nussbaum, and G. Baker, "Low Back Stresses When Learning to Use a Materials Handling Device," *Ergonomics,* **42,** 94–110 (1999).
14. M. A. Nussbaum and D. B. Chaffin, "Effects of Pacing When Using Material Handling Manipulators," *Human Factors,* **41,** 214–225 (1999).
15. I. Kingma, P. P. F. M. Kuijer, M. J. M. Hoozemans, J. H. van Dieën, A. J. van der Beek, and M. H. W. Frings-Dresen, "Effect of Design of Two-Wheeled Containers on Mechanical Loading," *Int. J. Ind. Ergon.,* **31,** 73–86 (2003).
16. P. P. F. M. Kuijer, M. J. M. Hoozemans, I. Kingma, J. H. van Dieën, W. H. K. de Vries, H. E. J. Veeger, A. J. van der Beek, B. Visser, and M. H. W. Frings-Dresen, "Effect of a Redesigned Two-Wheeled Container for Refuse Collecting on Mechanical Loading of Low Back and Shoulders," *Ergonomics,* **46,** 543–560 (2003).
17. P. P. F. M. Kuijer, M. H. W. Frings-Dresen, A. J. van der Beek, J. H. van Dieën, and B. Visser, "Effect of the Number of Two-Wheeled Containers at a Gathering Point on the Energetic Workload and Work Efficiency in Refuse Collecting," *Appl. Ergon.,* **33,** 571–577 (2002).
18. P. P. F. M. Kuijer, W. H. K. de Vries, A. J. van der Beek, J. H. van Dieën, B. Visser, and M. H. W. Frings-Dresen, "Effect of Job Rotation on Work Demands, Workload, and Recovery of Refuse Drivers and Collectors," *Human Factors,* **46,** 437–448 (2004).

BIBLIOGRAPHY

General

Chaffin, D. B., G. B. J. Andersson, and B. J. Martin, *Occupational Biomechanics,* 3rd ed., Wiley, New York, 1999.

Chengular, S. N., S. H. Rodgers, and T. E. Bernard, *Kodak's Ergonomic Design for People at Work,* 2nd ed., Wiley, New York, 2004.

Delleman, N. J., C. M. Haslegrave, and D. B. Chaffin, *Working Postures and Movements. Tools for Evaluation and Engineering,* CRC Press, Boca Raton, FL, 2004.

Konz, S., and S. Johnson, *Work Design: Occupational Ergonomics,* 6th ed., Holcomb Hathaway, Scottsdale, AZ, 2004.

Kroemer, K., H. Kroemer, and K. Kroemer-Elbert, *Ergonomics: How to Design for Ease and Efficiency,* 2nd ed., Prentice-Hall, Englewood Cliffs, NJ, 2001.

Mital, A., A. Kilbom, and S. Kumar, *Ergonomics Guidelines and Problem Solving,* Elsevier Ergonomics Book Series, Vol. 1, Elsevier, Amsterdam, 2000.

Anthropometry, Range-of-Motion, and Strength

Haber, R. N., and L. Haber, One size fits all? *Ergonomics in Design* **5,** 10–17 (1997).

Majoros, A. E., and S. A. Taylor, "Work Envelopes in Equipment Design," *Ergon. Design,* **5,** 18–24 (1997).

Roebuck, Jr., J. A., *Anthropometric Methods: Designing to Fit the Human Body,* Human Factors and Ergonomics Society, Santa Monica, CA, 1995.

Biomechanics

Hibbeler, R. C., *Engineering Mechanics: Dynamics,* Prentice-Hall, Singapore, 1997.

Kumar, S. (ed.), *Biomechanics in Ergonomics,* Taylor and Francis, London, 1999.

Robertson, D. G. E., G. E. Caldwell, J. Hamill, G. Kamen, and S. N. Whittlesey, *Research Methods in Biomechanics,* Human Kinetics, Champaign, IL, 2004.

Zatsiorsky, V. M., *Kinetics of Human Motion,* Human Kinetics, Champaign, IL, 2002.

Manual Material Handling

Ayoub, M. M., and A. Mital, *Manual Materials Handling,* Taylor and Francis, London, 1989.

Davis, K. G., and W. S. Marras, "The Effects of Motion on Trunk Biomechanics," *Clin. Biomechan.,* **15,** 703–717 (2000).

Hoozemans, M. J. M., A. J. van der Beek, M. H. W. Frings-Dresen, F. J. H. van Dijk, and L. H. V. van der Woude, "Pushing and Pulling in Relation to Musculoskeletal Disorders: A Review of Risk Factors," *Ergonomics,* **41,** 757–781 (1998).

J. H. van Dieën, M. J. M. Hoozemans, and H. M. Toussaint, "Stoop or Squat: A Review of Biomechanical Studies on Lifting Technique," *Clin. Biomechan.,* **14,** 685–696 (1999).

Fatigue

Chaffin, D. B., "Localized Muscle Fatigue. Definition and Measurement," *J. Occupat. Med.,* **15,** 346–354 (1973).

Enoka, R. M., and D. G. Stuart, "Neurobiology of Muscle Fatigue," *J. Appl. Physiol.,* **72,** 1631–1648 (1992).

Sjøgaard, G., "Muscle Fatigue," *Med. Sport Sci.,* **26,** 98–109 (1987).

Environmental Stress

Mechanical Vibration and Shock—Evaluation of Human Exposure to Whole-Body Vibrations, ISO 2631-1, International Organization for Standardization, Geneva, 1997.

Mechanical Vibration—Measurement and Evaluation of Human Exposure to Hand-Transmitted Vibration, ISO 5349-1, International Organization for Standardization, Geneva, 2001.

Parsons, K., *Human Thermal Environments,* 2nd ed., Taylor & Francis, London, 2003.

CHAPTER 23
ELECTRONIC MATERIALS AND PACKAGING

Warren C. Fackler, P.E.
Telesis Systems
Cedar Rapids, Iowa

1 INTRODUCTION

1.1 Scope

Electronic packaging is a multidisciplinary process consisting of the physical design, product development, and manufacture required to transform an electronic circuit schematic diagram into functional electronic equipment.

The categories of technical knowledge and design emphasis applicable to a given electronic product vary significantly in priority, depending on the intended product application (e.g., aerospace, automotive, computers, consumer goods, medical equipment, industrial equipment, agricultural equipment, military equipment, telephony, test equipment).

The key to successful electronic packaging is the ability to identify the applicable field(s) of technology and to select those design approaches most likely to offer solutions to a design problem.

1.2 Overview

There are four basic categories of concern to the electronic packaging engineer, working from the external environment inward:

1. *Exterior Conditions.* Equipment use requirements, service environments, and storage environments define equipment mounting needs, environmental conditions, electrical interconnections, power sources, heat sink availability, surface finishes, repair considerations, and ergonometric human factors requirements.
2. *Internal Conditions.* The equipment enclosure contains and protects internal electronic modules, subassemblies, components, and interconnections and provides thermal and electrical interfaces between the outer environment and the internal environment.

3. *Component Environments.* Module and subassembly housings define the interface between the individual electronic component subenvironment and the equipment's internal environment.
4. *Component Requirements.* Components in modules and subassemblies require protection, which includes limiting temperatures caused by internal heat generation, avoiding excessive mechanical stresses, reducing dynamic shock and vibration loads to below component fragility levels, avoidance of chemical and corrosive agents, and provision for correct mounting methods.

1.3 Approach

There are two major considerations in electronic packaging:

1. Correct selection and proper application of materials
2. Physical design of the equipment

Materials Selection

The process of materials selection includes the following steps:

1. *Identify* the desired or dominate material properties for the design application under consideration. See Section 2.1 for a listing of material properties.
2. *Define* overriding considerations: materials availability, manufacturing process limitations, customer directives, design legacy, and like issues.
3. *Review typical applications* similar to the design under consideration. See Section 2.2.
4. *Compile* a list of the most likely of candidate materials considering Section 2.3.
5. *Select* one or more materials that appear to offer the most likely solution.
6. *Verify suitability* by physically testing the resulting design.

Physical Design

The process of physical design includes the following steps:

1. *Develop* a list of design requirements and restrictions.
2. Using Section 3 and prior experience, *select* the most probable design approach.
3. *Apply* the selected design approach.
4. *Verify* the suitability of the resultant design.
5. *Identify* and solve the next lower level of design problems.
6. *Review* the net effect of the combination of solutions considering the interaction between the problem solutions.
7. *Iterate* until all design requirements and restrictions have been satisfied.

1.4 Design Techniques

Several computer-aided design programs address electronic packaging design and development tasks.[1] Most computer algorithms are adaptations of codes generated for other, related purposes (e.g., finite-element techniques for structures, stresses, thermal analysis, fluid flow analysis, and flow visualization; solid-modeling and drafting programs; printed circuit board design programs). In instances other than printed circuit programs and drafting programs,[2] the underlying computational assumptions inherent in the programs sometimes are not thoroughly documented and may thus produce questionable results. This will provide the opportunity to introduce errors when building a predictive model of a product. The electronic packaging engineer must possess a basic understanding of each physical phenomenon and the underlying assumptions implicit in each analytical model as applied to the specific equipment under analysis. It is always prudent to verify the predicted result by testing and by comparison of the results with one or more known successful designs of a similar nature.

2 ELECTRONIC PACKAGING MATERIALS

2.1 Material Properties

Electrical Conductivity

A material may be required to conduct electrical currents. This includes metals and some nonmetallic elements such as adhesives, greases, and other graphite or metal-powder-loaded compounds. When current flows through the material, electrical resistance creates voltage drop and heat generation, either of which may be a desirable or an undesirable consequence. In screening for electrical conductors, a maximum electrical resistance requirement must be defined; thus materials with equal or lower electrical resistance become candidates for selection. *Electrical resistance* is temperature sensitive; thus it is necessary to make sure that variances in the material electrical resistance are satisfactory over the temperature range of interest.

Conversely, it may be necessary to electrically isolate a component or current-carrying member to protect from undesired electrical short circuits. Materials used as electrical insulators (dielectrics), such as mica or glass or ceramics and many plastic materials, have very high electrical resistance, are poor conductors of electricity, and are employed to provide electrical isolation.[3] Some high-voltage dielectrics are liquids, which often act as heat transfer media in addition to providing electrical insulation. Examples include mineral oils, silicon oils, polybutanes, flourocarbons, and organic esters including castor oil. Problems with liquid dielectrics include limited temperature range, chemical decomposition over time, combustibility, and loss of dielectric strength.

Thermal Conductivity

All materials conduct heat; however, among materials there is a wide range of *thermal conductivity*. Good conductors of heat include metals, and poor conductors of heat such as ceramic and plastic foam materials are considered to be thermal insulators. The cooling of electronic components and the protection of electronic components from excessive temperatures are managed by employing mounting devices, heat sinks, and thermal insulators using materials of the appropriate thermal conductivity.

Thermal Emissivity

The *thermal emissivity* of a material is a measure of the efficiency by which a material will radiate or receive infrared energy. Thermal emissivity is a surface characteristic of a material and to some extent a function of material color. For example, a highly polished light-colored material will have low thermal emissivity and a dark-colored, high-surface-roughness material will exhibit a high thermal emissivity.

When an electronic component is mounted near a high-temperature surface, the heat absorbed by the component may be reduced by using materials and finishes with low thermal emissivity.

Emissivity control is of key importance for electronic equipment used in spacecraft and which may face the sun or be in shadow. Low-emissivity materials, such as plating an equipment case with polished gold, are used to limit the heating from the sun and also to reduce heat loss when the equipment is facing dark space.

If thermal emissivity is a design requirement, the selected material may be chosen for some other dominant criteria and conrol of thermal emissivity is achieved by surface finish selection, to include application of an electrodeposited highly emissive metal on the exposed surfaces of the part.

Painting a surface with a light-colored paint may slightly lower the emissivity; however, the roughness of a painted surface increases its effective area, which will impact more than the color of the paint. As a result, the emissivity of a painted surface is often essentially the same regardless of paint color, and thus the improvement is minor. Coating a dark, polished surface with a light-colored paint often leads to higher emissivity, not lower emissivity.

Thermal Expansion

The *coefficient of thermal expansion* is a material property which allows the electronic packaging engineer to predict the linear or volumetric change (expansion or contraction) of a material when the material is exposed to a change in temperature.

It is important to determine the coefficient of thermal expansion for materials which are bonded or mechanically secured together. For example, a ceramic component (very low coefficient of thermal expansion) soldered to a printed circuit board (a higher coefficient of thermal expansion) exposed to repeated temperature cycles, perhaps caused by turning the equipment on and off, can lead to solder joint failure. As another example, when two materials of different thermal expansion characteristics are mechanically joined and exposed to a temperature change, one material will be placed in compression and the other material will be placed in tension. This will result in the composite assembly bending in the direction of the material placed in tension. Also, depending on the relative strength of the materials, one or the other material may fracture or the joining method (adhesive joint, solder joint, weld, rivet, screw, etc.) may fracture or otherwise be degraded. Thermal expansion is a consideration when precise mechanical position is required. Thermal expansion will cause a change in the relative position of parts during a change in temperature. This movement may cause the loss of precise position and improper operation of the assembly.

Chemical Inertness

Materials may be exposed to potentially damaging chemicals during the life of a product. Such chemicals include fuels and lubricants, cleaning fluids, fluxes, and other chemicals used in industrial processes. Most tables which list material properties also provide limited information regarding *chemical resistance*. If the material is exposed to conditions not covered by information in the tables, then the designer must contact the provider of the material for additional information. It also may be necessary to experimentally determine and verify the suitability of a material in the presence of chemicals. It is important to note that chemical reactions are affected by temperature and the presence of other chemicals.

Chemical deterioration may be induced by ionizing radiation (which alters the molecular structure and thus the physical properties of materials) and ultraviolet radiation (such as from sunshine), which may cause depletion of plasticizers from elastomers, such as polyvinyl chloride (PVC) plastic, which is used as wire insulation material, thus eventually rendering the material brittle and weak.

Corrosion

Most tables of material properties provide information of corrosion resistance; however, this information may not be sufficiently complete to give confidence regarding the intended usage of a material. This may (for metals) include susceptibility to intergranular or stress–corrosion cracking in the presence of certain acids or bases. For all materials, and especially nonmetals, corrosion deterioration includes processes similar to those identified regarding the chemical inertness of a material.

The process of galvanic corrosion[4] is a primary concern to the electronic packaging engineer. Under certain conditions galvanic corrosion is a leading cause of loss of electrical conductivity or loss of strength in a material or at a riveted, bolted, or welded joint and as failure (perforation or delamination) of electroplated finishes.

Table 1 lists the galvanic series relative position of several materials used in electronic packaging.

Galvanic corrosion is the result of an electrochemical reaction which occurs when a design (which employs mating metals ranking at different levels on the galvanic chart) is in the presence of an electrolyte (i.e., saltwater, contaminated water, or other conducting chemicals). One material acts as an anode and the other as a cathode as in a simple electrical storage cell. The result is ionic migration during which the anodic element suffers sacrificial deterioration. This is a mechanism for leaching elements from grain boundaries and a corresponding weakening of the affected material and other undesirable effects. This process may be hastened or hindered by passing a small electrical current through the affected area.

Temperature Range

The operating and storage temperature ranges experienced by a material must be known and the properties of the material must be reviewed for stability over the range of temperatures. This is important for all materials, including metals when the temperatures are very high or very low. Metals and nonmetals will become brittle and stiff at lower temperatures and may soften, lose strength, and creep or flow at elevated temperatures. This is especially true with nonmetalic materials, where the operating temperature range is limited due to undesirable changes in material properties.

Table 1 Galvanic Series Position of Common Electronic Materials

Most anodic (corroded end) (+) *positive end*	Magnesium and alloys
	Zinc
	Aluminum 1100 alloys
	Cadmium
	Aluminum 2024 alloys
	Steel, iron, cast irons
	Type 304 and 316 CRES (corrosion-resistant steel) (active)
	Lead–tin solders
	Lead
	Tin
	Nickel (active)
	Inconel (active)
	Brasses, copper, bronze
	Monel and copper–nickel alloys
	Monel
	Silver solder
	Nickel (passive)
	Inconel (passive)
	Type 304 and 316 CRES (corrosion-resistant steel) (passive)
	Silver, gold, platinum
	Titanium
Most cathodic (noble end) (−) *negative end*	Graphite

Source: Reference 7.

Strength

A guide to the load-bearing capacity or strength of a material is its *tensile strength,* which for many metals is a function of alloy selection and heat treating (i.e., is the material annealed, half hard, or hard?). Strength of nonmetals (plastics) varies significantly with temperature and the rate at which a load is applied (impact strength). Weigh reduction is achieved in electronics by using smaller quantities of high-strength materials for forming brackets, structures, and fasteners.

Density

Density is the weight of a material for a given unit of volume. This property may be used as a guide to reducing weight in a design. Material density establishes the inertia and resonant frequency of mechanical elements and thus the degree to which they are affected by mechanical vibration and mechanical shock loading.

Another indicator of the density of a material is its *specific gravity*, which is a comparative ratio of the weight of a material compared with the weight of an equivalent volume of water.

Electromagnetic and Electrostatic Shielding

Electromagnetic shielding is employed to prevent the imposition of electrical waves from the environment (such as radio waves) on circuit elements, thus inducing currents and voltages and leading to circuit performance failure and possible component damage.

Electrostatic shielding is employed to prevent penetration of high-voltage charges on the surface of an electronic enclosure (such as caused by static discharge) from entering the enclosure and therein causing damage to components or circuit performance malfunction.

Protection of equipment from both of these sources of potential performance deterioration involves shielding the equipment, such as preventing radio waves or other electromagnetic noise from entering the enclosure, shielding the equipment to promote the collection of static charges on the surface of the equipment, and dissipating the unwanted energy by providing a conductive path to drain the energy to an appropriate sink (ground).

A circuit may be shielded by housing it in a conductive enclosure, thus creating a Faraday cage, where the imposing electromagnetic waves or static charges cause currents in the skin of the enclosure which do not penetrate into the enclosure. An example would be to shield a high-gain amplifier circuit and avoid malfunction of that circuit when the equipment is exposed to radio waves.

Conversely, essentially identical techniques are used to contain radio wave emissions from a circuit and prevent the radio waves from escaping the enclosure and causing malfunction in other equipment which may be susceptible to such emissions. High-speed digital circuits at speeds of 10 kHz and above are known to cause such emissions and often require electromagnetic shielding to meet the emission requirements set forth by Part 15 of the Federal Communications Commission Rules and Regulations.[5] Nonmetallic enclosures are sometimes used to shield internal circuits and components from the electromagnetic field by coating the enclosure with an electrically conductive paint.

The primary guide to the effectiveness of a material as an electrostatic shield is the *electrical conductivity* of the material.

Magnetic Shielding Properties

Magnetic shielding pertains to protecting a circuit from the presence of fluctuating magnetic field energy which may induce currents in wires and metal structural elements and lead to circuit malfunction or failure. A simple Faraday shield such as used for electromagnetic wave or static discharge protection will not suffice for protection from magnetic fields. It is usually necessary to restrict the shielding materials to alloys of iron. Iron-bearing alloys tend to block penetration of magnetic flux into the equipment. Magnetic flux may induce noise within a circuit, which may cause malfunction of a circuit, including erasure of memory from several classes of memory storage devices or the introduction of false pulses or blockage of pulses in digital circuits.

Fatigue Resistance

Fatigue is the loss of physical strength and eventual failure of a material due to the material being subjected to repeated loads. Such loads may be the result of thermal expansion stresses which occur during temperature cycles, mechanical loads induced during the operation of the device, shipping and handling stresses, exposure to vibration, and repeated mechanical impact shock loads.

Repetitive loads well below failure-inducing loads are known to cause small incremental damage, and that damage will accumulate until the material fails.[6] Fatigue resistance may be judged on a relative basis by referring to tables of *fatigue resistance,* which relate stress levels to time to failure for a specific material or alloy under ideal conditions such as rotating beam test data.

The actual ability of a component to withstand fatigue damage is highly sensitive to the loading history, the precise shape of the component (i.e., stress risers), equipment construction details, and variations in temperature. As a result, proof testing is usually required to verify the ability of the selected material and design configuration to avoid excessive fatigue damage.

Hardness

Hardness is taken as the ability of a material to resist denting as caused by a load bearing on the material using a dimensionally standardized probe. There are different scales by which hardness is measured, the most common are the *Brinell hardness number* and the *Rockwell hardness number* for a given *Rockwell scale.*

Material surfaces are sometimes hardened to reduce wear in mechanical latches, sliding surfaces, and electrical contacts. The lubricity properties of polymers such as nylon and Teflon are used to reduce friction in applications such as drawer slides where a plastic material is used to rub against a hard metal surface.

The hardness of a material is often established by the surface treatment of the material and is not always a bulk property of the base material. That is, the surface of steel and aluminum and many other materials may be chemically or physically hardened to improve the wear of the part under design. The ability of a material to be surface hardened varies widely among materials, with many materials lacking the ability to be surface hardened.

Ductility

The *ductility* of a material refers to the ability of the material to be deformed without breaking. Extreme examples would include tin (very ductile) and glass (brittle and thus not ductile). This is an important property for materials which are to be mechanically formed, which are to be subjected to bending stresses during use, and which must withstand impact shock loading without cracking or shattering.

A guide to ductility of a material is its *elongation,* which refers to the degree to which a material will deform under load without failing. Elongation is seen to vary with the extent to which a material is heat treated. Ductility in plastic materials is strongly affected by temperature. For example, a plastic gear may work well at room temperature, but at low temperatures it may become brittle and fracture under load. Material ductility often varies widely within a given general class of materials and each individual material must be reviewed for ductility.

Closely related to ductility is the material's *malleability,* which is the ability of the material (usually metals) to be shaped by stamping, spinning, cold heading, upsetting, rolling, or other physical process involving deformation of the basic material in solid form. A malleable material is one which may be physically formed without tearing, cracking, fracturing, or developing surface defects. Tables of materials properties often provide guidance relative to malleability, and subcontractors who provide metal-shaping services are a good source of information on malleability.

Wear Resistance

Wear resistance is a property of materials which does not often appear in tables of material properties. A guide to wear is *hardness*—the harder and smoother the material, the greater its resistance to wear. Ductility should be considered as a material may become brittle at low temperatures and exhibit surface fractures and fretting under load. Wear life is a function of the magnitude of load-induced localized compressive stresses and the methods, such as lubrication, used to reduce friction. Some nonmetallic materials, such as nylon, are self-lubricating and useful as bearing materials.

Sublimation

Sublimation is the loss of material when the material is exposed to low pressures (vacuum) and or elevated temperatures. It involves the conversion from solid form immediately to gaseous form apparently without passing through a liquid phase. Sublimation may occur to the surface of a pure material, or if the material is a more complex matrix, one or more constituent materials may sublimate, leaving a residual material possessing poor mechanical or electrical properties.

Sublimation may also result in the deposition of the sublimated material onto other, often cooler surfaces where the presence of the sublimated material is not desired (i.e., perhaps forming a conductive film causing circuit elements to short or reducing visibility through a window). Information on sublimation is limited and may require inquiry to the manufacturer of a material or to experimental test data.

Sublimation is of critical importance when the electronic packaging engineer selects materials which will be subjected to low-pressure and low-vacuum environments such as for electronic equipment for use in test equipment or in space vehicles.

Combustibility

Combustibility is the property of a material, under a given set of conditions, to oxidize or burn, which may range from the ability to burn but not sustain flame when the ignition source is removed to burning and sustaining flame after the ignition source is removed. Combustion may lead to the generation of noxious gases and destructive deterioration of the circuit elements in the vicinity of the combustion.

This property is most often associated with plastic and organic materials; however, some metals will burn in the presence of oxidizing gases or chemicals. Many tables of plastic material properties do not include information on combustibility; however, this information is usually available from the manufacturer of the material. If combustibility is to be prevented in a design, it is sometimes necessary to conduct tests to determine if a potential material is suitable for the intended application.

Creep

Creep is the ongoing deformation (flow) of a material when that material is subjected to a mechanical load

Failure due to creep is often very slow, and creep may occur only during certain conditions (such as a solder-sealed case which is sealed at sea level but operated in an aircraft or the same case operating at seal level but within which the internal pressure is increased due to heating during normal operation of the equipment). It is good design practice to never place such a soldered joint in sustained mechanical stress for the reason that the solder will creep until the stress is relieved or the joint fails. The design of a mechanical joint where solder is employed (e.g., to achieve an enduring electrical connection or to seal a joint) must have the mechanical load taken by structural members without the solder present. Mechanical joints or seals which utilize adhesives, elastomers, and other nonmetallic materials must be treated in the same way to avoid eventual failure.

A creep failure may not become evident for days, weeks, or even years, but it will happen if materials with the known ability to creep are improperly incorporated into a design.

Creep is not only a characteristic of some metals; it is also a property of several inorganic or organic materials. An example would the loosening of a mechanical joint where the joint consists of metallic elements bolted together and includes a nylon washer. This joint eventually loosens when the nylon washer creeps to relieve the initial clamping stresses. Another example would be when a compressible gasket made of a material which will creep is used in a flanged joint where the joint does not have mechanical stops to control the separation of the flanges when the bolts are tightened. In this case the joint will eventually suffer loosening and possible leakage when the overcompressed gasket creeps to relieve the joint compressive forces. A classic example of a creep-induced failure occurs when an insulated wire presses against a hard metallic object such as the edge of a bracket or panel. In this case, the insulation will creep until the inner conductor of the wire eventually makes contact with (shorts to) the metal object.

Moisture Absorption

In many instances, when a material is selected for use as an insulator or the material must have dimensional stability in humid environments, the *moisture absorption* characteristics of the material are important. For metals, moisture absorption is not a significant consideration; however, the presence of moisture will hasten galvanic corrosion when different metals are in intimate contact.

For nonmetals, porous sintered metal structures, and ceramics, moisture absorption is a significant consideration. The absorbed moisture may cause corrosion and lead to failure by expanding within the material at cold temperatures. For materials used as electrical insulators, the insulator may become electrically conductive when moist. Hydrophobic (water-resisting materials, the surface tension of which causes water to bead and not wet the surface of the material) materials are preferred for insulator applications.

Most nonmetallic materials are tested for moisture absorption and the results of these data are found in many listings of material properties. The majority of moisture absorption data relate the absorption of water over a specified period of time.

In some cases, particularly for organic materials, moisture absorption leads to a loss of strength and physical deterioration. If some materials become wet and are subsequently allowed to dry, they may exhibit permanent shrinkage and the inability to redevelop their original physical characteristics.

The chemical absorption properties of seals, paints, tubing, gasket materials, and plastic materials which are expected to be used in applications involving chemicals (oil, gasoline, fuels, alcohol, cleaning solutions, etc.) most likely have been investigated, and the materials supplier is usually able to provide test data and guidance in the selection of an appropriate material.

During conditions of high humidity and moisture in the presence of inorganic salts, conditions are created which encourage the growth of fungus. Fungus growth can lead to material degradation in appearance and loss of strength and the development of undesirable electrical shorts. Materials are generally categorized as fungus nutrient or fungus resistant. Materials such as metals, glass, ceramics, mica, silicone resins, asbestos, acrylics, diallyl phthalate, and a limited list of plastic resins are fungus resistant. Materials containing organic constituents, rubber, cellulose acetates, epoxy resins, many lubricants, medium- and low-density polyethylene, polyvinyl compounds, formaldehyde compounds, and others are fungus nutrient. The above listing is far from complete, and the designer is cautioned to verify the fungus-nutrient characteristics of any materials used in a design.

2.2 Typical Applications

General

An important guide to materials selection is found by considering how materials have been used for similar applications. Electronic packaging applications often fall into one or more of the following categories:

- Equipment attachment
- Equipment racks, frames, and mounting structures
- Equipment and module enclosures
- Temperature control
- Mechanical joints
- Finishes
- Position-sensitive assemblies
- Electrical contacts
- Harsh-environment endurance

Equipment Attachment

Other than desktop and portable equipment, electronic equipment is usually mechanically fastened to an equipment rack, frame, shelf, or mounting structure. These devices may sit on the floor in an equipment room or are bolted [using high-strength Society of Automotive Engineers (SAE) rated fasteners] or welded to the structure of the containing building or vehicle. The attachment may consist of, for relay rack-mounted devices, passing two or more bolts through holes in the equipment front panel into threaded fasteners in the rack frame. For equipment used in aircraft, ships, and vehicles, industry standard ARINC Air Transport Rack (ATR) modular cases are placed on an equipment shelf and are secured thereto. The ATR module mounting usually consists of an engagement device (pins fitting into holes) in the rear of the module and brackets with retained thumbscrews or latching hook arrangements at the lower edge at the front of the module, including a device fastened to the shelf into which the screws or latches engage. Larger equipment used in moving vehicles, such as tanks, battleships, and trucks, where the equipment must be protected from higher levels of vibration and shock loading, mounting often involves bolting the equipment to a frame, which incorporates energy-absorbing dampers, and bolting the frame to the structure of the vehicle. For consumer goods applications (car radios and CD players, motorcycle and bicycle electronics, etc.) a variety of mounting configurations are employed, some which feature quick release of the attached equipment.

Each application places similar requirements on the materials used for attachment. The dominant considerations are strength, wear resistance, corrosion avoidance, temperature range, and fatigue resistance. The materials are often selected from the steel family to achieve a favorable strength-to-weight ratio. For mounting bolts to fasten commercial goods and other equipment used in relatively benign environments, cadmium- or nickel-plated steel hardware is used. If the mechanical loads are severe, then higher strength SAE-grade steels must be used. For more corrosive environments, stainless steel (usually passivated) is employed. For aggressive high-stress applications where corrosion and wear are a concern, fastening devices and brackets are made from precipitation-hardened stainless steels which are passivated to prevent corrosion.

Equipment Racks, Frames, and Mounting Structures

Equipment racks, frames, and mounting structures such as an equipment shelf are designed to bear the weight of electronic equipment mounted thereto. The dominant considerations are strength, wear resistance, corrosion, electrical conductivity, and corrosion avoidance. The materials of preference include heat-treated high-strength aluminum (such as 6061-T6) and mild steel materials. The aluminum materials are almost always either anodized or treated (chromate dip or similar) and painted for appearance and protection from the elements. Steel materials are protected from rust by nickel, zinc, or other electroplated finish and often by a rust-resistant paint. Stainless steel parts may be passivated to improve resistance to corrosion.

Equipment and Module Enclosures

Enclosures are the cases or boxes which house electronic equipment. Dominant considerations include operating temperature range, electromagnetic and electrostatic shielding, magnetic shielding, strength, wear resistance, thermal conductivity, appearance, corrosion resistance, and chemical inertness.

Materials used for enclosures vary with the application and end-user market. They include both metal and plastic materials. Cost containment, ease of fabrication, and appearance are often the overriding considerations.

When electromagnetic, electrostatic, and magnetic shielding protection is not required, plastic materials are often used to fabricate enclosures. Aluminum and steel materials may be utilized to achieve electrostatic and magnetic shielding protection. If magnetic shielding protection is required, iron-based materials are required, and the material of choice is usually steel. Metal cases are almost always protected against corrosion and other environmental damage by a finish treatment, sometimes followed by paint to achieve a suitable appearance.

Plastic materials may be used to build cases for electromagnetic and electrostatic protecting enclosures if they are coated internally with a conductive paint which may bear either silver or nickel powder. Enclosures fabricated from plastic usually do not require application of paint, chemical treatment, or other surface protection. Another dominant consideration for plastic cases is combustibility because many plastics will burn and emit noxious or toxic fumes when exposed to high temperatures.

Temperature Control

Temperature control is directed at maintaining electronic component temperatures within prescribed limits. This is accomplished by the proper selection and application of materials which are used to conduct heat from or to the components.

The primary modes of heat transfer are conduction, convection, and radiation. The dominant considerations for heat transfer material selection include thermal conductivity, chemical inertness, resistance to corrosion, and sometimes wear resistance, sublimation, thermal expansion, and the ability to resist ionizing and ultraviolet radiation.

Conduction heat transfer involves the flow of thermal energy through a material from a heat source (component) to a heat sink. The design goal is to reduce the resistance to heat flow between the heat source and the heat sink. The effectiveness of heat transfer within a material is measured by its thermal conductivity. Metals such as silver, copper, and aluminum are good heat conductors; however, aluminum is most often used because it tends not to tarnish, is easy to form, and costs less. Copper or copper alloys are used when large amounts of heat are to be transferred and any small improvement in thermal conductivity is meaningful in the design.

When the design requires electrical insulation and good thermal conduction across a joint between two materials, thin layers of mica, silicon-impregnated cloth, and ceramic materials are often used. The thermal conductivity of these joints may be enhanced by applying a thermal grease to the joining parts to prevent the entrapment of air or other gas (a thermal insulator) between the mating surfaces.

Convection heat transfer involves the movement of heat between one material (usually metal) to a surrounding material such as a gas (air) or a liquid (water, Freon, etc.). A convection design seeks to conduct heat from a heat source into a physical configuration, such as fins, which presents a relatively large area of contact between the metal object and the gas or liquid with which the heat is to be exchanged. Good thermal conductivity is very important; so is strength and resistance to corrosion. Convection heat sinks are often made from high-strength aluminum alloys to meet these requirements. For resistance to corrosion, which may result in the formation of a thermally insulating film between the heat transfer media, the aluminum heat sink is sulfuric acid anodized black in color to also enhance heat transfer by radiation.

Convection heat transfer may be improved by mechanically moving cooling gas over the heat sink fins. This is called forced convection, and by using forced convection, the cooling effects of the heat sink may be greatly improved. Of course, forced convection is only effective when an atmosphere is present, and the ability of the forced convection design to handle a given heat load is strongly affected by the density and pressure of the cooling media.

Radiation heat transfer is the primary mode of heat transfer when considering space applications. Radiation at sea level can contribute 10% or more of the total heat transfer process depending on the temperature gradient and distance between the radiating and receiving objects. Radiation heat transfer applications feature use of thermally conductive materials to transfer heat to radiating surfaces. The effectiveness of the radiation surface emissivity is dependent on the finish applied to those surfaces. For materials which are to reduce the reception or reduce the ability to emit heat, the surfaces are smooth and shiny. Such surfaces are often electroplated with a material such as gold, chromium, or nickel which is chemically inert and may be buffed to a very high gloss. For materials for which the design requires effective heat absorption or emission, the material surface is not smooth (increasing surface area) and is usually black or another dark color. This is achieved on aluminum surfaces by application of a black sulfuric anodize finish such as used on the internal elements of solar collector devices.

Paint on a surface adds a layer of insulation and thus is an impediment to radiation heat transfer. Due to the surface roughness of painted surfaces, the color of the paint usually makes only small differences in surface radiation emissivity; however, a light-colored (white) paint will be somewhat less efficient at radiation heat transfer than a dark (black) paint under extreme thermal loads.

Plastic and ceramic materials are not effective radiation heat transfer materials and may be employed as surface insulators to reduce the effectiveness of radiation heat transfer.

Mechanical Joints

Mechanical joints are structural attachments within an electronic rack, enclosure, case, shelf, or housing and are material sensitive. Mechanical joints may be either permanent or semipermanent. A permanent mechanical joint could be brazed, welded, adhesively bonded, chemically joined, riveted, or mechanically upset. Semipermanent mechanical joints involve the application of mechanical fasteners and permit the mechanical joint to be attached and detached during service.

Mechanical joints may involve two or more metal elements, two or more nonmetallic elements, or a mixture of both metallic and nonmetallic elements.

For metallic fused joints the metals must be suitable to the joining method, and filler materials which are carefully selected achieve the joint, such as welding rod alloy materials or brazing materials. The dominant considerations include temperature range (will the material properties be altered by the temperatures required by the fusing process?), relative thermal expansion to avoid residual stresses, and avoidance of materials which in combination will lead to galvanic corrosion. If the metals gain their strength properties by being heat treated after forming, the joint materials may require annealing prior to fusing and heat treating after fusing.

Soldering metal parts together using conventional electronic lead–tin or lead-free solders is not a valid method by which to form a mechanical joint. Solder will creep under mechanical load, and if the mechanical load is continuously applied, the soldered joint will eventually creep to failure. Solder is a valid method for electrically bonding and environmentally sealing mechanical joints, such as riveted joints, which are designed to carry mechanical stress without the presence of the solder and where the solder is used for bonding or sealing and is not used to carry mechanical loads.

Nonmetallic (plastic) material joints may be formed by thermal bonding or ultrasonic welding identical formulations together (thermoplastics only) or by chemical fusing agents as recommended by the plastics manufacturer.

Mechanical joints involving a variety of materials (i.e., both metals and plastics) require careful materials selection, and the dominant characteristics of interest include thermal expansion, corrosion, creep, temperature range, and electrical conductivity.

Finishes

The majority of metallic materials and many plastic materials require the application of a surface finish to enhance resistance to corrosion and to provide an attractive appearance. Dominant considerations regarding application of material finishes include electrical conductivity, thermal conductivity, thermal expansion, chemical inertness, temperature range, wear resistance, emissivity, sublimation, and moisture absorption.

Aluminum materials used inside equipment are usually chemically cleaned and chromate conversion coated. Chromate conversion coating is also used as a surface treatment which promotes improved adhesion of paint to the aluminum. Chemical-film-coated aluminum passes electrical signals and may have an electrical conductivity greater that the base aluminum. These coatings are easily scratched and have poor abrasion resistance.

Anodizing of aluminum creates a wear-resistant and sometimes attractive finish that consists of a thin layer of aluminum oxide. This oxide film is an electrical insulator and may cause a problem if the part is to pass electricity to another component. Sulfuric acid anodize is used to provide protection from corrosion, causes very little thickness growth to the base metal, works well for dyed applications, and is avoided if the design has overlapping surfaces where the acid may be trapped during the anodize process. Chromic acid anodize is preferred when small-dimensional change is necessary for highly stressed parts or on assemblies that could entrap processing solution. Hard anodize provides an extremely hard finish with excellent abrasion resistance and poor electrical conductivity. Hard anodize coating is brittle and will crack should the anodized part be flexed. In addition, hard anodize causes a loss of strength because for 0.04 mm of film thickness, 0.02 mm of the underlying material is lost. In addition, the anodize thickness increases the part dimensions by about half the thickness of the film.

Aluminum may be electroplated to achieve a variety of properties, such as plating with a thin layer of gold over copper to achieve a low thermal emissivity or selective plating of the surface with tin to permit the solder attachment of an electrical conductor to the aluminum part. The entire part should be plated for maximum protection, and voids in the plating will expose the base material and thus may establish conditions for galvanic corrosion.

Steel parts must be protected from oxidizing (rusting) in the presence of even small amounts of moisture or humidity. Steel is usually protected by plating with nickel or zinc for fasteners and with tin when used as an electronic equipment chassis to which electrical connections are required. Nickel plating and chrome plating are used on steel parts to improve resistance to wear and to provide an attractive appearance. Cadmium plating is used extensively to protect steel against corrosion; however, it is limited to use below about 230° and is known to sublimate under vacuum conditions. Other treatments such as black oxide coating and phosphate coating offer alternative approaches to protection of steel surfaces.

Caution is required when a metal is electroplated with another metal. The two materials differ in coefficient of thermal expansion, which may promote delamination of the plating from the base metal when the composite is subjected to thermal cycles. Also, the further apart the materials are listed on the galvanic series, the greater the opportunity for galvanic corrosion to occur. This latter consideration will require the addition of protective coatings (paint, adhesives, etc.) at any location where the plating is removed and both the plating and the base material are exposed to the environment. Aggressive mechanical fastening techniques, such as the use of star-type (shakeproof) lock washers in a bolted joint, may cause local failure of the plating and eventually lead to joint failure (a loss of electrical continuity, mechanical strength, or both) by corrosion.

Stainless steel parts used in electronic applications are passivated to remove surface contamination and to form a uniform protective oxide film that assures that the material delivers maximum resistance to environmental contaminates.

Other commonly used metals, such as beryllium–copper, bronze, and brass, also must be protected and may be treated by processes similar to those used to protect steel.

Plastic materials may require finish treatments such as electroplating to achieve surface electrical conductivity or a metallic finish for appearance purposes such as applying chrome on a knob or bezel or other like application. Not all plastics may be painted due to the inability of commonly used paints to adhere to the plastic surface. The surface texture of thermoplastic parts may be given an attractive matte finish by vapor blasting.

Position-Sensitive Assemblies

Position-sensitive assemblies include devices the operation of which requires an invariant physical relationship between components over the range of temperatures to which the electronic product will be exposed. One example would be an antenna coupler or filter wherein the electrical circuit resonance frequency of the antenna coupler is determined by the length of protrusion of a cylinder (tuning element) into a cavity. The protrusion distance is varied by the equipment operator to be proper for each specific operational frequency and, once set, must be maintained within less than a millimeter over a significant temperature range. A second example would be a variable-frequency tank circuit which after being adjusted must maintain precise values of inductance, capacitance, and resistance over a range of temperature. A third example would be an electronic structure featuring panels or structural elements fabricated from different materials but attached together that must not be allowed to deform when subjected to a change in temperature.

The dominant consideration is the thermal expansion of the materials involved. Some materials, such as Invar, a nickel–steel alloy, and structural ceramics, have a very low coefficient of thermal expansion and are used to facilitate the design of critical positioning mechanisms.

Electrical Contacts

Electrical contacts in electronic applications usually consist of switch contacts and connector contacts. The requirement for each are similar; however, the usage varies in that the switch contacts may be frequently operated and the connector contacts are rarely engaged or disengaged.

Materials selected to make electrical contacts depend substantially upon the amount of electrical energy that the contacts must carry. The dominant considerations for selecting contact materials include electrical conductivity, chemical inertness, corrosion resistance, and wear resistance. Electrical contacts are usually made from metal materials; however, in unusual instances where a low electrical conductivity is not required, conductive plastics (plastics loaded with carbon or metallic particles) are employed. Sometimes a metal contact will ride upon a nonmetallic contact such as in a carbon composition potentiometer.

Connectors which must maintain reliable electrical contact over extended periods of time are often designed to produce high contact forces between the contacting elements. High contact forces result in significant mechanical erosion of the contacting materials whenever the connector is engaged or disengaged and can seriously decrease the number of uses of the connector. Connector engagement is beneficial because the mechanical action of the contacting materials sliding over each other tends to remove surface films and corrosion and thus lead to a lower contact resistance. This lowered contact resistance will last until the contact resistance is increased due to the formation of chemical films or corrosion products.

Contacts which are required to carry (make and break) very low electrical current, such as complementary metal–oxide–semiconductor (CMOS) level signals, are known as "dry contacts." Dry contacts are fabricated using inert metals or metals plated with an inert metal such as gold or rhodium upon which surface film formation is inhibited. Silver contacts and lead-based metal contacts are not used because they tarnish and the tarnish film is a poor conductor of electricity and thus induces resistance in the switched circuit. Long-term resistance to environmental chemicals is sometimes improved by coating the contacts with protective grease or other media.

Circuit contacts where the circuit carries higher levels of current provide greater freedom in selection of contact materials because the current flow is adequate to dispel the tarnish and other films that may form on the contacts. These contacts are often made from beryllium–copper, silver alloys, lead–tin alloys, nickel, rhodium, and other materials.

Dominant considerations for connectors include electrical conductivity, chemical inertness, prevention of galvanic corrosion, hardness, and wear resistance. Many of the same materials used for switch contacts are used for connector elements, but with more emphasis on wear resistance and chemical resistance. It is necessary to prove electrical contact performance for switches or connectors by extensive life testing with the primary variables being contact force and material hardness.

Encapsulation

Encapsulation or potting of an electronic component or subassembly is utilized to promote ruggedness, resist harsh environments, provide electrical isolation, and promote improved heat transfer for cooling of the encapsulated components. Encapsulation is also used to seal an assembly to protect proprietary design details from the curious.

A variety of material properties must be considered when selecting an encapsulant. Sometimes the requirements are inconsistent with the range of properties offered by any one material. Of interest are the ability of the material to flow freely, not capture or entrain air bubbles; be lightweight; offer acceptable electrical, thermal, and chemical properties; and resist cutting and abrasion after cure. During cure the material should not produce excessive temperatures or excessive stresses on the encapsulated parts. After cure, the material should not absorb moisture, not produce excessive forces during temperature cycles, be stiff enough to support components and connections, and be able to be removed for purposes of repair.

Some properties of the materials may vary during cure and for some duration following cure. An example would be dielectric strength which may not stabilize until several days after cure.

At least three primary groups of material formulations are used for encapsulation:

- *Epoxy-Resin-Based Materials.* These materials may be formulated for strength and rigidity. The addition of glass microspheres would reduce density, and addition of metal or other powders will increase thermal and electrical conductivity. Epoxy resins are usually two-part mixtures and require strict control of the mixing, pouring, and curing processes.
- *Silicon-Based Materials.* These materials are known for having an extended temperature range over which they do not degrade. They tend to remain flexible and are resistant to water. They are either one- or two-part formulations. Some formulations generate acetic acid during cure and may cause corrosion of parts and conductors. Cure time is a function of the thickness and mass of the encapsulant and may require more than one day to achieve full cure.
- *Polyurethane-Based Materials.* These materials generally consist of a catalyst and resin and require mixing prior to pouring. They tend to remain flexible over a useful temperature range and after cure are of a rubberlike consistency; however, they have less tendency to adhere to components, wires entering the potted volume, and potting box surfaces than an epoxy or silicone encapsulant.

Epoxy polymers are used to provide low-cost, high-performance reliability without hermeticity to provide environmental protection for bare circuit chips when used in chip-on-board (COB), ball grid array (BGA), multichip module (MCM), and chip scale packaging (CSP) assemblies.

Data describing the properties of encapsulants may be found on manufacturer's data sheets; however, this information may be incomplete and sometimes does not relate physical properties to the method of cure and the material thickness. It is often necessary to consult with the technical support personnel of the manufacturer and to perform testing to determine suitability for the intended design goals.

Harsh-Environment Endurance

Environmental endurance is the ability of the selected material(s) to resist degradation when subjected to the service conditions faced by the equipment or device constructed from those materials. Unusually harsh operating conditions may cause the material selection process to focus on the ability of the material to resist the service environment before the designer considers other important desired material properties.

High-temperature environments require the use of metals, ceramics, and temperature-resistant plastics.

Highly corrosive chemical environment possibly involving strong acids or strong alkali vapors or solutions is resisted by some glasses, some ceramics, stainless steels, and some plastics.

Solar radiation creates heating of materials and degrades some plastics, such as nylon and PVC, which lose weight and strength when exposed to ultraviolet radiation.

Glasses, ceramics, and many metals react well in the low pressures and vacuum conditions of outer space without sublimating or losing strength due to ultraviolet radiation exposure.

2.3 Candidate Materials

General

With the dominant and overriding design considerations defined, the electronics packaging engineer is ready to survey general classes of materials for the purpose of identifying one or more candidate material for the design task under consideration.

One primary materials selection consideration is often the conductivity of the material, that is, whether the material is a conductor or an insulator of electrical signals and currents. Conductivity is an important characteristic (printed circuit traces, wires, etc.) and is often the reason that metals are the materials of choice. Other important characteristics of metals include strength, durability, thermal conductivity, and ability to work over a wide temperature range.

Nonmetallic elements may exhibit many of the characteristics of metals; however, they are used to provide electrical and thermal insulation. Other properties such as inertness to chemicals, dielectric strength, appearance, low cost, and east of fabrication may lead to the use of nonmetallic elements.

Semiconductor materials may be employed to achieve, when properly formulated as in transistor junctions, the ability to become either an electrical conductor or an electrical insulator depending on the conditions of use.

Several major categories of materials exist from which the designer may identify subsets and specific materials for detailed review and selection.

Metals

A guide to the electrical conductivity of a metal is its electrical resistance. Table 2 provides examples of materials listed by volume resistivity, where:

- Insulators are considered to have a volume resistivity of 10×10^6 Ω·cm or greater.
- Semiconductors have volume resistivity in the range of 10×10^6 Ω·cm to about 10×10^{-3} Ω·cm
- Conductors are materials with volume resistivity of about 10×10^{-3} Ω·cm or lower.

Other selected properties of metals used in electronic packaging are provided by Table 3.

Table 2 Materials Categorized by Resistivity

Conductors	Semiconductors	Insulators
Basic metals	Silicon	Ceramics
Metal alloys	Germanium	Elastomers
Metal-bearing compounds	Gallium arsenide	Gases
Plasma	Selenium	Glass
Some liquids		Solvents
		Thermoplastics
		Thermosets

Table 3 Typical Properties of Selected Metals[a]

Metal	Melting Point (°C)	Density, 10×10^{-3} kg/m^3 (20°C)	Resistivity, 10×10^8 Ω·m (20°C)	Thermal Conductivity, W/m·K (0–100°C)	Thermal Expansion, 10×10^6 (1°C)
Aluminum	660	2.7	2.7	238	23.5
Antimony	630	6.7	42.0	17.6	8.11
Beryllium	1284	1.8	5.0	167	12.0
Bismuth	271	9.8	116	7.9	13.4
Cadmium	321	8.6	7.4	92.0	31.0
Chromium	1875	7.2	13.0	69.0	6.5
Copper	1491	8.9	1.7	393	17.0
Germanium	937	5.3	$>10 \times 10^6$	59.0	5.8
Gold	1064	19.3	2.3	263	5.8
Indium	156	7.3	9.1	81.9	25.0
Iron	1525	7.9	9.7	71.0	6.8
Lead	328	11.7	21.0	34.3	20.0
Lithium	179	0.5	9.4	71.1	55.0
Magnesium	650	1.7	4.0	167	26.0
Molybdenum	2595	10.2	5.6	142	5.2
Nickel	1254	8.9	6.8	88.0	13.4
Palladium	1550	12.0	10.9	71.0	10.9
Platinum	1770	21.5	10.7	71.0	9.1
Rhodium	1959	12.4	4.7	84.0	8.4
Silicon	1414	2.3	$>10 \times 10^{10}$	83.0	7.5
Silver	961	10.5	1.6	418	19.0
Tantalum	2985	17.0	14.0	54.0	6.5
Tin	232	7.3	12.9	65.0	12.0
Titanium	1670	4.5	55.0	17.0	8.9
Tungsten	3380	19.4	6.0	165	4.5
Uranium[a]	1130	19.0	29.0	29.3	23.0
Zinc	420	7.1	26.0	111	30.0
Zirconium	1860	6.5	45.0	20.0	5.9

[a] This table indicates the relative properties of various metals. Exact properties vary with alloy and temperature.

Iron and Its Alloys. Iron-bearing materials used in electronic packaging include various alloys of steel.[7] Steel materials[8] are strong and durable; however, they are subject to deterioration by oxidation (rust) and relative to other metals such as copper and aluminum are less efficient conductors of electricity and heat.

Steel alloys containing chromium, commonly referred to as "stainless steels" or CRES (corrosion-resistant steel),[9] resist oxidation and tarnishing and are often used in electronic packaging as fastener devices such as screws, nuts, washers, and other hardware. Some alloys of stainless steel are not magnetic, a property which is sometimes useful when used in conjunction with circuits which are sensitive to magnetic fields.

There are a wide variety of properties which may be developed for steel alloys, such as very high strength-to-weight ratios, hardness, wear, selective resistance to chemicals, the ability to be formed by mechanical cutting and shaping, and the ability to be joined by welding and brazing. Consumer goods electronic equipment chassis are sometimes constructed from tin-plated steel, which is a low-cost material that can be easily cut or formed and which may be soldered using conventional lead–tin electronic solders.

Aluminum and Its Alloys. Aluminum alloys[10] are commonly used in electronic equipment for manufacturing cases, chassis, covers, brackets, and other mechanical parts. Aluminum alloys range from being "dead soft" through alloys for which substantial strength and surface hardness improvements are achieved by heat treatment.

Aluminum alloys are subject to degradation by both acids and bases and are subject to galvanic corrosion when in contact with other metals. Aluminum is a good conductor of both electricity and heat. Common surface protection methods include iridite or chromate dip coating, anodizing, and painting.

Aluminum alloys are easily cut and formed by most mechanical methods. The softer aluminum alloys are very ductile and may be extruded, impact formed, rolled, or mechanically upset. Other alloys may be punched and formed to achieve a defined shape and then heat treated to achieve structural properties. Aluminum has a good strength-to-weight ratio and is used to achieve lightweight products. Commonly used aluminum alloy series include the following:

- 1100: Resists corrosion, good weldability, good conductivity, commonly heat treated, and used for sheetmetal parts, rivets, and other applications without high-strength requirements.
- 3003: Similar to 1100, except higher strength, more corrosion resistant, and used for heat exchanger fins and like applications.
- 5052: Excellent corrosion resistance, especially in marine environments, higher strength than 3003, and used for chassis parts, fan blades, and the like.
- 6061: A heat-treatable high-strength material used for structural parts.
- 356 series alloy: Used for cast aluminum parts. It may be heat treated and is used for sand, permanent mold, and investment casting processes.

Magnesium. Magnesium has similarities to aluminum. It is a good but slightly poorer electrical and thermal conductor with a high strength-to-weight ratio and is sometime used in place of aluminum to effect weight reduction. Care must be taken when using magnesium because magnesium corrodes readily, is highly reactive to oxidizing chemicals, and in shaving and powder form will oxidize (burn).

Copper and Its Alloys. The primary attributes of copper are its low electrical resistance and low resistance to the flow of heat (high thermal conductivity). Copper is used for conductive traces on printed circuit boards, in wiring, and in switches and relays. Copper may be machined, formed, and cast. It is easily soldered by conventional lead–tin electronic solders. Copper will tarnish and usually requires surface treatment to avoid tarnishing in service.

Copper is heavy, soft, and ductile and thus is not a good structural material unless used in alloys with other materials such as beryllium. Beryllium–copper alloys are used for electrical contacts (improved wear) and are good materials for constructing springs and spring-type electrical contacts.

Cadmium, Chromium, Nickel, and Zinc. These metals are often used as an electroplated finish on iron alloy parts, such as fasteners, to improve the appearance of the parts and to achieve protection against galvanic corrosion.

Gold and Silver. Gold and silver are precious metals that exhibit very low electrical resistance and very high thermal conductivity. Both are used to electroplate connector, relay, and switch contacts to achieve low contact resistance, durability, and resistance to corrosion.

Gold is an inert metal and not given to forming nonconductive surface films over time and thus is a good candidate for "dry circuit" switching. (A dry circuit is one in which very low currents are involved, such as signals in a CMOS circuit. The low currents are insufficient to penetrate and dispel the effect of environmental-induced nonconductive films on the contacts. The result is a detrimental increase in contact resistance over time.)

Care must be used when silver is employed in a design as silver will easily tarnish, and when used as a switch or other contacts, this tarnish will create a nonconductive surface film on the silver part. Sulfur, such as found in brown wrapping paper, will quickly cause silver to tarnish. The application of antioxidant grease is used to protect electrical contacts fabricated from silver.

Both gold and silver are easily soldered using conventional electronic solders. When soldering to contacts which are silver plated, the solder will leach the plated silver from the wetted area, resulting in degraded electrical performance. As a result, a "silver solder," containing approximately 2% silver, is be used to prevent leaching.

Lead and Tin. Lead is very heavy and may be used to shield electronic components which are subjected to ionizing radiation.

Tin is a soft and ductile material, is substantially inert to the environment, and has lubrication properties. Thus tin may be used to electroplate electrical contacts in connectors to improve the mechanical mating of connector elements.

Used together, lead and tin are the primary constituents in the majority of electronic solder alloys commonly used in the electronic industry. Lead–tin solders vary in the ratio of lead to tin in the alloy, with 63% lead and 37% tin being a eutectic mixture featuring the lowest melting point of solder, and when cooling the entire mixture "freezes" at the same temperature, which leads to improved joint integrity, particularly regarding the attachment of parts to a printed circuit board. Another common solder mixture consists of 60% lead and 40% tin. This mixture melts at a slightly higher temperature; however, it exhibits higher mechanical strength in the solidified form and thus is used to attach wires to connectors and in other general applications.

The presence of lead in solder is believed to cause health risks, and lead-free solders[11] and lead-free conductive adhesives are being used in sensitive applications where lead must be avoided.

Rhodium. This metal has very good wear resistance and is used to coat electrical contacts subjected to repetitive usage, such as high-performance relay contacts.

Titanium. Heavier than aluminum but lighter than steel, titanium has a melting point comparable to steel, offers a relatively inert structural material with a high strength-to-weight ratio, and is difficult to weld. Titanium electrical resistance is about 20 times that of aluminum, and its thermal conductivity is one-fourth that of aluminum. Its coefficient of thermal expansion is nearly the same as that of steel. Titanium is used for weight reduction when high strength is required for equipment attachments, mounting structures, and cases.

Plastics and Elastomers

Plastics or organic polymers find extensive use in electronic applications.[12] There are two basic types of plastics: thermoplastics and thermosetting plastics. The difference between thermoplastics and thermosets is whether or not the plastic is completely polymerized at the time of manufacture (i.e., a thermoplastic) or whether the plastic is provided as a resin plus a hardener which must be mixed and polymerized when the plastic is formed into a component (i.e., a thermosetting plastic).

Plastics are commonly used in applications where electrical insulation is required or acceptable. Plastics are unable to operate over the temperature range of metals and are subject to softening at warm temperatures and become brittle at cold temperatures. The physical properties vary widely; however, plastics are used in the construction of many electronic components and may be used for equipment housings and appearance hardware. Care must be taken when soldering near plastic parts, as touching plastic by a soldering iron tip can cause local melting and deformation. Resistance to chemicals varies widely, as does moisture absorption and mechanical load-bearing capacity. Most plastics are subject to creep and must be employed where permanent deformation due to sustained loading is controlled or may be tolerated.

Plastic materials may be formed into component parts by a wide variety of methods, and the cost per unit for large-volume production can be quite low. Typical plastics used in electronic packaging applications are identified in Table 4.

Thermoplastics. Thermoplastic materials are available in sheet, granules, and powder and may be fabricated by transfer molding, injection molding, laminating, pultrusion, compression molding, filament winding, and pouring or casting. Thermoplastic materials can be softened by heat and merged or shaped. The properties of thermoplastics may be modified by the addition of fillers or reinforcing fibers.

Acrylonitrile butadiene styrene (ABS) thermoplastics are tough and used to make enclosures, insulators, appearance sheet materials, and a variety of molded parts for electronic applications. They may be varied in color, but the temperature range is limited. Most applications of ABS thermoplastics are in consumer goods and other low-cost applications.

Acrylics may be used for conferral coating and in solid form are optically clear with very good weather resistance. They are temperature sensitive and will readily burn when exposed to flame.

Fluoropolymers are widely used in electronic applications due to the wide range of properties for which they may be formulated. They are high heat resistant, are inert to most chemicals, have essentially zero water absorption, have a low coefficient of friction, have high arc resistance, resist creep, and have low dielectric losses. Teflon, manufactured by DuPont, is an example of a fluoropolymer used for wire insulation, electrical insulation, low load bearings, and dielectric purposes. Fluoropolymers are expensive, relatively soft, and hard to process.

Nylon is a polyamide polymer which is tough and abrasion resistant with good mechanical strength. It is used in fiber form to make fabrics and in solid form to make gears,

Table 4 Plastics Used in Electronic Packaging

Thermoplastics	Thermosets	Elastomers
ABS	Allyl	Natural rubber
Acrylic	Epoxy	Isoprene
Fluoropolymer	Melamine	Chloroprene, neoprene
Nylon	Phenolic	Ethylene–propylene
Polycarbonate	Polyimide	Fluorinated copolymers
Polyethermide	Polyurethane	Butyl
Polyethylene	Silicone	Nitrile, Buna N
Polyimide		PVC
Polystyrene		Silicone copolymers
PVC		Polysulfide
		Polyurethane

standoffs, low-pressure unlubricated bearings, wire ties, connectors, wire jackets, and like items. Nylon absorbs water and its dielectric strength and resistivity vary with the humidity of the environment to which it is exposed. Nylon will degrade when subjected to sunlight and will exhibit creep or cold flow under sustained loads.

Polycarbonate materials exhibit high impact strength, are optically clear, are flame resistant, and will perform in higher temperature applications. This material is used to make windows and other functional and appearance-related components. These materials absorb water and when under mechanical stress are degraded by many chemicals and solvents.

Polyetherimide polymer finds usage as fuse blocks, circuit board insulators, chip carriers, equipment housings, switch and circuit breaker housings, and like applications. It exhibits low dielectric losses, functions over a wide temperature range, has high radiation resistance, and withstands boiling water.

Polyethylene is an inert, abrasion-resistant olefin with low weight per unit volume and has very low water absorption. This material is used in molded form for discrete-component enclosures, battery cases, and like applications. In sheet or film form it is used as a dielectric in the manufacture of capacitors. It also is used as a vapor barrier and as insulation and sheath for wires and cables. When exposed to flame, olefins will burn.

Polyimide plastic materials are used for radomes, antenna housings, films, printer circuit boards, and connectors. They have high radiation resistance, have low coefficient of thermal expansion, have high electrical resistance, are wear resistant, have a low dielectric loss factor, and resist high temperatures. They are chemically resistant but are degraded by hot caustic solutions and highly polar solvents. Hard to mold, polyimide shapes are formed by powder metallurgy and compression molding techniques.

Polystyrene has a very low dissipation factor of about 0.0001 and about a 2.4 dielectric constant and thus is used in foam insulation materials and in radio-frequency (RF) strip lines and similar applications. It is degraded by many solvents and chemicals and by modestly elevated temperatures.

PVC is used to build wire sleeves and jackets, tubing, and insulation due to its flexibility and resistance to most chemicals. It is sensitive to ultraviolet (sunlight) radiation, which causes it to become brittle and shrink. PVC use is restricted to temperatures below about 100°C, and it stiffens at temperatures below 0°C.

Thermosets. Thermoset plastic materials are supplied in the form of a resin and hardener, and the final polymerization of the plastic occurs during the final forming process. This final polymerization may be referred to as curing, hardening, vulcanizing, or crosslinking. Sometimes heat is required to complete or hasten the polymerization. The fabricator must be aware that the chemical process of curing involves the generation of heat by an exothermic reaction within the material, and this heat must be shunted away to prevent cracking and other deterioration. The material also shrinks up to 10% by volume during cure, thus having an impact on the configuration of the mold. The addition of fillers, pigments, and lubricants control shrinkage.

The allyl polymer family is characterized by good dimensional stability, having electrical insulation properties, the ability to withstand elevated temperatures, and good moisture resistance. A commonly used allyl is glass-filled diallyl phthalate, which is dimensionally stable and is used to make inserts in connectors, electronic parts, and housings and to form insulating standoff posts.

Epoxy materials adhere well and are used as adhesives as well as for potting, casting, and encapsulation purposes. The properties may be varied from flexible to rigid. The final weight, strength, thermal conduction, and thermal expansion may be modified by inclusion of filler materials. Epoxies are used to encapsulate circuits such as microcircuit chips in MCMs, COB, and conventional components in cordwood array packaging. Epoxy resins are used with glass fabrics and phenolic laminates to fabricate printed circuit boards and also for conformal coating, encapsulation, and adhesives and in forming durable paints and finishes.

Phenolic materials are used for high-temperature, high-mechanical-strength (relative to other plastics), low-cost applications. Phenolics are used to fabricate knobs, handles, connectors, and chip carriers an in laminate form to build small enclosures. Phenolics are not elastic and may fracture upon bending or impact loading.

Polyimide materials exhibit low outgassing, and good radiation resistance and have very good high- and low-temperature performance. They must be used with care because they are sensitive to moisture and are degraded by organic acids and alkalis. Applications include bearings, seals, insulators, flexible cable, tapes, sleeving, moldings, wire enamel, chip carriers, and laminates.

Polyurethane materials are often used as paints, wire enamels, conformal coatings, embedding compounds, and foam dielectric and thermal insulation. Some polyurethane materials may be poured and used as an encapsulant to protect circuits from the environment. They exhibit elastomeric properties and have tear resistance, abrasion resistance, and unusual toughness. They can be degraded by strong acids and bases. The operating temperature range is limited and should not exceed 120°C.

Silicone-based materials have a wide range of general-purpose applications in electronic equipment. They may be formulated for use as sealants, elastomers, adhesives, paints, conformal coatings, and encapsulants. They have a broad temperature range (−40 to +250°C), have excellent arc resistance, are nonburning, and are water resistant when fully cured.

A commonly used silicone family is known as the room temperature vulcanizing (RTV) adhesives and sealants often used to waterproof a component or electrical joint. These products polymerize with water when curing and in the process release acetic acid in some formulations or ethyl alcohol in other formulations. In applications where the presence of acetic acid vapor would degrade circuit elements, such as switch and relay contacts, the RTV should be allowed to fully cure before sealing it within an electronic enclosure. Alternatively, if the presence of acetic acid on a component is a problem and ethyl alcohol may be tolerated, an RTV which does not exude acetic acid may be used in the design.

Elastomers. Elastomers are substances such as natural rubber[13] and polymers which have material properties which resemble rubber.[14,15] These substances have the ability to rapidly return to nearly their original shape after sustaining substantial deformation due to the application of mechanical stress and the release of that stress. Elastomers are employed as sealing gaskets and drive belts and find other uses that involve bending and recovery of shape.

Sometimes used for gaskets and power train or drive belts, elastomers are much more vulnerable to aging and environmental degradation than other plastic materials and must be selected with care. The designer is cautioned to consider material properties such as compression set, creep, loss of elasticity, fatigue failure due to repetitive load, loss of elasticity at low temperatures, strength at elevated temperatures, tear resistance, and tensile strength that may be degraded by exposure to ultraviolet and ionizing radiation, elevated temperatures, exposure to some lubricants and chemicals, and continuous mechanical stress.

Ceramics and Glasses

Ceramics and glasses are used in similar ways in electronic packaging applications.[16] They have unique chemical, thermal, and mechanical properties. Both are primarily used as electrical insulators; however, glasses find a wider range of use to include substrates, capacitor and resistor bonding components, and equipment and component enclosure seals.

Ceramics. Electronic ceramic materials include electrical porcelains which have varied properties based on the amount of aluminum oxide incorporated into the material. Aluminum oxide is the additive of choice for ceramic materials for it improves mechanical strength, greatly improves thermal conductivity, and improves flexure strength. A concern, however, is that the use of aluminum oxide substantially increases the material coefficient of thermal expansion (about double that of silicon) and elevates the material dielectric constant. Ceramics are often bonded to metal substrates to perform useful electronic functions.

Other oxides may be used, such as beryllium (which is toxic) to achieve high thermal conductivity; aluminum nitride, which has a high thermal conductivity with thermal expansion nearly that of silicon; and boron nitride, which combines high thermal conductivity with machinability in a softer yet durable material.

Glass. Glass is an amorphous noncrystalline material that may be heated, even to a liquid phase, and formed into useful shapes. Glass performs like a liquid that has been sufficiently cooled to become substantially stiff and rigid. The inclusions of additives can substantially alter glass melting point, thermal expansion, and electrical properties. Glass may be optically clear, can be fused with other materials to make resistors and capacitor elements, and is vulnerable to impact mechanical loads and thermal shock loads that cause it to exhibit brittle fracture.

Soda–lime glasses are used to seal hermetic packages and to form insulator bushing for feedthrough devices.

Borosilicate glasses have excellent resistance to chemicals, high electrical resistance, and low dielectric constants and are used as binders for compounds associated with component construction.

Lead alkali borosilicate glasses have a lower melting point and are used for adhesive and sealing applications, including semiconductor processing.

Glass ceramics are machinable materials that may have strength twice that of ordinary silicon dioxide glasses. They have a crystalline structure with temperature stability lower than that of other glasses and may be shaped into a variety of useful configurations.

Adhesives

Adhesives are materials that, by adhesion, cohesion, or molecular bonding, cling to the surface of another material. Adhesive materials used for electronic applications are most often derived from elastomers and plastic polymers. Glasses are used to perform adhesive functions when the elastomers and plastics have inadequate temperature range or are insufficiently inert to the product environment. Lead–tin solders are used as conductive adhesives in the application of surface-mounted parts to printed circuit boards.

Adhesives are utilized for structural bonding, to seal enclosures, to bond and protect components from the environment, to seal mechanical joints and thus prevent galvanic corrosion, to join parts for less cost that with mechanical means, and to join materials where the temperatures developed by processes such as brazing or soldering would degrade the bonded elements.

The success of an adhesive application is at least as dependent upon proper joint design and application procedures as it is on the selection of a specific adhesive material.

The primary requirements for achieving a successful adhesive joint include the following:

- The area of the adhesive overlap must be adequate to withstand the mechanical loads imposed in service.

- The mechanical loading generally should not provide peel or cleavage stresses on the adhesive joint.
- The adhesive must be able to wet and bond to the substrate material to which it is applied.
- The surfaces to be joined must be chemically clean and may require a properly applied primer.
- If the adhesive mixture consists of multiple components, these components must be accurately measured and properly mixed.
- Pressure, elevated temperatures, and positioning fixtures may be needed while curing the adhesive.
- The cure cycle may require several hours for the adhesive to gain full strength.
- The adhesive material must withstand processing and usage environmental temperatures and chemicals.

There are three categories of adhesive materials:

1. Structural adhesives—intended to hold two or more parts together under conditions where the adhesive joint is subjected to high mechanical loads
2. Holding adhesives—intended to permanently or temporarily hold lightweight items in place
3. Sealing adhesives—intended to fill a space between two or more materials and provide a seal without the need to have high structural strength.

There are a wide variety of materials utilized as adhesives, including the adhesives identified in Table 5.

3 ELECTRONIC PACKAGING

3.1 Component Mounting

Discrete Components

Discrete components are circuit elements not incorporated into integrated-circuit-type packaging. Included are components and subassemblies that may or may not be mounted to a printed circuit board. Component specifications provided by the manufacturer usually provide a guide to mounting requirements.

Components not normally mounted to a printed circuit card include disk drives, liquid crystal displays, power relays, panel-mounted switches and controls, connectors, and devices that generate significant amounts of heat.

Table 5 Adhesives Used in Electronic Packaging

Thermosetting	Thermoplastic	Elastomeric	Hybrid
Cyanoacrylate	Polyvinyl acetate	Butyl	Epoxy phenolic
Polyester	Polyvinyl acetal	Styrene butadiene	Epoxy polysulfide
Epoxy	Polyamide	Phenolic	Neoprene
Phenolic	Acrylic	Polysulfide	Vinyl phenolic
Polyimide		Silicone	
		Neoprene	

Small discrete components are often mechanically attached to a structure, lead mounted and soldered to electrical terminals, or soldered to a printed circuit board. Examples include resistors and capacitors in leaded packages, individual transistors, rectifiers, bridges, relays, and light-emitting diodes (LEDs).

Printed Circuit Board Components

A printed circuit board may consist of one or more patterns of conductive traces separated by an insulator such as FR4 glass epoxy or other suitable rigid or flexible substrate.

Increasingly smaller components and integrated circuits have greater internal complexity with connection point counts which can exceed 400 points per device. Trace widths and spaces between traces of 0.010 in. or wider are common, as are fine-line board traces and spaces of from 0.010 to 0.006 in. Very fine line boards may have traces and spaces 0.001 in. or less; however, these densities are difficult to achieve in production.

The various types of components mounted to a printed circuit board may be classified as either leaded components or surface-mounted components.

Leaded components are mounted by inserting component leads through holes in the printed circuit board and soldering the leads into place followed by lead trimming and cleaning operations. This technology is mature. Leaded components include discrete components and leaded integrated circuits [dual in-line package (DIP) with two rows of pins and single in-line packages (SIPs)].

A variation of leaded components is the pin grid array (PGA), which features a matrix of up to 168 pins protruding from the bottom of the package; another variation is the quad flat pack, which features connection points on all four sides and must be mounted into a suitable chip carrier which is soldered to the circuit board.

Very large scale integrated (VLSI) circuits combine a multiplicity of circuit functions on an often custom-designed integrated circuit.

Surface-mount technology (SMT) consists of component circuit packages which feature conductive pads on the package with no leads protruding. These parts are attached to the printed circuit board by placing the SMT part onto a pattern of circuit board traces which have been coated with a solder paste mixture. Following placement, the assembly is heated to reflow the solder paste and to bond the components to the printed circuit board.

Converting from a through-hole design to an SMT design usually reduces the printed circuit board area to about 40% of the original size. The area reduction is highly dependent on the specific geometry of the components employed, interconnection requirements, and mechanical considerations.

SMT components include the following:

- Small-outline integrated circuits (SOICs), similar in appearance to DIP packages, except that the body of the component is smaller and the pins are replaced by gull wing or J-type lead configurations.
- Common SMT discrete packages known as Electronic Industry Association (EIA) sizes 1206, 0805, 0603, and 0402 for resistors, capacitors, and diodes; with EIA sizes A, B, C, D, and MELF packages for various types of capacitors and inductors.
- Plastic or ceramic leaded chip carriers (PLCC or CLCC), rectangular carriers with J-leads around all four edges. These components may be directly soldered to the printed circuit board or installed into a socket that in turn is soldered to the printed circuit board.
- Chip-on-board, which consists of adhesive bonding a basic silicon chip die to the printed circuit board, beam welding leads from the die to the printed circuit board, and encapsulating the die and leads in a drop of adhesive potting compound.

- Ball grid array packages, much like PGA packages, except that instead of an array of pins protruding from the bottom of the component, there is an array of solder balls, each affixed to a pad on the component. The component may be either a plastic (PBGA) or a ceramic (CBGA) package. The BGA package is placed onto a corresponding artwork pattern on the printed circuit board and the assembly is heated to reflow the solder balls, thus attaching the BGA to the printed circuit board.
- Flip-chip package, a component package manufactured with small solder balls placed directly on the circuit substrate where electric connections are required. The substrate is then "flipped," or turned over, so that the solder balls may be fused by reflow directly to pads on a printed circuit board.
- Multichip module, a component package, houses more than one interconnected silicon die within a subassembly. The subassembly is attached to the printed circuit board as a through-hole (DIP or SIP) component or as a SMT component.
- Silicon on silicon (SOS), a component package consisting of silicon die attached to a silicon substrate to create a custom integrated circuit assembly, which is attached to the printed circuit board like a conventional component.

3.2 Fastening and Joining

Mechanical Fastening

Conventional machine design techniques apply to the design of mechanical joints employing threaded fasteners, rivets, and pins; they apply when strength and deflection are the primary design criteria. Dynamic loads (shock and vibration) require additional consideration, as does the selection of fastener materials to avoid corrosion.

In many mechanical fastening applications in electronic equipment fastener strength is not an issue and fastener size is selected based on the need to reduce the number of screw sizes used in the design (cost issue), appearance, and space available for tool clearances. For instance, when material compatibility is required between the fastener and aluminum parts with exposure to moisture, weather, and corrosive environments, stainless steel fasteners are advised. In commercial applications where weather exposure and corrosive environments are not a significant issue, zinc- or cadmium-plated steel fasteners are employed. Brass fasteners are used when the electrical resistance must be low to facilitate the flow of electrical current through a joint. Where joint electrical insulation is necessary to prevent electrical current flow through a joint, nylon and machined Teflon fasteners or other nonmetal fasteners are used.

Screw head selection is important in electronic equipment applications. Phillips head screws are preferred over slotted head screws due to the ability to gain increased tightening torque and to avoid marring of nearby surfaces should the screwdriver blade slip out of the screw head slot. Pan-head screws are preferred over round-head screws due to the absence of sharp edges. Flat-head screws are used to bury the screw head within the material thickness of one of the structural elements; however, there is no allowance for tolerances which exist between flat-head screws in a multifastener joint.

When threaded fasteners are used, there is concern that the joint will loosen and become ineffective over time. Such loosening may be caused by thermal cycling, vibration, or repetitive bending of the joint. Techniques used to maintain threaded joint integrity include the following:

- Use of a lock washer between the nut and the base material. To avoid damage to the base material, a flat washer is placed between the lock washer and the base material.

- In the event that an electrical bond must be established through the threaded joint, a toothed-type lock washer without a flat washer must be employed.
- Use of a compression nut (formed as to cause friction between the nut and the screw threads). This device loses effectiveness if frequently removed and may require replacement.
- Use of a nut or screw with a compressible insert (a threaded insert pressed into the base material). This applies to screw sizes of #6 and larger. The same warning on reuse applies as for the compression nut.
- Use of a screw retention adhesive material on the threads prior to making the joint. The adhesive must be reapplied each time the joint is disassembled. Various degrees of hold are available.
- Use of an antirotation wire through a hole in the nut or in the head of the screw. Applicable to larger bolts only.
- Tooth-type lock washers should not be used in contact with printed circuit board (may damage or short to traces) or with nonmetallic base materials (will mar the mating surface).
- Joints where one or more elements are capable of cold flow (creep), e.g., nylon, plastics, and soft metal, require a retention method other than compression-type lock washers.

Rivets are used in electronics assembly to form a permanent joint and may be solid or tubular. One should not depend on a riveted joint to provide long-term electrical connectivity. Cold flow will lead to joint looseness when soft metals or plastic materials are involved. Rivet material must be compatible with other joint materials to avoid corrosion.

Pins pressed into holes in mating parts are sometimes used to make permanent joints. Pin joints may be disassembled; however, a larger diameter pin may be required to achieve full joint strength upon reassembly. Materials selection is important to avoid corrosion.

Welding and Soldering

Conventional spot welding, inert gas welding, torch welding, and brazing[17,18] are used in the construction of many metal chassis and other mechanical components. Such joints have consistent electrical conductivity. Material properties in the heat-affected zone are often altered and weakened and may cause early mechanical failure of highly stressed joints. Lap joints tend to draw liquids into themselves by capillary action. Lap joints must be carefully cleaned and protected from ingestion of contaminantes which may eventually cause corrosion, loss of electrical conductivity, and eventual mechanical failure of the joint.

Lead–tin and no-lead solders are used to make electrical joints,[19,20] and to bind components to printed circuit boards. Eutectic 63% lead–37% tin solder has a relatively low melting point and is commonly used, unless there is a no-lead requirement, for attachment of components to circuit boards. No-lead solder is also used; however, the flow temperature is higher by 20°C or more. The manufacturer of the solder should be contacted for recommendations and data on specific no-lead formulations.

In applications where a soldered electrical joint is needed and mechanical stresses will be present, the joint must be designed to accept the mechanical stresses without the solder present. Under load, a solder joint will creep until the loads are eliminated or when the joint fails. As a result, solder is generally used only for electrical connection purposes and not for carrying mechanical loads.

Solder often is the only means of mechanical and electrical support for surface-mounted parts on a circuit board assembly. Due to variations in the coefficient of thermal expansion between the circuit board substrate and the component materials, solder joints will be subjected to thermal cycling-induced stresses caused by environmental or operationally generated temperature changes. Successful surface-mount design requires that the mass of the individual parts is very small and that the circuit board is protected from bending stresses so that solder attachment points will not eventually fail as the result of creep or fatigue failure.

Adhesives

Adhesives[21] are used in electronic equipment for a variety of purposes, such as component attachment to circuit boards in preparation for wave soldering. Encapsulants are used to encase and protect components and circuits, and adhesives are used to seal mechanical joints to avoid liquid and gas leakage.

Adhesive joints withstand shear loads; however, adhesive joints are much weaker when subjected to peeling loads. The load-bearing properties of cured adhesive joints (creep, stiffness, modulus of elasticity, and shear stresses) may vary significantly over the temperature ranges often experienced in service. Successful joints using adhesives are designed to bear mechanical loads without the adhesive present, with the adhesive applied to achieve seal and environmental protection.

Adhesives may release chemicals and gases during cure which are corrosive to electrical contacts and components and materials used in the construction of electronic equipment. Such adhesives must be avoided or fully cured prior to introduction into a sealed electronic enclosure.

3.3 Interconnection

Discrete Wiring

Discrete wiring involves the connection from one component to another by use of electronic hookup wire which may or may not be insulated. In either case the individual connections are made mechanically by forming the component leads to fit the support terminals prior to applying solder to the connection. Care must be taken to route wires away from sharp objects and to avoid placing mechanical stresses on the electrical joints.

Board Level

Board-level interconnection is accomplished by soldering components to conductive patterns etched onto a printed circuit board. Panel- or bracket-mounted parts may require discrete wiring between the component and the printed circuit board. Board assemblies sometimes consist of two or more individual circuit boards where a smaller board assembly is soldered directly to a host circuit board.

Socket type connectors may be soldered to a circuit board to receive integrated circuits, relays, memory chips, and other discrete components. Care must be exercised to assure that the socket provides mechanical retention of the part to prevent the part from being dislodged by transportation and service environments.

Intramodule

Discrete components and circuit board assemblies located within an electronic subassembly, or module, are interconnected within the module. In addition, the module circuits and components are presented to an interface, such as one or more connectors, to facilitate interconnection with other modules or cable assemblies.

Intermodule

Individual modules are interconnected to achieve system-level functions. Modules may be plugged together directly using connectors mounted to each module, be interconnected by cable and wiring harness assemblies, or be plugged into connectors arrayed on a common interconnection circuit board sometimes called a "mother" board.

Interequipment

System-level interconnection between electronic equipment may consist of wiring harness assemblies, fiber-optic cables, or wireless interconnection.

Fiber-Optic Connections

Fiber-optic[20] links are sometimes employed to interconnect electronic systems instead of conventional metallic conductors. Fiber-optic communication consists of transmitting a modulated light beam through a small-diameter (100-μm) glass fiber to a receiver where the modulated signal is transformed into a corresponding electrical signal. Used extensively in communications equipment and systems, fiber-optic links are immune to radio and magnetic field interference. Design is centered on methods to provide connectors and splices without inducing signal reflection and attenuation. Also, the design must provide at least the manufacturer-recommended minimum bend radii to avoid reflections and distorting stresses on the fiber. It is important to mechanically support and control fiber cable runs to avoid excessive mechanical stresses.

3.4 Shock and Vibration

Fragility

Fragility refers to the vulnerability of a device to failure. It is necessary for the electronic packaging engineer to understand the most likely modes of failure under dynamic loads and adjust the design to protect against those failures. Fragility not only addresses loads which cause immediate mechanical or electrical malfunctions but also includes failures which result from accumulated damage (fatigue, fretting, or fatiguelike processes) leading to failure early in the expected useful life of a product.

Shock

Shock may be defined as a sudden change in momentum of a body. A shock pulse may range from a simple step function or haversine pulse to a brief but complex waveform composed of several frequencies. The shock pulse may result in operational malfunctions (in operating equipment), breakage of brittle materials, or bending displacement and subsequent (ringing) vibration of elements of the equipment receiving the shock pulse. Shock pulses near the fundamental or harmonic resonance frequency of the structure cause greatly magnified and destructive responses. Shock failures include the following:

- Permanent localized deformation at point of impact
- Permanent deformation of structural elements within an equipment such as distorted mounting brackets
- Secondary impact failures within an equipment should structural deformations cause components to strike adjacent surfaces
- Temporary or permanent malperformance of operating equipment
- Failure of fasteners, structural joints, and mounting attachment points
- Breakage of fragile components and fragile structural elements

Design techniques employed to avoid shock-induced damage[22–24] include the following:

- Characterization of the shock-producing event in terms of impulse waveform, energy, and point of application
- Computation or empirical determination of equipment responses to the shock pulse in terms of acceleration (or g level) vs. time
- Modifications of the equipment structure to avoid resonant frequencies which coincide with the frequency content of the shock pulse
- Assuring that the strength of structural elements is adequate to withstand the dynamic g loading without either permanent deformation or harmful displacements due to bending
- Selecting and using components which are known to withstand the internal shock environment to which they are subjected when the local mounting structure responds to the shock pulse transmitted inward from the external blow
- Employing protective measures such as energy-absorbing or resonance-modifying materials between the equipment and the point of shock application or within the equipment as applied to the mounting of fragile components

Vibration

The response of equipment to vibration may create damage if the equipment or elements thereof are resonant within the passband of the excitation spectra. Vibration failures include the following:

- Fretting, wear, and loosening of mechanical joints, thermal joints, and fasteners and within components such as connectors, switches, and potentiometers
- Fatigue-induced structural failure of brackets, circuit boards, and components
- Physical and operational failure should individual structural element bending displacements produce impact between adjacent objects
- Deviations in the electrical performance of components caused by the relative motion of elements within the component or by the relative motion between a component and other objects which modulate the local electrical and/or magnetic fields

Design techniques employed to avoid vibration-induced damage include the following:

- Characterization of the energy and frequency content of the source of vibration excitation
- Analytical and empirical determination of equipment primary and secondary structural responses and component sensitivity to vibration forces in the passband of the source vibration
- Control of individual resonance frequencies of an equipment structure and internal elements to avoid coincidence of resonance frequencies
- Employing materials which have adequate fatigue life to withstand the cumulative damage predicted to occur over the life of the equipment
- Use of energy-absorbing materials between the equipment and the excitation source and within the equipment for the protective mounting of sensitive components

Testing

The primary purposes of shock and vibration testing are to evaluate the dynamic response of an equipment and components to the dynamic excitation and to compare these responses with the fragility of the unit under test.

Basic characterization testing is usually performed on an electrodynamic vibration machine with the unit under test hard mounted to a vibration fixture which has no resonance in the passband of the excitation spectrum. The test input is a low-displacement-level sinusoid which is slowly varied in frequency (swept) over the frequency range of interest. Recorded results of sine sweep testing produces a history of the response (displacement or acceleration) as experienced by accelerometers located at selected points on the equipment.

Caution is advised when using a hard-mount vibration fixture because the fixture is very stiff and is capable of injecting more energy into the test specimen at specimen resonance than would be experienced in the actual service environment. For this reason the test input signal should be of low amplitude. In service, the reaction of a less stiff mounting structure to the specimen at specimen resonance could significantly reduce the energy injected into the specimen and lessen the damage thereto. If the specimen response history is known prior to testing, the test system may be programmed to control input levels vs. frequency to reproduce the actual response history. In this case, it is important to assure that the control accelerometer is placed at the exact location on the specimen where the accelerometer was located when the initial response spectrum was recorded.

Vibration test information is used to aid in adjusting the equipment design to avoid unfavorable responses to the serviced excitation, such as the occurrence of coupled resonance (i.e., a component having a resonance frequency coincident with the resonance frequency of its supporting structure or the structure having a significant resonance which coincides with the frequency of an input shock spectra). Individual components are often tested to determine and document the excitation levels and frequencies at which they occur. Finding correlation between the individual resonances and operational failures of the unit under test is fundamental to both shock and vibration design.

For more complex vibration service input spectra, such as multiple sinusoidal or random vibration spectra, additional testing is performed on individual parts of the equipment to ensure that the responses are predictable, which will allow the designer to "tune" the overall equipment to avoid coincident resonances and to limit the responses of sensitive components. The final equipment testing to the specified dynamic environments and durations, which may be different for each axis of excitation, may be combined with other variables such as temperature, humidity, and altitude environments.

3.5 Structural Design

General

Structural design of an electronic system[24,25] equipment structure, enclosure, module, or bracket involves analytical, empirical, and experimental techniques to predict and thus control mechanical stresses. Structural strength may be defined as the ability of a material to bear both static (sustained) and dynamic (time-varying) loads without significant permanent deformation.

Many nonferrous materials suffer permanent deformation under sustained loads (creep). Ductile materials withstand dynamic loads better than brittle materials, which may fracture under sudden load application. Materials such as plastics often exhibit significant changes in material properties over the temperature range encountered by a product.

Many structural elements require stiffness to control deflection but must be checked to assure that strength criteria are achieved.

Complexity and Mechanical Impedance

An equipment may be viewed as a collection of individual elements interconnected to achieve an overall systems function. Each element may be individually modeled, but the equipment model becomes complex when the elements are interconnected.

The concept of mechanical impedance[6,24] applies to dynamic environments and refers to the reaction between a structural element or component and its mounting points over a range of excitation frequencies. The reaction forces at the structural interface or mounting point are a function of the resonance response of an element and may have an amplifying effect or a dampening effect on the mounting structure. Mechanical impedance design involves control of element resonance frequencies and mounting structural resonance frequencies, so that they do not combine to cause failure.

Degree of Enclosure

The degree of enclosure is the extent to which the components within electronic equipment are isolated from the surrounding environment.

For vented enclosures the design must provide drain holes to facilitate elimination of induced liquids and condensation. Convection (natural or forced) cooled equipment, when used where dust is present, may require filtration of the air permitted to enter the equipment.

Partially sealed enclosures using permeable sealing materials (e.g., adhesives and plastics) are vulnerable to penetration by water vapor and other gases. Pachen's law states that the total pressure inside an enclosure is the sum of the partial pressures of the constituent gases. When the external partial pressure of a constituent gas is higher than the internal partial pressure of that gas, regardless of the total pressure inside the equipment, the constituent gas will enter through the permeable seal until the internal and external partial pressures are equalized. When water vapor is ingested, condensation may occur inside the enclosure during temperature cycling, resulting in corrosion and perhaps interruption of electrical signals. Permeable seals, which permit gases but not liquids to enter equipment, do not protect against internal moisture damage and corrosion.

Completely (hermetically) sealed equipment enclosures using metal or glass seals permit the internal humidity and pressure to be defined when the unit is sealed. It is necessary to control the dryness of internal gases introduced into the equipment to protect from condensation-induced corrosion. In addition, it is necessary to assure that internal pressures generated inside of the enclosure due to heating in combination with changes in external ambient pressure (e.g., due to changes in altitude or ambient air pressure) do not exceed the structural deformation limitations and stress-bearing capabilities of the enclosure.

Equipment which operates in the presence of explosive gases or combustible dust mixtures must incorporate components which will not cause ignition, and exposed circuits must operate at low-voltage and low-current conditions so that short-circuit heating and arcing are controlled or eliminated. Vented equipment used in explosive environments may require use of flame propagation barriers, such as a screen mesh, which demonstrate under test should ignition occur inside the equipment that the flame front will not propagate into the outer environment.

Thermal Expansion and Stresses

The coefficient of thermal expansion is a material property which defines the degree to which a material will expand or contract when that material is subjected to a change in temperature. The coefficient of thermal expansion varies widely among the materials available for use in the construction of electronic equipment. When two or more materials are bonded or fastened together and subjected to a change in temperature, they will attempt to expand or contract at different rates. This difference in expansion or contraction leads to bending and shear stresses which may be detrimental to the life and operation of the equipment.

Thermal cycling of mechanically joined materials may lead to failure such as loss of electrical contact in bolted joints, cracking and breaking of ceramic parts bonded to plastic or metal surfaces, solder joint failure, and accelerated fatigue failure when the thermal stresses add to the stresses normally expected during the life cycle of an electronic product. Thermal stresses may be reduced by selecting adjoining materials which have the least difference in coefficient of thermal expansion.

3.6 Thermal Design

Objectives

The purpose of thermal design[26–28] is to control component temperatures within acceptable limits and thus to achieve satisfactory product reliability.[29] Newer component fabrication techniques, such as the fabrication of metal–oxide–semiconductors (MOSs) and CMOSs, greatly reduce component input power requirements and thus greatly reduce heat generation within the component. However, components using these and similar techniques of construction are being reduced in size such that the power per unit volume of high-component-density electronic equipment is actually being increased as a result of using these newer fabrication technologies.

There are three primary levels of thermal design:

1. Equipment total heat generation and control of how that heat will be dissipated by the equipment enclosure to the local external environment
2. Equipment internal environment, which is the environment experienced by modules and subassemblies, and how heat from the modules and subassemblies will be passed to the equipment enclosure
3. Control of critical component temperatures and how heat is transferred between the components and the internal environment

In addition, the electronic packaging engineer must consider how materials, finishes, and lubricants will behave over the expected range of temperatures.

Heat flow and temperature are analogous to the flow of electrical current and voltage. The term *thermal resistance* (in degrees celsius per watt) is employed to relate temperature to the flow of thermal energy in the same manner as Ohm's law relates voltage to the flow of electrical current.

A material or device may be characterized by its thermal resistance. Examples include the thermal resistance of heat sinks, interfaces between components and mounting surfaces, thermal resistance of structural elements, and thermal resistance of mechanical joints.

Thermal design includes the definition of heat flow paths from the component to the ultimate heat sink; for each heat flow path thermal resistances are selected to ensure that component temperatures will be maintained within acceptable levels.

Basis Heat Transfer Modes

Conduction. Conduction is the transfer of thermal energy directly though a material medium, which may be a solid, liquid, or gas. Conduction of heat from a source to the ultimate heat sink includes as appropriate:

1. Component internal heat transfer from an internal feature, such as a semiconductor junction temperature, to the local mounting surface, liquid or air. Component specifications usually include a thermal resistance which relates an internal critical point temperature to a specified location on the component package.
2. Contact resistance between the component and its mounting surface. Contact resistance depends on contact area, contact pressure, and presence of thermal grease or

other materials used to lower contact resistance. Contact resistances are often determined experimentally.

3. Thermal resistance through structural elements (computed based on cross-sectional area and length) using thermal conductivities as defined in tables of material properties.
4. Interface resistance at each structural joint. These may be estimated by calculation and are often defined experimentally.
5. Thermal resistance of heat sinks[30] used to dissipate energy to gas or liquid coolants. Conduction heat sinks include liquid-cooled cold plates, convection (natural and forced) heat sinks, forced-convection heat exchangers, evaporative devices such as heat pipes, and thermoelectric (Peltier effect) cooling devices.

Free Convection. Free convection involves the rejection of thermal energy from a warm object to air or other gas surrounding the object in the absence of mechanically or environmentally induced motion of the gases. Free convection is a continuing process whereby the local warming of surrounding gases in immediate contact with a warm object produces buoyancy and movement of the heated gases away from the warm object. Cooler gases are drawn toward the warm object, replacing the escaping warm gases. In this manner, heat is passed from the warm object to the gas (or liquid) environment.

Free-convection thermal resistance depends on the orientation of warm surfaces relative to gravity, surface area, and surface finish. Free convection is enhanced by providing extended surfaces such as fins to increase the effective contact area between the warm surface and local gases. Manufacturers of commercially available heat sinks are able to provide thermal resistance information for the devices that they market.

Equipment which is to be cooled by natural convection often must be provided with vents to permit the escape of warm gases and the entrance of cooler gases. Warm components must be in the flow path of the cooling gases, and internal obstructions must be avoided which would impede flow of the cooling gases.

Natural-convection cooling is a mass-rate-of-flow process, and cooling effectiveness is decreased by reduction in ambient air pressure (thus lowering the density of the cooling gas) such as occurs at higher altitudes and during hot days when the density of surrounding gases is lower.

Forced Convection. Forced convection may be produced mechanically with fans, blowers, and pumps or movement of the equipment during use or by naturally occurring air movement (wind) over unsheltered equipment. Forced-convection thermal resistance values are much lower than those resulting from natural-convection cooling.

Within forced-convection cooled equipment, the cooling media is ducted first to the most temperature sensitive components and then to the less temperature sensitive components located downstream where local temperatures may be higher.

Forced-convection heat sinks[31] and heat exchangers are often used to cool heat-sensitive components and assemblies.

Forced-convection thermal resistance is sensitive to the mass rate of flow and velocity of the cooling medium. Mass rate of flow is pressure dependent and decreases with reductions in ambient pressure. Decreased mass rate of flow causes increased thermal resistance. Increased air velocity past the cooled object reduces the thickness of the boundary layer surrounding the object and thus decreases the effective thermal resistance;

Radiation. Radiation heat transfer is the transfer of energy from a warmer object to a cooler object by infrared radiation. Unlike convection cooling, radiation heat transfer is most effective in a vacuum and is not dependent upon the presence of a gaseous media between the objects.

The effectiveness of radiation heat transfer is dependent on the temperature differential between the objects, the distance between the objects, the projected area of each object, and the emissivity of each object. Radiation cooling (or heating) efficiency is reduced if smoke, moisture vapor, dust, or other particulate matter is suspended in the intervening gases.

Radiation is the primary mode of heat transfer in outer space; however, unless large temperature differences and short separation distances are experienced between earthbound objects, radiation involves a small fraction of the total heat flow.

Solar radiation causes heat buildup within an electronic enclosure when ultraviolet rays from the sun strike a closed equipment case adding to the heat load within the equipment. Should solar radiation pass into a device, as through glass or plastic windows or a lens which is transparent to ultraviolet rays, the radiation will heat internal surfaces, which in turn will radiate at infrared frequencies to which the transparent material is usually opaque and will not permit the infrared radiation to leave the enclosure. In this way thermal energy is trapped inside the enclosure and may lead to excessive internal temperatures. Solar-induced heat loading must be considered whenever electronic equipment is exposed to sunlight.

Evaporation. Evaporation cooling and condensation cooling techniques utilize the latent heat of vaporization of a heat transfer liquid such as water, alcohol, Freon, or other liquid to cause effective temperature control. Thus, when an object is submerged within a liquid, such as water, the temperature of the object is controlled to the boiling point of the liquid (100°C for water) within which it is submerged.

Heat pipe and evaporation chamber devices utilize evaporation by containing a liquid and providing an internal capillary structure to return condensed liquid from the cool end of the device to the heated end of the device.

3.7 Protective Packaging

Definition

Protective packaging includes the techniques employed to ensure that a product will survive handling, shipping, and storage environments without degradation.

Storage Environment Protection

Electronic equipment may be subjected to storage for extended periods of time. The storage environments may be more severe than the equipment will experience after being placed into service.

Protective packaging selected for storage must withstand the storage environments and offer protection to the enclosed product.

The materials from which storage containers and fillers are selected must remain chemically inert and not introduce detrimental effects on the stored equipment. For example, some paper products contain sulfur, the fumes of which accelerate tarnishing, thus increasing contact resistance of silver-plated contacts. Also, some adhesives and wrapping and cushioning materials outgas corrosive fumes at elevated temperatures.

Shipping Environment Protection

Protective packaging must withstand transportation environments and the handling associated with movement of the packaged equipment to and from carriers. The transportation and handling environments may include exposure to more severe shock and vibration than the protected product will experience after being placed into service.

The shipping containers must tolerate stacking and handling by mechanical lifting devices. When it is predicted that the transportation environment will be severe, the container will need to include packaging materials to cushion the product from mechanical shock and vibration loads.

REFERENCES

1. D. Agonafer and R. E. Fulton, *Computer Aided Design,* Marcel Dekker, EEP Vol. 3, ASME Press, New York, 1992.
2. G. L. Ginsberg, *Printed Circuits Design,* McGraw-Hill, New York, 1990.
3. W. T. Shugg, *Handbook of Electrical and Electronic Insulating Materials,* Van Nostrand Reinhold, New York, 1986.
4. B. S. Matisoff, *Handbook of Electronics Packaging,* Van Nostrand Reinhold, New York, 1982
5. Part 15, Federal Communications Rules and Regulations, www.fcc.gov.
6. W. C. Fackler, *Equivalent Techniques for Vibration Testing,* SVM-9, Naval Research Laboratory, Washington, DC, 1972.
7. *Alloy Cross Index,* Mechanical Properties Data Center, Battelle's Columbus Laboratories, Columbus, OH, 1981.
8. *Metals Handbook,* American Society for Metals, Metals Park, OH, 2000.
9. *Stainless Steel Handbook,* Allegheny Ludlum Steel Corporation, Pittsburgh, PA, 1956.
10. *Aluminum and Aluminum Alloys,* American Society for Metals, Metals Park, OH, 1999.
11. www.Pb-Free.com, an interactive website dedicated to providing information.
12. C. A. Harper, *Handbook of Plastics and Elastomers,* McGraw-Hill, New York, 1981.
13. M. Morton, *Rubber Technology,* Van Nostrand Reinhold, New York, 1987.
14. R. C. Buchanan, *Ceramic Materials for Electronics,* Marcel Dekker, New York, 1986.
15. C. A. Harper, *Electronic Packaging Handbook,* 3rd ed., McGraw-Hill Professional, New York, 2000.
16. J. B. Wachtman, *Mechanical Properties of Ceramics,* Wiley-Interscience, New York, 1986.
17. R. K. Wassink, *Soldering in Electronics,* 2nd ed., Electrochemical Publications, Isle of Man, British Isles, 1997.
18. A. Rahn, *The Basics of Soldering,* Wiley-Interscience, New York, 1993.
19. R. W. Woodgate, *Handbook of Machine Soldering,* Wiley-Interscience, New York, 1988.
20. M. G. Pecht, *Soldering Processes and Equipment,* Wiley-Interscience, Hew York, 1990.
21. A. J. Kinloch, *Adhesion and Adhesives,* Chapman and Hall, New York, 1987.
22. J. H. Williams, *Fundamentals of Applied Dynamics,* MIT Press, Cambridge, MA, 1995.
23. P. A. Engle, *Structural Analysis of Printed Circuit Board Systems,* Springer-Verlag, New York, 1993.
24. C. M. Harris and A. G. Piersol, *Harris' Shock and Vibration Handbook,* McGraw-Hill, New York, 2002.
25. G. R. Blackwell, *The Electronic Packaging Handbook,* CRC Press, Boca Raton, FL, 1999.
26. D. J. Dean, *Thermal Design of Electronic Circuit Boards and Packages,* Electrochemical Publications, Isle of Man, British Isles, 1997.
27. A. D. Kraus and A. Bar-Cohen, *Thermal Analysis and Control of Electronic Equipment,* McGraw-Hill, New York, 1993.
28. D. S. Steinberg, *Cooling Techniques for Electronic Equipment,* 2nd ed., Wiley-Interscience, New York, 1991.
29. F. Jensen, *Electronic Component Reliability,* Wiley-Interscience, New York, 1991.
30. A. D. Kraus and A. Bar-Cohen, *Design and Analysis of Heat Sinks,* Wiley, New York, 1995.
31. W. T. Kays and A. L. London, *Compact Heat Exchangers,* McGraw-Hill, New York, 1984.

CHAPTER 24

DESIGN OPTIMIZATION: AN OVERVIEW

A. Ravi Ravindran
Department of Industrial and Manufacturing Engineering
Pennsylvania State University
University Park, Pennsylvania

G. V. Reklaitis
School of Chemical Engineering
Purdue University
West Lafayette, Indiana

1 INTRODUCTION

This chapter presents an overview of optimization theory and its application to problems arising in engineering. In the most general terms, optimization theory is a body of mathematical results and numerical methods for finding and identifying the best candidate from a collection of alternatives without having to enumerate and evaluate explicitly all possible alternatives. The process of optimization lies at the root of engineering, since the classical function of the engineer is to design new, better, more efficient, and less expensive systems as well as to devise plans and procedures for the improved operation of existing systems. The power of optimization methods to determine the best case without actually testing all possible cases comes through the use of a modest level of mathematics and at the cost of performing iterative numerical calculations using clearly defined logical procedures or algorithms implemented on computing machines. Because of the scope of most engineering applications and the tedium of the numerical calculations involved in optimization algorithms, the techniques of optimization are intended primarily for computer implementation.

2 REQUIREMENTS FOR APPLICATION OF OPTIMIZATION METHODS

To apply the mathematical results and numerical techniques of optimization theory to concrete engineering problems, it is necessary to delineate clearly the boundaries of the engineering system to be optimized, to define the quantitative criterion on the basis of which candidates will be ranked to determine the "best," to select the system variables that will be used to characterize or identify candidates, and to define a model that will express the manner in which the variables are related. This composite activity constitutes the process of *formulating* the engineering optimization problem. Good problem formulation is the key to the success of an optimization study and is to a large degree an art. It is learned through practice and the study of successful applications and is based on the knowledge of strengths, weaknesses, and peculiarities of the techniques provided by optimization theory.

2.1 Defining System Boundaries

Before undertaking any optimization study it is important to define clearly the boundaries of the system under investigation. In this context a system is the restricted portion of the universe under consideration. The system boundaries are simply the limits that separate the system from the remainder of the universe. They serve to isolate the system from its surroundings, because, for purposes of analysis, all interactions between the system and its surroundings are assumed to be frozen at selected representative levels. Since interactions, nonetheless, always exist, the act of defining the system boundaries is the first step in the process of approximating the real system.

In many situations it may turn out that the initial choice of system boundary is too restrictive. To analyze a given engineering system fully, it may be necessary to expand the system boundaries to include other subsystems that strongly affect the operation of the system under study. For instance, suppose a manufacturing operation has a paint shop in which finished parts are mounted on an assembly line and painted in different colors. In an initial study of the paint shop we may consider it in isolation from the rest of the plant. However, we may find that the optimal batch size and color sequence we deduce for this system are strongly influenced by the operation of the fabrication department that produces the finished parts. A decision thus has to be made whether to expand the system boundaries to include the fabrication department. An expansion of the system boundaries certainly increases the size and complexity of the composite system and thus may make the study much more difficult. Clearly, to make our work as engineers more manageable, we would prefer as much as possible to break down large complex systems into smaller subsystems that can be dealt with individually. However, we must recognize that this decomposition is in itself a potentially serious approximation of reality.

2.2 Performance Criterion

Given that we have selected the system of interest and have defined its boundaries, we next need to select a criterion on the basis of which the performance or design of the system can be evaluated so that the best design or set of operating conditions can be identified. In many engineering applications an economic criterion is selected. However, there is a considerable choice in the precise definition of such a criterion: total capital cost, annual cost, annual net profit, return on investment, cost-to-benefit ratio, or net present worth. In other applications a criterion may involve some technology factors, for instance minimum production time, maximum production rate, minimum energy utilization, maximum torque, and minimum weight. Regardless of the criterion selected, in the context of optimization the best will always mean the candidate system with either the *minimum* or the *maximum* value of the performance index.

It is important to note that within the context of the optimization methods only *one* criterion or performance measure is used to define the optimum. It is not possible to find a solution that, say, simultaneously minimizes cost and maximizes reliability and minimizes energy utilization. This again is an important simplification of reality, because in many practical situations it would be desirable to achieve a solution that is best with respect to a number of different criteria. One way of treating multiple competing objectives is to select one criterion as primary and the remaining criteria as secondary. The primary criterion is then used as an optimization performance measure, while the secondary criteria are assigned acceptable minimum or maximum values and are treated as problem constraints. However, if careful consideration were not given while selecting the acceptable levels, a feasible design that satisfies all the constraints may not exist. This problem is overcome by a technique called *goal programming,* which is fast becoming a practical method for handling multiple-criteria. In this method, all the objectives are assigned target levels for achievement and a relative priority on achieving these levels. Goal programming treats these targets as goals to aspire for and not as absolute constraints. It then attempts to find an optimal solution that comes as "close" as possible" to the targets in the order of specified priorities. Readers interested in multiple-criteria optimizations are directed to recent specialized texts.[1,2]

2.3 Independent Variables

The third key element in formulating a problem for optimization is the selection of the independent variables that are adequate to characterize the possible candidate designs or operation conditions of the system. There are several factors that must be considered in selecting the independent variables. First, it is necessary to distinguish between variables whose values are amenable to change and variables whose values are fixed by external factors, lying outside the boundaries selected for the system in question. For instance, in the case of the paint shop, the types of parts and the colors to be used are clearly fixed by product specifications or customer orders. These are specified system parameters. On the other hand, the order in which the colors are sequenced is, within constraints imposed by the types of parts available and inventory requirements, an independent variable that can be varied in establishing a production plan.

Furthermore, it is important to differentiate between system parameters that can be treated as fixed and those that are subject to fluctuations which are influenced by external and uncontrollable factors. For instance, in the case of the paint shop equipment breakdown and worker absenteeism may be sufficiently high to influence the shop operations seriously. Clearly, variations in these key system parameters must be taken into account in the production planning problem formulation if the resulting optimal plan is to be realistic and operable.

Second, it is important to include in the formulation all of the important variables that influence the operation of the system or affect the design definition. For instance, if in the design of a gas storage system we include the height, diameter, and wall thickness of a cylindrical tank as independent variables but exclude the possibility of using a compressor to raise the storage pressure, we may well obtain a very poor design. For the selected fixed pressure we would certainly find the least cost tank dimensions. However, by including the storage pressure as an independent variable and adding the compressor cost to our performance criterion, we could obtain a design that has a lower overall cost because of a reduction in the required tank volume. Thus the independent variables must be selected so that all important alternatives are included in the formulation. Exclusion of possible alternatives, in general, will lead to suboptimal solutions.

Finally, a third consideration in the selection of variables is the level of detail to which the system is considered. While it is important to treat all of the key independent variables, it is equally important not to obscure the problem by the inclusion of a large number of fine details of subordinate importance. For instance, in the preliminary design of a process involving a number of different pieces of equipment—pressure vessels, towers, pumps, compressors, and heat exchangers—one would normally not explicitly consider all of the fine details of the design of each individual unit. A heat exchanger may well be characterized by a heat transfer surface area as well as shell-side and tube-side pressure drops. Detail design variables such as number and size of tubes, number of tube and shell passes, baffle spacing, header type, and shell dimensions would normally be considered in a separate design study involving that unit by itself. In selecting the independent variables, a good rule to follow is to include only those variables that have a significant impact on the composite system performance criterion.

2.4 System Model

Once the performance criterion and the independent variables have been selected, the next step in problem formulation is the assembly of the model that describes the manner in which the problem variables are related and the performance criterion is influenced by the independent variables. In principle, optimization studies may be performed by experimenting directly with the system. Thus, the independent variables of the system or process may be set to selected values, the system operated under those conditions, and the system performance index evaluated using the observed performance. The optimization methodology would then be used to predict improved choices of the independent variable values and experiments continued in this fashion. In practice, most optimization studies are carried out with the help of a model, a simplified mathematical representation of the real system. Models are used because it is too expensive or time consuming or risky to use the real system to carry out the study. Models are typically used in engineering design because they offer the cheapest and fastest way of studying the effects of changes in key design variables on system performance.

In general, the model will be composed of the basic material and energy balance equations, engineering design relations, and physical property equations that describe the physical phenomena taking place in the system. These equations will normally be supplemented by inequalities that define allowable operating ranges, specify minimum or maximum performance requirements, or set bounds on resource availabilities. In sum, the model consists of all of the elements that normally must be considered in calculating a design or in predicting the performance of an engineering system. Quite clearly the assembly of a model is a very time-consuming activity, and it is one that requires a thorough understanding of the system being considered. In simple terms, a model is a collection of equations and inequalities that define how the system variables are related and that constrain the variables to take on acceptable values.

From the preceding discussion, we observe that a problem suitable for the application of optimization methodology consists of a performance measure, a set of independent variables, and a model relating the variables. Given these rather general and abstract requirements, it is evident that the methods of optimization can be applied to a very wide variety of applications. We shall illustrate next a few engineering design applications and their model formulations.

3 APPLICATIONS OF OPTIMIZATION IN ENGINEERING

Optimization theory finds ready application in all branches of engineering in four primary areas:

1. Design of components of entire systems
2. Planning and analysis of existing operations
3. Engineering analysis and data reduction
4. Control of dynamic systems

In this section we briefly consider representative applications from the first three areas.

In considering the application of optimization methods in design and operations, the reader should keep in mind that the optimization step is but one step in the overall process of arriving at an optimal design or an efficient operation. Generally, that overall process will, as shown in Fig. 1, consist of an iterative cycle involving synthesis or definition of the structure of the system, model formulation, model parameter optimization, and analysis of the resulting solution. The final optimal design or new operating plan will be obtained only

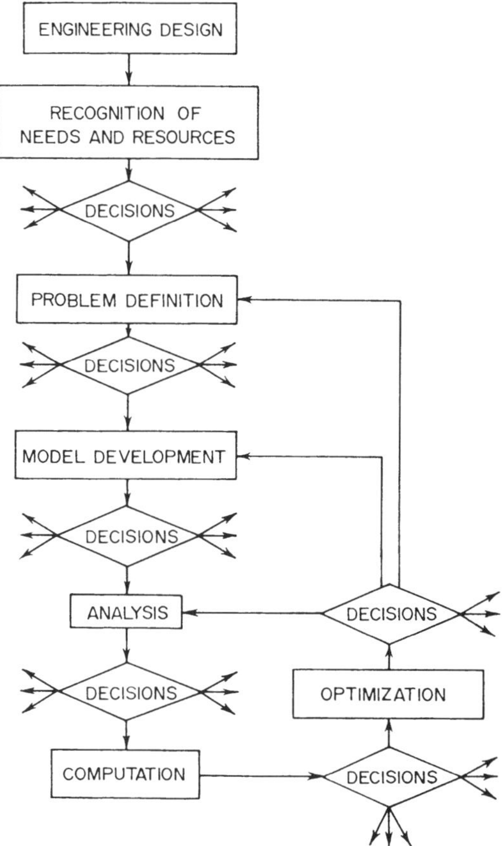

Figure 1 Optimal design process.

after solving a series of optimization problems, the solution to each of which will have served to generate new ideas for further system structures. In the interest of brevity, the examples in this section show only one pass of this iterative cycle and focus mainly on preparations for the optimization step. This focus should not be interpreted as an indication of the dominant role of optimization methods in the engineering design and systems analysis process. Optimization theory is but a very powerful tool that, to be effective, must be used skillfully and intelligently by an engineer who thoroughly understands the system under study. The primary objective of the following example is simply to illustrate the wide variety but common form of the optimization problems that arise in the design and analysis process.

3.1 Design Applications

Applications in engineering design range from the design of individual structural members to the design of separate pieces of equipment to the preliminary design of entire production facilities. For purposes of optimization the shape or structure of the system is assumed known and the optimization problem reduces to the selection of values of the unit dimensions and operating variables that will yield the best value of the selected performance criterion.

Example 1 Design of an Oxygen Supply System

Description. The basic oxygen furnace (BOF) used in the production of steel is a large fedbatch chemical reactor that employs pure oxygen. The furnace is operated in a cyclic fashion: Ore and flux are charged to the unit, are treated for a specified time period, and then are discharged. This cyclic operation gives rise to a cyclically varying demand rate for oxygen. As shown in Fig. 2, over each cycle there is a time interval of length t_1 of low

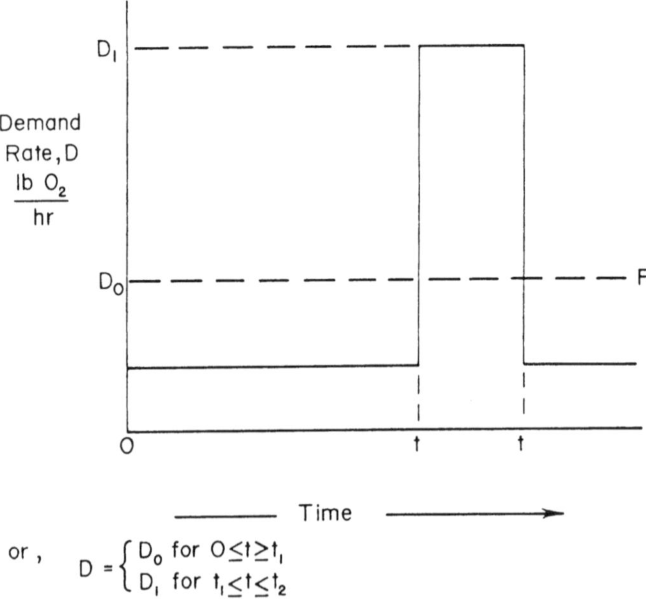

Figure 2 Oxygen demand cycle.

demand rate, D_0, and a time interval $(t_2 - t_1)$ of high demand rate, D_1. The oxygen used in the BOF is produced in an oxygen plant. Oxygen plants are standard process plants in which oxygen is separated from air using a combination of refrigeration and distillation. These are highly automated plants which are designed to deliver a fixed oxygen rate. To mesh the continuous oxygen plant with the cyclically operating BOF, a simple inventory system shown in Fig. 3 and consisting of a compressor and a storage tank must be designed. A number of design possibilities can be considered. In the simplest case, one could select the oxygen plant capacity to be equal to D_1, the high demand rate. During the low-demand interval the excess oxygen could just be vented to the air. At the other extreme, one could also select the oxygen plant capacity to be just enough to produce the amount of oxygen required by the BOF over a cycle. During the low-demand interval, the excess oxygen production would then be compressed and stored for use during the high-demand interval of the cycle. Intermediate designs could involve some combination of venting and storage of oxygen. The problem is to select the optimal design.

Formulation. The system of concern will consist of the O_2 plant, the compressor, and the storage tank. The BOF and its demand cycle are assumed fixed by external factors. A reasonable performance index for the design is the total annual cost, which consists of the oxygen production cost (fixed and variable), the compressor operating cost, and the fixed costs of the compressor and the storage vessel. The key independent variables are the oxygen plant production rate F (lb O_2/h), the compressor and storage tank design capacities H (hp) and V (ft^3), respectively, and the maximum tank pressure p (psia). Presumably the oxygen plant design is standard, so that the production rate fully characterizes the plant. Similarly, we assume that the storage tank will be of a standard design approved for O_2 service.

The model will consist of the basic design equations that relate the key independent variables. If I_{max} is the maximum amount of oxygen that must be stored, then using the corrected gas law we have

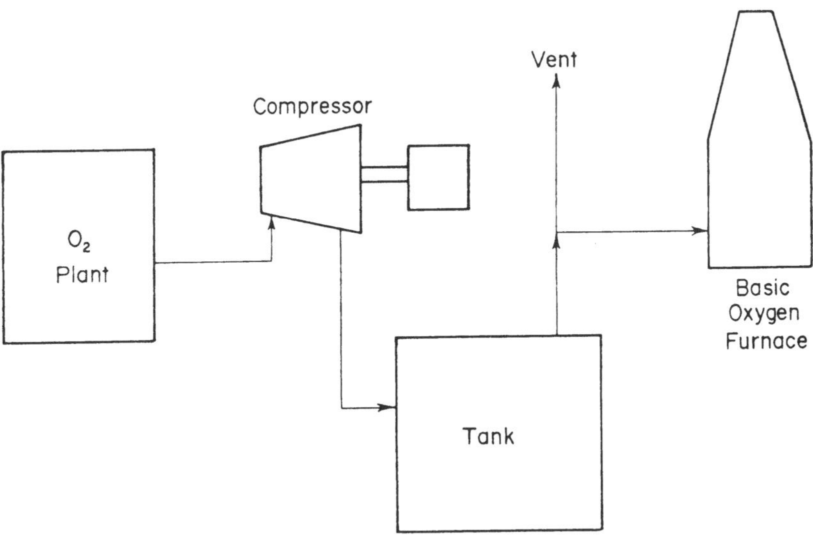

Figure 3 Design of oxygen production system.

$$V = \frac{I_{max}}{M} \frac{RT}{p} z \tag{1}$$

where R = gas constant
T = gas temperature (assume fixed)
z = compressability factor
M = molecular weight of O_2

From Fig. 1, the maximum amount of oxygen that must be stored is equal to the area under the demand curve between t_1 and t_2 and D_1 and F. Thus,

$$I_{max} = (D_1 - F)(t_2 - t_1) \tag{2}$$

Substituting Eq. (2) into Eq. (1), we obtain

$$V = \frac{(D_1 - F)(t_2 - t_1)}{M} \frac{RT}{p} z \tag{3}$$

The compressor must be designed to handle a gas flow rate of $(D_1 - F)(t_2 - t_1)/t_1$ and to compress it to the maximum pressure p. Assuming isothermal ideal gas compression,[3]

$$H = \frac{(D_1 - F)(t_2 - t_1)}{t_1} \frac{RT}{k_1 k_2} \ln\left(\frac{p}{p_0}\right) \tag{4}$$

where k_1 = unit conversion factor
k_2 = compression efficiency
p_0 = O_2 delivery pressure

In addition to Eqs. (3) and (4), the O_2 plant rate F must be adequate to supply the total oxygen demand, or

$$F \geq \frac{D_0 t + D_1 (t_2 - t_1)}{t_2} \tag{5}$$

Moreover, the maximum tank pressure must be greater than O_2 delivery pressure,

$$p \geq p_0 \tag{6}$$

The performance criterion will consist of the oxygen plant annual cost,

$$C_1(\$/\text{yr}) = a_1 + a_2 F \tag{7}$$

where a_1 and a_2 are empirical constants for plants of this general type and include fuel, water, and labor costs. The capital cost of storage vessels is given by a power law correlation,

$$C_2\ (\$) = b_1 V^{b_2} \tag{8}$$

where b_1 and b_2 are empirical constants appropriate for vessels of a specific construction.

The capital cost of compressors is similarly obtained from a correlation,

$$C_3\ (\$) = b_3 H^{b_4} \tag{9}$$

The compressor power cost will, as an approximation, be given by $b_5 t_1 H$, where b_5 is the cost of power.

The total cost function will thus be of the form

$$\text{Annual cost} = a_1 + a_2 F + d(b_1 V^{b_2} + b_3 H^{b_4}) + N b_5 t_1 H \tag{10}$$

where N = number of cycles per year
d = approximate annual cost factor

The complete design optimization problem thus consists of the problem of minimizing Eq. (10) by the appropriate choice of F, V, H, and p subject to Eqs. (3) and (4) as well as inequalities (5) and (6).

The solution of this problem will clearly be affected by the choice of the cycle parameters (N, D_0, D_l, t_1, and t_2), the cost parameters (a_1, a_2, $b_1 - b_5$, and d), as well as the physical parameters (T, p_0, k_2, z, and M).

In principle, we could solve this problem by eliminating V and H from Eq. (10) using Eqs. (3) and (4), thus obtaining a two-variable problem. We could then plot the contours of the cost function (10) in the plane of the two variables F and p, impose the inequalities (5) and (6), and determine the minimum point from the plot. However, the methods discussed in subsequent sections allow us to obtain the solution with much less work. For further details and a study of solutions for various parameter values the reader is invited to consult Ref. 4.

The preceding example presents a preliminary design problem formulation for a system consisting of several pieces of equipment. The next example illustrates a detailed design of a single structural element.

Example 2 Design of a Welded Beam

Description. A beam A is to be welded to a rigid support member B. The welded beam is to consist of 1010 steel and is to support a force F of 6000 lb. The dimensions of the beam are to be selected so that the system cost is minimized. A schematic of the system is shown in Fig. 4.

Formulation. The appropriate system boundaries are self-evident. The system consists of the beam A and the weld required to secure it to B. The independent or design variables in this case are the dimensions h, l, t, and b, as shown in Fig. 4. The length L is assumed to be specified at 14. in. For notational convenience we redefine these four variables in terms of the vector of unknowns $\mathbf{x}$,

$$\mathbf{x} = [x_1, x_2, x_3, x_4]^T = [h, l, t, b]^T$$

The performance index appropriate to this design is the cost of a weld assembly. The major cost components of such an assembly are (a) set-up labor cost, (b) welding labor cost, and (c) material cost:

$$F(x) = c_0 + c_1 + c_2 \tag{11}$$

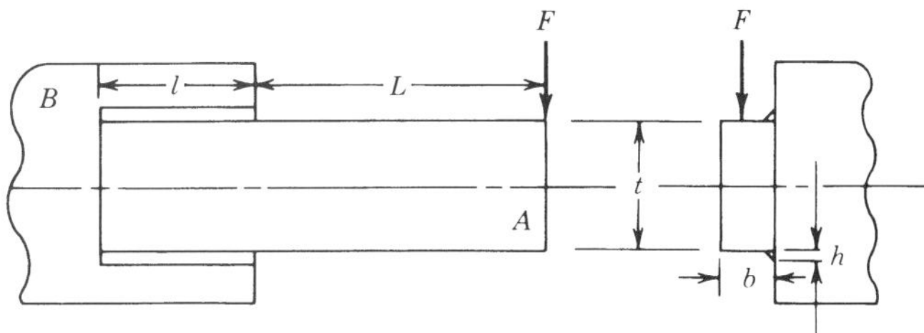

Figure 4 Welded beam.

where $F(x)$ = cost function
c_0 = set-up cost
c_1 = welding labor cost
c_2 = material cost

SET-UP COST: c_0. The company has chosen to make this component a weldment, because of the existence of a welding assembly line. Furthermore, assume that fixtures for set-up and holding of the bar during welding are readily available. The cost c_0 can, therefore, be ignored in this particular total cost model.

WELDING LABOR COST: c_1. Assume that the welding will be done by machine at a total cost of \$10 per hour (including operating and maintenance expense). Furthermore, suppose that the machine can lay down 1 in.^{3} of weld in 6 min. Therefore, the labor cost is

$$c_1 = \left(10\ \frac{\$}{\text{h}}\right)\left(\frac{1\ \text{h}}{60\ \text{min}}\right)\left(6\ \frac{\text{min}}{\text{in.}^3}\right) V_w = 1\left(\frac{\$}{\text{in.}^3}\right) V_w$$

where V_w is weld volume, in cubic inches.

MATERIAL COST: c_2

$$c_2 = c_3 V_w + c_4 V_B$$

where c_3 = \$/volume of weld material, =(0.37)(0.283) (\$/in.^{3})
c_4 = \$/volume of bar stock, =(017)(0.283) (\$/in.^{3})
V_B = volume of bar A (in.^{3})

From the geometry,

$$V_w = 2(\tfrac{1}{2}h^2 l) - h^2 l \qquad \text{and} \qquad V_B = tb(L + l)$$

so

$$c_2 = c_3 h^2 l + c_4 tb(L + l)$$

Therefore, the cost function becomes

$$F(x) = h^2 l + c_3 h^2 l + c_4 tb(L + l) \tag{12}$$

or, in terms of the x variables,

$$F(x) = (l + c_3) x_1^2 x_2 + c_4 x_3 x_4 (L + x_2) \tag{13}$$

Note all combinations of x_1, x_2, x_3, and x_4 can be allowed if the structure is to support the load required. Several functional relationships between the design variables that delimit the region of feasibility must certainly be defined. These relationships, expressed in the form of inequalities, represent the design model. Let us first define the inequalities and then discuss their interpretation:

$$g_1(x) = \tau_d - \tau(x) \geq 0 \tag{14}$$

$$g_2(x) = \sigma_d - \sigma(x) \geq 0 \tag{15}$$

$$g_3(x) = x_4 - x_1 \geq 0 \tag{16}$$

$$g_4(x) = x_2 \geq 0 \tag{17}$$

$$g_5(x) = x_3 \geq 0 \tag{18}$$

$$g_6(x) = P_c(x) - F \geq 0 \tag{19}$$

$$g_7(x) = x_1 - 0.125 \geq 0 \tag{20}$$

$$g_8(x) = 0.25 - \text{DEL}(x) \geq 0 \tag{21}$$

where τ_d = design shear stress of weld
$\tau(x)$ = maximum shear stress in weld; a function of x
σ_d = design normal stress for beam material
$\sigma(x)$ = maximum normal stress in beam; a function of x
$P_c(x)$ = bar buckling load; a function of x
$\text{DEL}(x)$ = bar end deflection; a function of x

To complete the model, it is necessary to define the important stress states.

WELD STRESS: $\tau(x)$. Following Shigley,[5] the weld shear stress has two components, τ' and τ'', where τ' is the primary stress acting over the weld throat area and τ'' is a secondary torsional stress:

$$\tau' = \frac{F}{\sqrt{2}\, x_1 x_2} \qquad \text{and} \qquad \tau'' = \frac{MR}{J}$$

with

$$M = F[L + (\tfrac{1}{2}x_2)]$$
$$R = \{(\tfrac{1}{4}x_2^2) + [\tfrac{1}{2}(x_3 + x_1)]^2\}^{1/2}$$
$$J = 2\{0.707x_1x_2[\tfrac{1}{12}x_2^2 + \tfrac{1}{4}(x_3 + x_1))^2]\}$$

where M = moment F about center of gravity of weld group
J = polar moment of inertia of weld group

Therefore, the weld stress τ becomes

$$\tau(x) = [(\tau')^2 + 2\,\tau'\,\tau''\cos\theta + (\tau'')^2]^{1/2}$$

where $\cos\theta = x_2/2R$.

BAR BENDING STRESS: $\sigma(x)$. The maximum bending stress can be shown to be equal to

$$\sigma(x) = \frac{6FL}{x_4 x_3^2}$$

BAR BUCKLING LOAD: $P_c(x)$. If the ratio $t/b = x_3/x_4$ grows large, there is a tendency for the bar to buckle. Those combinations of x_3 and x_4 that will cause this buckling to occur must be disallowed. It has been shown[6] that for narrow rectangular bars a good approximation to the buckling load is

$$P_c(x) = \frac{4.013\sqrt{EI\alpha}}{L^2}\left(1 - \frac{x_3}{2L}\sqrt{\frac{EI}{\alpha}}\right)$$

where E = Young's modulus, $=30 \times 10^6$ psi
$I = \frac{1}{12}\, x_3 x_4^3$
$\alpha = \frac{1}{3}\, G x_3 x_4^3$
G = shearing modulus, $=12 \times 10^6$ psi

BAR DEFLECTION: $\text{DEL}(x)$. To calculate the deflection assume the bar to be a cantilever of length L. Thus,

$$\text{DEL}(x) = \frac{4FL^3}{Ex_3^3 x_4}$$

The remaining inequalities are interpreted as follows: Inequality g_3 states that it is not practical to have the weld thickness greater than the bar thickness; g_4 and g_5 are nonnegativity restrictions on x_2 and x_3. Note that the nonnegativity of x_1 and x_4 are implied by g_3 and g_7. Constraint g_6 ensures that the buckling load is not exceeded. Inequality g_7 specifies that it is not physically possible to produce an extremely small weld.

Finally, the two parameters τ_d and σ_d in g_1 and g_2 depend on the material of construction. For 1010 steel τ_d = 13,600 psi and σ_d = 30,000 psi are appropriate.

The complete design optimization problem thus consists of the cost function (13) and the complex system of inequalities that results when the stress formulas are substituted into Eqs. (14)–(21). All of these functions are expressed in terms of four independent variables.

This problem is sufficiently complex that graphical solution is patently infeasible. However, the optimum design can readily be obtained numerically using the methods of subsequent sections. For a further discussion of this problem and its solution the reader is directed to Ref. 7.

3.2 Operations and Planning Applications

The second major area of engineering application of optimization is found in the tuning of existing operations. We shall discuss an application of goal programming model for machinability data optimization in metal cutting.[8]

Example 3 An Economic Machining Problem with Two Competing Objectives. Consider a single-point, single-pass turning operation in metal cutting wherein an optimum set of cutting speed and feed rate is to be chosen which balances the conflict between metal removal rate and tool life as well as being within the restrictions of horsepower, surface finish, and other cutting conditions. In developing the mathematical model of this problem, the following constraints will be considered for the machining parameters:

Constraint 1: Maximum Permissible Feed

$$f \leq f_{\max} \tag{22}$$

where f is the feed in inches per revolution and $f_{\max}$ is usually determined by a cutting force restriction or by surface finish requirements.[9]

Constraint 2: Maximum Cutting Speed Possible. If v is the cutting speed in surface feet per minute, then

$$v \leqq v_{\max} \tag{23}$$

where $v_{\max} = \pi D N_{\max}/12$

$N_{\max}$ = maximum spindle speed available on machine

Constraint 3: Maximum Horsepower Available. If $P_{\max}$ is the maximum horsepower available at the spindle, then

$$v f^{\alpha} \leq \frac{P_{\max}\,(33{,}000)}{c_t\, d_c^{\beta}}$$

where α, β, and c_t are constants[9] and d_c is the depth of cut in inches, which is fixed at a given value. For a given $P_{\max}$, c_t, β, and d_c, the right-hand side of the above constraint will be a constant. Hence, the horsepower constraint can be written simply as

$$v f^{\alpha} \leqq \text{const} \tag{24}$$

Constraint 4: Nonnegativity Restrictions on Feed Rate and Speed

$$v, f \geqq 0 \tag{25}$$

In optimizing metal cutting there are a number of optimality criteria that can be used. Suppose we consider the following objectives in our optimization: (i) maximize metal removal (MRR) and (ii) maximize tool life (TL). The expression for MRR is

$$\text{MRR} = 12vfd_c \qquad \text{in.}^3/\text{min} \tag{26}$$

TL for a given depth of cut is given by

$$\text{TL} = \frac{A}{v^{1/n} f^{1/n_1}} \tag{27}$$

where A, n, and n_1 are constants. We note that the MRR objective is directly proportional to feed and speed, while the TL objective is inversely proportional to feed and speed. In general, there is no single solution to a problem formulated in this way, since MRR and TL are competing objectives and their respective maxima must include some compromise between the maximum of MRR and the maximum of TL.

A Goal Programming Model

Goal programming is a technique specifically designed to solve problems involving complex, usually conflicting multiple objectives. Goal programming requires the user to select a set of goals (which may or may not be realistic) that ought to be achieved (if possible) for the various objectives. It then uses preemptive weights or priority factors to rank the different goals and tries to obtain an optimal solution satisfying as many goals as possible. For this, it creates a single objective function that minimizes the deviations from the stated goals according to their relative importance.

Before we discuss the goal programming formulation of the machining problem, we should discuss the difference between the terms *real constraint* and *goal constraint* (or simply *goal*) as used in goal programming models. The real constraints are absolute restrictions placed on the behavior of the design variables, while the goal constraints are conditions one would like to achieve but are not mandatory. For instance, a real constraint given by

$$x_1 + x_2 = 3$$

requires all possible values of $x_1 + x_2$ to always equal 3. As opposed to this, if we simply had a goal requiring $x_1 + x_2 = 3$, then this is not mandatory and we can choose values of x_1, x_2 such that $x_1 + x_2 \geq 3$ as well as $x_1 + x_2 \leq 3$. In a goal constraint positive and negative deviational variables are introduced as follows:

$$x_1 + x_2 + d_1^- - d_1^+ = 3 \qquad d_1^- \text{ and } d_1^+ \geq 0$$

Note that if $d_1^- > 0$, then $x_1 + x_2 < 3$, and if $d_1^+ > 0$, then $x_1 + x_2 > 3$. By assigning suitable preemptive weights on d_1^- and d_1^+, the model will try to achieve the sum $x_1 + x_2$ as close as possible to 3.

Returning to the machining problem with competing objectives, suppose that management considers that a given single-point, single-pass turning operation will be operating at an acceptable efficiency level if the following goals are met as closely as possible.

1. The MRR must be greater than or equal to a given rate M_1 (in.^{3}/min).
2. The tool life must equal T_1 (min).

In addition, management requires that a higher priority be given to achieving the first goal than the second.

The goal programming approach may be illustrated by expressing each of the goals as goal constraints as shown below. Taking the MRR goal first,

$$12vfd_c + d_1^- - d_1^+ = M_1$$

where d_1^- represents the amount by which the MRR goal is underachieved and d_1^+ represents any overachievement of the MRR goal. Similarly, the TL goal can be expressed as

$$\frac{A}{v^{1/n} f^{1/n1}} + d_2^- - d_2^+ = T_1$$

Since the objective is to have an MRR of at least M_1, the objective function must be set up so that a high penalty will be assigned to the underachievement variable d_1^-. No penalty will be assigned to d_1^+. To achieve a tool of life of T_1, penalties must be associated with both d_2^- and d_2^+ so that both of these variables are minimized to their fullest extent. The relative magnitudes of these penalties must reflect the fact that the first goal is considered to be more important than the second. Accordingly, the goal programming objective function for this problem is

$$\text{Minimize } z = P_1 d_1^- + P_2(d_2^- + d_2^+)$$

where P_1 and P_2 are nonnumerical preemptive priority factors such that $P_1 \gg P_2$ (i.e., P_1 is infinitely larger than P_2) . With this objective function every effort will be made to satisfy completely the first goal before any attempt is made to satisfy the second.

To express the problem as a linear goal programming problem, M_l is replaced by M_2, where

$$M_2 = \frac{M_1}{12d_c}$$

The goal T_1 is replaced by T_2, where

$$T_2 = \frac{A}{T_1}$$

and logarithms are taken of the goals and constraints. The problem can then be stated as follows:

$$\text{Minimize } z = P_1 d_1^- + P_2(d_2^- + d_2^+)$$

Subject to MRR goal $\quad \log v + \log f + d_1^- - d_1^+ = \log M_2$

TL goal $\quad (1/N) \log v + (1/n_1) \log f + d_2^- - d_2^+ = \log T_2$

f_{max} constraint $\quad \log f \leqq \log f_{max}$

V_{max} constraint $\quad \log v \leqq \log v_{max}$

Horsepower constraint $\quad \log v + \alpha \log f \leqq \log \text{const}$

$$\log v, \log f, d_1^-, d_1^+, d_2^-, d_2^+ \geqq 0$$

We would like to reemphasize here that the last three inequalities are real constraints on feed, speed, and horsepower that must be satisfied at all times, while the equations for MRR and TL are simply goal constraints. For a further discussion of this problem and its solution, see Ref. 8. An efficient algorithm and a computer code for solving linear goal programming problems are given in Ref. 10. Readers interested in other optimization models in metal cutting should see Ref. 11. The textbook by Lee[12] contains a good discussion of goal programming theory and its applications.

3.3 Analysis and Data Reduction Applications

A further fertile area for the application of optimization techniques in engineering can be found in nonlinear regression problems as well as in many analysis problems arising in engineering science. A very common problem arising in engineering model development is the need to determine the parameters of some semitheoretical model given a set of experimental data. This data reduction or regression problem inherently transforms to an optimization problem, because the model parameters must be selected so that the model fits the data as closely as possible.

Suppose some variable y is assumed to be dependent on an independent variable x and related to x through a postulated equation $y = f(x, \theta_1, \theta_2)$, which depends on two parameters θ_1 and θ_2. To establish the appropriate values of θ_1, θ_2, we run a series of experiments in which we adjust the independent variable x and measure the resulting y. As a result of a series of N experiments covering the range of x interest, a set of y and x values (y_i, x_i), $i = 1, \ldots, N$, is available. Using these data we now try to "fit" our function to the data by adjusting θ_1 and θ_2 until we get a "good fit." The most commonly used measure of a good fit is the *least-squares criterion,*

$$L(\theta_1, \theta_2) = \sum_{i=1}^{N} [y_i - f(x_i, \theta_1, \theta_2)]^2 \tag{28}$$

The difference $y_i - f(x_i, \theta_1, \theta_2)$ between the experimental value y_i and the predicted value $f(x_i, \theta_1, \theta_2)$ measures how close our model prediction is to the data and is called the *residual*. The sum of the squares of the residuals at all the experimental points gives an indication of goodness of fit. Clearly, if $L(\theta_1, \theta_2)$ is equal to zero, then the choice of θ_1, θ_2 has led to a perfect fit; the data points fall exactly on the predicted curve. The data-fitting problem can thus be viewed as an optimization problem in which $L(\theta_1, \theta_2)$ is minimized by appropriate choice of θ_1, θ_2.

Example 4 Nonlinear Curve Fitting

Description. The pressure–molar volume–temperature relationship of real gases is known to deviate from that predicted by the ideal gas relationship

$$Pv = RT$$

where P = pressure (atm)
v = molar volume ($\text{cm}^3/\text{g}\cdot\text{mol}$)
T = temperature (K)
R = gas constant ($82.06\ \text{atm}\cdot\text{cm}^3/\text{g}\cdot\text{mol K}$)

The semiempirical Redlich–Kwong equation

$$P = \frac{RT}{v - b} - \frac{a}{T^{1/2}\, v(v + b)} \tag{29}$$

is intended to direct for the departure from ideality but involves two empirical constants a and b whose values are best determined from experimental data. A series of PvT measurements listed in Table 1 are made for CO_2, from which a and b are to be estimated using nonlinear regression.

Formulation. Parameters a and b will be determined by minimizing the least-squares function (28). In the present case, the function will take the form

$$\sum_{i=1}^{8} \left[P_i - \frac{RT_i}{v_i - b} + \frac{a}{T^{1/2}\, v_i(v_i + b)} \right]^2 \tag{30}$$

where P_i is the experimental value at experiment i and the remaining two terms correspond to the value of P predicted from Eq. (29) for the conditions of experiment i for some selected value of the parameters a and b. For instance, the term corresponding to the first experimental point will be

$$\left(33 - \frac{82.06(273)}{500 - b} + \frac{a}{(273)^{1/2}\,(500)(500 + b)} \right)^2$$

Function (30) is thus a two-variable function whose value is to be minimized by appropriate choice of the independent variables a and b. If the Redlich–Kwong equation were to precisely match the data, then at the optimum the function (30) would be exactly equal to zero. In general, because of experimental error and because the equation is too simple to accurately model the CO_2 nonidealities, Eq. (30) will not be equal to zero at the optimum. For instance, the optimal values of $a = 6.377 \times 10^7$ and $b = 29.7$ still yield a squared residual of 9.7×10^{-2}.

4 STRUCTURE OF OPTIMIZATION PROBLEMS

Although the application problems discussed in the previous section originate from radically different sources and involve different systems, at the root they have a remarkably similar form. All four can be expressed as problems requiring the minimization of a real-valued function $f(x)$ of an N-component vector argument $x = (x_1, x_2, \text{and } x_N)$ whose values are restriced to satisfy a number of real-valued equations $h_k(x) = 0$, a set of inequalities $g_j(x) \geq 0$, and the variable bounds $x_i^{(U)} \geq x_i \geq x_i^{(L)}$. In subsequent discussions we will refer to the function $f(x)$ as the *objective function*, to the equations $h_k(x) = 0$ as *equality constraints,* and to the inequalities $g_j(x) \geq 0$ as the *inequality constraints.* For our purposes, these problem functions will always be assumed to be real valued, and their number will always be finite.

Table 1 Pressure–Molar Volume–Temperature Data for CO_2

Experiment Number	P (atm)	v (cm³/g·mol)	T (K)
1	33	500	273
2	43	500	323
3	45	600	373
4	26	700	273
5	37	600	323
6	39	700	373
7	38	400	273
8	63.6	400	373

The general problem

$$\begin{aligned} &\text{Minimize } f(x) \\ &\text{Subject to } h_k(x) = 0 \qquad k = 1, \ldots, K \\ &\qquad g_j(x) \geq 0 \qquad j = 1, \ldots, J \\ &\qquad x_i^{(L)} \geq x_i \geq x_i^{(L)} \qquad i = 1, \ldots, N \end{aligned}$$

is called a *constrained* optimization problem. For instance, Examples 1–3 are all constrained problems. The problems in which there are no constraints, that is,

$$J = K = 0$$

and

$$x_i^{(U)} = -x_i^{(L)} = \infty \qquad i = 1, \ldots, N$$

are called *unconstrained* optimization problems. Example 4 is an unconstrained problem. Optimization problems can be classified further based on the structure of the functions f, h_k, and g_j and on the dimensionality of x. Figure 5 illustrates one such classification. The basic subdivision is between unconstrained and constrained problems. There are two important classes of methods for solving the unconstrained problems. The direct-search methods require only that the objective function be evaluated at different points, at least through experimentation. Gradient-based methods require the analytical form of the objective function and its derivatives.

An important class of constrained optimization problems is *linear programming,* which requires both the objective function and the constraints to be linear functions. Out of all optimization models, linear programming models are the most widely used and accepted in practice. Professionally written software programs are available from all computer manufacturers for solving very large linear programming problems. Unlike the other optimization problems that require special solution methods based on the problem structure, linear programming has just one common algorithm, known as the "simplex method," for solving all types of linear programming problems. This essentially has contributed to the successful applications of linear programming model in practice. In 1984, Narendra Karmarkar,[13] an AT&T researcher, developed an interior point algorithm, which was claimed to be 50 times faster than the simplex method for solving linear programming problems. By 1990, Karmarkar's seminal work had spawned hundreds of research papers and a large class of interior point methods. It has become clear that while the initial claims are somewhat exaggerated, interior point methods do become competitive for very large problems. For a discussion of interior point methods, see Refs. 14–16.

Integer programming (IP) is another important class of linearly constrained problems where some or all of the design variables are restricted to be integers. But solutions of IP problems are generally difficult, time consuming, and expensive. Hence, a practical approach is to treat all the integer variables as continuous, solve the associated linear programming (LP) problem, and round off the fractional values to the nearest integers such as the constraints are not violated. This generally produces a good integer solution close to the optimal integer solution, particularly when the values of the variables are large. However, such an approach would fail when the values of the variables are small or binary valued (0 or 1). A good rule of thumb is to treat any integer variable whose value will be at least 20 as

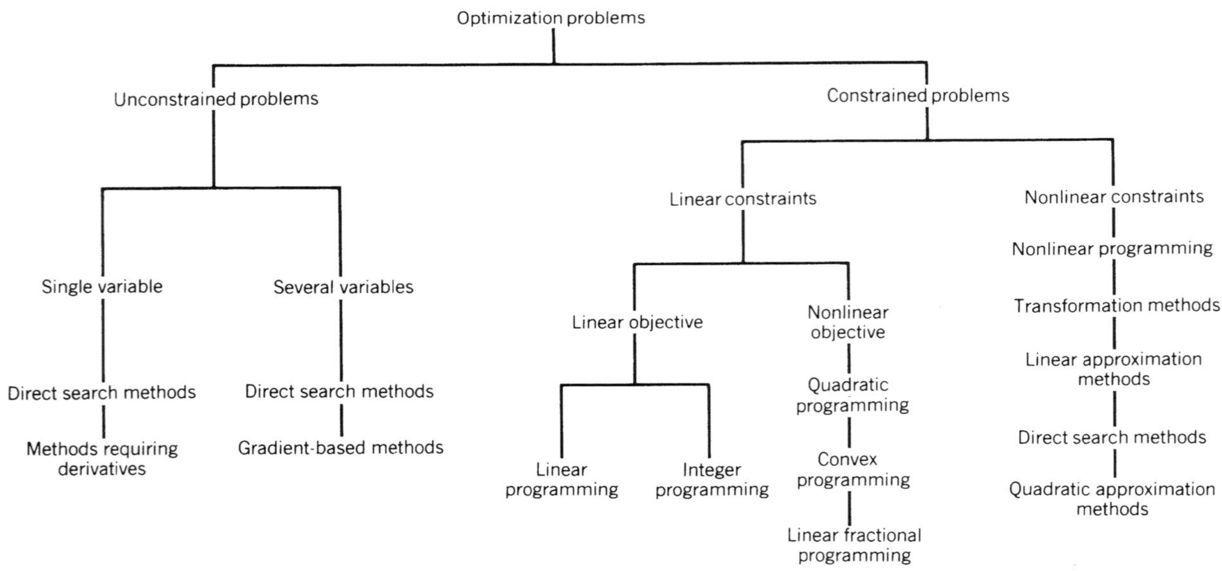

Figure 5 Classification of optimization problems.

continuous and use special-purpose IP algorithms for the rest. For a complete discussion of IP applications and algorithms, see Refs. 16–19.

The next class of optimization problems involves nonlinear objective functions and linear constraints. Under this class we have the following:

1. Quatratic programming, whose objective is a quadratic function
2. Convex programming, whose objective is a special nonlinear function satisfying an important mathematical property called *convexity*
3. Linear fractional programming, whose objective is the ratio of two linear functions

Special-purpose algorithms that take advantage of the particular form of the objective functions are available for solving these problems.

The most general optimization problems involve nonlinear objective functions and nonlinear constraints and are generally grouped under the term *nonlinear programming.* The majority of engineering design problems fall into this class. Unfortunately, there is no single method that is best for solving every nonlinear programming problems. Hence a host of algorithms are reviewed in the next section.

Nonlinear programming problems wherein the objective function and the constraints can be expressed as the sum of generalized polynomial functions are called *geometric programming* problems. A number of engineering design problems fall into the geometric programming framework. Since its earlier development in 1961, geometric programming has undergone considerable theoretical development, has experienced a proliferation of proposals for numerical solution techniques, and has enjoyed considerable practical engineering applications (see Refs. 20 and 21).

- Nonlinear programming problems where some of the design variables are restricted to be discrete or integer valued are called *mixed-integer nonlinear programming* (MINLP) problems. Such problems arise in process design, simulation optimization, industrial experimentation, and reliability optimization. MINLP problems are generally more difficult to solve since the problems have several local optima. Recently, *simulated annealing* and *genetic algorithms* have been emerging as powerful heuristic algorithms to solve MINLP problems. Simulated annealing has been successfully applied to solve problems in a variety of fields, including mathematics, engineering, and mathematical programming (see, e.g., Refs. 19 and 22–24).
- Genetic algorithms are heuristic search methods based on the two main principles of natural genetics, namely entities in a population reproduced to create offspring and the survival of the fittest (see, e.g., Refs. 23 and 25). For a discussion of the successful application of genetic algorithms and the areas of research in the field, See Ref. 26.

5 OVERVIEW OF OPTIMIZATION METHODS

Optimization methods can be viewed as nothing more than numerical hill-climbing procedures in which the objective function presenting the topology of the hill is searched to identify the highest point—or maximum—subject to constraining relations that might be equality constraints (stay on winding path) or inequality constraints (stay within fence boundaries). While the constraints do serve to reduce the area that must be searched, the numerical calculations required to ensure that the search stays on the path or within the fences generally do constitute a considerable burden. Accordingly, optimization methods for unconstrained problems and methods for linear constraints are less complex than those designed for nonlinear constraints. In this section, a selection of optimization techniques representative of the main families of methods will be discussed. For a more detailed presentation of individual methods the reader is invited to consult Ref. 27.

5.1 Unconstrained Optimization Methods

Methods for unconstrained problems are divided into those for single-variable functions and those appropriate for multivariable functions. The methods in the former class are important because single-variable optimization problems arise commonly as subproblems in the solution of multivariable problems. For instance, the problem of minimizing a function $f(x)$ for a point x^0 in a direction of d (often called a *line search*) can be posed as a minimization problem in the scalar variable α:

$$\text{Minimize } f(x^0 + \alpha d)$$

Single-Variable Methods

Single-variable methods are roughly divided into *region elimination methods* and *point estimation methods.* The former use comparison of function values at selected trial points to reject intervals within which the optimum of the function does not lie. The latter typically use polynomial approximating functions to estimate directly the location of the optimum. The simplest polynomial approximating function is the quadratic,

$$\bar{f}(x) = ax^2 + bx + c$$

whose coefficients a, b, c can be evaluated readily from those trial values of the actual function. The point at which the derivative of $\bar{f}$ is zero is used readily to predict the location of the optimum of the true function,

$$\tilde{x} = -\frac{b}{2a}$$

The process is repeated using successively improved trial values until the differences between successive estimates $\tilde{x}$ become sufficiently small.

Multivariable Unconstrained Methods

These algorithms can be divided into *direct-search methods* and *gradient-based methods*. The former methods only use direct function values to guide the search, while the latter also require the computation of function gradient and, in some cases, second-derivative values. Direct-search methods in widespread use in engineering applications include the simplex search, the pattern search method of Hooke and Jeeves, random-sampling-based methods, and the conjugate directions method of Powell (see Chapter 3 of Ref. 27). All but the last of these methods make no assumptions about the smoothness of the function contours and hence can be applied to both discontinuous and discrete-valued objective functions.

Gradient-based methods can be grouped into the classical methods of steepest descent (Cauchy) and Newton's method and the modern quasi-Newton methods such as the conjugate gradient, Davidson–Fletcher–Powell, and Broyden–Fletche–Shamo algorithms. All gradient-based methods employ the first derivative or gradient of the function at the current best solution estimate $\bar{x}$ to compute a direction in which the objective function value is guaranteed to decrease (a *descent* direction). For instance, Cauchy's classical method used the direction

$$d = -\nabla f(\bar{x})$$

followed by a line search from $\bar{x}$ in this direction. In Newton's method the gradient vector is premultiplied by the matrix of second derivatives to obtain an improved direction vector,

$$d = -[\nabla^2 f(\bar{x})^{-1}]\, \nabla f(\bar{x})$$

which in theory at least yields very good convergence behavior. However, the computation of $\nabla^2 f$ is often too burdensome for engineering applications. Instead, in recent years quasi-Newton methods have found increased application. In these methods, the direction vector is computed as

$$d = -H\, \nabla f(\bar{x})$$

where H is a matrix whose elements are updated as the iterations proceed using only values of gradient and function value difference from successive estimates. Quasi-Newton methods differ in the details of H updating, but all use the general form

$$H^{n+1} = H^n + C^n$$

where H^n is the previous value of H and C^n is a suitable correction matrix. The attractive feature of this family of methods is that the convergence rates approaching those of Newton's method are attained without the need for computing $\nabla^2 f$ or solving the linear equation set

$$\nabla^2 f(\bar{x})d = -\nabla f(\bar{x})$$

to obtain d. Recent developments in these methods have focused on strategies for eliminating the need for detailed line searching along the direction vectors and on enhancements for solving very large programs. For a detailed discussion of quasi-Newton methods the reader is directed to Refs. 27 and 28.

5.2 Constrained Optimization Methods

Constrained optimization methods can be classified into those applicable to totally linear or at least linearly constrained problems and those applicable to general nonlinear problems. The linear or linearly constrained problems can be well solved using methods of linear programming and extensions, as discussed earlier. The algorithms suitable for general nonlinear problems comprise four broad categories of methods:

1. Direct-search methods that use only objective and constraint function values
2. Transformation methods that use constructions that aggregate constraints with the original objective function to form a single composite unconstrained function
3. Linearization methods that use linear approximations of the nonlinear problem functions to produce efficient search directions
4. Successive quadratic programming methods that use quasi-Newton constructions to solve the general problem via a series of subproblems with quadratic objective functions and linear constraints

Direct Search

The direct-search methods essentially consist of extensions of unconstrained direct-search procedures to accommodate constraints. These extensions are generally only possible with inequality constraints or linear equality constraints. Nonlinear equalities must be treated by implicit or explicit variable elimination. That is, each equality constraint is either explicitly solved for a selected variable and used to eliminate that variable from the search or numerically solved for values of the dependent variables for each trial point in the space of the independent variables.

For example, the problem

$$\begin{aligned} \text{Minimize} \quad & f(x) = x_1 x_2 x_3 \\ \text{Subject to} \quad & h_1(x) = x_1 + x_2 + x_3 - 1 = 0 \\ & h_2(x) = x_1^2 x_3 + x_2 x_3^2 + x_2^{-1} - 2 = 0 \\ & 0 \le (x_1, x_3) \le \tfrac{1}{2} \end{aligned}$$

involves two equality constraints and hence can be viewed as consisting of two dependent variables and one independent variable. Clearly, h_1 can be solved for x_1 to yield

$$x_1 = 1 - x_2 - x_3$$

Thus, on substitution the problem reduces to

$$\begin{aligned} \text{Minimize} \quad & (1 - x_2 - x_3)x_2 x_3 \\ \text{Subject to} \quad & (1 - x_2 - x_3)^2 x_3 + x_2 x_3^2 + x_2^{-1} - 2 = 0 \\ & 0 \le 1 - x_2 - x_3 \le \tfrac{1}{2} \\ & 0 \le x_3 \le \tfrac{1}{2} \end{aligned}$$

Solution of the remaining equality constraint for one variable, say x_3, in terms of the other is very difficult. Instead, for each value of the independent variable x_2, the corresponding value of x_3 would have to be calculated numerically using some root-finding method.

Some of the more widely used direct-search methods include the adaptation of the simplex search due to Box (called the *complex method*), various direct random-sampling-type methods, and combined random sampling/heuristic procedures such as the combinatorial heuristic method[29] advanced for the solution of complex optimal mechanism design problems.

A typical direct sampling procedure is given by the formula

$$x_i p = \bar{x}_i \times z_i\,(2r - 1)^k \qquad \text{for each variable } x_i,\ i = 1, \ldots, n$$

where $\bar{x}$ = current best value of variable i
z_i = allowable range of variable i
r = random variable uniformly distributed on interval 0–1
k = adaptive parameter whose value is adjusted based on past successes or failures in search

For given $\bar{x}_i$, z, and k, r is sampled N times and the new point x^p evaluated. If x^p satisfies all constraints, it is retained; if it is infeasible, it is rejected and a new set of Nr values is generated. If x^p is feasible, $f(x^p)$ is compared to $f(\bar{x})$, and if improvement is found, x^p replaces $\bar{x}$. Otherwise x^p is rejected. The parameter k is an adaptive parameter whose value will regulate the contraction or expansion of the sampling region. A typical adjustment procedure for k might be to increase k by 2 whenever a specified number of improved points is found after a certain number of trials.

The general experience with direct-search and especially random-sampling-based methods for constrained problems is that they can be quite effective for severely nonlinear problems that involve multiple local minima but are of low dimensionality.

Transformation Methods

This family consists of strategies for converting the general constrained problem to a parametrized unconstrained problem that is solved repeatedly for successive values of the parameters. The approaches can be grouped into the penalty/barrier function constructions, exact penalty methods, and augmented Lagrangian methods. The classical penalty function approach is to transform the general constrained problem to the form

$$P(x,R) = f(x) + \Omega(R, g(x), h(x))$$

where R = penalty parameter
Ω = penalty term

The ideal penalty function will have the property that

$$P(x, R) = \begin{cases} f(x) & \text{if } x \text{ is feasible} \\ \infty & \text{if } x \text{ is infeasible} \end{cases}$$

Given this idealized construction, $P(x, R)$ could be minimized using any unconstrained optimization method, and hence, the underlying constrained problem would have been solved. In practice such radical discontinuities cannot be tolerated from a numerical point of view, and hence, practical penalty functions use penalty terms of the form

$$\Omega(R, g, h) = R\left(\sum_k h_k(x)\right)^2 + R\left(\sum [\min(0, g_j(x))]^2\right)$$

A series of unconstrained minimizations of $P(x, R)$ with different values of R are carried out beginning with a low value of R (say $R = 1$) and progressing to very large values of R. For low values of R, the unconstrained minima of $P(x,R)$ obtained will involve considerable constraint violations. As R increases, the violations decrease until, in the limit as $R \rightarrow \infty$, the violations will approach zero. A large number of the different forms of the Ω function have been proposed; however, all forms share the common feature that a sequence of problems must be solved and that, as the penalty parameter R becomes large, the penalty function becomes increasingly distorted and thus its minimization becomes increasingly more difficult. As a result the penalty function approach is best used for modestly sized problems (2–10 variables), few nonlinear equalities (2–5), and a modest number of inequalities. In engineering applications, the unconstrained subproblems are most commonly minimized using direct–search methods, although successful use of quasi-Newton methods is also reported.

The exact penalty function and augmented Lagrangian approaches have been developed in an attempt to circumvent the need to force convergence by using increasing values of the penalty parameter. One typical representative of this type of method is the so-called method of multipliers.[30] In this method, once a sufficiently large value of R is reached, further increases are not required. However, the method does involve additional finite parameters that must be updated between subproblem solutions. Computational evidence reported to date suggests that, while augmented Lagrangian approaches are more reliable than penalty function methods, they, as a class, are not suitable for larger dimensionality problems.

Linearization Methods

The common characteristic of this family of methods is the use of local linear approximations to the nonlinear problem functions to define suitable, preferably feasible, directions for search. Well-known members of this family include the method of feasible directions, the gradient projection method, and the generalized reduced gradient (GRG) method. Of these, the GRG method has seen the widest engineering application.

The key constructions of the GRG method are the following:

1. The calculation of the reduced objective function gradient $\nabla\tilde{f}$
2. The use of the reduced gradient to determine a direction vector in the space of the independent variables
3. The adjustment of the dependent variable values using Newton's method so as to achieve constraint satisfaction

Given a feasible point x^0, the gradients of the equality constraints are evaluated and used to form the constraint Jacobian matrix A. This matrix is partitioned into a square submatrix J and the residual rectangular matrix C where the variable associated with the columns of J are the dependent variables and those associated with C are the independent variables.

If J is selected to have nonzero determinant, then the reduced gradient is defined as

$$\nabla\tilde{f}(x^0) = \nabla\bar{f} - \nabla\hat{f}\, J^{-1}C$$

where $\nabla\bar{f}$ is the subvector of objective function partial derivatives corresponding to the (dependent) variables and $\hat{f}$ is the corresponding subvector whose components correspond to the independent variables. The reduced gradient $\nabla\tilde{f}$ provides an estimate of the rate of change of $f(x)$ with respect to the independent variables when the dependent variables are adjusted to satisfy the linear approximations to the constraints.

Given $\nabla\tilde{f}$, in the simplest version of the GRG algorithm, the direction subvector for the independent variables $\bar{d}$ is selected to be the reduced gradient descent direction

$$\bar{d} = -\nabla\tilde{f}$$

For a given step α in that direction, the constraints are solved iteratively to determine the value of the dependent variables $\hat{x}$ that will lead to a feasible point. Thus, the system

$$h_k\,(\bar{x}^0 + a\bar{d}, \hat{x}) = 0 \qquad k = 1, \ldots, K$$

is solved for the K unknown variables of $\hat{x}$. The new feasible point is checked to determine whether an improved objective value has been obtained and, if not, α is reduced and the solution for $\hat{x}$ repeated. The overall algorithm terminates when a point is reached at which the reduced gradient is sufficiently close to zero.

The GRG algorithm has been extended to accommodate inequality constraints as well as variable bound. Moreover, the use of efficient equation-solving procedures, line search procedures for α, and quasi-Newton formulas to generate improved direction vectors $\bar{d}$ has been investigated. A commercial-quality GRG code will incorporate such developments and thus will constitute a reasonably complex software package. Computational testing using such codes indicates that GRG implementations are among the most robust and efficient general-purpose nonlinear optimization methods currently available.[31] One of the particular advantages of this algorithm, which can be critical in engineering applications, is that it generates *feasible* intermediate points; hence, it can be interrupted prior to final convergence to yield a feasible solution. Of course, this attractive feature and the general efficiency of the method are attained at the price of providing (analytically or numerically) the values of the partial derivatives of all of the model functions.

Successive Quadratic Programming (SQP) Methods

This family of methods seeks to attain superior convergence rates by employing subproblems constructed using higher order approximating functions than those employed by the linearization methods. The SQP methods are still the subject of active research; hence, developments and enhancements are proceeding apace. However, the basic form of the algorithm is well established and can be sketched out as follows.

At a given point x^0, a direction-finding subproblem is constructed which takes the form of a quadratic programming problem:

$$\begin{aligned} \text{Minimize} \quad & \nabla^T f\, d + \tfrac{1}{2} d^T H d \\ \text{Subject to} \quad & h_k(x^0) + \nabla^T h_k\,(x^0) d = 0 \\ & g_j(x^0) + \nabla^T g_j\,(x^0) d \geq 0 \end{aligned}$$

The symmetric matrix H is a quasi-Newton approximation of the matrix of second derivatives of a composite function (the Lagrangian) containing terms corresponding to all of the functions f, h_k, and g_j. The matrix H is updated using only gradient differences as in the unconstrained case. The direction vector d is used to conduct a line search which seeks to minimize a penalty function of the type discussed earlier. The penalty function is required because, in general, the intermediate points produced by this method will be infeasible. Use of the penalty function ensures that improvements are achieved in either the objective function values or the constraint violations or both. One major advantage of the method is that very efficient methods are available for solving large quadratic programming problems and, hence, that the method is suitable for large-scale applications. Recent computational testing indicates that the SQP approach is very efficient, outperforming even the best GRG codes.[32] However, it is restricted to models in which infeasibilities can be tolerated and will produce feasible solutions only when the algorithm has converged.

5.3 Optimization Software

With the exception of the direct-search methods and the transformation-type methods, the development of computer programs implementing state-of-the-art optimization algorithms is a major effort requiring expertise in numerical methods in general and numerical linear algebra in particular. For that reason, it is generally recommended that engineers involved in design optimization studies take advantage of the number of good-quality implementations now available through various public sources.

Commercial computer codes for solving LP/IP/NLP problems are available from many computer manufacturers and private companies that specialize in marketing software for major computer systems. Depending on their capabilities, these codes vary in their complexity, ease of use, and cost (see, e.g., Ref. 33). LP models with a few hundred constraints can now be solved on personal computers (PCs). There are now at least a hundred small companies marketing LP software for PCs. For a 2003 survey of LP software, see Ref. 34.

Nash[35] presents a 1998 survey of nonlinear programming software that will run on PC compatibles, Macintosh systems, and UNIX-based workstations. Detailed product descriptions, prices, and capabilities of NLP software are included in the survey. There are now LP/IP/NLP solvers that can be invoked directly from inside spreadsheet packages. For example, Microsoft Excel and Microsoft Office for Windows and Macintosh contain a general-purpose optimizer for solving small-scale linear, integer, and nonlinear programming problems. Developed by Frontline Systems for Excel users, the LP optimizer is based on the simplex algorithm, while the NLP optimizer is based on the GRG algorithm. Frontline Systems also offers more powerful Premium Solver Platforms for Excel that can solve larger problems. For a review, see Ref. 36. A number of other spreadsheet add-ins are also available now for solving optimization problems. For a recent survey, see Ref. 37.

There are now modeling languages that allow the user to express a model in a very compact algebraic form, with whole classes of constraints and variables defined over index sets. Models with thousands of constraints and variables can be defined in a few pages in a syntax that is very close to standard algebraic notation. The algebraic form of the model is kept separate from the actual data for any particular instance of the model. The computer takes over the responsibility of transforming the abstract form of the model and the specific data into a specific constraint matrix. This has greatly simplified the building and, even more, the changing of the optimization models. There are several modeling languages available for PCs. The two high-end products are GAMS (General Algebraic Modeling System) and AMPL (A Mathematical Programming Language). For a discussion of GAMS, see Ref. 38. For a general introduction to modeling languages, see Refs. 33 and 39, and for an excellent discussion of AMPL, see Ref. 40.

For those who write their own programs, LINDO Systems offers a library of callable linear, quadratic, and MIP solvers.[41] Engineers familiar with MATLAB[42] can use its Optimization Toolbox for solving LP, QP, IP, and NLP problems. For a MATLAB software review, see Ref. 43. MATLAB and the Optimization Toolbox provide users an interactive environment to define models, gather data, and analyze results.

For optimization on the Internet,[44] users can get a complete list of optimization software available for LP, IP, and NLP problems at the following web site: www.ece.northwestern.edu/otc. Maintained by Argonne National Laboratory and Northwestern University, this OTC (Optimization Technology Center) site provides access not only to the NEOS (Network Enabled Optimization System) software guide but also to the other optimization-related sites that are continually updated. The NEOS guide on optimization software is based on the textbook by More and Wright,[45] an excellent resource for those interested in a broad review of the various optimization methods and their computer codes. The book is divided into two parts. Part I has an overview of algorithms for different optimization problems, categorized as unconstrained optimization, nonlinear least squares, nonlinear equations, linear programming, quadratic programming, bound-constrained optimization, network optimization, and integer programming. Part II includes product descriptions of 75 software packages that implement the algorithms described in Part I. Much of the software described in this book is in the public domain and can be obtained through the Internet. The NEOS guide also offers a number of engineering design optimization packages. Both MATLAB and the NEOS offer software for genetic algorithms also.

6 SUMMARY

In this chapter an overview was given of the elements and methods comprising design optimization methodology. The key element in the overall process of design optimization was seen to be the engineering model of the system constructed for this purpose. The assumptions and formulation details of the model govern the quality and relevance of the optimal design obtained. Hence, it is clear that design optimization studies cannot be relegated to optimization software specialists but are the proper domain of the well-informed design engineer.

The chapter also gave a structural classification of optimization problems and a broad-brush review of the main families of optimization methods. Clearly this review can only hope to serve as an entry point to this broad field. For a more complete discussion of optimization techniques with emphasis on engineering applications, guidelines for model formulation, and practical solution strategies, the readers are referred to the text by Reklaitis, Ravindran, and Ragsdell.[27]

The Design Automation Committee of the Design Engineering Division of the American Society of Mechanical Engineers (ASME) has been sponsoring conferences devoted to engineering design optimization. Several of these presentations have subsequently appeared in the *Journal of Mechanical Design, ASME Transactions.* Ragsdell[46] presents a review of the papers published up to 1977 in the areas of machine design applications and numerical methods in design optimization. ASME published, in 1981, a special volume entitled *Progress in Engineering Optimization,* edited by Mayne and Ragsdell.[47] It contains several articles pertaining to advances in optimization methods and their engineering applications in the areas of mechanism design, structural design, optimization of hydraulic networks, design of helical springs, optimization of hydrostatic journal bearing, and others. Finally, the persistent and mathematically oriented reader may wish to pursue the fine exposition given by Avrial,[48] which explores the theoretical properties and issues of nonlinear programming methods.

REFERENCES

1. M. Zeleny, *Multiple Criteria Decision Making,* McGraw-Hill, New York, 1982.
2. T. L. Vincent and W. J. Grantham, *Optimality in Parametric Systems,* Wiley, New York, 1981.
3. K. E. Bett, J. S. Rowlinson, and G. Saville, *Thermodynamics for Chemical Engineers,* MIT Press, Cambridge, MA 1975.
4. F. C. Jen, C. C. Pegels, and T. M. Dupuis, "Optimal Capacities of Production Facilities," *Manag. Sci.* **14B,** 570–580 (1968).
5. J. E. Shigley, *Mechanical Engineering Design,* McGraw-Hill, New York, 1973, p. 271.
6. S. Timoshenko and J. Gere, *Theory of Elastic Stability,* McGraw-Hill, New York, 1961, p. 257.
7. K. M. Ragsdell and D. T. Phillips, "Optimal Design of a Class of Welded Structures Using Geometric Programming," ASME *J. Eng. Ind. Ser. B,* **98**(3), 1021–1025 (1975).
8. R. H. Philipson and A. Ravindran, "Application of Goal Programming to Machinability Data Optimization," *J. Mech. Design, Trans.* ASME, **100,** 286–291 (1978).
9. E. J. A. Armarego and R. H. Brown, *The Machining of Metals,* Prentice-Hall, Englewood Cliffs, NJ, 1969.
10. J. L. Arthur and A. Ravindran, "PAGP-Partitioning Algorithm for (Linear) Goal Programming Problems," *ACM Trans. Math. Software,* **6,** 378–386 (1980).
11. R. H. Philipson and A. Ravindran, "Application of Mathematical Programming to Metal Cutting," *Math. Programming Study,* **11,** 116–134 (1979).
12. S. M. Lee, *Goal Programming for Decision Analysis,* Auerbach, Philadelphia, PA, 1972.
13. N. K. Karmarkar, "A New Polynomial Time Algorithm for Linear Programming," *Combinatorica,* **4,** 373–395 (1984).
14. A. Arbel, *Exploring Interior Point Linear Programming: Algorithms and Software,* MIT Press, Cambridge, MA, 1993.
15. S.-C. Fang and S. Puthenpura, *Linear Optimization and Extensions,* Prentice-Hall, Englewood Cliffs, NJ, 1993.
16. R. L. Rardin, *Optimization in Operations Research,* Prentice-Hall, Englewood Cliffs, NJ, 1998.
17. K. G. Murty, *Operations Research: Deterministic Optimization Models,* Prentice-Hall, Englewood Cliffs, NJ, 1995.
18. G. L. Nemhauser and L. A. Wolsey, *Integer and Combinatorial Optimization,* Wiley, New York, 1988.
19. A. D. Belegundu and T. R. Chandrupatla, *Optimization Concepts and Applications in Engineering,* Prentice-Hall, Englewood Cliffs, NJ, 1999.

20. C. S. Beightler and D. T. Phillips, *Applied Geometric Programming,* Wiley, New York, 1976.

21. M. J. Rijckaert, "Engineering Applications of Geometric Programming," in *Optimization and Design,* M. Avriel, M. J. Rijckaert, and D. J. Wilde (eds.), Prentice-Hall, Englewood Cliffs, NJ, 1974.

22. I. O. Bohachevsky, M. E. Johnson, and M. L. Stein, "Generalized Simulated Annealing for Function Optimization," *Technometrics,* **28,** 209–217 (1986).

23. L. Davis (ed.), *Genetic Algorithms and Simulated Annealing,* Pitman, London, 1987.

24. S. Kirkpatrick, C. D. Gelatt, and M. P. Vecchi, "Optimization by Simulated Annealing," *Science,* **220,** 670–680 (1983).

25. D. E. Goldberg, *Genetic Algorithm in Search, Optimization, and Machine Learning,* Addison-Wesley, Reading, MA, 1989.

26. A. Maria, "Genetic Algorithms for Multimodal Continuous Optimization Problems," Ph.D. Dissertation, University of Oklahoma, Norman, OK, 1995.

27. G. V. Reklaitis, A. Ravindran, and K. M. Ragsdell, *Engineering Optimization: Methods and Applications,* Wiley, New York 1983.

28. R. Fletcher, *Practical Methods of Optimization,* 2nd ed., Wiley, New York, 1987.

29. T. W. Lee and F. Fruedenstein, "Heuristic Combinatorial Optimization in the Kinematic Design of Mechanisms: Part 1: Theory," *J. Eng. Ind. Trans. ASME,* **98,** 1277–1280 (1976).

30. S. B. Schuldt, G. A. Gabriele, R. R. Root, E. Sandgren, and K. M. Ragsdell, "Application of a New Penalty Function Method to Design Optimization," *J. Eng. Ind. Trans. ASME,* **99,** 31–36 (1977).

31. E. Sandgren and K. M. Ragsdell, "The Utility of Nonlinear Programming Algorithms: A Comparative Study—Parts 1 and 2," *J. Mechan. Design Trans. ASME,* **102,** 540–541. (1980).

32. K. Schittkowski, *Nonlinear Programming Codes: Information, Tests, Performance,* Lecture Notes in Economics and Mathematical Systems, Vol. 183, Springer-Verlag, New York, 1980.

33. R. Sharda, *Linear and Discrete Optimization and Modeling Software: A Resource Handbook,* Lionheart, Atlanta, GA, 1993.

34. R. Fourer, "Linear Programming Software Survey," *OR/MS Today,* **30** (Dec. 2003).

35. S. G. Nash, "Nonlinear Programming Software Survey," *OR/MS Today,* **25** (June, 1998).

36. C. Albright, "Premium Solver Platform for Excel," *OR/MS Today,* **28,** 58–63 (June 2001).

37. T. A. Grossman, "Spreadsheet Add-Ins for OR/MS," *OR/MS Today,* **29,** (Aug. 2002).

38. A. Brooke, D. Kendrick, and A. Meeraus, *GAMS: A User's Guide,* Scientific Press, Redwood City, CA, 1988.

39. R. Sharda and G. Rampal, "Algebraic Modeling Languages on PC's", *OR/MS Today,* **22,** 58–63 (1995).

40. R. Fourer, D. M. Gay, and B. W. Kernighan, "A Modeling Language for Mathematical Programming," *Manag. Sci.* **36,** 519–554 (1990).

41. A. Felt, "LINDO API: Software Review," *OR/MS Today,* **29,** (Dec. 2002).

42. R. Biran and M. Breiner, *MATLAB for Engineers,* Addison-Wesley, Reading, MA, 1995.

43. J. J. Hutchinson, "MATLAB Software Review," *OR/MS Today,* **23** (Oct. 1996).

44. J. Czyzyk, J. H. Owen, and S. J. Wright, "Optimization on the Internet," *OR/MS Today,* **24** (Oct. 1997).

45. J. J. More and S. J. Wright, *Optimization Software Guide,* SIAM Publications, Philadelphia, PA, 1993.

46. K. M. Ragsdell, "Design and Automation," *J. Mech. Design Trans. ASME,* **102,** 424–429 (1980).

47. R. W. Mayne and K. M. Ragsdell (eds.), *Progress in Engineering Optimization,* American Society of Mechanical Engineers, New York, 1981.

48. M. Avriel, *Nonlinear Programming: Analysis and Methods,* Prentice-Hall, Englewood Cliffs, NJ, 1976.

CHAPTER 25

DESIGN FOR MANUFACTURE AND ASSEMBLY WITH PLASTICS

James A. Harvey
Under the Bridge Consulting, Inc.
Corvallis, Oregon

1 INTRODUCTION

This chapter is divided into three sections: plastic materials selection, plastics-joining techniques, and plastic part design. Our major focus will be on plastic materials selection. The information presented is based on both the lectures given and the information received in short courses taught to practicing engineers and scientists involved in all aspects of commercial plastics part designs and in graduate-school courses to budding new materials scientists and engineers.

In the open literature for material selection you will find articles with titles similar to "The Science of Material Selection" or "The Art of Material Selection." Hopefully this chapter will eliminate some of the mystery or confusion regarding material selection.

2 PLASTIC MATERIALS SELECTION

2.1 Polymers

In the selection of plastic materials for a commercial part design the first step is as in all technology development: to learn the basic definitions, concepts, and principles of that technology. The following terms will be defined as they are important in the selection of materials for plastics part design:

Reprinted from *Handbook of Materials Selection,* Wiley, New York, 2002, by permission of the publisher.

Polymer
Thermoplastics
Thermosets
Elastomers
Polymerization reactions
Molecular weight and distribution
Molecular structure of polymers
Five viscoelastic regions of polymers
Carothers equation
Additives

A *polymer* is a compound consisting of repeating structural units. A simple example of a repeat unit is the $-CH_2-$ chemical moiety. Two repeated units are equivalent to the organic compound ethane. Ethane is a gas at room temperature with a total molecular weight of 30 atomic mass units (amu). A polymer family with hundreds of thousands of these $-CH_2-$ repeat units represents the polyethylenes with molecular weights in the millions.

A *thermoplastic polymer* is a polymer that consists of linear polymers chains. Whenever you use a thermoplastic it is usually in its final molecular weight form. The major thermal event is to process it into the final part form. There are three types of thermoplastics: amorphous, semicrystalline, and liquid crystal.

A *thermosetting resin* is one that contains a highly crosslinked polymer network when processed. One has to "cook" or cure the resin before it can be formed into its final shape.

An *elastomer* is a very lightly crosslinked polymer with the ability to be extended to a high elongation and snap back to its original dimensions when the forces have been removed.

Polymerization reactions play a usual role in the process of material selection. From the name of the polymer and its polymerization reaction, one can make a reasonable first attempt to select a plastic material. But the reader must be cautioned that the preceding statement is a general rule. For example, polyethylene is named from the monomer from which it is made, ethylene. This monomer is polymerized through an addition reaction. Typically, addition polymers are water hating, or hydrophobic. For a first approximation, this type of material would be a good material to use in applications where exposure to water is required.

Now let us look at another polymer, polyethylene terephthalate. It is formed from the reaction of ethylene glycol with terephthalic acid or terephthalic acid ester. During the reaction, in order for the polymer, polyethylene terephthalate, to build up molecular weight, it loses either water or alcohol as a by-product. Polymers formed by adding two or more coreactants under conditions of time, temperature, and other reaction conditions with the formation of a byproduct such as water and alcohol are said to be formed by condensation and are named by the new chemical functional group formed. In general, these polymers are water loving, or hydrophilic.

As mentioned, polyesters are formed from the reaction of organic acids with organic alcohols with water as a byproduct. Nylons are formed from the condensation reaction of organic acids with organic amines with water as a byproduct. Polyimides are formed from the condensation reaction between acid anhydrides and organic amines with the release of water.

Organic chemistry plays a very important role in the selection of a polymer for a plastic part. Here we are only posing general rules, and to a first approximation one can make very reasonable selection in the early stages of plastic part design using these general observations.

For example, if you were assigned to design a plastic part that had to exist in a water environment, your first choice could be an addition polymer such as polyethylene rather than a condensation polymer such as polyethylene terephthalate. The polyethylene is water hating, or hydrophobic, and so should not be affected by water.

This writer can already imagine the "but what about this incident" remarks. Yes, water bottles are made from polyesters. The bottles are dated for lifetime and the companies that fill these bottles with their mountain spring fresh water want you to see how clear their water is. However, over time the polyester bottles will absorb water. Water (sport) bottles that are used over and over and filled by the consumer are made from the addition polymers. They are also opaque. Transparency in these sport plastic water containers is not important. This polymer character of being transparent or opaque will be discussed later.

Molecular weight and *molecular weight distributions* are other important parameters for a polymer. The polymerization reaction is complicated. The polymerization reaction does not result in a simple single molecule. The reaction yields many different sizes of polymer chains. The molecular mass of each chain refers to its molecular weight. And as mentioned, since many different sizes of polymer chains are formed thus, there will be a distribution of the molecular weights.

Knowledge of the molecular weight and its distribution aids in the selection of polymers for a plastics part and in the lifetime of plastic parts. The general techniques that can be used to determine molecular weights of polymers are achieved through viscosity measurements either in solution or using solid samples. Solution viscosities consist of timing the flow of polymer solutions of known concentrations through a fixed volume. The melt index or melt flow index (MFI) is derived from a standardized test in which a solid is used instead of a solution. A given amount of polymer is heated to a certain temperature, a known force is applied to the molten polymer, and its flow is timed. If all things are equal, the lower molecular weight polymer will flow through the given volume faster than a higher molecular weight polymer. Hence the higher molecular weight of a polymer, the lower the MFI value. For the members of a given polymer family this is a reasonable way to distinguish between low- and high-molecular-weight versions. The final technique is gel permeation (size exclusion) chromatography. The polymer is dissolved in a solvent. The solution is then passed through a series of tubes (columns) packed with different porous particles. As the solution passes through, the polymer chains with the highest molecular weights pass through the fastest. A detector measures the polymer chains as they exit the instrument. Thus one ends up with a chromatograph, which shows the distribution of the different molecular weights of the polymer chain in the sample. Solution viscosities are usually by the polymer manufacturer. Melt flow index is used as an initial tool for material selection and as a tool to help determine the molding process. Gel permeation (size exclusion) chromatography was in the past treated as a research tool, but lately it has gained a great deal of popularity as a quality control technique.

Another important parameter for the different polymers refers to their thermal behavior. A typical thermoplastic is a solid at ambient temperatures. As the material is heated, it starts to soften; then it flows and in some cases it melts. When it is cool, it solidifies. Depending on the container (mold) used, the thermoplastic will retain the shape of that mold. This process should be repeatable. Thus thermoplastics are recyclable. Another thermal property of thermoplastics is creep. This property refers to the ability of the material to flow under a load as a function of temperature.

Thermosetting resin systems are quite different. When one processes a thermoplastic into a particular plastic part, its molecular weight has already been established by the manufacturer. With thermosetting resin systems one starts with low-molecular-weight reactants, and to process these ingredients, the reactants are cured, or "cooked," into the desired final shape. If the reactants have been fully reacted, the result is one giant molecule. To process a thermosetting resin system into a part, the thermal events consist of heating the ingredients so they start to soften followed by some of the ingredients melting. As the temperature is raised, the system is totally liquid. As the temperature continues to rise, the onset of curing (crosslinking) occurs. As the reaction proceeds, the viscosity increases and the part hardens. At the end of the curing reaction the part is solid; then it is cooled to ambient temperatures. Once formed the part cannot be reheated to change its shape. If thermosetting resin system has been properly cured, it should not be affected by temperature or solvents. This behavior may be used to characterize the starting thermosetting material. Usually, as the material is curing, its solubility changes from soluble to insoluble. A good example are the liquid epoxy resins. These materials are usually eutectic mixtures of three or more reactive species (oligomers). Such analyses as the number and amount of reactive species or epoxy equivalent weights affect the quality of the cure or the reproducibility of the final cure structure.

The behavior of elastomers is somewhat different than that of thermoplastics and thermosets. As a first approximation, it behaves more like a thermoplastic. We all know that car tires soften in the hot months of summer. Most elastomers will swell when placed in a solvent.

Thermal analyses are a set of techniques used to characterize the thermal behavior of the different types of polymers. In addition to providing the thermal characteristics of polymers, they can assist in determining a processing cycle. Differential scanning calorimetry (DSC) yields the thermal events of a sample, i.e., melting points, onsets, maximums, and offsets of curing, decomposition temperatures, crystallization temperatures, and glass transition temperatures (this will be discussed later). Thermogravimetric analyses (TGAs) give the changes in mass of a sample as a function of temperature and environment. Thermal mechanical analyses (TMAs) reveal the changes in volume of a sample (warpage and shrinkage) and glass transition temperatures. Dynamic mechanical analyses (DMAs) provide the modulus and changes in modulus and glass transition temperature as a function of temperature, time, and oscillation (dynamic load).

The *molecular structure of the polymer* will determine if it is transparent or opaque. This internal structure is called polymer morphology. Thermoplastic polymers can be divided into amorphous, semicrystalline, crystalline, and liquid crystal polymers. This classification is only reserved for thermoplastics. Morphology refers to how the polymer chains are arranged, in an ordered or disordered manner. Amorphous refers to total disorder. Crystalline refers to total order. Semicrystalline is a combination of disorder with domains of order within the structure. The liquid crystal polymers are a special class of thermoplastics that retain their order in the melt. Based upon chemical principles, as a material goes from the solid state to the liquid state, it goes from a state of order to one of disorder. The liquid crystal polymers lack this transition, and this unique characteristic has an enhanced effect on the processing of these materials.

We can examine the internal structure of amorphous, semicrystalline, and crystalline thermoplastics another way by viewing the polymer chains as spaghetti. We have cooked spaghetti (disordered) as one extreme and uncooked (ordered) spaghetti as the other extreme. Except for the liquid crystalline polymers, most thermoplastic polymers are either amorphous or semicrystalline (ordered polymer chains with crystalline domains). Due to the presence of crystalline domains, the semicrystalline polymers have a melting point and light will be scattered as it hits these domains, thus giving the material an opaque appearance. Thus amorphous polymers do not have a melting point and are transparent.

The next important definition involves the *five viscoelastic regions of polymers.* If we plot the modulus of a thermoplastic material as a function of temperature, we obtain a graph such as the one shown in Fig. 1.

Region 1 represents the behavior of the material at low temperatures. It is in its glassy state. The mobility of the polymer chains has decreased. The material is hard. As it is heated, it reaches region 3. This region is known as the rubbery region and the material loses strength in the range of three orders of magnitude. As the sample is heated to an even higher temperature, the polymer (region 4) starts to decompose and finally, at region 5, decomposition occurs with loss of strength.

A semicrystalline thermoplastic has the appearance of the dashed line in Fig. 1. The drop in modulus from the glassy region to the rubbery region is not as drastic with semicrystalline polymers (region 2) as it is with amorphous polymers. As the semicrystalline thermoplastic reaches its melting point, its strength decreases sharply and, as expected, goes from a solid to a liquid.

The transition between the glassy region of a polymer to its rubbery region is known as its glass transition and the temperature at which it occurs at is its glass transition temperature.

Glass transition is defined as "the reversible change in an amorphous material or in amorphous regions of a partially crystalline material, from (or to) a viscous or rubbery condition to (or from) a hard and relatively brittle one" (Ref. 2).

Some individuals use the term *glass transition temperature* while discussing cured thermosetting resin systems. To this writer, if the thermosetting resin is completely cured, it should not have a glass transition temperature. If it is completely cured as the material is heated over a temperature range, it should be unaffected by temperature until it reaches its decomposition temperature. In an analysis, the presence of a glass transition temperature may be due to either the thermoset not being completely cured or the thermoplastic nature

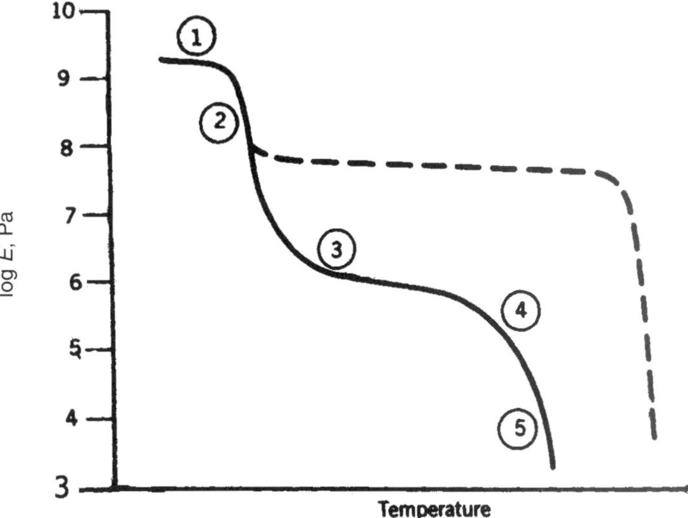

Figure 1 Five viscoelastic regions of a linear amorphous polymer. The dashed line represents the behavior of a semi-crystalline polymer. (From Ref. 1.)

of the crosslinked network. If you perform a thermal technique to determine glass transition temperature, cool the sample to ambient temperature and repeat the analysis on the same sample. If the material is not fully cured, the repeat run should indicate a higher apparent glass transition temperature and a lower drop in modulus.

Figure 2 shows the volume–temperature differences between the glass transition temperature and the crystalline melting point of a thermoplastic polymer. As the temperature is increased and a semicrystalline polymer (*b*) passes through its crystalline melting point (I_m), the change in volume is discontinuous—i.e., this is a step increase. As an amorphous polymer passes through its glass transition temperature (T_g), it goes from a rigid glass to an amorphous, softer material, which exhibits an increase in volume with increasing temperature. Thus in selecting a thermoplastic for a plastic part, it is always best to use the material below its T_g or T_m. There are exceptions to this rule. For example, polyethylenes are used in their rubbery region due to their subambient temperature glass transition temperatures.

The *Carothers equation* is included in our discussion of the selection of polymeric materials for plastic-molded parts because of its impact in thermoplastic polymerization:

$$\overline{X}_n = \frac{2}{2 - pr}$$

where $\overline{X}_n$ refers to the number-average degree of polymerization, p is the extent of the reaction, and r indicates either the ratio or the purity of the reactants. Basically what this equation teaches us is that one needs a high conversion of pure reactants and the correct stiochiometry to obtain the proper molecular weight. Small changes in purity, incomplete reaction, and incorrect ratio of reactants can have a drastic effect on the moldability of a part or its performance behavior. Thus the consistency and repeatability of a molded part shipped to a customer is highly dependent on the consistency and repeatability of the material from the polymer manufacturer compounder, and molder.

2.2 Plastics

Plastics are polymers with *additives*. These additives perform many different functions. Some refer to these materials as ''foo foo'' dust. The addition of additives to the polymer enhances the manufacture of the part, the product performance, the lifetime of the part, and the appearance of the part. A partial list of these additives includes antioxidants, light stabilizers, acid scavengers, lubricants, polymer-processing aids, antiblocking additives, slip additives, antifogging additives, antistatic additives, antimicrobials, flame retardants, chemical blowing

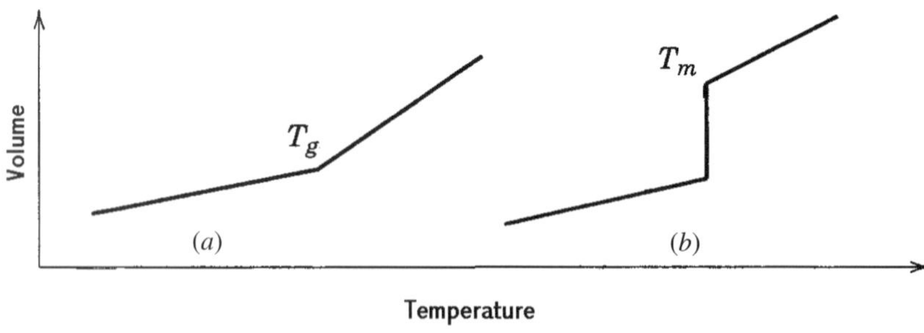

Figure 2 A comparison of the glass transition temperature: (*a*) to a melting point; (*b*) of a thermoplastic polymer. (From Ref. 1.)

agents, polyolefins, colorants, fluorescent whitening agents, fillers, nucleating agents, and plasticisizers.

From the name of an additive one can determine its function. There are exceptions to this, and in most cases we do not know what additive package compounders have put into a polymer. This information is treated as confidential. Thus one must be careful when one switches from one supplier to another. Even though the starting polymer may be the same, different additive packages will affect the performance behavior and lifetime of the designed plastic part. Thus once a material has been selected, this writer highly recommends to completely characterize that material in case the supplier changes the material or its consistency or you are involved in determining product failures. As a consultant, this writer has been involved in several failure analyses projects, and many problems are difficult and expensive to solve without baseline data.

Two types of additives will be defined in this section due to their importance: fillers and plasticizers. Fillers are added to polymers to affect the color and smoothness of the final molded part, to assist in the molding of the part by changing the flow behavior of the plastics, and to reduce the cost of the molded part.

Plasticizers are unique materials. They are added to a polymer to reduce the hardness of the polymer or make it more flexible. A good example is the ''smell'' we experience when we purchase a new car—a plasticiser causes this new-car smell as well as seats that are nice and soft. Over time the smell is gone, the seats become hard and brittle, and we have to clean the inside of the windshield. The reverse can occur with hydrophilic polymers in the presence of water. The polymer absorbs water, its glass transition temperature drops, and the material becomes softer. In the case of hydrophobic thermoplastic materials oils will have a similar effect.

Thus, it should be noted that these plastic systems are dynamic and are constantly changing. If you design a plastic part for a lifetime of five years, it would be ''nice'' to test the part for five years under the operating conditions of the plastic part assembly. However, based upon ''time to market,'' this is not an option. One can retain samples of production.

2.3 Reinforced Plastics

Reinforced plastics are plastics containing reinforcing elements within a plastic matrix that are long or short discontinued (chapped) fibers and the parts are manufactured by injection molding. There are many different types of reinforced plastics. Reinforced plastics are in essence forms of composite materials. A composite is a heterogeneous mixture of matrix resin, reinforcement, and other components that act in concert with each other. The matrix resin protects the reinforcement from wear and ''glues'' the reinforcement in place. The reinforcement provides strength and stiffness to the plastic part and enhances the properties of the matrix resin. Also the reinforcement helps dissipate the energy throughout the structure when impacted.

Properties of reinforced plastics should be obtained from the supplier. There are many variations of reinforcement forms and sizes.

3 PLASTIC MATERIALS SELECTION TECHNIQUES

Selecting plastic materials for a particular design may be a difficult process. It is wishful thinking to hope that anyone can design a plastic part. It would be helpful to have an individual in the organization who has had experience in plastic material selection or a materials engineer that knows the material science of plastics.

This writer's first experience in selecting materials was relatively simple. The criteria were set by the equipment available for the project. Then the temperature, chemical requirements, and number of parts needed were used to pick the material. The eight main criteria were functionality, chemical resistance, external processing (your supplier), internal processing, lifetime, design margins, cost, and greenness of the part. As the demands on the materials increased, subcriteria were developed. The criteria and subcriteria are presented in Table 1.

The above scheme works reasonably well, but it does not include the most important factor in material selection—the timing, schedule, or time to market. One factor that this treatment does give proper justification to is cost. However, sometimes cost decisions are not made by material scientists and engineers.

4 PLASTIC-JOINING TECHNIQUES

Table 2 contains a list of most of the techniques used to join two plastic parts together. Each technique has its own unique advantages and limitations. This writer has had more experience with adhesive bonding. In the selection of adhesives one can follow most of the criteria and subcriteria listed in Table 1.

In the adhesive bonding of plastic parts several issues should be considered. First is the failure mechanism that one should stride for in the design. There are three failure mechanisms for an adhesive joint: adhesive, cohesive, and substrate failures. An adhesive failure is failure within the adhesive joint. Cohesive failure is failure between the adhesive and substrate. Substrate failure is failure within the plastic parts.

The golden rule of design using adhesives is to design for cohesive or substrate failure.

A chemical principle that one should consider in the selection of adhesives is "like dissolves like": The more similar the adhesive is to the plastic one is trying to bond together, the stronger the bond.

Another consideration before using adhesives involves surface treatment. Whether one employs mechanical or chemical etching, preparing the surface is a critical procedure in the manufacturing step. Specific surface treatments of the various polymers can be found by searching the Internet.

5 PLASTIC PART DESIGN

This writer recommends the use of suppliers' design guides and the two design books listed in the Suggested Readings. However, there are tricks one can employ when performing failure analyses on molded plastic parts. One technique is to ash a plastic molded part. One can obtain information as to the flow of the plastic and fiber orientation if the part is reinforced.

One can also section the molded reinforced plastic into a smaller specimen that can be analyzed by TGA to determine percent resin and reinforcement contents, which in turn will show the consistency of the molded part.

Sometimes if the plastic has a long distance to flow in a mold, the polymer chains can separate. The smaller polymer chains travel faster than the larger chain. To verify the consistency of the polymer molecular weight throughout the molded plastic part, one can section the part and subject the specimen to gel permeation chromatography.

Table 1 Criteria for Selection of Materials (Thermoplastics, Thermosets, Elastomers, and Adhesives)

Main Criteria	Subcriteria
Functionality	Purpose of part
	Type and magnitude of normal service stresses
	Loading pattern and time under load
	Fatigue resistance
	Overloads and abuse
	Impact resistance
	Normal range of operating temperatures
	Maximum and minimum service temperatures
	Electrical resistivity
	Dielectric loss
	Antistatic properties
	Tracking resistance
	Flammability
	Surface finish
	Color matching and color retention
	Tolerances and dimensional stability
	Weight factors
	Space limitations
	Allowable deflections
Materials acceptance	Compatibility with chemicals
	Solvent and vapor attack
	Reactions with acids, bases, water, etc.
	Water absorption effects
	Ultraviolet light exposure and weathering
	Oxidation
	Chemical erosion and/or corrosion (electrochemical effects)
	Attack by fungi, bacteria, or insects
	Leaching of additives from the part material into its environment
	Absorption of components into the part from its environment
	Permeability of vapors and gases
	Normal range of operating temperatures
	Maximum and minimum service temperatures
Environmental concerns	Scrap rates
	Recyclability
	Chloro- and fluoropolymers
Lifetime	Product lifetime
	Reliability
	Product specifications
	Acceptance codes and specifications
Margin	Safety design factors
Internal process	Normal range of processing temperatures
	Maximum and minimum processing temperatures
	Choice of processes
	Method of assembly
	Secondary processes
	Finishing and decorating
	Quality control and inspection
	Contamination

Table 1 (*Continued*)

Main Criteria	Subcriteria
External process (supplier of parts)	Normal range of processing temperatures
	Maximum and minimum processing temperatures
	Choice of processes
	Method of assembly
	Secondary processes
	Finishing and decorating
	Quality control and inspection
	Contamination
	Timing for part design changes
	Timing for prototype molds
	Timing for production molds
	Technical support from supplier
	Contamination
Cost	Materials costs
	Materials availability
	Alternative material choices
	Supplier's availability
	Part costs
	Cost of capital plant: molds and processing machines
	Operation costs of component, including manufacturing and fuel consumption
	Capacity

6 PLASTIC PART MATERIAL SELECTION STRATEGY

After reading the first part of this chapter one may be either totally confused or have a headache. This writer understands perfectly. Materials selection is not an easy task. The following represents a series of suggestions and hints that hopefully make the task easier.

First, if you are part of a large organization, develop a team that may help you. This writer is in favor of at least a four-person team. The team should consist of a design engineer, a materials engineer (analytical chemists may be good substitutes if there are no materials engineers), an internal experience engineer, and a representative from the procurement department or an internal company buyer.

Table 2 Plastic-Joining Techniques

Adhesive bonding	Infrared welding
Electrofusion bonding	Laser welding
Friction welding	Mechanical fastening
Linear	Beading
Rotational	Hot stakes
Heated tool welding	Interface fits
Hot plate	Molded-in and ultrasonic inserts
Hot shoe	Molded-in threads: riveting, self-threading screws, snap fit
High-frequency welding	Solvent joining
Hot-gas welding	Thermal impulse welding
Induction welding	Ultrasonic welding

Some team selections are obvious. Each can handle part of the criteria and subcriteria listed in Table 1 or any other requirements that you develop.

The first step is probably the most difficult: selecting the polymer families to evaluate. Some consult the *Modern Plastic Encyclopedia*.[3] This author is partial to Domininghaus's book on plastics for engineers for the selection of a thermoplastic.[4] The *Modern Plastics Encyclopedia* is an excellent and well-respected source book. However, it only contains one data point within the total history of a particular thermoplastic. Domininghaus provides pressure–volume–temperature ($P–V–T$) graphs on the various thermoplastic families. These data are critical in the processing of materials.

The next action is to obtain samples and technical information from suppliers. As previously mentioned, most suppliers have design guides for their polymers. These are an excellent source of information that can be helpful in your efforts to design the plastic part. Also, obtain from the polymer supplier any analytical procedures as to how they characterize their materials. This will assist your internal analytical people to develop a material knowledge database.

Next you and your team should review all the available data on the polymer under consideration. A good literature search through a technical library may save you time, effort, and money. In addition, you may want to perform your own tests to fill in missing information. Polymer suppliers can be helpful in this area. They can provide molded, ASTM (American Society for Testing and Materials) test coupons to use for your own testing. If you are designing a plastic part to be in a certain chemical environment, you may wish to test your selection in a chemical-soak-type test. This can be achieved by soaking a test coupon in the chemical of concern. You may also want to soak the test coupon at different temperatures within the operating range of the designed plastic part or within a linear range of behavior of the polymer. If you can perform such a chemical soak at at least three different temperatures, you can predict the lifetime of the polymer if it can be related to a chemical failure. However, data of this sort must be obtained using the principles of chemical kinetics.

An example of aging a part can be found within the different outcomes (failure mechanisms) of an egg. Take an egg and set it on a shelf and leave it alone. After several months the egg becomes rotten. Take a similar egg and place it under a hen and after a while we have a cute little chick. Take another egg and place it in boiling water and after about 10 min you have a hard-boiled egg. Never pick temperatures throughout the viscoelastic region of a thermoplastic; you will obtain three different responses (at its glassy region, at its glass transition temperature, and at its rubbery region) of the material.[5]

Now we can proceed with the plastic(s) of choice. This is the polymer with the magical foo-foo dust that the compounder puts in it for various reasons. Chemical soak tests are extremely important in the cases where the plastic part is used in a chemical environment. We do not want anything from the plastic to be extracted into the "chemical environment," thus either affecting the properties of the plastic or contaminating the chemical environment. The reverse is also true: We do not want the plastic to absorb chemicals from the chemical environment. This could cause the properties of the plastics to be lower due to a plasticization effect.

The same chemical soak tests should be conducted on the final molded part with the chemical soaking to exposed portions of the design. In addition, to develop a material knowledge base for the particular part you are designing, you will need these data in the event you or others have to perform failure analysis on the molded part.

At this stage you should be dealing with the molded plastic part assembly. The next step is to design tests that reflect the functionality of the plastic molded part assembly after it leaves your facility and is in the field.

7 CONCLUSION

The goal of this chapter was to provide hints, suggestions, and tricks to assist in the selection of materials for plastic parts design. These hints, suggestions, and tricks have helped this writer in various industrial positions and in consulting projects that have been completed. We have all heard the phrase "if you don't have time to do it right the first time, you won't have time to redo it." It is always more productive to do a project right the first time!

This author has participated in projects in which millions were spent to develop a plastic assembly but the project failed because nothing was spent on an analytical test. Doing such a test is critical to the success of any project.

We often hear "mechanical engineers can pick materials." This may not be true. But, again, this author has participated in projects where simply consulting a polymer handbook would have saved thousands of dollars and many months of time. For example, polyethylene terephthalate was selected as the material of choice for a plastic part that had to withstand an internal processing step of being adhesively bonded to another part for 2 min at 150°C. The grade of polyethylene terephthalate used was a recycled grade with a glass transition temperature in the vicinity of 60°C. Placing the final assembly in an oven at 60°C to simulate an aging test failed all parts due to changes in the dimensions of the part. Subjecting a polyethylene terephthalate coupon to 60°C testing or checking the literature would have been helpful.

Another example is the use of polystyrene as a throw-away coffee cups. Several years ago a fast-food chain was sued for injuries a customer suffered for drinking coffee from one of these cups. Part of the injuries occurred because the cup was made from polystyrene. Polystyrene has a glass transition temperature in the vicinity of 100°C. This temperature is the same as the boiling point of water. Coffee that is extremely hot can reach temperatures close to its glass transition temperature or in the vicinity of when the material starts to transit from the glassy region to the rubbery region of the polymer. In one case it did, thus losing its structural integrity, causing the coffee to spill out of the cup, burning the customer, and resulting in legal action.

Acknowledgments

This author has been fortunate to have the benefit of good books, good teachers, a good network of suppliers, and good coworkers who have been part of my team as well as engineers who were willing to learn, and management that had faith in my methods. I am grateful to all of these.

REFERENCES

1. L. H. Sperling, *Polymeric Multicomponent Materials, an Introduction,* Wiley-Interscience, New York, 1997.
2. R. J. Seyler, "Opening Discussions," in *Assignment of the Glass Transition, STP 1249,* R. J. Seyler (ed.), American Society for Testing and Materials, Philadelphia, 1994.
3. *Modern Plastics Encyclopedia,* McGraw-Hill/Modern Plastics, various editions.
4. H. Domininghaus, *Plastics for Engineers—Materials, Properties, Applications,* Hanser/Gardner Publications, Cincinnati, OH, 1993.
5. K. T. Gillen, M. Celina, R. L. Clough, and J. Wise, "Extrapolation of Accelerated Aging Data—Arrhenius or Erroneous?," *Trends Polym. Sci.,* **5**(8), 250–257 (Aug. 1997).

SUGGESTED READINGS

Brostow, W. (ed.), *Performance of Plastics,* Hanser/Gardner Publications, Cincinnati, OH, 1995.

Brostow, W., and R. D. Corneliussen (eds.), *Failure of Plastics,* Hanser/Gardner Publications, Cincinnati, OH, 1986.

Dym, J. B., *Product Design with Plastics, a Practical Manual,* Industrial Press, New York, 1983.

Ezrin, M., *Plastics Failure Guide Cause and Prevention,* Hanser/Gardner Publications, Cincinnati, OH, 1996.

Handbook of Plastics Joining, a Practical Guide, Plastics Design Library Staff, Plastics Design Library, Norwich, NY, 1997.

Handbooks published by the American Society of Materials on polymers, adhesives and adhesion, and composites, ASM International, Materials Park, OH.

MacDermott, C. P., and A. V. Shenoy, *Selecting Thermoplastics for Engineering Application,* 2nd expanded ed., Marcel Dekker, New York, 1997.

Malloy, R. A., *Plastic Part Design for Injection Molding,* Hanser/Gardner Publications, Cincinnati, OH, 1994.

Morton-Jones, D. H., *Polymer Processing,* Chapman & Hall, New York, 1989.

Osswald, T. A., *Polymer Processing Fundamentials,* Hanser/Gardner Publications, Cincinnati, OH, 1998.

Osswald, T. A., and G. Menges, *Materials Science of Polymers for Engineers,* Hanser/Gardner Publications, Cincinnati, OH, 1995.

Rubin, A., *The Elements of Polymer Science and Engineering, an Introductory Text for Engineers and Chemists,* Academic, New York, 1982.

Young, R. J., and P. A. Lovell, *Introduction to Polymers,* 2nd ed., Chapman & Hall, New York, 1994.

CHAPTER 26

FAILURE MODES: PERFORMANCE AND SERVICE REQUIREMENTS FOR METALS

J. A. Collins
Department of Mechanical Engineering
The Ohio State University
Columbus, Ohio

S. R. Daniewicz
Department of Mechanical Engineering
Mississippi State University
Starkville, Mississippi

1 CRITERIA OF FAILURE

Any change in the size, shape, or material properties of a structure, machine, or machine part that renders it incapable of performing its intended function must be regarded as a mechanical failure of the device. It should be carefully noted that the key concept here is that *improper functioning* of a machine part constitutes failure. Thus, a shear pin that does *not* separate into two or more pieces upon the application of a preselected overload must be regarded as having failed as surely as a drive shaft has failed if it *does* separate into two pieces under normal expected operating loads.

Failure of a device or structure to function properly might be brought about by any one or a combination of many different responses to loads and environments while in service.

Reprinted from *Handbook of Materials Selection,* Wiley, New York, 2002, by permission of the publisher.

For example, too much or too little elastic deformation might produce failure. A fractured load-carrying structural member or a shear pin that does not shear under overload conditions each would constitute failure. Progression of a crack due to fluctuating loads or aggressive environment might lead to failure after a period of time if resulting excessive deflection or fracture interferes with proper machine function.

A primary responsibility of any mechanical designer is to ensure that his or her design functions as intended for the prescribed design lifetime and, at the same time, that it be competitive in the marketplace. Success in designing competitive products while averting premature mechanical failures can be achieved consistently only by recognizing and evaluating all potential modes of failure that might govern the design. To recognize potential failure modes, a designer must be acquainted with the array of failure modes observed in practice and with the conditions leading to these failures. The following section summarizes the mechanical failure modes most commonly observed in practice followed by a brief description of each one.

2 FAILURE MODES

A failure mode may be defined as the physical process or processes that take place or that combine their effects to produce a failure. In the following list of commonly observed failure modes, it may be noted that some failure modes are unilateral phenomena, whereas others are combined phenomena. For example, fatigue is listed as a failure mode, corrosion is listed as a failure mode, and corrosion fatigue is listed as still another failure mode. Such combinations are included because they are commonly observed, important, and often *synergistic*. In the case of corrosion fatigue, for example, the presence of active corrosion aggravates the fatigue process and at the same time the presence of a fluctuating load accelerates the corrosion process.

The following list is not presented in any special order, but it includes all commonly observed modes of mechanical failure[1]:

1. Force- and/or temperature-induced elastic deformation
2. Yielding
3. Brinnelling
4. Ductile rupture
5. Brittle fracture
6. Fatigue
   - **a.** High-cycle fatigue
   - **b.** Low-cycle fatigue
   - **c.** Thermal fatigue
   - **d.** Surface fatigue
   - **e.** Impact fatigue
   - **f.** Corrosion fatigue
   - **g.** Fretting fatigue
7. Corrosion
   - **a.** Direct chemical attack
   - **b.** Galvanic corrosion
   - **c.** Crevice corrosion

   - **d.** Pitting corrosion
   - **e.** Intergranular corrosion
   - **f.** Selective leaching
   - **g.** Erosion corrosion
   - **h.** Cavitation corrosion
   - **i.** Hydrogen damage
   - **j.** Biological corrosion
   - **k.** Stress corrosion

8. Wear
   - **a.** Adhesive wear
   - **b.** Abrasive wear
   - **c.** Corrosive wear
   - **d.** Surface fatigue wear
   - **e.** Deformation wear
   - **f.** Impact wear
   - **g.** Fretting wear

9. Impact
   - **a.** Impact fracture
   - **b.** Impact deformation
   - **c.** Impact wear
   - **d.** Impact fretting
   - **e.** Impact fatigue

10. Fretting
   - **a.** Fretting fatigue
   - **b.** Fretting wear
   - **c.** Fretting corrosion

11. Creep

12. Thermal relaxation

13. Stress rupture

14. Thermal shock

15. Galling and seizure

16. Spalling

17. Radiation damage

18. Buckling

19. Creep buckling

20. Stress corrosion

21. Corrosion wear

22. Corrosion fatigue

23. Combined creep and fatigue

As commonly used in engineering practice, the failure modes just listed may be defined and described briefly as follows. It should be emphasized that these failure modes only produce failure when they generate a set of circumstances that interfere with the proper functioning of a machine or device.

Force and/or temperature-induced elastic deformation failure occurs whenever the elastic (recoverable) deformation in a machine member, induced by the imposed operational loads or temperatures, becomes large enough to interfere with the ability of the machine to perform its intended function satisfactorily.

Yielding failure occurs when the plastic (unrecoverable) deformation in a ductile machine member, brought about by the imposed operational loads or motions, becomes large enough to interfere with the ability of the machine to perform its intended function satisfactorily.

Brinnelling failure occurs when the static forces between two curved surfaces in contact result in local yielding of one or both mating members to produce a permanent surface discontinuity of significant size. For example, if a ball bearing is statically loaded so that a ball is forced to permanently indent the race through local plastic flow, the race is brinnelled. Subsequent operation of the bearing might result in intolerably increased vibration, noise, and heating; and, therefore, failure would have occurred.

Ductile rupture failure occurs when the plastic deformation in a machine part that exhibits ductile behavior is carried to the extreme so that the member separates into two pieces. Initiation and coalescence of internal voids slowly propagate to failure, leaving a dull, fibrous rupture surface.

Brittle fracture failure occurs when the elastic deformation in a machine part that exhibits brittle behavior is carried to the extreme so that the primary interatomic bonds are broken and the member separates into two or more pieces. Preexisting flaws or growing cracks form initiation sites for very rapid crack propagation to catastrophic failure, leaving a granular, multifaceted fracture surface.

Fatigue failure is a general term given to the sudden and catastrophic separation of a machine part into two or more pieces as a result of the application of fluctuating loads or deformations over a period of time. Failure takes place by the initiation and propagation of a crack until it becomes unstable and propagates suddenly to failure. The loads and deformations that typically cause failure by fatigue are far below the static or monotonic failure levels. When loads or deformations are of such magnitude that more than about 10,000 cycles are required to produce failure, the phenomenon is usually termed *high-cycle fatigue*. When loads or deformations are of such magnitude that less than about 10,000 cycles are required to produce failure, the phenomenon is usually termed *low-cycle fatigue*. When load or strain cycling is produced by a fluctuating temperature field in the machine part, the process is usually termed *thermal fatigue*. *Surface fatigue* failure, usually associated with rolling surfaces in contact, manifests itself as pitting, cracking, and spalling of the contacting surfaces as a result of the cyclic Hertz contact stresses that result in maximum values of cyclic shear stresses slightly below the surface. The cyclic subsurface shear stresses generate cracks that propagate to the contacting surface, dislodging particles in the process to produce surface pitting. This phenomenon is often viewed as a type of wear. Impact fatigue, corrosion fatigue, and fretting fatigue are described later.

Corrosion failure, a very broad term, implies that a machine part is rendered incapable of performing its intended function because of the undesired deterioration of the material as a result of chemical or electrochemical interaction with the environment. Corrosion often interacts with other failure modes such as wear or fatigue. The many forms of corrosion include the following. *Direct chemical attack*, perhaps the most common type of corrosion, involves corrosive attack of the surface of the machine part exposed to the corrosive media, more or less uniformly over the entire exposed surface. *Galvanic corrosion* is an accelerated electrochemical corrosion that occurs when two dissimilar metals in electrical contact are made part of a circuit completed by a connecting pool or film of electrolyte or corrosive medium, leading to current flow and ensuing corrosion. *Crevice corrosion* is the accelerated corrosion process highly localized within crevices, cracks, or joints where small-volume regions of stagnant solution are trapped in contact with the corroding metal. *Pitting corrosion* is a very localized attack that leads to the development of an array of holes or pits that penetrate the metal. *Intergranular corrosion* is the localized attack occurring at grain boundaries of certain copper, chromium, nickel, aluminum, magnesium, and zinc alloys when they are improperly heat treated or welded. Formation of local galvanic cells that precipitate corrosion products at the grain boundaries seriously degrades the material strength because of the intergranular corrosive process. *Selective leaching* is a corrosion process in which one element of a solid alloy is removed, such as in dezincification of brass alloys or graphitization of gray cast irons. *Erosion corrosion* is the accelerated chemical attack that results when abrasive or viscid material flows past a containing surface, continuously baring fresh, unprotected material to the corrosive medium. *Cavitation corrosion* is the accelerated chemical corrosion that results when, because of differences in vapor pressure, certain bubbles and cavities within a fluid collapse adjacent to the pressure vessel walls, causing particles of the surface to be expelled, baring fresh, unprotected surface to the corrosive medium. *Hydrogen damage,* while not considered to be a form of direct corrosion, is induced by corrosion. Hydrogen damage includes hydrogen blistering, hydrogen embrittlement, hydrogen attack, and decarburization. *Biological corrosion* is a corrosion process that results from the activity of living organisms, usually by virtue of their processes of food ingestion and waste elimination, in which the waste products are corrosive acids or hydroxides. *Stress corrosion,* an extremely important type of corrosion, is described separately later.

Wear is the undesired cumulative change in dimensions caused by the gradual removal of discrete particles from contacting surfaces in motion, usually sliding, predominantly as a result of mechanical action. Wear is not a single process, but a number of different processes that can take place by themselves or in combination, resulting in material removal from contacting surfaces through a complex combination of local shearing, plowing, gouging, welding, tearing, and others. *Adhesive wear* takes place because of high local pressure and welding at asperity contact sites, followed by motion-induced plastic deformation and rupture of asperity functions, with resulting metal removal or transfer. *Abrasive wear* takes place when the wear particles are removed from the surface by the plowing, gouging, and cutting action of the asperities of a harder mating surface or by hard particles entrapped between the mating surfaces. When the conditions for either adhesive wear or abrasive wear coexist with conditions that lead to corrosion, the processes interact synergistically to produce *corrosive wear.* As described earlier, *surface fatigue wear* is a wear phenomenon associated with curved surfaces in rolling or sliding contact, in which subsurface cyclic shear stresses initiate microcracks that propagate to the surface to spall out macroscopic particles and form wear pits. *Deformation wear* arises as a result of repeated *plastic* deformation at the wearing surfaces, producing a matrix of cracks that grow and coalesce to form wear particles. Deformation wear is often caused by severe impact loading. *Impact wear* is impact-induced repeated *elastic* deformation at the wearing surface that produces a matrix of cracks that grows in accordance with the surface fatigue description just given. Fretting wear is described later.

Impact failure results when a machine member is subjected to nonstatic loads that produce in the part stresses or deformations of such magnitude that the member no longer is capable of performing its function. The failure is brought about by the interaction of stress or strain waves generated by dynamic or suddenly applied loads, which may induce local stresses and strains many times greater than would be induced by the static application of the same loads. If the magnitudes of the stresses and strains are sufficiently high to cause separation into two or more parts, the failure is called *impact fracture*. If the impact produces intolerable elastic or plastic deformation, the resulting failure is called *impact deformation*. If repeated impacts induce cyclic elastic strains that lead to initiation of a matrix of fatigue cracks, which grows to failure by the surface fatigue phenomenon described earlier, the process is called *impact wear*. If fretting action, as described in the next paragraph, is induced by the small lateral relative displacements between two surfaces as they impact together, where the small displacements are caused by Poisson strains or small tangential "glancing" velocity components, the phenomenon is called *impact fretting*. *Impact fatigue* failure occurs when impact loading is applied repetitively to a machine member until failure occurs by the nucleation and propagation of a fatigue crack.

Fretting action may occur at the interface between any two solid bodies whenever they are pressed together by a normal force and subjected to small-amplitude cyclic relative motion with respect to each other. Fretting usually takes place in joints that are not intended to move but, because of vibrational loads or deformations, experience minute cyclic relative motions. Typically, debris produced by fretting action is trapped between the surfaces because of the small motions involved. *Fretting fatigue* failure is the premature fatigue fracture of a machine member subjected to fluctuating loads or strains together with conditions that simultaneously produce fretting action. The surface discontinuities and microcracks generated by the fretting action act as fatigue crack nuclei that propagate to failure under conditions of fatigue loading that would otherwise be acceptable. Fretting fatigue failure is an insidious failure mode because the fretting action is usually hidden within a joint where it cannot be seen and leads to premature, or even unexpected, fatigue failure of a sudden and catastrophic nature. *Fretting wear* failure results when the changes in dimensions of the mating parts, because of the presence of fretting action, become large enough to interfere with proper design function or large enough to produce geometrical stress concentration of such magnitude that failure ensues as a result of excessive local stress levels. *Fretting corrosion* failure occurs when a machine part is rendered incapable of performing its intended function because of the surface degradation of the material from which the part is made, as a result of fretting action.

Creep failure results whenever the plastic deformation in a machine member accrues over a period of time under the influence of stress and temperature until the accumulated dimensional changes interfere with the ability of the machine part to satisfactorily perform its intended function. Three stages of creep are often observed: (1) transient or primary creep during which time the rate of strain decreases, (2) steady-state or secondary creep during which time the rate of strain is virtually constant, and (3) tertiary creep during which time the creep strain rate increases, often rapidly, until rupture occurs. This terminal rupture is often called creep rupture and may or may not occur, depending on the stress–time–temperature conditions.

Thermal relaxation failure occurs when the dimensional changes due to the creep process result in the relaxation of a prestrained or prestressed member until it no longer is able to perform its intended function. For example, if the prestressed flange bolts of a high-temperature pressure vessel relax over a period of time because of creep in the bolts, so that, finally, the peak pressure surges exceed the bolt preload to violate the flange seal, the bolts will have failed because of thermal relaxation.

Stress rupture failure is intimately related to the creep process except that the combination of stress, time, and temperature is such that rupture into two parts is ensured. In stress rupture failures the combination of stress and temperature is often such that the period of steady-state creep is short or nonexistent.

Thermal shock failure occurs when the thermal gradients generated in a machine part are so pronounced that differential thermal strains exceed the ability of the material to sustain them without yielding or fracture.

Galling failure occurs when two sliding surfaces are subjected to such a combination of loads, sliding velocities, temperatures, environments, and lubricants that massive surface destruction is caused by welding and tearing, plowing, gouging, significant plastic deformation of surface asperities, and metal transfer between the two surfaces. Galling may be thought of as a severe extension of the adhesive wear process. When such action results in significant impairment to intended surface sliding or in seizure, the joint is said to have failed by galling. *Seizure* is an extension of the galling process to such severity that the two parts are virtually welded together so that relative motion is no longer possible.

Spalling failure occurs whenever a particle is spontaneously dislodged from the surface of a machine part so as to prevent the proper function of the member. Armor plate fails by spalling, for example, when a striking missile on the exposed side of an armor shield generates a stress wave that propagates across the plate in such a way as to dislodge or spall a secondary missile of lethal potential on the protected side. Another example of spalling failure is manifested in rolling contact bearings and gear teeth because of the action of surface fatigue as described earlier.

Radiation damage failure occurs when the changes in material properties induced by exposure to a nuclear radiation field are of such a type and magnitude that the machine part is no longer able to perform its intended function, usually as a result of the triggering of some other failure mode and often related to loss in ductility associated with radiation exposure. Elastomers and polymers are typically more susceptible to radiation damage than are metals, whose strength properties are sometimes enhanced rather than damaged by exposure to a radiation field, although ductility is usually decreased.

Buckling failure occurs when, because of a critical combination of magnitude and/or point of load application, together with the geometrical configuration of a machine member, the deflection of the member suddenly increases greatly with only a slight change in load. This nonlinear response results in a buckling failure if the buckled member is no longer capable of performing its design function.

Creep buckling failure occurs when, after a period of time, the creep process results in an unstable combination of the loading and geometry of a machine part so that the critical buckling limit is exceeded and failure ensues.

Stress corrosion failure occurs when the applied stresses on a machine part in a corrosive environment generate a field of localized surface cracks, usually along grain boundaries, that render the part incapable of performing its function, often because of triggering some other failure mode. Stress corrosion is a very important type of corrosion failure mode because so many different metals are susceptible to it. For example, a variety of iron, steel, stainless steel, copper, and aluminum alloys are subject to stress corrosion cracking if placed in certain adverse corrosive media.

Corrosion wear failure is a combination failure mode in which corrosion and wear combine their deleterious effects to incapacitate a machine part. The corrosion process often produces a hard, abrasive corrosion product that accelerates the wear, while the wear process constantly removes the protective corrosion layer from the surface, baring fresh metal to the corrosive medium and thus accelerating the corrosion. The two modes combine to make the result more serious than either of the modes would have been otherwise.

Corrosion fatigue is a combination failure mode in which corrosion and fatigue combine their deleterious effects to cause failure of a machine part. The corrosion process often forms pits and surface discontinuities that act as stress raisers that in turn accelerate fatigue failure. Furthermore, cracks in the usually brittle corrosion layer also act as fatigue crack nuclei that propagate into the base material. On the other hand, the cyclic loads or strains cause cracking and flaking of the corrosion layer, which bares fresh metal to the corrosive medium. Thus, each process accelerates the other, often making the result disproportionately serious.

Combined creep and fatigue failure is a combination failure mode in which all of the conditions for both creep failure and fatigue exist simultaneously, each process influencing the other to produce failure. The interaction of creep and fatigue is probably synergistic but is not well understood.

3 ELASTIC DEFORMATION AND YIELDING

Small changes in the interatomic spacing of a material, induced by applied forces or changing temperatures, are manifested macroscopically as elastic strain. Although the maximum elastic strain in crystalline solids, including engineering metals, is typically very small, the force required to produce the small strain is usually large; hence, the accompanying stress is large. For uniaxial loading of a machine or structural element, the total elastic deformation of the member may be found by integrating the elastic strain over the length of the element. Thus, for a uniform bar subjected to uniaxial loading the total deformation of the bar in the axial direction is

$$\Delta l = l\varepsilon \tag{1}$$

where Δl is total axial deformation of the bar, l is the original bar length, and ε is the axial elastic strain. If Δl exceeds the design-allowable axial deformation, failure will occur. For example, if the axial deformation of an aircraft gas turbine blade, due to the centrifugal force field, exceeds the tip clearance gap, failure will occur because of force-induced elastic deformation. Likewise, if thermal expansion of the blade produces a blade-axial deformation that exceeds the tip clearance gap, failure will occur because of temperature-induced elastic deformation.

When the state of stress is more complicated, it becomes necessary to calculate the elastic strains induced by the multiaxial states of stress in three mutually perpendicular directions through the use of the generalized Hooke's law equations for isotropic materials:

$$\begin{aligned} \varepsilon_x &= \frac{1}{E}[\sigma_x - \nu(\sigma_y + \sigma_z)] \\ \varepsilon_y &= \frac{1}{E}[\sigma_y - \nu(\sigma_x + \sigma_z)] \\ \varepsilon_z &= \frac{1}{E}[\sigma_z - \nu(\sigma_x + \sigma_y)] \end{aligned} \tag{2}$$

where σ_x, σ_y, and σ_z are the normal stresses in the three coordinate directions, E and ν are Young's modulus and Poisson's ratio, respectively, and ε_x, ε_y, and ε_z are the elastic strains in the three coordinate directions. Again, total elastic deformation of a member in any of the coordinate directions may be found by integrating the strain over the member's length in that direction. If the change in length of the member in any direction exceeds the design-allowable deformation in that direction, failure will occur. The use of commercial finite-element analysis software packages is one commonly used means of determining both the elastic strains produced in a structural element and the subsequent elastic deformations produced.

If applied loads reach certain critical levels, the atoms within the microstructure may be moved into new equilibrium positions and the induced strains are not fully recovered upon release of the loads. Such permanent strains, usually the result of slip, are called plastic strains, and the macroscopic permanent deformation due to plastic strain is called yielding. If applied loads are increased even more, the plastic deformation process may be carried to the point of instability where *necking* begins: Internal voids form and slowly coalesce to finally produce a ductile rupture of the loaded member.

After plastic deformation has been initiated, Eqs. (2) are no longer valid and the predictions of plastic strains and deformations under multiaxial states of stress are more difficult. If a designer can tolerate a prescribed plastic deformation without experiencing failure, these plastic deformations may be determined using plasticity theory. Many commercial finite-element analysis software packages now possess the capability to compute both plastic strains and deformations for a prescribed nonlinear elastic–plastic constitutive relation.

For the case of simple uniaxial loading, the onset of yielding may be accurately predicted to occur when the uniaxial normal stress reaches a value equal to the yield strength of the material read from an engineering stress–strain curve. If the loading is more complicated and a multiaxial state of stress is produced by the loads, the onset of yielding may no longer be predicted by comparing any one of the normal stress components with uniaxial material yield strength, not even the maximum principal normal stress. Onset of yielding for multiaxially stressed critical points in a machine or structure is more accurately predicted through the use of a *combined stress theory of failure,* which has experimentally been validated for the prediction of yielding. The two most widely accepted theories for predicting the onset of yielding are the distortion energy theory (also called the octahedral shear stress theory or the von Mises criterion) and the maximum shearing stress theory (also called the Tresca criterion). The distortion energy theory is somewhat more accurate while the maximum shearing stress theory may be slightly easier to use and is more conservative.

In words, the distortion energy theory may be expressed as follows: *Failure is predicted to occur in the multiaxial state of stress when the distortion energy per unit volume becomes equal to or exceeds the distortion energy per unit volume at the time of failure in a simple uniaxial stress test using a specimen of the same material.*

Mathematically, the distortion energy theory may be formulated as: *Failure is predicted by the distortion energy theory to occur if*

$$\lfloor(\sigma_1 - \sigma_2)^2 + (\sigma_1 - \sigma_3)^2 + (\sigma_3 - \sigma_1)^2\rfloor \geq 2\sigma_f^2 \tag{3}$$

The maximum shearing stress theory may be stated in words as: *Failure is predicted to occur in the multiaxial state of stress when the maximum shearing stress magnitude becomes equal to or exceeds the maximum shearing stress magnitude at the time of failure in a simple uniaxial stress test using a specimen of the same material.*

Mathematically, the maximum shearing stress theory becomes: *Failure is predicted by the maximum shearing stress theory to occur if*

$$\sigma_1 - \sigma_3 \geq \sigma_f \tag{4}$$

where σ_1, σ_2, and σ_3 are the principal stresses at a point ordered such that $\sigma_1 \geq \sigma_2 \geq \sigma_3$ and σ_f is the uniaxial failure strength in tension.

Comparisons of these two failure theories with experimental data on yielding are shown in Fig. 1 for a variety of materials and different biaxial states of stress.

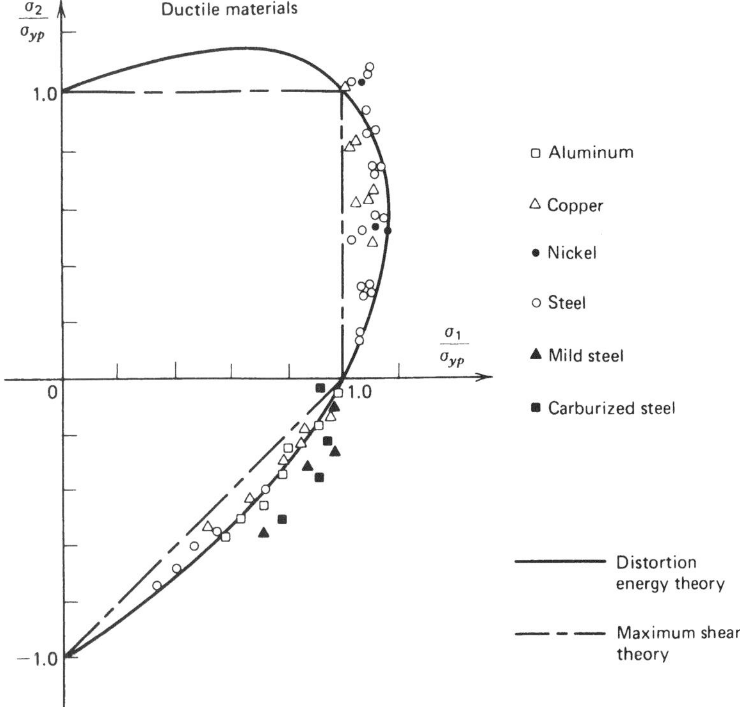

Figure 1 Comparison of biaxial yield strength data with theories of failure for a variety of ductile materials.

4 FRACTURE MECHANICS AND UNSTABLE CRACK GROWTH

When the material behavior is brittle rather than ductile, the mechanics of the failure process are very different. Instead of the slow coalescence of voids associated with ductile rupture, brittle fracture proceeds by the high-velocity propagation of a crack across the loaded member. If the material behavior is clearly brittle, fracture may be predicted with reasonable accuracy through use of the maximum normal stress theory of failure. In words, the maximum normal stress theory may be expressed as follows: *Failure is predicted to occur in the multiaxial state of stress when the maximum principal normal stress becomes equal to or exceeds the maximum normal stress at the time of failure in a simple uniaxial stress test using a specimen of the same material.*

Mathematically, the maximum normal stress theory becomes: *Failure is predicted by the maximum normal stress theory to occur if*

$$\sigma_1 \geq \sigma_t \qquad |\sigma_3| \geq \sigma_c \tag{5}$$

where σ_1, σ_2, and σ_3 are the principal stresses at a point ordered such that $\sigma_1 \geq \sigma_2 \geq \sigma_3$, σ_t is the uniaxial failure strength in tension, and σ_c is the uniaxial failure strength in compression. Comparison of this failure theory with experimental data on brittle fracture for different biaxial states of stress is shown in Fig. 2.

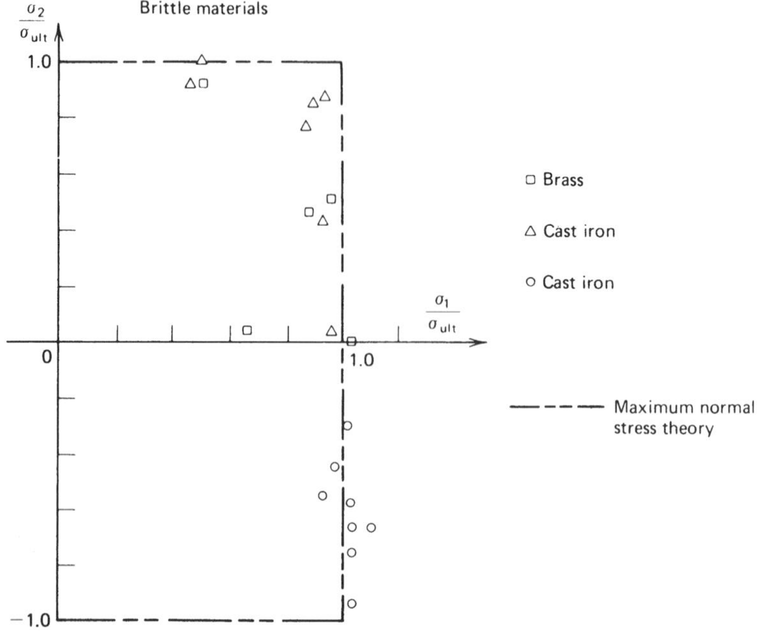

Figure 2 Comparison of biaxial brittle fracture strength data with maximum normal stress theory for several brittle materials.

On the other hand, more recent experience has led to the understanding that nominally ductile materials may also fail by a brittle fracture response in the presence of cracks or flaws if the combination of crack size, geometry of the part, temperature, and/or loading rate lies within certain critical regions. Furthermore, the development of higher strength structural alloys, the wider use of welding, and the use of thicker sections in some cases have combined their influence to reduce toward a critical level the capacity of some structural members to accommodate local plastic strain without fracture. At the same time, fabrication by welding, residual stresses due to machining, and assembly mismatch in production have increased the need for accommodating local plastic strain to prevent failure. Fluctuating service loads of greater severity and more aggressive environments have also contributed to unexpected fractures.

An important observation in studying fracture behavior is that the magnitude of the nominal applied stress that causes fracture is related to the size of the crack or cracklike flaw within the structure.[2–4] For example, observations of the behavior of central through-the-thickness cracks, oriented normal to the applied tensile stress in steel and aluminum plates, yielded the results shown in Figs. 3 and 4. In these tests, as the tensile loading on the precracked plates was slowly increased, the crack initially extended slowly for a time and then abruptly extended to failure by rapid crack propagation. The slow stable crack growth or tearing was characterized by speeds of the order of fractions of an inch per minute. The rapid crack propagation was characterized by speeds of the order of hundreds of feet per second. The data of Figs. 3 and 4 indicate that for longer initial crack lengths the fracture stress (the stress corresponding to the onset of stable tearing) was lower. For the aluminum alloy the fracture stress was less than the yield strength for cracks longer than about 0.75

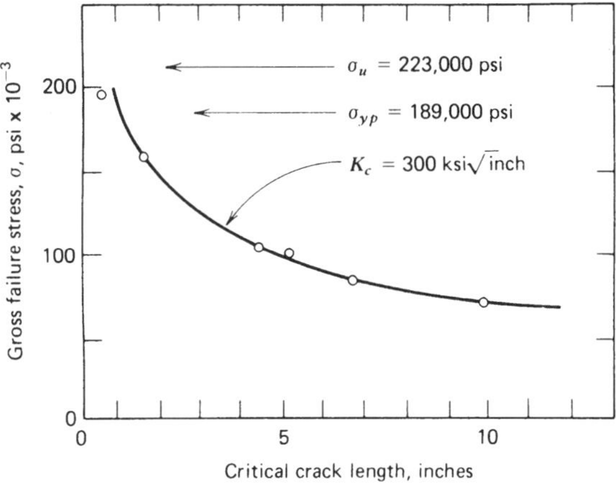

Figure 3 Influence of crack length on gross failure stress for center-cracked steel plate. (After Ref. 5. Copyright ASTM, adapted with permission.)

in. For the steel alloy the fracture stress was less than the yield strength for cracks longer than about 0.5 in. In both cases, for shorter cracks the fracture stress approaches the ultimate strength of the material determined from a conventional uniaxial tension test.

Experience has shown that the onset of stable tearing establishes an important material property termed *fracture toughness*. The fracture toughness may be used as a design criterion in fracture prevention, just as the yield strength is used as a design criterion in prevention of yielding of a ductile material under static loading. It should be noted that not all materials and/or specimen types exhibit stable tearing. Many materials and/or specimen configurations exhibit rapid crack propagation without any evidence of prior stable tearing.

In many cases slow crack propagation occurs by means other than stable tearing, especially under conditions of fluctuating loads and/or aggressive environments. In analyses and predictions involving fatigue failure phenomena, characterization of the rate of slow

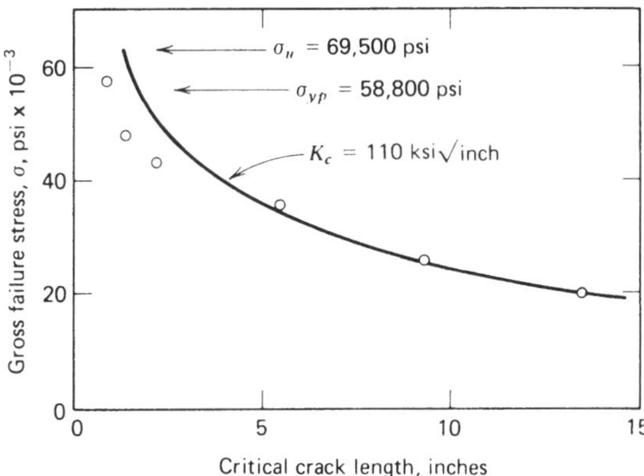

Figure 4 Influence of crack length on gross failure stress for center-cracked aluminum plate. (After Ref. 5. Copyright ASTM, adapted with permission.)

crack extension and the initial flaw size, together with critical crack size, are used to determine the useful life of a component or structure subjected to fluctuating loads. The topic of fatigue crack propagation is discussed further in Section 5.

The simplest useful model for the stresses near the tip of a crack is based on the assumptions of linear elastic material behavior and a two-dimensional analysis; thus, the procedure is often referred to as linear elastic fracture mechanics. Although the validity of the linear elastic assumption may be questioned in view of plastic zone formation at the tip of a crack in any real engineering material, as long as *small-scale yielding* occurs, that is, as long as the plastic zone size remains small compared to the dimensions of the crack, the linear elastic model gives good engineering results. Thus, the small-scale yielding concept implies that the small plastic zone is confined within a linear elastic field surrounding the crack tip. If the material properties, section size, loading conditions, and environment combine in such a way that large-scale plastic zones are formed, the basic assumptions of linear elastic fracture mechanics are violated, and elastic–plastic fracture mechanics methods must be employed.[4]

Three basic types of stress fields can be defined for crack-tip stress analysis, each one associated with a distinct mode of crack deformation, as illustrated in Fig. 5. The opening mode, mode I, is associated with local displacement in which the crack surfaces move directly apart, as shown in Fig. 5*a*. The sliding mode, mode II, is developed when crack surfaces slide over each other in a direction perpendicular to the leading edge of the crack, as shown in Fig. 5*b*. The tearing mode, mode III, is characterized by crack surfaces sliding with respect to each other in a direction parallel to the leading edge of the crack, as shown in Fig. 5*c*. Superposition of these three modes will fully describe the most general three-dimensional case of local crack-tip deformation and stress field, although mode I is most common.

In terms of the coordinates shown in Fig. 6, the stresses near the crack tip for mode I loading may be written as[2]

$$\sigma_x = \frac{K_\mathrm{I}}{\sqrt{2\pi r}} \cos\frac{\theta}{2}\left[1 - \sin\frac{\theta}{2}\sin\frac{3\theta}{2}\right] \tag{6}$$

$$\sigma_y = \frac{K_\mathrm{I}}{\sqrt{2\pi r}} \cos\frac{\theta}{2}\left[1 + \sin\frac{\theta}{2}\sin\frac{3\theta}{2}\right] \tag{7}$$

$$\tau_{xy} = \frac{K_\mathrm{I}}{\sqrt{2\pi r}} \sin\frac{\theta}{2}\cos\frac{\theta}{2}\cos\frac{3\theta}{2} \tag{8}$$

The parameter K_I is known as the mode I stress intensity factor. This parameter represents the strength of the stress field surrounding the tip of the crack. Since fracture is induced by the crack-tip stress field, the stress intensity factor is the primary correlation parameter used in current practice.

In general, the expressions for the stress intensity factor are of the form

$$K_\mathrm{I} = C\sigma\sqrt{\pi a} \tag{9}$$

where a is the crack size, σ is the gross-section stress, and C is dependent on the type of loading and the geometry away from the crack. Much work has been completed in determining values of C for a wide variety of conditions. (See, for example, Ref. 5.)

Many commercial finite-element analysis software packages possess special crack-tip elements allowing the numerical computation of stress intensity factors. A discussion of some

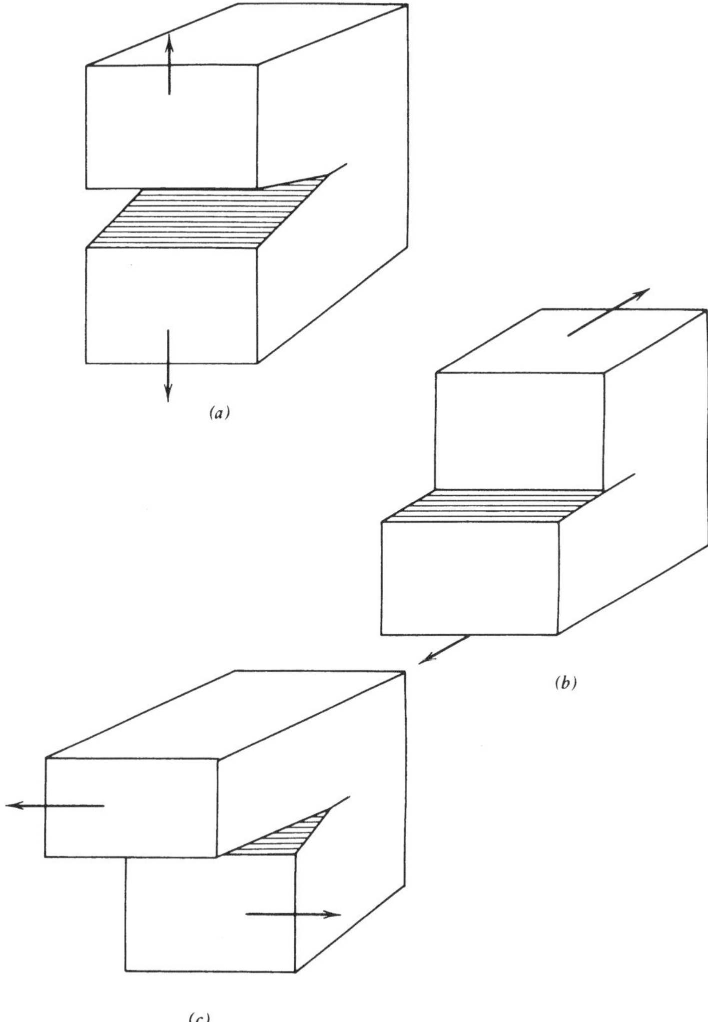

Figure 5 Basic modes of crack displacement: (*a*) mode I, (*b*) mode II, and (*c*) mode III.

of the techniques employed within these software packages is given by Anderson.[4] Through the use of weight functions,[6,35] stress intensity factors may also be computed easily using numerical integration and the stresses that would exist in the uncracked body.

From Eq. (9), the stress intensity factor increases proportionally with gross nominal stress σ and also is a function of the crack length a. The value of K_{I} associated with the onset of fracture has been designated the *critical stress intensity*, K_C. As noted earlier in Figs. 3 and 4, the fracture of specimens with different crack lengths occurs at different values of gross-section stress but at a constant value of K_C. Thus, K_C provides a single-parameter fracture criterion that allows the prediction of fracture based on (9). That is, *fracture is predicted to occur if*

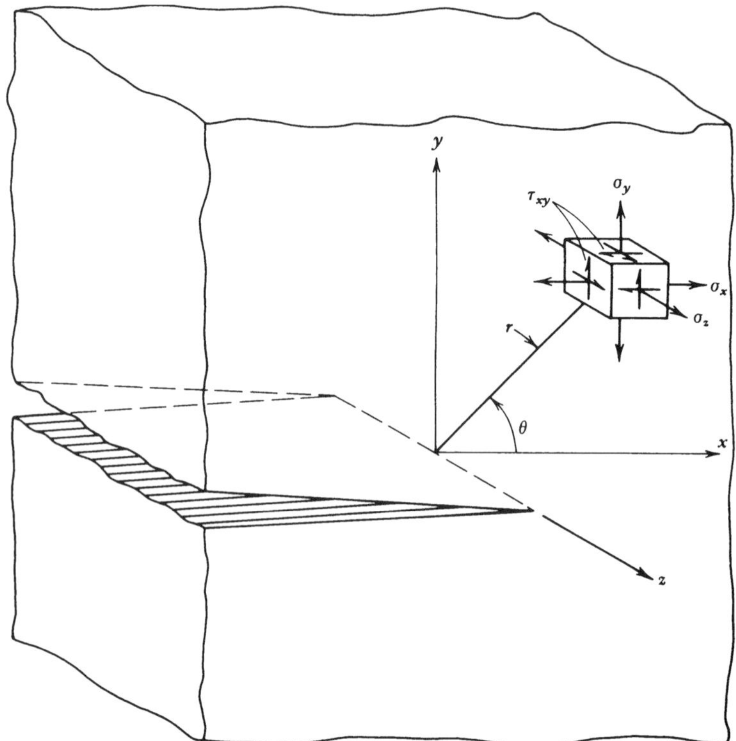

Figure 6 Coordinates measured from leading edge of a crack.

$$K_{\mathrm{I}} \geq K_C \tag{10}$$

In studying material behavior, one finds that for a given material, as the specimen thickness is increased, the critical stress intensity K_C decreases to a lower limiting value. This lower limiting value defines a basic material property $K_{\mathrm{I},C}$, the *plane-strain fracture toughness* for the material. Standard test methods have been established for the determination of $K_{\mathrm{I},C}$ values.[7] A few data are shown in Table 1. Useful compilations of fracture toughness values have been prepared by several organizations and individuals. These include Refs. 9–15.

For the plane-strain fracture toughness $K_{\mathrm{I},C}$ to be a valid failure prediction criterion for a specimen or a machine part, plane-strain conditions must exist at the crack tip; that is, the material must be thick enough to ensure plane-strain conditions. It has been estimated[7] empirically that for plane-strain conditions the minimum material thickness B must be

$$B \geq 2.5 \left(\frac{K_{\mathrm{I},C}}{\sigma_{\mathrm{yp}}} \right)^2 \tag{11}$$

where σ_{yp} is the material yield strength.

If the material is not thick enough to meet the criterion of (11), plane stress is a more likely state of stress at the crack tip; and K_C, the critical stress intensity factor for failure prediction under plane-stress conditions, may be estimated using a semiempirical relationship[16]:

Table 1 Yield Strength and Plane-Strain Fracture Toughness Data for Selected Engineering Alloys[8,9]

Alloy	Form	Test Temperature		σ_{yp}		$K_{I,C}$	
		°F	°C	ksi	MPa	ksi$\sqrt{\text{in.}}$	MPa$\sqrt{\text{m}}$
4340 (500°F temper) steel	Plate	70	21	217–238	1495–1640	45–57	50–63
4340 (800°F temper) steel	Forged	70	21	197–211	1360–1455	72–83	79–91
D6AC (1000°F temper) steel	Plate	170	21	217	1495	93	102
D6AC (1000°F temper) steel	Plate	−65	−54	228	1570	56	62
A 538 steel				250	1722	100	111
2014-T6 aluminum	Forged	75	24	64	440	28	31
2024-T351 aluminum	Plate	80	27	54–56	370–385	28–40	31–44
7075-T6 aluminum				85	585	30	33
7075-T651 aluminum	Plate	70	21	75–81	515–560	25–28	27–31
7075-T7351 aluminum	Plate	70	21	58–66	300–455	28–32	31–35
Ti-6Al-4V titanium	Plate	74	23	119	820	96	106

$$K_C = K_{I,C}\left[1 + \frac{1.4}{B^2}\left(\frac{K_{I,C}}{\sigma_{yp}}\right)^4\right]^{1/2} \tag{12}$$

As long as the crack-tip plastic zone remains in the regime of small-scale yielding, this estimation procedure provides a good design approach. For conditions that result in large crack-tip plastic zones (large applied stresses, large crack lengths), performing a failure assessment using linear elastic fracture mechanics (LEFM) is invalid and potentially nonconservative. A general rule of thumb is that plasticity effects become significant when the applied stresses approach 50% of the yield stress, but this is by no means a universal rule.[4] When small-scale yielding is not generated at the crack tip, a better design approach would involve the implementation of an appropriate elastic–plastic fracture mechanics (EPFM) procedure.

5 FATIGUE

Static or quasi-static loading is rarely observed in modern engineering practice, making it essential for the designer to address himself or herself to the implications of repeated loads, fluctuating loads, and rapidly applied loads. By far, the majority of engineering design projects involve machine parts subjected to fluctuating or cyclic loads. Such loading induces fluctuating or cyclic stresses that often result in failure by fatigue.

Fatigue failure investigations over the years have led to the observation that the fatigue process actually embraces two domains of cyclic stressing or straining that are significantly different in character and in each of which failure is probably produced by different physical mechanisms. One domain of cyclic loading is that for which significant plastic strain occurs during each cycle. This domain is associated with high loads and short lives, or low numbers of cycles to produce fatigue failure, and is commonly referred to as *low-cycle fatigue*. The other domain of cyclic loading is that for which the strain cycles are largely confined to the elastic range. This domain is associated with lower loads and long lives, or high numbers of cycles to produce fatigue failure, and is commonly referred to as *high-cycle fatigue*. Low-cycle fatigue is typically associated with cycle lives from 1 up to about 10^4 cycles. Fatigue may be characterized as a progressive failure phenomenon that proceeds by the *initiation* and *propagation* of cracks to an unstable size. Although there is not complete agreement on the microscopic details of the initiation and propagation of the cracks, processes of reversed slip and dislocation interaction appear to produce fatigue nuclei from which cracks may grow. Finally, the crack length reaches a critical dimension and one additional cycle then causes complete failure. The final failure region will typically show evidence of plastic deformation produced just prior to final separation. For ductile materials the final fracture area often appears as a shear lip produced by crack propagation along the planes of maximum shear.

Although designers find these basic observations of great interest, they must be even more interested in the macroscopic phenomenological aspects of fatigue failure and in avoiding fatigue failure during the design life. Some of the macroscopic effects and basic data requiring consideration in designing under fatigue loading include:

1. The effects of a simple, completely reversed alternating stress on the strength and properties of engineering materials
2. The effects of a steady stress with a superposed alternating component, that is, the effects of cyclic stresses with a nonzero mean
3. The effects of alternating stresses in a multiaxial state of stress
4. The effects of stress gradients and residual stresses, such as imposed by shot peening or cold rolling
5. The effects of stress raisers, such as notches, fillets, holes, threads, riveted joints, and welds.
6. The effects of surface finish, including the effects of machining, cladding, electroplating, and coating
7. The effects of temperature on fatigue behavior of engineering materials
8. The effects of size of the structural element
9. The effects of accumulating cycles at various stress levels and the permanence of the effect
10. The extent of the variation in fatigue properties to be expected for a given material
11. The effects of humidity, corrosive media, and other environmental factors
12. The effects of interaction between fatigue and other modes of failure, such as creep, corrosion, and fretting

5.1 Fatigue Loading and Laboratory Testing

Faced with the design of a fatigue-sensitive element in a machine or structure, a designer is very interested in the fatigue response of engineering materials to various loadings that might occur throughout the design life of the machine under consideration. That is, the designer is interested in the effects of various *loading spectra* and associated *stress spectra,* which will in general be a function of the design configuration and the operational use of the machine.

Perhaps the simplest fatigue stress spectrum to which an element may be subjected is a zero-mean sinusoidal stress–time pattern of constant amplitude and fixed frequency, applied for a specified number of cycles. Such a stress–time pattern, often referred to as a completely reversed cyclic stress, is illustrated in Fig. 7*a*. Utilizing the sketch of Fig. 7*b*, we can conveniently define several useful terms and symbols; these include

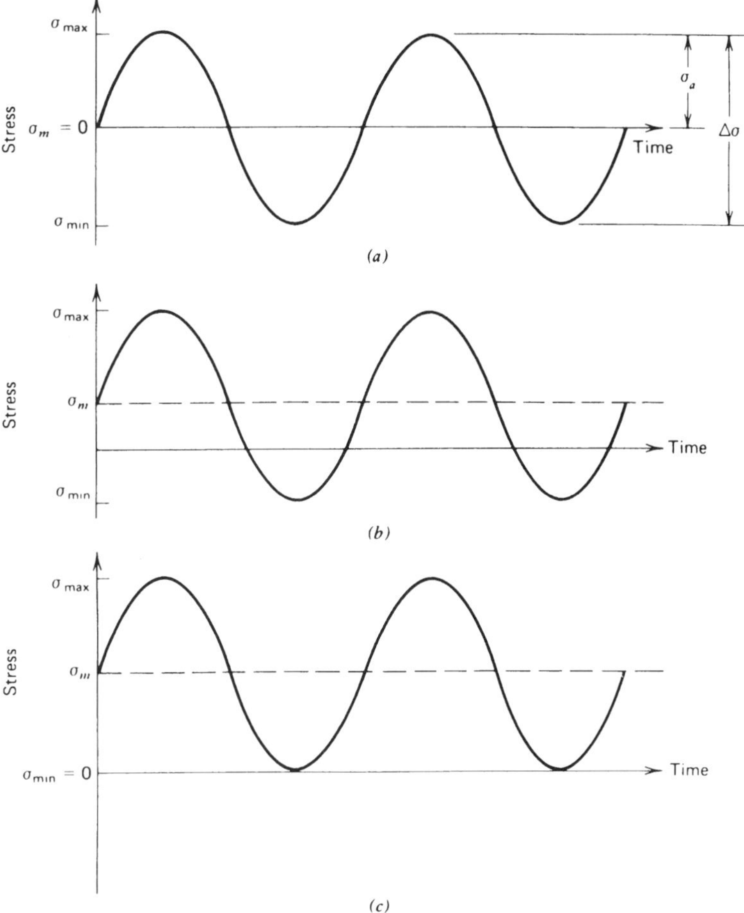

Figure 7 Several constant-amplitude stress–time patterns of interest: (*a*) completely reversed, $R = -1$, (*b*) nonzero mean stress, (*c*) released tension, $R = 0$.

$$\sigma_{max} = \text{maximum stress in the cycle}$$

$$\sigma_m = \text{mean stress, } = (\sigma_{max} + \sigma_{min})/2$$

$$\sigma_{min} = \text{minimum stress in the cycle}$$

$$\sigma_a = \text{alternating stress amplitude, } = (\sigma_{max} - \sigma_{min})/2$$

$$\Delta\sigma = \text{range of stress, } = \sigma_{max} - \sigma_{min}$$

$$R = \text{stress ratio, } = \sigma_{min}/\sigma_{max}$$

Any two of the quantities just defined, except the combinations σ_a and $\Delta\sigma$, are sufficient to describe completely the stress–time pattern.

More complicated stress–time patterns are produced when the mean stress, the stress amplitude, or both change during the operational cycle, as illustrated in Fig. 8. It may be noted that this stress–time spectrum is beginning to approach a degree of realism. Finally,

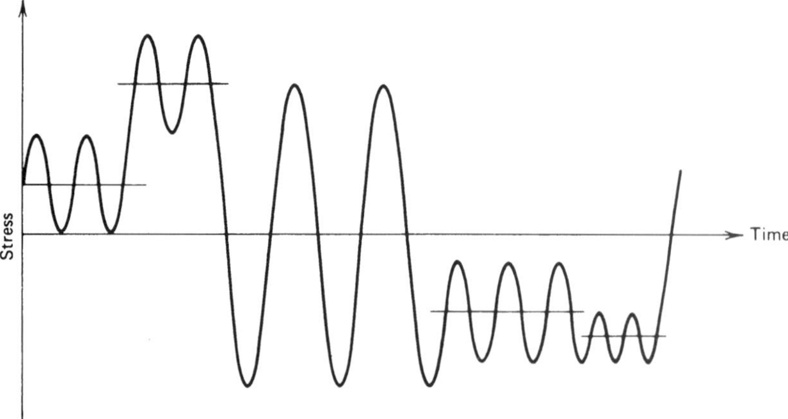

Figure 8 Stress–time pattern in which both mean and amplitude change to produce a more complicated stress spectrum.

in Fig. 9 a sketch of a realistic stress spectrum is given. This type of quasi-random stress–time pattern might be encountered in an airframe structural member during a typical mission including refueling, taxi, takeoff, gusts, maneuvers, and landing. The obtaining of useful, realistic data is a challenging task in itself. Instrumentation of existing machines, such as operational aircraft, provide some useful information to the designer if his or her mission is similar to the one performed by the instrumented machine. Recorded data from accelerometers, strain gauges, and other transducers may in any event provide a basis from which a

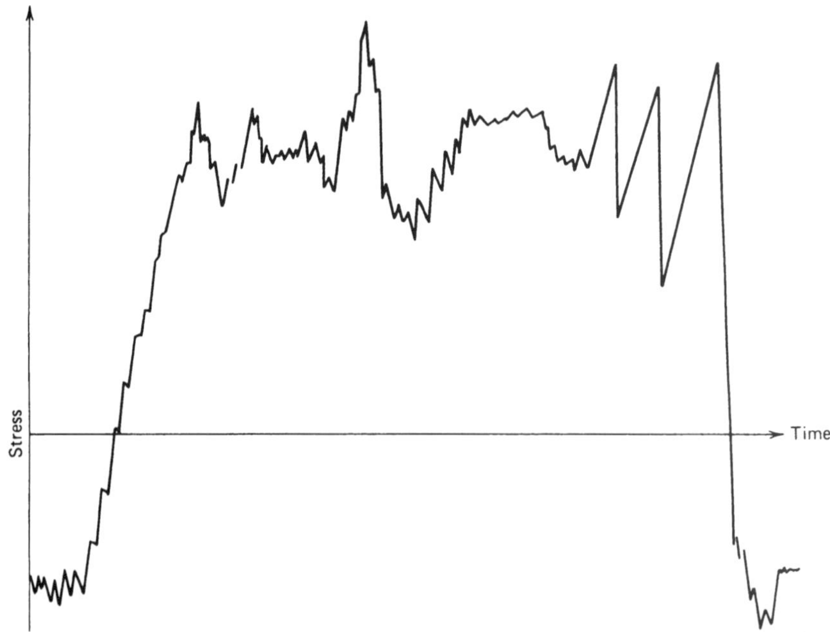

Figure 9 A quasi-random stress–time pattern typical of an aircraft during a given mission.

statistical representation can be developed and extrapolated to future needs if the fatigue processes are understood.

Basic data for evaluating the response of materials, parts, or structures are obtained from carefully controlled laboratory tests. Various types of testing machines and systems commonly used include:

1. Rotating-bending machines
   a. Constant bending moment type
   b. Cantilever bending type
2. Reciprocating-bending machines
3. Axial direct-stress machines
   a. Brute-force type
   b. Resonant type
4. Vibrating shaker machines
   a. Mechanical type
   b. Electromagnetic type
5. Repeated torsion machines
6. Multiaxial stress machines
7. Computer-controlled closed-loop machines
8. Component testing machines for special applications
9. Full-scale or prototype fatigue testing systems

Computer-controlled fatigue testing machines are widely used in all modern fatigue-testing laboratories. Usually such machines take the form of precisely controlled hydraulic systems with feedback to electronic controlling devices capable of producing and controlling virtually any strain–time, load–time, or displacement–time pattern desired. A schematic diagram of such a system is shown in Fig. 10.

Special testing machines for component testing and full-scale prototype testing are not found in the general fatigue-testing laboratory. These systems are built up especially to suit a particular need, for example, to perform a full-scale fatigue test of a commercial jet aircraft.

It may be observed that fatigue-testing machines range from very simple to very complex. The very complex testing systems, used, for example, to test a full-scale prototype, produce very specialized data applicable only to the particular prototype and test conditions used; thus, for the particular prototype and test conditions the results are very accurate, but extrapolation to other test conditions and other pieces of hardware is difficult, if not impossible. On the other hand, simple smooth-specimen laboratory fatigue data are very general and can be utilized in designing virtually any piece of hardware made of the specimen material. However, to use such data in practice requires a quantitative knowledge of many pertinent differences between the laboratory and the application, including the effects of nonzero mean stress, varying stress amplitude, environment, size, temperature, surface finish, residual stress pattern, and others. Fatigue testing is performed at the extremely simple level of smooth-specimen testing, the extremely complex level of full-scale prototype testing, and everywhere in the spectrum between. Valid arguments can be made for testing at all levels.

5.2 *S–N–P* Curves: A Basic Design Tool

Basic fatigue data in the high-cycle life range can be conveniently displayed on a plot of cyclic stress level versus the logarithm of life or, alternatively, on a log-log plot of stress

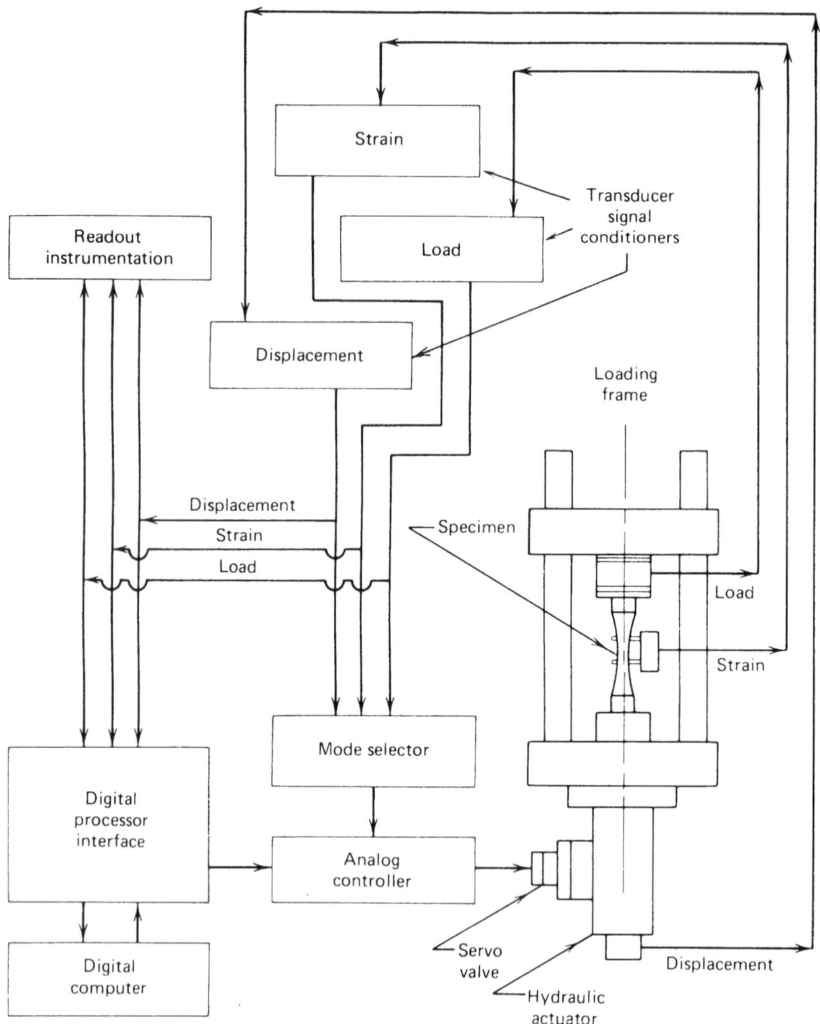

Figure 10 Schematic diagram of a computer-controlled closed-loop fatigue-testing machine.

versus life. These plots, called *S–N* curves, constitute design information of fundamental importance for machine parts subjected to repeated loading. Because of the scatter of fatigue life data at any given stress level, it must be recognized that there is not only one *S–N* curve for a given material, but a family of *S–N* curves with probability of failure as the parameter. These curves are called the *S–N–P* curves, or curves of constant probability of failure on a stress-versus-life plot. A representative family of *S–N–P* curves is illustrated in Fig. 11. It should also be noted that references to the "*S–N* curve" in the literature generally refer to the mean curve unless otherwise specified. Details regarding fatigue testing and the experimental generation of *S–N–P* curves may be found in Ref. 1.

The mean *S–N* curves sketched in Fig. 12 distinguish two types of material response to cyclic loading commonly observed. The ferrous alloys and titanium exhibit a steep branch in the relatively short life range, leveling off to approach a stress asymptote at longer lives.

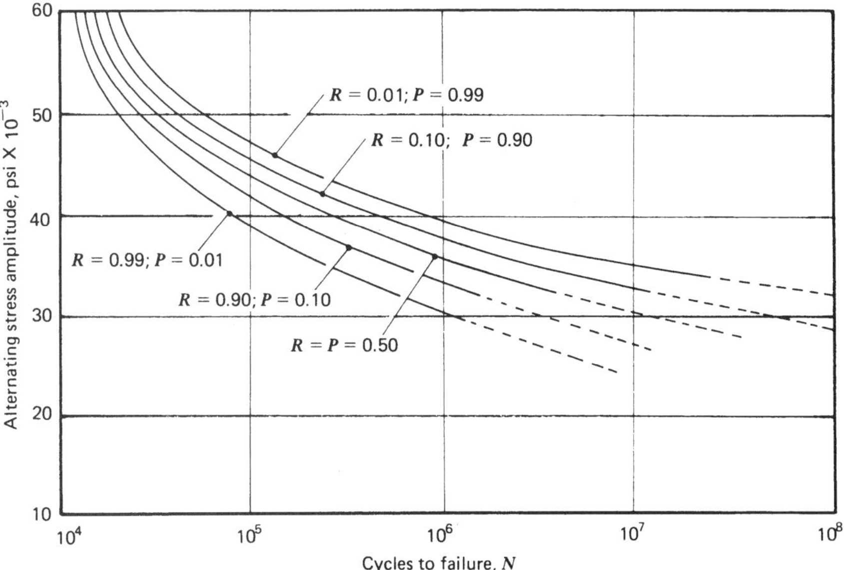

Figure 11 Family of *S–N–P* curves or *R–S–N* curves, for 7075-T6 aluminum alloy. Note: P = probability of failure; R = reliability = $1 - P$. (Adapted from Ref. 17, with permission from John Wiley & Sons, Inc.)

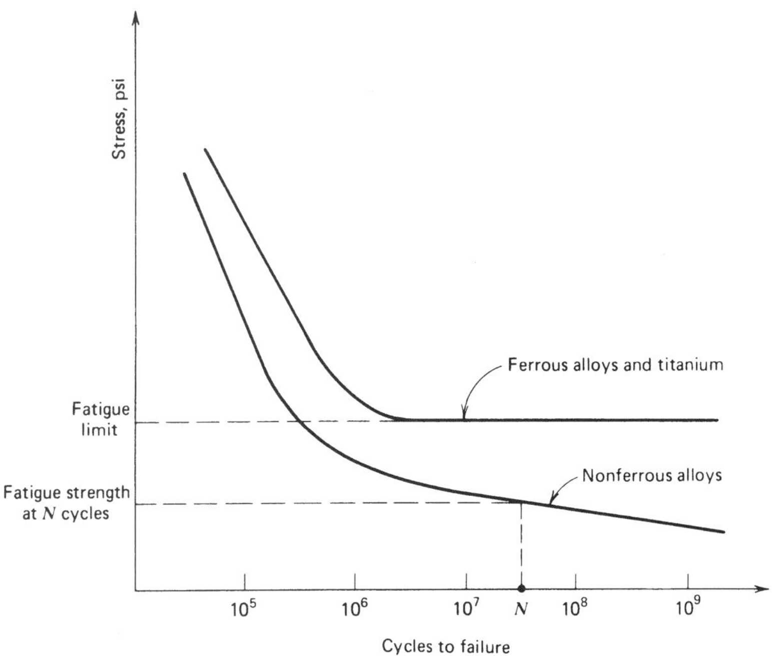

Figure 12 Two types of material response to cyclic loading.

This stress asymptote is called the *fatigue limit* or *endurance limit* and is the stress level below which an infinite number of cycles can be sustained without failure. The nonferrous alloys do not exhibit an asymptote, and the curve of stress versus life continues to drop off indefinitely. For such alloys there is no fatigue limit, and failure as a result of cyclic load is only a matter of applying enough cycles. All materials, however, exhibit a relatively flat curve in the long-life range.

To characterize the failure response of nonferrous materials and of ferrous alloys in the finite-life range, the term *fatigue strength at a specified life,* σ_N, is used. The term fatigue strength identifies the stress level at which failure will occur at the specified life. The specification of *fatigue strength* without specifying the corresponding life is meaningless. The specification of a *fatigue limit* always implies infinite life.

5.3 Factors That Affect *S–N–P* Curves

There are many factors that may influence the fatigue failure response of machine parts or laboratory specimens, including material composition, grain size and grain direction, heat treatment, welding, geometrical discontinuities, size effects, surface conditions, residual surface stresses, operating temperature, corrosion, fretting, operating speed, configuration of the stress–time pattern, nonzero mean stress, and prior fatigue damage. Typical examples of how some of these factors may influence fatigue response are shown in Figs. 13–19. It is usually necessary to search the literature and existing databases to find the information required for a specific application, and it may be necessary to undertake experimental testing programs to produce data where they are unavailable.

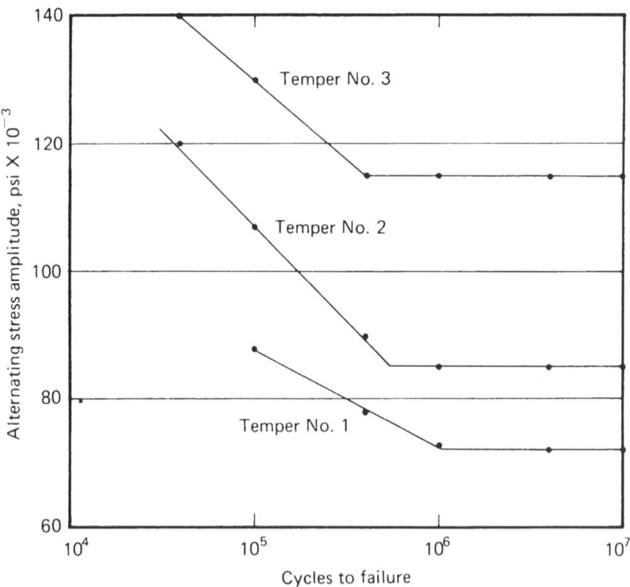

Figure 13 Effects of heat treatment on the *S–N* curve of oil-quenched SAE 4130 steel. Temper No. 1: S_u = 129 ksi; temper No. 2: S_u = 150 ksi; temper No. 3: S_u = 206 ksi.

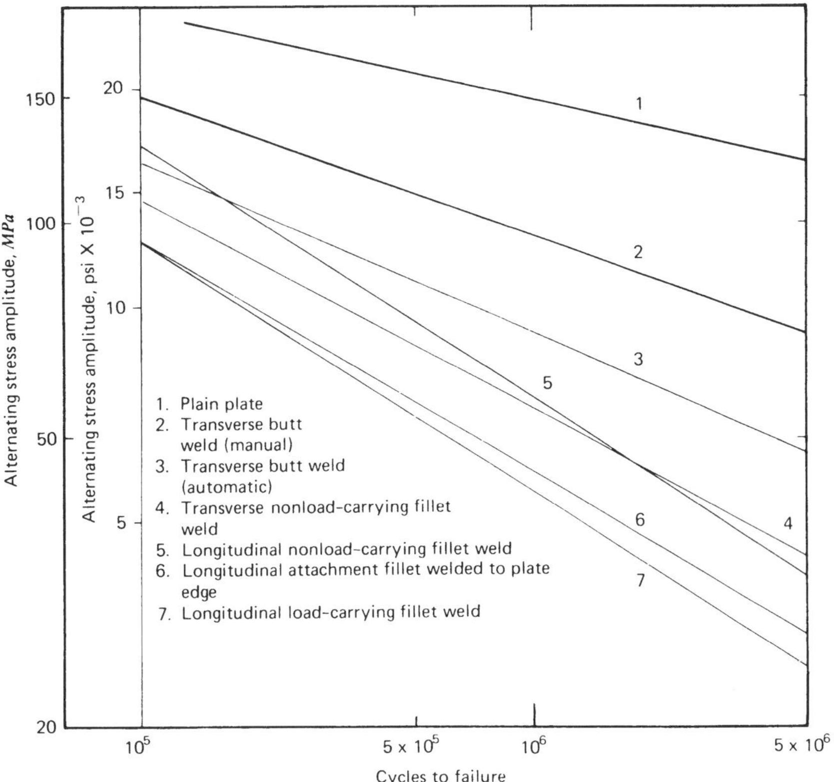

Figure 14 Effects of welding detail on the *S–N* curve of structural steel, with yield strength in the range of 30–52 ksi. Tests were released tension ($R = 0$). (Data from Ref. 18.)

5.4 Nonzero Mean Stress

Most basic fatigue data collected in the laboratory are for completely reversed alternating stresses, that is, zero mean cyclic stresses. Most service applications involve nonzero mean cyclic stresses. It is therefore very important to a designer to know the influence of mean stress on fatigue behavior so that he or she can utilize basic completely reversed laboratory data in designing machine parts subjected to nonzero mean cyclic stresses.

If a designer is fortunate enough to find test data for his or her proposed material under the mean stress conditions and design life of interest, the designer should, of course, use these data. Such data are typically presented on so-called *master diagrams* or *constant life diagrams* for the material. A master diagram for a 4340 steel alloy is shown in Fig. 20. An alternative means of presenting this type of fatigue data is illustrated in Fig. 21 for a 4130 steel alloy.

If data are not available to the designer, he or she may estimate the influence of nonzero mean stress by any one of several empirical relationships that relate failure at a given life under nonzero mean conditions to failure at the same life under zero mean cyclic stresses. Historically, the plot of alternating stress amplitude σ_a versus mean stress σ_m has been the

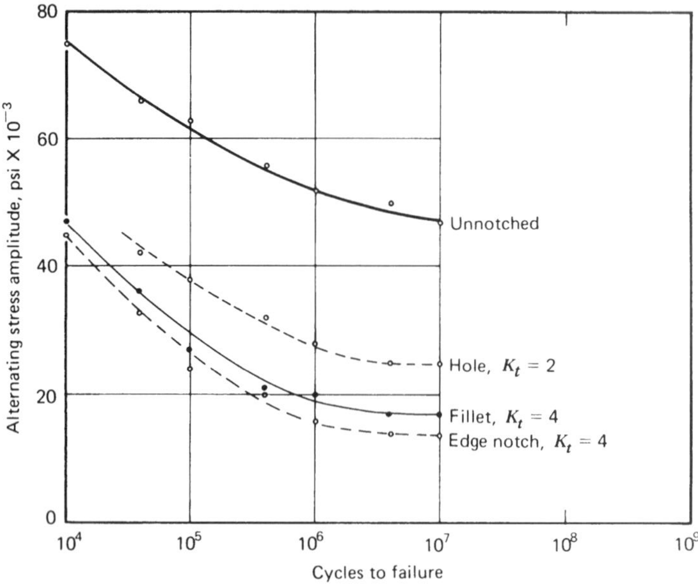

Figure 15 Effects of geometrical discontinuities on the *S–N* curve of SAE 4130 steel sheet, tested under completely reversed axial loading. Specimen dimensions (t = thickness, w = width, r = notch radius); unnotched: t = 0.075 in., w = 1.5 in.; hole: t = 0.075 in., w = 4.5 in., r = 1.5 in.; fillet: t = 0.075 in., w_{net} = 1.5 in., w_{gross} = 2.25 in., r = 0.0195 in.; edge notch: t = 0.075 in., w_{net} = 1.5 in., w_{gross} = 2.25 in., r = 0.057 in. (Data from Ref. 19.)

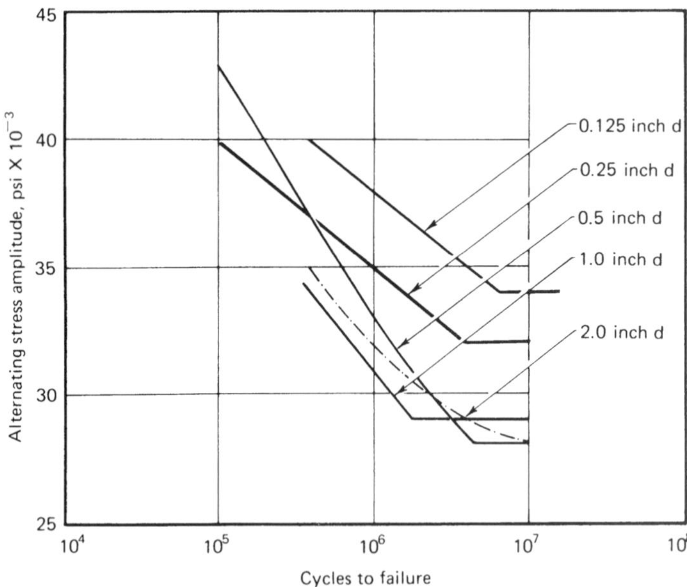

Figure 16 Size effects on the *S–N* curve of SAE 1020 steel specimens cut from a 3.5-in.-diameter hot-rolled bar, testing in rotating bending. (Data from Ref. 20.)

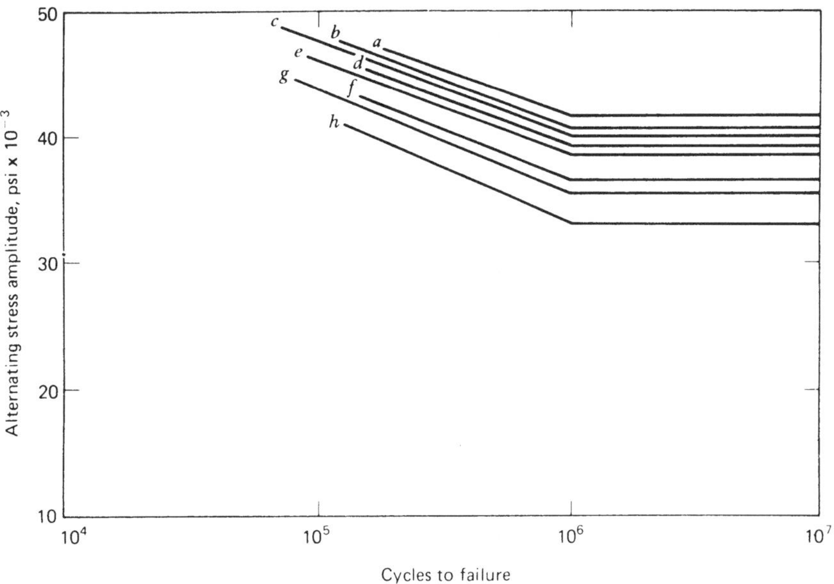

Figure 17 Effect of surface finish on the *S–N* curve of carbon steel specimens, testing in rotating bending: (*a*) high polish, longitudinal direction; (*b*) FF emery finish; (*c*) No. 1 emery finish; (*d*) coarse emery finish; (*e*) smooth file; (*f*) as-turned; (*g*) bastard file; (*h*) coarse file. (Data from Ref. 21.)

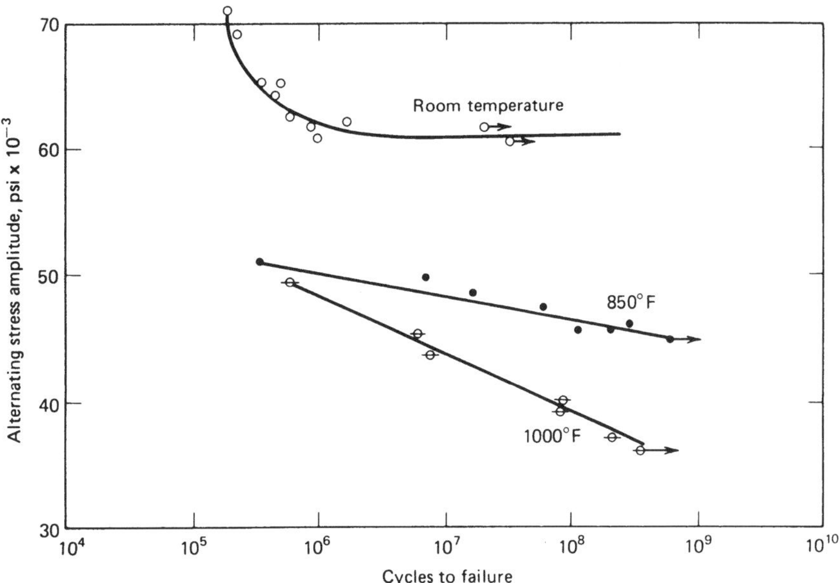

Figure 18 Effect of operating temperature on the *S–N* curve of a 12% chromium steel alloy. (Data from Ref. 22.)

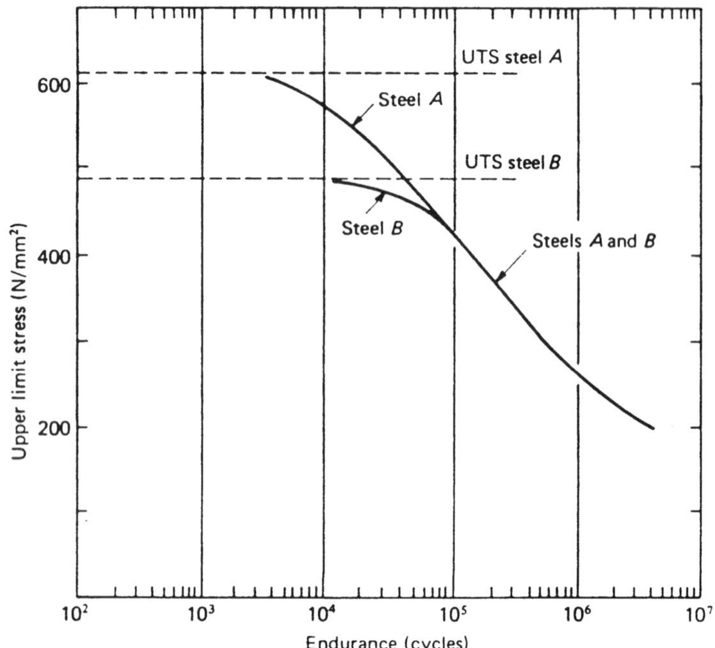

Figure 19 Effect of ultimate strength on the *S–N* curve for transverse butt welds in two steels. (Data from Ref. 23.)

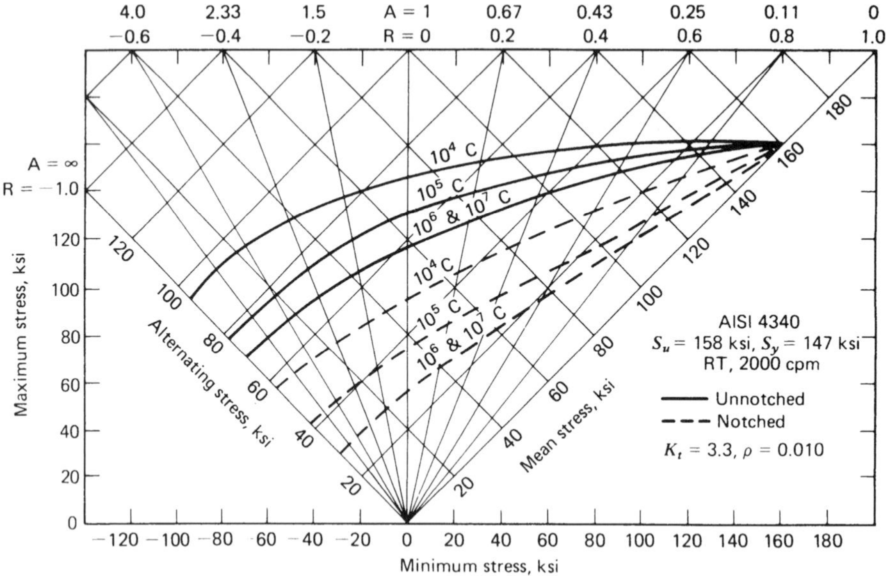

Figure 20 Master diagram for 4340 steel. (From Ref. 24, p. 37.)

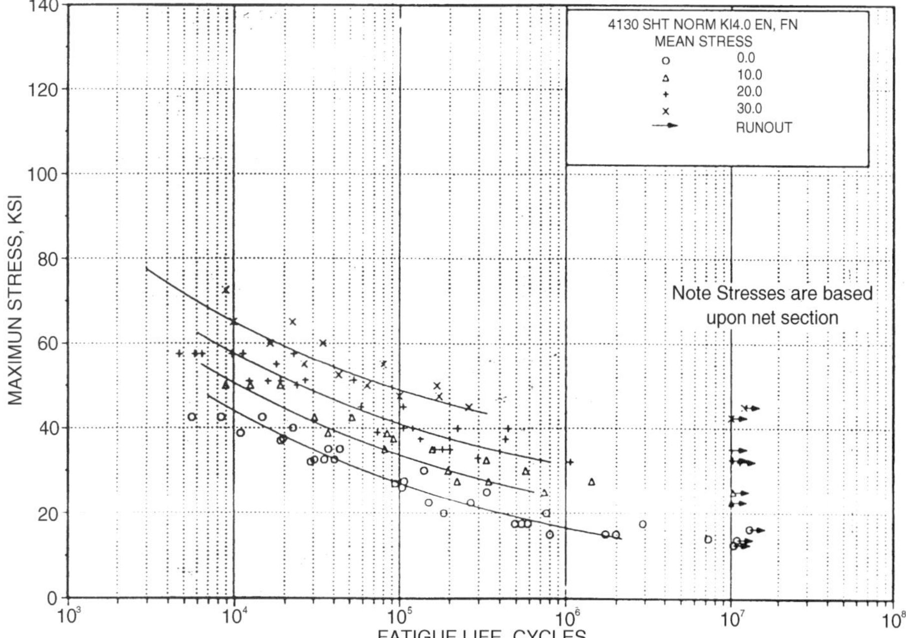

Figure 21 Best-fit *S–N* curves for notched 4130 alloy steel sheet, $K_t = 4.0$. (From Ref. 10.)

object of numerous empirical curve-fitting attempts. The more successful attempts have resulted in four different relationships:

1. Goodman's linear relationship
2. Gerber's parabolic relationship
3. Soderberg's linear relationship
4. The elliptic relationship

A modified form of the Goodman relationship is recommended for general use under conditions of high-cycle fatigue. For tensile mean stress ($\sigma_m > 0$), this relationship may be written

$$\frac{\sigma_a}{\sigma_N} + \frac{\sigma_m}{\sigma_u} = 1 \tag{13}$$

where σ_u is the material ultimate strength and σ_N is the zero mean stress fatigue strength for a given number of cycles N. For a given alternating stress, compressive mean stresses ($\sigma_m < 0$) have been empirically observed to increase fatigue resistance. However, for conservatism, it is typically assumed that compressive mean stress exerts no influence on fatigue life. Thus, for $\sigma_m < 0$, the fatigue response is identical to that for $\sigma_m = 0$ with $\sigma_a = \sigma_N$.

The modified Goodman relationship is illustrated in Fig. 22. This curve is a failure locus for the case of *uniaxial* fatigue stressing. Any cyclic loading that produces an alternating stress and mean stress that exceeds the bounds of the locus will cause failure in fewer than N cycles. Any alternating stress–mean stress combination that lies within the locus will result in more than N cycles without failure. Combinations on the locus produce failure in N cycles.

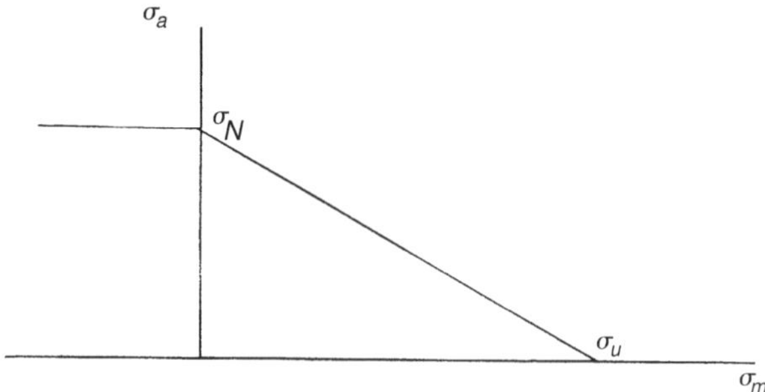

Figure 22 Modified Goodman relationship.

The modified Goodman relationship shown in Fig. 22 considers fatigue failure exclusively. The reader is cautioned to ensure that the maximum and minimum stresses produced by the cyclic loading do not exceed the material yield strength σ_{yp} such that failure by yielding would be predicted to occur.

5.5 Fatigue Crack Propagation

A fatigue crack that has been initiated by cyclic loading, or any other preexisting flaw in the structure or material, may be expected to grow under sustained cyclic loading until it reaches the critical size from which it will propagate rapidly to catastrophic failure in accordance with the principles of fracture mechanics. For many structures or machine elements, the time required for a fatigue-initiated crack or a preexisting flaw to grow to critical size is a significant portion of the total life.

The fatigue crack growth rate da/dN has been found to often correlate with the crack-tip stress intensity factor range such that

$$\frac{da}{dN} = g(\Delta K) \tag{14}$$

where ΔK is the mode I stress intensity factor range, computed using the maximum and minimum applied stresses with $\Delta K = K_{max} - K_{min}$. Most crack growth rate data produced have been characterized in terms of ΔK. For example, Fig. 23 illustrates indirectly the dependence of fatigue crack growth on the stress intensity factor. The crack growth rate, indicated by the slope of the a-versus-N curves, increases with both the applied load and crack length. Since the crack-tip stress intensity factor range also increases with applied load and crack length, it is clear that the crack growth rate is related to the applied stress intensity factor range.

To plot the data of Fig. 23 in terms of the stress intensity factor range and crack growth rate, the crack growth rate is estimated from a numerically determined slope of the a-versus-N curves between successive data points. Corresponding values of ΔK are then computed from the applied load range and mean crack length for each interval. The results of this procedure are shown in Fig. 24 for the data presented in Fig. 23. It should be noted that all the curves of Fig. 23 are incorporated into the single curve shown in Fig. 24 through use of the stress intensity factor, and the curve of Fig. 24 is therefore applicable to any combination of cyclic stress range and crack length for released loading ($R = 0$) on specimens of this geometry. Different geometries under different applied stresses will exhibit identical crack-

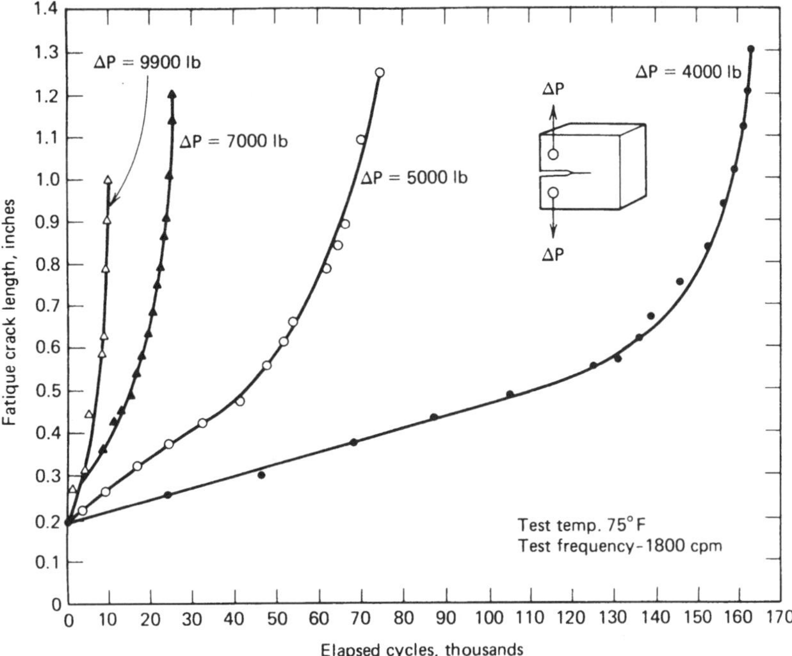

Figure 23 Effect of cyclic-load range on crack growth in Ni–Mo–V alloy steel for released tension loading. (From Ref. 37. Reprinted with permission of the Society for Experimental Mechanics, www.sem.org.)

tip stress fields if the stress intensity factors are equal. Thus, because the stress intensity factor characterizes the state of stress near the crack tip, the fatigue crack growth rate correlation shown in Fig. 24 is applicable to any cyclically loaded component with $R = 0$ manufactured using the same material. This allows crack growth data generated from simple laboratory specimens to be utilized for approximate crack growth predictions in more complex geometries.

Fatigue crack growth rate data similar to that shown in Fig. 24 have been reported for a wide variety of engineering metals. The linear behavior observed using log-log coordinates suggests that Eq. (14) may be generalized as follows:

$$\frac{da}{dN} = C(\Delta K)^n \tag{15}$$

where n is the slope of the plot of log da/dN versus log ΔK and C is the da/dN value found by extending the straight line to a ΔK value of unity. This relationship was first proposed by Paris and Erdogan.[27] The empirical parameters C and n are a function of material, R ratio, thickness, temperature, environment, and loading frequency. Standard methods have been established for conducting fatigue crack growth tests,[25] and fatigue crack growth rate data may be found in Refs. 9, 10, and 12–15. Many other fracture mechanics–based empirical correlations other than Eq. (15) have been proposed, some of which are discussed by Schijve.[26] An extensive overview of the fatigue crack propagation problem is provided by Pook.[32]

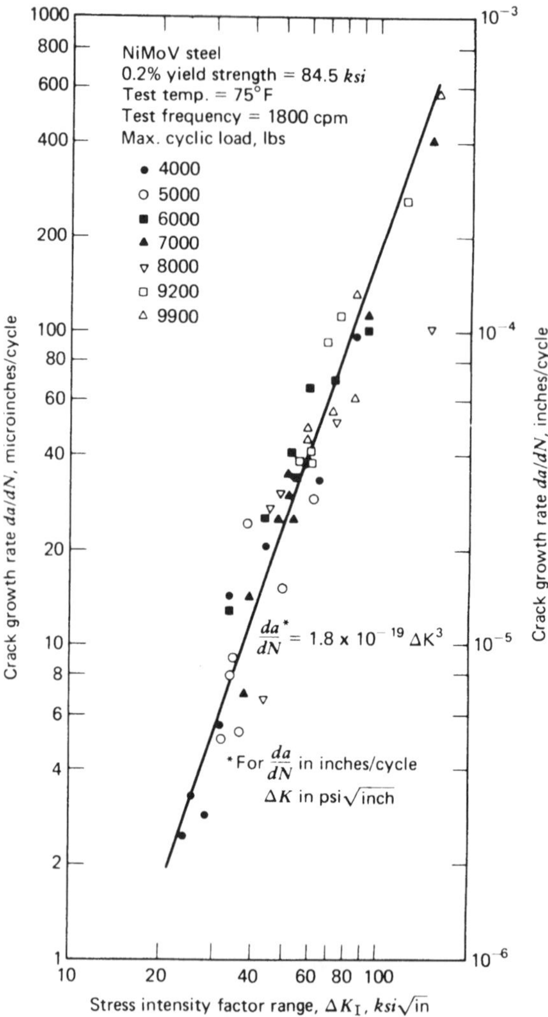

Figure 24 Crack growth rate as a function of stress–intensity range for Ni–Mo–V steel. (From Ref. 37. Reprinted with permission of the Society for Experimental Mechanics, www.sem.org.)

Given an initial crack of length a_i, Eq. (15) may be integrated to give the number of cycles N required to propagate a crack to a size a_N such that

$$N = \int_{a_i}^{a_N} \frac{da}{C(\Delta K)^n} \tag{16}$$

Given that ΔK is a function of crack length a, numerical integration techniques will in general be required to compute N. An approximate procedure and several idealized examples are presented by Parker.[35]

It must be emphasized that Eqs. (15) and (16) are applicable only to region II crack growth, as illustrated in Fig. 25. Region I of Fig. 25 exhibits a threshold ΔK_{th} below which

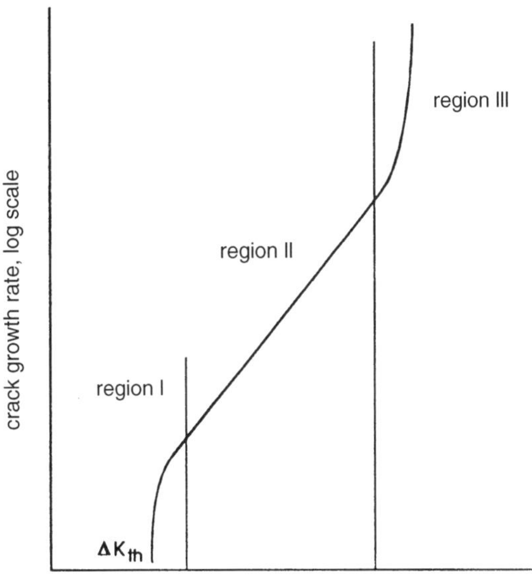

Figure 25 Schematic representation of fatigue crack growth rate data.

the crack will not propagate. Region III corresponds to the transition into the unstable regime of rapid crack extension. In this region, crack growth rates are large and the number of cycles associated with growth in this region is small.

With an initial crack of length a_i, from Eq. (16), the number of cycles required to grow a crack to a critical length a_c such that rapid crack extension would be predicted may be approximated as

$$N_p = \int_{a_i}^{a_c} \frac{da}{C(\Delta K)^n} \tag{17}$$

Assuming an initial crack that has been initiated by cyclic loading, the crack propagation life N_p given by Eq. (17) may then be added to the crack initiation life N_i to obtain an estimate of the total fatigue life N with

$$N = N_i + N_p \tag{18}$$

Such estimates are highly sensitive to the length of the initial crack a_i. While the local stress–strain approach[1,33] may be used to compute the number of cycles N_i required to initiate a crack, the corresponding length of this initiated crack is not defined specifically. No consensus has yet been reached regarding the length of this initiated crack. A size of between 0.25 and 5.0 mm has been suggested,[33] as cracks of this size normally exist at fracture in the small laboratory specimens used to generate the strain-versus-cycles-to-failure data required for the local stress–strain approach. An alternative approach would involve the assumption of a preexisting material or manufacturing defect such that $N_i \approx 0$. For example, such an assumption is often made during the analysis of welded joints.[28] If nondestructive techniques are employed, a reasonable assumption for the size of this initial defect would be the largest flaw that could avoid detection.

Crack growth rates determined from constant-amplitude cyclic loading tests are approximately the same as for random loading tests in which the maximum stress is held constant but mean and range of stress vary randomly. However, in random loading tests where the maximum stress is also allowed to vary, the sequence of loading cycles may have a marked effect on crack growth rate, with the overall crack growth being significantly higher for random loading spectra.

Many investigations have shown a significant delay in crack propagation following intermittent application of high stresses. That is, fatigue damage and crack extension are dependent on preceding cyclic load history. This dependence of crack extension on preceding history and the effects upon future damage increments are referred to as *interaction* effects. Most of the interaction studies conducted have dealt with *retardation* of crack growth as a result of the application of occasional tensile overload cycles. Retardation may be characterized as a period of reduced crack growth rate following the application of a peak load or loads higher and in the same direction as those peaks that follow.

The modeling of interaction effects requires consideration of crack-tip plasticity and its subsequent influence. In metals of all types, cracks will remain closed or partially closed for a portion of the applied cyclic load as a consequence of plastically deformed material left in the wake of the growing crack. Under cyclic loading, crack growth will occur during the loading portion of the cycle. Given that a plastic zone exists at the crack tip prior to crack extension, as the material at the crack tip separates, the newly formed crack surfaces will exhibit a layer of plastically deformed material along the newly formed crack faces. Subsequent unloading will compress this plastically deformed material, closing the crack while the applied stress remains tensile. This phenomenon is known as plasticity-induced fatigue crack closure and was first discussed by Elber.[30] Upon reloading during the following cycle, crack growth will not continue unless the applied load is sufficiently large such that the compressive stresses acting along the crack surfaces are overcome and the crack is fully opened. This load is known as the crack opening load and has been demonstrated to be a key parameter in determining fatigue crack growth rates under both constant amplitude and spectrum loading. Further information regarding crack closure may be found in Refs. 1, 4, 16, 26, and 31.

Discussion to this point has been limited to the growth of through-thickness cracks under mode I loading. While mode I loading is often dominant, under the most general circumstances the applied cyclic loads will generate stress intensity factor ranges ΔK_{I}, ΔK_{II}, and ΔK_{III} at the crack tip, and *mixed-mode* fatigue crack growth must be considered. Modeling methodologies for mixed-mode fatigue crack growth are discussed in Refs. 34 and 35. In addition, fatigue cracks in machine elements and structures are often not through-thickness cracks but rather surface cracks that extend partially through the thickness. Such surface cracks are often semielliptical in shape and the analysis of these cracks is considerably more complicated. Information regarding surface cracks may be found in Refs. 4 and 29.

Research has suggested that when fatigue cracks are small, crack growth rates are larger than would be predicted using Eq. (15) for a given ΔK.[36] Small-crack behavior is often important, as a significant portion of the fatigue life may be spent in the small-crack regime. The fatigue crack propagation life N_p will also be influenced by the presence of residual stresses such as might exist as a consequence of welding, heat treatment, carburizing, grinding, or shot peening. Compressive residual stresses are beneficial, decreasing the rate of fatigue crack growth and increasing propagation life. While approximate methodologies exist for incorporating the effects of residual stress within fatigue crack growth predictions,[35] residual stress distributions are often difficult to characterize.

Reasonable design estimates for the fatigue crack propagation life may be obtained using Eq. (17). However, the many uncertainties typically associated with fatigue life predictions emphasize the essential requirement to conduct full-scale fatigue tests to provide acceptable reliability.

6 CREEP AND STRESS RUPTURE

Creep in its simplest form is the progressive accumulation of plastic strain in a specimen or machine part under stress at elevated temperature over a period of time. Creep failure occurs when the accumulated creep strain results in a deformation of the machine part that exceeds the design limits. *Creep rupture* is an extension of the creep process to the limiting condition where the stressed member actually separates into two parts. *Stress rupture* is a term used interchangeably by many with creep rupture; however, others reserve the term stress rupture for the rupture termination of a creep process in which steady-state creep is never reached and use the term creep rupture for the rupture termination of a creep process in which a period of steady-state creep has persisted. Figure 26 illustrates these differences. The interaction of creep and stress rupture with cyclic stressing and the fatigue process has not yet been clearly understood but is of great importance in many modern high-performance engineering systems.

Creep strains of engineering significance are not usually encountered until the operating temperatures reach a range of approximately 35–70% of the melting point on a scale of absolute temperature. The approximate melting temperature for several substances is shown in Table 2.

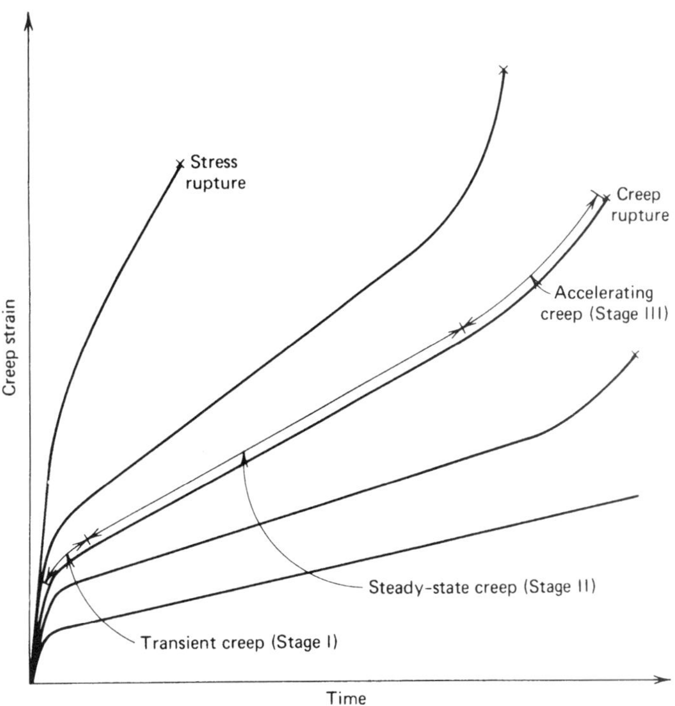

Figure 26 Illustration of creep and stress rupture.

Table 2 Melting Temperatures[38]

Material	°F	°C
Hafnium carbide	7030	3887
Graphite (sublimes)	6330	3500
Tungsten	6100	3370
Tungsten carbide	5190	2867
Magnesia	5070	2800
Molybdenum	4740	2620
Boron	4170	2300
Titanium	3260	1795
Platinum	3180	1750
Silica	3140	1728
Chromium	3000	1650
Iron	2800	1540
Stainless steels	2640	1450
Steel	2550	1400
Aluminum alloys	1220	660
Magnesium alloys	1200	650
Lead alloys	605	320

Not only is excessive deformation due to creep an important consideration, but other consequences of the creep process may also be important. These might include creep rupture, thermal relaxation, dynamic creep under cyclic loads or cyclic temperatures, creep and rupture under multiaxial states of stress, cumulative creep effects, and effects of combined creep and fatigue.

Creep deformation and rupture are initiated in the grain boundaries and proceed by sliding and separation. Thus, creep rupture failures are intercrystalline, in contrast, for example, to the transcrystalline failure surface exhibited by room temperature fatigue failures. Although creep is a plastic flow phenomenon, the intercrystalline failure path gives a rupture surface that has the appearance of brittle fracture. Creep rupture typically occurs without necking and without warning. Current state-of-the-art knowledge does not permit a reliable prediction of creep or stress rupture properties on a theoretical basis. Furthermore, there seems to be little or no correlation between the creep properties of a material and its room temperature mechanical properties. Therefore, test data and empirical methods of extending these data are relied on heavily for prediction of creep behavior under anticipated service conditions.

Metallurgical stability under long-time exposure to elevated temperatures is mandatory for good creep-resistant alloys. Prolonged time at elevated temperatures acts as a tempering process, and any improvement in properties originally gained by quenching may be lost. Resistance to oxidation and other corrosive media are also usually important attributes for a good creep-resistant alloy. Larger grain size may also be advantageous since this reduces the length of grain boundary, where much of the creep process resides.

6.1 Prediction of Long-Term Creep Behavior

Much time and effort has been expended in attempting to develop good short-time creep tests for accurate and reliable prediction of long-term creep and stress rupture behavior. It appears, however, that really reliable creep data can be obtained only by conducting long-term creep tests that duplicate actual service loading and temperature conditions as nearly as possible. Unfortunately, designers are unable to wait for years to obtain design data needed in creep failure analysis. Therefore, certain useful techniques have been developed for approximating long-term creep behavior based on a series of short-term tests. Data from creep testing may be plotted in a variety of different ways. The basic variables involved are stress, strain, time, temperature, and, perhaps, strain rate. Any two of these basic variables may be selected as plotting coordinates, with the remaining variables treated as parametric constants for a given curve. Three commonly used methods for extrapolating short-time creep data to long-term applications are the abridged method, the mechanical acceleration method, and the thermal acceleration method. In the abridged method of creep testing the tests are conducted at several different stress levels and at the contemplated operating temperature. The data are plotted as creep strain versus time for a family of stress levels, all run at constant temperature. The curves are plotted out to the laboratory test duration and then extrapolated to the required design life. In the mechanical acceleration method of creep testing, the stress levels used in the laboratory tests are significantly higher than the contemplated design stress levels, so the limiting design strains are reached in a much shorter time than in actual service. The data taken in the mechanical acceleration method are plotted as stress level versus time for a family of constant-strain curves all run at a constant temperature. The thermal acceleration method involves laboratory testing at temperatures much higher than the actual service temperature expected. The data are plotted as stress versus time for a family of constant temperatures where the creep strain produced is constant for the whole plot.

It is important to recognize that such extrapolations are not able to predict the potential of failure by creep rupture prior to reaching the creep design life. In any testing method it should be noted that creep-testing guidelines usually dictate that test periods of less than 1% of the expected life are not deemed to give significant results. Tests extending to at least 10% of the expected life are preferred where feasible.

Several different theories have been proposed to correlate the results of short-time elevated-temperature tests with long-term service performance at more moderate temperatures. One of the more accurate and useful of these proposals is the Larson–Miller theory.

The Larson–Miller theory[39] postulates that for each combination of material and stress level there exists a unique value of a parameter P that is related to temperature and time by the equation

$$P = (\theta + 460)(C + \log_{10} t) \tag{19}$$

where P = Larson–Miller parameter, constant for a given material and stress level
θ = temperature, °F
C = constant, usually assumed to be 20
t = time in hours to rupture or to reach a specified value of creep strain

This equation was investigated for both creep and rupture for some 28 different materials by Larson and Miller with good success. By using (19) it is a simple matter to find a short-term combination of temperature and time that is equivalent to any desired long-term service requirement. For example, for any given material at a specified stress level the test conditions listed in Table 3 should be equivalent to the operating conditions.

6.2 Creep under a Uniaxial State of Stress

Many relationships have been proposed to relate stress, strain, time, and temperature in the creep process. If one investigates experimental creep-strain-versus-time data, it will be observed that the data are close to linear for a wide variety of materials when plotted on log

Table 3 Equivalent Conditions Based on Larson–Miller Parameter

Operating Condition	Equivalent Test Condition
10,000 h at 1000°F	13 h at 1200°F
1,000 h at 1200°F	12 h at 1350°F
1,000 h at 1350°F	12 h at 1500°F
1,000 h at 300°F	2.2 h at 400°F

strain–log time coordinates. Such a plot is shown, for example, in Fig. 27 for three different materials. An equation describing this type of behavior is

$$\delta = At^a \tag{20}$$

where δ = true creep strain
t = time
A, a = empirical constants

Differentiating (20) with respect to time gives

$$\dot{\delta} = bt^{-n} \tag{21}$$

or, setting $aA = b$ and $1 - a = n$,

$$\dot{\delta} = bt^{-n} \tag{22}$$

This equation represents a variety of different types of creep-strain-versus-time curves, depending on the magnitude of the exponent n. If n is zero, the behavior, characteristic of high temperatures, is termed *constant creep rate*, and the creep strain is given as

$$\delta = b_1 t + C_1 \tag{23}$$

If n lies between 0 and 1, the behavior is termed *parabolic creep,* and the creep strain is given by

$$\delta = b_3 t^m + C_3 \tag{24}$$

This type of creep behavior occurs at intermediate and high temperatures. The coefficient b_3 increases exponentially with stress and temperature, and the exponent m decreases with stress and increases with temperature. The influence of stress level σ on creep rate can often be represented by the empirical expression

$$\dot{\delta} = b\sigma^N \tag{25}$$

Assuming the stress σ to be independent of time, we may integrate (25) to yield the creep strain

$$\delta = Bt\sigma^N + C' \tag{26}$$

If the constant C' is small compared with $Bt\sigma^N$, as it often is, the result is called the *log-log stress–time creep law,* given as

$$\sigma = Bt\sigma^N \tag{27}$$

As long as the instantaneous deformation on load application and the stage I transient creep are small compared to stage II steady-state creep, Eq. (27) is useful as a design tool.

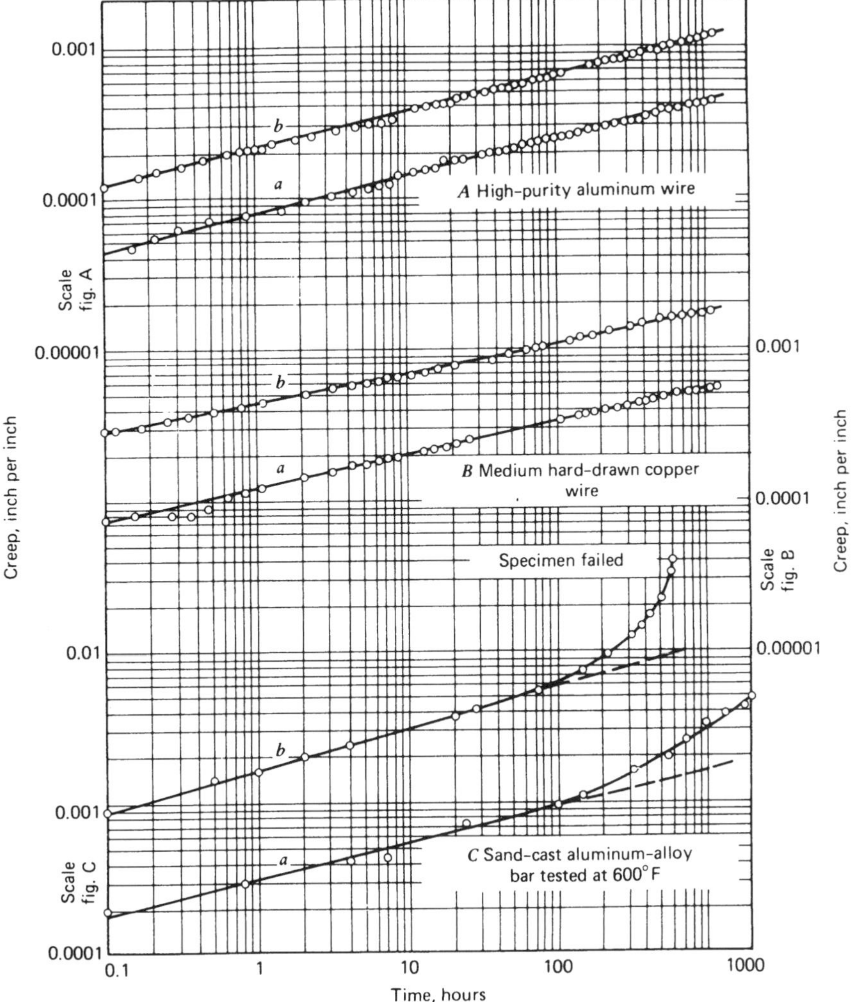

Figure 27 Creep curves for three materials. (From Ref. 40.)

If it is necessary to consider all stages of the creep process, the creep strain expression becomes much more complex. The most general expression for the creep process is[41]

$$\delta = \frac{\sigma}{E} + k_1\sigma^m + k_2(1 - e^{-qt})\sigma^n + k_3t\sigma^p \tag{28}$$

where
δ = total creep strain
σ/E = initial elastic strain
$k_1\sigma^m$ = initial plastic strain
$k_2(1 - e^{-qt})\sigma^n$ = anelastic strain
$k_3t\sigma^p$ = viscous strain
σ = stress

E = modulus of elasticity
m = reciprocal of strain-hardening exponent
k_1 = reciprocal of strength coefficient
q = reciprocal of Kelvin retardation time
k_2 = anelastic coefficient
n = empirical exponent
k_3 = viscous coefficient
p = empirical exponent
t = time

To utilize this empirical nonlinear expression in a design environment requires specific knowledge of the constants and exponents that characterize the material and temperature of the application. In all cases it must be recognized that stress rupture may intervene to terminate the creep process, and the prediction of this occurrence is difficult.

7 FRETTING AND WEAR

Fretting and wear share many common characteristics but, at the same time, are distinctly different in several ways. Basically, fretting action has, for many years, been defined as a combined mechanical and chemical action in which contacting surfaces of two solid bodies are pressed together by a normal force and are caused to execute oscillatory sliding relative motion, wherein the magnitude of normal force is great enough and the amplitude of the oscillatory sliding motion is small enough to significantly restrict the flow of fretting debris away from the originating site.[42] More recent definitions of fretting action have been broadened to include cases in which contacting surfaces periodically separate and then reengage as well as cases in which the fluctuating friction-induced surface tractions produce stress fields that may ultimately result in failure. The complexities of fretting action have been discussed by numerous investigators, who have postulated the combination of many mechanical, chemical, thermal, and other phenomena that interact to produce fretting. Among the postulated phenomena are plastic deformation caused by surface asperities plowing through each other, welding and tearing of contacting asperities, shear and rupture of asperities, friction-generated subsurface shearing stresses, dislodging of particles and corrosion products at the surfaces, chemical reactions, debris accumulation and entrapment, abrasive action, microcrack initiation, and surface delamination.[43–58]

Damage to machine parts due to fretting action may be manifested as corrosive surface damage due to fretting corrosion, loss of proper fit or change in dimensions due to fretting wear, or accelerated fatigue failure due to fretting fatigue. Typical sites of fretting damage include interference fits; bolted, keyed, splined, and riveted joints; points of contact between wires in wire ropes and flexible shafts; friction clamps; small-amplitude-of-oscillation bearings of all kinds; contacting surfaces between the leaves of leaf springs; and all other places where the conditions of fretting persist. Thus, the efficiency and reliability of the design and operation of a wide range of mechanical systems are related to the fretting phenomenon.

Wear may be defined as the undesired cumulative change in dimensions caused by the gradual removal of discrete particles from contacting surfaces in motion, due predominantly to mechanical action. It should be further recognized that corrosion often interacts with the wear process to change the character of the surfaces of wear particles through reaction with the environment. Wear is, in fact, not a single process but a number of different processes that may take place by themselves or in combination. It is generally accepted that there are at least five major subcategories of wear (see p. 120 of Ref. 59; see also Ref. 60), including adhesive wear, abrasive wear, corrosive wear, surface fatigue wear, and deformation wear. In addition, the categories of fretting wear and impact wear[55,61,62] have been recognized by wear specialists. Erosion and cavitation are sometimes considered to be categories of wear as well. Each of these types of wear proceeds by a distinctly different physical process and must be separately considered, although the various subcategories may combine their influence either by shifting from one mode to another during different eras in the operational lifetime of a machine or by simultaneous activity of two or more different wear modes.

7.1 Fretting Phenomena

Although fretting fatigue, fretting wear, and fretting corrosion phenomena are potential failure modes in a wide variety of mechanical systems, and much research effort has been devoted to the understanding of the fretting process, there are very few quantitative design data available, and no generally applicable design procedure has been established for predicting failure under fretting conditions. However, even though the fretting phenomenon is not fully understood and a good general model for prediction of fretting fatigue or fretting wear has not yet been developed, significant progress has been made in establishing an understanding of fretting and the variables of importance in the fretting process. It has been suggested that there may be more than 50 variables that play some role in the fretting process.[63] Of these, however, there are probably only 8 that are of major importance:

1. The magnitude of relative motion between the fretting surfaces
2. The magnitude and distribution of pressure between the surfaces at the fretting interface
3. The state of stress, including magnitude, direction, and variation with respect to time in the region of the fretting surfaces
4. The number of fretting cycles accumulated
5. The material, and surface condition, from which each of the fretting members is fabricated
6. Cyclic frequency of relative motion between the two members being fretted
7. Temperature in the region of the two surfaces being fretted
8. Atmospheric environment surrounding the surfaces being fretted

These variables interact so that a quantitative prediction of the influence of any given variable is very dependent on all the other variables in any specific application or test. Also, the combination of variables that produce a very serious consequence in terms of fretting fatigue damage may be quite different from the combinations of variables that produce serious fretting wear damage. No general techniques yet exist for quantitatively predicting the influence of the important variables of fretting fatigue and fretting wear damage, although many special cases have been investigated. However, it has been observed that certain trends usually exist when the variables just listed are changed. For example, fretting damage tends to increase with increasing contact pressure until a nominal pressure of a few thousand pounds per square inch is reached, and further increases in pressure seem to have relatively little direct effect. The state of stress is important, especially in fretting fatigue. Fretting damage accumulates with increasing numbers of cycles at widely different rates, depending on specific operating conditions. Fretting damage is strongly influenced by the material properties of the fretting pair-surface hardness, roughness, and finish. No clear trends have been established regarding frequency effects on fretting damage, and although both temperature and atmospheric environment are important influencing factors, their influences have not been clearly established. A clear presentation relative to these various parameters is given in Ref. 55.

Fretting fatigue is fatigue damage directly attributable to fretting action. It has been suggested that premature fatigue nuclei may be generated by fretting through either abrasive pit-digging action, asperity-contact microcrack initiation,[64] friction-generated cyclic stresses that lead to the formation of microcracks,[65] or subsurface cyclic shear stresses that lead to surface delamination in the fretting zone.[58] Under the abrasive pit-digging hypothesis, it is conjectured that tiny grooves or elongated pits are produced at the fretting interface by the asperities and abrasive debris particles moving under the influence of oscillatory relative motion. A pattern of tiny grooves would be produced in the fretted region with their longitudinal axes all approximately parallel and in the direction of fretting motion, as shown schematically in Fig. 28.

The asperity-contact microcrack initiation mechanism is postulated to proceed due to the contact force between the tip of an asperity on one surface and another asperity on the mating surface as the surfaces move back and forth. If the initial contact does not shear one or the other asperity from its base, the repeated contacts at the tips of the asperities give rise to cyclic or fatigue stresses in the region at the base of each asperity. It has been estimated[51] that under such conditions the region at the base of each asperity is subjected to large local stresses that probably lead to the nucleation of fatigue microcracks at these sites. As shown

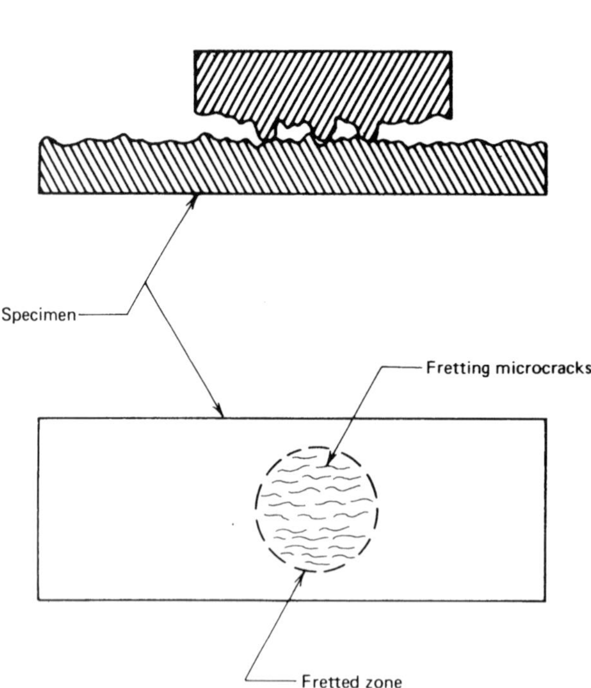

Figure 28 Idealized schematic illustration of the stress concentrations produced by the abrasive pit-digging mechanism.

schematically in Fig. 29, it would be expected that the asperity-contact mechanism would produce an array of microcracks whose longitudinal axes would be generally perpendicular to the direction of fretting motion.

The friction-generated cyclic stress fretting hypothesis[53] is based on the observation that when one member is pressed against the other and caused to undergo fretting motion, the tractive friction force induces a compressive tangential stress component in a volume of material that lies ahead of the fretting motion and a tensile tangential stress component in a volume of material that lies behind the fretting motion, as shown in Fig. 30*a*. When the fretting direction is reversed, the tensile and compressive regions change places. Thus, the volume of material adjacent to the contact zone is subjected to a cyclic stress that is postulated to generate a field of microcracks at these sites. Furthermore, the geometrical stress concentration associated with the clamped joint may contribute to microcrack generation at these sites.[54] As shown in Fig. 30*c*, it would be expected that the friction-generated microcrack mechanism would produce an array of microcracks whose longitudinal axes would be generally perpendicular to the direction of fretting motion. These cracks would lie in a region adjacent to the fretting contact zone.

In the delamination theory of fretting[58] it is hypothesized that the combination of normal and tangential tractive forces transmitted through the asperity-contact sites at the fretting interface produces a complex multiaxial state of stress, accompanied by a cycling deformation field, which produces subsurface peak shearing stress and subsurface crack nucleation sites. With further cycling, the cracks propagate approximately parallel to the surface, as in

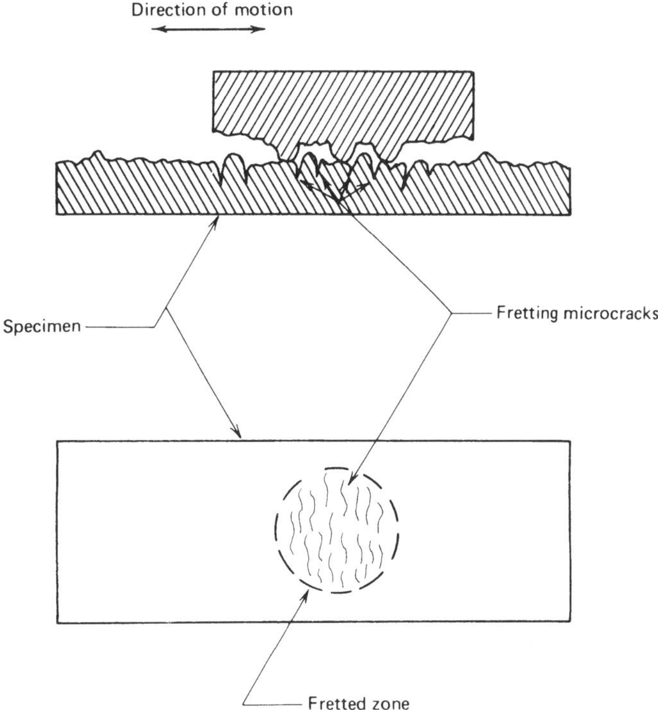

Figure 29 Idealized schematic illustration of the stress concentrations produced by the asperity-contact microcrack initiation mechanism.

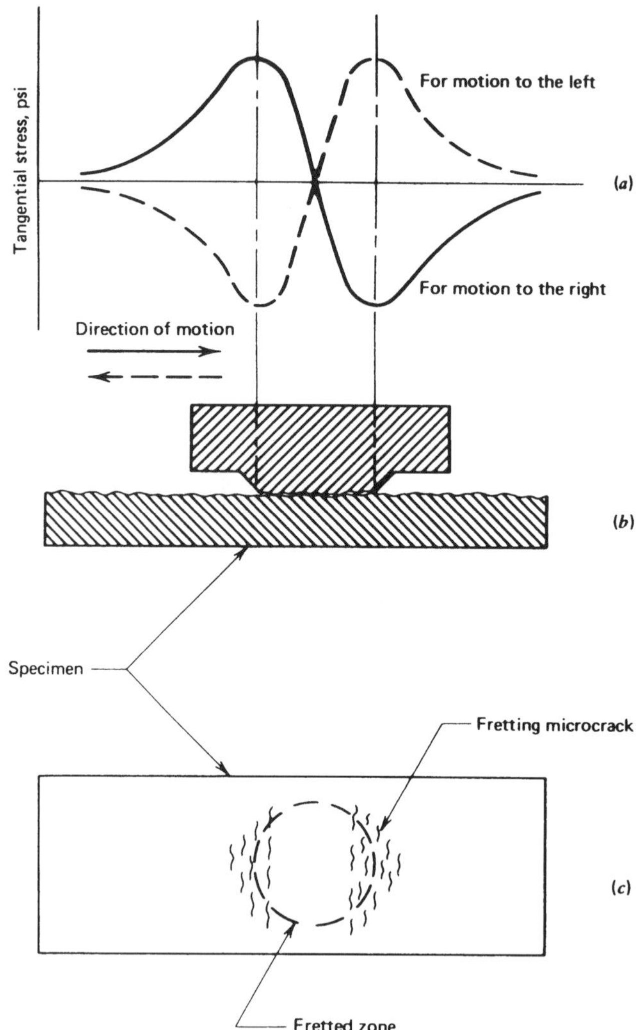

Figure 30 Idealized schematic illustration of the tangential stress components and microcracks produced by the friction-generated microcrack initiation mechanism.

the case of the surface fatigue phenomenon, finally propagating to the surface to produce a thin wear sheet, which "delaminates" to become a particle of debris.

Supporting evidence has been generated to indicate that under various circumstances each of the four mechanisms is active and significant in producing fretting damage.

The influence of the state of stress in the member during the fretting is shown for several different cases in Fig. 31, including static tensile and compressive mean stresses during fretting. An interesting observation in Fig. 31 is that fretting under conditions of compressive mean stress, either static or cyclic, produces a drastic reduction in fatigue properties. This, at first, does not seem to be in keeping with the concept that compressive stresses are beneficial in fatigue loading. However, it was deduced[66] that the compressive stresses during

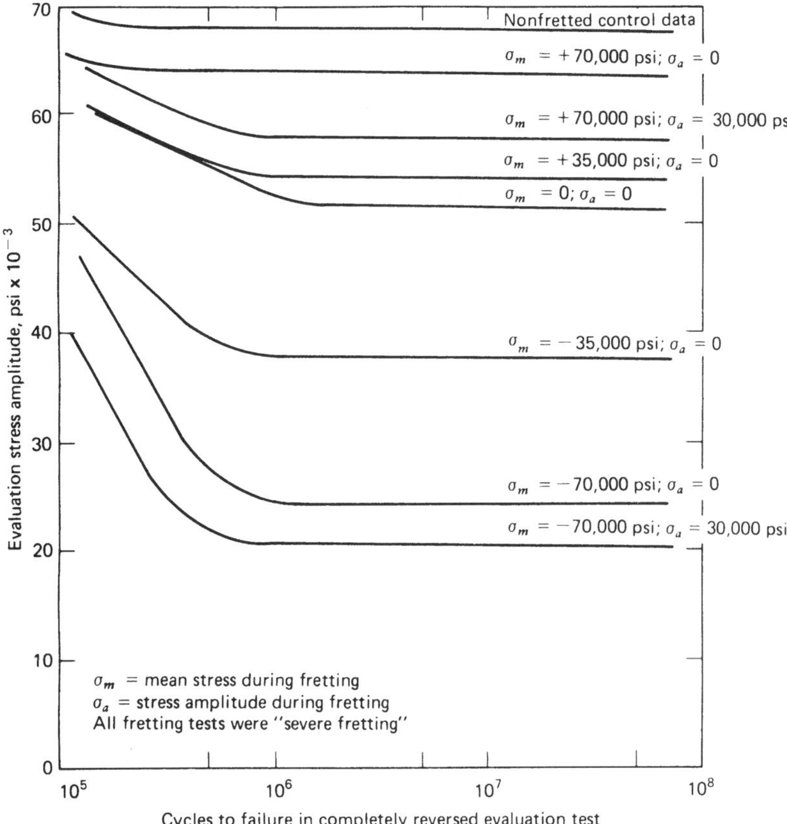

Figure 31 Residual fatigue properties subsequent to fretting under various states of stress.

fretting shown in Fig. 31 actually resulted in local residual tensile stresses in the fretted region. Likewise, the tensile stresses during fretting shown in Fig. 31 actually resulted in local residual compressive stresses in the fretted region. The conclusion, therefore, is that local compressive stresses are beneficial in minimizing fretting fatigue damage.

Further evidence of the beneficial effects of compressive residual stresses in minimizing fretting fatigue damage is illustrated in Fig. 32, where the results of a series of Prot (fatigue limit) tests are reported for steel and titanium specimens subjected to various combinations of shot peening and fretting or cold rolling and fretting. It is clear from these results that the residual compressive stresses produced by shot peening and cold rolling are effective in minimizing the fretting damage. The reduction in scatter of the fretted fatigue properties for titanium is especially important to a designer because design stress is closely related to the lower limit of the scatter band.

In the final analysis, it is necessary to evaluate the seriousness of fretting fatigue damage in any specific design by running simulated service tests on specimens or components. Within the current state-of-the-art knowledge in the area of fretting fatigue, there is no other safe course of action open to the designer.

Fretting wear is a change in dimensions through wear directly attributable to the fretting process between two mating surfaces. It is thought that the abrasive pit-digging mechanism,

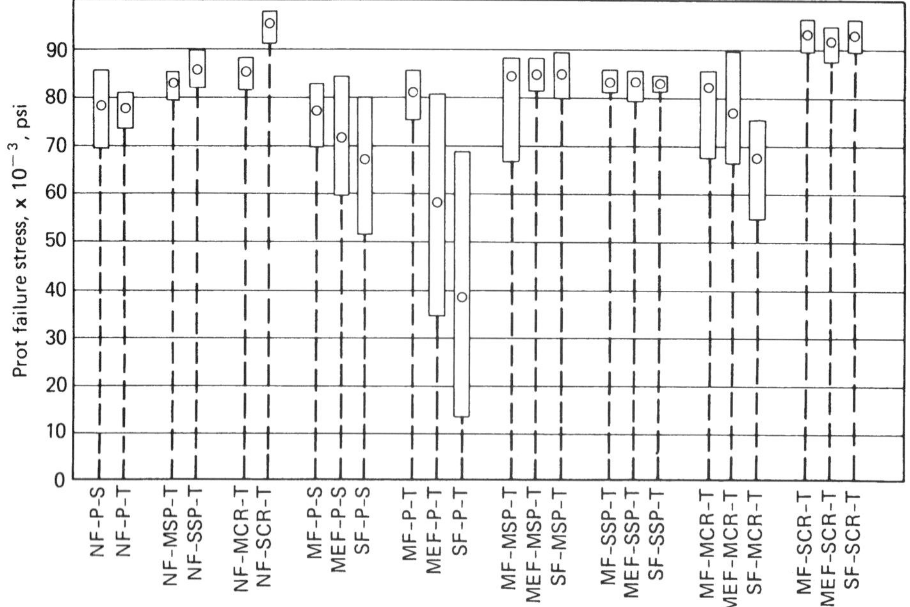

Test Condition Used	Code Designation	Sample Size	Mean Prot Failure Stress, psi	Unbiased Standard Deviation, psi
Nonfretted, polished, SAE 4340 steel	NF-P-S	15	78,200	5,456
Nonfretted, polished, Ti-140-A titanium	NF-P-T	15	77,800	2,454
Nonfretted, mildly shot-peened, Ti-140-A titanium	NF-MSP-T	15	83,100	1,637
Nonfretted, severely shot-peened, Ti-140-A titanium	NF-SSP-T	15	85,700	2,398
Nonfretted, mildly cold-rolled, Ti-140-A titanium	NF-MCR-T	15	85,430	1,924
Nonfretted, severely cold-rolled, Ti-140-A titanium	NF-SCR-T	15	95,400	2,120
Mildly fretted, polished, SAE 4340 steel	MF-P-S	15	77,280	4,155
Medium fretted, polished, SAE 4340 steel	MeF-P-S	15	71,850	5,492
Severely fretted, polished, SAE 4340 steel	SF-P-S	15	67,700	6,532
Mildly fretted, polished, Ti-140-A titanium	MF-P-T	15	81,050	3,733
Medium fretted, polished, Ti-140-A titanium	MeF-P-T	15	58,140	15,715
Severely fretted, polished, Ti-140-A titanium	SF-P-T	15	38,660	19,342
Mildly fretted, mildly shot-peened, Ti-140-A titanium	MF-MSP-T	15	84,520	5,239
Medium fretted, mildly shot-peened, Ti-140-A titanium	MeF-MSP-T	15	84,930	2,446
Severely fretted, mildly shot-peened, Ti-140-A titanium	SF-MSP-T	15	84,870	2,647
Mildly fretted, severely shot-peened, Ti-140-A titanium	MF-SSP-T	15	83,600	1,474
Medium fretted, severely shot-peened, Ti-140-A titanium	MeF-SSP-T	15	83,240	1,332
Severely fretted, severely shot-peened, Ti-140-A titanium	SF-SSP-T	15	83,110	1,280
Mildly fretted, mildly cold-rolled, Ti-140-A titanium	MF-MCR-T	15	82,050	4,313
Medium fretted, mildly cold-rolled, Ti-140-A titanium	MeF-MCR-T	15	76,930	8,305
Severely fretted, mildly cold-rolled, Ti-140-A titanium	SF-MCR-T	15	67,960	5,682
Mildly fretted, severely cold-rolled, Ti-140-A titanium	MF-SCR-T	15	93,690	1,858
Medium fretted, severely cold-rolled, Ti-140-A titanium	MeF-SCR-T	15	91,950	2,098
Severely fretted, severely cold-rolled, Ti-140-A titanium	SF-SCR-T	15	93,150	1,365

Figure 32 Fatigue properties of fretted steel and titanium specimens with various degrees of shot peening and cold rolling. (See Ref. 52.)

the asperity-contact microcrack initiation mechanism, and the wear-sheet delamination mechanism may all be important in most fretting wear failures. As in the case of fretting fatigue, there has been no good model developed to describe the fretting wear phenomenon in a way useful for design. An expression for weight loss due to fretting has been proposed[48] as

$$W_{\text{total}}\ (k_0 L^{1/2} - k_1 L)\frac{C}{F} + k_2 SLC \tag{29}$$

where W_{total} = total specimen weight loss
L = normal contact load
C = number of fretting cycles
F = frequency of fretting
S = peak-to-peak slip between fretting surfaces
k_0, k_1, k_2 = constants to be empirically determined

This equation has been shown to give relatively good agreement with experimental data over a range of fretting conditions using mild steel specimens. However, weight loss is not of direct use to a designer. Wear depth is of more interest. Prediction of wear depth in an actual design application must in general be based on simulated service testing.

Some investigators have suggested that estimates of fretting wear depth may be based on the classical adhesive or abrasive wear equations, in which wear depth is proportional to load and total distance slid, where the total distance slid is calculated by multiplying relative motion per cycle times number of cycles. Although there are some supporting data for such a procedure,[67] more investigation is required before it could be recommended as an acceptable approach for general application.

If fretting wear at a support interface, such as between tubes and support plates of a steam generator or heat exchanger or between fuel pins and support grids of a reactor core, produces loss of fit at a support site, impact fretting may occur. Impact fretting is fretting action induced by the small lateral relative displacements between two surfaces when they impact together, where the small displacements are caused by Poisson strains or small tangential "glancing" velocity components. Impact fretting has only recently been addressed in the literature,[68] but it should be noted that under certain circumstances impact fretting may be a potential failure mode of great importance.

Fretting corrosion may be defined as any corrosive surface involvement resulting as a direct result of fretting action. The consequences of fretting corrosion are generally much less severe than for either fretting wear or fretting fatigue. Note that the term *fretting corrosion* is not being used here as a synonym for fretting, as in much of the early literature on this topic. Perhaps the most important single parameter in minimizing fretting corrosion is proper selection of the material pair for the application. Table 4 lists a variety of material pairs grouped according to their resistance to fretting corrosion.[69] Cross comparisons from one investigator's results to another's must be made with care because testing conditions varied widely. The minimization or prevention of fretting damage must be carefully considered as a separate problem in each individual design application because a palliative in one application may significantly accelerate fretting damage in a different application. For example, in a joint that is designed to have no relative motion, it is sometimes possible to reduce or prevent fretting by increasing the normal pressure until all relative motion is arrested. However, if the increase in normal pressure does not completely arrest the relative motion, the result may be significantly increasing fretting damage instead of preventing it.

Nevertheless, there are several basic principles that are generally effective in minimizing or preventing fretting. These include:

Table 4 Fretting Corrosion Resistance of Various Material Pairs[69]

	Material Pairs Having Good Fretting Corrosion Resistance		
Sakmann and Rightmire	Lead	on	Steel
	Silver plate	on	Steel
	Silver plate	on	Silver plate
	"Parco-lubrized" steel	on	Steel
Gray and Jenny	Grit blasted steel plus lead plate	on	Steel (very good)
	$\frac{1}{16}$-in. nylon insert	on	Steel (very good)
	Zinc and iron phosphated (bonderizing) steel	on	Steel (good with thick coat)
McDowell	Laminated plastic	on	Gold plate
	Hardtool steel	on	Tool steel
	Cold-rolled steel	on	Cold-rolled steel
	Cast iron	on	Cast iron with phosphate coating
	Cast iron	on	Cast iron with rubber cement
	Cast iron	on	Cast iron with tungsten sulfide coating
	Cast iron	on	Cast iron with rubber insert
	Cast iron	on	Cast iron with Molykote lubricant
	Cast iron	on	Stainless steel with Molykote lubricant
	Material Pairs Having Intermediate Fretting Corrosion Resistance		
Sakmann and Rightmire	Cadmium	on	Steel
	Zinc	on	Steel
	Copper alloy	on	Steel
	Zinc	on	Aluminum
	Copper plate	on	Aluminum
	Nickel plate	on	Aluminum
	Silver plate	on	Aluminum
	Iron plate	on	Aluminum
Gray and Jenny	Sulfide-coated bronze	on	Steel
	Cast bronze	on	"Parco-lubrized" steel
	Magnesium	on	"Parco-lubrized" steel
	Grit-blasted steel	on	Steel
McDowell	Cast iron	on	Cast iron (rough or smooth surface)
	Copper	on	Cast iron
	Brass	on	Cast iron
	Zinc	on	Cast iron
	Cast iron	on	Silver plate
	Cast iron	on	Copper plate
	Magnesium	on	Copper plate
	Zirconium	on	Zirconium
Sakmann and Rightmire	Steel	on	Steel
	Nickel	on	Steel
	Aluminum	no	Steel
	Al–Si alloy	on	Steel
	Antimony plate	on	Steel
	Tin	on	Steel
	Aluminum	on	Aluminum
	Zinc plate	on	Aluminum

Table 4 (*Continued*)

Gray and Jenny	Grit blast plus silver plate	on	Steel[a]
	Steel	on	Steel
	Grit blast plus copper plate	on	Steel
	Grit blast plus in plate	on	Steel
	Grit blast and aluminum foil	on	Steel
	Be–Cu insert	on	Steel
	Magnesium	on	Steel
	Nitrided steel	on	Chromium-plated steel[b]
	Material Pairs Having Poor Fretting Corrosion Resistance		
McDowell	Aluminum	on	Cast iron
	Aluminum	on	Stainless steel
	Magnesium	on	Cast iron
	Cast iron	on	Chromium plate
	Laminated plastic	on	Cast iron
	Bakelite	on	Cast iron
	Hardtool steel	on	Stainless steel
	Chromium plate	on	Chromium plate
	Cast iron	on	Tin plate
	Gold plate	on	Gold plate

[a] Possibly effective with light loads and thick (0.005-in.) silver plate.
[b] Some improvement by heating chromium-plated steel to 538°C for 1 h.

1. Complete separation of the contacting surfaces.
2. Elimination of all relative motion between the contacting surfaces.
3. If relative motion cannot be eliminated, it is sometimes effective to superpose a large unidirectional relative motion that allows effective lubrication. For example, the practice of driving the inner or outer race of an oscillatory pivot bearing may be effective in eliminating fretting.
4. Providing compressive residual stresses at the fretting surface; this may be accomplished by shot peening, cold rolling, or interference fit techniques.
5. Judicious selection of material pairs.
6. Use of interposed low-shear-modulus shim material or plating, such as lead, rubber, or silver.
7. Use of surface treatments or coatings as solid lubricants.
8. Use of surface grooving or roughening to provide debris escape routes and differential strain matching through elastic action.

Of all these techniques, only the first two are completely effective in preventing fretting. The remaining concepts, however, may often be used to minimize fretting damage and yield an acceptable design.

7.2 Wear Phenomena

The complexity of the wear process may be better appreciated by recognizing that many variables are involved, including the hardness, toughness, ductility, modulus of elasticity, yield strength, fatigue properties, and structure and composition of the mating surfaces, as well as geometry, contact pressure, temperature, state of stress, stress distribution, coefficient of friction, sliding distance, relative velocity, surface finish, lubricants, contaminants, and ambient atmosphere at the wearing interface. Clearance versus contact-time history of the wearing surfaces may also be an important factor in some cases. Although the wear processes are complex, progress has been made toward development of quantitative empirical relationships for the various subcategories of wear under specified operating conditions.

Adhesive wear is often characterized as the most basic or fundamental subcategory of wear since it occurs to some degree whenever two solid surfaces are in rubbing contact and remains active even when all other modes of wear have been eliminated. The phenomenon of adhesive wear may be best understood by recalling that all real surfaces, no matter how carefully prepared and polished, exhibit a general waviness upon which is superposed a distribution of local protuberances or asperities. As two surfaces are brought into contact, therefore, only a relatively few asperities actually touch, and the *real* area of contact is only a small fraction of the *apparent* contact area. (See Chap. 1 of Ref. 44 and Chap. 2 of Ref. 70.) Thus, even under very small applied loads, the local pressures at the contact sites become high enough to exceed the yield strength of one or both surfaces, and local plastic flow ensues. If the contacting surfaces are clean and uncorroded, the very intimate contact generated by this local plastic flow brings the atoms of the two contacting surfaces close enough together to call into play strong adhesive forces. This process is sometimes called *cold welding*. Then if the surfaces are subjected to relative sliding motion, the cold-welded junctions must be broken. Whether they break at the original interface or elsewhere within the asperity depends on surface conditions, temperature distribution, strain-hardening characteristics, local geometry, and stress distribution. If the junction is broken away from the original interface, a particle of one surface is transferred to the other surface, marking one event in the adhesive wear process. Later sliding interactions may dislodge the transferred particles as loose wear particles or they may remain attached. If this adhesive wear process becomes severe and large-scale metal transfer takes place, the phenomenon is called *galling*. If the galling becomes so severe that two surfaces adhere over a large region so that the actuating forces can no longer produce relative motion between them, the phenomenon is called *seizure*. If properly controlled, however, the adhesive wear rate may be low and self-limiting, often being exploited in the "wearing-in" process to improve mating surfaces such as bearings or cylinders so that full film lubrication may be effectively used.

One quantitative estimate of the amount of adhesive wear is given as follows (see Ref. 59 and Chaps. 2 and 6 of Ref. 71):

$$d_{\text{adh}} = \frac{V_{\text{adh}}}{A_a} = \left(\frac{k}{9\sigma_{\text{yp}}}\right)\left(\frac{W}{A_a}\right) L_s \tag{30}$$

or

$$d_{\text{adh}} = k_{\text{adh}} p_m L_s \tag{31}$$

where d_{adh} is the average wear depth, A_a is the apparent contact area, L_s is the total sliding distance, V_{adh} is the wear volume, W is the applied load, $p_m = W/A_a$ is the mean nominal contact pressure between bearing surfaces, and $k_{\text{adh}} = k/9\sigma_{\text{yp}}$ is a wear coefficient that depends on the probability of formation of a transferred fragment and the yield strength (or hardness) of the softer material. Typical values of the wear constant k for several material pairs are shown in Table 5, and the influence of lubrication on the wear constant k is indicated in Table 6. Noting from (31) that

Table 5 Archard Adhesive Wear Constant k for Various Unlubricated Material Pairs in Sliding Contact

Material	Wear Constant k
Zinc on zinc	160×10^{-3}
Low-carbon steel on low-carbon steel	45×10^{-3}
Copper on copper	32×10^{-3}
Stainless steel on stainless steel	21×10^{-3}
Copper (on low-carbon steel)	1.5×10^{-3}
Low-carbon steel (on copper)	0.5×10^{-3}
Bakelite on bakelite	0.02×10^{-3}

Source: From Ref. 71, with permission of John Wiley & Sons.

$$k_{\text{adh}} = \frac{d_{\text{adh}}}{p_m L_s} \tag{32}$$

it may be observed that if the ratio $d_{\text{adh}}/p_m L_s$ is experimentally found to be constant, Eq. (31) should be valid. Experimental evidence has been accumulated (see pp. 124–125 of Ref. 59) to confirm that for a given material pair this ratio is constant up to mean nominal contact pressures approximately equal to the uniaxial yield strength. Above this level the adhesive wear coefficient increases rapidly, with attendant severe galling and seizure.

In the selection of metal combinations to provide resistance to adhesive wear, it has been found that the sliding pair should be composed of mutually insoluble metals and that at least one of the metals should be from the B subgroup of the periodic table. (See p. 31 of Ref. 72.) The reasons for these observations are that the number of cold-weld junctions formed is a function of the mutual solubility, and the strength of the junction bonds is a function of the bonding characteristics of the metals involved. The metals in the B subgroup of the periodic table are characterized by weak, brittle covalent bonds. These criteria have been verified experimentally, as shown in Table 7, where 114 of 123 pairs tested substantiated the criteria.

In the case of abrasive wear, the wear particles are removed from the surface by the plowing and gouging action of the asperities of a harder mating surface or by hard particles trapped between the rubbing surfaces. This type of wear is manifested by a system of surface grooves and scratches, often called *scoring*. The abrasive wear condition in which the hard

Table 6 Order-of-Magnitude Values for Adhesive Wear Constant k under Various Conditions of Lubrication

Lubrication Condition	Metal (on Metal)		Nonmetal
	Like	Unlike	(on Metal)
Unlubricated	5×10^{-3}	2×10^{-4}	5×10^{-6}
Poorly lubricated	2×10^{-4}	2×10^{-2}	5×10^{-6}
Average lubrication	2×10^{-5}	2×10^{-5}	5×10^{-6}
Excellent lubrication	2×10^{-6} to 10^{-7}	2×10^{-6} to 10^{-7}	2×10^{-6}

Source: From Ref. 71, with permission of John Wiley & Sons.

asperities of one surface wear away the mating surface is commonly called *two-body wear,* and the condition in which hard abrasive particles between the two surfaces cause the wear is called *three-body wear.*

An average abrasive wear depth d_{abr} may then be estimated as

$$d_{abr} = \frac{V_{abr}}{A_a} = \left(\frac{(\tan\theta)_m}{3\pi\sigma_{yp}}\right)\left(\frac{W}{A_a}\right) L_s \tag{33}$$

$$d_{abr} = k_{abr} p_m L_s \tag{34}$$

Table 7 Adhesive Wear Behavior of Various Pairs[a]

	Material Combination				
Description of Metal Pair	Al Disk	Steel Disk	Cu Disk	Ag Disk	Remarks
Soluble pairs with poor adhesive wear resistance	Be	Be	Be	Be	These pairs substantiate the criteria of solubility and B subgroup metals
	Mg	—	Mg	Mg	
	Al	Al	Al	—	
	Si	Si8	Si	Si	
	Ca	—	Ca	—	
	Ti	Ti	Ti	—	
	Cr	Cr	—	—	
	—	Mn	—	—	
	Fe	Fe	—	—	
	Co	Co	Co	—	
	Ni	Ni	Ni	—	
	Cu	—	Cu	—	
	—	Zn	Zn	—	
	Zr	Zr	Zr	Zr	
	Nb	Nb	Nb	—	
	Mo	Mo	Mo	—	
	Rh	Rh	Rh	—	
	—	Pd	—	—	
	Ag	—	Ag	—	
	—	—	Cd	Cd	
	—	—	In	In	
	Sn	—	Sn	—	
	Ce	Ce	Ce	—	
	Ta	Ta	Ta	—	
	W	W	W	—	
	—	Ir	—	—	
	Pt	Pt	Pt	—	
	Au	Au	Au	Au	
	Th	Th	Th	Th	
	U	U	U	U	
Soluble pairs with fair or good adhesive wear resistance. (F) = Fair	—	Cu(F)	—		These pairs do not substantiate the stated criteria
	Zn(F)	—	—		
	—	—	Sb(F)		
Insoluble pairs, neither from the B subgroup, with poor adhesive wear resistance		Li			These pairs substantiate the stated criteria
		Mg			
		Ca			
		Ba			

Table 7 (*Continued*)

Description of Metal Pair	Material Combination				Remarks
	Al Disk	Steel Disk	Cu Disk	Ag Disk	
Insoluble pairs, one from the B subgroup, with fair or good adhesive wear resistance. (F) = Fair	—	C(F)	—	—	These pairs substantiate the stated criteria
	—	—	—	Ti(F)	
	—	—	Cr(F)	Cr(F)	
	—	—	—	Fe(F)	
	—	—	—	Co(F)	
	—	—	Ge(F)	—	
	—	Se(F)	Se(F)	—	
	—	—	—	Nb(F)	
	—	Ag	—	—	
	Cd	Cd	—	—	
	Ln	In	—	—	
	—	Sn(F)	—	—	
	—	Sb(F)	Sb	—	
	Te(F)	Te(F)	Te(F)	—	
	Ti	Ti	Ti	—	
	Pb(F)	Pb	Pb	—	
	Bi(F)	Bi	Bi(F)	—	
Insoluble pairs, one from the B subgroup, with poor adhesive wear resistance	C	—	C	C	These pairs do not substantiate the stated criteria
	—	—	—	Ni	
	Se	—	—	—	
	—	—	—	Mo	

[a] See pp. 34–35 of Ref. 72.

where W is total applied load, θ is the angle a typical conical asperity makes with respect to the direction of sliding, $(\tan\,\theta)_m$ is a weighted mean value for all asperities, L_s is a total distance of sliding, σ_{yp} is the uniaxial yield point strength for the softer material, V_{abr} is abrasive wear volume, $p_m = W/A_a$ is the mean nominal contact pressure between bearing surfaces, and $k_{\mathrm{abr}} = (\tan\,\theta)_m/3\pi\sigma_{\mathrm{yp}}$ is an abrasive wear coefficient that depends on the roughness characteristics of the surface and the yield strength (or hardness) of the softer material.

Comparing (33) for abrasive wear volume with (30) for adhesive wear volume, we note that they are formally the same except the constant $k/3$ in the adhesive wear equation is replaced by $(\tan\,\theta)_m/\pi$ in the abrasive wear equation. Typical values of the wear constant $3(\tan\,\theta)_m/\pi$ for several materials are shown in Table 8. As indicated in Table 8, experimental evidence shows that k_{abr} for three-body wear is typically about an order of magnitude smaller than for the two-body case, probably because the trapped particles tend to roll much of the time and cut only a small part of the time.

In selecting materials for abrasive wear resistance, it has been established that both hardness and modulus of elasticity are key properties. Increasing wear resistance is associated with higher hardness and lower modulus of elasticity since both the amount of elastic deformation and the amount of elastic energy that can be stored at the surface are increased by higher hardness and lower modulus of elasticity.

Table 9 tabulates several materials in order of descending values of (hardness)/(modulus of elasticity). Well-controlled experimental data are not yet available, but general experience would provide an ordering of materials for decreasing wear resistance compatible with the

Table 8 Abrasive Wear Constant $3(\tan\theta)_m/\pi$ for Various Materials in Sliding Contact as Reported by Different Investigators

Materials	Wear Type	Particle Size (μm)	$3(\tan\theta)_m/\pi$
Many	Two body	—	180×10^{-3}
Many	Two body	110	150×10^{-3}
Many	Two body	40–150	120×10^{-3}
Steel	Two body	260	80×10^{-3}
Many	Two body	80	24×10^{-3}
Brass	Two Body	70	16×10^{-3}
Steel	Three body	150	6×10^{-3}
Steel	Three body	80	4.5×10^{-3}
Many	Three body	40	2×10^{-3}

Source: See p. 169 of Ref. 71. Reprinted with permission from John Wiley & Sons.

Table 9 Values of (Hardness/Modulus of Elasticity) for Various Materials

Material	Condition	BHN[a] ($E \times 10^{-6}$) (in mixed units)
Alundum (Al_2O_3)	Bonded	143
Chrome plate	Bright	83
Gray ion	Hard	33
Tungsten carbide	9% Co	22
Steel	Hard	21
Titanium	Hard	17
Aluminum alloy	Hard	11
Gray iron	As cast	10
Structural steel	Soft	5
Malleable iron	Soft	5
Wrought iron	Soft	3.5
Chromium metal	As cast	3.5
Copper	Soft	2.5
Silver	Pure	2.3
Aluminum	Pure	2.0
Lead	Pure	2.0
Tin	Pure	0.7

[a] Brinell hardness number.

Source: Reprinted from Ref. 59. Copyright 1957, with permission from Elsevier Science.

array of Table 9. When the conditions for adhesive or abrasive wear exist together with conditions that lead to corrosion, the two processes persist together and often interact synergistically. If the corrosion product is hard and abrasive, dislodged corrosion particles trapped between contacting surfaces will accelerate the abrasive wear process. In turn, the wear process may remove the "protective" surface layer of corrosion product to bare new metal to the corrosive atmosphere, thereby accelerating the corrosion process. Thus, the corrosion wear process may be self-accelerating and may lead to high rates of wear.

On the other hand, some corrosion products, for example, metallic phosphates, sulfides, and chlorides, form as soft lubricative films that actually improve the wear rate markedly, especially if adhesive wear is the dominant phenomenon.

Three major wear control methods have been defined, as follows (see p. 36 of Ref. 72): *principle of protective layers,* including protection by lubricant, surface film, paint, plating, phosphate, chemical, flame-sprayed, or other types of interfacial layers; *principle of conversion,* in which wear is converted from destructive to permissible levels through better choice of metal pairs, hardness, surface finish, or contact pressure; and *principle of diversion,* in which the wear is diverted to an economical replaceable wear element that is periodically discarded and replaced as "wear-out" occurs.

When two surfaces operate in rolling contact, the wear phenomenon is quite different from the wear of sliding surfaces just described, although the "delamination" theory[73] is very similar to the mechanism of wear between rolling surfaces in contact as described here. Rolling surfaces in contact result in Hertz contact stresses that produce maximum values of shear stress slightly below the surface. (See, for example, Ref. 74.) As the rolling contact zone moves past a given location on the surface, the subsurface peak shear stress cycles from zero to a maximum value and back to zero, thus producing a cyclic stress field. Such conditions may lead to fatigue failure by the initiation of a subsurface crack that propagates under repeated cyclic loading and that may ultimately propagate to the surface to spall out a macroscopic surface particle to form a wear pit. This action, called *surface fatigue wear,* is a common failure mode in antifriction bearings, gears, and cams and all machine parts that involve rolling surfaces in contact.

8 CORROSION AND STRESS CORROSION

Corrosion may be defined as the undesired deterioration of a material through chemical or electrochemical interaction with the environment or destruction of materials by means other than purely mechanical action. Failure by corrosion occurs when the corrosive action renders the corroded device incapable of performing its design function. Corrosion often interacts synergistically with another failure mode, such as wear or fatigue, to produce the even more serious combined failure modes, such as corrosion wear or corrosion fatigue. Failure by corrosion and protection against failure by corrosion has been estimated to cost in excess of $8 billion annually in the United States alone.[75]

The complexity of the corrosion process may be better appreciated by recognizing that many variables are involved, including environmental, electrochemical, and metallurgical aspects. For example, anodic reactions and rate of oxidation; cathodic reactions and rate of reduction; corrosion inhibition, polarization, or retardation; passivity phenomena; effect of oxidizers; effect of velocity; temperature; corrosive concentration; galvanic coupling; and metallurgical structure all influence the type and rate of the corrosion process.

Corrosion processes have been categorized in many different ways. One convenient classification divides corrosion phenomena into the following types[75,76]:

1. Direct chemical attack
2. Galvanic corrosion
3. Crevice corrosion
4. Pitting corrosion
5. Intergranular corrosion
6. Selective leaching
7. Erosion corrosion
8. Cavitation corrosion
9. Hydrogen damage
10. Biological corrosion
11. Stress corrosion cracking

Depending on the types of environment, loading, and mechanical function of the machine parts involved, any of the types of corrosion may combine their influence with other failure modes to produce premature failures. Of particular concern are interactions that lead to failure by corrosion wear, corrosion fatigue, fretting fatigue, and corrosion-induced fracture.

8.1 Types of Corrosion

Direct chemical attack is probably the most common type of corrosion. Under this type of corrosive attack the surface of the machine part exposed to the corrosive media is attacked more or less uniformly over its entire surface, resulting in a progressive deterioration and dimensional reduction of sound load-carrying net cross section. The rate of corrosion due to direct attack can usually be estimated from relatively simple laboratory tests in which small specimens of the selected material are exposed to a well-simulated actual environment, with frequent weight change and dimensional measurements carefully taken. The corrosion rate is usually expressed in mils per year (mpy) and may be calculated as[75]

$$R = \frac{534W}{\gamma A t} \tag{35}$$

where R is rate of corrosion penetration in mils (1 mil = 0.001 in.) per year (mpy), W is weight loss in milligrams, A is exposed area of the specimen in square inches, γ is density of the specimen in grams per cubic centimeter, and t is exposure time in hours. Use of this corrosion rate expression in predicting corrosion penetration in actual service is usually successful if the environment has been properly simulated in the laboratory. Corrosion rate data for many different combinations of materials and environments are available in the literature.[77–79] Figure 33 illustrates one presentation of such data.

Direct chemical attack may be reduced in severity or prevented by any one or a combination of several means, including selecting proper materials to suit the environment; using plating, flame spraying, cladding, hot dipping, vapor deposition, conversion coatings, and organic coatings or paint to protect the base material; changing the environment by using lower temperature or lower velocity, removing oxygen, changing corrosive concentration, or adding corrosion inhibitors; using cathodic protection in which electrons are supplied to the metal surface to be protected either by galvanic coupling to a sacrificial anode or by an external power supply; or adopting other suitable design modifications.

Galvanic corrosion is an accelerated electrochemical corrosion that occurs when two dissimilar metals in electrical contact are made part of a circuit completed by a connecting pool or film of electrolyte or corrosive medium. Under these circumstances, the potential

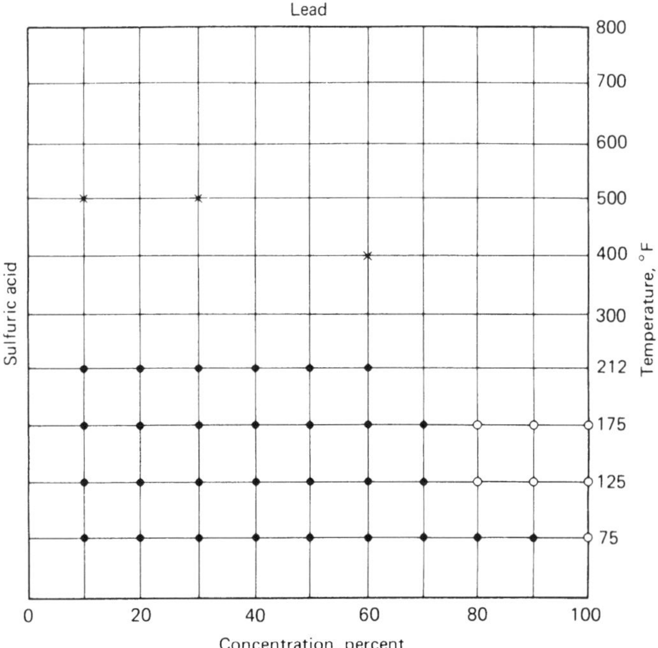

Figure 33 Nelson's method for summarizing corrosion rate data for lead in sulfuric acid environment as a function of concentration and temperature. (After Ref. 77; reprinted with permission of McGraw-Hill Book Company.)

difference between the dissimilar metals produces a current flow through the connecting electrolyte, which leads to corrosion, concentrated primarily in the more anodic or less noble metal of the pair. This type of action is completely analogous to a simple battery cell. Current must flow to produce galvanic corrosion, and, in general, more current flow means more serious corrosion. The relative tendencies of various metals to form galvanic cells and the probable direction of the galvanic action are illustrated for several commercial metals and alloys in seawater in Table 10.[75,76]

Ideally, tests in the actual service environment should be conducted; but, if such data are unavailable, the data of Table 10 should give a good indication of possible galvanic action. The farther apart the two dissimilar metals are in the galvanic series, the more serious the galvanic corrosion problem may be. Material pairs within any bracketed group exhibit little or no galvanic action. It should be noted, however, that there are sometimes exceptions to the galvanic series of Table 10, so wherever possible, corrosion tests should be performed with actual materials in the actual service environment.

The accelerated galvanic corrosion is usually most severe near the junction between the two metals, decreasing in severity at locations farther from the junction. The ratio of cathodic area to anodic area exposed to the electrolyte has a significant effect on corrosion rate. It is

Table 10 Galvanic Series of Several Commercial Metals and Alloys in Seawater

↑	Platinum
	Gold
Noble or cathodic	Graphite
(protected end)	Titanium
	Silver
	[Chlorimet 3 (62 Ni, 18 Cr, 18 Mo) Hastelloy C (62 Ni, 17 C, 15 Mo)]
	[18-8 Mo stainless steel (passive) 18-8 stainless steel (passive) Chromium stainless steel 11–30% Cr (passive)]
	[Inconel (passive) (80 Ni, 13 Cr, 7 Fe Nickel (passive)]
	Silver solder
	[Monel (70 Ni, 30 Cu) Cupronickels (60-90 Cu, 40-10 Ni) Bronzes (Cu–Sn) Copper Brasses (Cu–Zn)]
	[Chlorimet 2 (66 Ni, 32 Mo, 1 Fe) Hastelloy B (60 Ni, 30 Mo, 6 Fe, 1 Mn)]
	[Inconel (active) Nickel (active)]
	Tin
	Lead
	Lead–tin solders
	[18-8 Mo stainless steel (active) 18-8 stainless steel (active)]
	Ni-Resist (high-Ni cast iron)
	Chromium stainless steel, 13% Cr (active)
	[Cast iron Steel or iron]
Active or anodic	2024 aluminum (4.5 Cu, 1.5 Mg, 0.6 Mn)
(corroded end)	Cadmium
	Commercially pure aluminum (1100)
	Zinc
↓	Magnesium and magnesium alloys

Source: Ref. 75. Reprinted with permission of McGraw-Hill Book Company.

desirable to have a *small ratio* of cathode area to anode area. For this reason, if only *one* of two dissimilar metals in electrical contact is to be coated for corrosion protection, the more noble or more corrosion-resistant metal should be coated. Although this at first may seem the wrong metal to coat, the area effect, which produces anodic corrosion rate of 10^2–10^3 times cathodic corrosion rates for equal areas, provides the logic for this assertion.

Galvanic corrosion may be reduced in severity or prevented by one or a combination of several steps, including the selection of material pairs as close together as possible in the galvanic series, preferably in the same bracketed group; electrical insulation of one dissimilar metal from the other as completely as possible; maintaining as small a ratio of cathode area to anode area as possible; proper use and maintenance of coatings; the use of inhibitors to decrease the aggressiveness of the corroding medium; and the use of cathodic protection in which a third metal element anodic to both members of the operating pair is used as a sacrificial anode that may require periodic replacement.

Crevice corrosion is an accelerated corrosion process highly localized within crevices, cracks, and other small-volume regions of stagnant solution in contact with the corroding metal. For example, crevice corrosion may be expected in gasketed joints; clamped interfaces; lap joints; rolled joints; under bolt and rivet heads; and under foreign deposits of dirt, sand, scale, or corrosion product. Until recently, crevice corrosion was thought to result from differences in either oxygen concentration or metal ion concentration in the crevice compared to its surroundings. More recent studies seem to indicate, however, that the local oxidation and reduction reactions result in oxygen depletion in the stagnant crevice region, which leads to an excess positive charge in the crevice due to increased metal ion concentration. This, in turn, leads to a flow of chloride and hydrogen ions into the crevice, both of which accelerate the corrosion rate within the crevice. Such accelerated crevice corrosion is highly localized and often requires a lengthy incubation period of perhaps many months before it gets under way. Once started, the rate of corrosion accelerates to become a serious problem. To be susceptible to crevice corrosion attack, the stagnant region must be wide enough to allow the liquid to enter but narrow enough to maintain stagnation. This usually implies cracks and crevices of a few thousandths to a few hundredths of an inch in width.

To reduce the severity of crevice corrosion or prevent it, it is necessary to eliminate the cracks and crevices. This may involve caulking or seal welding existing lap joints; redesign to replace riveted or bolted joints by sound, welded joints; filtering foreign material from the working fluid; inspection and removal of corrosion deposits; or using nonabsorbent gasket materials.

Pitting corrosion is a very localized attack that leads to the development of an array of holes or pits that penetrate the metal. The pits may be widely scattered or so heavily concentrated that they simply appear as a rough surface. The mechanism of pit growth is virtually identical to that of crevice corrosion described, except that an existing crevice is not required to initiate pitting corrosion. The pit is probably initiated by a momentary attack due to a random variation in fluid concentration or a tiny surface scratch or defect. Some pits may become inactive because of a stray convective current, whereas others may grow large enough to provide a stagnant region of stable size, which then continues to grow over a long period of time at an accelerating rate. Pits usually grow in the direction of the gravity force field since the dense concentrated solution in a pit is required for it to grow actively. Most pits, therefore, grow downward from horizontal surfaces to ultimately perforate the wall. Fewer pits are formed on vertical walls, and very few pits grow upward from the bottom surface.

Measurement and assessment of pitting corrosion damage is difficult because of its highly local nature. Pit depth varies widely and, as in the case of fatigue damage, a statistical approach must be taken in which the probability of a pit of specified depth may be established in laboratory testing. Unfortunately, a significant size effect influences depth of pitting, and this must be taken into account when predicting service life of a machine part based on laboratory pitting corrosion data.

The control or prevention of pitting corrosion consists primarily of the wise selection of material to resist pitting or, since pitting is usually the result of stagnant conditions, imparting velocity to the fluid. Increasing its velocity may also decrease pitting corrosion attack.

Because of the atomic mismatch at the grain boundaries of polycrystalline metals, the stored strain energy is higher in the grain boundary regions than in the grains themselves. These high-energy grain boundaries are more chemically reactive than the grains. Under certain conditions depletion or enrichment of an alloying element or impurity concentration at the grain boundaries may locally change the composition of a corrosion-resistant metal, making it susceptible to corrosive attack. Localized attack of this vulnerable region near the grain boundaries is called intergranular corrosion. In particular, the austenitic stainless steels are vulnerable to intergranular corrosion if *sensitized* by heating into the temperature range from 950 to 1450°F, which causes depletion of the chromium near the grain boundaries as chromium carbide is precipitated at the boundaries. The chromium-poor regions then corrode because of local galvanic cell action, and the grains literally fall out of the matrix. A special case of intergranular corrosion, called "weld decay," is generated in the portion of the weld heat-affected zone, which is heated into the sensitizing temperature range.

To minimize the susceptibility of austenitic stainless steels to intergranular corrosion, the carbon content may be lowered to below 0.03%, stabilizers may be added to prevent depletion of the chromium near the grain boundaries, or a high-temperature solution heat treatment, called quench annealing, may be employed to produce a more homogeneous alloy.

Other alloys susceptible to intergranular corrosion include certain aluminum alloys, magnesium alloys, copper-based alloys, and die-cast zinc alloys in unfavorable environments.

The corrosion phenomenon in which one element of a solid alloy is removed is termed selective leaching. Although the selective leaching process may occur in any of several alloy systems, the more common examples are *dezincification* of brass alloys and *graphitization* of gray cast iron. Dezincification may occur as either a highly local "plug-type" or a broadly distributed layer-type attack. In either case, the dezincified region is porous, brittle, and weak. Dezincification may be minimized by adding inhibitors such as arsenic, antimony, or phosphorus to the alloy; by lowering oxygen in the environment; or by using cathodic protection.

In the case of graphitization of gray cast iron, the environment selectively leaches the iron matrix to leave the graphite network intact to form an active galvanic cell. Corrosion then proceeds to destroy the machine part. Use of other alloys, such as nodular or malleable cast iron, mitigates the problem because there is no graphite network in these alloys to support the corrosion residue. Other alloy systems in adverse environments that may experience selective leaching include aluminum bronzes, silicon bronzes, and cobalt alloys.

Erosion corrosion is an accelerated, direct chemical attack of a metal surface due to the action of a moving corrosive medium. Because of the abrasive wear action of the moving fluid, the formation of a protective layer of corrosion product is inhibited or prevented, and the corroding medium has direct access to bare, unprotected metal. Erosion corrosion is usually characterized by a pattern of grooves or peaks and valleys generated by the flow pattern of the corrosive medium. Most alloys are susceptible to erosion corrosion, and many different types of corrosive media may induce erosion corrosion, including flowing gases, liquids, and solid aggregates. Erosion corrosion may become a problem in such machine parts as valves, pumps, blowers, turbine blades and nozzles, conveyors, and piping and ducting systems, especially in the regions of bends and elbows.

Erosion corrosion is influenced by the velocity of the flowing corrosive medium, turbulence of the flow, impingement characteristics, concentration of abrasive solids, and characteristics of the metal alloy surface exposed to the flow. Methods of minimizing or preventing erosion corrosion include reducing the velocity, eliminating or reducing turbulence, avoiding sudden changes in the direction of flow, eliminating direct impingement where possible, filtering out abrasive particles, using harder and more corrosion-resistant alloys, reducing the temperature, using appropriate surface coatings, and using cathodic protection techniques.

Cavitation often occurs in hydraulic systems, such as turbines, pumps, and piping, when pressure changes in a flowing liquid give rise to the formation and collapse of vapor bubbles at or near the containing metal surface. The impact associated with vapor bubble collapse may produce high-pressure shock waves that may plastically deform the metal locally or destroy any protective surface film of corrosion product and locally accelerate the corrosion process. Furthermore, the tiny depressions so formed act as a nucleus for subsequent vapor bubbles, which continue to form and collapse at the same site to produce deep pits and pockmarks by the combined action of mechanical deformation and accelerated chemical corrosion. This phenomenon is called cavitation corrosion. Cavitation corrosion may be reduced or prevented by eliminating the cavitation through appropriate design changes. Smoothing the surfaces, coating the walls, using corrosion-resistant materials, minimizing pressure differences in the cycle, and using cathodic protection are design changes that may be effective.

Hydrogen damage, although not considered to be a form of direct corrosion, is often induced by corrosion. Any damage caused in a metal by the presence of hydrogen or the interaction with hydrogen is called hydrogen damage. Hydrogen damage includes hydrogen blistering, hydrogen embrittlement, hydrogen attack, and decarburization.

Hydrogen blistering is caused by the diffusion of hydrogen atoms into a void within a metallic structure where they combined to form molecular hydrogen. The hydrogen pressure builds to a high level that, in some cases, causes blistering, yielding, and rupture. Hydrogen blistering may be minimized by using materials without voids, by using corrosion inhibitors, or by using hydrogen-impervious coatings.

Hydrogen embrittlement is also caused by the penetration of hydrogen into the metallic structure to form brittle hydrides and pin dislocation movement to reduce slip, but the exact mechanism is not yet fully understood. Hydrogen embrittlement is more serious at the higher strength levels of susceptible alloys, which include most of the high-strength steels. Reduction and prevention of hydrogen embrittlement may be accomplished by ''baking out'' the hydrogen at relatively low temperatures for several hours, use of corrosion inhibitors, or use of less susceptible alloys.

Decarburization and hydrogen attack are both high-temperature phenomena. At high temperatures hydrogen removes carbon from an alloy, often reducing its tensile strength and increasing its creep rate. This carbon-removing process is called *decarburization*. It is also possible that the hydrogen may lead to the formation of methane in the metal voids, which may expand to form cracks, another form of hydrogen attack. Proper selection of alloys and coatings is helpful in prevention of these corrosion-related problems.

Biological corrosion is a corrosion process or processes that result from the activity of living organisms. These organisms may be microorganisms, such as aerobic or anaerobic bacteria, or they may be macroorganisms, such as fungi, mold, algae, or barnacles. The organisms may influence or produce corrosion by virtue of their processes of food ingestion and waste elimination. There are, for example, sulfate-reducing anaerobic bacteria, which produce iron sulfide when in contact with buried steel structures, and aerobic sulfur-oxidizing bacteria, which produce localized concentrations of sulfuric acid and serious corrosive attack on buried steel and concrete pipe lines. There are also iron bacteria, which ingest ferrous iron and precipitate ferrous hydroxide to produce local crevice corrosion attack. Other bacteria oxidize ammonia to nitric acid, which attacks most metals, and most bacteria produce carbon dioxide, which may form the corrosive agent carbonic acid. Fungi and mold assimilate organic matter and produce organic acids. Simply by their presence, fungi may provide the site for crevice corrosion attacks, as does the presence of attached barnacles and algae. Prevention or minimization of biological corrosion may be accomplished by altering the environment or by using proper coatings, corrosion inhibitors, bactericides or fungicides, or cathodic protection.

8.2 Stress Corrosion Cracking

Stress corrosion cracking is an extremely important failure mode because it occurs in a wide variety of different alloys. This type of failure results from a field of cracks produced in a metal alloy under the combined influence of tensile stress and a corrosive environment. The metal alloy is not attacked over most of its surface, but a system of intergranular or transgranular cracks propagates through the matrix over a period of time.

Stress levels that produce stress corrosion cracking are well below the yield strength of the material, and residual stresses as well as applied stresses may produce failure. The lower the stress level, the longer is the time required to produce cracking, and there appears to be a threshold stress level below which stress corrosion cracking does not occur.[75]

The chemical compositions of the environments that lead to stress corrosion cracking are highly specific and peculiar to the alloy system, and no general patterns have been observed. For example, austenitic stainless steels are susceptible to stress corrosion cracking in chloride environments but not in ammonia environments, whereas brasses are susceptible to stress corrosion cracking in ammonia environments but not in chloride environments. Thus, the "season cracking" of brass cartridge cases in the crimped zones was found to be stress corrosion cracking due to the ammonia resulting from decomposition of organic matter. Likewise, "caustic embrittlement" of steel boilers, which resulted in many explosive failures, was found to be stress corrosion cracking due to sodium hydroxide in the boiler water.

Stress corrosion cracking is influenced by stress level, alloy composition, type of environment, and temperature. Crack propagation seems to be intermittent, and the crack grows to a critical size, after which a sudden and catastrophic failure ensues in accordance with the laws of fracture mechanics. Stress corrosion crack growth in a statically loaded machine part takes place through the interaction of mechanical strains and chemical corrosion processes at the crack tip. The largest value of plane-strain stress intensity factor for which crack growth does not take place in a corrosive environment is designated K_{ISCC}. In many cases, corrosion fatigue behavior is also related to the magnitude of K_{ISCC}.[8]

Prevention of stress corrosion cracking may be attempted by lowering the stress below the critical threshold level, choice of a better alloy for the environment, changing the environment to eliminate the critical corrosive element, use of corrosion inhibitors, or use of cathodic protection. Before cathodic protection is implemented, care must be taken to ensure that the phenomenon is indeed stress corrosion cracking because hydrogen embrittlement is accelerated by cathodic protection techniques.

9 FAILURE ANALYSIS AND RETROSPECTIVE DESIGN

In spite of all efforts to design and manufacture machines and structures to function properly without failure, failures do occur. Whether the failure consequences simply represent an annoying inconvenience or a catastrophic loss of life and property, it is the responsibility of the designer to glean all of the information possible from the failure event so that similar events can be avoided in the future. Effective assessment of service failures usually requires the intense interactive scrutiny of a team of specialists, including at least a mechanical designer and a materials engineer trained in failure analysis techniques. The team might often include a manufacturing engineer and a field service engineer as well. The mission of the failure analysis team is to discover the initiating cause of failure, identify the best solution, and redesign the product to prevent future failures. Although the results of failure analysis investigations may often be closely related to product liability litigation, the legal issues will not be addressed in this discussion.

Techniques utilized in the failure analysis effort include the inspection and documentation of the event through direct examination, photographs, and eyewitness reports; preservation of all parts, especially failed parts; and pertinent calculations, analyses, and examinations that may help establish and validate the cause of failure. The materials engineer may utilize macroscopic examination, low-power magnification, microscopic examination, transmission or scanning electron microscopic techniques, energy-dispersive X-ray techniques, hardness tests, spectrographic analysis, metallographic examination, or other techniques of determining the failure type, failure location, material abnormalities, and potential causes of failure. The designer may perform stress and deflection analyses, examine geometry, assess service loading and environmental influences, reexamine the kinematics and dynamics of the application, and attempt to reconstruct the failure scenario. Other team members may examine the quality of manufacture, the quality of maintenance, the possibility of unusual or unconventional usage by the operator, or other factors that may have played a role in the service failure. Piecing all of this information together, it is the objective of the failure analysis team to identify as accurately as possible the probable cause of failure.

As undesirable as service failures may be, the results of a well-executed failure analysis may be transformed directly into improved product reliability by a designer who capitalizes on service failure data and failure analysis results. These techniques of retrospective design have become important working tools of the profession and are likely to continue to grow in importance.

REFERENCES

1. J. A. Collins, *Failure of Materials in Mechanical Design; Analysis, Prediction, Prevention,* 2nd ed., Wiley, New York, 1993.
2. K. Hellan, *Introduction to Fracture Mechanics,* McGraw-Hill, New York, 1984.
3. H. Tada, P. C. Paris, and G. R. Irwin, *The Stress Analysis of Cracks Handbook,* 2nd ed., Paris Productions, St. Louis, 1985.
4. T. L. Anderson, *Fracture Mechanics, Fundamentals and Applications,* 2nd ed., CRC Press, Boca Raton, FL, 1995.
5. "Progress in Measuring Fracture Toughness and Using Fracture Mechanics," *Materials Research and Standards,* March, 103–119 (1964).
6. X. Wu and A. J. Carlsson, *Weight Functions and Stress Intensity Factor Solutions,* Pergamon Press, Oxford, 1991.
7. E 399-90, "Standard Test Method for Plane-Strain Fracture Toughness of Metallic Materials," *Annual Book of ASTM Standards,* Vol. 03.01, ASTM, Philadelphia, 1991.
8. R. W. Hertzberg, *Deformation and Fracture Mechanics of Engineering Materials,* 3rd ed., Wiley, New York, 1989.
9. J. P. Gallagher (ed.), *Damage Tolerant Design Handbook,* 4 vols., Metals and Ceramics Information Center, Battelle Columbus Labs., Columbus, OH, 1983.

10. *Metallic Materials and Elements for Aerospace Vehicle Structures,* MIL-HDBK-5F, 2 vols., U.S. Dept. of Defense, Naval Publications and Forms Center, Philadelphia, 1987.
11. W. T. Matthews, *Plane Strain Fracture Toughness* (K_{IC}) *Data Handbook for Metals,* Report No. AMMRC MS 73-6, U.S. Army Materiel Command, NTIS, Springfield, VA, 1973.
12. C. M. Hudson and S. K. Seward, "A Compendium of Sources of Fracture Toughness and Fatigue Crack Growth Data for Metallic Alloys," *Int. J. Fracture,* **14,** R151 (1978).
13. C. M. Hudson and S. K. Seward, "A Compendium of Sources of Fracture Toughness and Fatigue Crack Growth Data for Metallic Alloys—Part II," *Int. J. Fracture,* **20,** R59 (1982).
14. C. M. Hudson and S. K. Seward, "A Compendium of Sources of Fracture Toughness and Fatigue Crack Growth Data for Metallic Alloys—Part III," *Int. J. Fracture,* **39,** R43 (1989).
15. C. M. Hudson and J. J. Ferrainolo, "A Compendium of Sources of Fracture Toughness and Fatigue Crack Growth Data for Metallic Alloys—Part IV," *Int. J. Fracture,* **48,** R19 (1991).
16. M. F. Kanninen and C. H. Popelar, *Advanced Fracture Mechanics,* Oxford University Press, New York, 1985.
17. A. F. Madayag, *Metal Fatigue, Theory and Design*, Wiley, New York, 1969.
18. *Proceedings of the Conference on Welded Structures,* Vols. I and II, The Welding Institute, Cambridge, England, 1971.
19. H. J. Grover, S. A. Gordon, and L. R. Jackson, *Fatigue of Metals and Structures,* GPO, Washington, DC, 1954.
20. H. F. Moore, "A Study of Size Effect and Notch Sensitivity in Fatigue Tests of Steel," *ASTM Proc.,* **45,** 507 (1945).
21. W. N. Thomas, "Effect of Scratches and Various Workshop Finishes upon the Fatigue Strength of Steel," *Engineering,* **116,** 449ff (1923).
22. G. V. Smith, *Properties of Metals at Elevated Temperatures,* McGraw-Hill, New York, 1950.
23. N. E. Frost and K. Denton, "The Fatigue Strength of Butt Welded Joints in Low Alloy Structural Steels," *Brit. Welding J.,* **14**(4), (1967).
24. H. J. Grover, *Fatigue of Aircraft Structures,* GPO, Washington, DC, 1966.
25. E 647-88a, "Standard Test Method for Measurement of Fatigue Crack Growth Rates," ASTM, Philadelphia, 1988.
26. J. Schijve, "Four Lectures on Fatigue Crack Growth," *Eng. Fracture Mech.,* **11,** 167–221 (1979).
27. P. C. Paris and F. Erdogan, "A Critical Analysis of Crack Propagation Laws," *Trans. ASME, J. Basic Eng.,* **85,** 528–534 (1963).
28. S. J. Maddox, *Fatigue Strength of Welded Structures,* 2nd ed., Abington, Cambridge, England, 1991.
29. W. G. Reuter, J. H. Underwood, and J. C. Newman (eds.), *Surface-Crack Growth: Models, Experiments and Structures,* STP-1060, ASTM, Philadelphia, 1990.
30. W. Elber, *Eng. Fracture Mech.,* **2,** 37–45 (1970).
31. J. C. Newman and W. Elber (eds.), *Mechanics of Fatigue Crack Closure,* STP-982, ASTM, Philadelphia, 1988.
32. L. P. Pook, *The Role of Crack Growth in Metal Fatigue,* The Metals Society, London, 1983.
33. H. O. Fuchs and R. I. Stephens, *Metal Fatigue in Engineering,* Wiley, New York, 1980.
34. D. Broek, *Elementary Engineering Fracture Mechanics,* 4th ed., Kluwer, London, 1986.
35. A. P. Parker, *The Mechanics of Fracture and Fatigue,* E. & F. N. Spon Ltd., New York, 1981.
36. R. D. Ritchie and J. Lankford (eds.), *Small Fatigue Cracks,* The Metallurgical Society, Warrendale, PA, 1986.
37. W. G. Clark, Jr., "Fracture Mechanics in Fatigue," *Experimental Mechanics,* Sept. (1971).
38. R. C. Juvinall, *Engineering Considerations of Stress, Strain, and Strength,* McGraw-Hill, New York, 1967.
39. F. R. Larson and J. Miller, "Time-Temperature Relationships for Rupture and Creep Stresses," *ASME Trans.,* **74,** 765 (1952).
40. R. G. Sturm, C. Dumont, and F. M. Howell, "A Method of Analyzing Creep Data," *ASME Trans.,* **58,** A62 (1936).
41. N. H. Polakowski and E. J. Ripling, *Strength and Structure of Engineering Materials*. Prentice-Hall, Englewood Cliffs, NJ, 1966.

42. J. A. Collins, "Fretting-Fatigue Damage-Factor Determination," *J. Eng. Industry,* **87**(8), 298–302 (1965).

43. D. Godfrey, "Investigation of Fretting by Microscopic Observation," NACA Report 1009, Cleveland, OH, 1951 (formerly TN-2039, Feb. 1950).

44. F. P. Bowden and D. Tabor, *The Friction and Lubrication of Solids,* Oxford University Press, Amen House, London, 1950.

45. D. Godfrey and J. M. Baily, "Coefficient of Friction and Damage to Contact Area during the Early Stages of Fretting; I-Glass, Copper, or Steel Against Copper," NACA TN-3011, Cleveland, OH, Sept. 1953.

46. M. E. Merchant, "The Mechanism of Static Friction," *J. Appl. Phys.,* **11**(3), 232 (1940).

47. E. E. Bisson, R. L. Johnson, M. A. Swikert, and D. Godfrey, "Friction, Wear, and Surface Damage of Metals as Affected by Solid Surface Films," NACA TN-3444, Cleveland, OH, May 1955.

48. H. H. Uhlig, "Mechanisms of Fretting Corrosion," *J. Appl. Mech.,* **76,** 401–407 (1954).

49. I. M. Feng and B. G. Rightmire, "The Mechanism of Fretting," *Lubrication Eng.,* **9,** 134ff (June 1953).

50. I. M. Feng, "Fundamental Study of the Mechanism of Fretting," Final Report, Lubrication Laboratory, Massachusetts Institute of Technology, Cambridge, MA, 1955.

51. H. T. Corten, "Factors Influencing Fretting Fatigue Strength," T. & A. M. Report No. 88, Department of Theoretical and Applied Mechanics, University of Illinois, Urbana, IL, June 1955.

52. W. L. Starkey, S. M. Marco, and J. A. Collins, "Effects of Fretting on Fatigue Characteristics of Titanium-Steel and Steel-Steel Joints," Paper 57-A-113, ASME, New York, 1957.

53. W. D. Milestone, "Fretting and Fretting-Fatigue in Metal-to-Metal Contacts," Paper 71-DE-38, ASME, New York, 1971.

54. G. P. Wright and J. J. O'Connor, "The Influence of Fretting and Geometric Stress Concentrations on the Fatigue Strength of Clamped Joints," *Proc. Inst. Mech. Eng.,* 186 (1972).

55. R. B. Waterhouse, *Fretting Corrosion,* Pergamon, New York, 1972.

56. "Fretting in Aircraft Systems," AGARD Conference Proceedings CP161, distributed through NASA, Langley Field, VA, 1974.

57. "Control of Fretting Fatigue," Report No. NMAB-333, National Academy of Sciences, National Materials Advisory Board, Washington, DC, 1977.

58. N. P. Suh, S. Jahanmir, J. Fleming, and E. P. Abrahamson, "The Delamination Theory of Wear-II," Progress Report, Materials Processing Lab, Mechanical Engineering Dept., MIT Press, Cambridge, MA, Sept. 1975.

59. J. T. Burwell, Jr., "Survey of Possible Wear Mechanisms," *Wear,* **1,** 119–141 (1957).

60. M. B. Peterson, M. K. Gabel, and M. J. Derine, "Understanding Wear"; K. C. Ludema, "A Perspective on Wear Models"; E. Rabinowicz, "The Physics and Chemistry of Surfaces"; J. McGrew, "Design for Wear of Sliding Bearings"; R. G. Bayer, "Design for Wear of Lightly Loaded Surfaces," *ASTM Standardization News,* **2**(9), 9–32 (1974).

61. P. A. Engel, "Predicting Impact Wear," *Machine Design,* May, 100–105 (1977).

62. P. A. Engel, *Impact Wear of Materials,* Elsevier, New York, 1976.

63. J. A. Collins, "A Study of the Phenomenon of Fretting-Fatigue with Emphasis on StressField Effects," Dissertation, Ohio State University, Columbus, OH, 1963.

64. J. A. Collins and F. M. Tovey, "Fretting Fatigue Mechanisms and the Effect of Direction of Fretting Motion on Fatigue Strength," *J. Materials,* **7**(4), (Dec. 1972).

65. W. D. Milestone, "An Investigation of the Basic Mechanism of Mechanical Fretting and Fretting-Fatigue at Metal-to-Metal Joints, with Emphasis on the Effects of Friction and Friction-Induced Stresses," Dissertation, Ohio State University, Columbus, 1966.

66. J. A. Collins and S. M. Marco, "The Effect of Stress Direction during Fretting on Subsequent Fatigue Life," *ASTM Proc.,* **64,** 547 (1964).

67. H. Lyons, "An Investigation of the Phenomenon of Fretting-Wear and Attendant Parametric Effects Towards Development of Failure Prediction Criteria," Ph.D. Dissertation, Ohio State University, Columbus, OH, 1978.

68. P. L. Ko, "Experimental Studies of Tube Fretting in Steam Generators and Heat Exchangers," ASME/CSME Pressure Vessels and Piping Conference, Nuclear and Materials Division, Montreal, Canada, June 1978.

69. R. B. Heywood, *Designing against Fatigue of Metals,* Reinhold, New York, 1962.

70. F. P. Bowden and D. Tabor, *Friction and Lubrication,* Methuen, London, 1967.

71. E. Rabinowicz, *Friction and Wear of Materials,* Wiley, New York, 1966.

72. C. Lipson, *Wear Considerations in Design,* Prentice-Hall, Englewood Cliffs, NJ, 1967.

73. N. P. Suh, "The Delamination Theory of Wear," *Wear,* **25,** 111–124 (1973).

74. J. E. Shigley and C. R. Mischke, *Mechanical Engineering Design,* 5th ed., McGraw-Hill, New York, 1989.

75. M. G. Fontana and N. D. Greene, *Corrosion Engineering,* 2nd ed., McGraw-Hill, New York, 1978.

76. L. S. Seabright and R. J. Fabian, "The Many Faces of Corrosion," *Materials in Design Eng.,* **57**(1) (1963).

77. G. Nelson, *Corrosion Data Survey,* National Association of Corrosion Engineers, Houston, TX, 1972.

78. H. H. Uhlig (ed.), *Corrosion Handbook,* Wiley, New York, 1948.

79. E. Rabald, *Corrosion Guide,* Elsevier, New York, 1951.

CHAPTER 27
FAILURE ANALYSIS OF PLASTICS

Vishu Shah
Diamond Bar, California

1 INTRODUCTION

The fundamental problem concerning the failures of parts made out of plastics materials is a lack of understanding of the difference between the nature of relatively new polymeric materials and traditional materials such as metal, wood, and ceramics. Designers, processors, and end users are all equally responsible in contributing to the problem. Merely copying a metal or wood product with some minor aesthetic changes can lead to premature and sometimes catastrophic failures. Designers are generally most familiar with metals and their behavior under load and varying conditions of temperature and environment. While designing metal parts, a designer can rely on instantaneous stress–strain properties and for most applications can disregard the effect of temperature, environment, and long-term load (creep). Plastics materials are viscoelastic in nature, and unlike other materials, properties can vary considerably under the influence of temperature, load, environment, and presence of chemicals. For example, a well-designed part may perform its intended function for a very long time at room temperature environment and under normal load. The same part may fall apart quickly when exposed to extreme cold or hot environment and the process can accelerate if it is subjected to mechanical loading or exposed to a chemical environment. Most often overlooked is the synergistic effect of all the conditions, such as temperature, creep, chemicals, ultraviolet (UV), and other environmental factors, on plastic parts. During processing, plastic materials are subjected to severe physical conditions involving elevated temperatures, high pressures, and high-shear-rate flow as well as chemical changes. The processor must follow the proper procedures and guidelines set forth by material suppliers and make sure that the material is processed optimally in well-maintained equipment. Lastly, end users must be educated by product manufacturers in the proper use of the plastics products and make sure that they are used for the intended purpose. Plastic parts often fail prematurely because of intentional or unintentional abuse by consumers.

Part failure is generally related to one of four key factors: material selection, design, process, and service conditions. Figure 1 shows reasons for plastic product failures.

1.1 Material Selection

Failures arising from hasty material selection are not uncommon in the plastics or any other industry. In an application that demands high-impact resistance, a high-impact material must be specified. If the material is to be used outdoors for a long period, a UV-resistant material must be specified. For proper material selection, careful planning, a thorough understanding of plastic materials, and reasonable prototype testing are required. The material selection should not be based solely on cost. A systematic approach to the material selection process is necessary to select the best material for any application. The proper material selection technique starts with carefully defining the application requirements in terms of mechanical, thermal, environmental, electrical, and chemical properties. In many instances, it makes sense to design a thinner wall part, taking advantage of the stiffness-to-weight ratio offered by higher priced, fast-cycling engineering materials. All special needs such as outdoor UV exposure, light transmission, fatigue, creep and stress relaxation, and regulatory requirements must be considered. Processing techniques and assembly methods play a key role in selecting the appropriate material and should be given consideration. Many plastic materials are susceptible to chemical attack and therefore the behavior of plastic material in a chemical environment is one of the most important considerations in selecting the material. No single property defines a material's ability to perform in a given chemical environment, and factors such as external or molded-in stresses, length of exposure, temperature, and chemical concentration should be carefully scrutinized. Many companies, including material suppliers, have developed software to assist in material selection simply by selecting the application requirement in the order of importance.

Some common pitfalls in the material selection process are relying on published material property data, misinterpretation of data sheets, and blindly accepting the material supplier's recommendations. Material property data sheets should only be used for screening various types and grades of materials and not for ultimate selection or engineering design. The reported data are generally derived from short-term tests and single-point measurements under laboratory conditions using standard test bars. The published values are generally

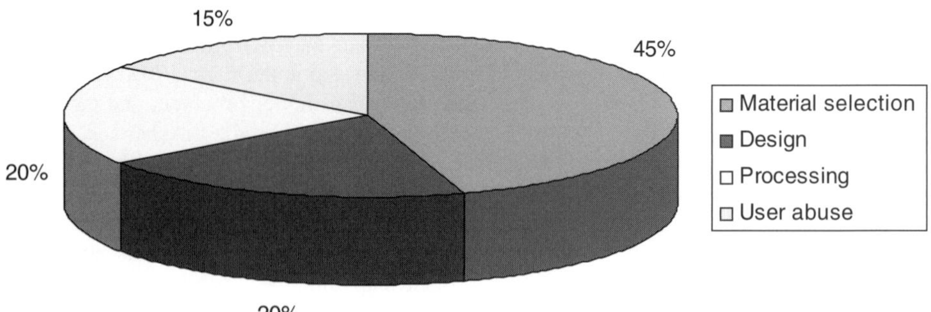

Figure 1 Major reasons for plastics product failures. (Reprinted with permission from Rapra Technology Ltd.)

higher and do not correlate well with actual-use conditions. Such data do not take into account the effect of time, temperature, environment, and chemicals. Figure 2 shows a typical failure arising from improper material selection.

1.2 Design

Proper material selection alone will not prevent a product from failing. While designing a plastic product, the designer must use the basic rules and guidelines provided by the material supplier for designing a particular part in that material. One must remember that with the exception of a few basic rules in designing plastic parts, the design criteria change from material to material as well as from application to application. The most common mistakes made by designers when working in plastics are related to wall thickness, sharp corners, creep, draft, environmental compatibility, and placement of ribs. Failure arising from designing parts with sharp corners (insufficient radius) by far exceeds all other reasons for part failures. Stresses build rapidly in internal sharp corners of the part, as shown in Fig. 3, which illustrates the influence of fillet radius on stress concentration. Maintaining uniform wall thickness is essential in keeping sink marks, voids, warpage, and more importantly areas of molded-in stresses to a minimum. The viscoelastic nature of plastic materials as opposed to metals requires designers to pay special attention to creep and stress relaxation data. Plastic parts will deform under load over time depending upon the type of material, amount of load, length of time, and temperature. Design guides for proper plastic part design are readily available from material suppliers. Table 1 shows a typical part design checklist. Figure 4 illustrates a typical part failure arising from improper design.

1.3 Process

After proper material selection and design, the responsibility shifts from the designer to the plastic processor. The most innovative design and a very careful material selection cannot make up for poor processing practices. Molded-in stresses, voids, weak weld lines, and moisture in the material are some of the most common causes for premature product failures.

The latest advancement in process control technology not only allows the processors to control the process with a high degree of reliability but also helps in recordkeeping should

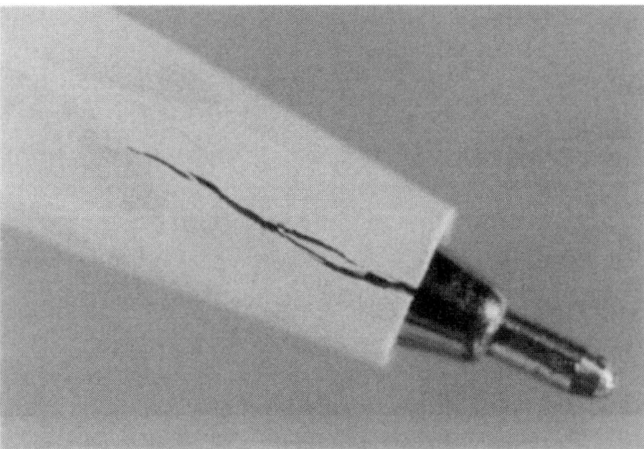

Figure 2 Failure arising from improper material selection.

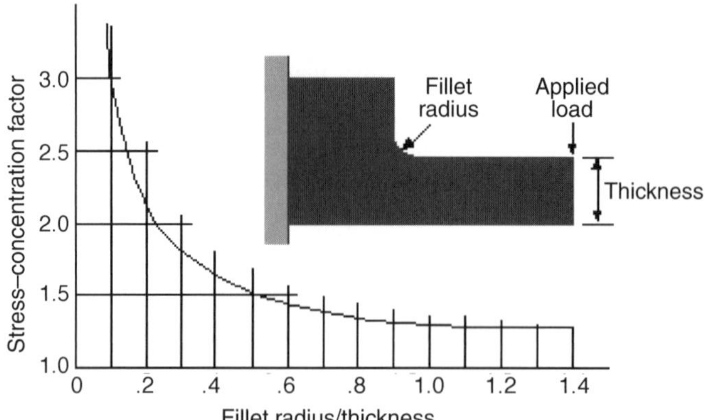

Figure 3 Influence of fillet radius on stress concentration.

a product fail at a later date. Such records of processing parameters are invaluable to a person conducting failure analysis. Any assembly or secondary operation on a processed part must be evaluated carefully to avoid premature failures. Failures arising from stress cracking around metal inserts, drilled holes, and welded joints are quite common. Part failure resulting from poor processing practices is shown in Fig. 5.

1.4 Service Conditions

In spite of warning labels and user's instructions, failures arising from service conditions is quite common in the plastics industry. Five categories of unintentional service conditions are as follows:

1. Reasonable misuse
2. Use beyond intended lifetime
3. Unstable service conditions
4. Beyond reasonable misuse
5. Simultaneous application of two stresses operating synergistically

Figure 6 shows a failed nut due to overtightening by the consumer.

Most stresses imposed on plastic products in service can be grouped under thermal, chemical, physical, biological, mechanical, and electrical.

2 TYPES OF FAILURES

2.1 Mechanical Failure

Mechanical failure arises from applied external forces which, when they exceed the yield strength of the material, cause the product to deform, crack, or break into pieces. The force may have been applied in tension, compression, and impact for a short or a long period of time at varying temperatures and humidity conditions.

Table 1 Part Design Checklist

Part Design Checklist

for injection molded engineering thermoplastics

MILES

Polymers Division

Material Selection Requirements

☐ LOADS	___ Magnitude	___ Duration	___ Impact	___ Fatigue	___ Wear
☐ ENVIRONMENT	___ Temperature	___ Chemicals	___ Humidity	___ Cleaning agents	
	___ Lubricants	___ U.V. light			
☐ SPECIAL	___ Transparency	___ Paintability	___ Warpage/Shrinkage		
	___ Plateability	___ Flammability	___ Cost	___ Agency test	

Part Details Review

☐ RADII	___ Sharp corners	___ Ribs	___ Bosses	___ Lettering
☐ WALL THICKNESS				
• Material	___ Strength	___ Electrical	___ Flammability	
• Flow	___ Flow length	___ Too thin	___ Thin to thick	
	___ Picture framing	___ Orientation		
• Uniformity	___ Thick areas	___ Thin areas	___ Abrupt changes	
☐ RIBS	___ Radii	___ Draft	___ Height	___ Spacing
	___ Base thickness			
☐ BOSSES	___ Radii	___ Draft	___ Inside diameter/outside diameter	
	___ Base thickness	___ Length/diameter		
☐ WELD LINES	___ Proximity to load	___ Strength vs. load	___ Visual area	
☐ DRAFT	___ Draw polish	___ Texture depth	___ ½ degree (minimum)	
☐ TOLERANCE	___ Part geometry	___ Material		
	___ Tool design (across parting line, slides)			

Assembly Considerations

☐ PRESS FIT	___ Tolerances	___ Long-term retention	___ Hoop stress
☐ SNAP FIT	___ Allowable strain	___ Assembly force	
	___ Tapered beam	___ Multiple assembly	
☐ SCREWS	___ Thread-cutting vs. forming	___ Avoid countersinks	
☐ MOLDED THREAD	___ Avoid feather-edges, sharp corners and pipe threads		
☐ ULTRASONICS	___ Energy director	___ Shear joint interference	
	___ Wall thickness	___ Hermetic seal	
☐ ADHESIVE &	___ Shear vs. butt joint	___ Compatibility	
SOLVENT BONDS	___ Trapped vapors		
☐ GENERAL	___ Interfit tolerances	___ Stack tolerances	___ Thermal expansion
	___ Care with rivets and molded-in inserts		___ Component compatibility

Mold Concerns

☐ WARPAGE	___ Cooling (corners)	___ Ejector placement	
☐ GATES	___ Type	___ Size	___ Location
☐ RUNNERS	___ Size and shape	___ Sprue size	___ Balanced flow
	___ Cold slug well	___ Sharp corners	
☐ GENERAL	___ Draft	___ Part ejection	___ Avoid thin/long cores

Figure 4 Brittle failure due to sharp corner (lack of radius) on part.

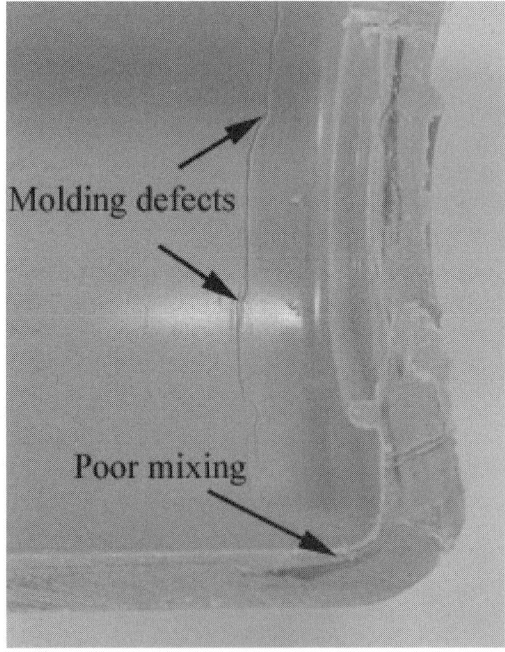

Figure 5 Failure as result of poor processing practices. (Courtesy of The Madison Group.)

Figure 6 Failed nut due to overtightening nut from a water supply line. (Courtesy of The Madison Group.)

2.2 Thermal Failure

Thermal failures occur from exposing products to an extremely hot or extremely cold environment. At abnormally high temperatures the product may warp, twist, melt, or even burn. Plastics tend to get brittle at low temperatures. Even the slightest amount of load may cause the product to crack or even shatter.

2.3 Chemical Failure

Very few plastics are totally impervious to all chemicals. Failure occurring from exposing the products to certain chemicals is quite common. Residual or molded-in stress, high temperatures, and external loading tend to aggravate the problem. Figure 7 illustrates failure of a product from chemical attack.

2.4 Environmental Failure

Plastics exposed to outdoor environments are susceptible to many types of detrimental factors. Ultraviolet rays, humidity, microorganisms, ozone, heat, and pollution are major environmental factors that seriously affect plastics. The effect can be a mere loss of color, slight crazing and cracking, or complete breakdown of the polymer structure. A typical product failure arising from weathering effects is shown in Fig. 8

3 ANALYZING FAILURES

The first step in analyzing any type of failure is to determine the cause of the failure. Before proceeding with elaborate tests, some basic information regarding the product must be gathered. If the product is returned from the field, the district manager or consumer should be asked for basic information such as the date of purchase, date of installation, date when the first failure encountered, geographic location, types of chemicals used with or around the product, and whether the product was used indoors or outdoors. This information is vital if one is to analyze the defective product proficiently. For example, if the report from the field indicates a certain type of chemical was used with the product, one can easily check the chemical compatibility of the product or go one step further and simulate the actual-use condition with that chemical. Recordkeeping also simplifies the task of failure analysis. A

Figure 7 Polystyrene (PS) degraded by prolonged contact with gasoline. (Courtesy of The Madison Group.)

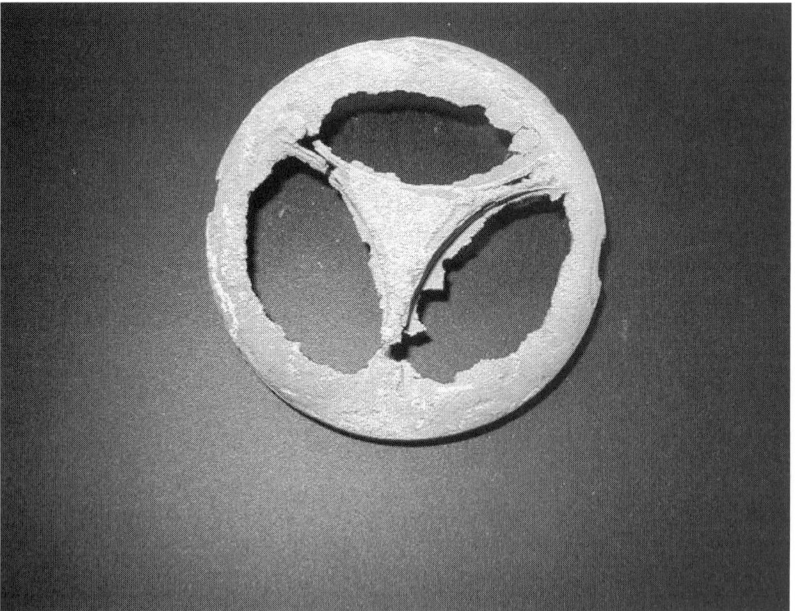

Figure 8 Failure occurring in plastics part due to prolonged exposure and resulting degradation from UV rays.

simple date code or cavity identification number will enhance the traceability. Many types of checklists to help analyze the failures have been developed. Ten basic methods are employed to analyze product failure:

1. Visual examination
2. Identification analysis
3. Stress analysis
4. Heat reversion technique
5. Microtoming (microstructural analysis)
6. Mechanical testing
7. Thermal analysis
8. Nondestructive testing (NDT) techniques
9. Fractography
10. Simulation testing

By zeroing in on the type of failure, one can easily select the appropriate method of failure analysis.

3.1 Visual Examination

A careful visual examination of the returned part can reveal many things. Excessive splay marks indicate that the materials were not adequately dried before processing. The failure to remove moisture from hydroscopic materials can lower the overall physical properties of the molded article and in some cases even cause them to become brittle. The presence of foreign material and other contaminants is also detrimental and could have caused the part to fail. Burn marks on molded articles are easy to detect. They are usually brown streaks and black spots. These marks indicate the possibility of material degradation during processing, causing the breakdown of molecular structure and leading to overall reduction in the physical properties. Sink marks and weak weld lines, readily visible on molded parts, represent poor processing practices and may contribute to part failure.

A careful visual examination will also reveal the extent of consumer abuse. The presence of unusual chemicals, grease, pipe dope, and other substances may give some clues. Heavy marks and gouges could be the sign of excessively applied external force.

The defective part should also be cut in half using a sharp saw blade. The object here is to look for voids caused by trapped gas and excessive shrinkage, especially in thick sections during molding. A reduction in wall thickness caused by such voids could be less than adequate for supporting compressive or tensile force or withstanding impact load and may cause a part to fail. Lastly, if the product has failed because of exposure to UV rays and other environmental factors, a slight chalking, microscopic cracks, large readily visible cracks, or loss of color will be evident.

3.2 Identification Analysis

One of the main reasons for product failure is use of the wrong material. When a defective product is returned from the field, material identification tests must be carried out to verify that the material used in the defective product is, in fact, the material specified on the product drawing. However, identifying the type of material is simply not enough. Since all plastic materials are supplied in a variety of grades with a broad range of properties, the grade of material must also be determined. A simple technique such as the melt index test can be carried out to confirm the grade of a particular type of material. The percentage of regrind material mixed with virgin material has a significant effect on the physical properties. Generally, the higher the level of regrind material mixed with virgin, the lower the physical properties. If during processing higher than recommended temperature and long residence time are used, chances are the material will degrade. This degraded material, when reground and mixed with virgin material, can cause a significant reduction in overall properties. Unfortunately, the percentage of regrind used with virgin material is almost impossible to determine by performing tests on the molded parts. However, a correlation between the melt index value and the part failure rate can be established by conducting a series of tests to determine the minimum or maximum acceptable melt index value.

Part failures due to impurities and contamination of virgin material are quite common. Material contamination usually occurs during processing. A variety of purging materials are used to purge the previous material from the extruder barrel before using the new material. Not all of these purging materials are compatible. Such incompatibility can cause the loss of properties, brittleness, and delamination. In the vinyl compounding operation, failure to add key ingredients, such as an impact modifier, can result in premature part failure. Simple laboratory techniques cannot identify such impurities, contamination, or the absence of a key ingredient. More sophisticated techniques, such as Fourier transform infrared (FTIR) analysis and gel permeation chromatography (GPC), must be employed. These methods can not only identify the basic material but also point out the type and level of impurities in most cases. Plastic materials are rarely manufactured without some type of additives. These additives play an important role in thermally or environmentally stabilizing the base polymer. Antioxidants, flame retardants, UV stabilizers, heat stabilizers, and lubricants are routinely added for enhancement. If for some reason these important additives are left out or depleted due to poor processing practices, plastic parts may fail prematurely. Techniques such as deformulation or reverse engineering a product are employed for separation, identification, and quantification of ingredients.

3.3 Stress Analysis

Once the part failure resulting from poor molding practices or improper material usage through visual examination and material identification is ruled out, the next logical step is to carry out an experimental stress analysis. Experimental stress analysis is one of the most versatile methods for analyzing parts for possible failure. The part can be externally stressed or can have residual or molded-in stresses. External stresses or molded-in stresses or a combination of both can cause a part to fail prematurely. Stress analysis is an important part of failure identification. Detection of residual stresses has a different meaning than evaluation of stresses due to applied forces. It is possible of course to see failure due to poor design or underestimating forces. These failures are usually detected in proof testing or in early production. Residual stresses are different: A molding process can generate residual stress just about anywhere, anytime. Here, ongoing photoelastic inspection can prove extremely helpful, allowing detection of defective molded parts or identification of failures in clear plastic products. Experimental stress analysis can be conducted to determine the actual levels of stress in the part. Five methods are used to conduct stress analysis: photoelastic method, brittle-coating method, strain gauge method, chemical method, and heat reversion.

Photoelastic Method

The photoelastic method for experimental stress analysis is quite popular among design engineers and has proven to be an extremely versatile, yet simple technique.

If the parts to be analyzed are made of one of the transparent materials, stress analysis is simple. All transparent plastics, being birefringent, lend themselves to photoelastic stress analysis. The transparent part is placed between two polarizing media and viewed from the opposite side of the light source. The fringe patterns are observed without applying external stress. This allows the observer to study the molded-in or residual stresses in the part. High fringe order indicates the area of high stress level whereas low fringe order represents an unstressed area. Also, close spacing of fringes represents a high stress gradient. A uniform color indicates uniform stress in the part. Next, the part should be stressed by applying external force and simulating actual-use conditions. The areas of high stress concentration can be easily pinpointed by observing changes in fringe patterns brought forth by external stress. Figure 9 illustrates a typical stress pattern in a part. This type of evaluation is useful as a regular part of product inspection for quality control of transparent parts for any manufacturer to maintain product quality and consistency and to prevent failures. The method often reveals problems associated with many other process control parameters, such as temperature, material, fill rate, design, etc.

Another technique known as the photoelastic coating technique can be used to photoelastically stress analyze opaque plastic parts. The part to be analyzed is coated with a photoelastic coating, service loads are applied to the part, and the coating is illuminated by polarized light from the reflection polariscope. Molded-in or residual stresses cannot be observed with this technique. However, the same part can be fabricated using one of the transparent plastic materials. In summary, photoelastic techniques can be used successfully for failure analysis of a defective product.

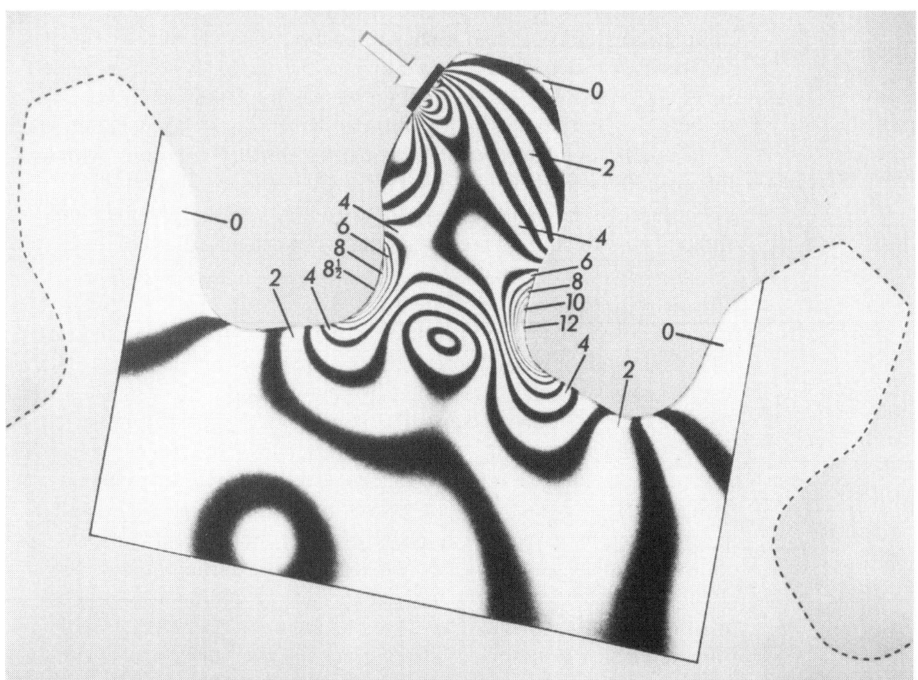

Figure 9 Typical photoelastic stress pattern. (Courtesy of Measurements Group, Inc.)

Brittle-Coating Method

The brittle-coating method is another technique of conveniently measuring the localized stresses in a part. Brittle coatings are specially prepared lacquers that are usually applied by spraying on the actual part. The part is subjected to stress after air drying the coating. The location of maximum strain and the direction of the principal strain is indicated by the small cracks that appear on the surface of the part as a result of external loading. Thus, the technique offers valuable information regarding the overall picture of the stress distribution over the surface of the part. The data obtained from the brittle coating method can be used to determine the exact areas for strain gauge location and orientation, allowing precise measurement of the strain magnitude at points of maximum interest. They are also useful for the determination of stresses at stress concentration points that are too small or inconveniently located for installation of strain gauges. The brittle-coating technique, however, is not suitable for detailed quantitative analysis such as photoelasticity. Sometimes it is necessary to apply an undercoating prior to the brittle coating to promote adhesion and to minimize compatibility problems. Further discussion on this subject is found in the literature (see the Bibliography).

Strain Gauge Method

The electrical resistance strain gauge method is the most popular and widely accepted method for strain measurements. The strain gauge consists of a grid of strain-sensitive metal foil bonded to a plastic backing material. When a conductor is subjected to a mechanical deformation, its electrical resistance changes proportionally. This principle is applied in the operation of a strain gauge. For strain measurements, the strain gauge is bonded to the surface of a part with a special adhesive and then connected electrically to a measuring instrument. When the test part is subjected to a load, the resulting strain produced on the surface of the part is transmitted to the foil grid. The strain in the grid causes a change in its length and cross section and produces a change in the resistivity of the grid material. This change in grid resistance, which is proportional to the strain, is then measured with a strain gauge recording instrument. In using strain gauges for failure analysis, care must be taken to test the adhesives for compatibility with particular plastics to avoid stress cracking problems.

Residual or molded-in stresses can be directly measured with strain gauges using the hole-drilling method. This method involves measuring a stress at a particular location, drilling a hole through the part to relieve the frozen-in stresses, and then remeasuring the stress. The difference between the two measurements is calculated as residual stress.

Chemical Method

When exposed to certain chemicals while under stress, most plastics show stress cracking. This phenomenon is used in the stress analysis of molded parts. The level of molded-in or residual stress can be determined by this method. The part is immersed in a mixture of glacial acetic acid and water for 2 min at 73°F and later inspected for cracks that occur where tensile stress at the surface is greater than the critical stress. The part may also be externally stressed to a predetermined level and sprayed with the chemical to determine critical stresses. Stress cracking curves for many types of plastics have been developed by material suppliers. If a defective product returned from the field appears to have stress cracked, similar tests should be carried out to determine molded-in stresses as well as the effect of external loading by simulating end-use conditions. Failures of such types are seen often in parts where metal inserts are molded in or inserted after molding. Three other tests—stain resistance test, solvent stress cracking resistance, and environmental stress cracking resistance (ESCR)—are also employed to analyze failed parts. The acetone immersion test to determine the quality of rigid polyvinyl chloride (PVC) pipe and fittings as indicated by their reaction to immersion in anhydrous acetone is very useful. An unfused PVC compound attacked by an anhydrous acetone causes the material to swell, flake, or completely disintegrate. A properly fused PVC compound is impervious to anhydrous acetone and only a minor swelling, if any, is observed. Defective PVC pipe or fittings returned from the field are subjected to this test for failure analysis.

Solvent stress analysis provides a quantitative means of determining stress levels in parts molded from various plastic materials. The test is primarily used in evaluating residual molded-in stress. The test is also very useful in determining the effect of external stress such as the ones induced by the screw, rivet, or insert used in assembly. Inherent stresses develop in molded parts as a result of injecting hot molten plastic material into a relatively cold mold. Such molecular orientation develops during the mold-filling phase as the melt is injected through the nozzle, runners, gates, and cavity. Internal stresses related to molecular orientation can lead to warpage or remain as internal stresses, reducing the durability and environmental stress crack resistance of the molded part. Highly stressed parts are attacked by solvents above its critical stress level. The net effect is crazing and cracking, which can be visually seen on the part. Figure 10 illustrates the result of solvent attack in terms of crazing and cracking around the highly stressed mounting holes.

3.4 Heat Reversion (ASTM F1057)

All plastics manufacturing processes introduce some degree of stress in the finished product. The stresses in molded parts are commonly referred to as molded-in (residual) stresses. By reversing the process, by reheating the molded or extruded product, the presence of stress

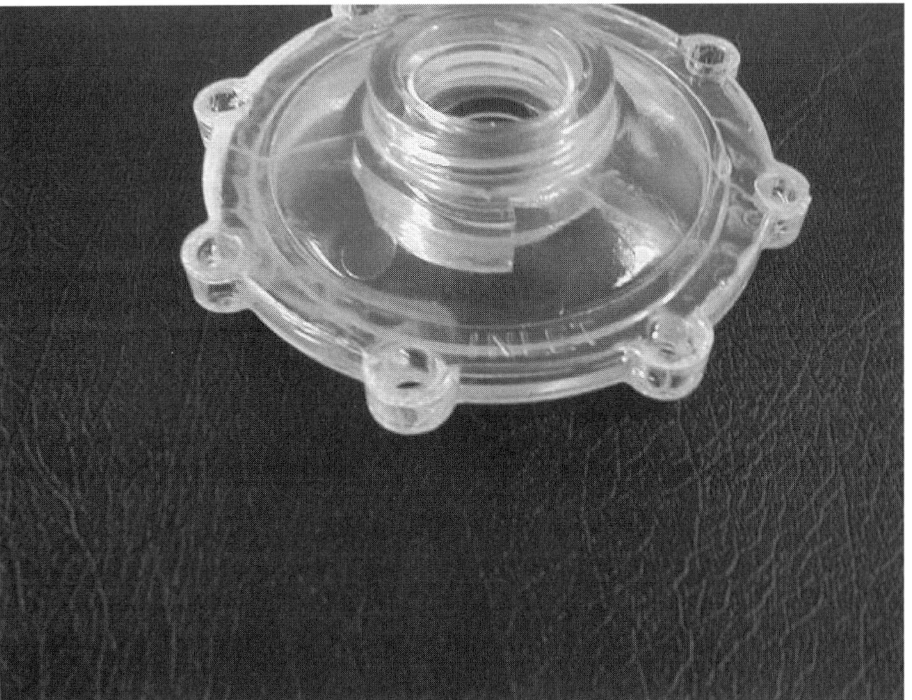

Figure 10 Result of solvent attack in terms of crazing and cracking around highly stressed mounting holes.

can be determined. The test is conducted by simply placing the entire specimen or a portion of the specimen in a thermostatically controlled, circulating air oven and subjecting to a predetermined temperature for a specified time. The specimens are visually examined for a variety of attributes. The degree and severity of warpage, blistering, wall separation, fish scaling, and distortion in the gate area of molded parts indicate the stress level. Stresses and molecular orientation effects in the plastic material are relieved, and the plastic starts to revert to a more stable form. The temperature at which this begins to occur is important. If changes start below the heat distortion temperature of the material, high levels of stress and flow orientation are indicated. The test has been significantly improved by new methods, including the attachment of strain gauges to critical regions of the part to carefully monitor initial changes in the shape. ASTM F1057 describes the standard practice for estimating the quality of extruded PVC pipe by heat reversion.

3.5 Microtoming (Microstructural Analysis)

Microtoming is a technique of slicing an ultrathin section from a molded plastic part for microscopic examination. This technique has been used by biologists and metallurgists for years, but only in the last decade has this technique been used successfully as a valuable failure analysis tool.

Microtoming begins with the skillful slicing of an 8–10-μm-thick section from a part and mounting the slice on a transparent glass slide. The section is then examined under a light transmission microscope equipped with a polarizer for photoelastic analysis. A high-power (1000×) microscope which will permit photographic recording of the structure in color is preferred. By examining the microstructure of a material, much useful information can be derived. For example, microstructural examination of a finished part that is too brittle may show that the melt temperature was either nonuniform or too low. The presence of unmelted particles is usually evident in such cases. Another reason for frequent failures of the injection-molded part is failure to apply sufficient time and pressure to freeze the gates. This causes the parts to be underpacked, which creates center-wall shrinkage voids. Figure 11 illustrates such shrinkage voids. Voids tend to reduce the load-bearing capabilities and toughness of a part through the concentration of stress in a weak area. Contamination, indicated by abnormality in the microstructure, almost always creates some problems. Contamination caused by the mixing of different polymers can be detected through such analysis by carefully studying the differences in polymer structures. Quite often, a poor pigment dispersion also causes parts to be brittle. This is readily observable through the microtoming technique. To achieve optimum properties, additives such as glass fibers and fillers must disperse properly. Microtoming a glass-fiber-reinforced plastic part reveals the degree of bonding of the glass fiber to the resin matrix as well as the dispersion and orientation of glass fibers. Molded-in stresses as well as stresses resulting from external loading are readily observed under cross-polarized light because of changes in birefringence when the molecular structure is strained. Microtoming can also be applied to check the integrity of spin and ultrasonic or vibration welds.

3.6 Mechanical Testing

Defective product returned from the field is often subjected to a variety of mechanical tests to determine the integrity of the product. Two basic methods are employed. The first one involves conducting mechanical tests such as tensile, impact, or compression on the actual

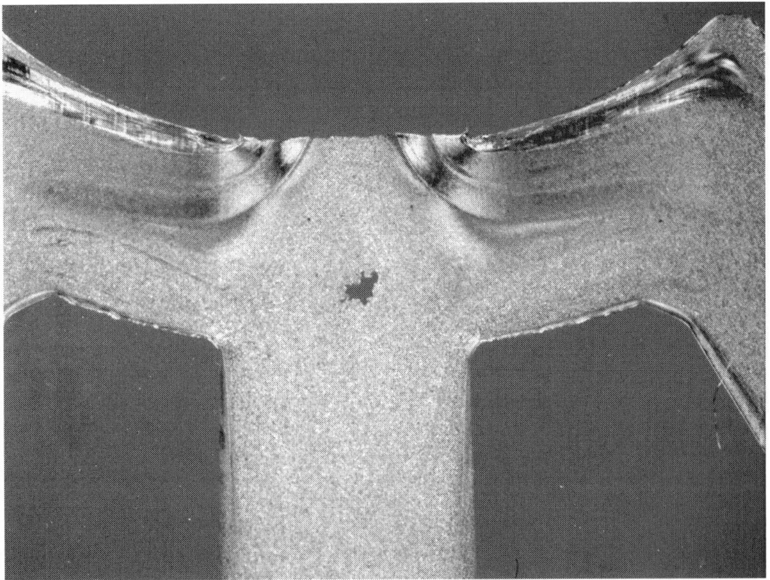

Figure 11 Shrinkage voids created by insufficient time and pressure to freeze the gates during injection-molding process. (Courtesy of BASF Corp.)

part or a small sample cut from the part. The test results are then compared to the test results obtained from the retained samples. The second method requires grinding the defective parts and either compression or injection molding standard test bars and conducting mechanical tests. The test results are compared to the published data for the virgin material. The amount of material available for molding the test bars quite often precludes injection molding. The test data obtained from compression-molded samples are generally lower than for injection-molded samples. Fatigue failure tests such as flexural fatigue or tensile fatigue can be employed to determine premature failure from cyclic loading.

3.7 Thermal Analysis

Thermal analysis techniques are used extensively in failure analysis. Three of the most popular techniques are differential scanning calorimetry (DSC), thermogravimetric analysis (TGA), and thermomechanical analysis (TMA). DSC is a technique in which the heat flow rate to the sample (differential power) is measured while the temperature of the sample, in a specified atmosphere, is programmed. Because all materials have a finite heat capacity, heating or cooling a sample specimen results in a flow of heat in or out of the sample. TGA is a test procedure in which changes in weight of a specimen are monitored as the specimen is progressively heated. The sample weight is continuously monitored as the temperature is increased either at a constant rate or through a series of steps. The components of a polymer or elastomer formulation volatilize or decompose at different temperatures. This leads to a series of weight-loss steps that allow the components to be quantitatively measured. TMA is a thermal analysis technique consisting of measuring physical expansion or contraction of a material or changes in its modulus of viscosity as a function of temperature.

3.8 Nondestructive Testing Techniques

NDT techniques are useful in determining the flaws, discontinuities, and joints. In its simplest form, NDT involves measurement, weighing, visual examination, and looking for surface imperfections, weak knit lines, etc. Ultrasonic testing and acoustic emission are the most common techniques used for failure analysis.

3.9 Fractography (ASTM C1145)

Fractography is one of the most valuable tools available for failure analysis. It differs from microstructural analysis techniques using a microtome, an instrument which slices a very

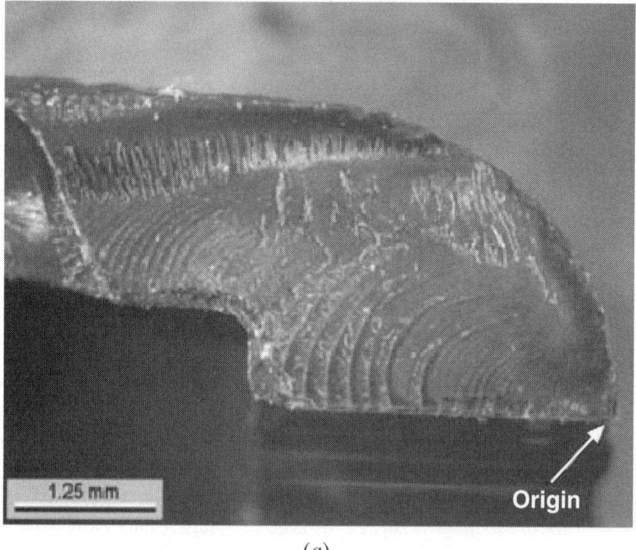

(*a*)

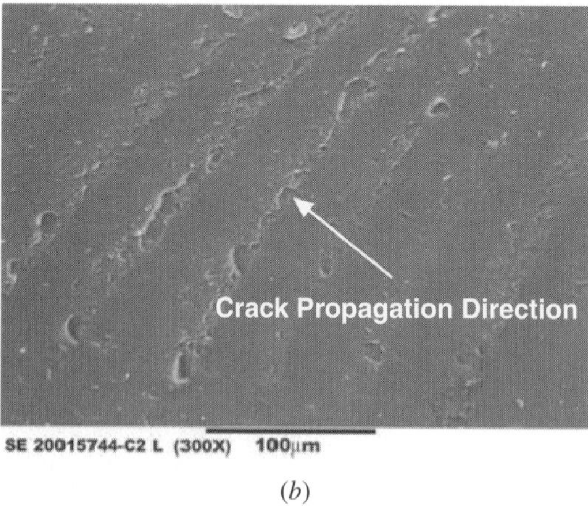

(*b*)

Figure 12 (*a*) Fatigue striations emanating from fracture origin of polycarbonate latch handle. (*b*) SEM micrograph of fatigue striations shown in (*a*). (Courtesy of IMR Test Labs.)

thin section from a part for microscopic examination. Fractography is defined as the means and methods for characterizing a fractured specimen or component. By examining the exposed fractured surface and interpreting the crack pattern and fracture markings, one can determine the origin of the crack, direction of the crack propagation, failure mechanism, and stresses involved. On the microscopic scale, all fractures fall into one of the two categories: ductile and brittle. Ductile fractures are characterized by material tearing and exhibit gross plastic deformation. Brittle fractures display little or no microscopically visible plastic deformation and require less energy to form. Fracture analysis usually involves the unexpected brittle failure of normally ductile material. The scanning electron microscope (SEM) is the most widely used instrument for fractographic analysis. Figure 12 shows a micrograph of a failed specimen.

3.10 Simulation Testing

In many instances, by simulating certain conditions such as exposing a part similar to the one that has failed to chemicals and/or other environment, valuable knowledge concerning possible cause of failure can be gained.

BIBLIOGRAPHY

Durelli, A. J., E. A. Phillips, and C. H. Tsao, *Introduction to the Theoretical and Experimental Analysis of Stress and Strain,* McGraw-Hill, New York, 1958.

Ezrin, M., *Plastics Failure Guide Causes and Prevention,* Hanser Gardner, 1996, p. 154.

Holman, J. P., *Experimental Methods for Engineers,* McGraw-Hill, New York, 1971, pp. 333–334.

"Introduction to Stress Analysis by the Photoelastic Coating Technique," Tech. Bull. IDCA-I, Photoelastic Division, Vishay Intertechnology, Raleigh, NC.

Levy, S., "Product Testing, Insurance against Failure," *Plastics Design Forum,* July–Aug. 1984, p. 83.

Measurements Group, Tech. Bull., Vishay Intertechnology, Raleigh, NC.

Miller, E., *Plastics Products Design Handbook,* Marcel Dekker, New York, 1981, p. 4

Morita, D. R., "QC Tests That Can Help Pin Point Material or Design Problems," *Plastics Design Forum,* May–June 1980, pp. 51–55.

Perrington, R., "Fractography of Metals and Plastics," www.imrtest.com.

"Reflection Polariscope," Tech. Bull. S-103-A, Photoelastic Division, Vishay Intertechnology, Raleigh, NC.

Sessions, M. L., *Microtoming, Engineering Design with Dupont Plastics,* E. I. Dupont Co., Wilmington, DE, Spring 1977, p. 12.

Spier, I. M., "The Most Common Mistakes Made by Design Engineer Working in Plastics," *Plastic Design Forum,* Mar./Apr. 1986, p. 24.

Tobin, W., "Why Products Fail," *Plastic Design Forum,* Jan./Feb. 1987, p. 48.

Vogt, J. P., "Testing for Mechanical Integrity Assures Service Life of Plastic Parts," *Plast. Design Process.,* Mar. 1976, pp. 12–13.

CHAPTER 28

FAILURE MODES: PERFORMANCE AND SERVICE REQUIREMENTS FOR CERAMICS

Dietrich Munz
Institut für Zuverlässigkeit und Schadenskunde im Maschinenbau
University of Karlsruhe
Karlsruhe, Germany

1 OVERVIEW

A reliable design of components requires the knowledge of the possible failure modes under service conditions and knowledge of the material properties describing these failure modes. These material properties are characterized by specific material parameters, such as strength, fracture toughness, hardness, and relations between loading parameters and materials response, e.g., the stress–strain curve or the *S–N* curve for fatigue loading. A reliable design has to incorporate the uncertainties in the loading parameters and in the material response. Therefore, a correct description of the scatter of the material properties is an important part in any design procedure. Material properties usually are determined under simple loading conditions: uniaxial loading, homogeneous stress state, constant temperature, simple load–time variations, such as constant load, cyclic loading, or constant loading rate. The loading conditions of components are more complex: multiaxial and inhomogeneous stress state, complex load, and temperature variation. Therefore, methods to transform results from simple loading conditions to complex situations in components have to be developed and applied.

For metals, a good knowledge of failure modes, design rules following from these failure modes, and criteria for materials selection exists. Ceramics are brittle materials that fail

Reprinted from *Handbook of Materials Selection,* Wiley, New York, 2002, by permission of the publisher.

without preceding plastic deformation. Only at high temperatures may creep deformation occur before fracture. The brittle failure mode requires specific design rules and criteria for material selection. Whereas in metals high local stresses lead to local plastic deformation only and not to failure of the component in ceramics, these high local stresses can lead to failure at relatively low global stresses. High local stresses can occur under different situations: sharp notches, thermal shock, contact loading. Tension stresses under these conditions are dangerous because of the lower tensile strength compared to compressive strength. From this behavior general design rules follow:

- Avoid sharp notches.
- Avoid surface damage, e.g., scratches.
- Avoid sudden changes in temperature.
- Avoid point or line loading of ceramic components.

The failure in ceramics starts from flaws introduced during fabrication or surface treatment. These flaws may be pores, inclusions, cracks, or surface scratches. Due to the scatter in the size of the flaws, also the strength shows a large scatter, which is much larger than in metals. This requires special statistical design criteria. This large scatter is associated with a large effect of the size of the component on the strength, which has to be included in the statistical design criteria.

Under mechanical or thermal loading, different failure modes are possible:

- Unstable fracture
- Subcritical extension of flaws under constant loading (static fatigue)
- Subcritical extension of flaws under cyclic loading (cyclic fatigue)
- Creep deformation and creep fracture at high temperatures
- Corrosion, especially oxidation
- Wear in components under sliding contact

The different failure modes are characterized by material properties and relations between the applied loading in terms of stresses or stress intensity factors and the materials response in terms of lifetime or plastic deformation. The material properties are:

- Tensile strength σ_c or, to be more specific, σ_{ct}
- Compressive strength σ_{cc}
- Fracture toughness K_{IC}
- Hardness H

The failure relations are:

- Relation between stress σ and lifetime t_f
- Relation between stress amplitude σ_a or stress range $\Delta\sigma = 2\sigma_a$ and number of cycles to failure N_f
- Relation between creep strain and time

The flaws leading to failure can be described as cracks. Therefore, the methods of fracture mechanics can be applied to describe the flaw extension behavior. The corresponding relations are:

- Relation between flaw size and strength
- Relation between crack growth rate and stress intensity factor for static fatigue
- Relation between crack extension for one load cycle and the range of the stress intensity factor for cyclic fatigue

Because of the scatter in strength and in the lifetime, the failure cannot be described by a unique strength or a fixed lifetime. For a given load, only the probability of failure can be specified. This failure probability depends not only on the maximum load in the component but also on the size of the component and the stress distribution in the component.

2 FLAWS

Looking on a fracture surface of a ceramic, one very often can find the origin of the fracture. An example is shown in Fig. 1. It can distinguish between intrinsic and extrinsic flaws. Intrinsic flaws are generated during the fabrication process. They can be pores, shrinkage cracks, or areas of low density. Inclusions of other materials and specific grain configurations, especially large grains, may lead to high microstructural residual stresses after cooling from the sintering temperature. These stresses may cause cracks during the application of a small mechanical load. Phase transformation during cooling or under an applied load, accompanied by a change in volume, e.g., in zirconia, is another cause of crack nucleation. All these flaws are primarily volume flaws that, however, may extend to the surface. Extrinsic flaws are surface flaws caused by surface treatment, such as machining or grinding. Oxidation at high temperature may introduce surface pits.

3 FRACTURE MECHANICS

The flaws that cause failure are described as cracks. Consequently, the methods of linear elastic fracture mechanics can be applied. The relation between the size of the flaw a and the strength σ_c is

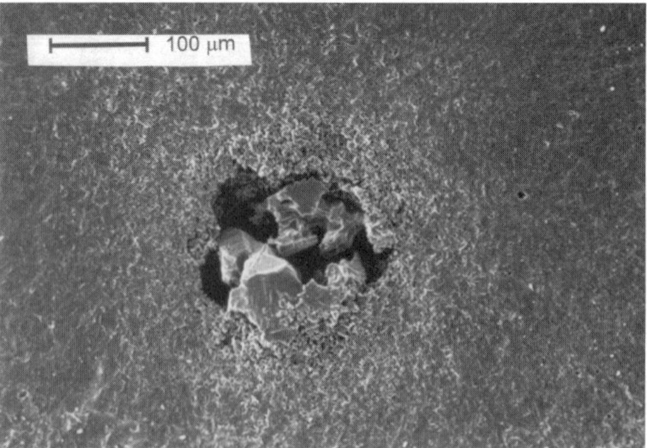

Figure 1 Flaw origin in a silicon carbide.

$$\sigma_c = \frac{K_{IC}}{\sqrt{a}Y} \tag{1}$$

where K_{IC} is the fracture toughness and Y a constant that depends on the shape of the crack. For a semicircular surface flaw $Y = 1.28$.

The fracture toughness is a material property. In Table 1 typical values of fracture toughness are listed. They range from less than 1 MPa·$\sqrt{\text{m}}$ for glass to 8 MPa·$\sqrt{\text{m}}$ for silicon nitride and 12 MPa·$\sqrt{\text{m}}$ for zirconia. For a given material the fracture toughness depends on the specific microstructure and density. Therefore, a large variation of K_{IC} is observed.

In Fig. 2 the strength σ_c according to Eq. (1) is plotted versus the crack size a for different values of K_{IC}. A flaw of a size of 100 μm leads to a strength of 391 MPa for a material with $K_{IC} = 5$ MPa·$\sqrt{\text{m}}$ (e.g., alumina) and to a strength of 781 MPa for a material with $K_{IC} = 10$ MPa·$\sqrt{\text{m}}$ (e.g., a zirconia). For a flaw size of 20 μm the corresponding values are 873 and 1747 MPa. In view of this large effect of the flaw size on the strength, a process technique that leads to a reduction of the processing flaws is desirable. Fracture may also be caused by surface flaws, introduced by grinding. Therefore, polishing the surface may also increase the strength.

The type of flaws also affects the strength. Applying Eq. (1) requires the existence of a crack. Three-dimensional flaws such as pores are described as spheres with a surrounding annular crack (Fig. 3). The stress intensity factor then depends on the ratio of the size of the crack a to the radius of the sphere R. The size of the crack may be on the order of the grain size.

4 STRENGTH

Equation (1) can be applied to predict the strength if the crack size is known. In metals this relation is used, for instance, to predict the remaining strength of a component with a fatigue

Table 1 Fracture Toughness and Strength of Different Ceramics

	Fracture Toughness (MPa$\sqrt{\text{m}}$)	Tensile Strength (MPa)	Compressive Strength (MPa)
AlN	3–4.5	300–500	
Al_2O_3	3–4.5	300–500	2500
Al_2TiO_5		10–50	
B_4C	3.5	300–500	2900
BeO		200–400	1500
$MoSi_2$		300	2500
SiC hot pressed	4.6	500–850	460
Si_3N_4 dense reaction bonded	4–8 1.5–2.8	500–1200 150–350	
SiO_2 fused quartz	0.8	50–100	1300 2000
TiB_2		300–400	1600
TiC		400	
ZrB_2		300	
ZrO_2	8–12	800–1000	2000
Glass	0.6–1.0		

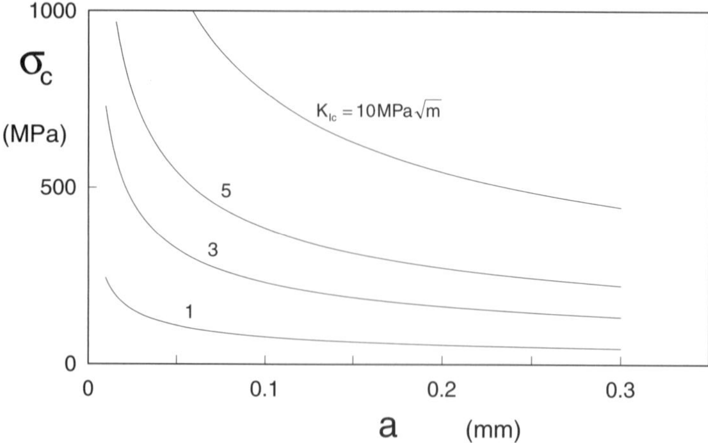

Figure 2 Strength as a function of flaw size for materials with different fracture toughness.

crack, whose size has been determined by a nondestructive testing method and is on the order of millimeters. In ceramic materials the flaws are small, on the order of 10–100 μm, and difficult to detect. Therefore Eq. (1) has not yet been used to predict the strength or to reject components after a nondestructive testing procedure. However, the relation shows the potential of a material. A material with a high fracture toughness and a low strength may have a potential for higher strength by a reduction in the flaw size.

In Fig. 4 the strength of SiC is plotted versus the crack size in a log-log plot. For cracks larger than about 100 μm a straight line with a slope of $-\frac{1}{2}$ is obtained according to Eq. (1). For smaller cracks a deviation with an upper limit is observed. The equivalent crack size is a length transformed to an internal crack in a plate. The reasons for the deviation from Eq. (1) for small flaw sizes will not be discussed here.

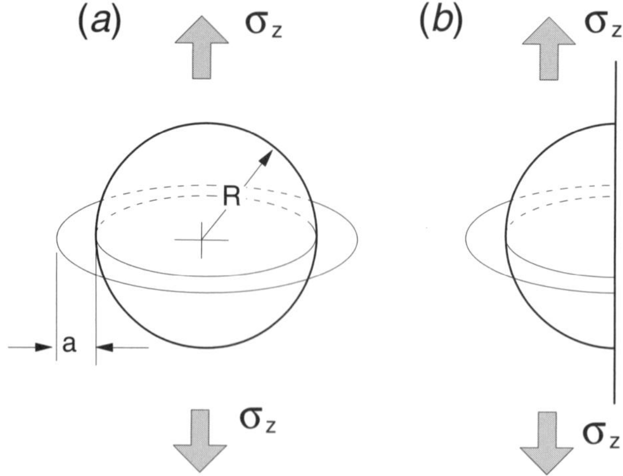

Figure 3 Pores with annular crack.

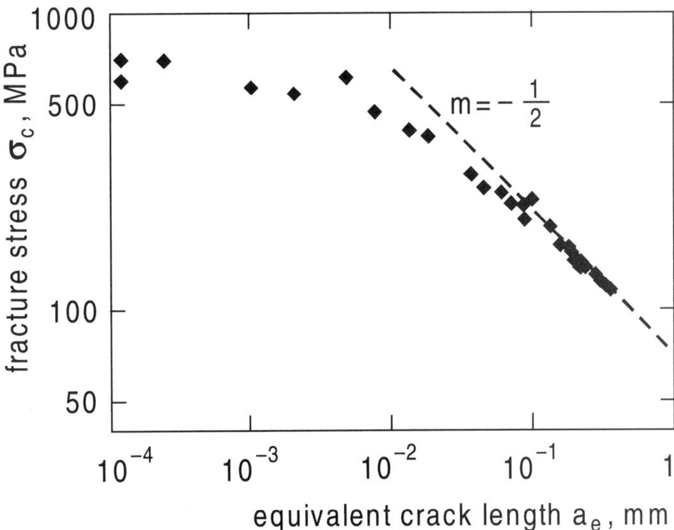

Figure 4 Strength of silicon carbide vs. crack size. (From Ref. 1.)

Because of the large dependence of the strength on the flaw size for a given material, the strength exhibits a large variation, depending on the processing method of the material. Strength also can be varied for a given material by varying the grain size and the microstructure in multiphase materials because this may change the fracture toughness and the processing flaw size. The values given in Table 1 are typical values of dense materials. Generally, the tensile strengths of ceramics are lower than those of high-strength metals. Silicon nitride and some zirconia materials, however, have strength values of up to 1000 MPa.

5 DELAYED FAILURE

Under cyclic loading, but also under constant loading, ceramic materials may fail after some time. This is called static fatigue for constant loading and cyclic fatigue for cyclic loading. This designation contrasts with that in metals, where the designation of fatigue is restricted to the damage under cyclic loading. The failure behavior is described by an *S–N* curve, where the stress is plotted versus the lifetime under static loading and the stress range $\Delta\sigma$ (twice the amplitude) versus the number of cycles to failure under cyclic loading. In a log-log plot straight lines are observed, which correspond to the relations

$$t_f = \frac{A}{\sigma^n} \tag{2a}$$

$$N_f = \frac{A}{(\Delta\sigma)^n} \tag{2b}$$

The phenomenon of delayed failure is caused by subcritical extension of the flaws. For static loading the crack growth rate depends on the stress intensity factor K and for cyclic loading on the range of the stress intensity factor ΔK. In a large range of crack growth rate, the subcritical crack extension can be described by the relations

$$\frac{da}{dt} = CK^n \tag{3a}$$

$$\frac{da}{dN} = C(\Delta K)^n \tag{3b}$$

where da/dN is the crack extension for one load cycle. Equation (3) can be obtained by integration of Eq. (4) from an initial crack size a_i to a critical crack size a_c. Because of the high values of n, the critical crack size can be set to infinity. The values of n in Eqs. (3) and (4) are the same for corresponding loading situations (static or cyclic). The parameters A, C, and n are different for static and cyclic loading, n_{cyclic} being always somewhat smaller than n_{static}. For cyclic loading A and C depend on the mean stress, which conveniently is expressed by the ratio $R = \sigma_{\min}/\sigma_{\max} = K_{\min}/K_{\max}$.

The following relation was found:

$$\frac{da}{dN} = \frac{C}{(1 - R)^{n-p}} (\Delta K)^n \tag{4}$$

Integration leads to

$$N_f = C* \frac{(1 - R)^{n-p}}{(\Delta\sigma)^n} = C* \frac{1}{(\Delta\sigma)^p(\sigma_m + \Delta\sigma/2)^{n-p}} \tag{5}$$

Some values are given in Table 2. High values of n mean a large effect of the stress or the stress range on the lifetime.

Summary

- Delayed failure is described by a power law relation between crack growth rate and the stress intensity factor for constant loading and by a power law relation between crack extension for one load cycle and the range of the stress intensity factor for cyclic loading.

6 SCATTER OF STRENGTH AND LIFETIME

6.1 Scatter of Strength

Because of the statistical distribution of the flaw size, the strength is subjected to considerable scatter. As a consequence, the design of ceramic components cannot be based on an average strength and the application of a safety factor, as is usually done for metallic components. The scatter is associated with an effect of the size of a component and the stress distribution

Table 2 Some Values of n and p^2

Material	n	p^2
ZrO_2 (Y-TZP)	19	3
Si_3N_4	29	1.3
SiC_w/Al_2O_3	10.2	4.8

in the component on the strength. Therefore, a statistical treatment of the scatter and a design according to prescribed failure probability is necessary. In this section the method of applying Weibull statistics to uniaxial loading is described. The design applying multiaxial Weibull statistics will be presented in Section 7.

The scatter in the strength is described by the two-parameter Weibull distribution. The failure probability, i.e., the probability of failure occurring below the stress $\sigma = \sigma_c$, is given by

$$F = 1 - \exp\left[-\left(\frac{\sigma_c}{\sigma_0}\right)^m\right] \tag{6}$$

The Weibull parameter m is called the Weibull modulus and is a measure of the scatter. The Weibull parameter σ_0 is related to the strength. Whereas m is a material parameter, σ_0 depends on the material as well as the size of the component and the stress distribution in the component.

From Eq. (6) it follows that failure is possible also at very low stresses. Therefore, a three-parameter Weibull distribution with a lower limit σ_u is considered sometimes:

$$F = 1 - \exp\left[-\left(\frac{\sigma_c - \sigma_u}{\sigma_0}\right)^m\right] \qquad \sigma_c > \sigma_u \tag{7}$$

It is, however, not possible to determine σ_u with sufficient accuracy when applying a limited number of test specimens. Hence, the two-parameter distribution is used usually.

Failure can be caused by surface flaws or volume flaws. For a homogeneous stress distribution in a component and failure by volume flaws, Eq. (6) can be replaced by

$$F = 1 - \exp\left[-\frac{V}{V_0}\left(\frac{\sigma_c}{\sigma_V}\right)^m\right] \tag{8}$$

where V is the volume and V_0 is the unit volume, which is introduced for normalization. Now σ_V is a quantity independent of the size of the component. An inhomogeneous stress distribution in a component is described by the relation

$$\sigma(x, y, z) = \sigma^* g(x, y, z) \tag{9}$$

where σ^* is a characteristic stress in the component and $g(x, y, z)$ a geometry function. The failure probability then is given by

$$F = 1 - \exp\left[-\frac{1}{V_0}\left(\frac{\sigma_c^*}{\sigma_V}\right)^m \int g^m \, dV\right] = 1 - \exp\left[-\frac{V_{\text{eff}}}{V_0}\left(\frac{\sigma_c^*}{\sigma_V}\right)^m\right] \tag{10}$$

where σ_c^* is the value of σ^* at fracture and

$$V_{\text{eff}} = \int g^m \, dV \tag{11}$$

is the effective volume.

For surface flaws the volume in Eqs. (8)–(11) has to be replaced by the surface area.

If Eq. (6) is used to calculate the failure probability for a component, $\sigma_{0,C}$ of the component can be calculated from $\sigma_{0,\text{SP}}$ of the specimen by

$$\sigma_{0,C} = \sigma_{0,\text{SP}}\left(\frac{V_{\text{eff, SP}}}{V_{\text{eff},C}}\right)^{1/m} \tag{12}$$

Experimental results are presented in a Weibull diagram, where $\ln \ln 1/(1 - F)$ is plotted versus $\ln \sigma_c$. To correlate the measured strength values with the failure probabilities, the strength values are ranked in increasing order and numbered from 1 to n. Then, the strength values σ_{ci} are related to the failure probability F_i according to

$$F_i = \frac{i - 0.5}{n} \tag{13}$$

The slope of the Weibull plot is m; σ_0 is the strength $F = 0.632$. Examples are shown in Fig. 5.

The average strength $\overline{\sigma_c}$ and the median strength for σ_{med} (σ_c for $F = 0.5$) are related to m and σ_0 by

$$\overline{\sigma_c} = \sigma_0\Gamma\left(1 + \frac{1}{m}\right) \qquad \sigma_{\text{med}} = (0.693)^{1/m} \tag{14}$$

where Γ is the gamma function. In Table 3 the ratios $\overline{\sigma_c}/\sigma_0$ and $\sigma_{\text{med}}/\sigma_0$ are given for different values of m.

Equations (8) and (10) lead to a large effect of the size of the component on the strength, which depends on the Weibull modulus m. They also show a large effect of m on the failure probability. This effect is illustrated in Fig. 6 by the strength at different failure probabilities, Weibull parameters m, and different effective volumes. For a material with a mean strength $\overline{\sigma_c} = 500$ MPa determined with four-point specimens having an inner span $L = 20$ mm, width $W = 4.5$ mm, and height $H = 3.5$ mm, the effective volume is

$$V_{\text{eff}} = \frac{LWH}{2(m + 1)} \tag{15}$$

In Table 4 the parameter σ_0 and the effective volume are given for different failure probabilities and different values of m.

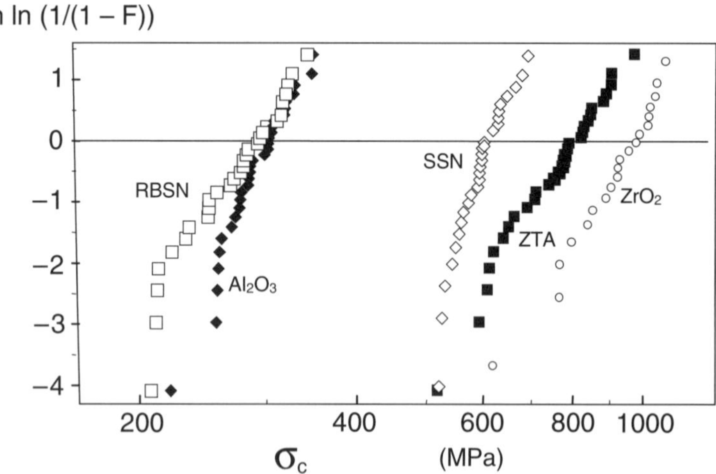

Figure 5 Weibull plots of strength for several materials (RBSN: reaction-bonded silicon nitride, SSN: sintered silicon nitride, ZTA: zirconia-toughend alumina).

Table 3 Ratios $\overline{\sigma_c}/\sigma_0$ and σ_{med}/σ_0

	$m = 5$	$m = 10$	$m = 20$	$m = 30$	$m = 50$
$\overline{\sigma_c}/\sigma_0$	0.918	0.951	0.974	0.982	0.989
σ_{med}/σ_0	0.929	0.964	0.982	0.988	0.993

6.2 Scatter of Lifetime

The scatter of the lifetime under constant or cyclic loading is much larger than that of the strength and can also be described by a Weibull distribution. For static fatigue the failure probability is

$$F = 1 - \exp\left[-\left(\frac{t_f}{t_0}\right)^{m^*}\right] \quad (16a)$$

and for cyclic fatigue

$$F = 1 - \exp\left[-\left(\frac{N_f}{N_0}\right)^{m^*}\right] \quad (16b)$$

If the strength and lifetime are caused by the same flaws, then the Weibull parameters m^*, t_0, and N_0 can be related to the Weibull parameters m and σ_0 of the strength:

$$m^* = \frac{m}{n-2} \quad (17)$$

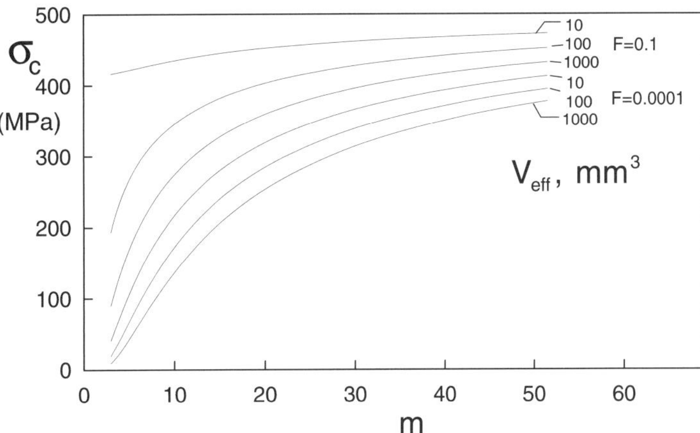

Figure 6 Effect of Weibull modulus m and effective volume on the strength for two different failure probabilities for materials with the same average strength of 500 MPa.

Table 4 Strength at Different Failure Probabilities and Effective Volumes (Materials with the Same Average Strength in Bending Tests)

m	V_{eff} (mm^3)	σ_0	$F = 0.0001$	$F = 0.001$	$F = 0.01$	$F = 0.1$	$F = 0.5$
5	10	661	105	166	264	4222	615
	100	417	66	105	168	266	388
	1000	263	42	66	195	168	245
10	10	545	217	273	435	435	526
	100	433	172	217	346	346	417
	1000	349	137	172	275	275	332
20	10	506	319	358	402	452	496
	100	451	284	319	358	403	442
	1000	402	253	284	319	359	394
50	10	495	411	431	451	473	491
	100	473	393	412	431	452	470
	1000	452	376	394	412	432	449

$$t_0 = \frac{B\sigma_0^{n-2}}{\sigma^n} \tag{18a}$$

$$N_0 = \frac{B\sigma_0^{n-2}}{(\Delta\sigma)^n} \tag{18b}$$

where n is the crack growth exponent for static and cyclic crack extension, respectively; B is given by

$$B = \frac{2}{(n-2)CY^2K_{IC}^{n-2}} \tag{19}$$

with C from Eq. (4).

In Eq. (18) $\sigma_{0,C}$ of the component has to be used. If $\sigma_{0,\text{SP}}$ as obtained from the specimens is known, the parameters t_0 and N_0 in Eqs. (16) are

$$t_{0,C} = \frac{B\sigma_{0,\text{SP}}^{n-2}}{\sigma^n}\left(\frac{V_{\text{eff, SP}}}{V_{\text{eff}, C}}\right)^{(n-2)/m} \tag{20a}$$

$$N_{0,C} = \frac{B\sigma_{0,\text{SP}}^{n-2}}{(\Delta\sigma)^n}\left(\frac{V_{\text{eff, SP}}}{V_{\text{eff}, C}}\right)^{(n-2)/m} \tag{20b}$$

Summary

- The scatter of the strength is described by the two-parameter Weibull distribution with the Weibull modulus m.
- The average strength and the strength at given failure probability depend on the size of the component and the stress distribution in the component.
- The lifetime is described by a Weibull distribution; the parameters can be obtained from the Weibull parameters of the strength, if strength and lifetime are caused by the same flaws.

7 DESIGN APPLYING MULTIAXIAL WEIBULL STATISTICS

7.1 Strength under Compression Loading

The strength of ceramic materials under compressive loading is much higher than under tension loading. Usually compression tests are performed applying cylindrical specimens. The cylinders are loaded with parallel stamps. Due to the different lateral expansion of the specimen and the stamp under compressive loading, radial stresses occur in the loaded specimen, which may lead to premature failure. Therefore the compressive strength measured may depend on the loading conditions. A detailed description of the compressive test is given in Ref. 2. Some compressive strength results are presented in Table 1. The ratio between compressive and tensile strength ranges between 2.5 and 18 depending on the material and the porosity. Dense materials have a higher ratio than porous materials.

7.2 Global Multiaxial Fracture Criterion

Whereas the strength of materials is measured under uniaxial tension or compression loading, multiaxial stresses very often occur in components. This is the case for uniaxial loading of notched components, components under internal pressure, or components under inhomogeneous temperature distribution. The degree of multiaxiality is described by the ratios

$$\alpha = \frac{\sigma_2}{\sigma_1} \qquad \beta = \frac{\sigma_3}{\sigma_1} \tag{21}$$

where σ_1, σ_2, and σ_3 are the principal stresses. Two different approaches exist to assess multiaxial stresses: global multiaxiality and local multiaxiality criteria.

A global multiaxiality criterion is a relation between the principal stresses and one or two material properties. The most reliable criterion is Mohr's hypothesis, which is presented here for a plane-stress situation. It is based on the tensile strength σ_{ct} and compressive strength σ_{cc}. The criterion is shown in Fig. 7 and reads

$$\begin{aligned} \sigma_1 &= \sigma_{ct} && \text{for } \sigma_1 > 0,\ \sigma_2 > 0 \\ \sigma_2 &= -\sigma_{cc} && \text{for } \sigma_1 < 0,\ \sigma_2 < 0 \\ \sigma_1 &= \sigma_{ct}\left(1 + \frac{\sigma_2}{\sigma_{cc}}\right) && \text{for } \sigma_1 > 0,\ \sigma_2 < 0 \end{aligned} \tag{22}$$

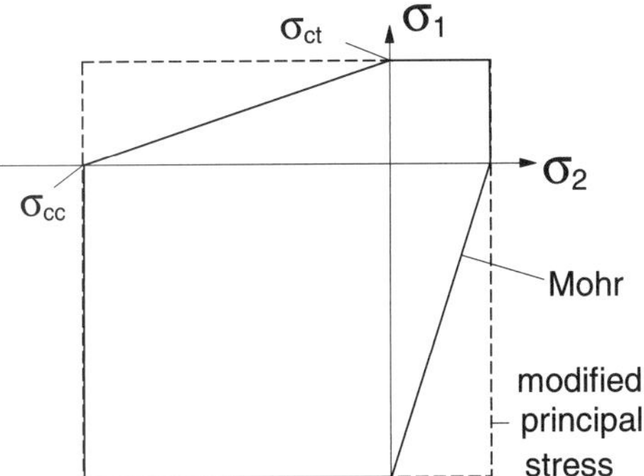

Figure 7 Mohr's hypothesis for biaxial loading.

7.3 Local Multiaxiality Criterion

The local multiaxiality criterion starts with a flaw type that may be a pore or a crack. The local fracture criterion is a critical tensile stress for a pore or a critical stress intensity factor for a crack. Details can be found in Ref. 3. In the pore models only the tensile strength appears as a material property. For a spherical pore the failure relation is

$$\sigma_1 + \frac{15\nu - 3}{27 - 15\nu}\sigma_2 - \frac{3 + 15\nu}{27 - 15\nu}\sigma_3 = \sigma_{ct} \tag{23}$$

From this relation the compressive strength follows as

$$\sigma_{cc} = \frac{27 - 15\nu}{3 + 15\nu}\sigma_{ct} = \begin{cases} 4\sigma_{ct} & \text{for } \nu = 0.2 \\ 3\sigma_{ct} & \text{for } \nu = 0.3 \end{cases} \tag{24}$$

Under triaxial compression, no failure is expected because the stresses at the pore are negative. In Fig. 8 the failure diagram for plane stress is shown. Results for an ellipsoidal pore are represented in Fig. 9. From these results it can be concluded that the ratio of compressive to tensile strength may vary between about 3 and 6, depending on the shape of the pore and the Poisson ratio. It also shows that Mohr's hypothesis is a good approximation to the more detailed pore models.

For a given pore geometry (spherical or ellipsoidal with a given ratio of b/a), the strength is independent of the size of the pore. Consequently, it is not possible to derive a statistical multiaxial failure relation. Such a relation can be obtained for a crack model. In this model the flaws are described as cracks with a given shape, e.g., as circular cracks or semicircular surface cracks. Then, a statistical multiaxial failure criterion can be developed in the following steps. The cracks are assumed to be randomly orientated and have a size distribution that leads to the Weibull distribution of the strength under uniaxial loading. Depending on the orientation of the flaw with respect to the principal stress axis and the ratios of the principal stresses α and β, the stresses at the crack tip are described by the mixed-mode stress intensity factors K_I, K_{II}, and K_{III}: K_I is caused by the normal stress on the crack area σ_n; K_{II} and K_{III} are caused by the shear stress τ acting on the crack plane.

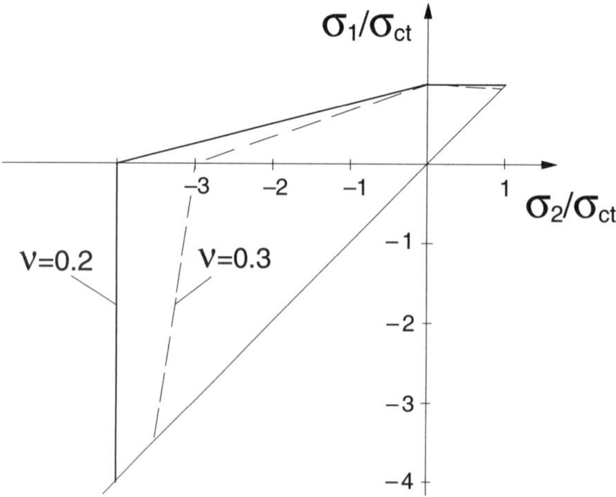

Figure 8 Failure diagram for a spherical pore.

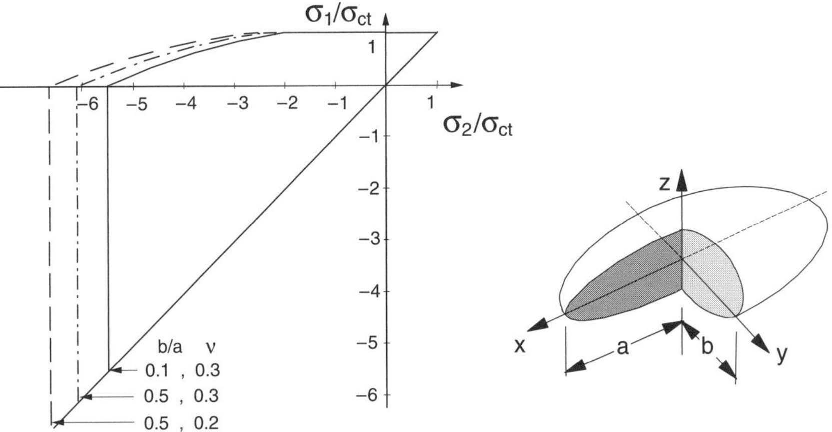

Figure 9 Failure diagram for an elliptical pore. (From Ref. 3.)

The local failure criterion is a relation between these three stress intensity factors and the fracture toughness K_{IC}. Different relations are discussed in literature, some of them also include the fracture toughness $K_{\mathrm{II},C}$ for mode II loading. An example is the criterion of constant-energy release rate:

$$K_{\mathrm{I}}^2 + K_{\mathrm{II}}^2 + \frac{1}{1-\nu} K_{\mathrm{III}}^2 = K_{IC}^2 \tag{25}$$

For any local stress state a critical orientation of the crack exists. If the normal stress is compressive ($\sigma_n < 0$), then $K = 0$ and K_{II} and K_{III} have to be calculated with an effective shear stress:

$$\tau_{\mathrm{eff}} = \begin{cases} |\tau| + \mu\sigma_n & \text{for } |\mu\sigma_n| < \tau \\ 0 & \text{for } |\mu\sigma_n| > \tau \end{cases} \tag{26}$$

with the friction coefficient μ. As an example, the resulting failure criterion is shown for circular cracks and a specific mixed-mode failure criterion in Fig. 10.

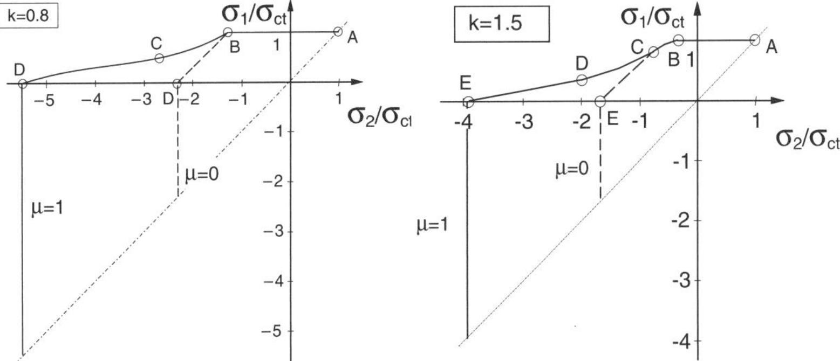

Figure 10 Failure diagram for circular cracks (k characterizes the mixed-mode criterion). (From Ref. 3.)

If the scatter of the strength is included in the calculation procedure, the resulting failure criterion also depends on the Weibull parameter m. For details see Ref. 3.

For an inhomogeneous stress distribution the failure probability of a component is given by the integral

$$F = 1 - \exp\left[-\frac{1}{2\pi} \left(\frac{\sigma_c^*}{\sigma_{I0}} \right)^m \iint h^m \sin \Phi \, d\Phi \, d\Psi \int g^m \, dV \right] \tag{27}$$

The angles Φ and Ψ characterize the orientation of the crack normal to the principal stress axis. The quantity h depends on the stress ratios α and β and on the mixed-mode fracture criterion; σ_{I0} is a material-dependent critical stress.

The flowchart for the computation of the failure probability is presented in Fig. 11. It shows the necessary input information:

- The stress state in the component (usually from finite-element calculation)
- A crack model
- A local failure criterion
- The material properties m and σ_{I0}

Postprocessor programs are available for the calculation of failure properties.

Summary

- The failure probability of components under multiaxial loading can be calculated applying multiaxial Weibull theory, where the flaws are described as randomly orientated cracks.
- As a global multiaxial fracture criterion Mohr's hypothesis can be applied.

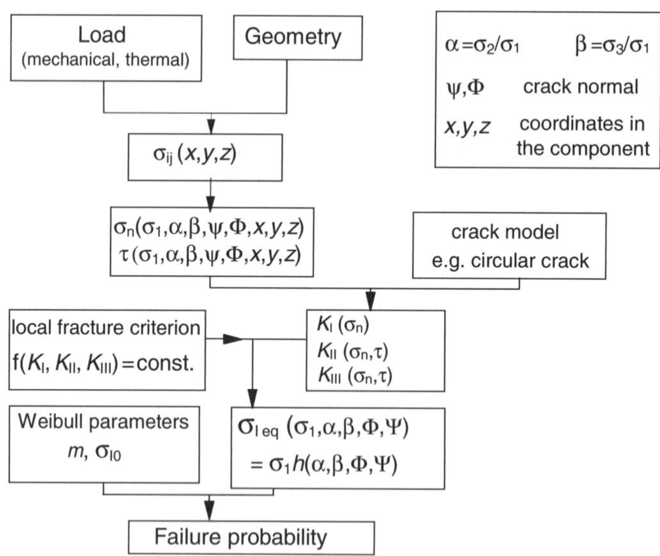

Figure 11 Flowchart for computation of failure probability for multiaxially loaded components. (From Ref. 3.)

8 MATERIALS SELECTION FOR THERMAL SHOCK CONDITIONS

Ceramic materials may fail during rapid cooling or rapid heating. Due to the time-dependent transient temperature distribution in a component, the thermal expansion depends on the location and has to be compensated for by mechanical deformation. Whereas in metals these strains lead to small local plastic deformation, they may cause high local stresses in ceramics that may exceed the local strength. The failure probability of a component under thermal shock conditions can be calculated applying the multiaxial Weibull statistics described in Section 7.

The resistance against thermal shock varies for different materials. It depends on the following material properties: thermal expansion coefficient α, elastic properties Young's modulus E, Poisson's ratio ν, thermal conductivity λ, density ρ, head capacity C_p, and strength σ_c. Under different thermal conditions different thermal shock parameters have been defined which characterize the thermal shock sensitivity.

For perfect heat transfer from the medium to the component the thermal stress at the surface after a thermal shock by ΔT is

$$\sigma_{\max} = \frac{E\alpha\,\Delta T}{1 - \nu} \tag{28}$$

The critical temperature for fracture under thermal shock therefore is

$$R = \frac{\sigma_c(1 - \nu)}{E\alpha} \tag{29}$$

For cooling with a constant heat transfer coefficient h the maximum stress at the surface increases with time and decreases after passing a maximum. The maximum stress is given by

$$\sigma_{\max} = \frac{E\alpha\,\Delta T}{1 - \nu} f(B) \tag{30}$$

where f is a function of the Biot modulus

$$B = \frac{hd}{\lambda} \tag{31}$$

where λ is the thermal conductivity and d a characteristic size parameter of the component. For small values of B the maximum stress is proportional to B. Hence the critical temperature for failure is proportional to

$$R' = \frac{\sigma_c\lambda(1 - \nu)}{E\alpha} \tag{32}$$

For a constant heating rate at the surface the maximum stress is

$$\sigma_{\max} = -C\frac{\alpha E\rho C_p}{\lambda(1 - \nu)}\frac{dT}{dt} \tag{33}$$

where ρ is the density, C_p the heat capacity, and C a constant depending on the geometry. The critical heating rate therefore is proportional to

$$R'' = \frac{\sigma_c\lambda(1 - \nu)}{E\alpha\rho C_p} \tag{34}$$

Whereas under mechanical loading a crack extends unstably after initiation, a crack stops after some extension under thermal shock conditions. The remaining strength, however, decreases considerably. If the strength of the component is not of importance, a limited amount of crack extension may be tolerable. Then, the condition for materials selection may be a small extension of an existing crack. It can be shown that the amount of crack extension during thermal shock increases with increasing crack size. Consequently a material with a large initial crack may be of advantage. The initial crack size follows from

$$\alpha_i = \left(\frac{K_{IC}}{\sigma_c Y}\right)^2 \tag{35}$$

leading to another thermal shock parameter

$$R'''' = \frac{K_{IC}^2}{\sigma_c^2} \tag{36}$$

(R''' is not considered here).

In Table 5 typical values of the thermal shock parameters are given.

Summary

- The sensitivity to failure by thermal shock increases with increasing Young's modulus, increasing thermal expansion coefficient, decreasing strength, and decreasing thermal conductivity.
- For a constant heating rate at the surface the sensitivity decreases further with increasing density and increasing heat capacity.

9 FAILURE AT HIGH TEMPERATURES

Failure at high temperatures is caused by creep elongation and creep rupture. Another failure mode is oxidation. Different mechanisms may contribute to creep and creep damage. In

Table 5 Thermal Shock Parameters for Different Materials

					Si_3N_4			
	Al_2O_3	MgO	ZrO_2	SiC	HPSN	RBSN	BeO	Al_2TiO_5
α (10^{-6} K^{-1})	8	12	11	4	3.2	2.5	8	1.8
E (GPa)	400	270	200	350	300	180	360	30
ν	0.22	0.17	0.25	0.2	0.28	0.23	0.25	0.2
λ (W m^{-1} K^{-1})	30	30	2.5	100	35	10	300	2.5
ρ (g cm^{-3})	3.9	3.5	6.0	1.0	0.7	0.7	1.3	0.7
C (J g^{-1} K^{-1})	1.0	1.0	0.5	1.0	0.7	0.7	1.3	0.7
σ_c(MPa)	300	180	950	360	660	200	180	65
K_{IC} (MPa·$m^{1/2}$)	4.5	3.0	10	4	7	2	4.8	—
R (K)	73	46	324	206	495	342	47	962
R' (kW m^{-1})	2.19	1.38	0.81	20.6	17.3	3.42	14.1	2.41
R'' (W cm^2 g^{-1} K)	5.6	3.9	2.7	66	75	20	36	9.6
R'''' (mm)	0.23	0.28	0.11	0.12	0.11	0.10	0.71	

many cases, creep is caused by grain boundary effects, such as viscous flow of an amorphous grain boundary phase, diffusion of vacancies along grain boundaries, formation and extension of cavities at grain boundaries or dissolution, and reprecipitation of material through the glassy phase. Creep within the grain is caused by the motion of dislocations or vacancies.

Damage mechanisms leading to the rupture of a component can be the formation of grain boundary cavities, coalescence to cracks, and growth of the cracks until final failure.

9.1 Creep Strain

After application of a load to a specimen, the same variation of strain with time as in metals is observed. After an instantaneous strain ε_0, the creep curve can be subdivided into three stages. In stage I—the primary creep—the strain rate decreases and reaches in stage II—secondary creep—a constant value. Afterwards, in stage III—tertiary creep—the strain rate increases.

Primary creep can be described by one of the following relations:

$$\varepsilon_p = At^m \quad \text{with } m < 1 \tag{36a}$$

$$\varepsilon_p = A[1 - \exp(-mt)] \tag{36b}$$

The effect of the stress on the creep rate is often described by Norton's law:

$$\dot{\varepsilon} = B\sigma^n \tag{37}$$

Other proposed relations are

$$\dot{\varepsilon} = B[\exp(\alpha\sigma) - 1] \tag{38}$$

$$\dot{\varepsilon} = B \sinh(\alpha\sigma) \tag{39}$$

These relations can be applied for the primary and secondary creep range with possibly different parameters α and n.

Creep under compression differs from creep in tension, especially for secondary creep, where the creep rate in compression is lower than under tension. An example is shown in Fig. 12. In this case the exponent n in Eq. (37) is similar in tension and compression; the constant B, however, is different. Figure 12 also shows that Eq. (37) cannot be applied for

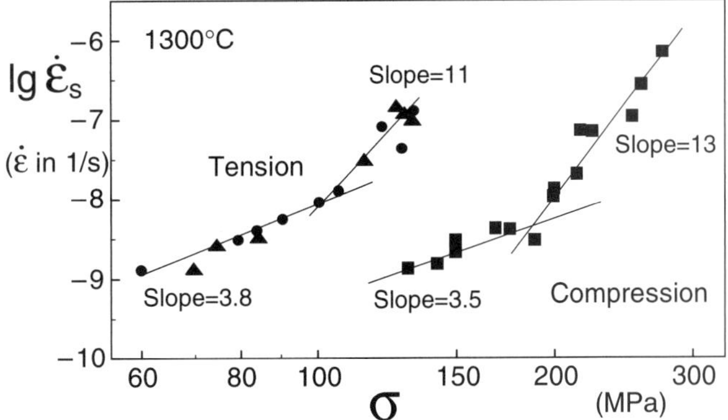

Figure 12 Stress dependence of the secondary creep rate for reaction-bonded siliconized silicon carbide tested under tension and compression (from Ref. 4).

the whole range of the stress: The exponent n for small stresses is lower than at higher stresses.

The effect of temperature usually can be described by

$$\dot{\varepsilon} = C \exp\left(-\frac{Q}{RT}\right) \tag{40}$$

where R is the gas constant and Q the activation energy for the leading creep process. An example of Eq. (40) is shown in Fig. 13.

The problem of these relations is that the lowest creep rates measured are usually higher than the allowable creep rate in a component. It is obvious from Fig. 12 that the lowest creep rates measured in these investigations are on the order of 10^{-9}/s. For this creep rate the creep strain is 3% in one year. Therefore an extrapolation to lower creep rates often is necessary.

9.2 Creep Rupture

The mechanism of creep rupture differs from the mechanism of subcritical crack extension at low temperatures. This can be seen in Fig. 14, where the stress is plotted versus the lifetime for an alumina at 1100°C in a log-log plot. Two ranges can be distinguished. At high stresses, the failure is caused by the extension of preexisting flaws. The relation between stress and failure strain can be described by Eq. (3a) with $n = 12$ [the slightly curved line in Fig. 14 is obtained by applying Eq. (3a) for the bending tests taking into account small stress redistribution due to creep]. At lower stresses, the observed lifetime is much lower than predicted from Eq. (3).

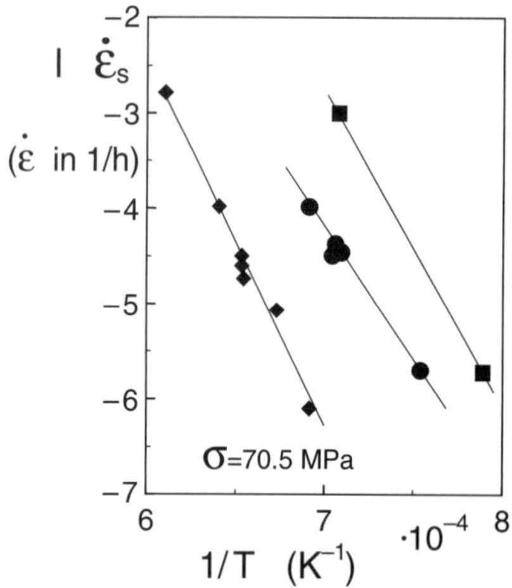

Figure 13 Secondary creep rate as a function of temperature for three hot-pressed silicon nitrides. (From Ref 5.)

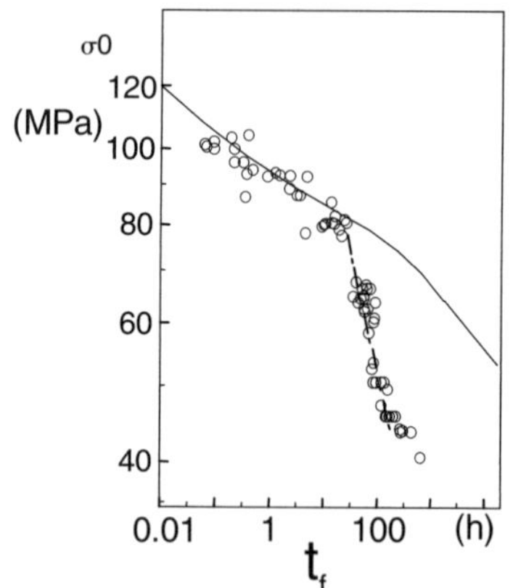

Figure 14 Lifetime as a function of the (elastically calculated) outer fiber bending stress for static bending tests on Al_2O_3 with 4% glass content. Solid line predictions from subcritical crack growth.

Summary

- Creep at high temperatures is described by relations between creep rate, stress, and temperature.
- The time to failure in the creep range is less stress dependent than at lower temperatures.

REFERENCES

1. S. Usami, I. Takahashi, and T. Machida, "Static Fatigue Limit of Ceramic Materials Containing Small Flaws," *Eng. Fract. Mech.,* **25,** 483–495 (1986).
2. G. Sines and M. Adams, *Compression Testing of Ceramics, Fracture Mechanics of Ceramics,* Vol. 3, Plenum, New York, 1978, pp. 403–434.
3. D. Munz and T. Fett, *Ceramics—Mechanical Properties, Failure Behaviour, Materials Selection,* Springer, Berlin, 1999.
4. D. F. Caroll and S. M. Wiederthom, "High Temperature Creep Testing of Ceramics," in *Mechanical Testing of Engineering Ceramics at High Temperatures,* Elsevier Applied Science, London, 1989, pp. 135–149.
5. R. Kossowsky, D. G. Miller, and E. S. Diaz, "Tensile Creep Strength of Hot-Pressed Si_3N_4," *J. Mater. Sci.,* **10,** 983–997 (1975).

CHAPTER 29

MECHANICAL RELIABILITY AND LIFE PREDICTION FOR BRITTLE MATERIALS

G. S. White, E. R. Fuller, Jr., and S. W. Freiman
Ceramics Division
Materials Science and Engineering Laboratory
National Institute of Standards and Technology
Gaithersburg, Maryland

1 SCOPE

This chapter is intended to provide the reader with a general understanding of why brittle materials fail in a time-dependent manner in service and how to estimate the lifetimes that can be expected for such materials. In addition, we describe procedures to evaluate the confidence with which these lifetime predictions can be applied. Throughout this chapter, we assume that the material under investigation can be treated as isotropic and homogeneous and that microstructural influences on properties can be ignored. Therefore, more complicated issues, such as crystallographic texture, grain boundary phases, and *R*-curve behavior, lie outside the scope of this discussion and will not be addressed herein.

2 INTRODUCTION

Techniques to determine reliability of components fabricated from brittle materials (e.g., ceramics and glasses) have been extensively developed over the last 30 years.[1–7] Reliability is generally defined as the *probability* that a component, or system, will perform its intended function for a specified *period of time*.[8] Accordingly, the two overarching principles influencing reliability are the statistical nature of component strength and its time-dependent, environmentally enhanced degradation under stress. The statistical aspect of strength derives from the distribution of the most severe defects in the components (i.e., the strength-

Reprinted from *Handbook of Materials Selection,* Wiley, New York, 2002, by permission of the publisher.

determining flaws).[9–15] The time-dependent aspect of strength results from the growth of defects under stress and environment, resulting in time-dependent component failure.[16–19] These concepts have led to a lifetime prediction formalism that incorporates strength and crack growth as a function of stress. Predicted reliability, or lifetime, is only meaningful, however, when coupled with a confidence estimate. Therefore, the final step in the lifetime prediction process must be a statistical analysis of the experimental results.[2,20–22]

While phenomenological, the reliability methodology has been useful in predicting lifetimes for myriad applications, including all-glass aircraft windows,[22] spacecraft[23,24] windows, flat panel displays,[25] optical glass fibers,[26] porcelain insulators,[27] vitreous grinding wheels,[28] and electronic substrates.[29] Moreover, although developed for isotropic and homogeneous materials, the methodology appears to be generally valid for most fine-grain ceramic materials[7,30–32] and perhaps for single-crystal materials.

The lifetime prediction approach, described below, represents the currently accepted procedures of reliability assessment for homogeneous, brittle materials. There are, of course, a number of assumptions built into any technical procedure and, frequently, these assumptions cannot be tested. The assumptions inherent in steps associated with lifetime prediction are clearly stated in this chapter, and implications that arise if the assumptions are violated are discussed at the end of each section.

The structure of this chapter has been chosen to allow the user to access easily regions of particular interest. It has been arranged to provide a general overview of the processes involved in lifetime prediction of brittle materials followed by a series of appendixes that describe the technical details and requirements for each of the different stipulated experimental procedures. The overview describes why the various measurement procedures are necessary and how they fit into an overall lifetime prediction model. For details associated with the derivation of the models, the user is directed to the references at the end of the chapter.

3 OVERVIEW

3.1 General Considerations

The most basic assumption made in this chapter is that the material whose lifetime is of interest is truly brittle; that means there are no energy dissipation mechanisms (e.g., plastic deformation, internal friction, phase transformations, creep) other than catastrophic bond rupture occurring during mechanical failure. It has been well documented that brittle materials fail from flaws that locally amplify the magnitude of stresses to which the material is subjected.[9,10,14,15,33] These flaws, e.g., scratches, pores, pits, inclusions, or cracks, result from processing, handling, and use conditions. For a given applied or residual stress, the initial flaw distribution determines whether the material will survive application of the stress or will immediately fail. Similarly, the evolution of the flaw population with time determines how long the surviving material will remain intact.

Other key assumptions in reliability predictions for brittle materials are that the experiments used to determine inert strength distributions do not alter the initial flaw population and that the flaw population in the test pieces mimics that in the final component. In addition, we assume that the flaws change only through the crack growth processes described in the model. It is essential, therefore, that the test specimens that are used to determine the initial flaw distribution be fabricated by the same processing conditions and prepared with the identical surface preparation as applied to the final components.[4,34] It is also essential that care be taken to avoid damaging test specimens during fabrication, storage, and handling.

The care taken often includes, but is not limited to, chamfering test specimens to reduce failures resulting from edge defects, individually wrapping test specimens to prevent damage during storage and transportation, and avoiding contact with the tensile surface of the specimens.

3.2 Strength

Because the flaw population under stress defines the initial strength of a brittle material, it is necessary to characterize this distribution or, equivalently, the distribution of initial strengths. Almost always, the initial strength distribution is easier to characterize. This can be done in three ways: (1) a statistical characterization of the initial strength distribution, (2) an overload proof test to establish a minimum strength, or (3) direct nondestructive flaw detection from which strength can be calculated. Each of these characterization techniques, described below, has specific advantages. However, each of the techniques also contains stringent requirements and limitations. It is essential to bear in mind that if these requirements are not met or the limitations are exceeded, the techniques will lead to erroneous lifetime predictions.

Statistical Strength Distribution

The first strength characterization technique to be discussed is the statistical strength characterization approach. This approach is particularly useful to evaluate the suitability of different materials, processing procedures, and surface treatments for a given application. However, because the approach is statistical, it cannot guarantee that components will have strengths greater than some minimum value; rather, it provides a failure (or survival) *probability* for a given load level.

Because it is the most critical flaw in a component or test specimen that leads to failure, it is not the distribution of flaws itself that is important but rather the distribution of the extremes of the flaw distribution. Fortunately, distributions of extremes, or extreme-value statistics, have been thoroughly studied.[11–13,35] Strength distributions are typically described by the Weibull distribution, one of three extreme-value distributions. Procedures for estimating the parameters of the Weibull distribution are well documented.[36–38] The generally accepted procedure is to use maximum-likelihood estimators. Procedures for the two-parameter Weibull distribution are described in the ASTM standard practice guide ASTM C 1239-95.[36] Occasionally, the strength data exhibit threshold-strength behavior and must be fitted to a three-parameter Weibull distribution.[22] In such cases a number of difficulties are associated with the numerical techniques used to estimate the Weibull parameters.[37,38] Typically a Newton–Raphson iteration procedure is used (see Appendix F of Ref. 37), but a sequential search technique with subsequent bisectional bracketing also works well.[22]

The strength distribution obtained is unequivocal provided that it is determined on actual components (for each batch of material) under conditions that simulate the service conditions. However, this requirement often poses an intolerable economic constraint, because so many tests must be performed to achieve an acceptable confidence level. Consequently, tests are often conducted on small coupons whose processing history and surface treatment mimic those of the final component.

Some care must be exercised in collecting the strength data. Measurements should be conducted at a high loading rate and in an inert environment, e.g., a dry gaseous nitrogen environment, to avoid environmental effects discussed below. In addition, the stress state, σ, used for the Weibull test must represent as closely as possible the stress conditions the component will see in service. For example, conducting the Weibull tests in uniaxial tension will not necessarily provide meaningful data if the actual component will experience biaxial loading.

After the Weibull strength distribution has been obtained and analyzed, a particular probability of failure, F, can then be associated with a corresponding initial strength, S. The relationship between F and S, i.e., the spread of initial strengths, is controlled by the distribution of initial flaw sizes in the material.

Finally, if the tests have been conducted on test coupons rather than final components, it is essential to relate the area under load in the coupon to the area under load in the component (see Appendix 1) because the probability of encompassing a more severe flaw is proportional to the area under load. Therefore, the test specimen area under load must be normalized to the component area under load. It should be borne in mind that, if failure results from subsurface defects, the important parameter to normalize is the volume under load rather than the surface area.

Minimum Strength Overload Proof Test

An overload proof test is a procedure for establishing the upper limit on the most critical flaw in a component or, equivalently, *the lower limit on the initial strength distribution*.[4,7,39–43] The technique is expensive since real components are tested. However, a properly conducted proof test completely eliminates the problem of failure through statistical outliers that is unavoidable when the statistical strength distribution technique is used. Therefore, if the statistical strength distribution analysis indicates that there is an unacceptably high probability of failure at the required load, the overload proof test procedure can be used to eliminate the weak components and truncate the failure distribution. It should be noted that, because the overload proof test is applied to finished components, it is generally not a technique that is appropriate for material or design selection. In addition, for a final component with a complicated geometry, it can be difficult to apply controlled stresses that are equivalent in orientation to those experienced in service.

In concept, the overload proof test is straightforward. One simply applies an overload stress to each component typically two to three times the service stress. The ratio of the overload stress to the service stress is the *proof-test ratio*. Components with critical flaws larger than a predetermined size or with strengths less than a predetermined minimum (namely, the proof-test stress) will break. Accordingly, such components are automatically eliminated from the distribution.

However, the requirements for a properly conducted proof test must be stringently met. Specifically, the proof test must be devised such that the stress at each location of a component exceeds the service stress in that same location by an amount at least as large as the recommended proof-test ratio. The proof test must be performed under controlled conditions of environment and proof-test load cycle, the most critical aspect being to have conditions as inert as possible and to unload from the proof stress as rapidly as possible. If these conditions are satisfied, then proof testing is undoubtedly the best life prediction procedure, because no aspects of the analysis are unresolved. If the conditions are not met, then crack growth (discussed below) could occur during unloading, obviating the desired truncation of the initial flaw population and leaving the components weaker than predicted by the proof-test analysis.

Nondestructive Flaw Detection

In principle, a good nondestructive evaluation (NDE) test for flaw detection offers several benefits: 100% inspection, on-line inspection, in-field health monitoring, and quantitative critical flaw detection. In practice, the approach is almost never used for brittle materials. The difficulties associated with using NDE involve both flaw detection and flaw quantification. Since the toughness of most ceramics and glasses is on the order of 1–5 $MPa \cdot \sqrt{m}$, critical flaw sizes are less than approximately 50 μm for even low-strength materials. This is currently beyond the detection limit of most available NDE procedures. However, measurement techniques are rapidly improving and, within the foreseeable future, instruments may be commonly available that will detect defects of micrometer and submicrometer size. A more formidable task will then be relating the NDE measurements to quantifiable reliability predictions.

As discussed above, brittle materials fail from the most severe, not the average, flaws. The phrase "most severe flaw" means the flaw that generates the largest stress intensity in the expected stress field. Consequently, the flaw size, shape, and orientation relative to the stress field must be quantified before results from an NDE measurement can be used to predict strength.

Unlike metals, in which flaws can grow to many times their original size without component rupture while NDE techniques monitor the flaw growth or the concomitant material property degradation, brittle materials fail with very limited flaw extension and almost no property degradation. Consequently, unless a specific NDE test has been clearly shown to predict failure strength accurately for the same brittle material with the same surface finish and under the same loading conditions of interest, it is better to use the statistical strength distribution or the overload proof-test techniques described previously to determine strength information.

3.3 Environmentally Enhanced Fracture

The second property influencing the reliability of brittle materials is the growth of flaws in the presence of stress. If the flaws remained constant with time,* there would be no need to proceed beyond the initial strength analysis. Based upon the Weibull distribution, the probability of failure would be known for any stress condition and would be constant. Alternatively, based on an overload proof test or NDE, a minimum strength is defined such that failure would not occur for stresses below these values. However, it is well known that flaws grow under stress in most brittle materials.[1,16–19,44,45] As the flaws grow, the material weakens, leading to a time-dependent mechanical failure. This process is known by several names; three of the most common are "environmentally enhanced fracture," "delayed failure," and "static fatigue."

Environmentally enhanced fracture is a process that is still not completely understood. However, it is known that flaws in materials under stress can react with certain environments, resulting in bond rupture and flaw growth at stresses well below the nominal bond strength of the material. Water is particularly effective at enhancing fracture in many brittle materials. In addition, water is almost ubiquitous in either liquid or vapor phase. Therefore, it is essential that the effects of water-enhanced fracture be incorporated into any lifetime prediction model.†

Because the process of environmentally enhanced fracture is only partially understood, expressions relating crack growth velocity V to stress intensity factor K_I are typically phenomenological rather than truly theoretical. Although several crack growth rate expressions

---

*Throughout this discussion, we consider only the *growth* of preexisting flaws over time and ignore the possibility of generating a new flaw population during use.[4,34]

†Few environments are more effective than water in enhancing crack growth for most brittle materials. An exception is very basic solutions for silica-based glasses.[45] For such cases, the following discussion still applies, but the tests must be conducted in the new environment.

have been proposed, the one used in almost all lifetime predictions for brittle materials is given by[2–4,6,7,39]

$$V = V_C \left(\frac{K_I}{K_{IC}} \right)^N \qquad (1)$$

where V_C is a constant and N is called the environmentally enhanced crack growth parameter, sometimes denoted the inverse crack growth susceptibility. K_{IC} is the critical stress intensity factor at which fracture ensues catastrophically but, for the present context, can be considered as a scaling parameter for the stress intensity factor, K_I. N denotes the extent to which the environment interacts with the stressed crack to cause enhanced crack growth. For a given V_C and K_{IC}, the larger N is, the less effective the environment is at enhancing crack growth.

The fact that Eq. (1) is phenomenological rather than rigorously based on theory should not be a concern as long as the equation describes the observed behavior, *particularly in the low-stress region*. Guaranteeing this condition is difficult, however, since the crack growth rates for a component in use are usually several orders of magnitude smaller than the lowest crack growth rates that can be determined experimentally. As long as any deviation from Eq. (1) is of the form $V_{\text{real}}(K_I) < V_{\text{Eq. (1)}}(K_I)$, that is, the real velocity is less than the velocity predicted from Eq. (1), lifetime predictions based upon Eq. (1) are conservative. However, if $V_{\text{real}}(K_I) > V_{\text{Eq. (1)}}(K_I)$, lifetime predictions based on Eq. (1) are no longer conservative.

At this point, it should be noted that there are a number of other possible crack growth expressions that are statistically indistinguishable from Eq. (1) over the measurable crack velocity range.[46] Some of these expressions are more conservative, i.e., predict higher crack growth rates, at low K_I than the power law. A particularly important example is the expression $V = B \exp[bK_I]$, which is derived from chemical rate theory and has been used by several authors to parameterize environmentally enhanced fracture.[3,24,46–48] Although having a stronger scientific basis than the power law [Eq. (1)], the exponential expression for crack growth rates has never developed the following obtained by the power law expression. The dominance of the power law expression derives from the fact that it allows closed-form, analytical lifetime solutions to be obtained even when the critical flaws are surrounded by a residual stress field, as will be discussed in the section on constant-loading-rate experiments. With the advent of powerful desktop computing, it is now possible to use numerical techniques, rather than analytical derivations, to predict lifetimes. The approach outlined above is still required, i.e., determination of the initial strength distribution through Weibull statistics[36–38] or proof-test[4,39] measurements, evaluation of crack growth parameters through dynamic fatigue experiments,[4,49] and estimation of confidence limits through statistical analysis, e.g., bootstrap methods[21,22,50–52] as discussed below. Consequently, it is now possible to make lifetime predictions using crack growth relationships like the exponential expression. However, it is much easier to demonstrate how the different measurements are combined to obtain an expression for the lifetime if analytical expressions are used. Therefore, for the purposes of this chapter, we will focus exclusively on the power law expression for crack growth.

Constant-Loading-Rate Experiments

Conducting fracture mechanics crack growth experiments to evaluate the parameters governing environmentally enhanced crack growth is a process that requires both a large amount of time and a thorough understanding of issues associated with environmentally enhanced fracture. Consequently, a faster evaluation method that is less prone to experimental subtleties is desirable. Equation (2) defines the constant-loading-rate or dynamic fatigue measurement relationship (Appendix 2). In Eq. (2), λ is a fitting parameter related to V_C and K_{IC} and $N' = (3N + 2)/4$, where V_C, K_{IC}, and N are parameters in Eq. (1):

$$\sigma_f^{N'-1} = \lambda \frac{d\sigma}{dt} \tag{2}$$

Although there are various possible geometries and several critical experimental requirements for dynamic fatigue measurements (see Appendix 2), the measurements are conceptually straightforward. A series of specimens are subjected to a load that increases linearly with time. The log of the failure strength, σ_f, plotted as a function of stressing rate, σ, gives the parameters N' and λ. To obtain conservative values of N' and λ, dynamic fatigue measurements need to be carried out in the harshest environmental conditions under which the final component will be used. For example, if water vapor enhances crack growth, liquid water could provide the limiting case of water-enhanced environmental fracture. However, it is not necessary that the initial flaw distribution be maintained. A change in flaw distribution will change σ_f, but not the slope, $\lambda/(N - 1)$. Therefore, dynamic fatigue experiments are usually conducted on indented specimens; the indentations provide a uniform and repeatable critical flaw.[53–57] Typically, the resultant data have much less scatter than those obtained from the initial flaw distribution, thereby resulting in a more precise determination of the slope.

Inert Strength for Indented Specimens

The inert strength for indented specimens, S_v, is determined using the same procedure as that used to obtain the Weibull distribution: high loading rate and an inert atmosphere. However, instead of using the natural flaw population as the strength-limiting defects, S_v uses the same level of indentations that were used in the dynamic fatigue experiments. These data provide an upper limit on the dynamic fatigue strength data.

3.4 Lifetime Prediction

Once the Weibull distribution, the dynamic fatigue data, and the inert strength for the indented specimens have been obtained, the lifetime, t_f, for any initial strength, S, and any applied stress, σ, can be calculated using Eq. (3)[22]:

$$t_f = \frac{\lambda}{N' + 1} \left(\frac{S}{S_v}\right)^{N'-2} \sigma^{-N'} \tag{3}$$

Although Eq. (3) predicts a time to failure for a component under load, there is still no way to estimate how statistically significant the value is. To obtain the most information from the lifetime prediction model, it is necessary to estimate the confidence limits surrounding t_f.

3.5 Confidence Limits

As with any experimentally determined quantities, there are uncertainties associated with each of the parameters in Eq. (3). Some of the parameters may not be amenable to standard error propagation analysis. For example, for large N', e.g., $N' \geq 60$, the uncertainties in evaluating the slope of the dynamic fatigue curve are much larger than the calculated slope itself. Under such conditions, standard error propagation techniques are invalid. An alternative is the bootstrap technique.[50–52] It is a statistical procedure that takes advantage of modern computing power and makes it relatively straightforward to determine confidence limits (Appendix 3). It applies to any data set that is *independent and identically distributed* (iid). The major assumption with the bootstrap technique is that the data used to evaluate the parameters in Eq. (3) adequately represent the statistical scatter in the experiments. If this condition is met, combining the bootstrap technique with the data used to evaluate t_f permits the user to estimate the time to failure at whatever confidence level is desired.

3.6 Life Prediction Process

The individual steps required for lifetime prediction have been laid out above. If the steps are followed and careful attention is paid to the requirements and restrictions at each step in the process, a lifetime prediction with a specific failure probability and known confidence limits for a specific applied load is obtained. However, frequently, other questions are of interest. For example, the needed lifetime may be an input parameter, and the limiting stress value may be the unknown parameter of interest. As another example, a required lifetime at a known load may be the input parameters, and the unknown may be whether a given material can be used and, if so, with what probability of failure. The formalism outlined above is adequate to address any of these questions and, in fact, can be used in an iterative manner to determine how variations in one group of parameters affect others. In this section, failure diagrams will be introduced. These diagrams, constructed from the data obtained using the techniques outlined above, graphically relate the parameters to show quantitatively how modifications in one parameter, e.g., stress or lifetime, affect the others.

Figure 1*a* is derived directly from Eq. (3). Each of the solid lines maps the applied stress, σ, onto the time to failure, t_f, for a different value of the initial strength or, equivalently, different probability of immediate failure upon the application of the σ (see Appendix 1). The lines are all parallel with slope of $-N'$. The bands about each line represent the confidence limit at which the position of the line is known and can be determined by a statistical analysis tool such as the bootstrap method. For conservative calculations, the left-hand confidence limit should be used instead of the centerline. Figure 1*a* immediately shows, in a quantitative way, how all the parameters affecting lifetime predictions are coupled if a statistical strength distribution is being used. Known or required values of some of the parameters can be located on the plot and the range of possible values for the remaining parameters can be taken directly from the graph.

If the overload proof-test, rather than the statistical strength distribution, approach is being used, Fig. 1*b* maps the applied stress onto the lifetime. This figure is obtained from Eq. (3) by replacing the initial strength *S* divided by σ by the proof-test ratio, *R*. This gives a linear relationship between $\log(\sigma)$ and $\log(t_f)$ with a slope of -2. The different parallel lines in Fig. 1*b* are now determined by changes in *R*. The larger *R* is, the larger t_f is for a given σ, but at the cost of more components being broken. To estimate the fraction of components that will fail during the proof test (one refers again to the probability diagram), the proportion of components likely to fail in the proof test can be estimated by determining the intersection of the stress coordinates of the proof stress, $\sigma_p = R\sigma_a$, and the time coordinate for the time at the peak proof load, t_p. If this proportion is unacceptable, e.g., most of the components fail in the proof test, then the material, including flaw distribution, is inadequate,

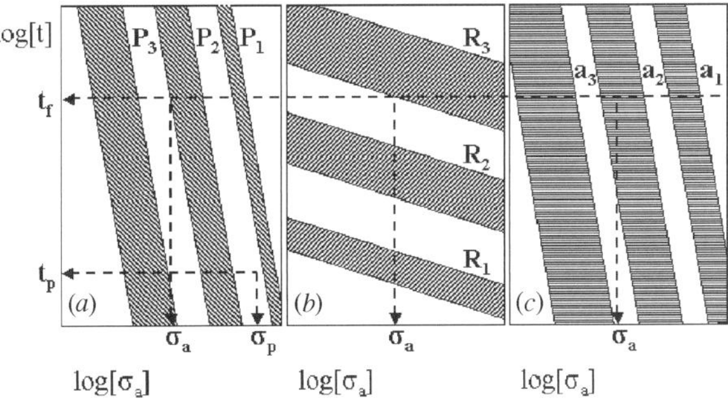

Figure 1 Design approach.

and an alternate material should be explored. This process should proceed until a material with a satisfactory proof-test failure proportion is identified.

Finally, if proof testing is unsuitable, for example, as a result of the complexity of the component, NDE techniques could be explored. The approach taken at this juncture is strongly material dependent. Figure 1*c* relates the lifetime, t_f, to the applied stress, σ_a, and the flaw size, a. To use Fig. 1*c*, an approximate value for the flaw size that must be detected to ensure integrity is first obtained from the graph, without accounting for the extension of the confidence limits. Then, the ability to detect such flaw types should be evaluated using available NDE techniques. At this stage, confidence limits on flaw detection in the test material can be established for each NDE technique, and the flaw size detection requirement can be more closely defined. The suitability of the material for this application is determined by the capability of detecting *all* flaws in the critical parts of the component that exceed the specified size. As discussed previously, use of this approach requires that NDE techniques are available that provide quantitative size and shape data regarding flaws.

4 SUMMARY

In summary, the following procedure is required to predict lifetimes in isotropic, homogeneous, brittle materials:

1. Fast fracture tests in an inert environment to determine the Weibull distribution of the initial flaws/strengths and the probability of failure for any arbitrary initial strength, *or* a proof test to weed out specimens below a predetermined initial strength, *or* an NDE technique that guarantees a lower limit on strength
2. Dynamic fatigue measurements on indented specimens to determine susceptibility to environmentally enhanced fracture
3. Fast fracture tests in an inert environment on indented specimens to determine the inert strength (strength in the absence of environmental effects) of the indented material
4. Bootstrap Eq. (3), using the data in steps 1–3, to predict lifetimes and determine confidence levels for predetermined initial strength

It is essential for the user to remember that this lifetime prediction methodology assumes that the flaw/strength distribution determined from the Weibull analysis is the same distribution to be found in the final components (for the statistical strength distribution approach), that the tests do not alter the flaw distribution, and that flaws change over time only through environmentally enhanced fracture as described by Eq. (1).

APPENDIX 1: WEIBULL TESTS

Since brittle failure is expected to occur from the most severe defect in the component, the strengths are assumed to come from an extreme-value distribution. A critical assumption in the following discussion is that the strength distribution in the strength test is the same as the distribution in the final components. Typically, either a two- or a three-parameter Weibull is used to evaluate the distribution. Equation (A1.1) gives the form of the three-parameter Weibull expression[8,22,35]

$$(1 - F) = \exp\left[-\left(\frac{\sigma(x) - S_t}{S_0 - S_t}\right)^m\right] \tag{A1.1}$$

Two of the three parameters in Eq. (A1.1) are the shape parameter, m, which determines the shape of the probability distribution function, and the scale parameter, S_0, which defines the scale of the strength axis. The shape parameter, m, defines how quickly the probability of survival drops from a value of 1.0 as the stress on the material increases. Low values of m correspond to a broad, relatively slowly varying decrease in the probability of survival as a function of increasing σ. In contrast, high m values generate a survival probability curve approaching a step shape; the survival probability remains near 1.0 until σ nears S_0 and then drops rapidly to zero. As implied by the previous sentence, S_0 is related to the value of stress at which the probability of survival is near the midpoint between 1.0 and 0. For $S_t = 0$, the probability of survival at $\sigma = S_0$ is $1/e$. In fact, in the limit of m approaching infinity and $S_t = 0$, S_0 is identically the value of stress for which the probability of survival drops from 1.0 to 0.

The third Weibull parameter, S_t, represents a "threshold" value for strength below which there is zero probability of failure. The form of the two-parameter Weibull expression is identical with Eq. (A1.1) with S_t equal to zero; i.e., there is no lower strength threshold. The decision as to whether S_t is zero or finite is not always easy to make. The question may become moot if it is borne in mind that the two-parameter distribution is a special case of the three-parameter distribution. The decision whether to include a finite value for S_t could be addressed statistically by determining the significance of a threshold value and uncertainties related to its inclusion (i.e., with what confidence can a finite value be attributed to S_t?) by the bootstrap simulation (Appendix 3). However, the implications of assuming a two-parameter or a three-parameter Weibull distribution should be kept in mind. A two-parameter distribution is always more conservative and, therefore, safer, since it implicitly assumes that there is a possibility for failure at any load. However, if a threshold does exist, use of a two-parameter distribution will have economic consequences because it will predict a finite failure probability at loads for which no failures will occur.

Based on Eq. (A1.1), a set of test specimens can be broken rapidly in bending (Appendix 2) in an inert environment and the strength of each specimen plotted against the survivability ranking, $1 - F$:

$$\ln\left[\ln\left(\frac{1}{1 - F}\right)\right] = m \ln[\sigma(x) - S_t] - m \ln(S_0 - S_t) \tag{A1.2}$$

This results in a line of slope m and intercept of $-m \ln(S_0 - S_t)$. Therefore, the survival probability of a given *test specimen* could be predicted for a given load. However, usually, the area under load on a component in use is not identical to the area sampled in the bend test. Because the flaws are assumed to be distributed randomly in both the final components and the test specimens, if the area under load in the final component is larger than the area sampled by the Weibull test, the component is likely to sample larger flaws than those sampled during the tests. Therefore, the probability of failure at a given load will be greater for the component than for the test specimens. The converse is true if the area sampled in the final component is smaller than that sampled by the test specimens. Thus, a weighting factor is incorporated into the analysis to account for differences in sampling area between the test specimens and the final components.

This is handled in the following way. If an elemental area, dA, of a component surface at position x experiences a tensile stress $\sigma(x)$ and a survival probability $1 - F$, then the survival probability of the component, $1 - F_c$, is the joint survival probability of all the individual surface elements. In the limit of infinitesimal elemental areas, this survival probability is

$$[1 - F_c(S)] = \exp\left[-k(m, S_t)\left(\frac{S - S_t}{S_0 - S_t}\right)^m\right] \tag{A1.3}$$

where F_c is the failure probability of the component, S is the maximum tensile stress in the component (i.e., the strength), and $k(m, S_t)$ is an area-scaling factor given by

$$k(m, S_t) = \frac{1}{A_0}\iint\left(\frac{\sigma(x) - S_t}{S - S_t}\right)^m dA \tag{A1.4}$$

If A_0 is chosen to make the k factor for the laboratory specimen unity, $k(m, S_t)$ is the relative area-scaling ratio between the components and the laboratory specimens.

Best Practices

- Unless there is clear evidence that a threshold exists in the Weibull distribution, it is better (i.e., conservative) to assume a two-parameter Weibull distribution.
- A good rule of thumb is that 30 or more specimens should be used to generate a Weibull plot.
- It is frequently valuable to identify the failure source of specimens under an optical microscope.[58,59] This procedure distinguishes failures from machining damage, inclusions, handling accidents, etc. A description of fractography is outside the range of this document, but an excellent discussion can be found at the website http://www.ceramics.nist.gov/webbook/fracture/fracture.htm.
- Measurements should be conducted in as inert an environment as possible. Dry flowing N_2 gas in a glove bag is usually a good choice.
- Measurements should be made at the highest loading rate for which the testing machine provides reliable data.

Dangers

- Modifying the flaw distribution on the specimen tensile surfaces through improper handling procedures
- Testing in such a manner (speed or environment) that allows slow crack growth to occur during the test

APPENDIX 2: STRENGTH AND DYNAMIC FATIGUE TESTS

Strength tests are tests in which the strength of specimens (final components or specially fabricated test specimens) is measured at a constant loading rate. For inert strength measurements, the loading rate is high and the specimens are tested in an inert environment to minimize slow crack growth. In contrast, dynamic fatigue tests[4,49] are strength tests that are conducted in an active environment over a wide range of loading rates to quantify parameters that describe slow crack growth. Setting aside for the moment issues of environment, loading rate, and natural flaw distribution versus indentation flaws, all of which are addressed below, the mechanics associated with measuring strength are the same for inert strength, Weibull distribution, or dynamic fatigue.

Bend Tests. Strength tests are usually conducted under uniaxial[60–63] or biaxial[64–68] bending conditions. The stress states associated with both of these bending geometries are well known, so data analysis is straightforward. A brief discussion of the advantages and disadvantages of each of these stress tests follows.

The most commonly used uniaxial bend test is the four-point bend test. This test, which is thoroughly described in the ASTM standard C 1161-94,[63] consists of a rectangular parallelepiped specimen in a loading fixture that consists of two parallel cylindrical loading pins in contact with one face of the specimen and two more widely spaced cylindrical loading pins in contact with the opposite face. The loading pins are perpendicular to and are spaced symmetrically about the center point of the specimen's long axis. This loading geometry has the advantage of generating a uniform, uniaxial stress longitudinally on the tensile side of the bend bar; the size of the uniform tensile region matches the spacing of the narrowly spaced loading cylinders. The large region of uniform tensile stress makes the four-point bend geometry much preferable to the three-point bend geometry, also described in ASTM standard C 1161-94.[63] In the three-point bend geometry, the tensile stress is largest at the center of the specimen but drops off immediately as a function of distance away from the center. Therefore, in the four-point bend geometry, a much larger population of initial defects is sampled at maximum stress than is sampled in the three-point bend geometry.

Edge failures pose a problem in uniaxial testing. The process of machining specimen material into bars unavoidably generates machining damage along the edges of the specimens. If the stress intensity at one of the machining flaws is greater than the stress intensity values at the flaws of interest, the specimen will fail from the machining flaw. This gives rise to a couple of difficulties. On the simplest level, if specimens that fail from the edge are ignored in the analysis, edge failures still involve time and material that are wasted. On a more subtle level, the question arises whether it is legitimate to ignore data from specimens that fail from the edge? How can it be determined whether a specimen failed from a flaw that resulted from specimen manufacture or whether it failed from a flaw of interest that happened to lie near or at an edge? Certainly, in the latter case, discarding the data is not appropriate. Clearly, it would be better to avoid edge failures altogether. Two ways of accomplishing this for uniaxial bending involve either machine chamfering or hand rounding specimen edges on the tensile surface or use of a standard flaw large enough to be the source of failure. Edge modification requires that the 90° corners along the longitudinal tensile edges of the specimen be machined away or rounded the length of the specimen. This approach greatly reduces the probability of edge failures by eliminating the high stress concentrating effect of the corner. However, it reduces the tensile surface area of the specimen and adds both machining expense and time to the specimen preparation. In addition, a sloppy chamfering or rounding job can cause more damage than it relieves. Nevertheless, if carefully done, chamfering or hand rounding corners can nearly eliminate edge failures. The second approach to reduce edge failures through the use of large standard sized flaws will be discussed below for both uniaxial and biaxial bend tests. At this point, it is sufficient to say that the use of standard flaws clearly is not a solution to edge failures if the purpose of the strength tests is to obtain a Weibull distribution (see Appendix 1). However, standard flaws can be very useful in dynamic fatigue experiments.

Biaxial tests generate an equi-biaxial stress state on the tensile surface of disk-shaped specimens.[64–68] This is accomplished through the use of a test geometry in which a ring composed of small balls* (e.g., tungsten carbide balls) is placed in contact with one side of the specimen. This ring has a radius somewhat smaller than the radius of the specimen disk. On the other side of the disk, a ring or circular flat (e.g., a tungsten carbide ball that has had a flat machined onto it) is pressed against the specimen. When the flat is pressed against the specimen, an equi-biaxial tensile stress with a radius equal to the radius of the small flat is generated on the tensile surface of the specimen. The literature contains descriptions of biaxial test configurations in which a ball, rather than a flat or ring, is used to apply the load.[69] Like the three-point bend test, point loading under biaxial conditions generates tensile stresses that fall monotonically from the center of the specimen; there is no region of high uniform tensile stress like that which occurs when a flat is used. Therefore, flat-on-ring rather than ball-on-ring loading is recommended.

Biaxial loading has several major advantages. First, because the stress is equi-biaxial, the most severe flaw will cause failure regardless of its orientation in the specimen. Second, because the loading geometry is circularly symmetric, alignment in the plane of the disk is less of a problem than for uniaxial bend tests. Third, because the high-stress region is in the center of the disk and the stress at the edge of the specimen is only about 30% of the maximum stress, edge failures are almost completely eliminated. However, these benefits can be offset by the fact that biaxial testing requires much more material that uniaxial testing.

Both uniaxial and biaxial bend tests are subject to errors resulting from loading train misalignment and friction. The ASTM standards address both of these issues. Here we only mention the existence of the problems and commend the standards to the attention of the reader.

Standard Flaws. If it is not necessary to measure strengths from flaws representative of the flaw population in the final components; i.e., to determine the Weibull distribution, it is usually beneficial to generate a large, uniform flaw from which the specimen will fail. Such flaws can greatly reduce scatter in the strength data, allowing trends to be observed that might be obscured by the scatter resulting from the natural flaw population. Typically, such flaws are generated by use of a Vickers indenter to place an indentation impression in the center of what will be the high-tensile-stress region of the specimen. When indentations are used, their depths should remain fairly shallow, since the tensile stress in the bend test decreases with depth into the specimen. A rule of thumb is that the indentation impression, including cracks, should be less than one-fourth the thickness of the specimen. Examining the fracture surface of a bend specimen after testing can verify this.

It is important to be aware that indentations leave a residual stress field around the indentation impression. Therefore, strength tests on indented specimens must take into account both the applied far-field stresses as well as the local residual stresses at the indentation. In the dynamic fatigue measurements, the residual stress field results in an expression usually expressed in terms of $N' = (3N + 2)/4$ rather than N [see Eq. (2)].[6] A residual field also typically surrounds natural flaws.[53] The residual stress associated with point flaws is identical in form to that around an indentation. Even many surface scratches can be treated as a series of point flaws, due to the nonuniform crack formation that occurs along the length of the scratch.

Dynamic Fatigue Measurements. As stated at the beginning of this appendix, dynamic fatigue tests are simply strength tests that have been made in the environment of interest as a function of stressing rate. Therefore, the discussion regarding uniaxial and biaxial strength

---

*A solid ring of a somewhat compliant material may also be used.[22,68]

tests applies directly to dynamic fatigue tests. The purpose of dynamic fatigue measurements is to quantify crack growth parameters. For the power law crack growth expression, it is particularly necessary to determine N. As mentioned above, dynamic fatigue experiments are usually conducted on indented specimens to minimize scatter in the strength. From Eq. (2), we see that evaluating N' requires that the independent axis be plotted as $\log(\dot{\sigma})$. For even moderate values of N, the slope of the dynamic fatigue curve will be shallow [e.g., for $N = 30$, $1/(N' - 1) \approx 0.05$]. Consequently, to obtain accurate values for N', it is necessary to take the dynamic fatigue data over as wide a stressing rate as possible. We recommend taking the data over five orders of magnitude in stressing rate, if possible. Data taken over less than three orders of magnitude should not be trusted. An analysis of the confidence with which N' (or N) is known (see Appendix 3) will give guidance as to whether the stressing rate range was large enough.

It should be mentioned that recent work suggests that dynamic fatigue measurements made at too high a velocity could give values for N that are somewhat too low.[70,71] The reported work reflects ongoing research, and a discussion of it lies outside the scope of this chapter; the upper velocity at which this effect occurs is dependent upon both material and environment. In addition, it is not clear that the variation in N is generally large enough to detect experimentally. At any rate, any proposed error in N that would result from measuring dynamic fatigue at too high a stressing rate would result in a conservative lifetime prediction. Therefore, at this time, we recommend as wide a stressing rate range as possible.

Indented Inert Strength. Indented inert strength measurements are made with indented specimens rather than specimens that will fail from their natural flaw population. As stated earlier in this appendix, inert strength measurements need to be made at a high loading rate in an inert (e.g., N_2 gas) environment. It is worth mentioning that regular dry nitrogen gas is sufficient; there is no measurable advantage in going to a superdry gas as long as the loading rate is high.

Best Practices

- For either uniaxial or biaxial bend tests, articulated, self-aligning fixtures should be used to minimize extraneous stresses (see ASTM standard C 1161-94 for uniaxial testing[62,63] and Refs. 64, 65, 67, and 68 for a discussion of biaxial testing).
- Inert strength tests require high loading rate and inert environment. Enclosing the test rig in a glove bag and flowing N_2 gas through the bag is a simple way to achieve an inert environment.
- To minimize edge failures in bend bars, the edges can be rounded or chamfered.
- Vickers indentations are useful in reducing scatter in bend tests. In uniaxial tests, the indentation should be oriented so that one set of diagonal cracks will be perpendicular to the stress axis.
- Dynamic fatigue tests should be made over as wide a range of stressing rates as possible. Three orders of magnitude is the minimum range that should be considered.

Dangers

- Three-point bend or ball-on-circle tests should be avoided in favor of four-point bend or flat-on-circle tests.
- Do not use indentations for Weibull strength tests.
- In biaxial tests, do not use a rigid punch as the flat.

APPENDIX 3: CONFIDENCE LIMITS

The lifetime analysis presented above is based on three sets of experimental data: (1) dynamic fatigue data for indented specimens, (2) inert strength data for indented specimens, and (3) inert strength distribution for the strength-limiting flaws (Weibull data). To estimate the confidence in a lifetime prediction based on Eq. (3) and these three data sets, i.e., the lifetime at a particular *level of confidence,* we need to determine the distribution of lifetimes that results from the experimental uncertainties in these data sets. The *nonparametric bootstrap technique*[21,22,50–52] is ideal for this purpose. The nonparametric bootstrap technique involves the creation of a collection of simulated data sets based upon the original data. If the original data set contained n data points, the simulated data sets also contain n values and are generated by the random selection of values from the original data set. After each value is selected to provide a simulated data point, it is returned to the original data pool so that every value selected for the simulated sets is selected out of the full collection of n values (i.e., a select-and-replace procedure). The new set of data is called a *bootstrap sample*. This procedure is repeated a large number of times and the statistic of interest is determined.

The requirement for the original data set is that the data be iid. Clearly, if the data for the Weibull distribution and the indented inert strength are taken properly, they meet this criterion. Equally clearly, the dynamic fatigue data do not; the strength values depend upon the stressing rate. To use the bootstrap technique to simulate dynamic fatigue data, the following steps are taken. First, a linear regression analysis is performed on the dynamic fatigue data, regressing the logarithm of the fracture stress versus the logarithm of the stressing rate. The residuals of this regression analysis, namely, the failure stress minus the best-fit value, do meet the iid requirement. Therefore, for each loading rate, residuals are chosen using the select-and-replace procedure previously described. A fatigue stress is calculated from the selected residual and the fitted line and associated with the loading rate to simulate a dynamic fatigue data set. A new estimate of lifetime can be made from each simulated set of Weibull, indented inert strength and dynamic fatigue data. This procedure can be repeated thousands of times to generate a distribution of lifetimes. The value of the lifetime at any arbitrary confidence level, $y\%$ (e.g., 99%), is the lifetime such that $y\%$ of the lifetime estimates are greater than this value.

Best Practice

- The bootstrap technique should be used to evaluate the confidence limits for the parameters from *each* of the three data sets as well as the confidence limits for the overall lifetime. This can alert the user to problems in a particular experimental procedure (e.g., a wide enough loading rate range in the dynamic fatigue experiment).

Acknowledgment

The authors gratefully acknowledge the critical reviews of the manuscript and the most helpful suggestions of S. J. Dapkunas and S. M. Wiederhorn.

REFERENCES

1. S. M. Wiederhorn and L. H. Bolz, "Stress Corrosion and Static Fatigue of Glass," *J. Am. Ceram. Soc.,* **53,** 543 (1970).

2. S. M. Wiederhorn, "Reliability, Life Prediction, and Proof Testing of Ceramics," in *Ceramics for High Performance Applications,* J. J. Burke, A. G. Gorum, and R. N. Katz (eds.), Brook Hill, Chestnut Hill, MA, 1974, pp. 633–663.

3. S. M. Wiederhorn, E. R. Fuller, Jr., J. Mandel, and A. G. Evans, "An Error Analysis of Failure Prediction Techniques Derived from Fracture Mechanics," *J. Am. Ceram. Soc.,* **59**(9–10), 403–411 (1976).

4. J. E. Ritter, Jr., "Engineering Design and Fatigue Failure of Brittle Materials," in *Fracture Mechanics of Ceramics,* Vol. 4, R. C. Bradt, D. P. H. Hasselman, and F. F. Lange (eds.), Plenum, New York, 1978, pp. 667–686.

5. J. E. Ritter, Jr., S. M. Wiederhorn, N. J. Tighe, and E. R. Fuller, Jr., "Application of Fracture Mechanics in Assuring against Fatigue Failure of Ceramic Components," in *Ceramics for High Performance Applications, III, Reliability,* Vol. 6, E. M. Lenoe, R. N. Katz and J. J. Burke (eds.), Plenum, New York, 1981, pp. 49–59.

6. E. R. Fuller, Jr., B. R. Lawn, and R. F. Cook, "Theory of Fatigue for Brittle Flaws Originating from Residual Stress Concentrations," *J. Am. Ceram. Soc.,* **66**(5), 314–321 (1983).

7. S. M. Wiederhorn and E. R. Fuller, Jr., "Structural Reliability of Ceramic Materials," *Mater. Sci. Eng.,* **71**(1–2), 169–186 (1985).

8. K. C. Kapur and L. R. Lamberson, *Reliability in Engineering Design,* Wiley, New York, 1977.

9. A. A. Griffith, "The Phenomena of Rupture and Flow in Solids," *Phil. Trans. Roy. Soc. London A,* **221,** 163–198 (1921).

10. A. A. Griffith, "The Theory of Rupture," in *Proc. 1st Intl. Cong. Appl. Mech.,* C. B. Biezeno and J. M. Burgers (eds.), Technische Bockhaudel en Drukkerij, Delft, The Netherlands, 1924, pp. 55–63.

11. W. Weibull, "A Statistical Theory of the Strength of Materials," *Ing. Vetenskaps Akad. Handl.,* **151,** 45 (1939).

12. W. Weibull, "Phenomenon of Rupture in Solids," *Ing. Vetenskaps Akad. Handl.,* **153,** 55 (1939).

13. W. Weibull, "A Statistical Distribution Function of Wide Applicability," *J. Appl. Mech.,* **18**(3), 293–297 (1951).

14. F. W. Preston, "The Mechanical Properties of Glass," *J. Appl. Phys.,* **13,** 623–634 (1942).

15. R. E. Mould and S. D. Southwick, "Strength and Static Fatigue of Abraded Glass under Controlled Ambient Conditions: II, Effect of Various Abrasions and the Universal Fatigue Curve," *J. Am. Ceram. Soc.,* **42**(12), 582–592 (1959).

16. S. M. Wiederhorn, "Influence of Water Vapor on Crack Propagation in Soda-Lime Glass," *J. Am. Ceram. Soc.,* **50,** 407 (1967).

17. S. M. Wiederhorn, "Subcritical Crack Growth in Ceramics," in *Fracture Mechanics of Ceramics,* Vol. 2, R. C. Bradt, D. P. H. Hasselman, and F. F. Lange (eds.), Plenum, New York, 1974, pp. 613–646.

18. S. W. Freiman, "Stress-Corrosion Cracking of Glasses and Ceramics," in *Stress-Corrosion Cracking,* R. H. Jones (ed.), ASM International, Materials Park, OH, 1992, pp. 337–344.

19. G. S. White, "Environmental Effects on Crack Growth," in *Mechanical Testing Methodology for Ceramic Design and Reliability,* D. C. Cranmer and D. W. Richerson (eds.), Marcel Dekker, New York, 1998, pp. 17–42.

20. D. F. Jacobs and J. E. Ritter, Jr., "Uncertainty in Minimum Lifetime Predictions," *J. Am. Ceram. Soc.,* **59**(11–12), 481–487 (1976).

21. C. A. Johnson and W. T. Tucker, "Advanced Statistical Concepts of Fracture in Brittle Materials," in *Engineered Materials Handbook,* Vol. 4, *Ceramics and Glasses,* ASM International, Materials Park, OH, 1992, pp. 709–715.

22. E. R. Fuller, Jr., S. W. Freiman, J. B. Quinn, G. D. Quinn, and W. C. Carter, "Fracture Mechanics Approach to the Design of Glass Aircraft Windows: A Case Study," *SPIE Proc.,* **2286,** 419–430 (1994).

23. S. M. Wiederhorn, A. G. Evans, and D. E. Roberts, "A Fracture Mechanics Study of the Skylab Windows," in *Fracture Mechanics of Ceramics,* Vol. 2, R. C. Bradt, D. P. H. Hasselman, and F. F. Lange (eds.), Plenum, New York, 1974, pp. 829–841.

24. S. M. Wiederhorn, A. G. Evans, E. R. Fuller, Jr., and H. Johnson, "Application of Fracture Mechanics to Space-Shuttle Windows," *J. Am. Ceram. Soc.,* **57**(7), 319–323 (1974).

25. S. T. Gulati and J. D. Helfinstine, "Long-Term Durability of Flat Panel Displays for Automotive Applications," in *Vehicle Displays '96,* SID, Ypsilanti, MI, 1996, pp. 49–56.

26. J. E. Ritter, Jr., K. Jakus, and R. C. Babinski, "Effect of Temperature and Humidity on Delayed Failure of Optical Glass Fibers," in *Methods for Assessing the Structural Reliability of Brittle Materials,* ASTM STP 884, S. W. Freiman and C. M. Hudson (eds.), American Society for Testing and Materials, West Conshohocken, PA, 1984, pp. 131–141.

27. A. G. Evans, S. M. Wiederhorn, M. Linzer, and E. R. Fuller, Jr., "Proof Testing of Porcelain Insulators and Application of Acoustic Emission," *Am. Ceram. Soc. Bull.,* **54**(6), 576–581 (1975).

28. J. E. Ritter, Jr. and S. A. Wulf, " Evaluation of Proof Testing to Assure Against Delayed Failure," *Am. Ceram. Soc. Bull.,* **57,** 186–189 (1978).

29. J. N. Humenik and J. E. Ritter, Jr., "Susceptibility of Alumina Substrates to Stress Corrosion Cracking during Wet Processing," *Am. Ceram. Soc. Bull.,* **59,** 1205 (1981).

30. A. C. Gonzalez, H. Multhopp, R. F. Cook, B. R. Lawn, and S. W. Freiman, "Fatigue Properties of Ceramics with Natural and Controlled Flaws: A Study of Alumina," in *Methods for Assessing the Structural Reliability of Brittle Materials,* ASTM STP 884, S. W. Freiman and C. M. Hudson (eds.), American Society for Testing and Materials, West Conshohocken, PA, 1984, pp. 43–56.

31. T. Fett and D. Munz, "Lifetime Prediction for Hot-Presses Silicon Nitride at High Temperatures," in *Methods for Assessing the Structural Reliability of Brittle Materials,* ASTM STP 884, S. W. Freiman and C. M. Hudson (eds.), American Society for Testing and Materials, West Conshohocken, PA, 1984, pp. 154–176.

32. G. D. Quinn, "Static Fatigue in High-Performance Ceramics," in *Methods for Assessing the Structural Reliability of Brittle Materials,* ASTM STP 884, S. W. Freiman and C. M. Hudson (eds.), American Society for Testing and Materials, West Conshohocken, PA, 1984, pp. 177–193.

33. A. Zimmermann, M. Hoffman, B. D. Flinn, R. K. Bordia, T.-J. Chuang, E. R. Fuller, Jr., and J. Rödel, "Fracture of Alumina with Controlled Pores," *J. Am. Ceram. Soc.,* **81**(9), 2449–2457 (1998).

34. R. W. Davidge, J. R. McLaren, and G. Tappin, *J. Mater. Sci.,* **8,** 1699 (1973).

35. E. J. Gumbell, *Statistics of Extremes,* Columbia University Press, New York, 1958.

36. ASTM Standard Practice, ASTM C 1239-95: Standard Practice for Reporting Uniaxial Strength Data and Estimating Weibull Distribution Parameters for Advanced Ceramics, Annual Book of ASTM Standards, Vol. 15.01.

37. J. F. Lawless, *Statistical Models and Methods for Lifetime Data,* Wiley, New York, 1982, Section 4.4.1, Appendix E, and Appendix F.

38. H. Rockette, C. Antle, and L. A. Kimko, "Maximum Likelihood Estimation with the Weibull Model," *J. Am. Stat. Assoc.,* **69**(345), 246–249 (1974).

39. A. G. Evans and S. M. Wiederhorn, "Proof Testing of Ceramic Materials—An Analytical Basis for Failure Prediction," *Int. J. Fract. Mech.,* **10**(3), 379 (1974).

40. A. G. Evans and E. R. Fuller, Jr., "Proof-Testing—The Effects of Slow Crack Growth," *Mater. Sci. Eng.,* **19**(1), 69–77 (1975).

41. J. E. Ritter, Jr., P. B. Oates, E. R. Fuller, Jr., and S. M. Wiederhorn, "Proof Testing of Ceramics: Part 1. Experiment," *J. Mater. Sci.,* **15,** 2275–2281 (1980).

42. E. R. Fuller, Jr., S. M. Wiederhorn, J. E. Ritter, Jr., and P. B. Oates, "Proof Testing of Ceramics: Part 2. Theory," *J. Mater. Sci.,* **15,** 2282–2295 (1980).

43. S. M. Wiederhorn, S. W. Freiman, E. R. Fuller, Jr., and H. Richter, "Effects of Multiregion Crack Growth on Proof Testing," in *Methods for Assessing the Structural Reliability of Brittle Materials,* ASTM STP 884, S. W. Freiman and C. M. Hudson (eds.), American Society for Testing and Materials, West Conshohocken, PA, 1984, pp. 95–116.

44. S. M. Wiederhorn, S. W. Freiman, E. R. Fuller, Jr., and C. J. Simmons, "Effect of Water and Other Dielectrics on Crack Growth," *J. Mater. Sci.,* **17,** 3460–3478 (1982).

45. G. S. White, S. W. Freiman, S. M. Wiederhorn, and T. D. Coyle, "Effects of Counterions on Crack Growth in Vitreous Silica," *J. Am. Ceram. Soc.,* **70**(12), 891–895 (1987).

46. S. M. Wiederhorn, "Dependence of Lifetime Predictions on the Form of the Crack Propagation Equation," in *Fracture 1977,* Vol. 3, ICF4, D. M. R. Taplin (ed.), University of Waterloo Press, Waterloo, Ontario, Canada, 1977, pp. 893–902.

47. S. M. Wiederhorn and J. E. Ritter, Jr., "Application of Fracture Mechanics Concepts to Structural Ceramics," in *Fracture Mechanics Applied to Brittle Materials,* ASTM STP 678, S. W. Freiman (ed.), American Society for Testing and Materials, West Conshohocken, PA, 1979, pp. 202–214.

48. K. Jakus, J. E. Ritter, Jr., and J. M. Sullivan, "Dependence of Fatigue Predictions on the Form of the Crack Velocity Equation," *J. Am. Ceram. Soc.,* **64**(6), 372–374 (1981).

49. ASTM Standard Practice, ASTM C 1368-97: Standard Test Method for Determination of Slow Crack Growth Parameters of Advanced Ceramics by Constant Stress-Rate Flexural Testing at Ambient Temperature, Annual Book of ASTM Standards, Vol. 15.01.

50. P. Diaconis and B. Efron, "Computer-Intensive Methods in Statistics," *Sci. Am.,* **248,** 116–130 (1983).

51. B. Efron and R. Tibshirani, "Bootstrap Methods for Standard Errors, Confidence Intervals, and Other Measures of Statistical Accuracy," *Stat. Sci.,* **1,** 54–77 (1986).

52. B. Efron and R. J. Tibshirani, *An Introduction to the Bootstrap,* Chapman & Hall, New York, 1993.

53. B. R. Lawn and D. B. Marshall, "Residual Stress Effects in Failure from Flaws," *J. Am. Ceram. Soc.,* **62**(1–2), 106–108 (1979).

54. B. R. Lawn and T. R. Wilshaw, "Indentation Fracture: Principles and Applications," *J. Mater. Sci.,* **10**(6), 1049–1081 (1975).

55. B. R. Lawn and E. R. Fuller, "Equilibrium Penny-Like Cracks in Indentation Fracture," *J. Mater. Sci.,* **10**(12), 2016–2024 (1975).

56. B. R. Lawn and D. B. Marshall, "Hardness, Toughness, and Brittleness: An Indentation Analysis," *J. Am. Ceram. Soc.,* **62**(7–8), 347–350 (1979).

57. B. R. Lawn, A. G. Evans, and D. B. Marshall, "Elastic/Plastic Indentation Damage in Ceramics," *J. Am. Ceram. Soc.,* **63**(9–10), 574–581 (1980).

58. J. J. Mecholsky, R. W. Rice, and S. W. Freiman, "Prediction of Fracture Energy and Flaw Size in Glasses from Measurements of Mirror Size," *J. Am. Ceram. Soc.,* **57**(10), 440–443 (1974).

59. V. D. Frechette, *Failure Analysis of Brittle Materials,* Advances in Ceramics, Vol. 28, American Ceramic Society, Westerville, OH, 1990.

60. W. H. Duckworth, "Precise Tensile Properties of Ceramic Bodies," *J. Am. Ceram. Soc.,* **34**(1), 1–9 (1951).

61. F. I. Baratta, "Requirements for Flexure Testing of Brittle Materials," in *Methods for Assessing the Structural Reliability of Brittle Materials,* ASTM STP 884, S. W. Freiman and C. M. Hudson (eds.), American Society for Testing and Materials, West Conshohocken, PA, 1984, pp. 194–222.

62. F. I. Baratta, W. T. Matthews, and G. D. Quinn, "Errors Associated with Flexure Testing of Brittle Materials," U.S. Army MTL TR 87-35, 1987.

63. ASTM Standard Practice, ASTM C 1161-94: Standard Test Method for Flexural Strength of Advanced Ceramics at Ambient Temperature, Annual Book of ASTM Standards, Vol. 15.01.

64. J. E. Ritter, Jr., K. Jakus, A. Batakis, and N. Bandyopadhyay, "Appraisal of Biaxial Strength Testing," *J. Non-Cryst. Sol.,* **38,** 419–424 (1980).

65. D. K. Shetty, A. R. Rosenfield, P. McGuire, G. K. Bansal, and W. H. Duckworth, "Biaxial Flexure Tests for Ceramics," *J. Am. Ceram. Soc.,* **59**(12), 1193–1197 (1980).

66. H. Fessler, D. C. Fricker, and D. J. Godfrey, "A Comparative Study of the Mechanical Strength of Reaction-Bonded Silicon Nitride," in *Ceramics for High Performance Applications III,* E. M. Lenoe, R. N. Katz, and J. J. Burke (eds.), Plenum, New York, 1983, pp. 705–736.

67. H. Fessler and D. C. Fricker, "A Theoretical Analysis of the Ring-On-Ring Loading Disk Test," *J. Am. Ceram. Soc.,* **67**(9), 582–588 (1984).

68. W. F. Adler and D. J. Mihora, "Biaxial Flexure Testing: Analysis and Experimental Results," in *Fracture Mechanics of Ceramics,* Vol. 10, R. C. Bradt, D. P. H. Hasselman, D. Munz, M. Sakai, and V. Ya Shevchenko (eds.), Plenum, New York, 1992, pp. 227–246.

69. J. B. Wachtman, Jr., W. Capps, and J. Mandel, "Biaxial Flexure Tests of Ceramic Substrates, *J. Mater.,* **7**(2), 188–194 (1972).

70. F. Sudreau, C. Olagnon, and G. Fantozzi, "Lifetime Prediction of Ceramics: Importance of Test Method," *Ceram. Int.,* **10,** 125–135 (1994).

71. J. A. Salem and M. G. Jenkins, "The Effect of Stress Rate on Slow Crack Growth Parameters," in *Fracture Resistance Testing of Monolithic and Composite Brittle Materials,* ASTM STP 1409, J. A. Salem, G. D. Quinn, and M. G. Jenkins (eds.), American Society for Testing and Materials, West Conshohocken, PA, 2001.

CHAPTER 30

TOTAL QUALITY MANAGEMENT IN MECHANICAL DESIGN

B. S. Dhillon
Department of Mechanical Engineering
University of Ottawa
Ottawa, Ontario, Canada

1 INTRODUCTION

In today's competitive environment, the age-old belief of many companies that "the customer is always right" has a new twist. To survive, companies are focusing their entire organization on customer satisfaction. The approach followed for ensuring customer satisfaction is known as total quality management (TQM). The challenge is to "manage" so that the "total" and the "quality" are experienced in an effective manner[1].

The history of quality-related efforts may be traced back to ancient times. For example, Egyptian wall paintings of around 1450 BC show some evidence of measurement and inspection activity.[2] In the modern times, total quality movement appears to have started in the early 1900s in the time-and-motion study works of Frederick W. Taylor, the father of scientific management.[3–5]

During the 1940s, the efforts of people such as W. E. Deming, J. Juran, and A. V. Feigenbaum greatly helped to strengthen the TQM movement.[4] In 1951, the Japanese Union of Scientists and Engineers established a prize named after W. E. Deming to be awarded to the organization that implements the most successful quality policies.[6] In 1987, on similar lines, the U.S. government established the Malcolm Baldrige award and, Nancy Warren and American behavioral scientist, coined the term *total quality management.*[7]

Quality cannot be inspected out of a product; it must be built in. The consideration of quality in design begins during the specification-writing phase. Many factors contribute to the success of the quality consideration in engineering or mechanical design. TQM is a useful tool for application during the design phase. It should be noted that the material presented in this chapter does not deal specifically with mechanical design but with design in general. However, the same material is equally applicable to the design of mechanical items and some important aspects of TQM considered useful to mechanical design are presented.

2 TERMS AND DEFINITIONS

This section presents terms and definitions used in TQM that are considered useful for mechanical design.[8–14]

- *Quality.* The totality of characteristics and features of an item or service that bear on its ability to satisfy stated requirements.
- *Total Quality Management.* A philosophy, set of methods or approaches, and process whose output leads to satisfied customers and continuous improvement.
- *Design.* The process of originating a conceptual solution to a need and expressing it in a form from which an item may be manufactured/produced or a service delivered.
- *Quality Measure.* A quantitative measure of the features and characteristics of a product or a service.
- *Design Review.* A formal documented and systematic critical study of a given design by individuals other that the designer.
- *Relative Quality.* The degree of excellence of a product/service.
- *Design Rationale.* The justification to explain why specific design implementation approaches were or were not selected.
- *Quality Conformance.* The extent to which the item/service conforms with the stated requirements.
- *Quality Control.* The operational methods and activities used to satisfy requirements for quality.
- *Quality Program.* The documented set of plans for implementing the quality system.
- *Quality Problem.* The difference between the specified quality and the achieved quality.
- *Quality Requirements.* Those requirements that pertain to the characteristics and features of an item/service and are required to be satisfied to meet a specified need.
- *Quality Performance Reporting System.* The system used to collect and report performance statistics of the product and service.

3 TQM IN GENERAL

This section presents some basic and general aspects of TQM.

3.1 Comparison between TQM and Traditional Quality Assurance Program

Table 1 presents a comparison between TQM and traditional quality assurance programs with respect to seven factors: objective, quality defined, customer, quality responsibility, decision making, cost, and definitions.[6,14–16]

3.2 TQM Elements

Seven important elements of the TQM are as follows[17]:

- *Team Approach.* The main goal is to get everyone involved with TQM, including customers, subcontractors, and vendors. The quality team size may vary from 3 to 15 members and the team membership is voluntary. However, team members possess sufficient knowledge in areas such as cost–benefit analysis, planning and controlling projects, brainstorming, public relations, flow charting, and statistic presentation techniques. Furthermore, the leader of the team usually has a management background and possesses qualities such as group leadership skills, communication skills, skill in statistical methods and techniques, skill in group dynamics, and presentation skills.
- *Management Commitment and Leadership.* These are crucial to the success of TQM and are normally achieved only after senior management personnel has effectively understood the TQM concept. Consequently, management establishes organization goals and directions and plays an instrumental role in achieving them.
- *Cost of Quality.* This is a quality measurement tool. It is used to monitor the TQM process effectiveness, choose quality improvement projects, and justify the cost to doubters. The cost of quality is expressed by[18]

$$C_q = \text{QMC} + \text{DC} = \text{PC} + \text{AC} + \text{DC} \tag{1}$$

where C_q = cost of quality
QMC = quality management cost
AC = appraisal cost
PC = prevention cost
DC = deviation cost

Table 1 Comparison between TQM and Traditional Quality Assurance Program

Factor	Traditional Quality Assurance Program	TQM
Objective	Discovers errors	Prevents errors
Quality defined	Creates manufactured goods that satisfy specifications	Produces goods suitable for consumer/customer use
Customer	Ambiguous comprehension of customer needs	Well-defined mechanisms to understand and meet customer needs
Quality responsibility	Quality control/inspection department	Involves all people in the organization
Decision making	Follows top-down management approach	Follows a team approach with worker groups
Cost	Better quality results in higher cost	Better quality lowers cost and enhances productivity
Definitions	Product driven	Customer driven

- *Supplier Participation.* This is important because an organization's ability to provide quality goods or services largely depends upon the types of relationships that exist among the parties involved in the process (e.g., the customer, the supplier, and the processor). Nowadays, companies increasingly require their suppliers to have well-established TQM programs as a precondition for potential business.[19]
- *Customer Service.* The application of the TQM concept to this area in the form of joint teams usually leads to customer satisfaction. These teams are useful to determine customer satisfaction through interactions with customers and they develop joint plans, goals, and controls.
- *Statistical Methods.* These are used to, for example, identify and separate causes of quality problem; verify, reproduce, and repeat measurements based on data; and make decisions on facts based on actual data instead of opinions of various people.[17,20] Some examples of statistical methods are Pareto diagrams, control charts, scatter diagrams, and cause-and-effect diagrams.[19,21,22]
- *Training.* A Japanese axiom states, "Quality begins with training and ends with training."[19] Under TQM, quality is the responsibility of all company employees; the training effort must be targeted to all hierarchy levels of the organization. Furthermore, it should cover areas such as fundamentals of TQM, team problem solving, and interpersonal communication.

3.3 TQM Principles and Barriers to TQM Success

The concept of TQM is based upon many principles. The important ones are shown in Fig. 1.[23]

Over the years, professionals and researchers working in TQM have identified many barriers to success with TQM. The most frequently occurring barriers are as follows[24,25]:

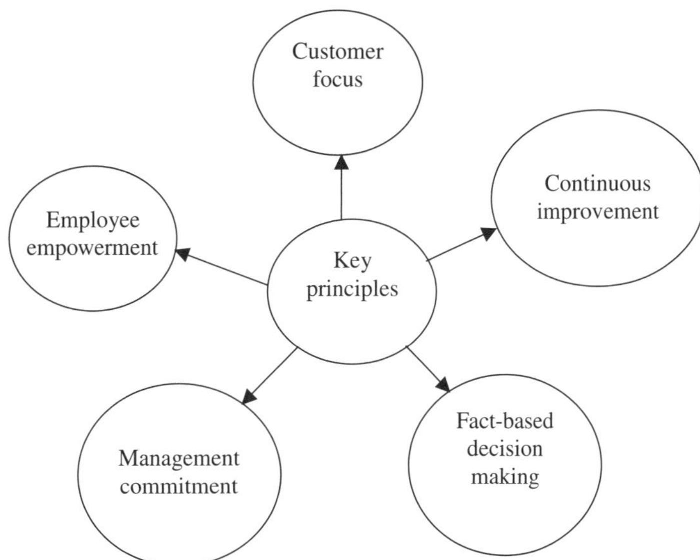

Figure 1 Important TQM principles.

1. Inadequate management commitment
2. Incorrect planning
3. Poor continuous training and education
4. Poor or improper usage of empowerment and teamwork
5. Inability to make changes to organizational culture
6. Giving poor attention to external and internal customers
7. Poor measurement methods and inadequate access to data and results
8. Incompatible organizational setup and isolated individuals, groups, and departments

4 THREE GENERAL APPROACHES TO TQM

W. E. Deming, J. M. Juran, and P. B. Crosby played instrumental roles in the development of TQM. Their approaches to TQM are presented below.[6,24,26–30]

4.1 Deming's Approach to TQM

Deming's approach is composed of the following:

1. Establish consistency of purpose for improving services.
2. Adopt the new philosophy for making the accepted levels of defects, delays, or mistakes unwanted.
3. Stop reliance on mass inspection as it neither improves nor guarantees quality. Remember that teamwork between the firm and its suppliers improves the process.
4. Stop awarding business with respect to the price.
5. Discover problems. Management must work continually to improve the system.
6. Take advantage of modern methods used for training. In developing a training program, take into consideration such items as:
   - Identification of company objectives
   - Identification of the training goals
   - Understanding of goals by everyone involved
   - Orientation of new employees
   - Training of supervisors in statistical thinking
   - Team building
   - Analysis of the teaching need
7. Institute modern supervision approaches.
8. Eradicate fear so that everyone involved may work to his or her full capacity.
9. Tear down departmental barriers so that everyone can work as a team member.
10. Eliminate items such as goals, posters, and slogans that call for new productivity levels without the improvement of methods.
11. Make your organization free of work standards prescribing numeric quotas.
12. Eliminate factors that inhibit employee workmanship pride.
13. Establish an effective education and training program.

14. Develop a program that will encourge the above 13 steps every day for never-ending improvement.

4.2 Crosby's Approach to TQM

Crosby's approach to TQM is composed of the following:

1. Ensure management commitment to quality.
2. Establish quality improvement teams/groups having representatives from all departments.
3. Identify locations of current and potential quality-related problems.
4. Establish the quality cost and describe its application as a management tool.
5. Promote the quality awareness and personal concern of all employees in the organization.
6. Take appropriate measures to rectify problems highlighted through previous steps.
7. Form a group or committee for the zero-defects program.
8. Train all supervisors to actively perform their part of the quality improvement program.
9. Hold a "zero-defects day" to let all organization personnel realize the change.
10. Encourage all employees to develop improvement goals for themselves and their respective groups.
11. Encourage all individuals to inform management about the obstacles faced by them to attain their set improvement goals.
12. Recognize appropriately those who participate.
13. Form quality councils for communicating on a regular basis.
14. Repeat steps 1–13 to emphasize that the quality improvement program is a never ending process.

4.3 Juran's Approach to TQM

Juran's approach is composed of the following:

1. Promote awareness of the requirement and opportunity for improvement.
2. Establish appropriate goals for improvement.
3. Organize for attaining the set goals (i.e., form a quality council, highlight problems, choose projects, form teams, and designate facilitators).
4. Provide appropriate training.
5. Carry out projects for finding solutions to problems.
6. Report progress.
7. Give appropriate recognitions.
8. Communicate results to all concerned individuals.
9. Keep records of scores.
10. Maintain momentum by making yearly improvements part of the regular company systems and processes.

5 QUALITY IN THE DESIGN PHASE

Although TQM will generally help to improve design quality, specific quality-related steps are also necessary during the design phase. These additional steps will further enhance product design.

An informal review during specification writing may be regarded as the beginning of quality assurance in the design phase. As soon as the first draft of the specification is complete, the detailed analysis begins.

This section presents topics considered directly or indirectly useful to improving quality in the design phase.[31–33]

5.1 Quality Design Characteristics

There are seven quality design characteristics[31,32]: performance, durability, conformance, features, serviceability, reliability, and aesthetics. For example, with a manufactured product (e.g., stereo amplifier), these characteristics would involve the signal-to-noise ratio (power), useful life, workmanship, remote control, ease of repair, mean time to failure, and oak cabinet, respectively. Similarly, for a service product (e.g., checking account) these are time to process customer requests, keeping pace with industry trends, accuracy, automatic bill paying, resolution of errors, variability of time to process, and appearance of bank lobby, respectively.

5.2 Steps for Controlling Design

Effective control of the design activity can help to improve product quality. Any design activity can be controlled by following the steps listed below during the design process[8]:

- Establish customer requirements.
- Translate the customer requirements into a definitive specification document of the requirements.
- Perform a feasibility study to find if the requirements can be accomplished.
- Plan for accomplishing the requirements.
- Organize materials and resources for accomplishing the requirements.
- Perform a project definition study to find the most suitable solution out of many possible solutions.
- Develop a specification document that describes the product's/service's features and characteristics.
- Develop a prototype/model of the proposed design.
- Conduct trials to determine the degree to which the product/service under development meets the design requirements and customer needs.
- Feed data back into the design and repeat the process until the product or service satisfies the specified requirements in an effective manner.

5.3 Product Design Review

Various types of design reviews are conducted during the product design phase. One reason for performing these reviews is to improve quality. Design reviews conducted during the design phase include preliminary design review, detailed design reviews (the number of which may vary from one project to another), critical design review (the purpose of this review is to approve the final design), preproduction design review (this review is performed after the prototype tests), postproduction design review, and operations and support design review.

The consideration of quality begins at the preliminary design review and becomes stronger as the design develops. The role of quality assurance in preliminary design review is to ensure that the new design is free of quality problems of similar existing designs. This requires a good knowledge of the strengths and weaknesses of the competing products.

5.4 Quality Function Deployment, Quality Loss Function, and Benchmarking

These three approaches are quite useful in ensuring quality during the design phase. Quality function deployment (QFD) is a value analysis tool used during product and process development. It is an extremely useful concept for developing test strategies and translating needs to specification.

QFD was developed in Japan. In the case of new product development, it is simply a matrix of consumer/customer requirements versus design requirements. Some sources for the input are market surveys, interviews, and brainstorming. To use the example of an automobile, customer needs include price, expectations at delivery (safety, perceived quality, service ability, performance, workmanship, etc.), and expectations over time (including customer support, durability, reliability, performance, repair part availability, low preventive maintenance and maintenance cost, mean time between failures within prediction, etc.).

Finally, QFD helps to turn needs into design engineering requirements.

The basis for the quality loss function is that if all parts are produced close to their specified values, then it is fair to expect the best product performance and lower cost to society. According to Taguchi, quality cost goes up not only when the finished product is outside given specifications but also when it deviates from the set target value within the specifications.[4]

One important point to note, using Taguchi's philosophy, is that a product's final quality and cost are determined to a large extent by its design and manufacturing processes. It may be said that the loss function concept is simply the application of a life-cycle cost model to quality assurance. Taguchi expresses the loss function as

$$L(x) = c(x - T_v)^2 \tag{2}$$

where x = variable
$L(x)$ = loss at x
T_v = targeted value of variable at which product is expected to show its best performance
c = proportionality constant
$x - T_v$ = deviation from target value

In the formulation of the loss function, assumptions are made, such as zero loss at the target value and that the dissatisfaction of a customer is proportional only to the deviation from the target value. The value of the proportionality constant c can be determined by estimating the loss value for an unacceptable deviation, such as the tolerance limit. Thus, the following relationship can be used to estimate the value of c:

$$c = \frac{L_a}{\Delta^2} \tag{3}$$

where L_a = amount of loss expressed in dollars
Δ = deviation amount from target value T_v

Example Assume that the estimated loss for the Rockwell hardness number beyond 56 is \$150 and the targeted value of the hardness number is 52. Estimate the value of the proportionality constant. Substituting the given data into Eq. (3), we get

$$c = \frac{150}{(56 - 52)^2} = 9.375$$

Thus, the value of the proportionality constant is 9.375.

Benchmarking is a process of comparing in-house products and processes with the most effective ones in the field and setting objectives for gaining a competitive advantage. The following steps are associated with benchmarking[34]:

- Identify items and their associated key features to benchmark during product planning.
- Select companies, industries, or technologies to benchmark against. Determine existing strengths of the items to benchmark against.
- Determine the best-in-class target from each selected benchmark item.
- Evaluate, as appropriate, in-house processes and technologies with respect to benchmarks.
- Set improvement targets, remembering that the best-in-class target is always a moving target.

5.5 Process Design Review

Soon after the approval of a preliminary design, a process flowchart is prepared. So that proper consideration is given to quality, the quality engineer should work with process and reliability engineers. In addition, the quality engineer's expertise in variation control provides important input.

Lack of integration between quality assurance and manufacturing is one of the main reasons for the failure of the team effort. Process failure mode and effect analysis (FMEA) helps this integration to take place early. Consideration of the total manufacturing process during FMEA, rather than just of the equipment, is also a useful step in this regard. For FMEA to produce promising results, quality and manufacturing engineers have to work as a team. In addition FMEA is a useful tool for performing analysis of a new process, including the analysis of receiving, handling, and storing materials and tools. Also, the participation of suppliers in FMEA studies enhances FMEA's value. The following steps are associated with FMEA[35]:

- Develop a process flowchart that includes all process inputs: materials, storage and handling, transportation, etc.
- List all components/elements of the process.
- Write down each component/element description and identify all possible failure modes.
- Assign failure rate/probability to each component/element failure mode.
- Describe each failure mode cause and effect.

- Enter remarks for each failure mode.
- Review each critical failure mode and take corrective measures.

5.6 Plans for Acquisition and Process Control

The development of quality assurance plans for procurement and process control during the design phase is useful for improving product quality. One immediate advantage is the smooth transition from design to production. The equipment procurement plan should include such items as equipment performance verification, statistical tolerance analysis, testing for part interchangeability, and pilot runs. Similarly, the component procurement quality plans should address concerns and cooperation on such areas as component qualification, closed-loop failure management, process control implementation throughout the production lines, and standard and special screening test.

Prior to embarking on product manufacturing, there is a need to identify the critical points where the probability of occurrence of a serious defect is quite high. Thus, process control plans should be developed by applying QFD and process FMEA. These plans should include the following:

- Acceptance of standard definitions
- Procedures to monitor defects
- Approaches for controlling critical process points

5.7 Taguchi's Quality Philosophy Summary and Kume's Approach for Process Improvement

Taguchi's approach was discussed earlier, but because of its importance, this section summarizes his quality philosophy again[36,37]:

1. A critical element of a manufactured item's quality is the total loss generated by that item to society as whole.
2. In today's market, continuous cost reduction and quality improvement are critical for companies to stay in business.
3. Design and its associated manufacturing processes determine, to a large extent, the ultimate quality and cost of a manufactured product.
4. Unceasing reduction in a product performance characteristic's variation from its target values is part of a continuous quality improvement effort.
5. The loss of customers due to variation in an item's performance is frequently almost proportional to the square of the performance characteristic's deviation from its target value.
6. The identification of the product and process parameter settings that reduce performance variation can be accomplished through statistically designed experiments.
7. Reduction in the performance variation of a product or process can be achieved by exploiting the product or process parameter nonlinear effects on performance characteristics.

To improve a process, Kume outlined a seven-step approach[38]:

- Select project.
- Observe the process under consideration.

- Perform process analysis.
- Take corrective measures.
- Evaluate effectiveness of corrective measures.
- Standardize the change.
- Review and make appropriate modifications, if applicable, in future plans.

5.8 Design for Six Sigma

The Design for Six Sigma (DFSS) approach is quite useful to improve product or service quality and is composed of the following steps[39]:

- Identify the new product or service to be designed and establish a project plan and a project team.
- Plan and conduct research for understanding customer and other related requirements.
- Develop alternative design approaches, choose an approach for high-level design, and assess the design capability to satisfy the specified requirements.
- Develop the design, determine its capability, and plan a pilot test study.
- Perform the pilot test study, analyze the associated results, and make changes to the design as required.

This approach is described in detail in Ref. 40.

5.9 Design Guidelines for Quality Improvement

Design guidelines for quality improvement with their advantages in parentheses are as follows[32,41]:

- Minimize number of part/component numbers (less variations of like parts or components).
- Select parts/components that can withstand process operations (less degradation of parts/components and less damage to parts/components).
- Eradicate adjustments altogether (elimination of assembly adjustment errors and adjustment parts with high failure rates).
- Utilize repeatable, well-understood processes (easy-to-control part quality and assembly quality).
- Eradicate all engineering changes on released items (less errors due to change-over and multiple revisions/versions).
- Design for robustness (low sensitivity to component/part variability).
- Design for effective, i.e., efficient and adequate, testing (less mistaking "good" for "bad" item/product and vice-versa).
- Minimize number of parts/components (less parts/components to fail, fewer part and assembly drawings, fewer parts/components to hold to required quality characteristics, and less complicated assemblies).
- Make assembly easy, straightforward, and foolproof (self-securing parts, impossible to assemble parts incorrectly, no "force fitting" of parts, easy to spot missing parts, and assembly tooling designed into part).

- Lay out parts/components for reliable process completion (less damage to parts/components during handling and assembly).

6 TQM METHODS

Over the years many methods that can be used to improve product or service quality have been developed. Effective application of these approaches can help to improve mechanical design/product quality directly or indirectly.

Reference 41 presents 100 TQM methods classified into four areas: idea generation, analytical, management, and data collection analysis and display:

- The idea generation area has 17 methods, including brainstorming, buzz groups, idea writing, lateral thinking, list reduction, morphological forced connections, opportunity analysis, and snowballing.
- The analytical area has 19 methods, including cause-and-effect analysis, failure-modes-and-effect analysis, Taguchi methods, force field analysis, tolerance design, solution effect analysis, and process cost of quality.
- The management area contains 31 methods, including Deming wheel, benchmarking, Kaizen, Pareto analysis, quality circles, QFD, error proofing (poka-yoke), and zero defects.
- The data collection analysis and display area has 33 methods, including check sheets, histograms, Hoshin Kanri, scatter diagrams, statistical process control, pie charts, spider web diagrams, and matrix diagram.

Thirteen methods belonging to the above four areas are presented in Sections 16.1–16.13.[30,42]

6.1 Affinity Diagram

This diagram is used to organize large volumes of data into groups according to some kind of natural affinity.[42,43] The affinity diagram is used when a group of people are ascertaining customer needs for the purpose of translating them into design requirements.

Each member of the group writes ideas about customer needs on file cards and then lays the cards on a table without discussing the ideas with other team members. The team members are asked to arrange the requirements on their cards into groups that have an affinity for each other.

The main advantage of the affinity diagram approach is that organizing data in the form of affinity can illustrate natural associations between customer needs rather than just logical connections. This approach is described in detail in Ref. 43.

6.2 Deming Wheel

This is a management concept proposed by W. E. Deming for meeting the customer quality requirements by using a cycle made up of four elements: plan, do, check, and action.[26,42] More specifically, the wheel can be used when developing a new product based on customer requirements.

The approach calls for teamwork among various groups of the company: product development, manufacturing, sales, and market research. The cycle, or wheel, is considered constantly rotating and its four elements are described below.

- *Plan.* This is concerned with determining the cause of the problem after its detection in the product development and planning corrective actions based on facts.
- *Do.* This calls for a quality improvement team to take responsibility for taking necessary steps to correct the problem after its detection.
- *Check.* This is concerned with determining if the improvement process was successful.
- *Action.* This is concerned with accepting the new and better quality level if the improvement process is successful. If it is not, it (i.e., action) calls for repeating the Deming cycle.

This approach is described in detail in Ref. 26.

6.3 Fishbone Diagram

This approach, also known as the cause-and-effect or Ishikawa diagram, was originally developed by K. Ishikawa in Japan. The diagram serves as a useful tool in quality-related studies to perform cause-and-effect analysis for generating ideas and finding the root cause of a problem for investigation. The diagram resembles a fishbone; thus the name.

Major steps for developing a fishbone diagram are as follow[30]:

- Establish problem statement.
- Brainstorm to highlight possible causes.
- Categorize major causes into natural groupings and stratify by steps of the process.
- Insert the problem or effect in the "fish head" box on the right-hand side. Develop the diagram by unifying the causes following the necessary process steps.
- Refine categories by asking questions such as "What causes this?" and "Why does this condition exist?"

This method is described in detail in Ref. 30.

6.4 Pareto Diagram

An Italian economist, Vilfredo Pareto (1848–1923), developed a formula in 1897 to show that the distribution of income is uneven.[44,45] In 1907, a similar theory was put forward in a diagram by M. C. Lorenz, a U.S. economist. In later years, J. M. Juran applied Lorenz's diagram to quality problems and called it Pareto analysis.[46]

In quality control work, Pareto analysis means there are always a few defects in hardware manufacture that loom large in occurrence, frequency, and severity. Economically, these defects are costly and thus of great significance. It may simply be stated that on average about 80% of the costs occur due to 20% of the defects.

The Pareto diagram, derived from the above reasoning, is helpful in identifying where to concentrate one's effort. The Pareto diagram is a frequency chart with bars arranged in descending order from left to right, visually highlighting the major problem areas. The Pareto principle can be quite instrumental in TQM, particularly in improving quality of product designs.

6.5 Kaizen Method

Kaizen means "improvement" in Japanese, and the Kaizen philosophy maintains that the current way of life deserves to be improved on a continuous basis. This philosophy is broader than TQM because it calls for ongoing improvement as workers, leaders, managers, and so on. Thus, Kaizen includes TQM, quality circles, zero defects, new product design, continuous quality improvement, customer service agreements, and so on. Kaizen is often referred to as "the improvement movement" because it encompasses constant improvement in social life, working life, and the home life of everyone. This method is described in detail in Ref. 30.

6.6 Force Field Analysis

This method was developed by Kurk Lewin to identify forces existing in a situation. It calls first for clear understanding of the driving and restraining forces and then for developing plans to implement change.[30]

The change is considered a dynamic process and the result of a struggle between driving forces (i.e., those forces seeking to upset the status quo) and restraining forces (i.e., those forces attempting to maintain the status quo). A change occurs only when the driving forces are stronger than the restraining forces.

The group brainstorming method serves as a useful tool to identify the driving forces and restraining forces in a given situation.

With respect to improving engineering design quality, the force field analysis facilitates change in the following way:

- It forces concerned personnel to identify and to think through the certain facets of an acquired change.
- It highlights the priority order of the involved driving and restraining forces.
- It leads to the establishment of a priority action plan.

An example of the force field analysis technique is given in Ref. 30.

6.7 Customer Needs Mapping Method

This approach is used to identify consumer requirements and then the ability of in-house processes to meet those requirements satisfactorily. The two major customers of a process are the external customer (the purchaser of the product or service) and the internal customer (the next step in the process of receiving the output). Past experience has shown that the internal customer is often overlooked by groups such as inventory control, accounting, facilities, and computer.

The wants or requirements of both external and internal customers can be identified through brainstorming, customer interviews, and so on.

Some advantages of customer needs mapping are as follows:

- It enhances the understanding of the customer background.
- It highlights customer wants.
- It translates customer needs into design features.
- It focuses attention on process steps important to customers.
- It highlights overlooked customer needs.

6.8 Control Charts

Control charts were developed by Walter A. Shewhart of Bell Telephone Laboratories in 1924 for analyzing discrete or continuous data collected over a period of time.[47] A control chart is a graphical tool used for assessing the states of control of a given process. The variations or changes are inherent in all processes, but their magnitude may be large or very small. Minimizing or eliminating such variations will help to improve the quality of a product or service.

Physically, a control chart is composed of a center line (CL), upper control limit (UCL), and lower control limit (LCL). In most cases, a control chart uses control limits of plus or minus three standard deviations from the mean of the quality or other characteristic under study.

Following are some of the reasons for using control charts for improving product design quality[47]:

- To provide a visual display of a process
- To determine if a process is in statistical control
- To stop unnecessary process-related adjustments
- To provide information on trends over time
- To take appropriate corrective measures

Prior to developing a control chart, the following factors must be considered:

- Sample size, frequency, and the approach to be followed for selecting them
- Objective of the control chart under consideration
- Characteristics to be examined
- Required gauges or testing devices

6.9 Poka-Yoke Method

This is a mistake-proofing approach. It calls for designing a process or product so that the occurrence of any anticipated defect is eliminated. This is accomplished through the use of automatic test equipment that "inspects" the operations performed in manufacturing a product and then allows the product to proceed only when everything is correct. Poka-yoke makes it possible to achieve the goal of zero defects in the production process. The method is described in detail in Ref. 30.

6.10 Hoshin Planning Method

This method, also known as the "seven management tools," helps to tie quality improvement activities to the long-term plans of the organization.[48] Hoshin planning focuses on policy development issues: planning objective identification, management and employee action identification, and so on. The following are the three basic Hoshin planning processes:

- General planning begins with the study of consumers for the purpose of focusing the organization's attention on satisfying their needs.
- Intermediate planning begins after the general planning is over. It breaks down the general planning premises into various segments for the purpose of addressing them individually.
- Detailed planning begins after the completion of the intermediate planning and is assisted by the arrow diagram and by the process decision program chart.

The seven management tools related to or used in each of the above three areas are as follows:

General

- Interrelationship diagram
- Affinity chart

Intermediate

- Matrix data analysis
- Tree diagram
- Matrix diagram

Detailed

- Arrow diagram
- Process decision program chart

These management tools are discussed below:

- The interrelationship diagram is used to identify cause-and-effect links among ideas produced. It is particularly useful in situations where there is a requirement to identify root causes. One important limitation of the interrelationship diagram is the overwhelming attempt to identify linkages between all generated ideas.
- The affinity chart is used to sort related ideas into groups and then label each similar group. The affinity chart is extremely useful in handling large volumes of ideas, including the requirement to identify broad issues.
- The matrix data analysis is used to show linkages between two variables, particularly when there is a requirement to show visually the strength of their relationships. The main drawback of this approach is that only two relationships can be compared at a time.
- The tree diagram is used to map required tasks into detailed groupings. This method is extremely useful when there is a need to divide broad tasks or general objectives into subtasks.
- The matrix diagram is used to show relationships between activities, such as tasks and people. It is an extremely useful tool for showing relationships clearly.
- The arrow diagram is used as a detailed planning and scheduling tool and helps to identify time requirements and relationships among activities. The arrow diagram is an extremely powerful tool in situations requiring detailed planning and control of complex tasks with many interrelationships.
- The process decision program chart is used to map contingencies along with countermeasures. The process decision program chart is an advantage in implementing a new plan with potential problems so that the countermeasures can be thought through.

6.11 Gap Analysis Method

This method is used to understand services offered from different perspectives. The method considers five major gaps that are evaluated so that when differences are highlighted between perceptions, corrective measures can be initiated to narrow the gap or difference:

- Consumer expectation and management perception gap
- Management perception of consumer expectation and service quality specification gap

- Service quality specifications and service delivery gap
- External communication and service delivery gap
- Consumer expectation concerning the service and the actual service received gap

This method is described in detail in Ref. 30.

6.12 Stratification

This method is used in defining a problem by identifying where it does and does not occur.[43] It is a technique of splitting data according to meeting or not meeting a set of criteria. More specifically, the stratification approach highlights patterns in the data. It is used before and after collecting data for designing the way to collect data and the way of focusing the analysis, respectively. The following steps are associated with the stratification approach[42]:

- Brainstorm to identify characteristics that could cause systematic differences in the data.
- Design data collection forms for incorporating these characteristics.
- Collect data and then examine them to identify any trends or patterns.

The main advantage of the stratification approach is that it ensures the collection of all necessary data the first time; thus it avoids wasting time and effort.

6.13 Opportunity Analysis

This is an efficient method used to evaluate a long list of options against desired goals and available resources. It can be used by an individual or a group. The method is composed of the following steps[42]:

- List all your desired goals in the situation under consideration.
- Rank each goal's importance with respect to satisfying customers and rate your capability to accomplish them.
- Determine if there are sufficient required resources available to accomplish these goals.
- Select the most promising options.

REFERENCES

1. C. R. Farquhar and C. G. Johnston, *Total Quality Management: A Competitive Imperative,* Report No. 60-90-E, Conference Board of Canada, Ottawa, Ontario, Canada, 1990.
2. D. C. Dague, "Quality: Historical Perspective," in *Quality Control in Manufacturing,* Society of Automotive Engineers, Warrendale, PA, 1981, pp. 23–29.
3. A. Rao, A. Naido, L. Leo, and K. Arthur, *Total Quality Management: A Cross Functional Perspective,* Wiley, New York, 1996.
4. C. D. Gevirtz, *Developing New Products with TQM,* McGraw-Hill, New York, 1994.
5. D. L. Goetsch and S. Davis, *Implementing Total Quality,* Prentice-Hall, Englewood Cliffs, New Jersey, 1995.
6. W. H. Schmidt, and J. P. Finnigan, *The Race without a Finish Line: America's Quest for Total Quality,* Jossey-Bass, San Francisco, CA, 1992.
7. M. Walton, *Deming Management at Work,* Putnam, New York, 1990.
8. D. Hoyle, *ISO 9000 Quality Systems Handbook,* Butterworth-Heinemann, Boston, 1998.

9. T. P. Omdahl (ed.), *Reliability, Availability, and Maintainability (RAM) Dictionary,* ASQC Quality Press, Milwaukee, WI, 1988.

10. *Quality Systems Terminology,* ANSI/ASQC A3-1978, American Society for Quality Control (ASQC), Milwaukee, WI, 1978.

11. *Glossary and Tables for Statistical Quality Control,* American Society for Quality Control, Milwaukee, WI, 1973.

12. J. L. Hradesky, *Total Quality Management Handbook,* McGraw-Hill, New York, 1995.

13. B. S. Dhillon, *Quality Control, Reliability, and Engineering Design,* Marcel Dekker, New York, 1985.

14. B. S. Dhillon, *Engineering and Technology Management Tools and Applications,* Artech House, Norwood, MA, 2002.

15. C. N. Madu and K. Chu-Hua, "Strategic Total Quality Management (STQM)," in *Management of New Technologies for Global Competitiveness,* C. N. Madu (ed.), Quorum Books, Westport, CT, 1993, pp. 3–25.

16. B. S. Dhillon, *Advanced Design Concepts for Engineers,* Technomic Publishing, Lancaster, PA, 1998.

17. J. L. Burati, M. F. Matthews, and S. N. Kalidindi, "Quality Management Organizations and Techniques," *Construct. Eng. Manag.,* **118,** 112–128 (Mar. 1992).

18. W. B. Ledbetter, *Measuring the Cost of Quality in Design and Construction,* Publication No. 10–2, Construction Industry Institute, Austin, TX, 1989.

19. M. Imai, *Kaizen: The Key to Japan's Competitive Success,* Random House, New York, 1986.

20. J. Perisco, "Team Up for Quality Improvement," *Quality Prog.,* **22**(1), 33–37 (1989).

21. H. Kume, *Statistical Methods for Quality Improvement,* Association for Overseas Technology Scholarship, Tokyo, 1985.

22. K. Ishikawa, *Guide to Quality Control,* Asian Productivity Organization, Tokyo, 1982.

23. L. Martin, "Total Quality Management in the Public Sector," *Nat. Productivity Rev.,* **10,** 195–213 (1993).

24. G. M. Smith, *Statistical Process Control and Quality Improvement,* Prentice-Hall, Inc., Upper Saddle River, NJ, 2001.

25. R. J. Masters, "Overcoming the Barriers to TQM Success," *Quality Prog.,* May 1996, pp. 50–52.

26. W. E. Deming, *Out of the Crisis,* Center for Advanced Engineering Study, Massachusetts Institute of Technology, Cambridge, MA, 1986.

27. P. B. Crosby, *Quality without Tears,* McGraw-Hill, New York, 1984.

28. J. S. Oakland, *Total Quality Management,* Butterworth-Heinemann, Boston, 2003.

29. J. Heizer and B. Render, *Production and Operations Management,* Prentice-Hall, Upper Saddle River, NJ, 1995.

30. P. Mears, *Quality Improvement Tools and Techniques,* McGraw-Hill, New York, 1995.

31. P. E. Pisek, "Defining Quality at the Marketing/Development Interface," *Quality Prog.,* June 1987, pp. 28–36.

32. J. R. Evans and W. M. Lindsay, *The Management and Control of Quality,* West Publishing, New York, 1989.

33. D. G. Raheja, *Assurance Technologies,* McGraw-Hill, New York, 1991.

34. *Total Quality Management: A Guide for Implementation,* Document No. DOD 5000.51.6 (draft), U.S. Department of Defense, Washington, DC, March 23, 1989.

35. B. S. Dhillon and C. Singh, *Engineering Reliability: New Techniques and Applications,* Wiley, New York, 1981.

36. R. H. Lochner and J. E. Matar, *Designing for Quality,* ASQC Quality, Milwaukee, WI, 1990.

37. R. N. Kackar, "Taguchi's Quality Philosophy: Analysis Commentary," *Quality Prog.,* **19,** 21–29 (1986).

38. H. Kume, *Statistical Methods for Quality Improvement,* Japanese Quality, Tokyo, 1987.

39. G. J. Hahn, N. Doganaksoy, and R. Hoerl, "The Evolution of Six Sigma," *Quality Eng.,* **12**(3), 317–326 (2000).

40. F. M. Gryna, *Quality Planning and Analysis,* McGraw-Hill, New York, 2001.

41. D. Daetz, "The Effect of Product Design on Product Quality and Product Cost," *Quality Prog.,* June 1987, pp. 63–67.

42. G. K. Kanji and M. Asher, *100 Methods for Total Quality Management,* Sage Publications, London, 1996.

43. B. Bergman and B. Klefsjo, *Quality: From Customer Needs to Customer Satisfaction,* McGraw-Hill, New York, 1994.

44. J. A. Burgess, *Design Assurance for Engineers and Managers,* Marcel Dekker, New York, 1984.

45. B. S. Dhillon, *Quality Control, Reliability and Engineering Design,* Marcel Dekker, New York, 1985.

46. J. M. Juran, F. M. Gryna, and R. S. Bingham (ed.), *Quality Control Handbook,* McGraw-Hill, New York, 1979.

47. *Statistical Quality Control Handbook,* AT&T Technologies, Indianapolis, IN, 1956.

48. B. King, *Hoshin Planning: The Developmental Approach,* Methuen, Boston, MA, 1969.

BIBLIOGRAPHY

TQM

Baker, W., *TQM: A Philosophy and Style of Managing,* Faculty of Administration, University of Ottawa, Ottawa, Ontario, 1992.

Besterfield, D. H., *Total Quality Management,* Prentice-Hall, Upper Saddle River, NJ, 2003.

Evans, J. R., *Total Quality: Management, Organization, and Strategy,* Thomson/South-Western, Mason, OH, 2003.

Farquhar, C. R., and C. G. Johnston, *Total Quality Management: A Competitive Imperative,* Report No. 60-90-E, Conference Board of Canada, Ottawa, Ontario, 1990

Feigenbaum, A. V., *Total Quality Control,* McGraw-Hill, New York, 1983.

Gevirtz, C. D., *Developing New Products with TQM,* McGraw-Hill, New York, 1994.

Mears, P., *Quality Improvement Tools and Techniques,* McGraw-Hill, New York, 1995, pp. 229–246.

Oakland, J. S., *Total Quality Management: Text with Cases,* Butterworth-Heinemann, Boston, MA, 2003.

Oakland, J. S. (ed.), *Total Quality Management Proceedings of the Second International Conference,* IFS Publications, Kempston, Bedford, UK, 1989.

Ross, J. E., *Total Quality Management: Text, Cases, and Readings,* St. Lucie, Boca Raton, FL, 1999.

Shores, A. R., *Survival of the Fittest: Total Quality Control and Management,* ASQC Quality, Milwaukee, WI, 1988.

Stein, R. E., *The Next Phase of Total Quality Management,* Marcel Dekker, New York, 1994.

Tenner, R. R., and I. J. Detoro, *Total Quality Management: Three Steps to Continuous Improvement,* Addison-Wesley, Reading, MA, 1992.

Design Quality

Burgess, J. A., "Assuring the Quality of Design," *Machine Design,* Feb. 1982, pp. 65–69.

Chaparian, A. P., "Teammates: Design and Quality Engineers," *Quality Prog.,* **10**(4), 16–17 (Apr. 1977).

Colloquium on Management of Design Quality Assurance, IEE Colloquium Digest No. 1988/6, Institution of Electrical Engineers, London, 1988.

Daetz, D., "The Effect of Product Design on Product Quality and Product Cost," *Quality Prog.,* **20,** 63–67 (June 1987).

Evans, J. R., and W. M. Lindsay, *The Management and Control of Quality,* West, New York, 1982, pp. 188–221.

Gryna, F. M., "Designing for Quality," in *Quality Planning and Analysis,* F. M. Gryna (ed.), McGraw-Hill, New York, 2001, pp. 332–372.

Juran, J. J., *Juran on Quality by Design,* Free Press, New York, 1992.

Lockner, R. H., and J. E. Matar, *Designing for Quality,* ASQC Quality, Milwaukee, WI, 1990.

Michalek, J. M., and R. K. Holmes, "Quality Engineering Techniques in Product Design/Process," in *Quality Control in Manufacturing,* SP-483, Society of Automotive Engineers, Warrendale, PA, pp. 17–22.

Phadke, M. S., *Quality Engineering Using Robust Design,* Prentice-Hall, Englewood Cliffs, NJ, 1986.

Pignatiello, J. J., and J. S. Ramberg, "Discussion on Off-Line Quality Control, Parameter Design, and the Taguchi Method," *J. Quality Technol.,* **17,** 198–206 (1985).

Quality Assurance in the Design of Nuclear Power Plants: A Safety Guide, Report No. 50-SG-QA6, International Atomic Energy Agency, Vienna, 1981.

Quality Assurance in the Procurement, Design, and Manufacture of Nuclear Fuel Assemblies: A Safety Guide, Report No. 50-SG-QA 11, International Atomic Energy Agency, Vienna, 1983.

Revelle, J. B., J. W. Moran, and C. A. Cox, *The QFD Handbook,* Wiley, New York, 1998.

Ross, P. J., *Taguchi Techniques for Quality Engineering,* McGraw-Hill, New York, 1988.

Turmel, J., and L. Gartz, "Designing in Quality Improvement: A Systematic Approach to Designing for Six Sigma," *Proceedings of the Annual Quality Congress,* 1997, pp. 391–398.

Vonderembse, M., and T. Van Fossen, "Is Quality Function Deployment Good for Product Development? Forty Companies Say Yes," *Quality Manag. J.,* **4**(3), 65–79 (1997).

CHAPTER 31

RELIABILITY IN THE MECHANICAL DESIGN PROCESS

B. S. Dhillon
Department of Mechanical Engineering
University of Ottawa
Ottawa, Ontario, Canada

1 INTRODUCTION

The history of reliability may be traced back to the early 1930s when probability concepts were applied to problems related to electric power systems.[1–7] During World War II, German researchers applied the basic reliability concepts to improve reliability of their V1 and V2 rockets. During the period 1945–1950, the U.S. Department of Defense (DOD) conducted various studies that revealed a definite need to improve equipment reliability. Consequently, the DOD formed an ad hoc committee on reliability in 1950. In 1952, this committee was transformed to a permanent body: Advisory Group on the Re,-liability of Electronic Equipment (AGREE).[8] The group released its report in 1957.

In 1951, W. Weibull proposed a function to represent time to failure of various engineering items.[9] Subsequently, this function became known as the Weibull distribution and is regarded as the starting point of mechanical reliability along with the works of A. M. Freudenthal.[10–11]

In the early 1960s, the National Aeronautics and Space Administration (NASA) played an important role in the development of mechanical reliability, basically due to the following three factors[12]:

- The loss of Syncom I in space in 1963 due to a bursting high-pressure gas tank
- The loss of Mariner III in 1964 due to a mechanical failure
- The frequent failure of components such as valves, regulators, and pyrotechnics in the Gemini spacecraft systems

Consequently, NASA initiated and completed many projects concerned with mechanical reliability. A detailed history of mechanical reliability is given in Refs. 13–15 along with a comprehensive list of publications on the subject up to 1992.

2 STATISTICAL DISTRIBUTIONS AND HAZARD RATE MODELS

Various types of statistical distributions and hazard rate models are used in mechanical reliability to represent failure times of mechanical items. This section presents some of these distributions and models considered useful to perform various types of mechanical reliability analyses.

2.1 Statistical Distributions

This section presents three statistical or probability distributions: exponential, Weibull, and normal.

Exponential Distribution

This is probably the most widely used distribution in reliability work to represent the failure behavior of various engineering items.[16] Moreover, it is relatively easy to handle in performing reliability analysis in the industrial sector. Its probability density function is expressed by[14,16]

$$f(t) = \lambda e^{-\lambda t} \qquad \text{for} \quad \lambda > 0 \qquad t \geq 0 \tag{1}$$

where $f(t)$ = probability density function

λ = distribution parameter; in reliability work, it is known as the constant failure rate

t = time.

The cumulative distribution function is given by[14,16]

$$\begin{aligned} F(t) &= \textstyle\int_0^t f(t)\,dt = \int_0^t \lambda e^{-\lambda t}\,dt \\ &= 1 - e^{-\lambda t} \end{aligned} \tag{2}$$

where $F(t)$ is the cumulative distribution function.

Weibull Distribution

This distribution was developed by W. Weibull in the early 1950s and can be used to represent many different physical phenomena.[9] The distribution probability density function is expressed by[14,16]

$$f(t) = \frac{\alpha t^{\alpha-1}}{\theta^{\alpha}} e{-(t/\theta)^{\alpha}} \qquad \text{for} \quad t \geq 0 \qquad \alpha > 0 \qquad \theta > 0 \tag{3}$$

where σ and α are the distribution scale and shape parameters, respectively. The cumulative distribution function is given by[14,16]

$$F(t) = \int_0^t f(t)\, dt = \int_0^t \frac{\alpha t^{\alpha-1}}{\theta^{\alpha}} e^{-(t/\theta)^{\alpha}}\, dt$$

$$= 1 - e^{-(t/\theta)^{\alpha}} \tag{4}$$

For $\alpha = 1$ and $\alpha = 2$, the Weibull distribution becomes the exponential and Rayleigh distributions, respectively.

Normal Distribution

This is one of the most widely known distributions. In mechanical reliability, it is often used to represent an item's stress and strength. The probability density function of the distribution is expressed by

$$f(t) = \frac{1}{\sigma\sqrt{2\pi}} \exp\left[-\frac{(t-\mu)^2}{2\sigma^2}\right] \qquad -\infty < t < +\infty \tag{5}$$

where μ and σ are the distribution parameters (i.e., mean and standard deviation, respectively). The cumulative distribution function is given by[14,16]

$$F(t) = \int_{-\infty}^t f(t)\, dt = \frac{1}{\sigma\sqrt{2\pi}} \int_{-\infty}^t \exp\left[-\frac{(x-\mu)^2}{2\sigma^2}\right] dx \tag{6}$$

2.2 Hazard Rate Models

In reliability studies, the term *hazard rate* is often used. It simply means the constant or nonconstant failure rate of an item. Thus, the hazard rate of an item is expressed by

$$h(t) = \frac{f(t)}{1 - F(t)} \tag{7}$$

where $h(t)$ is the item hazard rate.

This section presents four hazard rate models considered useful to perform various types of mechanical reliability studies: exponential, Weibull, normal, and general.

Exponential Distribution

By substituting Eqs. (1) and (2) into Eq. (7), we get the following equation for the exponential distribution hazard rate function:

$$h(t) = \frac{\lambda e^{-\lambda t}}{1 - (1 - e^{-\lambda t})} = \lambda \tag{8}$$

As the right-hand side of Eq. (8) is independent of time, λ is called the failure rate.

Weibull Distribution

By substituting Eqs. (3) and (4) into Eq. (7), we get the following equation for the Weibull distribution hazard rate function:

$$h(t) = \frac{[(\alpha t^{\alpha-1}/\theta^{\alpha})\, e^{-(t/\theta)^{\alpha}}]}{1 - [1 - e^{-(t/\theta)^{\alpha}}]} = \frac{\alpha t^{\alpha-1}}{\theta^{\alpha}} \tag{9}$$

For $\alpha = 1$ and $\alpha = 2$, Eq. (9) becomes the hazard rate function for the exponential and Rayleigh distributions, respectively.

Normal Distribution

By substituting Eqs. (5) and (6) into Eq. (7), we get the following equation for the normal distribution hazard rate function:

$$h(t) = \frac{1/(\sigma\sqrt{2\pi}) \exp\,[-(t-\mu)^2/2\sigma^2]}{1 - 1/(\sigma\sqrt{2\pi}) \int_{-\infty}^{t} \exp\,[-(x-\mu)^2/2\sigma^2]\, dx} \tag{10}$$

General Distribution

The distribution hazard rate function is defined by[17]

$$h(t) = c\lambda\alpha t^{\alpha-1} + (1-c)mt^{m-1}\,\theta e^{\theta t^m} \qquad \text{for} \quad 0 \le c \le 1 \qquad \alpha, \theta, m, \lambda > 0 \tag{11}$$

where θ, λ = scale parameters
α, m = shape parameters
t = time

The following distribution hazard rate functions are the special cases of Eq. (11):

- Bathtub; for $m = 1$, $\alpha = 0.5$
- Makeham; for $m = 1$, $\alpha = 1$
- Extreme value; for $c = 0$, $m = 1$
- Weibull; for $c = 1$
- Rayleigh; for $c = 1$, $\alpha = 2$
- Exponential; for $c = 1$, $\alpha = 1$

3 COMMON RELIABILITY NETWORKS

Components of a mechanical system can form configurations such as series, parallel, series–parallel, parallel–series, k out of m, standby, and bridge. Often, these configurations are referred to as the standard configurations. Sometime during the design process, it might be desirable to determine the reliability or the values of other related parameters of systems forming such configurations. All these configurations or networks are described below.[14,16]

3.1 Series Network

The block diagram of an "m"-unit series network or configuration is shown in Fig. 1. Each block represents a system unit or component. If any one of the components fails, the system fails; that is, all of the series units must work normally for the system to succeed. For independent units, the reliability of the system shown in Fig. 1 is

$$R_S = R_1 R_2 R_3 \cdots R_m \tag{12}$$

where R_S = series system reliability
m = number of units
R_i = reliability of unit i for $i = 1, 2, 3, \ldots, m$

For constant unit failure rates of the units, Eq. (12) becomes[14]

$$R_S(t) = e^{-\lambda_1 t} \cdot e^{-\lambda_2 t} \cdot e^{-\lambda_3 t} \cdots e^{-\lambda_m t} = \exp\left(-\sum_{i=1}^{m} \lambda_i t\right) \tag{13}$$

where $R_S(t)$ = series system reliability at time t
λ_i = constant failure rate of unit i for $i = 1, 2, 3, \ldots, m$

The system hazard rate is given by[14]

$$\lambda_S(t) = \frac{f_S(t)}{1 - F_S(t)} = -\frac{1}{R_S(t)} \frac{dR_S(t)}{dt}$$

$$= \sum_{i=1}^{m} \lambda_i \tag{14}$$

where $\lambda_S(t)$ = series system hazard rate or total failure rate
$f_S(t)$ = series system probability density function
$F_S(t)$ = series system cumulative distribution function

It is to be noted that the system total failure rate given by Eq. (14) is the sum of the failure rates of all of the units. It means that whenever the failure rates of units are added, it is automatically assumed that the units are acting in series (i.e., if one unit fails, the system fails). This is the worst-case assumption often practiced in the design of engineering systems.

The system mean time to failure is given by[14]

$$\text{MTTF}_S = \int_0^\infty R_S(t)\, dt = \int_0^\infty \exp\left(-\sum_{i=1}^{m} \lambda_i t\right) dt$$

$$= \frac{1}{\sum_{i=1}^{m} (1/\lambda_i)} \tag{15}$$

where MTTF_S is the series system mean time to failure.

Figure 1 Series system block diagram.

3.2 Parallel Network

This type of configuration can be used to improve a mechanical system's reliability during the design phase. The block diagram of an "m"-unit parallel network is shown in Fig. 2. Each block in the diagram represents a unit. This configuration assumes that all of its units are active and at least one unit must work normally for the system to succeed. For independently failing units, the reliability of the parallel network shown in Fig. 2 is expressed by[14]

$$R_p = 1 - \prod_{i=1}^{m} (1 - R_i) \tag{16}$$

where R_p = reliability of parallel network
R_i = reliability of unit i for $i = 1, 2, 3, \ldots, m$

For constant failure rates of the units, Eq. (16) becomes[14]

$$R_p(t) = 1 - \prod_{i=1}^{m} (1 - e^{-\lambda_i t}) \tag{17}$$

where $R_p(t)$ = parallel network reliability at time t
λ_i = constant failure rate of unit i for $i = 1, 2, 3, \ldots, m$

For identical units, the network mean time to failure is given by[14]

$$\text{MTTF}_p = \int_0^{\infty} R_p(t)\, dt = \frac{1}{\lambda} \sum_{i=1}^{m} \frac{1}{i} \tag{18}$$

where MTTF_p = parallel network mean time to failure
λ = unit constant failure rate

3.3 Series–Parallel Network

The network block diagram is shown in Fig. 3. Each block in the diagram represents a unit. This network represents a system having m number of subsystems in series. In turn, each subsystem contains n number of active units in parallel. All subsystems must operate normally for the system to succeed. For independent units, the reliability of the series–parallel network shown in Fig. 3 is given by[18,19]

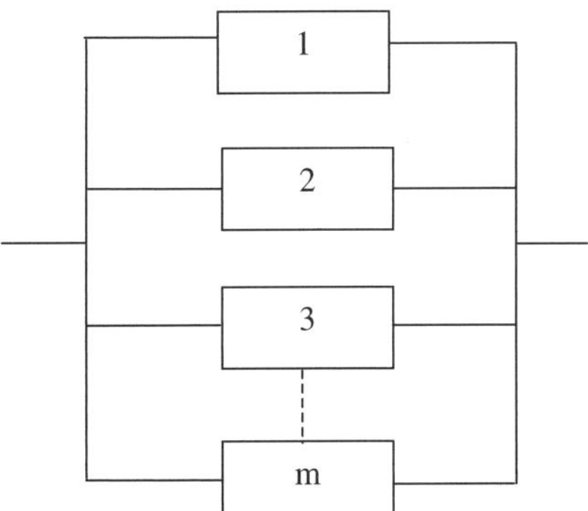

Figure 2 Parallel system block diagram.

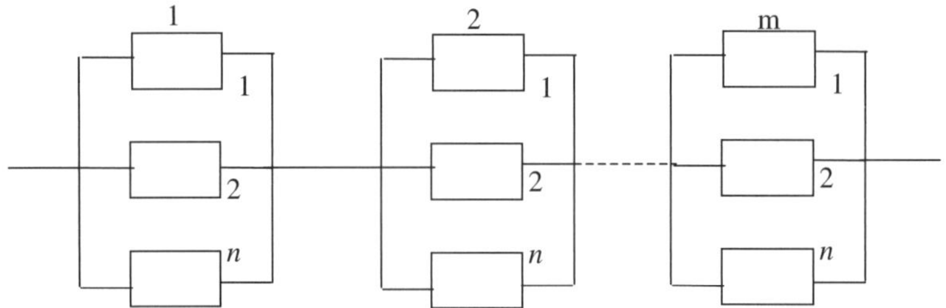

Figure 3 Series–parallel network block diagram.

$$R_{Sp} = \prod_{i=1}^{m}\left(1 - \prod_{j=1}^{n} F_{ij}\right) \tag{19}$$

where R_{Sp} = series–parallel network reliability
m = number of subsystems
n = number of units
F_{ij} = failure probability of the ith subsystem's jth unit

For constant unit failure rates, Eq. (19) becomes[18,19]

$$R_{Sp}(t) = \prod_{i=1}^{m}\left[1 - \prod_{j=1}^{n}(1 - e^{-\lambda_{ij}t})\right] \tag{20}$$

where $R_{Sp}(t)$ = series–parallel network reliability at time t
λ_{ij} = constant failure rate of unit ij

For identical units, the network mean time to failure is given by[19]

$$\begin{aligned}\text{MTTF}_{Sp} &= \int_0^{\infty} R_{Sp}(t)\,dt \\ &= \frac{1}{\lambda}\sum_{i=1}^{m}\left[(-1)^{i+1}\binom{m}{i}\sum_{j=1}^{in}\frac{1}{j}\right]\end{aligned} \tag{21}$$

where MTTF_{Sp} = series–parallel network mean time to failure
λ = unit failure rate

3.4 Parallel–Series Network

This network represents a system having m number of subsystems in parallel. In turn, each subsystem contains n number of units in series. At least one subsystem must function normally for the system to succeed. The network block diagram is shown in Fig. 4. Each block in the diagram represents a unit. For independent units, the reliability of the parallel–series network shown in Fig. 4 is expressed by[18,19]

$$R_{pS} = 1 - \prod_{i=1}^{m}\left(1 - \prod_{j=1}^{n} R_{ij}\right) \tag{22}$$

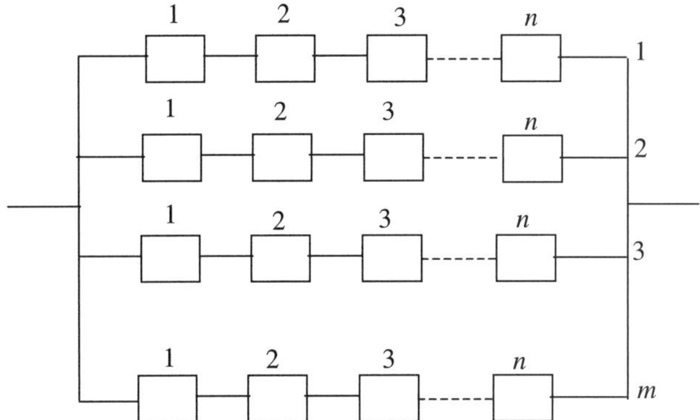

Figure 4 Parallel–series network block diagram.

where R_{pS} = parallel–series network reliability
m = number of subsystems
n = number of units
R_{ij} = reliability of the jth unit in the ith subsystem

For constant unit failure rates, Eq. (23) becomes[18,19]

$$R_{pS}(t) = 1 - \prod_{i=1}^{m}\left(1 - \prod_{j=1}^{n} e^{-\lambda_{ij}t}\right) \tag{23}$$

where $R_{pS}(t)$ = parallel–series network reliability at time t
λ_{ij} = constant failure rate of unit ij

For identical units, the network mean time to failure is given by[19]

$$\text{MTTF}_{pS} = \int_0^\infty R_{pS}(t)\,dt = \frac{1}{n\lambda}\sum_{i=1}^{m}\frac{1}{i} \tag{24}$$

where MTTF_{pS} is the parallel–series network mean time to failure.

3.5 *K*-out-of-*m*-Unit Network

This network is sometimes referred to as a partially redundant network. It is a parallel network with a condition that at least K units out of the total of m units must operate normally for the system to succeed.

For independent and identical units, the network reliability is given by[14,16]

$$R_{K/m} = \sum_{i=K}^{m}\binom{m}{i}R^i(1-R)^{m-i} \tag{25}$$

where

$$\binom{m}{i} = \frac{m!}{i!(m-i)!}$$

$R_{K/m}$ is the K-out-of-m-unit network reliability, and R is the unit reliability. For $K = 1$ and $K = m$, Eq. (25) becomes the reliability expression for parallel and series networks, respectively. More specifically, parallel and series networks are the special cases of the K-out-of-m-unit network.

For constant failure rates of the units, Eq. (25) becomes[14,16]

$$R_{K/m} = \sum_{i=K}^{m} \binom{m}{i} e^{-i\lambda t}(1 - e^{-\lambda t})^{m-i} \tag{26}$$

where $R_{K/m}$ = K-out-of-m-unit network reliability at time t
λ = unit constant failure rate

The network mean time to failure is given by[14,16]

$$\text{MTTF}_{K/m} = \int_0^{\infty} R_{k/m}(t)\, dt = \frac{1}{\lambda} \sum_{i=k}^{m} \frac{1}{i} \tag{27}$$

where $\text{MTTF}_{K/m}$ is the K-out-of-m-unit network mean time to failure.

3.6 Standby System

This is another configuration used to improve reliability. The block diagram of an $(m + 1)$-unit standby system is shown in Fig. 5. Each block in the diagram represents a unit. In this configuration, one unit operates and m units are kept on standby. As soon as the operating unit fails, it is replaced by one of the standbys. The system fails when all of its units fail (i.e., operating plus all standbys). For perfect switching, independent and identical units, and as-good-as-new standby units, the standby system reliability is given by[14,16]

$$R_{SS}(t) = \sum_{i=0}^{m} \frac{[\int_0^t \lambda(t)\, dt]^i \exp(-\int_0^t \lambda(t)\, dt)}{i!} \tag{28}$$

where $R_{SS}(t)$ = standby system reliability at time t
$\lambda(t)$ = unit hazard rate or time-dependent failure rate
m = number of standbys

For constant failure rates of the units [i.e., $\lambda(t) = \lambda$] Eq. (28) becomes

$$R_{SS}(t) = \sum_{i=0}^{m} \frac{(\lambda\, t)^i\, e^{-\lambda t}}{i!} \tag{29}$$

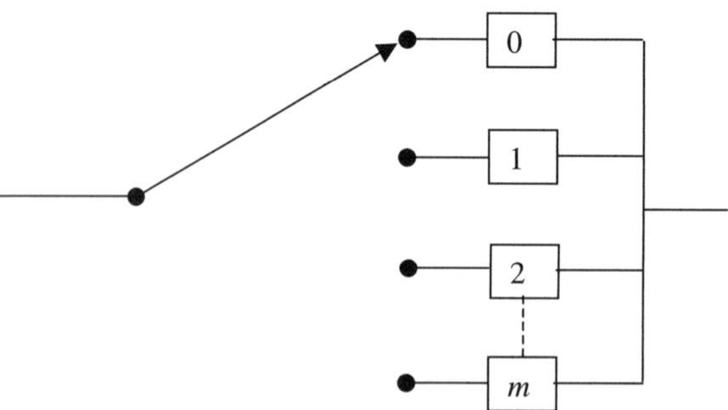

Figure 5 Block diagram of an $(m + 1)$-unit standby system.

where λ is the unit constant failure rate. The system mean time to failure is given by[14]

$$\begin{aligned}\mathrm{MTTF}_{\mathrm{SS}} &= \int_0^\infty R_{\mathrm{SS}}(t)\,dt \\ &= \int_0^\infty \sum_{i=0}^{m} \frac{(\lambda t)^i e^{-\lambda t}}{i!}\,dt \\ &= \frac{m+1}{\lambda} \qquad (30)\end{aligned}$$

where $\mathrm{MTTF}_{\mathrm{SS}}$ is the standby system mean time to failure.

3.7 Bridge Network

The block diagram of a bridge network is shown in Fig. 6. Each block in the diagram represents a unit. Mechanical components sometimes can form this type of configuration. For independent units, the reliability of the bridge network shown in Fig. 6 is[20]

$$\begin{aligned}R_{\mathrm{bn}} &= 2\prod_{i=1}^{5} R_i + \prod_{i=1}^{4} R_i + R_1 R_3 R_5 + R_1 R_4 + R_2 R_5 \\ &\quad - \prod_{i=2}^{5} R_i - \prod_{i=1}^{4} R_i - R_5 \prod_{i=1}^{3} R_i - R_1 \prod_{i=3}^{5} R_i - R_1 R_2 R_4 R_5 \qquad (31)\end{aligned}$$

where R_{bn} = bridge network reliability
R_i = unit i reliability for i = 1, 2, 3, 4, 5

For identical units and constant failure rates of the units, Eq. (31) becomes[14]

$$R_{\mathrm{bn}}(t) = 2e^{-5\lambda t} - 5e^{-4\lambda t} + 2e^{-3\lambda t} + 2e^{-2\lambda t} \qquad (32)$$

where λ is the unit constant failure rate. The network mean time to failure is given by

$$\mathrm{MTTF}_{\mathrm{bn}} = \int_0^\infty R_{\mathrm{bn}}(t)\,dt = \frac{49}{60\lambda} \qquad (33)$$

where $\mathrm{MTTF}_{\mathrm{bn}}$ is the bridge network mean time to failure.

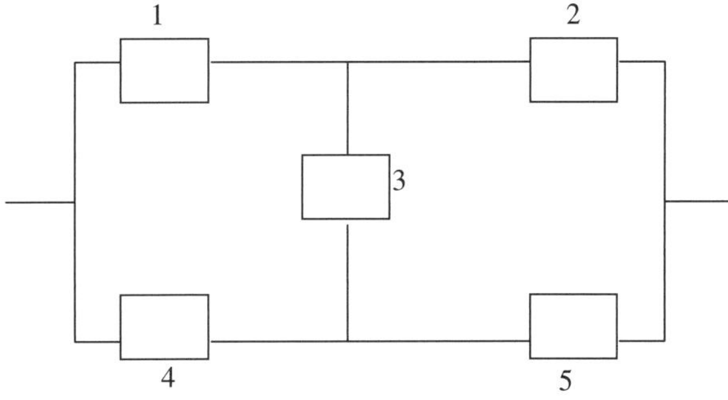

Figure 6 Block diagram of a five-unit bridge network.

4 MECHANICAL FAILURE MODES AND CAUSES OF GENERAL AND GEAR FAILURES

A mechanical failure may be defined as any change in the shape, size, or material properties of a structure, piece of equipment, or equipment part that renders it unfit to carry out its specified mission adequately.[13] Thus, there are many different types of failure modes associated with mechanical items. Good design practices can reduce or eliminate altogether the occurrence of these failure modes. Some of these failure modes are as follows[21–23]:

- Fatigue failure
- Material flaw failure
- Bending failure
- Metallurgical failure
- Bearing failure
- Instability failure
- Shear loading failure
- Compressive failure
- Creep/rupture failure
- Tensile yield strength failure
- Ultimate tensile strength failure
- Stress concentration failure

There are many causes of product failures. Some of these are as follows[24]:

- Defective design
- Wear-out
- Defective manufacturing
- Wrong application
- Incorrect installation
- Failure of other parts

A study performed over a period of 35 years reported a total of 931 gear failures.[24] They were classified under four categories: breakage (61.2%), surface fatigue (20.3%), wear (3.2%), and plastic flow (5.3%). The causes of these failures were grouped into the five categories shown in Fig. 7.

These five categories were further divided into various elements. The elements of the service-related classification were continual overloading (25%), improper assembly (21.2%), impact loading (13.9%), incorrect lubrication (11%), foreign material (1.4%), abusive handling (1.2%), bearing failure (0.7%), and operator errors (0.3%). The elements of the heat-treatment-related classification were incorrect hardening (5.9%), inadequate case depth (4.8%), inadequate core hardness (2%), excessive case depth (1.8%), improper tempering (1%), excessive core hardness (0.5%), and distortion (0.2%). The elements of the design-related classification were wrong design (2.8%), specification of suitable heat treatment (2.5%), and incorrect material selection (1.6%). The elements of the manufacturing-related classification were grinding burns (0.7%) and tool marks or notches (0.7%). Finally, the three elements of the material-related classification were steel defects (0.5%), mixed steel or incorrect composition (0.2%), and forging defects (0.1%).

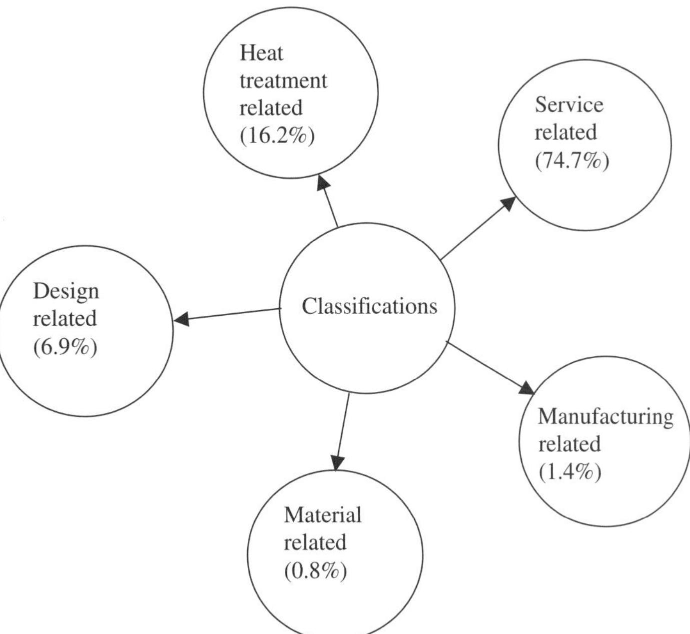

Figure 7 Classifications of gear failure causes.

5 RELIABILITY-BASED DESIGN AND DESIGN-BY-RELIABILITY METHODOLOGY

It would be unwise to expect a system to perform to a desired level of reliability unless it is specifically designed for that level. Desired system/equipment/part reliability design specifications due to factors such as well-publicized failures (e.g., the space shuttle Challenger disaster and the Chernobyl nuclear accident) have increased the importance of reliability-based design. The starting point for reliability-based design is during the writing of the design specification. In this phase, all reliability needs and specifications are entrenched into the design specification. Examples of these requirements might include item mean time to failure (MTTF), mean time to repair (MTTR), test or demonstration procedures to be used, and applicable documents.

Over the years the DOD has developed various reliability documents for use during the design and development of an engineering item. Many times, such documents are entrenched into the item design specification document. Table 1 presents some of these documents. Many professional bodies and other organizations have also developed documents on various aspects of reliability.[15,22,25] References 22 and 26 provide descriptions of documents developed by the DOD.

Reliability is an important consideration during the design phase. According to Ref. 27, as many as 60% of failures can be eliminated through design changes. There are many strategies the designer could follow to improve design:

- Eliminate failure modes.
- Focus design for fault tolerance.
- Focus design for fail safe.
- Focus design to include mechanism for early warnings of failure through fault diagnosis.

Table 1 Selected Design-Related Documents Developed by the U.S. Department of Defense

Document No.	Document Title
MIL-STD-721	Definitions of Terms for Reliability and Maintainability
MIL-STD-217	Reliability Prediction of Electronic Equipment
MIL-STD-781	Reliability Design, Qualification, and Production Acceptance Tests: Exponential Distribution
MIL-STD-756	Reliability Modeling and Prediction
MIL-STD-785	Reliability Program for Systems and Equipment
MIL-HDBK-251	Reliability/Design Thermal Applications
MIL-STD-1629	Procedures for Performing a Failure Mode, Effects, and Criticality Analysis
RADC-TR-75-22	Non-Electronic Reliability Notebook
MIL-STD-965	Parts Control Program
MIL-STD-2074	Failure Classification for Reliability Testing

During the design phase of a product, various types of reliability and maintainability analyses can be performed, including reliability evaluation and modeling, reliability allocation, maintainability evaluation, human factors/reliability evaluation, reliability testing, reliability growth modeling, and life-cycle cost. In addition, some of the design improvement strategies are zero-failure design, fault-tolerant design, built-in testing, derating, design for damage detection, modular design, design for fault isolation, and maintenance-free design. During design reviews, reliability and maintainability-related actions recommended/taken are to be thoroughly reviewed from desirable aspects.

A systematic series of steps are taken to design a reliable mechanical item. The "design by methodology" is composed of such steps[14,28,29]:

- Define the design problem under consideration.
- Identify and list all associated design variables and parameters.
- Perform failure mode, effect, and criticality analyses according to MIL-STD-1629.[30]
- Verify critical design parameter selection.
- Establish appropriate relationships between the failure-governing criteria and the critical parameters.
- Determine the failure-governing stress and strength functions and then the most appropriate failure-governing stress and strength distributions.
- Estimate the reliability utilizing the failure-governing stress and strength distributions for all critical failure modes.
- Iterate the design until reliability goals are achieved.
- Optimize design with respect to factors such as safety, reliability, cost, performance, maintainability, weight, and volume.
- Repeat the design optimization process for all critical components.
- Estimate item reliability.
- Iterate the design until item reliability goals are fully satisfied.

6 DESIGN RELIABILITY ALLOCATION AND EVALUATION METHODS

Over the years, many reliability allocation and evaluation methods have been developed for use during the design phase.[14,29] This section presents some that are considered useful, particularly in designing mechanical items.

6.1 Failure Rate Allocation Method

This method is used to allocate failure rates to system components when the overall system required failure rate is given. The method is based on the following three assumptions[14]:

- All system components fail independently.
- Component failure rates are constant.
- System components form a series network.

Thus, the system failure rate using Eq. (14) is

$$\lambda_S = \sum_{i=1}^{m} \lambda_i \tag{34}$$

where λ_S = system failure rate
m = total number of system components
λ_i = failure rate of component i for $i = 1, 2, 3, \ldots, m$

If the specified failure rate of the system is λ_{Sp}, then the component failure rate is allocated such that

$$\sum_{i=1}^{m} \lambda_i^* \leq \lambda_{Sp} \tag{35}$$

where λ_i^* is the failure rate allocated to component i for $i = 1, 2, 3, \ldots, m$. The following three steps are associated with this approach:

- Estimate failure rates of the system components (i.e., λ_i for $i = 1, 2, 3, \ldots, m$) using the field data.
- Calculate the relative weight α_i of component i using the preceding step failure rate data and the expression

$$\alpha_i = \frac{\lambda_i}{\sum_{i=1}^{m} \lambda_i} \qquad \text{for} \quad i = 1, 2, 3, \ldots, m \tag{36}$$

  It is to be noted that α_i represents the relative failure vulnerability of component i and

$$\sum_{i=1}^{m} \alpha_i = 1 \tag{37}$$

- Allocate the failure rate to part or component i by using the equation

$$\lambda_i^* = \alpha_i \lambda_{Sp} \qquad \text{for} \quad i = 1, 2, 3, \ldots, m \tag{38}$$

A solved example in Ref. 14 demonstrates the application of this method.

6.2 Hybrid Reliability Allocation Method

This method combines two reliability allocation methods: similar familiar systems and factors of influence. The method is more attractive because it incorporates the benefits of both systems.[14,29]

The basis for the similar-familiar-systems approach is the familiarity of the designer with similar systems as well as the utilization of failure data collected on similar systems from various sources during the allocation process. The principal disadvantage of the similar-familiar-systems method is the assumption that the reliability and life-cycle cost of similar systems are satisfactory or adequate.

The factors-of-influence method is based upon four factors that are considered to affect item reliability: failure criticality, environment, complexity/time, and state of the art.

The failure criticality factor is concerned with how critical is the failure of an item (e.g., the failure of some auxiliary equipment in an aircraft may not be as critical as the failure of an engine). The environmental factor takes into account the susceptibility of items to conditions such as vibration, humidity, and temperature.

The complexity/time factor relates to the number of item/subsystem components/parts and the operating time of the item under consideration during the total system operating period. Finally, the state-of-the-art factor relates to the advancement in the state of the art for an item under consideration.

During the reliability allocation process, each item is rated with respect to each of these four factors by assigning a number from 1 to 10. The assignment of 1 means the item is least affected by the factor under consideration and 10 means the item is most affected by the same factor. Subsequently, reliability is determined by weighing these numbers for all four factors.

6.3 Safety Factor and Safety Margin

The safety factor and safety margin are arbitrary multipliers used to ensure the reliability of mechanical items during the design phase. These indexes can provide satisfactory design if they are established using considerable past experiences and data.

Safety Factor

A safety factor can be defined in many different ways.[13,31–35] Two commonly used definitions follow.

Definition I. The safety factor is defined by[36]

$$\mathrm{SF} = \frac{m_{\mathrm{Sh}}}{m_{\mathrm{SS}}} \geq 1 \tag{39}$$

where SF = safety factor
m_{Sh} = mean failure governing strength
m_{SS} = mean failure governing stress

This index is a good measure of safety when both stress and strength are normally distributed. However, when the spread of both strength and/or stress is large, the index becomes meaningless because of positive failure rate.[14]

Definition II. The safety factor is defined by[32–37]

$$\mathrm{SF} = \frac{\mathrm{US}}{\mathrm{WS}} \tag{40}$$

where SF = safety factor
WS = working stress expressed in pounds per square inch (psi)
US = ultimate strength expressed in psi

Safety Margin
The safety margin (SM) is defined as[14,31]

$$\text{SM} = \text{SF} - 1 \tag{41}$$

The negative value of this measure means that the item under consideration will fail. Thus, its value must always be greater than zero.

The safety margin for normally distributed stress and strength is expressed by[14,31]

$$\text{SM} = \frac{\mu_{\text{as}} - \mu_{\text{ms}}}{\sigma_{\text{Sh}}} \tag{42}$$

where μ_{as} = average strength
μ_{ms} = maximum stress
σ_{Sh} = strength standard deviation

In turn the maximum stress μ_{ms} is expressed by

$$\mu_{\text{ms}} = \mu_{\text{SS}} + C\sigma_{\text{SS}} \tag{43}$$

where μ_{SS} = mean stress
σ_{SS} = stress standard deviation
C = factor between 3 and 6

6.4 Stress–Strength Interference Theory Method

This method is used to determine the reliability of a mechanical item when its associated stress and strength probability density functions are known. The item reliability is defined by[13,14,29]

$$R = P(y < x) = P(x > y) \tag{44}$$

where R = item reliability
P = probability
x = strength random variable
y = stress random variable

Equation (44) is rewritten in the form[13,14,29]

$$R = \int_{-\infty}^{\infty} f(y)\left[\int_{y}^{\infty} f(x)\,dx\right] dy \tag{45}$$

where $f(x)$ = strength probability density function
$f(y)$ = stress probability density function

Special Case Model: Exponentially Distributed Stress and Strength
In this case, an item's stress and strength are defined by

$$f(y) = \theta e^{-\theta y} \qquad 0 \le y < \infty \tag{46}$$

and

$$f(x) = \lambda e^{-\lambda x} \qquad 0 \leq x < \infty \tag{47}$$

where σ and λ are the reciprocals of the mean values of stress and strength, respectively. Using Eqs. (46) and (47) in Eq. (45) yields

$$\begin{aligned} R &= \int_0^\infty \theta e^{-\theta y} \left(\int_y^\infty \lambda e^{-\lambda x} \, dx \right) dy \\ &= \frac{\theta}{\theta + \lambda} \end{aligned} \tag{48}$$

For $\theta = 1/\bar{y}$, and $\lambda = 1/\bar{x}$, Eq. (48) becomes

$$R = \frac{\bar{x}}{\bar{x} + \bar{y}} \tag{49}$$

where $\bar{x}$ = mean strength
$\bar{y}$ = mean stress

Similarly, models for other stress and strength probability distributions can be developed. A number of such models are presented in Ref. 13.

6.5 Failure Modes and Effect Analysis (FMEA)

FMEA is a vital tool for evaluating system design from the point of view of reliability. It was developed in the early 1950s to evaluate the design of various flight control systems.[38]

The difference between the FMEA and the failure mode, effect, and criticality analysis (FMECA) is that FMEA is a qualitative technique used to evaluate a design, whereas FMECA is composed of FMEA and criticality analysis (CA). Criticality analysis is a quantitative method used to rank critical failure mode effects by taking into account their occurrence probabilities.

As FMEA is a widely used method in industry, there are many standards/documents written about it. In Ref. 30 45 such publications were collected and evaluated, prepared by organizations such as the DOD, NASA, and the Institute of Electrical and Electronic Engineers (IEEE). These documents include[39]:

- DOD: MIL-STD-785A (1969), MIL-STD-1629 (draft) (1980), MIL-STD-2070 (AS) (1977), MIL-STD-1543 (1974), AMCP-706-196 (1976)
- NASA: NHB 5300.4 (1A) (1970), ARAC Proj. 79-7 (1976)
- IEEE: ANSI N 41.4 (1976)

Details of the above documents as well as a list of publications on FMEA are given in Ref. 24.

The main steps involved in performing FMEA are as follows[29]:

- Define carefully all system boundaries and detailed requirements.
- List all parts/subsystems in the system under consideration.
- Identify and describe each part and list all its associated failure modes.
- Assign failure rate/probability to each failure mode.
- List effects of each failure mode on subsystem/system/plant.

- Enter remarks for each failure mode.
- Review each critical failure mode and take appropriate measures.

This method is described in detail in Ref. 14.

6.6 Fault Tree Analysis (FTA)

This method, so called because it arranges fault events in a tree-shaped diagram, is one of the most widely used techniques for performing system reliability analysis. In particular, it is probably the most widely used method in the nuclear power industry. The technique is well suited for determining the combined effects of multiple failures.

The fault tree technique is more costly to use than the FMEA approach. It was developed in the early 1960s at Bell Telephone Laboratories to evaluate the reliability of the Minuteman Launch Control System. Since that time, hundreds of publications on the method have appeared.[15]

The FTA begins by identifying an undesirable event, called the "top event," associated with a system. Fault events that could cause the occurrence of the top event are generated and connected by AND and OR logic gates. The construction of a fault tree proceeds by generation of fault events (by asking the question "How could this event occur?") in a successive manner until the fault events need not be developed further. These events are known as primary or elementary events. In simple terms, the fault tree may be described as the logic structure relating the top event to the primary events. This method is described in detail in Ref. 14.

7 HUMAN ERROR AND RELIABILITY CONSIDERATION IN MECHANICAL DESIGN

As in the reliability of any other system, human reliability and error play an important role in the reliability of mechanical systems. Over the years many times mechanical systems/equipment have failed due to human error rather than hardware failure. Careful consideration of human error and reliability during the design of mechanical systems can help to eliminate or reduce the occurrence of non-hardware-related failures during the operation of such systems. Human errors may be classified under the following seven distinct categories[40–42]:

- Design errors
- Operator errors
- Assembly errors
- Inspection errors
- Maintenance errors
- Installation errors
- Handling errors

Each of these categories is described in detail in Ref. 14.

There are numerous causes for the occurrence of human error including poor equipment design, complex tasks, poor work layout, poorly written operating and maintenance procedures, poor job environment (e.g., poor lighting, high/low temperature, crowded work space, high noise level), inadequate work tools, poor skill of involved personnel, and poor motivation of involved personnel.[40–42]

Human reliability of time-continuous tasks such as aircraft maneuvering, scope monitoring, and missile countdown can be calculated by using the equation[42,43]

$$R_h(t) = \exp\left(-\int_0^t \lambda(t)\, dt\right) \tag{50}$$

where $R_h(t)$ = human reliability at time t
$\lambda(t)$ = time-dependent human error rate

For constant human error rate [i.e., $\lambda(t) = \lambda$], Eq. (50) becomes

$$R_h(t) = \exp\left(-\int_0^t \lambda\, dt\right) = e^{-\lambda t} \tag{51}$$

where λ is the constant human error rate. The subject of human reliability and error is discussed in detail in Ref. 42.

Example. A person is performing a certain time-continuous task. Assume his or her error rate is 0.004 per hour. Calculate the person's probability of performing the task correctly during a 5-h period.

By using the values in Eq. (51), we get

$$R_h\,(5) = e^{-(0.004)5} = 0.98$$

So, there is approximately a 98% chance that the person will perform the task correctly during the specified period.

8 FAILURE RATE ESTIMATION MODELS FOR VARIOUS MECHANICAL ITEMS

Many mathematical models available in the literature can be used to estimate failure rates of items such as bearings, pumps, brakes, filters, compressors, and seals.[44–46] This section presents some of these models.

8.1 Brake System Failure Rate Estimation Model

The brake system failure rate is expressed by[44]

$$\lambda_{\text{brs}} = \sum_{i=1}^{6} \lambda_i \tag{52}$$

where λ_{brs} = brake system failure rate, expressed in failures/10^6 h
λ_1 = brake housing failure rate
λ_2 = total failure rate of actuators
λ_3 = total failure rate of seals
λ_4 = total failure rate of bearings
λ_5 = total failure rate of springs
λ_6 = total failure rate of brake friction materials

The values of λ_i for i = 1, 2, ... , 6 are obtained through various means.[44,47]

8.2 Compressor System Failure Rate Estimation Model

The compressor system failure rate is expressed by[45]

$$\lambda_{\text{comp}} = \sum_{i=1}^{6} \lambda_i \tag{53}$$

where λ_{comp} = compressor system failure rate, expressed in failures/10^6 h
λ_1 = failure rate of all compressor bearings
λ_2 = compressor casing failure rate
λ_3 = failure rate due to design configuration
λ_4 = failure rate of valve assay (if any)
λ_5 = failure rate of all compressor seals
λ_6 = failure rate of all compressor shafts

Procedures for calculating λ_1, λ_2, λ_3, λ_4, λ_5, and λ_6 are presented in Ref. 45.

8.3 Filter Failure Rate Estimation Model

The filter failure rate is expressed by[46]

$$\lambda_{\text{ft}} = \lambda_b \prod_{i=1}^{6} \theta_i \tag{54}$$

where λ_{ft} = filter failure rate, expressed in failures/10^6 h
λ_{b} = filter base failure rate
θ_{i} = ith modifying factor; $i = 1$ is for temperature effects, $i = 2$ for water contamination effects, $i = 3$ for cyclic flow effects, $i = 4$ for differential pressure effects, $i = 5$ for cold-start effects, and $i = 6$ for vibration effects

Procedures for estimating λ_b, θ_1, θ_2, θ_3, θ_4, θ_5, and θ_6 are given in Ref. 46.

8.4 Pump Failure Rate Estimation Model

The pump failure rate is expressed by[46]

$$\lambda_{\text{pm}} = \sum_{i=1}^{5} \lambda_i \tag{55}$$

where λ_{pm} = pump failure rate, expressed in failures/10^6 cycles
λ_1 = pump fluid driver failure rate
λ_2 = pump casing failure rate
λ_3 = pump shaft failure rate
λ_4 = failure rate of all pump seals
λ_5 = failure rate of all pump bearings

Procedures for calculating λ_1, λ_2, λ_3, λ_4, and λ_5 are presented in Ref. 46.

9 FAILURE DATA AND FAILURE DATA COLLECTION SOURCES

Failure data provide invaluable information to reliability engineers, design engineers, management, and so on, concerning the product performance. These data are the final proof of

Table 2 Failure Rates for Selected Mechanical Items

Item Description	Failure Rate 10^{-6} h
Roller bearing	8.323
Bellows (general)	13.317
Filter (liquid)	6.00
Compressor (general)	33.624
Pipe	0.2
Hair spring	1.0
Pump (vacuum)	10.610
Gear (spur)	3.152
Seal (O-ring)	0.2
Nut or bolt	0.02
Brake (electromechanical)	16.00
Knob (general)	2.081
Washer (lock)	0.586
Washer (flat)	0.614
Duct (general)	2.902

Note: Use environment: ground fixed or general.

the success or failure of the effort expended during the design and manufacture of a product used under designed conditions. During the design phase of a product, past information concerning its failures plays a critical role in the reliability analysis of that product. Failure data can be used to estimate item failure rate, perform effective design reviews, predict reliability and maintainability of redundant systems, conduct trade-off and life-cycle cost studies, and perform preventive maintenance and replacement studies. Table 2 presents failure rates for selected mechanical items.[13,48,49]

There are many different ways to collect failure data. For example, during the equipment life cycle, there are eight identifiable data sources: repair facility reports, development testing of the item, previous experience with similar or identical items, customer's failure-reporting systems, inspection records generated by quality control and manufacturing groups, tests conducted during field demonstration, environmental qualification approval, and field installation, acceptance testing, and warranty claims.[50] Table 3 presents some sources for collecting failure data for use during the design phase.[13]

Table 3 Selected Failure Data Sources for Mechanical Items

Source	Developed By
Ref. 48	Reliability Analysis Center, Rome Air Development Center, Griffis Air Force Base, Rome, New York
Component Reliability Data for Use in Probabilistic Safety Assessment (1998)	International Atomic Energy Agency, Vienna, Austria
R. G. Arno, Non-Electronic Parts Reliability Data (Rept. No. NPRD-2, 1981)	Reliability Analysis Center, Rome Air Development Center, Griffis Air Force Base, Rome, New York
Government Industry Data Exchange Program (GIDEP)	GIDEP Operations Center, U.S. Dept. of Navy, Seal Beach, Corona, California.
Ref. 49	Reliability Analysis Center, Rome Air Development Center, Griffis Air Force Base, Rome, New York

REFERENCES

1. W. J. Layman, "Fundamental Considerations in Preparing a Master System Plan," *Electrical World,* **101,** 778–792 (1933).
2. S. A. Smith, "Spare Capacity Fixed by Probabilities of Outage," *Electrical World,* **103,** 222–225 (1934).
3. S. A. Smith, "Probability Theory and Spare Equipment," *Edison Electric Inst. Bull.,* Mar. 1934, pp. 310–314.
4. S. A. Smith, "Service Reliability Measured by Probabilities of Outage," *Electrical World,* **103,** 371–374 (1934).
5. P. E. Benner, "The Use of the Theory of Probability to Determine Spare Capacity," *General Electric Rev.,* **37,** 345–348 (1934).
6. S. M. Dean, "Considerations Involved in Making System Investments for Improved Service Reliability," *Edison Electric Inst. Bull.,* **6,** 491–496 (1938).
7. B. S. Dhillon, *Power System Reliability, Safety, and Management,* Ann Arbor Science, Ann Arbor, MI, 1983.
8. A. Coppola, "Reliability Engineering of Electronic Equipment: A Historical Perspective," *IEEE Trans. Reliabil.,* **33,** 29–35 (1984).
9. W. Weibull, "A Statistical Distribution Function of Wide Applicability," *J. Appl. Mech.,* **18,** 293–297 (1951).
10. A. M. Freudenthal, and E. J. Gumbel, "Failure and Survival in Fatigue," *J. Appl. Phys.,* **25,** 110–120 (1954).
11. A. M. Freudenthal, "Safety and Probability of Structural Failure," *Trans. Am. Soc. Civil Eng.,* **121,** 1337–1397 (1956).
12. W. M. Redler, "Mechanical Reliability Research in the National Aeronautics and Space Administration," *Proceedings of the Reliability and Maintainability Conference,* 1966, pp. 763–768.
13. B. S. Dhillon, *Mechanical Reliability: Theory, Models, and Applications,* American Institute of Aeronautics and Astronautics, Washington, DC, 1988.
14. B. S. Dhillon, *Design Reliability: Fundamentals and Applications,* CRC, Boca Raton, FL, 1999.
15. B. S. Dhillon, *Reliability and Quality Control: Bibliography on General and Specialized Areas,* Beta Publishers, Gloucester, Ontario, Canada, 1992.
16. P. Kales, *Reliability: For Technology, Engineering, and Management,* Prentice-Hall, Upper Saddle River, NJ, 1998.
17. B. S. Dhillon, "A Hazard Rate Model," *IEEE Trans. Reliabil.,* **29,** 150–151 (1979).
18. B. S. Dhillon, *Systems Reliability, Maintainability, and Management,* Petrocelli Books, New York, 1983.
19. B. S. Dhillon, *Reliability, Quality, and Safety for Engineers,* CRC, Boca Raton, FL, 2005.
20. J. P. Lipp, "Topology of Switching Elements Versus Reliability," *Trans. IRE Reliabil. Quality Control,* **7,** 21–34 (1957).
21. J. A. Collins, *Failure of Materials in Mechanical Design,* Wiley, New York, 1981.
22. W. Grant Ireson, C. F. Coombs, and R. Y. Moss (eds.), *Handbook of Reliability Engineering and Management,* McGraw-Hill, New York, 1996.
23. R. L. Doyle, "Mechanical System Reliability, Tutorial Notes," *Annual Reliability and Maintainability Symposium,* Las Vegas, NV, 1992.
24. C. Lipson, "Analysis and Prevention of Mechanical Failures," Course Notes No. 8007, University of Michigan, Ann Arbor, MI, June 1980.
25. S. S. Rao, *Reliability-Based Design,* McGraw-Hill, New York, 1992.
26. J. W. Wilbur, and N. B. Fuqua, *A Primer for DOD Reliability, Maintainability, and Safety,* Standards Document No. PRIM 1, Rome Air Development Center, Griffiss Air Force Base, Rome, NY, 1988.
27. D. G. Raheja, *Assurances Technologies,* McGraw-Hill, New York, 1991.
28. D. Kececioglu, "Reliability Analysis of Mechanical Components and Systems," *Nucl. Eng. Design,* **19,** 259–290 (1972).
29. B. S. Dhillon, and C. Singh, *Engineering Reliability: New Techniques and Applications,* Wiley, New York, 1981.

30. *Procedures for Performing Failure Mode, Effects, and Criticality Analysis,* MIL-STD-1629, Department of Defense, Washington, DC, 1980.
31. D. Kececioglu, and E. B. Haugen, "A Unified Look at Design Safety Factors, Safety Margin, and Measures of Reliability," *Proceedings of the Annual Reliability and Maintainability Conference,* 1968, pp. 522–530.
32. G. M. Howell, "Factors of Safety," *Machine Design,* July 12, 1956, pp. 76–81.
33. R. B. McCalley, "Nomogram for Selection of Safety Factors," *Design News,* Sept. 1957, pp. 138–141.
34. R. Schoof, "How Much Safety Factor?" *Allis-Chalmers Elec. Rev.,* 1960, pp. 21–24.
35. J. E. Shigley, and L. D. Mitchell, *Mechanical Engineering Design,* McGraw-Hill, New York, 1983, pp. 610–611.
36. J. H. Bompass-Smith, *Mechanical Survival: The Use of Reliability Data,* McGraw-Hill, London, 1973.
37. V. M. Faires, *Design of Machine Elements,* Macmillan, New York, 1955.
38. J. S. Countinho, "Failure Effect Analysis," *Trans. N.Y. Acad. Sci.,* **26,** 564–584 (1964).
39. B. S. Dhillon, "Failure Modes and Effects Analysis: Bibliography," *Microelectron. Reliabil.,* **32,** 719–732 (1992).
40. D. Meister, "The Problem of Human-Initiated Failures," *Proceedings of the Eighth National Symposium on Reliability and Quality Control,* 1962, pp. 234–239.
41. J. I. Cooper, "Human-Initiated Failures and Man-Function Reporting," *IRE Trans. Human Factors,* **10,** 104–109 (1961).
42. B. S. Dhillon, *Human Reliability: With Human Factors,* Pergamon, New York, 1986.
43. T. L. Regulinski, and W. B. Askern, "Mathematical Modeling of Human Performance Reliability," in *Proceedings of the Annual Symposium on Reliability,* 1969, pp. 5–11.
44. S. Rhodes, J. J. Nelson, J. D. Raze, and M. Bradley, "Reliability Models for Mechanical Equipment," *Proceedings of the Annual Reliability and Maintainability Symposium,* 1988, pp. 127–131.
45. J. D. Raze, J. J. Nelson, D. J. Simard, and M. Bradley, "Reliability Models for Mechanical Equipment," *Proceedings of the Annual Reliability and Maintainability Symposium,* 1987, pp. 130–134.
46. J. J. Nelson, J. D. Raze, J. Bowman, G. Perkins, and A. Wannamaker, "Reliability Models for Mechanical Equipment," *Proceedings of the Annual Reliability and Maintainability Symposium,* 1989, pp. 146–153.
47. T. D. Boone, *Reliability Prediction Analysis for Mechanical Brake Systems,* NAVAIR/SYSCOM Report, Department of Navy, Department of Defense, Washington, DC, Aug. 1981.
48. M. J. Rossi, *Non-Electronic Parts Reliability Data,* Report No. NRPD-3, Reliability Analysis Center, Rome Air Development Center, Griffiss Air Force Base, NY, 1985.
49. R. E. Schafer, J. E. Angus, J. M. Finkelstein, M. Yerasi, and D. W. Fulton, *RADC Non-electronic Reliability Notebook,* Report No. RADC-TR-85-194, Reliability Analysis Center, Rome Air Development Center, Griffiss Air Force Base, NY, 1985.
50. B. S. Dhillon, and H. C. Viswanath, "Bibliography of Literature on Failure Data," *Microelectron. Reliabil.,* **30,** 723–750 (1990).

BIBLIOGRAPHY

Bompas-Smith, J. H., *Mechanical Survival,* McGraw-Hill, London, 1973.

Carter, A. D. S., *Mechanical Reliability,* Macmillan Education, London, 1986.

Carter, A. D. S., *Mechanical Reliability and Design,* Wiley, New York, 1997.

Dhillon, B. S., *Robot Reliability and Safety,* Springer-Verlag, New York, 1991.

Frankel, E. G., *Systems Reliability and Risk Analysis,* Martinus Nijhoff, The Hague, 1984.

Haugen, E. B., *Probabilistic Mechanical Design,* Wiley, New York, 1980.

Kapur, K. C., and L. R. Lamberson, *Reliability in Engineering Design,* Wiley, New York, 1977.

Kivenson, G., *Durability and Reliability in Engineering Design,* Hayden, New York, 1971.

Little, A., *Reliability of Shell Buckling Predictions,* MIT Press, Cambridge, MA, 1964.

Little, R. E., *Mechanical Reliability Improvement: Probability and Statistics for Experimental Testing,* Marcel Dekker, New York, 2003.

Mechanical Reliability Concepts, American Society of Mechanical Engineers, New York, 1965.

Middendorf, W. H., *Design of Devices and Systems,* Marcel Dekker, New York, 1990.

Milestone, W. D. (ed.), *Reliability, Stress Analysis and Failure Prevention Methods in Mechanical Design,* American Society of Mechanical Engineers, New York, 1980.

Shooman, M. L., *Probabilistic Reliability: An Engineering Approach,* Krieger, Melbourne, FL, 1990.

Siddell, J. N., *Probabilistic Engineering Design,* Marcel Dekker, New York, 1983.

Additional publications on mechanical design reliability may be found in Refs. 13 and 15.

CHAPTER 32
LUBRICATION OF MACHINE ELEMENTS

Bernard J. Hamrock
Department of Mechanical Engineering
The Ohio State University
Columbus, Ohio

By the middle of this century two distinct regimes of lubrication were generally recognized. The first of these was hydrodynamic lubrication. The development of the understanding of this lubrication regimen began with the classical experiments of Tower,[1] in which the existence of a film was detected from measurements of pressure within the lubricant, and of Petrov,[2] who reached the same conclusion from friction measurements. This work was closely followed by Reynolds' celebrated analytical paper[3] in which he used a reduced form of the Navier–Stokes equations in association with the continuity equation to generate a second-order differential equation for the pressure in the narrow, converging gap of a bearing contact. Such a pressure enables a load to be transmitted between the surfaces with very low friction since the surfaces are completely separated by a film of fluid. In such a situation it is the physical properties of the lubricant, notably the dynamic viscosity, that dictate the behavior of the contact.

The second lubrication regimen clearly recognized by 1950 was boundary lubrication. The understanding of this lubrication regimen is normally attributed to Hardy and Doubleday,[4,5] who found that very thin films adhering to surfaces were often sufficient to assist relative sliding. They concluded that under such circumstances the chemical composition of the fluid is important, and they introduced the term "boundary lubrication." Boundary lubrication is at the opposite end of the lubrication spectrum from hydrodynamic lubrication. In boundary lubrication it is the physical and chemical properties of thin films of molecular proportions and the surfaces to which they are attached that determine contact behavior. The lubricant viscosity is not an influential parameter.

In the last 30 years research has been devoted to a better understanding and more precise definition of other lubrication regimens between these extremes. One such lubrication regimen occurs in nonconformal contacts, where the pressures are high and the bearing surfaces deform elastically. In this situation the viscosity of the lubricant may rise considerably, and this further assists the formation of an effective fluid film. A lubricated contact in which such effects are to be found is said to be operating elastohydrodynamically. Significant progress has been made in our understanding of the mechanism of elastohydrodynamic lubrication, generally viewed as reaching maturity.

This chapter describes briefly the science of these three lubrication regimens hydrodynamic, elastohydrodynamic, and boundary) and then demonstrates how this science is used in the design of machine elements.

SYMBOLS

A_p	total projected pad area, m^2
a_b	groove width ratio
a_f	bearing-pad load coefficient
B	total conformity of ball bearing
b	semiminor axis of contact, m; width of pad, m
$\bar{b}$	length ratio, b_s/b_r
b_g	length of feed groove region, m
b_r	length of ridge region, m
b_s	length of step region, m
C	dynamic load capacity, N
C_l	load coefficient, F/p_aRl
c	radial clearance of journal bearing, m
c'	pivot circle clearance, m
c_b	bearing clearance at pad minimum film thickness (Fig. 16), m
c_d	orifice discharge coefficient
D	distance between race curvature centers, m
$\tilde{D}$	material factor
D_x	diameter of contact ellipse along x axis, m
D_y	diameter of contact ellipse along y axis, m
d	diameter of rolling element or diameter of journal, m
d_a	overall diameter of ball bearing (Fig. 76), m
d_b	bore diameter of ball bearing, m
d_c	diameter of capillary tube, m
d_i	inner-race diameter of ball bearing, m
d_o	outer-race diameter of ball bearing, m

$\bar{d}_o$	diameter of orifice, m
E	modulus of elasticity, N/m^2
E'	effective elastic modulus, $2\left(\frac{1-\nu_a^2}{E_a}+\frac{1-\nu_b^2}{E_b}\right)^{-1}$, N/m^2
$\tilde{E}$	metallurgical processing factor
$\mathcal{E}$	elliptic integral of second kind
e	eccentricity of journal bearing, m
F	applied normal load, N
F'	load per unit length, N/m
$\tilde{F}$	lubrication factor
$\mathcal{F}$	elliptic integral of first kind
F_c	pad load component along line of centers (Fig. 41), N
F_e	rolling-element bearing equivalent load, N
F_r	applied radial load, N
F_s	pad load component normal to line of centers (Fig. 41), N
F_t	applied thrust load, N
f	race conformity ratio
f_c	coefficient dependent on materials and rolling-element bearing type (Table 19)
G	dimensionless materials parameter
$\tilde{G}$	speed effect factor
G_f	groove factor
g_e	dimensionless elasticity parameter, $W^{8/3}/U^2$
g_v	dimensionless viscosity parameter, GW^3/U^2
H	dimensionless film thickness, h/R_x
$\tilde{H}$	misalignment factor
H_a	dimensionless film thickness ratio, h_s/h_r
H_b	pad pumping power, N·m/s
H_c	power consumed in friction per pad, W
H_f	pad power coefficient
$H_{\min}$	dimensionless minimum film thickness, $h_{\min}/R_x$
$\hat{H}_{\min}$	dimensionless minimum film thickness, $H_{\min}(W/U)^2$
H_p	dimensionless pivot film thickness, h_p/c
H_t	dimensionless trailing-edge film thickness, h_t/c
h	film thickness, m
$\bar{h}_i$	film thickness ratio, h_i/h_o
h_i	inlet film thickness, m
h_l	leading-edge film thickness, m
$h_{\min}$	minimum film thickness, m
h_o	outlet film thickness, m
h_p	film thickness at pivot, m
h_r	film thickness in ridge region, m
h_s	film thickness in step region, m

h_t film thickness at trailing edge, m
h_0 film constant, m
J number of stress cycles
K load deflection constant
$\bar{K}$ dimensionless stiffness coefficient, cK_p/p_aRl
K_a dimensionless stiffness, $-c\ \partial\bar{W}/\partial c$
K_p film stiffness, N/m
K_1 load–deflection constant for a roller bearing
$K_{1.5}$ load–deflection constant for a ball bearing
$\bar{K}_\infty$ dimensionless stiffness, cK_p/p_aRl
k ellipticity parameter, D_y/D_x
k_c capillary tube constant, m^3
k_o orifice constant, $m^4/N^{1/2}\cdot s$
L fatigue life
L_a adjusted fatigue life
L_{10} fatigue life where 90% of bearing population will endure
L_{50} fatigue life where 50% of bearing population will endure
l bearing length, m
l_c length of capillary tube, m
l_r roller effective length, m
l_t roller length, m
l_v length dimension in stress volume, m
l_1 total axial length of groove, m
M probability of failure
$\bar{M}$ stability parameter, $\bar{m}p_ah_r^5/2R^5/\eta^2$
m number of rows of rolling elements
$\bar{m}$ mass supported by bearing, $N\cdot s^2/m$
m_p preload factor
N rotational speed, rps
N_R Reynolds number
n number of rolling elements or number of pads or grooves
P dimensionless pressure, p/E'
P_d diametral clearance, m
P_e free endplay, m
p pressure, N/m^2
p_a ambient pressure, N/m^2
p_l lift pressure, N/m^2
p_{max} maximum pressure, N/m^2
p_r recess pressure, N/m^2
p_s bearing supply pressure, N/m^2
Q volume flow of lubricant, m^3/s
$\bar{Q}$ dimensionless flow, $3\eta Q/\pi p_a h_r^3$

Q_c volume flow of lubricant in capillary, m^3/s
Q_o volume flow of lubricant in orifice, m^3/s
Q_s volume side flow of lubricant, m^3/s
q constant, $\pi/2 - 1$
q_f bearing-pad flow coefficient
R curvature sum on shaft or bearing radius, m
$\overline{R}$ groove length fraction, $(R_o - R_g)/R_o - R_i)$
R_g groove radius (Fig. 60), m
R_o orifice radius, m
R_x effective radius in x direction, m
R_y effective radius in y direction, m
R_1 outer radius of sector thrust bearing, m
R_2 inner radius of sector thrust bearing, m
r race curvature radius, m
r_c roller corner radius, m
S probability of survival
Sm Sommerfeld number for journal bearings, $\eta N d^3 l/2Fc^2$
Sm_t Sommerfeld number for thrust bearings, $\eta u b l^2/F h_0^2$
s shoulder height, m
T tangential force, N
$\overline{T}$ dimensionless torque, $6\, T_r/\pi p_a/(R_1^2 + R_2^2)\, h_r \Lambda_c$
T_c critical temperature
T_r torque, N·m
U dimensionless speed parameter, $u\eta_0/E'R_x$
u mean surface velocity in direction of motion, m/s
v elementary volume, m^3
N dimensionless load parameter, $F/E'R_x^2$
$\overline{W}$ dimensionless load capacity, $F/p_a l(b_r + b_s + b_g)$
$\overline{W}_\infty$ dimensionless load, $1.5 G_f F/\pi p_a (R_1^2 - R_2^2)$
X, Y factors for calculation of equivalent load
x,y,z coordinate system
$\overline{x}$ distance from inlet edge of pad to pivot, m
α radius ratio, R_y/R_x
α_a offset factor
α_b groove width ratio, $b_s/(b_r + b_s)$
α_p angular extent of pad, deg
α_r radius ratio, R_2/R_1
β contact angle, deg
β' iterated value of contact angle, deg
β_a groove angle, deg
β_f free or initial contact angle, deg

β_p	angle between load direction and pivot, deg
Γ	curvature difference
γ	groove length ratio, l_1/l
Δ	rms surface finish, m
δ	total elastic deformation, m
ϵ	eccentricity ratio, e/c
η	absolute viscosity of lubricant, $N \cdot s/m^2$
η_k	kinematic viscosity, ν/ρ, m^2/s
η_0	viscosity at atmospheric pressure, $N \cdot s/m^2$
θ	angle used to define shoulder height, deg
$\bar{\theta}$	dimensionless step location, $\theta_i/(\theta_i + \theta_o)$
θ_g	angular extent of lubrication feed groove, deg
θ_i	angular extent of ridge region, deg
θ_o	angular extent of step region, deg
Λ	film parameter ratio of minimum film thickness to composite surface roughness)
Λ_c	dimensionless bearing number, $3\eta\omega(R_1^2 - R_2^2)/p_a h_r^2$
Λ_j	dimensionless bearing number, $6\eta\omega R^2/p_a c^2$
Λ_t	dimensionless bearing number, $6\eta ul/p_a h_r^2$
λ	length-to-width ratio
λ_a	length ratio, $(b_r + b_s + b_g)/l$
λ_b	$(1 + 2/3\alpha)^{-1}$
μ	coefficient of friction, T/F
ν	Poisson's ratio
ξ	pressure–viscosity coefficient of lubricant, m^2/N
ξ_p	angle between line of centers and pad leading edge, deg
ρ	lubricant density, $N \cdot s^2/m^4$
ρ_0	density at atmospheric pressure, $N \cdot s^2/m^4$
σ_{max}	maximum Hertzian stress, N/m^2
τ	shear stress, N/m^2
τ_0	maximum shear stress, N/m^2
ϕ	attitude angle in journal bearings, deg
ϕ_p	angle between pad leading edge and pivot, deg
ψ	angular location, deg
ψ_l	angular limit of ψ, deg
ψ_s	step location parameter, $b_s/(b_r + b_s + b_g)$
ω	angular velocity, rad/s
ω_B	angular velocity of rolling-element race contact, rad/s
ω_b	angular velocity of rolling element about its own center, rad/s
ω_c	angular velocity of rolling element about shaft center, rad/s
ω_d	rotor whirl frequency, rad/s
$\bar{\omega}_d$	whirl frequency ratio, ω_d/ω_j
ω_j	journal rotational speed, rad/s

Subscripts

a	solid a
b	solid b
EHL	elastohydrodynamic lubrication
e	elastic
HL	hydrodynamic lubrication
i	inner
iv	isoviscous
o	outer
pv	piezoviscous
r	rigid
x,y,z	coordinate system

1 LUBRICATION FUNDAMENTALS

A lubricant is any substance that is used to reduce friction and wear and to provide smooth running and a satisfactory life for machine elements. Most lubricants are liquids (like mineral oils, the synthetic esters and silicone fluids, and water), but they may be solids (such as polytetrafluoroethylene) for use in dry bearings, or gases (such as air) for use in gas bearings. An understanding of the physical and chemical interactions between the lubricant and the tribological surfaces is necessary if the machine elements are to be provided with satisfactory life. To help in the understanding of this tribological behavior, the first section describes some lubrication fundamentals.

1.1 Conformal and Nonconformal Surfaces

Hydrodynamic lubrication is generally characterized by surfaces that are conformal; that is, the surfaces fit snugly into each other with a high degree of geometrical conformity (as shown in Fig. 1), so that the load is carried over a relatively large area. Furthermore, the load-carrying surface remains essentially constant while the load is increased. Fluid-film journal bearings (as shown in Fig. 1) and slider bearings exhibit conformal surfaces. In journal bearings the radial clearance between the shaft and bearing is typically one-thousandth of the shaft diameter; in slider bearings the inclination of the bearing surface to the runner is typically one part in a thousand. These converging surfaces, coupled with the fact that there is relative motion and a viscous fluid separating the surfaces, enable a positive pressure to be developed and exhibit a capacity to support a normal applied load. The magnitude of the pressure developed *is not* generally large enough to cause significant elastic deformation of the surfaces. The minimum film thickness in a hydrodynamically lubricated

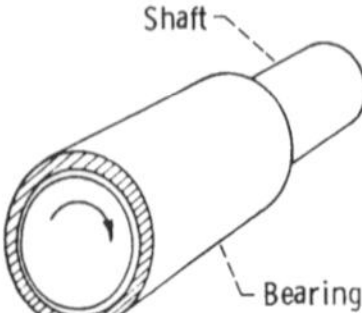

Figure 1 Conformal surfaces. (From Ref. 6.)

bearing is a function of applied load, speed, lubricant viscosity, and geometry. The relationship between the minimum film thickness $h_{\min}$ and the speed u and applied normal load F is given as

$$(h_{\min})_{\mathrm{HL}} \propto \left(\frac{u}{F}\right)^{1/2} \tag{1}$$

More coverage of hydrodynamic lubrication can be found in Section 2.

Many machine elements have contacting surfaces that *do not* conform to each other very well, as shown in Fig. 2 for a rolling-element bearing. The full burden of the load must then be carried by a very small contact area. In general, the contact areas between nonconformal surfaces enlarge considerably with increasing load, but they are still smaller than the contact areas between conformal surfaces. Some examples of nonconformal surfaces are mating gear teeth, cams and followers, and rolling-element bearings (as shown in Fig. 2). The mode of lubrication normally found in these nonconformal contacts is elastohydrodynamic lubrication. The requirements necessary for hydrodynamic lubrication (converging surfaces, relative motion, and viscous fluid) are also required for elastohydrodynamic lubrication.

The relationship between the minimum film thickness and normal applied load and speed for an elastohydrodynamically lubricated contact is

$$(h_{\min})_{\mathrm{EHL}} \propto F^{-0.073} \tag{2}$$

$$(h_{\min})_{\mathrm{EHL}} \propto u^{0.68} \tag{3}$$

Comparing the results of Eqs. (2) and (3) with that obtained for hydrodynamic lubrication expressed in Eq. (1) indicates that:

1. The exponent on the normal applied load is nearly seven times larger for hydrodynamic lubrication than for elastohydrodynamic lubrication. This implies that in elastohydrodynamic lubrication the film thickness is only slightly affected by load while in hydrodynamic lubrication it is significantly affected by load.
2. The exponent on mean velocity is slightly higher for elastohydrodynamic lubrication than that found for hydrodynamic lubrication.

More discussion of elastohydrodynamic lubrication can be found in Section 3.

The load per unit area in conformal bearings is relatively low, typically averaging only 1 MN/m^2 and seldom over 7 MN/m^2. By contrast, the load per unit area in nonconformal contacts will generally exceed 700 MN/m^2 even at modest applied loads. These high pressures result in elastic deformation of the bearing materials such that elliptical contact areas are formed for oil-film generation and load support. The significance of the high contact pressures is that they result in a considerable increase in fluid viscosity. Inasmuch as viscosity is a measure of a fluid's resistance to flow, this increase greatly enhances the lubricant's

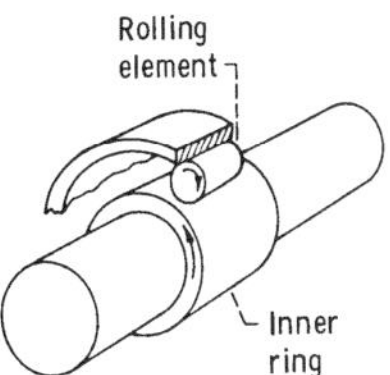

Figure 2 Nonconformal surfaces. (From Ref. 6.)

ability to support load without being squeezed out of the contact zone. The high contact pressures in nonconforming surfaces therefore result in both an elastic deformation of the surfaces and large increases in the fluid's viscosity. The minimum film thickness is a function of the parameters found for hydrodynamic lubrication with the addition of an effective modulus of elasticity parameter for the bearing materials and a pressure–viscosity coefficient for the lubricant.

1.2 Bearing Selection

Ball bearings are used in many kinds of machines and devices with rotating parts. The designer is often confronted with decisions on whether a nonconformal bearing such as a rolling-element bearing or a conformal bearing such as a hydrodynamic bearing should be used in a particular application. The following characteristics make rolling-element bearings *more desirable* than hydrodynamic bearings in many situations:

1. Low starting and good operating friction
2. The ability to support combined radial and thrust loads
3. Less sensitivity to interruptions in lubrication
4. No self-excited instabilities
5. Good low-temperature starting

Within reasonable limits changes in load, speed, and operating temperature have but little effect on the satisfactory performance of rolling-element bearings.

The following characteristics make nonconformal bearings such as rolling-element bearings *less desirable* than conformal (hydrodynamic) bearings:

1. Finite fatigue life subject to wide fluctuations
2. Large space required in the radial direction
3. Low damping capacity
4. High noise level
5. More severe alignment requirements
6. Higher cost

Each type of bearing has its particular strong points, and care should be taken in choosing the most appropriate type of bearing for a given application.

The Engineering Services Data Unit documents[7,8] provide an excellent guide to the selection of the type of journal or thrust bearing most likely to give the required performance when considering the load, speed, and geometry of the bearing. The following types of bearings were considered:

1. Rubbing bearings, where the two bearing surfaces rub together (e.g., unlubricated bushings made from materials based on nylon, polytetrafluoroethylene, also known as PTFE, and carbon)
2. Oil-impregnated porous metal bearings, where a porous metal bushing is impregnated with lubricant and thus gives a self-lubricating effect (as in sintered-iron and sintered-bronze bearings)
3. Rolling-element bearings, where relative motion is facilitated by interposing rolling elements between stationary and moving components (as in ball, roller, and needle bearings)

4. Hydrodynamic film bearings, where the surfaces in relative motion are kept apart by pressures generated hydrodynamically in the lubricant film

Figure 3, reproduced from the Engineering Sciences Data Unit publication,[7] gives a guide to the typical load that can be carried at various speeds for a nominal life of 10,000 h at room temperature by journal bearings of various types on shafts of the diameters quoted. The heavy curves indicate the preferred type of journal bearing for a particular load, speed, and diameter and thus divide the graph into distinct regions. From Fig. 3 it is observed that rolling-element bearings are preferred at lower speeds and hydrodynamic oil-film bearings are preferred at higher speeds. Rubbing bearings and oil-impregnated porous metal bearings are not preferred for any of the speeds, loads, or shaft diameters considered. Also, as the shaft diameter is increased, the transitional point at which hydrodynamic bearings are preferred over rolling-element bearings moves to the left.

The applied load and speed are usually known, and this enables a preliminary assessment to be made of the type of journal bearing most likely to be suitable for a particular appli-

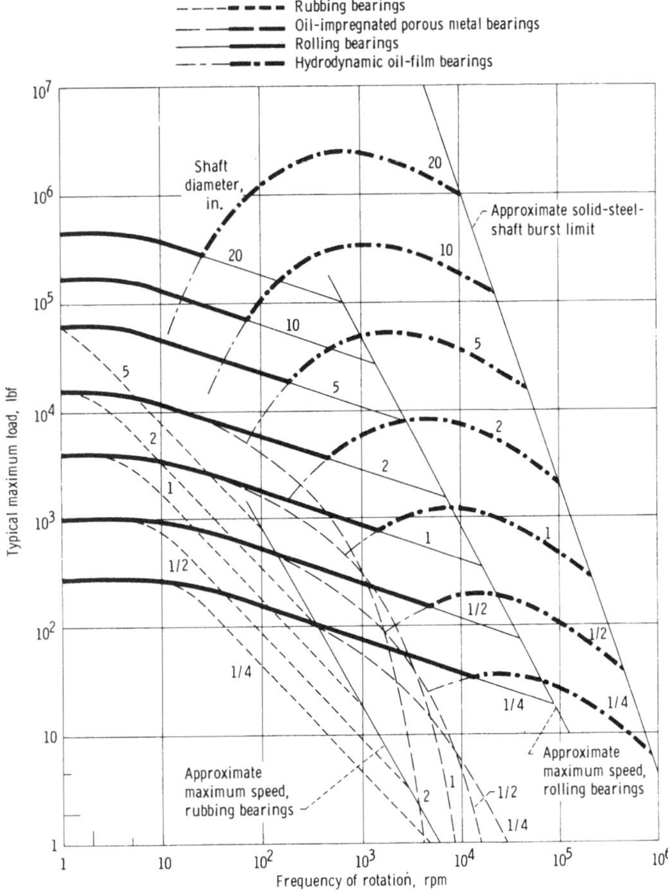

Figure 3 General guide to journal bearing type. (Except for roller bearings, curves are drawn for bearings with width equal to diameter. A medium-viscosity mineral oil lubricant is assumed for hydrodynamic bearings.) (From Ref. 7.)

cation. In many cases the shaft diameter will have been determined by other considerations, and Fig. 3 can be used to find the type of journal bearing that will give adequate load capacity at the required speed. These curves are based upon good engineering practice and commercially available parts. Higher loads and speeds or smaller shaft diameters are possible with exceptionally high engineering standards or specially produced materials. Except for rolling-element bearings the curves are drawn for bearings with a width equal to the diameter. A medium-viscosity mineral oil lubricant is assumed for the hydrodynamic bearings.

Similarly, Fig. 4, reproduced from the Engineering Sciences Data Unit publication,[8] gives a guide to the typical maximum load that can be carried at various speeds for a nominal life of 10,000 h at room temperature by thrust bearings of the various diameters quoted. The heavy curves again indicate the preferred type of bearing for a particular load, speed, and diameter and thus divide the graph into major regions. As with the journal bearing results (Fig. 3) the hydrodynamic bearing is preferred at lower speeds. A difference between Figs. 3 and 4 is that at very low speeds there is a portion of the latter figure in which the rubbing bearing is preferred. Also, as the shaft diameter is increased, the transitional point at which

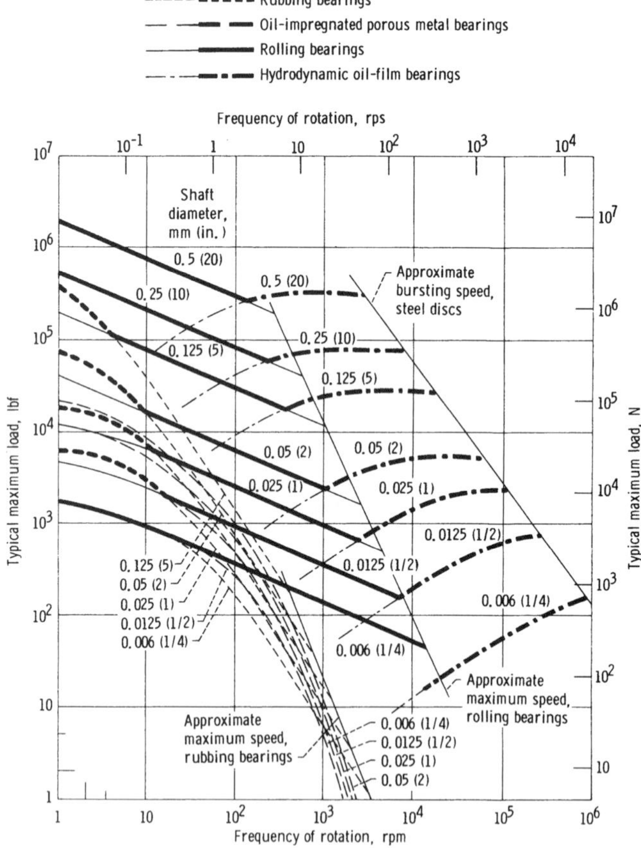

Figure 4 General guide to thrust bearing type. (Except for roller bearings, curves are drawn for typical ratios of inside-to-outside diameter. A medium-viscosity mineral oil lubricant is assumed for hydrodynamic bearings.) (From Ref. 8.)

hydrodynamic bearings are preferred over rolling-element bearings moves to the left. Note also from this figure that oil-impregnated porous metal bearings are not preferred for any of the speeds, loads, or shaft diameters considered.

1.3 Lubricants

Both oils and greases are extensively used as lubricants for all types of machine elements over wide a range of speeds, pressures, and operating temperatures. Frequently, the choice is determined by considerations other than lubrication requirements. The requirements of the lubricant for successful operation of nonconformal contacts such as in rolling-element bearings and gears are considerably more stringent than those for conformal bearings and therefore will be the primary concern in this section.

Because of its fluidity, oil has several advantages over grease: It can enter the loaded conjunction most readily to flush away contaminants, such as water and dirt, and, particularly, to transfer heat from heavily loaded machine elements. Grease, however, is extensively used because it permits simplified designs of housings and enclosures, which require less maintenance, and because it is more effective in sealing against dirt and contaminants.

Viscosity

In hydrodynamic and elastohydrodynamic lubrication the most important physical property of a lubricant is its viscosity. The viscosity of a fluid may be associated with its resistance to flow, that is, with the resistance arising from intermolecular forces and internal friction as the molecules move past each other. Thick fluids, like molasses, have relatively high viscosity; they do not flow easily. Thinner fluids, like water, have lower viscosity; they flow very easily.

The relationship for internal friction in a viscous fluid (as proposed by Newton)[9] can be written as

$$\tau = \eta \frac{du}{dz} \tag{4}$$

where τ = internal shear stress in the fluid in the direction of motion
η = coefficient of absolute or dynamic viscosity or coefficient of internal friction
du/dz = velocity gradient perpendicular to the direction of motion (i.e., shear rate)

It follows from Eq. (4) that the unit of dynamic viscosity must be the unit of shear stress divided by the unit of shear rate. In the newton-meter-second system the unit of shear stress is the newton per square meter while that of shear rate is the inverse second. Hence the unit of dynamic viscosity will be newton per square meter multiplied by second, or $\text{N}\cdot\text{s/m}^2$. In the SI system the unit of pressure or stress (N/m^2) is known as the pascal, abbreviated Pa, and it is becoming increasingly common to refer to the SI unit of viscosity as the pascal-second (Pa·s). In the cgs system, where the dyne is the unit of force, dynamic viscosity is expressed as dyne-second per square centimeter. This unit is called the poise, with its submultiple the centipoise (1 cP = 10^{-2} P) of a more convenient magnitude for many lubricants used in practice.

Conversion of dynamic viscosity from one system to another can be facilitated by Table 1. To convert from a unit in the column on the left-hand side of the table to a unit at the top of the table, multiply by the corresponding value given in the table. For example, η = 0.04 $\text{N}\cdot\text{s/m}^2$ = 0.04 × 1.45 × 10^{-4} $\text{lbf}\cdot\text{s/in.}^2$ = 5.8 × 10^{-6} $\text{lbf}\cdot\text{sec/in.}^2$. One English and three metric systems are presented—all based on force, length, and time. Metric units are the centipoise, the kilogram force-second per square meter, and the newton-second per square

Table 1 Viscosity Conversion

To Convert From—	To— cP	kgf·s/m²	N·s/m²	lbf·s/in.²
	Multiply By—			
cP	1	1.02×10^{-4}	10^{-3}	1.45×10^{-7}
kgf·s/m²	9.807×10^{3}	1	9.807	1.422×10^{-3}
N·s/m²	10^{3}	1.02×10^{-1}	1	1.45×10^{-4}
lbf·s/in²	6.9×10^{6}	7.034×10^{2}	6.9×10^{3}	1

meter (or Pa·s). The English unit is pound force-second per square inch, or reyn, in honor of Osborne Reynolds.

In many situations it is convenient to use the *kinematic viscosity* rather than the dynamic viscosity. The kinematic viscosity η_k is equal to the dynamic viscosity η divided by the density ρ of the fluid ($\eta_k = \eta/\rho$). The ratio is literally kinematic, all trace of force or mass cancelling out. The unit of kinematic viscosity may be written in SI units as square meters per second or in English units as square inches per second or, in cgs units, as square centimeters per second. The name stoke, in honor of Sir George Gabriel Stokes, was proposed for the cgs unit by Max Jakob in 1928. The centistoke, or one-hundredth part, is an everyday unit of more convenient size, corresponding to the centipoise.

The viscosity of a given lubricant varies within a given machine element as a result of the nonuniformity of pressure or temperature prevailing in the lubricant film. Indeed, many lubricated machine elements operate over ranges of pressure or temperature so extensive that the consequent variations in the viscosity of the lubricant may become substantial and, in turn, may dominate the operating characteristics of machine elements. Consequently, an adequate knowledge of the viscosity–pressure and viscosity–pressure–temperature relationships of lubricants is indispensable.

Oil Lubrication

Except for a few special requirements, petroleum oils satisfy most operating conditions in machine elements. High-quality products, free from adulterants that can have an abrasive or lapping action, are recommended. Animal or vegetable oils or petroleum oils of poor quality tend to oxidize, to develop acids, and to form sludge or resinlike deposits on the bearing surfaces. They thus penalize bearing performance or endurance.

A composite of recommended lubricant kinematic viscosities at 38°C (100°F) is shown in Fig. 5. The ordinate of this figure is the speed factor, which is bearing bore size measured in millimeters multiplied by the speed in revolutions per minute. In many rolling-element bearing applications an oil equivalent to an SAE-10 motor oil [4×10^{-6} m²/s, or 40 cS, at 38°C (100°F)] or a light turbine oil is the most frequent choice.

For a number of military applications where the operational requirements span the temperature range −54–204°C (−65–400°F), synthetic oils are used. Ester lubricants are most frequently employed in this temperature range. In applications where temperatures exceed 260°C (500°F), most synthetics will quickly break down, and either a solid lubricant (e.g., MoS_2) or a polyphenyl ether is recommended. A more detailed discussion of synthetic lubricants can be found in Bisson and Anderson.[11]

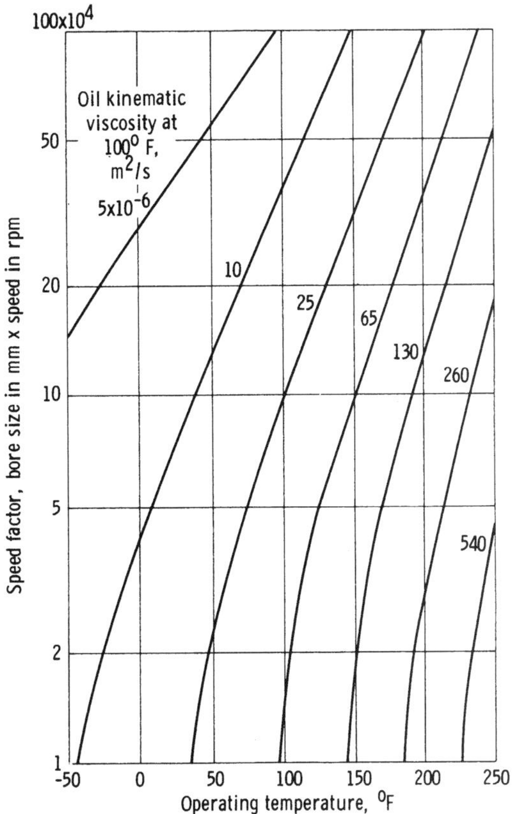

Figure 5 Recommended lubricant viscosities for ball bearings. (From Ref. 10.)

Grease Lubrication

The simplest method of lubricating a bearing is to apply grease because of its relatively nonfluid characteristics. The danger of leakage is reduced, and the housing and enclosure can be simpler and less costly than those used with oil. Grease can be packed into bearings and retained with inexpensive enclosures, but packing should not be excessive and the manufacturer's recommendations should be closely adhered to.

The major limitation of grease lubrication is that it is not particularly useful in high-speed applications. In general, it is not employed for speed factors over 200,000, although selected greases have been used successfully for higher speed factors with special designs.

Greases vary widely in properties depending on the type and grade or consistency. For this reason few specific recommendations can be made. Greases used for most bearing operating conditions consist of petroleum, diester, polyester, or silicone oils thickened with sodium or lithium soaps or with more recently developed nonsoap thickeners. General characteristics of greases are as follows:

1. Petroleum oil greases are best for general-purpose operation from −34 to 149°C (−30–300°F).
2. Diester oil greases are designed for low-temperature service down to −54°C (−65°F).

3. Ester-based greases are similar to diester oil greases but have better high-temperature characteristics, covering the range from −73 to 177°C (−100–350°F).
4. Silicone oil greases are used for both high- and low-temperature operation, over the widest temperature range of all greases [−73–232°C (−100–450°F)], but have the disadvantage of low load-carrying capacity.
5. Fluorosilicone oil greases have all of the desirable features of silicone oil greases plus good load capacity and resistance to fuels, solvents, and corrosive substances. They have a very low volatility in vacuum down to 10^{-7} Torr, which makes them useful in aerospace applications.
6. Perfluorinated oil greases have a high degree of chemical inertness and are completely nonflammable. They have good load-carrying capacity and can operate at temperatures as high as 280°C (550°F) for long periods, which makes them useful in the chemical processing and aerospace industries, where high reliability justifies the additional cost.

Grease consistency is important since grease will slump badly and churn excessively when too soft and fail to lubricate when too hard. Either condition causes improper lubrication, excessive temperature rise, and poor performance and can shorten machine element life. A valuable guide to the estimation of the useful life of grease in rolling-element bearings has been published by the Engineering Sciences Data Unit.[12]

It has recently been demonstrated by Aihara and Dowson[13] and by Wilson[14] that the film thickness in grease-lubricated components can be calculated with adquate accuracy by using the viscosity of the base oil in the elastohydrodynamic equation (see Section 3). This enables the elastohydrodynamic lubrication film thickness formulas to be applied with confidence to grease-lubricated machine elements.

1.4 Lubrication Regimes

If a machine element is adequately designed and lubricated, the lubricated surfaces are separated by a lubricant film. Endurance testing of ball bearings, as reported by Tallian et al.,[15] has demonstrated that when the lubricant film is thick enough to separate the contacting bodies, fatigue life of the bearing is greatly extended. Conversely, when the film is not thick enough to provide full separation between the asperities in the contact zone, the life of the bearing is adversely affected by the high shear resulting from direct metal-to-metal contact.

To establish the effect of film thickness on the life of the machine element, we first introduce a relevant parameter Λ. The relationship between Λ and the minimum film thickness $h_{\min}$ is defined to be

$$\Lambda = \frac{h_{\min}}{(\Delta_a^2 + \Delta_b^2)^{1/2}} \tag{5}$$

where Δ_a = rms surface finish of surface a
Δ_b = rms surface finish of surface b

Hence Λ is just the minimum film thickness in units of the composite roughness of the two bearing surfaces.

Hydrodynamic Lubrication Regime

Hydrodynamic lubrication occurs when the lubricant film is sufficiently thick to prevent the opposite solids from coming into contact. This condition is often referred to as the ideal form of lubrication since it provides low friction and a high resistance to wear. The lubrication of the contact is governed by the bulk physical properties of the lubricant, notably viscosity, and the frictional characteristics arise purely from the shearing of the viscous lubricant. The pressure developed in the oil film of hydrodynamically lubricated bearings is due to two factors:

1. The geometry of the moving surfaces produces a convergent film shape.
2. The viscosity of the liquid results in a resistance to flow.

The lubricant films are normally many times thicker than the surface roughness so that the physical properties of the lubricant dictate contact behavior. The film thickness normally exceeds 10^{-6} m. For hydrodynamic lubrication the film parameter Λ, defined in Eq. (5), is an excess of 10 and may even rise to 100. Films of this thickness are clearly also insensitive to chemical action in surface layers of molecular proportions.

For normal load support to occur in bearings, positive-pressure profiles must develop over the length of the bearing. Three different forms of hydrodynamic lubrication are presented in Fig. 6. Figure 6*a* shows a slider bearing. For a positive load to be developed in the slider bearing shown in Fig. 6*a* the lubricant film thickness must be decreasing in the direction of sliding.

A squeeze film bearing is another mechanism of load support of hydrodynamic lubrication, and it is illustrated in Fig. 6*b*. The squeeze action is the normal approach of the bearing surfaces. The squeeze mechanism of pressure generation provides a valuable cushioning effect when the bearing surfaces tend to be pressed together. Positive pressures will be generated whcn thc film thickness is diminishing.

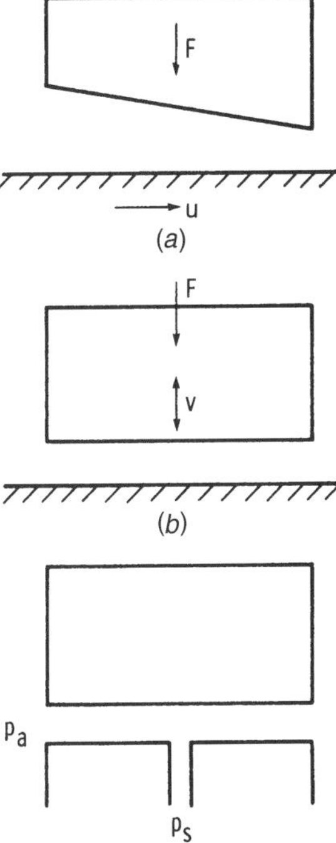

Figure 6 Mechanisms of load support for hydrodynamic lubrication. (*a*) Slider bearing. (*b*) Squeeze film bearing. (*c*) Externally pressurized bearing.

An externally pressurized bearing is yet a third mechanism of load support of hydrodynamic lubrication, and it is illustrated in Fig. 6*c*. The pressure drop across the bearing is used to support the load. The load capacity is independent of the motion of the bearing and the viscosity of the lubricant. There is no problem of contact at starting and stopping as with the other hydrodynamically lubricated bearings because pressure is applied before starting and is maintained until after stopping.

Hydrodynamically lubricated bearings are discussed further in Section 2.

Elastohydrodynamic Lubrication Regime

Elastohydrodynamic lubrication is a form of hydrodynamic lubrication where elastic deformation of the bearing surfaces becomes significant. It is usually associated with highly stressed machine components of low conformity. There are two distinct forms of elastohydrodynamic lubrication (EHL).

Hard EHL. Hard EHL relates to materials of *high* elastic modulus, such as metals. In this form of lubrication both the elastic deformation and the pressure–viscosity effects are equally important. Engineering applications in which elastohydrodynamic lubrication is important for high-elastic-modulus materials include gears and rolling-element bearings.

Soft EHL. Soft EHL relates to materials of *low* elastic modulus, such as rubber. For these materials the elastic distortions are large, even with light loads. Another feature of the elastohydrodynamics of low-elastic-modulus materials is the negligible effect of the relatively low pressures on the viscosity of the lubricating fluid. Engineering applications in which elastohydrodynamic lubrication is important for low-elastic-modulus materials include seals, human joints, tires, and a number of lubricated elastomeric material machine elements.

The common factors in hard and soft EHL are that the local elastic deformation of the solids provides coherent fluid films and that asperity interaction is largely prevented. Elastohydrodynamic lubrication normally occurs in contacts where the minimum film thickness is in the range $0.1\ \mu\text{m} < h_{\min} \leq 10\ \mu\text{m}$ and the film parameter Λ is in the range $3 \leq \Lambda < 10$. Elastohydrodynamic lubrication is discussed further in Section 3.

Boundary Lubrication Regime

If in a lubricated contact the pressures become too high, the running speeds too low, or the surface roughness too great, penetration of the lubricant film will occur. Contact will take place between the asperities. The friction will rise and approach that encountered in dry friction between solids. More importantly, wear will take place. Adding a small quantity of certain active organic compounds to the lubricating oil can, however, extend the life of the machine elements. These additives are present in small quantities (<1%) and function by forming low-shear-strength surface films strongly attached to the metal surfaces. Although they are sometimes only one or two molecules thick, such films are able to prevent metal-to-metal contact.

Some boundary lubricants are long-chain molecules with an active end group, typically an alcohol, an amine, or a fatty acid. When such a material, dissolved in a mineral oil, meets a metal or other solid surface, the active end group attaches itself to the solid and gradually builds up a surface layer. The surface films vary in thickness from 5×10^{-9} to 10^{-8} m depending on molecular size, and the film parameter Λ is less than unity ($\Lambda < 1$). Boundary lubrication is discussed further in Section 4.

Figure 7 illustrates the film conditions existing in hydrodynamic, elastohydrodynamic, and boundary lubrication. The surface slopes in this figure are greatly distorted for the purpose of illustration. To scale, real surfaces would appear as gently rolling hills rather than sharp peaks.

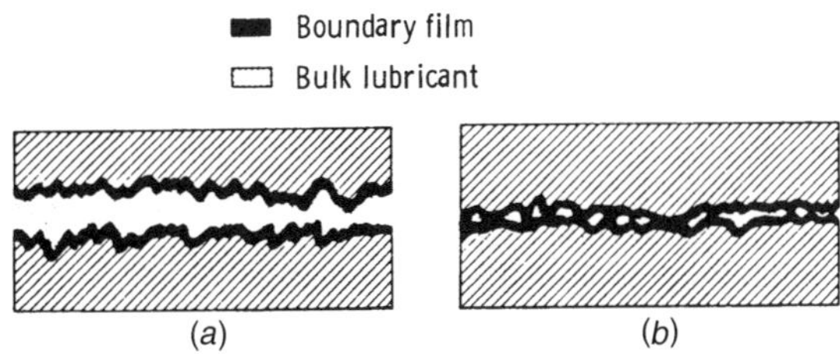

Figure 7 Film conditions of lubrication regimes. (*a*) Hydrodynamic and elastohydrodynamic lubrication—surfaces separated by bulk lubricant film. (*b*) Boundary lubrication—performance essentially dependent on boundary film.

1.5 Relevant Equations

This section presents the equations frequently used in hydrodynamic and elastohydrodynamic lubrication theory. They are not relevant to boundary lubrication since in this lubrication regime bulk fluid effects are negligible. The differential equation governing the pressure distribution in hydrodynamically and elastohydrodynamically lubricated machine elements is known as the Reynolds equation. For steady-state hydrodynamic lubrication the Reynolds equation normally appears as

$$\frac{\partial}{\partial x}\left(h^3 \frac{\partial p}{\partial x}\right) + \frac{\partial}{\partial y}\left(h^3 \frac{\partial p}{\partial y}\right) = 12\eta u \frac{\partial h}{\partial x} \tag{6}$$

where h = film shape measured in the z direction, m
p = pressure, N/m^2
η = lubricant viscosity, $N \cdot s/m^2$
u = mean velocity, $(u_a + u_b)/2$, m/s

Solutions of Eq. (6) are rarely achieved analytically, and approximate numerical solutions are sought.

For elastohydrodynamic lubrication the steady-state form of the Reynolds equation normally appears as

$$\frac{\partial}{\partial x}\left(\frac{\rho h^3}{\eta} \frac{\partial p}{\partial x}\right) + \frac{\partial}{\partial y}\left(\frac{\rho h^3}{\eta} \frac{\partial p}{\partial y}\right) = 12u \frac{\partial(\rho h)}{\partial x} \tag{7}$$

where ρ is lubricant density in $N \cdot s^2/m^2$. The essential difference between Eqs. (6) and (7) is that Eq. (7) allows for variation of viscosity and density in the x and y directions. Equations (6) and (7) allow for the bearing surfaces to be of finite length in the y direction. Side leakage, or flow in the y direction, is associated with the second term in Eqs. (6) and (7). The solution of Eq. (7) is considerably more difficult than that of Eq. (6); therefore, only numerical solutions are available.

The viscosity of a fluid may be associated with the resistance to flow, with the resistance arising from the intermolecular forces and internal friction as the molecules move past each other. Because of the much larger pressure variation in the lubricant conjunction, the viscosity of the lubricant for elastohydrodynamic lubrication does not remain constant as is approximately true for hydrodynamic lubrication.

As long ago as 1893, Barus[16] proposed the following formula for the isothermal viscosity–pressure dependence of liquids:

$$\eta = \eta_0 e^{\xi p} \tag{8}$$

where η_0 = viscosity at atmospheric pressure
ξ = pressure–viscosity coefficient of lubricant

The pressure–viscosity coefficient ξ characterizes the liquid considered and depends in most cases only on temperature, not on pressure.

Table 2 lists the absolute viscosities of 12 lubricants at atmospheric pressure and three temperatures as obtained from Jones et al.[17] These values would correspond to η_0 to be used in Eq. (8) for the particular fluid and temperature to be used. The 12 fluids with manufacturer and manufacturer's designation are shown in Table 3. The pressure–viscosity coefficients ξ, expressed in square meters per newton, for these 12 fluids at three different temperatures are shown in Table 4.

For a comparable change in pressure the relative density change is smaller than the viscosity change. However, very high pressures exist in elastohydrodynamic films, and the liquid can no longer be considered as an incompressible medium. From Dowson and Higginson[18] the density can be written as

$$\rho = \rho_0 \left(1 + \frac{0.6p}{1 + 1.7p}\right) \tag{9}$$

where p is given in gigapascals.

The film shape appearing in Eq. (7) can be written with sufficient accuracy as

$$h = h_0 + \frac{x^2}{2R_x} + \frac{y^2}{2R_y} + \delta(x,y) \tag{10}$$

where h_0 = constant, m
$\delta(x,y)$ = total elastic deformation, m
R_x = effective radius in x direction, m
R_y = effective radius in y direction, m

The elastic deformation can be written, from standard elasticity theory, in the form

Table 2 Absolute Viscosities of Test Fluids at Atmospheric Pressure and Three Temperatures (From Ref. 17)

	Temperature, °C		
	38	99	149
Test Fluid	Absolute Viscosity, η, cP		
Advanced ester	25.3	4.75	2.06
Formulated advanced ester	27.6	4.96	2.15
Polyalkyl aromatic	25.5	4.08	1.80
Polyalkyl aromatic + 10 wt % heavy resin	32.2	4.97	2.03
Synthetic paraffinic oil (lot 3)	414	34.3	10.9
Synthetic paraffinic oil (lot 4)	375	34.7	10.1
Synthetic paraffinic oil (lot 4) + antiwear additive	375	34.7	10.1
Synthetic paraffinic oil (lot 2) + antiwear additive	370	32.0	9.93
C-ether	29.5	4.67	2.20
Superrefined naphthenic mineral oil	68.1	6.86	2.74
Synthetic hydrocarbon (traction fluid)	34.3	3.53	1.62
Fluorinated polyether	181	20.2	6.68

Table 3 Fluids with Manufacturer and Manufacturer's Designation (From Ref. 17)

Test Fluid	Manufacturer	Designation
Advanced ester	Shell Oil Co.	Aeroshell® turbine oil 555 (base oil)
Formulated advanced ester	Shell Oil Co.	Aeroshell® turbine oil 555 (WRGL-358)
Polyalkyl aromatic	Continental Oil Co.	DN-600
Synthetic paraffinic oil (lot 3)	Mobil Oil Corp.	XRM 109F3
Synthetic paraffinic oil (lot 4)	↓	XRM 109F4
Synthetic paraffinic oil + antiwear additive (lot 2)	↓	XRM 177F2
Synthetic paraffinic oil + antiwear additive (lot 4)	↓	XRM 177F4
C-ether	Monsanto Co.	MCS-418
Superrefined naphthenic mineral oil	Humble Oil and Refining Co.	FN 2961
Synthetic hydrocarbon (traction fluid)	Monsanto Co.	MCS-460
Fluorinated polyether	DuPont Co.	PR 143 AB (Lot 10)

$$\delta(x,y) = \frac{2}{\pi E'} \iint_{\Lambda} \frac{p(x,y)\, dx_1\, dy_1}{[(x - x_1)^2 + (y - y_1)^2]^{1/2}} \tag{11}$$

where

$$E' = 2\left(\frac{1 - \nu_a^2}{E_a} + \frac{1 - \nu_b^2}{E_b}\right)^{-1} \tag{12}$$

Table 4 Pressure–Viscosity Coefficients for Test Fluids at Three Temperatures (From Ref. 17)

	Temperature, °C		
	38	99	149
Test Fluid	Pressure–Viscosity Coefficient, ξ, m^2/N		
Advanced ester	1.28×10^{-8}	0.987×10^{-8}	0.851×10^{-8}
Formulated advanced ester	1.37	1.00	0.874
Polyalkyl aromatic	1.58	1.25	1.01
Polyalkyl aromatic + 10 wt % heavy resin	1.70	1.28	1.06
Synthetic paraffinic oil (lot 3)	1.77	1.51	1.09
Synthetic paraffinic oil (lot 4)	1.99	1.51	1.29
Synthetic paraffinic oil (lot 4) + antiwear additive	1.96	1.55	1.25
Synthetic paraffinic oil (lot 2) + antiwear additive	1.81	1.37	1.13
C-ether	1.80	0.980	0.795
Superrefined naphthenic mineral oil	2.51	1.54	1.27
Synthetic hydrocarbon (traction fluid)	3.12	1.71	0.939
Fluorinated polyether	4.17	3.24	3.02

and ν = Poisson's ratio
E = modulus of elasticity, N/m^2

Therefore, Eq. (6) is normally involved in hydrodynamic lubrication situations, while Eqs. (7)–(11) are normally involved in elastohydrodynamic lubrication situations.

2 HYDRODYNAMIC AND HYDROSTATIC LUBRICATION

Surfaces lubricated hydrodynamically are normally conformal, as pointed out in Section 1.1. The conformal nature of the surfaces can take its form either as a thrust bearing or as a journal bearing, both of which will be considered in this section. Three features must exist for hydrodynamic lubrication to occur:

1. A viscous fluid must separate the lubricated surfaces.
2. There must be relative motion between the surfaces.
3. The geometry of the film shape must be larger in the inlet than at the outlet so that a convergent wedge of lubricant is formed.

If feature 2 is absent, lubrication can still be achieved by establishing relative motion between the fluid and the surfaces through external pressurization. This is discussed further in Section 2.3.

In hydrodynamic lubrication the entire friction arises from the shearing of the lubricant film so that it is determined by the viscosity of the oil: the thinner (or less viscous) the oil, the lower the friction. The great advantages of hydrodynamic lubrication are that the friction can be very low ($\mu \simeq 0.001$) and, in the ideal case, there is no wear of the moving parts. The main problems in hydrodynamic lubrication are associated with starting or stopping since the oil-film thickness theoretically is zero when the speed is zero.

The emphasis in this section is on hydrodynamic and hydrostatic lubrication. This section is not intended to be all inclusive but rather to typify the situations existing in hydrodynamic and hydrostatic lubrication. For additional information the reader is recommended to investigate Gross et al.,[19] Reiger,[20] Pinkus and Sternlicht,[21] and Rippel.[22]

2.1 Liquid-Lubricated Hydrodynamic Journal Bearings

Journal bearings, as shown in Fig. 8, are used to support shafts and to carry radial loads with minimum power loss and minimum wear. The bearing can be represented by a plain cylindrical bush wrapped around the shaft, but practical bearings can adopt a variety of forms. The lubricant is supplied at some convenient point through a hole or a groove. If the bearing extends around the full 360° of the shaft, the bearing is described as a full journal bearing. If the angle of wrap is less than 360°, the term "partial journal bearing" is employed.

Plain

Journal bearings rely on the motion of the shaft to generate the load-supporting pressures in the lubricant film. The shaft does not normally run concentric with the bearing center. The distance between the shaft center and the bearing center is known as the eccentricity. This eccentric position within the bearing clearance is influenced by the load that it carries. The amount of eccentricity adjusts itself until the load is balanced by the pressure generated in the converging portion of the bearing. The pressure generated, and therefore the load capacity of the bearing, depends on the shaft eccentricity e, the frequency of rotation N, and the

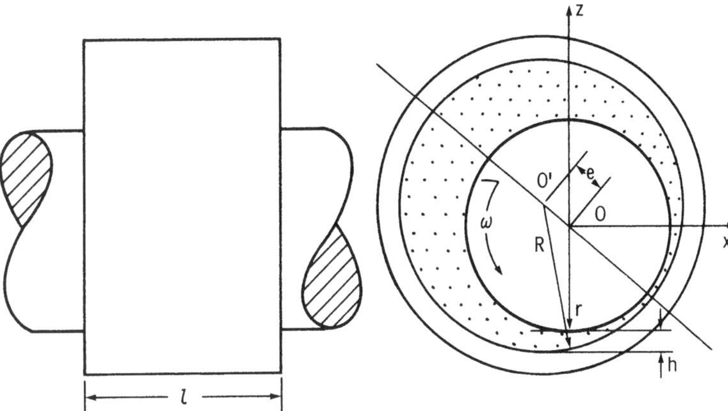

Figure 8 Journal bearing.

effective viscosity of the lubricant η in the converging film, as well as the bearing dimensions l and d and the clearance c. The three dimensionless groupings normally used for journal bearings are:

1. The eccentricity ratio, $\epsilon = e/c$
2. The length-to-diameter ratio, $\lambda = l/d$
3. The Sommerfeld number, $\text{Sm} = \eta N d^3 l / 2Fc^2$

When designing a journal bearing, the first requirement to be met is that it should operate with an adequate minimum film thickness, which is directly related to the eccentricity ($h_{\min} = c - e$). Figures 9, 10, and 11 show the eccentricity ratio, the dimensionless minimum film thickness, and the dimensionless Sommerfeld number for, respectively, a full journal bearing and partial journal bearings of 180° and 120°. In these figures a recommended op-

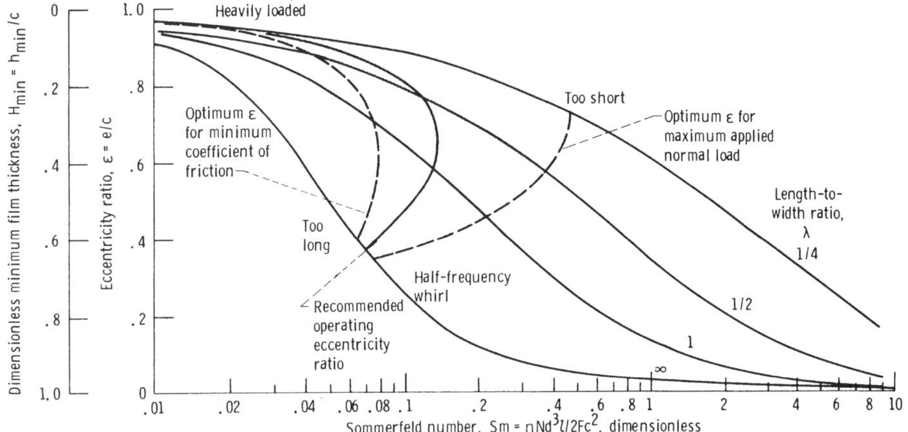

Figure 9 Design figure showing eccentricity ratio, dimensionless minimum film thickness, and Sommerfeld number for full journal bearings. (Adapted from Ref. 23.)

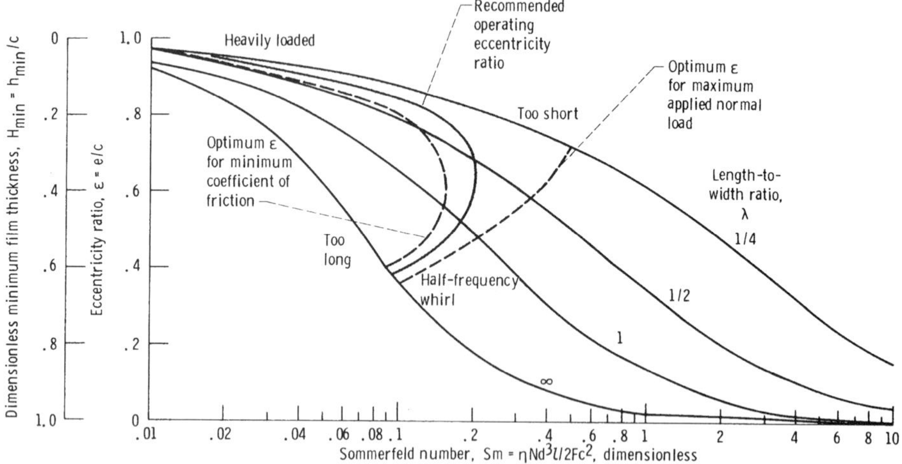

Figure 10 Design figure showing eccentricity ratio, dimensionless minimum film thickness, and Sommerfeld number for 180° partial journal bearings, centrally loaded. (Adapted from Ref. 23.)

erating eccentricity ratio is indicated as well as a preferred operational area. The left boundary of the shaded zone defines the optimum eccentricity ratio for minimum coefficient of friction, and the right boundary is the optimum eccentricity ratio for maximum load. In these figures it can be observed that the shaded area is significantly reduced for the partial bearings as compared with the full journal bearing. These plots were adapted from results given in Raimondi and Boyd.[23]

Figures 12, 13, and 14 show a plot of attitude angle ϕ (angle between the direction of the load and a line drawn through the centers of the bearing and the journal) and the bearing characteristic number for various length-to-diameter ratios for, respectively, a full journal bearing and partial journal bearings of 180° and 120°. This angle establishes where the

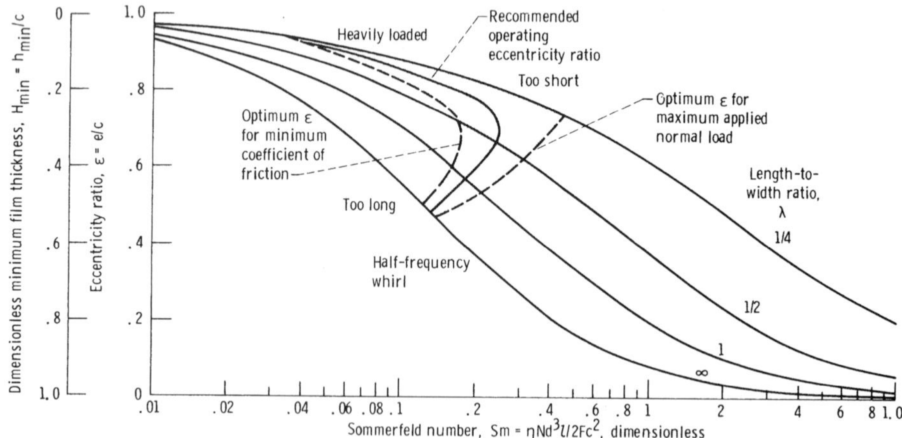

Figure 11 Design figure showing eccentricity ratio, dimensionless minimum film thickness, and Sommerfeld number for 120° partial journal bearings, centrally loaded. (Adapted from Ref. 23.)

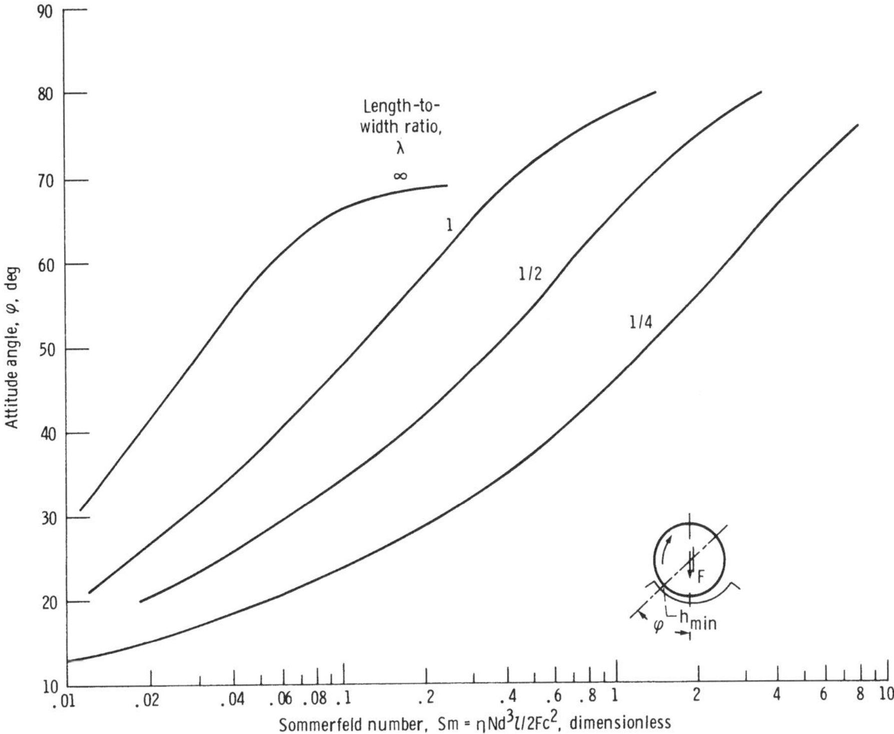

Figure 12 Design figure showing attitude angle (position of minimum film thickness) and Sommerfeld number for full journal bearings, centrally loaded. (Adapted from Ref. 23.)

minimum and maximum film thicknesses are located within the bearing. These plots were also adapted from results given in Raimondi and Boyd,[23] where additional information about the coefficient of friction, the flow variable, the temperature rise, and the maximum film pressure ratio for a complete range of length-to-diameter ratios as well as for full or partial journal bearings can be found.

Nonplain

As applications have demanded higher speeds, vibration problems due to critical speeds, imbalance, and instability have created a need for journal bearing geometries other than plain journal bearings. These geometries have various patterns of variable clearance so as to create pad film thicknesses that have more strongly converging and diverging regions. Figure 15 shows elliptical, offset half, three-lobe, and four-lobe bearings—bearings different from the plain journal bearing. An excellent discussion of the performance of these bearings is provided in Allaire and Flack,[24] and some of their conclusions are presented here. In Fig. 15, each pad is moved in toward the center of the bearing some fraction of the pad clearance in order to make the fluid-film thickness more converging and diverging than that which occurs in a plain journal bearing. The pad center of curvature is indicated by a cross. Generally, these bearings give good suppression of instabilities in the system but can be subject to subsynchronous vibration at high speeds. Accurate manufacturing of these bearings is not always easy to obtain.

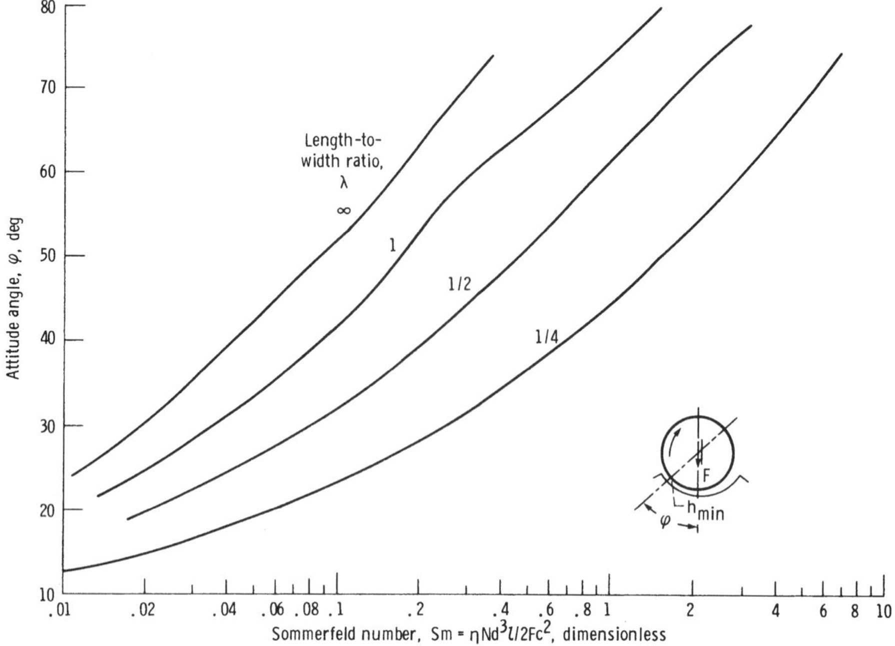

Figure 13 Design figure showing attitude angle (position of minimum film thickness) and Sommerfeld number for 180° partial journal bearings, centrally loaded. (Adapted from Ref. 23.)

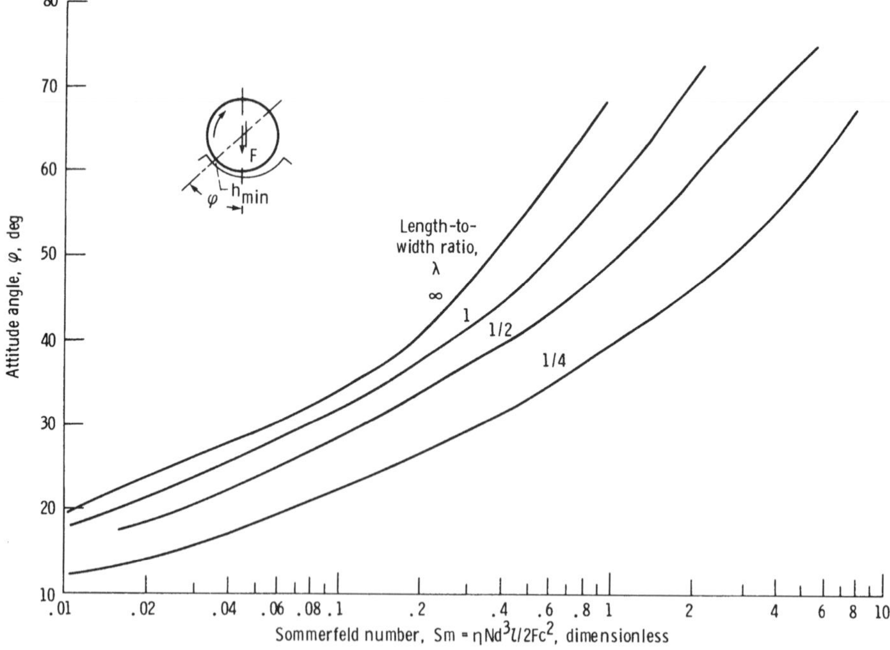

Figure 14 Design figure showing attitude angle (position of minimum film thickness) and Sommerfeld number for 120° partial journal bearings, centrally loaded. (Adapted from Ref. 23.)

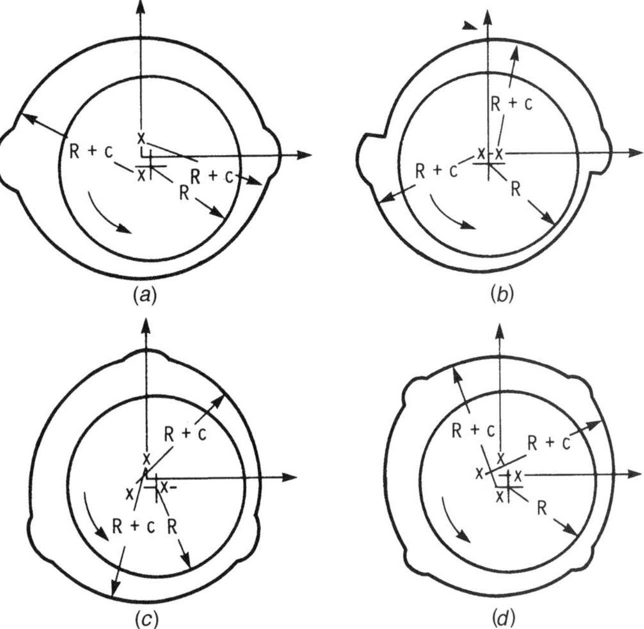

Figure 15 Types of fixed-incline pad preloaded journal bearings. (From Ref. 24.) (*a*) Elliptical bore bearing ($\alpha_a = 0.5$, $m_p = 0.4$). (*b*) Offset half bearing ($\alpha_a = 1.125$, $m_p = 0.4$). (*c*) Three-lobe bearing ($\alpha_a = 0.5$, $m_p = 0.4$). (*d*) Four-lobe bearing ($\alpha_a = 0.5$, $m_p = 0.4$).

A key parameter used in describing these bearings is the fraction of length in which the film thickness is converging to the full pad length, called the offset factor and defined as

$$\alpha_a = \frac{\text{length of pad with converging film thickness}}{\text{full pad length}}$$

The elliptical bearing, shown in Fig. 15, indicates that the two pad centers of curvature are moved along the y axis. This creates a pad with one-half of the film shape converging and the other half diverging (if the shaft were centered), corresponding to an offset factor $\alpha_a = 0.5$. The offset half bearing in Fig. 15*b* consists of a two-axial-groove bearing that is split by moving the top half horizontally. This results in low vertical stiffness.

Generally, the vibration characteristics of this bearing are such as to avoid the previously mentioned oil whirl, which can drive a machine unstable. The offset half bearing has a purely converging film thickness with a converged pad arc length of 160° and the point opposite the center of curvature at 180°. Both the three-lobe and four-lobe bearings shown in Figs. 15*c* and 15*d* have an offset factor of $\alpha_a = 0.5$.

The fractional reduction of the film clearance when the pads are brought in is called the preload factor m_p. Let the bearing clearance at the pad minimum film thickness (with the shaft center) be denoted by c_b. Figure 16*a* shows that the largest shaft that can be placed in the bearing has a radius $R + c_b$, thereby establishing the definition of c_b. The preload factor m_p is given by

$$m_p = \frac{c - c_b}{c}$$

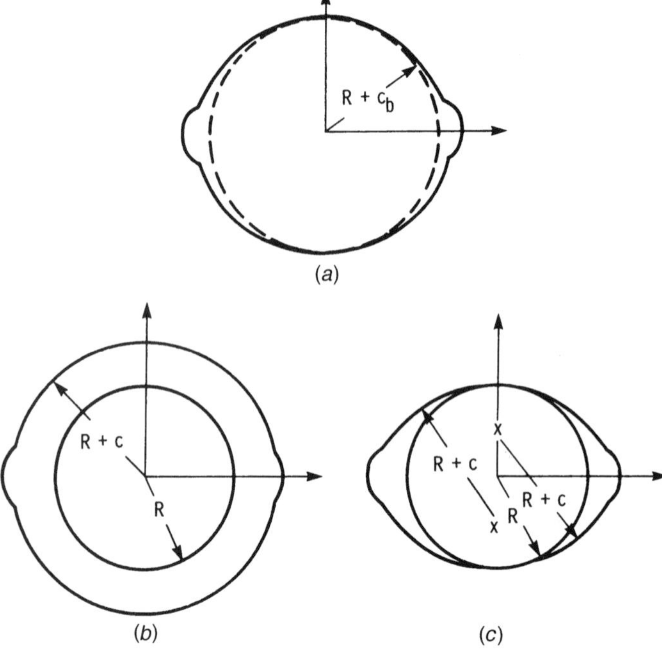

Figure 16 Effect of preload on two-lobe bearings. (From Ref. 24.) (*a*) Largest shaft that fits in bearing. (*b*) $m = 0$, largest shaft $= R + c$, bearing clearance $c_b = (c)$. (*c*) $m = 1.0$, largest shaft $= R$, bearing clearance $c_b = 0$.

A preload factor of zero corresponds to having all of the pad centers of curvature coinciding at the center of the bearing; a preload factor of 1.0 corresponds to having all of the pads touching the shaft. Figures 16*b* and 16*c* illustrate these extreme situations. Values of the preload factor are indicated in the various types of fixed journal bearings shown in Fig. 15.

Figure 17 shows the variation of the whirl ratio with Sommerfeld number at the threshold of instability for the four bearing types shown in Fig. 15. It is evident that a definite relationship exists between the stability and whirl ratio such that the more stable bearing distinctly whirls at a lower speed ratio. With the exception of the elliptical bearing, all bearings whirl at speeds less than 0.48 of the rotor speed. The offset bearing attains a maximum whirl ratio of 0.44 at a Sommerfeld number of about 0.4 and decreases to a steady value of 0.35 at higher Sommerfeld numbers. This observation corresponds to the superior stability with the offset bearing at high-speed and light-load operations.

The whirl ratios with the three-lobe and four-lobe bearings share similar characteristics. They both rise sharply at low Sommerfeld numbers and remain fairly constant for most portions of the curves. Asymptotic whirl ratios of 0.47 and 0.48, respectively, are reached at high Sommerfeld numbers. In comparison with the four-lobe bearing, the three-lobe bearing always has the lower whirl ratio.

The elliptical bearing is the least desirable for large Sommerfeld numbers. At $Sm > 1.3$ the ratio exceeds 0.5.

2.2 Liquid-Lubricated Hydrodynamic Thrust Bearings

In a thrust bearing, a thrust plate attached to or forming part of the rotating shaft is separated from the sector-shaped bearing pads by a film of lubricant. The load capacity of the bearing

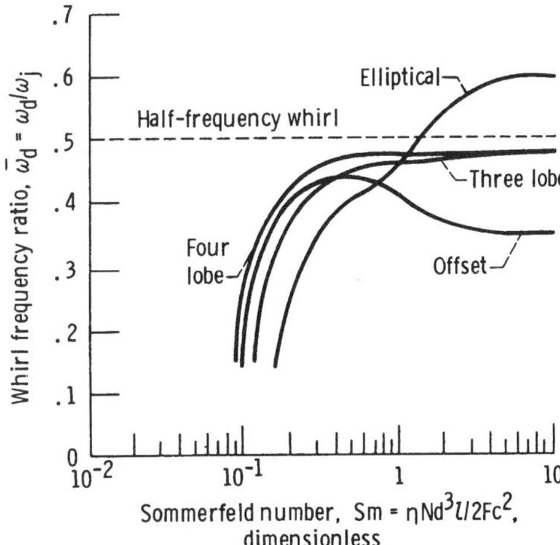

Figure 17 Chart for determining whirl frequency ratio. (From Ref. 24.)

arises entirely from the pressure generated by the motion of the thrust plate over the bearing pads. This action is achieved only if the clearance space between the stationary and moving components is convergent in the direction of motion. The pressure generated in, and therefore the load capacity of, the bearing depends on the velocity of the moving slider $u = (R_1 + R_2)\omega/2 = \pi(R_1 + R_2)N$, the effective viscosity, the length of the pad l, the width of the pad b, the normal applied load F, the inlet film thickness h_i, and the outlet film thickness h_0. For thrust bearings three dimensionless parameters are used:

1. $\lambda = l/b$, pad length-to-width ratio
2. $\mathrm{Sm}_t = \eta u b l^2/F h_o^2$, Sommerfeld number for thrust bearings
3. $\bar{h}_i = h_i/h_o$, film thickness ratio

It is important to recognize that the total thrust load F is equal to nF, where n is the number of pads in a thrust bearing. In this section three different thrust bearings will be investigated. Two fixed-pad types, a fixed incline and a step sector, and a pivoted-pad type will be discussed.

Fixed-Incline Pad

The simplest form of fixed-pad thrust bearing provides only straight-line motion and consists of a flat surface sliding over a fixed pad or land having a profile similar to that shown in Fig. 18. The fixed-pad bearing depends for its operation on the lubricant being drawn into a wedge-shaped space and thus producing pressure that counteracts the load and prevents contact between the sliding parts. Since the wedge action only takes place when the sliding surface moves in the direction in which the lubricant film converges, the fixed-incline bearing, shown in Fig. 18, can only carry load for this direction of operation. If reversibility is desired, a combination of two or more pads with their surfaces sloped in opposite direction is required. Fixed-incline pads are used in multiples as in the thrust bearing shown in Fig. 19.

The following procedure assists in the design of a fixed-incline pad thrust bearing:

1. Choose a pad width-to-length ratio. A square pad ($\lambda = 1$) is generally felt to give good performance. From Fig. 20, if it is known whether maximum load or minimum

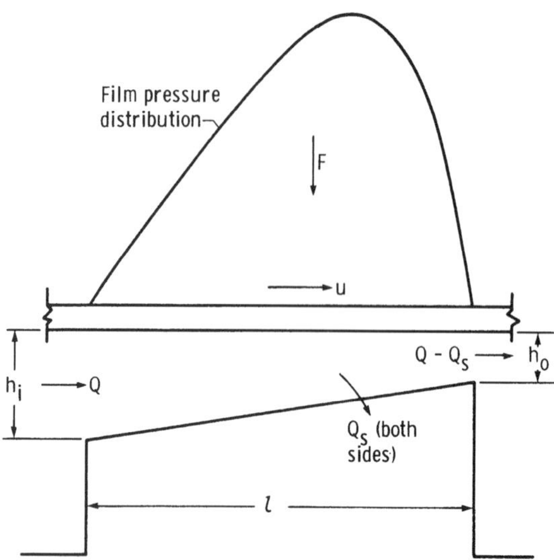

Figure 18 Configuration of fixed-incline pad bearing. (From Ref. 25. Reprinted by permission of ASME.)

power is most important in the particular application, a value of the film thickness ratio can be determined.

2. Within the terms in the Sommerfeld number the term least likely to be preassigned is the outlet film thickness. Therefore, determine h_o from Fig. 21. Since $\overline{h}_i$ is known from Fig. 20, h_i can be determined ($h_i = \overline{h}_i h_o$).
3. Check Table 5 to see if minimum (outlet) film thickness is sufficient for the preassigned surface finish. If not:
   a. Increase the fluid viscosity or speed of the bearing.
   b. Decrease the load or the surface finish. Upon making this change return to step 1.
4. Once an adequate minimum film thickness has been determined, use Figs. 22–24 to obtain, respectively, the coefficient of friction, the power consumed, and the flow.

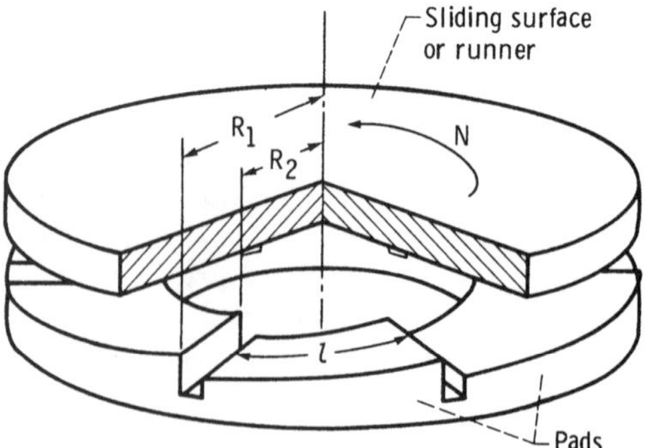

Figure 19 Configuration of fixed-incline pad thrust bearing. (From Ref. 25.)

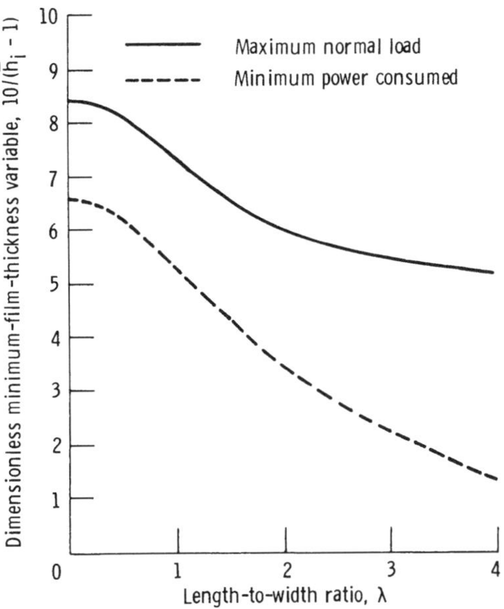

Figure 20 Chart for determining minimum film thickness corresponding to maximum load or minimum power for various pad proportions—fixed-incline pad bearings. (From Ref. 25. Reprinted by permission of ASME.)

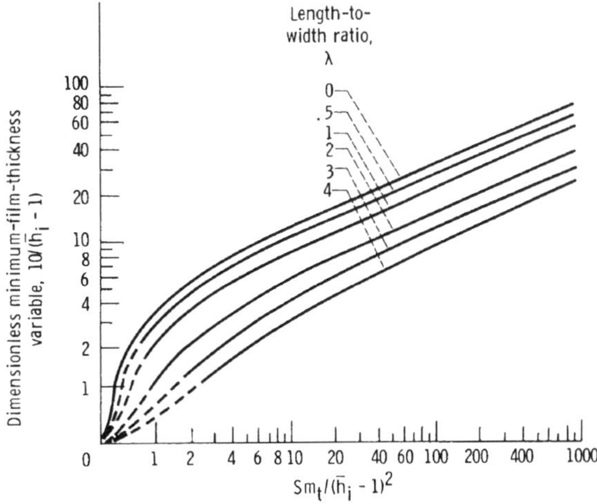

Figure 21 Chart for determining minimum film thickness for fixed-incline pad thrust bearings. (From Ref. 25. Reprinted by permission of ASME.)

Table 5 Allowable Minimum Outlet Film Thickness for a Given Surface Finish (From Ref. 8)

Surface Finish[a]					Allowable Minimum Outlet Film Thickness,[b] h_0	
Familiar British Units, μin., CLA	SI Units, μin., CLA	Description of Surface	Examples of Manufacturing Methods	Approximate Relative Costs	Familiar British Units, in.	SI Units, m
4–8	0.1–0.2	Mirrorlike surface without toolmarks, close tolerances	Grind, lap, and superfinish	17–20	0.00010	0.0000025
8–16	0.2–0.4	Smooth surface without scratches, close tolerances	Grind and lap	17–20	0.00025	0.0000062
16–32	0.4–0.8	Smooth surface, close tolerances	Grind, file, and lap	10	0.00050	0.0000125
32–63	0.8–1.6	Accurate bearing surface without toolmarks	Grind, precision mill, and file	7	0.00100	0.000025
63–125	1.6–3.2	Smooth surface without objectionable toolmarks, moderate tolerances	Shape, mill, grind, and turn	5	0.00200	0.000050

[a] CLA = centerline average. μm = micrometer; 40 μin. (microinch) = 1 μm.

[b] The values of film thickness are given only for guidance. They indicate the film thickness required to avoid metal-to-metal contact under clean-oil conditions with no misalignment. It may be necessary to take a larger film thickness than that indicated (e.g., to obtain an acceptable temperature rise). It has been assumed that the average surface finish of the pads is the same as that of the runner.

Pivoted Pad

The simplest form of pivoted-pad bearing provides only for straight-line motion and consists of a flat surface sliding over a pivoted pad as shown in Fig. 25. If the pad is assumed to be in equilibrium under a given set of operating conditions, any change in these conditions, such as a change in load, speed, or viscosity, will alter the pressure distribution and thus momentarily shift the center of pressure and create a moment that causes the pad to change its inclination until a new position of equilibrium is established. It can be shown that if the position of that pivot, as defined by the distance $\bar{x}$, is fixed by choosing $\bar{x}/l$, the ratio of the inlet film thickness to the outlet film thickness, h_i/h_o, also becomes fixed and is independent of load, speed, and viscosity. Thus the pad will automatically alter its inclination so as to maintain a constant value of h_i/h_o.

Pivoted pads are sometimes used in multiples as pivoted-pad thrust bearings, shown in Fig. 26. Calculations are carried through for a single pad, and the properties for the complete bearing are found by combining these calculations in the proper manner.

Normally, a pivoted pad will only carry load if the pivot is placed somewhere between the center of the pad and the outlet edge ($0.5 < \bar{x}/l \leq 1.0$). With the pivot so placed, the pad therefore can only carry load for one direction of rotation.

The following procedure helps in the design of pivoted-pad thrust bearings:

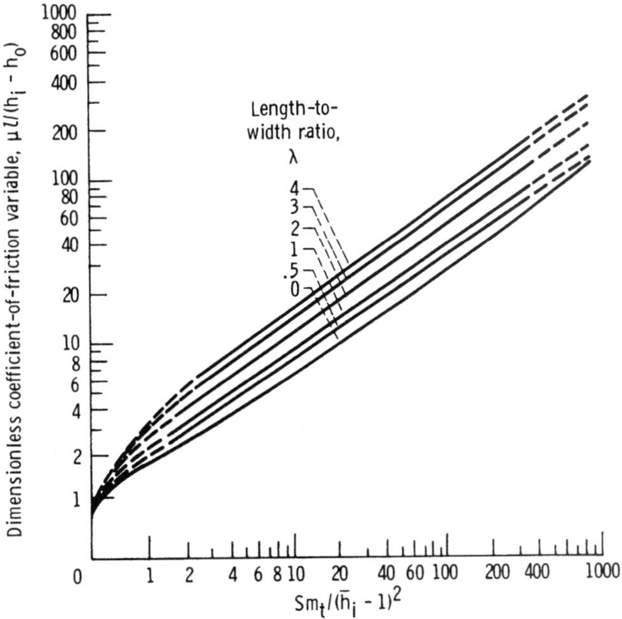

Figure 22 Chart for determining coefficient of friction for fixed-incline pad thrust bearings. (From Ref. 25. Reprinted by permission of ASME.)

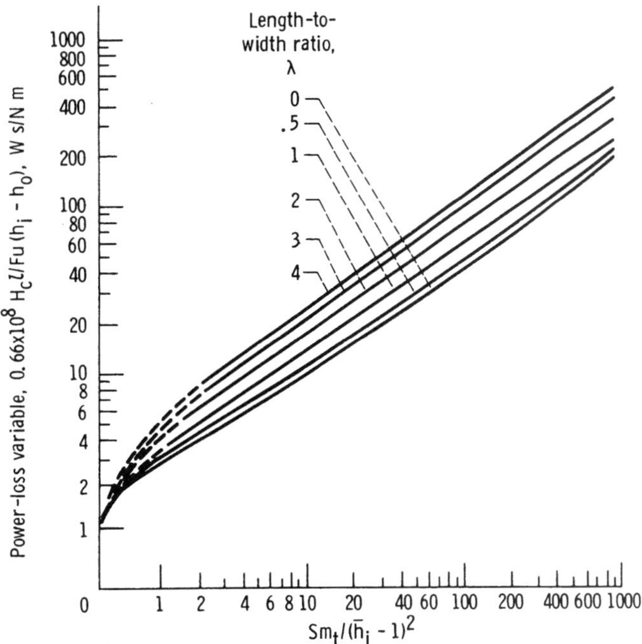

Figure 23 Chart for determining power loss for fixed-incline pad thrust bearings. (From Ref. 25. Reprinted by permission of ASME.)

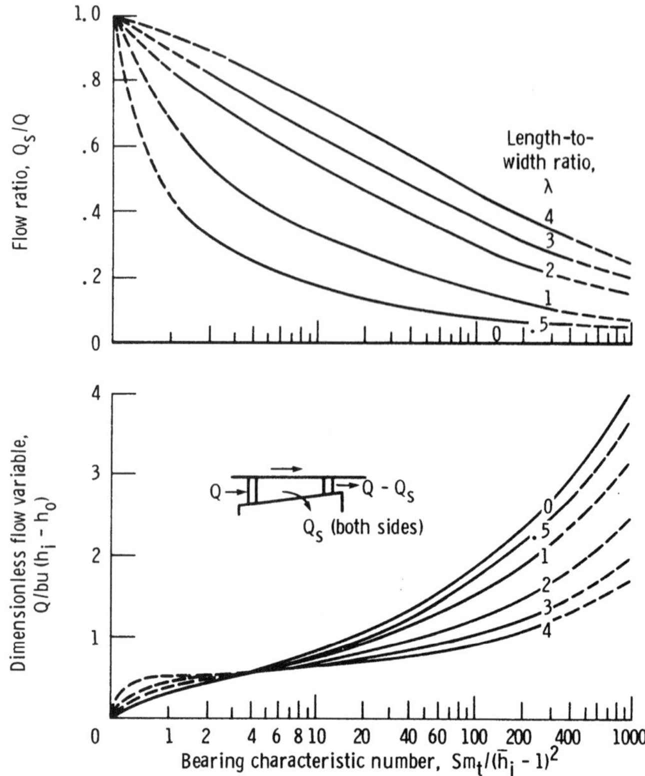

Figure 24 Charts for determining lubricant flow for fixed-incline pad thrust bearings. (From Ref. 25. Reprinted by permission of ASME.)

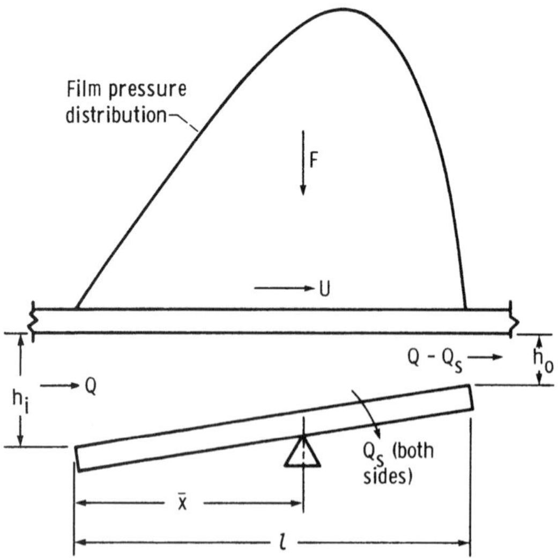

Figure 25 Configuration of pivoted-pad bearings. (From Ref. 25. Reprinted by permission of ASME.)

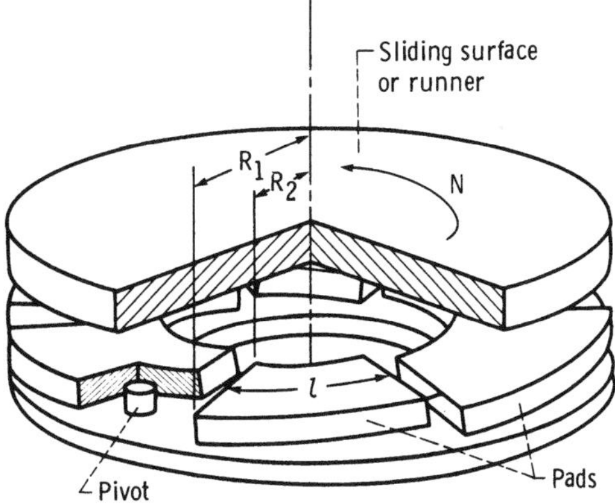

Figure 26 Configuration of pivoted-pad thrust bearings. (From Ref. 25. Reprinted by permission of ASME.)

1. Having established if minimum power or maximum load is more critical in the particular application and chosen a pad length-to-width ratio, establish the pivot position from Fig. 27.
2. In the Sommerfeld number for thrust bearings the unknown parameter is usually the outlet or minimum film thickness. Therefore, establish the value of h_o from Fig. 28.
3. Check Table 5 to see if the outlet film thickness is sufficient for the preassigned surface finish. If sufficient, go on to step 4. If not, consider:
   a. Increasing the fluid viscosity
   b. Increasing the speed of the bearing
   c. Decreasing the load of the bearing
   d. Decreasing the surface finish of the bearing lubrication surfaces

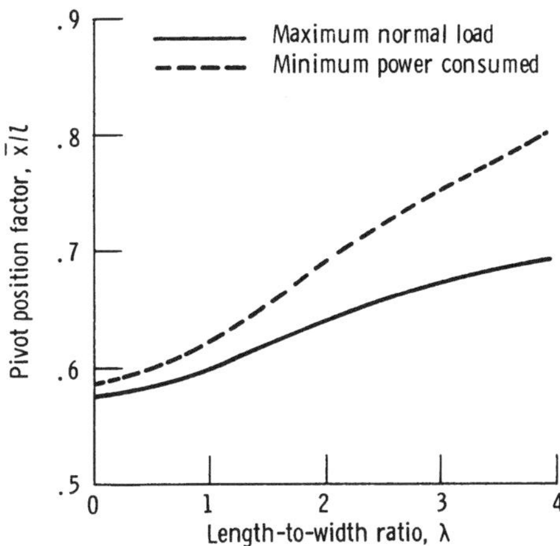

Figure 27 Chart for determining pivot position corresponding to maximum load or minimum power loss for various pad proportions—pivoted-pad bearings. (From Ref. 25. Reprinted by permission of ASME.)

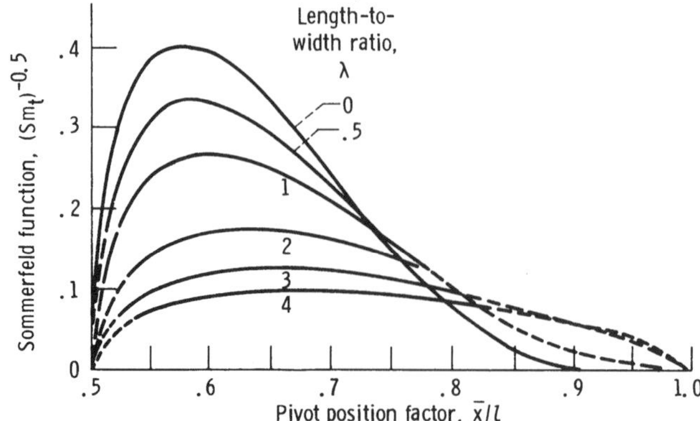

Figure 28 Chart for determining outlet film thickness for pivoted-pad thrust bearings. (From Ref. 25. Reprinted by permission of ASME.)

Upon making this change return to step 1.

4. Once an adequate outlet film thickness is established, determine the film thickness ratio, power loss, coefficient of friction, and flow from Figs. 29–32.

Step Sector

The configuration of a step-sector thrust bearing is shown in Fig. 33. The parameters used to define the dimensionless load and stiffness are:

1. $\bar{h}_i = h_i/h_o$, film thickness ratio
2. $\bar{\theta} = \theta_i/(\theta_i + \theta_o)$, dimensionless step location
3. n, number of sectors

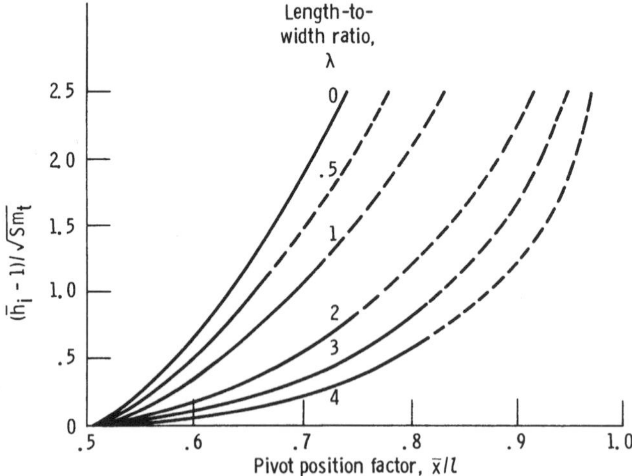

Figure 29 Chart for determining film thickness ratio $\bar{h}_i$ for pivoted-pad thrust bearings. (From Ref. 25. Reprinted by permission of ASME.)

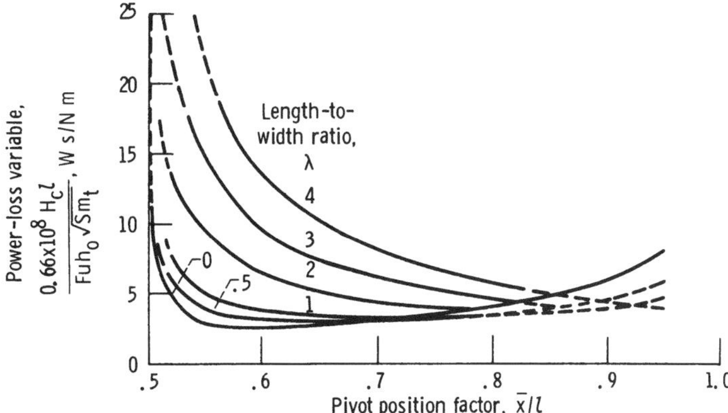

Figure 30 Chart for determining power loss for pivoted-pad thrust bearings. (From Ref. 25. Reprinted by permission of ASME.)

4. $\alpha_r = R_2R_1$, radius ratio

5. θ_g, angular extent of lubrication feed groove

Note that the first four parameters are dimensionless and the fifth is dimensional and expressed in radians.

The optimum parallel step-sector bearing for *maximum load capacity* for a given α_r and θ_g is

$$\bar{\theta}_{\text{opt}} = 0.558 \qquad (\bar{h}_i)_{\text{opt}} = 1.668 \qquad n_{\text{opt}} = \frac{2\pi}{\theta_g + 2.24(1 - \alpha_r)/(1 + \alpha_r)}$$

where n_{opt} is rounded off to the nearest integer and its minimum value is 3. For *maximum stiffness,* results are identical to the above with the exception that $(\bar{h}_i)_{\text{opt}} = 1.467$. These results are obtained from Hamrock.[26]

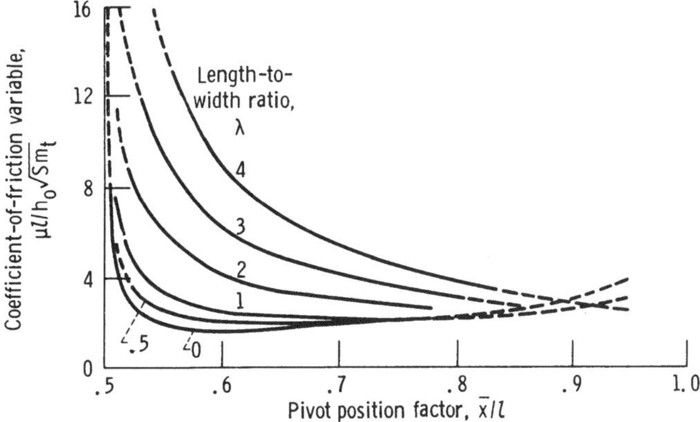

Figure 31 Chart for determining coefficient of friction for pivot-pad thrust bearings. (From Ref. 25. Reprinted by permission of ASME.)

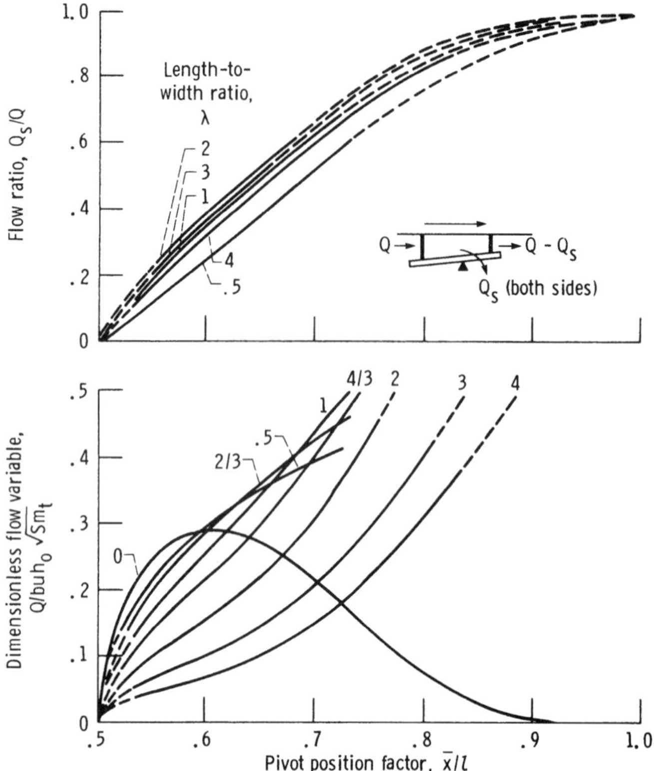

Figure 32 Chart for determining lubricant flow for pivot-pad thrust bearings. (From Ref. 25. Reprinted by permission of ASME.)

2.3 Hydrostatic Bearings

In Sections 2.1 and 2.2 the load-supporting fluid pressure is generated by relative motion between the bearing surfaces. Thus its load capacity depends on the relative speeds of the surfaces. When the relative speeds of the bearing are low or the loads are high, the liquid-lubricated journal and thrust bearings may not be adequate. If full-film lubrication with no metal-to-metal contact is desired under such conditions, another technique, called hydrostatic or externally pressurized lubrication, may be used.

The one salient feature that distinguishes hydrostatic from hydrodynamic bearings is that the fluid is pressurized externally to the bearings and the pressure drop across the bearing is used to support the load. The load capacity is independent of the motion of bearing surfaces or the fluid viscosity. There is no problem of contact of the surfaces at starting and stopping as with conventional hydrodynamically lubricated bearings because pressure is applied before starting and maintained until after stopping. Hydrostatic bearings can be very useful under conditions of little or no relative motion and under extreme conditions of temperature or corrosivity, where it may be necessary to use bearing materials with poor boundary lubricating properties. Surface contact can be avoided completely, so material properties are much less important than in hydrodynamic bearings. The load capacity of a hydrostatic bearing is proportional to the available pressure.

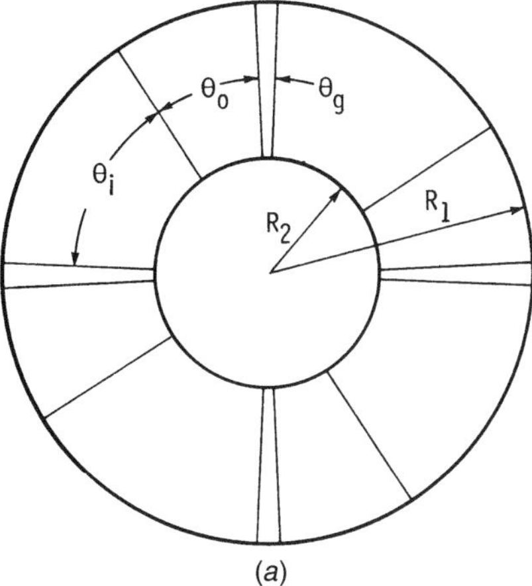

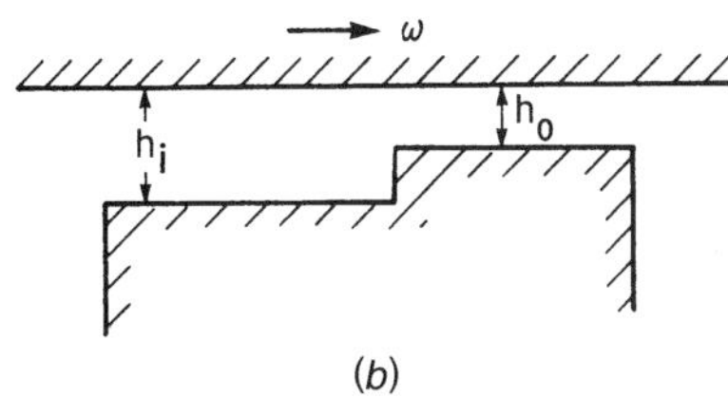

Figure 33 Configuration of step-sector thrust bearing. (From Ref. 26.) (*a*) Top view. (*b*) Section through a sector.

Hydrostatic bearings do, however, require an external source of pressurization such as a pump. This represents an additional system complication and cost.

The chief advantage of hydrostatic bearings is their ability to support extremely heavy loads at slow speeds with a minimum of driving force. For this reason they have been successfully applied in rolling mills, machine tools, radio and optical telescopes, large radar antennas, and other heavily loaded, slowly moving equipment.

The formation of a fluid film in a hydrostatic bearing system is shown in Fig. 34. A simple bearing system with the pressure source at zero pressure is shown in Fig. 34*a*. The runner under the influence of a load F is seated on the bearing pad. As the source pressure builds up, Fig. 34*b*, the pressure in the pad recess also increases. The pressure in the recess is built up to a point, Fig. 34*c*, where the pressure on the runner over an area equal to the pad recess area is just sufficient to lift the load. This is commonly called the lift pressure. Just after the runner separates from the bearing pad, Fig. 34*d*, the pressure in the recess is less than that required to lift the bearing runner ($p_r < p_l$). After lift, flow commences through the system. Therefore, a pressure drop exists between the pressure source and the bearing (across the restrictor) and from the recess to the exit of the bearing.

If more load is added to the bearing, Fig. 34*e*, the film thickness will decrease and the recess pressure will rise until pressure within the bearing clearance and the recess is sufficient to carry the increased load. If the load is now decreased to less than the original, Fig. 34*f*, the film thickness will increase to some higher value and the recess pressure will decrease accordingly. The maximum load that can be supported by the pad will be reached, theoretically, when the pressure in the recess is equal to the pressure at the source. If a load greater than this is applied, the bearing will seat and remain seated until the load is reduced and can again be supported by the supply pressure.

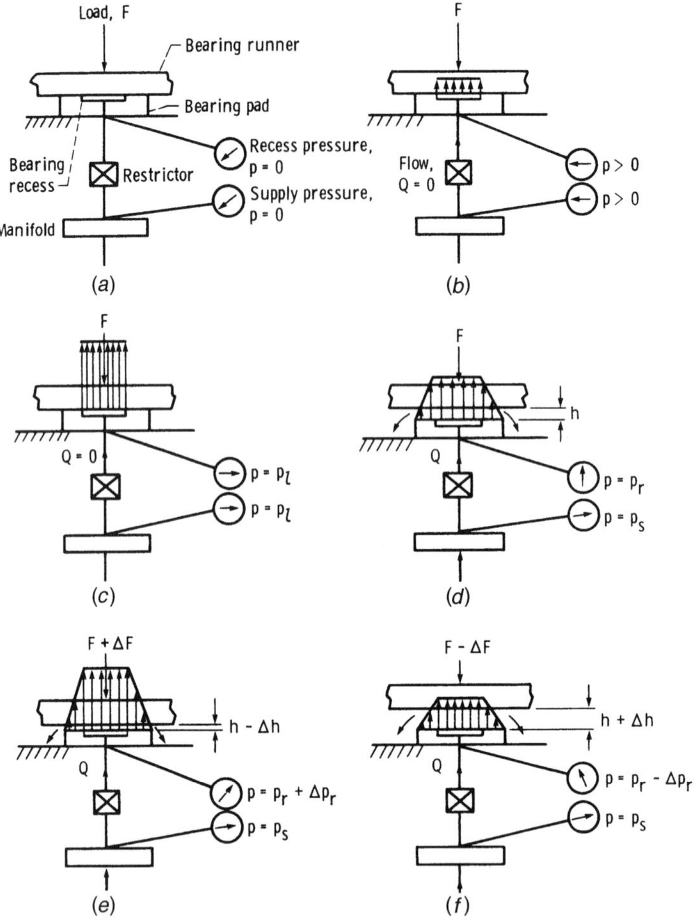

Figure 34 Formation of fluid film in hydrostatic bearing system. (From Ref. 22.) (*a*) Pump off. (*b*) Pressure building up. (*c*) Pressure × recess area = *F*. (*d*) Bearing operating. (*e*) Increased load. (*f*) Decreased load.

Pad Coefficients

To find the load-carrying capacity and flow requirements of any given hydrostatic bearing pad, it is necessary to determine certain pad coefficients. Since the selection of pad and recess geometries is up to the designer, the major design problem is the determination of particular bearing coefficients for particular geometries.

The load-carrying capacity of a bearing pad, regardless of its shape or size, can be expressed as

$$F = a_f A_p p_r \tag{13}$$

where a_f = bearing-pad load coefficients
A_p = total projected pad area, m^2
p_r = recess pressure, N/m^2

The amount of lubricant flow across a pad and through the bearing clearance is

$$Q = q_f \frac{F}{A_p} \frac{h^3}{\eta} \tag{14}$$

where q_f = pad flow coefficient
h = film thickness, m
η = lubricant absolute viscosity, N·s/m^2

The pumping power required by the hydrostatic pad can be evaluated by determining the product of recess pressure and flow:

$$H_b = p_r Q = H_f \left(\frac{F}{A_p}\right)^2 \frac{h^3}{\eta} \tag{15}$$

where $H_f = q_f / a_f$ is the bearing-pad power coefficient. Therefore, in designing hydrostatic bearings the designer is primarily concerned with the bearing coefficients (a_f, q_f, and H_f) expressed in Eqs. (13)–(15).

Bearing coefficients are dimensionless quantities that relate performance characteristics of load, flow, and power to physical parameters. The bearing coefficients for two types of bearing pads will be considered, both of which exhibit pure radial flow and are flat, thrust-loaded types of bearings. For other types of hydrostatic bearings the reader is referred to Rippel.[22]

Circular Step Bearing Pad. The bearing coefficients for this type of pad are expressed as

$$a_f = \frac{1}{2}\left[\frac{1 - (R_o/R)^2}{\log_e(R/R_o)}\right] \tag{16}$$

$$q_f = \frac{\pi}{3}\left[\frac{1}{1 - (R_o/R)^2}\right] \tag{17}$$

$$H_f = \frac{2\pi \log_e(R/R_o)}{3[1 - (R_o/R)^2]^2} \tag{18}$$

For this type of pad the total projected bearing pad area A_p is equal to πR^2.

Figure 35 shows the three bearing-pad coefficients for various ratios of recess radius to bearing radius for a circular step thrust bearing. The bearing-pad load coefficient a_f varies from zero for extremely small recesses to unity for bearings having large recesses with respect to pad dimensions. In a sense, a_f is a measure of how efficiently the bearing uses the recess pressure to support the applied load.

In Fig. 35 we see that the pad flow coefficient q_f varies from unity for pads with relatively small recesses to a value approaching infinity for bearings with extremely large recesses. Physically, as the recess becomes larger with respect to the bearing, the hydraulic resistance to fluid flow decreases, and thus flow increases.

From Fig. 35, the power coefficient H_f approaches infinity for very small recesses, decreases to a minimum value as the recess size increases, and approaches infinity again for very large recesses. For this particular bearing the minimum value of H_f occurs at a ratio of recess radius to bearing radius R_o/R of 0.53. All bearing-pad configurations exhibit minimum values of H_f when their ratios of recess length to bearing length are approximately 0.4–0.6.

Annular Thrust Bearing. Figure 36 shows an annular thrust pad bearing. In this bearing the lubricant flows from the annular recess over the inner and outer sills. For this type of bearing the pad coefficients are

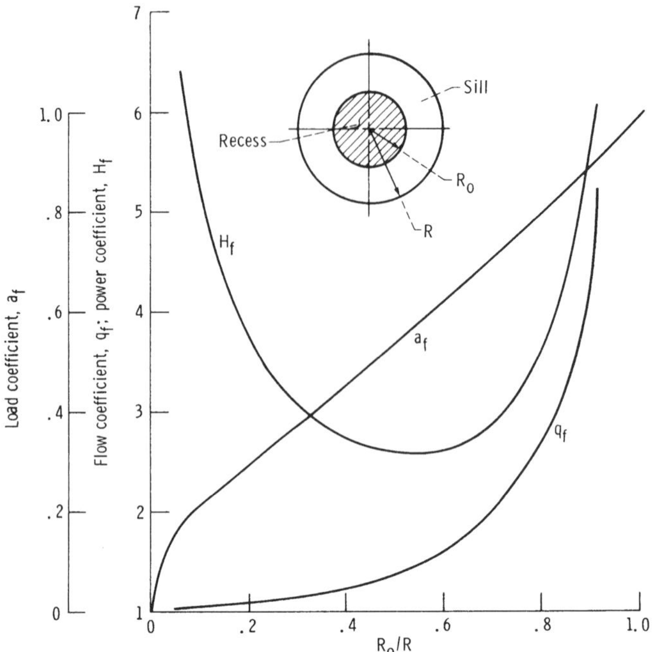

Figure 35 Chart for determining bearing pad coefficients for circular step thrust bearing. (From Ref. 22.)

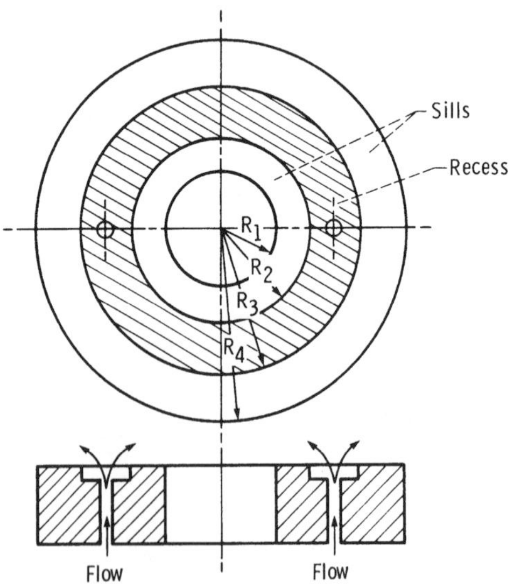

Figure 36 Configuration of annular thrust pad bearing. (From Ref. 22.)

$$a_f = \frac{1}{2(R_4^2 - R_1^2)} \left[\frac{R_4^2 - R_3^2}{\log_e(R_4/R_3)} - \frac{R_2^2 - R_1^2}{\log_e(R_2/R_1)} \right] \tag{19}$$

$$q_f = \frac{\pi}{6q_f} \left[\frac{1}{\log_e(R_4/R_3)} - \frac{1}{\log_e(R_2/R_1)} \right] \tag{20}$$

$$H_f = \frac{q_f}{a_f} \tag{21}$$

For this bearing the total projected bearing-pad area is

$$A_p = \pi(R_4^2 - R_1^2) \tag{22}$$

Figure 37 shows the bearing-pad load coefficient for an annular thrust pad bearing as obtained from Eqs. (19)–(21). For this figure it is assumed that the annular recess is centrally located within the bearing width; this therefore implies that $R_1 + R_4 = R_2 + R_3$. The curve for a_f applies for all R_1/R_4 ratios.

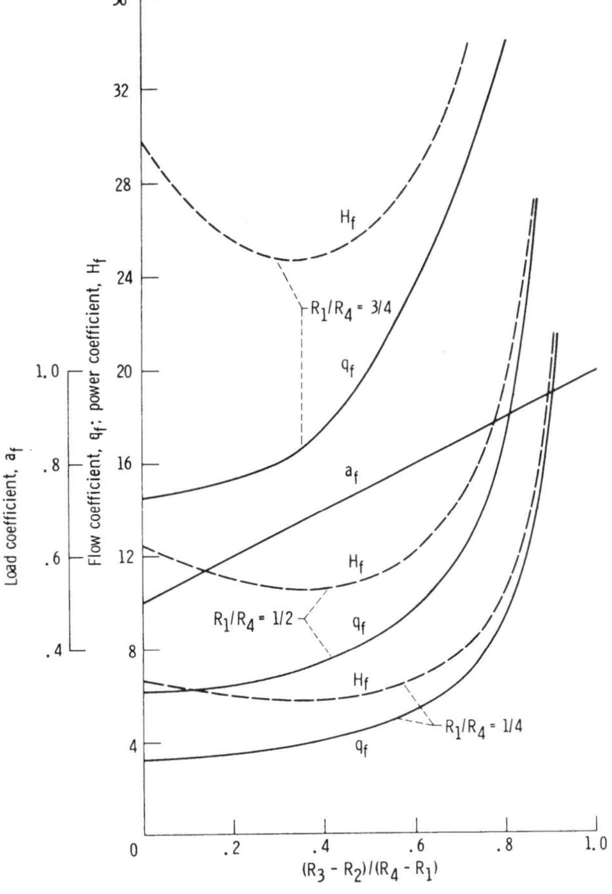

Figure 37 Chart for determining bearing pad coefficients for annular thrust pad bearings. (From Ref. 22.)

The hydrostatic bearings considered in this section have been limited to flat thrust-loaded bearings. Design information about other pad configurations can be obtained from Rippel.[22] The approach used for the simple, flat, thrust-loaded pad configuration is helpful in considering the more complex geometries covered by Rippel.[22]

Compensating Elements

As compared with common bearing types, hydrostatic bearings are relatively complex systems. In addition to the bearing pad, the system includes a pump and compensating elements. Three common types of compensating elements for hydrostatic bearings are the capillary tube, the sharp-edged orifice, and constant-flow valve.

Capillary Compensation. Figure 38 shows a capillary-compensated hydrostatic bearing as obtained from Rippel.[22] The small diameter of the capillary tube provides restriction and resultant pressure drop in the bearing pad. The characteristic feature of capillary compensation is a long tube or a relatively small diameter ($l_c > 20d_c$). The laminar flow of fluid through such a tube, while neglecting entrance and exit effects and viscosity changes due to temperature and pressure effects, can be expressed as

$$Q_c = \frac{k_c(p_s - p_r)}{\eta} \tag{23}$$

where

$$k_c = \frac{\pi d_c^4}{128 l_c} \tag{24}$$

For a given capillary tube, k_c is a constant expressed in cubic meters. Thus, from Eq. (23) the flow through a capillary tube is related linearly to the pressure drop across it. In a hydrostatic bearing with capillary compensation and a fixed supply pressure, the flow through the bearing will decrease with increasing load since the pocket pressure p_r is proportional to the load. To satisfy the assumption of laminar flow, the Reynolds number must be less than 2000 when expressed as

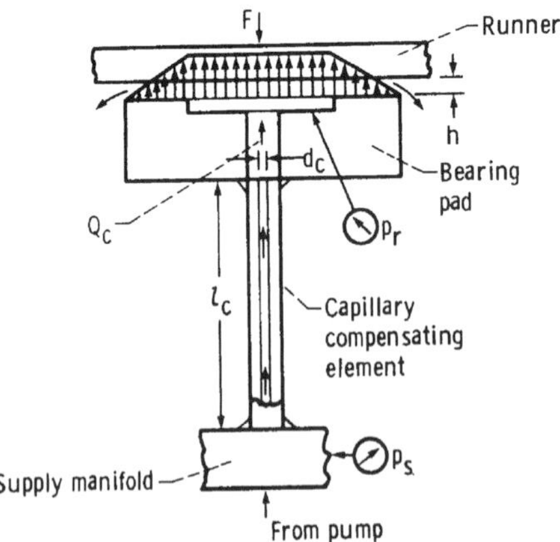

Figure 38 Capillary-compensated hydrostatic bearing. (From Ref. 22.)

$$N_R = \frac{4\rho Q_c}{\pi d_c \eta} < 2000 \tag{25}$$

where ρ is the mass density of the lubricant in $N \cdot s^2/m^4$. Hypodermic needle tubing serves quite well as capillary tubing for hydrostatic bearings. Although very small diameter tubing is available, diameters less than 6×10^{-4} m should not be used because of their tendency to clog.

Orifice Compensation. Orifice compensation is illustrated in Fig. 39. The flow of an incompressible fluid through a sharp-edged orifice can be expressed as

$$Q_o = k_o(p_s - p_r)^{1/2} \tag{26}$$

where

$$k_o = \frac{\pi c_d d_o^2}{\sqrt{8\rho}}$$

and c_d is the orifice discharge coefficient. For a given orifice size and given lubricant, k_o is a constant expressed in $m^4/s \cdot N^{1/2}$. Thus, from Eq. (26) flow through an orifice is proportional to the square root of the pressure difference across the orifice.

The discharge coefficient c_d is a function of the Reynolds number. For an orifice the Reynolds number is

$$N_R = \frac{d_o}{\eta}\,[2\rho(p_s - p_r)]^{1/2} \tag{27}$$

For a Reynolds number greater than approximately 15, which is the usual case in orifice-compensated hydrostatic bearings, c_d is about 0.6 for $d_o/D < 0.1$. For a Reynolds number less than 15, the discharge coefficient is approximately

$$c_d = 0.20\sqrt{N_R} \tag{28}$$

The pipe diameter D at the orifice should be at least 10 times the orifice diameter d_o. Sharp-edged orifices, depending on their diameters, have a tendency to clog; therefore orifice diameters d_o less than 5×10^{-4} m should be avoided.

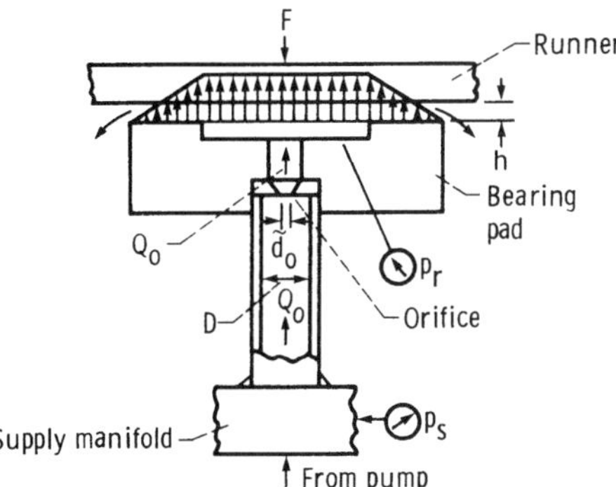

Figure 39 Orifice-compensated hydrostatic bearing. (From Ref. 22.)

Constant-Flow-Valve Compensation. Constant-flow-valve compensation is illustrated in Fig. 40. This type of restrictor has a constant flow regardless of the pressure difference across the valve. Hence, the flow is independent of recess pressure.

The relative ranking of the three types of compensating elements with regard to a number of considerations is given in Table 6. A rating of 1 in this table indicates best or most desirable. This table should help in deciding which type of compensation is most desirable in a particular application.

Basically, any type of compensating element can be designed into a hydrostatic bearing system if loads on the bearing never change. But if stiffness, load, or flow vary, the choice of the proper compensating element becomes more difficult and the reader is again referred to Rippel.[22]

2.4 Gas-Lubricated Hydrodynamic Bearings

A relatively recent (within the last 30 years) extension of hydrodynamic lubrication that is of growing importance is gas lubrication. It consists of using air or some other gas as a lubricant rather than a mineral oil. The viscosity of air is 1000 times smaller than that of very thin mineral oils. Consequently, the viscous resistance is very much less. However, the distance of nearest approach (i.e., the closest distance between the shaft and the bearing) is also correspondingly smaller, so that special precautions must be taken. To obtain full benefits from gas lubrication, the following should be observed:

1. Surfaces must have a very fine finish.
2. Alignment must be very good.
3. Dimensions and clearances must be very accurate.
4. Speeds must be high.
5. Loading must be relatively low.

Another main difference between the behavior of similar gas and liquid films besides that of viscosity is the compressibility of the gas. At low relative speeds it is reasonable to expect the gas-film density to remain nearly constant and the film therefore to behave as if it were incompressible. At high speeds, however, the density change is likely to become of

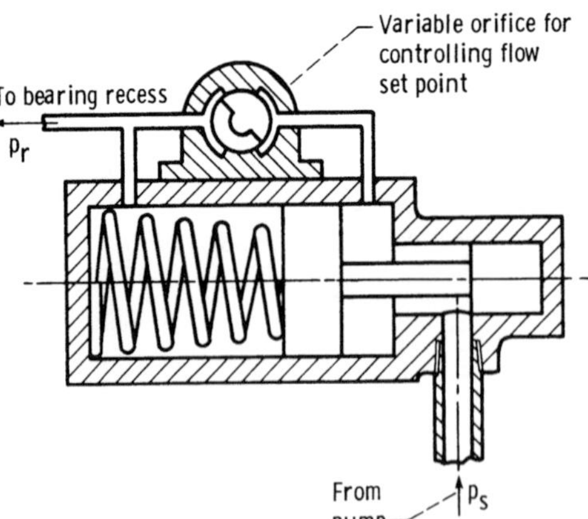

Figure 40 Constant-flow-valve compensation in hydrostatic bearing. (From Ref. 22.)

Table 6 Compensating-Element Considerations[a] (From Ref. 22)

Consideration	Compensating Element		
	Capillary	Orifice	Constant-Flow Valve
Initial cost	2	1	3
Cost to fabricate and install	2	3	1
Space required	2	1	3
Reliability	1	2	3
Useful life	1	2	3
Commercial availability	2	3	1
Tendency to clog	1	2	3
Serviceability	2	1	3
Adjustability	3	2	1

[a] Rating of 1 is best or most desirable.

primary importance so that such gas-film properties must differ appreciably from those of similar liquid films.

Gas-lubricated bearings can also operate at very high temperatures since the lubricant will not degrade chemically. Furthermore, if air is used as the lubricant, it costs nothing. Gas bearings are finding increasing use in gas cycle machinery where the cycle gas is used in the bearings, thus eliminating the need for a conventional lubrication system; in gyros, where precision and constancy of torque are critical; in food and textile processing machinery, where cleanliness and absence of contaminants are critical; and also in the magnetic recording tape industry.

Journal Bearings

Plain gas-lubricated journal bearings are of little interest because of their poor stability characteristics. Lightly loaded bearings that operate at low eccentricity ratios are subjected to fractional frequency whirl, which can result in bearing destruction. Two types of gas-lubricated journal bearings find wide-spread use, namely, the pivoted pad and the herringbone groove.

Pivoted Pad. Pivoted-pad journal bearings are most frequently used as shaft supports in gas-bearing machinery because of their excellent stability characteristics. An individual pivot pad and shaft are shown in Fig. 41, and a three-pad pivoted-pad bearing assembly is shown in Fig. 42. Generally, each pad provides pad rotation degrees of freedom about three orthogonal axes (pitch, roll, and yaw). Pivoted-pad bearings are complex because of the many geometric variables involved in their design. Some of these variables are:

1. Number of pads
2. Circumferential extent of pads, α_p
3. Aspect ratio of pad, R/l
4. Pivot location, ϕ_p/α_p
5. Machined-in clearance ratio, c/R

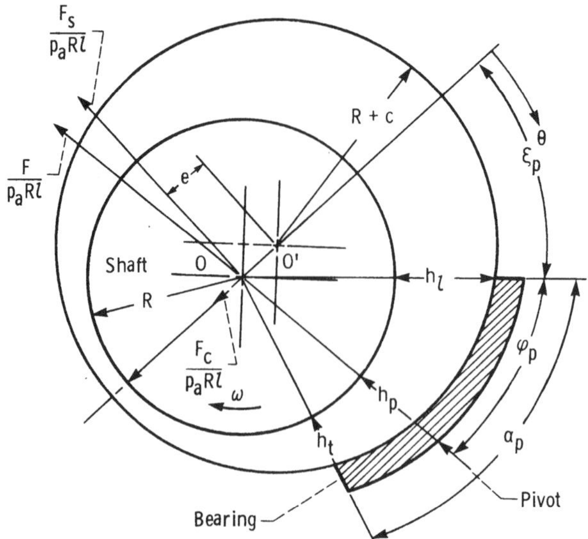

Figure 41 Geometry of individual shoe-shaft bearing. (From Ref. 27.)

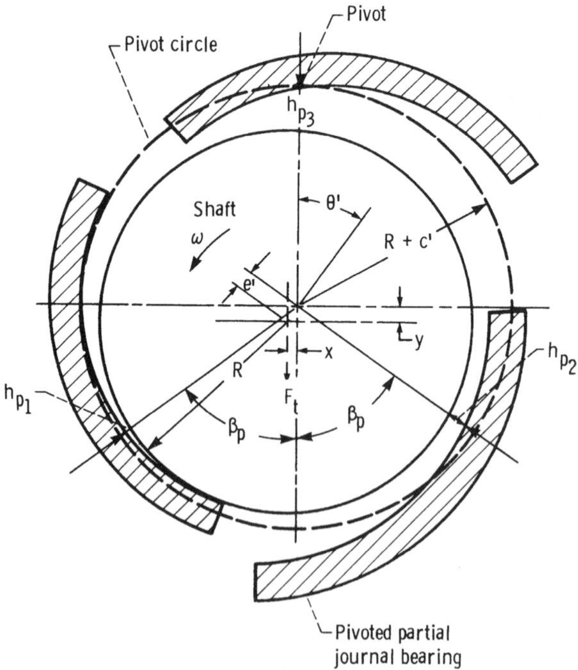

Figure 42 Geometry of pivoted-pad journal bearing with three shoes. (From Ref. 27.)

6. Pivot circle clearance ratio, c'/R
7. Angle between line of centers and pad leading edge, ξ_p

Analysis is accomplished by first determining the characteristics of an individual pad. Both geometric and operating parameters influence the design of a pivoted pad. The operating parameter of importance is the dimensionless bearing number Λ_j, where

$$\Lambda_j = \frac{6\eta\omega R^2}{p_a c^2}$$

The results of computer solutions obtained from Gunter et al.[27] for the performance of a single pad are shown in Figs. 43–45. These figures illustrate load coefficient, pivot film thickness, and trailing-edge film thickness as functions of pivot location and eccentricity ratio. These field maps apply for a pad with a radius-to-width ratio of 0.606, a circumferential extent of 94.5° (an aspect ratio of 1), and $\Lambda_j = 3.5$. For other geometries and Λ values similar maps must be generated. Additional maps are given in Gunter et al.[27]

Figures 46–48 show load coefficient and stiffness coefficient for a range of Λ_j values up to 4. These plots are for a pivot position of ⅔.

When the individual pad characteristics are known, the characteristics of the multipad bearing can be determined by using a trial-and-error approach. With the arrangement shown in Fig. 42, the load is directed between the two lower pivots. For this case the load carried by each of the lower pads is initially assumed to be $F \cos \beta$. The pivot film thicknesses h_{p_1} and h_{p_2} are then calculated. The upper-pad pivot film thickness h_{p_3}, eccentricity ratio ε, and load coefficient C_{l_3} can be determined. The additional load on the shaft due to the reaction of pad 3 is added to the system load. Calculations are repeated until the desired accuracy is achieved.

Pivoted-pad journal bearings are usually assembled with a pivot circle clearance c' somewhat less than the machined-in clearance c. When $c'/c < 1$, the bearing is said to be preloaded. Preload is usually given in terms of a preload coefficient, which is equal to $(c - c')/c$. Preloading is used to increase bearing stiffness and to prevent complete unloading of one or more pads. The latter condition can lead to pad flutter and possible contact of the pad leading edge and the shaft, which, in turn, can result in bearing failure.

Herringbone Groove. A fixed-geometry bearing that has demonstrated good stability characteristics and thus promise for use in high-speed gas bearings is the herringbone bearing. It consists of a circular journal and bearing sleeve with shallow, herringbone-shaped grooves cut into either member. Figure 49 illustrates a partially grooved herringbone journal bearing. In this figure the groove and bearing parameters are also indicated. Figures 50–54 were obtained from Hamrock and Fleming[28] and are design charts that present curves for optimizing the design parameters for herringbone journal bearings for maximum radial load. The (*a*) portion of these figures is for the grooved member rotating and the (*b*) portion is for the smooth member rotating. The only groove parameter not represented in these figures is the number of grooves to be used. From Hamrock and Fleming[28] it was found that the *minimum* number of grooves to be placed around the journal can be represented by $n \geq \Lambda_j/5$.

More than any other factors, self-excited whirl instability and low-load capacity limit the usefulness of gas-lubricated journal bearings. The whirl problem is the tendency of the journal center to orbit the bearing center at an angular speed less than or equal to half that of the journal about its own center. In many cases the whirl amplitude is large enough to cause destructive contact of the bearing surfaces.

Figure 55, obtained from Fleming and Hamrock,[29] shows the stability attained by the optimized herringbone bearings. In this figure the stability parameter $\overline{M}$ is introduced, where

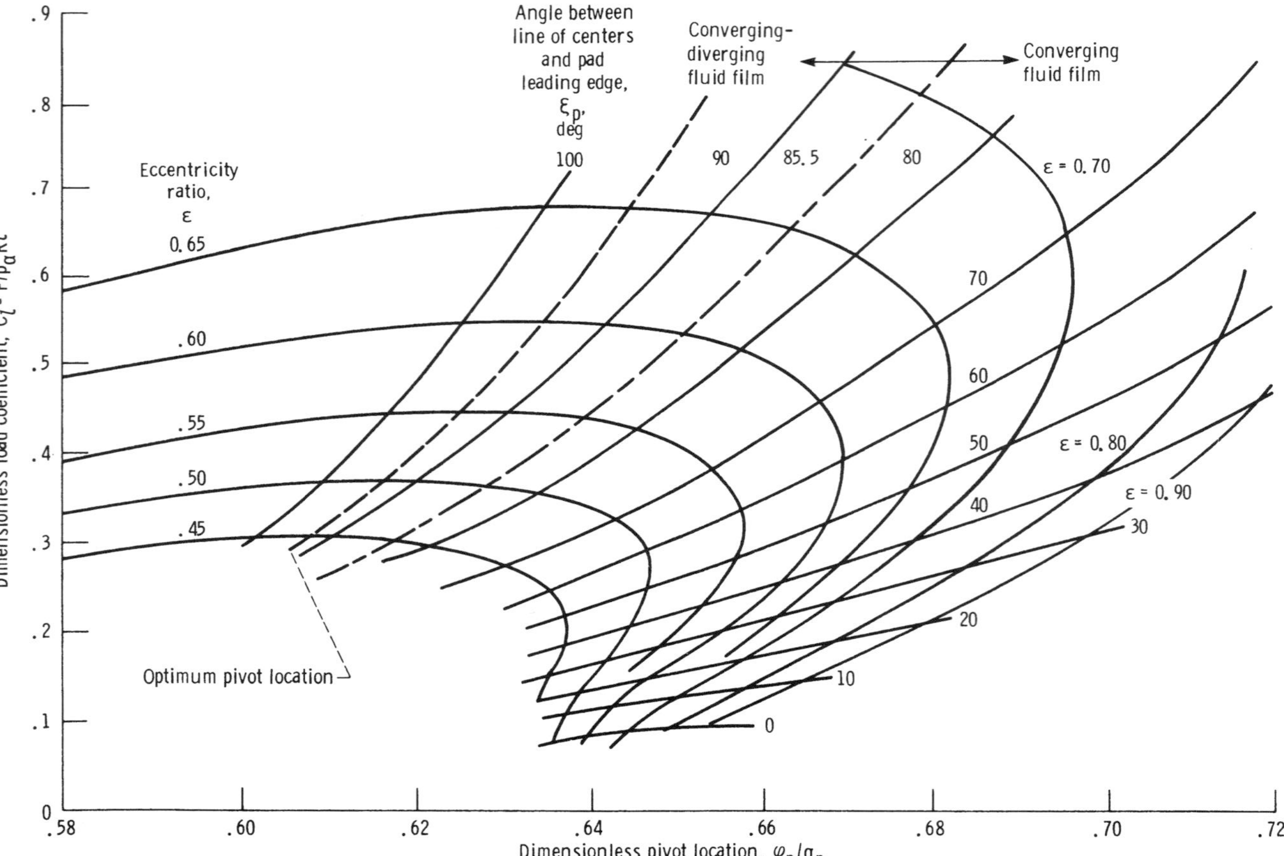

Figure 43 Chart for determining load coefficient. Bearing radius-to-length ratio, R/l, 0.6061; angular extent of pad, α_p, 94.5°; dimensionless bearing number, A_j, 3.5. (From Ref. 27.)

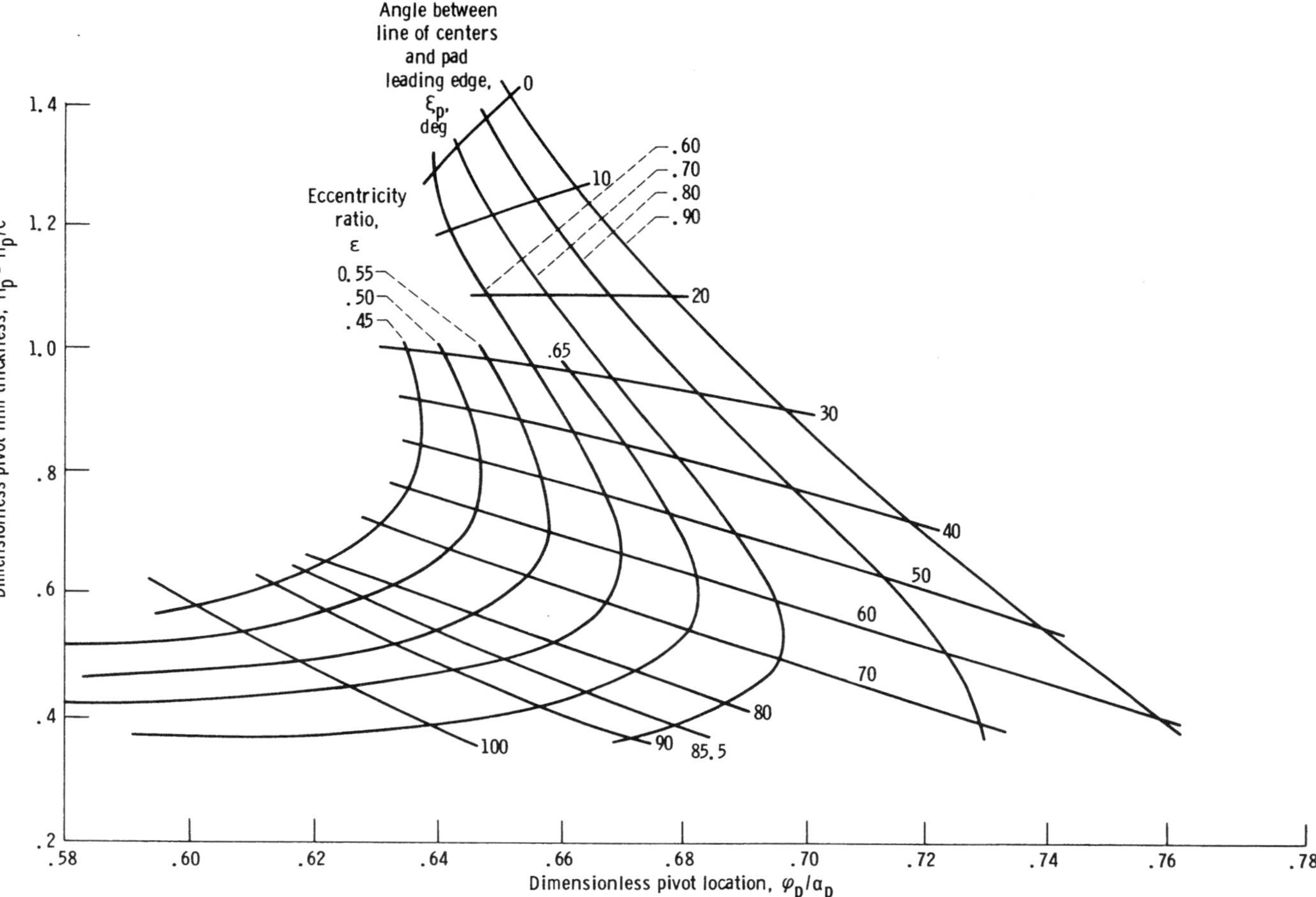

Figure 44 Chart for determining pivot film thickness. Bearing radius-to-length ratio, R/l, 0.6061; angular extent of pad, α_p, 94.5°; dimensionless bearing number, A_j, 3.5. (From Ref. 27.)

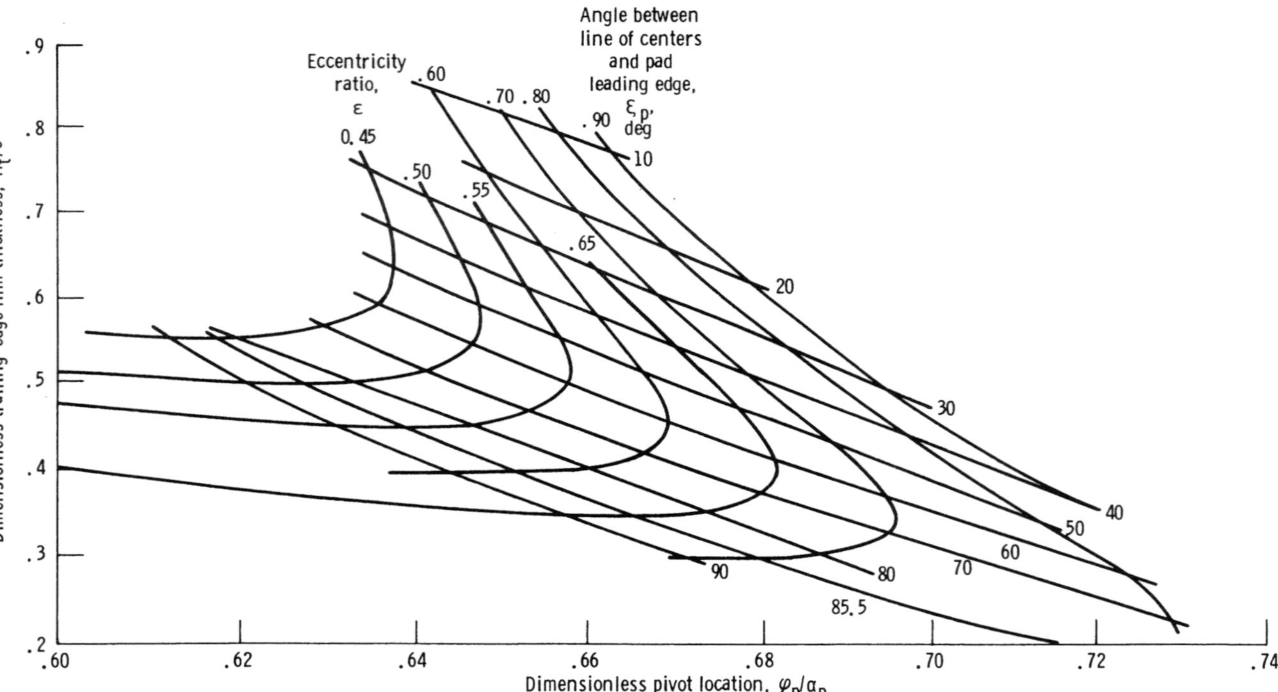

Figure 45 Chart for determining trailing-edge film thickness. Bearing radius-to-length ratio, R/l, 0.6061; angular extent of pad, α_p, 94.5°, dimensionless bearing number, A_j, 3.5. (From Ref. 27.)

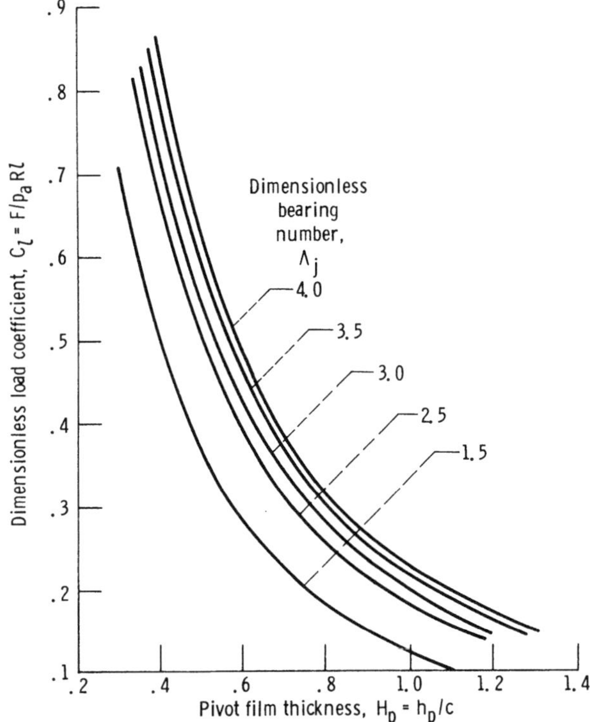

Figure 46 Chart for determining load coefficient. Angular extent of pad, α_p, 94.5°; ratio of angle between pad leading edge and pivot to α_p, φ_p/α_p, $\frac{2}{3}$; length-to-width ratio, λ, 1.0. (From Ref. 27.)

$$\overline{M} = \frac{\overline{m} p_a h_r^5}{2R^5 l \eta^2}$$

and $\overline{m}$ is the mass supported by the bearing.

In Fig. 55, the bearings with the grooved member rotating are substantially more stable than those with the smooth member rotating, especially at high compressibility numbers.

Thrust Bearings

Two types of gas-lubricated thrust bearings have found the widest use in practical applications. These are the Rayleigh step and the spiral- or herringbone-groove bearings.

Rayleigh Step Bearing. Figure 56 shows a Rayleigh step thrust bearing. In this figure the ridge region is where the film thickness is h_r and the step region is where the film thickness is h_s. The feed groove is the deep groove separating the end of the ridge region and the beginning of the next step region. Although not shown in the figure, the feed groove is orders of magnitude deeper than the film thickness h_r. A pad is defined as the section that includes ridge, step, and feed groove regions. The length of the feed groove is small relative to the length of the pad. It should be noted that each pad acts independently since the pressure profile is broken at the lubrication feed groove.

The load capacity and stiffness of a Rayleigh step thrust bearing are functions of the following parameters:

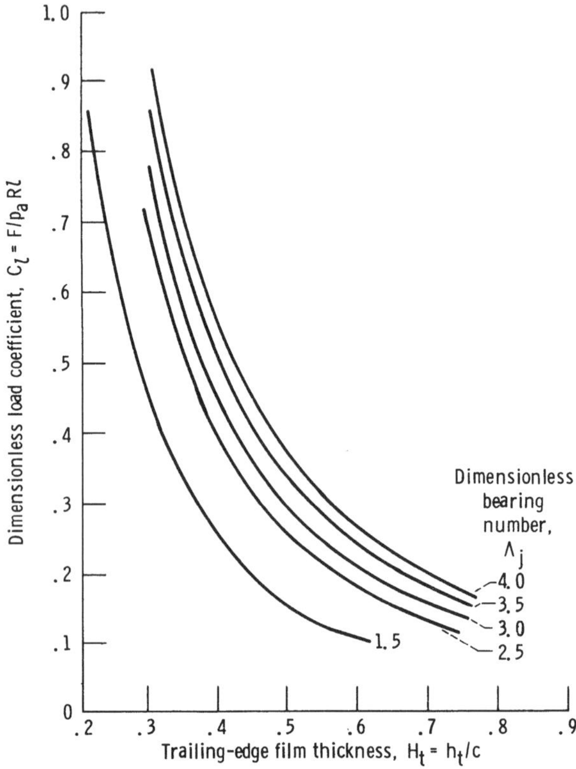

Figure 47 Chart for determining load coefficient. Angular extent of pad, α_p, 94.5°; ratio of angle between pad loading edge and pivot to α_p, φ_p/α_p, $\frac{2}{3}$; length-to-width ratio, λ, 1.0. (From Ref. 27.)

1. $\Lambda_t = 6\eta u l / p_a h_r^2$, the dimensionless bearing number
2. $\lambda_a = (b_r + b_s + b_g)/l$, length ratio
3. $H_a = h_s/h_r$, film thickness ratio
4. $\psi_s = b_s/(b_s + b_r + b_g)$, step location parameter

Figure 57*a* shows the effect of Λ_t on λ_a, H_a, and ψ_s for the *maximum-load-capacity condition.* The optimal step parameters Λ_a, H_a, and ψ_s approach an asymptote as the dimensionless bearing number Λ_t becomes small. This asymptotic condition corresponds to the incompressible solution or $\lambda_a = 0.918$, $\psi_s = 0.555$, $H_a = 1.693$. For $\Lambda_t > 1$ it is observed that there is a different optimum value of λ_a, H_a, and ψ_s for each value of Λ_t.

Figure 57*b* shows the effect of Λ_t on λ_a, H_a, and ψ_s for the *maximum-stiffness condition.* As in Fig. 57*a* the optimal step parameters approach asymptotes as the incompressible solution is reached. The asymptotes are $\lambda_a = 0.915$, $\psi_s = 0.557$, and $H_a = 1.470$. Note that there is a difference in the asymptote for the film thickness ratio but virtually no change in λ_a and ψ_s when compared with the results obtained for the maximum-load-capacity condition.

Figure 58 shows the effect of dimensionless bearing number Λ_t on dimensionless load capacity and stiffness. The difference in these figures is that the optimal step parameters are obtained in Fig. 58*a* for maximum load capacity and in Fig. 58*b* for maximum stiffness.

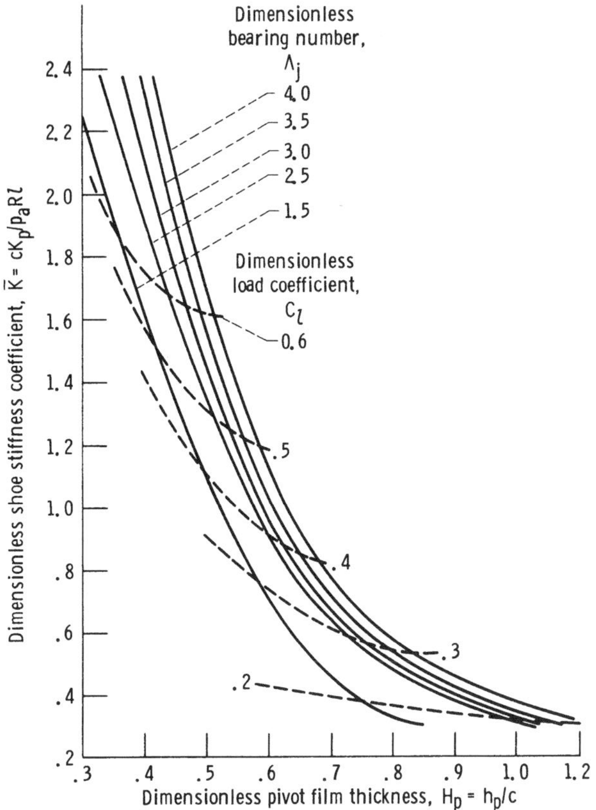

Figure 48 Chart for determining shoe stiffness coefficient. (From Ref. 27.)

For optimization of a step-sector thrust bearing, parameters for the sector must be found that are analogous to those for the rectangular step bearing. The following substitutions accomplish this transformation:

$$l \rightarrow R_1 - R_2$$

$$n(b_s + b_r + b_g) \rightarrow \pi(R_1 + R_2)$$

$$u \rightarrow \frac{\omega}{2}(R_1 + R_2)$$

where n is the number of pads placed in the step sector. By making use of these equations, the dimensionless bearing number can be rewritten as

$$\Lambda_c = \frac{3\eta\omega(R_1^2 - R_2^2)}{p_a h_r^2}$$

The optimal number of pads to be placed in the sector is obtained from the formula

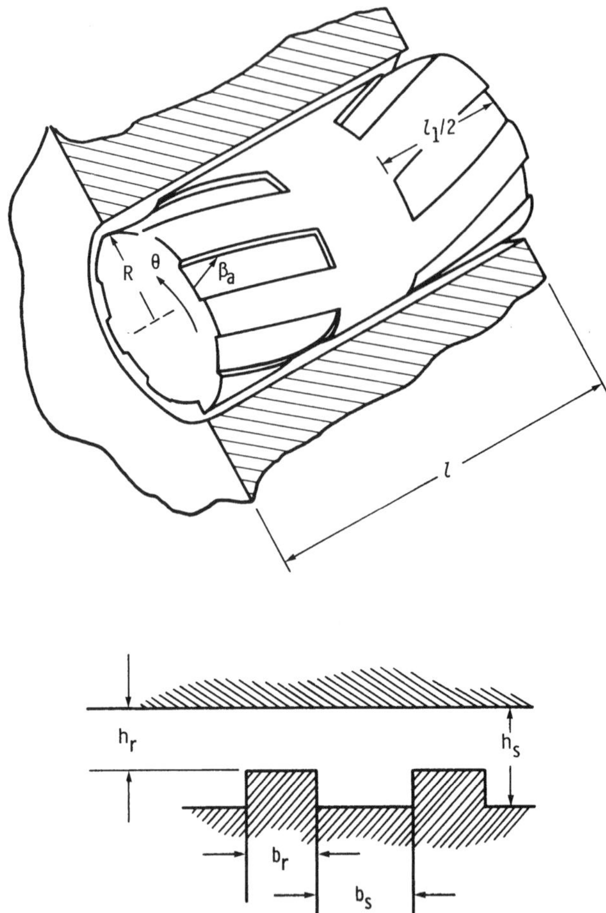

Figure 49 Configuration of concentric herringbone-groove journal bearing. Bearing parameters; $\lambda = l/2R$; $\Lambda_j = 6\mu UR/p_a h_r^2$. Groove parameters; $H_a = h_s/h_r$; $\alpha_b = b_s/(b_r + b_s)$; β_a, $\gamma = l_1/l$; n. (From Ref. 28.)

$$n = \frac{\pi(R_1 + R_2)}{(\lambda_a)_{\text{opt}}(R_1 - R_2)}$$

where $(\lambda_a)_{\text{opt}}$ is obtained from Fig. 57*a* or 57*b* for a given dimensionless bearing number Λ_t. Since n will not normally be an integer, rounding it to the nearest integer is required. Therefore, through the parameter transformation discussed above, the results presented in Figs. 57 and 58 are directly usable in designing optimal step-sector gas-lubricated thrust bearings.

Spiral-Groove Thrust Bearings. An inward-pumping spiral-groove thrust bearing is shown in Fig. 59. An inward-pumping thrust bearing is somewhat more efficient than an outward-pumping thrust bearing and therefore is the only type considered here.

The dimensionless parameters normally associated with a spiral-groove thrust bearing are:

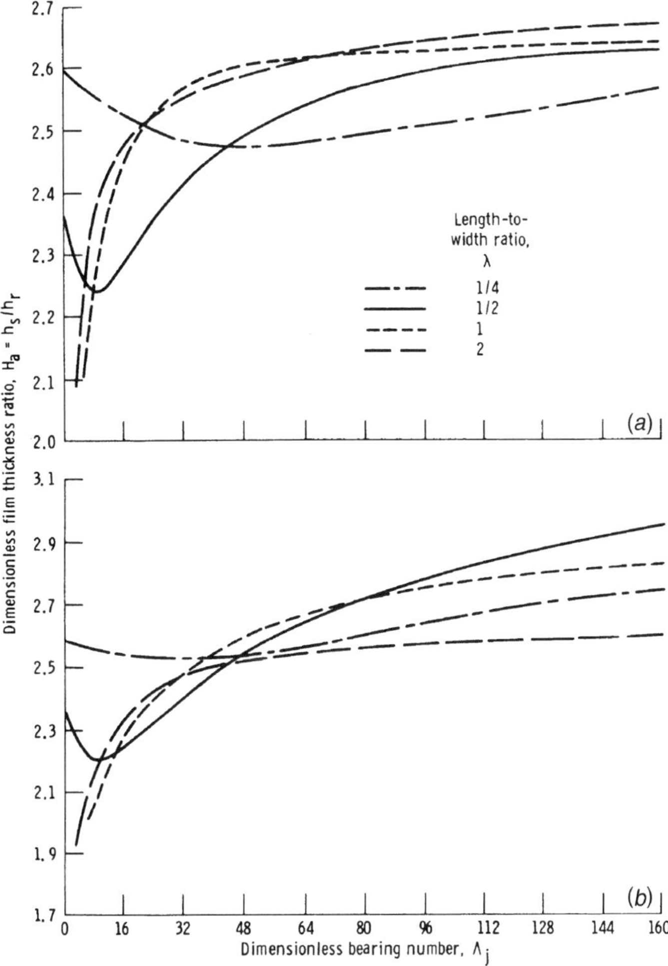

Figure 50 Chart for determining optimal film thickness. (From Ref. 28.) (*a*) Grooved member rotating. (*b*) Smooth member rotating.

1. Angle of inclination, β_a
2. Width ratio, $\bar{b} = b_s/b_r$
3. Film ratio, $H_a = h_s/h_r$
4. Radius ratio, $\alpha_r = R_2/R_1$
5. Groove length fraction, $\bar{R} = (R_1 - R_g)/(R_1 - R_2)$
6. Number of grooves, n
7. Dimensionless bearing number, $\Lambda_c = 3\eta\omega(R_1^2 - R_2^2)/p_a h_r^2$

The first six parameters are geometrical parameters and the last parameter is an operating parameter.

The performance of spiral-groove thrust bearings is represented by the following dimensionless parameters:

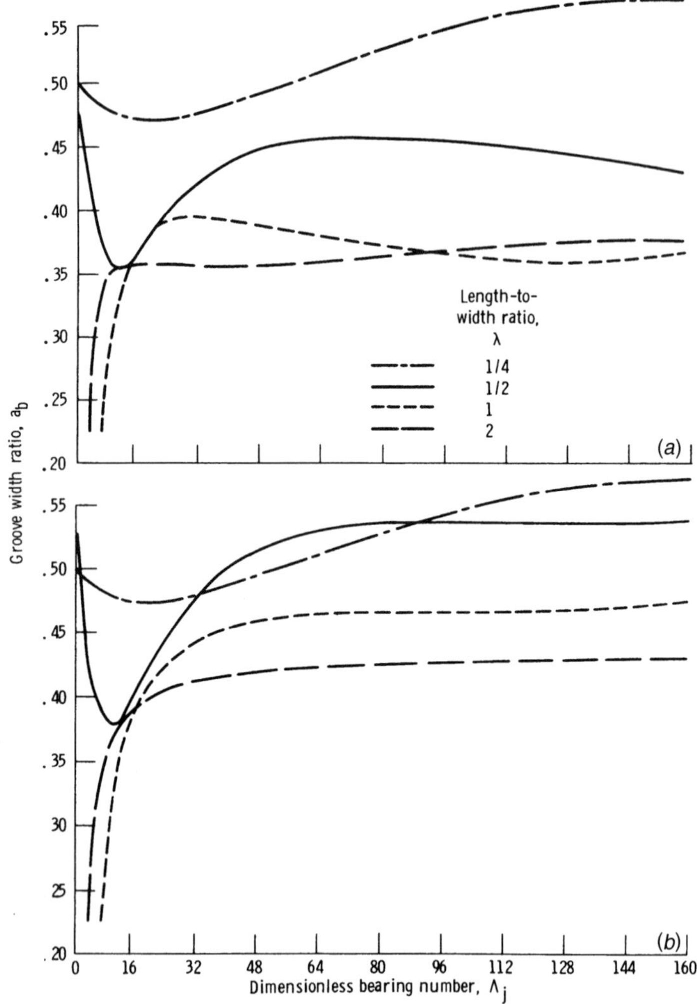

Figure 51 Chart for determining optimal groove width ratio. (From Ref. 28.) (*a*) Grooved member rotating. (*b*) Smooth member rotating.

Load

$$\overline{W}_\infty = \frac{1.5G_fF}{\pi p_a(R_1^2 - R_2^2)} \tag{29}$$

Stiffness

$$\overline{K}_\infty = \frac{1.5h_rG_fK_p}{\pi p_a(R_1^2 - R_2^2)} \tag{30}$$

Flow

$$\overline{Q} = \frac{3\eta Q}{\pi p_a h_r^3} \tag{31}$$

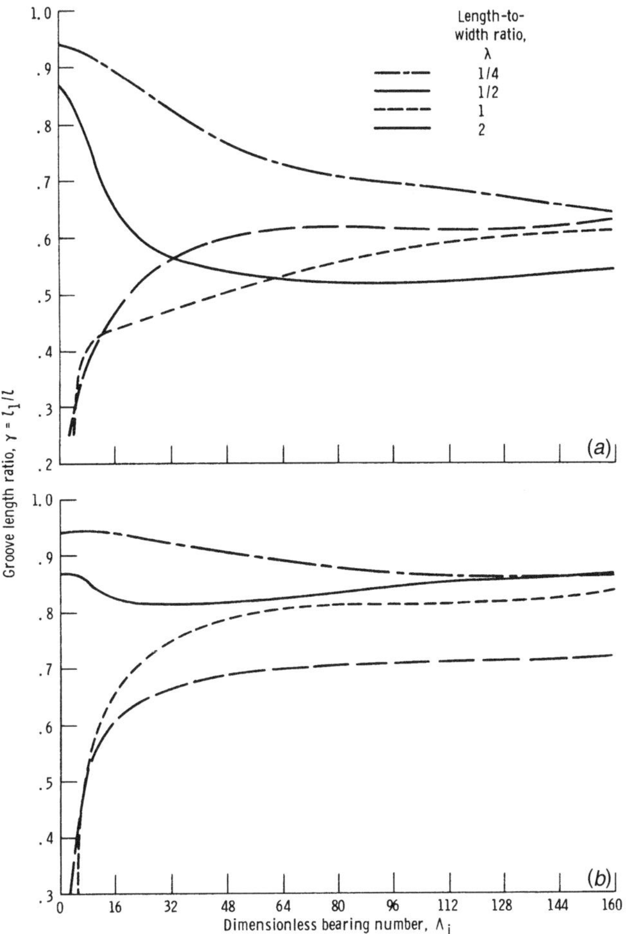

Figure 52 Chart for determining optimal groove length ratio. (From Ref. 28.) (*a*) Grooved member rotating. (*b*) Smooth member rotating.

Torque

$$\overline{T} = \frac{6T_r}{\pi p_a(R_1^2 + R_2^2)h_r\Lambda_c} \tag{32}$$

When the geometrical and operating parameters are specified, the load, stiffness, flow, and torsion can be obtained.

The design charts of Reiger[20] are reproduced as Figs. 60–66. Figure 60 shows the dimensionless load for various radius ratios as a function of dimensionless bearing number Λ_c. This figure can be used to calculate the dimensionless load for a finite number of grooves; Fig. 61 can be used to determine the value of the groove factor. Figure 62 shows curves of dimensionless stiffness; Fig. 63 shows curves of dimensionless flow; and Fig. 64 shows curves of dimensionless torque. Optimized groove geometry parameters can be obtained from Fig. 65. Finally, Fig. 66 is used to calculate groove radius R_g (shown in Fig. 59). Figure 66

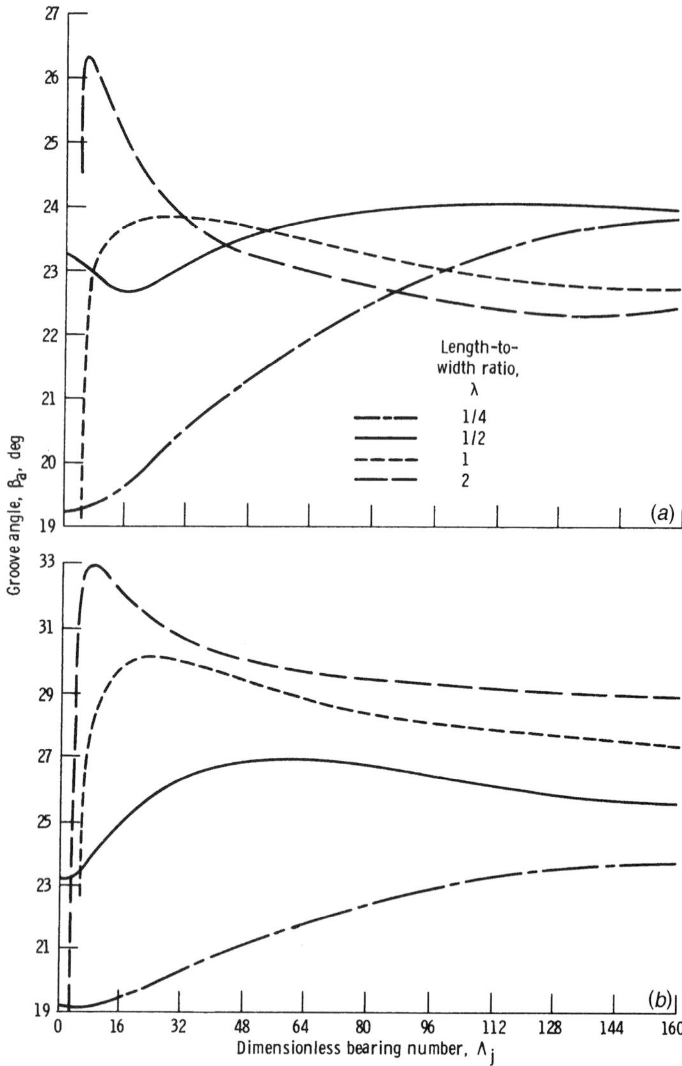

Figure 53 Chart for determining optimal groove angle. (From Ref. 28.) (*a*) Grooved member rotating. (*b*) Smooth member rotating.

shows the required groove length fraction $\overline{R} = (R_o - R_g)/(R_o - R_i)$ to ensure stability from self-excited oscillations.

In a typical design problem the given factors are load, speed, bearing envelope, gas viscosity, ambient pressure, and an allowable radius-to-clearance ratio. The maximum value of the radius-to-clearance ratio is usually dictated by the distortion likely to occur to the bearing surfaces. Typical values are 5000–10,000. The procedure normally followed in designing a spiral-groove thrust bearing while using the design curves given in Figs. 60–66 is as follows:

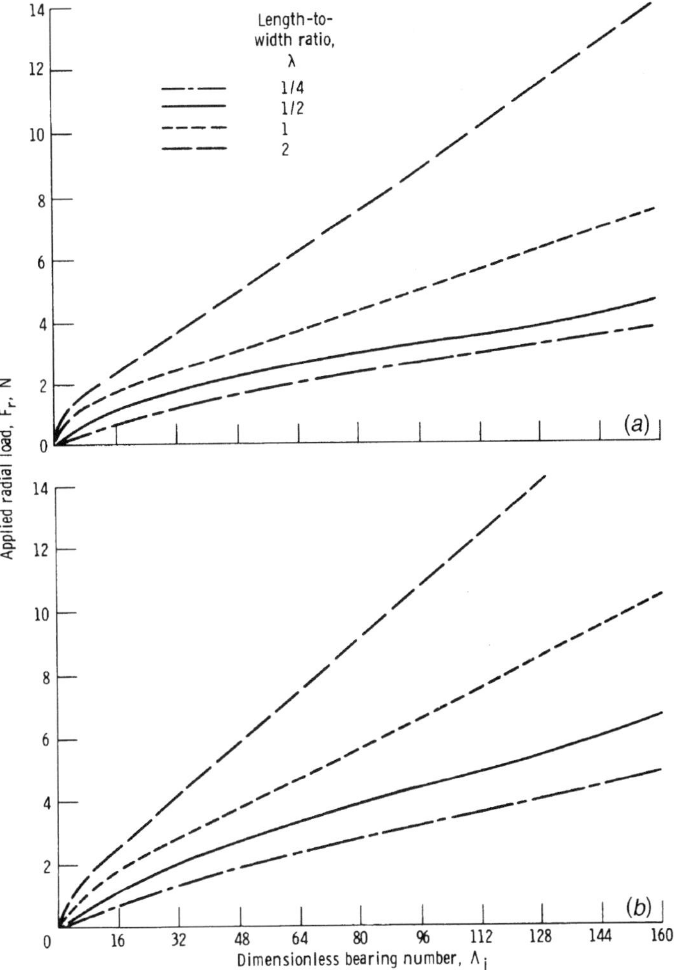

Figure 54 Chart for determining maximum radial load capacity. (From Ref. 28.) (*a*) Grooved member rotating. (*b*) Smooth member rotating.

1. Select the number of grooves n.
2. From Fig. 61 determine the groove factor G_f for given $\alpha_r = R_i/R_o$ and n.
3. Calculate $\overline{W}_\infty = 1.5G_fF/\pi p_a(R_1^2 - R_2^2)$.
4. If $\overline{W}_\infty < 0.8$, R_1 must be increased. Return to step 2.
5. From Fig. 60, given $\overline{W}_\infty$ and α_r, establish Λ_c.
6. Calculate

$$\frac{R_1}{h_r} = \left\{ \frac{\Lambda_c p_a}{3\eta(\omega_h - \omega_o)[1 - (R_2/R_1)^2]} \right\}^{1/2}$$

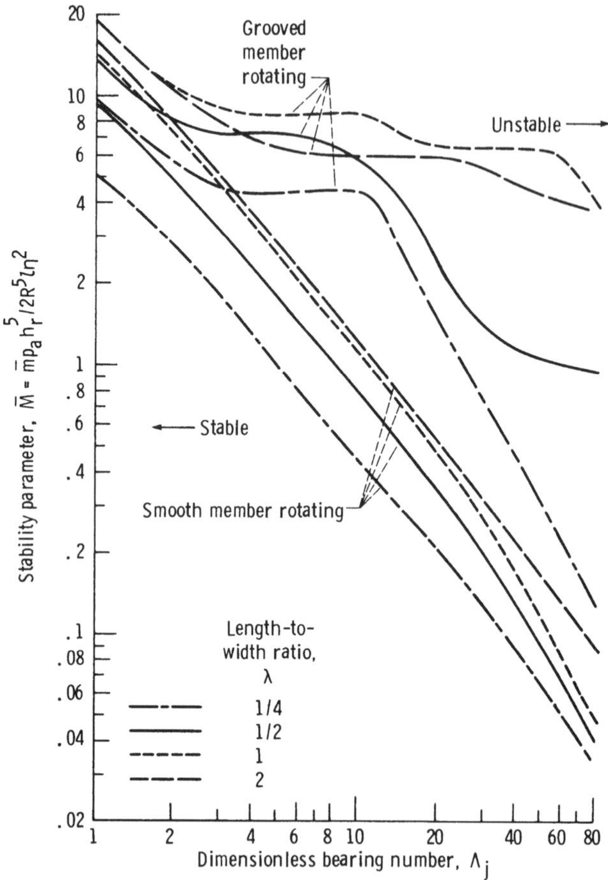

Figure 55 Chart for determining maximum stability of herringbone-groove bearings. (From Ref. 29.)

If $R_1/h_r > 10{,}000$ (or whatever preassigned radius-to-clearance ratio), a larger bearing or higher speed is required. Return to step 2. If these changes cannot be made, an externally pressurized bearing must be used.

7. Having established what α_r and Λ_c should be, obtain values of $\overline{K}_\infty$, $\overline{Q}$, and $\overline{T}$ from Figs. 62, 63, and 64, respectively. From Eqs. (29), (30), and (31) calculate K_p, Q, and T_r.
8. From Fig. 65 obtain groove geometry (b, β_a, and H_a) and from Fig. 66 obtain R_g.

3 ELASTOHYDRODYNAMIC LUBRICATION

Downson[31] defines elastohydrodynamic lubrication (EHL) as "the study of situations in which elastic deformation of the surrounding solids plays a significant role in the hydrodynamic lubrication process." Elastohydrodynamic lubrication implies complete fluid-film lubrication and no asperity interaction of the surfaces. There are two distinct forms of elastohydrodynamic lubrication.

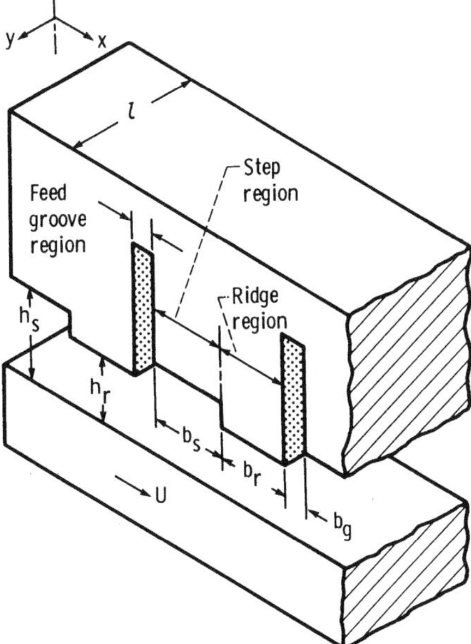

Figure 56 Configuration of rectangular step thrust bearing. (From Ref. 30.)

1. *Hard EHL.* Hard EHL relates to materials of high elastic modulus, such as metals. In this form of lubrication not only are the elastic deformation effects important, but the pressure–viscosity effects are equally as important. Engineering applications in which this form of lubrication is dominant include gears and rolling-element bearings.

2. *Soft EHL.* Soft EHL relates to materials of low elastic modulus, such as rubber. For these materials elastic distortions are large, even with light loads. Another feature is the negligible pressure–viscosity effect on the lubricating film. Engineering applications in which soft EHL is important include seals, human joints, tires, and a number of lubricated elastomeric material machine elements.

The recognition and understanding of elastohydrodynamic lubrication present one of the major developments in the field of tribology in this century. The revelation of a previously unsuspected regimen of lubrication is clearly an event of importance in tribology. Elastohydrodynamic lubrication not only explained the remarkable physical action responsible for the effective lubrication of many machine elements, but it also brought order to the understanding of the complete spectrum of lubrication regimens, ranging from boundary to hydrodynamic.

A way of coming to an understanding of elastohydrodynamic lubrication is to compare it to hydrodynamic lubrication. The major developments that have led to our present understanding of hydrodynamic lubrication[1,3] predate the major developments of elastohydrodynamic lubrication[32,33] by 65 years. Both hydrodynamic and elastohydrodynamic lubrication are considered as fluid-film lubrication in that the lubricant film is sufficiently thick to prevent the opposing solids from coming into contact. Fluid-film lubrication is often referred to as the ideal form of lubrication since it provides low friction and high resistance to wear.

This section highlights some of the important aspects of elastohydrodynamic lubrication while illustrating its use in a number of applications. It is not intended to be exhaustive but to point out the significant features of this important regimen of lubrication. For more details the reader is referred to Hamrock and Dowson.[10]

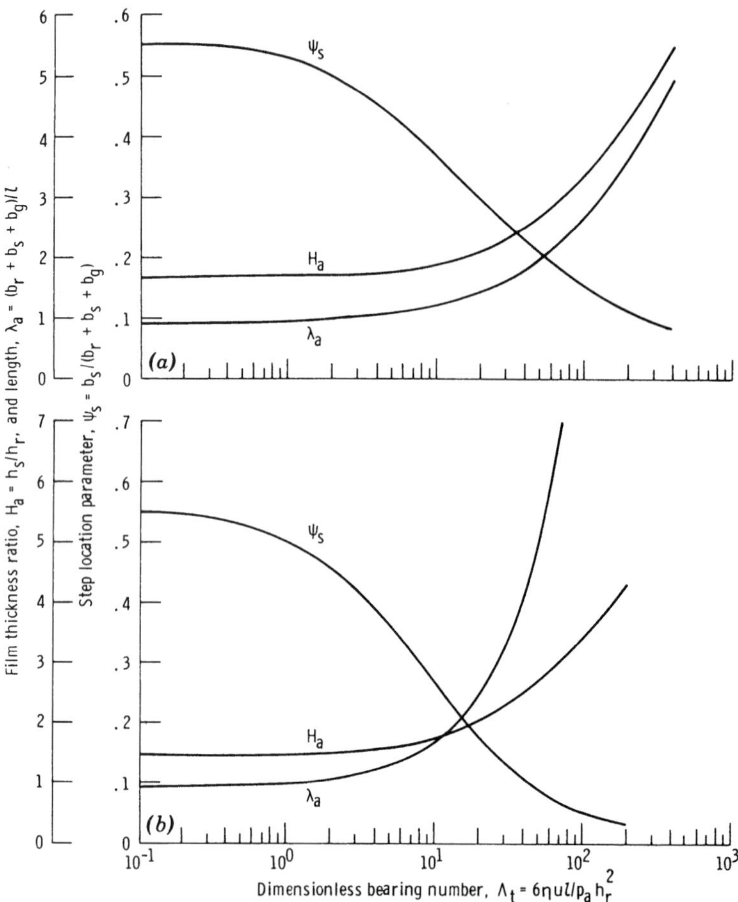

Figure 57 Chart for determining optimal step parameters. (From Ref. 30.) (*a*) Maximum dimensionless load. (*b*) Maximum dimensionless stiffness.

3.1 Contact Stresses and Deformations

As was pointed out in Section 1.1, elastohydrodynamic lubrication is the mode of lubrication normally found in nonconformal contacts such as rolling-element bearings. A load–deflection relationship for nonconformal contacts is developed in this section. The deformation within the contact is calculated from, among other things, the ellipticity parameter and the elliptic integrals of the first and second kinds. Simplified expressions that allow quick calculations of the stresses and deformations to be made easily from a knowledge of the applied load, the material properties, and the geometry of the contacting elements are presented in this section.

Elliptical Contacts

The undeformed geometry of contacting solids in a nonconformal contact can be represented by two ellipsoids. The two solids with different radii of curvature in a pair of principal planes

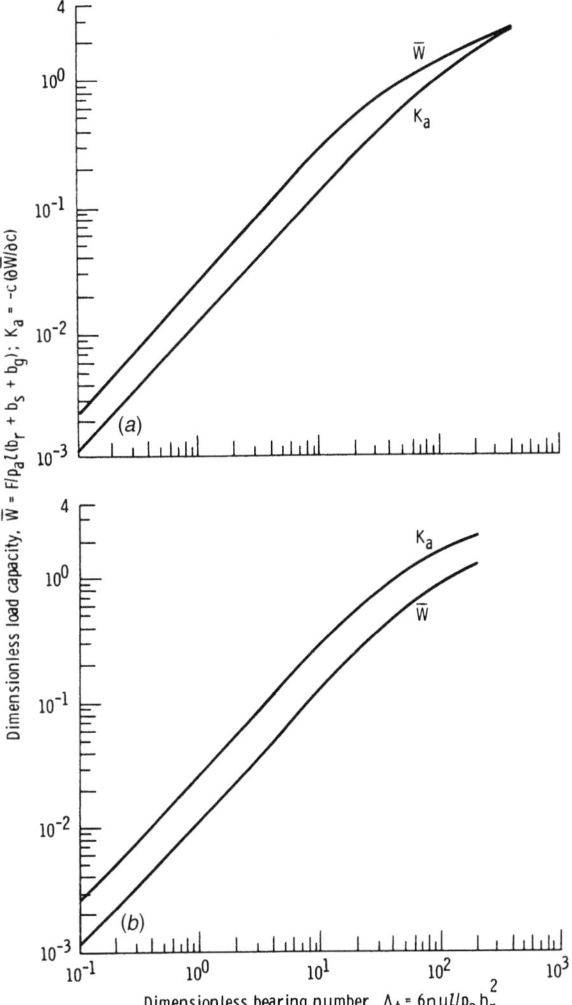

Figure 58 Chart for determining dimensionless load capacity and stiffness. (From Ref. 30.) (*a*) Maximum dimensionless load capacity. (*b*) Maximum stiffness.

(x and y) passing through the contact between the solids make contact at a single point under the condition of zero applied load. Such a condition is called point contact and is shown in Fig. 67, where the radii of curvature are denoted by r's. It is assumed that convex surfaces, as shown in Fig. 67, exhibit positive curvature and concave surfaces exhibit negative curvature. Therefore if the center of curvature lies within the solids, the radius of curvature is positive; if the center of curvature lies outside the solids, the radius of curvature is negative. It is important to note that if coordinates x and y are chosen such that

$$\frac{1}{r_{ax}} + \frac{1}{r_{bx}} > \frac{1}{r_{ay}} + \frac{1}{r_{by}} \tag{33}$$

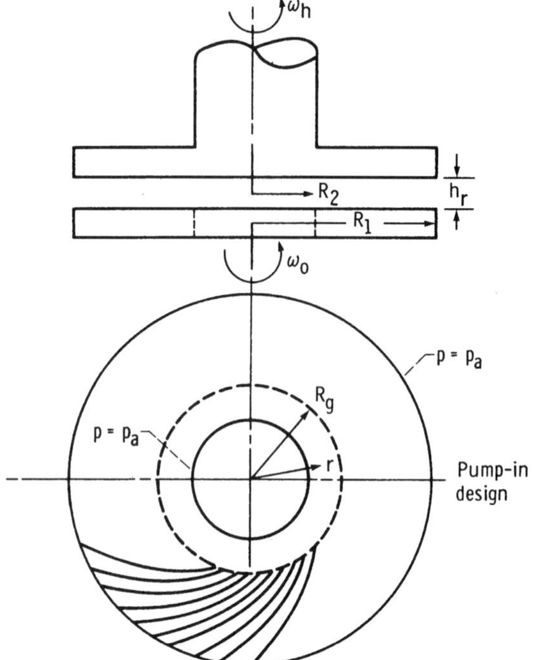

Figure 59 Configuration of spiral-groove thrust bearing. (From Ref. 20.)

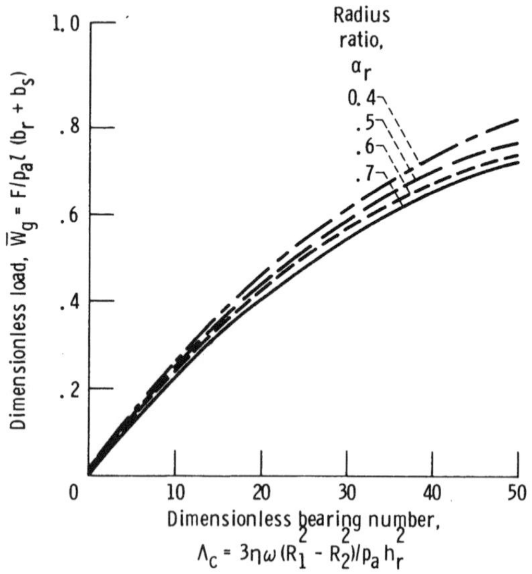

Figure 60 Chart for determining load for spiral-groove thrust bearings. (From Ref. 20.)

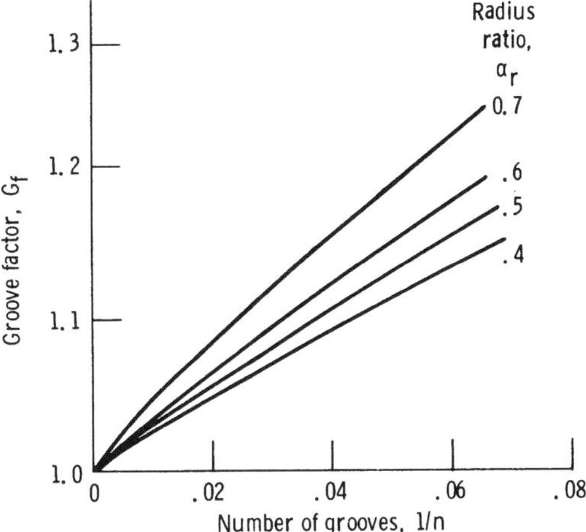

Figure 61 Chart for determining groove factor for spiral-groove thrust bearings. (From Ref. 20.)

coordinate x then determines the direction of the semiminor axis of the contact area when a load is applied and y determines the direction of the semimajor axis. The direction of motion is always considered to be along the x axis.

The curvature sum and difference, which are quantities of some importance in the analysis of contact stresses and deformations, are

$$\frac{1}{R} = \frac{1}{R_x} + \frac{1}{R_y} \tag{34}$$

$$\Gamma = R\left(\frac{1}{R_x} - \frac{1}{R_y}\right) \tag{35}$$

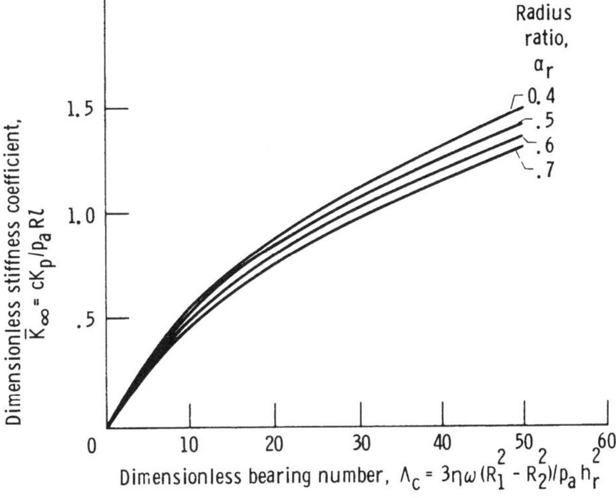

Figure 62 Chart for determining stiffness for spiral-groove thrust bearings. (From Ref. 20.)

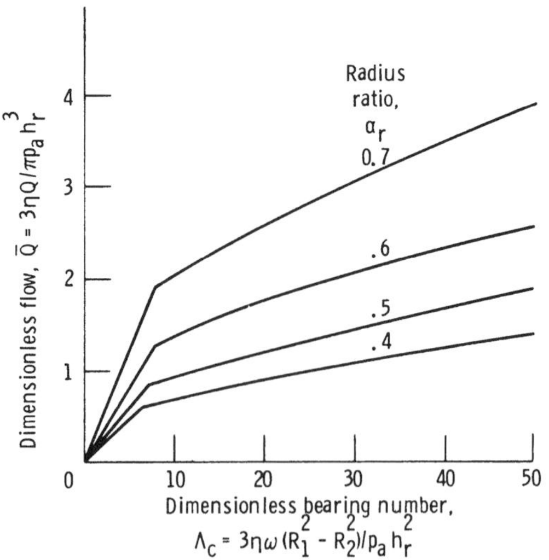

Figure 63 Chart for determining flow for spiral-groove thrust bearings. (From Ref. 20.)

where

$$\frac{1}{R_x} = \frac{1}{r_{ax}} + \frac{1}{r_{bx}} \tag{36}$$

$$\frac{1}{R_y} = \frac{1}{r_{ay}} + \frac{1}{r_{by}} \tag{37}$$

$$\alpha = \frac{R_y}{R_x} \tag{38}$$

Equations (36) and (37) effectively redefine the problem of two ellipsoidal solids approaching one another in terms of an equivalent ellipsoidal solid of radii R_x and R_y approaching a plane.

The ellipticity parameter k is defined as the elliptical-contact diameter in the y direction (transverse direction) divided by the elliptical-contact diameter in the x direction (direction of motion) or $k = D_y/D_x$. If Eq. (33) is satisfied and $\alpha \geq 1$, the contact ellipse will be

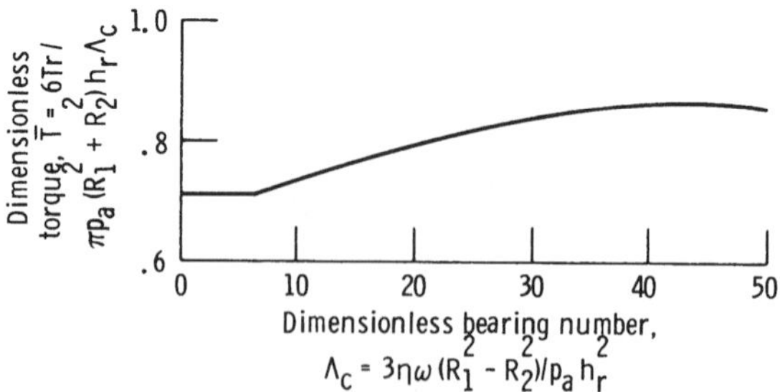

Figure 64 Chart for determining torque for spiral-groove thrust bearings. Curve is for all radius ratios. (From Ref. 20.)

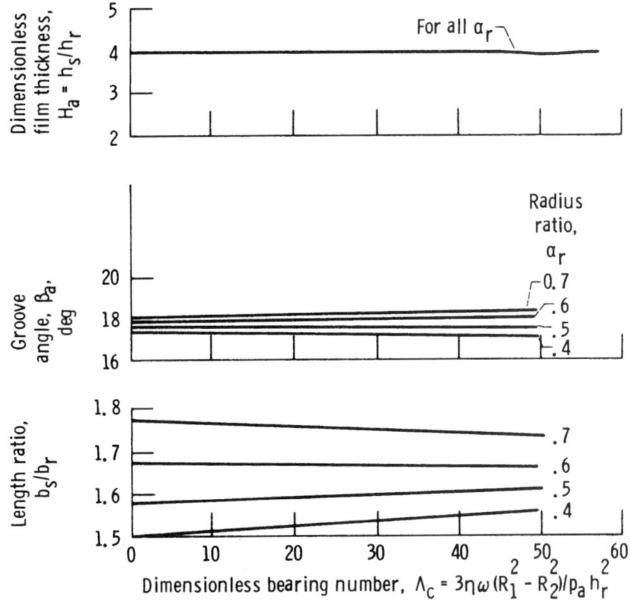

Figure 65 Chart for determining optimal groove geometry for spiral-groove thrust bearings. (From Ref. 20.)

oriented so that its major diameter will be transverse to the direction of motion, and, consequently, $k \geq 1$. Otherwise, the major diameter would lie along the direction of motion with both $\alpha \leq 1$ and $k \leq 1$. Figure 68 shows the ellipticity parameter and the elliptic integrals of the first and second kinds for a range of curvature ratios ($\alpha = R_y/R_x$) usually encountered in concentrated contacts.

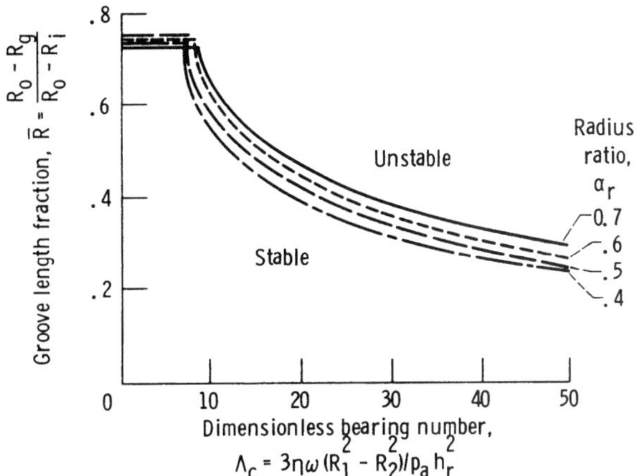

Figure 66 Chart for determining groove length fraction for spiral-groove thrust bearings. (From Ref. 20.)

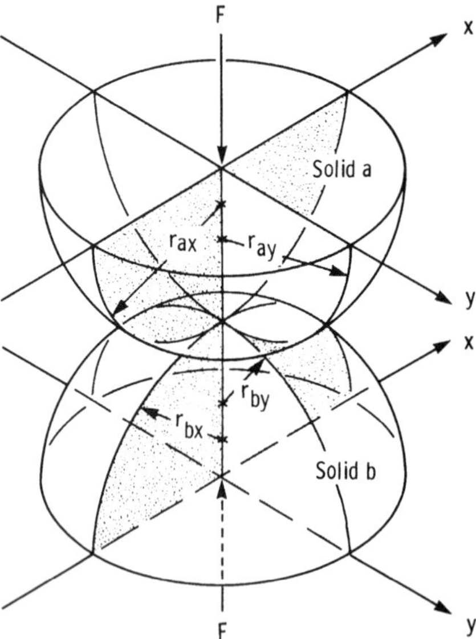

Figure 67 Geometry of contacting elastic solids. (From Ref. 10.)

Simplified Solutions for $\alpha > 1$. The classical Hertzian solution requires the calculation of the ellipticity parameter k and the complete elliptic integrals of the first and second kinds $\mathcal{F}$ and $\mathcal{E}$. This entails finding a solution to a transcendental equation relating k, $\mathcal{F}$, and $\mathcal{E}$ to the geometry of the contacting solids. Possible approaches include an iterative numerical procedure, as described, for example, by Hamrock and Anderson,[35] or the use of charts, as shown by Jones.[36] Hamrock and Brewe[34] provide a shortcut to the classical Hertzian solution for the local stress and deformation of two elastic bodies in contact. The shortcut is accomplished by using simplified forms of the ellipticity parameter and the complete elliptic in-

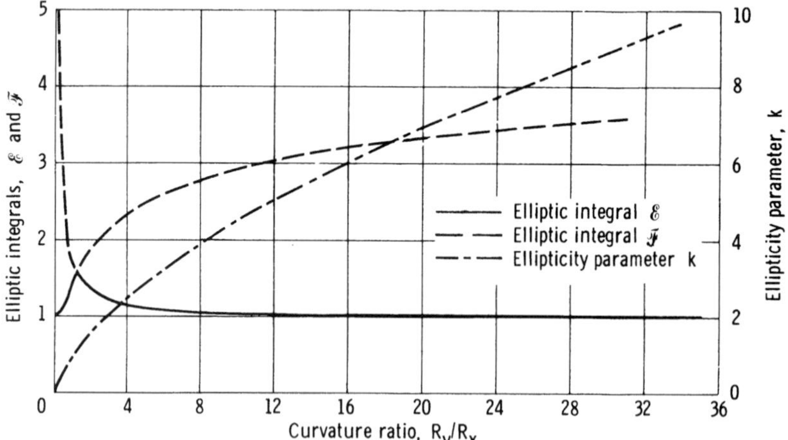

Figure 68 Chart for determining ellipticity parameter and elliptic integrals of first and second kinds. (From Ref. 34.)

tegrals, expressing them as functions of the geometry. The results of Hamrock and Brewe's work[34] are summarized here.

A power fit using linear regression by the method of least squares resulted in the following expression for the ellipticity parameter:

$$k = \alpha^{2/\pi} \qquad \text{for} \quad \alpha \geq 1 \tag{39}$$

The asymptotic behavior of $\mathcal{E}$ and $\mathcal{F}$ ($\alpha \to 1$ implies $\mathcal{E} \to \mathcal{F} \to \pi/2$ and $\alpha \to \infty$ implies $\mathcal{F} \to \infty$ and $\mathcal{E} \to 1$) was suggestive of the type of functional dependence that $\mathcal{E}$ and $\mathcal{F}$ might follow. As a result, an inverse and a logarithmic fit were tried for $\mathcal{E}$ and $\mathcal{F}$, respectively. The following expressions provided excellent curve fits:

$$\mathcal{E} = 1 + \frac{q}{\alpha} \qquad \text{for} \quad \alpha \geq 1 \tag{40}$$

$$\mathcal{F} = \frac{\pi}{2} + q \ln \alpha \qquad \text{for} \quad \alpha \geq 1 \tag{41}$$

where

$$q = \frac{\pi}{2} - 1 \tag{42}$$

When the ellipticity parameter k [Eq. (39)], the elliptic integrals of the first and second kinds [Eqs. (40) and (41)], the normal applied load F, Poisson's ratio ν, and the modulus of elasticity E of the contacting solids are known, we can write the major and minor axes of the contact ellipse and the maximum deformation at the center of the contact, from the analysis of Hertz,[37] as

$$D_y = 2\left(\frac{6k^2\mathcal{E}FR}{\pi E'}\right)^{1/3} \tag{43}$$

$$D_x = 2\left(\frac{6\mathcal{E}FR}{\pi k E'}\right)^{1/3} \tag{44}$$

$$\delta = F\left[\left(\frac{9}{2\mathcal{E}R}\right)\left(\frac{F}{\pi k E'}\right)^2\right]^{1/3} \tag{45}$$

where [as in Eq. (12)]

$$E' = 2\left(\frac{1-\nu_a^2}{E_a} + \frac{1-\nu_b^2}{E_b}\right)^{-1} \tag{46}$$

In these equations D_y and D_x are proportional to $F^{1/3}$ and δ is proportional to $F^{2/3}$.

The maximum Hertzian stress at the center of the contact can also be determined by using Eqs. (42) and (44),

$$\sigma_{\max} = \frac{6F}{\pi D_x D_y} \tag{47}$$

Simplified Solutions for $\alpha \leq 1$. Table 7 gives the simplified equations for $\alpha < 1$ as well as for $\alpha \geq 1$. Recall that $\alpha \geq 1$ implies $k \geq 1$ and Eq. (33) is satisfied and $\alpha < 1$ implies $k < 1$ and Eq. (33) is not satisfied. It is important to make the proper evaluation of α, since it has a great significance in the outcome of the simplified equations.

Table 7 Simplified Equations (From Ref. 6)

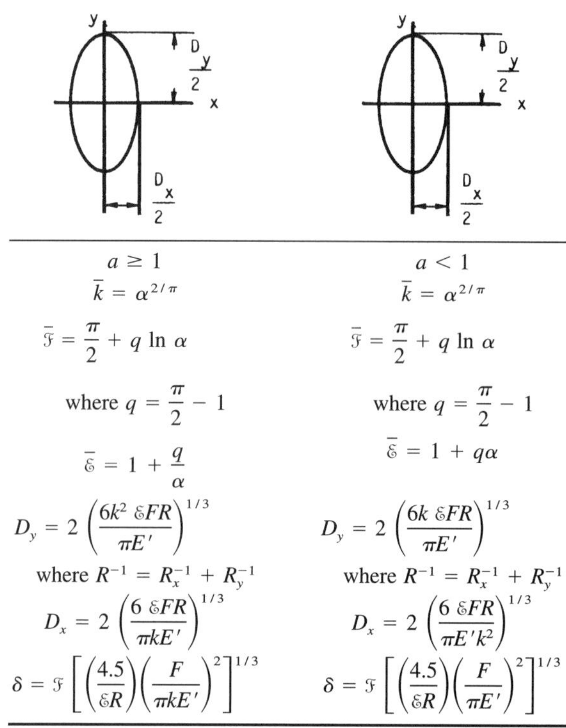

$a \geq 1$	$a < 1$
$\bar{k} = \alpha^{2/\pi}$	$\bar{k} = \alpha^{2/\pi}$
$\bar{\mathcal{F}} = \dfrac{\pi}{2} + q \ln \alpha$	$\bar{\mathcal{F}} = \dfrac{\pi}{2} + q \ln \alpha$
where $q = \dfrac{\pi}{2} - 1$	where $q = \dfrac{\pi}{2} - 1$
$\bar{\mathcal{E}} = 1 + \dfrac{q}{\alpha}$	$\bar{\mathcal{E}} = 1 + q\alpha$
$D_y = 2\left(\dfrac{6k^2\,\mathcal{E}FR}{\pi E'}\right)^{1/3}$	$D_y = 2\left(\dfrac{6k\,\mathcal{E}FR}{\pi E'}\right)^{1/3}$
where $R^{-1} = R_x^{-1} + R_y^{-1}$	where $R^{-1} = R_x^{-1} + R_y^{-1}$
$D_x = 2\left(\dfrac{6\,\mathcal{E}FR}{\pi k E'}\right)^{1/3}$	$D_x = 2\left(\dfrac{6\,\mathcal{E}FR}{\pi E' k^2}\right)^{1/3}$
$\delta = \mathcal{F}\left[\left(\dfrac{4.5}{\mathcal{E}R}\right)\left(\dfrac{F}{\pi k E'}\right)^2\right]^{1/3}$	$\delta = \mathcal{F}\left[\left(\dfrac{4.5}{\mathcal{E}R}\right)\left(\dfrac{F}{\pi E'}\right)^2\right]^{1/3}$

Figure 69 shows three diverse situations in which the simplified equations can be usefully applied. The locomotive wheel on a rail (Fig. 69*a*) illustrates an example in which the ellipticity parameter k and the radius ratio α are less than 1. The ball rolling against a flat plate (Fig. 69*b*) provides pure circular contact (i.e., $\alpha = k = 1.0$). Figure 69*c* shows how the contact ellipse is formed in the ball–outer-race contact of a ball bearing. Here the semimajor axis is normal to the direction of rolling and, consequently, α and k are greater than 1. Table 8 shows how the degree of conformity affects the contact parameters for the various cases illustrated in Fig. 69.

Rectangular Contacts

For this situation the contact ellipse discussed in the preceding section is of infinite length in the transverse direction ($D_y \to \infty$). This type of contact is exemplified by a cylinder loaded against a plate, a groove, or another parallel cylinder or by a roller loaded against an inner or outer ring. In these situations the contact semiwidth is given by

$$b = R_x\left(\frac{8W}{\pi}\right)^{1/2} \tag{48}$$

where

$$W = \frac{F'}{E'R_x} \tag{49}$$

and F' is the load per unit length along the contact.

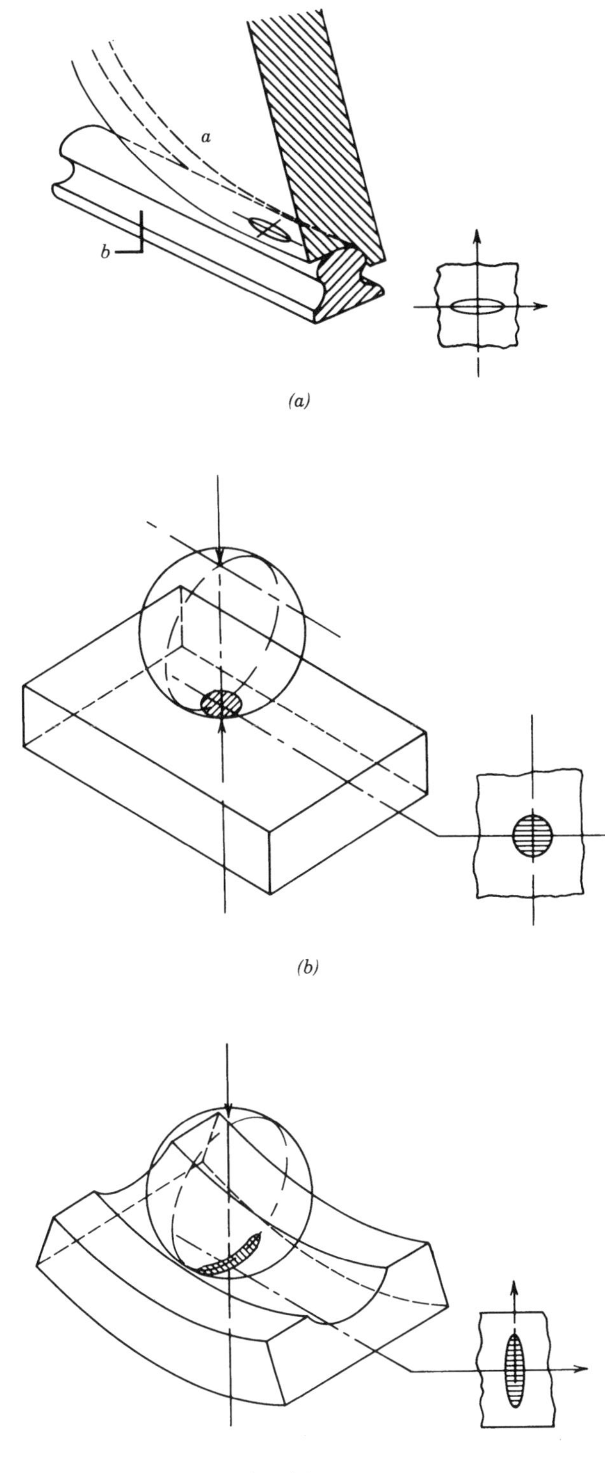

Figure 69 Three degrees of conformity. (From Ref. 34.) (*a*) Wheel on rail. (*b*) Ball on plane. (*c*) Ball–outer-race contact.

Table 8 Practical Applications for Differing Conformities[a] (From Ref. 34)

Contact Parameters	Wheel on Rail	Ball on Plane	Ball–Outer-Race Contact
F	1.00×10^5 N	222.4111 N	222.4111 N
r_{ax}	50.1900 cm	0.6350 cm	0.6350 cm
r_{ay}	∞	0.6350 cm	0.6350 cm
r_{bx}	∞	∞	−3.8900 cm
r_{by}	30.0000 cm	∞	−0.6600 cm
u	0.5977	1.0000	22.0905
k	0.7206	1.0000	7.1738
$\mathcal{E}$	1.3412	1.5708	1.0258
$\mathcal{F}$	1.8645	1.5708	3.3375
D_y	1.0807 cm	0.0426 cm	0.1810 cm
D_x	1.4997 cm	0.0426 cm	0.0252 cm
δ	0.0108 cm	7.13×10^{-4} cm	3.57×10^{-4} cm
$\sigma_{\max}$	1.1784×10^5 N/cm^2	2.34×10^5 N/cm^2	9.30×10^4 N/cm^2

[a] $E' = 2.197 \times 10^7$ N/cm^2.

The maximum deformation due to the approach of centers of two cylinders can be written as[12]

$$\delta = \frac{2WR_x}{\pi}\left[\frac{2}{3} + \ln\left(\frac{2r_{ax}}{b}\right) + \ln\left(\frac{2r_{bx}}{b}\right)\right] \tag{50}$$

The maximum Hertzian stress in a rectangular contact can be written as

$$\sigma_{\max} = E'\left(\frac{W}{2\pi}\right)^{1/2} \tag{51}$$

3.2 Dimensionless Grouping

The variables appearing in elastohydrodynamic lubrication theory are

E' = effective elastic modulus, N/m^2

F = normal applied load, N

h = film thickness, m

R_x = effective radius in x (motion) direction, m

R_y = effective radius in y (transverse) direction, m

u = mean surface velocity in x direction, m/s

ξ = pressure–viscosity coefficient of fluid, m^2/N

η_0 = atmospheric viscosity, N·s/m^2;

From these variables the following five dimensionless groupings can be established.

Dimensionless film thickness

$$H = \frac{h}{R_x} \tag{52}$$

Ellipticity parameter

$$k = \frac{D_y}{D_x} = \left(\frac{R_y}{R_x}\right)^{2/\pi} \tag{53}$$

Dimensionless load parameter

$$W = \frac{F}{E'R_x^2} \tag{54}$$

Dimensionless speed parameter

$$U = \frac{\eta_0 u}{E'R_x} \tag{55}$$

Dimensionless materials parameter

$$G = \xi E' \tag{56}$$

The dimensionless minimum film thickness can now be written as a function of the other parameters involved:

$$H = f(k, U, W, G)$$

The most important practical aspect of elastohydrodynamic lubrication theory becomes the determination of this function f for the case of the minimum film thickness within a conjunction. Maintaining a fluid-film thickness of adequate magnitude is clearly vital to the efficient operation of machine elements.

3.3 Hard-EHL Results

By using the numerical procedures outlined in Hamrock and Dowson,[38] the influence of the ellipticity parameter and the dimensionless speed, load, and materials parameters on minimum film thickness was investigated by Hamrock and Dowson.[39] The ellipticity parameter k was varied from 1 (a ball-on-plate configuration) to 8 (a configuration approaching a rectangular contact). The dimensionless speed parameter U was varied over a range of nearly two orders of magnitude and the dimensionless load parameter W over a range of one order of magnitude. Situations equivalent to using materials of bronze, steel, and silicon nitride and lubricants of paraffinic and naphthenic oils were considered in the investigation of the role of the dimensionless materials parameter G. Thirty-four cases were used in generating the minimum-film-thickness formula for hard EHL given here:

$$H_{\min} = 3.63U^{0.68}G^{0.49}W^{-0.073}(1 - e^{-0.68k}) \tag{57}$$

In this equation the dominant exponent occurs on the speed parameters, while the exponent on the load parameter is very small and negative. The materials parameter also carries a significant exponent, although the range of this variable in engineering situations is limited.

In addition to the minimum-film-thickness formula, contour plots of pressure and film thickness throughout the entire conjunction can be obtained from the numerical results. A representative contour plot of dimensionless pressure is shown in Fig. 70 for k = 1.25, U = 0.168×10^{-11}, and G = 4522. In this figure and in Fig. 71, the + symbol indicates the

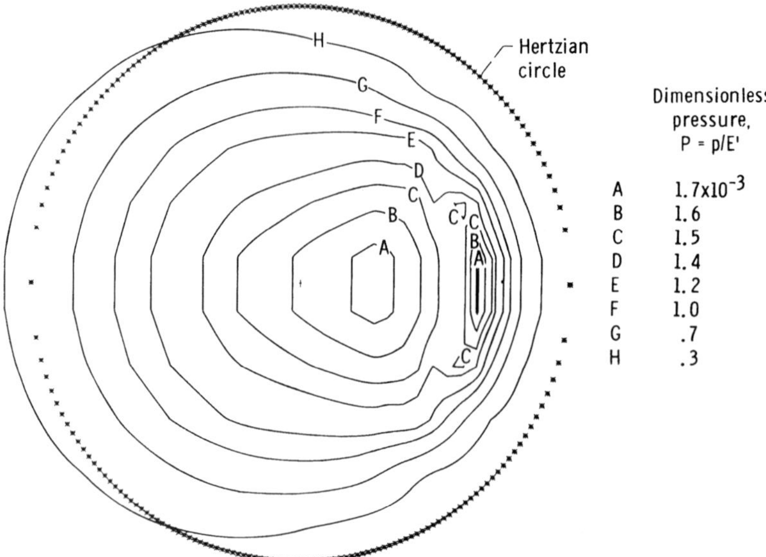

Figure 70 Contour plot of dimensionless pressure. $k = 1.25$; $U = 0.168 \times 10^{-11}$; $W = 0.111 \times 10^{-6}$; $G = 4522$. (From Ref. 39.)

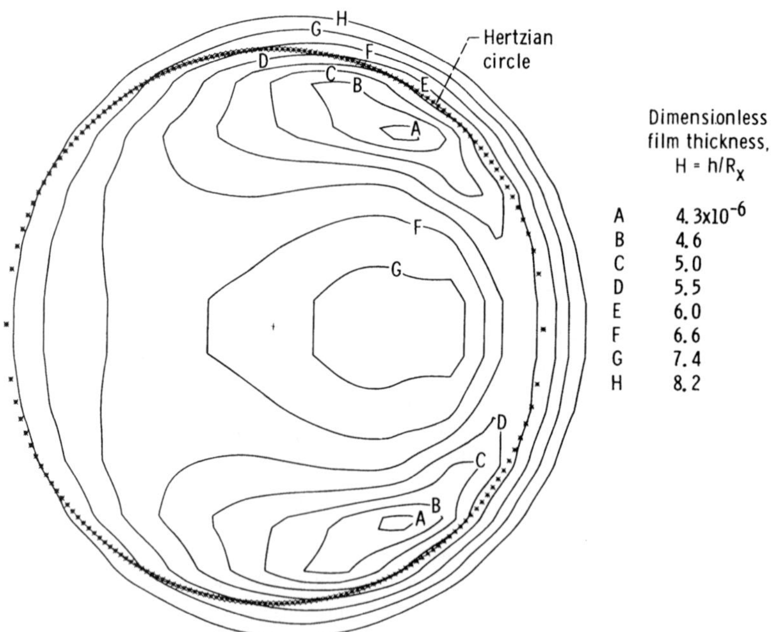

Figure 71 Contour plot of dimensionless film thickness. $k = 1.25$; $U = 0.168 \times 10^{-11}$; $W = 0.111 \times 10^{-6}$; $G = 4522$. (From Ref. 39.)

center of the Hertzian contact zone. The dimensionless representation of the X and Y coordinates causes the actual Hertzian contact ellipse to be a circle regardless of the value of the ellipticity parameter. The Hertzian contact circle is shown by asterisks. On this figure is a key showing the contour labels and each corresponding value of dimensionless pressure. The inlet region is to the left and the exit region is to the right. The pressure gradient at the exit end of the conjunction is much larger than that in the inlet region. In Fig. 70 a pressure spike is visible at the exit of the contact.

Contour plots of the film thickness are shown in Fig. 71 for the same case as Fig. 70. In this figure two minimum regions occur in well-defined lobes that follow, and are close to, the edge of the Hertzian contact circle. These results contain all of the essential features of available experimental observations based on optical interferometry.[40]

3.4 Soft-EHL Results

In a similar manner, Hamrock and Dowson[41] investigated the behavior of soft-EHL contacts. The ellipticity parameter was varied from 1 (a circular configuration) to 12 (a configuration approaching a rectangular contact), while U and W were varied by one order of magnitude and there were two different dimensionless materials parameters. Seventeen cases were considered in obtaining the dimensionless minimum-film-thickness equation for soft EHL:

$$H_{\min} = 7.43U^{0.65}W^{-0.21}(1 - 0.85e^{-0.31k}) \tag{58}$$

The powers of U in Eqs. (57) and (58) are quite similar, but the power of W is much more significant for soft-EHL results. The expression showing the effect of the ellipticity parameter is of exponential form in both equations, but with quite different constants.

A major difference between Eqs. (57) and (58) is the absence of the materials parameter in the expression for soft EHL. There are two reasons for this: one is the negligible effect of the relatively low pressures on the viscosity of the lubricating fluid, and the other is the way in which the role of elasticity is automatically incorporated into the prediction of conjunction behavior through the parameters U and W. Apparently the chief effect of elasticity is to allow the Hertzian contact zone to grow in response to increases in load.

3.5 Film Thickness for Different Regimens of Fluid-Film Lubrication

The types of lubrication that exist within nonconformal contacts like that shown in Fig. 70 are influenced by two major physical effects: the elastic deformation of the solids under an applied load and the increase in fluid viscosity with pressure. Therefore, it is possible to have four regimens of fluid-film lubrication, depending on the magnitude of these effects and on their relative importance. In this section, because of the need to represent the four fluid-film lubrication regimens graphically, the dimensionless grouping presented in Section 3.2 will need to be recast. That is, the set of dimensionless parameters given in Section 3.2—$\{H, U, W, G, \text{and } k\}$—will be reduced by one parameter without any loss of generality. Thus the dimensionless groupings to be used here are:

Dimensionless film parameter

$$\hat{H} = H\left(\frac{W}{U}\right)^2 \tag{59}$$

Dimensionless viscosity parameter

$$g_v = \frac{GW^3}{U^2} \tag{60}$$

Dimensionless elasticity parameter

$$g_e = \frac{W^{8/3}}{U^2} \tag{61}$$

The ellipticity parameter remains as discussed in Section 3.1, Eq. (39). Therefore the reduced dimensionless group is $\{\hat{H}, g_v, g_e, k\}$.

Isoviscous-Rigid Regimen

In this regimen the magnitude of the elastic deformation of the surfaces is such an insignificant part of the thickness of the fluid film separating them that it can be neglected, and the maximum pressure in the contact is too low to increase fluid viscosity significantly. This form of lubrication is typically encountered in circular-arc thrust bearing pads; in industrial processes in which paint, emulsion, or protective coatings are applied to sheet or film materials passing between rollers; and in very lightly loaded rolling bearings.

The influence of conjunction geometry on the isothermal hydrodynamic film separating two rigid solids was investigated by Brewe et al.[42] The effect of geometry on the film thickness was determined by varying the radius ratio R_y/R_x from 1 (circular configuration) to 36 (a configuration approaching a rectangular contact). The film thickness was varied over two orders of magnitude for conditions representative of steel solids separated by a paraffinic mineral oil. It was found that the computed minimum film thickness had the same speed, viscosity, and load dependence as the classical Kapitza solution,[43] so that the new dimensionless film thickness H is constant. However, when the Reynolds cavitation condition ($\partial p/\partial n = 0$ and $p = 0$) was introduced at the cavitation boundary, where n represents the coordinate normal to the interface between the full film and the cavitation region, an additional geometrical effect emerged. According to Brewe et al.,[42] the dimensionless minimum-film-thickness parameter for the isoviscous-rigid regime should now be written as

$$(\hat{H}_{\min})_{\text{ir}} = 128\alpha\lambda_b^2 \left[0.131 \tan^{-1}\left(\frac{\alpha}{2}\right) + 1.683\right]^2 \tag{62}$$

where

$$\alpha = \frac{R_y}{R_x} \approx (k)^{\pi/2} \tag{63}$$

and

$$\lambda_b = \left(1 + \frac{2}{3\alpha}\right)^{-1} \tag{64}$$

In Eq. (62) the dimensionless film thickness parameter $\hat{H}$ is shown to be strictly a function only of the geometry of the contact described by the ratio $\alpha = R_y/R_x$.

Piezoviscous-Rigid Regime

If the pressure within the contact is sufficiently high to increase the fluid viscosity within the conjunction significantly, it may be necessary to consider the pressure–viscosity characteristics of the lubricant while assuming that the solids remain rigid. For the latter part of this assumption to be valid, it is necessary that the deformation of the surfaces remain an insignificant part of the fluid-film thickness. This form of lubrication may be encountered on roller end-guide flanges, in contacts in moderately loaded cylindrical tapered rollers, and between some piston rings and cylinder liners.

From Hamrock and Dowson[44] the minimum-film-thickness parameter for the piezoviscous-rigid regimen can be written as

$$(\hat{H}_{\min})_{\text{pvr}} = 1.66\, g_v^{2/3}\, (1 - e^{-0.68k}) \tag{65}$$

Note the absence of the dimensionless elasticity parameter g_e from Eq. (65).

Isoviscous-Elastic (Soft-EHL) Regimen

In this regimen the elastic deformation of the solids is a significant part of the thickness of the fluid film separating them, but the pressure within the contact is quite low and insufficient to cause any substantial increase in viscosity. This situation arises with materials of low elastic modulus (such as rubber), and it is a form of lubrication that may be encountered in seals, human joints, tires, and elastomeric material machine elements.

If the film thickness equation for soft EHL [Eq. (58)] is rewritten in terms of the reduced dimensionless grouping, the minimum-film-thickness parameter for the isoviscous-elastic regimen can be written as

$$(\hat{H}_{\min})_{\text{ie}} = 8.70 g_e^{0.67}(1 - 0.85e^{-0.31k}) \tag{66}$$

Note the absence of the dimensionless viscosity parameter g_v from Eq. (66).

Piezoviscous-Elastic (Hard-EHL) Regime

In fully developed elastohydrodynamic lubrication the elastic deformation of the solids is often a significant part of the thickness of the fluid film separating them, and the pressure within the contact is high enough to cause a significant increase in the viscosity of the lubricant. This form of lubrication is typically encountered in ball and roller bearings, gears, and cams.

Once the film thickness equation [Eq. (57)] has been rewritten in terms of the reduced dimensionless grouping, the minimum film parameter for the piezoviscous-elastic regimen can be written as

$$(\hat{H}_{\min})_{\text{pve}} = 3.42 g_v^{0.49} g_e^{0.17}(1 - e^{-0.68k}) \tag{67}$$

An interesting observation to make in comparing Eqs. (65)–(67) is that in each case the sum of the exponents on g_v and g_e is close to the value of ⅔ required for complete dimensional representation of these three lubrication regimens: piezoviscous-rigid, isoviscous-elastic, and piezoviscous-elastic.

Contour Plots

Having expressed the dimensionless minimum-film-thickness parameter for the four fluid-film regimens in Eqs. (62)–(67), Hamrock and Dowson[44] used these relationships to develop a map of the lubrication regimens in the form of dimensionless minimum-film-thickness parameter contours. Some of these maps are shown in Figs. 72–74 on a log-log grid of the dimensionless viscosity and elasticity parameters for ellipticity parameters of 1, 3, and 6, respectively. The procedure used to obtain these figures can be found in Ref. 44. The four lubrication regimens are clearly shown in Figs. 72–74. By using these figures for given values of the parameters k, g_v, and g_e, the fluid-film lubrication regimen in which any elliptical conjunction is operating can be ascertained and the approximate value of $\hat{H}_{\min}$ can be determined. When the lubrication regimen is known, a more accurate value of $\hat{H}_{\min}$ can

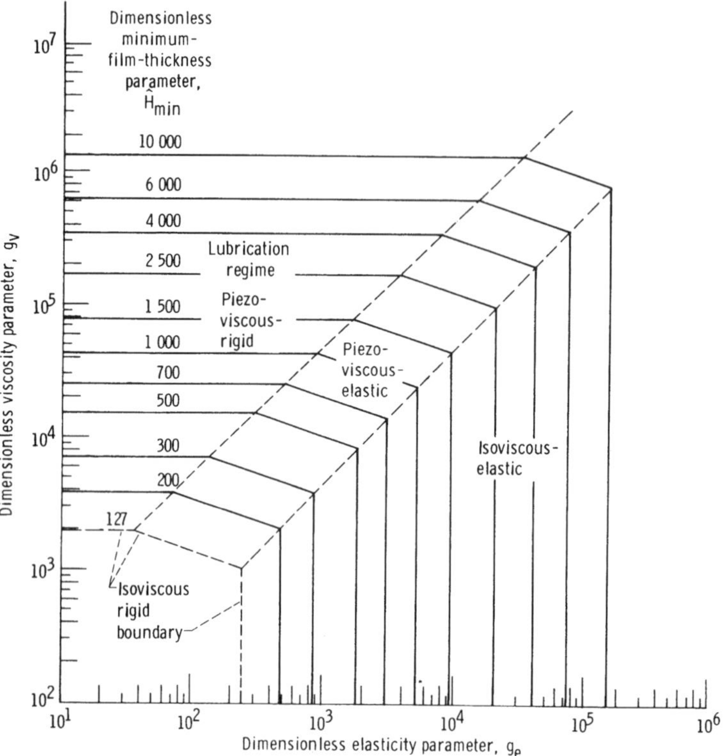

Figure 72 Map of lubrication regimens for ellipticity parameter k of 1. (From Ref. 44.)

be obtained by using the appropriate dimensionless minimum-film-thickness equation. These results are particularly useful in initial investigations of many practical lubrication problems involving elliptical conjunctions.

3.6 Rolling-Element Bearings

Rolling-element bearings are precision, yet simple, machine elements of great utility, whose mode of lubrication is elastohydrodynamic. This section describes the types of rolling-element bearings and their geometry, kinematics, load distribution, and fatigue life and demonstrates how elastohydrodynamic lubrication theory can be applied to the operation of rolling-element bearings. This section makes extensive use of the work by Hamrock and Dowson[10] and by Hamrock and Anderson.[6]

Bearing Types

A great variety of both design and size range of ball and roller bearings are available to the designer. The intent of this section is not to duplicate the complete descriptions given in manufacturers' catalogs, but rather to present a guide to representative bearing types along with the approximate range of sizes available. Tables 9–17 illustrate some of the more widely used bearing types. In addition, there are numerous types of specialty bearings available for

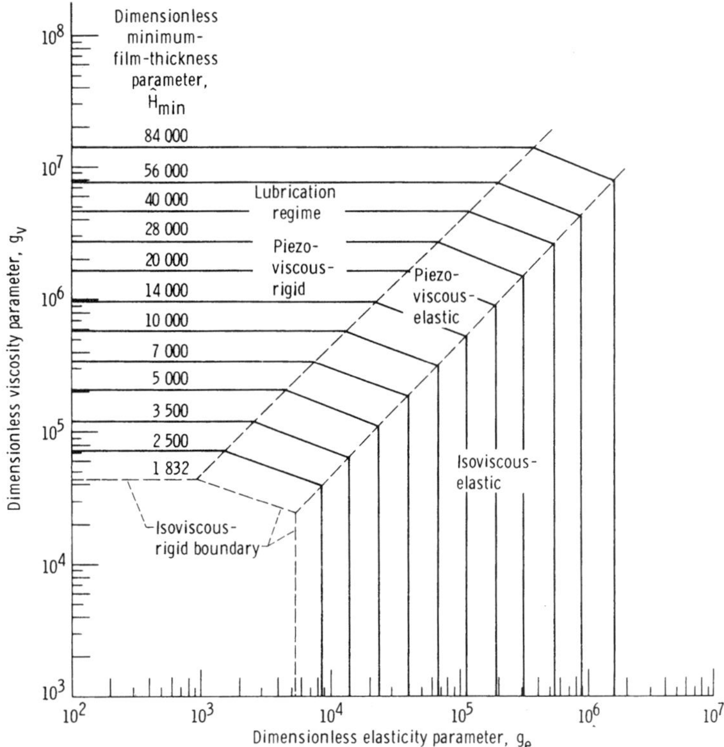

Figure 73 Map of lubrication regimens for ellipticity parameter k of 3. (From Ref. 44.)

which space does not permit a complete cataloging. Size ranges are given in metric units. Traditionally, most rolling-element bearings have been manufactured to metric dimensions, predating the efforts toward a metric standard. In addition to bearing types and approximate size ranges available, Tables 9–17 also list approximate relative load-carrying capabilities, both radial and thrust, and, where relevant, approximate tolerances to misalignment.

Rolling bearings are an assembly of several parts–an inner race, an outer race, a set of balls or rollers, and a cage or separator. The cage or separator maintains even spacing of the rolling elements. A cageless bearing, in which the annulus is packed with the maximum rolling-element complement, is called a full-complement bearing. Full-complement bearings have high load capacity but lower speed limits than bearings equipped with cages. Tapered-roller bearings are an assembly of a cup, a cone, a set of tapered rollers, and a cage.

Ball Bearings. Ball bearings are used in greater quantity than any other type of rolling bearing. For an application where the load is primarily radial with some thrust load present, one of the types in Table 9 can be chosen. A Conrad, or deep-groove, bearing has a ball complement limited by the number of balls that can be packed into the annulus between the inner and outer races with the inner race resting against the inside diameter of the outer race. A stamped and riveted two-piece cage, piloted on the ball set, or a machined two-piece cage, ball piloted or race piloted, is almost always used in a Conrad bearing. The only exception is a one-piece cage with open-sided pockets that is snapped into place. A filling-

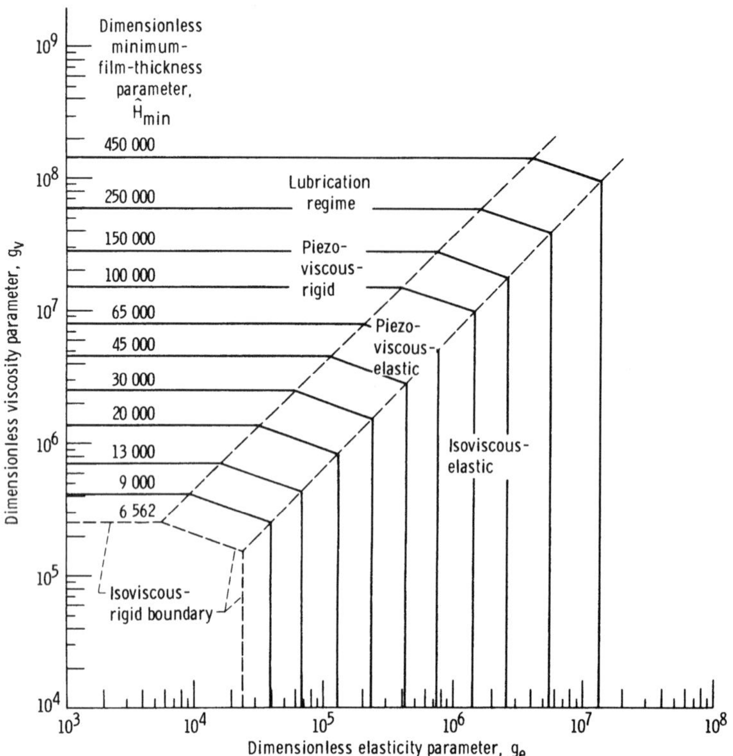

Figure 74 Map of lubrication regimens for ellipticity parameter k of 6. (From Ref. 44.)

notch bearing has both inner and outer races notched so that a ball complement limited only by the annular space between the races can be used. It has low thrust capacity because of the filling notch.

The self-aligning internal bearing shown in Table 9 has an outer-race ball path ground in a spherical shape so that it can accept high levels of misalignment. The self-aligning external bearing has a multipiece outer race with a spherical interface. It too can accept high misalignment and has higher capacity than the self-aligning internal bearing. However, the external self-aligning bearing is somewhat less self-aligning than its internal counterpart because of friction in the multipiece outer race.

Representative angular-contact ball bearings are illustrated in Table 10. An angular-contact ball bearing has a two-shouldered ball groove in one race and a single-shouldered ball groove in the other race. Thus it is capable of supporting only a unidirectional thrust load. The cutaway shoulder allows assembly of the bearing by snapping over the ball set after it is positioned in the cage and outer race. This also permits use of a one-piece, machined, race-piloted cage that can be balanced for high-speed operation. Typical contact angles vary from 15° to 25°.

Angular-contact ball bearings are used in duplex pairs mounted either back to back or face to face as shown in Table 10. Duplex bearing pairs are manufactured so that they "preload" each other when clamped together in the housing and on the shaft. The use of preloading provides stiffer shaft support and helps prevent bearing skidding at light loads.

Table 9 Characteristics of Representative Radial Ball Bearings (From Ref. 10)

Type		Approximate Range of Bore Sizes, mm		Relative Capacity		Limiting Speed Factor	Tolerance to Misalignment
		Minimum	Maximum	Radial	Thrust		
Conrad or deep groove		3	1060	1.00	0.7[a]	1.0	±0°15′
Maximum capacity or filling notch		10	130	1.2–1.4	0.2[a]	1.0	±0°3′
Magneto or counterbored outer		3	200	0.9–1.3	0.5–0.9[b]	1.0	±0°5′
Airframe or aircraft control		4.826	31.75	High static capacity	0.5[a]	0.2	0°
Self-aligning, internal		5	120	0.7	0.2[b]	1.0	±2°30′
Self-aligning, external		—	—	1.0	0.7[a]	1.0	High
Double row, maximum		6	110	1.5	0.2[a]	1.0	±0°3′
Double row, deep groove		6	110	1.5	1.4[a]	1.0	0°

[a] Two directions.
[b] One direction.

Proper levels of preload can be obtained from the manufacturer. A duplex pair can support bidirectional thrust load. The back-to-back arrangement offers more resistance to moment or overturning loads than does the face-to-face arrangement.

Where thrust loads exceed the capability of a simple bearing, two bearings can be used in tandem, with both bearings supporting part of the thrust load. Three or more bearings are occasionally used in tandem, but this is discouraged because of the difficulty in achieving good load sharing. Even slight differences in operating temperature will cause a maldistribution of load sharing.

The split-ring bearing shown in Table 10 offers several advantages. The split ring (usually the inner) has its ball groove ground as a circular arc with a shim between the ring halves. The shim is then removed when the bearing is assembled so that the split-ring ball groove has the shape of a gothic arch. This reduces the axial play for a given radial play and results in more accurate axial positioning of the shaft. The bearing can support bidirectional thrust loads but must not be operated for prolonged periods of time at predominantly

Table 10 Characteristics of Representative Angular-Contact Ball Bearings (From Ref. 10)

Type	Approximate Range of Bore Sizes, mm		Relative Capacity		Limiting Speed Factor	Tolerance to Misalignment
	Minimum	Maximum	Radial	Thrust		
One-directional thrust	10	320	1.00–1.15[b]	1.5–2.3[a,b]	1.1–3.0[b]	±0°2′
Duplex, back to back	10	320	1.85	1.5[c]	3.0	0°
Duplex, face to face	10	320	1.85	1.5[c]	3.0	0°
Duplex, tandem	10	320	1.85	2.4[a]	3.0	0°
Two-directional or split ring	10	110	1.15	1.5[c]	3.0	±0°2′
Double row	10	140	1.5	1.85[c]	0.8	0°
Double row, maximum	10	110	1.65	0.5[a] 1.5[d]	0.7	0°

[a] One direction.
[b] Depends on contact angle.
[c] Two directions.
[d] In other direction.

Table 11 Characteristics of Representative Thrust Ball Bearings (From Ref. 10)

Type	Approximate Range of Bore Sizes, mm		Relative Capacity		Limiting Speed Factor	Tolerance to Misalignment
	Minimum	Maximum	Radial	Thrust		
One directional, flat race	6.45	88.9	0	0.7[a]	0.10	0°[b]
One directional, grooved race	6.45	1180	0	1.5[a]	0.30	0°
Two directional, grooved race	15	220	0	1.5[c]	0.30	0°

[a] One direction.
[b] Accepts eccentricity.
[c] Two directions.

Table 12 Characteristics of Representative Cylindrical Roller Bearings (From Ref. 10)

Type	Approximate Range of Bore Sizes, mm		Relative Capacity		Limiting Speed Factor	Tolerance to Misalignment
	Minimum	Maximum	Radial	Thrust		
Separable outer ring, nonlocating (RN, RIN)	 10	320	1.55	0	1.20	±0°5′
Separable inner ring, nonlocating (RU, RIU)	 12	500	1.55	0	1.20	±0°5′
Separable outer ring, one-direction locating (RF, RIF)	 40	177.8	1.55	Locating[a]	1.15	±0°5′
Separable inner ring, one-direction locating (RJ, RIJ)	 12	320	1.55	Locating[a]	1.15	±0°5′
Self-contained, two-direction locating	 12	100	1.35	Locating[b]	1.15	±0°5′
Separable inner ring, two-direction locating (RT, RIT)	 20	320	1.55	Locating[b]	1.15	±0°5′
Nonlocating, full complement (RK, RIK)	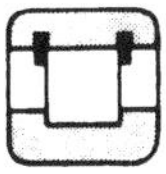 17	75	2.10	0	0.20	±0°5′
Double row, separable outer ring, nonlocating (RD)	 30	1060	1.85	0	1.00	0°
Double row, separable inner ring, nonlocating	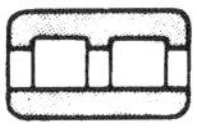 70	1060	1.85	0	1.00	0°

[a] One direction.
[b] Two directions.

Table 13 Characteristics of Representative Spherical Roller Bearings (From Ref. 10)

Type	Approximate Range of Bore Sizes, mm		Relative Capacity		Limiting Speed Factor	Tolerance to Misalignment
	Minimum	Maximum	Radial	Thrust		
Single row, barrel or convex	20	320	2.10	0.20	0.50	±2°
Double row, barrel or convex	25	1250	2.40	0.70	0.50	±1°30′
Thrust	85	360	0.10[a] 0.10[b]	1.80[a] 2.40[b]	0.35–0.50	±3°
Double row, concave	50	130	2.40	0.70	0.50	±1°30′

[a] Symmetric rollers.
[b] Asymmetric rollers.

radial loads. This results in three-point ball–race contact and relatively high frictional losses. As with the conventional angular-contact bearing, a one-piece precision-machined cage is used.

Ball thrust bearings (90° contact angle), Table 11, are used almost exclusively for machinery with vertical oriented shafts. The flat-race bearing allows eccentricity of the fixed and rotating members. An additional bearing must be used for radial positioning. It has low load capacity because of the very small ball–race contacts and consequent high Hertzian stress. Grooved-race bearings have higher load capacities and are capable of supporting low-magnitude radial loads. All of the pure thrust ball bearings have modest speed capability

Table 14 Characteristics of Standardized Double-Row, Spherical Roller Bearings (From Ref. 10)

Type	Roller Design	Retainer Design	Roller Guidance	Roller-Race Contact
SLB	Symmetric	Machined, roller piloted	Retainer pockets	Modified line, both races
SC	Symmetric	Stamped, race piloted	Floating guide ring	Modified line, both races
SD	Asymmetric	Machined, race piloted	Inner-ring center rib	Line contact, outer; point contact, inner

Table 15 Characteristics of Spherical Roller Bearings (From Ref. 10)

Series	Types	Approximate Range of Bore Sizes, mm		Approximate Relative Capacity[a]		Limiting Speed Factor
		Minimum	Maximum	Radial	Thrust	
202	Single-row barrel	20	320	1.0	0.11	0.5
203	Single-row barrel	20	240	1.7	0.18	0.5
204	Single-row barrel	25	110	2.1	0.22	0.4
212	SLB	35	75	1.0	0.26	0.6
213	SLB	30	70	1.7	0.53	↓
22, 22K	SLB, SC, SD	30	320	1.7	0.46	↓
23, 23K	SLB, SC, SD	40	280	2.7	1.0	↓
30, 30K	SLB, SC, SD	120	1250	1.2	0.29	0.7
31, 31K	SLB, SC, SD	110	1250	1.7	0.54	0.6
32, 32K	SLB, SC, SD	100	850	2.1	0.78	0.6
39, 39K	SD	120	1250	.7	0.18	0.7
40, 40K	SD	180	250	1.5	—	0.7

[a]Load capacities are comparative within the various series of spherical roller bearings only. For a given envelope size, a spherical roller bearing has a radial capacity approximately equal to that of a cylindrical roller bearing.

because of the 90° contact angle and the consequent high level of ball spinning and frictional losses.

Roller Bearings. Cylindrical roller bearings, Table 12, provide purely radial load support in most applications. An N or U type of bearing will allow free axial movement of the shaft relative to the housing to accommodate differences in thermal growth. An F or J type of bearing will support a light thrust load in one direction; and a T type of bearing will support a light bidirectional thrust load.

Cylindrical roller bearings have moderately high radial load capacity as well as high-speed capability. Their speed capability exceeds that of either spherical or tapered-roller bearings. A commonly used bearing combination for support of a high-speed rotor is an angular-contact ball bearing or duplex pair and a cylindrical roller bearing.

As explained in the following section on bearing geometry, the rollers in cylindrical roller bearings are seldom pure cylinders. They are crowned or made slightly barrel shaped to relieve stress concentrations of the roller ends when any misalignment of the shaft and housing is present.

Cylindrical roller bearings may be equipped with one- or two-piece cages, usually race piloted. For greater load capacity, full-complement bearings can be used, but at a significant sacrifice in speed capability.

Spherical roller bearings, Tables 13–15, are made as either single- or double-row bearings. The more popular bearing design uses barrel-shaped rollers. An alternative design employs hourglass-shaped rollers. Spherical roller bearings combine very high radial load capacity with modest thrust load capacity (with the exception of the thrust type) and excellent tolerance to misalignment. They find widespread use in heavy-duty rolling mill and industrial gear drives, where all of these bearing characteristics are requisite.

Tapered-roller bearings, Table 16, are also made as single- or double-row bearings with combinations of one- or two-piece cups and cones. A four-row bearing assembly with two-

Table 16 Characteristics of Representative Tapered Roller Bearings (From Ref. 10)

Type	Subtype	Approximate Range of Bore Sizes, mm	
		Minimum	Maximum
Single row (TS)	TST—tapered bore	8	1690
	TSS—steep angle	24	430
	TS—pin cage	16	1270
	TSE, TSK—keyway cones		—
	TSF, TSSF—flanged cup	12	380
	TSG—steering gear (without cone)	8	1070
			—
Two row, double cone, single cups (TDI)	TDIK, TDIT, TDITP—tapered bore	30	1200
		30	860
	TDIE, TDIKE—slotted double cone	24	690
		55	520
	TDIS—steep angle		
Two row, double cup, single cones, adjustable (TDO)	TDO	8	1830
	TDOS—steep angle	20	1430
Two row, double cup, single cones, nonadjustabe (TNA)	TNA	20	60
	TNASW—slotted cones	30	260
	TNASWE—extended cone rib	20	305
		8	70
	TNASWH—slotted cones, sealed	—	—
	TNADA, TNHDADX—self-aligning cup AD		
Four row, cup adjusted (TQO)		70	1500
	TQO, TQOT—tapered bore	250	1500
Four row, cup adjusted (TQI)	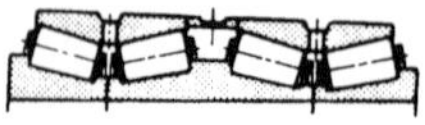 TQIT—tapered bore	—	—

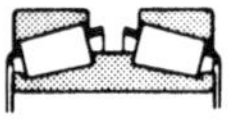

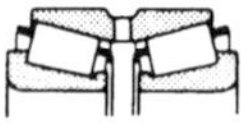

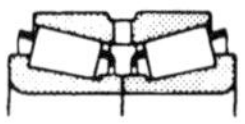

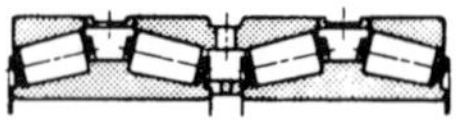

or three-piece cups and cones is also available. Bearings are made with either a standard angle for applications in which moderate thrust loads are present or with a steep angle for high thrust capacity. Standard and special cages are available to suit the application requirements.

Single-row tapered-roller bearings must be used in pairs because a radially loaded bearing generates a thrust reaction that must be taken by a second bearing. Tapered-roller bearings are normally set up with spacers designed so that they operate with some internal play. Manufacturers' engineering journals should be consulted for proper setup procedures.

Needle roller bearings, Table 17, are characterized by compactness in the radial direction and are frequently used without an inner race. In the latter case the shaft is hardened and ground to serve as the inner race. Drawn cups, both open and closed end, are frequently used for grease retention. Drawn cups are thin walled and require substantial support from

Table 17 Characteristics of Representative Needle Roller Bearings (From Ref. 10)

Type		Bore Sizes, mm		Relative Load Capacity		Limiting Speed Factor	Misalignment Tolerance
		Minimum	Maximum	Dynamic	Static		
Drawn cup, needle	Open end; Closed end	3	185	High	Moderate	0.3	Low
Drawn cup, needle, grease retained		4	25	High	Moderate	0.3	Low
Drawn cup, roller	Open end; Closed end	5	70	Moderate	Moderate	0.9	Moderate
Heavy-duty roller		16	235	Very high	Moderate	1.0	Moderate
Caged roller		12	100	Very high	High	1.0	Moderate
Cam follower		12	150	Moderate to high	Moderate to high	0.3–0.9	Low
Needle thrust		6	105	Very high	Very high	0.7	Low

the housing. Heavy-duty roller bearings have relatively rigid races and are more akin to cylindrical roller bearings with long-length-to-diameter-ratio rollers.

Needle roller bearings are more speed limited than cylindrical roller bearings because of roller skewing at high speeds. A high percentage of needle roller bearings are full-complement bearings. Relative to a caged needle bearing, these have higher load capacity but lower speed capability.

There are many types of specialty bearings available other than those discussed here. Aircraft bearings for control systems, thin-section bearings, and fractured-ring bearings are some of the more widely used bearings among the many types manufactured. A complete coverage of all bearing types is beyond the scope of this chapter.

Angular-contact ball bearings and cylindrical roller bearings are generally considered to have the highest speed capabilities. Speed limits of roller bearings are discussed in conjunction with lubrication methods. The lubrication system employed has as great an influence on limiting bearing speed as does the bearing design.

Geometry

The operating characteristics of a rolling-element bearing depend greatly on the diametral clearance of the bearing. This clearance varies for the different types of bearings discussed in the preceding section. In this section, the principal geometrical relationships governing the operation of unloaded rolling-element bearings are developed. This information will be of vital interest when such quantities as stress, deflection, load capacity, and life are considered in subsequent sections. Although bearings rarely operate in the unloaded state, an understanding of this section is vital to the appreciation of the remaining sections.

Geometry of Ball Bearings

PITCH DIAMETER AND CLEARANCE. The cross section through a radial, single-row ball bearing shown in Fig. 75 depicts the radial clearance and various diameters. The pitch diameter d_e is the mean of the inner- and outer-race contact diameters and is given by

$$d_e = d_i + \tfrac{1}{2}(d_o - d_i) \qquad \text{or} \qquad d_e = \tfrac{1}{2}(d_o + d_i) \tag{68}$$

Also from Fig. 75, the diametral clearance denoted by P_d can be written as

$$P_d = d_o - d_i - 2d \tag{69}$$

Diametral clearance may therefore be thought of as the maximum distance that one race can move diametrally with respect to the other when no measurable force is applied and both races lie in the same plane. Although diametral clearance is generally used in connection with single-row radial bearings, Eq. (69) is also applicable to angular-contact bearings.

RACE CONFORMITY. Race conformity is a measure of the geometrical conformity of the race and the ball in a plane passing through the bearing axis, which is a line passing through the center of the bearing perpendicular to its plane and transverse to the race. Figure 76 is a cross section of a ball bearing showing race conformity, expressed as

$$f = \frac{r}{d} \tag{70}$$

For perfect conformity, where the radius of the race is equal to the ball radius, f is equal to ½. The closer the race conforms to the ball, the greater the frictional heat within the contact. On the other hand, open-race curvature and reduced geometrical conformity, which reduce friction, also increase the maximum contact stresses and, consequently, reduce the bearing fatigue life. For this reason, most ball bearings made today have race conformity ratios in the range $0.51 \le f \le 0.54$, with $f = 0.52$ being the most common value. The race conformity

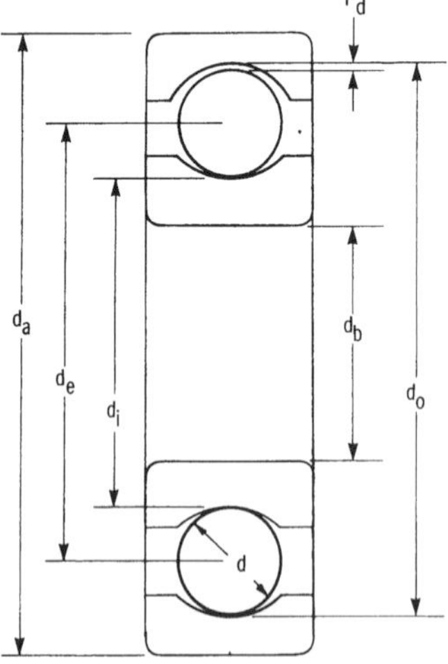

Figure 75 Cross section through radial, single-row ball bearing. (From Ref. 10.)

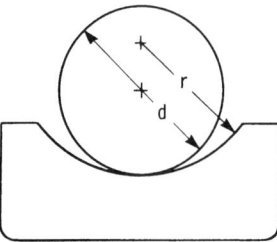

Figure 76 Cross section of ball and outer race, showing race conformity. (From Ref. 10.)

ratio for the outer race is usually made slightly larger than that for the inner race to compensate for the closer conformity in the plane of the bearing between the outer race and ball than between the inner race and ball. This tends to equalize the contact stresses at the inner- and outer-race contacts. The difference in race conformities does not normally exceed 0.02.

Contact Angle. Radial bearings have some axial play since they are generally designed to have a diametral clearance, as shown in Fig. 77. This implies a free-contact angle different from zero. Angular-contact bearings are specifically designed to operate under thrust loads. The clearance built into the unloaded bearing, along with the race conformity ratio, determines the bearing free-contact angle. Figure 77 shows a radial bearing with contact due to the axial shift of the inner and outer races when no measurable force is applied.

Before the free-contact angle is discussed, it is important to define the distance between the centers of curvature of the two races in line with the center of the ball in both Figs. 77*a* and 77*b*. This distance—denoted by x in Fig. 77*a* and by D in Fig. 77*b*—depends on race radius and ball diameter. Denoting quantities referred to the inner and outer races by subscripts i and o, respectively, we see from Figs. 77*a* and 77*b* that

$$\frac{P_d}{4} + d + \frac{P_d}{4} = r_o - x + r_i$$

or

$$x = r_o + r_i - d - \frac{P_d}{2}$$

and

$$d = r_o - D + r_i$$

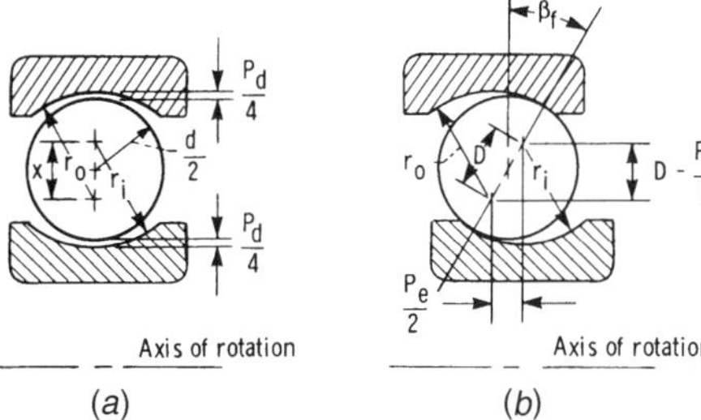

Figure 77 Cross section of radial ball bearing, showing ball–race contact due to axial shift of inner and outer rings. (From Ref. 10.) (*a*) Initial position. (*b*) Shifted position.

or

$$D = r_o + r_i - d \tag{71}$$

From these equations, we can write

$$x = D - \frac{P_d}{2}$$

This distance, shown in Fig. 77, will be useful in defining the contact angle.

By using Eq. (70), we can write Eq. (71) as

$$D = Bd \tag{72}$$

where

$$B = f_o + f_i - 1 \tag{73}$$

The quantity B in Eq. (72) is known as the total conformity ratio and is a measure of the combined conformity of both the outer and inner races to the ball. Calculations of bearing deflection in later sections depend on the quantity B.

The free-contact angle β_f (Fig. 77) is defined as the angle made by a line through the points of contact of the ball and both races with a plane perpendicular to the bearing axis of rotation when no measurable force is applied. Note that the centers of curvature of both the outer and inner races lie on the line defining the free-contact angle. From Fig. 77, the expression for the free-contact angle can be written as

$$\cos \beta_f = \frac{D - P_d/2}{D} \tag{74}$$

By using Eqs. (69) and (71), we can write Eq. (74) as

$$\beta_f = \cos^{-1}\left[\frac{r_o + r_i - \tfrac{1}{2}(d_o - d_i)}{r_o + r_i - d}\right] \tag{75}$$

Equation (75) shows that if the size of the balls is increased and everything else remains constant, the free-contact angle is decreased. Similarly, if the ball size is decreased, the free-contact angle is increased.

From Eq. (74) the diametral clearance P_d can be written as

$$P_d = 2D(1 - \cos \beta_f) \tag{76}$$

This is an alternative definition of the diametral clearance given in Eq. (69).

Endplay. Free endplay P_e is the maximum axial movement of the inner race with respect to the outer race when both races are coaxially centered and no measurable force is applied. Free endplay depends on total curvature and contact angle, as shown in Fig. 77, and can be written as

$$P_e = 2D \sin \beta_f \tag{77}$$

The variation of free-contact angle and free endplay with the ratio $P_d/2d$ is shown in Fig. 78 for four values of total conformity normally found in single-row ball bearings. Eliminating β_f in Eqs. (76) and (77) enables the establishment of the following relationships between free endplay and diametral clearance:

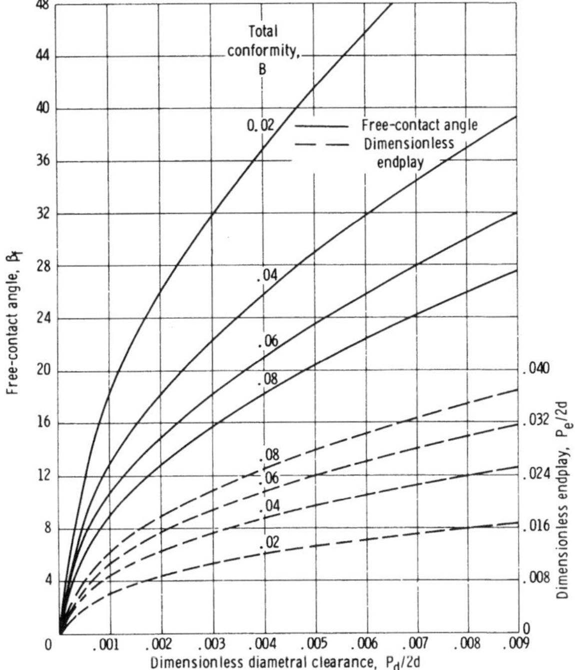

Figure 78 Chart for determining free-contact angle and endplay. (From Ref. 10.)

$$P_d = 2D - [(2D)^2 - P_c^2]^{1/2}$$

$$P_e = (4DP_d - P_d^2)^{1/2}$$

Shoulder Height. The shoulder height of ball bearings is illustrated in Fig. 79. Shoulder height, or race depth, is the depth of the race groove measured from the shoulder to the bottom of the groove and is denoted by s in Fig. 79. From this figure the equation defining the shoulder height can be written as

$$s = r(1 - \cos \theta) \tag{78}$$

The maximum possible diametral clearance for complete retention of the ball–race contact within the race under zero thrust load is given by

$$(P_d)_{\max} = \frac{2Ds}{r}$$

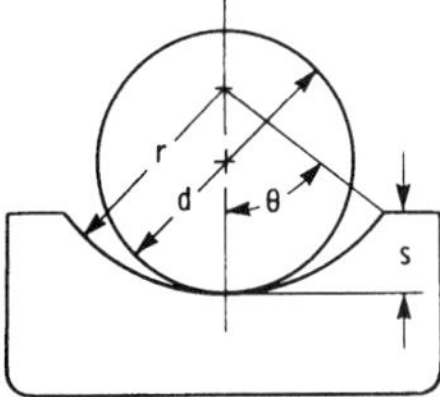

Figure 79 Shoulder height in ball bearing. (From Ref. 10.)

Curvature Sum and Difference. A cross section of a ball bearing operating at a contact angle β is shown in Fig. 80. Equivalent radii of curvature for both inner- and outer-race contacts in, and normal to, the direction of rolling can be calculated from this figure. The radii of curvature for the ball–inner-race contact are

$$r_{ax} = r_{ay} = \frac{d}{2} \tag{79}$$

$$r_{bx} = \frac{d_c - d \cos \beta}{2 \cos \beta} \tag{80}$$

$$r_{by} = -f_i d = -r_i \tag{81}$$

The radii of curvature for the ball–outer-race contact are

$$r_{ax} = r_{ay} = \frac{d}{2} \tag{82}$$

$$r_{bx} = \frac{d_c + d \cos \beta}{2 \cos \beta} \tag{83}$$

$$r_{by} = -f_o d = -r_o \tag{84}$$

In Eqs. (80) and (81), β is used instead of β_f since these equations are also valid when a load is applied to the contact. By setting $\beta = 0°$, Eqs. (79)–(84) are equally valid for radial ball bearings. For thrust ball bearings, $r_{bx} = \infty$ and the other radii are defined as given in the preceding equations.

Equations (36) and (37) effectively redefine the problem of two ellipsoidal solids approaching one another in terms of an equivalent ellipsoidal solid of radii R_x and R_y approaching a plane. From the radius-of-curvature expressions, the radii R_x and R_y for the contact example discussed earlier can be written for the ball–inner-race contact as

$$R_x = \frac{d(d_e - d \cos \beta)}{2d_e} \tag{85}$$

$$R_y = \frac{f_i d}{2f_i - 1} \tag{86}$$

and for the ball–outer-race contact as

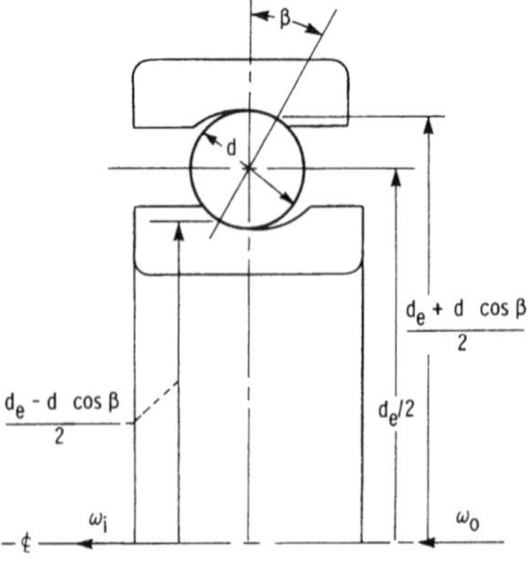

Figure 80 Cross section of ball bearing. (From Ref. 10.)

$$R_x = \frac{d(d_e + d \cos \beta)}{2d_e} \tag{87}$$

$$R_y = \frac{f_o d}{2f_o - 1} \tag{88}$$

Roller Bearings. The equations developed for the pitch diameter d_e and diametral clearance P_d for ball bearings in Eqs. (68) and (69), respectively, are directly applicable for roller bearings.

CROWNING. To prevent high stresses at the edges of the rollers in cylindrical roller bearings, the rollers are usually crowned as shown in Fig. 81. A fully crowned roller is shown in Fig. 81*a* and a partially crowned roller in Fig. 81*b*. In this figure the crown curvature is greatly exaggerated for clarity. The crowning of rollers also gives the bearing protection against the effects of slight misalignment. For cylindrical rollers, $r_{ay}/d \approx 10^2$. In contrast, for spherical rollers in spherical roller bearings, as shown in Fig. 81, $r_{ay}/d \approx 4$. In Fig. 81 it is observed that the roller effective length l_r is the length presumed to be in contact with the races under loading. Generally, the roller effective length can be written as

$$l_r = l_t - 2r_c$$

where r_c is the roller corner radius or the grinding undercut, whichever is larger.

RACE CONFORMITY. Race conformity applies to roller bearings much as it applies to ball bearings. It is a measure of the geometrical conformity of the race and the roller. Figure 82 shows a cross section of a spherical roller bearing. From this figure the race conformity can be written as

$$f = \frac{r}{2r_{ay}}$$

In this equation if subscript i or o is added to f and r, we obtain the values for the race conformity for the inner- and outer-race contacts.

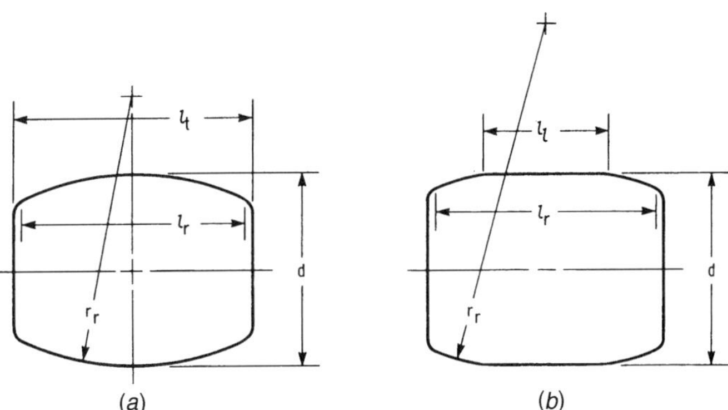

Figure 81 Spherical and cylindrical rollers. (From Ref. 6.) (*a*) Spherical roller (fully crowned). (*b*) Cylindrical roller (partially crowned).

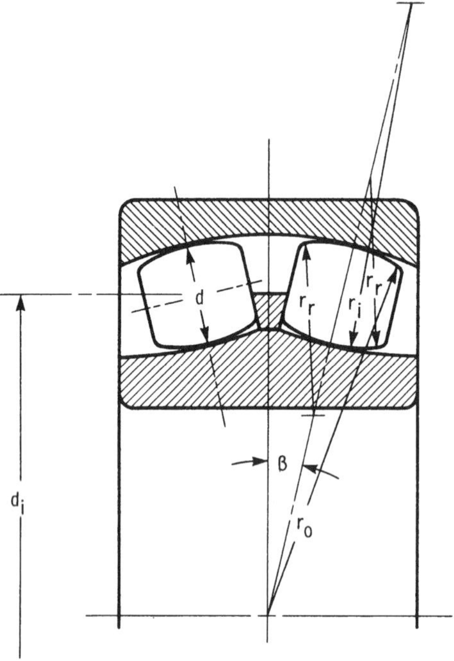

Figure 82 Spherical roller bearing geometry. (From Ref. 6.)

Free Endplay and Contact Angle. Cylindrical roller bearings have a contact angle of zero and may take thrust load only by virtue of axial flanges. Tapered-roller bearings must be subjected to a thrust load or the inner and outer races (the cone and cup) will not remain assembled; therefore, tapered-roller bearings do not exhibit free diametral play. Radial spherical roller bearings are, however, normally assembled with free diametral play and, hence, exhibit free endplay. The diametral play P_d for a spherical roller bearing is the same as that obtained for ball bearings as expressed in Eq. (69). This diametral play as well as endplay is shown in Fig. 83 for a spherical roller bearing. From this figure we can write that

$$r_o \cos \beta = \left(r_o - \frac{P_d}{2} \right) \cos \gamma$$

or

$$\beta = \cos^{-1} \left[\left(1 - \frac{P_d}{2r_o} \right) \cos \gamma \right]$$

Also from Fig. 83 the free endplay can be written as

$$P_e = 2r_o(\sin \beta - \sin \gamma) + P_d \sin \gamma$$

Curvature Sum and Difference. The same procedure will be used for defining the curvature sum and difference for roller bearings as was used for ball bearings. For spherical roller bearings, as shown in Fig. 82, the radii of curvature for the roller–inner-race contact can be written as

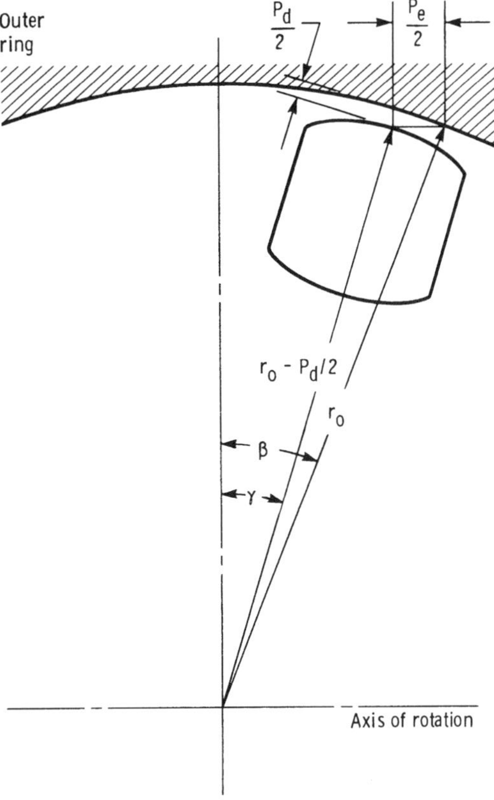

Figure 83 Schematic diagram of spherical roller bearing, showing diametral play and endplay. (From Ref. 6.)

$$r_{ax} = \frac{d}{2} \qquad r_{ay} = f_i \left(\frac{r_i}{2}\right)$$

$$r_{bx} = \frac{d_e - d \cos \beta}{2 \cos \beta} \qquad r_{by} = -2f_i r_{ay}$$

For the spherical roller bearing shown in Fig. 82 the radii of curvature for the roller–outer-race contact can be written as

$$r_{ax} = \frac{d}{2} \qquad r_{ay} = f_o \left(\frac{r_o}{2}\right)$$

$$r_{bx} = -\frac{d_e + d \cos \beta}{2 \cos \beta} \qquad r_{by} = -2f_o r_{ay}$$

Knowing the radii of curvature for the contact condition, we can write the curvature sum and difference directly from Eqs. (34) and (35). Furthermore, the radius-of-curvature expressions R_x and R_y for spherical roller bearings can be written for the roller–inner-race contact as

$$R_x = \frac{d(d_c - d\cos\beta)}{2d_c} \tag{89}$$

$$R_y = \frac{2r_{ay}f_i}{2f_i - 1} \tag{90}$$

and for the roller–outer-race contact as

$$R_x = \frac{d(d_c + d\cos\beta)}{2d_c} \tag{91}$$

$$R_y = \frac{2r_{ay}f_o}{2f_o - 1} \tag{92}$$

Kinematics

The relative motions of the separator, the balls or rollers, and the races of rolling-element bearings are important to understanding their performance. The relative velocities in a ball bearing are somewhat more complex than those in roller bearings, the latter being analogous to the specialized case of a zero- or fixed-value contact-angle ball bearing. For that reason the ball bearing is used as an example here to develop approximate expressions for relative velocities. These are useful for rapid but reasonably accurate calculation of elastohydrodynamic film thickness, which can be used with surface roughnesses to calculate the lubrication life factor.

When a ball bearing operates at high speeds, the centrifugal force acting on the ball creates a difference between the inner- and outer-race contact angles, as shown in Fig. 84, in order to maintain force equilibrium on the ball. For the most general case of rolling and spinning at both inner- and outer-race contacts, the rolling and spinning velocities of the ball are as shown in Fig. 85.

The equations for ball and separator angular velocity for all combinations of inner- and outer-race rotation were developed by Jones.[45] Without introducing additional relationships to describe the elastohydrodynamic conditions at both ball–race contacts, however, the ball–spin-axis orientation angle ϕ cannot be obtained. As mentioned, this requires a long nu-

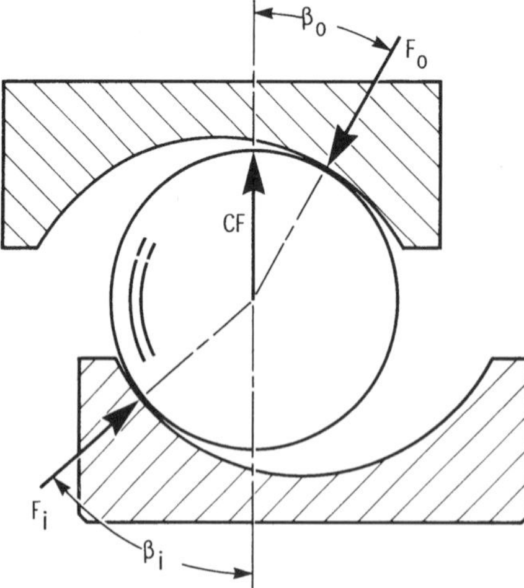

Figure 84 Contact angles in a ball bearing at appreciable speeds. (From Ref. 6.)

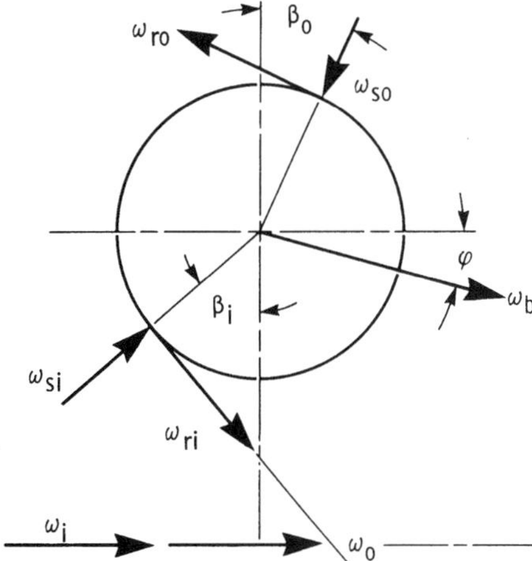

Figure 85 Angular velocities of a ball. (From Ref. 6.)

merical solution except for the two extreme cases of outer- or inner-race control. These are illustrated in Fig. 86.

Race control assumes that pure rolling occurs at the controlling race, with all of the ball spin occurring at the other race contact. The orientation of the ball rotational axis is then easily determinable from bearing geometry. Race control probably occurs only in dry bearings or dry-film-lubricated bearings where Coulomb friction conditions exist in the ball–race contact ellipses. Pure rolling will occur at the race contact with the higher magnitude spin-opposing moment. This is usually the inner race at low speeds and the outer race at high speeds.

In oil-lubricated bearings in which elastohydrodynamic films exist in both ball–race contacts, rolling with spin occurs at both contacts. Therefore, precise ball motions can only be determined through use of a computer analysis. We can approximate the situation with a reasonable degree of accuracy, however, by assuming that the ball rolling axis is normal to the line drawn through the centers of the two ball–race contacts.This is shown in Fig. 80.

The angular velocity of the separator or ball set ω_c about the shaft axis can be shown to be

$$\omega_c = \frac{(v_i + v_o)/2}{d_e/2}$$

$$= \frac{1}{2}\left[\omega_i\left(1 - \frac{d\cos\beta}{d_e}\right) + \omega_o\left(1 + \frac{d\cos\beta}{d_e}\right)\right] \tag{93}$$

where v_i and v_o are the linear velocities of the inner and outer contacts. The angular velocity of a ball ω_b about its own axis is

$$\omega_b = \frac{v_i - v_o}{d_e/2}$$

$$= \frac{d_e}{2d}\left[\omega_i\left(1 - \frac{d\cos\beta}{d_e}\right) - \omega_o\left(1 + \frac{d\cos\beta}{d_e}\right)\right] \tag{94}$$

To calculate the velocities of the ball–race contacts, which are required for calculating elastohydrodynamic film thicknesses, it is convenient to use a coordinate system that rotates

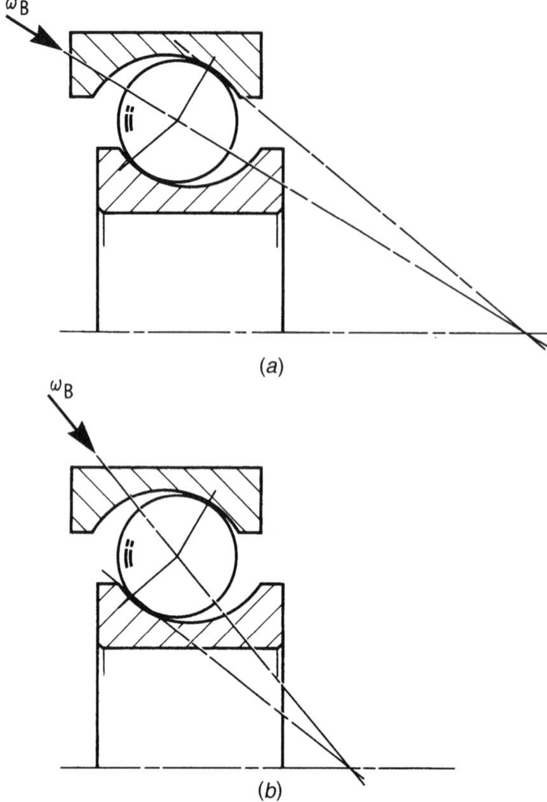

Figure 86 Ball spin axis orientations for outer- and inner-race control. (From Ref. 6.) (*a*) Outer-race control. (*b*) Inner-race control.

at ω_c. This fixes the ball–race contacts relative to the observer. In the rotating coordinate system the angular velocities of the inner and outer races become

$$\omega_{ir} = \omega_i - \omega_c = \left(\frac{\omega_i - \omega_o}{2}\right)\left(1 + \frac{d \cos \beta}{d_e}\right)$$

$$\omega_{or} = \omega_o - \omega_c = \left(\frac{\omega_o - \omega_i}{2}\right)\left(1 - \frac{d \cos \beta}{d_e}\right)$$

The surface velocities entering the ball–inner-race contact for pure rolling are

$$u_{ai} = u_{bi} = \left(\frac{d_e - d \cos \beta}{2}\right) \omega_{ir} \tag{95}$$

or

$$u_{ai} = u_{bi} = \frac{d_e(\omega_i - \omega_o)}{4}\left(1 - \frac{d^2 \cos^2 \beta}{d_e^2}\right) \tag{96}$$

and those at the ball–outer-race contact are

$$u_{ao} = u_{bo} = \left(\frac{d_e + d \cos \beta}{2}\right) \omega_{or}$$

or

$$u_{ao} = u_{bo} = \frac{d_e(\omega_o - \omega_i)}{4}\left(1 - \frac{d^2 \cos^2 \beta}{d_e^2}\right) \tag{97}$$

For a cylindrical roller bearing $\beta = 0°$ and Eqs. (92), (94), (96), and (97) become, if d is roller diameter,

$$\begin{aligned}
\omega_c &= \frac{1}{2}\left[\omega_i\left(1 - \frac{d}{d_e}\right) + \omega_o\left(1 + \frac{d}{d_e}\right)\right] \\
\omega_R &= \frac{d_e}{2d}\left[\omega_i\left(1 - \frac{d}{d_e}\right) + \omega_o\left(1 + \frac{d}{d_e}\right)\right] \\
u_{ai} = u_{bi} &= \frac{d_e(\omega_i - \omega_o)}{4}\left(1 - \frac{d^2}{d_e^2}\right) \\
u_{ao} = u_{bo} &= \frac{d_e(\omega_o - \omega_i)}{4}\left(1 - \frac{d^2}{d_e^2}\right)
\end{aligned} \tag{98}$$

For a tapered-roller bearing, equations directly analogous to those for a ball bearing can be used if d is the average diameter of the tapered roller, d_e is the diameter at which the geometric center of the rollers is located, and ω is the angle as shown in Fig. 87.

Static Load Distribution

Having defined a simple analytical expression for the deformation in terms of load in Section 3.1, it is possible to consider how the bearing load is distributed among the rolling elements. Most rolling-element bearing applications involve steady-state rotation of either the inner or outer race or both; however, the speeds of rotation are usually not so great as to cause ball or roller centrifugal forces or gyroscopic moments of significant magnitudes. In analyzing

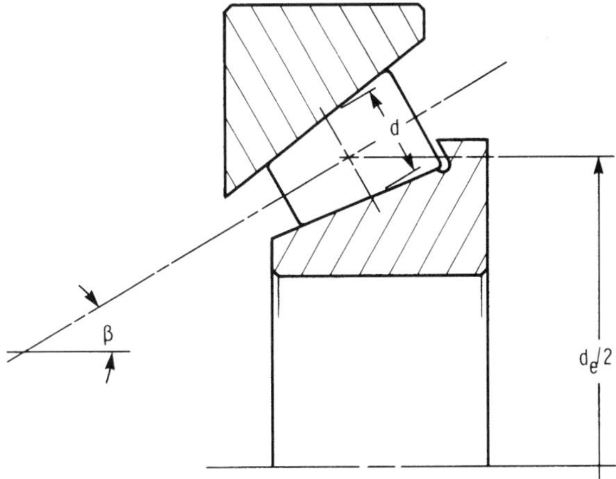

Figure 87 Simplified geometry for tapered-roller bearing. (From Ref. 6.)

the loading distribution on the rolling elements, it is usually satisfactory to ignore these effects in most applications. In this section the load–deflection relationships for ball and roller bearings are given, along with radial and thrust load distributions of statically loaded rolling elements.

Load–Deflection Relationships. For an elliptical contact the load–deflection relationship given in Eq. (45) can be written as

$$F = K_{1.5}\delta^{3/2} \tag{99}$$

where

$$K_{1.5} = \pi k E' \left(\frac{2\delta R}{9\mathcal{F}^3}\right)^{1/2} \tag{100}$$

Similarly for a rectangular contact, Eq. (50) gives

$$F = K_1 \delta$$

where

$$K_1 = \left(\frac{\pi l E'}{2}\right)\left[\frac{1}{2/3 + \ln(2r_{ax}/b) + \ln(2r_{bx}/b)}\right] \tag{101}$$

In general, then,

$$F = K_j \delta^j \tag{102}$$

in which j is 1.5 for ball bearings and 1.0 for roller bearings. The total normal approach between two races separated by a rolling element is the sum of the deformations under load between the rolling element and both races. Therefore

$$\delta = \delta_o + \delta_i \tag{103}$$

where

$$\delta_o = \left[\frac{F}{(K_j)_o}\right]^{1/j} \tag{104}$$

$$\delta_i = \left[\frac{F}{(K_j)_i}\right]^{1/j} \tag{105}$$

Substituting Eqs. (103)–(105) into Eq. (102) gives

$$K_j = \frac{1}{\{[1/(K_j)_o]^{1/j} + [1/(K_j)_i]^{1/j}\}^j}$$

Recall that $(K_j)_o$ and $(K_j)_i$ are defined by Eq. (100) or (101) for an elliptical or rectangular contact, respectively. From these equations we observe that $(K_j)_o$ and $(K_j)_i$ are functions of only the geometry of the contact and the material properties. The radial and thrust load analyses are presented in the following two sections and are directly applicable for radially loaded ball and roller bearings and thrust-loaded ball bearings.

Radially Loaded Ball and Roller Bearings. A radially loaded rolling element with radial clearance P_d is shown in Fig. 88. In the concentric position shown in Fig. 88*a*, a uniform radial clearance between the rolling element and the races of $P_d/2$ is evident. The application of a small radial load to the shaft causes the inner race to move a distance $P_d/2$ before

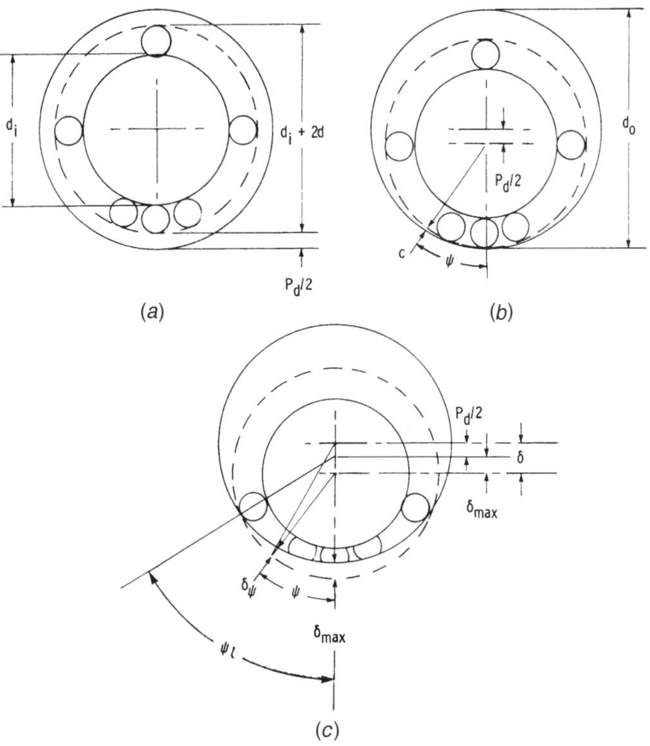

Figure 88 Radially loaded rolling-element bearing. (From Ref. 10.) (*a*) Concentric arrangement. (*b*) Initial contact. (*c*) Interference.

contact is made between a rolling element located on the load line and the inner and outer races. At any angle there will still be a radial clearance c that, if P_d is small compared with the radius of the tracks, can be expressed with adequate accuracy by

$$c = \tfrac{1}{2}(1 - \cos \psi)P_d$$

On the load line where $\psi = 0$ the clearance is zero, but when $\psi = 90°$, the clearance retains its initial value of $P_d/2$.

The application of further load will cause elastic deformation of the balls and the elimination of clearance around an arc $2\psi_c$. If the interference or total elastic compression on the load is δ_{max}, the corresponding elastic compression of the ball δ_ψ along a radius at angle ψ to the load line will be given by

$$\delta_\psi = (\delta_{max} \cos \psi - c) = (\delta_{max} + \tfrac{1}{2}P_d) \cos \psi - \tfrac{1}{2}P_d$$

This assumes that the races are rigid. Now, it is clear from Fig. 88 that $\delta_{max} + P_d/2$ represents the total relative radial displacement of the inner and outer races. Hence,

$$\delta_\psi = \delta \cos \psi - \tfrac{1}{2}P_d \tag{106}$$

The relationship between load and elastic compression along the radius at angle ψ to the load vector is given by Eq. (102) as

$$F_\psi = K_j \delta_\psi^j$$

Substituting Eq. (106) into this equation gives

$$F_\psi = K_j(\delta \cos \psi - \tfrac{1}{2}P_d)^j$$

For static equilibrium the applied load must equal the sum of the components of the rolling-element loads parallel to the direction of the applied load:

$$F_r = \sum F_\psi \cos \psi$$

Therefore

$$F_r = K_j \sum \left(\delta \cos \psi - \frac{P_d}{2}\right)^j \cos\psi \tag{107}$$

The angular extent of the bearing arc $2\psi_l$ in which the rolling elements are loaded is obtained by setting the root expression in Eq. (107) equal to zero and solving for ψ:

$$\psi_i = \cos^{-1}\left(\frac{P_d}{2\delta}\right)$$

The summation in Eq. (107) applies only to the angular extent of the loaded region. This equation can be written for a roller bearing as

$$F_r = \left(\psi_l - \frac{P_d}{2\delta} \sin \psi_l\right) \frac{nK_1\delta}{2\pi} \tag{108}$$

and similarly in integral form for a ball bearing as

$$F_r = \frac{n}{\pi} K_{1.5} \delta^{3/2} \int_0^{\psi_l} \left(\cos \psi - \frac{P_d}{2\delta}\right)^{3/2} \cos \psi \, d\psi$$

The integral in the equation can be reduced to a standard elliptic integral by the hypergeometric series and the beta function. If the integral is numerically evaluated directly, the following approximate expression is derived:

$$\int_0^{\psi_l} \left(\cos \psi - \frac{P_d}{2\delta}\right)^{3/2} \cos \psi \, d\psi = 2.491 \left\{\left[1 + \left(\frac{P_d/2\delta - 1}{1.23}\right)^2\right]^{1/2} - 1\right\}$$

This approximate expression fits the exact numerical solution to within $\pm 2\%$ for a complete range of $P_d/2\delta$.

The load carried by the most heavily loaded ball is obtained by substituting $\psi = 0°$ in Eq. (107) and dropping the summation sign:

$$F_{\max} = K_j \delta^j \left(1 - \frac{P_d}{2\delta}\right)^j$$

Dividing the maximum ball load [Eq. (109)] by the total radial load for a roller bearing [Eq. (108)] gives

$$F_r = \frac{[\psi_l - (P_d/2\delta) \sin \psi_l] \, nF_{\max}}{2\pi(1 - P_d/2\delta)} \tag{109}$$

and similarly for a ball bearing

$$F_r = \frac{nF_{\max}}{Z} \tag{110}$$

where

$$Z = \frac{\pi(1 - P_d/2\delta)^{3/2}}{2.491 \left\{ \left[1 + \left(\frac{1 - P_d/2\delta}{1.23} \right)^2 \right]^{1/2} - 1 \right\}} \tag{111}$$

For *roller bearings* when the diametral clearance P_d is zero, Eq. (105) gives

$$F_r = \frac{nF_{\max}}{4} \tag{112}$$

For *ball bearings* when the diametral clearance P_d is zero, the value of Z in Eq. (110) becomes 4.37. This is the value derived by Stribeck[46] for ball bearings of zero diametral clearance. The approach used by Stribeck was to evaluate the finite summation for various numbers of balls. He then derived the celebrated Stribeck equation for static load-carrying capacity by writing the more conservative value of 5 for the theoretical value of 4.37:

$$F_r = \frac{nF_{\max}}{5} \tag{113}$$

In using Eq. (113), it should be remembered that Z was considered to be a constant and that the effects of clearance and applied load on load distribution were not taken into account. However, these effects were considered in obtaining Eq. (110).

Thrust-Loaded Ball Bearings. The static-thrust-load capacity of a ball bearing may be defined as the maximum thrust load that the bearing can endure before the contact ellipse approaches a race shoulder, as shown in Fig. 89, or the load at which the allowable mean compressive stress is reached, whichever is smaller. Both the limiting shoulder height and the mean compressive stress must be calculated to find the static-thrust-load capacity.

The contact ellipse in a bearing race under a load is shown in Fig. 89. Each ball is subjected to an identical thrust component F_t/n, where F_t is the total thrust load. The initial contact angle before the application of a thrust load is denoted by β_f. Under load, the normal ball thrust load F acts at the contact angle β and is written as

$$F = \frac{F_t}{n} \sin \beta \tag{114}$$

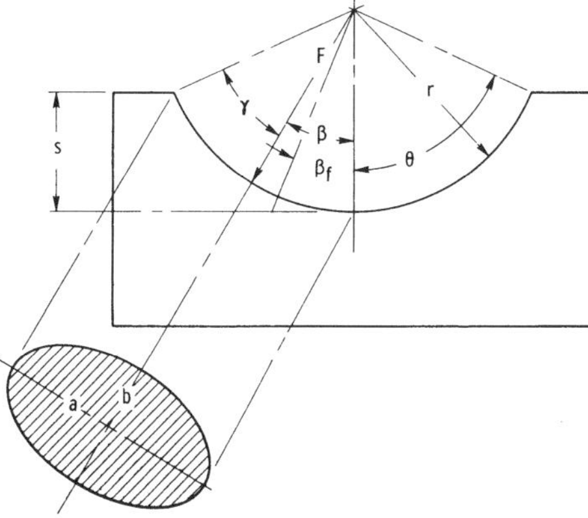

Figure 89 Contact ellipse in bearing race. (From Ref. 10.)

A cross section through an angular-contact bearing under a thrust load F_t is shown in Fig. 90. From this figure the contact angle after the thrust load has been applied can be written as

$$\beta = \cos^{-1}\left(\frac{D - P_d/2}{D + \delta}\right) \tag{115}$$

The initial contact angle was given in Eq. (74). Using that equation and rearranging terms in Eq. (115) give, solely from geometry (Fig. 90),

$$\delta = D\left(\frac{\cos\beta_f}{\cos\beta} - 1\right) = \delta_o + \delta_i$$

$$= \left[\frac{F}{(K_j)_o}\right]^{1/j} + \left[\frac{F}{(K_j)_i}\right]^{1/j}$$

$$K_j = 1 \Bigg/ \left\{\left[\frac{1}{(K_j)_o}\right]^{1/j} + \left[\frac{1}{(K_j)_i}\right]^{1/j}\right\}^j$$

$$K_j = 1 \Bigg/ \left\{\left[\frac{4.5\,\mathcal{F}_o^3}{\pi k_o E_o'(R_o\,\mathcal{E}_o)^{1/2}}\right]^{2/3} + \left[\frac{4.5\,\mathcal{F}_i^3}{\pi k_i E_i(R_i\,\mathcal{E}_i)^{1/2}}\right]^{2/3}\right\} \tag{116}$$

$$F = K_j D^{3/2}\left(\frac{\cos\beta_f}{\cos\beta} - 1\right)^{3/2} \tag{117}$$

where

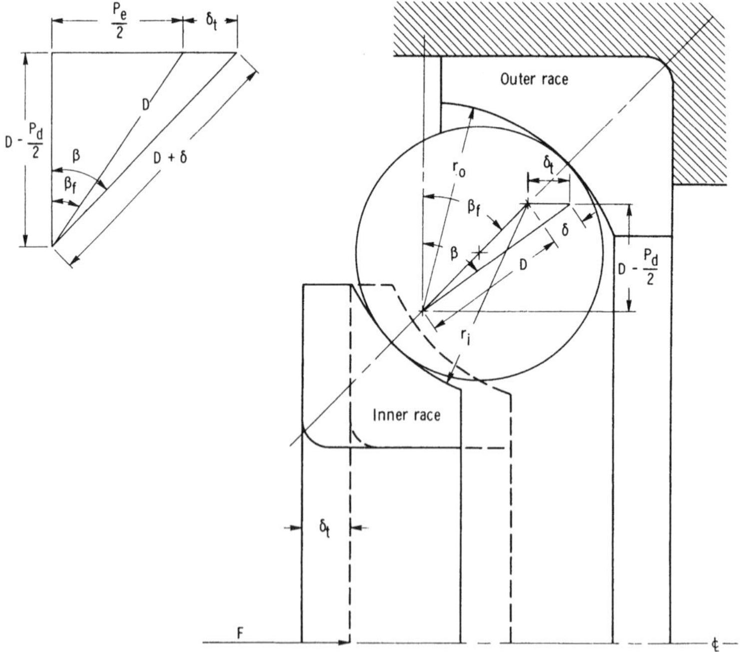

Figure 90 Angular-contact ball bearing under thrust load. (From Ref. 10.)

$$K_{1.5} = \pi k E' \left(\frac{R\,\mathcal{E}}{4.5\,\mathcal{F}^3} \right)^{1/2} \tag{118}$$

and k, $\mathcal{E}$, and $\mathcal{F}$ are given by Eqs. (39), (40), and (41), respectively.

From Eqs. (114) and (117), we can write

$$\frac{F_t}{n \sin \beta} = F$$
$$\frac{F_t}{nK_f D^{3/2}} = \sin \beta \left(\frac{\cos \beta_f}{\cos \beta} - 1 \right)^{3/2} \tag{119}$$

This equation can be solved numerically by the Newton–Raphson method. The iterative equation to be satisfied is

$$\beta' - \beta = \frac{\dfrac{F_t}{nK_{1.5}D^{3/2}} - \sin \beta \left(\dfrac{\cos \beta_f}{\cos \beta} - 1 \right)^{3/2}}{\cos \beta \left(\dfrac{\cos \beta_f}{\cos \beta} - 1 \right)^{3/2} + \dfrac{3}{2} \cos \beta_f \tan^2 \beta \left(\dfrac{\cos \beta_f}{\cos \beta} - 1 \right)^{1/2}} \tag{120}$$

In this equation convergence is satisfied when $\beta' - \beta$ becomes essentially zero.

When a thrust load is applied, the shoulder height is limited to the distance by which the pressure–contact ellipse can approach the shoulder. As long as the following inequality is satisfied, the pressure–contact ellipse will not exceed the shoulder height limit:

$$\theta > \beta + \sin^{-1} \left(\frac{D_y}{fd} \right)$$

From Fig. 79 and Eq. (68), the angle used to define the shoulder height θ can be written as

$$\theta = \cos^{-1} \left(\frac{1 - s}{fd} \right)$$

From Fig. 77 the axial deflection δ_t corresponding to a thrust load can be written as

$$\delta_t = (D + \delta) \sin \beta - D \sin \beta_f \tag{121}$$

Substituting Eq. (116) into Eq. (121) gives

$$\delta_t = \frac{D \sin(\beta - \beta_f)}{\cos \beta}$$

Having determined β from Eq. (120) and β_f from Eq. (103), we can easily evaluate the relationship for δ_t.

Preloading. The use of angular-contact bearings as duplex pairs preloaded against each other is discussed in the first subsection in Section 3.6. As shown in Table 10 duplex bearing pairs are used in either back-to-back or face-to-face arrangements. Such bearings are usually preloaded against each other by providing what is called "stickout" in the manufacture of the bearing. This is illustrated in Fig. 91 for a bearing pair used in a back-to-back arrangement. The magnitude of the stickout and the bearing design determine the level of preload on each bearing when the bearings are clamped together as in Fig. 91. The magnitude of preload and the load–deflection characteristics for a given bearing pair can be calculated by using Eqs. (74), (99), (114), and (116)–(119).

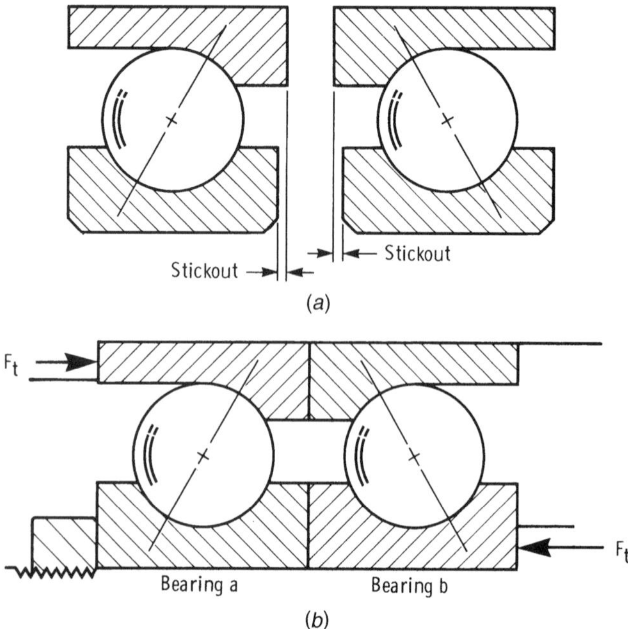

Figure 91 Angular-contact bearings in back-to-back arrangement, shown individually as manufactured and as mounted with preload. (From Ref. 6.) (*a*) Separated. (*b*) Mounted and preloaded.

The relationship of initial preload, system load, and final load for bearings *a* and *b* is shown in Fig. 92. The load–deflection curve follows the relationship $\delta = KF^{2/3}$. When a system thrust load F_t is imposed on the bearing pairs, the magnitude of load on bearing *b* increases while that on bearing *a* decreases until the difference equals the system load. The physical situation demands that the change in each bearing deflection be the same ($\Delta a = \Delta b$ in Fig. 92). The increments in bearing load, however, are not the same. This is important because it always requires a system thrust load far greater than twice the preload before one bearing becomes unloaded. Prevention of bearing unloading, which can result in skidding and early failure, is an objective of preloading.

Rolling Bearing Fatigue Life

Contact Fatigue Theory. Rolling fatigue is a material failure caused by the application of repeated stresses to a small volume of material. It is a unique failure type. It is essentially a process of seeking out the weakest point at which the first failure will occur. A typical spall is shown in Fig. 93. We can surmise that on a microscale there will be a wide dispersion in material strength or resistance to fatigue because of inhomogeneities in the material. Because bearing materials are complex alloys, we would not expect them to be homogeneous or equally resistant to failure at all points. Therefore, the fatigue process can be expected to be one in which a group of supposedly identical specimens exhibit wide variations in failure time when stressed in the same way. For this reason it is necessary to treat the fatigue process statistically.

To be able to predict how long a typical bearing will run under a specific load, we must have the following two essential pieces of information:

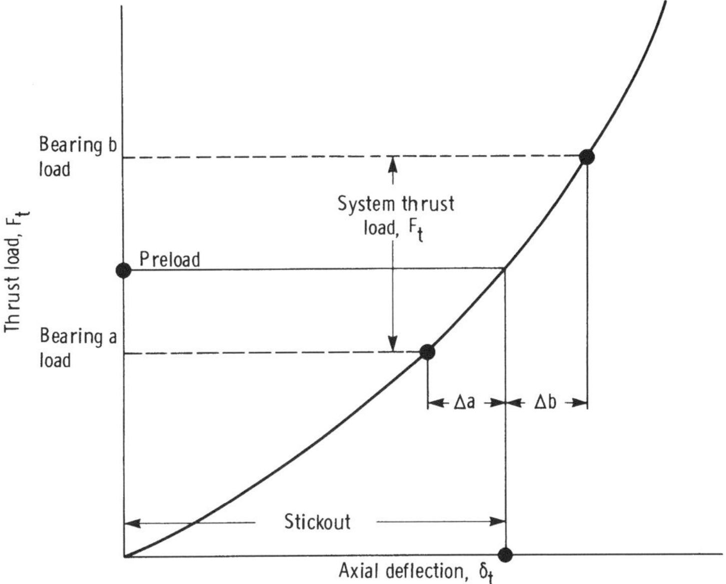

Figure 92 Thrust load–axial deflection curve for a typical ball bearing. (From Ref. 6.)

1. An accurate, quantitative estimate of the life dispersion or scatter.
2. The life at a given survival rate or reliability level. This translates into an expression for the "load capacity," or the ability of the bearing to endure a given load for a stipulated number of stress cycles or revolutions. If a group of supposedly identical bearings are tested at a specific load and speed, there will be a wide scatter in bearing lives, as shown in Fig. 94.

Figure 93 Typical fatigue spall.

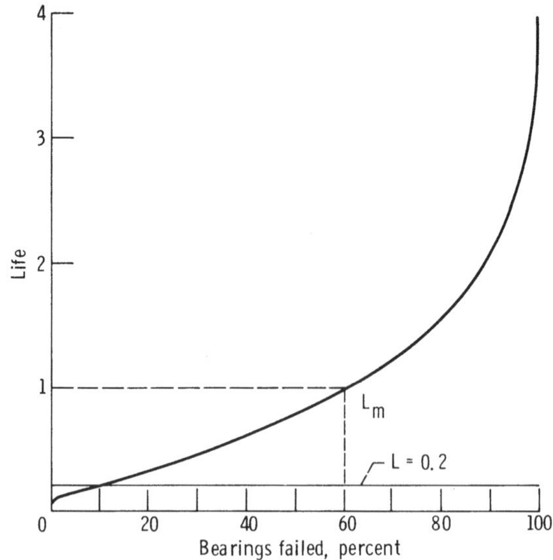

Figure 94 Distribution of bearing fatigue failures. (From Ref. 6.)

The Weibull Distribution. Weibull[47] postulates that the fatigue lives of a homogeneous group of rolling-element bearings are dispersed according to the following relation:

$$\ln \ln \frac{1}{S} = e_1 \ln L/A$$

where S is the probability of survival, L is the fatigue life, and e_1 and A are constants. The Weibull distribution results from a statistical theory of strength based on probability theory, where the dependence of strength on volume is explained by the dispersion in material strength. This is the "weakest link" theory.

Consider a volume being stressed that is broken up into m similar volumes:

$S_1 = 1 - M_1$ $\quad$ $S_2 = 1 - M_2$ $\quad$ $S_3 = 1 - M_3$ $\quad \cdots \quad$ $S_m = 1 - M_m$

The M's represent the probability of failure and the S's represent the probability of survival. For the entire volume we can write

$$S = S_1 \cdot S_2 \cdot S_3 \cdots\cdots S_m$$

Then

$$1 - M = (1 - M_1)(1 - M_2)(1 - M_3)\cdots(1 - M_m)$$

$$1 - M = \prod_{i=1}^{m} (1 - M_i)$$

$$S = \prod_{i=1}^{m} (1 - M_i)$$

The probability of a crack starting in the ith volume is

$$M_i = f(x)v_i$$

where $f(x)$ is a function of the stress level, the number of stress cycles, and the depth into the material where the maximum stress occurs and v_i is the elementary volume. Therefore,

$$S = \prod_{i=1}^{m} [1 - f(x)v_i]$$

$$\ln S = \sum_{i=1}^{m} \ln[1 - f(x)v_i]$$

Now if $f(x)v_i \ll 1$, then $\ln[1 - f(x)v_i] = -f(x)v_i$ and

$$\ln S = -\sum_{i=1}^{m} f(x)v_i$$

Let $v_i \longrightarrow 0$; then

$$\sum_{i=1}^{m} f(x)v_i = \int (x)\, dv = f(x)V$$

Lundberg and Palmgren[48] assume that $f(x)$ could be expressed as a power function of shear stress τ_0, number of stress cycles J, and depth to the maximum shear stress Z_0:

$$f(x) = \frac{\tau_0^{c_1} J^{c_2}}{Z_0^{c_3}} \tag{122}$$

They also choose as the stressed volume

$$V = D_y Z_0 l_v$$

Then

$$\ln S = -\frac{\tau_0^{c_1} J^{c_2} D_y l_v}{Z_0^{c_3 - 1}}$$

or

$$\ln \frac{1}{S} = \frac{\tau_0^{c_1} J^{c_2} D_y l_v}{Z_0^{c_3 - 1}}$$

For a specific bearing and load (e.g., stress) τ_0, D_y, l_v, and Z_0 are all constant, so that

$$\ln \frac{1}{S} \approx J^{c_2}$$

Designating J as life L in stress cycles gives

$$\ln \frac{1}{S} = \left(\frac{L}{A}\right)^{c_2}$$

or

$$\ln \ln \frac{1}{S} = c_2 \ln \left(\frac{L}{A}\right) \tag{123}$$

This is the Weibull distribution, which relates probability of survival and life. It has two principal functions. First, bearing fatigue lives plot as a straight line on Weibull coordinates (log log vs. log), so that the life at any reliability level can be determined. Of most interest are the L_{10} life ($S = 0.9$) and the L_{50} life ($S = 0.5$). Bearing load ratings are based on the L_{10} life. Second, Eq. (123) can be used to determine what the L_{10} life must be to obtain a required life at any reliability level. The L_{10} life is calculated, from the load on the bearing and the bearing dynamic capacity or load rating given in manufacturers' catalogs and engineering journals, by using the equation

$$L = \left(\frac{C}{F_e}\right)^m$$

where C = basic dynamic capacity or load rating
F_e = equivalent bearing load
m = 3 for elliptical contacts and 10/3 for rectangular contacts

A typical Weibull plot is shown in Fig. 95.

Lundberg–Palmgren Theory. The Lundberg–Palmgren theory, on which bearing ratings are based, is expressed by Eq. (122). The exponents in this equation are determined experimentally from the dispersion of bearing lives and the dependence of life on load, geometry, and bearing size. As a standard of reference, all bearing load ratings are expressed in terms of the specific dynamic capacity C, which, by definition, is the load that a bearing can carry for 10^6 inner-race revolutions with a 90% chance of survival.

Factors on which specific dynamic capacity and bearing life depend are:

1. Size of rolling element
2. Number of rolling elements per row
3. Number of rows of rolling elements
4. Conformity between rolling elements and races

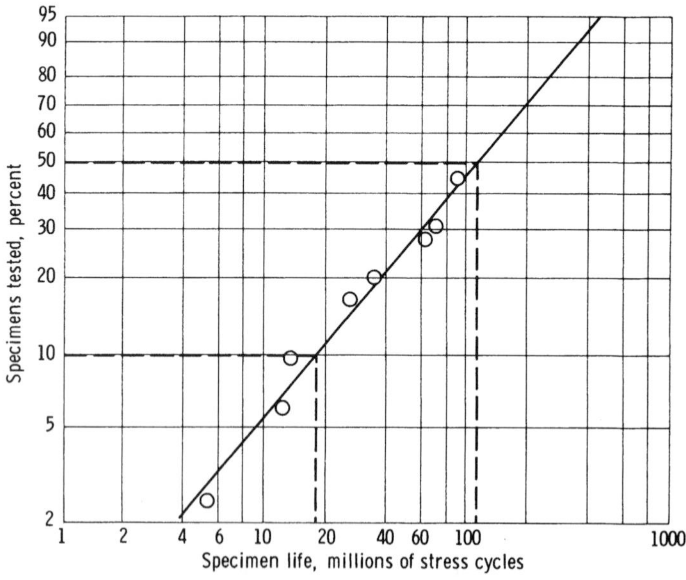

Figure 95 Typical Weibull plot of bearing fatigue failures. (From Ref. 10.)

5. Contact angle under load
6. Material properties
7. Lubricant properties
8. Operating temperature
9. Operating speed

Only factors 1–5 are incorporated in bearing dynamic capacities developed from the Lundberg–Palmgren theory. The remaining factors must be taken into account in the life adjustment factors discussed later.

The formulas for specific dynamic capacity as developed by Lundberg and Palmgren[48,49] are as follows:

For radial ball bearings with $d \leq 25$ mm,

$$C = f_c(i \cos \beta)^{0.7} n^{2/3} \left(\frac{d}{0.0254}\right)^{1.8}$$

where d = diameter of rolling element, m
i = number of rows of rolling elements
n = number of rolling elements per row
β = contact angle
f_c = coefficient dependent on material and bearing type

For radial ball bearings with $d \geq 25$ mm,

$$C = f_c(i \cos \beta)^{0.7} n^{2/3} \left(\frac{d}{0.0254}\right)^{1.4}$$

For radial roller bearings,

$$C = f_c(i \cos \beta)^{0.78} n^{3/4} \left(\frac{d}{0.0254}\right)^{1.07} \left(\frac{l_t}{0.0254}\right)^{0.78}$$

where l_t is roller length in meters.

For thrust ball bearings with $\beta \neq 90°$,

$$C = f_c(i \cos \beta)^{0.7}(\tan \beta) n^{2/3} \left(\frac{d}{0.0254}\right)^{1.8}$$

For thrust roller bearings with $\beta \neq 90°$,

$$C = f_c(i \cos \beta)^{0.78}(\tan \beta) n^{3/4} \left(\frac{l_t}{0.0254}\right)^{0.78}$$

For thrust ball bearings with $\beta = 90°$,

$$C = f_c i^{0.7} n^{2/3} \left(\frac{d}{0.0254}\right)^{1.8}$$

For thrust roller bearings with $\beta = 90°$,

$$C = f_c i^{0.78} n^{3/4} \left(\frac{d}{0.0254}\right)^{1.07} \left(\frac{l_t}{0.0254}\right)^{0.78}$$

For ordinary bearing steels such as SAE 52100 with mineral oil lubrication, f_c can be evaluated by using Tables 18 and 19, but a more convenient method is to use tabulated values from the most recent Antifriction Bearing Manufacturers Association (AFBMA) doc-

Table 18 Capacity Formulas for Rectangular and Elliptic Contacts[a] (From Ref. 6)

Function	Elliptical Contact of Ball Bearings			Rectangular Contact of Roller Bearings		
c	$f_c f_a i^{0.7} N^{2/3} d^{1.8}$			$f_c f_a i^{7/9} N^{3/4} d^{29/27} l_{t,i}^{7/9}$		
f_c	$g_c f_1 f_2 \left(\frac{d_i}{d_i - d}\right)^{0.41}$			$g_c f_1 f_2$		
g_c	$\left[1 + \left(\frac{c_i}{c_o}\right)^{10/8}\right]^{-0.8}$			$\left[1 + \left(\frac{c_i}{c_o}\right)^{9/2}\right]^{-2/9}$		
c_i/c_o	$f_3 \left[\frac{d_i(d_o - d)}{d_o(d_i - d)}\right]^{0.41}$			$f_3 \left(\frac{l_{t,i}}{l_{t,o}}\right)^{7/9}$		
	Radial	Thrust		Radial	Thrust	
		$\beta \neq 90°$	$\beta = 90°$		$\beta \neq 90°$	$\beta = 90°$
γ	$\frac{d \cos \beta}{d_e}$		$\frac{d}{d_e}$	$\frac{d \cos \beta}{d_e}$		$\frac{d}{d_e}$
f_a	$(\cos \beta)^{0.7}$	$(\cos \beta)^{0.7} \tan \beta$	1	$(\cos \beta)^{7/9}$	$(\cos \beta)^{7/9} \tan \beta$	1
f_1	3.7–4.1	6–10		18–25	36–60	
f_2	$\frac{\gamma^{0.3}(1 - \gamma)^{1.39}}{(1 + \gamma)^{1/3}}$		$\gamma^{0.3}$	$\frac{\gamma^{2/9}(1 - \gamma)^{29/27}}{(1 + \gamma)^{1/3}}$		$\gamma^{2/9}$
f_3	$104 f_4$	f_4	1	$1.14\ f_4$	f_4	1
f_4	$\left(\frac{1 - \gamma}{1 + \gamma}\right)^{1.72}$			$\left(\frac{1 - \gamma}{1 + \gamma}\right)^{38/37}$		

[a] Units in kg and mm.

uments on dynamic load ratings and life.[50] The value of C is calculated or determined from bearing manufacturers' catalogs. The equivalent load F_e can be calculated from the equation

$$F_e = XF_r + YF_t$$

Factors X and Y are given in bearing manufacturers' catalogs for specific bearings.

In addition to specific dynamic capacity C, every bearing has a specific static capacity, usually designated as C_0. Specific static capacity is defined as the load that, under static conditions, will result in a permanent deformation of 0.0001 times the rolling-element diameter. For some bearings C_0 is less than C, so it is important to avoid exposing a bearing to a static load that exceeds C_0. Values of C_0 are also given in bearing manufacturers' catalogs.

The AFBMA Method. Shortly after publication of the Lundberg–Palmgren theory, the AFBMA began efforts to standardize methods for establishing bearing load ratings and making life predictions. Standardized methods of establishing load ratings for ball bearings[51] and roller bearings[52] were devised, based essentially on the Lundberg–Palmgren theory. These early standards are published in their entirety in Jones.[45] In recent years significant advances have been made in rolling-element bearing material quality and in our understanding of the role of lubrication in bearing life through the development of elastohydrodynamic theory. Therefore the original AFBMA standards in AFBMA[51,52] have been updated with life ad-

Table 19 Capacity Formulas for Mixed Rectangular and Elliptical Contacts[a] (From Ref. 6)

Function	Radial Bearing	Thrust Bearing $\beta \neq 90°$	Thrust Bearing $\beta = 90°$	Radial Bearing	Thrust Bearing $\beta \neq 90°$	Thrust Bearing $\beta = 90°$
	Inner Race			Outer Race		
γ	$\dfrac{d \cos \beta}{d_e}$		$\dfrac{d}{d_e}$	$\dfrac{d \cos \beta}{d_e}$		$\dfrac{d}{d_e}$
	Rectangular Contact c_i			Elliptical Contact c_o		
c_i or c_o	$f_1 f_2 f_a i^{7/9} N^{3/4} D^{29/27} l_{t,i}^{7/9}$			$f_1 f_2 f_a \left(\dfrac{2R}{D}\dfrac{r_o}{r_o - R}\right)^{0.41} i^{0.7} N^{2/3} D^{1.8}$		
f_a	$(\cos \beta)^{7/9}$	$(\cos \beta)^{7/9} \tan \beta$	1	$(\cos \beta)^{0.7}$	$(\cos \beta)^{0.7} \tan \beta$	1
f_1	18–25	36–60		3.5–3.9	6–10	
f_2	$\dfrac{\gamma^{2/9}(1-\gamma)^{29/27}}{(1+\gamma)^{1/3}}$		$\gamma^{3/9}$	$\dfrac{\gamma^{0.3}(1+\gamma)^{1.39}}{(1-\gamma)^{1/3}}$		$\gamma^{0.3}$
	Point Contact c_i			Line Contact c_o		
c_i or c_o	$f_1 f_2 f_a \left(\dfrac{2R}{D}\dfrac{r_i}{r_i - R}\right)^{0.41} i^{0.7} n^{2/3} d^{1.8}$			$f_1 f_2 f_a i^{7/9} n^{3/4} d^{29/27} l_{t,o}^{7.9}$		
f_a	$(\cos \alpha)^{0.7}$	$(\cos \alpha)^{0.7} \tan \alpha$	1	$(\cos \alpha)^{7/9}$	$(\cos \alpha)^{7/9} \tan \alpha$	1
f_1	3.7–4.1	6–10		15–22	36–60	
f_2	$\dfrac{\gamma^{0.3}(1-\gamma)^{1.39}}{(1+\gamma)^{1/3}}$		$\gamma^{0.3}$	$\dfrac{\gamma^{2/9}(1+\gamma)^{29/27}}{(1-\gamma)^{1/3}}$		$\gamma^{2/9}$

[a] $C = C_i\,[1 + (C_i/C_o)^4]^{1/4}$ units in kg and mm.

justment factors. These factors have been incorporated into ISO,[50] which is discussed in the following section.

Life Adjustment Factors. A comprehensive study of the factors affecting the fatigue life of bearings, which were not taken into account in the Lundberg–Palmgren theory, is reported in Bamberger et al.[53] In that reference it was assumed that the various environmental or bearing design factors are multiplicative in their effect on bearing life. The following equation results:

$$L_A = (\tilde{D})(\tilde{E})(\tilde{F})(\tilde{G})(\tilde{H})L_{10}$$

or

$$L_A = (\tilde{D})(\tilde{E})(\tilde{F})(\tilde{G})(\tilde{H})(C/F_e)^m$$

where $\tilde{D}$ = materials factor
$\tilde{E}$ = metallurgical processing factor
$\tilde{F}$ = lubrication factor
$\tilde{G}$ = speed effect factor
$\tilde{H}$ = misalignment factor
F_e = bearing equivalent load
m = load-life exponent; either 3 for ball bearings or 10/3 for roller bearings

Factors, $\tilde{D}$, $\tilde{E}$, and $\tilde{F}$ are briefly reviewed here. The reader is referred to Bamberger et al.[53] for a complete discussion of all five life adjustment factors.

MATERIALS FACTORS $\tilde{D}$ AND $\tilde{E}$. For over a century, AISI 52100 steel has been the predominant material for rolling-element bearings. In fact, the basic dynamic capacity as defined by AFBMA in 1949 is based on an air-melted 52100 steel, hardened to at least Rockwell C 58. Since that time, better control of air-melting processes and the introduction of vacuum remelting processes have resulted in more homogeneous steels with fewer impurities. Such steels have extended rolling-element bearing fatigue lives to several times the AFBMA or catalog life. Life improvements of 3–8 times are not uncommon. Other steel compositions, such as AISI M-1 and AISI M-50, chosen for their higher temperature capabilities and resistance to corrosion, also have shown greater resistance to fatigue pitting when vacuum melting techniques are employed. Case-hardened materials, such as AISI 4620, AISI 4118, and AISI 8620, used primarily for roller bearings, have the advantage of a tough, ductile steel core with a hard, fatigue-resistant surface.

The recommended $\tilde{D}$ factors for various alloys processed by air melting are shown in Table 20. Insufficient definitive life data were found for case-hardened materials to recommended $\tilde{D}$ factors for them. It is recommended that the user refer to the bearing manufacturer for the choice of a specific case-hardened material.

The metallurgical processing variables considered in the development of the $\tilde{E}$ factor included melting practice (air and vacuum melting) and metal working (thermomechanical working). Thermomechanical working of M-50 has also been shown to result in improved life, but it is costly and still not fully developed as a processing technique. Bamberger et al.[53] recommended an $\tilde{E}$ factor of 3 for consumable-electrode-vacuum-melted materials.

The translation of factors into a standard[50] is discussed later.

LUBRICATION FACTOR $\tilde{F}$. Until approximately 1960 the role of the lubricant between surfaces in rolling contact was not fully appreciated. Metal-to-metal contact was presumed to occur in all applications with attendant required boundary lubrication. The development of elastohydrodynamic lubrication theory showed that lubricant films of thickness of the order of microinches and tens of microinches occur in rolling contact. Since surface finishes are

Table 20 Material Factor for Through-Hardened Bearing Materials[a] (From Ref. 53)

Material	$\tilde{D}$ Factor
52100	2.0
M-1	0.6
M-2	0.6
M-10	2.0
M-50	2.0
T-1	0.6
Halmo	2.0
M-42	0.2
WB 49	0.6
440C	0.6–0.8

[a] Air-melted materials assumed.

of the same order of magnitude as the lubricant film thicknesses, the significance of rolling-element bearing surface roughnesses to bearing performance became apparent. Tallian[54] first reported on the importance on bearing life of the ratio of elastohydrodynamic lubrication film thickness to surface roughness. Figure 96 shows life as a percentage of calculated L_{10} life as a function of Λ, where

$$\Lambda = \frac{h_{\text{min}}}{(\Delta_a^2 + \Delta_b^2)^{1/2}}$$

Figure 97, from Bamberger et al.,[53] presents a curve of the recommended $\tilde{F}$ factor as a function of the Λ parameter. A mean of the curves presented in Tallian[54] for ball bearings and in Skurka[55] for roller bearings is recommended for use. A formula for calculating the minimum film thickness h_{min} in the hard-EHL regimen is given in Eq. (57).

The results of Bamberger et al.[53] have not been fully accepted into the current AFBMA standard represented by ISO.[50] The standard presents the following:

1. Life and dynamic load rating formulas for radial and thrust ball bearings and radial and thrust roller bearings
2. Tables of f_c for all cases
3. Tables of X and Y factors for calculating equivalent loads
4. Load rating formulas for multirow bearings
5. Life correction factors for high-reliability levels a_1, materials a_2, and lubrication or operating conditions a_3

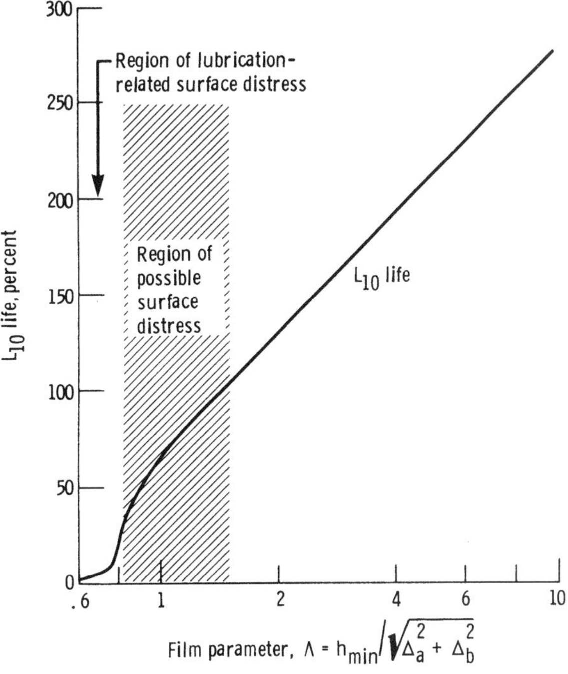

Figure 96 Chart for determining group fatigue life L_{10}. (From Ref. 54.)

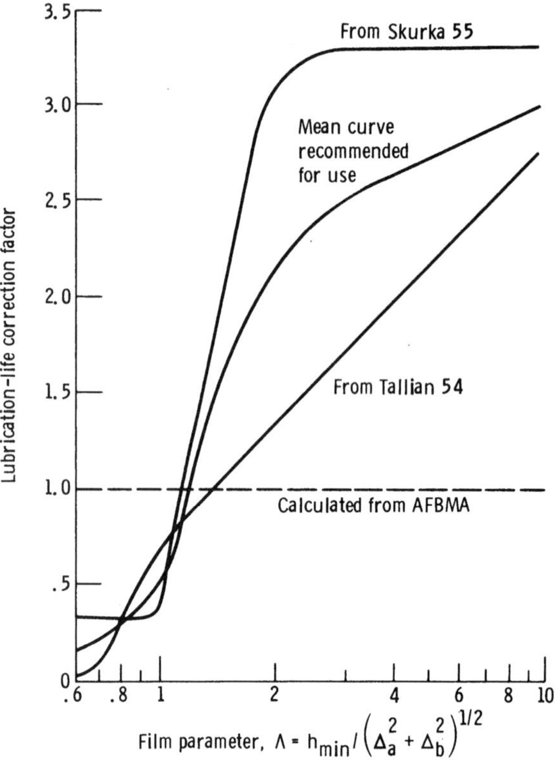

Figure 97 Chart for determining lubrication-life correction factor. (From Ref. 53.)

Procedures for calculating a_2 and a_3 are less than definitive, reflecting the need for additional research, life data, and operating experience.

Applications

In this section two applications of the film thickness equations developed throughout this chapter are presented to illustrate how the fluid-film lubrication conditions in machine elements can be analyzed. Specifically, a typical roller bearing and a typical ball bearing problem are considered.

Cylindrical-Roller-Bearing Problem. The equations for elastohydrodynamic film thickness that have been developed earlier relate primarily to elliptical contacts, but they are sufficiently general to allow them to be used with adequate accuracy in line-contact problems, as would be found in a cylindrical roller bearing. Therefore, the minimum elastohydrodynamic film thicknesses on the inner and outer races of a cylindrical roller bearing with the following dimensions are calculated:

Inner-race diameter, d_i, mm (m)	65 (0.064)
Outer-race diameter, d_o, mm (m)	96 (0.096)
Diameter of cylindrical rollers, d, mm (m)	16 (0.016)
Axial length of cylindrical rollers, l, mm (m)	16 (0.016)
Number of rollers in complete bearing, n	9

A bearing of this kind might well experience the following operating conditions:

Radial load, F_r, N	10,800
Inner-race angular velocity, ω_i, rad/s	524
Outer-race angular velocity, ω_o, rad/s	0
Lubricant viscosity at atmospheric pressure at operating temperature of bearings, η_o, N·s/m^2	0.01
Viscosity–pressure coefficient, ξ, m^2/N	2.2×10^{-8}
Modulus of elasticity for both rollers and races, E, N/m^2	2.075×10^{11}
Poisson's ratio, ν	0.3

CALCULATION. From Eq. (124), the most heavily loaded roller can be expressed as

$$F_{\max} = \frac{4F_r}{n} = \frac{4(10{,}800\ \text{N})}{9} = 4800\ \text{N} \tag{124}$$

Therefore, the radial load per unit length on the most heavily loaded roller is

$$F'_{\max} = \frac{4800\ \text{N}}{0.016\ \text{m}} = 0.3\ \text{MN/m} \tag{125}$$

From Fig. 98 we can write the radii of curvature as

$$r_{ax} = 0.008\ \text{m} \qquad r_{ay} = \infty$$

$$r_{bx,i} = 0.032\ \text{m} \qquad r_{by,i} = \infty$$

$$r_{bx,o} = 0.048\ \text{m} \qquad r_{by,o} = \infty$$

Then

$$\frac{1}{R_{x,i}} = \frac{1}{0.008} + \frac{1}{0.032} = \frac{5}{0.032}$$

giving $R_{x,i} = 0.0064$ m,

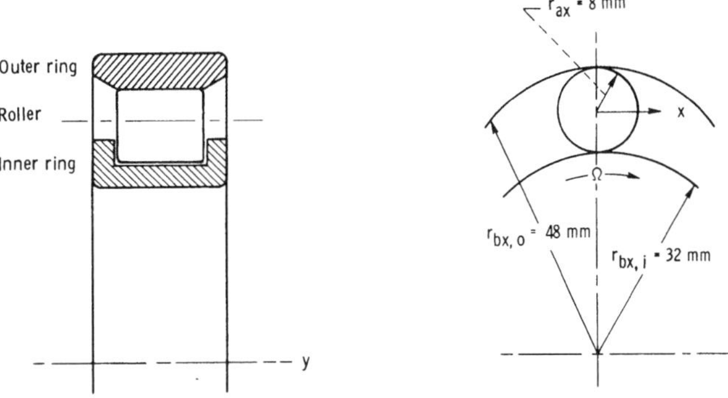

Figure 98 Roller bearing example: $r_{ay} = r_{by,f} = r_{by,o} = \infty$.

$$\frac{1}{R_{x,o}} = \frac{1}{0.008} - \frac{1}{0.048} = \frac{5}{0.048} \tag{126}$$

giving $R_{x,o} = 0.0096$ m, and

$$\frac{1}{R_{y,i}} = \frac{1}{R_{y,o}} = \frac{1}{\infty} + \frac{1}{\infty} = 0 \tag{127}$$

giving $R_{y,i} = R_{y,o} = \infty$.

From the input information, the effective modulus of elasticity can be written as

$$E' = 2\left(\frac{1 - \nu_a^2}{E_a} + \frac{1 - \nu_b^2}{E_b}\right) = 2.28 \times 10^{11}\ \text{N/m}^2 \tag{128}$$

For pure rolling, the surface velocity u relative to the lubricated conjunctions for a cylindrical roller is

$$u = |\omega_i - \omega_o| \frac{d_e^2 - d^2}{4d_e} \tag{129}$$

where d_e is the pitch diameter and d is the roller diameter:

$$d_e = \frac{d_o + d_i}{2} = \frac{0.096 + 0.064}{2} = 0.08\ \text{m} \tag{130}$$

Hence,

$$u = \frac{0.08^2 - 0.16^2}{4 \times 0.08} |524 - 0| = 10.061\ \text{m/s} \tag{131}$$

The dimensionless speed, materials, and load parameters for the inner- and outer-race conjunctions thus become

$$U_i = \frac{\eta_0 u}{E'R_{x,i}} = \frac{0.01 \times 10.061}{2.28 \times 10^{11} \times 0.0064} = 6.895 \times 10^{-11} \tag{132}$$

$$G_i = \xi E' = 5016 \tag{133}$$

$$W_i = \frac{F}{E'(R_{x,i})^2} = \frac{4800}{2.28 \times 10^{11} \times (0.0064)^2} = 5.140 \times 10^{-4} \tag{134}$$

$$U_o = \frac{\eta_o u}{E'R_{x,o}} = \frac{0.01 \times 10.061}{2.28 \times 10^{11} \times 0.0096} = 4.597 \times 10^{-11} \tag{135}$$

$$G_o = \xi E' = 5016 \tag{136}$$

$$W_o = \frac{F}{E'(R_{x,o})^2} = \frac{4800}{2.28 \times 10^{11} \times (0.0096)^2} = 2.284 \times 10^{-4} \tag{137}$$

The appropriate elliptical-contact elastohydrodynamic film thickness equation for a fully flooded conjunction is developed in Section 3.3 and recorded as Eq. (138):

$$H_{\min} = \frac{h_{\min}}{R_x} = 3.63U^{0.68}G^{0.49}W^{-0.073}(1 - e^{-0.68k}) \tag{138}$$

For a roller bearing, $k = \infty$ and this equation reduces to

$$H_{\min} = 3.63U^{0.68}G^{0.49}W^{-0.073}$$

The dimensionless film thickness for the roller–inner-race conjunction is

$$H_{\min} = \frac{h_{\min}}{R_{x,i}} = 3.63 \times 1.231 \times 10^{-7} \times 65.04 \times 1.783 = 50.5 \times 10^{-6}$$

and hence

$$h_{\min} = 0.0064 \times 50.5 \times 10^{-6} = 0.32\ \mu\text{m}$$

The dimensionless film thickness for the roller–outer-race conjunction is

$$H_{\min} = \frac{h_{\min}}{R_{x,o}} = 3.63 \times 9.343 \times 10^{-8} \times 65.04 \times 1.844 = 40.7 \times 10^{-6}$$

and hence

$$h_{\min} = 0.0096 \times 40.7 \times 10^{-6} = 0.39\ \mu\text{m}$$

It is clear from these calculations that the smaller minimum film thickness in the bearing occurs at the roller–inner-race conjunction, where the geometrical conformity is less favorable. It was found that if the ratio of minimum film thickness to composite surface roughness is greater than 3, an adequate elastohydrodynamic film is maintained. This implies that a composite surface roughness of $<0.1\ \mu$m is needed to ensure that an elastohydrodynamic film is maintained.

Radial Ball Bearing Problem. Consider a single-row, radial, deep-groove ball bearing with the following dimensions:

Inner-race diameter, d_i, m	0.052291
Outer-race diameter, d_o, m	0.077706
Ball diameter, d, m	0.012700
Number of balls in complete bearing, n	9
Inner-groove radius, r_i, m	0.006604
Outer-groove radius, r_o, m	0.006604
Contact angle, β, deg	0
rms surface finish of balls, Δ_b, μm	0.0625
rms surface finish of races, Δ_a, μm	0.175

A bearing of this kind might well experience the following operating conditions:

Radial load, F_r, N	8900
Inner-race angular velocity, ω_i, rad/s	400
Outer-race angular velocity, ω_o, rad/s	0
Lubricant viscosity at atmospheric pressure and effective operating temperature of bearing, η_0, N·s/m^2	0.04
Viscosity–pressure coefficient, ξ, m^2/N	2.3×10^{-8}
Modulus of elasticity for both balls and races, E, N/m^2	2×10^{11}
Poisson's ratio for both balls and races, ν	0.3

The essential features of the geometry of the inner and outer conjunctions (Figs. 75 and 76) can be ascertained as follows:

Pitch diameter [*Eq.* (*68*)]:

$$d_e = 0.5(d_o + d_i) = 0.065 \text{ m}$$

Diametral clearance [*Eq.* (*69*)]:

$$P_d = d_o - d_i - 2d = 1.5 \times 10^{-5} \text{ m}$$

Race conformity [*Eq.* (*70*)]:

$$f_i = f_o = \frac{r}{d} = 0.52$$

Equivalent radius [*Eq.* (*85*)]:

$$R_{x,i} = \frac{d(d_e - d)}{2d_e} = 0.00511 \text{ m}$$

Equivalent radius [*Eq.* (*87*)]:

$$R_{x,o} = \frac{d(d_e + d)}{2d_e} = 0.00759 \text{ m}$$

Equivalent radius [*Eq.* (*86*)]:

$$R_{y,i} = \frac{f_i d}{2f_i - 1} = 0.165 \text{ m}$$

Equivalent radius [*Eq.* (*88*)]:

$$R_{y,o} = \frac{f_o d}{2f_o - 1} = 0.165 \text{ m}$$

The curvature sum

$$\frac{1}{R_i} = \frac{1}{R_{x,i}} + \frac{1}{R_{y,i}} = 201.76 \tag{139}$$

gives $R_i = 4.956 \times 10^{-3}$ m, and the curvature sum

$$\frac{1}{R_o} = \frac{1}{R_{x,o}} + \frac{1}{R_{y,o}} = 137.81 \tag{140}$$

gives $R_o = 7.256 \times 10^{-3}$ m. Also, $\alpha_i = R_{y,i}/R_{x,i} = 32.35$ and $\alpha_o = R_{y,o}/R_{x,o} = 21.74$.

The nature of the Hertzian contact conditions can now be assessed:

Ellipticity parameters:

$$k_i = \alpha_i^{2/\pi} = 9.42 \qquad k_o = \alpha_o^{2/\pi} = 7.09 \tag{141}$$

Elliptic integrals:

$$q = \frac{\pi}{2} - 1$$

$$\mathcal{E}_i = 1 + \frac{q}{\alpha_i} = 1.0188 \qquad \mathcal{E}_o = 1 + \frac{q}{\alpha_o} = 1.0278 \tag{142}$$

$$\mathcal{F}_i = \frac{\pi}{2} + q \ln \alpha_i = 3.6205 \qquad \mathcal{F}_o = \frac{\pi}{2} + q \ln \alpha_o = 3.3823 \tag{143}$$

The effective elastic modulus E' is given by

$$E' = 2\left(\frac{1 - v_a^2}{E_a} + \frac{1 - v_b^2}{E_b}\right) - 1 = 2.198 \times 10^{11}\ \mathrm{N/m^2}$$

To determine the load carried by the most heavily loaded ball in the bearing, it is necessary to adopt an iterative procedure based on the calculation of local static compression and the analysis presented in the fourth subsection in Section 3.6. Stribeck[46] found that the value of Z was about 4.37 in the expression

$$F_{\max} = \frac{ZF_r}{n}$$

where $F_{\max}$ = load on most heavily loaded ball
F_r = radial load on bearing
n = number of balls

However, it is customary to adopt a value of $Z = 5$ in simple calculations in order to produce a conservative design, and this value will be used to begin the iterative procedure.

Stage 1. Assume $Z = 5$. Then

$$F_{\max} = \frac{5F_r}{9} = \frac{5}{9} \times 8900 = 4944\ \mathrm{N} \tag{144}$$

The maximum local elastic compression is

$$\delta_i = \mathcal{F}_i \left[\left(\frac{9}{2\mathcal{E}_i R_i}\right)\left(\frac{F_{\max}}{\pi k_i E'}\right)^2\right]^{1/3} = 2.902 \times 10^{-5}\ \mathrm{m}$$

$$\delta_o = \mathcal{F}_o \left[\left(\frac{9}{2\mathcal{E}_o R_o}\right)\left(\frac{F_{\max}}{\pi k_o E'}\right)^2\right]^{1/3} = 2.877 \times 10^{-5}\ \mathrm{m} \tag{145}$$

The sum of the local compressions on the inner and outer races is

$$\delta = \delta_i + \delta_o = 5.799 \times 10^{-5}\ \mathrm{m}$$

A better value for Z can now be obtained from

$$Z = \frac{\pi(1 - P_d/2\delta)^{3/2}}{2.491\left\{\left[1 + \left(\frac{1 - P_d/2\delta}{1.23}\right)^2\right]^{1/2} - 1\right\}}$$

since $P_d/2\delta = (1.5 \times 10^{-5})/(5.779 \times 10^{-5}) = 0.1298$. Thus

$$Z = 4.551$$

Stage 2.

$$Z = 4.551$$

$$F_{\max} = (4.551 \times 8900)/9 = 4500\ \mathrm{N}$$

$$\delta_i = 2.725 \times 10^{-5}\ \mathrm{m} \qquad \delta_o = 2.702 \times 10^{-5}\ \mathrm{m}$$

$$\delta = 5.427 \times 10^{-5}\ \mathrm{m}$$

$$\frac{P_d}{2\delta} = 0.1382$$

Thus

$$Z = 4.565$$

Stage 3.

$$Z = 4.565$$

$$F_{\max} = \frac{4.565 \times 8900}{9} = 4514 \text{ N}$$

$$\delta_i = 2.731 \times 10^{-5} \text{ m} \qquad \delta_o = 2.708 \times 10^{-5} \text{ m}$$

$$\delta = 5.439 \times 10^{-5} \text{ m}$$

$$\frac{P_d}{2\delta} = 0.1379$$

and hence

$$Z = 4.564$$

This value is very close to the previous value from stage 2 of 4.565, and a further iteration confirms its accuracy.

Stage 4.

$$Z = 4.564$$

$$F_{\max} = \frac{4.564 \times 8900}{9} = 4513 \text{ N}$$

$$\delta_i = 2.731 \times 10^{-5} \text{ m} \qquad \delta_o = 2.707 \times 10^{-5} \text{ m}$$

$$\delta = 5.438 \times 10^{-5} \text{ m}$$

$$\frac{P_d}{2\delta} = 0.1379$$

and hence

$$Z = 4.564$$

The load on the most heavily loaded ball is thus 4513 N.

Elastohydrodynamic Minimum Film Thickness. For pure rolling

$$u = |\omega_o - \omega_i| \frac{d_e^2 - d^2}{4d_e} = 6.252 \text{ m/s} \tag{146}$$

The dimensionless speed, materials, and load parameters for the inner- and outer-race conjunctions thus become

$$U_i = \frac{\eta_0 u}{E' R_{x,i}} = \frac{0.04 \times 6.252}{2.198 \times 10^{11} \times 5.11 \times 10^{-3}} = 2.227 \times 10^{-10} \tag{147}$$

$$G_i = \xi E' = 2.3 \times 10^{-8} \times 2.198 \times 10^{11} = 5055 \tag{148}$$

$$W_i = \frac{F}{E'(R_{x,i})^2} = \frac{4513}{2.198 \times 10^{11} \times (5.11)^2 \times 10^{-6}} = 7.863 \times 10^{-4} \tag{149}$$

$$U_o = \frac{\eta_0 u}{E' R_{x,o}} = \frac{0.04 \times 6.252}{2.198 \times 10^{11} \times 7.59 \times 10^{-3}} = 1.499 \times 10^{-10} \tag{150}$$

$$G_o = \xi E' = 2.3 \times 10^{-8} \times 2.198 \times 10^{11} = 5055 \tag{151}$$

$$W_o = \frac{\mathrm{F}}{E'(R_{x,o})^2} = \frac{4513}{2.198 \times 10^{11} \times (7.59)^2 \times 10^{-6}} = 3.564 \times 10^{-4} \tag{152}$$

The dimensionless minimum elastohydrodynamic film thickness in a fully flooded elliptical contact is given by

$$H_{\min} = \frac{h_{\min}}{R_x} = 3.63 U^{0.68} G^{0.49} W^{-0.073} (1 - e^{-0.68k}) \tag{153}$$

For the ball–inner-race conjunction it is

$$\begin{aligned}(H_{\min})_i &= 3.63 \times 2.732 \times 10^{-7} \times 65.29 \times 1.685 \times 0.9983 \\ &= 1.09 \times 10^{-4}\end{aligned} \tag{154}$$

Thus

$$(h_{\min})_i = 1.09 \times 10^{-4} \, R_{x,i} = 0.557 \ \mu\text{m}$$

The lubrication factor Λ discussed in the fifth subsection of Section 3.6 was found to play a significant role in determining the fatigue life of rolling-element bearings. In this case

$$\Lambda_i = \frac{(h_{\min})_i}{(\Delta_a^2 + \Delta_b^2)^{1/2}} = \frac{0.557 \times 10^{-6}}{[(0.175)^2 + (0.06225)^2]^{1/2} \times 10^{-6}} = 3.00 \tag{155}$$

Ball–outer-race conjunction is given by

$$\begin{aligned}(H_{\min})_o &= \frac{(h_{\min})_o}{R_{x,o}} = 3.63 U_o^{0.68} G^{0.49} W^{-0.073} \, (1 - e^{-0.68k_o}) \\ &= 3.63 \times 2.087 \times 10^{-7} \times 65.29 \times 1.785 \times 0.9919 \\ &= 0.876 \times 10^{-4}\end{aligned} \tag{156}$$

Thus

$$(h_{\min})_o = 0.876 \times 10^{-4} \, R_{x,o} = 0.665 \ \mu\text{m}$$

In this case, the lubrication factor Λ is given by

$$\Lambda_o = \frac{0.665 \times 10^{-6}}{[(0.175)^2 + (0.0625)^2]^{1/2} \times 10^{-6}} = 3.58 \tag{157}$$

Once again, it is evident that the smaller minimum film thickness occurs between the most heavily loaded ball and the inner race. However, in this case the minimum elastohydrodynamic film thickness is about three times the composite surface roughness, and the bearing lubrication can be deemed to be entirely satisfactory. Indeed, it is clear from Fig. 97 that very little improvement in the lubrication factor $\tilde{F}$ and thus in the fatigue life of the bearing could be achieved by further improving the minimum film thickness and hence Λ.

4 BOUNDARY LUBRICATION

If the pressures in fluid-film-lubricated machine elements are too high, the running speeds are too low, or if the surface roughness is too great, penetration of the lubricant film will occur. Contact will take place between asperities, leading to a rise in friction and wear rate. Figure 99 (obtained from Bowden and Tabor[56]) shows the behavior of the coefficient of friction in the different lubrication regimens. It is to be noted in this figure that in boundary lubrication, although the friction is much higher than in the hydrodynamic regime, it is still much lower than for unlubricated surfaces. As the running conditions are made more severe, the amount of lubricant breakdown increases, until the system scores or seizes so badly that the machine element can no longer operate successfully.

Figure 100 shows the wear rate in the different lubrication regimens as determined by the operating load. In the hydrodynamic and elastohydrodynamic lubrication regimens, since there is no asperity contact, there is little or no wear. In the boundary lubrication regimen the degree of asperity interaction and wear rate increases as the load increases. The transition from boundary lubrication to an unlubricated condition is marked by a drastic change in wear rate. Machine elements cannot operate successfully in the unlubricated region. Together Figs. 99 and 100 show that both friction and wear can be greatly decreased by providing a boundary lubricant to unlubricated surfaces.

Understanding boundary lubrication depends first on recognizing that bearing surfaces have asperities that are large compared with molecular dimensions. On the smoothest machined surfaces these asperities may be 25 nm (0.025 μm) high; on rougher surfaces they may be ten to several hundred times higher. Figure 101 illustrates typical surface roughness as a random distribution of hills and valleys with varying heights, spacing, and slopes. In the absence of hydrodynamic or elastohydrodynamic pressures these hills or asperities must support all of the load between the bearing surfaces. Understanding boundary lubrication also depends on recognizing that bearing surfaces are often covered by boundary lubricant films such as are idealized in Fig. 101. These films separate the bearing materials and, by shearing preferentially, provide some control of friction, wear, and surface damage.

Many mechanisms, such as door hinges, operate totally under conditions (high load, low speed) of boundary lubrication. Others are designed to operate under full hydrodynamic or elastohydrodynamic lubrication. However, as the oil-film thickness is a function of speed, the film will be unable to provide complete separation of the surfaces during startup and rundown, and the condition of boundary lubrication will exist. The problem from the boundary lubrication standpoint is to provide a boundary film with the proper physical characteristics to control friction and wear. The work of Bowden and Tabor,[56] Godfrey,[59] and Jones[60] was relied upon in writing the sections that follow.

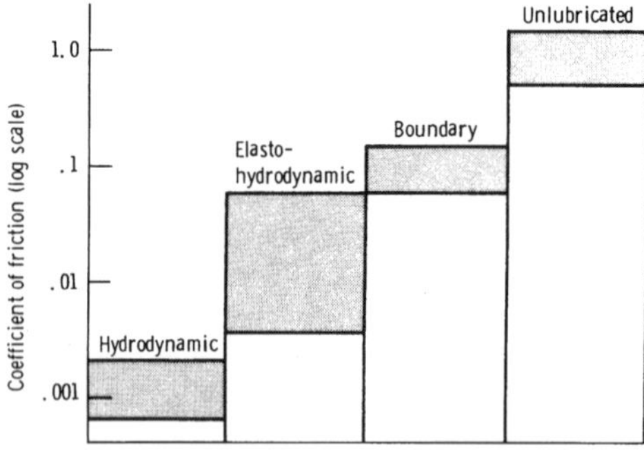

Figure 99 Schematic drawing showing how type of lubrication shifts from hydrodynamic to elastohydrodynamic to boundary lubrication as the severity of running conditions is increased. (From Ref. 56.)

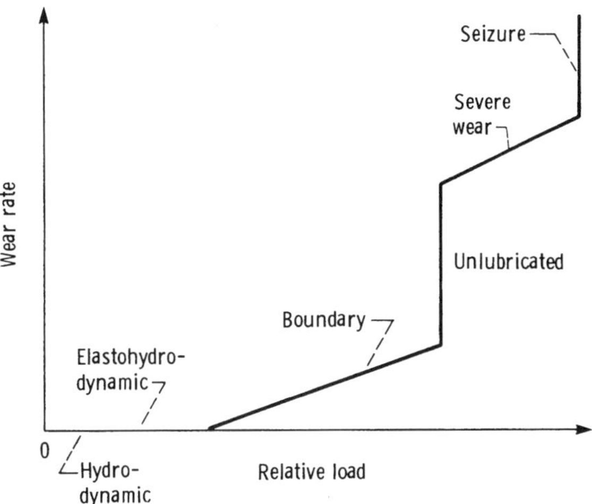

Figure 100 Chart for determining wear rate for various lubrication regimens. (From Ref. 57.)

4.1 Formation of Films

The most important aspect of boundary lubrication is the formation of surface films that will protect the contacting surfaces. There are three ways of forming a boundary lubricant film: physical adsorption, chemisorption, and chemical reaction. The surface action that determines the behavior of boundary lubricant films is the energy binding the film molecules to the surface, a measure of the film strength. The formation of films is presented in the order of such a film strength, the weakest being presented first.

Physical Adsorption

Physical adsorption involves intermolecular forces analogous to those involved in condensation of vapors to liquids. A layer of lubricant one or more molecules thick becomes

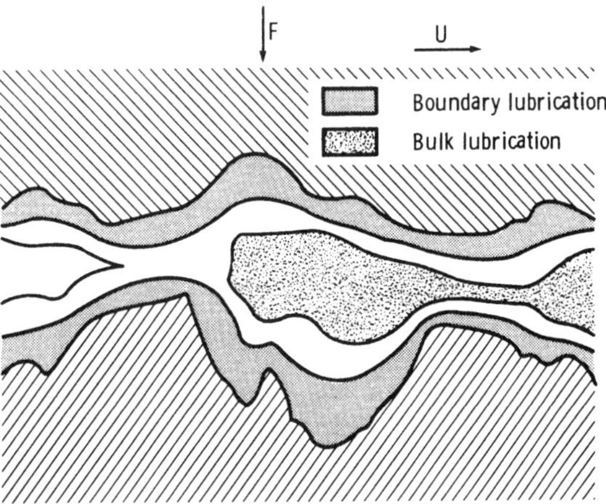

Figure 101 Lubricated bearing surfaces. (From Ref. 58.)

attached to the surfaces of the solids, and this provides a modest protection against wear. Physical adsorption is usually rapid, reversible, and nonspecific. Energies involved in physical adsorption are in the range of heats of condensations. Physical adsorption may be monomolecular or multilayer. There is no electron transfer in this process. An idealized example of physical adsorption of hexadecanol on an unreactive metal is shown in Fig. 102. Because of the weak bonding energies involved, physically adsorbed species are usually not very effective boundary lubricants.

Chemical Adsorption

Chemically adsorbed films are generally produced by adding animal and vegetable fats and oils to the base oils. These additives contain long-chain fatty acid molecules, which exhibit great affinity for metals at their active ends. The usual configuration of these polar molecules resembles that of a carpet pile with the molecules standing perpendicular to the surface. Such fatty acid molecules form metal soaps that are low-shear-strength materials with coefficients of friction in the range 0.10–0.15. The soap film is dense because of the preferred orientation of the molecules. For example, on a steel surface stearic acid will form a monomolecular layer of iron stearate, a soap containing 10^{14} molecules/cm^2 of surface. The effectiveness of these layers is limited by the melting point of the soap (180°C for iron stearate). It is clearly essential to choose an additive that will react with the bearing metals, so that less reactive, inert metals like gold and platinum are not effectively lubricated by fatty acids.

Examples of fatty acid additives are stearic, oleic, and lauric acid. The soap films formed by these acids might reduce the coefficient of friction to 50% of that obtained by a straight mineral oil. They provide satisfactory boundary lubrication at moderate loads, temperatures, and speeds and are often successful in situations showing evidence of mild surface distress.

Chemisorption of a film on a surface is usually specific, may be rapid or slow, and is not always reversible. Energies involved are large enough to imply that a chemical bond has formed (i.e., electron transfer has taken place). In contrast to physical adsorption, chemi-

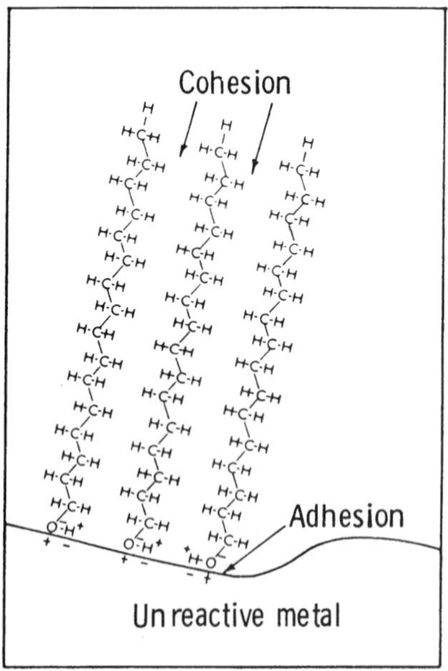

Figure 102 Physical adsorption of hexadecanol. (From Ref. 59.)

sorption may require an activation energy. A film may be physically adsorbed at low temperatures and chemisorbed at higher temperatures. In addition, physical adsorption may occur on top of a chemisorbed film. An example of a film of stearic acid chemisorbed on an iron oxide surface to form iron stearate is shown in Fig. 103.

Chemical Reaction

Films formed by chemical reaction provide the greatest film strength and are used in the most severe operating conditions. If the load and sliding speeds are high, significant contact temperatures will be developed. It has already been noted that films formed by physical and chemical adsorption cease to be effective above certain transition temperatures, but some additives start to react and form new high-melting-point inorganic solids at high temperatures. For example, sulfur will start to react at about 100°C to form sulfides with melting points of over 1000°C. Lubricants containing additives like sulfur, chlorine, phosphorus, and zinc are often referred to as extreme-pressure (EP) lubricants, since they are effective in the most arduous conditions.

The formation of a chemical reaction film is specific; may be rapid or slow (depending on temperature, reactivity, and other conditions); and is irreversible. An idealized example of a reacted film of iron sulfide on an iron surface is shown in Fig. 104.

4.2 Physical Properties of Boundary Films

The two physical properties of boundary films that are most important in determining their effectiveness in protecting surfaces are melting point and shear strength. It is assumed that the film thicknesses involved are sufficient to allow these properties to be well defined.

Melting Point

The melting point of a surface film appears to be one discriminating physical property governing failure temperature for a wide range of materials including inorganic salts. It is

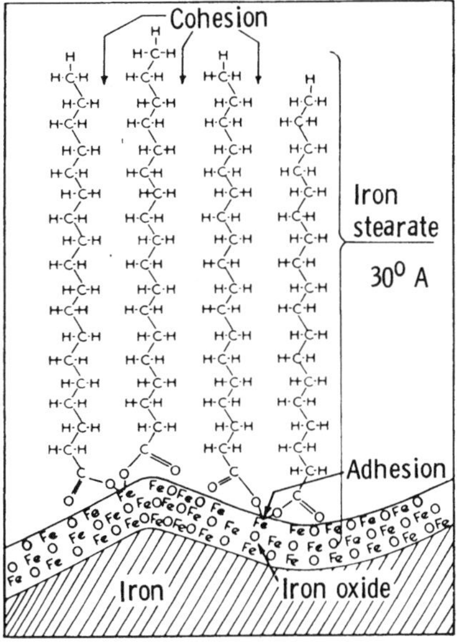

Figure 103 Chemisorption of stearic acid on iron surface to form iron stearate. (From Ref. 59.)

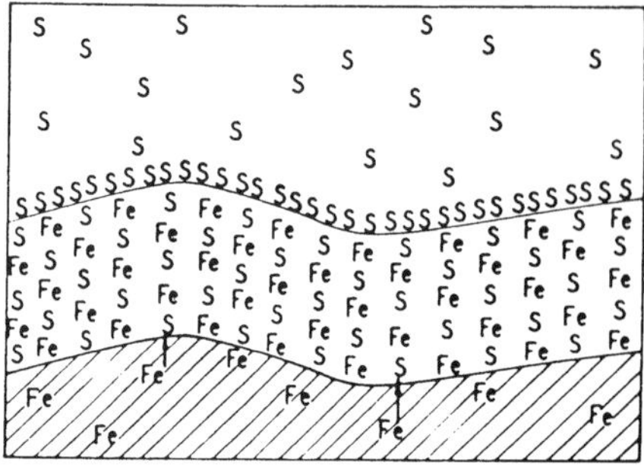

Figure 104 Formation of inorganic film by reaction of sulfur with iron to form iron sulfide. (From Ref. 59.)

based on the observation that only a surface film that is solid can properly interfere with potentially damaging asperity contacts. Conversely, a liquid film allows high friction and wear. Under practical conditions, physically adsorbed additives are known to be effective only at low temperatures and chemisorbed additives at moderate temperatures. High-melting-point inorganic materials are used for high-temperature lubricants.

The correlation of melting point with failure temperature has been established for a variety of organic films. An illustration is given in Fig. 105 (obtained from Russell et al.[61]) showing the friction transition for copper lubricated with pure hydrocarbons. Friction data for two hydrocarbons (mesitylene and dotriacontane) are given in Fig. 105 as a function of temperature. In this figure the boundary film failure occurs at the melting point of each hydrocarbon.

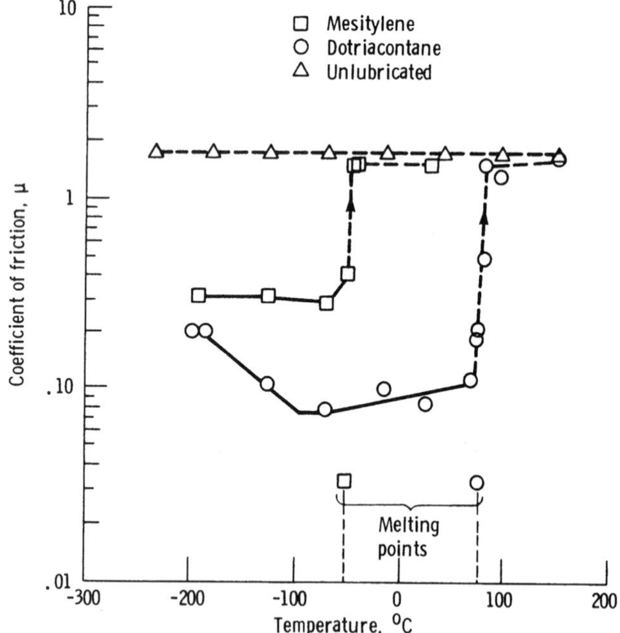

Figure 105 Chart for determining friction of copper lubricated with hydrocarbons in dry helium. (From Ref. 61.)

In contrast, chemisorption of fatty acids on reactive metals yields failure temperature based on the softening point of the soap rather than the melting point of the parent fatty acid.

Shear Strength

The shear strength of a boundary lubricating film should be directly reflected in the friction coefficient. In general, this is true with low-shear-strength soaps yielding low friction and high-shear-strength salts yielding high friction. However, the important parameter in boundary friction is the ratio of shear strength of the film to that of the substrate. This relationship is shown in Fig. 106, where the ratio is plotted on the horizontal axis with a value of 1 at the left and zero at the right. These results are in agreement with experience. For example, on steel an MoS_2 film gives low friction and Fe_2O_3 gives high friction. The results from Fig. 106 also indicate how the same friction value can be obtained with various combinations provided that the ratio is the same. It is important to recognize that shear strength is also affected by pressure and temperature.

4.3 Film Thickness

Boundary film thickness can vary from a few angstroms (adsorbed gas) to thousands of angstroms (chemical reaction films). In general, as the thickness of a boundary film increases, the coefficient of friction decreases. This effect is shown in Fig. 107*a*, which shows the coefficient of friction plotted against oxide film thickness formed on a copper surface. However, continued increases in thickness may result in an increase in friction. This effect is shown in Fig. 107*b*, which shows the coefficient of friction plotted against indium film thickness on copper surface. It should also be pointed out that the shear strengths of all boundary films decrease as their thicknesses increase, which may be related to the effect seen in Fig. 107*b*.

For physically adsorbed or chemisorbed films, surface protection is usually enhanced by increasing film thickness. The frictional transition temperature of multilayers also increases with increasing number of layers.

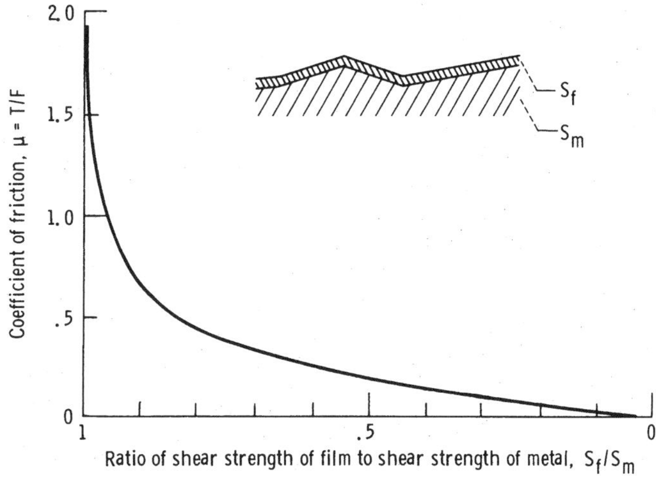

Figure 106 Chart for determining friction as function of shear strength ratio. (From Ref. 59.)

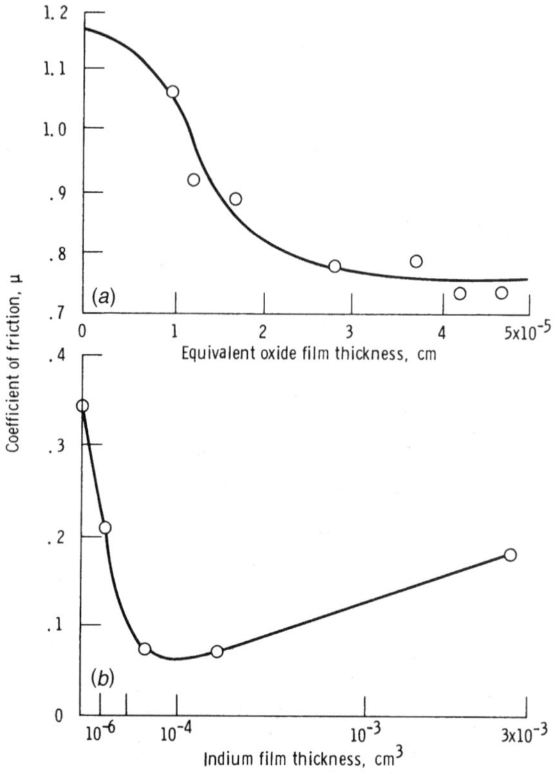

Figure 107 Chart for determining relationship of friction and thickness of films on copper surfaces. (From Ref. 62.)

For thick chemically reacted films there is an optimum thickness for minimum wear that depends on temperature, concentration, or load conditions. The relationship between wear and lubricant (or additive) reactivity is shown in Fig. 108. Here, if reactivity is not great enough to produce a thick enough film, adhesion wear occurs. On the other hand, if the material is too reactive, very thick films are formed and corrosive wear ensues.

4.4 Effect of Operating Variables

The effect of load, speed, temperature, and atmosphere can be important for the friction and wear of boundary lubrication films. Such effects are considered in this section.

On Friction

Load. The coefficient of friction is essentially constant with increasing load.

Speed. In general, in the absence of viscosity effects, friction changes little with speed over a sliding speed range of 0.005–1.0 cm/s. When viscosity effects do come into play, two types of behavior are observed, as shown in Fig. 109. In this figure relatively nonpolar

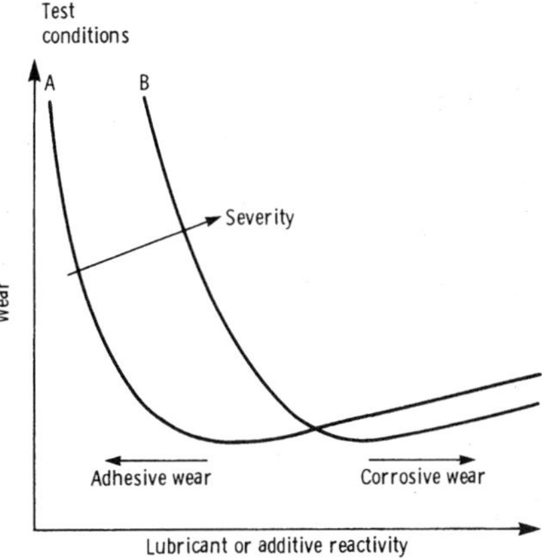

Figure 108 Relationship between wear and lubricant reactivity. (From Ref. 63.)

materials such as mineral oils show a decrease in friction with increasing speed, while polar fatty acids show the opposite trend. At higher speeds viscous effects will be present, and increases in friction are normally observed.

Temperature. It is difficult to make general comments on the effect of temperature on boundary friction since so much depends on the other conditions and the type of materials present. Temperature can cause disruption, desorption, or decomposition of boundary films. It can also provide activation energy for chemisorption or chemical reactions.

Atmosphere. The presence of oxygen and water vapor in the atmosphere can greatly affect the chemical processes that occur in the boundary layer. These processes can, in turn, affect the friction coefficient.

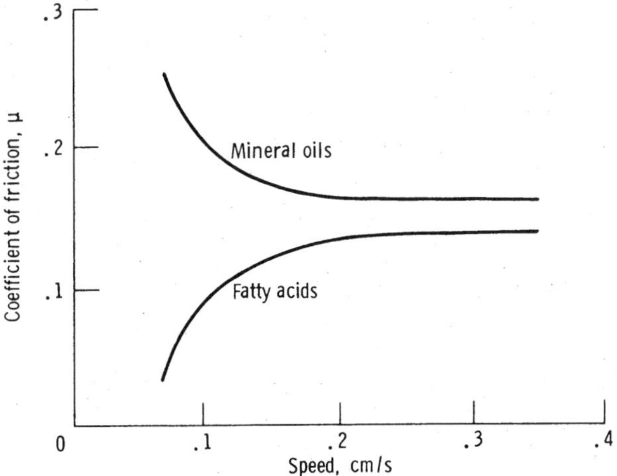

Figure 109 Effect of speed on coefficient of friction. (From Ref. 64.)

On Wear

Load. It is generally agreed that wear increases with increasing load, but no simple relationship seems to exist, at least before the transition to severe wear occurs. At this point a discontinuity of wear versus load is often like that illustrated in Fig. 100.

Speed. For practical purposes, wear rate in a boundary lubrication regime is essentially independent of speed. This assumes no boundary film failure due to contact temperature rise.

Temperature. As was the case for friction, there is no way to generalize the effect of temperature on wear. The statement that pertains to friction also pertains to wear.

Atmosphere. Oxygen has been shown to be an important ingredient in boundary lubrication experiments involving load-carrying additives. The presence of oxygen or moisture in the test atmosphere has a great effect on the wear properties of lubricants containing aromatic species.

4.5 Extreme-Pressure Lubricants

The best boundary lubricant films cease to be effective above 200–250°C. At these high temperatures the lubricant film may iodize. For operation under more severe conditions, EP lubricants might be considered.

Extreme-pressure lubricants usually consist of a small quantity of an EP additive dissolved in a lubricating oil, usually referred to as the base oil. The most common additives used for this purpose contain phosphorus, chlorine, or sulfur. In general, these materials function by reacting with the surface to form a surface film that prevents metal-to-metal contact. If, in addition, the surface film formed has a low shear strength, it will not only protect the surface, but it will also give a low coefficient of friction. Chloride films give a lower coefficient of friction ($\mu = 0.2$) than sulfide films ($\mu = 0.5$). Sulfide films, however, are more stable, are unaffected by moisture, and retain their lubricating properties to very high temperatures.

Although EP additives function by reacting with the surface, they must not be too reactive; otherwise chemical corrosion may be more troublesome than frictional wear. They should only react when there is a danger of seizure, usually noted by a sharp rise in local or global temperature. For this reason it is often an advantage to incorporate in a lubricant a small quantity of a fatty acid that can provide effective lubrication at temperatures below those at which the additive becomes reactive. Bowden and Tabor[56] describe this behavior in Fig. 110, where the coefficient of friction is plotted against temperature. Curve *A* is for paraffin oil (the base oil) and shows that the friction is initially high and increases as the temperature is raised. Curve *B* is for a fatty acid dissolved in the base oil: It reacts with the surface to form a metallic soap, which provides good lubrication from room temperature up to the temperature at which the soap begins to soften. Curve *C* is for a typical EP additive in the base oil; this reacts very slowly below the temperature T_c, so that in this range the lubrication is poor, while above T_c the protective film is formed and effective lubrication is provided to a very high temperature. Curve *D* is the result obtained when the fatty acid is added to the EP solution. Good lubrication is provided by the fatty acid below T_c, while above this temperature the greater part of the lubrication is due to the additive. At still higher temperatures, a deterioration of lubricating properties will also occur for both curves *C* and *D*.

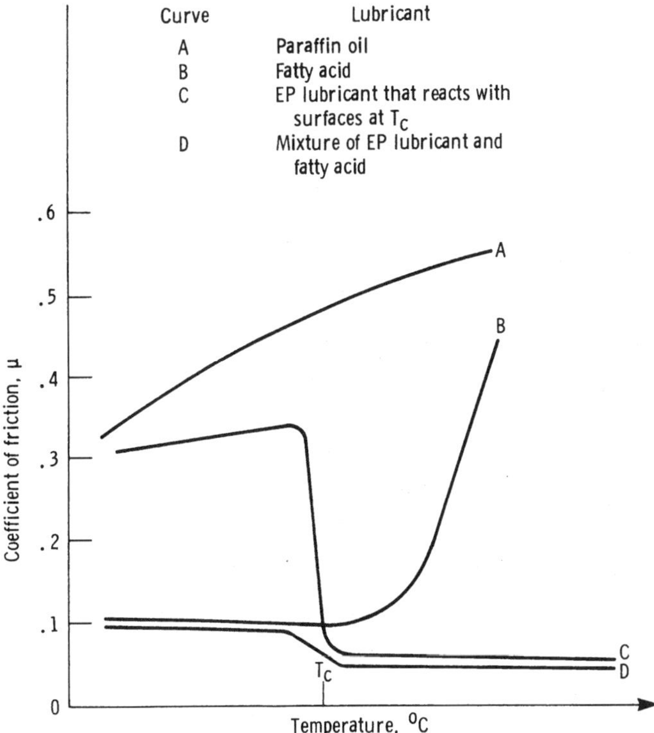

Figure 110 Graph showing frictional behavior of metal surfaces with various lubricants. (From Ref. 56.)

REFERENCES

1. B. Tower, "First Report on Friction Experiments (Friction of Lubricated Bearings)," *Proc. Inst. Mech. Eng., London,* 632–659 (1883).
2. N. P. Petrov, "Friction in Machines and the Effect of the Lubricant," *Inzh. Zh. St-Petreb.,* **1,** 71–140 (1883); **2,** 227–279 (1883); **3,** 377–436 (1883); **4,** 535–564 (1883).
3. O. Reynolds, "On the Theory of Lubrication and Its Application to Mr. Beauchamp Tower's Experiments, Including an Experimental Determination of the Viscosity of Olive Oil," *Philos. Trans. R. Soc. London,* **177,** 157–234 (1886).
4. W. B. Hardy and I. Doubleday, "Boundary Lubrication—The Temperature Coefficient," *Proc. R. Soc.,* **A101,** 487–492 (1922).
5. W. B. Hardy and I. Doubleday, "Boundary Lubrication—The Paraffin Series," *Proc. R. Soc.,* **A104,** 25–39 (1922).
6. B. J. Hamrock and W. J. Anderson, *Rolling-Element Bearings,* NASA RP-1105, 1983.
7. ESDU, "General Guide to the Choice of Journal Bearing Type," Engineering Sciences Data Unit, Item 65007, Institution of Mechanical Engineers, London, 1965.
8. ESDU, "General Guide to the Choice of Thrust Bearing Type," Engineering Sciences Data Unit, Item 67033, Institution of Mechanical Engineers, London, 1967.
9. I. Newton, *Philosophiae Naturalis Principia Mathematica.* 1687. Imprimature S. Pepys. Reg. Soc. Praess, 5 Julii 1866. Revised and supplied with a historical and explanatory appendix by F. Cajori,

edited by R. T. Crawford, 1934. Published by the University of California Press, Berkeley and Los Angeles, 1966.

10. B. J. Hamrock and D. Dowson, *Ball Bearing Lubrication—The Elastohydrodynamics of Elliptical Contacts,* Wiley, New York, 1981.
11. E. E. Bisson and W. J. Anderson, *Advanced Bearing Technology,* NASA SP-38, 1964.
12. ESDU, "Contact Stresses," Engineering Sciences Data Unit, Item 78035, Institution of Mechanical Engineers, London, 1978.
13. S. Aihara and D. Dowson, "A Study of Film Thickness in Grease Lubricated Elastohydrodynamic Contacts," in *Proceedings of Fifth Leeds–Lyon Symposium on Tribology on "Elastohydrodynamics and Related Topics,"* D. Dowson, C. M. Taylor, M. Godet, and D. Berthe (eds.), Mechanical Engineering Publications, Bury St. Edmunds, Suffolk, 1979, pp. 104–115.
14. A. R. Wilson, "The Relative Thickness of Grease and Oil Films in Rolling Bearings," *Proc. Inst. Mech. Eng., London,* **193**(17), 185–192 (1979).
15. T. Tallian, L. Sibley, and R. Valori, "Elastohydrodynamic Film Effect on the Load-Life Behavior of Rolling Contacts," ASME Paper 65-LUB-11, 1965.
16. C. Barus, "Isotherms, Isopeistics and Isometrics Relative to Viscosity," *Am. J. Sci.,* **45,** 87–96 (1893).
17. W. R. Jones, R. L. Johnson, W. O. Winer, and D. M. Sanborn, "Pressure-Viscosity Measurements for Several Lubricants to 5.5×10^8 Newtons per Square Meter (8×10^4 psi) and 149°C (300°F)," *ASLE Trans.,* **18**(4), 249–262 (1975).
18. D. Dowson and G. R. Higginson, *Elastohydrodynamic Lubrication, the Fundamentals of Roller and Gear Lubrication,* Pergamon, Oxford, 1966.
19. W. A. Gross, L. A. Matsch, V. Castelli, A. Eshel, and M. Wildmann, *Fluid Film Lubrication,* Wiley, New York, 1980.
20. N. F. Reiger, *Design of Gas Bearings,* Mechanical Technology, Latham, NY, 1967.
21. O. Pinkus and B. Sternlicht, *Theory of Hydrodynamic Lubrication,* McGraw-Hill, New York, 1961.
22. H. C. Rippel, *Cast Bronze Hydrostatic Bearing Design Manual,* Cast Bronze Bearing Institute, Cleveland, OH, 1963.
23. A. A. Raimondi and J. Boyd, "A Solution for the Finite Journal Bearing and Its Application to Analysis and Design; III," *Trans. ASLE,* **1**(1), 194–209 (1959).
24. P. E. Allaire and R. D. Flack, "Journal Bearing Design for High Speed Turbomachinery," *Bearing Design—Historical Aspects, Present Technology and Future Problems,* W. J. Anderson (ed.), American Society of Mechanical Engineers, New York, 1980, pp. 111–160.
25. A. A. Raimondi and J. Boyd, "Applying Bearing Theory to the Analysis and Design of Pad-Type Bearings," *Trans. ASME,* 287–309 (April 1955).
26. B. J. Hamrock, "Optimum Parallel Step-Sector Bearing Lubricated with an Incompressible Fluid," NASA TM-83356, 1983.
27. E. J. Gunter, J. G. Hinkle, and D. D. Fuller, "Design Guide for Gas-Lubricated Tilting-Pad Journal and Thrust Bearings with Special Reference to High Speed Rotors," Franklin Institute Research Laboratories Report I-A2392-3-1, 1964.
28. B. J. Hamrock and D. P. Fleming, "Optimization of Self-Acting Herringbone Grooved Journal Bearings for Minimum Radial Load," in *Proceedings of Fifth International Gas Bearing Symposium,* University of Southampton, Southampton, England, 1971, Paper 13.
29. D. P. Fleming and B. J. Hamrock, "Optimization of Self-Acting Herringbone Journal Bearings for Maximum Stability," paper presented at the Sixth International Gas Bearing Symposium, University of Southampton, Southampton, England, 1974, Paper c1, pp. 1–11.
30. B. J. Hamrock, "Optimization of Self-Acting Step Thrust Bearings for Load Capacity and Stiffness," *ASLE Trans.,* **15**(3), 159–170 (1972).
31. D. Dowson, "Elastohydrodynamic Lubrication—An Introduction and a Review of Theoretical Studies," *Inst. Mech. Eng., London, Proc.,* **180**(Pt. 3B), 7–16 (1965).
32. A. N. Grubin, "Fundamentals of the Hydrodynamic Theory of Lubrication of Heavily Loaded Cylindrical Surfaces," in *Investigation of the Contact Machine Components,* Kh. F. Ketova (ed.), translation of Russian Book No. 30, Central Scientific Institute of Technology and Mechanical Engineering, Moscow, 1949, Chap. 2. (Available from Dept. of Scientific and Industrial Research, Great Britain, Transl. CTS-235, and from Special Libraries Association, Chicago, Transl. R-3554.)
33. A. J. Petrusevich, "Fundamental Conclusion from the Contact-Hydrodynamic Theory of Lubrication," *Zv. Akad, Nauk, SSSR* (*OTN*), **2,** 209 (1951).

34. B. J. Hamrock and D. Brewe, "Simplified Solution for Stresses and Deformation," *J. Lubr. Technol.,* **105**(2), 171–177 (1983).

35. B. J. Hamrock and W. J. Anderson, "Analysis of an Arched Outer-Race Ball Bearing Considering Centrifugal Forces," *J. Lubr. Techn.,* **95**(3), 265–276 (1973).

36. A. B. Jones, "Analysis of Stresses and Deflections," New Departure Engineering Data, General Motors Corporation, Bristol, CT, 1946.

37. H. Hertz, "The Contact of Elastic Solids," *J. Reine Angew. Math.,* **92,** 156–171 (1881).

38. B. J. Hamrock and D. Dowson, "Isothermal Elastohydrodynamic Lubrication of Point Contacts, Part I—Theoretical Formulation," *J. Lubr. Technol.,* **98**(2), 223–229 (1976).

39. B. J. Hamrock and D. Dowson, "Isothermal Elastohydrodynamic Lubrication of Point Contacts, Part III—Fully Flooded Results," *J. Lubr. Technol.,* **99**(2), 264–276 (1977).

40. A. Cameron and R. Gohar, "Theoretical and Experimental Studies of the Oil Film in Lubricated Point Contacts," *Proc. R. Soc. London, Ser. A,* **291,** 520–536 (1966).

41. B. J. Hamrock and D. Dowson, "Elastohydrodynamic Lubrication of Elliptical Contacts for Materials of Low Elastic Modulus, Part I—Fully Flooded Conjunction," *J. Lubr. Technol.,* **100**(2), 236–245 (1978).

42. D. E. Brewe, B. J. Hamrock, and C. M. Taylor, "Effect of Geometry on Hydrodynamic Film Thickness," *J. Lubr. Technol.,* **101**(2), 231–239 (1979).

43. P. L. Kapitza, "Hydrodynamic Theory of Lubrication during Rolling," *Zh. Tekh. Fig.,* **25**(4), 747–762 (1955).

44. B. J. Hamrock and D. Dowson, "Minimum Film Thickness in Elliptical Contacts for Different Regimes of Fluid-Film Lubrication," in *Proceedings of Fifth Leeds–Lyon Symposium on Tribology on Elastohydrodynamics and Related Topics,* D. Dowson, C. M. Taylor, M. Godet, and D. Berthe (eds.), Mechanical Engineering Publications, Bury St. Edmunds, Suffolk, 1979, pp. 22–27.

45. A. B. Jones, "The Mathematical Theory of Rolling Element Bearings," in *Mechanical Design and Systems Handbook,* H. A. Rothbart (ed.), McGraw-Hill, New York, 1964, pp. 13-1–13-76.

46. R. Stribeck, "Kugellager fur beliebige Belastungen," *Z. VDI-Zeitschrift,* **45**(3), 73–125 (1901).

47. W. Weibull, "A Statistical Representation of Fatigue Failures in Solids," *Trans. Roy. Inst. Tech., Stockholm,* **27** (1949).

48. G. Lundberg and A. Palmgren, "Dynamic Capacity of Rolling Bearings," *Acta Polytech.,* Mechanical Engineering Series, **I**(3) (1947).

49. G. Lundberg and A. Palmgren, "Dynamic Capacity of Rolling Bearings," *Acta Polytech.,* Mechanical Engineering Series, **II**(4) (1952).

50. ISO, "Rolling Bearings, Dynamic Load Ratings and Rating Life," ISO/TC4/JC8, Revision of ISOR281. Issued by International Organization for Standardization, Technical Committee ISO/TC4, 1976.

51. AFBMA, *Method of Evaluating Load Ratings for Ball Bearings,* AFBMA Standard System No. 9, Revision No. 4, Anti-Friction Bearing Manufacturers Association, Arlington, VA, 1960.

52. AFBMA, *Method of Evaluating Load Ratings for Roller Bearings,* AFBMA Standard System No. 11, Anti-Friction Bearing Manufacturers Association, Arlington, VA, 1960.

53. E. N. Bamberger, T. A. Harris, W. M. Kacmarsky, C. A. Moyer, R. J. Parker, J. J. Sherlock, and E. V. Zaretsky, *Life Adjustment Factors for Ball and Roller Bearings—An Engineering Design Guide,* American Society for Mechanical Engineers, New York, 1971.

54. T. E. Tallian, "On Competing Failure Modes in Rolling Contact," *Trans. ASLE,* **10,** 418–439 (1967).

55. J. C. Skurka, "Elastohydrodynamic Lubrication of Roller Bearings," *J. Lubr. Technol.,* **92**(2), 281–291 (1970).

56. F. P. Bowden and D. Tabor, *Friction—An Introduction to Tribology,* Heinemann, London, 1973.

57. A. Beerbower, "Boundary Lubrication," *Scientific and Technical Application Forecasts,* DAC-19-69-C-0033, Department of the Army, 1972.

58. R. S. Fein and F. J. Villforth, "Lubrication Fundamentals," in *Lubrication,* Texaco, New York, 1973, pp. 77–88.

59. D. Godfrey, "Boundary Lubrication," in *Interdisciplinary Approach to Friction and Wear,* P. M. Ku (ed.), NASA SP-181, 1968, pp. 335–384.

60. W. R. Jones, *Boundary Lubrication—Revisited,* NASA TM-82858, 1982.

61. J. A. Russell, W. C. Campbell, R. A. Burton, and P. M. Ku, "Boundary Lubrication Behavior of Organic Films at Low Temperatures," *ASLE Trans.,* **8**(1), 48 (1965).

62. I. V. Kragelski, *Friction and Wear,* Butterworth, London, 1965, pp. 158–163.

63. C. N. Rowe, "Wear Corrosion and Erosion," in *Interdisciplinary Approach to Liquid Lubricant Technology,* P. M. Ku (ed.), NASA SP-237, 1973, pp. 469–527.

64. D. Clayton, "An Introduction to Boundary and Extreme Pressure Lubrication," *Phys. Lubr., Br. J. Appl. Phys.,* **2** (Suppl. 1), 25 (1951).

CHAPTER 33
SEAL TECHNOLOGY

Bruce M. Steinetz
NASA Glenn Research Center at Lewis Field
Cleveland, Ohio

1 INTRODUCTION

Seals are required to fulfill critical needs in meeting the ever-increasing system-performance requirements of modern machinery. Approaching a seal design, one has a wide range of available seal choices. This chapter aids the practicing engineer in making an initial seal selection and provides current reference material to aid in the final design and application.

This chapter provides design insight and application for both static and dynamic seals. Static seals reviewed include gaskets, O-rings, and selected packings. Dynamic seals reviewed include mechanical face, labyrinth, honeycomb, and brush seals. For each of these seals, typical configurations, materials, and applications are covered. Where applicable, seal flow models are presented.

2 STATIC SEALS

2.1 Gaskets

Gaskets are used to effect a seal between two mating surfaces subjected to differential pressures. Gasket types and materials are limited only by one's imagination. Table 1 lists some common gasket materials and Table 2 lists common elastomer properties.[1] The following gasket characteristics are considered important for good sealing performance.[2] Selecting the gasket material that has the best balance of the following properties will result in the best practical gasket design.

- Chemical compatibility
- Heat resistance
- Compressibility

Table 1 Common Gasket Materials, Gasket Factors (m), and Minimum Design Seating Stress (y) (Table 2-5.1 ASME Code for Pressure Vessels, 1995)

Gasket Material	Gasket Factor m	Min. Design Seating Stress y, psi	Sketches
Self-energizing types (O-rings, metallic, elastomer, other gasket types considered as self-sealing)	0	0	...
Elastomers without fabric or high percent of asbestos fiber:			
Below 75A Shore Durometer	0.50	0	
75A or higher Shore Durometer	1.00	200	
Asbestos with suitable binder for operating conditions:			
⅛ in. thick	2.00	1600	
$\frac{1}{16}$ in. thick	2.75	3700	
$\frac{1}{32}$ in. thick	3.50	6500	
Elastomers with cotton fabric insertion	1.25	400	
Elastomers with asbestos fabric insertion (with or without wire reinforcement):			
3-ply	2.25	2200	
2-ply	2.50	2900	
1-ply	2.75	3700	
Vegetable fiber	1.75	1100	
Spiral-wound metal, asbestos filled:			
Carbon	2.50	10,000	
Stainless, Monel, and nickel-base alloys	3.00	10,000	
Corrugated metal, asbestos inserted, or corrugated metal, jacketed asbestos filled:			
Soft aluminum	2.50	2900	
Soft copper or brass	2.75	3700	
Iron or soft steel	3.00	4500	
Monel or 4%–6% chrome	3.25	5500	
Stainless steels and nickel-base alloys	3.50	6500	

Table 1 (*Continued*)

Gasket Material	Gasket Factor m	Min. Design Seating Stress y, psi	Sketches
Corrugated metal:			
Soft aluminum	2.75	3700	
Soft copper or brass	3.00	4500	
Iron or soft steel	3.25	5500	
Monel or 4%–6% chrome	3.50	6500	
Stainless steels and nickel-base alloys	3.75	7600	
Flat metal, jacketed asbestos filled:			
Soft aluminum	3.25	5500	
Soft copper or brass	3.50	6500	
Iron or soft steel	3.75	7600	
Monel	3.50	8000	
4%–6% chrome	3.75	9000	
Stainless steels and nickel-base alloys	3.75	9000	
Grooved metal:			
Soft aluminum	3.25	5500	
Soft copper or brass	3.50	6500	
Iron or soft steel	3.75	7600	
Monel or 4%–6% chrome	3.75	9000	
Stainless steels and nickel-base alloys	4.25	10,100	
Solid flat metal:			
Soft aluminum	4.00	8800	
Soft copper or brass	4.75	13,000	
Iron or soft steel	5.50	18,000	
Monel or 4%–6% chrome	6.00	21,800	
Stainless steels and nickel-base alloys	6.50	26,000	
Ring joint:			
Iron or soft steel	5.50	18,000	
Monel or 4%–6% chrome	6.00	21,800	
Stainless steels and nickel-base alloys	6.50	26,000	

- Microconformability (asperity sealing)
- Recovery
- Creep relaxation
- Erosion resistance
- Compressive strength (crush resistance)
- Tensile strength (blowout resistance)

Table 2 The Most Important Elastomers and Their Properties

Elastomer	Composition	Working temperature range, °C	Tensile strength, bar	Elongation, %	Hardness, °Shore	Water	Steam	Hydraulic fluids, non-flammable (ester-based)	Mineral fats and oils	Vegetable and animal fats and oils	Ozone	Hydrocarbons: Aliphatic	Hydrocarbons: Aromatic	Hydrocarbons: Halogenated	Alcohols	Ketones	Esters	Dilute acids	Concentrated acids	Dilute alkalis	Concentrated alkalis	Saline solutions
Natural rubber	Rubber, K. W. Coil Refining-type polymerisate	−30–120	50–280	1000	30–98	x	x	—	—	—	—	—	—	—	x	x	—	o	—	x	o	x
S.B.R.	Butadiene–styrene copolymer	−30–130	50–240	700	40–95	x	x	—	—	—	o	—	—	—	x	x	—	x	o	x	x	x
Nitrile N	Butadiene–acrylonitrile copolymer	−30–130	50–240	700	40–95	x	o	—	x	x	o	x	o	—	x	—	—	o	—	o	o	x
Neoprene	Chlorinated–butadiene polymerisate	−40–140	50–270	800	40–95	x	x	—	o	o	x	o	—	—	x	—	—	x	o	x	x	x
Butyl	Isobutylene–isoprene copolymer	−50–150	40–170	900	40–90	x	x	o	—	o	x	—	o	—	x	o	o	x	o	x	x	x
Hypalon	Chlorosulfonated polyethylene	−40–140	40–200	600	40–95	x	o	—	—	o	x	—	—	—	x	—	—	x	o	x	x	x
Silicone rubber	Polycondensates of dialkylsiloxanes	−100–200	20–80	500	40–80	o	—	—	o	x	x	o	—	—	x	o	—	x	o	x	o	o
Thiokol	Alkylopolysulfide	−40–80	10–60	200	65–80	x	—	x	x	x	x	x	o	o	x	o	x	x	o	x	x	x
Polyacrylic	Polyacrylate	−30–120	20–70	700	70–85	o	—	x	x	x	x	x	o	o	o	—	x	—	—	o	—	o
Vulcollan	Polyurethane	−30–80	200–320	600	70–95	o	—	—	x	x	x	x	—	—	o	—	—	—	—	—	—	—
Adiprene	Polyurethane	−40–120	80–300	700	70–95	x	o	—	x	x	x	x	—	—	o	—	—	—	—	—	—	—
Kel-F	Copolymer of chlorotriethylene and vinylidene fluoride	−50–180	30–120	700	60–90	x	x	—	—	o	x	o	—	—	x	—	—	x	x	x	x	x
Viton	Vinylidene fluoride–hexafluoropropylene copolymer	−60–200	80–160	300	60–95	x	o	o	x	x	x	x	x	o	x	—	—	x	o	o	—	x
PTFE	Polytetrafluoroethylene (PTFE)	−200–280	140–310	200	55D	x	x	x	x	x	x	x	x	x	x	x	x	x	x	x	x	x
E.P.R.	Ethylene–propylene	−55–200	50–160	400	70–95	x	x	x	—	—	x	—	—	—	x	x	o	x	o	x	x	x
F.S.R.	Fluorosilicone rubber	−60–230	55–85	400	40–80	o	o	o	x	x	x	x	x	o	x	—	o	o	—	x	o	x

Note: From Ref. 1. x, stable; o, stable under certain conditions; —, unstable.

- Shear strength (flange shearing movement)
- Removal, or "Z," strength
- Antistick
- Heat conductivity
- Acoustic isolation
- Dimensional stability

Nonmetallic Gaskets. Most *nonmetallic gaskets* consist of a fibrous base held together with some form of an elastomeric binder. A gasket is formulated to provide the best load-bearing properties while being compatible with the fluid being sealed.

Nonmetallic gaskets are often reinforced to improve torque retention and blowout resistance for more severe service requirements. Some types of reinforcements include perforated cores, solid cores, perforated skins, and solid skins, each suited for specific applications. After a gasket material has been reinforced by either material additions or laminating, manufacturers can emboss the gasket raising a sealing lip, which increases localized pressures, thereby increasing sealability.

Metallic Gaskets. Metallic gaskets are generally used where either the joint temperature or load is extreme or in applications where the joint might be exposed to particularly caustic chemicals. A good seal capable of withstanding very high temperature is possible if the joint is designed to yield locally over a narrow location with application of bolt load. Some of the most common metallic gaskets range from soft varieties, such as copper, aluminum, brass, and nickel, to highly alloyed steels. Noble metals, such as platinum, silver, and gold, also have been used in difficult locations.

Metallic gaskets are available in both standard and custom designs. Since there is such a wide variety of designs and materials used, it is recommended that the reader directly contact metallic gasket suppliers for design and sealing information.

Required Bolt Load

ASME Method. The ASME Code for Pressure Vessels, Section VIII, Div. 1, App. 2, is the most commonly used design method for gasketed joints where important joint properties, including flange thickness and bolt size and pattern, are specified. An integral part of the AMSE Code revolves around two gasket factors:

1. An m factor, often called the gasket maintenance factor, is associated with the hydrostatic end force and the operation of the joint.
2. The y factor is a rough measure of the minimum seating stress associated with a particular gasket material. The y factor pertains only to the initial assembly of the joint.

The ASME Code makes use of two basic equations to calculate bolt load, with the larger calculated load being used for design:

$$W_{m1} = H + H_p = \frac{\pi}{4} G^2 P + 2\pi bGmP$$

$$W_{m2} = H_y = \pi bGy$$

where W_{m1} = minimum required bolt load from maximum operating or working conditions, lb

W_{m2} = minimum required initial bolt load for gasket seating (atmospheric-temperature conditions) without internal pressure, lb

H = total hydrostatic end force, lb $[(\pi/4)G^2P]$

H_p = total joint–contact–surface compression load, lb

H_y = total joint–contact–surface seating load, lb

G = diameter at location of gasket load reaction; generally defined as follows: when $b_0 < 1/4$ in., G = mean diameter of gasket contact face, in.; when $b_0 > 1/4$ in., G = outside diameter of gasket contact face less $2b$, in.

P = maximum internal design pressure, psi

b = effective gasket or joint–contact–surface *seating* width, in; $= b_0$ when $b_0 \leq 1/4$ in., $= 0.5\sqrt{b_0}$ when $b_0 > 1/4$ in. (see also ASME Table 2-5.2)

$2b$ = effective gasket or joint–contact–surface *pressure* width, in.

b_0 = basic gasket seating width

m = gasket factor per ASME Table 2-5.1 (repeated here as Table 1).

y = gasket or joint–contact–surface unit seating load, per ASME Table 2-5.1 (repeated here as Table 1), psi

The factor m provides a margin of safety to be applied when the hydrostatic end force becomes a determining factor. Unfortunately, this value is difficult to obtain experimentally since it is not a constant. The equation for W_{m2} assumes that a certain unit stress is required on a gasket to make it conform to the sealing surfaces and be effective. The second empirical constant y represents the gasket yield–stress value and is very difficult to obtain experimentally.

Practical Considerations

Flange Surfaces. Preparing the flange surfaces is paramount for effecting a good gasket seal. Surface finish affects the degree of sealability. The rougher the surface, the more bolt load required to provide an adequate seal. Extremely smooth finishes can cause problems for high operating pressures, as lower frictional resistance leads to a higher tendency for blowout. Surface finish lay is important in certain applications to mitigate leakage. Orienting finish marks transverse to the normal leakage path will generally improve sealability.

Flange Thickness. Flange thickness must also be sized correctly to transmit bolt clamping load to the area between the bolts. Maintaining seal loads at the midpoint between the bolts must be kept constantly in mind. Adequate thickness is also required to minimize the bowing of the flange. If the flange is too thin, the bowing will become excessive and no bolt load will be carried to the midpoint, preventing sealing.

Bolt Pattern. Bolt pattern and frequency are critical in effecting a good seal. The best bolt clamping pattern is invariably a combination of the maximum practical number of bolts, optimum spacing, and positioning.

One can envision the bolt loading pattern as a series of straight lines drawn from bolt to adjacent bolt until the circuit is completed. If the sealing areas lie on either side of this pattern, it will likely be a potential leakage location. Figure 1 shows an example of the various conditions.[2] If bolts cannot be easily repositioned on a problematic flange, Fig. 2 illustrates techniques to improve gasket effectiveness through reducing gasket face width where bolt load is minimum. Note that gasket width is retained in the vicinity of the bolt to support local bolt loads and minimize gasket tearing.

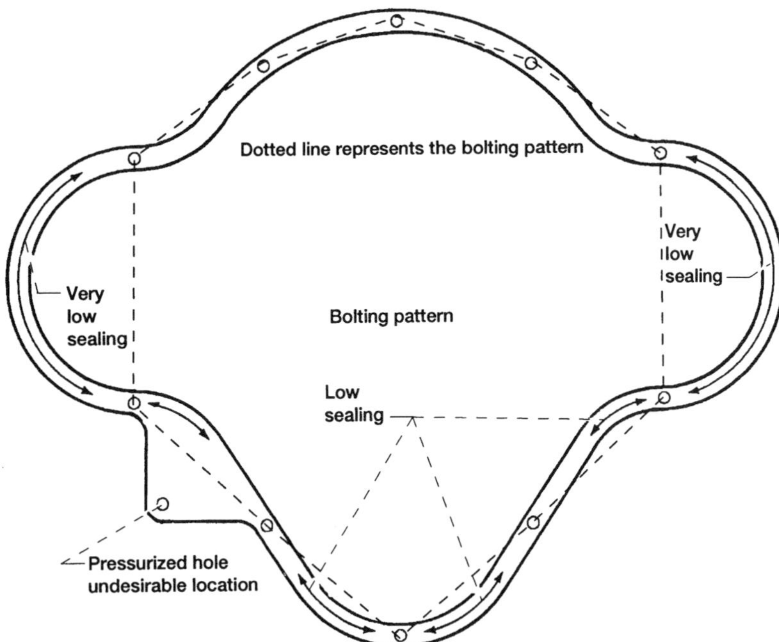

Figure 1 Bolting pattern indicating poor sealing areas. (From Ref. 2.)

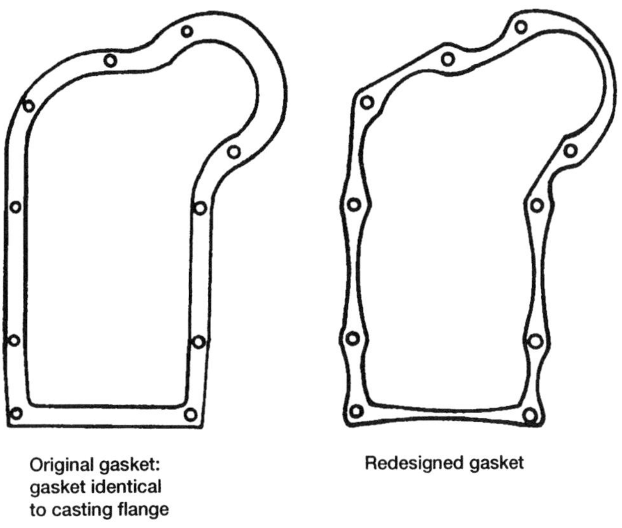

Figure 2 Original versus redesigned gasket for improved sealing. (From Ref. 2.)

Gasket Thickness and Compressibility. Gasket thickness and compressibility must be matched to the rigidity, roughness, and unevenness of the mating flanges. An effective gasket seal is achieved only if the stress level imposed on the gasket at installation is adequate for the specific gasket and joint requirements.

Gaskets made of compressible materials should be as thin as possible. Adequate gasket thickness is required to seal and conform to the unevenness of the mating flanges, including surface finish, flange flatness, and flange warpage during use. A gasket that is too thick can compromise the seal during pressurization cycles and is more likely to exhibit creep relaxation over time.

Elevated-Temperature Service. Use of gaskets at elevated temperatures results in some additional challenges. The Pressure Vessel Research Council of the Welding Research Council has published several bulletins in this area (see Refs. 3 and 4).

2.2 O-Rings

O-ring seals are perhaps one of the most common forms of seals. Following relatively straightforward design guidelines, a designer can be confident of a high-quality seal over a wide range of operating conditions. This section provides useful insight to designers approaching an O-ring seal design, including the basic sealing mechanism, preload, temperature effects, common materials, and chemical compatibility with a range of working fluids. The reader is directed to manufacturer's design manuals for detailed information on the final selection and specification.[5]

Basic Sealing Mechanism

O-rings are compressed between the two mating surfaces and are retained in a seal gland. The initial compression provides initial sealing critical to successful sealing. Upon increase of the pressure differential across the seal, the seal is forced to flow to the lower pressure side of the gland (see Fig. 3). As the seal moves, it gains greater area and force of sealing contact. At the pressure limit of the seal, the O-ring just begins to extrude into the gap between the inner and outer member of the gap. If this pressure limit is exceeded, the O-ring will fail by extruding into the gap. The shear strength of the seal material is no longer sufficient to resist flow and the seal material extrudes (flows) out of the open passage. Back-up rings are used to prevent seal extrusion for high-pressure static and for dynamic applications.

Preload

The tendency of an O-ring to return to its original shape after the cross section is compressed is the basic reason why O-rings make such excellent seals. The maximum linear compression suggested by manufacturers is 30% for static applications and 16% for dynamic seals (up to 25% for small cross-sectional diameters). Compression less than these values is acceptable, within reason, if assembly problems are an issue. Manufacturers recommend[5] a minimum amount of initial linear compression to overcome the compression set that O-rings exhibit.

O-ring compression force depends principally on the hardness of the O-ring, its cross-sectional dimension, and the amount of compression. Figure 4 illustrates the range of compressive force per linear inch of seal for typical linear percent compressions (0.139 in. cross-sectional diameter) and compound hardness (Shore A hardness scale). Softer compounds provide better sealing ability, as the rubber flows more easily into the grooves. Harder compounds are specified for high pressures, to limit chance of extruding into the groove,

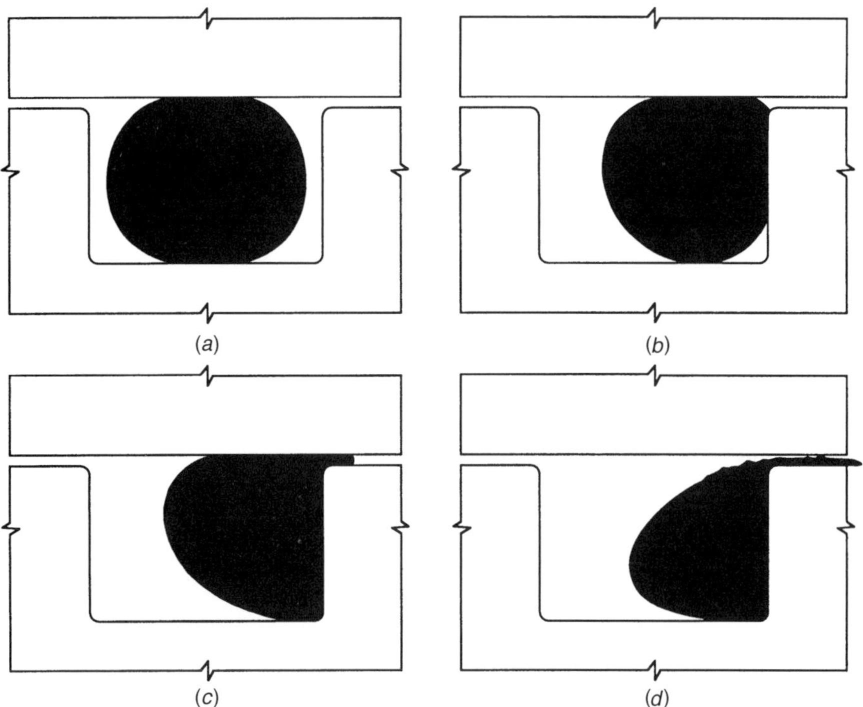

Figure 3 Basic O-ring sealing mechanism: (*a*) O-ring installed; (*b*) O-ring under pressure; (*c*) O-ring extruding; (*d*) O-ring failure. (From Ref. 5.)

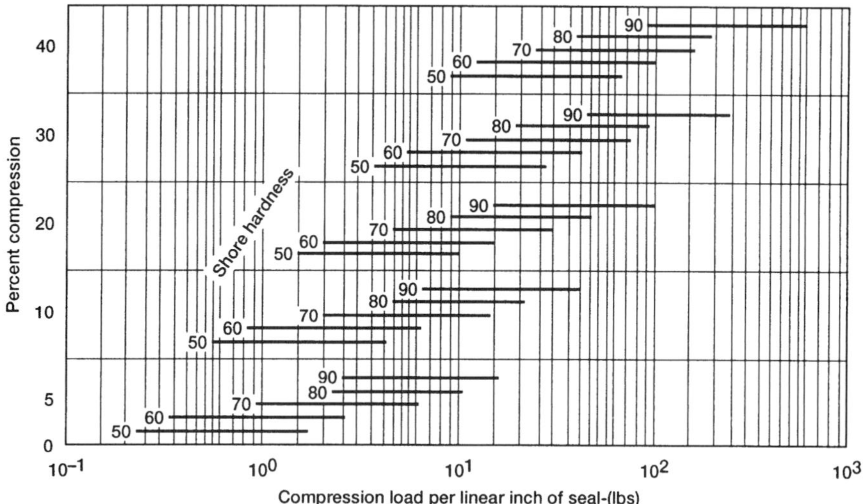

Figure 4 Effect of percent compression and material Shore hardness on seal compression load, 0.139 in. cross section. (From Ref. 5.)

and to improve wear life for dynamic service. For most applications, compounds having a type A durometer hardness of 70–80 are the most suitable compromise.[5]

Thermal Effects

O-ring seals respond to temperature changes. Therefore, it is critical to ensure the correct material and hardness are selected for the application. High temperatures soften compounds. This softening can negatively affect the seal's extrusion resistance at temperature. Over long periods of time at high temperature, chemical changes occur. These generally cause an increase in hardness, along with volume and compression set changes.

O-ring compounds harden and contract at cold temperatures. These effects can both lead to a loss of seal if initial compression is not set properly. Because the compound is harder, it does not flow into the mating surface irregularities as well. Just as important, the more common O-ring materials have a coefficient of thermal expansion (CTE) 10 times greater than that of steel (i.e., nitrile CTE is 6.2×10^{-5} per °F).

Groove dimensions must be sized correctly to account for this dimensional change. Manufacturers' design charts[5] are devised such that proper O-ring sealing is ensured for the temperature ranges for standard elastomeric materials. However, the designer may want to modify gland dimensions for a given application that experiences only high or low temperatures in order to maintain a particular squeeze on the O-ring. Martini[6] gives several practical examples showing how to tailor groove dimensions to maintain a given squeeze for the operating temperature.

Material Selection/Chemical Compatibility

Seal compounds must work properly over the required temperature range, have the proper hardness to resist extrusion while effectively sealing, and must resist chemical attack and resultant swelling caused by the operating fluids. Table 2 summarizes the most important elastomers, their working temperature range, and their resistance to a range of common working fluids.

Rotary Applications

O-rings are also used to seal rotary shafts where surface speeds and pressures are relatively low. One factor that must be carefully considered when applying O-ring seals to rotary applications is the Gow–Joule effect.[6] When a rubber O-ring is stretched slightly around a rotating shaft (e.g., put in tension), friction between the ring and shaft generates heat, causing the ring to contract, exhibiting a negative expansion coefficient. As the ring contracts, friction forces increase, generating additional heat and further contraction. This positive-feedback cycle causes rapid seal failures. Similar failures in reciprocating applications and static applications are unusual because surface speeds are too low to initiate the cycle. Further, in reciprocating applications the seal is moved into contact with cooler adjacent material. To prevent the failure cycle, O-rings are not stretched over shafts but are oversized slightly (circumferentially) and compressed into the sealing groove. The precompression of the cross section results in O-ring stresses that oppose the contraction stress, preventing the failure cycle described. Martini[6] provides guidelines for specifying the O-ring seal. Following appropriate techniques O-ring seals have run for significant periods of time at speeds up to 750 fpm and pressures up to 200 psi.

2.3 Packings and Braided Rope Seals

Rope packings used to seal stuffing boxes and valves and prevent excessive leakage can be traced back to the early days of the Industrial Revolution. An excellent summary of types of rope seal packings is given in Ref. 7. Novel adaptations of these seal packings have been required as temperatures have continued to rise to meet modern system requirements. New ceramic materials are being investigated to replace asbestos in a variety of gasket and rope-packing constructions.

Materials

Packing materials are selected for the intended temperature and the chemical environment. Graphite-based packing/gaskets are rated for up to 1000°F for oxidizing environments and up to 5400°F for reducing environments.[8] Used within its recommended temperature, graphite will provide a good seal with acceptable ability to track joint movement during temperature/pressure excursions. Graphite can be laminated with itself to increase thickness or with metal/plastic to improve handling and mechanical strength. Table 2 provides working temperatures for conventional (e.g., nitrile, PTFE, neoprene, among others) gasket/packings. Table 3 provides typical maximum working temperatures for high-temperature gasket/packing materials.

Packings and Braided Rope Seals for High-Temperature Service

High-temperature packings and rope seals are required for a variety of applications, including sealing: furnace joints and locations within continuous casting units (gate seals, mold seals, runners, spouts, etc.), among others. High-temperature packings are used for numerous aerospace applications, including turbine casing and turbine engine locations, Space Shuttle thermal protection systems, and nozzle joint seals.

Aircraft engine turbine inlet temperatures and industrial system temperatures continue to climb to meet aggressive cycle thermal efficiency goals. Advanced material systems, including monolithic/composite ceramics, intermetallic alloys (i.e., nickel aluminide), and carbon–carbon composites, are being explored to meet aggressive temperature, durability, and weight requirements. Incorporating these materials in the high-temperature locations in

Table 3 Gasket/Rope Seal Materials

Fiber Material	Maximum Working Temperature, °F
Graphite	
Oxidizing environment	1000
Reducing	5400
Fiberglass (glass dependent)	1000
Superalloy metals (depending on alloy)	1300–1600
Oxide ceramics (Ref. Tompkins, 1995)[a]	
62% Al_2O_3 24% SiO_2 14% B_2O_3 (Nextel 312)	1800[b]
70% Al_2O_3 28% SiO_2 2% B_2O_3 (Nextel 440)	2000[b]
73% Al_2O_3 27% SiO_2 (Nextel 550)	2100[b]

[a] Tompkins, T. L. "Ceramic Oxide Fibers: Building Blocks for New Applications," Ceramic Industry Publications, Business News Publishing, Apr. 1995.

[b] Temperature at which fiber retains 50% (nominal) room temperature strength. Materials can be used at higher temperatures than these for short-term. (Consult the manufacturer for guidance.)

the system, designers must overcome materials issues, such as differences in thermal expansion rates and lack of material ductility.

Designers are finding that one way to avoid cracking and buckling of the high-temperature brittle components rigidly mounted in their support structures is to allow relative motion between the primary and supporting components.[9] Often this joint occurs in a location where differential pressures exist, requiring high-temperature seals. These seals or packings must exhibit the following important properties: operate hot ($\geq$1300°F); exhibit low leakage; resist mechanical scrubbing caused by differential thermal growth and acoustic loads; seal complex geometries; retain resilience after cycling; and support structural loads.

In an industrial seal application, a high-temperature all-ceramic seal is being used to seal the interface between a low-expansion-rate primary structure and the surrounding support structure. The seal consists of a dense uniaxial fiber core overbraided with two two-dimensional braided sheath layers.[9] Both core and sheath are composed of 8-μm alumina–silica fibers (Nextel 550) capable of withstanding 2000+°F temperatures. In this application over a heat/cool cycle, the support structure moves 0.3 in. relative to the primary structure, precluding normal fixed-attachment techniques. Leakage flows for the all-ceramic seal are shown in Fig. 5 for three temperatures after simulated scrubbing[9] (10 cycles $\times$ 0.3-in. at 1300°F). Studies[9] have shown the benefits of high sheath braid angle and double-stage seals for reducing leakage. Increasing hybrid seal sheath braid angle and increasing core coverage led to increased compressive force (for the same linear seal compression) and one-third the leakage of the conventional hybrid design. Adding a second seal stage reduced seal leakage 30% relative to a single stage.

In a turbine vane application, the conventional braze joint is replaced with a floating-seal arrangement incorporating a small-diameter (1/16-in.) rope seal (Fig. 6). The seal is designed to serve as a seal and a compliant mount, allowing relative thermal growth between the high-temperature turbine vane and the lower temperature support structure, preventing

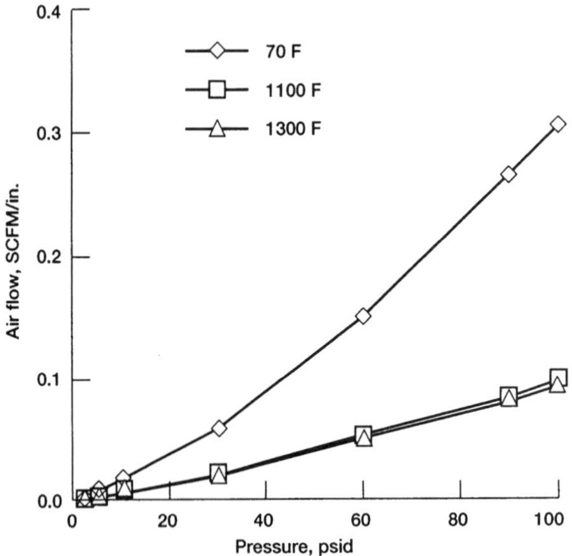

Figure 5 Flow versus pressure data for three temperatures, 1/16-in.-diameter all-ceramic seal, 0.022 in. seal compression, after scrubbing. (From Ref. 9.)

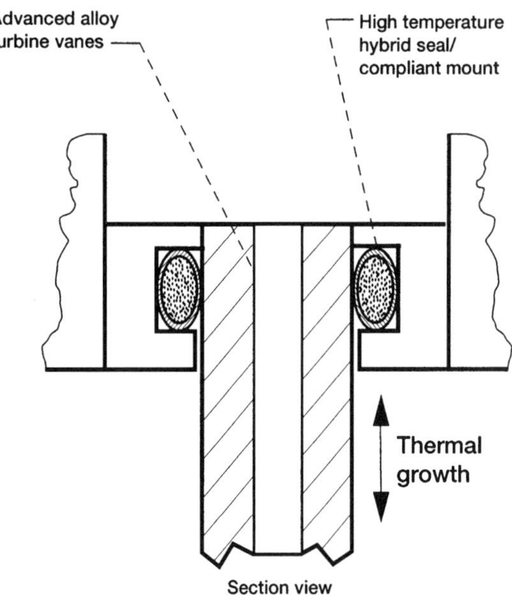

Figure 6 Schematic of turbine vane seal. (From Ref. 10.)

thermal strains and stresses. A hybrid seal consisting of a dense uniaxial ceramic core (8-μm alumina–silica Nextel 550 fibers) overbraided with a superalloy wire (0.0016-in.-diameter Haynes 188 alloy) abrasion-resistant sheath has proven successful for this application.[10] Leakage flows for the hybrid seal are shown in Fig. 7 for two temperatures and pressures under two preload conditions after simulated scrubbing (10 cycles × 0.3 in. at 1300°F). Researchers at NASA Glenn Research Center continue to strive for higher operating hybrid seals (metallic sheath over a ceramic fiber core). Recent oxidation studies[11] showed that wires made from alumina forming scale base alloys (e.g., Plansee PM2000) could resist oxidation at temperatures to 2200°F (1200°C) for test times up to 70 h. Tests showed that alumina-forming alloys with reactive element additions performed best at 2200°F under all test conditions in the presence of oxygen, moisture, and temperature cycling. Wire samples exhibited slow-growing oxide and adherent scales.

Space Shuttle rocket motor designers have found success in implementing braided carbon ropes as thermal barriers to protect temperature-sensitive downstream O-ring seals.[12–18] In this application, carbon fiber thermal barriers are used to effectively block the 5500°F gas temperatures from reaching the downstream Viton O-ring pressure seals that are rated to only 500°F (800°F short term). Carbon can be used in this application because of the reducing environment. The braided rope seals are unique in that they effectively block the high temperatures but do not impede the system pressures (900 psi) from seating the O-rings that form the final pressure seal. In the full-scale solid rocket motor tests performed, motor manufacturer ATK-Thiokol measured gas temperatures upstream (hot side) and downstream of the thermal barriers. ATK-Thiokol observed a significant drop in gas temperature across the two thermal barriers, permitting only cool (<200°F) gas to reach the downstream Viton O-rings seals. Full-scale motor tests have certified the carbon thermal barrier for Space Shuttle flight. The NASA Glenn thermal barrier is also being used by other solid rocket manufacturers. Attributes of the thermal barrier include the following:

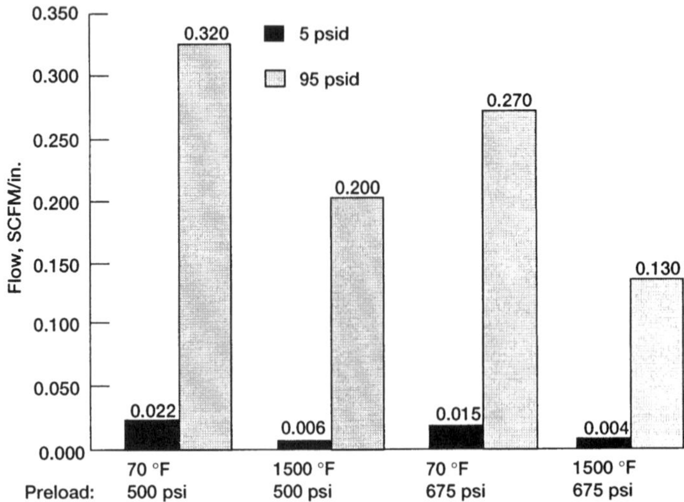

Figure 7 The effect of temperature, pressure, and representative compression on seal flow after cycling for 0.060-in. hybrid vane seal. (From Ref. 10.)

- *Unique Structure.* The unique braided structure[18] permits designers to tailor the thermal barrier/seal's properties. Tighter, denser braids form a more effective flow restriction. Looser braids offer more flexibility and allow tighter bend radii.
- *Flexibility/Resiliency.* The carbon thermal barrier provides much-needed flexibility and resiliency to accommodate either joint closings or openings during rocket pressurization and launch not afforded by competing approaches.
- *Self-Seating Feature.* Tests have shown that, upon joint pressurization, the thermal barrier seats itself in the groove to provide a more effective barrier to hot-gas flow.
- *Gas Jet Diffusion.* The thermal barrier diffuses and spreads the incoming high-pressure (900-psi) combustion gas jets, preventing damage to downstream O-rings.
- *Burn Resistance.* NASA Glenn tests showed that the carbon thermal barrier exhibits burn resistance over 60 times greater than similarly constructed ceramic thermal barriers.
- *Slag Block.* The thermal barrier also blocks molten alumina (3700°F) slag (products of combustion) from impinging on temperature-sensitive O-rings.
- *Simplified Installation.* The thermal barrier installs easily into joints in one-sixth the time, eliminating current laborious, time-consuming steps of applying the formerly used joint fill compound, checking joint fill integrity, and replacing/repairing joint fill.

3 DYNAMIC SEALS

3.1 Initial Seal Selection

An engineer approaching a dynamic seal design has a wide range of seals to choose from. A partial list of seals available ranges from the mechanical face seal through the labyrinth and brush seal, as indicated in Fig. 8. To aid in the initial seal selection, a "decision tree"

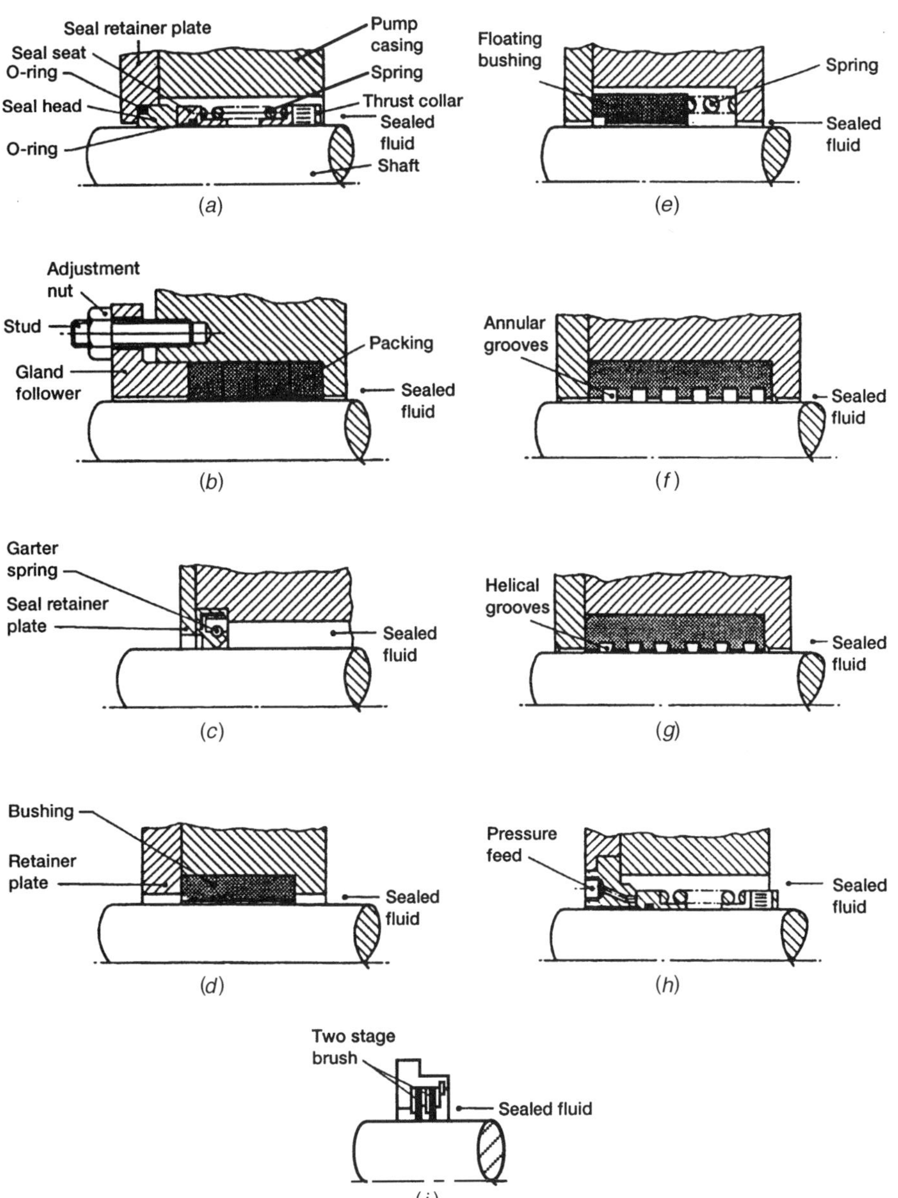

Figure 8 Examples of the main types of rotary seal: (*a*) mechanical face seal; (*b*) stuffing box; (*c*) lip seal; (*d*) fixed bushing; (*e*) floating bushing; (*f*) labyrinth; (*g*) viscoseal; (*h*) hydrostatic seal; (*i*) brush seal. [(*a*)–(*h*) From Ref. 19.]

has been proposed by Fern and Nau.[19] The decision tree (see Fig. 9) has been updated for the current work to account for the emergence of brush seals. In this chart, a majority of answers either "yes" or "no" to the questions at each stage leads the designer to an appropriate seal starting point. If answers are equally divided, both alternatives should be explored using other design criteria, such as performance, size, and cost.

The scope of this chapter does not permit treatment of every entry in the decision tree. However, several examples are given below to aid in understanding its use.

Radial lip seals are used to prevent fluids, normally lubricated, from leaking around shafts and their housings. They are also used to prevent dust, dirt, and foreign contaminants from entering the lubricant chamber. Depending on conditions, lip seals have been designed to operate at very high shaft speeds (6000–12,000 rpm) with light oil mist and no pressure in a clean environment. Lip seals have replaced mechanical face seals in automotive water pumps at pressures to 30 psi, temperatures of −45°F–350°F, and shaft speeds to 8000 sfpm (American Variseal, 1994). Lip seals are also used in completely flooded low-speed applications or in muddy environments. A major advantage of the radial lip seal is its compactness. A 0.32-in. × 0.32-in. lip seal provides a very good seal for a 2-in.-diameter shaft.

Mechanical face seals are capable of handling much higher pressures and a wider range of fluids. Mechanical face seals are recommended over brush seals where very high pressures must be sealed in a single stage. Mechanical face seals have a lower leakage than brush seals because their effective clearances are several times smaller. However, the mechanical face seal requires much better control of dimensions and tolerates less shaft misalignment and runout, thereby increasing costs.

Turbine Engine Seals. Readers interested particularly in turbine engine seals are referred to Steinetz and Hendricks,[20] who review in greater depth the trade-offs in selecting seals for turbine engine applications. Technical factors increasing seal design complexity for aircraft engines include high temperatures (≥1000°F), high surface speeds (up to 1500 fps), rapid thermal/structural transients, small-space claim, maneuver and landing loads, and the requirement to be lightweight.

3.2 Mechanical Face Seals

The primary elements of a conventional spring-loaded mechanical face seal are the primary seal (the main sealing faces), the secondary seal (seals shaft leakage), and the spring or bellows element that keep the primary seal surfaces in contact, shown in Fig. 8. The primary seal faces are generally lapped to demanding surface flatness, with surface flatness of 40 μin. (1 μm) not uncommon. Surface flatness this low is required to make a good seal, since the running clearances are small. Conventional mechanical face seals operate with clearances of 40–200 μin. Dry-running, noncontacting gas face seals that use spiral-groove face geometry reliably run at pressures of 1800 psig and speeds up to 590 fps.[20a]

Seal Balance

Seal balancing is a technique whereby the primary seal front and rear areas are used to minimize the contact pressure between the mating seal faces to reduce wear and to increase the operating pressure capability. The concept of seal balancing is illustrated in Fig. 10.[21] The front and rear faces of the seal in Fig. 10*a* are identical and the full fluid pressure exerted on A′ is carried on the seal face A. By modifying the geometry of the primary seal head ring to establish a smaller frontal area A′ (Fig. 10*b*) and to provide a shoulder on the opposite side of the seal ring to form a front face B′, the hydraulic pressure counteracts part

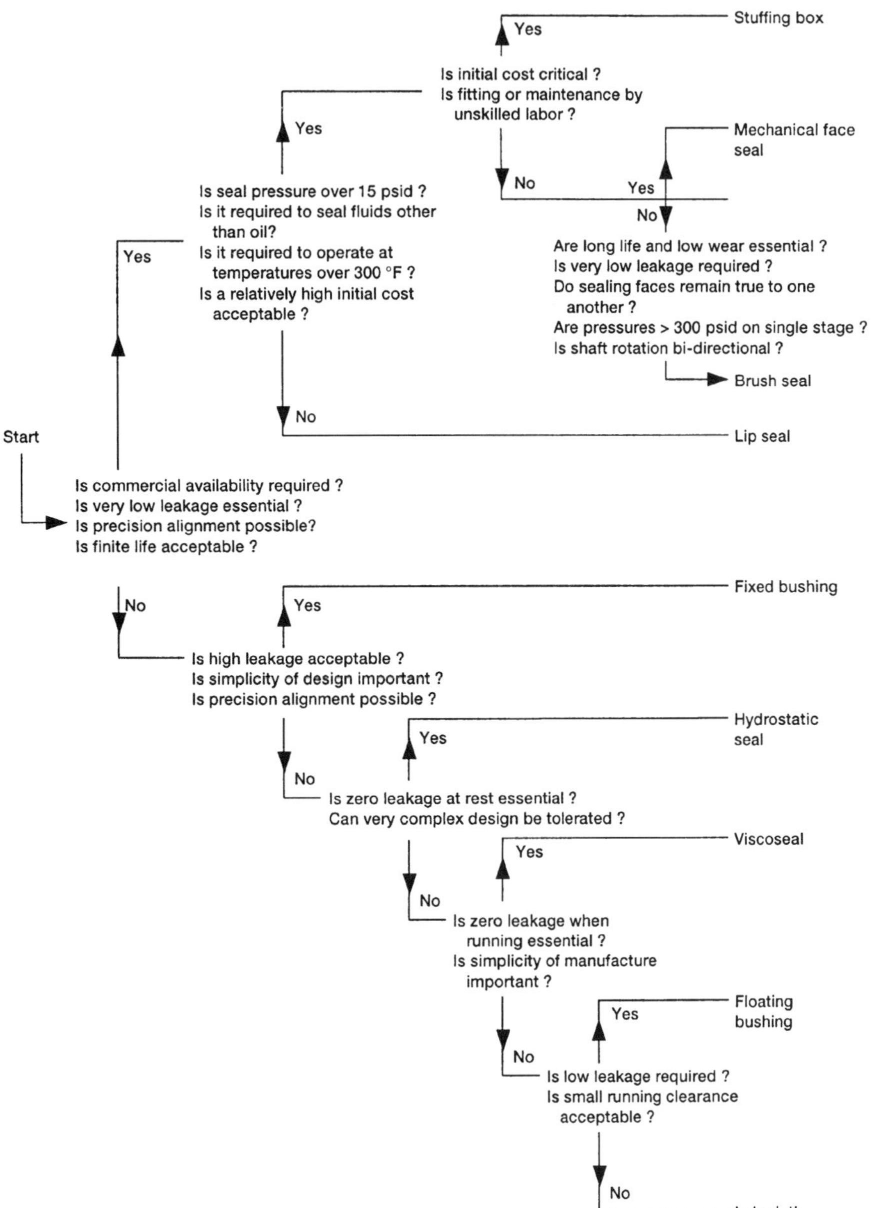

Figure 9 Seal selection chart (a majority answer of "yes" or "no" to the question at each stage leads the reader to the appropriate decision; if answers are equally divided, both alternatives should be explored). (Adapted from Ref. 19.)

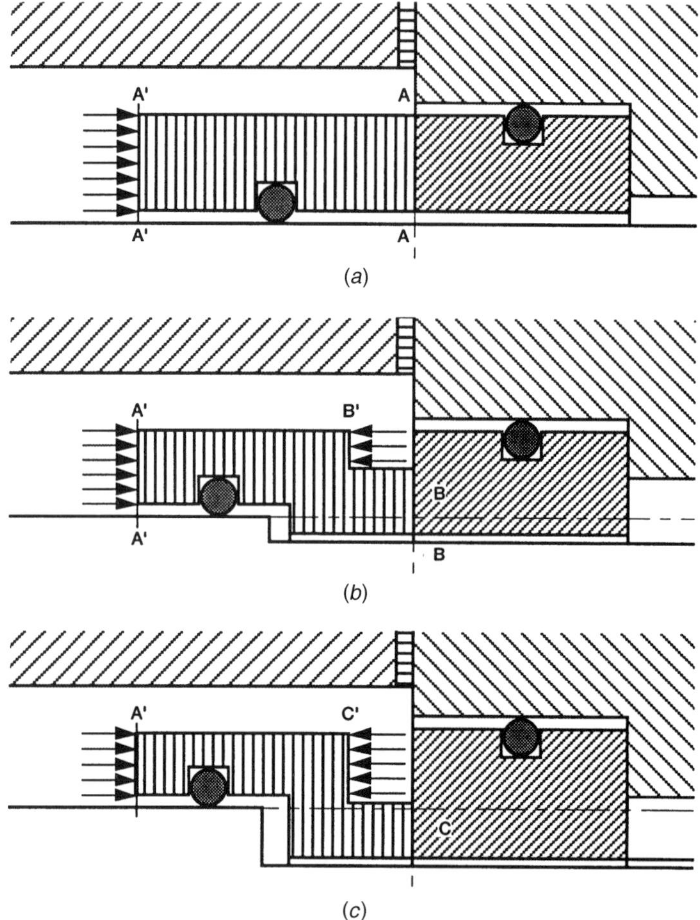

Figure 10 Illustration of face seal balance conditions: (*a*) unbalanced; (*b*) partially balanced; (*c*) fully balanced. (From Ref. 21.)

of the hydraulic loading from A′. Consequently, the remaining face pressure in the contact interface is significantly reduced. Depending on the relative sizes of surfaces A′ and B′, the seal is either partially balanced (Fig. 10*b*) or fully balanced (Fig. 10*c*). In fully balanced seals, there is no net hydraulic load exerted on the seal face. Seals are generally run with a partial balance, however, to minimize face loads and wear while keeping the seal closed during possible transient overpressure conditions. Partially balanced seals can run at pressures greater than six times unbalanced seals can for the same speed and temperature conditions.

Mechanical Face Seal Leakage

Liquid Flow. Minimizing leakage between seal faces is possible only through maintaining small clearances. Volumetric flow (Q) can be determined for the following two conditions[22]:

For coned faces:

$$Q = \frac{\pi \phi r_m}{3\mu} \left(\frac{P_o - P_i}{1/h_o^2 - 1/h_i^2} \right)$$

For parallel faces:

$$Q = \frac{-\pi r_m h^3}{6\mu} \frac{(P_o - P_i)}{(r_o - r_i)} \qquad h_o = h_i \quad \text{and} \quad (r_o - r_i)/r_m < 0.1$$

where ϕ (radians) is the cone angle (positive if faces are convergent travelling inward radially); r_o, r_i (in.) outer and inner radii; r_m (in.) mean radius (in.); h_o, h_i (in.) outer and inner film thicknesses; P_o, P_i (psi) outer and inner pressures; μ (lbf·s/in.^{2}) viscosity. The need for small clearances is demonstrated by noting that doubling the film clearance, h, increases the leakage flow eight fold.

Gas Flow. Closed-form equations for gas flow through parallel faces can be written only for conditions of laminar flow (Reynolds number < 2300). For laminar flow with a parabolic pressure distribution across the seal faces, the mass flow is given as[22]

$$\dot{m} = \frac{\pi r_m h^3}{12\mu RT} \frac{(P_i^2 - P_o^2)}{(r_o - r_i)} \qquad (r_o - r_i)/r_m < 0.1$$

where R is the gas constant (53.3 lbf·ft/lb$_m$·°R for air) and T (°R) is the gas temperature (isothermal throughout).

In cases where flow is both laminar and turbulent, iterative schemes must be employed. See Refs. 22 and 23 for numerical algorithms to use in solving for the seal leakage rates. Reference 24 treats the most general case of two-phase flow through the seal faces.

Seal Face Flatness

In addition to lapping faces to the 40-μin. flatness, there are several other points to consider. The lapped rings should be mounted on surfaces that are themselves flat. The ring must be stiff enough to resist distortions caused either by thermal or fluid pressure stresses.

The primary mode of distortion of a mechanical seal face under combined fluid and thermal stresses is solid-body rotation about the seal's neutral axis.[19] If the sum of the moments M (in.-lb/in.) per unit of circumference around the neutral axis can be calculated, then the angular deflection θ (radians) of the sealing face can be obtained from

$$\theta = \frac{M r_m^2}{EI}$$

where E (psi) = Young's modulus
I (in.^{4}) = the second moment of areas about the neutral axis
r_m (in.) = the mean radius of the seal ring

Face Seal Materials

Selecting the correct materials for a given seal application is critical to ensuring desired performance and durability. Seal components for which material selection is important from a tribology standpoint are the stationary nosepiece (or primary seal ring) and the mating ring (or seal seat). Properties considered ideal for the primary seal ring are shown below.[25]

1. Mechanical:
   (a) High modulus of elasticity
   (b) High tensile strength
   (c) Low coefficient of friction
   (d) Excellent wear characteristics and hardness
   (e) Self-lubrication

2. Thermal:
   (a) Low coefficient of expansion
   (b) High thermal conductivity
   (c) Thermal shock resistance
   (d) Thermal stability
3. Chemical:
   (a) Corrosion resistance
   (b) Good wetability
4. Miscellaneous:
   (a) Dimensional stability
   (b) Good machinability and ease of manufacture
   (c) Low cost and ready availability

Carbon–graphite is often the first choice for one of the running seal surfaces because of its superior dry-running (i.e., startup) behavior. It can run against itself, metals, or ceramics without galling or seizing. Carbon–graphite is generally impregnated with resin or with a metal to increase thermal conductivity and bearing characteristics. In cases where the seal will see considerable abrasives, carbon may wear excessively and then it is desirable to select very hard face seal materials. A preferred combination for very long wear (subject to other constraints) is tungsten carbide running on tungsten carbide. For a comprehensive coverage of face seal material selection, including chemical compatibility, see Ref. 26. Secondary seals are either O-rings or bellows. Temperature ranges and chemical compatibility for common O-ring secondary seals such as nitrile, fluorocarbon (Viton), and PTFE (Teflon) are provided in Table 2.

A mechanical seal is considered to have failed if the seal leakage exceeds the plant site operating or environmental limits. Seal failures can be a major contributor to rotary equipment failure and downtime. The Fluid Sealing Association has compiled an excellent guide for troubleshooting mechanical seals.[27] Seals can fail from one or more of the following reasons: (1) incorrect selection of the seal design or materials for the intended application; (2) abuse of the seal before installation; (3) erroneous installation; (4) improper startup, including dry running or failure of the environmental controls; (5) improper equipment operation; (6) contamination of the sealing fluid with either abrasive or corrosive materials; (7) equipment-induced excessive shaft run-out, deflection, vibration, or worn bearings; and (8) a worn-out seal. The effect of bearing performance on seal life is detailed in the *Pump and Systems Handbook*.[28]

3.3 Emission Concerns

Mechanical face seals have played and will continue to play a major role for many years in minimizing emissions to the atmosphere. New federal, state, and local environmental regulations have intensified the focus on mechanical face seal performance in terms of emissions. Within a short time, regulators have gone from little or no concern about fugitive hazardous emissions to a position of severely restricting all hazardous emissions. For instance, under the authority of Title III of the 1990 Clean Air Act Amendment (CAAA), the U.S. Environmental Protection Agency (EPA) adopted the National Emission Standards for Hazardous Air Pollutants (NESHAP) for the control of emissions of volatile hazardous air pollutants

(Ref. STLE, 1994).[29] Leak definitions per the regulation [EPA HON Subpart H (5)] are as follows:

Phase I: 10,000 parts per million volumetric (ppmv), beginning on compliance date

Phase II: 5000 ppmv, 1 year after compliance date

Phase III: 1000–5000 ppmv, depending on application, 2½ years after compliance date

The Clean Air Act regulations require U.S. plants to reduce emissions of 189 hazardous air pollutants by 80% in the next several years.[30] The American Petroleum Industry (API) has responded with a standard of its own, known as API 682, that seeks to reduce maintenance costs and control volatile organic compound (VOC) emissions on centrifugal and rotary pumps in heavy service. API 682, a pump shaft sealing standard, is designed to help refinery pump operators and similar users to comply with environmental emissions regulations. These regulations will continue to have a major impact on users of valves, pumps, compressors, and other processing devices. Seal users are cautioned to check with their state and local air quality control authorities for specific information.

Sealing Approaches for Emissions Controls

The Society of Tribologists and Lubrication Engineers published a guideline of mechanical seals for meeting the fugitive emissions requirements.[29] Seal technology available meets approximately 95% of current and anticipated federal, state, and local emission regulations. Applications not falling within the guidelines include food, pharmaceutical, and monomer-type products where dual seals cannot be used because of product purity requirements and chemical reaction of dual-seal buffer fluids with the sealed product. Bowden[31] compares various seal arrangements and factors to consider in a design when targeting low fugitive emissions based on operational experience in industrial applications.

Three sealing approaches for meeting the new regulatory requirements are discussed below: single seals, tandem seals, and double seals.[29]

Single Seals. The most economical approach available is the single seal mounted inside a stuffing box (Fig. 11). Generally, this type of seal uses the pumped product for lubrication. Due to some finite clearance between the faces, there is a small amount of leakage to the atmosphere. Using current technology in the design of a single seal, emissions can be con-

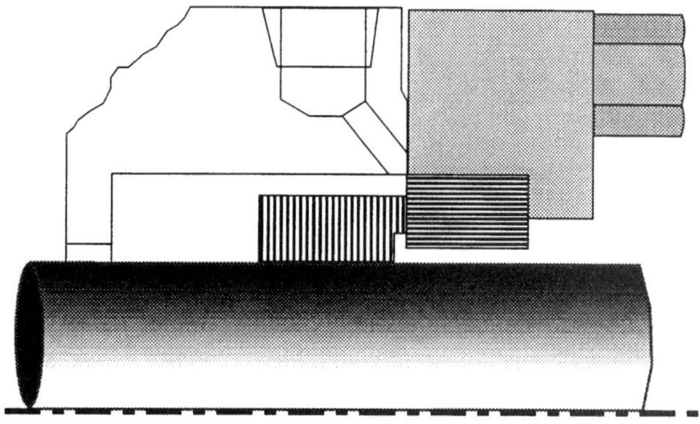

Figure 11 Single seal. (From Ref. 29.)

trolled to 500 ppm based on both laboratory and field test data. Emissions to the atmosphere can be eliminated by venting the atmospheric side to a vapor recovery or disposal system. Using this approach, emission readings approaching zero can be achieved. Since single seals have a minimum of contacting parts and normally require minimum support systems, they are considered highly reliable.

Tandem Seals. Tandem seals consist of two seal assemblies between which a barrier fluid operates at a pressure less than the pumped process pressure. The inboard primary seal seals the full pumped product pressure, and the outboard seal typically seals a nonpressurized barrier fluid (Fig. 12). Tandem seal system designs are available that provide zero emission of the pumped product to the environment, provided the vapor pressure of the product is higher than that of the barrier fluid and the product is immiscible in the barrier fluid. The barrier fluid isolates the pumped product from the atmosphere and is maintained by a support system. This supply system generally includes a supply tank assembly and optional cooling system and means for drawing off the volatile component (generally at the top of the supply tank). Examples of common barrier fluids are found in Table 4.

Tandem seal systems also provide a high level of sealing and reliability and are simple systems to maintain, due to the typical use of nonpressurized barrier fluid. Pumped product contamination by the barrier fluid is avoided since the barrier fluid is at a lower pressure than the pumped product.

Double Seals. Double seals differ from tandem seals in that the barrier fluid between the primary and outboard seal is pressurized (Fig. 13). Double seals can be either externally or internally pressurized. An externally pressurized system requires a lubrication unit to pressurize the barrier fluid above the pumped product pressure and to provide cooling. An internally pressurized double seal refers to a system that internally pressurizes the fluid film at the inboard faces as the shaft rotates. In this case, the barrier fluid in the seal chamber is normally at atmospheric pressure. This results in less heat generation from the system.

Application Guide. The areas of application based on emissions to the atmosphere for the three types of seals discussed are illustrated in Fig. 14. The scope of this chart is for seals less than 6 in. in diameter, for pressures 600 psig and less, and for surface speeds up to 5600 fpm. Waterbury[30] provides a modern overview of several commercial products aimed at achieving zero leakage or leak-free operation in compliance with current regulations.

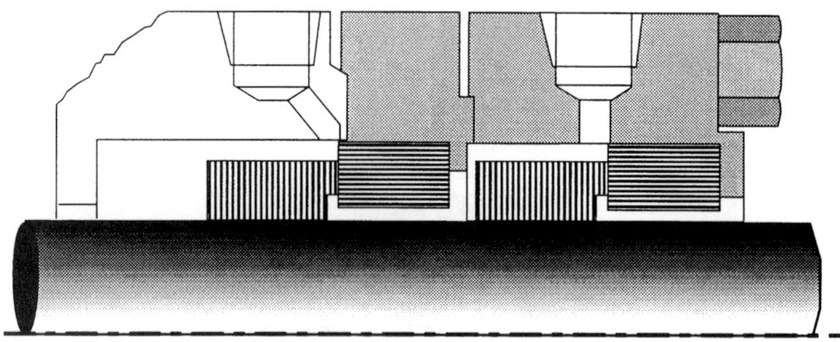

Figure 12 Tandem seal. (From Ref. 29.)

Table 4 Properties of Common Barrier Fluids for Tandem or Double Seals[a]

Barrier Fluid	Temperature Limits, °F		Comments
	Lower	Upper	
Water	40	180	Use corrosion-resistant materials Protect from freezing
Propylene glycol	−76	368	Consult seal manufacturer for proper mixture with water to avoid excessive viscosity
n–Propyl alcohol	−147	157	
Kerosene	0	300	
No. 2 diesel fuel	10	300	Contains additives

[a] STLE, "Guidelines for Meeting Emission Regulations for Rotating Machinery with Mechanical Seals," Special Publication SP-30, Society of Tribologists and Lubrication Engineers, Park Ridge, IL, 1990.

3.4 Noncontacting Seals for High-Speed/Aerospace Applications

For very high speed turbomachinery, including gas turbines, seal runner speeds may reach speeds greater than 1300 fps, requiring novel seal arrangements to overcome wear and pressure limitations of conventional face seals. Two classes of seals are used that rely on a thin film of air to separate the seal faces. Hydrostatic face seals port high pressure fluid to the sealing face to induce opening force and maintain controlled face separation (see Fig. 15). The fluid pressure developed between the faces is dependent upon the gap dimension and the pressure varies between the lower and upper limits shown in the figure. Any change in the design clearance results in an increase or decrease of the opening force in a stabilizing sense. Of the four configurations shown, the coned seal configuration is the most popular. Converging faces are used to provide seal stability. Hydrostatic face seals suffer from contact during startup. To overcome this, the seals can be externally pressurized, but this adds cost and complexity.

The aspirating hydrostatic face seal (Fig. 15*d*) under development by GE and Stein Seal for turbine engine applications provides a unique failsafe feature.[32–34] The seal is designed

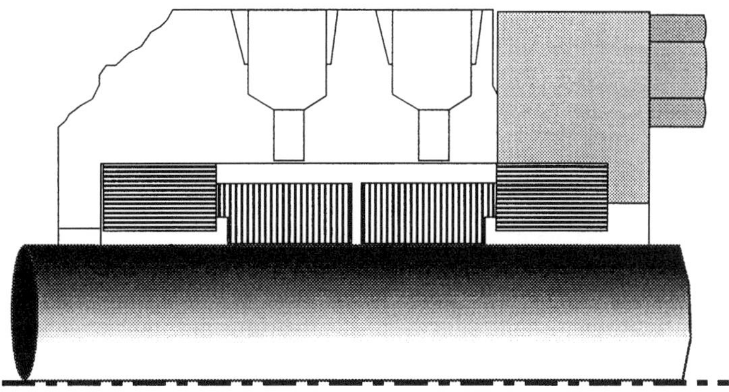

Figure 13 Double seal. (From Ref. 29.)

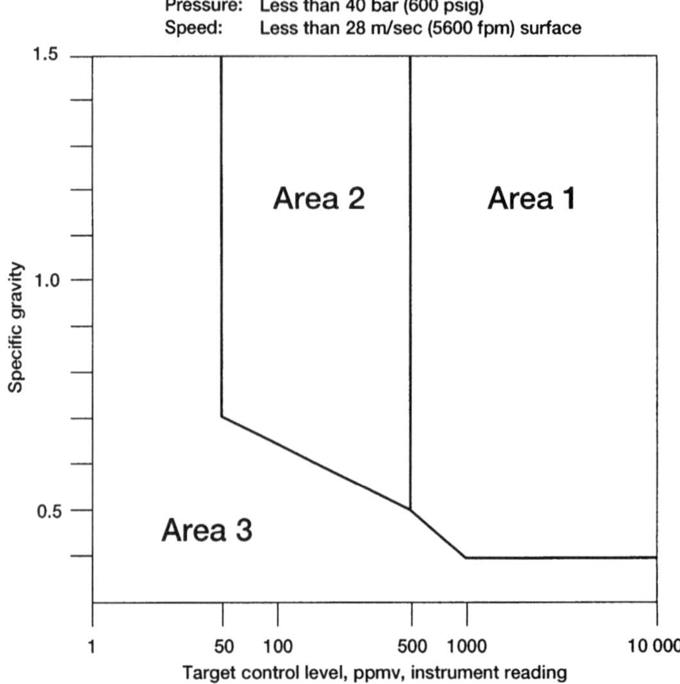

Chart area	Recommended technology
1	General purpose single seals, or dual (double and tandem) seals
2	Special purpose single seals, or dual (double and tandem) seals
3	Dual pressurized (double) seals Single or dual non-pressurized (tandem) seals vented to a closed vent system, above 0.4 specific gravity

Figure 14 Application guide to control emissions. (From Ref. 29.)

to be open during initial rotation and after system shutdown—the two periods during which potentially damaging rubs are most common. Upon system pressurization, the aspirating teeth set up an initial pressure drop across the seal (6 psi nominal) that generates a closing force to overcome the retraction spring force F_s, causing the seal to close to its operating clearance (nominal 0.0015–0.0025 in.). System pressure is ported to the face seal to prevent touchdown and provide good film stiffness during operation. At engine shutdown, the pressure across the seal drops and the springs retract the seal away from the rotor, preventing contact.

Hydrodynamic or self-acting face seals incorporate lift pockets to generate a hydrodynamic film between the two faces to prevent seal contact. A number of lift pocket configurations are employed, including shrouded Rayleigh step, spiral groove, circular groove, and annular groove (Fig. 16). In these designs, hydrodynamic lift is independent of the seal

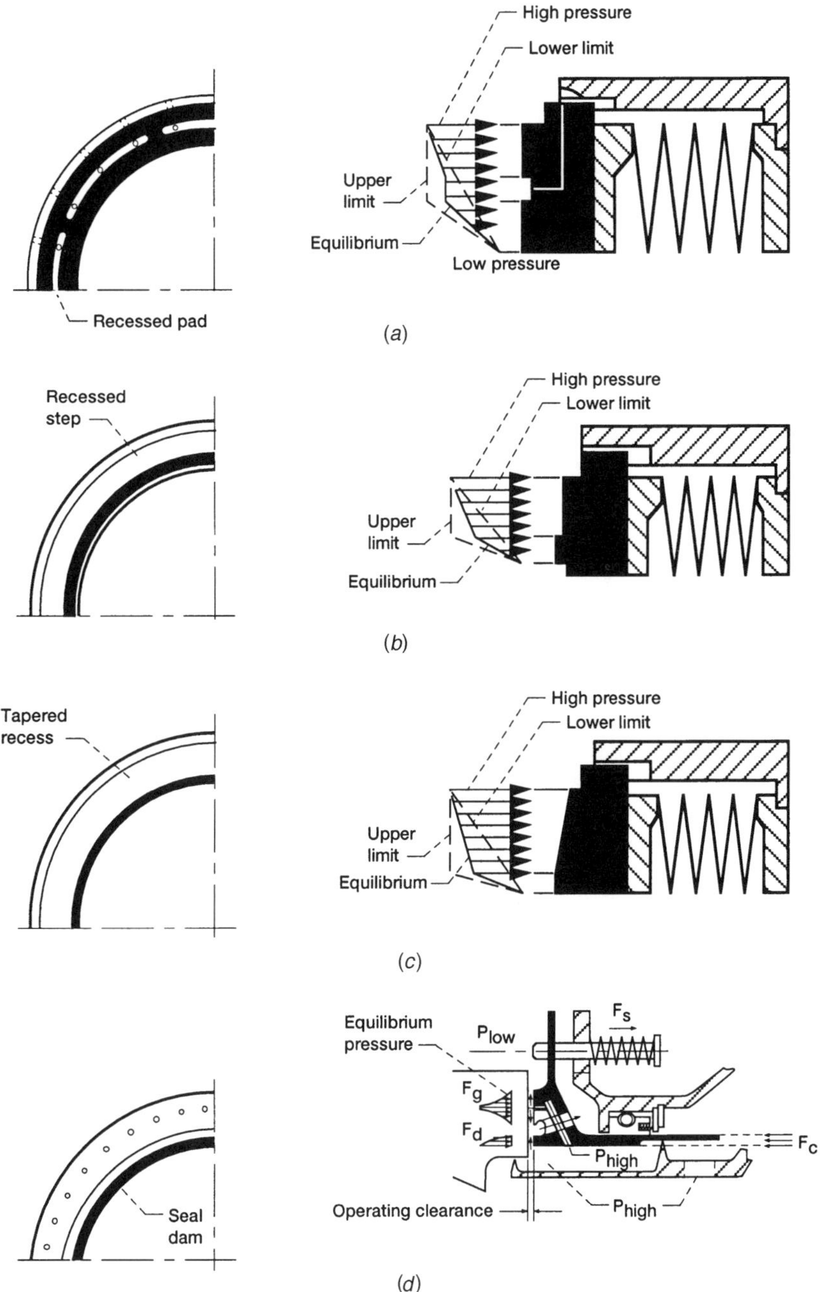

Figure 15 Self-energized hydrostatic noncontacting mechanical face seals: (*a*) recessed pads with orifice compensation; (*b*) recessed step; (*c*) convergent tapered face; (*d*) aspirating seal. [(*a*)–(*c*) from Ref. 1; (*d*) from Ref. 32.]

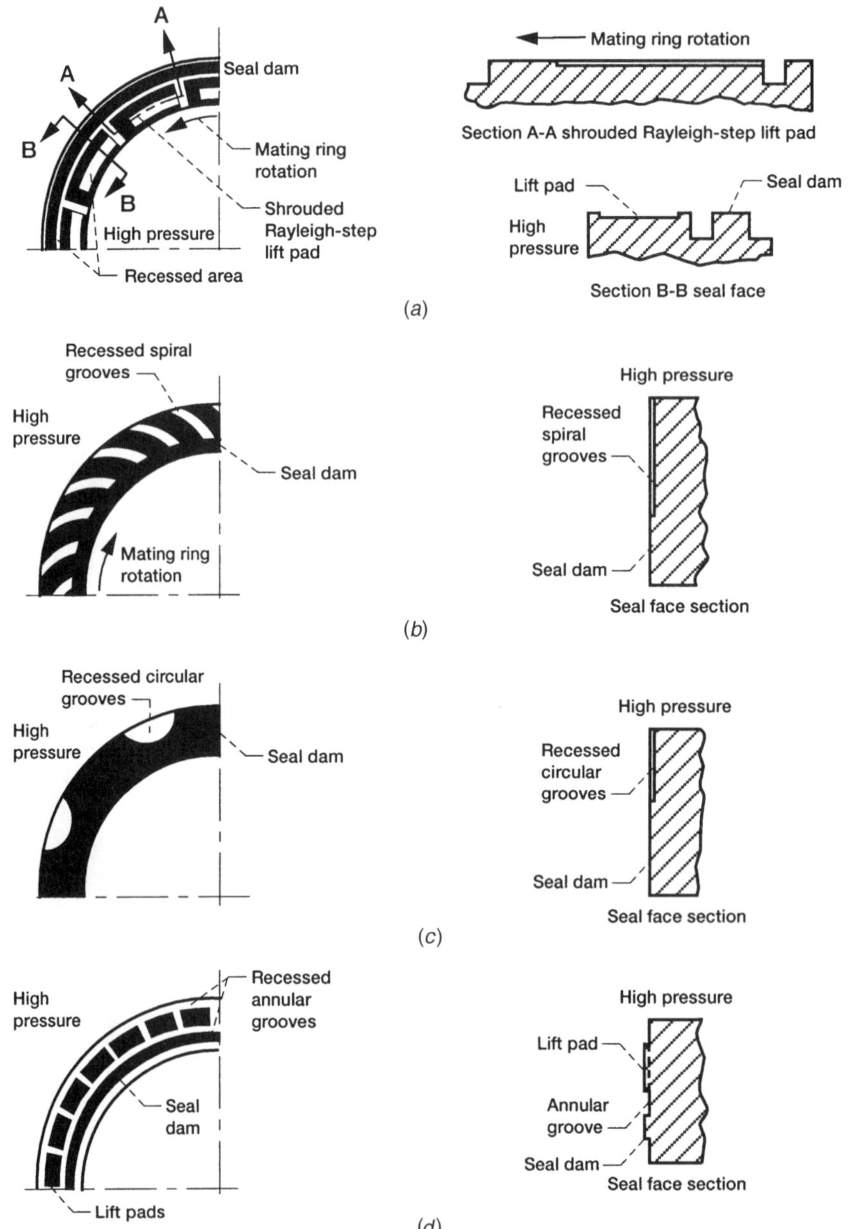

Figure 16 Various types of hydrodynamic noncontacting mechanical face seals: (*a*) shrouded Rayleigh step; (*b*) spiral groove; (*c*) circular groove; (*d*) annular groove. (From Ref. 1.)

pressure; it is proportional to the rotation speed and to the fluid viscosity. Therefore a minimum speed is required to develop sufficient lift force for face separation. Hydrodynamic seals operate on small (≤0.0005 in. nominal) clearances, resulting in very low leakage compared to labyrinth or brush seals, as shown in Fig. 17.[35] Because rubbing occurs during startup and shutdown, seal face materials must be selected for good rubbing characteristics for low wear (see Face Seal Materials, above).

Computer Analysis Tools: Face/Annular Seals

To aid aerospace and industrial seal designers alike, NASA sponsored the development of computer codes to predict the seal performance under a variety of conditions.[36] NASA seal design computer codes are available to U.S. persons through Open Channel Software.[37] Codes were developed to treat both incompressible (e.g., liquid) and compressible (e.g., gas) flow conditions. In general, the codes assess seal performance characteristics, including load capacity, leakage flow, power requirements, and dynamic characteristics in the form of stiffness and damping coefficients. These performance characteristics are computed as functions of seal and groove geometry, loads or film thicknesses, running speed, fluid viscosity, and boundary pressures. The GFACE code predicts performance for the following face seal geometries: hydrostatic, hydrostatic recess, radial and circumferential Rayleigh step, and radial and circumferential tapered land. The GCYLT code predicts performance for both hydrodynamic and hydrostatic cylindrical seals, including the following geometries: circumferential multilobe and Rayleigh step, Rayleigh step in direction of flow, tapered and self-energized hydrostatic. A description of these codes and their validation is given by Shapiro.[38] The SPIRALG/SPIRALI codes predict characteristics of gas-lubricated (SPIRALG) and liquid-lubricated (SPIRALI) spiral-groove, cylindrical, and face seals.[39]

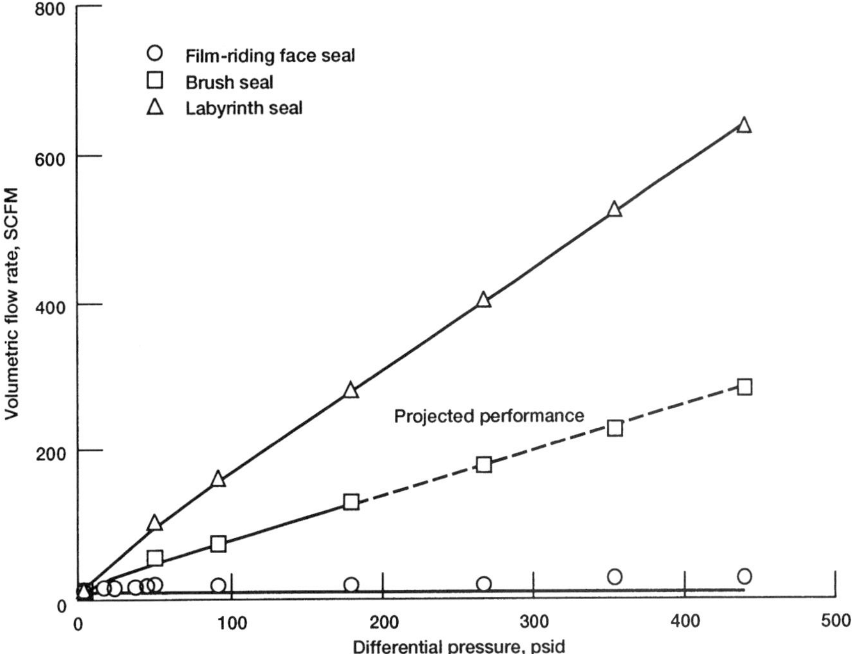

Figure 17 Comparison of brush, labyrinth and self-acting, film-riding face seal leakage rates as a function of differential pressure. Seal diameter, 5.84 in. (From Ref. 35.)

Dynamic response of seal rings to rotor motions is an important consideration in seal design. For contact seals, dynamic motion can impose significant interfacial forces, resulting in high wear and reduction in useful life. For fluid-film seals, the rotor excursions are generally greater than the film thickness, and if the ring does not track, contact and failure may occur. The computer code DYSEAL predicts the tracking capability of fluid-film seals and can be used for parametric geometric variations to find acceptable configurations.[40]

3.5 Labyrinth Seals

By their nature, labyrinth seals are clearance seals that also permit shaft excursions without potentially catastrophic rub-induced rotor instability problems. By design, labyrinth seals restrict leakage by dissipating the kinetic energy of fluid flow through a series of flow constrictions and cavities that sequentially accelerate and decelerate the fluid flow or change its direction abruptly to create the maximum flow friction and turbulence. The ideal labyrinth seal would transform all kinetic energy at each throttling into internal energy (heat) in each cavity. However, in practical labyrinth seals, a considerable amount of kinetic energy is transferred from one passage to the next. The advantage of labyrinth seals is that the speed and pressure capability is limited only by the structural design. One disadvantage, however, is a relatively high leakage rate. Labyrinth seals are used in so many gas sealing applications because of their very high running speed (1500 ft/s), pressure (250 psi), and temperature (≥1300°F) and the need to accommodate shaft excursions caused by transient loads. Labyrinth seal leakage rates have been reduced over the years through novel design concepts but are still higher than desired because labyrinth seal leakage is clearance dependent and this clearance opens due to periodic transient rubs.

Seal Configurations

Labyrinth seals can be configured in many ways (Fig. 18). The labyrinth seal configurations typically used are straight, angled-teeth straight, stepped, staggered, and abradable or wear-in. Optimizing labyrinth seal geometry depends on the given application and greatly affects the labyrinth seal leakage. Stepped labyrinth seals have been used extensively as turbine interstage air seals. Leakage flow through inclined, stepped labyrinths is about 40% that of straight labyrinths for similar conditions (Fig. 19). Performance benefits of stepped labyrinths must be balanced with other design issues. They require more radial space, are more difficult to manufacture, and may produce an undesirable thrust load because of the stepped area.

Leakage Flow Modeling

Leakage flow through labyrinth seals is generally modeled as a sequential series of throttlings through the narrow blade tip clearances. Ideally, the kinetic energy increase across each annular orifice would be completely dissipated in the cavity. However, dissipation is not complete. Various authors handle this in different ways: Egli[43] introduced the concept of "carryover" to account for the incomplete dissipation of kinetic energy in straight labyrinth seals. Vermes[44] introduced the residual energy factor, α, to account for the residual energy in the flow as it passes from one stage to the next:

$$W = 5.76K \frac{A_g}{[RT_o]^{1/2}} \frac{P_o}{[1 - \alpha]^{1/2}} \beta \qquad \text{where} \quad \beta = \left[\frac{1 - \left[\dfrac{P_N}{P_o}\right]^2}{N - \ln\left[\dfrac{P_N}{P_o}\right]} \right]^{1/2}$$

and the residual energy factor

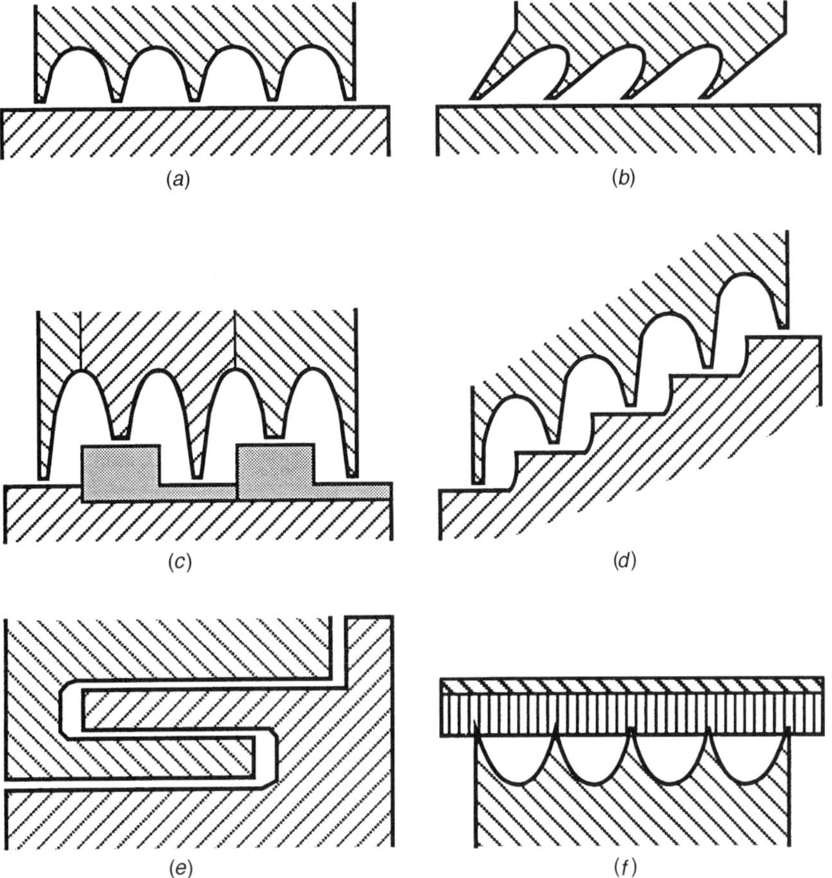

Figure 18 Labyrinth seal configurations: (*a*) straight labyrinth; (*b*) inclined- or angled-teeth straight labyrinth; (*c*) staggered labyrinth; (*d*) stepped labyrinth; (*e*) interlocking labyrinth; (*f*) abradable (wear-in) labyrinth. (From Ref. 41.)

$$\alpha = \frac{8.52}{\left[\dfrac{\mathrm{TP} - L}{c}\right] + 7.23}$$

where A_g = flow area of single annular orifice (in.^{2})
c = clearance (in.)
g_c = gravitational constant (32.2 ft/s^2)
G = mass flux (lb$_m$/ft^2 · s)
K = clearance factor for annular orifice (see Fig. 20)
L = tooth width at sealing point (in.)
N = number of teeth (in.)
N_{Re} = Reynolds number, defined as $G(c/12)/\mu g_c$
TP = tooth pitch (in.)
P_o, P_N = inlet pressure, pressure at tooth N

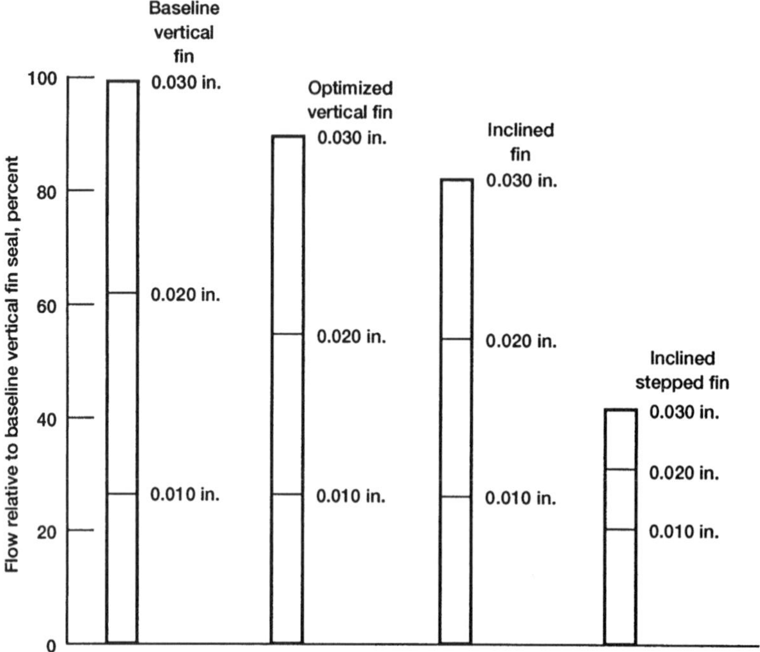

Figure 19 Labyrinth seal leakage flow performance for typical designs and clearances, relative to a baseline five-finned straight labyrinth seal of various gaps at pressure ratio of 2. (From Ref. 42.)

R = gas constant (lbf · ft/lbm · °R)
T_o = gas inlet temperature (°R)
W = weight flow lb/s
μ = gas viscosity ($\text{lb}_f \cdot \text{s/ft}^2$)

The clearance factor is plotted in Fig. 20 for a range of Reynolds numbers and tooth width-to-clearance ratios. Since K is a function of N_{Re} and since N_{Re} is a function of the unknown mass flow, the necessary first approximation can be made with $K = 0.67$. Vermes[44] also presented methods for calculating mass flow for stepped labyrinth seals and for off-design conditions (e.g., the stepped seal teeth are offset from their natural lands). Tooth shape also plays a role in leakage resistance. Mahler[45] showed that sharp corners provide the highest leakage resistance.

Applications

There are innumerable applications of labyrinth seals in the field. They are used to seal rolling-element bearings, machine spindles, and other applications where some leakage can be tolerated. Since the development of the gas turbine engine, the labyrinth seal has been perhaps the most common seal applied to sealing both primary and secondary airflow.[20] Its combined pressure–speed–life limits have for many years exceeded those of its rubbing-contact seal competitors. Labyrinth seals are also used extensively in cryogenic rocket turbopump applications.

Seals and Rotordynamic Stability

Although the primary function of a seal is to control leakage, a secondary but equally important purpose is not to negatively affect rotordynamic stability, especially in high-speed

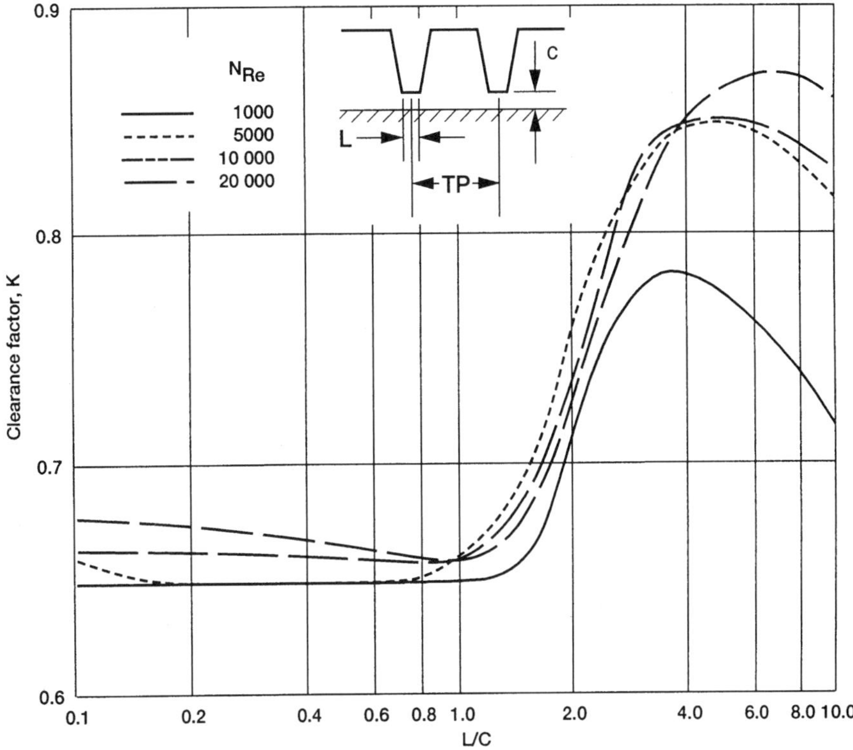

Figure 20 Clearance factor (K) versus ratio of tooth width (L) to tooth clearance (c). [For various Reynolds numbers (N_{Re}).] (From Ref. 44.)

turbomachinery. When a clearance gap such as in either annular or labyrinth seals changes with time, lateral forces occur that act out of phase with case distortion. Depending on changes in gap in the flow direction, excitation or damping of the lateral forces occurs. If the lateral forces become too large, they can contribute to shaft instability problems. These problems and several solutions, including swirl brakes, are further discussed in Hendricks,[46] Bently,[47] Alford,[48–50] and Benckert and Wachter,[51] among others.

Computer Analysis Tools: Labyrinth Seals

The computer code KTK calculates the leakage and pressure distribution through a labyrinth seal based on a detailed knife-to-knife (KTK) analysis. This code was developed by Allison Gas Turbines for the Air Force[52] and is also documented in Shapiro et al.[40] Rhode and Nail[53] present recent work in the application of a Reynolds-averaged computer code to generic labyrinth seals operating in the compressible region Mach number ≥ 0.3.

3.6 Honeycomb Seals

Honeycomb seals are used extensively in mating contact with labyrinth knife edges machined onto the rotor in applications where there are significant shaft movements. After brazing the honeycomb material to the case, the inner diameter is machined to seal tolerance requirements. Properly designed honeycomb seals, in extensive tests performed by Stocker et al.[54]

under a NASA contract, showed dramatic leakage reductions under select gap and honeycomb cell size combinations.

For applications where low leakage is paramount, designers will specify a small radial clearance between the labyrinth teeth and abradable surface (honeycomb or sprayed abradable). Designers will take advantage of normal centrifugal growth of the rotor to reduce this clearance to line-to-line and often to a wear-in condition, making an effective labyrinth seal. A "green" slow-speed-ramp wear-in cycle is recommended.

Materials. Honeycomb elements are often fabricated of Hastelloy X,[55] a nickel-base alloy. Honeycomb seals provide for low-energy rubs when transient conditions cause the labyrinth knife edges to wear into the surface (low-energy rubs minimize potentially damaging shaft vibrations). In very high surface speed applications and where temperatures are high the labyrinth teeth are "tipped" with a hard abrasive coating, increasing cutting effectiveness and reducing the thermal stresses in the labyrinth teeth during rubs.

Honeycomb Annular Seals. Honeycomb seals are also being considered now as annular seals to greatly improve damping over either smooth surfaces or labyrinth seals. Childs et al.[56] showed that honeycombs properly applied in annular seals control leakage, have good stiffness, and exhibit damping characteristics six times those of labyrinth seals alone.

3.7 Brush Seals

As described by Ferguson,[42] the brush seal is the first simple, practical alternative to the finned labyrinth seal that offers extensive performance improvements. Benefits of brush seals over labyrinth seals include the following:

- Reduced leakage compared to labyrinth seals. If properly applied, leakage reductions upward of 50% are possible.
- Flexible brush seal accommodates shaft excursions due to stop/start operations and other transient conditions. Labyrinth seals often incur permanent clearance increases under such conditions, degrading seal and machine performance.
- Requires significantly less axial space than labyrinth seal.
- More stable leakage characteristics over long operating periods.

Brush seals have matured significantly over the past 20 years. Typical operating conditions of state-of-the-art brush seals include the following[57]:

Differential pressure	Up to 300 psid per stage; higher pressures (e.g., 1000 psid) possible with multiple stages
Surface speed	Up to 1200 ft/s
Operating temperature	Up to 1200°F
Diameter range	Up to 120 in.

Basic brush seal construction is quite simple, as shown in cross section in Fig. 21. A dense pack of fine-diameter wire bristles is sandwiched and welded between a backing ring (downstream side) and a sideplate (upstream side). The wire bristles protrude radially inward and are machined to form a brush bore fit around a mating rotor, with a slight interference. Brush seal interferences and preload must be properly selected to prevent potentially catastrophic overheating of the rotor and excessive rotor thermal growths. The weld on the seal outer diameter is machined to form a close-tolerance outer diameter sealing surface that is fitted into a suitable seal housing.

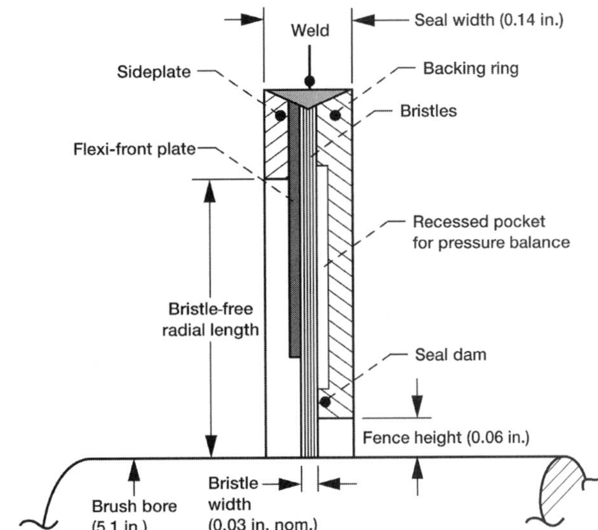

Figure 21 Brush seal cross section with typical dimensions. (From Ref. 20.)

To accommodate anticipated radial shaft movements, the bristles must bend. To allow the bristles to bend without buckling, the wires are oriented at an angle (typically 45°–55°) to a radial line through the shaft. The bristles point in the direction of rotation. The angled construction also greatly facilitates seal installation, considering the slight inner diameter interference with the rotors. The backing ring provides structural support to the otherwise flexible bristles and assists the seal in limiting leakage. To minimize brush seal hysteresis caused by brush bristle binding on the back plate, new features have been added to the backing ring. These include reliefs of various forms. An example design is shown in Fig. 21 and includes the recessed pocket and seal dam. The recessed pocket assists with pressure balancing of the seal and the relatively small contact area at the seal dam minimizes friction, allowing the bristles to follow the speed-dependent shaft growths. Bristle free radial length and packing pattern are selected to accommodate anticipated shaft radial movements while operating within the wire's elastic range at temperature. A number of brush seal manufacturers[57,58] include some form of flow deflector (e.g., see flexi-front plate in Fig. 21) on the high-pressure side of the wire bristles. This element aids in mitigating the radial pressure closing loads (e.g., sometimes known as "pressure closing") caused by air forces urging the bristles against the shaft. This element can also aid in reducing installation damage, bristle flutter in highly turbulent flow fields, and foreign-object damage. The backing ring clearance is sized slightly larger than anticipated rotor radial excursions and relative thermal and mechanical growth to ensure that the rotor never contacts the ring, causing rotor and casing damage. An abradable rub surface added to the backing ring has been proposed to mitigate this problem by allowing tighter backing-plate clearances.

Brush Seal Design Considerations

To properly design and specify brush seals for an application, many design factors must be considered and traded off. A comprehensive brush seal design algorithm was proposed by Holle and Krishan.[59] An iterative process must be followed to satisfy seal basic geometry, stress, thermal (especially during transient rub conditions), leakage, and life constraints to arrive at an acceptable design. Table 5 illustrates many of the characteristics that must be

Table 5 Brush Seal Characteristics Evaluated during Design for Successful Application

Pressure capability	Seal upstream protection
Frequency	Seal high- and low-cycle fatigue (HCF, LCF) analysis
Seal leakage	Seal oxidation
Seal stiffness	Seal creep
Seal blow-down (e.g., pressure-closing effect)	Seal wear
Bristle tip forces and pressure-stiffening effect	Solid-particle erosion
Seal heat generation	Reverse rotation
Bristle tip temperature	Seal life/long-term considerations
Rotor dynamics	Performance predictions
Rotor thermal stability	Oil sealing
Secondary flow and cavity flow (including swirl flow)	Shaft considerations (e.g., coating)

Source: From Ref. 60.

considered and understood for a successful brush seal design.[60] Design criteria are required for each of the different potential failure modes, including stress, fatigue life, creep life, wear life, and oxidation life, among others.

Implementation Issues. Improper design and implementation of brush seals can result in premature seal failures. The following is a partial list of potential pitfalls to be mindful of in specifying brush seals[60]:

- *Excessive Interference.* Specifying too tight of a fit at assembly can lead to excessive frictional heat and wear of bristles. In the worst case, seals can become thermally unstable: Frictional heating can cause the rotor to grow radially into the seal, thereby increasing frictional heating and leading to additional rotor growth. If left unchecked, the rotor can grow into the backing plate, leading to seal and possibly equipment failure. Designers need to consider the relative seal-to-shaft closure across the entire operating speed/temperature range. In large ground-based turbine designs, brush seals are often assembled with a clearance to preclude excessive interference and heating during thermal and speed transients.
- *Excessive Operating Speed.* Brush seals have been run successfully to 1200 fps.[57,58] In some limited applications they have run faster than this. However, excessive surface speeds combined with high unit pressures can lead to excessive heat generation and premature failure.
- *Inadequate Understanding of Flow Fields.* The design of upstream cavities is key to reducing brush seal flutter, especially in high-swirl-flow fields.[60] Poor designs lead to bristle aerodynamic instability and high-cycle fatigue. Furthermore, brush seals can restrict leakage so much better than labyrinth seals that inadequate flow may be supplied to expensive downstream components (e.g., turbine vanes, blades, buckets, and/or wheel cavities), resulting in life-limiting conditions for those components. A clear understanding of the flow fields is essential for successful implementation.
- *Improper Brush Pack Design.* Designers should consult with brush seal manufacturers to aid in specifying the brush parameters. Manufacturers can aid in selecting the correct wire diameter, brush pack width, bristle free height, and fence height for the speed, pressure, and transient conditions anticipated. Improper brush pack width, for instance, can result in excessive bending of the bristles under the backing ring, leading to excessive bristle wear.

Brush Pack Considerations. Depending on required sealing pressure differentials and life, wire bristle diameters are chosen in the range of 0.0028–0.006 in.[61] Better load and wear properties are found with larger bristle diameters. Bristle pack widths also vary depending on application: The higher the pressure differential, the greater the pack width. Higher pressure applications require bristle packs with higher axial stiffness to prevent the bristles from blowing under the backing ring. Dinc et al[60] and his group have developed brush seals that have operated at air pressures up to 400 psid in a single stage. Brush seals have been made in very large diameters. Large brush seals, especially for ground power applications, are often made segmented to allow easy assembly and disassembly, especially on machines where the shaft stays in place during refurbishment.

Other Considerations. If not properly considered, brush seals can exhibit three other phenomena deserving some discussion: *seal hysteresis, bristle stiffening,* and *pressure closing.* As described in Short et al.[61] and Basu et al.,[62] after the rotor moves into the bristle pack (due to radial excursions or thermal growths), the displaced bristles do not immediately recover against the frictional forces between them and the backing ring. As a result, a significant leakage increase (more than double) was observed[62] following rotor movement. This leakage hysteresis exists until after the pressure load is removed (e.g., after the engine is shut down). Furthermore, if the bristle pack is not properly designed, the seal can exhibit a considerable stiffening effect with application of pressure. This phenomenon results from interbristle friction loads, making it more difficult for the brush bristles to flex during shaft excursions. Air leaking through the seal also exerts a radially inward force on the bristles, resulting in what has been termed pressure closing or bristle "blow-down." This extra contact load, especially on the upstream side of the brush, affects the life of the seal (upstream bristles are worn in either a scalloped or coned configuration) and higher interface contact pressure. Because of these and other considerations, designers should consult with brush seal manufacturers[57,58] for application assistance.

Multiple brush seals are generally used where large pressure drops must be accommodated. The primary reason for using multiple seals is not to improve sealing but to reduce pressure-induced distortions in the brush pack, namely axial brush distortions under the backing ring, that cause wear. Researchers have noticed greater wear on the downstream brush if the flow jet coming from the upstream brush is not deflected away from the downstream brush–rotor contact.

Leakage Performance Comparisons

Ferguson[42] compared brush seal leakage with that of traditional five-finned labyrinth seals of various configurations. The results of this study (Fig. 22) indicate that the flow of a *new* brush seal is only 4% that of a vertical finned seal with a 0.03-in. radial gap and one-fifth that of an inclined-fin labyrinth seal with a step up and a 0.01-in. gap.

Addy et al.[63] showed similar large reductions in leakage testing a 5.1-in. bore seal across a wide temperature and speed range. Table 6 compares air leakage between a new brush seal and similarly sized labyrinth seals.

Effects of Speed. Proctor and Delgado studied the effects of speed (up to 1200 ft/s), temperature (up to 1200°F), and pressure (up to 75 psid) on brush seal and finger seal leakage and power loss.[64] They determined that leakage generally decreased with increasing speed. It is believed that leakage decreases with speed since the rotor diameter increases, causing both a decrease in the effective seal clearance and an increase in contact stresses.

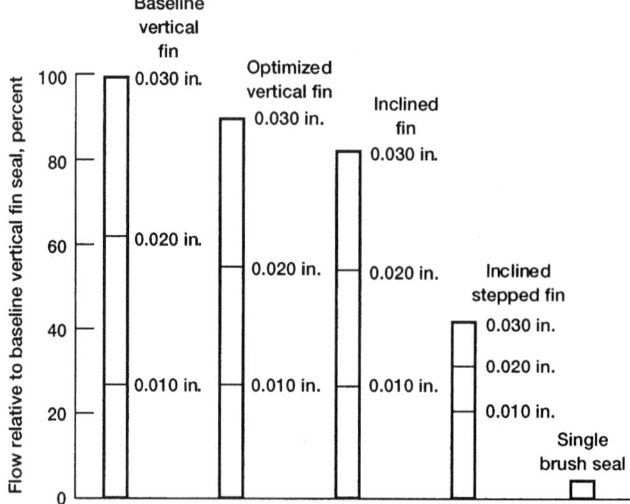

Figure 22 Sealing performance of new brush seals relative to baseline five-finned labyrinth seals of various radial gaps at pressure ratio of 2. (From Ref. 42.)

Table 6 Comparison of New Labyrinth Seal (Smooth and Honeycomb Lands) versus New Brush Seal Leakage Rates for Comparable Conditions

Seal	Rotor Diameter (in.)	Seal Clearance (in.)	Brush Seal Interference (in.)	Pressure Ratio, P_i/P_o	ϕ-Flow Parameter, $\left(\frac{lb_m \cdot R^{1/2}}{lb_f \cdot s}\right)$	Mass Flow, (lb_m/s)
4-Tooth labyrinth vs. smooth land[a]	6.0	0.010	—	3.0	0.36	0.141
4-Tooth labyrinth vs. honeycomb land 0.062-in. cell size[a]	6.0	0.010	—	3.0	0.35	0.137
Brush seal[b]	5.1	—	0.004	3.0	0.0053	0.0099

Labyrinth seal: 4-tooth labyrinth; 0.11 in. pitch; 0.11 in. knife height.
Brush seal: 0.028 in. brush width; 0.0028-in.-diameter bristles; 0.06 in. fence clearance.
Static, 0 rpm.

Flow parameter, $\phi = \dfrac{m\sqrt{T_i}}{P_i A}$

[a] Ref. 54.
[b] Ref. 63.

Aircraft Turbine Engine Performance. Mahler and Boyes[65] have made leakage comparisons of new and aircraft engine-tested brush seals. They concluded that performance did not deteriorate significantly for periods approaching one engine overhaul cycle (3000 h). Of the three brush seals examined, the "worst-case" brush seal's leakage rates doubled compared to a new brush seal. Even so, brush seal leakage was still less than half the leakage of the labyrinth seal.

Cryogenic Brush Seals. The long life and low leakage of brush seals make them candidates for use in rocket engine turbopumps. Brush seals 2 in. in diameter with nominal 0.005–in. radial interference were tested in liquid nitrogen (LN2) at shaft speeds up to 35,000 and 65,000 rpm, respectively, and at pressure drops up to 175 psid per brush.[66] A labyrinth seal was also tested in liquid nitrogen to provide a baseline. The LN2 leakage rate of a single brush seal with an initial radial shaft interference of 0.005 in. measured one-half to one-third the leakage rate of a 12-tooth labyrinth seal with a radial clearance of 0.005 in.

Brush seals are not a solution for all seal problems. However, when they are applied within design limits, brush seal leakage will be lower than that of competing labyrinth seals and remain closer to design goals even after transient rub conditions.

Brush Seal Flow Modeling

Brush seal flow modeling is complicated by several factors unique to porous structures, in that the leakage depends on the seal porosity, which depends on the pressure drop across the seal. Flow through the brush travels perpendicular to the brush pack through the annulus formed by the backing ring inner diameter and the shaft diameter, radially inward at successive layers within the brush, and between the bristle tips and the shaft.

A flow model proposed by Holle et al.[67] uses a single parameter, effective brush thickness, to correlate the flows through the seal. Variation in seal porosity with pressure difference is accounted for by normalizing the varying brush thicknesses by a minimum or ideal brush thickness. Maximum seal flow rates are computed by using an iterative procedure that has converged when the difference in successive iterations for the flow rate is less than a preset tolerance.

Flow models proposed by Hendricks et al.[68,69] are based on a bulk average flow through the porous media. These models account for brush porosity, bristle loading and deformation, brush geometry parameters, and multiple flow paths. Flow through a brush configuration is simulated by using an electrical analog that has driving potential (pressure drops), current (mass flow), and resistance (flow losses, friction, and momentum) as the key variables. All of the above models require some empirical data to establish the correlating constants. Once these are established the models predict seal flow reasonably well.

A number of researchers (for example see Refs. 70–73) have applied numerical techniques to model brush seal flows and bristle pressure loadings. Though these models are more complex, they permit a more detailed investigation of the subtleties of flow and stresses within the brush pack.

Brush Seal Materials

Brush wire bristles range in diameter from 0.0028 in. for low pressures to 0.006 in. for high pressures. The most commonly used material for brush seals is the cobalt-base alloy Haynes 25. Brush seals are generally run against a smooth, hard-face coating to minimize shaft wear and minimize chances of wear-induced cracks from affecting the structural integrity of the rotor. The usual coatings selected are ceramic, including chromium carbide and aluminum oxide. Selecting the correct mating wire and shaft surface finish for a given application can reduce friction heating and extend seal life through reduced oxidation and wear. For extreme operating temperatures to above 1300°F, Derby[74] has shown low wear and friction for the nickel-based superalloy Haynes 214 (heat treated for high strength) running against a solid-film lubricated hard-face coating Triboglide. Fellenstein et al.[75,76] investigated a number of bristle/rotor coating material pairs corroborating the benefits of Haynes 25 wires run against chrome carbide but observed Haynes 214 bristle flairing when run against chrome carbide and zirconia coatings.

Nonmetallic Bristles. High-speed turbine designers have long wondered if brush seals could replace labyrinth seals in bearing sump locations. Brush seals would mitigate traditional labyrinth seal clearance opening and corresponding increased leakage. Issues slowing early application of brush seals in these locations included coking (carburization of oil particles at excessively high temperatures), metal particle damage of precision rolling-element bearings, and potential for fires. GE-Global research has found success in applying aramid bristles for certain bearing sump locations.[77,78] Advantages of the aramid bristles include stable properties up to 300°F (150°C) operating temperatures, negligible amount of shrinkage and moisture absorption, lower wear than Haynes 25 up to 300°F, lower leakage (due to smaller 12-μm diameters), and resistance to coking.[77] Based on laboratory demonstration, the aramid fiber seals were installed in a GE 7EA frame (#1) inlet bearing sealing location. Preliminary field data showed that the nonmetallic brush seal maintained a higher pressure difference between the air and bearing drain cavities and enhanced the effectiveness of the sealing system, allowing less oil particles to migrate out of the bearing.

Aircraft Turbine Engines

Applications and Benefits. Brush seals are seeing extensive service in both commercial and military turbine engines. Lower leakage brush seals permit better management of cavity flows and significant reductions in specific fuel consumption when compared to competing labyrinth seals. Allison Engines has implemented brush seals in engines for the Saab 2000, Cessna Citation-X, and V-22 Osprey. GE has implemented a number of brush seals in the balance piston region of the GE90 engine for the Boeing 777 aircraft. PW has entered revenue service with brush seals in three locations[65] on the PW4168 for Airbus aircraft and on the PW4084 for the Boeing 777.

Ground-Based Turbine Engines. Brush seals are being retrofitted into ground-based turbines both individually and combined with labyrinth seals to greatly improve turbine power output and heat rate (see Refs. 60 and 79–84). Dinc et al.[60] report that incorporating brush seals in a GE Frame 7EA turbine in the high-pressure packing location increased output by 1.0% and decreased heat rate by 0.5%. Using brush seals in the interstage location resulted in similar improvements. Brush seals have proven effective for service lives of up to 40,000 hr![60]

3.8 Ongoing Developments

Long life and durability under very high temperature (≥1300°F) conditions are hurdles to overcome to meet goals of advanced turbine engines under development for next-generation commercial subsonic, supersonic, and military fighter engine requirements. The tribology phenomena are complex and installation specific. In order to extend engine life and bring down maintenance costs, research and development are continuing in this area. To extend brush seal lives at high temperature, Addy et al.,[63] Hendricks et al.,[85] and Howe[86] have investigated approaches to replace metallic bristles with ceramic fibers. Ceramic fibers offer the potential for operating above 815°C (1500°F) and for reducing bristle wear rates and increasing seal lives while maintaining good flow resistance. Though early results indicate rotor coating wear, ceramic brush leakage rates were less than half those of labyrinth seal (0.007-in. clearance) and bristle wear was low.[63]

Designers continue to pursue seal designs that address the wear of brush seals. A sample of some of the seal designs being pursued includes the following. Justak has developed a hybrid floating brush seal that combines hydrodynamic seal shoes with a secondary brush seal.[87,88] Gail and Klemens[89] have patented a seal that combines a brush with a slide ring to improve sealing effect and reduce wear. Proctor and Steinetz[90] and Braun et al.[91] are developing an innovative noncontacting finger seal. In this seal, the brush bristles are replaced with precision-machined upstream/downstream finger laminates. Shaft movement is accommodated by bending of the finger elements. Noncontact operation is afforded by hydrodynamic or hydrostatic lift pads located on the downstream laminate. These pads cause the seal to lift during shaft transients. Grondahl has patented a pressure-actuated leaf seal that is designed to overcome wear during transient startup/shutdown conditions.[92]

REFERENCES

1. I. E. Etsion and B. M. Steinetz "Seals," in *Mechanical Design Handbook,* H. A. Rothbart (ed.), McGraw-Hill, New York, 1996, Section 17.
2. R. V. Brink, D. E. Czernik, and L. A. Horve, *Handbook of Fluid Sealing,* McGraw-Hill, New York, 1993.
3. A. Bazergui and L. Marchand, "Development of Tightness Test Procedures for Gaskets in Elevated Temperature Service," *Welding Res. Council Bull.* **339,** Dec. 1988.
4. M. Derenne, L. Marchand, J. R. Payne, and A. Bazergui, "Elevated Temperature Testing of Gaskets for Bolted Flanged Connections," *Welding Res. Council Bull.,* **391,** May, 1994.
5. *Parker O-ring Handbook,* Cleveland, OH, 2001.
6. L. J. Martini, *Practical Seal Design,* Marcel Dekker, 1984.
7. A. Mathews and G. R. McKillop, "Compression Packings," in *Machine Design Seals Reference Issue,* Penton, Mar. 1967, Chap. 8.
8. R. A. Howard, P. S. Petrunich, and K. C. Schmidt, *Grafoil Engine Design Manual,* Vol. 1, Union Carbide Corp., 1987.
9. B. M. Steinetz and M. L. Adams, "Effects of Compression, Staging and Braid Angle on Braided Rope Seal Performance," *J. Propulsion and Power,* **14**(6), 934–940, (Dec. 1998). See also 33rd AIAA/ASME/SAE/ASEE Joint Propulsion Conference & Exhibit, AIAA-97-2872, Seattle, WA, July 1997, NASA TM, Vol. 107504, July 1997.
10. B. M. Steinetz et al., *High Temperature Braided Rope Seals for Static Sealing Applications,* NASA TM-107233; also *AIAA J. Propulsion and Power,* **13**(5), 1997.
11. E. J. Opila, J. A. Lorincz, and J. J. DeMange, "Oxidation of High-Temperature Alloy Wires in Dry Oxygen and Water Vapor," in *High Temperature Corrosion and Materials Chemistry,* Vol. 5, E. Opila, J. Fergus, T. Maruyama, J. Mizusaki, T. Narita, D. Shifler, and E. Wuchina (eds.), Electrochemical Society, Pennington, NJ, 2005.
12. P. Bauer, *Development of an Enhanced Thermal Barrier for RSRM Nozzle Joints* AIAA-2000-3566, July, 2000.
13. P. Bauer, P., *MNASA as a Test Bed for Carbon Fiber Thermal Barrier Development,* AIAA-2001-3454, July, 2001.
14. P. Totman, A. Prince, D. Frost, and P. Himebaugh, *Alternatives to Silicon Rubber Thermal Barrier in RSRM Nozzle Joints,* AIAA-99-2796, July, 1999.
15. M. Ewing, J. R. McGuire, B. B. McWhorter, and D. L. Frost, *Performance Enhancement of the Space Shuttle RSRM Nozzle-to-Case Joint Using a Carbon Rope Barrier,* AIAA-99-2899, July, 1999.

16. B. M. Steinetz, and P. H. Dunlap, "Development of Thermal Barriers for Solid Rocket Motor Nozzle Joints," *J. Propulsion and Power,* **17**(5), 1023–1034 (Sept./Oct., 2001); also NASA TM-209278, June 1999.

17. B. M. Steinetz and P. H. Dunlap, "Feasibility Assessment of Thermal Barrier Seals for Extreme Transient Temperatures," *J. Propulsion and Power,* **16**(2), 347–356 (Mar./Apr. 2000); also NASA TM-208484, July 1998.

18. B. M. Steinetz and P. H. Dunlap, "Rocket Motor Joint Construction Including Thermal Barrier," U.S. Patent No. 6,446,979 B1, Sept. 10, 2002.

19. A. G. Fern and B. S. Nau, *Seals,* Engineering Design Guide 15, published for Design Council, British Standards Institution and Council of Engineering Institutions, Oxford University Press, 1976.

20. B. M. Steinetz and R. C. Hendricks, "Aircraft Engine Seals," Chapter 9 of *Tribology for Aerospace Applications,* STLE Special Publication SP-37, 1997.

20a. J. Crane, "Dry Running Noncontacting Gas Seal," Bulletin No. S-3030, 1993.

21. H. Buchter, *Industrial Sealing Technology,* Wiley, New York, 1979.

22. A. O. Lebeck, *Principles and Design of Mechanical Face Seals,* Wiley, New York, 1991.

23. J. Zuk and P. J. Smith, *Quasi-One Dimensional Compressible Flow across Face Seals and Narrow Slots—II. Computer Program,* NASA TN D-6787, 1972.

24. W. F. Hughes et al., *Dynamics of Face and Annular Seals with Two-Phase Flow,* NASA CR-4256, 1989.

25. P. F. Brown, "Status of Understanding for Seal Materials," *Tribology in the 80's,* NASA CP-23000, Vol. 2, 1984, pp. 811–829.

26. J. C. Dahlheimer, *Mechanical Face Seal Handbook,* Chilton Book Co., Philadelphia, 1972.

27. *Mechanical Seal Handbook,* Fluid Sealing Association Wayne, PA, 2000.

28. *Pumps and Systems Handbook,* Cahaba Media Group, Tuscaloosa, AL, 2003.

29. *Guidelines for Meeting Emission Regulations for Rotating Machinery with Mechanical Seals,* Special Publication SP-30, Society of Tribologists and Lubrication Engineers, Park Ridge, IL, revised 1994.

30. R. C. Waterbury, "Zero-Leak Seals Cut Emissions," *Pumps and Systems Magazine,* AES Marketing, Fort Collins, CO, July 1996.

31. P. E. Bowden, "Design and Selection of Mechanical Seals to Minimize Emissions," *Pro. Inst. Mech. Eng.,* **213**(Pt J) (1999).

32. H. Hwang, T. Tseng, B. Shucktis, and B. Steinetz, *Advanced Seals for Engine Secondary Flowpath,* AIAA-95-2618, presented at the 1995 AIAA/ASME/SAE/ASEE Joint Propulsion Conference, San Diego, CA, 1995.

33. C. E. Wolfe et al., *Full Scale Testing and Analytical Validation of an Aspirating Face Seal,* AIAA Paper 96-2802, 1996.

34. B. Bagepalli et al., *Dynamic Analysis of an Aspirating Face Seal for Aircraft-Engine Applications,* AIAA Paper 96-2803, 1996.

35. J. Munson, *Testing of a High Performance Compressor Discharge Seal,* AIAA Paper 93-1997, 1993.

36. R. C. Hendricks, *Seals Code Development*—'95, NASA CP-10181, 1995.

37. Open Channel Software, Chicago, IL, phone: (773) 334-8177.

38. W. Shapiro, *Numerical, Analytical, Experimental Study of Fluid Dynamic Forces in Seals,* Volume *2–Description of Gas Seal Codes GCYLT and GFACE,* NASA Contract Report for Contract NAS3-25644, Sept. 1995.

39. J. Walowit and W. Shapiro, *Numerical, Analytical, Experimental Study of Fluid Dynamic Forces in Seals,* Volume 3—*Description of Spiral-Groove Codes SPIRALG and SPIRALI,* NASA Contract Report for Contract NAS3-25644, Sept. 1995.

40. W. Shapiro et al., *Numerical, Analytical, Experimental Study of Fluid Dynamic Forces in Seals,* Volume 5—*Description of Seal Dynamics Code DYSEAL and Labyrinth Seals Code KTK,* NASA Contract Report for Contract NAS3-25644, Sept. 1995.

41. R. E. Burcham and R. B. Keller, Jr., *Liquid Rocket Engine Turbopump Rotating-Shaft Seals,* NASA SP-8121, 1979.

42. J. G. Ferguson, *Brushes as High Performance Gas Turbine Seals,* ASME Paper 88-GT-182, 1988.

43. A. Egli, "The Leakage of Steam through Labyrinth Seals," *ASME Trans.* **57**(3), 115–122 (1935).

44. G. Vermes, "A Fluid Mechanics Approach to the Labyrinth Seal Leakage Problem," *J. Eng. Power,* **83**(2), 161–169, 1961.

45. F. H. Mahler, *Advanced Seal Technology,* Report PWA-4372, Contract F33615-71-C-1534, Pratt and Whitney Aircraft Co., East Hartford, CT, 1972.

46. R. C. Hendricks, L. T. Tam, and A. Muszynska, *Turbomachine Sealing and Secondary Flows, Part 2–Review of Rotordynamic Issues in Inherently Unsteady Flow Systems with Small Clearances,* NASA TM-2004-211,991, July 2004.

47. D. E. Bently, C. T. Hatch, and B. Grissom (eds.), *Fundamentals of Rotating Machinery Diagnostics,* Bently Pressurized Bearings, Minden, NV, 2002.

48. J. S. Alford, *Protection of Labyrinth Seals from Flexural Vibration,* ASME Paper 63-AHGT-9, 1963; also *J. Eng. Power,* Apr. 1964, pp. 141–148.

49. J. S. Alford, "Protecting Turbomachinery from Self-Excited Rotor Whirl," *J. Eng. for Power,* Series A, **87,** 333–344 (Oct. 1965).

50. J. S. Alford, "Protecting Turbomachinery from Unstable and Oscillatory Flows," *J. Eng. for Power,* Series A, **89,** 513–528 (Oct. 1967).

51 H. Benckert and J. Wachter, "Studies on Vibrations Stimulated by Lateral Forces in Sealing Gaps," paper presented at the AGARD Power, Energetics, and Propulsion Meeting on Seal Technology in Gas Turbine Engines, AGARD CP-237 (AGARD AR-123), Paper 9, 1978.

52. D. L. Tipton, T. E. Scott, and R. E. Vogel, *Labyrinth Seal Analysis:* Volume III—*Analytical and Experimental Development of a Design Model for Labyrinth Seals,* AFWAL TR-85-2103, Allison Gas Turbine Division, General Motors, Indianapolis, IN, 1986.

53. D. L. Rhode and G. H. Nail, "Computation of Cavity-by-Cavity Flow Development in Generic Labyrinth Seals," *J. Tribol.* **14,** 47–51, 1992.

54. H. L. Stocker, D. M. Cox, and G. F. Holle, *Aerodynamic Performance of Conventional and Advanced Design Labyrinth Seals with Solid Smooth, Abradable, and Honeycomb Lands—Gas Turbine Engines,* NASA CR-135307, 1977.

55. Z. Galel, F. Brindisi, and D. Norstrom, *Chemical Stripping of Honeycomb Airseals, Overview and Update,* ASME Paper 90-GT-318, 1990.

56. D. W. Childs, D. Elrod, and K. Hale, "Annular Honeycomb Seals: Test Results for Leakage and Rotordynamic Coefficients—Comparison to Labyrinth and Smooth Configurations," in *Rotordynamic Instability Problems in High-Performance Turbomachinery,* NASA CP-3026, 1989, pp. 143–159.

57. Perkin Elmer Fluid Sciences product literature. http://fluidsciences.perkinelemer.com/turbomachinery.

58. Cross Manufacturing product literature, www.crossmanufacturing.com.

59. G. F. Holle and M. R. Krishnan *Gas Turbine Engine Brush Seal Applications,* AIAA Paper 90-2142, 1990.

60. S. Dinc, M. Demiroglu, N. Turnquist, G. Toetze, J. Maupin, J. Hopkins, C. Wolfe, and M. Florin, "Fundamental Design Issues of Brush Seals for Industrial Applications," *J. Turbomachinery,* **124** (Apr. 2002).

61. J. F. Short et al., *Advanced Brush Seal Development,* AIAA Paper 96-2907, 1996.

62. P. Basu et al., *Hysteresis and Bristle Stiffening Effects of Conventional Brush Seals,* AIAA Paper 93-1996, 1993.

63. H. E. Addy et al., *Preliminary Results of Silicon Carbide Brush Seal Testing at NASA Lewis Research Center,* AIAA Paper 95-2763, 1995.

64. M. P. Proctor and I. R. Delgado, "Leakage and Power Loss Test Results for Competing Turbine Engine Seals," GT2004-53935, in *Proc. of ASME Turbo Expo, Power for Land, Sea, and Air,* Vienna, Austria, June 2004.

65. F. Mahler and E. Boyes, *The Application of Brush Seals in Large Commercial Jet Engines,* AIAA Paper 95-2617, 1995.

66. M. P. Proctor, J. F. Walker, H. D. Perkins, J. F. Hoopes, and G. S. Williamson, *Brush Seals for Cryogenic Applications: Performance, Stage Effects, and Preliminary Wear Results in LN2 and LH2,* NASA Technical Paper 3536, Oct. 1996.

67. G. F. Holle, R. E. Chupp, and C. A. Dowler, "Brush Seal Leakage Correlations Based on Effective Thickness," paper presented at the Fourth International Symposium on Transport Phenomena and Dynamics of Rotating Machinery, preprint Vol. A., 1992, pp. 296–304.

68. R. C. Hendricks et al., *A Bulk Flow Model of a Brush Seal System,* ASME Paper 91-GT-325, 1991.

69. R. C. Hendricks et al., "Investigation of Flows in Bristle and Fiberglass Brush Seal Configurations," paper presented at the Fourth International Symposium on Transport Phenomena and Dynamics of Rotating Machinery, preprint Vol. A, 1992, pp. 315–325.

70. M. J. Braun and V. V. Kudriavtsev, "A Numerical Simulation of Brush Seal Section and Some Experimental Results." *J. Turbomachinery* (1995).

71. M. T. Turner, J. W. Chew, and C. A. Long, "Experimental Investigation and Mathematical Modeling of Clearance Brush Seals," ASME 97-GT-282, presented at the International Gas Turbine and Aeroengine Congress and Exhibition, Orlando, FL, 1997.

72. L. H. Chen, P. E. Wood, T. V. Jones, and J. W. Chew, "An Iterative CFD and Mechanical Brush Seal Model and Comparisons with Experimental Results," ASME 98-GT-372, presented at the International Gas Turbine and Aeroengine Congress and Exhibition, Stockholm, Sweden, 1998.

73. M. F. Aksit, "Analysis of Brush Seal Bristle Stresses with Pressure Friction Coupling," ASME GT2003-38718, presented at the ASME/IGTI Turbo Expo, Atlanta, GA, June, 2003.

74. J. Derby and R. England, *Tribopair Evaluation of Brush Seal Applications,* AIAA Paper 92-3715, 1992.

75. J. Fellenstein, C. Della Corte, K. D. Moore, and E. Boyes, *High Temperature Brush Seal Tuft Testing of Metallic Bristles vs Chrome Carbide,* NASA TM-107238, AIAA-96-2908, 1996.

76. J. A. Fellenstein, C. Della Corte, K. A. Moore, and E. Boyes, *High Temperature Brush Seal Tuft Testing of Selected Nickel–Chrome and Cobalt–Chrome Superalloys,* NASA TM-107497, AIAA-97-2634, 1997.

77. N. Bhate, A. C. Thermos, M. F. Aksit, M. Demiroglu, and H. Kizil, "Non-Metallic Brush Seals for Gas Turbine Bearings," GT2004-54296, in *Proc. of ASME Turbo Expo, Power for Land, Sea, and Air,* Vienna, Austria, June 2004.

78. M. F. Aksit, Y. Dogu, and M. Gursoy, "Hydrodynamic Lift of Brush Seals in Oil Sealing Applications," AIAA-2004-3721, presented at the 40th AIAA/ASME/SAE/ASEE Joint Propulsion Conference and Exhibit, Ft. Lauderdale, FL, July 2004.

79. R. E. Chupp, R. P. Johnson, and R. G. Loewenthal, *Brush Seal Development for Large Industrial Gas Turbines,* AIAA Paper 95-3146, 1995.

80. R. E. Chupp, R. J. Prior, and R. G. Loewenthal, *Update on Brush Seal Development for Large Industrial Gas Turbines,* AIAA Paper 96-3306, 1996.

81. R. E. Chupp, M. F. Aksit, F. Ghasripoor, N. A. Turnquist, and M. Demiroglu, *Advanced Seals for Industrial Turbine Applications,* AIAA Paper 2001-3626, 2001.

82. E. Bancalari, I. S. Diakunchak, and G. McQuiggan, "A Review of W501G Engine Design, Development and Field Operating Experience," GT2003-38843, in *Proc. of ASME Turbo Expo, Power for Land, Sea, and Air,* Atlanta, GA, June 2003.

83. I. S. Diakunchak, G. R. Gaul, G. McQuiggan, and L. R. Southall, "Siemens Westinghouse Advanced Turbine Systems Program Final Summary," GT2002-30654, in *Proc. of ASME Turbo Expo, Power for Land, Sea, and Air,* Amsterdam, The Netherlands, June 2002.

84. S. Ingistov, "Compressor Discharge Brush Seal for Gas Turbine Model 7EA," *ASME J. of Turbomachinery,* **124**(Apr. 2002).

85. R. C. Hendricks, R. Flower, and H. Howe, "Development of a Brush Seals Program Leading to Ceramic Brush Seals," in *Seals Flow Code Development—'93,* NASA CP-10136, 1994, pp. 99–117.

86. H. Howe, "Ceramic Brush Seals Development," in *Seals Flow Code Development—'93,* NASA CP-10136, pp. 133–150, 1994.

87. J. Justak, "Robust Hydrodynamic Brush Seal," U.S. Patent No. 6,428,009, 2002.

88. A. Delgado, L. S. Andres, and J. Justak, "Analysis of Performance and Rotordynamic Force Coefficients of Brush Seals with Reverse Rotation Ability," ASME GT 2004-53614, presented at the ASME Turbo Expo 2004 Power for Land, Seal and Air, Vienna, Austria, June 2004.

89. A. Gail and W. Klemens, "Brush Seal," U.S. Patent No. 6,695,314, 2004.

90. M. P. Proctor and B. M. Steinetz, "Non-Contacting Finger Seal," U.S. Patent No. 6,811,154, 2004.

91. M. J. Braun, H. Pierson, D. Deng, and F. Choi, "Non-Contacting Finger Seal Investigations," in *Conference Proceedings of the NASA Seal/Secondary Air Flow System Workshop,* Cleveland, OH, Nov. 2004.

92. C. M. Grondahl, "Seal Assembly and Rotary Machine Containing Such Seal," U.S. Patent No. 6,644,667, 2003.

BIBLIOGRAPHY

American Society of Mechanical Engineers, *Code for Pressure Vessels,* Sec. VIII, Div. 1, App. 2, 2004.

American Variseal, *Variseal™ Design Guide,* AVDG394 American Variseal Co., Broomfield, CO, 2005.

Howard, R. A., *Grafoil Engineering Design Manual,* Union Carbide, Cleveland, OH, 1987.

CHAPTER 34

VIBRATION AND SHOCK

Singiresu S. Rao
Department of Mechanical and Aerospace Engineering
University of Miami
Coral Gables, Florida

1 INTRODUCTION

The subject of vibrations is a specialized area of dynamics. Dynamics is concerned with the motion of physical systems. Vibrations are concerned with the motion of a system in the form of oscillation or repetitive motion relative to a reference state. The reference state, in most cases, denotes the static equilibrium position of the system (as in the case of buildings). The reference state, in some cases, denotes the steady motion of the system (as in the case of spinning of a satellite). In vibrations, the motion is induced by a disturbing force. If the disturbing force acts only initially (at time zero), the resulting vibration is known as free vibration. On the other hand, if the disturbing force acts for some time (even after time zero), the resulting vibration is called forced vibration. Shock is a somewhat loosely defined term that implies a degree of suddenness and severity. It denotes a nonperiodic excitation in the form of a pulse of short duration. For the analysis and design of mechanical and structural systems, it is essential that the analyst or the designer has a thorough understanding of the basic principles of vibration and shock.

Some common sources of vibration and shock to equipment include handling; transportation by trucks, trains, airplanes, ships, and other vehicles; and blasts from explosions. Vibrations can be observed in the following typical situations: an automobile traveling over a rough surface, earthquakes, motion of an airplane in turbulence, and a washing machine during spin cycle. The objectives of study of vibration and shock are as follows:

To explain, predict, and control the oscillatory behavior or response of systems

To avoid fatigue damage in structures and machines

To avoid physical discomfort to humans operating/working near vibrating equipment

To safely design buildings, bridges, mechanical equipment, and machinery

To avoid noise generated by vibration

The study of vibration requires the following mathematical and computational tools:

1. *Linear Algebra.* Matrix operations, solution of simultaneous equations, and solution of algebraic eigenvalue problem.
2. *Complex Algebra.* Representation and manipulation of harmonic and periodic functions in complex form.
3. *Ordinary Differential Equations.* Solution by Laplace transforms and other techniques.
4. *Partial Differential Equations.* Solution by method of separation of variables and other techniques.
5. *Approximate Analytical Methods.* Solution by Rayleigh, Rayleigh–Ritz, Galerkin, least-squares, and other variational and weighted residual approaches.
6. *Numerical Methods.* Solution of continuous systems as equivalent discrete systems using finite-difference and finite-element methods.

2 MODELING OF PHYSICAL SYSTEMS

A physical system is to be replaced by a mathematical model in order to predict its vibration behavior. The accuracy of the predicted behavior depends on the level of difficulty associated with the mathematical model. The model must account for the four basic phenomena associated with the physical system, namely, the elasticity, inertia, excitation or input energy, and damping or dissipation of energy. The mathematical model should not be too complex and overly sophisticated to include more details of the system than are necessary. An overly sophisticated model will require more computational effort than necessary. In some cases, it may not be possible to find the solution of an overly complicated model. Based on the nature of the mathematical model used, the system may be called a discrete (or lumped) system or a continuous (or distributed) system. In the discrete model, the physical system is assumed to consist of several rigid bodies (usually considered as point masses) connected by springs and dampers. The springs denote restoring forces that tend to return the masses to their respective undisturbed (or equilibrium) states. The dampers provide resistance to velocity and dissipate the energy of the system. In the continuous model, the mass, elasticity, and damping are assumed to be distributed throughout the system. The equations of motion of a discrete system are in the form of a system of n coupled second-order ordinary differential equations, where n denotes the number of masses (discrete masses or rigid bodies). The number of independent coordinates needed to describe the configuration of a system at any time during vibration defines the degrees of freedom of the system. For example, Figs. 1, 2 and 3 denote typical one-, two-, and three-degree-of-freedom systems, respectively. A point mass can have three translational degrees of freedom while a rigid body can have three translational and three rotational degrees of freedom. Many mechanical and structural components and systems such as bars, beams, plates, and shells have distributed mass, elasticity, and damping. The equation of motion of a continuous system is in the form of a partial

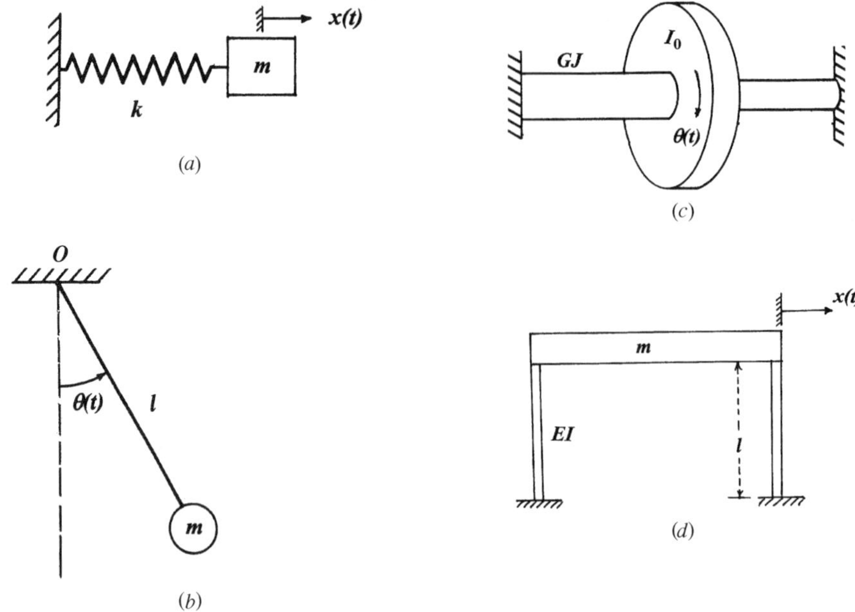

Figure 1 Typical one-degree-of-freedom systems.

differential equation. A continuous system can be modeled either as a discrete- or lumped-parameter system with varying number of degrees of freedom or as a continuous system with infinite number of degrees of freedom, as illustrated for a cantilever beam in Fig. 4.

The oscillatory motion of a body may be harmonic, periodic, or nonperiodic in nature. If the time variation of the displacement of the mass is sinusoidal, the motion will be harmonic. The number of cycles of motion per unit time defines the frequency, and the maxi-

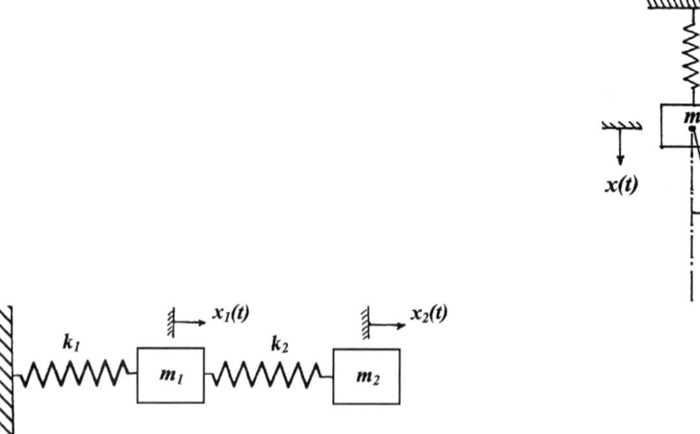

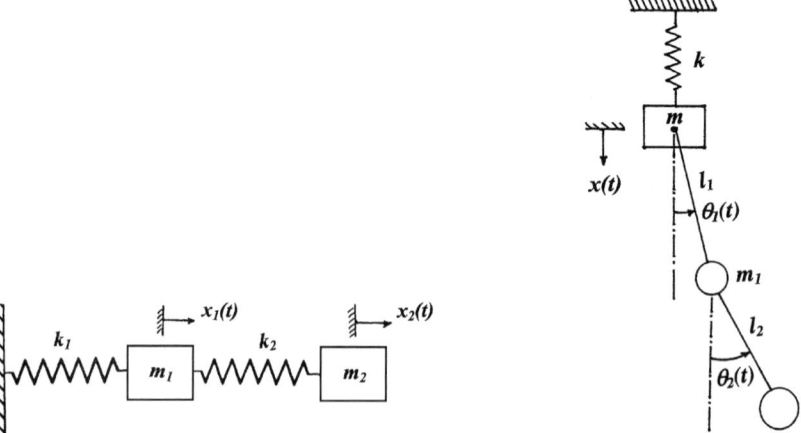

Figure 2 Two-degree-of-freedom system.

Figure 3 Three-degree-of-freedom system.

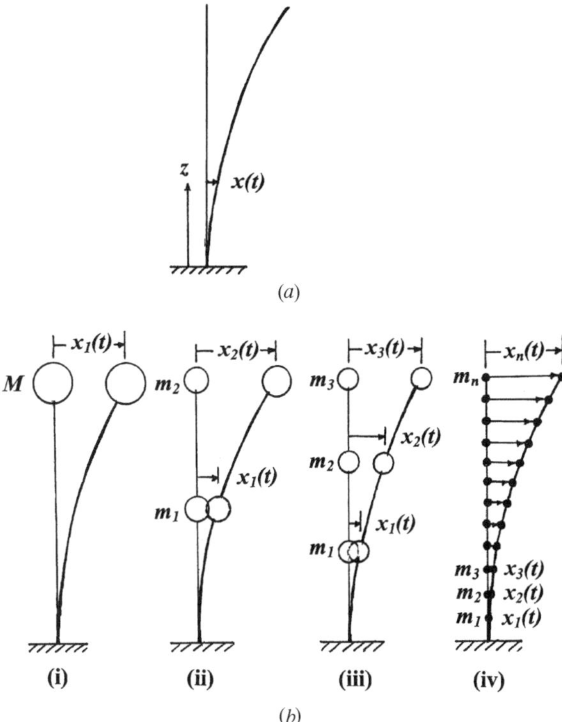

Figure 4 Modeling of a continuous system: (*a*) continuous system; (*b*) discrete system.

mum magnitude of motion is called the amplitude of vibration. If the periodic variation of motion is not harmonic, the motion will be periodic. In this case, the periodic motion can be expressed as a sum of harmonic motions of different frequencies. If the time variation of the displacement of the mass is arbitrary (nonperiodic), the motion is said to be nonperiodic. If the nonperiodic motion can be described either by an equation or by a set of tabulated values, the motion is considered to be deterministic. On the other hand, if the motion cannot be described by any equation or tabulated values, it is said to be random or probabilistic. When an external force or excitation is applied to a mechanical or structural system, the amplitude of the resulting vibration can become very large when a frequency component of the applied force or excitation approaches one of the natural frequencies of the system, particularly the fundamental one. Such a condition, known as resonance, and the attendant stresses and strains might cause a failure of the system. Because of this, designers should have a means of determining the natural frequencies of mechanical and structural systems using analytical or experimental approaches.

3 SINGLE-DEGREE-OF-FREEDOM SYSTEM

A study of the vibration characteristics of a single-degree-of-freedom-system is extremely important in the study of vibration and shock because the approximate or qualitative response of most systems can be determined by using a single-degree-of-freedom model for the system. A general single-degree-of-freedom system consists of a mass m, a spring of stiffness k, and a viscous damper with a damping constant c, as shown in Fig. 5. The significance of the quantities m, c and k for different types of systems is given in Table 1.

3.1 Equation of Motion

The equation of motion is given by

$$m\ddot{x} + c\dot{x} + kx = f(t) \tag{1}$$

where a dot (two dots) above x denotes the first (second) derivative with respect to time.

3.2 Free Vibration

The free-vibration response of the system, corresponding to $f(t) = 0$ in Eq. (1), is given by

$$x(t) = e^{-\zeta\omega_n t}\left(C_1 e^{\omega_n\sqrt{\zeta^2-1}} + C_2 e^{-\omega_n\sqrt{\zeta^2-1}}\right) \tag{2}$$

where C_1 and C_2 are constants, ω_n is the undamped natural frequency of vibration

$$\omega_n = \sqrt{\frac{k}{m}} \tag{3}$$

and ζ is the damping ratio

$$\zeta = \frac{c}{c_{\text{cri}}} = \frac{c}{2m\omega_n} = \frac{c}{2\sqrt{km}} \tag{4}$$

with c_{cri} denoting critical damping:

$$c_{\text{cri}} = 2m\omega_n = 2\sqrt{km} \tag{5}$$

Equation (2) indicates that the physical behavior of a system depends primarily upon the magnitude of the damping ratio ζ. There are three distinct damping cases: (i) $\zeta > 1$, (ii) $\zeta = 1$, and (iii) $\zeta < 1$.

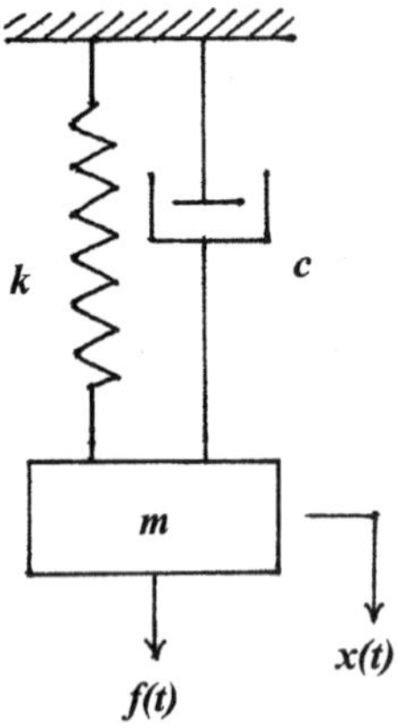

Figure 5 Mass–spring–damper system.

Table 1 Significance of m, c, and k in Different Systems

Vibrating System	m	c	k	Variable x
1. Translatory spring–mass–damper system, Fig. 5	Mass (kg)	Viscous damping constant (N·s/m)	Spring stiffness (N/m)	Linear displacement (m)
2. Rotational spring–mass–damper system, Fig. 1c	Mass moment of inertia (kg·m^2)	Torsional damping constant (m·N·s/rad)	Torsional spring stiffness (m·N/rad)	Angular displacement (rad)
3. Swinging pendulum, Fig. 1b	Moment of inertia of bob (kg·m^2)	Damping constant of surrounding medium (m·N·s/rad)	Angular stiffness constant due to gravity (N·m/rad)	Angular displacement (rad)
4. Transversely vibrating cantilever beam, Fig. 4a	Mass at end of beam (kg)	Damping constant due to surrounding medium (N·s/m)	Flexural stiffness of beam (N/m)	Transverse displacement of mass at end of cantilever (m)

Case (i)

When $\zeta > 1$, the system is said to be overdamped and will not oscillate when displaced from its static equilibrium position. A typical displacement–time curve of the system when the mass is given an initial displacement x_0 from its static equilibrium position and released with nonzero initial velocity is shown in Fig. 6*a*. It can be seen that an overdamped system takes an extremely long time (theoretically infinite time) to return to its static equilibrium position.

Case (ii)

When $\zeta = 1$, the system is said to be critically damped and the free-vibration solution becomes

$$x(t) = (C_1 + C_2 t)e^{-\omega_n t} \tag{6}$$

In this case, the system will not oscillate when given an initial displacement but returns to its static equilibrium position similar to that of an overdamped system. Here also it takes an infinite time for the system to return to its static equilibrium position, as shown in Fig. 6*b*. However, the critically damped system returns to its static equilibrium position in minimum possible time without oscillation.

Case (iii)

When $\zeta < 1$, the system is said to be underdamped and the free-vibration solution can be expressed as

$$x(t) = e^{-\zeta\omega_n t} (C_1 e^{i\omega_d t} + C_2 e^{-i\omega_d t}) \tag{7}$$

or

$$x(t) = X_0 e^{-\zeta\omega_n t} \sin(\omega_d t + \phi) \tag{8}$$

where

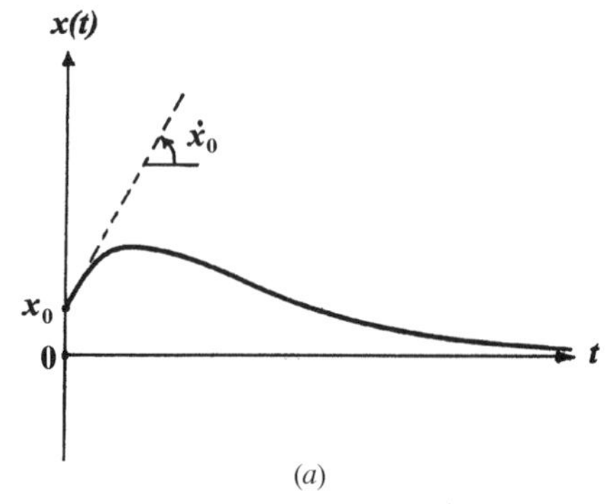

(a)

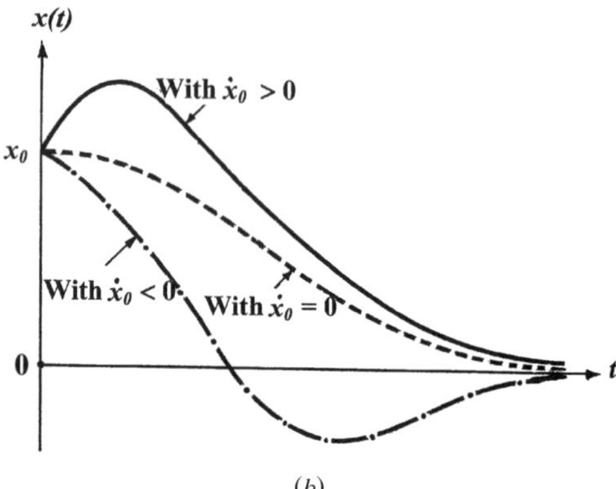

(b)

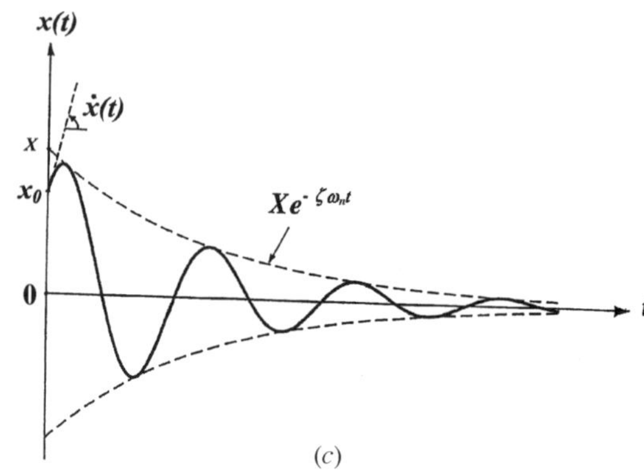

(c)

Figure 6 Free-vibration response: (a) $\zeta > 1$; (b) $\zeta = 1$; (c) $\zeta < 1$.

$$\omega_d = \omega_n \sqrt{1 - \zeta^2} \tag{9}$$

denotes the damped frequency of vibration and C_1 and C_2 or X_0 and ϕ are constants which can be determined from the initial conditions. If the initial displacement and initial velocity of the system are specified as

$$x(t = 0) = x_0 \qquad \dot{x}(t = 0) = \dot{x}_0 \tag{10}$$

Eqs. (7) and (8) can be expressed in equivalent form as

$$x(t) = e^{-\zeta\omega_n t}\left(x_0 \cos \omega_d t + \frac{\dot{x}_0 + \zeta\omega_n x_0}{\omega_d} \sin \omega_d t\right) \tag{11}$$

A typical response of the system when given an initial displacement x_0 with nonzero initial velocity is shown in Fig. 6*c*.

3.3 Forced-Harmonic Vibration

When a single-degree-of-freedom system is subjected to a harmonic force in addition to the initial conditions, the equation of motion and its solution are given by

$$m\ddot{x} + c\dot{x} + kx = f_0 \sin \omega t \tag{12}$$

and

$$x(t) = e^{-\zeta\omega_n t}(C_1 \cos \omega_d t + C_2 \sin \omega_d t) + X \sin(\omega t - \phi) \tag{13}$$

where f_0 is the magnitude and ω is the frequency of the applied force. In Eq. (13), the first term denotes the transient vibration (x_h) while the second term indicates the steady-state solution (x_p), with X indicating the amplitude and ϕ the phase angle. The amplitude X and the phase angle ϕ are given by

$$X = \frac{f_0/k}{\sqrt{[1 - (\omega/\omega_n)^2]^2 + [2\zeta(\omega/\omega_n)]^2}} = \frac{\delta_{st}}{\sqrt{(1 - r^2)^2 + (2\zeta r)^2}} \tag{14}$$

and

$$\phi = \tan^{-1}\left(\frac{2\zeta(\omega/\omega_n)}{1 - (\omega/\omega_n)^2}\right) = \tan^{-1}\left(\frac{2\zeta r}{1 - r^2}\right) \tag{15}$$

with

$$\delta_{st} = \frac{f_0}{k} = \text{static deflection under force } f_0 \tag{16}$$

$$r = \frac{\omega}{\omega_n} = \text{frequency ratio} \tag{17}$$

3.4 Forced Nonharmonic Vibration

When $f(t)$ is nonharmonic, the steady-state response or the particular integral of Eq. (1), $x_p(t)$, is given by the Duhamel or convolution integral:

$$x_P(t) = \frac{1}{m\omega_d} \int_0^t f(\eta) e^{-\zeta\omega_n(t-\eta)} \sin \omega_d(t - \eta)\, d\eta \tag{18}$$

Unless the forcing function $f(t)$ is a simple analytical function, such as a step function, the integral in Eq. (18) needs to be evaluated by some numerical method. The solution given by Eq. (18) for a step function $f(t) = f_0$, $t > 0$, can be expressed as

$$x_P(t) = \frac{f_0}{k} \left\{ 1 - \frac{e^{-(\zeta 2\pi t/\tau_n)}}{\sqrt{1 - \zeta^2}} \left[\cos\left(\frac{2\pi\sqrt{1 - \zeta^2}\, t}{\tau_n} - \phi \right) \right] \right\} \tag{19}$$

where

$$\phi = \tan^{-1} \frac{\zeta}{\sqrt{1 - \zeta^2}} \tag{20}$$

and

$$\tau_n = \frac{2\pi}{\omega_n} = 2\pi \sqrt{\frac{m}{k}} \tag{21}$$

If the initial conditions of the system are not zero, the homogeneous solution of the differential equation (1), x_h(t) is to be added to the particular integral so that the complete solution becomes

$$\begin{aligned} x(t) &= x_h(t) + x_p(t) \\ &= e^{-\zeta\omega_n t} \left(x_0 \cos \omega_d t + \frac{\dot{x}_0 + \zeta\omega_n x_0}{\omega_d} \sin \omega_d t \right) \\ &\quad + \frac{1}{m\omega_d} \int_0^t f(\eta) e^{-\zeta\omega_n(t-\eta)} \sin \omega_d(t - \eta)\, d\eta \end{aligned} \tag{22}$$

4 SHOCK SPECTRUM

Many mechanical and structural systems are subjected, on occasion, to sudden large forces acting for short periods of time which are small compared to the natural time period of the system. Such forces, known as shock, can cause damage to the system and can induce undesirable vibration of the system. Hence the prediction of the response of a system to shock is of great importance in mechanical and structural design. Generally, the severity of a shock is measured in terms of the maximum value of the response. It is customary to use the response of an undamped single-degree-of-freedom system for comparison purposes. A plot of the maximum response of an undamped single-degree-of-freedom system to a given shock as a function of the natural time period (or natural frequency) of the system is known as the shock spectrum or response spectrum. These plots help in finding the ratio of the maximum dynamic stress to the corresponding static stress in a mechanical or structural system. In most practical cases, the time at which the maximum response of the system occurs is also of interest; hence such plots are also given along with the shock spectrum. It is known that the maximum response of a single-degree-of-freedom system subjected to a force depends upon the characteristics of the system as well as the nature of the applied force. For an undamped system, the natural period or frequency is the characteristic that determines the response for any specified forcing function. The shape and duration of the force $f(t)$ also plays an important role in determining the response.

For example, assume the half sine wave shown in Fig. 7 denotes a shock pulse where f_0 denotes the peak value of the force and t_0 indicates the duration of the pulse. This force can be regarded as the superposition of two one-sided sine functions—one starting at $t = 0$ and the other starting at $t = t_0 = \pi/\omega$, as shown in Fig. 7*b*. Defining the unit step function applied at $x = a$ as $u(t - a)$,

$$u(t-a) = \begin{cases} 0 & t < a \\ 1 & t > a \end{cases} \tag{23}$$

The shock pulse of Fig. 7*a* can be described as

$$f(t) = f_0[\sin \omega t u(t) + \sin \omega(t - t_0)u(t - t_0)] \tag{24}$$

The response of an undamped spring–mass system to the forcing function given by Eq. (24) can be expressed as

$$x(t) = \frac{f_0}{k}\frac{1}{1-(\omega/\omega_n)^2} \times \left\{ \left(\sin \omega t - \frac{\omega}{\omega_n}\sin \omega_n t \right) u(t) + [\sin \omega(t - t_0) - \frac{\omega}{\omega_n} \sin \omega_n(t - t_0)]u(t - t_0) \right\} \tag{25}$$

It is to be noted that the first term in the curly brackets on the right side of Eq. (25) gives the response of the system for $0 < t < \pi/\omega$ (during the time of application of the pulse) while the response of the system for any time after the termination of the pulse ($t > \pi/\omega$) can be expressed in compact form as

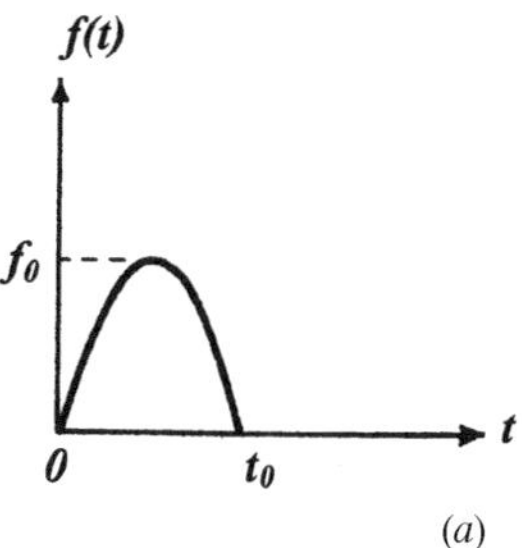

(*a*)

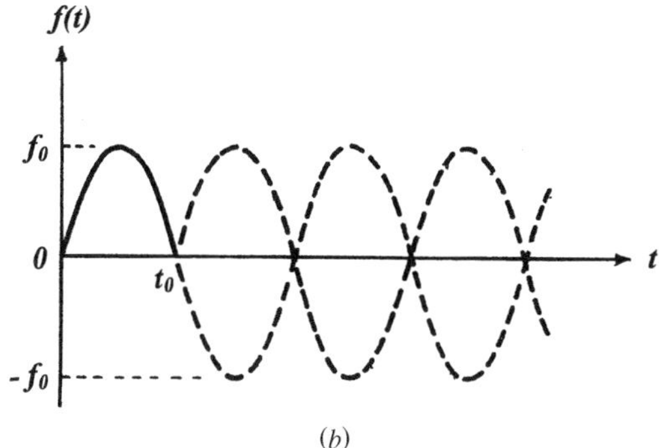

(*b*)

Figure 7 Half sine wave.

$$x(t) = \frac{f_0\,(\omega_n/\omega)}{k[1 - (\omega_n/\omega)^2]}\,[\sin \omega_n t + \sin \omega_n(t - t_0)] \tag{26}$$

To determine the maximum response of the system, we need to first find the time $t_{\max}$ at which the maximum response $x_{\max}$ occurs by solving the equation $\dot{x} = dx/dt = 0$ and then substitute the value of $t_{\max}$ for t in the expression of $x(t)$ to find $x_{\max}$. The time derivatives of $x(t)$ are given by

$$\dot{x}(t) = \frac{f_0\omega}{k[1 - (\omega/\omega_n)^2]}\,(\cos \omega t - \cos \omega_n t) \qquad 0 < t < \frac{\pi}{\omega} \tag{27}$$

and

$$\dot{x}(t) = \frac{f_0(\omega_n^2/\omega)}{k[1 - (\omega_n/\omega)^2]}\,[\cos \omega_n t + \cos \omega_n(t - t_0)] \qquad t > \frac{\pi}{\omega} \tag{28}$$

By setting $\dot{x} = 0$, Eqs. (27) and (28) yield

$$t_{\max} = \frac{2j\pi}{\omega_n + \omega} \qquad j = 1, 2, 3, \ldots \tag{29}$$

$$x_{\max} = \frac{f_0(\omega_n/\omega)}{k[(\omega_n/\omega) - 1]} \sin \frac{2j\pi}{1 + (\omega_n/\omega)} \qquad j < \frac{1}{2}\left(1 + \frac{\omega_n}{\omega}\right) \qquad \text{for} \quad 0 < t < \frac{\pi}{\omega} \tag{30}$$

and

$$t_{\max} = (2j - 1)\frac{\pi}{2\omega_n} + \frac{t_0}{2} \qquad j = 1, 2, 3, \ldots \tag{31}$$

$$x_{\max} = \frac{2f_0(\omega_n/\omega)}{k[1 - (\omega_n/\omega)^2]} \cos \frac{\pi\omega_n}{2\omega} \qquad \text{for} \quad t > \frac{\pi}{\omega} \tag{32}$$

Thus the shock spectrum can be obtained by plotting $x_{\max}$ versus ω_n/ω using both Eqs. (30) and (32). The shock spectrum is shown in nondimensional form in Fig. 8.

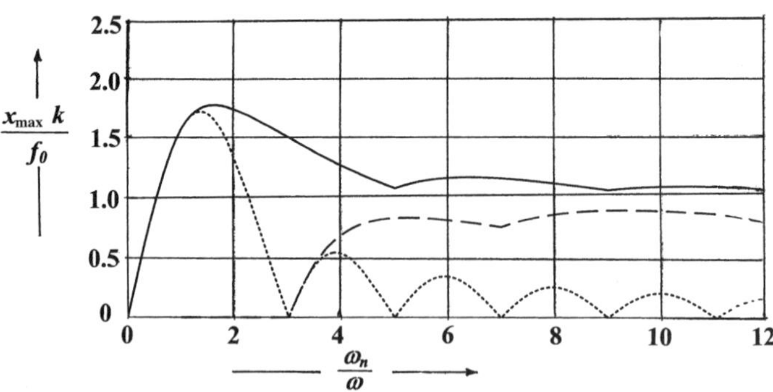

Figure 8 Shock spectrum.

5 MULTI-DEGREE-OF-FREEDOM SYSTEM

Most mechanical and structural systems have distributed mass, elasticity, and damping. These systems are modeled as multi- (n-) degree-of-freedom systems to facilitate analysis of their vibration behavior. Several methods are available to construct an n-degree-of-freedom model from a continuous system. These include the physical lumping or modeling method, finite-element method, finite-difference method, modal analysis method, Rayleigh–Ritz method, Galerkin method, and many others. In most cases, the number of degrees of freedom (n) to be used in the model depends on the frequency range. If the system is expected to undergo significant deformations at higher frequencies, the model should include enough number of degrees of freedom to cover all the important frequencies. Most vibration characteristics of a n-degree-of-freedom system are similar to those of a single-degree-of-freedom system. An n-degree-of-freedom system will have n natural frequencies, its free vibrations denote exponentially decaying motions, its forced vibrations exhibit resonance behavior, etc. However, there are some vibration characteristics that are unique to an n-degree-of-freedom system which are absent in single-degree-of-freedom systems. For example, the existence of normal modes, orthogonality of normal modes, and decomposition of the response of the system (free or forced) in terms of normal modes are unique to multi-degree-of-freedom systems.

5.1 Equations of Motion

Consider a three-degree-of-freedom system consisting of three masses m_1, m_2, and m_3; three springs with stiffnesses k_1, k_2, and k_3; and three viscous dampers with damping constants c_1, c_2, and c_3, as shown in Fig. 9. The mass m_i subjected to the force $f_i(t)$ undergoes a displacement $x_I(t)$, $i = 1, 2, 3$. The equations of motion, derived from the free-body diagrams of the masses, can be expressed in matrix form as

$$\begin{bmatrix} m_1 & 0 & 0 \\ 0 & m_2 & 0 \\ 0 & 0 & m_3 \end{bmatrix} \begin{Bmatrix} \ddot{x}_1 \\ \ddot{x}_2 \\ \ddot{x}_3 \end{Bmatrix} + \begin{bmatrix} c_1 + c_2 & -c_2 & 0 \\ -c_2 & c_2 + c_3 & -c_3 \\ 0 & -c_3 & c_3 \end{bmatrix} \begin{Bmatrix} \dot{x}_1 \\ \dot{x}_2 \\ \dot{x}_3 \end{Bmatrix}$$
$$+ \begin{bmatrix} k_1 + k_2 & -k_2 & 0 \\ -k_2 & k_2 + k_3 & -k_3 \\ 0 & -k_3 & k_3 \end{bmatrix} \begin{Bmatrix} x_1 \\ x_2 \\ x_3 \end{Bmatrix} + \begin{Bmatrix} f_1(t) \\ f_2(t) \\ f_3(t) \end{Bmatrix} \tag{33}$$

Equation (33) can be generalized to an n-degree-of-freedom system as

$$[m]\,\ddot{\mathbf{x}} + [c]\,\dot{\mathbf{x}} + [k]\,\mathbf{x} = \mathbf{f}(t) \tag{34}$$

where $[m]$, $[c]$, and $[k]$ are the mass, damping, and stiffness matrices, respectively;

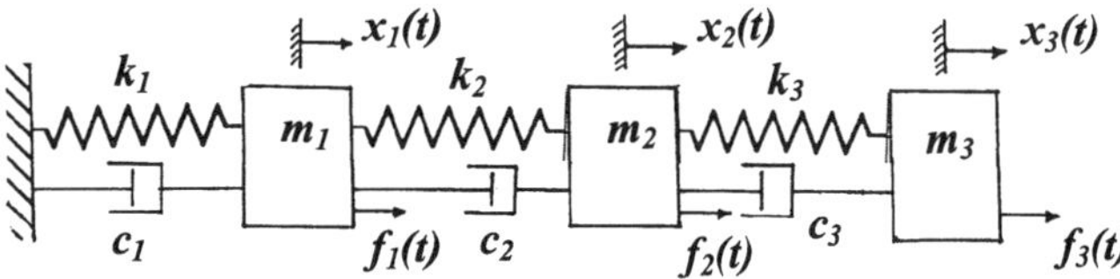

Figure 9 Three-degree-of-freedom system.

$\mathbf{x}$, $\dot{\mathbf{x}}$ and $\ddot{\mathbf{x}}$ are the displacement, velocity, and acceleration vectors, respectively; and $\mathbf{f}(t)$ is the vector of forces with

$$\mathbf{x}(t) = \begin{Bmatrix} x_1(t) \\ x_2(t) \\ \vdots \\ x_n(t) \end{Bmatrix} \qquad \mathbf{f}(t) = \begin{Bmatrix} f_1(t) \\ f_2(t) \\ \vdots \\ f_n(t) \end{Bmatrix} \tag{35}$$

The mass matrix $[m]$ is a symmetric positive-definite matrix, $[c]$ is a symmetric nonnegative matrix, and $[k]$ is a symmetric nonnegative matrix. The symmetry of the matrices $[m]$, $[c]$, and $[k]$ implies that the Maxwell–Betti reciprocity theorem[1] is valid for the system.

5.2 Free-Vibration Response

When $\mathbf{f}(t) = 0$, Eq. (34) gives the free-vibration equation. For the undamped free vibration of the system, the harmonic solution is assumed as

$$\mathbf{x}(t) = \mathbf{X} \cos \omega t \tag{36}$$

where $\mathbf{X}$ is the vector of amplitudes of the masses (or mode shape) and ω is the natural frequency. By substituting $[c] = [0]$ and Eq. (36) into the free-vibration equation, we obtain

$$\omega^2[m]\mathbf{X} = [k]\mathbf{X} \tag{37}$$

The solution of the eigenvalue problem of Eq. (37) gives the natural frequencies ω_1, ω_2, . . . , ω_n and the corresponding mode shapes, natural modes, or eigenvectors $\mathbf{X}^{(1)}$, $\mathbf{X}^{(2)}$, . . . , $\mathbf{X}^{(n)}$. The mode shapes $\mathbf{X}^{(i)}$ are orthogonal with respect to the mass and stiffness matrices. When the mode shapes are normalized with respect to the mass matrix, we obtain

$$\mathbf{X}^{(i)T}\,[m]\,\mathbf{X}^{(j)} = \delta_{ij} \tag{38}$$

$$\mathbf{X}^{(i)T}\,[k]\,\mathbf{X}^{(i)} = [\omega_i^2] \tag{39}$$

where δ_{ij} is the Kronecker delta defined by

$$\delta_{ij} = \begin{cases} 1 & \text{if } i = j \\ 0 & \text{if } i \neq j \end{cases} \tag{40}$$

and $[\omega_i^2]$ is a diagonal matrix consisting of ω_1^2, ω_2^2, . . . , ω_n^2 in the diagonal locations. For the free vibration of the undamped system, a procedure known as modal analysis can be used. In this procedure, the free-vibration solution is assumed as

$$\mathbf{x}(t) = \sum_{i=1}^{n} \eta_i(t)\mathbf{X}^{(i)} \tag{41}$$

where η_i(t) is the generalized coordinate corresponding to the mode i. Using the initial conditions

$$\mathbf{x}(t = 0) = \mathbf{x}_0 \qquad \dot{\mathbf{x}}(t = 0) = \dot{\mathbf{x}}_0 \tag{42}$$

the generalized coordinates can be found as

$$\eta_i(t) = \eta_i(0) \cos \omega_i t + \frac{\dot{\eta}_i(0)}{\omega_i} \sin \omega_i t \qquad i = 1, 2, \ldots, n \tag{43}$$

where the initial values of the generalized displacements $\eta_i(0)$ and the generalized velocities

$\dot{\eta}_i(0)$ can be found using Eqs. (42). Finally the free-vibration solution of the multi-degree-of-freedom system can be determined using Eq. (41).

5.3 Forced-Vibration Response

The forced vibration of the system, given by the solution of Eq. (34), can be found using modal analysis. For simplicity, we assume the damping to be proportional and the matrix $[c]$ is assumed to be given by a linear combination of the mass and stiffness matrices:

$$[c] = \alpha[m] + \beta[k] \tag{44}$$

where α and β are constants. By using a solution of the form of Eq. (42) and the orthogonality properties of the normal modes, the generalized coordinates can be found as

$$\eta_i(t) = a_i e^{-\zeta_i \omega_i t} \sin(\omega_{d_i} t + \phi_i) + \frac{1}{\omega_{d_i}} e^{-\zeta_i \omega_i t} \int_0^t F_i(\tau) e^{\zeta_i \omega_i \tau} \sin \omega_{d_i} (t - \tau)\, d\tau \tag{45}$$

where

$$\omega_{d_i} = \omega_i \sqrt{1 - \zeta_i^2} \tag{46}$$

a_i and ϕ_i can be determined from the initial values of the modal coordinates $\eta_i(0)$ and $\dot{\eta}_i(0)$, and ζ_i is the damping ratio determined from the relation

$$\zeta_i = \frac{1}{2\omega_i}(\alpha + \beta\omega_i^2) \qquad i = 1, 2, \ldots, n \tag{47}$$

where ω_i denotes the ith natural frequency of the system. Once the generalized coordinates are known, the forced-vibration response of the multi-degree-of-freedom system can be found from Eq. (42).

6 VIBRATION OF CONTINUOUS SYSTEMS

Some practical mechanical and structural systems with uniform or regular geometry can be modeled as simple continuous vibratory systems. In these models, the inertial and elastic properties are distributed in terms of mass density and elastic moduli. A continuous vibratory system will have infinite degrees of freedom. Examples of continuous systems include strings, bars, shafts, beams, membranes, plates, and shells. The vibration of simple undamped continuous systems due to only initial conditions is considered in this section. The responses of undamped continuous systems due only to nonzero initial conditions are sinusoidal transient responses and denote free vibrations.

6.1 Transverse Vibration of a String

A string is geometrically a line and hence will support only tensile load. Each point of the string is assumed to have only transverse motion (Fig. 10). We assume that the transverse displacement of the string (w) is small so that the tension $T(x)$ may be assumed to be a constant independent of the position along the curve of the string. In the free-body diagram of a differential length dx of the string shown in Fig. 10b, the horizontal components of the tension, $T(x)$ and $T(x + dx)$, are approximately equal to each other but of opposite in sign:

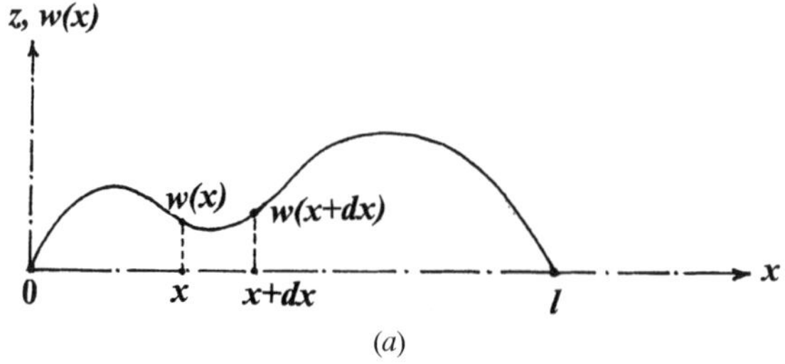

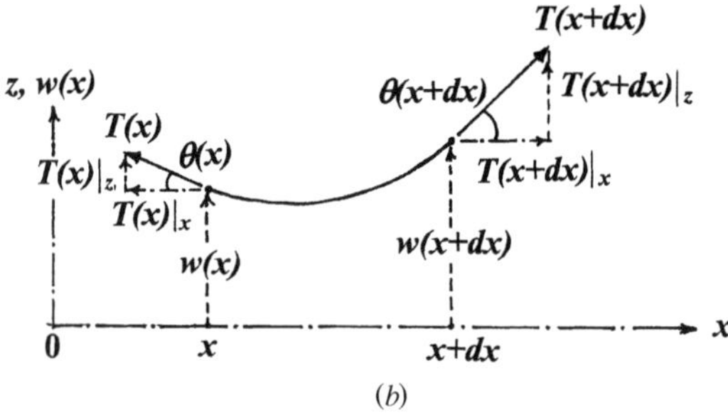

Figure 10 Vibrating string.

$$T(x)|_x = -T(x + dx)|_x \tag{48}$$

According to the Newton's second law, the sum of the vertical components of the tensile forces causes the time rate of change of the linear momentum of the differential element of the string in the transverse direction:

$$T(x)|_z + T(x + dx)|_z = \mu \, dx \, \frac{\partial^2 w}{\partial t^2} \tag{49}$$

where μ is the mass per unit length and $\partial^2 w / \partial t^2$ is the acceleration of the differential element of the string. Using

$$T(x)|_z = T \sin \theta(x) \qquad \sin \theta(x) \cong -\frac{\partial w(x)}{\partial x}$$

$$T(x + dx)|_z = T\left(\frac{\partial w}{\partial x} + \frac{\partial^2 w}{\partial x^2} \, dx\right) \tag{50}$$

Eq. (49) yields the equation of motion as

$$\frac{\partial^2 w}{\partial x^2} = \frac{1}{\alpha^2} \frac{\partial^2 w}{\partial t^2} \tag{51}$$

where

$$\alpha = \sqrt{\frac{T}{\mu}} \tag{52}$$

Equation (51) is called the wave equation and α denotes the wave velocity.

6.2 Longitudinal Vibration of a Bar

By proceeding as in the case of the transverse vibration of a string, the free-body diagram of an element of the bar yields the equation of motion for the longitudinal vibration of a bar as

$$\frac{\partial^2 u}{\partial x^2} = \frac{1}{\alpha^2} \frac{\partial^2 u}{\partial t^2} \tag{53}$$

where u is the axial displacement of the cross section of the bar at x and

$$\alpha = \sqrt{\frac{E}{\mu}} \tag{54}$$

with E denoting Young's modulus and μ the mass density of the bar. If the bar is nonuniform, the cross-sectional area $A(x)$ varies with x and the equation of motion becomes

$$\frac{\partial^2 u}{\partial x^2} + \frac{1}{A(x)} \frac{\partial A}{\partial x} \frac{\partial u}{\partial x} = \frac{\mu}{E} \frac{\partial^2 u}{\partial t^2} \tag{55}$$

6.3 Torsional Vibration of a Shaft

Denoting the angular displacement of the cross section of the shaft at x as $\theta(x,t)$, the equation of motion can be derived as

$$\frac{\partial^2 \theta}{\partial x^2} = \frac{1}{\alpha^2} \frac{\partial^2 \theta}{\partial t^2} \tag{56}$$

where

$$\alpha = \sqrt{\frac{G}{\mu}} \tag{57}$$

with μ denoting the mass density and G the shear modulus of the material. If the cross-sectional area of the shaft $A(x)$ varies with x, the equation of motion becomes

$$G\left(J \frac{\partial^2 \theta}{\partial x^2} + \frac{\partial J}{\partial x} \frac{\partial \theta}{\partial x}\right) = \mu J \frac{\partial^2 \theta}{\partial t^2} \tag{58}$$

It is to be noted that the equations of motion of a string, bar, and shaft [Eqs. (51), (53), and (56)] have the same form and hence their solutions will be similar.

6.4 Flexural Vibration of a Beam

The bending moment–deflection relation of the beam is given by

$$M(x) = EI \frac{\partial^2 w}{\partial x^2} \tag{59}$$

where E is Young's modulus and I is the area moment of inertia of the cross section of the beam. The equation of motion of the beam is given by

$$EI \frac{\partial^4 w}{\partial x^4} + \mu A \frac{\partial^2 w}{\partial t^2} = 0 \tag{60}$$

where μ is the density and A is the area of cross section of the beam. If an external force $f(x,t)$ per unit length acts on the beam, the equation of motion takes the form

$$EI \frac{\partial^4 w}{\partial x^4} + \mu A \frac{\partial^2 w}{\partial t^2} = f(x,t) \tag{61}$$

For a nonuniform beam, the cross section varies along the length and the equation of motion for forced vibration becomes

$$\frac{\partial^2}{\partial x^2}\left(EI \frac{\partial^2 w}{\partial x^2}\right) + \mu A \frac{\partial^2 w}{\partial t^2} = f(x,t) \tag{62}$$

6.5 Free-Vibration Solution

The free-vibration solution of a continuous system depends on its boundary conditions and initial conditions. The free-vibration solution can be determined either in the form of a wave solution or in the form of superposition of normal modes of vibration. These methods are presented in the following sections.

Wave Solution

We shall consider the free-vibration equation of strings, bars, and shafts to illustrate the wave solution. The one-dimensional wave equation can be expressed in the general form

$$\frac{\partial^2 f}{\partial x^2} = \frac{1}{\alpha^2}\frac{\partial^2 f}{\partial t^2} \tag{63}$$

where α indicates the wave propagation velocity and $f(x,t)$ denotes the transverse displacement in the case of strings, longitudinal displacement in the case of bars, and angular displacement in the case of shafts. The solution of Eq. (63) is

$$f(x,t) = f_1(x - \alpha t) + f_2(x + \alpha t) \tag{64}$$

where f_1 and f_2 are arbitrary functions of the arguments $x - \alpha t$ and $x + \alpha t$, respectively. Here, $f_1(x - \alpha t)$ denotes a displacement wave of arbitrary shape f_1 traveling in the positive x direction with a constant velocity α and without a change in shape. Similarly, $f_2(x + \alpha t)$ represents a displacement wave of arbitrary shape f_2 traveling in the negative x direction. Thus a general type of motion consists of a superposion of two waves of arbitrary shape traveling in opposite directions as shown in Fig. 11.

Normal-Mode Solution

We shall consider the free-vibration equation of beams to illustrate the normal-mode solution. The governing equation is

$$\alpha^2 \frac{\partial^4 w}{\partial x^4} + \frac{\partial^2 w}{\partial t^2} = 0 \tag{65}$$

where

$$\alpha = \sqrt{\frac{EI}{\rho A}} \tag{66}$$

The solution of Eq. (65) is assumed to be harmonic with frequency ω as

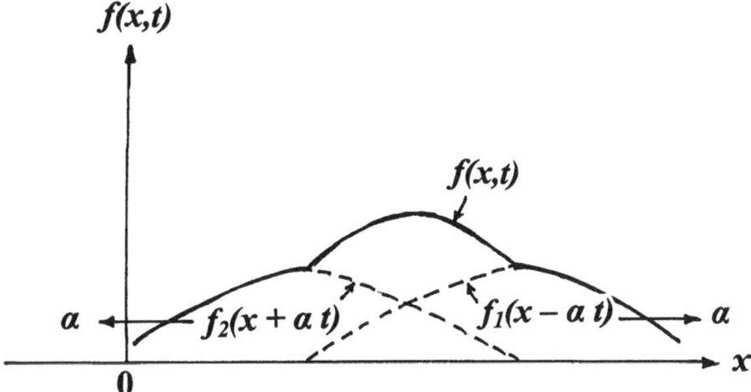

Figure 11 Wave solution.

$$w(x,t) = W(x) \cos \omega t \tag{67}$$

where $W(x)$ is a function of x indicating the deflection (normal mode) of the beam during vibration at frequency ω. Substitution of Eq. (67) into Eq. (65) yields

$$\frac{d^4W}{dx^4} - \beta^4 W = 0 \tag{68}$$

where

$$\beta^4 = \frac{\omega^2}{\alpha^2} \tag{69}$$

The solution of Eq. (68) can be expressed as

$$W(x) = C_1 \sinh \beta x + C_2 \cosh \beta x + C_3 \sin \beta x + C_4 \cos \beta x \tag{70}$$

where the constants C_1, C_2, C_3, and C_4 are evaluated using the boundary conditions of the beam. To indicate the procedure, consider a beam simply supported at both ends. The boundary conditions can be expressed as

$$w(0,t) = \frac{\partial^2 w(0,t)}{\partial x^2} = 0 \qquad \text{or} \qquad W(0) = \frac{d^2W(0)}{dx^2} = 0 \tag{71}$$

$$w(l,t) = \frac{\partial^2 w(l,t)}{\partial \chi^2} = 0 \qquad \text{or} \qquad W(l) = \frac{d^2W(l)}{dx^2} = 0 \tag{72}$$

Substitution of the solution [Eq. (70)] into Eqs. (71) and (72) leads to

$$C_2 + C_4 = 0 \tag{73}$$

$$\beta^2(C_2 - C_4) = 0 \tag{74}$$

$$C_1 \sinh \beta l + C_2 \cosh \beta l + C_3 \sin \beta l + C_4 \cos \beta l = 0 \tag{75}$$

$$\beta^2 (C_1 \sinh \beta l + C_2 \cosh \beta l - C_3 \text{ sini } \beta l - C_4 \cos \beta l) = 0 \tag{76}$$

By setting the determinant of the coefficient matrix of C_1, C_2, C_3 and C_4 in Eqs. (73)–(76) equal to zero, we obtain the frequency equation

$$\sin \beta l = 0 \tag{77}$$

The roots of Eq. (77) are given by

$$\beta l = n\pi \qquad n = 1, 2, \ldots \tag{78}$$

which, in view of Eq. (69), gives

$$\omega_n = (\beta_n l)^2 \sqrt{\frac{EI}{\rho A l^4}} \tag{79}$$

The free-vibration response of the simply supported beam can be determined by superposing the normal modes as

$$w(x,t) = \sum_{n=1}^{\infty} \sin \frac{n\pi x}{l} (A_n \sin \omega_n t + B_n \cos \omega_n t) \tag{80}$$

where $A_n = C_2 A$ and $B_n = C_2 B$. The constants A_n and B_n are evaluated by applying the initial conditions. The frequency equations and the first three natural frequencies of beams with some simple boundary conditions are given in Table 2.

7 VIBRATION AND SHOCK ISOLATION

Vibration and shock isolation denotes the storage of energy temporarily and its release in a different time relation. Isolation may be either for force or for motion. Isolation of motion involves the reduction of the deflections and stresses in systems whose supports experience motion due to vibration or shock. The concept of isolation of motion can be explained with reference to the system shown in Fig. 12. First consider a system that consists of a precision instrument of mass m connected to a larger mass M by a spring k. The mass M rests on a support (without the spring K) that is subjected to vibration and shock. To reduce the motion of mass m relative to M, an isolator in the form of a spring K is inserted between the mass M and the support S, as shown in Fig. 12. The characteristics of the isolator K depend on the type of system or equipment to be protected and the nature of the disturbing motion $y(t)$. Force isolation involves the forces developed by the operation of machinery. For example, in an internal combustion engine, forces are generated by the firing of the fuel–air mixture. Additional forces result from the inertia of the unbalanced rotating and reciprocationg members. All these forces are transmitted to the support or base or foundation of the engine. These forces may cause unacceptably high stresses in the support or the motion of the support may cause annoyance to the personnel working nearby. The forces created by the operation

Table 2 Frequency Equations and Natural Frequencies[a] of Uniform Beams

Boundary Conditions of Beam	Frequency Equation	$\beta_1 l$	$\beta_2 l$	$\beta_3 l$
Simply supported at both ends	$\sin \beta l = 0$	3.1416	6.2832	9.4248
Fixed at both ends	$\cos \beta l \cosh \beta l = 1$	4.7300	7.8532	10.9956
Free at both ends	$\cos \beta l \cosh \beta l = 1$	4.7300	7.8532	10.9956
Fixed–free beam	$\cos \beta l \cosh \beta l = -1$	1.8751	4.6941	7.8548
Fixed–simply supported beam	$\tan \beta l = \tanh \beta l$	3.9266	7.0686	10.2102
Simply supported–free beam	$\tan \beta l = \tanh \beta l$	3.9266	7.0686	10.2102

[a] $\omega_n = (\beta_n l)^2 \sqrt{EI/(\rho A l^4)}$.

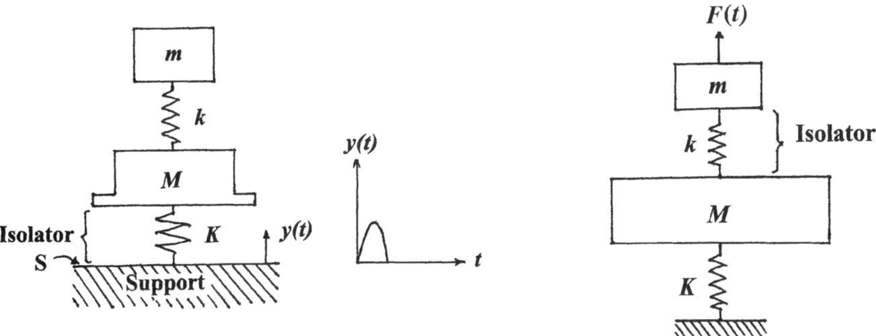

Figure 12 Isolation of motion.

Figure 13 Isolation of force.

of the engine can be isolated from the support by mounting the engine on a properly designed isolation system.

The concept of force isolation can be explained with reference to the system shown in Fig. 13, where the engine of mass m is placed on a support or foundation that can be represented by a mass M and a spring K (with no spring k). Here $f(t)$ indicates the force developed due to gas pressure and inertia force. Since the engine is directly attached to the foundation, both the engine and the foundation experience a common deflection in response to the force $f(t)$. By inserting a properly designed isolator k in between the engine and its support as shown in Fig. 13, the motion of the support M can be reduced. This will reduce the stress in the spring K and will reduce the disturbance experienced by the personnel working on or near the engine support.

7.1 Vibration Isolation

As stated earlier, two aspects of vibration isolation are to be considered—one for the isolation of forces and the other for the isolation of motions. The isolation of forces is required for rotating and reciprocating machinery such as fans, electric motors, compressors, and diesel engines. In these cases, the objective of isolation is to reduce the magnitude of the force transmitted to the support, base, or foundation of the machine. The isolation of motion is required for automobiles, trucks, trains, airplanes, and ships. In this case, the objective of isolation is to reduce the amplitude of vibration so that the mounted equipment is subjected to less severe vibration compared to the supporting structure, base, or foundation.

Consider a system consisting of a concentrated mass m supported by a spring of stiffness k and that moves only in the vertical direction. When a harmonic force $F(t) = F_0 \sin \omega t$ is applied to the mass, as shown in Fig. 14, the steady-state response or displacement of the mass m is given by

$$x(t) = \frac{F_0/k}{1 - \omega^2/\omega_n^2} \sin \omega t \tag{81}$$

where $\omega_n = \sqrt{k/m}$ is the natural frequency of the system. The applied force $F(t)$ is often due to the unbalanced rotating or reciprocating members. The spring k acts as an isolator. The purpose of the isolator is to ensure that the force transmitted to the support is less than the force $F(t)$ applied to the mass m. The force transmitted to the support is equal to $kx(t)$. The effectiveness of the isolator is measured in terms of a quantity known as force trans-

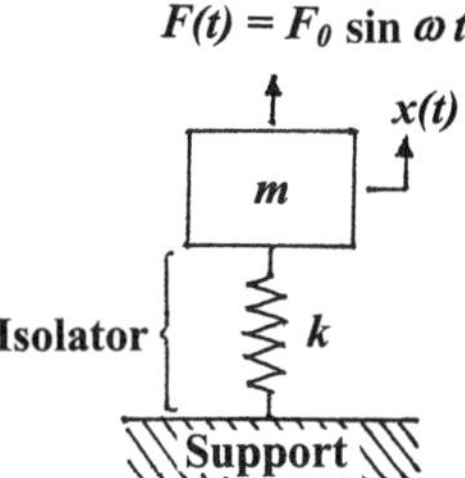

Figure 14 Foundation isolation.

missibility T_F, which is defined as the ratio of the force transmitted by the isolator to the force applied to the mass:

$$T_F = \frac{kx}{F} = \frac{1}{1 - \omega^2/\omega_n^{\,2}} \tag{82}$$

It can be seen that the transmissibility T_F becomes negative when the frequency ratio ω/ω_n is larger than unity. Thus, for $\omega/\omega_n < 1$, the mass m moves downward if $F(t)$ acts downward. The maximum downward force exerted on the support occurs at the same instant when the downward force $F(t)$ is maximum. On the other hand, for $\omega/\omega_n > 1$, the maximum downward force on the support occurs at the same instant when the force $F(t)$ is maximum upward.

Next, consider a system consisting of a concentrated mass m supported by a spring of stiffness k in which the support is subjected to a harmonic disturbance (or motion), as shown in Fig. 15:

$$y(t) = y_0 \sin \omega t \tag{83}$$

In this case also the spring acts as an isolator. The purpose of the isolator is that the displacement (motion) transmitted to the mass m is less than the displacement applied to the support. The displacement transmissibility T_D is defined as the ratio of the maximum displacement of the mass m to the maximum displacement of the support:

$$T_D = \frac{x_0}{y_0} = \frac{1}{1 - \omega^2/\omega_n^{\,2}} \tag{84}$$

It can be seen that the expression of the displacement transmissibility is identical to that of the force transmissibility. In the design of vibration isolators, a quantity, isolation, is used

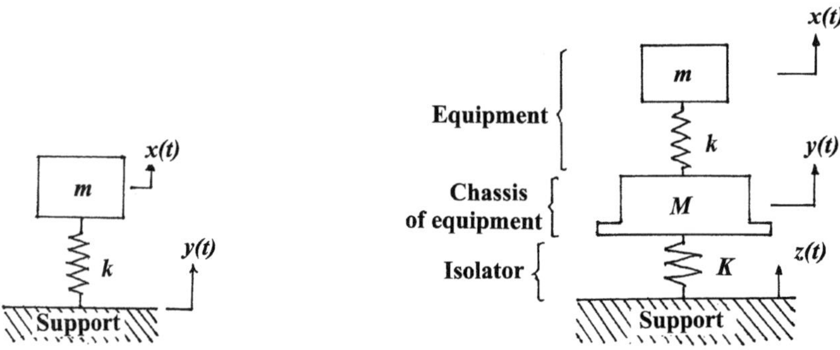

Figure 15 Displacement isolation of mass.

Figure 16 Shock isolation.

to denote the complement of transmissibility. Thus a system with a transmissibility of 0.25 (or 25%) is said to have an isolation of 0.75 (or 75%).

7.2 Shock Isolation

Consider a shock to equipment that occurs due to a sudden change in velocity caused, for example, by the dropping of a package from a height. The equipment can be modeled, for a simple analysis, by a mass m and a spring of stiffness k, as shown in Fig. 16. Here the equipment is originally assumed to be attached to a chassis of mass M, which in turn is secured rigidly to a support S (with no spring K). Under a velocity shock, the support experiences a sudden velocity

$$\ddot{z} = \frac{dz}{dt} = V = \sqrt{2gh} \tag{85}$$

where h is the height from which the chassis and equipment are dropped and g is the acceleration due to gravity. Since the chassis is rigidly attached to the support, it experiences the same velocity so that

$$\dot{y} = \dot{z} = V \tag{86}$$

and hence the displacement of the mass M is given by

$$y(t) = Vt \tag{87}$$

The displacement of the equipment (mass m) can be expressed as

$$x(t) = V\left(t - \frac{1}{\omega_n} \sin \omega_n t\right) \tag{88}$$

where

$$\omega_n = \sqrt{\frac{k}{m}} \tag{89}$$

is the natural frequency of the equipment. The severity of the shock is usually measured by the maximum deformation of the spring, $\delta_{\max}$. The deformation of the spring, δ, is given by

$$\delta = y - x = \frac{V}{\omega_n} \sin \omega_n t \tag{90}$$

and hence

$$\delta_{\max} = \frac{V}{\omega_n} \tag{91}$$

In some cases, the severity of the shock experienced by the equipment or one of its components is measured by the acceleration experienced by the mass m. The acceleration of the equipment, $\ddot{x}(t)$, is given by

$$\ddot{x}(t) = V\omega_n \sin \omega_n t \tag{92}$$

and hence

$$\ddot{x}(t)\big|_{\max} = V\omega_n \tag{93}$$

To isolate the equipment from shock, an isolator, in the form of a spring of stiffness K, is inserted between the chassis of the equipment and the support, as shown in Fig. 16. When the support is subjected to a velocity shock $\dot{z} = V$, the displacement of the support will be

$$z(t) = Vt \tag{94}$$

The equation of motion of the chassis is given by

$$M\ddot{y} + K(y - z) = 0 \tag{95}$$

with the solution

$$y(t) = V\left(t - \frac{1}{\Omega_n} \sin \Omega_n t\right) \tag{96}$$

where

$$\Omega_n = \sqrt{\frac{K}{M}} \tag{97}$$

is the natural frequency of the chassis–isolator system. The equation of motion of the equipment (mass m) can be expressed as

$$m\ddot{x} + k(x - y) = 0 \tag{98}$$

Using Eq. (96), the solution of Eq. (98) can be determined as

$$x(t) = V\left[t - \frac{\sin \omega_n t}{\omega_n(1 - \omega_n^2/\Omega_n^2)} - \frac{\sin \Omega_n t}{\Omega_n(1 - \Omega_n^2/\omega_n^2)}\right] \tag{99}$$

where ω_n is the natural frequency of the equipment:

$$\omega_n = \sqrt{\frac{k}{m}} \tag{100}$$

The maximum acceleration of the equipment, m, is given by

$$\ddot{x}|_{\max} = \frac{V\omega_n}{1 - \omega_n/\Omega_n} \tag{101}$$

The shock transmissibility T_s is defined as the ratio of maximum accelerations of the equipment with and without the isolator:

$$T_s = \frac{\ddot{x}|_{\max} \text{ with isolator, Eq. (101)}}{\ddot{\mathbf{x}}_{\max} \text{ without isolator, Eq. (93)}} = \frac{1}{1 - \omega_n/\Omega_n} \tag{102}$$

The effectiveness of the shock isolator is measured in terms of the shock transmissibility, which can be seen to be a function of the characteristics of both the isolator and the equipment.

8 STANDARDS FOR VIBRATION OF MACHINES

Standards are established for vibration of machines by international organizations such as the International Organization for Standardization (ISO), which works in cooperation with national organizations such as American National Standards Institute (ANSI) in the United States and the British Standards Institute (BSI) in the United Kingdom, the Deutches Institut für Normung (DIN) in Germany, the Danish Standards Association (DS) in Denmark, and the Indian Standards Institute (ISI) in India. Vibration standards are intended to establish criteria for rating or classifying the performance of machines and equipment, provide a basis for selection of machines, and establish procedures for the calibration of machines. Vibration standards are developed by classifying machinery into four categories as follows:

1. Reciprocating machinery having both rotating and reciprocating components such as diesel engines, compressors, and pumps. In these machines, the amplitude of vibration is measured on the main structure of the machine at low frequencies in order to establish the severity of vibration.
2. Rotating machinery having rigid rotors such as electric motors and single-stage pumps. For these machines, vibration is measured on the main structure such as a pedestal or bearing cap to determine the severity of vibration.
3. Rotating machinery having flexible rotors such as large steam turbine generators, compressors, and multistage pumps. For these machines, vibration is measured on the shaft (rotor) directly to assess the severity of vibration.
4. Rotating machinery having quasi-rigid rotors such as low-pressure steam turbines, fans, and axial-flow compressors. In these cases, vibration amplitude is measured on the bearing cap of the flexible rotor to establish the severity of vibration.

The classification of severity of machine vibration is based on the amplitude of the motion variable, namely, the displacement, velocity, or acceleration. The particular motion variable used depends on the type of standard, frequency range, and other factors. For example, the rms (root-mean-square) value of velocity in the frequency range 10–1000 Hz is commonly used in many standards. The vibration velocity can be converted to vibration displacement for a single-frequency component as

$$S_f = \frac{V_f}{\omega_f}\sqrt{2} = \frac{V_f}{2\pi f}\sqrt{2} = 0.225\,\frac{V_f}{f} \tag{103}$$

where S_f is the peak displacement amplitude, V_f is the rms value of the vibration velocity, f is the linear frequency (in hertz), and ω_f is the angular frequency (radian per second). The ISO standards, for example, specifying the allowable vibration severity of rotating machinery are given in the Bibliography.

SYMBOLS

A	Cross-sectional area
A_n, B_n	Constants
c	Damping constant
c_{cri}	Critical damping constant
$C_1, \ldots, C_4$	Constants
E	Young's modulus
$f(t)$	External force
f_0	Amplitude of force
f_1, f_2	Arbitrary functions
G	Shear modulus
I	Area moment of inertia
k, K	Spring stiffness
l	Length of beam

m	Mass
M	Mass, bending moment
r	Frequency ratio
t	Time
T	Tension in string
T_D (T_F)	Displacement (force) transmissibility
T_s	Shock transmissibility
u	Axial displacement of bar, unit step function
V	Velocity
w	Transverse displacement of string or beam
x	Displacement
x_h (x_p)	Homogeneous solution (particular integral)
x_0 ($\dot{x}_0$)	Initial displacement (velocity)
X, X_0	Amplitude of displacement
$\mathbf{X}^{(i)}$	ith modal vector
α	Constant, wave velocity
β	Constant, frequency parameter in beam vibrations
ϕ	Phase angle
θ	Angular displacement of shaft
ζ	Damping ratio
μ	Mass density, mass per unit length
ω	Frequency of applied force
ω_n, Ω_n	Natural frequency of vibration
$\eta_i(t)$	ith generalized (modal) coordinate
δ_{st}	Static deflection
δ_{ij}	Kronecker delta
ω_n	Natural frequency, nth frequency
ω_d	Damped frequency
$(\cdot)$	Maximum value of $(\cdot)$
$[\cdot]$	Matrix

$\dot{x} = dx/dt$, $\ddot{x} = d^2x/dt^2$

REFERENCE

1. Kelly, S. G., *Fundamentals of Mechanical Vibrations,* McGraw-Hill, New York, 2000.

BIBLIOGRAPHY

Braun, S. G., D. J. Ewins, and S. S. Rao (eds.), *Encyclopedia of Vibration,* 3 vol., Academic, San Diego, CA, 2002.

Crede, C. E. *Vibration and Shock Isolation,* Wiley, New York, 1951.

Donaldson, B. K., *Analysis of Aircraft Structures: An Introduction,* McGraw-Hill, New York, 1993.

Harris, C. M., *Shock and Vibration Handbook,* 4th ed., McGraw-Hill, New York, 1996.

ISO 2372, "Mechanical Vibration of Machines with Operating Speeds from 10 to 200 rps—Basis for Specifying Evaluation Standards," International Organization for Standardization.

ISO 2954, "Mechanical Vibration of Rotating and Reciprocating Machinery—Requirements for Instruments for Measuring Vibration Severity," International Organization for Standardization.

ISO 3945, "The Measurement and Evaluation of Vibration Severity of Large Rotating Machines, in Situ; Operating at Speeds from 10 to 200 rps," International Organization for Standardization.

ISO/DIS 10816 Series, ISO/DIS 10816/1, "General Guidelines, Mechanical Vibration—Evaluation of Machine Vibration by Measurements on Non-Rotaing Parts," International Organization for Standardization.

ISO Standards Handbook—Mechanical Vibration and Shock, Vol. 1: *Terminology and Symbols; Tests and Test Equipment; Balancing and Balancing Equipment,* Vol. 2: *Human Exposure to Vibration and Shock, Vibration in Relation to Vehicles, Specific Equipment and Machines, Buildings.*

Meirovitch, L. *Fundamentals of Vibrations,* McGraw-Hill, New York, 2001.

Rao, S. S., *Mechanical Vibrations,* 4th ed., Pearson Prentice-Hall, Upper Saddle River, NJ, 2004.

Tedesco, J. W., W. G. McDougal, and C. A. Ross, *Structural Dynamics: Theory and Applications,* Addison-Wesley, Menlo Park, CA, 1999.

Vierck, R. K., *Vibration Analysis,* 2nd ed., Thomas Y. Crowell Harper & Row, New York, 1979.

Volterra, E. and E. C. Zachmanoglou, *Dynamics of Vibrations,* C. E. Merrill Books, Columbus, OH, 1965.

CHAPTER 35
NOISE MEASUREMENT AND CONTROL

George M. Diehl
Machinery Acoustics
Phillipsburg, New Jersey

1 SOUND CHARACTERISTICS

Sound is a compressional wave. The particles of the medium carrying the wave vibrate longitudinally, or back and forth, in the direction of travel of the wave, producing alternating regions of compression and rarefaction. In the compressed zones the particles move forward in the direction of travel, whereas in the rarefied zones they move opposite to the direction of travel. Sound waves differ from light waves in that light consists of transverse waves, or waves that vibrate in a plane normal to the direction of propagation.

2 FREQUENCY AND WAVELENGTH

Wavelength, the distance from one compressed zone to the next, is the distance the wave travels during one cycle. Frequency is the number of complete waves transmitted per second. Wavelength and frequency are related by the equation

$$v = f\lambda$$

where v = velocity of sound, m/s
f = frequency, cycles/s, Hz
λ = wavelength, m

3 VELOCITY OF SOUND

The velocity of sound in air depends on the temperature and is equal to

$$v = 20.05\sqrt{273.2 + {}^\circ\mathrm{C}}\ \mathrm{m/s}$$

where °C is the temperature in degrees Celsius.

The velocity in the air may also be expressed as

$$v = 49.03\sqrt{459.7 + {}^\circ\mathrm{F}}\ \mathrm{ft/s}$$

where °F is the temperature in degrees Fahrenheit.

The velocity of sound in various materials is shown in Tables 1, 2, and 3.

4 SOUND POWER AND SOUND PRESSURE

Sound power is measured in watts. It is independent of distance from the source and independent of the environment. Sound intensity, or watts per unit area, is dependent on distance. Total radiated sound power may be considered to pass through a spherical surface surrounding the source. Since the radius of the sphere increases with distance, the intensity, or watts per unit area, must also decrease with distance from the source.

Microphones, sound-measuring instruments, and the ear of a listener respond to changing pressures in a sound wave. Sound power, which cannot be measured directly, is proportional to the mean-square sound pressure, p^2, and can be determined from it.

5 DECIBELS AND LEVELS

In acoustics, sound is expressed in decibels instead of watts. By definition, a decibel is 10 times the logarithm, to the base 10, of a ratio of two powers, or powerlike quantities. The reference power is 1 pW, or 10^{-12} W. Therefore,

$$L_W = 10 \log \frac{W}{10^{-12}} \tag{1}$$

where L_W = sound power level, dB
W = sound power, W
log = logarithm to base 10

Sound pressure level is 10 times the logarithm of the pressure ratio squared, or 20 times

Table 1 Velocity of Sound in Solids

Material	Longitudinal Bar Velocity		Plate (Bulk) Velocity	
	cm/s	fps	cm/s	fps
Aluminum	5.24×10^5	1.72×10^4	6.4×10^5	2.1×10^4
Antimony	3.40×10^5	1.12×10^4	—	—
Bismuth	1.79×10^5	5.87×10^3	2.18×10^5	7.15×10^3
Brass	3.42×10^5	1.12×10^4	4.25×10^5	1.39×10^4
Cadmium	2.40×10^5	7.87×10^3	2.78×10^5	9.12×10^3
Constantan	4.30×10^5	1.41×10^4	5.24×10^5	1.72×10^4
Copper	3.58×10^5	1.17×10^4	4.60×10^5	1.51×10^4
German silver	3.58×10^5	1.17×10^4	4.76×10^5	1.56×10^4
Gold	2.03×10^5	6.66×10^3	3.24×10^5	1.06×10^4
Iridium	4.79×10^5	1.57×10^4	—	—
Iron	5.17×10^5	1.70×10^4	5.85×10^5	1.92×10^4
Lead	1.25×10^5	4.10×10^3	2.40×10^5	7.87×10^3
Magnesium	4.90×10^5	1.61×10^4	—	—
Manganese	3.83×10^5	1.26×10^4	4.66×10^5	1.53×10^4
Nickel	4.76×10^5	1.56×10^4	5.60×10^5	1.84×10^4
Platinum	2.80×10^5	9.19×10^3	3.96×10^5	1.30×10^4
Silver	2.64×10^5	8.66×10^3	3.60×10^5	1.18×10^4
Steel	5.05×10^5	1.66×10^4	6.10×10^5	2.00×10^4
Tantalum	3.35×10^5	1.10×10^4	—	—
Tin	2.73×10^5	8.96×10^3	3.32×10^5	1.09×10^4
Tungsten	4.31×10^5	1.41×10^4	5.46×10^5	1.79×10^4
Zinc	3.81×10^5	1.25×10^4	4.17×10^5	1.37×10^4
Cork	5.00×10^4	1.64×10^3	—	—
Crystals				
Quartz X-cut	5.44×10^5	1.78×10^4	5.72×10^5	1.88×10^4
Rock salt X-cut	4.51×10^5	1.48×10^4	4.78×10^5	1.57×10^4
Glass				
Heavy flint	3.49×10^5	1.15×10^4	3.76×10^5	1.23×10^4
Extra heavy flint	4.55×10^5	1.49×10^4	4.80×10^5	1.57×10^4
Heaviest crown	4.71×10^5	1.55×10^4	5.26×10^5	1.73×10^4
Crown	5.30×10^5	1.74×10^4	5.66×10^5	1.86×10^4
Quartz	5.37×10^5	1.76×10^4	5.57×10^5	1.81×10^4
Granite	3.95×10^5	1.30×10^4	—	—
Ivory	3.01×10^5	9.88×10^3	—	—
Marble	3.81×10^5	1.25×10^4	—	—
Slate	4.51×10^5	1.48×10^4	—	—
Wood				
Elm	1.01×10^5	3.31×10^3	—	—
Oak	4.10×10^5	1.35×10^4	—	—

Table 2 Velocity of Sound in Liquids

Material	Temperature °C	Temperature °F	Velocity cm/s	Velocity fps
Alcohol, ethyl	12.5	54.5	1.21×10^5	3.97×10^3
	20	68	1.17×10^5	3.84×10^3
Benzene	20	68	1.32×10^5	4.33×10^3
Carbon bisulfide	20	68	1.16×10^5	3.81×10^3
Chloroform	20	68	1.00×10^5	3.28×10^3
Ether, ethyl	20	68	1.01×10^5	3.31×10^3
Glycerine	20	68	1.92×10^5	6.30×10^3
Mercury	20	68	1.45×10^5	4.76×10^3
Pentane	20	68	1.02×10^5	3.35×10^3
Petroleum	15	59	1.33×10^5	4.36×10^3
Turpentine	3.5	38.3	1.37×10^5	4.49×10^3
	27	80.6	1.28×10^5	4.20×10^3
Water, fresh	17	62.6	1.43×10^5	4.69×10^3
Water, sea	17	62.6	1.51×10^5	4.95×10^3

Table 3 Velocity of Sound in Gases

Material	Temperature °C	Temperature °F	Velocity cm/s	Velocity fps
Air	0	32	3.31×10^4	1.09×10^3
	20	68	3.43×10^4	1.13×10^3
Ammonia gas	0	32	4.15×10^4	1.48×10^3
Carbon dioxide	0	32	2.59×10^4	8.50×10^2
Carbon monoxide	0	32	3.33×10^4	1.09×10^3
Chlorine	0	32	2.06×10^4	6.76×10^2
Ethane	10	50	3.08×10^4	1.01×10^3
Ethylene	0	32	3.17×10^4	1.04×10^3
Hydrogen	0	32	1.28×10^5	4.20×10^3
Hydrogen chloride	0	32	2.96×10^4	9.71×10^2
Hydrogen sulfide	0	32	2.89×10^4	9.48×10^2
Methane	0	32	4.30×10^4	1.41×10^3
Nitric oxide	10	50	3.24×10^4	1.06×10^3
Nitrogen	0	32	3.34×10^4	1.10×10^3
	20	68	3.51×10^4	1.15×10^3
Nitrous oxide	0	32	2.60×10^4	8.53×10^2
Oxygen	0	32	3.16×10^4	1.04×10^3
	20	68	3.28×10^4	1.08×10^3
Sulfur dioxide	0	32	2.13×10^4	6.99×10^2
Water vapor	0	32	1.01×10^4	3.31×10^2
	100	212	1.05×10^4	3.45×10^2

the logarithm of the pressure ratio. The reference sound pressure is 20 μPa, or 20×10^{-6} Pa. Therefore,

$$L_p = 20 \log \frac{p}{20 \times 10^{-6}} \tag{2}$$

where L_p = sound pressure level, dB
p = root-mean-square sound pressure, Pa
log = logarithm to base 10

6 COMBINING DECIBELS

It is often necessary to combine sound levels from several sources. For example, it may be desired to estimate the combined effect of adding another machine in an area where other equipment is operating. The procedure for doing this is to combine the sounds on an energy basis, as follows:

$$L_p = 10 \log[10^{0.1L_1} + 10^{0.1L_2} + \cdots + 10^{0.1L_n}] \tag{3}$$

where L_p = total sound pressure level, dB
L_1 = sound pressure level of source 1
L_n = sound pressure level of source n
log = logarithm to base 10

7 SOUND PRODUCED BY SEVERAL MACHINES OF THE SAME TYPE

The total sound produced by a number of machines of the same type can be determined by adding 10 log n to the sound produced by one machine alone. That is,

$$L_p(n) = L_p + 10 \log n$$

where $L_p(n)$ = sound pressure level of n machines
L_p = sound pressure level of one machine
n = number of machines of the same type

In practice, the increase in sound pressure level measured at any location seldom exceeds 6 dB, no matter how many machines are operating. This is because of the necessary spacing between machines and the fact that sound pressure level decreases with distance.

8 AVERAGING DECIBELS

There are many occasions when the average of a number of decibel readings must be calculated. One example is when sound power level is to be determined from a number of sound-pressure-level readings. In such cases the average may be calculated as follows:

$$\overline{L_p} = 10 \log \left\{ \frac{1}{n} [10^{0.1L_1} + 10^{0.1L_2} + \cdots + 10^{0.1L_n}] \right\} \tag{4}$$

where $\overline{L_p}$ = average sound pressure level, dB
L_1 = sound pressure level at location 1

L_n = sound pressure level at location n
n = number of locations
log = logarithm to base 10

The calculation may be simplified if the difference between maximum and minimum sound pressure levels is small. In such cases arithmetic averaging may be used instead of logarithmic averaging, as follows:

If the difference between the maximum and minimum of the measured sound pressure levels is 5 dB or less, average the levels arithmetically.

If the difference between maximum and minimum sound pressure levels is between 5 and 10 dB, average the levels arithmetically and add 1 dB.

The results will usually be correct within 1 dB when compared to the average calculated by Eq. (4).

9 SOUND-LEVEL METER

The basic instrument in all sound measurements is the sound-level meter. It consists of a microphone, a calibrated attenuator, an indicating meter, and weighting networks. The meter reading is in terms of root-mean-square sound pressure level.

The A-weighting network is the one most often used. Its response characteristics approximate the response of the human ear, which is not as sensitive to low-frequency sounds as it is to high-frequency sounds. A-weighted measurements can be used for estimating annoyance caused by noise and for estimating the risk of noise-induced hearing damage. Sound levels read with the A-network are referred to as dBA.

10 SOUND ANALYZERS

The octave-band analyzer is the most common analyzer for industrial noise measurements. It separates complex sounds into frequency bands one octave in width and measures the level in each of the bands.

An octave is the interval between two sounds having a frequency ratio of 2. That is, the upper cutoff frequency is twice the lower cutoff frequency. The particular octaves read by the analyzer are identified by the center frequency of the octave. The center frequency of each octave is its geometric mean, or the square root of the product of the lower and upper cutoff frequencies. That is,

$$f_0 = \sqrt{f_1 f_2}$$

where f_0 = center frequency, Hz
f_1 = lower cutoff frequency, Hz
f_2 = upper cutoff frequency, Hz

The frequencies f_1 and f_2 can be determined from the center frequency. Since $f_2 = 2f_1$, it can be shown that $f_1 = f_0/\sqrt{2}$ and $f_2 = \sqrt{2}\, f_0$.

Third-octave-band analyzers divide the sound into frequency bands one-third octave in width. The upper cutoff frequency is equal to $2^{1/3}$, or 1.26, times the lower cutoff frequency.

When unknown frequency components must be identified for noise control purposes, narrow-band analyzers must be used. They are available with various bandwidths.

11 CORRECTION FOR BACKGROUND NOISE

The effect of ambient or background noise should be considered when measuring machine noise. Ambient noise should preferably be at least 10 dB below the machine noise. When the difference is less than 10 dB, adjustments should be made to the measured levels, as shown in Table 4.

If the difference between machine octave-band sound pressure levels and background octave-band sound pressure levels is less than 6 dB, the accuracy of the adjusted sound pressure levels will be decreased. Valid measurements cannot be made if the difference is less than 3 dB.

12 MEASUREMENT OF MACHINE NOISE

The noise produced by a machine may be evaluated in various ways, depending on the purpose of the measurement and the environmental conditions at the machine. Measurements are usually made in overall A-weighted sound pressure levels, plus either octave-band or third-octave-band sound pressure levels. Sound power levels are calculated from sound-pressure-level measurements.

13 SMALL MACHINES IN A FREE FIELD

A free field is one in which the effects of the boundaries are negligible, such as outdoors or in a very large room. When small machines are sound tested in such locations, measurements at a single location are often sufficient. Many sound test codes specify measurements at a distance of 1 m from the machine.

Sound power levels, octave band, third-octave band, or A-weighted, may be determined by the following equation:

$$L_W = L_p + 20 \log r + 7.8 \tag{5}$$

where L_W = sound power level, dB
L_p = sound pressure level, dB
r = distance from source, m
log = logarithm to base 10

Table 4 Correction for Background Sound

Level Increase due to the Machine (dB)	Value to Be Subtracted from Measured Level (dB)
3	3.0
4	2.2
5	1.7
6	1.3
7	1.0
8	0.8
9	0.6
10	0.5

14 MACHINES IN SEMIREVERBERANT LOCATIONS

Machines are almost always installed in semireverberant environments. Sound pressure levels measured in such locations will be greater than they would be in a free field. Before sound power levels are calculated, adjustments must be made to the sound-pressure-level measurements.

There are several methods for determining the effect of the environment. One uses a calibrated reference sound source, with known sound power levels, in octave or third-octave bands. Sound pressure levels are measured on the machine under test, at predetermined microphone locations. The machine under test is then replaced by the reference sound source, and measurements are repeated. Sound power levels can then be calculated as follows:

$$L_{Wx} = \overline{L_{px}} + (L_{Ws} - \overline{L_{ps}}) \tag{6}$$

where L_{Wx} = band sound power level of the machine under test
$\overline{L_{px}}$ = average sound pressure level measured on the machine under test
L_{Ws} = band sound power level of the reference source
$\overline{L_{ps}}$ = average sound pressure level on the reference source

Another procedure for qualifying the environment uses a reverberation test. High-speed recording equipment and a special noise source are used to measure the time for the sound pressure level, originally in a steady state, to decrease 60 dB after the special noise source is stopped. This reverberation time must be measured for each frequency or each frequency band of interest.

Unfortunately, neither of these two laboratory procedures is suitable for sound tests on large machinery, which must be tested where it is installed. This type of machinery usually cannot be shut down while tests are being made on a reference sound source, and reverberation tests cannot be made in many industrial areas because ambient noise and machine noise interfere with reverberation time measurements.

15 TWO-SURFACE METHOD

A procedure that can be used in most industrial areas to determine sound pressure levels and sound power levels of large operating machinery is called the two-surface method. It has definite advantages over other laboratory-type tests. The machine under test can continue to operate. Expensive, special instrumentation is not required to measure reverberation time. No calibrated reference source is needed; the machine is its own sound source. The only instrumentation required is a sound-level meter and an octave-band analyzer. The procedure consists of measuring sound pressure levels on two imaginary surfaces enclosing the machine under test. The first measurement surface, S_1, is a rectangular parallelepiped 1 m away from a reference surface. The reference surface is the smallest imaginary rectangular parallelepiped that will just enclose the machine and terminate on the reflecting plane or floor. The area, in square meters, of the first measurement surface is given by the formula

$$S_1 = ab + 2ac + 2bc \tag{7}$$

where $a = L + 2$
$b = W + 2$
$c = H + 1$

and L, W, and H are the length, width, and height of the reference parallelepiped in meters.

The second measurement surface, S_2, is a similar but larger, rectangular parallelepiped located at some greater distance from the reference surface. The area, in square meters, of the second measurement surface is given by the formula

$$S_2 = de + 2df + 2ef \tag{8}$$

where $d = L + 2x$
$e = W + 2x$
$f = H + x$

and x is the distance in meters from the reference surface to S_2.

Microphone locations are usually those shown on Fig. 1.

First, the measured sound pressure levels should be corrected for background noise as shown in Table 4. Next, the average sound pressure levels in each octave band of interest should be calculated as shown in Eq. (4).

Octave-band sound pressure levels, corrected for both background noise and the semi-reverberant environment, may then be calculated by the equations

$$\overline{L_p} = \overline{L_{p1}} - C \tag{9}$$

$$C = 10 \log \left\{ \left[\frac{K}{K - 1} \right] \left[1 - \frac{S_1}{S_2} \right] \right\} \tag{10}$$

$$K = 10^{0.1(\overline{L_{p1}} - \overline{L_{p2}})} \tag{11}$$

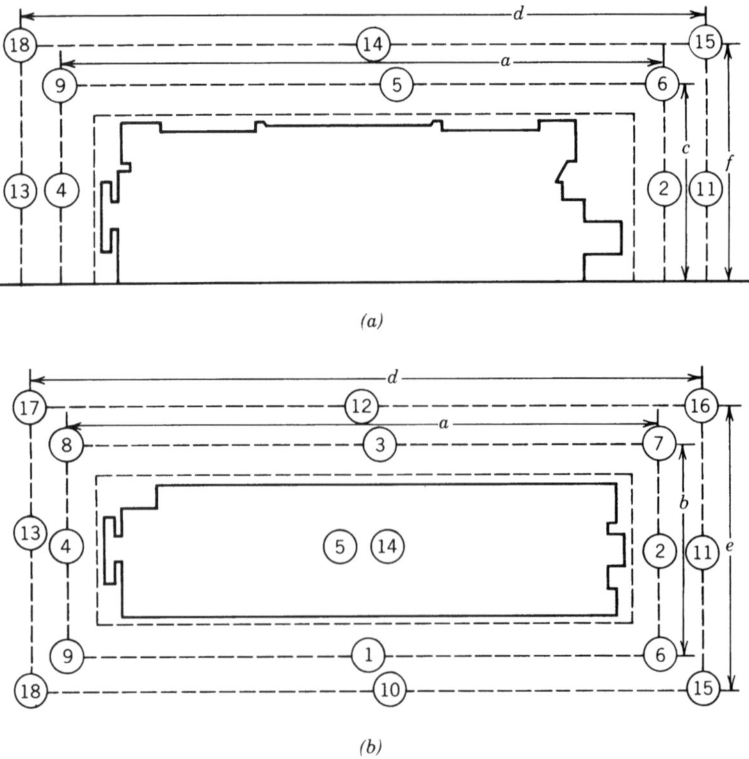

Figure 1 Microphone locations: (*a*) side view; (*b*) plan view.

where $\overline{L_p}$ = average octave-band sound pressure level over area S_1 corrected for both background sound and environment

$\overline{L_{p1}}$ = average octave-band sound pressure level over area S_1 corrected for background sound only

C = environmental correction

$\overline{L_{p2}}$ = average octave-band sound pressure level over area S_2 corrected for background sound

As an alternative, the environmental correction C may be obtained from Fig. 2.

Sound power levels in each octave band of interest may be calculated by the equation

$$L_W = \overline{L_p} + 10 \log \left[\frac{S_1}{S_0}\right] \tag{12}$$

where L_W = octave-band sound power level, dB

$\overline{L_p}$ = average octave-band sound pressure level over area S_1 corrected for both background sound and environment

S_1 = area of measurement surface S_1, m^2

S_0 = 1 m^2

For simplicity, this equation can be written

$$L_W = \overline{L_p} + 10 \log S_1$$

16 MACHINERY NOISE CONTROL

There are five basic methods used to reduce noise: sound absorption, sound isolation, vibration isolation, vibration damping, and mufflers. In most cases several of the available methods are used in combination to achieve a satisfactory solution. Actually, most sound-absorbing materials provide some isolation, although it may be very small, and most sound-isolating

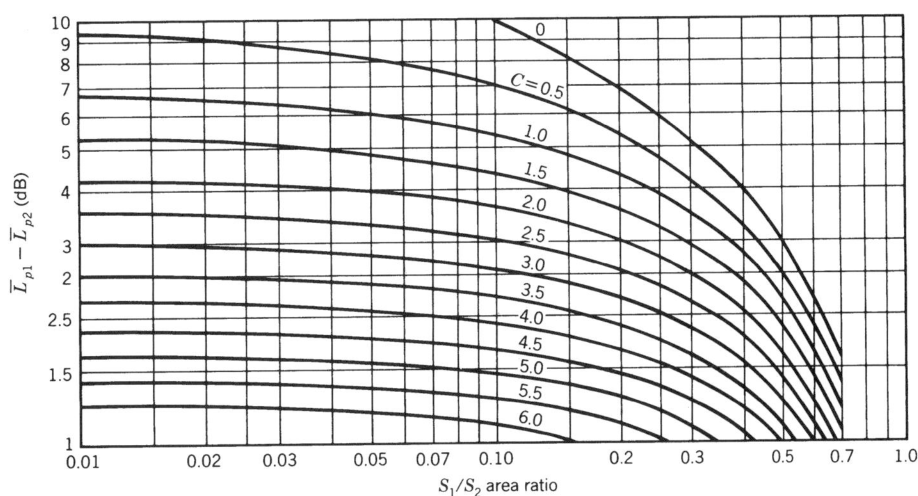

Figure 2 S_1S_2 area ratio.

materials provide some absorption, even though it may be negligible. Many mufflers rely heavily on absorption, although they are classified as a separate means of sound control.

17 SOUND ABSORPTION

The sound-absorbing ability of a material is given in terms of an absorption coefficient, designated by α. The absorption coefficient is defined as the ratio of the energy absorbed by the surface to the energy incident on the surface. Therefore, α can be anywhere between 0 and 1. When $\alpha = 0$, all the incident sound energy is reflected; when $\alpha = 1$, all the energy is absorbed.

The value of the absorption coefficient depends on the frequency. Therefore, when specifying the sound-absorbing qualities of a material, either a table or a curve showing α as a function of frequency is required. Sometimes, for simplicity, the acoustical performance of a material is stated at 500 Hz only, or by a noise reduction coefficient (NRC) that is obtained by averaging, to the nearest multiple of 0.05, the absorption coefficients at 250, 500, 1000, and 2000 Hz.

The absorption coefficient varies somewhat with the angle of incidence of the sound wave. Therefore, for practical use, a statistical average absorption coefficient at each frequency is usually measured and stated by the manufacturer. It is often better to select a sound-absorbing material on the basis of its characteristics for a particular noise rather than by its average sound-absorbing qualities.

Sound absorption is a function of the length of the path relative to the wavelength of the sound, and not the absolute length of the path of sound in the material. This means that at low frequencies the thickness of the material becomes important, and absorption increases with thickness. Low-frequency absorption can be improved further by mounting the material at a distance of one-quarter wavelength from a wall, instead of directly on it.

Table 5 shows absorption coefficients of various materials used in construction.

Table 5 Absorption Coefficients

Material	125 cps	250 cps	500 cps	1000 cps	2000 cps	4000 cps
Brick, unglazed	0.03	0.03	0.03	0.04	0.05	0.07
Brick, unglazed, painted	0.01	0.01	0.02	0.02	0.02	0.03
Concrete block	0.36	0.44	0.31	0.29	0.39	0.25
Concrete block, painted	0.10	0.05	0.06	0.07	0.09	0.08
Concrete	0.01	0.01	0.015	0.02	0.02	0.02
Wood	0.15	0.11	0.10	0.07	0.06	0.07
Glass, ordinary window	0.35	0.25	0.18	0.12	0.07	0.04
Plaster	0.013	0.015	0.02	0.03	0.04	0.05
Plywood	0.28	0.22	0.17	0.09	0.10	0.11
Tile	0.02	0.03	0.03	0.03	0.03	0.02
6 lb/ft^2 fiberglass	0.48	0.82	0.97	0.99	0.90	0.86

The sound absorption of a surface, expressed in either square feet of absorption or sabins, is equal to the area of the surface, in square feet, times the absorption coefficient of the material on the surface.

The average absorption coefficient $\bar{\alpha}$ is calculated as follows:

$$\bar{\alpha} = \frac{\alpha_1 S_1 + \alpha_2 S_2 + \cdots + \alpha_n S_n}{S_1 + S_2 + \cdots + S_n} \tag{13}$$

where $\bar{\alpha}$ = average absorption coefficient
$\alpha_1, \alpha_2, \alpha_n$ = absorption coefficients of materials on various surfaces
S_1, S_2, S_n = areas of various surfaces

18 NOISE REDUCTION DUE TO INCREASED ABSORPTION IN ROOM

A machine in a large room radiates noise that decreases at a rate inversely proportional to the square of the distance from the source. Soon after the machine is started the sound wave impinges on a wall. Some of the sound energy is absorbed by the wall and some is reflected. The sound intensity will not be constant throughout the room. Close to the machine the sound field will be dominated by the source, almost as though it were in a free field, while farther away the sound will be dominated by the diffuse field, caused by sound reflections. The distance where the free-field and diffuse-field conditions control the sound depends on the average absorption coefficient of the surfaces of the room and the wall area. This critical distance can be calculated by the following equation:

$$r_c = 0.2\sqrt{R} \tag{14}$$

where r_c = distance from source, m
R = room constant of the room, m^2

The room constant is equal to the product of the average absorption coefficient of the room and the total internal area of the room divided by the quantity 1 minus the average absorption coefficient. That is,

$$R = \frac{S_t \bar{\alpha}}{1 - \bar{\alpha}} \tag{15}$$

where R = room constant, m^2
$\bar{\alpha}$ = average absorption coefficient
S_t = total area of the room, m^2

Essentially free-field conditions exist farther from a machine in a room with a large room constant than they do in a room with a small room constant.

The distance r_c determines where absorption will reduce noise in the room. An operator standing close to a noisy machine will not benefit by adding sound-absorbing material to the walls and ceiling. Most of the noise heard by the operator is radiated directly by the machine, and very little is reflected noise. On the other hand, listeners farther away, at distances greater than r_c, will benefit from the increased absorption.

The noise reduction in those areas can be estimated by the following equation:

$$\text{NR} = 10 \log \frac{\bar{\alpha}_2 S}{\bar{\alpha}_1 S} \tag{16}$$

where NR = far-field noise reduction, dB
$\overline{\alpha}_1 S$ = room absorption before treatment
$\overline{\alpha}_2 S$ = room absorption after treatment

Equation (16) shows that doubling the absorption will reduce noise by 3 dB. It requires another doubling of the absorption to get another 3-dB reduction. This is much more difficult than getting the first doubling and considerably more expensive.

19 SOUND ISOLATION

Noise may be reduced by placing a barrier or wall between a noise source and a listener. The effectiveness of such a barrier is described by its transmission coefficient.

The sound transmission coefficient of a partition is defined as the fraction of incident sound transmitted through it.

Sound transmission loss is a measure of the sound-isolating ability and is equal to the number of decibels by which sound energy is reduced in transmission through a partition. By definition, it is 10 times the logarithm to the base 10 of the reciprocal of the sound transmission coefficient. That is,

$$\mathrm{TL} = 10 \log \frac{1}{\tau} \tag{17}$$

where TL = transmission loss, dB
τ = transmission coefficient

Transmission of sound through a rigid partition or solid wall is accomplished mainly by the forced vibration of the wall. That is, the partition is forced to vibrate by the pressure variations in the sound wave.

Under certain conditions porous materials can be used to isolate high-frequency sound, and, in general, the loss provided by a uniform porous material is directly proportional to the thickness of the material. For most applications, however, sound-absorbing materials are very ineffective sound isolators because they have the wrong characteristics. They are porous, instead of airtight, and they are lightweight, instead of heavy. The transmission loss of nonporous materials is determined by weight per square foot of surface area and how well all cracks and openings are sealed. Transmission loss is affected also by dynamic bending stiffness and internal damping. Table 6 shows the transmission loss of various materials used in construction.

20 SINGLE PANEL

The simplest type of sound-isolating barrier is a single, homogeneous, nonporous partition. In general, the transmission loss of a single wall of this type is proportional to the logarithm of the mass. Its isolating ability also increases with frequency, and the approximate relationship is given by the following equation:

$$\mathrm{TL} = 20 \log W + 20 \log f - 33 \tag{18}$$

where TL = transmission loss, dB
W = surface weight, $\mathrm{lb/ft^2}$
f = frequency, Hz

Table 6 Transmission Loss of Building Materials

Item	TL
Hollow-core door (3⁄16-in. panels)	15
1¾-in. solid-core oak door	20
2½-in. heavy wood door	25–30
4-in. cinder block	20–25
4-in. cinder block, plastered	40
4-in. cinder slab	40–45
4-in. slab-suspended concrete, plastered	50
Two 4-in. cinder blocks—4-in. air space	55
4-in. brick	45
4-in. brick, plastered	47
8-in. brick, plastered	50
Two 8-in. cinder block—4-in. air space	57

This means that the transmission loss increases 6 dB each time the weight is doubled and 6 dB each time the frequency is doubled. In practice, these numbers are each about 5 dB, instead of 6 dB.

In general, a single partition or barrier should not be counted on to provide noise reduction of more than about 10 dB.

21 COMPOSITE PANEL

Many walls or sound barriers are made of several different materials. For example, machinery enclosures are commonly constructed of sheet steel, but they may have a glass window to observe instruments inside the enclosure. The transmission coefficients of the two materials are different. Another example is when there are necessary cracks or openings in the enclosure where it fits around a rotating shaft. In this case, the transmission loss of the opening is zero, and the transmission coefficient is 1.0.

The effectiveness of such a wall is related to both the transmission coefficients of the materials in the wall and the areas of the sections. A large area transmits more noise than a small one made of the same material. Also, more noise can be transmitted by a large area with a relatively small transmission coefficient than by a small one with a comparatively high transmission coefficient. On the other hand, a small area with a high transmission coefficient can ruin the effectiveness of an otherwise excellently designed enclosure. Both transmission coefficients and areas must be controlled carefully.

The average transmission coefficient of a composite panel is

$$\bar{\tau} = \frac{\tau_1 S_1 + \tau_2 S_2 + \tau_3 S_3 + \cdots + \tau_n S_n}{S_1 + S_2 + S_3 + \cdots + S_n} \tag{19}$$

where $\bar{\tau}$ = average transmission coefficient
$\tau_1, \ldots, \tau_n$ = transmission coefficients of the various areas
$S_1, \ldots, S_n$ = various areas

Figure 3 shows how the transmission loss of a composite wall or panel may be deter-

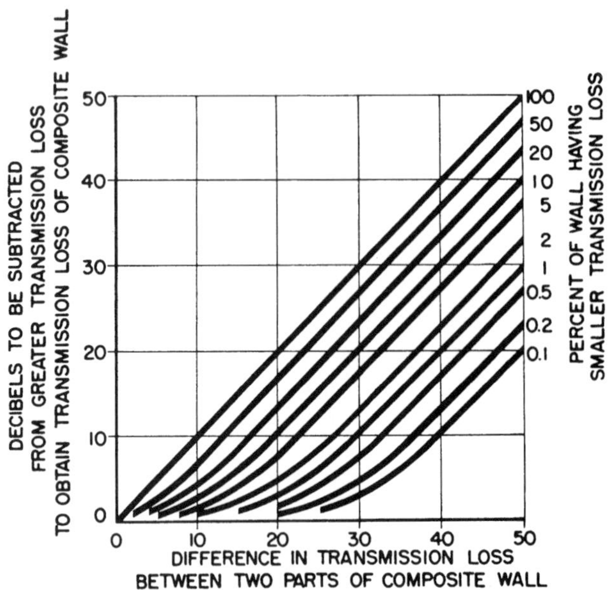

Figure 3 Decibels to be subtracted from greater transmission loss to obtain transmission loss of composite wall. The difference in transmission loss between two parts of composite wall. Percentage of wall having smaller transmission loss.

mined from the transmission loss values of its parts. It also shows the damaging effect of small leaks in the enclosure. In this case the transmission loss of the leak opening is zero.

A leak of only 0.1% in an expensive, high-quality door or barrier constructed of material with a transmission loss of 50 dB would reduce the transmission loss by 20 dB, resulting in a TL of only 30 dB instead of 50 dB. A much less expensive 30-dB barrier would be reduced by only 3 dB, resulting in a TL of 27 dB. This shows that small leaks are more damaging to a high-quality enclosure than to a lower quality one.

22 ACOUSTIC ENCLOSURES

When machinery noise must be reduced by 20 dB or more, it is usually necessary to use complete enclosures. It must be kept in mind that the actual decrease in noise produced by an enclosure depends on other things as well as the transmission loss of the enclosure material. Vibration resonances must be avoided, or their effects must be reduced by damping; structural and mechanical connections must not be permitted to short circuit the enclosure; and the enclosure must be sealed as well as possible to prevent acoustic leaks. In addition, the actual noise reduction depends on the acoustic properties of the room in which the enclosure is located. For this reason, published data on transmission loss of various materials should not be assumed to be the same as the noise reduction that will be obtained when using those materials in enclosures. How materials are used in machinery enclosures is just as important as which materials are used.

A better description of the performance of an acoustic enclosure is given by its noise reduction NR, which is defined as the difference in sound pressure level between the enclosure and the receiving room. The relation between noise reduction and transmission loss is given by the following equation:

$$NR = TL - 10 \log \left[\frac{1}{4} + \frac{S_{wall}}{R_{room}} \right] \tag{20}$$

where NR = difference in sound pressure level, in dB, between the enclosure and the receiving room
TL = transmission loss of the enclosure walls, dB
S_{wall} = area of the enclosure walls, ft^2
R_{room} = room constant of the receiving room, ft^2

Equation (20) indicates that if the room constant is very large, like it would be outdoors, or in a very large area with sound-absorbing material on the walls and ceiling, the NR could exceed the TL by almost 6 dB. In most industrial areas the NR is approximately equal to or several decibels less than the TL.

If there is no absorption inside the enclosure, and it is highly reverberant, like smooth sheet steel, sound reflects back and forth many times. As the noise source continues to radiate noise, with none of it being absorbed, the noise continues to increase without limit. Theoretically, with zero absorption, the sound level will increase to such a value that no enclosure can contain it.

Practically, this condition cannot exist; there will always be some absorption present, even though it may be very little. However, the sound level inside the enclosure will be greater than it would be without the enclosure. For this reason, the sound level inside the enclosure should be assumed to equal the actual noise source plus 10 dB, unless a calculation shows it to be otherwise.

When absorbing material is added to the inside of the enclosure, sound energy decreases each time it is reflected.

An approximate method for estimating the noise reduction of an enclosure is

$$NR = 10 \log \left[1 + \frac{\overline{\alpha}}{\overline{\tau}} \right] \tag{21}$$

where NR = noise reduction, dB
$\overline{a}$ = average absorption coefficient of the inside of the enclosure
$\overline{\tau}$ = average transmission coefficient of the enclosure

Equation (21) shows that in the theoretical case where there is no sound absorption, there is no noise reduction.

23 DOUBLE WALLS

A 4-in.-thick brick wall has a transmission loss of about 45 dB. An 8-in.-thick brick wall with twice as much weight has a transmission loss of about 50 dB. After a certain point has been reached it is found to be impractical to try to obtain higher isolation values simply by doubling the weight, since both the weight and the cost become excessive, and only a 5-dB improvement is gained for each doubling of weight.

An increase can be obtained, however, by using double-wall construction. That is, two 4-in.-thick walls separated by an air space are better than one 8-in. wall. However, noise radiated by the first panel can excite vibration of the second one and cause it to radiate noise. If there are any mechanical connections between the two panels, vibration of one directly couples to the other, and much of the benefit of double-wall construction is lost.

There is another factor that can reduce the effectiveness of double-wall construction. Each of the walls represents a mass, and the air space between them acts as a spring. This mass–spring–mass combination has a series of resonances that greatly reduce the transmission loss at the corresponding frequencies. The effect of the resonances can be reduced by adding sound-absorbing material in the space between the panels.

24 VIBRATION ISOLATION

There are many instances where airborne sound can be reduced substantially by isolating a vibrating part from the rest of the structure. A vibration isolator in its simplest form is some type of resilient support. The purpose of the isolator may be to reduce the magnitude of force transmitted from a vibrating machine or part of a machine to its supporting structure. Conversely, its purpose may be to reduce the amplitude of motion transmitted from a vibrating support to a part of the system that is radiating noise due to its vibration.

Vibration isolators can be in the form of steel springs, cork, felt, rubber, plastic, or dense fiberglass. Steel springs can be calculated quite accurately and can do an excellent job of vibration isolation. However, they also can have resonances, and high-frequency vibrations can travel through them readily, even though they are effectively isolating the lower frequencies. For this reason, springs are usually used in combination with elastomers or similar materials. Elastomers, plastics, and materials of this type have high internal damping and do not perform well below about 15 Hz. However, this is below the audible range, and, therefore, it does not limit their use in any way for effective sound control.

The noise reduction that can be obtained by installing an isolator depends on the characteristics of the isolator and the associated mechanical structure. For example, the attenuation that can be obtained by spring isolators depends not only on the spring constant, or spring stiffness (the force necessary to stretch or compress the spring one unit of length), but also on the mass load on the spring, the mass and stiffness of the foundation, and the type of excitation.

If the foundation is very massive and rigid, and if the mounted machine vibrates at constant amplitude, the reduction in force on the foundation is independent of frequency. If the machine vibrates at a constant force, the reduction in force depends on the ratio of the exciting frequency to the natural frequency of the system.

When a vibrating machine is mounted on an isolator, the ratio of the force applied to the isolator by the machine to the force transmitted by the isolator to the foundation is called the "transmissibility." That is,

$$\text{Transmissibility} = \frac{\text{transmitted force}}{\text{impressed force}}$$

Under ideal conditions this ratio would be zero. In practice, the objective is to make it as small as possible. This can be done by designing the system so that the natural frequency of the mounted machine is very low compared to the frequency of the exciting force.

If no damping is present, the transmissibility can be expressed by the following equation:

$$T = \frac{1}{1 - (\omega/\omega_n)^2} \tag{22}$$

where T = transmissibility, expressed as a fraction
ω = circular frequency of the exciting force, rad/s
ω_n = the circular frequency of the mounted system, rad/s

When $\omega/\omega_n = 0$, the transmissibility equals 1.0. That is, there is no benefit obtained from the isolator.

If ω/ω_n is greater than zero but less than 1.41, the isolator actually increases the magnitude of the transmitted force. This is called the "region of amplification." In fact, when ω/ω_n equals 1.0, the theoretical amplitude of the transmitted force goes to infinity, since this is the point where the frequency of the disturbing force equals the system natural frequency.

Equation (22) indicates that the transmissibility becomes negative when ω/ω_n is greater than 1.0. The negative number is simply due to the phase relation between force and motion, and it can be disregarded when considering only the amount of transmitted force.

Since vibration isolation is achieved only when ω/ω_n is greater than 1.41, the equation for transmissibility can be written so that T is positive:

$$T = \frac{1}{(\omega/\omega_n)^2 - 1}$$

Also, since $\omega = 2\pi f$,

$$T = \frac{1}{(f/f_n)^2 - 1} \tag{23}$$

The static deflection of a spring when stretched or compressed by a weight is related to its natural frequency by the equation

$$f_n = 3.14\sqrt{\frac{1}{d}} \tag{24}$$

where f_n = natural frequency, Hz
d = deflection, in.

When this is substituted in the equation for transmissibility, Eq. (23), it can be shown that

$$d = \left(\frac{3.14}{f}\right)^2\left(\frac{1}{T} + 1\right) \tag{25}$$

This shows that the transmissibility can be determined from the deflection of the isolator due to its supported load.

Equations (23) and (25) can be plotted, as shown in Fig. 4, for convenience in selecting isolator natural frequencies or deflections.

For critical applications, the natural frequency of the isolator should be about one-tenth to one-sixth of the disturbing frequency. That is, the transmissibility should be between 1 and 3%. For less critical conditions, the natural frequency of the isolator should be about one-sixth to one-third of the driving frequency, with transmissibility between 3 and 12%.

25 VIBRATION DAMPING

Complex mechanical systems have many resonant frequencies, and whenever an exciting frequency is coincident with one of the resonant frequencies, the amplitude of vibration is limited only by the amount of damping in the system. If the exciting force is wide band, several resonant vibrations can occur simultaneously, thereby compounding the problem. Damping is one of the most important factors in noise and vibration control.

There are three kinds of damping. Viscous damping is the type that is produced by viscous resistance in a fluid, for example, a dashpot. The damping force is proportional to

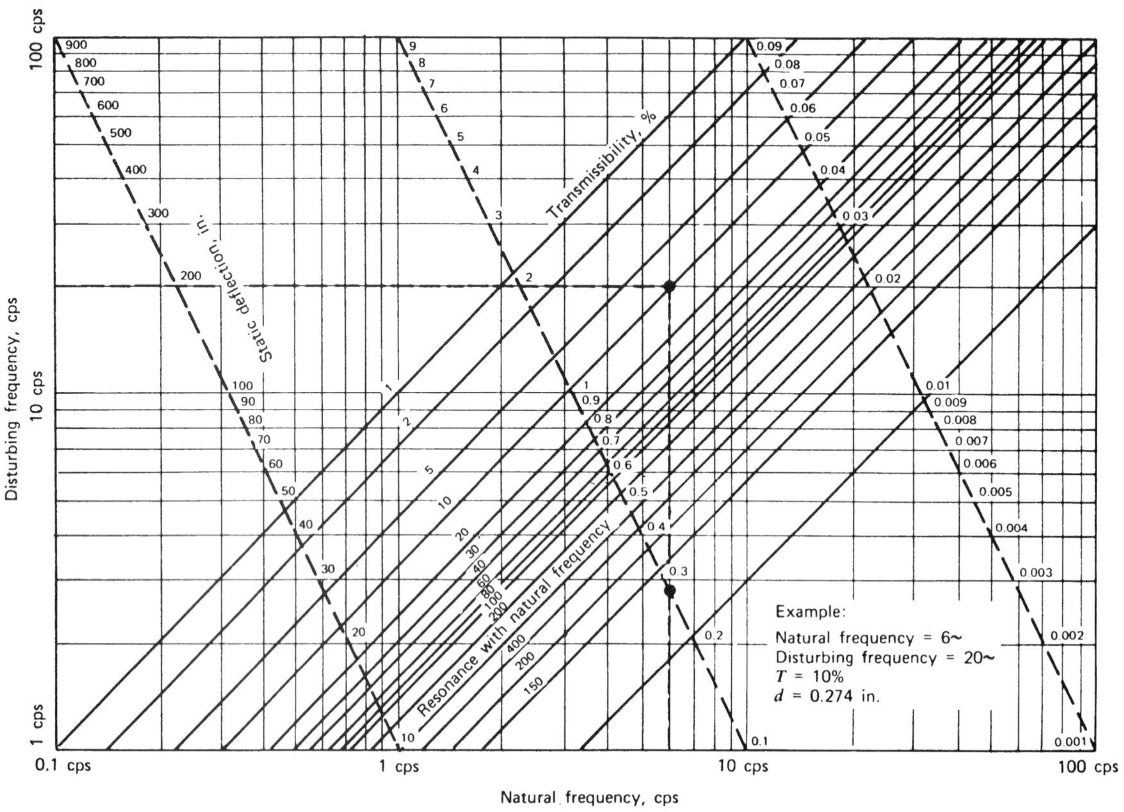

Figure 4 Transmissibility of flexible mountings: $T = 1/[(\omega/\omega)^2 - 1]$.

velocity. Dry friction, or Coulomb damping, produces a constant damping force, independent of displacement and velocity. The damping force is produced by dry surfaces rubbing together, and it is opposite in direction to that of the velocity. Hysteresis damping, also called material damping, produces a force that is in phase with the velocity but is proportional to displacement. This is the type of damping found in solid materials, such as elastomers, widely used in sound control.

A large amount of noise radiated from machine parts comes from vibration of large areas or panels. These parts may be integral parts of the machine or attachments to the machine. They can be flat or curved, and vibration can be caused by either mechanical or acoustic excitation. The radiated noise is a maximum when the parts are vibrating in resonance.

When the excitation is mechanical, vibration isolation may be all that is needed. In other instances, the resonant response can be reduced by bonding a layer of energy-dissipating polymeric material to the structure. When the structure bends, the damping material is placed alternately in tension and compression, thus dissipating the energy as heat.

This extensional or free-layer damping is remarkably effective in reducing resonant vibration and noise in relatively thin, lightweight structures such as panels. It becomes less effective as the structure stiffness increases, because of the excessive increase in thickness of the required damping layer.

In a vibrating structure, the amount of energy dissipated is a function of the amount of energy necessary to deflect the structure, compared to that required to deflect the damping material. If 99% of the vibration energy is required to deflect the structure, and 1% is required to deflect the damping layer, then only 1% of the vibration energy is dissipated.

Resonant vibration amplitude in heavier structures can be controlled effectively by using constrained-layer damping. In this method, a relatively thin layer of viscoelastic damping material is constrained between the structure and a stiff cover plate. Vibration energy is removed from the system by the shear motion of the damping layer.

26 MUFFLERS

Silencers, or mufflers, are usually divided into two categories: absorptive and reactive. The absorptive type, as the name indicates, removes sound energy by the use of sound-absorbing materials. They have relatively wide-band noise reduction characteristics and are usually applied to problems associated with continuous spectra, such as fans, centrifugal compressors, jet engines, and gas turbines. They are also used in cases where a narrow-band noise predominates but the frequency varies because of a wide range of operating speed.

A variety of sound-absorbing materials are used in many different configurations, determined by the level of the unsilenced noise and its frequency content, the type of gas being used, the allowable pressure drop through the silencer, the gas velocity, gas temperature and pressure, and the noise criterion to be met.

Fiberglass or mineral wool with density approximately 0.5–6.0 lb/ft^3 is frequently used in absorptive silencers. These materials are relatively inexpensive and have good sound-absorbing characteristics. They operate on the principle that sound energy causes the material fibers to move, converting the sound energy into mechanical vibration and heat. The fibers do not become very warm since the sound energy is actually quite low, even at fairly high decibel levels.

The simplest kind of absorptive muffler is a lined duct, where either the absorbing material is added to the inside of the duct walls or the duct walls themselves are made of sound-absorbing material. The attenuation depends on the duct length, thickness of the lining, area of the air passage, type of absorbing material, and frequency of the sound passing through.

The acoustical performance of absorptive mufflers is improved by adding parallel or annular baffles to increase the amount of absorption. This also increases pressure drop through the muffler, so that spacing and area must be carefully controlled.

Reactive mufflers have a characteristic performance that does not depend to any great extent on the presence of sound-absorbing material but utilizes the reflection characteristics and attenuating properties of conical connectors, expansion chambers, side branch resonators, tail pipes, and so on, to accomplish sound reduction.

Expansion chambers operate most efficiently in applications involving discrete frequencies rather than broadband noise. The length of the chamber is adjusted so that reflected waves cancel the incident waves, and since wavelength depends on frequency, expansion chambers should be tuned to some particular frequency. When a number of discrete frequencies must be attenuated, several expansion chambers can be placed in series, each tuned to a particular wavelength.

An effective type of reactive muffler, called a Helmholtz resonator, consists of a vessel containing a volume of air that is connected to a noise source, such as a piping system. When a pure-tone sound wave is propagated along the pipe, the air in the vessel expands and contracts. By proper design of the area and length of the neck and volume of the chamber, sound wave cancellation can be obtained, thereby reducing the tone. This type of resonator produces maximum noise reduction over a very narrow frequency range, but it is possible to combine several Helmholtz resonators on a piping system so that not only will each cancel out at its own frequency but they can be made to overlap so that noise is attenuated over a wider range instead of at sharply tuned points.

Helmholtz resonators are normally located in side branches, and for this reason they do not affect flow in the main pipe.

The resonant frequency of these devices can be calculated by the equation

$$f = \frac{C}{2\pi}\sqrt{\frac{A}{LV}} \tag{26}$$

where f = resonant frequency, Hz
C = speed of sound in the fluid, ft/s
A = cross-sectional area of the neck, ft^2
L = length of the neck, ft
V = volume of the chamber, ft^3

The performance of all types of mufflers can be stated in various ways. Not everyone uses the same terminology, but in general, the following definitions apply:

INSERTION LOSS is defined as the difference between two sound pressure levels measured at the same point in space before and after a muffler is inserted in the system.

DYNAMIC INSERTION LOSS is the same as insertion loss, except that it is measured when the muffler is operating under rated flow conditions. Therefore, the dynamic insertion loss is of more interest than ratings based on no-flow conditions.

TRANSMISSION LOSS is defined as the ratio of sound power incident on the muffler to the sound power transmitted by the muffler. It cannot be measured directly, and it is difficult to calculate analytically. For these reasons the transmission loss of a muffler has little practical application.

ATTENUATION is used to describe the decrease in sound power as sound waves travel through the muffler. It does not convey information about how a muffler performs in a system.

NOISE REDUCTION is defined as the difference between sound pressure levels measured at the inlet of a muffler and those at the outlet.

27 SOUND CONTROL RECOMMENDATIONS

Sound control procedures should be applied during the design stages of a machine, whenever possible. A list of recommendations for noise reduction follows:

1. Reduce horsepower. Noise is proportional to horsepower. Therefore, the machine should be matched to the job. Excess horsepower means excess noise.
2. Reduce speed. Slow-speed machinery is quieter than high-speed machinery.
3. Keep impeller tip speeds low. However, it is better to keep the rpm low and the impeller diameter large than to keep the rpm high and the impeller diameter small, even though the tip speeds are the same.
4. Improve dynamic balance. This decreases rotating forces, structure-borne sound, and the excitation of structural resonances.

5. Reduce the ratio of rotating masses to fixed masses.
6. Reduce mechanical run-out of shafts. This improves the initial static and dynamic balance.
7. Avoid structural resonances. These are often responsible for many unidentified components in the radiated sound. In addition to being excited by sinusoidal forcing frequencies, they can be excited by impacting parts and sliding and rubbing contacts.
8. Eliminate or reduce impacts. Reduce either the mass of impacting parts or their striking velocities.
9. Reduce peak acceleration. Reduce the rate of change of velocity of moving parts by using the maximum time possible to produce the required velocity change and by keeping the acceleration as nearly constant as possible over the available time period.
10. Improve lubrication. Inadequate lubrication is often the cause of bearing noise, structure-borne noise due to friction, and the excitation of structural resonances.
11. Maintain closer tolerances and clearances in bearings and moving parts.
12. Install bearings correctly. Improper installation accounts for approximately half of bearing noise problems.
13. Improve alignment. Improper alignment is a major source of noise and vibration.
14. Use center-of-gravity mounting whenever feasible. When supports are symmetrical with respect to the center of gravity, translational modes of vibration do not couple to rotational modes.
15. Maintain adequate separation between operating speeds and lateral and torsional resonant speeds.
16. Consider the shape of impeller vanes from an acoustic standpoint. Some configurations are noisier than others.
17. Keep the distance between impeller vanes and cutwater or diffuser vanes as large as possible. Close spacing is a major source of noise.
18. Select combinations of rotating and stationary vanes that are not likely to excite strong vibration and noise.
19. Design turning vanes properly. They are a source of self-generated noise.
20. Keep the areas of inlet passages as large as possible and their length as short as possible.
21. Remove or keep at a minimum any obstructions, bends, or abrupt changes in fluid passages.
22. Pay special attention to inlet design. This is extremely important in noise generation.
23. Item 22 applies also to the discharge, but the inlet is more important than the discharge from an acoustic standpoint.
24. Maintain gradual, not abrupt, transition from one area to the next in all fluid passages.
25. Reduce flow velocities in passages, pipes, and so on. Noise can be reduced substantially by reducing flow velocities.
26. Reduce jet velocities. Jet noise is proportional to the eighth power of the velocity.
27. Reduce large radiating areas. Surfaces radiating certain frequencies can often be divided into smaller areas with less radiating efficiency.

28. Disconnect possible sound-radiating parts from other vibrating parts by installing vibration breaks to eliminate metal-to-metal contact.
29. Provide openings or air leaks in large radiating areas so that air can move through them. This reduces pressure build-up and decreases radiated noise.
30. Reduce clearances, piston weights, and connecting rod weights in reciprocating machinery to reduce piston impacts.
31. Apply additional sound control devices, such as inlet and discharge silencers and acoustic enclosures.
32. When acoustic enclosures are used, make sure that all openings are sealed properly.
33. Install machinery on adequate mountings and foundations to reduce structure-borne sound and vibration.
34. Take advantage of all directivity effects whenever possible by directing inlet and discharge openings away from listeners or critical areas.
35. When a machine must meet a particular sound specification, purchase driving motors, turbines, gears, and auxiliary equipment that produce 3–5-dB lower sound levels than the machine alone. This ensures that the combination is in compliance with the specification.

CHAPTER 36
NONDESTRUCTIVE INSPECTION

Robert L. Crane and Jeremy S. Knopp
Air Force Research Laboratory
Materials Directorate
Wright Patterson Air Force Base
Dayton, Ohio

1 INTRODUCTION

This chapter deals with the nondestructive inspection of materials, components, and structures. The term *nondestructive inspection* (NDI) or *nondestructive evaluation* (NDE) is defined as that class of physical and chemical tests that permit the detection and/or measurement of significant properties or the detection of defects in a material without impairing its usefulness. The inspection process is often complicated by the fact that many materials are anisotropic and most NDI techniques were developed for isotropic materials such as metals. The added complication due to the anisotropy usually means that an inspection is more complicated than it would be with isotropic materials.

Inspection of complex materials and structures is frequently carried out by comparing the expected inspection data with a standard and noting any significant deviations. This means a well-defined standard must be available for calibration of the inspection instrumentation. Furthermore, standards also must contain implanted flaws that mimic those that naturally occur in the material or structure to be inspected. Without a well-defined standard to calibrate the inspection process, the analysis of NDI results can be significantly in error. For example, to estimate the amount of porosity in a cast component from ultrasonic measurements, standard calibration specimens with calibrated levels of porosity must be available to calibrate the instrumentation. Without such standards, estimation of porosity from ultrasonic data is a highly speculative process.

This chapter covers some important and some less well-known NDI tests. Since information on the less frequently used tests is not generally in standard texts, additional sources of information are listed in the References.

Inspection instrumentation must possess four qualities in order to receive widespread acceptance in the NDI community:

1. *Accuracy.* The instrument must accurately measure a property of the material or structure that can be used to infer either its properties or the presence of flaws.
2. *Reliability.* The instrument must be highly reliable, i.e., it must consistently detect and quantify flaws or a property with a high degree of reliability. If an instrument is not reliable, then it may not detect flaws that can lead to failure of the component, or it may indicate the presence of a flaw where none exists. The detection of a phantom flaw can mean that an adequate component is rejected, which is a costly error.
3. *Simplicity.* The most frequently used instruments are those used by factory or repair technicians. The inspection community rarely uses highly skilled operators due to the cost constraints.
4. *Low Cost.* An instrument need not be low cost in an absolute sense. Instead, it must be inexpensive relative either to the value of the component under test or to the cost of a failure or aborted mission. For example, in the aircraft industry as much as 12% of the value of the component may be spent on inspection of a flight-critical aircraft component.

1.1 Information on Inspection Methods

To the engineer confronted by a new inspection requirement, there may arise the question of where to find pertinent information regarding an inspection procedure and its interpretation. Fortunately, many potential sources of information about instrumentation and techniques are available for NDI, and a brief examination of this literature is presented here. Many of these references were generated because of the demands of materials used in flight-critical aerospace structures. In this chapter, we will refer only to scientific and engineering books and journals that one would reasonably expect to find in a well-provisioned library. With the rise of the Internet, there are now many electronic sources of information available on the World Wide Web. These include library catalogues, societal home pages, online journals devoted to inspection, home pages of instrument manufacturers with online demonstrations of their capabilities and inspection services, inspection software, and online forums devoted to solving inspection problems. The References provides many such sources. However, with new electronic sources appearing daily, it is only a brief snapshot of those available at the beginning of the twenty-first century. For those new to the technology, American Society for Testing and Materials (ASTM) standards are particularly valuable because they give very detailed directions on many NDE techniques. More importantly, they are widely accepted standards for inspections. The References also provides sources for those situations where standard inspection methods are not sufficient to detect the material condition of interest.

General NDE Reference Books
General overviews to NDE techniques are provided in Refs. 1–22. The reader will note that some of these citations are not recent, but they are included because of their value to the engineer who does not possess formal training in the latest inspection technologies. Additionally, some older works were included because of their clarity of presentation, completeness, or usefulness to the inspection of complex structures.

NDE Journals
The periodical literature is often a source of the latest research results for new or modified inspection methodologies.[5,23–28] Some excellent journals are no longer available but are still a valuable source of information or may contain data available nowhere else. Whenever possible, World Wide Web addresses are provided to give the reader ready access to this material.

1.2 Electronic References

There are many useful electronic references for those working in NDE technology. Only a few of the many useful sites on the World Wide Web, are included here. Many sites contain links to other sites that contain information on a special topic of interest to the reader. Because the Web is constantly being updated, the list in the References represents a very brief snapshot of the information available to the NDE community. Some useful sites associated with government agencies were not included due to space limitations. The Web addresses provided are associated with NDE societies,[5,6,27,30–36] institutes,[7,8,10,38–40] government agencies,[25,38,41] and general interest sites.[26,33,35,42–44] There are also many references for the reader interested in using or modifying existing NDE techniques.[31,45–47]

1.3 Future NDE Capabilities

At this point, the reader might be tempted to ask if there are new technologies on the horizon that will enable more cost-effective, anticipatory inspection or monitoring of materials and structures. The answer to this is an emphatic yes. There are new developments in solid-state detectors that should significantly affect both inspection capability and cost. For example, optical and X-ray detectors now give the inspector the ability to rapid scan large areas of structures for defects. Many new developments in these areas are the outgrowth of advances in noninvasive medical imaging. By coupling this technology with computer algorithms that search an image, the inspection of large areas can be automated, providing more accurate inspections with much less operator fatigue. Hopefully this technological advance will remove much of the drudgery of detecting the rather small number of flaws in an otherwise large population of satisfactory components.

The area of data fusion is just beginning to be explored in the NDE field. This means that data collected with one technique can be combined with another technique to detect a range of flaws not detected when either is used independently. Data from several techniques can then be coupled at the basic physics level to provide a more complete description of the microstructural details of a material than is now possible.

Finally, the development of new semiconductor-based devices microelectromechanical systems (MEMS) and radio-frequency identification (RFID) allows the implantation of monitoring devices into a material at the time of manufacture to enable real-time structural health monitoring. These devices will permit the inspector to detect and quantify material or structural degradation remotely. This should also enable management of the components and structures for optimum usage over their lifetimes. Remote inspection and tracking of material degradation should reduce the burden of inspection while giving the inspector the ability to examine areas of structure that are now called "hidden." For more information about this rapidly evolving area the reader is referred to the literature.[29,33,48–50]

This is a brief review of the commonly used NDI methods listed in Table 1 along with types of flaws that each method detects and the advantages and disadvantages of each technique. For detailed information regarding the capabilities of any particular method, the reader is referred to the literature. A good place to start any search for the latest NDE technology is the home page of the American Society for Nondestructive Testing.[6]

Table 1 Capabilities of the Common NDI Methods

Method	Typical Flaws Detected	Typical Application	Advantages	Disadvantages
Radiography	Voids, porosity, inclusions, and cracks	Castings, forging, weldments, and structural assemblies	Detects internal flaws; useful on a wide variety of geometric shapes; portable; provides a permanent record	High cost; insensitive to thin laminar flaws, such as tight fatigue cracks and delaminations; potential health hazard
Liquid penetrants technique	Cracks, gouges, porosity, laps, and seams open to a surface	Castings, forging, weldments, and components subject to fatigue or stress–corrosion cracking	Inexpensive; easy to apply; portable; easily interpreted	Flaw must be open to an accessible surface, level of detectability operator dependent
Eddy current inspection	Cracks and variations in alloy composition or heat treatment, wall thickness, dimensions	Tubing, local regions of sheet metal, alloy sorting, and coating thickness measurement	Moderate cost; readily automated; portable	Detects flaws which change conductivity of metals; shallow penetration; geometry sensitive
Magnetic particles method	Cracks, laps, voids, porosity, and inclusions	Castings, forging, and extrusions	Simple; inexpensive; detects shallow subsurface flaws as well as surface flaws	Useful on ferromagnetic materials only; surface preparation required; irrelevant indications often occur; operator dependent
Thermal testing	Voids or disbonds in both metallic and nonmetallic materials, location of hot or cold spots in thermally active assemblies	Laminated structures, honeycomb, and electronic circuit boards	Produces a thermal image that is easily interpreted	Difficult to control surface emissivity and poor discrimination between flaw types
Ultrasonic methods	Cracks, voids, porosity, inclusions and delaminations, and lack of bonding between dissimilar materials	Composites, forgings, castings, and weldments and pipes	Excellent depth penetration; good sensitivity and resolution; can provide permanent record	Requires acoustic coupling to component; slow; interpretation of data is often difficult

2 LIQUID PENETRANTS

Liquid penetrants are used to detect surface-connected discontinuities, such as cracks, porosity, and laps, in solid, nonporous materials.[51] The method uses a brightly colored visible or fluorescent penetrating liquid which is applied to the surface of a cleaned part. During a specified "dwell time" the liquid enters the discontinuity and is then removed from the surface of the part in a separate step. The penetrant is drawn from the flaw to the surface by a developer to provide an indication of surface-connected defects. This process is depicted schematically in Figs. 1–4. A penetrant indication of a flaw in a turbine blade is shown in Fig. 5.

2.1 Penetrant Process

Both technical societies and military specifications require a classification system for penetrants. Society documents (typically ASTM E165)[51] categorize penetrants into visible and fluorescent, depending on the type of dye used. In each category, there are three types, depending on how the excess penetrant is removed from the part. These are water washable, postemulsifiable, and solvent removable.

The first step in penetrant testing (PT) or inspection is to clean the part. This critical step is one of the most neglected phases of the PT procedure. Since PT only detects flaws that are open to the surface, the flaw and part surface must be free of dirt, grease, oil, water, chemicals, and other foreign materials that might block the penetrant's entrance into a defect. Typical cleaning procedures use vapor degreasers, ultrasonic cleaners, alkaline cleaners, or solvents.

After the surface is clean, a liquid penetrant is applied to the part by dipping, spraying, or brushing. In this step, the penetrant on the surface is wicked into the flaw. In the case of tight or narrow surface openings, such as fatigue cracks, the penetrant must be allowed to remain on the part for a minimum of 30 min to completely fill the flaw. High-sensitivity fluorescent dye penetrants are used for this type of inspection.

After the dwell time, excess penetrant is removed by one of the processes mentioned previously. For water-based penetrants an emulsifier is sprayed onto the part and again a dwell time is observed. Water is then used to remove the penetrant from the surface of the part. In some cases, the emulsifier is included in the penetrant, so one only needs to wash the part after the penetrant has had time to penetrate the flaw. These penetrants are therefore called "water washable." Of course, the emulsifier reduces the brightness of any flaw indication because it dilutes the penetrant. Ideally, only the surface penetrant is removed with the penetrant in the flaw left undisturbed.

The final step in a basic penetrant inspection is the application of a fine powder developer. This may be applied either wet or dry. The developer aids in wicking the penetrant

Figure 1 Schematic representation of a part surface before cleaning for penetrant inspection.

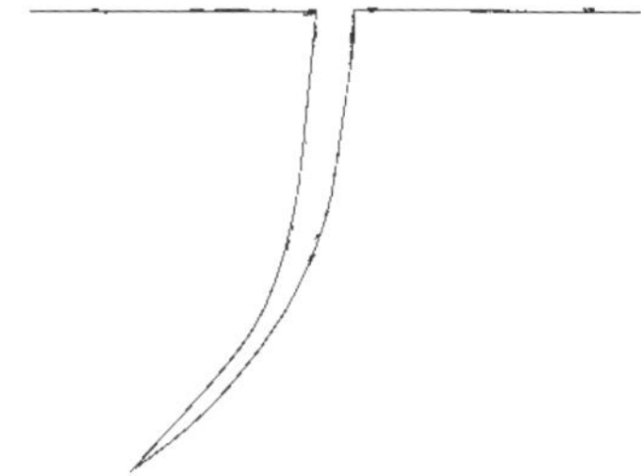

Figure 2 Part surface after cleaning and before penetrant application.

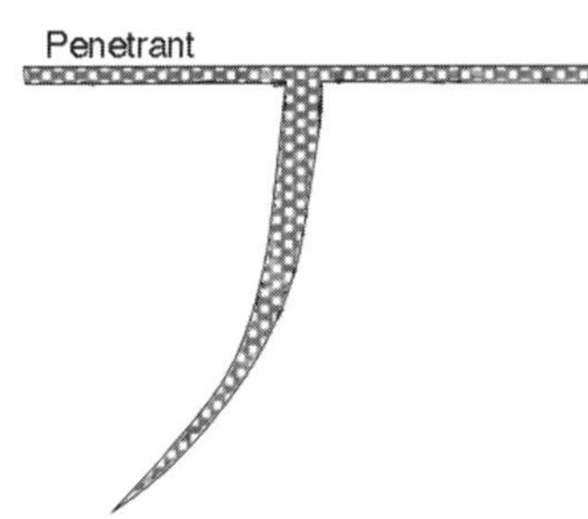

Figure 3 Part after penetrant application.

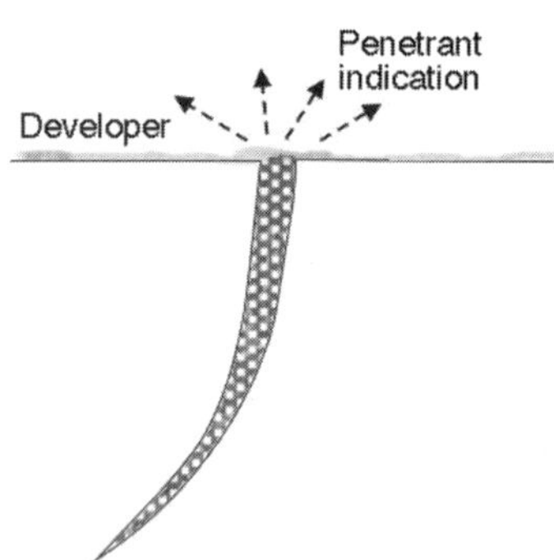

Figure 4 Schematic representation of part after excess penetrant has been removed and the developer has been applied.

from the flaw and provides a suitable background for its detection. The part is then viewed under a suitable illumination—either an ultraviolet or a visible source. A typical fluorescent penetrant indication for a crack in a jet engine turbine blade is shown in Fig. 5.

2.2 Reference Standards

Several reference standards are used to check the effectiveness of liquid penetrant systems. One of the oldest and most often used methods involves applying penetrant to hard chromium-plated brass panels. The panel is bent to place the chromium in tension, producing a series of cracks in the plating. These panels are available in sets containing fine, medium, and coarse cracks. The panels are used to classify penetrant materials by sensitivity and to detect degrading changes in the penetrant process.

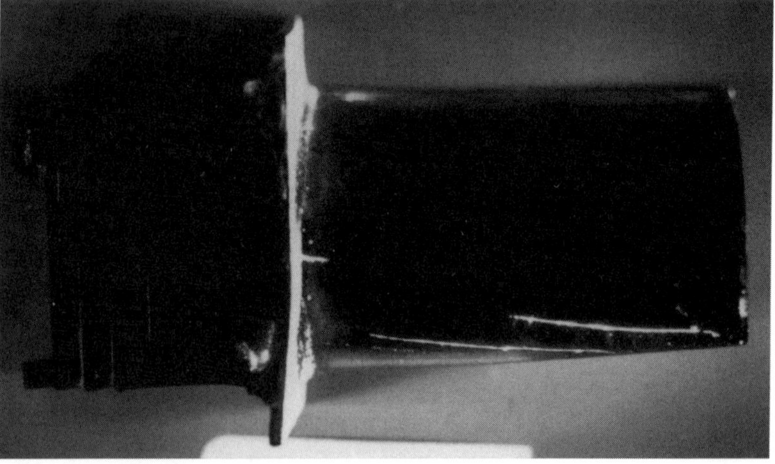

Figure 5 Penetrant indication of a crack running along the edge of a jet engine turbine blade. Ultraviolet illumination causes the extracted penetrant to fluoresce.

2.3 Limitations of Penetrant Inspections

The major limitation of liquid penetrant inspection is that it can only detect flaws that are open to the surface. Other inspection methods must be used to detect subsurface defects. A factor that may inhibit the effectiveness of liquid penetrant inspection is surface roughness. Rough surfaces are likely to produce false indications by trapping penetrant, therefore, PT is not suited to the inspection of porous materials. Other penetrant-like methods are available for porous components—see the discussion of filtered particle inspection in Ref. 51.

3 RADIOGRAPHY

In radiography used in NDE, the projected X-ray attenuations of a multitude of paths through a specimen are recorded as a two-dimensional image on a recording media, usually film. One might ask if the newer solid-state X-ray imaging technologies used in medicine also apply to NDE. The answer is yes, as will be discussed in the latter part of this section. The most used recording medium is still film because it is the simplest to apply and provides a resolution of subtle details not currently available with solid-state detectors. However, this situation may not be the case much longer as rapid progress is being made in the development of solid-state detectors with significantly enhanced resolution capabilities. Therefore, since this chapter is written at the beginning of the twenty-first century, when film usage for inspection is still commonplace, this portion of the chapter approaches radiography from the standpoint of film-based recording. Since most quantitative relationships for film also apply to solid-state detectors, the material presented should be applicable for the near future.

The radiography testing (RT) process is shown schematically in Figure 6. RT records any feature that changes the attenuation of the X-ray beam as it traverses the component. This local change in attenuation produces a change in the intensity of the X-ray beam, which translates into a change in the density, or darkness, on a film. This change in brightness may appear as a distinct shadow or in some cases a delicate shadow on the radiograph. The inspector is greatly aided in detecting a flaw or discrepancy in a part by his or her knowledge of part shape and its influence on the radiographic image. Flaws, which do not change the attenuation of the X-ray beam on passage through the part, are not recorded. For example, a delamination in a laminated specimen is not visible because there is no local change in attenuation of the X-ray beam as it transverses the part. Conversely, flaws that are oriented parallel to the X-ray path do not attenuate the beam as much, allowing more radiation to

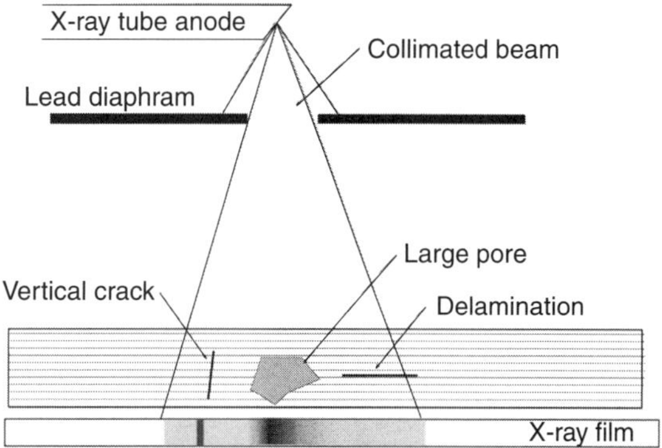

Figure 6 Schematic radiograph of a typical with typical flaws.

expose the film and appearing darker than the surrounding image. An example of a crack in the correct orientation to be visible on a radiograph of a piece of tubing is shown in Fig. 7.

3.1 Generation and Absorption of X-Radiation

X-radiation can be produced from a number of processes. The most common method of generating X-rays is with an electron tube in which a beam of energetic electrons impacts a metal target. As the electrons are rapidly decelerated by this collision, a wide band of X-radiation is produced, analogous to white light. This band of radiation is referred to as *Bremsstrahlung* or breaking radiation. These high-energy electrons produce short-wavelength energetic X-rays. The relationship between the shortest wavelength radiation and the highest voltage applied to the tube is given by

$$\lambda = \frac{12{,}336}{\text{voltage}}$$

where λ is the wavelength in angstroms and is the shortest wavelength of the X-radiation produced. The more energetic the radiation, the more penetrating powers it possesses, and very high energy radiation is used on dense materials such as metals. While it is possible to analytically predict what X-ray energy would provide the best image for a specific material and geometry, a simpler method of determining the optimum X-ray energy is shown in Fig. 8. Note that high-energy X-ray beams are used for dense materials, e.g., steels, or for thick low-density materials, e.g., large plastic parts. An alternative method to using this figure is to use the radiographic equivalence factors given in Table 2.[52] Aluminum is the standard material for X-ray tube voltages below 100 KeV, while steel is the standard above this voltage. When radiographing another material, its thickness is multiplied by the factor in this table to obtain the equivalent thickness of the standard material. The radiographic parameters are set up for this thickness of aluminum or steel. When used in this manner, good radiographs can be obtained for most parts. For example, assume that one must radiograph a 0.75-in.-thick piece of brass with a 400-keV X-ray source. The inspector should multiply the 0.75 in. of brass by the factor of 1.3 to obtain 0.98. This means that an acceptable radiograph of the brass plates would be obtained with the same exposure parameters as would be used for 0.98 in. (approximately 1 in.) of steel.

Radiation for RT can also be obtained from the decay of radioactive sources. In this case, the process is usually referred to as gamma radiography. These radiation sources have several characteristics that differ from X-ray tubes. First, gamma radiation is very nearly monochromatic; that is, the spectrum of radiation contains only one or two dominant energies. Second, the energies of most sources are on the order of millions-of-volts range, making

Figure 7 A radiograph of a crack in end of aluminum tubing.

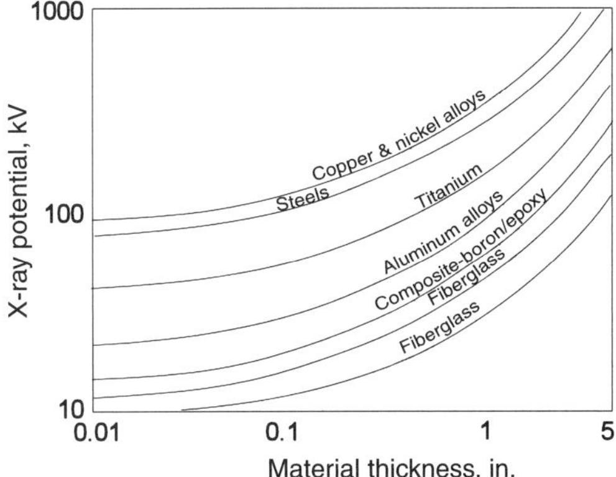

Figure 8 A plot of the X-ray tube voltage versus thickness of several industrial materials.

this source ideal for inspecting highly attenuating materials or very large structures. Third, the small size of these sources permits them to be used in tight locations where an X-ray tube could not fit. Fourth, since the gamma-ray source is continually decaying, adjustments to the exposure time must be made in order to achieve consistent results over time. Finally, the operator must always remember that the source is continually on and is therefore a persistent safety hazard! Aside from these differences, gamma radiography differs little from standard practice, so no further distinction between the two will be given.

3.2 Neutron Radiography

Neutron radiography[53] may useful to inspect some materials and structures. Because the attenuation of neutrons is not related to the elemental composition of the part, some elements

Table 2 Approximate Radiographic Equivalence Factors

Energy Level	100 kV	150 kV	220 kV	250 kV	400 kV	1 MeV	2 MeV	4–25 MeV	^{192}Ir	^{60}Co
Metal										
Magnesium	0.05	0.05	0.08							
Aluminum	0.08	0.12	0.18						0.35	0.35
Aluminum alloy	0.10	0.14	0.18						0.35	0.35
Titanium		0.54	0 54		0 71	0.9	0.9	0.9	0.9	0.9
Iron/all steels	1.0	1.0	1.0	1.0	1.0	1.0	1.0	1.0	1.0	1.0
Copper	1.5	1.6	1.4	1.4	1.4	1.1	1.1	1.2	1.1	1.1
Zinc		1.4	1.3		1.3			1.2	1.1	1.0
Brass		1.4	1.3		1.3	1.2	1.1	1.0	1.1	1.0
Inconel X		1.4	1.3		1.3	1.3	1.3	1.3	1.3	1 3
Monel	1.7		1.2							
Zirconium	2.4	2.3	2.0	1.7	1.5	1.0	1.0	1.0	1.2	1.0
Lead	14.0	14.0	12.0			5.0	2.5	2.7	4.0	2.3
Halfnium			14.0	12.0	9.0	3.0				
Uranium			20.0	16.0	12.0	4.0		3.9	12.6	3.4

Source: From Ref. 52.

can be more easily detected than others. While X-rays are most heavily absorbed by high-atomic-number elements, this is not true of neutrons, as shown in Fig. 9. In Fig. 10 two aluminum panels are bonded with an epoxy adhesive. The reader can discern that hydrogen adsorbs neutrons more than aluminum does, and thus the missing adhesive is easily detectable.

Neutron radiography, however, does have several constraints. First, neutrons do not expose radiographic film and therefore a fluorescing medium is often used to produce light, which exposes the film. The image produced in this manner is not as sharp and well defined as that from X-rays. Second, at present there is no portable high-flux, portable source of neutrons. This means that a nuclear reactor is most often used to supply the neutron radiation. Although neutron radiography has these severe restrictions, at times there is no alternative, and the utility of this method outweighs its expense and complexity.

3.3 Attenuation of X-Radiation

An appreciation of how radiographs are interpreted requires a fundamental understanding of X-ray absorption. The relationship governing this phenomenon is de Beer's law:

$$I = I_0 e^{-\mu x}$$

where I, I_0 = transmitted and incident X-ray beam intensities, respectively
μ = attenuation coefficient of material, cm^{-1}
x = thickness of specimen, cm

Since the attenuation coefficient is a function of both the composition of the specimen and the wavelength of the X-rays, it would be necessary to calculate or measure it for each wavelength used in RT. However, it is possible to calculate the attenuation coefficient of a material for a specific X-ray energy using the mass absorption coefficient μ_m as defined below. The mass absorption coefficients for most elements are readily available for a variety of X-ray energies,[54]

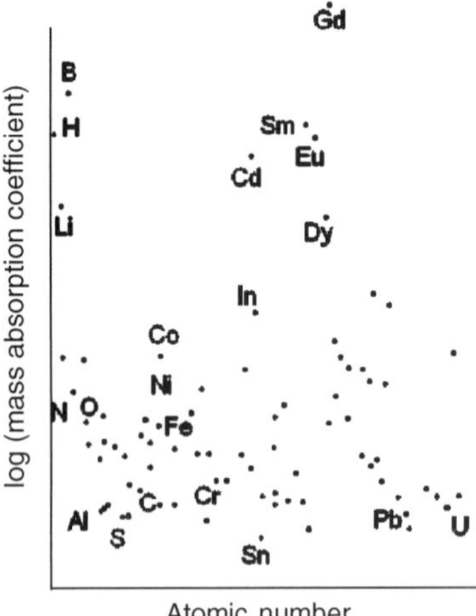

Figure 9 A plot of mass absorption coefficient for neutron radiography versus atomic number.

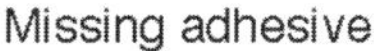

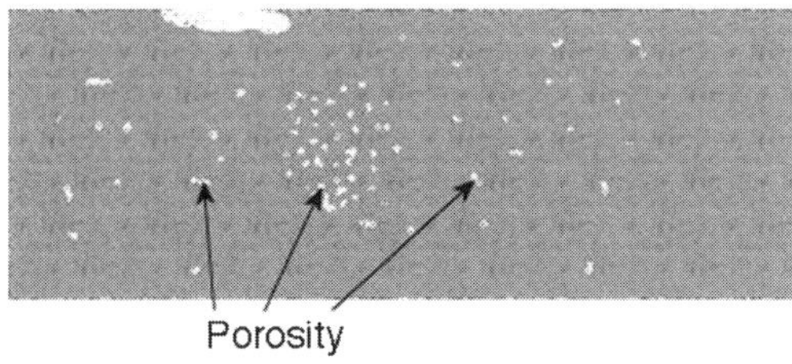

Figure 10 A schematic representation of a neutron radiograph showing flaws in an adhesive bond.

$$\mu_m = \frac{\mu}{\rho}$$

where μ = attenuation coefficient of an element, cm^{-1}
ρ = density, g/cm^3

The mass absorption coefficient for the material is obtained, at a specific X-ray energy, by multiplying the μ_m of each element by its weight fraction in a material and summing these quantities. Multiplying this sum by the density of the material yields its attenuation coefficient for the material. This procedure is often not used in practice because the results are valid only for a narrow band of wavelengths. Radiographic equivalency factors are used instead. This process points out that each element in a material contributes to the attenuation coefficient by an amount proportional to its amount in the material.

3.4 Film-Based Radiography

The classical method of recording an X-ray image is with film. Because of the continued importance of this medium of recording and the fact that much of the technology associated with it is applicable to newer solid-state recording methods, this section explores film radiography in some detail.

The relationship between the darkness produced on an X-ray film and the quantity of radiation impinging on it is shown by log-log plots of darkness, or film density, and relative exposure (Figs. 11 and 12). Varying the time of exposure, intensity of the beam, or specimen thickness changes the density, or darkness, of the image. The slope of the curve along its linear portion is referred to as the film gamma, γ. Film has characteristics that are analogous to electronic devices: The greater the gamma or amplification capability of the film, the smaller its dynamic range—the range of exposures over which density is linearly related to thickness. If it is necessary to use a high-gamma film to detect very subtle flaws in a part with a wide range of thicknesses, then it is necessary to use several different film types in the same cassette or package. In this way, each film will be optimized for flaw detection in a narrow thickness range of the part.

Using this information, one may calculate the minimum detectable flaw size for a specific RT inspection. A simple method is available to check the radiographic procedure to determine if this detectability has been achieved on the film. This method does not ensure that the radiograph was taken with the specimen in the proper orientation; it merely provides a method of checking for proper execution of a radiographic procedure; see Section 3.5.

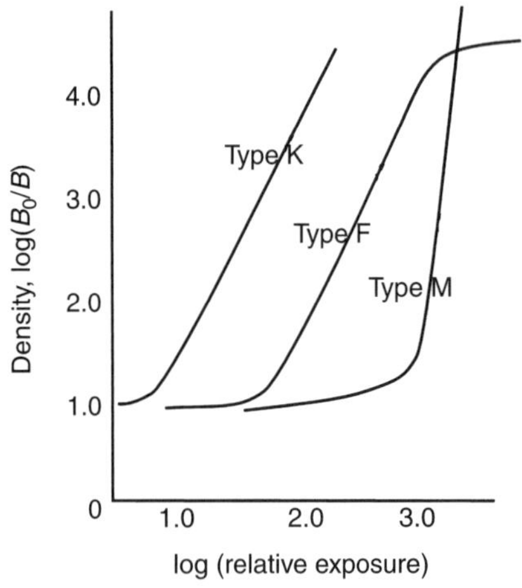

Figure 11 Density or darkness of X-ray film versus relative exposure for three common films.

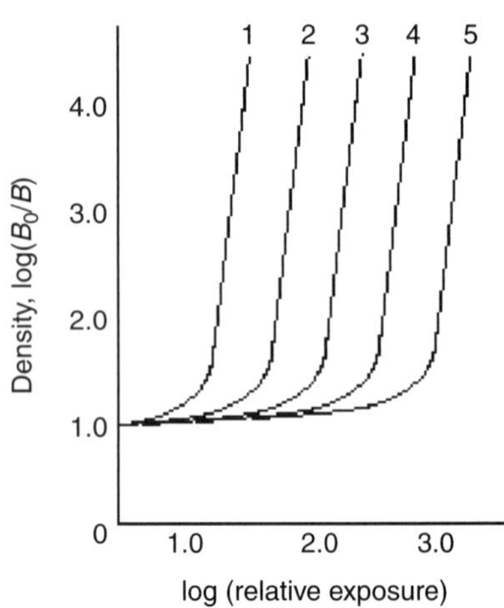

Figure 12 Density versus relative exposure for films that could be used in a multiple film exposure to obtain optimum flaw delectability in a complex part.

Using a knowledge of the minimum density difference that is detectable by the average radiographic inspector, the following equation relates the radiographic sensitivity, S, to radiographic parameters:

$$S = \frac{2.3}{\gamma \mu x}$$

where S is the radiographic sensitivity in percent, γ is the film gamma, μ is the attenuation coefficient of the specimen material, and x is the maximum thickness of the part associated with a particular radiographic film. The radiographer to uses a penetrameter to determine if this sensitivity was achieved. Table 3 give the sensitivity S in percent and the expected RT performance in penetrameter values (see Section 3.5).

3.5 Penetrameter

An example of a penetrameter is shown schematically in Fig. 13, while its image on a radiograph is shown in Fig. 14. While there are many types of penetrameters, this one was

Table 3 Radiographic Sensitivity with the Thinnest Penetrameter and Smallest Hole Visible on Radiograph

Sensitivity, S (%)	Quality Level (%T – Hole Diameter)
0.7	1 – 1T
1.0	1 – 2T
1.4	2 – 1T
2.0	2 – 2T
2.8	2 – 4T
4.0	4 – 2T

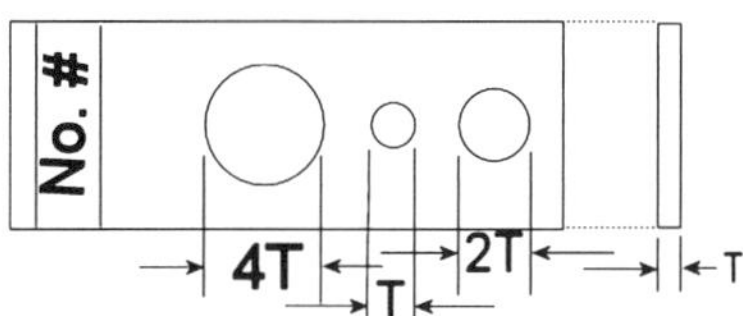

Figure 13 Schematic of typical film penetrameter.

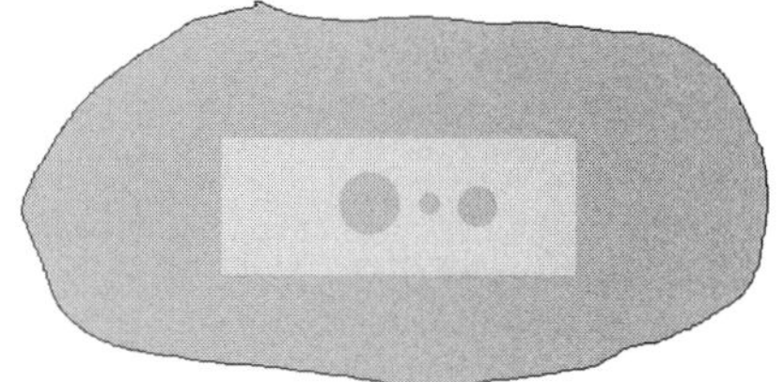

Figure 14 A radiograph of the penetrameter shown in previous figure. The 1T hole is just visible indicating the resolution obtained in the radiograph.

chosen because it is easily related to radiographic sensitivity. The penetrameter is simply a thin strip of metal or polymeric material[55] in which three holes of varying sizes are drilled or punched. It is composed of the same material as the specimen and has a thickness 1%, 2%, or 4% of maximum part thickness. The holes in the penetrameter have diameters that are 1, 2, and 4 T. The sensitivity achieved for each radiographic is determined by noting the smallest hole just visible in the thinnest penetrameter on a film and using Table 3 to determine the sensitivity achieved. By calculating the radiographic sensitivity and then noting the level achieved in practice, the radiographic process can be quantitatively evaluated. While this procedure does not offer any guarantee of flaw detection, it is useful in evaluating the effectiveness of the RT process.

Almost all variables of the radiographic process may be easily and rapidly changed with the aid of tables, graphs, and nomograms, which are usually provided by film manufacturers free of charge. For more information, the reader is referred to the commercial literature.

3.6 Real-Time Radiography

While film radiography represents the bulk of radiographic NDE performed at this time (the beginning of the twenty-first century), new methods of both recording the data and analyzing it are coming into widespread usage. For example, filmless radiography (FR) and real-time radiography (RTR) use solid-state detectors and digital signal processing (DSP) software instead of film to record and enhance the radiographic image. These methods have many advantages, along with some disadvantages. For example, FR permits viewing a radiographic image while the specimen is being moved. This often permits the detection of flaws that would normally be missed in conventional film radiography because of the limited number of views or exposures taken—remember that the X-ray beam must pass along a crack or void for it to be detectable. Additionally, the motion of some flaws enhances their delectability because they present the inspector with a different image as a function of time. Additionally, image enhancement techniques can now be economically and rapidly applied to these images because of the availability of inexpensive, fast computing hardware. The disadvantage of RTR is its lower resolution compared to film. Typical resolution capabilities of RTR or FR systems are in the range of 4 to perhaps 20 line pairs/mm, while film resolution capabilities are in the range of 10–100 line pairs/mm. This means some very fine flaws may not be detectable with FR and the inspector must resort to film. However, in cases where resolution is not the limiting factor, the benefits of software image enhancement can be significant. While the images on film may also be enhanced using the image processing schemes, they cannot be performed in real or near real time, as can be done with an electronic system.

3.7 Computed Tomography

Another advance in industrial radiography has been the incorporation of computed tomography (CT) into the repertoire of the radiographer. Unfortunately, CT has not been exploited to it fullest extent principally due to the high cost of instrumentation. The capability of CT to link NDE measurements with engineering design and analysis gives this inspection a unique ability to provide quanitative estimates of performance not associated with NDE.

The principal advantage of this method is that it produces an image of a thin slice of the specimen under examination. This slice is parallel to the path of the X-ray beam that passes through the specimen, in contrast to the shadowgraph image produced by traditional radiography shown in Figure 7. Whereas the shadowgraph image can be difficult to interpret, the computed CT image does not contain information from planes outside the thin slice.

A comparison between CT and traditional film radiography is best made with images from these two modalities. Figure 7 shows a typical radiograph where one can easily see the image of the top and bottom surfaces of the tube under inspection. The reader can contrast this with the image in Fig. 15, a CT image of a flashlight. The individual components of the flashlight are easily visible and any misplacement of its components or defects in its assembly can be easily detected. An image with a finer scale that reveals the microstructural details of a pencil is shown in Fig. 16. Clearly visible are not only the key features and even the growth rings of the wood. In fact, the details of the growth during each season are visible as rings within rings. The information in the CT image contrasted with conventional radiographs is striking. First, the detectability of a defect is independent of its position in the image. This is not the case with the classical radiograph, where the defect detectability decreases significantly with depth in the specimen, because the defect represents a smaller change in the attenuation of the X-ray beam as the depth increases. Second, the defect detectability is very nearly independent of its orientation. This again is clearly not the case with classical radiography. New applications for CT are constantly being discovered. For example, with a digital CT image it is possible to search for various flaw conditions using

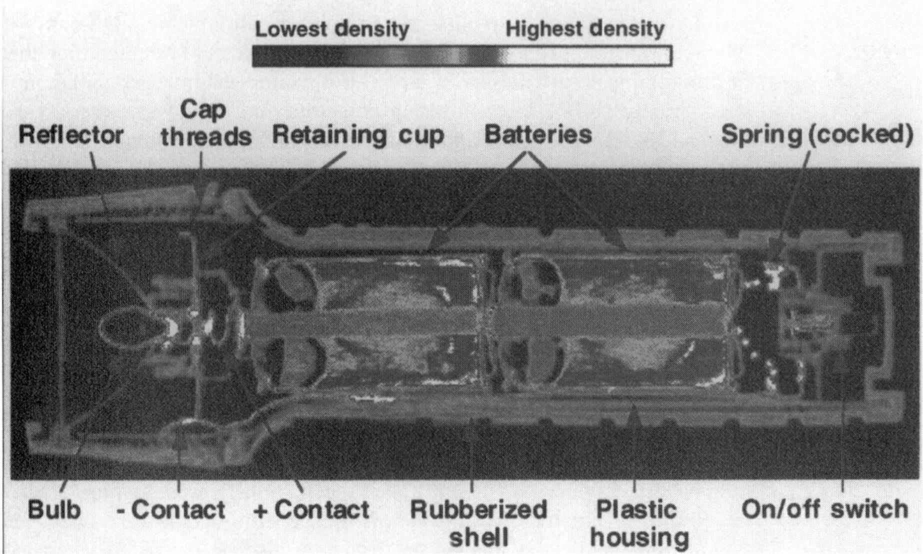

Figure 15 A computed tomography image of flashlight showing details of its internal structure.

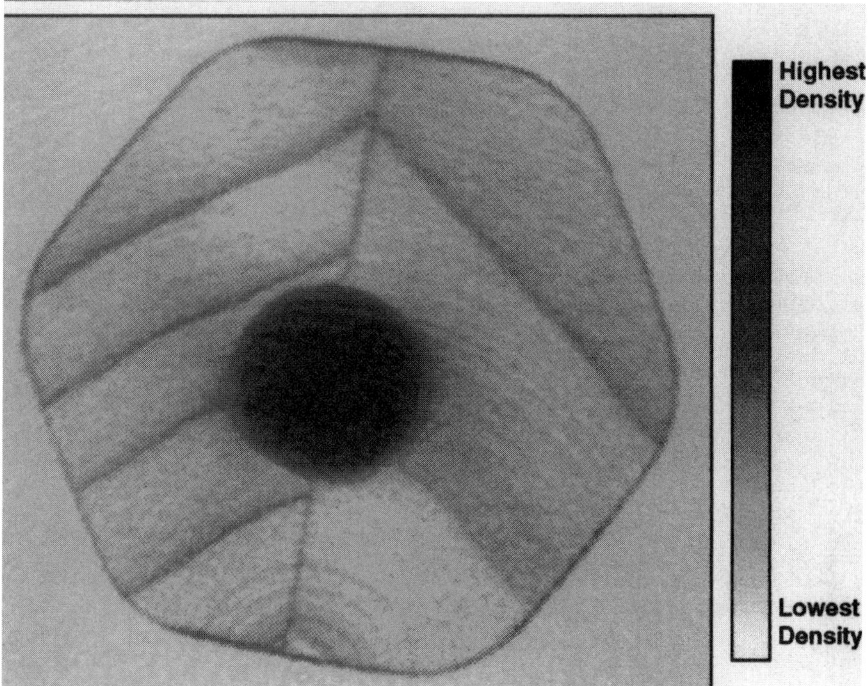

Figure 16 A computed tomograph of pencil. The reader will note the yearly growth rings and even the growth variations within a single growing season.

computer analysis and relieve the inspector of much of the tedium of examining structures for the odd flaw. In addition, it is possible to link the digital CT image with finite-element analysis software to examine precisely how the flaws present will affect such parameters as stress distribution, heat flow, etc. With little effort, one could analyze the full three-dimensional performance of many engineering structures.

4 ULTRASONIC METHODS

Ultrasonic inspection methods utilize high-frequency sound waves to inspect the interior of solid parts. Sound waves are mechanical or elastic disturbances or waves that propagate in fluid and solid media. Ultrasonic testing (UT) or inspection is similar to the angler who uses sonar to detect fish.[56] The government and various technical societies have developed standard practice specifications for UT. These include ASTM specifications 214-68, 428-71, and 494-75 and military specification MIL-1-8950H. Acoustic and ultrasonic testing can take many forms, from simple coin tapping to the transmission and reception of very high frequency or ultrasonic waves into a part to analyze its internal structure.

UT instruments operating in the frequency range between 20 and 500 kHz are referred to as *sonic* instruments, while those that operate above 500 kHz are called *ultrasonic*. To generate and receive ultrasonic waves, a piezoelectric transducer is employed to convert electrical signals to sound waves and back again. The usual form of a transducer is a piezoelectric crystal mounted in a waterproof housing that is electrically connected to a pulsar (transmitter) and a receiver. In the transmit mode a high-voltage, short-duration electrical spike is applied to the crystal, causing it to rapidly change shape and emit an acoustic pulse. In the receive mode, sound waves or returning echoes compress the piezoelectric crystal, producing an electrical signal that is amplified and processed by the receiver. This process is shown schematically in Fig. 17.

4.1 Sound Waves

Ultrasonic waves have physical characteristics such as wavelength (λ), frequency (f), velocity (v), pressure (p), and amplitude (a). The following relationship between wavelength, frequency, and sound velocity is valid for all sound waves:

$$f\lambda = v$$

For example, the wavelength of longitudinal ultrasonic waves of frequency 2 MHz propagating in steel is 3 mm and the wavelength of shear waves is about half this value, 1.6 mm. The relation between the sound pressure and the particle amplitude is

$$p = 2\pi f \rho v a$$

where p is density, f is the frequency of the sound wave, v is its velocity, and a the amplitude.

Ultrasonic waves are reflected from boundaries between different materials or media. Each medium has characteristic acoustic impedance and reflections occur in a manner similar to those observed with electrical signals. The acoustic impedance Z of any media capable of supporting sound waves is defined by

$$Z = \rho v$$

where ρ = density of medium, g/cm^3
v = velocity of sound along direction of propagation

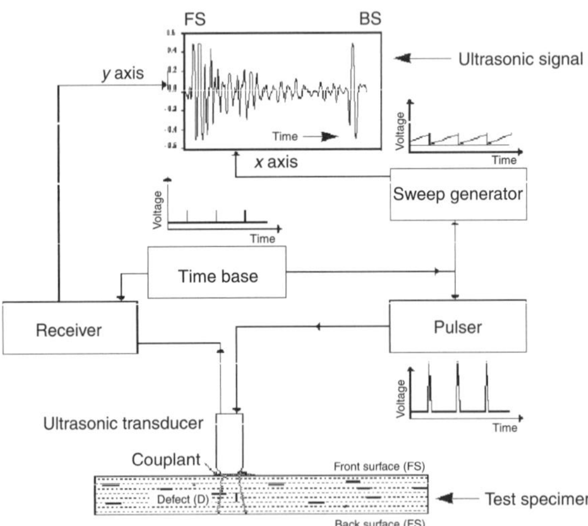

Figure 17 A schematic of ultrasonic data collection and display in the A-scan mode.

Materials with high acoustic impedance are often referred to as sonically hard, in contrast to sonically soft materials with low impedances. For example, steel ($Z = 7.7\ \text{g/cm}^3 \times 5.9\ \text{km/s} = 45.4 \times 10^6\ \text{kg/m}^2\cdot\text{s}$) is sonically harder than aluminum ($Z = 2.7\ \text{g/cm}^3 \times 6.3\ \text{km/s} = 17 \times 10^6\ \text{kg/m}^2\cdot\text{s}$). The Appendix at the end of this chapter lists the acoustic properties of many common materials.

4.2 Reflection and Transmission of Sound

Almost all acoustic energy incident on air–solid interfaces is reflected because of the large impedance mismatch between air and most solids. For this reason, a medium with impedance close to that of the part is used to couple the sonic energy from the transducer into the part. A liquid couplant has obvious advantages for parts with a complex geometry, and water is the couplant of choice for most inspection situations. The receiver, in addition to amplifying the returning echoes, also time gates the returning echoes between the front surface and rear surfaces of the component. Thus, any unusually occurring echo is displayed separately or used to set off an alarm, as shown in Fig. 17. This method of displaying the voltage amplitude of the returning pulse versus time or depth (if acoustic velocity is known) at a single point in the specimen is known as an A-scan. In this figure, the first signal corresponds to a reflection from the front surface (FS) of the part and the last signal corresponds to the reflection from its back surface (BS). The signal or echo between the FS and BS is from the defect in the middle of the part.

The portion of sound energy that is either reflected from or transmitted through each interface is a function of the impedances of the medium on each side of that interface. The reflection coefficient R (ratio of the sound pressures or intensities of the reflected and incident waves) and the power reflection coefficient R_{pwr} (ratio of the power in the reflected and incident sound waves) for normally incident waves onto an interface are given as

$$R = \frac{p_r}{p_i} = \frac{Z_1 - Z_2}{Z_1 + Z_2}$$

$$R_{\text{pwr}} = \frac{I_r}{I_i} = \left(\frac{Z_1 - Z_2}{Z_1 + Z_2}\right)^2$$

Likewise, the transmission coefficients T and T_{pwr} are defined as

$$T = \frac{p_t}{p_i} = \frac{2Z_2}{Z_1 + Z_2}$$

$$T_{\text{pwr}} = \frac{I_t}{I_i} = \frac{4(Z_2/Z_1)}{[1 + (Z_2/Z_1)]^2}$$

where I_i, I_r and I_t are the incident, reflected, and transmitted acoustic field intensities, respectively; Z_1 is the acoustic impedance of the medium from which the sound wave is incident; and Z_2 is impedance of the medium into which the wave is transmitted. From these equations one can calculate the reflection and transmission coefficients for a planar flaw containing air, $Z_1 = 450\ \text{kg/cm}^2\cdot\text{s}$, located in a steel part, $Z_2 = 45.4 \times 10^6\ \text{kg/m}^2\cdot\text{s}$. In this case, the reflection coefficient for the flaw is virtually -1.0. The minus sign indicates a phase change of 180° for the reflected pulse (note that the defect signal in Fig. 17 is inverted or phase shifted by 180° from the FS signal). Effectively no acoustic energy is transmitted across an air gap, necessitating the use of water as a coupling media in ultrasonic testing.

Using the acoustic properties of common materials given in the Appendix the reader can make a number of simple, yet informative, calculations.

Thus far, our discussion has involved only longitudinal waves. This is the only wave that travels through fluids such as air and water. The particle motion in this wave, if one could see it, is similar to the motion of a spring, or a Slinky toy, where the displacement and wave motion are collinear (the oscillations occur along the direction of propagation). The wave is called compressional or dilatational since both compressional and dilatational forces are active in it. Audible sound waves that we hear are compressional waves. This wave propagates in liquids and gases as well as in solids. However, a solid medium can also support additional types of waves such as shear and Rayleigh or surface waves. Shear or transverse waves have a particle motion that is analogous to what one sees in an oscillating rope. That is, the displacement of the rope is perpendicular to the direction of wave propagation. The velocity of this wave is about half that of compressional waves and is only found in solid media, as indicated in the Appendix. Shear waves are often generated when a longitudinal wave is incident on a fluid–solid interface at angles of incidence other than 90°. Rayleigh or surface waves have elliptical wave motion, as shown in Fig. 18, and penetrate the surface for about one wavelength; therefore, they can be used to detect surface and very near surface flaws. The velocity of Rayleigh waves is about 90% of the shear wave velocity. Their generation requires a special device, or wedge as shown in Figure 18, which enables an incident ultrasonic wave on the sample at a specific angle that is characteristic of the material (Rayleigh angle). The reader can find more details in the scientific literature.[57–59]

4.3 Refraction of Sound

The direction of propagation of acoustic waves is governed by the acoustic equivalent of Snell's law. Referring to Fig. 19, the direction of propagation is determined with the equation

$$\frac{\sin \theta_i}{c_{\mathrm{I}}} = \frac{\sin \theta_r}{c_{\mathrm{I}}} = \frac{\sin \gamma_r}{b_{\mathrm{I}}} = \frac{\sin \theta_t}{c_{\mathrm{II}}} = \frac{\sin \gamma_t}{b_{\mathrm{II}}}$$

where c_{I} is the velocity of the incident longitudinal wave, c_{I} and b_{I} are the velocities of the longitudinal and shear reflected waves, and c_{II} and b_{II} are the velocities of the longitudinal and shear transmitted waves in solid II. In the water–steel interface, there is no reflected shear wave because these waves do not propagate in fluids such as water. In this case, the above relationship is simplified. Since the water has a lower longitudinal wave speed than

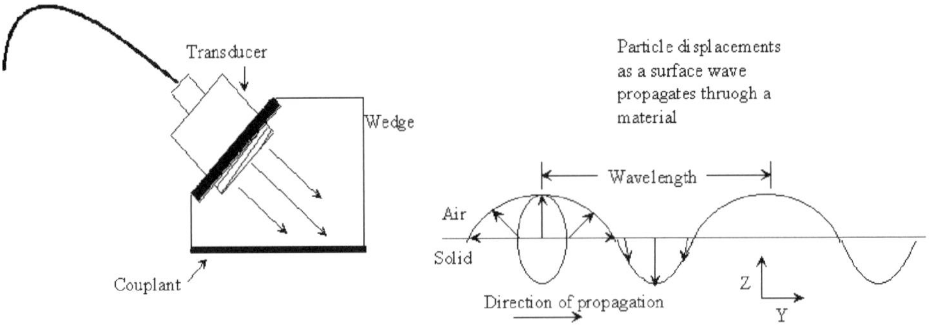

Figure 18 Generation and propagation of surface waves in a material.

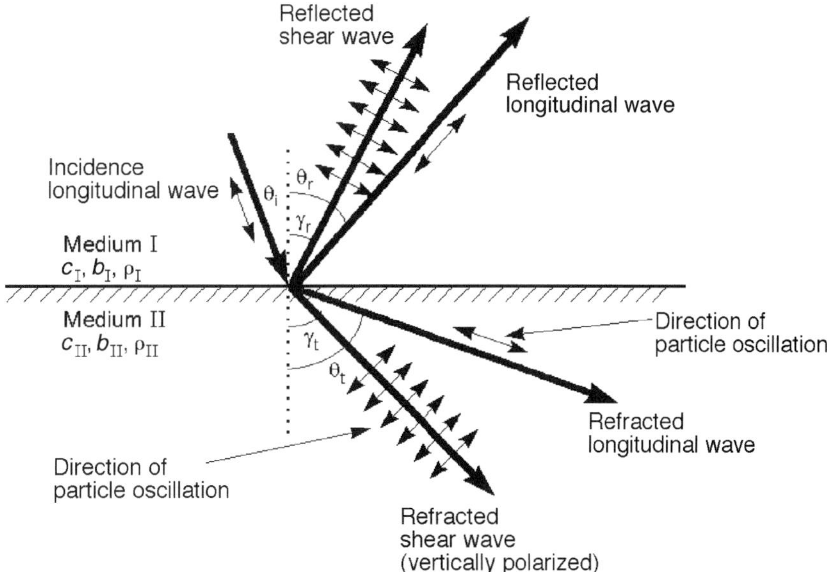

Figure 19 Schematic representation of Snell's law and the mode conversion of a longitudinal wave incident on a solid/solid interface.

either the longitudinal or shear wave speeds of the steel, the transmitted acoustic waves are refracted away from the normal. If the incident wave approaches the interface at increasing angles, there will be an angle above which there will be no transmitted acoustic wave in the higher wave speed material. This angle is refered to as a critical angle. At this angle, the refracted wave travels along the interface and does not enter the solid. A computer-generated curve is shown in Fig. 20 in which the normalized acoustic energy that is reflected and refracted at a water–steel interface is plotted as a function of the angle of the incident longitudinal wave. Note that a longitudinal or first critical angle for steel occurs at 14.5°. Likewise, the shear or second critical angle occurs at about 30°. If the angle of incidence is increased above the first critical angle but less than the second critical angle, only the shear wave is generated in the metal and travels at an angle of refraction described by Snell's law. Angles of incidence above the second critical angle produce a complete reflection of the incident acoustic wave; that is, no acoustic energy enters the solid. At a specific angle of incidence (Rayleigh angle) surface acoustic waves are generating on the material. The Rayleigh angle can be easily calculated from Snell's law by assuming that the refracted angle is 90°. The Rayleigh angle for steel occurs at 29.5°. In the region between the two critical angles, only the shear wave is generated and is referred to as *shear wave testing.* There are two distinct advantages to inspecting parts with this type of shear wave. First, with only one type of wave present, the ambiguity that would exist concerning which type of wave is reflected from a defect does not occur. Second, the lower wave speed of the shear wave means that it is easier to resolve distances within the part. For these reasons, shear wave inspection is often chosen for inspection of thin metallic structures such as those in aircraft.

Using the relationships for the reflection and transmission coefficients, a great deal of information can be deduced about any ultrasonic inspection situation when the acoustic wave is incident at 90° to the surface. For other angles of incidence, computer software is often used to analyze the acoustic interactions. Analytic predictions of ultrasonic performance in

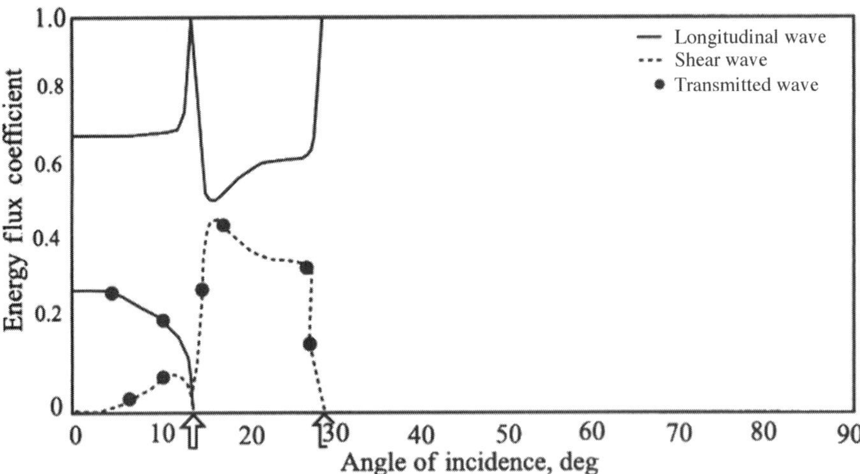

Figure 20 Amplitude (energy flux) and phase of the reflected coefficient and transmitted amplitude versus angle of incidence for a longitudinal wave incident on a water-steel interface. The arrows indicate the critical angles for the interface.

complex materials such as fiber-reinforced composites require the use of more complex algorithms, because more complicate modes of wave propagation can occur. Examples of these include Lamb waves (plate waves), Stoneley waves (interface waves), Love waves (guided in layers of a solid material coated onto another one), and others.

4.4 Inspection Process

Once the type of ultrasonic inspection has been chosen and the optimum experimental parameters determined, one must choose the mode of presentation of the data. If the principal dimension of the flaw is less than the diameter of the transducer, then the A-scan method may be chosen, as shown in Fig. 17. The acquisition of a series of A-scans obtained by scanning the transducer in one direction across the specimen and displaying the data as distance versus depth is referred to as a B-scan. This is the mode most often used by medical ultrasound instrumentation. In the A-scan mode, the size of the flaw may be inferred by comparing the amplitude of the defect signal to a set of standard calibration blocks. Each block has a flat bottom hole (FBH) drilled from one end. Calibration blocks have FBH diameters that vary in $\frac{1}{64}$-in. increments, for example, a number 5 block has a $\frac{5}{64}$-in. FBH. By comparing the amplitude of the signal from a calibration block with one from a defect, the inspector may specify a defect size as equivalent to a certain size FBH. The equivalent size is meaningful only for smooth flaws that are nearly perpendicular to the path of the ultrasonic beam and is used in many industrial situations where a reference size is required by a UT procedure.

If the flaw size is larger than the transducer diameter, then the C-scan mode is usually selected. In this mode, shown in Fig. 21, the transducer is rastered back and forth across the part. In normal operation, a line is traced on a computer monitor or piece of paper. When a flaw signal is detected between the front and back surfaces, the line drawing ceases and a blank place appears on the paper or monitor. Using this mode of presentation, a planar

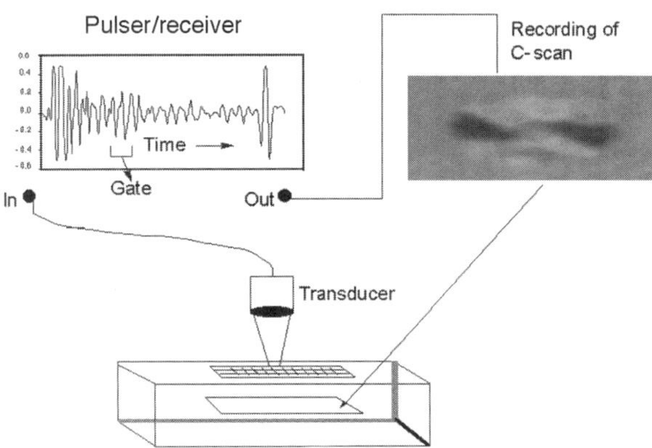

Figure 21 A schematic representation of ultrasonic data collection. The data are displayed using the C-scan mode. The image shows a defect located at a certain depth in the material.

projection of each flaw is presented to the viewer and its positional relationship to other flaws and to the component boundaries is easily ascertained. Unfortunately, the C-scan mode does not show depth information, unless an electronic gate is set to capture only information from within a specified time window or time gate in the part. With current computer capability, it is a rather simple matter to store all of the returning A-scan data and display only the data in a C-scan mode for a specific depth.

Depending on the structural complexity and the attenuation of the signal, cracklike flaws as small as 0.015 in. in diameter may be reliably detected and quantified with this method. An example of a typical C-scan printout of an adhesively bonded test panel is shown in Fig. 22. This panel was fabricated with a void-simulating Teflon implant and the numerous additional white areas indicate the presence of a great deal of porosity in the part.

Through-Transmission versus Pulse Echo

Thus far, the discussion of ultrasonic inspection methods has been concerned with the setup that uses a single transducer to send a signal into the part and to receive any returning echoes. This method is variously referred to as pulse-echo or pitch-catch inspection and is shown schematically in Fig. 23. The other frequently used inspection setup for many structures is called through-transmission. With this setup two transducers are used, one to send ultrasonic pulses and the other placed on the opposite side of the part to receive the trans-

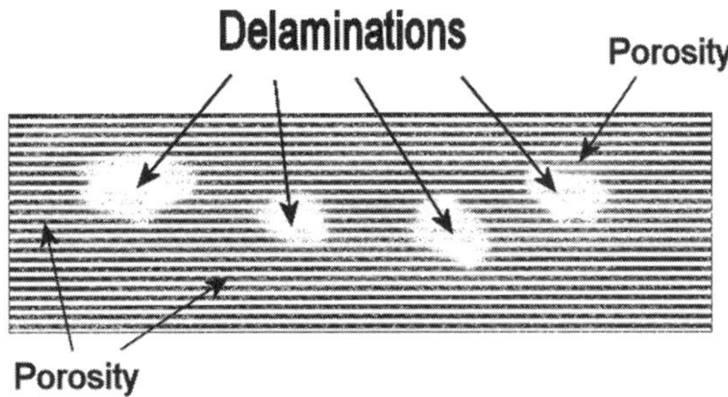

Figure 22 A typical C-scan image of a composite specimen showing delaminations and porosity.

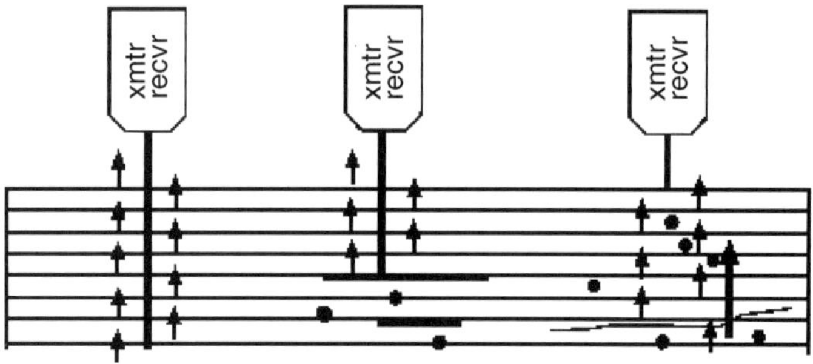

Figure 23 A schematic representation of the pulse-echo mode of ultrasonic inspection.

mitted signals, as shown schematically in Fig. 24. In Fig. 23 and 24 a large number of reflections occur for the many individual layers of in a composite part which can obscure subtle reflections from inclusions whose reflectivity is similar to that of the layered materials. An inclusion with an acoustic impedance very close to that of the part, e.g., paper or peal-plys in polymer-based composites, is very difficult to detect with the through-transmission mode of inspection. In this case, the pulse-echo inspection mode is often used to detect these flaws. On the other hand, reflections from distributed flaws such as porosity, as shown on the right-hand side of Figs. 23 and 24, can be obscured by the general background noise present in an acoustic signal. Therefore, it is the loss in signal strength of the transmitted signal of the through-transmission method that is most often used to detect this type of flaw. While porosity is detectable in this manner, its location may not be determined. In this situation, the pulse-echo mode is required because the distance from the front or back surface to the flaw can be determined by the relative position of the reflections of the scattered porosity with respect to the surface reflection. Because each method supplies important information about potential flaws and its location, modern ultrasonic instrumentation is frequently equipped to perform both types of inspection nearly simultaneously.[60,61] In such a setup, two transducers are used to conduct a through-transmission test and then each is used

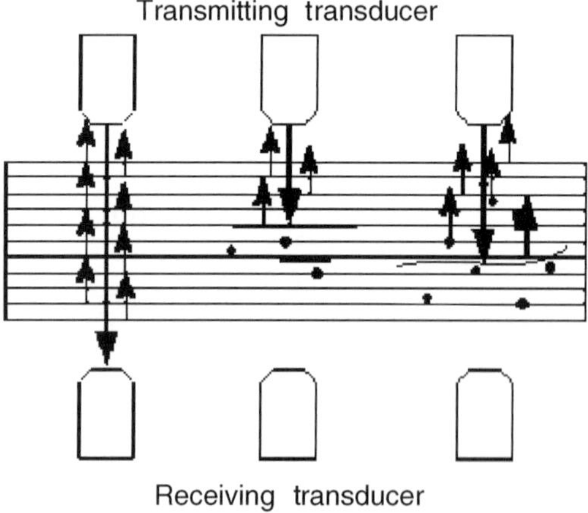

Figure 24 A schematic representation of the through-transmission mode of inspection.

separately to conduct pulse-echo tests from opposite sides of the part. This method also helps ensure that a large flaw does not shadow a smaller one, as shown in Figs. 23 and 24.

Portable Ultrasonic Systems

This ability to image defects on specific levels within a layered component, e.g., a composite, is so important that C-scan instrumentation has been miniaturized for usage in the field. An example of one such system developed for aircraft inspection is shown in Fig. 25. The heart of this system is a computer that records the position of a hand-held transducer and the complete A-scan wave train at each point of the scan. Since the equipment tracks the motion of the transducer as it is scanned manually across a structure, the inspector can see which areas have been scanned. If areas are missed, he or she can return to "color them in," as shown in the image on the monitor. Additionally, computer manipulation of ultrasonic image data allows the inspector to select either one or a small number of layers for evaluation. In this way, an orderly assessment of the flaws in critical structures can be accomplished. This process of selecting flaws on a layer-by-layer basis for evaluation is shown schematically in Fig. 26 for the instrument depicted in Fig. 25.

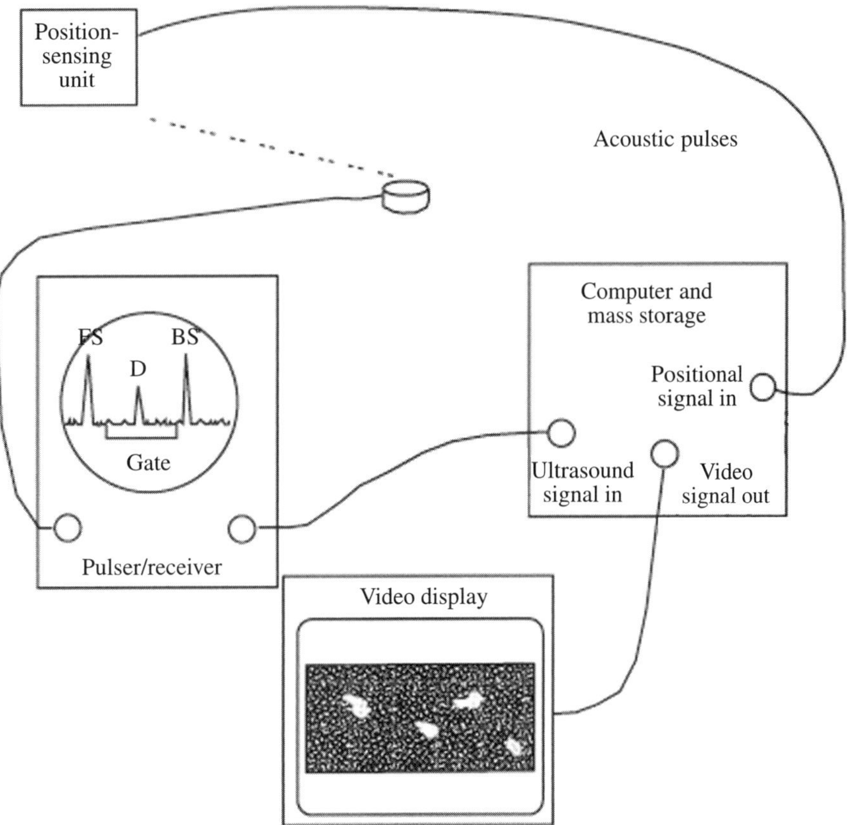

Figure 25 Field level C-scan instrumentation that is capable of simultaneously tracking the motion of a hand held transducer, and recording the ultrasonic information.

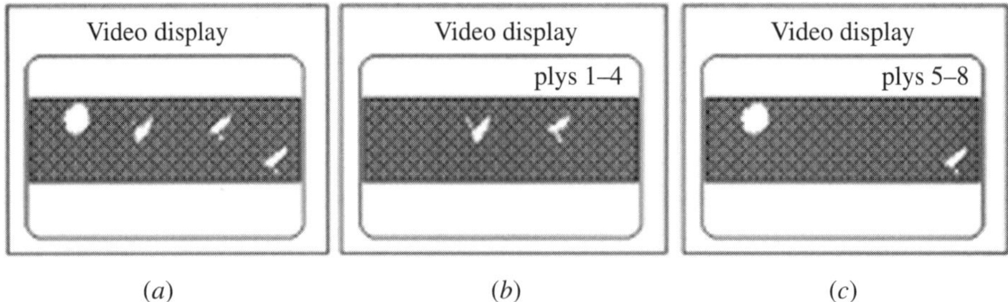

(*a*) (*b*) (*c*)

Figure 26 Three different displays of delaminations in a 16-ply composite obtained from a field level C-scan system. The image at "a" is a projection all the flaws through the specimen. Display's b and c are images from selected depths within the specimen.

4.5 Bond Testers

A great deal of the ultrasonic inspection literature is devoted to instruments that test adhesive bonds. There has been a recent resurgence of interest in the inspection of adhesive bonds due to concerns about the viability of bonded patches on our aging aircraft.[62] For an extensive treatment of most of the currently used instruments, the reader is referred to review articles.[63–67] However, while there may seem to be a large number or instruments, some with exaggerated claims of performance, most operate on the same physical principles.

Bond-testing instruments use a variety of means to excite sonic or low-frequency sound waves into the part. In these methods, a low-frequency acoustic transducer is attached to the structure through a couplant. As the driving frequency of the transducer is varied, the amplitude and phase of the transducer oscillations change dramatically as it passes through a resonance. The phase and amplitude of these vibrations change very rapidly and reach a maximum as the driving frequency passes through the resonance frequency of the transducer. The effect of the structure is to dampen the resonant response of the transducer–block combination because of the transfer of acoustic energy into it. Defects such as delaminations and porosity in the adhesive bond layer increase the stiffness of the structure and lower the resonant frequency of the combination. The amplitude of the resonance is increased since there is less material to adsorb the sound energy. These changes in the sharpness of the resonant response are easily detectable electronically.

An alternate method of detecting flaws in bonded components is with a low-frequency or sonic instrument that senses the change in the time of flight for sound waves in the layered structure due to the presence of planar delamination. In such instruments, the increased time of traveling from a transmitting to a receiving transducer is detected electronically. Several commercially available bond-testing instruments successfully exploit this principle. A clever adaptation of a commercial version of this instrument has recently been used to successfully test the joints of structures made from sheet molding compounds.[64]

Probably the most often used method of detecting delaminations in laminated structures is with a coin or tap hammer. This simple instrument is surprisingly effective in trained hands at detecting flaws since an exceedingly complex computer interprets the output signal, i.e., the human brain. Consider, for a moment, that most parents can easily hear their child playing a musical instrument at a school concert. They can perform this task even though their child may have a minor part to play and all the other instruments are much louder than the one that their child is playing. With this powerful real-time signal-processing capability, inspectors can often detect flaws that cannot be detectable with current instrumentation and computers.

5 MAGNETIC PARTICLE METHOD

The magnetic particle method of nondestructive testing is used to locate surface and subsurface discontinuities in ferromagnetic materials.[68] An excellent reference for this NDE method is Ref. 21, especially Chapters 10–16. Magnetic particle inspection is based on the principle that magnetic lines of force, when present in a ferromagnetic material, are distorted by changes in material continuity, such as cracks or inclusions, as shown schematically in Fig. 27. If the flaw is open to the surface or close to it, the flux lines escape the surface at the site of the discontinuity. Near-surface flaws, such as nonmagnetic inclusions, cause the same bulging of the lines of force above the surface. These distorted fields, usually referred to as a leakage fields, reveals the presence of the discontinuity when fine magnetic particles are attracted to them during magnetic particle inspection. If these particles are fluorescent, their presence at a flaw will be visible under ultraviolet light, much like penetrant indications. Magnetic particle inspection is used for steel components because it is fast and easily implemented and has rather simple flaw indications. The part is usually magnetized with an electric current and then a solution containing fluorescent particles is applied by flowing it over the part. The particles stick to the part, forming the indication of the flaw.

5.1 Magnetizing Field

The magnetizing field may be applied to a component by a number of methods. Its function is to generate a residual magnetic field at the surface of the part. The application of a magnetizing force (H) generates a magnetic flux (B) in the component, as shown schematically in Fig. 28. In this figure, the magnetic flux density B has units of newtons per ampere or webers per square meter and the strength of the magnetic field or magnetic flux intensity H has units of oersteds or amperes per meter. Starting at the origin, a magnetizing force is applied and the magnetic field internal to the part increases in a nonlinear fashion along the path shown by the arrows. If the force is reversed, the magnetic field does not return to zero but follows the arrows around the curve as shown. The reader will note that once the magnetizing force is removed, the flux density does not return to zero but remains at an elevated value called the material's remanence. This is the point at which most magnetic particle inspections are performed. The reader will also note that an appreciable reverse magnetic force H must be applied before the internal field density is again zero. This point is referred to as the coercivity of the material. If the magnetizing force is applied and reversed, the material will respond by continually moving around this hysteresis loop.

Selection of the type of magnetizing current depends primarily on whether the defects are open to the surface or are wholly below it. Alternating-current (AC) magnetization is best for the detection of surface discontinuities because the current is concentrated in the near-surface region of the part. Direct-currents (DC) magnetization is best suited for subsurface discontinuities because of its deeper penetration of the part. While DC can be obtained from batteries or DC generators, it is usually produced by half-wave or full-wave rectification of commercial power. Rectified current is classified as half-wave direct current (HWDC) or full-wave direct current (FWDC). Alternating-current fields are usually obtained

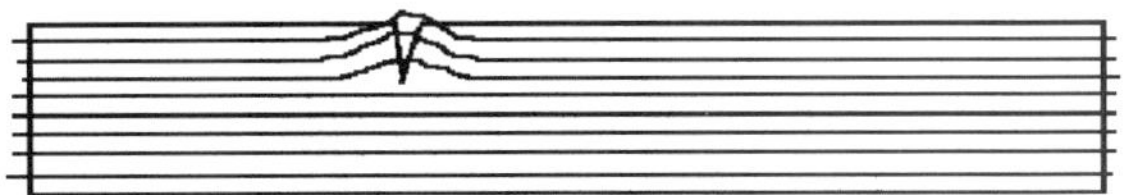

Figure 27 Schematic representation of the magnetic lines of flux in a ferromagnetic metal near a flaw. Small magnetic particles are attracted to the leakage field associated with the flaw.

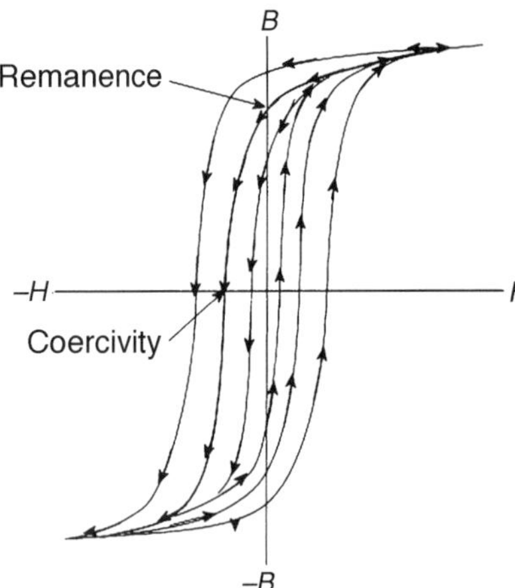

Figure 28 A magnetic flux intensity, H, versus magnetic flux density, B hysteresis curve for a typical steel. Initial magnetization starts at the origin and progresses as shown by the arrows. Demagnetization follows the arrows of the smaller hysteresis loops.

from conventional power mains, but it is supplied to the part at reduced voltage for reasons of safety and the high-current requirements of the magnetizing process.

Two general types of magnetic particles are available to highlight flaws. One type is low-carbon steel with high-permeability and low-retentivity particles, which are used dry and consist of different sizes and shapes to respond to leakage fields. The other type of is very fine particles of magnetic iron oxide that are suspended in a liquid (either a petroleum distillate or water). These particles are smaller and have a lower permeability than the dry particles. Their small mass permits them to be held by the weak leakage fields at very fine surface cracks. Magnetic particles are available in several colors to increase their contrast against different surfaces or backgrounds. Dry powders are typically gray, red, yellow, and black, while wet particles are usually red, black, or fluorescent.

5.2 Continuous versus Noncontinuous Fields

Because the field is always stronger while the magnetizing current is on, the continuous magnetizing method is generally preferred. Additionally, for specimens with low retentivity this continuous method is often preferred. In the continuous method, the current can be applied in short pulses, typically 0.5 s. The magnetic particles are applied to the surface during this interval and are free to move to the site of the leakage fields. Liquid suspended fluorescent particles produces the most sensitive indications. For field inspections, the magnetizing current is often continuously applied during the test to give time for the powder to migrate to the defect site. In the residual method, the particles are applied after the magnetizing current is removed. This method is particularly well suited for production inspection of multiple parts.

The choice of direction of the magnetizing field within the part involves the nature of the flaw and its direction with respect to the surface and the major axis of the part. In circular magnetization, the field runs circumferentially around the part. It is induced into the part by passing current through it between two contacting electrodes. Since flaws perpendicular to the magnetizing lines are readily detectable, circular magnetization is used to detect flaws that are parallel or less than 45° to the surface of the long, circular specimens. Placing the specimen inside a coil to create a field running lengthwise through the part produces longitudinal magnetization. This induction method is used to detect transverse discontinuities to the axis of the part.

5.3 Inspection Process

The surface of the part to be examined should be clean, dry, and free of contaminants such as oil, grease, loose rust, loose sand, loose scale, lint, thick paint, welding flux, and weld splatter. Cleaning of the specimen may be accomplished with detergents, organic solvents, or mechanical means, such as scrubbing or grit blasting.

Portable and stationary equipment are available for this inspection process. Selection of the specific type of equipment depends on the nature and location of testing. Portable equipment is available in lightweight units (35–90 lb) which can be readily taken to the inspection site. Generally, these units operate at 115, 230, or 460 V AC and supply current outputs of 750–1500 amps in half-wave AC.

5.4 Demagnetizing the Part

Once the inspection process is complete, the part must be demagnetized. This is done by one of several ways depending on the subsequent usage of the component. A simple method of demagnetizing to remove residual magnetism from small tools is to draw it through the loop shaped coil tip of a soldering iron. This has the effect of retracing the hysteresis loop a large number of times, each time with a smaller magnetizing force. When completely withdrawn, the tool will then have a very small remnant magnetic field, which for all practical purposes is zero. This same process is accomplished with an industrial part by slowly reducing and reversing the magnetizing current until it is essentially zero, as shown schematically by the arrows in Fig. 28. Another method of demagnetizing a part is to heat it above its Curie temperature (about 550°C for iron), at which point all residual magnetism disappears. This last process is the best means of removing all residual magnetism, but it requires the expense and time of an elevated heat treatment.

6 THERMAL METHODS

Thermal nondestructive inspection methods involve the detection of infrared energy emitted from the surface of a test object.[69] This technique is used to detect the flow of thermal energy either into or out of a specimen and the effect of anomalies have on the surface temperature distribution. The material properties that influence this method are heat capacity, density, thermal conductivity, and emissivity. Defects that are usually detected include porosity, cracks, and delaminations that are parallel to the surface. The sensitivity of any thermal method is greatest for near-surface flaws that impede heat flow and degrades rapidly for deeply buried flaws in high-conductivity materials. Materials with lower thermal conductivity yield better resolution because they allow larger thermal gradients.

6.1 Infrared Cameras

All objects emit infrared (IR) radiation with a temperature above absolute zero. At room temperature, the thermal radiation is predominately IR with a wavelength of approximately 10 μm. IR cameras are available that can produce images from this radiation and are capable of viewing large areas by scanning. Since the IR images are usually captured and stored in digital form, image processing is easily performed and the enhanced images are stored on magnetic or optical media. For many applications, an uncalibrated thermal image of a specimen is sufficient to detect flaws. However, if absolute temperatures are required, the IR instrumentation must be calibrated to account for the surface emissivity of the test subject.

The ability of thermography to detect flaws is often affected by the type of flaw and its orientation with respect to the surface of the object. To have a maximum effect on the surface temperature, the flaw must interrupt heat flow to the surface. Since a flaw can occur at any angle to the surface, the important parameter is its projected area to the camera. Subsurface flaws such as cracks parallel to the surface of the object, porosity, and debonding of a surface layer are easily detected. Cracks that are perpendicular to the surface can be very difficult or impossible to detect using thermography.

Most thermal NDE methods do not have good spatial resolution due to spreading of thermal energy as it diffuses to the surface. The greatest advantage of thermography is that it can be a noncontact, remote-viewing technique requiring only line-of-sight access to one side of a test specimen. Large areas can be viewed rapidly, since scan rates for IR cameras run between 16 and 30 frames per second. Temperature differences of 0.02°C or less can be detected in a controlled environment.

6.2 Thermal Paints

A number of contact thermal methods are available for inspection purposes. These usually involve applying a coating to the sample and observing a color change as the specimen is thermally cycled. Several different types of coatings are available that cover a wide temperature ranges. Temperature-sensitive pigments in the form of paints have been made to cover a temperature range of 40–1600°C. Thermal phosphors emit visible light when exposed to UV radiation. (The amount of visible light is inversely proportional to temperature.) Thermochromic compounds and cholesteric liquid crystals change color over large temperature ranges. The advantages of these approaches are the simplicity of application and relatively low cost if only small areas are scanned.

6.3 Thermal Testing

Excellent results may be achieved for thermographic inspections performed in dynamic environments where the transient effects of heat flow in the test object can be monitored. This enhances detection of areas where different heat transfer rates are present. Applications involving steady-state conditions are more limited. Thermography has been successfully used in several different areas of testing. In medicine, it is used to detect subsurface tumors. In aircraft manufacturing and maintenance, it may be used to detect debonding in layered materials and structures. In the electronics industry, it is used to detect poor thermal performance of circuit board components. Recently thermography has been used to detect stress-induced thermal gradients around defects in dynamically loaded test samples. For more information on thermal NDE methods, the reader is referred to Refs. 69-71.

7 EDDY CURRENT METHODS

7.1 Eddy Current Inspection

Eddy current (EC) methods are used to inspect electrically conducting components for flaws. Flaws that cause a change in electrical conductivity or magnetic permeability such as surface-breaking cracks, subsurface cracks, voids, and errors in heat treatment are detectable using

EC methods. Thickness measurements and the thickness of nonconducting coatings on metal substrates can also be determined with EC methods.[72] Quite often, several of these conditions can be monitored simultaneously if instrumentation capable of measuring the phase of the EC signal is used.

This inspection method is based on the principle that eddy currents are induced in a conducting material when a coil (probe) is excited with an alternating or transient electric current that is placed in close proximity to the surface of a conductor. The induced currents create an electromagnetic field that opposes the field of the inducing coil in accordance with Lenz's law. The eddy currents circulate in the part in closed, continuous paths, and their magnitude depends on many variables. These include the magnitude and frequency of the current in the inducing coil, the coil's shape and position relative to the surface of the part, electrical conductivity, magnetic permeability, shape of the part, and presence of discontinuities or inhomogeneities within the material. Therefore, EC inspection is useful for measuring the electrical properties of materials and detecting discontinuities or variations in the geometry of components.

Skin Effect

Eddy current inspections are limited to the near-surface region of the conductor by the skin effect. Within the material, the EC density decreases with the depth. The density of the EC field falls off exponentially with depth and diminishes to a value of about 37% of the surface value at a depth referred to as the standard depth of penetration (SDP). The SDP in meters is calculated with the formula

$$\text{SDP} = \frac{1}{\sqrt{\pi f \sigma \mu}}$$

where f = test frequency, Hz
σ = test material's electrical conductivity, mho/m
μ = permeability, H/m

The latter quantity is the product of the relative permeability of the specimen, 1.0 for nonmagnetic materials, and the permeability of free space, $4\pi \times 10^{-7}$ H/m.

Impedance Plane

While the SDP is used to give an indication of the depth from which useful information can be obtained, the choice of the independent variables in most test situations is usually made using the impedance plane diagram suggested by Förster.[73] It is theoretically possible to calculate the optimum inspection parameters from numerical codes based on Maxwell's equations, but this is a laborious task that is justified in special situations.

The eddy currents induced at the surface of a material are time varying and have amplitude and phase. The complex impedance of the coil used in the inspection of a specimen is a function of a number of variables. The effect of changes in these variables can be conveniently displayed with the impedance diagram, which shows the variations in amplitude and phase of the coil impedance as functions of the dependent variables specimen conductivity, thickness, and distance between the coil and specimen, or lift-off. For the case of an encircling coil on a solid cylinder, shown schematically in Fig. 29, the complex impedance plane is displayed in Fig. 30. The reader will note that the ordinate and abscissa are normalized by the inductive reactance of the empty coil. This eliminates the effect of the geometry of the coil and specimen. The numerical values shown on the large curve, which are called reference numbers, are used to combine the effects of the conductivity, size of the test specimen, and frequency of the measurement into a single parameter. This yields a diagram that is useful for most test conditions. The reference numbers shown on the outermost curve are obtained with the following relationship for nonmagnetic materials:

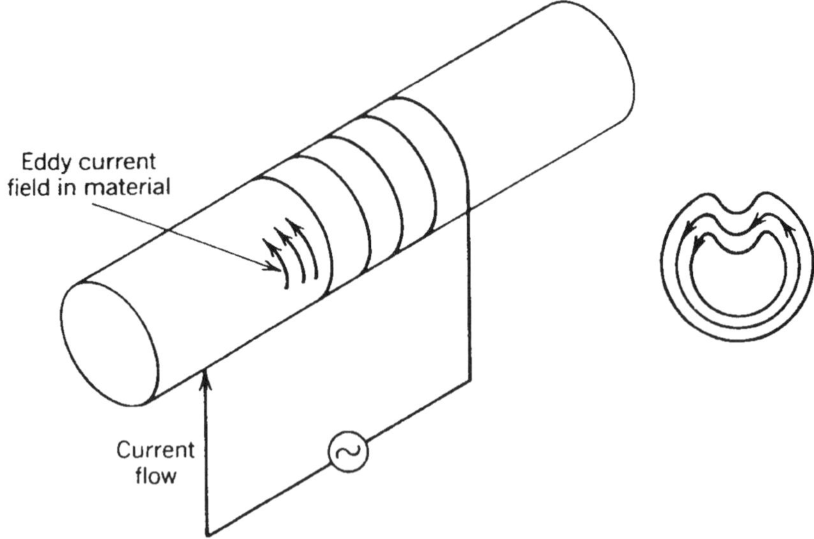

Figure 29 Schematic representations for an eddy current inspection of a solid cylinder. Also shown are the eddy current paths within the cross-section of the cylinder near the crack.

$$\text{Reference number} = r\sqrt{2\pi f \mu \sigma}$$

where r = radius of bar, m
f = frequency of test, Hz
m = magnetic permeability of free space, $4\pi \times 10^{-7}$ H/m
σ = conductivity of specimen in, mho/m

The outer curve in Figures 30 and 31 is useful only for the case where the coil is the same size as the solid cylinder, which can never happen. For those cases where the coil is larger than the test specimen, which is usually the case, a coil-filling factor is calculated. This is quite easily accomplished with the formula

$$N = \frac{\text{diameter}_{\text{specimen}}}{\text{diameter}_{\text{coil}}}$$

Figure 30 shows the impedance plane with a curve for specimen/coil inspection geometry with a fill factor of 0.75. Note that the reference numbers on the curves representing the different fill factors can be determined by projecting a straight line from the point 1.0 on the ordinate to the reference number of interest, as is shown for the reference number 5.0. Both the fill factor and the reference number change when the size of either the specimen or coil changes. Assume that a reference number of 5.0 is appropriate to a specific test with $N = 1.0$; if the coil diameter is changed so that the fill factor becomes 0.75, then the new reference number will be equal to approximately 7. While the actual change in reference number for this case follows the path indicated by the dashed line in Fig. 30, we have estimated the change along a straight line. This yields a small error in optimizing the test setup but is sufficient for most purposes. For a more detailed treatment of the impedance plane, the reader is referred to Ref. 72. The inspection geometry discussed thus far has been for a solid cylinder. The other geometry of general interest is the thin-walled tube. In this case the skin effect limits the thickness of the metal that may be effectively inspected.

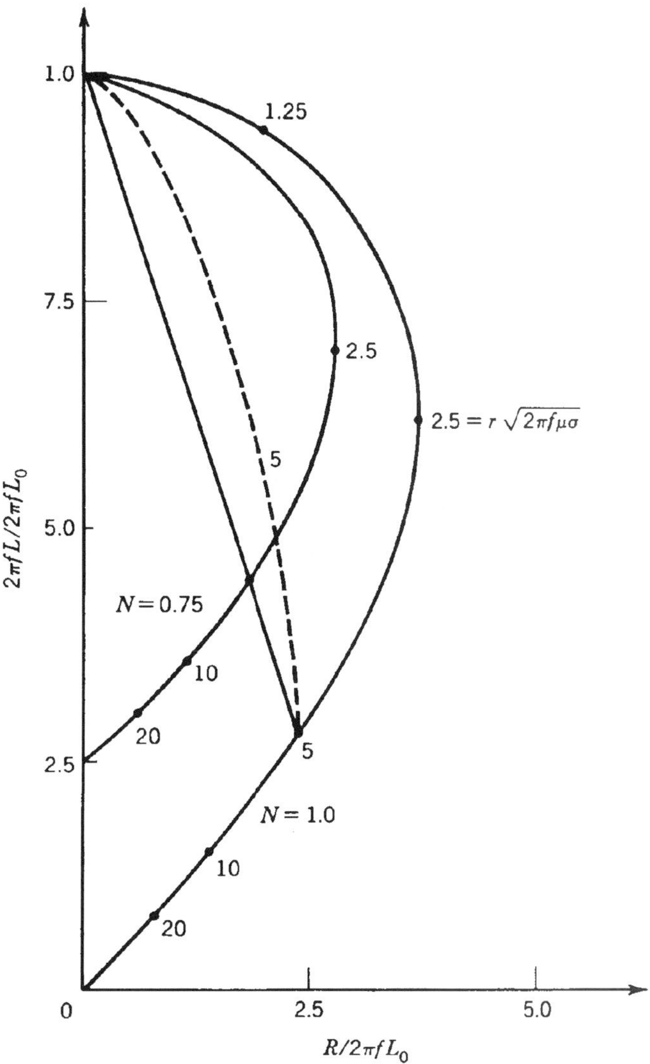

Figure 30 The normalized impedance diagram for a long encircling coil on a solid, nonferromagnetic cylinder. For N = 1 the coil and cylinder have the same diameter, while for N = 0.75 the coil is approximately 1.155 times larger than the cylinder.

For an infinitely thin-walled tube, the impedance plane is shown in Fig. 31, which includes the curve for a solid cylinder. The dashed lines that connect these two cases are for thin-walled cylinders of varying thicknesses. The semicircular curve for the thin cylinder is used in the same manner as described above for the solid cylinder.

Lift-off of Inspection Coil from Specimen

In most inspection situations, the only independent variables are frequency and lift-off. High-frequency excitations are frequently used for detecting defects such as surface-connected cracks or corrosion, while low frequencies are used to detect subsurface flaws. It is also

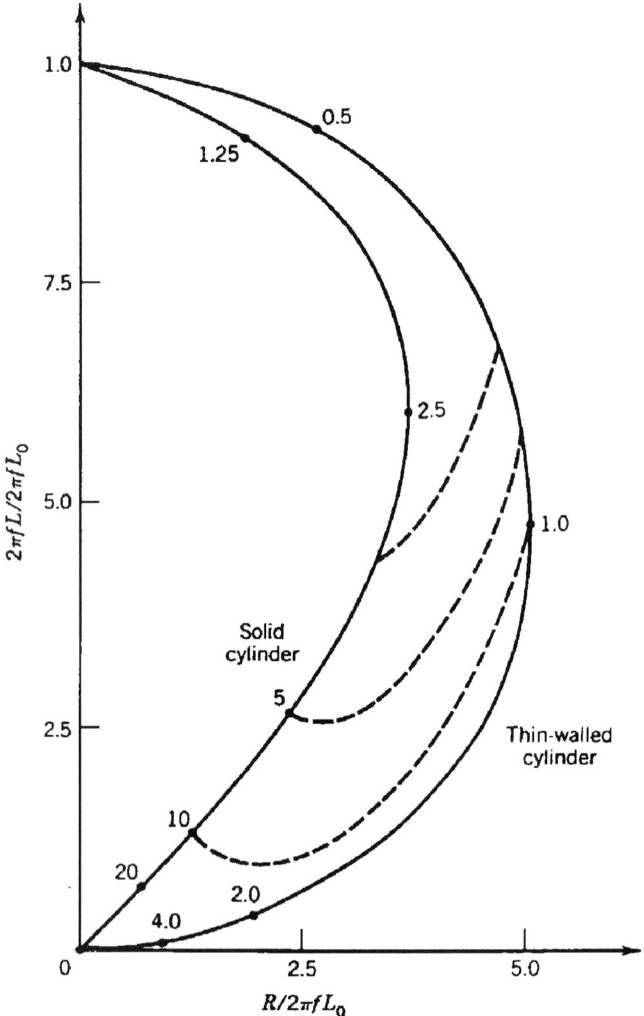

Figure 31 The normalized impedance diagram for a long encircling coil on both a solid and thin-walled conductive but nonferromagnetic cylinders. The dashed lines represent the effects of varying wall thicknesses.

possible to change the coil shape and measurement configuration to enhance detectability, but the discussion of these more complex parameters is beyond the scope of this chapter and the reader is referred to the literature. The relationships discussed so far may be applied by examining Fig. 32, where changes in thickness, lift-off, and conductivity are represented by vectors. These vectors all point in different directions representing the phases of the different possible signals. Instrumentation with phase discrimination circuitry can differentiate between these signals and therefore is often capable of detecting two changes in specimen condition at once. Changes in conductivity can arise from several different conditions. For example, aluminum alloys can have different conductivities depending on their heat treatment. Changes in apparent conductivity are also due to the presence of cracks or voids.

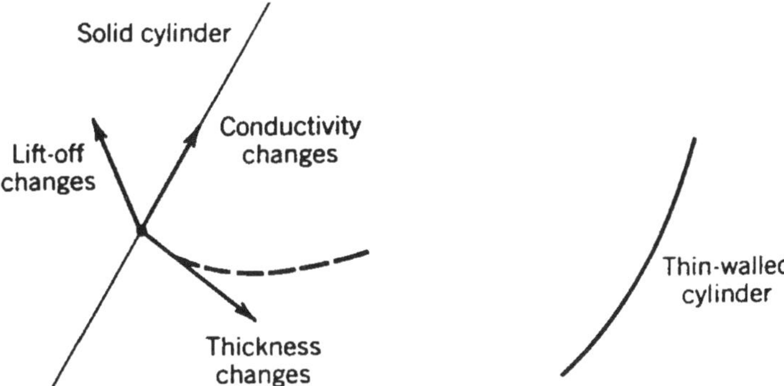

Figure 32 Effects of various changes in inspection conditions on local signal changes in the impedance plane of previous figure. Phase differentiation is relatively easily accomplished with current instrumentation.

A crack decreases the apparent conductivity of the specimen because the eddy currents must travel a longer distance to complete their circuit within the material. Lift-off and wall thinning are also shown in Fig. 32. Thus, two different flaw conditions can be rapidly detected. There are situations where changes in wall thickness and lift-off result in signals that are very nearly out of phase and therefore the net change is not detectable. If this situation is suspected, then inspection at two different frequencies is warranted. There are other inspection situations that cannot be covered in this brief description. These include the inspection of ferromagnetic alloys, plate, and sheet stock and the measurement of film thicknesses on metal substrates. For a treatment of these and other special applications of EC inspection, the reader is referred to Ref. 72.

7.2 Probes and Sensors

In some situations, it may be advantageous to have a core with a high magnetic permeability inside the coil. Magnetic fields will pass through the medium with the highest permeability if possible. Therefore, materials with high permeability can be placed in different geometric configurations to enhance the sensitivity of a probe. One example of this is the cup-core probe where the coil has a ferrite core, shield, and cap.[74]

For low-frequency or transient tests, using the inductive coil as a sensor is not sufficient since it responds to the time change in magnetic field and not the direct magnetic field. It is necessary to induce a magnetic field sensor that responds well at the lower frequencies. Sensors such as the Hall effect sensor and giant magnetoresistive (GMR) sensors have been used to accomplish this.[75]

There are numerous methods of making eddy current NDE measurements. Two of the more common methods are shown schematically in Fig. 33. In the absolute coil arrangement, very accurate measurements can be made of the differences between the two samples. In the differential coil method, it is the differences between the two variables at two slightly different locations that may be detected. For this arrangement, slightly varying changes in dimensions and conductivity are not sensed, while singularities such as cracks or voids are highlighted, even in the presence of other slowly changing variables. Since the specific electronic circuitry used to accomplish this task can vary dramatically, depending on the specific inspection situation, the reader is referred to the current NDE and instrumentation Refs. 73, 76, and 77.

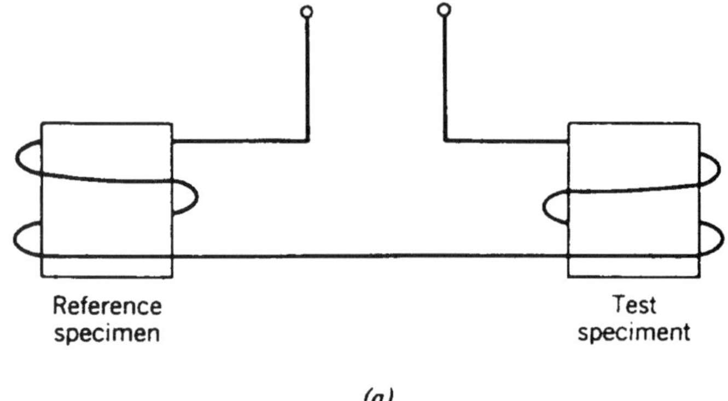

(a)

Reference
coil

Inspection
coil

(b)

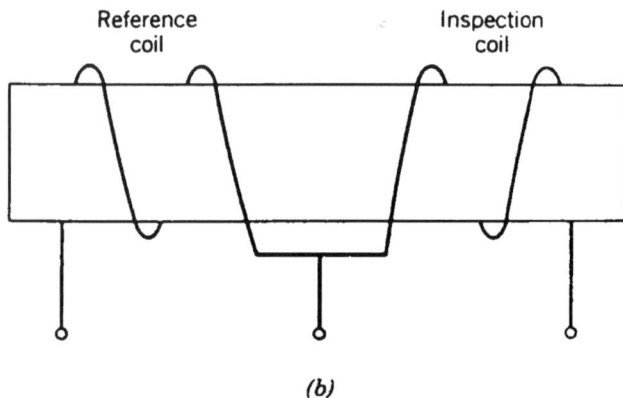

Figure 33 A schematic representation of an absolute (a) versus (b) a differential coil configurations used in eddy current testing.

APPENDIX: ULTRASONIC PROPERTIES OF COMMON MATERIALS

Material	Comments	Density (g/cm^3)	V_l (km/s)	V_s (km/s)	Impedance (MRayl)	Attn (dB/cm/MHz)	Attn (dB/cm at 5 MHz)
Gases							
Alcohol vapor			0.23				
Air	25 atm		0.33				
	50 atm		0.34				
	100 atm		0.35				
			0.34				
			0.39				
			0.55				
Ammonia			0.42				
Argon		0.00178	0.32				
Carbon monoxide		0.34					

Material	Comments	Density (g/cm^3)	V_1 (km/s)	V_s (km/s)	Impedance (MRayl)	Attn (dB/cm/MHz)	Attn (dB/cm at 5 MHz)
Carbon dioxide			0.26				
Carbon disulfide			0.19				
Chlorine			0.21				
Ether vapor			0.18				
Ethylene			0.31				
Helium		0.00018	0.97				
Hydrogen		0.00009	1.28				
Methane		0.00074	0.43				
Neon		0.0009	0.43				
Nitric oxide			0.33				
Nitrogen		0.00125	0.33				
		0.00116	0.35				
Nitrous oxide			0.26				
Oxygen		0.00142	0.32				
		0.00132	0.33				
Water vapor			0.4				
			0.41				
			0.42				
Gases—Cryogenic							
Argon		1.404	0.84			1.18	
		1.424	0.86			1.23	
Helium		0.125	0.18			0.023	
		0.146	0.23			0.034	
Helium-4	Liquid at 2 K	0.15	0.18			0.027	
Hydrogen	Liquid at 20 K	0.07	1.19			0.08	
		0.355	1.13			0.401	
Nitrogen		0.815	0.87			0.708	
		0.843	0.93			0.783	
Oxygen		1.143	0.97			1.04	
		1.272	1.13			1.44	
		1.149	0.95			1.09	
Liquids							
Acetate butyl (*n*)		0.871	1.17			1.02	
Acetate ethyl		0.9	1.18			1.06	
Acetate methyl		0.928	1.15			1.07	
Acetate propyl		0.891	1.18			1.05	
Acetone		0.79	1.17			0.92	
		0.791	1.16		0.92		0.0469
Acetonitrile		0.783	1.29			1.01	
Acetonyl acetone		0.729	1.4			1.36	
Acetylene dichloride		1.26	1.02			1.29	
Adiprene	CW-520	0.79	1.68		1.33		0.0469
Alcohol, butyl		0.81	1.24			1	
Alcohol, ethyl		0.789	1.18			0.93	
Alcohol, furfuryl		1.135	1.45			1.65	
Alcohol, isopropyl		0.79	1.17		0.92		0.08
Alcohol, methyl		0.792	1.12			0.89	
Alcohol, propyl (*i*)		0.786	1.17			0.92	

Material	Comments	Density (g/cm³)	V_l (km/s)	V_s (km/s)	Impedance (MRayl)	Attn (dB/cm/MHz)	Attn (dB/cm at 5 MHz)
Alcohol, propyl (*n*)		0.804	1.22			0.98	
Alcohol, *t*-amyl		0.81	1.2			0.97	
Alkazene 13		0.86	1.32			1.14	
Analine		1.022	1.69			1.68	
A-Spirit	Ethanol>96%	0.79	1.18			0.93	
Benzene		0.87	1.3			1.13	
	C_6H_6	0.88	1.31			1.15	
Benzol		0.878	1.33			1.17	
Benzol ethyl		0.868	1.34			1.16	
Bromo-benzene	C_6H_5Br	1.52	1.17			1.78	
Bromoform		2.89	0.92			2.66	
Butanol	Butyl	0.71	1.27		0.9		
		0.81	1.27			1.03	
Butoxyethanol	(2*n*-)		1.31				
tert-Butyl chloride		0.84	0.98			0.82	
Butylene glycol (2.3)		1.019	1.48			1.51	
Butyrate ethyl		0.877	1.17			1.03	
Carbitol		0.988	1.46			1.44	
Carbon disulfide			1.16				
		1.26	1.15			1.45	
Carbon tetrachloride	CCl_4	1.595	0.93			1.48	
Cerechlor 42		1.26	1.43			1.8	
Chinolin		1.09	1.57			1.71	
Chlorobenzene	C_6H_5Cl	1.1	1.3			1.43	
Chloroform		1.49	0.99			1.47	
Chlorohexanol	>98%	0.95	1.42			1.35	
Cyclohexanol		0.962	1.45			1.39	
	Freon		1.2				
	DTE 21 oil		1.39				
	Glycerol		1.52				
Decahydronaphtaline		0.948	1.42			1.39	
	$C_{10}H_{18}$	0.89	1.42			1.27	
	Paraffin		1.41				
Diacetyl		0.99	1.24			1.22	
Diamine propane	(1.3) >99%	0.89	1.66			1.47	
Dichloro isobutane (1.3)		1.14	1.22			1.39	
Diethylamine	$(C_2H_5)_2NH$	0.7	1.13			0.8	
Diethylene glycol		1.116	1.58			1.76	
Diethyl ketone		0.813	1.31			1.07	
Dimethyl phthalate		1.2	1.46			1.75	
Dioxane		1.033	1.38			1.43	
Diphenyl	Diphenyl oxide		1.5				
Dodecanol		0.83	1.41			1.16	
DTE 21	Mobil		1.39				
DTE 24	Mobil		1.42				
DTE 26	Mobil		1.43				
Dubanol	Shell		1.43				
			1.44				
	Water		1.55				

Material	Comments	Density (g/cm^3)	V_1 (km/s)	V_s (km/s)	Impedance (MRayl)	Attn (dB/cm/MHz)	Attn (dB/cm at 5 MHz)
Ethanol	C_2H_4OH	0.79	1.13		0.89		0.0421
Ethanol amide		1.018	1.72			1.75	
Ethyl acetate		0.9	1.19			1.07	
Ethylenclycolol	Sodium benzoate		1.67				
Ethylene diamine	>99.5%	0.9	1.69			1.52	
Ethylene glycol	>99.5%	1.11	1.69			1.88	
	1.2-Ethanediol	1.112	1.67			1.86	
		1.113	1.66			1.85	
	H_2O 1:4		1.6				
	H_2O 2:3		1.68				
	H_2O 3:2		1.72				
	H_2O 4:1		1.72				
Ethyl ether		0.713	0.99			0.7	
Fluorinert	H_2O-72	1.68	0.51			0.86	
	FC-104	1.76	0.58			1.01	
	FC-75	1.76	0.59			1.02	
	FC-77	1.78	0.6			1.05	
	FC-43	1.85	0.66			1.21	
	FC-40	1.86	0.64			1.19	
	FC-70	1.94	0.69			1.33	
Fluoro-benzene	C_6H_5F	1.024	1.18			1.21	
Formamide		1.134	1.62			1.84	
Freon			0.68				
Freon MF 21.1		1.485	0.8			1.19	
Freon TF 21.1		1.574	0.97			1.52	
Furfural		1.157	1.45			1.68	
Gasoline		0.803	1.25			1	
Glycerine	$CH_2OHCHOHCH_2OH$	1.23	1.9			2.34	
	Glycerol >98%	1.26	1.88			2.37	
		1.26	1.92			2.42	
Glycerol	Water	1.22	1.88			2.29	
	Butanol		1.45				
	Ethanol		1.52				
			1.56				
	Isopropanol		1.57				
Glycerol trioleate		0.91	1.44			1.31	
Glycol	Polyethylene	1.06	1.62			1.71	
		1.087	1.62			1.75	
	Ethylene	1.108	1.59			1.76	
		1.112	1.67			1.86	
Hexane	$(n\text{-})C_6H_{14}$	0.659	1.1			0.727	
n-Hexanol		0.819	1.3			1.06	
Honey	Sue Bee Orange	1.42	2.03			2.89	
Iodobenzene	C_6H_5I	1.183	1.1			2.01	
Isopentane		0.62	0.99			0.62	
Isopropanol			1.14				
Isopropyl alcohol		0.786	1.17			0.92	
	Propylene glycol	0.79	1.14			0.9	0
		0.84	1.21			1.01	0.14
		0.88	1.24			1.09	0.4

Material	Comments	Density (g/cm³)	V_1 (km/s)	V_s (km/s)	Impedance (MRayl)	Attn (dB/cm/MHz)	Attn (dB/cm at 5 MHz)
		0.88	1.28			1.13	0.23
		0.92	1.3			1.2	0.44
		0.94	1.35			1.27	0.33
		0.96	1.36			1.31	0.53
		0.97	1.36			1.32	1.12
		0.99	1.43			1.41	0.47
		1	1.42			1.42	1.08
		1	1.43			1.43	0.67
		1.03	1.48			1.52	1.13
		1.03	1.51			1.56	0.72
		1.04	1.54			1.64	1.2
		1.07	1.58			1.7	0.93
		1.09	1.6			1.75	1.4
		1.13	1.65			1.86	1.68
		1.16	1.75			2.03	2.57
		1.2	1.82			2.19	2.12
		1.25	1.92			2.39	4.55
Jeffox WL-1400			1.53				
Kerosene		0.81	1.32			1.07	
Linalool		0.884	1.4			1.24	
Mercury	Hg	13.6	1.45			19.7	
Mercury, 20°C			1.42			19.7	
Mesityloxide		0.85	1.31			1.11	
Methanol	CH_3OH	0.796	1.09		0.87	0.87	0.0262
Methyl acetate		0.934	1.21			1.13	
Methylene iodide			0.98				
Methylethyl ketone		0.805	1.21			0.97	
Methyl napthalene		1.09	1.51			1.65	
Methyl salicylate		1.16	1.38			1.6	
Modinet P40		1.06	1.38			1.47	
Monochlorobenzene		1.107	1.27			1.41	
Morpholine		1	1.44			1.44	
M-xylol		0.864	1.32			1.14	
NaK	Mix	0.64	1.66				
		0.713	1.72				
		0.714	1.84				
		0.73	1.95				
		0.736	1.77				
		0.738	1.89				
		0.754	2				
		0.759	1.82				
		0.761	1.99				
		0.778	2.05				
		0.781	1.88				
		0.784	1.99				
		0.801	2.1				
		0.804	1.93				
		0.807	2.04				
		0.825	2.15				
		0.826	1.98				

Material	Comments	Density (g/cm³)	V_1 (km/s)	V_s (km/s)	Impedance (MRayl)	Attn (dB/cm/MHz)	Attn (dB/cm at 5 MHz)
		0.83	2.09				
		0.848	2.2				
		0.849	2.04				
		0.853	2.14				
		0.871	2.25				
		0.876	2.19				
		0.893	2.31				
Nicotine	$C_{10}H_{14}N_2$	1.01	1.49			1.51	
Nitrobenzene		1.2	1.46			1.75	
Nitrogen	N_2	0.8	0.86			0.68	
Nitromethane		1.13	1.33			1.5	
Oil, baby		0.821	1.43			1.17	
Oil, castor	Jeffox WL-1400		1.52				
	Castor	0.95	1.54			1.45	
	Ricinus oil	0.969	1.48			1.43	
Oil, corn		0.922	1.46			1.34	
Oil, cutting	64 AS (red)		1.4				
Oil, diesel			1.25				
Oil, fluorosilicone	Dow FS-1265		0.76				
Oil, grape seed	Cerechlor	0.92	1.43				
	Castor oil	0.936	1.44			1.35	
Oil, gravity fuel AA		0.99	1.49			1.48	
Oil, linseed		0.922	1.77			1.63	
		0.94	1.46			1.34	
Oil, mineral (heavy)		0.843	1.46			1.37	
Oil, mineral (light)		0.825	1.44			1.19	
Oil, motor (2-cycle)			1.43				
Oil, motor (SAE 20)		0.87	1.74			1.51	
Oil, motor (SAE 30)		0.88	1.7			1.5	
Oil, olive			1.43				
		0.918	1.45			1.32	
		0.948	1.43			1.39	
Oil, paraffin			1.28				
			1.43				
		0.835	1.42			1.86	
Oil, peanut		0.914	1.44			1.31	
		0.936	1.46			1.37	
Oil, safflower		0.92	1.45			1.34	
Oil, silicone	Dow 710 fluid		1.35				
	Silicone 200	0.818	0.96			0.74	
		0.94	0.97			0.91	
		0.972	0.99			0.96	
	30 cP	0.993	0.99			0.983	
		1.1	1.37			1.5	
Oil, soybean		0.93	1.43			1.32	
Oil, sperm		0.88	1.44			1.27	
Oil, sun	Nivea		1.41				
Oil, sunflower		0.92	1.45			1.34	
Oil, synthetic		0.98	1.27			1.33	
Oil, transformer		0.92	1.39			1.28	

Material	Comments	Density (g/cm³)	V_1 (km/s)	V_s (km/s)	Impedance (MRayl)	Attn (dB/cm/MHz)	Attn (dB/cm at 5 MHz)
Oil, transmission	Dexron (red)		1.42				
Oil, velocite	Mobil		1.3				
Oil, wheat germ		0.94	1.49			1.39	
Paraffin		1.5	1.5			2.3	
d-Penchone		0.94	1.32			1.24	
Pentane		0.621	1.01			0.63	
	(*n*-) C_5H_{12}	0.626	1.03			0.64	
Petroleum		0.825	1.29			1.06	
Polypropylene glycol	Polyglycol P-400		1.3				
	Polyglycol P-1200		1.3				
	Polyglycol E-200		1.57				
Polypropylene oxide	Ambiflo		1.37				
Potassium		0.662	1.49				
		0.685	1.55				
		0.707	1.6				
		0.729	1.65				
		0.751	1.71				
		0.773	1.76				
		0.796	1.81				
		0.818	1.86				
Propane diol	(1.3) >97%	1.05	1.62			1.7	
Pyridine		0.982	1.41			1.38	
Sodium		0.759	2.15				
		0.784	2.21				
		0.809	2.26				
		0.833	2.31				
		0.857	2.37				
		0.881	2.42				
		0.904	2.48				
		0.926	2.53				
Solvesso #3		0.877	1.37			0.201	
Sonotrack	Coupling gel	1.4	1.62			1.68	
Span 20			1.48				
Span 85			1.46				
Tallow			0.39				
Tetraethylene glycol		1.12	1.58			1.77	
Tetrahydronaphtaline	(1.2.3.4)	0.97	1.47			1.42	
Trichloroethylene		1.05	1.05			0.41	
Triethylene glycol		1.12	1.61			1.81	
		1.123	1.61			1.98	
Trithylamine	$(C_2H_5)_3N$	0.73	1.12			0.81	
Turpentine		0.87	1.25			1.11	
		0.893	1.28			1.14	
			1.27				
Ucon 75H450			1.54				
Univis 800		0.87	1.35			1.19	
Water–salt solution	10%		1.47				
	15%		1.53				
	20%		1.6				

Material	Comments	Density (g/cm^3)	V_1 (km/s)	V_s (km/s)	Impedance (MRayl)	Attn (dB/cm/MHz)	Attn (dB/cm at 5 MHz)
Water, sea		1.025	1.53			1.57	
		1.026	1.5			1.54	
Water		1	1.51			1.51	
		1	1.55			1.55	
	Propylene glycol	1	1.5			1.5	
		1.01	1.61			1.63	0.021
		1.02	1.69			1.72	0.038
		1.03	1.51			1.56	0.669
		1.03	1.62			1.66	0.213
		1.03	1.69			1.73	0.088
		1.05	1.6			1.69	
		1.06	1.69			1.79	0.059
		1.07	1.58			1.7	1.025
		1.07	1.66			1.78	0.395
		1.07	1.71			1.83	0.174
		1.07	1.73			1.84	0.112
		1.11	1.71			1.89	0.086
		1.11	1.75			1.95	0.321
		1.11	1.76			1.94	0.117
		1.11	1.77			1.97	0.182
		1.12	1.66			1.86	1.744
		1.12	1.71			1.91	0.582
		1.16	1.75			2.03	2.57
		1.16	1.78			2.06	1.242
		1.16	1.8			2.09	0.175
		1.16	1.81			2.09	0.648
		1.16	1.82			2.11	0.241
		1.16	1.82			2.11	0.397
		1.2	1.82			2.19	2.12
		1.2	1.85			2.23	1.469
		1.2	1.85			2.22	2.033
		1.2	1.86			2.24	1.023
		1.2	1.87			2.24	0.731
		1.2	1.88			2.25	0.544
		1.25	1.92			2.39	4.55
Water	UCON 50HB400	0.79	1.16			0.91	0.11
		0.83	1.25			1.04	0.06
		0.83	1.27			1.06	0.06
		0.83	1.28			1.07	0.07
		0.83	1.29			1.07	0.07
		0.84	1.21			1.01	0.15
		0.84	1.24			1.03	0.08
		0.87	1.41			1.23	0.3
		0.88	1.31			1.16	0.15
		0.88	1.34			1.18	0.15
		0.88	1.37			1.2	0.2
		0.88	1.4			1.22	0.24
		0.89	1.26			1.12	0.26
		0.91	1.54			1.4	0.55

Material	Comments	Density (g/cm^3)	V_1 (km/s)	V_s (km/s)	Impedance (MRayl)	Attn (dB/cm/MHz)	Attn (dB/cm at 5 MHz)
		0.92	1.52		1.4	0.74	
		0.93	1.44		1.33	0.4	
		0.93	1.48		1.38	0.57	
		0.94	1.32		1.24	0.35	
		0.94	1.4		1.31	0.35	
		0.96	1.63		1.57	1.38	
		0.96	1.64		1.57	0.03	
		0.97	1.54		1.5	1.06	
		0.97	1.59		1.55	1.54	
		0.99	1.38		1.37	0.52	
		0.99	1.49		1.46	0.7	
		1	1.48		1.48		
		1	1.5		1.5	0	
		1	1.5		1.5	0.04	
		1.01	1.61		1.63	0.13	
		1.01	1.61		1.63	0.13	
		1.01	1.63		1.65	0	
		1.01	1.69		1.71	0.4	
		1.02	1.64		1.69	2.72	
		1.02	1.66		1.7	2.11	
		1.02	1.66		1.7	2.11	
		1.02	1.66		1.7	2.11	
		1.02	1.68		1.72	0.84	
		1.02	1.69		1.72	1.5	
		1.02	1.69		1.73	0.17	
		1.02	1.69		1.73	0.29	
		1.02	1.7		1.73	0.08	
		1.02	1.71		1.74	0.44	
		1.03	1.5		1.55	0.92	
		1.03	1.5		1.55	0.92	
		1.03	1.53		1.58	0.72	
		1.03	1.53		1.58	0.72	
		1.03	1.54		1.6	0.72	
		1.03	1.54		1.6	0.72	
		1.03	1.56		1.61	0.73	
		1.03	1.56		1.61	0.73	
		1.03	1.57		1.63	2.13	
		1.03	1.57		1.61	0.92	
		1.03	1.57		1.62	0.58	
		1.03	1.57		1.63	2.13	
		1.03	1.57		1.61	0.92	
		1.03	1.57		1.62	0.58	
		1.03	1.57		1.63	2.13	
		1.03	1.58		1.63	1.25	
		1.03	1.58		1.63	1.25	
		1.03	1.6		1.65	0.47	
		1.03	1.6		1.65	0.47	
		1.03	1.61		1.65	0.56	
		1.03	1.61		1.65	0.56	

Material	Comments	Density (g/cm³)	V_l (km/s)	V_s (km/s)	Impedance (MRayl)	Attn (dB/cm/MHz)	Attn (dB/cm at 5 MHz)
		1.03	1.62			1.67	0.38
		1.03	1.62			1.67	0.38
		1.03	1.63			1.68	0.9
		1.03	1.63			1.68	0.9
		1.03	1.64			1.69	2.72
		1.03	1.65			1.7	1.53
		1.03	1.65			1.7	0.46
		1.03	1.65			1.7	0.32
		1.03	1.65			1.7	1.53
		1.03	1.65			1.7	0.46
		1.03	1.65			1.7	0.32
		1.04	1.44			1.5	0.94
		1.04	1.44			1.5	0.94
		1.04	1.46			1.52	1.05
		1.04	1.48			1.53	1.02
		1.04	1.51			1.57	0.85
Water, D_2O		1.104	1.4			1.55	
Xylene hexaflouride		1.37	0.88			1.21	
Solids (Metals and Alloys)							
Aluminum		2.7	6.32	3.1		17.1	
	Duraluminum	2.71	6.32	3.1		17.1	
Al 1100-0	2S0	2.71	6.35	3.1	2.9	17.2	
Al 2014	14S	2.8	6.32	3.1		17.7	
Al 2024 T4	24ST	2.77	6.37	3.2	2.95	17.6	
Al 2117 T4	17ST	2.8	6.5	3.1		18.2	
Antimony	Sb		3.4				
Bearing Babbit		10.1	2.3			23.2	
Beryllium		1.82	12.9	8.9	7.87	23.5	
Bismuth		9.8	2.18	1.1		21.4	
Brass	70% Cu-30% Zn	8.64	4.7	2.1		40.6	
		8.56	4.28	2		36.6	
	Half Hard	8.1	3.83	2.1		31.0	
	Naval	8.42	4.43	2.1	1.95	37.3	
Bronze	Phospho	8.86	3.53	2.2	2.01	31.3	
Cadmium	Cd	8.6	2.8	1.5		42.0	
		8.64	2.78	1.5		24.0	
Cesium		1.88	0.97			1.82	
Columbium		8.57	4.92	2.1		42.2	
Constantan		8.88	5.24	2.6		46.5	
Copper		8.93	4.66	2.3	1.93	41.6	
Copper, rolled	Cu	8.9	5.01	2.3		44.6	
E-Solder		2.71	1.9	1		5.14	
Gallium		5.95	2.74			16.3	
Germanium		5.47	5.41			29.6	
Gold	Hard drawn	19.32	3.24	1.2		62.6	
Hafnium			3.84				
Inconel		8.25	5.72	3	2.79	64.5	

Material	Comments	Density (g/cm³)	V_l (km/s)	V_s (km/s)	Impedance (MRayl)	Attn (dB/cm/MHz)	Attn (dB/cm at 5 MHz)
Indium		7.3	2.22			16.2	
Iron		7.7	5.9	3.2	2.79	45.4	
	Cast	7.22	4.6	2.6		33.2	
Lead		11.4	2.16	0.7	0.63	24.6	
	5% Antimony	10.9	2.17	0.8	0.74	23.7	
Magnesium		1.74	6.31			11.0	
	AM-35	1.74	5.79	3.1	2.87	10.1	
	FS-1	1.69	5.47	3		9.2	
	J-1	1.7	5.67	3		9.6	
	M	1.75	5.76	3.1		10.1	
	O-1	1.82	5.8	3		10.6	
	ZK-60A-TS	1.83	5.71	3.1		10.4	
		1.72	5.8	3		10.0	
Manganese		7.39	4.66	2.4		34.4	
Molybdenum		10.2	6.29	3.4	3.11	64.2	
Monel		8.83	6.02	2.7	1.96	53.2	
Nickel		8.88	5.63	3	2.64	50.0	
Nickel–silver		11.2	3.58	2.2		40.0	
Platinum		21.4	3.96	1.7		84.7	
Plutonium			1.79			28.2	
	1% Gallium		1.82			28.6	
Potassium		0.83	1.82			1.51	
Rubidium		1.53	1.26			1.93	
Silver		10.5	3.6	1.6		37.8	
	Nickel	8.75	4.62	2.3	1.69	40.4	
	Germanium	8.7	4.76			41.4	
Steel	302 Cres	8.03	5.66	3.1	3.12	45.4	
	347 Cres	7.91	5.74	3.1		45.4	
	410 Cres	7.67	7.39	3	2.16	56.7	
	1020	7.71	5.89	3.2		45.4	
	1095	7.8	5.9	3.2		51.0	
	4150	7.84	5.86	2.8		45.9	
		7.82	5.89	3.2		46.1	
		7.81	5.87	3.2		45.8	
		7.8	5.82	2.8		45.4	
	4340	7.8	5.85	3.2		51.0	
	Mild	7.8	5.9	3.2		46.00	
	Stainless 347	7.89	5.79	3.1		45.70	
Tantalum		16.6	4.1	2.9		54.8	
Thallium		11.9	1.62			19.3	
Thorium		11.3	2.4	1.6		33.2	
Tin		7.3	3.3	1.7		24.1	
Titanium		4.5	6.07	3.1		27.3	
		4.48	6.1	3.1		27.3	
Tungsten		19.25	5.18	2.9	2.65	99.7	
Uranium		18.5	3.4	2		63.0	
Vanadium		6.03	6	2.8		36.2	
Zinc		7.1	4.17	2.4		29.6	
Zircalloy			4.72	2.4		44.2	
Zirconium		6.48	4.65	2.3		30.1	

Material	Comments	Density (g/cm^3)	V_1 (km/s)	V_s (km/s)	Impedance (MRayl)	Attn (dB/cm/MHz)	Attn (dB/cm at 5 MHz)
Solids (Ceramics)							
Ammonium dihydrogen phosphate (ADP)	502/118.9:1	1.35	2.73		3.69		
	502/118.5:1	1.35	2.67		3.60		
			3.28				
Arsenic trisulfide		3.2	2.58	1.4		8.25	
Barium titanate		5.55	5.64	2.9		33.5	
Boron carbide		2.4	11			26.4	
Brick		1.7	4.3			7.40	
		3.6	3.65	2.6		15.3	
Calcium fluoride	CaFl. X-cut		6.74				
Clay rock		2.5	3.48	3.4		14.2	
Concrete		2.6	3.1			8.00	
Flint		3.6	4.26	3		18.9	
Glass	Crown	2.24	5.1	2.8		11.4	
	205 Sheet	2.49	5.66			14.1	
	FK3	2.26	4.91	2.9		11.1	
	FK6	2.28	4.43	2.5		10.1	
	Flint	3.6	4.5			16.0	
	Macor	2.54	5.51			14.0	
	Plate	2.75	5.71			10.7	
	Pyrex	2.24	5.64	3.3		13.1	
	Quartz	2.2	5.57	3.4		14.5	
	Silica	2.2	5.9			13.0	
	Soda lime	2.24	6			13.4	
	T1K	2.38	4.38			10.5	
	Window		6.79	3.4			
Glass crown	Reg.	2.6	5.66	3.5		14.5	
Granite		4.1	6.5			26.8	
Graphite	Pyrolytic	1.46	4.6			6.60	
	Pressed	1.8	2.4			4.10	
Hydrogen	Solid at 4.2 K	0.089	2.19			0.19	
Ice		0.92	3.6			3.20	
		2.65	3.99	3.3		16.4	
Ivory		2.17	3.01			10.4	
Leadmeta niobate	$PbNbO_3$	6.2	3.3			20.5	
	K-81	6.2	3.3			20.5	
	K-83	4.3	5.33			22.9	
	K-85	5.5	3.35			18.4	
Lead zirconate titanate	$PbZrTiO_3$	7.75	3.28			29.3	
		7.5	4			30.0	
		7.45	4.2			31.3	
		7.43	4.44			33.0	
		7.95	4.72			37.5	
Lithium niobate	46 Rot. Y-cut	4.7	7.08			33.0	
	Z-cut	4.64	7.33			34.0	
	Y-cut		6.88				

Material	Comments	Density (g/cm^3)	V_1 (km/s)	V_s (km/s)	Impedance (MRayl)	Attn (dB/cm/MHz)	Attn (dB/cm at 5 MHz)
Lithium sulfate	Y-cut	2.06	5.46			11.2	
Marble		2.8	3.8			10.5	
Porcelain		2.3	5.9			13.50	
Potassium bromide			3.38				
Potassium chloride			4.14				
Potassium sodium niobate		4.46	6.94			31.0	
PZT-2		7.6	4.41	1.7		31.3	
PZT-4		7.5	4.6	1.9		34.5	
PZT-5A		7.75	4.35	1.7		33.7	
PZT-5H		7.5	4.56	1.8		34.2	
Quartz		6.82	5.66			15.2	
	X-cut	2.65	5.75			15.3	
Salt	NaCl	2.17	4.85			10.5	
Salt, rochelle		2.2	5.36	3.8		13.1	
	$KNaC_4H4_{(6}$		2.47				
Salt, rock	X dir		4.78				
Sapphire		2.6	9.8			11.7	
	Al_2O_3	3.98	11.2			44.5	
Silica, fused		2.2	5.96	3.8		13.1	
Silicon	Anisotropic	2.33	9			21.0	
Silicon carbide		13.8	6.66			91.8	
Silicon nitride		3.27	11	6.3		36.0	
Slate			4.5				
		3	4.5			13.5	
Sodium bismuth titanate		6.5	4.06			26.4	
Sodium bromide	NaBr		2.79				
Sulfur			1.35				
Titanium carbide		5.15	8.27	5.2		42.6	
Tourmaline	Z-cut	3.1	7.54			23.4	
Uranium oxide	UO_2		5.18			56.7	
Zinc oxide		5.68	6.4	3		36.4	
Solids (Polymer)							
ABS	Acrylonitrile	1.04	2.11		2.20		
Acrylic		1.2	2.7		3.24		
Acrylic resin		1.18	2.67	1.1		3.15	
Araldite	502/956	1.16	2.62		4.04		
Bakelite		1.4	2.59			3.63	
		1.9	1.9			4.80	
Butyl rubber		1.11	1.8			2.00	
Carbon, pyrolytic	Soft	2.21	3.31			7.31	
Carbon, vitreous		1.47	4.26	2.7		6.26	
Celcon	Acetal copolymer	1.41	2.51			3.54	
Cellulose acetate		1.3	2.45			3.19	
Cycolac	Acrylonitrile–butadiene–styrene		2.27			2.49	

Material	Comments	Density (g/cm^3)	V_l (km/s)	V_s (km/s)	Impedance (MRayl)	Attn (dB/cm/MHz)	Attn (dB/cm at 5 MHz)
ECHOGEL 1265	100PHA of B	9.19	1.32			12.2	
		1.4	1.7			2.38	
		1.1	1.71			1.90	
EPON 828	MPDA	1.21	2.83	1.2		3.40	
EPOTEK 301		1.08	2.64			2.85	
EPOTEK 330		1.14	2.57			2.94	
EPOTEK H70S		1.68	2.91			4.88	
EPOTEK V6	10PHA of B	1.23	2.55			3.14	
		1.23	2.61			3.21	
		1.26	2.55			3.22	
		1.25	2.6			3.25	
Epoxy	Silver	3.098	1.89			5.85	
		3.383	1.87			6.31	
EPX-1 or EPX-2	100PHA of B	1.1	2.44			2.68	
Ethyl vinyl acetate		0.94	1.8			1.69	
		0.95	1.68			1.60	
		0.93	1.86			1.72	
Delrin		1.36	2.47			3.36	
	Acetal homopolymer	1.42	2.52			3.57	
DER317	10.5PHR DEH20	1.18	2.75			3.25	
		2.23	2.07			4.61	
	13.5PHR MPDA	1.6	2.4			3.84	
		2.03	2.19			4.44	
		3.4	1.86	0.9		6.40	
	9PHR DEH20	7.27	1.5			10.9	
		2.23	2.03	1		4.53	
		2.37	1.93			4.58	
DER332	10PHR DEH20	1.76	3.18	1.6		5.58	
		1.2	2.6			3.11	
	10.5PHR DEH20	1.29	2.65			3.41	
		1.26	2.61			3.29	
		1.37	2.75			3.78	
	11PHR DEH20	1.72	2.35			4.05	
		1.29	2.71			3.49	
	14PHR MPDA	1.25	2.59			3.24	
	15PHR MPDA	1.54	2.78	1.5		4.27	
		1.49	2.8	1.4		4.18	
		1.24	2.66			3.30	
		1.24	2.55	1.2		3.16	
		2.15	3.75			8.06	
		2.24	3.9			8.74	
		6.45	1.75			11.3	
	64PHR V140	1.13	2.36			2.65	
	75PHR V140	1.12	2.35			2.62	
	100PHR V140	1.1	2.32			2.55	
		1.13	2.27			2.55	
		1.16	2.36			2.74	
Glucose		1.56	3.2			5.00	
Hysol	C8-4143/3404	1.58	2.85			4.52	

Material	Comments	Density (g/cm^3)	V_l (km/s)	V_s (km/s)	Impedance (MRayl)	Attn (dB/cm/MHz)	Attn (dB/cm at 5 MHz)
	C9-4183/3561	3.17	2.16			7.04	
		2.14	2.49			5.33	
		1.8	2.62			4.70	
		1.48	2.92			4.30	
		2.66	2.3			6.10	
	C8-4412	1.68	2.02			3.39	
		1.5	2.32			3.49	
Hysol	R9-2039/3404						
Ivory			3.01				
Kel-F			1.79				
Kydex		1.35	2.22			2.99	
Lucite	Polymethylacrylate	1.29	2.72			3.50	
		1.18	2.68	1.3		3.16	
		1.15	2.7	1.1		3.10	
Marlex 5003	High-density polyethylene	0.95	2.56			2.43	
Melopas		1.7	2.9			4.93	
Micarta	Linen base		3				
Mylar		1.18	2.54			3.00	
Neoprene		1.31	1.6			2.10	
Noryl	Polyphenylene oxide	1.08	2.27			2.45	
Nylon 6-6		1.12	2.6	1.1		2.90	
Penton	Chlorinated polyether	1.4	2.57			3.60	
	Syntactic foam (33 lb/ft^3)	0.53	2.57			1.36	
Phenolic		1.34	1.42			1.90	
Plexiglas	UVA	1.27	2.76			3.51	
	UVAII	1.18	2.73	1.4		3.22	
Polyamide			2.6			2.90	
Polycarbonate	Lexan	1.18	2.3			2.71	
Polyester	Casting Resin	1.07	2.29			2.86	
Polyethylene	Low density	0.92	2.06		1.90	22	26.5
		1.1	2.67			2.80	
	TCI		1.6				
	HD. LB-861	0.96	2.43			2.33	
Polyisobutylene			1.49				
	mol. wt. 200		1.85				
Polypropylene	Profax 6423	0.901	2.49			2.24	
		0.88	2.74			2.40	
Polysulfone		1.24	2.24			2.78	
Polystyrene		1.1	2.67			2.80	
	Styron 666	1.05	2.4			2.52	
Polyurethane	RP-6400	1.04	1.5			1.56	
	RP-6401	1.07	1.71		1.83	35	73
		1.07	1.63			1.74	
	RP-6402	1.08	1.77			1.91	
	RP-6403	1.1	1.87			2.05	
	RP-6405	1.3	2.09			2.36	
	RP-6410	1.04	1.71		1.78	36	73
		1.04	1.33			1.38	

Material	Comments	Density (g/cm^3)	V_l (km/s)	V_s (km/s)	Impedance (MRayl)	Attn (dB/cm/MHz)	Attn (dB/cm at 5 MHz)
	RP-6413	1.04	1.71		1.78	21	35.2
		1.04	1.65			1.66	
	RP-6414	1.05	1.78			1.86	
		1.05	1.85		1.94	18	35.2
	RP-6422	1.04	1.6			1.66	
	EN-9	1.01	1.68			1.70	
	REN plastic	1.07	1.71			1.83	35
		1.04	1.49			1.55	36
		1.04	1.62			1.69	15
		1.04	1.71			1.78	21
		1.05	1.85			1.92	18
	RP6422	1.04	1.62		1.69	14	27.6
Polyvinyl chloride (PVC)		1.45	2.27			3.31	
Polyvinylbutyral	Butracite	1.11	2.35			2.60	
Polyvinylidene difluoride		1.79	2.3			4.20	
Profax	Polypropylene	0.9	2.79			2.51	
Refrasil		1.73	3.75			6.49	
Rubber	BFG#6063-19-71	0.97	1.53			1.56	
	BFG#35080						
	Hard	1.1	1.45			2.64	
	Rho-C	1	1.55			1.55	
	Soft	0.95	0.07			1.00	
Scotchcast	XR2535	1.49	2.48			3.70	
Scotchply XP241	Syntactic foam (42 lb/ft^3)	0.65	2.84			1.84	
	Syntactic foam (38 lb/ft^3)	0.61	2.81			1.71	
Scotchply	XP241	0.65	2.84			1.84	
	SP1002	1.94	3.25			6.24	
Scotch tape	2.5 mils thick	1.16	1.9			2.08	
Silicon rubber	Sylgard 170	1.38	0.97			1.34	
	Sylgard 182	1.05	1.03			1.07	
		1.12	1.03			1.15	
	Sylgard 184	1.03	1.03			1.04	
	RTV-11	1.18	1.05			1.24	
	RTV-21	1.31	1.01			1.32	
	RTV-30	1.45	0.97			1.41	
	RTV-41	1.31	1.01			1.32	
	RTV-60	1.47	0.96			1.41	
	RTV-77	1.33	1.02			1.36	
	RTV-90	1.5	0.96			1.44	
	RTV-112	1.05	0.94			0.99	
	RTV-511	1.18	1.11			1.31	
	RTV-116	1.1	1.02			1.12	
	RTV-118	1.04	1.03			1.07	
	RTV-577	1.35	1.08			1.46	
	RTV-560	1.42	1.03			1.46	
	RTV-602	1.02	1.16			1.18	

Material	Comments	Density (g/cm^3)	V_1 (km/s)	V_s (km/s)	Impedance (MRayl)	Attn (dB/cm/MHz)	Attn (dB/cm at 5 MHz)
	RTV-615	1.02	1.08			1.10	
	RTV-616	1.22	1.06			1.29	
	RTV-630	1.24	1.05			1.30	
	PRC 1933-2	1.48	0.95			1.40	
Silly Putty		1	1			1.00	
Stycast	1251-40	1.67	2.9	1.5		4.83	
		1.63	2.95			4.82	
		1.57	2.88			4.53	
		1.5	2.77			4.16	
	1264	1.19	2.22			2.64	
	2741	1.17	2.29			2.68	
	CPC-41	1.01	1.52			1.54	
	CPC-39	1.06	1.53			1.63	
Styrene 50D	Polystyrene	1.04	2.33			2.43	
Styron	Modified polystyrene	1.03	2.24			2.31	
Surlyn	1555 Ionomer	0.95	1.91			1.81	
Tapox	Epoxy	1.11	2.48			2.76	
Techform	EA700	1.2	2.63			3.14	
Teflon		2.14	1.39			2.97	
		2.2	1.35			2.97	
TPX	DX845	0.83	2.22		1.84	4.2	5.8
Tracon	2135 D	1.03	2.45			1.52	
	2143 D	1.05	2.37			2.50	
	2162 D	1.19	2.02			2.41	
	3011	1.2	2.12			2.54	
	401 ST	1.62	2.97			4.82	
Uvex			2.11				
WR 106-1	Fluoro elastomer		0.87				
Zytel-101	Nylon-101	1.14	2.71			3.08	
Solids (Natural)							
Ash	Along fiber		4.67				
Beech	Along fiber		3.34				
Beef			1.55			1.68	
Brain			1.49			1.55	
Cork			0.5				
Douglas Fir	Cross grain		1.4				
	With grain		4.8				
Elm			1.4			0.798	
Human			1.47			1.58	
Kidney			1.54			1.62	
Liver			1.54			1.65	
Maple	Along fiber		4.11				
Oak			4.47			3.60	
Pine	Along fiber		3.32				
Poplar	Along fiber		4.28				
Spleen			1.5			1.60	
Sycamore	Along fiber		4.46				
Water		0.88	4	2		3.50	
Wood	Cork	0.24	0.5			0.12	

Material	Comments	Density (g/cm^3)	V_1 (km/s)	V_s (km/s)	Impedance (MRayl)	Attn (dB/cm/MHz)	Attn (dB/cm at 5 MHz)
	Elm		4.1				
	Oak	0.72	4			1.57	
	Pine	0.45	3.5			1.57	

REFERENCES

1. *Metals Handbook,* 3rd ed., Vol. 17: *Nondestructive Evaluation and Quality Control,* ASM International, Metals Park, OH, 1989.
2. M. R. Mitchell and O. Buck (eds.), *Cyclic Deformation, Fracture, and Nondestructive Evaluation of Advanced Materials,* American Society for Testing and Materials, Philadephlia, PA, 1992.
3. H. J. Shapuk (ed.), *Annual Book of ASTM Standards: E-7, Nondestructive Testing,* American Society for Testing and Materials, West Conshohocken, PA, 1997.
4. R. E. Green, Jr. (ed.), *Nondestructive Characterization of Materials,* Vol. 8, International Symposium on Nondestructive Characterization of Materials, Plenum, New York, 1998.
5. *British Journal of Nondestructive Testing,* no longer published.
6. American Society for Nondestructive Testing, *http://www.asnt.org.*
7. Center for Nondestructive Evaluation, *http://www.cnde.iastate.edu.*
8. Center for Quality Engineering & Failure Prevention, *http://www.cqe.nwu.edu.*
9. *Nondestructive Evaluation System Reliability Assessment, http://www.ihserc.com.*
10. Nondestructive Testing Information Analysis Center, *http://www.ntiac.com.*
11. W. Altergott and E. Henneke (eds.), *Characterization of Advanced Materials,* Plenum, New York, 1990.
12. R. Halmshaw, *Nondestructive Testing Handbook,* Chapman & Hall, London, 1991.
13. R. A. Kline, *Nondestructive Characterization of Materials,* Technomic Publishing, Lancaster, PA, 1992.
14. P. K. Mallick, "Nondestructive Tests," in *Composites Engineering Handbook,* P. K. Mallick (ed.), Marcel Dekker, New York, 1997.
15. W. McGonnagle, *Nondestructive Testing,* Gordon Breach, New York, 1961.
16. C. O. Ruud et al. (eds.), *Nondestructive Characterization of Materials,* Vols. I–IV, Plenum, New York, 1986.
17. R. S. Sharpe, *Research Techniques in Nondestructive Testing,* Academic, New York, 1984.
18. J. Summerscales, "Manufacturing Defects in Fibre-Reinforced Plastic Composites," *Insight,* **36**(12), 936–942 (1994).
19. D. O. Thompson and D. E. Chimenti (eds.), *Review of Progress in Quantitative Nondestructive Evaluation,* Plenum, New York, 1982–2000.
20. J. Boogaard and G. M. van Dijk (eds.), *Nondestructive Testing: Proceedings of the 12th World Conference on Nondestructive Testing,* Elsevier Science, New York, 1989.
21. D. E. Bray and R. K. Stanley, *Nondestructive Evaluation, A Tool for Design, Manufacturing, and Service,* McGraw-Hill, New York, 1989.
22. M. H. Geier, *Quality Handbook for Composite Materials,* Chapman & Hall, London, 1994.
23. Online Journal Publication Service, *http://ojps.aip.org.*
24. *Journal of Composite Materials,* 2004.
25. Electronic journals, *http://lib-www.lanl.gov/cgi-bin/ejrnlsrch.cgi.*
26. Elsevier Science, *http://www.elsevier.com/homepage/elecserv.htt.*
27. *Japanese Journal of Nondestructive Inspection, http://sparc5.kid.ee.cit.nihon-u.ac.jp/homepage_Eng.html.*
28. *Journal of Nondestructive Evaluation,* Kluwer Academic, 2004,

29. *Journal of Micromechanics and Microengineering,* 2004.
30. British Institute of Non-Destructive Testing, http://www.bindt.org/, 1999.
31. IFANT, International Foundation for the Advancement of Nondestructive Testing, *http://www.ifant.org*.
32. Japan JSNDI, *http://sparc5.kid.ee.cit.nihonu.ac.jp/homepage_Eng.html.*
33. SPIE, *http://spie.org/.*
34. Institute of Electrical and Electronic Engineers, *http://www.ieee.org/.*
35. *IEEE-ASME, Journal of Microelecromechanical Systems,* Vol. 2000, 2004.
36. American Society of Mechanical Engineers, *http://www.asme.org/.*
37. Center for Nondestructive Evaluation, *http://www.cnde.com.*
38. Airport and Aircraft Safety Research & Development, *http://www.asp.tc.faa.gov.*
39. Fraunhofer IZFP, *http://www.fhg.de/english/profile/institute/izfp/index.html.*
40. Stasuk Testing & Inspection, *http://www.nde.net.*
41. AFRL electronic journals, *http://www.wrs.afrl.af.mil/infores/library/ejournals.htm.*
42. Link, Springer Verlag, *http://link.springer-ny.com/.*
43. IBM Intellectual Property Network, *http://www.patents.ibm.com.*
44. Lavender International NDT, in *Lavender International,* 2004.
45. *Trends in NDE Science and Technology, Proceedings of the 14th World Conferences on Nondestructive Testing,* Brookfield VT, Ashgate Publishing, 1997.
46. J. L. Rose and A. A. Tseng (eds.), *New Directions in Nondestructive Evaluation of Advanced Materials,* American Society of Mechanical Engineers, New York, 1988.
47. *Journal of Intelligent Material Systems and Structures, http://www.techpub.com.*
48. Smart Structures, *http://www.adaptive-ss.com/.*
49. Smart Materials and Structures, http://www.adaptive-ss.com/, 2001.
50. Smart Structures—Harvard, *http://iti.acns.nwu.edu/clear/infr/imat_smart.html.*
51. N. Tracy (ed.), *Liquid Penetrant Testing,* 3rd ed., Vol. 2 of *Nondestructive Testing Handbook,* P. Moore (ed.), American Society for Nondsetructive Testing, Columbus, OH, 1999.
52. R. A. Quinn, *Industrial Radiography—Theory and Practice,* Eastman Kodak, Rochester, NY, 1980.
53. H. Burger, *Neutron Radiography; Methods, Capabilities and Applications,* Elsevier Science, New York, 1965.
54. R. H. Bossi, F. A. Iddings, and G. C. Wheeler (eds.), *Radiographic Testing,* 3rd ed., *Nondestructive Testing Handbook,* P. Moore (ed.), American Society for Nondsetructive Testing, Columbus, OH, 2002.
55. R. H. Fassbender and D. J. Hagemaier, "Low-Kilovoltage Radiography of Composites," *Mater. Eval.,* **41**(7), 381–838 (1983).
56. A. S. Birks and J. Green, *Ultrasonic Testing,* 2nd ed., Vol. 7 of *Nondestructive Testing Handbook,* P. Intire (ed.), American Society for Nondestructive Testing, Columbus, OH, 1991.
57. E. A. Ash and E. G. S. Paige, *Rayleigh Wave Theory and Application,* Springer Series on Wave Phenomena, Vols. 1 and 2, Springer-Verlag, Berlin, 1985.
58. I. A. Viktorov, *Rayleigh and Lamb Waves,* Plenum Press, New York, 1967.
59. J. Krautkramer and H. Krautkramer, *Ultrasonic Testing of Materials,* 3rd ed., Springer-Verlag, New York, 1983.
60. R. B. Jones and D. E. W. Stone, "Toward an Ultrasonic-Attenuation Technique to Measure Void Content in Carbon-Fibre Composites," *Nondestructive Testing,* **9**(3), 71–79 (1976).
61. T. S. Jones, "Inspection of Composites Using the Automated Ultrasonic Scanning System (AUSS)," *Mater. Eval.,* **43**(5), 746–753 (1985).
62. D. K. Hsu and T. C. Patton, "Development of Ultrasonic Inspection for Adhesive Bonds in Aging Aircraft," *Mater. Eval.,* **51**(12), 1390–1397 (1993).
63. R. N. Swamy and A. M. A. H. Ali, "Assessment of In Situ Concrete Strength by Various Non-Destructive Tests," *NDT Int.,* **17**(3), 139–146 (1984).
64. E. P. Papadakis and G. B. Chapman II, "Modification of a Commercial Ultrasonic Bond Tester for Quantitative Measurements in Sheet-Molding Compound Lap Joints," *Mater. Eval.,* **51**(4), 496–500 (1993).

65. D. J. Hagemier, "Bonded Joints and Nondestructive Testing—1," *Nondestructive Testing,* **4**(12), 401–406 (1971).

66. D. J. Hagemier, "Bonded Joints and Nondestructive Testing—2," *Nondestructive Testing,* **5**(2), 38–47 (1972).

67. D. J. Hagemier, "Nondestructive Testing of Bonded Metal-to-Metal Joints—2," *Nondestructive Testing,* **5**(6), 144–153 (1972).

68. J. T. Schmidt and K. Skeie (eds.), *Magnetic Particle Testing,* 2nd ed., Vol. 6 of *Nondestructive Testing Handbook,* P. McIntire (ed.), American Society for Nondestructive Testing, Columbus, OH, 2001.

69. X. P. V. Maldague (ed.), *Infrared and Thermal Testing,* 3rd ed., Vol. 3 of *Nondestructive Testing Handbook,* P. Moore (ed.), American Society for Nondestructive Testing, Columbus, OH, 2001.

70. R. K. Stanley (ed.), *Special Nondestructive Testing Methods,* 2nd ed., Vol. 9 of *Nondestructive Testing Handbook,* P. O. Moore and P. McIntire (eds.), American Society for Nondestructive Testing, Columbus, OH, 1995.

71. T. B. Zorc (ed.), *Nondestructive Evaluation and Quality Control,* 9th ed., Vol. 17 of *Metals Handbook,* ASM International, Metals Park, OH, 1989.

72. S. S. Udpa (ed.), *Electromagnetic Testing,* 3rd ed., Vol. 5 of *Nondestructive Testing Handbook,* P. Moore (ed.), American Society for Nondestructive Testing, Columbus, OH, 2004.

73. F. Förster, "Theoretische und experimentalle Grundlagen der zerstörungfreien Werkstoffprufung mit Wirbelstromverfahren, I. Das Tastpulverfahern," *Zeitschrift fur Metallkunde,* **43,** 163–171 (1952).

74. S. N. Vernon, "Parametric Eddy Current Defect Depth Model and Its Application to Graphite Epoxy," *NDT Int.,* **22**(3), 139–148 (1989).

75. Wincheski et al., "Development of Giant Magnetoresistive Inspection System for Detection of Deep Fatigue Cracks under Airframe Fastners," in *Review of Progress in Quantitative Nondestructive Evaluation,* 2002.

76. J. Blitz, *Electrical and Magnetic Method of Non-Destructive Testing,* Chapman & Hall, 1997.

77. H. L. Libby, *Introduction to Electromagnetic Nondestructive Test Methods,* Wiley-Interscience, New York, 1971.

INDEX

A

B

C

D

F

G

H

I

J

K

M

O

P

Q

T

U

V

W

X

Y